# 11 Years' SOLVED PAPERS

## 2010-2020

# 11 Years'

# SOLVED PAPERS

## 2010-2020

# IIT JEE

## (JEE Main & Advanced)

**PHYSICS | CHEMISTRY**
**MATHEMATICS**

arihant
ARIHANT PRAKASHAN (Series), MEERUT

**ARIHANT PRAKASHAN (Series), MEERUT**



卐 **Administrative & Production Offices**

**Regd. Office**

'Ramchhaya' 4577/15, Agarwal Road, Darya Ganj, New Delhi -110002
Tele: 011- 47630600, 43518550

**Head Office**

Kalindi, TP Nagar, Meerut (UP) - 250002
Tel: 0121-7156203, 7156204

卐 **Sales & Support Offices**

Agra, Ahmedabad, Bengaluru, Bareilly, Chennai, Delhi, Guwahati, Hyderabad, Jaipur, Jhansi, Kolkata, Lucknow, Nagpur & Pune.

卐 **ISBN** 978-93-25293-28-1

卐 **PRICE** ₹590.00

**PO No :** TXT-XX-XXXXXXX-X-XX

Published by Arihant Publications (India) Ltd.

For further information about the books published by Arihant, log on to www.arihantbooks.com or e-mail at info@arihantbooks.com

Follow us on   

# Contents

JEE Main & Advanced, IIT JEE

## Solved Papers (2010-2020)

# Contents

JEE Main & Advanced, IIT JEE

Solved Papers (2010-2020)

# JEE ADVANCED

# Solved Paper 2020

## Paper 1

### Section 1 (Maximum Marks : 18)

- This section contains **SIX (06)** questions.
- Each question has **FOUR** options. **ONLY ONE** of these four options is the correct answer.
- For each question, choose the option corresponding to the correct answer.
- Answer to each question will be evaluated according to the following marking scheme:

| | | |
|---|---|---|
| *Full Marks* | : | **+3** If ONLY the correct options is chosen. |
| *Zero Marks* | : | **0** If none of the options is chosen. (i.e., the question is unanswered) |
| *Negative Marks* | : | **−1** In all other cases. |

**1.** A football of radius $R$ is kept on a hole of radius $r$ $(r < R)$ made on a plank kept horizontally. One end of the plank is now lifted so that it gets tilted making an angle $\theta$ from the horizontal as shown in the figure below. The maximum value of $\theta$ so that the football does not start rolling down the plank satisfies (figure is schematic and not drawn to scale)

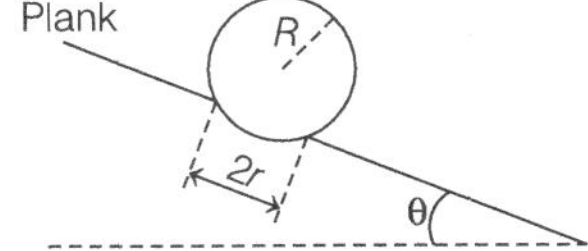


(a) $\sin\theta = \frac{r}{R}$ (b) $\tan\theta = \frac{r}{R}$

(c) $\sin\theta = \frac{r}{2R}$ (d) $\cos\theta = \frac{r}{2R}$

**2.** A light disc made of aluminium (a nonmag- netic material) is kept horizontally and is free to rotate about its axis as shown in the figure. A strong magnet is held vertically at a point above the disc away from its axis. On revolving the magnet about the axis of the disc, the disc will (figure is schematic and not drawn to scale)

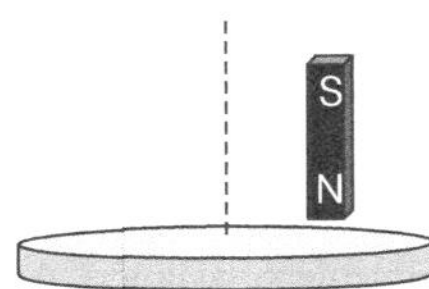


(a) rotate in the direction opposite to the direction of magnet's motion

(b) rotate in the same direction as the direction of magnet's motion

(c) not rotate and its temperature will remain unchanged

(d) not rotate but its temperature will slowly rise.

**3.** A small roller of diameter 20 cm has an axle of diameter 10 cm (see figure below on the left). It is on a horizontal floor and a meter scale is positioned horizontally on its axle with one edge of the scale on top of the axle (see figure on the right). The scale is now pushed slowly on the axle so that it moves without slipping on the axle, and the roller starts rolling without slipping. After the roller has moved 50 cm, the position of the scale will look like (figures are schematic and not drawn to scale)

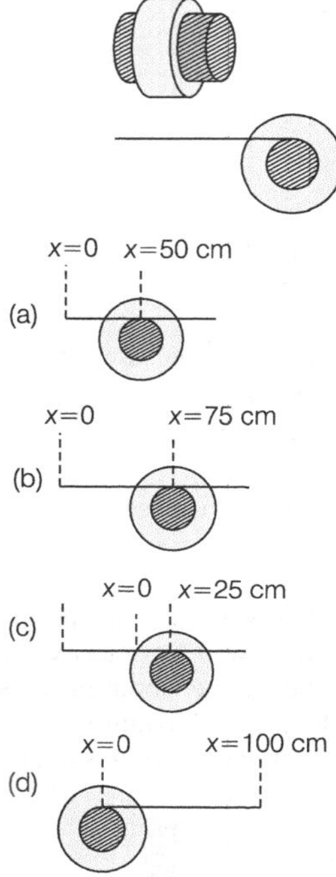


**4.** A circular coil of radius $R$ and $N$ turns has negligible resistance. As shown in the schematic figure, its two ends are connected to two wires and it is hanging by those wires with its plane being vertical. The wires are connected to a capacitor with charge $Q$ through a switch. The coil is in a horizontal uniform magnetic field $B_0$ parallel to the plane of the coil. When the switch is closed, the capacitor gets discharged through the coil in a very short time. By the time the capacitor is discharged fully, magnitude of the angular momentum gained by the coil will be (assume that the discharge time is so short that the coil has hardly rotated during this time)

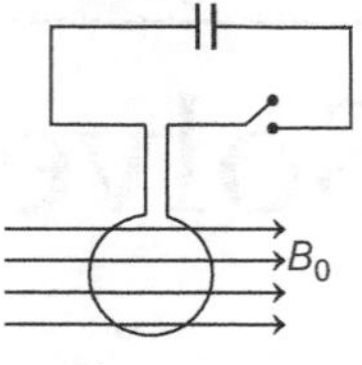


(a) $\frac{\pi}{2}NQB_0R^2$ (b) $\pi NQB_0R^2$

(c) $2\pi NQB_0R^2$ (d) $4\pi NQB_0R^2$

**5.** A parallel beam of light strikes a piece of transparent glass having cross section as shown in the figure below. Correct shape of the emergent wavefront will be (figures are schematic and not drawn to scale)

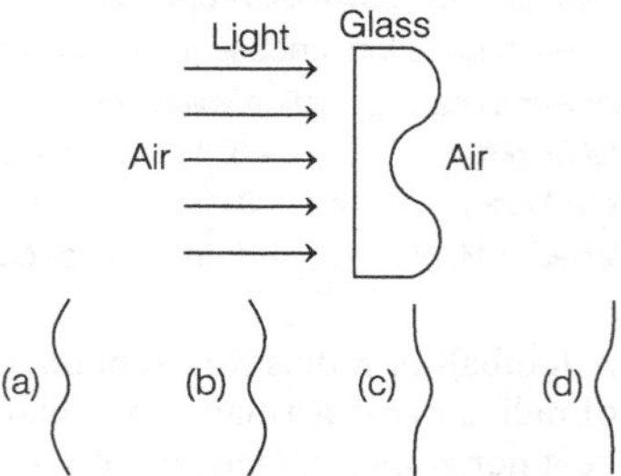


**6.** An open-ended U-tube of uniform cross-sectional area contains water (density $10^3$ kg m$^{-3}$). Initially the water level stands at 0.29 m from the bottom in each arm. Kerosene oil (a water-immiscible liquid) of density 800 kg m$^{-3}$ is added to the left arm until its length is 0.1 m, as shown in the schematic figure below. The ratio $\left(\frac{h_1}{h_2}\right)$ of the heights of the liquid in the two arms is

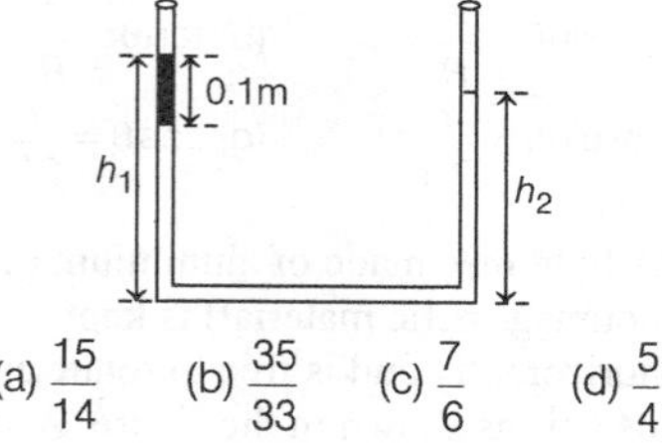


(a) $\frac{15}{14}$ (b) $\frac{35}{33}$ (c) $\frac{7}{6}$ (d) $\frac{5}{4}$

## Section 2 (Maximum Marks : 24)

- This section contains **SIX (06)** questions.
- Each question has **FOUR** options. **ONE OR MORE THAN ONE** of these four option(s) is (are) correct options(s).
- For each question, choose the options(s) corresponding to (all) the correct answer(s).
- Answer to each question will be evaluated according to the following marking scheme:

| | | | |
|---|---|---|---|
| *Full Marks* | : | **+4** | If only (all) the correct option(s) is (are) chosen. |
| *Partial Marks* | : | **+3** | If all the four options are correct but ONLY three options are chosen. |
| *Partial Marks* | : | **+2** | If three or more options are correct but ONLY two options are chosen, both of which are correct. |
| *Partial Marks* | : | **+1** | If two or more options are correct but ONLY one option is chosen and it is a correct option; |
| *Zero Marks* | : | **0** | If none of the option is chosen (i.e., the question is unanswered); |
| *Negative Marks* | : | **−1** | In all other cases. |

**7.** A particle of mass $m$ moves in circular orbits with potential energy $V(r) = Fr$, where $F$ is a positive constant and $r$ is its distance from the origin. Its energies are calculated using the Bohr model. If the radius of the particle's orbit is denoted by $R$ and its speed and energy are denoted by $v$ and $E$, respectively, then for the $n^{th}$ orbit (here $h$ is the Planck's constant)

(a) $R \propto n^{1/3}$ and $v \propto n^{2/3}$

(b) $R \propto n^{2/3}$ and $v \propto n^{1/3}$

(c) $E = \frac{3}{2}\left(\frac{n^2h^2F^2}{4\pi^2m}\right)^{1/3}$

(d) $E = 2\left(\frac{n^2h^2F^2}{4\pi^2m}\right)^{1/3}$

**8.** The filament of a light bulb has surface area 64 mm$^2$. The filament can be considered as a black body at temperature 2500 K emitting radiation like a point source when viewed from far. At night the light bulb is observed from a distance of 100 m. Assume the pupil of the eyes of the observer to be circular with radius 3 mm. Then (Take Stefan-Boltzmann constant $= 5.67 \times 10^{-8}$ Wm$^{-2}$K$^{-4}$, Wien's displacement constant $= 2.90 \times 10^{-3}$ m-K, Planck's constant $= 6.63 \times 10^{-34}$ Js, speed of light in vacuum $= 3.00 \times 10^8$ ms$^{-1}$)

(a) power radiated by the filament is in the range 642 W to 645 W

(b) radiated power entering into one eye of the observer is in the range $3.15 \times 10^{-8}$ W to $3.25 \times 10^{-8}$ W

(c) the wavelength corresponding to the maximum intensity of light is 1160 nm

(d) taking the average wavelength of emitted radiation to be 1740 nm, the total number of photons entering per second into one eye of the observer is in the range $2.75 \times 10^{11}$ to $2.85 \times 10^{11}$

**9.** Sometimes it is convenient to construct a system of units so that all quantities can be expressed in terms of only one physical quantity. In one such system, dimensions of different quantities are given in terms of a quantity $X$ as follows: [position] = $[X^{\alpha}]$; [speed] = $[X^{\beta}]$; [acceleration] = $[X^{p}]$; [linear momentum] = $[X^{q}]$; [force] = $[X^{r}]$. Then

(a) $\alpha + p = 2\beta$ (b) $p + q - r = \beta$

(c) $p - q + r = \alpha$ (d) $p + q + r = \beta$

**10.** A uniform electric field, $\mathbf{E} = -400\sqrt{3}\hat{\mathbf{y}}$ NC$^{-1}$ is applied in a region. A charged particle of mass m carrying positive charge $q$ is projected in this region with an initial speed of $2\sqrt{10} \times 10^6$ ms$^{-1}$. This particle is aimed to hit a target $T$, which is 5 m away from its entry point into the field as shown schematically in the figure.

Take $\frac{q}{m} = 10^{10}$ $\text{Ckg}^{-1}$. Then

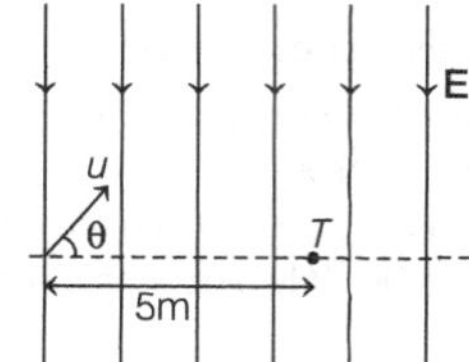


(a) the particle will hit $T$ if projected at an angle 45° from the horizontal

(b) the particle will hit $T$ if projected either at an angle 30° or 60° from the horizontal

(c) time taken by the particle to hit $T$ could be $\sqrt{\frac{5}{6}}$ μs as well as $\sqrt{\frac{5}{2}}$ μs

(d) time taken by the particle to hit $T$ is $\sqrt{\frac{5}{3}}$ μs

**11.** Shown in the figure is a semicircular metallic strip that has thickness $t$ and resistivity $\rho$. Its inner radius is $R_1$ and outer radius is $R_2$. If a voltage $V_0$ is applied between its two ends, a current $I$ flows in it. In addition, it is observed that a transverse voltage $\Delta V$ develops between its inner and outer surfaces due to purely kinetic effects of moving electrons (ignore any role of the magnetic field due to the current). Then (figure is schematic and not drawn to scale)

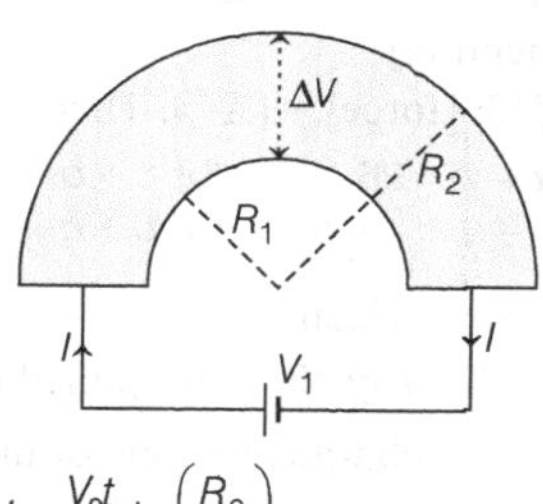


(a) $I = \frac{V_0 t}{\pi \rho} \ln\left(\frac{R_2}{R_1}\right)$

(b) the outer surface is at a higher voltage than the inner surface

(c) the outer surface is at a lower voltage than the inner surface

(d) $\Delta V \propto I^2$

**12.** As shown schematically in the figure, two vessels contain water solutions (at temperature $T$) of potassium permanganate ($KMnO_4$) of different concentrations $n_1$ and $n_2$ ($n_1 > n_2$) molecules per unit volume with $\Delta n = (n_1 - n_2) << n_1$. When they are connected by a tube of small length $l$ and cross-sectional area $S$, $KMnO_4$ starts to diffuse from the left to the right vessel through the tube. Consider the collection of molecules to behave as dilute ideal gases and the difference in their partial pressure in the two vessels causing the diffusion. The speed $v$ of the molecules is limited by the viscous force $-\beta v$ on each molecule, where $\beta$ is a constant. Neglecting all terms of the order $(\Delta n)^2$, which of the following is/are correct? ($k_B$ is the Boltzmann constant)

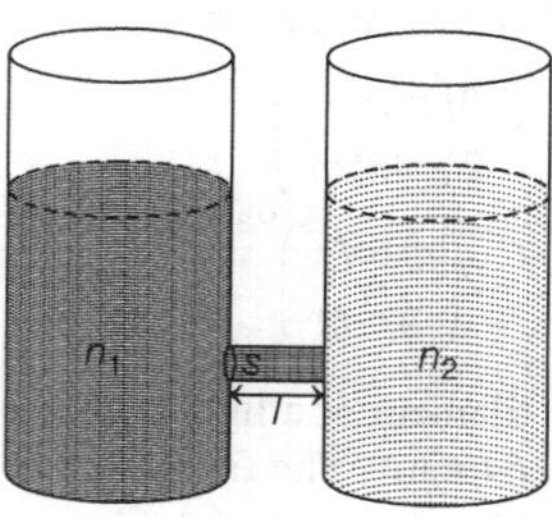


(a) the force causing the molecules to move across the tube is $\Delta n k_B T S$

(b) force balance implies $n_1 \beta v l = \Delta n k_B T$

(c) total number of molecules going across the tube per sec is $\left(\frac{\Delta n}{l}\right)\left(\frac{k_B T}{\beta}\right) S$

(d) rate of molecules getting transferred through the tube does not change with time

## Section 3 (Maximum Marks : 24)

- This section contains **SIX (06)** questions. The answer to each question is a **NUMERICAL VALUE.**
- Four each question, enter the correct numerical value of the answer using the mouse and the on-screen virtual numeric keypad in the place designated to enter the answer. If the numerical value has more than two decimal places, truncate/round-off the value to TWO decimal places.
- Answer to each question will be evaluated according to the following marking scheme:

*Full Marks* : **+4** If ONLY the correct numerical value is entered.
*Zero Marks* : **0** In all other cases.

**13.** Put a uniform meter scale horizontally on your extended index fingers with the left one at 0.00 cm and the right one at 90.00 cm. When you attempt to move both the fingers slowly towards the center, initially only the left finger slips with respect to the scale and the right finger does not. After some distance, the left finger stops and the right one starts slipping. Then the right finger stops at a distance $x_R$ from the center (50.00 cm) of the scale and the left one starts slipping again. This happens because of the difference in the frictional forces on the two fingers. If the coefficients of static and dynamic friction between the fingers and the scale are 0.40 and 0.32, respectively, the value of $x_R$ (in cm) is ......... .

**14.** When water is filled carefully in a glass, one can fill it to a height $h$ above the rim of the glass due to the surface tension of water. To calculate $h$ just before water starts flowing, model the shape of the water above the rim as a disc of thickness $h$ having semicircular edges, as shown schematically in the figure. When the pressure of water at the bottom of this disc exceeds what can be withstood due to the surface tension, the water surface breaks near the rim and water starts flowing from there. If the density of water, its surface tension and the acceleration due to gravity are $10^3$ kg m$^{-3}$, 0.07 Nm$^{-1}$ and 10 ms$^{-2}$, respectively, the value of $h$ (in mm) is ......... .

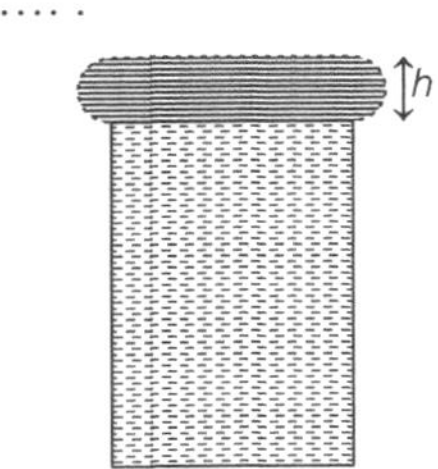


**15.** One end of a spring of negligible unstretched length and spring constant $k$ is fixed at the origin (0,0). A point particle of mass $m$ carrying a positive charge $q$ is attached at its other end. The entire system is kept on a smooth horizontal surface. When a point dipole **p** pointing towards the charge $q$ is fixed at the origin, the spring gets stretched to a length $l$ and attains a new equilibrium position (see figure below). If the point mass is now displaced slightly by $\Delta l << l$ from its equilibrium position and released, it is found to oscillate at frequency $\frac{1}{\delta}\sqrt{\frac{k}{m}}$.

The value of $\delta$ is ______.

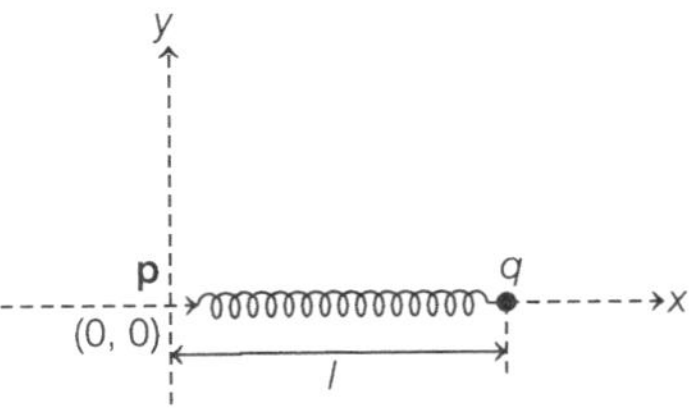

**16.** Consider one mole of helium gas enclosed in a container at initial pressure $P_1$ and volume $V_1$. It expands isothermally to volume $4V_1$. After this, the gas expands adiabatically and its volume becomes $32V_1$. The work done by the gas during isothermal and adiabatic expansion processes are $W_{iso}$ and $W_{adia}$, respectively. If the ratio $\frac{W_{iso}}{W_{adia}} = f \ln 2$, then $f$ is________.

**17.** A stationary tuning fork is in resonance with an air column in a pipe. If the tuning fork is moved with a speed of $2\ ms^{-1}$ in front of the open end of the pipe and parallel to it, the length of the pipe should be changed for the resonance to occur with the moving tuning fork. If the speed of sound in air is320 $ms^{-1}$, the smallest value of the percentage change required in the length of the pipe is __________.

**18.** A circular disc of radius $R$ carries surface charge density $\sigma(r) = \sigma_0\left(1 - \frac{r}{R}\right)$, where $\sigma_0$ is a constant and $r$ is the distance from the center of the disc. Electric flux through a large spherical surface that encloses the charged disc completely is $\phi_0$. Electric flux through another spherical surface of radius $\frac{R}{4}$ and concentric with the disc is $\phi$. Then the ratio $\frac{\phi_0}{\phi}$ is_________.

# CHEMISTRY

## Section 1 (Maximum Marks : 18)

- This section contains **SIX (06)** questions.
- Each question has **FOUR** options. **ONLY ONE** of these four options is the correct answer.
- For each question, choose the option corresponding to the correct answer.
- Answer to each question will be evaluated according to the following marking scheme:

*Full Marks* : **+3** If only the correct option is chosen.
*Zero Marks* : **0** If none of the options is chosen. (i.e. the question is unanswered);
*Negative Marks* : **−1** In all other cases.

**19.** If the distribution of molecular speeds of a gas is as per the figure shown below, then the ratio of the most probable, the average, and the root mean square speeds, respectively, is

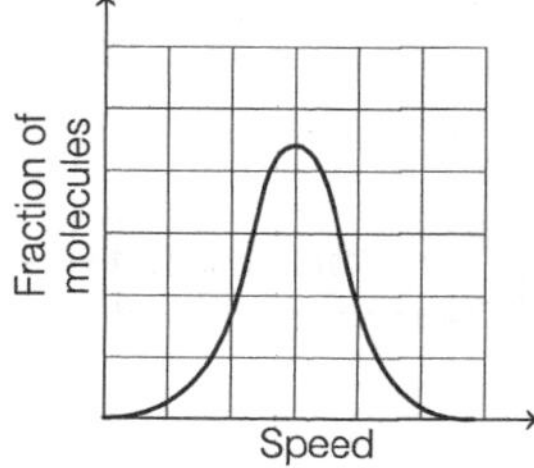


(a) 1 : 1 : 1 (b) 1 : 1 : 1.224
(c) 1 : 1.128 : 1.224 (d) 1 : 1.128 : 1

**20.** Which of the following liberates $O_2$ upon hydrolysis?

(a) $Pb_3O_4$ (b) $KO_2$
(c) $Na_2O_2$ (d) $Li_2O_2$

**21.** A colourless aqueous solution contains nitrates of two metals, $X$ and $Y$. When it was added to an aqueous solution of NaCl, a white precipitate was formed. This precipitate was found to be partly soluble in hot water to give a residue $P$ and a solution $Q$. The residue $P$ was soluble in aqueous $NH_3$ and also in excess sodium thiosulphate. The hot solution $Q$ gave a yellow precipitate with KI. The metals $X$ and $Y$, respectively, are

(a) Ag and Pb
(b) Ag and Cd
(c) Cd and Pb
(d) Cd and Zn

**22.** Newman projections $P$, $Q$, $R$ and $S$ are shown below :

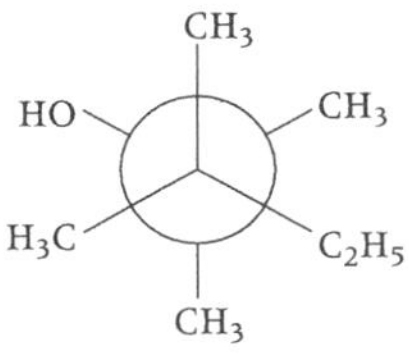

(P): front $CH_3$ (top), $CH_3$ (upper right), $C_2H_5$ (lower right); back HO (upper left), $H_3C$ (lower left), $CH_3$ (bottom)

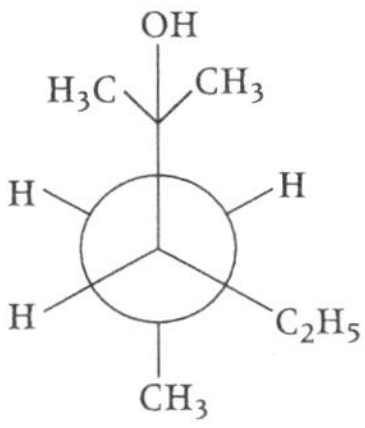

(Q): OH, $H_3C$, $CH_3$ (top); H (upper left), H (upper right), H (lower left), $C_2H_5$ (lower right), $CH_3$ (bottom)

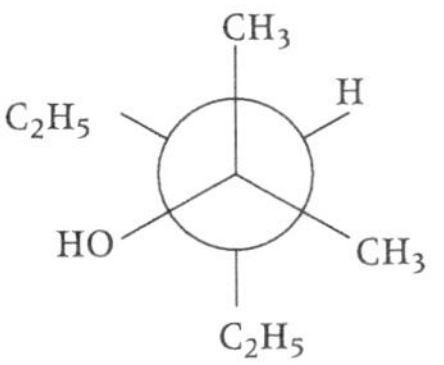

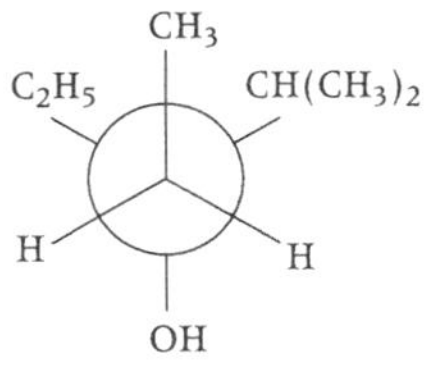

(R): $CH_3$ (top), $C_2H_5$, H, HO, $CH_3$, $C_2H_5$ (bottom)

(S): $CH_3$ (top), $C_2H_5$, $CH(CH_3)_2$, H, H, OH (bottom)

Which one of the following options represents identical molecules?

(a) $P$ and $Q$
(b) $Q$ and $S$
(c) $Q$ and $R$
(d) $R$ and $S$

**23.** Which one of the following structures has the IUPAC name 3-ethynyl-2-hydroxy-4-methylhex-3-en-5-ynoic acid?

(a) 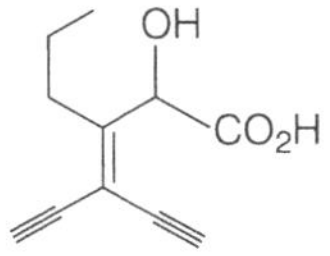 [structure: OH, $CO_2H$]
(b) 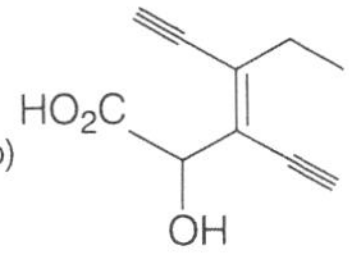 [structure: $HO_2C$, OH]
(c) 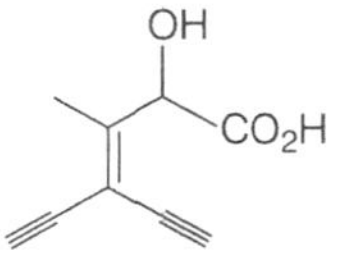 [structure: OH, $CO_2H$]
(d) 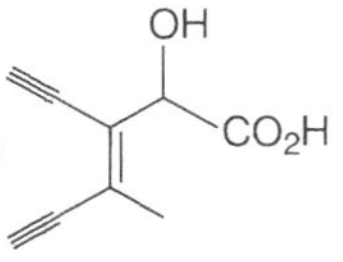 [structure: OH, $CO_2H$]

**24.** The Fischer projection of D-erythrose is shown below :

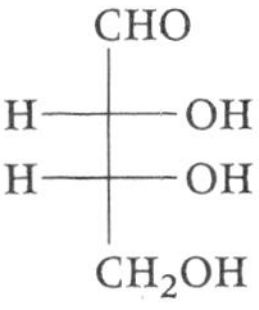

CHO
H—C—OH
H—C—OH
$CH_2OH$
D-erythrose

D-erythrose and its isomers are listed as P, Q, R, and S in Column-I. Choose the correct relationship of P, Q, R, and S with D-erythrose from Column II.

| | Column I | | Column II |
|---|---|---|---|
| P. | 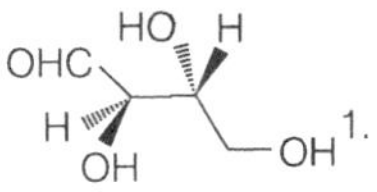 OHC, HO, H, H, OH, OH | 1. | Diastereomer |
| Q. | 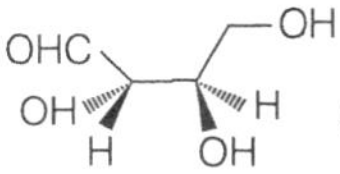 OHC, OH, OH, H, H, OH | 2. | Identical |
| R. | 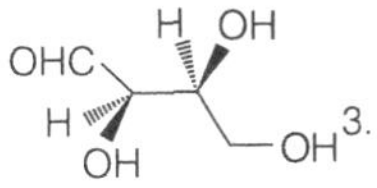 OHC, H, OH, H, OH, OH | 3. | Enantiomer |
| S. | 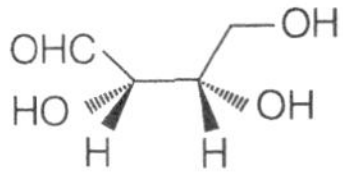 OHC, OH, HO, OH, H, H | | |

**Codes**

| | P | Q | R | S |
|---|---|---|---|---|
| (a) | 2 | 3 | 2 | 2 |
| (b) | 3 | 1 | 1 | 2 |
| (c) | 2 | 1 | 1 | 3 |
| (d) | 2 | 3 | 3 | 1 |

## Section 2 (Maximum Marks : 24)

- This section contains **SIX (06)** questions.
- Each question has **FOUR** options. **ONE OR MORE THAN ONE** of these four option(s) is (are) correct answer(s).
- For each question, choose the option(s) corresponding to (all) the correct answer(s).
- Answer to each question will be evaluated according to the following marking scheme:

*Full Marks* : **+4** If only (all) the correct option(s) is (are) chosen.
*Partial Marks* : **+3** If all the four options are correct but ONLY three options are chosen.
*Partial Marks* : **+2** If three or more options are correct but ONLY two options are chosen, both of which are correct.
*Partial Marks* : **+1** If two or more options are correct but ONLY one option is chosen and it is a correct option;
*Zero Marks* : **0** If none of the option is chosen (i.e., the question is unanswered);

***25.*** In thermodynamics, the $p$-$V$ work done is given by

$$w = -\int dV\, p_{\text{ext}}$$

For a system undergoing a particular process, the work done is

$$w = -\int dV\left(\frac{RT}{V-b} - \frac{a}{V^2}\right)$$

This equation is applicable to a

(a) system that satisfies the van der Waals' equation of state
(b) process that is reversible and isothermal
(c) process that is reversible and adiabatic
(d) process that is irreversible and at constant pressure

***26.*** With respect to the compounds I-V. Choose the correct statement(s).

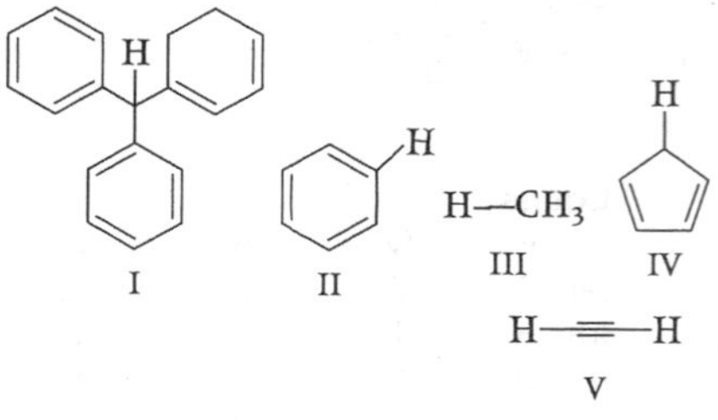

I II H—CH$_3$ III IV H—≡—H V

(a) The acidity of compound I is due to delocalisation in the conjugate base.
(b) The conjugate base of compound IV is aromatic.
(c) Compound II becomes more acidic, when it has a —NO$_2$ substituent.
(d) The acidity of compounds follows the order

I > IV > V > II > III.

***27.*** In the reaction scheme shown below $Q$, $R$ and $S$ are the major products.

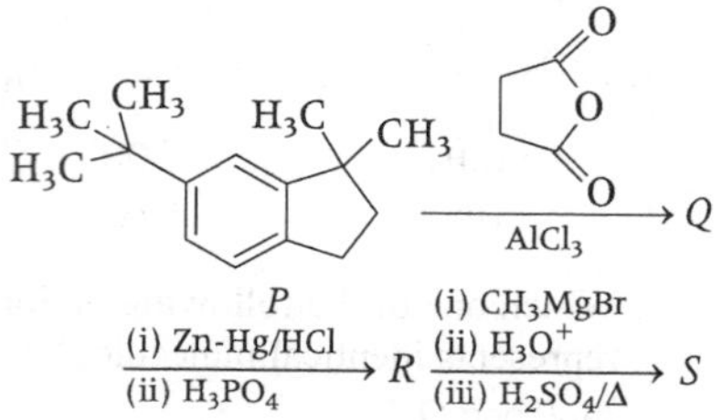

$P$ $\xrightarrow{\text{AlCl}_3}$ $Q$

$\xrightarrow[\text{(ii) H}_3\text{PO}_4]{\text{(i) Zn-Hg/HCl}}$ $R$ $\xrightarrow[\text{(iii) H}_2\text{SO}_4/\Delta]{\text{(i) CH}_3\text{MgBr, (ii) H}_3\text{O}^+}$ $S$

The correct structure of

(a) $S$ is

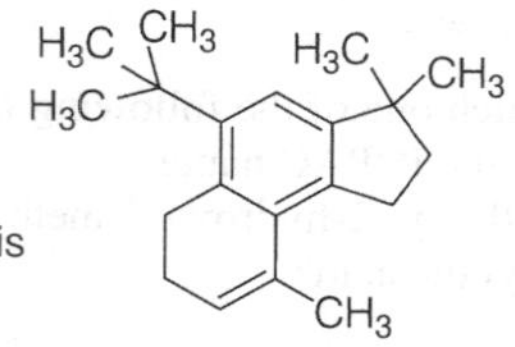

(b) $Q$ is

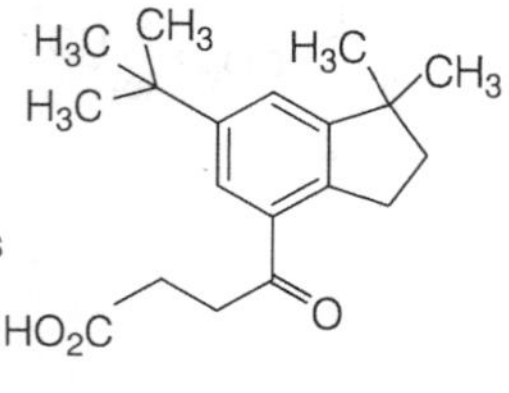

(c) $R$ is

(d) $S$ is

**28.** Choose the correct statement(s) among the following:

(a) $[FeCl_4]^-$ has tetrahedral geometry.

(b) $[Co(en)(NH_3)_2Cl_2]^+$ has 2 geometrical isomers.

(c) $[FeCl_4]^-$ has higher spin-only magnetic moment than $[Co(en)(NH_3)_2Cl_2]^+$.

(d) The cobalt ion in $[Co(en)(NH_3)_2Cl_2]^+$ has $sp^3d^2$ hybridisation.

**29.** With respect to hypochlorite, chlorate and perchlorate ions, choose the correct statement(s).

(a) The hypochlorite ion is the strongest conjugate base.

(b) The molecular shape of only chlorate ion is influenced by the lone pair of electrons of Cl.

(c) The hypochlorite and chlorate ions disproportionate to give rise to identical set of ions.

(d) The hypochlorite ion oxidises the sulphite ion.

**30.** The cubic unit cell structure of a compound containing cation $M$ and anion $X$ is shown below. When compared to the anion, the cation has smaller ionic radius. Choose the correct statement(s).

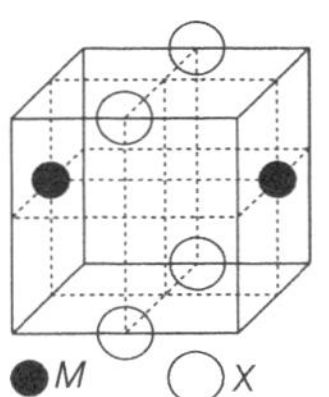


(a) The empirical formula of the compound is $MX$.

(b) The cation $M$ and anion $X$ have different coordination geometries.

(c) The ratio of $M - X$ bond length of the cubic unit cell edge length is 0.866.

(d) The ratio of the ionic radii of cation $M$ to anion $X$ is 0.414.

## Section 3 (Maximum Marks : 24)

- This section contains **SIX (06)** questions. The answer to each question is a **NUMERICAL VALUE.**
- Four each question, enter the correct numerical value of the answer using the mouse and the on-screen virtual numeric keypad in the place designated to enter the answer. If the numerical value has more than two decimal places, truncate/round-off the value to TWO decimal places.
- Answer to each question will be evaluated according to the following marking scheme:

Full Marks : **+4** If ONLY the correct numerical value is entered.

Zero Marks : **0** In all other cases.

**31.** 5.00 mL of 0.10 M oxalic acid solution taken in a conical flask is titrated against NaOH from a burette using phenolphthalein indicator. The volume of NaOH required for the appearance of permanent faint pink color is tabulated below for five experiments. What is the concentration, in molarity, of the NaOH solution?

| **Exp. No.** | **Vol. of NaOH** (mL) |
|---|---|
| 1 | 12.5 |
| 2 | 10.5 |
| 3 | 9.0 |
| 4 | 9.0 |
| 5 | 9.0 |

**32.** Consider the reaction, $A \rightleftharpoons B$ at 1000 K. At time $t'$, the temperature of the system was increased to 2000 K and the system was allowed to reach equilibrium. Throughout this experiment the partial pressure of $A$ was maintained at 1 bar. Given, below is the plot of the partial pressure of $B$ with time. What is the ratio of the standard Gibbs energy of the reaction at 1000 K to that at 2000 K?

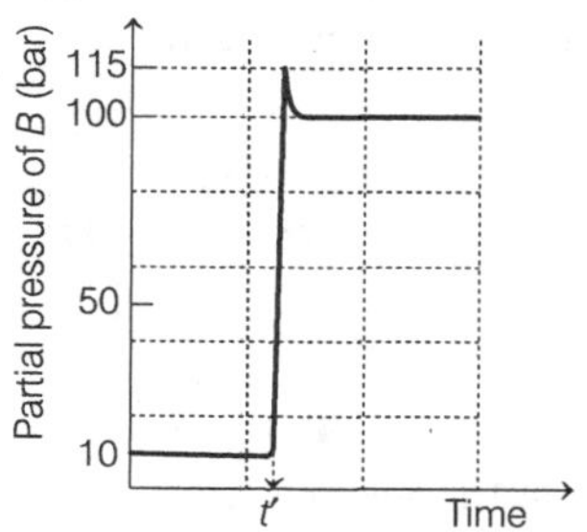


**33.** Consider a 70% efficient hydrogen-oxygen fuel cell working under standard conditions at 1 bar and 298 K. Its cell reaction is

$$H_2(g) + \frac{1}{2}O_2(g) \longrightarrow H_2O(l)$$

The work derived from the cell on the consumption of $1.0 \times 10^{-3}$ mole of $H_2(g)$ is used to compress 1.00 mole of a monoatomic ideal gas in a thermally insulated container. What is the change in the temperature (in K) of the ideal gas?

The standard reduction potentials for the two half-cells are given below:

$$O_2(g) + 4H^+(aq) + 4e^- \longrightarrow 2H_2O(l), \quad E° = 1.23 \text{ V},$$

$$2H^+(aq) + 2e^- \longrightarrow H_2(g), \quad E° = 0.00 \text{ V}$$

Use, $F = 96500$ C mol$^{-1}$, $R = 8.314$ J mol$^{-1}$ K$^{-1}$.

**34.** Aluminium reacts with sulphuric acid to form aluminium sulphate and hydrogen. What is the volume of hydrogen gas in litre (L) produced at 300 K and 1.0 atm pressure, when 5.4 g of aluminium and 50.0 mL of 5.0 M sulphuric acid are combined for the reaction?

(Use molar mass of aluminium as 27.0 g mol$^{-1}$, $R = 0.082$ atm L mol$^{-1}$ K$^{-1}$)

**35.** $^{238}_{92}U$ is known to undergo radioactive decay to form $^{206}_{82}Pb$ by emitting alpha and beta particles. A rock initially contained $68 \times 10^{-6}$ g of $^{238}_{92}U$. If the number of alpha particles that it would emit during its radioactive decay of $^{238}_{92}U$ to $^{206}_{82}Pb$ in three half-lives is $Z \times 10^{18}$, then what is the value of $Z$?

**36.** In the following reaction, compound $Q$ is obtained from compound $P$ *via* an ionic intermediate.

$CO_2CH_3$

$C_6H_5$ $C_6H_5$

$C_6H_5$

($P$)

$\xrightarrow[\text{}]{\text{Conc. } H_2SO_4}$ $Q$ (A coloured compound)

What is the degree of unsaturation of $Q$?

# MATHEMATICS

## Section 1 (Maximum Marks : 18)

- This section contains **SIX (06)** questions.
- Each question has **FOUR** options. **ONLY ONE** of these four options is the correct answer.
- For each question, choose the option corresponding to the correct answer.
- Answer to each question will be evaluated according to the following marking scheme:

| | | |
|---|---|---|
| *Full Marks* | : | **+3** If ONLY the correct options is chosen. |
| *Zero Marks* | : | **0** If none of the options is chosen. (i.e. the question is unanswered) |
| *Negative Marks* | : | **−1** In all other cases. |

***37.*** Suppose $a, b$ denote the distinct real roots of the quadratic polynomial $x^2 + 20x - 2020$ and suppose $c, d$ denote the distinct complex roots of the quadratic polynomial $x^2 - 20x + 2020$. Then the value of $ac(a-c) + ad(a-d) + bc(b-c) + bd(b-d)$ is

(a) 0 (b) 8000
(c) 8080 (d) 16000

***38.*** If the function $f: \mathbf{R} \to \mathbf{R}$ is defined by $f(x) = |x|(x - \sin x)$, then which of the following statements is TRUE?

(a) $f$ is one-one, but NOT onto
(b) $f$ is onto, but NOT one-one
(c) $f$ is BOTH one-one and onto
(d) $f$ is NEITHER one-one NOR onto

***39.*** Let the functions $f: \mathbf{R} \to \mathbf{R}$ and $g: \mathbf{R} \to \mathbf{R}$ be defined by

$$f(x) = e^{x-1} - e^{-|x-1|}$$

and $g(x) = \frac{1}{2}(e^{x-1} + e^{1-x})$.

Then the area of the region in the first quadrant bounded by the curves $y = f(x)$, $y = g(x)$ and $x = 0$ is

(a) $(2 - \sqrt{3}) + \frac{1}{2}(e - e^{-1})$

(b) $(2 + \sqrt{3}) + \frac{1}{2}(e - e^{-1})$

(c) $(2 - \sqrt{3}) + \frac{1}{2}(e + e^{-1})$

(d) $(2 + \sqrt{3}) + \frac{1}{2}(e + e^{-1})$

***40.*** Let $a, b$ and $\lambda$ be positive real numbers. Suppose $P$ is an end point of the latus rectum of the parabola $y^2 = 4\lambda x$, and suppose the ellipse $\frac{x^2}{a^2} + \frac{y^2}{b^2} = 1$ passes through the point $P$. If the tangents to the parabola and the ellipse at the point $P$ are perpendicular to each other, then the eccentricity of the ellipse is

(a) $\frac{1}{\sqrt{2}}$ (b) $\frac{1}{2}$
(c) $\frac{1}{3}$ (d) $\frac{2}{5}$

***41.*** Let $C_1$ and $C_2$ be two biased coins such that the probabilities of getting head in a single toss are $\frac{2}{3}$ and $\frac{1}{3}$, respectively. Suppose $\alpha$ is the number of heads that appear when $C_1$ is tossed twice, independently, and suppose $\beta$ is the number of heads that appear when $C_2$ is tossed twice, independently. Then the probability that the roots of the quadratic polynomial $x^2 - \alpha x + \beta$ are real and equal, is

(a) $\frac{40}{81}$ (b) $\frac{20}{81}$
(c) $\frac{1}{2}$ (d) $\frac{1}{4}$

**42.** Consider the rectangles lying in the region

$$\left\{(x, y) \in R \times R : 0 \le x \le \frac{\pi}{2} \text{ and } 0 \le y \le 2\sin(2x)\right\}$$

and having one side on the $X$-axis. The area of the rectangle which has the maximum perimeter among all such rectangles, is

(a) $\frac{3\pi}{2}$ (b) $\pi$ (c) $\frac{\pi}{2\sqrt{3}}$ (d) $\frac{\pi\sqrt{3}}{2}$

## Section 2 (Maximum Marks : 24)

- This section contains **SIX (06)** questions.
- Each question has **FOUR** options. **ONE OR MORE THAN ONE** of these four option(s) is (are) correct answer(s).
- For each question, choose the options(s) corresponding to (all) the correct answer(s).
- Answer to each question will be evaluated according to the following marking scheme:

| | | |
|---|---|---|
| *Full Marks* | : | **+4** If only (all) the correct option(s) is (are) chosen. |
| *Partial Marks* | : | **+3** If all the four options are correct but ONLY three options are chosen. |
| *Partial Marks* | : | **+2** If three or more options are correct but ONLY two options are chosen, both of which are correct. |
| *Partial Marks* | : | **+1** If two or more options are correct but ONLY one option is chosen and it is a correct option; |
| *Zero Marks* | : | **0** If none of the options is chosen (i.e., the question is unanswered); |
| *Negative Marks* | : | **−2** In all other cases. |

**43.** Let the function $f : \mathbf{R} \to \mathbf{R}$ be defined by $f(x) = x^3 - x^2 + (x-1)\sin x$ and let $g : \mathbf{R} \to \mathbf{R}$ be an arbitrary function. Let $fg : \mathbf{R} \to \mathbf{R}$ be the product function defined by $(fg)(x) = f(x)g(x)$. Then which of the following statements is/are TRUE?

(a) If $g$ is continuous at $x = 1$, then $fg$ is differentiable at $x = 1$

(b) If $fg$ is differentiable at $x = 1$, then $g$ is continuous at $x = 1$

(c) If $g$ is differentiable at $x = 1$, then $fg$ is differentiable at $x = 1$

(d) If $fg$ is differentiable at $x = 1$, then $g$ is differentiable at $x = 1$

**44.** Let $M$ be a $3 \times 3$ invertible matrix with real entries and let $I$ denote the $3 \times 3$ identity matrix. If $M^{-1} = \text{adj}(\text{adj}\, M)$, then which of the following statements is/are ALWAYS TRUE?

(a) $M = I$ (b) $\det M = 1$

(c) $M^2 = I$ (d) $(\text{adj } M)^2 = I$

**45.** Let $S$ be the set of all complex numbers $z$ satisfying $|z^2 + z + 1| = 1$. Then which of the following statements is/are TRUE?

(a) $\left|z + \frac{1}{2}\right| \le \frac{1}{2}$ for all $z \in S$

(b) $|z| \le 2$ for all $z \in S$

(c) $\left|z + \frac{1}{2}\right| \ge \frac{1}{2}$ for all $z \in S$

(d) The set $S$ has exactly four elements

**46.** Let $x, y$ and $z$ be positive real numbers. Suppose $x, y$ and $z$ are the lengths of the sides of a triangle opposite to its angles $X, Y$ and $Z$, respectively. If

$$\tan\frac{X}{2} + \tan\frac{Z}{2} = \frac{2y}{x + y + z},$$

then which of the following statements is/are TRUE?

(a) $2Y = X + Z$ (b) $Y = X + Z$

(c) $\tan\frac{X}{2} = \frac{x}{y + z}$ (d) $x^2 + z^2 - y^2 = xz$

**47.** Let $L_1$ and $L_2$ be the following straight lines.

$$L_1 : \frac{x-1}{1} = \frac{y}{-1} = \frac{z-1}{3} \text{ and}$$

$$L_2 : \frac{x-1}{-3} = \frac{y}{-1} = \frac{z-1}{1}.$$

Suppose the straight line

$$L : \frac{x-\alpha}{l} = \frac{y-1}{m} = \frac{z-\gamma}{-2}$$

lies in the plane containing $L_1$ and $L_2$ and passes through the point of intersection of $L_1$ and $L_2$. If the line $L$ bisects the acute angle between the lines $L_1$ and $L_2$, then which of the following statements is/are TRUE?

(a) $\alpha - \gamma = 3$ (b) $l + m = 2$
(c) $\alpha - \gamma = 1$ (d) $l + m = 0$

**48.** Which of the following inequalities is/are TRUE?

(a) $\int_0^1 x\cos x dx \geq \frac{3}{8}$

(b) $\int_0^1 x\sin x dx \geq \frac{3}{10}$

(c) $\int_0^1 x^2 \cos x dx \geq \frac{1}{2}$

(d) $\int_0^1 x^2 \sin x dx \geq \frac{2}{9}$

## Section 3 (Maximum Marks : 24)

- This section contains **SIX (06)** questions. The answer to each question is a **NUMERICAL VALUE.**
- For each question, enter the correct numerical value of the answer using the mouse and the on-screen virtual numeric keypad in the place designated to enter the answer. If the numerical value has more than two decimal places, **truncate/round-off** the value to **TWO** decimal places.
- Answer to each question will be evaluated according to the following marking scheme:

Full Marks : **+4** If ONLY the correct numerical value is entered.
Zero Marks : **0** In all other cases.

**49.** Let $m$ be the minimum possible value of $\log_3(3^{y_1} + 3^{y_2} + 3^{y_3})$, where $y_1, y_2, y_3$ are real numbers for which $y_1 + y_2 + y_3 = 9$. Let $M$ be the maximum possible value of $(\log_3 x_1 + \log_3 x_2 + \log_3 x_3)$, where $x_1, x_2, x_3$ are positive real numbers for which $x_1 + x_2 + x_3 = 9$. Then the value of $\log_2(m^3) + \log_3(M^2)$ is...... .

**50.** Let $a_1, a_2, a_3, ....$ be a sequence of positive integers in arithmetic progression with common difference 2. Also, let $b_1, b_2, b_3, ....$ be a sequence of positive integers in geometric progression with common ratio 2. If $a_1 = b_1 = c$, then the number of all possible values of $c$, for which the equality $2(a_1 + a_2 + ... + a_n) = b_1 + b_2 + ... + b_n$ holds for some positive integer $n$, is ......

**51.** Let $f : [0, 2] \to \mathbf{R}$ be the function defined by

$$f(x) = (3 - \sin(2\pi x))\sin\left(\pi x - \frac{\pi}{4}\right) - \sin\left(3\pi x + \frac{\pi}{4}\right).$$

If $\alpha, \beta \in [0, 2]$ are such that $\{x \in [0,2] : f(x) \geq 0\} = [\alpha, \beta]$, then the value of $\beta - \alpha$ is ......

**52.** In a triangle $PQR$, let $\mathbf{a} = \mathbf{QR}$, $\mathbf{b} = \mathbf{RP}$ and $\mathbf{c} = \mathbf{PQ}$. If $|\mathbf{a}| = 3$, $|\mathbf{b}| = 4$ and $\frac{\mathbf{a}.(\mathbf{c} - \mathbf{b})}{\mathbf{c}.(\mathbf{a} - \mathbf{b})} = \frac{|\mathbf{a}|}{|\mathbf{a}| + |\mathbf{b}|}$, then the value of $|\mathbf{a} \times \mathbf{b}|^2$ is ...... .

**53.** For a polynomial $g(x)$ with real coefficients, let $m_g$ denote the number of distinct real roots of $g(x)$.

Suppose $S$ is the set of polynomials with real coefficients defined by
$$S = \{(x^2-1)^2(a_0 + a_1x + a_2x^2 + a_3x^3) : a_0, a_1, a_2, a_3 \in \mathbf{R}\};$$
For a polynomial $f$, let $f'$ and $f''$ denote its first and second order derivatives, respectively. Then the minimum possible value of $(m_{f'} + m_{f''})$, where $f \in S$, is .........

**54.** Let $e$ denote the base of the natural logarithm. The value of the real number $a$ for which the right hand limit
$$\lim_{x \to 0^+} \frac{(1-x)^{\frac{1}{x}} - e^{-1}}{x^a}$$
is equal to a non zero real number, is ...... .

# Paper 2

## PHYSICS

### Section 1 (Maximum Marks : 18)

- This section contains **SIX (06)** questions.
- The answer to each question is a **SINGLE DIGIT INTEGER** ranging from 0 to 9, **BOTH INCLUSIVE**.
- For each question, enter the correct integer corresponding to the answer using the mouse and the on-screen virtual numeric keypad in the place designated to enter the answer.
- Answer to each question will be evaluated according to the following marking scheme:

*Full Marks* : **+3** If only (all) the correct option(s) is (are) chosen.
*Zero Marks* : **0** If the question is unanswered).
*Negative Marks* : **−1** In all other cases.

**1.** A large square container with thin transparent vertical walls and filled with water (refractive index $\frac{4}{3}$) is kept on a horizontal table. A student holds a thin straight wire vertically inside the water 12 cm from one of its corners, as shown schematically in the figure. Looking at the wire from this corner, another student sees two images of the wire, located symmetrically on each side of the line of sight as shown. The separation (in cm) between these images is ___________.

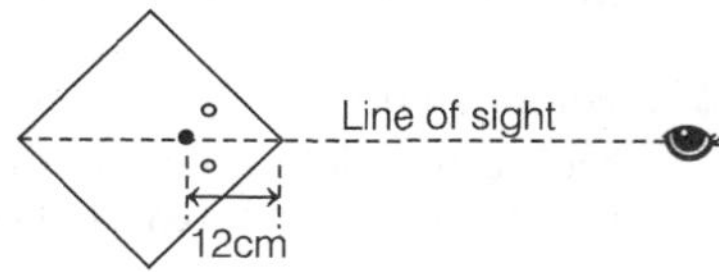


**2.** A train with cross-sectional area $S_t$ is moving with speed $v_0$ inside a long tunnel of cross-sectional area $S_0(S_0 = 4S_t)$. Assume that almost all the air (density $\rho$) in front of the train flows back between its sides and the walls of the tunnel. Also, the air flow with respect to the train is steady and laminar. Take the ambient pressure and that inside the train to be $p_0$. If the pressure in the region between the sides of the train and the tunnel walls is $p$ then $p_0 - p = \frac{7}{2N}\rho v_t^2$. The value of $N$ is ________.

**3.** Two large circular discs separated by a distance of 0.01 m are connected to a battery via a switch as shown in the figure. Charged oil drops of density 900 kg m$^{-3}$ are released through a tiny hole at the center of the top disc. Once some oil drops achieve terminal velocity, the switch is closed to apply a voltage of 200 V across the discs. As a result, an oil drop of radius $8 \times 10^{-7}$ m stops moving vertically and floats between the discs. The number of electrons present in this oil drop is ________. (neglect the buoyancy force, take acceleration due to gravity $= 10\ ms^{-2}$ and charge on an electron $(e)$ $= 1.6 \times 10^{-19}$ C)

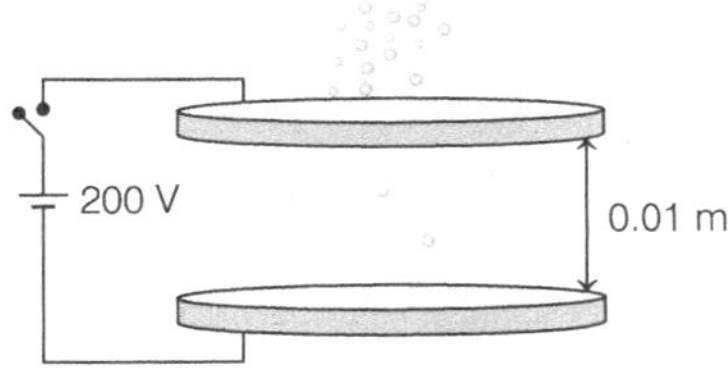


**4.** A hot air balloon is carrying some passengers, and a few sandbags of mass 1 kg each so that its total mass is 480 kg. Its effective volume giving the balloon its buoyancy is $V$. The balloon is floating at an equilibrium height of 100 m. When $N$ number of sandbags are thrown out, the balloon rises to a new equilibrium height close to 150 m with its volume $V$ remaining unchanged. If the variation of the density of air with height h from the ground is $pho(h) = \rho_0 e^{-\frac{h}{h_0}}$, where $\rho_0 = 1.25$ kg $m^{-3}$ and $h_0 = 6000$ m, the value of $N$ is _________.

**5.** A point charge $q$ of mass $m$ is suspended vertically by a string of length $l$. A point dipole of dipole moment **p** is now brought towards $q$ from infinity so that the charge moves away. The final equilibrium position of the system including the direction of the dipole, the angles and distances is shown in the figure below. If the work done in bringing the dipole to this position is $N \times (mgh)$, where $g$ is the acceleration due to gravity, then the value of $N$ is _________. (Note that for three coplanar forces keeping a point mass in equilibrium, $\frac{F}{\sin\theta}$ is the same for all forces, where $F$ is any one of the forces and $\theta$ is the angle between the other two forces)

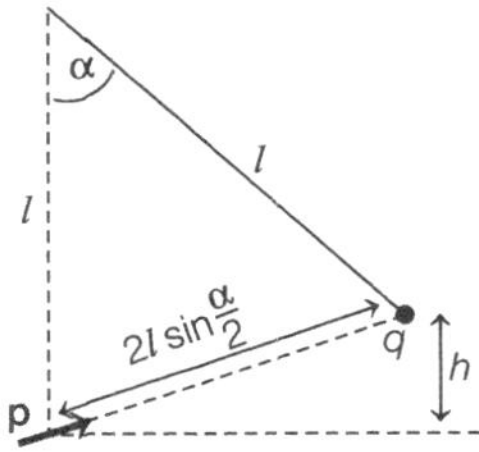


**6.** A thermally isolated cylindrical closed vessel of height 8 m is kept vertically. It is divided into two equal parts by a diathermic (perfect thermal conductor) frictionless partition of mass 8.3 kg. Thus the partition is held initially at a distance of 4 m from the top, as shown in the schematic figure below. Each of the two parts of the vessel contains 0.1 mole of an ideal gas at temperature 300 K. The partition is now released and moves without any gas leaking from one part of the vessel to the other. When equilibrium is reached, the distance of the partition from the top (in m) will be ________ (take the acceleration due to gravity $= 10\ ms^{-2}$ and the universal gas constant$= 9.3$ J $mol^{-1}$ $K^{-1}$).

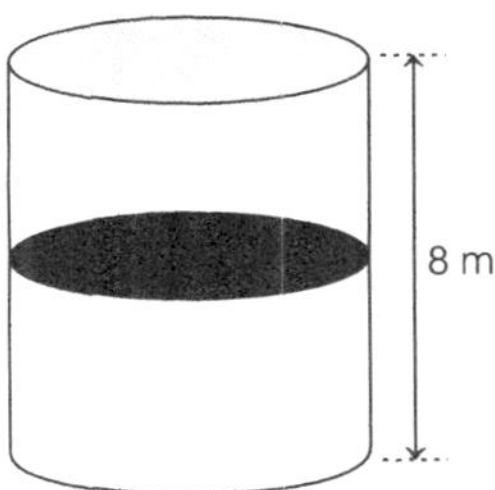

## Section 2 (Maximum Marks : 24)

- This section contains **SIX (06)** questions.
- Each question has **FOUR** options. **ONE OR MORE THAN ONE** of these four option(s) is (are) correct options(s).
- For each question, choose the options(s) corresponding to (all) the correct answer(s).
- Answer to each question will be evaluated according to the following marking scheme:

*Full Marks* : **+4** If only (all) the correct option(s) is (are) chosen.
*Partial Marks* : **+3** If all the four options are correct but ONLY three options are chosen.
*Partial Marks* : **+2** If three or more options are correct but ONLY two options are chosen, both of which are correct.
*Partial Marks* : **+1** If two or more options are correct but ONLY one option is chosen and it is a correct option;
*Zero Marks* : **0** If none of the option is chosen (i.e., the question is unanswered);
*Negative Marks* : **−2** In all other cases.

**7.** A beaker of radius $r$ is filled with water$\left(\text{refractive indexd } \frac{4}{3}\right)$ up to a height $H$ as shown in the figure on the left. The beaker is kept on a horizontal table rotating with angular speed $\omega$. This makes the water surface curved so that the difference in the height of water level at the center and at the circumference of the beaker is $h(h << H, h << r)$, as shown in the figure on the right. Take this surface to be approximately spherical with a radius of curvature $R$. Which of the following is/are correct? ($g$ is the acceleration due to gravity)

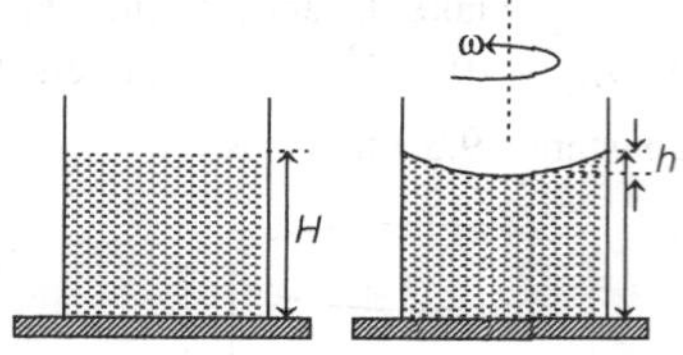


(a) $R = \frac{h^2 + r^2}{2h}$

(b) $R = \frac{3r^2}{2h}$

(c) Apparent depth of the bottom of the beaker is close to $\frac{3H}{2}\left(1 + \frac{\omega^2 H}{2g}\right)^{-1}$

(d) Apparent depth of the bottom of the beaker is close to $\frac{3H}{4}\left(1+ \frac{\omega^2 H}{4g}\right)^{-1}$

**8.** A student skates up a ramp that makes an angle 30° with the horizontal. He/she starts (as shown in the figure) at the bottom of the ramp with speed $v_0$ and wants to turn around over a semicircular path $xyz$ of radius $R$ during which he/she reaches a maximum height $h$ (at point $y$) from the ground as shown in the figure. Assume that the energy loss is negligible and the force required for this turn at the highest point is provided by his/her weight only. Then ($g$ is the acceleration due to gravity)

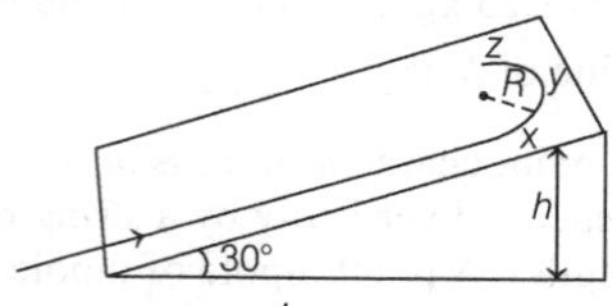


(a) $v_0^2 - 2gh = \frac{1}{2}gR$

(b) $v_0^2 - 2gh = \frac{\sqrt{3}}{2}gR$

(c) the centripetal force required at points $x$ and $z$ is zero

(d) the centripetal force required is maximum at points $x$ and $z$

**9.** A rod of mass $m$ and length $L$, pivoted at one of its ends, is hanging vertically. A bullet of the same mass moving at speed $v$ strikes the rod horizontally at a distance $x$ from its pivoted end and gets embedded in it. The combined system now rotates with angular speed $\omega$ about the pivot. The maximum angular speed $\omega_M$ is achieved for $x = x_M$. Then

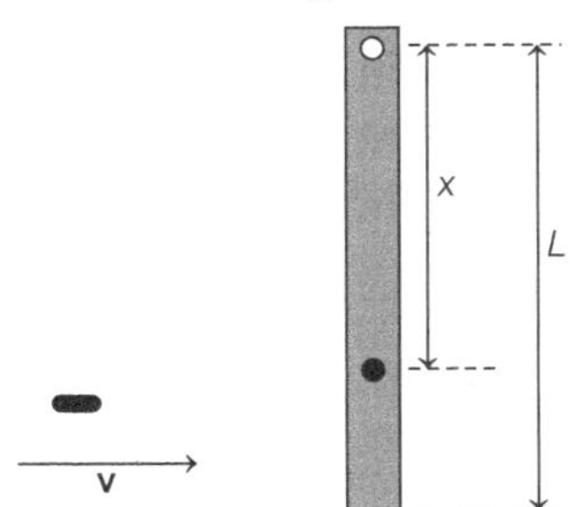


(a) $\omega = \dfrac{3vx}{L^2 + 3x^2}$
(b) $\omega = \dfrac{12vx}{L^2 + 12x^2}$
(c) $x_M = \dfrac{L}{\sqrt{3}}$
(d) $\omega_M = \dfrac{v}{2L}\sqrt{3}$

**10.** In an $X$-ray tube, electrons emitted from a filament (cathode) carrying current $I$ hit a target (anode) at a distance $d$ from the cathode. The target is kept at a potential $V$ higher than the cathode resulting in emission of continuous and characteristic $X$-rays. If the filament current $I$ is decreased to $\dfrac{I}{2}$, the potential difference $V$ is increased to $2V$, and the separation distance $d$ is reduced to $\dfrac{d}{2}$, then

(a) the cut-off wavelength will reduce to half, and the wavelengths of the characteristic $X$-rays will remain the same
(b) the cut-off wavelength as well as the wavelengths of the characteristic $X$-rays will remain the same
(c) the cut-off wavelength will reduce to half, and the intensities of all the $X$-rays will decrease
(d) the cut-off wavelength will become two times larger, and the intensity of all the $X$-rays will decrease

**11.** Two identical non-conducting solid spheres of same mass and charge are suspended in air from a common point by two non-conducting, massless strings of same length. At equilibrium, the angle between the strings is $\alpha$. The spheres are now immersed in a dielectric liquid of density 800 kg m$^{-3}$ and dielectric constant 21. If the angle between the strings remains the same after the immersion, then

(a) electric force between the spheres remains unchanged
(b) electric force between the spheres reduces
(c) mass density of the spheres is 840 kg m$^{-3}$
(d) the tension in the strings holding the spheres remains unchanged

**12.** Starting at time $t = 0$ from the origin with speed 1 ms$^{-1}$, a particle follows a two-dimensional trajectory in the $x$-$y$ plane so that its coordinates are related by the equation $y = \dfrac{x^2}{2}$. The $x$ and $y$ components of its accel- eration are denoted by $a_x$ and $a_y$, respectively. Then

(a) $a_x = 1\,\text{ms}^{-2}$ implies that when the particle is at the origin, $a_y = 1\,\text{ms}^{-2}$
(b) $a_x = 0$ implies $a_y = 1\,\text{ms}^{-2}$ at all times
(c) at $t = 0$, the particle's velocity points in the $x$-direction
(d) $a_x = 0$ implies that at $t = 1$ s, the angle between the particle's velocity and the $x$ axis is 45°

## Section 3 (Maximum Marks : 24)

- This section contains **SIX (06)** questions. The answer to each question is a **NUMERICAL VALUE.**
- Four each question, enter the correct numerical value of the answer using the mouse and the on-screen virtual numeric keypad in the place designated to enter the answer. If the numerical value has more than two decimal places, truncate/round-off the value to TWO decimal places.
- Answer to each question will be evaluated according to the following marking scheme:

*Full Marks* : **+4** If ONLY the correct numerical value is entered.
*Zero Marks* : **0** In all other cases.

**13.** A spherical bubble inside water has radius $R$. Take the pressure inside the bubble and the water pressure to be $p_0$. The bubble now gets compressed radially in an adiabatic manner so that its radius becomes $(R - a)$. For $a << R$ the magnitude of the work done in the process is given by $(4\pi p_0 Ra^2)X$, where $X$ is a constant and $y = C_p/C_V = 41/30$. The value of $X$ is________.

**14.** In the balanced condition, the values of the resistances of the four arms of a Wheatstone bridge are shown in the figure below. The resistance $R_3$ has temperature coefficient $0.0004°C^{-1}$. If the temperature of $R_3$ is increased by $100°C$, the voltage developed between $S$ and $T$ will be ______ volt.

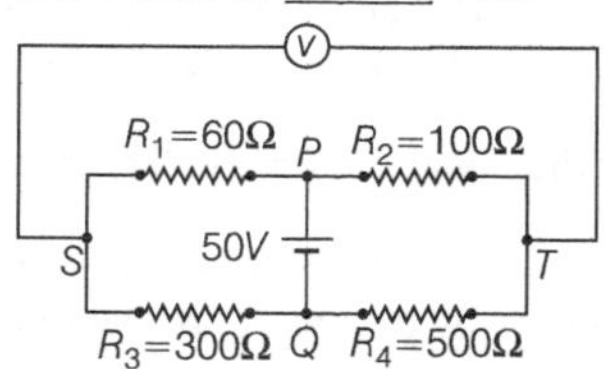


**15.** Two capacitors with capacitance values $C_1 = 2000 \pm 10$pF and $C_2 = 3000 \pm 15$ pF are connected in series. The voltage applied across this combination is $C = 5.00 \pm 0.02$ V. The percentage error in the calculation of the energy stored in this combination of capacitors is _______.

**16.** A cubical solid aluminium (bulk modulus $= -V\dfrac{dP}{dV} = 70\,\text{GPa}$) block has an edge length of 1 m on the surface of the earth. It is kept on the floor of a 5 km deep ocean. Taking the average density of water and the acceleration due to gravity to be $10^3$ kg m$^{-3}$ and 10 ms$^{-2}$, respectively, the change in the edge length of the block in mm is _____.

**17.** The inductors of two $LR$ circuits are placed next to each other, as shown in the figure. The values of the self-inductance of the inductors, resistances, mutual-inductance and applied voltages are specified in the given circuit. After both the switches are closed simultaneously, the total work done by the batteries against the induced $EMF$ in the inductors by the time the currents reach their steady state values is_______ mJ.

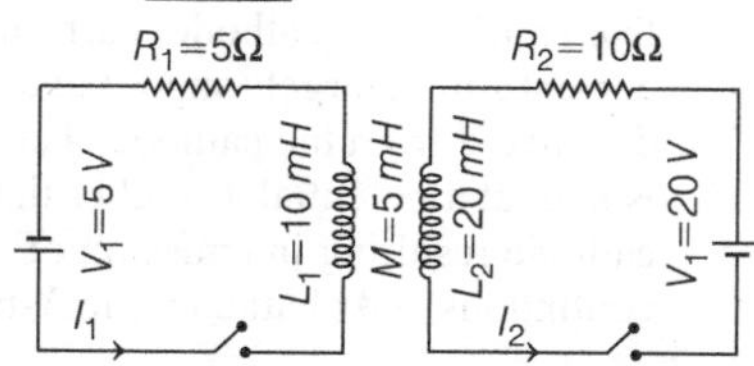


**18.** A container with 1 kg of water in it is kept in sunlight, which causes the water to get warmer than the surroundings. The average energy per unit time per unit area received due to the sunlight is 700 Wm$^{-2}$ and it is absorbed by the water over an effective area of $0.05$ m$^2$. Assuming that the heat loss from the water to the surroundings is governed by Newton's law of cooling, the difference (in °C) in the temperature of water and the surroundings after a long time will be ______. (Ignore effect of the container and take constant for Newton's law of cooling = 0.001 s$^{-1}$, Heat capacity of water = 4200 J kg$^{-1}$K$^{-1}$)

# CHEMISTRY

## Section 1 (Maximum Marks : 18)

- This section contains **SIX (06)** questions.
- The answer to each question is a **SINGLE DIGIT INTEGER** ranging from 0 TO 9. **BOTH INCLUSIVE**.
- For each question, enter the correct integer corresponding to the answer using the mouse and the on-screen virtual numeric keypad in the place designated to enter the answer.
- Answer to each question will be evaluated according to the following marking scheme:

*Full Marks* : **+3** If ONLY the correct integer is entered.

*Zero Marks* : **0** If the question is unanswered.

*Negative Marks* : **−1** In all other cases.

***19.*** The 1st, 2nd and 3rd ionisation enthalpies, $I_1$, $I_2$, and $I_3$, of four atoms with atomic numbers $n$, $n + 1$, $n + 2$, and $n + 3$, where $n < 10$, are tabulated below. What is the value of $n$?

| Atomic number | Ionisation enthalpy (kJ/mol) | | |
|---|---|---|---|
| | $I_1$ | $I_2$ | $I_3$ |
| $n$ | 1681 | 3374 | 6050 |
| $n + 1$ | 2081 | 3952 | 6122 |
| $n + 2$ | 496 | 4562 | 6910 |
| $n + 3$ | 738 | 1451 | 7733 |

***20.*** Consider the following compounds in the liquid form :

$O_2$, $HF$, $H_2O$, $NH_3$, $H_2O_2$, $CCl_4$, $CHCl_3$, $C_6H_6$, $C_6H_5Cl$

When a charged comb is brought near their flowing stream, how many of them show deflection as per the following figure?

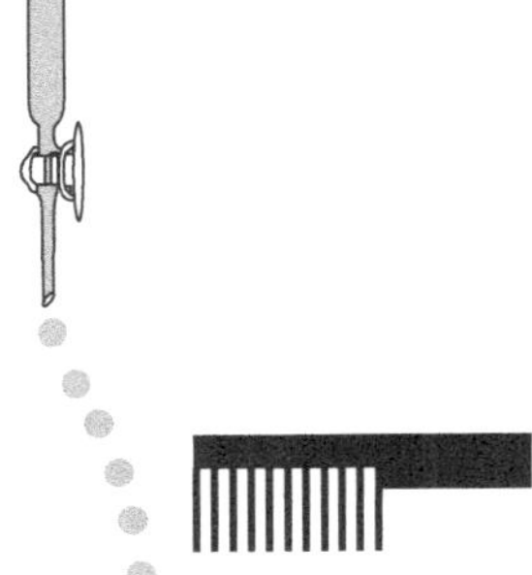

***21.*** In the chemical reaction between stoichiometric quantities of $KMnO_4$ and KI in weakly basic solution, what is the number of moles of $I_2$ released for 4 moles of $KMnO_4$ consumed?

***22.*** An acidified solution of potassium chromate was layered with an equal volume of amyl alcohol. When it was shaken after the addition of 1 mL of 3% $H_2O_2$, a blue alcohol layer was obtained. The blue colour is due to the formation of a chromium (VI) compound '$X$'. What is the number of oxygen atoms bonded to chromium through only single bond in a molecule of $X$?

***23.*** The structure of a peptide is given below.

If the absolute values of the net charge of the peptide at pH = 2, pH = 6, and pH = 11 are and $|Z_3|$, respectively, then what is $|Z_1| + |Z_2| + |Z_3|$?

***24.*** An organic compound ($C_8H_{10}O_2$) rotates plane-polarised light. It produces pink color with neutral $FeCl_3$ solution. What is the total number of all the possible isomers for this compound?

## Section 2 (Maximum Marks : 24)

- This section contains **SIX (06)** questions.
- Each question has **FOUR** options. **ONE OR MORE THAN ONE** of these four option(s) is (are) correct asnswer(s).
- For each question, choose the option(s) corresponding to (all) the correct answer(s).
- Answer to each question will be evaluated according to the following marking scheme:

| | | |
|---|---|---|
| *Full Marks* | : | **+4** If only (all) the correct option(s) is (are) chosen. |
| *Partial Marks* | : | **+3** If all the four options are correct but ONLY three options are chosen. |
| *Partial Marks* | : | **+2** If three or more options are correct but ONLY two options are chosen, both of which are correct. |
| *Partial Marks* | : | **+1** If two or more options are correct but ONLY one option is chosen and it is a correct option. |
| *Zero Marks* | : | **0** If none of the options is chosen (i.e. the question is unanswered). |
| *Negative Marks* | : | **−2** In all other cases. |

**25.** In an experiment, *m* grams of a compound *X* (gas/liquid/solid) taken in a container is loaded in a balance as shown in figure I below.

In the presence of a magnetic field, the pan with *X* is either deflected upwards (figure II), or deflected downwards (figure III), depending on the compound *X*. Identify the correct statement(s).

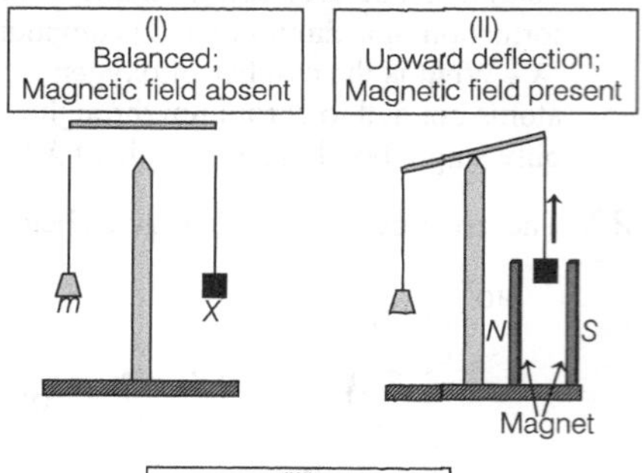


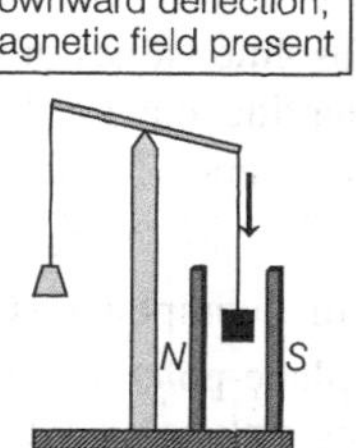


(a) If *X* is $H_2O(l)$, deflection of the pan is upwards.

(b) If *X* is $K_4[Fe(CN)_6](s)$, deflection of the pan is upwards.

(c) If *X* is $O_2(g)$, deflection of the pan is downwards.

(d) If *X* is $C_6H_6(l)$, deflection of the pan is downwards.

**26.** Which of the following plots is(are) correct for the given reaction?

($[P]_0$ is the initial concentration of *P*)

$$\underset{(P)}{(CH_3)_3C\text{—}Br} + NaOH \longrightarrow \underset{(Q)}{(CH_3)_3C\text{—}OH} + NaBr$$

(a)

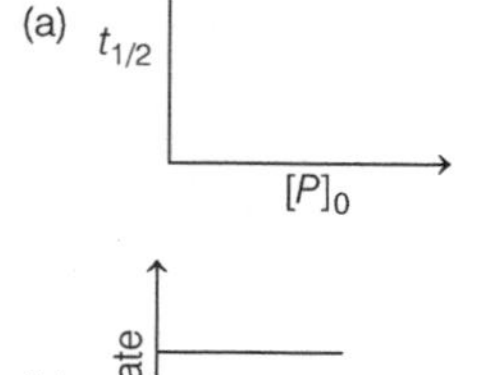


(b)

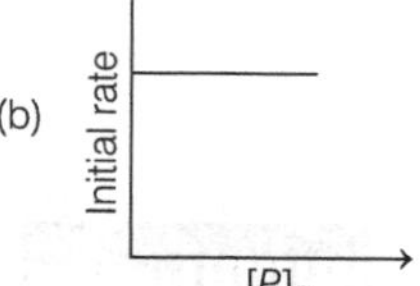

(c)

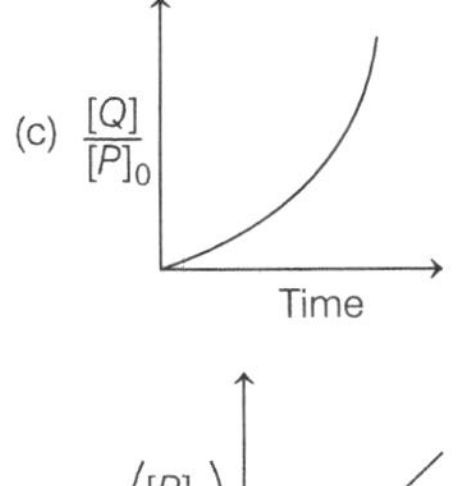


(d) $\ln\left(\frac{[P]}{[P]_0}\right)$

Time

**27.** Which among the following statement(s) is(are) true for the extraction of aluminium from bauxite?

(a) Hydrated $Al_2O_3$ precipitates, when $CO_2$ is bubbled through a solution of sodium aluminate.

(b) Addition of $Na_3AlF_6$ lowers the melting point of alumina.

(c) $CO_2$ is evolved at the anode during electrolysis.

(d) The cathode is a steel vessel with a lining of carbon.

**28.** Choose the correct statement(s) among the following:

(a) $SnCl_2 \cdot 2H_2O$ is a reducing agent.

(b) $SnO_2$ reacts with KOH to form $K_2[Sn(OH)_6]$.

(c) A solution of $PbCl_2$ in HCl contains $Pb^{2+}$ and $Cl^-$ ions.

(d) The reaction of $Pb_3O_4$ with hot dilute nitric acid to give $PbO_2$ is a redox reaction.

**29.** Consider the following four compounds, I, II, III, and IV.

$NH_2$ — I

$H_3C$ N $CH_3$ — II

$NH_2$, $O_2N$, $NO_2$, $NO_2$ — III

$H_3C$ N $CH_2$, $O_2N$, $NO_2$, $NO_2$ — IV

Choose the correct statement(s).

(a) The order of basicity is II > I > III > IV.

(b) The magnitude of $pK_b$ difference between I and II is more than that between III and IV.

(c) Resonance effect is more in III than in IV.

(d) Steric effect makes compound IV more basic than III.

**30.** Consider the following transformations of a compound *P*.

$R$ (Optically active) $\xleftarrow[\text{(ii) } C_6H_5COCH_3 \quad \text{(iii) } H_3O^+/\Delta]{\text{(i) } NaNH_2}$ $P$ $(C_9H_{12})$ $\xrightarrow[\text{(ii) } KMnO_4/H_2SO_4/\Delta]{\text{(i) } X \text{ (reagent)}}$ $Q$ $(C_8H_{12}O_6)$ (Optically active acid)

$P \xrightarrow{Pt/H_2}$ (cyclohexyl–$CH_2CH_2CH_3$)

Choose the correct option(s).

(a) *P* is

(b) *X* is Pd-C/quinoline/$H_2$

(c) *P* is

(d) *R* is

## Section 3 (Maximum Marks : 24)

- This section contains **SIX (06)** questions. The answer to each question is a **NUMERICAL VALUE.**
- For each question, enter the correct numerical value of the answer using the mouse and the on-screen virtual numeric keypad in the place designated to enter the answer. If the numerical value has more than two decimal places, **truncate/round-off** the value to **TWO** decimal places.
- Answer to each question will be evaluated according to the following marking scheme:

*Full Marks* : **+4** If only the correct numerical value is entered as answer.

*Zero Marks* : **0** In all other cases.

---

**31.** A solution of 0.1 M weak base ($B$) is titrated with 0.1 M of a strong acid (H$A$). The variation of pH of the solution with the volume of H$A$ added is shown in the figure below. What is the p$K_b$ of the base? The neutralisation reaction is given by

$$B + HA \longrightarrow BH^+ + A^-.$$

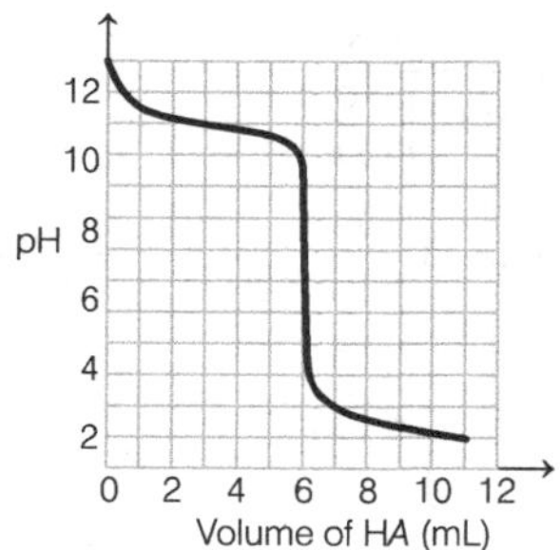


**32.** Liquids $A$ and $B$ form ideal solution for all compositions of $A$ and $B$ at 25°C. Two such solutions with 0.25 and 0.50 mole fractions of $A$ have the total vapour pressure of 0.3 and 0.4 bar, respectively. What is the vapour pressure of pure liquid $B$ in bar?

**33.** The figure below is the plot of potential energy *versus* internuclear distance ($d$) of $H_2$ molecule in the electronic ground state. What is the value of the net potential energy $E_0$ (as indicated in the figure) in kJ mol$^{-1}$, for $d = d_0$ at which the electron-electron repulsion and the nucleus-nucleus repulsion energies are absent? As reference, the potential energy of H atom is taken as zero when its electron and the nucleus are infinitely far apart. Use Avogardo constant as $6.023 \times 10^{23}$ mol$^{-1}$.

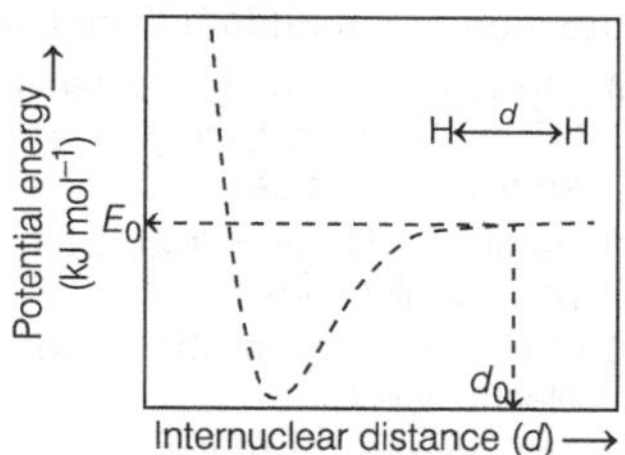


**34.** Consider the reaction sequence from $P$ to $Q$ shown below. The overall yield of the major product $Q$ from $P$ is 75%. What is the amount in grams of $Q$ obtained from 9.3 mL of $P$? (Use density of $P$ = 1.00 g mL$^{-1}$; Molar mass of C = 12.0, H = 1.0, O = 16.0 and N = 14.0 g mol$^{-1}$)

$-NH_2$ (i) $NaNO_2$+HCl / 0-5°C; (ii) OH; (iii) $CH_3CO_2H/H_2O$ $\longrightarrow$

$P$

$Q$ + NaOH

**35.** Tin is obtained from cassiterite by reduction with coke. Use the data given below to determine the minimum temperature (in K) at which the reduction of cassiterite by coke would take place.

At 298 K : $\Delta_f H°[SnO_2(s)] = -581.0$ kJ mol$^{-1}$,

$$\Delta_f H^\circ[(CO_2)(g)] = -394.0 \text{ kJ mol}^{-1}$$
$$S^\circ[SnO_2(s)] = 56.0 \text{J K}^{-1}\text{mol}^{-1}$$
$$S^\circ[Sn(s)] = 52.0 \text{JK}^{-1}\text{ mol}^{-1}$$
$$S^\circ[C(s)] = 6.0 \text{J K}^{-1}\text{ mol}^{-1}$$
$$S^\circ[CO_2(g)] = 210.0 \text{JK}^{-1}\text{ mol}^{-1}$$

Assume that, the enthalpies and the entropies are temperature independent.

**36.** An acidified solution of 0.05 M $Zn^{2+}$ is saturated with 0.1 M $H_2S$. What is the minimum molar concentration ($M$) of $H^+$ required to prevent the precipitation of ZnS?

Use $K_{sp}(\text{ZnS}) = 1.25 \times 10^{-22}$ and overall dissociation constant of $H_2S$, $K_{net} = K_1K_2 = 1 \times 10^{-21}$.

# MATHEMATICS

## Section 1 (Maximum Marks : 18)

- This section contains **SIX (06)** questions.
- The answer to each question is a **SINGLE DIGIT INTEGER** ranging from 0 TO 9. **BOTH INCLUSIVE**.
- For each question, enter the correct integer corresponding to the answer using the mouse and the on-screen virtual numeric keypad in the place designated to enter the answer.
- Answer to each question will be evaluated according to the following marking scheme:

*Full Marks* : **+3** If ONLY the correct integer is entered.
*Zero Marks* : **0** If the question is unanswered.
*Negative Marks* : **−1** In all other cases.

**37.** For a complex number $z$, let $\text{Re}(z)$ denote the real part of $z$. Let $S$ be the set of all complex numbers $z$ satisfying $z^4 - |z|^4 = 4iz^2$, where $i = \sqrt{-1}$. Then the minimum possible value of $|z_1 - z_2|^2$, where $z_1, z_2 \in S$ with $\text{Re}(z_1) > 0$ and $\text{Re}(z_2) < 0$ is …… .

**38.** The probability that a missile hits a target successfully is 0.75. In order to destroy the target completely, at least three successful hits are required. Then the minimum number of missiles that have to be fired so that the probability of completely destroying the target is **NOT** less than 0.95, is ……… .

**39.** Let $O$ be the centre of the circle $x^2 + y^2 = r^2$, where $r > \frac{\sqrt{5}}{2}$. Suppose $PQ$ is a chord of this circle and the equation of the line passing through $P$ and $Q$ is $2x + 4y = 5$. If the centre of the circumcircle of the triangle $OPQ$ lies on the line $x + 2y = 4$, then the value of $r$ is ……… .

**40.** The trace of a square matrix is defined to be the sum of its diagonal entries. If $A$ is a $2 \times 2$ matrix such that the trace of $A$ is 3 and the trace of $A^3$ is $-18$, then the value of the determinant of $A$ is ……… .

**41.** Let the functions $f : (-1, 1) \to \mathbf{R}$ and $g : (-1, 1) \to (-1, 1)$ be defined by

$$f(x) = |2x - 1| + |2x + 1| \text{ and}$$
$$g(x) = x - [x],$$

where $[x]$ denotes the greatest integer less than or equal to $x$. Let $f \circ g$: $(-1, 1) \to R$ be the composite function defined by $(f \circ g)(x) = f(g(x))$. Suppose $c$ is the number of points in the interval $(-1, 1)$ at which $f \circ g$ is **NOT** continuous, and suppose $d$ is the number of points in the interval $(-1, 1)$ at which $f \circ g$ is NOT differentiable. Then the value of $c + d$ is ......... .

**42.** The value of the limit

$$\lim_{x \to \frac{\pi}{2}} \frac{4\sqrt{2}(\sin 3x + \sin x)}{\left(2\sin 2x \sin \frac{3x}{2} + \cos \frac{5x}{2}\right) - \left(\sqrt{2} + \sqrt{2}\cos 2x + \cos \frac{3x}{2}\right)}$$

is ......... .

## Section 2 (Maximum Marks : 24)

- This section contains **SIX (06)** questions.
- Each question has **FOUR** options. **ONE OR MORE THAN ONE** of these four option(s) is (are) correct asnswer(s).
- For each question, choose the option(s) corresponding to (all) the correct answer(s).
- Answer to each question will be evaluated according to the following marking scheme:

*Full Marks* : **+4** If only (all) the correct option(s) is (are) chosen.
*Partial Marks* : **+3** If all the four options are correct but ONLY three options are chosen.
*Partial Marks* : **+2** If three or more options are correct but ONLY two options are chosen, both of which are correct.
*Partial Marks* : **+1** If two or more options are correct but ONLY one option is chosen and it is a correct option.
*Zero Marks* : **0** If none of the options is chosen (i.e. the question is unanswered).
*Negative Marks* : **−2** In all other cases.

**43.** Let $b$ be a nonzero real number. Suppose $f : \mathbf{R} \to \mathbf{R}$ is a differentiable function such that $f(0) = 1$. If the derivative $f'$ of $f$ satisfies the equation

$$f'(x) = \frac{f(x)}{b^2 + x^2}$$

for all $x \in \mathrm{R}$, then which of the following statements is/are TRUE?

(a) If $b > 0$, then $f$ is an increasing function
(b) If $b < 0$, then $f$ is a decreasing function
(c) $f(x)f(-x) = 1$ for all $x \in \mathrm{R}$
(d) $f(x) - f(-x) = 0$ for all $x \in \mathrm{R}$

**44.** Let $a$ and $b$ be positive real numbers such that $a > 1$ and $b < a$. Let $P$ be a point in the first quadrant that lies on the hyperbola $\frac{x^2}{a^2} - \frac{y^2}{b^2} = 1$. Suppose the tangent to the hyperbola at $P$ passes through the point $(1, 0)$, and suppose the normal to the hyperbola at $P$ cuts off equal intercepts on the coordinate axes. Let $\Delta$ denote the area of the triangle formed by the tangent at $P$, the normal at $P$ and the X-axis. If $e$ denotes the eccentricity of the hyperbola, then which of the following statements is/are TRUE?

(a) $1 < e < \sqrt{2}$
(b) $\sqrt{2} < e < 2$
(c) $\Delta = a^4$
(d) $\Delta = b^4$

**45.** Let $f : \mathrm{R} \to \mathrm{R}$ and $g : \mathrm{R} \to \mathrm{R}$ be functions satisfying

$f(x + y) = f(x) + f(y) + f(x)f(y)$ and $f(x) = xg(x)$

for all $x, y \in \mathrm{R}$. If $\lim_{x \to 0} g(x) = 1$, then which of the following statements is/are TRUE?

(a) $f$ is differentiable at every $x \in \mathbf{R}$
(b) If $g(0) = 1$, then $g$ is differentiable at every $x \in \mathbf{R}$
(c) The derivative $f'(1)$ is equal to 1
(d) The derivative $f'(0)$ is equal to 1

**46.** Let $\alpha, \beta, \gamma, \delta$ be real numbers such that $\alpha^2 + \beta^2 + \gamma^2 \neq 0$ and $\alpha + \gamma = 1$. Suppose the point $(3, 2, -1)$ is the mirror image of the point $(1, 0, -1)$ with respect to the plane $\alpha x + \beta y + \gamma z = \delta$. Then which of the following statements is/are TRUE?

(a) $\alpha + \beta = 2$ (b) $\delta - \gamma = 3$
(c) $\delta + \beta = 4$ (d) $\alpha + \beta + \gamma = \delta$

**47.** Let $a$ and $b$ be positive real numbers. Suppose $\mathbf{PQ} = a\hat{\mathbf{i}} + b\hat{\mathbf{j}}$ and $\mathbf{PS} = a\hat{\mathbf{i}} - b\hat{\mathbf{j}}$ are adjacent sides of a parallelogram $PQRS$. Let $\mathbf{u}$ and $\mathbf{v}$ be the projection vectors of $\mathbf{w} = \hat{\mathbf{i}} + \hat{\mathbf{j}}$ along $\mathbf{PQ}$ and $\mathbf{PS}$, respectively. If $|\mathbf{u}| + |\mathbf{v}| = |\mathbf{w}|$ and if the area of the parallelogram $PQRS$ is 8, then which of the following statements is/are TRUE?

(a) $a + b = 4$
(b) $a - b = 2$
(c) The length of the diagonal $PR$ of the parallelogram $PQRS$ is 4
(d) $\mathbf{w}$ is an angle bisector of the vectors $\mathbf{PQ}$ and $\mathbf{PS}$

**48.** For non-negative integers $s$ and $r$, let

$$\binom{s}{r} = \begin{cases} \dfrac{s!}{r!(s-r)!} & \text{if } r \leq s, \\ 0 & \text{if } r > s \end{cases}$$

For positive integers $m$ and $n$, let

$$g(m, n) = \sum_{p=0}^{m+n} \frac{f(m, n, p)}{\binom{n+p}{p}}$$

where for any non-negative integer $p$,

$$f(m, n, p) = \sum_{i=0}^{p} \binom{m}{i} \binom{n+i}{p} \binom{p+n}{p-i}.$$

Then which of the following statements is/are TRUE?

(a) $g(m, n) = g(n, m)$ for all positive integers $m, n$
(b) $g(m, n+1) = g(m+1, n)$ for all positive integers $m, n$
(c) $g(2m, 2n) = 2g(m, n)$ for all positive integers $m, n$
(d) $g(2m, 2n) = (g(m, n))^2$ for all positive integers $m, n$

## Section 3 (Maximum Marks : 24)

- This section contains **SIX (06)** questions. The answer to each question is a **NUMERICAL VALUE.**
- For each question, enter the correct numerical value of the answer using the mouse and the on-screen virtual numeric keypad in the place designated to enter the answer. If the numerical value has more than two decimal places, **truncate/round-off** the value to **TWO** decimal places.
- Answer to each question will be evaluated according to the following marking scheme:

*Full Marks* : **+4** If ONLY the correct numerical value is entered.
*Zero Marks* : **0** In all other cases.

**49.** An engineer is required to visit a factory for exactly four days during the first 15 days of every month and it is mandatory that no two visits take place on consecutive days. Then the number of all possible ways in which such visits to the factory can be made by the engineer during 1-15 June 2021 is .........

**50.** In a hotel, four rooms are available. Six persons are to be accommodated in these four rooms in such a way that each of these rooms contains at least one person and at most two persons. Then the number of all possible ways in which this can be done is .........

**51.** Two fair dice, each with faces numbered 1, 2, 3, 4, 5 and 6, are rolled together and the sum of the numbers on the faces is observed. This process is repeated till the sum is either a prime number or a perfect square. Suppose the sum turns out to be a perfect square before it turns out to be a prime number. If $p$ is the probability that this perfect square is an odd number, then the value of $14\,p$ is ......... .

**52.** Let the function $f:[0,1] \to \text{R}$ be defined by

$$f(x) = \frac{4^x}{4^x + 2}$$

Then the value of

$$f\left(\frac{1}{40}\right) + f\left(\frac{2}{40}\right) + f\left(\frac{3}{40}\right) + \dots + f\left(\frac{39}{40}\right) - f\left(\frac{1}{2}\right)$$

is .........

**53.** Let $f:\mathbf{R} \to \mathbf{R}$ be a differentiable function such that its derivative $f'$ is continuous and $f(\pi) = -6$.
If $F:[0,\pi] \to \mathbf{R}$ is defined by $F(x) = \int_0^x f(t)dt$, and if

$$\int_0^\pi (f'(x) + F(x))\cos x\, dx = 2$$

then the value of $f(0)$ is ......... .

**54.** Let the function $f:(0,\pi) \to \mathbf{R}$ be defined by
$f(\theta) = (\sin\theta + \cos\theta)^2 + (\sin\theta - \cos\theta)^4$
Suppose the function $f$ has a local minimum at $\theta$ precisely when $\theta \in \{\lambda_1\pi, \dots, \lambda_r\pi\}$, where $0 < \lambda_1 < \dots \lambda_r < 1$. Then the value of $\lambda_1 + \dots + \lambda_r$ is ....... .

# Answers

## Paper 1

| | | | | | | | | | | | | | | | | | | | |
|---|---|---|---|---|---|---|---|---|---|---|---|---|---|---|---|---|---|---|---|
| 1. | a | 2. | b | 3. | b | 4. | b | 5. | a | 6. | b | 7. | b,c | 8. | b,c,d | 9. | a,b | 10. | b,c |
| 11. | a,c,d | 12. | a,b,c | 13. | 25.60 | 14. | 3.74 | 15. | 3.14 | 16. | 1.77 to 1.78 | 17. | 0.62 to 0.63 | 18. | 6.40 | 19. | b | 20. | b |
| 21. | a | 22. | c | 23. | d | 24. | c | 25. | a,b,c | 26. | a,b,c | 27. | b,d | 28. | a,c | 29. | a,b,d | 30. | a,c |
| 31. | 0.11 | 32. | 0.25 | 33. | 13.32 | 34. | 6.15 | 35. | 1.2 | 36. | 18 | 37. | d | 38. | c | 39. | a | 40. | a |
| 41. | b | 42. | c | 43. | a,c | 44. | b,c,d | 45. | b,c | 46. | b,c | 47. | a,b | 48. | a,b,d | 49. | 8.00 | 50. | 3.00 |
| 51. | 1.00 | 52. | 108.00 | 53. | 5.00 | 54. | 1.00 | | | | | | | | | | | | |

## Paper 2

| | | | | | | | | | | | | | | | | | | | |
|---|---|---|---|---|---|---|---|---|---|---|---|---|---|---|---|---|---|---|---|
| 1. | * | 2. | 9 | 3. | 6 | 4. | 4 | 5. | 2 | 6. | * | 7. | a,d | 8. | a,d | 9. | a,c,d | 10. | a,c |
| 11. | b,c | 12. | a,b,c,d | 13. | 2.05 | 14. | 0.27 | 15. | 1.3 | 16. | 0.24 | 17. | 55.00 | 18. | 8.33 | 19. | 9 | 20. | 6 |
| 21. | 6 | 22. | 4 | 23. | 5 | 24. | 6 | 25. | a,b,c | 26. | a | 27. | a,b,c,d | 28. | a,b | 29. | c,d | 30. | b,c |
| 31. | 3.00 | 32. | 0.2 | 33. | −5242.41 | 34. | 18.60 | 35. | 935 | 36. | 0.2 | 37. | 8 | 38. | 6 | 39. | 2 | 40. | 5 |
| 41. | 4 | 42. | 8 | 43. | a,c | 44. | a,d | 45. | a,b,d | 46. | a,b,c | 47. | a,c | 48. | a,b,d | 49. | 495.00 | 50. | 1080.00 |
| 51. | 8.00 | 52. | 19.00 | 53. | 4.00 | 54. | 0.50 | | | | | | | | | | | | |

**Note** (*) None of the option is correct.

# Answer *with* Explanations

## Paper 1

### Physics

**1.** *(a)*

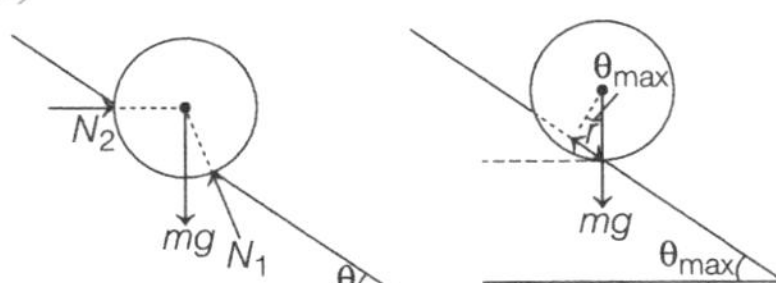


For $\theta_{max}$, the football is about to roll, then $N_2 = 0$ and all the forces ($mg$ and $N_1$) must pass through contact point.

$\therefore \cos(90° - \theta_{max}) = \frac{r}{R} \Rightarrow \sin\theta_{max} = \frac{r}{R}$

**2.** *(b)* When the magnet is moved, it creates a state, where the plate moves through the magnetic flux, due to which an electromotive force is generated in the plate and eddy currents are induced. These currents are such that it opposes the relative motion $\Rightarrow$ disc will rotate in the direction of rotation of magnet.

**3.** *(b)* For no slipping at ground,

$$v_c = \omega R$$

(where, $r$ is the radius of roller)

$\therefore$ Velocity of scale $= (v_c + \omega r)$ (where, $r$ is the radius of axle)

Given, $v_c \cdot t = 50\,\text{cm}$

$\therefore$ Distance moved by scale $= (v_c + \omega r)t$

$$= \left(v_c + \frac{v_c r}{r}\right)t = \frac{3v_c}{2} \cdot t = 75\,\text{cm}$$

Therefore, relative displacement (with respect to centre of roller) is (75 – 50) cm = 25 cm.

**4.** *(b)* Torque experienced by circular loop $= \mathbf{M} \times \mathbf{B}$

where, $\mathbf{M}$ is magnetic moment and $\mathbf{B}$ is magnetic field.

$$\therefore \quad \tau = i\pi R^2 NB_0$$

(at the instant shown $\theta = \pi/2$)

$$\therefore \quad \tau dt = d\mathbf{L} = i\pi R^2 NB_0 dt$$

$$= Q\pi R^2 NB_0 \qquad (\because \; idt = Q)$$

**5.** *(a)*

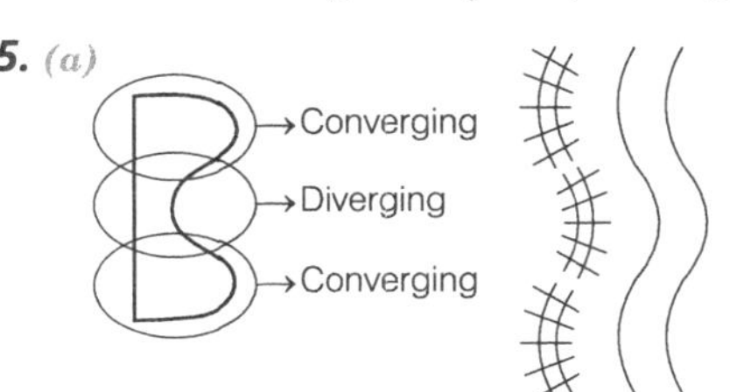


**6.** *(b)* We have, $h_1 + h_2 = 0.29 \times 2 + 0.1$

$$h_1 + h_2 = 0.68 \qquad \ldots \text{(i)}$$

$$\Rightarrow \quad p_0 + \rho_k g(0.1) + \rho_w g(h_1 - 0.1) - \rho_w g h_2 = p_0$$

where, $\rho_k$ = density of kerosene and $\rho_w$ = density of water.

$$\Rightarrow \rho_k g(0.1) + \rho_w g h_1 - \rho_w g \times (0.1) = \rho_w g h_2$$

$$\Rightarrow \quad 800 \times 10 \times 0.1 + 1000 \times 10 \times h_1 - 1000 \times 10 \times 0.1 = 1000 \times 10 \times h_2$$

$$\Rightarrow \quad 10000(h_1 - h_2) = 200$$

$$\Rightarrow \quad h_1 - h_2 = 0.02 \qquad \ldots \text{(ii)}$$

Solving Eqs. (i) and (ii), we get

$$h_1 = 0.35$$

and $\quad h_2 = 0.33$

So, $\quad \frac{h_1}{h_2} = \frac{35}{33}$

**7.** *(b, c)* $|\text{Force}| = \left|-\frac{dV}{dr}\right| \Rightarrow F = \frac{mv^2}{r} \qquad \ldots \text{(i)}$

$$mvr = \frac{nh}{2\pi} \qquad \ldots \text{(ii)}$$

From Eqs. (i) and (ii), we get

$$\therefore \quad F = \frac{m}{r}\left(\frac{nh}{2\pi mr}\right)^2 \quad \left(\because \; v = \frac{nh}{2\pi mr}\right)$$

or $\quad F = \frac{mn^2h^2}{4\pi^2 m^2 r^3}$

or $\quad F = \frac{n^2h^2}{4\pi^2 m r^3}$

$$r^3 \propto n^2 \qquad (\text{as } F = \text{constant})$$

$$r \propto n^{2/3}$$
$$v = \frac{nh}{2\pi mr}$$
$$v \propto \frac{n}{n^{2/3}}$$
or $\quad v \propto n^{1/3}$

$\therefore \quad \text{KE} = \frac{1}{2}mv^2 = \frac{Fr}{2}$

$\because \quad \text{TE} = \text{PE} + \text{KE} = Fr + \frac{Fr}{2}$

$\Rightarrow \quad \frac{3Fr}{2} = \text{TE}$

or $\quad \text{TE} = E = \frac{3F}{2} \times \left(\frac{n^2h^2}{4\pi^2 rnF}\right)^{1/3}$

$$= \frac{3}{2}\left(\frac{n^2h^2F^2}{4\pi^2 m}\right)^{1/3}$$

**8.** *(b, c, d)* Here, $A = 64\ \text{mm}^2$, $T = 2500\ \text{K}$ (where, $A$ = surface area of filament, $T$ = temperature of filament, $d$ is distance of bulb from observer and $R_e$ = radius of pupil of eye)

Point source, $d = 100\ \text{m}$, $R_e = 3\ \text{mm}$

(a) $P = \sigma AeT^{-4}$

$= 5.67 \times 10^{-8} \times 64 \times 10^{-6} \times 1 \times (2500)^4$

($\because e = 1$ for black body)

$= 141.75\ \text{W}$

(b) Power reaching to the eye

$$= \frac{P}{4\pi d^2} \times (\pi R_e^2)$$
$$= \frac{141.75}{4\pi \times (100)^2} \times \pi \times (3 \times 10^{-3})^2$$
$$= 3.189375 \times 10^{-8}\ \text{W}$$

(c) $\lambda_m T = b$

$\lambda_m \times 2500 = 2.9 \times 10^{-3}$

$\Rightarrow \quad \lambda_m = 1.16 \times 10^{-6} = 1160\ \text{nm}$

(d) Power received by one eye of observer

$$= \left(\frac{hc}{\lambda}\right) \times \frac{\Delta N}{\Delta t}$$

where, $\frac{\Delta N}{\Delta t}$ = number of photons entering into eye per second.

$\Rightarrow 3.189375 \times 10^{-8}$

$$= \frac{6.63 \times 10^{-34} \times 3 \times 10^8}{1740 \times 10^{-9}} \times \frac{\Delta N}{\Delta t}$$

$\Rightarrow \quad \frac{\Delta N}{\Delta t} = 2.79 \times 10^{11}$

**9.** *(a, b)* $[x] = x^\alpha \Rightarrow \frac{p}{v} = \text{time}$

$[v] = x^\beta$

$[a] = x^p \quad \Rightarrow \quad \frac{v}{a} = \text{time}$

$[p] = x^q$

$[F] = x^r$

$$\Rightarrow \left(\frac{\text{position}}{\text{speed}}\right) = \left(\frac{\text{speed}}{\text{acceleration}}\right) = \left(\frac{p}{F}\right)$$

$$\Rightarrow \quad \frac{x^\alpha}{x^\beta} = \frac{x^\beta}{x^p} = \frac{x^q}{x^r}$$

$$\Rightarrow \quad x^{\alpha-\beta} = x^{\beta-p} = x^{q-r}$$

$$\alpha - \beta = \beta - p = q - r$$
$$\alpha + p = 2\beta$$
$$q + p = \beta + r$$
$$p + q - r = \beta$$

**10.** *(b, c)*

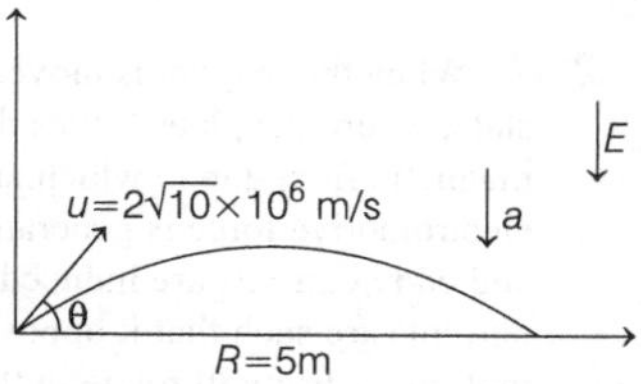

$u = 2\sqrt{10} \times 10^6$ m/s, $\theta$, $R = 5$m, $E$, $a$

$$a = \frac{qE}{m} = 10^{10} \times 400\sqrt{3}\ \text{m/s}^2$$

or $\quad a = 4\sqrt{3} \times 10^{12}\ \text{m/s}^2$

$\therefore \quad R = \frac{u^2 \sin 2\theta}{a}$

$\Rightarrow \quad \sin 2\theta = \frac{Ra}{u^2}$

or $\quad \sin 2\theta = \frac{5 \times 4\sqrt{3} \times 10^{12}}{4 \times 10 \times 10^{12}}$

$\Rightarrow \quad \sin 2\theta = \frac{\sqrt{3}}{2}$

$\Rightarrow \quad 2\theta = 60°, 120°$

$\Rightarrow \quad \theta = 30°$ or $60°$ for same range

$\therefore \quad T = \frac{2u\sin\theta}{a}$

$$= \frac{2 \times 2\sqrt{10} \times 10^6}{2 \times 4\sqrt{3} \times 10^{12}}$$

$$\Rightarrow T = \sqrt{\frac{10}{12}} \times 10^{-6} \Rightarrow T = \sqrt{\frac{5}{6}} \times 10^{-6}$$

**11.** (a, c, d)

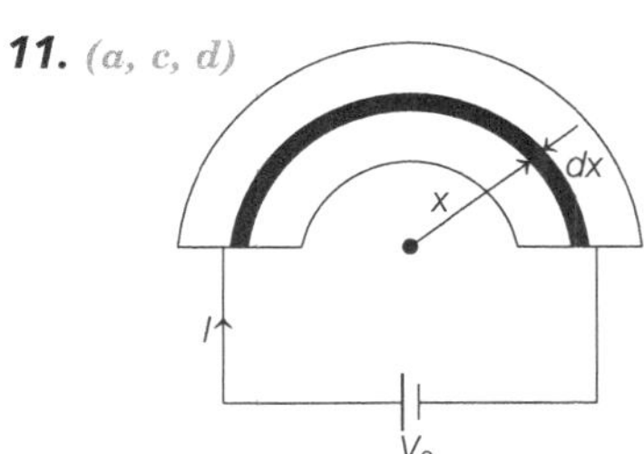


All the elements are in parallel.

$$\therefore \quad \int \frac{1}{dr} = \int_{R_1}^{R_2} \frac{t\,dx}{\rho\,\pi x}$$

$$\frac{1}{r} = \frac{t}{\pi\rho} \ln\left(\frac{R_2}{R_1}\right)$$

$$\text{Resistance} = \frac{\pi\rho}{t\ln\left(\frac{R_2}{R_1}\right)}$$

$$\therefore \quad I = \frac{V}{\text{Resistance}}$$

$$= \frac{Vt}{\pi\rho} \ln\left(\frac{R_2}{R_1}\right)$$

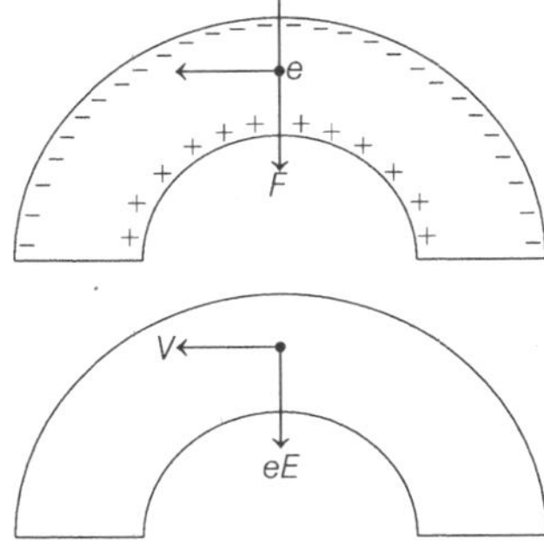


$$eE = \frac{mv^2}{r}$$

$$I = neA\,V_d$$

$$\frac{V_0 t dr}{\rho\pi r} = ne(drt)V$$

$$\Rightarrow \quad V = \sqrt{\frac{V_0}{\rho\pi ner}}$$

$$\therefore \quad E = \frac{m}{er} \frac{V_0^2 t^2}{\rho^2\pi^2 n^2 e^2 r^2}$$

$$\Rightarrow \quad E = \frac{mV_0^2}{e^3 r^2 \rho^2 \pi^2 n^2}$$

$$dV = -Edr \Rightarrow dV = -\frac{K}{r^3}dr$$

$$V = K\int_{R_1}^{R_2} r^{-3}dr$$

$$\Delta V = \frac{mV_0^2}{e^3\rho^2\pi^2 n^2}\left(\frac{1}{r_1^2} - \frac{1}{r_2^2}\right)$$

$$\Delta V \propto V_0^2 \quad \Rightarrow \quad I \propto V_0$$

**12.** (a, b, c) $n_1 >> (n_1 - n_2) = \Delta n$

$$\because \quad p_1 = \frac{n_1 RT}{N_A} \quad \text{and} \quad p_2 = \frac{n_2 RT}{N_A}$$

$$F = (n_1 - n_2)k_B TS = \Delta n k_B TS$$

$$V = \frac{\Delta n k_B TS}{\beta}$$

$$\Delta n k_B TS = l n_1 S\beta v$$

$$\Rightarrow \quad n_1 \beta v l = \Delta n\, k_B T$$

Total number of molecules/second

$$= \frac{(n_1 v dt)S}{dt}$$

$$= n_1 vS = \frac{\Delta n\, k_B T\, vS}{\beta v l} = \left(\frac{\Delta n}{l}\right)\left(\frac{k_B T}{\beta}\right)S$$

As $\Delta n$ will decrease with time, therefore rate of molecules getting transfer decreases with time.

**13.** (25.60)

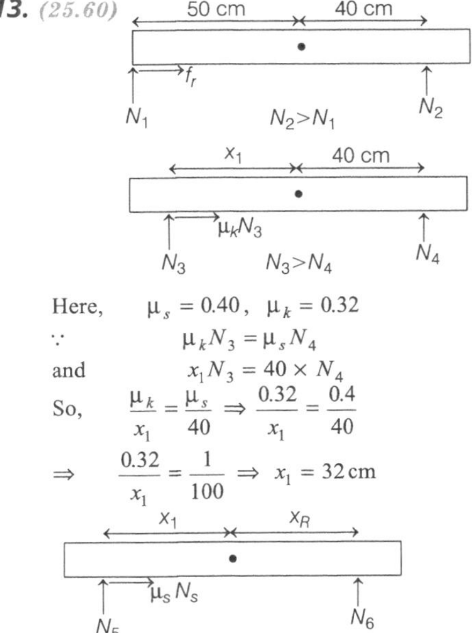


Here, $\mu_s = 0.40$, $\mu_k = 0.32$

$$\because \quad \mu_k N_3 = \mu_s N_4$$

and $\quad x_1 N_3 = 40 \times N_4$

So, $\quad \frac{\mu_k}{x_1} = \frac{\mu_s}{40} \Rightarrow \frac{0.32}{x_1} = \frac{0.4}{40}$

$$\Rightarrow \quad \frac{0.32}{x_1} = \frac{1}{100} \Rightarrow x_1 = 32\,\text{cm}$$

$$\frac{\mu_S N_5}{x_1 N_5} = \frac{\mu_k N_6}{x_R N_6} \Rightarrow \frac{0.40}{10 \times 32} \times \frac{0.32}{x_R}$$

$$\Rightarrow \quad x_R = 25.6$$

**14.** *(3.74)*

$h/2$

$R_2 = h/2$

Pressure at the bottom of disc = Pressure due to surface tension

$$\rho g h = T\left(\frac{1}{R_1} + \frac{1}{R_2}\right)$$

$$R_1 >>> R_2$$

So, $\frac{1}{R_1} <<< \frac{1}{R_2}$ and $R_2 = h/2$

$$\therefore \; \rho g h = T\left(\frac{1}{R_1} + \frac{1}{R_2}\right) = T\left(0 + \frac{1}{h/2}\right)$$

$$h^2 = \frac{2T}{\rho g} \Rightarrow h = \sqrt{\frac{2T}{\rho g}}$$

$$\Rightarrow \quad = \sqrt{\frac{2 \times 0.07}{10^3 \times 10}} = \sqrt{\frac{14 \times 100}{10^4 \times 100}}$$

$$h = \sqrt{14}\,\text{mm} = 3.741$$

**15.** *(3.14)* Electrostatic force = Spring force

At $l$, $Fe = F_{sp}$ $kl = \frac{2\alpha pq}{l^3}$ $\left(\text{Here, } \alpha = \frac{1}{4\pi\varepsilon_0}\right)$

$y$ $p$ $q$ $Fe$ $(0, 0)$ $F_{sp}$ $l$

Now, the mass $m$ is displaced by $\Delta l = x$ from the mean position

$$F_{net} = F_{sp} - Fe = k(l + x) - \frac{q(2\alpha p)}{(l + x)^3}$$

$$= k(x + l) - \frac{q(2\alpha p)}{l^3(1 + x/l)^3}$$

$$= kx + kl - q\left(\frac{2\alpha p}{l^3}\right)\left(1 - \frac{3x}{l}\right)$$

$$= kx + kl - q\left(\frac{2\alpha p}{l^3}\right) + \frac{2\alpha pq}{l^3} \cdot \frac{3x}{l}$$

Substituting $\frac{2\alpha pq}{l^3} = kl$, we get

$$F_{net} = kx + kl\left(\frac{3x}{l}\right) = 4kx$$

This is restoring in nature.

Hence, $k_{eq} = 4k$

or $\quad T = 2\pi\sqrt{\frac{m}{4k}} = \pi\sqrt{\frac{m}{k}}$

$\therefore \quad f = \frac{1}{\pi}\sqrt{\frac{k}{m}}$

So, $\quad \delta = \pi = 3.14$

**16.** *(1.77 to 1.78)*

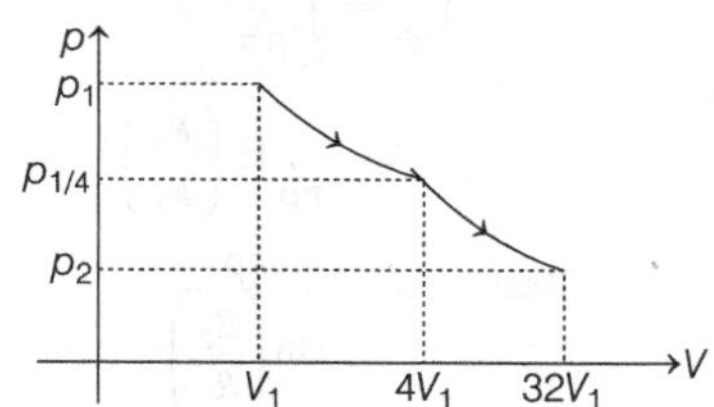


$$\frac{p_1}{4}(4V_1)^{5/3} = p_2(32V_1)^{5/3}$$

$$p_2 = \frac{p_1}{4}\left(\frac{1}{8}\right)^{5/3} = \frac{p_1}{128}$$

$$W_{adi} = \frac{p_1V_1 - p_2V_2}{\gamma - 1}$$

$$= \frac{p_1V_1 - \frac{p_1}{128}(32\,V_1)}{\frac{5}{3} - 1}$$

$$= \frac{p_1V_1(3/4)}{2/3} = \frac{9}{8}p_1V_1$$

$$W_{iso} = p_1V_1 \ln\left(\frac{4V_1}{V_1}\right) = 2p_1V_1 \ln 2$$

$$\therefore \quad \frac{W_{iso}}{W_{adi}} = \frac{16}{9}\ln 2$$

$$\Rightarrow \quad f = \frac{16}{9} = 1.7778 \approx 1.78$$

**17.** *(0.62 to 0.63)*

$$f \propto \frac{1}{l_1} \Rightarrow f = \frac{k}{l_1} \quad \ldots \text{(i)}$$

(where, $l_1 \Rightarrow$ initial length of pipe)

$$\left(\frac{v}{v - v_T}\right) f = \frac{k}{l_1} \quad \ldots \text{(ii)}$$

(where, $v_T$ = speed of tuning fork, $l_2$ = new length of pipe)

Dividing Eq. (i) by Eq. (ii), we get

$$\frac{v - v_T}{c} = \frac{l_2}{l_1} \Rightarrow \frac{l_2}{l_1} - 1 = \frac{v - v_T}{v} - 1$$

$$\frac{l_2 - l_1}{l_1} = \frac{-v_T}{v}$$

$$\frac{l_2 - l_1}{l_1} \times 100 = \frac{-2}{320} \times 100 = -0.625$$

Therefore, smallest value of percentage change required in the length of pipe is 0.625.

**18.** (6.40)

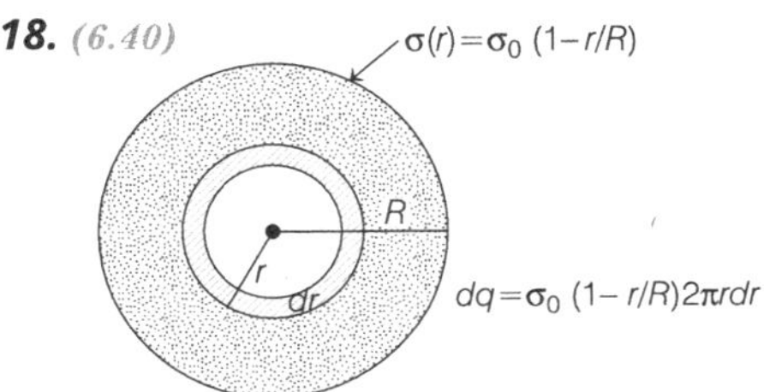


$$\phi_0 = \frac{\int dq}{\varepsilon_0} = \frac{\int_0^R \sigma_0\left(1 - \frac{r}{R}\right) 2\pi r\, dr}{\varepsilon_0}$$

$$\phi_0 = \frac{\int dq}{\varepsilon_0} = \frac{\int_0^{R/4} \sigma_0\left(1 - \frac{r}{R}\right) 2\pi r\, dr}{\varepsilon_0}$$

$$\therefore \quad \frac{\phi_0}{\phi} = \frac{\sigma_0 2\pi \int_0^R \left(r - \frac{r^2}{R}\right) dr}{\sigma_0 2\pi \int_0^{R/4} \left(r - \frac{r^2}{R}\right) dr}$$

$$= \frac{\frac{R^2}{2} - \frac{R^2}{3}}{\frac{R^2}{32} - \frac{R^2}{3 \times 64}} = \frac{32}{5} = 6.40$$

# Chemistry

**19.** (b) Fraction of molecules *vs* velocity graph is Maxwell distribution curve.

This curve is slightly unsymmetrical as shown below.

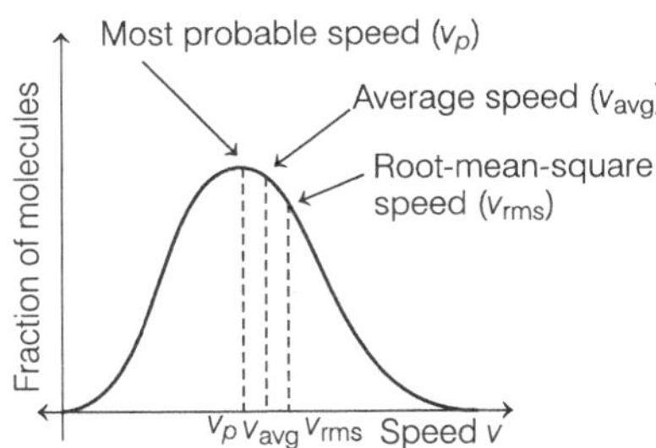


The ratio of most probable, the average and the root mean square speeds for this graph is 1 : 1·128 : 1·224

But the graph in question is completly symmetrical. Therefore, the most probable and the average speed will be same here but root mean square speed will be greater than average speed. So, the correct ratio is 1 : 1 : 1.224.

**20.** (b) (a) $Pb_3O_4 + H_2O \longrightarrow$ No reaction

$Pb_3O_4$ is insoluble in water or do not react with water.

(b) $KO_2 + 2H_2O \longrightarrow KOH + H_2O_2 + 1/2\,O_2$

Potassium superoxide is a strong oxidant, able to convert oxides into peroxides or molecular oxygen. Hydrolysis gives oxygen gas, hydrogen peroxide and potassium hydroxide.

(c) $Na_2O_2 + 2H_2O \longrightarrow 2NaOH + H_2O_2$

When sodium peroxide dissolves in water, it is hydrolysed and forms sodium hydroxide and hydrogen peroxide. The reaction is highly exothermic.

(d) $Li_2O_2 + 2H_2O \longrightarrow 2LiOH + H_2O_2$

The reactivity of $Li_2O_2$ toward water differs from $LiO_2$, in $Li_2O_2$ results in $H_2O_2$ as a product.

Hence, the correct option is (b).

**21.** *(a)* $X$ : Ag $Y$ : Pb $P$ : AgCl $Q$ : $PbCl_2$

$$AgNO_3 + Pb(NO_3)_2 \xrightarrow{NaCl} \underset{(P)}{AgCl\,(\downarrow)} + \underset{(Q)}{PbCl_2}$$

Heated and filtered

AgCl $(P)$ (White ppt.) PbCl$_2$ $(Q)$ (Hot solution) $\xrightarrow{KI}$ PbI$_2$ (Yellow ppt.)

$$\underset{(P)}{AgCl(s)} + \underset{(Excess)}{2NH_3} \longrightarrow [Ag(NH_3)_2]^+Cl^-$$

$AgCl(s)$ $(P)$ $\xrightarrow{2N_2S_2O_3}$ $Na_3[Ag(S_2O_3)_2] + NaCl$
Soluble complex salt

Hence, the correct option is (a).

**22.** *(c)* IUPAC name of $P$, $Q$, $R$ and $S$ are :

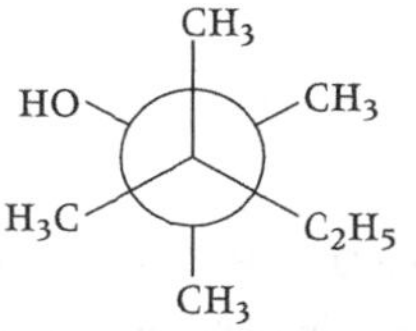

Newman projection $(P)$: front: CH$_3$ (top), HO, H$_3$C; back: CH$_3$, C$_2$H$_5$, CH$_3$ (bottom)

$(P)$

$\Rightarrow$ $C_2H_5—C(CH_3)_2—C(OH)(CH_3)—CH_3$

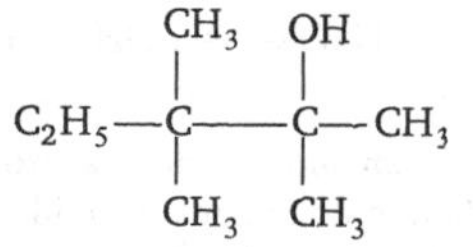

(2,3,3-trimethylpentan-2-ol)

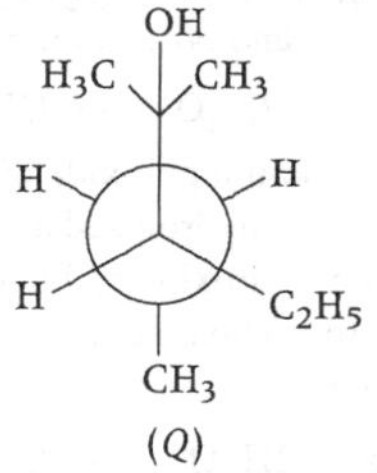

Newman projection $(Q)$: OH, H$_3$C, CH$_3$ (top); H, H, H, C$_2$H$_5$; CH$_3$ (bottom)

$(Q)$

$\Rightarrow$ $C_2H_5—CH(CH_2CH_3)—C(CH_3)(CH_3)—OH$

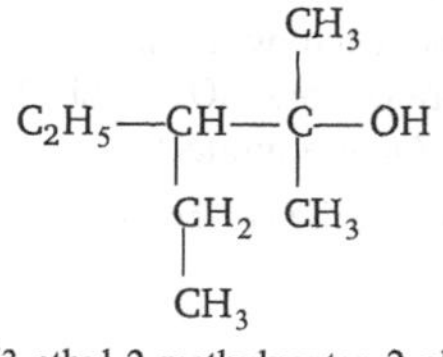

(3-ethyl-2-methylpentan-2-ol)

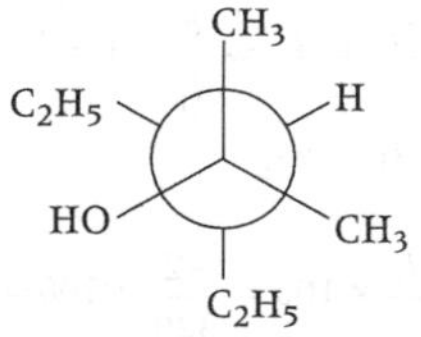

Newman projection $(R)$: CH$_3$ (top), C$_2$H$_5$, H, HO, CH$_3$, C$_2$H$_5$ (bottom)

$(R)$

$\Rightarrow$ $C_2H_5—CH(C_2H_5)—C(CH_3)(OH)—CH_3$

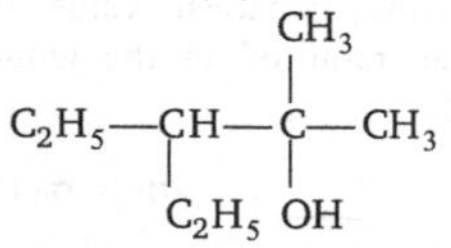

(3-ethyl-2-methylpentan-2-ol)

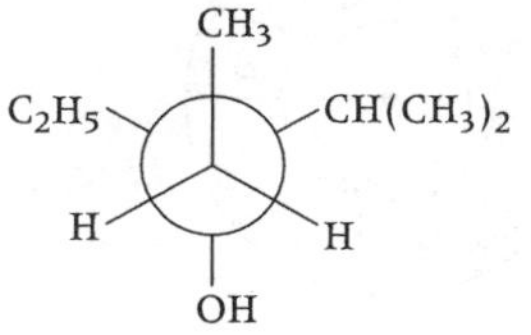

Newman projection $(S)$: CH$_3$ (top), C$_2$H$_5$, CH(CH$_3$)$_2$, H, H, OH (bottom)

$(S)$

$\Rightarrow$ $C_2H_5—C(CH_2CH_3)(OH)—CH(CH_3)—CH_3$

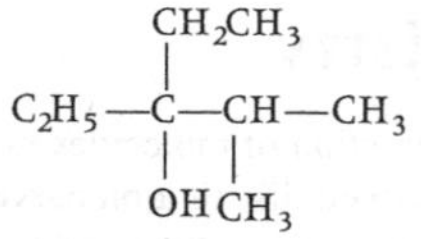

(3-ethyl-2-methylpentan-3-ol)

IUPAC name of compounds $Q$ and $R$ are same, thus the correct option is (c).

**23.** *(d)* According to the IUPAC rules, first of all select the longest carbon chain (6-carbon atoms) then numbering should be start from most prior functional group, which is carboxylic acid in the given compound, thus

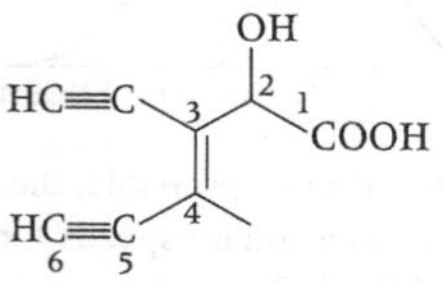

HC≡C–C$^3$(=C$^4$(CH$_3$)–C$^5$≡C$^6$H)–C$^2$H(OH)–C$^1$OOH

3-ethynyl-2-hydroxy-4-methylhex-3-en-5-ynoic acid

**24.** *(c)* First of all, we need to convert wedge-dash formula into Fischer projection formula.

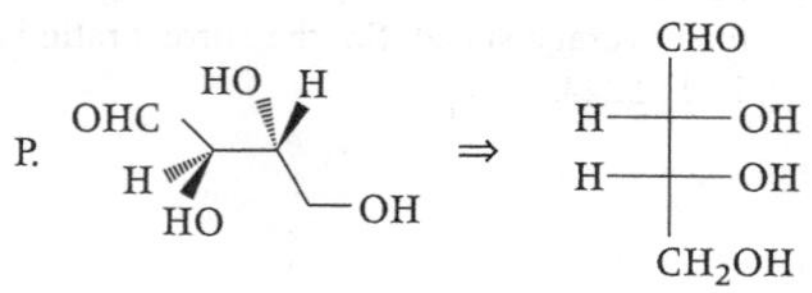

P. (wedge-dash formula: OHC, HO, H, H, HO, OH) $\Rightarrow$ Fischer projection: CHO (top); H—C—OH; H—C—OH; CH$_2$OH (bottom)

Q. 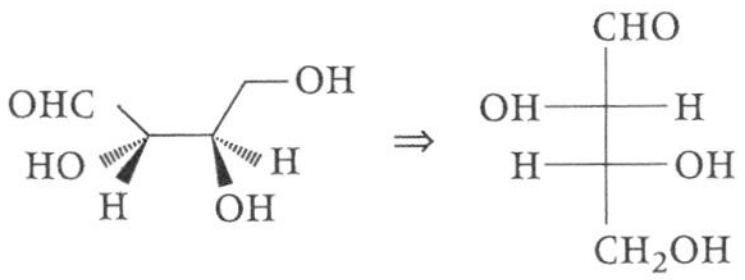 OHC, HO, H, OH, H, OH ⇒ CHO; OH—H; H—OH; $CH_2OH$

R. 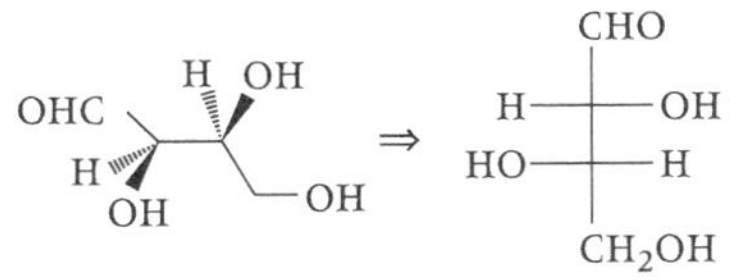 OHC, H, OH, H OH, OH ⇒ CHO; H—OH; HO—H; $CH_2OH$

S. 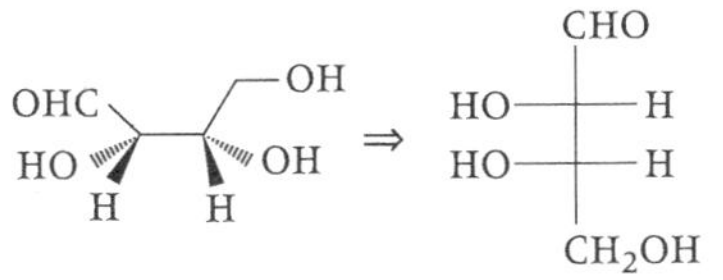 OHC, HO, H, OH, H, OH ⇒ CHO; HO—H; HO—H; $CH_2OH$

Now we can clearly see, that compound P is same as given compound, thus compound P is identical to D-erythrose.

Compound Q and D-erythrose are not mirror images of one another and are non-superimposable on one another. Thus, they are diastereomers.

Compound R and D - erythrose are also not mirror images of one another and are non-superimposable on one another. Thus, they are diastereomers.

Compound S and D-erythrose are chiral molecules. They are mirror images of one another. Furthermore, the molecules are non-superimosable on one another. This means that the molecules cannot be placed on top of one another and give the same molecule. Chiral molecules with one or more stereocenters can be enantiomers.

Hence, the correct option is (c).

**25.** *(a, b, c)* Given, $w = -\int p_{\text{ext}}\, dV$

For 1 mole van der Waals' gas

$$p = \left(\frac{RT}{V-b} - \frac{a}{V^2}\right)$$

For reversible process, $p_{\text{ext}} = p_{\text{gas}}$

$$w = -\int\left(\frac{RT}{V-b} - \frac{a}{V^2}\right) dV$$

But, it is not applicable for irreversible process which are carried out very fast. So, work done is calculated assuming final pressure remains constant throughout the process. Thus, statement (a), (b) and (c) correct while statement (d) is incorrect.

**26.** *(a, b, c)* Triphenylmethane (I) is acidic because its conjugate base is stabilised by resonance.

$$(Ph)_3CH \longrightarrow (Ph)_3\overset{\ominus}{C} + H^+$$

Triphenylmethyl carbanion

Cyclopentadiene (IV) is acidic because its conjugate base is aromatic.

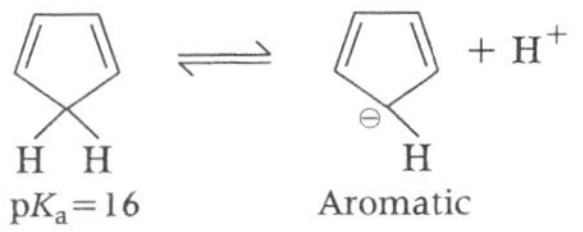

H H, $pK_a = 16$ ⇌ H, Aromatic $+ H^+$

Nitrobenzene is more acidic than benzene because nitro group is electron withdrawing. It will stabilise the conjugate base of benzene by $-R$ and $-I$ effect.

The acidic strength order on the basis of $pK_a$ data is

$$IV > V > I > II > III.$$

Hence, the correct options are (a), (b) and (c) only.

**27.** *(b, d)* Compound (*P*) undergoes Friedel-Craft acylation reaction in presence of succinic anhydride and $AlCl_3$ to gives an acid (*Q*).

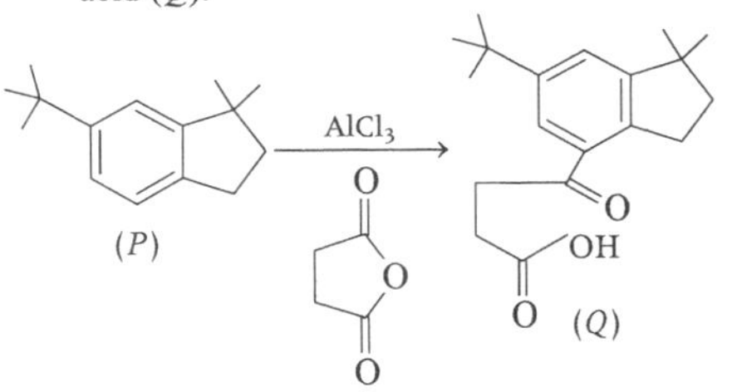

(*P*) $\xrightarrow{AlCl_3}$ (succinic anhydride) (*Q*)

Compound (*Q*) on treatment with Zn-Hg / HCl undergoes Clemmensen reduction, where ketone group change into alkane and on further reaction with dehydrating agent $H_3PO_4$ gives compound (*R*).

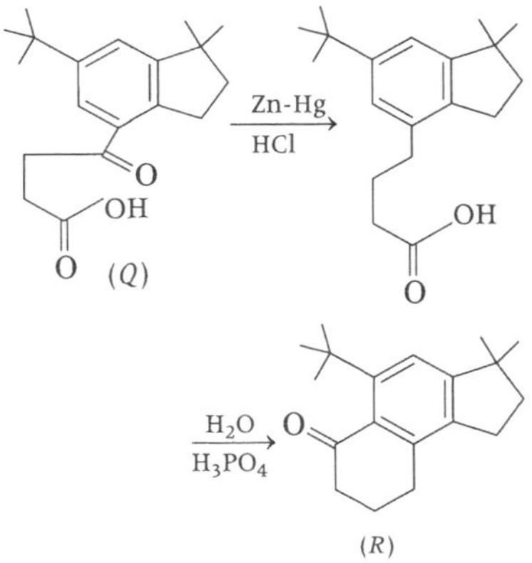

(*Q*) $\xrightarrow[HCl]{Zn\text{-}Hg}$ $\xrightarrow[H_3PO_4]{H_2O}$ (*R*)

Compound ($R$) on reaction with Grignard reagent and heated in presence of acid gives dehydrating product ($S$).

$(R)$ —(i) $CH_3MgBr$, (ii) $H_3O^+$→ [HO, $CH_3$ alcohol] —(iii) $H_2SO_4$, $-H_2O\ \Delta$→ $(S)$

Thus, the option (b) and (d) are correct.

**28.** *(a, c)*

(a) In $[FeCl_4]^-$, oxidation number of Fe atom = + 3

Electronic configuration of Fe in ground state = $3d^6 4s^2$

Electronic configuration of $Fe^{3+} = 3d^5 4s^0 4p^0$

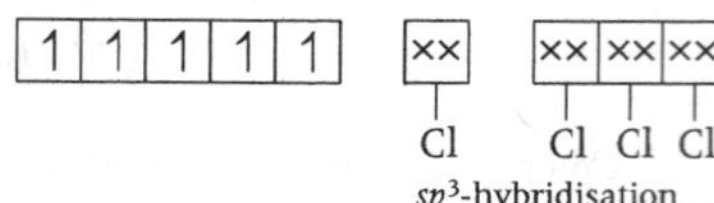


Thus, $[FeCl_4]^-$ has tetrahedral geometry.

(b) $[Co(en)(NH_3)_2Cl_2]^+$ have three geometrical isomers. Thus, statement (b) is incorrect.

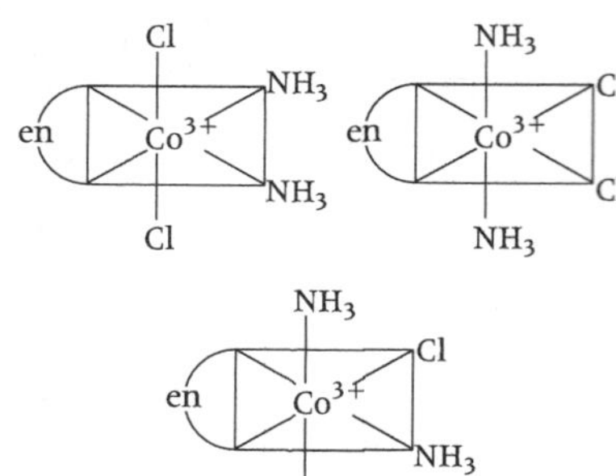


(c) $Fe^{3+}$ in $[FeCl_4]^-$ is $sp^3$-hybridised with 5 unpaired electrons. (higher spin-only magnetic moment $= \sqrt{n(n+2)} = 5.92$ BM). While $Co^{3+}$ in $[Co(en)(NH_3)_2Cl_2]^+$ is $d^2sp^3$-hybridised with zero unpaired electrons (low spin-only magnetic moment $= \sqrt{n(n+2)} = 0$ BM).

Thus, the statement (c) is correct.

(d) $Co^{3+}$ $[Co(en)(NH_3)_2Cl_2]^+$

$Co^{3+}$ : $[Ar]3d^6$ — $e_g$ [ ][ ], $t_{2g}$ [⇅][⇅][⇅], $\Delta_0 > P$

$Co^{3+}$ in $[Co(en)(NH_3)_2Cl_2]^+$ is $d^2sp^3$-hybridised and has octahedral geometry with 0 unpaired electron. Thus, statement (d) is incorrect.

**29.** *(a, b, d)*

(a) Order of acid strength different oxyacids of chlorine are :

$$\underset{\text{(Hypochlorous acid)}}{\overset{(+1)}{HClO}} < \underset{\text{(Chloric acid)}}{\overset{(+5)}{HClO_3}} < \underset{\text{(Perchloric acid)}}{\overset{(+7)}{HClO_4}}$$

Weak acid have strong conjugate base thus hypochlorite ion has strongest conjugate base. Therefore, statement (a) is correct.

(b) Hypochlorite ion is linear and perchlorate ion is tetrahedral and there is no effect of lone pair on hypochlorite ion. Thus statement (b) is correct.

$Cl—O^-$ (Hypochlorite ion) $\quad$ $^-O—Cl(=O)_2$ with lone pair (Chlorate ion) $\quad$ $^-O—Cl(=O)_3$ (Perchlorate ion)

(c) In the disproportionation reaction, chlorate ion Cl(+ 5) is oxidised to perchlorate, Cl(+ 7) and reduced to chloride, Cl (−1).

$$4ClO_3^- \longrightarrow 3ClO_4^- + Cl^-$$

While in hypochlorite ion, chlorite ion Cl(+ 1) is oxidised to chlorate, Cl(+ 5) and reduced to chloride, Cl(− 1) ion.

$$3ClO^- \longrightarrow ClO_3^- + 2Cl^-$$

Thus, statement (c) is incorrect.

(d) The hypochlorite ion oxidises the sulphite ion to sulphate ion, because HOCl is the strongest oxidising Cloxyacids,

$$ClO^- + SO_3^{2-} \longrightarrow SO_4^{2-} + Cl^-$$

Thus, statement (d) is correct.

**30.** *(a, c)*

(a) The empirical formula of the compound: Contribution of $M$ and $X$:

$M_{\left(2\times\frac{1}{2}\right)} X_{\left(4\times\frac{1}{4}\right)} \Rightarrow MX$

(b) Coordination number of both $M$ and $X$ is 8.

(c) Distance between $M$ and $X$

$= \sqrt{\frac{a^2}{4} + \frac{a^2}{2}} = \sqrt{\frac{3}{4}}a = \frac{\sqrt{3}}{2}a$

$\Rightarrow 0.866a$

(d) $r_M : r_X = (\sqrt{3} - 1) : 1 \Rightarrow 0.732:1$, thus statement (d) is incorrect.

**31.** *(0.11)* Oxalic acid solution titrated with NaOH solution using phenolphthalein as an indicator.

$$H_2C_2O_4 + 2NaOH \longrightarrow Na_2C_2O_4 + 2H_2O$$

Equivalent of $H_2C_2O_4$ reacted = Equivalent of NaOH reacted

$$= \frac{5 \times 2 \times 0.1}{1000} = \frac{9 \times M_{(NaOH)} \times 1}{1000}$$

$$M_{(NaOH)} = \frac{1}{9} = 0.11$$

**32.** *(0.25)* Given : $A \rightleftharpoons B$ $(p_A = 1 \text{ bar})$

Using $\Delta G = \Delta G^\circ + RT \ln K_p$

At equilibrium : $\Delta G^\circ = - RT \ln K_p$

$$\Delta G_1^\circ = - RT_1 \ln K_{p_1} \quad \ldots \text{(i)}$$

$$\Delta G_2^\circ = - RT_2 \ln K_{p_2} \quad \ldots \text{(ii)}$$

From Eqs. (i) and (ii),

$$\frac{\Delta G_1^\circ}{\Delta G_2^\circ} = \frac{T_1}{T_2} \times \frac{\ln K_{p_1}}{\ln K_{p_2}}$$

$$= \frac{1000}{2000} \times \frac{\ln(10)}{\ln(100)} = \frac{1}{4} = 0.25$$

**33.** *(13.32)* Vessel is insulated, thus $q = 0$

For the given reaction :

$$H_2(g) + 1/2\, O_2(g) \longrightarrow H_2O(l);$$

$$E^\circ = 1.23 - 0.00 = 1.23 \text{ V}$$

$$\Delta G^\circ = - nFE^\circ$$

$$= - 2 \times 96500 \times 1.23 \text{ J / mol}$$

Therefore, work derived from this fuel cell using 70% efficiency and on consumption of $1.0 \times 10^{-3}$ mol of $H_2(g)$

$$= 2 \times 96500 \times 1.23 \times 0.7 \times 1 \times 10^{-3}$$

$$= 166.17 \text{ J}$$

This work done = change in internal energy (for monoatomic gas,

$$C_{V,m} = 3R / 2 ),$$

$$166.17 = nC_{V,m}\Delta T$$

$$\Rightarrow \quad \Delta T = \frac{166.17 \times 2}{1 \times 3 \times 8.314}$$

$\Rightarrow 13.32$ K

**34.** *(6.15)* Aluminium reacts with sulphuric acid to form aluminium sulphate and hydrogen.

$$\underset{\left(\frac{5.4}{27} = 0.2 \text{ mol}\right)}{2Al} + \underset{\left(\frac{50 \times 5}{1000} = 0.25 \text{ mol}\right)}{3H_2SO_4} \longrightarrow Al_2(SO_4)_3 + 3H_2$$

$H_2SO_4$ is limiting reagent and moles of $H_2(g)$ produced = 0.25 mol

Using ideal gas equation,

$$pV = nRT$$

$$\Rightarrow V = \frac{0.25 \times 0.082 \times 300}{1 \text{ atm}} \Rightarrow 6.15 \text{ L}$$

**35.** *(1.2)* ${}^{238}_{92}U \longrightarrow {}^{206}_{82}Pb + 8\,{}^{4}_{2}He + 6\,{}^{0}_{-1}\beta$

Number of moles of ${}^{238}_{92}U$ present initially

$$= \frac{68 \times 10^{-6}}{238}$$

After three half-lifes, moles of ${}^{238}_{92}U$ decayed

$$= \frac{68 \times 10^{-6}}{238} \times \left(1 - \frac{1}{2^3}\right)$$

$$= \frac{68 \times 10^{-6}}{238} \times \frac{7}{8}$$

Therefore, number of $\alpha$-particles emitted

$$= \frac{68 \times 10^{-6}}{238} \times \frac{7}{8} \times 8 \times 6.023 \times 10^{23}$$

$$= 1.204 \times 10^{18}$$

$$\approx 1.2 \times 10^{18}$$

Thus, the correct answer is 1.2.

**36.** *(18)* Compound (P) on treatment with concentrated $H_2SO_4$, gives :

$H^+$ Conc. $H_2SO_4$ $-CH_3OH$

This intermediate compound, on delocalisation, gives a coloured compound $Q$.

$-\overset{\oplus}{H}$ $ArS_E2$ (Intramolecular)

(Q)

Number of rings = 4 + 1 = 5

$\pi$-bonds = $4 \times 3 + 1 = 13$

Thus, the degree of unsaturation of $Q$ is 18.

## Mathematics

**37.** *(d)* Given quadratic polynomials,

$x^2 + 20x - 2020$ and $x^2 - 20x + 2020$

having $a, b$ distinct real and $c, d$ distinct complex roots respectively.

So, $a + b = -20, ab = -2020$

and $c + d = 20, cd = 2020$

Now, $ac(a-c) + ad(a-d) + bc(b-c) + bd(b-d)$

$= a^2(c+d) - a(c^2+d^2) + b^2(c+d) - b(c^2+d^2)$

$= (c+d)(a^2+b^2) - (c^2+d^2)(a+b)$

$= (c+d)[(a+b)^2 - 2ab] - (a+b)[(c+d)^2 - 2cd]$

$= 20[(20)^2 + 4040] + 20[(20)^2 - 4040]$

$= 2 \times 20 \times (20)^2 = 40 \times 400 = 16000$

**38.** *(c)* The given function $f : R \to R$ is

$$f(x) = |x|(x - \sin x) \quad \dots \text{(i)}$$

$\because$ The function '$f$' is a odd and continuous function and as $\lim_{x \to \infty} f(x) = \infty$ and $\lim_{x \to -\infty} f(x) = -\infty$, so range is $R$, therefore '$f$' is a onto function.

$$\because f(x) = \begin{cases} x(x - \sin x), & x \geq 0 \\ -x(x - \sin x), & x < 0 \end{cases}$$

$$\therefore f'(x) = \begin{cases} 2x - \sin x - x\cos x, & x > 0 \\ -2x + \sin x + x\cos x, & x < 0 \end{cases}$$

$$\begin{cases} (x - \sin x) + x(1 - \cos x), & x > 0 \\ (-x + \sin x) - x(1 - \cos x), & x < 0 \end{cases}$$

$\because$ for $x > 0$, $x - \sin x > 0$ and $x(1 - \cos x) > 0$

$\therefore f'(x) > 0 \; \forall \; x \in (0, \infty)$

$\Rightarrow f$ is strictly increasing function, $\forall x \in (0, \infty)$.

Similarly, for $x < 0$, $-x + \sin x > 0$ and $(-x)(1 - \cos x) > 0$, therefore, $f'(x) > 0 \; \forall x \in (-\infty, 0)$

$\Rightarrow f$ is strictly increasing function, $\forall x \in (0, \infty)$

Therefore '$f$' is a strictly increasing function for $x \in R$ and it implies that $f$ is one-one function.

**39.** *(a)* The given functions $f : R \to R$ and $g : R \to R$ be defined by

$$f(x) = e^{x-1} - e^{-|x-1|}$$

$$= \begin{cases} e^{x-1} - e^{1-x}, & x \geq 1 \\ 0, & x < 1 \end{cases}$$

and $g(x) = \frac{1}{2}(e^{x-1} + e^{1-x})$

For point of intersection of curves $f(x)$ and $g(x)$ put $f(x) = g(x)$

for $x \geq 1$, $e^{x-1} - e^{1-x} = \frac{1}{2}(e^{x-1} + e^{1-x})$

$\Rightarrow \quad e^{x-1} = 3e^{1-x} \Rightarrow e^{2x} = 3e^2$

$\Rightarrow \quad x = \frac{1}{2}\log_e^3 + 1$

So, required area is

$$\int_0^{1/2\log_e^3 + 1} (g(x) - f(x))dx$$

$$= \int_0^{1/2\log_e^3 + 1} g(x)dx - \int_1^{1/2\log_e^3 + 1} f(x)dx$$

$$= \frac{1}{2}\int_0^{1/2\log_e^3 + 1} (e^{x-1} + e^{1-x})dx - \int_1^{1/2\log_e^3 + 1} (e^{x-1} - e^{1-x})dx$$

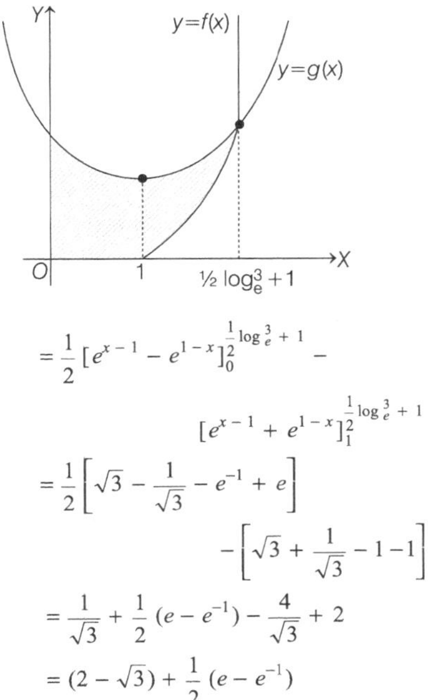


$$= \frac{1}{2}[e^{x-1} - e^{1-x}]_0^{\frac{1}{2}\log_e^3 + 1} - [e^{x-1} + e^{1-x}]_1^{\frac{1}{2}\log_e^3 + 1}$$

$$= \frac{1}{2}\left[\sqrt{3} - \frac{1}{\sqrt{3}} - e^{-1} + e\right] - \left[\sqrt{3} + \frac{1}{\sqrt{3}} - 1 - 1\right]$$

$$= \frac{1}{\sqrt{3}} + \frac{1}{2}(e - e^{-1}) - \frac{4}{\sqrt{3}} + 2$$

$$= (2 - \sqrt{3}) + \frac{1}{2}(e - e^{-1})$$

**40.** *(a)* Equation of given parabola is

$$y^2 = 4\lambda x \quad \dots \text{(i)}$$

So the end point of the latus rectum of the parabola (i), $P(\lambda, 2\lambda)$ and the given ellipse $\frac{x^2}{a^2} + \frac{y^2}{b^2} = 1$, passes through point $P(\lambda, 2\lambda)$.

On differentiating the equation of parabola, w.r.t. '$x$', we get

$$\frac{dy}{dx} = \frac{2\lambda}{y}$$

$\therefore$ Slope of tangent to the parabola at point $P$ is $m_1 = 1$

Similarly, on differentiating the equation of given ellipse, $\frac{x^2}{a^2} + \frac{y^2}{b^2} = 1$, w.r.t $x$, we get

$$\frac{2x}{a^2} + \frac{2y}{b^2}\frac{dy}{dx} = 0 \Rightarrow \frac{dy}{dx} = -\frac{b^2x}{a^2y}$$

$\therefore$ Slope of tangent to the ellipse at point $P$ is

$$m_2 = -\frac{b^2}{2a^2}$$

$\because$ It is given that the tangents are perpendicular to each other. So,

$$m_1m_2 = -1$$

$$\Rightarrow (1)\left(-\frac{b^2}{2a^2}\right) = -1$$

$$\Rightarrow \frac{b^2}{a^2} = 2 \Rightarrow b = \sqrt{2}a$$

$\therefore$ Eccentricity of ellipse $\frac{x^2}{a^2} + \frac{y^2}{b^2} = 1$ will be

$$e = \sqrt{1 - \frac{a^2}{b^2}} = \sqrt{1 - \frac{1}{2}} = \frac{1}{\sqrt{2}} \{\because b > a\}$$

**41.** *(b)* It is given that $\alpha$ is the number of heads that appear when $C_1$ is tossed twice, the probability distribution of random variable $\alpha$ is

| $\alpha$ | 0 | 1 | 2 |
|---|---|---|---|
| $P(\alpha)$ | $\left(\frac{1}{3}\right)^2$ | $2\left(\frac{2}{3}\right)\left(\frac{1}{3}\right)$ | $\left(\frac{2}{3}\right)^2$ |

Similarly, it is given that $\beta$ is the number of heads that appear when $C_2$ is tossed twice, so probability distribution of random variable $\beta$ is

| $\beta$ | 0 | 1 | 2 |
|---|---|---|---|
| $P(\beta)$ | $\left(\frac{2}{3}\right)^2$ | $\frac{4}{9}$ | $\frac{1}{9}$ |

Now, as the roots of quadratic polynomial $x^2 - \alpha x + \beta$ are real and equal, so $D = \alpha^2 - 4\beta = 0$ and it is possible if $(\alpha, \beta) = (0, 0)$ or $(2, 1)$

$\therefore$ Required probability

$$= \left(\frac{1}{3}\right)^2\left(\frac{2}{3}\right)^2 + \left(\frac{2}{3}\right)^2\left(\frac{4}{9}\right)$$

$$= \frac{4}{81} + \frac{16}{81} = \frac{20}{81}$$

**42.** *(c)* Given region is

$$\{(x, y) \in R \times R : 0 \le x \le \frac{\pi}{2}$$

and $\quad 0 \le y \le 2\sin(2x)$

On drawing the diagram,

Let the side $PS$ on the $X$-axis, such that $P(x, 0)$, and $Q(x, 2\sin(2x))$, so length of the sides $PS = QR = 2\left(\frac{\pi}{4} - x\right)$ and

$PQ = RS = 2\sin 2x$.

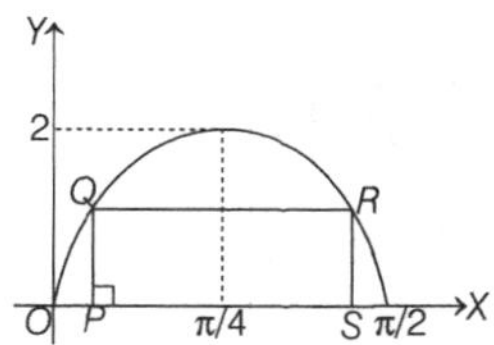


$\therefore$ Perimeter of the rectangle

$$y = 4\left[\frac{\pi}{4} - x + \sin 2x\right]$$

For maximum, $\frac{dy}{dx} = 0$

$$\Rightarrow -1 + 2\cos 2x = 0 \Rightarrow \cos 2x = \frac{1}{2}$$

$$\Rightarrow \quad 2x = \frac{\pi}{3} \Rightarrow \left\{x = \frac{\pi}{6} \in \left[0, \frac{\pi}{2}\right]\right\}$$

and $\left.\frac{d^2y}{dx^2}\right|_{x=\frac{\pi}{6}} = -4\sin 2x\big|_{x=\frac{\pi}{6}} < 0$

$\therefore$ At $x = \frac{\pi}{6}$, the rectangle $PQRS$ have maximum perimeter.

So length of sides

$$PS = QR = 2\left(\frac{\pi}{4} - \frac{\pi}{6}\right) = \frac{\pi}{6}$$

and $\quad PQ = RS = 2\sin\left(\frac{\pi}{3}\right) = \sqrt{3}$

$\therefore$ Required area $= \frac{\pi}{6} \times \sqrt{3} = \frac{\pi}{2\sqrt{3}}$

**43.** *(a, c)* Given functions $f : R \to R$ be defined by $f(x) = x^3 - x^2 + (x-1)\sin x$ and $g : R \to R$ be an arbitrary function.

Now, let $g$ is continuous at $x = 1$, then

$$\lim_{x \to 1^-} \frac{(fg)(x) - (fg)(1)}{x - 1}$$

$$= \lim_{h \to 0} \frac{(fg)(1-h) - (fg)(1)}{1 - h - 1}$$

$\because \quad (fg)(x) = f(x) \cdot g(x)$ (given)

$$= \lim_{h \to 0} \frac{f(1-h) \cdot g(1-h) - f(1) \cdot g(1)}{-h}$$

$$= \lim_{h \to 0} \frac{f(1-h) \cdot g(1)}{-h}$$

$\{\because f(1) = 0$ and $g$ is continuous at $x = 1$, so $g(1-h) = g(1)\}$

$$= g(1) \lim_{h \to 0} \frac{(1-h)^2(-h) + (-h)\sin(1-h)}{-h}$$

$$= (1 + \sin 1)g(1)$$

Similarly, $\lim_{x \to 1^+} \frac{(fg)(x) - (fg)(1)}{x - 1}$

$$= \lim_{h \to 0} \frac{f(1+h) \cdot g(1)}{h}$$

$$= g(1) \lim_{h \to 0} \frac{(1+h)^2(h) + h\sin(1+h)}{h}$$

$$= (1 + \sin 1)g(1)$$

$\because$ RHD and LHD of function $fg$ at $x = 1$ is finitely exists and equal, so $fg$ is differentiable at $x = 1$

Now, let $(fg)(x)$ is differentiable at $x = 1$, so

$$\lim_{x \to 1^-} \frac{(fg)(x) - (fg)(1)}{x - 1} = \lim_{x \to 1^+} \frac{(fg)(x) - (fg)(1)}{x - 1}$$

$$\Rightarrow \lim_{x \to 1^-} \frac{f(x)g(x) - f(1)g(1)}{x - 1} = \lim_{x \to 1^+} \frac{f(x)g(x) - f(1)g(1)}{x - 1}$$

$$\Rightarrow \lim_{x \to 1^-} \frac{f(x)g(x)}{x - 1} = \lim_{x \to 1^+} \frac{f(x)g(x)}{x - 1}$$

$\{\because f(1) = 0\}$

$$\Rightarrow \lim_{h \to 0} \frac{f(1-h)g(1-h)}{-h} = \lim_{h \to 0} \frac{f(1+h)g(1+h)}{h}$$

$$\lim_{h \to 0} \frac{[(1-h)^2(-h) + (-h)\sin(1-h)]g(1-h)}{-h}$$

$$= \lim_{h \to 0} \frac{[(1+h)^2(h) + (h)\sin(1+h)]g(1+h)}{h}$$

$$\lim_{h \to 0} [(1-h)^2 + \sin(1-h)]g(1-h) = \lim_{h \to 0} [(1+h)^2 + \sin(1+h)]g(1+h)$$

It does not mean that $g(x)$ is continuous or differentiable at $x = 1$.

But if $g$ is differentiable at $x = 1$, then it must be continuous at $x = 1$ and so $fg$ is differentiable at $x = 1$.

**44.** *(b, c, d)* It is given that matrix $M$ be a $3 \times 3$ invertible matrix, such that

$$M^{-1} = \text{adj}(\text{adj}\, M) \Rightarrow M^{-1} = |M|\, M$$

$(\because$ for a matrix $A$ of order '$n$' $\text{adj}(\text{adj}A) = |A|^{n-2}A\}$

$$\Rightarrow M^{-1}M = |M|M^2 \Rightarrow M^2\,|M| = I \quad \ldots \text{(i)}$$

$$\because \quad \det(M^2|M|) = \det(I) = 1$$

$$\Rightarrow \quad |M|^3\,|M|^2 = 1 \Rightarrow |M| = 1 \quad \ldots \text{(ii)}$$

from Eqs. (i) and (ii), we get

$$M^2 = I$$

As, adj $M = |M| M^{-1} = M$

$\Rightarrow$ $(\text{adj } M)^2 = M^2 \Rightarrow (\text{adj } M)^2 = I$

**45.** *(b, c)* It is given that the complex number $Z$, satisfying

$$|z^2 + z + 1| = 1 \Rightarrow \left|\left(z + \frac{1}{2}\right)^2 + \frac{3}{4}\right| = 1$$

$\because$ $|z_1 - z_2| \geq ||z_1| - |z_2||$

$\therefore$ $\left|\left(z + \frac{1}{2}\right)^2 - \left(-\frac{3}{4}\right)\right| \geq \left|\left|z + \frac{1}{2}\right|^2 - \left|-\frac{3}{4}\right|\right|$

$\Rightarrow$ $\left|\left|z + \frac{1}{2}\right|^2 - \frac{3}{4}\right| \leq 1$

$\Rightarrow$ $-1 \leq \left|z + \frac{1}{2}\right|^2 - \frac{3}{4} \leq 1$

$\Rightarrow$ $-\frac{1}{4} \leq \left|z + \frac{1}{2}\right|^2 \leq \frac{7}{4}$

$\Rightarrow$ $0 \leq \left|z + \frac{1}{2}\right|^2 \leq \frac{7}{4}$ $\{\because |z| \geq 0\}$

$\Rightarrow$ $0 \leq \left|z + \frac{1}{2}\right| \leq \frac{\sqrt{7}}{2}$ ... (i)

$\because$ $\left|z + \frac{1}{2}\right| \leq \frac{\sqrt{7}}{2}$

$\because$ $|z_1 + z_2| \geq ||z_1| - |z_2||$

$\therefore$ $\left|z + \frac{1}{2}\right| \geq \left||z| - \frac{1}{2}\right|$

$\Rightarrow$ $\left||z| - \frac{1}{2}\right| \leq \left|z + \frac{1}{2}\right| \leq \frac{\sqrt{7}}{2}$

$\Rightarrow$ $-\frac{\sqrt{7}}{2} \leq |z| - \frac{1}{2} \leq \frac{\sqrt{7}}{2}$

$\Rightarrow$ $\frac{1 - \sqrt{7}}{2} \leq |z| \leq \frac{\sqrt{7} + 1}{2}$

$\Rightarrow$ $|z| \leq \frac{1 + \sqrt{7}}{2}$, $\therefore |z| \leq 2$

$\because$ $\left|\left(z + \frac{1}{2}\right)^2 + \frac{3}{4}\right| \leq \left|\left|z + \frac{1}{2}\right|^2 + \frac{3}{4}\right|$

$\Rightarrow$ $1 \leq \left|\left|z + \frac{1}{2}\right|^2 + \frac{3}{4}\right| \Rightarrow \left|z + \frac{1}{2}\right|^2 + \frac{3}{4} \geq 1$

$\Rightarrow$ $\left|z + \frac{1}{2}\right|^2 \geq \frac{1}{4} \Rightarrow \left|z + \frac{1}{2}\right| \geq \frac{1}{2}$ ... (ii)

from Eqs. (i) and (ii), we get

$$\frac{1}{2} \leq \left|z + \frac{1}{2}\right| \leq \frac{\sqrt{7}}{2}$$

**46.** *(b, c)* For a $\Delta XYZ$, it is given that

$$\tan\frac{X}{2} + \tan\frac{Z}{2} = \frac{2y}{x + y + z}$$

$\Rightarrow$ $\frac{\Delta}{s(s-x)} + \frac{\Delta}{s(s-z)} = \frac{y}{s}$

$\Rightarrow$ $\Delta\frac{(s - z + s - x)}{(s - x)(s - z)} = y$

$\Rightarrow$ $\Delta = (s - x)(s - z)$

$\Rightarrow s(s-x)(s - y)(s - z) = (s - x)^2 (s - z)^2$

$\Rightarrow$ $s^2 - sy = s^2 - (x + z)s + xz$

$\Rightarrow$ $s(x + z - y) = xz$

$\Rightarrow$ $(x + z)^2 - y^2 = 2xz$

$\Rightarrow$ $x^2 + z^2 = y^2 \Rightarrow y = \frac{\pi}{2}$

$\because$ $X + Y + Z = \pi \Rightarrow X + Z = \frac{\pi}{2} = Y$

$\Rightarrow$ $X + Z = Y$

$\because$ $\tan\frac{X}{2} = \sqrt{\frac{1 - \cos x}{1 + \cos x}} = \sqrt{\frac{1 - \frac{z}{y}}{1 + \frac{z}{y}}}$

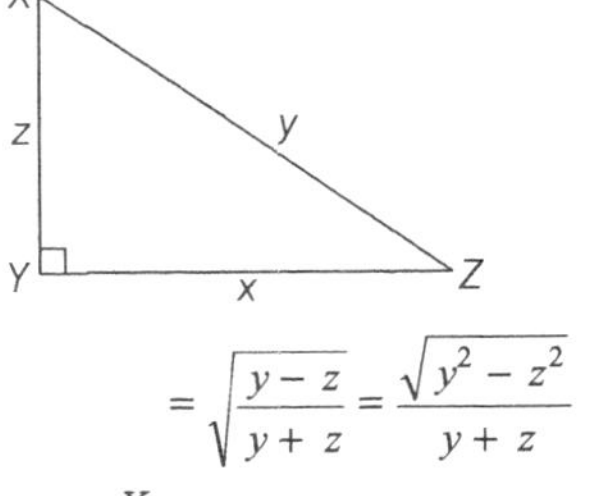


$$= \sqrt{\frac{y - z}{y + z}} = \frac{\sqrt{y^2 - z^2}}{y + z}$$

$\tan\frac{X}{2} = \frac{x}{y + z}$ $\{\because x^2 + z^2 = y^2\}$

**47.** *(a, b)* Equation of given straight lines

$$L_1 : \frac{x - 1}{1} = \frac{y}{-1} = \frac{z - 1}{3}$$

and $L_2 : \frac{x - 1}{-3} = \frac{y}{-1} = \frac{z - 1}{1}$

having vector form respectively, are

$$\mathbf{r} = (\hat{\mathbf{i}} + \hat{\mathbf{k}}) + \lambda(\hat{\mathbf{i}} - \hat{\mathbf{j}} + 3\hat{\mathbf{k}})$$

and $\mathbf{r} = (\hat{\mathbf{i}} + \hat{\mathbf{k}}) + \nu(-3\hat{\mathbf{i}} - \hat{\mathbf{j}} + \hat{\mathbf{k}})$

$\because$ $(\hat{\mathbf{i}} - \hat{\mathbf{j}} + 3\hat{\mathbf{k}}) . (-3\hat{\mathbf{i}} - \hat{\mathbf{j}} + \hat{\mathbf{k}})$

$= -3 + 1 + 3 = 1$ is positive,

$\therefore$ Angle between supporting line vectors of lines $L_1$ and $L_2$ is acute, and point of intersection of given lines $L_1$ and $L_2$ is (1, 0, 1).

Now, vector along the acute angle bisector of vectors $(\hat{\mathbf{i}} - \hat{\mathbf{j}} + 3\hat{\mathbf{k}})$ and $(-3\hat{\mathbf{i}} - \hat{\mathbf{j}} + \hat{\mathbf{k}})$ is $(-\hat{\mathbf{i}} - \hat{\mathbf{j}} + 2\hat{\mathbf{k}})$ or $(\hat{\mathbf{i}} + \hat{\mathbf{j}} - 2\hat{\mathbf{k}})$.

It is given that line

$L: \frac{x-\alpha}{l} = \frac{y-1}{m} = \frac{z-\gamma}{-2}$ is the bisector of the acute angle between the lines $L_1$ and $L_2$, so

$l = 1$ and $m = 1$

and $\quad \frac{1-\alpha}{1} = \frac{0-1}{1} = \frac{1-\gamma}{-2}$

$\Rightarrow \quad \alpha = 2, \gamma = -1$

$\therefore \quad \alpha - \gamma = 3,\ l + m = 2$

**48.** *(a, b, d)*

$\because \quad \cos x = 1 - \frac{x^2}{2!} + \frac{x^4}{4!} - \frac{x^6}{6!} + \dots$

and $\quad \sin x = x - \frac{x^3}{3!} + \frac{x^5}{5!} - \frac{x^7}{7!} + \dots$

$\therefore \quad \int_0^1 x\cos x\,dx \ge \int_0^1 \left(x - \frac{x^3}{2!}\right) dx$

$$= \left[\frac{x^2}{2} - \frac{x^4}{8}\right]_0^1 = \frac{1}{2} - \frac{1}{8} = \frac{3}{8}$$

$\Rightarrow \quad \int_0^1 x\cos x\,dx \ge \frac{3}{8}$

and, $\quad \int_0^1 x \sin\,dx \ge \int_0^1 \left(x^2 - \frac{x^4}{6}\right) dx$

$$= \left[\frac{x^3}{3} - \frac{x^5}{30}\right]_0^1 = \frac{1}{3} - \frac{1}{30} = \frac{9}{30} = \frac{3}{10}$$

$\Rightarrow \quad \int_0^1 x \sin x\,dx \ge \frac{3}{10}$

and, $\quad \int_0^1 x^2 \cos x\,dx \ge \int_0^1 \left(x^3 - \frac{x^5}{2}\right) dx$

$$= \left[\frac{x^4}{4} - \frac{x^6}{12}\right]_0^1 = \frac{1}{4} - \frac{1}{12} = \frac{2}{12} = \frac{1}{6}$$

$\therefore \quad \int_0^1 x^2 \cos x\,dx \ge \frac{1}{6}$

and, $\quad \int_0^1 x^2 \sin x\,dx \ge \int_0^1 \left(x^3 - \frac{x^5}{6}\right) dx$

$$= \left[\frac{x^4}{4} - \frac{x^6}{36}\right]_0^1 = \frac{1}{4} - \frac{1}{36} = \frac{8}{36} = \frac{2}{9}$$

$\therefore \quad \int_0^1 x^2 \sin x\,dx \ge \frac{2}{9}$

**49.** *(8.00)* For real numbers $y_1, y_2, y_3$, the quantities $3^{y_1}, 3^{y_2}$ and $3^{y_3}$ are positive real numbers, so according to the AM-GM inequality, we have

$$\frac{3^{y_1} + 3^{y_2} + 3^{y_3}}{3} \ge (3^{y_1} . 3^{y_2} . 3^{y_3})^{\frac{1}{3}}$$

$$\Rightarrow 3^{y_1} + 3^{y_2} + 3^{y_3} \ge 3\,(3^{y_1} . 3^{y_2} . 3^{y_3})^{\frac{1}{3}}$$

On applying logarithm with base '3', we get

$$\log_3(3^{y_1} + 3^{y_2} + 3^{y_3}) \ge \left[1 + \frac{1}{3}(y_1 + y_2 + y_3)\right]$$

$$= 1 + 3$$
$$= 4$$

$\{\because y_1 + y_2 + y_3 = 9\}$

$\therefore \quad m = 4$

Now, for positive real numbers $x_1, x_2$ and $x_3$, according to AM-GM inequality, we have

$$\frac{x_1 + x_2 + x_3}{3} \ge (x_1x_2x_3)^{\frac{1}{3}}$$

On applying logarithm with base '3', we get

$$\log_3\left(\frac{x_1 + x_2 + x_3}{3}\right) \ge \frac{1}{3}$$

$(\log_3 x_1 + \log_3 x_2 + \log_3 x_3)$

$\Rightarrow$

$1 \ge \frac{1}{3}(\log_3 x_1 + \log_3 x_2 + \log_3 x_3)$

$\{\because x_1 + x_2 + x_3 = 9\}$

$\therefore \quad M = 3$

Now, $\log_2(m^3) + \log_3(M^2)$

$$= 3\log_2(4) + 2\log_3(3) = 6 + 2 = 8$$

**50.** *(3.00)* Given arithmetic progression of positive integers terms $a_1, a_2, a_3, \dots$ having common difference '2' and geometric progression of positive integers terms $b_1, b_2, b_3, \dots$ having common ratio '2' with $a_1 = b_1 = c$, such that $2(a_1 + a_2 + a_3 + \dots + a_n) = b_1 + b_2 + b_3 + \dots + b_n$

$\Rightarrow \quad 2 \times \frac{n}{2}[2C + (n-1)2] = C\left(\frac{2^n - 1}{2 - 1}\right)$

$\Rightarrow \quad 2nC + 2n^2 - 2n = 2^n.C - C$

$\Rightarrow \quad C[2^n - 2n - 1] = 2n^2 - 2n$

$\because \quad C \in N \Rightarrow 2n^2 - 2n \ge 2^n - 2n - 1$

$\Rightarrow \quad 2n^2 + 1 \ge 2^n \Rightarrow n \le 6$

and, also $C > 0 \Rightarrow n > 2$

$\therefore$ The possible values of $n$ are 3, 4, 5, 6

So, at $n = 3$, $C = \dfrac{(2 \times 9) - 6}{8 - 6 - 1} = 12$

at, $n = 4$, $C = \dfrac{32 - 8}{16 - 8 - 1} = \dfrac{24}{9} = \dfrac{8}{3} \notin N$

at, $n = 5$, $C = \dfrac{50 - 10}{32 - 10 - 1} = \dfrac{40}{21} \notin N$

and at, $n = 6$, $C = \dfrac{72 - 12}{64 - 12 - 1} = \dfrac{60}{51} \notin N$

$\therefore$ The required value of $C = 12$ for $n = 3$.

**51.** (1.00) The given function $f : [0,2] \to R$ defined by

$$f(x) = (3 - \sin(2\pi x))\sin\left(\pi x - \frac{\pi}{4}\right) - \sin\left(3\pi x + \frac{\pi}{4}\right)$$

$$= (3 - \sin(2\pi x))\left[\frac{\sin \pi x}{\sqrt{2}} - \frac{\cos \pi x}{\sqrt{2}}\right] - \left\{\frac{\sin 3\pi x}{\sqrt{2}} + \frac{\cos(3\pi x)}{\sqrt{2}}\right\}$$

$$= (3 - \sin(2\pi x))\frac{[\sin(\pi x) - \cos(\pi x)]}{\sqrt{2}} - \frac{1}{\sqrt{2}}[3\sin(\pi x) - 4\sin^3(\pi x) + 4\cos^3(\pi x) - 3\cos(\pi x)]$$

$$= \frac{\sin(\pi x) - \cos(\pi x)}{\sqrt{2}}[3 - \sin(2\pi x) - 3 + 4\{\sin^2(\pi x) + \cos^2(\pi x) + \sin(\pi x)\cos(\pi x)\}]$$

$$= \frac{\sin(\pi x) - \cos(\pi x)}{\sqrt{2}}[4 + \sin(2\pi x)]$$

As, $f(x) \geq 0 \ \forall \in [\alpha, \beta]$, where $\alpha, \beta \in [0, 2]$, so

$$\sin(\pi x) - \cos(\pi x) \geq 0$$

as $\quad 4 + \sin(2\pi x) > 0 \quad \forall x \in R.$

$\Rightarrow \quad \pi x \in \left[\frac{\pi}{4}, \frac{5\pi}{4}\right] \Rightarrow x \in \left[\frac{1}{4}, \frac{5}{4}\right]$

$\therefore \quad \alpha = \frac{1}{4}$ and $\beta = \frac{5}{4}$

Therefore the value of $(\beta - \alpha) = 1$

**52.** (108.00) It is given that, in a $\Delta PQR$

$\mathbf{PQ} = \mathbf{c}$, $\mathbf{QR} = \mathbf{a}$ and $\mathbf{RP} = \mathbf{b}$, so

$\mathbf{a} + \mathbf{b} + \mathbf{c} = 0$

$\therefore \quad |\mathbf{a}|^2 + \mathbf{a} \cdot \mathbf{b} + \mathbf{a} \cdot \mathbf{c} = 0 \quad \ldots$ (i)

and $\quad \mathbf{a} \cdot \mathbf{b} + |\mathbf{b}|^2 + \mathbf{c} \cdot \mathbf{b} = 0 \quad \ldots$ (ii)

and, also given that $|\mathbf{a}| = 3$, $|\mathbf{b}| = 4$

and $\dfrac{\mathbf{a} \cdot \mathbf{c} - \mathbf{a} \cdot \mathbf{b}}{\mathbf{c} \cdot \mathbf{a} - \mathbf{c} \cdot \mathbf{b}} = \dfrac{|\mathbf{a}|}{|\mathbf{a}| + |\mathbf{b}|} = \dfrac{3}{7}$

$\Rightarrow 7[(\mathbf{a} \cdot \mathbf{c}) - (\mathbf{a} \cdot \mathbf{b})] = 3[(\mathbf{c} \cdot \mathbf{a}) - (\mathbf{c} \cdot \mathbf{b})]$

from Eqs. (i) and (ii), on putting the values of $\mathbf{a} \cdot \mathbf{c}$ and $\mathbf{c} \cdot \mathbf{b}$, we get

$\Rightarrow \quad 7[-9 - (\mathbf{a} \cdot \mathbf{b}) - (\mathbf{a} \cdot \mathbf{b})] = 3[-9 - (\mathbf{a} \cdot \mathbf{b}) + 16 + (\mathbf{a} \cdot \mathbf{b})]$

$\Rightarrow \quad 7[-9 - 2(\mathbf{a} \cdot \mathbf{b})] = 3 \times 7$

$\Rightarrow \quad \mathbf{a} \cdot \mathbf{b} = -6$

$\because |\mathbf{a} \times \mathbf{b}|^2 = (|\mathbf{a}||\mathbf{b}|)^2 - (\mathbf{a} \cdot \mathbf{b})^2$

$= 144 - 36 = 108$

**53.** (5.00) Given set $S$ of polynomials with real coefficients

$S = \{(x^2 - 1)^2(a_0 + a_1x + a_2x^2 + a_3x^3) : a_0, a_1, a_2, a_3 \in R\}$

and for a polynomial $f \in S$, Let

$$f(x) = (x^2 - 1)^2 (a_0 + a_1x + a_2x^2 + a_3x^3)$$

it have $-1$ and 1 as repeated roots twice, so graph of $f(x)$ touches the $X$-axis at $x = -1$ and $x = 1$, so $f'(x)$ having at least three roots $x = -1, 1$ and $\alpha$. Where $\alpha \in (-1, 1)$ and $f''(x)$ having at least two roots in interval $(-1, 1)$

So, $\quad m_{f'} = 3$ and $m_{f''} = 2$

$\therefore$ Minimum possible value of $(m_{f'} + m_{f''}) = 5$

**54.** (1.00) The right hand limit

$$\lim_{x \to 0^+} \frac{(1 - x)^{1/x} - e^{-1}}{x^a}$$

$$= \lim_{x \to 0^+} \frac{e^{\left\{\frac{1}{x}\log_e(1 - x)\right\}} - e^{-1}}{x^a}$$

$$= \lim_{x \to 0^+} \frac{e^{\frac{1}{x}\left(-x - \frac{x^2}{2} - \frac{x^3}{3} - \ldots\right)} - e^{-1}}{x^a}$$

$$= \lim_{x \to 0^+} \frac{e^{-1} \cdot e^{\left(-\frac{x}{2} - \frac{x^2}{3} - \frac{x^3}{4} - \ldots\ldots\right)} - e^{-1}}{x^a}$$

$$= e^{-1} \lim_{x \to 0^+} \frac{e^{-\left(\frac{x}{2} + \frac{x^2}{3} + \frac{x^3}{4} + \ldots\right)} - 1}{x^a}$$

The above limit will be non-zero, if $a = 1$. And at $a = 1$, the value of the limit is

$$= e^{-1}\left(-\frac{1}{2}\right) = -\frac{1}{2e}$$

# Paper 2

## Physics

**1.** Question was wrong. According to official website of IIT-JEE, marks were given to all students in this question.

**2.** *(9)*

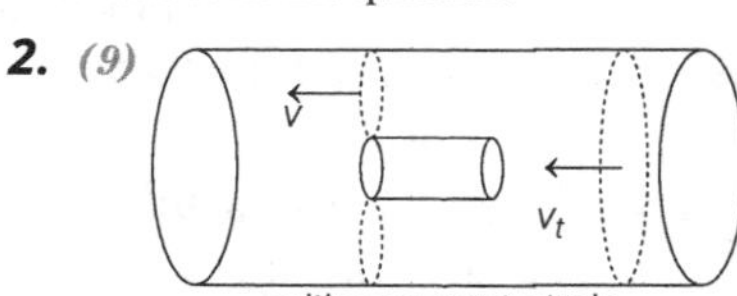


Applying Bernoulli's equation,

$$p_0 + \frac{1}{2}\rho v_1^2 = p + \frac{1}{2}\rho v^2$$

$$p_0 - p = \frac{1}{2}\rho(v^2 - v_1^2) \quad \dots \text{(i)}$$

From equation of continuity,

$$4S_t v_t = v \times 3S_t$$

$$\Rightarrow \quad v = \frac{4}{3}v_t \quad \dots \text{(ii)}$$

From Eqs. (i) and (ii), we get

$$p_0 - p = \frac{1}{2}\rho\left(\frac{16}{9}v_t^2 - v_t^2\right) = \frac{1}{2}\rho\frac{7v_t^2}{9}$$

$$\therefore \quad N = 9$$

**3.** *(6)* $qE = Mg \Rightarrow \quad q = \frac{mg}{E} = \left(\frac{mg}{V/d}\right)$

$$= \frac{900 \times \frac{4}{5}\pi(8 \times 10^{-7})^3 \times 10}{\frac{200}{0.01}}$$

$$N = \frac{q}{e} = \frac{900 \times 4\pi \times 8^3 \times 10^{-21} \times 10 \times 0.01}{200 \times 1.6 \times 10^{-19}}$$

$$N = 6$$

**4.** *(4)* Weight = Upthrust

$$mg = F_u \Rightarrow 480 \times 10 = \rho V g$$

$$480 \times 10 = \rho_0 e^{-\frac{h}{h_0}} \Rightarrow 480 \times 10 = \rho_0 e^{-\frac{100}{6000}} Vg \quad \dots \text{(i)}$$

$$(480 - N \times 1)\, 10 = \rho' Vg$$

$$(480 - N)\, 10 = \rho_0 e^{-\frac{150}{6000}} Vg \dots \text{(ii)}$$

Dividing Eq. (i) by Eq. (ii), we get

$$\frac{480}{480 - N} = e^{\left(\frac{150-100}{6000}\right)}$$

$$\frac{480}{480 - N} = e^{\frac{50}{6000}} \Rightarrow \frac{480}{480 - N} = e^{\frac{1}{120}}$$

$$N = 4$$

**5.** *(2)*

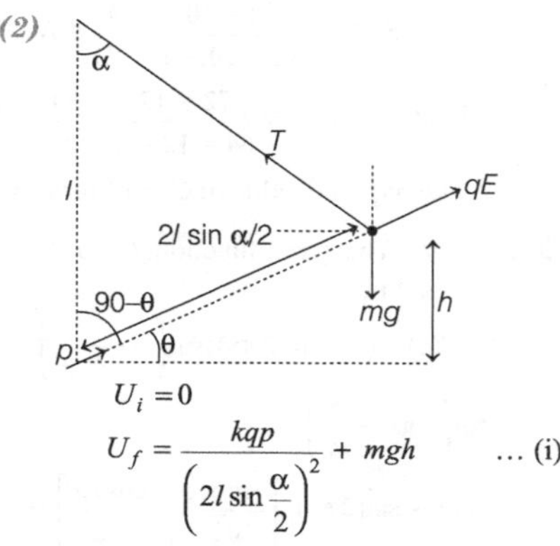


$$U_i = 0$$

$$U_f = \frac{kqp}{\left(2l\sin\frac{\alpha}{2}\right)^2} + mgh \quad \dots \text{(i)}$$

Now, from $\Delta OAB$,

$$\alpha + 90 - \theta + 90 - \theta = 180$$

$$\Rightarrow \quad \alpha = 2\theta$$

From $\Delta ABC, h = 2l\sin\left(\frac{\alpha}{2}\right)\sin\theta$

$$h = 2l\sin\left(\frac{\alpha}{2}\right)\sin\left(\frac{\alpha}{2}\right) \Rightarrow h = 2l\sin^2\left(\frac{\alpha}{2}\right)$$

Now, charge is in equilibrium at point $B$. So, using sine rule,

$$\frac{mg}{\sin\left(90 + \frac{\alpha}{2}\right)} = \frac{qE}{\sin 1(80 - 2\theta)}$$

$$\Rightarrow \quad \frac{mg}{\cos\frac{\alpha}{2}} = \frac{qE}{\sin 2\theta}$$

$$\Rightarrow \frac{mg}{\cos\frac{\alpha}{2}} = \frac{qE}{\sin\alpha} \Rightarrow \frac{mg}{\cos\frac{\alpha}{2}} = \frac{qE}{2\sin\frac{\alpha}{2}\cos\frac{\alpha}{2}}$$

$$\Rightarrow \quad qE = mg\, 2\sin\left(\frac{\alpha}{2}\right)$$

$$\Rightarrow \quad \frac{q2kp}{\left(2l\sin\frac{\alpha}{2}\right)^3} = mg\, 2\sin\left(\frac{\alpha}{2}\right)$$

$$\Rightarrow \quad \frac{kpq}{\left(2l\sin\frac{\alpha}{2}\right)^2} = mg\sin\left(\frac{\alpha}{2}\right) \times \left(2l\sin\frac{\alpha}{2}\right)$$

$$\Rightarrow \quad \frac{kpq}{\left(2l\sin\frac{\alpha}{2}\right)^2} = mgh$$

Substituting this in Eq. (i), we get

$$U_f = mgh + \frac{kpq}{\left(2l\sin\frac{\alpha}{2}\right)^2} \Rightarrow U_f = 2mgh$$

$$W = \Delta U = Nmgh = N = 2$$

**6.** Question was wrong. According to official website of IIT-JEE, marks were given to all students in this question.

**7.** *(a, d)*

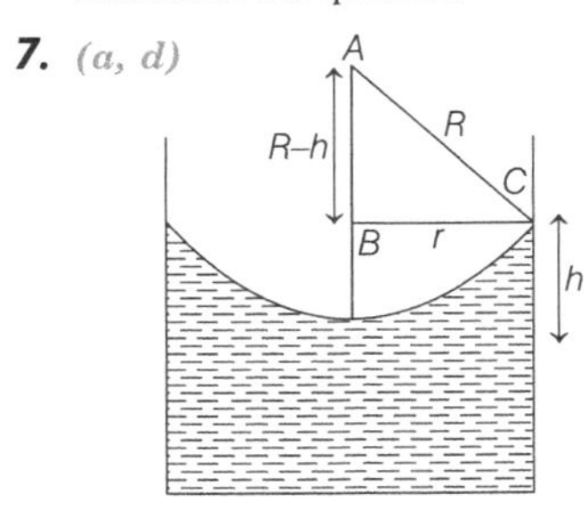


$$(R-h)^2 + r^2 = R^2$$

$$R^2 + h^2 - 2Rh + r^2 = R^2$$

$$h^2 + r^2 = 2Rh \Rightarrow \frac{h^2+r^2}{2h} = R \quad \ldots \text{(i)}$$

When $r >>> h$, then

$$R = \frac{r^2}{2h} \Rightarrow h = \frac{\omega^2 r^2}{2g}$$

$$R = \frac{r^2 2g}{2\omega^2 r^2} \Rightarrow R = \frac{g}{\omega^2} \quad \ldots \text{(ii)}$$

$$\frac{\mu_1}{v} - \frac{\mu_2}{u} = \frac{\mu_1 - \mu_2}{R} \Rightarrow \frac{1}{v} + \frac{4}{3u} = \frac{1-\frac{4}{3}}{R}$$

$$\frac{1}{v} = -\left(\frac{1}{3R} + \frac{4}{3(H-h)}\right)$$

$$\frac{1}{v} = -\left(\frac{1}{3R} + \frac{4}{3H}\right)$$

Putting the value of Eq. (ii) $h << H$

$$\frac{1}{v} = -\left[\frac{4}{3H}\left(1 + \frac{3H}{4}\frac{\omega^2}{39}\right)\right]$$

$$v = -\left[\frac{3H}{4}\left(1 + \frac{\omega^2 H}{4}\right)^{-1}\right]$$

**8.** *(a, d)* Apply energy conservation,

$$\frac{1}{2}mv_0^2 = mgh + \frac{1}{2}mv^2$$

$$\frac{1}{2}mv^2 = \frac{1}{2}mv_0^2 - mgh \quad \ldots \text{(i)}$$

At top point $y$,

$$mg\sin\theta = \text{centripetal force} = \frac{mv^2}{R}$$

$$\frac{mv^2}{R} = mg\sin\theta$$

Putting this value in Eq. (i), we get

$$\frac{mgR\sin\theta}{2} = \frac{1}{2}mv_0^2 - mgh$$

$$\frac{mgR}{2}\frac{1}{2} = \frac{1}{2}mv_0^2 - mgh$$

$$\frac{yR}{2} = v_0^2 - 2gh$$

**9.** *(a, c, d)* According to conservation of angular momentum about suspension point,

$$mvx = \left(\frac{mL^2}{3} + mx^2\right)\omega$$

$$vx = \left(\frac{L^2}{3} + x^2\right)\omega \Rightarrow \omega = \frac{3vx}{L^2 + 3x^2}$$

For maximum,

$$\frac{d\omega}{dx} = 0 \Rightarrow x = \frac{L}{\sqrt{3}} \Rightarrow \omega = \frac{\sqrt{3}\,v}{2L}$$

**10.** *(a, c)* $\lambda \propto \frac{1}{V}$

So, when $V$ double, then cut-off wavelength become half. Characteristic does not depend on voltage. When $I$ is halved, then intensity will decrease.

**11.** *(b, c)* In air,

$$T\cos\theta = mg \quad \ldots \text{(i)}$$

$$T\sin\theta = qE \quad \ldots \text{(ii)}$$

$$qE = mg\tan\theta \quad \ldots \text{(iii)}$$

In liquid,

$$T'\cos\theta = mg - u$$

$$T'\cos\theta = mg\left(1 + \frac{\rho_l}{\rho_s}\right) \quad \ldots \text{(iv)}$$

$$T'\sin\theta = \frac{qE}{K} \quad \ldots \text{(v)}$$

Dividing Eq. (v) by Eq.(ii), we get

$$\frac{T'}{T} = \frac{1}{K} \Rightarrow T' = \frac{T}{K} = \frac{T}{21} \quad \ldots \text{(vi)}$$

Dividing Eq. (iv) by Eq. (v), we get

$$\frac{qE}{K} = mg\left(1 - \frac{\rho_l}{\rho_s}\right)\tan\theta$$

$$qE = Kmg\left(1 - \frac{\rho_l}{\rho_s}\right)\tan\theta \quad \text{...(vii)}$$

From Eqs. (vii) and (iii), we get

$$\rho_s = 840 \text{ kg/m}^3$$

**12.** *(a, b, c, d)* Given, equation $y = \frac{x^2}{2}$

$$\frac{dy}{dt} = \frac{2x}{2}\frac{dx}{dt} \Rightarrow v_y = xv_x$$

$$\frac{dv_y}{dt} = x\frac{dv_x}{dt} + \frac{dx}{dt}v_x \Rightarrow a_y = xa_x + v_x^2$$

Put the condition of option you get the answer.

**13.** *(2.05)* $W = (\Delta p)_{\text{avg}} \times 4\pi R^2 a$

$$\approx \left|\frac{dp}{2} \cdot 4\pi R^2 a\right|$$

[for small change $(\Delta p)_{\text{avg}} < p >$ arithmetic mean]

For adiabatic process, $\frac{dp}{dV} = -\gamma\left(\frac{p}{V}\right)$

$$\therefore \quad dp = -\gamma\frac{p}{V}dV = -\frac{\gamma p_0}{V}4\pi R^2 a$$

$$= \frac{\gamma p_0}{2V} \times 4\pi R^2 a \times 4\pi R^2 a$$

$$= \frac{\gamma p_0}{2 \times 4\pi R^3} 4\pi R^2 a \times 4\pi R^2 a$$

$$= (4\pi Rp \times a^2)\frac{3\gamma}{2} \Rightarrow x \approx 2.05$$

**14.** *(0.27)* $R_3' = 300\,(1 + \alpha\Delta T) = 312\,\Omega$

Now,

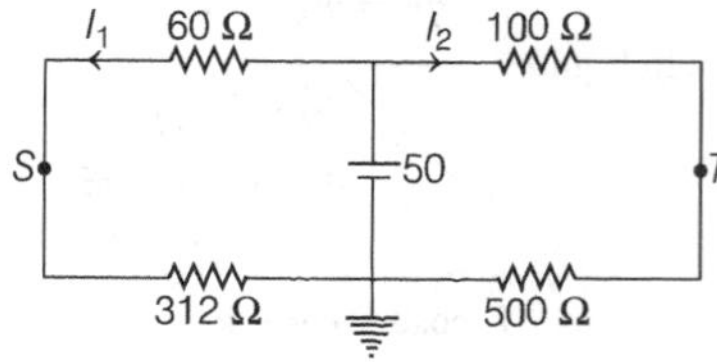


$$I_1 = \frac{50}{372} \text{ and } I_2 = \frac{50}{600}$$

$$V_S - V_T = 312I_1 - 500I_2$$

$$= 41.94 - 41.67 = 0.27 \text{ V}$$

**15.** *(1.3)* $\because \quad \frac{1}{C_{\text{eq}}} = \frac{1}{C_1} + \frac{1}{C_2}$

$$\Rightarrow \quad C_{\text{eq}} = \frac{2000 \times 3000}{5000} = 1200 \text{ pF}$$

$$\frac{-dC_{\text{eq}}}{C_{\text{eq}}^2} = \frac{-dC_1}{C_1^2} - \frac{dC_2}{C_2^2}$$

$$\Rightarrow \quad dC_{\text{eq}} = 6 \text{ pF} \Rightarrow \varepsilon = \frac{1}{2}CV^2$$

$$\frac{d\varepsilon}{q} \times 100 = \left(\frac{dC}{C} + \frac{2dV}{V}\right) \times 100$$

$$= \left(\frac{6}{1200} + 2 \times \frac{0.02}{5}\right) \times 100 = 1.3\%$$

**16.** *(0.24)* $\frac{dV}{V} = -\frac{dp}{B}$ (where, $B$ = bulk modulus)

$$V = l^3 \Rightarrow \frac{\Delta V}{V} = 3\frac{\Delta l}{l}$$

$$3\frac{\Delta l}{l} = \left|-\frac{\Delta p}{B}\right| = \frac{\rho g h}{B} \Rightarrow \Delta T = \frac{\rho g h l}{3B}$$

Substituting the given values, we get $\Delta l = 0.24$ mm

**17.** *(55.00)* Mutual inductance is producing flux in same direction as self-inductance.

$$\therefore \quad U = \frac{1}{2}L_1 I_1^2 + \frac{1}{2}L_2 I_2^2 + MI_1 I_2$$

$$\Rightarrow \quad U = \frac{1}{2} \times (10 \times 10^{-3})1^2$$

$$+ \frac{1}{2} \times (20 \times 10^{-3}) \times 2^2 + (5 \times 10^{-3}) \times 1 \times 2$$

$$= 55 \text{ mJ}$$

**18.** *(8.33)* $\frac{dQ}{dt} = e\sigma A(T^4 - T_0^4)$

For small temperature change,

$$\frac{dQ}{dt} = e\sigma AT^3\Delta T \quad \text{...(i)}$$

$$\frac{mCdT}{dt} = e\sigma AT^3\Delta T \Rightarrow \frac{dT}{dt} = \frac{e\sigma AT^3}{mC}\Delta T$$

$\frac{e\sigma AT^3}{mC} \rightarrow$ constant for Newton law of cooling

$$\frac{e\sigma AT^3}{mC} = 0.001 \Rightarrow e\sigma AT^3 = mC \times 0.001$$

$$= 1 \times 4200 \times 0.001$$

$$e\sigma AT^3 = 4.2 \quad \text{...(ii)}$$

$$\frac{dQ}{dt} = 700 \times 0.05 = 35 \text{ W} \quad \text{...(iii)}$$

Putting the value of Eqs. (ii) and (iii) in Eq. (i), we get

$$35 = 4.2\,\Delta T \Rightarrow \frac{35}{4.2} = \Delta T \Rightarrow \Delta T = 8.33$$

## Chemistry

**19.** (9) By observing the values of different ionisation energies, $I_1$, $I_2$ and $I_3$ for atomic number $(n + 2)$, it is observed that there is very large difference between the second ionisation energy and first ionisation energy $(I_2 >> I_1)$.

This indicates that number of valence shell electrons is 1 and atomic number $(n + 2)$ should be an alkali metal.

Also for atomic number $(n + 3)$, $I_3 >> I_2$.

This indicates that it will be an alkaline earth metal which suggests that atomic number $(n + 1)$ should be a noble gas and atomic number $(n)$ should belong to halogen family. Since, $n < 10$; hence, $n = 9$ (F atom)

**20.** (6) Only polar liquid will be attracted towards charged comb due to the formation of electrically charged droplets in the polar liquid stream, induced by a nearby charged object. Hence, liquid showing deflection are $HF, H_2O, NH_3, H_2O_2, CHCl_3, C_6H_5Cl$.

**21.** (6) In alkaline medium : Iodide is oxidised to iodate

$$2MnO_4^- + H_2O + I^- \longrightarrow 2MnO_2 + 2OH^- + IO_3^-$$

But, in weakly basic solution :

$$K\overset{+7}{Mn}O_4 + K\overset{-1}{I} \longrightarrow \overset{+4}{Mn}O_2 + \overset{0}{I_2}$$

$$\text{Eq. of } KMnO_4 = \text{Eq. of } I_2$$

$$4 \times 3 = n \times 2 \Rightarrow n = 6$$

**22.** (4) When a solution of $K_2CrO_4$ is treated with amyl alcohol and acidified $H_2O_2$, the layer of amyl alcohol turns blue because acidified $H_2O_2$ converts $K_2CrO_4$ to $CrO_5$ to given the blue colouration,

$$CrO_4^{2-} + 2H + 2H_2O_2 \longrightarrow \underset{\text{(Blue coloured compound)}}{CrO_5} + 3H_2O$$

$CrO_5 \Rightarrow$ [structure: Cr with one O double bonded and two O–O peroxo groups]

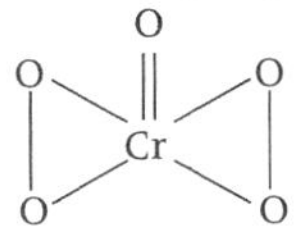

Number of oxygen atom bonded with chromium with single bond is (4).

**23.** (5) At pH = 2,

There are two $—NH_2$ group, and + 1 charge on each group because all amino groups exist in the form of $— NH_3^{\oplus}$.

Therefore, $|Z_1| = 2$.

At pH = 6,

$NH_2$ of lysine (+ 1) (pH = 9.47) and $COOH(-1)$ of glutamic (pH = 3.08) acid, so because of dipolar ion exists, therefore $|Z_2| = 0$.

At pH = 11,

COOH of glutamic acid has (– 1), COOH of lysine (– 1) and OH of phenol (– 1).

Therefore, $|Z_3| = |-3| = 3$ (All COOH and OH exist in the form of $—COO^-$ and $—O^-$).

$$\therefore \quad |Z_1| + |Z_2| + |Z_3| = 2 + 0 + 3 = 5$$

**24.** (6) Phenolic (–OH) group gives positive test with neutral $FeCl_3$ solution, means organic compound $(C_8H_{10}O_2)$ is a phenol derivative.

Compound $(C_8H_{10}O_2)$ also rotate plane polarised light means compound is an optically active and chiral carbon atom is present.

So, the possible structures which are optically active and have phenolic group are as followed :

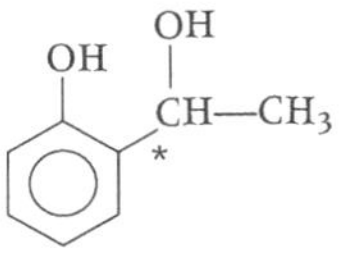

[o-isomer: OH on ring, *CH(OH)–CH₃] $(d+l)=2$

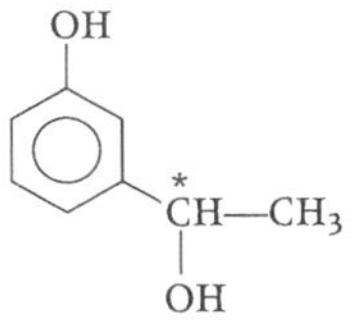

[m-isomer: OH on ring, *CH(OH)–CH₃] $(d+l)=2$

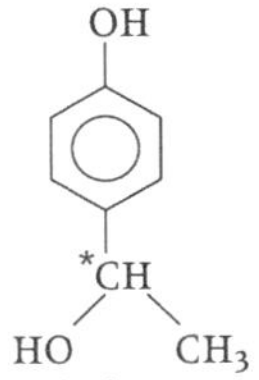

[p-isomer: OH on ring, *CH(OH)CH₃] $(d+l)=2$

Therefore, total optically active isomers will be 6.

**25.** *(a, b, c)* Paramagnetism is a form of magnetism whereby some materials are attracted by an externally applied magnetic field, and form internal, induced magnetic fields in the direction of the applied magnetic field. So, magnetic balance shows downward deflection. While diamagnetic substance shows repulsion in magnetic field and magnetic balance shows upward deflection.

(a) $X = H_2O(l)$
(Water has no unpaired electrons and is thus diamagnetic). Hence, statement (a) is correct.

(b) $X = K_4[Fe(CN)_6](s)$
(CN is a strong field ligand which forces the $d$-orbital electrons to pair up $(t_{2g}^6 e_g^0)$ and making it diamagnetic. Hence, statement (b) is correct)

(c) $X = O_2$ (g)
[Here, $O_2(g)$ is paramagnetic due to two-unpaired electrons present in $\pi^*$ (antibonding orbitals). Hence, statement (c) is correct.

(d) $X = C_6H_6(l)$
(Here, $C_6H_6$ is diamagnetic due to presence of 0 unpaired electrons. Hence, statement (d) is incorrect.

**26.** *(a)* $(CH_3)_3C—Br + NaOH \xrightarrow[\text{(First o rder reaction)}]{S_N1} (CH_3)_3C—OH + NaBr$

This is first order reaction and for first order reaction $t_{1/2} = \dfrac{0.693}{k}$

So, half-life is independent of initial concentration.

Therefore, the plot (a) correct.

For first order reaction,
$(r) = k[(CH_3)_3C—Br]$;

$$\ln\left(\frac{P_0}{P}\right) = k\ x\ t$$

Or $$\ln\left(\frac{P}{P_0}\right) = -\ kxt$$

Hence, plot (b) and (d) are incorrect.

For first order reaction,

$$Q = [P_0](1 - e^{kt})$$

Or $$\frac{[Q]}{[P_0]} = (1 - e^{kt})$$

Hence, plot (c) is incorrect.

**27.** *(a, b, c, d)*

(a) $2Na[Al(OH)_4](aq) + CO_2 \longrightarrow Na_2CO_3 + H_2O + 2Al(OH)_3 \downarrow$ or $Al_2O_3.2H_2O$ (ppt.)
Hence, the statement (a) is true.

(b) The electrolysis of alumina by Hall-Heroult's process is carried by using a fused mixture of alumina and cryolite ($Na_2AlF_6$) along with minor quantities of aluminium fluoride and fluorspar. The addition of cryolite and fluorspar increases the electrical conductivity of alumina and lowers the fusion temperature. Hence, the statement (b) is true.

(c) At anode, alumina reacts with fluorine to liberate oxygen and evolved oxygen reacts with carbon to form carbon dioxide and hence, statement (c) is true.

(d) Steel cathode with carbon lining and graphite anode are used and hence,the statement (d) is true.

**28.** *(a, b)*

(a) $Sn^{2+}$ of stannous chloride dihydrate $(SnCl_2 \cdot 2H_2O)$ tends to convert into $Sn^{4+}$.
Hence, statement (a) is correct.

(b) $SnO_2$ reacts with KOH and gives $K_2SnO_3.3H_2O$ or $K_2[Sn(OH)_6]$ because it is amphoteric in nature.

$$SnO_2 + KOH \longrightarrow K_2SnO_3 + H_2O$$
or $$K_2[Sn(OH)_6]$$

Hence, statement (b) is correct.

(c) In conc. HCl, $PbCl_2$ exists as chloroplumbous acid, $H_2[PbCl_4]$

$$\overset{II}{Pb}Cl_2 + 2HCl \longrightarrow H_2[\overset{II}{Pb}Cl_4]$$

Hence, statement (c) is incorrect.

(d) $Pb_3O_4$ is a mixture of $(2\overset{+2}{Pb}O + \overset{+4}{Pb}O_2)$

$$Pb_3O_4 + 4HNO_3 \longrightarrow 2\overset{+2}{Pb}(NO_3)_2 + \overset{+4}{Pb}O_2 + 2H_2O$$

It is not a redox reaction. Thus, the statement (d) is incorrect.

**29.** *(c, d)*

(a) The correct basic strength order is
(IV) > (II) > (I) > (III);
(IV) is strongest base due to steric inhibition to resonance effect.
(III) is weakest base due to $-M$ group of three nitro groups present at *ortho* and *para* positions.
(II) is stronger than (I) since (III) is tertiary and (I) primary aromatic amine.
So, option (a) is incorrect.

(b) $pK_b$ different between I and II is 0.53 and that of III and IV is 4.6. So, option (b) is incorrect.

(c) and (d) In 2, 4, 6-trinitro aniline (III) due to strong $-R$ effect of — $NO_2$ groups, the lone pair of — $NH_2$ is more involved with benzene ring hence it has least basic strength. Whereas (IV) N, N-dimethyl 2, 4, 6-trinitro aniline, due to steric inhibition to resonance (SIR) effect; the lone pair of nitrogen is not in the plane of benzene, hence makes it lone pair more free to protonate.

**30.** *(b,c)* Compound (*P*) on reaction with Lindlar catalyst (*X*) followed by oxidation in presence of acidified $KMnO_4$, gives an optically active compound (*Q*) 3-carboxy-heptane-1, 7-dioic acid.

(*P*) $\xrightarrow[(X)]{Pd\text{–}C/H_2}$ [cyclohexene ring C5, C6, C4, C3 with side chain 2, $\overset{1}{C}H=CH_2$]

$\xrightarrow[\Delta - CO_2]{KMnO_4/H_2SO_4}$ $\overset{7}{C}OOH$–6–5–4–$\overset{3*}{C}H(COOH)$–2–$\overset{1}{C}OOH$

(*Q*)

Compound (*P*) also react with strong base ($NaNH_2$), gives alkyaline which one treatment with $C_6H_5COCH_3 / H_3O^+ / \Delta$, gives an optically active compound (*R*).

(*P*) $\xrightarrow{\text{(i) } NaNH_2}$ cyclohexenyl–$CH_2$–$C\equiv\overset{\ominus}{C}\overset{\oplus}{Na}$

$\downarrow$ (ii) $C_6H_5COCH_3$ (iii) $H_3O^+/\Delta$, $-H_2O$

cyclohexenyl*–$CH_2$–$C\equiv C$–$C(=CH_2)$–$C_6H_5$

(*R*)
Optically active

**31.** *(3.00)* From the given diagram, 6 mL volume of H*A* used till equivalene point. At half of equivalence point, solution will be basic buffer with *B* and $BH^+$.

$$\because \quad pOH = pK_b + \log\frac{[BH^+]}{[B]}$$

At half equivalence point :

$$[BH^+] = [B] \qquad (\because pH = 11)$$

Therefore, $pOH = pK_b = 14 - 11 = 3$

$\because \quad pK_b = 3.00$

**32.** *(0.2)* Using Raoult's law equation for a mixture of volatile liquids.

$$p_T = p_A^\circ \chi_A + p_B^\circ \chi_B$$

$$0.3 = 0.25 \chi p_A^\circ + 0.75 \chi p_B^\circ \qquad \ldots \text{(i)}$$

$$0.4 = 0.5 \chi p_A^\circ + 0.5 \chi p_B^\circ \qquad \ldots \text{(ii)}$$

By solving equation (i) and (ii)

$p_A^\circ = 0.6$ bar and $p_B^\circ = 0.2$ bar

Thus, the vapour pressure of pure liquid *B* in bar is 0.2.

**33.** *(−5242.41)* Given that, electrons and nucleus are at infinite distance, so potential energy of H-atom is taken as zero.

Therefore, according to Bohr's model, potential energy of a H-atom with electron in its ground state = − 27.2 eV

At $d = d_0$, nucleus-nucleus and electron-electron repulsion is absent.

Hence, potential energy will be calculated for 2 H atoms = − 2 × 27.2 eV= − 54.4 eV

Potential enery of 1 mol H atoms in kJ

$$= \frac{54.4 \times 6.02 \times 10^{23} \times 1.6 \times 10^{-19}}{1000}$$

$$= -5242.4192 \text{ kJ/mol}$$

**34.** *(18.60)* Aniline ($P$) is treated with $NaNO_2$ and HCl in cold condition to form benzene diazonium chloride. The process of conversion of primary aromatic amines into its diazonium salt is called diazotisation.

$NH_2$ (aniline) $\xrightarrow[0-5°C]{HCl + NaNO_2}$ $N_2^+Cl^-$ (benzene diazonium chloride)

β-naphthol couples with phenyldiazonium electrophile to produce an intense orange-red dye ($Q$) as major product.

$N_2^+Cl^-$ + β-naphthol (2-naphthol) (OH) $\xrightarrow{0-5°C}$ Ph–N=N–(naphthol, OH) + HCl

2-naphthol aniline dye
(Orange-red dye)

Given that, volume of aniline ($P$) $= 9.3$ mL
(density of $P = 1.00$ g/mL)
So, mass of aniline $= 9.3$ g
Molecular mass of aniline ($C_6H_7N$) $= 93$ g/mol
Therefore, moles of aniline $= 9.3 / 93 = 0.1$ mol of $P$.
Molecular mass of 2 napthol aniline orange dye ($Q$) $= 248$ g/mol
$\Rightarrow$ 0.1 mol of aniline ($P$) will produce 0.1 mol of compound ($Q$).
But, according to the question the major product $Q$ from $P$ is 75%.
Therefore, mass of '$Q$' produced

$$= (0.1 \times 248 \times 0.75)\text{g} = 18.60\,\text{g}$$

**35.** *(935)* Tin is obtained from cassiterite by reduction with coke and the balanced chemical reaction is

$$SnO_2(s) + C(s) \longrightarrow Sn(s) + CO_2(g)$$

Standard enthalpy of reaction,

$$\Delta H^\circ_{R^n} = (\Delta H^\circ_f\ CO_2(g)) - (\Delta H^\circ_f\ SnO_2(s))$$

$$\Delta H^\circ_{R^n} = -394 - (-581) \Rightarrow 187\text{ kJ}$$

Standard entropy of reaction,

$$\Delta S^\circ_{R^n} = \Delta S^\circ_{\text{Products}} - \Delta S^\circ_{\text{Reactants}}$$

$$\Delta S^\circ_{R^n} = [S^\circ(Sn(s)) + S^\circ(CO_2(g))] - [S^\circ(SnO_2(s)) + S^\circ(C(s))]$$

$$\Delta S^\circ_{R^n} = [52 + 210] - [56 + 6]$$

$$\Rightarrow 200\text{ JK}^{-1}\text{ mol}^{-1}$$

We know that, $\Delta H^\circ = T\Delta S^\circ$

$$\therefore \quad T = \frac{\Delta H^\circ}{\Delta S^\circ} = \frac{187 \times 1000}{200} \Rightarrow 935\text{K}$$

For the reaction to be spontaneous, the temperature should be greater than 935 K.

**36.** *(0.2)* $ZnS(s) \longrightarrow \underset{(0.05\text{ M})}{Zn^{2+}}(aq) + S^{2-}(aq)$

$$K_{sp}(ZnS) = [Zn^{2+}][S^{2-}] = 1.25 \times 10^{-22}$$

$$0.05 \times [S^{2-}] = 1.25 \times 10^{-22}$$

$$\Rightarrow [S^{2-}] = \frac{1.25 \times 10^{-22}}{0.05} \Rightarrow 25 \times 10^{-22}\text{ M}$$

For $H_2S$, $\underset{(0.1\text{ M})}{H_2S} \longrightarrow 2H^+ + \underset{(25 \times 10^{-22}\text{ M})}{S^{-2}}$

$$K_{net} = 1 \times 10^{-21} = \frac{[H^+]^2[S^{2-}]}{[H_2S]}$$

$$1 \times 10^{-21} = \frac{[H^+]^2 \times 25 \times 10^{-22}}{[0.1]}$$

$$[H^+]^2 = \frac{1}{25}$$

$$[H^+] = \frac{1}{5} \Rightarrow 0.2\text{ M}$$

## Mathematics

**37.** *(8)* For a complex number $z$, it is given that

$$z^4 - |z|^4 = 4iz^2$$

$$\Rightarrow \quad z^4 - z^2\bar{z}^2 = 4iz^2$$

$$\Rightarrow \quad z^2(z - \bar{z})(z + \bar{z}) = 4iz^2$$

So, either $z^2 = 0$ or $(z - \bar{z})(z + \bar{z}) = 4i$

Now, **Case-I**, if $z^2 = 0$ and $z = x + iy$

So, $\quad x^2 - y^2 + 2ixy = 0$

$\Rightarrow \quad x^2 - y^2 = 0$

and $\quad xy = 0$

$\Rightarrow \quad x = y = 0$

$\Rightarrow z = 0$, which is not possible according to given conditions.

**Case-II**, if $(z - \bar{z})(z + \bar{z}) = 4i$ and $z = x + iy$

So, $(2iy)(2x) = 4i$

$\Rightarrow xy = 1$ is an equation of rectangular hyperbola and for minimum value of $|z_1 - z_2|^2$, the $z_1$ and $z_2$ must be vertices of the rectangular hyperbola.

Therefore, $z_1 = 1 + i$ and $z_2 = -1 - i$

$\therefore$ Minimum value of $|z_1 - z_2|^2$

$= (1+1)^2 + (1+1)^2 = 4 + 4 = 8.$

**38.** (6) It is given that the probability, a missile hits a target successfully $p = \frac{3}{4}$, so the probability to not hits the target is $q = \frac{1}{4}$.

And it is also given that to destroy the target completely, at least three successful hits are required.

Now, according to the question, let the minimum number of missiles required to fired is $n$, so

$${}^nC_3 p^3 q^{n-3} + {}^nC_4 p^4 q^{n-4} + \ldots + {}^nC_n p^n \geq 0.95$$

$$\Rightarrow 1 - \left\{ {}^nC_0 \left(\frac{1}{4}\right)^n + {}^nC_1 \left(\frac{3}{4}\right)\left(\frac{1}{4}\right)^{n-1} + {}^nC_2 \left(\frac{3}{4}\right)^2 \left(\frac{1}{4}\right)^{n-2} \right\} \geq 0.95$$

$$\Rightarrow 1 - \frac{95}{100} \geq \frac{1}{4^n} + \frac{3n}{4^n} + \frac{n(n-1)}{2}\frac{9}{4^n}$$

$$\Rightarrow \frac{4^n}{20} \geq \frac{2 + 6n + 9n^2 - 9n}{2}$$

$$\Rightarrow 10(9n^2 - 3n + 2) \leq 4^n$$

Now, at $n = 3$, LHS = 720, RHS = 64

at $n = 4$, LHS = 1340, RHS = 256

at $n = 5$, LHS = 2120, RHS = 1024

at $n = 6$, LHS = 3080, RHS = 4096

Hence, $n = 6$ missiles should be fired.

**39.** (2) As we know that the equation of family of circles passes through the points of intersection of given circle $x^2 + y^2 = r^2$ and line $PQ: 2x + 4y = 5$ is,

$$(x^2 + y^2 - r^2) + \lambda(2x + 4y - 5) = 0 \quad \ldots \text{(i)}$$

Since, the circle (i) passes through the centre of circle

$$x^2 + y^2 = r^2,$$

So, $-r^2 - 5\lambda = 0$

or $5\lambda + r^2 = 0 \quad \ldots \text{(ii)}$

and the centre of circle (i) lies on the line $x + 2y = 4$, so centre $(-\lambda, -2\lambda)$ satisfy the line $x + 2y = 4$.

Therefore, $-\lambda - 4\lambda = 4$

$\Rightarrow \quad -5\lambda = 4$

$\Rightarrow \quad r^2 = 4 \quad$ {from Eq. (ii)}

$\Rightarrow \quad r = 2$

**40.** (5) Let a square matrix '$A$' of order $2 \times 2$, such that $t_r(A) = 3$, is

$$A = \begin{bmatrix} x & y \\ z & 3-x \end{bmatrix}$$

So, $$A^2 = \begin{bmatrix} x & y \\ z & 3-x \end{bmatrix}\begin{bmatrix} x & y \\ z & 3-x \end{bmatrix}$$

$$= \begin{bmatrix} x^2 + yz & xy + 3y - xy \\ xz + 3z - xz & yz + (3-x)^2 \end{bmatrix}$$

$$\therefore \quad A^3 = \begin{bmatrix} x^2 + yz & 3y \\ + 3z & yz + 9 + x^2 - 6x \end{bmatrix} \begin{bmatrix} x & y \\ z & 3-x \end{bmatrix}$$

$$\because \quad t_r(A^3) = x^3 + xyz + 3yz + 3yz + 3yz - xyz + 27 - 9x + 3x^2 - x^3 - 18x + 6x^2$$

$$= 9yz + 27 - 27x + 9x^2 = -18 \text{ (given)}$$

$\Rightarrow \quad yz + 3 - 3x + x^2 = -2$

$\Rightarrow \quad 3x - x^2 - yz = 5$

$$\because \quad |A| = \begin{vmatrix} x & y \\ z & 3-x \end{vmatrix}$$

$$= 3x - x^2 - yz = 5.$$

**41.** (4) The given functions $f: (-1, 1) \longrightarrow R$ and $g: (-1, 1) \longrightarrow (-1, 1)$ be defined by

$$f(x) = |2x - 1| + |2x + 1|$$

and $\quad g(x) = x - [x].$

As, we know the composite function $(fog)(x)$ is discontinuous at the points, where $g(x)$ is discontinuous for given domain. And, since $g(x)$ is discontinuous at $x = 0$ lies in interval $(-1, 1)$, so value of $c = 1$.

And, since $(fog)(x)$ is not differentiable at the point where $g(x)$ is not differentiable as well as at those points also where $g(x)$ attains the values so that $f(g(x))$ is non-differentiable.

Since $g(x)$ is not continuous at $x = 0 \in (-1, 1)$ so $fog(x)$ is not differentiable and as $f(x) = |2x-1| + |2x+1|$ is not differentiable at $x = -1/2$ and $1/2$, so $(fog)(x)$ is not differentiable for those $x$, for which $g(x) = -1/2$ or $1/2$.

But $g(x) \geq 0$, so $g(x)$ can be 1/2 only and for $x = -1/2$ and $1/2$, $g(x) = \frac{1}{2}$.

So, $(fog)(x)$ is not differentiable at $x = -1/2, 0, 1/2$, therefore value of $d = 3$

$\therefore \quad c + d = 1 + 3 = 4.$

**42.** *(8)* The limit

$$\lim_{x \to \frac{\pi}{2}} \frac{4\sqrt{2}(\sin 3x + \sin x)}{\left(2\sin 2x \sin \frac{3x}{2} + \cos \frac{5x}{2}\right) - \left(\sqrt{2} + \sqrt{2}\cos 2x + \cos \frac{3x}{2}\right)}$$

$$= \lim_{x \to \frac{\pi}{2}} \frac{4\sqrt{2}(2\sin 2x \cos x)}{2\sin 2x \sin \frac{3x}{2} + \left(\cos \frac{5x}{2} - \cos \frac{3x}{2}\right) - \sqrt{2}(1 + \cos 2x)}$$

$$= \lim_{x \to \frac{\pi}{2}} \frac{8\sqrt{2}\sin 2x \cos x}{2\sin 2x \sin \frac{3x}{2} - 2\sin 2x \sin \frac{x}{2} - \sqrt{2}(2\cos^2 x)}$$

$$= \lim_{x \to \frac{\pi}{2}} \frac{4\sqrt{2}\sin 2x \cos x}{\sin 2x\left(\sin \frac{3x}{2} - \sin \frac{x}{2}\right) - \sqrt{2}\cos^2 x}$$

$$= \lim_{x \to \frac{\pi}{2}} \frac{4\sqrt{2}\sin 2x \cos x}{2\sin 2x \cos x \sin \frac{x}{2} - \sqrt{2}\cos^2 x}$$

$$= \lim_{x \to \frac{\pi}{2}} \frac{4\sqrt{2}\sin 2x}{2\sin 2x \sin \frac{x}{2} - \sqrt{2}\cos x}$$

$$= \lim_{x \to \frac{\pi}{2}} \frac{8\sqrt{2}\sin x}{4\sin x \sin \frac{x}{2} - \sqrt{2}}$$

$$= \frac{8\sqrt{2}}{\frac{4}{\sqrt{2}} - \sqrt{2}} = \frac{16}{4-2} = 8.$$

**43.** *(a, c)* Given differential equation

$$f'(x) = \frac{f(x)}{b^2 + x^2}$$

$$\Rightarrow \int \frac{f'(x)}{f(x)}\,dx = \int \frac{dx}{b^2 + x^2}$$

$$\Rightarrow \log_e |f(x)| = \frac{1}{b}\tan^{-1}\left(\frac{x}{b}\right) + c$$

$\because \quad f(0) = 1$, so $c = 0$

$\therefore \quad |f(x)| = e^{\frac{1}{b}\tan^{-1}\left(\frac{x}{b}\right)}$

$\Rightarrow f(x) = e^{\frac{1}{b}\tan^{-1}\left(\frac{x}{b}\right)}$ or $-e^{\frac{1}{b}\tan^{-1}\left(\frac{x}{b}\right)}$

$\because \quad f(0) = 1 \Rightarrow f(x) = e^{\frac{1}{b}\tan^{-1}\left(\frac{x}{b}\right)}$

and $f'(x) = \dfrac{1}{b^2}\dfrac{e^{\frac{1}{b}\tan^{-1}\left(\frac{x}{b}\right)}}{1 + \left(\frac{x}{b}\right)^2} = \dfrac{e^{\frac{1}{b}\tan^{-1}\left(\frac{x}{b}\right)}}{b^2 + x^2}$

$\therefore f'(x) > 0 \; \forall x \in R$ and $b \in R_0$.

Therefore, $f(x)$ is an increasing function $\forall \, b \in R_0$.

and $\quad f(x)\,f(-x)$

$$= e^{\frac{1}{b}\tan^{-1}\left(\frac{x}{b}\right)} \cdot e^{-\frac{1}{b}\tan^{-1}\left(\frac{x}{b}\right)} = 1$$

**44.** *(a, d)* Equation of given hyperbola is

$$\frac{x^2}{a^2} - \frac{y^2}{b^2} = 1, \; a > b \text{ and } a > 1$$

Let point $P(a\sec\theta, b\tan\theta)$ on the hyperbola in first quadrant i.e., $Q \in \left(0, \frac{\pi}{2}\right)$.

Now equation of tangent to the hyperbola at point $P$ is

$$\frac{x\sec\theta}{a} - \frac{y\tan\theta}{b} = 1 \qquad \ldots\text{(i)}$$

$\because$The tangent (i) passes through point $A(1, 0)$

So, $\quad a = \sec\theta > 1. \qquad \ldots$ (ii)

and equation of normal to the hyperbola at point $P$ having slope is $\frac{b\sec\theta}{a\tan\theta}$, as normal cuts off equal intercepts on the coordinate axes, so slope must be $-1$.

Therefore, $\quad \dfrac{b\sec\theta}{a\tan\theta} = -1$

$\Rightarrow \quad b = -\tan\theta \quad$ {from Eq. (ii)}

$\because$ The eccentricity of hyperbola

$$e = \sqrt{1 + \frac{b^2}{a^2}} = \sqrt{1 + \sin^2\theta}$$

$\therefore \quad 1 < e < \sqrt{2}$, as $\theta \in \left(0, \frac{\pi}{2}\right)$

And the area of required triangle is, which is isosceles is

$$\Delta = \frac{1}{2}(AP)^2 = \frac{1}{2}[(1 - \sec^2\theta)^2 + \tan^4\theta]$$

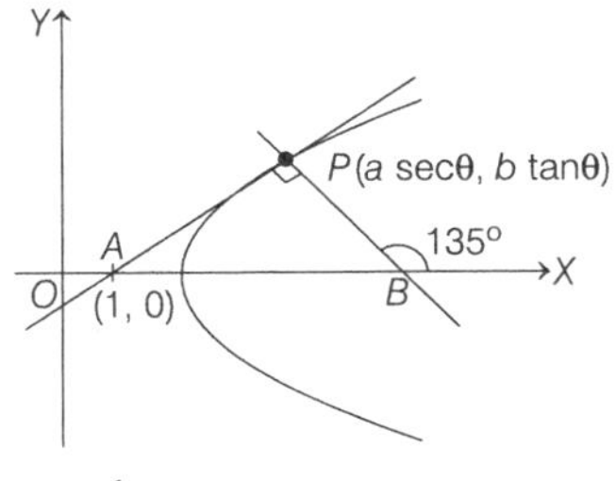


$$= \frac{1}{2}(2\tan^4\theta) = \tan^4\theta = b^4.$$

**45.** (a, b, d) The given function $f : R \to R$ is satisfying as

$$f(x + y) = f(x) + f(y) + f(x)\, f(y).$$

So, $f'(x) = \lim_{h \to 0} \frac{f(x + h) - f(x)}{h}$

$$= \lim_{h \to 0} \frac{f(x) + f(h) + f(x)\, f(h) - f(x)}{h}$$

$$= \lim_{h \to 0} \frac{f(h)}{h}(1 + f(x))$$

$\because \quad f(x) = xg(x) \Rightarrow g(x) = \frac{f(x)}{x}$

$\therefore \quad \lim_{x \to 0} \frac{f(x)}{x} = \lim_{x \to 0} g(x) = 1$ (given)

$\therefore \quad f'(x) = 1 + f(x) \Rightarrow \frac{f'(x)}{1 + f(x)} = 1$

$\Rightarrow \quad \log_e(1 + f(x)) = x + c$

$\Rightarrow \quad 1 + f(x) = e^{x + c}$

$\Rightarrow \quad f(x) = e^{x + c} - 1$

$\because \quad f(0) = 0 \Rightarrow c = 0$

Therefore, $f(x) = e^x - 1$ is differentiable at every $x \in R$.

And $\quad f'(x) = e^x \Rightarrow f'(0) = 1$

Now, $g(x) = \frac{f(x)}{x} = \frac{e^x - 1}{x}$ and if $g(0) = 1$

LHD (at $x = 0$) of

$$g(x) = \lim_{h \to 0} \frac{g(0 - h) - g(0)}{-h}$$

$$= \lim_{h \to 0} \frac{\frac{e^{-h} - 1}{-h} - 1}{-h}$$

$$= \lim_{h \to 0} \frac{e^{-h} - 1 + h}{h^2} = \frac{1}{2}$$

and, RHD (at $x = 0$) of

$$g(x) = \lim_{h \to 0} \frac{g(0 + h) - g(0)}{h}$$

$$= \lim_{h \to 0} \frac{e^h - 1 - h}{h^2} = \frac{1}{2}$$

So, if $g(0) = 1$, then $g$ is differentiable at every $x \in R$.

**46.** (a, b, c) Since, the point $A(3, 2, -1)$ is the mirror image of the point $B(1, 0, -1)$ with respect to the plane $\alpha x + \beta y + \gamma z = \delta$, then

$$\frac{\alpha}{3 - 1} = \frac{\beta}{2 - 0} = \frac{\gamma}{(-1) - (-1)}$$

$$\Rightarrow \quad \frac{\alpha}{1} = \frac{\beta}{1} = \frac{\gamma}{0} \Rightarrow \gamma = 0$$

it is given that $\alpha + \gamma = 1 \Rightarrow \alpha = 1$, so $\beta = 1$.

And, the mid-point of $AB$, $M(2, 1, -1)$ lies on the given plane, so

$$2\alpha + \beta - \gamma = \delta$$

$\Rightarrow \quad 2(1) + (1) - 0 = \delta \Rightarrow \delta = 3$

$\therefore \alpha + \beta = 2, \delta - \gamma = 0, \delta + \beta = 4.$

**47.** (a, c) Given vectors $\mathbf{PQ} = a\hat{\mathbf{i}} + b\hat{\mathbf{j}}$ and $\mathbf{PS} = a\hat{\mathbf{i}} - b\hat{\mathbf{j}}$ are adjacent sides of a parallelogram $PQRS$, so area of parallelogram $PQRS = |\mathbf{PQ} \times \mathbf{PS}|$

$= 2ab = 8$ (given)

$\therefore \quad ab = 4 \quad \ldots$(i)

According to the question,

$|\mathbf{u}| = |$projection vector of $\mathbf{w} = \hat{\mathbf{i}} + \hat{\mathbf{j}}$ along $\mathbf{PQ}|$

$$= \left| \frac{(\hat{\mathbf{i}} + \hat{\mathbf{j}}) \cdot (a\hat{\mathbf{i}} + b\hat{\mathbf{j}})}{\sqrt{a^2 + b^2}} \right| = \frac{|a + b|}{\sqrt{a^2 + b^2}}$$

and, similarly, $|\mathbf{v}| = |$projection vector of $\mathbf{w} = \hat{\mathbf{i}} + \hat{\mathbf{j}}$ along $\mathbf{PS}|$

$$= \frac{|(\hat{\mathbf{i}} + \hat{\mathbf{j}}) \cdot (a\hat{\mathbf{i}} + b\hat{\mathbf{j}})|}{\sqrt{a^2 + b^2}}$$

$$= \frac{|a - b|}{\sqrt{a^2 + b^2}} \Rightarrow |\mathbf{u}| + |\mathbf{v}| = |\mathbf{w}|$$

$$\Rightarrow \quad \frac{|a+b|}{\sqrt{a^2+b^2}} + \frac{|a-b|}{\sqrt{a^2+b^2}} = \sqrt{2}$$

$$\Rightarrow |a+b| + |a-b| = \sqrt{2}\sqrt{a^2+b^2}$$

If $a \geq b > 0$, then $2a = \sqrt{2}\sqrt{a^2+b^2}$

$$\Rightarrow \quad 4a^2 = 2a^2 + 2b^2 \Rightarrow a^2 = b^2$$

$$\Rightarrow \quad a = b \quad \dots \text{(ii)}$$

From Eqs. (i) and (ii), we get

$$a = 2 = b \Rightarrow a + b = 4$$

and the length of diagonal $PR$

$$= |a\hat{\mathbf{i}} + b\hat{\mathbf{j}} + a\hat{\mathbf{i}} - b\hat{\mathbf{j}}|$$
$$= |2a\hat{\mathbf{i}}| = 2a = 4$$

And, the angle bisector of vector **PQ** and **PS** is along the vector

$$\pm \frac{\lambda}{\sqrt{a^2+b^2}}((a\hat{\mathbf{i}} + b\hat{\mathbf{j}}) \pm (a\hat{\mathbf{i}} - b\hat{\mathbf{j}}))$$

**48.** *(a, b, d)* Since,

$$f(m, n, p) = \sum_{i=0}^{p} \binom{m}{i}\binom{n+i}{p}\binom{p+n}{p-i}$$

$$= \sum_{i=0}^{p} \frac{m!}{i!(m-i)!} \times \frac{(n+i)!}{p!(n+i-p)!} \times \frac{(p+n)!}{(p-i)!(n+i)!}$$

$$= \frac{(n+p)!}{p!} \sum_{i=0}^{p} {}^mC_i \frac{1}{(p-i)!(n+i-p)!}$$

$$= \frac{(n+p)!}{n!p!} \sum_{i=0}^{p} {}^mC_i \times {}^nC_{p-i}$$

$$= {}^{n+p}C_p \; {}^{m+n}C_p$$

Since $g(m, n) = \sum_{p=0}^{m+n} \frac{f(m, n, p)}{\binom{n+p}{p}}$

$$= \sum_{p=0}^{m+n} \frac{{}^{n+p}C_p \times {}^{m+n}C_p}{{}^{n+p}C_p}$$

$$= \sum_{p=0}^{m+n} {}^{m+n}C_p = 2^{m+n}$$

$\therefore \quad g(m, n) = g(n, m) = 2^{m+n}$, and

$$g(m, n+1) = 2^{m+n+1} = g(m+1, n)$$

and $\quad g(2m; 2n) = 2^{2(m+n)} = (2^{(m+n)})^2$
$$= (g(m, n))^2.$$

**49.** *(495.00)* Let the engineer visits the factory first time after $x_1$ days to 1 June, second time after $x_2$ days to first visit and so on.

$$\therefore \quad x_1 + x_2 + x_3 + x_4 + x_5 = 11$$

where $x_1, x_5 \geq 0$ and $x_2, x_3, x_4 \geq 1$ according to the requirement of the question.

Now, let $x_2 = a + 1, x_3 = b + 1$ and $x_4 = c + 1$ where $a, b, c \geq 0$

$\therefore$ New equation will be

$$x_1 + a + b + c + x_5 = 8$$

Now, the number of all possible ways in which the engineer can made visits is equals to the non-negative integral solution of equation

$x_1 + a + b + c + x_5 = 8$, and it is equal to

$${}^{8+5-1}C_{5-1} = {}^{12}C_4 = \frac{12 \times 11 \times 10 \times 9}{4 \times 3 \times 2}$$
$$= 495.$$

**50.** *(1080.00)* The groups of persons can be made only in 2, 2, 1, 1

$\therefore$ So the number of required ways is equal to number of ways to distribute the 6 distinct objects in group sizes 1, 1, 2 and 2

$$= \frac{6!}{(2!)^2 (1!)^2 (2!) (2!)} (4!)$$
$$= 360 \times 3 = 1080.$$

**51.** *(8.00)* Let an event $E$ of sum of outputs are perfect square (i.e., 4 or 9), so

$E = \{(1, 3), (2, 2), (3, 1), (3, 6), (4, 5), (5, 4), (6, 3)\}$

and an event $F$ of sum of outputs are prime numbers

(i.e., 2, 3, 5, 7, 11) so

$F = \{(1, 1), (1, 2), (2, 1), (1, 4), (2, 3), (3, 2), (4, 1), (1, 6), (2, 5), (3, 4), (4, 3), (5, 2), (6, 1), (5, 6), (6, 5)\}$

and event $T$ of sum of outputs are odd numbers

(i.e., 3, 5, 7, 9, 11)

$T = \{(1, 2), (2, 1), (1, 4), (2, 3), (3, 2), (4, 1), (1, 6), (2, 5), (3, 4), (4, 3), (5, 2), (6, 1), (3, 6), (4, 5), (5, 4), (6, 3), (5, 6), (6, 5)\}$

Now, required probability $p = P(T / E)$

$$= \frac{P(T \cap E)}{P(E)}$$

where, $P(T \cap E)$ = probability of occuring perfect square odd number before prime

$$= \left(\frac{4}{36}\right) + \left(\frac{14}{36}\right)\left(\frac{4}{36}\right) + \left(\frac{14}{36}\right)^2\left(\frac{4}{36}\right) + \dots\infty$$

$$= \frac{\frac{4}{36}}{1 - \frac{14}{36}} = \frac{4}{22} = \frac{2}{11}$$

and $P(E)$ = probability of occuring perfect square before prime

$$= \left(\frac{7}{36}\right) + \left(\frac{14}{36}\right)\left(\frac{7}{36}\right) + \left(\frac{14}{36}\right)^2\left(\frac{7}{36}\right) + \dots\infty$$

$$= \frac{\frac{7}{36}}{1 - \frac{14}{36}} = \frac{7}{22}$$

$$\therefore P(T / E) = \frac{\frac{2}{11}}{\frac{7}{22}} = \frac{4}{7} = p \Rightarrow 14p = 8$$

**52.** *(19.00)* The given function $f : [0, 1] \to R$ be define by

$$f(x) = \frac{4^x}{4^x + 2}$$

$$\because \quad f(1 - x) = \frac{4^{1-x}}{4^{1-x} + 2} = \frac{2}{2 + 4^x}$$

$$\therefore f(x) + f(1 - x) = \frac{4^x}{4^x + 2} + \frac{2}{2 + 4^x}$$

$$= \frac{4^x + 2}{4^x + 2}$$

So, $\quad f(x) + f(1 - x) = 1 \quad \dots (i)$

$$\therefore f\left(\frac{1}{40}\right) + f\left(\frac{2}{40}\right) + f\left(\frac{3}{40}\right) + \dots + f\left(\frac{39}{40}\right) - f\left(\frac{1}{2}\right)$$

$$= \left[f\left(\frac{1}{40}\right) + f\left(\frac{39}{40}\right)\right] + \left[f\left(\frac{2}{40}\right) + f\left(\frac{38}{40}\right)\right] + \dots + \left[f\left(\frac{18}{40}\right) + f\left(\frac{22}{40}\right)\right]$$

$$+ \left[f\left(\frac{19}{40}\right) + f\left(\frac{21}{40}\right)\right] + f\left(\frac{20}{40}\right) - f\left(\frac{1}{2}\right)$$

$$= \{1 + 1 + \dots + 1 + 1\} + f\left(\frac{1}{2}\right) - f\left(\frac{1}{2}\right)$$

{from Eq. (i)}

19 – times

= 19.

**53.** *(4.00)* It is given that, for functions $f : R \to R$ and $F : [0, \pi] \to R$,

$$F(x) = \int_0^x f(t)dt, \text{ where } f(\pi) = -6$$

$$\Rightarrow \quad F'(\pi) = f(\pi) = -6 \quad \dots (i)$$

Now, $\int_0^\pi (f'(x) + F(x))\cos x \, dx$

$$= \int_0^\pi f'(x) \cos x dx + \int_0^\pi F(x)\cos x dx$$

$$= [\cos x \, f(x)]_0^\pi + \int_0^\pi f(x) \sin x \, dx + \int_0^\pi F(x) \cos x \, dx$$

{by integration by parts}

$$= (-1)(-6) - f(0) + [(\sin x) F(x)]_0^\pi - \int_0^\pi F(x)\cos x \, dx + \int_0^\pi F(x)\cos x \, dx$$

$$= 6 - f(0) = 2 \quad \text{(given)}$$

$$\Rightarrow \quad f(0) = 4$$

**54.** *(0.50)* The given function $f : R \to R$ be defined by

$$f(\theta) = (\sin\theta + \cos\theta)^2 + (\sin\theta - \cos\theta)^4$$

$$= 1 + \sin 2\theta + (1 - \sin 2\theta)^2$$

$$= 1 + \sin 2\theta + 1 + \sin^2 2\theta - 2\sin 2\theta$$

$$= \sin^2 2\theta - \sin 2\theta + 2$$

$$= \left(\sin 2\theta - \frac{1}{2}\right)^2 + \frac{7}{4}$$

The local minimum of function '$f$' occurs when

$$\sin 2\theta = \frac{1}{2}$$

$$\Rightarrow \quad 2\theta = \frac{\pi}{6}, \frac{5\pi}{6}, \frac{13\pi}{6} \dots$$

$$\Rightarrow \quad \theta = \frac{\pi}{12}, \frac{5\pi}{12}, \frac{13\pi}{12}, \dots$$

but $\quad \theta \in \{\lambda_1\pi, \lambda_2\pi, \dots \lambda_r\pi\}$,

where $\quad 0 < \lambda_1 < \dots < \lambda_r < 1$.

$$\therefore \quad \theta = \frac{\pi}{12}, \frac{5\pi}{12}$$

So $\lambda_1 + \dots + \lambda_r = \frac{1}{12} + \frac{5}{12} = 0.50$

# JEE ADVANCED
## SOLVED PAPER 2019

**Duration** : 3 Hours **Max. Marks : 360**

## Paper 1

### Section 1 (Maximum Marks : 12)

- This section contains **FOUR (04)** questions.
- Each question has **FOUR** options. **ONLY ONE** of these four options is the correct answer.
- For each question, choose the option corresponding to the correct answer.
- Answer to each question will be evaluated according to the following marking scheme:

*Full Marks* : **+3** If ONLY the correct option is chosen.
*Zero Marks* : **0** If none of the options is chosen. (i.e., the question is unanswered)
*Negative Marks* : **−1** In all other cases.

**1.** In a radioactive sample, ${}^{40}_{19}\text{K}$ nuclei either decay into stable ${}^{40}_{20}\text{Ca}$ nuclei with decay constant $4.5 \times 10^{-10}$ per year or into stable ${}^{40}_{18}\text{Ar}$ nuclei with decay constant $0.5 \times 10^{-10}$ per year. Given that in this sample all the stable ${}^{40}_{20}\text{Ca}$ and ${}^{40}_{18}\text{Ar}$ nuclei are produced by the ${}^{40}_{19}\text{K}$ nuclei only. In time $t \times 10^9$ years, if the ratio of the sum of stable ${}^{40}_{20}\text{Ca}$ and ${}^{40}_{18}\text{Ar}$ nuclei to the radioactive ${}^{40}_{19}\text{K}$ nuclei is 99, the value of $t$ will be
[Given: ln 10 = 2.3]

(a) 9.2 (b) 1.15 (c) 4.6 (d) 2.3

**2.** A current carrying wire heats a metal rod. The wire provides a constant power ($P$) to the rod. The metal rod is enclosed in an insulated container. It is observed that the temperature ($T$) in the metal rod changes with time ($t$) as

$$T(t) = T_0\left(1 + \beta t^{\frac{1}{4}}\right),$$ where, $\beta$ is a constant with appropriate dimension while $T_0$ is a constant with dimension of temperature. The heat capacity of the metal is

(a) $\dfrac{4P(T(t) - T_0)^4}{\beta^4 T_0^5}$ (b) $\dfrac{4P(T(t) - T_0)^3}{\beta^4 T_0^4}$

(c) $\dfrac{4P(T(t) - T_0)}{\beta^4 T_0^2}$ (d) $\dfrac{4P(T(t) - T_0)^2}{\beta^4 T_0^3}$

**3.** A thin spherical insulating shell of radius $R$ carries a uniformly distributed charge such that the potential at its surface is $V_0$. A hole with a small area $\alpha 4\pi R^2 (\alpha << 1)$ is made on the shell without affecting the rest of the shell. Which one of the following statements is correct?

(a) The ratio of the potential at the center of the shell to that of the point at $\frac{1}{2}R$ from center towards the hole will be $\frac{1-\alpha}{1-2\alpha}$.

(b) The potential at the center of the shell is reduced by $2\alpha V_0$.

(c) The magnitude of electric field at the center of the shell is reduced by $\frac{\alpha V_0}{2R}$.

(d) The magnitude of electric field at a point, located on a line passing through the hole and shell's centre, on a distance $2R$ from the center of the spherical shell will be reduced by $\frac{\alpha V_0}{2R}$.

**4.** Consider a spherical gaseous cloud of mass density $\rho(r)$ in free space where $r$ is the radial distance from its center. The gaseous cloud is made of particles of equal mass $m$ moving in circular orbits about the common center with the same kinetic energy $K$. The force acting on the particles is their mutual gravitational force. If $\rho(r)$ is constant in time, the particle number density $n(r) = \rho(r)/m$ is [$G$ is universal gravitational constant]

(a) $\frac{K}{6\pi r^2 m^2 G}$ (b) $\frac{K}{\pi r^2 m^2 G}$

(c) $\frac{3K}{\pi r^2 m^2 G}$ (d) $\frac{K}{2\pi r^2 m^2 G}$

## Section 2 (Maximum Marks : 32)

- This section contains **EIGHT (08)** questions.
- Each question has **FOUR** options. **ONE OR MORE THAN ONE** of these four option(s) is (are) correct options(s).
- For each question, choose the options(s) corresponding to (all) the correct answer(s).
- Answer to each question will be evaluated according to the following marking scheme:

*Full Marks* : **+4** If only (all) the correct option(s) is (are) chosen.

*Partial Marks* : **+3** If all the four options are correct but ONLY three options are chosen.

*Partial Marks* : **+2** If three or more options are correct but ONLY two options are chosen, both of which are correct.

*Partial Marks* : **+1** If two or more options are correct but ONLY one option is chosen and it is a correct option;

*Zero Marks* : 0 If none of the option is chosen (i.e., the question is unanswered);

*Negative Marks* : **−1** In all other cases.

- For example, in a question, if (A), (B) and (D) are the ONLY three options corresponding to correct answers, then
choosing ONLY (A), (B) and (D) will get + 4 marks
choosing ONLY (A) and (B) will get + 2 marks
choosing ONLY (A) and (D) will get + 2 marks
choosing ONLY (B) and (D) will get + 2 marks
choosing ONLY (A) will get + 1 marks
choosing ONLY (B) will get + 1 marks
choosing ONLY (D) will get + 1 marks
choosing no option (i.e., the question is unanswered) will get 0 marks; and
choosing any other combination of options will get – 1 mark.

**5.** A cylindrical capillary tube of 0.2 mm radius is made by joining two capillaries $T_1$ and $T_2$ of different materials having water contact angles of $0^\circ$ and $60^\circ$, respectively. The capillary tube is dipped vertically in water in two different configurations, case I and II as shown in figure. Which of the following option(s) is(are) correct?

[Surface tension of water $= 0.075 \text{ N/m}$, density of water $= 1000 \text{ kg/m}^3$, take $g = 10 \text{ m/s}^2$]

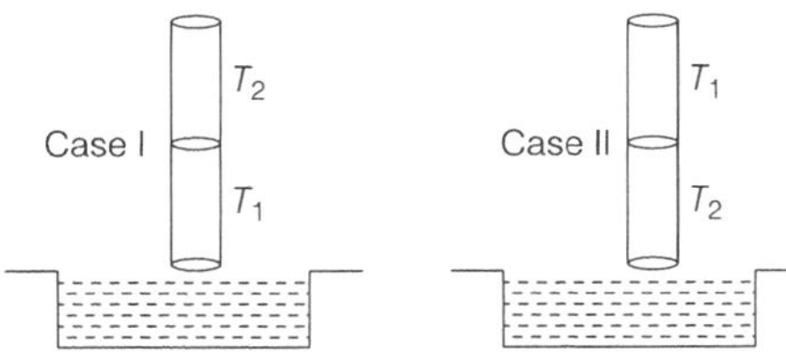


(a) For case I, if the joint is kept at 8 cm above the water surface, the height of water column in the tube will be 7.5 cm. (Neglect the weight of the water in the meniscus).

(b) For case I, if the capillary joint is 5 cm above the water surface, the height of water column raised in the tube will be more than 8.75 cm. (Neglect the weight of the water in the meniscus).

(c) The correction in the height of water column raised in the tube, due to weight of water contained in the meniscus, will be different for both cases.

(d) For case II, if the capillary joint is 5 cm above the water surface, the height of water column raised in the tube will be 3.75 cm. (Neglect the weight of the water in the meniscus).

**6.** A conducting wire of parabolic shape, initially $y = x^2$, is moving with velocity $\mathbf{v} = v_0\hat{\mathbf{i}}$ in a non-uniform magnetic field

$$\mathbf{B} = B_0\left(1 + \left(\frac{y}{L}\right)^{\beta}\right)\hat{\mathbf{k}},$$ as shown in figure. If $V_0$, $B_0$, $L$ and $\beta$ are positive constants and $\Delta\phi$ is the potential difference developed between the ends of the wire, then the correct statement(s) is/are

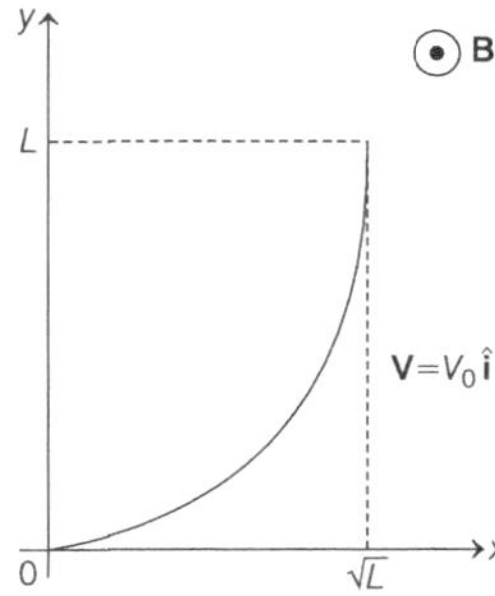


(a) $|\Delta\phi| = \frac{4}{3}B_0V_0L$ for $\beta = 2$

(b) $|\Delta\phi|$ remains the same if the parabolic wire is replaced by a straight wire, $y = x$ initially, of length $\sqrt{2}L$

(c) $|\Delta\phi| = \frac{1}{2}B_0V_0L$ for $\beta = 0$

(d) $|\Delta\phi|$ is proportional to the length of the wire projected on the $y$-axis.

**7.** A thin convex lens is made of two materials with refractive indices $n_1$ and $n_2$, as shown in the figure. The radius of curvature of the left and right spherical surfaces are equal. $f$ is the focal length of the lens when $n_1 = n_2 = n$. The focal length is $f + \Delta f$ when $n_1 = n$ and $n_2 = n + \Delta n$. Assuming $\Delta n << (n - 1)$ and $1 < n < 2$, the correct statement(s) is/are

(a) If $\frac{\Delta n}{n} < 0$ then $\frac{\Delta f}{f} > 0$

(b) For $n = 1.5$, $\Delta n = 10^{-3}$ and $f = 20$ cm, the value of $|\Delta f|$ will be 0.02 cm (round off to 2nd decimal place).

(c) $\left|\frac{\Delta f}{f}\right| < \left|\frac{\Delta n}{n}\right|$

(d) The relation between $\frac{\Delta f}{f}$ and $\frac{\Delta n}{n}$ remains unchanged if both the convex surfaces are replaced by concave surfaces of the same radius of curvature.

**8.** One mole of a monatomic ideal gas goes through a thermodynamic cycle, as shown in the volume versus temperature ($V$ - $T$) diagram. The correct statement(s) is/are [$R$ is the gas constant]

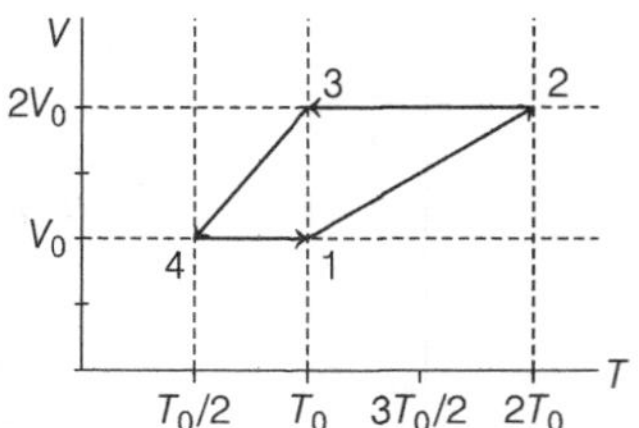


(a) Work done in this thermodynamic cycle $(1 \to 2 \to 3 \to 4 \to 1)$ is $|W| = \frac{1}{2} RT_0$.

(b) The ratio of heat transfer during processes $1 \to 2$ and $2 \to 3$ is $\left|\frac{Q_{1 \to 2}}{Q_{2 \to 3}}\right| = \frac{5}{3}$.

(c) The above thermodynamic cycle exhibits only isochoric and adiabatic processes.

(d) The ratio of heat transfer during processes $1 \to 2$ and $3 \to 4$ is $\left|\frac{Q_{1 \to 2}}{Q_{3 \to 4}}\right| = \frac{1}{2}$.

**9.** In the circuit shown, initially there is no charge on the capacitors and keys $S_1$ and $S_2$ are open. The values of the capacitors are $C_1 = 10\ \mu F$, $C_2 = 30\ \mu F$ and $C_3 = C_4 = 80\ \mu F$.

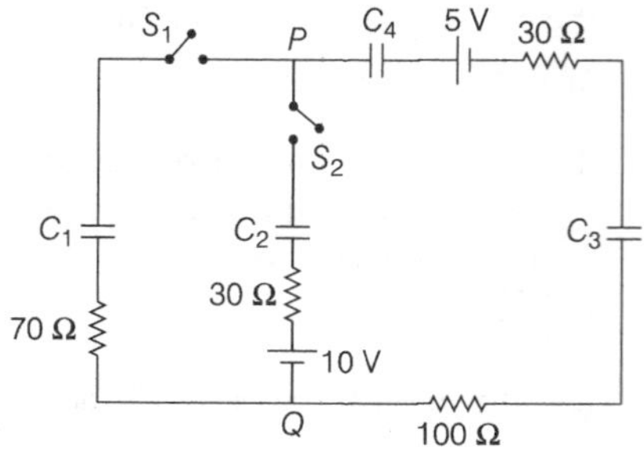


Which of the statement(s) is/are correct?

(a) The key $S_1$ is kept closed for long time such that capacitors are fully charged. Now, key $S_2$ is closed, at this time, the instantaneous current across $30\ \Omega$ resistor (between points $P$ and $Q$) will be 0.2 A (round off to 1st decimal place).

(b) If key $S_1$ is kept closed for long time such that capacitors are fully charged, the voltage across the capacitor $C_1$ will be 4V.

(c) At time $t = 0$, the key $S_1$ is closed, the instantaneous current in the closed circuit will be 25 mA.

(d) If key $S_1$ is kept closed for long time such that the capacitors are fully charged, the voltage difference between points $P$ and $Q$ will be 10 V.

**10.** A charged shell of radius $R$ carries a total charge $Q$. Given $\phi$ as the flux of electric field through a closed cylindrical surface of height $h$, radius $r$ and with its center same as that of the shell. Here, center of the cylinder is a point on the axis of the cylinder which is equidistant from its top and bottom surfaces. Which of the following option(s) is/are correct? [$\epsilon_0$ is the permittivity of free space]

(a) If $h > 2R$ and $r = 4R/5$ then $\phi = Q/5\epsilon_0$

(b) If $h > 2R$ and $r = 3R/5$ then $\phi = Q/5\epsilon_0$

(c) If $h < 8R/5$ and $r = 3R/5$ then $\phi = 0$

(d) If $h > 2R$ and $r > R$ then $\phi = Q/\epsilon_0$

**11.** Two identical moving coil galvanometers have $10\ \Omega$ resistance and full scale deflection at $2\mu A$ current. One of them is converted into a voltmeter of 100 mV full scale reading and the other into an ammeter of 1 mA full scale current using appropriate resistors. These are then used to measure the voltage and current in the Ohm's law experiment with $R = 1000\ \Omega$ resistor by using an ideal cell. Which of the following statement(s) is/are correct?

(a) The resistance of the voltmeter will be $100\,k\Omega$.

(b) The resistance of the ammeter will be $0.02\,\Omega$.
(round off to 2nd decimal place).

(c) If the ideal cell is replaced by a cell having internal resistance of $5\Omega$ then the measured value of $R$ will be more than $1000\,\Omega$.

(d) The measured value of $R$ will be $978\,\Omega < R < 982\,\Omega$.

**12.** Let us consider a system of units in which mass and angular momentum are dimensionless. If length has dimension of $L$, which of the following statement(s) is/are correct?

(a) The dimension of force is $[L]^{-3}$.

(b) The dimension of power is $[L]^{-5}$.

(c) The dimension of energy is $[L]^{-2}$.

(d) The dimension of linear momentum is $[L]^{-1}$.

## Section 3 (Maximum Marks : 18)

- This section contains **SIX (06)** questions. The answer to each question is a **NUMERICAL VALUE.**
- Four each question, enter the correct numerical value of the answer using the mouse and the on-screen virtual numeric keypad in the place designated to enter the answer. If the numerical value has more than two decimal places, truncate/round-off the value to TWO decimal places.
- Answer to each question will be evaluated according to the following marking scheme:

**13.** A parallel plate capacitor of capacitance $C$ has spacing $d$ between two plates having area $A$. The region between the plates is filled with $N$ dielectric layers, parallel to its plates, each with thickness, $\delta = \frac{d}{N}$. The dielectric constant of the $m^{th}$ layer is $K_m = K\left(1 + \frac{m}{N}\right)$. For a very large $N (> 10^3)$, the capacitance $C$ is $\alpha\left(\frac{K\varepsilon_0 A}{d \ln 2}\right)$.
The value of $\alpha$ will be .............. .

[$\epsilon_0$ is the permittivity of free space.]

**14.** A planar structure of length $L$ and width $W$ is made of two different optical media of refractive indices $n_1 = 1.5$ and $n_2 = 1.44$ as shown in figure. If $L >> W$, a ray entering from end $AB$ will emerge from end $CD$ only if the total internal reflection condition is met inside the structure. For $L = 9.6$ m, if the incident angle $\theta$ is varied, the maximum time taken by a ray to exit the plane $CD$ is $t \times 10^{-9}$ s, where, $t$ is .............

[Speed of light, $c = 3 \times 10^8$ m/s]



**15.** A block of weight 100 N is suspended by copper and steel wires of same cross-sectional area $0.5\,cm^2$ and length $\sqrt{3}$ m and 1 m, respectively. Their other ends are fixed on a ceiling as shown in figure. The angles subtended by copper and steel wires with ceiling are $30^\circ$ and $60^\circ$, respectively.

If elongation in copper wire is $(\Delta l_C)$ and elongation in steel wire is $(\Delta l_S)$, then the ratio $\frac{\Delta l_C}{\Delta l_S}$ is ......... .

[Young's modulus for copper and steel are $1\times10^{11}$ N / m$^2$ and $2\times10^{11}$ N/m$^2$ respectively.]

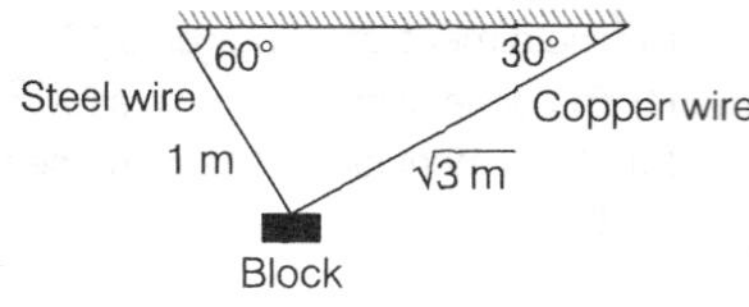


**16.** A liquid at 30°C is poured very slowly into a Calorimeter that is at temperature of 110°C. The boiling temperature of the liquid is 80°C. It is found that the first 5 gm of the liquid completely evaporates. After pouring another 80 gm of the liquid the equilibrium temperature is found to be 50°C. The ratio of the latent heat of the liquid to its specific heat will be ............ °C.

[Neglect the heat exchange with surrounding]

**17.** A particle is moved along a path AB-BC-CD-DE-EF-FA, as shown in figure, in presence of a force $\mathbf{F} = (\alpha y\hat{\mathbf{i}} + 2\alpha x\hat{\mathbf{j}})$ N, where $x$ and $y$ are in meter and $\alpha = -1\,\text{Nm}^{-1}$. The work done on the particle by this force **F** will be ............. Joule.

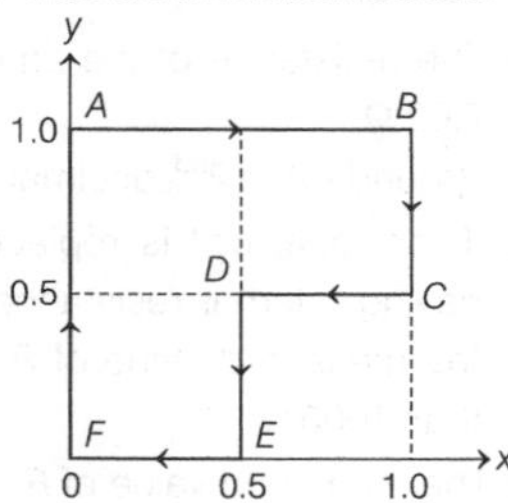


**18.** A train S1, moving with a uniform velocity of 108 km/h, approaches another train S2 standing on a platform. An observer O moves with a uniform velocity of 36 km/h towards S2, as shown in figure.

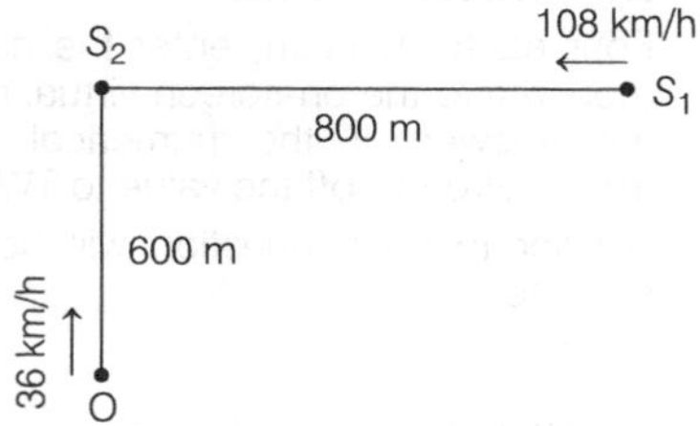


Both the trains are blowing whistles of same frequency 120 Hz. When O is 600 m away from S2 and distance between S1 and S2 is 800 m, the number of beats heard by O is ......... . [Speed of the sound = 330 m / s ......... .]

# CHEMISTRY

## Section 1 (Maximum Marks : 12)

***Instructions:*** Same as given in Physics.

**19.** The correct order of acid strength of the following carboxylic acids is

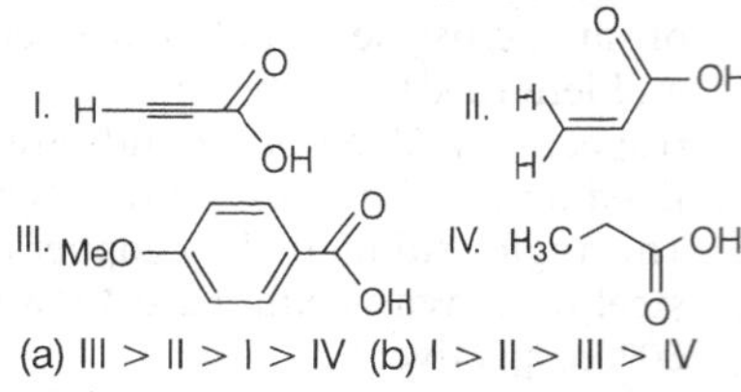


(a) III > II > I > IV (b) I > II > III > IV
(c) II > I > IV > III (d) I > III > II > IV

**20.** The green colour produced in the borax bead test of a chromium (III) salt is due to

(a) $Cr_2O_3$ (b) CrB
(c) $Cr(BO_2)_3$ (d) $Cr_2(B_4O_7)_3$

**21.** Calamine, malachite, magnetite and cryolite, respectively, are

(a) $ZnCO_3$, $CuCO_3$, $Fe_2O_3$, $Na_3AlF_6$
(b) $ZnSO_4$, $CuCO_3$, $Fe_2O_3$, $AlF_3$
(c) $ZnSO_4$, $Cu(OH)_2$, $Fe_3O_4$, $Na_3AlF_6$
(d) $ZnCO_3$, $CuCO_3\cdot Cu(OH)_2$, $Fe_3O_4$, $Na_3AlF_6$

**22.** Molar conductivity ($\Lambda_m$) of aqueous solution of sodium stearate, which behaves as a strong electrolyte, is recorded at varying concentrations ($C$) of sodium stearate. Which one of the following plots provides the correct representation of micelle formation in the solution?

(critical micelle concentration (CMC) is marked with an arrow in the figures)

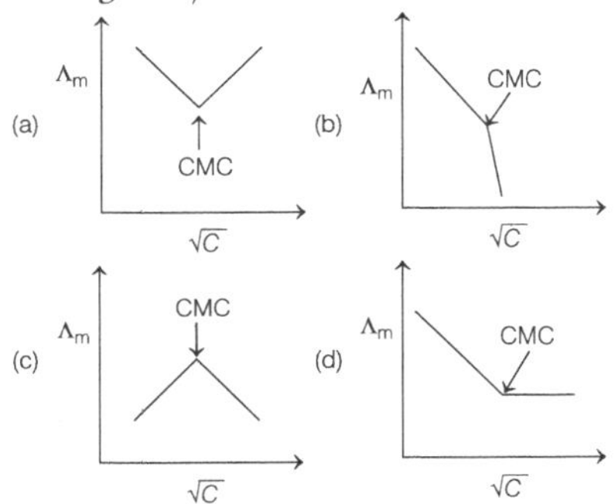


## Section 2 (Maximum Marks : 32)

***Instructions:*** Same as given in Physics.

**23.** Choose the reaction(s) from the following options, for which the standard enthalpy of reaction is equal to the standard enthalpy of formation.

(a) $2C(g) + 3H_2(g) \longrightarrow C_2H_6(g)$

(b) $2H_2(g) + O_2(g) \longrightarrow 2H_2O(l)$

(c) $\frac{3}{2}O_2(g) \longrightarrow O_3(g)$

(d) $\frac{1}{8}S_8(s) + O_2(g) \longrightarrow SO_2(g)$

**24.** A tin chloride $Q$ undergoes the following reactions (not balanced)

$$Q + Cl^- \longrightarrow X$$

$$Q + Me_3N \longrightarrow Y$$

$$Q + CuCl_2 \longrightarrow Z + CuCl$$

$X$ is a monoanion having pyramidal geometry. Both $Y$ and $Z$ are neutral compounds.

Choose the correct option(s).

(a) There is a coordinate bond in $Y$

(b) The central atom in $Z$ has one lone pair of electrons

(c) The oxidation state of the central atom in $Z$ is + 2

(d) The central atom in $X$ is $sp^3$ hybridised

**25.** In the decay sequence.

$${}^{238}_{92}U \xrightarrow{-x_1} {}^{234}_{90}Th \xrightarrow{-x_2} {}^{234}_{91}Pa \xrightarrow{-x_3} {}^{234}Z \xrightarrow{-x_4} {}^{230}_{90}Th$$

$x_1$, $x_2$, $x_3$ and $x_4$ are particles/radiation emitted by the respective isotopes. The correct option(s) is(are)

(a) $Z$ is an isotope of uranium

(b) $x_2$ is $\beta^-$

(c) $x_1$ will deflect towards negatively charged plate

(d) $x_3$ is $\gamma$-ray

**26.** Which of the following statement(s) is(are) correct regarding the root mean square speed ($U_{rms}$) and average translational kinetic energy ($E_{av}$) of a molecule in a gas at equilibrium?

(a) $U_{rms}$ is inversely proportional to the square root of its molecular mass

(b) $U_{rms}$ is doubled when its temperature is increased four times

(c) $E_{av}$ is doubled when its temperature is increased four times

(d) $E_{av}$ at a given temperature does not depend on its molecular mass

**27.** Choose the correct option(s) for the following set of reactions.

$$C_6H_{10}O \xrightarrow[\text{(ii) } H_2O]{\text{(i) MeMgBr}} Q \xrightarrow{\text{Conc. HCl}} S \text{ (major)}$$

$$Q \xrightarrow{20\%\ H_3PO_4,\ 360\ K} R \text{ (major)}$$

$$T \text{ (major)} \xleftarrow[\text{(ii) } Br_2,\ h\nu]{\text{(i) } H_2,\ \text{Ni}} R \xrightarrow[\Delta]{\text{HBr, benzoyl peroxide}} U \text{ (major)}$$

(a) S: Cl, $CH_3$ (cyclohexene) T: $H_3C$, Br (cyclohexane)

(b) U: $H_3C$, Br (cyclohexane) S: $CH_3$, Cl (cyclohexane)

(c) S: $H_3C$, Cl (cyclohexane) U: $CH_3$, Br (cyclohexane)

(d) U: $CH_3$, Br (cyclohexane) T: $H_3C$, Br (cyclohexane)

**28.** Which of the following statement(s) is(are) true?

(a) The two six-membered cyclic hemiacetal forms of D-(+)-glucose are called anomers

(b) Oxidation of glucose with bromine water gives glutamic acid

(c) Monosaccharides cannot be hydrolysed to given polyhydroxy aldehydes and ketones

(d) Hydrolysis of sucrose gives dextrorotatory glucose and laevorotatory fructose

**29.** Fusion of $MnO_2$ with KOH in presence of $O_2$ produces a salt $W$. Alkaline solution of $W$ upon electrolytic oxidation yields another salt $X$. The manganese containing ions present in $W$ and $X$, respectively, are $Y$ and $Z$. Correct statement(s) is (are)

(a) Both $Y$ and $Z$ are coloured and have tetrahedral shape

(b) $Y$ is diamagnetic in nature while $Z$ is paramagnetic

(c) In both $Y$ and $Z$, $\pi$-bonding occurs between $p$-orbitals of oxygen and $d$-orbitals of manganese

(d) In aqueous acidic solution, $Y$ undergoes disproportionation reaction to give $Z$ and $MnO_2$

**30.** Each of the following options contains a set of four molecules. Identify the option(s) where all four molecules posses permanent dipole moment at room temperature.

(a) $SO_2$, $C_6H_5Cl$, $H_2Se$, $BrF_5$ (b) $BeCl_2$, $CO_2$, $BCl_3$, $CHCl_3$

(c) $NO_2$, $NH_3$, $POCl_3$, $CH_3Cl$ (d) $BF_3$, $O_3$, $SF_6$, $XeF_6$

## Section 3 (Maximum Marks : 18)

***Instructions:*** Same as given in Physics.

**31.** Among $B_2H_6$, $B_3N_3H_6$, $N_2O$, $N_2O_4$, $H_2S_2O_3$ and $H_2S_2O_8$, the total number of molecules containing covalent bond between two atoms of the same kind is ............

**32.** On dissolving 0.5 g of a non-volatile non-ionic solute to 39 g of benzene, its vapour pressure decreases from 650 mmHg to 640 mmHg. The depression of freezing point of benzene (in K) upon addition of the solute is ............
(Given data : Molar mass and the molal freezing point depression constant of benzene are 78 g $mol^{-1}$ and 5.12 K kg $mol^{-1}$, respectively).

**33.** Consider the kinetic data given in the following table for the reaction
$A + B + C \longrightarrow$ Product

| Experiment No. | $[A]$ (mol $dm^{-3}$) | $[B]$ (mol $dm^{-3}$) | $[C]$ (mol $dm^{-3}$) | Rate of reaction (mol $dm^{-3}s^{-1}$) |
|---|---|---|---|---|
| 1 | 0.2 | 0.1 | 0.1 | $6.0 \times 10^{-5}$ |
| 2 | 0.2 | 0.2 | 0.1 | $6.0 \times 10^{-5}$ |
| 3 | 0.2 | 0.1 | 0.2 | $1.2 \times 10^{-4}$ |
| 4 | 0.3 | 0.1 | 0.1 | $9.0 \times 10^{-5}$ |

The rate of the reaction for $[A] = 0.15$ mol $dm^{-3}$, $[B] = 0.25$ mol $dm^{-3}$ and $[C] = 0.15$ mol $dm^{-3}$ is found to be $Y \times 10^{-5}$ mol $dm^{-3}s^{-1}$. The value of $Y$ is ............

**34.** For the following reaction, the equilibrium constant $K_c$ at 298 K is $1.6 \times 10^{17}$.

$$Fe^{2+}(aq) + S^{2-}(aq) \rightleftharpoons FeS(s)$$

When equal volumes of 0.06 M $Fe^{2+}(aq)$ and 0.2 M $S^{2-}(aq)$ solutions are mixed, the equilibrium concentration of $Fe^{2+}(aq)$ is found by $Y \times 10^{-17}$ M. The value of $Y$ is ..........

**35.** At 143 K, the reaction of $XeF_4$ with $O_2F_2$ produces a xenon compound $Y$. The total number of lone pair(s) of electrons present on the whole molecule of $Y$ is ...........

**36.** Schemes 1 and 2 describe the conversion of $P$ to $Q$ and $R$ to $S$, respectively. Scheme 3 describes the synthesis of $T$ from $Q$ and $S$. The total number of Br atoms in a molecule of $T$ is .............

**Scheme 1**

$P$ ($NH_2$) $\xrightarrow[\text{(iv) } H_3O^+, \Delta \quad \text{(v) } SOCl_2, \text{ pyridine}]{\text{(i) } Br_2\text{(excess)}, H_2O \quad \text{(ii) } NaNO_2, HCl, 273\text{ K} \quad \text{(iii) CuCN/KCN}}$ $Q$ (major)

**Scheme 2**

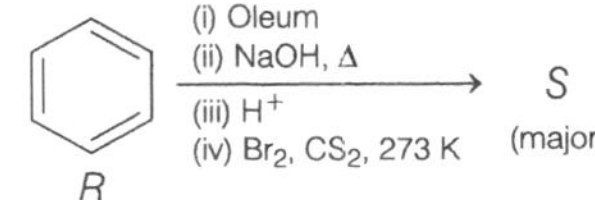


**Scheme 3**

$S \xrightarrow[\text{(ii) } Q]{\text{(i) NaOH}, \Delta} T$ (major)

# MATHEMATICS

## Section 1 (Maximum Marks : 12)

***Instructions:*** Same as given in Physics.

**37.** Let $S$ be the set of all complex numbers $z$ satisfying $|z - 2 + i| \geq \sqrt{5}$. If the complex number $z_0$ is such that $\frac{1}{|z_0 - 1|}$ is the maximum of the set $\left\{\frac{1}{|z-1|} : z \in S\right\}$, then the principal argument of $\frac{4 - z_0 - \bar{z}_0}{z_0 - \bar{z}_0 + 2i}$ is

(a) $\frac{\pi}{4}$ (b) $\frac{3\pi}{4}$ (c) $-\frac{\pi}{2}$ (d) $\frac{\pi}{2}$

**38.** Let $M = \begin{bmatrix} \sin^4\theta & -1 - \sin^2\theta \\ 1 + \cos^2\theta & \cos^4\theta \end{bmatrix}$ $= \alpha I + \beta M^{-1}$,

where $\alpha = \alpha(\theta)$ and $\beta = \beta(\theta)$ are real numbers, and $I$ is the $2 \times 2$ identity matrix. If $\alpha^*$ is the minimum of the set $\{\alpha(\theta) : \theta \in [0, 2\pi)\}$ and $\beta^*$ is the minimum of the set $\{\beta(\theta) : \theta \in [0, 2\pi)\}$, then the value of $\alpha^* + \beta^*$ is

(a) $-\frac{17}{16}$ (b) $-\frac{31}{16}$ (c) $-\frac{37}{16}$ (d) $-\frac{29}{16}$

**39.** A line $y = mx + 1$ intersects the circle $(x-3)^2 + (y+2)^2 = 25$ at the points $P$ and $Q$. If the midpoint of the line segment $PQ$ has $x$-coordinate $-\frac{3}{5}$, then which one of the following options is correct?

(a) $6 \leq m < 8$ (b) $-3 \leq m < -1$
(c) $4 \leq m < 6$ (d) $2 \leq m < 4$

**40.** The area of the region $\{(x, y) : xy \leq 8, 1 \leq y \leq x^2\}$ is

(a) $8\log_e 2 - \frac{14}{3}$ (b) $8\log_e 2 - \frac{7}{3}$
(c) $16\log_e 2 - \frac{14}{3}$ (d) $16\log_e 2 - 6$

## Section 2 (Maximum Marks : 32)

***Instructions:*** Same as given in Physics.

**41.** Let $\Gamma$ denote a curve $y = y(x)$ which is in the first quadrant and let the point $(1, 0)$ lie on it. Let the tangent to $\Gamma$ at a point $P$ intersect the $y$-axis at $Y_P$. If $PY_P$ has length 1 for each point $P$ on $\Gamma$, then which of the following options is/are correct?

(a) $xy' + \sqrt{1 - x^2} = 0$

(b) $xy' - \sqrt{1 - x^2} = 0$

(c) $y = \log_e\left(\frac{1 + \sqrt{1 - x^2}}{x}\right) - \sqrt{1 - x^2}$

(d) $y = -\log_e\left(\frac{1 + \sqrt{1 - x^2}}{x}\right) + \sqrt{1 - x^2}$

**42.** Define the collections $\{E_1, E_2, E_3, \dots\}$ of ellipses and $\{R_1, R_2, R_3, \dots\}$ of rectangles as follows :

$$E_1 : \frac{x^2}{9} + \frac{y^2}{4} = 1$$

$R_1$ : rectangle of largest area, with sides parallel to the axes, inscribed in $E_1$;

$E_n$ : ellipse $\frac{x^2}{a_n^2} + \frac{y^2}{b_n^2} = 1$ of the largest area inscribed in $R_{n-1}$, $n > 1$;

$R_n$ : rectangle of largest area, with sides parallel to the axes, inscribed in $E_n$, $n > 1$.

Then which of the following options is/are correct?

(a) The eccentricities of $E_{18}$ and $E_{19}$ are NOT equal.

(b) The distance of a focus from the centre in $E_9$ is $\frac{\sqrt{5}}{32}$.

(c) $\sum_{n=1}^{N}$(area of $R_n$) $< 24$, for each positive integer $N$.

(d) The length of latusrectum of $E_9$ is $\frac{1}{6}$

**43.** In a non-right-angled triangle $\Delta PQR$, let $p, q, r$ denote the lengths of the sides opposite to the angles at $P, Q, R$ respectively. The median from $R$ meets the side $PQ$ at $S$, the perpendicular from $P$ meets the side $QR$ at $E$, and $RS$ and $PE$ intersect at $O$. If $p = \sqrt{3}$, $q = 1$, and the radius of the circumcircle of the $\Delta PQR$ equals 1, then which of the following options is/are correct?

(a) Length of $OE = \frac{1}{6}$

(b) Length of $RS = \frac{\sqrt{7}}{2}$

(c) Area of $\Delta SOE = \frac{\sqrt{3}}{12}$

(d) Radius of incircle of $\Delta PQR = \frac{\sqrt{3}}{2}(2 - \sqrt{3})$

**44.** Let $\alpha$ and $\beta$ be the roots of $x^2 - x - 1 = 0$, with $\alpha > \beta$. For all positive integers $n$, define

$$a_n = \frac{\alpha^n - \beta^n}{\alpha - \beta}, n \geq 1,$$

$b_1 = 1$ and $b_n = a_{n-1} + a_{n+1}$, $n \geq 2$

Then which of the following options is/are correct?

(a) $\sum_{n=1}^{\infty} \frac{b_n}{10^n} = \frac{8}{89}$

(b) $b_n = \alpha^n + \beta^n$ for all $n \geq 1$

(c) $a_1 + a_2 + a_3 + \dots + a_n = a_{n+2} - 1$ for all $n \geq 1$

(d) $\sum_{n=1}^{\infty} \frac{a_n}{10^n} = \frac{10}{89}$

**45.** Let $L_1$ and $L_2$ denote the lines

$$\mathbf{r} = \hat{\mathbf{i}} + \lambda(-\hat{\mathbf{i}} + 2\hat{\mathbf{j}} + 2\hat{\mathbf{k}}), \lambda \in R$$

and $\mathbf{r} = \mu(2\hat{\mathbf{i}} - \hat{\mathbf{j}} + 2\hat{\mathbf{k}}), \mu \in R$

respectively. If $L_3$ is a line which is perpendicular to both $L_1$ and $L_2$ and cuts both of them, then which of the following options describe(s) $L_3$?

(a) $\mathbf{r} = \frac{2}{9}(2\hat{\mathbf{i}} - \hat{\mathbf{j}} + 2\hat{\mathbf{k}}) + t(2\hat{\mathbf{i}} + 2\hat{\mathbf{j}} - \hat{\mathbf{k}})$, $t \in R$

(b) $\mathbf{r} = \frac{1}{3}(2\hat{\mathbf{i}} + \hat{\mathbf{k}}) + t(2\hat{\mathbf{i}} + 2\hat{\mathbf{j}} - \hat{\mathbf{k}}), t \in R$

(c) $\mathbf{r} = \frac{2}{9}(4\hat{\mathbf{i}} + \hat{\mathbf{j}} + \hat{\mathbf{k}}) + t(2\hat{\mathbf{i}} + 2\hat{\mathbf{j}} - \hat{\mathbf{k}}), t \in R$

(d) $\mathbf{r} = t(2\hat{\mathbf{i}} + 2\hat{\mathbf{j}} - \hat{\mathbf{k}}), t \in R$

**46.** There are three bags $B_1$, $B_2$ and $B_3$. The bag $B_1$ contains 5 red and 5 green balls, $B_2$ contains 3 red and 5 green balls, and $B_3$ contains 5 red and 3 green balls. Bags $B_1$, $B_2$ and $B_3$ have probabilities $\frac{3}{10}, \frac{3}{10}$ and $\frac{4}{10}$ respectively of being chosen. A bag is selected at random and a ball is chosen at random from the bag. Then which of the following options is/are correct?

(a) Probability that the chosen ball is green, given that the selected bag is $B_3$, equals $\frac{3}{8}$.

(b) Probability that the selected bag is $B_3$, given that the chosen ball is green, equals $\frac{5}{13}$.

(c) Probability that the chosen ball is green equals $\frac{39}{80}$.

(d) Probability that the selected bag is $B_3$ and the chosen ball is green equals $\frac{3}{10}$.

**47.** Let $M = \begin{bmatrix} 0 & 1 & a \\ 1 & 2 & 3 \\ 3 & b & 1 \end{bmatrix}$ and

$\text{adj } M = \begin{bmatrix} -1 & 1 & -1 \\ 8 & -6 & 2 \\ -5 & 3 & -1 \end{bmatrix}$

where $a$ and $b$ are real numbers. Which of the following options is/are correct?

(a) $\det(\text{adj } M^2) = 81$

(b) If $M \begin{bmatrix} \alpha \\ \beta \\ \gamma \end{bmatrix} = \begin{bmatrix} 1 \\ 2 \\ 3 \end{bmatrix}$, then $\alpha - \beta + \gamma = 3$

(c) $(\text{adj } M)^{-1} + \text{adj } M^{-1} = -M$

(d) $a + b = 3$

**48.** Let $f : R \to R$ be given by

$$f(x) = \begin{cases} x^5 + 5x^4 + 10x^3 + 10x^2 + 3x + 1, & x < 0; \\ x^2 - x + 1, & 0 \le x < 1; \\ \frac{2}{3}x^3 - 4x^2 + 7x - \frac{8}{3}, & 1 \le x < 3; \\ (x-2)\log_e(x-2) - x + \frac{10}{3}, & x \ge 3; \end{cases}$$

Then which of the following options is/are correct?

(a) $f$ is increasing on $(-\infty, 0)$

(b) $f'$ is NOT differentiable at $x = 1$

(c) $f$ is onto

(d) $f'$ has a local maximum at $x = 1$

## Section 3 (Maximum Marks : 18)

***Instructions:*** Same as given in Physics.

**49.** Let $S$ be the sample space of all $3 \times 3$ matrices with entries from the set $\{0, 1\}$. Let the events $E_1$ and $E_2$ be given by

$E_1 = \{A \in S : \det A = 0\}$ and

$E_2 = \{A \in S :$ sum of entries of $A$ is $7\}$.

If a matrix is chosen at random from $S$, then the conditional probability $P(E_1 \mid E_2)$ equals ...........

**50.** Let the point $B$ be the reflection of the point $A(2, 3)$ with respect to the line $8x - 6y - 23 = 0$. Let $\Gamma_A$ and $\Gamma_B$ be circles of radii 2 and 1 with centres $A$ and $B$ respectively. Let $T$ be a common tangent to the circles $\Gamma_A$ and $\Gamma_B$ such that both the circles are on the same side of $T$. If $C$ is the point of intersection of $T$ and the line passing through $A$ and $B$, then the length of the line segment $AC$ is ...........

**51.** Let $\omega \neq 1$ be a cube root of unity. Then the minimum of the set $\{|a + b\omega + c\omega^2|^2 : a, b, c \text{ distinct non-zero integers}\}$ equals ............

**52.** If $I = \frac{2}{\pi} \int_{-\pi/4}^{\pi/4} \frac{dx}{(1 + e^{\sin x})(2 - \cos 2x)}$

then $27I^2$ equals .........

**53.** Three lines are given by $\mathbf{r} = \lambda\hat{\mathbf{i}}, \lambda \in R$, $\mathbf{r} = \mu(\hat{\mathbf{i}} + \hat{\mathbf{j}}), \mu \in R$ and $\mathbf{r} = \nu(\hat{\mathbf{i}} + \hat{\mathbf{j}} + \hat{\mathbf{k}}), \nu \in R$

Let the lines cut the plane $x + y + z = 1$ at the points $A, B$ and $C$ respectively. If the area of the triangle $ABC$ is $\Delta$ then the value of $(6\Delta)^2$ equals ..........

**54.** Let $AP(a\,; d)$ denote the set of all the terms of an infinite arithmetic progression with first term $a$ and common difference $d > 0$. If $AP(1\,; 3) \cap AP(2\,; 5) \cap AP(3\,; 7) = AP(a\,; d)$, then $a + d$ equals .........

# Paper 2

## PHYSICS

### Section 1 (Maximum Marks : 32)

- This section contains **EIGHT (08)** questions.
- Each question has **FOUR** options for correct answer(s). **ONE OR MORE THAN ONE** of these four option(s) is (are) correct options(s).
- For each question, choose the correct options(s) to answer the question.
- Answer to each question will be evaluated according to the following marking scheme:

  *Full Marks* : **+4** If only (all) the correct option(s) is (are) chosen.
  *Partial Marks* : **+3** If all the four options are correct but ONLY three options are chosen.
  *Partial Marks* : **+2** If three or more options are correct but ONLY two options are chosen, both of which are correct options.
  *Partial Marks* : **+1** If two or more options are correct but ONLY one option is chosen and it is a correct option.
  *Zero Marks* : **0** If none of the options is chosen (i.e. the question is unanswered).
  *Negative Marks* : **−1** In all other cases.
- **For example:** in a question, if (A), (B) and (D) are the ONLY three options corresponding to correct answer, then
  choosing ONLY (A), (B) and (D) will get +4 marks;
  choosing ONLY (A) and (B) will get +2 marks;
  choosing ONLY (A) and (D) will get +2 marks;
  choosing ONLY (B) and (D) will get +2 marks;
  choosing ONLY (A) will get +1 mark;
  choosing ONLY (B) will get +1 mark;
  choosing ONLY (D) will get +1 mark;
  choosing no option (i.e. the question is unanswered) will get 0 marks; and
  choosing any other combination of options will −1 mark.

**1.** An electric dipole with dipole moment $\frac{p_0}{\sqrt{2}}(\hat{\mathbf{i}}+\hat{\mathbf{j}})$ is held fixed at the origin $O$ in the presence of a uniform electric field of magnitude $E_0$.

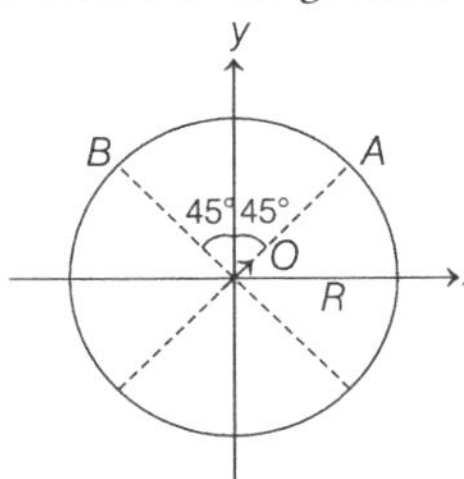


If the potential is constant on a circle of radius $R$ centered at the origin as shown in figure, then the correct statement(s) is/are,
($\epsilon_0$ is the permittivity of the free space, $R >>$ dipole size)

(a) The magnitude of total electric field on any two points of the circle will be same.

(b) Total electric field at point $B$ is $\vec{\mathbf{E}}_B = 0$

(c) $R = \left(\frac{p_0}{4\pi \epsilon_0 E_0}\right)^{1/3}$

(d) Total electric field at point $A$ is $\vec{\mathbf{E}}_A = \sqrt{2}E_0(\hat{\mathbf{i}}+\hat{\mathbf{j}})$

**2.** A small particle of mass $m$ moving inside a heavy, hollow and straight tube along the tube axis undergoes elastic collision at two ends. The tube has no friction and it is closed at one end by a flat surface while the other end is fitted with a heavy movable flat piston as shown in figure.

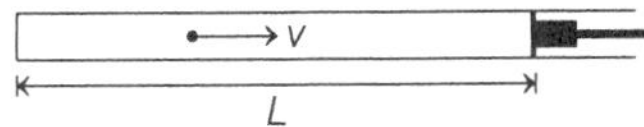


When the distance of the piston from closed end is $L = L_0$, the particle speed is $v = v_0$. The piston is moved inward at a very low speed $V$ such that $V << \frac{dL}{L} v_0$, where $dL$ is the infinitesimal displacement of the piston. Which of the following statement(s) is/are correct?

(a) After each collision with the piston, the particle speed increases by $2V$.

(b) If the piston moves inward by $dL$, the particle speed increases by $2v\frac{dL}{L}$.

(c) The particle's kinetic energy increases by a factor of 4 when the piston is moved inward from $L_0$ to $1/2\ L_0$.

(d) The rate at which the particle strikes the piston is $v/L$.

**3.** Three glass cylinders of equal height $H = 30$ cm and same refractive index $n = 1.5$ are placed on a horizontal surface as shown in figure.

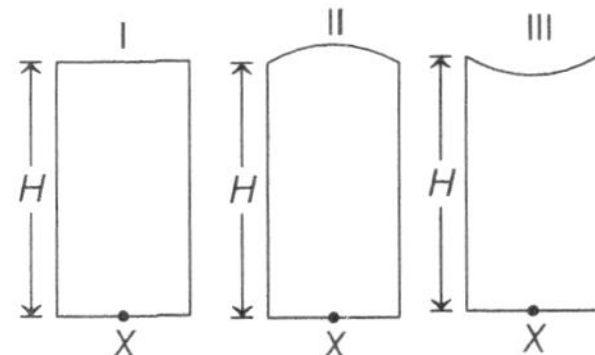


Cylinder I has a flat top, cylinder II has s convex top and cylinder III has a concave top. The radii of curvature of the two curved tops are same ($R = 3$ m). If $H_1$, $H_2$, and $H_3$ are the apparent depths of a point $X$ on the bottom of the three cylinders, respectively, the correct statement(s) is/are

(a) $H_2 > H_1$

(b) $H_3 > H_1$

(c) 0.85 cm $< (H_2 - H_1) <$ 0.9 cm

(d) $H_2 > H_3$

**4.** A block of mass $2M$ is attached to a massless spring with spring-constant $k$.

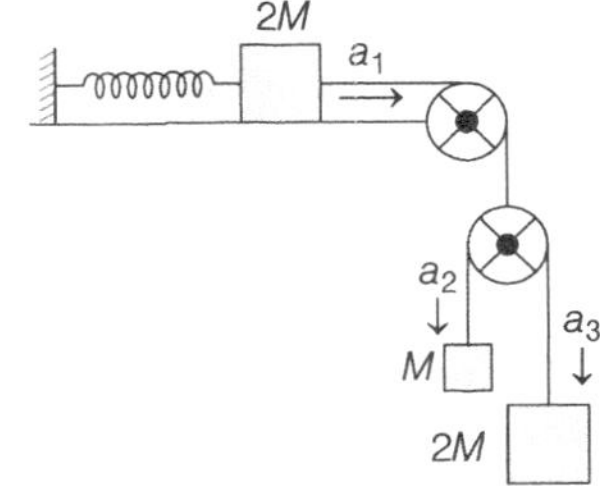

This block is connected to two other blocks of masses $M$ and $2M$ using two massless pulleys and strings. The accelerations of the blocks are $a_1$, $a_2$ and $a_3$ as shown in the figure. The system is released from rest with the spring in its unstretched state. The maximum extension of the spring is $x_0$. Which of the following option(s) is/are correct?

[$g$ is the acceleration due to gravity. Neglect friction]

(a) $a_2 - a_1 = a_1 - a_3$

(b) At an extension of $\frac{x_0}{4}$ of the spring, the magnitude of acceleration of the block connected to the spring is $\frac{3g}{10}$.

(c) $x_0 = \frac{4Mg}{k}$

(d) When spring achieves an extension of $\frac{x_0}{2}$ for the first time, the speed of the block connected to the spring is $3g\sqrt{\frac{M}{5k}}$.

**5.** A thin and uniform rod of mass $M$ and length $L$ is held vertical on a floor with large friction. The rod is released from rest so that it falls by rotating about its contact-point with the floor without slipping. Which of the following statement(s) is/are correct, when the rod makes an angle 60° with vertical?
[$g$ is the acceleration due to gravity]

(a) The angular acceleration of the rod will be $\frac{2g}{L}$.

(b) The normal reaction force from the floor on the rod will be $\frac{Mg}{16}$.

(c) The radial acceleration of the rod's center of mass will be $\frac{3g}{4}$.

(d) The angular speed of the rod will be $\sqrt{\frac{3g}{2L}}$.

**6.** In a Young's double slit experiment, the slit separation $d$ is 0.3 mm and the screen distance $D$ is 1 m. A parallel beam of light of wavelength 600 nm is incident on the slits at angle $\alpha$ as shown in figure.

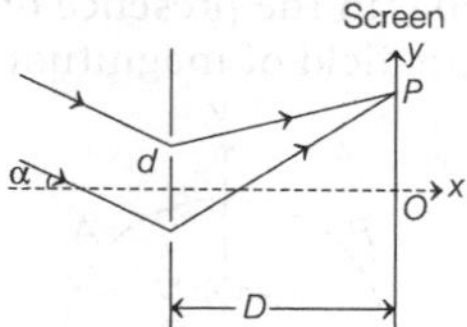


On the screen, the point $O$ is equidistant from the slits and distance $PO$ is 11.0 mm. Which of the following statement(s) is/are correct?

(a) For $\alpha = 0$, there will be constructive interference at point $P$.

(b) For $\alpha = \frac{0.36}{\pi}$ degree, there will be destructive interference at point $P$.

(c) For $\alpha = \frac{0.36}{\pi}$ degree, there will be destructive interference at point $O$.

(d) Fringe spacing depends on $\alpha$.

**7.** A free hydrogen atom after absorbing a photon of wavelength $\lambda_a$ gets excited from the state $n = 1$ to the state $n = 4$. Immediately after that the electron jumps to $n = m$ state by emitting a photon of wavelength $\lambda_e$. Let the change in momentum of atom due to the absorption and the emission be $\Delta p_a$ and $\Delta p_e$, respectively. If $\frac{\lambda_a}{\lambda_e} = \frac{1}{5}$, which of the option(s) is/are correct?
[Use $hc = 1242$ eVnm; 1 nm $= 10^{-9}$ m, $h$ and $c$ are Planck's constant and speed of light in vacuum, respectively]

(a) The ratio of kinetic energy of the electron in the state $n = m$ to the state, $n = 1$ is $\frac{1}{4}$

(b) $m = 2$

(c) $\frac{\Delta p_a}{\Delta p_e} = \frac{1}{2}$

(d) $\lambda_e = 418$ nm

**8.** A mixture of ideal gas containing 5 moles of monatomic gas and 1 mole of rigid diatomic gas is initially at pressure $P_0$, volume $V_0$, and temperature $T_0$. If the gas mixture is adiabatically compressed to a volume $\frac{V_0}{4}$, then the correct statement(s) is/are

(Given, $2^{1.2} = 2.3$; $2^{3.2} = 9.2$; $R$ is a gas constant)

(a) The final pressure of the gas mixture after compression is in between $9P_0$ and $10P_0$.

(b) The average kinetic energy of the gas mixture after compression is in between $18\,RT_0$ and $19\,RT_0$.

(c) Adiabatic constant of the gas mixture is 1.6.

(d) The work $|W|$ done during the process is $13RT_0$.

## Section 2 (Maximum Marks : 18)

- This section contains **SIX (06)** questions. The answer to each question is a **NUMERICAL VALUE.**
- For each question, enter the correct numerical value of the answer using the mouse and the on-screen virtual numeric keypad in the place designated to enter the answer. If the numerical value has more than two decimal places, **truncate/round-off** the value to **TWO** decimal places.
- Answer to each question will be evaluated according to the following marking scheme:

*Full Marks* : **+3** If ONLY the correct numerical value is entered as answer.

*Zero Marks* : **0** In all other cases.

**9.** Suppose a $^{226}_{88}\text{Ra}$ nucleus at rest and in ground state undergoes $\alpha$-decay to a $^{222}_{86}\text{Rn}$ nucleus in its excited state. The kinetic energy of the emitted $\alpha$ particle is found to be 4.44 MeV. $^{222}_{86}\text{Rn}$ nucleus then goes to its ground state by $\gamma$-decay. The energy of the emitted $\gamma$ photon is ......... keV.

[Given : atomic mass of $^{226}_{88}\text{Ra} = 226.005$ u, atomic mass of $^{222}_{86}\text{Rn} = 222.000$ u, atomic mass of $\alpha$ particle $= 4.000$ u, $1\text{ u} = 931\text{ MeV}/c^2$, c is speed of the light]

**10.** A monochromatic light is incident from air on a refracting surface of a prism of angle 75° and refractive index $n_0 = \sqrt{3}$. The other refracting surface of the prism is coated by a thin film of material of refractive index $n$ as shown in figure. The light suffers total internal reflection at the coated prism surface for an incidence angle of $\theta \le 60°$. The value of $n^2$ is ......... .

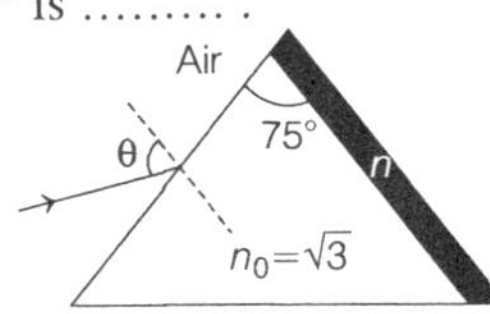


**11.** A perfectly reflecting mirror of mass $M$ mounted on a spring constitutes a spring-mass system of angular frequency $\Omega$ such that $\frac{4\pi M\Omega}{h} = 10^{24}$ m$^{-2}$ with $h$ as Planck's constant. $N$ photons of wavelength $\lambda = 8\pi \times 10^{-6}$ m strike the mirror simultaneously at normal incidence such that the mirror gets displaced by 1 $\mu$m.

If the value of $N$ is $x \times 10^{12}$, then the value of $x$ is ......... .
[Consider the spring as massless]

**12.** An optical bench has 1.5 m long scale having four equal divisions in each cm. While measuring the focal length of a convex lens, the lens is kept at 75 cm mark of the scale and the object pin is kept at 45 cm mark. The image of the object pin on the other side of the lens overlaps with image pin that is kept at 135 cm mark. In this experiment, the percentage error in the measurement of the focal length of the lens is .........

**13.** A ball is thrown from ground at an angle $\theta$ with horizontal and with an initial speed $u_0$. For the resulting projectile motion, the magnitude of average velocity of the ball up to the point when it hits the ground for the first time is $V_1$. After hitting the ground, the ball rebounds at the same angle $\theta$ but with a reduced speed of $\frac{u_0}{\alpha}$. Its motion continues for a long time as shown in figure. If the magnitude of average velocity of the ball for entire duration of motion is $0.8\ V_1$, the value of $\alpha$ is ............ .

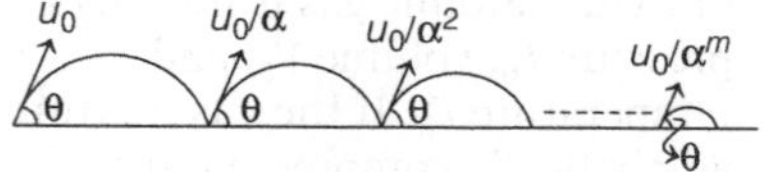


**14.** A 10 cm long perfectly conducting wire $PQ$ is moving with a velocity 1 cm/s on a pair of horizontal rails of zero resistance. One side of the rails is connected to an inductor $L = 1$ mH and a resistance $R = 1\ \Omega$ as shown in figure. The horizontal rails, $L$ and $R$ lie in the same plane with a uniform magnetic field $B = 1$ T perpendicular to the plane. If the key $S$ is closed at certain instant, the current in the circuit after 1 millisecond is $x \times 10^{-3}$ A, where the value of $x$ is ...... .
[Assume the velocity of wire $PQ$ remains constant (1 cm/s) after key $S$ is closed. Given, $e^{-1} = 0.37$, where $e$ is base of the natural logarithm]

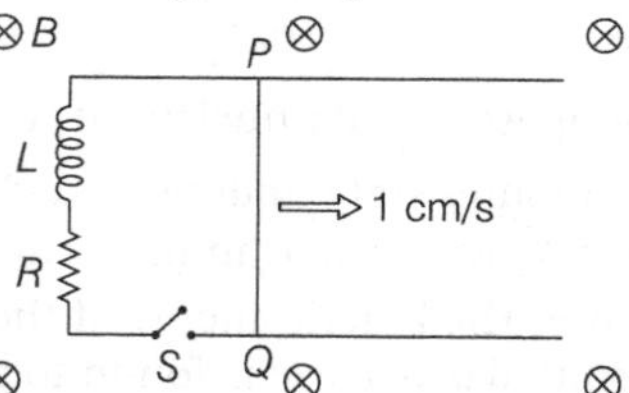


## Section 3 (Maximum Marks : 12)

- This section contains **TWO (02)** List-Match sets.
- Each List-Match set has **TWO (02)** Multiple Choice Questions.
- Each List-Match set has two lists : List-I and List-II
- List-I has **Four** entries (I), (II), (III) and (IV) and **List-II** has **Six** entries (P), (Q), (R), (S), (T) and (U).
- **FOUR** options are given in each Multiple Choice Question based on **List-I** and **List-II** and **ONLY ONE** of these four options satisfies the condition asked in the Multiple Choice Questions.
- Answer to each question will be evaluated according to the following marking scheme :

*Full Marks* : **+3** If ONLY the option corresponding to the correct combination is chosen:

*Zero Marks* : **0** If none of the options is chosen (i.e. the question is unanswered).

*Negative Marks* : **−1** In all other cases.

*Answer the following by appropriately matching the lists based on the information given in the paragraph*

A musical instrument is made using four different metal strings, 1, 2, 3 and 4 with mass per unit length $\mu$, $2\mu$, $3\mu$ and $4\mu$ respectively. The instrument is played by vibrating the strings by varying the free length in between the range $L_0$ and $2L_0$. It is found that in string-1 ($\mu$) at free length $L_0$ and tension $T_0$ the fundamental mode frequency is $f_0$.

List-I gives the above four strings while list-II lists the magnitude of some quantity.

| List-I | List-II |
|---|---|
| (I) String-1 ($\mu$) | (P) 1 |
| (II) String-2 ($2\mu$) | (Q) 1/2 |
| (III) String-3 ($3\mu$) | (R) $1/\sqrt{2}$ |
| (IV) String-4 ($4\mu$) | (S) $1/\sqrt{3}$ |
| | (T) 3/16 |
| | (U) 1/16 |

**15.** If the tension in each string is $T_0$, the correct match for the highest fundamental frequency in $f_0$ units will be

(a) I $\rightarrow$ P, II $\rightarrow$ Q, III $\rightarrow$ T, IV $\rightarrow$ S
(b) I $\rightarrow$ P, II $\rightarrow$ R, III $\rightarrow$ S, IV $\rightarrow$ Q
(c) I $\rightarrow$ Q, II $\rightarrow$ S, III $\rightarrow$ R, IV $\rightarrow$ P
(d) I $\rightarrow$ Q, II $\rightarrow$ P, III $\rightarrow$ R, IV $\rightarrow$ T

**16.** The length of the strings 1, 2, 3 and 4 are kept fixed at $L_0, \dfrac{3L_0}{2} \cdot \dfrac{5L_0}{4}$ and $\dfrac{7L_0}{4}$, respectively. Strings 1, 2, 3 and 4 are vibrated at their 1st, 3rd, 5th and 14th harmonics, respectively such that all the strings have same frequency.

The correct match for the tension in the four strings in the units of $T_0$ will be

(a) I $\rightarrow$ P, II $\rightarrow$ R, III $\rightarrow$ T, IV $\rightarrow$ U
(b) I $\rightarrow$ P, II $\rightarrow$ Q, III $\rightarrow$ R, IV $\rightarrow$ T
(c) I $\rightarrow$ P, II $\rightarrow$ Q, III $\rightarrow$ T, IV $\rightarrow$ U
(d) I $\rightarrow$ T, II $\rightarrow$ Q, III $\rightarrow$ R, IV $\rightarrow$ U

*Answer the following by appropriately matching the lists based on the information given in the paragraph*

In a thermodynamic process on an ideal monatomic gas, the infinitesimal heat absorbed by the gas is given by $T\Delta X$. where, $T$ is temperature of the system and $\Delta X$ is the infinitesimal change in a thermodynamic quantity $X$ of the system. For a mole of monatomic ideal gas,

$$X = \frac{3}{2} R \ln\left(\frac{T}{T_A}\right) + R \ln\left(\frac{V}{V_A}\right)$$

Here, $R$ is gas constant, $V$ is volume of gas, $T_A$ and $V_A$ are constants.

The List-I below gives some quantities involved in a process and List-II gives some possible values of these quantities.

| List-I | List-II |
|---|---|
| (I) Work done by the system in process $1 \rightarrow 2 \rightarrow 3$ | (P) $\frac{1}{3} RT_0 \ln 2$ |
| (II) Change in internal energy in process $1 \rightarrow 2 \rightarrow 3$ | (Q) $\frac{1}{3} RT_0$ |
| (III) Heat absorbed by the system in process $1 \rightarrow 2 \rightarrow 3$ | (R) $RT_0$ |
| (IV) Heat absorbed by the system in process $1 \rightarrow 2$ | (S) $\frac{4}{3} RT_0$ |
| | (T) $\frac{1}{3} RT_0(3 + \ln 2)$ |
| | (U) $\frac{5}{6} RT_0$ |

**17.** If the process on one mole of monatomic ideal gas is as shown in the *TV*-diagram with $P_0V_0 = \frac{1}{3}RT_0$, the correct match is,

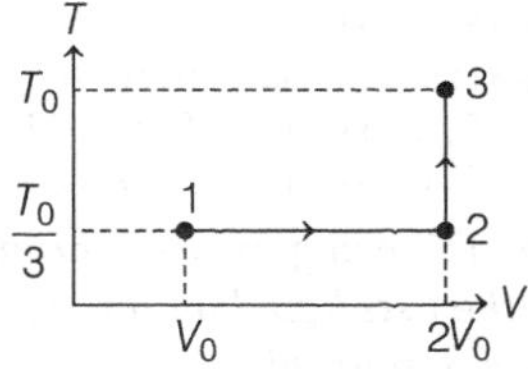


(a) I → P, II → R, III → T, IV → S
(b) I → P, II → T, III → Q, IV → T
(c) I → S, II → T, III → Q, IV → U
(d) I → P, II → R, III → T, IV → P

**18.** If the process carried out on one mole of monoatomic ideal gas is as shown in the *PV*-diagram with $p_0V_0 = \frac{1}{3}RT_0$, the correct match is,

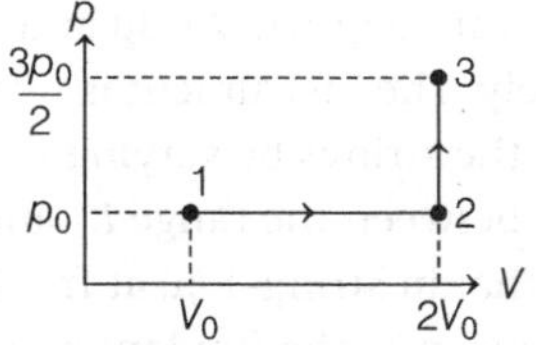


(a) I → S, II → R, III → Q, IV → T
(b) I → Q, II → R, III → P, IV → U
(c) I → Q, II → S, III → R, IV → U
(d) I → Q, II → R, III → S, IV → U

# CHEMISTRY

## Section 1 (Maximum Marks : 32)

***Instructions:*** Same as given in Physics.

**19.** Choose the correct option(s) for the following reaction sequence

MeO, CHO; (i) $Hg^{2+}$, dil. $H_2SO_4$ (ii) $AgNO_3$, $NH_4OH$ (iii) Zn-Hg, conc. HCl → *Q*

(i) $SOCl_2$ pyridine (ii) $AlCl_3$ → *R*; Zn-Hg conc. HCl → *S*

Consider *Q*, *R* and *S* as major products.

(a) MeO, *R*, O; MeO, *S*

(b) MeO, $CO_2H$, *Q*; MeO, *R*, O

(c) MeO, *R*, O; MeO, *S*

(d) OH, MeO, $CO_2H$, *Q*; MeO, *S*

**20.** Choose the correct option(s) that give(s) an aromatic compound as the major product.

(a) Br; NaOEt →

(b) + $Cl_2$ (excess) $\xrightarrow{\text{UV, 500 K}}$

(c) $H_3C$–CH(Br)–CH$_2$–Br $\xrightarrow[\text{(iii) Red hot iron tube, 873 K}]{\text{(i) Alc. KOH (ii) NaNH}_2}$

(d) (cyclopentadiene) $\xrightarrow{\text{NaOMe}}$

**21.** The ground state energy of hydrogen atom is $-13.6$ eV. Consider an electronic state $\Psi$ of $He^+$ whose energy, azimuthal quantum number and magnetic quantum number are $-3.4$ eV, 2 and 0, respectively.

Which of the following statement(s) is(are) true for the state $\Psi$?

(a) It is a 4*d* state

(b) The nuclear charge experienced by the electron in this state is less than 2*e*, where *e* is the magnitude of the electronic charge

(c) It has 2 angular nodes

(d) It has 3 radial nodes

**22.** Consider the following reactions (unbalanced).

$$Zn + \text{Hot conc. } H_2SO_4 \longrightarrow G + R + X$$

$$Zn + \text{conc. NaOH} \longrightarrow T + Q$$

$$G + H_2S + NH_4OH \longrightarrow Z \text{ (a precipitate)} + X + Y$$

Choose the correct option(s).

(a) The oxidation state of Zn in *T* is +1

(b) *R* is a V-shaped molecule

(c) Bond order of *Q* is 1 in its ground state

(d) *Z* is dirty white in colour

**23.** With reference to *aqua-regia*, choose the correct option(s).

(a) *Aqua-regia* is prepared by mixing conc. HCl and conc. $HNO_3$ in 3 : 1 (*v* / *v*) ratio

(b) The yellow colour of *aqua-regia* is due to the presence of NOCl and $Cl_2$

(c) Reaction of gold with *aqua-regia* produces an anion having Au in +3 oxidation state

(d) Reaction of gold with *aqua regia* produces $NO_2$ in the absence of air

**24.** Choose the correct option(s) from the following.

(a) Teflon is prepared by heating tetrafluoroethene in presence of a persulphate catalyst at high pressure

(b) Natural rubber is polyisoprene containing *trans* alkene units

(c) Cellulose has only $\alpha$-D-glucose units that are joined by glycosidic linkages

(d) Nylon-6 has amide linkages

**25.** The cyanide process of gold extraction involves leaching out gold from its ore with $CN^-$ in the presence of *Q* in water to form *R*. Subsequently, *R* is treated with *T* to obtain Au and *Z*. Choose the correct option(s).

(a) *Q* is $O_2$ (b) *Z* is $[Zn(CN)_4]^{2-}$

(c) *T* is Zn (d) *R* is $[Au(CN)_4]^-$

**26.** Which of the following reactions produce(s) propane as a major product?

(a) $H_3C$–CH$_2$–CH$_2$–Cl $\xrightarrow{\text{Zn, dil. HCl}}$

(b) $H_3C$–CH(Br)–CH$_2$–Br $\xrightarrow{\text{Zn}}$

(c) $H_3C$–CH$_2$–CH$_2$–COONa $\xrightarrow{\text{NaOH, CaO, }\Delta}$

(d) $H_3C$–CH$_2$–CH$_2$–COONa + $H_2O$ $\xrightarrow{\text{Electrolysis}}$

## Section 2 (Maximum Marks : 18)

***Instructions:*** Same as given in Physics.

**27.** The decomposition reaction

$2N_2O_5(g) \xrightarrow{\Delta} 2N_2O_4(g) + O_2(g)$ is started in a closed cylinder under isothermal isochoric condition at an initial pressure of 1 atm. After $Y \times 10^3$ s, the pressure inside the cylinder is found to be 1.45 atm. If the rate constant of the reaction is $5 \times 10^{-4}\ s^{-1}$, assuming ideal gas behaviour, the value of *Y* is ........

**28.** The mole fraction of urea in an aqueous urea solution containing 900 g of water is 0.05. If the density of the solution is 1.2 g $cm^{-3}$, then molarity of urea solution is ......
(Given data : Molar masses of urea and water are 60 g $mol^{-1}$ and 18 g $mol^{-1}$, respectively)

**29.** Total number of hydroxyl groups present in a molecule of the major product $P$ is .........

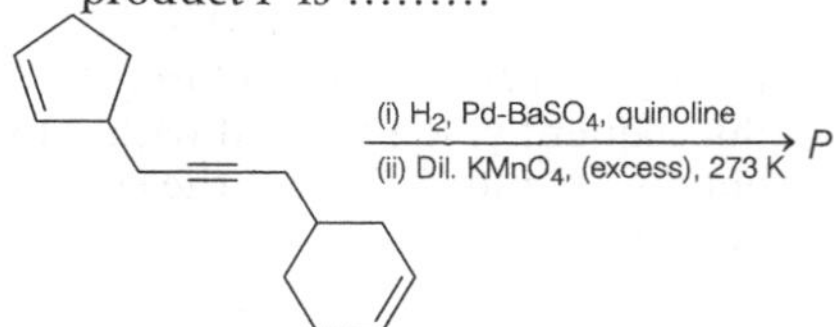


**30.** Total number of *cis* N—Mn—Cl bond angles (that is Mn—N and Mn—Cl bonds in *cis* positions) present in a molecule of *cis* [$Mn(en)_2Cl_2$] complex is .........
(en = $NH_2CH_2CH_2NH_2$)

**31.** The amount of water produced (in g) in the oxidation of 1 mole of rhombic sulphur by conc. $HNO_3$ to a compound with the highest oxidation state of sulphur is ......
(Given data : Molar mass of water = 18 g $mol^{-1}$)

**32.** Total number of isomers considering both structural and stereoisomers of cyclic ethers with the molecular formula $C_4H_8O$ is .........

## Section 3 (Maximum Marks : 12)

***Instructions:*** Same as given in Physics.

Answer the following by appropriately matching the lists based on the information given in the paragraph.

**33.** Consider the Bohr's model of a one-electron atom where the electron moves around the nucleus. In the following List-I contains some quantities for the *n*th orbit of the atom and List-II contains options showing how they depend on *n*.

| | List-I | | List-II |
|---|---|---|---|
| (I) | Radius of the *n* th orbit | (P) | $\propto n^{-2}$ |
| (II) | Angular momentum of the electron in the *n*th orbit | (Q) | $\propto n^{-1}$ |
| (III) | Kinetic energy of the electron in the *n*th orbit | (R) | $\propto n^{0}$ |
| (IV) | Potential energy of the electron in the *n*th orbit | (S) | $\propto n^{1}$ |
| | | (T) | $\propto n^{2}$ |
| | | (U) | $\propto n^{1/2}$ |

Which of the following options has the correct combination considering List-I and List-II?

(a) (III), (P) (b) (III), (S)
(c) (IV), (U) (d) (IV), (Q)

**34.** Consider the Bohr's model of a one-electron atom where the electron moves around the nucleus. In the following List-I contains some quantities for the *n*th orbit of the atom and List-II contains options showing how they depend on *n*.

| | List-I | | List-II |
|---|---|---|---|
| (I) | Radius of the *n*th orbit | (P) | $\propto n^{-2}$ |
| (II) | Angular momentum of the electron in the *n*th orbit | (Q) | $\propto n^{-1}$ |
| (III) | Kinetic energy of the electron in the *n*th orbit | (R) | $\propto n^{0}$ |
| (IV) | Potential energy of the electron in the *n*th orbit | (S) | $\propto n^{1}$ |
| | | (T) | $\propto n^{2}$ |
| | | (U) | $\propto n^{1/2}$ |

Which of the following options has the correct combination considering List-I and List-II?

(a) (II), (R) (b) (I), (P)
(c) (I), (T) (d) (II), (Q)

**35.** List-I includes starting materials and reagents of selected chemical reactions. List-II gives structures of compounds that may be formed as intermediate products and/or final products from the reactions of List-I.

| | List-I | | List-II |
|---|---|---|---|
| (I) | 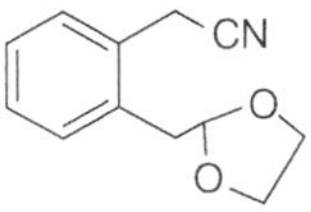 CN, O, O; (i) DIBAL-H (ii) Dil. HCl (iii) $NaBH_4$ (iv) Conc. $H_2SO_4$ | (P) | 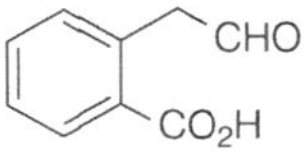 CHO, $CO_2H$ |
| (II) | 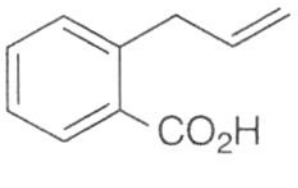 $CO_2H$; (i) $O_3$ (ii) Zn, $H_2O$ (iii) $NaBH_4$ (iv) Conc. $H_2SO_4$ | (Q) | 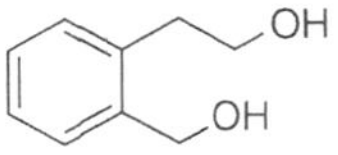 OH, OH |
| (III) | 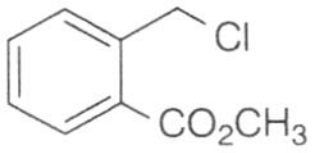 Cl, $CO_2CH_3$; (i) KCN (ii) $H_3O^+$,Δ (iii) $LiAlH_4$ (iv) Conc. $H_2SO_4$ | (R) | 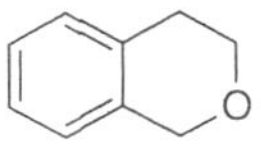 O |
| (IV) | 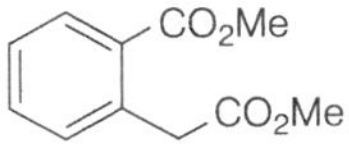 $CO_2Me$, $CO_2Me$; (iii) $LiAlH_4$ (iv) Conc. $H_2SO_4$ | (S) | 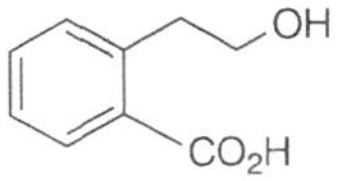 OH, $CO_2H$ |
| | | (T) | 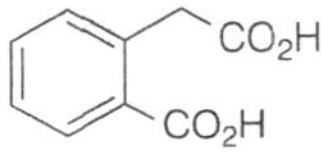 $CO_2H$, $CO_2H$ |
| | | (U) | 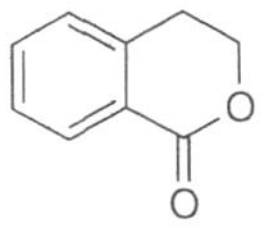 O, O |

Which of the following options has correct combination considering List-I and List-II?

(a) (III), (S), (R) (b) (IV), (Q), (R) (c) (III), (T), (U) (d) (IV), (Q), (U)

**36.** List-I includes starting materials and reagents of selected chemical reactions. List-II gives structures of compounds that may be formed as intermediate products and/or final products from the reactions of List-I.

| | List-I | | List-II |
|---|---|---|---|
| (I) | 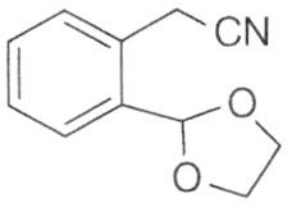 CN, O, O; (i) DIBAL-H (ii) Dil. HCl (iii) $NaBH_4$ (iv) Conc. $H_2SO_4$ | (P) | 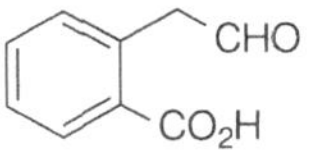 CHO, $CO_2H$ |
| (II) | 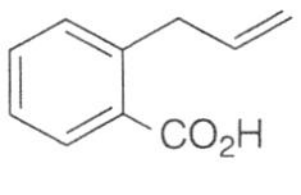 $CO_2H$; (i) $O_3$ (ii) Zn, $H_2O$ (iii) $NaBH_4$ (iv) Conc. $H_2SO_4$ | (Q) | 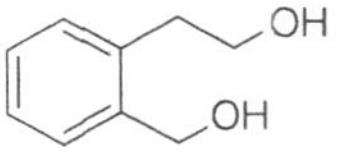 OH, OH |

| List-I | | List-II | |
|---|---|---|---|
| (III) | 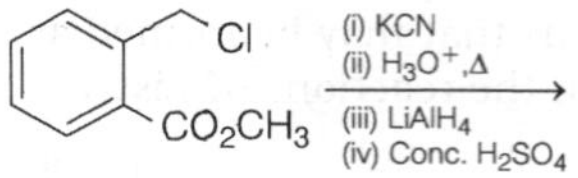 Cl, $CO_2CH_3$ (i) KCN (ii) $H_3O^+, \Delta$ (iii) $LiAlH_4$ (iv) Conc. $H_2SO_4$ | (R) | 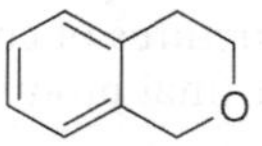 O |
| (IV) | 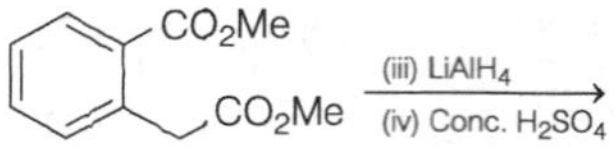 $CO_2Me$, $CO_2Me$ (iii) $LiAlH_4$ (iv) Conc. $H_2SO_4$ | (S) | 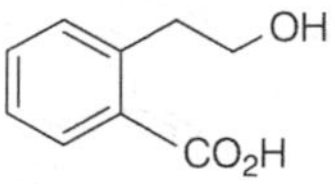 OH, $CO_2H$ |
| | | (T) | 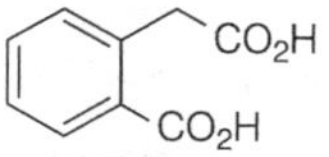 $CO_2H$, $CO_2H$ |
| | | (U) | 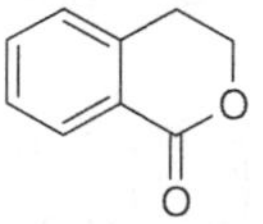 O, O |

Which of the following options has correct combination considering List-I and List-II?

(a) (II), (P), (S), (U) (b) (I), (Q), (T), (U) (c) (II), (P), (S), (T) (d) (I), (S), (Q), (R)

# MATHEMATICS

## Section 1 (Maximum Marks : 32)

***Instructions:*** Same as given in Physics.

**37.** For non-negative integers $n$, let

$$f(n) = \frac{\sum_{k=0}^{n} \sin\left(\frac{k+1}{n+2}\pi\right)\sin\left(\frac{k+2}{n+2}\pi\right)}{\sum_{k=0}^{n} \sin^2\left(\frac{k+1}{n+2}\pi\right)}$$

Assuming $\cos^{-1} x$ takes values in $[0, \pi]$, which of the following options is/are correct?

(a) If $\alpha = \tan(\cos^{-1} f(6))$, then $\alpha^2 + 2\alpha - 1 = 0$

(b) $f(4) = \frac{\sqrt{3}}{2}$

(c) $\sin(7\cos^{-1} f(5)) = 0$

(d) $\lim_{n\to\infty} f(n) = \frac{1}{2}$

**38.** Let $f : R \to R$ be given by $f(x) = (x-1)(x-2)(x-5)$. Define

$$F(x) = \int_0^x f(t)dt, \ x > 0$$

Then which of the following options is/are correct?

(a) $F(x) \neq 0$ for all $x \in (0, 5)$

(b) $F$ has a local maximum at $x = 2$

(c) $F$ has two local maxima and one local minimum in $(0, \infty)$

(d) $F$ has a local minimum at $x = 1$

**39.** Let, $f(x) = \frac{\sin \pi x}{x^2}$, $x > 0$

Let $x_1 < x_2 < x_3 < \ldots < x_n < \ldots$ be all the points of local maximum of $f$ and $y_1 < y_2 < y_3 < \ldots < y_n < \ldots$ be all the points of local minimum of $f$.

Then which of the following options is/are correct?

(a) $|x_n - y_n| > 1$ for every $n$

(b) $x_{n+1} - x_n > 2$ for every $n$

(c) $x_1 < y_1$

(d) $x_n \in \left(2n, 2n + \frac{1}{2}\right)$ for every $n$

**40.** Three lines $L_1 : \mathbf{r} = \lambda\hat{\mathbf{i}}, \lambda \in R$,

$L_2 : \mathbf{r} = \hat{\mathbf{k}} + \mu\hat{\mathbf{j}}, \mu \in R$ and

$L_3 : \mathbf{r} = \hat{\mathbf{i}} + \hat{\mathbf{j}} + \nu\hat{\mathbf{k}}, \nu \in R$ are given.

For which point(s) $Q$ on $L_2$ can we find a point $P$ on $L_1$ and a point $R$ on $L_3$ so that $P$, $Q$ and $R$ are collinear?

(a) $\hat{\mathbf{k}}$ (b) $\hat{\mathbf{k}} + \hat{\mathbf{j}}$ (c) $\hat{\mathbf{k}} + \frac{1}{2}\hat{\mathbf{j}}$ (d) $\hat{\mathbf{k}} - \frac{1}{2}\hat{\mathbf{j}}$

**41.** For $a \in R$, $|a| > 1$, let $\lim_{n\to\infty}\left(\dfrac{1 + \sqrt[3]{2} + \ldots + \sqrt[3]{n}}{n^{7/3}\left(\dfrac{1}{(an+1)^2} + \dfrac{1}{(an+2)^2} + \ldots + \dfrac{1}{(an+n)^2}\right)}\right) = 54$

Then the possible value(s) of $a$ is/are

(a) −6 (b) 7 (c) 8 (d) −9

**42.** Let $f : R$ be a function. We say that $f$ has PROPERTY 1 if $\lim_{h\to 0}\dfrac{f(h) - f(0)}{\sqrt{|h|}}$ exists and is finite, and PROPERTY 2 if $\lim_{h\to 0}\dfrac{f(h) - f(0)}{h^2}$ exists and is finite.

Then which of the following options is/are correct?

(a) $f(x) = \sin x$ has PROPERTY 2 (b) $f(x) = x^{2/3}$ has PROPERTY 1

(c) $f(x) = |x|$ has PROPERTY 1 (d) $f(x) = x|x|$ has PROPERTY 2

**43.** Let $x \in R$ and let $P = \begin{bmatrix} 1 & 1 & 1 \\ 0 & 2 & 2 \\ 0 & 0 & 3 \end{bmatrix}$, $Q = \begin{bmatrix} 2 & x & x \\ 0 & 4 & 0 \\ x & x & 6 \end{bmatrix}$ and $R = PQP^{-1}$, the which of the following options is/are correct?

(a) There exists a real, number $x$ such that $PQ = QP$

(b) For $x = 0$, if $R\begin{bmatrix} 1 \\ a \\ b \end{bmatrix} = 6\begin{bmatrix} 1 \\ a \\ b \end{bmatrix}$, then $a + b = 5$

(c) For $x = 1$, there exists a unit vector $\alpha\hat{\mathbf{i}} + \beta\hat{\mathbf{j}} + \gamma\hat{\mathbf{k}}$ for which $R\begin{bmatrix} \alpha \\ \beta \\ \gamma \end{bmatrix} = \begin{bmatrix} 0 \\ 0 \\ 0 \end{bmatrix}$

(d) $\det R = \det\begin{bmatrix} 2 & x & x \\ 0 & 4 & 0 \\ x & x & 5 \end{bmatrix} + 8$, for all $x \in R$

**44.** Let $P_1 = I = \begin{bmatrix} 1 & 0 & 0 \\ 0 & 1 & 0 \\ 0 & 0 & 1 \end{bmatrix}$, $P_2 = \begin{bmatrix} 1 & 0 & 0 \\ 0 & 0 & 1 \\ 0 & 1 & 0 \end{bmatrix}$, $P_3 = \begin{bmatrix} 0 & 1 & 0 \\ 1 & 0 & 0 \\ 0 & 0 & 1 \end{bmatrix}$, $P_4 = \begin{bmatrix} 0 & 1 & 0 \\ 0 & 0 & 1 \\ 1 & 0 & 0 \end{bmatrix}$,

$P_5 = \begin{bmatrix} 0 & 0 & 1 \\ 1 & 0 & 0 \\ 0 & 1 & 0 \end{bmatrix}$, $P_6 = \begin{bmatrix} 0 & 0 & 1 \\ 0 & 1 & 0 \\ 1 & 0 & 0 \end{bmatrix}$ and $X = \sum_{k=1}^{6} P_k \begin{bmatrix} 2 & 1 & 3 \\ 1 & 0 & 2 \\ 3 & 2 & 1 \end{bmatrix} P_k^T$

where, $P_k^T$ denotes the transpose of the matrix $P_k$. Then which of the following option is/are correct?

(a) $X$ is a symmetric matrix

(b) The sum of diagonal entries of $X$ is 18

(c) $X - 30I$ is an invertible matrix

(d) If $X\begin{bmatrix}1\\1\\1\end{bmatrix} = \alpha\begin{bmatrix}1\\1\\1\end{bmatrix}$, then $\alpha = 30$

## Section 2 (Maximum Marks : 18)

***Instructions:*** Same as given in Physics.

**45.** Let $\mathbf{a} = 2\hat{\mathbf{i}} + \hat{\mathbf{j}} - \hat{\mathbf{k}}$ and $\mathbf{b} = \hat{\mathbf{i}} + 2\hat{\mathbf{j}} + \hat{\mathbf{k}}$ be two vectors. Consider a vector $\mathbf{c} = \alpha\mathbf{a} + \beta\mathbf{b}$, $\alpha, \beta \in R$. If the projection of $\mathbf{c}$ on the vector $(\mathbf{a} + \mathbf{b})$ is $3\sqrt{2}$, then the minimum value of $(\mathbf{c} - (\mathbf{a} \times \mathbf{b}))\cdot\mathbf{c}$ equals .........

**46.** Let $|X|$ denote the number of elements in a set $X$. Let $S = \{1, 2, 3, 4, 5, 6\}$ be a sample space, where each element is equally likely to occur. If $A$ and $B$ are independent events associated with $S$, then the number of ordered pairs $(A, B)$ such that $1 \le |B| < |A|$, equals .........

**47.** Suppose

$$\det\begin{bmatrix}\sum_{k=0}^{n} k & \sum_{k=0}^{n} {}^nC_k k^2 \\ \sum_{k=0}^{n} {}^nC_k k & \sum_{k=0}^{n} {}^nC_k 3^k\end{bmatrix} = 0 \text{ holds}$$

for some positive integer $n$. Then $\sum_{k=0}^{n} \frac{{}^nC_k}{k+1}$ equals .........

**48.** Five persons $A, B, C, D$ and $E$ are seated in a circular arrangement. If each of them is given a hat of one of the three colours red, blue and green, then the number of ways of distributing the hats such that the persons seated in adjacent seats get different coloured hats is .........

**49.** The value of the integral

$$\int_0^{\pi/2} \frac{3\sqrt{\cos\theta}}{(\sqrt{\cos\theta} + \sqrt{\sin\theta})^5}\, d\theta \text{ equals}$$

............

**50.** The value of

$$\sec^{-1}\left(\frac{1}{4}\sum_{k=0}^{10} \sec\left(\frac{7\pi}{12} + \frac{k\pi}{2}\right) \sec\left(\frac{7\pi}{12} + \frac{(k+1)\pi}{2}\right)\right)$$

in the interval $\left[-\frac{\pi}{4}, \frac{3\pi}{4}\right]$ equals......

## Section 3 (Maximum Marks : 12)

***Instructions:*** Same as given in Physics.

*Answer the following by appropriately matching the lists based on the information given in the paragraph.*

**51.** Let $f(x) = \sin(\pi\cos x)$ and $g(x) = \cos(2\pi\sin x)$ be two functions defined for $x > 0$. Define the following sets whose elements are written in the increasing order :

$X = \{x : f(x) = 0\}$, $Y = \{x : f'(x) = 0\}$

$Z = \{x : g(x) = 0\}$, $W = \{x : g'(x) = 0\}$

List-I contains the sets $X, Y, Z$ and $W$. List-II contains some information regarding these sets.

| | List-I | | List-II |
|---|---|---|---|
| (I) | $X$ | (P) | $\supseteq \left\{\frac{\pi}{2}, \frac{3\pi}{2}, 4\pi, 7\pi\right\}$ |
| (II) | $Y$ | (Q) | an arithmetic progression |
| (III) | $Z$ | (R) | NOT an arithmetic progression |
| (IV) | $W$ | (S) | $\supseteq \left\{\frac{\pi}{6}, \frac{7\pi}{6}, \frac{13\pi}{6}\right\}$ |
| | | (T) | $\supseteq \left\{\frac{\pi}{3}, \frac{2\pi}{3}, \pi\right\}$ |
| | | (U) | $\supseteq \left\{\frac{\pi}{6}, \frac{3\pi}{4}\right\}$ |

Which of the following is the only CORRECT combination?

(a) (IV), (P), (R), (S)
(b) (III), (P), (Q), (U)
(c) (III), (R), (U)
(d) (IV), (Q), (T)

**52.** Let $f(x) = \sin(\pi \cos x)$ and $g(x) = \cos(2\pi \sin x)$ be two functions defined for $x > 0$. Define the following sets whose elements are written in the increasing order :

$X = \{x : f(x) = 0\}$, $Y = \{x : f'(x) = 0\}$

$Z = \{x : g(x) = 0\}$, $W = \{x : g'(x) = 0\}$

List-I contains the sets $X, Y, Z$ and $W$. List-II contains some information regarding these sets.

| | List-I | | List-II |
|---|---|---|---|
| (I) | $X$ | (P) | $\supseteq \left\{\frac{\pi}{2}, \frac{3\pi}{2}, 4\pi, 7\pi\right\}$ |
| (II) | $Y$ | (Q) | an arithmetic progression |
| (III) | $Z$ | (R) | NOT an arithmetic progression |
| (IV) | $W$ | (S) | $\supseteq \left\{\frac{\pi}{6}, \frac{7\pi}{6}, \frac{13\pi}{6}\right\}$ |
| | | (T) | $\supseteq \left\{\frac{\pi}{3}, \frac{2\pi}{3}, \pi\right\}$ |
| | | (U) | $\supseteq \left\{\frac{\pi}{6}, \frac{3\pi}{4}\right\}$ |

Which of the following is the only CORRECT combination?

(a) (II), (Q), (T) (b) (II), (R), (S)
(c) (I), (P), (R) (d) (I), (Q), (U)

**53.** Let the circles $C_1 : x^2 + y^2 = 9$ and $C_2 : (x-3)^2 + (y-4)^2 = 16$, intersect at the points $X$ and $Y$. Suppose that another circle $C_3 : (x-h)^2 + (y-k)^2 = r^2$ satisfies the following conditions :

(i) Centre of $C_3$ is collinear with the centres of $C_1$ and $C_2$.

(ii) $C_1$ and $C_2$ both lie inside $C_3$ and

(iii) $C_3$ touches $C_1$ at $M$ and $C_2$ at $N$.

Let the line through $X$ and $Y$ intersect $C_3$ at $Z$ and $W$, and let a common tangent of $C_1$ and $C_3$ be a tangent to the parabola $x^2 = 8\alpha y$.

There are some expression given in the List-I whose values are given in List-II below.

| | List-I | | List-II |
|---|---|---|---|
| (I) | $2h + k$ | (P) | 6 |
| (II) | $\frac{\text{Length of } ZW}{\text{Length of } XY}$ | (Q) | $\sqrt{6}$ |
| (III) | $\frac{\text{Area of triangle } MZN}{\text{Area of triangle } ZMW}$ | (R) | $\frac{5}{4}$ |
| (IV) | $\alpha$ | (S) | $\frac{21}{5}$ |
| | | (T) | $2\sqrt{6}$ |
| | | (U) | $\frac{10}{3}$ |

Which of the following is the only INCORRECT combination?

(a) (III), (R) (b) (IV), (S)
(c) (I), (P) (d) (IV), (U)

**54.** Let the circle $C_1 : x^2 + y^2 = 9$ and $C_2 : (x-3)^2 + (y-4)^2 = 16$, intersect at the points $X$ and $Y$. Suppose that another circle $C_3 : (x-h)^2 + (y-k)^2 = r^2$ satisfies the following conditions :

(i) centre of $C_3$ is collinear with the centers of $C_1$ and $C_2$.

(ii) $C_1$ and $C_2$ both lie inside $C_3$, and

(iii) $C_3$ touches $C_1$ at $M$ and $C_2$ at $N$.

Let the line through $X$ and $Y$ intersect $C_3$ at $Z$ and $W$, and let a common tangent of $C_1$ and $C_3$ be a tangent to the parabola $x^2 = 8\alpha y$.

There are some expression given in the List-I whose values are given in List-II below.

| List-I | List-II |
|---|---|
| (I) $2h + k$ | (P) 6 |
| (II) $\frac{\text{Length of } ZW}{\text{Length of } XY}$ | (Q) $\sqrt{6}$ |
| (III) $\frac{\text{Area of triangle } MZN}{\text{Area of triangle } ZMW}$ | (R) $\frac{5}{4}$ |
| (IV) $\alpha$ | (S) $\frac{21}{5}$ |
| | (T) $2\sqrt{6}$ |
| | (U) $\frac{10}{3}$ |

Which of the following is the only CORRECT combination?

(a) (II), (T) (b) (I), (S) (c) (II), (Q) (d) (I), (U)

# Answers

## Paper 1

| | | | | | | | | | |
|---|---|---|---|---|---|---|---|---|---|
| 1. | a | 2. | b | 3. | a | 4. | d | 5. | a,c,d |
| 6. | a,b,d | 7. | a,b,d | 8. | a,b | 9. | b,c | 10. | b,c,d |
| 11. | b,d | 12. | a,c,d | 13. | 1.00 | 14. | 50.00 | 15. | 2.00 |
| 16. | 270.00 | 17. | 0.75 | 18. | 8.13 | 19. | b | 20. | c |
| 21. | d | 22. | b | 23. | c,d | 24. | a,d | 25. | a,b,c |
| 26. | a,b,d | 27. | c,d | 28. | a,c,d | 29. | a,c,d | 30. | a,c |
| 31. | 4.00 | 32. | 1.02 | 33. | 6.75 | 34. | 8.9 | 35. | 19 |
| 36. | 4 | 37. | c | 38. | d | 39. | d | 40. | c |
| 41. | a,c | 42. | c,d | 43. | a,b,d | 44. | b,c,d | 45. | a,b,c |
| 46. | a,c | 47. | b,c,d | 48. | b,c,d | 49. | 0.50 | 50. | 10 |
| 51. | 3.00 | 52. | 4.0 | 53. | 0.75 | 54. | 157.00 | | |

## Paper 2

| | | | | | | | | | |
|---|---|---|---|---|---|---|---|---|---|
| 1. | c,d | 2. | a,c | 3. | a,d | 4. | a | 5. | b,c,d |
| 6. | b | 7. | a,b | 8. | a,c,d | 9. | 135.00 | 10. | 1.50 |
| 11. | 1.00 | 12. | 1.38 & 1.39 | 13. | 4.00 | 14. | 0.63 | 15. | b |
| 16. | c | 17. | d | 18. | d | 19. | a,b | 20. | c,d |
| 21. | a,c | 22. | b,c,d | 23. | a,b,c | 24. | a,d | 25. | a,b,c |
| 26. | a,c | 27. | 2.3 | 28. | 2.98 mole | 29. | 6 | 30. | 6 |
| 31. | 288 | 32. | 10.0 | 33. | a | 34. | c | 35. | b |
| 36. | a | 37. | a,b,c | 38. | a,b,d | 39. | a,b,d | 40. | c,d |
| 41. | c,d | 42. | b,c | 43. | b,d | 44. | a,b,d | 45. | 18 |
| 46. | 1523 | 47. | 6.20 | 48. | 30 | 49. | 0.5 | 50. | 0 |
| 51. | a | 52. | a | 53. | b | 54. | c | | |

# Answer *with* Explanations

## Paper 1

**1.** *(a)*

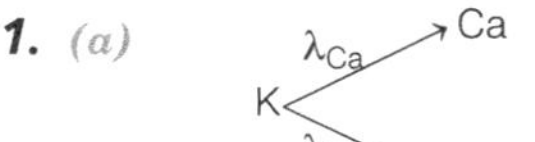


at $t = 0$ dissipated energy,

$$\frac{dN}{dt} = -(\lambda_1 + \lambda_2) \times N$$

$$\Rightarrow \quad \frac{dN}{N} = -(\lambda_1 + \lambda_2)\, dt$$

Integration on both sides, we get

$$\log_e\left(\frac{N}{N_0}\right) = -(\lambda_1 + \lambda_2)t$$

$$2.3 \times \log_{10}\left(\frac{N_0}{N_0/100}\right) = 5 \times 10^{-10} t$$

$$\frac{2.303 \times 2}{5 \times 10^{-10}} = t$$

$$2.303 \times 0.4 \times 10^{10} = t$$

$$\Rightarrow \quad t = 9.2 \times 10^9 \text{ Yr}$$

**2.** *(b)* Heat capacity, $\frac{dQ}{dt} = H\frac{dT}{dt}$

Power of the rod, $P = H.T_0.\beta.\frac{1}{4}.t^{-3/4}$

$$\frac{4P}{T_0 \cdot \beta} = t^{-3/4} \cdot H$$

$$\Rightarrow \quad H = \left(\frac{4P}{T_0\beta}\right) t^{3/4} \quad \ldots \text{(i)}$$

Now, $T - T_0 = T_0\beta t^{1/4}$

So, $t^{3/4} = \left(\frac{T - T_0}{T_0\beta}\right)^3$

Substituting this value of $t^{3/4}$ in equation (i) we get,

$$H = \frac{4P(T - T_0)^3}{T_0^4\beta^4}$$

**3.** *(a)*

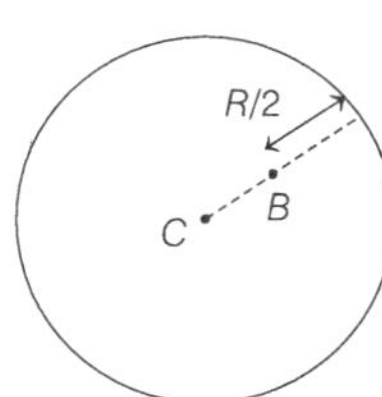


Given, $V$ at surface of the sphere

$$V_0 = \frac{KQ}{R}$$

Here, $K = \frac{1}{4\pi\varepsilon_0} = \text{constant}$

$V$ at point $C$,

$$V_C = \frac{KQ}{R} - \frac{K\alpha Q}{R} = V_0(1 - \alpha)$$

$V$ at point $B$,

$$V_B = \frac{KQ}{R} - \frac{K(\alpha Q)}{R/2} = V_0(1 - 2\alpha)$$

$$\therefore \quad \frac{V_C}{V_B} = \frac{1 - \alpha}{1 - 2\alpha}$$

**4.** *(d)*

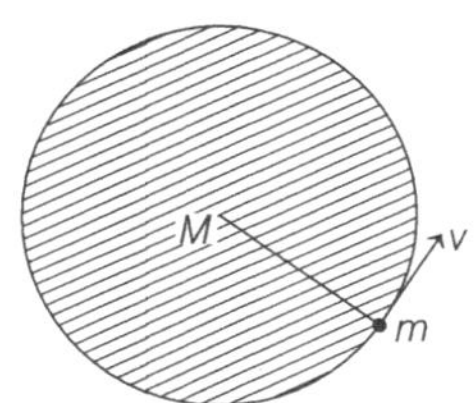


Gravitational force = Centripetal force of the earth

$$\frac{GMm}{r^2} = \frac{mv^2}{r}$$

($\because M$ = total mass from 0 to $r$)

$$= \frac{2}{r}\left(\frac{1}{2}mv^2\right)$$

$$\Rightarrow \quad \frac{GMm}{r^2} = \frac{2K}{r} \Rightarrow M = \frac{2Kr}{Gm}$$

$$\left(\because \frac{1}{2}mv^2 = K\right)$$

Differentiate on both sides, we get

$$\Rightarrow \quad dM = \frac{2K}{Gm}dr$$

$$\Rightarrow \quad 4\pi r^2 dr\rho = \frac{2K}{Gm}dr$$

($\because$ volume = mass × density)

$$\therefore \quad \rho = \frac{K}{2\pi Gmr^2}$$

$$\therefore \quad \frac{\rho}{m} = \frac{K}{2\pi r^2 m^2 G} \quad \left(\because \frac{\rho}{m} = \text{volume}\right)$$

**5.** *(a, c, d)* Heights if only single material tubes are used of sufficient length,

$$h_1 = \frac{2R\cos\theta}{\rho rg}$$

$$= \frac{2\times 0.075\times \cos 0^\circ}{1000\times 2\times10^{-4}\times10} = 7.5\,\text{cm}$$

$$h_2 = \frac{2T\cos\theta'}{\rho rg}$$

$$= \frac{2\times 0.075\times \cos 60^\circ}{1000\times 2\times10^{-4}\times10} = 3.75\,\text{cm}$$

**Case- I**

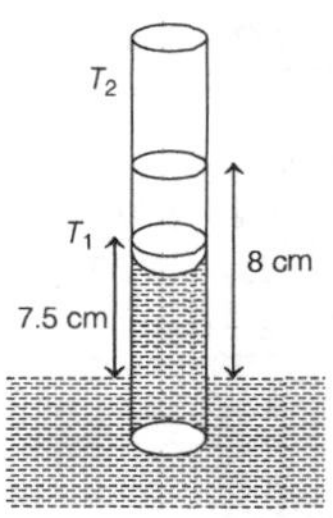


**Case-II**

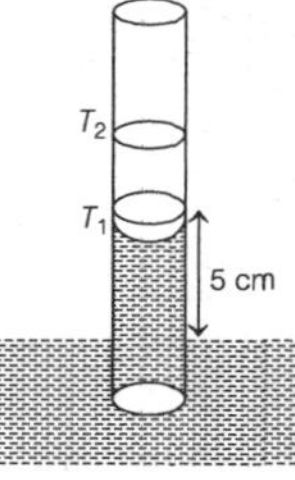


2. Liquid will rise only upto height of 5 cm and meniscus will adjust by changing is radius of curvature. If the liquid goes up in tube 2 then it will not be able to support the weight of the liquid.
3. Weight of water in meniscus will be different in two cases because angle of contact is different.

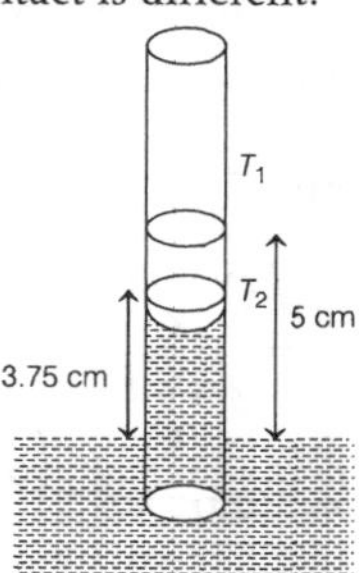


**6.** *(a, b, d)*

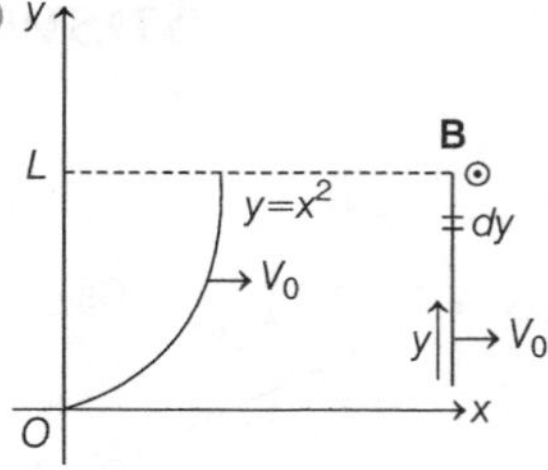


Motional emf across the length $dy$ is,

$$d\varepsilon = BV_0\,dy = B_0\left[1+\left(\frac{y}{L}\right)^\beta\right]V_0dy$$

$$\therefore\quad \varepsilon = \int_0^L B_0\left(1+\left(\frac{y}{L}\right)^\beta\right)V_0dy$$

$$= B_0V_0\left[1+\frac{1}{\beta+1}\right]$$

emf in loop is proportional to $L$ for given value of $\beta$,

$$\beta = 0,\ \varepsilon = 2B_0V_0L$$

$$\beta = 2,\ \varepsilon = B_0V_0L\left[1+\frac{1}{3}\right] = \frac{4}{3}B_0V_0L$$

The length of projection of the wire $Y = X$ of length $\sqrt{2}L$ on the $y$-axis is $L$ thus, the answer remain unchanged.

Therefore, correct options are are (a), (b) and (d).

**7.** *(a, b, d)* $\frac{1}{f} = (n-1)\left(\frac{1}{R}-\frac{1}{\infty}\right)$

$$\Rightarrow\quad \frac{1}{f_0} = \frac{2(n-1)}{R} \qquad \ldots(1)$$

$$\frac{1}{\Delta f} = (n+\Delta n-1)\left(\frac{1}{R}-\frac{1}{\infty}\right)$$

$$\frac{1}{f_0+\Delta f_0} = \frac{(n-1)}{R} + (n+\Delta n-1)\left(\frac{1}{R}\right)$$

$$\frac{1}{f_0+\Delta f_0} = \frac{2n+\Delta n-2}{R} \qquad \ldots(2)$$

From Eqs. (i) and (ii), we get

$$\frac{f_0+\Delta f_0}{f_0} = \frac{\frac{2(n-1)}{R}}{\frac{2n+\Delta n-2}{R}}$$

$$1+\frac{\Delta f_0}{f_0} = \frac{2(n-1)}{(2n+\Delta n-2)}$$

$$\frac{\Delta f_0}{f_0} = \frac{-\Delta n}{(2n+\Delta n-2)}$$

$$\frac{\Delta f_0}{20} = \frac{10^{-3}}{3+10^{-3}-2}$$

$$\Rightarrow \quad \Delta f_0 = -2\times10^{-2}$$

$$|\Delta f_0| = 0.02\,\text{cm}$$

**8.** *(a, b)* $P$-$V$ graph of the given $V$-$T$ graph is shown below.

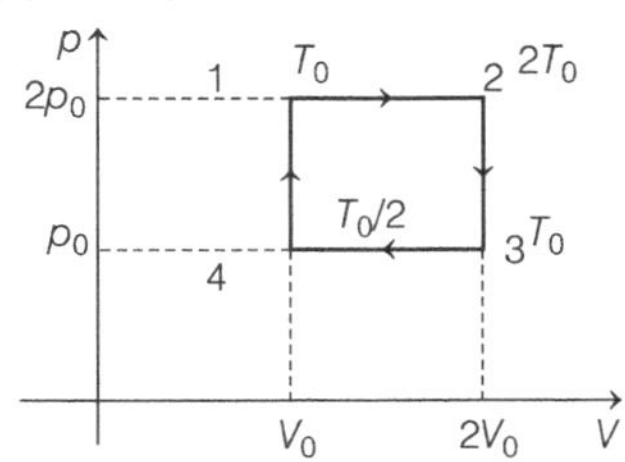


(d) $\left|\frac{\Delta Q_{1\to2}}{\Delta Q_{3\to4}}\right| = \left|\frac{NC_p\Delta T_{1\to2}}{NC_p\Delta T_{3\to4}}\right| = \frac{T_0}{T_0/2} = 2$

(b) $\left|\frac{\Delta Q_{1\to2}}{\Delta Q_{2\to3}}\right| = \left|\frac{NC_p\Delta T_{1\to2}}{NC_p\Delta T_{3\to4}}\right| = \frac{C_p}{C_V} = \frac{5}{3}$

(a) $W_{\text{cycle}} = p_0V_0 = nR\left[\frac{T_0}{2}\right]$

***Note*** *For ideal gas equation,*

$$pV = nRT$$

$\therefore \quad (pV)_4 = (nRT)_4$

*or* $\quad p_0V_0 = nR\frac{T_0}{2} = R\frac{T_0}{2}$

*as* $\quad n = 1$

(c) Wrong as no adiabatic process is involved.

**9.** *(b, c)* Just after closing of switch charge on any capacitor is zero.

$\therefore$ Replace all capacitors by conducting wires.

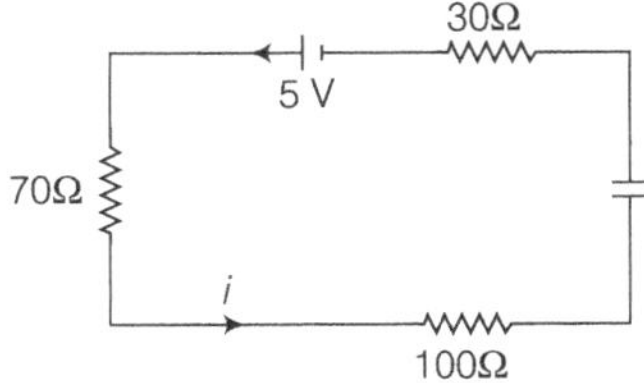


Current flow in the circuit,

$$i = \frac{5}{70+100+30} = \frac{5}{200} = 25\,\text{mA}$$

Now $S_1$ is kept closed for long time circuit is in steady state.

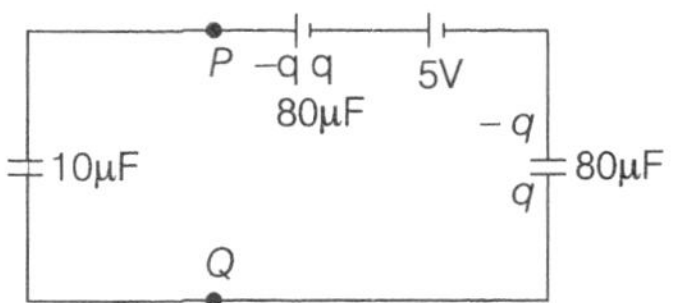


$$\frac{q}{10} + \frac{q}{80} + \frac{q}{80} - 5 = 0 \Rightarrow \frac{10q}{80} = 5$$

$\therefore \quad q = 40\,\mu\text{C}$

$\therefore \quad V$ across $C_1 = \frac{40}{10} = 4\,\text{V}$

Now just after closing $S_2$ charge on each capacitor remains same.

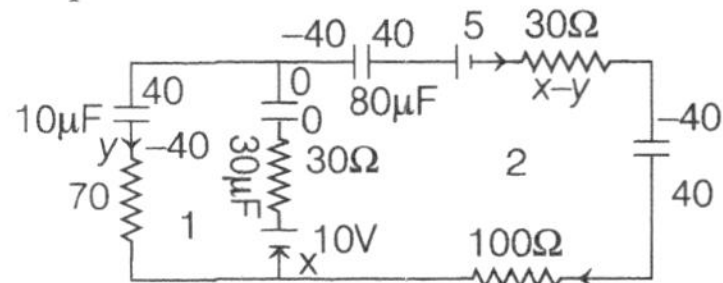


Applying KVL,

$$-10 + x\times30 + \frac{40}{10} + y\times70 = 0$$

$$30x + 70y = 6 \quad \ldots(1)$$

$$-\frac{40}{80} + 5 + (x-y)\,30 - \frac{40}{80} + (x+y)$$

$$\times100 - 10 + x\times30 = 0$$

$$160x - 130y - 6 = 0 \quad \ldots(2)$$

$$y = \frac{96}{1510}$$

$$x = 0.05\,\text{A}$$

**10.** *(b, c, d)*

(a) $h > 2R$ and $r > R$

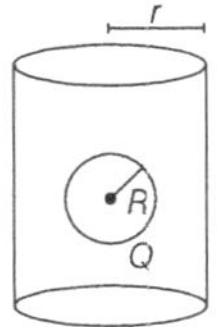


$\phi = \frac{Q}{\varepsilon_0}$, clearly from Gauss' Law

(b) suppose $h = \frac{8R}{5}$ and $r = \frac{3R}{5}$

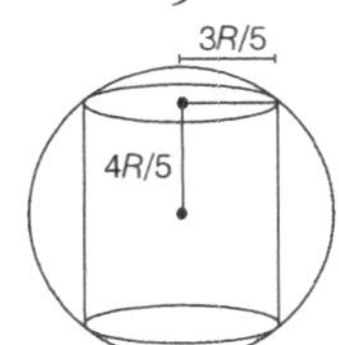

$\therefore \quad \phi = 0$

so, for $h < \frac{8R}{5}$ then $\phi = 0$.

(c) For $h = 2R$ and $r = \frac{4R}{5}$

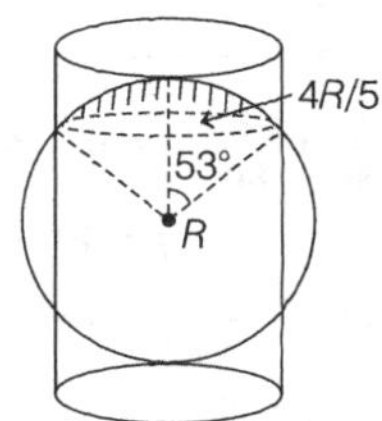


Shaded charge $= 2\pi\,(1 - \cos 53^\circ) \times \frac{Q}{4\pi} = \frac{Q}{5}$

$$\therefore \quad q_{\text{enclosed}} = \frac{2Q}{5}$$

$$\therefore \quad \phi = \frac{2Q}{5\varepsilon_0}$$

$\therefore$ for $h > 2R$ and $r = \frac{4R}{5}$

$$\therefore \quad \phi = \frac{2Q}{5\varepsilon_0}$$

(d) like option $c$ for $h = 2R$ and $r = \frac{3R}{5}$

$$q_{\text{enclosed}} = 2 \times 2\pi\,(1 - \cos 37^\circ)\,\frac{Q}{4\pi} = \frac{Q}{5}$$

$\therefore$ Electric flux, $\phi = \frac{Q}{5\varepsilon_0}$

**11.** *(b, d)*

$I_g$ (G) —— ($R_g$)

$$V = 100 \times 10^{-3} = 10^{-1}\text{ V}$$

$$V = I_g(R_g + R_V)$$

$$\frac{10^{-1}}{2 \times 10^{-6}} = R_g + R_V$$

$$5 \times 10^4\,\Omega \approx R_V \qquad (\because R_V < 10^5\,\Omega)$$

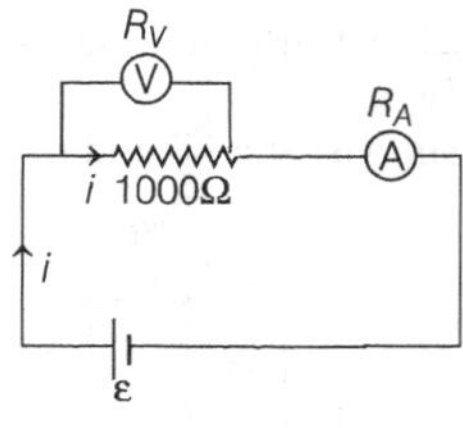


$$I_g R_g = (I - I_g)S$$

$$S = \frac{2 \times 10^{-6} \times 10}{10^{-3} - 2 \times 10^{-6}}$$

$$S = 2 \times 10^{-5} \times 10^3 = 2 \times 10^{-2} = 20\text{m}\Omega$$

$$R_A = \frac{SR_g}{S + R_g} = \frac{20 \times 10^{-3} \times 10}{10 + 20 \times 10^{-3}}$$

$$\approx 20 \times 10^{-3}\,\Omega$$

$$i = \frac{\varepsilon}{\left(\frac{1000 \times 50 \times 10^3}{51 \times 10^3}\right)} = \frac{51\varepsilon}{5 \times 10^4}$$

$$(\because R_A \to 0)$$

$$i' = i\left(\frac{R_V}{51 \times 10^3}\right) = \frac{\varepsilon}{1000}$$

Measured resistance,

$$\therefore R_m = \frac{i' \times 1000}{i} = \frac{\varepsilon}{51\varepsilon} \times 5 \times 10 = 980.4\,\Omega$$

If the voltmeter shows full scale deflection then

$$\frac{\varepsilon}{980} \times \left(\frac{1000}{51 \times 10^3}\right) \times 5 \times 10^4 = 10^{-1}$$

$$\therefore \quad \varepsilon = 999.6\text{ mV}$$

Since, $i_A = 1$ mA so maximum reading of $R$ can be

$$\frac{999.6\text{ mV}}{1\text{ mA}} = 999.6\,\Omega$$

**12.** *(a,c,d)* $[M] = [\text{Mass}] = [M^0L^0T^0]$

$[J] = [\text{Angular momentum}] = [ML^2T^{-1}]$

$[L] = [\text{Length}]$

Now, $[ML^2T^{-1}] = [M^0L^0T^0]$

$\therefore \quad [L^2] = [T]$

Power $[P] = [MLT^{-2}.LT^{-1}] = [ML^2T^{-3}]$

$= [L^2 L^{-6}]$

$[P] = [L^{-4}]$

Energy/Work $[W] = [MLT^{-2}.L]$

$= [L^2 L^{-4}] = [L^{-2}]$

Force $[F] = [MLT^{-2}] = [L.L^{-4}] = [L^{-3}]$

Linear momentum $[p] = [MLT^{-1}] = [L.L^{-2}]$

$[p] = [L^{-1}]$

**13.** *(1.00)* Parallel plate capacitor,

$$\frac{x}{m}=\frac{d}{N}$$

$$d\left(\frac{1}{C}\right)=\frac{dx}{K_m\varepsilon_0 A}=\frac{dx}{K\varepsilon_0 A\left(1+\frac{m}{N}\right)}$$

$$=\frac{dx}{K\varepsilon_0 A\left(1+\frac{x}{d}\right)}$$

Integration on both sides, we get

$$\frac{1}{C_{eq}}=\int d\left(\frac{1}{C}\right)=\int_0^D \frac{d\,dx}{K\varepsilon_0 A(d+x)}$$

$$\frac{1}{C_{eq}}=\int\frac{d}{K\varepsilon_0 A}\ln 2 \Rightarrow C_{eq}=\frac{K\varepsilon_0 A}{d\ln 2}$$

Therefore, $\alpha = 1$.

**14.** *(50.00)*

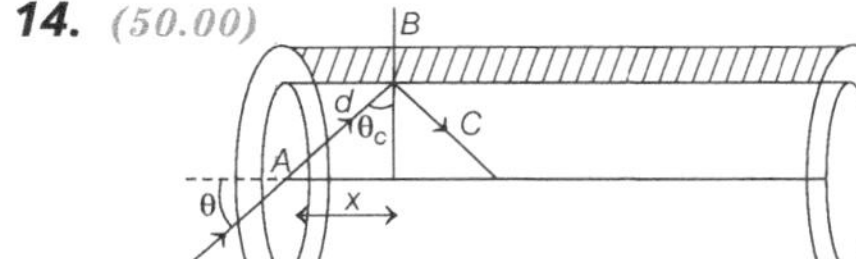


According to total internal reflection (TIR),

$$1.5\sin\theta_c = 1.44\sin 90^\circ$$

$$\sin\theta_c=\frac{1.44}{1.50}=\frac{24}{25}$$

$$\therefore \quad \sin\theta_c=\frac{x}{d}=\frac{24}{25}\Rightarrow d=\frac{25x}{24}$$

$\therefore$ Total length travelled by light,

$$\therefore\ t=\frac{S}{\left(\frac{c}{n_1}\right)}=\frac{10}{\frac{3\times10^8}{1.5}}=\frac{1}{2}\times10^{-7}=5\times10^{-8}$$

$$t = 50\,\text{ns} \Rightarrow t = 50\times10^{-9}$$

**15.** *(2.00)*

$T_s$ $T_c$ 60° 30° 100 N

$$\frac{T_S}{2}=T_C\frac{\sqrt{3}}{2}$$

$$T_S=\sqrt{3}\,T_C$$

$$\Delta l=\frac{Tl}{Ay}$$

$$\therefore\ \frac{\Delta l_C}{\Delta l_S}=\left(\frac{T_C}{T_S}\right)\left(\frac{l_C}{l_S}\right)\left(\frac{Y_S}{Y_C}\right)$$

$$=\left(\frac{1}{\sqrt{3}}\right)\left(\frac{\sqrt{3}}{1}\right)\left(\frac{2\times10^{11}}{1\times10^{11}}\right)=2.00$$

**16.** *(270.00)*

**Case-I** $5C\times50+5L=C_2\times30$ ...(i)

**Case-II** $80C[50-30]=C_2[80-50]$ ...(ii)

By Eq. (i) and (ii)

$$1600C=250+5L$$

$$\therefore\ \frac{L}{C}=\frac{1350}{5}=270^\circ\text{C}$$

**17.** *(0.75)* $d=\mathbf{F}\cdot\mathbf{dr}$

$$d=\alpha y dx+2\alpha x dy$$

$A\to B$, $y=1$, $dy=0$

then $W_{A\to B}=\int\alpha y dx=\alpha 1\int_0^1 dx=\alpha$

$B\to C$, $x=1$, $dx=0$

then $W_{B\to C}=2\alpha 1\int_1^{0.5}dy=-2\alpha(0.5)=-\alpha$

$C\to D$, $y=0.5$, $dy=0$

then $W_{C\to D}=\int_1^{0.5}\alpha y dx=\alpha.\frac{1}{2}\int_1^{0.5}dx=-\frac{\alpha}{4}$

$D\to E$, $x=0.5$, $dx=0$

then $W_{D\to E}=2\alpha\int xdy=2\alpha.\frac{1}{2}\int_{0.5}^{0}dy=-\frac{\alpha}{2}$

$E\to F$, $y=0$, $dy=0$ then $W_{EF}=0$

$F\to A$, $x=0$, $dx=0$ then $W_{F\to A}=0$

$$\therefore\quad W=\alpha-\alpha-\frac{\alpha}{4}-\frac{\alpha}{2}=-\frac{3\alpha}{4}$$

Given, $\alpha=-1\Rightarrow W=\frac{3}{4}\text{J}=0.75\,\text{J}$

**18.** *(8.13)* $V_{\text{sound}}=330$ m/s

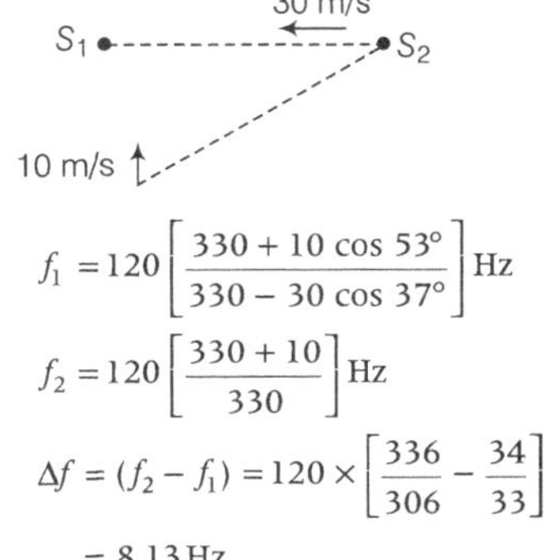


$$f_1=120\left[\frac{330+10\cos 53^\circ}{330-30\cos 37^\circ}\right]\text{Hz}$$

$$f_2=120\left[\frac{330+10}{330}\right]\text{Hz}$$

$$\Delta f=(f_2-f_1)=120\times\left[\frac{336}{306}-\frac{34}{33}\right]$$

$$=8.13\,\text{Hz}$$

**19.** *(b)* Acidic nature depends upon nature of electron withdrawing group and electronegativity. Electronegativity further depends on % $s$ character. Higher the $s$-character, greater will be the electronegativity and hence tendency to loose H increases thus acidic character also increases.

(I) $H-C\equiv C-COOH$
$sp$- hybridisation
(50% $s$ character)
($pK_a = 1.86$)

(II) $H_2C=CH-COOH$
$sp^2$ -hybridisation
(30-33% $s$ character)
($pK_a = 4.3$)

(III) $MeO-C_6H_4-COOH$
$sp^2$ -hybridisation
(Resonance effect)
($pK_a = 4.5$)

(IV) $H_3C-CH_2-COOH$
$sp^3$ -hybridisation
(25% $s$-character)
($pK_a = 4.8$)

Hence, acidic order I > II > III > IV.

II is more acidic than III since electron donating group ($-OCH_3$) is attached to benzene ring in III which decreases the acidic character.

On the other hand, $pK_a$ value also determined acidic nature, lower $pK_a$ value gives maximum acidic character.

Hence, option (b) is correct.

**20.** *(c)* Borax bead test is performed only for coloured salt. Borax (sodium pyroborate), $Na_2B_4O_7 \cdot 10H_2O$ on heating gets fused and lose water of crystallisation. It swells up into fluffy white porous mass which melts into a colourless liquid which later form a clear transparent glassy bead consisting of boric anhydride and sodium metaborate.

$$Na_2B_4O_7 \cdot 10H_2O \xrightarrow{\Delta} Na_2B_4O_7 + 10H_2O \uparrow$$

$$Na_2B_4O_7 \xrightarrow{\Delta} \underbrace{\underset{\text{Boric anhydride}}{B_2O_3} + \underset{\text{Sodium metaborate}}{2NaBO_2}}_{\text{glassy bead}}$$

Boric anhydride is non-volalite. When it react with Cr(III) salt then deep green complex is formed.

$$2Cr^{3+} + 3B_2O_3 \longrightarrow \underset{\text{Deep green}}{2Cr(BO_2)_3}$$

Hence, option (c) is correct.

**21.** *(d)* $ZnCO_3$-Calamine (zinc ore)

$CuCO_3 \cdot Cu(OH)_2$-Malachite (copper ore)

$Fe_3O_4$-Magnetite (iron ore)

$Na_3AlF_6$-Cryolite (aluminium ore)

Thus, option (d) is correct.

**22.** *(b)* **Key Idea** The aqueous solution of ionic surfactant, i.e. sodium stearate ($C_{17}H_{35}CO\bar{O}N\overset{+}{a}$) acts as a strong univalent type of electrolyte in the concentration range below the CMC and the linear function of dependence of $\Lambda_m$ on $\sqrt{C}$ has a small negative slope.

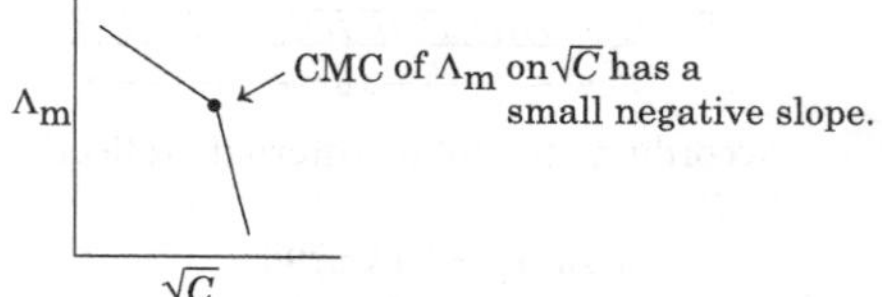


At normal or low concentration, sodium stearate $[CH_3(CH_2)_{16}COO^-Na^+]$ behaves as strong electrolyte and for strong electrolyte, molar conductance ($\Lambda_m$) decreases with increase in concentration. Above particular concentration, sodium stearate forms aggregates known as micelles. The concentration is called as CMC. Since, number of ions decreases and hence $\Lambda_m$ also decreases.

Hence, option (b) is correct.

**23.** *(c, d)* The standard enthalpy of formation is defined as standard enthalpy change for formation of 1 mole of a substance from its elements, present in their most stable state of aggregation.

$$\frac{3}{2}O_2(g) \longrightarrow O_3(g);$$

$$\frac{1}{8}S_8(s) + O_2(g) \longrightarrow SO_2(g)$$

In the above two reactions standard enthalpy of reaction is equal to standard enthalpy of formation.

**24.** *(a, d)*

$$\underset{\substack{\text{Tin chloride}\\(Q)}}{SnCl_2} + Cl^- \longrightarrow \underset{(X)}{SnCl_3^-}$$

$\left[\ddot{S}n(Cl)_3\right]^{\ominus}$ $sp^3$ (pyramidal)

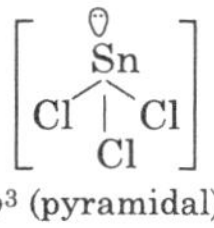

$SnCl_3^-$ has $(3\sigma + 1lp)$ and exist in pyramidal structure.

$$\underset{\substack{(3°\text{amine})\\(Q)}}{SnCl_2 + Me_3N} \longrightarrow \underset{(Y)}{SnCl_2 \cdot NMe_3}$$

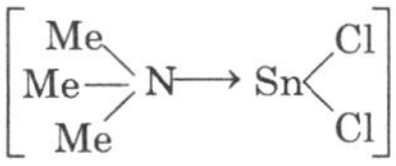

$[Me_3N \rightarrow SnCl_2]$

$Y$ complex has coordinate bond in between nitrogen and Sn metal.

Reduction

$$\underset{(Q)}{\overset{+2}{Sn}Cl_2} + 2\overset{+2}{Cu}Cl_2 \longrightarrow \underset{(Z)}{\overset{+4}{Sn}Cl_4} + 2\overset{+1}{Cu}Cl$$

Oxidation

$Z$ is oxidised product and oxidation state of Sn is +4 in $Z$ compound. Structure of $SnCl_4(Z)$ is

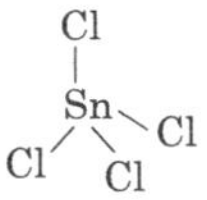

$SnCl_4$ (Sn bonded to four Cl)

Thus, options (a, d) are correct.

**25.** *(a,b,c)* **Key Idea** The lose of one α-particle will decrease the mass number by 4 and atomic number by 2. On the other hand, loss of β-particle will increase the atomic number by 1.

In decay sequence,

$$_{92}U^{238} \longrightarrow {_{90}Th^{234}} + \underset{X_1 \text{ particle}}{_2He^4 \text{ (or } \alpha)}$$

$$\underset{\substack{Z^{234}\text{ is isotope}\\\text{of uranium}}}{_{92}U^{234}} + \underset{X_3\text{ particle}}{(\beta^- \text{ or } {_{-1}e^0})} \longleftarrow {_{91}Pa^{234}} + \underset{X_2 \text{ particle}}{(\beta^- \text{ or } {_{-1}e^0})}$$

$$\longrightarrow {_{90}Th^{230}} + \underset{X_4\text{ particle}}{_2He^4 \text{ (or } \alpha)}$$

$X_1$ particle will deflect towards negatively charged plate due to presence of positive charge on α- particles.

Hence, options (a, b, c) are correct.

**26.** *(a, b, d)* The explanation of given statements are as follows:

(a) $U_{rms}$ is inversely proportional to the square root of its molecular mass.

$$U_{rms} = \sqrt{\frac{3RT}{M}}$$

Hence, option (a) is correct.

(b) When temperature is increased four times then $U_{rms}$ become doubled.

$$U_{rms} = \sqrt{\frac{3R}{M} \times 4T}$$

$$U_{rms} = 2 \times \sqrt{\frac{3RT}{M}}$$

Hence, option (b) is correct.

(c) and (d) $E_{av}$ is directly proportional to temperature but does not depends on its molecular mass at a given temperature as $E_{av} = \frac{3}{2}KT$. If temperature raised four times than $E_{av}$ becomes four time multiple.

Thus, option (c) is incorrect and option (d) is correct.

**27.** *(c, d)* The given road map problem is

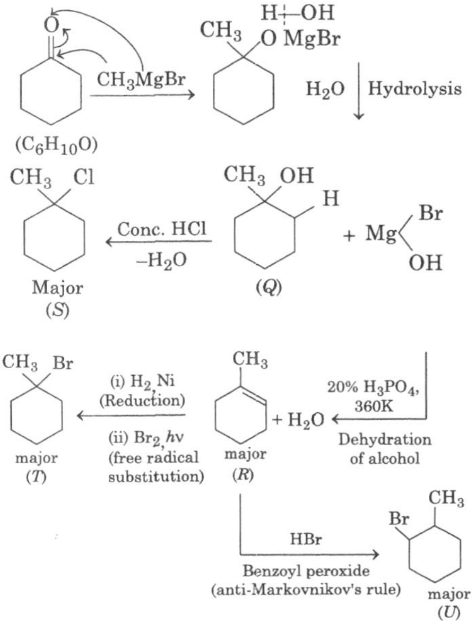

Cyclohexanone $(C_6H_{10}O)$ $\xrightarrow{CH_3MgBr}$ 1-methylcyclohexyl-OMgBr (H–OH) $\xrightarrow[\text{Hydrolysis}]{H_2O}$ 1-methylcyclohexanol $(Q)$ + $Mg(Br)(OH)$

$(Q)$ $\xrightarrow[-H_2O]{\text{Conc. HCl}}$ 1-chloro-1-methylcyclohexane, Major $(S)$

$(Q)$ + $H_2O$ ← $\xleftarrow[\text{Dehydration of alcohol}]{20\%\ H_3PO_4,\ 360K}$ 1-methylcyclohexene, major $(R)$

$(R)$ $\xrightarrow[\text{(ii) } Br_2, h\nu \text{ (free radical substitution)}]{\text{(i) } H_2, Ni \text{ (Reduction)}}$ 1-bromo-1-methylcyclohexane, major $(T)$

$(R)$ $\xrightarrow[\text{Benzoyl peroxide (anti-Markovnikov's rule)}]{HBr}$ 1-bromo-2-methylcyclohexane, major $(U)$

Hence, options (c, d) are correct.

**28.** *(a, c, d)* The explanation of given statements are as follows:

(a) Two six membered cyclic hemiacetal form of D-(+)- glucose are called anomers.

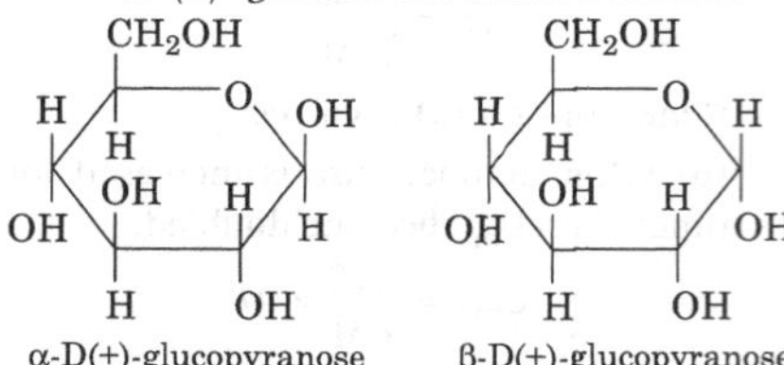

α-D(+)-glucopyranose β-D(+)-glucopyranose

Both are anomers.

(b) Oxidation of glucose in presence of $Br_2$ water gives gluconic acid.

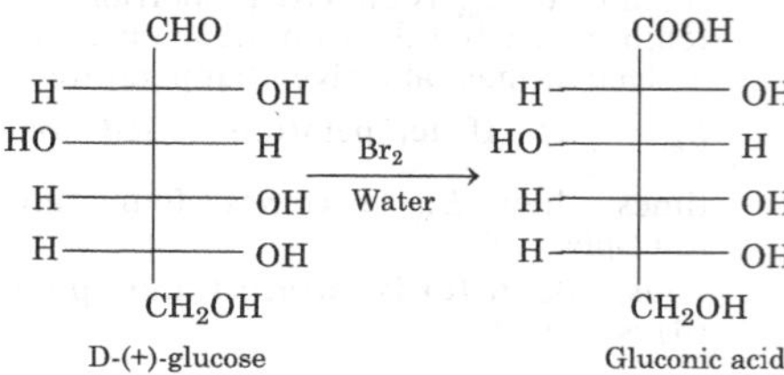

D-(+)-glucose Gluconic acid

(c) Monosaccharides can not be hydrolysed into polyhydroxy aldehydes and ketones.

(d) Hydrolysis of sucrose gives D-glucose and L-fructose.

$$C_{12}H_{22}O_{11} + H_2O \xrightarrow{\text{Invertase}} \underset{\substack{\text{D-glucose} \\ \text{or dextrorotatory}}}{C_6H_{12}O_6} + \underset{\substack{\text{L-fructose} \\ \text{or laevorotatory}}}{C_6H_{12}O_6}$$

Hence, options (a, c, d) are correct.

**29.** *(a, c, d)*

$$MnO_2 + 2KOH + \frac{1}{2}O_2 \xrightarrow{\Delta} \underset{\substack{(W)\text{ potassium} \\ \text{manganate}}}{K_2MnO_4} + H_2O$$

$$\underset{(W)}{K_2MnO_4\,(aq)} \rightleftharpoons 2K^+(aq) + \underset{(Y)}{MnO_4^{2-}(aq)}$$

$\left[ O^{\ominus}\text{–}Mn(=O)(O)_2 \right]$ $sp^3$ hybridisation, tetrahedral (manganate ion) Green coloured complex

$MnO_4^{2-}$ ion has one unpaired electrons, therefore it gives *d-d* transition to form green colour. *Y* complex has paramagnetic nature due to presence of one unpaired electron.

In aqueous solution,

$$\underset{(W)}{K_2MnO_4} + H_2O \xrightarrow{\text{Electrolytic oxidation}} H_2 + KOH + \underset{(X)}{KMnO_4}$$

$$KMnO_4(aq) \xrightarrow{\Delta} K^+ + \underset{(Z)}{MnO_4^-}$$

$\left[ O^{\ominus}\text{–}Mn(=O)(O)_2 \right]$ $sp^3$, tetrahedral (purple coloured complex ion)

$MnO_4^-$ ions gives charge transfer spectrum in which a fraction of electronic charge is transferred between the molecular entities.

$$\because \quad \underset{(Y)}{MnO_4^{2-}} \xrightarrow[\text{oxidation}]{\text{Electrolytic}} \underset{(Z)}{MnO_4^-} + e^-$$

In acidic medium, *Y* undergoes disproportionation reaction.

$$\underset{(Y)}{3MnO_4^{2-}(aq)} + 4H^+ \longrightarrow \underset{(Z)}{2MnO_4^-} + MnO_2 + 2H_2O$$

$\underset{(Y)}{MnO_4^{2-}}$ and $\underset{(Z)}{MnO_4^-}$ both ions form π-bonding between *p*-orbitals of oxygen and *d*-orbitals of manganese.

Thus, options (a, c, d) are correct.

**30.** *(a,c)* **Key Idea** Dipole moment of a bond depends on the difference in the electronegativities of bonded atoms. More is the difference in the electronegativities, greater will be the dipole moment. Also,

For symmetrical molecule, $\mu = 0$

For unsymmetrical molecule, $\mu \neq 0$

The molecules which gives permanent dipole moment are polar in nature.

$SO_2$: $\mu \neq 0$ (polar), $C_6H_5Cl$: $\mu \neq 0$ (polar), $SeH_2$: $\mu \neq 0$ (polar)

$BrF_5$: $\mu \neq 0$ (polar), $Cl \overset{\mu_1}{\longleftarrow} Be \overset{\mu_2}{\longrightarrow} Cl$: $\mu_{net} \approx 0$ (non-polar)

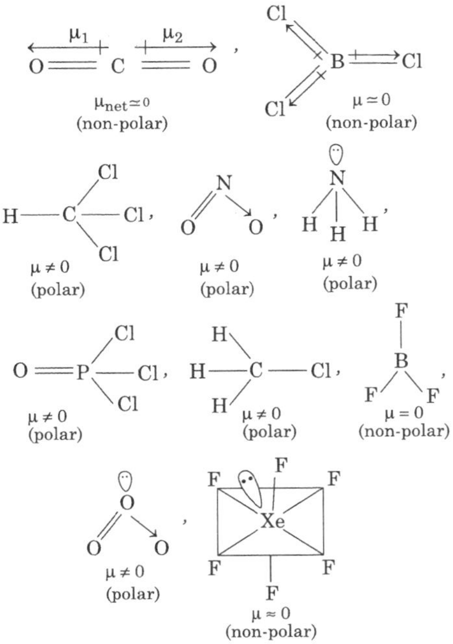

$\overset{\mu_1}{\longleftarrow}$ $\overset{\mu_2}{\longrightarrow}$ O=C=O, $\mu_{net} \simeq 0$ (non-polar); $BCl_3$, $\mu = 0$ (non-polar)

$CHCl_3$, $\mu \neq 0$ (polar); $NO_2$, $\mu \neq 0$ (polar); $NH_3$, $\mu \neq 0$ (polar)

$POCl_3$, $\mu \neq 0$ (polar); $CH_2Cl_2$ (H—C—Cl), $\mu \neq 0$ (polar); $BF_3$, $\mu = 0$ (non-polar)

$O_3$, $\mu \neq 0$ (polar); $XeF_5$ (Xe), $\mu \approx 0$ (non-polar)

Thus, options (a, c) are correct.

**31.** (4.00) $N_2O$, $N_2O_4$, $H_2S_2O_3$ and $H_2S_2O_8$ molecules are containing covalent bond between two atoms.

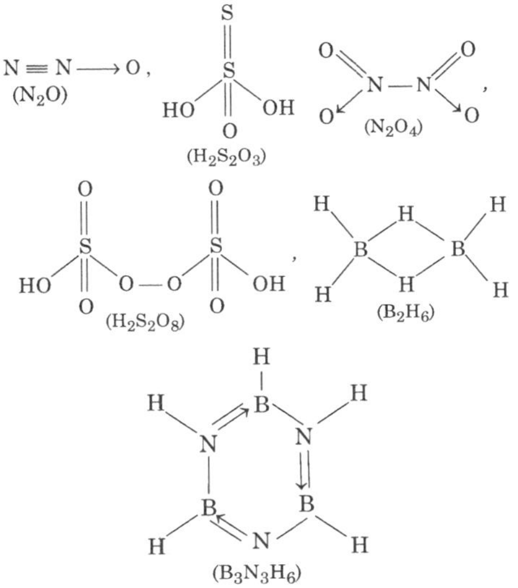

$N \equiv N \longrightarrow O$ ($N_2O$), ($H_2S_2O_3$), ($N_2O_4$), ($H_2S_2O_8$), ($B_2H_6$), ($B_3N_3H_6$)

$B_2H_6$ and $B_3N_6H_6$ have polar bond, but do not have same kind of atom.

**32.** (1.02) **Key Idea** First calculate, molar mass of solute using the formula,

$$\frac{p^\circ - p_s}{p^\circ} = \frac{n_{solute}}{n_{solute} + n_{solvent}}$$ and then calculate $\Delta T_f$ by applying the formula; $\Delta T_f = K_f \times m$.

When 0.5 g of non-volatile solute dissolve into 39 gm of benzene then relative lowering of vapour pressure occurs. Hence, vapour pressure decreases from 650 mmHg to 640 mmHg.

Given, vapour pressure of solvent $(p^\circ) = 650$ mmHg

Vapour pressure of solution $(p_s) = 640$ mmHg

Weight of non-volatile solute = 0.5 g

Weight of solvent (benzene) = 39 g

From relative lowering of vapour pressure,

$$\frac{p^\circ - p_s}{p^\circ} = x_{Solute} = \frac{n_{solute}}{n_{solute} + n_{solvent}}$$

$$\frac{650 - 640}{650} = \frac{\frac{0.5}{\text{molar mass}}}{\frac{0.5}{\text{molar mass}} + \frac{39}{78}}$$

$$\frac{10}{650} = \frac{\frac{0.5}{\text{molar mass}}}{\frac{0.5}{\text{molar mass}} + 0.5}$$

$0.5 + 0.5 \times$ molar mass $= 65 \times 0.5$

$\therefore$ Molar mass of solute = 64 g

From molal depression of freezing point,

$$\Delta T_f = K_f \times \text{molality} = \frac{K_f \times w_{solute}}{(MW)_{solute} \times w_{solvent}}$$

$$\Delta T_f = 5.12 \times \frac{0.5 \times 1000}{64 \times 39} \Rightarrow \Delta T_f = 1.02\,\text{K}$$

**33.** (6.75) Rate $= k[A]^x[B]^y[C]^z$

$$\frac{(\text{Rate})_1}{(\text{Rate})_2} = \frac{[0.2]^x\,[0.1]^y\,[0.1]^z}{[0.2]^x\,[0.2]^y\,[0.1]^z} = \frac{6 \times 10^{-5}}{6 \times 10^{-5}}$$

$\Rightarrow y = 0$

$$\frac{(\text{Rate})_1}{(\text{Rate})_3} = \frac{[0.2]^x\,[0.1]^y\,[0.1]^z}{[0.2]^x\,[0.1]^y\,[0.2]^z} = \frac{6 \times 10^{-5}}{1.2 \times 10^{-4}}$$

$\Rightarrow z = 1$

$$\frac{(\text{Rate})_1}{(\text{Rate})_4} = \frac{[0.2]^x\,[0.1]^y\,[0.1]^z}{[0.3]^x\,[0.1]^y\,[0.1]^z} = \frac{6 \times 10^{-5}}{9 \times 10^{-5}}$$

$\Rightarrow x = 1$

So, rate $= k[A]^1 [C]^1$

From exp-Ist,

$$\text{Rate} = 6.0 \times 10^{-5} \text{ mol dm}^{-3} \text{ s}^{-1}$$

$$6.0 \times 10^{-5} = k[0.2]^1 [0.1]^1$$

$$k = 3 \times 10^{-3}$$

Given, $[A] = 0.15 \text{ mol dm}^{-3}$

$[B] = 0.25 \text{ mol dm}^{-3}$

$[C] = 0.15 \text{ mol dm}^{-3}$

$$\therefore \text{ Rate} = (3 \times 10^{-3}) \times [0.15]^1 [0.25]^0 [0.15]^1$$

$$= 3 \times 10^{-3} \times 0.15 \times 0.15$$

$$\text{Rate} = 6.75 \times 10^{-5} \text{ mol dm}^{-3} \text{ s}^{-1}$$

Thus, $Y = 6.75$

**34.** *(8.9)* Given, equilibrium constant ($K_c$) at 298 K $= 1.6 \times 10^{17}$

| | $Fe^{2+}(aq)$ + | $S^{2-}(aq)$ $\rightleftharpoons$ | $FeS(s)$ |
|---|---|---|---|
| At initial concentration (Before mixing) | 0.06 M | 0.2 M | – |
| At initial concentration (After mixing) | 0.03 M | 0.1 M | – |
| At equilibrium | 0.03-$X$ | 0.1 – 0.03 = 0.07 | – |

[Here, $K_c >> 10^3$, thus limiting reagent will be consumed almost completely, $0.03 - X = 0$ $\therefore$ $X = 0.03$]

From equilibrium constant,

$$K_C = \frac{[FeS]}{[Fe^{2+}][S^{2-}]}$$

$$K_C = \frac{1}{X \times 0.07}$$

[For FeS(s) $= 1 \text{ mol L}^{-1}$]
(Pure solid)

$$1.6 \times 10^{17} = \frac{1}{X \times 0.07}$$

$$X = \frac{1}{1.6 \times 10^{17} \times 0.07}$$

$$= 8.9 \times 10^{-17}$$

Given, $X = Y \times 10^{-17} = 8.9 \times 10^{-17}$

$\therefore$ $Y = 8.9$

**35.** *(19)* $XeF_4$ reacts with $O_2F_2$ to form $XeF_6$ · $O_2F_2$ is fluoronating reagent.

$$XeF_4 + O_2F_2 \xrightarrow{143 \text{ K}} \underset{(Y)}{XeF_6} + O_2$$

The structore of $XeF_6$ is

$Y$ compound ($XeF_6$) has 3 lone pair in each fluorine and one lone pair in xenon. Hence, total number of lone pairs electrons is 19.

**36.** *(4)* **Scheme -1**

Aniline ($P$) $\xrightarrow[H_2O]{\text{(i) } Br_2\text{(excess)}}$ 2,4,6-tribromoaniline $\xrightarrow[\text{273K (diazotisation of aniline)}]{\text{(ii) } NaNO_2 + HCl}$ 2, 4, 6 tribromo diazonium chloride salt ($N_2^+Cl^-$) $\xrightarrow[\text{Cl}^- \text{ is displace by CN}^- \text{ ion}]{\text{(iii) CuCN/KCN}}$ 2,4,6-tribromobenzonitrile (CN) $+ N_2\uparrow$

(iv) $H_3O^+$ (complete hydrolysis of —CN group which convert into —COOH) → 2,4,6-tribromobenzoic acid (COOH) $\xrightarrow{\text{(v) } SOCl_2, \text{ Pyridine}}$ 2,4,6-tribromobenzoyl chloride (C(=O)—Cl), Major ($Q$)

**Scheme-2**

Benzene ($R$) $\xrightarrow[\substack{(H_2SO_4 + SO_3) \\ \text{(Sulphonation)}}]{\text{(i) Oleum } H_2S_2O_7}$ $SO_3H$ $\xrightarrow{\text{(ii) NaOH, }\Delta}$ $SO_3Na$

$\xrightarrow{\text{(iii) } H^+ \text{ Hydrolysis}}$ OH (phenol) $+ NaHSO_3$

$\xrightarrow[\substack{CS_2, \text{ 273K} \\ \text{(Bromination)}}]{Br_2}$ 2- bromophenol (minor) + 4- bromophenol, major ($S$)

**Scheme-3**

OH, Br (S) $\xrightarrow{\text{(i) NaOH, }\Delta}$ $O^-Na^+$, Br $\xrightarrow{(Q)}$ (Q: O=C–Cl with Br, Br, Br on the ring) → (T) (Major) + NaCl

(S) (T) (Major)

$T$ compound has total number of Br atom =4

**37.** (c) The complex number $z$ satisfying $|z-2+i| \geq \sqrt{5}$, which represents the region outside the circle (including the circumference) having centre $(2,-1)$ and radius $\sqrt{5}$ units.

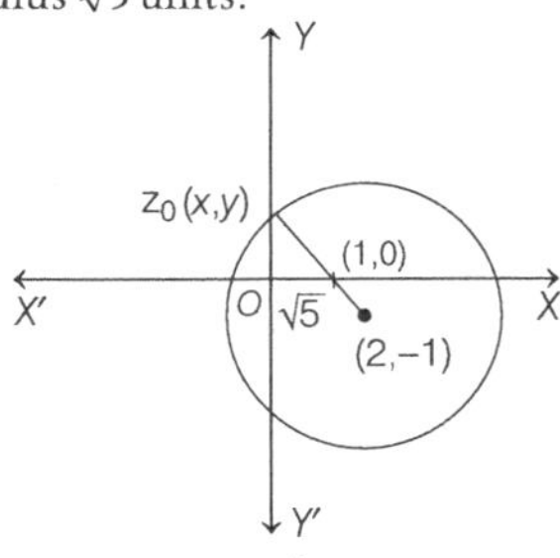


Now, for $z_0 \in S \dfrac{1}{|z_0-1|}$ is maximum.

When $|z_0-1|$ is minimum. And for this it is required that $z_0 \in S$, such that $z_0$ is collinear with the points $(2,-1)$ and $(1,0)$ and lies on the circumference of the circle $|z-2+i| = \sqrt{5}$.

So let $z_0 = x+iy$, and from the figure $0 < x < 1$ and $y > 0$.

$$\text{So, } \frac{4-z_0-\bar{z}_0}{z_0-\bar{z}_0+2i} = \frac{4-x-iy-x+iy}{x+iy-x+iy+2i}$$

$$= \frac{2(2-x)}{2i(y+1)} = -i\left(\frac{2-x}{y+1}\right)$$

$\because \dfrac{2-x}{y+1}$ is a positive real number, so $\dfrac{4-z_0-\bar{z}_0}{z_0-\bar{z}_0+2i}$ is purely negative imaginary number.

$$\Rightarrow \arg\left(\frac{4-z_0-\bar{z}_0}{z_0-\bar{z}_0+2i}\right) = -\frac{\pi}{2}$$

**38.** (d) It is given that matrix

$$M = \begin{bmatrix} \sin^4\theta & -1-\sin^2\theta \\ 1+\cos^2\theta & \cos^4\theta \end{bmatrix} = \alpha I + \beta M^{-1},$$

where

$\alpha = \alpha(\theta)$ and $\beta = \beta(\theta)$ are real numbers and $I$ is the $2\times 2$ identity matrix.

Now,

$$\det(M) = |M| = \sin^4\theta\cos^4\theta + 1 + \sin^2\theta + \cos^2\theta + \sin^2\theta\cos^2\theta$$

$$= \sin^4\theta\cos^4\theta + \sin^2\theta\cos^2\theta + 2$$

$$\text{and} \begin{bmatrix} \sin^4\theta & -1-\sin^2\theta \\ 1+\cos^2\theta & \cos^4\theta \end{bmatrix} = \begin{bmatrix} \alpha & 0 \\ 0 & \alpha \end{bmatrix} + \frac{\beta}{|M|}(\text{adj } M)$$

$$\left[\because M^{-1} = \frac{\text{adj } M}{|M|}\right]$$

$$\Rightarrow \begin{bmatrix} \sin^4\theta & -1-\sin^2\theta \\ 1+\cos^2\theta & \cos^4\theta \end{bmatrix} = \begin{bmatrix} \alpha & 0 \\ 0 & \alpha \end{bmatrix} + \frac{\beta}{|M|}\begin{bmatrix} \cos^4\theta & 1+\sin^2\theta \\ -1-\cos^2\theta & \sin^4\theta \end{bmatrix}$$

$$\left\{\because \text{adj}\begin{bmatrix} a & b \\ c & d \end{bmatrix} = \begin{bmatrix} d & -b \\ -c & a \end{bmatrix}\right\}$$

$$\Rightarrow \beta = -|M| \text{ and } \alpha = \sin^4\theta + \cos^4\theta$$

$$\Rightarrow \alpha = \alpha(\theta) = 1 - \frac{1}{2}\sin^2(2\theta), \text{ and}$$

$$\beta = \beta(\theta) = -\left\{\left(\sin^2\theta\cos^2\theta + \frac{1}{2}\right)^2 + \frac{7}{4}\right\}$$

$$= -\left\{\left(\frac{\sin^2(2\theta)}{4} + \frac{1}{2}\right)^2 + \frac{7}{4}\right\}$$

Now, $\alpha^* = \alpha_{\min} = \frac{1}{2}$

and $\beta^* = \beta_{\min} = -\frac{37}{16}$

$\because$ $\alpha$ is minimum at $\sin^2(2\theta) = 1$ and $\beta$ is minimum at $\sin^2(2\theta) = 1$

So, $\alpha^* + \beta^* = \frac{1}{2} - \frac{37}{16} = -\frac{29}{16}$

**39.** *(d)* **Key Idea** Firstly find the centre of the given circle and write the coordinates of mid point (*A*), of line segment *PQ*, since $AC \perp PQ$ therefore use (slope of *AC*) × (slope of *PQ*) = − 1

It is given that points *P* and *Q* are intersecting points of circle

$$(x-3)^2 + (y+2)^2 = 25 \quad \dots(\text{i})$$

Line $\quad y = mx + 1 \quad \dots(\text{ii})$

And, the mid-point of *PQ* is *A* having *x*-coordinate $\frac{-3}{5}$

so *y*-coordinate is $1 - \frac{3}{5}m$.

So, $A\left(-\frac{3}{5}, 1 - \frac{3}{5}m\right)$

From the figure,

$\because$ $AC \perp PQ$

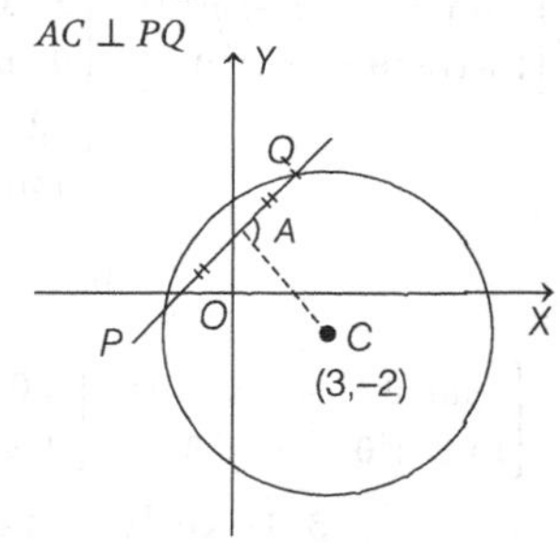


$\Rightarrow$ (slope of *AC*) × (slope of *PQ*) = −1

$$\Rightarrow \left(\frac{-2-1+\frac{3}{5}m}{3+\frac{3}{5}}\right) \times m = -1$$

$$\Rightarrow \frac{(3/5)m - 3}{18/5} m = -1$$

$$\Rightarrow \left(\frac{3m-15}{18}\right) m = -1$$

$$\Rightarrow 3m^2 - 15m + 18 = 0$$

$$\Rightarrow m^2 - 5m + 6 = 0$$

$$\Rightarrow m = 2 \text{ or } 3$$

**40.** *(c)* The given region $\{(x, y) : xy \le 8, 1 \le y \le x^2\}$.

From the figure, region *A* and *B* satisfy the given region, but only *A* is bounded region, so area of bounded region

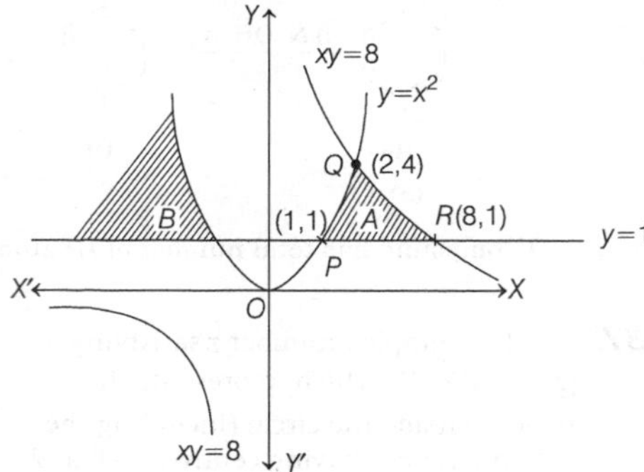


$$A = \int_1^2 (x^2 - 1)\,dx + \int_2^8 \left(\frac{8}{x} - 1\right) dx$$

[$\because$ Points *P*(1, 1), *Q* (2, 4) and *R*(8, 1)]

$$= \left[\frac{x^3}{3} - x\right]_1^2 + [8\log|x| - x]_2^8$$

$$= \left(\frac{8}{3} - 2 - \frac{1}{3} + 1\right) + 8\log 8 - 8 - 8\log 2 + 2$$

$$= -\frac{14}{3} + 16\log 2$$

$$= 16\log 2 - \frac{14}{3}$$

**41.** *(a,c)* Let a point *P*(*h*, *k*) on the curve $y = y(x)$, so equation of tangent to the curve at point *P* is

$$y - k = \left(\frac{dy}{dx}\right)_{h,k} (x - h) \quad \dots(\text{i})$$

Now, the tangent (i) intersect the *Y*-axis at $Y_p$, so coordinates $Y_p$ is $\left(0, k - h\frac{dy}{dx}\right)$, where $\frac{dy}{dx} = \left(\frac{dy}{dx}\right)_{(h,k)}$

So, $PY_p = 1$ (given)

$$\Rightarrow \sqrt{h^2 + h^2\left(\frac{dy}{dx}\right)^2} = 1$$

$$\Rightarrow \frac{dy}{dx} = \pm\frac{\sqrt{1-x^2}}{x}$$

[on replacing *h* by *x*]

$$\Rightarrow dy = \pm\frac{\sqrt{1-x^2}}{x}dx$$

On putting $x = \sin\theta$, $dx = \cos\theta\, d\theta$, we get

$$dy = \pm \frac{\sqrt{1-\sin^2\theta}}{\sin\theta}\cos\theta d\theta$$

$$= \pm \frac{\cos^2\theta}{\sin\theta} d\theta$$

$$= \pm(\text{cosec}\theta - \sin\theta)d\theta$$

$$\Rightarrow \quad y = \pm[\ln(\text{cosec}\theta - \cot\theta) + \cos\theta] + C$$

$$\Rightarrow \quad y = \pm\left[\ln\left(\frac{1-\cos\theta}{\sin\theta}\right) + \cos\theta\right] + C$$

$$\Rightarrow y = \pm\left[\ln\left(\frac{1-\sqrt{1-\sin^2\theta}}{\sin\theta}\right) + \sqrt{1-\sin^2\theta}\right] + C$$

$$\Rightarrow y = \pm\left[\ln\left(\frac{1-\sqrt{1-x^2}}{x}\right) + \sqrt{1-x^2}\right] + C$$

[$\because x = \sin\theta$]

$$= \pm\left[-\ln\frac{1+\sqrt{1-x^2}}{x} + \sqrt{1-x^2}\right] + C$$

[on rationalization]

$\because$ The curve is in the first quadrant so $y$ must be positive, so

$$y = \ln\left(\frac{1+\sqrt{1-x^2}}{x}\right) - \sqrt{1-x^2} + C$$

As curve passes through $(1, 0)$, so

$0 = 0 - 0 + c \Rightarrow c = 0$, so required curve is

$$y = \ln\left(\frac{1+\sqrt{1-x^2}}{x}\right) - \sqrt{1-x^2}$$

and required differential equation is

$$\frac{dy}{dx} = -\frac{\sqrt{1-x^2}}{x}$$

$$\Rightarrow \quad xy' + \sqrt{1-x^2} = 0$$

Hence, options (a) and (c) are correct.

**42.** *(c,d)* Given equation of ellipse

$$E_1 : \frac{x^2}{9} + \frac{y^2}{4} = 1 \qquad \ldots(\text{i})$$

Now, let a vertex of rectangle of largest area with sides parallel to the axes, inscribed in $E_1$ be $(3\cos\theta, 2\sin\theta)$.

So, area of rectangle

$R_1 = 2(3\cos\theta) \times 2(2\sin\theta) = 12\sin(2\theta)$

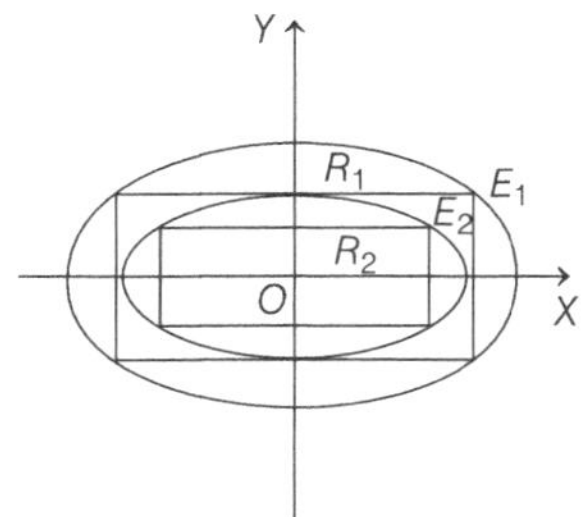


The area of $R_1$ will be maximum, if $\theta = \frac{\pi}{4}$ and maximum area is 12 square units and length of sides of rectangle $R_1$ are $2a\cos\theta = \sqrt{2}\,a = 3\sqrt{2} =$ length of major axis of ellipse $E_2$ and $2b\sin\theta = \sqrt{2}\,b = 2\sqrt{2} =$ length of minor axis of ellipse $E_2$.

So, $E_2 : \dfrac{x^2}{\left(\frac{a}{\sqrt{2}}\right)^2} + \dfrac{y^2}{\left(\frac{b}{\sqrt{2}}\right)^2} = 1$ and

maximum area of rectangle

$R_2 = 2\left(\frac{a}{\sqrt{2}}\right)\left(\frac{b}{\sqrt{2}}\right)$ and so on.

So, $E_n = \dfrac{x^2}{\left(\frac{a}{(\sqrt{2})^{n-1}}\right)^2} + \dfrac{y^2}{\left(\frac{b}{(\sqrt{2})^{n-1}}\right)^2} = 1,$

and maximum area of rectangle

$R_n = 2\left(\dfrac{a}{(\sqrt{2})^{n-1}}\right)\left(\dfrac{b}{(\sqrt{2})^{n-1}}\right)$

Now option (a),

Since, eccentricity of ellipse

$$E_n = e'_n = \sqrt{1 - \frac{(b_n)^2}{(a_n)^2}}$$

$$= \sqrt{1 - \frac{\left(\frac{b}{(\sqrt{2})^{n-1}}\right)^2}{\left(\frac{a}{(\sqrt{2})^{n-1}}\right)^2}}$$

$$= \sqrt{1 - \frac{b^2}{a^2}} = \sqrt{1 - \frac{4}{9}} = \frac{\sqrt{5}}{3}$$

is independent of '$n$', so eccentricity of $E_{18}$ and $E_{19}$ are equal.

Option (b),

Distance between focus and centre of

$$E_9 = e \cdot a_9 = \frac{a}{(\sqrt{2})^8}(e)$$

$$= \frac{3}{2^4} \times \frac{\sqrt{5}}{3} = \frac{\sqrt{5}}{16} \text{ unit}$$

Option (c),

$$\because \sum_{n=1}^{N} (\text{area of } R_n) < (\text{area of } R_1) + (\text{area of } R_2) + \ldots\ldots \infty$$

$$< 2ab + 2\frac{ab}{2} + 2\frac{ab}{2^2} + \ldots\ldots.$$

$$< 2ab\left(1 + \frac{1}{2} + \frac{1}{2^2} + \ldots..\right)$$

$$< 12\left(\frac{1}{1-1/2}\right)$$

$\Rightarrow \sum_{n=1}^{N} (\text{area of } R_n) < 24$, for each positive integer $N$.

Option (d),

$$\text{Length of latusrectum } E_9 = \frac{2b_9^2}{a_9} = \frac{2b^2}{a(\sqrt{2})^8}$$

$$= \frac{2 \times 4}{3 \times 16} = \frac{1}{6} \text{ units}$$

Hence, options (c) and (d) are correct.

**43.** *(a, b, d)* Let a non-right angled $\Delta PQR$.

Now, by sine rule

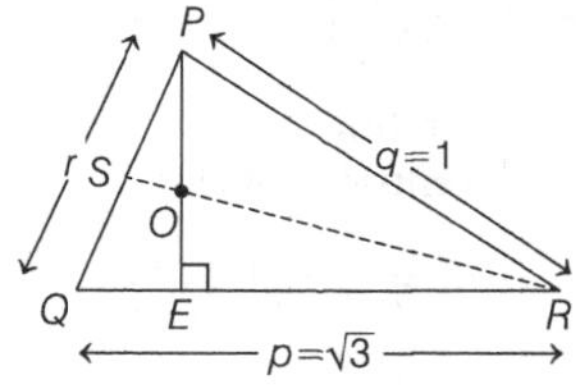


$$\frac{p}{\sin P} = \frac{q}{\sin Q} = \frac{r}{\sin R} = 2 \times \text{circumradius}$$

$$\Rightarrow \frac{\sqrt{3}}{\sin P} = \frac{1}{\sin Q} = \frac{r}{\sin R} = 2 \times 1$$

[circumradius = 1 unit]

$$\Rightarrow \sin P = \frac{\sqrt{3}}{2} \text{ and } \sin Q = \frac{1}{2}$$

$$\Rightarrow P = 120^\circ \text{ and } Q = 30^\circ$$

($\because \Delta PQR$ is non-right angled triangle)

So, $R = 30^\circ$

$\Rightarrow r = 1$, so $\Delta PQR$ is an isosceles triangle. And, since $RS$ and $PE$ are the median of $\Delta PQR$, so '$O$' is centroid of the $\Delta PQR$.

Now,

Option (a),

From Apollonius theorem,

$$2(PE^2 + QE^2) = PQ^2 + PR^2$$

$$\Rightarrow \quad 2\left(PE^2 + \frac{3}{4}\right) = 1 + 1$$

$$\Rightarrow \quad PE^2 = 1 - \frac{3}{4} \Rightarrow PE^2 = \frac{1}{4}$$

$$\Rightarrow \quad PE = \frac{1}{2} \text{ units}$$

$$\text{and} \quad OE = \frac{1}{3} PE = \frac{1}{6} \text{ units}$$

[$\because O$ divides $PE$ is 2 : 1]

Option (b),

Again from Apollonius theorem,

$$2(PS^2 + RS^2) = PR^2 + QR^2$$

$$\Rightarrow \quad 2\left(\frac{1}{4} + RS^2\right) = 1 + 3$$

$$\Rightarrow \quad RS^2 = 2 - \frac{1}{4} \Rightarrow RS^2 = \frac{7}{4}$$

$$\Rightarrow \quad RS = \frac{\sqrt{7}}{2} \text{ units}$$

Option (c),

$$\text{Area of } \Delta SOE = \frac{1}{2}(OE)(ST)$$

$$= \frac{1}{2} \times \frac{1}{6}[(PS)\sin 60^\circ]$$

$$= \frac{1}{12} \times \frac{1}{2} \times \frac{\sqrt{3}}{2}$$

$$= \frac{\sqrt{3}}{48} \text{ square units}$$

Option (d),

$\because$ Inradius of

$$\Delta PQR = \frac{\Delta}{s} = \frac{\frac{1}{2} pq \sin R}{\frac{1}{2}(p + q + r)} = \frac{\frac{1}{2}(\sqrt{3})(1)\frac{1}{2}}{\frac{1}{2}(\sqrt{3} + 1 + 1)}$$

$$= \frac{\sqrt{3}}{2}(2 - \sqrt{3}) \text{ units}$$

Hence, options (a), (b) and (d) are correct.

**44.** *(b,c,d)* Given quadratic equation $x^2 - x - 1 = 0$ having roots $\alpha$ and $\beta$, $(\alpha > \beta)$

So, $\alpha = \frac{1+\sqrt{5}}{2}$ and $\beta = \frac{1-\sqrt{5}}{2}$

and $\alpha + \beta = 1, \alpha\beta = -1$

$\because \quad a_n = \frac{\alpha^n - \beta^n}{\alpha - \beta}, n \ge 1$

So, $a_{n+1} = \frac{\alpha^{n+1} - \beta^{n+1}}{\alpha - \beta}$

$= \alpha^n + \alpha^{n-1}\beta + \alpha^{n-2}\beta^2 + \dots + \alpha\beta^{n-1} + \beta^n$

$= \alpha^n - \alpha^{n-2} - \alpha^{n-3}\beta - \dots - \beta^{n-2} + \beta^n$ [as $\alpha\beta = -1$]

$= \alpha^n + \beta^n - (\alpha^{n-2} + \alpha^{n-3}\beta + \dots + \beta^{n-2})$

$= \alpha^n + \beta^n - a_{n-1}$

$\left[\text{as } a_{n-1} = \frac{\alpha^{n-1} - \beta^{n-1}}{\alpha - \beta} = \alpha^{n-2} + \alpha^{n-3}\beta + \dots + \beta^{n-2}\right]$

$\Rightarrow a_{n+1} + a_{n-1} = \alpha^n + \beta^n = b_n, \forall n \ge 1$

So, option (b) is correct.

Now, $\sum_{n=1}^{\infty} \frac{b_n}{10^n} = \sum_{n=1}^{\infty} \frac{\alpha^n + \beta^n}{10^n}$ [as, $b_n = \alpha^n + \beta^n$]

$= \sum_{n=1}^{\infty}\left(\frac{\alpha}{10}\right)^n + \sum_{n=1}^{\infty}\left(\frac{\beta}{10}\right)^n$ $\left[\because \left|\frac{\alpha}{10}\right| < 1 \text{ and } \left|\frac{\beta}{10}\right| < 1\right]$

$= \frac{\frac{\alpha}{10}}{1 - \frac{\alpha}{10}} + \frac{\frac{\beta}{10}}{1 - \frac{\beta}{10}} = \frac{\alpha}{10-\alpha} + \frac{\beta}{10-\beta}$

$= \frac{10\alpha - \alpha\beta + 10\beta - \alpha\beta}{(10-\alpha)(10-\beta)}$

$= \frac{10(\alpha+\beta) - 2\alpha\beta}{100 - 10(\alpha+\beta) + \alpha\beta}$

$= \frac{10(1) - 2(-1)}{100 - 10(1) - 1}$ [as $\alpha + \beta = 1$ and $\alpha\beta = -1$]

$= \frac{12}{89}$

So, option (a) is not correct.

$\because \quad \alpha^2 = \alpha + 1$ and $\beta^2 = \beta + 1$

$\Rightarrow \alpha^{n+2} = \alpha^{n+1} + \alpha^n$ and $\beta^{n+2} = \beta^{n+1} + \beta^n$

$\Rightarrow (\alpha^{n+2} + \beta^{n+2}) = (\alpha^{n+1} + \beta^{n+1}) + (\alpha^n + \beta^n)$

$\Rightarrow \quad a_{n+2} = a_{n+1} + a_n$

Similarly, $a_{n+1} = a_n + a_{n-1}$

$a_n = a_{n-1} + a_{n-2}$

.........

.........

$a_3 = a_2 + a_1$

On adding, we get

$a_{n+2} = (a_n + a_{n-1} + a_{n-2} + \dots + a_2 + a_1) + a_2$

$\left[\because a_2 = \frac{\alpha^2 - \beta^2}{\alpha - \beta} = \alpha + \beta = 1\right]$

So, $a_{n+2} - 1 = a_1 + a_2 + a_3 + \dots + a_n$

So, option (c) is also correct.

And, now $\sum_{n=1}^{\infty} \frac{a_n}{10^n} = \sum_{n=1}^{\infty} \frac{\alpha^n - \beta^n}{(\alpha - \beta)10^n}$

$= \frac{1}{\alpha - \beta}\left[\sum_{n=1}^{\infty}\left(\frac{\alpha}{10}\right)^n - \sum_{n=1}^{\infty}\left(\frac{\beta}{10}\right)^n\right]$

$= \frac{1}{\alpha - \beta}\left[\frac{\frac{\alpha}{10}}{1 - \frac{\alpha}{10}} - \frac{\frac{\beta}{10}}{1 - \frac{\beta}{10}}\right],$

$\left[\text{as } \left|\frac{\alpha}{10}\right| < 1 \text{ and } \left|\frac{\beta}{10}\right| < 1\right]$

$= \frac{1}{\alpha - \beta}\left(\frac{\alpha}{10-\alpha} - \frac{\beta}{10-\beta}\right) = \frac{1}{\alpha - \beta}$

$\left[\frac{10\alpha - \alpha\beta - 10\beta + \alpha\beta}{100 - 10(\alpha+\beta) + \alpha\beta}\right]$

$= \frac{10(\alpha - \beta)}{(\alpha - \beta)[100 - 10(\alpha+\beta) + \alpha\beta]}$

$= \frac{10}{100 - 10 - 1} = \frac{10}{89}$

Hence, options (b), (c) and (d) are correct.

**45.** *(a,b,c)* Given lines

$L_1$: $\mathbf{r} = \hat{\mathbf{i}} + \lambda(-\hat{\mathbf{i}} + 2\hat{\mathbf{j}} + 2\hat{\mathbf{k}}), \lambda \in R$ and

$L_2$: $\mathbf{r} = \mu(2\hat{\mathbf{i}} - \hat{\mathbf{j}} + 2\hat{\mathbf{k}}), \mu \in R$

and since line $L_3$ is perpendicular to both lines $L_1$ and $L_2$.

Then a vector along $L_3$ will be,

$\begin{vmatrix} \hat{\mathbf{i}} & \hat{\mathbf{j}} & \hat{\mathbf{k}} \\ -1 & 2 & 2 \\ 2 & -1 & 2 \end{vmatrix} = \hat{\mathbf{i}}(4+2) - \hat{\mathbf{j}}(-2-4) + \hat{\mathbf{k}}(1-4)$

$= 6\hat{\mathbf{i}} + 6\hat{\mathbf{j}} - 3\hat{\mathbf{k}} = 3(2\hat{\mathbf{i}} + 2\hat{\mathbf{j}} - \hat{\mathbf{k}})$ ...(i)

Now, let a general point on line $L_1$.
$P(1-\lambda, 2\lambda, 2\lambda)$ and on line $L_2$.
as $Q(2\mu, -\mu, 2\mu)$ and let $P$ and $Q$ are point of intersection of lines $L_1, L_3$ and $L_2, L_3$, so direction ratio's of $L_3$

$$(2\mu+\lambda-1, -\mu-2\lambda, 2\mu-2\lambda) \quad \ldots\text{(ii)}$$

Now, $\dfrac{2\mu+\lambda-1}{2}=\dfrac{-\mu-2\lambda}{2}=\dfrac{2\mu-2\lambda}{-1}$

[from Eqs.(i) and (ii)]

$\Rightarrow \quad \lambda=\dfrac{1}{9}$ and $\mu=\dfrac{2}{3}$

So, $P\left(\dfrac{8}{9}, \dfrac{2}{9}, \dfrac{2}{9}\right)$ and $Q\left(\dfrac{4}{9}, -\dfrac{2}{9}, \dfrac{4}{9}\right)$

Now, we can take equation of line $L_3$ as $\mathbf{r}=\mathbf{a}+t(2\hat{\mathbf{i}}+2\hat{\mathbf{j}}-\hat{\mathbf{k}})$, where $\mathbf{a}$ is position vector of any point on line $L_3$ and possible vector of $\mathbf{a}$ are

$\left(\dfrac{8}{9}\hat{\mathbf{i}}+\dfrac{2}{9}\hat{\mathbf{j}}+\dfrac{2}{9}\hat{\mathbf{k}}\right)$ or $\left(\dfrac{4}{9}\hat{\mathbf{i}}-\dfrac{2}{9}\hat{\mathbf{j}}+\dfrac{4}{9}\hat{\mathbf{k}}\right)$

or $\left(\dfrac{2}{3}\hat{\mathbf{i}}+\dfrac{1}{3}\hat{\mathbf{k}}\right)$

Hence, options (a), (b) and (c) are correct.

**46.** *(a,c)* **Key Idea:** Use conditional probability, total probability and Baye's theorem.

It is given that there are three bags $B_1, B_2$ and $B_3$ and probabilities of being chosen $B_1, B_2$ and $B_3$ are respectively.

$\therefore \quad P(B_1)=\dfrac{3}{10}, P(B_2)=\dfrac{3}{10}$ and $P(B_3)=\dfrac{4}{10}$.

| $B_1$ | $B_2$ | $B_3$ |
|---|---|---|
| 5 R | 3 R | 5 R |
| 5 G | 5 G | 3 G |

Now, probability that the chosen ball is green, given that selected bag is

$$B_3=P\left(\frac{G}{B_3}\right)=\frac{3}{8}$$

Now, probability that the selected bag is $B_3$, given that the chosen ball is green

$$=P\left(\frac{B_3}{G}\right)$$

$$=\frac{P\left(\frac{G}{B_3}\right)P(B_3)}{P\left(\frac{G}{B_1}\right)P(B_1)+P\left(\frac{G}{B_2}\right)P(B_2)+P\left(\frac{G}{B_3}\right)P(B_3)}$$

[by Baye's theorem]

$$=\frac{\left(\frac{3}{8}\times\frac{4}{10}\right)}{\left(\frac{5}{10}\times\frac{3}{10}\right)+\left(\frac{5}{8}\times\frac{3}{10}\right)+\left(\frac{3}{8}\times\frac{4}{10}\right)}$$

$$=\frac{\frac{1}{2}}{\frac{1}{2}+\frac{5}{8}+\frac{1}{2}}=\frac{4}{13}$$

Now, probability that the chosen ball is green $=P(G)$

$$=P(B_1)P\left(\frac{G}{B_1}\right)+P(B_2)P\left(\frac{G}{B_2}\right)+P(B_3)P\left(\frac{G}{B_3}\right)$$

[By using theorem of total probability]

$$=\left(\frac{3}{10}\times\frac{5}{10}\right)+\left(\frac{3}{10}\times\frac{5}{8}\right)+\left(\frac{4}{10}\times\frac{3}{8}\right)$$

$$=\frac{3}{20}+\frac{3}{16}+\frac{3}{20}=\frac{12+15+12}{80}=\frac{39}{80}$$

Now, probability that the selected bag is $B_3$ and the chosen ball is green

$$=P(B_3)\times P\left(\frac{G}{B_3}\right)=\frac{4}{10}\times\frac{3}{8}=\frac{3}{20}$$

Hence, options (a) and (c) are correct.

**47.** *(b,c,d)* Given square matrix

$$M=\begin{bmatrix}0 & 1 & a\\ 1 & 2 & 3\\ 3 & b & 1\end{bmatrix}$$

and $\operatorname{adj}(M)=\begin{bmatrix}-1 & 1 & -1\\ 8 & -6 & 2\\ -5 & 3 & -1\end{bmatrix}$

$\because \quad |\operatorname{adj}(M)|=|M|^2=\begin{vmatrix}-1 & 1 & -1\\ 8 & -6 & 2\\ -5 & 3 & -1\end{vmatrix}$

$\Rightarrow$

$$|M|^2=-1(6-6)-1(-8+10)-1(24-30)$$
$$=-2+6=4$$

$\Rightarrow \quad |M|=\pm 2$

$\therefore \det(\text{adj } M^2) = |M^2|^2 = |M|^4 = 16$

As we know $A(\text{adj } A) = |A| I$

$\Rightarrow \quad M = |M| (\text{adj } M)^{-1} \quad \ldots(\text{i})$

$$\because (\text{adj } M)^{-1} = \frac{1}{|\text{adj } M|}\begin{bmatrix} 0 & -2 & -6 \\ -2 & -4 & -2 \\ -4 & -6 & -2 \end{bmatrix}^T$$

$$= \frac{1}{4}\begin{bmatrix} 0 & -2 & -4 \\ -2 & -4 & -6 \\ -6 & -2 & -2 \end{bmatrix}$$

$$\text{So} \begin{bmatrix} 0 & 1 & a \\ 1 & 2 & 3 \\ 3 & b & 1 \end{bmatrix} = \frac{|M|}{4}\begin{bmatrix} 0 & -2 & -4 \\ -2 & -4 & -6 \\ -6 & -2 & -2 \end{bmatrix}$$

$\Rightarrow |M| = -2, a = 2$ and $b = 1$

$$\Rightarrow M = \begin{bmatrix} 0 & 1 & 2 \\ 1 & 2 & 3 \\ 3 & 1 & 1 \end{bmatrix}$$

$$\text{Now, If} \quad M\begin{bmatrix} \alpha \\ \beta \\ \gamma \end{bmatrix} = \begin{bmatrix} 1 \\ 2 \\ 3 \end{bmatrix}$$

$$\Rightarrow \begin{bmatrix} 0 & 1 & 2 \\ 1 & 2 & 3 \\ 3 & 1 & 1 \end{bmatrix}\begin{bmatrix} \alpha \\ \beta \\ \gamma \end{bmatrix} = \begin{bmatrix} 1 \\ 2 \\ 3 \end{bmatrix}$$

$\Rightarrow \quad B + 2\gamma = 1, \alpha + 2\beta + 3\gamma = 2$

and $3\alpha + \beta + \gamma = 3$

$\Rightarrow \quad \alpha = 1, \beta = -1$ and $\gamma = 1$

$\therefore \alpha - \beta + \gamma = 3$

And $(adj\ M)^{-1} + adj\ (M^{-1}) = 2(adj\ M)^{-1}$

$[\because adj\ (M^{-1}) = (adj\ M)^{-1}]$

$$= 2\left(-\frac{M}{2}\right) = -M$$

$\left[\because (adj\ M)^{-1} = \dfrac{M}{|M|} \text{ from Eq. (i)}\right]$

and $\because a = 2$ and $b = 1$, so $a + b = 3$

Hence, options (b), (c) and (d) are correct.

**48.** *(b,c,d)* Given function $f: R \longrightarrow R$ is

$$f(x) = \begin{cases} x^5 + 5x^4 + 10x^3 + 10x^2 + 3x + 1, & x < 0 \\ x^2 - x + 1, & 0 \le x < 1 \\ \frac{2}{3}x^3 - 4x^2 + 7x - \frac{8}{3}, & 1 \le x < 3 \\ (x-2)\log_e(x-2) - x + \frac{10}{3}, & x \ge 3 \end{cases}$$

So,

$$f'(x) = \begin{cases} 5x^4 + 20x^3 + 30x^2 + 20x + 3, & x < 0 \\ 2x - 1, & 0 < x < 1 \\ 2x^2 - 8x + 7, & 1 < x < 3 \\ \log_e(x-2), & x > 3 \end{cases}$$

At $x = 1$, $f''(1^-) = 2 > 0$ and

$f''(1^+) = 4 - 8 = -4 < 0$

$\therefore f'(x)$ is not differentiable at $x = 1$ and $f'(x)$ has a local maximum at $x = 1$.

For $x \in (-\infty, 0)$

$f'(x) = 5x^4 + 20x^3 + 30x^2 + 20x + 3$

and since

$f'(-1) = 5 - 20 + 30 - 20 + 3 = -2 < 0$

So, $f(x)$ is not increasing on $x \in (-\infty, 0)$.

Now, as the range of function $f(x)$ is $R$, so $f$ is onto function.

Hence, options (b), (c) and (d) are correct.

**49.** *(0.50)* Given sample space $(S)$ of all $3 \times 3$ matrices with entries from the set $\{0, 1\}$ and events

$E_1 = \{A \in S : \det(A) = 0\}$ and

$E_2 = \{A \in S : \text{sum of entries of } A \text{ is } 7\}$.

For event $E_2$, means sum of entries of matrix $A$ is 7, then we need seven 1s and two 0s.

$\therefore$ Number of different possible matrices

$$= \frac{9!}{7!\,2!}$$

$\Rightarrow \quad n(E_2) = 36$

For event $E_1$, $|A| = 0$, both the zeroes must be in same row/column.

$\therefore$ Number of matrices such that their determinant is zero $= 6 \times \dfrac{3!}{2!} = 18$

$= n(E_1 \cap E_2)$

$\therefore$ Required probability,

$$P\left(\frac{E_1}{E_2}\right) = \frac{n(E_1 \cap E_2)}{n(E_2)}$$

$$= \frac{18}{36}$$

$$= \frac{1}{2} = 0.50$$

**50.** *(10)* According to given informations the figure is as following

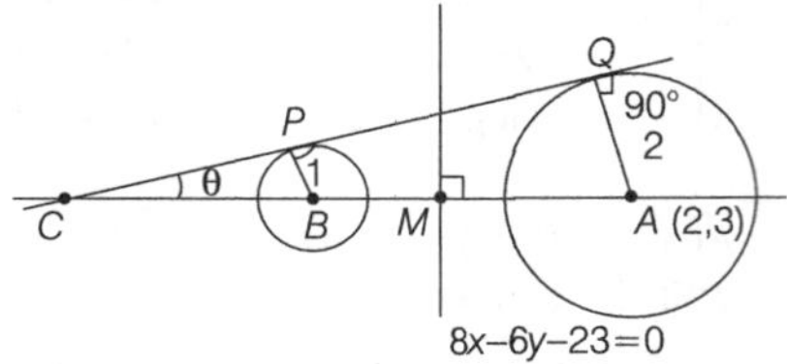


From the figure,

$$AC = \frac{2}{\sin\theta} \quad \ldots(\text{i})$$

$\because \quad \sin\theta = \frac{1}{CB}$ (from $\Delta CPB$)...(ii)

and $\quad \sin\theta = \frac{2}{AC} = \frac{2}{CB + AB}$

(from $\Delta CQA$)...(iii)

$\because \quad AB = AM + MB = 2AM \quad [\because AM = MB]$

$$= 2\frac{|(8 \times 2) - (6 \times 3) - 23|}{\sqrt{64 + 36}}$$

$$= \frac{2 \times 25}{10} = 5.00$$

From Eqs. (ii) and (iii), we get

$$\sin\theta = \frac{1}{CB} = \frac{2}{CB + AB}$$

$\Rightarrow \quad \frac{1}{CB} = \frac{2}{CB + 5} \quad [\because AB = 5]$

$\Rightarrow \quad CB + 5 = 2CB \Rightarrow CB = 5 = \frac{1}{\sin\theta}$

From the Eq. (i), we get

$$AC = \frac{2}{\sin\theta} = 2 \times 5 = 10.00$$

**51.** *(3.00)* Given, $\omega \neq 1$ be a cube root of unity, then $|a + b\omega + c\omega^2|^2$

$= (a + b\omega + c\omega^2)\,\overline{(a + b\omega + c\omega^2)}$,

$[\because z\bar{z} = |z|^2]$

$= (a + b\omega + c\omega^2)\,(a + b\bar{\omega} + 2c\bar{\omega}^2)$

$= (a + b\omega + c\omega^2)\,(a + b\omega^2 + c\omega)$

$[\because \bar{\omega} = \omega^2 \text{ and } \bar{\omega}^2 = \omega]$

$= a^2 + b^2 + c^2 + ab(\omega^2 + \omega) + bc(\omega^2 + \omega^4) + ac(\omega + \omega^2)$

$[\text{as } \omega^3 = 1]$

$= a^2 + b^2 + c^2 + ab(-1) + bc(-1) + ac(-1)$

$[\text{as } \omega + \omega^2 = -1, \omega^4 = \omega]$

$= a^2 + b^2 + c^2 - ab - bc - ca$

$= \frac{1}{2}\{(a - b)^2 + (b - c)^2 + (c - a)^2\}$

$\because a, b$ and $c$ are distinct non-zero integers.

For minimum value $a = 1$, $b = 2$ and $c = 3$

$\therefore \quad |a + b\omega + c\omega^2|^2_{\min} = \frac{1}{2}\{1^2 + 1^2 + 2^2\}$

$$= \frac{6}{2} = 3.00$$

**52.** *(4.0)* Given,

$$I = \frac{2}{\pi}\int_{-\pi/4}^{\pi/4} \frac{dx}{(1 + e^{\sin x})\,(2 - \cos 2x)} \quad \ldots(\text{i})$$

On applying property

$\int_a^b f(x)dx = \int_a^b f(a + b - x)\,dx$, we get

$$I = \frac{2}{\pi}\int_{-\pi/4}^{\pi/4} \frac{e^{\sin x}\,dx}{(1 + e^{\sin x})\,(2 - \cos 2x)} \quad \ldots(\text{ii})$$

On adding integrals (i) and (ii), we get

$$2I = \frac{2}{\pi}\int_{-\pi/4}^{\pi/4} \frac{dx}{2 - \cos 2x}$$

$$\Rightarrow \quad I = \frac{1}{\pi}\int_{-\pi/4}^{\pi/4} \frac{dx}{2 - \dfrac{1 - \tan^2 x}{1 + \tan^2 x}}$$

$$\left[\text{as } \cos 2x = \frac{1 - \tan^2 x}{1 + \tan^2 x}\right]$$

$$= \frac{2}{\pi}\int_0^{\pi/4} \frac{\sec^2 x}{1 + 3\tan^2 x}\,dx$$

$$\left[\because \frac{\sec^2 x}{1 + 3\tan^2 x} \text{ is even function}\right]$$

Put $\sqrt{3}\tan x = t \Rightarrow \sqrt{3}\sec^2 dx = dt$, and at $x = 0$, $t = 0$ and at $x = \sqrt{3}$, $t = \sqrt{3}$

So, $I = \frac{2}{\pi}\int_0^{\sqrt{3}} \frac{1}{\sqrt{3}}\frac{dt}{1 + t^2} = \frac{2}{\sqrt{3}\pi}[\tan^{-1} t]_0^{\sqrt{3}}$

$$= \frac{2}{\sqrt{3}\pi}\left(\frac{\pi}{3}\right) = \frac{2}{3\sqrt{3}} \Rightarrow 27I^2 = 4.00$$

**53.** *(0.75)* Given three lines

$\mathbf{r} = \lambda\hat{\mathbf{i}}, \lambda \in R,$

$\mathbf{r} = \mu(\hat{\mathbf{i}} + \hat{\mathbf{j}}), \mu \in R$

and $\quad \mathbf{r} = v(\hat{\mathbf{i}} + \hat{\mathbf{j}} + \hat{\mathbf{k}}), v \in R$

cuts the plane $x + y + z = 1$ at the points $A$, $B$ and $C$, respectively. So, for point $A$, put $(\lambda, 0, 0)$ in the plane, we get $\lambda + 0 + 0 = 1 \Rightarrow \lambda = 1 \Rightarrow A \equiv (1,0,0)$.

Similarly, for point $B$, put $(\mu, \mu, 0)$ in the plane, we get $\mu + \mu + 0 = 1 \Rightarrow \mu = \frac{1}{2}$

$$\Rightarrow B \equiv \left(\frac{1}{2}, \frac{1}{2}, 0\right).$$

and for point $C$, put $(v, v, v)$ in the plane we get

$$v + v + v = 1 \Rightarrow v = \frac{1}{3} \Rightarrow C \equiv \left(\frac{1}{3}, \frac{1}{3}, \frac{1}{3}\right)$$

Now, area of $\Delta ABC = \frac{1}{2}|\mathbf{AB} \times \mathbf{AC}| = \Delta$

$$\because \quad \mathbf{AB} = -\frac{1}{2}\hat{\mathbf{i}} + \frac{1}{2}\hat{\mathbf{j}},$$

and $\mathbf{AC} = -\frac{2}{3}\hat{\mathbf{i}} + \frac{1}{3}\hat{\mathbf{j}} + \frac{1}{3}\hat{\mathbf{k}}$

$$\therefore \mathbf{AB} \times \mathbf{AC} = \begin{vmatrix} \hat{\mathbf{i}} & \hat{\mathbf{j}} & \hat{\mathbf{k}} \\ -1/2 & 1/2 & 0 \\ -2/3 & 1/3 & 1/3 \end{vmatrix}$$

$$= \hat{\mathbf{i}}\left(\frac{1}{6}\right) - \hat{\mathbf{j}}\left(-\frac{1}{6}\right) + \hat{\mathbf{k}}\left(-\frac{1}{6} + \frac{2}{6}\right)$$

$$= \frac{1}{6}(\hat{\mathbf{i}} + \hat{\mathbf{j}} + \hat{\mathbf{k}})$$

$$\Rightarrow |\mathbf{AB} \times \mathbf{AC}| = \frac{1}{6}\sqrt{3} = \frac{1}{2\sqrt{3}}$$

$$\Rightarrow \quad \Delta = \frac{1}{4\sqrt{3}}$$

$$\Rightarrow \quad (6\Delta)^2 = 36\frac{1}{16 \times 3} = \frac{3}{4} = 0.75$$

**54.** *(157.00)* Given that, $AP(a; d)$ denote the set of all the terms of an infinite arithmetic progression with first term '$a$' and common difference $d > 0$.

Now, let $m^{\text{th}}$ term of first progression

$$AP(1\,;\,3) = 1 + (m-1)3 = 3m - 2 \quad \ldots\text{(i)}$$

and $n^{\text{th}}$ term of progression

$$AP(2; 5) = 2 + (n-1)5 = 5n - 3 \quad \ldots\text{(ii)}$$

and $r^{\text{th}}$ term of third progression $AP\ (3; 7)$

$$= 3 + (r-1)7 = 7r - 4 \quad \ldots\text{(iii)}$$

are equal.

Then, $3m - 2 = 5n - 3 = 7r - 4$

Now, for $AP(1\,;\,3) \cap AP(2\,;\,5) \cap AP(3; 7)$, the common terms of first and second progressions, $m = \frac{5n-1}{3}$

$\Rightarrow n = 2, 5, 11, \ldots$ and the common terms of second and the third progressions,

$$r = \frac{5n+1}{7} \Rightarrow n = 4, 11, \ldots$$

Now, the first common term of first, second and third progressions (when $n = 11$), so $a = 2 + (11-1)5 = 52$

and $d = \text{LCM}(3, 5, 7) = 105$

So, $AP(1\,;\,3) \cap AP(2\,;\,5) \cap AP(3\,;\,7)$

$$= AP(52\,;105)$$

So, $a = 52$ and $d = 105$

$$\Rightarrow \quad a + d = 157.00$$

# Paper 2

**1.** *(c, d)* $R >>$ Dipole size.

Circle is equipotential.

So, $E_{\text{net}}$ should be perpendicular to surface hence,

$$\frac{kp_0}{r^3} = E_0$$

$$\Rightarrow \quad R = \left(\frac{kp_0}{E_0}\right)^{1/3}$$

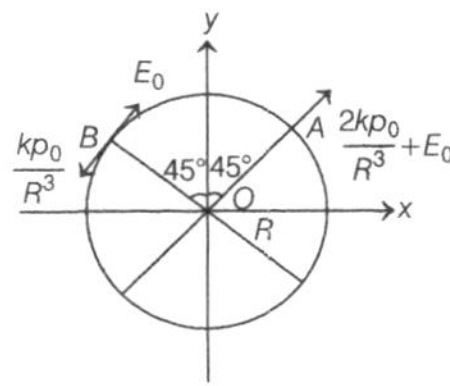


At point $B$, net electric field will be zero.

$$E_B = 0$$

$$(E_A)_{\text{net}} = \frac{2kp_0}{R^3} + E_0 = 3E_0$$

Electric field at point $A$, $\mathbf{E}_A = \frac{3}{\sqrt{2}}E_0[\hat{\mathbf{i}} + \hat{\mathbf{j}}]$

$$(E_B)_{\text{net}} = 0$$

**2.** *(a, c)*

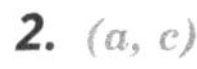

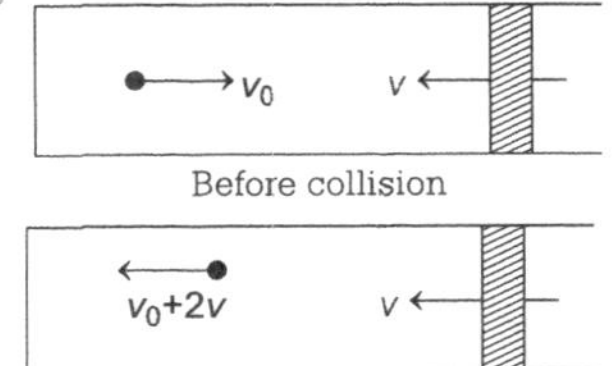


Before collision

After Collision

Change in speed $= (2v + v_0 - v_0) = 2v$

In every collision it acquires $2v$,

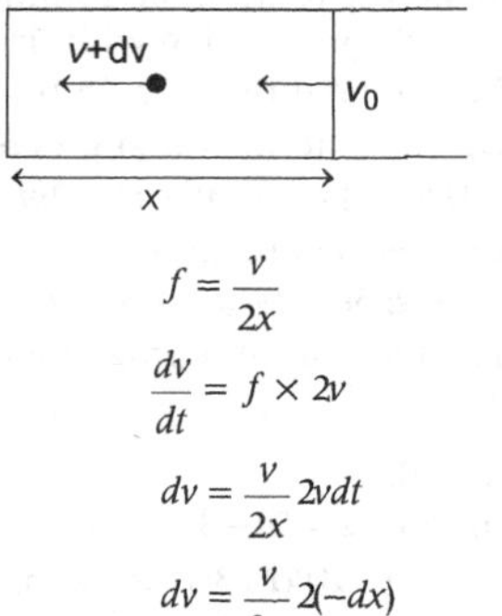


$$f = \frac{v}{2x}$$

$$\frac{dv}{dt} = f \times 2v$$

$$dv = \frac{v}{2x} 2v dt$$

$$dv = \frac{v}{2x} 2(-dx)$$

Integration on both sides limits $v_0$ to $v$, we get

$$\int_{v_0}^{v} \frac{dv}{v} = \int_{\ell}^{x} \frac{-dx}{x}$$

$\Rightarrow$ $$\ln \frac{v}{v_0} = -\ln \frac{x}{\ell}$$

$\Rightarrow$ $$v = \frac{v_0 \ell}{x}$$

where, $x = \frac{\ell}{2}$, $v = 2v_0$

so, $$f = \frac{2v_0}{2\frac{\ell}{2}} = \frac{2v_0}{\ell}$$

$\therefore$ $$\frac{k_f}{k_i} = 4$$

**3.** *(a, d)* **Case-I**

$H = 30$ cm

$n = \frac{3}{2}$

$H_1 = \frac{H}{n}$

$\Rightarrow$ $\frac{30 \times 2}{3} = 20$ cm

**Case-II**

$R = 300$ cm

$$\frac{n_2}{v} - \frac{n_1}{u} = \frac{n_2 - n_1}{R}$$

$$\frac{1}{-H_2} - \frac{3}{-2 \times 30} = \frac{1 - \frac{3}{2}}{-300}$$

$$H_2 = \frac{600}{29} = 20.684 \text{ cm}$$

**Case-III**

$$\frac{n_2}{v} - \frac{n_1}{u} = \frac{n_2 - n_1}{R};$$

$$\frac{1}{-H_3} - \frac{3}{-2 \times 30} = \frac{1 - \frac{3}{2}}{300}$$

$$H_3 = \frac{600}{31} = 19.354 \text{ cm}$$

**4.** *(a)*

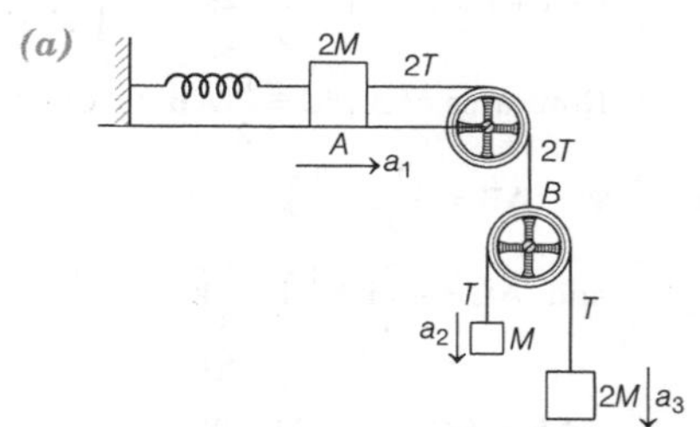


In the frame of pulley $B$,

the hanging masses have accelerations :

$M \rightarrow (a_2 - a_1)$, $2M \rightarrow (a_3 - a_1)$ : downward.

$\therefore$ $(a_2 - a_1) = -(a_3 - a_1)$ [constant]

Assuming that the extension of the spring is $x$

We consider the FBD of $A$ :

$$2M \cdot \frac{d^2x}{dt^2} = 2T - kx$$

where, $a_1 \equiv \frac{d^2x}{dt^2}$ ...(i)

and the FBD of the rest of the system in the frame of pulley $B$ :

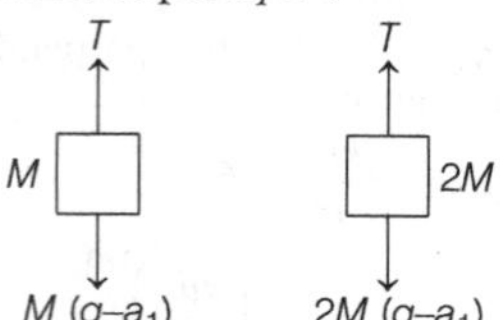


Upward acceleration of block $M$ w.r.t. the pulley $B$

= Downward acceleration of block $2M$ w.r.t the pulley,

$$\frac{T - M(g - a_1)}{M} = \frac{2M(g - a_1) - T}{2M}$$

$\Rightarrow$ $$T = \frac{4M}{3}(g - a_1) \quad \text{...(ii)}$$

Substituting in Eq. (i), we get

$$2M \cdot a_1 = \frac{8M}{3}(g - a_1) - kx$$

or $$\frac{14M}{3} a_1 = \frac{8Mg}{3} - kx \qquad \ldots \text{(iii)}$$

This is equation of SHM

Maximum extension = 2 × amplitude

i.e., $$x_0 = 2 \times \frac{8Mg}{3k}$$

$$\text{Amplitude} = \frac{x_0}{2} = \frac{3Mg}{3k} \text{ and } \omega = \sqrt{\frac{3k}{14M}}$$

At $\frac{x_0}{4}$, acceleration is easily found from Eq. (iii),

$$\frac{14M}{3} a_1 = \frac{8Mg}{3} - \frac{4Mg}{3}$$

$$a_1 = \frac{2g}{7}$$

At $\frac{x_0}{2}$, speed of the block

$(2M) = \omega \times$ amplitude

$$= \sqrt{\frac{3k}{14M}} \times \frac{8Mg}{3k}$$

∴ option (a) is correct

**5.** (b, c, d)

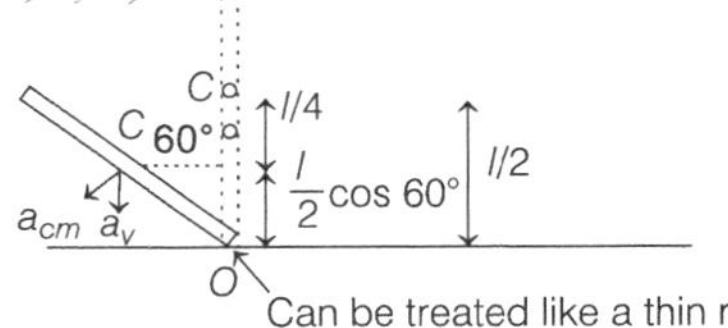


$$\Delta K + \Delta U = 0$$

$$\frac{1}{2} I_0 \omega^2 = -\Delta U$$

$$\frac{1}{2}\frac{m\ell^2}{3}\omega^2 = -\left(-mg\frac{\ell}{4}\right)$$

$$\omega = \sqrt{\frac{3g}{2\ell}}$$

$$\Rightarrow \quad a_{\text{radial}} = \omega^2 \frac{\ell}{2} = \frac{3g}{2\ell}\frac{\ell}{2} = \frac{3g}{4} \Rightarrow \tau = I_0 \alpha$$

$$\alpha = \frac{mg\frac{\ell}{2}\sin 60^\circ}{m\frac{\ell^2}{3}} = \frac{3\sqrt{3}g}{4\ell}$$

$$\Rightarrow \quad a_v = \left(\alpha \frac{\ell}{2}\right)\sin 60^\circ + \omega^2 \frac{\ell}{2}\cos 60^\circ$$

$$a_v = \frac{3\sqrt{3}g}{8}\frac{\sqrt{3}}{2} + \frac{3g}{8} \Rightarrow a_v = \frac{9g}{16} + \frac{6g}{16}$$

$$mg - N = ma_v \Rightarrow N = \frac{mg}{16}$$

**6.** (b)

$$\Delta x = d\sin\alpha + d\sin\theta$$

θ and α are small angles

∴ $$\Delta x = d\alpha + \frac{dy}{D}$$

(a) $\alpha = 0$

$$\therefore \quad \Delta x = \frac{dy}{D} = \frac{0.3 \times 11}{1000} = 33 \times 10^{-4} \text{ mm}$$

$$\Delta x \text{ in terms of } \lambda = \frac{33 \times 10^{-4}}{600 \times 10^{-6}}\lambda = \frac{11\lambda}{2}$$

as $$\Delta x = (2n - 1)\frac{\lambda}{2}$$

There will be destructive interference.

(b) $$\Delta x = 0.3\,\text{mm} \times \frac{0.36}{\pi} \times \frac{\pi}{180} + \frac{0.3\text{ mm} \times 11\text{ mm}}{1000}$$

$$= 39 \times 10^{-4} \text{ mm}$$

$$39 \times 10^{-4} = (2n - 1) \times \frac{600 \times 10^{-9} \times 10^3}{2}$$

$$n = 7$$

So, there will be destruction interference.

(c) $$\Delta x = 3\,\text{mm} \times \frac{0.36}{\pi} \times \frac{\pi}{180} + 0 = 600\,\text{nm}$$

$$600\text{ nm} = n\lambda \Rightarrow n = 1$$

So, there will be construction interference.

(d) Fringe width does not depend on α.

**7.** (a, b) $$\frac{\lambda_a}{\lambda_e} = \frac{E_4 - E_1}{E_4 - E_m} = \frac{1 - \frac{1}{16}}{\frac{1}{m^2} - \frac{1}{16}} = \frac{1}{5}$$

On solving we get,

$$m = 2$$

$$\lambda_e = \frac{12400 \times 4}{13.6} = 3647$$

$\frac{K_2}{K_1} = \frac{1^2}{2^2} = \frac{1}{4}$ as kinetic energy is proportional to $\frac{1}{n^2}$.

**8.** *(a, c, d)* $\gamma_{mix} = \dfrac{n_1 C_{P1} + n_2 C_{P2}}{n_1 C_{V1} + n_2 C_{V2}} = \dfrac{8}{5}$

$$W = \frac{P_1V_1 - P_2V_2}{\gamma - 1}$$

$$P_0 V_0^{8/5} = P_2\left(\frac{V_0}{4}\right)^{8/5}$$

$$P_2 = 9.2P_0$$

$$W = \frac{P_0V_0 - 9.2P_0\frac{V_0}{4}}{3/5} = -13RT_0$$

$\therefore \quad |W| = 13RT_0$

(b) $T_1V_1^{\gamma-1} = T_2V_2^{\gamma-1}$

$$T_2 = T_1(2)^{6/5} = 2.3T_0$$

Average kinetic energy of gas mixture

$$= nC_{V_{mix}}T_2 = 23RT_0$$

**9.** *(135.00)* Mass defect

$\Delta m = 226.005 - 222.000 - 4.000$

$= 0.005\,\text{amu}$

$\therefore$ $Q$ value $= 0.005 \times 931.5 = 4.655\,\text{MeV}$

Momentum $P = \sqrt{2km}$ = constant

Also $\dfrac{KE_\alpha}{KE_{Rn}} = \dfrac{m_{Rn}}{m_\alpha}$

$\Rightarrow \quad KE_{Rn} = \dfrac{m_\alpha}{m_{Rn}} \cdot KE_\alpha$

$= \dfrac{4}{222} \times 4.44 = 0.08\,\text{MeV}$

$\therefore$ Energy of $\gamma$ – Photon

$= 4.655 - (4.44 + 0.08)$

$= 0.135\,\text{MeV} = 135\,\text{keV}$

**10.** *(1.50)* For TIR at coating,

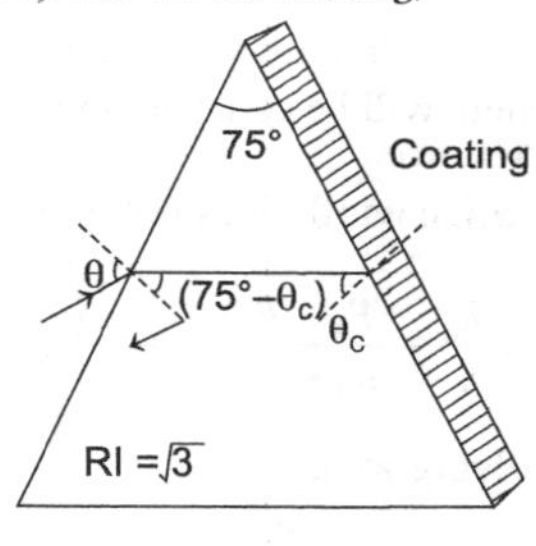


$$\sin\theta_c = \frac{n}{\sqrt{3}}$$

Applying snell's law at first surface

$\sin\theta = \sqrt{3}\sin(75° - \theta_c)$

For limiting condition, at $\theta = 60°$

$$\sin 60° = \sqrt{3}\sin(75° - \theta_c)$$

$$\frac{\sqrt{3}}{2} = \sqrt{3}\sin(75° - \theta_c)$$

$$\frac{1}{2} = \sin(75° - \theta_c)$$

$$\Rightarrow \quad \sin 30° = \sin(75° - \theta_c)$$

$$30° = 75° - \theta_c \Rightarrow \theta_c = 45°$$

$$\frac{n}{\sqrt{3}} = \frac{1}{\sqrt{2}} \Rightarrow n^2 = \frac{3}{2} = 1.50$$

**11.** *(1.00)* Momentum transferred on mirror $= \dfrac{2Nh}{\lambda}$

$$\frac{2Nh}{\lambda} = MV_{(\text{mean position})}$$

$V_{(\text{mean position})} = \Omega A \quad$ (where, $A = 1\,\mu\text{m}$)

$$\frac{2Nh}{\lambda} = M\Omega A$$

(where, $\lambda = 8\pi \times 10^{-6}$)

$$N = \frac{M\Omega(10^{-6})\lambda}{2h}$$

$$= \frac{M\Omega 8\pi \times 10^{-6} \times 10^{-6}}{2h}$$

$$N = \frac{4\pi M\Omega}{h} \times 10^{-12}$$

$$= 10^{24} \times 10^{-12}$$

$$N = 1 \times 10^{12} \Rightarrow x = 1$$

**12.** *(1.38 & 1.39)*

$u = (x_2 - x_1) = 75 - 45 = 30\,\text{cm}$

$\Delta u = \Delta x_2 + \Delta x_1 = \dfrac{1}{4} + \dfrac{1}{4} = \dfrac{1}{2}\,\text{cm}$

$v = (x_3 - x_2) = 135 - 75 = 60\,\text{cm}$

$\Delta v = \Delta x_3 + \Delta x_2 = \dfrac{1}{4} + \dfrac{1}{4} = \dfrac{1}{2}\,\text{cm}$

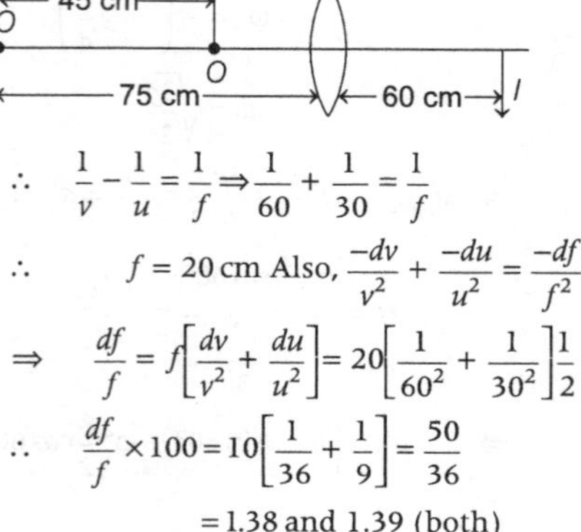


$\therefore \quad \dfrac{1}{v} - \dfrac{1}{u} = \dfrac{1}{f} \Rightarrow \dfrac{1}{60} + \dfrac{1}{30} = \dfrac{1}{f}$

$\therefore \quad f = 20\,\text{cm}$ Also, $\dfrac{-dv}{v^2} + \dfrac{-du}{u^2} = \dfrac{-df}{f^2}$

$\Rightarrow \quad \dfrac{df}{f} = f\left[\dfrac{dv}{v^2} + \dfrac{du}{u^2}\right] = 20\left[\dfrac{1}{60^2} + \dfrac{1}{30^2}\right]\dfrac{1}{2}$

$\therefore \quad \dfrac{df}{f} \times 100 = 10\left[\dfrac{1}{36} + \dfrac{1}{9}\right] = \dfrac{50}{36}$

$= 1.38$ and $1.39$ (both)

**13.** *(4.00)* For first projectile,

$$< V > = \frac{R}{T} = U_x = v_1$$

For journey,

$$< v >_{1 \text{ to } n} = \frac{R_1 + R_2 + \ldots + R_n}{T_1 + T_2 + \ldots + T_n}$$

$$= \frac{\frac{2u_{x_1}u_{y_1}}{g} + \frac{2u_{x_2}u_{y_2}}{g} + \ldots + \frac{2u_{x_n}u_{y_n}}{g}}{\frac{2u_{y_1}}{g} + \frac{2u_{y_2}}{g} + \ldots \frac{2u_{y_n}}{g}}$$

$$u_x\left[\frac{1 + \frac{1}{\alpha^2} + \frac{1}{\alpha^4} + \ldots \frac{1}{\alpha^{2n}}}{1 + \frac{1}{\alpha} + \frac{1}{\alpha^2} + \ldots + \frac{1}{\alpha^n}}\right] = 0.8v_1$$

$$\frac{v_0\left[\frac{1}{1 - \frac{1}{\alpha^2}}\right]}{\left[\frac{1}{1 - \frac{1}{\alpha}}\right]} = 0.8v_1$$

$$\Rightarrow \quad \frac{\alpha}{1 + \alpha} = 0.8$$

$$\Rightarrow \quad \alpha = 4$$

**14.** *(0.63)* Motional emf,

$$e = (\mathbf{v} \times \mathbf{B})\mathbf{d}l = 10^{-2} \times 1 \times 10^{-1}$$

$$e = 10^{-3}\text{ V}$$

$$\tau_L = LR = (10^{-3})(1) = 10^{-3}\text{ s} = 1\text{ ms}$$

$$i = i_0(1 - e^{-t/\tau_L}) = \frac{10^{-3}}{1}(1 - e^{-1})$$

$$i = 10^{-3}(1 - 0.37)$$

$$l = 0.63\,\text{mA}$$

**15.** *(b)* Fundamental frequency is maximum when length is minimum i.e. $L_0$,

**Case 1.** $L = L_0, T = T_0, f = f_0;$

$$f_1 = \frac{1}{2L_0}\sqrt{\frac{T_0}{\mu}}$$

**Case 2.** $f_2 = \frac{1}{L_0}\sqrt{\frac{T_2}{2\mu}} = \frac{f_0}{\sqrt{2}}$

**Case 3.** $f_3 = \frac{1}{L_0}\sqrt{\frac{T_2}{3\mu}} = \frac{f_0}{\sqrt{3}}$

**Case 4.** $f_4 = \frac{1}{L_0}\sqrt{\frac{T_2}{4\mu}} = \frac{f_0}{2}$

**16.** *(c)*

**Case 1.** $L = L_0, T = T_0, f = f_0$

$$f_1 = \frac{1}{2L_0}\sqrt{\frac{T_0}{\mu}}$$

**Case 2.** $L = \frac{3L_0}{2}$

$$f_2 = \frac{3}{2 \times \frac{3L_0}{2}}\sqrt{\frac{T_2}{2\mu}} = f_0$$

$$\Rightarrow \quad f_0 = \frac{1}{2L_0}\sqrt{\frac{T_2}{\mu}} \Rightarrow T_2 = \frac{T_0}{2}$$

**Case 3.** $L = \frac{5L_0}{4}$

$$f_3 = \frac{5}{2 \times \frac{5L_0}{4}}\sqrt{\frac{T_3}{3\mu}} = f_0$$

$$\Rightarrow \quad f_0 = \frac{2}{\sqrt{3}L_0}\sqrt{\frac{T_3}{\mu}} \Rightarrow T_3 = \frac{3T_0}{16}$$

**Case 4.** $L = \frac{7L_0}{4} \Rightarrow f_4 = \frac{14}{2 \times \frac{7L_0}{4}}\sqrt{\frac{T_4}{4\mu}} = f_0$

$$\Rightarrow \quad f_0 = \frac{2}{L_0}\sqrt{\frac{T_4}{\mu}} \Rightarrow T_4 = \frac{T_0}{16}$$

**17.** *(d)* 1-2 process is isothermal and 2-3 process is isochoric.

(I) $W_{1\to 2} = nRI\ln\frac{V_f}{V_i} = 1 \times R\frac{T_0}{3}\ln\frac{V_2}{V_1}$

$$= \frac{RT_0}{3}\ln\frac{2V_0}{V_0} = \frac{RT_0}{3}\ln 2$$

$W_{2\to 3} = 0$ (Isochoric process)

$$W_{1\to 2\to 3} = W_{1\to 2} + W_{2\to 3}$$

$$= \frac{RT_0}{3}\ln 2\,(I \to P)$$

(II) $\Delta U = \frac{f}{2}nR(T_f - T_i)$

$$\Delta U_{1\to 2\to 3} = \frac{3}{2}\left[1 \times R\left(T_0 - \frac{T_0}{3}\right)\right]$$

$= RT_0$ (II $\to R$)

(III) $Q_{1\to 2\to 3} = \Delta U_{1\to 2\to 3} + W_{1\to 2\to 3}$

(First law of thermodynamics)

$$= RT_0 + \frac{RT_0}{3}\ln 2$$

$= \frac{RT_0}{3}[3 + \ln 2]$ (III $\to T$)

(IV) $Q_{1\to2} = \Delta U_{1\to2} + W_{1\to2} = 0 + \frac{RT_0}{3}\ln 2 = \frac{RT_0}{3}\ln 2$ (IV → P)

**18.** (d) (I) $W_{1\to2\to3} = W_{1\to2} + W_{2\to3} = P_0[2V_0 - V_0] + 0 = P_0V_0$

$$W_{1\to2\to3} = P_0V_0 = \frac{RT_0}{3} \qquad (\text{I} \to Q)$$

(II) $U_{1\to2\to3} = \frac{3}{2}\left[\frac{3P_0}{2} \times 2V_0 - P_0V_0\right] = \frac{3}{2} \times 2P_0V_0 = 3P_0V_0 = RT_0$ (II → R)

(III) $Q_{1\to2\to3} = U_{1\to2\to3} + W_{1\to2\to3} = RT_0 + \frac{RT_0}{3} = \frac{4RT_0}{3}$ (III → S)

(IV) $Q_{1\to2} = nC_P\Delta T = n\frac{5}{2}R(T_2 - T_1) = \frac{5}{2}[P_0 2V_0 - P_0V_0] = \frac{5}{2}P_0V_0 = \frac{5}{2}\frac{RT_0}{3} = \frac{5}{6}R_0T_0$

(IV → U)

**19.** (a, b)

$CH_3O-C_6H_4-C\equiv C-CH_2-CHO$ —(i) $Hg^{2+}$, dil. $H_2SO_4$ (Hydration of Alkyne)→ $CH_3O-C_6H_4-C(OH)=CH-CH_2-CHO$

—Keto–enol Tautomerism→ $CH_3O-C_6H_4-CO-CH_2-CH_2-CHO$

—(ii) $AgNO_3+NH_4OH$ (Tollen's reagent) (—CHO converts into $COO^-$)→ $CH_3O-C_6H_4-CO-CH_2-CH_2-COO^-$

—(iii) Zn – Hg + Conc. HCl (Clemmensen reduction) (C=O reduce to —$CH_2$)→ $CH_3O-C_6H_4-CH_2-CH_2-CH_2-COOH$ (Q)

—(i) $SOCl_2$+Pyridine (ii) $AlCl_3$→ Electrophilic substitution reaction → cyclic ketone (R)

—Zn-Hg/Conc. HCl Clemmensen reduction→ (S)

**20.** (c, d) **Key Idea** An aromatic compound must be cyclic and planar. It must follow $(4n+2)e^-$ rule and have the conjugated system in it.

(a) Bromocyclobutene —NaOEt→ Substitution product (OEt) + Ellimination product (cyclobutadiene) —Dimerise→ (Non-aromatic)

(b) Benzene react with $Cl_2$ (excess) in presence of UV light and 500 K of temperature to form benzene hexachloride (non-aromatic).

Benzene + $Cl_2$ (excess) —UV, 500 K→ $C_6H_6Cl_6$

(Non-aromatic)

(c) $CH_3CH(Br)CH_2Br \xrightarrow[\text{(ii) } NaNH_2]{\text{(i) Alc. KOH}} CH_3C \equiv CH$

$\xrightarrow[\text{873 K (Trimerisation)}]{\text{(iii) Red hot iron tube}}$ 1,3,5-trimethylbenzene ($CH_3$, $CH_3$, $CH_3$)

(Aromatic)

(d) Cyclopentadiene $\xrightarrow[-MeOH]{NaOMe}$ cyclopentadienyl anion $\ominus Na^+$

(Aromatic ion)

Thus, (c) and (d) options are correct.

**21.** (a, c) Given, ground state energy of hydrogen atom = − 13.6 eV

Energy of $He^+$ = − 3.4 eV, $Z = 2$

$$\text{Energy of } He^+, E = -\frac{13.6 \times Z^2}{n^2} \text{ eV}$$

$$-3.4 \text{ eV} = \frac{-13.6 \times (2)^2}{n^2}$$

$$\Rightarrow \quad n = \sqrt{\frac{13.6 \times 4}{3.4}} \Rightarrow n = 4$$

Given, azimuthal quantum number ($l$) = 2 ($d$ – subshell

Magnetic quantum number ($m$) = 0

∴ Angular nodes ($l$) = 2

Radial node = $n - l - 1 = 4 - 2 - 1 = 1$

$nl = 4d$ state

Hence, options (a), (c) are correct.

**22.** (b, c, d) When Zn react with hot conc. $H_2SO_4$ then $SO_2$ is released and $ZnSO_4$ is obtained.

$$\underset{\text{(Hot + Conc.)}}{Zn + 2H_2SO_4} \longrightarrow \underset{(G)}{ZnSO_4} + \underset{(R)}{SO_2 \uparrow} + \underset{(X)}{2H_2O}$$

$R(SO_2)$ molecule is V-Shaped

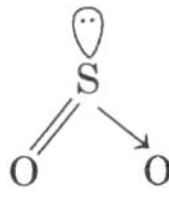


Thus, option (b) is correct.

When Zn is react with conc. NaOH then $H_2$ gas is evolved and $Na_2ZnO_2$ is obtained.

$$Zn + 2NaOH(\text{conc.}) \longrightarrow \underset{(T)}{Na_2ZnO_2} + \underset{(Q)}{H_2\uparrow}$$

In ground state, H—H (Q) (bond order = 1)

Thus, option (c) is correct.

The oxidation state of Zn in $T(Na_2ZnO_2)$ is +2

Thus, option (a) is incorrect.

$$\underset{(G)}{ZnSO_4} + H_2S + NH_4OH \longrightarrow \underset{(Z)}{ZnS\downarrow} + \underset{(X)}{2H_2O} + \underset{(Y)}{(NH_4)_2SO_4}$$

ZnS ($Z$) compound is dirty white coloured.

Thus, option (d) is correct.

**23.** (a, b, c) The explanation of given statements are as follows:

(a) *Aqua-regia* is prepared by mixing conc. HCl and conc. $HNO_3$ in 3:1 ($v/v$) ratio and is used in oxidation of gold and platinum. Hence, option (a) is correct.

(b) Yellow colour of *aqua-regia* is due to its decomposition into NOCl (orange yellow) and $Cl_2$ (greenish yellow). Hence, option (b) is correct.

(c) When gold reacts with *aqua-regia* then it produces $AuCl_4^-$ anion complex in which Au has +3 oxidation state.

$$\overset{0}{Au} + HNO_3 + 4HCl \longrightarrow \overset{+3}{AuCl_4^-} + H_3O^+ + NO + H_2O$$

Oxidation

Hence, option (c) is correct.

(d) Reaction of gold with *aqua-regia* produces NO gas in absence of air. Hence, option (d) is incorrect.

**24.** (a, d) The explanation of given statements are as follows: (a) Teflon is prepared by heating tetrafluoroethene in presence of persulphate catalyst at higher pressure.

$$nCF_2{=}CF_2 \xrightarrow[\text{catalyst}]{\text{Persulphate}} {-\!\!\left(CF_2{-}CF_2\right)\!\!-}$$

Thermoplastic Polymer (PTFE)

Thus, option (a) is correct.

(b) Natural rubber is polyisoprene containing *cis* alkene units.

$$nCH_2{=}C(CH_3){-}CH{=}CH_2 \longrightarrow \left(CH_2(CH_3)C{=}C(H)CH_2\right)_n$$

Isoprene $\qquad$ *cis*-ployisoprene unit (natural rubber)

Thus, option (b) is incorrect.

(c) Cellulose has only β-D-glucose units that are joined together by glycosidic linkages as shown in the following structure:

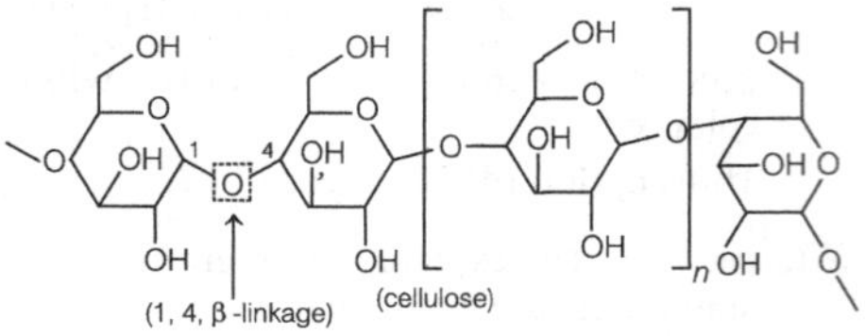

(1, 4, β-linkage) (cellulose)

Thus, option (c) is incorrect.

(d) Nylon-6 has amide linkages.

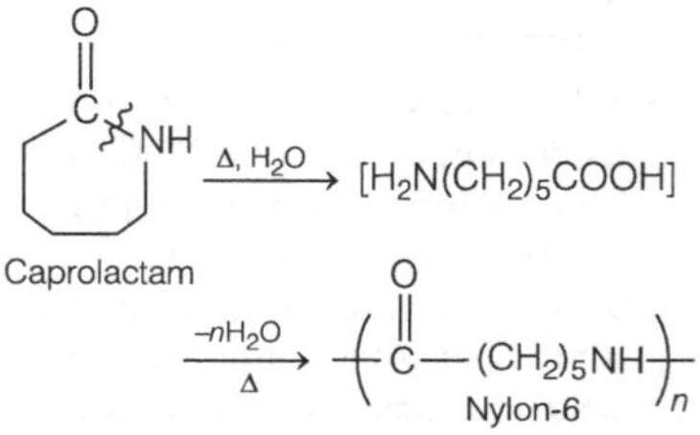

Caprolactam $\xrightarrow{\Delta, H_2O}$ $[H_2N(CH_2)_5COOH]$ $\xrightarrow[\Delta]{-nH_2O}$ $\left[ -\overset{O}{\overset{\|}{C}}-(CH_2)_5NH- \right]_n$ Nylon-6

Thus, option (d) is correct.

**25.** *(a, b, c)* Cyanide process of gold extraction involves leaching out gold from its ore with $CN^-$ in the presence of $O_2$ (*Q*) in water to form $[Au(CN)_2]^-$ (*R*).

When $[Au(CN)_2]^-$ reacts with Zn (*T*), it froms $[Zn(CN)_4]^{2-}$ (*Z*) and Au.

The corresponding reactions are as follows :

$$4Au(s) + 8CN^-(aq) \xrightarrow{H_2O + O_2(Q)} \underset{(R)}{4[Au(CN)_2]^-} + 4OH^-(aq)$$

$$\underset{(R)}{2[Au(CN)_2]^-(aq)} + \underset{(T)}{Zn(s)} \longrightarrow [Zn(CN)_4]^{2-}(aq) + 2Au(s)$$

Hence, options (a, b, c) are correct.

**26.** *(a, c)* The given reactions takes place as follows:

(a) 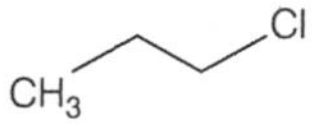 $CH_3CH_2CH_2Cl \xrightarrow{Zn, \text{ dil. } HCl} \underset{\text{(Propane)}}{CH_3—CH_2—CH_3}$

(b) 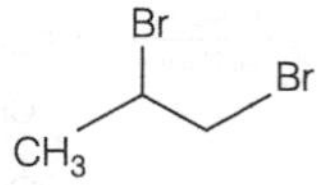 $CH_3CH(Br)CH_2Br \xrightarrow[\Delta]{Zn} \underset{\text{(Propene)}}{CH_3—CH=CH_2} + ZnBr_2$

(Elimination reaction)

(c) 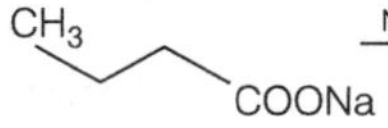 $CH_3CH_2CH_2COONa \xrightarrow[\text{Electrolysis}]{NaOH, CaO, \Delta} \underset{\text{(Propane)}}{CH_3—CH_2—CH_3} + CO_2\uparrow$

(d) 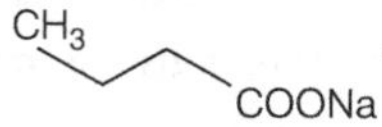 $CH_3CH_2CH_2COONa \xrightarrow{\text{Electrolysis}} \underset{n\text{-hexane}}{CH_3—(CH_2)_4—CH_3}$

Thus, options (a, c) are correct.

**27.** *(2.3)* At constant *V*, *T*

| | $2N_2O_5(g) \xrightarrow{\Delta}$ | $2N_2O_4(g)$ | + $O_2(g)$ |
|---|---|---|---|
| At initial $t = 0$ | 1 | 0 | 0 |
| $t = Y \times 10^3$ sec | $1 - 2p$ | $2p$ | $p$ |

$$p_{\text{Total}} = 1 - 2p + 2p + p$$
$$1.4 = 1 + p$$
$$p = 0.45 \text{ atm}$$

According to first order reaction,

$$k = \frac{2303}{t} \log \frac{p_i}{p_i - 2p}$$
$$p_i = 1 \text{ atm (given)}$$
$$2p = 2 \times 0.45 = 0.9 \text{ atm}$$

On substituting the values in above equation,

$$2k \cdot t = 2.303 \log \frac{1}{1 - 0.9}$$
$$2 \times 5 \times 10^{-4} \times y \times 10^3 = 2.303 \log \frac{1}{0.1}$$
$$y = 2.303 = 2.3$$

**Note** Unit of rate constant (*k*), i.e. $s^{-1}$ represents that it is a first order reaction.

**28.** *(2.98 mole)*

> **Key Idea** Molarity
> $$(M) = \frac{\text{Number of moles of solute} \times 1000}{\text{Volume of solution (in mL)}}$$

Also, volume $= \dfrac{\text{Mass}}{\text{Density}}$

Given, mole fraction of urea ($\chi_{urea}$) = 0.05

Mass of water = 900 g

Density = 1.2 g/cm$^3$

$$\chi_{urea} = \frac{n_{urea}}{n_{urea} + 50}$$

$$\left[\because \text{Moles of water} = \frac{900}{18} = 50\right]$$

$$0.05 = \frac{n_{urea}}{n_{urea} + 50}$$

$$\Rightarrow \quad 19n_{urea} = 50$$

$$n_{urea} = 2.6315 \text{ moles}$$

$$w_{urea} = n_{urea} \times (M \cdot wt)_{urea}$$

$$= (2.6315 \times 60) \text{ g}$$

$$V = \frac{2.6315 \times 60 + 900}{1.2}$$

$$\left[\because \text{Density} = \frac{\text{Mass of solution}}{\text{Volume of solution}}\right]$$

$$= 881.57 \text{ mL}$$

Now, molarity

$$= \text{Number of moles of solute} \times \frac{1000}{\text{Volume of solution (mL)}}$$

$$= \frac{2.6315 \times 1000}{881.57}$$

$$= 2.98 \text{ M}$$

**29.** *(6)*

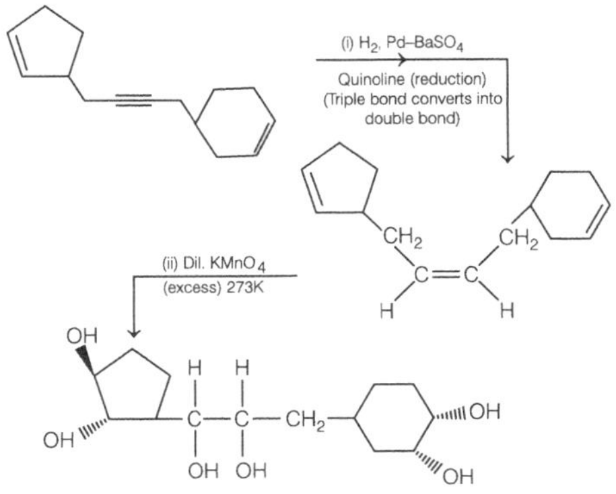

(i) $H_2$, Pd–$BaSO_4$, Quinoline (reduction) (Triple bond converts into double bond)

(ii) Dil. $KMnO_4$ (excess) 273K

(*P*) (Major product)

Compound (*P*) has total number of hydroxyl groups = 6

**30.** *(6)* The structure of *cis*-$[Mn(en)_2Cl_2]$ complex is

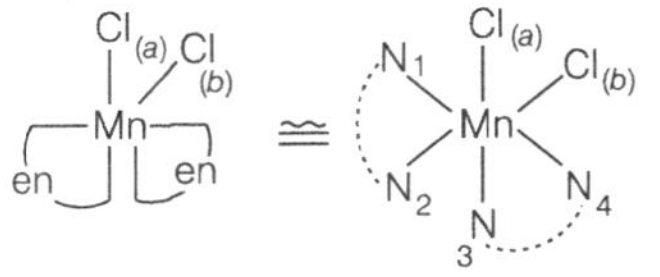

Bond angles (Mn—N and Mn—Cl bond in *cis* positions)

Cl (*a*) —— Mn —— $N_{(1)}$

Cl (*a*) —— Mn —— $N_{(2)}$

Cl (a) —— Mn —— $N_{(4)}$

Cl (*b*) —— Mn —— $N_1$

Cl (*b*) —— Mn —— $N_3$

Cl (*b*) —— Mn —— $N_4$

Number of *cis* Cl—Mn—N = 6

**31.** *(288)* **Key Idea** Rhombic sulphur ($S_8$) gets oxidised into sulphuric acid and water, $NO_2$ gas is released on reaction with conc. $HNO_3$.

When rhombic sulphur ($S_8$) is oxidised by conc. $HNO_3$ then $H_2SO_4$ is obtained and $NO_2$ gas is released.

$$S_8 + 48HNO_3 \longrightarrow 8H_2SO_4 + 48NO_2 + 16\,H_2O$$

1 mole of rhombic sulphur produces = 16 moles of $H_2O$

∴ Mass of water = 16 × 18 (molar mass of $H_2O$)

$$= 288 \text{ g}$$

**32.** *(10.0)* The structure of cyclic ether with molecular formula, $C_4H_8O$ are as follows:

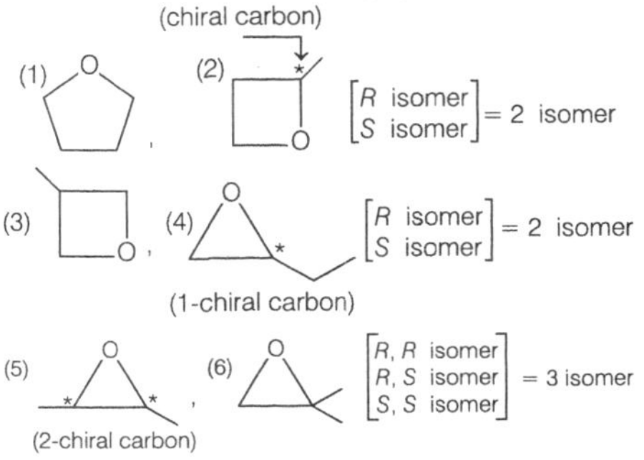

(chiral carbon)

(1) (2) $\begin{bmatrix} R \text{ isomer} \\ S \text{ isomer} \end{bmatrix}$ = 2 isomer

(3) (4) $\begin{bmatrix} R \text{ isomer} \\ S \text{ isomer} \end{bmatrix}$ = 2 isomer

(1-chiral carbon)

(5) (6) $\begin{bmatrix} R, R \text{ isomer} \\ R, S \text{ isomer} \\ S, S \text{ isomer} \end{bmatrix}$ = 3 isomer

(2-chiral carbon)

Total number of isomers of cyclic ether with molecular formula, $C_4H_8O$ are 10.

**33.** *(a)* (III) Kinetic energy of the electron in $n$th orbit,

$$\text{K.E.} = +13.6 \times \frac{Z^2}{n^2} \quad \text{or} \quad \text{K.E.} \propto \frac{1}{n^2} \quad \text{or} \quad \text{K.E.} \propto n^{-2}$$

From list-II, correct match is (III $P$).

(IV) Potential energy of the electron in the $n$th orbit,

$$\text{P.E.} = -2 \times 13.6 \times \frac{Z^2}{n^2} \Rightarrow \text{P.E.} \propto \frac{1}{n^2} \Rightarrow \text{P.E.} \propto n^{-2}$$

From List II, correct match is (IV $P$).

Hence, correct matching from list-I and list-II on the basis of given option is (III, $P$).

**34.** *(c)* (I) Radius of the $n$th orbit, $r = 0.529 \times \frac{n^2}{Z}$

Here, $r \propto n^2$

From list-II, correct match is (I, $T$)

(II) Angular momentum of the electron, $mvr = \frac{nh}{2\pi}$ or $mvr \propto n$

From list-II, correct match (II, $S$)

Hence, correct matching from list-I and list-II on the basis of given option is (I, $T$).

**35.** (b)

(III) [o-(chloromethyl)benzoate, $C_6H_4(CH_2Cl)(COOCH_3)$] $\xrightarrow[\text{Nucleophilic substitution reaction (Cl}^- \text{ replaces by CN}^- \text{ ion)}]{\text{(i) KCN}}$ $C_6H_4(CH_2CN)(COOCH_3)$

$\xrightarrow[\Delta]{\text{(ii) } H_3O^+}$ (Complete hydrolysis of —CN give —COOH and ester also get hydrolysed into —COOH and alcohol) $CH_3OH + C_6H_4(CH_2COOH)(COOH)$ $(T)$

$\xrightarrow{LiAlH_4}$ $C_6H_4(CH_2CH_2OH)(CH_2OH)$ $(Q)$ $\xrightarrow[\Delta]{\text{(iv) conc. } H_2SO_4}$ $H_2O$ + (isochroman) $(R)$

Hence, correct match of (III) are $T$, $Q$, $R$.

(IV) $C_6H_4(COOMe)(CH_2COOMe)$ $\xrightarrow{\text{(i) } LiAlH_4}$ $C_6H_4(CH_2CH_2OH)(CH_2OH)$ $(Q)$ $\xrightarrow[\text{Dehydration}]{\text{(ii) conc. } H_2SO_4}$ (isochroman) $(R)$ $+ H_2O$

Hence, correct match of IV is $Q$, $R$.

Hence, correct matching from list-I and list II on the basis of given option is (IV), Q, R.

**36.** (a)

(I) [2-(1,3-dioxolan-2-yl)phenyl]acetonitrile $\xrightarrow[\text{(ii) Dil. HCl}]{\text{(i) DIBAL–H}}$ benzene ring with $CH_2—\overset{O}{\overset{\|}{C}}—H$ and $CH\langle^{OH}_{OH}$

$\downarrow -H_2O$

benzene ring with $CH_2—\overset{O}{\overset{\|}{C}}H$ and $\underset{O}{\underset{\|}{C}}—H$

$\xrightarrow[\text{(reduction)}]{\text{(iii) NaBH}_4}$ benzene ring with $CH_2OH$ and $CH_2OH$ $(Q)$

$(Q) \xrightarrow[\text{(dehydration)}]{\text{(iv) Conc. } H_2SO_4}$ $(R)$ (isochroman)

Hence, correct match of (I) are $(Q, R)$

(II) $(S)$ 2-allylbenzoic acid (benzene ring with $CH_2CH=CH_2$ and COOH) $\xrightarrow[\text{(Ozonolysis)}]{\text{(i) } O_3\text{, (ii) Zn + } H_2O}$ $(P)$ benzene ring with $CH_2—\overset{O}{\overset{\|}{C}}—H$ and COOH

$\downarrow$ (iii) $NaBH_4$

benzene ring with $CH_2—OH$ and $\underset{O}{\underset{\|}{C}}—OH$ $\xrightarrow[-H_2O]{\text{(iv) Conc. } H_2SO_4}$ $(U)$ (isochroman-1-one)

Hence, correct match of II is $(P, S, U)$.

Hence, correct matching from list-I and list-II on the basis of given option is (II), $P, S, U$.

**37.** (a, b, c) It is given, that for non-negative integers '$n$',

$$f(n)=\frac{\sum_{k=0}^{n}\sin\left(\frac{k+1}{n+2}\pi\right)\sin\left(\frac{k+2}{n+2}\pi\right)}{\sum_{k=0}^{n}\sin^2\left(\frac{k+1}{n+2}\pi\right)}=\frac{\sum_{k=0}^{n}\left(\cos\frac{\pi}{n+2}-\cos\left(\frac{2k+3}{n+2}\pi\right)\right)}{\sum_{k=0}^{n}\left(1-\cos\left(\frac{2k+2}{n+2}\pi\right)\right)}$$

[$\because 2\sin A\sin B=\cos(A-B)-\cos(A+B)$ and $2\sin^2 A=1-\cos 2A$]

$$=\frac{\left(\cos\left(\frac{\pi}{n+2}\right)\right)\sum_{k=0}^{n}1-\left\{\cos\frac{3\pi}{n+2}+\cos\frac{5\pi}{n+2}+\cos\frac{7\pi}{n+2}+\ldots..+\cos\left(\frac{2n+3}{n+2}\pi\right)\right\}}{\sum_{k=0}^{n}1-\left\{\cos\frac{2\pi}{n+2}+\cos\frac{4\pi}{n+2}+\cos\frac{6\pi}{n+2}+\ldots..+\cos\left(\frac{2n+2}{n+2}\pi\right)\right\}}$$

$$= \frac{(n+1)\cos\left(\frac{\pi}{n+2}\right) - \frac{\sin\left(\frac{n\pi}{n+2}\right)}{\sin\left(\frac{\pi}{n+2}\right)}\cos\left(\frac{n+3}{n+2}\pi\right)}{(n+1) - \frac{\sin\left(\frac{n\pi}{n+2}\right)}{\sin\left(\frac{\pi}{n+2}\right)}\cos\left(\frac{n+2}{n+2}\pi\right)}$$

$$\left[\because \cos(\alpha) + \cos(\alpha+\beta) + \cos(\alpha+2\beta) + \ldots + \cos(\alpha+(n-1)\beta) = \frac{\sin\left(\frac{n\beta}{2}\right)}{\sin\left(\frac{\beta}{2}\right)}\cos\left(\frac{2\alpha+(n-1)\beta}{2}\right)\right]$$

$$= \frac{(n+1)\cos\left(\frac{\pi}{n+2}\right) - \frac{\sin\left(\pi - \frac{\pi}{n+2}\right)}{\sin\left(\frac{\pi}{n+2}\right)}\cos\left(\pi + \frac{\pi}{n+2}\right)}{(n+1) - \frac{\sin\left(\pi - \frac{\pi}{n+2}\right)}{\sin\left(\frac{\pi}{n+2}\right)}\cos(\pi)}$$

$$= \frac{(n+1)\cos\left(\frac{\pi}{n+2}\right) + \frac{\sin\left(\frac{\pi}{n+2}\right)}{\sin\left(\frac{\pi}{n+2}\right)}\cos\left(\frac{\pi}{n+2}\right)}{(n+1) + \frac{\sin\left(\frac{\pi}{n+2}\right)}{\sin\left(\frac{\pi}{n+2}\right)}}$$

$$= \frac{(n+2)\cos\left(\frac{\pi}{n+2}\right)}{(n+2)} = \cos\left(\frac{\pi}{n+2}\right)$$

$\Rightarrow \quad f(n) = \cos\left(\frac{\pi}{n+2}\right)$

Now, $f(6) = \cos\left(\frac{\pi}{8}\right)$

$\because \quad \alpha = \tan(\cos^{-1} f((6))) = \tan\left(\frac{\pi}{8}\right)$ $\quad \left\{\because \cos^{-1}\cos x = x \text{ if } x \in \left(0, \frac{\pi}{2}\right)\right\}$

$= \sqrt{2} - 1$

$\Rightarrow \; (\alpha + 1) = \sqrt{2} \Rightarrow (\alpha+1)^2 = 2 \Rightarrow \alpha^2 + 2\alpha + 1 = 2$

$\Rightarrow \; \alpha^2 + 2\alpha - 1 = 0$

Now, $f(4) = \cos\left(\frac{\pi}{4+2}\right) = \cos\left(\frac{\pi}{6}\right) = \frac{\sqrt{3}}{2}$,

Now,

$$\sin(7\cos^{-1} f(5)) = \sin\left(7\cos^{-1}\left(\cos\left(\frac{\pi}{5+2}\right)\right)\right)$$
$$= \sin\left(7\left(\frac{\pi}{7}\right)\right) = \sin\pi = 0$$

and Now,

$$\lim_{n\to\infty} f(x) = \lim_{n\to\infty} \cos\frac{\pi}{n+2} = \cos 0 = 1$$

Hence, options (a), (b) and (c) are correct.

**38.** (a, b, d) Given, $f: R \to R$ and

$f(x) = (x-1)(x-2)(x-5)$

Since, $F(x) = \int_0^x f(t)\,dt,\ x > 0$

So, $F'(x) = f(x) = (x-1)(x-2)(x-5)$

According to wavy curve method

$$\begin{array}{ccccccc} - & & + & & - & & + \\ \hline & 1 & & 2 & & 5 & \end{array}$$

$F'(x)$ changes, it's sign from negative to positive at $x = 1$ and 5, so, $F(x)$ has minima at $x = 1$ and 5 and as $F'(x)$ changes, it's sign from positive to negative at $x = 2$, so $F(x)$ has maxima at $x = 2$.

$$\because F(2) = \int_0^2 f(t)\,dt = \int_0^2 (t^3 - 8t^2 + 17t - 10)dt$$
$$= \left[\frac{t^4}{4} - 8\frac{t^3}{3} + 17\frac{t^2}{2} - 10t\right]_0^2$$
$$= 4 - \frac{64}{3} + 34 - 20 = 38 - \frac{124}{3} = -\frac{10}{3}$$

$\because$ At the point of maxima $x = 2$, the functional value $F(2) = -\frac{10}{3}$, is negative for the interval $x \in (0, 5)$, so $F(x) \neq 0$ for any value of $x \in (0, 5)$,

Hence, options (a), (b) and (d) are correct.

**39.** (a, b, d) Given, $f(x) = \frac{\sin(\pi x)}{x^2}$, $x > 0$

$$\Rightarrow f'(x) = \frac{x^2\pi\cos(\pi x) - 2x\sin(\pi x)}{x^4}$$
$$= \frac{2x\cos(\pi x)\left[\frac{x\pi}{2} - \tan(\pi x)\right]}{x^4}$$
$$= \frac{2\cos(\pi x)\left[\frac{x\pi}{2} - \tan(\pi x)\right]}{x^3}$$

Since, for maxima and minima of $f(x)$, $f'(x) = 0$

$\Rightarrow \cos(\pi x) = 0$ or $\tan(\pi x) = \frac{\pi x}{2}$, (as $x > 0$)

$\because \cos(\pi x) \neq 0 \Rightarrow \tan(\pi x) = \frac{\pi x}{2}$

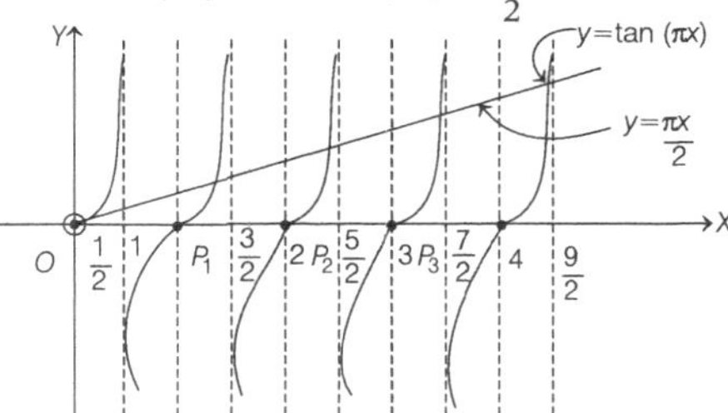


$\because f'(P_1^-) < 0$ and $f'(P_1^+) > 0$

$\Rightarrow x = P_1 \in \left(1, \frac{3}{2}\right)$

is point of local minimum.

$\because f'(P_2^-) > 0$ and $f'(P_2^+) < 0$

$\Rightarrow x = P_2 \in \left(2, \frac{5}{2}\right)$

is point of local maximum.

From the graph, for points of maxima $x_1, x_2, x_3$ ...... it is clear that

$\frac{5}{2} - x_1 > \frac{9}{2} - x_2 > \frac{13}{2} - x_3 > \frac{17}{2} - x_4$ ......

$\Rightarrow x_{n+1} - x_n > 2, \forall n.$

From the graph for points of minima $y_1, y_2, y_3$ ....., it is clear that

$\frac{3}{2} - y_1 > \frac{5}{2} - x_1 > \frac{7}{2} - y_2 > \frac{9}{2} - x_2$ ......

$|x_n - y_n| > 1, \forall n$ and $x_1 > (y_1 + 1)$

And $x_1 \in \left(2, \frac{5}{2}\right)$, $x_2 \in \left(4, \frac{9}{2}\right)$, $x_3 \in \left(6, \frac{13}{2}\right)$

........

$\Rightarrow x_n \in \left(2n, 2n + \frac{1}{2}\right), \forall n.$

Hence, options (a), (b) and (d) are correct.

**40.** (c, d)

> ***Key Idea*** Points, *A*, *B*, *C* are collinear $\Rightarrow$ **AB**, **BC** are collinear vectors $\Rightarrow$ **AB** $= \lambda$**BC** for some non-zero scalar $\lambda$.

Given lines,

$L_1 : \mathbf{r} = \lambda\hat{\mathbf{i}}, \lambda \in R$ ... (i)

$L_2 : \mathbf{r} = \mu\hat{\mathbf{j}} + \hat{\mathbf{k}}, \mu \in R$ ... (ii)

and $L_3 : \mathbf{r} = \hat{\mathbf{i}} + \hat{\mathbf{j}} + v\hat{\mathbf{k}}, v \in R$ ... (iii)

Now, let the point $P$ on $L_1 = (\lambda, 0, 0)$

the point $Q$ on $L_2 = (0, \mu, 1)$, and

the point $R$ on $L_3 = (1, 1, v)$

For collinearity of points $P, Q$ and $R$, there should be a non-zero scalar '$m$', such that

$\mathbf{PQ} = m\,\mathbf{PR}$

$$\Rightarrow (-\lambda\hat{\mathbf{i}} + \mu\hat{\mathbf{j}} + \hat{\mathbf{k}}) = m[(1-\lambda)\hat{\mathbf{i}} + \hat{\mathbf{j}} + v\hat{\mathbf{k}}]$$

$$\Rightarrow \frac{\lambda}{\lambda - 1} = \frac{\mu}{1} = \frac{1}{v}$$

$$\Rightarrow v = \frac{1}{\mu} \text{ and } \lambda = \frac{\mu}{\mu - 1}$$

where, $\mu \neq 0$ and $\mu \neq 1$

$$\Rightarrow Q \neq \hat{\mathbf{k}} \text{ and } Q \neq \hat{\mathbf{k}} + \hat{\mathbf{j}}$$

Hence, $Q$ can not have coordinater (0, 0, 1) and (0, 1, 1)

Hence, options (c) and (d) are correct.

**41.** *(c, d)* Since,

$$\lim_{n \to \infty} \left[ \frac{1 + \sqrt[3]{2} + \sqrt[3]{3} + \dots + \sqrt[3]{n}}{n^{7/3}\left(\frac{1}{(an+1)^2} + \frac{1}{(an+2)^2} + \dots + \frac{1}{(an+n)^2}\right)} \right],$$

$a \in R, |a| > 1$

$$= \lim_{n \to \infty} \frac{\sum_{r=1}^{n} (r^{1/3})}{n^{7/3} \sum_{r=1}^{n} \frac{1}{(an+r)^2}}$$

$$= \lim_{n \to \infty} \frac{\sum_{r=1}^{n} \left(\frac{r}{n}\right)^{1/3} \frac{1}{n}}{\sum_{r=1}^{n} \frac{1}{\left(a + \frac{r}{n}\right)^2} \frac{1}{n}}$$

$$= \frac{\int_0^1 x^{1/3} dx}{\int_0^1 \frac{dx}{(a+x)^2}} = 54, \quad \text{(given)}$$

$$\Rightarrow \frac{\frac{3}{4}[x^{4/3}]_0^1}{\left[-\frac{1}{x+a}\right]_0^1} = 54$$

$$\Rightarrow \frac{3/4}{-\frac{1}{a+1} + \frac{1}{a}} = 54$$

$$\Rightarrow \frac{3}{4 \times 54} = \frac{1}{a(a+1)} \Rightarrow a^2 + a = 72$$

$$\Rightarrow a^2 + 9a - 8a - 72 = 0$$

$$\Rightarrow a(a+9) - 8(a+9) = 0$$

$$\Rightarrow (a-8)(a+9) = 0$$

$$\Rightarrow a = 8 \text{ or } -9$$

Hence, options (c) and (d) are correct.

**42.** *(b, c)* It is given, that $f : R \to R$ and

Property 1 : $\lim_{h \to 0} \frac{f(h) - f(0)}{\sqrt{|h|}}$ exists and finite, and

Property 2 : $\lim_{h \to 0} \frac{f(h) - f(0)}{h^2}$ exists and finite.

**Option a,**

$$\text{P2} : \lim_{h \to 0} \frac{\sin h - \sin 0}{h^2} = \lim_{h \to 0} \frac{1}{h}\left(\frac{\sin h}{h}\right)$$

= doesn't exist.

**Option b,**

$$P1 : \lim_{h \to 0} \frac{h^{2/3} - 0}{\sqrt{|h|}} = \lim_{h \to 0} h^{2/3 - 1/2}$$

$$= \lim_{h \to 0} h^{1/6} = 0$$

exists and finite.

**Option c,**

$$P1 : \lim_{h \to 0} \frac{|h| - 0}{\sqrt{|h|}} = \lim_{h \to 0} \sqrt{|h|} = 0, \text{ exists}$$

and finite.

**Option d,**

$$P2 : \lim_{h \to 0} \frac{h|h| - 0}{h^2} = \lim_{h \to 0} \frac{|h|}{h}$$

$$= \begin{cases} 1, & \text{if } h \to 0^+ \\ -1, & \text{if } h \to 0^- \end{cases}$$

So $\lim_{h \to 0} \frac{f(h) - f(0)}{h^2}$ does not exist.

Hence, options (b) and (c) are correct.

**43.** *(b, d)* It is given, that matrices

$$P = \begin{bmatrix} 1 & 1 & 1 \\ 0 & 2 & 2 \\ 0 & 0 & 3 \end{bmatrix}, Q = \begin{bmatrix} 2 & x & x \\ 0 & 4 & 0 \\ x & x & 6 \end{bmatrix}$$

$$\therefore \quad P^{-1} = \frac{\text{adj}(P)}{|P|}$$

as $|P| = 6$ and adj $P = \begin{bmatrix} 6 & 0 & 0 \\ -3 & 3 & 0 \\ 0 & -2 & 2 \end{bmatrix}^T$

$$\Rightarrow \quad P^{-1} = \frac{1}{6}\begin{bmatrix} 6 & -3 & 0 \\ 0 & 3 & -2 \\ 0 & 0 & 2 \end{bmatrix}$$

$\therefore \; |R| = |PQP^{-1}| \quad [\because R = PQP^{-1} \text{ (given)}]$

$\Rightarrow |R| = |P|\,|Q|\,|P^{-1}| = |Q|$

$[\because |P|\,|P^{-1}| = |I| = 1]$

$$= \begin{vmatrix} 2 & x & x \\ 0 & 4 & 0 \\ x & x & 6 \end{vmatrix} = \begin{vmatrix} 2 & x & x \\ 0 & 4 & 0 \\ x & x & 5 \end{vmatrix} + \begin{vmatrix} 2 & x & 0 \\ 0 & 4 & 0 \\ x & x & 1 \end{vmatrix}$$

$$= \begin{vmatrix} 2 & x & x \\ 0 & 4 & 0 \\ x & x & 5 \end{vmatrix} + 2(4-0) - x(0-0) + 0(0-4x)$$

$$= \begin{vmatrix} 2 & x & x \\ 0 & 4 & 0 \\ x & x & 5 \end{vmatrix} + 8, \text{ for all } x \in R$$

$$\because \quad PQ = \begin{bmatrix} 1 & 1 & 1 \\ 0 & 2 & 2 \\ 0 & 0 & 3 \end{bmatrix}\begin{bmatrix} 2 & x & x \\ 0 & 4 & 0 \\ x & x & 6 \end{bmatrix}$$

$$= \begin{bmatrix} 2+x & 4+2x & x+6 \\ 2x & 2x+8 & 12 \\ 3x & 3x & 18 \end{bmatrix}$$

$$\text{and } QP = \begin{bmatrix} 2 & x & x \\ 0 & 4 & 0 \\ x & x & 6 \end{bmatrix}\begin{bmatrix} 1 & 1 & 1 \\ 0 & 2 & 2 \\ 0 & 0 & 3 \end{bmatrix}$$

$$= \begin{bmatrix} 2 & 2+2x & 2+5x \\ 0 & 8 & 8 \\ x & 3x & 3x+18 \end{bmatrix}$$

There is no common value of '$x$', for which each corresponding element of matrices $PQ$ and $QP$ is equal.

$$\text{For } x = 0, Q = \begin{bmatrix} 2 & 0 & 0 \\ 0 & 4 & 0 \\ 0 & 0 & 6 \end{bmatrix}$$

$$\text{then, if } R\begin{bmatrix} 1 \\ a \\ b \end{bmatrix} = 6\begin{bmatrix} 1 \\ a \\ b \end{bmatrix}$$

$$\Rightarrow PQP^{-1}\begin{bmatrix} 1 \\ a \\ b \end{bmatrix} = 6\begin{bmatrix} 1 \\ a \\ b \end{bmatrix} \quad [\because R = PQP^{-1}]$$

$$\Rightarrow \frac{1}{6}\begin{bmatrix} 1 & 1 & 1 \\ 0 & 2 & 2 \\ 0 & 0 & 3 \end{bmatrix}\begin{bmatrix} 2 & 0 & 0 \\ 0 & 4 & 0 \\ 0 & 0 & 6 \end{bmatrix}\begin{bmatrix} 6 & -3 & 0 \\ 0 & 3 & -2 \\ 0 & 0 & 2 \end{bmatrix}\begin{bmatrix} 1 \\ a \\ b \end{bmatrix} = 6\begin{bmatrix} 1 \\ a \\ b \end{bmatrix}$$

$$\Rightarrow \quad \frac{1}{6}\begin{bmatrix} 2 & 4 & 6 \\ 0 & 8 & 12 \\ 0 & 0 & 18 \end{bmatrix}\begin{bmatrix} 6 & -3 & 0 \\ 0 & 3 & -2 \\ 0 & 0 & 2 \end{bmatrix}\begin{bmatrix} 1 \\ a \\ b \end{bmatrix} = 6\begin{bmatrix} 1 \\ a \\ b \end{bmatrix}$$

$$\Rightarrow \quad \begin{bmatrix} 12 & 6 & 4 \\ 0 & 24 & 8 \\ 0 & 0 & 36 \end{bmatrix}\begin{bmatrix} 1 \\ a \\ b \end{bmatrix} = 36\begin{bmatrix} 1 \\ a \\ b \end{bmatrix}$$

$$\Rightarrow \quad \begin{bmatrix} 12+6a+4b \\ 0+24a+8b \\ 0+0+36b \end{bmatrix} = \begin{bmatrix} 36 \\ 36a \\ 36b \end{bmatrix}$$

$\Rightarrow \quad 6a + 4b = 24 \quad \text{and} \quad 12a = 8b$

$\Rightarrow \quad 3a + 2b = 12 \quad \text{and} \quad 3a = 2b$

$\Rightarrow \quad a = 2 \text{ and } b = 3$

So $a + b = 5$.

$$\text{Now, } R\begin{bmatrix} \alpha \\ \beta \\ \gamma \end{bmatrix} = \begin{bmatrix} 0 \\ 0 \\ 0 \end{bmatrix}$$

and $\alpha\hat{i} + \beta\hat{j} + \gamma\hat{k}$ is a unit vector,

so det $(R) = 0$

$\Rightarrow \det(Q) = 0 \quad [\because R = PQP^{-1} \text{ So, } |R| = |Q|]$

$$\Rightarrow \begin{vmatrix} 2 & x & x \\ 0 & 4 & 0 \\ x & x & 6 \end{vmatrix} = 0$$

$\Rightarrow \; 2(24-0) - x(0-0) + x(0-4x) = 0$

$\Rightarrow \quad 48 - 4x^2 = 0$

$\Rightarrow \quad x^2 = 12 \Rightarrow x = \pm 2\sqrt{3}$

So, for $x = 1$, there does not exist a unit vector $\alpha\hat{i} + \beta\hat{j} + \gamma\hat{k}$, for which $R\begin{bmatrix} \alpha \\ \beta \\ \gamma \end{bmatrix} = \begin{bmatrix} 0 \\ 0 \\ 0 \end{bmatrix}$

Hence, options (b) and (d) are correct.

**44.** *(a, b, d)* Given matrices,

$$P_1 = I = \begin{bmatrix} 1 & 0 & 0 \\ 0 & 1 & 0 \\ 0 & 0 & 1 \end{bmatrix}, P_2 = \begin{bmatrix} 1 & 0 & 0 \\ 0 & 0 & 1 \\ 0 & 1 & 0 \end{bmatrix},$$

$$P_3 = \begin{bmatrix} 0 & 1 & 0 \\ 1 & 0 & 0 \\ 0 & 0 & 1 \end{bmatrix}, P_4 = \begin{bmatrix} 0 & 1 & 0 \\ 0 & 0 & 1 \\ 1 & 0 & 0 \end{bmatrix},$$

$$P_5 = \begin{bmatrix} 0 & 0 & 1 \\ 1 & 0 & 0 \\ 0 & 1 & 0 \end{bmatrix}, P_6 = \begin{bmatrix} 0 & 0 & 1 \\ 0 & 1 & 0 \\ 1 & 0 & 0 \end{bmatrix}$$

and $X = \sum_{K=1}^{6} P_k \begin{bmatrix} 2 & 1 & 3 \\ 1 & 0 & 2 \\ 3 & 2 & 1 \end{bmatrix} P_K^T$

$\because \quad P_1^T = P_1, P_2^T = P_2, P_3^T = P_3,$

$P_4^T = P_5, P_5^T = P_4$

and $P_6^T = P_6$

and Let $Q = \begin{bmatrix} 2 & 1 & 3 \\ 1 & 0 & 2 \\ 3 & 2 & 1 \end{bmatrix}$

and $\because Q^T = Q$

Now,

$$X = (P_1QP_1^T) + (P_2QP_2^T) + (P_3QP_3^T) + (P_4QP_4^T) + (P_5QP_5^T) + (P_6QP_6^T)$$

So, $X^T = (P_1QP_1^T)^T + (P_2QP_2^T)^T + (P_3QP_3^T)^T + (P_4QP_4^T)^T + (P_5QP_5^T)^T + (P_6QP_6^T)^T$

$$= P_1QP_1^T + P_2QP_2^T + P_3QP_3^T + P_4QP_4^T + P_5QP_5^T + P_6QP_6^T$$

$[\because (ABC)^T = C^TB^TA^T$ and $(A^T)^T = A$ and $Q^T = Q]$

$\Rightarrow X^T = X$

$\Rightarrow X$ is a symmetric matrix.

The sum of diagonal entries of $X = Tr(X)$

$$= \sum_{i=1}^{6} T_r(P_i QP_i^T) = \sum_{i=1}^{6} T_r(QP_i^TP_i)$$

$[\because T_r(ABC) = T_r(BCA)]$

$$= \sum_{i=1}^{6} T_r(QI)$$

[$\because P_i$'s are orthogonal matrices]

$$= \sum_{i=1}^{6} T_r(Q) = 6\,T_r(Q) = 6 \times 3 = 18$$

Now, Let $R = \begin{bmatrix} 1 \\ 1 \\ 1 \end{bmatrix}$, then

$$XR = \sum_{K=1}^{6} (P_KQP_K^T)R = \sum_{K=1}^{6} (P_KQP_K^TR)$$

$$= \sum_{K=1}^{6} (P_KQR) \qquad [\because P_K^T R = R]$$

$$= \sum_{K=1}^{6} P_K \begin{bmatrix} 6 \\ 3 \\ 6 \end{bmatrix}$$

$$= \sum_{K=1}^{6} P_K \begin{bmatrix} 6 \\ 3 \\ 6 \end{bmatrix} = \begin{bmatrix} 2 & 2 & 2 \\ 2 & 2 & 2 \\ 2 & 2 & 2 \end{bmatrix} \begin{bmatrix} 6 \\ 3 \\ 6 \end{bmatrix}$$

$$\Rightarrow XR = \begin{bmatrix} 30 \\ 30 \\ 30 \end{bmatrix} \Rightarrow XR = 30R$$

$$\Rightarrow \quad X \begin{bmatrix} 1 \\ 1 \\ 1 \end{bmatrix} = 30 \begin{bmatrix} 1 \\ 1 \\ 1 \end{bmatrix}$$

$\Rightarrow (X - 30I)\,R = 0 \Rightarrow |X - 30I| = 0$

So, $(X - 30I)$ is not invertible and value of $\alpha = 30$.

Hence, options (a), (b) and (d) are correct.

**45.** *(18)* Given vectors $\mathbf{a} = 2\hat{\mathbf{i}} + \hat{\mathbf{j}} - \hat{\mathbf{k}}$

and $\mathbf{b} = \hat{\mathbf{i}} + 2\hat{\mathbf{j}} + \hat{\mathbf{k}}$

So, $\mathbf{a} + \mathbf{b} = 3\hat{\mathbf{i}} + 3\hat{\mathbf{j}} \Rightarrow |\mathbf{a} + \mathbf{b}| = 3\sqrt{2}$

Since, it is given that projection of $\mathbf{c} = \alpha\mathbf{a} + \beta\mathbf{b}$ on the vector $(\mathbf{a} + \mathbf{b})$ is $3\sqrt{2}$, then

$$\frac{(\mathbf{a} + \mathbf{b}).\mathbf{c}}{|\mathbf{a} + \mathbf{b}|} = 3\sqrt{2}$$

$\Rightarrow \quad (\mathbf{a} + \mathbf{b}) . (\alpha\mathbf{a} + \beta\mathbf{b}) = 18$

$\Rightarrow \alpha(\mathbf{a}.\mathbf{a}) + \beta(\mathbf{a}.\mathbf{b}) + \alpha(\mathbf{b}.\mathbf{a}) + \beta(\mathbf{a}.\mathbf{b}) = 18$

$\Rightarrow 6\alpha + 3\beta + 3\alpha + 6\beta = 18$

$\Rightarrow 9\alpha + 9\beta = 18 \Rightarrow (\alpha + \beta) = 2 \qquad \dots$ (i)

Now, for minimum value of $(\mathbf{c} - (\mathbf{a} \times \mathbf{b})) . \mathbf{c}$

$= (\alpha\mathbf{a} + \beta\mathbf{b} - (\mathbf{a} \times \mathbf{b})) . (\alpha\mathbf{a} + \beta\mathbf{b})$

$= \alpha^2 (\mathbf{a}.\mathbf{a}) + \alpha\beta(\mathbf{a}.\mathbf{b}) + \alpha\beta(\mathbf{a}.\mathbf{b}) + \beta^2(\mathbf{b}.\mathbf{b})$

$[\because (\mathbf{a} \times \mathbf{b}) . \mathbf{a} = 0 = (\mathbf{a} \times \mathbf{b}) . \mathbf{b}]$

$= 6\alpha^2 + 6\alpha\beta + 6\beta^2 = 6(\alpha^2 + \beta^2 + \alpha\beta)$

$= 6[(\alpha + \beta)^2 - \alpha\beta] = 6[4 - \alpha\beta]$

$= 6[4 - \alpha(2 - \alpha)]$

$= 6[4 - 2\alpha + \alpha^2]$

The minimum value of

$$6(4 - 2\alpha + \alpha^2) = 6(3) = 18$$

[As minimum value of $ax^2 + bx + c = -\frac{D}{4a}$, if $a > 0$]

**46.** (1523) Given sample space $S = \{1, 2, 3, 4, 5, 6\}$ and let there are $i$ elements in set $A$ and $j$ elements in set $B$.

Now, according to information $1 \le j < i \le 6$. So, total number of ways of choosing sets $A$ and

$$B = \sum_{1 \le j < i \le 6} \sum {}^6C_i \, {}^6C_j$$

$$= \frac{\left(\sum_{r=1}^{6} {}^6C_r\right)^2 - \sum_{r=1}^{6} ({}^6C_r)^2}{2}$$

$$= \frac{(2^6 - 1)^2 - ({}^{12}C_6 - 1)}{2}$$

$$= \frac{(63)^2 - \frac{12!}{6!6!} + 1}{2}$$

$$= \frac{3969 - 924 + 1}{2} = \frac{3046}{2} = 1523$$

**47.** (6.20) It is given that

$$\begin{vmatrix} \sum_{k=0}^{n} k & \sum_{k=0}^{n} {}^nC_k k^2 \\ \sum_{k=0}^{n} {}^nC_k . k & \sum_{k=0}^{n} {}^nC_k 3^k \end{vmatrix} = 0$$

$$\Rightarrow \begin{vmatrix} \frac{n(n+1)}{2} & n(n+1)2^{n-2} \\ n.2^{n-1} & 4^n \end{vmatrix} = 0$$

$$\left[\because \sum_{k=0}^{n} k = \frac{n(n+1)}{2}, \quad \sum_{k=0}^{n} {}^nC_k k = n.2^{n-1}, \quad \sum_{k=0}^{n} {}^nC_k k^2 = n(n+1)2^{n-2} \text{ and } \sum_{k=0}^{n} {}^nC_k 3^k = 4^n\right]$$

$$\Rightarrow \frac{n(n+1)}{2} 4^n - n^2(n+1) 2^{2n-3} = 0$$

$$\Rightarrow \frac{4^n}{2} - n\frac{4^{n-1}}{2} = 0$$

$$\Rightarrow n = 4$$

$$\therefore \sum_{k=0}^{n} \frac{{}^nC_k}{k+1} = \sum_{k=0}^{4} \frac{{}^4C_k}{k+1}$$

$$= \frac{1}{5} \sum_{k=0}^{4} {}^5C_{k+1} = \frac{1}{5}(2^5 - 1)$$

$$= \frac{1}{5}(32 - 1) = \frac{31}{5} = 6.20$$

**48.** (30) Given that, no two persons sitting adjacent in circular arrangement, have hats of same colour. So, only possible combination due to circular arrangement is $2 + 2 + 1$.

So, there are following three cases of selecting hats are

$2R + 2B + 1G$ or $2B + 2G + 1R$ or $2G + 2R + 1B$.

To distribute these 5 hats first we will select a person which we can done in ${}^5C_1$ ways and distribute that hat which is one of it's colour. And, now the remaining four hats can be distributed in two ways.

So, total ways will be $3 \times {}^5C_1 \times 2$

$$= 3 \times 5 \times 2 = 30$$

**49.** (0.5) **Key Idea** Use property $\int_0^a f(x)\,dx = \int_0^a f(a - x)\,dx$

The given integral

$$I = \int_0^{\pi/2} \frac{3\sqrt{\cos\theta}}{(\sqrt{\cos\theta} + \sqrt{\sin\theta})^5} d\theta \quad \ldots \text{(i)}$$

$$\Rightarrow I = \int_0^{\pi/2} \frac{3\sqrt{\sin\theta}}{(\sqrt{\sin\theta} + \sqrt{\cos\theta})^5} d\theta \quad \ldots \text{(ii)}$$

[Using the property $\int_0^a f(x)\,dx = \int_0^a f(a - x)\,dx$]

Now, on adding integrals (i) and (ii), we get

$$2I = \int_0^{\pi/2} \frac{3}{(\sqrt{\sin\theta} + \sqrt{\cos\theta})^4} d\theta$$

$$= \int_0^{\pi/2} \frac{3\sec^2\theta}{(1 + \sqrt{\tan\theta})^4} d\theta$$

Now, let $\tan\theta = t^2 \Rightarrow \sec^2\theta \, d\theta = 2t\,dt$

and at $\theta = \frac{\pi}{2}$, $t \to \infty$

and at $\theta = 0$, $t \to 0$

So, $2I = \int_0^\infty \frac{6t\,dt}{(1 + t)^4} = 6\int_0^\infty \frac{t + 1 - 1}{(t + 1)^4} dt$

$$\Rightarrow I = 3\left[\int_0^\infty \frac{dt}{(t + 1)^3} - \int_0^\infty \frac{dt}{(t + 1)^4}\right]$$

$$= 3\left[-\frac{1}{2(t + 1)^2} + \frac{1}{3(t + 1)^3}\right]_0^\infty$$

$$\Rightarrow I = 3\left[\frac{1}{2} - \frac{1}{3}\right] = 3\left(\frac{1}{6}\right) = \frac{1}{2} \Rightarrow I = 0.5$$

**50.** *(0)* $\because \sum_{k=0}^{10} \sec\left(\frac{7\pi}{12}+\frac{k\pi}{2}\right)\sec\left(\frac{7\pi}{12}+\frac{(k+1)\pi}{2}\right)$

$$= \sum_{k=0}^{10} \frac{1}{\cos\left(\frac{7\pi}{12}+\frac{k\pi}{2}\right)\cos\left(\frac{7\pi}{12}+\frac{(k+1)\pi}{2}\right)}$$

$$= \sum_{k=0}^{10} \frac{\sin\left[\left(\frac{7\pi}{12}+\frac{(k+1)\pi}{2}\right)-\left(\frac{7\pi}{12}+\frac{k\pi}{2}\right)\right]}{\cos\left(\frac{7\pi}{12}+\frac{k\pi}{2}\right)\cos\left(\frac{7\pi}{12}+\frac{(k+1)\pi}{2}\right)}$$

$$\left[\because \frac{7\pi}{12}+\frac{(k+1)\pi}{2}-\left(\frac{7\pi}{12}+\frac{k\pi}{2}\right)=\frac{\pi}{2} \text{ and } \sin\frac{\pi}{2}=1\right]$$

$$= \sum_{k=0}^{10} \frac{\sin\left(\frac{7\pi}{12}+\frac{(k+1)\pi}{2}\right)\cos\left(\frac{7\pi}{12}+\frac{k\pi}{2}\right) - \sin\left(\frac{7\pi}{12}+\frac{k\pi}{2}\right)\cos\left(\frac{7\pi}{12}+\frac{(k+1)\pi}{2}\right)}{\cos\left(\frac{7\pi}{12}+\frac{k\pi}{2}\right)\cos\left(\frac{7\pi}{12}+\frac{(k+1)\pi}{2}\right)}$$

$$= \sum_{k=0}^{10}\left[\tan\left(\frac{7\pi}{12}+\frac{(k+1)\pi}{2}\right)-\tan\left(\frac{7\pi}{12}+\frac{k\pi}{2}\right)\right]$$

$$= \tan\left(\frac{7\pi}{12}+\frac{\pi}{2}\right)-\tan\left(\frac{7\pi}{12}\right) + \tan\left(\frac{7\pi}{12}+\frac{2\pi}{2}\right)-\tan\left(\frac{7\pi}{12}+\frac{\pi}{2}\right)$$

$$\vdots$$

$$+ \tan\left(\frac{7\pi}{12}+\frac{11\pi}{2}\right)-\tan\left(\frac{7\pi}{12}+\frac{10\pi}{2}\right)$$

$$= \tan\left(\frac{7\pi}{12}+\frac{11\pi}{2}\right)-\tan\frac{7\pi}{12} = \tan\frac{\pi}{12}+\cot\frac{\pi}{12}$$

$$= \frac{1}{\sin\frac{\pi}{12}\cos\frac{\pi}{12}} = \frac{2}{\sin\frac{\pi}{6}} = 4$$

So, $\sec^{-1}\left(\frac{1}{4}\sum_{k=0}^{10}\sec\left(\frac{7\pi}{12}+\frac{k\pi}{2}\right)\sec\left(\frac{7\pi}{12}+\frac{(k+1)\pi}{2}\right)\right)$

$= \sec^{-1}(1) = 0$

**51.** *(a)* For $Z = \{x : g(x) = 0\}$, $x > 0$

$\because \quad g(x) = \cos(2\pi \sin x) = 0$

$\Rightarrow 2\pi \sin x = (2n+1)\frac{\pi}{2}$, $n \in$ Integer

$\Rightarrow \quad \sin x = \frac{2n+1}{4}$

$\Rightarrow \quad \sin x = -\frac{3}{4}, -\frac{1}{4}, \frac{1}{4}, \frac{3}{4}$

$[\because \sin x \in [-1, 1]]$

here values of $\sin x$, $-\frac{3}{4}, -\frac{1}{4}, \frac{1}{4}, \frac{3}{4}$ are in an A.P. but corresponding values of $x$ are not in an AP so, (iii) $\rightarrow$ R.

For $\quad W = \{x : g'(x) = 0\}$, $x > 0$

so, $\quad g'(x) = -2\pi \cos x \sin(2\pi \sin x) = 0$

$\Rightarrow$ either $\cos x = 0$ or $\sin(2\pi \sin x) = 0$

$\Rightarrow$ either $x = (2n+1)\frac{\pi}{2}$ or $2\pi \sin x = n\pi$, $n \in$ Integers.

$\because 2\pi \sin x = n\pi$

$\Rightarrow \quad \sin x = \frac{n}{2} = -1, -\frac{1}{2}, 0, \frac{1}{2}, 1$

$\{\because \sin x \in [-1, 1)\}$

$\therefore x = n\pi, (2n+1)\frac{\pi}{2}$ or $n\pi + (-1)^n\left(\pm\frac{\pi}{6}\right)$

$\Rightarrow$ (iv) $\rightarrow$ P, R, S

Hence, option (a) is correct.

**52.** *(a)* For, $X = \{x : f(x) = 0\}$, $x > 0$

Now, $f(x) = 0$

$\Rightarrow \sin(\pi \cos x) = 0$, $x > 0$

$\Rightarrow \pi \cos x = n\pi$, $n \in$ Integer.

$\Rightarrow \quad \cos x = n$

$\Rightarrow \quad \cos x = -1, 0, 1$

$\{\because \cos x \in [-1, 1]\}$

$\Rightarrow x = n\pi$ or $(2n+1)\frac{\pi}{2}$, $n$ is an integer. so, (i) $\rightarrow$ (P), (Q)

For, $\quad Y = \{x : f'(x) = 0\}$, $x > 0$

Now, $\quad f'(x) = 0$

$\Rightarrow \quad -\pi \sin x \cos(\pi \cos x) = 0$

$\Rightarrow$ either $\sin x = 0 \Rightarrow x = n\pi$, $n$ is an integer, or $\cos(\pi \cos x) = 0$

$\Rightarrow \pi \cos x = (2n+1)\frac{\pi}{2}$, $n$ is an integer

$\Rightarrow \quad \cos x = \frac{2n+1}{2}$

$\Rightarrow \cos x = \pm \frac{1}{2}$, $\{\because \cos x \in [-1, 1]\}$

$\Rightarrow x = 2n\pi \pm \frac{\pi}{3}$ or $2n\pi \pm \frac{2\pi}{3}$, $n$ is an integer.

So, (ii) $\rightarrow$ (Q), (T)

Hence, option (a) is correct.

**53.** *(b)* It is given that, the centres of circles $C_1$, $C_2$ and $C_3$ are co-linear,

$$\therefore \begin{vmatrix} 0 & 0 & 1 \\ 3 & 4 & 1 \\ h & k & 1 \end{vmatrix} = 0 \Rightarrow 4h = 3k \quad \ldots \text{(i)}$$

and $MN$ is the length of diameter of circle $C_3$, so

$$MN = 3 + \sqrt{(3-0)^2 + (4-0)^2} + 4$$
$$= 3 + 5 + 4 = 12$$

So, radius of circle $C_3$, $r = 6$ ... (ii)

Since, the circle $C_3$ touches $C_1$ at $M$ and $C_2$ at $N$, so

$$|C_1 C_3| = |r - 3|$$
$$\Rightarrow \sqrt{h^2 + k^2} = 3$$
$$\Rightarrow h^2 + k^2 = 9 \quad \ldots \text{(iii)}$$

From Eqs. (i) and (iii), we get

$$h^2 + \frac{16h^2}{9} = 9 \Rightarrow 25h^2 = 81$$
$$\Rightarrow h = +\frac{9}{5} \text{ and } k = +\frac{12}{5}$$

So, $2h + k = \frac{18}{5} + \frac{12}{5} = 6$

Now, equation common chord $XY$ of circles $C_1$ and $C_2$ is

$$C_1 - C_2 = 0$$
$$\Rightarrow 6x + 8y = 18$$
$$\Rightarrow 3x + 4y = 9 \quad \ldots \text{(iv)}$$

Now,

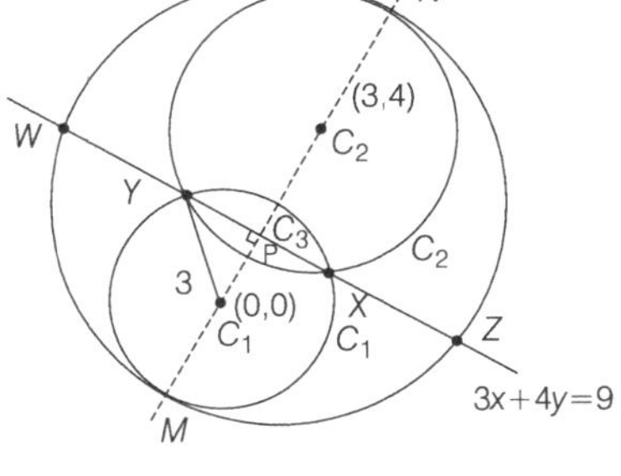

$$C_1P = \frac{9}{5}$$

Now, $PY^2 = GY^2 - GP^2$

$$= 9 - \frac{81}{25} = \frac{144}{25}$$
$$\Rightarrow PY = \frac{12}{5}$$
$$\because XY = 2PY = 2 \times \frac{12}{5} = \frac{24}{5}$$

Now, $C_3P = \dfrac{\left|3\left(\frac{9}{5}\right) + 4\left(\frac{12}{5}\right) - 9\right|}{5} = \dfrac{6}{5}$

So, $PW^2 = C_3W^2 - C_3P^2 = 36 - \frac{36}{25} = \frac{864}{25}$ $\{\because C_3W = r = 6\}$

$$\Rightarrow PW = \frac{12\sqrt{6}}{5}$$
$$\because ZW = 2PW = \frac{24\sqrt{6}}{5}$$
$$\therefore \frac{\text{length of } ZW}{\text{length of } XY} = \sqrt{6}$$

Now, area of

$$\Delta MZN = \frac{1}{2}(MN)(PZ) = \frac{1}{2} \times (12)\left(\frac{1}{2} WZ\right) \quad \{\because MN = 12\}$$
$$= 3WZ = 3\,\frac{24\sqrt{6}}{5} = \frac{72\sqrt{6}}{5}$$

and area of $\Delta ZMW = \frac{1}{2}(ZW)(MP)$

$$= \frac{1}{2}\left(\frac{24\sqrt{6}}{5}\right)(MG + GP)$$
$$= \frac{12\sqrt{6}}{5}\left(3 + \frac{9}{5}\right) \quad \left\{\because MG = 3 \text{ and } GP = \frac{9}{5}\right\}$$
$$= \frac{12\sqrt{6}}{5}\left(\frac{24}{5}\right)$$
$$= \frac{288\sqrt{6}}{25}$$

$$\therefore \frac{\text{Area of } \Delta MZN}{\text{Area of } \Delta ZMW} = \frac{\frac{72\sqrt{6}}{5}}{\frac{288\sqrt{6}}{25}} = \frac{5}{4}$$

$\because$ Common tangent of circles $C_1$ and $C_3$ is $C_1 - C_3 = 0$

$$\Rightarrow \quad (x^2 + y^2 - 9) - \left[\left(x - \frac{9}{5}\right)^2 + \left(y - \frac{12}{5}\right)^2 - 36\right] = 0$$

$$\Rightarrow \quad \frac{18}{5}x + \frac{24}{5}y + 18 = 0$$

$$\Rightarrow \quad 3x + 4y + 15 = 0 \quad \ldots \text{(v)}$$

$\because$ Tangent (v) is also touches the parabola $x^2 = 8\alpha y$,

$$\therefore \quad -2\alpha\left(-\frac{3}{4}\right)^2 = -\frac{15}{4}$$

$$\Rightarrow \quad \alpha = \frac{10}{3}$$

So combination (iv), (S) is only incorrect.
Hence, option (b) is correct.

**54.** *(c)* $\because \dfrac{\text{length of } ZW}{\text{length of } XY} = \sqrt{6}$

So, combination (ii), Q is only correct.
Hence, option (c) is correct.

# JEE Main

# Solved Paper 2018

## INSTRUCTIONS

1. This test consists of 90 questions.
2. Each question is allotted 4 marks for correct response.
3. Candidates will be awarded marks as stated above in instruction no. 2 for correct response of each question. 1 mark will be deducted for indicating incorrect response of each question. No deduction from the total score will be made if no response is indicated for an item in the answer sheet.
4. There is only one correct response for each question. Filling up more than one response in any question will be treated as wrong response and marks for wrong response will be deducted according as per instructions.

## Physics

**1.** The density of a material in the shape of a cube is determined by measuring three sides of the cube and its mass. If the relative errors in measuring the mass and length are respectively 1.5% and 1%, the maximum error in determining the density is

(a) 2.5% (b) 3.5%
(c) 4.5% (d) 6%

**2.** All the graphs below are intended to represent the same motion. One of them does it incorrectly. Pick it up.

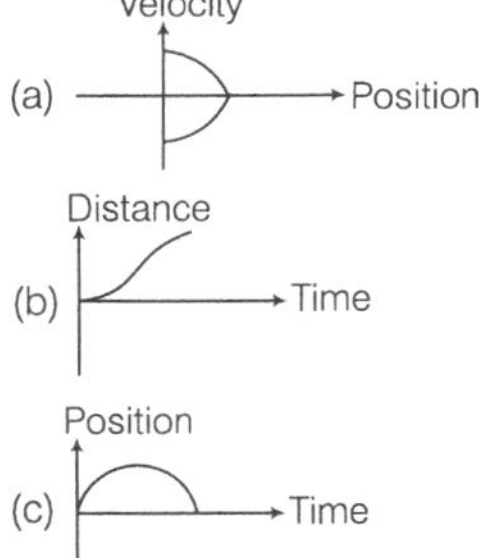


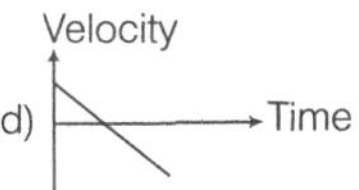


**3.** Two masses $m_1 = 5$ kg and $m_2 = 10$ kg connected by an inextensible string over a frictionless pulley, are moving as shown in the figure. The coefficient of friction of horizontal surface is 0.15. The minimum weight $m$ that should be put on top of $m_2$ to stop the motion is

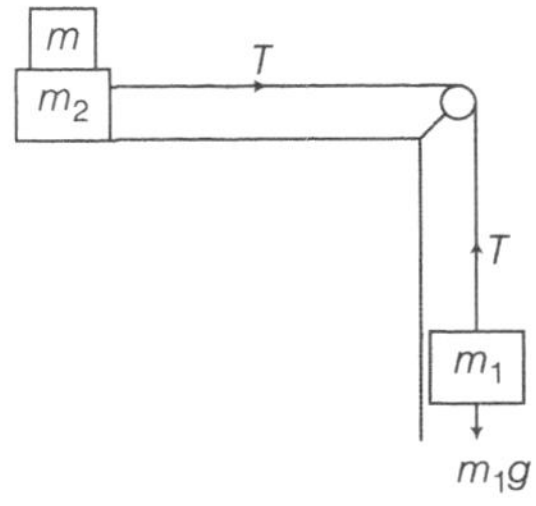


(a) 18.3 kg (b) 27.3 kg
(c) 43.3 kg (d) 10.3 kg

**4.** A particle is moving in a circular path of radius $a$ under the action of an attractive potential $U = -\frac{k}{2r^2}$. Its total energy is

(a) $-\frac{k}{4a^2}$ (b) $\frac{k}{2a^2}$

(c) zero (d) $-\frac{3}{2}\frac{k}{a^2}$

**5.** In a collinear collision, a particle with an initial speed $v_0$ strikes a stationary particle of the same mass. If the final total kinetic energy is 50% greater than the original kinetic energy, the magnitude of the relative velocity between the two particles after collision, is

(a) $\frac{v_0}{4}$ (b) $\sqrt{2}\, v_0$

(c) $\frac{v_0}{2}$ (d) $\frac{v_0}{\sqrt{2}}$

**6.** Seven identical circular planar discs, each of mass $M$ and radius $R$ are welded symmetrically as shown in the figure. The moment of inertia of the arrangement about the axis normal to the plane and passing through the point $P$ is

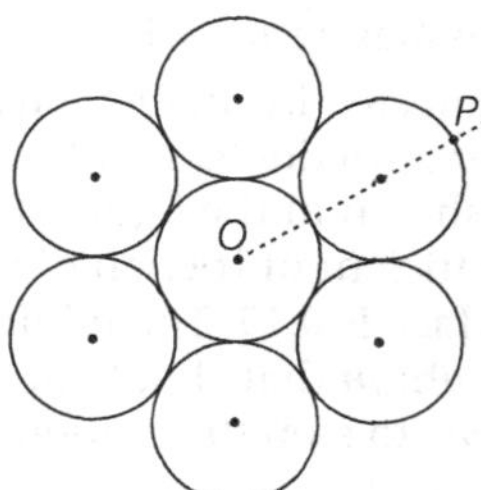


(a) $\frac{19}{2}MR^2$ (b) $\frac{55}{2}MR^2$

(c) $\frac{73}{2}MR^2$ (d) $\frac{181}{2}MR^2$

**7.** From a uniform circular disc of radius $R$ and mass $9M$, a small disc of radius $\frac{R}{3}$ is removed as shown in the figure. The moment of inertia of the remaining disc about an axis perpendicular to the plane of the disc and passing through centre of disc is

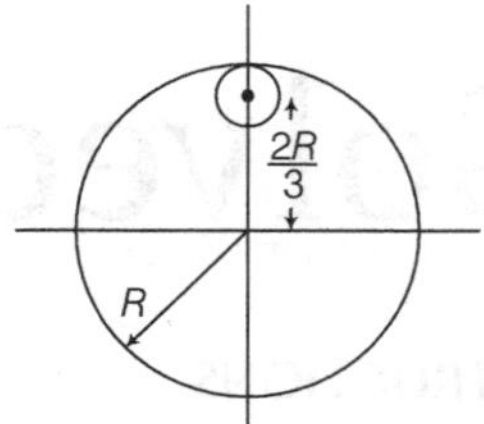


(a) $4MR^2$ (b) $\frac{40}{9}MR^2$

(c) $10MR^2$ (d) $\frac{37}{9}MR^2$

**8.** A particle is moving with a uniform speed in a circular orbit of radius $R$ in a central force inversely proportional to the $n$th power of $R$. If the period of rotation of the particle is $T$, then :

(a) $T \propto R^{3/2}$ for any $n$

(b) $T \propto R^{\frac{n}{2}+1}$

(c) $T \propto R^{(n+1)/2}$

(d) $T \propto R^{n/2}$

**9.** A solid sphere of radius $r$ made of a soft material of bulk modulus $K$ is surrounded by a liquid in a cylindrical container. A massless piston of area $a$ floats on the surface of the liquid, covering entire cross-section of cylindrical container. When a mass $m$ is placed on the surface of the piston to compress the liquid, the fractional decrement in the radius of the sphere, $\left(\frac{dr}{r}\right)$ is

(a) $\frac{Ka}{mg}$ (b) $\frac{Ka}{3mg}$

(c) $\frac{mg}{3Ka}$ (d) $\frac{mg}{Ka}$

**10.** Two moles of an ideal monoatomic gas occupies a volume $V$ at 27°C. The gas expands adiabatically to a volume 2 $V$. Calculate (i) the final temperature of the gas and (ii) change in its internal energy.

(a) (i) 189 K (ii) 2.7 kJ
(b) (i) 195 K (ii) −2.7 kJ
(c) (i) 189 K (ii) −2.7 kJ
(d) (i) 195 K (ii) 2.7 kJ

**11.** The mass of a hydrogen molecule is $3.32\times 10^{-27}$ kg. If $10^{23}$ hydrogen molecules strike per second, a fixed wall of area 2 $cm^2$ at an angle of 45° to the normal and rebound elastically with a speed of $10^3$ m/s, then the pressure on the wall is nearly

(a) $2.35\times10^3$ N/m$^2$ (b) $4.70\times10^3$ N/m$^2$
(c) $2.35\times10^2$ N/m$^2$ (d) $4.70\times10^2$ N/m$^2$

**12.** A silver atom in a solid oscillates in simple harmonic motion in some direction with a frequency of $10^{12}$ per second. What is the force constant of the bonds connecting one atom with the other?
(Take, molecular weight of silver $=108$ and Avogadro number $=6.02\times10^{23}$ g mol$^{-1}$)

(a) 6.4 N/m (b) 7.1 N/m
(c) 2.2 N/m (d) 5.5 N/m

**13.** A granite rod of 60 cm length is clamped at its middle point and is set into longitudinal vibrations. The density of granite is $2.7\times10^3$ kg/m$^3$ and its Young's modulus is $9.27\times10^{10}$ Pa. What will be the fundamental frequency of the longitudinal vibrations?

(a) 5 kHz (b) 2.5 kHz
(c) 10 kHz (d) 7.5 kHz

**14.** Three concentric metal shells $A$, $B$ and $C$ of respective radii $a$, $b$ and $c$ $(a<b<c)$ have surface charge densities $+\sigma$, $-\sigma$ and $+\sigma$, respectively. The potential of shell $B$ is

(a) $\frac{\sigma}{\varepsilon_0}\left[\frac{a^2-b^2}{a}+c\right]$ (b) $\frac{\sigma}{\varepsilon_0}\left[\frac{a^2-b^2}{b}+c\right]$
(c) $\frac{\sigma}{\varepsilon_0}\left[\frac{b^2-c^2}{b}+a\right]$ (d) $\frac{\sigma}{\varepsilon_0}\left[\frac{b^2-c^2}{c}+a\right]$

**15.** A parallel plate capacitor of capacitance 90 pF is connected to a battery of emf 20 V. If a dielectric material of dielectric constant $K=\frac{5}{3}$ is inserted between the plates, the magnitude of the induced charge will be

(a) 1.2 nC (b) 0.3 nC
(c) 2.4 nC (d) 0.9 nC

**16.** In an AC circuit, the instantaneous emf and current are given by

$$e=100\sin 30\,t,\ i=20\sin\left(30\,t-\frac{\pi}{4}\right)$$

In one cycle of AC, the average power consumed by the circuit and the wattless current are, respectively

(a) 50 , 10 (b) $\frac{1000}{\sqrt{2}}$ , 10
(c) $\frac{50}{\sqrt{2}}$ , 0 (d) 50 , 0

**17.** Two batteries with emf 12 V and 13 V are connected in parallel across a load resistor of 10 Ω. The internal resistances of the two batteries are 1 Ω and 2 Ω, respectively. The voltage across the load lies between

(a) 11.6 V and 11.7 V
(b) 11.5 V and 11.6 V
(c) 11.4 V and 11.5 V
(d) 11.7 V and 11.8 V

**18.** An electron, a proton and an alpha particle having the same kinetic energy are moving in circular orbits of radii $r_e$, $r_p$, $r_\alpha$ respectively, in a uniform magnetic field $B$. The relation between $r_e$, $r_p$, $r_\alpha$ is

(a) $r_e>r_p=r_\alpha$ (b) $r_e<r_p=r_\alpha$
(c) $r_e<r_p<r_\alpha$ (d) $r_e<r_\alpha<r_p$

**19.** The dipole moment of a circular loop carrying a current $I$, is $m$ and the magnetic field at the centre of the loop is $B_1$. When the dipole moment is doubled by keeping the current constant, the magnetic field at the centre of the loop is $B_2$. The ratio $\frac{B_1}{B_2}$ is

(a) 2 (b) $\sqrt{3}$

(c) $\sqrt{2}$ (d) $\frac{1}{\sqrt{2}}$

**20.** For an $R$-$L$-$C$ circuit driven with voltage of amplitude $v_m$ and frequency $\omega_0 = \frac{1}{\sqrt{LC}}$, the current exhibits resonance. The quality factor, $Q$ is given by

(a) $\frac{\omega_0 L}{R}$ (b) $\frac{\omega_0 R}{L}$

(c) $\frac{R}{\omega_0 C}$ (d) $\frac{CR}{\omega_0}$

**21.** An EM wave from air enters a medium. The electric fields are $\mathbf{E}_1 = E_{01}\hat{\mathbf{x}}\cos\left[2\pi\nu\left(\frac{z}{c} - t\right)\right]$ in air and $\mathbf{E}_2 = E_{02}\hat{\mathbf{x}}\cos[k(2z - ct)]$ in medium, where the wave number $k$ and frequency $\nu$ refer to their values in air. The medium is non-magnetic.

If $\varepsilon_{r_1}$ and $\varepsilon_{r_2}$ refer to relative permittivities of air and medium respectively, which of the following options is correct?

(a) $\frac{\varepsilon_{r_1}}{\varepsilon_{r_2}} = 4$ (b) $\frac{\varepsilon_{r_1}}{\varepsilon_{r_2}} = 2$

(c) $\frac{\varepsilon_{r_1}}{\varepsilon_{r_2}} = \frac{1}{4}$ (d) $\frac{\varepsilon_{r_1}}{\varepsilon_{r_2}} = \frac{1}{2}$

**22.** Unpolarised light of intensity $I$ passes through an ideal polariser $A$. Another identical polariser $B$ is placed behind $A$. The intensity of light beyond $B$ is found to be $\frac{I}{2}$. Now, another identical polariser $C$ is placed between $A$ and $B$. The intensity beyond $B$ is now found to be $\frac{1}{8}$. The angle between polariser $A$ and $C$ is

(a) 0° (b) 30°

(c) 45° (d) 60°

**23.** The angular width of the central maximum in a single slit diffraction pattern is 60°. The width of the slit is 1 μm. The slit is illuminated by monochromatic plane waves. If another slit of same width is made near it, Young's fringes can be observed on a screen placed at a distance 50 cm from the slits. If the observed fringe width is 1 cm, what is slit separation distance? (i.e. distance between the centres of each slit.)

(a) 25 μm (b) 50 μm

(c) 75 μm (d) 100 μm

**24.** An electron from various excited states of hydrogen atom emit radiation to come to the ground state. Let $\lambda_n, \lambda_g$ be the de-Broglie wavelength of the electron in the $n$th state and the ground state, respectively. Let $\Lambda_n$ be the wavelength of the emitted photon in the transition from the $n$th state to the ground state. For large $n$, ($A, B$ are constants)

(a) $\Lambda_n \approx A + \frac{B}{\lambda_n^2}$ (b) $\Lambda_n \approx A + B\lambda_n^2$

(c) $\Lambda_n^2 \approx A + B\lambda_n^2$ (d) $\Lambda_n^2 \approx \lambda$

**25.** If the series limit frequency of the Lyman series is $\nu_L$, then the series limit frequency of the Pfund series is

(a) $25\,\nu_L$ (b) $16\,\nu_L$

(c) $\frac{\nu_L}{16}$ (d) $\frac{\nu_L}{25}$

**26.** It is found that, if a neutron suffers an elastic collinear collision with deuterium at rest, fractional loss of its energy is $P_d$; while for its similar collision with carbon nucleus at rest, fractional loss of energy is $P_c$. The values of $P_d$ and $P_c$ are respectively

(a) (.89, .28) (b) (.28, .89)
(c) (0, 0) (d) (0, 1)

**27.** The reading of the ammeter for a silicon diode in the given circuit is

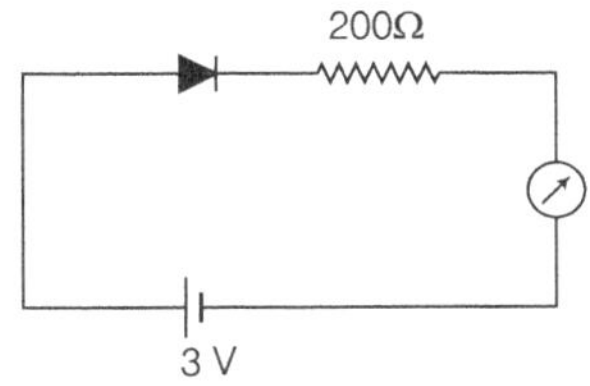


(a) 0 (b) 15 mA
(c) 11.5 mA (d) 13.5 mA

**28.** A telephonic communication service is working at carrier frequency of 10 GHz. Only 10% of it is utilised for transmission. How many telephonic channels can be transmitted simultaneously, if each channel requires a bandwidth of 5 kHz?

(a) $2 \times 10^3$ (b) $2 \times 10^4$
(c) $2 \times 10^5$ (d) $2 \times 10^6$

**29.** In a potentiometer experiment, it is found that no current passes through the galvanometer when the terminals of the cell are connected across 52 cm of the potentiometer wire. If the cell is shunted by a resistance of 5 Ω, a balance is found when the cell is connected across 40 cm of the wire. Find the internal resistance of the cell.

(a) 1 Ω (b) 1.5 Ω
(c) 2 Ω (d) 2.5 Ω

**30.** On interchanging the resistances, the balance point of a meter bridge shifts to the left by 10 cm. The resistance of their series combination is 1 kΩ. How much was the resistance on the left slot before interchanging the resistances?

(a) 990 Ω (b) 505 Ω
(c) 550 Ω (d) 910 Ω

# Chemistry

**31.** The ratio of mass per cent of C and H of an organic compound $(C_xH_yO_z)$ is 6 : 1. If one molecule of the above compound $(C_xH_yO_z)$ contains half as much oxygen as required to burn one molecule of compound $C_xH_y$ completely to $CO_2$ and $H_2O$. The empirical formula of compound $C_xH_yO_z$ is

(a) $C_3H_6O_3$ (b) $C_2H_4O$
(c) $C_3H_4O_2$ (d) $C_2H_4O_3$

**32.** Which type of 'defect' has the presence of cations in the interstitial sites?

(a) Schottky defect
(b) Vacancy defect
(c) Frenkel defect
(d) Metal deficiency defect

**33.** According to molecular orbital theory, which of the following will not be a viable molecule?

(a) $He_2^{2+}$ (b) $He_2^+$
(c) $H_2^-$ (d) $H_2^{2-}$

**34.** Which of the following lines correctly show the temperature dependence of equilibrium constant, $K$, for an exothermic reaction?

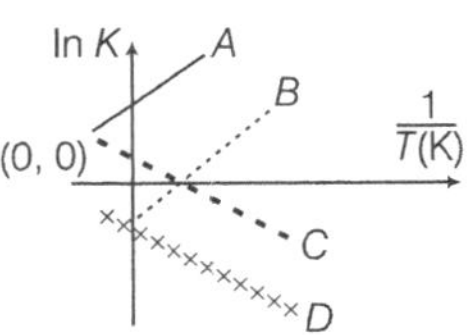


(a) $A$ and $B$ (b) $B$ and $C$
(c) $C$ and $D$ (d) $A$ and $D$

**35.** The combustion of benzene ($l$) gives $CO_2(g)$ and $H_2O(l)$. Given that heat of combustion of benzene at constant volume is $-3263.9$ kJ $mol^{-1}$ at 25° C; heat of combustion (in kJ $mol^{-1}$) of benzene at constant pressure will be ($R = 8.314$ $JK^{-1}mol^{-1}$)

(a) 4152.6 (b) −452.46
(c) 3260 (d) −3267.6

**36.** For 1 molal aqueous solution of the following compounds, which one will show the highest freezing point?

(a) $[Co(H_2O)_6]Cl_3$
(b) $[Co(H_2O)_5Cl]Cl_2 \cdot H_2O$
(c) $[Co(H_2O)_4Cl_2]Cl \cdot 2H_2O$
(d) $[Co(H_2O)_3Cl_3] \cdot 3H_2O$

**37.** An aqueous solution contains 0.10 M $H_2S$ and 0.20 M HCl. If the equilibrium constants for the formation of $HS^-$ from $H_2S$ is $1.0 \times 10^{-7}$ and that of $S^{2-}$ from $HS^-$ ions is $1.2 \times 10^{-13}$ then the concentration of $S^{2-}$ ions in aqueous solution is

(a) $5 \times 10^{-8}$ (b) $3 \times 10^{-20}$
(c) $6 \times 10^{-21}$ (d) $5 \times 10^{-19}$

**38.** An aqueous solution contains an unknown concentration of $Ba^{2+}$. When 50 mL of a 1 M solution of $Na_2SO_4$ is added, $BaSO_4$ just begins to precipitate. The final volume is 500 mL. The solubility product of $BaSO_4$ is $1 \times 10^{-10}$. What is the original concentration of $Ba^{2+}$?

(a) $5 \times 10^{-9}$ M (b) $2 \times 10^{-9}$ M
(c) $1.1 \times 10^{-9}$ M (d) $1.0 \times 10^{-10}$ M

**39.** At 518°C, the rate of decomposition of a sample of gaseous acetaldehyde, initially at a pressure of 363 Torr, was 1.00 Torr $s^{-1}$ when 5% had reacted and 0.5 Torr $s^{-1}$ when 33% had reacted. The order of the reaction is

(a) 2 (b) 3 (c) 1 (d) 0

**40.** How long (approximate) should water be electrolysed by passing through 100 amperes current so that the oxygen released can completely burn 27.66 g of diborane? (Atomic weight of $B = 10.8$ $\mu$)

(a) 6.4 hours (b) 0.8 hours
(c) 3.2 hours (d) 1.6 hours

**41.** The recommended concentration of fluoride ion in drinking water is up to 1 ppm as fluoride ion is required to make teeth enamel harder by converting $[3Ca_3(PO_4)_2 \cdot Ca(OH)_2]$ to :

(a) $[CaF_2]$
(b) $[3(CaF_2) \cdot Ca(OH)_2]$
(c) $[3Ca_3(PO_4)_2 \cdot CaF_2]$
(d) $[3\{Ca_3(PO_4)_2\} \cdot CaF_2]$

**42.** Which of the following compounds contain(s) no covalent bond(s)?

$KCl, PH_3, O_2, B_2H_6, H_2SO_4$

(a) $KCl, B_2H_6, PH_3$ (b) $KCl, H_2SO_4$
(c) KCl (d) $KCl, B_2H_6$

**43.** Which of the following are Lewis acids?

(a) $PH_3$ and $BCl_3$ (b) $AlCl_3$ and $SiCl_4$
(c) $PH_3$ and $SiCl_4$ (d) $BCl_3$ and $AlCl_3$

**44.** Total number of lone pair of electron in $I_3^-$ ion is

(a) 3 (b) 6 (c) 9 (d) 12

**45.** Which of the following salts is the most basic in aqueous solution?

(a) $Al(CN)_3$ (b) $CH_3COOK$
(c) $FeCl_3$ (d) $Pb(CH_3COO)_2$

**46.** Hydrogen peroxide oxidises $[Fe(CN)_6]^{4-}$ to $[Fe(CN)_6]^{3-}$ in acidic medium but reduces $[Fe(CN)_6]^{3-}$ to $[Fe(CN)_6]^{4-}$ in alkaline medium. The other products formed are, respectively.

(a) $(H_2O + O_2)$ and $H_2O$
(b) $(H_2O + O_2)$ and $(H_2O + OH^-)$
(c) $H_2O$ and $(H_2O + O_2)$
(d) $H_2O$ and $(H_2O + OH^-)$

**47.** The oxidation states of Cr, in $[Cr(H_2O)_6]Cl_3$, $[Cr(C_6H_6)_2]$, and $K_2[Cr(CN)_2(O)_2(O_2)(NH_3)]$ respectively are

(a) +3, +4 and +6 (b) +3, +2 and +4
(c) +3, 0 and +6 (d) +3, 0 and +4

**48.** The compound that does not produce nitrogen gas by the thermal decomposition is

(a) $Ba(N_3)_2$ (b) $(NH_4)_2Cr_2O_7$
(c) $NH_4NO_2$ (d) $(NH_4)_2SO_4$

**49.** When metal '*M*' is treated with NaOH, a white gelatinous precipitate '*X*' is obtained, which is soluble in excess of NaOH. Compound '*X*' when heated strongly gives an oxide which is used in chromatography as an adsorbent. The metal '*M*' is

(a) Zn (b) Ca
(c) Al (d) Fe

**50.** Consider the following reaction and statements :

$$[Co(NH_3)_4Br_2]^+ + Br^- \longrightarrow [Co(NH_3)_3Br_3] + NH_3$$

I. Two isomers are produces if the reactant complex ion is a *cis*-isomer.
II. Two isomers are produced if the reactant complex ion is a *trans*-isomer.
III. Only one isomer is produced if the reactant complex ion is a *trans*-isomer.
IV. Only one isomer is produced if the reactant complex ion is a *cis*-isomer.

The correct statements are

(a) (I) and (II) (b) (I) and (III)
(c) (III) and (IV) (d) (II) and (IV)

**51.** Glucose on prolonged heating with HI gives

(a) *n*-hexane (b) 1-hexene
(c) Hexanoic acid (d) 6-iodohexanal

**52.** The *trans*-alkenes are formed by the reduction of alkynes with

(a) $H_2$-Pd/C, $BaSO_4$ (b) $NaBH_4$
(c) Na/liq. $NH_3$ (d) Sn-HCl

**53.** Which of the following compounds will be suitable for Kjeldahl's method for nitrogen estimation?

(a) N (pyridine ring)
(b) $NH_2$
(c) $NO_2$
(d) $N_2^+Cl^-$

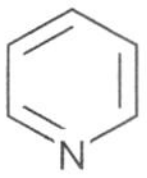

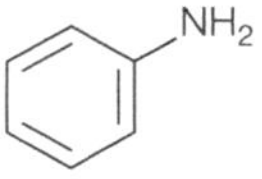

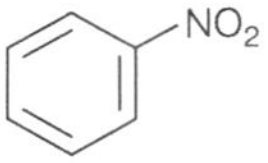

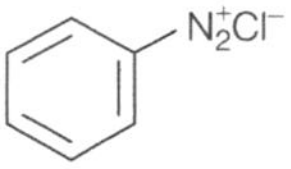

**54.** Phenol on treatment with $CO_2$ in the presence of NaOH followed by acidification produces compound *X* as the major product. *X* on treatment with $(CH_3CO)_2O$ in the presence of catalytic amount of $H_2SO_4$ produces:

(a) O, O, $CH_3$, $CO_2H$
(b) O, O, $CH_3$, $CO_2H$
(c) O, C, O, $CH_3$, O, OH
(d) $CO_2H$, $CO_2H$, O, $CH_3$, O

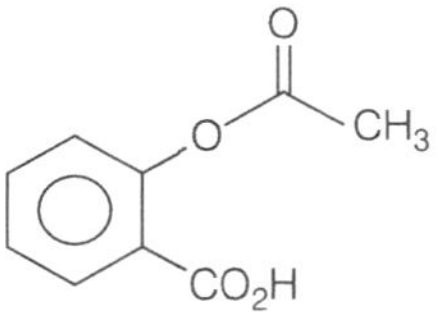

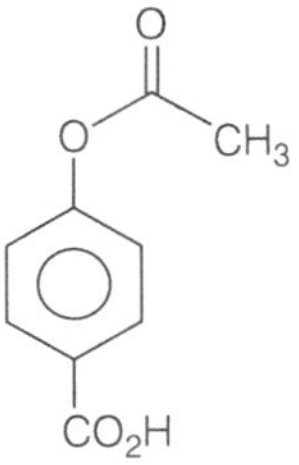

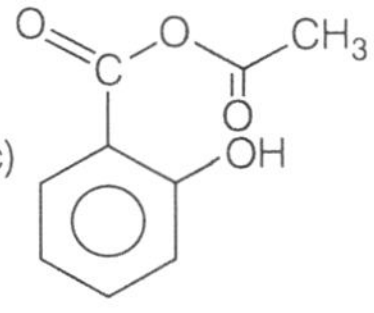

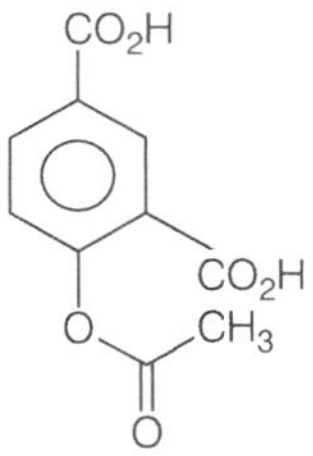

**55.** An alkali is titrated against an acid with methyl orange as indicator, which of the following is a correct combination?

| | Base | Acid | End point |
|---|---|---|---|
| (a) | Weak | Strong | Colourless to pink |
| (b) | Strong | Strong | Pinkish red to yellow |
| (c) | Weak | Strong | Yellow to pinkish red |
| (d) | Strong | Strong | Pink to colourless |

**56.** The predominant form of histamine present in human blood is ($pK_a$, Histidine = 6.0)

(a) 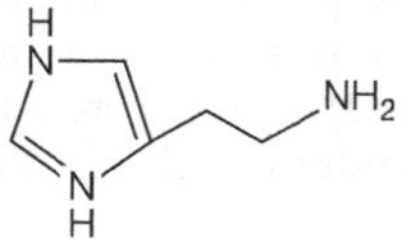 H N ... NH$_2$ ... N H

(b) 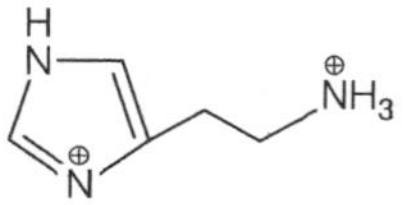 H N ... $\overset{\oplus}{N}H_3$ ... $\overset{\oplus}{N}$

(c) 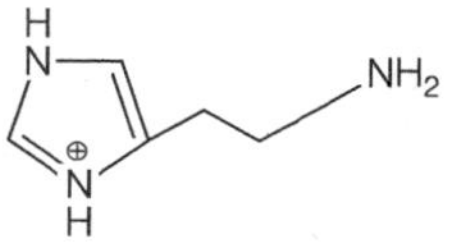 H N ... NH$_2$ ... $\overset{\oplus}{N}$ H

(d) 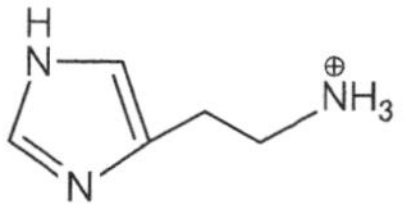 H N ... $\overset{\oplus}{N}H_3$ ... N

**57.** Phenol reacts with methyl chloroformate in the presence of NaOH to form product *A*. *A* reacts with $Br_2$ to form product *B*. *A* and *B* are respectively

(a) 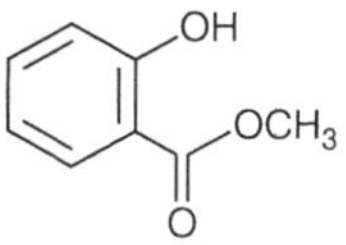 OH, OCH$_3$, O and Br, OH, OCH$_3$, O

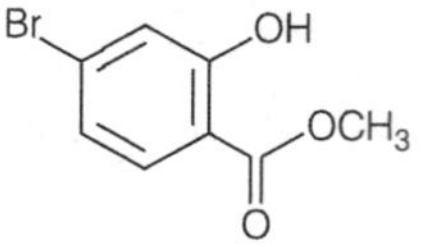

(b) 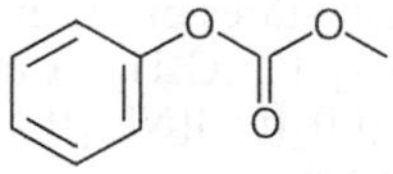 O, O, O and O, O, O, Br

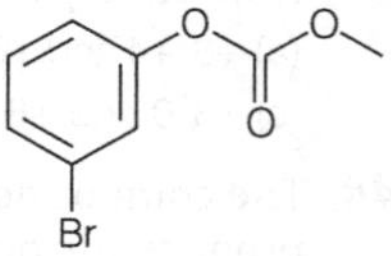

(c) 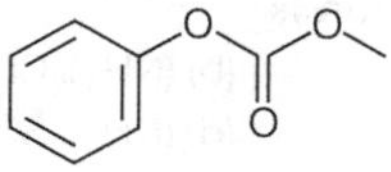 O, O, O and O, O, O, Br

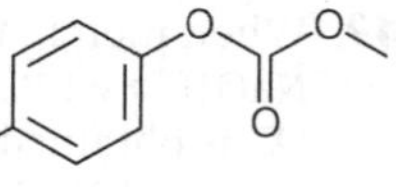

(d) 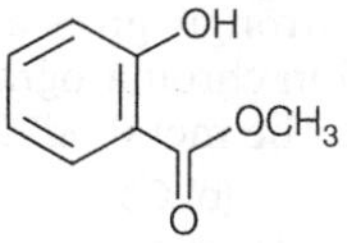 OH, OCH$_3$, O and OH, OCH$_3$, Br, O

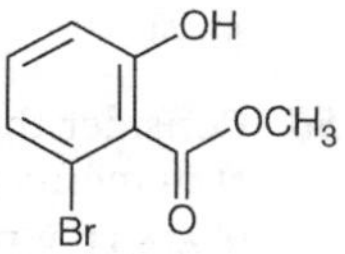

**58.** The increasing order of basicity of the following compounds is

I.  NH$_2$ II. 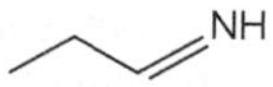 NH

III. 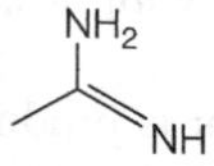 NH$_2$, NH IV. 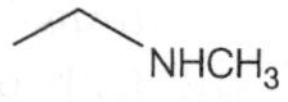 NHCH$_3$

(a) (I) < (II) < (III) < (IV)

(b) (II) < (I) < (III) < (IV)

(c) (II) < (I) < (IV) < (III)

(d) (IV) < (II) < (I) < (III)

**59.** The major product formed in the following reaction is

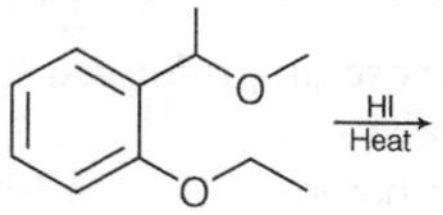

O, O $\xrightarrow[\text{Heat}]{\text{HI}}$

(a) 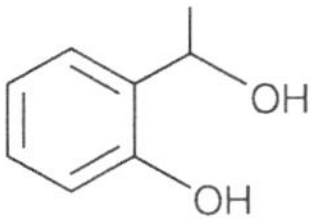 (b)

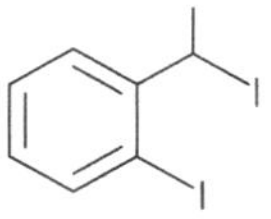

(c) 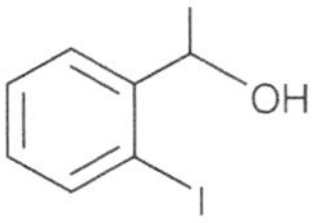 (d)

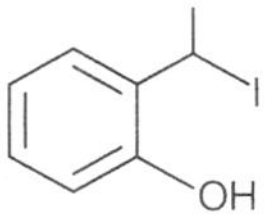

**60.** The major product of the following reaction is

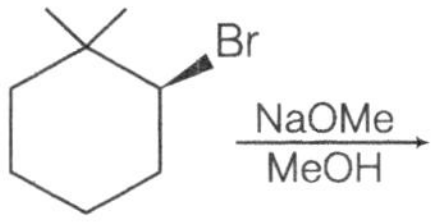

NaOMe / MeOH

(a) 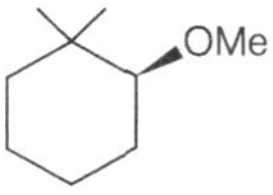 (b)

(c) 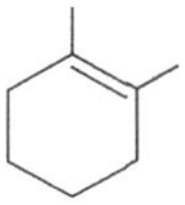 (d)

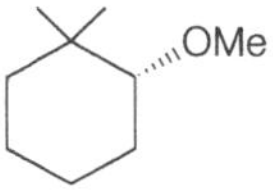

# Mathematics

**61.** Two sets $A$ and $B$ are as under $A=\{(a,b)\in\mathbf{R}\times\mathbf{R}:|a-5|<1$ and $|b-5|<1\}$; $B=\{(a,b)\in\mathbf{R}\times\mathbf{R}:4(a-6)^2+9(b-5)^2\le 36\}$. Then,

(a) $B\subset A$
(b) $A\subset B$
(c) $A\cap B=\phi$ (an empty set)
(d) neither $A\subset B$ nor $B\subset A$

**62.** Let $S=\{x\in\mathbf{R}:x\ge 0$ and $2|\sqrt{x}-3|+\sqrt{x}(\sqrt{x}-6)+6=0$. Then, $S$

(a) is an empty set
(b) contains exactly one element
(c) contains exactly two elements
(d) contains exactly four elements

**63.** If $\alpha,\beta\in\mathbf{C}$ are the distinct roots of the equation $x^2-x+1=0$, then $\alpha^{101}+\beta^{107}$ is equal to

(a) –1 (b) 0
(c) 1 (d) 2

**64.** If $\begin{vmatrix} x-4 & 2x & 2x \\ 2x & x-4 & 2x \\ 2x & 2x & x-4 \end{vmatrix}$ $=(A+Bx)(x-A)^2$, then the ordered pair $(A,B)$ is equal to

(a) (–4, – 5) (b) (–4, 3)
(c) (–4, 5) (d) (4, 5)

**65.** If the system of linear equations

$$x+ky+3z=0$$
$$3x+ky-2z=0$$
$$2x+4y-3z=0$$

has a non-zero solution $(x,y,z)$, then $\frac{xz}{y^2}$ is equal to

(a) –10 (b) 10
(c) –30 (d) 30

**66.** From 6 different novels and 3 different dictionaries, 4 novels and 1 dictionary are to be selected and arranged in a row on a shelf, so that the dictionary is always in the middle. The number of such arrangements is

(a) atleast 1000
(b) less than 500
(c) atleast 500 but less than 750
(d) atleast 750 but less than 1000

**67.** The sum of the coefficients of all odd degree terms in the expansion of $\left(x+\sqrt{x^3-1}\right)^5+\left(x-\sqrt{x^3-1}\right)^5$, $(x>1)$ is

(a) –1 (b) 0
(c) 1 (d) 2

**68.** Let $a_1, a_2, a_3, \ldots, a_{49}$ be in AP such that $\sum_{k=0}^{12} a_{4k+1} = 416$ and $a_9 + a_{43} = 66$. If $a_1^2 + a_2^2 + \ldots + a_{17}^2 = 140$ m, then $m$ is equal to

(a) 66 (b) 68
(c) 34 (d) 33

**69.** Let $A$ be the sum of the first 20 terms and $B$ be the sum of the first 40 terms of the series

$1^2 + 2\cdot 2^2 + 3^2 + 2\cdot 4^2 + 5^2 + 2\cdot 6^2 + \ldots$

If $B - 2A = 100\lambda$, then $\lambda$ is equal to

(a) 232 (b) 248
(c) 464 (d) 496

**70.** For each $t \in \mathbf{R}$, let $[t]$ be the greatest integer less than or equal to $t$. Then,

$$\lim_{x \to 0^+} x\left(\left[\frac{1}{x}\right] + \left[\frac{2}{x}\right] + \ldots + \left[\frac{15}{x}\right]\right)$$

(a) is equal to 0
(b) is equal to 15
(c) is equal to 120
(d) does not exist (in **R**)

**71.** Let $S = (t \in \mathbf{R} : f(x) = |x - \pi| \cdot (e^{|x|} - 1)\sin|x|$ is not differentiable at $t\}$. Then, the set $S$ is equal to

(a) $\phi$ (an empty set) (b) $\{0\}$
(c) $\{\pi\}$ (d) $\{0, \pi\}$

**72.** If the curves $y^2 = 6x, 9x^2 + by^2 = 16$ intersect each other at right angles, then the value of $b$ is

(a) 6 (b) $\frac{7}{2}$
(c) 4 (d) $\frac{9}{2}$

**73.** Let $f(x) = x^2 + \frac{1}{x^2}$ and $g(x) = x - \frac{1}{x}$, $x \in \mathbf{R} - \{-1, 0, 1\}$. If $h(x) = \frac{f(x)}{g(x)}$, then the local minimum value of $h(x)$ is

(a) 3 (b) $-3$
(c) $-2\sqrt{2}$ (d) $2\sqrt{2}$

**74.** The integral

$$\int \frac{\sin^2 x \cos^2 x}{(\sin^5 x + \cos^3 x \sin^2 x + \sin^3 x \cos^2 x + \cos^5 x)^2} dx$$

is equal to

(a) $\frac{1}{3(1+\tan^3 x)} + C$ (b) $\frac{-1}{3(1+\tan^3 x)} + C$
(c) $\frac{1}{1+\cot^3 x} + C$ (d) $\frac{-1}{1+\cot^3 x} + C$

(where $C$ is a constant of integration)

**75.** The value of $\int_{-\pi/2}^{\pi/2} \frac{\sin^2 x}{1+2^x} dx$ is

(a) $\frac{\pi}{8}$ (b) $\frac{\pi}{2}$
(c) $4\pi$ (d) $\frac{\pi}{4}$

**76.** Let $g(x) = \cos x^2$, $f(x) = \sqrt{x}$ and $\alpha, \beta$ $(\alpha < \beta)$ be the roots of the quadratic equation $18x^2 - 9\pi x + \pi^2 = 0$. Then, the area (in sq units) bounded by the curve $y = (gof)(x)$ and the lines $x = \alpha$, $x = \beta$ and $y = 0$, is

(a) $\frac{1}{2}(\sqrt{3} - 1)$ (b) $\frac{1}{2}(\sqrt{3} + 1)$
(c) $\frac{1}{2}(\sqrt{3} - \sqrt{2})$ (d) $\frac{1}{2}(\sqrt{2} - 1)$

**77.** Let $y = y(x)$ be the solution of the differential equation

$\sin x \frac{dy}{dx} + y\cos x = 4x$, $x \in (0, \pi)$.

If $y\left(\frac{\pi}{2}\right) = 0$, then $y\left(\frac{\pi}{6}\right)$ is equal to

(a) $\frac{4}{9\sqrt{3}}\pi^2$ (b) $\frac{-8}{9\sqrt{3}}\pi^2$
(c) $-\frac{8}{9}\pi^2$ (d) $-\frac{4}{9}\pi^2$

**78.** A straight line through a fixed point (2, 3) intersects the coordinate axes at distinct points $P$ and $Q$. If $O$ is the origin and the rectangle $OPRQ$ is completed, then the locus of $R$ is

(a) $3x + 2y = 6$ (b) $2x + 3y = xy$
(c) $3x + 2y = xy$ (d) $3x + 2y = 6xy$

**79.** Let the orthocentre and centroid of a triangle be $A(-3, 5)$ and $B(3, 3)$, respectively. If $C$ is the circumcentre of this triangle, then the radius of the circle having line segment $AC$ as diameter, is

(a) $\sqrt{10}$ (b) $2\sqrt{10}$

(c) $3\sqrt{\frac{5}{2}}$ (d) $\frac{3\sqrt{5}}{2}$

**80.** If the tangent at $(1, 7)$ to the curve $x^2 = y - 6$ touches the circle $x^2 + y^2 + 16x + 12y + c = 0$, then the value of $c$ is

(a) 195 (b) 185

(c) 85 (d) 95

**81.** Tangent and normal are drawn at $P(16, 16)$ on the parabola $y^2 = 16x$, which intersect the axis of the parabola at $A$ and $B$, respectively. If $C$ is the centre of the circle through the points $P$, $A$ and $B$ and $\angle CPB = \theta$, then a value of $\tan\theta$ is

(a) $\frac{1}{2}$ (b) 2

(c) 3 (d) $\frac{4}{3}$

**82.** Tangents are drawn to the hyperbola $4x^2 - y^2 = 36$ at the points $P$ and $Q$. If these tangents intersect at the point $T(0, 3)$, then the area (in sq units) of $\Delta PTQ$ is

(a) $45\sqrt{5}$ (b) $54\sqrt{3}$

(c) $60\sqrt{3}$ (d) $36\sqrt{5}$

**83.** If $L_1$ is the line of intersection of the planes $2x - 2y + 3z - 2 = 0$, $x - y + z + 1 = 0$ and $L_2$ is the line of intersection of the planes $x + 2y - z - 3 = 0, 3x - y + 2z - 1 = 0$, then the distance of the origin from the plane, containing the lines $L_1$ and $L_2$ is

(a) $\frac{1}{4\sqrt{2}}$ (b) $\frac{1}{3\sqrt{2}}$

(c) $\frac{1}{2\sqrt{2}}$ (d) $\frac{1}{\sqrt{2}}$

**84.** The length of the projection of the line segment joining the points $(5, -1, 4)$ and $(4, -1, 3)$ on the plane, $x + y + z = 7$ is

(a) $\frac{2}{\sqrt{3}}$ (b) $\frac{2}{3}$

(c) $\frac{1}{3}$ (d) $\sqrt{\frac{2}{3}}$

**85.** Let $\mathbf{u}$ be a vector coplanar with the vectors $\mathbf{a} = 2\hat{\mathbf{i}} + 3\hat{\mathbf{j}} - \hat{\mathbf{k}}$ and $\mathbf{b} = \hat{\mathbf{j}} + \hat{\mathbf{k}}$. If $\mathbf{u}$ is perpendicular to $\mathbf{a}$ and $\mathbf{u} \cdot \mathbf{b} = 24$, then $|\mathbf{u}|^2$ is equal to

(a) 336 (b) 315

(c) 256 (d) 84

**86.** A bag contains 4 red and 6 black balls. A ball is drawn at random from the bag, its colour is observed and this ball along with two additional balls of the same colour are returned to the bag. If now a ball is drawn at random from the bag, then the probability that this drawn ball is red, is

(a) $\frac{3}{10}$ (b) $\frac{2}{5}$

(c) $\frac{1}{5}$ (d) $\frac{3}{4}$

**87.** If $\sum_{i=1}^{9}(x_i - 5) = 9$ and $\sum_{i=1}^{9}(x_i - 5)^2 = 45$, then the standard deviation of the 9 items $x_1, x_2, \ldots, x_9$ is

(a) 9 (b) 4

(c) 2 (d) 3

**88.** If sum of all the solutions of the equation

$$8\cos x \cdot \left(\cos\left(\frac{\pi}{6} + x\right) \cdot \cos\left(\frac{\pi}{6} - x\right) - \frac{1}{2}\right) = 1$$

in $[0, \pi]$ is $k\pi$, then $k$ is equal to

(a) $\frac{2}{3}$ (b) $\frac{13}{9}$

(c) $\frac{8}{9}$ (d) $\frac{20}{9}$

**89.** $PQR$ is a triangular park with $PQ = PR = 200$ m. A TV tower stands at the mid-point of $QR$. If the angles of elevation of the top of the tower at $P$, $Q$ and $R$ are respectively 45°, 30° and 30°, then the height of the tower (in m) is

(a) 100 (b) 50

(c) $100\sqrt{3}$ (d) $50\sqrt{2}$

**90.** The boolean expression $\sim (p \vee q) \vee (\sim p \wedge q)$ is equivalent to

(a) $\sim p$ (b) $p$

(c) $q$ (d) $\sim q$

# Answers

## Physics

| 1. | (c) | 2. | (b) | 3. | (b) | 4. | (c) | 5. | (b) | 6. | (d) | 7. | (a) | 8. | (c) | 9. | (c) | 10. | (c) |
|---|---|---|---|---|---|---|---|---|---|---|---|---|---|---|---|---|---|---|---|
| 11. | (a) | 12. | (b) | 13. | (a) | 14. | (b) | 15. | (a) | 16. | (b) | 17. | (b) | 18. | (b) | 19. | (c) | 20. | (a) |
| 21. | (c) | 22. | (c) | 23. | (a) | 24. | (a) | 25. | (d) | 26. | (a) | 27. | (c) | 28. | (c) | 29. | (b) | 30. | (c) |

## Chemistry

| 31. | (d) | 32. | (c) | 33. | (d) | 34. | (a) | 35. | (d) | 36. | (d) | 37. | (b) | 38. | (c) | 39. | (a) | 40. | (c) |
|---|---|---|---|---|---|---|---|---|---|---|---|---|---|---|---|---|---|---|---|
| 41. | (c) | 42. | (c) | 43. | (d) | 44. | (c) | 45. | (b) | 46. | (c) | 47. | (c) | 48. | (d) | 49. | (c) | 50. | (b) |
| 51. | (a) | 52. | (c) | 53. | (b) | 54. | (a) | 55. | (c) | 56. | (d) | 57. | (c) | 58. | (c) | 59. | (d) | 60. | (b) |

## Mathematics

| 61. | (b) | 62. | (c) | 63. | (c) | 64. | (c) | 65. | (b) | 66. | (a) | 67. | (d) | 68. | (c) | 69. | (b) | 70. | (c) |
|---|---|---|---|---|---|---|---|---|---|---|---|---|---|---|---|---|---|---|---|
| 71. | (a) | 72. | (d) | 73. | (d) | 74. | (b) | 75. | (d) | 76. | (a) | 77. | (c) | 78. | (c) | 79. | (c) | 80. | (d) |
| 81. | (b) | 82. | (a) | 83. | (b) | 84. | (d) | 85. | (a) | 86. | (b) | 87. | (c) | 88. | (b) | 89. | (a) | 90. | (a) |

# Answer *with* Explanations

## Physics

**1.** (c) ∴ Density, $\rho = \frac{\text{Mass}}{\text{Volume}} = \frac{M}{L^3}$

or $\rho = \frac{M}{L^3}$

$\Rightarrow$ Error in density $\frac{\Delta\rho}{\rho} = \frac{\Delta M}{M} + \frac{3\Delta L}{L}$

So, maximum % error in measurement of $\rho$ is

$$\frac{\Delta\rho}{\rho} \times 100 = \frac{\Delta M}{M} \times 100 + \frac{3\Delta L}{L} \times 100$$

or % error in density $= 1.5 + 3 \times 1$

% error $= 4.5\%$

**2.** (b) **Key Idea** *If velocity versus time graph is a straight line with negative slope, then acceleration is constant and negative.*

With a negative slope distance-time graph will be parabolic $\left(s = ut - \frac{1}{2}at^2\right)$.

So, option (b) will be incorrect.

**3.** (b) Motion stops when pull due to $m_1 \leq$ force of friction between $m$ and $m_2$ and surface.

$\Rightarrow \quad m_1 g \leq \mu(m_2 + m)g$

$\Rightarrow \quad 5 \times 10 \leq 0.15(10 + m) \times 10$

$\Rightarrow \quad m \geq 23.33\,\text{kg}$

Here, nearest value is 27.3 kg

So, $m_{\text{min}} = 27.3\,\text{kg}$

**4.** (c) ∴ Force $= -\frac{dU}{dr}$

$\Rightarrow \quad F = -\frac{d}{dr}\left(\frac{-k}{2r^2}\right) = -\frac{k}{r^3}$

As particle is on circular path, this force must be centripetal force.

$$\Rightarrow \quad |F| = \frac{mv^2}{r}$$

So, $\frac{k}{r^3} = \frac{mv^2}{r} \Rightarrow \frac{1}{2}mv^2 = \frac{k}{2r^2}$

$\therefore$ Total energy of particle = KE + PE

$$= \frac{k}{2r^2} - \frac{k}{2r^2} = 0$$

Total energy = 0

**5.** (b) ***Key Idea*** *Momentum is conserved in all type of collisions,*

Final kinetic energy is 50% more than initial kinetic energy

$$\Rightarrow \quad \frac{1}{2}mv_2^2 + \frac{1}{2}mv_1^2 = \frac{150}{100} \times \frac{1}{2}mv_0^2 \quad \dots\text{(i)}$$

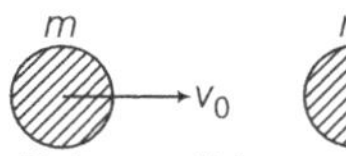

Before collision

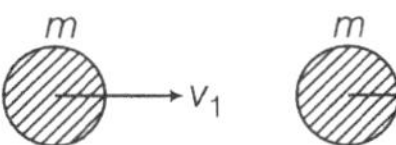

After collision

Conservation of momentum gives,

$$mv_0 = mv_1 + mv_2$$

$$v_0 = v_2 + v_1 \quad \dots\text{(ii)}$$

From Eqs. (i) and (ii), we have

$$v_1^2 + v_2^2 + 2v_1v_2 = v_0^2$$

$$\Rightarrow \quad 2v_1v_2 = \frac{-v_0^2}{2}$$

$$\therefore \quad (v_1 - v_2)^2 = (v_1 + v_2)^2 - 4v_1v_2 = 2v_0^2$$

or $\quad v_{rel} = \sqrt{2}v_0$

**6.** (d) ***Key Idea*** *First we found moment of inertia (MI) of system using parallel axis theorem about centre of mass, then we use it to find moment of inertia about given axis.*

Moment of inertia of an outer disc about the axis through centre is

$$= \frac{MR^2}{2} + M(2R)^2 = MR^2\left(4 + \frac{1}{2}\right) = \frac{9}{2}MR^2$$

For 6 such discs,

moment of inertia $= 6 \times \frac{9}{2}MR^2 = 27MR^2$

So, moment of inertia of system

$$= \frac{MR^2}{2} + 27MR^2 = \frac{55}{2}MR^2$$

Hence, $\quad I_P = \frac{55}{2}MR^2 + (7M \times 9R^2)$

$$\Rightarrow \quad I_P = \frac{181}{2}MR^2$$

$$I_{\text{system}} = \frac{181}{2}MR^2$$

**7.** (a) Moment of inertia of remaining solid
= Moment of intertia of complete solid
− Moment of inertia of removed portion

$$\therefore \quad I = \frac{9MR^2}{2} - \left[\frac{M\left(\frac{R}{3}\right)^2}{2} + M\left(\frac{2R}{3}\right)^2\right]$$

$$\Rightarrow \quad I = 4MR^2$$

**8.** (c) $\therefore$ Force = Mass × Acceleration $= m\omega^2R$

and given, $\quad F \propto \frac{1}{R^n} \Rightarrow F = \frac{k}{R^n}$

So, we have

$$\frac{k}{R^n} = m\left(\frac{2\pi}{T}\right)^2 R$$

$$\Rightarrow \quad T^2 = \frac{4\pi^2 m}{k} \cdot R^{n+1} \quad \Rightarrow \quad T \propto R^{\frac{n+1}{2}}$$

**9.** (c) $\therefore$ Bulk modulus,

$$K = \frac{\text{Volumetric stress}}{\text{Volumetric strain}} = \frac{\Delta p}{\frac{\Delta V}{V}}$$

$$\Rightarrow K = \frac{mg}{a\left(\frac{3\Delta r}{r}\right)} \left[\because V = \frac{4}{3}\pi r^3, \text{ so } \frac{\Delta V}{V} = \frac{3\Delta r}{r}\right]$$

$$\Rightarrow \quad \frac{\Delta r}{r} = \frac{mg}{3aK}$$

**10.** (c) For adiabatic process relation of temperature and volume is,

$$T_2V_2^{\gamma-1} = T_1V_1^{\gamma-1}$$

$$\Rightarrow \quad T_2(2V)^{2/3} = 300(V)^{2/3}$$

$$\left[\gamma = \frac{5}{3} \text{ for monoatomic gases}\right]$$

$$\Rightarrow \quad T_2 = \frac{300}{2^{2/3}} \approx 189\text{K}$$

Also, in adiabatic process,

$$\Delta Q = 0, \Delta U = -\Delta W$$

or $\Delta U = \dfrac{-nR(\Delta T)}{\gamma - 1}$

$$= -2 \times \frac{3}{2} \times \frac{25}{3}(300 - 189)$$

$$\approx -2.7 \text{ kJ}$$

$T_2 \approx 189K, \Delta U \approx 2.7\text{kJ}$

**11.** (a)

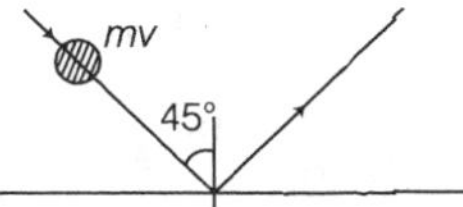


Momentum imparted due to first collision

$= 2mv\sin 45° = \sqrt{2}mv$ $\left[\because \sin 45° = \dfrac{1}{\sqrt{2}}\right]$

$\therefore$ Pressure on surface

$$= \frac{n\sqrt{2}mv}{\text{Area}} = \frac{10^{23} \times \sqrt{2} \times 3.32 \times 10^{-27} \times 10^3}{(2 \times 10^{-2})^2}$$

$p = 2.35 \times 10^3 \text{ N/m}^2$

**12.** (b) For a harmonic oscillator,

$T = 2\pi\sqrt{\dfrac{m}{k}}$, where $k$ = force constant and

$T = \dfrac{1}{f}$

$\therefore \quad k = 4\pi^2 f^2 m$

$$= 4 \times \left(\frac{22}{7}\right)^2 \times (10^{12})^2 \times \frac{108 \times 10^{-3}}{6.02 \times 10^{23}}$$

$\Rightarrow \quad k = 7.1 \text{ N/m}$

**13.** (a)

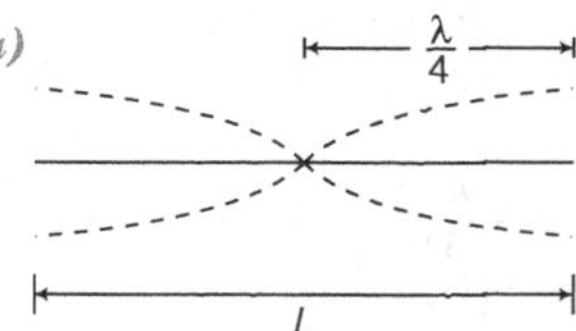


From vibration mode,

$$\frac{\lambda}{2} = L \Rightarrow \lambda = 2L$$

$\therefore$ Wave speed, $v = \sqrt{\dfrac{Y}{\rho}}$

So, frequency $f = \dfrac{v}{\lambda}$

$\Rightarrow \quad f = \dfrac{1}{2L}\sqrt{\dfrac{Y}{\rho}}$

$$= \frac{1}{2 \times 60 \times 10^{-2}}\sqrt{\frac{9.27 \times 10^{10}}{2.7 \times 10^3}}$$

$$\approx 5000 \text{ Hz}$$

$$f = 5 \text{ k Hz}$$

**14.** (b) ***Key Idea*** *Potential of B = Potential due to charge on A + Potential due to charge on B + Potential due to charge on C.*

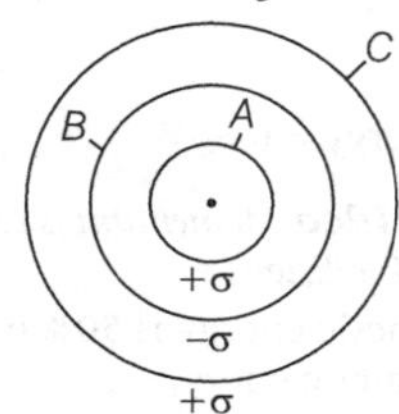


$$\therefore V_B = \frac{k(Q_A + Q_B)}{b} + \frac{kQ_C}{c}$$

$$= \frac{1}{4\pi\varepsilon_0}\left[\frac{\sigma 4\pi a^2}{b} - \frac{\sigma 4\pi b^2}{b} + \frac{\sigma 4\pi c^2}{c}\right]$$

$$= \frac{\sigma\varepsilon}{\varepsilon_0}\left(\frac{a^2 - b^2}{b} + \frac{c^2}{c}\right)$$

$$= \frac{\sigma}{\varepsilon_0}\left(\frac{a^2 - b^2}{b} + c\right)$$

$$V_B = \frac{\sigma}{\varepsilon_0}\left(\frac{a^2 - b^2}{b} + c\right)$$

**15.** (a) Magnitude of induced charge is given by $Q' = (K - 1)\, CV_0$

$$= \left(\frac{5}{3} - 1\right) 90 \times 10^{-12} \times 20$$

$$= 1.2 \times 10^{-9}\text{C}$$

$\Rightarrow \quad Q' = 1.2\text{nC}$

**16.** (b) Given, $e = 100\sin 30t$

and $\quad i = 20\sin\left(30t - \dfrac{\pi}{4}\right)$

$\therefore$ Average power ,

$$P_{av} = V_{rms} I_{rms} \cos\phi = \frac{100}{\sqrt{2}} \times \frac{20}{\sqrt{2}} \times \cos\frac{\pi}{4}$$

$$= \frac{1000}{\sqrt{2}} \text{ watt}$$

Wattless current is,

$$I = I_{rms}\sin\phi = \frac{20}{\sqrt{2}} \times \sin\frac{\pi}{4} = \frac{20}{2} = 10\text{A}$$

$\therefore \quad P_{av} = \dfrac{1000}{\sqrt{2}}$ watt and $I_{wattless} = 10\text{A}$

**17.** (b)

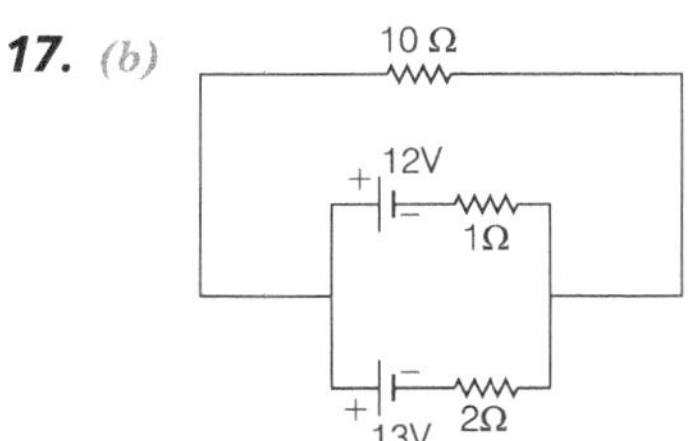


For parallel combination of cells,

$$E_{eq} = \frac{\frac{E_1}{r_1} + \frac{E_2}{r_2}}{\frac{1}{r_1} + \frac{1}{r_2}}$$

$$\therefore \quad E_{eq} = \frac{\frac{12}{1} + \frac{13}{2}}{\frac{1}{1} + \frac{1}{2}} = \frac{37}{3} \text{ V}$$

Potential drop across 10 Ω resistance,

$$V = \left(\frac{E}{R_{total}}\right) \times 10$$

$$= \frac{\frac{37}{3}}{\left(10 + \frac{2}{3}\right)} \times 10 = 11.56 \text{V}$$

$$\therefore \quad V = 11.56 \text{ V}$$

**Alternative Method**

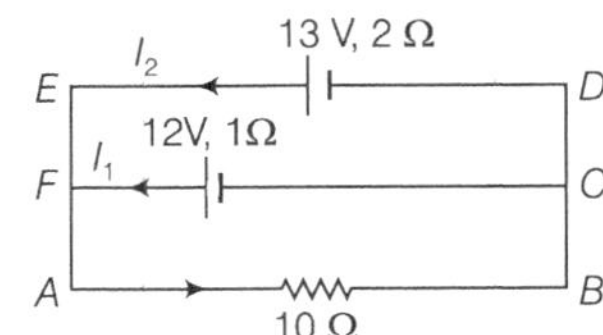


Applying KVL,

in loop *ABCFA*,

$$-12 + 10(I_1 + I_2) + 1 \times I_1 = 0$$

$$\Rightarrow \quad 12 = 11I_1 + 10I_2 \quad \ldots \text{(i)}$$

Similarly,

in loop *ABDEA*,

$$-13 + 10(I_1 + I_2) + 2 \times I_2 = 0$$

$$\Rightarrow \quad 13 = 10I_1 + 12I_2 \quad \ldots \text{(ii)}$$

Solving Eqs. (i) and (ii), we get

$$I_1 = \frac{7}{16}\text{A}, \; I_2 = \frac{23}{32}\text{A}$$

∴ Voltage drop across 10 Ω resistance is,

$$V = 10\left(\frac{7}{16} + \frac{23}{32}\right) = 11.56 \text{ V}$$

**18.** (b) From $Bqv = \frac{mv^2}{r}$, we have

$$r = \frac{mv}{Bq} = \frac{\sqrt{2mK}}{Bq}$$

where, $K$ is the kinetic energy.

As, kinetic energies of particles are same;

$$r \propto \frac{\sqrt{m}}{q}$$

$$\Rightarrow r_e : r_p : r_\alpha = \frac{\sqrt{m_e}}{e} : \frac{\sqrt{m_p}}{e} : \frac{\sqrt{4m_p}}{2e}$$

Clearly, $r_p = r_\alpha$ and $r_e$ is least $[\because m_e < m_p]$

So, $\quad r_p = r_\alpha > r_e$

**19.** (c) **Key Idea** *As m = IA, so to change dipole moment (current is kept constant), we have to change radius of loop.*

Initially, $\quad m = I\pi R^2$ and $B_1 = \frac{\mu_0 I}{2R_1}$

Finally, $\quad m' = 2m = I\pi R_2^2$

$$\Rightarrow \quad 2I\pi R_1^2 = I\pi R_2^2 \quad \text{or} \quad R_2 = \sqrt{2}R_1$$

So, $\quad B_2 = \frac{\mu_0 I}{2(R_2)} = \frac{\mu_0 I}{2\sqrt{2}R_1}$

Hence, $\quad \text{ratio } \frac{B_1}{B_2} = \frac{\left(\frac{\mu_0 I}{2R_1}\right)}{\left(\frac{\mu_0 I}{2\sqrt{2}R_1}\right)} = \sqrt{2}$

$$\therefore \quad \text{Ratio } \frac{B_1}{B_2} = \sqrt{2}$$

**20.** (a) Sharpness of resonance of a resonant *L-C-R* circuit is determined by the ratio of resonant frequency with the selectivity of circuit. This ratio is also called "Quality Factor" or *Q*-factor.

$$Q\text{-factor} = \frac{\omega_0}{2\Delta\omega} = \frac{\omega_0 L}{R} = \frac{1}{\omega_0 CR}$$

**21.** (c) **Key Idea** *Speed of progressive wave is given by,* $v = \frac{\omega}{k}$

As electric field in air is,

$$E_1 = E_{01}\,\mathbf{x}\cos\left(\frac{2\pi\nu z}{c} - 2\pi\nu t\right)$$

$$\therefore \quad \text{Speed in air} = \frac{2\pi\nu}{\left(\frac{2\pi\nu}{c}\right)} = c$$

Also, $\quad c = \frac{1}{\sqrt{\mu_0 \varepsilon_{r_1} \varepsilon_0}} \quad \ldots \text{(i)}$

In medium, $E_2 = E_{02}\mathbf{x} \cos(2kz - kct)$

$\therefore$ Speed in medium $= \frac{kc}{2k} = \frac{c}{2}$

Also, $\frac{c}{2} = \frac{1}{\sqrt{\mu_0 \varepsilon_{r_2} \varepsilon_0}}$ ...(ii)

As medium is non-magnetic medium, $\mu_{\text{medium}} = \mu_{\text{air}}$

On dividing Eq. (i) by Eq. (ii), we have

$$2 = \sqrt{\frac{\varepsilon_{r_2}}{\varepsilon_{r_1}}} \Rightarrow \frac{\varepsilon_{r_1}}{\varepsilon_{r_2}} = \frac{1}{4}$$

**22.** *(c)*

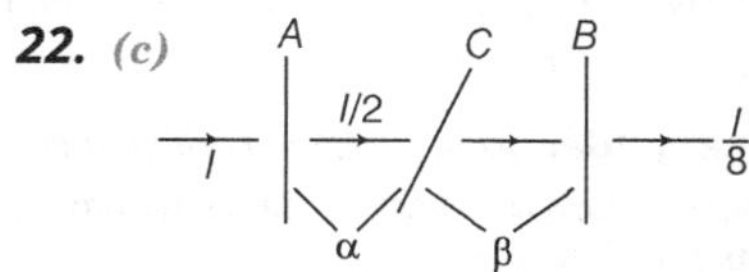


Using Malus law, intensity available after $C = \frac{I}{2} \times \cos^2 \alpha$

and intensity available after

$$B = \frac{I}{2}\cos^2 \alpha \times \cos^2 \beta = \frac{I}{8} \quad \text{(given)}$$

So, $\frac{I}{2} \times \cos^2 \alpha \cdot \cos^2 \beta = \frac{I}{8}$

$\Rightarrow \cos^2 \alpha \cdot \cos^2 \beta = \frac{1}{4}$

This is satisfied with $\alpha = 45°$ and $\beta = 45°$

So, angle between $A$ and $C$ is 45°.

**23.** *(a)* Angular width of diffraction pattern = 60°

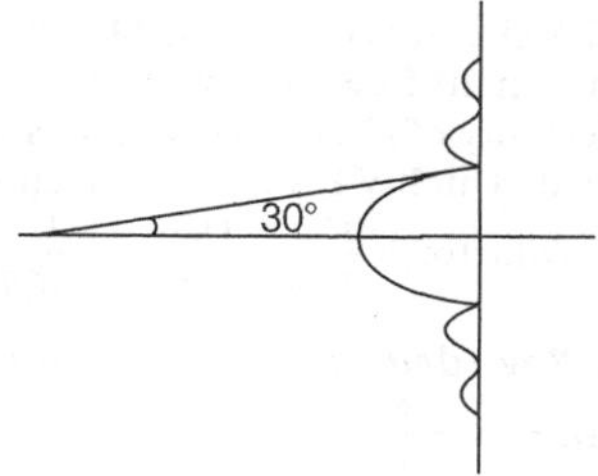


For first minima,

$\frac{a}{2}\sin\theta = \frac{\lambda}{2}$, [here, $a = 10^{-6}$ m, $\theta = 30°$]

$\Rightarrow \lambda = 10^{-6} \times \sin 30° \Rightarrow \lambda = \frac{10^{-6}}{2}$ m

Now, in case of interference caused by bringing second slit,

$\therefore$ Fringe width, $\beta = \frac{\lambda D}{d}$

[here, $\lambda = \frac{10^{-6}}{2}$ m, $\beta = 1$ cm $= \frac{1}{100}$ m,

$d = ?$ and $D = 50$ cm $= \frac{50}{100}$ m]

So, $d = \frac{\lambda D}{\beta} = \frac{10^{-6} \times 50}{2 \times \frac{1}{100} \times 100} = 25 \times 10^{-6}$ m

or $d = 25\,\mu\text{m}$

**24.** *(a)* If wavelength of emitted photon in de-excitation is $\Lambda_n$;

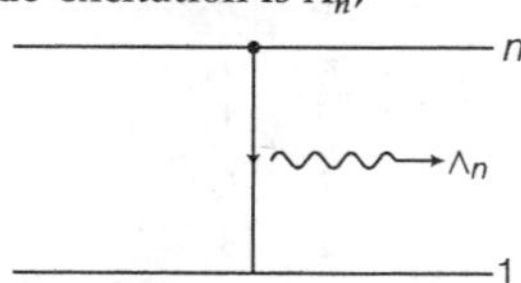


Then, $\frac{hc}{\Lambda_n} = E_n - E_g$

$$\frac{hc}{\Lambda_n} = \frac{p_n^2}{2m} - \frac{p_g^2}{2m}, \quad \left[\because E = \frac{p^2}{2m}\right]$$

As energies are negative, we get

$$\frac{hc}{\Lambda_n} = \frac{p_g^2}{2m} - \frac{p_n^2}{2m}$$

$$\Rightarrow \quad = \frac{p_g^2}{2m}\left(1 - \left(\frac{p_n}{p_g}\right)^2\right)$$

$$= \frac{h^2}{2m\lambda_g^2}\left(1 - \frac{\lambda_g^2}{\lambda_n^2}\right) \quad [\because p \propto \lambda^{-1}, p = \frac{h}{\lambda}]$$

$$\Rightarrow \Lambda_n = \frac{2m\lambda_g^2 c}{h}\left(1 - \frac{\lambda_g^2}{\lambda_n^2}\right)^{-1}$$

$$\Rightarrow \Lambda_n = \frac{2m\lambda_g^2 c}{h}\left(1 + \frac{\lambda_g^2}{\lambda_n^2}\right)$$

$[\because (1 - x)^{-n} = 1 + nx]$

$$\Rightarrow \Lambda_n \simeq A + \frac{B}{\lambda_n^2}$$

where, $A = \left[\frac{2mc\lambda_g^2}{h}\right]$ and $B = \left[\frac{2mc\lambda_g^4}{h}\right]$

are constants.

**25.** *(d)* **Key Idea** *Series limit occurs in the transition $n_2 = \infty$ to $n_1 = 1$ in Lyman series* and $n_2 = \infty$ to $n_1 = 5$ in *p*fund series. For Lyman series,

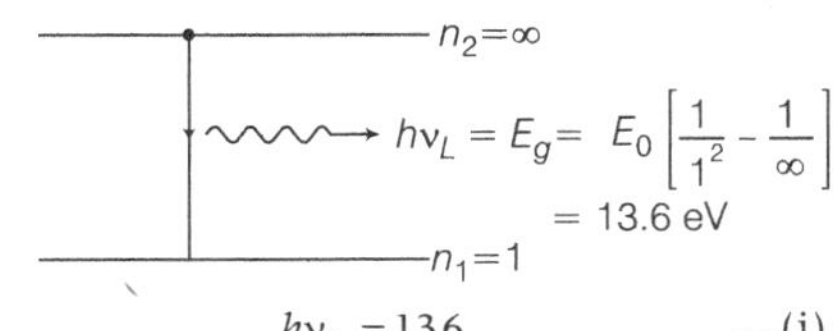


$$h\nu_L = 13.6 \quad \ldots(\text{i})$$

In *pfund* series

$n_2=\infty$

$$h\nu_p = E_0\left[\frac{1}{5^2}-\frac{1}{\infty}\right]=\frac{13.6}{5^2}$$

$n_2 = 5$

$$h\nu_p = \frac{13.6}{5^2} \quad \ldots(\text{ii})$$

From Eqs. (i) and (ii), we get

$$25h\nu_p = h\nu_L$$

$$\therefore \quad \nu_p = \frac{\nu_L}{25}$$

**26.** *(a)* Neutron-Deuterium collision;

$m \rightarrow 2m \rightarrow m \rightarrow 2m \rightarrow$

$v_0 \quad v=0 \quad v_1 \quad v_2$

Momentum conservation gives;

$$mv_0 = mv_1 + 2mv_2$$

$$\Rightarrow \quad v_0 = v_1 + 2v_2 \quad \ldots(\text{i})$$

Collision given is elastic .

So, coefficient of restitution, $e = 1$

$$\therefore \quad e = 1 = \frac{\text{Velocity of separation}}{\text{Velocity of approach}}$$

$$\Rightarrow \quad 1 = \frac{v_2 - v_1}{v_0 - 0} \quad \Rightarrow \quad v_0 = v_2 - v_1 \quad \ldots(\text{ii})$$

On adding Eqs. (i) and (ii), we get

$$2v_0 = 3v_2$$

$$\Rightarrow \quad \frac{2v_0}{3} = v_2$$

So, from Eq. (i), we get

$$v_1 = v_0 - 2v_2 = v_0 - \frac{4v_0}{3} \Rightarrow v_1 = -\frac{v_0}{3}$$

Fractional loss of energy of neutron

$$= \left(\frac{-K_f + K_i}{K_i}\right)_{\text{for neutron}}$$

$$= \frac{-\frac{1}{2}mv_1^2 + \frac{1}{2}mv_0^2}{\frac{1}{2}mv_0^2} = \frac{-\frac{v_0^2}{9} + v_0^2}{v_0^2}$$

$$= \left(-\frac{1}{9}+1\right) = \frac{8}{9} = 0.8\overline{8} = 0.89$$

$m \xrightarrow{v_0} 12m \rightarrow m \xrightarrow{v_1} + 12m \rightarrow v_2$

Similarly, for neutron-carbon atom collision;

Momentum conservation gives;

$$v_0 = v_1 + 12v_2 \text{ and } e = 1$$

$$\Rightarrow \quad v_0 = v_2 - v_1$$

$$\text{So,} \quad v_1 = \frac{11}{13}v_0$$

$$\therefore \text{Loss of energy} = \left(-\frac{121}{169}+1\right) = 0.28$$

So, $P_d = 0.89$ and $P_c = 0.28$

**27.** *(c)* ***Key Idea*** *Potential drop in a silicon diode in forward bias is around 0.7 V.*

In given circuit, potential drop across 200 Ω resistor is

$$I = \frac{\Delta V_{\text{net}}}{R} = \frac{3-0.7}{200}$$

$$\Rightarrow I = 0.0115\text{A} \quad \Rightarrow \quad I = 11.5 \text{ mA}$$

**28.** *(c)* Only 10% of 10 GHz is utilised for transmission.

$\therefore$ Band available for transmission

$$= \frac{10}{100} \times 10 \times 10^9 \text{ Hz} = 10^9 \text{ Hz}$$

Now, if there are $n$ channels each using 5 kHz then, $n \times 5 \times 10^3 = 10^9$

$$\Rightarrow \quad n = 2 \times 10^5$$

**29.** *(b)* With only the cell,

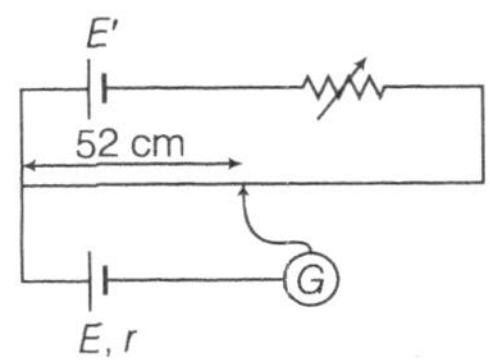


On balancing, $E = 52 \times x \quad \ldots(\text{i})$

where, $x$ is the potential gradient of the wire.

When the cell is shunted,

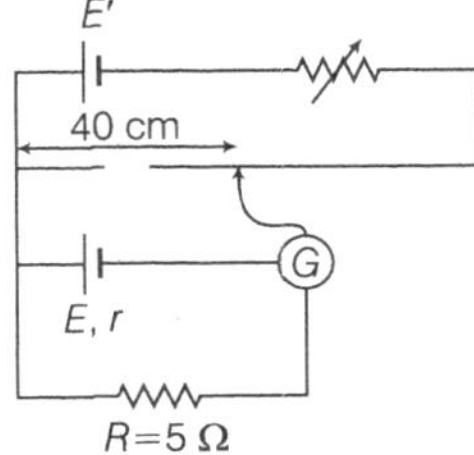

Similarly, on balancing,

$$V = E - \frac{Er}{(R+r)} = 40 \times x \quad \dots(\text{ii})$$

Solving Eqs. (i) and (ii), we get

$$\frac{E}{V} = \frac{1}{1 - \frac{r}{R+r}} = \frac{52}{40}$$

$$\Rightarrow \quad \frac{E}{V} = \frac{R+r}{R} = \frac{52}{40}$$

$$\Rightarrow \quad \frac{5+r}{5} = \frac{52}{40}$$

$$\Rightarrow \quad r = \frac{3}{2}\Omega \qquad r = 1.5\,\Omega$$

**30.** *(c)* We have, $X + Y = 1000\,\Omega$

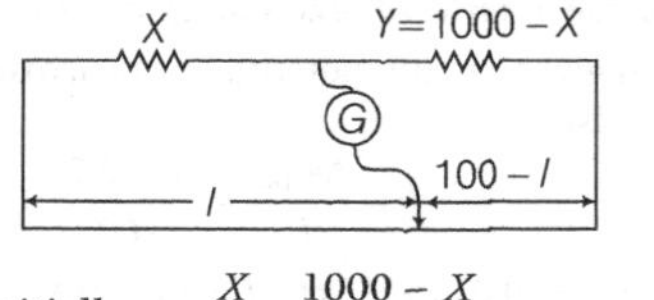


$$\text{Initially,} \quad \frac{X}{l} = \frac{1000 - X}{100 - l} \quad \dots(\text{i})$$

When $X$ and $Y$ are interchanged, then

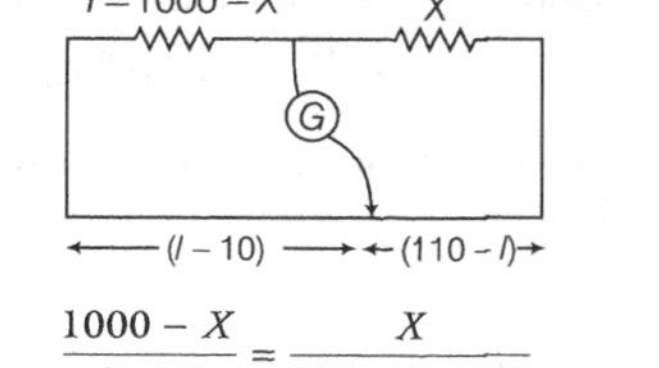


$$\frac{1000 - X}{l - 10} = \frac{X}{100 - (l - 10)}$$

$$\text{or} \quad \frac{1000 - X}{l - 10} = \frac{X}{110 - l} \quad \dots(\text{ii})$$

From Eqs. (i) and (ii), we get

$$\frac{100 - l}{l} = \frac{l - 10}{110 - l}$$

$$(100 - l)(110 - l) = (l - 10)l$$

$$11000 - 100l - 110l + l^2 = l^2 - 10l$$

$$\Rightarrow \quad 11000 = 200l$$

$$\therefore \quad l = 55\,\text{cm}$$

Substituting the value of $l$ in Eq. (i), we get

$$\frac{X}{55} = \frac{1000 - 55}{100 - 55}$$

$$\Rightarrow \quad 20X = 11000$$

$$\therefore \quad X = 550\,\Omega$$

# Chemistry

**31.** *(d)* We can calculate the simplest whole number ratio of C and H from the data given, as

| Element | Relative mass | Molar mass | Relative mole | Simplest whole number ratio |
|---|---|---|---|---|
| C | 6 | 12 | $\frac{6}{12} = 0.5$ | $\frac{0.5}{0.5} = 1$ |
| H | 1 | 1 | $\frac{1}{1} = 1$ | $\frac{1}{0.5} = 2$ |

Alternatively this ratio can also be calculated directly in the terms of $x$ and $y$ as

$\frac{12x}{y} = \frac{6}{1}$ (given and molar mass of C = 12, H = 1)

Now, after calculating this ratio look for condition 2 given in the question i.e. quantity of oxygen is half of the quantity required to burn one molecule of compound $C_xH_y$ completely to $CO_2$ and $H_2O$. We can calculate number of oxygen atoms from this as consider the equation.

$$C_xH_y + \left[x + \frac{y}{4}\right]O_2 \longrightarrow xCO_2 + \frac{y}{2}H_2O$$

Number of oxygen atoms required

$$= 2 \times \left[x + \frac{y}{4}\right] = \left[2x + \frac{y}{2}\right]$$

$$\text{Now given, } z = \frac{1}{2}\left[2x + \frac{y}{2}\right] = \left[x + \frac{y}{4}\right]$$

Here we consider $x$ and $y$ as simplest ratios for C and H so now putting the values of $x$ and $y$ in the above equation.

$$z = \left[x + \frac{y}{4}\right]$$

$$= \left[1 + \frac{2}{4}\right] = 1.5$$

Thus, the simplest ratio figures for $x$, $y$ and $z$ are $x = 1$, $y = 2$ and $z = 1.5$

Now, put these values in the formula given i.e. $C_xH_yO_z = C_1H_2O_{1.5}$

So, empirical formula will be

$[C_1H_2O_{1.5}] \times 2 = C_2H_4O_3$

**32.** (c) It is the "Frenkel defect" in which cations leave their original site and occupy interstitial site as shown below.

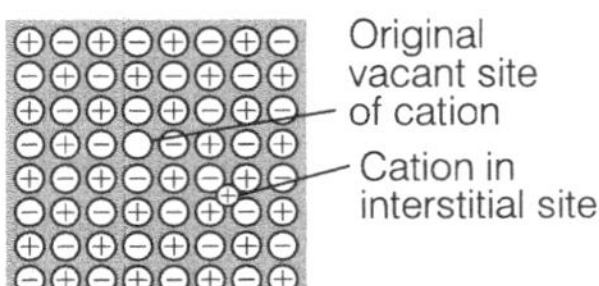


**33.** (d) ***Key Idea*** *According to M.O.T, the viability of any molecule can be judged through the calculation of bond order.*

| Electronic | Configuration | Bond order |
|---|---|---|
| $He_2^+$ | $\sigma_{1s^2}\sigma^*_{1s^1}$ | $\frac{2-1}{2} = 0.5$ |
| $H_2^-$ | $\sigma_{1s^2}\sigma^*_{1s^1}$ | $\frac{2-1}{2} = 0.5$ |
| $H_2^{2-}$ | $\sigma_{1s^2}\sigma^*_{1s^2}$ | $\frac{2-2}{2} = 0$ |
| $He_2^{2+}$ | $\sigma_{1s^2}$ | $\frac{2-0}{2} = 1$ |

The molecule having zero bond order will not be viable hence, $H_2^{2-}$ (option d) is the correct answer.

**34.** (a) From thermodynamics,

$$\ln k = \frac{-\Delta H^\circ}{RT} + \frac{\Delta S^\circ}{R} \qquad \ldots\text{(i)}$$

Mathematically, the equation of straight line is

$$y = c + mx \qquad \ldots\text{(ii)}$$

After comparing Eq. (ii) with (i) we get,

slope $= \frac{-\Delta H^\circ}{R}$ and intercept $= \frac{\Delta S^\circ}{R}$

Now, we know for exothermic reaction $\Delta H$ is negative (−)ve. But here,

$$\text{Slope} = \frac{-\Delta H^\circ}{R} \text{ is positive}$$

So, lines $A$ and $B$ in the graph represent temperature dependence of equilibrium constant $K$ for an exothermic reaction as shown below

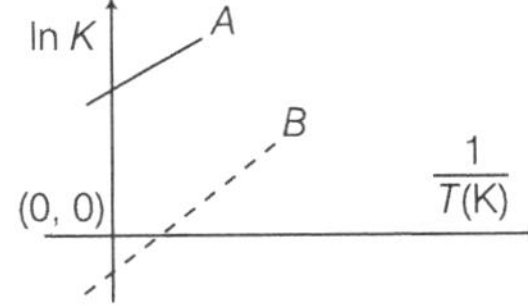


**35.** (d) ***Key idea*** *Calculate the heat of combustion with the help of following formula*

$$\Delta H_p = \Delta U + \Delta n_g RT$$

*where, $\Delta H_p$ = Heat of combustion at constant pressure*

*$\Delta U$ = Heat at constant volume (It is also called $\Delta E$)*

*$\Delta n_g$ = Change in number of moles (In gaseous state).*

*$R$ = Gas constant; $T$ = Temperature.*

From the equation,

$$C_6H_6(l) + \frac{15}{2}O_2(g) \longrightarrow 6CO_2(g) + 3H_2O(l)$$

Change in the number of gaseous moles i.e.

$$\Delta n_g = 6 - \frac{15}{2} = -\frac{3}{2} \text{ or } -1.5$$

Now we have $\Delta n_g$ and other values given in the question are

$$\Delta U = -3263.9 \text{ kJ/mol}$$

$$T = 25^\circ\text{C} = 273 + 25 = 298 \text{ K}$$

$$R = 8.314 \text{JK}^{-1}\text{mol}^{-1}$$

So, $\Delta H_p = (-3263.9) + (-1.5) \times 8.314 \times 10^{-3} \times 298$

$$= -3267.6 \text{ kJ mol}^{-1}$$

**36.** (d) ***Key idea*** *"Addition of solute particles to a pure solvent results to depression in its freezing point."*

*All the compounds given in question are ionic in nature so, consider their van't Hoff factor ($i$) to reach at final conclusion.*

The solution with maximum freezing point must have minimum number of solute particles. This generalisation can be done with the help of van't Hoff factor ($i$) i.e.

Number of solute particles $\propto$ van't Hoff factor ($i$)

Thus, we can say directly

Solution with maximum freezing point will be the one in which solute with minimum van't Hoff factor is present

Now, for

$$Co(H_2O)_6]Cl_3 \rightleftharpoons [Co(H_2O)_6]^{3+} + 3Cl^-$$

van't Hoff factor ($i$) is 4. Similarly for,

$$[Co(H_2O)_5Cl]Cl_2 \cdot H_2O \rightleftharpoons [Co(H_2O)_5Cl]^{2+} + 2Cl^- \quad \text{'}i\text{' is } 3$$

$[Co(H_2O)_4Cl_2]Cl \cdot 2H_2O \rightleftharpoons$
$[Co(H_2O)_4Cl_2]^+ + Cl^-$ '$i$' is 2

and for $[Co(H_2O)_3Cl_3] \cdot 3H_2O$, '$i$' is 1 as it does not show ionisation. Hence, $[Co(H_2O)_3Cl_3] \cdot 3H_2O$ have minimum number of particles in the solution.

So, freezing point of its solution will be maximum.

**37.** *(b)* Given $[H_2S] = 0.10\,M$

$[HCl] = 0.20\,M$

So, $[H^+] = 0.20\,M$

$$H_2S \rightleftharpoons H^+ + HS^-, \quad K_1 = 1.0 \times 10^{-7}$$

$$HS^- \rightleftharpoons H^+ + S^{2-}, \quad K_2 = 1.2 \times 10^{-13}$$

It means for,

$$H_2S \rightleftharpoons 2H^+ + S^{2-}$$

$$K = K_1 \times K_2$$
$$= 1.0 \times 10^{-7} \times 1.2 \times 10^{-13}$$
$$= 1.2 \times 10^{-20}$$

Now $[S^{2-}] = \dfrac{K \times [H_2S]}{[H^+]^2}$ [according to the final equation]

$$= \frac{1.2 \times 10^{-20} \times 0.1\,M}{(0.2M)^2}$$

$$= \frac{1.2 \times 10^{-20} \times 1 \times 10^{-1}\,M}{4 \times 10^{-2} M}$$

$$= 3 \times 10^{-20} M$$

**38.** *(c)* Its given that the final volume is 500 mL and this final volume was arrived when 50 mL of 1 M $Na_2SO_4$ was added to unknown $Ba^{2+}$ solution.

So, we can interpret the volume of unknown $Ba^{2+}$ solution as 450 mL i.e.

$$\underset{\substack{Ba^{2+}\\ \text{solution}}}{450\text{mL}} + \underset{\substack{Na_2SO_4\\ \text{solution}}}{50\text{mL}} \longrightarrow \underset{\substack{BaSO_4\\ \text{solution}}}{500\text{mL}}$$

From this we can calculate the concentration of $SO_4^{2-}$ ion in the solution *via*

$$M_1V_1 = M_2V_2$$
$$1 \times 50 = M_2 \times 500$$

(as $1M\,Na_2SO_4$ is taken into consideration)

$$M_2 = \frac{1}{10} = 0.1\,M$$

Now for just precipitation,

Ionic product = Solubility product ($K_{sp}$)

i.e. $[Ba^{2+}][SO_4^{2-}] = K_{sp}$ of $BaSO_4$

Given $K_{sp}$ of $BaSO_4 = 1 \times 10^{-10}$

So, $[Ba^{2+}][0.1] = 1 \times 10^{-10}$

or $[Ba^{2+}] = 1 \times 10^{-9}$ M

***Remember*** This is the concentration of $Ba^{2+}$ ions in final solution. Hence, for calculating the $[Ba^{2+}]$ in original solution we have to use

$$M_1V_1 = M_2V_2$$

as $M_1 \times 450 = 10^{-9} \times 500$

so, $M_1 = 1.1 \times 10^{-9}$M

**39.** *(a)* For the reaction,

$$CH_3CHO(g) \xrightarrow{\text{Decomposes}} CH_4 + CO$$

Let order of reaction with respect to $CH_3CHO$ is $m$.

Its given, $r_1 = 1$ torr/sec. when $CH_3CHO$ is 5% reacted i.e. 95% unreacted. Similarly, $r_2 = 0.5$ torr/sec when $CH_3CHO$ is 33% reacted i.e., 67% unreacted.

Use the formula, $r \propto (a - x)^m$

where $(a - x)$ = amount unreacted

so, $\dfrac{r_1}{r_2} = \dfrac{(a - x_1)^m}{(a - x_2)^m}$ or $\dfrac{r_1}{r_2} = \left[\dfrac{a - x_1}{a - x_2}\right]^m$

Now putting the given values

$$\frac{1}{0.5} = \left(\frac{0.95}{0.67}\right)^m$$

$\Rightarrow$ $2 = (1.41)^m$

or $m = 2$

**40.** *(c)* Given that, $i = 100$ amp. also, 27.66 g of diborane ($B_2H_6$)

Molecular mass of $B_2H_6$
$= 10.8 \times 2 + 6 = 27.6$

Number of moles of $B_2H_6$ in 27.66 g

$$= \frac{\text{Given mass}}{\text{Molar mass}} = \frac{27.66}{27.6} \approx 1$$

Now consider the equation

$$B_2H_6 + 3O_2 \longrightarrow B_2O_3 + 3H_2O$$

From the equation we can interpret that 3 moles of oxygen is required to burn 1 mole (i.e. 27.6 g) $B_2H_6$ completely.

Also consider the electrolysis reaction of water i.e.

$$H_2O \rightleftharpoons 2H^+ + O^{--}$$

$$2H^+ \xrightarrow[\text{Cathode}]{+2e^-} 2H \longrightarrow H_2 \uparrow$$

$$O^{--} \xrightarrow[-2e^-]{\text{Anode}} O \xrightarrow[\text{atoms}]{\text{2 such}} O_2 \uparrow$$

From the above equation it can be easily interpreted that in electrolysis of water for the production of 1 mole of oxygen from 1 mole of $H_2O$ at anode 4 moles electrons are required.

Likewise for the production of 3 moles of $O_2$ 12(3 × 4) moles of electrons will be needed.

So, the total amount of charge required to produce 3 moles of oxygen will be $12 \times F$ or $12 \times 96500$

We know $\quad Q = it$

So, $\quad 12 \times 96500 = 100 \times t$ in seconds

or $\quad \dfrac{12 \times 96500}{100 \times 3600} = t$ in hours $= 3.2$ hours

**41.** (c) Fluoride ions help in making teeth enamel harder by converting $[3Ca_3(PO_4)_2 \cdot Ca(OH)_2]$ i.e. Hydroxy apatite to $[3Ca_3(PO_4)_2 \cdot CaF_2]$ i.e., Fluorapatite (Harder teeth enamel) *via* following reaction:

$$[3Ca_3(PO_4)_2 \cdot Ca(OH)_2] + \underset{\text{From drinking water}}{2F^-} \longrightarrow [3Ca_3(PO_4)_2 \cdot CaF_2] + 2OH^-$$

**42.** (c) KCl is the only ionic compound. The structure of $PH_3$, $O_2$, $B_2H_6$ and $H_2SO_4$ are given below

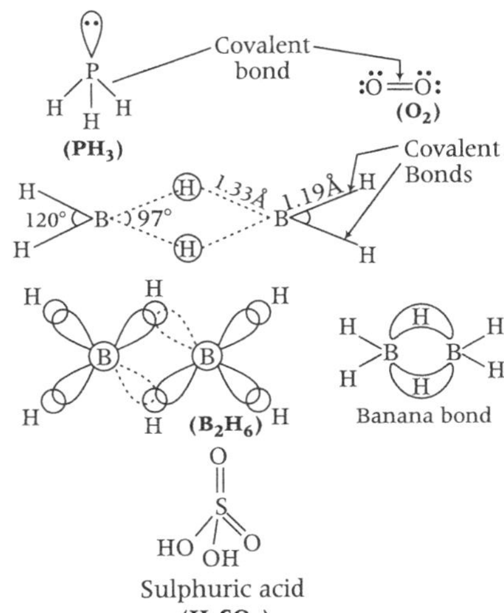


All bond between S and O atom are covalent bonds.

**43.** (d) **Key Idea** *Lewis acids are defined as, "Electron deficient compounds which have the ability to accept atleast one lone pair."*

The compound given are

**$PH_3$**-Octet complete although P has vacant 3*d*-orbital but does not have the tendency to accept lone pair in it. Hence, it cannot be considered as Lewis acid.

**$BCl_3$**-Incomplete octet with following orbital picture.

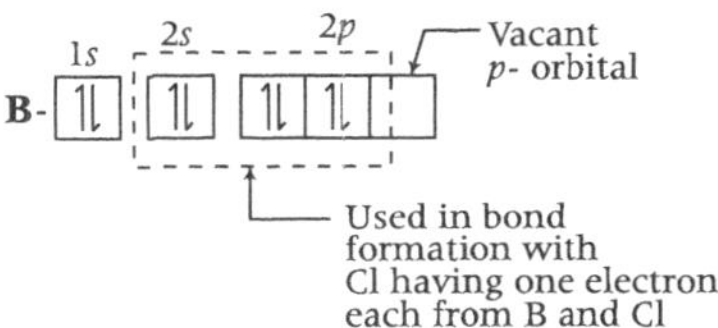


Hence, vacant *p*-orbital of B can accept one lone pair thus it can be considered as Lewis acid.

**$AlCl_3$**-Similar condition is visible in $AlCl_3$ as well i.e.

Al ( Valence orbital only) =

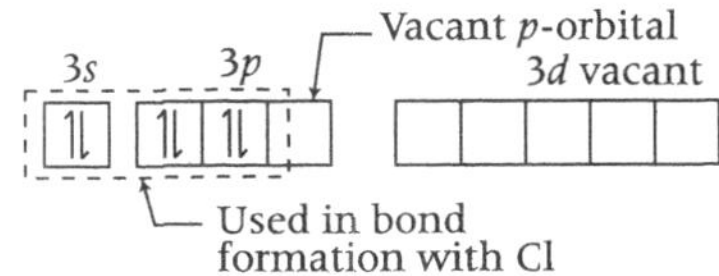


Hence this compound can also be considered as Lewis acid.

**$SiCl_4$** - Although this compound does not have incomplete octet but it shows the tendency to accept lone pair of electrons in its vacant *d*-orbital. This tendency of $SiCl_4$ is visible in following reaction.

$$SiCl_4 + H_2\ddot{O}: \longrightarrow Cl_4Si \leftarrow OH_2 \longrightarrow Cl_3Si{-}OH + HCl$$

Lone pair acceptance in *d*-orbital

Thus option (b) and (d) both appear as correct but most suitable answer is (d) as the condition of a proper Lewis acid is more well defined in $BCl_3$ and $AlCl_3$.

**44.** *(c)* The structure of $I_3^-$ ion is

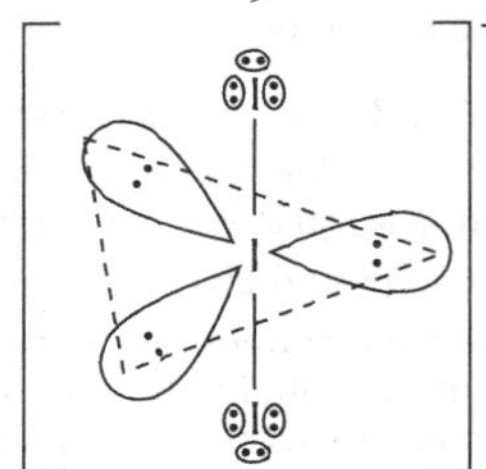

Hence, 9 is the correct answer.

**45.** *(b)* Among the given salts

**$FeCl_3$** is acidic in nature i.e., have acidic solution as it is the salt of weak base and strong acid.

**$Al(CN)_3$** and **$Pb(CH_3COO)_2$** are the salts of weak acid and weak base.

**$CH_3COOK$** is the salt of strong base and weak acid.

Hence, the solution of $CH_3COOK$ will be most basic because of the following reaction.

$$CH_3COOK + H_2O \rightleftharpoons \underset{\text{(Weak acid)}}{CH_3COOH} + \underset{\text{(Strong base)}}{KOH}$$

**46.** *(c)* Both reactions in their complete format are written below

(i) In acidic medium,

$$[Fe^{2+}(CN)_6]^{4-} + H_2\overset{-1}{O_2} + 2H^+ \longrightarrow [Fe^{3+}(CN)_6]^{3-} + 2H_2\overset{-2}{O}$$

(ii) In alkaline medium,

$$[Fe^{3+}(CN)_6]^{3-} + H_2\overset{-1}{O_2} + 2OH^- \longrightarrow [Fe^{2+}(CN)_6]^{4-} + O_2 + 2H_2O$$

Hence, $H_2O$ (for reaction (i)) and $O_2 + H_2O$ (for reaction (ii)) are produced as by product.

**47.** *(c)* Let the oxidation state of Cr in all cases is '$x$'

(i) Oxidation state of Cr in $[Cr(H_2O)_6]Cl_3$

$$x + (0 \times 6) + (-1 \times 3) = 0$$

or $x + 0 - 3 = 0$

or $x = +3$

(ii) Oxidation state of Cr in $[Cr(C_6H_6)_2]$

$$x + (2 \times 0) = 0$$

or $x = 0$

(iii) Oxidation state of Cr in

$$K_2[Cr(CN)_2(O)_2(O_2)(NH_3)]$$

$$1 \times 2 + x + (-1 \times 2) + (-2 \times 2) + (-2) + 0 = 0$$

or $2 + x - 2 - 4 - 2 = 0$

or $x - 6 = 0$

hence $x = +6$

Thus, +3, 0 and +6 is the answer.

**48.** *(d)* The thermal decomposition of given compounds is shown below

$$(NH_4)_2Cr_2O_7 \xrightarrow{\Delta} N_2 + 4H_2O + Cr_2O_3$$

$$NH_4NO_2 \xrightarrow{\Delta} N_2 + 2H_2O$$

$$(NH_4)_2SO_4 \xrightarrow{\Delta} 2NH_3 + H_2SO_4$$

$$Ba(N_3)_2 \longrightarrow Ba + 3N_2$$

Thus, only $(NH_4)_2SO_4$ does not gives $N_2$ on heating (It give $NH_3$). While rest of the given compounds gives $N_2$ on their thermal decomposition.

**49.** *(c)* Among the given metals Al forms white gelatinous ppt. with NaOH. Hence, the probable metal can be Al. This ppt. is dissolved in excess of NaOH due to the formation of sodium metal Aluminate. Both the reactions are shown below.

$$Al^{3+} \xrightarrow{NaOH} \underset{\substack{\text{White gelatinous ppt.} \\ \text{(X) of aluminium} \\ \text{hydroxide}}}{Al(OH)_3} \xrightarrow{\text{Excess of NaOH}} \underset{\substack{\text{Sodium} \\ \text{metaaluminate)} \\ \text{soluble}}}{NaAlO_2}$$

Aluminium hydroxide on strong heating gives alumina ($Al_2O_3$) which is used as an adsorbent in chromatography. This reaction can be seen as :

$$2Al(OH)_3 \xrightarrow{\Delta} Al_2O_3 + 3H_2O$$

Thus, metal $M$ is Al.

**Ca**, being below sodium in electrochemical reactivity series, cannot displaces Na from its aqueous solution.

**Zn** reacts with NaOH to form sodium zincate which is a soluble compound.

**Fe** reacts with sodium hydroxide to form tetrahydroferrate (II) sodium which is again a soluble complex.

**50.** *(b)* If the reactant is *cis* isomer than following reaction takes place.

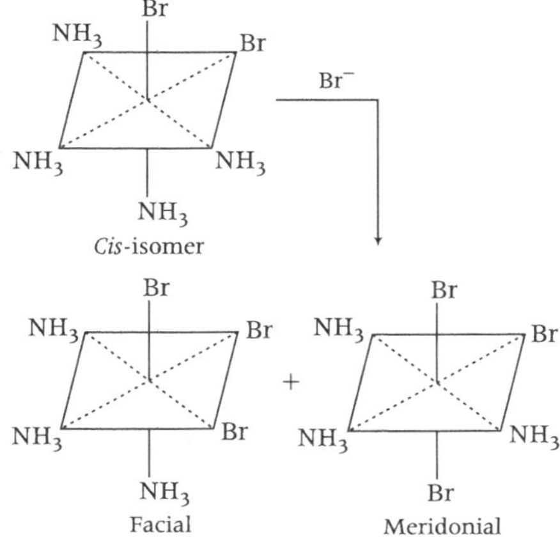


i.e. two isomers are produced. If the reactant is *trans* isomer than following reaction takes place.

Br, $NH_3$, $NH_3$, $NH_3$, $NH_3$, Br — *Trans* $\xrightarrow{Br^-}$ Br, $NH_3$, $NH_3$, $NH_3$, Br, Br — Meridonial

i.e. only 1 isomer is produced. Thus, statement (I) and (III) are correct resulting to option (b) as the correct answer.

**51.** *(a)* HI is a strong reducing agent. It reduces both primary and secondary alcoholic groups of glucose along with the carbonyl group to produce *n*-hexane as

CHO ← Carbonyl group
$(CHOH)_4$ ← Secondary alcoholic group
$CH_2OH$ ← Primary alcoholic group
Glucose

$\xrightarrow[\Delta\ \text{(Prolonged)}]{HI}$ $CH_3$–$(CH_2)_4$–$CH_3$ *n*-hexane

**52.** *(c)* Sodium metal in liquid ammonia reduces alkynes with anti stereochemistry to give *trans* alkenes. The reduction is selectively *anti* since the vinyl radical formed during reduction is more stable in *trans* configuration.

**Mechanism**

$RC{\equiv}CR' \longrightarrow$ R(•)C=C(−)R' $\xrightarrow[-NH_2]{H-NH_2}$ R–C=C(H)R' (Vinylic radical)
$\dot{Na} \rightarrow Na^+$

$\dot{Na} \rightarrow Na^+$ ↓

R–C(−)=C(H)R' (Vinylic anion) $\xrightarrow[-NH_2]{H-NH_2}$ R(H)C=C(H)R'

*Sodium atom donates an electron to alkyne which after H-abstraction from $NH_3$ forms vinylic radical. Transfer of another electron gives a vinylic anion, which is more stable in trans form. This in turn gives trans-alkene after H-abstraction from $NH_3$.*

**53.** *(b)* Estimation of nitrogen through Kjeldahl's method is not suitable for organic compounds containing nitrogen in ring or nitrogen in nitro or azo groups. It is because of the fact that nitrogen of these compounds does not show conversion to Ammonium sulphate ($(NH_4)_2SO_4$) during the process. Hence, among the given compounds only aniline can be used suitably for estimation of nitrogen by Kjeldahl's method.

**54.** *(a)*

$C_6H_5OH + CO_2 + NaOH \xrightarrow[\text{acidification}]{\text{Followed by}} X \xrightarrow[\text{(Catalytic amount)}]{(CH_3CO)_2O,\ \text{conc. } H_2SO_4} ?$

The very first reaction in the above road map looks like Kolbe's reaction which results to salicylic acid as

$C_6H_5OH \xrightarrow[\text{(ii) Acidification}]{\text{(i) } CO_2,\ NaOH}$ o-HO–$C_6H_4$–COOH

Salicylic acid
(*X*)

The salicylic acid with acetic anhydride [$(CH_3CO)_2O$] in the presence of catalytic amount of conc. $H_2SO_4$ undergoes acylation to produce aspirin as

$$\text{C}_6\text{H}_4(\text{OH})(\text{COOH}) + (\text{CH}_3\text{CO})_2\text{O} \xrightarrow{\text{Conc. } H_2SO_4} \text{C}_6\text{H}_4(\text{O—CO—CH}_3)(\text{COOH}) + CH_3COOH$$

Acetyl salicylic acid (Aspirin)

Aspirin is a non-narcotic analgesic (Pain killer).

**55.** *(c)* Methyl orange show Pinkish colour towards more acidic medium and yellow orange colour towards basic or less acidic media. Its working pH range is

Pinkish Red ← 3.9 –4.5 → Yellow orange

Weak base have the pH range greater than 7. When methyl orange is added to this weak base solution it shows yellow orange colour.

Now when this solution is titrated against strong acid the pH move towards more acidic range and reaches to end point near 3.9 where yellow orange colour of methyl orange changes to Pinkish red resulting to similar change in colour of solution as well.

**56.** *(d)* Our blood is slightly basic in nature with pH range from 7.35-7.4.

The structure of histamine is given below:

Imidazole ring; Aliphatic amino group ($NH_2$); Basic nitrogen of imidazole ring

It is produced by decarboxylation of histidine having following structure. It is clearly visible from the above structure that histamine has two basic centres namely aliphatic amino group and basic nitrogen of imidazole ring.

The aliphatic amino group has $pK_a$ around 9.4. In blood with pH around 7.4 the aliphatic amino group of histamine become protonated to give a single charged cation as shown below

(Imidazole ring—$CH_2CH_2$—$\overset{\oplus}{N}H_3$)

**57.** *(c)* Given

$$\text{C}_6\text{H}_5\text{OH} + \text{Methyl chloroformate} \xrightarrow{NaOH} A \xrightarrow{Br_2} B$$

In the above road map, first reaction appears as acid base reaction followed by $S_NAE$ (Nucleophilic substitution through Addition and Elimination). Both the steps are shown below

(i) **Acid base reaction**

$$\text{C}_6\text{H}_5\text{OH} \xrightarrow{^-OH} \text{C}_6\text{H}_5\text{O}^- + H_2O$$

(ii) $\mathbf{S_NAE}$

$$\text{C}_6\text{H}_5\text{O}^- + \text{Cl—C(=O)—OCH}_3 \longrightarrow \text{C}_6\text{H}_5\text{—O—C(=O)—O—CH}_3$$

In the product of $S_NAE$ the attached group is *ortho* and *para*-directing due to following cross conjugation

(Ring positions 1–6, δ– at 2, 4, 6; $\ddot{O}_1$—C(=O)—$\ddot{O}_2$—$CH_3$)

Cross conjugation due to which lone pair of oxygen 1 will be easily available to ring resulting to higher electron density at 2, 4,6 position with respect to group. However from the stability point of view *ortho* positions are not preferred by substituents as group —O—C(=O)—O—$CH_3$ is bulky.

Hence, on further bromination of $S_NAE$ product *para* bromo derivative will be the preferred product i.e.

$$\text{C}_6\text{H}_5\text{—O—C(=O)—O—CH}_3 + Br_2 \longrightarrow p\text{-Br—C}_6\text{H}_4\text{—O—C(=O)—OCH}_3 + HBr$$

**58.** *(c)* ***Key Idea*** *Among the given compounds the basic nature depends upon their tendency to donate electron pair.*

Among the given compounds in $CH_3CH_2CH{=}NH$, Nitrogen is $sp^2$-hybridised. This marginally increases the electronegativity of nitrogen which in turn decreases the electron donation tendency of nitrogen. Thus making compound least basic.

Among the rest $CH_3C(=NH)NH_2$ is totally different from others as in this compound lone pair of one nitrogen are in conjugation with $\pi$ bond i.e. As a result of this conjugation the cation formed after protonation becomes resonance stabilised

In conjugation

Protonation

Equivalent resonance

This equivalent resonance in cation makes $CH_3C(=\ddot{N}H)\ddot{N}H_2$ most basic among all.

Categorisation is very simple between rest two as

$CH_2{=}CHCH_2NH_2$ (primary amine) is less basic than $CH_3CH_2$—N(H)—$CH_3$ (secondary amine)

Hence, the correct order is

(II) < (I) < (IV) < (III) i.e. option *(c)* is correct.

**59.** *(d)* ***Key idea*** *The reaction given is a nucleophilic substitution reaction in which cleavage at C—O bond is visible. The product formation can be visualised with the help of following analysis.*

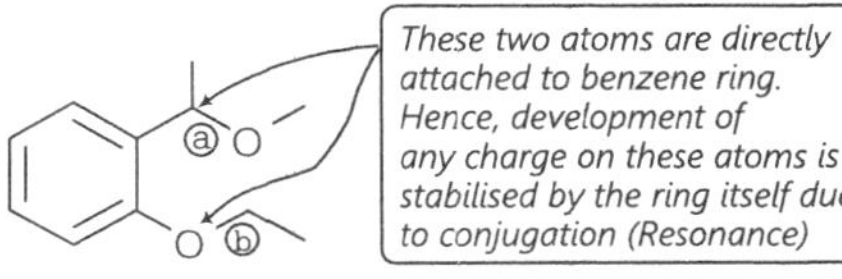


*If any one properly visualise the fact written with figure above, than a conclusion can be made that C—O bonds marked (a) and (b) in the figure will undergo heterolysis during the reaction.*

The reaction can be represented as

HI, Heat

$\longrightarrow$ + $CH_3CH_2I$ + $CH_3OH$

### Mechanism

*Step* I The reaction begins with the attack of $H^+$ of HI on oxygen to form oxonium ion as

Oxonium ion

*Step* II This oxonium ion undergoes lysis and addition of $I^-$ to form two products as

+ $CH_3OH$

*Step* III Similar pathway is followed at the other oxygen atom, which can be visualised as

+ $CH_3CH_2I$

***Note*** *Mechanism of a reaction is always a logical sequencing of events which may occur simultaneously as well.*

**60.** *(b)* Complete reaction can be represented as

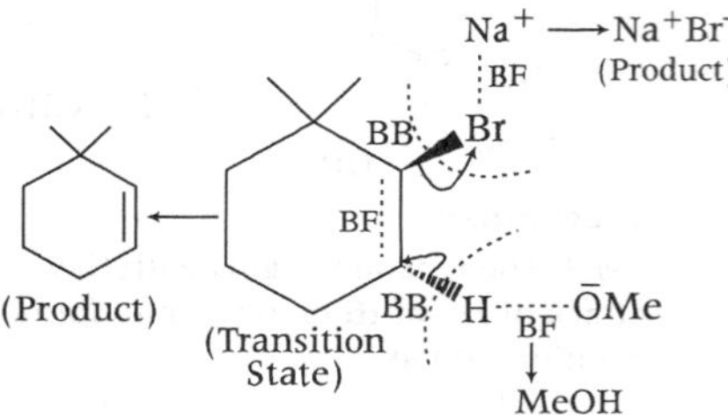

Thus, the given reaction is dehydrohalogenation which is a β-elimination proceeding through $E_2$ mechanism.

***Mechanism*** The reaction proceeds through the formation of following transition state with simultaneous removal of Br and H atoms.

Here, BB = Bond breakage
BF = Bond formation

# Mathematics

**61.** *(b)* We have,

$$|a-5|<1 \text{ and } |b-5|<1$$

$\therefore \quad -1<a-5<1$ and $-1<b-5<1$

$\Rightarrow \quad 4<a<6$ and $4<b<6$

Now, $4(a-6)^2+9(b-5)^2 \le 36$

$$\Rightarrow \quad \frac{(a-6)^2}{9}+\frac{(b-5)^2}{4} \le 1$$

Taking axes as $a$-axis and $b$-axis

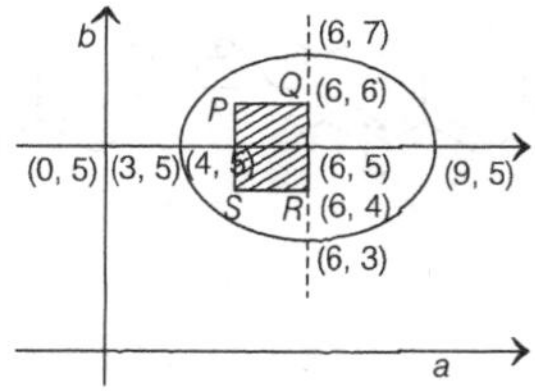


The set $A$ represents square $PQRS$ inside set $B$ representing ellipse and hence $A \subset B$.

**62.** *(c)* We have,

$$2|\sqrt{x}-3|+\sqrt{x}(\sqrt{x}-6)+6=0$$

Let $\sqrt{x}-3=y$

$\Rightarrow \quad \sqrt{x}=y+3$

$\therefore \quad 2|y|+(y+3)(y-3)+6=0$

$\Rightarrow \quad 2|y|+y^2-3=0$

$\Rightarrow \quad |y|^2+2|y|-3=0$

$\Rightarrow \quad (|y|+3)(|y|-1)=0$

$\Rightarrow \quad |y| \ne -3 \Rightarrow |y|=1$

$\Rightarrow \quad y=\pm 1 \Rightarrow \sqrt{x}-3=\pm 1$

$\Rightarrow \quad \sqrt{x}=4, 2 \Rightarrow x=16, 4$

**63.** *(c)* We have,

$\alpha, \beta$ are the roots of $x^2-x+1=0$

$\because$ Roots of $x^2-x+1=0$ are $-\omega, -\omega^2$

$\therefore$ Let $\alpha=-\omega$ and $\beta=-\omega^2$

$\Rightarrow \alpha^{101}+\beta^{107}=(-\omega)^{101}+(-\omega^2)^{107}$

$=-(\omega^{101}+\omega^{214})$

$=-(\omega^2+\omega) \quad (\because \omega^3=1)$

$=-(-1) \quad [\because 1+\omega+\omega^2=0]$

$=1$

**64.** *(c)* Given,

$$\begin{vmatrix} x-4 & 2x & 2x \\ 2x & x-4 & 2x \\ 2x & 2x & x-4 \end{vmatrix}=(A+Bx)(x-A)^2$$

$\Rightarrow$ Apply $C_1 \to C_1+C_2+C_3$

$$\begin{vmatrix} 5x-4 & 2x & 2x \\ 5x-4 & x-4 & 2x \\ 5x-4 & 2x & x-4 \end{vmatrix}=(A+Bx)(x-A)^2$$

Taking common $(5x-4)$ from $C_1$, we get

$$(5x-4)\begin{vmatrix} 1 & 2x & 2x \\ 1 & x-4 & 2x \\ 1 & 2x & x-4 \end{vmatrix}=(A+Bx)(x-A)^2$$

Apply $R_2 \to R_2-R_1$ and $R_3 \to R_3-R_1$

$$\therefore (5x-4)\begin{vmatrix} 1 & 2x & 0 \\ 0 & -x-4 & 0 \\ 0 & 0 & -x-4 \end{vmatrix}$$

$$=(A+Bx)(x-A)^2$$

Expanding along $C_1$, we get

$$(5x-4)(x+4)^2=(A+Bx)(x-A)^2$$

Equating, we get

$$A = -4 \text{ and } B = 5$$

**65.** *(b)* We have,

$x + ky + 3z = 0;\ 3x + ky - 2z = 0;$
$2x + 4y - 3z = 0$

System of equation has non-zero solution, if

$$\begin{vmatrix} 1 & k & 3 \\ 3 & k & -2 \\ 2 & 4 & -3 \end{vmatrix} = 0$$

$\Rightarrow (-3k + 8) - k(-9 + 4) + 3(12 - 2k) = 0$

$\Rightarrow \quad -3k + 8 + 9k - 4k + 36 - 6k = 0$

$\Rightarrow \quad -4k + 44 = 0 \Rightarrow k = 11$

Let $\quad z = \lambda$, then we get

$$x + 11y + 3\lambda = 0 \quad \ldots(\text{i})$$

$$3x + 11y - 2\lambda = 0 \quad \ldots(\text{ii})$$

and $\quad 2x + 4y - 3\lambda = 0 \quad \ldots(\text{iii})$

Solving Eqs. (i) and (ii), we get

$$x = \frac{5\lambda}{2}, y = \frac{-\lambda}{2}, z = \lambda$$

$$\Rightarrow \quad \frac{xz}{y^2} = \frac{5\lambda^2}{2 \times \left(-\frac{\lambda}{2}\right)^2} = 10$$

**66.** *(a)* Given 6 different novels and 3 different dictionaries.

Number of ways of selecting 4 novels from 6 novels is ${}^6C_4 = \frac{6!}{2!4!} = 15$

Number of ways of selecting 1 dictionary is from 3 dictionaries is ${}^3C_1 = \frac{3!}{1!2!} = 3$

$\therefore$ Total number of arrangement of 4 novels and 1 dictionary where dictionary is always in the middle, is

$$15 \times 3 \times 4! = 45 \times 24 = 1080$$

**67.** *(d)* ***Key idea*** $= (a + b)^n + (a - b)^n$

$= 2({}^nC_0a^n + {}^nC_2a^{n-2}b^2 + {}^nC_4a^{n-4}b^4 \ldots)$

We have,

$(x + \sqrt{x^3 - 1})^5 + (x - \sqrt{x^3 - 1})^5, x > 1$

$= 2({}^5C_0x^5 + {}^5C_2x^3(\sqrt{x^3 - 1})^2$

$\quad + {}^5C_4x(\sqrt{x^3 - 1})^4)$

$= 2(x^5 + 10x^3(x^3 - 1) + 5x(x^3 - 1)^2)$

$= 2(x^5 + 10x^6 - 10x^3 + 5x^7 - 10x^4 + 5x)$

Sum of coefficients of all odd degree terms is $2(1 - 10 + 5 + 5) = 2$

**68.** *(c)* We have, $a_1, a_2, a_3, \ldots a_{49}$ are in AP.

$$\sum_{k=0}^{12} a_{4k+1} = 416 \text{ and } a_9 + a_{43} = 66$$

Let $a_1 = a$ and $d =$ common difference

$\because \quad a_1 + a_5 + a_9 + \cdots + a_{49} = 416$

$\therefore a + (a + 4d) + (a + 8d) + \ldots (a + 48d) = 416$

$\Rightarrow \quad \frac{13}{2}(2a + 48d) = 416$

$\Rightarrow \quad a + 24d = 32 \quad \ldots(\text{i})$

Also, $\quad a_9 + a_{43} = 66$

$\therefore a + 8d + a + 42d = 66$

$\Rightarrow \quad 2a + 50d = 66$

$\Rightarrow \quad a + 25d = 33 \quad \ldots(\text{ii})$

Solving Eqs. (i) and (ii), we get

$$a = 8 \text{ and } d = 1$$

Now, $a_1^2 + a_2^2 + a_3^2 + \cdots + a_{17}^2 = 140m$

$$8^2 + 9^2 + 10^2 + \ldots + 24^2 = 140m$$

$\Rightarrow (1^2 + 2^2 + 3^2 + \ldots + 24^2) - (1^2 + 2^2$

$\quad + 3^2 + \ldots + 7^2) = 140m$

$\Rightarrow \quad \frac{24 \times 25 \times 49}{6} - \frac{7 \times 8 \times 15}{6} = 140m$

$\Rightarrow \quad \frac{3 \times 7 \times 8 \times 5}{6}(7 \times 5 - 1) = 140m$

$\Rightarrow \quad 7 \times 4 \times 5 \times 34 = 140m$

$\Rightarrow \quad 140 \times 34 = 140m \Rightarrow \quad m = 34$

**69.** *(b)* We have,

$1^2 + 2 \cdot 2^2 + 3^2 + 2 \cdot 4^2 + 5^2 + 2 \cdot 6^2 + \ldots$

$A =$ sum of first 20 terms

$B =$ sum of first 40 terms

$\therefore \quad A = 1^2 + 2 \cdot 2^2 + 3^2 + 2 \cdot 4^2 + 5^2$

$\quad + 2 \cdot 6^2 + \ldots + 2 \cdot 20^2$

$A = (1^2 + 2^2 + 3^2 + \ldots + 20^2)$

$\quad + (2^2 + 4^2 + 6^2 + \ldots + 20^2)$

$A = (1^2 + 2^2 + 3^2 + \ldots + 20^2)$

$\quad + 4(1^2 + 2^2 + 3^2 + \ldots + 10^2)$

$$A = \frac{20 \times 21 \times 41}{6} + \frac{4 \times 10 \times 11 \times 21}{6}$$

$$A = \frac{20 \times 21}{6}(41 + 22) = \frac{20 \times 41 \times 63}{6}$$

Similarly, $B = (1^2 + 2^2 + 3^2 + \ldots + 40^2)$

$\quad + 4(1^2 + 2^2 + \ldots + 20^2)$

$$B = \frac{40 \times 41 \times 81}{6} + \frac{4 \times 20 \times 21 \times 41}{6}$$

$$B = \frac{40 \times 41}{6}(81 + 42) = \frac{40 \times 41 \times 123}{6}$$

Now, $B - 2A = 100\lambda$

$$\therefore \quad \frac{40 \times 41 \times 123}{6} - \frac{2 \times 20 \times 21 \times 63}{6} = 100\lambda$$

$$\Rightarrow \quad \frac{40}{6}(5043 - 1323) = 100\lambda$$

$$\Rightarrow \quad \frac{40}{6} \times 3720 = 100\lambda$$

$$\Rightarrow \quad 40 \times 620 = 100\lambda$$

$$\Rightarrow \quad \lambda = \frac{40 \times 620}{100} = 248$$

**70.** *(c)* ***Key idea*** *Use property of greatest integer function* $[x] = x - \{x\}$.

We have,

$$\lim_{x \to 0^+} x\left(\left[\frac{1}{x}\right] + \left[\frac{2}{x}\right] + \ldots + \left[\frac{15}{x}\right]\right)$$

We know, $[x] = x - \{x\}$

$$\therefore \quad \left[\frac{1}{x}\right] = \frac{1}{x} - \left\{\frac{1}{x}\right\}$$

Similarly, $\left[\frac{n}{x}\right] = \frac{n}{x} - \left\{\frac{n}{x}\right\}$

$\therefore$ Given limit

$$= \lim_{x \to 0^+} x\left(\frac{1}{x} - \left\{\frac{1}{x}\right\} + \frac{2}{x} - \left\{\frac{2}{x}\right\} + \ldots \frac{15}{x} - \left\{\frac{15}{x}\right\}\right)$$

$$= \lim_{x \to 0^+} (1 + 2 + 3 + \ldots + 15) - x\left(\left\{\frac{1}{x}\right\} + \left\{\frac{2}{x}\right\} + \ldots + \left\{\frac{15}{x}\right\}\right)$$

$$= 120 - 0 = 120$$

$$\left[\begin{array}{l} \because 0 \le \left\{\frac{n}{x}\right\} < 1, \text{therefore} \\ 0 \le x\left\{\frac{n}{x}\right\} < x \Rightarrow \lim_{x \to 0^+} x\left\{\frac{n}{x}\right\} = 0 \end{array}\right]$$

**71.** *(a)* We have,

$$f(x) = |x - \pi| \cdot (e^{|x|} - 1)\sin|x|$$

$$f(x) = \begin{cases} (x - \pi)(e^{-x} - 1)\sin x, & x < 0 \\ -(x - \pi)(e^{x} - 1)\sin x, & 0 \le x < \pi \\ (x - \pi)(e^{x} - 1)\sin x, & x \ge \pi \end{cases}$$

We check the differentiability at $x = 0$ and $\pi$.

We have,

$$f'(x) = \begin{cases} (x - \pi)(e^{-x} - 1)\cos x + (e^{-x} - 1)\sin x \\ \quad + (x - \pi)\sin x e^{-x}(-1), x < 0 \\ -[(x - \pi)(e^{x} - 1)\cos x + (e^{x} - 1)\sin x \\ \quad + (x - \pi)\sin x e^{x}], 0 < x < \pi \\ (x - \pi)(e^{x} - 1)\cos x + (e^{x} - 1)\sin x \\ \quad + (x - \pi)\sin x e^{x}, x > \pi \end{cases}$$

Clearly, $\lim_{x \to 0^-} f'(x) = 0 = \lim_{x \to 0^+} f'(x)$

and $\lim_{x \to \pi^-} f'(x) = 0 = \lim_{x \to \pi^+} f'(x)$

$\therefore$ $f$ is differentiable at $x = 0$ and $x = \pi$

Hence, $f$ is differentiable for all $x$.

**72.** *(d)* We have, $y^2 = 6x$

$$\Rightarrow \quad 2y\frac{dy}{dx} = 6 \Rightarrow \frac{dy}{dx} = \frac{3}{y}$$

Slope of tangent at $(x_1, y_1)$ is $m_1 = \frac{3}{y_1}$

Also, $\quad 9x^2 + by^2 = 16$

$$\Rightarrow \quad 18x + 2by\frac{dy}{dx} = 0 \quad \Rightarrow \quad \frac{dy}{dx} = \frac{-9x}{by}$$

Slope of tangent at $(x_1, y_1)$ is $m_2 = \frac{-9x_1}{by_1}$

Since, these are intersection at right angle.

$$\therefore \quad m_1 m_2 = -1 \Rightarrow \frac{27x_1}{by_1^2} = 1$$

$$\Rightarrow \quad \frac{27x_1}{6bx_1} = 1 \quad [\because y_1^2 = 6x_1]$$

$$\Rightarrow \quad b = \frac{9}{2}$$

**73.** *(d)* We have,

$$f(x) = x^2 + \frac{1}{x^2} \text{ and } g(x) = x - \frac{1}{x}$$

$$\Rightarrow h(x) = \frac{f(x)}{g(x)}$$

$$\therefore \quad h(x) = \frac{x^2 + \frac{1}{x^2}}{x - \frac{1}{x}} = \frac{\left(x - \frac{1}{x}\right)^2 + 2}{x - \frac{1}{x}}$$

$$\Rightarrow \quad h(x) = \left(x - \frac{1}{x}\right) + \frac{2}{x - \frac{1}{x}}$$

$$x - \frac{1}{x} > 0,$$

$$\left(x - \frac{1}{x}\right) + \frac{2}{x - \frac{1}{x}} \in [2\sqrt{2}, \infty)$$

$$x - \frac{1}{x} < 0,$$

$$\left(x - \frac{1}{x}\right) + \frac{2}{x - \frac{1}{x}} \in (-\infty, 2\sqrt{2}]$$

$\therefore$ Local minimum value is $2\sqrt{2}$.

**74.** (b) We have,

$$I = \int \frac{\sin^2 x \cdot \cos^2 x}{(\sin^5 x + \cos^3 x \cdot \sin^2 x + \sin^3 x \cdot \cos^2 x + \cos^5 x)^2} dx$$

$$= \int \frac{\sin^2 x \cos^2 x}{\{\sin^3 x(\sin^2 x + \cos^2 x) + \cos^3 x(\sin^2 x + \cos^2 x)\}^2} dx$$

$$= \int \frac{\sin^2 x \cos^2 x}{(\sin^3 x + \cos^3 x)^2} dx$$

$$= \int \frac{\sin^2 x \cos^2 x}{\cos^6 x(1 + \tan^3 x)^2} dx$$

$$= \int \frac{\tan^2 x \sec^2 x}{(1 + \tan^3 x)^2} dx$$

Put $\tan^3 x = t \Rightarrow 3\tan^2 x \sec^2 x dx = dt$

$$\therefore \quad I = \frac{1}{3}\int \frac{dt}{(1 + t)^2}$$

$$\Rightarrow \quad I = \frac{-1}{3(1 + t)} + C$$

$$\Rightarrow \quad I = \frac{-1}{3(1 + \tan^3 x)} + C$$

**75.** (d) ***Key idea***

Use property $= \int_a^b f(x)dx = \int_a^b f(a + b - x)dx$

Let $I = \int_{-\pi/2}^{\pi/2} \frac{\sin^2 x}{1 + 2^x} dx$

$$\Rightarrow I = \int_{-\pi/2}^{\pi/2} \frac{\sin^2\left(-\frac{\pi}{2} + \frac{\pi}{2} - x\right)}{1 + 2^{-\frac{\pi}{2} + \frac{\pi}{2} - x}} dx$$

$$\left[\because \int_a^b f(x)dx = \int_a^b f(a + b - x)dx\right]$$

$$\Rightarrow \quad I = \int_{-\pi/2}^{\pi/2} \frac{\sin^2 x}{1 + 2^{-x}} dx$$

$$\Rightarrow \quad I = \int_{-\pi/2}^{\pi/2} \frac{2^x \sin^2 x}{2^x + 1} dx$$

$$\Rightarrow \quad 2I = \int_{-\pi/2}^{\pi/2} \sin^2 x\left(\frac{2^x + 1}{2^x + 1}\right)dx$$

$$\Rightarrow \quad 2I = \int_{-\pi/2}^{\pi/2} \sin^2 x \, dx$$

$$\Rightarrow \quad 2I = 2\int_0^{\pi/2} \sin^2 x \, dx$$

[$\because \sin^2 x$ is an even function]

$$\Rightarrow \quad I = \int_0^{\pi/2} \sin^2 x dx$$

$$\Rightarrow \quad I = \int_0^{\pi/2} \cos^2 x dx$$

$$\left[\because \int_0^a f(x)dx = \int_0^a f(a - x)dx\right]$$

$$\Rightarrow \quad 2I = \int_0^{\pi/2} dx$$

$$\Rightarrow \quad 2I = [x]_0^{\pi/2} \Rightarrow I = \frac{\pi}{4}$$

**76.** (a) We have,

$$\Rightarrow \quad 18x^2 - 9\pi x + \pi^2 = 0$$

$$\Rightarrow \quad 18x^2 - 6\pi x - 3\pi x + \pi^2 = 0$$

$$(6x - \pi)(3x - \pi) = 0$$

$$\Rightarrow \quad x = \frac{\pi}{6}, \frac{\pi}{3}$$

Now, $\alpha < \beta \alpha = \frac{\pi}{6}, \beta = \frac{\pi}{3}$

Given, $g(x) = \cos x^2$ and $f(x) = \sqrt{x}$

$$y = gof(x)$$

$$\therefore \quad y = g(f(x)) = \cos x$$

Area of region bounded by $x = \alpha$, $x = \beta$, $y = 0$ and curve $y = g(f(x))$ is

$$A = \int_{\pi/6}^{\pi/3} \cos x \, dx$$

$$A = [\sin x]_{\pi/6}^{\pi/3}$$

$$A = \sin\frac{\pi}{3} - \sin\frac{\pi}{6} = \frac{\sqrt{3}}{2} - \frac{1}{2}$$

$$A = \left(\frac{\sqrt{3} - 1}{2}\right)$$

**77.** (c) We have,

$$\sin x \frac{dy}{dx} + y\cos x = 4x$$

$$\Rightarrow \quad \frac{dy}{dx} + y\cot x = 4x \operatorname{cosec} x$$

This is a linear differential equation of form

$$\frac{dy}{dx} + Py = Q$$

where $P = \cot x$, $Q = 4x \operatorname{cosec} x$

Now, $IF = e^{\int P dx} = e^{\int \cot x dx} = e^{\log \sin x} = \sin x$

Solution of the differential equation is

$$y \cdot \sin x = \int 4x \operatorname{cosec} x \sin x dx + C$$

$$\Rightarrow \quad y\sin x = \int 4x dx + C = 2x^2 + C$$

Put $x = \frac{\pi}{2}, y = 0$, we get

$$C = -\frac{\pi^2}{2}$$

$$\Rightarrow \quad y\sin x = 2x^2 - \frac{\pi^2}{2}$$

Put $\quad x = \frac{\pi}{6}$

$$\therefore \quad y\left(\frac{1}{2}\right) = 2\left(\frac{\pi^2}{36}\right) - \frac{\pi^2}{2}$$

$$\Rightarrow \quad y = \frac{\pi^2}{9} - \pi^2$$

$$\Rightarrow \quad y = -\frac{8\pi^2}{9}$$

**Alternate Method**

We have, $\sin x \frac{dy}{dx} + y\cos x = 4x$, which can be written as $\frac{d}{dx}(\sin x \cdot y) = 4x$

On integrating both sides, we get

$$\int \frac{d}{dx}(\sin x \cdot y) \cdot dx = \int 4x \cdot dx$$

$$\Rightarrow \quad y \cdot \sin x = \frac{4x^2}{2} + C$$

$$\Rightarrow \quad y \cdot \sin x = 2x^2 + C$$

Now, as $y = 0$ when $x = \frac{\pi}{2}$

$$\therefore \quad C = -\frac{\pi^2}{2}$$

$$\Rightarrow \quad y \cdot \sin x = 2x^2 - \frac{\pi^2}{2}$$

Now, putting $x = \frac{\pi}{6}$, we get

$$y\left(\frac{1}{2}\right) = 2\left(\frac{\pi^2}{36}\right) - \frac{\pi^2}{2}$$

$$\Rightarrow \quad y = \frac{\pi^2}{9} - \pi^2 = -\frac{8\pi^2}{9}$$

**78.** (c)

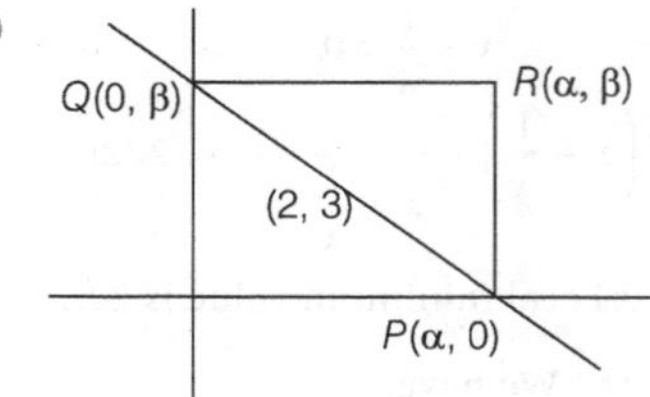


Equation of line $PQ$ is $\quad \frac{x}{\alpha} + \frac{y}{\beta} = 1$

Since this line is passes through fixed point (2, 3).

$$\therefore \quad \frac{2}{\alpha} + \frac{3}{\beta} = 1$$

$\therefore$ Locus of $R$ is $2\beta + 3\alpha = \alpha\beta$

i.e. $\quad 2y + 3x = xy$

$\Rightarrow \quad 3x + 2y = xy$

**79.** (c) **Key idea** *Otrhocentre, centroid and circumcentre are collinear and centroid divide orthocentre and circumcentre in 2 : 1 ratio.*

We have orthocentre and centroid of a triangle be $A(-3, 5)$ and $B(3, 3)$ respectively and $C$ circumcentre.

A(−3, 5) —— B(3,3) —— C

Clearly, $AB = \sqrt{(3+3)^2 + (3-5)^2}$
$= \sqrt{36+4} = 2\sqrt{10}$

We know that, $AB : BC = 2 : 1$

$\Rightarrow \quad BC = \sqrt{10}$

Now,
$AC = AB + BC = 2\sqrt{10} + \sqrt{10} = 3\sqrt{10}$

Since, $AC$ is a diameter of circle.

$$\therefore \quad r = \frac{AC}{2}$$

$$\Rightarrow \quad r = \frac{3\sqrt{10}}{2} = 3\sqrt{\frac{5}{2}}$$

**80.** (d) **Key idea** *Equation of tangent to the curve*
*$x^2 = 4ay$ at $(x_1, y_1)$ is* $xx_1 = 4a\left(\frac{y + y_1}{2}\right)$

Tangent to the curve $x^2 = y - 6$ at $(1, 7)$ is

$$x = \frac{y+7}{2} - 6$$

$\Rightarrow \quad 2x - y + 5 = 0 \quad \ldots(i)$

Equation of circle is

$x^2 + y^2 + 16x + 12y + c = 0$

Centre $(-8, -6)$

$$r = \sqrt{8^2 + 6^2 - c} = \sqrt{100 - c}$$

Since, line $2x - y + 5 = 0$ also touches the circle.

$$\therefore \quad \sqrt{100 - c} = \left|\frac{2(-8) - (-6) + 5}{\sqrt{2^2 + 1^2}}\right|$$

$$\Rightarrow \quad \sqrt{100 - c} = \left|\frac{-16 + 6 + 5}{\sqrt{5}}\right|$$

$$\Rightarrow \quad \sqrt{100 - c} = |-\sqrt{5}|$$

$$\Rightarrow \quad 100 - c = 5 \Rightarrow c = 95$$

**81.** (b) Equation of tangent and normal to the curve $y^2 = 16x$ at $(16, 16)$ is $x - 2y + 16 = 0$ and $2x + y - 48 = 0$, respectively.

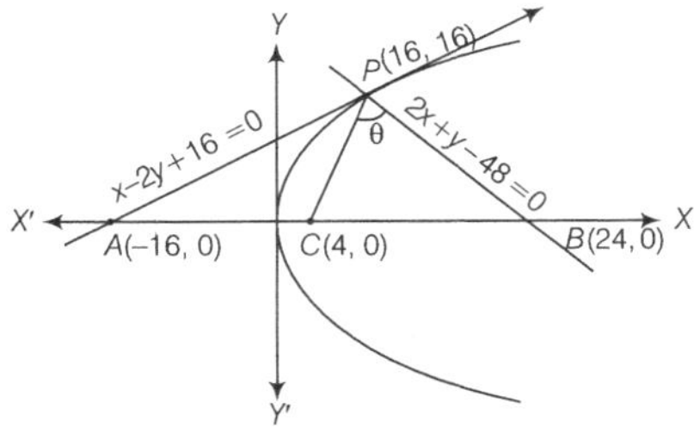


$A = (-16, 0)$ ; $B = (24, 0)$

$C$ is the centre of circle passing through $PAB$

i.e. $\quad C = (4, 0)$

Slope of $PC = \dfrac{16 - 0}{16 - 4} = \dfrac{16}{12} = \dfrac{4}{3} = m_1$

Slope of $PB = \dfrac{16 - 0}{16 - 24} = \dfrac{16}{-8} = -2 = m_2$

$$\tan\theta = \left|\frac{m_1 - m_2}{1 + m_1 m_2}\right|$$

$$\Rightarrow \quad \tan\theta = \left|\frac{\frac{4}{3} + 2}{1 - \left(\frac{4}{3}\right)(2)}\right|$$

$\Rightarrow \quad \tan\theta = 2$

**82.** (a) Tangents are drawn to the hyperbola $4x^2 - y^2 = 36$ at the point $P$ and $Q$.

Tangent intersects at point $T(0, 3)$

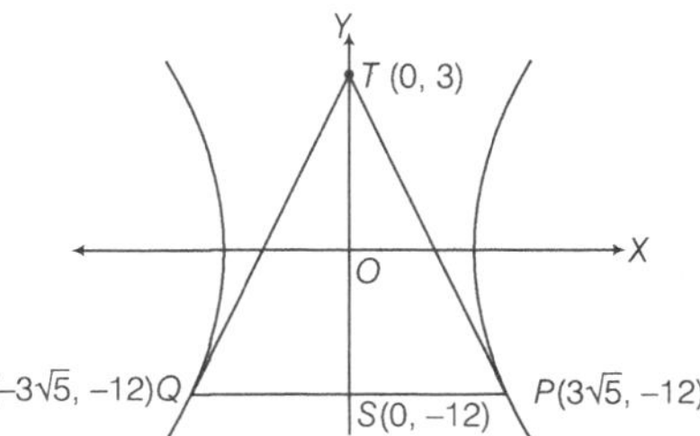


Clearly, $PQ$ is chord of contact.

$\therefore$ Equation of $PQ$ is $-3y = 36$

$\Rightarrow \quad y = -12$

Solving the curve $4x^2 - y^2 = 36$ and $y = -12$,

we get $\quad x = \pm 3\sqrt{5}$

Area of $\Delta PQT = \dfrac{1}{2} \times PQ \times ST = \dfrac{1}{2}(6\sqrt{5} \times 15)$

$= 45\sqrt{5}$

**83.** (b) $L_1$ is the line of intersection of the plane $2x - 2y + 3z - 2 = 0$ and $x - y + z + 1 = 0$ and $L_2$ is the line of intersection of the plane $x + 2y - z - 3 = 0$ and $3x - y + 2z - 1 = 0$

Since $L_i$ is parallel to $\begin{vmatrix} \hat{i} & \hat{j} & \hat{k} \\ 2 & -2 & 3 \\ 1 & -1 & 1 \end{vmatrix} = \hat{i} + \hat{j}$

$L_2$ is parallel to $\begin{vmatrix} \hat{i} & \hat{j} & \hat{k} \\ 1 & 2 & -1 \\ 3 & -1 & 2 \end{vmatrix} = 3\hat{i} - 5\hat{j} - 7\hat{k}$

Also, $L_2$ passes through $\left(\dfrac{5}{7}, \dfrac{8}{7}, 0\right)$.

[put $z = 0$ in last two planes]

So, equation of plane is

$$\begin{vmatrix} x - \frac{5}{7} & y - \frac{8}{7} & z \\ 1 & 1 & 0 \\ 3 & -5 & -7 \end{vmatrix} = 0$$

$$\Rightarrow \quad 7x - 7y + 8z + 3 = 0$$

Now, perpendicular distance from origin is

$$\left|\frac{3}{\sqrt{7^2 + 7^2 + 8^2}}\right| = \frac{3}{\sqrt{162}} = \frac{1}{3\sqrt{2}}$$

**84.** (d) **Key idea** *length of projection of the line segment joining* $\mathbf{a}_1$ *and* $\mathbf{a}_2$ *on the plane* $\mathbf{r}\cdot\mathbf{n}$

$= d$ *is* $\left|\dfrac{(\mathbf{a}_2-\mathbf{a}_1)\times\mathbf{n}}{|\mathbf{n}|}\right|$

Length of projection the line segment joining the points $(5, -1, 4)$ and $(4, -1, 3)$ on the plane $x + y + z = 7$ is

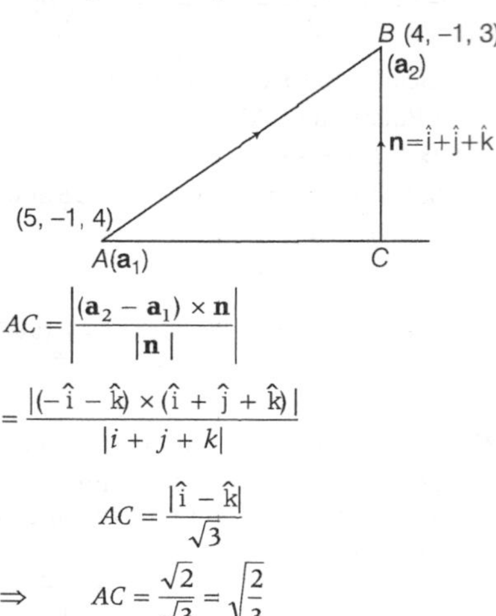


$$AC = \left|\frac{(\mathbf{a}_2-\mathbf{a}_1)\times\mathbf{n}}{|\mathbf{n}|}\right|$$

$$= \frac{|(-\hat{i}-\hat{k})\times(\hat{i}+\hat{j}+\hat{k})|}{|i+j+k|}$$

$$AC = \frac{|\hat{i}-\hat{k}|}{\sqrt{3}}$$

$$\Rightarrow \quad AC = \frac{\sqrt{2}}{\sqrt{3}} = \sqrt{\frac{2}{3}}$$

**Alternative Method**

Clearly, DR's of $AB$ are $4-5, -1+1, 3-4$, i.e. $-1, 0, -1$

and DR's of normal to plane are 1, 1, 1.

Now, let $\theta$ be the angle between the line and plane, then $\theta$ is given by

$$\sin\theta = \frac{|-1+0-1|}{\sqrt{(-1)^2+(-1)^2}\sqrt{1^2+1^2+1^2}}$$

$$= \frac{2}{\sqrt{2}\sqrt{3}} = \sqrt{\frac{2}{3}}$$

B (4, –1, 3)
A (5, –1, 4)
θ
C
θ
Plane :
x + y + z = 7

$$\Rightarrow \sin\theta = \sqrt{\frac{2}{3}} \Rightarrow \cos\theta = \sqrt{1-\sin^2\theta}$$

$$= \sqrt{1-\frac{2}{3}} = \frac{1}{\sqrt{3}}$$

Clearly, length of projection

$$= AB\cos\theta = \sqrt{2}\frac{1}{\sqrt{3}} \qquad [\because AB = \sqrt{2}]$$

$$= \sqrt{\frac{2}{3}}$$

**85.** (a) **Key idea** *If any vector* $\mathbf{x}$ *is coplanar with the vector* $\mathbf{y}$ *and* $\mathbf{z}$, *then* $\mathbf{x} = \lambda\mathbf{y} + \mu\mathbf{z}$

Here, $\mathbf{u}$ is coplanar with $\mathbf{a}$ and $\mathbf{b}$.

$\therefore \quad \mathbf{u} = \lambda\mathbf{a} + \mu\mathbf{b}$

Dot product with $\mathbf{a}$, we get

$$\mathbf{u}\cdot\mathbf{a} = \lambda(\mathbf{a}\cdot\mathbf{a}) + \mu(\mathbf{b}\cdot\mathbf{a})$$

$$\Rightarrow \quad 0 = 14\lambda + 2\mu \qquad \ldots(\text{i})$$

$$[\because \mathbf{a} = 2\hat{\mathbf{i}} + 3\hat{\mathbf{j}} - \hat{\mathbf{k}}, \mathbf{b} = \hat{\mathbf{j}} + \hat{\mathbf{k}}, \mathbf{u}\cdot\mathbf{a} = 0]$$

Dot product with $\mathbf{b}$, we get

$$\mathbf{u}\cdot\mathbf{b} = \lambda(\mathbf{a}\cdot\mathbf{b}) + \mu(\mathbf{b}\cdot\mathbf{b})$$

$$24 = 2\lambda + 2\mu \ldots(\text{ii}) \quad [\because \mathbf{u}\cdot\mathbf{b} = 24]$$

Solving Eqs. (i) and (ii), we get

$$\lambda = -2, \mu = 14$$

Dot product with $\mathbf{u}$, we get

$$|\mathbf{u}|^2 = \lambda(\mathbf{u}\cdot\mathbf{a}) + \mu(\mathbf{u}\cdot\mathbf{b})$$

$$|\mathbf{u}|^2 = -2(0) + 14(24)$$

$$\Rightarrow \quad |\mathbf{u}|^2 = 336$$

**86.** (b) **Key idea** *Use the theorem of total probability*

Let $E_1$ = Event that first ball drawn is red

$E_2$ = Event that first ball drawn is black

$A$ = Event that second ball drawn is red

$$P(E_1) = \frac{4}{10}, P\left(\frac{A}{E_1}\right) = \frac{6}{12}$$

$$\Rightarrow \quad P(E_2) = \frac{6}{10}, P\left(\frac{A}{E_2}\right) = \frac{4}{12}$$

By law of total probability

$$P(A) = P(E_1)\times P\left(\frac{A}{E_1}\right) + P(E_2)\times P\left(\frac{A}{E_2}\right)$$

$$= \frac{4}{10}\times\frac{6}{12} + \frac{6}{10}\times\frac{4}{12}$$

$$= \frac{24+24}{120} = \frac{48}{120} = \frac{2}{5}$$

**87.** (c) **Key idea** *Standard deviation is remain unchanged, if observations are added or subtracted by a fixed number*

We have,

$$\sum_{i=1}^{9}(x_1 - 5) = 9 \text{ and } \sum_{i=1}^{9}(x_1 - 5)^2 = 45$$

$$SD = \sqrt{\frac{\sum_{i=1}^{9}(x_1 - 5)^2}{9} - \left(\frac{\sum_{i=1}^{9}(x_1 - 5)}{9}\right)^2}$$

$$\Rightarrow \quad SD = \sqrt{\frac{45}{9} - \left(\frac{9}{9}\right)^2}$$

$$\Rightarrow \quad SD = \sqrt{5-1} = \sqrt{4} = 2$$

**88.** (b) ***Key idea*** *Apply the identity*

$\cos(x + y)\cos(x - y) = \cos^2 x - \sin^2 y$

and $\cos 3x = 4\cos^3 x - 3\cos x$

We have,

$$8\cos x\left(\cos\left(\frac{\pi}{6} + x\right)\cos\left(\frac{\pi}{6} - x\right) - \frac{1}{2}\right) = 1$$

$$\Rightarrow \quad 8\cos x\left(\cos^2\frac{\pi}{6} - \sin^2 x - \frac{1}{2}\right) = 1$$

$$\Rightarrow \quad 8\cos x\left(\frac{3}{4} - \sin^2 x - \frac{1}{2}\right) = 1$$

$$\Rightarrow \quad 8\cos x\left(\frac{3}{4} - \frac{1}{2} - 1 + \cos^2 x\right) = 1$$

$$\Rightarrow \quad 8\cos x\left(\frac{-3 + 4\cos^2 x}{4}\right) = 1$$

$$\Rightarrow \quad 2(4\cos^3 x - 3\cos x) = 1$$

$$\Rightarrow \quad 2\cos 3x = 1 \Rightarrow \cos 3x = \frac{1}{2}$$

$$\Rightarrow \quad 3x = \frac{\pi}{3}, \frac{5\pi}{3}, \frac{7\pi}{3} \qquad [0 \le 3x \le 3\pi]$$

$$\Rightarrow \quad x = \frac{\pi}{9}, \frac{5\pi}{9}, \frac{7\pi}{9}$$

$$\text{Sum} = \frac{\pi}{9} + \frac{5\pi}{9} + \frac{7\pi}{9} = \frac{13\pi}{9} \Rightarrow k\pi = \frac{13\pi}{9}$$

Hence, $\quad k = \frac{13}{9}$

**89.** (a)



Let height of tower $TM$ be $h$.

In $\quad \Delta PMT$, $\tan 45° = \frac{TM}{PM}$

$$\Rightarrow \quad 1 = \frac{h}{PM}$$

$$\Rightarrow \quad PM = h$$

In $\Delta TQM$, $\tan 30° = \frac{h}{QM}$;

$$QM = \sqrt{3}h$$

In $\Delta PMQ$, $PM^2 + QM^2 = PQ^2$

$$h^2 + (\sqrt{3}h)^2 = (200)^2$$

$$\Rightarrow \quad 4h^2 = (200)^2$$

$$\Rightarrow \quad h = 100\text{ m}$$

**90.** (a) ***Key idea*** *Use De-Morgan's and distributive law.*

We have, $\sim(p \vee q) \vee (\sim p \wedge q)$

$\equiv (\sim p \wedge \sim q) \vee (\sim p \wedge q)$

[$\because$ By De-Morgan's law $\sim(p \vee q) = (\sim p \wedge \sim q)$]

$\equiv \sim p \wedge (\sim q \vee q)$

[By distributive law]

$\equiv \sim p \wedge t \qquad [\sim q \vee q = t]$

$\equiv \sim p$

# JEE ADVANCED

# SOLVED PAPER 2018

## Paper 1

## PHYSICS

### Section 1 (Maximum Marks : 24)

- This section contains **SIX (06)** questions.
- Each question has **FOUR** options for correct answer(s). **ONE OR MORE THAN ONE** of these four option(s) is (are) correct options(s).
- For each question, choose the correct options(s) to answer the question.
- Answer to each question will be evaluated according to the following marking scheme:

| | | |
|---|---|---|
| *Full Marks* | : **+4** | If only (all) the correct option(s) is (are) chosen. |
| *Partial Marks* | : **+3** | If all the four options are correct but ONLY three options are chosen. |
| *Partial Marks* | : **+2** | If three or more options are correct but ONLY two options are chosen, both of which are correct options. |
| *Partial Marks* | : **+1** | If two or more options are correct but ONLY one option is chosen and it is a correct option. |
| *Zero Marks* | : **0** | If none of the options is chosen (i.e. the question is unanswered). |
| *Negative Marks* | : **−2** | In all other cases. |

- **For example:** If first, third and fourth are the ONLY three correct options for a question with second option being an incorrect option; selecting only all the three correct options will result in + 4 marks. Selecting only two of the three correct options (e.g. the first and fourth options), without selecting any incorrect option (second option in this case), will result in +2 marks. Selecting only one of the three correct options (either first or third or fourth option), without selecting any incorrect option (second option in this case), will result in +1 marks. Selecting any incorrect option(s) (second option in this case), with or without selection of any correct option(s) will result in −2 marks.

**1.** The potential energy of mass $m$ at a distance $r$ from a fixed point $O$ is given by $V(r) = kr^2/2$, where $k$ is a positive constant of appropriate dimensions. This particle is moving in a circular orbit of radius $R$ about the point $O$. If $v$ is the speed of the particle and $L$ is the magnitude of its angular momentum about $O$, which of the following statements is (are) true ?

(a) $v = \sqrt{\frac{k}{2m}}R$ (b) $v = \sqrt{\frac{k}{m}}R$

(c) $L = \sqrt{mk}R^2$ (d) $L = \sqrt{\frac{mk}{2}}R^2$

**2.** Consider a body of mass 1.0 kg at rest at the origin at time $t = 0$ . A force

$\mathbf{F} = (\alpha t\,\hat{\mathbf{i}} + \beta\,\hat{\mathbf{j}})$ is applied on the body, where $\alpha = 1.0\text{Ns}^{-1}$ and $\beta = 1.0\text{N}$. The torque acting on the body about the origin at time $t = 1.0\text{s}$ is $\tau$. Which of the following statements is (are) true ?

(a) $|\tau| = \frac{1}{3}\text{N-m}$

(b) The torque $\tau$ is in the direction of the unit vector $+\hat{\mathbf{k}}$

(c) The velocity of the body at $t = 1$ s is $\mathbf{v} = \frac{1}{2}(\hat{\mathbf{i}} + 2\hat{\mathbf{j}})\,\text{ms}^{-1}$

(d) The magnitude of displacement of the body at $t = 1$ s is $\frac{1}{6}$m

**3.** A uniform capillary tube of inner radius $r$ is dipped vertically into a beaker filled with water. The water rises to a height $h$ in the capillary tube above the water surface in the beaker. The surface tension of water is $\sigma$. The angle of contact between water and the wall of the capillary tube is $\theta$. Ignore the mass of water in the meniscus. Which of the following statements is (are) true ?

(a) For a given material of the capillary tube, $h$ decreases with increase in $r$

(b) For a given material of the capillary tube, $h$ is independent of $\sigma$

(c) If this experiment is performed in a lift going up with a constant acceleration, then $h$ decreases

(d) $h$ is proportional to contact angle $\theta$

**4.** In the figure below, the switches $S_1$ and $S_2$ are closed simultaneously at $t = 0$ and a current starts to flow in the circuit. Both the batteries have the same magnitude of the electromotive force (emf) and the polarities are as indicated in the figure. Ignore mutual inductance between the inductors. The current $I$ in the middle wire reaches its maximum magnitude $I_{max}$ at time $t = \tau$. Which of the following statements is (are) true ?

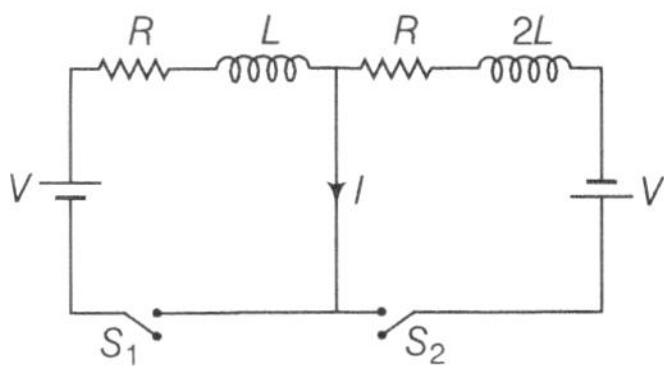


(a) $I_{max} = \frac{V}{2R}$

(b) $I_{max} = \frac{V}{4R}$

(c) $\tau = \frac{L}{R}\ln 2$

(d) $\tau = \frac{2L}{R}\ln 2$

**5.** Two infinitely long straight wires lie in the $xy$-plane along the lines $x = \pm R$. The wire located at $x = +R$ carries a constant current $I_1$ and the wire located at $x = -R$ carries a constant current $I_2$. A circular loop of radius $R$ is suspended with its centre at $(0,0,\sqrt{3}R)$ and in a plane parallel to the $xy$-plane. This loop carries a constant current $I$ in the clockwise direction as seen from above the loop. The current in the wire is taken to be positive, if it is in the $+\hat{\mathbf{j}}$-direction. Which of the following statements regarding the magnetic field $\mathbf{B}$ is (are) true?

(a) If $I_1 = I_2$, then $\mathbf{B}$ cannot be equal to zero at the origin (0,0,0)

(b) If $I_1 > 0$ and $I_2 < 0$, then $\mathbf{B}$ can be equal to zero at the origin (0,0,0)

(c) If $I_1 < 0$ and $I_2 > 0$, then $\mathbf{B}$ can be equal to zero at the origin (0,0,0)

(d) If $I_1 = I_2$, then the $z$-component of the magnetic field at the centre of the loop is $\left(-\frac{\mu_0 I}{2R}\right)$

**6.** One mole of a monoatomic ideal gas undergoes a cyclic process as shown in the figure (where, $V$ is the volume and $T$ is the temperature). Which of the statements below is (are) true ?

(a) Process I is an isochoric process

(b) In process II, gas absorbs heat

(c) In process IV, gas releases heat

(d) Processes I and III are not isobaric

## Section 2 (Maximum Marks : 24)

- This section contains **EIGHT (08)** questions. The answer to each question is a **NUMERICAL VALUE.**
- Four each question, enter the correct numerical value (in decimal notation, truncated/rounded-off to the **second decimal place;** e.g. 6.25, 7.00.– 0.33, –.30, 30.27, –127.30) using the mouse and the on-screen virtual numeric keypad in the place designated to enter the answer.
- Answer to each question will be evaluated according to the following marking scheme:

*Full Marks* : **+3** If ONLY the correct numerical value is entered as answer.

*Zero Marks* : **0** In all other cases.

**7.** Two vectors **A** and **B** are defined as $\mathbf{A} = a\hat{\mathbf{i}}$ and $\mathbf{B} = a(\cos\omega t\hat{\mathbf{i}} + \sin\omega t\hat{\mathbf{j}})$, where $a$ is a constant and $\omega = \pi/6$ rad $s^{-1}$. If $|\mathbf{A}+\mathbf{B}| = \sqrt{3}|\mathbf{A}-\mathbf{B}|$ at time $t = \tau$ for the first time, the value of $\tau$, in seconds, is ............ .

**8.** Two men are walking along a horizontal straight line in the same direction. The main in front walks at a speed $1.0\,ms^{-1}$ and the man behind walks at a speed $2.0\,ms^{-1}$. A third man is standing at a height 12 m above the same horizontal line such that all three men are in a vertical plane. The two walking men are blowing identical whistles which emit a sound of frequency 1430 Hz. The speed of sound in air $330\,ms^{-1}$. At the instant, when the moving men are 10 m apart, the stationary man is equidistant from them. The frequency of beats in Hz, heard by the stationary man at this instant, is ............ .

**9.** A ring and a disc are initially at rest, side by side, at the top of an inclined plane which makes an angle 60° with the horizontal. They start to roll without slipping at the same instant of time along the shortest path. If the time difference between their reaching the ground is $(2-\sqrt{3})/\sqrt{10}$ s, then the height of the top of the inclined plane, in metres, is .......... . (Take, $g = 10\,ms^{-2}$)

**10.** A spring block system is resting on a frictionless floor as shown in the figure. The spring constant is $2.0\,Nm^{-1}$ and the mass of the block is 2.0 kg . Ignore the mass of the spring. Initially, the spring is in an unstretched condition. Another block of mass 1.0 kg moving with a speed of $2.0\,ms^{-1}$ collides elastically with the first block. The collision is such that the 2.0 kg block does not hit the wall. The distance, in metres, between the two blocks when the spring returns to its unstretched position for the first time after the collision is ........... .

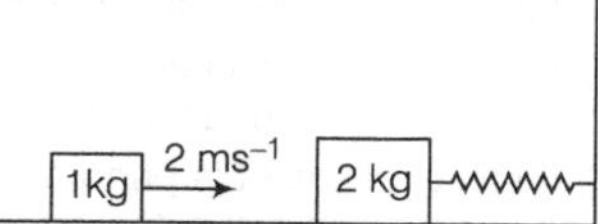


**11.** Three identical capacitors $C_1$, $C_2$ and $C_3$ have a capacitance of 1.0 μF each and they are uncharged initially. They are connected in a circuit as shown in the figure and $C_1$ is then filled completely with a dielectric material of relative permittivity $\varepsilon_r$. The cell electromotive force (emf) $V_0 = 8\,V$. First the switch $S_1$ is closed while the switch $S_2$ is kept open. When the capacitor $C_3$ is fully charged, $S_1$ is opened and $S_2$ is closed simultaneously. When all the capacitors reach equilibrium, the acitors reach equilibrium, the charge on $C_3$ is found to be $5\mu C$. The value of $\varepsilon_r$ = .............

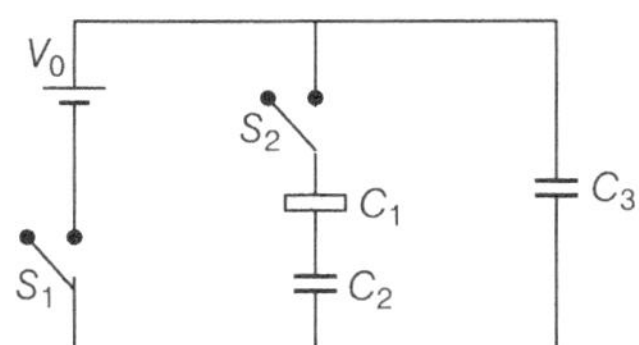


**12.** In the $xy$-plane, the region $y > 0$ has a uniform magnetic field $B_1\hat{\mathbf{k}}$ and the region $y < 0$ has another uniform magnetic field $B_2\hat{\mathbf{k}}$. A positively charged particle is projected from the origin along the positive $Y$-axis with speed $v_0 = \pi \text{ ms}^{-1}$ at $t = 0$, as shown in figure. Neglect gravity in this problem. Let $t = T$ be the time when the particle crosses the $X$-axis from below for the first time. If $B_2 = 4B_1$, the average speed of the particle, in $\text{ms}^{-1}$, along the $X$-axis in the time interval $T$ is............ .

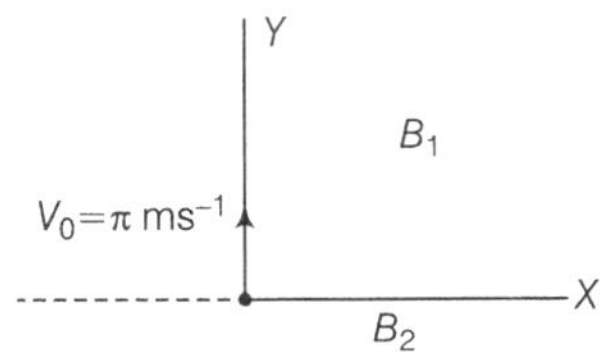


**13.** Sunlight of intensity $1.3 \text{ kWm}^{-2}$ is incident normally on a thin convex lens of focal length 20 cm. Ignore the energy loss of light due to the lens and assume that the lens aperture size is much smaller than its focal length. The average intensity of light, in $\text{kW m}^{-2}$, at a distance 22 cm from the lens on the other side is ............ .

**14.** Two conducting cylinders of equal length but different radii are connected in series between two heat baths kept at temperatures $T_1 = 300\text{ K}$ and $T_2 = 100\text{ K}$, as shown in the figure. The radius of the bigger cylinder is twice that of the smaller one and the thermal conductivities of the materials of the smaller and the larger cylinders are $K_1$ and $K_2$, respectively. If the temperature at the junction of the cylinders in the steady state is $200\text{ K}$, then $K_1/K_2$ =.......... .

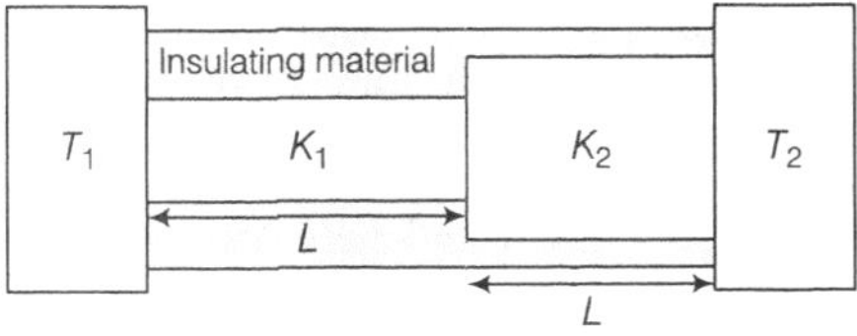


## Section 3 (Maximum Marks : 12)

- This section contains **TWO (02)** paragraphs. Based on each paragraph, there are **TWO(02)** questions.
- Each question has **FOUR** options. **ONLY ONE** of these four options corresponds to the correct answer.
- Four each question, choose the option corresponding to the correct answer.
- Answer to each question will be evaluated according to the following marking scheme:

*Full Marks* : **+3** If ONLY the correct option is chosen.
*Zero Marks* : **0** If none of the options is chosen (i.e. the question is unanswered).
*Negative Marks* : **−1** In all other cases.

### Paragraph X

In electromagnetic theory, the electric and magnetic phenomena are related to each other. Therefore, the dimensions of electric and magnetic quantities must also be related to each other. In the questions below, [E] and [B] stand for dimensions of electric and magnetic fields respectively, while $[\varepsilon_0]$ and $[\mu_0]$ stand for dimensions of the permittivity and permeability of free space, respectively. [L] and [T] are dimensions of length and time, respectively. All the quantities are given in SI units.

*(There are two questions based on PARAGRAPH "X", the question given below is one of them)*

**15.** The relation between $[E]$ and $[B]$ is

(a) [E] = [B] [L] [T]

(b) $[E] = [B][L]^{-1}[T]$

(c) $[E] = [B][L][T]^{-1}$

(d) $[E] = [B][L]^{-1}[T]^{-1}$

**16.** The relation between $[\varepsilon_0]$ and $[\mu_0]$ is

(a) $[\mu_0] = [\varepsilon_0][L]^2[T]^{-2}$

(b) $[\mu_0] = [\varepsilon_0][L]^{-2}[T]^2$

(c) $[\mu_0] = [\varepsilon_0]^{-1}[L]^2[T]^{-2}$

(d) $[\mu_0] = [\varepsilon_0]^{-1}[L]^{-2}[T]^2$

## Paragraph A

If the measurement errors in all the independent quantities are known, then it is possible to determine the error in any dependent quantity. This is done by the use of series expansion and truncating the expansion at the first power of the error. For example, consider the relation $z = x/y$. If the errors in $x$, $y$ and $z$ are $\Delta x$, $\Delta y$ and $\Delta z$ respectively, then

$$z \pm \Delta z = \frac{x \pm \Delta x}{y \pm \Delta y}$$

$$= \frac{x}{y}\left(1 \pm \frac{\Delta x}{x}\right)\left(1 \pm \frac{\Delta y}{y}\right)^{-1}$$

The series expansion for $\left(1 \pm \frac{\Delta y}{y}\right)^{-1}$, to first power in $\Delta y/y$, is $1 \mp (\Delta y/y)$. The relative errors in independent variables are always added. So, the error in $z$ will be

$$\Delta z = z\left(\frac{\Delta x}{x} + \frac{\Delta y}{y}\right)$$

The above derivation makes the assumption that $\Delta x/x \ll 1$, $\Delta y/y \ll 1$. Therefore, the higher powers of these quantities are neglected.

*(There are two questions based on PARAGRAPH "A", the question given below is one of them)*

**17.** Consider the ratio $r = \frac{(1-a)}{(1+a)}$ to be determined by measuring a dimensionless quantity $a$. If the error in the measurement of $a$ is $\Delta a$ $(\Delta a/a \ll 1)$, then what is the error $\Delta r$ in determining $r$?

(a) $\frac{\Delta a}{(1+a)^2}$

(b) $\frac{2\Delta a}{(1+a)^2}$

(c) $\frac{2\Delta a}{(1-a)^2}$

(d) $\frac{2a\Delta a}{(1-a^2)}$

**18.** In an experiment, the initial number of radioactive nuclei is 3000. It is found that 1000 ± 40 nuclei decayed in the first 1.0s. For $|x| \ll 1$, $\ln(1+x) = x$ up to first power in $x$. The error $\Delta\lambda$, in the determination of the decay constant $\lambda$ in $s^{-1}$, is

(a) 0.04

(b) 0.03

(c) 0.02

(d) 0.01

# CHEMISTRY

## Section 1 (Maximum Marks : 24)

- This section contains **SIX (06)** questions.
- Each question has **FOUR** options for correct answer(s). **ONE OR MORE THAN ONE** of these four option(s) is (are) correct options(s).
- For each question, choose the correct options(s) to answer the question.
- Answer to each question will be evaluated according to the following marking scheme:

*Full Marks* : **+4** If only (all) the correct option(s) is (are) chosen.
*Partial Marks* : **+3** If all the four options are correct but ONLY three options are chosen.
*Partial Marks* : **+2** If three or more options are correct but ONLY two options are chosen, both of which are correct options.
*Partial Marks* : **+1** If two or more options are correct but ONLY one option is chosen and it is a correct option.
*Zero Marks* : **0** If none of the options is chosen (i.e. the question is unanswered).
*Negative Marks* : **–2** In all other cases.

- **For example:** If first, third and fourth are the ONLY three correct options for a question with second option being an incorrect option; selecting only all the three correct options will result in + 4 marks. Selecting only two of the three correct options (e.g. the first and fourth options), without selecting any incorrect option (second option in this case), will result in +2 marks. Selecting only one of the three correct options (either first or third or fourth option), without selecting any incorrect option (second option in this case), will result in +1 marks. Selecting any incorrect option(s) (second option in this case), with or without selection of any correct option(s) will result in –2 marks.

**1.** The compound(s) which generate (s) $N_2$ gas upon thermal decomposition below 300°C is (are)

(a) $NH_4NO_3$ (b) $(NH_4)_2Cr_2O_7$
(c) $Ba(N_3)_2$ (d) $Mg_3N_2$

**2.** The correct statement (s) regarding the binary transition metal carbonyl compounds is (are) (Atomic numbers : Fe = 26, Ni = 28)

(a) Total number of valence shell electrons at metal centre in $Fe(CO)_5$ or $Ni(CO)_4$ is 16
(b) These are predominantly low spin in nature
(c) Metal-carbon bond strengthens when the oxidation state of the metal is lowered
(d) The carbonyl C—O bond weakens when the oxidation state of the metal is increased

**3.** Based on the compounds of group 15 elements, the correct statement(s) is (are)

(a) $Bi_2O_5$ is more basic than $N_2O_5$
(b) $NF_3$ is more covalent than $BiF_3$
(c) $PH_3$ boils at lower temperature than $NH_3$
(d) The N—N single bond is stronger than the P—P single bond

**4.** In the following reaction sequence, the correct structure (s) of $X$ is (are)

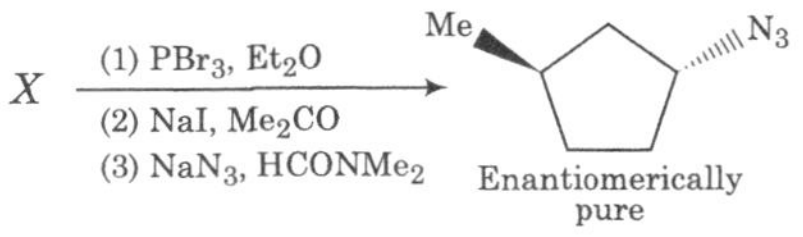

$$X \xrightarrow[\text{(2) NaI, Me}_2\text{CO} \quad \text{(3) NaN}_3\text{, HCONMe}_2]{\text{(1) PBr}_3\text{, Et}_2\text{O}}$$

Me, $N_3$ (cyclopentane) Enantiomerically pure

(a) Me, OH (cyclopentane)

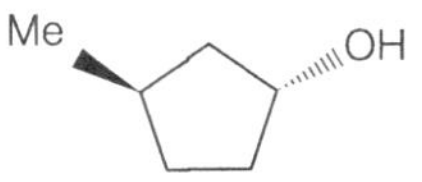

(b) Me, OH (cyclopentane)

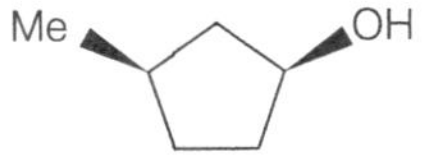

(c) Me (cyclopentane ring with Me and OH)

(d) Me (cyclopentane ring with Me and OH)

**5.** The reaction(s) leading to the formation of 1,3,5-trimethylbenzene is (are)

(a) Acetone $\xrightarrow[\Delta]{\text{Conc. } H_2SO_4}$

(b) Me—≡—H $\xrightarrow[873\text{ K}]{\text{Heated iron tube}}$

(c) 1,3,5-triacetylbenzene $\xrightarrow[(3)\ \text{Sodalime, } \Delta]{(1)\ Br_2,\ NaOH\quad (2)\ H_3O^+}$

(d) Benzene-1,3,5-tricarbaldehyde (CHO, OHC, CHO) $\xrightarrow{\text{Zn/Hg, HCl}}$

**6.** A reversible cyclic process for an ideal gas is shown below. Here, $p, V$ and $T$ are pressure, volume and temperature, respectively. The thermodynamic parameters $q, w, H$ and $U$ are heat, work, enthalpy and internal energy, respectively.

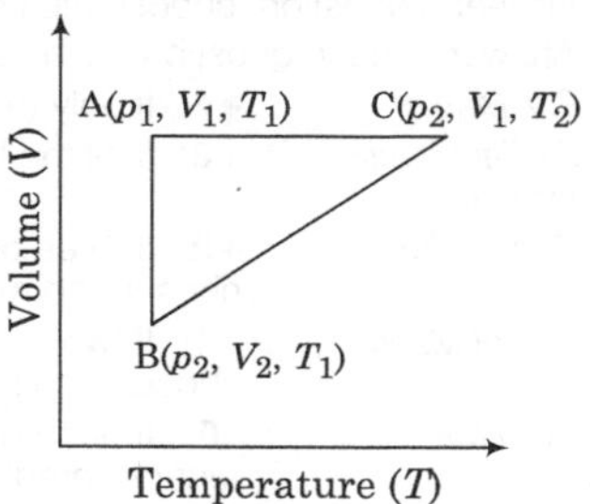


The correct options is (are)

(a) $q_{AC} = \Delta U_{BC}$ and $w_{AB} = p_2(V_2 - V_1)$

(b) $w_{BC} = p_2(V_2 - V_1)$ and $q_{BC} = \Delta H_{AC}$

(c) $\Delta H_{CA} < \Delta U_{CA}$ and $q_{AC} = \Delta U_{BC}$

(d) $q_{BC} = \Delta H_{AC}$ and $\Delta H_{CA} > \Delta U_{CA}$

## Section 2 (Maximum Marks : 24)

- This section contains **EIGHT (08)** questions. The answer to each questions is a **NUMERICAL VALUE.**
- Four each question, enter the correct numerical value (in decimal notation, truncated/rounded-off to the **second decimal place;** e.g. 6.25, 7.00.– 0.33, –.30, 30.27, –127.30) using the mouse and the on screen virtual numeric keypad in the place designated to enter the answer.
- Answer to each question will be evaluated according to the following marking scheme

*Full Marks* : +3 If ONLY the correct numerical value is entered as answer.

*Zero Marks* : 0 In all other cases.

**7.** Among the species given below, the total number of diamagnetic species is____

H atom, $NO_2$ monomer, $O_2^-$ (superoxide), dimeric sulphur in vapour phase,

$Mn_3O_4$, $(NH_4)_2[FeCl_4]$, $(NH_4)_2[NiCl_4]$, $K_2MnO_4$, $K_2CrO_4$

**8.** The ammonia prepared by treating ammonium sulphate with calcium hydroxide is completely used by $NiCl_2 \cdot 6H_2O$ to form a stable coordination compound. Assume that both the reactions are 100% complete. If 1584 g of ammonium sulphate and 952 g of $NiCl_2 \cdot 6H_2O$ are used in the preparation, the combined weight (in grams) of gypsum and the nickel-ammonia coordination compound thus produced is____

(Atomic weights in g $mol^{-1}$ :
H = 1, N = 14, O = 16, S = 32,
Cl = 35.5, Ca = 40, Ni = 59)

**9.** Consider an ionic solid $MX$ with NaCl structure. Construct a new structure ($Z$) whose unit cell is constructed from the unit cell of $MX$ following the sequential instruction given below. Neglect the charge balance.

(a) Remove all the anions ($X$) except the central one
(b) Replace all the face centered cations ($M$) by anions ($X$)
(c) Remove all the corner cations ($M$)
(d) Replace the central anion ($X$) with cation ($M$)

The value of $\left(\frac{\text{Number of anions}}{\text{Number of cations}}\right)$ in $Z$ is ___

**10.** For the electrochemical cell,

$Mg(s)|Mg^{2+}(aq, 1\,M)|\,|Cu^{2+}(aq, 1\,M)|Cu(s)$

The standard emf of the cell is 2.70 V at 300 K. When the concentration of $Mg^{2+}$ is changed to $x$ M, the cell potential changes to 2.67 V at 300 K. The value of $x$ is ______.

(Given, $\frac{F}{R} = 11500\ K\,V^{-1}$, where $F$ is the Faraday constant and $R$ is the gas contant, In (10) = 2.30)

**11.** A closed tank has two compartments $A$ and $B$, both filled with oxygen (assumed to be ideal gas). The partition separating the two compartments is fixed and is a perfect heat insulator (Fig. 1). If the old partition is replaced by a new partition which can slide and conduct heat but does not allow the gas to leak across (Fig. 2), the volume (in $m^3$) of the compartment $A$ after the system attains equilibrium is ____.

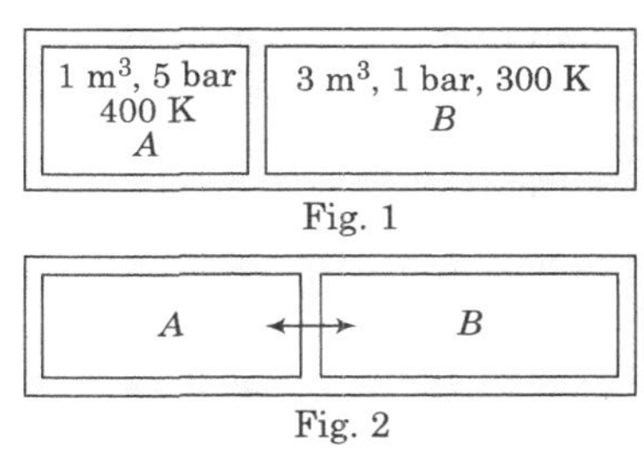

Fig. 1

Fig. 2

**12.** Liquids $A$ and $B$ form ideal solution over the entire range of composition. At temperature $T$, equimolar binary solution of liquids $A$ and $B$ has vapour pressure 45 torr. At the same temperature, a new solution of $A$ and $B$ having mole fractions $x_A$ and $x_B$, respectively, has vapour pressure of 22.5 torr. The value of $x_A / x_B$ in the new solution is ____.

(Given that the vapour pressure of pure liquid $A$ is 20 Torr at temperature $T$)

**13.** The solubility of a salt of weak acid ($AB$) at pH 3 is $Y \times 10^{-3}$ mol $L^{-1}$. The value of $Y$ is_____

(Given that the value of solubility product of $AB$ $(K_{sp}) = 2 \times 10^{-10}$ and the value of ionisation constant of HB $(K_a) = 1 \times 10^{-8}$

**14.** The plot given below shows $p - T$ curves (where $p$ is the pressure and $T$ is the temperature) for two solvents $X$ and $Y$ and isomolal solution of NaCl in these solvents. NaCl completely dissociates in both the solvents.

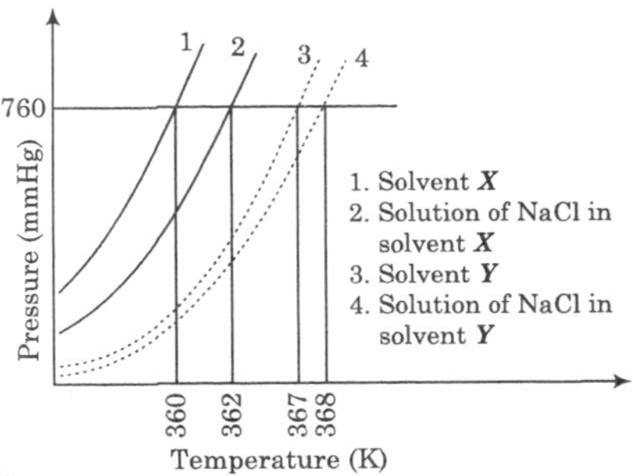

On addition of equal number of moles of a non-volatile solute $S$ in equal amount (in kg) of these solvents, the elevation of boiling point of solvent $X$ is three times that of solvent $Y$. Solute S is known to undergo dimerisation in these solvents. If the degree of dimerisation is 0.7 in solvent $Y$, the degree of dimerisation in solvent $X$ is ____.

## Section 3 (Maximum Marks : 12)

- This section contains **TWO (02)** paragraphs. Based on each paragraph, there are **TWO(02)** questions.
- Each question has **FOUR** options. **ONLY ONE** of these four options corresponds to the correct answer.
- Four each question, choose the option corresponding to the correct answer.
- Answer to each question will be evaluated according to the following marking scheme:

*Full Marks* : +3 If ONLY the correct option is chosen.
*Zero Marks* : 0 If none of the options is chosen (i.e. the question is unanswered).
*Negative Marks* : −1 In all other cases.

### Paragraph X

Treatment of benzene with $CO/HCl$ in the presence of anhydrous $AlCl_3/CuCl$ followed by reaction with $Ac_2O/NaOAc$ gives compound $X$ as the major product. Compound $X$ upon reaction with $Br_2/Na_2CO_3$ followed by heating at 473 K with moist KOH furnishes $Y$ as the major product. Reaction of $X$ with $H_2/Pd$-C, followed by $H_3PO_4$ treatment gives $Z$ as the major product.

**15.** The compound $Y$ is

(a) 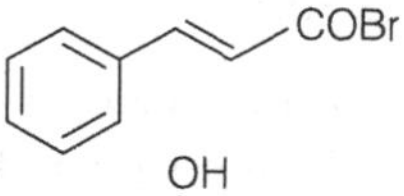 COBr

(b) 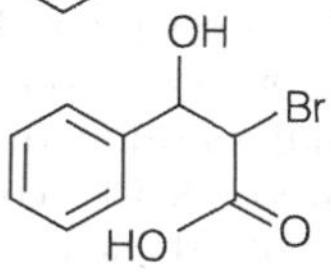 OH, Br, HO, O

(c)

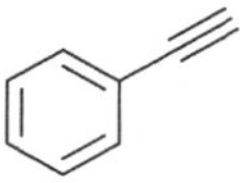

(d) 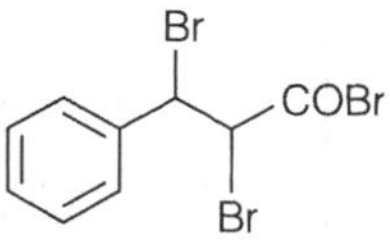 Br, COBr, Br

**16.** The compound $Z$ is

(a) 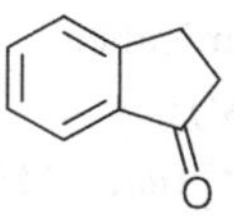 O

(b) 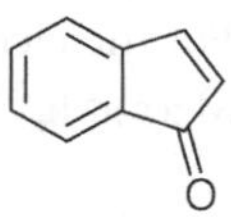 O

(c) 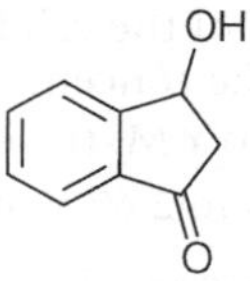 OH, O

(d)

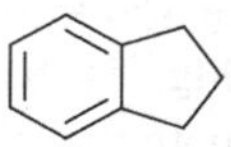

### Paragraph A

An organic acid $P(C_{11}H_{12}O_2)$ can easily be oxidised to a dibasic acid which reacts with ethylene glycol to produce a polymer dacron. Upon ozonolysis, $P$ gives an aliphatic ketone as one of the products. $P$ undergoes the following reaction sequences to furnish $R$ via $Q$. The compound $P$ also undergoes another set of reactions to produce $S$.

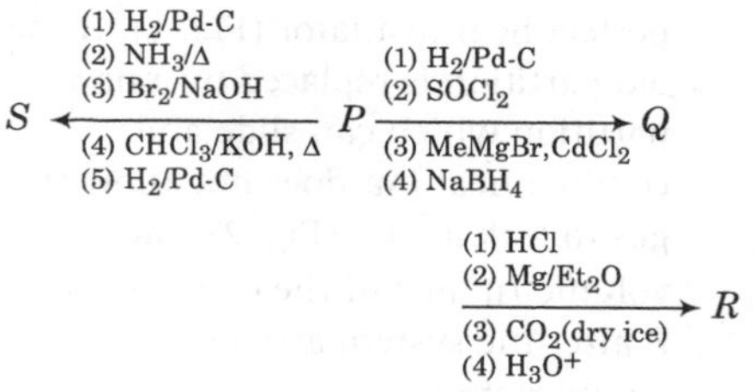

$S \xleftarrow[\text{(4) } CHCl_3/KOH, \Delta \text{ (5) } H_2/Pd\text{-}C]{\text{(1) } H_2/Pd\text{-}C \text{ (2) } NH_3/\Delta \text{ (3) } Br_2/NaOH} P \xrightarrow[\text{(3) } MeMgBr, CdCl_2 \text{ (4) } NaBH_4]{\text{(1) } H_2/Pd\text{-}C \text{ (2) } SOCl_2} Q \xrightarrow[\text{(3) } CO_2\text{(dry ice) (4) } H_3O^+]{\text{(1) } HCl \text{ (2) } Mg/Et_2O} R$

**17.** The Compound $R$ is

(a)

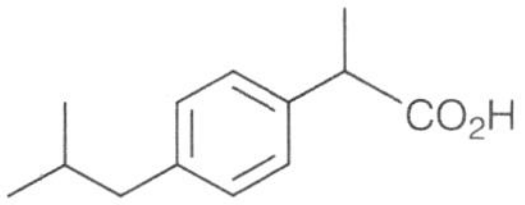

(b)

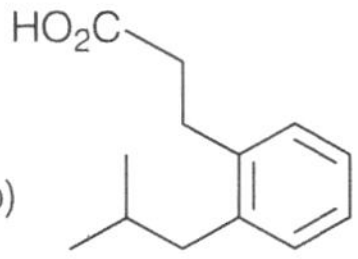

(c)

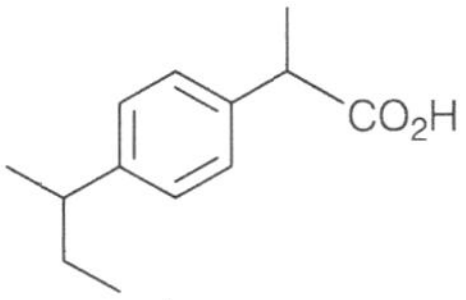

(d)

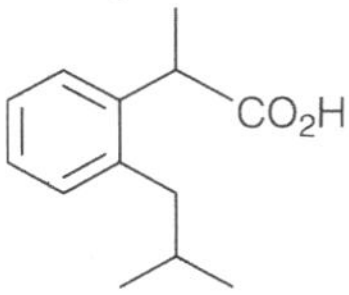

**18.** The compound $S$ is

(a)

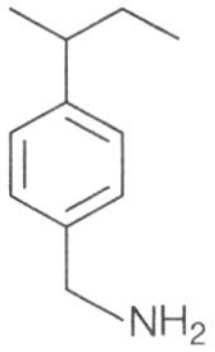

(b)

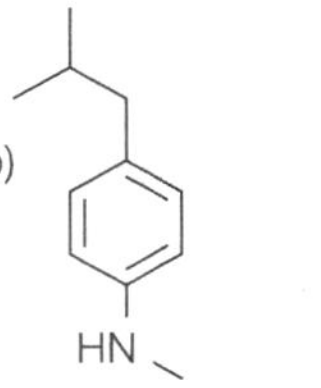

(c)

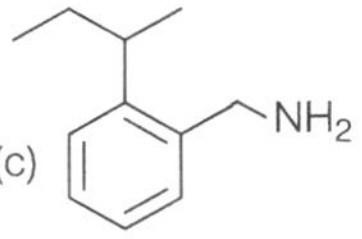

(d)

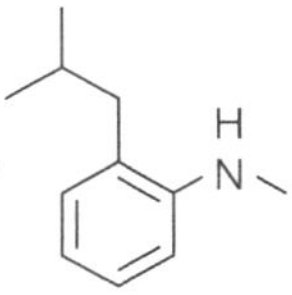

# MATHEMATICS

## Section 1 (Maximum Marks : 24)

- This section contains **SIX (06)** questions.
- Each question has **FOUR** options for correct answer(s). **ONE OR MORE THAN ONE** of these four option(s) is (are) correct options(s).
- For each question, choose the correct options(s) to answer the question.
- Answer to each question will be evaluated according to the following marking scheme:

*Full Marks* : **+4** If only (all) the correct option(s) is (are) chosen.
*Partial Marks* : **+3** If all the four options are correct but ONLY three options are chosen.
*Partial Marks* : **+2** If three or more options are correct but ONLY two options are chosen, both of which are correct options.
*Partial Marks* : **+1** If two or more options are correct but ONLY one option is chosen and it is a correct option.
*Zero Marks* : **0** If none of the options is chosen (i.e. the question is unanswered).
*Negative Marks* : **–2** In all other cases.

- **For example:** If first, third and fourth are the ONLY three correct options for a question with second option being an incorrect option; selecting only all the three correct options will result in + 4 marks. Selecting only two of the three correct options (e.g. the first and fourth options), without selecting any incorrect option (second option in this case), will result in +2 marks. Selecting only one of the three correct options (either first or third or fourth option), without selecting any incorrect option (second option in this case), will result in +1 marks. Selecting any incorrect option(s) (second option in this case), with or without selection of any correct option(s) will result in –2 marks.

**1.** For a non-zero complex number $z$, let $\arg(z)$ denote the principal argument with $-\pi < \arg(z) \le \pi$. Then, which of the following statement(s) is (are) FALSE ?

(a) $\arg(-1-i) = \frac{\pi}{4}$, where $i = \sqrt{-1}$

(b) The function $f : R \to (-\pi, \pi]$, defined by $f(t) = \arg(-1 + it)$ for all $t \in R$, is continuous at all points of $R$, where $i = \sqrt{-1}$.

(c) For any two non-zero complex numbers $z_1$ and $z_2$, $\arg\left(\frac{z_1}{z_2}\right) - \arg(z_1) + \arg(z_2)$ is an integer multiple of $2\pi$.

(d) For any three given distinct complex numbers $z_1$, $z_2$ and $z_3$, the locus of the point $z$ satisfying the condition $\arg\left(\frac{(z - z_1)(z_2 - z_3)}{(z - z_3)(z_2 - z_1)}\right) = \pi$, lies on a straight line.

**2.** In a $\Delta PQR$, let $\angle PQR = 30°$ and the sides $PQ$ and $QR$ have lengths $10\sqrt{3}$ and 10, respectively. Then, which of the following statement(s) is (are) TRUE?

(a) $\angle QPR = 45°$

(b) The area of the $\Delta PQR$ is $25\sqrt{3}$ and $\angle QRP = 120°$

(c) The radius of the incircle of the $\Delta PQR$ is $10\sqrt{3} - 15$

(d) The area of the circumcircle of the $\Delta PQR$ is $100\pi$

**3.** Let $P_1 : 2x + y - z = 3$ and $P_2 : x + 2y + z = 2$ be two planes. Then, which of the following statement(s) is (are) TRUE?

(a) The line of intersection of $P_1$ and $P_2$ has direction ratios $1, 2, -1$

(b) The line $\frac{3x-4}{9} = \frac{1-3y}{9} = \frac{z}{3}$ is perpendicular to the line of intersection of $P_1$ and $P_2$

(c) The acute angle between $P_1$ and $P_2$ is $60°$

(d) If $P_3$ is the plane passing through the point $(4, 2, -2)$ and perpendicular to the line of intersection of $P_1$ and $P_2$, then the distance of the point $(2, 1, 1)$ from the plane $P_3$ is $\frac{2}{\sqrt{3}}$

**4.** For every twice differentiable function $f : R \to [-2, 2]$ with $(f(0))^2 + (f'(0))^2 = 85$, which of the following statement(s) is (are) TRUE ?

(a) There exist $r, s \in R$, where $r < s$, such that $f$ is one-one on the open interval $(r, s)$

(b) There exists $x_0 \in (-4, 0)$ such that $|f'(x_0)| \le 1$

(c) $\lim_{x \to \infty} f(x) = 1$

(d) There exists $\alpha \in (-4, 4)$ such that $f(\alpha) + f''(\alpha) = 0$ and $f'(\alpha) \ne 0$

**5.** Let $f : R \to R$ and $g : R \to R$ be two non-constant differentiable functions. If $f'(x) = (e^{(f(x) - g(x))})\, g'(x)$ for all $x \in R$ and $f(1) = g(2) = 1$, then which of the following statement(s) is (are) TRUE?

(a) $f(2) < 1 - \log_e 2$

(b) $f(2) > 1 - \log_e 2$

(c) $g(1) > 1 - \log_e 2$

(d) $g(1) < 1 - \log_e 2$

**6.** Let $f : [0, \infty) \to R$ be a continuous function such that $f(x) = 1 - 2x + \int_0^x e^{x-t} f(t)\, dt$ for all $x \in [0, \infty)$. Then, which of the following statement(s) is (are) TRUE?

(a) The curve $y = f(x)$ passes through the point $(1, 2)$

(b) The curve $y = f(x)$ passes through the point $(2, -1)$

(c) The area of the region $\{(x, y) \in [0, 1] \times R : f(x) \le y \le \sqrt{1 - x^2}\}$ is $\frac{\pi - 2}{4}$

(d) The area of the region $\{(x, y) \in [0, 1] \times R : f(x) \le y \le \sqrt{1 - x^2}\}$ is $\frac{\pi - 1}{4}$

## Section 2 (Maximum Marks : 24)

- This section contains **EIGHT (08)** questions. The answer to each question is a **NUMERICAL VALUE.**
- Four each question, enter the correct numerical value (in decimal notation, truncated/rounded-off to the **second decimal place;** e.g. 6.25, 7.00.– 0.33, –.30, 30.27, –127.30) using the mouse and the on-screen virtual numeric keypad in the place designated to enter the answer.
- Answer to each question will be evaluated according to the following marking scheme:

*Full Marks* : **+3** If ONLY the correct numerical value is entered as answer.

*Zero Marks* : **0** In all other cases.

**7.** The value of

$$\left((\log_2 9)^2\right)^{\frac{1}{\log_2 (\log_2 9)}} \times (\sqrt{7})^{\frac{1}{\log_4 7}}$$ is

............. .

**8.** The number of 5 digit numbers which are divisible by 4, with digits from the set $\{1, 2, 3, 4, 5\}$ and the repetition of digits is allowed, is ..................... .

**9.** Let $X$ be the set consisting of the first 2018 terms of the arithmetic progression 1, 6, 11, ...., and $Y$ be the set consisting of the first 2018 terms of the arithmetic progression 9, 16, 23, ..... Then, the number of elements in the set $X \cup Y$ is .................. .

**10.** The number of real solutions of the equation $\sin^{-1}\left(\sum_{i=1}^{\infty} x^{i+1} - x\sum_{i=1}^{\infty}\left(\frac{x}{2}\right)^i\right)$

$$= \frac{\pi}{2} - \cos^{-1}\left(\sum_{i=1}^{\infty}\left(-\frac{x}{2}\right)^i - \sum_{i=1}^{\infty}(-x)^i\right)$$

lying in the interval $\left(-\frac{1}{2}, \frac{1}{2}\right)$ is .................. .

(Here, the inverse trigonometric functions $\sin^{-1} x$ and $\cos^{-1} x$ assume values in $\left[-\frac{\pi}{2}, \frac{\pi}{2}\right]$ and $[0, \pi]$, respectively.)

**11.** For each positive integer $n$, let

$$y_n = \frac{1}{n}\left((n+1)(n+2)\ldots(n+n)\right)^{\frac{1}{n}}.$$

For $x \in R$, let $[x]$ be the greatest integer less than or equal to $x$. If $\lim_{n\to\infty} y_n = L$, then the value of $[L]$ is ................ .

**12.** Let **a** and **b** be two unit vectors such that $\mathbf{a} \cdot \mathbf{b} = 0$. For some $x, y \in R$, let $\mathbf{c} = x\mathbf{a} + y\mathbf{b} + (\mathbf{a} \times \mathbf{b})$. If $|\mathbf{c}| = 2$ and the vector **c** is inclined at the same angle $\alpha$ to both **a** and **b**, then the value of $8\cos^2\alpha$ is ............... .

**13.** Let $a, b, c$ be three non-zero real numbers such that the equation $\sqrt{3}\, a\cos x + 2b\sin x = c$, $x \in \left[-\frac{\pi}{2}, \frac{\pi}{2}\right]$, has two distinct real roots $\alpha$ and $\beta$ with $\alpha + \beta = \frac{\pi}{3}$. Then, the value of $\frac{b}{a}$ is ................. .

**14.** A farmer $F_1$ has a land in the shape of a triangle with vertices at $P(0, 0)$, $Q(1, 1)$ and $R(2, 0)$. From this land, a neighbouring farmer $F_2$ takes away the region which lies between the sides $PQ$ and a curve of the form $y = x^n$ $(n > 1)$. If the area of the region taken away by the farmer $F_2$ is exactly 30% of the area of $\Delta PQR$, then the value of $n$ is ................... .

## Section 3 (Maximum Marks : 12)

- This section contains **TWO (02)** paragraphs. Based on each paragraph, there are **TWO(02)** questions.
- Each question has **FOUR** options. **ONLY ONE** of these four options corresponds to the correct answer.
- Four each question, choose the option corresponding to the correct answer.
- Answer to each question will be evaluated according to the following marking scheme:

*Full Marks* : **+3** If ONLY the correct option is chosen.
*Zero Marks* : **0** If none of the options is chosen (i.e. the question is unanswered).
*Negative Marks* : **−1** In all other cases.

### Paragraph X

Let $S$ be the circle in the $XY$-plane defined by the equation $x^2 + y^2 = 4$.

*(There are two questions based on Paragraph "X", the question given below is one of them)*

**15.** Let $E_1E_2$ and $F_1F_2$ be the chords of $S$ passing through the point $P_0$ (1, 1) and parallel to the $X$-axis and the $Y$-axis, respectively. Let $G_1G_2$ be the chord of $S$ passing through $P_0$ and having slope −1. Let the tangents to $S$ at $E_1$ and $E_2$ meet at $E_3$, then tangents to $S$ at $F_1$ and $F_2$ meet at $F_3$, and the tangents to $S$ at $G_1$ and $G_2$ meet at $G_3$. Then, the points $E_3$, $F_3$ and $G_3$ lie on the curve

(a) $x + y = 4$
(b) $(x-4)^2 + (y-4)^2 = 16$
(c) $(x-4)(y-4) = 4$
(d) $xy = 4$

**16.** Let $P$ be a point on the circle $S$ with both coordinates being positive. Let the tangent to $S$ at $P$ intersect the coordinate axes at the points $M$ and $N$. Then, the mid-point of the line segment $MN$ must lie on the curve

(a) $(x+y)^2 = 3xy$
(b) $x^{2/3} + y^{2/3} = 2^{4/3}$
(c) $x^2 + y^2 = 2xy$
(d) $x^2 + y^2 = x^2y^2$

### Paragraph A

There are five students $S_1, S_2, S_3, S_4$ and $S_5$ in a music class and for them there are five seats $R_1, R_2, R_3, R_4$ and $R_5$ arranged in a row, where initially the seat $R_i$ is allotted to the student $S_i$, $i = 1, 2, 3, 4, 5$. But, on the examination day, the five students are randomly allotted the five seats.

*(There are two questions based on Paragraph "A", the question given below is one of them)*

**17.** The probability that, on the examination day, the student $S_1$ gets the previously allotted seat $R_1$, and NONE of the remaining students gets the seat previously allotted to him/her is

(a) $\frac{3}{40}$ (b) $\frac{1}{8}$
(c) $\frac{7}{40}$ (d) $\frac{1}{5}$

**18.** For $i = 1, 2, 3, 4$, let $T_i$ denote the event that the students $S_i$ and $S_{i+1}$ do NOT sit adjacent to each other on the day of the examination. Then, the probability of the event $T_1 \cap T_2 \cap T_3 \cap T_4$ is

(a) $\frac{1}{15}$ (b) $\frac{1}{10}$
(c) $\frac{7}{60}$ (d) $\frac{1}{5}$

# Paper 2

## PHYSICS

### Section 1 (Maximum Marks : 24)

- This section contains **SIX (06)** questions.
- Each question has **FOUR** options for correct answer(s). **ONE OR MORE THAN ONE** of these four option(s) is (are) correct options(s).
- For each question, choose the correct options(s) to answer the question.
- Answer to each question will be evaluated according to the following marking scheme:

| | | |
|---|---|---|
| *Full Marks* | : **+4** | If only (all) the correct option(s) is (are) chosen. |
| *Partial Marks* | : **+3** | If all the four options are correct but ONLY three options are chosen. |
| *Partial Marks* | : **+2** | If three or more options are correct but ONLY two options are chosen, both of which are correct options. |
| *Partial Marks* | : **+1** | If two or more options are correct but ONLY one option is chosen and it is a correct option. |
| *Zero Marks* | : **0** | If none of the options is chosen (i.e. the question is unanswered). |
| *Negative Marks* | : **−2** | In all other cases. |

- **For example:** If first, third and fourth are the ONLY three correct options for a question with second option being an incorrect option; selecting only all the three correct options will result in + 4 marks. Selecting only two of the three correct options (e.g. the first and fourth options), without selecting any incorrect option (second option in this case), will result in +2 marks. Selecting only one of the three correct options (either first or third or fourth option), without selecting any incorrect option (second option in this case), will result in +1 marks. Selecting any incorrect option(s) (second option in this case), with or without selection of any correct option(s) will result in −2 marks.

**1.** A particle of mass $m$ is initially at rest at the origin. It is subjected to a force and starts moving along the $X$-axis. Its kinetic energy $K$ changes with time as $dK / dt = \gamma t$, where $\gamma$ is a positive constant of appropriate dimensions. Which of the following statements is (are) true?

(a) The force applied on the particle is constant

(b) The speed of the particle is proportional to time

(c) The distance of the particle from the origin increases linearly with time

(d) The force is conservative

**2.** Consider a thin square plate floating on a viscous liquid in a large tank. The height $h$ of the liquid in the tank is much less than the width of the tank. The floating plate is pulled horizontally with a constant velocity $u_0$. Which of the following statements is (are) true?

(a) The resistive force of liquid on the plate is inversely proportional to $h$

(b) The resistive force of liquid on the plate is independent of the area of the plate

(c) The tangential (shear) stress on the floor of the tank increases with $u_0$

(d) The tangential (shear) stress on the plate varies linearly with the viscosity $\eta$ of the liquid

**3.** An infinitely long thin non-conducting wire is parallel to the $Z$-axis and carries a uniform line charge density $\lambda$. It pierces a thin non-conducting spherical shell of radius $R$ in such a way that the arc $PQ$ subtends an angle $120°$ at the centre $O$ of the spherical shell, as shown in the figure. The permittivity of free space is $\varepsilon_0$. Which of the following statements is (are) true?

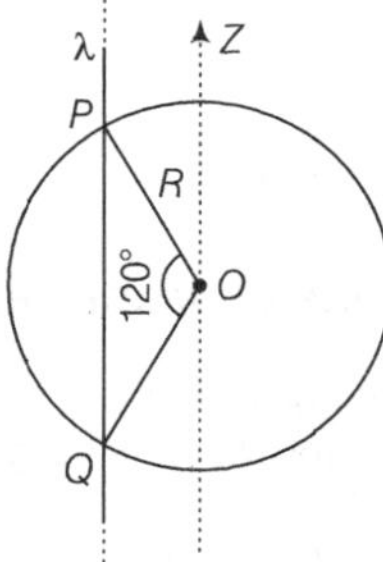


(a) The electric flux through the shell is $\sqrt{3}\, R\lambda / \varepsilon_0$

(b) The $z$-component of the electric field is zero at all the points on the surface of the shell

(c) The electric flux through the shell is $\sqrt{2}\, R\lambda / \varepsilon_0$

(d) The electric field is normal to the surface of the shell at all points

**4.** A wire is bent in the shape of a right angled triangle and is placed in front of a concave mirror of focal length $f$ as shown in the figure. Which of the figures shown in the four options qualitatively represent(s) the shape of the image of the bent wire? (These figures are not to scale.)

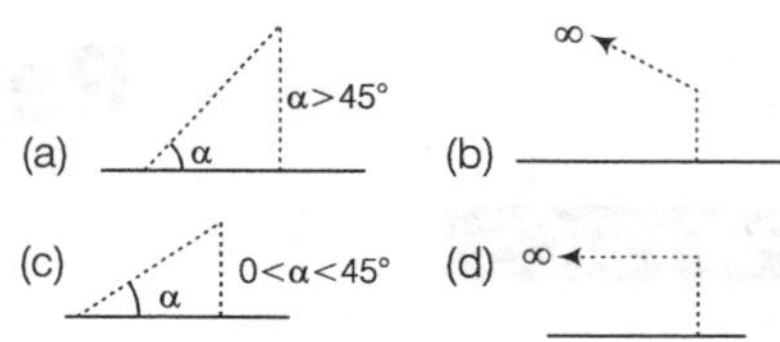


**5.** In a radioactive decay chain, ${}^{232}_{90}\text{Th}$ nucleus decays to ${}^{212}_{82}\text{Pb}$ nucleus. Let $N_\alpha$ and $N_\beta$ be the number of $\alpha$ and $\beta$- particles respectively, emitted in this decay process. Which of the following statements is (are) true?

(a) $N_\alpha = 5$ (b) $N_\alpha = 6$

(c) $N_\beta = 2$ (d) $N_\beta = 4$

**6.** In an experiment to measure the speed of sound by a resonating air column, a tuning fork of frequency 500 Hz is used. The length of the air column is varied by changing the level of water in the resonance tube. Two successive resonances are heard at air columns of length 50.7 cm and 83.9 cm. Which of the following statements is (are) true?

(a) The speed of sound determined from this experiment is 332 m s$^{-1}$

(b) The end correction in this experiment is 0.9 cm

(c) The wavelength of the sound wave is 66.4 cm

(d) The resonance at 50.7 cm corresponds to the fundamental harmonic

## Section 2 (Maximum Marks : 24)

- This section contains **EIGHT (08)** questions. The answer to each question is a **NUMERICAL VALUE.**
- Four each question, enter the correct numerical value (in decimal notation, truncated/rounded-off to the **second decimal place;** e.g. 6.25, 7.00.– 0.33, –.30, 30.27, –127.30) using the mouse and the on-screen virtual numeric keypad in the place designated to enter the answer.
- Answer to each question will be evaluated according to the following marking scheme:

*Full Marks* : **+3** If ONLY the correct numerical value is entered as answer.

*Zero Marks* : **0** In all other cases.

**7.** A solid horizontal surface is covered with a thin layer of oil. A rectangular block of mass $m = 0.4$ kg is at rest on this surface. An impulse of 1.0 N s is applied to the block at time $t = 0$, so that it starts moving along the $X$-axis with a velocity $v(t) = v_0 e^{-t/\tau}$, where $v_0$ is a constant and $\tau = 4$ s. The displacement of the block, in metres, at $t = \tau$ is ............. . (Take, $e^{-1} = 0.37$).

**8.** A ball is projected from the ground at an angle of 45° with the horizontal surface. It reaches a maximum height of 120 m and returns to the ground. Upon hitting the ground for the first time, it loses half of its kinetic energy. Immediately after the bounce, the velocity of the ball makes an angle of 30° with the horizontal surface. The maximum height it reaches after the bounce, in metres, is ........ .

**9.** A particle of mass $10^{-3}$kg and charge 1.0 C is initially at rest. At time $t = 0$, the particle comes under the influence of an electric field $\mathbf{E}(t) = E_0 \sin \omega t\, \hat{\mathbf{i}}$, where $E_0 = 1.0\ \text{NC}^{-1}$ and $\omega = 10^3\ \text{rad s}^{-1}$. Consider the effect of only the electrical force on the particle. Then, the maximum speed in m $\text{s}^{-1}$, attained by the particle at subsequent times is ........ .

**10.** A moving coil galvanometer has 50 turns and each turn has an area $2 \times 10^{-4}\text{m}^2$. The magnetic field produced by the magnet inside the galvanometer is 0.02 T. The torsional constant of the suspension wire is $10^{-4}\text{N - m rad}^{-1}$. When a current flows through the galvanometer, a full scale deflection occurs, if the coil rotates by 0.2 rad. The resistance of the coil of the galvanometer is 50 Ω. This galvanometer is to be converted into an ammeter capable of measuring current in the range 0 – 1.0 A. For this purpose, a shunt resistance is to be added in parallel to the galvanometer. The value of this shunt resistance in ohms, is ............ .

**11.** A steel wire of diameter 0.5 mm and Young's modulus $2 \times 10^{11}\ \text{N m}^{-2}$ carries a load of mass $m$. The length of the wire with the load is 1.0 m. A vernier scale with 10 divisions is attached to the end of this wire. Next to the steel wire is a reference wire to which a main scale, of least count 1.0 mm, is attached. The 10 divisions of the vernier scale correspond to 9 divisions of the main scale. Initially, the zero of vernier scale coincides with the zero of main scale. If the load on the steel wire is increased by 1.2 kg, the vernier scale division which coincides with a main scale division is ......... . Take, $g = 10\ \text{ms}^{-2}$ and $\pi = 3.2$).

**12.** One mole of a monoatomic ideal gas undergoes an adiabatic expansion in which its volume becomes eight times its initial value. If the initial temperature of the gas is 100 K and the universal gas constant $R = 8.0\ \text{j mol}^{-1}\ \text{K}^{-1}$, the decrease in its internal energy in joule, is .......... .

**13.** In a photoelectric experiment, a parallel beam of monochromatic light with power of 200 W is incident on a perfectly absorbing cathode of work function 6.25 eV. The frequency of light is just above the threshold frequency, so that the photoelectrons are emitted with negligible kinetic energy. Assume that the photoelectron emission efficiency is 100%. A potential difference of 500 V is applied between the cathode and the anode. All the emitted electrons are incident normally on the anode and are absorbed. The anode experiences a force $F = n \times 10^{-4}$N due to the impact of the electrons. The value of $n$ is ........... . (Take mass of the electron, $m_e = 9\times10^{-31}$kg and eV $=1.6\times10^{-19}$J)

**14.** Consider a hydrogen-like ionised atom with atomic number $Z$ with a single electron. In the emission spectrum of this atom, the photon emitted in the $n = 2$ to $n = 1$ transition has energy 74.8 eV higher than the photon emitted in the $n = 3$ to $n = 2$ transition. The ionisation energy of the hydrogen atom is 13.6 eV. The value of $Z$ is ............ .

## Section 3 (Maximum Marks : 12)

- This section contains **TWO (02)** paragraphs. Based on each paragraph, there are **TWO(02)** questions.
- Each question has **FOUR** options. **ONLY ONE** of these four options corresponds to the correct answer.
- Four each question, choose the option corresponding to the correct answer.
- Answer to each question will be evaluated according to the following marking scheme:
  *Full Marks* : **+3** If ONLY the correct option is chosen.
  *Zero Marks* : **0** If none of the options is chosen (i.e. the question is unanswered).
  *Negative Marks* : **−1** In all other cases.

**15.** The electric field $E$ is measured at a point $P\,(0, 0, d)$ generated due to various charge distributions and the dependence of $E$ on $d$ is found to be different for different charge distributions. List-I contains different relations between $E$ and $d$. List-II describes different electric charge distributions, along with their locations. Match the functions in List-I with the related charge distributions in List-II.

| | List-I | | List-II |
|---|---|---|---|
| P. | $E$ is independent of $d$ | 1. | A point charge $Q$ at the origin |
| Q. | $E \propto \frac{1}{d}$ | 2. | A small dipole with point charges $Q$ at $(0, 0, l)$ and $-Q$ at $(0, 0, -1)$. (Take, $2l \ll d$) |
| R. | $E \propto \frac{1}{d^2}$ | 3. | An infinite line charge coincident with the $X$-axis, with uniform linear charge density $\lambda$. |
| S. | $E \propto \frac{1}{d^3}$ | 4. | Two infinite wires carrying a uniform linear charge density parallel to the $X$- axis. The one along $(y = 0, z = l)$ has a charge density $+\lambda$ and the one along $(y = 0, z = -l)$ has a charge density $-\lambda$. (Take, $2l \ll d$). |
| | | 5. | Infinite plane charge coincident with the $xy$-plane with uniform surface charge density. |

(a) P $\to$5; Q $\to$3,4; R $\to$1; S $\to$2
(b) P $\to$5; Q $\to$3; R $\to$1, 4; S $\to$2
(c) P $\to$5; Q $\to$3; R $\to$1, 2; S $\to$4
(d) P $\to$4; Q $\to$2, 3; R $\to$1; S $\to$5

**16.** A planet of mass $M$, has two natural satellites with masses $m_1$ and $m_2$. The radii of their circular orbits are $R_1$ and $R_2$, respectively. Ignore the gravitational force between the satellites. Define $v_1, L_1, K_1$ and $T_1$ to be respectively, the orbital speed, angular momentum, kinetic energy and time period of revolution of satellite 1; and $v_2, L_2, K_2$ and $T_2$ to be the corresponding quantities of satellite 2. Given, $m_1 / m_2 = 2$ and $R_1 / R_2 = 1/4$, match the ratios in List-I to the numbers in List-II.

| List-I | List-II |
|---|---|
| P. $\frac{v_1}{v_2}$ | 1. $\frac{1}{8}$ |
| Q. $\frac{L_1}{L_2}$ | 2. 1 |
| R. $\frac{K_1}{K_2}$ | 3. 2 |
| S. $\frac{T_1}{T_2}$ | 4. 8 |

(a) P → 4; Q → 2; R → 1; S → 3
(b) P → 3; Q → 2; R → 4; S → 1
(c) P → 2; Q → 3; R → 1; S → 4
(d) P → 2; Q → 3; R → 4; S → 1

**17.** One mole of a monoatomic ideal gas undergoes four thermodynamic processes as shown schematically in the $pV$-diagram below. Among these four processes, one is isobaric, one is isochoric, one is isothermal and one is adiabatic. Match the processes mentioned in List-I with the corresponding statements in List-II.

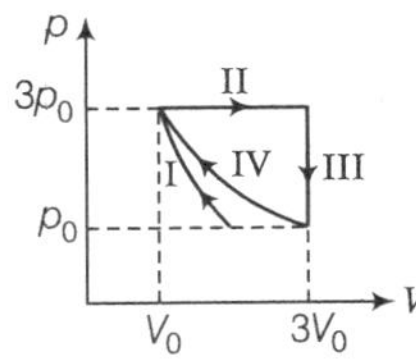


| List-I | List-II |
|---|---|
| P. In process I | 1. Work done by the gas is zero |
| Q. In process II | 2. Temperature of the gas remains unchanged |
| R. In process III | 3. No heat is exchanged between the gas and its surroundings |
| S. In process IV | 4. Work done by the gas is $6p_0V_0$ |

(a) P → 4; Q → 3; R → 1; S → 2
(b) P → 1; Q → 3; R → 2; S → 4
(c) P → 3; Q → 4; R → 1; S → 2
(d) P → 3; Q → 4; R → 2; S → 1

**18.** In the List-I below, four different paths of a particle are given as functions of time. In these functions, α and β are positive constants of appropriate dimensions and $\alpha \neq \beta$. In each case, the force acting on the particle is either zero or conservative. In List-II, five physical quantities of the particle are mentioned: **p** is the linear momentum, **L** is the angular momentum about the origin, $K$ is the kinetic energy, $U$ is the potential energy and $E$ is the total energy. Match each path in List-I with those quantities in List-II, which are conserved for that path.

| List-I | List-II |
|---|---|
| P. $\mathbf{r}(t) = \alpha t\,\mathbf{i} + \beta t\,\mathbf{j}$ | 1. **p** |
| Q. $\mathbf{r}(t) = \alpha \cos \omega t\,\mathbf{i} + \beta \sin \omega t\,\mathbf{j}$ | 2. **L** |
| R. $\mathbf{r}(t) = \alpha(\cos \omega t\,\mathbf{i} + \sin \omega t\,\mathbf{j})$ | 3. $K$ |
| S. $\mathbf{r}(t) = at\,\mathbf{i} + \frac{\beta}{2} t^2\,\mathbf{j}$ | 4. $U$ |
| | 5. $E$ |

(a) P → 1, 2, 3, 4, 5; Q → 2, 5; R → 2, 3, 4, 5; S → 5
(b) P → 1, 2, 3, 4, 5; Q → 3, 5; R → 2, 3, 4, 5; S → 2, 5
(c) P → 2, 3, 4; Q → 5; R → 1, 2, 4; S → 2, 5
(d) P → 1, 2, 3, 5; Q → 2, 5; R → 2, 3, 4, 5; S → 2, 5

# CHEMISTRY

## Section 1 (Maximum Marks : 24)

- This section contains **SIX (06)** questions.
- Each question has **FOUR** options for correct answers. **ONE OR MORE THAN ONE** of these four options is are correct options(s).
- For each question, choose the correct options to answer the question.
- Answer to each question will be evaluated according to the following marking scheme:

*Full Marks* : + 4 If only all the correct options is are chosen.

*Partial Marks* : + 3 If all the four options are correct but ONLY three options are chosen.

*Partial Marks* : +2 If three or more options are correct but ONLY two options are chosen, both of which are correct options

*Partial Marks* : + 1 If two or more options are correct but ONLY one option is chosen and it is a correct option.

*Zero Marks* : 0 If none of the options is chosen (i.e. the question is unanswered).

*Negative Marks* : −2 In all other cases.

- **For example:** If first, third and fourth are the ONLY three correct options for a question with second option being an incorrect option; selecting only all the three correct options will result in +4 marks. Selecting only two of the three correct options (e.g. the first and fourth options), without selecting any incorrect options (second option in this case), will result in +2 marks. Selecting only one of the three correct options (either first or third or fourth option), without selecting any incorrect optin (second option in this case), will result in +1 marks. Selecting any incorrect options (second option in this case), with or without selection of any correct option(s) will result in −2 marks.

**1.** The correct option(s) regarding the complex $[Co(en)(NH_3)_3(H_2O)]^{3+}$ ($en = H_2NCH_2CH_2NH_2$) is (are)

(a) It has two geometrical isomers

(b) It will have three geometrical isomers, if bidentate 'en' is replaced by two cyanide ligands

(c) It is paramagnetic

(d) It absorbs light at longer wavelength as compared to $[Co(en)(NH_3)_4]^{3+}$

**2.** The correct option(s) to distinguish nitrate salts to $Mn^{2+}$ and $Cu^{2+}$ taken separately is (are)

(a) $Mn^{2+}$ shows the characteristic green colour in the flame test

(b) Only $Cu^{2+}$ shows the formation of precipitate by passing $H_2S$ in acidic medium

(c) Only $Mn^{2+}$ shows the formation of precipitate by passing $H_2S$ in faintly basic medium

(d) $Cu^{2+}/Cu$ has higher reduction potential than $Mn^{2+}/Mn$ (measured under similar conditions)

**3.** Aniline reacts with mixed acid (conc. $HNO_3$ and conc. $H_2SO_4$) at 288 K to give P (51%), Q (47%) and R (2%). The major product(s) of the following sequence is (are)

$$R \xrightarrow[\substack{(3)\ H_3O^+ \\ (4)\ NaNO_2,\ HCl/273\text{-}278\ K \\ (5)\ EtOH,\ \Delta}]{\substack{(1)\ Ac_2O,\ pyridine \\ (2)\ Br_2,\ CH_3CO_2H}} S$$

$$\xrightarrow[\substack{(3)\ NaNO_2,\ HCl/273\text{-}278\ K \\ (4)\ H_3PO_2}]{\substack{(1)\ Sn/HCl \\ (2)\ Br_2/H_2O\ (excess)}}$$

Major product(s)

(a) (b) (c) (d)

**4.** The Fischer presentation of D-glucose is given below.

| | CHO | |
|---|---|---|
| H | | OH |
| HO | | H |
| H | | OH |
| H | | OH |
| | $CH_2OH$ | |

D-glucose

The correct structure(s) of β-L-glucopyranose is (are)

(a) (b) (c) (d)

**5.** For a first order reaction $A(g) \longrightarrow 2B(g) + C(g)$ at constant volume and 300 K, the total pressure at the beginning $(t = 0)$ and at time $t$ are $p_0$ and $p_t$, respectively.

Initially, only A is present with concentration $[A]_0$, and $t_{1/3}$ is the time required for the partial pressure of A to reach $1/3^{rd}$ of its initial value. The correct option(s) is (are)

(Assume that all these gases behave as ideal gases)

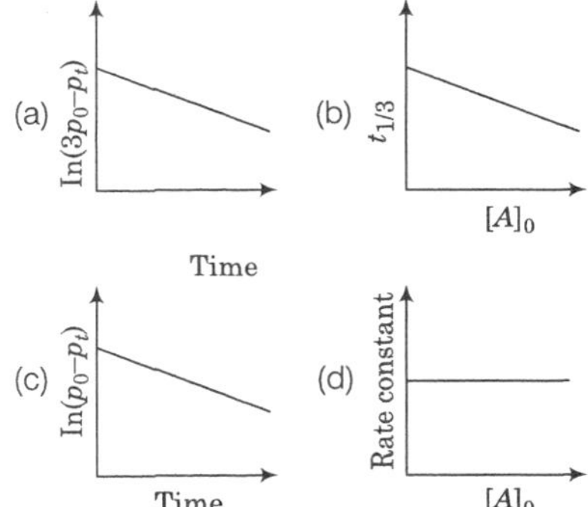


**6.** For a reaction, $A \rightleftharpoons P$, the plots of $[A]$ and $[P]$ with time at temperatures $T_1$ and $T_2$ are given below.

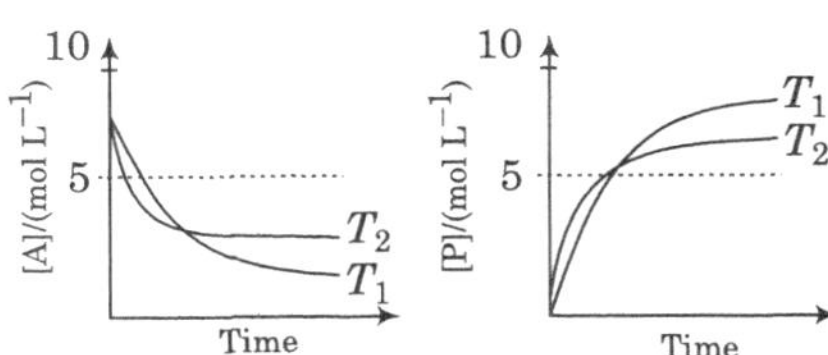


If $T_2 > T_1$, the correct statement(s) is are

(Assume $\Delta H^\ominus$ and $\Delta S^\ominus$ are independent of temperature and ratio of $\ln K$ at $T_1$ to $\ln K$ at $T_2$ is greater than $T_2 / T_1$. Here $H, S, G$ and $K$ are enthalpy, entropy, Gibbs energy and equilibrium constant, respectively.)

(a) $\Delta H^\ominus < 0, \Delta S^\ominus < 0$

(b) $\Delta G^\ominus < 0, \Delta H^\ominus > 0$

(c) $\Delta G^\ominus < 0, \Delta S^\ominus < 0$

(d) $\Delta G^\ominus < 0, \Delta S^\ominus > 0$

## Section 2 (Maximum Marks : 24)

- This section contains **EIGHT (08)** questions. The answer to each questions is a **NUMERICAL VALUE.**
- Four each question, enter the correct numerical value (in decimal notation, truncated/rounded-off to the **second decimal place;** e.g. 6.25, 7.00.– 0.33, –.30, 30.27, –127.30) using the mouse and the on screen virtual numeric keypad in the place designated to enter the answer.
- Answer to each question will be evaluated according to the following marking scheme:

*Full Marks* : +3 If ONLY the correct numerical value is entered as answer.

*Zero Marks* : 0 In all other cases.

**7.** The total number of compounds having at least one bridging oxo group among the molecules given below is ......... .

$N_2O_3, N_2O_5, P_4O_6, P_4O_7, H_4P_2O_5, H_5P_3O_{10}, H_2S_2O_3, H_2S_2O_5$

**8.** Galena (an ore) is partially oxidised by passing air through it at high temperature. After some time, the passage of air is stopped, but the heating is continued in a closed furnace such that the content undergo self-reduction. The weight (in kg) of Pb produced per kg of $O_2$ consumed is ......... .

(Atomic weights in g $mol^{-1}$ : O = 16, S = 32, Pb = 207)

**9.** To measure the quantity of $MnCl_2$ dissolved in an aqueous solution, it was completely converted to $KMnO_4$ using the reaction,

$$MnCl_2 + K_2S_2O_8 + H_2O \longrightarrow KMnO_4 + H_2SO_4 + HCl$$

(equation not balanced).

Few drops of concentrated HCl were added to this solution and gently warmed. Further, oxalic acid (225 mg) was added in portions till the colour of the permanganate ion disappeared. The quantity of $MnCl_2$ (in mg) present in the initial solution is ......... .

(Atomic weights in g $mol^{-1}$: Mn = 55, Cl = 35.5)

**10.** For the given compound *X*, the total number of optically active stereoisomers is ......... .

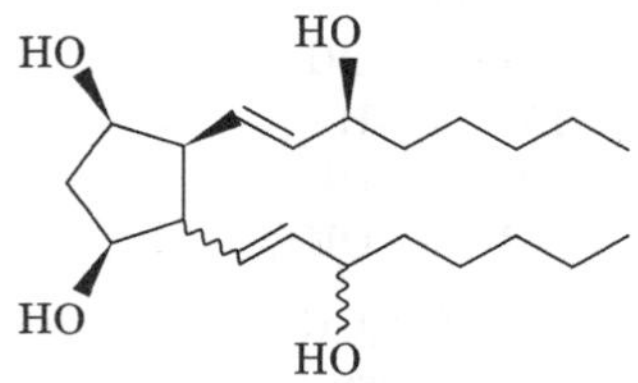


(wedge bond) This type of bond indicates that the configuration at the specific carbon and the geometry of the double bond is fixed

(wavy bond) This type of bond indicates that the configuration at the specific carbon and the geometry of the double bond is not fixed

**11.** In the following reaction sequence, the amount of *D* (in gram) formed from 10 moles of acetophenone is .......

(Atomic weights in g $mol^{-1}$ : H = 1, C = 12, N = 14, O = 16, Br = 80. The yield (%) corresponding to the product in each step is given in the parenthesis)

$$\text{Acetophenone} \xrightarrow[H_3O^+]{NaOBr} \underset{(60\%)}{A} \xrightarrow{NH_3, \Delta} \underset{(50\%)}{B} \xrightarrow{Br_2/KOH} \underset{(50\%)}{C} \xrightarrow[AcOH]{Br_2(3\text{ equivalent})} \underset{(100\%)}{D}$$

**12.** The surface of copper gets tarnished by the formation of copper oxide. $N_2$ gas was passed to prevent the oxide formation during heating of copper at 1250 K. However, the $N_2$ gas contains 1 mole % of water vapour as impurity. The water vapour oxidises copper as per the reaction given below

$$2Cu(g) + H_2O(g) \longrightarrow Cu_2O(s) + H_2(g)$$

$p_{H_2}$ is the minimum partial pressure of $H_2$ (in bar) needed to prevent the oxidation at 1250 K. The value of $\ln(p_{H_2})$ is ......... .

(Given : total pressure = 1 bar, $R$ (universal gas constant) = 8 J $K^{-1}$ $mol^{-1}$, ln(10) = 2.30 Cu(s) and $Cu_2O(s)$ are mutually immiscible.)

At 1250 K :

$$2Cu(s) + 1/2\, O_2(g) \longrightarrow Cu_2O(s); \quad \Delta G^\ominus = -78{,}000 \text{ J mol}^{-1}$$

$$H_2(g) + \frac{1}{2}O_2(g) \longrightarrow H_2O(g);$$

$\Delta G^\ominus = -1{,}78{,}000$ J $mol^{-1}$;

$G$ is the Gibbs energy

**13.** Consider the following reversible reaction,

$$A(g) + B(g) \rightleftharpoons AB(g)$$

The activation energy of the backward reaction exceeds that of the forward reaction by $2RT$ (in J $mol^{-1}$). If the pre-exponential factor of the forward reaction is 4 times that of the reverse reaction, the absolute value of $\Delta G^\ominus$ (in J $mol^{-1}$) for the reaction at 300 K is ......... .

(Given ; ln(2) = 0.7 $RT$ = 2500 J $mol^{-1}$ at 300 K and $G$ is the Gibbs energy)

**14.** Consider an electrochemical cell : $A(s) \mid A^{n+}(aq, 2M) \parallel B^{2n+}(aq, 1M) \mid B(s)$.

The value of $\Delta H^\ominus$ for the cell reaction is twice of $\Delta G^\ominus$ at 300 K. If the emf of the cell is zero, the $\Delta S^\ominus$ (in J $K^{-1}$ $mol^{-1}$) of the cell reaction per mole of B formed at 300 K is .... .

(Given : ln(2) = 0.7, $R$ (universal gas constant) = 8.3 J $K^{-1}$ $mol^{-1}$. $H, S$ and $G$ are enthalpy, entropy and Gibbs energy, respectively.)

## Section 3 (Maximum Marks : 12)

- This section contains **FOUR (04)** questions.
- Each question has **TWO (02)** matching lists : **LIST-I** and **LIST-II**.
- **FOUR** options are given representing matching of elements from **LIST-I** and **LIST-II. ONLY ONE** of these four options corresponds to a correct matching.
- For each question, choose the option corresponding to the correct matching.
- For each question, marks will be awarded according to the following marking scheme:

| | | |
|---|---|---|
| *Full Marks* | : +3 | If ONLY the option corresponding to the correct matching is chosen. |
| *Zero Marks* | : 0 | If none of the options is chosen (i.e. the question is unanswered). |
| *Negative Marks* | : −1 | In all other cases. |

**15.** Match each set of hybrid orbitals from List–I with complexes given in List-II.

| List–I | List–II |
|---|---|
| P. $dsp^2$ | 1. $[FeF_6]^{4-}$ |
| Q. $sp^3$ | 2. $[Ti(H_2O)_3Cl_3]$ |
| R. $sp^3d^2$ | 3. $[Cr(NH_3)_6]^{3+}$ |
| S. $d^2sp^3$ | 4. $[FeCl_4]^{2-}$ |
| | 5. $[Ni(CO)_4]$ |
| | 6. $[Ni(CN)_4]^{2-}$ |

The correct option is

(a) $P \rightarrow 5$; $Q \rightarrow 4, 6$; $R \rightarrow 2, 3$; $S \rightarrow 1$

(b) $P \rightarrow 5,6$; $Q \rightarrow 4$; $R \rightarrow 3$; $S \rightarrow 1,2$

(c) $P \rightarrow 6$; $Q \rightarrow 4, 5$; $R \rightarrow 1$; $S \rightarrow 2, 3$

(d) $P \rightarrow 4,6$; $Q \rightarrow 5, 6$; $R \rightarrow 1,2$; $S \rightarrow 3$

**16.** The desired product $X$ can be prepared by reacting the major product of the reactions in List-I with one or more appropriate reagents in List-II.

(given, order of migratory aptitude : aryl > alkyl > hydrogen)

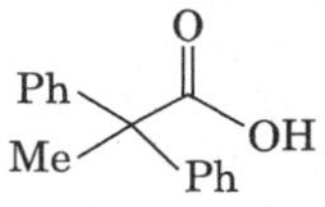

O
Ph
Me
OH
Ph

$X$

| List–I | List–II |
|---|---|
| P. 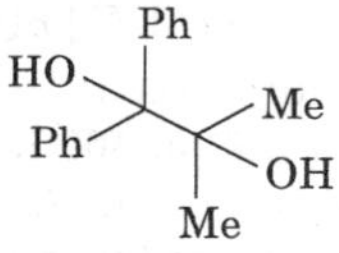 Ph, HO, Ph, Me, OH, Me + $H_2SO_4$ | 1. $I_2$, NaOH |
| Q. 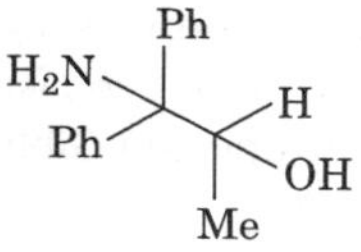 Ph, $H_2N$, Ph, H, OH, Me + $HNO_2$ | 2. $[Ag(NH_3)_2]$ OH |
| R. 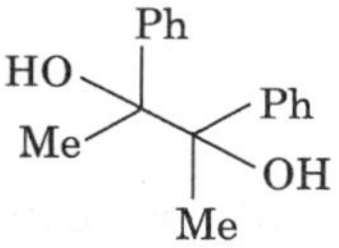 Ph, HO, Me, Ph, OH, Me + $H_2SO_4$ | 3. Fehling solution |
| S. 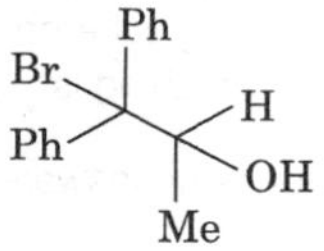 Ph, Br, Ph, H, OH, Me + $AgNO_3$ | 4. HCHO, NaOH |
| | 5. NaOBr |

The correct option is

(a) P $\rightarrow$ 1; Q $\rightarrow$ 2, 3; R $\rightarrow$ 1, 4; S $\rightarrow$ 2,4
(b) P $\rightarrow$ 1, 5; Q $\rightarrow$ 3,4; R $\rightarrow$ 4, 5; S $\rightarrow$ 3
(c) P $\rightarrow$ 1, 5; Q $\rightarrow$ 3,4; R $\rightarrow$ 5; S $\rightarrow$ 2,4
(d) P $\rightarrow$ 1, 5; Q $\rightarrow$ 2, 3; R $\rightarrow$ 1,5; S $\rightarrow$ 2,3

**17.** List-I contains reactions and List-II contains major products.

| List-I | List-II |
|---|---|
| P. 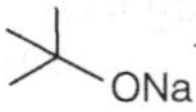 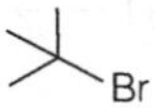 ONa + Br $\rightarrow$ | 1. 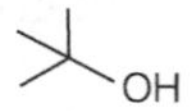 OH |
| Q. 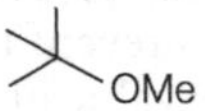 OMe + HBr $\rightarrow$ | 2. 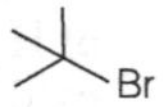 Br |
| R. 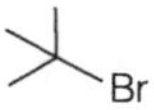 Br + NaOMe $\rightarrow$ | 3. 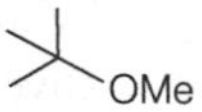 OMe |
| S. 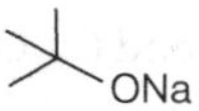 ONa + MeBr $\rightarrow$ | 4. 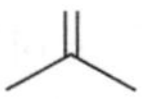 |
| | 5. 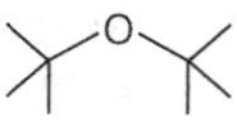 O |

Match each reaction in List-I with one or more products in List-II and choose the correct option.

(a) P $\rightarrow$ 1, 5; Q $\rightarrow$ 2; R $\rightarrow$ 3; S $\rightarrow$ 4
(b) P $\rightarrow$ 1, 4; Q $\rightarrow$ 2; R $\rightarrow$ 4; S $\rightarrow$ 3
(c) P $\rightarrow$ 1, 4; Q $\rightarrow$ 1,2; R $\rightarrow$ 3,4; S $\rightarrow$ 4
(d) P $\rightarrow$ 4, 5; Q $\rightarrow$ 4; R $\rightarrow$ 4; S $\rightarrow$ 3,4

**18.** Dilution processes of different aqueous solutions, with water, are given in List-I. The effects of dilution of the solution on $[H^+]$ are given in List-II.

(**Note** Degree of dissociation ($\alpha$) of weak acid and weak base is << 1; degree of hydrolysis of salt << 1; $[H^+]$ represents the concentration of $H^+$ ions)

| List-I | List-II |
|---|---|
| P. (10 mL of 0.1 M NaOH + 20 mL of 0.1 M acetic acid) diluted to 60 mL | 1. the value of $[H^+]$ does not change on dilution |
| Q. (20 mL of 0.1 M NaOH + 20 mL of 0.1 M acetic acid) diluted to 80 mL | 2. the value of $[H^+]$ changes to half of its initial value on dilution |

| List-I | | List-II | |
|---|---|---|---|
| R. | (20 mL of 0.1M HCl + 20 mL of 0.1 M ammonia solution) diluted to 80 mL | 3. | the value of $[H^+]$ changes to two times of its initial value on dilution. |
| S. | 10 mL saturated solution of $Ni(OH)_2$ in equilibrium with exces solid $Ni(OH)_2$ is diluted to 20 mL (solid $Ni(OH)_2$ is still present after dilution). | 4. | the value of $[H^+]$ changes to $\frac{1}{\sqrt{2}}$ times of its initial value on dilution |
| | | 5. | the value of $[H^+]$ changes to $\sqrt{2}$ times of its initial value on dilution |

Match each process given in List-I with one or more effect(s) in List-II. The correct option is

(a) $P \to 4$; $Q \to 2$; $R \to 3$; $S \to 1$
(b) $P \to 4$; $Q \to 3$; $R \to 2$; $S \to 3$
(c) $P \to 1$; $Q \to 4$; $R \to 5$; $S \to 3$
(d) $P \to 1$; $Q \to 5$; $R \to 4$; $S \to 1$

# MATHEMATICS

## Section 1 (Maximum Marks : 24)

- This section contains **SIX (06)** questions.
- Each question has **FOUR** options for correct answer(s). **ONE OR MORE THAN ONE** of these four option(s) is (are) correct options(s).
- For each question, choose the correct options(s) to answer the question.
- Answer to each question will be evaluated according to the following marking scheme:

| | | |
|---|---|---|
| *Full Marks* | : **+4** | If only (all) the correct option(s) is (are) chosen. |
| *Partial Marks* | : **+3** | If all the four options are correct but ONLY three options are chosen. |
| *Partial Marks* | : **+2** | If three or more options are correct but ONLY two options are chosen, both of which are correct options. |
| *Partial Marks* | : **+1** | If two or more options are correct but ONLY one option is chosen and it is a correct option. |
| *Zero Marks* | : **0** | If none of the options is chosen (i.e. the question is unanswered). |
| *Negative Marks* | : **−2** | In all other cases. |

- **For example:** If first, third and fourth are the ONLY three correct options for a question with second option being an incorrect option; selecting only all the three correct options will result in + 4 marks. Selecting only two of the three correct options (e.g. the first and fourth options), without selecting any incorrect option (second option in this case), will result in +2 marks. Selecting only one of the three correct options (either first or third or fourth option), without selecting any incorrect option (second option in this case), will result in +1 marks. Selecting any incorrect option(s) (second option in this case), with or without selection of any correct option(s) will result in −2 marks.

**1.** For any positive integer $n$, define $f_n:(0,\infty)\to R$ as

$$f_n(x)=\sum_{j=1}^{n}\tan^{-1}\left(\frac{1}{1+(x+j)(x+j-1)}\right)$$

for all $x\in(0,\infty)$. (Here, the inverse trigonometric function $\tan^{-1}x$ assumes values in $\left(-\frac{\pi}{2},\frac{\pi}{2}\right)$). Then, which of the following statement(s) is (are) TRUE?

(a) $\sum_{j=1}^{5}\tan^2(f_j(0))=55$

(b) $\sum_{j=1}^{10}(1+f'_j(0))\sec^2(f_j(0))=10$

(c) For any fixed positive integer $n$, $\lim_{x\to\infty}\tan(f_n(x))=\frac{1}{n}$

(d) For any fixed positive integer $n$, $\lim_{x\to\infty}\sec^2(f_n(x))=1$

**2.** Let $T$ be the line passing through the points $P(-2,7)$ and $Q(2,-5)$. Let $F_1$ be the set of all pairs of circles $(S_1, S_2)$ such that $T$ is tangent to $S_1$ at $P$ and tangent to $S_2$ at $Q$, and also such that $S_1$ and $S_2$ touch each other at a point, say $M$. Let $E_1$ be the set representing the locus of $M$ as the pair $(S_1, S_2)$ varies in $F_1$. Let the set of all straight line segments joining a pair of distinct points of $E_1$ and passing through the point $R(1,1)$ be $F_2$. Let $E_2$ be the set of the mid-points of the line segments in the set $F_2$. Then, which of the following statement(s) is (are) TRUE?

(a) The point $(-2, 7)$ lies in $E_1$

(b) The point $\left(\frac{4}{5}, \frac{7}{5}\right)$ does NOT lie in $E_2$

(c) The point $\left(\frac{1}{2}, 1\right)$ lies in $E_2$

(d) The point $\left(0, \frac{3}{2}\right)$ does NOT lie in $E_1$

**3.** Let $S$ be the set of all column matrices $\begin{bmatrix} b_1 \\ b_2 \\ b_3 \end{bmatrix}$ such that $b_1, b_2, b_3 \in R$ and the system of equations (in real variables)

$$-x+2y+5z=b_1$$
$$2x-4y+3z=b_2$$
$$x-2y+2z=b_3$$

has at least one solution. Then, which of the following system(s) (in real variables) has (have) at least one solution for each $\begin{bmatrix} b_1 \\ b_2 \\ b_3 \end{bmatrix} \in S$?

(a) $x+2y+3z=b_1$, $4y+5z=b_2$ and $x+2y+6z=b_3$

(b) $x+y+3z=b_1$, $5x+2y+6z=b_2$ and $-2x-y-3z=b_3$

(c) $-x+2y-5z=b_1$, $2x-4y+10z=b_2$ and $x-2y+5z=b_3$

(d) $x+2y+5z=b_1$, $2x+3z=b_2$ and $x+4y-5z=b_3$

**4.** Consider two straight lines, each of which is tangent to both the circle $x^2+y^2=(1/2)$ and the parabola $y^2=4x$. Let these lines intersect at the point $Q$. Consider the ellipse whose centre is at the origin $O(0,0)$ and whose semi-major axis is $OQ$. If the length of the minor axis of this ellipse is $\sqrt{2}$, then which of the following statement(s) is (are) TRUE?

(a) For the ellipse, the eccentricity is $1/\sqrt{2}$ and the length of the latus rectum is 1

(b) For the ellipse, the eccentricity is 1/2 and the length of the latus rectum is 1/2

(c) The area of the region bounded by the ellipse between the lines $x=\frac{1}{\sqrt{2}}$ and $x=1$ is $\frac{1}{4\sqrt{2}}(\pi-2)$

(d) The area of the region bounded by the ellipse between the lines $x=\frac{1}{\sqrt{2}}$ and $x=1$ is $\frac{1}{16}(\pi-2)$

**5.** Let $s, t, r$ be non-zero complex numbers and $L$ be the set of solutions $z=x+iy$ $(x, y \in R, i=\sqrt{-1})$ of the equation $sz+t\bar{z}+r=0$, where $\bar{z}=x-iy$. Then, which of the following statement(s) is (are) TRUE?

(a) If $L$ has exactly one element, then $|s| \neq |t|$

(b) If $|s|=|t|$, then $L$ has infinitely many elements

(c) The number of elements in $L \cap \{z: |z-1+i|=5\}$ is at most 2

(d) If $L$ has more than one element, then $L$ has infinitely many elements

**6.** Let $f:(0,\pi)\to R$ be a twice differentiable function such that $\lim_{t\to x}\frac{f(x)\sin t-f(t)\sin x}{t-x}=\sin^2 x$ for all $x\in(0,\pi)$.

If $f\left(\frac{\pi}{6}\right)=-\frac{\pi}{12}$, then which of the following statement(s) is (are) TRUE?

(a) $f\left(\frac{\pi}{4}\right)=\frac{\pi}{4\sqrt{2}}$

(b) $f(x)<\frac{x^4}{6}-x^2$ for all $x\in(0,\pi)$

(c) There exists $\alpha\in(0,\pi)$ such that $f'(\alpha)=0$

(d) $f''\left(\frac{\pi}{2}\right)+f\left(\frac{\pi}{2}\right)=0$

## Section 2 (Maximum Marks : 24)

- This section contains **EIGHT (08)** questions. The answer to each question is a **NUMERICAL VALUE.**
- Four each question, enter the correct numerical value (in decimal notation, truncated/rounded-off to the **second decimal place;** e.g. 6.25, 7.00.– 0.33, –.30, 30.27, –127.30) using the mouse and the on-screen virtual numeric keypad in the place designated to enter the answer.
- Answer to each question will be evaluated according to the following marking scheme:

*Full Marks* : **+3** If ONLY the correct numerical value is entered as answer.

*Zero Marks* : **0** In all other cases.

**7.** The value of the integral $\int_0^{1/2}\frac{1+\sqrt{3}}{((x+1)^2(1-x)^6)^{1/4}}dx$ is .......

**8.** Let $P$ be a matrix of order $3\times 3$ such that all the entries in $P$ are from the set $\{-1,0,1\}$. Then, the maximum possible value of the determinant of $P$ is ......... .

**9.** Let $X$ be a set with exactly 5 elements and $Y$ be a set with exactly 7 elements. If $\alpha$ is the number of one-one functions from $X$ to $Y$ and $\beta$ is the number of onto functions from $Y$ to $X$, then the value of $\frac{1}{5!}(\beta-\alpha)$ is ......... .

**10.** Let $f:R\to R$ be a differentiable function with $f(0)=0$. If $y=f(x)$ satisfies the differential equation $\frac{dy}{dx}=(2+5y)(5y-2)$, then the value of $\lim_{x\to-\infty}f(x)$ is ...... .

**11.** Let $f:R\to R$ be a differentiable function with $f(0)=1$ and satisfying the equation $f(x+y)=f(x)f'(y)+f'(x)f(y)$ for all $x,y\in R$.

Then, the value of $\log_e(f(4))$ is ....... .

**12.** Let $P$ be a point in the first octant, whose image $Q$ in the plane $x+y=3$ (that is, the line segment $PQ$ is perpendicular to the plane $x+y=3$ and the mid-point of $PQ$ lies in the plane $x+y=3$) lies on the $Z$-axis. Let the distance of $P$ from the $X$-axis be 5. If $R$ is the image of $P$ in the $XY$-plane, then the length of $PR$ is ......... .

**13.** Consider the cube in the first octant with sides $OP$, $OQ$ and $OR$ of length 1, along the $X$-axis, $Y$-axis and $Z$-axis, respectively, where $O(0,0,0)$ is the origin. Let $S\left(\frac{1}{2},\frac{1}{2},\frac{1}{2}\right)$ be the centre of the cube and $T$ be the vertex of the cube opposite to the origin $O$ such that $S$ lies on the diagonal $OT$. If $\mathbf{p}=\mathbf{SP}$, $\mathbf{q}=\mathbf{SQ}$, $\mathbf{r}=\mathbf{SR}$ and $\mathbf{t}=\mathbf{ST}$, then the value of $|(\mathbf{p}\times\mathbf{q})\times(\mathbf{r}\times\mathbf{t})|$ is ..... .

**14.** Let $X = ({}^{10}C_1)^2 + 2({}^{10}C_2)^2 + 3({}^{10}C_3)^2 + \ldots + 10({}^{10}C_{10})^2$, where ${}^{10}C_r$, $r \in \{1, 2, \ldots, 10\}$ denote binomial coefficients. Then, the value of $\frac{1}{1430}X$ is .......... .

## Section 3 (Maximum Marks : 12)

- This section contains **FOUR (04)** questions.
- Each question has **TWO (02)** matching lists : **LIST-I** and **LIST-II.**
- **FOUR** options are given representing matching of elements from **LIST-I** and **LIST-II. ONLY ONE** of these four options corresponds to a correct matching.
- For each question, choose the option corresponding to the correct matching.
- For each question, marks will be awarded according to the following marking scheme.

*Full Marks* : **+3** If ONLY the option corresponding to the correct matching is chosen.
*Zero Marks* : **0** If none of the options is chosen (i.e. the question is unanswered).
*Negative Marks* : **−1** In all other cases.

**15.** Let $E_1 = \{x \in R : x \neq 1 \text{ and } \frac{x}{x-1} > 0\}$ and

$$E_2 = \left\{ x \in E_1 : \sin^{-1}\left(\log_e\left(\frac{x}{x-1}\right)\right) \text{ is a real number} \right\}$$

(Here, the inverse trigonometric function $\sin^{-1} x$ assumes values in $\left[-\frac{\pi}{2}, \frac{\pi}{2}\right]$.). Let $f : E_1 \to R$ be the function defined by $f(x) = \log_e\left(\frac{x}{x-1}\right)$ and $g : E_2 \to R$ be the function defined by $g(x) = \sin^{-1}\left(\log_e\left(\frac{x}{x-1}\right)\right)$.

| | List-I | | List-II |
|---|---|---|---|
| P. | The range of $f$ is | 1. | $\left(-\infty, \frac{1}{1-e}\right] \cup \left[\frac{e}{e-1}, \infty\right)$ |
| Q. | The range of $g$ contains | 2. | $(0, 1)$ |
| R. | The domain of $f$ contains | 3. | $\left[-\frac{1}{2}, \frac{1}{2}\right]$ |
| S. | The domain of $g$ is | 4. | $(-\infty, 0) \cup (0, \infty)$ |
| | | 5. | $\left(-\infty, \frac{e}{e-1}\right]$ |
| | | 6. | $(-\infty, 0) \cup \left(\frac{1}{2}, \frac{e}{e-1}\right]$ |

The correct option is
(a) P → 4; Q → 2; R → 1; S → 1
(b) P → 3; Q → 3; R → 6; S → 5
(c) P → 4; Q → 2; R → 1; S → 6
(d) P → 4; Q → 3; R → 6; S → 5

**16.** In a high school, a committee has to be formed from a group of 6 boys $M_1, M_2, M_3, M_4, M_5, M_6$ and 5 girls $G_1, G_2, G_3, G_4, G_5$.

(i) Let $\alpha_1$ be the total number of ways in which the committee can be formed such that the committee has 5 members, having exactly 3 boys and 2 girls.

(ii) Let $\alpha_2$ be the total number of ways in which the committee can be formed such that the committee has at least 2 members, and having an equal number of boys and girls.

(iii) Let $\alpha_3$ be the total number of ways in which the committee can be formed such that the committee has 5 members, at least 2 of them being girls.

(iv) Let $\alpha_4$ be the total number of ways in which the committee can be formed such that the committee has 4 members, having at least 2 girls such that both $M_1$ and $G_1$ are NOT in the committee together.

| List-I | | List-II |
|---|---|---|
| P. The value of $\alpha_1$ is | 1. | 136 |
| Q. The value of $\alpha_2$ is | 2. | 189 |
| R. The value of $\alpha_3$ is | 3. | 192 |
| S. The value of $\alpha_4$ is | 4. | 200 |
| | 5. | 381 |
| | 6. | 461 |

The correct option is

(a) P → 4; Q → 6; R → 2; S → 1
(b) P → 1; Q → 4; R → 2; S → 3
(c) P → 4; Q → 6; R → 5; S → 2
(d) P → 4; Q → 2; R → 3; S → 1

**17.** Let $H: \frac{x^2}{a^2} - \frac{y^2}{b^2} = 1$, where $a > b > 0$, be a hyperbola in the $XY$-plane whose conjugate axis $LM$ subtends an angle of $60°$ at one of its vertices $N$. Let the area of the $\Delta LMN$ be $4\sqrt{3}$.

| | List-I | | List-II |
|---|---|---|---|
| P. | The length of the conjugate axis of $H$ is | 1. | 8 |
| Q. | The eccentricity of $H$ is | 2. | $\frac{4}{\sqrt{3}}$ |
| R. | The distance between the foci of $H$ is | 3. | $\frac{2}{\sqrt{3}}$ |
| S. | The length of the latus rectum of $H$ is | 4. | 4 |

The correct option is

(a) P → 4; Q → 2; R → 1; S → 3
(b) P → 4; Q → 3; R → 1; S → 2
(c) P → 4; Q → 1; R → 3; S → 2
(d) P → 3; Q → 4; R → 2; S → 1

**18.** Let $f_1 : R \to R$, $f_2 : \left(-\frac{\pi}{2}, \frac{\pi}{2}\right) \to R$, $f_3 : (-1, e^{\pi/2} - 2) \to R$ and $f_4 : R \to R$ be functions defined by

(i) $f_1(x) = \sin(\sqrt{1 - e^{-x^2}})$,

(ii) $f_2(x) = \begin{cases} \frac{|\sin x|}{\tan^{-1} x} & \text{if } x \neq 0 \\ 1 & \text{if } x = 0 \end{cases}$, where the inverse trigonometric function $\tan^{-1} x$ assumes values in $\left(-\frac{\pi}{2}, \frac{\pi}{2}\right)$,

(iii) $f_3(x) = [\sin(\log_e(x + 2))]$, where for $t \in R$, $[t]$ denotes the greatest integer less than or equal to $t$,

(iv) $f_4(x) = \begin{cases} x^2 \sin\left(\frac{1}{x}\right) & \text{if } x \neq 0 \\ 0 & \text{if } x = 0 \end{cases}$

| List-I | | List-II |
|---|---|---|
| P. The function $f_1$ is | 1. | NOT continuous at $x = 0$ |
| Q. The function $f_2$ is | 2. | continuous at $x = 0$ and NOT differentiable at $x = 0$ |
| R. The function $f_3$ is | 3. | differentiable at $x = 0$ and its derivative is NOT continuous at $x = 0$ |
| S. The function $f_4$ is | 4. | differentiable at $x = 0$ and its derivative is continuous at $x = 0$ |

The correct option is

(a) P → 2; Q → 3; R → 1; S → 4
(b) P → 4; Q → 1; R → 2; S → 3
(c) P → 4; Q → 2; R → 1; S → 3
(d) P → 2; Q → 1; R → 4; S → 3

# Answers

## Paper 1

### Physics

1. (b, c)  2. (a, c)  3. (a, c)  4. (b, d)  5. (a, b, d)  6. (b, c, d)  7. 2.00 s
8. 5.00 Hz  9. 0.75 m  10. 2.09 m  11. 1.50  12. 2.00 m/s  13. 130 $kW/m^2$
14. 4.00  15. (c)  16. (d)  17. (b)  18. (c)

### Chemistry

1. (b, c)  2. (b, c)  3. (a, b, c)  4. (b)  5. (a, b, d)  6. (b, c)  7. (1)
8. (2992)  9. (3)  10. (10)  11. (2.22)  12. (19)  13. (4.47)  14. (0.05)  15. (c)
16. (a)  17. (a)  18. (b)

### Mathematics

1. (a, b, d)  2. (b, c, d)  3. (c, d)  4. (a, b, d)  5. (b,c)
6. (b, c)  7. (8)  8. (625)  9. (3748)  10. (2)  11. (1)  12. (3)
13. (0.5)  14. (4)  15. (a)  16. (d)  17. (a)  18. (c)

## Paper 2

### Physics

1. (a, b)  2. (a, c, d)  3. (a, b)  4. (d)  5. (a, c)  6. (a, c)  7. 6.30 m
8. 30.00 m  9. 2.00 $ms^{-1}$  10. 5.55 Ω  11. 3.00  12. 900.00 J  13. 24.00
14. 3.00  15. (b)  16. (b)  17. (c)  18. (a)

### Chemistry

1. (a, b, d)  2. (b, d)  3. (d)  4. (d)  5. (a, d)  6. (a, c)  7. (6)
8. (6.47 kg)  9. (126 mg)  10. (7)  11. (495 g)  12. (– 14.6)
13. (+ 8500 J/mol)  14. (– 11.62 $JK^{-1}$ $mol^{-1}$)  15. (c)  16. (d)  17. (b)  18. (d)

### Mathematics

1. (d)  2. (a, d)  3. (a, d)  4. (a, c)  5. (a, c, d)  6. (b, c, d)  7. (2)  8. (4)
9. (119)  10. (0.4)  11. (2)  12. (8)  13. (0.5)  14. (646)  15. (a)
16. (c)  17. (b)  18. (d)

# Answer *with* Explanations

## Paper 1

### Physics

**1.** *(b, c)* $V = \dfrac{Kr^2}{2}$

$$F = -\frac{dV}{dr} = -Kr \text{ (towards centre)}$$

$$\left[F = -\frac{dV}{dr}\right]$$

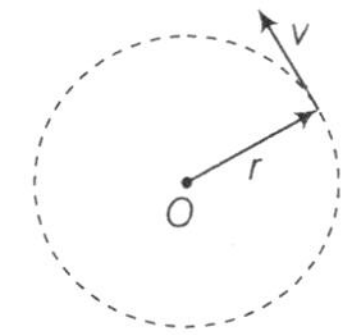


$$kR = \frac{mv^2}{R} \text{ (Centripetal force)}$$

$$v = \sqrt{\frac{kR^2}{m}} = \sqrt{\frac{k}{m}}R$$

$$\Rightarrow \quad L = mvR = \sqrt{\frac{k}{m}}R^2$$

**2.** *(a, c)* $\mathbf{F} = (\alpha t)\hat{\mathbf{i}} + \beta\hat{\mathbf{j}}$ $\quad$ [at $t=0, v=0$, $\mathbf{r}=0$]

$$\alpha = 1, \beta = 1$$

$$\mathbf{F} = t\hat{\mathbf{i}} + \hat{\mathbf{j}}$$

$$m\frac{d\mathbf{v}}{dt} = t\hat{\mathbf{i}} + \hat{\mathbf{j}}$$

On integrating, $m\mathbf{v} = \dfrac{t^2}{2}\hat{\mathbf{i}} + t\hat{\mathbf{j}}$ $\quad$ [$m = 1$ kg]

$$\frac{d\mathbf{r}}{dt} = \frac{t^2}{2}\hat{\mathbf{i}} + t\hat{\mathbf{j}} \qquad [\mathbf{r} = 0 \text{ at } t = 0]$$

Again, on integrating,

$$\mathbf{r} = \frac{t^3}{6}\hat{\mathbf{i}} + \frac{t^2}{2}\hat{\mathbf{j}}$$

At $t = 1$ s, $\tau = (\mathbf{r} \times \mathbf{F}) = \left(\dfrac{1}{6}\hat{\mathbf{i}} + \dfrac{1}{2}\hat{\mathbf{j}}\right) \times (\hat{\mathbf{i}} + \hat{\mathbf{j}})$

$$\tau = \frac{1}{3}\hat{\mathbf{k}}$$

$$v = \frac{t^3}{2}\hat{\mathbf{i}} + t\hat{\mathbf{j}}$$

At $t = 1$ s, $\mathbf{v} = \left(\dfrac{1}{2}\hat{\mathbf{i}} + \hat{\mathbf{j}}\right) = \dfrac{1}{2}(\hat{\mathbf{i}} + 2\hat{\mathbf{j}})$ m/s

At $t = 1$ s, $\mathbf{r}_1 - \mathbf{r}_0 = \left[\dfrac{1}{6}\hat{\mathbf{i}} + \dfrac{1}{2}\hat{\mathbf{j}}\right] - [0]$

$$\mathbf{s} = \frac{1}{6}\hat{\mathbf{i}} + \frac{1}{2}\hat{\mathbf{j}}$$

$$|\mathbf{s}| = \sqrt{\left(\frac{1}{6}\right)^2 + \left(\frac{1}{2}\right)^2} \Rightarrow \frac{\sqrt{10}}{6} \text{ m}$$

**3.** *(a, c)* $h = \dfrac{2\sigma\cos\theta}{r\rho g}$

(a) $\rightarrow h \propto \dfrac{1}{r}$

(b) $h$ depends upon $\sigma$.

(c) If lift is going up with constant acceleration.

$$g_{\text{eff}} = (g + a) \quad \Rightarrow h = \frac{2\sigma\cos\theta}{r\rho(g + a)}$$

It means $h$ decreases.

(d) $h$ is proportional to $\cos\theta$.

**4.** *(b, d)* $I_1 = \dfrac{V}{R}\left(1 - e^{-\frac{tR}{L}}\right)$

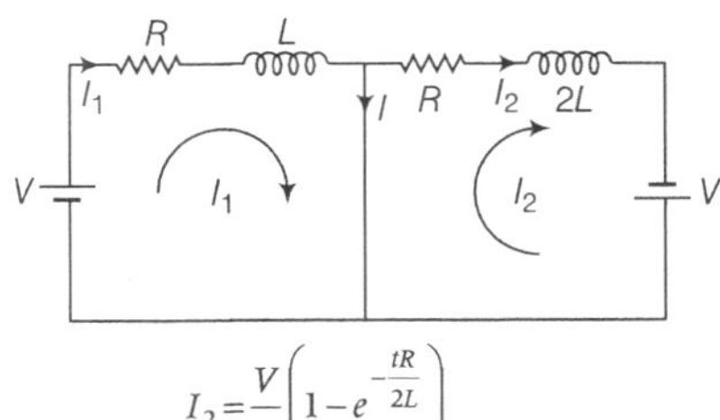


$$I_2 = \frac{V}{R}\left(1 - e^{-\frac{tR}{2L}}\right)$$

From principle of superposition,

$$I = I_1 - I_2 \Rightarrow I = \frac{V}{R}e^{-\frac{tR}{2L}}\left(1 - e^{-\frac{tR}{2L}}\right) \quad \ldots\text{(i)}$$

$I$ is maximum when $\dfrac{dI}{dt} = 0$, which gives

$$e^{-\frac{tR}{2L}} = \frac{1}{2} \text{ or } t = \frac{2L}{R}\ln 2$$

Substituting this time in Eq. (i), we get

$$I_{\max} = \frac{V}{4R}$$

**5.** *(a,b,d)*

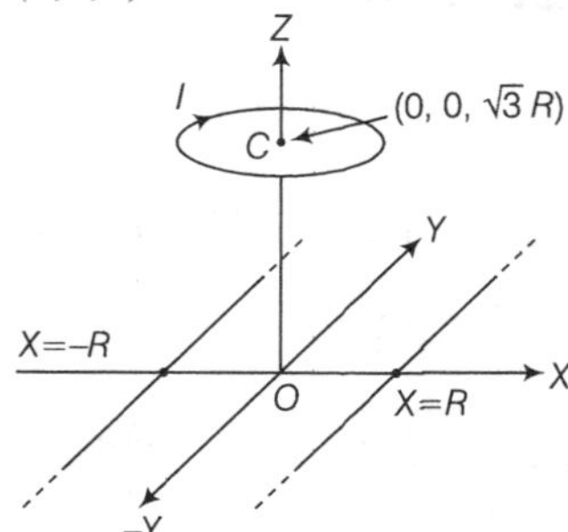


(a) At origin, $\mathbf{B}=0$ due to two wires if $I_1=I_2$, hence $(\mathbf{B}_{net})$ at origin is equal to $\mathbf{B}$ due to ring. which is non-zero.

(b) If $I_1>0$ and $I_2<0$, $\mathbf{B}$ at origin due to wires will be along $+\hat{\mathbf{k}}$. Direction of $\mathbf{B}$ due to ring is along $-\hat{\mathbf{k}}$ direction and hence $\mathbf{B}$ can be zero at origin.

(c) If $I_1<0$ and $I_2>0$, $\mathbf{B}$ at origin due to wires is along $-\hat{\mathbf{k}}$ and also along $-\hat{\mathbf{k}}$ due to ring, hence $\mathbf{B}$ cannot be zero.

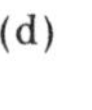

(d)

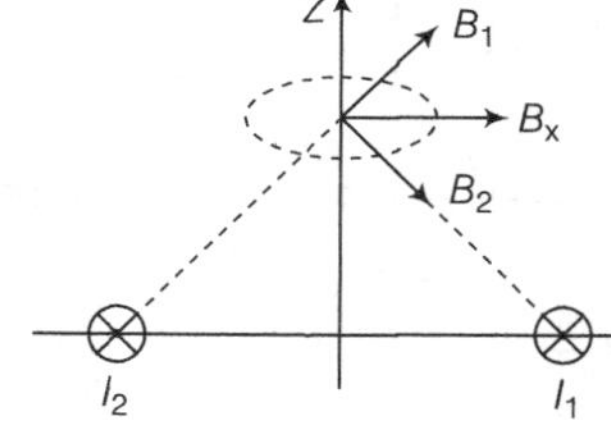


At centre of ring, $\mathbf{B}$ due to wires is along $x$-axis.

Hence z-component is only because of ring which $B=\dfrac{\mu_0 i}{2R}(-\hat{\mathbf{k}})$.

**6.** *(b,c,d)* (b) Process-II is isothermal expansion,

$$\Delta U=0,\ W>0$$
$$\Delta Q=W>0$$

(c) Process-IV is isothermal compression,

$$\Delta U=0,\ W<0$$
$$\Delta Q=W<0$$

(d) Process-I and III are not isobaric because in isobaric process $T\propto V$ hence, $T$-$V$ graph should be a straight line passing through origin.

**7.** *(2.0)* $\mathbf{A}=a\hat{\mathbf{i}}$ and $\mathbf{B}=a\cos\omega\hat{\mathbf{i}}+a\sin\omega t\hat{\mathbf{j}}$

$$\mathbf{A}+\mathbf{B}=(a+a\cos\omega t)\hat{\mathbf{i}}+a\sin\omega t\hat{\mathbf{j}}$$
$$\mathbf{A}-\mathbf{B}=(a-a\cos\omega t)\hat{\mathbf{i}}+a\sin\omega t\hat{\mathbf{j}}$$
$$|\mathbf{A}+\mathbf{B}|=\sqrt{3}|\mathbf{A}-\mathbf{B}|$$
$$\sqrt{(a+a\cos\omega t)^2+(a\sin\omega t)^2}=\sqrt{3}\sqrt{(a-a\cos\omega t)^2+(a\sin\omega t)^2}$$
$$\Rightarrow\quad 2\cos\frac{\omega t}{2}=\pm\sqrt{3}\times 2\sin\frac{\omega t}{2}$$
$$\tan\frac{\omega t}{2}=\pm\frac{1}{\sqrt{3}}$$
$$\frac{\omega t}{2}=n\pi\pm\frac{\pi}{6}$$
$$\frac{\pi}{12}t=n\pi\pm\frac{\pi}{6}$$
$$t=(12n\pm 2)\,\text{s}$$
$$=2\text{s},10\text{s},14\text{s} \text{ and so on.}$$

**8.** *(5 Hz)*

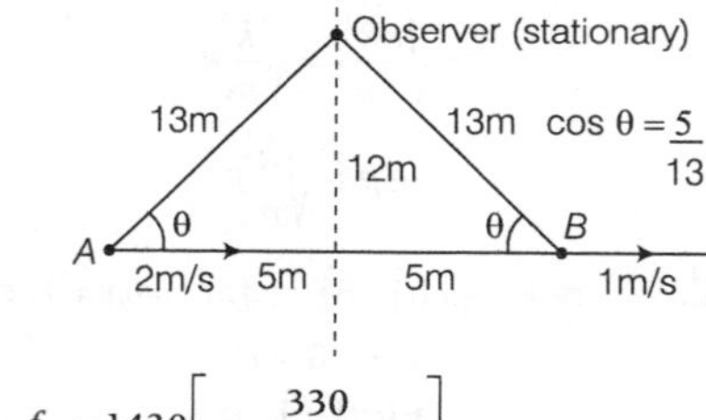


$$f_A=1430\left[\frac{330}{330-2\cos\theta}\right]$$
$$=1430\left[\frac{1}{1-\frac{2\cos\theta}{330}}\right]\approx 1430\left[1+\frac{2\cos\theta}{330}\right]$$

[from binomial expansion]

$$f_B=1430\left[\frac{330}{330+1\cos\theta}\right]$$
$$\approx 1430\left[1-\frac{\cos\theta}{330}\right]$$

Beat frequency $=f_A-f_B$

$$=1430\left[\frac{3\cos\theta}{330}\right]=13\cos\theta$$
$$=13\left(\frac{5}{13}\right)=5.00\,\text{Hz}$$

**9.** *(0.75 m)* $a=\dfrac{g\sin\theta}{1+\dfrac{I}{MR^2}}$

$$a_{\text{ring}} = \frac{g\sin\theta}{2} \qquad (I = MR^2)$$

$$a_{\text{disc}} = \frac{2g\sin\theta}{3} \qquad \left(I = \frac{MR^2}{2}\right)$$

$$s = \frac{h}{\sin\theta} = \frac{1}{2}at^2$$

$$= \frac{1}{2}\left(\frac{g\sin\theta}{2}\right)t_1^2$$

$$\Rightarrow \quad t_1 = \sqrt{\frac{4h}{g\sin^2\theta}} = \sqrt{\frac{16h}{3g}}$$

$$s = \frac{h}{\sin\theta} = \frac{1}{2}at^2 = \frac{1}{2}\left(\frac{2g\sin\theta}{3}\right)t_2^2$$

$$\Rightarrow \quad t_2 = \sqrt{\frac{3h}{g\sin^2\theta}} = \sqrt{\frac{4h}{g}}$$

$$t_2 - t_1 = \sqrt{\frac{16h}{3g}} - \sqrt{\frac{4h}{g}}$$

$$= \frac{2-\sqrt{3}}{\sqrt{10}}$$

$$\sqrt{h}\left[\frac{4}{\sqrt{3}} - 2\right] = 2 - \sqrt{3}$$

Soving this equation we get, $h = 0.75\,\text{m}$.

**10.** (2.09 **m**) **Just Before Collision,**

**Just After Collision**

Let velocities of 1 kg and 2 kg blocks just after collision be $v_1$ and $v_2$ respectively.

From momentum conservation principle,

$$1 \times 2 = 1v_1 + 2v_2 \qquad \ldots\text{(i)}$$

Collision is elastic. Hence $e = 1$ or relative velocity of separation = relation velocity of approach.

$$v_2 - v_1 = 2 \qquad \ldots\text{(ii)}$$

From Eqs. (i) and (ii),

$$v_2 = \frac{4}{3}\text{m/s}, \; v_1 = \frac{-2}{3}\text{m/s}$$

2 kg block will perform SHM after collision,

$$t = \frac{T}{2} = \pi\sqrt{\frac{m}{k}} = 3.14\,\text{s}$$

$$\text{Distance} = |v_1|t = \frac{2}{3} \times 3.14$$

$$= 2.093 = 2.09\text{ m}$$

**11.** (1.50)

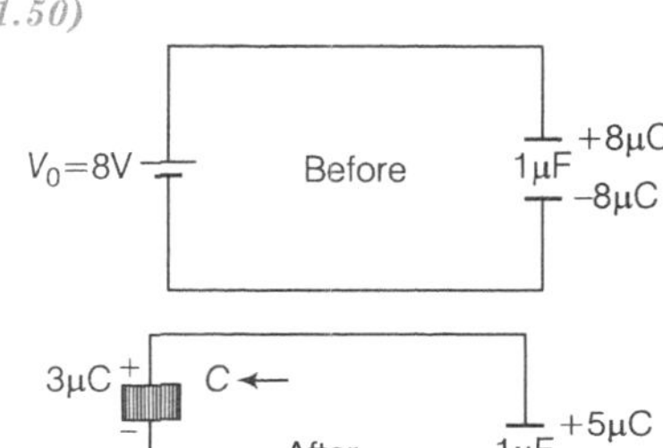


$$C = \varepsilon_r C_1 = (\varepsilon_r)\mu\text{F}$$

Applying loop rule,

$$\frac{5}{1} - \frac{3}{\varepsilon_r} - \frac{3}{1} = 0 \Rightarrow \frac{3}{\varepsilon_r} = 2$$

$$\varepsilon_r = 1.50$$

**12.** (2 **m / s**) If average speed is considered along $x$-axis,

$$R_1 = \frac{mv_0}{qB_1}, \; R_2 = \frac{mv_0}{qB_2} = \frac{mv_0}{4qB_1}$$

$$R_1 > R_2$$

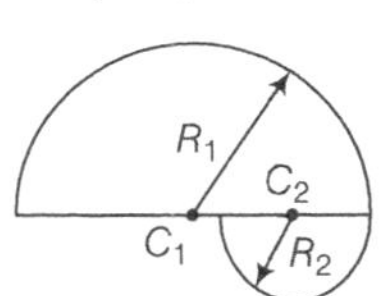


Distance travelled along $x$-axis, $\Delta x = 2(R_1 + R_2) = \dfrac{5mv_0}{2qB_1}$

$$\text{Total time} = \frac{T_1}{2} + \frac{T_2}{2} = \frac{\pi m}{qB_1} + \frac{\pi m}{qB_2}$$

$$= \frac{\pi m}{qB_1} + \frac{\pi m}{4qB_1} = \frac{5\pi m}{4qB_1}$$

$$\text{Magnitude of average speed} = \frac{\dfrac{5mv_0}{2qB_1}}{\dfrac{5\pi m}{4qB_1}}$$

$$= 2\text{m/s}$$

**13.** *(130.0* **kW / m²***)*

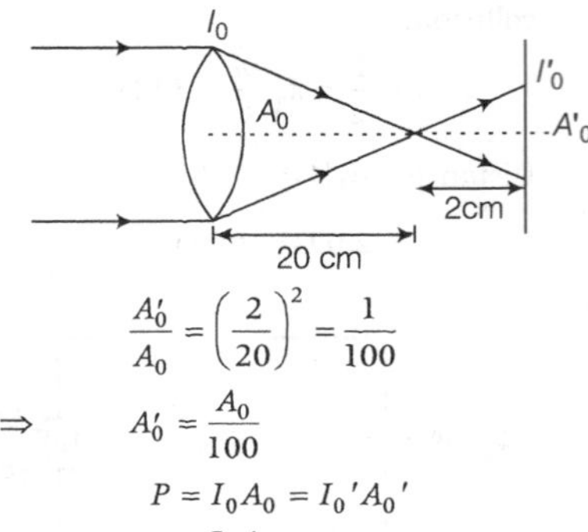


$$\frac{A_0'}{A_0} = \left(\frac{2}{20}\right)^2 = \frac{1}{100}$$

$$\Rightarrow \quad A_0' = \frac{A_0}{100}$$

$$P = I_0 A_0 = I_0' A_0'$$

$$\Rightarrow \quad I_0' = \frac{I_0 A_0}{\frac{A_0}{100}} = 100 I_0 = 130\,\text{kW/m}^2$$

**14.** *(4)* Rate of heat flow will be same,

$$\therefore \quad \frac{300-200}{R_1} = \frac{200-100}{R_2}$$

$$\left(\text{as } H = \frac{dQ}{dt} = \frac{T \cdot D}{R}\right)$$

$$\therefore \quad R_1 = R_2 \Rightarrow \frac{L_1}{K_1 A_1} = \frac{L_2}{K_2 A_2}$$

$$\Rightarrow \quad \frac{K_1}{K_2} = \frac{A_2}{A_1} = 4$$

**15.** *(c)* In terms of dimension, $F_e = F_m$

$$\Rightarrow \quad qE = qvB \quad \text{or} \quad E = vB$$

$$[E] = [B][\text{LT}^{-1}]$$

**16.** *(d)* $c = \dfrac{1}{\sqrt{\mu_0 \varepsilon_0}}$

$$c^2 = \frac{1}{\mu_0 \varepsilon_0}$$

$$\mu_0 = \varepsilon_0^{-1} \cdot c^{-2}$$

$$[\mu_0] = [\varepsilon_0]^{-1}[\text{L}^{-2}\text{T}^2]$$

**17.** *(b)* $r = \dfrac{1-a}{1+a}$

$$\ln r = \ln(1-a) - \ln(1+a)$$

Differentiating, we get

$$\frac{dr}{r} = -\frac{da}{1-a} - \frac{da}{1+a}$$

or, we can write

$$\frac{\Delta r}{r} = -\left[\frac{\Delta a}{1-a} + \frac{\Delta a}{1+a}\right]$$

$$\frac{\Delta r}{r} = \frac{-2\Delta a}{1-a^2}$$

or $$\Delta r = -\left(\frac{2\Delta a}{1-a^2}\right)(r) = \frac{-2\Delta a}{(1+a)^2}$$

**18.** *(c)* $N = N_0 e^{-\lambda t}$

$$\ln N = \ln N_0 - \lambda t$$

Differentiating w.r.t $\lambda$, we get

$$\frac{1}{N} \cdot \frac{dN}{d\lambda} = 0 - t$$

$$\Rightarrow \quad |d\lambda| = \frac{dN}{Nt} = \frac{40}{2000 \times 1} = 0.02$$

## Chemistry

**1.** *(b, c)* Among the given compounds, those which generate $N_2$ on thermal decomposition below 300°C are **ammonium dichromate** i.e., $(NH_4)_2Cr_2O_7$ and **barium azide** or nitride i.e., $Ba(N_3)_2$. Reactions of their thermal decomposition are given below

(i) $$(NH_4)_2Cr_2O_7 \xrightarrow[\text{Below } 300°\text{C}]{\Delta} N_2\uparrow + Cr_2O_3 + 4H_2O$$

(ii) $$Ba(N_3)_2 \xrightarrow[\text{Around } 160° \text{ and above}]{\Delta} Ba + 3N_2\uparrow$$

**Ammonium nitrate** ($NH_4NO_3$) on heating below 300°C gives $N_2O$ as

$$NH_4NO_3 \xrightarrow[\text{below } 300°\text{C}]{\Delta} N_2O + 2H_2O$$

However, on rapid heating or explosion

(i.e. above 300°C) it gives off nitrogen as

$$2NH_4NO_3 \xrightarrow[\text{or explosion}]{\text{Rapid heating}} 2N_2 + O_2 + 4H_2O$$

**Magnesium nitride** ($Mg_3N_2$) does not decompose at lower temperatures being comparatively more stable. Its thermal decomposition requires a minimum temperature of 700°C and proceeds as

$$Mg_3N_2 \xrightarrow{700-1500°\text{C}} 3Mg + N_2\uparrow$$

**2.** *(b, c)* Statement wise explanation is

(i) **Statement (a)** The total number of valence shell electrons at metal centre in $Fe(CO)_5$ or $Ni(CO)_4$ is 8 instead of 16 as shown below

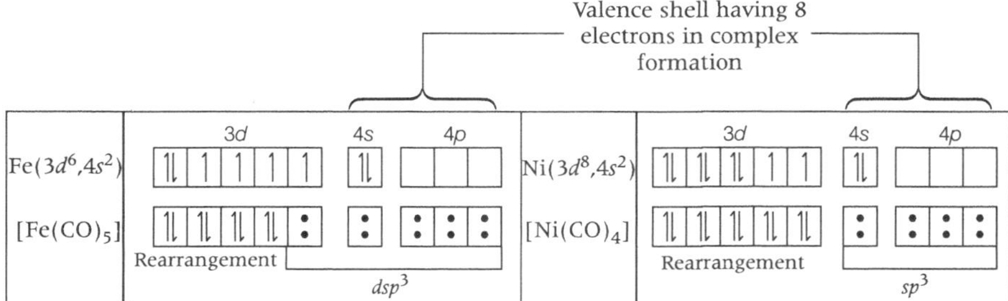


Hence, this statement is incorrect.

(ii) **Statement (b)** Carbonyl complexes are predominantly low spin complexes due to strong ligand fields. Hence, this statement is correct.

(iii) **Statement (c)** For central metal lowering of oxidation state results to increase in electron density on it. This in turn results to increase in extent of synergic bonding. Thus, we can say "metal carbonyl bond strengthens, when oxidation state of metal is lowered".

Hence, it is a correct statement.

(iv) **Statement(d)** Increase in positive charge on metal (i.e., increase in oxidation state) results to decrease in synergic bonding strength.

This in turn makes C—O bond stronger instead of weaker. Hence this statement is also incorrect.

**3.** ***(a,b,c)*** Statement wise explanation is

(i) **Statement(a)** $Bi_2O_5$ is a metallic oxide while $N_2O_5$ is a non-metallic oxide.

Metallic oxides being **ionic** are **basic in nature** while non metallic oxides being **covalent** are **acidic in nature**. This confirms more basic nature of $Bi_2O_5$ in comparison to $N_2O_5$. Hence, this is a correct statement.

(ii) **Statement (b)** The electronegativity difference between N(3) and F(4) is less as compared to the electronegativity difference between Bi(1.7) and F(4). More electronegativity difference leads to ionic compounds. Thus, $NF_3$ must be more covalent in nature as compared to $BiF_3$. Hence, this statement is also correct.

(iii) **Statement (c)** In $NH_3$ intermolecular hydrogen bonding is present, which is altogether absent in $PH_3$. Thus, $PH_3$ boils at lower temperature than $NH_3$. Hence, this is also a correct statement.

(iv) **Statement (d)** Due to smaller size of N the lone pair-lone pair repulsion is more in N—N single bond as compared to P—P single bond. This results to weaker N—N single bond as compared to P—P single bond. Hence, this statement is incorrect.

**4.** *(b)* ***Key idea*** *All the reactions involved in the problem are Nucleophilic substitution i.e., $S_N2$ which biomolecular results in inversion of configuration at the chiral carbon.*

**Reaction 1** in its generalised format is seen as

$$ROH \xrightarrow[\text{In diethyl ether } (Et_2O)]{PBr_3} RBr$$

**Reaction 2** is simple halogen exchange reaction called **Finkelstein reaction**. Its generalised format is

$$RX + NaI \xrightarrow{\text{In acetone}(Me_2CO)} RI + NaX$$

where $X$ = Cl or Br

**Reaction 3** in its generalised format seen as

$$RI + NaN_3 \xrightarrow{HCONMe_2} RN_3 + NaI$$

Now if the given product is

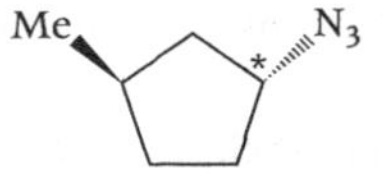

and which is too enantiomerically pure i.e. 100% either dextrorotatory or leavorotatory form, then the 'X' must be

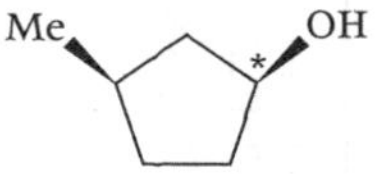

**Note** *The configuration at carbon * atom in 'X' becomes inverted due to $S_N2$ mechanism which is visible in the product as well.*

Thus, the probable reactions will be

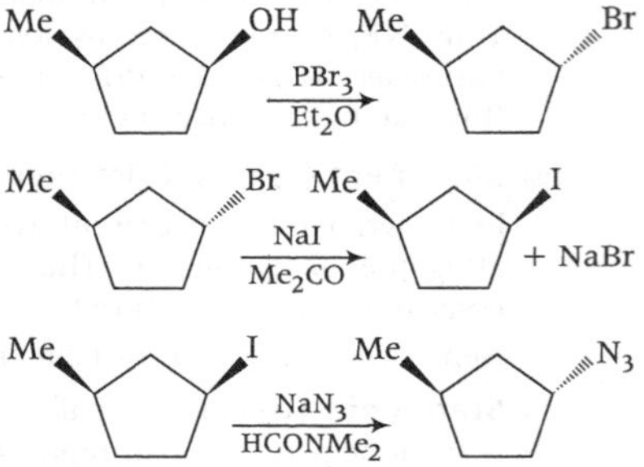

$\xrightarrow[Et_2O]{PBr_3}$

$\xrightarrow[Me_2CO]{NaI}$ + NaBr

$\xrightarrow[HCONMe_2]{NaN_3}$

**5.** *(a, b, d)* Reaction shown in option (a) is aldol condensation in the presence of conc. $H_2SO_4$ at high temperature.

In summerised way the formation of mesitylene through this can be visualised as

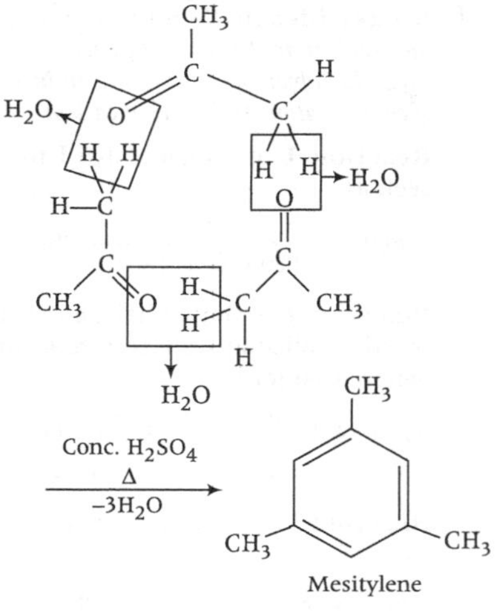

$\xrightarrow[-3H_2O]{\text{Conc. } H_2SO_4, \Delta}$ Mesitylene

Reaction given in option (b) is simple polymerisation (trimerisation) reaction of alkyne i.e.,

Me—≡—H or $CH_3—C≡CH$ when passed through heated iron tube at 873K then mesitylene is formed as

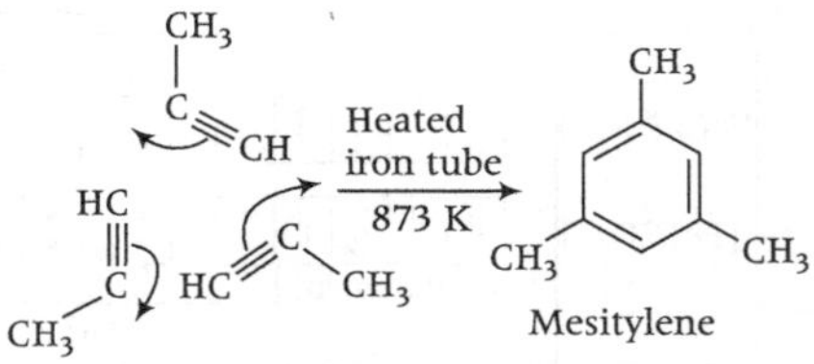

$\xrightarrow[873\text{ K}]{\text{Heated iron tube}}$ Mesitylene

This reaction is also called aromatisation.

(1) and (2) reactions of option (c) combined to give haloform reaction while (3) reaction given in this option is decarboxylation reaction i.e.,

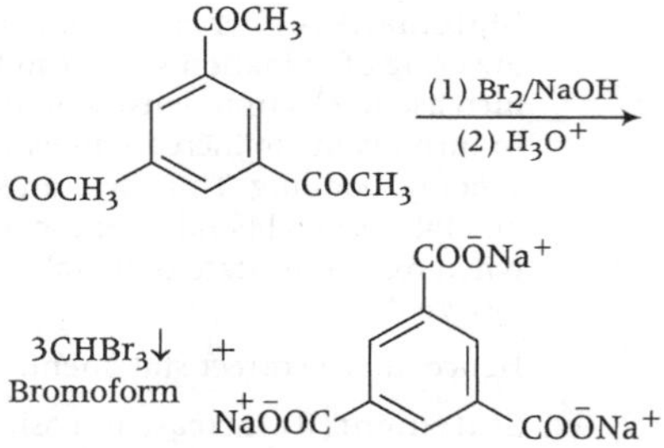

$\xrightarrow[(2)\ H_3O^+]{(1)\ Br_2/NaOH}$ $3CHBr_3\downarrow$ + (Bromoform)

The above product of haloform reaction on decarboxylation gives benzene as

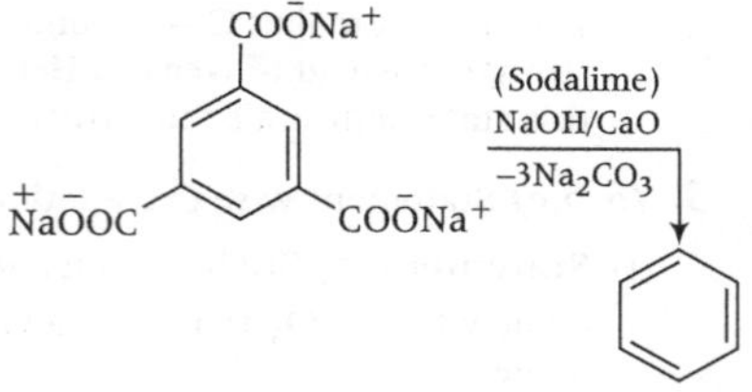

$\xrightarrow[-3Na_2CO_3]{\text{(Sodalime) } NaOH/CaO}$ Benzene

The reaction given in option (d) is Clemmensen reduction i.e.,

$$>C{=}O \xrightarrow[HCl]{Zn-Hg} >CH_2 + H_2O$$

Hence, the final product of this reaction is also mesitylene which can be seen as

Zn-Hg/HCl: CHO-substituted benzene (OHC, CHO, CHO) → CH$_3$, H$_3$C, CH$_3$ substituted benzene

Mesitylene

**6.** *(b, c)* In the given curve *AC* represents **isochoric process** as volume at both the points is same i.e., $V_1$

Similarly, *AB* represents **isothermal process** (as both the points are at $T_1$ temperature) and *BC* represents **isobaric process** as both the points are at $p_2$ pressure.

Now (i) for option (a)

$$q_{AC} = \Delta U_{BC} = nC_V(T_2 - T_1)$$

where, $n=$ number of moles

$C_v=$ specific heat capacity at constant volume

However, $W_{AB} \neq p_2(V_2 - V_1)$ instead

$$W_{AB} = nRT_1 \ln\left(\frac{V_2}{V_1}\right)$$

So, this option is incorrect.

(ii) For option (b)

$$q_{BC} = \Delta H_{AC} = nC_p\,(T_2 - T_1)$$

where, $C_p$ = specific heat capacity at constant pressure

Likewise,

$$W_{BC} = -p_2(V_1 - V_2)$$

Hence, this option is correct.

(iii) For option (c)

as $nC_p\,(T_2 - T_1) < nC_V\,(T_2 - T_1)$

so $\Delta H_{CA} < \Delta U_{CA}$

and $q_{AC} = \Delta U_{BC}$

Hence, this option is also correct.

(iv) For option (d)

Although $q_{BC} = \Delta H_{AC}$

but $\Delta H_{CA} \not> \Delta U_{CA}$

Hence, this option is incorrect.

**7.** *(1)* Among the given species only $K_2CrO_4$ is diamagnetic as central metal atom Cr in it has [Ar]$3d^0$ electronic configuration i.e., all paired electrons. The structure and oxidation state of central metal atom of this compound are as follows

**Structure** $K^+$ [Cr with =O, =O, O$^-$, O$^-$],

**Oxidation state** $Cr^{6+}$

Rest all the compounds are **paramagnetic**. Reasons for their paramagnetism are given below

(i) H-atom have $1s^1$ electronic configuration i.e., 1 unpaired electron.

(ii) $NO_2$ i.e. O=Ṅ–O (with ⊙ on N) in itself is **an odd electron species.**

(iii) $O_2^-$ (Superoxide) has one unpaired electron in $\pi^*$ molecular orbital.

(iv) $S_2$ in vapour phase has $O_2$ like electronic configuration i.e., have 2 unpaired electrons in $\pi^*$ molecular orbitals.

(v) $Mn_3O_4$ has following structure

$\overset{+2}{Mn}$ (O)(O) $\overset{+4}{Mn}$ (O)(O) $\overset{+2}{Mn}$

Thus, Mn is showing +2 and +4 oxidation states. The outermost electronic configuration of elemental Mn is $3d^5 4s^2$. Hence, in both the above oxidation states it has unpaired electrons as

$Mn^{2+}$— 3d: [1][1][1][1][1] 4s: [ ]

5 unpaired electrons

$Mn^{4+}$— 3d: [1][1][1][ ][ ] 4s: [ ]

3 unpaired electrons

(vi) $(NH_4)_2FeCl_4$ has Fe as central metal atom with +2 oxidation state. The electronic configuration of $Fe^{2+}$ in the complex is

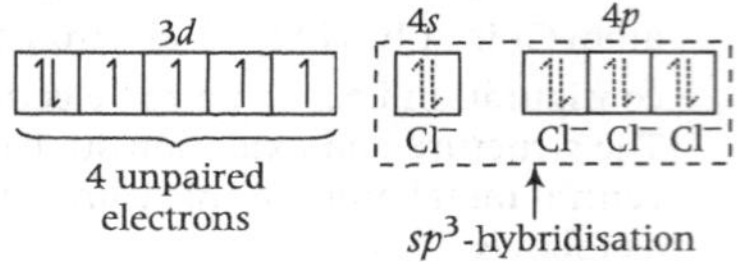


(vii) $(NH_4)_2NiCl_4$ has Ni as central metal atom with +2 oxidation state. The electronic configuration of $Ni^{2+}$ in the complex is

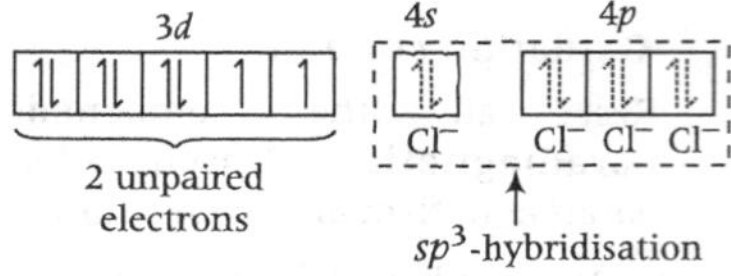


(viii) In $K_2MnO_4$ central metal atom Mn has +6 oxidation state with following structure

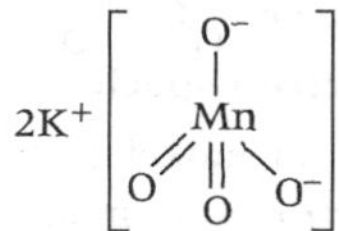


Electronic configuration of $Mn^{6+}$ is

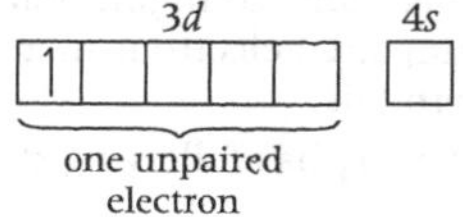


**8.** *(2992)* Balanced equations of reactions used in the problem are as follows

(i) $\underset{\substack{1\text{ mol}\\132\text{ g}}}{(NH_4)_2SO_4} + Ca(OH)_2 \longrightarrow \underset{\substack{1\text{ mol}\\172\text{ g}}}{CaSO_4 \cdot 2H_2O} + \underset{\substack{2\text{ mol}\\(2\times17)=34\text{ g}}}{2NH_3}$

(ii) $\underset{\substack{1\text{ mol}\\238\text{ g}}}{NiCl_2 \cdot 6H_2O} + \underset{\substack{6\text{ mol}\\102\text{ g}}}{6NH_3} \longrightarrow \underset{\substack{1\text{ mol}\\232\text{ g}}}{[Ni(NH_3)_6]Cl_2} + 6H_2O$

Now, in Eq. (i)

if, 1584 g of ammonium sulphate is used.

i.e., 1584 g $(NH_4)_2SO_4 = \frac{1584}{132} = 12\text{mol}$

So, according to the Eq. (i) given above 12 moles of $(NH_4)_2SO_4$ produces

(a) 12 moles of gypsum

(b) 24 moles of ammonia

Here, 12 moles of gypsum $= 12\times172$
$= 2064\text{g}$

and 24 moles of $NH_3 = 24\times17 = 408$ g

Further, as given in question,

24 moles of $NH_3$ produced in reaction (i) is completly utilised by 952g or 4 moles of $NiCl_2 \cdot 6H_2O$ to produce 4 moles of $[Ni(NH_3)_6]Cl_2$.

So 4 moles of $[Ni(NH_3)_6]Cl_2 = 4\times232$
$= 928\text{gms}$

Hence, total mass of gypsum and nickel ammonia coordination compound $[Ni(NH_3)_6]Cl_2$

$= 2064 + 928 = 2992$

**9.** *(3)* The unit cell of initial structure of ionic solid *MX* looks like

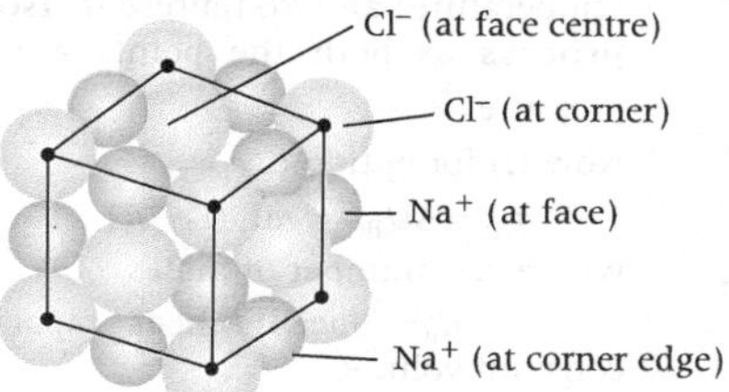


In NaCl type of solids cations ($Na^+$) occupy the octahedral voids while anions ($Cl^-$) occupy the face centre positions.

However, as per the demand of problem the position of cations and anions are swapped.

We also know that (for 1 unit cell)

(A) Total number of atoms at FCC = 4

(B) Total number of octahedral voids = 4

(as no. of atoms at FCC = No. of octahedral voids)

Now taking the conditions one by one

(i) If we remove all the anions except the central one than number of left anions.
$= 4 - 3 = 1$

(ii) If we replace all the face centred cations by anions than effective number of cations will be $= 4 - 3 = 1$
Likewise effective number of anions will be $= 1 + 3 = 4$

(iii) If we remove all the corner cations then effective number of cations will be $1 - 1 = 0$

(iv) If we replace central anion with cation then effective number of cations will be $0+1=1$
Likewise effective number of anions will be $4-1=3$

Thus, as the final outcome, total number of cations present in $Z$ after fulfilling all the four sequential instructions = 1

Likewise, total number of anions = 3

Hence, the value of

$$\frac{\text{Number of anions}}{\text{Number of cations}}=\frac{3}{1}=3$$

**10.** (10) Equation of cell reaction according to the cell notation given, is

$$Mg(s) + Cu^{2+}(aq) \longrightarrow Mg^{2+}(aq) + Cu(s)$$

(Reduction: $Cu^{2+}(aq) \to Cu(s)$; Oxidation: $Mg(s) \to Mg^{2+}(aq)$)

Given, $E^{\circ}_{\text{cell}}=2.70\text{ V}$, $T=300\text{ K}$

with $[Mg^{2+}(aq)]=1\text{ M}$ and $[Cu^{2+}(aq)]=1\text{ M}$ and $n=2$

Further, $E_{\text{cell}}=2.67\text{V}$

with $[Cu^{2+}(aq)]=1\text{M}$ and $[Mg^{2+}(aq)]=x\text{M}$

and $\frac{F}{R}=11500\text{KV}^{-1}$

where $F=$ Faraday constant, $R=$ gas constant

From the formula,

$$E_{\text{cell}}=E^{\circ}_{\text{cell}}-\frac{RT}{nF}\ln\frac{[Mg^{2+}(aq)]}{[Cu^{2+}(aq)]}$$

After putting the given values

$$2.67=2.70-\frac{RT}{2F}\ln\frac{x}{1}$$

or $$2.67=2.70-\frac{R\times300}{2F}\times\ln x$$

$$-0.03=\frac{-R\times300}{2F}\times\ln x$$

or $$\ln x=\frac{0.03\times2}{300}\times\frac{F}{R}$$

$$=\frac{0.03\times2\times11500}{300}=2.30$$

So, $\ln x=2.30$

or $x=10$ (as given $\ln(10)=2.30$)

**11.** (2.22) Given $p_1=5\text{bar}$, $V_1=1\text{ m}^3$, $T_1=400\text{ K}$

So, $$n_1=\frac{5}{400R} \quad (\text{from } pV=nRT)$$

Similarly, $p_2=1$ bar, $V_2=3\text{ m}^3$, $T_2=300$ K,

$$n_2=\frac{3}{300R}$$

Let at equilibrium the new volume of $A$ will be $(1+x)$

So, the new volume of $B$ will be $(3-x)$

Now, from the ideal gas equation.

$$\frac{p_1V_1}{n_1RT_1}=\frac{p_2V_2}{n_2RT_2}$$

and at equilibrium (due to conduction of heat)

$$\frac{p_1}{T_1}=\frac{p_2}{T_2}$$

So, $$\frac{V_1}{n_1}=\frac{V_2}{n_2} \text{ or } V_1n_2=V_2n_1$$

After putting the values

$$(1+x)\times\frac{3}{300R}=(3-x)\times\frac{5}{400R}$$

or $(1+x)=\frac{(3-x)5}{4}$ or $4(1+x)=15-5x$

or $4+4x=15-5x$ or $x=\frac{11}{9}$

Hence, new volume of $A$ i.e., $(1+x)$ will comes as $1+\frac{11}{9}=\frac{20}{9}$ or 2.22.

**12.** (19) ***Key Idea*** *Use the formula*

$$p_{\text{Total}}=p^{\circ}_A\times\chi_A+p^{\circ}_B\times\chi_B$$

*and for equimolar solutions* $\chi_A=\chi_B=1/2$

Given, $p_{\text{Total}}=45\text{torr}$ for equimolar solution

$$p^{\circ}_A=20\text{ torr}$$

So, $$45=p^{\circ}_A\times\frac{1}{2}+p^{\circ}_B\times\frac{1}{2}=\frac{1}{2}(p^{\circ}_A+p^{\circ}_B)$$

or $p^{\circ}_A+p^{\circ}_B=90\text{torr}$ ...(i)

But we know $p^{\circ}_A=20$ torr

so, $p^{\circ}_B=90-20=70\text{torr}$ (From Eq. (i))

Now, for the new solution from the same formula

$$p_{\text{Total}}=22.5\text{torr}$$

$$22.5=20\chi_A+70(1-\chi_A) \quad (\text{As } \chi_A+\chi_B=1)$$

or $22.5=70-50\chi_A$

So, $$\chi_A=\frac{70-22.5}{50}=0.95$$

Thus $\chi_B=1-0.95=0.05$ (as $\chi_A+\chi_B=1$)

Hence, the ratio

$$\frac{\chi_A}{\chi_B}=\frac{0.95}{0.05}=19$$

**13.** *(4.47)* Let solubility of $AB$ in the buffer of pH 3 = $x$

$$AB(s) \rightleftharpoons A^+(aq) + B^-(aq) \quad K_1 = K_{sp}$$

$$B^-(aq) + H^+(aq) \rightleftharpoons HB(aq) \quad K_2 = \frac{1}{K_a}$$

$$AB(s) + H^+(aq) \rightleftharpoons \underset{x}{HB(aq)} + \underset{x}{A^+(aq)}$$

$$K_3 = \frac{K_{sp}}{K_a}$$

$$K_3 = \frac{[HB][A^+]}{[H^+]} = \frac{K_{sp}}{K_a}$$

$$\therefore \quad \frac{x^2}{(10^{-3})} = \frac{2 \times 10^{-10}}{1 \times 10^{-8}}$$

$$\therefore \quad x = 4.47 \times 10^{-3}\ M = y \times 10^{-3}\ M$$

$$\therefore \quad y = \mathbf{4.47}$$

**14.** *(0.05)* From the graph we can note $\Delta T_b$ for solution $X$ i.e.,

$$\Delta T_{b(X)} = 362 - 360 = 2$$

Likewise, $\Delta T_b$ for solution $Y$ i.e.,

$$\Delta T_{b(Y)} = 368 - 367 = 1$$

Now by using the formula

$$\Delta T_b = i \times \text{molality of solution} \times K_b$$

For solution $X$

$$2 = i \times m_{NaCl} \times K_{b(X)} \quad \text{...(i)}$$

Similarly for solution $y$

$$1 = i \times m_{NaCl} \times K_{b(Y)} \quad \text{...(ii)}$$

from Eq. (i) and (ii) above

$$\frac{K_{b(X)}}{K_{b(Y)}} = \frac{2}{1} \text{ or } 2$$

or $\quad K_{b(X)} = 2K_{b(Y)}$

For solute $S$

$$2S \longrightarrow S_2$$

(given due to dimerisation)

| | $2S$ | $S_2$ |
|---|---|---|
| Initial | $\alpha$ | 0 |
| Final | $(1-\alpha)$ | $\alpha/2$ |

So, here $\quad i = \left(1 - \frac{\alpha}{2}\right)$

$$\Delta T_{b[X](s)} = \left(1 - \frac{\alpha_1}{2}\right) K_{b(X)}$$

$$\Delta T_{b[Y](s)} = \left(1 - \frac{\alpha_2}{2}\right) K_{b(Y)}$$

Given, $\Delta T_{b(X)(s)} = 3\Delta T_{b(Y)(s)}$

$$\left(1 - \frac{\alpha_1}{2}\right) K_{b(X)} = 3 \times \left(1 - \frac{\alpha_2}{2}\right) \times K_{b(Y)}$$

or $\quad 2\left(1 - \frac{\alpha_1}{2}\right) = 3\left(1 - \frac{\alpha_2}{2}\right) \quad [\because K_{b(X)} = 2K_{b(Y)}]$

or $\quad 2\left(1 - \frac{\alpha_1}{2}\right) = 3\left(1 - \frac{0.7}{2}\right)$ (as given, $\alpha_2 = 0.7$)

or $\quad 4 - 2\alpha_1 = 6 - 2.1$ or $2\alpha_1 = 0.1$

so, $\quad \alpha_1 = 0.05$

**15.** *(c)* Given,

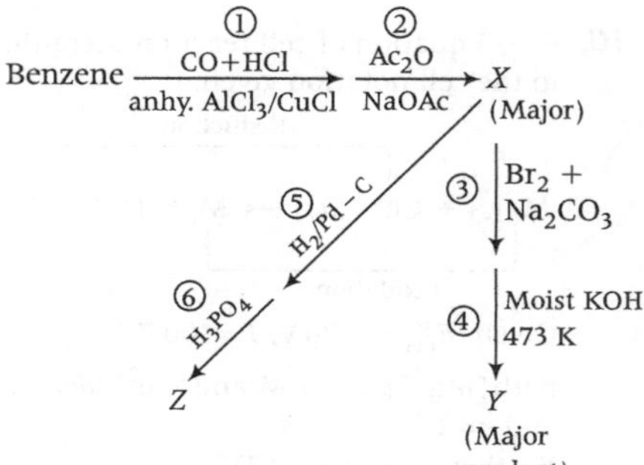


For this question we require only reaction 1 to 4 written above.

Let us explore them one by one.

**Reaction 1** It is called formylation or Gatterman Koch reaction. A —CHO group is introduced to benzene ring through this reaction as

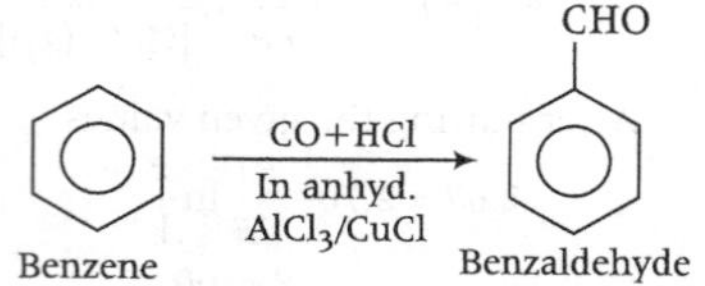


The attacking electrophile is $H—\overset{+}{C}{=}O$ which is generated as

(i) $CO + HCl \rightleftharpoons H—\underset{\underset{O}{\|}}{C}—Cl$

(ii) $H—\underset{\underset{O}{\|}}{C}—Cl + AlCl_3$

$$\rightleftharpoons H—\underset{\underset{O}{\|}}{\overset{+}{C}} + AlCl_4^-$$

**Reaction 2** It is Perkin condensation which results in α, β unsaturated acid as

$$C_6H_5CHO + (CH_3CO)_2O \xrightarrow[180°C]{CH_3COO^-Na^+} C_6H_5CH{=}CHCOOH + CH_3COOH$$

Cinnamic acid

**Note** *Besides* $CH_3COO^-Na^+$, *quinoline, pyridine,* $Na_2CO_3$, *triethylamine can also be used as bases in this reaction.*

**Reaction 3** It is simple addition of bromine to unsaturated acid formed through reaction 2.

$$C_6H_5CHO + (CH_3CO)_2O \xrightarrow[180°C]{CH_3COO^-Na^+} C_6H_5CH{=}CHCOOH + CH_3COOH$$

Cinnamic acid

$Na_2CO_3$ works as a base in the reaction to trap $H^+$ to be released by — COOH of the reactant. $C_6H_5CH(OH)—CH(Br)COO^-Na^+$ is also formed in the reaction as the minor product.

**Reaction 4** It is decarboxylation and dehydrohalogenation of product produced by reaction 3 as

$$C_6H_5CHBr—CHBrCOO^-Na^+ \xrightarrow[437\ K]{\text{Moist KOH}} C_6H_5C{\equiv}CH + Na_2CO_3 + KBr$$

Hence, $Y$ is $C_6H_5C{\equiv}CH$ i.e., (c) is the correct answer.

**16.** (a) **Reaction 5** The Perkin condensation product $X$ is

$C_6H_5CH{=}CHCOOH$ i.e., cinnamic acid (Ref. Q. 15)

This compound on hydrogenation with $H_2$ in the presence of Pd activated with charcoal (Pd–C) gives

$$C_6H_5CH{=}CHCOOH + H_2 \xrightarrow{Pd\text{-}C} C_6H_5CH_2CH_2COOH$$

3-Phenyl propanoic acid

**Reaction 6** The product of reaction 5 on heating with $H_3PO_4$ dehydrates to give

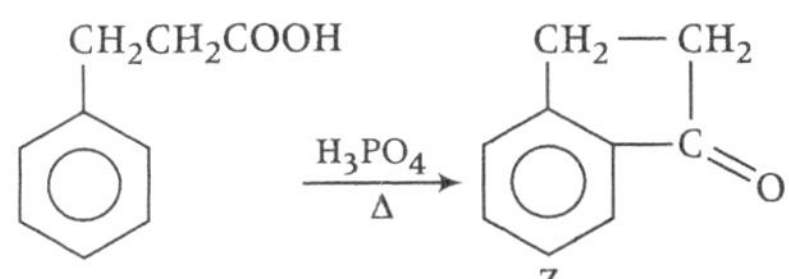


**17.** (a) Given,

(i) $C_{11}H_{12}O_2 \xrightarrow{\text{Oxidation}}$ Dibasic acid
(An organic acid '$P$')

This indicates the presence of alkyl or alkenyl branch in $P$ along with –COOH group.

(ii) Dibasic acid produced by oxidation of *P*

$$\xrightarrow[\text{(Ethylene glycol)}]{\underset{OH\quad\ OH}{\overset{CH_2-CH_2}{|\qquad\ |}}} \left[-O-\overset{O}{\overset{\|}{C}}-C_6H_4-\underset{O}{\underset{\|}{C}}-O-(CH_2)_2-O-\right]$$

Dacron

This indicates presence of benzene ring in *P*; as concluded from the structure of dacron given below.

$$\left[-\underbrace{O-\overset{O}{\overset{\|}{C}}-C_6H_4-\underset{O}{\underset{\|}{C}}}_{\text{Dicarboxylic acid component}}-\underbrace{O-CH_2-CH_2-O}_{\text{Ethylene glycol component}}-\right]$$

Dacron

Attachment of —COO group in dacron also confirm the *para* position of branch with respect to —COOH group in *P*.

(iii) $P \xrightarrow{\text{Ozonolysis}}$ Aliphatic ketone + other oxidised products.

This reaction confirms the presence of multiple bonded branch i.e., alkenyl group in *P*.

Thus *P* can be

$$HOOC-C_6H_4-CH=C(CH_3)_2$$

IUPAC name : 4-(2-methyl) prop-l-enyl benzoic acid

Now look for the reactions

$$\underset{(P)}{HOOC-C_6H_4-CH=C(CH_3)_2} \xrightarrow{\text{Oxidation}} \underset{\text{Terephthalic acid}}{HOOC-C_6H_4-COOH}$$

$$\underset{(P)}{HOOC-C_6H_4-CH=C(CH_3)_2} \xrightarrow{\text{Ozonolysis}} \underset{\substack{\text{Aliphatic}\\ \text{ketone}\\ \text{(Acetone)}}}{CH_3COCH_3} + \underset{\text{Glyoxal}}{\underset{CHO}{\overset{CHO}{|}}} + \underbrace{\underset{CHO}{\overset{COOH}{C=O}} + \underset{CHO}{\overset{CHO}{C=O}}}_{\text{other oxidised products}}$$

Given, $P \xrightarrow[\substack{3.\ MeMgBr,\ CdCl_2 \\ 4.\ NaBH_4}]{\substack{1.\ H_2/Pd-C \\ 2.\ SOCl_2}} Q$

So,

(P) $\xrightarrow{H_2/Pd—C}$ $\xrightarrow{SOCl_2}$ $\xrightarrow{MeMgBr,\ CdCl_2}$ $\xrightarrow{NaBH_4}$ (Q)

COOH, COOH, Cl–CO, C(=O)—Me, H—C(Me)—OH

Further, $Q \xrightarrow[\substack{3.\ CO_2\ (Dry\ ice) \\ 4.\ H_3O^+}]{\substack{1.\ HCl \\ 2.\ Mg/Et_2O}} R$

(Q) H—C(Me)—OH $\xrightarrow{HCl}$ H—C(Me)—Cl $\xrightarrow[\text{(Dry ether)}]{Mg,\ Et_2O}$ H—C(Me)—MgCl (Grignard's reagent) $\xrightarrow[(4)\ H_3O^+]{(3)\ CO_2\ \text{(Carboxylation)}}$ H—C(Me)—COOH (R) + Mg(OH)Cl

**18.** *(b)* Given (In connection with Q. 17)

$P \xrightarrow[\substack{4.\ CHCl_3/KOH,\Delta \\ 5.\ H_2/\ Pd-C}]{\substack{1.\ H_2/Pd-C \\ 2.\ NH_3/\Delta \\ 3.\ Br_2/NaOH}} S$

So, COOH (P) $\xrightarrow{H_2/Pd–C}$ COOH $\xrightarrow{NH_3/\Delta}$ $CONH_2$ $\xrightarrow[\text{(Hofmann bromamide)}]{Br_2/NaOH}$ $NH_2$ $\xrightarrow[\text{(Carbylamine)}]{CHCl_3/KOH}$ N≡C $\xrightarrow{H_2/Pd–C}$ $CH_3$–NH (S)

# Mathematics

**1.** *(a, b, d)*

(a) Let $z = -1 - i$ and $\arg(z) = \theta$

Now, $\tan\theta = \left|\dfrac{\text{im}(z)}{\text{Re}(z)}\right| = \left|\dfrac{-1}{-1}\right| = 1$

$\Rightarrow \quad \theta = \dfrac{\pi}{4}$

Since, $x < 0$, $y < 0$

$\therefore \quad \arg(z) = -\left(\pi - \dfrac{\pi}{4}\right) = -\dfrac{3\pi}{4}$

(b) We have, $f(t) = arg(-1 + it)$

$$\arg(-1 + it) = \begin{cases} \pi - \tan^{-1} t, & t \geq 0 \\ -(\pi + \tan^{-1} t), & t < 0 \end{cases}$$

This function is discontinuous at $t = 0$.

(c) We have,

$\arg\left(\dfrac{z_1}{z_2}\right) - arg(z_1) + arg(z_2)$

Now, $arg\left(\dfrac{z_1}{z_2}\right) = arg(z_1) - arg(z_2) + 2n\pi$

$\therefore arg\left(\dfrac{z_1}{z_2}\right) - arg(z_1) + arg(z_2)$

$= arg(z_1) - arg(z_2) + 2n\pi - arg(z_1) + arg(z_2)$

$= 2n\pi$

So, given expression is multiple of $2\pi$.

(d) We have, $\arg\left(\dfrac{(z - z_1)(z_2 - z_3)}{(z - z_3)(z_2 - z_1)}\right) = \pi$

$\Rightarrow \left(\dfrac{z - z_1}{z - z_3}\right)\left(\dfrac{z_2 - z_3}{z_2 - z_1}\right)$ is purely real

Thus, the points $A(z_1)$, $B(z_2)$, $C(z_3)$ and $D(z)$ taken in order would be concyclic if $[(z - z_1)(z_2 - z_3)/(z_2 - z_1)(z - z_3)]$ purely real.

Hence, it is a circle.

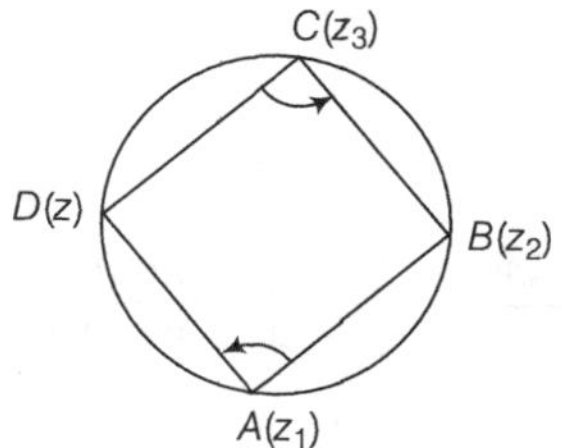


$\therefore$ (a), (b), (d) are false statement.

Hence, option (a), (b), (d) are correct answer.

**2.** *(b, c, d)* We have,

In $\Delta PQR$

$\angle PQR = 30°$

$PQ = 10\sqrt{3}$

$QR = 10$

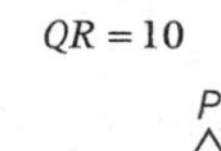

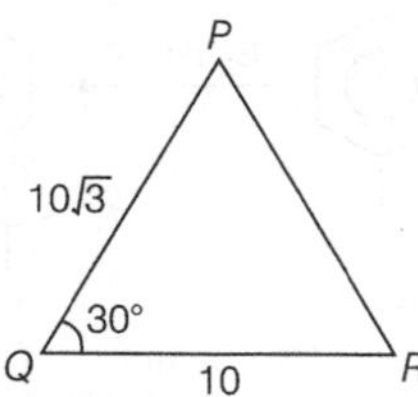


By cosine rule

$\cos 30° = \dfrac{PQ^2 + QR^2 - PR^2}{2PQ \cdot QR}$

$\Rightarrow \quad \dfrac{\sqrt{3}}{2} = \dfrac{300 + 100 - PR^2}{200\sqrt{3}}$

$\Rightarrow \quad 300 = 300 + 100 - PR^2$

$\Rightarrow \quad PR = 10$

Since, $PR = QR = 10$

$\therefore \quad \angle QPR = 30°$ and $\angle QRP = 120°$

Area of $\Delta PQR = \dfrac{1}{2} PQ \cdot QR \cdot \sin 30°$

$= \dfrac{1}{2} \times 10\sqrt{3} \times 10 \times \dfrac{1}{2} = 25\sqrt{3}$

Radius of incircle of

$\Delta PQR = \dfrac{\text{Area of } \Delta PQR}{\text{Semi-perimetre of } \Delta PQR}$

i.e. $r = \dfrac{\Delta}{s} = \dfrac{25\sqrt{3}}{\dfrac{10\sqrt{3} + 10 + 10}{2}} = \dfrac{25\sqrt{3}}{5(\sqrt{3} + 2)}$

$\Rightarrow \quad r = 5\sqrt{3}(2 - \sqrt{3}) = 10\sqrt{3} - 15$

and radius of circumcircle

$(R) = \dfrac{abc}{4\Delta} = \dfrac{10\sqrt{3} \times 10 \times 10}{4 \times 25\sqrt{3}} = 10$

$\therefore$ Area of circumcircle of

$\Delta PQR = \pi R^2 = 100\pi$

Hence, option (b), (c) and (d) are correct answer.

**3.** *(c, d)* We have,

$P_1 : 2x + y - z = 3$

and $\quad P_2 : x + 2y + z = 2$

Here, $\vec{n_1} = 2\hat{i} + \hat{j} - \hat{k}$

and $\vec{n_2} = \hat{i} + 2\hat{j} + \hat{k}$

(a) Direction ratio of the line of intersection of $P_1$ and $P_2$ is $\theta\, \vec{n_1} \times \vec{n_2}$

i.e. $\begin{vmatrix} \hat{i} & \hat{j} & \hat{k} \\ 2 & 1 & -1 \\ 1 & 2 & 1 \end{vmatrix}$

$= (1 + 2)\hat{i} - (2 + 1)\hat{j} + (4 - 1)\hat{k}$

$= 3(\hat{i} - \hat{j} + \hat{k})$

Hence, statement *a* is false.

(b) We have,

$$\frac{3x - 4}{9} = \frac{1 - 3y}{9} = \frac{z}{3}$$

$$\Rightarrow \frac{x - \frac{4}{3}}{3} = \frac{\left(y - \frac{1}{3}\right)}{-3} = \frac{z}{3}$$

This line is parallel to the line of intersection of $P_1$ and $P_2$.

Hence, statement (b) is false.

(c) Let acute angle between $P_1$ and $P_2$ be $\theta$.

We know that,

$$\cos\theta = \frac{\vec{n_1} \cdot \vec{n_2}}{|\vec{n_1}||\vec{n_2}|}$$

$$= \frac{(2\hat{i} + \hat{j} - \hat{k}) \cdot (\hat{i} + 2\hat{j} + \hat{k})}{|2\hat{i} + \hat{j} - \hat{k}||\hat{i} + 2\hat{j} + \hat{k}|}$$

$$= \frac{2 + 2 - 1}{\sqrt{6} \times \sqrt{6}} = \frac{1}{2}$$

$\theta = 60°$

Hence, statement (c) is true.

(d) Equation of plane passing through the point $(4, 2, -2)$ and perpendicular to the line of intersection of $P_1$ and $P_2$ is

$$3(x - 4) - 3(y - 2) + 3(z + 2) = 0$$

$$\Rightarrow 3x - 3y + 3z - 12 + 6 + 6 = 0$$

$$\Rightarrow x - y + z = 0$$

Now, distance of the point $(2, 1, 1)$ from the plane $x - y + z = 0$ is

$$D = \left|\frac{2 - 1 + 1}{\sqrt{1 + 1 + 1}}\right| = \frac{2}{\sqrt{3}}$$

Hence, statement (d) is true.

**4.** *(a, b, d)* We have,

$$(f(0))^2 + (f'(0))^2 = 85$$

and $f : R \to [-2, 2]$

(a) Since, $f$ is twice differentiable function, so $f$ is continuous function.

$\therefore$ This is true for every continuous function.

Hence, we can always find $x \in (r, s)$, where $f(x)$ is one-one.

$\therefore$ This statement is true.

(b) By L.M.V.T

$$f'(c) = \frac{f(b) - f(a)}{b - a}$$

$$\Rightarrow |f'(c)| = \left|\frac{f(b) - f(a)}{b - a}\right|$$

$$\Rightarrow |f'(x_0)| = \left|\frac{f(0) - f(-4)}{0 + 4}\right|$$

$$= \left|\frac{f(0) - f(-4)}{4}\right|$$

Range of $f$ is $[-2, 2]$

$\therefore \quad -4 \le f(0) - f(-4) \le 4$

$$\Rightarrow 0 \le \left|\frac{f(0) - f(-4)}{4}\right| \le 1$$

Hence, $|f'(x_0)| = 1$.

Hence, statement is true.

(c) As no function is given, then we assume $f(x) = 2\sin\left(\frac{\sqrt{85}x}{2}\right)$

$$\therefore \quad f'(x) = \sqrt{85}\cos\left(\frac{\sqrt{85}\,x}{2}\right)$$

Now, $(f(0))^2 + (f'(0))^2 = (2\sin 0)^2 + (\sqrt{85}\cos 0)^2$

$$(f(0))^2 + (f'(0))^2 = 85$$

and $\lim_{x \to \infty} f(x)$ does not exists.

Hence, statement is false.

(d) From option *b*, $|f'(x_0)| \le 1$

$$x_0 \in (-4, 0)$$

$$\therefore \quad (f'(x_0))^2 \le 1$$

Hence,

$$g(x_0) = (f(x_0))^2 + (f'(x_0))^2 \le 4 + 1$$

$$[\because f(x_0) \in [-2, 2]]$$

$$\Rightarrow \quad g(x_0) \le 5$$

Now, let $p \in (-4, 0)$ for which $g(p) = 5$

Similarly, let $q$ be smallest positive number $q \in (0, 4)$

such that $g(q) = 5$

Hence, by Rolle's theorem is $(p, q)$

$g'(c) = 0$ for $\alpha \in (-4, 4)$ and since $g(x)$ is greater than 5 as we move form $x = p$ to $x = q$

and $\quad f(x))^2 \le 4$

$\Rightarrow (f'(x))^2 \ge 1$ in $(p, q)$

Thus, $\quad g'(c) = 0$

$\Rightarrow \quad f'f + f'f'' = 0$

So, $f(\alpha) + f''(\alpha) = 0$ and $f'(\alpha) \neq 0$

Hence, statement is true.

**5.** *(b, c)* We have,

$f'(x) = e^{(f(x) - g(x))} g'(x) \; \forall \; x \in R$

$$\Rightarrow \quad f'(x) = \frac{e^{f(x)}}{e^{g(x)}} g'(x)$$

$$\Rightarrow \quad \frac{f'(x)}{e^{f(x)}} = \frac{g'(x)}{e^{g(x)}}$$

$$\Rightarrow \quad e^{-f(x)} f'(x) = e^{-g(x)} g'(x)$$

On integrating both side, we get

$$e^{-f(x)} = e^{-g(x)} + C$$

At $x = 1$

$$e^{-f(1)} = e^{-g(1)} + C$$

$$e^{-1} = e^{-g(1)} + C \; [\because f(1) = 1] \quad \ldots(\text{i})$$

At $x = 2$

$$e^{-f(2)} = e^{-g(2)} + C$$

$$\Rightarrow \quad e^{-f(2)} = e^{-1} + C \; [\because g(2) = 1] \ldots(\text{ii})$$

From Eqs. (i) and (ii)

$$e^{-f(2)} = 2e^{-1} - e^{-g(1)} \quad \ldots(\text{iii})$$

$$\Rightarrow \quad e^{-f(2)} > 2e^{-1}$$

We know that, $e^{-x}$ is decreasing

$$\therefore \quad -f(2) < \log_e 2 - 1$$

$$f(2) > 1 - \log_e 2$$

$$\Rightarrow \quad e^{-g(1)} + e^{-f(2)} = 2e^{-1} \; [\text{from Eq. (iii)}]$$

$$\Rightarrow \quad e^{-g(1)} < 2e^{-1}$$

$$-g(1) < \log_e 2 - 1$$

$$\Rightarrow \quad g(1) > -\log_e 2$$

**6.** *(b, c)* We have,

$$f(x) = 1 - 2x + \int_0^x e^{x-t} f(t)\, dt$$

On multiplying $e^{-x}$ both sides, we get

$$e^{-x} f(x) = e^{-x} - 2xe^{-x} + \int_0^x e^{-t} f(t)\, dt$$

On differentiating both side w.r.t. $x$, we get

$$e^{-x} f'(x) - e^{-x} f(x) = -e^{-x} - 2e^{-x} + 2xe^{-x} + e^{-x} f(x)$$

$$\Rightarrow \quad f'(x) - 2f(x) = 2x - 3$$

[dividing both sides by $e^{-x}$]

Let $\quad f(x) = y$

$$\Rightarrow \quad f'(x) = \frac{dy}{dx}$$

$$\therefore \quad \frac{dy}{dx} - 2y = 2x - 3$$

which is linear differential equation of the form $\frac{dy}{dx} + Py = Q$. Here, $P = -2$ and $Q = 2x - 3$.

Now, $\text{IF} = e^{\int P\, dx} = e^{\int -2\, dx} = e^{-2x}$

$\therefore$ Solution of the given differential equation is

$$y \cdot e^{-2x} = \int \underset{\text{I}}{(2x - 3)} \underset{\text{II}}{e^{-2x}}\, dx + C$$

$$y \cdot e^{-2x} = \frac{-(2x - 3) \cdot e^{-2x}}{2} + 2\int \frac{e^{-2x}}{2}\, dx + C$$

[by using integration by parts]

$$\Rightarrow y \cdot e^{-2x} = \frac{-(2x - 3)\, e^{-2x}}{2} - \frac{e^{-2x}}{2} + C$$

$$\Rightarrow \quad y = (1 - x) + Ce^{2x}$$

On putting $x = 0$ and $y = 1$, we get

$$1 = 1 + C \Rightarrow C = 0$$

$$\therefore \quad y = 1 - x$$

$y = 1 - x$ passes through $(2, -1)$

Now, area of region bounded by curve $y = \sqrt{1 - x^2}$ and $y = 1 - x$ is shows as

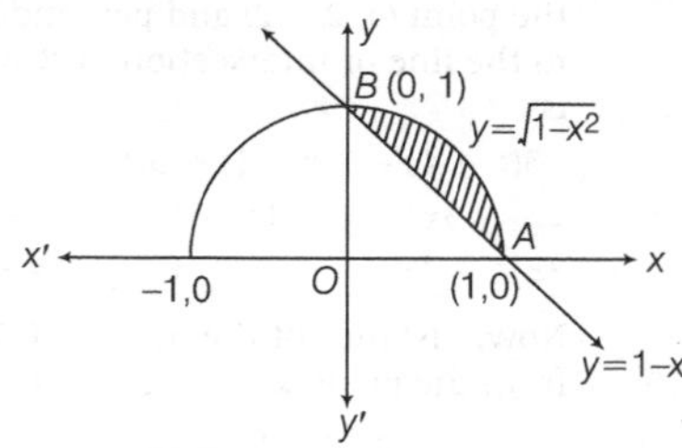


$\therefore$ Area of shaded region

= Area of 1st quadrant of a circle – Area of $\Delta\, OAB$

$= \frac{\pi}{4}(1)^2 - \frac{1}{2} \times 1 \times 1$

$= \frac{\pi}{4} - \frac{1}{2} = \frac{\pi - 2}{4}$

Hence, options *b* and *c* are correct.

**7.** *(8)* $((\log_2 9)^2)^{\frac{1}{\log_2(\log_2 9)}} \times (\sqrt{7})^{\frac{1}{\log_4 7}}$

$= (\log_2 9)^{\frac{2 \cdot \log 2}{\log_2 9}} \times 7^{\frac{1}{2} \cdot \log_7 4}$

$= (\log_2 9)^{\log_{\log_2 9} 2^2} \times 7^{\log_7 2}$

$= 2^2 \times 2 = 8$

**8.** *(625)* A number is divisible by 4 if last 2 digit number is divisible by 4.

$\therefore$ Last two digit number divisible by 4 from (1, 2, 3, 4, 5) are

12, 24, 32, 44, 52

$\therefore$ The number of 5 digit number which are divisible by 4, from the digit (1, 2, 3, 4, 5) and digit is repeated is

$5 \times 5 \times 5 \times (5 \times 1) = 625$

**9.** *(3748)* Here, $X = \{1, 6, 11, \ldots, 10086\}$

$[\because a_n = a + (n-1)d]$

and $Y = \{9, 16, 23, \ldots, 14128\}$

$X \cap Y = \{16, 51, 86, \ldots\}$

$t_n$ of $X \cap Y$ is less than or equal to 10086

$\therefore \quad t_n = 16 + (n-1)\, 35 \le 10086$

$\Rightarrow \quad n \le 288.7$

$\therefore \quad n = 288$

$\because \quad n(X \cap Y) = n(X) + n(Y) - n(X \cap Y)$

$\therefore \quad n(X \cap Y) = 2018 + 2018 - 288$

$= 3748$

**10.** *(2)* We have,

$$\sin^{-1}\left(\sum_{i=1}^{\infty} x^{i+1} - x \sum_{i=1}^{\infty} \left(\frac{x}{2}\right)^i\right) = \frac{\pi}{2} - \cos^{-1}\left(\sum_{i=1}^{\infty} \left(\frac{-x}{2}\right)^i - \sum_{i=1}^{\infty} (-x)^i\right)$$

$$\Rightarrow \sin^{-1}\left[\frac{x^2}{1-x} - \frac{x \cdot \frac{x}{2}}{1 - \frac{x}{2}}\right] = \frac{\pi}{2} - \cos^{-1}\left[\frac{\frac{-x}{2}}{1 + \frac{x}{2}} - \frac{(-x)}{1+x}\right]$$

$\left[\because \sum_{i=1}^{\infty} x^{i+1} = x^2 + x^3 + x^4 + \ldots = \frac{x^2}{1-x}\right.$ using sum of infinite terms of GP $\left.\right]$

$$\Rightarrow \sin^{-1}\left[\frac{x^2}{1-x} - \frac{x^2}{2-x}\right] = \frac{\pi}{2} - \cos^{-1}\left[\frac{x}{1+x} - \frac{x}{2+x}\right]$$

$$\Rightarrow \sin^{-1}\left[\frac{x^2}{1-x} - \frac{x^2}{2-x}\right] = \sin^{-1}\left(\frac{x}{1+x} - \frac{x}{2+x}\right)$$

$\left[\because \sin^{-1} x = \frac{\pi}{2} - \cos^{-1} x\right]$

$$\Rightarrow \frac{x^2}{1-x} - \frac{x^2}{2-x} = \frac{x}{1+x} - \frac{x}{2+x}$$

$$\Rightarrow x^2\left(\frac{2 - x - 1 + x}{(1-x)(2-x)}\right) = x\,\frac{(2 + x - 1 - x)}{(1+x)(2+x)}$$

$$\Rightarrow \frac{x}{2 - 3x + x^2} = \frac{1}{2 + 3x + x^2} \text{ or } x = 0$$

$\Rightarrow \quad x^3 + 3x^2 + 2x = x^2 - 3x + 2$

$\Rightarrow x^3 + 2x^2 + 5x - 2 = 0$ or $x = 0$

Let $f(x) = x^3 + 2x^2 + 5x - 2$

$f'(x) = 3x^2 + 4x + 5$

$f'(x) > 0, \forall\, x \in R$

$\therefore$ $x^3 + 2x^2 + 5x - 2$ has only one real roots

Therefore, total number of real solution is 2.

**11.** *(1)* We have,

$y_n = \frac{1}{n}[(n+1)(n+2)\ldots(n+n)]^{1/n}$

and $\lim_{n \to \infty} y_n = L$

$\Rightarrow \quad L = \lim_{n \to \infty} \frac{1}{n}[(n+1)(n+2)(n+3) \ldots (n+n)]^{1/n}$

$$\Rightarrow L = \lim_{n \to \infty}\left[\left(1 + \frac{1}{n}\right)\left(1 + \frac{2}{n}\right)\left(1 + \frac{3}{n}\right)\ldots\left(1 + \frac{n}{n}\right)\right]^{\frac{1}{n}}$$

$$\Rightarrow \log L = \lim_{n \to \infty} \frac{1}{n}\left[\log\left(1 + \frac{1}{n}\right) + \log\left(1 + \frac{2}{n}\right) \ldots \log\left(1 + \frac{n}{n}\right)\right]$$

$$\Rightarrow \log L = \lim_{n\to\infty} \frac{1}{n} \sum_{r=1}^{n} \log\left(1 + \frac{r}{n}\right)$$

$$\Rightarrow \log L = \int_0^1 \underset{\text{II}}{1} \times \underset{\text{I}}{\log(1+x)}\, dx$$

$$\Rightarrow \log L = (x \cdot \log(1+x))_0^1 - \int_0^1 \left[\frac{d}{dx}(\log(1+x)\int dx\right] dx$$

[by using integration by parts]

$$\Rightarrow \log L = [x \log(1+x)]_0^1 - \int_0^1 \frac{x}{1+x}\, dx$$

$$\Rightarrow \log L = \log 2 - \int_0^1 \left(\frac{x+1}{x+1} - \frac{1}{x+1}\right) dx$$

$$\Rightarrow \log L = \log 2 - [x]_0^1 + [\log(x+1)]_0^1$$

$$\Rightarrow \log L = \log 2 - 1 + \log 2 - 0$$

$$\Rightarrow \log L = \log 4 - \log e = \log \frac{4}{e}$$

$$\Rightarrow \quad L = \frac{4}{e}$$

$$\Rightarrow \quad [L] = \left[\frac{4}{e}\right] = 1$$

**12.** *(3)* We have,

$$\vec{c} = x\vec{a} + y\vec{b} + \vec{a} \times \vec{b} \text{ and } \vec{a} \cdot \vec{b} = 0$$

$$|\vec{a}| = |\vec{b}| = 1$$

and $\quad |\vec{c}| = 2$

Also, given $\vec{c}$ is inclined on $\vec{a}$ and $\vec{b}$ with same angle $\alpha$.

$$\therefore \quad \vec{a} \cdot \vec{c} = x|\vec{a}|^2 + y(\vec{a} \cdot \vec{b}) + \vec{a} \cdot (\vec{a} \times \vec{b})$$

$$|\vec{a}||\vec{c}| \cos\alpha = x + 0 + 0$$

$$x = 2\cos\alpha$$

Similarly,

$$|\vec{b}||\vec{c}| \cos\alpha = 0 + y + 0$$

$$\Rightarrow \quad y = 2\cos\alpha$$

$$|\vec{c}|^2 = x^2 + y^2 + |\vec{a} \times \vec{b}|^2$$

$$4 = 8\cos^2\alpha + |a|^2 |b|^2 \sin^2 90°$$

$$4 = 8\cos^2\alpha + 1$$

$$\Rightarrow \quad 8\cos^2\alpha = 3$$

**13.** *(0.5)* We have, $\alpha, \beta$ are the roots of

$$\sqrt{3}a \cos x + 2b \sin x = c$$

$$\therefore \quad \sqrt{3}a\cos\alpha + 2b\sin\alpha = c \quad \ldots(\text{i})$$

and $\quad \sqrt{3}a\cos\beta + 2b\sin\beta = c \quad \ldots(\text{ii})$

On subtracting Eq. (ii) from Eq. (i), we get

$$\sqrt{3}a(\cos\alpha - \cos\beta) + 2b(\sin\alpha - \sin\beta) = 0$$

$$\Rightarrow \sqrt{3}a\left(-2\sin\left(\frac{\alpha+\beta}{2}\right)\right)\sin\left(\frac{\alpha-\beta}{2}\right) + 2b\left(2\cos\left(\frac{\alpha+\beta}{2}\right)\right)\sin\left(\frac{\alpha-\beta}{2}\right) = 0$$

$$\Rightarrow \sqrt{3}a \sin\left(\frac{\alpha+\beta}{2}\right) = 2b\cos\left(\frac{\alpha+\beta}{2}\right)$$

$$\Rightarrow \tan\left(\frac{\alpha+\beta}{2}\right) = \frac{2b}{\sqrt{3}a}$$

$$\Rightarrow \tan\left(\frac{\pi}{6}\right) = \frac{2b}{\sqrt{3}a} \quad \left[\because \alpha + \beta = \frac{\pi}{3}, \text{ given}\right]$$

$$\Rightarrow \frac{1}{\sqrt{3}} = \frac{2b}{\sqrt{3}a} \Rightarrow \frac{b}{a} = \frac{1}{2}$$

$$\Rightarrow \frac{b}{a} = 0.5$$

**14.** *(4)* We have,

$$y = x^n, n > 1$$

$\because$ $P(0,0)$ $Q(1,1)$ and $R(2,0)$ are vertices of $\Delta PQR$.

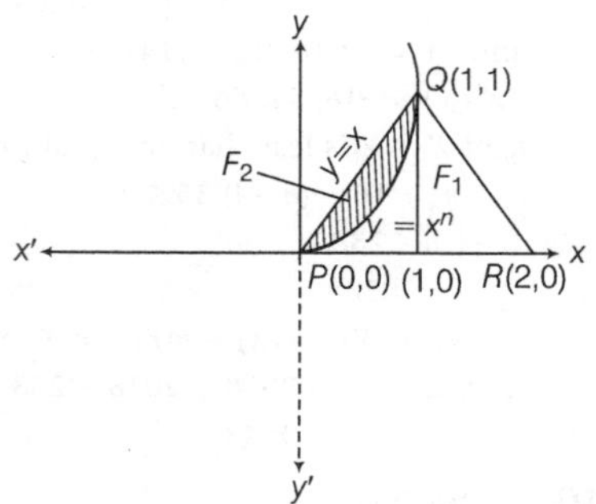


$\therefore$ Area of shaded region = 30% of area of $\Delta PQR$

$$\Rightarrow \int_0^1 (x - x^n)\, dx = \frac{30}{100} \times \frac{1}{2} \times 2 \times 1$$

$$\Rightarrow \left[\frac{x^2}{2} - \frac{x^{n+1}}{n+1}\right]_0^1 = \frac{3}{10}$$

$$\Rightarrow \left(\frac{1}{2} - \frac{1}{n+1}\right) = \frac{3}{10}$$

$$\Rightarrow \frac{1}{n+1} = \frac{1}{2} - \frac{3}{10} = \frac{2}{10} = \frac{1}{5}$$

$$\Rightarrow n + 1 = 5 \Rightarrow n = 4$$

**15.** *(a)*

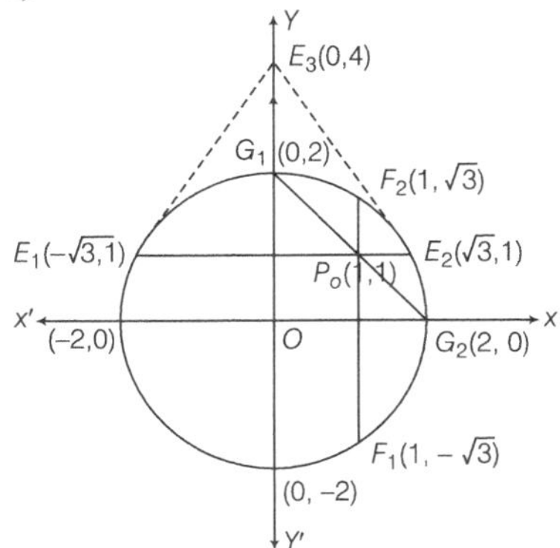


Equation of tangent at $E_1(-\sqrt{3}, 1)$ is

$-\sqrt{3}x + y = 4$ and at $E_2(\sqrt{3}, 1)$ is

$$\sqrt{3}x + y = 4$$

Intersection point of tangent at $E_1$ and $E_2$ is (0, 4)

$\therefore$ Coordinates of $E_3$ is (0, 4)

Similarly, equation of tangent at $F_1(1, -\sqrt{3})$ and $F_2(1, \sqrt{3})$ are $x - \sqrt{3}y = 4$ and $x + \sqrt{3}y = 4$, respectively and intersection point is (4, 0), i.e., $F_3(4, 0)$ and equation of tangent at $G_1(0, 2)$ and $G_2(2, 0)$ are $2y = 4$ and $2x = 4$, respectively and intersection point is (2, 2) i.e., $G_3(2, 2)$.

Point $E_3(0, 4)$, $F_3(4, 0)$ and $G_3(2, 2)$ satisfies the line $x + y = 4$.

**16.** *(d)* We have,

$$x^2 + y^2 = 4$$

Let $P(2\cos\theta, 2\sin\theta)$ be a point on a circle.

$\therefore$ Tangent at $P$ is

$$2\cos\theta\, x + 2\sin\theta\, y = 4$$

$$\Rightarrow \quad x\cos\theta + y\sin\theta = 2$$

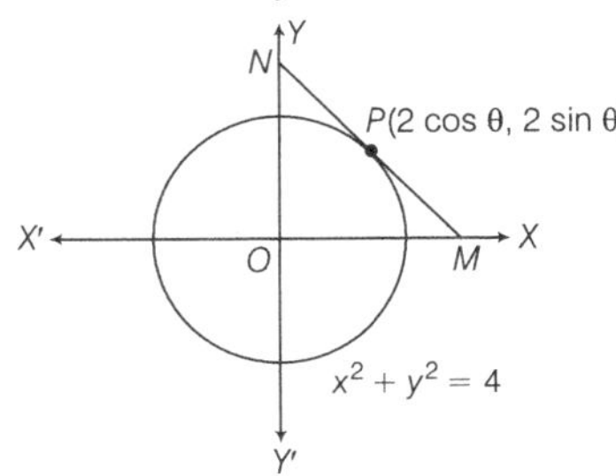


$\therefore$ The coordinates at $M\left(\dfrac{2}{\cos\theta}, 0\right)$ and $N\left(0, \dfrac{2}{\sin\theta}\right)$

Let $(h, k)$ is mid-point of $MN$

$$\therefore \quad h = \frac{1}{\cos\theta} \text{ and } k = \frac{1}{\sin\theta}$$

$$\Rightarrow \quad \cos\theta = \frac{1}{h}$$

$$\text{and} \quad \sin\theta = \frac{1}{k}$$

$$\Rightarrow \quad \cos^2\theta + \sin^2\theta = \frac{1}{h^2} + \frac{1}{k^2}$$

$$\Rightarrow \quad 1 = \frac{h^2 + k^2}{h^2 \cdot k^2}$$

$$\Rightarrow \quad h^2 + k^2 = h^2 k^2$$

$\therefore$ Mid-point of $MN$ lie on the curve

$$x^2 + y^2 = x^2 y^2$$

**17.** *(a)* Here, five students $S_1, S_2, S_3, S_4$ and $S_5$ and five seats $R_1, R_2, R_3, R_4$ and $R_5$

$\therefore$ Total number of arrangement of sitting five students is $5! = 120$

Here, $S_1$ gets previously alloted seat $R_1$

$\therefore S_2, S_3, S_4$ and $S_5$ not get previously seats.

Total number of way $S_2, S_3, S_4$ and $S_5$ not get previously seats is

$$4!\left(1 - \frac{1}{1!} + \frac{1}{2!} - \frac{1}{3!} + \frac{1}{4!}\right)$$

$$= 24\left(1 - 1 + \frac{1}{2} - \frac{1}{6} + \frac{1}{24}\right)$$

$$= 24\left(\frac{12 - 4 + 1}{24}\right) = 9$$

$$\therefore \quad \text{Required probability} = \frac{9}{120} = \frac{3}{40}$$

**18.** *(c)* Here, $n(T_1 \cap T_2 \cap T_3 \cap T_4) = \text{Total} - n(\overline{T_1} \cup \overline{T_2} \cup \overline{T_3} \cup \overline{T_4})$

$$\Rightarrow n(T_1 \cap T_2 \cap T_3 \cap T_4)$$

$$= 5! - [{}^4C_1\ 4!\ 2! - ({}^3C_1\ 3!\ 2! + {}^3C_1\ 3!\ 2!\ 2!) + ({}^2C_1\ 2!\ 2! + {}^4C_1\ 2 \cdot 2!) - 2]$$

$$\Rightarrow \quad n(T_1 \cap T_2 \cap T_3 \cap T_4)$$

$$= 120 - [192 - (36 + 72) + (8 + 16) - 2]$$

$$= 120 - [192 - 108 + 24 - 2] = 14$$

$$\therefore \quad \text{Required probability} = \frac{14}{120} = \frac{7}{60}$$

# Paper 2

## Physics

**1.** *(a, b)* $K = \frac{1}{2}mv^2 \Rightarrow \frac{dK}{dt} = mv\frac{dv}{dt}$

Given, $\frac{dK}{dt} = \gamma t \Rightarrow mv\frac{dv}{dt} = \gamma t$

$$\Rightarrow \int_0^v v\,dv = \int_0^t \frac{\gamma}{m} t\,dt \Rightarrow \frac{v^2}{2} = \frac{\gamma}{m}\frac{t^2}{2}$$

$$\Rightarrow \quad v = \sqrt{\frac{\gamma}{m}}\,t \Rightarrow a = \frac{dv}{dt} = \sqrt{\frac{\gamma}{m}}$$

$$\therefore \quad F = ma = \sqrt{\gamma m} = \text{constant}$$

$$\therefore \quad V = \frac{ds}{dt} = \sqrt{\frac{\gamma}{m}}\,t \Rightarrow s = \sqrt{\frac{\gamma}{m}}\frac{t^2}{2}$$

**Note** *Force is constant. In the website of IIT, option (d) is given correct. In the opinion of author all constant forces are not necessarily conservative. For example : viscous force at terminal velocity is a constant force but it is not conservative.*

**2.** *(a, c, d)*

A

$F_v$ ← [hatched plate] → $u_0$

Plate

$$F_v = -\eta A\left(\frac{dv}{dy}\right)$$

Since, height $h$ of the liquid in tank is very small.

$$\Rightarrow \quad \frac{dv}{dy} = \frac{\Delta v}{\Delta y} = \left(\frac{u_0}{h}\right)$$

$$F_v = -(\eta)A\left(\frac{u_0}{h}\right)$$

$$F_v \propto \left(\frac{1}{h}\right), F_v \propto u_0, F \propto A, F_v \propto \eta$$

**3.** *(a, b)* $PQ = (2)\,R\sin 60°$

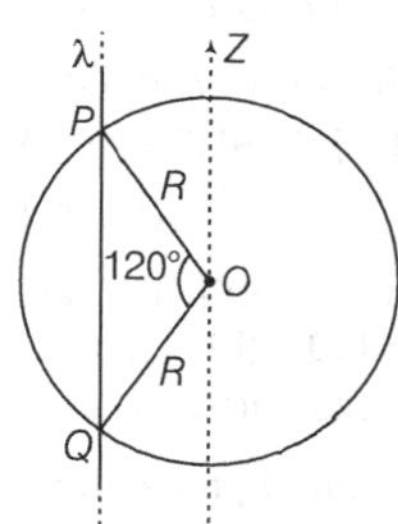


$$= (2R)\frac{\sqrt{3}}{2} = (\sqrt{3}R)$$

$$q_{\text{enclosed}} = \lambda(\sqrt{3}R)$$

We have, $\phi = \frac{q_{\text{enclosed}}}{\varepsilon_0}$

$$\Rightarrow \quad \phi = \left(\frac{\sqrt{3}\lambda R}{\varepsilon_0}\right)$$

Also, electric field is perpendicular to wire, so $Z$-component will be zero.

**4.** *(d)* Image of point $A$

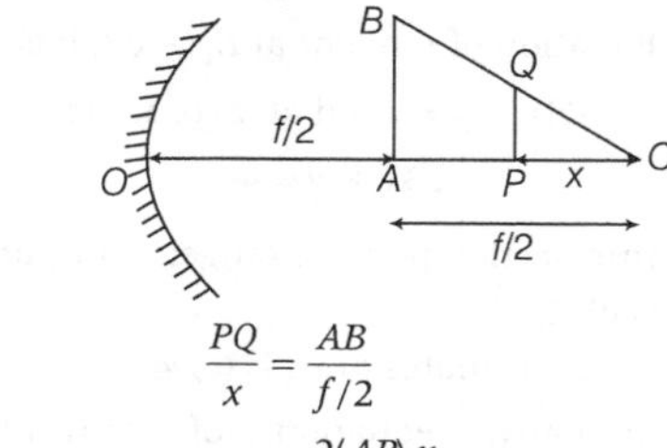


$$\frac{PQ}{x} = \frac{AB}{f/2}$$

$$\Rightarrow \quad PQ = \frac{2(AB)x}{f}$$

**For $A$ :**

$$\frac{1}{v} + \frac{1}{[-(f/2)]} = \frac{1}{-f} \Rightarrow v = f$$

$$\Rightarrow \quad \frac{I_{AB}}{AB} = -\frac{v}{u} = -\frac{f}{\left(-\frac{f}{2}\right)}$$

$$\Rightarrow \quad I_{AB} = 2\,AB$$

For height of $PQ$,

$$\frac{1}{v} + \frac{1}{-[(f-x)]} = \frac{1}{-f}$$

$$\Rightarrow \quad \frac{1}{v} = \frac{1}{(f-x)} - \frac{1}{f}$$

$$\Rightarrow \quad v = \frac{f(f-x)}{x}$$

$$\Rightarrow \quad \frac{I_{PQ}}{PQ} = -\frac{v}{u} = \frac{f(f-x)}{x[(f-x)]} = \left(\frac{f}{x}\right)$$

$$\Rightarrow \quad I_{PQ} = \frac{f}{x}PQ = \left(\frac{f}{x}\right)\left(\frac{2(AB)x}{f}\right)$$

$$\left[\because PQ = \frac{2(AB)x}{f}\right]$$

$$I_{PQ} = 2AB$$

(Size of image is independent of $x$. So, final image will be of same height terminating at infinity)

**5.** *(a, c)* $^{232}_{90}Th$ is converting into $^{212}_{82}Pb$.

Change in mass number $(A) = 20$

$\therefore$ Number of $\alpha$-particle emitted $= \frac{20}{4} = 5$

Due to 5 $\alpha$-particles, $Z$ will change by 10 units.

Since, given change is 8, therefore number of $\beta$-particles emitted is 2.

**6.** *(a, c)* Let $n$ th harmonic is corresponding to 50.7 cm and $(n+1)$th harmonic is corresponding 83.9 cm.

$\therefore$ Their difference is $\frac{\lambda}{2}$.

$$\therefore \quad \frac{\lambda}{2} = (83.9 - 50.7) \text{ cm}$$

$$\text{or} \quad \lambda = 66.4 \text{ cm}$$

$$\therefore \quad \frac{\lambda}{4} = 16.6 \text{cm}$$

Length corresponding to fundamental mode must be close to $\frac{\lambda}{4}$ and 50.7 cm must be an odd multiple of this length. $16.6 \times 3 = 49.8$ cm. Therefore, 50.7 is 3rd harmonic.

If end correction is $e$, then

$$e + 50.7 = \frac{3\lambda}{4}$$

$$e = 49.8 - 50.7 = -0.9 \text{ cm}$$

$\therefore$ Speed of sound, $v = f\lambda$

$$\Rightarrow \quad v = 500 \times 66.4 \text{ cm/s} = 332 \text{ m/s}$$

**7.** *(6.30)* Linear impulse, $J = mv_0$

$$\therefore \quad v_0 = \frac{J}{m} = 2.5 \text{ m/s}$$

$$\therefore \quad v = v_0 e^{-t/\tau}$$

$$\frac{dx}{dt} = v_0 e^{-t/\tau}$$

$$\int_0^x dx = v_0 \int_0^\tau e^{-t/\tau} dt$$

$$x = v_0 \left[\frac{e^{-t/\tau}}{-1/\tau}\right]_0^\tau$$

$$x = 2.5(-4)(e^{-1} - e^0)$$

$$= 2.5(-4)(0.37 - 1)$$

$$x = 6.30 \text{ m}$$

**8.** *(30)* $\because H = \frac{u^2 \sin^2 45^\circ}{2g} = 120$ m

$$\Rightarrow \quad \frac{u^2}{4g} = 120 \text{ m}$$

If speed is $v$ after the first collision, then speed should remain $\frac{1}{\sqrt{2}}$ times, as kinetic energy has reduced to half.

$$\Rightarrow \quad v = \frac{u}{\sqrt{2}}$$

$$\therefore \quad h_{max} = \frac{v^2 \sin^2 30^\circ}{2g}$$

$$= \frac{(u/\sqrt{2})^2 \sin^2 30^\circ}{2g}$$

$$= \left(\frac{u^2/4g}{4}\right) = \frac{120}{4} = 30 \text{ m}$$

**9.** *(2)* $a = \frac{F}{m} = \frac{qE}{m} = 10^3 \sin(10^3 t)$

$$\frac{dv}{dt} = 10^3 \sin(10^3 t)$$

$$\Rightarrow \quad \int_0^v dv = \int_0^t 10^3 \sin(10^3 t)\, dt$$

$$\therefore \quad v = \frac{10^3}{10^3}[1 - \cos(10^3 t)]$$

Velocity will be maximum when $\cos(10^3 t) = -1$

$$v_{max} = 2 \text{ m/s}$$

**10.** *(5.55)* Given, $N = 50$,

$$A = 2 \times 10^{-4} \text{ m}^2, C = 10^{-4}, R = 50\Omega,$$

$$B = 0.02 \text{ T}, \theta = 0.2 \text{ rad}$$

$$\therefore N i_g AB = C\theta$$

$$\Rightarrow \quad i_g = \frac{C\theta}{NAB}$$

$$= \frac{10^{-4} \times 0.2}{50 \times 2 \times 10^{-4} \times 0.02} = 0.1 \text{ A}$$

$$\therefore \quad V_{ab} = i_g \times G = (i - i_g)S$$

$$0.1 \times 50 = (1 - 0.1) \times S$$

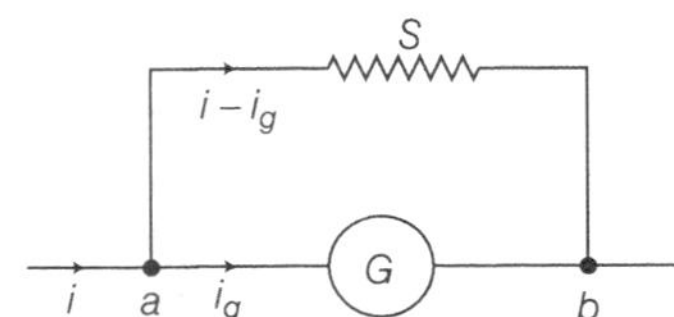

$$5 = 0.9 \times S$$

$$\therefore \quad S = \frac{50}{9}\Omega = 5.55\Omega$$

**11.** *(3)* Given, $d = 0.5\,\text{mm}$,

$Y = 2\times10^{11}\ \text{Nm}^{-2}$,

$l = 1\ \text{m}$

$$\Delta l = \frac{Fl}{AY} = \frac{mgl}{\frac{\pi d^2}{4}Y}$$

$$= \frac{1.2\times10\times1}{\frac{\pi}{4}\times(5\times10^{-4})^2\times2\times10^{11}}$$

$= 0.3\ \text{mm}$

$$\text{LC of vernier} = \left(1-\frac{9}{10}\right)\text{mm} = 0.1\ \text{mm}$$

So, 3rd division of vernier scale will coincide with main scale.

**12.** *(900)* Given, $n = 1, \gamma = \frac{5}{3}$

$T$-$V$ equation in adiabatic process is

$TV^{\gamma-1} = \text{constant}$

$T_1V_1^{\gamma-1} = T_2V_2^{\gamma-1}$

$$\Rightarrow \quad T_2 = T_1\left(\frac{V_1}{V_2}\right)^{\gamma-1} = 100\times\left(\frac{1}{8}\right)^{\frac{2}{3}}$$

$\Rightarrow \quad T_2 = 25\,\text{K}$

$C_V = \frac{3}{2}R$ for monoatomic gas

$$\therefore \quad \Delta U = nC_V\Delta T = n\times\left(\frac{3R}{2}\right)(T_2 - T_1)$$

$$= 1\times\frac{3}{2}\times8\times(25-100)$$

$$= -\ 900\ \text{J}$$

$\therefore$ Decrease in internal energy = 900 J

**13.** *(24)* $\therefore$ Power $= nhf$

(where, $n$ = number of photons incident per second)

Since, KE = 0, $hf$ = work-funcition $W$

$$200 = nW = n\,[6.25\times1.6\times10^{-19}]$$

$$\Rightarrow \quad n = \frac{200}{1.6\times10^{-19}\times6.25}$$

As photon is just above threshold frequency $KE_{\text{max}}$ is zero and they are accelerated by potential difference of 500 V.

$$\therefore \quad \text{KE}_f = q\Delta V$$

$$\frac{P^2}{2m} = q\Delta V$$

$$\Rightarrow \quad P = \sqrt{2mq\,\Delta V}$$

Since, efficiency is 100%, number of electrons emitted per second = number of photons incident per second.

As, photon is completely absorbed, force exerted

$$= n(mV) = nP = n\sqrt{2mq\Delta V}$$

$$= \frac{200}{6.25\times1.6\times10^{-19}}\times\sqrt{2\,(9\times10^{-31})\times1.6\times10^{-19}\times500}$$

$$= 24$$

**14.** *(3)*

$$\Delta E_{2-1} = 13.6\times Z^2\left[1-\frac{1}{4}\right] = 13.6\times Z^2\left[\frac{3}{4}\right]$$

$$\Delta E_{3-2} = 13.6\times Z^2\left[\frac{1}{4}-\frac{1}{9}\right] = 13.6\times Z^2\left[\frac{5}{36}\right]$$

$$\therefore \quad \Delta E_2 = \Delta E_{3-2} + 74.8$$

$$13.6\times Z^2\left[\frac{3}{4}\right] = 13.6\times Z^2\left[\frac{5}{36}\right] + 74.8$$

$$13.6\times Z^2\left[\frac{3}{4}-\frac{5}{36}\right] = 74.8$$

$$Z^2 = 9$$

$$\therefore \quad Z = 3$$

**15.** *(b)* **List-II**

(1) $E = \frac{1}{4\pi\varepsilon_0}\frac{Q}{d^2} \quad \Rightarrow \quad E \propto \frac{1}{d^2}$

(2) $E_{\text{axis}} = \frac{1}{4\pi\varepsilon_0}\frac{2Q(2l)}{d^3} \quad \Rightarrow \quad E \propto \frac{1}{d^3}$

(3) $E = \frac{\lambda}{2\pi\varepsilon_0 d} \quad \Rightarrow \quad E \propto \frac{1}{d}$

(4) $E = \frac{\lambda}{2\pi\varepsilon_0(d-l)} - \frac{\lambda}{2\pi\varepsilon_0(d+l)} = \frac{\lambda(2l)}{2\pi\varepsilon_0 d^2}$

$$\Rightarrow E \propto \frac{1}{d^2}$$

(5) $E = \frac{\sigma}{2\varepsilon_0} \quad \Rightarrow$ E is independent of $d$

**16.** *(b)* $v = \sqrt{\frac{GM}{R}}$

Let $R_1 = R$, then $R_2 = 4R$

If $m_2 = m$, then $m_1 = 2m$

**List-I**

(P) $\frac{v_1}{v_2}=\sqrt{\frac{R_2}{R_1}}=\sqrt{\frac{4R}{R}}=2:1$

(Q) $L=mvR$

$\frac{L_1}{L_2}=\frac{R(2m)v_1}{4R(m)v_2}=\frac{1}{2}(2)=1:1$

(R) $\frac{K_1}{K_2}=\frac{\frac{1}{2}(2m)v_1^2}{\frac{1}{2}(m)v_2^2}=2(4)=8:1$

(S) $\frac{T_1}{T_2}=\left(\frac{R_1}{R_2}\right)^{3/2}=\left(\frac{1}{4}\right)^{3/2}=1:8$

**17.** *(c)* $\left(\frac{dp}{dV}\right)_{\text{adiabatic}}=\gamma\left(\frac{dp}{dV}\right)_{\text{isothermal}}$

**List-I**

(P) Process I $\Rightarrow$ Adiabatic $\Rightarrow Q=0$

(Q) Process II $\Rightarrow$ Isobaric

$\therefore\ W=p\Delta V=3p_0[3V_0-V_0]=6p_0V_0$

(R) Process III $\Rightarrow$ Isochoric $\Rightarrow W=0$

(S) Process (IV) $\Rightarrow$ Isothermal

$\Rightarrow$ Temperature = Constant

**18.** *(a)* When force $F=0\Rightarrow$ potential energy $U=$ constant

$F\neq 0\Rightarrow$ force is conservative $\Rightarrow$ Total energy $E=$ constant

**List-I**

(P) $\mathbf{r}(t)=\alpha t\hat{\mathbf{i}}+\beta t\hat{\mathbf{j}}$

$\frac{d\mathbf{r}}{dt}=\mathbf{v}=\alpha\hat{\mathbf{i}}+\beta\hat{\mathbf{j}}=\text{constant}$

$\Rightarrow\ \mathbf{p}=\text{constant}$

$|\mathbf{v}|=\sqrt{\alpha^2+\beta^2}=\text{constant}$

$\Rightarrow\ K=\text{constant}$

$\frac{d\mathbf{v}}{dt}=\mathbf{a}=0\Rightarrow F=0\Rightarrow U=\text{constant}$

$E=U+K=\text{constant}$

$\mathbf{L}=m(\mathbf{r}\times\mathbf{v})=0$

$\mathbf{L}=\text{constant}$

$P\rightarrow 1,2,3,4,5$

(Q) $\mathbf{r}(t)=\alpha\cos\omega t\hat{\mathbf{i}}+\beta\sin\omega t\hat{\mathbf{j}}$

$\frac{d\mathbf{r}}{dt}=\mathbf{v}=\alpha\omega\sin\omega t(-\hat{\mathbf{i}})+\beta\omega\cos\omega t\hat{\mathbf{j}}$

$\neq$constant

$\Rightarrow\ \mathbf{p}\neq\text{constant}$

$|\mathbf{v}|=\omega\sqrt{(\alpha\sin\omega t)^2+(\beta\cos\omega t)^2}\neq\text{constant}$

$\Rightarrow K\neq\text{constant}$

$\mathbf{a}=\frac{d\mathbf{v}}{dt}=-\omega^2\mathbf{r}\neq 0$

$\Rightarrow\ E=\text{constant}=K+U$

But $K\neq$ constant $\Rightarrow U\neq$ constant

$\mathbf{L}=m(\mathbf{r}\times\mathbf{v})=m\omega\alpha\beta\,(\hat{\mathbf{k}})=\text{constant}$

$Q\rightarrow 2,5$

(R) $\mathbf{r}(t)=\alpha(\cos\omega t\hat{\mathbf{i}}+\sin\omega t\hat{\mathbf{j}})$

$\frac{d\mathbf{r}}{dt}=\mathbf{v}=\alpha\omega[\sin\omega t(-\hat{\mathbf{i}})+\cos\omega t\hat{\mathbf{j}}]$

$\neq$ constant

$\Rightarrow\mathbf{p}\neq\text{constant}$

$|\mathbf{v}|=\alpha\omega=\text{constant}\Rightarrow K=\text{constant}$

$\mathbf{a}=\frac{d\mathbf{v}}{dt}=-\omega^2\mathbf{r}\neq 0\Rightarrow E=\text{constant},$

$U=\text{constant}$

$\mathbf{L}=m(\mathbf{r}\times\mathbf{v})=m\omega\alpha^2\hat{\mathbf{k}}=\text{constant}$

$R\rightarrow 2,3,4,5$

(S) $\mathbf{r}(t)=\alpha t\hat{\mathbf{i}}+\frac{\beta}{2}t^2\hat{\mathbf{j}}$

$\frac{d\mathbf{r}}{dt}=\mathbf{v}=\alpha\hat{\mathbf{i}}+\beta t\hat{\mathbf{j}}\neq\text{constant}$

$\Rightarrow\ \mathbf{p}\neq\text{constant}$

$|\mathbf{v}|=\sqrt{\alpha^2+(\beta t)^2}\neq\text{constant}$

$\Rightarrow\ K\neq\text{constant}$

$\mathbf{a}=\frac{d\mathbf{v}}{dt}=\beta\hat{\mathbf{j}}\neq 0$

$\Rightarrow\ E=\text{constant}=K+U$

But $K\neq$ constant

$\therefore\ U\neq$ constant

$\mathbf{L}=m(\mathbf{r}\times\mathbf{v})=\frac{1}{2}\alpha\beta t^2\hat{\mathbf{k}}\neq\text{constant}$

$S\rightarrow 5$

# Chemistry

**1.** *(a, b, d)* Statement wise explanation is

**Statement (a)**

$[Co(en)(NH_3)_3H_2O]^{3+}$ have following 2 geometrical isomers.

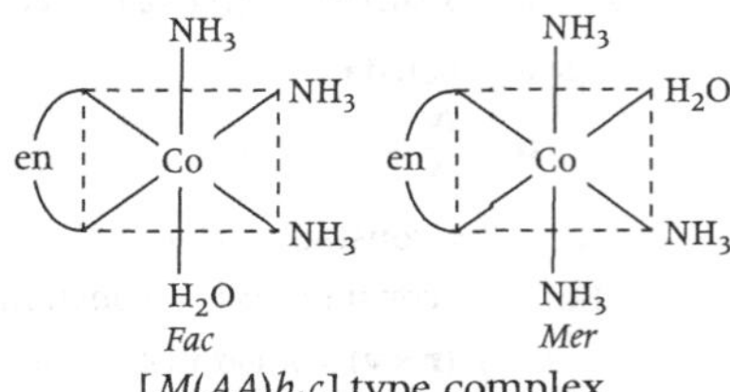


$[M(AA)b_3c]$ type complex

Hence, this is correct statement.

**Statement (b)**

If bidentate ligand 'en' is replaced by two cyanide ligands then $[Co(NH_3)_3(H_2O)(CN)_2]^+$ is formed.

It is $[Ma_3b_2c]$ type complex which has following 3 geometrical isomers.

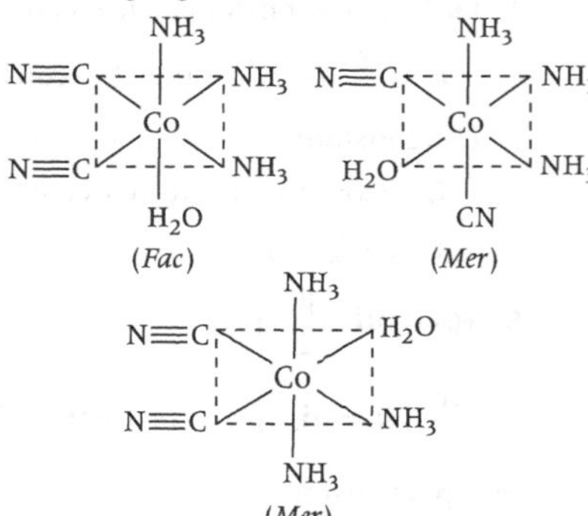


Hence, this statement is also correct.

**Statement (c)**

Co metal has $[Ar]3d^7 4s^2$ configuration while in $[Co(en)(NH_3)_3(H_2O)]^{3+}$ it is in +3 oxidation state. Thus, $Co^{3+}$ has $[Ar]3d^6$ configuration, and $d^2sp^3$ in complex.

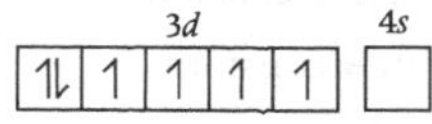


As *en* is a strong ligand, so pairing will occur

| 3d | | | | | 4s |
|---|---|---|---|---|---|
| ⇅ | ⇅ | ⇅ | | | |

Due to the presence of all paired electrons it show diamagnetic behaviour rather than paramagnetic.

Hence, this statement is incorrect.

**Statement (d)**

According to CFT, absorption of light by coordination complexes depends upon CFSE i.e., crystal field splitting energy $(\Delta_0)$as

$$\Delta_0 \propto \frac{1}{\lambda}$$

Among the complexes given $[Co(en)(NH_3)_4]^{3+}$ has more $\Delta_0$ value as compared to complex $[Co(en)(NH_3)_3(H_2O)]^{3+}$. Thus, $[Co(en)(NH_3)_3(H_2O)]^{3+}$ absorbs the light at longer wavelength for *d-d* transition.

Hence, this statement is also correct.

***Note :*** *For any complex, the value of $\Delta_0$ can be calculated via the difference or gap between $e_g$ and $t_{2g}$ values.*

**2.** *(b,d)* Statement wise explanation is

**Statement (a)** $Mn^{2+}$ produces yellow-green colour in flame test while $Cu^{2+}$ produces bluish-green colour in flame test. Thus, due to the presence of green colour in both the cases, flame test is not the suitable method to distinguish between nitrate salts of $Cu^{2+}$ and $Mn^{2+}$.

Hence this statement is wrong.

**Statement (b)** $Cu^{2+}$ belong to group II of cationic or basic radicals. It gives black ppt. of CuS if $H_2S$ is passed through it in the presence of acid (e.g HCl). $Mn^{2+}$ does not show this property hence this can be considered as a suitable method to distinguish between $Mn^{2+}$ and $Cu^{2+}$.

Hence, this statement is correct.

**Statement (c)** In faintly basic medium when $H_2S$ is passed both $Cu^{2+}$ and $Mn^{2+}$ forms precipitates. Thus, it is not suitable method to distinguish between them.

Hence, this statement is incorrect

**Statement (d)** The standard reduction potential of $Cu^{2+}/Cu$ is +0.34 V while that of $Mn^{2+}/Mn$ is –1.18V. This can be used to distinguish between $Cu^{2+}$ and $Mn^{2+}$. In general less electropositive metals have higher SRP.

Hence, this statement is correct.

**3.** *(d)* Given, Aniline $\xrightarrow[\text{Conc. } H_2SO_4]{\text{Conc. } HNO_3+}$ $P$ (51%) + $Q$ (47%) + $R$ (2%)

Then $P$, $Q$ and $R$ will be

Aniline $\xrightarrow[\text{Conc. } H_2SO_4]{\text{Conc. } HNO_3+}$ p-nitroaniline + m-nitroaniline + o-nitroaniline

$P$ (51%) $Q$ (47%) $R$ (2%)

Its given

$$R \xrightarrow[\text{(iii) } H_3O^+ \text{ (iv) } NaNO_2, \text{ HCl/273-278 K (v) EtOH, } \Delta]{\text{(i) } Ac_2O, \text{ Pyridine (ii) } Br_2, CH_3COOH} S$$

o-nitroaniline ($R$) $\xrightarrow[\text{Pyridine}]{Ac_2O}$ $HN-\overset{O}{\overset{||}{C}}-CH_3$ (o-nitroacetanilide) $\xrightarrow[CH_3COOH]{Br_2}$ 4-bromo-2-nitroacetanilide $\xrightarrow[\text{(Hydrolysis)}]{-CH_3COOH,\ H_3O^+}$ 4-bromo-2-nitroaniline $\xrightarrow[\text{(Diazotisation)}]{NaNO_2/HCl,\ 273\text{-}278K}$ $\overset{+}{N_2}\overset{-}{Cl}$ (4-bromo-2-nitrobenzenediazonium chloride) $\xrightarrow[\Delta]{EtOH}$ m-bromonitrobenzene ($S$)

Now from S to major products its given.

$$S \xrightarrow[\text{(iii) } NaNO_2, \text{ HCl/273-278K (iv) } H_3PO_2]{\text{(i) Sn/ HCl (ii) } Br_2/H_2O \text{ (Excess)}} \text{Major product}$$

So,

$\text{(S) 1-bromo-3-nitrobenzene} \xrightarrow{Sn/HCl} \text{3-bromoaniline} \xrightarrow[H_2O\ (Excess)]{Br_2} \text{2,3,4,6-tetrabromoaniline} \xrightarrow[273\text{-}278K]{NaNO_2+HCl} \text{Ar}N_2^+Cl^- \xrightarrow{H_3PO_2}$

Major product
1, 3, 4, 5-tetrabromobenzene

Hence, only (d) is the correct answer.

**4.** *(d)*

| D-glucose | | L-glucose |
|---|---|---|
| $H-C_1=O$ | | $H-C_1=O$ |
| $H-C_2-OH$ | | $HO-C_2-H$ |
| $HO-C_3-H$ | | $H-C_3-OH$ |
| $H-C_4-OH$ | | $HO-C_4-H$ |
| $H-C_5-OH$ | | $HO-C_5-H$ |
| $_6CH_2OH$ | | $_6CH_2OH$ |

So, β-L glucopyranose is formed as

β-L-glucopyranose

The α-L-glucopyranose has configurational change at $C_1$ only and looks like

Configuration at this carbon atom is opposite in α and β-forms

α-L-glucopyranose

**5.** *(a, d)* Given for the reaction (at $T = 300$ K and constant volume $= V$)

$$A(g) \longrightarrow 2B(g) + C(g)$$

| | $A(g)$ | $2B(g)$ | $C(g)$ |
|---|---|---|---|
| at $t = 0$ | $p_0$ | – | – |
| at $t = t$ | $p_0 - x$ | $2x$ | $x$ |
| at $t = t_{1/3}$ | $\left[p_0 - \frac{2p_0}{3}\right] = \frac{p_0}{3}$ | $\frac{4p_0}{3}$ | $\frac{2p_0}{3}$ |

We can calculate,

$$p_t = p_0 - x + 2x + x = p_0 + 2x$$

or $\quad 2x = p_t - p_0 \quad$ or $\quad x = \frac{p_t - p_0}{2}$

Now for first order reaction,

$$t = \frac{1}{k} \ln \frac{p_0}{(p_0 - x)}$$

Putting the value of $x$ in the equation,

$$t = \frac{1}{k} \ln \frac{p_0}{p_0 - \left(\frac{p_t - p_0}{2}\right)}$$

$$= \frac{1}{k} \ln \frac{2p_0}{2p_0 - p_t + p_0}$$

or $\quad kt = \ln \frac{2p_0}{(3p_0 - p_t)}$

or $\quad kt = \ln 2p_0 - \ln (3p_0 - p_t)$

or $\quad \boxed{\ln (3p_0 - p_t) = -kt + \ln 2p_0}$

It indicates graph between $\ln (3p_0 - p_t)$ *vs* '$t$' will be a straight line with negative slope , so option (a) is correct

$$t_{1/3} = \frac{1}{k} \ln \frac{p_0}{p_0/3} = \frac{1}{k} \ln 3$$

It indicates $t_{1/3}$ is independent of initial concentration so, option (b) is incorrect.

Likewise, rate constant also does not show its dependence over initial concentration. Thus, graph between rate constant and $[A]_0$ will be a straight line parallel to $X$-axis .

$$(a - x) \text{ at time } t = \frac{yp_i - p_t}{(y - 1)}$$

**6.** *(a,c)* For the reaction,

$$A \rightleftharpoons P$$

Given, $\quad T_1 < T_2$

$$\frac{\ln K_1}{\ln K_2} > \frac{T_2}{T_1} \quad \ldots \text{(i)}$$

It shows,

On increasing the temperature, $K$ decreases so reaction is exothermic i.e.,

$$\Delta H^\circ < 0$$

Besides, graph shows $K > 1$

So $\quad \Delta G^\circ < 0$

Now from equation (i)

$$T_1 \ln K_1 > T_2 \ln K_2$$

$$-\Delta G^\circ_1 > -\Delta G^\circ_2$$

Likewise

$$(-\Delta H^\circ + T_1 \Delta S^\circ) > (-\Delta H^\circ + T_2 \Delta S^\circ)$$

or simply

$$T_1 \Delta S^\circ > T_2 \Delta S^\circ$$

So, $(T_2 - T_1) \Delta S^\circ < 0$

$\therefore \quad \Delta S^\circ < 0$

In other words, increase of $\Delta G$ with increase in temperature is possible only when $\Delta S^\circ < 0$.

Hence, options (a) and (c) are correct.

**7.** *(6)* The structures of various molecules given in problem are discussed below—

1. $\mathbf{N_2O_3}$ It is the tautomeric mixture of following two structures—

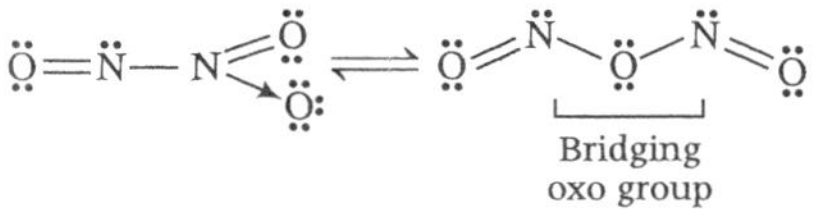

**Conclusion** 1 bridging oxo group is present in the compound.

2. $\mathbf{N_2O_5}$ It has following structure.

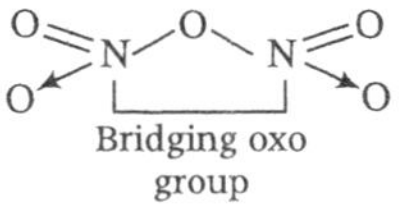

**Conclusion** 1 bridging oxo group is present in the compound.

where, $p_i$ = initial pressure, $p_t$ = total pressure

$y$ = number of gaseous products per mole of reactant

3. $\mathbf{P_4O_6}$

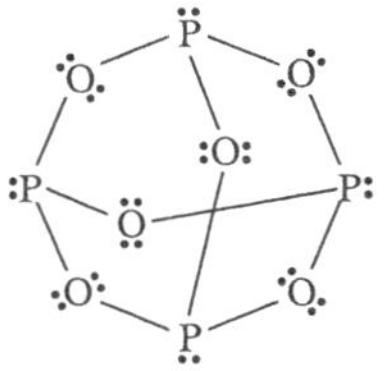

**Conclusion** 6 bridging oxo groups are present in the compound.

4. $\mathbf{P_4O_7}$

**Conclusion** 6 bridging oxo groups are present in the compound.

5. $\mathbf{H_4P_2O_5}$

**Conclusion** 1 bridging oxo group is present in the compound.

6. $H_5P_3O_{10}$

**Conclusion** 2 bridging oxo groups are present in the compound.

7. $\mathbf{H_2S_2O_3}$

**Conclusion** This compound does not contain any bridging oxo group.

8. $\mathbf{H_2S_2O_5}$

**Conclusion** This compound also does not contain any bridging oxo group.

**8.** *(6.47 kg)* The equations of chemical reactions occurring during the process are—

**In the presence of oxygen**

$$2PbS + 3O_2 \longrightarrow 2PbO + 2SO_2 \quad \ldots(i)$$

By self reduction

$$2PbO + PbS \longrightarrow 3Pb + SO_2$$

Thus 3 moles of $O_2$ produces 3 moles of Pb i.e.,

$32 \times 3 = 96$ g of $O_2$ produces

$3 \times 207 = 621$ g of Pb

So 1000 g (1kg) of oxygen will produce $\frac{621}{96} \times 1000 = 6468.15$ g

$$= 6.4687 \text{ kg} \approx 6.47 \text{ kg}$$

**Alternatively**

From the direct equation,

$$\underset{}{PbS} + \underset{32\text{ g}}{O_2} \longrightarrow \underset{207\text{ g}}{Pb} + SO_2$$

So, 32 g of $O_2$ gives 207 g of Pb

1 g of $O_2$ will give $\frac{207}{32}$ g of Pb

1000g of $O_2$ will give $\frac{207}{32} \times 1000$

$$= 6468.75\text{ g} = 6.46875 \text{ kg}$$
$$\approx 6.47\text{kg}$$

**9.** *(126 mg)*

$$Mn^{2+} \longrightarrow MnO_4^- + 5e^-$$
$$S_2O_8^{2-} + 2e^- \longrightarrow 2SO_4^{2-}$$

$$\therefore \; 2Mn^{2+} + 5S_2O_8^{2-} \longrightarrow 2MnO_4^- + 10SO_4^{2-}$$

Also,

$$2MnO_4^- + 5C_2O_4^{2-} \longrightarrow 2Mn^{2+} + 10CO_2$$

Hence, $2Mn^{2+} \equiv 5C_2O_4^{2-}$

$$1\ MnCl_2 \equiv 2.5\,H_2C_2O_4$$

Oxalic acid taken $= 225$ mg

$= \frac{225}{90} = 2.5$ millimoles

Hence, $MnCl_2 = 1$ millimole

$= (55 + 71) = \mathbf{126\ mg}$

**10.** *(7)* As given in the question 3 stereocentres are visible, i.e.,

Hence, the total number of stereoisomers $= 2^3 = 8$

But out of these the following one is optically inactive due to symmetry

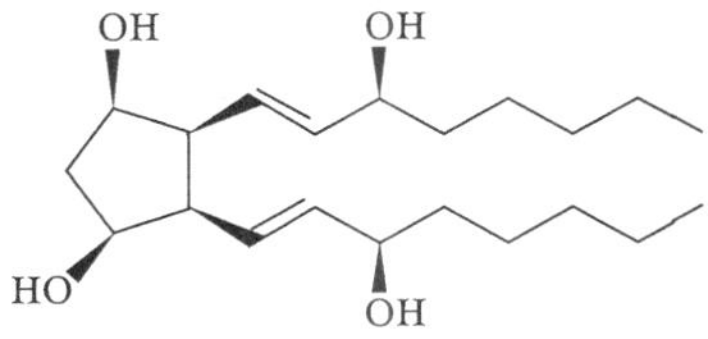

Hence, total number of optically active stereoisomers = 7

**11.** *(495 g)* Given,

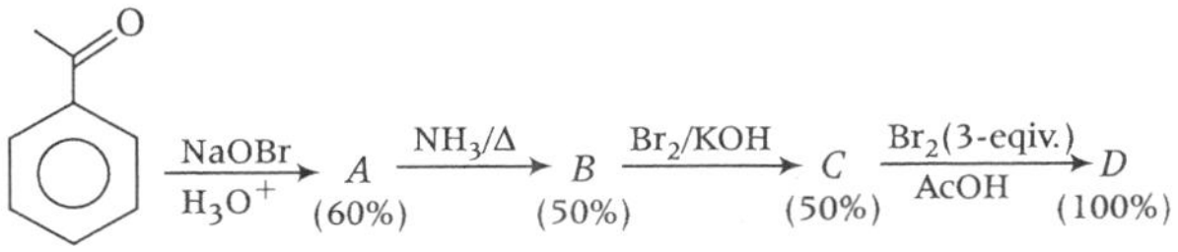

$$\text{Acetophenone} \xrightarrow[H_3O^+]{NaOBr} \underset{(60\%)}{A} \xrightarrow{NH_3/\Delta} \underset{(50\%)}{B} \xrightarrow{Br_2/KOH} \underset{(50\%)}{C} \xrightarrow[AcOH]{Br_2(3\text{-eqiv.})} \underset{(100\%)}{D}$$

The products formed are

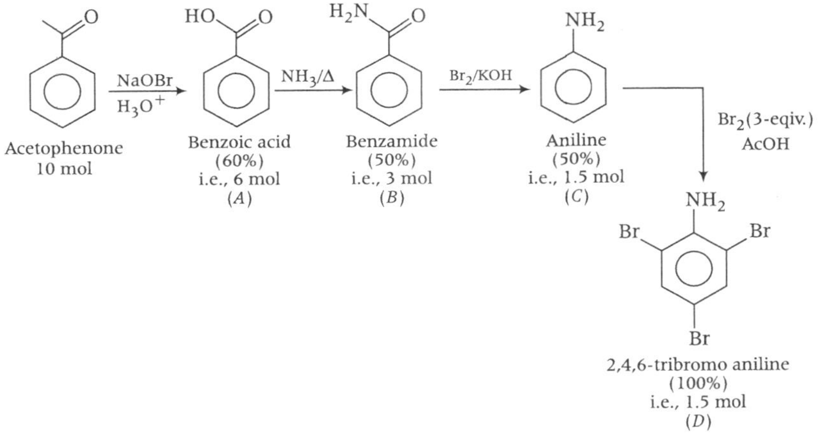

Acetophenone 10 mol $\xrightarrow[H_3O^+]{NaOBr}$ Benzoic acid (60%) i.e., 6 mol (*A*) $\xrightarrow{NH_3/\Delta}$ Benzamide (50%) i.e., 3 mol (*B*) $\xrightarrow{Br_2/KOH}$ Aniline (50%) i.e., 1.5 mol (*C*) $\xrightarrow[AcOH]{Br_2(3\text{-eqiv.})}$ 2,4,6-tribromo aniline (100%) i.e., 1.5 mol (*D*)

So, 1.5 mol of 2,4,6-tribromo aniline are produced from 10 moles of acetophenone.

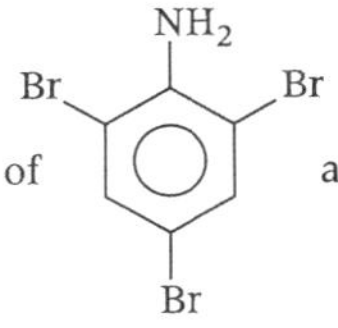

Molar mass of 2,4,6-tribromo aniline $= 240 + 14 + 4 + 72 = 330$

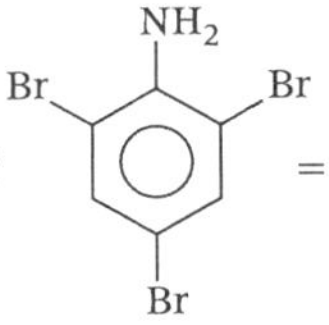

Hence, amount of 2,4,6-tribromo aniline produced is $330 \times 1.5 = 495$ g

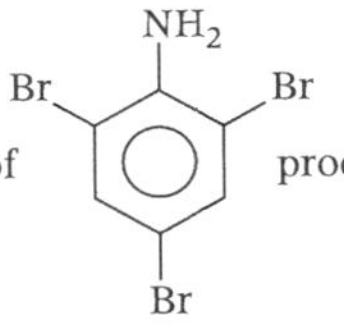

**12.** *(–14.6)* Given

(i) $2Cu(s)+\frac{1}{2}O_2(g) \longrightarrow Cu_2O(s);$

$$\Delta G^o = -78000 \text{ J mol}^{-1}$$
$$= -78 \text{ kJ mol}^{-1}$$

(ii) $H_2(g)+\frac{1}{2}O_2(g) \longrightarrow H_2O(g);$

$$\Delta G^o = -178000 \text{ J mol}^{-1}$$
$$= -178 \text{ kJ mol}^{-1}$$

So, net reaction is (By (i)-(ii))

$$2Cu(s)+H_2O(g) \longrightarrow Cu_2O(s)+H_2(g);$$
$$\Delta G = 100000 \text{ J/mol or } 10^5 \text{ J/mol}$$
$$= 100 \text{ kJ mol}^{-1}$$

Now , for the above reaction

$$\Delta G = \Delta G^o + RT \ln\left[\frac{p_{H_2}}{p_{H_2O}}\right]$$

and to prevent above reaction,

$$\Delta G \geq 0$$

So,

$$\Delta G^o + RT \ln\left[\frac{p_{H_2}}{p_{H_2O}}\right] \geq 0$$

After putting the values,

$$10^5 + 8\times 1250 \ln\left[\frac{p_{H_2}}{p_{H_2O}}\right] \geq 0$$

or $\quad 10^5 + 10^4 \ln\left[\frac{p_{H_2}}{p_{H_2O}}\right] \geq 0$

or $\quad 10^4(\ln p_{H_2} - \ln p_{H_2O}) \geq -10^5$

or $\quad \ln p_{H_2} \geq -10 + \ln p_{H_2O}$

or $\quad \ln p_{H_2} \geq -10 + 2.3 \log(0.01)$ (as $p_{H_2O} = 1\%$)

$$\geq -10 - 4.6$$

so $\quad \ln p_{H_2} \geq -14.6$

**13.** *(+ 8500 J/mol)*

For the reaction,

$$A(g) + B(g) \rightleftharpoons AB(g)$$

Given $\quad E_{a_b} = E_{a_f} + 2RT$

or $\quad E_{a_b} - E_{a_f} = 2RT$

Further

$$A_f = 4A_b \quad \text{or} \quad \frac{A_f}{A_b} = 4$$

Now, rate constant for forward reaction,

$$k_f = A_f e^{-E_{a_f}/RT}$$

Likewise, rate constant for backward reaction,

$$k_b = A_b e^{-E_{a_b}/RT}$$

At equilibrium,

Rate of forward reaction = Rate of backward reaction

i.e., $\quad k_f = k_b$

or $\quad \frac{k_f}{k_b} = k_{eq}$

so $\quad k_{eq} = \frac{A_f e^{-E_{a_f}/RT}}{A_b e^{-E_{a_b}/RT}}$

$$= \frac{A_f}{A_b} e^{-(E_{a_f} - E_{a_b})/RT}$$

After putting the given values

$$k_{eq} = 4e^2$$

(as $E_{a_b} - E_{a_f} = 2RT$ and $\frac{A_f}{A_b} = 4$)

Now, $\quad \Delta G^\circ = -RT \ln K_{eq}$

$$= -2500 \ln(4e^2)$$
$$= -2500 (\ln 4 + \ln e^2)$$
$$= -2500 (1.4 + 2)$$
$$= -2500 \times 3.4$$
$$= -8500 \text{ J/mol}$$

Absolute value = 8500 J/mol

**14.** *(– 11.62 JK$^{-1}$ mol$^{-1}$)*

Given,

$$A(s)\,|\,A^{n+}(aq, 2\text{ M})\,||\,B^{2n+}(aq, 1\text{ M})\,|\,B(s)$$

So, reactions at respective electrode will be

**Anode** $\quad A(s) \longrightarrow A^{n+} + ne^- \times 2$

**Cathode** $\quad B^{2n+} + 2ne^- \longrightarrow B(s)$

**Overall reaction**

$$2A(s) + B^{2n+}(aq) \longrightarrow 2A^{n+}(aq) + B(s)$$

Further,

$\Delta H^\circ = 2\Delta G^\circ$ and $E_{cell} = 0$ is also given

Now by using the Nernst equation

$$E_{cell} = E^\circ_{cell} - \frac{RT}{nF}\ln\frac{[\text{Product}]}{[\text{Reactant}]}$$

After putting the values

$$0 = E^\circ_{cell} - \frac{RT}{2nF}\ln\frac{[A^{n+}]^2}{[B^{2n+}]}$$

or $\quad E^\circ = \frac{RT}{2nF}\ln\frac{[2]^2}{[1]} = \frac{RT}{2nF}\ln 4 \qquad \ldots(i)$

Further from the formula,

$$\Delta G^\circ = -nFE^\circ$$

$$\Delta G^\circ = -2nFE^\circ$$

Now putting the value of $E^\circ$ from eq. (i)

$$\Delta G^\circ = -2nF \times \frac{RT}{2nF}\ln 4 \quad \ldots(ii)$$

$$\Delta G^\circ = -RT\ln 4$$

Finally, using the formula

$$\Delta G^\circ = \Delta H^\circ - T\Delta S^\circ$$

$$\Delta G^\circ = 2\Delta G^\circ - T\Delta S^\circ$$

(as $\Delta H^\circ = 2\Delta G^\circ$, given)

$$\Delta G^\circ = T\Delta S^\circ$$

or $\Delta S^\circ = \frac{\Delta G^\circ}{T} = \frac{-RT\ln 4}{T}$

(from eq. (ii), $\Delta G^\circ = -RT\ln 4$)

$= -R\ln 4 = -8.3 \times 2 \times 0.7$

(as all values given)

$= -11.62$ J/K-mol

**15.** *(c)* **For, *P* i.e. $dsp^2$,** It is seen in $[Ni(CN)_4]^{2-}$

$$Ni—[Ar]3d^8 4s^2$$

$$Ni^{2+}—[Ar]3d^8$$

$Ni^{2+}$— 3d: ⥮ ⥮ ⥮ ↿ ↿ | 4s: □ | 4p: □ □ □

as $CN^-$ is a strong ligand so when it approaches towards central metal pairing of unpaired electrons takes place.

Thus, in $[Ni(CN)_4]^{2-}$

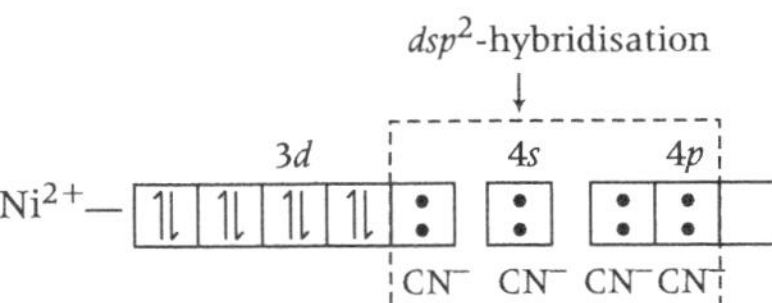


**Structure** : Square planar

So correct match for P is 6.

**For *Q* i.e., $sp^3$**

It is seen in $[FeCl_4]^{2-}$ and $Ni(CO)_4$

$$Fe - [Ar]3d^6 4s^2$$

$$Fe^{2+} - [Ar]3d^6$$

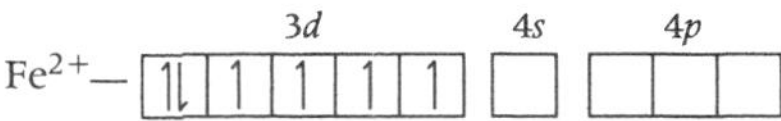


As $Cl^-$ is a weak ligand so when it approaches towards central metal pairing of unpaired electrons does not take place.

Thus, in $[FeCl_4]^{2-}$

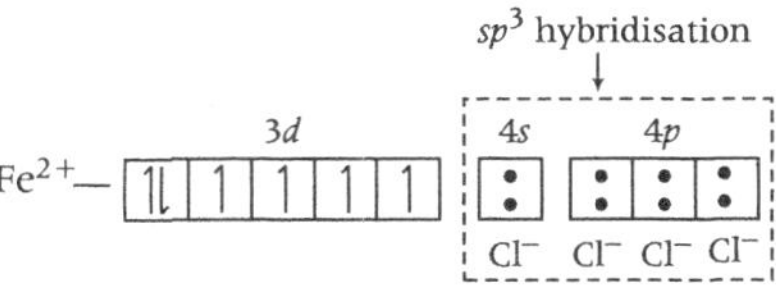


**Structure**-Tetrahedral

Likewise in $Ni(CO)_4$

$$Ni—[Ar]3d^8 4s^2$$

As CO is a strong ligand, hence when it approaches towards central metal atom pairing of unpaired electron of central atom takes place.

Thus, in $Ni(CO)_4$

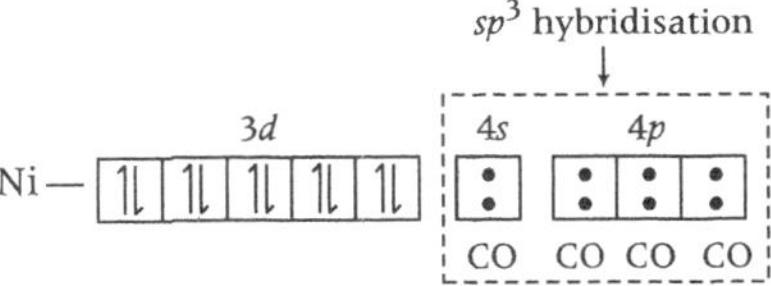


**Structure** Tetrahedral

So, for *Q*-4 and 5 are correct match.

**For *R* i.e., $sp^3d^2$**

It is seen in $[FeF_6]^{4-}$

$$Fe—[Ar]3d^6 4s^2$$

$$Fe^{2+}—[Ar]3d^6$$

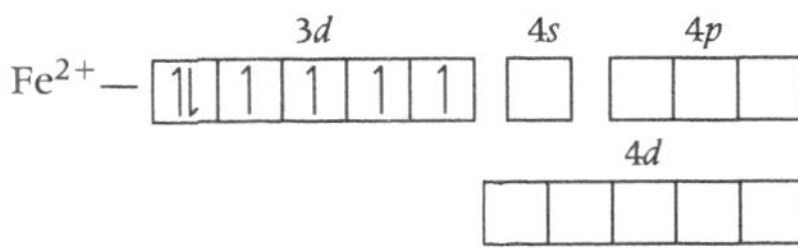


As $F^-$ is a weak field ligand hence, when it approaches towards central metal atom, pairing of its electrons does not take place.

Thus, in $[FeF_6]^{4-}$

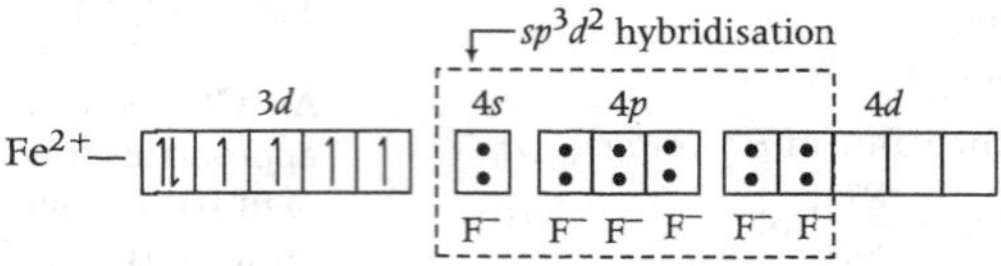


**Structure** : Octahedral

So, 1 is the correct match for *R*.

**For *S* i.e., $d^2sp^3$**

It is seen in $[Ti(H_2O)_3Cl_3]$ and $[Cr(NH_3)_6]^{3+}$

$Ti—[Ar]3d^2 4s^2$

$Ti^{3+}—[Ar]3d^1$

$Ti^{3+}$ — [3d: 1 _ _ _ _] [4s: _] [4p: _ _ _]

Here, both $H_2O$ and Cl are weak ligands

So, in $[Ti(H_2O)_3Cl_3]$

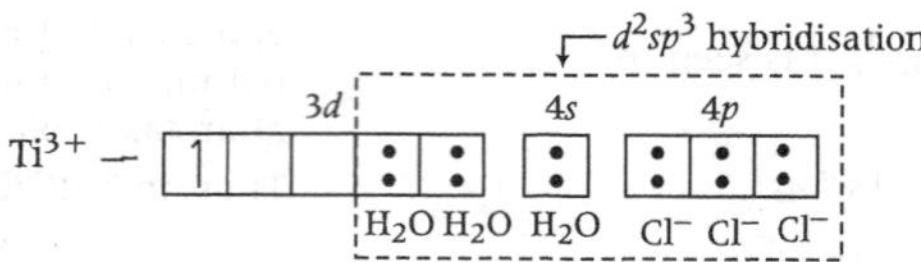


**Structure** Octahedral

Likewise in $[Cr(NH_3)_6]^{3+}$

$Cr—[Ar]3d^5 4s^1$

$Cr^{3+}—[Ar]3d^3 4s^0$

$Cr^{3+}$ — [3d: 1 1 1 _ _] [4s: _] [4p: _ _ _]

Here, $NH_3$ is also a weak field ligand so due to its approach no pairing takes place in Cr.

Thus, In $[Cr(NH_3)_6]^{3+}$

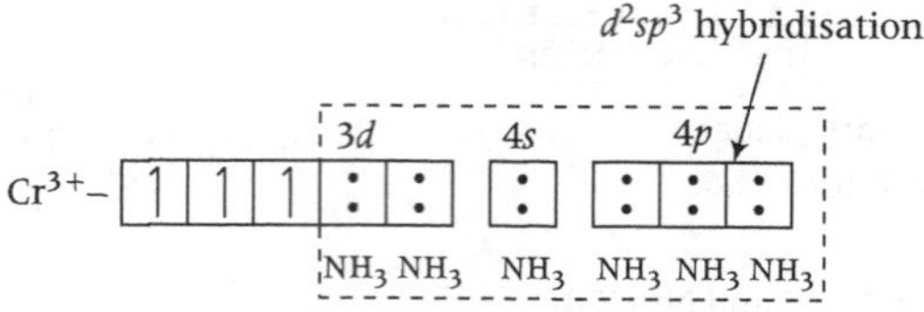


So for, S-2 and 3 are the correct match.

**16.** *(d)*

**For *P*** i.e., (Ph)₂C(OH)–C(Me)₂(OH) + $H_2SO_4$

The correct match is 1 i.e., $I_2$, NaOH and 5 i.e., NaOBr

The reactions proceed as

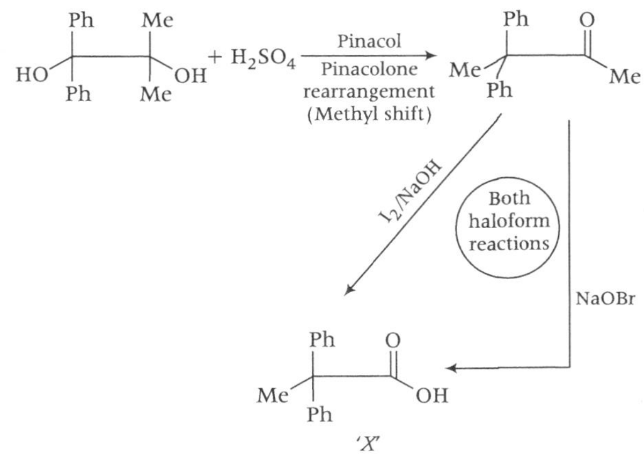


**For *Q*** i.e., $NH_2$–C(Ph)(Ph)–C(H)(Me)–OH + $HNO_2$

The correct match is 2 i.e. $[Ag(NH_3)_2]OH$ and 3 i.e., Fehling's solution.

The reactions proceed as

$H_2N$–C(Ph)(Ph)–C(H)(Me)–OH + $HNO_2$ $\xrightarrow[\text{rearrangement (Methyl shift)}]{\text{Deamination}}$ Me–C(Ph)(Ph)–CHO

Fehling's solution / $[Ag(NH_3)_2]OH$ (Tollens's reagent) — Both oxidation Reactions

→ Me–C(Ph)(Ph)–COOH 'X'

**For *R*** i.e., HO–C(Ph)(Me)–C(Ph)(Me)–OH + $H_2SO_4$

The correct match is 1, 5 again.

The reaction proceed as

$$\text{HO–C(Ph)(Me)–C(Ph)(Me)–OH} + H_2SO_4 \xrightarrow[\text{rearrangement (Phenyl shift)}]{\text{Pinacol Pinacolone}} \text{Me–C(Ph)}_2\text{–CO–Me}$$

$\xrightarrow{I_2/NaOH}$ / $\xrightarrow{NaOBr}$ (Both haloform reactions) $\text{Me–C(Ph)}_2\text{–COOH}$ ($X$)

**For *S*** i.e., $\text{Br–C(Ph)}_2\text{–C(H)(Me)–OH} + AgNO_3$

The correct match is 2, 3.

The reaction proceed as

$$\text{Br–C(Ph)}_2\text{–C(H)(Me)–OH} + AgNO_3 \xrightarrow{\text{dibromination rearrangement}} \text{Me–C(Ph)}_2\text{–CHO}$$

Rest procedure is same as seen for *Q* above i.e., via oxidation.

**17.** *(b)* **For *P*,** i.e. $(CH_3)_3C\overset{-}{O}\overset{+}{Na} + (CH_3)_3C\text{–Br}$

For this reaction 1 and 4 are probable products.

Product 1 i.e., $(CH_3)_3C\text{–OH}$ is formed due to substitution while product 4 i.e., $(CH_3)_2C{=}CH_2$ is formed due to elimination. A tertiary carbocation i.e, $(CH_3)_3C^+$ formed during the reaction. Remember for 3° carbocation ions elimination product predominates.

**For *Q*,** i.e. $(CH_3)_3C\text{–OMe} + HBr$

Correctly matched product for this reaction is 2 i.e., $(CH_3)_3C\text{–Br}$.

The reaction proceeds as

$$(CH_3)_3C\text{–OMe} + HBr \longrightarrow MeOH + (CH_3)_3C\text{–Br}$$

**For *R*** i.e., $(CH_3)_3C\text{–Br} + NaOMe$

Correctly matched product is 4 i.e., $(CH_3)_2C{=}CH_2$. It is a normal elimination reaction and proceeds as

$\text{(CH}_3)_3\text{C–Br} + \overset{+}{\text{Na}}\overset{-}{\text{O}}\text{Me}$ (Alkali in alcohol) $\longrightarrow$ $\text{CH}_2=\text{C(CH}_3)_2$

3° alkyl halide preferes elimination.

**For** $S$ i.e., $\text{(CH}_3)_3\text{C–ONa}$ + MeBr

The correct match is 3 i.e., $\text{(CH}_3)_3\text{C–OMe}$.
The reaction proceeds as

$\text{(CH}_3)_3\text{C–}\overset{-}{\text{O}}\overset{+}{\text{Na}} + \text{MeBr} \longrightarrow \text{(CH}_3)_3\text{C–OMe} + \text{NaBr}$

**18.** *(d)* **For** $P$ **i.e., (10 mL of 0.1 M NaOH + 20 mL of 0.1 M acetic acid) is diluted to 60 mL**

The correct match is 1 i.e., the value of $[H^+]$ does not change on dilution due to the formation of following buffer.

$$NaOH + CH_3COOH \rightleftharpoons CH_3COO^-Na^+ + H_2O$$

Initial millimol 1 2
Final millimol
1 1

Final volume – 30 mL (20 + 10) in which millimoles of $CH_3COOH$ and $CH_3COO^-Na^+$ are counted.

**For** $Q$ **i.e., (20 mL of 0.1 M NaOH + 20 mL of 0.1 M $CH_3COOH$) is diluted to 80 mL**

The correct match is 5 i.e., the value of $[H^+]$ changes to
$\sqrt{2}$ times of its initial value on dilution.

As per the condition given in $Q$ the resultant solution before dilution contain 2 millimoles of $CH_3COO^-Na^+$ is 40 mL solution. Hence, it is the salt of weak acid and strong base. So,

$$[H^+]_{initial} = \sqrt{\frac{K_W K_a}{C}}$$

After dilution to 80 mL, the new '$C$' becomes $\frac{C}{2}$, So,

$$[H^+]_{new} = \sqrt{\frac{K_W K_a}{C/2}} \text{ or } [H^+]_{initial} \times \sqrt{2}$$

**For** $R$ **i.e., (20 mL of 0.1 M HCl + 20 mL of 0.1 M $NH_3$) is diluted to 80 mL**

The correct match is 4 i.e., the value of $[H^+]$ changes to $\frac{1}{\sqrt{2}}$ times of its initial value of dilution.

As per the condition given in $R$ the resultant solution before dilution contains 2 millimoles of $NH_4Cl$ in 40 mL of solution. Hence, a salt of strong acid and weak base is formed.

For this,

$$[H^+]_{initial} = \sqrt{\frac{K_w \times C}{K_b}}$$

Now on dilution upto 80 mL new conc. becomes $C/2$

So, $$[H^+]_{new} = \sqrt{\frac{K_w \times \frac{C}{2}}{K_b}}$$

or $$[H^+]_{new} = [H^+]_{initial} \times \frac{1}{\sqrt{2}}$$

**For** $S$ **i.e., 10 mL saturated solution of $Ni(OH)_2$ in equilibrium with excess solid $Ni(OH)_2$ is diluted to 20 mL and solid $Ni(OH)_2$ is still present after dilution.**

The correct match is 1.

$$Ni(OH)_2(s) \rightleftharpoons Ni^{2+} + 2OH^-$$

as per the condition given it is a sparingly soluble salt. Hence, on dilution the concentration of $OH^-$ ions remains constant in saturated solution.

So for this solution,

$$[H^+]_{new} = [H^+]_{initial}$$

# Mathematics

**1.** *(d)* We have,

$$f_n(x) = \sum_{j=1}^{n} \tan^{-1}\left(\frac{1}{1+(x+j)(x+j-1)}\right)$$

for all $x \in (0, \infty)$

$$\Rightarrow f_n(x) = \sum_{j=1}^{n} \tan^{-1}\left(\frac{(x+j)-(x+j-1)}{1+(x+j)(x+j-1)}\right)$$

$$\Rightarrow f_n(x) = \sum_{j=1}^{n} [\tan^{-1}(x+j) - \tan^{-1}(x+j-1)]$$

$\Rightarrow f_n(x) = (\tan^{-1}(x+1) - \tan^{-1}x)$
$+ (\tan^{-1}(x+2) - \tan^{-1}(x+1))$
$+ (\tan^{-1}(x+3) - \tan^{-1}(x+2))$
$+ \ldots + (\tan^{-1}(x+n) - \tan^{-1}(x+n-1))$

$\Rightarrow f_n(x) = \tan^{-1}(x+n) - \tan^{-1}x$

This statement is false as $x \neq 0$. i.e., $x \in (0, \infty)$

(b) This statement is also false as $0 \notin (0, \infty)$

(c) $f_n(x) = \tan^{-1}(x+n) - \tan^{-1}x$

$$\lim_{x\to\infty} \tan(f_n(x)) = \lim_{x\to\infty} \tan(\tan^{-1}(x+n) - \tan^{-1}x)$$

$$\Rightarrow \lim_{x\to\infty} \tan(f_n(x)) = \lim_{x\to\infty} \tan\left(\tan^{-1}\frac{n}{1+nx+x^2}\right)$$

$$= \lim_{x\to\infty} \frac{n}{1+nx+x^2} = 0$$

$\therefore$ (c) statement is false.

(d) $\lim_{x\to\infty} \sec^2(f_n(x)) = \lim_{x\to\infty} (1+\tan^2 f_n(x))$

$= 1 + \lim_{x\to\infty} \tan^2(f_n(x))$

$= 1 + 0 = 1$

$\therefore$ (d) statement is true

**2.** *(a, d)* It is given that $T$ is tangents to $S_1$ at $P$ and $S_2$ at $Q$ and $S_1$ and $S_2$ touch externally at $M$.

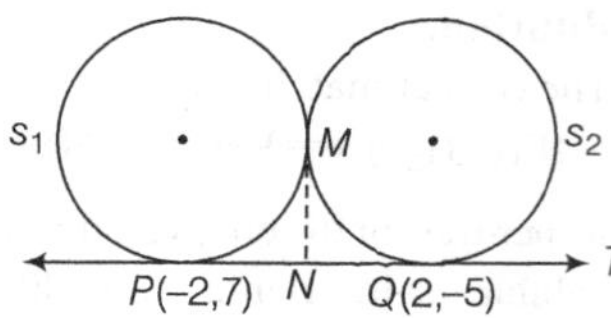


$\therefore \quad MN = NP = NQ$

$\therefore$ Locus of $M$ is a circle having $PQ$ as its diameter of circle

$\therefore$ Equation of circle

$(x-2)(x+2) + (y+5)(y-7) = 0$

$\Rightarrow \quad x^2 + y^2 - 2y - 39 = 0$

Hence,

$E_1 : x^2 + y^2 - 2y - 39 = 0, x \neq \pm 2$

Locus of mid-point of chord $(h,k)$ of the circle $E_1$ is

$xh + yk - (y+k) - 39 = h^2 + k^2 - 2k - 39$

$\Rightarrow \quad xh + yk - y - k = h^2 + k^2 - 2k$

Since, chord is passing through $(1, 1)$.

$\therefore$ Locus of mid-point of chord $(h, k)$ is

$h + k - 1 - k = h^2 + k^2 - 2k$

$\Rightarrow \quad h^2 + k^2 - 2k - h + 1 = 0$

Locus is $E_2 : x^2 + y^2 - x - 2y + 1 = 0$

Now, after checking options, (a) and (d) are correct.

**3.** *(a, d)* We have,

$$-x + 2y + 5z = b_1$$
$$2x - 4y + 3z = b_2$$
$$x - 2y + 2z = b_3$$

has at least one solution.

$$\therefore \quad D = \begin{vmatrix} -1 & 2 & 5 \\ 2 & -4 & 3 \\ 1 & -2 & 2 \end{vmatrix}$$

and $D_1 = D_2 = D_3 = 0$

$$\Rightarrow \quad D_1 = \begin{vmatrix} b_1 & 2 & 5 \\ b_2 & -4 & 3 \\ b_3 & -2 & 2 \end{vmatrix}$$

$$= -2b_1 - 14b_2 + 26b_3 = 0$$

$\Rightarrow b_1 + 7b_2 = 13b_3$ ...(i)

$$\text{(a) } D = \begin{vmatrix} 1 & 2 & 3 \\ 0 & 4 & 5 \\ 1 & 2 & 6 \end{vmatrix} = 1(24-10) + 1(10-12)$$

$$= 14 - 2 = 12 \neq 0$$

Here, $D \neq 0 \Rightarrow$ unique solution for any $b_1$, $b_2$, $b_3$.

$$\text{(b) } D = \begin{vmatrix} 1 & 1 & 3 \\ 5 & 2 & 6 \\ -2 & -1 & -3 \end{vmatrix}$$

$= 1(-6+6) - 1(-15+12)$
$+ 3(-5+4) = 0$

For atleast one solution

$D_1 = D_2 = D_3 = 0$

$$\text{Now, } D_1 = \begin{vmatrix} b_1 & 1 & 3 \\ b_2 & 2 & 6 \\ b_3 & -1 & -3 \end{vmatrix}$$

$= b_1(-6+6) - b_2(-3+3) + b_3(6-6)$
$= 0$

$$D_2 = \begin{vmatrix} 1 & b_1 & 3 \\ 5 & b_2 & 6 \\ -2 & b_3 & -3 \end{vmatrix}$$

$= -b_1(-15+12) + b_2(-3+6) - b_3(6-15)$

$= 3b_1 + 3b_2 + 9b_3 = 0 \Rightarrow b_1 + b_2 + 3b_3 = 0$
not satisfies the Eq. (i)

$\therefore$ It has no solution.

(c) $D = \begin{vmatrix} -1 & 2 & -5 \\ 2 & -4 & 10 \\ 1 & -2 & 5 \end{vmatrix}$

$= -1(-20 + 20) - 2(10 - 10) - 5(-4 + 4)$
$= 0$

Here, $b_2 = -2b_1$ and $b_3 = -b_1$ satisfies the Eq. (i)

Planes are parallel.

(d) $D = \begin{vmatrix} 1 & 2 & 5 \\ 2 & 0 & 3 \\ 1 & 4 & -5 \end{vmatrix}$

$= 1(0 - 12) - 2(-10 - 3) + 5(8 - 0) = 54$

$D \neq 0$

$\therefore$ It has unique solution for any $b_1$, $b_2$, $b_3$.

**4.** *(a, c)* We have,

Equations of circle

$$x^2 + y^2 = \frac{1}{2}$$

and Equations of parabola

$$y^2 = 4x$$

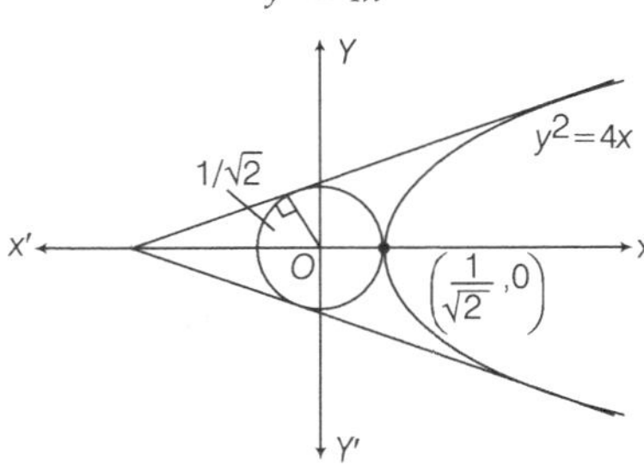


Let the equation of common tangent of parabola and circle is

$$y = mx + \frac{1}{m}$$

Since, radius of circle $= \frac{1}{\sqrt{2}}$

$$\therefore \quad \frac{1}{\sqrt{2}} = \left| \frac{0 + 0 + \frac{1}{m}}{\sqrt{1 + m^2}} \right|$$

$\Rightarrow m^4 + m^2 - 2 = 0$

$\Rightarrow \quad m = \pm 1$

$\therefore$ Equation of common tangents are

$$y = x + 1 \text{ and } y = -x - 1$$

Intesection point of common tangent at $Q(-1, 0)$

$\therefore$ Equation of ellipse

$$\frac{x^2}{1} + \frac{y^2}{1/2} = 1$$

where, $a^2 = 1$, $b^2 = 1/2$

Now, eccentricity

$$(e) = \sqrt{1 - \frac{b^2}{a^2}} = \sqrt{1 - \frac{1}{2}} = \frac{1}{\sqrt{2}}$$

and length of latusrectum

$$= \frac{2b^2}{a} = \frac{2\left(\frac{1}{2}\right)}{1} = 1$$

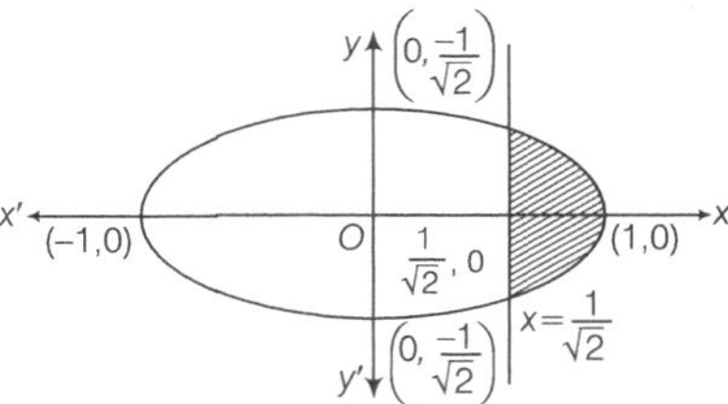


$\therefore$ Area of shaded region

$$= 2\int_{1/\sqrt{2}}^{1} \frac{1}{\sqrt{2}} \sqrt{1 - x^2}\, dx$$

$$= \sqrt{2}\left[\frac{x}{2}\sqrt{1 - x^2} + \frac{1}{2}\sin^{-1} x\right]_{1/\sqrt{2}}^{1}$$

$$= \sqrt{2}\left[\left(0 + \frac{\pi}{4}\right) - \left(\frac{1}{4} + \frac{\pi}{8}\right)\right]$$

$$= \sqrt{2}\left(\frac{\pi}{8} - \frac{1}{4}\right) = \frac{\pi - 2}{4\sqrt{2}}$$

**5.** *(a, c, d)* We have,

$$sz + t\bar{z} + r = 0 \qquad \text{...(i)}$$

On taking conjugate

$$\bar{s}\bar{z} + \bar{t}z + \bar{r} = 0 \qquad \text{...(ii)}$$

On solving Eqs. (i) and (ii), we get

$$z = \frac{\bar{r}t - r\bar{s}}{|s|^2 - |t|^2}$$

(a) For unique solutions of $z$

$|s|^2 - |t|^2 \neq 0 \Rightarrow |s| \neq |t|$

It is true

(b) If $|s| = |t|$, then $\bar{r}t - r\bar{s}$ may or may not be zero. So, $z$ may have no solutions.

$\therefore$ $L$ may be an empty set.

It is false.

(c) If elements of set $L$ represents line, then this line and given circle intersect at maximum two point. Hence, it is true.

(d) In this case locus of $z$ is a line, so $L$ has infinite elements. Hence, it is true.

**6.** *(b, c, d)* Given,

$$\lim_{t \to x} \frac{f(x) \sin t - f(t) \sin x}{t - x} = \sin^2 x$$

Using $L'$Hospital rules

$$\lim_{t \to x} \frac{f(x) \cos t - f'(t) \sin x}{1} = \sin^2 x$$

$$\Rightarrow f(x) \cos x - f'(x) \sin x = \sin^2 x$$

$$\Rightarrow f'(x) \sin x - f(x) \cos x = -\sin^2 x$$

$$\Rightarrow \frac{f'(x) \sin x - f(x) \cos x}{\sin^2 x} = -1$$

$$\Rightarrow \quad d\left(\frac{f(x)}{\sin x}\right) = -1$$

On integrating, we get

$$\frac{f(x)}{\sin x} = -x + C$$

$$\Rightarrow \quad f(x) = -x \sin x + C \sin x$$

It is given that $x = \frac{\pi}{6}, f\left(\frac{\pi}{6}\right) = -\frac{\pi}{12}$

$$\therefore \quad f\left(\frac{\pi}{6}\right) = -\frac{\pi}{6}\sin\frac{\pi}{6} + C\sin\frac{\pi}{6}$$

$$= -\frac{\pi}{12} = -\frac{\pi}{12} + \frac{1}{2}C$$

$\Rightarrow C = 0$

$\therefore \; f(x) = -x \sin x$

(a) $f(x) = -x \sin x$

$$f\left(\frac{\pi}{4}\right) = -\frac{\pi}{4}\sin\frac{\pi}{4} = -\frac{\pi}{4\sqrt{2}} \text{ false}$$

(b) $f(x) = -x \sin x$

$$\sin x > x - \frac{x^3}{6}, \forall x \in (0, \pi)$$

$$\Rightarrow \quad -x \sin x < -x^2 + \frac{x^4}{6}$$

$$\Rightarrow \quad f(x) < \frac{x^4}{6} - x^2, \forall x \in (0, \pi)$$

It is true

(c) $f(x) = -x \sin x$

$$f'(x) = -\sin x - x \cos x$$

$$f'(x) = 0$$

$$\Rightarrow -\sin x - x \cos x = 0$$

$$\tan x = -x$$

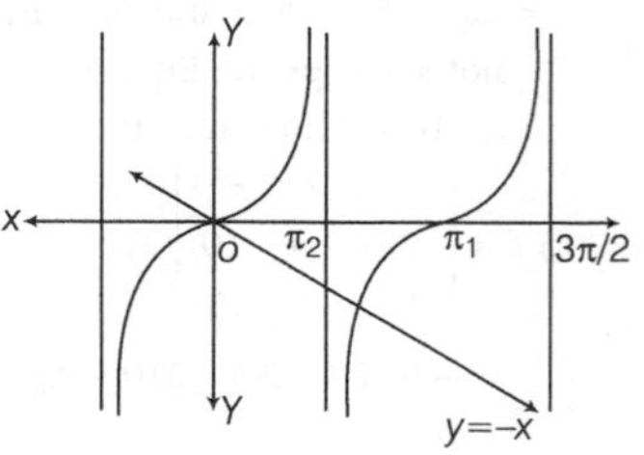


$\Rightarrow$ Their exists $\alpha \in (0, \pi)$ for which $f'(\alpha) = 0$

It is true

(d) $f(x) = -x \sin x$

$$f'(x) = -\sin x - x \cos x$$

$$f''(x) = -2 \cos x + x \sin x$$

$$f''\left(\frac{\pi}{2}\right) = \frac{\pi}{2}, \; f\left(\frac{\pi}{2}\right) = -\frac{\pi}{2}$$

$$\therefore \quad f''\left(\frac{\pi}{2}\right) + f\left(\frac{\pi}{2}\right) = 0$$

It is true.

**7.** *(2)* Let $I = \int_0^{1/2} \frac{1 + \sqrt{3}}{[(x+1)^2 (1-x)^6]^{1/4}} dx$

$$\Rightarrow \quad I = \int_0^{1/2} \frac{1 + \sqrt{3}}{(1 - x)^2 \left[\left(\frac{1 - x}{1 + x}\right)^6\right]^{1/4}} dx$$

Put $\frac{1-x}{1+x} = t \Rightarrow \frac{-2\,dx}{(1+x)^2} = dt$

when $x = 0, t = 1, x = \frac{1}{2}, t = \frac{1}{3}$

$$\therefore \quad I = \int_1^{1/3} \frac{(1 + \sqrt{3})\,dt}{-2(t)^{6/4}}$$

$$\Rightarrow I = \frac{-(1 + \sqrt{3})}{2}\left[\frac{-2}{\sqrt{t}}\right]_1^{1/3}$$

$$\Rightarrow \quad I = (1 + \sqrt{3})(\sqrt{3} - 1) \Rightarrow I = 3 - 1 = 2$$

**8.** *(4)* Let $\text{Det}(P) = \begin{vmatrix} a_1 & b_1 & c_1 \\ a_2 & b_2 & c_2 \\ a_3 & b_3 & c_3 \end{vmatrix}$

$$= a_1(b_2c_3 - b_3c_2) - a_2(b_1c_3 - b_3c_1) + a_3(b_1c_2 - b_2c_1)$$

Now, maximum value of Det $(P) = 6$

If $a_1 = 1, a_2 = -1, a_3 = 1,$

$$b_2c_3 = b_1c_3 = b_1c_2 = 1$$

and $\quad b_3c_2 = b_3c_1 = b_2c_1 = -1$

But it is not possible as

$$(b_2c_3)(b_3c_1)(b_1c_2) = -1$$

and $(b_1c_3)(b_3c_2)(b_2c_1) = 1$
i.e., $b_1b_2b_3c_1c_2c_3 = 1$ and $-1$
Similar contradiction occurs when
$a_1 = 1, a_2 = 1, a_3 = 1, b_2c_1 = b_3c_1 = b_1c_2 = 1$
and $b_3c_2 = b_1c_3 = b_1c_2 = -1$
Now, for value to be 5 one of the terms must be zero but that will make 2 terms zero which means answer cannot be 5

Now,

$$\begin{vmatrix} 1 & 1 & 1 \\ -1 & 1 & 1 \\ 1 & -1 & 1 \end{vmatrix} = 4$$

Hence, maximum value is 4.

**9.** *(119)* Given, $X$ has exactly 5 elements and $Y$ has exactly 7 elements.

$\therefore \quad n(X) = 5$

and $\quad n(Y) = 7$

Now, number of one-one functions from $X$ to $Y$ is

$$\alpha = {}^7P_5 = {}^7C_5 \times 5!$$

Number of onto functions from $Y$ to $X$ is $\beta$

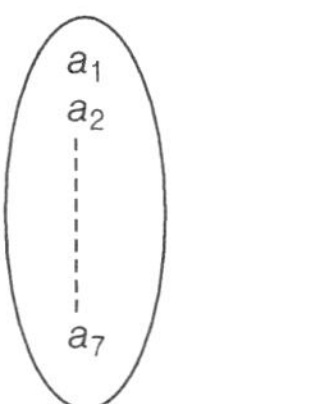


1, 1, 1, 1, 3 or 1, 1, 1, 2, 2

$$\therefore \quad \beta = \frac{7!}{3!\,4!} \times 5! + \frac{7!}{(2!)^3 3!} \times 5!$$

$$= ({}^7C_3 + 3\,{}^7C_3)\,5! = 4 \times {}^7C_3 \times 5!$$

$$\therefore \quad \frac{\beta - \alpha}{5!} = \frac{(4 \times {}^7C_3 - {}^7C_5)\,5!}{5!}$$

$$= 4 \times 35 - 21 = 140 - 21 = 119$$

**10.** *(0.4)* We have,

$$\frac{dy}{dx} = (2 + 5y)(5y - 2)$$

$$\Rightarrow \quad \frac{dy}{25y^2 - 4} = dx$$

$$\Rightarrow \quad \frac{1}{25}\left(\frac{dy}{y^2 - \frac{4}{25}}\right) = dx$$

On integrating both sides, we get

$$\frac{1}{25}\int \frac{dy}{y^2 - \left(\frac{2}{5}\right)^2} = \int dx$$

$$\Rightarrow \quad \frac{1}{25} \times \frac{1}{2 \times 2/5} \log\left|\frac{y - 2/5}{y + 2/5}\right| = x + C$$

$$\Rightarrow \quad \log\left|\frac{5y - 2}{5y + 2}\right| = 20(x + C)$$

$$\Rightarrow \quad \left|\frac{5y - 2}{5y + 2}\right| = Ae^{20x} \qquad [\because e^{20C} = A]$$

when $x = 0 \Rightarrow y = 0$, then $A = 1$

$$\therefore \quad \left|\frac{5y - 2}{5y + 2}\right| = e^{20x}$$

$$\lim_{x \to -\infty}\left|\frac{5f(x) - 2}{5f(x) + 2}\right| = \lim_{x \to -\infty} e^{20x}$$

$$\Rightarrow \lim_{n \to -\infty}\left|\frac{5f(x) - 2}{5f(x) + 2}\right| = 0$$

$$\Rightarrow \quad \lim_{n \to -\infty} 5f(x) - 2 = 0$$

$$\Rightarrow \quad \lim_{n \to -\infty} f(x) = \frac{2}{5} = 0.4$$

**11.** *(2)* Given,

$$f(x + y) = f(x)\,f'(y) + f'(x)\,f(y), \forall x, y \in R$$

and $f(0) = 1$

Put $x = y = 0$, we get

$$f(0) = f(0)\,f'(0) + f'(0)\,f(0)$$

$$\Rightarrow \quad 1 = 2f'(0) \Rightarrow f'(0) = \frac{1}{2}$$

Put $x = x$ and $y = 0$, we get

$$f(x) = f(x)\,f'(0) + f'(x)\,f(0)$$

$$\Rightarrow \quad f(x) = \frac{1}{2}f(x) + f'(x)$$

$$\Rightarrow \quad f'(x) = \frac{1}{2}f(x) \Rightarrow \frac{f'(x)}{f(x)} = \frac{1}{2}$$

On integrating, we get

$$\log f(x) = \frac{1}{2}x + C$$

$$\Rightarrow \quad f(x) = Ae^{\frac{1}{2}x}, \text{ where } e^C = A$$

If $f(0) = 1$, then $A = 1$

Hence, $\quad f(x) = e^{\frac{1}{2}x}$

$$\Rightarrow \quad \log_e f(x) = \frac{1}{2}x$$

$$\Rightarrow \quad \log_e f(4) = \frac{1}{2} \times 4 = 2$$

**12.** *(8)* Let $P(\alpha, \beta, \gamma)$ and $R$ is image of $P$ in the $XY$-plane.

$\therefore R(\alpha, \beta, -\gamma)$

Also, $Q$ is the image of $P$ in the plane $x + y = 3$

$$\therefore \quad \frac{x-\alpha}{1} = \frac{y-\beta}{1} = \frac{z-\gamma}{0} = \frac{-2(\alpha+\beta-3)}{2}$$

$$x = 3-\beta,\ y = 3-\alpha,\ z = \gamma$$

Since, $Q$ is lies on $Z$-axis

$\therefore \quad \beta = 3, \alpha = 3, z = \gamma$

$\therefore P(3, 3, \gamma)$

Given, distance of $P$ from $X$-axis be 5

$$\therefore \quad 5 = \sqrt{3^2 + \gamma^2}$$

$$25 - 9 = \gamma^2$$

$$\Rightarrow \quad \gamma = \pm 4$$

Then, $\quad PR = |2\gamma| = |2 \times 4| = 8$

**13.** *(0.5)* Here, $P(1, 0, 0)$, $Q(0, 1, 0)$, $R(0, 0, 1)$, $T = (1, 1, 1)$ and $S = \left(\frac{1}{2}, \frac{1}{2}, \frac{1}{2}\right)$.

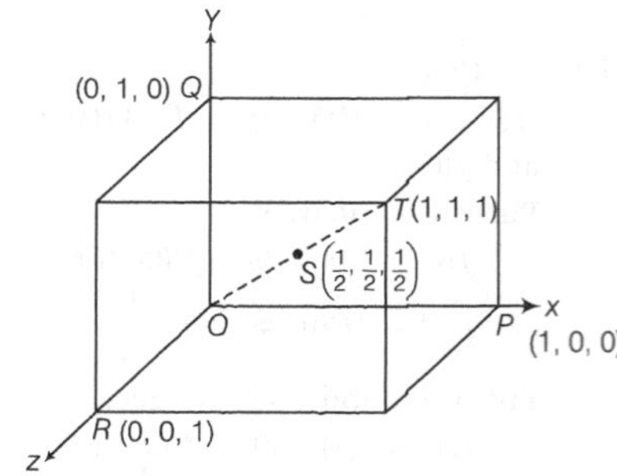


Now, $\vec{p} = \overrightarrow{SP} = \overrightarrow{OP} - \overrightarrow{OS}$

$$= \left(\frac{1}{2}\hat{i} - \frac{1}{2}\hat{j} - \frac{1}{2}\hat{k}\right) = \frac{1}{2}(\hat{i} - \hat{j} - \hat{k})$$

$$\vec{q} = \overrightarrow{SQ} = \frac{1}{2}(-\hat{i} + \hat{j} - \hat{k})$$

$$\vec{r} = \overrightarrow{SR} = \frac{1}{2}(-\hat{i} - \hat{j} + \hat{k})$$

and $\quad \vec{t} = \overrightarrow{ST} = \frac{1}{2}(\hat{i} + \hat{j} + \hat{k})$

$$\vec{p} \times \vec{q} = \frac{1}{4}\begin{vmatrix} \hat{i} & \hat{j} & \hat{k} \\ 1 & -1 & -1 \\ -1 & 1 & -1 \end{vmatrix} = \frac{1}{4}(2\hat{i} + 2\hat{j})$$

and $\vec{r} \times \vec{t} = \frac{1}{4}\begin{vmatrix} \hat{i} & \hat{j} & \hat{k} \\ -1 & -1 & 1 \\ 1 & 1 & 1 \end{vmatrix} = \frac{1}{4}(-2\hat{i} + 2\hat{j})$

Now, $(\vec{p} \times \vec{q}) \times (\vec{r} \times \vec{t}) = \frac{1}{16}\begin{vmatrix} \hat{i} & \hat{j} & \hat{k} \\ 2 & 2 & 0 \\ -2 & 2 & 0 \end{vmatrix}$

$$= \frac{1}{16}(8\hat{k}) = \frac{1}{2}\hat{k}$$

$$\therefore |(p \times q) \times (\vec{r} \times \vec{t})| = \left|\frac{1}{2}\hat{k}\right| = \frac{1}{2} = 0.5$$

**14.** *(646)* We have,

$$X = ({}^{10}C_1)^2 + 2({}^{10}C_2)^2 + 3({}^{10}C_3)^2 + \ldots + 10({}^{10}C_{10})^2$$

$$\Rightarrow X = \sum_{r=1}^{10} r({}^{10}C_r)^2$$

$$\Rightarrow X = \sum_{r=1}^{10} r\ {}^{10}C_r\ {}^{10}C_r$$

$$\Rightarrow X = \sum_{r=1}^{10} r \times \frac{10}{r}\ {}^{9}C_{r-1}\ {}^{10}C_r \left[\because {}^{n}C_r = \frac{n}{r}\ {}^{n-1}C_{r-1}\right]$$

$$\Rightarrow X = 10\sum_{r=1}^{10} {}^{9}C_{r-1}\ {}^{10}C_r$$

$$\Rightarrow X = 10\sum_{r=1}^{10} {}^{9}C_{r-1}\ {}^{10}C_{10-r} \quad [\because {}^{n}C_r = {}^{n}C_{n-r}]$$

$$\Rightarrow X = 10 \times {}^{19}C_9$$

$$[\because {}^{n-1}C_{r-1}\ {}^{n}C_{n-r} = {}^{2n-1}C_{n-1}]$$

Now, $\frac{1}{1430}X = \frac{10 \times {}^{19}C_9}{1430} = \frac{{}^{19}C_9}{143} = \frac{{}^{19}C_9}{11 \times 13}$

$$= \frac{19 \times 17 \times 16}{8} = 19 \times 34 = 646$$

**15.** *(a)* We have,

$$E_1 = \left\{x \in R : x \neq 1 \text{ and } \frac{x}{x-1} > 0\right\}$$

$$\therefore E_1 = \frac{x}{x-1} > 0$$

+    –    +
0    1

$$E_1 = x \in (-\infty, 0) \cup (1, \infty)$$

and $E_2 = \left\{\begin{matrix} x \in E_1 : \sin^{-1}\left(\log_e\left(\frac{x}{x-1}\right)\right) \\ \text{is a real number} \end{matrix}\right\}$

$$E_2 = -1 \le \log_e \frac{x}{x-1} \le 1$$

$$\Rightarrow \quad e^{-1} \le \frac{x}{x-1} \le e$$

Now, $\frac{x}{x-1} \ge e^{-1} \Rightarrow \frac{x}{x-1} - \frac{1}{e} \ge 0$

$$\Rightarrow \frac{ex - x + 1}{e(x-1)} \ge 0 \Rightarrow \frac{x(e-1)+1}{(x-1)e} \ge 0$$

+ – +

$\frac{-1}{e-1}$ 1

$$\Rightarrow \quad x \in \left(-\infty, \frac{1}{1-e}\right] \cup (1, \infty)$$

Also, $\frac{x}{x-1} \le e$

$$\Rightarrow \quad \frac{(e-1)x - e}{x-1} \ge 0$$

+ – +

1 $\frac{e}{e-1}$

$$\Rightarrow \quad x \in (-\infty, 1) \cup \left[\frac{e}{e-1}, \infty\right)$$

So, $E_2 = \left(-\infty, \frac{1}{1-e}\right] \cup \left[\frac{e}{e-1}, \infty\right)$

$\therefore$ The domain of $f$ and $g$ are

$$\left(-\infty, \frac{1}{1-e}\right] \cup \left[\frac{e}{e-1}, \infty\right)$$

and Range of $\frac{x}{x-1}$ is $R^+ - \{1\}$

$\Rightarrow$ Range of $f$ is $R - \{0\}$ or $(-\infty, 0) \cup (0, \infty)$

Range of $g$ is $\left[-\frac{\pi}{2}, \frac{\pi}{2}\right] - \{0\}$ or $\left[-\frac{\pi}{2}, 0\right) \cup \left(0, \frac{\pi}{2}\right]$

Now, $P \to 4$, $Q \to 2$, $R \to 1$, $S \to 1$

Hence, option (a) is correct answer.

**16.** (c) Given 6 boys $M_1, M_2, M_3, M_4, M_5, M_6$ and 5 girls $G_1, G_2, G_3, G_4, G_5$

(i) $\alpha_1 \to$ Total number of ways of selecting 3 boys and 2 girls from 6 boys and 5 girls.

i..e, ${}^6C_3 \times {}^5C_2 = 20 \times 10 = 200$

$\therefore \quad \alpha_1 = 200$

(ii) $\alpha_2 \to$ Total number of ways selecting at least 2 member and having equal number of boys and girls

i.e., ${}^6C_1{}^5C_1 + {}^6C_2{}^5C_2 + {}^6C_3{}^5C_3 + {}^6C_4{}^5C_4 + {}^6C_5{}^5C_5$

$= 30 + 150 + 200 + 75 + 6 = 461$

$\Rightarrow \quad \alpha_2 = 461$

(iii) $\alpha_3 \to$ Total number of ways of selecting 5 members in which at least 2 of them girls

i.e., ${}^5C_2{}^6C_3 + {}^5C_3{}^6C_2 + {}^5C_4{}^6C_1 + {}^5C_5{}^6C_0$

$= 200 + 150 + 30 + 1 = 381$

$\alpha_3 = 381$

(iv) $\alpha_4 \to$ Total number of ways of selecting 4 members in which at least two girls such that $M_1$ and $G_1$ are not included together.

$G_1$ is included $\to$

${}^4C_1 \cdot {}^5C_2 + {}^4C_2 \cdot {}^5C_1 + {}^4C_3$

$= 40 + 30 + 4 = 74$

$M_1$ is included $\to$

${}^4C_2 \cdot {}^5C_1 + {}^4C_3 = 30 + 4 = 34$

$G_1$ and $M_1$ both are not included

${}^4C_4 + {}^4C_3 \cdot {}^5C_1 + {}^4C_2 \cdot {}^5C_2$

$1 + 20 + 60 = 81$

$\therefore$ Total number $= 74 + 34 + 81 = 189$

$\alpha_4 = 189$

Now, $P \to 4$; $Q \to 6$; $R \to 5$; $S \to 2$

Hence, option (c) is correct.

**17.** (b) We have,

Equation of hyperbola

$$\frac{x^2}{a^2} - \frac{y^2}{b^2} = 1$$

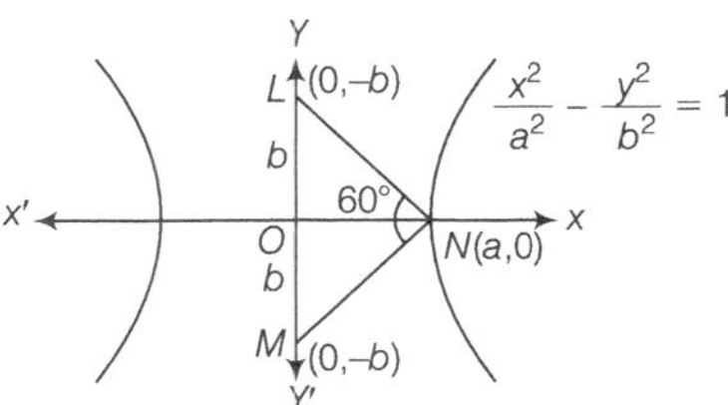


It is given,

$$\angle LNM = 60°$$

and Area of $\Delta LMN = 4\sqrt{3}$

Now, $\Delta LNM$ is an equilateral triangle whose sides is $2b$ $\quad [\because \Delta LON \cong \Delta MOL;$
$\therefore \quad \angle NLO = \angle NMO = 60°]$

$\therefore$ Area of $\Delta LMN = \frac{\sqrt{3}}{4}(2b)^2$

$\Rightarrow \quad 4\sqrt{3} = \sqrt{3}b^2 \Rightarrow b = 2$

Also, area of $\Delta LMN = \frac{1}{2}a(2b) = ab$

$\Rightarrow \quad 4\sqrt{3} = a(2)$

$\Rightarrow \quad a = 2\sqrt{3}$

(P) Length of conjugate axis = $2b = 2(2) = 4$

(Q)

Eccentricity $(e) = \sqrt{1 + \frac{b^2}{a^2}} = \sqrt{1 + \frac{4}{12}}$

$= \frac{4}{2\sqrt{3}} = \frac{2}{\sqrt{3}}$

(R) Distance between the foci

$= 2ae = 2 \times 2\sqrt{3} \times \frac{2}{\sqrt{3}} = 8$

(S) The length of latusrectum

$= \frac{2b^2}{a} = \frac{2(4)}{2\sqrt{3}} = \frac{4}{\sqrt{3}}$

$P \to 4; Q \to 3; R \to 1; S \to 2$

Hence, option (b) is correct.

**18.** *(d)*

(i) Given,

$f_1 : R \to R$ and $f_1(x) = \sin(\sqrt{1 - e^{-x^2}})$

$\therefore$ $f_1(x)$ is continuous at $x = 0$

Now,

$$f_1'(x) = \cos\sqrt{1 - e^{-x^2}} \cdot \frac{1}{2\sqrt{1 - e^{-x^2}}}(2xe^{-x^2})$$

At $x = 0$

$f_1'(x)$ does not exists.

$\therefore$ $f_1(x)$ is not differential at $x = 0$

Hence, option (2) for $P$.

(ii) Given, $f_2(x) = \begin{cases} \frac{|\sin x|}{\tan^{-1} x}, & \text{if } x \neq 0 \\ 1, & \text{if } x = 0 \end{cases}$

$$\Rightarrow \quad f_2(x) = \begin{cases} \frac{-\sin x}{\tan^{-1} x} & x < 0 \\ \frac{\sin x}{\tan^{-1} x} & x > 0 \\ 1 & x = 0 \end{cases}$$

Clearly, $f_2(x)$ is not continuous at $x = 0$.

$\therefore$ Option (1) for $Q$.

(iii) Given, $f_3(x) = [\sin(\log_e(x + 2))]$, where [ ] is G.I.F.

and $f_3 : (-1, e^{\pi/2} - 2) \to R$

It is given

$$-1 < x < e^{\pi/2} - 2$$

$\Rightarrow \quad -1 + 2 < x + 2 < e^{\pi/2} - 2 + 2$

$\Rightarrow \quad 1 < x + 2 < e^{\pi/2}$

$\Rightarrow \quad \log_e 1 < \log_e(x + 2) < \log_e e^{\pi/2}$

$\Rightarrow \quad 0 < \log_e(x + 2) < \frac{\pi}{2}$

$\Rightarrow \quad \sin 0 < \sin\log_e(x + 2) < \sin\frac{\pi}{2}$

$\Rightarrow \quad 0 < \sin\log_e(x + 2) < 1$

$\therefore \quad [\sin\log_e(x + 2)] = 0$

$\therefore \quad f_3(x) = 0, f'_3(x) = f_3''(x) = 0$

It is differentiable and continuous at $x = 0$.

$\therefore$ Option (4) for $R$

(iv) Given, $f_4(x) = \begin{cases} x^2\sin\left(\frac{1}{x}\right), & \text{if } x \neq 0 \\ 0, & \text{if } x = 0 \end{cases}$

Now, $\lim_{x \to 0} f_4(x) = \lim_{x \to 0} x^2\sin\left(\frac{1}{x}\right) = 0$

$$f_4'(x) = 2x\sin\left(\frac{1}{x}\right) - \cos\left(\frac{1}{x}\right)$$

For $x = 0$, $f_4'(x) = \lim_{h \to 0}\frac{f(0 + h) - f(0)}{h}$

$\Rightarrow f_4'(x) = \lim_{h \to 0}\frac{h^2\sin\left(\frac{1}{h}\right) - 0}{h}$

$\Rightarrow f_4'(x) = \lim_{h \to 0} h\sin\left(\frac{1}{h}\right) = 0$

Thus,

$$f_4'(x) = \begin{cases} 2x\sin\left(\frac{1}{x}\right) - \cos\left(\frac{1}{x}\right), & x \neq 0 \\ 0, & x = 0 \end{cases}$$

Again, $\lim_{x \to 0}$

$f'(x) = \lim_{x \to 0}\left(2x\sin\left(\frac{1}{x}\right) - \cos\left(\frac{1}{x}\right)\right)$

does not exists.

Since, $\lim_{x \to 0}\cos\left(\frac{1}{x}\right)$ does not exists.

Hence, $f'(x)$ is not continuous at $x = 0$.

$\therefore$ Option (3) for $S$.

# SOLVED PAPER 2017

# JEE Main

### Instructions

1. This test consists of 90 questions.
2. There are three parts in the question paper consisting of Physics, Chemistry & Mathematics having 30 questions in each part of equal weightage. Each question is allotted 4 marks for correct response.
3. Candidates will be awarded marks as stated above in instruction no. 2 for correct response of each question. 1 mark will be deducted for indicating incorrect response of each question. No deduction from the total score will be made if no response is indicated for an item in the answer sheet.
4. There is only one correct response for each question. Filling up more than one response in any question will be treated as wrong response and marks for wrong response will be deducted according as per instructions.

## Physics

1. An observer is moving with half the speed of light towards a stationary microwave source emitting waves at frequency 10 GHz. What is the frequency of the microwave measured by the observer? (speed of light $= 3 \times 10^8 \text{ ms}^{-1}$)

(a) 12.1 GHz

(b) 17.3 GHz

(c) 15.3 GHz

(d) 10.1 GHz

2. The following observations were taken for determining surface tension $T$ of water by capillary method. Diameter of capillary, $d = 1.25 \times 10^{-2}$ m rise of water, $h = 1.45 \times 10^{-2}$m. Using $g = 9.80$ m/s$^2$ and the simplified relation $T = \frac{rhg}{2} \times 10^3 \text{N/m}$, the possible error in surface tension is closest to

(a) 1.5% (b) 2.4%

(c) 10% (d) 0.15%

3. Some energy levels of a molecule are shown in the figure. The ratio of the wavelengths $r = \lambda_1 / \lambda_2$ is given by

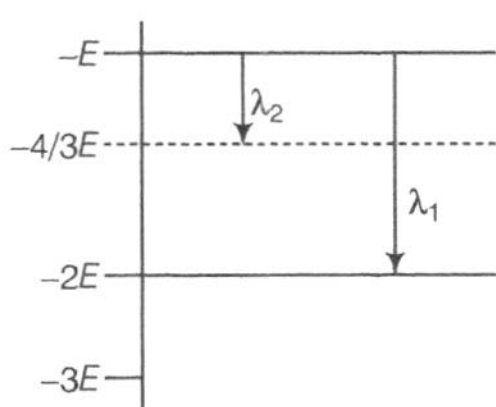


(a) $r = \frac{2}{3}$ (b) $r = \frac{3}{4}$

(c) $r = \frac{1}{3}$ (d) $r = \frac{4}{3}$

4. A body of mass $m = 10^{-2}$kg is moving in a medium and experiences a frictional force $F = -kv^2$. Its initial speed is $v_0 = 10$ ms$^{-1}$. If, after 10 s, its energy is $\frac{1}{8} mv_0^2$, the value of $k$ will be

(a) $10^{-3}$kgs$^{-1}$ (b) $10^{-4}$kgm$^{-1}$

(c) $10^{-1}$kgm$^{-1}$s$^{-1}$ (d) $10^{-3}$kgm$^{-1}$

5. $C_p$ and $C_V$ are specific heats at constant pressure and constant volume, respectively. It is observed that $C_p - C_V = a$ for hydrogen gas $C_p - C_V = b$ for nitrogen gas. The correct relation between $a$ and $b$ is

(a) $a = b$ (b) $a = 14b$

(c) $a = 28b$ (d) $a = \frac{1}{14}b$

6. The moment of inertia of a uniform cylinder of length $l$ and radius $R$ about its perpendicular bisector is $I$. What is the ratio $l/R$ such that the moment of inertia is minimum?

(a) $\frac{\sqrt{3}}{2}$ (b) 1 (c) $\frac{3}{\sqrt{2}}$ (d) $\sqrt{\frac{3}{2}}$

7. A radioactive nucleus $A$ with a half-life $T$, decays into a nucleus $B$. At $t = 0$, there is no nucleus $B$. After sometime $t$, the ratio of the number of $B$ to that of $A$ is 0.3. Then, $t$ is given by

(a) $t = T\frac{\log 1.3}{\log_e 2}$ (b) $t = T \log 1.3$

(c) $t = \frac{T}{\log 1.3}$ (d) $t = \frac{T \log_e 2}{2 \log 1.3}$

8. Which of the following statements is false?

(a) In a balanced Wheatstone bridge, if the cell and the galvanometer are exchanged, the null point is disturbed

(b) A rheostat can be used as a potential divider

(c) Kirchhoff's second law represents energy conservation

(d) Wheatstone bridge is the most sensitive when all the four resistances are of the same order of magnitude

9. A capacitance of 2 μF is required in an electrical circuit across a potential difference of 1kV. A large number of 1 μF capacitors are available which can withstand a potential difference of not more than 300 V. The minimum number of capacitors required to achieve this is

(a) 16 (b) 24 (c) 32 (d) 2

10. In the given circuit diagram, when the current reaches steady state in the circuit, the charge on the capacitor of capacitance $C$ will be

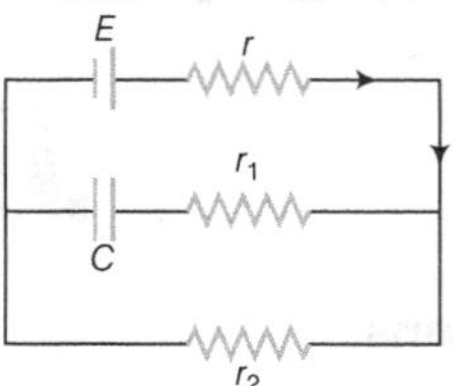


(a) $CE\frac{r_1}{(r_2 + r)}$

(b) $CE\frac{r_2}{(r + r_2)}$

(c) $CE\frac{r_1}{(r_1 + r)}$

(d) $CE$

11. In the below circuit, the current in each resistance is

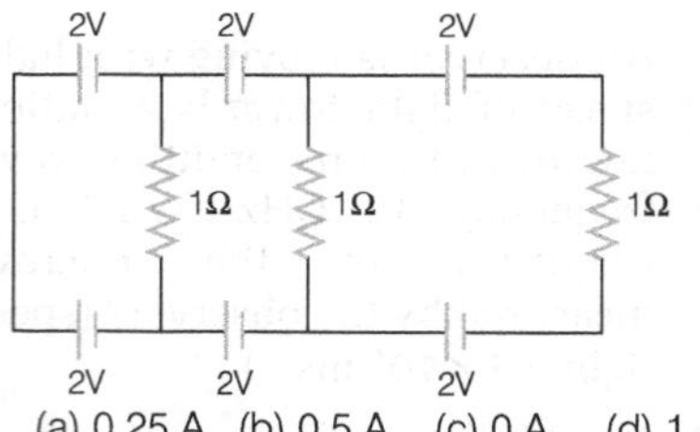


(a) 0.25 A (b) 0.5 A (c) 0 A (d) 1 A

12. In amplitude modulation, sinusoidal carrier frequency used is denoted by $\omega_c$ and the signal frequency is denoted by $\omega_m$. The bandwidth $(\Delta\omega_m)$ of the signal is such that $\Delta\omega_m << \omega_c$. Which of the following frequencies is not contained in the modulated wave?

(a) $\omega_c$ (b) $\omega_m + \omega_c$

(c) $\omega_c - \omega_m$ (d) $\omega_m$

13. In a common emitter amplifier circuit using an $n$-$p$-$n$ transistor, the phase difference between the input and the output voltages will be

(a) 90° (b) 135°

(c) 180° (d) 45°

14. A copper ball of mass 100 g is at a temperature $T$. It is dropped in a copper calorimeter of mass 100 g, filled with 170 g of water at room temperature. Subsequently, the temperature of the system is found to be 75°C. $T$ is (Given, room temperature = 30°C, specific heat of copper = 0.1 cal/g°C)

(a) 885°C (b) 1250°C (c) 825°C (d) 800°C

15. In a Young's double slit experiment, slits are separated by 0.5 mm and the screen is placed 150 cm away. A beam of light consisting of two wavelengths, 650 nm and 520 nm, is used to obtain interference fringes on the screen. The least distance from the common central maximum to the point where the bright fringes due to both the wavelengths coincide, is

(a) 7.8 mm (b) 9.75 mm
(c) 15.6 mm (d) 1.56 mm

16. An electric dipole has a fixed dipole moment $\mathbf{p}$, which makes angle $\theta$ with respect to $X$-axis. When subjected to an electric field $\mathbf{E}_1 = E\hat{\mathbf{i}}$, it experiences a torque $\mathbf{T}_1 = \tau\hat{\mathbf{k}}$. When subjected to another electric field $\mathbf{E}_2 = \sqrt{3}E_1\hat{\mathbf{j}}$, it experiences a torque $\mathbf{T}_2 = -\mathbf{T}_1$. The angle $\theta$ is

(a) 45° (b) 60° (c) 90° (d) 30°

17. A slender uniform rod of mass $M$ and length $l$ is pivoted at one end so that it can rotate in a vertical plane (see the figure). There is negligible friction at the pivot. The free end is held vertically above the pivot and then released. The angular acceleration of the rod when it makes an angle $\theta$ with the vertical, is

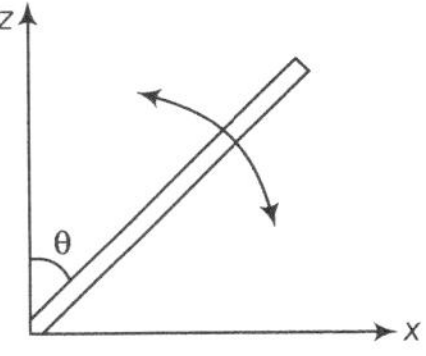


(a) $\frac{2g}{3l}\sin\theta$ (b) $\frac{3g}{2l}\cos\theta$
(c) $\frac{2g}{3l}\cos\theta$ (d) $\frac{3g}{2l}\sin\theta$

18. An external pressure $P$ is applied on a cube at 0°C so that it is equally compressed from all sides. $K$ is the bulk modulus of the material of the cube and $\alpha$ is its coefficient of linear expansion. Suppose we want to bring the cube to its original size by heating. The temperature should be raised by

(a) $\frac{P}{\alpha K}$ (b) $\frac{3\alpha}{PK}$ (c) $3PK\alpha$ (d) $\frac{P}{3\alpha K}$

19. A diverging lens with magnitude of focal length 25 cm is placed at a distance of 15 cm from a converging lens of magnitude of focal length 20 cm. A beam of parallel light falls on the diverging lens. The final image formed is

(a) virtual and at a distance of 40 cm from convergent lens
(b) real and at a distance of 40 cm from the divergent lens
(c) real and at a distance of 6 cm from the convergent lens
(d) real and at a distance of 40 cm from convergent lens

20. An electron beam is accelerated by a potential difference $V$ to hit a metallic target to produce $X$-rays. It produces continuous as well as characteristic $X$-rays. If $\lambda_{min}$ is the smallest possible wavelength of $X$-rays in the spectrum, the variation of $\log \lambda_{min}$ with $\log V$ is correctly represented in

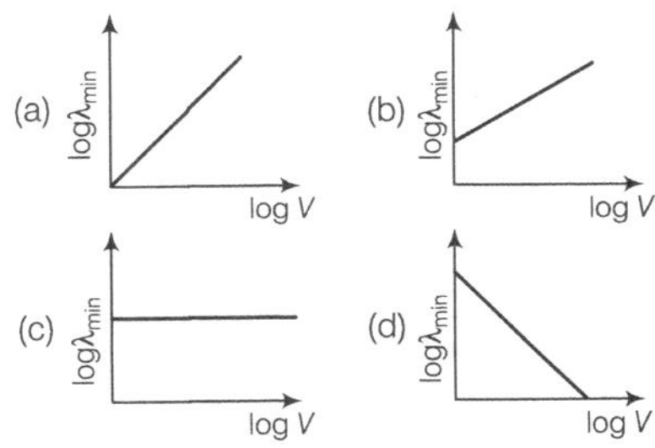

21. The temperature of an open room of volume 30 $m^3$ increases from 17°C to 27°C due to the sunshine. The atmospheric pressure in the room remains $1 \times 10^5$ Pa. If $n_i$ and $n_f$ are the number of molecules in the room before and after heating, then $n_f - n_i$ will be

(a) $1.38 \times 10^{23}$

(b) $2.5 \times 10^{25}$

(c) $-2.5 \times 10^{25}$

(d) $-1.61 \times 10^{23}$

22. In a coil of resistance 100 Ω, a current is induced by changing the magnetic flux through it as shown in the figure. The magnitude of change in flux through the coil is

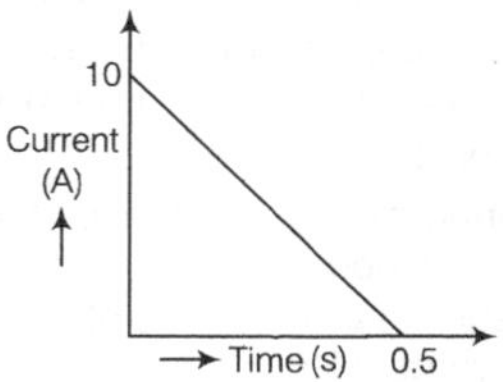


(a) 225 Wb (b) 250 Wb

(c) 275 Wb (d) 200 Wb

23. When a current of 5 mA is passed through a galvanometer having a coil of resistance 15 Ω, it shows full scale deflection. The value of the resistance to be put in series with the galvanometer to convert it into a voltmeter of range 0-10 V is

(a) $2.045 \times 10^3$ Ω

(b) $2.535 \times 10^3$ Ω

(c) $4.005 \times 10^3$ Ω

(d) $1.985 \times 10^3$ Ω

24. A time dependent force $F = 6t$ acts on a particle of mass 1 kg. If the particle starts from rest, the work done by the force during the first 1 s will be

(a) 22 J (b) 9 J

(c) 18 J (d) 4.5 J

25. A magnetic needle of magnetic moment $6.7 \times 10^{-2}$ $Am^2$ and moment of inertia $7.5 \times 10^{-6}$ kg $m^2$ is performing simple harmonic oscillations in a magnetic field of 0.01 T. Time taken for 10 complete oscillations is

(a) 8.89 s (b) 6.98 s

(c) 8.76 s (d) 6.65 s

26. The variation of acceleration due to gravity $g$ with distance $d$ from centre of the Earth is best represented by ($R$ = Earth's radius)

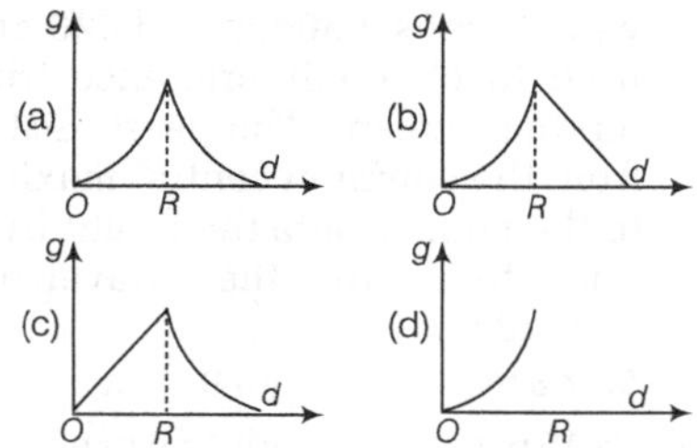


27. A body is thrown vertically upwards. Which one of the following graphs correctly represent the velocity *vs* time?

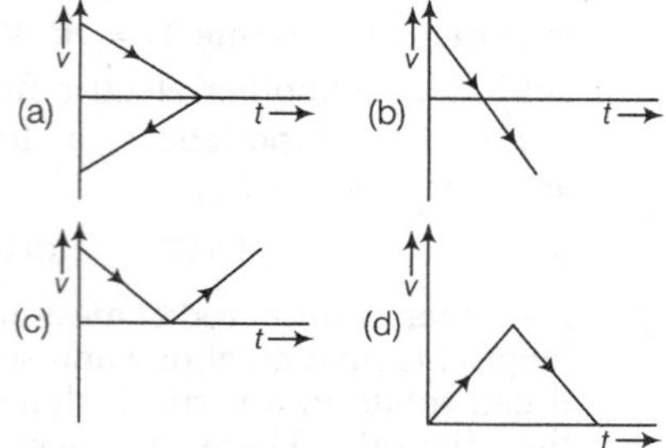


28. A particle $A$ of mass $m$ and initial velocity $v$ collides with a particle $B$ of mass $\frac{m}{2}$ which is at rest. The collision is head on, and elastic. The ratio of the de-Broglie wavelengths $\lambda_A$ to $\lambda_B$ after the collision is

(a) $\frac{\lambda_A}{\lambda_B} = 2$ (b) $\frac{\lambda_A}{\lambda_B} = \frac{2}{3}$

(c) $\frac{\lambda_A}{\lambda_B} = \frac{1}{2}$ (d) $\frac{\lambda_A}{\lambda_B} = \frac{1}{3}$

29. A particle is executing simple harmonic motion with a time period $T$. At time $t = 0$, it is at its position of equilibrium. The kinetic energy-time graph of the particle will look, like

(a)
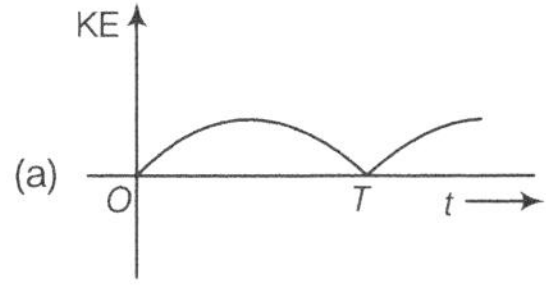


(b)
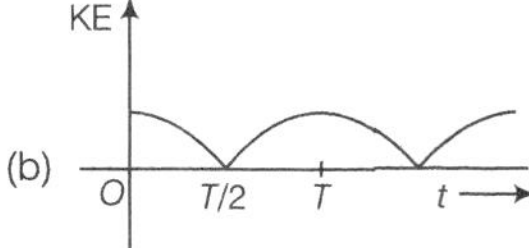


(c)
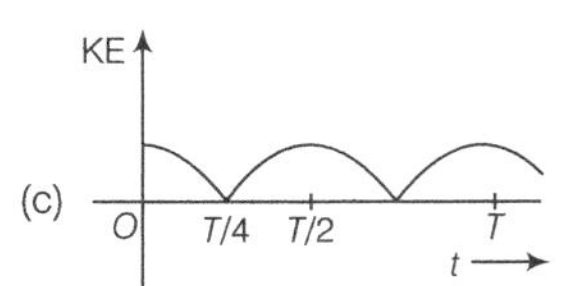


(d)
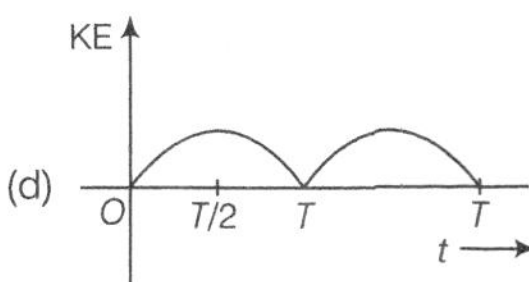


30. A man grows into a giant such that his linear dimensions increase by a factor of 9. Assuming that his density remains same, the stress in the leg will change by a factor of

(a) $\frac{1}{9}$ (b) 81 (c) $\frac{1}{81}$ (d) 9

# Chemistry

31. Which of the following compounds will give significant amount of *meta*-product during mononitration reaction?

(a)
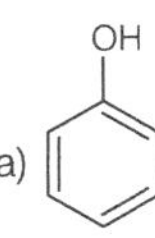


(b)
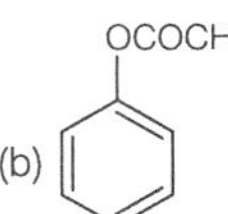


(c)
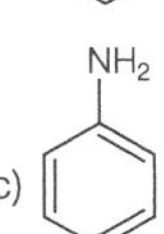


(d) NHCOCH$_3$

32. $\Delta U$ is equal to

(a) isochoric work (b) isobaric work
(c) adiabatic work (d) isothermal work

33. The increasing order of reactivity of the following halides for the $S_N1$ reaction is

I. $CH_3CH(Cl)CH_2CH_3$
II. $CH_3CH_2CH_2Cl$
III. $p$-$H_3CO$— $C_6H_4$— $CH_2Cl$

(a) (III) < (II) < (I) (b) (II) < (I) < (III)
(c) (I) < (III) < (II) (d) (II) < (III) < (I)

34. The radius of the second Bohr orbit for hydrogen atom is (Planck's constant $(h) = 6.6262 \times 10^{-34}$ Js; mass of electron $= 9.1091 \times 10^{-31}$ kg ; charge of electron $(e) = 1.60210 \times 10^{-19}$ C; permitivity of vacuum $(\epsilon_0) = 8.854185 \times 10^{-12}$ $kg^{-1}m^{-3}A^2$)

(a) 1.65 Å
(b) 4.76 Å
(c) 0.529 Å
(d) 2.12 Å

35. $pK_a$ of a weak acid ($HA$) and $pK_b$ of a weak base ($BOH$) are 3.2 and 3.4, respectively. The pH of their salt ($AB$) solution is

(a) 7.2 (b) 6.9 (c) 7.0 (d) 1.0

36. The formation of which of the following polymers involves hydrolysis reaction?

(a) Nylon-6 (b) Bakelite
(c) Nylon-6, 6 (d) Terylene

37. The most abundant elements by mass in the body of a healthy human adult are Oxygen (61.4%); Carbon (22.9%), Hydrogen (10.0 %); and Nitrogen (2.6%). The weight which a 75 kg person would gain if all $^1H$ atoms are replaced by $^2H$ atoms is

(a) 15 kg (b) 37.5 kg (c) 7.5 kg (d) 10 kg

38. Which of the following, upon treatment with *tert*-BuONa followed by addition of bromine water, fails to decolourise the colour of bromine?

(a) 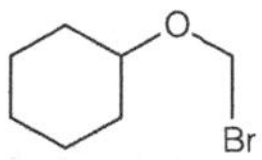 O Br (b) $C_6H_5$ Br

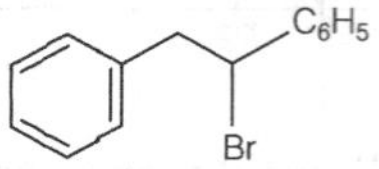

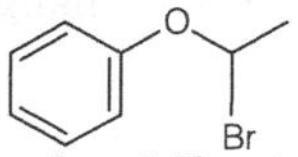

(c) O Br (d) 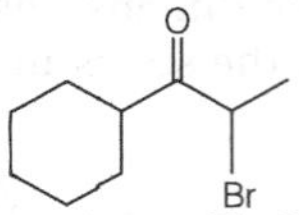 O Br

39. In the following reactions, ZnO is respectively acting as a/an

(i) $ZnO + Na_2O \longrightarrow Na_2ZnO_2$

(ii) $ZnO + CO_2 \longrightarrow ZnCO_3$

(a) base and acid (b) base and base

(c) acid and acid (d) acid and base

40. Both lithium and magnesium display several similar properties due to the diagonal relationship; however, the one which is incorrect is

(a) Both form basic carbonates

(b) Both form soluble bicarbonates

(c) Both form nitrides

(d) nitrates of both Li and Mg yield $NO_2$ and $O_2$ on heating

41. 3-methyl-pent-2-ene on reaction with HBr in presence of peroxide forms an addition product. The number of possible stereoisomers for the product is

(a) six (b) zero (c) two (d) four

42. A metal crystallises in a face centred cubic structure. If the edge length of its unit cell is '*a*', the closest approach between two atoms in metallic crystal will be

(a) $2a$ (b) $2\sqrt{2}a$ (c) $\sqrt{2}a$ (d) $\frac{a}{\sqrt{2}}$

43. Two reactions $R_1$ and $R_2$ have identical pre- exponential factors. Activation energy of $R_1$ exceeds that of $R_2$ by 10 kJ $mol^{-1}$. If $k_1$ and $k_2$ are rate constants for reactions $R_1$ and $R_2$, respectively at 300 K, then $\ln\left(\frac{k_2}{k_1}\right)$ is equal to

$(R = 8.314 \text{ J mol}^{-1}\text{K}^{-1})$

(a) 8 (b) 12 (c) 6 (d) 4

44. The correct sequence of reagents for the following conversion will be

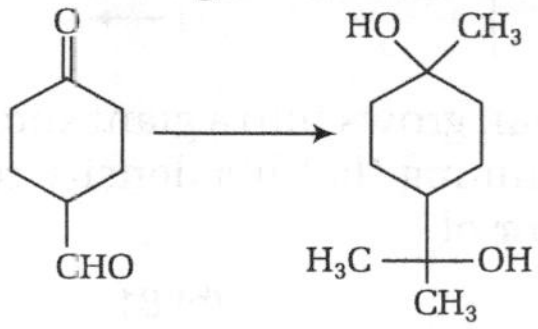

O, CHO ⟶ HO CH₃, $H_3C$—OH, $CH_3$

(a) $[Ag(NH_3)_2]^+ OH^-$, $H^+/CH_3OH$, $CH_3MgBr$

(b) $CH_3MgBr$, $H^+/CH_3OH$, $[Ag(NH_3)_2]^+ OH^-$

(c) $CH_3MgBr$, $[Ag(NH_3)_2]^+ OH^-$, $H^+/CH_3OH$

(d) $[Ag(NH_3)_2]^+ OH^-$, $CH_3MgBr$, $H^+/CH_3OH$

45. The Tyndall effect is observed only when following conditions are satisfied

1. The diameter of the dispersed particles is much smaller than the wavelength of the light used.
2. The diameter of the dispersed particle is not much smaller than the wavelength of the light used.
3. The refractive indices of the dispersed phase and dispersion medium are almost similar in magnitude.
4. The refractive indices of the dispersed phase and dispersion medium differ greatly in magnitude.

(a) 1 and 4 (b) 2 and 4

(c) 1 and 3 (d) 2 and 3

46. Which of the following compounds will behave as a reducing sugar in an aqueous KOH solution?

(a)
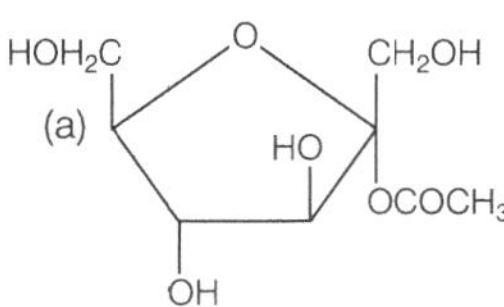


(b)
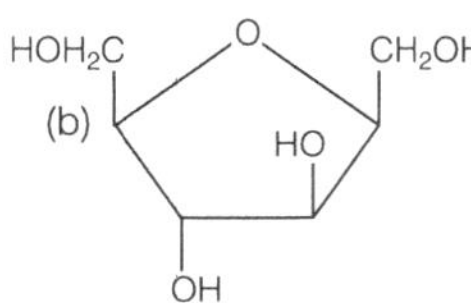


(c) HOH$_2$C, O, CH$_2$OH, HO, OCH$_3$, OH

(d) HOH$_2$C, O, CH$_2$OCH$_3$, OH, OH, OH

47. Given,

$C_{(graphite)} + O_2(g) \longrightarrow CO_2(g);$
$\Delta_r H° = -393.5 \text{ kJ mol}^{-1}$

$H_2(g) + \frac{1}{2}O_2(g) \longrightarrow H_2O(l);$

$\Delta_r H° = -285.8 \text{ kJ mol}^{-1}$

$CO_2(g) + 2H_2O(l) \longrightarrow CH_4(g) + 2O_2(g);$

$\Delta_r H° = +890.3 \text{ kJ mol}^{-1}$

Based on the above thermochemical equations, the value of $\Delta_r H°$ at 298 K for the reaction,

$C_{(graphite)} + 2H_2(g) \longrightarrow CH_4(g)$ will be

(a) $+78.8 \text{ kJ mol}^{-1}$ (b) $+144.0 \text{ kJ mol}^{-1}$

(c) $-74.8 \text{ kJ mol}^{-1}$ (d) $-144.0 \text{ kJ mol}^{-1}$

48. Which of the following reactions is an example of a redox reaction?

(a) $XeF_4 + O_2F_2 \longrightarrow XeF_6 + O_2$

(b) $XeF_2 + PF_5 \longrightarrow [XeF]^+PF_6^-$

(c) $XeF_6 + H_2O \longrightarrow XeOF_4 + 2HF$

(d) $XeF_6 + 2H_2O \longrightarrow XeO_2F_2 + 4HF$

49. The products obtained when chlorine gas reacts with cold and dilute aqueous NaOH are

(a) $ClO^-$ and $ClO_3^-$ (b) $ClO_2^-$ and $ClO_3^-$

(c) $Cl^-$ and $ClO^-$ (d) $Cl^-$ and $ClO_2^-$

50. The major product obtained in the following reaction is

$C_6H_5CH(Br)(H)CH_2C_6H_5 \xrightarrow[\Delta]{^tBuOK}$

(a) $(\pm)$ $C_6H_5CH(O^tBu)CH_2C_6H_5$

(b) $C_6H_5CH = CHC_6H_5$

(c) $(+)$ $C_6H_5CH(O^tBu)CH_2C_6H_5$

(d) $(-)C_6H_5CH(O^tBu)CH_2C_6H_5$

51. Sodium salt of an organic acid '$X$' produces effervescence with conc. $H_2SO_4$. '$X$' reacts with the acidified aqueous $CaCl_2$ solution to give a white precipitate which decolourises acidic solution of $KMnO_4$.'$X$' is

(a) $C_6H_5COONa$ (b) $HCOONa$

(c) $CH_3COONa$ (d) $Na_2C_2O_4$

52. Which of the following species is not paramagnetic?

(a) NO (b) CO

(c) $O_2$ (d) $B_2$

53. The freezing point of benzene decreases by 0.45°C when 0.2 g of acetic acid is added to 20 g of benzene. If acetic acid associates to form a dimer in benzene, percentage association of acetic acid in benzene will be ($K_f$ for benzene = 5.12 K kg mol$^{-1}$)

(a) 64.6 % (b) 80.4 %

(c) 74.6 % (d) 94.6 %

54. Which of the following molecules is least resonance stabilised?

(a) 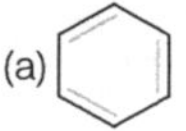 (b) 

(c) 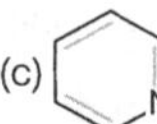 (d)

55. On treatment of 100 mL of 0.1 M solution of $CoCl_3.6H_2O$ with excess of $AgNO_3$; $1.2 \times 10^{22}$ ions are precipitated. The complex is

(a) $[Co(H_2O)_4Cl_2]\ Cl.2H_2O$
(b) $[Co(H_2O)_3Cl_3].3H_2O$
(c) $[Co(H_2O)_6]Cl_3$
(d) $[Co(H_2O)_5Cl]\ Cl_2.H_2O$

56. The major product obtained in the following reaction is

DIBAL-H

COOH

(a) OH CHO COOH (b) OH CHO CHO

(c) CHO COOH (d) CHO CHO

57. A water sample has ppm level concentration of following anions

$F^- = 10;\ SO_4^{2-} = 100;\ NO_3^- = 50$

the anion/anions that make/makes the water sample unsuitable for drinking is/are

(a) Only $NO_3^-$ (b) Both $SO_4^{2-}$ and $NO_3^-$
(c) Only $F^-$ (d) Only $SO_4^{2-}$

58. 1 g of a carbonate $(M_2CO_3)$ on treatment with excess HCl produces 0.01186 mole of $CO_2$. The molar mass of $M_2CO_3$ in g $mol^{-1}$ is

(a) 1186 (b) 84.3 (c) 118.6 (d) 11.86

59. Given, $E^\circ_{Cl_2/Cl^-} = 1.36$ V,

$E^\circ_{Cr^{3+}/Cr} = -0.74$ V

$E^\circ_{Cr_2O_7^{2-}/Cr^{3+}} = 1.33$ V,

$E^\circ_{MnO_4^-/Mn^{2+}} = 1.51$ V

Among the following, the strongest reducing agent is

(a) Cr (b) $Mn^{2+}$ (c) $Cr^{3+}$ (d) $Cl^-$

60. The group having isoelectronic species is

(a) $O^{2-}, F^-, Na^+, Mg^{2+}$
(b) $O^-, F^-, Na, Mg^+$
(c) $O^{2-}, F^-, Na, Mg^{2+}$
(d) $O^-, F^-, Na^+, Mg^{2+}$

# Mathematics

61. If $S$ is the set of distinct values of $b$ for which the following system of linear equations

$$x + y + z = 1, \Rightarrow x + ay + z = 1$$
and $\quad ax + by + z = 0$

has no solution, then $S$ is

(a) an infinite set
(b) a finite set containing two or more elements
(c) singleton set
(d) a empty set

62. The statement $(p \to q) \to [(\sim p \to q) \to q]$ is

(a) a tautology
(b) equivalent to $\sim p \to q$
(c) equivalent to $p \to \sim q$
(d) a fallacy

63. If $5(\tan^2 x - \cos^2 x) = 2\cos 2x + 9$, then the value of $\cos 4x$ is

(a) $-\frac{3}{5}$ (b) $\frac{1}{3}$ (c) $\frac{2}{9}$ (d) $-\frac{7}{9}$

64. For three events $A, B$ and $C$, if $P$(exactly one of $A$ or $B$ occurs) $= P$(exactly one of $B$ or $C$ occurs) $= P$ (exactly one of $C$ or $A$ occurs) $= \frac{1}{4}$ and $P$ (all the three events occur simultaneously) $= \frac{1}{16}$, then the probability that atleast one of the events occurs, is

(a) $\frac{7}{32}$ (b) $\frac{7}{16}$ (c) $\frac{7}{64}$ (d) $\frac{3}{16}$

65. Let $\omega$ be a complex number such that $2\omega + 1 = z$, where $z = \sqrt{-3}$. If

$$\begin{vmatrix} 1 & 1 & 1 \\ 1 & -\omega^2 - 1 & \omega^2 \\ 1 & \omega^2 & \omega^7 \end{vmatrix} = 3k, \text{ then } k \text{ is}$$

equal to

(a) $-z$ (b) $z$ (c) $-1$ (d) 1

66. Let $k$ be an integer such that the triangle with vertices $(k, -3k)$, $(5, k)$ and $(-k, 2)$ has area 28 sq units. Then, the orthocentre of this triangle is at the point

(a) $\left(2, -\frac{1}{2}\right)$ (b) $\left(1, \frac{3}{4}\right)$

(c) $\left(1, -\frac{3}{4}\right)$ (d) $\left(2, \frac{1}{2}\right)$

67. If 20 m of wire is available for fencing off a flower-bed in the form of a circular sector, then the maximum area (in sq m) of the flower-bed is

(a) 12.5 (b) 10 (c) 25 (d) 30

68. The area (in sq units) of the region $\{(x, y) : x \geq 0, x + y \leq 3, x^2 \leq 4y$ and $y \leq 1 + \sqrt{x}\}$ is

(a) $\frac{59}{12}$ (b) $\frac{3}{2}$ (c) $\frac{7}{3}$ (d) $\frac{5}{2}$

69. If the image of the point $P(1, -2, 3)$ in the plane $2x + 3y - 4z + 22 = 0$ measured parallel to the line $\frac{x}{1} = \frac{y}{4} = \frac{z}{5}$ is $Q$, then $PQ$ is equal to

(a) $3\sqrt{5}$ (b) $2\sqrt{42}$ (c) $\sqrt{42}$ (d) $6\sqrt{5}$

70. For $x \in \left(0, \frac{1}{4}\right)$, if the derivative of $\tan^{-1}\left(\frac{6x\sqrt{x}}{1 - 9x^3}\right)$ is $\sqrt{x} \cdot g(x)$, then $g(x)$ equals

(a) $\frac{9}{1 + 9x^3}$ (b) $\frac{3x\sqrt{x}}{1 - 9x^3}$

(c) $\frac{3x}{1 - 9x^3}$ (d) $\frac{3}{1 + 9x^3}$

71. If $(2 + \sin x)\frac{dy}{dx} + (y + 1)\cos x = 0$ and $y(0) = 1$, then $y\left(\frac{\pi}{2}\right)$ is equal to

(a) $\frac{1}{3}$ (b) $-\frac{2}{3}$ (c) $-\frac{1}{3}$ (d) $\frac{4}{3}$

72. Let a vertical tower $AB$ have its end $A$ on the level ground. Let $C$ be the mid-point of $AB$ and $P$ be a point on the ground such that $AP = 2AB$. If $\angle BPC = \beta$, then $\tan \beta$ is equal to

(a) $\frac{6}{7}$ (b) $\frac{1}{4}$

(c) $\frac{2}{9}$ (d) $\frac{4}{9}$

73. If $A = \begin{bmatrix} 2 & -3 \\ -4 & 1 \end{bmatrix}$, then adj $(3A^2 + 12A)$ is equal to

(a) $\begin{bmatrix} 72 & -84 \\ -63 & 51 \end{bmatrix}$ (b) $\begin{bmatrix} 51 & 63 \\ 84 & 72 \end{bmatrix}$

(c) $\begin{bmatrix} 51 & 84 \\ 63 & 72 \end{bmatrix}$ (d) $\begin{bmatrix} 72 & -63 \\ -84 & 51 \end{bmatrix}$

74. For any three positive real numbers $a, b$ and $c$, if $9(25a^2 + b^2) + 25(c^2 - 3ac) = 15b(3a + c)$, then

(a) $b, c$ and $a$ are in GP
(b) $b, c$ and $a$ are in AP
(c) $a, b$ and $c$ are in AP
(d) $a, b$ and $c$ are in GP

75. The distance of the point $(1, 3, -7)$ from the plane passing through the point $(1, -1, -1)$ having normal perpendicular to both the lines

$\frac{x-1}{1}=\frac{y+2}{-2}=\frac{z-4}{3}$ and
$\frac{x-2}{2}=\frac{y+1}{-1}=\frac{z+7}{-1}$, is

(a) $\frac{20}{\sqrt{74}}$ units (b) $\frac{10}{\sqrt{83}}$ units

(c) $\frac{5}{\sqrt{83}}$ units (d) $\frac{10}{\sqrt{74}}$ units

76. Let $I_n=\int \tan^n x\,dx\ (n>1)$. If $I_4+I_6=a\tan^5 x+bx^5+C$, where $C$ is a constant of integration, then the ordered pair $(a,b)$ is equal to

(a) $\left(-\frac{1}{5},1\right)$ (b) $\left(\frac{1}{5},0\right)$

(c) $\left(\frac{1}{5},-1\right)$ (d) $\left(-\frac{1}{5},0\right)$

77. The eccentricity of an ellipse whose centre is at the origin is 1/2. If one of its directrices is $x=-4$, then the equation of the normal to it at $\left(1,\frac{3}{2}\right)$ is

(a) $2y-x=2$ (b) $4x-2y=1$

(c) $4x+2y=7$ (d) $x+2y=4$

78. If a hyperbola passes through the point $P(\sqrt{2},\sqrt{3})$ and has foci at $(\pm 2,0)$, then the tangent to this hyperbola at $P$ also passes through the point

(a) $(3\sqrt{2},2\sqrt{3})$ (b) $(2\sqrt{2},3\sqrt{3})$

(c) $(\sqrt{3},\sqrt{2})$ (d) $(-\sqrt{2},-\sqrt{3})$

79. The function $f:R\to\left[-\frac{1}{2},\frac{1}{2}\right]$ defined as $f(x)=\frac{x}{1+x^2}$ is

(a) invertible

(b) injective but not surjective

(c) surjective but not injective

(d) neither injective nor surjective

80. $\lim\limits_{x\to\pi/2}\frac{\cot x-\cos x}{(\pi-2x)^3}$ equals

(a) $\frac{1}{24}$ (b) $\frac{1}{16}$ (c) $\frac{1}{8}$ (d) $\frac{1}{4}$

81. Let $\mathbf{a}=2\hat{\mathbf{i}}+\hat{\mathbf{j}}-2\hat{\mathbf{k}}$, $\mathbf{b}=\hat{\mathbf{i}}+\hat{\mathbf{j}}$ and $\mathbf{c}$ be a vector such that $|\mathbf{c}-\mathbf{a}|=3$, $|(\mathbf{a}\times\mathbf{b})\times\mathbf{c}|=3$ and the angle between $\mathbf{c}$ and $\mathbf{a}\times\mathbf{b}$ is 30°. Then, $\mathbf{a}\cdot\mathbf{c}$ is equal to

(a) $\frac{25}{8}$ (b) 2 (c) 5 (d) $\frac{1}{8}$

82. The normal to the curve $y(x-2)(x-3)=x+6$ at the point, where the curve intersects the $Y$-axis passes through the point

(a) $\left(-\frac{1}{2},-\frac{1}{2}\right)$ (b) $\left(\frac{1}{2},\frac{1}{2}\right)$

(c) $\left(\frac{1}{2},-\frac{1}{3}\right)$ (d) $\left(\frac{1}{2},\frac{1}{3}\right)$

83. If two different numbers are taken from the set {0, 1, 2, 3, ..., 10}, then the probability that their sum as well as absolute difference are both multiple of 4, is

(a) $\frac{6}{55}$ (b) $\frac{12}{55}$ (c) $\frac{14}{45}$ (d) $\frac{7}{55}$

84. A man $X$ has 7 friends, 4 of them are ladies and 3 are men. His wife $Y$ also has 7 friends, 3 of them are ladies and 4 are men. Assume $X$ and $Y$ have no common friends. Then, the total number of ways in which $X$ and $Y$ together can throw a party inviting 3 ladies and 3 men, so that 3 friends of each of $X$ and $Y$ are in this party, is

(a) 485 (b) 468 (c) 469 (d) 484

85. The value of $({}^{21}C_1-{}^{10}C_1)+({}^{21}C_2-{}^{10}C_2)+({}^{21}C_3-{}^{10}C_3)+({}^{21}C_4-{}^{10}C_4)+\ldots+({}^{21}C_{10}-{}^{10}C_{10})$ is

(a) $2^{21}-2^{11}$ (b) $2^{21}-2^{10}$

(c) $2^{20}-2^{9}$ (d) $2^{20}-2^{10}$

86. A box contains 15 green and 10 yellow balls. If 10 balls are randomly drawn one-by-one with replacement, then the variance of the number of green balls drawn is

(a) $\frac{12}{5}$ (b) 6 (c) 4 (d) $\frac{6}{25}$

87. Let $a, b, c \in R$. If $f(x) = ax^2 + bx + c$ be such that $a + b + c = 3$ and $f(x+y) = f(x) + f(y) + xy$, $\forall\ x, y \in R$, then $\sum_{n=1}^{10} f(n)$ is equal to

(a) 330 (b) 165 (c) 190 (d) 255

88. The radius of a circle having minimum area, which touches the curve $y = 4 - x^2$ and the lines $y = |x|$, is

(a) $2(\sqrt{2}+1)$ (b) $2(\sqrt{2}-1)$
(c) $4(\sqrt{2}-1)$ (d) $4(\sqrt{2}+1)$

89. For a positive integer $n$, if the quadratic equation,
$x(x+1) + (x+1)(x+2) + \ldots + (x+\overline{n-1})(x+n) = 10n$
has two consecutive integral solutions, then $n$ is equal to

(a) 12 (b) 9
(c) 10 (d) 11

90. $\int_{\pi/4}^{3\pi/4} \frac{dx}{1+\cos x}$ is equal to

(a) $-2$ (b) 2
(c) 4 (d) $-1$

# Answer with Explanation

## Physics

1. *(b)* As the observer is moving towards the source, so frequency of waves emitted by the source will be given by the formula

$$f_{\text{observed}} = f_{\text{actual}} \cdot \left(\frac{1 + v/c}{1 - v/c}\right)^{1/2}$$

Here, frequency $\frac{v}{c} = \frac{1}{2}$

So, $$f_{\text{observed}} = f_{\text{actual}} \left(\frac{3/2}{1/2}\right)^{1/2}$$

$\therefore$ $$f_{\text{observed}} = 10 \times \sqrt{3} = 17.3 \text{ GHz}$$

2. *(a)* By ascent formula, we have surface tension,

$$T = \frac{rhg}{2} \times 10^3 \ \frac{\text{N}}{\text{m}}$$
$$= \frac{dhg}{4} \times 10^3 \ \frac{\text{N}}{\text{m}} \qquad \left(\because r = \frac{d}{2}\right)$$

$\Rightarrow \frac{\Delta T}{T} = \frac{\Delta d}{d} + \frac{\Delta h}{h}$ [given, $g$ is constant]

So, percentage

$$= \frac{\Delta T}{T} \times 100 = \left(\frac{\Delta d}{d} + \frac{\Delta h}{h}\right) \times 100$$
$$= \left(\frac{0.01 \times 10^{-2}}{1.25 \times 10^{-2}} + \frac{0.01 \times 10^{-2}}{1.45 \times 10^{-2}}\right) \times 100 = 1.5\%$$

$$\therefore \frac{\Delta T}{T} \times 100 = 1.5\%$$

3. *(c)* We have, $\lambda = \frac{hc}{\Delta E}$

So, ratio of wave lengths

$$\frac{\lambda_1}{\lambda_2} = \frac{hc/\Delta E_1}{hc/\Delta E_2} = \frac{\Delta E_2}{\Delta E_1} = \frac{\left(\frac{4}{3}E - E\right)}{2E - E} = \frac{1}{3}$$

4. *(b)* Given, force, $F = -kv^2$

$\therefore$ Acceleration, $a = \frac{-k}{m} v^2$

or $$\frac{dv}{dt} = \frac{-k}{m} v^2 \Rightarrow \frac{dv}{v^2} = -\frac{k}{m} \cdot dt$$

Now, with limits, we have

$$\int_{10}^{v} \frac{dv}{v^2} = -\frac{k}{m} \int_0^t dt$$

$$\Rightarrow \left(-\frac{1}{v}\right)_{10}^{v} = -\frac{k}{m} t \Rightarrow \frac{1}{v} = 0.1 + \frac{kt}{m}$$

$$\Rightarrow v = \frac{1}{0.1 + \frac{kt}{m}} = \frac{1}{0.1 + 1000k}$$

$$\Rightarrow \frac{1}{2} \times m \times v^2 = \frac{1}{8} m v_0^2$$

$$\Rightarrow v = \frac{v_0}{2} = 5$$

$$\Rightarrow \frac{1}{0.1 + 1000k} = 5 \Rightarrow 1 = 0.5 + 5000k$$

$$\Rightarrow k = \frac{0.5}{5000} \Rightarrow k = 10^{-4} \text{ kg/m}$$

5. *(b)* By Mayor's relation, for 1 g mole of a gas,

$$C_p - C_V = R$$

So, when $n$ gram moles are given,

$$C_p - C_V = \frac{R}{n}$$

As per given question,

$$a = C_p - C_V = \frac{R}{2}; \text{ for } H_2 \quad \ldots \text{(i)}$$

$$b = C_p - C_V = \frac{R}{28}; \text{ for } N_2 \quad \ldots \text{(ii)}$$

From Eqs. (i) and (ii), we get

$$a = 14b$$

6. *(d)* MI of a solid cylinder about its perpendicular bisector of length is

$$I = m\left(\frac{l^2}{12} + \frac{R^2}{4}\right)$$

$$\Rightarrow \quad I = \frac{mR^2}{4} + \frac{ml^2}{12} = \frac{m^2}{4\pi\rho l} + \frac{ml^2}{12}$$

$$[\because \rho\pi R^2 l = m]$$

For $I$ to be maximum,

$$\frac{dI}{dl} = -\frac{m^2}{4\pi\rho}\left(\frac{1}{l^2}\right) + \frac{ml}{6} = 0$$

$$\Rightarrow \quad \frac{m^2}{4\pi\rho} = \frac{ml^3}{6}$$

Now, putting $m = \rho\pi R^2 l$

$$\therefore \quad l^3 = \frac{3}{2\pi\rho} \cdot \rho\pi R^2 l$$

$$\frac{l^2}{R^2} = \frac{3}{2}$$

$$\therefore \quad \frac{l}{R} = \sqrt{\frac{3}{2}}$$

7. *(a)* Decay scheme is ,

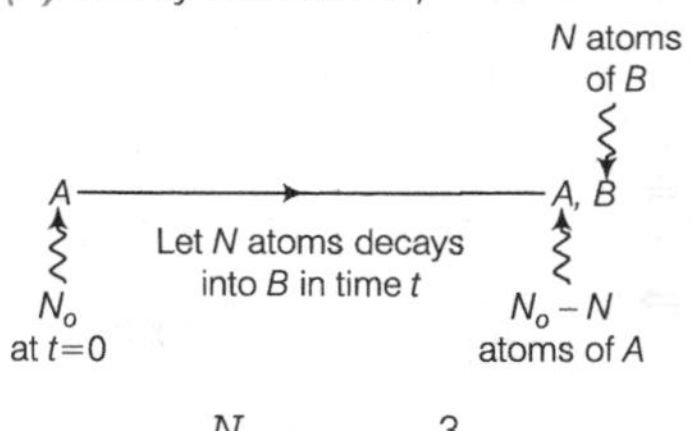


Given, $\quad \frac{N_B}{N_A} = 0.3 = \frac{3}{10}$

$$\Rightarrow \quad \frac{N_B}{N_A} = \frac{30}{100}$$

So, $\quad N_0 = 100 + 30 = 130$ atoms

By using $\quad N = N_0 e^{-\lambda t}$

We have, $\quad 100 = 130\, e^{-\lambda t}$

$$\Rightarrow \quad \frac{1}{1.3} = e^{-\lambda t} \Rightarrow \log 1.3 = \lambda t$$

If $T$ is half-life, then $\lambda = \frac{\log_e 2}{T}$

$$\Rightarrow \quad \log 1.3 = \frac{\log_e 2}{T} \cdot t$$

$$\therefore \quad t = \frac{T \cdot \log(1.3)}{\log_e 2}$$

8. *(a)* In a balanced Wheatstone bridge, there is no effect on position of null point, if we exchange the battery and galvanometer. So, option (a) is incorrect.

9. *(c)* As each capacitors cannot withstand more than 300 V, so there should be four capacitors in each row became in this condition 1 kV i.e. 1000 V will be divided by 4 (i.e. 250 not more than 300 V).

Now, equivalent capacitance of one row

$$= \frac{1}{4} \times 1\mu F = 0.25\mu F$$

$$\left[\because \text{in series combination, } C_{eq} = \frac{c}{n}\right]$$

Now, we need equivalent of 2μF, so let we need $n$ such rows

$$\therefore \quad n \times 0.25 = 2\mu F$$

$$[\because \text{in parallel combination } C_{eq} = nc]$$

$$n = \frac{2}{0.25}$$

$$= 8$$

$\therefore$ Total number of capacitors
= number of rows
× number of capacitors in each row
$= 8 \times 4 = 32$

10. *(b)* In steady state, no current flows through the capacitor. So, resistance $r_p$ becomes ineffective.

So, the current in circuit,

$$I = \frac{E}{r + r_2 \text{ (Total Resistance)}}$$

$\because$ Potential drop across capacitor

$$= \text{Potential drop across } r_2 = Ir_2 = \frac{Er_2}{r + r_2}$$

$\therefore$ Stored charge of capacitor, $Q = CV$

$$= CE\frac{r_2}{r + r_2}$$

**11.** *(c)* Each resistance is converted with two cells combined in opposite direction, so potential drop across each resistor is zero. Hence the current through each of resistor is zero.

**12.** *(d)* Frequency spectrum of modulated wave is

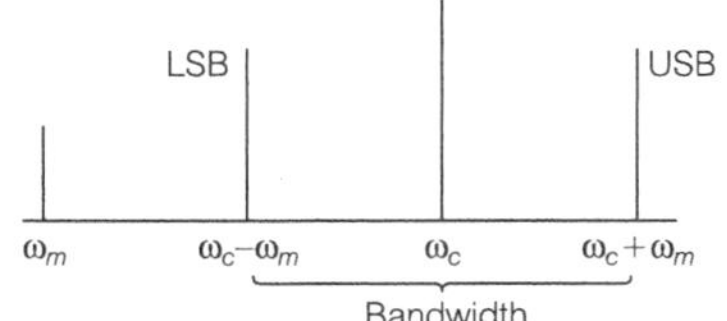


Clearly, $\omega_m$ is not included in the spectrum.

**13.** *(c)*

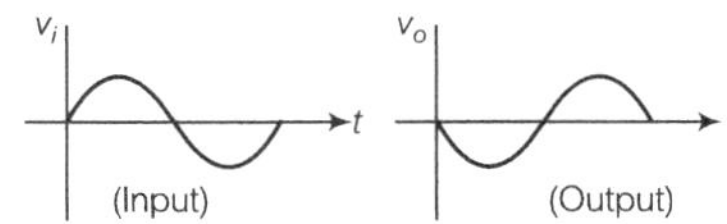


In a CE *npn* transistor amplifier output is 180° out of phase with input.

**14.** *(a)* Heat gained (water + calorimeter) = Heat lost by copper ball

$$\Rightarrow \quad m_w s_w \Delta T + m_c s_c \Delta T = m_B s_B \Delta T$$

$$\Rightarrow \quad 170 \times 1 \times (75-30) + 100 \times 0.1 \times (75-30) = 100 \times 0.1 \times (T-75)$$

$$\therefore \quad T = 885°\text{C}$$

**15.** *(a)* Let $n_1$th fringe formed due to first wavelength and $n_2$th fringe formed due to second wavelength coincide i.e. their distance from common central maxima will be same

i.e. $\quad y_{n_1} = y_{n_2}$

$$\Rightarrow \quad \frac{n_1 \lambda_1 D}{d} = \frac{n_2 \lambda_2 D}{d}$$

$$\Rightarrow \quad \frac{n_1}{n_2} = \frac{\lambda_1}{\lambda_2} = \frac{520}{650} = \frac{4}{5}$$

Hence, distance of the point of coincidence from the central maxima is

$$y = \frac{n_1 \lambda_1 D}{d} = \frac{n_2 \lambda_2 D}{d} = \frac{4 \times 650 \times 10^{-9} \times 1.5}{0.5 \times 10^{-3}} = 7.8 \text{ mm}$$

**16.** *(b)*

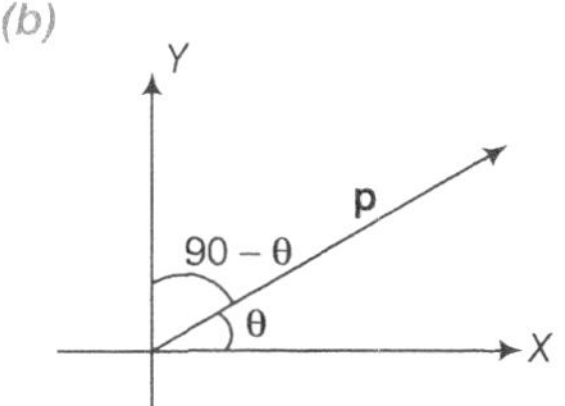


Torque applied on a dipole $\tau = pE \sin\theta$

where $\theta$ = angle between axis of dipole and electric field.

For electric field $E_1 = E\hat{\mathbf{i}}$

it means field is directed along positive $X$ direction, so angle between dipole and field will remain $\theta$, therefore torque in this direction

$$E_1 = pE_1 \sin\theta$$

In electric field $E_2 = \sqrt{3}\, E\hat{\mathbf{j}}$, it means field is directed along positive $Y$-axis, so angle between dipole and field will be $90 - \theta$.

Torque in this direction

$$T_2 = pE \sin(90 - \theta)$$
$$= p\sqrt{3}\, E_1 \cos\theta$$

According to question

$$\tau_2 = -\tau_1 \Rightarrow |\tau_2| = |\tau_1|$$

$$\therefore \quad pE_1 \sin\theta = p\sqrt{3}\, E_1 \cos\theta$$

$$\tan\theta = \sqrt{3}$$

$$\Rightarrow \quad \tan\theta = \tan 60°$$

$$\therefore \quad \theta = 60°$$

**17.** *(d)* As the rod rotates in vertical plane so a torque is acting on it, which is due to the vertical component of weight of rod.

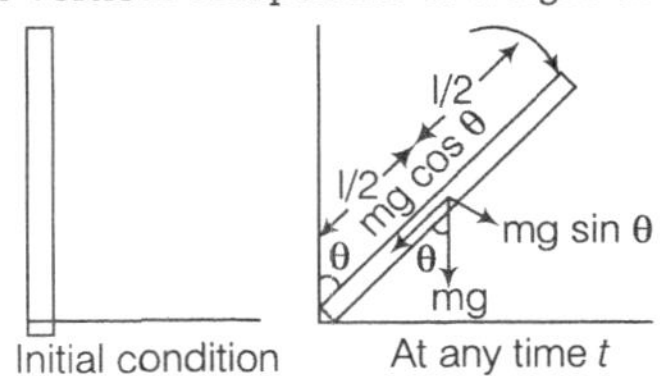


Now, Torque $\tau$ = force × perpendicular distance of line of action of force from axis of rotation

$$= mg \sin\theta \times \frac{l}{2}$$

Again, Torque $\tau = I\alpha$

Where, $I$ = moment of inertia $= \frac{ml^2}{3}$

[Force and Torque frequency along axis of rotation passing through in end]

$\alpha$ = angular acceleration

$$\therefore \quad mg\sin\theta \times \frac{l}{2} = \frac{ml^2}{3}\alpha$$

$$\therefore \quad \alpha = \frac{3g\sin\theta}{2l}$$

**18.** *(d)* $K = \dfrac{P}{(-\Delta V/V)}$

$$\Rightarrow \quad -\frac{\Delta V}{V} = \frac{P}{K}$$

$$\Rightarrow \quad -\Delta V = \frac{PV}{K}$$

Change in volume $\Delta V = \gamma v \Delta T$

Where $r$ = coefficient of volume expansion

Again, $\gamma = 3\alpha$

$\alpha$ is coefficient of linear expansion

$$\therefore \quad \Delta V = V(3\alpha)\Delta T$$

$$\therefore \quad \frac{PV}{K} = V(3\alpha)\Delta T$$

$$\therefore \quad \Delta T = \frac{P}{3\alpha K}$$

**19.** *(d)* Focal length of diverging lens is 25 cm.

As the rays are coming parallel, so the image ($I_1$) will be formed at the focus of diverging lens i.e. at 25 cm towards left of diverging lens.

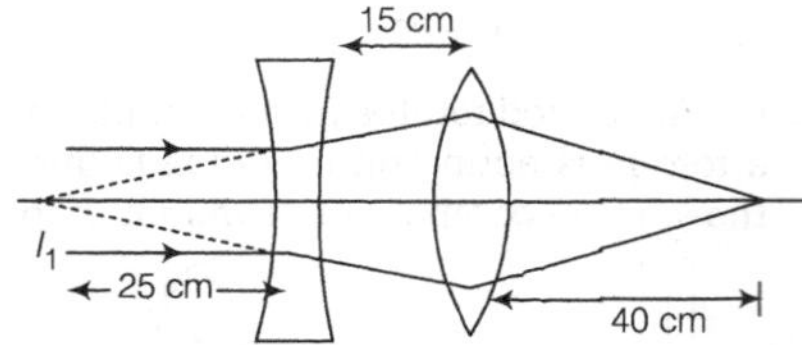


Now, the image ($I_1$) will work as object for converging lens.

For convergying lens, distance of object $u$ (i.e. distance of $I_1$) = $-(25 + 15)$

$= -40$ cm

$f = 20$ cm

$\therefore$ From len's formula $\dfrac{1}{f} = \dfrac{1}{v} - \dfrac{1}{u}$

$$\frac{1}{20} = \frac{1}{v} - \frac{1}{-40} \Rightarrow \frac{1}{v} = \frac{1}{20} - \frac{1}{40}$$

$$\frac{1}{v} = \frac{1}{90} \Rightarrow v = 40 \text{ cm}$$

$v$ is positive so image will be real and will form at right side of converging lens at 40 cm.

**20.** *(d)* $\lambda_{min} = \dfrac{hc}{eV}$

$$\log(\lambda_{min}) = \log\left(\frac{hc}{e}\right) - \log V$$

$$y = c - mx$$

So, the required graph is given in option (d).

**21.** *(c)* From $pV = nRT = \dfrac{N}{N_A}RT$

We have, $n_f - n_i = \dfrac{pVN_A}{RT_f} - \dfrac{pVN_A}{RT_i}$

$$\Rightarrow \quad n_f - n_i = \frac{10^5 \times 30}{8.3} \times 6.02 \times 10^{23} \cdot \left(\frac{1}{300} - \frac{1}{290}\right)$$

$$= -2.5 \times 10^{25}$$

$$\therefore \quad \Delta n = -2.5 \times 10^{25}$$

**22.** *(b)* Induced constant, $I = \dfrac{e}{R}$

Here, $e$ = induced emf

$$= \frac{d\phi}{dt}$$

$$I = \frac{e}{R} = \left(\frac{d\phi}{dt}\right) \cdot \frac{1}{R}$$

$$d\phi = IRdt$$

$$\phi = \int IRdt$$

$\therefore$ Here, $R$ is constant

$$\therefore \quad \phi = R\int Idt$$

$\int I \cdot dt$ = Area under $I-t$ graph

$$= \frac{1}{2} \times 10 \times 0.5 = 2.5$$

$$\therefore \quad \phi = R \times 2.5$$

$= 100 \times 2.5$

$= 250$ Wb.

**23.** *(d)* Suppose a resistance $R_s$ is connected in series with galvanometer to convert it into voltmeter.

$I_g$ G $R_s$ V

$$I_g(G + R_s) = V$$

$$\Rightarrow \quad R = \frac{V}{I_g} - G$$

$\Rightarrow R = 1985 = 1.985$ **kΩ**

or $R = 1.985 \times 10^3$ Ω

24. *(d)* From Newton's second law, $\frac{\Delta p}{\Delta t} = F$

$\Rightarrow \quad \Delta p = F\Delta t$

$\therefore \quad p = \int dp = \int_0^1 F\,dt$

$\Rightarrow \quad p = \int_0^1 6t\,dt = 3 \text{ kg}\left(\frac{\text{m}}{\text{s}}\right)$

Also, change in kinetic energy

$$\Delta k = \frac{\Delta p^2}{2m} = \frac{3^2}{2 \times 1}$$
$$= 4.5$$

From work-energy theorem, work done = change in kinetic energy.

So, work done = $\Delta k = 4.5$ J

25. *(d)* Time period of oscillation is

$$T = 2\pi\sqrt{\frac{I}{MB}}$$

$$\Rightarrow T = 2\pi\sqrt{\frac{7.5 \times 10^{-6}}{6.7 \times 10^{-2} \times 0.01}}$$
$$= 0.665 \text{ s}$$

Hence, time for 10 oscillations is $t = 6.65$ s.

26. *(c)* Inside the earth surface $g = \frac{GM}{R^3} r$

i.e. $\quad g \propto r$

Out the earth surface $g = \frac{Gm}{r^2}$

i.e. $\quad g \propto \frac{1}{r^2}$

So, till earth surface '$g$' increases linearly with distance $r$, shown only in graph (c).

27. *(b)* Initially velocity keeps on decreasing at a constant rate, then it increases in negative direction with same rate.

28. *(a)* For elastic collision,

$$p_{\text{before collision}} = p_{\text{after collision}}.$$
$$mv = mv_A + \frac{m}{2}v_B$$
$$2v = 2v_A + v_B \quad \ldots\text{(i)}$$

Now, coefficient of restitution,

$$e = \frac{v_B - v_A}{u_A - v_B}$$

Here, $u_B = 0$ (Particle at rest) and for elastic collisione = 1

$$\therefore \quad 1 = \frac{v_B - v_A}{v} \Rightarrow v = v_B - v_A \quad \ldots\text{(ii)}$$

From Eq. (i) and Eq. (ii)

$$v_A = \frac{v}{3} \text{ and } v_B = \frac{4v}{3}$$

Hence, $\frac{\lambda_A}{\lambda_B} = \frac{\left(\frac{h}{mV_A}\right)}{\frac{h}{\frac{m}{2}.V_B}} = \frac{V_B}{2V_A} = \frac{4/3}{2/3} = 2$

29. *(c)* KE is maximum at mean position and minimum at extreme position $\left(\text{at } t = \frac{T}{4}\right)$.

30. *(d)* Stress $= \frac{\text{Weight}}{\text{Area}}$

Volume will become $(9^3)$ times.

So weight = volume × density × $g$ will also become $(9)^3$ times.

Area of cross section will become $(9)^2$ times.

$$= \frac{9^3 \times W_0}{9^2 \times A_0} = 9\left(\frac{W_0}{A_0}\right)$$

Hence, the stress increases by a factor of 9.

# Chemistry

31. *(c)* Aniline in presence of nitrating mixture (conc. $HNO_3$ + conc. $H_2SO_4$) gives significance amount ($\approx 47\%$) of meta-product because in presence of $H_2SO_4$ its protonation takes place and anilinium ion is formed.

$NH_2$ (Aniline) $\xrightarrow{H_2SO_4}$ $\overset{\oplus}{N}H_3\overset{\ominus}{HSO_4}$ (Anilinium ion)

Here, anilinium ion is strongly deactivating group and *meta*-directing in nature. So, it gives *meta*-nitration product.

$NH_2$ (benzene) $\xrightarrow[\text{+Conc.}HNO_3]{\text{Conc.}H_2SO_4}$ $\overset{\oplus}{N}H_3\overset{\ominus}{HSO_4}$ (with $NO_2$ at meta position)

32. *(c)* According to first law of thermodynamics,

$$\Delta U = q + W = q - p\Delta V$$

In isochoric process ($\Delta V = 0$),

$$\Delta U = q$$

In isobaric process ($\Delta p = 0$),

$$\Delta U = q$$

In adiabatic process ($q = 0$)

$$\Delta U = W$$

In isothermal process ($\Delta T = 0$) and $\Delta U = 0$

$\therefore$ $\Delta U$ is equal to adiabatic work.

**33.** *(b)* (i) The rate of $S_N1$ reaction depends only upon the concentration of the alkyl halide.

(ii) $S_N1$ reaction proceeds through the formation of carbocation.

The reactivity is decided by ease of dissociation of alkyl halide.

$$R—X \rightleftharpoons R^{\oplus} + X^{\ominus}$$

Higher the stability of $R^+$ (carbocation), higher would be the reactivity towards $S_N1$ reaction.

$p$-$H_3CO—C_6H_4—CH_2^{\oplus}$ is the most stable carbocation due to resonance and then $CH_3\overset{\oplus}{C}HCH_2CH_3$ (2° carbocation) while $CH_3CH_2\overset{\oplus}{C}H_2$ (1°) is least stable.

Thus, the correct increasing order of the reactivity of the given halides towards the $S_N1$ reaction is

$$\underset{\text{(II)}}{CH_3CH_2CH_2Cl} < \underset{\text{(I)}}{CH_3\underset{\underset{Cl}{|}}{C}HCH_2CH_3} < \underset{\text{(III)}}{p\text{-}H_3CO—C_6H_4—CH_2Cl}$$

**34.** *(d)* Bohr radius $(r_n) = \epsilon_0\, n^2h^2$

$$r_n = \frac{n^2h^2}{4\pi^2 me^2 kZ}$$

$$k = \frac{1}{4\pi\, \epsilon_0}$$

$$\therefore \quad r_n = \frac{n^2h^2\, \epsilon_0}{\pi me^2 Z}$$

$$= n^2\frac{a_0}{Z}$$

where, $m$ = mass of electron

$e$ = charge of electron

$h$ = Planck's constant

$k$ = Coulomb constant

$$r_n = \frac{n^2 \times 0.53}{Z}\, \text{Å}$$

Radius of $n^{th}$ Bohr orbit for H-atom

$= 0.53\, n^2$ Å

[$Z = 1$ for H-atom]

$\therefore$ Radius of $2^{nd}$ Bohr orbit for H-atom

$= 0.53 \times (2)^2 = 2.12$ Å

**35.** *(b)* For a salt of weak acid and weak base,

$$pH = 7 + \frac{1}{2}pK_a - \frac{1}{2}pK_b$$

Given, $pK_a(HA) = 3.2$, $pK_a(BOH) = 3.4$

$$\therefore \quad pH = 7 + \frac{1}{2}(3.2) - \frac{1}{2}(3.4)$$

$$= 7 + 1.6 - 1.7 = 6.9$$

**36.** *(a)* Nylon-6 or perlon is prepared by polymerisation of amino caproic acid at high temperature. Caprolactam is first hydrolysed with water to form amino acid which on heating undergoes polymerisation to give nylon-6.

Caprolactam $\xrightarrow{\text{Hydrolysis}}$ $\overset{+}{H_3N}—(CH_2)_5—\overset{\overset{O}{||}}{C}—O^-$ $\xrightarrow[\text{Polymerisation}]{\Delta}$ $\underset{\text{Nylon-6}}{\left[HN—(CH_2)_5—\overset{\overset{O}{||}}{C}\right]_n}$

**37.** *(c)* Given, abundance of elements by mass

oxygen = 61.4%, carbon = 22.9%, hydrogen = 10% and nitrogen = 2.6%

Total weight of person = 75 kg

$$\text{Mass due to } ^1H = \frac{75 \times 10}{100} = 7.5 \text{ kg}$$

$^1H$ atoms are replaced by $^2H$ atoms,

Mass due to $^2H = (7.5 \times 2)$ kg

$\therefore$ Mass gain by person = 7.5 kg

38. *(a)* To show decolourisation, compound must be unsaturated.

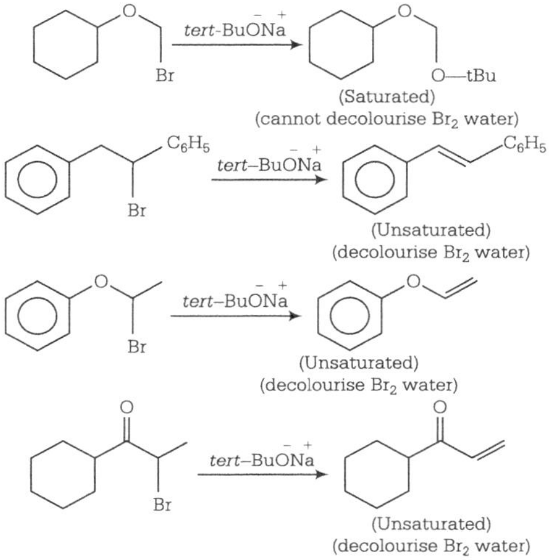

tert-BuONa → (Saturated) (cannot decolourise $Br_2$ water)

tert–BuONa → (Unsaturated) (decolourise $Br_2$ water)

tert–BuONa → (Unsaturated) (decolourise $Br_2$ water)

tert–BuONa → (Unsaturated) (decolourise $Br_2$ water)

39. *(d)* Zinc oxide (ZnO) when react with $Na_2O$ it act as acid while with $CO_2$ it act as base. Therefore, it is an amphoteric oxide.

$$\underset{\text{Acid}}{ZnO} + \underset{\text{Base}}{Na_2O} \longrightarrow \underset{\text{Salt}}{Na_2ZnO_2}$$

$$\underset{\text{Base}}{ZnO} + \underset{\text{Acid}}{CO_2} \longrightarrow \underset{\text{Salt}}{ZnCO_3}$$

40. *(a)* **Mg** can form basic carbonate while **Li** cannot.

$$5\,Mg^{2+} + 6\,CO_3^{2-} + 7H_2O \longrightarrow 4MgCO_3 \cdot Mg(OH)_2 \cdot 5H_2O + 2\,HCO_3^-$$

41. *(d)* The number of stereoisomers in molecules which are not divisible into two equal halves and have *n* number of asymmetric C-atoms $= 2^n$.

3-methyl-pent-2-ene on reaction with HBr in presence of peroxide forms an addition product i.e. 2-bromo-3-methyl pentane. It has two chiral centres. Therefore, 4 stereoisomers are possible

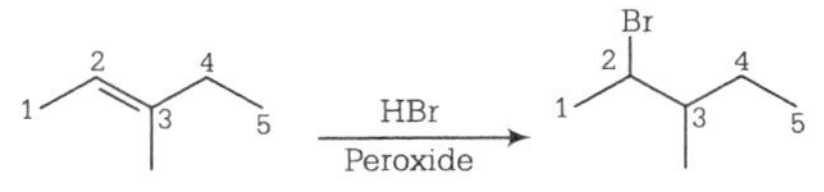

3-methylpent-2-ene → HBr / Peroxide →

Anti-Markownikoff's addition Four stereoisomers are possible (As molecule has two chiral centres and asymmetric).

42. *(d)* For fcc arrangement, $4r = \sqrt{2}a$

where, $r$ = radius and $a$ = edge length

$\therefore$ Closest distance $= 2r = \dfrac{\sqrt{2}\,a}{2} = \dfrac{a}{\sqrt{2}}$

43. *(d)* According to Arrhenius equation

$$k = Ae^{-E_a/RT}$$

where, $A$ = collision number or pre-exponential factor.

$R$ = gas constant

$T$ = absolute temperature

$E_a$ = energy of activation

For reaction $R_1$, $k_1 = Ae^{-E_{a_1}/RT}$ ...(i)

For reaction $R_2$, $k_2 = Ae^{-E_{a_2}/RT}$ ...(ii)

On dividing Eq. (ii) by Eq. (i), we get

$$\frac{k_2}{k_1} = e^{-\frac{(E_{a_2} - E_{a_1})}{RT}} \quad \text{...(iii)}$$

[$\because$ Pre-exponential factor '$A$' is same for both reactions]

Taking ln on both the sides of Eq. (iii), we get

$$\ln\left(\frac{k_2}{k_1}\right) = \frac{E_{a_1} - E_{a_2}}{RT}$$

Given, $E_{a_1} = E_{a_2} + 10\text{ kJ mol}^{-1}$

$= E_{a_2} + 10{,}000\text{ J mol}^{-1}$

$$\therefore \ln\frac{k_2}{k_1} = \frac{10{,}000\text{ J mol}^{-1}}{8.314\text{ J mol}^{-1}\text{K}^{-1} \times 300\text{ K}} = 4$$

44. *(a)*

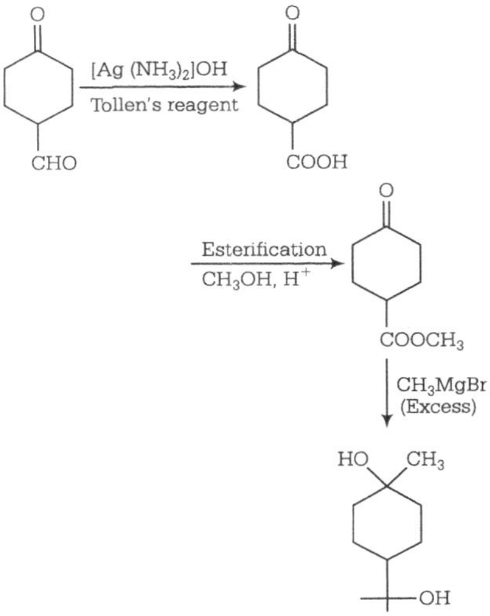

$[Ag(NH_3)_2]OH$ / Tollen's reagent: CHO → COOH

Esterification ($CH_3OH$, $H^+$) → $COOCH_3$

$CH_3MgBr$ (Excess) → HO, $CH_3$ ... OH

Before final product is formed, intermediate is

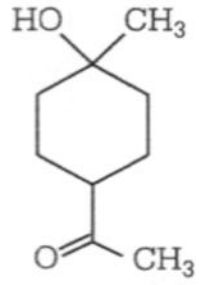

HO CH$_3$

O CH$_3$

45. *(b)* Colloidal solutions show Tyndall effect due to scattering of light by colloidal particles in all directions in space. It is observed only under the following conditions.
   (i) The diameter of the colloids should not be much smaller than the wavelength of light used.
   (ii) The refractive indices of the dispersed phase and dispersion medium should differ greatly in magnitude.

46. *(a)* Sugars that have an aldehyde, a ketone, a hemiacetal or a hemiketal group is able to reduce an oxidising agent. These sugars are classified as reducing sugars.

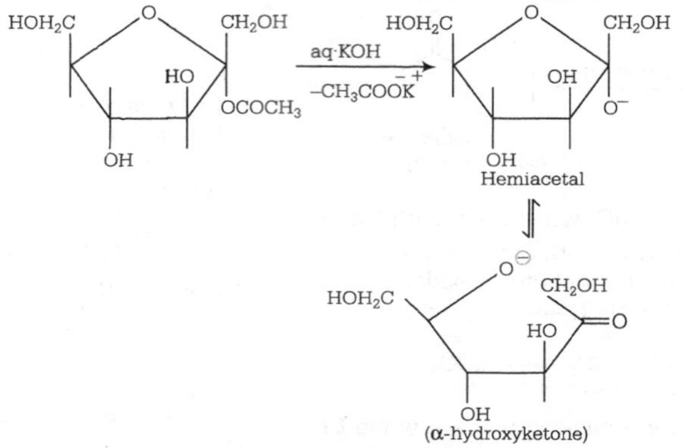

HOH$_2$C O CH$_2$OH HO OCOCH$_3$ OH $\xrightarrow[-CH_3\overset{-}{C}OO\overset{+}{K}]{aq \cdot KOH}$ HOH$_2$C O CH$_2$OH OH O$^-$ OH

Hemiacetal

⇅

HOH$_2$C O$^\ominus$ CH$_2$OH HO =O OH

(α-hydroxyketone)

Hemiacetal can be easily reduced by oxidising agent such as Tollen's reagent.

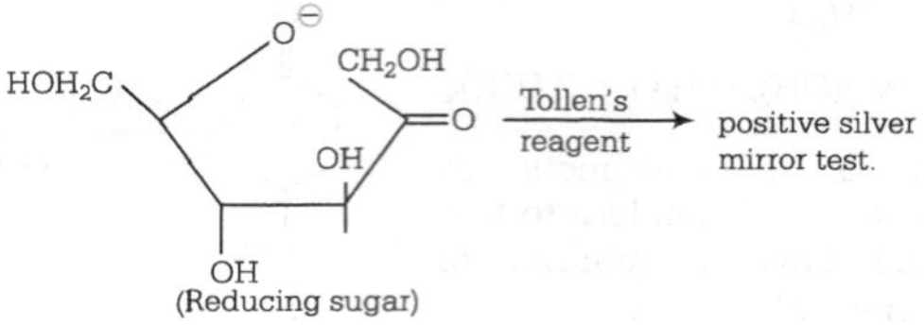

HOH$_2$C O$^\ominus$ CH$_2$OH =O OH OH $\xrightarrow[\text{reagent}]{\text{Tollen's}}$ positive silver mirror test.

(Reducing sugar)

47. *(c)* Based on given $\Delta_r H^\circ$

$\Delta_f H^\circ = H^\circ_{CO_2} = -393.5 \text{ kJ mol}^{-1}$ ...(i)

$\Delta_f H^\circ = H^\circ_{H_2O} = -285.8 \text{ kJ mol}^{-1}$ ...(ii)

$\Delta_f H^\circ = H^\circ_{O_2} = 0.00$ (elements) ...(iii)

Required thermal reaction is for $\Delta_f H^\circ$ of $CH_4$

Thus, from III

$890.3 = [\Delta_f H^\circ(CH_4) + 2\Delta_f H^\circ(O_2)]$

$- [\Delta_f H^\circ(CO_2) + 2\Delta_f H^\circ(H_2O)]$

$= \Delta_f H^\circ(CH_4) + 0] - [-393.5 - 2 \times 285.5]$

$\therefore \quad \Delta_f H^\circ(CH_4) = -74.8 \text{ kJ/mol}$

48. *(a)* The reaction in which oxidation and reduction occur simultaneously are termed as redox reaction.

$$\overset{+4}{(X)}eF_4 + \overset{+1}{O_2}(F_2) \longrightarrow \overset{+6}{X}eF_6 + \overset{0}{O}_2$$

Since, Xe undergoes oxidation while O undergoes reduction. So, it is an example of redox reaction.

49. *(c)* $Cl_2$, $Br_2$ and $I_2$ form a mixture of halide and hypohalites when react with cold dilute alkalies while a mixture of halides and haloate when react with concentrated cold alkalies.

$$\underset{\text{Cold and dilute}}{Cl_2 + 2NaOH} \longrightarrow NaCl + NaClO + H_2O$$

$\therefore$ $Cl^-$ and $ClO^-$ are obtained as products when chlorine gas reacts with cold and dilute aqueous NaOH.

50. *(b)* An alkyl halide in presence of a bulkier base removes a proton from a carbon adjacent to the carbon bonded to the halogen. This reaction is called E2 ($\beta$-elimination reaction).

$C_6H_5$–CH(Br)–CH(H)–$C_6H_5$ $\xrightarrow{t\text{-BuO}^-K^+}$ $C_6H_5$(H)C=C(H)$C_6H_5$ + *tert*-BuOH + $Br^-$

51. *(d)* The reaction takes place as follows

$$\underset{(X)}{Na_2C_2O_4} + \underset{\text{(conc.)}}{H_2SO_4} \longrightarrow Na_2SO_4 + H_2O + \underset{\text{Effervescence}}{CO\uparrow + CO_2\uparrow}$$

$$\underset{(X)}{Na_2C_2O_4} + CaCl_2 \longrightarrow \underset{\text{White ppt.}}{CaC_2O_4} + 2NaCl$$

$$5CaC_2O_4 + \underset{\text{Purple}}{2KMnO_4} + 8H_2SO_4 \longrightarrow K_2SO_4 + 5CaSO_4 + \underset{\text{Colourless}}{2MnSO_4} + 10CO_2 + 8H_2O$$

Hence, $X$ is $Na_2C_2O_4$.

52. *(b)* To identify the magnetic nature we need to check the molecular orbital configuration. If all orbitals are fully occupied, species is diamagnetic while when one or more molecular orbitals is/are singly occupied, species is paramagnetic.

(a) NO $(7 + 8 = 15) - \sigma 1s^2, \sigma^* 1s^2, \sigma 2s^2,$

$$\sigma^* 2s^2, \pi 2p_x^2 = \pi 2p_y^2, \pi 2p_z^2, \pi^* 2p_x^1 = \pi^* 2p_y^0$$

One unpaired electron is present. Hence, it is paramagnetic.

(b) CO $(6 + 8 = 14) - \sigma 1s^2, \sigma^* 1s^2, \sigma 2s^2,$

$$\sigma^* 2s^2, \pi 2p_x^2 = \pi 2p_y^2, \sigma 2p_z^2$$

No unpaired electron is present. Hence, it is diamagnetic.

(c) $O_2$ $(8 + 8 = 16) - \sigma 1s^2, \sigma^* 1s^2, \sigma 2s^2,$

$$\sigma^* 2s^2, \sigma 2p_z^2, \pi 2p_x^2 = \pi 2p_x^2, \pi^* 2p_x^1 = \pi^* 2p_y^1$$

Two unpaired electrons are present. Hence, it is paramagnetic.

(d) $B_2(5 + 5) - \sigma 1s^2, \sigma^* 1s^2, \sigma 2s^2,$

$$\sigma^* 2s^2, \pi 2p_x^1 = \pi 2p_y^1$$

Two unpaired electrons are present. Hence, it is paramagnetic.

53. *(d)* Let the degree of association of acetic acid ($CH_3COOH$) in benzene is $\alpha$, then

| | $2CH_3COOH \rightleftharpoons$ | $(CH_3COOH)_2$ |
|---|---|---|
| Initial moles | 1 | 0 |
| Moles at equilibrium | $1 - \alpha$ | $\frac{\alpha}{2}$ |

$\therefore$ Total moles $= 1 - \alpha + \frac{\alpha}{2} = 1 - \frac{\alpha}{2}$

or $\quad i = 1 - \frac{\alpha}{2}$

Now, depression in freezing point $(\Delta T_f)$ is given as

$$\Delta T_f = i K_f m \quad \ldots (i)$$

where, $K_f$ = molal depression constant or cryoscopic constant.

$m$ = molality

$$\text{Molality} = \frac{\text{number of moles of solute}}{\text{weight of solvent (in kg)}} = \frac{0.2}{60} \times \frac{1000}{20}$$

Putting the values in Eq. (i)

$$\therefore \quad 0.45 = \left[1 - \frac{\alpha}{2}\right](5.12)\left[\frac{0.2}{60} \times \frac{1000}{20}\right]$$

$$1 - \frac{\alpha}{2} = \frac{0.45 \times 60 \times 20}{5.12 \times 0.2 \times 1000}$$

$$\Rightarrow \quad 1 - \frac{\alpha}{2} = 0.527 \quad \Rightarrow \frac{\alpha}{2} = 1 - 0.527$$

$$\therefore \quad \alpha = 0.946$$

Thus, percentage of association = 94.6%

**54.** *(d)* Aromatic compounds are stable due to resonance while non-aromatics are not. According to Huckel's rule (or $4n+2$ rule), "For a planar, cyclic compound to be aromatic, its $\pi$ cloud must contain $(4n+2)\pi$ electrons, where, $n$ is any whole number." Thus,

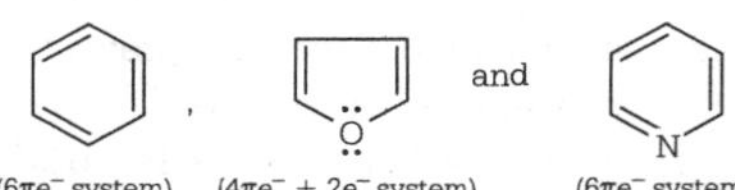


are aromatic and stabilised by resonance. They follow Huckel's rule.

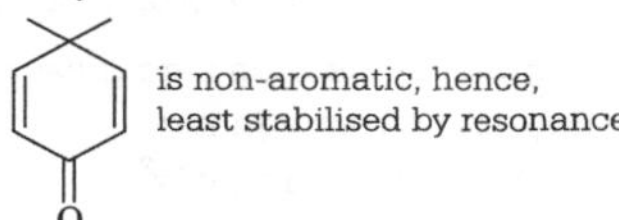

is non-aromatic, hence, least stabilised by resonance.

**55.** *(d)* Molarity $(M)$

$$= \frac{\text{Number of moles of solute}}{\text{Volume of solution (in L)}}$$

$\therefore$ Number of moles of complex

$$= \frac{\text{Molarity} \times \text{volume (in mL)}}{1000}$$

$$= \frac{0.1 \times 100}{1000} = 0.01 \text{ mole}$$

Number of moles of ions precipitate

$$= \frac{12 \times 10^{22}}{6.02 \times 10^{23}} = 0.02 \text{ moles}$$

$\therefore$ Number of $Cl^-$ present in ionisation sphere

$$= \frac{\text{Number of moles of ions precipitated}}{\text{Number of moles of complex}}$$

$$= \frac{0.02}{0.01} = 2$$

$\therefore$ 2 $Cl^-$ are present outside the square brackets, i.e. in ionisation sphere. Thus, the formula of complex is $[Co(H_2O)_5Cl]Cl_2 \cdot H_2O$.

**56.** *(a)* DIBAL-H (Di-isobutyl aluminium hydride) is a reducing agent with formula. This is generally used for the preparation of aldehydes. Using DIBAL —H, Lactones are reduced directly to aldehydes.

**57.** *(c)* $\mathbf{NO_3^-}$ The maximum limit of nitrate $(NO_3^-)$ in drinking water is 50 ppm and its source is fertilisers. If the maximum limit is increased in water it will cause methemoglobinemia (blue baby syndrome.)

$\mathbf{SO_4^{2-}}$ The maximum limit of sulphate $(SO_4^{2-})$ according to WHO is 500 pm and its sources are acid rain, industries. Excess $SO_4^{2-}$ has laxative effect.

$\mathbf{F^-}$ The maximum limit of fluoride $(F^-)$ is about 1.5 ppm. Its higher concentration converts enamel to more harder fluorapatite. Concentration (>2ppm) causes brown mottling of teeth and high concentration (>10 ppm) are harmful for bones and teeth.

$\therefore$ $SO_4^{2-}$ (100 ppm) and $NO_3^-$ (50 ppm) in water is suitable for drinking but the concentration of $F^-$ (10 ppm) makes water unsuitable for drinking.

**58.** *(b)*

$$\underset{1\,g}{M_2CO_3} + 2HCl \longrightarrow 2MCl + H_2O + \underset{0.01186\ \text{mole}}{CO_2}$$

Number of moles of $M_2CO_3$ reacted = Number of moles of $CO_2$ evolved

$$\frac{1}{M} = 0.01186 \ [M = \text{molar mass of } M_2CO_3]$$

$$M = \frac{1}{0.01186} = 84.3 \text{ g mol}^{-1}$$

**59.** *(a)* The substances which have lower reduction potentials are stronger reducing agents. Therefore, Cr ($E^\circ_{Cr^{3+}/Cr} = -0.74$ V) is the strongest reducing agent among all the other given options.

**60.** *(a)* Isoelectronic species are those which contains same number of electrons.

| Species | Atomic number | Number of electrons |
|---|---|---|
| $O^{2-}$ | 8 | 10 |
| $F^-$ | 9 | 10 |
| $Na^+$ | 11 | 10 |
| $Mg^{2+}$ | 12 | 10 |
| $O^-$ | 8 | 9 |
| Na | 11 | 11 |
| $Mg^+$ | 12 | 11 |

$\therefore$ Option (a) is correct which contains isoelectronic species $O^{2-}$, $F^-$, $Na^+$, $Mg^{2+}$.

# Mathematics

**61.** *(d)* $\because\ \Delta = \begin{vmatrix} 1 & 1 & 1 \\ 1 & a & 1 \\ a & b & 1 \end{vmatrix} = 1(a-b) - 1(1-a) + 1(b-a^2)$

$= -(a-1)^2$

$\Delta_1 = \begin{vmatrix} 1 & 1 & 1 \\ 1 & a & 1 \\ 0 & b & 1 \end{vmatrix} = 1(a-b) - 1(1) + 1(b)$

$= -(a-1)$

$\Delta_2 = \begin{vmatrix} 1 & 1 & 1 \\ 1 & 1 & 1 \\ a & 0 & 1 \end{vmatrix} = 1(1) - 1(1-a) + 1(0-a) = 0$ and $\Delta_3 = \begin{vmatrix} 1 & 1 & 1 \\ 1 & a & 1 \\ a & b & 0 \end{vmatrix}$

$= 1(-b) - 1(-a) + 1(b-a^2)$

$= -a(a-1)$

For $a = 1$

$\Delta = \Delta_1 = \Delta_2 = \Delta_3 = 0$

$\Delta$ for $b = 1$ only

$x + y + z = 1,$

$x + y + z = 1$

and $x + y + z = 0$

i.e. no solution (∵ RHS is not equal)

Hence, for no solution $b = 1$ only

**62.** *(a)* The truth table of the given expression is given below :

| p | q | x ≡ p → q | ~ p | ~p → q | y ≡ (~ p → q) → q | x → y |
|---|---|---|---|---|---|---|
| T | T | T | F | T | T | T |
| T | F | F | F | T | F | T |
| F | T | T | T | T | T | T |
| F | F | T | T | F | T | T |

Hence, it is a tautology.

**63.** *(d)* Given, $5(\tan^2 x - \cos^2 x) = 2\cos 2x + 9$

$\Rightarrow 5\left(\frac{1-\cos 2x}{1+\cos 2x} - \frac{1+\cos 2x}{2}\right) = 2\cos 2x + 9$

Put $\cos 2x = y$, we have

$$5\left(\frac{1-y}{1+y} - \frac{1+y}{2}\right) = 2y + 9$$

$\Rightarrow 5(2 - 2y - 1 - y^2 - 2y) = 2(1+y)(2y+9)$

$\Rightarrow 5(1 - 4y - y^2) = 2(2y + 9 + 2y^2 + 9y)$

$\Rightarrow 5 - 20y - 5y^2 = 22y + 18 + 4y^2$

$\Rightarrow 9y^2 + 42y + 13 = 0$

$\Rightarrow 9y^2 + 3y + 39y + 13 = 0$

$\Rightarrow 3y(3y+1) + 13(3y+1) = 0$

$\Rightarrow (3y+1)(3y+13) = 0$

$$\Rightarrow \quad y = -\frac{1}{3}, -\frac{13}{3}$$

$$\therefore \quad \cos 2x = -\frac{1}{3}, -\frac{13}{3}$$

$$\Rightarrow \quad \cos 2x = -\frac{1}{3} \quad \left[\because \cos 2x \neq -\frac{13}{3}\right]$$

Now, $\cos 4x = 2\cos^2 2x - 1$

$$= 2\left(-\frac{1}{3}\right)^2 - 1$$

$$= \frac{2}{9} - 1 = -\frac{7}{9}$$

**64.** *(b)* We have, $P$ (exactly one of $A$ or $B$ occurs)

$$= P(A \cup B) - P(A \cap B)$$
$$= P(A) + P(B) - 2P(A \cap B)$$

According to the question,

$$P(A) + P(B) - 2P(A \cap B) = \frac{1}{4} \quad \ldots\text{(i)}$$

$$P(B) + P(C) - 2P(B \cap C) = \frac{1}{4} \quad \ldots\text{(ii)}$$

and $P(C) + P(A) - 2P(C \cap A) = \frac{1}{4}$ ...(iii)

On adding Eqs. (i), (ii) and (iii), we get

$$2\,[P(A) + P(B) + P(C) - P(A \cap B) - P(B \cap C) - P(C \cap A)] = \frac{3}{4}$$

$$\Rightarrow P(A) + P(B) + P(C) - P(A \cap B) - P(B \cap C) - P(C \cap A) = \frac{3}{8}$$

$\therefore$ $P$ (atleast one event occurs)

$$= P(A \cup B \cup C)$$
$$= P(A) + P(B) + P(C) - P(A \cap B) - P(B \cap C) - P(C \cap A) + P(A \cap B \cap C)$$

$$= \frac{3}{8} + \frac{1}{16} = \frac{7}{16} \quad \left[\because P(A \cap B \cap C) = \frac{1}{16}\right]$$

**65.** *(a)* Given, $2\omega + 1 = z$

$$\Rightarrow \quad 2\omega + 1 = \sqrt{-3} \quad [\because z = \sqrt{-3}]$$

$$\Rightarrow \quad \omega = \frac{-1 + \sqrt{3}i}{2}$$

Since, $\omega$ is cube root of unity.

$$\therefore \quad \omega^2 = \frac{-1 - \sqrt{3}i}{2}$$

and $\omega^{3n} = 1$

Now, $\begin{vmatrix} 1 & 1 & 1 \\ 1 & -\omega^2 - 1 & \omega^2 \\ 1 & \omega^2 & \omega^7 \end{vmatrix} = 3k$

$$\Rightarrow \quad \begin{vmatrix} 1 & 1 & 1 \\ 1 & \omega & \omega^2 \\ 1 & \omega^2 & \omega \end{vmatrix} = 3k$$

$[\because 1 + \omega + \omega^2 = 0 \text{ and } \omega^7 = (\omega^3)^2 \cdot \omega = \omega]$

On applying $R_1 \to R_1 + R_2 + R_3$, we get

$$\begin{vmatrix} 3 & 1 + \omega + \omega^2 & 1 + \omega + \omega^2 \\ 1 & \omega & \omega^2 \\ 1 & \omega^2 & \omega \end{vmatrix} = 3k$$

$$\Rightarrow \begin{vmatrix} 3 & 0 & 0 \\ 1 & \omega & \omega^2 \\ 1 & \omega^2 & \omega \end{vmatrix} = 3k$$

$$\Rightarrow \quad 3(\omega^2 - \omega^4) = 3k$$

$$\Rightarrow \quad (\omega^2 - \omega) = k$$

$$\therefore \quad k = \left(\frac{-1 - \sqrt{3}i}{2}\right) - \left(\frac{-1 + \sqrt{3}i}{2}\right)$$

$$= -\sqrt{3}i = -z$$

**66.** *(d)* Given, vertices of triangle are $(k, -3k)$, $(5, k)$ and $(-k, 2)$.

$$\therefore \quad \frac{1}{2}\begin{vmatrix} k & -3k & 1 \\ 5 & k & 1 \\ -k & 2 & 1 \end{vmatrix} = \pm 28$$

$$\Rightarrow \quad \begin{vmatrix} k & -3k & 1 \\ 5 & k & 1 \\ -k & 2 & 1 \end{vmatrix} = \pm 56$$

$$\Rightarrow k(k - 2) + 3k(5 + k) + 1(10 + k^2) = \pm 56$$

$$\Rightarrow k^2 - 2k + 15k + 3k^2 + 10 + k^2 = \pm 56$$

$$\Rightarrow \quad 5k^2 + 13k + 10 = \pm 56$$

$$\Rightarrow \quad 5k^2 + 13k - 66 = 0$$

$$\text{or} \quad 5k^2 + 13k - 46 = 0$$

$$\Rightarrow \quad k = 2 \quad [\because k \in I]$$

Thus, the coordinates of vertices of triangle are $A(2, -6)$, $B(5, 2)$ and $C(-2, 2)$.

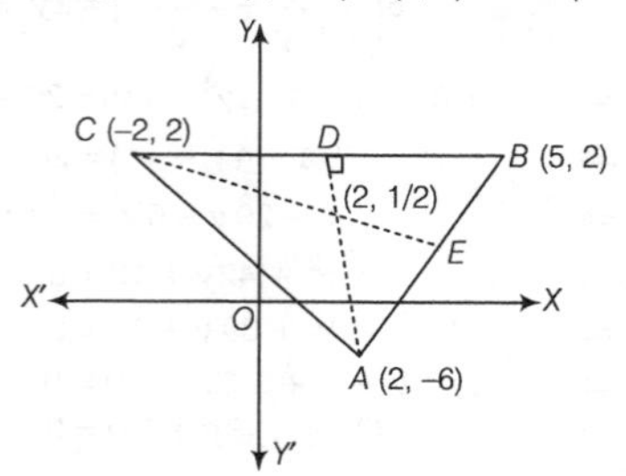

Now, equation of altitude from vertex $A$ is

$$y-(-6)=\frac{-1}{\left(\frac{2-2}{-2-5}\right)}(x-2) \Rightarrow x=2 \quad \ldots\text{(i)}$$

Equation of altitude from vertex $C$ is

$$y-2=\frac{-1}{\left[\frac{2-(-6)}{5-2}\right]}[x-(-2)]$$

$$\Rightarrow \quad 3x+8y-10=0 \quad \ldots\text{(ii)}$$

On solving Eqs. (i) and (ii), we get $x=2$ and $y=\frac{1}{2}$.

$\therefore$ Orthocentre $=\left(2, \frac{1}{2}\right)$

**67.** *(c)* Total length $=2r+r\theta=20$

$$\Rightarrow \quad \theta=\frac{20-2r}{r}$$

$r\theta$

Now, area of flower-bed,

$$A=\frac{1}{2}r^2\theta \Rightarrow A=\frac{1}{2}r^2\left(\frac{20-2r}{r}\right)$$

$$\Rightarrow \quad A=10r-r^2$$

$$\therefore \quad \frac{dA}{dr}=10-2r$$

For maxima or minima, put $\frac{dA}{dr}=0$.

$$\Rightarrow \quad 10-2r=0 \Rightarrow r=5$$

$$\therefore \quad A_{\max}=\frac{1}{2}(5)^2\left[\frac{20-2(5)}{5}\right]$$

$$=\frac{1}{2}\times 25\times 2=25 \text{ sq m}$$

**68.** *(d)* Required area

$$=\int_0^1 (1+\sqrt{x})dx+\int_1^2 (3-x)dx-\int_0^2 \frac{x^2}{4}dx$$

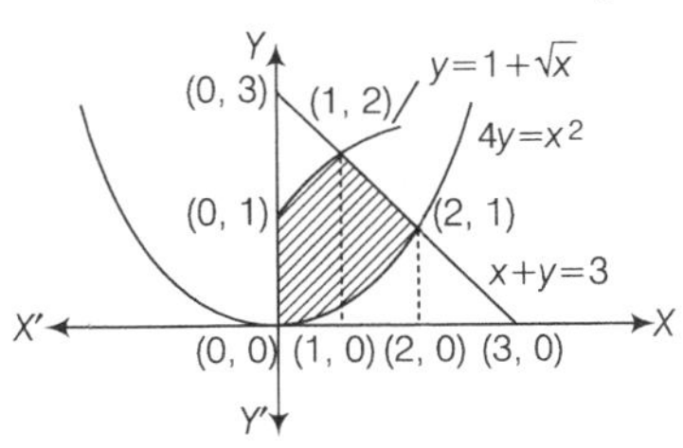


$$=\left[x+\frac{x^{3/2}}{3/2}\right]_0^1+\left[3x-\frac{x^2}{2}\right]_1^2-\left[\frac{x^3}{12}\right]_0^2$$

$$=\left(1+\frac{2}{3}\right)+\left(6-2-3+\frac{1}{2}\right)-\left(\frac{8}{12}\right)$$

$$=\frac{5}{3}+\frac{3}{2}-\frac{2}{3}=1+\frac{3}{2}=\frac{5}{2} \text{ sq units}$$

**69.** *(b)* Any line parallel to $\frac{x}{1}=\frac{y}{4}=\frac{z}{5}$ and passing through $P(1,-2,3)$ is

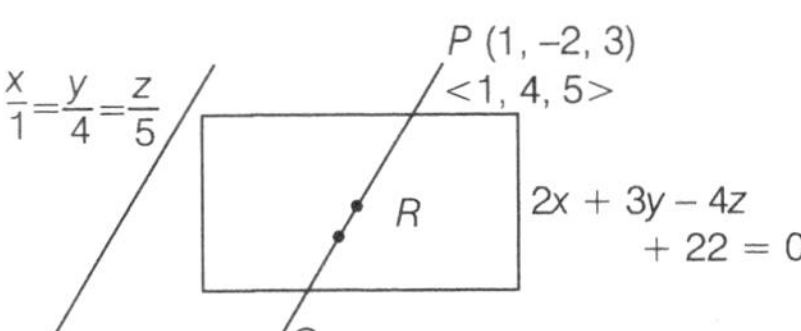


$$\frac{x-1}{1}=\frac{y+2}{4}=\frac{z-3}{5}=\lambda \text{ (say)}$$

Any point on above line can be written as $(\lambda+1, 4\lambda-2, 5\lambda+3)$.

$\therefore$ Coordinates of $R$ are $(\lambda+1, 4\lambda-2, 5\lambda+3)$.

Since, point $R$ lies on the above plane.

$$\therefore 2(\lambda+1)+3(4\lambda-2)-4(5\lambda+3)+22=0$$

$$\Rightarrow \quad \lambda=1$$

So, point $R$ is $(2, 2, 8)$.

Now, $PR=\sqrt{(2-1)^2+(2+2)^2+(8-3)^2}$

$$=\sqrt{42}$$

$$\therefore \quad PQ=2PR=2\sqrt{42}$$

**70.** *(a)* Let $y=\tan^{-1}\left(\frac{6x\sqrt{x}}{1-9x^3}\right)$

$$=\tan^{-1}\left[\frac{2\cdot(3x^{3/2})}{1-(3x^{3/2})^2}\right]$$

$$=2\tan^{-1}(3x^{3/2})$$

$$\left[\because 2\tan^{-1}x=\tan^{-1}\frac{2x}{1-x^2}\right]$$

$$\therefore \quad \frac{dy}{dx}=2\cdot\frac{1}{1+(3x^{3/2})^2}\cdot 3\times\frac{3}{2}(x)^{1/2}$$

$$=\frac{9}{1+9x^3}\cdot\sqrt{x}$$

$$\therefore \quad g(x)=\frac{9}{1+9x^3}$$

**71.** *(a)* We have,

$$(2+\sin x)\frac{dy}{dx}+(y+1)\cos x=0$$

$$\Rightarrow \quad \frac{dy}{dx}+\frac{\cos x}{2+\sin x}y=\frac{-\cos x}{2+\sin x}$$

which is a linear differential equation.

$$\therefore \quad \text{IF}=e^{\int\frac{\cos x}{2+\sin x}dx}=e^{\log(2+\sin x)}$$

$$=2+\sin x$$

$\therefore$ Required solution is given by

$$y\cdot(2+\sin x)=\int\frac{-\cos x}{2+\sin x}\cdot(2+\sin x)dx+C$$

$$\Rightarrow \quad y(2+\sin x)=-\sin x+C$$

Also, $\quad y(0)=1$

$$\therefore \quad 1(2+\sin 0)=-\sin 0+C$$

$$\Rightarrow \quad C=2$$

$$\therefore \quad y=\frac{2-\sin x}{2+\sin x}$$

$$\Rightarrow \quad y\left(\frac{\pi}{2}\right)=\frac{2-\sin\frac{\pi}{2}}{2+\sin\frac{\pi}{2}}=\frac{1}{3}$$

**72.** *(c)* Let $AB=h$, then $AD=2h$ and $AC=BC=\frac{h}{2}$

Again, let $\angle CPA=\alpha$

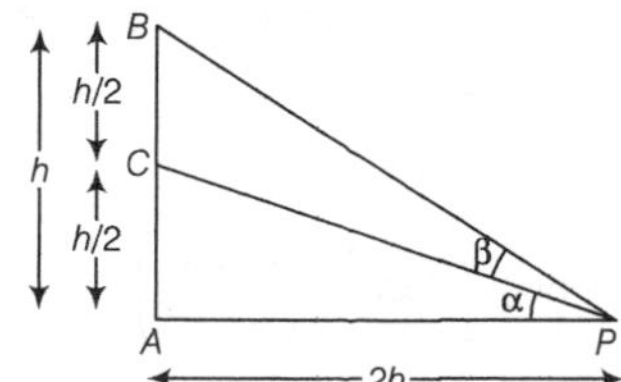


Now, in $\Delta ABP$,

$$\tan(\alpha+\beta)=\frac{AB}{AP}=\frac{h}{2h}=\frac{1}{2}$$

Also, in $\Delta ACP$, $\tan\alpha=\frac{AC}{AP}=\frac{\frac{h}{2}}{2h}=\frac{1}{4}$

Now, $\tan\beta=\tan[(\alpha+\beta)-\alpha]$

$$=\frac{\tan(\alpha+\beta)-\tan\alpha}{1+\tan(\alpha+\beta)\tan\alpha}$$

$$=\frac{\frac{1}{2}-\frac{1}{4}}{1+\frac{1}{2}\times\frac{1}{4}}=\frac{\frac{1}{4}}{\frac{9}{8}}=\frac{2}{9}$$

**73.** *(b)* We have, $A=\begin{bmatrix}2 & -3\\ -4 & 1\end{bmatrix}$

$$\therefore \quad A^2=A\cdot A=\begin{bmatrix}2 & -3\\ -4 & 1\end{bmatrix}\begin{bmatrix}2 & -3\\ -4 & 1\end{bmatrix}$$

$$=\begin{bmatrix}4+12 & -6-3\\ -8-4 & 12+1\end{bmatrix}$$

$$=\begin{bmatrix}16 & -9\\ -12 & 13\end{bmatrix}$$

Now,

$$3A^2+12A=3\begin{bmatrix}16 & -9\\ -12 & 13\end{bmatrix}+12\begin{bmatrix}2 & -3\\ -4 & 1\end{bmatrix}$$

$$=\begin{bmatrix}48 & -27\\ -36 & 39\end{bmatrix}+\begin{bmatrix}24 & -36\\ -48 & 12\end{bmatrix}$$

$$=\begin{bmatrix}72 & -63\\ -84 & 51\end{bmatrix}$$

$$\therefore \text{adj}(3A^2+12A)=\begin{bmatrix}51 & 63\\ 84 & 72\end{bmatrix}$$

**74.** *(c)* We have,

$$225a^2+9b^2+25c^2-75ac-45ab-15bc=0$$

$$\Rightarrow (15a)^2+(3b)^2+(5c)^2-(15a)(5c)-(15a)(3b)-(3b)(5c)=0$$

$$\Rightarrow \frac{1}{2}[(15a-3b)^2+(3b-5c)^2+(5c-15a)^2]=0$$

$$\Rightarrow \quad 15a=3b, 3b=5c \text{ and } 5c=15a$$

$$\therefore \quad 15a=3b=5c$$

$$\Rightarrow \quad \frac{a}{1}=\frac{b}{5}=\frac{c}{3}=\lambda \text{ (say)}$$

$$\Rightarrow \quad a=\lambda, b=5\lambda, c=3\lambda$$

Hence, $a, b$ and $c$ are in AP.

**75.** *(b)* Given, equations of lines are

$$\frac{x-1}{1}=\frac{y+2}{-2}=\frac{z-4}{3}$$

and $$\frac{x-2}{2}=\frac{y+1}{-1}=\frac{z+7}{-1}$$

Let $\mathbf{n}_1=\hat{\mathbf{i}}-2\hat{\mathbf{j}}+3\hat{\mathbf{k}}$ and $\mathbf{n}_2=2\hat{\mathbf{i}}-\hat{\mathbf{j}}-\hat{\mathbf{k}}$

$\therefore$ Any vector $\mathbf{n}$ perpendicular to both $\mathbf{n}_1, \mathbf{n}_2$ is given by

$$\mathbf{n}=\mathbf{n}_1\times\mathbf{n}_2$$

$$\Rightarrow \quad \mathbf{n} = \begin{vmatrix} \hat{\mathbf{i}} & \hat{\mathbf{j}} & \hat{\mathbf{k}} \\ 1 & -2 & 3 \\ 2 & -1 & -1 \end{vmatrix}$$

$$= 5\hat{\mathbf{i}} + 7\hat{\mathbf{j}} + 3\hat{\mathbf{k}}$$

$\therefore$ Equation of a plane passing through $(1, -1, -1)$ and perpendicular to $\mathbf{n}$ is given by

$$5(x-1) + 7(y+1) + 3(z+1) = 0$$

$$\Rightarrow \quad 5x + 7y + 3z + 5 = 0$$

$\therefore$ Required distance $= \left| \dfrac{5 + 21 - 21 + 5}{\sqrt{5^2 + 7^2 + 3^2}} \right|$

$$= \frac{10}{\sqrt{83}} \text{ units}$$

**76.** *(b)* We have, $I_n = \int \tan^n x \, dx$

$$\therefore \quad I_n + I_{n+2} = \int \tan^n x \, dx + \int \tan^{n+2} x \, dx$$

$$= \int \tan^n x (1 + \tan^2 x) \, dx$$

$$= \int \tan^n x \sec^2 x \, dx$$

$$= \frac{\tan^{n+1} x}{n+1} + C$$

Put $n = 4$, we get $I_4 + I_6 = \dfrac{\tan^5 x}{5} + C$

$$\therefore \quad a = \frac{1}{5} \text{ and } b = 0$$

**77.** *(b)* We have, $e = \dfrac{1}{2}$ and $\dfrac{a}{e} = 4$

$$\therefore \quad a = 2$$

Now, $b^2 = a^2(1 - e^2) = (2)^2 \left[ 1 - \left(\dfrac{1}{2}\right)^2 \right]$

$$= 4\left(1 - \frac{1}{4}\right) = 3 \Rightarrow b = \sqrt{3}$$

$\therefore$ Equation of the ellipse is

$$\frac{x^2}{(2)^2} + \frac{y^2}{(\sqrt{3})^2} = 1$$

$$\Rightarrow \quad \frac{x^2}{4} + \frac{y^2}{3} = 1$$

Now, the equation of normal at $\left(1, \dfrac{3}{2}\right)$ is

$$\frac{a^2 x}{x_1} - \frac{b^2 y}{y_1} = a^2 - b^2$$

$$\Rightarrow \quad \frac{4x}{1} - \frac{3y}{(3/2)} = 4 - 3$$

$$\Rightarrow \quad 4x - 2y = 1$$

**78.** *(b)* Let the equation of hyperbola be $\dfrac{x^2}{a^2} - \dfrac{y^2}{b^2} = 1.$

$$\therefore \quad ae = 2 \Rightarrow a^2 e^2 = 4$$

$$\Rightarrow \quad a^2 + b^2 = 4$$

$$\Rightarrow \quad b^2 = 4 - a^2$$

$$\therefore \quad \frac{x^2}{a^2} - \frac{y^2}{4 - a^2} = 1$$

Since, $(\sqrt{2}, \sqrt{3})$ lie on hyperbola.

$$\therefore \quad \frac{2}{a^2} - \frac{3}{4 - a^2} = 1$$

$$\Rightarrow \quad 8 - 2a^2 - 3a^2 = a^2(4 - a^2)$$

$$\Rightarrow \quad 8 - 5a^2 = 4a^2 - a^4$$

$$\Rightarrow \quad a^4 - 9a^2 + 8 = 0$$

$$\Rightarrow \quad (a^4 - 8)(a^4 - 1) = 0$$

$$\Rightarrow \quad a^4 = 8, a^4 = 1$$

$$\therefore \quad a = 1$$

Now, equation of hyperbola is $\dfrac{x^2}{1} - \dfrac{y^2}{3} = 1.$

$\therefore$ Equation of tangent at $(\sqrt{2}, \sqrt{3})$ is given by

$$\sqrt{2}x - \frac{\sqrt{3}y}{3} = 1$$

$$\Rightarrow \quad \sqrt{2}x - \frac{y}{\sqrt{3}} = 1$$

which passes through the point $(2\sqrt{2}, 3\sqrt{3})$.

**79.** *(c)* We have, $f(x) = \dfrac{x}{1 + x^2}$

$$\therefore \quad f\left(\frac{1}{x}\right) = \frac{\frac{1}{x}}{1 + \frac{1}{x^2}} = \frac{x}{1 + x^2} = f(x)$$

$$\therefore \quad f\left(\frac{1}{2}\right) = f(2) \text{ or } f\left(\frac{1}{3}\right) = f(3) \text{ and so on.}$$

So, $f(x)$ is many-one function.

Again, let $\quad y = f(x)$

$$\Rightarrow \quad y = \frac{x}{1 + x^2}$$

$$\Rightarrow \quad y + x^2 y = x$$

$$\Rightarrow \quad yx^2 - x + y = 0$$

As, $\quad x \in R$

$$\therefore \quad (-1)^2 - 4(y)(y) \geq 0$$

$$\Rightarrow \quad 1 - 4y^2 \geq 0$$

$$\Rightarrow \quad y \in \left[ \frac{-1}{2}, \frac{1}{2} \right]$$

$\therefore$ Range = Codomain $= \left[\frac{-1}{2}, \frac{1}{2}\right]$

So, $f(x)$ is surjective.

Hence, $f(x)$ is surjective but not injective.

**80.** *(b)* $\lim_{x \to \pi/2} \frac{\cot x - \cos x}{(\pi - 2x)^3}$

$$= \lim_{x \to \pi/2} \frac{1}{8} \cdot \frac{\cos x(1 - \sin x)}{\sin x\left(\frac{\pi}{2} - x\right)^3}$$

$$= \lim_{h \to 0} \frac{1}{8} \cdot \frac{\cos\left(\frac{\pi}{2} - h\right)\left[1 - \sin\left(\frac{\pi}{2} - h\right)\right]}{\sin\left(\frac{\pi}{2} - h\right)\left(\frac{\pi}{2} - \frac{\pi}{2} + h\right)^3}$$

$$= \frac{1}{8} \lim_{h \to 0} \frac{\sin h\,(1 - \cos h)}{\cos h \cdot h^3}$$

$$= \frac{1}{8} \lim_{h \to 0} \frac{\sin h\left(2\sin^2 \frac{h}{2}\right)}{\cos h \cdot h^3}$$

$$= \frac{1}{4} \lim_{h \to 0} \frac{\sin h \cdot \sin^2\left(\frac{h}{2}\right)}{h^3 \cos h}$$

$$= \frac{1}{4} \lim_{h \to 0} \left(\frac{\sin h}{h}\right)\left(\frac{\sin \frac{h}{2}}{\frac{h}{2}}\right)^2 \cdot \frac{1}{\cos h} \cdot \frac{1}{4}$$

$$= \frac{1}{4} \times \frac{1}{4} = \frac{1}{16}$$

**81.** *(b)* We have, $\mathbf{a} = 2\hat{\mathbf{i}} + \hat{\mathbf{j}} - 2\hat{\mathbf{k}}$

$\Rightarrow \quad |\mathbf{a}| = \sqrt{4 + 1 + 4} = 3$

and $\quad \mathbf{b} = \hat{\mathbf{i}} + \hat{\mathbf{j}} \Rightarrow |\mathbf{b}| = \sqrt{1 + 1} = \sqrt{2}$

Now, $|\mathbf{c} - \mathbf{a}| = 3 \Rightarrow |\mathbf{c} - \mathbf{a}|^2 = 9$

$\Rightarrow \quad (\mathbf{c} - \mathbf{a}) \cdot (\mathbf{c} - \mathbf{a}) = 9$

$\Rightarrow |\mathbf{c}|^2 + |\mathbf{a}|^2 - 2\mathbf{c} \cdot \mathbf{a} = 9 \quad \ldots$(i)

Again, $|(\mathbf{a} \times \mathbf{b}) \times \mathbf{c}| = 3$

$\Rightarrow \quad |\mathbf{a} \times \mathbf{b}||\mathbf{c}| \sin 30^\circ = 3 \Rightarrow |\mathbf{c}| = \frac{6}{|\mathbf{a} \times \mathbf{b}|}$

But $\mathbf{a} \times \mathbf{b} = \begin{vmatrix} \hat{\mathbf{i}} & \hat{\mathbf{j}} & \hat{\mathbf{k}} \\ 2 & 1 & -2 \\ 1 & 1 & 0 \end{vmatrix} = 2\hat{\mathbf{i}} - 2\hat{\mathbf{j}} + \hat{\mathbf{k}}$

$\therefore \quad |\mathbf{c}| = \frac{6}{\sqrt{4 + 4 + 1}} = 2 \quad \ldots$(ii)

From Eqs. (i) and (ii), we get

$(2)^2 + (3)^2 - 2\mathbf{c} \cdot \mathbf{a} = 9$

$\Rightarrow \quad 4 + 9 - 2\mathbf{c} \cdot \mathbf{a} = 9$

$\Rightarrow \quad \mathbf{c} \cdot \mathbf{a} = 2$

**82.** *(b)* Given curve is

$$y(x - 2)(x - 3) = x + 6 \quad \ldots\text{(i)}$$

Put $x = 0$ in Eq. (i), we get

$$y(-2)(-3) = 6 \Rightarrow y = 1$$

So, point of intersection is (0, 1).

Now, $y = \frac{x + 6}{(x - 2)(x - 3)}$

$$\Rightarrow \frac{dy}{dx} = \frac{1(x - 2)(x - 3) - (x + 6)(x - 3 + x - 2)}{(x - 2)^2(x - 3)^2}$$

$$\Rightarrow \left(\frac{dy}{dx}\right)_{(0, 1)} = \frac{6 + 30}{4 \times 9} = \frac{36}{36} = 1$$

$\therefore$ Equation of normal at (0, 1) is given by

$$y - 1 = \frac{-1}{1}(x - 0)$$

$\Rightarrow \quad x + y - 1 = 0$

which passes through the point $\left(\frac{1}{2}, \frac{1}{2}\right)$.

**83.** *(a)* Total number of ways of selecting 2 different numbers from {0, 1, 2, ..., 10} $= {}^{11}C_2 = 55$

Let two numbers selected be $x$ and $y$.

Then, $\quad x + y = 4m \quad \ldots$(i)

and $\quad x - y = 4n \quad \ldots$(ii)

$\Rightarrow \quad 2x = 4(m + n)$ and $2y = 4(m - n)$

$\Rightarrow \quad x = 2(m + n)$

and $\quad y = 2(m - n)$

$\therefore$ $x$ and $y$ both are even numbers.

| $x$ | $y$ |
|---|---|
| 0 | 4, 8 |
| 2 | 6, 10 |
| 4 | 0, 8 |
| 6 | 2, 10 |
| 8 | 0, 4 |
| 10 | 2, 6 |

$\therefore$ Required probability $= \frac{6}{55}$

**84.** *(a)* Given, $X$ has 7 friends, 4 of them are ladies and 3 are men while $Y$ has 7 friends, 3 of them are ladies and 4 are men.

$\therefore$ Total number of required ways

$$= {}^3C_3 \times {}^4C_0 \times {}^4C_0 \times {}^3C_3 + {}^3C_2 \times {}^4C_1 \times {}^4C_1 \times {}^3C_2 + {}^3C_1 \times {}^4C_2 \times {}^4C_2 \times {}^3C_1 + {}^3C_0 \times {}^4C_3 \times {}^4C_3 \times {}^3C_0$$

$= 1 + 144 + 324 + 16 = 485$

85. (d) $({}^{21}C_1 - {}^{10}C_1) + ({}^{21}C_2 - {}^{10}C_2) + ({}^{21}C_3 - {}^{10}C_3) + \dots + ({}^{21}C_{10} - {}^{10}C_{10})$

$$= ({}^{21}C_1 + {}^{21}C_2 + \dots + {}^{21}C_{10}) - ({}^{10}C_1 + {}^{10}C_2 + \dots + {}^{10}C_{10})$$

$$= \frac{1}{2}({}^{21}C_1 + {}^{21}C_2 + \dots + {}^{21}C_{20}) - (2^{10} - 1)$$

$$= \frac{1}{2}({}^{21}C_1 + {}^{21}C_2 + \dots + {}^{21}C_{21} - 1) - (2^{10} - 1)$$

$$= \frac{1}{2}(2^{21} - 2) = 2^{20} - 1 - 2^{10} + 1 - (2^{10} - 1)$$

$$= 2^{20} - 2^{10}$$

86. (a) Given box contains 15 green and 10 yellow balls.

$\therefore$ Total number of balls = 15 + 10 = 25

$P(\text{green balls}) = \frac{15}{25} = \frac{3}{5} = p$ = Probability of success

$P(\text{yellow balls}) = \frac{10}{25} = \frac{2}{5} = q$ = Probability of unsuccess and $n = 10$ = Number of trials.

$\therefore$ Variance $= npq = 10 \times \frac{3}{5} \times \frac{2}{5} = \frac{12}{5}$

87. (a) We have, $f(x) = ax^2 + bx + c$

Now, $f(x + y) = f(x) + f(y) + xy$

Put $y = 0 \Rightarrow f(x) = f(x) + f(0) + 0$

$\Rightarrow f(0) = 0 \Rightarrow c = 0$

Again, put $y = -x$

$\therefore f(0) = f(x) + f(-x) - x^2$

$\Rightarrow 0 = ax^2 + bx + ax^2 - bx - x^2$

$\Rightarrow 2ax^2 - x^2 = 0$

$\Rightarrow a = \frac{1}{2}$

Also, $a + b + c = 3$

$\Rightarrow \frac{1}{2} + b + 0 = 3$

$\Rightarrow b = \frac{5}{2}$

$\therefore f(x) = \frac{x^2 + 5x}{2}$

Now, $f(n) = \frac{n^2 + 5n}{2} = \frac{1}{2}n^2 + \frac{5}{2}n$

$$\therefore \sum_{n=1}^{10} f(n) = \frac{1}{2}\sum_{n=1}^{10} n^2 + \frac{5}{2}\sum_{n=1}^{10} n$$

$$= \frac{1}{2} \cdot \frac{10 \times 11 \times 21}{6} + \frac{5}{2} \times \frac{10 \times 11}{2}$$

$$= \frac{385}{2} + \frac{275}{2} = \frac{660}{2} = 330$$

88. (c) Let the radius of circle with least area be $r$.

Then, then coordinates of centre $= (0, 4 - r)$.

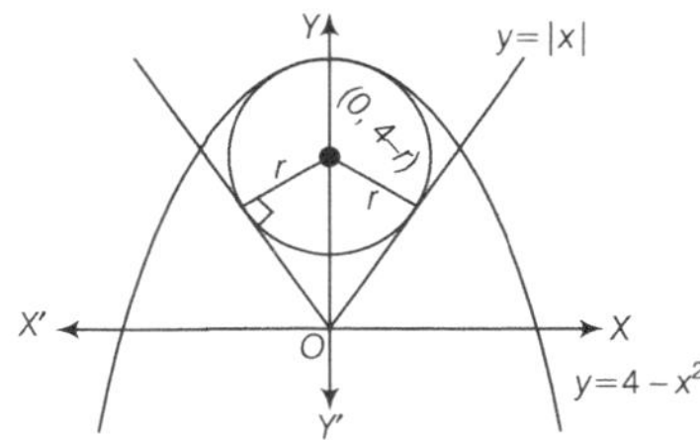


Since, circle touches the line $y = x$ in first quadrant.

$\therefore \left|\frac{0 - (4 - r)}{\sqrt{2}}\right| = r$

$\Rightarrow r - 4 = \pm r\sqrt{2}$

$\Rightarrow r = \frac{4}{\sqrt{2} + 1}$ or $\frac{4}{1 - \sqrt{2}}$

But $r \neq \frac{4}{1 - \sqrt{2}}$ $\left[\because \frac{4}{1 - \sqrt{2}} < 0\right]$

$\therefore r = \frac{4}{\sqrt{2} + 1} = 4(\sqrt{2} - 1)$

89. (d) Given quadratic equation is

$x(x + 1) + (x + 1)(x + 2) + \dots + (x + \overline{n - 1})(x + n) = 10n$

$\Rightarrow (x^2 + x^2 + \dots + x^2) + [(1 + 3 + 5 + \dots + (2n - 1)]x + [(1 \cdot 2 + 2 \cdot 3 + \dots + (n - 1)n] = 10n$

$\Rightarrow nx^2 + n^2x + \frac{n(n^2 - 1)}{3} - 10n = 0$

$\Rightarrow x^2 + nx + \frac{n^2 - 1}{3} - 10 = 0$

$\Rightarrow 3x^2 + 3nx + n^2 - 31 = 0$

Let $\alpha$ and $\beta$ be the roots.

Since, $\alpha$ and $\beta$ are consecutive.

$\therefore |\alpha - \beta| = 1$

$\Rightarrow (\alpha - \beta)^2 = 1$

Again, $(\alpha - \beta)^2 = (\alpha + \beta)^2 - 4\alpha\beta$

$\Rightarrow \quad 1 = \left(\frac{-3n}{3}\right)^2 - 4\left(\frac{n^2 - 31}{3}\right)$

$\Rightarrow \quad 1 = n^2 - \frac{4}{3}(n^2 - 31)$

$\Rightarrow \quad 3 = 3n^2 - 4n^2 + 124$

$\Rightarrow \quad n^2 = 121$

$\Rightarrow \quad n = \pm 11$

$\therefore \quad n = 11 \qquad [\because n > 0]$

**90.** *(b)* Let $I = \int_{\pi/4}^{3\pi/4} \frac{dx}{1 + \cos x}$

$= \int_{\pi/4}^{3\pi/4} \frac{1 - \cos x}{1 - \cos^2 x} dx$

$= \int_{\pi/4}^{3\pi/4} \frac{1 - \cos x}{\sin^2 x} dx$

$= \int_{\pi/4}^{3\pi/4} (\text{cosec}^2 x - \text{cosec } x \cot x) dx$

$= [-\cot x + \text{cosec } x]_{\pi/4}^{3\pi/4}$

$= [(1 + \sqrt{2}) - (-1 + \sqrt{2})] = 2$

# SOLVED PAPER 2017

# JEE Advanced

## Paper 1

## Physics

### Section 1 (Maximum Marks : 28)

- This section contains **SEVEN** questions.
- Each question has **FOUR** options (a), (b), (c) and (d). **ONE OR MORE THAN ONE** of these four options is (are) correct.
- For each question, darken the bubble(s) corresponding to all the correct option(s) in the ORS.
- For each question, marks will be awarded in one of the following categories:

*Full Marks* : + 4 If only the bubble(s) corresponding to all the correct option(s) is (are) darkened.

*Partial Marks* : + 1 For darkening a bubble corresponding **to each correct option**, provided NO incorrect option is darkened

*Zero Marks* : 0 If none of the bubbles is darkened.

*Negative Marks* : −2 In all other cases.

**1.** In the circuit shown, $L = 1\ \mu\text{H}$, $C = 1\ \mu\text{F}$ and $R = 1\ \text{k}\Omega$. They are connected in series with an AC source $V = V_0 \sin \omega t$ as shown. Which of the following options is/are correct?

$L$=1μH $C$=1μF $R$=1kΩ

$V_0 \sin \omega t$

(a) At $\omega \sim 0$ the current flowing through the circuit becomes nearly zero

(b) The frequency at which the current will be in phase with the voltage is independent of $R$

(c) The current will be in phase with the voltage if $\omega = 10^4\ \text{rads}^{-1}$

(d) At $\omega >> 10^6\ \text{rads}^{-1}$, the circuit behaves like a capacitor

**2.** For an isosceles prism of angle $A$ and refractive index $\mu$, it is found that the angle of minimum deviation $\delta_m = A$. Which of the following options is/are correct?

(a) For the angle of incidence $i_1 = A$, the ray inside the prism is parallel to the base of the prism

(b) At minimum deviation, the incident angle $i_1$ and the refracting angle $r_1$ at the first refracting surface are related by $r_1 = \left(\frac{i_i}{2}\right)$

(c) For this prism, the emergent ray at the second surface will be tangential to the surface when the angle of incidence at the first surface is $i_1 = \sin^{-1}\left[\sin A \sqrt{4\cos^2 \frac{A}{2} - 1} - \cos A\right]$

(d) For this prism, the refractive index $\mu$ and the angle prism $A$ are related as $A = \frac{1}{2}\cos^{-1}\left(\frac{\mu}{2}\right)$

3. A circular insulated copper wire loop is twisted to form two loops of area $A$ and $2A$ as shown in the figure. At the point of crossing the wires remain electrically insulated from each other. The entire loop lies in the plane (of the paper). A uniform magnetic field **B** points into the plane of the paper. At $t = 0$, the loop starts rotating about the common diameter as axis with a constant angular velocity $\omega$ in the magnetic field. Which of the following options is/are correct?

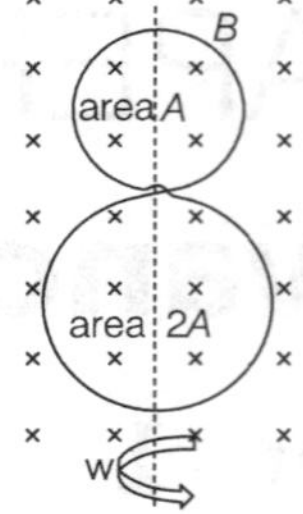


(a) the emf induced in the loop is proportional to the sum of the areas of the two loops
(b) The rate of change of the flux is maximum when the plane of the loops is perpendicular to plane of the paper
(c) The net emf induced due to both the loops is proportional to $\cos \omega t$
(d) The amplitude of the maximum net emf induced due to both the loops is equal to the amplitude of maximum emf induced in the smaller loop alone

4. A flat plane is moving normal to its plane through a gas under the action of a constant force $F$. The gas is kept at a very low pressure. The speed of the plate $v$ is much less than the average speed $u$ of the gas molecules. Which of the following options is/are true?
(a) At a later time the external force $F$ balances the resistive force
(b) The plate will continue to move with constant non-zero acceleration, at all times
(c) The resistive force experienced by the plate is proportional to $v$
(d) The pressure difference between the leading and trailing faces of the plate is proportional to $uv$

5. A block of mass $M$ has a circular cut with a frictionless surface as shown. The block rests on the horizontal frictionless surfaced of a fixed table. Initially the right edge of the block is at $x = 0$, in a coordinate system fixed to the table. A point mass $m$ is released from rest at the topmost point of the path as shown and it slides down. When the mass loses contact with the block, its position is $x$ and the velocity is $v$. At that instant, which of the following option is/are correct?

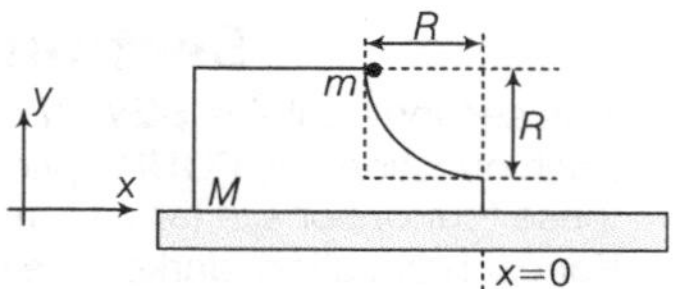


(a) The velocity of the point mass $m$ is $v = \sqrt{\dfrac{2gR}{1+\dfrac{m}{M}}}$
(b) The $x$ component of displacement of the centre of mass of the block $M$ is $-\dfrac{mR}{M+m}$
(c) The position of the point mass is $x = -\sqrt{2}\dfrac{mR}{M+m}$
(d) The velocity of the block $M$ is $V = -\dfrac{m}{M}\sqrt{2gR}$

6. A block $M$ hangs vertically at the bottom end of a uniform rope of constant mass per unit length. The top end of the rope is attached to a fixed rigid support at $O$. A transverse wave pulse (Pulse 1) of wavelength $\lambda_0$ is produced at point $O$ on the rope. The pulse takes time $T_{OA}$ to reach point $A$. If the wave pulse of wavelength $\lambda_0$ is produced at point $A$ (Pulse 2) without disturbing the position of $M$ it takes time $T_{AO}$ to reach point $O$. Which of the following options is/are correct?

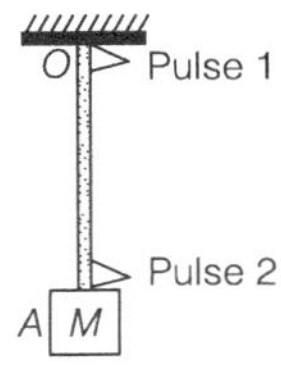


(a) The time $T_{AO} = T_{OA}$
(b) The wavelength of Pulse 1 becomes longer when it reaches point $A$
(c) The velocity of any pulse along the rope is independent of its frequency and wavelength
(d) The velocities of the two pulses (Pulse 1 and Pulse 2) are the same at the mid-point of rope

7. A human body has a surface area of approximately 1 $m^2$. The normal body temperature is 10K above the surrounding room temperature $T_0$. Take the room temperature to be $T_0 = 300$ K. For $T_0 = 300$ K, the value of $\sigma T_0^4 = 460 \text{ Wm}^{-2}$ (where $\sigma$ is the Stefan Boltzmann constant). Which of the following options is/are correct?

(a) If the body temperature rises significantly, then the peak in the spectrum of electromagnetic radiation emitted by the body would shift to longer wavelengths
(b) If the surrounding temperature reduces by a small amount $\Delta T_0 << T_0$, then to maintain the same body temperature the same (living) human being needs to radiate $\Delta W = 4\sigma T_0^3 \Delta T_0$ more energy per unit time
(c) The amount of energy radiated by the body in 1s is close to 60 J
(d) Reducing the exposed surface area of the body (e.g. by curling up) allows humans to maintain the same body temperature while reducing the energy lost by radiation

## Section 2 (Maximum Marks : 15)

- This section contains **FIVE** questions
- The answer to each question is a **SINGLE DIGIT INTEGER** ranging from 0 to 9, both inclusive.
- For each question, darken the bubble corresponding to the correct integer in the ORS.
- For each question, marks will be awarded in one of the following categories:

*Full Marks* : + 3 If only the bubble corresponding to all the correct answer is darkened.

*Zero Marks* : 0 In all other cases.

8. An electron in a hydrogen atom undergoes a transition from an orbit with quantum number $n_i$ to another with quantum number $n_f$. $v_i$ and $v_f$ are respectively the initial and final potential energies of the electron. If $\frac{v_i}{v_f} = 6.25$, then the smallest possible $n_f$ is

9. A drop of liquid of radius $R = 10^{-2}$ m having surface tension $S = \frac{0.1}{4\pi} \text{ Nm}^{-1}$ divides itself into $K$ identical drops. In this process the total change in the surface energy $\Delta U = 10^{-3}$ J. If $K = 10^{\alpha}$, then the value of $\alpha$ is

10. A stationary source emits sound of frequency $f_0 = 492$ Hz. The sound is reflected by a large car approaching the source with a speed of 2 $ms^{-1}$. The reflected signal is received by the source and superposed with the original. What will be the beat frequency of the resulting signal in Hz? (Given that the speed of sound in air is 330 ms $^{-1}$ and the car reflects the sound at the frequency it has received).

**11.** $^{131}I$ is an isotope of Iodine that $\beta$ decays to an isotope of Xenon with a half-life of 8 days. A small amount of a serum labelled with $^{131}I$ is injected into the blood of a person. The activity of the amount of $^{131}I$ injected was $2.4 \times 10^5$ Becquerel (Bq). It is known that the injected serum will get distributed uniformly in the blood stream in less than half an hour. After 11.5 h, 2.5 ml of blood is drawn from the person's body, and gives an activity of 115 Bq. The total volume of blood in the person's body, in litres is approximately (you may use $e^2 \approx 1 + x$ for $|x| << 1$ and $\ln 2 \approx 0.7$).

**12.** A monochromatic light is travelling in a medium of refractive index $n = 1.6$. It enters a stack of glass layers from the bottom side at an angle $\theta = 30°$. The interfaces of the glass layers are parallel to each other. The refractive indices of different glass layers are monotonically decreasing as $n_m = n - m\Delta n$, where $n_m$ is the refractive index of the $m$th slab and $\Delta n = 0.1$ (see the figure). The ray is refracted out parallel to the interface between the $(m-1)$th and $m$th slabs from the right side of the stack. What is the value of $m$?

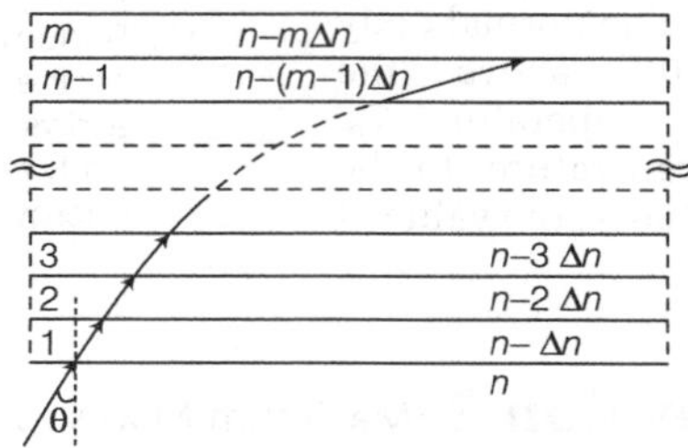


## Section 3 (Maximum Marks : 18)

- This section contains **SIX** questions of matching type.
- The section contains **TWO** tables (each having 3 columns and 4 rows)
- Based on each table, there are **THREE** questions.
- Each question has **FOUR** options [a], [b], [c] and [d]. **ONLY ONE** of these four options is correct.
- For each question, darken the bubble corresponding to the correct option in the ORS.
- For each question, Marks will be awarded in one of the following categories:

*Full Marks* : + 3 If only the bubble corresponding to the correct option is darkened

*Zero Marks* : 0 If none of the bubbles is darkened

*Negative Marks* : –1 In all other cases

**Directions** (Q.Nos. 13-15) Matching the information given in the three columns of the following table.

A charged particle (electron or proton) is introduced at the origin ($x = 0, y = 0, z = 0$) with a given initial velocity **v**. A uniform electric field **E** and a uniform magnetic field **B** exist everywhere. The velocity **v**, electric field **E** and magnetic field **B** are given in columns 1, 2 and 3, respectively. The quantities $E_0$, $B_0$ are positive in magnitude.

| | Column 1 | | Column 2 | | Column 3 |
|---|---|---|---|---|---|
| (I) | Electron with $\mathbf{v} = 2\frac{E_0}{B_0}\hat{x}$ | (i) | $\mathbf{E} = E_0\hat{z}$ | (P) | $\mathbf{B} = -B_0\hat{x}$ |
| (II) | Election with $\mathbf{v} = \frac{E_0}{B_0}\hat{y}$ | (ii) | $\mathbf{E} = -E_0\hat{y}$ | (Q) | $\mathbf{B} = B_0\hat{x}$ |
| (III) | Proton with $\mathbf{v} = 0$ | (iii) | $\mathbf{E} = -E_0\hat{x}$ | (R) | $\mathbf{B} = B_0\hat{y}$ |
| (IV) | Proton with $\mathbf{v} = 2\frac{E_0}{B_0}\hat{x}$ | (iv) | $\mathbf{E} = E_0\hat{x}$ | (S) | $\mathbf{B} = B_0\hat{z}$ |

**13.** In which case would the particle move in a straight line along the negative direction of $Y$-axis (i.e. move along $-\hat{y}$)?

(a) (IV) (ii) (S) (b) (II) (iii) (Q) (c) (III), (ii) (R) (d) (III) (ii) (P)

**14.** In which case will the particle move in a straight line with constant velocity?

(a) (II) (iii) (S) (b) (III) (iii) (P) (c) (IV) (i) (S) (d) (III) (ii) (R)

**15.** In which case will the particle describe a helical path with axis along the positive $z$-direction?

(a) (II) (ii) (R) (b) (III) (iii) (P) (c) (IV) (i) (S) (d) (IV) (ii) (R)

**Directions** (Q.Nos. 16-18) Matching the information given in the three columns of the following table.

An ideal gas is undergoing a cyclic thermodynamic process in different ways as shown in the corresponding $p$-$V$ diagrams in column 3 of the table. Consider only the path from state 1 to state 2. $W$ denotes the corresponding work done on the system. The equations and plots in the table have standards notations and used in thermodynamic processes. Here $\gamma$ is the ratio of heat capacities at constant pressure and constant volume. The number of moles in the gas is $n$.

| | Column 1 | | Column 2 | | Column 3 |
|---|---|---|---|---|---|
| (I) | $W_{1\to 2} = \frac{1}{\gamma - 1}(p_2V_2 - p_1V_1)$ | (i) | Isothermal | (P) | 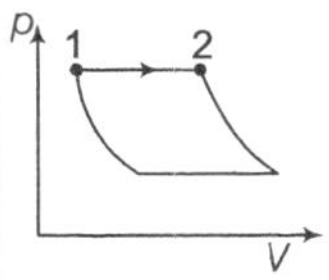 |
| (II) | $W_{1\to 2} = -pV_2 + pV_1$ | (ii) | Isochoric | (Q) | 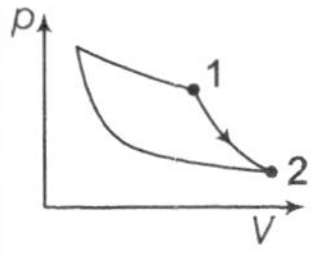 |
| (III) | $W_{1\to 2} = 0$ | (iii) | Isobaric | (R) | 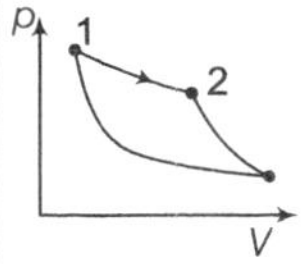 |
| (IV) | $W_{1\to 2} = -nRT\ln\left(\frac{V_2}{V_1}\right)$ | (iv) | Adiabatic | (S) | 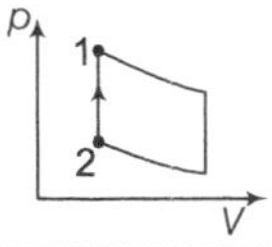 |

16. Which one of the following options correctly represents a thermodynamic process that is used as a correction in the determination of the speed of sound in an ideal gas?

(a) (IV) (ii) (R)
(b) (I) (ii) (Q)
(c) (I), (iv) (Q)
(d) (III) (iv) (R)

17. Which of the following options is the only correct representation of a process in which $\Delta U = \Delta Q - p\Delta V$ ?

(a) (II) (iii) (S) (b) (II) (iii) (P)
(c) (III) (iii) (P) (d) (II) (iv) (R)

18. Which one of the following options is the correct combination?

(a) (II) (iv) (P) (b) (III) (ii) (S)
(c) (II) (iv) (R) (d) (IV) (ii) (S)

# Chemistry

## Section 1 (Maximum Marks : 28)

- This section contains **SEVEN** questions.
- Each question has **FOUR** options (a), (b), (c) and (d). **ONE OR MORE THAN ONE** of these four option(s) is(are) correct.
- For each question, darken the bubble(s) corresponding to all the correct option(s) in the ORS.
- For each question, marks will be awarded in **one of the following categories** :

| | | |
|---|---|---|
| Full Marks | : + 4 | If only the bubble(s) corresponding to all the correct option(s) is(are) darkened. |
| Partial Marks | : +1 | For darkening a bubble corresponding to each correct option, provided NO incorrect option is darkened. |
| Zero Marks | : 0 | If none of the bubbles is darkened. |
| Negative Marks | : −2 | In all other cases. |

- For example, if (a), (c) and (d) are all the correct options for a question, darkening all these three will get + 4 marks; darkening only (a) and (d) will get +2 marks and darkening (a) and (b) will get − 2 marks, as a wrong option is also darkened.

1. An ideal gas is expanded form $(p_1, V_1, T_1)$ to $(p_2, V_2, T_2)$ under different conditions. The correct statement(s) among the following is (are)

(a) The work done by the gas is less when it is expanded reversibly from $V_1$ to $V_2$ under adiabatic conditions as compared to that when expanded reversibly form $V_1$ to $V_2$ under isothermal conditions.

(b) The change in internal energy of the gas is (i) zero, if it is expanded reversibly with $T_1 = T_2$, and (ii) positive, if it is expanded reversibly under adiabatic conditions with $T_1 \neq T_2$

(c) If the expansion is carried out freely, it is simultaneously both isothermal as well as adiabatic

(d) The work done on the gas is maximum when it is compressed irrversibly from $(p_2, V_2)$ to $(p_1, V_1)$ against constant pressure $p_1$

2. The IUPAC name(s) of the following compound is (are)

$H_3C$—$C_6H_4$—Cl

(a) 4-methylchlorobenzene
(b) 4-chlorotoluene
(c) 1-chloro-4-methylbenzene
(d) 1-methyl-4-chlorobenzene

3. For a solution formed by mixing liquids $L$ and $M$, the vapour pressure of $L$ plotted against the mole fraction of $M$ in solution is shown in the following figure. Here $x_L$ and $x_M$ represent mole fractions of $L$ and $M$,

respectively, in the solution. The correct statement(s) applicable to this system is (are)

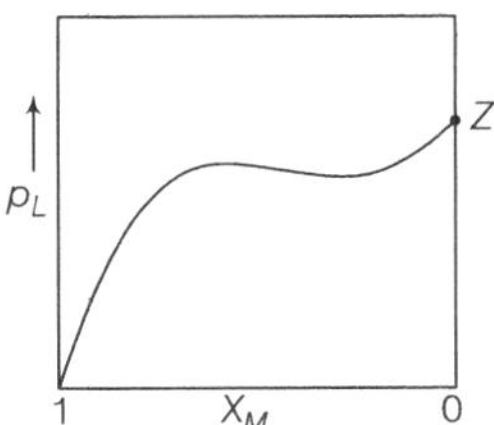


(a) The point $Z$ represents vapour pressure of pure liquid $M$ and Raoult's law is obeyed from $x_L = 0$ to $x_L = 1$

(b) Attractive intermolecular interactions between $L$ - $L$ in pure liquid $L$ and $M$ - $M$ in pure liquid $M$ are stronger than those between $L$ - $M$ when mixed in solution

(c) The point $Z$ represents vapour pressure of pure liquid $M$ and Raoult's law is obeyed when $x_L \to 0$

(d) The point $Z$ represents vapour pressure of pure liquid $L$ and Raoult's law is obeyed when $x_L \to 1$

4. The correct statement(s) for the following addition reactions is (are)

(i) $H_3C$, H / H, $CH_3$ $\xrightarrow{Br_2/CHCl_3}$ $M$ and $N$

(i) $H_3C$, H / H, $CH_3$ $\xrightarrow{Br_2/CHCl_3}$ $O$ and $P$

(a) ($M$ and $O$) and ($N$ and $P$) are two pairs of enantiomers

(b) Bromination proceeds through *trans*-addition in both the reactions

(c) $O$ and $P$ are identical molecules

(d) ($M$ and $O$) and ($N$ and $P$) two pairs of diastereomers

5. Addition of excess aqueous ammonia to a pink coloured aqueous solution of $MCl_2 \cdot 6H_2O(X)$ and $NH_4Cl$ gives an octahedral complex $Y$ in the presence of air. In aqueous solution, complex $Y$ behaves as 1 : 3 electrolyte. The reaction of $X$ with excess HCl at room temperature results in the formation of a blue colured complex $Z$. The calculated spin only magnetic moment of $X$ and $Z$ is 3.87 B.M., whersas it is zero for complex $Y$.

Among the following options, which statement(s) is (are) correct?

(a) The hybridisation of the central metal ion in $Y$ is $d^2sp^3$

(b) Addition of silver nitrate to $Y$ given only two equivalents of silver chloride

(c) When $X$ and $Y$ are in equilibrium at 0°C, the colour of the solution is pink

(d) $Z$ is a tetrahedral complex

6. The correct statement(s) about the oxoacids, $HClO_4$ and HClO, is (are)

(a) The central atom in both $HClO_4$ and HClO is $sp^3$-hybridised

(b) $HClO_4$ is formed in the reaction between $Cl_2$ and $H_2O$

(c) The conjugate base of $HClO_4$ is weaker base than $H_2O$

(d) $HClO_4$ is more acidic than HClO because of the resonance stabilisation of its anion

7. The colour of the $X_2$ molecules of group 17 elements changes gradually from yellow to violet down the group. This is due to

(a) decrease in $\pi^* - \sigma^*$ gap down the group

(b) decrease in ionisation energy down the group

(c) the physical state of $X_2$ at room temperature changes from gas to solid down the group

(d) decreases in HOMO-LUMO gap down the group

## Section 2 (Maximum Marks : 15)

- This section contains **FIVE** questions.
- The answer to each question is a **SINGLE DIGIT INTEGER** ranging from 0 to 9, both inclusive.
- For each question, darken the bubble corresponding to the correct integer in the ORS.
- For each question, marks will be awarded in one of th following categories :

Full Marks : + 3 If only the bubble corresponding to all the correct answer is darkened.

Zero Marks : 0 In all other cases.

**8.** The sum of the number of lone pairs of electrons on each central atom in the following species is

$[TeBr_6]^{2-}$, $[BrF_2]^+$, $SNF_3$ and $[XeF_3]^-$

(Atomic numbers : N = 7, F = 9, S = 16, Br = 35, Te = 52, Xe = 54)

**9.** Among the following, the number of aromatic compound(s) is

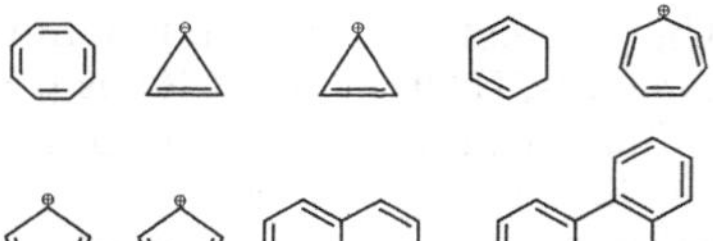

**10.** Among $H_2, He_2^+, Li_2, Be_2, B_2, C_2, N_2,$ and $F_2O_2^-$, the number of diamagnetic species is

(Atomic numbers : H = 1, He = 2, Li = 3, Be = 4, B = 5, C = 6, N = 7, O = 8, F = 9)

**11.** A crystalline solid of a pure substance has a face-centred cubic structure with a cell edge of 400 pm. If the density of the substance in the crystal is 8 g $cm^{-3}$, then the number of atoms present in 256 g of the crystal is $N \times 10^{24}$. The value of $N$ is

**12.** The conductance of a 0.0015 M aqueous solution of a weak monobasic acid was determined by using a conductivity cell consisting of platinised Pt electrodes. The distance between the electrodes is 120 cm with an area of cross section of 1 $cm^2$. The conductance of this solution was found to be $5 \times 10^{-7}$ S. The pH of the solution is 4. The value of limiting molar conductivity ($\Lambda_m^\circ$) of this weak monobasic acid in aqueous solution is $Z \times 10^2$ S $cm^{-1}$ $mol^{-1}$. The value of $Z$ is

## Section 3 (Maximum Marks : 18)

- This section contains **SIX** question of matching type.
- This section contains **TWO** tables (each having 3 columns and 4 rows)
- Based on each table, there are **THREE** questions.
- Each question has **FOUR** options (a), (b), (c) and (d). **ONLY ONE** of these options is correct.
- For each question, darken the bubble corresponding to the correct option is the ORS.
- For each question, marks will be awarded in **one of the following categories**:

Full Marks : + 3 If only the bubble corresponding to the correct option is darkened

Zero Marks : 0 If none of the bubbles is darkened

Negative Marks : –1 In all othe cases

**Answer Q. 13, Q. 14 and Q. 15 by appropriately matching the information given in the three columns of the following table.**

The wave function, $\psi_{n}, l, m_l$ is a mathematical function whose value depends upon spherical polar coordinates $(r, \theta, \phi)$ of the electron and characterised by the quantum number $n$, $l$ and $m_l$. Here $r$ is distance from nucleus, $\theta$ is colatitude and $\phi$ is azimuth. In the mathematical functions given in the Table, $Z$ is atomic number and $a_0$ is Bohr radius.

| Column 1 | Column 2 | Column 3 |
|---|---|---|
| (I) 1s-orbital | (i) $\psi_{n, l, m_l} \propto \left(\frac{Z}{a_0}\right)^{\frac{3}{2}} e^{-\left(\frac{Zr}{a_0}\right)}$ | (P) $\psi_{n, l, m_l}(r)$ vs $r/a_0$ graph (0) |
| (II) 2s-orbital | (ii) One radial node | (Q) Probability density at nucleus $\propto \frac{1}{a_0^3}$ |
| (III) $2p_z$-orbital | (iii) $\psi_{n, l, m_l} \propto \left(\frac{Z}{a_0}\right)^{\frac{5}{2}} re^{-\left(\frac{Zr}{a_0}\right)} \cos\theta$ | (R) Probability density is maximum at nucleus |
| (IV) $3d_z^2$-orbital | (iv) $xy$-plane is a nodal plane | (S) Energy needed to excite electron from $n = 2$ state to $n = 4$ state is $\frac{27}{32}$ times the energy needed to excite electron from $n = 2$ state to $n = 6$ state |

**13.** For $He^+$ ion, the only **INCORRECT** combination is

(a) (I) (i) (S) (b) (II) (ii) (Q) (c) (I) (iii) (R) (d) (I) (i) (R)

**14.** For the given orbital in Column 1, the Only **CORRECT** combination for any hydrogen-like species is

(a) (II) (ii) (P) (b) (I) (ii) (S) (c) (IV) (iv) (R) (d) (III) (iii) (P)

**15.** For hydrogen atom, the only **CORRECT** combination is

(a) (I) (i) (P) (b) (I) (iv) (R) (c) (II) (i) (Q) (d) (I) (i) (S)

**Answer Q. 16, Q. 17 and Q. 18 by appropriately matching the information given in the three columns of the following table.**

Column 1, 2 and 3 contain starting materials, reaction conditions, and type of reactions, respectively.

| Column 1 | Column 2 | Column 3 |
|---|---|---|
| (I) Toluene | (i) $NaOH/Br_2$ | (P) Condensation |
| (II) Acetophenone | (ii) $Br_2/h\nu$ | (Q) Carboxylation |
| (III) Benzaldehyde | (iii) $(CH_3CO)_2O/CH_3COOK$ | (R) Substitution |
| (IV) Phenol | (iv) $NaOH/CO_2$ | (S) Haloform |

16. The only **CORRECT** combination in which the reaction proceeds through radical mechanism is

(a) (IV) (i) (Q) (b) (III) (ii) (P)
(c) (II) (iii) (R) (d) (I) (ii) (R)

17. For the synthesis of benzoic acid, the only CORRECT combination is

(a) (II) (i) (S) (b) (I) (iv) (Q)
(c) (IV) (ii) (P) (d) (III) (iv) (R)

18. The only **CORRECT** combination that gives two different carboxylic acids is

(a) (IV) (iii) (Q) (b) (II) (iv) (R)
(c) (I) (i) (S) (d) (III) (iii) (P)

# Mathematics

## Section 1 (Maximum Marks : 28)

- This section contains **SEVEN** questions.
- Each question has **FOUR** options (a), (b), (c) and (d). **ONE OR MORE THAN ONE** of these four option(s) is(are) correct.
- For each question, darken the bubble(s) corresponding to all the correct option(s) in the ORS.
- For each question, marks will be awarded in one of the following categories:

*Full Marks* : +4 If only the bubble(s) corresponding to all the correct option(s) is(are) darkened.
*Partial Marks* : +1 For darkening a bubble corresponding to each correct option, provided NO incorrect option is darkened.
*Zero Marks* : 0 If none of the bubbles is darkened.
*Negative Marks* : −2 In all other cases.

- For example, if (a), (c) and (d) are all the correct options for a question, darkening all these three will get +4 marks; darkening only (a) and (d) will get +2 marks and darkening (a) and (b) will get −2 marks, as a wrong option is also darkened.

1. Let $X$ and $Y$ be two events such that $P(X) = \frac{1}{3}$, $P(X|Y) = \frac{1}{2}$ and $P(Y|X) = \frac{2}{5}$. Then

(a) $P(Y) = \frac{4}{15}$ (b) $P(X'|Y) = \frac{1}{2}$
(c) $P(X \cup Y) = \frac{2}{5}$ (d) $P(X \cap Y) = \frac{1}{5}$

2. Let $f : R \to (0, 1)$ be a continuous function. Then, which of the following function(s) has (have) the value zero at some point in the interval (0, 1) ?

(a) $e^x - \int_0^x f(t) \sin t \, dt$
(b) $f(x) + \int_0^{\frac{\pi}{2}} f(t) \sin t \, dt$
(c) $x - \int_0^{\frac{\pi}{2} - x} f(t) \cos t \, dt$
(d) $x^9 - f(x)$

3. Let $a, b, x$ and $y$ be real numbers such that $a - b = 1$ and $y \neq 0$. If the complex number $z = x + iy$ satisfies $\text{Im}\left(\frac{az + b}{z + 1}\right) = y$, then which of the following is(are) possible value(s) of $x$?

(a) $1 - \sqrt{1 + y^2}$
(b) $-1 - \sqrt{1 - y^2}$
(c) $1 + \sqrt{1 + y^2}$
(d) $-1 + \sqrt{1 - y^2}$

4. If $2x - y + 1 = 0$ is a tangent to the hyperbola $\frac{x^2}{a^2} - \frac{y^2}{16} = 1$ then which of the following CANNOT be sides of a right angled triangle?

(a) $a$, 4, 1 (b) $2a$, 4, 1
(c) $a$, 4, 2 (d) $2a$, 8, 1

5. Let $[x]$ be the greatest integer less than or equals to $x$. Then, at which of the following point(s) the function $f(x) = x\cos(\pi(x+[x]))$ is discontinuous ?

(a) $x = -1$ (b) $x = 1$
(c) $x = 0$ (d) $x = 2$

6. Which of the following is(are) NOT the square of a $3 \times 3$ matrix with real entries?

(a) $\begin{bmatrix} 1 & 0 & 0 \\ 0 & 1 & 0 \\ 0 & 0 & -1 \end{bmatrix}$ (b) $\begin{bmatrix} 1 & 0 & 0 \\ 0 & -1 & 0 \\ 0 & 0 & -1 \end{bmatrix}$

(c) $\begin{bmatrix} -1 & 0 & 0 \\ 0 & -1 & 0 \\ 0 & 0 & -1 \end{bmatrix}$ (d) $\begin{bmatrix} 1 & 0 & 0 \\ 0 & 1 & 0 \\ 0 & 0 & 1 \end{bmatrix}$

7. If a chord, which is not a tangent, of the parabola $y^2 = 16x$ has the equation $2x + y = p$, and mid-point $(h, k)$, then which of the following is(are) possible value(s) of $p, h$ and $k$?

(a) $p = -1, h = 1, k = -3$
(b) $p = 2, h = 3, k = -4$
(c) $p = -2, h = 2, k = -4$
(d) $p = 5, h = 4, k = -3$

## Section 2 (Maximum Marks : 15)

- This section contains **FIVE** questions.
- The answer to each question is a **SINGLE DIGIT INTEGER** ranging from 0 to 9, both inclusive.
- For each question, darken the bubble corresponding to the correct integer in the ORS.
- For each question, marks will be awarded in one of the following categories:

Full Marks : +3 If only the bubble corresponding to the correct answer is darkened.
Zero Marks : 0 In all other cases.

8. For a real number $\alpha$, if the system

$$\begin{bmatrix} 1 & \alpha & \alpha^2 \\ \alpha & 1 & \alpha \\ \alpha^2 & \alpha & 1 \end{bmatrix} \begin{bmatrix} x \\ y \\ z \end{bmatrix} = \begin{bmatrix} 1 \\ -1 \\ 1 \end{bmatrix}$$

of linear equations, has infinitely many solutions,

then $1 + \alpha + \alpha^2 =$

9. The sides of a right angled triangle are in arithmetic progression. If the triangle has area 24, then what is the length of its smallest side?

10. Let $f : R \to R$ be a differentiable function such that $f(0) = 0, f\left(\frac{\pi}{2}\right) = 3$ and $f'(0) = 1$.

If $g(x) = \int_x^{\frac{\pi}{2}} [f'(t)\ \text{cosec}\ t - \cot t\ \text{cosec}\ t\ f(t)]\,dt$

for $x \in \left(0, \frac{\pi}{2}\right]$, then $\lim_{x \to 0} g(x) =$

11. For how many values of $p$, the circle $x^2 + y^2 + 2x + 4y - p = 0$ and the coordinate axes have exactly three common points?

12. Words of length 10 are formed using the letters A, B, C, D, E, F, G, H, I, J. Let $x$ be the number of such words where no letter is repeated; and let $y$ be the number of such words where exactly one letter is repeated twice and no other letter is repeated. Then, $\frac{y}{9x} =$

## Section 3 (Maximum Marks : 18)

- This section contains **SIX** questions of matching type.
- This section contains **TWO** tables (each having 3 columns and 4 rows).
- Based on each table, there are **THREE** questions.
- Each question has **FOUR** options (a), (b), (c) and (d). **ONLY ONE** of these four options is correct.
- For each question, darken the bubble corresponding to the correct option in the ORS.
- For each question, marks will be awarded in one of the following categories:
  *Full Marks* : + 3 If only the bubble corresponding to the correct option is darkened.
  *Zero Marks* : 0 If none of the bubbles is darkened.
  *Negative Marks* : – 1 In all other cases.

**Directions** (Q.Nos. 13-15) by appropriately matching the information given in the three columns of the following table.

Columns 1, 2 and 3 contain conics, equations of tangents to the conics and points of contact, respectively.

| | Column-1 | | Column-2 | | Column-3 |
|---|---|---|---|---|---|
| (I) | $x^2 + y^2 = a^2$ | (i) | $my = m^2x + a$ | (P) | $\left(\frac{a}{m^2}, \frac{2a}{m}\right)$ |
| (II) | $x^2 + a^2y^2 = a^2]$ | (ii) | $y = mx + a\sqrt{m^2 + 1}$ | (Q) | $\left(\frac{-ma}{\sqrt{m^2+1}}, \frac{a}{\sqrt{m^2+1}}\right)$ |
| (III) | $y^2 = 4ax$ | (iii) | $y = mx + \sqrt{a^2m^2 - 1}$ | (R) | $\left(\frac{-a^2m}{\sqrt{a^2m^2+1}}, \frac{1}{\sqrt{a^2m^2+1}}\right)$ |
| (IV) | $x^2 - a^2y^2 = a^2$ | (iv) | $y = mx + \sqrt{a^2m^2 + 1}$ | (S) | $\left(\frac{-a^2m}{\sqrt{a^2m^2-1}}, \frac{-1}{\sqrt{a^2m^2-1}}\right)$ |

**13.** For $a = \sqrt{2}$, if a tangent is drawn to a suitable conic (Column 1) at the point of contact $(-1, 1)$, then which of the following options is the only CORRECT combination for obtaining its equation?

(a) (I) (ii) (Q) (b) (I) (i) (P)
(c) (III) (i) (P) (d) (II) (ii) (Q)

**14.** The tangent to a suitable conic (Column 1) at $\left(\sqrt{3}, \frac{1}{2}\right)$ is found to be $\sqrt{3}x + 2y = 4$, then which of the following options is the only CORRECT combination?

(a) (IV) (iv) (S) (b) (II) (iv) (R)
(c) (IV) (iii) (S) (d) (II) (iii) (R)

**15.** If a tangent to a suitable conic (Column 1) is found to be $y = x + 8$ and its point of contact is (8, 16), then which of the following options is the only CORRECT combination?

(a) (III) (i) (P) (b) (I) (ii) (Q)
(c) (II) (iv) (R) (d) (III) (ii) (Q)

**Directions** (Q.Nos. 16-18) by appropriately matching the information given in the three columns of the following table.

Let $f(x) = x + \log_e x - x \log_e x, x \in (0, \infty)$

Column 1 contains information about zeros of $f(x)$, $f'(x)$ and $f''(x)$.

Column 2 contains information about the limiting behaviour of $f(x)$, $f'(x)$ and $f''(x)$ at infinity.

Column 3 contains information about increasing/decreasing nature of $f(x)$ and $f'(x)$.

| | Column-1 | | Column-2 | | Column-3 |
|---|---|---|---|---|---|
| (I) | $f(x) = 0$ for some $x \in (1, e^2)$ | (i) | $\lim_{x \to \infty} f(x) = 0$ | (P) | $f$ is increasing in (0, 1) |
| (II) | $f'(x) = 0$ for some $x \in (1, e)$ | (ii) | $\lim_{x \to \infty} f(x) = -\infty$ | (Q) | $f$ is decreasing in $(e, e^2)$ |
| (III) | $f'(x) = 0$ for some $x \in (0, 1)$ | (iii) | $\lim_{x \to \infty} f'(x) = -\infty$ | (R) | $f'$ is increasing in (0, 1) |
| (IV) | $f''(x) = 0$ for some $x \in (1, e)$ | (iv) | $\lim_{x \to \infty} f''(x) = 0$ | (S) | $f'$ is decreasing in $(e, e^2)$ |

16. Which of the following options is the only INCORRECT combination?

(a) (I) (iii) (P)
(b) (II) (iv) (Q)
(c) (II) (iii) (P)
(d) (III) (i) (R)

17. Which of the following options is the only CORRECT combination?

(a) (I) (ii) (R)
(b) (III) (iv) (P)
(c) (II) (iii) (S)
(d) (IV) (i) (S)

18. Which of the following options is the only CORRECT combination?

(a) (III) (iii) (R)
(b) (IV) (iv) (S)
(c) (II) (ii) (Q)
(d) (I) (i) (P)

# Paper 2

## Physics

### Section 1 (Maximum Marks : 21)

- This section contains **SEVEN** questions.
- Each question has **FOUR** options (a), (b), (c) and (d). **ONLY ONE** of these four options is correct.
- For each question, darken the bubble corresponding to the correct option in the ORS.
- For each question, marks will be awarded in one of the following categories.

*Full Marks* : + 3 If only the bubble corresponding to the correct option is darkened.

*Zero Marks* : 0 If none of the bubbles is darkened.

*Negative Marks* : −1 In all other cases

**1.** Consider regular polygons with number of sides $n = 3, 4, 5$ ...... as shown in the figure. The centre of mass of all the polygons is at height $h$ from the ground. They roll on a horizontal surface about the leading vertex without slipping and sliding as depicted. The maximum increase in height of the locus of the centre of mass for each each polygon is $\Delta$. Then, $\Delta$ depends on $n$ and $h$ as

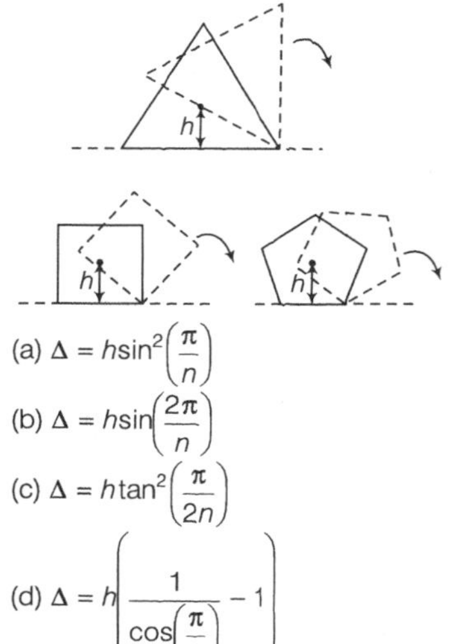


(a) $\Delta = h\sin^2\left(\frac{\pi}{n}\right)$

(b) $\Delta = h\sin\left(\frac{2\pi}{n}\right)$

(c) $\Delta = h\tan^2\left(\frac{\pi}{2n}\right)$

(d) $\Delta = h\left(\frac{1}{\cos\left(\frac{\pi}{n}\right)} - 1\right)$

**2.** Consider an expanding sphere of instantaneous radius $R$ whose total mass remains constant. The expansion is such that the instantaneous density $\rho$ remains uniform throughout the volume. The rate of fractional change in density $\left(\frac{1}{\rho}\frac{d\rho}{dt}\right)$ is constant. The velocity $v$ of any point of the surface of the expanding sphere is proportional to

(a) $R$ (b) $\frac{1}{R}$

(c) $R^3$ (d) $R^{\frac{2}{3}}$

**3.** A photoelectric material having work-function $\phi_0$ is illuminated with light of wavelength $\lambda \left(\lambda < \frac{hc}{\phi_0}\right)$. The fastest photoelectron has a de-Broglie wavelength $\lambda_d$. A change in wavelength of the incident light by $\Delta\lambda$ results in a change $\Delta\lambda_d$ in $\lambda_d$. Then, the ratio $\frac{\Delta\lambda_d}{\Delta\lambda}$ is proportional to

(a) $\frac{\lambda_d^2}{\lambda^2}$ (b) $\frac{\lambda_d}{\lambda}$

(c) $\frac{\lambda_d^3}{\lambda}$ (d) $\frac{\lambda_d^3}{\lambda^2}$

4. Three vectors **P**, **Q** and **R** are shown in the figure. Let $S$ be any point on the vector **R**. The distance between the points $P$ and $S$ is $b[\mathbf{R}]$. The general relation among vectors **P**, **Q** and **S** is

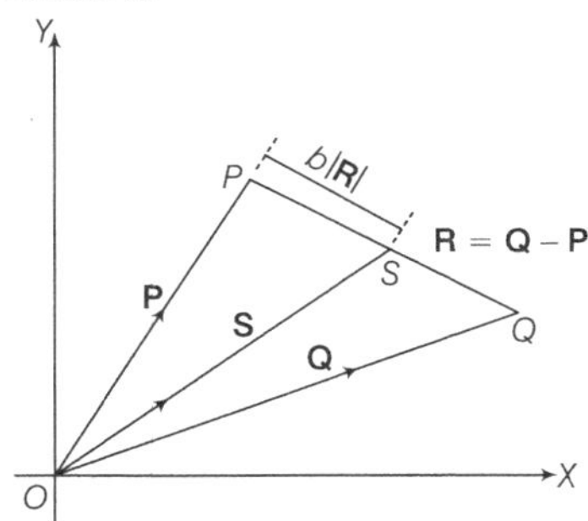


(a) $\mathbf{S} = (1 - b^2)\mathbf{P} + b\mathbf{Q}$
(b) $\mathbf{S} = (b - 1)\mathbf{P} + b\mathbf{Q}$
(c) $\mathbf{S} = (1 - b)\mathbf{P} + b\mathbf{Q}$
(d) $\mathbf{S} = (1 - b)\mathbf{P} + b^2\mathbf{Q}$

5. A person measures the depth of a well by measuring the time interval between dropping a stone and receiving the sound of impact with the bottom of the well. The error in his measurement of time is $\delta T = 0.01\text{s}$ and he measures the depth of the well to be $L = 20$ m. Take the acceleration due to gravity $g = 10\ \text{ms}^{-2}$ and the velocity of sound is $300\ \text{ms}^{-1}$. Then the fractional error in the measurement, $\frac{\delta L}{L}$, is closest to

(a) 1% (b) 5%
(c) 3% (d) 0.2%

6. A rocket is launched normal to the surface of the Earth, away from the Sun, along the line joining the Sun and the Earth. The Sun is $3 \times 10^5$ times heavier than the Earth and is at a distance $2.5 \times 10^4$ times larger than the radius of Earth. The escape velocity from Earth's gravitational field is $v_e = 11.2$ km $\text{s}^{-1}$. The minimum initial velocity $(v_s)$ required for the rocket to be able to leave the Sun-Earth system is closest to

(Ignore the rotation and revolution of the Earth and the presence of any other planet)

(a) $v_s = 72\ \text{km s}^{-1}$
(b) $v_s = 22\ \text{km s}^{-1}$
(c) $v_s = 42\ \text{km s}^{-1}$
(d) $v_s = 62\ \text{km s}^{-1}$

7. A symmetric star shaped conducting wire loop is carrying a steady state current $I$ as shown in the figure. The distance between the diametrically opposite vertices of the star is $4a$.

The magnitude of the magnetic field at the center of the loop is

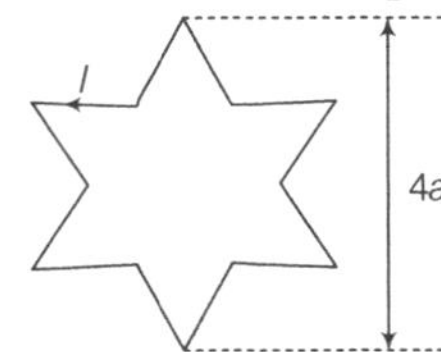


(a) $\frac{\mu_0 I}{4\pi a} 6[\sqrt{3} - 1]$
(b) $\frac{\mu_0 I}{4\pi a} 6[\sqrt{3} + 1]$
(c) $\frac{\mu_0 I}{4\pi a} 3[\sqrt{3} - 1]$
(d) $\frac{\mu_0 I}{4\pi a} 3[2 - \sqrt{3}]$

## Section 2 (Maximum Marks : 28)

- This section contains **SEVEN** questions.
- Each question has **FOUR** options (a), (b) (c) and (d). **ONE OR MORE THAN ONE** of these four option(s) is (are) correct.
- For each question, darken the bubble(s) corresponding to all the correct option(s) in the ORS.
- For each questions, marks will be awarded in one of the following categories :

  *Full Marks* : + 4 If only the bubble(s) corresponding to all the correct option(s) is (are) darkened.

  *Partial Marks* : + 1 For darkening a bubble corresponding **to each correct option,** provided NO incorrect option is darkened.

  *Zero Marks* : 0 If none of the bubbles is darkened.

  *Negative Marks* : −2 In all other cases.
- For example, if (a), (c) and (d) are all the correct options for a question, darkening all these three will get in + 4 marks ; darkening only (a) and (d) will get + 2 marks and darkening (a) and (b) will result in −2 marks, as a wrong option is also darkened.

**8.** A wheel of radius $R$ and mass $M$ is placed at the bottom of a fixed step of height $R$ as shown in the figure. A constant force is continuously applied on the surface of the wheel so that it just climbs the step without slipping. Consider the torque $\tau$ about an axis normal to the plane of the paper passing through the point $Q$. Which of the following options is/are correct?

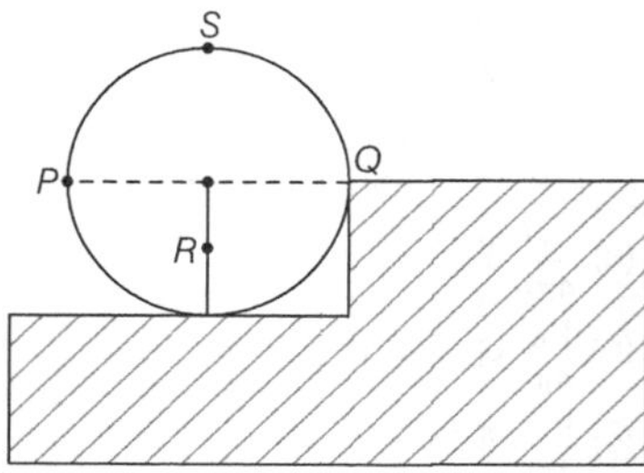


(a) If the force is applied normal to the circumference at point $P$, then $\tau$ is zero

(b) If the force is applied tangentially at point $S$, then $\tau \neq 0$ but the wheel never climbs the step

(c) If the force is applied at point $P$ tangentially, then $\tau$ decreases continuously as the wheel climbs

(d) If the force is applied normal to the circumference at point $X$, then $\tau$ is constant

**9.** Two coherent monochromatic point sources $S_1$ and $S_2$ of wavelength $\lambda = 600$ nm are placed symmetrically on either side of the centre of the circle as shown. The sources are separated by a distance $d = 1.8$ mm. This arrangement produces interference fringes visible as alternate bright and dark spots on the circumference of the circle. The angular separation between two consecutive bright spots is $\Delta\theta$. Which of the following options is/are correct?

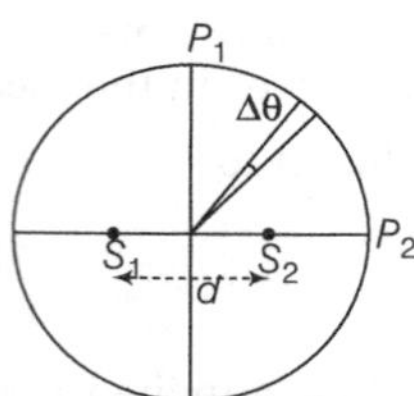


(a) The angular separation between two consecutive bright spots decreases as we move from $P_1$ to $P_2$ along the first quadrant

(b) A dark spot will be formed at the point $P_2$

(c) The total number of fringes produced between $P_1$ and $P_2$ in the first quadrant is close to 3000

(d) At $P_2$ the order of the fringe will be maximum

**10.** A point charge $+Q$ is placed just outside an imaginary hemispherical surface of radius $R$ as shown in the figure. Which of the following statements is/are correct?

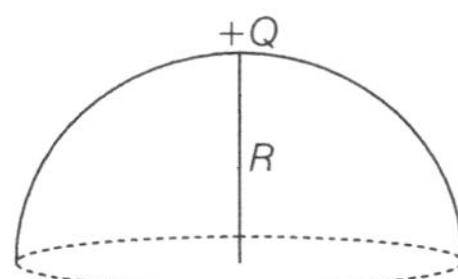


(a) The electric flux passing through the curved surface of the hemisphere is $-\frac{Q}{2\varepsilon_0}\left(1-\frac{1}{\sqrt{2}}\right)$.

(b) The component of the electric field normal to the flat surface is constant over the surface

(c) Total flux through the curved and the flat surfaces is $\frac{Q}{\varepsilon_0}$

(d) The circumference of the flat surface is an equipotential

**11.** A rigid uniform bar $AB$ of length $L$ is slipping from its vertical position on a frictionless floor (as shown in the figure). At some instant of time, the angle made by the bar with the vertical is $\theta$. Which of the following statements about its motion is/are correct?

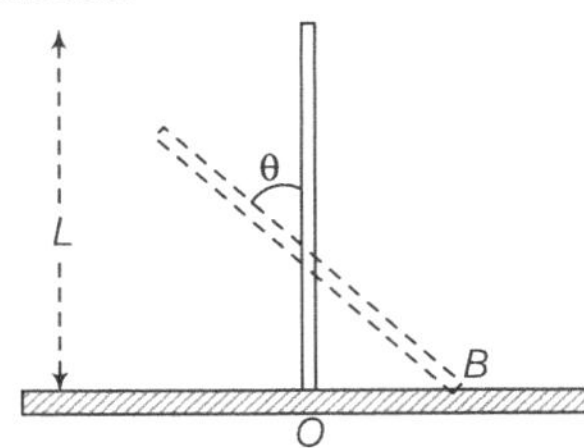


(a) Instantaneous torque about the point in contact with the floor is proportional to $\sin\theta$

(b) The trajectory of the point $A$ is parabola

(c) The mid-point of the bar will fall vertically downward

(d) When the bar makes an angle $\theta$ with the vertical, the displacement of its mid-point from the initial position is proportional to $(1-\cos\theta)$

**12.** A source of constant voltage $V$ is connected to a resistance $R$ and two ideal inductors $L_1$ and $L_2$ through a switch $S$ as shown. There is no mutual inductance between the two inductors. The switch $S$ is initially open. At $t = 0$, the switch is closed and current begins to flow. Which of the following options is/are correct?

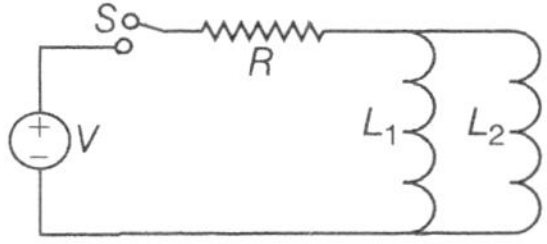


(a) After a long time, the current through $L_1$ will be $\frac{V}{R}\frac{L_2}{L_1+L_2}$

(b) After a long time, the current through $L_2$ will be $\frac{V}{R}\frac{L_1}{L_1+L_2}$

(c) The ratio of the currents through $L_1$ and $L_2$ is fixed at all times ($t > 0$)

(d) At $t = 0$, the current through the resistance $R$ is $\frac{V}{R}$

**13.** A uniform magnetic field $B$ exists in the region between $x = 0$ and $x = \frac{3R}{2}$ (region 2 in the figure) pointing normally into the plane of the paper. A particle with charge $+Q$ and momentum $p$ directed along $X$-axis enters region 2 from region 1 at point $P_1$ $(y = -R)$.

Which of the following option(s) is/are correct?

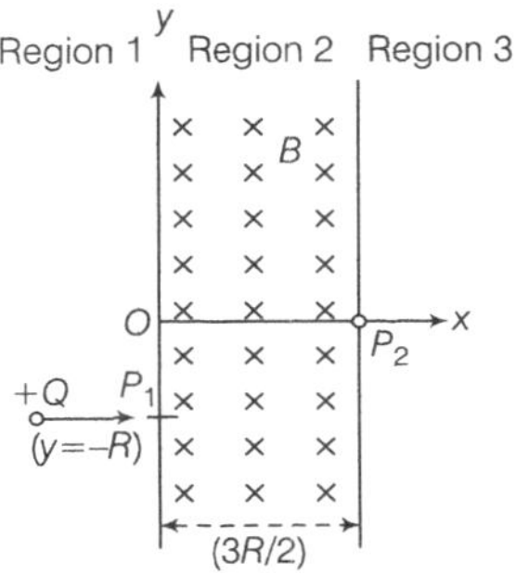

(a) When the particle re-enters region 1 through the longest possible path in region 2, the magnitude of the change in its linear momentum between point $P_1$ and the farthest point from $Y$-axis is $\frac{p}{\sqrt{2}}$

(b) For $B = \frac{8}{13}\frac{p}{QR}$, the particle will enter region 3 through the point $P_2$ on $X$-axis

(c) For $B > \frac{2}{3}\frac{p}{QR}$, the particle will re-enter region 1

(d) For a fixed $B$, particles of same charge $Q$ and same velocity $v$, the distance between the point $P_1$ and the point of re-entry into region 1 is inversely proportional to the mass of the particle

**14.** The instantaneous voltages at three terminals marked $X, Y$ and $Z$ are given by $V_X = V_0 \sin \omega t$,

$$V_Y = V_0 \sin\left(\omega t + \frac{2\pi}{3}\right) \text{ and}$$

$$V_Z = V_0 \sin\left(\omega t + \frac{4\pi}{3}\right).$$

An ideal voltmeter is configured to read rms value of the potential difference between its terminals. It is connected between points $X$ and $Y$ and then between $Y$ and $Z$. The reading(s) of the voltmeter will be

(a) $V_{YZ}^{rms} = V_0\sqrt{\frac{1}{2}}$

(b) $V_{XY}^{rms} = V_0\sqrt{\frac{3}{2}}$

(c) independent of the choice of the two terminals

(d) $V_{XY}^{rms} = V_0$

## Section 3 (Maximum Marks : 12)

- This section contains **TWO** paragraphs.
- Based on each paragraph, there are **TWO** questions.
- Each question has **FOUR** options (a), (b) (c) and (d). **ONLY ONE** of these four options is correct.
- For each question, darken the bubble corresponding to all the correct option in the ORS.
- For each questions, marks will be awarded in one of the following categories.

*Full Marks* : + 3 If only the bubble corresponding to all the correct answer is darkened.

*Zero Marks* : 0 In all other cases.

### Paragraph 1

Consider a simple $RC$ circuit as shown in Figure 1.

**Process 1** In the circuit the switch $S$ is closed at $t = 0$ and the capacitor is fully charged to voltage $V_0$ (i.e. charging continues for time $T >> RC$). In the process some dissipation ($E_D$) occurs across the resistance $R$. The amount of energy finally stored in the fully charged capacitor is $E_c$.

**Process 2** In a different process the voltage is first set to $\frac{V_0}{3}$ and maintained for a charging time $T >> RC$. Then, the voltage is raised to $\frac{2V_0}{3}$ without discharging the capacitor and again maintained for a time $T >> RC$. The process is repeated one more time by raising the voltage to $V_0$ and the capacitor is charged to the same final voltage $V_0$ as in Process 1.

These two processes are depicted in Figure 2.

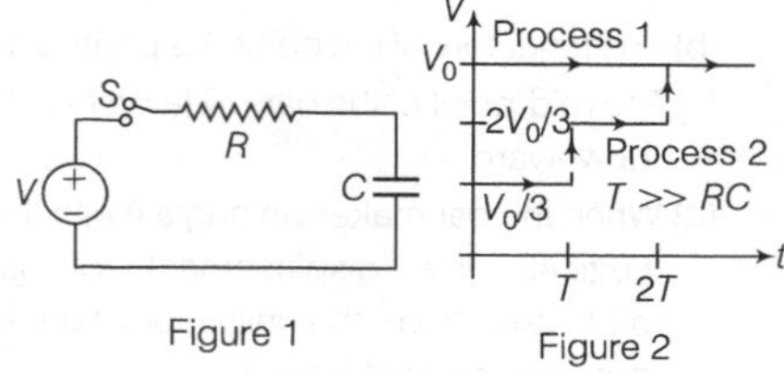

Figure 1 Figure 2

15. In Process 1, the energy stored in the capacitor $E_C$ and heat dissipated across resistance $E_D$ are related by

(a) $E_C = E_D \ln 2$ (b) $E_C = E_D$
(c) $E_C = 2E_D$ (d) $E_C = \frac{1}{2} E_D$

16. In Process 2, total energy dissipated across the resistance $E_D$ is

(a) $E_D = \frac{1}{3}\left(\frac{1}{2}CV_0^2\right)$

(b) $E_D = 3\left(\frac{1}{2}CV_0^2\right)$

(c) $E_D = 3CV_0^2$

(d) $E_D = \frac{1}{2}CV_0^2$

## Paragraph 2

One twirls a circular ring (of mass $M$ and radius $R$) near the tip of one's finger as shown in Figure 1. In the process the finger never loses contact with the inner rim of the ring. The finger traces out the surface of a cone, shown by the dotted line. The radius of the path traced out by the point where the ring and the finger is in contact is $r$. The finger rotates with an angular velocity $\omega_0$. The rotating ring rolls without slipping on the outside of a smaller circle described by the point where the ring and the finger is in contact (Figure 2). The coefficient of friction between the ring and the finger is $\mu$ and the acceleration due to gravity is $g$.

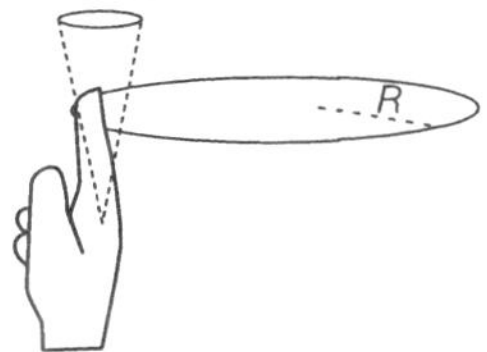

Figure 1

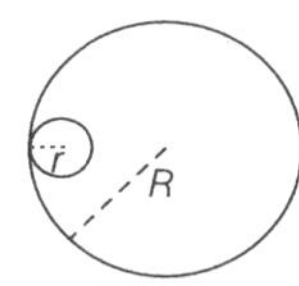

Figure 2

17. The total kinetic energy of the ring is

(a) $M\omega_0^2(R-r)^2$

(b) $\frac{1}{2}M\omega_0^2(R-r)^2$

(c) $M\omega_0^2R^2$

(d) $\frac{3}{2}M\omega_0^2(R-r)^2$

18. The minimum value of $\omega_0$ below which the ring will drop down is

(a) $\sqrt{\frac{g}{2\mu(R-r)}}$

(b) $\sqrt{\frac{3g}{2\mu(R-r)}}$

(c) $\sqrt{\frac{g}{\mu(R-r)}}$

(d) $\sqrt{\frac{2g}{\mu(R-r)}}$

# Chemistry

## Section 1 (Maximum Marks : 21)

- This section contain **SEVEN** questions
- Each question has FOUR options (a), (b), (c) and (d). **ONLY ONE** of these four options is correct.
- For each question, darken the bubble corresponding to the correct option in the ORS.
- For each question, marks will be awarded in one of the following categories :

| | | |
|---|---|---|
| Full Marks | : | + 3 If only the bubble corresponding to the correct option is darkened. |
| Partial Marks | : | +1 For darkening a bubble corresponding to each correct option, provided NO incorrect option is darkened. |
| Zero Marks | : | 0 If none of the bubbles is darkened. |
| Negative Marks | : | −2 In all other cases. |

**1.** The order of basicity among the following compounds is

(I) $H_3C-C(=O)-NH_2$ (II) (III) (IV) $H_2N-C(=NH)-NH_2$

(a) II>I>IV>III (b) I>IV>III>I (c) IV>I>II>I (d) IV>I>II>III

**2.** Pure water freezes at 273 K and 1 bar. The addition of 34.5 g of ethanol to 500 g of water changes the freezing point of the solution. Use the freezing point depression constant of water as 2 K kg $mol^{-1}$. The figures shown below represent plots of vapour pressure (V.P.) *versus* temperature ($T$). [Molecular weight of ethanol is 46 g $mol^{-1}$]

Among the following, the option representing change in the freezing point is

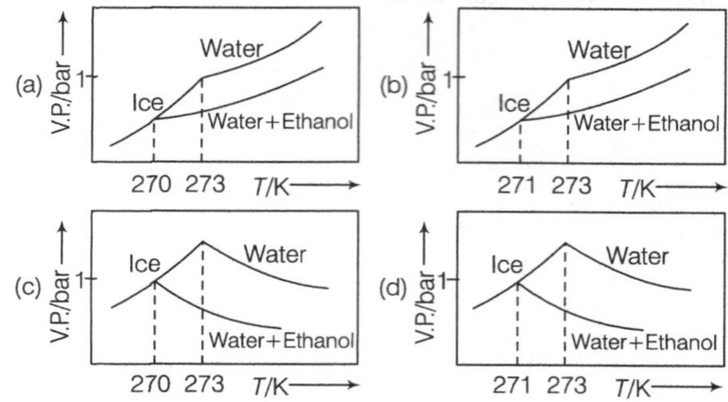


**3.** The major product of the following reaction is

OH, $NH_2$ (biphenyl) $\xrightarrow[\text{(ii) } (aq.\ NaOH)]{\text{(i) } NaNO_2/HCl/0°C}$

(a) 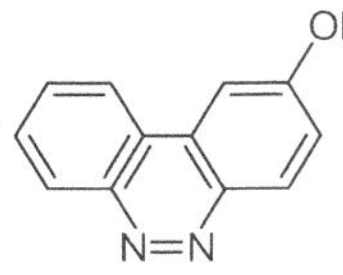 OH, N=N

(c) 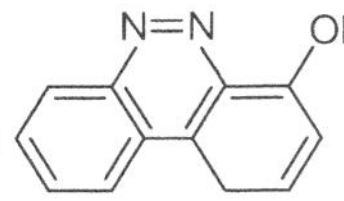 N=N, OH

(b) 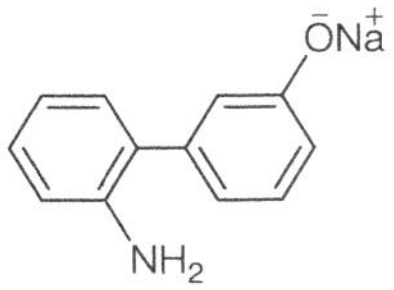 $\bar{O}\overset{+}{Na}$, $NH_2$

(d) 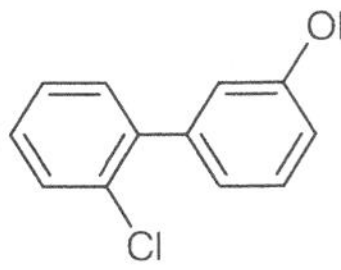 OH, Cl

4. For the following cell,

$Zn(s) \mid ZnSO_4(aq) \parallel CuSO_4(aq) \mid Cu(s)$

when the concentration of $Zn^{2+}$ is 10 times the concentration of $Cu^{2+}$, the expression for $\Delta G$ (in J $mol^{-1}$) is

[F is Faraday constant; $R$ is gas constant; $T$ is temperature; $E°$ (cell) = 1.1 V]

(a) 2.303 $RT$ +1.1 F (b) 1.1 F
(c) 2.303 $RT$ − 2.2 F (d) −2.2 F

5. The order of the oxidation state of the phosphorus atom in $H_3PO_2$, $H_3PO_4$, $H_3PO_3$ and $H_4P_2O_6$ is

(a) $H_3PO_4 > H_3PO_2 > H_3PO_3 > H_4P_2O_6$
(b) $H_3PO_4 > H_4P_2O_6 > H_3PO_3 > H_3PO_2$
(c) $H_3PO_2 > H_3PO_3 > H_4P_2O_6 > H_3PO_4$
(d) $H_3PO_3 > H_3PO_2 > H_3PO_4 > H_4P_2O_6$

6. The standard state Gibbs free energies of formation of C(graphite) and C(diamond) at $T$ = 298 K are

$\Delta_f G°$ [C(graphite)] = 0 kJ $mol^{-1}$

$\Delta_f G°$ [C(diamond)] = 2.9 kJ $mol^{-1}$

The standard state means that the pressure should be 1 bar, and substance should be pure at a given temperature. The conversion of graphite [C(graphite)] to diamond [C(diamond)] reduces its volume by $2 \times 10^{-6}$ $m^3$ $mol^{-1}$. If C(graphite) is converted to C(diamond) isothermally at $T$ = 298K, the pressure at which C(graphite) is in equilibrium with C(diamond), is

[Useful information : 1 J = 1 kg $m^2 s^{-2}$,
1 Pa = 1 kg $m^{-1}$ $s^{-2}$; 1 bar = $10^5$ Pa]

(a) 58001 bar
(b) 1450 bar
(c) 14501 bar
(d) 29001 bar

7. Which of the following combination will produce $H_2$ gas?

(a) Fe metal and conc. $HNO_3$
(b) Cu metal and conc. $HNO_3$
(c) Au metal and NaCN ($aq$) in the presence of air
(d) Zn metal and NaOH ($aq$)

## Section 2 (Maximum Marks : 28)

- This section contains **SEVEN** questions.
- Each question has FOUR option (a), (b), (c) and (d). **ONE OR MORE THAN ONE** of these four option(s) is(are) correct.
- For each question, darken the bubble(s) corresponding to all correct option(s) in the ORS.
- For each question, marks will be awarded in one of the following categories :

Full Marks : +4 If only the bubble(s) corresponding to all the correct option(s) is(are)

Partial Marks : +1 For darkening a bubble corresponding to each correct option, provided NO incorrect option is darkened.

Zero Marks : 0 If none of the bubbles is darkened.

Negative Marks : −2 In all other cases

- For example, if (a), (c) and (d) are all the correct option for a question, darkening all these three will get +4 marks; darkening only (a) and (d) will get +2 marks and darkening (a) and (b) will get −2 marks, as a wrong option is also darkened.

**8.** **For a reaction taking place in a container in equilibrium with its surroundings, the effect of temperature on its equilibrium constant K in terms of change in entropy is described by**

(a) With increase in temperature, the value of K for endothermic reaction increases because unfavourable change in entropy of the surroundings decreases

(b) With increase in temperature, the value of K for exothermic reaction decreases because favourable change in entropy of the surrounding decreases

(c) With increase in temperature, the value of K for endothermic reaction increases because the entropy change of the system is negative

(d) With increase in temperature, the value of K for exothermic reaction decreases because the entropy change of the system is positive

**9.** **In a bimolecular reaction, the steric factor $P$ was experimentally determined to be 4.5. the correct option(s) among the following is(are)**

(a) The activation energy of the reaction is unaffected by the value of the steric factor

(b) Experimentally determined value of frequency factor is higher than that predicted by Arrhenius equation

(c) The value of frequency factor predicted by Arrhenius equation is higher than that determined experimentally

(d) Since $P = 4.5$, the reaction will not proceed unless an effective catalyst is used

**10.** **For the following compounds, the correct statement(s) with respect to nucleophilic substitution reaction is(are)**

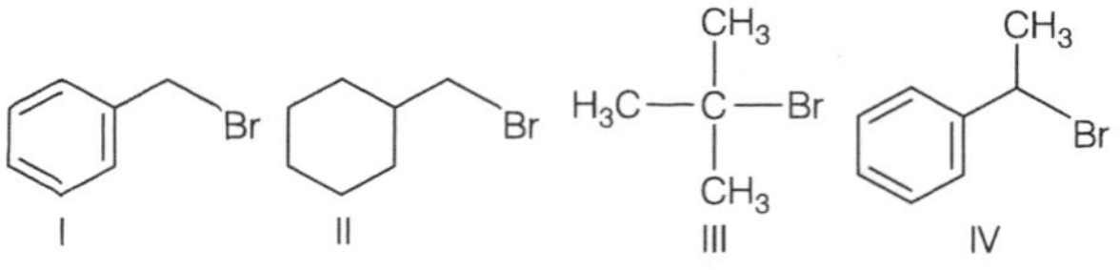

I, II, III, IV

(a) Compound IV undergoes inversion of configuration

(b) The order of reactivity for I, III and IV is : IV > I > III

(c) I and III follow $S_N1$ mechanism

(d) I and II follow $S_N1$ mechanism

11. Among the following, the correct statement(s) is(are)

(a) $Al(CH_3)_3$ has the three-centre two-electron bonds in its dimeric structure
(b) The Lewis acidity of $BCl_3$ is greater than that of $AlCl_3$
(c) $AlCl_3$ has the three-centre two-electron bonds in its dimeric structure
(d) $BH_3$ has the three-centre two-electron bonds in its dimeric structure

12. The option(s) with only amphoteric oxides is(are)

(a) NO, $B_2O_3$, PbO, $SnO_2$ (b) $Cr_2O_3$, CrO, SnO, PbO
(c) $Cr_2O_3$, BeO, SnO, $SnO_2$ (d) ZnO, $Al_2O_3$, PbO, $PbO_2$

13. The correct statement(s) about surface properties is(are)

(a) The critical temperatures of ethane and nitrogen are 563 K and 126 K, respectively. The adsorption of ethane will be more than that of nitrogen of same amount of activated charcoal at a given temperature
(b) Cloud is an emulsion type of colloid in which liquid is dispersed phase and gas is dispersion medium
(c) Adsorption is accompanied by decrease in enthalpy and decrease in entropy of the system
(d) Brownian motion of colloidal particles does not depend on the size of the particles but depends on viscosity of the solution

14. Compound $P$ and $R$ upon ozonolysis produce $Q$ and $S$, respectively. The molecular formula of $Q$ and $S$ is $C_8H_8O$. $Q$ undergoes Cannizzaro reaction but not haloform reaction, whereas $S$ undergoes haloform reaction but not Cannizzaro reaction.

(i) $P \xrightarrow[\text{(ii) Zn/H}_2\text{O}]{\text{(i) O}_3/\text{CH}_2\text{Cl}_2} \underset{(C_8H_8O)}{Q}$ (ii) $R \xrightarrow[\text{(ii) Zn/H}_2\text{O}]{\text{(i) O}_3/\text{CH}_2\text{Cl}_2} \underset{(C_8H_8O)}{S}$

The option(s) with suitable combination of $P$ and $R$, respectively, is(are)

(a) 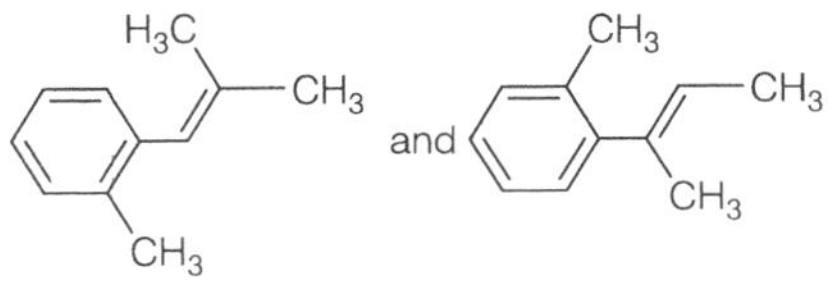 $H_3C$, $CH_3$, $CH_3$ and $CH_3$, $CH_3$, $CH_3$

(b) 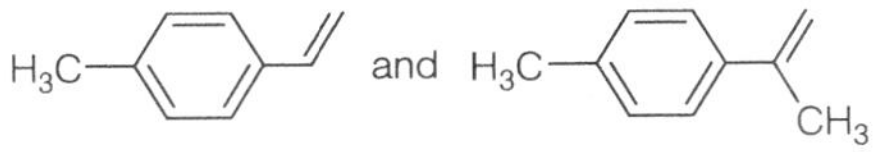 $H_3C-$ and $H_3C-$ $CH_3$

(c) 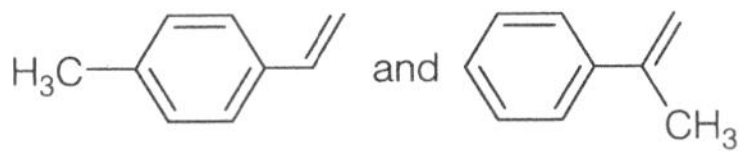 $H_3C-$ and $CH_3$

(d) 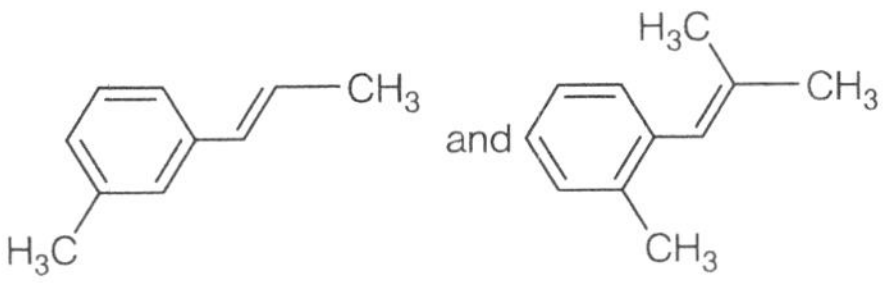 $CH_3$, $H_3C$ and $H_3C$, $CH_3$, $CH_3$

## Section 3 (Maximum Marks : 12)

- This section contains **TWO** paragraphs.
- Based on each paragraph, there are **TWO** questions.
- Each question has **FOUR** options (a), (b), (c) and (d). **ONLY ONE** of these four options is correct.
- For each question, darken the bubble corresponding to the correct integer in the ORS.
- For each question, marks will be awarded in one of the following categories :

Full Marks : + 3 If only the bubble corresponding to the correct option is darkened

Zero Marks : 0 If none of the bubbles is darkened

### Paragraph 1

Upon heating $KClO_3$ in presence of catalytic amount of $MnO_2$, a gas *W* is formed. Excess amount of *W* reacts with white phosphorus to give *X*. The reaction of *X* with pure $HNO_3$ gives *Y* and *Z*.

**15.** *Y* and *Z* are, respectively

(a) $N_2O_4$ and $HPO_3$ (b) $N_2O_4$ and $H_3PO_3$
(c) $N_2O_3$ and $H_3PO_4$ (d) $N_2O_5$ and $HPO_3$

**16.** *W* and *X* are, respectively

(a) $O_2$ and $P_4O_{10}$ (b) $O_2$ and $P_4O_6$
(c) $O_3$ and $P_4O_6$ (d) $O_3$ and $P_4O_{10}$

### Paragraph 2

The reaction of compound *P* with $CH_3MgBr$ (excess) in $(C_2H_5)_2O$ followed by addition of $H_2O$ gives *Q*. The compound *Q* on treatment with $H_2SO_4$ at 0° C gives *R*. The reaction of *R* with $CH_3COCl$ in the presence of anhydrous $AlCl_3$ in $CH_2Cl_2$ followed by treatment with $H_2O$ produces compound *S*. [Et in compound *P* is ethyl group]

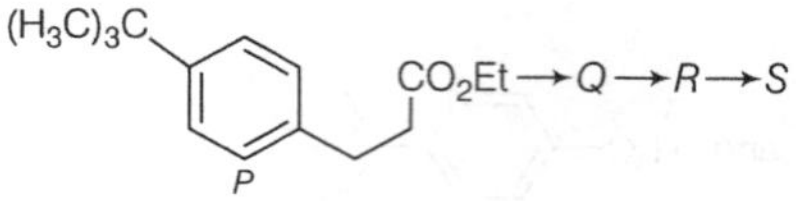

$(H_3C)_3C$ — [benzene ring] — $CH_2CH_2CO_2Et \rightarrow Q \rightarrow R \rightarrow S$

*P*

**17.** The product *S* is

(a) $(H_3C)_3C$, $COCH_3$, $CH_3$ [indane structure]

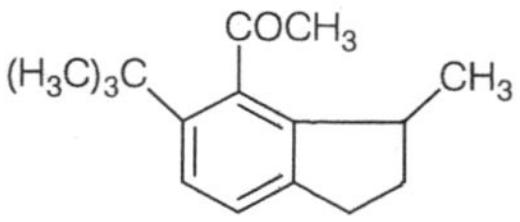

(b) $(H_3C)_3C$, $H_3C$ $CH_3$, $COCH_3$ [indane structure]

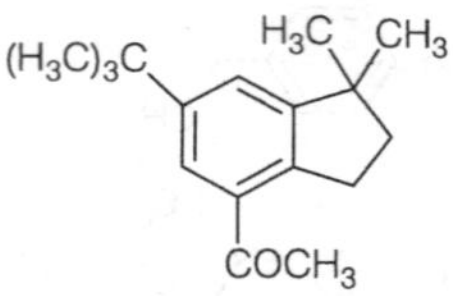

(c) $(H_3C)_3C$, $HO_3S$, $O$, $CH_3$, $COCH_3$

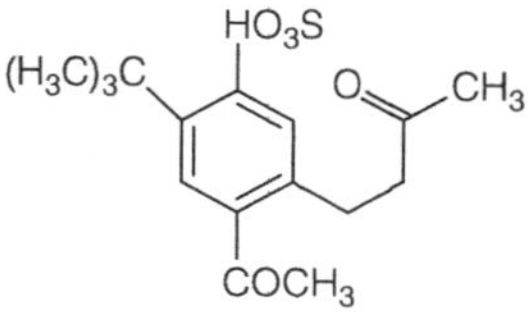

(d) $(H_3C)_3C$, $H_3COC$, $H_3C$, $CH_3$

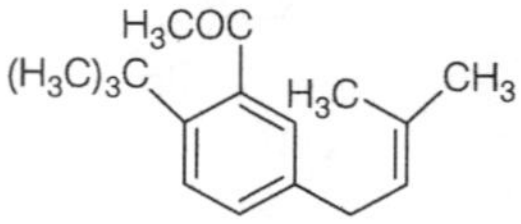

18. The reactions, $Q$ to $R$ and $R$ to $S$, are

(a) Aromatic sulfonation and Friedel-Crafts acylation
(b) Friedel-Crafts alkylation and Friedel-Crafts acylation
(c) Friedel-Crafts alkylation, dehydration and Friedel-Crafts acylation
(d) Dehydration and Friedel-Crafts acylation

# Mathematics

## Section 1 (Maximum Marks : 21)

- This section contains **SEVEN** questions.
- Each question has **FOUR** options (a), (b), (c) and (d). **ONLY ONE** of these four options is correct.
- For each question, darken the bubble corresponding to the correct option in the ORS.
- For each question, marks will be awarded in one of the following categories :
  *Full Marks* : +3 If only the bubble corresponding to the correct option is darkened.
  *Zero Marks* : 0 If none of the bubbles is darkened.
  *Negative Marks* : −1 In all other cases.

1. If $f : R \to R$ is a twice differentiable function such that $f''(x) > 0$ for all $x \in R$, and $f\left(\frac{1}{2}\right) = \frac{1}{2}$, $f(1) = 1$, then

(a) $f'(1) \le 0$
(b) $f'(1) > 1$
(c) $0 < f'(1) \le \frac{1}{2}$
(d) $\frac{1}{2} < f'(1) \le 1$

2. If $y = y(x)$ satisfies the differential equation

$$8\sqrt{x}\left(\sqrt{9+\sqrt{x}}\right)dy = \left(\sqrt{4+\sqrt{9+\sqrt{x}}}\right)^{-1} dx,\ x > 0$$

and $y(0) = \sqrt{7}$, then $y(256) =$

(a) 16 (b) 3 (c) 9 (d) 80

3. How many $3 \times 3$ matrices $M$ with entries from $\{0, 1, 2\}$ are there, for which the sum of the diagonal entries of $M^T M$ is 5 ?

(a) 198 (b) 162
(c) 126 (d) 135

4. Three randomly chosen nonnegative integers $x$, $y$ and $z$ are found to satisfy the equation $x + y + z = 10$. Then the probability that $z$ is even, is

(a) $\frac{1}{2}$ (b) $\frac{36}{55}$ (c) $\frac{6}{11}$ (d) $\frac{5}{11}$

5. Let $S = \{1, 2, 3, \dots\dots, 9\}$. For $k = 1, 2, \dots\dots 5$, let $N_k$ be the number of subsets of $S$, each containing five elements out of which exactly $k$ are odd. Then $N_1 + N_2 + N_3 + N_4 + N_5 =$

(a) 210 (b) 252
(c) 126 (d) 125

6. Let $O$ be the origin and let $PQR$ be an arbitrary triangle. The point $S$ is such that

$$\mathbf{OP}\cdot\mathbf{OQ} + \mathbf{OR}\cdot\mathbf{OS} = \mathbf{OR}\cdot\mathbf{OP} + \mathbf{OQ}\cdot\mathbf{OS} = \mathbf{OQ}\cdot\mathbf{OR} + \mathbf{OP}\cdot\mathbf{OS}$$

Then the triangle $PQR$ has $S$ as its

(a) centroid
(b) orthocentre
(c) incentre
(d) circumcentre

**7.** The equation of the plane passing through the point (1, 1, 1) and perpendicular to the planes $2x + y - 2z = 5$ and $3x - 6y - 2z = 7$ is

(a) $14x + 2y - 15z = 1$
(b) $-14x + 2y + 15z = 3$
(c) $14x - 2y + 15z = 27$
(d) $14x + 2y + 15z = 31$

## Section 2 (Maximum Marks : 28)

- This section contains **SEVEN** questions.
- Each question has **FOUR** options (a), (b), (c) and (d). **ONE OR MORE THAN ONE** of these four option(s) is(are) correct.
- For each question, darken the bubble(s) corresponding to all the correct option(s) in the ORS.
- For each question, marks will be awarded in one of the following categories :
  *Full Marks* : +4 If only the bubble(s) corresponding to all the correct option(s) is(are) darkened.
  *Partial Marks* : +1 For darkening a bubble corresponding to each correct option, provided NO incorrect option is darkened.
  *Zero Marks* : 0 If none of the bubbles is darkened.
  *Negative Marks* : −2 In all other cases.
- For example, if (a), (c) and (d) are all the correct options for a question, darkening all these three will get + 4 marks; darkening only (a) and (d) will get +2 marks and darkening (a) and (b) will get − 2 marks, as a wrong option is also darkened.

**8.** If $f : R \to R$ is a differentiable function such that $f'(x) > 2f(x)$ for all $x \in R$, and $f(0) = 1$ then

(a) $f(x) > e^{2x}$ in $(0, \infty)$
(b) $f'(x) < e^{2x}$ in $(0, \infty)$
(c) $f(x)$ is increasing in $(0, \infty)$
(d) $f(x)$ is decreasing in $(0 \infty)$

**9.** If $I = \sum_{k=1}^{98} \int_k^{k+1} \frac{k+1}{x(x+1)} dx$, then

(a) $I > \log_e 99$
(b) $I < \log_e 99$
(c) $I < \frac{49}{50}$
(d) $I > \frac{49}{50}$

**10.** If the line $x = \alpha$ divides the area of region $R = \{(x, y) \in R^2 : x^3 \le y \le x, 0 \le x \le 1\}$ into two equal parts, then

(a) $2\alpha^4 - 4\alpha^2 + 1 = 0$
(b) $\alpha^4 + 4\alpha^2 - 1 = 0$
(c) $\frac{1}{2} < \alpha < 1$
(d) $0 < \alpha \le \frac{1}{2}$

**11.** Let $\alpha$ and $\beta$ be non zero real numbers such that $2(\cos\beta - \cos\alpha) + \cos\alpha \cos\beta = 1$. Then which of the following is/are true?

(a) $\sqrt{3}\tan\left(\frac{\alpha}{2}\right) - \tan\left(\frac{\beta}{2}\right) = 0$
(b) $\tan\left(\frac{\alpha}{2}\right) - \sqrt{3}\tan\left(\frac{\beta}{2}\right) = 0$
(c) $\tan\left(\frac{\alpha}{2}\right) + \sqrt{3}\tan\left(\frac{\beta}{2}\right) = 0$
(d) $\sqrt{3}\tan\left(\frac{\alpha}{2}\right) + \tan\left(\frac{\beta}{2}\right) = 0$

**12.** Let

$$f(x) = \frac{1 - x(1 + |1 - x|)}{|1 - x|} \cos\left(\frac{1}{1 - x}\right)$$

for $x \ne 1$. Then

(a) $\lim_{x \to 1^+} f(x) = 0$
(b) $\lim_{x \to 1^-} f(x)$ does not exist
(c) $\lim_{x \to 1^-} f(x) = 0$
(d) $\lim_{x \to 1^+} f(x)$ does not exist

13. If $g(x) = \int_{\sin x}^{\sin(2x)} \sin^{-1}(t)\, dt$, then

(a) $g'\left(-\frac{\pi}{2}\right) = 2\pi$

(b) $g'\left(-\frac{\pi}{2}\right) = -2\pi$

(c) $g'\left(\frac{\pi}{2}\right) = 2\pi$

(d) $g'\left(\frac{\pi}{2}\right) = -2\pi$

14. If $f(x) = \begin{vmatrix} \cos(2x) & \cos(2x) & \sin(2x) \\ -\cos x & \cos x & -\sin x \\ \sin x & \sin x & \cos x \end{vmatrix}$, then

(a) $f(x)$ attains its minimum at $x = 0$

(b) $f(x)$ attains its maximum at $x = 0$

(c) $f'(x) = 0$ at more than three points in $(-\pi, \pi)$

(d) $f'(x) = 0$ at exactly three points in $(-\pi, \pi)$

## Section 3 (Maximum Marks : 12)

- This section contains **TWO** paragraphs.
- Based on each paragraph, there are **TWO** questions.
- Each question has **FOUR** options (a), (b), (c) and (d). **ONLY ONE** of these four options is correct.
- For each question, darken the bubble corresponding to the correct option in the ORS.
- For each question, marks will be awarded in one of the following categories:

*Full Marks* : + 3 If only the bubble corresponding to the correct option is darkened.

*Zero Marks* : 0 In all other cases.

### Paragraph 1

Let $O$ be the origin and **OX**, **OY**, **OZ** be three unit vectors in the directions of the sides **QR**, **RP**, **PQ** respectively, of a triangle $PQR$.

15. If the triangle $PQR$ varies, then the minimum value of $\cos(P+Q) + \cos(Q+R) + \cos(R+P)$ is

(a) $-\frac{3}{2}$ (b) $\frac{3}{2}$

(c) $\frac{5}{3}$ (d) $-\frac{5}{3}$

16. $|\mathbf{OX} \times \mathbf{OY}| =$

(a) $\sin(P+Q)$ (b) $\sin(P+R)$

(c) $\sin(Q+R)$ (d) $\sin 2R$

### Paragraph 2

Let $p, q$ be integers and let $\alpha, \beta$ be the roots of the equation, $x^2 - x - 1 = 0$ where $\alpha \neq \beta$. For $n = 0, 1, 2, \ldots\ldots$, let $a_n = p\alpha^n + q\beta^n$.

**FACT** : If $a$ and $b$ are rational numbers and $a + b\sqrt{5} = 0$, then $a = 0 = b$.

17. $a_{12} =$

(a) $a_{11} + 2a_{10}$

(b) $2a_{11} + a_{10}$

(c) $a_{11} - a_{10}$

(d) $a_{11} + a_{10}$

18. If $a_4 = 28$, then $p + 2q =$

(a) 14 (b) 7

(c) 21 (d) 12

# Answer *with* Explanations

## Paper 1

### Physics

1. *(a,b)* At $\omega \approx 0$, $X_C = \frac{1}{\omega C} = \infty$. Therefore, current is nearly zero.

Further at resonance frequency, current and voltage are in phase. This resonance frequency is given by,

$$\omega_r = \frac{1}{\sqrt{LC}} = \frac{1}{\sqrt{10^{-6} \times 10^{-6}}} = 10^6 \text{ rad/s}$$

We can see that this frequency is independent of $R$.

Further, $X_L = \omega L$, $X_C = \frac{1}{\omega C}$

At, $\omega = \omega_r = 10^6$ rad/s, $X_L = X_C$.

For $\omega > \omega_r$, $X_L > X_C$. So, circuit is inductive.

2. *(a,b,c)* The minimum deviation produced by a prism

$$\delta_m = 2i - A = A$$

$\therefore$ $i_1 = i_2 = A$ and $r_1 = r_2 = A/2$

$\therefore$ $r_1 = i_1/2$

Now, using Snell's law

$\sin A = \mu \sin A/2$

$\Rightarrow \mu = 2\cos(A/2)$

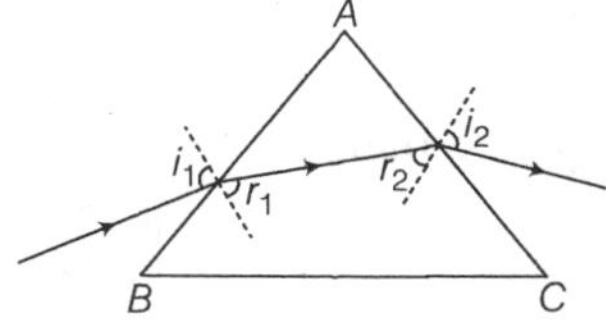


For this prism when the emergent ray at the second surface is tangential to the surface

$i_2 = \pi/2$

$\therefore$ $r_2 = \theta_c$

$\therefore$ $r_1 = A - \theta_c$

so, $\sin i_1 = \mu \sin(A - \theta_c)$

so,

$$i_1 = \sin^{-1}\left[\sin A\sqrt{4\cos^2\frac{A}{2} - 1} - \cos A\right]$$

For minimum deviation through isosceles prism, the ray inside the prism is parallel to the base of the prism if $\angle B = \angle C$.

But it is not necessarily parallel to the base if, $\angle A = \angle B$ or $\angle A = \angle C$

3. *(b,d)* The net magnetic flux through the loops at time $t$ is

$$\phi = B(2A - A)\cos \omega t = BA\cos \omega t$$

so, $\left|\frac{d\phi}{dt}\right| = B\omega A \sin\omega t$

$\therefore$ $\left|\frac{d\phi}{dt}\right|$ is maximum when $\phi = \omega t = \pi/2$

The emf induced in the smaller loop,

$$\varepsilon_{\text{smaller}} = -\frac{d}{dt}(BA\cos\omega t) = B\omega A \sin\omega t$$

$\therefore$ Amplitude of maximum net emf induced in both the loops

= Amplitude of maximum emf induced in the smaller loop alone.

4. *(a,c,d)*

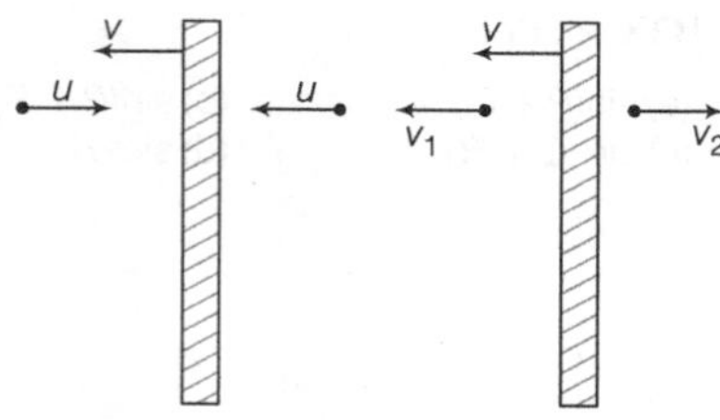


Just before the collision    Just after the collision

$$v_1 = u + 2v$$
$$\Delta v_1 = (2u + 2v)$$
$$F_1 = \frac{dp_1}{dt}$$

$$= \rho A(u + v)(2u + 2v)$$
$$= 2\rho A(u + v)^2$$
$$v_2 = (u - 2v)$$
$$\Delta v_2 = (2u - 2v)$$
$$F_2 = \frac{dp_2}{dt}$$
$$= \rho A(u - v)(2u - 2v)$$
$$= 2\rho A(u - v)^2$$

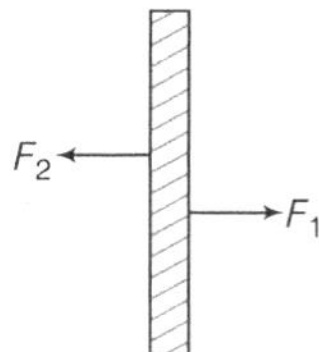


[$\Delta F$ is the net force due to the air molecules on the plate]

$$\Delta F = 2\rho A(4uv) = 8\rho Auv$$
$$P = \frac{\Delta F}{A} = 8\rho(uv)$$
$$F_{net} = (F - \Delta F) = ma$$

[$m$ is mass of the plate]

$$F - (8\rho Au)v = ma$$

5. *(a,b)* $\Delta x_{cm}$ of the block and point mass system = 0

$\therefore \quad m(x + R) + Mx = 0$

where, $x$ is displacement of the block.

Solving this equation, we get

$$x = -\frac{mR}{M + m}$$

From conservation of momentum and mechanical energy of the combined system

$$0 = mv - MV$$
$$mgR = \frac{1}{2}mv^2 + \frac{1}{2}MV^2$$

Solving these two equations, we get

$$v = \sqrt{\frac{2gR}{1 + \frac{m}{M}}}$$

6. *(a,c,d or a,c)* $v = \sqrt{\frac{T}{\mu}}$, so speed at any position will be same for both pulses, therefore time taken by both pulses will be same.

$$\lambda f = v$$
$$\Rightarrow \quad \lambda = \frac{v}{f}$$
$$\Rightarrow \quad \lambda \propto v \propto T$$

since when pulse 1 reaches at $A$ tension and hence speed decreases therefore $\lambda$ decreases.

**Note** *If we refer velocity by magnitude only, then option (a, c, d) will be correct, else only (a, c) will be correct.*

7. *(b,c,d or d)* **Assumption** $e = 1$

[black body radiation]

$$P = \sigma A(T^4 - T_0^4)$$

(c) $P_{rad} = \sigma AT^4 = \sigma \cdot 1 \cdot (T_0 + 10)^4$

$$= \sigma \cdot T_0^4 \left(1 + \frac{10}{T_0}\right)^4$$

[$T_0 = 300$ K given]

$$= \sigma \cdot (300)^4 \cdot \left(1 + \frac{40}{300}\right) \approx 460 \times \frac{17}{15} \approx 520 \text{ J}$$
$$P_{net} = 520 - 460 \approx 60 \text{ W}$$

$\Rightarrow$ Energy radiated in 1 s = 60 J

(b) $P = \sigma A(T^4 - T_0^4)$

$$dP = \sigma A(0 - 4T_0^3 \cdot dT)$$

and $\quad dT = -\Delta T$

$\Rightarrow \quad dP = 4\sigma AT_0^3 \Delta T$

(d) If surface area decreases, then energy radiation also decreases.

**Note** *While giving answer (b) and (c) it is assumed that energy radiated refers the net radiation. If energy radiated is taken as only emission, then (b) and (c) will not be included in answer.*

8. *(5)* Potential energy of hydrogen atom ($Z = 1$) in $n$th orbit (in eV)

$$\text{P, E} = -\frac{27.2}{n^2}$$
$$\frac{v_f}{v_i} = \frac{-\frac{27.2}{n_f^2}}{-\frac{27.2}{n_i^2}} = \frac{1}{6.25}$$
$$6.25 = \frac{n_f^2}{n_i^2}$$
$$\frac{n_f}{n_i} = 2.5 = \frac{5}{2}$$

Hence the answer is 5.

9. *(6)* From mass conservation,

$$\rho \cdot \frac{4}{3}\pi R^3 = \rho \cdot K \cdot \frac{4}{3}\pi r^3$$
$$\Rightarrow \quad R = K^{1/3} r$$

$\therefore \quad \Delta U = T\Delta A = T(K \cdot 4\pi r^2 - 4\pi R^2)$

$= T(K \cdot 4\pi R^2 K^{-2/3} - 4\pi R^2)$

$\Delta U = 4\pi R^2 T[K^{1/3} - 1]$

Putting the values, we get

$$10^{-3} = \frac{10^{-1}}{4\pi} \times 4\pi \times 10^{-4}[K^{1/3} - 1]$$

$$100 = K^{1/3} - 1$$

$$\Rightarrow \quad K^{1/3} \cong 100 = 10^2$$

Given that $K = 10^{\alpha}$

$$\therefore \quad 10^{\alpha/3} = 10^2$$

$$\Rightarrow \quad \frac{\alpha}{3} = 2$$

$$\Rightarrow \quad \alpha = 6$$

**10.** *(6)*

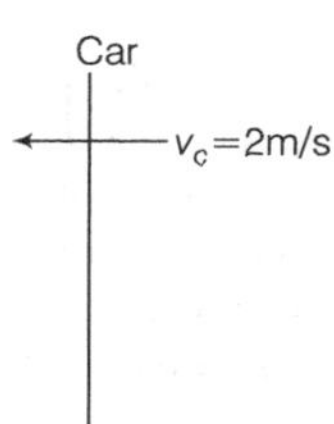


Frequency observed at car

$$f_1 = f_0\left(\frac{v + v_C}{v}\right)$$

($v$ = speed of sound)

Frequency of reflected sound as observed at the source

$$f_2 = f_1\left(\frac{v}{v - v_C}\right) = f_0\left(\frac{v + v_C}{v - v_C}\right)$$

Beat frequency $= f_2 - f_0$

$$= f_0\left[\frac{v + v_C}{v - v_C} - 1\right] = f_0\left[\frac{2v_C}{v - v_C}\right]$$

$$= 492 \times \frac{2 \times 2}{328} = 6 \text{ Hz}$$

**11.** *(5)* $I^{131} \xrightarrow[T_{1/2} = 8 \text{ Days}]{} Xe^{131} + \beta$

$A_0 = 2.4 \times 10^5 \text{ Bq} = \lambda N_0$

Let the volume is $V$,

$$t = 0$$

$$A_0 = \lambda N_0$$

$$t = 11.5 \text{ h}$$

$$A = \lambda N$$

$$115 = \lambda\left(\frac{N}{V} \times 2.5\right)$$

$$115 = \frac{\lambda}{V} \times 2.5 \times (N_0 e^{-\lambda t})$$

$$115 = \frac{(N_0\lambda)}{V} \times (2.5) \times e^{-\frac{\ln 2}{8 \text{ day}}(11.5 \text{ h})}$$

$$115 = \frac{(2.4 \times 10^5)}{V} \times (2.5) \times e^{-1/24}$$

$$V = \frac{2.4 \times 10^5}{115} \times 2.5\left[1 - \frac{1}{24}\right]$$

$$= \frac{2.4 \times 10^5}{115} \times 2.5\left[\frac{23}{24}\right]$$

$$= \frac{10^5 \times 23 \times 25}{115 \times 10^2}$$

$$= 5 \times 10^3 \text{ ml} = 5 \text{ L}$$

**12.** *(8)* But this value of refractive index is not possible.

$$1.6 \sin\theta = (n - m\Delta n)\sin 90°$$

$$1.6 \sin\theta = n - m\Delta n$$

$$1.6 \times \frac{1}{2} = 1.6 - m(0.1)$$

$$0.8 = 1.6 - m(0.1)$$

$$m \times 0.1 = 0.8$$

$$m = 8$$

**13.** *(c)* For particle to move in negative $y$-direction, either its velocity must be in negative $y$-direction (if initial velocity $\neq 0$) and force should be parallel to velocity or it must experience a net force in negative $y$-direction only (if initial velocity $= 0$)

**14.** *(a)* $\mathbf{F}_{\text{net}} = \mathbf{F}_e + \mathbf{F}_B$

$$= q\mathbf{E} + q\mathbf{v} \times \mathbf{B}$$

For particle to move in straight line with constant velocity, $\mathbf{F}_{\text{net}} = 0$

$$\therefore \quad q\mathbf{E} + q\mathbf{v} \times \mathbf{B} = 0$$

**15.** *(c)* For path to be helix with axis along positive $z$-direction, particle should experience a centripetal acceleration in $xy$-plane.

For the given set of options only option (c) satisfy the condition. Path is helical with increasing pitch.

**16.** *(c)* $V_{\text{sound}} = \sqrt{\frac{\gamma RT}{M}}$. As the sound wave propagates, the air in a chamber undergoes compression and rarefaction very fastly, hence undergo a adiabatic process. So, curves are steeper than isothermal.

$$\left(\frac{dp}{dV}\right)_{\text{Adi}} = -\gamma\left(\frac{p}{V}\right) \quad \ldots\text{(i)}$$

$$\left(\frac{dp}{dV}\right)_{\text{Iso}} = -\left(\frac{P}{V}\right) \quad \ldots\text{(ii)}$$

Graph 'Q' satisfies Eq. (i)

17. *(b)* $\Delta U = \Delta Q - p\Delta V$, $\Delta U + p\Delta V = \Delta Q$

As $\Delta U \neq 0$, $W \neq 0$, $\Delta Q \neq 0$. The process represents, isobaric process

$$W_{\text{gas}} = -p(\Delta V)$$

$$= -p(V_2 - V_1)$$

$$= -pV_2 + pV_1$$

Graph '*P*' satisfies isobaric process.

18. *(b)* Work done in isochoric process is zero.

$W_{12} = 0$ as $\Delta V = 0$

Graph '*S* represents isochoric process.

# Chemistry

1. *(a, c, d)*

(a)

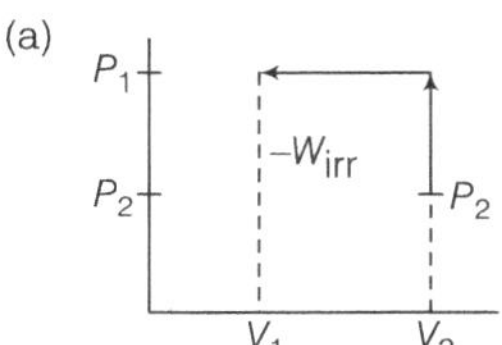


Irreversible compression

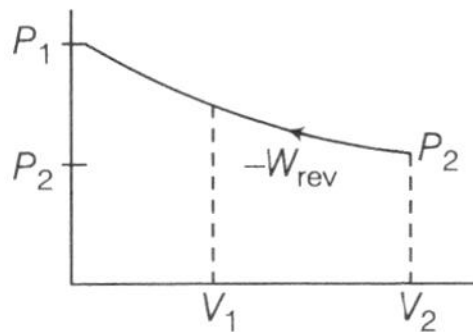


Reversible compression

Maximum work is done on the system when compression occur irreversibly and minimum work is done is reversible compression.

(b)

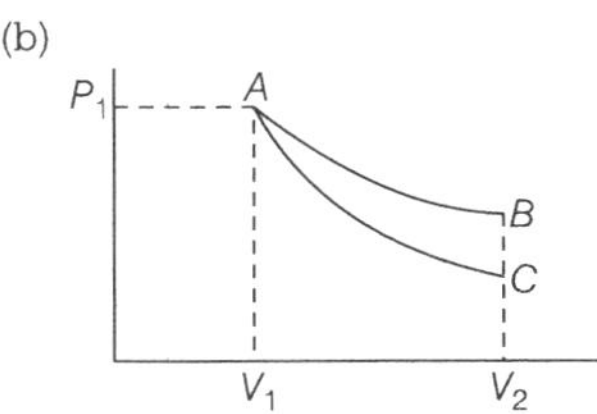


*AB* is isothermal and *AC* is adiabatic path. Work done is area under the curve. Hence, less work is obtained in adiabatic process than in isothermal process.

(c) It is incorrect. In adiabatic expansion cooling is observed, hence $\Delta U = nC_v\Delta T < 0$.

(d) $q = 0$ (adiabatic), $W = 0$ (Free expansion)

Hence, $\Delta U = 0$, $\Delta T = 0$ (Isothermal)

2. *(b,c)* Since, there is no principal functional groups, numbering of disubstituted benzene is done in alphabetical order as

Cl
①②③④⑤⑥
$CH_3$

Hence IUPAC name of this compound is 1-chloro-4-methyl benzene.

Also, toluene is an acceptable name in IUPAC, hence this compound can also be named as 4-chloro toluene.

3. *(b,d)* The graph shown indicates that there is positive deviation because the observed vapour pressure of *L* is greater than the ideal pressure

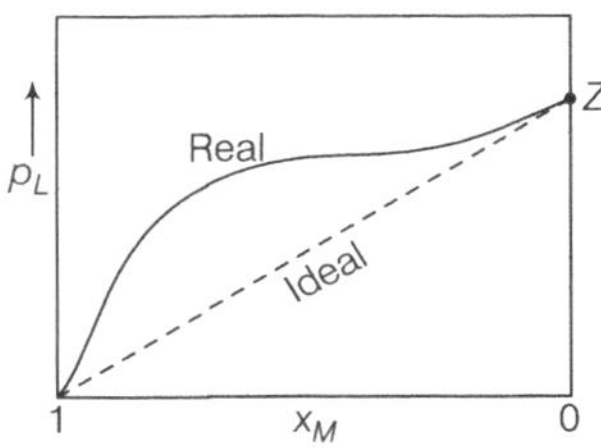


Since, deviation is positive, the intermolecular force between *L* and *M* is smaller than the same in pure *L* and pure *M*.

Also as $x_L \to 1$, $x_M \to 0$, the real curve approaching ideal curve where Raoult's law will be obeyed.

4. *(b,d)* Addition of halogen at double bond occur in antiorientation *via* cyclic halonium ion intermediate.

(i)

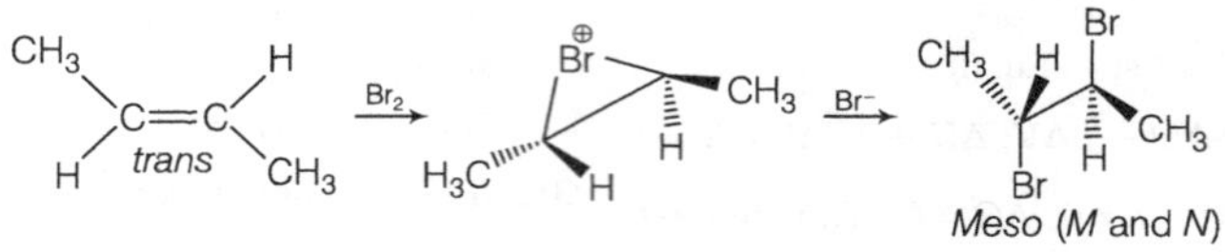

*trans* $\xrightarrow{Br_2}$ $\xrightarrow{Br^-}$ *Meso* (*M* and *N*)

(ii)

$\xrightarrow{Br_2}$

$\xrightarrow{Br^-}$ *O* + *P*

Enantiomers

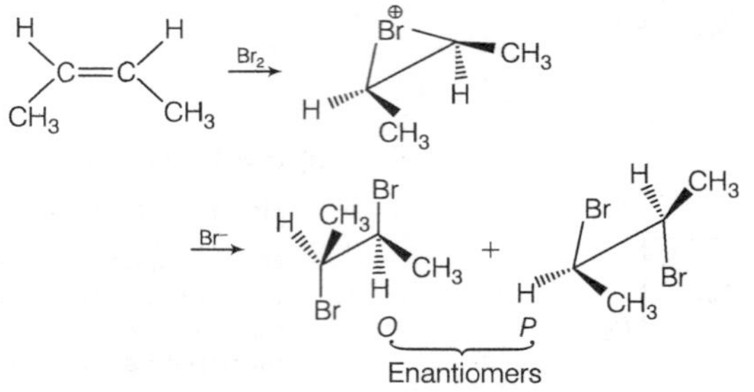

Here (*M* + *O*) and (*N* + *P*) are pair of diastereomers.

5. *(a, c, d)* $\underset{\text{Pink }(X)}{[Co(H_2O)_6]Cl_2} \xrightarrow[O_2\ (Air)]{\text{Excess } NH_4OH/NH_4Cl} \underset{Y}{Co(NH_3)_6]Cl_3}$

$$\underset{X}{[Co(H_2O)_6]^{2+}} + \underset{(Excess)}{4Cl^-} \longrightarrow \underset{\text{blue } Z}{[CoCl_4]^{2-}}$$

(a) Since $NH_3$ is moderately strong ligand, hybridisation of cobalt in *Y* is $d^2sp^3$.

(b) Cobalt is $sp^3$-hybridised in $[CoCl_4]^{2-}$.

(c) $\underset{Y}{[Co(NH_3)_6]Cl_3} + 3AgNO_3(aq) \longrightarrow 3AgCl\downarrow$

(d) $\underset{\text{Blue}}{[CoCl_4]^{2-}} + 6H_2O \rightleftharpoons \underset{\text{Pink}}{[Co(H_2O)_6]^{2+}} + 4Cl^-;\ \Delta H < 0$

6. *(a, c, d)*

(a) $ClO_4^-$ is more stable than $ClO^-$.

(b) Incorrect : $Cl_2 + H_2O \longrightarrow HCl + HOCl$

(c)

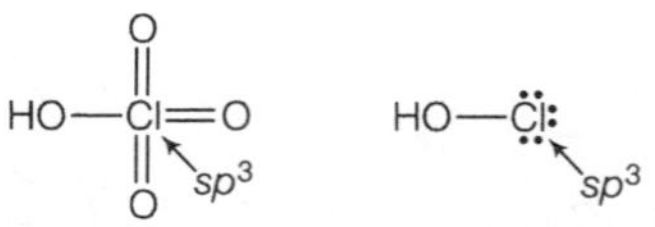

HO—Cl(=O)₃ ($sp^3$)    HO—Cl: ($sp^3$)

(d) $HClO_4$ is stronger acid than $H_2O$.

7. *(a,d)* Colour of halogen arises due to transition from HOMO to LUMO in the visible region. On moving down a group, the difference in energy between HOMO and LUMO decreases electronic transition occur more easily and colour intensity increases.

8. (6)

| S.N. | Species | No. of $\sigma$-bonds with central atom | No. of L.P at central atom |
|---|---|---|---|
| (i) | In $[TeBr_6]^{2-}$ | 6 | 1 |
| (ii) | In $[BrF_2]^+$ | 2 | 2 |
| (iii) | In $SNF_3$ | 4 | 0 |
| (iv) | In $[XeF_3]^-$ | 3 | 3 |

9. (5) The aromatic systems are

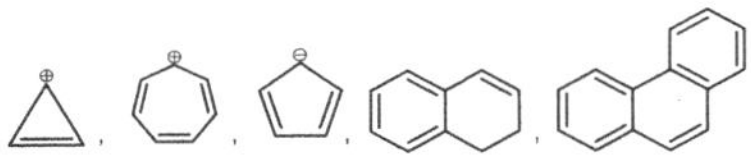

10. (6) $H_2, Li_2, Be_2, C_2, N_2$ and $F_2$ are diamagnetic according to molecular orbital theory.

11. (2) Density $(\rho) = 8 = \dfrac{4 \times M}{N_A (4 \times 10^{-8} \text{cm})^3}$

$$\Rightarrow \quad M = 128 \times 10^{-24} N_A$$

$$\Rightarrow \quad \text{No of atoms} = \frac{256}{M} \times N_A$$

$$= \frac{256}{128 \times 10^{-24} N_A} \times N_A$$

$$= 2 \times 10^{24}$$

12. (6) $\text{pH} = C\alpha = 10^{-4}$

$$\Rightarrow \quad \alpha = \frac{10^{-4}}{0.0015}$$

Also, conductance $(G) = \kappa\left(\dfrac{A}{l}\right)$

$$\Rightarrow \kappa = G\left(\frac{l}{A}\right) = 5 \times 10^{-7} \times \frac{120}{1} = 6 \times 10^{-5}$$

$$\Rightarrow \quad \Lambda^c = \frac{\kappa \times 1000}{C}$$

$$= \frac{6 \times 10^{-5} \times 1000}{0.0015}$$

$$\Rightarrow \Lambda^\infty = \frac{\Lambda^c}{\alpha} = \frac{6 \times 10^{-5} \times 1000}{0.0015} \times \frac{0.0015}{10^{-4}}$$

$$= 600 = 6 \times 10^2 \text{ S cm}^{-1}\text{mol}^{-1}$$

13. (c) In the wave function $(\psi)$ expression for 1s-orbital of $He^+$, there should be no angular part. Hence (iii) can't be true for $\psi_{1s}$ of $He^+$.

14. (a) Correct : 2s orbital has one radial node.

No of radial node $= n - l - 1 = 2 - 0 - 1 = 1$

Also, when radial part of wave function $(\psi)$ is plotted against "$r$", wave function changes its sign at node.

15. (d) $i$ is the correct expression of wave function for 1s-orbital of hydrogenic system.

Also, $E(2-4) = kZ^2\left(\dfrac{1}{4} - \dfrac{1}{16}\right)$

$$= kZ^2\left(\frac{3}{16}\right) \quad \ldots\text{(i)}$$

$$E(2-6) = kZ^2\left(\frac{1}{4} - \frac{1}{36}\right) = kZ^2\left(\frac{2}{9}\right) \ldots\text{(ii)}$$

$$\Rightarrow \quad E(2-6) \times \frac{27}{32} = kZ^2 \times \frac{2}{9} \times \frac{27}{32}$$

$$= kZ^2\left(\frac{3}{16}\right) = E(2-4)$$

16. (d) $C_6H_5{-}CH_3 \xrightarrow[hv]{Br_2}$

$C_6H_5{-}CH_2Br + HBr$
(Free radical bromination)

17. (a) $C_6H_5{-}\overset{O}{\overset{\|}{C}}{-}CH_3 \text{ (II)} \xrightarrow[\text{(i)}]{NaOH/Br_2}$

$C_6H_5COONa + CHBr_3(P)$
Bromoform

18. (d)

$C_6H_5CHO + CH_3{-}\overset{O}{\overset{\|}{C}}{-}O{-}\overset{O}{\overset{\|}{C}}{-}CH_3 \xrightarrow[\text{Perkin's condensation}]{CH_3COOK}$

$C_6H_5{-}CH{=}CH{-}COOH$
(Cinnamic acid)

Cinnamic acid shows *cis-trans* isomerism.

# Mathematics

1. (a,b) $P(X)=\frac{1}{3}$

$$P\left(\frac{X}{Y}\right)=\frac{P(X\cap Y)}{P(Y)}=\frac{1}{2}$$

$$P\left(\frac{Y}{X}\right)=\frac{P(X\cap Y)}{P(X)}=\frac{2}{5}$$

$$P(X\cap Y)=\frac{2}{15}$$

$$P(Y)=\frac{4}{15}$$

$$P\left(\frac{X'}{Y}\right)=\frac{P(Y)-P(X\cap Y)}{P(Y)}$$

$$=\frac{\frac{4}{15}-\frac{2}{15}}{\frac{4}{15}}=\frac{1}{2}$$

$$P(X\cup Y)=\frac{1}{3}+\frac{4}{15}-\frac{2}{15}$$

$$=\frac{7}{15}=\frac{7}{15}$$

2. (c,d)

(a) $\because$ $e^x\in(1,e)$ in $(0,1)$ and

$\int_0^x f(t)\sin t\,dt\in(0,1)$ in $(0,1)$

$\therefore$ $e^x-\int_0^x f(t)\sin t\,dt$ cannot be zero.

So, option (a) is incorrect.

(b) $f(x)+\int_0^{\frac{\pi}{2}} f(t)\sin t\,dt$ always positive

$\therefore$Option (b) is incorrect.

(c) Let $h(x)=x-\int_0^{\frac{\pi}{2}-x} f(t)\cos t\,dt$,

$$h(0)=-\int_0^{\frac{\pi}{2}} f(t)\cos t\,dt<0$$

$$h(1)=1-\int_0^{\frac{\pi}{2}-1} f(t)\cos t\,dt>0$$

$\therefore$ Option (c) is correct.

(d) Let $g(x)=x^9-f(x)$

$g(0)=-f(0)<0,\ g(1)=1-f(1)>0$

$\therefore$ Option (d) is correct.

3. (b,d) $\frac{az+b}{z+1}=\frac{ax+b+aiy}{(x+1)+iy}$

$$=\frac{(ax+b+aiy)((x+1)-iy)}{(x+1)^2+y^2}$$

$\therefore$ $\text{Im}\left(\frac{az+b}{z+1}\right)$

$$=\frac{-(ax+b)y+ay(x+1)}{(x+1)^2+y^2}$$

$\Rightarrow$ $\frac{(a-b)y}{(x+1)^2+y^2}=y$

$\because$ $a-b=1$

$\therefore$ $(x+1)^2+y^2=1$

$\therefore$ $x=-1\pm\sqrt{1-y^2}$

4. (a,b,c) Tangent $\equiv 2x-y+1=0$

Hyperbola $\equiv\frac{x^2}{a^2}-\frac{y^2}{16}=1$

It point $\equiv(a\sec\theta, 4\tan\theta)$,

tangent $\equiv\frac{x\sec\theta}{a}-\frac{y\tan\theta}{4}=1$

On comparing, we get

$$\sec\theta=-2a$$

$$\tan\theta=-4$$

$\Rightarrow$ $4a^2-16=1$

$\therefore$ $a=\frac{\sqrt{17}}{2}$

Substitute the value of $a$ in option (a), (b), (c) and (d).

5. (a,b,d) $f(x)=x\cos(\pi(x+[x]))$

At $x=0$

$$\lim_{x\to0} f(x)=\lim_{x\to0} x\cos(\pi(x+[x])=0$$

and $f(x)=0$

$\therefore$ It is continuous at $x=0$ and clearly discontinuous at other integer points.

6. (a,c) For a matrix to be square of other matrix its determinant should be positive.

(a) and (c) $\rightarrow$ Correct

(b) and (d) $\rightarrow$ Incorrect

7. (b) Equation of chord with mid-point $(h, k)$.

$$T = s_1$$

$$yk - 8x - 8h = k^2 - 16h$$

$$2x - \frac{yk}{4} = 2h - \frac{k^2}{4}$$

$\because \quad 2x + y = p$

$\therefore \quad k = -4$ and $p = 2h - 4$

8. (1) $\begin{vmatrix} 1 & \alpha & \alpha^2 \\ \alpha & 1 & \alpha \\ \alpha^2 & \alpha & 1 \end{vmatrix} = 0$

$\Rightarrow \quad \alpha^4 - 2\alpha^2 + 1 = 0$

$\Rightarrow \quad \alpha^2 = 1$

$\Rightarrow \quad \alpha = \pm 1$

But $\quad \alpha = 1$ not possible

[Not satisfying equation]

$\therefore \quad \alpha = -1$

Hence, $1 + \alpha + \alpha^2 = 1$

9. (6) Let the sides are $a-d$, $a$ and $a+d$.

Then, $\quad a(a-d) = 48$

and $a^2 - 2ad + d^2 + a^2 = a^2 + 2ad + d^2$

$\Rightarrow \quad a^2 = 4ad$

$\Rightarrow \quad a = 4d$

Thus, $\quad a = 8, d = 2$

Hence, $\quad a - d = 6$

10. (2) $g(x)$

$= \int_x^{\frac{\pi}{2}} f'(t)\text{cosec } t - \cot t \text{ cosec } t f(t))dt$

$\therefore \quad g(x) = f\left(\frac{\pi}{2}\right) \text{cosec } \frac{\pi}{2} - f(x) \text{cosec } x$

$\Rightarrow \quad g(x) = 3 - \frac{f(x)}{\sin x}$

$$\lim_{x \to 0} g(x) = \lim_{x \to 0}\left(\frac{3\sin x - f(x)}{\sin x}\right)$$

$$= \lim_{x \to 0} \frac{3\cos x - f'(x)}{\cos x}$$

$$= \frac{3-1}{1} = 2$$

11. (2) The circle and coordinate axes can have 3 common points, if it passes through origin. $[p = 0]$

If circle is cutting one axis and touching other axis.

Only possibility is of touching $X$-axis and cutting $Y$-axis. $[p = -1]$

12. (5) $x = 10!$

$$y = {}^{10}C_1 \times {}^{9}C_8 \times \frac{10!}{2!}$$

$$= 10 \times 9 \times \frac{10!}{2}$$

$$\frac{y}{9x} = \frac{10}{2} = 5$$

**Solutions.** (Q.Nos. 13-15)

Using standard equations

I-ii-Q

II-iv-R

III-i-P

IV-iii-S

Comparing with equations in given questions, we get

13. (a) Either option (a) or (c) is correct. But $a = \sqrt{2}$ and $(-1, 1)$ satisfy $x^2 + y^2 = a^2$.

So, option (a) is correct.

14. (b) Either option (b) or (c) is correct.

Satisfying the point $\left(\sqrt{3}, \frac{1}{2}\right)$ in (II), we get

$$a^2 = 4.$$

$\therefore$ The conic is $x^2 + 4y^2 = 4$

Now, equation of tangent at $(\sqrt{3}, \frac{1}{2})$ is $\sqrt{3}x + 2y = 4$, which is the given equation.

So, option (b) is correct.

15. (a) Either option (a), (b) or (c) is correct.

Satisfying the point (8, 16) in (III), we get

$\therefore$ The conic is $y^2 = 32x = 16(2x)$

Now, equation of tangent at (8, 16) is $y \cdot 16 = 16(x + 8)$

$\Rightarrow$ $y = x + 8$, which is the given in equation.

So, option (a) is correct.

**Solutions.** (Q.Nos. 16-18)

$f(x) = x + \ln x - x\ln x$

$f(1) = 1 > 0$

$f(e^2) = e^2 + 2 - 2e^2 = 2 - e^2 < 0$

$\Rightarrow$ $f(x) = 0$ for some $x \in (1, e^2)$

$\therefore$ I is correct

$f'(x) = 1 + \frac{1}{x} - \ln x - 1$

$= \frac{1}{x} - \ln x$

$f'(x) > 0$ for $(0, 1)$

$f'(x) < 0$ for $(e, \infty)$

$\therefore$ $P$ and $Q$ are correct, II is correct, III is incorrect.

$f''(x) = \frac{-1}{x^2} - \frac{1}{x}$

$f''(x) < 0$ for $(0, \infty)$

$\therefore$ $S$, is correct, $R$ is incorrect.

IV is incorrect.

$\lim_{x\to\infty} f(x) = -\infty$

$\lim_{x\to\infty} f'(x) = -\infty$

$\lim_{x\to\infty} f''(x) = 0$

$\therefore$ ii, iii, iv are correct.

16. *(d)*

17. *(c)*

18. *(c)*

# Paper 2

## Physics

1. *(d)*

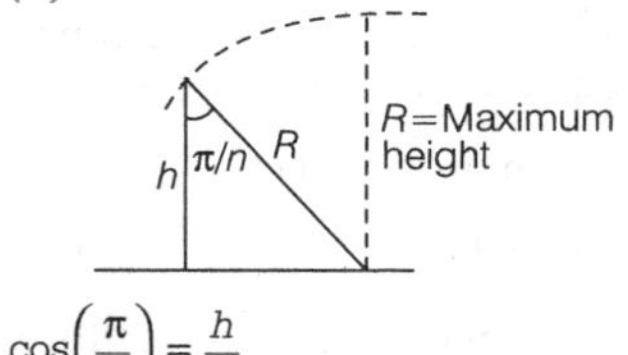


$\cos\left(\frac{\pi}{n}\right) = \frac{h}{R}$

$$\Delta = R - h = \frac{h}{\cos\left(\frac{\pi}{n}\right)} - h = h\left[\frac{1}{\cos\left(\frac{\pi}{n}\right)} - 1\right]$$

2. *(a)* $m = \frac{4\pi R^3}{3} \times \rho$

On taking log both sides, we have

$\ell n(m) = \ell n\left(\frac{4\pi}{3}\right) + \ell n(\rho) + 3\ell n(R)$

On differentiating with respect to time,

$0 = 0 + \frac{1}{\rho}\frac{d\rho}{dt} + \frac{3}{R}\frac{dR}{dt}$

$\Rightarrow$ $\left(\frac{dR}{dt}\right) = v \propto -R \times \frac{1}{\rho}\left(\frac{d\rho}{dt}\right)$

$v \propto R$

3. *(d)* According to photoelectric effect equation

$$KE_{max} = \frac{hc}{\lambda} - \phi_0$$

$$\frac{p^2}{2m} = \frac{hc}{\lambda} - \phi_0$$

$[KE = p^2/2m]$

$$\frac{(h/\lambda_d)^2}{2m} = \frac{hc}{\lambda} - \phi_0$$

$[p = h/\lambda]$

Assuming small changes, differentiating both sides,

$$\frac{h^2}{2m}\left(-\frac{2d\lambda_d}{\lambda_d^3}\right) = -\frac{hc}{\lambda^2}d\lambda$$

$$\frac{d\lambda_d}{d\lambda} \propto \frac{\lambda_d^3}{\lambda^2}$$

4. *(c)* $\mathbf{S} = \mathbf{P} + b\mathbf{R} = \mathbf{P} + b(\mathbf{Q} - \mathbf{P})$

$= \mathbf{P}(1 - b) + b\mathbf{Q}$

5. *(a)* $t = \sqrt{\frac{L}{5}} + \frac{L}{300}$

$$dt = \frac{1}{\sqrt{5}}\frac{1}{2}L^{-1/2}dL + \left(\frac{1}{300}dL\right)$$

$$dt = \frac{1}{2\sqrt{5}}\frac{1}{\sqrt{20}}\,dL + \frac{dL}{300} = 0.01$$

$$dL\left(\frac{1}{20} + \frac{1}{300}\right) = 0.01$$

$$dL\left[\frac{15}{300}\right] = 0.01$$

$$dL = \frac{3}{16}$$

$$\frac{dL}{L} \times 100 = \frac{3}{16} \times \frac{1}{20} \times 100 = \frac{15}{16} \simeq 1\%$$

6. *(c)* Given, $v_e = 11.2\ \text{km/s} = \sqrt{\frac{2GM_e}{R_e}}$

From energy conservation,

$$K_i + U_i = K_f + U_f$$

$$\frac{1}{2}mv_s^2 - \frac{GM_s m}{r} - \frac{GM_e m}{R_e} = 0 + 0$$

Here, $r$ = distane of rocket from sun

$$\Rightarrow \quad v_s = \sqrt{\frac{2GM_e}{R_e} + \frac{2GM_s}{r}}$$

Given, $M_s = 3 \times 10^5\ M_e$

and $r = 2.5 \times 10^4\ R_e$

$$\Rightarrow \quad v_s = \sqrt{\frac{2GM_e}{R_e} + (2G)\left(\frac{3 \times 10^5 M_e}{2.5 \times 10^4 R_e}\right)}$$

$$= \sqrt{\frac{2GM_e}{R_e}\left(1 + \frac{3 \times 10^5}{2.5 \times 10^4}\right)}$$

$$= \sqrt{\frac{2GM_e}{R_e} \times 13}$$

$$\Rightarrow \quad v_s \simeq 42\ \text{km/s}$$

7. *(a)*

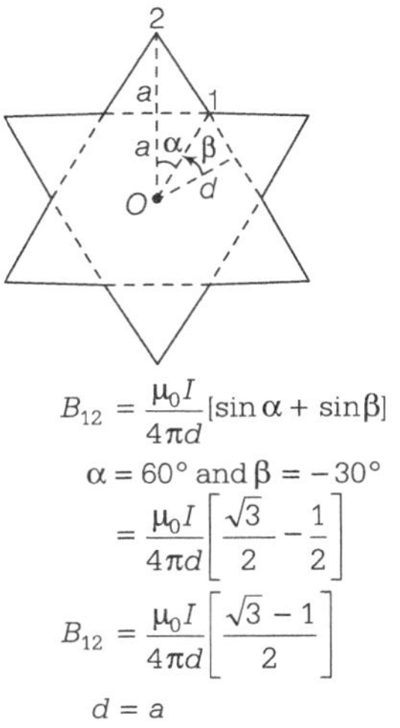


$$B_{12} = \frac{\mu_0 I}{4\pi d}[\sin\alpha + \sin\beta]$$

$$\alpha = 60° \text{ and } \beta = -30°$$

$$= \frac{\mu_0 I}{4\pi d}\left[\frac{\sqrt{3}}{2} - \frac{1}{2}\right]$$

$$B_{12} = \frac{\mu_0 I}{4\pi d}\left[\frac{\sqrt{3} - 1}{2}\right]$$

$$d = a$$

$$B_0 = 12\ B_{12}$$

$$= 12 \times \frac{\mu_0 I}{4\pi d}\left[\frac{\sqrt{3} - 1}{2}\right]$$

$$= \frac{\mu_0 I}{4\pi a} 6\ [\sqrt{3} - 1]$$

8. *(a,c)*

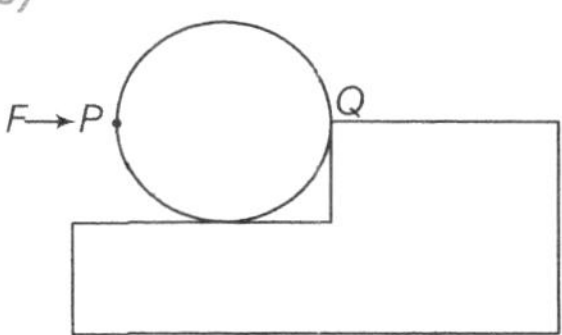


(a) If force is applied normal to surface at $P$, then line of action of force will pass from $Q$ and thus $\tau = 0$.

(b) Wheel can climb.

(c) $\tau = F(2R\cos\theta) - mgR\cos\theta$

$$\tau \propto \cos\theta$$

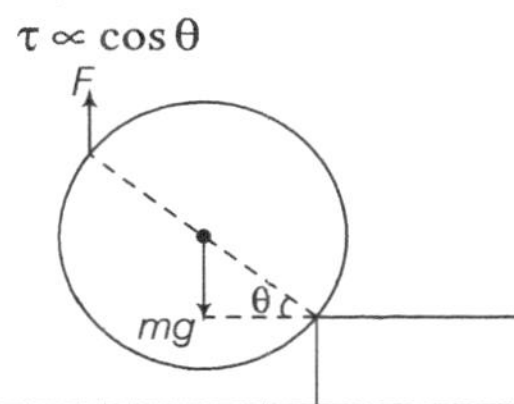


Hence, as $\theta$ increases, $\tau$ decreases. So its correct.

(d)

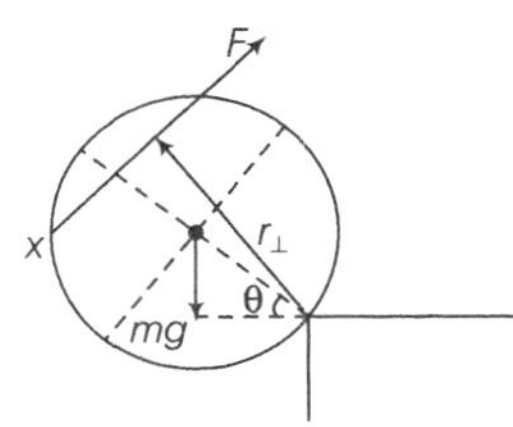


$\tau = Fr_\perp - mg\cos\theta$ ; $\tau$ increases with $\theta$.

9. *(c,d)*

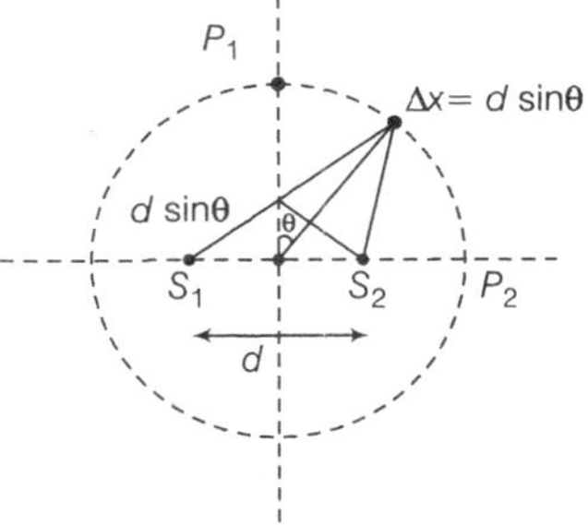


$\lambda = 600$ nm

at $P_1$ $\quad \Delta x = 0$

at $P_2$ $\quad \Delta x = 1.8 \text{ mm} = n\lambda$

Number of maximas will be

$$n = \frac{\Delta x}{\lambda} = \frac{1.8 \text{ mm}}{600 \text{ nm}} = 3000$$

at $P_2$ $\quad \Delta x = 3000\lambda$

Hence, bright fringe will be formed.

At $P_2$, 3000 th maxima is formed.

For (a) option

$$\Delta x = d\sin\theta$$

$$d\Delta x = d\cos d\theta$$

$$R\lambda = d\cos\theta R d\theta$$

$$R d\theta = \frac{R\lambda}{d\cos\theta}$$

As we move from $P_1$ to $P_2$

$$\theta \uparrow \cos\theta \downarrow R d\theta \uparrow$$

**10.** *(a,d)*

(a) $\Omega = 2\pi(1 - \cos\theta)$; $\theta = 45°$

$$\phi = -\frac{\Omega}{4\pi} \times \frac{Q}{\varepsilon_0} = -\frac{2\pi(1-\cos\theta)}{4\pi}\frac{Q}{\varepsilon_0}$$

$$= -\frac{Q}{2\varepsilon_0}\left(1 - \frac{1}{\sqrt{2}}\right)$$

(b) The component of the electric field perpendicular to the flat surface will decrease so we move away from the centre as the distance increases (magnitude of electric field decreases) as well as the angle between the normal and electric field will increase. Hence, the component of the electric field normal to the flat surface is not constant.

**Alternate solution**

$$x = \frac{R}{\cos\theta}$$

$$E = \frac{KQ}{x^2} = \frac{KQ\cos^2\theta}{R^2}$$

$$\Rightarrow \quad E_\perp = \frac{KQ\cos^3\theta}{R^2}$$

As we move away from centre

$$\theta \uparrow \cos\theta \downarrow \text{ so } E_\perp \downarrow$$

(c) Total flux $\phi$ due to charge $Q$ is $\frac{Q}{\varepsilon_0}$.

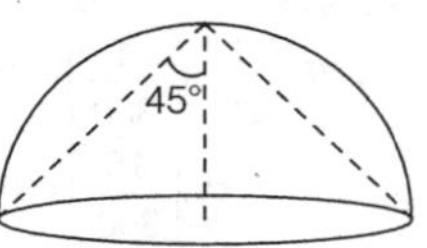


So, $\phi$ through the curved and flat surface will be less than $\frac{Q}{\varepsilon_0}$.

(d) Since, the circumference is equidistant from $Q$ it will be equipotential

$$V = \frac{KQ}{\sqrt{2}R}.$$

**11.** *(a,c,d)* When the bar makes an angle $\theta$, the height of its COM (mid-point) is $\frac{L}{2}\cos\theta$.

$\therefore$ Displacement

$$= L - \frac{L}{2}\cos\theta = \frac{L}{2}(1 - \cos\theta)$$

Since, force on COM is only along the vertical direction, hence COM is falling vertically downward. Instantaneous torque about point of contact is

$$\tau = mg \times \frac{L}{2}\sin\theta \quad \text{or} \quad \tau \propto \sin\theta$$

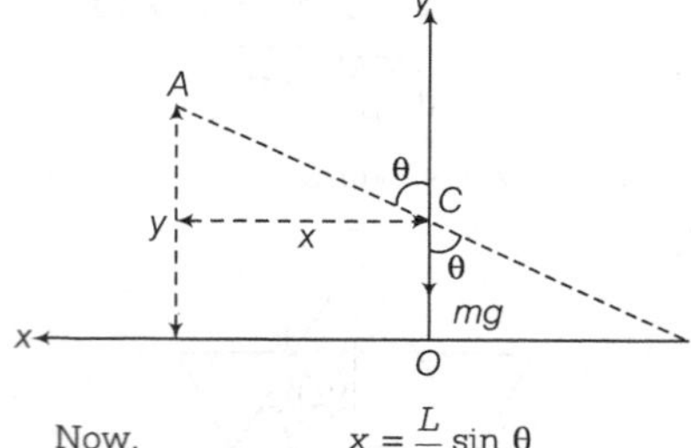


Now, $\quad x = \frac{L}{2}\sin\theta$

$$y = L\cos\theta$$

$$\frac{x^2}{(L/2)^2} + \frac{y^2}{L^2} = 1$$

Path of $A$ is an ellipse.

**12.** *(a,b,c)*

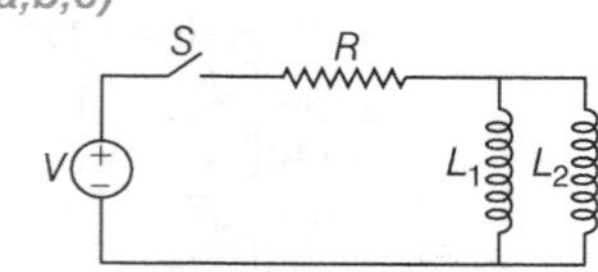

Since inductors are connected in parallel

$$V_{L_1} = V_{L_2}$$

$$L_1 \frac{dI_1}{dt} = L_2 \frac{dI_2}{dt}$$

$$L_1 I_1 = L_2 I_2$$

$$\frac{I_1}{I_2} = \frac{L_2}{L_1}$$

Current through resistor at any time $t$ is given by

$$I = \frac{V}{R}(1 - e^{-\frac{RT}{L}}), \text{ where } L = \frac{L_1 L_2}{L_1 + L_2}$$

After long time $I = \frac{V}{R}$

$$I_1 + I_2 = I \qquad \text{...(i)}$$

$$L_1 I_1 = L_2 I_2 \qquad \text{...(ii)}$$

From Eqs. (i) and (ii), we get

$$I_1 = \frac{V}{R}\frac{L_2}{L_1 + L_2}, \quad I_2 = \frac{V}{R}\frac{L_1}{L_1 + L_2}$$

(d) Value of current is zero at $t = 0$

Value of current is $V/R$ at $t = \infty$

Hence option (d) is incorrect.

13. *(b,c)* (a)

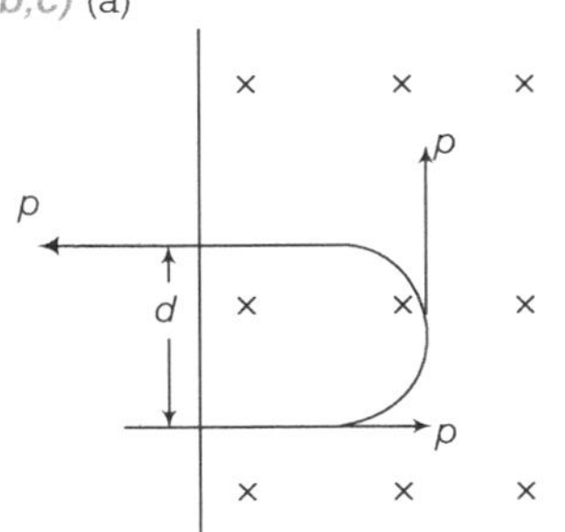


$$|\Delta \mathbf{P}| = \sqrt{2}p$$

(b) $r(1 - \cos\theta) = R$

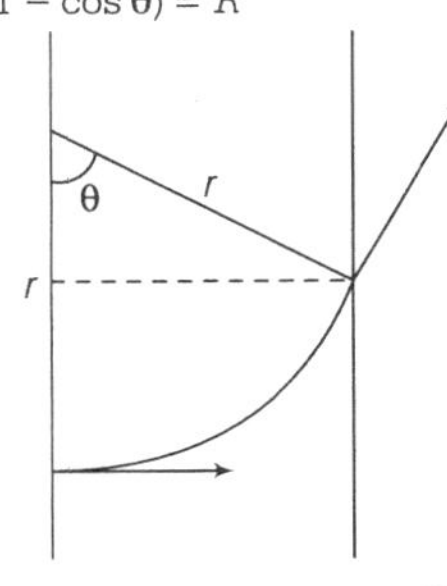


$$r \sin\theta = \frac{3R}{2}$$

$$\frac{\sin\theta}{1 - \cos\theta} = \frac{3}{2}$$

$$\frac{2\sin\frac{\theta}{2}\cos\frac{\theta}{2}}{2\sin^2\frac{\theta}{2}} = \frac{3}{2}$$

$$\cot\frac{\theta}{2} = \frac{3}{2}$$

$$\Rightarrow \quad \tan\frac{\theta}{2} = \frac{2}{3}$$

$$\Rightarrow \quad \tan\theta = \frac{2\left(\frac{2}{3}\right)}{1 - \frac{4}{9}} = \frac{\frac{4}{3}}{\frac{5}{9}} = \frac{12}{5}$$

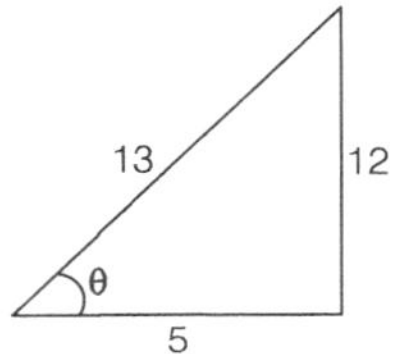


$$\sin\theta = \frac{12}{13}$$

$$r\left(\frac{12}{13}\right) = \frac{3R}{2}; \quad r = \frac{13R}{8} = \frac{P}{QB}; \quad B = \frac{8P}{13QR}$$

(c) $\frac{P}{QB} < \frac{3R}{2}, B > \frac{2P}{3QR}$

(d) $r = \frac{mv}{QB}, d = 2r = \frac{2mv}{QB} \quad \Rightarrow \quad d \propto m$

14. *(b,c)* $V_{XY} = V_0 \sin\left(\omega t + \frac{2\pi}{3}\right) - V_0 \sin \omega t$

$$= V_0 \sin\left(\omega t + \frac{2\pi}{3}\right) + V_0 \sin(\omega t + \pi)$$

$$\Rightarrow \quad \phi = \pi - \frac{2\pi}{3} = \frac{\pi}{3}$$

$$\Rightarrow \quad V_0' = 2V_0 \cos\left(\frac{\pi}{6}\right)$$

$$= \sqrt{3}\, V_0$$

$$\Rightarrow \quad V_{XY} = \sqrt{3}V_0 \sin(\omega t + \phi)$$

$$\Rightarrow \quad (V_{XY})_{\text{rms}} = (V_{YZ})_{\text{rms}} = \sqrt{3}\frac{V_0}{\sqrt{2}}$$

15. *(b)* When switch is closed for a very long time capacitor will get fully charged and charge on capacitor will be $q = CV$

Energy stored in capacitor,

$$E_C = \frac{1}{2}CV^2 \qquad \text{...(i)}$$

R
S
C

Work done by a battery,
$W = Vq = VCV = CV^2$

Energy dissipated across resistance

$$E_D = \text{(work done by battery)} - \text{(energy stored)}$$

$$E_D = CV^2 - \frac{1}{2}CV^2 = \frac{1}{2}CV^2 \quad \ldots\text{(ii)}$$

From Eqs. (i) and (ii)

$$E_D = E_C$$

16. *(a)* **For process (1)**

Charge on capacitor $= \frac{CV_0}{3}$

Energy stored in capacitor

$$= \frac{1}{2}C\frac{V_0^2}{9} = \frac{CV_0^2}{18}$$

Work done by battery $= \frac{CV_0}{3} \times \frac{V}{3} = \frac{CV_0^2}{9}$

Heat loss $= \frac{CV_0^2}{9} - \frac{CV_0^2}{18} = \frac{CV_0^2}{18}$

**For process (2)**

Charge on capacitor $= \frac{2CV_0}{3}$

Extra charge flow through battery $= \frac{CV_0}{3}$

Work done by battery $= \frac{CV_0}{3} \cdot \frac{2V_0}{3} = \frac{2CV_0^2}{9}$

Final energy stored in capacitor

$$= \frac{1}{2}C\left(\frac{2V_0}{3}\right)^2 = \frac{4CV_0^2}{18}$$

Energy stored in process

$$2 = \frac{4CV_0^2}{18} - \frac{CV_0^2}{18} = \frac{3CV_0^2}{18}$$

Heat loss in process (2) = work done by battery in process (2)
– energy store in capacitor process (2)

$$= \frac{2CV_0^2}{9} - \frac{3CV_0^2}{18}$$

$$= \frac{CV_0^2}{18}$$

**For process (3)**

Charge on capacitor $= CV_0$

Extra charge flown through battery =

$$CV_0 - \frac{2CV_0}{3} = \frac{CV_0}{3}$$

Work done by battery in this process

$$= \left(\frac{CV_0}{3}\right)(V_0) = \frac{CV_0^2}{3}$$

Final energy stored in capacitor $= \frac{1}{2}CV_0^2$

Energy stored in this process

$$= \frac{1}{2}CV_0^2 - \frac{4CV_0^2}{18} = \frac{5CV_0^2}{18}$$

Heat loss in process (3)

$$= \frac{CV_0^2}{3} - \frac{5CV_0^2}{18} = \frac{CV_0^2}{18}$$

Now, total heat loss $(E_D)$

$$= \frac{CV_0^2}{18} + \frac{CV_0^2}{18} + \frac{CV_0^2}{18} = \frac{CV_0^2}{6}$$

Final energy stored in capacitor $= \frac{1}{2}CV_0^2$

So we can say that $E_D = \frac{1}{3}\left(\frac{1}{2}CV_0^2\right)$

17. *(*)* Question is not very clear.

18. *(c)* If height of the cone $h >> r$

Then, $\mu N = mg$

$$\mu m(R - r)\omega_0^2 = mg$$

$$\omega_0 = \sqrt{\frac{g}{\mu(R - r)}}$$

## Chemistry

1. *(d)* IV is most basic as conjugate acid is stabilised by resonance of two —$NH_2$.

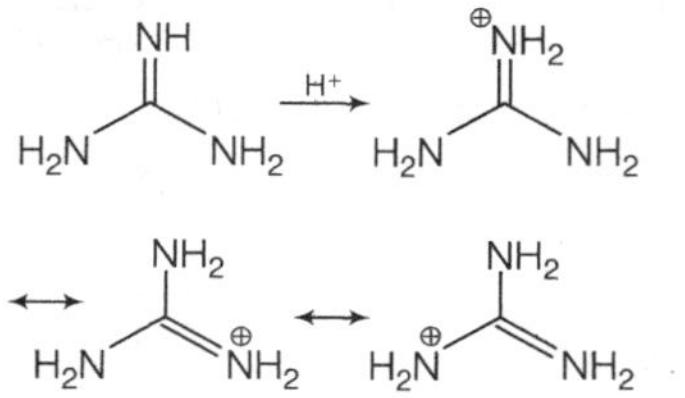

$$H_2N{-}C({=}NH){-}NH_2 \xrightarrow{H^+} H_2N{-}C({=}\overset{\oplus}{N}H_2){-}NH_2 \longleftrightarrow H_2N{-}C(NH_2){=}\overset{\oplus}{N}H_2 \longleftrightarrow \overset{\oplus}{H_2N}{=}C(NH_2){-}NH_2$$

III is least basic as

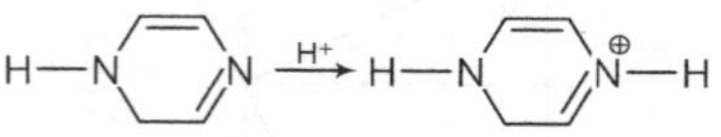

$$H{-}N\langle\text{ring}\rangle N \xrightarrow{H^+} H{-}N\langle\text{ring}\rangle \overset{\oplus}{N}{-}H$$

Destablised by $-I$-effect of $sp^2$-carbons.

2. *(a)* $-\Delta T_f = ik_f m_2$

$$= 1 \times 2 \times \frac{34.5}{46 \times 500} \times 1000 = 3$$

Vapour pressure curves shown in *(b)* is in agreement with the calculated value of $-\Delta T_f$. *(a)* is wrong, vapour pressure decreases on cooling.

3. *(a)* *Diazo* coupling occur at *para*-position of phenol.

4. *(c)* The redox reaction is :

$Zn(s) + Cu^{2+} \longrightarrow Zn^{2+} + Cu$

The Nernst equation is

$$E = E^{\circ} - \frac{2.303RT}{2F}\log 10$$
$$= 1.1 - \frac{2.303RT}{2F}$$

$$\text{Also, } \Delta G = -nEF = -2F\left(1.1 - \frac{2.303RT}{2F}\right)$$
$$= -2.2F + 2.303RT$$
$$= 2.303RT - 2.2F$$

5. *(b)* $H_3\overset{+5}{P}O_4 > H_4\overset{+4}{P_2}O_6 > H_3\overset{+3}{P}O_3 > H_3\overset{+1}{P}O_2$

6. *(c)* $G = H - TS = U + pV - TS$

$\Rightarrow dG = dU + pdV + Vdp - TdS - SdT = Vdp - SdT$

$[\because dU + pdV = dq = TdS]$

$\Rightarrow dG = Vdp$ if isothermal process $(dT = 0)$

$\Rightarrow \Delta G = V\Delta p$

Now taking initial state as standard state

$$G_{gr} - G_{gr}{}^{\circ} = V_{gr}\Delta p \quad \ldots(i)$$
$$G_d - G_d{}^{\circ} = V_d\Delta p \quad \ldots(ii)$$

Now (ii)-(i) gives,

$$(V_d - V_{gr})\Delta p = G_d - G_{gr} + (G_{gr}^{\circ} - G_d^{\circ}$$

At equilibrium, $G_d = G_{gr}$

$$\Rightarrow \quad (V_{gr} - V_d)\Delta p = G_d{}^{\circ} - G_{gr}{}^{\circ}$$
$$= 2.9 \times 10^3 \text{ J}$$

$$\Rightarrow \quad \Delta p = \frac{2.9 \times 10^3}{2 \times 10^{-6}}\text{Pa}$$
$$= \frac{29}{2} \times 10^8 \text{ Pa} = \frac{29000}{2}\text{ bar}$$
$$p = p_0 + \frac{29000}{2}$$
$$= 1 + \frac{29000}{2} = 14501 \text{ bar}$$

7. *(d)* $\underset{\text{Amphoteric}}{Zn} + 2NaOH \longrightarrow Na_2ZnO_2 + H_2$

8. *(a, b)* $\Delta S_{surr} = \dfrac{-\Delta H}{T_{surr}}$

For endothermic reaction, if $T_{surr}$ increases, $\Delta S_{surr}$ will increase. For exothermic reaction, if $T_{surr}$ increases, $\Delta S_{surr}$ will decrease.

9. *(a, b)* If steric factor is considered, the corrected Arrhenius equation will be

$k = pAe^{\frac{-E_a}{RT}}$ where $A$ = frequency factor by Arrhenius.

$\because$ $p > 1$, $pA > A$ hence, (a) is correct.

Activation energy is not related to steric factor.

10. *(c, d)*

(a) Both I and II are 1° halide, undergos $S_N2$ reaction.

(d) III is a tertiary halide, undergoes $S_N2$ reaction. I is benzylic bromide, it is very reactive in $S_N1$ also as if produces stable benzylic carbonation.

11. *(a, b, d)*

(a)

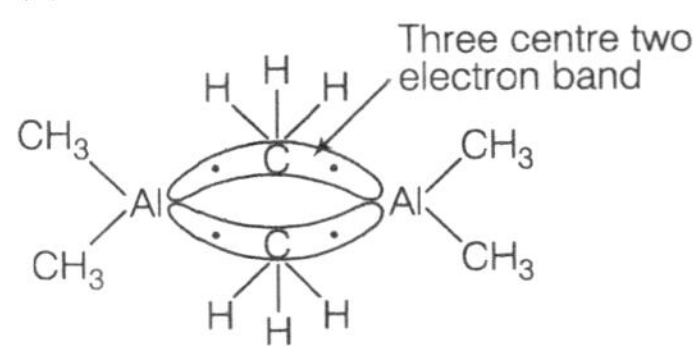


(b)

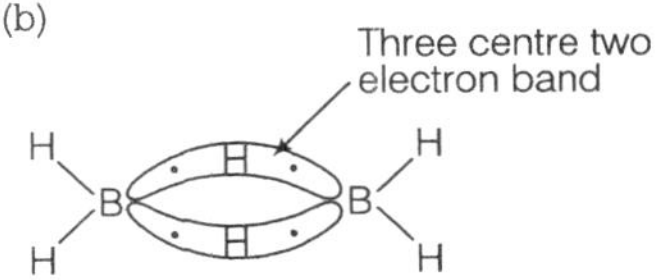

(c) $BCl_3$ is stronger Lewis acid than $AlCl_3$ due to greater extent of $p_\pi - p_\pi$ back bonding in $AlCl_3$.

(d)

Three centre four electron bond

Cl, Cl — Al ← :Cl: → Al — Cl, Cl (dimer $Al_2Cl_6$ with bridging Cl)

**12.** *(c, d)*

(c) is incorrect because NO is neutral oxide.

(d) is incorrect because CrO is basic oxide.

**13.** *(a, c)*

(b) Higher the critical temperature, greater the extent of adsorption.

(c) $\underset{\text{Adsorbent}}{P(s)} + \underset{\text{Adsorbate}}{Q(g)} \longrightarrow PQ(s)$

As gaseous adsorbate is adsorbed on solid surface, entropy decreases. Also formation of bond between $P$ and $Q$ results in release of energy, hence $\Delta H < 0$.

**14.** *(c, d)*

(b) $\underset{(P)}{CH_3-C_6H_4-CH{=}CH_2} \xrightarrow[\text{Zn-}H_2O]{O_3} \underset{(R)}{CH_3-C_6H_4-CHO} + H_2CO$

Gives Cannizzaro but not haloform reaction

$\underset{(Q)}{C_6H_5-C(CH_3){=}CH_2} \xrightarrow[\text{Zn-}H_2O]{O_3} \underset{(S)}{C_6H_5-\overset{O}{\overset{\|}{C}}-CH_3} + H_2CO$

Gives haloform reaction but not Cannizzaro reaction

(c) $\underset{P}{m\text{-}CH_3-C_6H_4-CH{=}CH-CH_3} \xrightarrow[\text{Zn-}H_2O]{O_3} \underset{(R)}{m\text{-}CH_3-C_6H_4-CHO} + CH_3CHO$

Gives Cannizzaro but not haloform reaction

$\underset{Q}{C_6H_5-C(CH_3){=}C(CH_3)_2} \xrightarrow[\text{Zn-}H_2O]{O_3} \underset{(S)}{C_6H_5-\overset{O}{\overset{\|}{C}}-CH_3} + CH_3-\overset{O}{\overset{\|}{C}}-CH_3$

Gives haloform reaction but not Cannizzaro reaction

## Paragraph-1

$$KClO_3 \xrightarrow[\Delta]{MnO_2} KCl + \underset{W}{O_2}$$

$$O_2 + P_4 \xrightarrow{\Delta} \underset{X}{P_4O_{10}} \xrightarrow{HNO_3} \underset{Y}{N_2O_5} + \underset{Z}{HPO_3}$$

**15.** *(d)* **16.** *(a)*

**Paragraph-2**

OEt+ (i) $CH_3MgBr(ex)$ (ii) $H_2O$

OH

Q

$H^+/0°C$

$CH_3COCl$ $AlCl_3$

R(F.C. Alkylation)

S(F.C. Acylation)

17. (b) 18. (b)

## Mathematics

1. (b) $f'(x)$ is increasing

For some $x$ in $\left(\frac{1}{2}, 1\right)$

$f'(x) = 1$ [LMVT]

$\therefore \quad f'(1) > 1$

2. (b) $\frac{dy}{dx} = \frac{1}{8\sqrt{x}\sqrt{9+\sqrt{x}}\sqrt{4+\sqrt{9+\sqrt{x}}}}$

$\Rightarrow \quad y = \sqrt{4+\sqrt{9+\sqrt{x}}} + c$

Now, $y(0) = \sqrt{7} + c$

$\Rightarrow \quad c = 0$

$y(256) = \sqrt{4+\sqrt{9+16}} = \sqrt{4+5} = 3$

3. (a) Sum of diagonal entries of $M^T M$ is $\sum a_i^2$.

$$\sum_{i=1}^{9} a_i^2 = 5$$

Possibilities

I. 2, 1, 0, 0, 0, 0, 0, 0, 0, which gives $\frac{9!}{7!}$ matrices

II. 1, 1, 1, 1, 1, 0, 0, 0, 0, which gives $\frac{9!}{4! \times 5!}$ matrices

Total matrices $= 9 \times 8 + 9 \times 7 \times 2 = 198$

4. (c) Sample space $\rightarrow {}^{12}C_2$

Number of possibilities for $z$ is even.

$z = 0 \Rightarrow {}^{11}C_1$

$z = 2 \Rightarrow {}^{9}C_1$

$z = 4 \Rightarrow {}^{7}C_1$

$z = 6 \Rightarrow {}^{5}C_1$

$z = 8 \Rightarrow {}^{3}C_1$

$z = 10 \Rightarrow {}^{1}C_1$

Total $= 36$

$\therefore$ Probability $= \frac{36}{66} = \frac{6}{11}$

5. (c) $N_i = {}^5C_k \times {}^4C_{5-k}$

$N_1 = 5 \times 1$

$N_2 = 10 \times 4$

$N_3 = 10 \times 6$

$N_4 = 5 \times 4$

$N_5 = 1$

$N_1 + N_2 + N_3 + N_4 + N_5 = 126$

6. (b) $\mathbf{OP} \cdot \mathbf{OQ} + \mathbf{OR} \cdot \mathbf{OS}$

$= \mathbf{OR} \cdot \mathbf{OP} + \mathbf{OQ} \cdot \mathbf{OS}$

$\Rightarrow \quad \mathbf{OP}(\mathbf{OQ} - \mathbf{OR}) + \mathbf{OS}(\mathbf{OR} - \mathbf{OQ}) = 0$

$\Rightarrow \quad (\mathbf{OP} - \mathbf{OS})(\mathbf{OQ} - \mathbf{OR}) = 0$

$\Rightarrow \quad \mathbf{SP} \cdot \mathbf{RQ} = 0$

Similarly $\mathbf{SR} \cdot \mathbf{PQ} = 0$ and $\mathbf{SQ} \cdot \mathbf{PR} = 0$

$\therefore$ $S$ is orthocentre.

7. *(d)* Let the equation of plane be $ax + by + cz = 1$. Then

$$a + b + c = 1$$
$$2a + b - 2c = 0$$
$$3a - 6b - 2c = 0$$
$$\Rightarrow \quad a = 7b$$
$$c = \frac{15b}{2}$$
$$b = \frac{2}{31}, a = \frac{14}{31}, c = \frac{15}{31}$$
$$\therefore \quad 14x + 2y + 15z = 31$$

8. *(a,c)* $f'(x) > 2f(x)$

$$\Rightarrow \quad \frac{dy}{y} > 2dx$$
$$\Rightarrow \quad \int_1^{f(x)} \frac{dy}{y} > 2\int_0^x dx$$
$$\ln(f(x)) > 2x$$
$$\therefore \quad f(x) > e^{2x}$$

Also, as $f'(x) > 2f(x)$

$\therefore f'(x) > 2c^{2x} > 0$

9. *(b,d)* $I = \sum_{k=1}^{98} \int_k^{k+1} \frac{(k+1)}{x(x+1)} dx$

Clearly, $I = \sum_{k=1}^{98} \int_k^{k+1} \frac{(k+1)}{x(x+1)^2} dx$

$$\Rightarrow \quad I > \sum_{k=1}^{98} (k+1) \int_k^{k+1} \frac{1}{(x+1)^2} dx$$
$$\Rightarrow \quad I > \sum_{k=1}^{98} (-(k+1)) \left[\frac{1}{k+2} - \frac{1}{k+1}\right]$$
$$\Rightarrow \quad I > \sum_{k=1}^{98} \frac{1}{k+2}$$
$$\Rightarrow \quad I > \frac{1}{3} + \ldots + \frac{1}{100} > \frac{98}{100}$$
$$\Rightarrow \quad I > \frac{49}{50}$$

Also, $I < \sum_{k=1}^{98} \int_k^{k+1} \frac{k+1}{x(k+1)} dx$

$$= \sum_{k=1}^{98} [\log_e(k+1) - \log_e k]$$
$$I < \log_e 99$$

10. *(a,c)* $\int_0^1 (x - x^3) dx = 2\int_0^\alpha (x - x^3) dx$

$$\frac{1}{4} = 2\left(\frac{\alpha^2}{2} - \frac{\alpha^4}{4}\right)$$
$$2\alpha^4 - 4\alpha^2 + 1 = 0$$
$$\alpha^2 = \frac{4 - \sqrt{16 - 8}}{4} \quad (\because \alpha \in (0, 1))$$
$$\because \quad \alpha^2 = 1 - \frac{1}{\sqrt{2}}$$

11. *(b,c)* We have,

$$2(\cos\beta - \cos\alpha) + \cos\alpha\cos\beta = 1$$
$$\text{or} \quad 4(\cos\beta - \cos\alpha) + 2\cos\alpha\cos\beta = 2$$
$$\Rightarrow 1 - \cos\alpha + \cos\beta - \cos\alpha\cos\beta$$
$$= 3 + 3\cos\alpha - 3\cos\beta - 3\cos\alpha\cos\beta$$
$$\Rightarrow \quad (1 - \cos\alpha)(1 + \cos\beta)$$
$$= 3(1 + \cos\alpha)(1 - \cos\beta)$$
$$\Rightarrow \quad \frac{(1 - \cos\alpha)}{(1 + \cos\alpha)} = \frac{3(1 - \cos\beta)}{1 + \cos\beta}$$
$$\Rightarrow \quad \tan^2\frac{\alpha}{2} = 3\tan^2\frac{\beta}{2}$$
$$\therefore \quad \tan\frac{\alpha}{2} \pm \sqrt{3}\tan\frac{\beta}{2} = 0$$

12. *(c,d)* $f(x) = \frac{1 - x(1 + |1 - x|)}{|1 - x|} \cos\left(\frac{1}{1 - x}\right)$

Now, $\lim_{x \to 1^-} f(x)$

$$= \lim_{x \to 1^-} \frac{1 - x(1 + 1 - x)}{1 - x} \cos\left(\frac{1}{1 - x}\right)$$
$$= \lim_{x \to 1^-} (1 - x)\cos\left(\frac{1}{1 - x}\right) = 0$$

and $\lim_{x \to 1^+} f(x) = \lim_{x \to 1^+}$

$$\frac{1 - x(1 - 1 + x)}{x - 1} \cos\left(\frac{1}{1 - x}\right)$$
$$= \lim_{x \to 1^+} -(x+1) \cdot \cos\left(\frac{1}{x+1}\right), \text{ which}$$

does not exist.

13. *(*)* $g(x) = \int_{\sin x}^{\sin 2x} \sin^{-1}(t) dt$

$$g'(x) = 2\cos 2x \sin^{-1}(\sin 2x)$$
$$- \cos x \sin^{-1}(\sin x)$$

$$g'\left(\frac{\pi}{2}\right) = -2\sin^{-1}(0) = 0$$

$$g'\left(-\frac{\pi}{2}\right) = -2\sin^{-1}(0) = 0$$

No option is matching.

**14.** *(b,c)* $f(x) = \begin{vmatrix} \cos 2x & \cos 2x & \sin 2x \\ -\cos x & \cos x & -\sin x \\ \sin x & \sin x & \cos x \end{vmatrix}$

$\cos 2x(\cos^2 x + \sin^2 x) - \cos 2x$

$(-\cos^2 x + \sin^2 x) + \sin 2x(-\sin 2x)$

$= \cos 2x + \cos 4x$

$f'(x) = -2\sin 2x - 4\sin 4x$

$= -2\sin 2x(1 + 4\cos 2x)$

At $x = 0$

$f'(x) = 0$ and $f(x) = 2$

Also, $f'(x) = 0$ $\sin 2x = 0$ or $\cos 2x = \frac{-1}{4}$

$\Rightarrow$ $x = \frac{n\pi}{2}$ or $\cos 2x = -\frac{1}{4}$

**15.** *(a)* $\cos(P + Q) + \cos(Q + R) + \cos(R + P)$

$= -(\cos R + \cos P + \cos Q)$

Max. of $\cos P + \cos Q + \cos R = \frac{3}{2}$

Min. of $\cos(P + Q)$

$+ \cos(Q + R) + \cos(R + P)$ is

$= -\frac{3}{2}$

**16.** *(a)* $\sin R = \sin(P + Q)$

**17.** *(d)* $\alpha^2 = \alpha + 1$

$\beta^2 = \beta + 1$

$a_n = p\alpha^n + q\beta^n$

$= p(\alpha^{n-1} + \alpha^{n-2}) + q(\beta^{n-1} + \beta^{n-2})$

$= a_{n-1} + a_{n-2}$

$\therefore$ $a_{12} = a_{11} + a_{10}$

**18.** *(d)* $\alpha = \frac{1 + \sqrt{5}}{2}$,

$\beta = \frac{1 - \sqrt{5}}{2}$

$a_4 = a_3 + a_2$

$= 2a_2 + a_1$

$= 3a_1 + 2a_0$

$28 = p(3\alpha + 2) + q(3\beta + 2)$

$28 = (p + q)\left(\frac{3}{2} + 2\right) + (p - q)\left(\frac{3\sqrt{5}}{2}\right)$

$\therefore$ $p - q = 0$

and $(p + q) \times \frac{7}{2} = 28$

$\Rightarrow$ $p + q = 8$

$\Rightarrow$ $p = q = 4$

$\therefore$ $p + 2q = 12$

# Solved Paper 2016

# JEE Main

## Joint Entrance Examination

**Instructions**

1. This test consists of 90 questions.
2. There are three parts in the question paper A, B, C consisting of Physics, Chemistry & Mathematics having 30 questions in each part of equal weightage. Each question is allotted 4 marks for correct response.
3. Candidates will be awarded marks as stated above in instruction no. 2 for correct response of each question. 1 mark will be deducted for indicating incorrect response of each question. No deduction from the total score will be made if no response is indicated for an item in the answer sheet.
4. There is only one correct response for each question. Filling up more than one response in any question will be treated as wrong response and marks for wrong response will be deducted according as per instructions.

## A. Physics

**1.** A student measures the time period of 100 oscillations of a simple pendulum four times. The data set is 90s, 91s, 92s and 95s. If the minimum division in the measuring clock is 1s, then the reported mean time should be

(a) (92 ± 2s)
(b) (92 ± 5s)
(c) (92 ± 1.8s)
(d) (92 ± 3s)

**2.** A particle of mass $m$ is moving along the side of a square of side $a$, with a uniform speed $v$ in the $XY$-plane as shown in the figure.

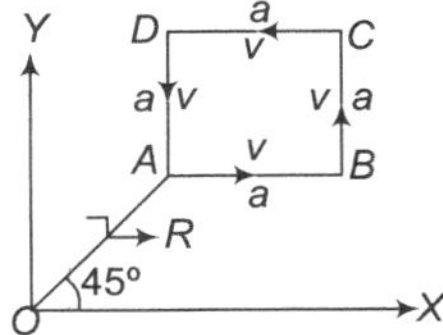


Which of the following statements is **false** for the angular momentum **L** about the origin?

(a) $\mathbf{L} = \frac{-mv}{\sqrt{2}} R\hat{\mathbf{k}}$ when the particle is moving from $A$ to $B$

(b) $\mathbf{L} = mv\left(\frac{R}{\sqrt{2}} - a\right)\hat{\mathbf{k}}$ when the particle is moving from $C$ to $D$

(c) $\mathbf{L} = mv\left(\frac{R}{\sqrt{2}} + a\right)\hat{\mathbf{k}}$ when the particle is moving from $B$ to $C$

(d) $\mathbf{L} = \frac{mv}{\sqrt{2}} R\hat{\mathbf{k}}$ when the particle is moving from $D$ to $A$

**3.** A point particle of mass m, moves along the uniformly rough track $PQR$ as shown in the figure. The coefficient of friction, between the particle and the rough track equals $\mu$. The particle is released from rest, from the point $P$ and it comes to rest at a point $R$. The energies lost by the ball, over the parts $PQ$ and $QR$ of the track, are equal to each other, and no energy is lost when particle changes direction from $PQ$ to $QR$.

The values of the coefficient of friction $\mu$ and the distance $x$ (= $QR$), are respectively close to

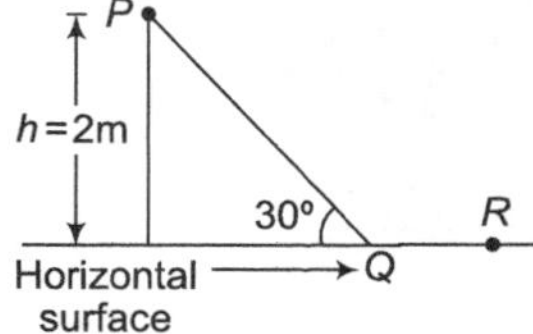


(a) 0.2 and 6.5 m (b) 0.2 and 3.5 m
(c) 0.29 and 3.5 m (d) 0.29 and 6.5 m

**4.** A person trying to lose weight by burning fat lifts a mass of 10 kg upto a height of 1 m 1000 times. Assume that the potential energy lost each time he lowers the mass is dissipated. How much fat will he use up considering the work done only when the weight is lifted up? Fat supplies $3.8 \times 10^7$ J of energy per kg which is converted to mechanical energy with a 20% efficiency rate. (Take, $g = 9.8 \text{ ms}^{-2}$)

(a) $2.45 \times 10^{-3}$ kg (b) $6.45 \times 10^{-3}$ kg
(c) $9.89 \times 10^{-3}$ kg (d) $12.89 \times 10^{-3}$ kg

**5.** A roller is made by joining together two corners at their vertices $O$. It is kept on two rails $AB$ and $CD$ which are placed asymmetrically (see the figure), with its axis perpendicular to $CD$ and its centre $O$ at the centre of line joining $AB$ and $CD$ (see the figure). It is given a light push, so that it starts rolling with its centre $O$ moving parallel to $CD$ in the direction shown. As it moves, the roller will tend to

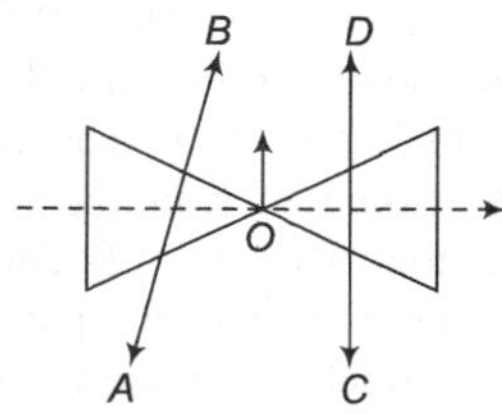


(a) turn left (b) turn right
(c) go straight
(d) turn left and right alternately

**6.** A satellite is revolving in a circular orbit at a height $h$ from the Earth's surface (radius of earth $R, h << R$). The minimum increase in its orbital velocity required, so that the satellite could escape from the earth's gravitational field, is close to (neglect the effect of atmosphere)

(a) $\sqrt{2gR}$ (b) $\sqrt{gR}$
(c) $\sqrt{gR/2}$ (d) $\sqrt{gR}\,(\sqrt{2}-1)$

**7.** A pendulum clock loses 12 s a day if the temperature is 40°C and gains 4 s a day if the temperature is 20°C. The temperature at which the clock will show correct time, and the coefficient of linear expansion $\alpha$ of the metal of the pendulum shaft are, respectively

(a) 25°C, $\alpha = 1.85 \times 10^{-5}$/°C
(b) 60°C, $\alpha = 1.85 \times 10^{-4}$/°C
(c) 30°C, $\alpha = 1.85 \times 10^{-3}$/°C
(d) 55°C, $\alpha = 1.85 \times 10^{-2}$/°C

**8.** An ideal gas undergoes a quasi static, reversible process in which its molar heat capacity $C$ remains constant. If during this process the relation of pressure $p$ and volume $V$ is given by $pV^n$ = constant, then $n$ is given by (here, $C_p$ and $C_V$ are molar specific heat at constant pressure and constant volume, respectively)

(a) $n = \dfrac{C_p}{C_V}$ (b) $n = \dfrac{C - C_p}{C - C_V}$
(c) $n = \dfrac{C_p - C}{C - C_V}$ (d) $n = \dfrac{C - C_V}{C - C_p}$

**9.** $n$ moles of an ideal gas undergoes a process $A$ and $B$ as shown in the figure. The maximum temperature of the gas during the process will be

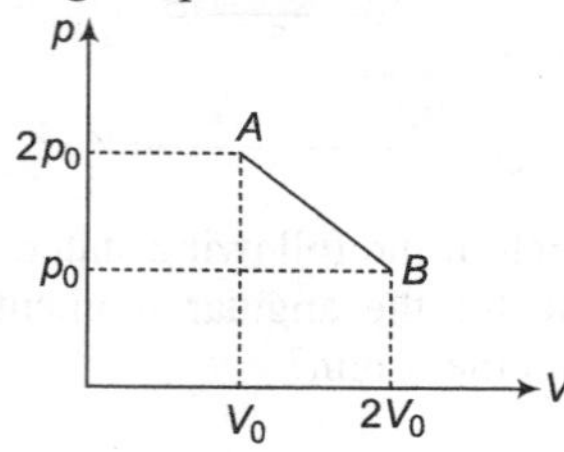

(a) $\frac{9}{4}\frac{p_0V_0}{nR}$ (b) $\frac{3}{2}\frac{p_0V_0}{nR}$
(c) $\frac{9}{2}\frac{p_0V_0}{nR}$ (d) $\frac{9p_0V_0}{nR}$

**10.** A particle performs simple harmonic motion with amplitude $A$. Its speed is trebled at the instant that it is at a distance $\frac{2}{3}A$ from equilibrium position. The new amplitude of the motion is

(a) $\frac{A}{3}\sqrt{41}$ (b) $3A$ (c) $A\sqrt{3}$ (d) $\frac{7}{3}A$

**11.** A uniform string of length 20 m is suspended from a rigid support. A short wave pulse is introduced at its lowest end. It starts moving up the string. The time taken to reach the support is (Take, $g = 10\ \text{ms}^{-2}$)

(a) $2\pi\sqrt{2}$ s (b) 2 s
(c) $2\sqrt{2}$ s (d) $\sqrt{2}$ s

**12.** The region between two concentric spheres of radii $a$ and $b$, respectively (see the figure), has volume charge density $\rho = \frac{A}{r}$, where, $A$ is a constant and $r$ is the distance from the centre. At the centre of the spheres is a point charge $Q$. The value of $A$, such that the electric field in the region between the spheres will be constant, is

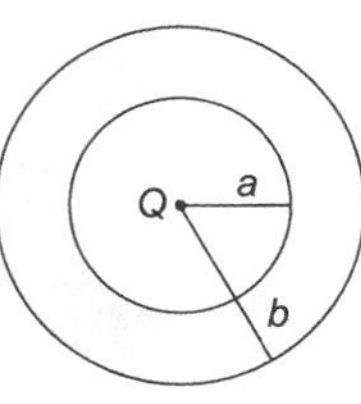


(a) $\frac{Q}{2\pi a^2}$ (b) $\frac{Q}{2\pi(b^2 - a^2)}$
(c) $\frac{2Q}{\pi(a^2 - b^2)}$ (d) $\frac{2Q}{\pi a^2}$

**13.** A combination of capacitors is set-up as shown in the figure. The magnitude of the electric field, due to a point charge $Q$ (having a charge equal to the sum of the charges on the 4 μF and 9μF capacitors), at a point distant 30 m from it, would equal to

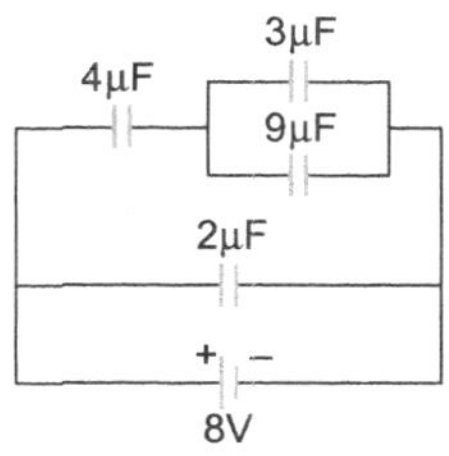


(a) 240 N/C (b) 360 N/C
(c) 420 N/C (d) 480 N/C

**14.** The temperature dependence of resistances of Cu and undoped Si in the temperature range 300-400 K, is best described by

(a) linear increase for Cu, linear increase for Si
(b) linear increase for Cu, exponential increase for Si
(c) linear increase for Cu, exponential decrease for Si
(d) linear decrease for Cu, linear decrease for Si

**15.** Two identical wires $A$ and $B$, each of length $l$, carry the same current $I$. Wire $A$ is bent into a circle of radius $R$ and wire $B$ is bent to form a square of side $a$. If $B_A$ and $B_B$ are the values of magnetic field at the centres of the circle and square respectively, then the ratio $\frac{B_A}{B_B}$ is

(a) $\frac{\pi^2}{8}$ (b) $\frac{\pi^2}{16\sqrt{2}}$ (c) $\frac{\pi^2}{16}$ (d) $\frac{\pi^2}{8\sqrt{2}}$

**16.** Hysteresis loops for two magnetic materials $A$ and $B$ are as given below:

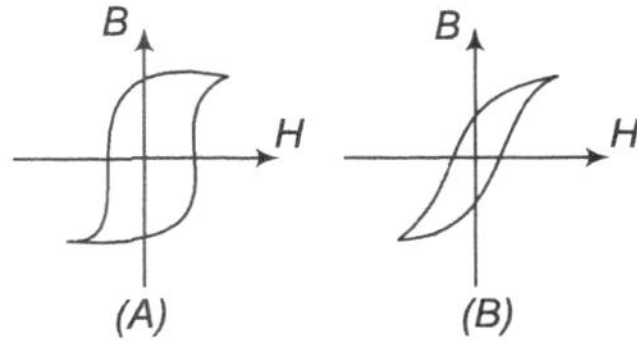


These materials are used to make magnets for electric generators, transformer core and electromagnet core. Then, it is proper to use

(a) $A$ for electric generators and transformers

(b) $A$ for electromagnets and $B$ for electric generators
(c) $A$ for transformers and $B$ for electric generators
(d) $B$ for electromagnets and transformers

**17.** An arc lamp requires a direct current of 10 A at 80 V to function. If it is connected to a 220 V (rms), 50 Hz AC supply, the series inductor needed for it to work is close to

(a) 80 H (b) 0.08 H
(c) 0.044 H (d) 0.065 H

**18.** Arrange the following electromagnetic radiations in the order of increasing energy.

A. Blue light B. Yellow light
C. X-ray D. Radio wave

(a) D, B, A, C (b) A, B, D, C
(c) C, A, B, D (d) B, A, D, C

**19.** An observer looks at a distant tree of height 10 m with a telescope of magnifying power of 20. To observer the tree appears

(a) 10 times taller (b) 10 times nearer
(c) 20 times taller (d) 20 times nearer

**20.** The box of a pin hole camera of length $L$, has a hole of radius $a$. It is assumed that when the hole is illuminated by a parallel beam of light of wavelength $\lambda$ the spread of the spot (obtained on the opposite wall of the camera) is the sum of its geometrical spread and the spread due to diffraction. The spot would then have its minimum size (say $b_{min}$) when

(a) $a = \frac{\lambda^2}{L}$ and $b_{min} = \left(\frac{2\lambda^2}{L}\right)$

(b) $a = \sqrt{\lambda L}$ and $b_{min} = \left(\frac{2\lambda^2}{L}\right)$

(c) $a = \sqrt{\lambda L}$ and $b_{min} = \sqrt{4\lambda L}$

(d) $a = \frac{\lambda^2}{L}$ and $b_{min} = \sqrt{4\lambda L}$

**21.** Radiation of wavelength $\lambda$, is incident on a photocell. The fastest emitted electron has speed $v$. If the wavelength is changed to $\frac{3\lambda}{4}$, the speed of the fastest emitted electron will be

(a) $> v\left(\frac{4}{3}\right)^{1/2}$ (b) $< v\left(\frac{4}{3}\right)^{1/2}$

(c) $= v\left(\frac{4}{3}\right)^{1/2}$ (d) $= v\left(\frac{3}{4}\right)^{1/2}$

**22.** Half-lives of two radioactive elements $A$ and $B$ are 20 min and 40 min, respectively. Initially, the samples have equal number of nuclei. After 80 min, the ratio of decayed numbers of $A$ and $B$ nuclei will be

(a) 1 : 16 (b) 4 : 1 (c) 1 : 4 (d) 5 : 4

**23.** If $a$, $b$, $c$ and $d$ are inputs to a gate and $x$ is its output, then, as per the following time graph, the gate is

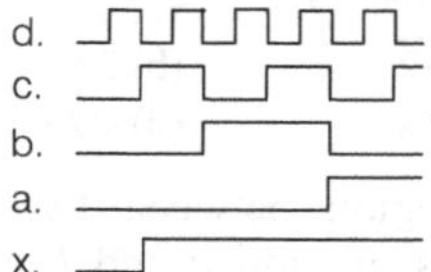


(a) NOT (b) AND (c) OR (d) NAND

**24.** Choose the correct statement.

(a) In amplitude modulation, the amplitude of the high frequency carrier wave is made to vary in proportion to the amplitude of the audio signal
(b) In amplitude modulation, the frequency of the high frequency carrier wave is made to vary in proportion to the amplitude of the audio signal
(c) In frequency modulation, the amplitude of the high frequency carrier wave is made to vary in proportion to the amplitude of the audio signal
(d) In frequency modulation, the amplitude of the high frequency carrier wave is made to vary in proportion to the frequency of the audio signal

**25.** A screw gauge with a pitch of 0.5 mm and a circular scale with 50 divisions is used to measure the thickness of a thin sheet of Aluminium. Before starting the measurement, it is found that when the two jaws of the screw gauge are brought in contact, the 45th division coincides with the main scale line and that the zero of the main scale is barely visible. What is the thickness of the sheet, if the main scale reading is 0.5 mm and the 25th division coincides with the main scale line?

(a) 0.75 mm (b) 0.80 mm
(c) 0.70 mm (d) 0.50 mm

**26.** A pipe open at both ends has a fundamental frequency $f$ in air. The pipe is dipped vertically in water, so that half of it is in water. The fundamental frequency of the air column is now

(a) $\frac{f}{2}$ (b) $\frac{3f}{4}$ (c) $2f$ (d) $f$

**27.** A galvanometer having a coil resistance of 100 Ω gives a full scale deflection when a current of 1 mA is passed through it. The value of the resistance which can convert this galvanometer into ammeter giving a full scale deflection for a current of 10 A, is

(a) 0.01 Ω (b) 2 Ω (c) 0.1 Ω (d) 3 Ω

**28.** In an experiment for determination of refractive index of glass of a prism by $i$-$\delta$ plot, it was found that a ray incident at an angle 35° suffers a deviation of 40° and that it emerges at an angle 79°. In that case, which of the following is closest to the maximum possible value of the refractive index?

(a) 1.5 (b) 1.6 (c) 1.7 (d) 1.8

**29.** Identify the semiconductor devices whose characteristics are as given below, in the order (a),(b),(c),(d).

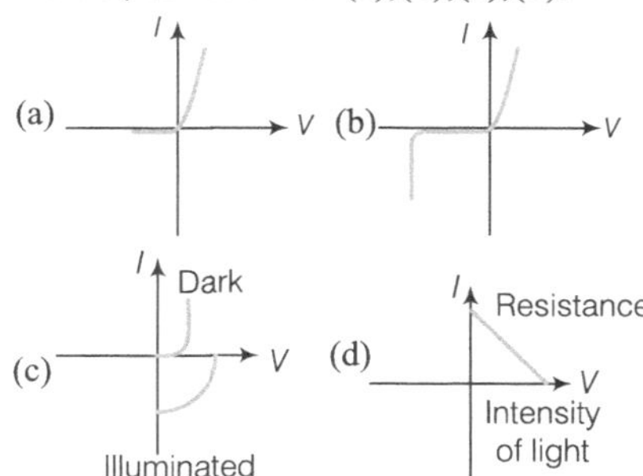


(a) Simple diode, Zener diode, Solar cell, Light dependent resistance
(b) Zener diode, Simple diode, Light dependent resistance, Solar cell
(c) Solar cell, Light dependent resistance, Zener diode, Simple diode
(d) Zener diode, Solar cell, Simple diode, Light dependent resistance

**30.** For a common-emitter configuration, if $\alpha$ and $\beta$ have their usual meanings, the **incorrect** relationship between $\alpha$ and $\beta$ is

(a) $\frac{1}{\alpha}=\frac{1}{\beta}+1$ (b) $\alpha=\frac{\beta}{1-\beta}$
(c) $\alpha=\frac{\beta}{1+\beta}$ (d) $\alpha=\frac{\beta^2}{1+\beta^2}$

# B. Chemistry

**1.** A stream of electrons from a heated filament was passed between two charged plates kept at a potential difference $V$ esu. If $e$ and $m$ are charge and mass of an electron, respectively, then the value of $h/\lambda$ (where, $\lambda$ is wavelength associated with electron wave) is given by

(a) $2meV$ (b) $\sqrt{meV}$
(c) $\sqrt{2meV}$ (d) $meV$

**2.** 2-chloro-2-methylpentane on reaction with sodium methoxide in methanol yields

I. $C_2H_5CH_2\overset{CH_3}{\underset{CH_3}{C}}—OCH_3$

II. $C_2H_5CH_2\underset{CH_3}{C}=CH_2$

III. $C_2H_5CH{=}C(CH_3){-}CH_3$

(a) Both I and III (b) Only III
(c) Both I and II (d) All of these

**3.** Which of the following compounds is metallic and ferromagnetic?

(a) $CrO_2$ (b) $VO_2$ (c) $MnO_2$ (d) $TiO_2$

**4.** Which of the following statements about low density polythene is false?

(a) It is a poor conductor of electricity
(b) Its synthesis required dioxygen or a peroxide initiator as a catalyst
(c) It is used in the manufacture of buckets, dustbins etc.
(d) Its synthesis requires high pressure

**5.** For a linear plot of log $(x/m)$ *versus* log $p$ in a Freundlich adsorption isotherm, which of the following statements is correct? ($k$ and $n$ are constants)

(a) $1/n$ appears as the intercept
(b) Only $1/n$ appears as the slope
(c) $\log\left(\frac{1}{n}\right)$ appears as the intercept
(d) Both $k$ and $1/n$ appear in the slope term

**6.** The heats of combustion of carbon and carbon monoxide are $-393.5$ and $-283.5$ kJ mol$^{-1}$, respectively. The heat of formation (in kJ) of carbon monoxide per mole is

(a) 676.5 (b) $-676.5$
(c) $-110.5$ (d) 110.5

**7.** The hottest region of Bunsen flame shown in the figure given below is

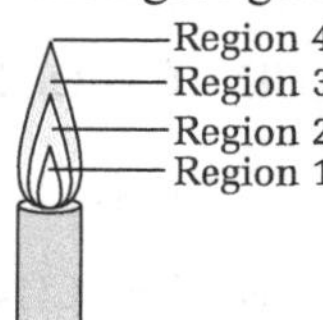


(a) region 2 (b) region 3
(c) region 4 (d) region 1

**8.** Which of the following is an anionic detergent?

(a) Sodium lauryl sulphate
(b) Cetyltrimethyl ammonium bromide
(c) Glyceryl oleate
(d) Sodium stearate

**9.** 18 g of glucose ($C_6H_{12}O_6$) is added to 178.2 g water. The vapour pressure of water (in torr) for this aqueous solution is

(a) 76.0 (b) 752.4 (c) 759.0 (d) 7.6

**10.** The distillation technique most suited for separating glycerol from spent lye in the soap industry is

(a) fractional distillation
(b) steam distillation
(c) distillation under reduced pressure
(d) simple distillation

**11.** The species in which the N-atom is in a state of *sp* hybridisation is

(a) $NO_2^-$ (b) $NO_3^-$ (c) $NO_2$ (d) $NO_2^+$

**12.** Decomposition of $H_2O_2$ follows a first order reaction. In 50 min, the concentration of $H_2O_2$ decreases from 0.5 to 0.125 M in one such decomposition. When the concentration of $H_2O_2$ reaches 0.05 M, the rate of formation of $O_2$ will be

(a) $6.93\times10^{-4}$ mol min$^{-1}$
(b) 2.66 L min$^{-1}$ at STP
(c) $1.34\times10^{-2}$ mol min$^{-1}$
(d) $6.93\times10^{-2}$ mol min$^{-1}$

**13.** The pair having the same magnetic moment is [at. no. Cr = 24, Mn = 25, Fe = 26 and Co = 27]

(a) $[Cr(H_2O)_6]^{2+}$ and $[Fe(H_2O)_6]^{2+}$
(b) $[Mn(H_2O)_6]^{2+}$ and $[Cr(H_2O)_6]^{2+}$
(c) $[CoCl_4]^{2-}$ and $[Fe(H_2O)_6]^{2+}$
(d) $[Cr(H_2O)_6]^{2+}$ and $[CoCl_4]^{2-}$

**14.** The absolute configuration of

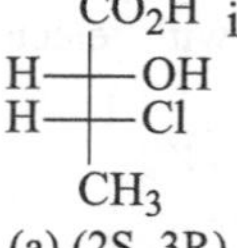


is

(a) (2S, 3R) (b) (2S, 3S)
(c) (2R, 3R) (d) (2R, 3S)

**15.** The equilibrium constant at 298 K for a reaction, $A + B \rightleftharpoons C + D$ is 100. If the initial concentrations of all the four species were 1 M each, then equilibrium concentration of $D$ (in mol $L^{-1}$) will be

(a) 0.818 (b) 1.818
(c) 1.182 (d) 0.182

**16.** Which one of the following ores is best concentrated by froth floatation method?

(a) Siderite (b) Galena
(c) Malachite (d) Magnetite

**17.** At 300 K and 1 atm, 15 mL of a gaseous hydrocarbon requires 375 mL air containing 20% $O_2$ by volume for complete combustion. After combustion, the gases occupy 330 mL. Assuming that the water formed is in liquid form and the volumes were measured at the same temperature and pressure, the formula of the hydrocarbon is

(a) $C_3H_8$ (b) $C_4H_8$
(c) $C_4H_{10}$ (d) $C_3H_6$

**18.** The pair in which phosphorus atoms have a formal oxidation state of +3 is

(a) pyrophosphorous and hypophosphoric acids
(b) orthophosphorous and hypophosphoric acids
(c) pyrophosphorous and pyrophosphoric acids
(d) orthophosphorous and pyrophosphorous acids

**19.** Which one of the following complexes shows optical isomerism?

(a) *cis* $[Co(en)_2Cl_2]Cl$
(b) *trans* $[Co(en)_2Cl_2]Cl$
(c) $[Co(NH_3)_4Cl_2]Cl$
(d) $[Co(NH_3)_3Cl_3]$

**20.** The reaction of zinc with dilute and concentrated nitric acid, respectively, produce

(a) $NO_2$ and NO (b) NO and $N_2O$
(c) $NO_2$ and $N_2O$ (d) $N_2O$ and $NO_2$

**21.** Which one of the following statements about water is false?

(a) Water can act both as an acid and as a base
(b) There is extensive intramolecular hydrogen bonding in the condensed phase
(c) Ice formed by heavy water sinks in normal water
(d) Water is oxidised to oxygen during photosynthesis

**22.** The concentration of fluoride, lead, nitrate and iron in a water sample from an underground lake was found to be 1000 ppb, 40 ppb, 100 ppm and 0.2 ppm, respectively. This water is unsuitable for drinking due to high concentration of

(a) lead (b) nitrate
(c) iron (d) fluoride

**23.** The main oxides formed on combustion of Li, Na and K in excess of air respectively are

(a) $LiO_2$, $Na_2O_2$ and $K_2O$
(b) $Li_2O_2$, $Na_2O_2$ and $KO_2$
(c) $Li_2O$, $Na_2O_2$ and $KO_2$
(d) $Li_2O$, $Na_2O$ and $KO_2$

**24.** Thiol group is present in

(a) cystine (b) cysteine
(c) methionine (d) cytosine

**25.** Galvanisation is applying a coating of

(a) Cr (b) Cu (c) Zn (d) Pb

**26.** Which of the following atoms has the highest first ionisation energy?

(a) Na (b) K (c) Sc (d) Rb

**27.** In the Hofmann-bromamide degradation reaction, the number of moles of NaOH and $Br_2$ used per mole of amine produced are

(a) four moles of NaOH and two moles of $Br_2$
(b) two moles of NaOH and two moles of $Br_2$
(c) four moles of NaOH and one mole of $Br_2$
(d) one mole of NaOH and one mole of $Br_2$

**28.** Two closed bulbs of equal volume ($V$) containing an ideal gas initially at pressure $p_i$ and temperature $T_1$ are connected through a narrow tube of negligible volume as shown in the figure below. The temperature of one of the bulbs is then raised to $T_2$. The final pressure $p_f$ is

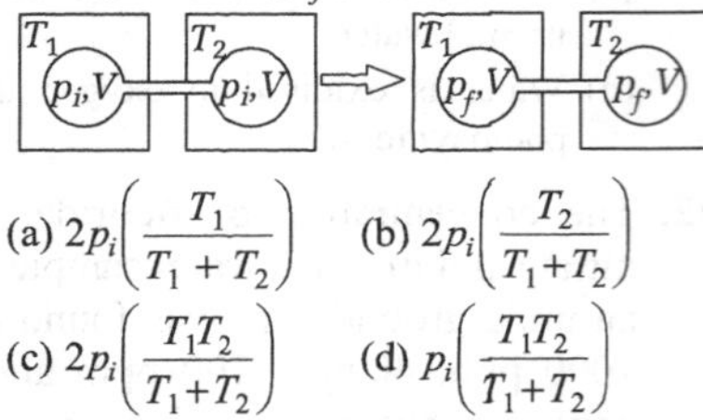

(a) $2p_i\left(\frac{T_1}{T_1+T_2}\right)$ (b) $2p_i\left(\frac{T_2}{T_1+T_2}\right)$

(c) $2p_i\left(\frac{T_1T_2}{T_1+T_2}\right)$ (d) $p_i\left(\frac{T_1T_2}{T_1+T_2}\right)$

**29.** The reaction of propene with HOCl $(Cl_2 + H_2O)$ proceeds through the intermediate

(a) $CH_3-\overset{+}{C}H-CH_2-Cl$

(b) $CH_3-CH(OH)-\overset{+}{C}H_2$

(c) $CH_3-CHCl-\overset{+}{C}H_2$

(d) $CH_3-\overset{+}{C}H-CH_2-OH$

**30.** The product of the reaction given below is

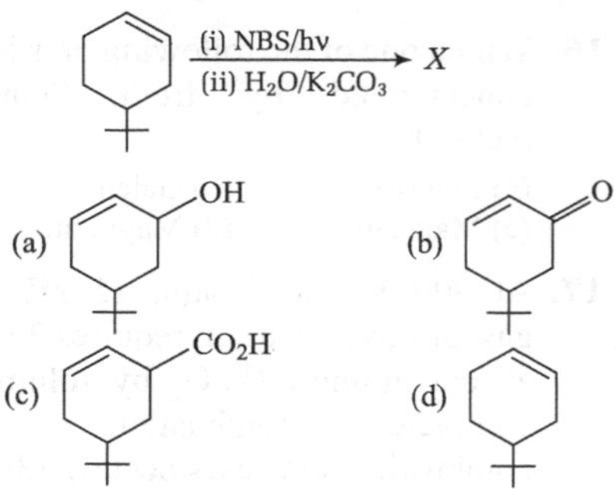

# C. Mathematics

**1.** If $f(x)+2f\left(\frac{1}{x}\right)=3x, x \neq 0$ and

$S=\{x \in R: f(x)=f(-x)\}$; then $S$

(a) is an empty set
(b) contains exactly one element
(c) contains exactly two elements
(d) contains more than two elements

**2.** A value of $\theta$ for which $\frac{2+3i\sin\theta}{1-2i\sin\theta}$ is purely imaginary, is

(a) $\frac{\pi}{3}$ (b) $\frac{\pi}{6}$

(c) $\sin^{-1}\left(\frac{\sqrt{3}}{4}\right)$ (d) $\sin^{-1}\left(\frac{1}{\sqrt{3}}\right)$

**3.** The sum of all real values of $x$ satisfying the equation $(x^2-5x+5)^{x^2+4x-60}=1$ is

(a) 3 (b) $-4$ (c) 6 (d) 5

**4.** If $A=\begin{bmatrix} 5a & -b \\ 3 & 2 \end{bmatrix}$ and $A$ adj $A = AA^T$, then $5a+b$ is equal to

(a) $-1$ (b) 5 (c) 4 (d) 13

**5.** The system of linear equations

$x+\lambda y-z=0$; $\lambda x-y-z=0$;
$x+y-\lambda z=0$

has a non-trivial solution for

(a) infinitely many values of $\lambda$
(b) exactly one value of $\lambda$
(c) exactly two values of $\lambda$
(d) exactly three values of $\lambda$

**6.** If all the words (with or without meaning) having five letters, formed using the letters of the word SMALL and arranged as in a dictionary, then the position of the word SMALL is

(a) 46th (b) 59th
(c) 52nd (d) 58th

**7.** If the number of terms in the expansion of $\left(1-\frac{2}{x}+\frac{4}{x^2}\right)^n$, $x \neq 0$, is 28, then the sum of the coefficients of all the terms in this expansion, is

(a) 64 (b) 2187
(c) 243 (d) 729

**8.** If the 2nd, 5th and 9th terms of a non-constant AP are in GP, then the common ratio of this GP is

(a) $\frac{8}{5}$ (b) $\frac{4}{3}$ (c) 1 (d) $\frac{7}{4}$

**9.** If the sum of the first ten terms of the series

$$\left(1\frac{3}{5}\right)^2 + \left(2\frac{2}{5}\right)^2 + \left(3\frac{1}{5}\right)^2 + 4^2 + \left(4\frac{4}{5}\right)^2$$

$+ \dots$, is $\frac{16}{5}m$, then $m$ is equal to

(a) 102 (b) 101
(c) 100 (d) 99

**10.** Let $p = \lim_{x \to 0^+} (1 + \tan^2 \sqrt{x})^{1/2x}$, then $\log p$ is equal to

(a) 2 (b) 1 (c) $\frac{1}{2}$ (d) $\frac{1}{4}$

**11.** For $x \in R$, $f(x) = |\log 2 - \sin x|$ and $g(x) = f(f(x))$, then

(a) $g$ is not differentiable at $x = 0$
(b) $g'(0) = \cos(\log 2)$
(c) $g'(0) = -\cos(\log 2)$
(d) $g$ is differentiable at $x = 0$ and $g'(0) = -\sin(\log 2)$

**12.** Consider $f(x) = \tan^{-1}\left(\sqrt{\frac{1 + \sin x}{1 - \sin x}}\right)$, $x \in \left(0, \frac{\pi}{2}\right)$. A normal to $y = f(x)$ at $x = \frac{\pi}{6}$ also passes through the point

(a) $(0, 0)$ (b) $\left(0, \frac{2\pi}{3}\right)$
(c) $\left(\frac{\pi}{6}, 0\right)$ (d) $\left(\frac{\pi}{4}, 0\right)$

**13.** A wire of length 2 units is cut into two parts which are bent respectively to form a square of side $= x$ units and a circle of radius $= r$ units. If the sum of the areas of the square and the circle so formed is minimum, then

(a) $2x = (\pi + 4)r$ (b) $(4 - \pi)x = \pi r$
(c) $x = 2r$ (d) $2x = r$

**14.** The integral $\int \frac{2x^{12} + 5x^9}{(x^5 + x^3 + 1)^3}\, dx$ is equal to

(a) $\frac{-x^5}{(x^5 + x^3 + 1)^2} + C$
(b) $\frac{x^{10}}{2(x^5 + x^3 + 1)^2} + C$
(c) $\frac{x^5}{2(x^5 + x^3 + 1)^2} + C$
(d) $\frac{-x^{10}}{2(x^5 + x^3 + 1)^2} + C$

where, $C$ is an arbitrary constant.

**15.** $\lim_{n \to \infty} \left[\frac{(n+1)(n+2)\dots 3n}{n^{2n}}\right]^{1/n}$ is equal to

(a) $\frac{18}{e^4}$ (b) $\frac{27}{e^2}$
(c) $\frac{9}{e^2}$ (d) $3\log 3 - 2$

**16.** The area (in sq units) of the region $\{(x, y) : y^2 \geq 2x$ and $x^2 + y^2 \leq 4x$, $x \geq 0, y \geq 0\}$ is

(a) $\pi - \frac{4}{3}$ (b) $\pi - \frac{8}{3}$
(c) $\pi - \frac{4\sqrt{2}}{3}$ (d) $\frac{\pi}{2} - \frac{2\sqrt{2}}{3}$

**17.** If a curve $y = f(x)$ passes through the point $(1, -1)$ and satisfies the differential equation, $y(1 + xy)\,dx = x\,dy$, then $f\left(-\frac{1}{2}\right)$ is equal to

(a) $-\frac{2}{5}$ (b) $-\frac{4}{5}$
(c) $\frac{2}{5}$ (d) $\frac{4}{5}$

**18.** Two sides of a rhombus are along the lines, $x - y + 1 = 0$ and $7x - y - 5 = 0$. If its diagonals intersect at $(-1, -2)$, then which one of the following is a vertex of this rhombus?

(a) $(-3, -9)$ (b) $(-3, -8)$
(c) $\left(\frac{1}{3}, -\frac{8}{3}\right)$ (d) $\left(-\frac{10}{3}, -\frac{7}{3}\right)$

**19.** The centres of those circles which touch the circle, $x^2 + y^2 - 8x - 8y - 4 = 0$, externally and also touch the $X$-axis, lie on

(a) a circle
(b) an ellipse which is not a circle
(c) a hyperbola
(d) a parabola

**20.** If one of the diameters of the circle, given by the equation, $x^2 + y^2 - 4x + 6y - 12 = 0$, is a chord of a circle $S$, whose centre is at $(-3, 2)$, then the radius of $S$ is

(a) $5\sqrt{2}$ (b) $5\sqrt{3}$ (c) 5 (d) 10

**21.** Let $P$ be the point on the parabola, $y^2 = 8x$, which is at a minimum distance from the centre $C$ of the circle, $x^2 + (y+6)^2 = 1$. Then, the equation of the circle, passing through $C$ and having its centre at $P$ is

(a) $x^2 + y^2 - 4x + 8y + 12 = 0$
(b) $x^2 + y^2 - x + 4y - 12 = 0$
(c) $x^2 + y^2 - \frac{x}{4} + 2y - 24 = 0$
(d) $x^2 + y^2 - 4x + 9y + 18 = 0$

**22.** The eccentricity of the hyperbola whose length of the latusrectum is equal to 8 and the length of its conjugate axis is equal to half of the distance between its foci, is

(a) $\frac{4}{3}$ (b) $\frac{4}{\sqrt{3}}$ (c) $\frac{2}{\sqrt{3}}$ (d) $\sqrt{3}$

**23.** The distance of the point $(1, -5, 9)$ from the plane $x - y + z = 5$ measured along the line $x = y = z$ is

(a) $3\sqrt{10}$ (b) $10\sqrt{3}$ (c) $\frac{10}{\sqrt{3}}$ (d) $\frac{20}{3}$

**24.** If the line, $\frac{x-3}{2} = \frac{y+2}{-1} = \frac{z+4}{3}$ lies in the plane, $lx + my - z = 9$, then $l^2 + m^2$ is equal to

(a) 26 (b) 18 (c) 5 (d) 2

**25.** Let $\hat{\mathbf{a}}, \hat{\mathbf{b}}$ and $\hat{\mathbf{c}}$ be three unit vectors such that $\hat{\mathbf{a}} \times (\hat{\mathbf{b}} \times \hat{\mathbf{c}}) = \frac{\sqrt{3}}{2}(\hat{\mathbf{b}} + \hat{\mathbf{c}})$. If $\hat{\mathbf{b}}$ is not parallel to $\hat{\mathbf{c}}$, then the angle between $\hat{\mathbf{a}}$ and $\hat{\mathbf{b}}$ is

(a) $\frac{3\pi}{4}$ (b) $\frac{\pi}{2}$ (c) $\frac{2\pi}{3}$ (d) $\frac{5\pi}{6}$

**26.** If the standard deviation of the numbers 2, 3, $a$ and 11 is 3.5, then which of the following is true?

(a) $3a^2 - 26a + 55 = 0$
(b) $3a^2 - 32a + 84 = 0$
(c) $3a^2 - 34a + 91 = 0$
(d) $3a^2 - 23a + 44 = 0$

**27.** Let two fair six-faced dice $A$ and $B$ be thrown simultaneously. If $E_1$ is the event that die $A$ shows up four, $E_2$ is the event that die $B$ shows up two and $E_3$ is the event that the sum of numbers on both dice is odd, then which of the following statements is not true?

(a) $E_1$ and $E_2$ are independent
(b) $E_2$ and $E_3$ are independent
(c) $E_1$ and $E_3$ are independent
(d) $E_1$, $E_2$ and $E_3$ are independent

**28.** If $0 \le x < 2\pi$, then the number of real values of $x$, which satisfy the equation $\cos x + \cos 2x + \cos 3x + \cos 4x = 0$, is

(a) 3 (b) 5 (c) 7 (d) 9

**29.** A man is walking towards a vertical pillar in a straight path, at a uniform speed. At a certain point $A$ on the path, he observes that the angle of elevation of the top of the pillar is 30°. After walking for 10 min from $A$ in the same direction, at a point $B$, he observes that the angle of elevation of the top of the pillar is 60°. Then, the time taken (in minutes) by him, from $B$ to reach the pillar, is

(a) 6 (b) 10
(c) 20 (d) 5

**30.** The Boolean expression $(p \wedge \sim q) \vee q \vee (\sim p \wedge q)$ is equivalent to

(a) $\sim p \wedge q$ (b) $p \wedge q$
(c) $p \vee q$ (d) $p \vee \sim q$

# Answers

## Paper 1

### Physics

| | | | | | | |
|---|---|---|---|---|---|---|
| **1.** (a) | **2.** (b,d) | **3.** (c) | **4.** (d) | **5.** (a) | **6.** (d) | **7.** (a) |
| **8.** (b) | **9.** (a) | **10.** (d) | **11.** (c) | **12.** (a) | **13.** (c) | **14.** (c) |
| **15.** (d) | **16.** (d) | **17.** (d) | **18.** (a) | **19.** (d) | **20.** (c) | **21.** (a) |
| **22.** (d) | **23.** (c) | **24.** (a) | **25.** (b) | **26.** (d) | **27.** (a) | **28.** (a) |
| **29.** (a) | **30.** (b,d) | | | | | |

### Chemistry

| | | | | | | |
|---|---|---|---|---|---|---|
| **1.** (c) | **2.** (d) | **3.** (a) | **4.** (c) | **5.** (b) | **6.** (c) | **7.** (a) |
| **8.** (a) | **9.** (b) | **10.** (c) | **11.** (d) | **12.** (a) | **13.** (a) | **14.** (a) |
| **15.** (b) | **16.** (b) | **17.** (*) | **18.** (d) | **19.** (a) | **20.** (d) | **21.** (b) |
| **22.** (b) | **23.** (c) | **24.** (b) | **25.** (a) | **26.** (c) | **27.** (c) | **28.** (b) |
| **29.** (a) | **30.** (a) | | | | | |

**Note** '*' Means, none options are not correct.

### Mathematics

| | | | | | | |
|---|---|---|---|---|---|---|
| **1.** (c) | **2.** (d) | **3.** (a) | **4.** (b) | **5.** (d) | **6.** (d) | **7.** (d) |
| **8.** (b) | **9.** (b) | **10.** (c) | **11.** (b) | **12.** (b) | **13.** (c) | **14.** (b) |
| **15.** (b) | **16.** (b) | **17.** (d) | **18.** (c) | **19.** (d) | **20.** (b) | **21.** (a) |
| **22.** (c) | **23.** (b) | **24.** (d) | **25.** (d) | **26.** (b) | **27.** (d) | **28.** (c) |
| **29.** (d) | **30.** (c) | | | | | |

# Answer *with* Explanations

## Physics

1. *(a)* True value $= \dfrac{90+91+95+92}{4} = 92$

Mean absolute error

$$= \frac{|92-90|+|92-91|+|92-95|+|92-92|}{4}$$

$$= \frac{2+1+3+0}{4} = 1.5$$

$$\text{Value} = (92 \pm 1.5)$$

Since, least count is 1 s

$$\therefore \quad \text{Value} = (92 \pm 2\text{s})$$

2. *(b, d)* We can apply $\mathbf{L} = m(\mathbf{r} \times \mathbf{v})$ for different parts.
For example :

In part (a), coordinates of $A$ are $\left(\dfrac{R}{\sqrt{2}}, \dfrac{R}{\sqrt{2}}\right)$

Therefore, $\mathbf{r} = \dfrac{R}{\sqrt{2}}\hat{\mathbf{i}} + \dfrac{R}{\sqrt{2}}\hat{\mathbf{j}}$ and $\mathbf{v} = v\hat{\mathbf{i}}$

So, substituting in $\mathbf{L} = m(\mathbf{r} \times \mathbf{v})$ we get,

$$\mathbf{L} = -\frac{mvR}{\sqrt{2}}\hat{\mathbf{k}}$$

Hence, option (a) is correct. Similarly, we can check other options also.

3. *(c)* As energy loss is same, thus

$$\mu mg\cos\theta \cdot (PQ) = \mu mg \cdot (QR)$$

$$\therefore \quad QR = (PQ)\cos\theta$$

$$\Rightarrow \quad QR = 4 \times \frac{\sqrt{3}}{2}$$

$$= 2\sqrt{3} \approx 3.5 \text{ m}$$

Further, decrease in potential energy = loss due to friction

$$\therefore \quad mgh = (\mu mg\cos\theta)d_1 + (\mu mg)d_2$$

$$m \times 10 \times 2 = \mu \times m \times 10 \times \frac{\sqrt{3}}{2} \times 4 + \mu \times m \times 10 \times 2\sqrt{3}$$

$$\Rightarrow \quad 4\sqrt{3}\,\mu = 2$$

$$\Rightarrow \quad \mu = \frac{1}{2\sqrt{3}}$$

$$= 0.288 = 0.29$$

4. *(d)* Work done in lifting mass

$$= (10 \times 9.8 \times 1) \times 1000$$

If $m$ is mass of fat burnt, then energy

$$= m \times 3.8 \times 10^7 \times \frac{20}{100}$$

Equating the two, we get

$$\therefore \quad m = \frac{49}{3.8} \approx 12.89 \times 10^{-3}\text{kg}$$

5. *(a)*

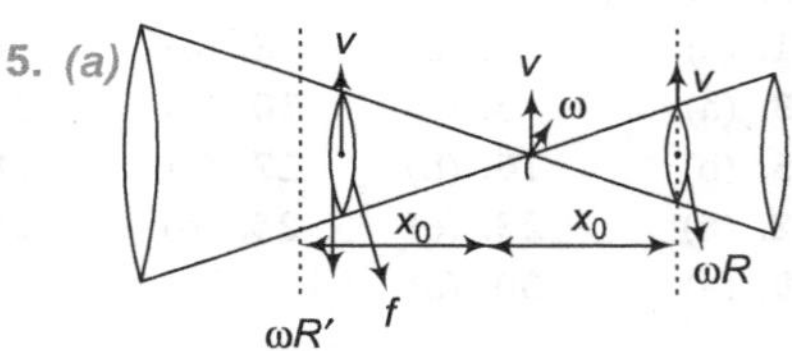


At distance $x_0$ from $O$, $v = \omega R$

distance less than $x_0$, $v > \omega R$

Initially, there is pure rolling at both the contacts. As the cone moves forward, slipping at $AB$ will start in forward direction, as radius at left contact decreases.

Thus, the cone will start turning towards left. As it moves further slipping at $CD$ will start in backward direction which will also turn the cone towards left.

6. *(d)* $v_{\text{orbital}} = \sqrt{\dfrac{GM}{R}} = \sqrt{gR}$

$$v_{\text{escape}} = \sqrt{2gR}$$

$\therefore$ Extra velocity required

$$= v_{\text{escape}} - v_{\text{orbital}} = \sqrt{gR}\,(\sqrt{2}-1)$$

7. *(a)* $T_0 = 2\pi\sqrt{\dfrac{L}{g}}$

$$T' = T_0 + \Delta T = 2\pi\sqrt{\frac{L+\Delta L}{g}}$$

$$\therefore T' = T_0 + \Delta T = 2\pi\sqrt{\frac{L(1+\alpha\Delta\theta)}{g}}$$

$$= \left\{2\pi\sqrt{\frac{L}{g}}\right\}(1+\alpha\Delta\theta)^{\frac{1}{2}} \approx T_0\left(1+\frac{\alpha\Delta\theta}{2}\right)$$

$$\therefore \quad \Delta T = T' - T_0 = \frac{\alpha \Delta\theta T_0}{2} \quad \ldots(i)$$

$$\text{or} \quad \frac{\Delta T_1}{\Delta T_2} = \frac{\alpha \Delta\theta_1 T_0}{\alpha \Delta\theta_2 T_0}$$

$$\Rightarrow \quad \frac{12}{4} = \frac{40° - \theta}{\theta - 20°}$$

$$\Rightarrow \quad 3(\theta - 20°) = 40° - \theta$$

$$\Rightarrow \quad 4\theta = 100°$$

$$\Rightarrow \quad \theta = 25° \text{ C}$$

Time gained or lost is given by

$$\Delta T = \left(\frac{\Delta T}{T_0 + \Delta T}\right) t \approx \frac{\Delta t}{T_0} t$$

From Eq. (i), $\frac{\Delta T}{T_0} = \frac{\alpha \Delta\theta}{2}$

$$\therefore \Delta t = \frac{\alpha (\Delta\theta) t}{2}$$

$$12 = \frac{\alpha (40° - 25°)(24 \times 3600)}{2}$$

$$\therefore \alpha = 1.85 \times 10^{-5} /° \text{C}$$

8. (b) $\Delta Q = \Delta U + \Delta W$

In the process $pV^n$ = constant, molar heat capacity is given by

$$C = \frac{R}{\gamma - 1} + \frac{R}{1-n} = C_V + \frac{R}{1-n}$$

$$C - C_V = \frac{R}{1-n}$$

$$\Rightarrow 1 - n = \frac{C_p - C_V}{C - C_V}$$

$$\therefore \quad n = 1 - \left(\frac{C_p - C_V}{C - C_V}\right)$$

$$= \frac{(C - C_V) - (C_p - C_V)}{C - C_V}$$

$$= \frac{C - C_p}{C - C_V}$$

9. (a) $p$-$V$ equation for path $AB$

$$p = -\left(\frac{p_0}{V_0}\right) V + 3p_0$$

$$\Rightarrow \quad pV = 3p_0 V - \frac{p_0}{V_0} V^2$$

$$\text{or} \quad T = \frac{pV}{nR} = \frac{1}{nR}\left(3p_0 V - \frac{p_0}{V_0} V^2\right)$$

For maximum temperature,

$$\frac{dT}{dV} = 0$$

$$\Rightarrow \quad 3p_0 - \frac{2p_0 V}{V_0} = 0 \Rightarrow V = \frac{3}{2} V_0$$

$$\text{and} \quad p = 3p_0 - \frac{p_0}{V_0} = \frac{3p_0}{2}$$

Therefore, at these values

$$T_{\max} = \frac{\left(\frac{3p_0}{2}\right)\left(\frac{3V_0}{2}\right)}{nR} = \frac{9p_0 V_0}{4nR}$$

10. (d) $v = \omega\sqrt{A^2 - x^2}$

At, $x = \frac{2A}{3}$

$$v = \omega\sqrt{A^2 - \left(\frac{2A}{3}\right)^2} = \frac{\sqrt{5}}{3} \omega A$$

As, velocity is trebled, hence $v' = \sqrt{5} A\omega$

This leads to new amplitude $A'$

$$\therefore \quad \omega\sqrt{A'^2 - \left(\frac{2A}{3}\right)^2} = \sqrt{5} A\omega$$

$$\Rightarrow \quad \omega^2\left[A'^2 - \frac{4A^2}{9}\right] = 5A^2\omega^2$$

$$\Rightarrow A'^2 = 5A^2 + \frac{4}{9}A^2 = \frac{49}{9}A^2$$

$$A' = \frac{7}{3}A$$

11. (c) At distance $x$ from the bottom

$$v = \sqrt{\frac{T}{\mu}} = \sqrt{\frac{\left(\frac{mgx}{L}\right)}{\left(\frac{m}{L}\right)}} = \sqrt{gx}$$

$$\therefore \quad \frac{dx}{dt} = \sqrt{x}\sqrt{g}$$

$$\Rightarrow \quad \int_0^L x^{-1/2} dx = \sqrt{g}\int_0^t dt$$

$$\Rightarrow \quad \left[\frac{x^{1/2}}{(1/2)}\Bigg|_0^L\right] = \sqrt{g} \cdot t$$

$$\Rightarrow \quad t = \frac{2\sqrt{L}}{\sqrt{g}}$$

$$\Rightarrow \quad t = 2\sqrt{\frac{20}{10}} = 2\sqrt{2} \text{ s}$$

12. *(a)* As $E$ is constant,

Hence, $E_a = E_b$

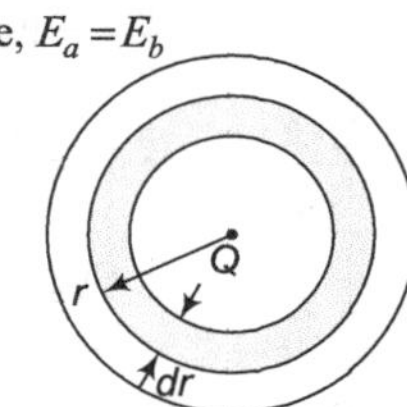


As per Guass theorem, only $Q$ contributes in electric field.

$$\therefore \quad \frac{kQ}{a^2} = \frac{k\left[Q + \int_a^b 4\pi r^2 dr \cdot \frac{A}{r}\right]}{b^2}$$

Here, $k = \frac{1}{4\pi\varepsilon_0}$

$$\Rightarrow \quad Q\frac{b^2}{a^2} = Q + 4\pi A\left[\frac{r^2}{2}\Big|_a^b\right]$$

$$= Q + 4\pi A \cdot \left(\frac{b^2 - a^2}{2}\right)$$

$$\Rightarrow Q\left(\frac{b^2}{a^2}\right) = Q + 2\pi A\,(b^2 - a^2)$$

$$\Rightarrow Q\left(\frac{b^2 - a^2}{a^2}\right) = 2\pi A\,(b^2 - a^2)$$

$$\Rightarrow \quad A = \frac{Q}{2\pi a^2}$$

13. *(c)* $3\,\mu$ and $9\mu = 12\mu F$

$$4\mu\text{F and } 12\mu\text{F} = \frac{4 \times 12}{4 \quad 12} = 3\mu\text{F}$$

$Q = CV = 3 \times 8 = 24\mu C$ (on $4\mu$F and $3\mu$F)

Now, this $24\mu$C distributes in direct ratio of capacity between $3\mu$F and $9\mu$F.

Therefore, $Q_{9\mu F} = 18\mu C$

$\therefore Q_{4\mu F} + Q_{9\mu F} = 24 + 18$

$= 42\mu C = Q$ (say)

$$E = \frac{kQ}{R^2}$$

$$= \frac{9 \times 10^9 \times 42 \times 10^{-6}}{30^2} = 420\text{ N/C}$$

14. *(c)* As, we know Cu is conductor, so increase in temperature, resistance will increases. Then, Si is semiconductor, so with increase in temperature, resistance will decreases.

15. *(d)* $B$ at centre of a circle $= \frac{\mu_0 I}{2R}$

$B$ at centre of a square

$$= 4 \times \frac{\mu I}{4\pi \cdot \frac{l}{2}}[\sin 45^\circ + \sin 45^\circ]$$

$$= 4\sqrt{2}\,\frac{\mu_0 I}{2\pi l}$$

Now, $R = \frac{L}{2\pi}$ and $l = \frac{L}{4}$

(as $L = 2\pi R = 4l$)

where, $L$ = length of wire.

$$\therefore \quad B_A = \frac{\mu_0 I}{2 \cdot \frac{L}{2\pi}} = \frac{\pi\mu_0 I}{L} = \pi\left[\frac{\mu_0 I}{L}\right]$$

$$B_B = 4\sqrt{2}\,\frac{\mu_0 I}{2\pi\left(\frac{L}{4}\right)}$$

$$= \frac{8\sqrt{2}\mu_0 I}{\pi L} = \frac{8\sqrt{2}}{\pi}\left[\frac{\mu_0 I}{L}\right]$$

$$\therefore \quad \frac{B_A}{B_B} = \pi^2 : 8\sqrt{2}$$

16. *(d)* We need high retentivity and high coercivity for electromagnets and small area of hysteresis loop for transformers.

17. *(d)* $V^2 = V_R^2 + V_L^2 \Rightarrow 220^2 = 80^2 + V_L^2$

Solving we get,

$$V_L \approx 205\text{ V}$$

$$X_L = \frac{V_L}{I} = \frac{205}{10} = 20.5\ \Omega = \omega L$$

$$\therefore \quad L = \frac{20.5}{2\pi \times 50} = 0.065\text{ H}$$

18. *(a)* Theoretical question. Therefore, no solution is required.

19. *(d)* Telescope resolves and brings objects closer. Hence, telescope with magnifying power of 20, the tree appears 20 times nearer.

20. *(c)*

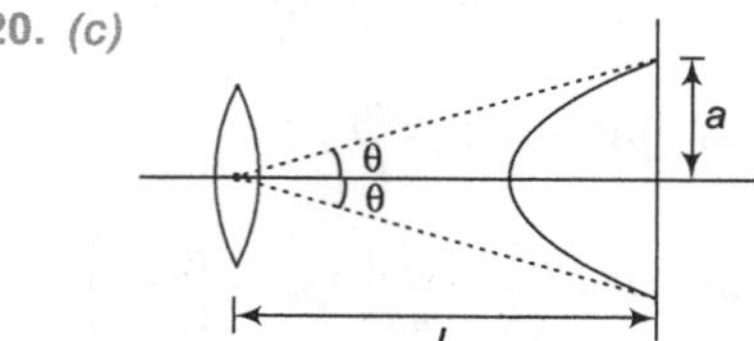

$\because \quad a\sin\theta \approx \lambda$

$\Rightarrow \quad a\left(\frac{a}{L}\right) \approx \lambda$

$\Rightarrow \quad a = \sqrt{\lambda L}$

$\therefore \quad \text{Spread} = 2a = \sqrt{4\lambda L}$

21. *(a)* According to the law of conservation of energy, i.e. Energy of a photon $(h\nu)$ = Work function $(\phi)$ + Kinetic energy of the photoelectron $\left(\frac{1}{2}mv_{max}^2\right)$

According to Einstein's photoelectric emission of light

i.e. $\quad E = (KE)_{max} + \phi$

As, $\quad \frac{hc}{\lambda} = (KE)_{max} + \phi$

If the wavelength of radiation is changed to $\frac{3\lambda}{4}$, then

$\Rightarrow \quad \frac{4}{3}\frac{hc}{\lambda} = \left(\frac{4}{3}(KE)_{max} + \frac{\phi}{3}\right) + \phi$

$(KE)_{max.}$ for fastest emitted electron $= \frac{1}{2}mv'^2 + \phi$

$\Rightarrow \quad \frac{1}{2}mv'^2 = \frac{4}{3}\left(\frac{1}{2}mv^2\right) + \frac{\phi}{3}$

i.e $\quad v' > v\left(\frac{4}{3}\right)^{1/2}$

22. *(d)* **A :** Numbers left :

$$N \rightarrow \frac{N}{2} \rightarrow \frac{N}{4} \rightarrow \frac{N}{8} \rightarrow \frac{N}{16}$$

$\therefore$ Number decayed,

$$N_A = N - \frac{N}{16} = \frac{15}{16}N$$

**B :** Numbers left : $N \rightarrow \frac{N}{2} \rightarrow \frac{N}{4}$

$\therefore$ Numbers decayed,

$$N_B = N - \frac{N}{4} = \frac{3}{4}N$$

Ratio : $\quad \frac{N_A}{N_B} = \frac{(15/16)N}{(3/4)N} = \frac{5}{4}$

23. *(c)* Output of OR gate is 0 when all inputs are 0 and output is 1 when atleast one of the input is 1.

**Observing output $x$** It is 0 when all inputs are 0 and it is 1 when atleast one of the input is 1.

$\therefore$ The gate is OR.

**Alternative Method**

**OR Gate**

| A | B | C | D | X |
|---|---|---|---|---|
| 0 | 0 | 0 | 0 | 0 |
| 0 | 0 | 0 | 1 | 1 |
| 0 | 0 | 1 | 0 | 1 |
| 0 | 0 | 1 | 1 | 1 |
| 0 | 1 | 0 | 0 | 1 |
| 0 | 1 | 0 | 1 | 1 |
| 0 | 1 | 1 | 0 | 1 |
| 0 | 1 | 1 | 1 | 1 |
| 1 | 0 | 0 | 0 | 1 |

24. *(a)* In amplitude modulation, $\mu = \frac{A_m}{A_c}$

25. *(b)* Least count

$$= \frac{\text{pitch}}{\text{number of divisions on circular scale}}$$

$$= \frac{0.5 \text{ mm}}{50}$$

$\therefore \quad LC = 0.01$

Negative zero error $= -5 \times LC$

$= -0.005$ mm

Measured value = main scale reading + screw gauge reading − zero error

$= 0.5 \text{ mm} + \{25 \times 0.01 - (-0.05)\}$ mm

$= 0.8$ mm

26. *(d)* Fundamental frequency of open pipe.

$$f = \frac{v}{2l}$$

Now, after half filled with water it becomes a closed pipe of length $\frac{l}{2}$.

Fundamental frequency of this closed pipe,

$$f' = \frac{v}{4(l/2)}$$

$$= \frac{v}{2l} = f$$

27. (a)

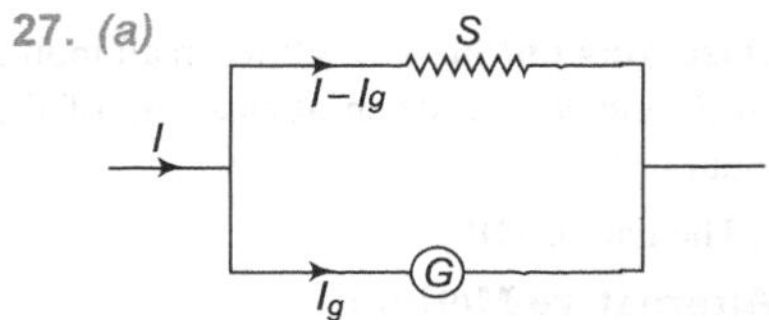


In parallel, current distributes in inverse ratio of resistance.

Hence, $\frac{I-I_g}{I_g}=\frac{G}{S} \Rightarrow S=\frac{GI_g}{I-I_g}$

As $I_g$ is very small, hence $S=\frac{GI_g}{I}$

$$b=\frac{(100)(1\times10^{-3})}{10}=0.01\,\Omega$$

28. (a) $\delta=(i_1+i_2)-A$

$\Rightarrow \quad 40^\circ=(35^\circ+79^\circ)-A$

$\Rightarrow \quad A=74^\circ$

Now, we know that $\mu=\frac{\sin\left(\frac{A+\delta_m}{2}\right)}{\sin\left(\frac{A}{2}\right)}$

It we take the given deviation as the minimum deviation then,

$$\mu=\frac{\sin\left(\frac{74^\circ+40^\circ}{2}\right)}{\sin\left(\frac{74^\circ}{2}\right)}=1.51$$

The given deviation may or may not be the minimum deviation. Rather it will be less than this value. Therefore, $\mu$ will be less than 1.51.

Hence, maximum possible value of refractive index is 1.51.

29. (a) Theoretical question. Therefore, no solution is required.

30. (b, d) $I_b+I_c=I_e$

$\Rightarrow \quad \frac{I_b}{I_c}+1=\frac{I_e}{I_c}$

$\Rightarrow \quad \frac{1}{\beta}+1=\frac{1}{\alpha} \Rightarrow \frac{1}{\alpha}=\frac{1+\beta}{\beta}$

$\Rightarrow \quad \alpha=\frac{\beta}{1+\beta}$

# Chemistry

1. (c) ***Plan*** As you can see in options, energy term is mentioned hence, we have to find out relation between $h/\lambda$ and energy. For this, we shall use de-Broglie wavelength and kinetic energy term in eV.

de-Broglie wavelength for an electron

$$(\lambda)=\frac{h}{p} \Rightarrow p=\frac{h}{\lambda} \quad \ldots\text{(i)}$$

Kinetic energy of an electron = eV

As we know that, $KE=\frac{p^2}{2m}$

$$\therefore \quad eV=\frac{p^2}{2m} \quad \text{or} \quad p=\sqrt{meV} \quad \ldots\text{(ii)}$$

From Eqs. (i) and (ii), we get

$$\frac{h}{\lambda}=\sqrt{meV}$$

2. (d) ***Key concept*** Strong nucleophile

($\overline{O}Me$) in polar solvent (MeOH) gives elimination products over substitution products but all products are possible in different yields.

$$CH_3-\underset{CH_3}{\overset{Cl}{C}}-CH_2CH_2CH_3 \xrightarrow[MeOH]{\overset{-}{Me}\overset{+}{ONa}} CH_3-\underset{CH_3}{\overset{OCH_3}{C}}-CH_2-CH_2-CH_3 + $$

(Less yield)

$$CH_3-\underset{CH_3}{C}=CH-CH_2-CH_3 + CH_2=\underset{CH_3}{C}-CH_2-CH_2-CH_3$$

(More yield)

3. (a) Only three elements iron (Fe), cobalt (Co) and nickel (Ni) show ferromagnetism at room temperature. $CrO_2$ is also a metallic and ferromagnetic compound which is used to make magnetic tapes for cassette recorders.

4. *(c)* High density polythene is used in the manufacture of buckets, dustbins etc.

5. *(b)* According to Freundlich adsorption isotherm, $\frac{x}{m} = kp^{1/n}$

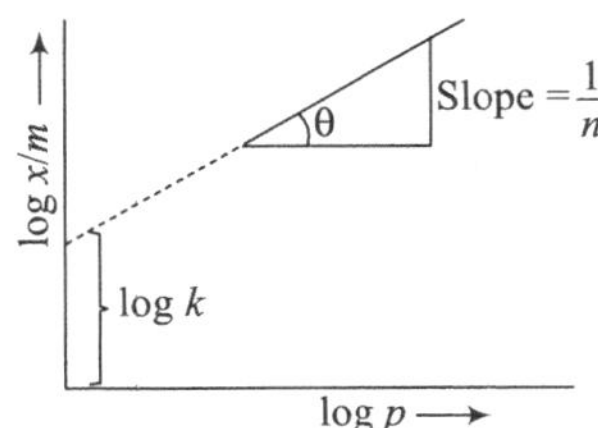


On taking logarithm of both sides, we get

$$\log \frac{x}{m} = \log k + \log p^{1/n} \text{ or}$$

$$\log \frac{x}{m} = \log k + \frac{1}{n} \log p$$

$$y = c + mx$$

$$y = \log \frac{x}{m}, c = \text{intercept} = \log k$$

$$m = \text{slope} = \frac{1}{n}$$

and $x = \log p$

6. *(c)* $C(s) + O_2(g) \longrightarrow CO_2(g)$;

$$\Delta H = -393.5 \text{ kJ mol}^{-1} \quad \ldots\text{(i)}$$

$$CO \quad - O_2 \longrightarrow CO_2\ g;$$

$$\Delta H = -283.5 \text{ kJ mol}^{-1} \quad \ldots\text{(ii)}$$

On subtracting Eq. (ii) from Eq. (i), we get

$$C(s) + \frac{1}{2}O_2(g) \longrightarrow CO(g);$$

$$\Delta H = (-393.5 + 283.5) \text{ kJ mol}^{-1} = -110 \text{ kJmol}^{-1}$$

(approx.)

7. *(a)* Region 1 (Pre-heating zone)

Region 2 (Primary combustion zone, hottest zone)

Region 3 (Internal zone)

Region 4 (Secondary reaction zone)

8. *(a)* Sodium lauryl sulphate $[(CH_3(CH_2)_{10}CH_2OSO_3^- Na^+)]$

= Anionic detergent

Cetyltrimethyl ammonium bromide

$$\left[ CH_3(CH_2)_{15} - \overset{\overset{CH_3}{|}}{\underset{\underset{CH_3}{|}}{N}} - CH_3 \right]^+ Br^-$$

= Cationic detergent

Glyceryl oleate $[(C_{17}H_{32}COO)_3C_3H_5]$

= Non-ionic detergent

Sodium stearate $[C_{17}H_{35}COO^- Na^+]$

= Anionic soap

9. *(b)* ***Key concept*** Vapour pressure of water ($p°$) = 760 torr

Number of moles of glucose

$$= \frac{\text{Mass (g)}}{\text{Molecular mass (g/mol)}}$$

$$= \frac{18 \text{ g}}{180 \text{ gmol}^{-1}}$$

$$= 0.1 \text{ mol}$$

Molar mass of water = 18 g/mol

Mass of water (given) = 178.2g

Number of moles of water

$$= \frac{\text{Mass of water}}{\text{Molar mass of water}}$$

$$= \frac{178.2\text{g}}{18 \text{ g/mol}} = 9.9 \text{ mol}$$

Total number of moles

= (0.1 + 9.9) moles = 10 moles

Now, mole fraction of glucose in solution = Change in pressure with respect to initial pressure

i.e. $\frac{\Delta p}{p°} = \frac{0.1}{10}$

or $\Delta p = 0.01 p°$

$= 0.01 \times 760 = 7.6$ torr

∴ Vapour pressure of solution

= (760 − 7.6) torr

= 752.4 torr

10. *(c)* Glycerol with high boiling point (290°C) can be separated from spent lye by distillation under reduced pressure. This process is used to purify liquids having very high boiling points. By this process, liquid is made to boil at lower temperature than its boiling point by lowering the pressure on its surface.

**11.** *(d)* Species Hybridisation

| Species | Hybridisation |
|---|---|
| $NO_2$ (N with lone pair, bonded to two O) | $sp^2$ |
| $NO_3^-$ | $sp^2$ |
| $NO_2^-$ | $sp^2$ |
| $O{=}\overset{+}{N}{=}O$ | $sp$ |

**12.** *(a)* For first order reaction, $k = \frac{2.303}{t} \log \frac{a}{a-x}$

Given, $t = 50$ min, $a = 0.5$ M,

$a - x = 0.125$ M

$\therefore \quad k = \frac{2.303}{50} \log \frac{0.5}{0.125} = 0.0277 \text{ min}^{-1}$

Now, as per reaction

$$2H_2O_2 \longrightarrow 2H_2O + O_2$$

$$-\frac{1}{2}\frac{d[H_2O_2]}{dt} = \frac{1}{2}\frac{d[H_2O]}{dt} = \frac{d[O_2]}{dt}$$

Rate of reaction, $-\frac{d[H_2O_2]}{dt} = k[H_2O_2]$

$$\therefore \frac{d[O_2]}{dt} = -\frac{1}{2}\frac{d[H_2O_2]}{dt} = \frac{1}{2}k[H_2O_2] \quad \ldots(i)$$

When the concentration of $H_2O_2$ reaches 0.05 M,

$$\frac{d[O_2]}{dt} = \frac{1}{2} \times 0.0277 \times 0.05 \text{ [from Eq. (i)]}$$

or $\frac{d[O_2]}{dt} = 6.93 \times 10^{-4} \text{ mol min}^{-1}$

**Alternative Method**

In 50 min, the concentration of $H_2O_2$ decreases from 0.5 to 0.125 M or in one half-life, concentration of $H_2O_2$ decreases from 0.5 to 0.25 M. In two half-lives, concentration of $H_2O_2$ decreases from 0.5 to 0.125 M or $2t_{1/2} = 50$ min

$t_{1/2} = 25$ min

$\therefore \quad k = \left(\frac{0.693}{25}\right) \text{min}^{-1}$

or $\frac{d[O_2]}{dt} = -\frac{1}{2}\frac{d[H_2O_2]}{dt} = \frac{k[H_2O_2]}{2}$

$= 6.93 \times 10^{-4} \text{mol min}^{-1}$

**13.** *(a)*

| Complex ion | Electronic configuration of metal ion | Number of unpaired electrons (*n*) |
|---|---|---|
| $[Cr(H_2O)_6]^{2+}$ | $Cr^{2+}$; [Ar] $3d^4$ | ↑ ↑ ↑ ↑ _ ; 4 |
| $[Fe(H_2O)_6]^{2+}$ | $Fe^{2+}$; [Ar] $3d^6$ | ⇅ ↑ ↑ ↑ ↑ ; 4 |
| $[Mn(H_2O)_6]^{2+}$ | $Mn^{2+}$; [Ar] $3d^5$ | ↑ ↑ ↑ ↑ ↑ ; 5 |
| $[CoCl_4]^{2-}$ | $Co^{2+}$; [Ar] $3d^7$ | ⇅ ⇅ ↑ ↑ ↑ ; 3 |

**14.** *(a)*

```
        1
       COOH
   H —2— OH
   H —3— Cl
       4
       CH3
```

For C-2, order of priority of substituents is

$$OH > CH(Cl)(CH_3) > COOH$$

For C-3, order of priority of substituents is

$$Cl > CH(OH)COOH > CH_3$$

Hence, according to CIP rules,

```
       3
       COOH
   H —— 1OH   2S
   H —2 1Cl   3R
       3
       CH3
```

**15.** *(b)*

| | $A$ | $+$ | $B$ | $\rightleftharpoons$ | $C$ | $+$ | $D$ |
|---|---|---|---|---|---|---|---|
| Initially at $t = 0$ | 1 | | 1 | | 1 | | 1 |
| At equilibrium | $1-x$ | | $1-x$ | | $1+x$ | | $1+x$ |

$$K_{eq} = \frac{[C][D]}{[A][B]} = \frac{(1+x)(1+x)}{(1-x)(1-x)} = \frac{(1+x)^2}{(1-x)^2}$$

or $\quad 100 = \left(\frac{1+x}{1-x}\right)^2$ or

$$10 = \frac{1+x}{1-x}$$

or $\quad 10 - 10x = 1 + x$

$10 - 1 = x + 10x$

$9 = 11x$

$x = \frac{9}{11} = 0.818$

$\therefore \; [D] = 1 + x = 1 + 0.818 = 1.818$

16. *(b)* Sulphide ores are concentrated by froth floatation method e.g. Galena (PbS).

17. *(None)*

$$C_xH_y(g) + \left(x + \frac{y}{4}\right) O_2(g) \longrightarrow xCO_2(g) + \frac{y}{2}H_2O(l)$$

(75 mL of $O_2$; 30 mL of $CO_2$)

$O_2$ used $= 20\%$ of $375 = 75$ mL

Inert part of air $= 80\%$ of $375 = 300$ mL

Total volume of gases

$= CO_2 +$ Inert part of air

$= 30 + 300 = 330$ mL

$$\frac{x}{1} = \frac{30}{15}$$

$$\Rightarrow \quad x = 2$$

$$\frac{x + \frac{y}{4}}{1} = \frac{75}{15}$$

$$\Rightarrow \quad x + \frac{y}{4} = 5$$

$$\Rightarrow \quad x = 2,\ y = 12 \Rightarrow C_2H_{12}$$

18. *(d)* Orthophosphorous acid, $H_3PO_3$ :

O
||
HO—P—OH
|
H

$$H_3\overset{x}{P}O_3 = 3 + x + 3(-2) = 0$$

or $\quad x = +3$

Pyrophosphorous acid, $H_4P_2O_5$ :

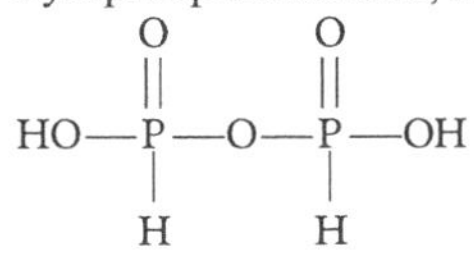


$$H_4\overset{x}{P_2}O_5 = 4 + 2x + 5(-2) = 0$$

$$4 + 2x - 10 = 0,\ x = +3$$

19. *(a)*

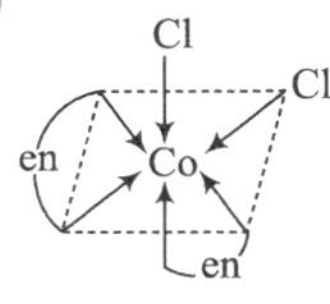


*cis*-$[Co(en)_2Cl_2]Cl$ (optically active)

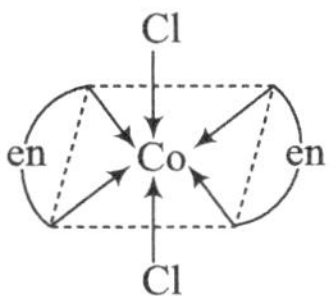


*trans*-$[Co(en)_2Cl_2]Cl$ (optically inactive due to plane of symmetry)

$[Co(NH_3)_4Cl_2]Cl$ can exist in both *cis* and *trans* forms that are given below:

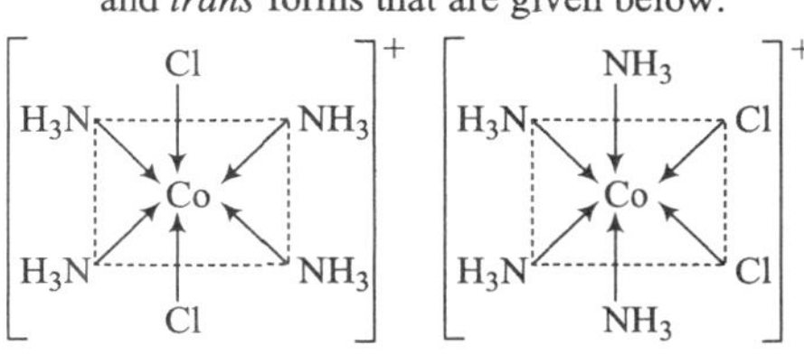


*trans*-$[Co(NH_3)_4Cl_2]Cl$ (optically inactive)

*cis*-$[Co(NH_3)_4Cl_2]Cl$ (optically inactive)

$[Co(NH_3)_3Cl_3]$ exists in *fac* and *mer*-isomeric forms and both are optically inactive.

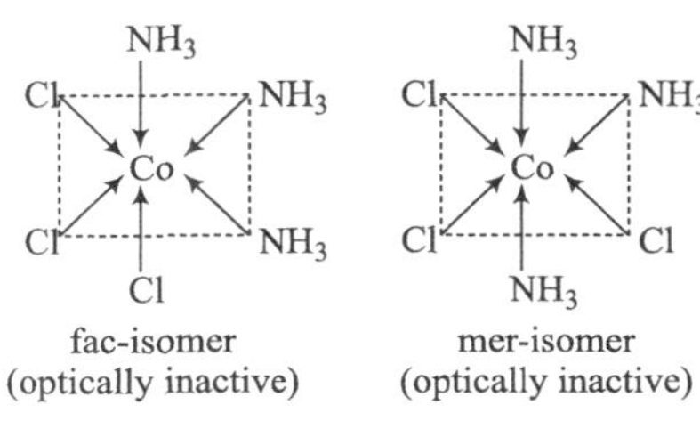


fac-isomer (optically inactive)

mer-isomer (optically inactive)

20. *(d)*

$$\underset{\text{(Conc.)}}{Zn + 4HNO_3} \longrightarrow Zn(NO_3)_2 + 2H_2O + 2NO_2$$

$$\underset{\text{(Dil.)}}{4Zn + 10HNO_3} \longrightarrow 4Zn(NO_3)_2 + N_2O + 5H_2O$$

21. *(b)* There is extensive intermolecular H-bonding in the condensed phase.

22. *(b)* This water is unsuitable for drinking due to high concentration of nitrate. In drinking water, maximum permissible concentration of

Lead $\simeq$ 50 ppb, Nitrate $\simeq$ 50 ppb

Iron $\simeq$ 0.2 ppm, Fluoride $\simeq$ 1 ppm

23. *(c)*

$$2Li + \frac{1}{2}\underset{\text{(Excess)}}{O_2(g)} \longrightarrow Li_2O$$

$$2Na + \underset{\text{(Excess)}}{O_2(g)} \longrightarrow Na_2O_2;$$

$$K + \underset{\text{(Excess)}}{O_2(g)} \longrightarrow KO_2$$

24. *(b)*

Cystine: HO—C(=O)—CH($NH_2$)—$CH_2$—S—S—$CH_2$—CH($NH_2$)—COOH

Cysteine: HS—$CH_2$—CH($NH_2$)—COOH

Methionine $H_3C-S-CH_2-CH_2-CH(NH_2)-C(=O)-OH$,

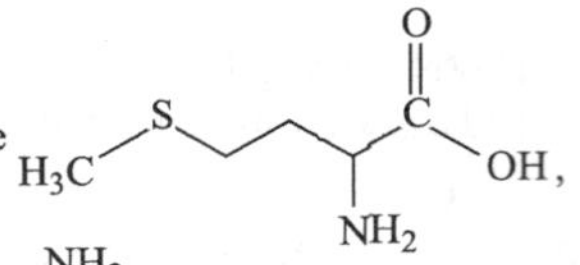

Cytosine (ring structure with $NH_2$, N, O, N–H)

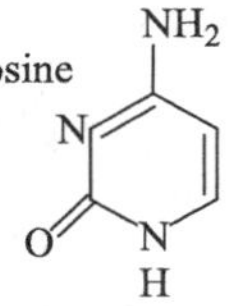

Thiol group (SH) is present in cysteine.

25. *(c)* Zinc metal is the most stable metal to cover iron surfaces. The process of coating the iron surface by zinc is called galvanisation.

26. *(c)* Order of first ionisation energy is Sc > Na > K > Rb.

Due to poor shielding effect, removal of one electron from 4*s* orbital is difficult as compared to 3*s*-orbital.

27. *(c)* Hofmann-bromamide degradation reaction is given as:

$$RCONH_2 + 4NaOH + Br_2 \longrightarrow \underset{(1^\circ \text{ amine})}{RNH_2} + Na_2CO_3 + 2NaBr + 2H_2O$$

Hence, four moles of NaOH and one mole of are used.

28. *(b)* Initially, Number of moles of gases in each container $= \frac{p_iV}{RT_1}$

Total number of moles of gases in both containers $= 2\frac{p_iV}{RT_1}$

After mixing, number of moles in left chamber $= \frac{p_fV}{RT_1}$

Number of moles in right chamber $= \frac{p_fV}{RT_2}$

Total number of moles $= \frac{p_fV}{RT_1} + \frac{p_fV}{RT_2}$

$$= \frac{p_fV}{R}\left(\frac{1}{T_1} + \frac{1}{T_2}\right)$$

As total number of moles remains constant.

Hence, $\frac{2p_iV}{RT_1} = \frac{p_fV}{RT_1} + \frac{p_fV}{RT_2}$

$$\Rightarrow \quad p_f = 2p_i\left(\frac{T_2}{T_1 + T_2}\right)$$

29. *(a)*

$$CH_3-CH=CH_2 \xrightarrow[\text{(Electrophilic addition)}]{\overset{\delta^-}{HO}-\overset{\delta^+}{Cl}} \underset{\text{(Intermediate)}}{CH_3-\overset{+}{C}H-CH_2-Cl} \xrightarrow{\bar{O}H} CH_3-\underset{OH}{CH}-CH_2-Cl$$

30. *(a)*

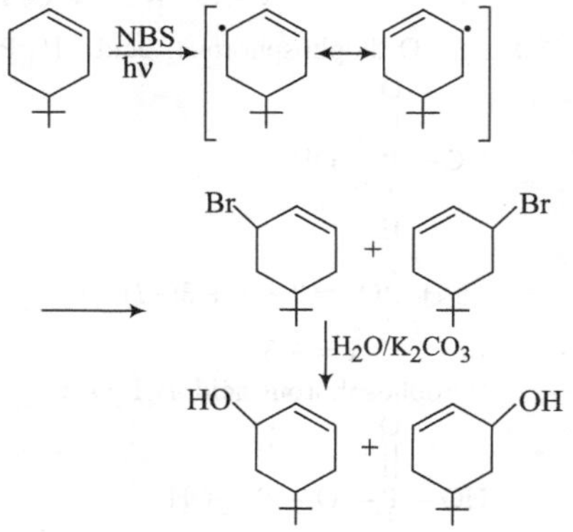

4-*tert*-butylcyclohexene $\xrightarrow[h\nu]{NBS}$ [allylic radical resonance structures] $\longrightarrow$ bromo-*tert*-butylcyclohexenes (Br) + (Br) $\xrightarrow{H_2O/K_2CO_3}$ (HO) + (OH) *tert*-butylcyclohexenols

# Mathematics

1. *(c)* We have, $f(x) + 2f\left(\frac{1}{x}\right) = 3x$,

$$x \neq 0 \quad \ldots(i)$$

On replacing $x$ by $\frac{1}{x}$ in the above Eq. (i),

we get $f\left(\frac{1}{x}\right) + 2f(x) = \frac{3}{x}$

$$\Rightarrow \quad 2f(x) + f\left(\frac{1}{x}\right) = \frac{3}{x} \quad \ldots(ii)$$

On multiplying Eq. (ii) by 2 and subtracting Eq. (i) from Eq. (ii), we get

$$4f(x) + 2f\left(\frac{1}{x}\right) = \frac{6}{x}$$

$$f(x) + 2f\left(\frac{1}{x}\right) = 3x$$

$$- \quad - \quad -$$

$$3f(x) = \frac{6}{x} - 3x$$

$$\Rightarrow \quad f(x) = \frac{2}{x} - x$$

Now, consider $f(x) = f(-x)$

$$\Rightarrow \quad \frac{2}{x} - x = -\frac{2}{x} + x$$

$$\Rightarrow \quad \frac{4}{x} = 2x$$

$$\Rightarrow \quad 2x^2 = 4 \Rightarrow x^2 = 2$$

$$\Rightarrow \quad x = \pm\sqrt{2}$$

Hence, $S$ contains exactly two elements.

2. *(d)* Let $z = \dfrac{2 + 3i \sin\theta}{1 - 2i \sin\theta}$ is purely imaginary. Then, we have $\text{Re}(z) = 0$

Now, consider $z = \dfrac{2 + 3i \sin\theta}{1 - 2i \sin\theta}$

$$= \frac{(2 + 3i \sin\theta)(1 + 2i \sin\theta)}{(1 - 2i \sin\theta)(1 + 2i \sin\theta)}$$

$$= \frac{2 + 4i \sin\theta + 3i \sin\theta + 6i^2 \sin^2\theta}{1^2 - (2i \sin\theta)^2}$$

$$= \frac{2 + 7i \sin\theta - 6\sin^2\theta}{1 + 4\sin^2\theta}$$

$$= \frac{2 - 6\sin^2\theta}{1 + 4\sin^2\theta} + i\,\frac{7\sin\theta}{1 + 4\sin^2\theta}$$

$$\because \quad \text{Re}(z) = 0$$

$$\therefore \quad \frac{2 - 6\sin^2\theta}{1 + 4\sin^2\theta} = 0$$

$$\Rightarrow \quad 2 = 6\sin^2\theta$$

$$\Rightarrow \quad \sin^2\theta = \frac{1}{3}$$

$$\Rightarrow \quad \sin\theta = \pm\frac{1}{\sqrt{3}}$$

$$\Rightarrow \theta = \sin^{-1}\left(\pm\frac{1}{\sqrt{3}}\right) = \pm\sin^{-1}\left(\frac{1}{\sqrt{3}}\right)$$

3. *(a)* Given, $(x^2 - 5x + 5)^{x^2 + 4x - 60} = 1$

Clearly, this is possible when

I. $x^2 + 4x - 60 = 0$

and $x^2 - 5x + 5 \neq 0$

or

II. $x^2 - 5x + 5 = 1$

or

III. $x^2 - 5x + 5 = -1$

and $x^2 + 4x - 60 =$ Even integer.

***Case* I** When $\mathbf{x^2 + 4x - 60 = 0}$

$$\Rightarrow \quad x^2 + 10x - 6x - 60 = 0$$

$$\Rightarrow \quad x(x + 10) - 6(x + 10) = 0$$

$$\Rightarrow \quad (x + 10)(x - 6) = 0$$

$$\Rightarrow \quad x = -10 \text{ or } x = 6$$

Note that, for these two values of $x$, $x^2 - 5x + 5 \neq 0$

***Case* II** When $x^2 - 5x + 5 = 1$

$$\Rightarrow \quad x^2 - 5x + 4 = 0$$

$$\Rightarrow \quad x^2 - 4x - x + 4 = 0$$

$$\Rightarrow \quad x(x - 4) - 1(x - 4) = 0$$

$$\Rightarrow \quad (x - 4)(x - 1) = 0$$

$$\Rightarrow \quad x = 4 \text{ or } x = 1$$

***Case* III** When $x^2 - 5x + 5 = -1$

$$\Rightarrow \quad x^2 - 5x + 6 = 0$$

$$\Rightarrow \quad x^2 - 2x - 3x + 6 = 0$$

$$\Rightarrow \quad x(x - 2) - 3(x - 2) = 0$$

$$\Rightarrow \quad (x - 2)(x - 3) = 0$$

$$\Rightarrow \quad x = 2 \text{ or } x = 3$$

Now, when $x = 2$, $x^2 + 4x - 60$ $= 4 + 8 - 60 = -48$, which is an even integer.

When $x = 3$,

$x^2 + 4x - 60 = 9 + 12 - 60 = -39$, which is not an even integer.

Thus, in this case, we get $x = 2$.

Hence, the sum of all real values of $x = -10 + 6 + 4 + 1 + 2 = 3$

4. *(b)* Given, $A = \begin{bmatrix} 5a & -b \\ 3 & 2 \end{bmatrix}$ and

$A \text{ adj } A = AA^T$

Clearly, $A(\text{adj } A) = |A| I_2$

[$\because$ if $A$ is square matrix of order $n$, then $A(\text{adj } A) = (\text{adj } A) \cdot A = |A| I_n$]

$$= \begin{vmatrix} 5a & -b \\ 3 & 2 \end{vmatrix} I_2 = (10a + 3b) I_2$$

$$= (10a + 3b)\begin{bmatrix} 1 & 0 \\ 0 & 1 \end{bmatrix}$$

$$= \begin{bmatrix} 10a + 3b & 0 \\ 0 & 10a + 3b \end{bmatrix} \quad \text{...(i)}$$

and $AA^T = \begin{bmatrix} 5a & -b \\ 3 & 2 \end{bmatrix}\begin{bmatrix} 5a & 3 \\ -b & 2 \end{bmatrix}$

$= \begin{bmatrix} 25a^2 + b^2 & 15a - 2b \\ 15a - 2b & 13 \end{bmatrix}$ ...(ii)

$\because A(\text{adj } A) = AA^T$

$\therefore \begin{bmatrix} 10a + 3b & 0 \\ 0 & 10a + 3b \end{bmatrix} =$

$\begin{bmatrix} 25a^2 + b^2 & 15a - 2b \\ 15a - 2b & 13 \end{bmatrix}$

[using Eqs. (i) and (ii)]

$\Rightarrow \quad 15a - 2b = 0 \Rightarrow a = \frac{2b}{15}$ ...(iii)

and $\quad 10a + 3b = 13$ ...(iv)

On substituting the value of '$a$' from Eq. (iii) in Eq. (iv), we get

$10 \cdot \left(\frac{2b}{15}\right) + 3b = 13$

$\Rightarrow \quad \frac{20b + 45b}{15} = 13$

$\Rightarrow \quad \frac{65b}{15} = 13$

$\Rightarrow \quad b = 3$

Now, substituting the value of $b$ in Eq. (iii), we get $5a = 2$

Hence, $\quad 5a + b = 2 + 3 = 5$

5. *(d)* Given, system of linear equation is

$x + \lambda y - z = 0;\ \lambda x - y - z = 0;$
$x + y - \lambda z = 0$

Note that, given system will have a non-trivial solution only if determinant of coefficient matrix is zero, i.e.

$$\begin{vmatrix} 1 & \lambda & -1 \\ \lambda & -1 & -1 \\ 1 & 1 & -\lambda \end{vmatrix} = 0$$

$\Rightarrow 1(\lambda + 1) - \lambda(-\lambda^2 + 1) - 1(\lambda + 1) = 0$

$\Rightarrow \quad \lambda + 1 + \lambda^3 - \lambda - \lambda - 1 = 0$

$\Rightarrow \quad \lambda^3 - \lambda = 0 \Rightarrow \lambda(\lambda^2 - 1) = 0$

$\Rightarrow \quad \lambda = 0 \text{ or } \lambda = \pm 1$

Hence, given system of linear equation has a non-trivial solution for exactly three values of $\lambda$.

6. *(d)* Clearly, number of words start with A $= \frac{4!}{2!} = 12$

Number of words start with $= 4! = 24$

Number of words start with $= \frac{4!}{2!} = 12$

Number of words start with $= \frac{3!}{2!} = 3$

Number of words start with $= 3! = 6$

Note that, next word will be "SMALL".

Hence, the position of word "SMALL" is 58th.

7. *(d)* Clearly, number of terms in the expansion of

$\left(1 - \frac{2}{x} + \frac{4}{x^2}\right)^n$ is $\frac{(n + 2)(n + 1)}{2}$ or ${}^{n+2}C_2$.

[assuming $\frac{1}{x}$ and $\frac{1}{x^2}$ distinct]

$\therefore \quad \frac{(n + 2)(n + 1)}{2} = 28$

$\Rightarrow (n + 2)(n + 1) = 56 = (6 + 1)(6 + 2)$

$\Rightarrow \quad n = 6$

Hence, sum of coefficients

$= (1 - 2 + 4)^6 = 3^6 = 729$

**Note** As $\frac{1}{x}$ and $\frac{1}{x^2}$ are functions of same variables, therefore number of dissimilar terms will be $2n + 1$, i.e. odd, which is not possible. Hence, it contains error.

8. *(b)* Let $a$ be the first term and $d$ be the common difference. Then, we have $a + d$, $a + 4d$, $a + 8d$ in GP,

i.e. $(a + 4d)^2 = (a + d)(a + 8d)$

$\Rightarrow a^2 + 16d^2 + 8ad$

$= a^2 + 8ad + ad + 8d^2$

$\Rightarrow \quad 8d^2 = ad$

$\Rightarrow \quad 8d = a \quad [\because d \neq 0]$

Now, common ratio,

$r = \frac{a + 4d}{a + d}$

$= \frac{8d + 4d}{8d + d}$

$= \frac{12d}{9d} = \frac{4}{3}$

9. *(b)* Let $S_{10}$ be the sum of first ten terms of the series. Then, we have

$$S_{10} = \left(1\frac{3}{5}\right)^2 + \left(2\frac{2}{5}\right)^2 + \left(3\frac{1}{5}\right)^2 + 4^2 + \left(4\frac{4}{5}\right)^2 + \dots \text{ to 10 terms}$$

$$= \left(\frac{8}{5}\right)^2 + \left(\frac{12}{5}\right)^2 + \left(\frac{16}{5}\right)^2 + 4^2 + \left(\frac{24}{5}\right)^2 + \dots \text{ to 10 terms}$$

$$= \frac{1}{5^2}(8^2 + 12^2 + 16^2 + 20^2 + 24^2 + \dots \text{ to 10 terms}$$

$$= \frac{4^2}{5^2}(2^2 + 3^2 + 4^2 + 5^2 + \dots \text{ to 10 terms})$$

$$= \frac{4^2}{5^2}(2^2 + 3^2 + 4^2 + 5^2 + \dots + 11^2)$$

$$= \frac{16}{25}((1^2 + 2^2 + \dots + 11^2) - 1^2)$$

$$= \frac{16}{25}\left[\frac{11 \cdot (11+1)(2 \cdot 11 + 1)}{6} - 1\right]$$

$$= \frac{16}{25}(506 - 1) = \frac{16}{25} \times 505$$

$$\Rightarrow \quad \frac{16}{5} m = \frac{16}{25} \times 505 = 101$$

10. *(c)* Given, $p = \lim\limits_{x \to 0^+} (1 + \tan^2 \sqrt{x})^{\frac{1}{2x}}$ [$1^\infty$ form]

$$= e^{\lim\limits_{x \to 0^+} \frac{\tan^2 \sqrt{x}}{2x}} = e^{\frac{1}{2}\lim\limits_{x \to 0^+}\left(\frac{\tan\sqrt{x}}{\sqrt{x}}\right)^2} = e^{\frac{1}{2}}$$

$$\therefore \quad \log p = \log e^{\frac{1}{2}} = \frac{1}{2}$$

11. *(b)* We have, $f(x) = |\log 2 - \sin x|$ and

$$g(x) = f(f(x)), x \in R$$

Note that, for $x \to 0$, $\log 2 > \sin x$

$\therefore \quad f(x) = \log 2 - \sin x$

$\Rightarrow \quad g(x) = \log 2 - \sin(f(x))$

$\quad = \log 2 - \sin(\log 2 - \sin x)$

Clearly, $g(x)$ is differentiable at $x = 0$ as $\sin x$ is differentiable.

Now,

$$g'(x) = -\cos(\log 2 - \sin x)(-\cos x)$$

$$= \cos x \cdot \cos(\log 2 - \sin x)$$

$\Rightarrow g'(0) = 1 \cdot \cos(\log 2)$

12. *(b)* We have,

$$f(x) = \tan^{-1}\sqrt{\frac{1 + \sin x}{1 - \sin x}}, x \in \left(0, \frac{\pi}{2}\right)$$

$$\Rightarrow \quad f(x) = \tan^{-1}\sqrt{\frac{\left(\cos\frac{x}{2} + \sin\frac{x}{2}\right)^2}{\left(\cos\frac{x}{2} - \sin\frac{x}{2}\right)^2}}$$

$$= \tan^{-1}\left(\frac{\cos\frac{x}{2} + \sin\frac{x}{2}}{\cos\frac{x}{2} - \sin\frac{x}{2}}\right)$$

$$\left(\because \cos\frac{x}{2} > \sin\frac{x}{2} \text{ for } 0 < \frac{x}{2} < \frac{\pi}{4}\right)$$

$$= \tan^{-1}\left(\frac{1 + \tan\frac{x}{2}}{1 - \tan\frac{x}{2}}\right)$$

$$= \tan^{-1}\left[\tan\left(\frac{\pi}{4} + \frac{x}{2}\right)\right] = \frac{\pi}{4} + \frac{x}{2}$$

$$\Rightarrow f'(x) = \frac{1}{2} \Rightarrow f'\left(\frac{\pi}{6}\right) = \frac{1}{2}$$

Now, equation of normal at $x = \frac{\pi}{6}$ is given by

$$\left(y - f\left(\frac{\pi}{6}\right)\right) = -2\left(x - \frac{\pi}{6}\right)$$

$$\Rightarrow \quad \left(y - \frac{\pi}{3}\right) = -2\left(x - \frac{\pi}{6}\right)$$

$$\left[\because f\left(\frac{\pi}{6}\right) = \frac{\pi}{4} + \frac{\pi}{12} = \frac{4\pi}{12} = \frac{\pi}{3}\right]$$

which passes through $\left(0, \frac{2\pi}{3}\right)$.

13. *(c)* According to given information, we have

Perimeter of square + Perimeter of circle = 2 units

$$\Rightarrow 4x + 2\pi r = 2 \Rightarrow r = \frac{1 - 2x}{\pi} \quad \dots\text{(i)}$$

Now, let $A$ be the sum of the areas of the square and the circle. Then,

$$A = x^2 + \pi r^2$$

$$= x^2 + \pi\frac{(1 - 2x)^2}{\pi^2}$$

$$\Rightarrow \quad A(x) = x^2 + \frac{(1 - 2x)^2}{\pi}$$

Now, for minimum value of $A(x)$, $\frac{dA}{dx} = 0$

$$\Rightarrow 2x + \frac{2(1-2x)}{\pi} \cdot (-2) = 0$$

$$\Rightarrow \quad x = \frac{2-4x}{\pi}$$

$$\Rightarrow \quad \pi x + 4x = 2$$

$$\Rightarrow \quad x = \frac{2}{\pi + 4} \quad \text{...(ii)}$$

Now, from Eq. (i), we get

$$r = \frac{1 - 2 \cdot \frac{2}{\pi + 4}}{\pi}$$

$$= \frac{\pi + 4 - 4}{\pi(\pi + 4)} = \frac{1}{\pi + 4} \quad \text{...(iii)}$$

From Eqs. (ii) and (iii),

we get $x = 2r$

**14.** *(b)* Let $I = \int \frac{2x^{12} + 5x^9}{(x^5 + x^3 + 1)^3}\, dx$

$$= \int \frac{2x^{12} + 5x^9}{x^{15}(1 + x^{-2} + x^{-5})^3}\, dx$$

$$= \int \frac{2x^{-3} + 5x^{-6}}{(1 + x^{-2} + x^{-5})^3}\, dx$$

Now, put $1 + x^{-2} + x^{-5} = t$

$$\Rightarrow \quad (-2x^{-3} - 5x^{-6})\, dx = dt$$

$$\Rightarrow \quad (2x^{-3} + 5x^{-6})\, dx = -dt$$

$$\therefore \quad I = -\int \frac{dt}{t^3} = -\int t^{-3}\, dt$$

$$= -\frac{t^{-3+1}}{-3+1} + C = \frac{1}{2t^2} + C$$

$$= \frac{x^{10}}{2(x^5 + x^3 + 1)^2} + C$$

**15.** *(b)* Let

$$l = \lim_{n \to \infty} \left[\frac{(n+1)\cdot(n+2)\dots(3n)}{n^{2n}}\right]^{\frac{1}{n}}$$

$$= \lim_{n \to \infty} \left[\frac{(n+1)\cdot(n+2)\dots(n+2n)}{n^{2n}}\right]^{\frac{1}{n}}$$

$$= \lim_{n \to \infty} \left[\left(\frac{n+1}{n}\right)\left(\frac{n+2}{n}\right)\dots\left(\frac{n+2n}{n}\right)\right]^{\frac{1}{n}}$$

On taking log both sides, we get

$$\log l = \lim_{n \to \infty} \frac{1}{n} \left[\log\left\{\left(1 + \frac{1}{n}\right)\left(1 + \frac{2}{n}\right)\dots\left(1 + \frac{2n}{n}\right)\right\}\right]$$

$$\Rightarrow \quad \log l = \lim_{n \to \infty} \frac{1}{n} \left[\log\left(1 + \frac{1}{n}\right) + \log\left(1 + \frac{2}{n}\right) + \dots + \log\left(1 + \frac{2n}{n}\right)\right]$$

$$\Rightarrow \quad \log l = \lim_{n \to \infty} \frac{1}{n} \sum_{r=1}^{2n} \log\left(1 + \frac{r}{n}\right)$$

$$\Rightarrow \quad \log l = \int_0^2 \log(1 + x)\, dx$$

$$\Rightarrow \quad \log l = \left[\log(1+x)\cdot x - \int \frac{1}{1+x} \cdot x\, dx\right]_0^2$$

$$\Rightarrow \quad \log l = [\log(1+x)\cdot x]_0^2 - \int_0^2 \frac{x + 1 - 1}{1 + x}\, dx$$

$$\Rightarrow \quad \log l = 2 \cdot \log 3 - \int_0^2 \left(1 - \frac{1}{1+x}\right) dx$$

$$\Rightarrow \quad \log l = 2 \cdot \log 3 - [x - \log|1 + x|]_0^2$$

$$\Rightarrow \quad \log l = 2 \cdot \log 3 - [2 - \log 3]$$

$$\Rightarrow \quad \log l = 3 \cdot \log 3 - 2$$

$$\Rightarrow \quad \log l = \log 27 - 2$$

$$\therefore \quad l = e^{\log 27 - 2} = 27 \cdot e^{-2}$$

$$= \frac{27}{e^2}$$

**16.** *(b)* Given equations of curves are

$$y^2 = 2x \quad \text{...(i)}$$

which is a parabola with vertex (0, 0) and axis parallel to $X$-axis.

and $\quad x^2 + y^2 = 4x$

which is a circle with centre (2, 0) and radius = 2 ...(ii)

On substituting $y^2 = 2x$ in Eq. (ii), we get

$$x^2 + 2x = 4x$$

$\Rightarrow \quad x^2 = 2x$

$\Rightarrow \quad x = 0$

or $\quad x = 2$

$\Rightarrow \quad y = 0$ or $y = \pm 2$ [using Eq. (i)]

Now, the required area is the area of shaded region, i.e.

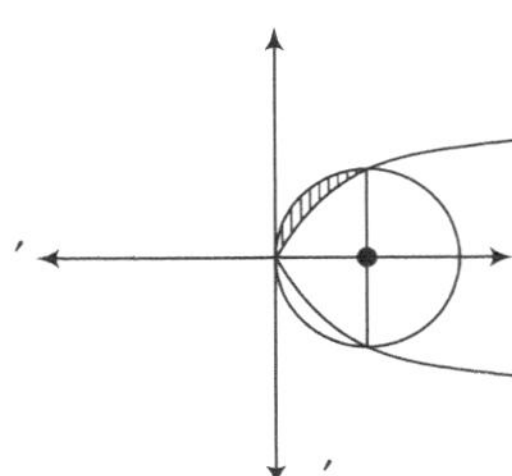

$$\text{Required area} = \frac{\text{Area of a circle}}{4} - \int_0^2 \sqrt{2x}\, dx$$

$$= \frac{\pi(2)^2}{4} - \sqrt{2}\int_0^2 x^{1/2} dx$$

$$= \pi - \sqrt{2}\left[\frac{x^{3/2}}{3/2}\right]_0^2$$

$$= \pi - \frac{2\sqrt{2}}{3}[2\sqrt{2} - 0]$$

$$= \left(\pi - \frac{8}{3}\right) \text{ sq unit}$$

17\. (d) Given differential equation is

$$y(1 + xy)\, dx = x\, dy$$

$$\Rightarrow \quad y\, dx + xy^2\, dx = x\, dy$$

$$\Rightarrow \quad \frac{x\, dy - y\, dx}{y^2} = x\, dx$$

$$\Rightarrow \quad -\frac{(y\, dx - x\, dy)}{y^2} = x\, dx$$

$$\Rightarrow \quad -d\left(\frac{x}{y}\right) = x\, dx$$

On integrating both sides, we get

$$-\frac{x}{y} = \frac{x^2}{2} + C \quad \ldots(\text{i})$$

$\because$ It passes through $(1, -1)$.

$$\therefore \quad 1 = \frac{1}{2} + C$$

$$\Rightarrow \quad C = \frac{1}{2}$$

Now, from Eq. (i)

$$-\frac{x}{y} = \frac{x^2}{2} + \frac{1}{2}$$

$$\Rightarrow \quad x^2 + 1 = -\frac{2x}{y}$$

$$\Rightarrow \quad y = -\frac{2x}{x^2 + 1}$$

$$\therefore \quad f\left(-\frac{1}{2}\right) = \frac{4}{5}$$

18\. (c) As the given lines $x - y + 1 = 0$ and $7x - y - 5 = 0$ are not parallel, therefore they represent the adjacent sides of the rhombus.

On solving $x - y + 1 = 0$ and $7x - y - 5 = 0$, we get $x = 1$ and $y = 2$. Thus, one of the vertex is $A(1, 2)$.

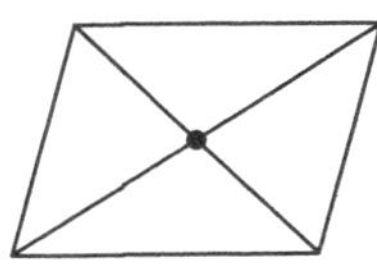

Let the coordinates of point $C$ be $(x, y)$.

Then, $\quad -1 = \frac{x + 1}{2}$ and $-2 = \frac{y + 2}{2}$

$\Rightarrow \quad x + 1 = -2$ and $y = -4 - 2$

$\Rightarrow \quad x = -3$ and $y = -6$

Hence, coordinates of $C = (-3, -6)$

Note that, vertices $B$ and $D$ will satisfy $x - y + 1 = 0$ and $7x - y - 5 = 0$, respectively.

Since, option (c) satisfies $7x - y - 5 = 0$, therefore coordinates of vertex $D$ is $\left(\frac{1}{3}, \frac{-8}{3}\right)$.

19\. (d) Given equation of circle is $x^2 + y^2 - 8x - 8y - 4 = 0$, whose centre is $C(4, 4)$ and radius

$$= \sqrt{4^2 + 4^2 + 4}$$

$$= \sqrt{36} = 6$$

Let the centre of required circle be $C_1(x, y)$. Now, as it touch the $X$-axis, therefore its radius $= |y|$.

Also, it touch the circle

$x^2 + y^2 - 8x - 8y - 4 = 0,$

therefore $CC_1 = 6 + |y|$

$\Rightarrow \sqrt{(x-4)^2 + (y-4)^2} = 6 + |y|$

$\Rightarrow \quad x^2 + 16 - 8x + y^2 + 16 - 8y$

$= 36 + y^2 + 12|y|$

$\Rightarrow \quad x^2 - 8x - 8y + 32 = 36 + 12|y|$

$\Rightarrow \quad x^2 - 8x - 8y - 4 = 12|y|$

***Case* I** If $y > 0$, then we have

$x^2 - 8x - 8y - 4 = 12y$

$\Rightarrow \quad x^2 - 8x - 20y - 4 = 0$

$\Rightarrow \quad x^2 - 8x - 4 = 20y$

$\Rightarrow \quad (x-4)^2 - 20 = 20y$

$\Rightarrow \quad (x-4)^2 = 20(y+1)$

which is a parabola.

***Case* II** If $y < 0$, then we have

$x^2 - 8x - 8y - 4 = -12y$

$\Rightarrow x^2 - 8x - 8y - 4 + 12y = 0$

$\Rightarrow \quad x^2 - 8x + 4y - 4 = 0$

$\Rightarrow \quad x^2 - 8x - 4 = -4y$

$\Rightarrow \quad (x-4)^2 = 20 - 4y$

$\Rightarrow (x-4)^2 = -4(y-5)$, which is again a parabola.

**20.** *(b)* Given equation of a circle is $x^2 + y^2 - 4x + 6y - 12 = 0$, whose centre is $(2, -3)$ and radius

$= \sqrt{2^2 + (-3)^2 + 12} = \sqrt{4 + 9 + 12} = 5$

Now, according to given information, we have the following figure.

$x^2 + y^2 - 4x + 6y - 12 = 0$

Clearly, $AO \perp BC$, as $O$ is mid-point of the chord.

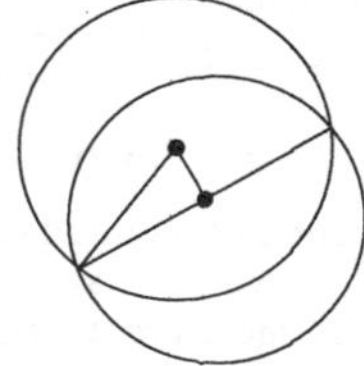

Now, in $\Delta AOB$, we have

$OA = \sqrt{(-3-2)^2 + (2+3)^2}$

$= \sqrt{25 + 25} = \sqrt{50} = 5\sqrt{2}$

and $\quad OB = 5$

$\therefore \quad AB = \sqrt{OA^2 + OB^2}$

$= \sqrt{50 + 25}$

$= \sqrt{75} = 5\sqrt{3}$

**21.** *(a)* Centre of circle $x^2 + (y+6)^2 = 1$ is $C(0, -6)$.

Let the coordinates of point $P$ be $(2t^2, 4t)$.

Now, let $D = CP$

$= \sqrt{(2t^2)^2 + (4t + 6)^2}$

$\Rightarrow \quad D = \sqrt{4t^4 + 16t^2 + 36 + 48t}$

Squaring on both sides

$\Rightarrow \quad D^2(t) = 4t^4 + 16t^2 + 48t + 36$

Let $\quad F(t) = 4t^4 + 16t^2 + 48t + 36$

For minimum, $F'(t) = 0$

$\Rightarrow \quad 16t^3 + 32t + 48 = 0$

$\Rightarrow \quad t^3 + 2t + 3 = 0$

$\Rightarrow (t+1)(t^2 - t + 3) = 0$

$\Rightarrow \quad t = -1$

Thus, coordinate of point $P$ are $(2, -4)$.

Now ,

$CP = \sqrt{2^2 + (-4+6)^2}$

$= \sqrt{4 + 4}$

$= 2\sqrt{2}$

Hence, the required equation of circle is

$(x-2)^2 + (y+4)^2 = (2\sqrt{2})^2$

$\Rightarrow x^2 + 4 - 4x + y^2 + 16 + 8y = 8$

$\Rightarrow \quad x^2 + y^2 - 4x + 8y + 12 = 0$

**22.** *(c)* We have, $\dfrac{2b^2}{a} = 8$ and $2b = ae$

$\Rightarrow \quad b^2 = 4a$ and $2b = ae$

Consider $\quad 2b = ae$

$\Rightarrow \quad 4b^2 = a^2e^2$

$\Rightarrow \quad 4a^2(e^2 - 1) = a^2e^2$

$\Rightarrow \quad 4e^2 - 4 = e^2 \quad [\because a \neq 0]$

$\Rightarrow \quad 3e^2 = 4$

$\Rightarrow \quad e = \frac{2}{\sqrt{3}} \quad [\because e > 0]$

23. *(b)* Equation of line passing through the point (1, – 5, 9) and parallel to $x = y = z$ is

$$\frac{x-1}{1} = \frac{y+5}{1} = \frac{z-9}{1} = \lambda \text{ (say)}$$

Thus, any point on this line is of the form $(\lambda + 1, \lambda - 5, \lambda + 9)$.

Now, if $P(\lambda + 1, \lambda - 5, \lambda + 9)$ is the point of intersection of line and plane, then

$(\lambda + 1) - (\lambda - 5) + \lambda + 9 = 5$

$\Rightarrow \quad \lambda + 15 = 5$

$\Rightarrow \quad \lambda = -10$

$\therefore$ Coordinates of point $P$ are $(-9, -15, -1)$.

Hence, required distance

$$= \sqrt{(1+9)^2 + (-5+15)^2 + (9+1)^2}$$

$$= \sqrt{10^2 + 10^2 + 10^2} = 10\sqrt{3}$$

24. *(d)* Since, the line $\frac{x-3}{2} = \frac{y+2}{-1} = \frac{z+4}{3}$ lies in the plane $lx + my - z = 9$, therefore we have $2l - m - 3 = 0$

[$\because$ normal will be perpendicular to the line]

$\Rightarrow \quad 2l - m = 3 \quad$ ...(i)

and $\quad 3l - 2m + 4 = 9$

[$\because$ point (3, – 2, – 4) lies on the plane]

$\Rightarrow \quad 3l - 2m = 5 \quad$ ...(ii)

On solving Eqs. (i) and (ii), we get

$l = 1$ and $m = -1$

$\therefore \quad l^2 + m^2 = 2$

25. *(d)* Given, $|\hat{\mathbf{a}}| = |\hat{\mathbf{b}}| = |\hat{\mathbf{c}}| = 1$

and $\quad \hat{\mathbf{a}} \times (\hat{\mathbf{b}} \times \hat{\mathbf{c}}) = \frac{\sqrt{3}}{2}(\hat{\mathbf{b}} + \hat{\mathbf{c}})$

Now, consider $\hat{\mathbf{a}} \times (\hat{\mathbf{b}} \times \hat{\mathbf{c}}) = \frac{\sqrt{3}}{2}(\hat{\mathbf{b}} + \hat{\mathbf{c}})$

$\Rightarrow (\hat{\mathbf{a}} \cdot \hat{\mathbf{c}})\,\hat{\mathbf{b}} - (\hat{\mathbf{a}} \cdot \hat{\mathbf{b}})\,\hat{\mathbf{c}} = \frac{\sqrt{3}}{2}\hat{\mathbf{b}} + \frac{\sqrt{3}}{2}\hat{\mathbf{c}}$

On comparing, we get

$$\hat{\mathbf{a}} \cdot \hat{\mathbf{b}} = -\frac{\sqrt{3}}{2}$$

$\Rightarrow |\hat{\mathbf{a}}|\,|\hat{\mathbf{b}}| \cos\theta = -\frac{\sqrt{3}}{2}$

$\Rightarrow \quad \cos\theta = -\frac{\sqrt{3}}{2} \quad [\because |\hat{\mathbf{a}}| = |\hat{\mathbf{b}}| = 1]$

$\Rightarrow \quad \cos\theta = \cos\left(\pi - \frac{\pi}{6}\right)$

$\Rightarrow \quad \theta = \frac{5\pi}{6}$

26. *(b)* We know that, if $x_1, x_2, \ldots, x_n$ are $n$ observations, then their standard deviation is given by

$$\sqrt{\frac{1}{n}\Sigma x_i^2 - \left(\frac{\Sigma x_i}{n}\right)^2}$$

We have, $(3.5)^2 = \frac{(2^2 + 3^2 + a^2 + 11^2)}{4} - \left(\frac{2+3+a+11}{4}\right)^2$

$\Rightarrow \quad \frac{49}{4} = \frac{4 + 9 + a^2 + 121}{4} - \left(\frac{16+a}{4}\right)^2$

$\Rightarrow \quad \frac{49}{4} = \frac{134 + a^2}{4} - \frac{256 + a^2 + 32a}{16}$

$\Rightarrow \quad \frac{49}{4} = \frac{4a^2 + 536 - 256 - a^2 - 32a}{16}$

$\Rightarrow \quad 49 \times 4 = 3a^2 - 32a + 280$

$\Rightarrow \quad 3a^2 - 32a + 84 = 0$

27. *(d)* Clearly,

$E_1 = \{(4,1),(4,2),(4,3),(4,4),(4,5),(4,6)\}$

$E_2 = \{(1,2),(2,2),(3,2),(4,2),(5,2),(6,2)\}$

and $E_3 = \{(1,2),(1,4),(1,6),$
$(2,1),(2,3),(2,5),(3,2),(3,4),(3,6),$
$(4,1),(4,3),(4,5),(5,2),(5,4),(5,6),$
$(6,1),(6,3),(6,5)\}$

$\Rightarrow \quad P(E_1) = \frac{6}{36} = \frac{1}{6}$

$P(E_2) = \frac{6}{36} = \frac{1}{6}$

and $\quad P(E_3) = \frac{18}{36} = \frac{1}{2}$

Now, $\quad P(E_1 \cap E_2) = P$
(getting 4 on die $A$ and 2 on die $B$)

$$= \frac{1}{36} = P(E_1) \cdot P(E_2)$$

$P(E_2 \cap E_3) = P$ (getting 2 on die $B$ and sum of numbers on both dice is odd)

$$= \frac{3}{36} = P(E_2) \cdot P(E_3)$$

$P(E_1 \cap E_3) = P$ (getting 4 on die $A$ and sum of numbers on both dice is odd)

$$= \frac{3}{36} = P(E_1) \cdot P(E_3)$$

and $P(E_1 \cap E_2 \cap E_3) = P$

[getting 4 on die $A$, 2 on die $B$ and sum of numbers is odd]

$= P$ (impossible event) $= 0$

Hence, $E_1$, $E_2$ and $E_3$ are not independent.

**28.** *(c)* Given equation is

$$\cos x + \cos 2x + \cos 3x + \cos 4x = 0$$

$$\Rightarrow (\cos x + \cos 3x) + (\cos 2x + \cos 4x) = 0$$

$$\Rightarrow \quad 2 \cos 2x \cos x + 2 \cos 3x \cos x = 0$$

$$\Rightarrow \quad 2 \cos x (\cos 2x + \cos 3x) = 0$$

$$\Rightarrow \quad 2 \cos x \left(2 \cos \frac{5x}{2} \cos \frac{x}{2}\right) = 0$$

$$\Rightarrow \quad \cos x \cdot \cos \frac{5x}{2} \cdot \cos \frac{x}{2} = 0$$

$$\Rightarrow \quad \cos x = 0 \text{ or } \cos \frac{5x}{2} = 0 \text{ or } \cos \frac{x}{2} = 0$$

Now, $\cos x = 0$

$$\Rightarrow \quad x = \frac{\pi}{2}, \frac{3\pi}{2} \quad [\because 0 \le x < 2\pi]$$

$$\cos \frac{5x}{2} = 0$$

$$\Rightarrow \quad \frac{5x}{2} = \frac{\pi}{2}, \frac{3\pi}{2}, \frac{5\pi}{2}, \frac{7\pi}{2}, \frac{9\pi}{2}, \frac{11\pi}{2}, \dots,$$

$$\Rightarrow \quad x = \frac{\pi}{5}, \frac{3\pi}{5}, \pi, \frac{7\pi}{5}, \frac{9\pi}{5}$$

$$[\because 0 \le x < 2\pi]$$

and $\cos \frac{x}{2} = 0$

$$\Rightarrow \quad \frac{x}{2} = \frac{\pi}{2}, \frac{3\pi}{2}, \frac{5\pi}{2}, \dots$$

$$\Rightarrow \quad x = \pi \quad [\because 0 \le x < 2\pi]$$

Hence, $x = \frac{\pi}{2}, \frac{3\pi}{2}, \pi, \frac{\pi}{5}, \frac{3\pi}{5}, \frac{7\pi}{5}, \frac{9\pi}{5}$

**29.** *(d)* According to given information, we have the following figure

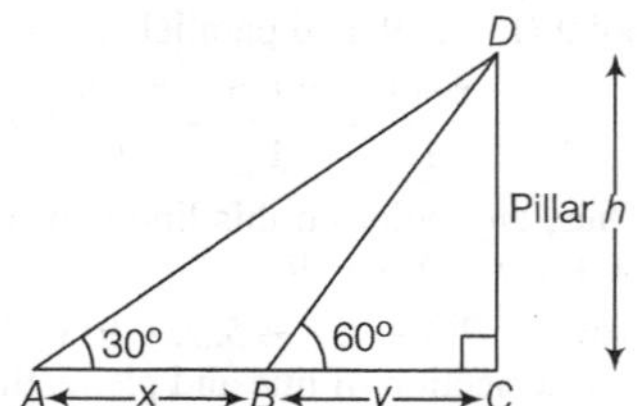


Now, from $\Delta ACD$ and $\Delta BCD$, we have

$$\tan 30^\circ = \frac{h}{x + y}$$

and $$\tan 60^\circ = \frac{h}{y}$$

$$\Rightarrow \quad h = \frac{x + y}{\sqrt{3}} \quad \text{...(i)}$$

and $$h = \sqrt{3}\, y \quad \text{...(ii)}$$

From Eqs. (i) and (ii),

$$\frac{x + y}{\sqrt{3}} = \sqrt{3}\, y$$

$$\Rightarrow \quad x + y = 3y$$

$$\Rightarrow \quad x - 2y = 0 \Rightarrow y = \frac{x}{2}$$

$\because$ Speed is uniform.

$\therefore$ Distance $y$ will be cover in 5 min.

$\because$ Distance $x$ covered in 10 min.

$\therefore$ Distance $\frac{x}{2}$ will be cover in 5 min.

**30.** *(c)* Consider, $(p \wedge \sim q) \vee q \vee (\sim p \wedge q)$

$$\equiv [(p \wedge \sim q) \vee q] \vee (\sim p \wedge q)$$
$$\equiv [(p \vee q) \wedge (\sim q \vee q)] \vee (\sim p \wedge q)$$
$$\equiv [(p \vee q) \wedge t] \vee (\sim p \wedge q)$$
$$\equiv (p \vee q) \vee (\sim p \wedge q)$$
$$\equiv (p \vee q \vee \sim p) \wedge (p \vee q \vee q)$$
$$\equiv (q \vee t) \wedge (p \vee q)$$
$$\equiv t \wedge (p \vee q) \equiv p \vee q$$

# Solved Paper 2016

# JEE Advanced

## Paper 1

## Physics

### Section 1 (Maximum Marks : 15)

- This section contains **FIVE** questions.
- Each question has **FOUR** options (a), (b), (c) and (d). **ONLY ONE** of these four options is correct.
- Four each question, darken the bubble corresponding to the correct option in the ORS.
- For each question, marks will be awarded in one of the following categories.

| | |
|---|---|
| Full Marks | : + 3 If only the bubble corresponding to the correct option is darkened. |
| Zero Marks | : 0 If none of the bubbles is darkened. |
| Negative Marks | : −1 In all other cases. |

**1.** A parallel beam of light is incident from air at an angle $\alpha$ on the side $PQ$ of a right angled triangular prism of refractive index $n=\sqrt{2}$. Light undergoes total internal reflection in the prism at the face $PR$ when $\alpha$ has a minimum value of 45°. The angle $\theta$ of the prism is

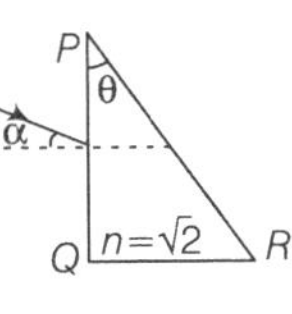


(a) 15° (b) 22.5° (c) 30° (d) 45°

**2.** In a historical experiment to determine Planck's constant, a metal surface was irradiated with light of different wavelengths. The emitted photoelectron energies were measured by applying a stopping potential. The relevant data for the wavelength ($\lambda$) of incident light and the corresponding stopping potential ($V_0$) are given below:

| $\lambda$ (μm) | $V_0$ (Volt) |
|---|---|
| 0.3 | 2.0 |
| 0.4 | 1.0 |
| 0.5 | 0.4 |

Given that $c=3\times10^8\,\text{ms}^{-1}$ and $e=1.6\times10^{-19}$ C, Planck's constant (in units of J-s) found from such an experiment is)

(a) $6.0\times10^{-34}$
(b) $6.4\times10^{-34}$
(c) $6.6\times10^{-34}$
(d) $6.8\times10^{-34}$

**3.** A water cooler of storage capacity 120 litres can cool water at a constant rate of $P$ watts. In a closed circulation system (as shown schematically in the figure), the water from the cooler is used to cool an external device that generates constantly 3 kW of heat (thermal load).

The temperature of water fed into the device cannot exceed 30°C and the entire stored 120 litres of water is initially cooled to 10°C. The entire system is thermally insulated. The minimum value of $P$ (in watts) for which the device can be operated for 3 hours is

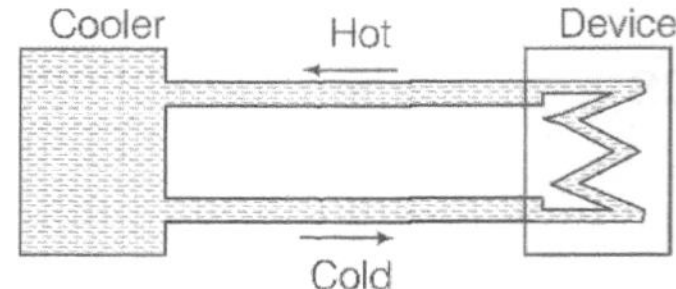


(Specific heat of water is $4.2 \text{ kJ kg}^{-1}\text{K}^{-1}$ and the density of water is $1000 \text{kg m}^{-3}$)

(a) 1600 (b) 2067
(c) 2533 (d) 3933

**4.** A uniform wooden stick of mass 1.6 kg of length $l$ rests in an inclined manner on a smooth, vertical wall of height $h (< l)$ such that a small portion of the stick extends beyond the wall. The reaction force of the wall on the stick is perpendicular to the stick. The stick makes an angle of 30° with the wall and the bottom of the stick is on a rough floor. The reaction of the wall on the stick is equal in magnitude to the reaction of the floor on the stick. The ratio $\frac{h}{l}$ and the frictional force $f$ at the bottom of the stick are ($g = 10 \text{ms}^{-2}$)

(a) $\frac{h}{l} = \frac{\sqrt{3}}{16}, f = \frac{16\sqrt{3}}{3}$ N

(b) $\frac{h}{l} = \frac{3}{16}, f = \frac{16\sqrt{3}}{3}$ N

(c) $\frac{h}{l} = \frac{3\sqrt{3}}{16}, f = \frac{8\sqrt{3}}{3}$ N

(d) $\frac{h}{l} = \frac{3\sqrt{3}}{16}, f = \frac{16\sqrt{3}}{3}$ N

**5.** An infinite line charge of uniform electric charge density $\lambda$ lies along the axis of an electrically conducting infinite cylindrical shell of radius $R$. At time $t = 0$, the space inside the cylinder is filled with a material of permittivity $\varepsilon$ and electrical conductivity $\sigma$. The electrical conduction in the material follows Ohm's law. Which one of the following graphs best describes the subsequent variation of the magnitude of current density $j$ (t) at any point in the material?

(a)
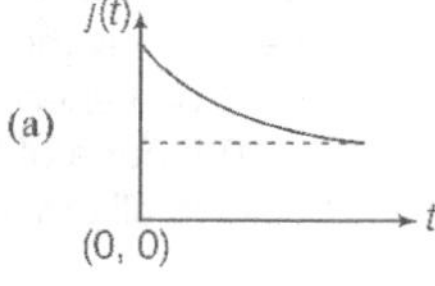


(b)
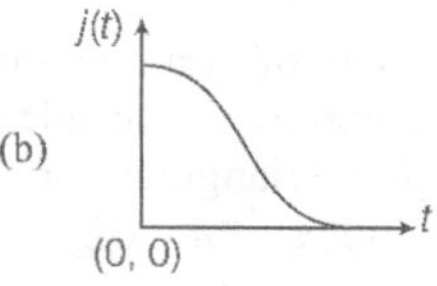


(c)
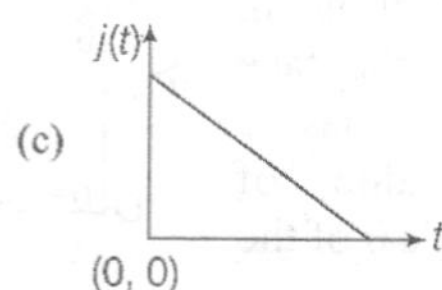


(d)
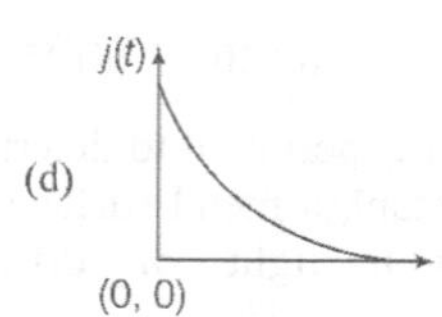

## Section 2 (Maximum Marks : 32)

- This section contains **EIGHT** questions.
- Each question has **FOUR** options (a), (b) (c) and (d). **ONE OR MORE THAN ONE** of these four option(s) is (are) correct.
- For each question, darken the bubble(s) corresponding to all the correct option(s) in the ORS.
- For each questions, marks will be awarded in one of the following categories :

| | | |
|---|---|---|
| Full Marks | : + 4 | If only the bubble(s) corresponding to all the correct option(s) is (are) darkened. |
| Partial Marks | : + 1 | For darkening a bubble corresponding to **each correct option**, provided NO incorrect option is darkened. |
| Zero Marks | : 0 | If none of the bubbles is darkened. |
| Negative Marks | : −2 | In all other cases. |

**6.** A plano-convex lens is made of material of refractive index $n$. When a small object is placed 30 cm away in front of the curved surface of the lens, an image of double the size of the object is produced. Due to reflection from the convex surface of the lens, another faint image is observed at a distance of 10 cm away from the lens. Which of the following statement(s) is (are) true?

(a) The refractive index of the lens is 2.5
(b) The radius of curvature of the convex surface is 45 cm
(c) The faint image is erect and real
(d) The focal length of the lens is 20 cm

**7.** A conducting loop in the shape of a right angled isosceles triangle of height 10 cm is kept such that the 90° vertex is very close to an infinitely long conducting wire (see the figure). The wire is electrically insulated from the loop. The hypotenuse of the triangle is parallel to the wire. The current in the triangular loop is in counterclockwise direction and increased at a constant rate of $10\,\text{As}^{-1}$. Which of the following statement(s) is (are) true?

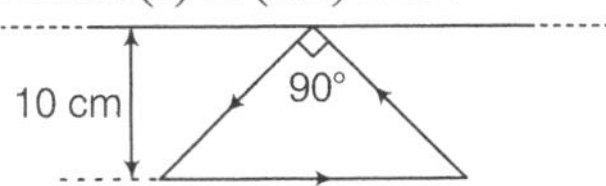


(a) There is a repulsive force between the wire and the loop
(b) If the loop is rotated at a constant angular speed about the wire, an additional emf of $\left(\frac{\mu_0}{\pi}\right)$ volt is induced in the wire
(c) The magnitude of induced emf in the wire is $\left(\frac{\mu_0}{\pi}\right)$ volt
(d) The induced current in the wire is in opposite direction to the current along the hypotenuse

**8.** The position vector $\mathbf{r}$ of particle of mass $m$ is given by the following equation $\mathbf{r}(t) = \alpha t^3 \hat{\mathbf{i}} + \beta t^2 \hat{\mathbf{j}}$

where, $\alpha = \frac{10}{3}\,\text{ms}^{-3}$, $\beta = 5\,\text{ms}^{-2}$ and $m = 0.1$ kg. At $t = 1\text{s}$, which of the following statement(s) is (are) true about the particle?

(a) The velocity $\mathbf{v}$ is given by $\mathbf{v} = (10\hat{\mathbf{i}} + 10\hat{\mathbf{j}})\,\text{ms}^{-1}$
(b) The angular momentum $\mathbf{L}$ with respect to the origin is given by $\mathbf{L} = (5/3)\hat{\mathbf{k}}$ Nms
(c) The force $\mathbf{F}$ is given by $\mathbf{F} = (\hat{\mathbf{i}} + 2\hat{\mathbf{j}})$N
(d) The torque $\tau$ with respect to the origin is given by $\tau = -\frac{20}{3}\hat{\mathbf{k}}$ Nm

**9.** A length-scale $(l)$ depends on the permittivity $(\varepsilon)$ of a dielectric material, Boltzmann constant $(k_B)$, the absolute temperature $(T)$, the number per unit volume $(n)$ of certain charged particles, and the charge ($q$) carried by each of the particles. Which of the following expression (s) for $l$ is (are) dimensionally correct?

(a) $l=\sqrt{\left(\frac{nq^2}{\varepsilon k_B T}\right)}$ (b) $l=\sqrt{\left(\frac{\varepsilon k_B T}{nq^2}\right)}$

(c) $l=\sqrt{\left(\frac{q^2}{\varepsilon n^{2/3} k_B T}\right)}$ (d) $l=\sqrt{\left(\frac{q^2}{\varepsilon n^{1/3} k_B T}\right)}$

**10.** Two loudspeakers $M$ and $N$ are located 20m apart and emit sound at frequencies 118 Hz and 121 Hz, respectively. A car in initially at a point $P$, 1800 m away from the midpoint $Q$ of the line $MN$ and moves towards $Q$ constantly at 60 km/h along the perpendicular bisector of $MN$. It crosses $Q$ and eventually reaches a point $R$, 1800 m away from $Q$.

Let $v(t)$ represent the beat frequency measured by a person sitting in the car at time $t$. Let $\nu_P$, $\nu_Q$ and $\nu_R$ be the beat frequencies measured at locations $P$, $Q$ and $R$ respectively. The speed of sound in air is $330\ \text{ms}^{-1}$. Which of the following statement(s) is (are) true regarding the sound heard by the person?

(a) The plot below represents schematically the variation of beat frequency with time

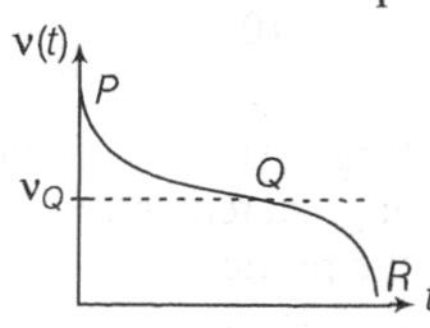


(b) The rate of change in beat frequency is maximum when the car passes through $Q$

(c) $\nu_P + \nu_R = 2\nu_Q$

(d) The plot below represents schematically the variations of beat frequency with time

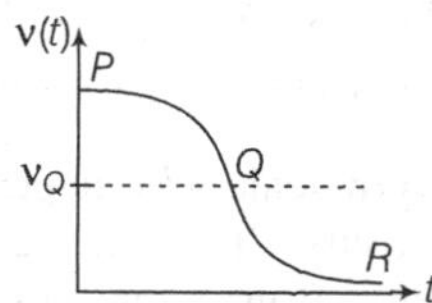


**11.** A transparent slab of thickness $d$ has a refractive index $n(z)$ that increases with $z$. Here, $z$ is the vertical distance inside the slab, measured from the top. The slab is placed between two media with uniform refractive indices $n_1$ and $n_2 (> n_1)$, as shown in the figure. A ray of light is incident with angle $\theta_i$ from medium 1 and emerges in medium 2 with refraction angle $\theta_f$ with a lateral displacement $l$.

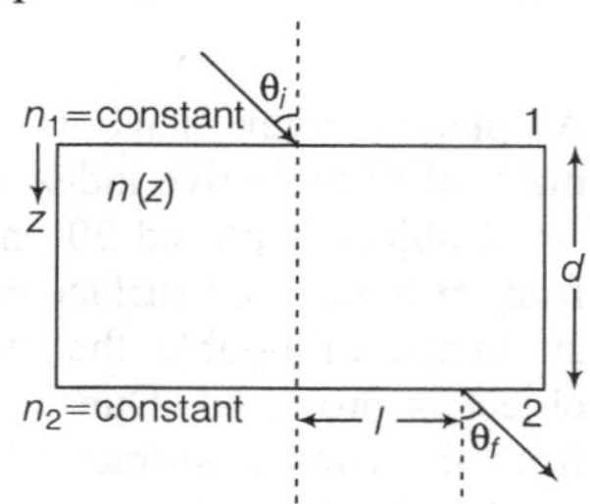


Which of the following statement(s) is (are) true?

(a) $l$ is independent on $n(z)$

(b) $n_1 \sin\theta_i = (n_2 - n_1)\sin\theta_f$

(c) $n_1 \sin\theta_i = n_2 \sin\theta_f$

(d) $l$ is independent of $n_2$

**12.** Highly excited states for hydrogen-like atoms (also called Rydberg states) with nuclear charge $Ze$ are defined by their principle quantum number $n$, where $n >> 1$. Which of the following statement(s) is (are) true?

(a) Relative change in the radii of two consecutive orbitals does not depend on $Z$

(b) Relative change in the radii of two consecutive orbitals varies as $1/n$

(c) Relative change in the energy of two consecutive orbitals varies as $1/n^3$

(d) Relative change in the angular momenta of two consecutive orbitals varies as $1/n$

**13.** An incandescent bulb has a thin filament of tungsten that is heated to high temperature by passing an electric current. The hot filament emits black-body radiation. The filament is observed to break up at random locations after a sufficiently long time of operation due to non-uniform evaporation of tungsten from the filament. If the bulb is powered at constant voltage, which of the following statement(s) is (are) true?

(a) The temperature distribution over the filament is uniform

(b) The resistance over small sections of the filament decreases with time

(c) The filament emits more light at higher band of frequencies before it breaks up

(d) The filament consumes less electrical power towards the end of the life of the bulb

## Section 3 (Maximum Marks : 15)

- This section contains **FIVE** questions.
- The answer to each question is a **SINGLE DIGIT INTEGER** ranging from 0 to 9, both inclusive.
- For each question, darken the bubble corresponding to the correct integer in the ORS.
- For each questions, marks will be awarded in one of the following categories.

Full Marks : + 3 If only the bubble corresponding to all the correct answer is darkened.

Zero Marks : 0 In all other cases.

**14.** A hydrogen atom in its ground state is irradiated by light of wavelength 970Å.

Taking $hc/e = 1.237 \times 10^{-6}$ eVm and the ground state energy of hydrogen atom as – 13.6 eV, the number of lines present in the emission spectrum is

**15.** The isotopes ${}^{12}_{5}B$ having a mass 12.014 u undergoes β-decay to ${}^{12}_{6}C$. ${}^{12}_{6}C$ has an excited state of the nucleus (${}^{12}_{6}C^*$) at 4.041 MeV above its ground state. If ${}^{12}_{5}B$ decays to ${}^{12}_{5}C^*$, the maximum kinetic energy of the β-particle in units of MeV is (1u = 931.5 MeV/$c^2$, where $c$ is the speed of light in vacuum)

**16.** Consider two solid spheres $P$ and $Q$ each of density 8 gm cm$^{-3}$ and diameters 1 cm and 0.5 cm, respectively. Sphere $P$ is dropped into a liquid of density 0.8 gm cm$^{-3}$ and viscosity $\eta = 3$ poiseulles. Sphere $Q$ is dropped into a liquid of density 1.6 gm cm$^{-3}$ and viscosity $\eta = 2$ poiseulles. The ratio of the terminal velocities of $P$ and $Q$ is

**17.** Two inductors $L_1$ (inductance 1mH, internal resistance 3 Ω) and $L_2$ (inductance 2 mH, internal resistance 4 Ω), and a resistor $R$ (resistance 12 Ω) are all connected in parallel across a 5V battery. The circuit is switched on at time $t = 0$. The ratio of the maximum to the minimum current $(I_{max}/I_{min})$ drawn from the battery is

**18.** A metal is heated in a furnace where a sensor is kept above the metal surface to read the power radiated $(P)$ by the metal. The sensor has a scale that displays $\log_2(P/P_0)$, where $P_0$ is a constant. When the metal surface is at a temperature of 487°C, the sensor shows a value 1. Assume that the emissivity of the metallic surface remains constant. What is the value displayed by the sensor when the temperature of the metal surface is raised to 2767°C?

# Chemistry

## Section 1 (Maximum Marks : 15)

- This section contains **FIVE** questions.
- Each question has **FOUR** options (a), (b), (c) and (d). Only one of these four options is correct.
- For each question, darken the bubble corresponding to the correct option in the ORS.
- For each question, marks will be awarded in one of the following categories.
  Full Marks : + 3, If only the bubble corresponding to the correct option is darkened.
  Zero Marks : 0, If none of the bubbles is darkened.
  Negative Marks : – 1, In all other cases.

**19.** *P* is the probability of finding the 1*s* electron of hydrogen atom in a spherical shell of infinitesimal thickness, *dr*, at a distance *r* from the nucleus. The volume of this shell is $4\pi r^2 dr$. The qualitative sketch of the dependence of *P* on *r* is

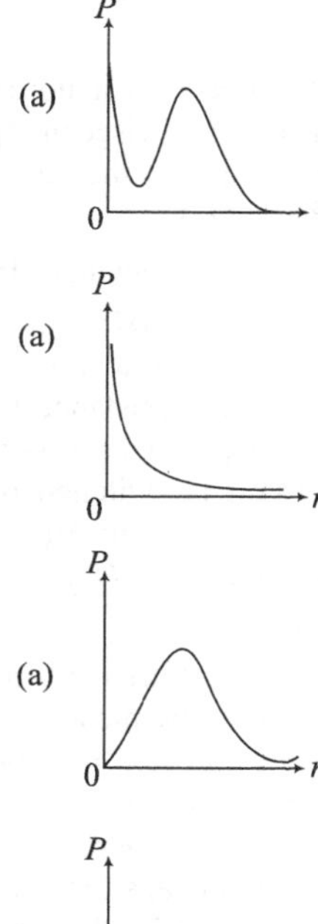


**20.** One mole of an ideal gas at 300 K in thermal contact with surroundings expands isothermally from 1.0 L to 2.0 L against a constant pressure of 3.0 atm. In this process, the change in entropy of surroundings ($\Delta S_{surr}$) in J $K^{-1}$ is (1 L atm = 101.3 J)

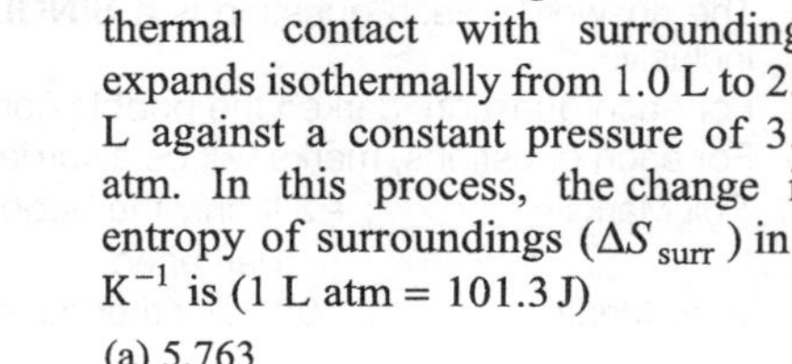

(a) 5.763
(b) 1.013
(c) – 1.013
(d) – 5.763

**21.** Among $[Ni(CO)_4]$, $[NiCl_4]^{2-}$, $[Co(NH_3)_4Cl_2]Cl$, $Na_3[CoF_6]$, $Na_2O_2$ and $CsO_2$, the total number of paramagnetic compounds is

(a) 2
(b) 3
(c) 4
(d) 5

**22.** The increasing order of atomic radii of the following Group 13 elements is

(a) Al < Ga < In < Tl
(b) Ga < Al < In < Tl
(c) Al < In < Ga < Tl
(d) Al < Ga < Tl < In

**23.** On complete hydrogenation, natural rubber produces

(a) ethylene-propylene copolymer
(b) vulcanised rubber
(c) polypropylene
(d) polybutylene

## Section 2 (Maximum Marks : 32)

- This section contains **EIGHT** questions.
- Each question has **FOUR** options (a), (b), (c) and (d). **ONE OR MORE THAN ONE** of these four options is (are) correct.
- For each question, darken the bubble(s) corresponding to all the correct option(s) in the ORS.
- For each question, marks will be awarded in **one of the following categories**:

| | | |
|---|---|---|
| Full Marks | : + 4, | If only the bubble(s) corresponding to all the correct option(s) is (are) darkened. |
| Partial Marks | : + 1, | For darkening a bubble corresponding **to each correct option** provided **NO** incorrect option is darkened. |
| Zero Marks | : 0, | If none of the bubbles is darkened. |
| Negative Marks | : − 2, | In all other cases. |

For example, if (a), (c) and (d) are all the correct options for a question, darkening all these three will result in + 4 marks; darkening only (a) and (d) will result in + 2 marks; and darkening (a) and (b) will result in − 2 marks, as a wrong option is also darkened.

**24.** The product(s) of the following reaction sequence is (are)

$NH_2$ (aniline) $\xrightarrow[\text{(iii) } H_3O^+\text{, Heat; (iv) } NaNO_2/HCl\text{, 273-278 K; (v) Cu/HBr}]{\text{(i) Acetic anhydride/Pyridine; (ii) } KBrO_3/Br}$

(a) 1,3-dibromobenzene (Br, Br)

(b) 1,4-dibromobenzene (Br, Br)

(c) 1,2,3-tribromobenzene (Br, Br, Br)

(d) 1,2,3,5-tetrabromobenzene (Br, Br, Br, Br)

**25.** The correct statement(s) about the following reaction sequence is (are)

Cumene $(C_9H_{12}) \xrightarrow[\text{(ii) } H_3O^+]{\text{(i) } O_2} P \xrightarrow{CHCl_3/NaOH} Q\text{(major)} + R\text{(minor)},$

$Q \xrightarrow[PhCH_2Br]{NaOH} S$

(a) $R$ is steam volatile

(b) $Q$ gives dark violet colouration with 1% aqueous $FeCl_3$ solution

(c) $S$ gives yellow precipitate with 2, 4-dinitrophenylhydrazine

(d) $S$ gives dark violet colouration with 1% aqueous $FeCl_3$ solution

**26.** The crystalline form of borax has

(a) tetranuclear $[B_4O_5(OH)_4]^{2-}$ unit

(b) all boron atoms in the same plane

(c) equal number of $sp^2$ and $sp^3$ hybridised boron atoms

(d) one terminal hydroxide per boron atom

**27.** The reagent(s) that can selectively precipitate $S^{2-}$ from a mixture of $S^{2-}$ and $SO_4^{2-}$ in aqueous solution is (are)

(a) $CuCl_2$ (b) $BaCl_2$
(c) $Pb(OOCCH_3)_2$ (d) $Na_2[Fe(CN)_5NO]$

**28.** A plot of the number of neutrons ($n$) against the number of protons ($p$) of stable nuclei exhibits upward deviation from linearity for atomic number, $Z > 20$. For an unstable nucleus having $n/p$ ratio less than 1, the possible mode(s) of decay is (are)

(a) $\beta^-$- decay ($\beta$- emission)
(b) orbital or $K$-electron capture
(c) neutron emission
(d) $\beta^+$-decay (positron emission)

**29.** Positive Tollen's test is observed for

(a) H, H, O, H, Br (b) CHO

(c) OH, Ph, Ph, O (d) O, Ph, Ph

**30.** The compound(s) with two lone pairs of electrons on the central atom is (are)

(a) $BrF_5$
(b) $ClF_3$
(c) $XeF_4$
(d) $SF_4$

**31.** According to the Arrhenius equation,

(a) a high activation energy usually implies a fast reaction
(b) rate constant increases with increase in temperature. This is due to a greater number of collisions whose energy exceeds the activation energy
(c) higher the magnitude of activation energy, stronger is the temperature dependence of the rate constant
(d) the pre-exponential factor is a measure of the rate at which collisions occur, irrespective of their energy

## Section 3 (Maximum Marks : 15)

- This section contains **FIVE** questions.
- The answer to each question is a **SINGLE DIGIT INTEGER** ranging from 0 to 9, both inclusive.
- For each question, darken the bubble corresponding to the correct integer in the ORS.
- For each question, marks will be awarded in **one of the following categories**:

Full Marks : +3 If only the bubble corresponding to the correct answer is darkened.

Zero marks : 0 In all other cases.

**32.** In the following monobromination reaction, the number of possible chiral product(s) is (are)...

$CH_2CH_2CH_3$
H—|—Br $\xrightarrow[300°C]{Br_2\,(1.0\ mole)}$
$CH_3$
(1.0 mole) (Enantiomerically pure)

**33.** The mole fraction of a solute in a solution is 0.1. At 298 K, molarity of this solution is the same as its molality. Density of this solution at 298 K is 2.0 g $cm^{-3}$. The ratio of the molecular weights of the solute and solvent, $\left(\frac{m_{solute}}{m_{solvent}}\right)$ is ... .

**34.** The possible number of geometrical isomers for the complex $[CoL_2Cl_2]^-$ ($L = H_2NCH_2CH_2O^-$) is (are) ...

**35.** In neutral or faintly alkaline solution, 8 moles of permanganate anion quantitative oxidise thiosulphate anions to produce $X$ moles of a sulphur containing product. The magnitude of $X$ is ... .

**36.** The diffusion coefficient of an ideal gas is proportional to its mean free path and mean speed. The absolute temperature of an ideal gas is increased 4 times and its pressure is increased 2 times. As a result, the diffusion coefficient of this gas increases $x$ times. The value of $x$ is ...

# Mathematics

## Section 1 (Maximum Marks : 15)

- This section contains **FIVE** questions.
- Each question has **FOUR** options (a), (b), (c) and (d). **ONLY ONE** of these four options is correct.
- For each question, darken the bubble corresponding to the correct option in the ORS.
- For each question, marks will be awarded in one of the following categories:

| | | |
|---|---|---|
| Full Marks | : +3 | If only the bubble corresponding to the correct option is darkened. |
| Zero Marks | : 0 | If none of the bubbles is darkened. |
| Negative Marks | : −1 | In all other cases. |

**37.** A debate club consists of 6 girls and 4 boys. A team of 4 members is to be selected from this club including the selection of a captain (from among these 4 members) for the team. If the team has to include atmost one boy, the number of ways of selecting the team is

(a) 380 (b) 320
(c) 260 (d) 95

**38.** The least value of $\alpha \in R$ for which $4\alpha x^2 + \frac{1}{x} \geq 1$, for all $x > 0$, is

(a) $\frac{1}{64}$ (b) $\frac{1}{32}$
(c) $\frac{1}{27}$ (d) $\frac{1}{25}$

**39.** Let $-\frac{\pi}{6} < \theta < -\frac{\pi}{12}$. Suppose $\alpha_1$ and $\beta_1$ are the roots of the equation $x^2 - 2x\sec\theta + 1 = 0$, and $\alpha_2$ and $\beta_2$ are the roots of the equation $x^2 + 2x\tan\theta - 1 = 0$. If $\alpha_1 > \beta_1$ and $\alpha_2 > \beta_2$, then $\alpha_1 + \beta_2$ equals

(a) $2(\sec\theta - \tan\theta)$
(b) $2\sec\theta$
(c) $-2\tan\theta$
(d) 0

**40.** Let $S = \left\{x \in (-\pi, \pi) : x \neq 0, \pm\frac{\pi}{2}\right\}$. The sum of all distinct solutions of the equation $\sqrt{3}\sec x + \operatorname{cosec} x + 2(\tan x - \cot x) = 0$ in the set $S$ is equal to

(a) $-\frac{7\pi}{9}$ (b) $-\frac{2\pi}{9}$
(c) 0 (d) $\frac{5\pi}{9}$

**41.** A computer producing factory has only two plants $T_1$ and $T_2$. Plant $T_1$ produces 20% and plant $T_2$ produces 80% of the total computers produced. 7% of computers produced in the factory turn out to be defective. It is knownthat $P$ (computer turns out to be defective, given that it is produced in plant $T_1$) $= 10P$ (computer turns out to be defective, given that it is produced in plant $T_2$), where $P(E)$ denotes the probability of an event $E$. A computer produced in the factory is randomly selected and it does not turn out to be defective. Then, the probability that it is produced in plant $T_2$, is

(a) $\frac{36}{73}$ (b) $\frac{47}{79}$ (c) $\frac{78}{93}$ (d) $\frac{75}{83}$

## Section 2 (Maximum Marks : 32)

- This section contains **EIGHT** questions.
- Each question has **FOUR** options (a), (b), (c) and (d). **ONE OR MORE THAN ONE** of these four option (s) is (are) correct.
- For each question, darken the bubble (s) corresponding to all the correct option(s) in the ORS.
- For each question, marks will be awarded in one of the following categories:

Full Marks : +4 If only the bubble (s) corresponding to all correct option(s) is (are) darkened.

Partial Marks : +1 For darkening a bubble corresponding **to each correct option**, provided no incorrect option is darkened.

Zero Marks : 0 If none of the dubbles is darkened

Negative Marks : −2 In all other cases.For example, if (a), (c) and (d) are all the correct options for a question, darkening all these three will result in +4 marks; darkening only (a) and (d) will result in +2 marks and darkening (a) and (b) will result

**42.** A solution curve of the differential equation

$$(x^2 + xy + 4x + 2y + 4)\frac{dy}{dx} - y^2 = 0,$$

$x > 0$, passes through the point $(1, 3)$. Then, the solution curve

(a) intersects $y = x + 2$ exactly at one point

(b) intersects $y = x + 2$ exactly at two points

(c) intersects $y = (x + 2)^2$

(d) does not intersect $y = (x + 3)^2$

**43.** Consider a pyramid $OPQRS$ located in the first octant $(x \geq 0,\ y \geq 0,\ z \geq 0)$ with $O$ as origin, and $OP$ and $OR$ along the $X$-axis and the $Y$-axis, respectively. The base $OPQR$ of the pyramid is a square with $OP = 3$. The point $S$ is directly above the mid-point $T$ of diagonal $OQ$ such that $TS = 3$.

Then,

(a) the acute angle between $OQ$ and $OS$ is $\frac{\pi}{3}$

(b) the equation of the plane containing the $\Delta OQS$ is $x - y = 0$

(c) the length of the perpendicular from $P$ to the plane containing the $\Delta OQS$ is $\frac{3}{\sqrt{2}}$

(d) the perpendicular distance from $O$ to the straight line containing $RS$ is $\sqrt{\frac{15}{2}}$

**44.** In a $\Delta XYZ$, let $x, y, z$ be the lengths of sides opposite to the angles $X, Y, Z$ respectively and $2s = x + y + z$. If $\frac{s-x}{4} = \frac{s-y}{3} = \frac{s-z}{2}$ and area of incircle of the $\Delta XYZ$ is $\frac{8\pi}{3}$

then

(a) area of the $\Delta XYZ$ is $6\sqrt{6}$

(b) the radius of circumcircle of the $\Delta XYZ$ is $\frac{35}{6}\sqrt{6}$

(c) $\sin\frac{X}{2}\sin\frac{Y}{2}\sin\frac{Z}{2} = \frac{4}{35}$

(d) $\sin^2\left(\frac{X+Y}{2}\right) = \frac{3}{5}$

**45.** Let $RS$ be the diameter of the circle $x^2 + y^2 = 1$, where $S$ is the point $(1, 0)$. Let $P$ be a variable point (other than $R$ and $S$) on the circle and tangents to the circle at $S$ and $P$ meet at the point $Q$. The normal to the circle at $P$

intersects a line drawn through $Q$ parallel to $RS$ at point $E$. Then, the locus of $E$ passes through the point(s)

(a) $\left(\frac{1}{3}, \frac{1}{\sqrt{3}}\right)$

(b) $\left(\frac{1}{4}, \frac{1}{2}\right)$

(c) $\left(\frac{1}{3}, -\frac{1}{\sqrt{3}}\right)$

(d) $\left(\frac{1}{4}, -\frac{1}{2}\right)$

**46.** The circle $C_1 : x^2 + y^2 = 3$ with centre at $O$ intersects the parabola $x^2 = 2y$ at the point $P$ in the first quadrant. Let the tangent to the circle $C_1$ at $P$ touches other two circles $C_2$ and $C_3$ at $R_2$ and $R_3$, respectively. Suppose $C_2$ and $C_3$ have equal radii $2\sqrt{3}$ and centres $Q_2$ and $Q_3$, respectively. If $Q_2$ and $Q_3$ lie on the $Y$-axis, then

(a) $Q_2Q_3 = 12$

(b) $R_2R_3 = 4\sqrt{6}$

(c) area of the $\Delta OR_2R_3$ is $6\sqrt{2}$

(d) area of the $\Delta PQ_2Q_3$ is $4\sqrt{2}$

**47.** Let $f: R \to R, g: R \to R$ and $h: R \to R$ be differentiable functions such that $f(x) = x^3 + 3x + 2$, $g(f(x)) = x$ and $h(g(g(x))) = x$ for all $x \in R$. Then,

(a) $g'(2) = \frac{1}{15}$

(b) $h'(1) = 666$

(c) $h(0) = 16$

(d) $h(g(3)) = 36$

**48.** Let $f: (0, \infty) \to R$ be a differentiable function such that $f'(x) = 2 - \frac{f(x)}{x}$ for all $x \in (0, \infty)$ and $f(1) \neq 1$.

Then

(a) $\lim_{x \to 0+} f'\left(\frac{1}{x}\right) = 1$

(b) $\lim_{x \to 0+} x\, f\left(\frac{1}{x}\right) = 2$

(c) $\lim_{x \to 0+} x^2 f'(x) = 0$

(d) $|f(x)| \le 2$ for all $x \in (0, 2)$

**49.** Let $P = \begin{bmatrix} 3 & -1 & -2 \\ 2 & 0 & \alpha \\ 3 & -5 & 0 \end{bmatrix}$, where $\alpha \in R$.

Suppose $Q = [q_{ij}]$ is a matrix such that $PQ = kI$, where $k \in R, k \neq 0$ and $I$ is the identity matrix of order 3. If $q_{23} = -\frac{k}{8}$

and $\det(Q) = \frac{k^2}{2}$

then

(a) $\alpha = 0, k = 8$

(b) $4\alpha - k + 8 = 0$

(c) $\det(P \operatorname{adj}(Q)) = 2^9$

(d) $\det(Q \operatorname{adj}(P)) = 2^{13}$

## Section 3 (Maximum Marks : 12)

- This section contains **FIVE** questions.
- The answer to each question is a **SIGNLE DIGIT INTEGER** ranging from 0 to 9, both inclusive.
- For each question, darken the bubble corresponding to the correct option in the ORS.
- For each question, marks will be awarded in **one of the following categories**:

Full Marks : + 3 If only the bubble corresponding to the correct option is darkened.

Zero Marks : 0 In all other cases.

**50.** Let $m$ be the smallest positive integer such that the coefficient of $x^2$ in the expansion of

$$(1+x)^2 + (1+x)^3 + \dots + (1+x)^{49} + (1+mx)^{50}$$

is $(3n+1)\ {}^{51}C_3$ for some positive integer $n$. Then, the value of $n$ is

**51.** The total number of distincts $x \in R$ for which $\begin{vmatrix} x & x^2 & 1+x^3 \\ 2x & 4x^2 & 1+8x^3 \\ 3x & 9x^2 & 1+27x^3 \end{vmatrix} = 10$ is

**52.** Let $z = \dfrac{-1+\sqrt{3}i}{2}$, where $i = \sqrt{-1}$, and $r, s \in \{1, 2, 3\}$.

Let $P = \begin{bmatrix} (-z)^r & z^{2s} \\ z^{2s} & z^r \end{bmatrix}$ and $I$ be the identity matrix of order 2. Then, the total number of ordered pairs $(r, s)$ for which $P^2 = -I$ is

**53.** The total number of distincts $x \in [0, 1]$ for which

$$\int_0^x \frac{t^2}{1+t^4}\,dt = 2x - 1 \text{ is}$$

**54.** Let $\alpha, \beta \in R$ be such that $\lim\limits_{x \to 0} \dfrac{x^2 \sin(\beta x)}{\alpha x - \sin x} = 1$. Then, $6(\alpha + \beta)$ equals

# Paper 2

## Physics

### Section 1 (Maximum Marks : 32)

- This section contains **SIX** questions.
- Each question has **FOUR** options (a), (b), (c) and (d). **ONLY ONE** of these four options is correct.
- Four each question, darken the bubble corresponding to the correct option in the ORS.
- For each question, marks will be awarded in one of the following categories.

Full Marks : + 3 If only the bubble corresponding to the correct option is darkened.
Zero Marks : 0 If none of the bubbles is darkened.
Negative Marks : – 1 In all other cases.

**1.** An accident in a nuclear laboratory resulted in deposition of a certain amount of radioactive material of half-life 18 days inside the laboratory. Tests revealed that the radiation was 64 times more than the permissible level required for safe operation of the laboratory. What is the minimum number of days after which the laboratory can be considered safe for use?

(a) 64 (b) 90 (c) 108 (d) 120

**2.** The electrostatic energy of $Z$ protons uniformly distributed throughout a spherical nucleus of radius $R$ is given by

$$E = \frac{3}{5}\frac{Z(Z-1)e^2}{4\pi\varepsilon_0 R}$$

The measured masses of the neutron, ${}^1_1\text{H}$, ${}^{15}_7\text{N}$ and ${}^{15}_8\text{O}$ are 1.008665u, 1.007825u, 15.000109u and 15.003065 u, respectively. Given that the radii of both the ${}^{15}_7\text{N}$ and ${}^{15}_8\text{O}$ nuclei are same, $1\ \text{u} = 931.5\ \text{MeV}/c^2$ ($c$ is the speed of light) and $e^2/(4\pi\varepsilon_0) = 1.44$ MeV fm. Assuming that the difference between the binding energies of ${}^{15}_7\text{N}$ and ${}^{15}_8\text{O}$ is purely due to the electrostatic energy, the radius of either of the nuclei is ($1\ \text{fm} = 10^{-15}$ m)

(a) 2.85 fm (b) 3.03 fm
(c) 3.42 fm (d) 3.80 fm

**3.** A gas is enclosed in a cylinder with a movable frictionless piston. Its initial thermodynamic state at pressure $P_i = 10^5$ Pa and volume $V_1 = 10^{-3}\ \text{m}^3$ changes to a final state at $P_f = (1/32) \times 10^5$ Pa and $V_f = 8 \times 10^{-3}\ \text{m}^3$ in an adiabatic quasi-static process, such that $P^3V^5 =$ constant. Consider another thermodynamic process that brings the system from the same initial state to the same final state in two steps : an isobaric expansion at $P_i$ followed by an isochoric (isovolumetric) process at volume $V_f$. The amount of heat supplied to the system in the two-step process is approximately

(a) 112 J (b) 294 J (c) 588 J (d) 813 J

**4.** There are two Vernier calipers both of which have 1 cm divided into 10 equal divisions on the main scale. The Vernier scale of one of the calipers ($C_1$) has 10 equal divisions that correspond to 9 main scale divisions. The Vernier scale of the other caliper ($C_2$) has 10 equal divisions that correspond to 11 main scale divisions.

The readings of the two calipers are shown in the figure. The measured values (in cm) by calipers $C_1$ and $C_2$ respectively, are

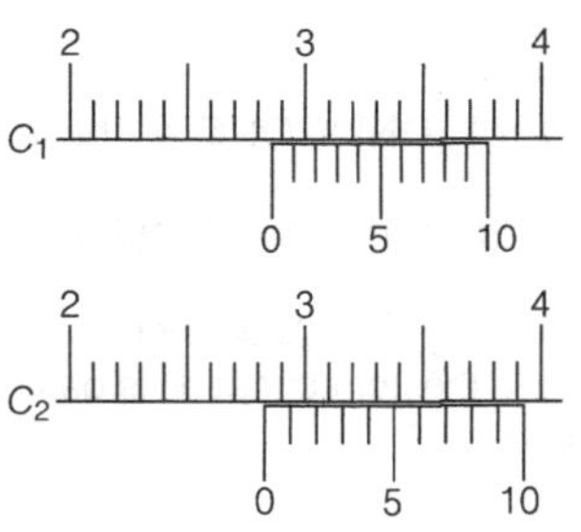


(a) 2.87 and 2.87 (b) 2.87 and 2.83
(c) 2.85 and 2.82 (d) 2.87 and 2.86

**5.** A small object is placed 50 cm to the left of a thin convex lens of focal length 30 cm. A convex spherical mirror of radius of curvature 100 cm is placed to the right of the lens at a distance of 50 cm. The mirror is tilted such that the axis of the mirror is at an angle $\theta = 30°$ to the axis of the lens, as shown in the figure.

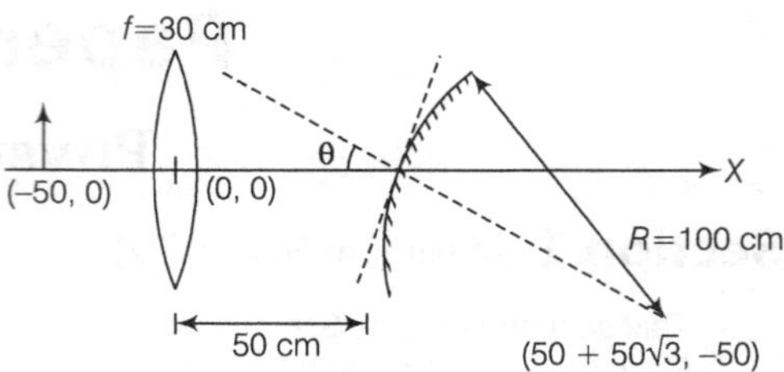


If the origin of the coordinate system is taken to be at the centre of the lens, the coordinates (in cm) of the point $(x, y)$ at which the image is formed are

(a) $(125/3, 25/\sqrt{3})$ (b) $(50 - 25\sqrt{3}, 25)$
(c) $(0, 0)$ (d) $(25, 25\sqrt{3})$

**6.** The ends $Q$ and $R$ of two thin wires, $PQ$ and $RS$, are soldered (joined) together. Initially, each of the wire has a length of 1 m 10°C. Now, the end $P$ is maintained at 10°C, while the end $S$ is heated and maintained at 400°C. The system is thermally insulated from its surroundings. If the thermal conductivity of wire $PQ$ is twice that of the wire $RS$ and the coefficient of linear thermal expansion of $PQ$ is $1.2 \times 10^{-5}\ K^{-1}$, the change in length of the wire $PQ$ is

(a) 0.78 mm (b) 0.90 mm
(c) 1.56 mm (d) 2.34 mm

## Section 2 (Maximum Marks : 32)

- This section contains **EIGHT** questions.
- Each question has **FOUR** options (a), (b) (c) and (d). **ONE OR MORE THAN ONE** of these four option(s) is (are) correct.
- For each question, darken the bubble(s) corresponding to all the correct option(s) in the ORS.
- For each questions, marks will be awarded in one of the following categories :

  Full Marks : + 4 If only the bubble(s) corresponding to all the correct option(s) is (are) darkened.

  Partial Marks : + 1 For darkening a bubble corresponding to **each correct option,** provided NO incorrect option is darkened.

  Zero Marks : 0 If none of the bubbles is darkened.

  Negative Marks : −2 In all other cases.
- For example, if (a), (c) and (d) are all the correct options for a question, darkening all these three will result in + 4 marks ; darkening only (a) and (d) will result in + 2 marks and darkening (a) and (b) will result in −2 marks, as a wrong option is also darkened.

**7.** Two thin circular discs of mass $m$ and $4m$, having radii of $a$ and $2a$, respectively, are rigidly fixed by a massless, rigid rod of length $l = \sqrt{24}\, a$ through their centers. This assembly is laid on a firm and flat surface and set rolling without slipping on the surface so that the angular speed about the axis of the rod is $\omega$. The angular momentum of the entire assembly about the point '$O$' is **L** (see the figure). Which of the following statement(s) is (are) true?

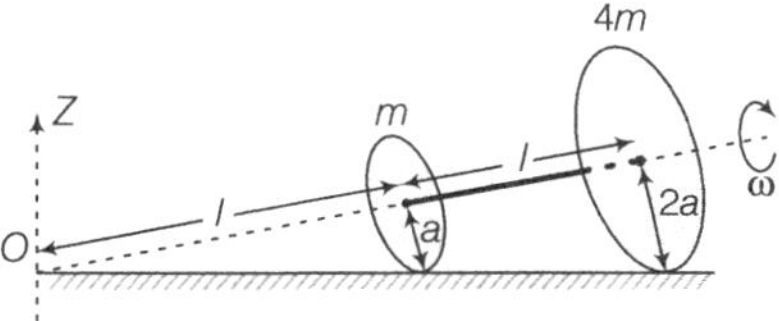


(a) The magnitude of the $z$-component of **L** is $55\, ma^2\, \omega$

(b) The magnitude of angular momentum of centre of mass of the assembly about the point $O$ is $81\, ma^2\, \omega$

(c) The centre of mass of the assembly rotates about the $Z$-axis with an angular speed of $\frac{\omega}{5}$

(d) The magnitude of angular momentum of the assembly about its centre of mass is $17\, ma^2 \frac{\omega}{2}$

**8.** Consider two identical galvanometers and two identical resistors with resistance $R$. If the internal resistance of the galvanometers $R_c < R/2$, which of the following statement(s) about anyone of the galvanometers is (are) true?

(a) The maximum voltage range is obtained when all the components are connected in series

(b) The maximum voltage range is obtained when the two resistors and one galvanometer are connected in series, and the second galvanometer is connected in parallel to the first galvanometer

(c) The maximum current range is obtained when all the components are connected in parallel

(d) The maximum current range is obtained when the two galvanometers are connected in series, and the combination is connected in parallel with both the resistors

**9.** A rigid wire loop of square shape having side of length $L$ and resistance $R$ is moving along the $X$-axis with a constant velocity $v_0$ in the plane of the paper . At $t = 0$, the right edge of the loop enters a region of length $3L$ where there is a uniform magnetic field $B_0$ into the plane of the paper, as shown in the figure. For sufficiently large $v_0$, the loop eventually crosses the region. Let $x$ be the location of the right edge of the loop. Let $v(x)$, $I(x)$ and $F(x)$ represent the velocity of the loop, current in the loop, and force on the loop, respectively, as a function of $x$. Counter-clockwise current is taken as positive.

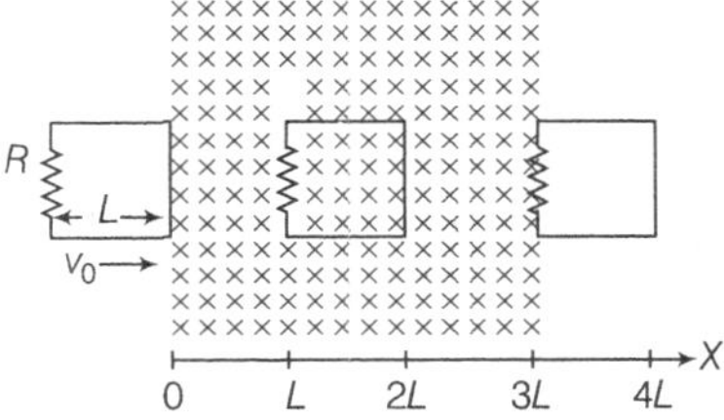


Which of the following schematic plot(s) is (are) correct? (Ignore gravity)

(a)

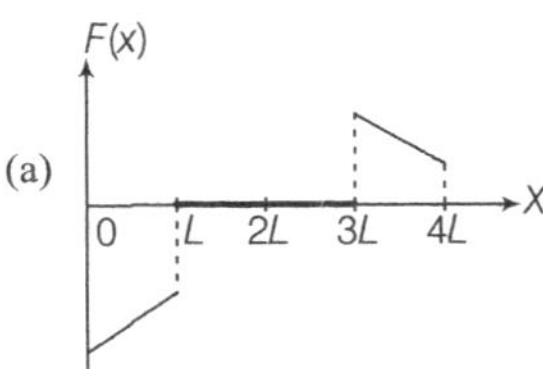

(b)

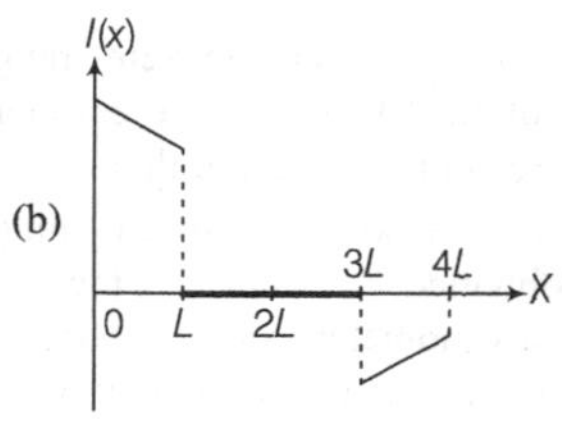


(c)

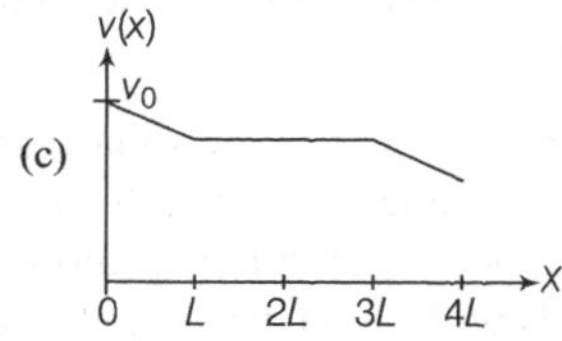


(d)

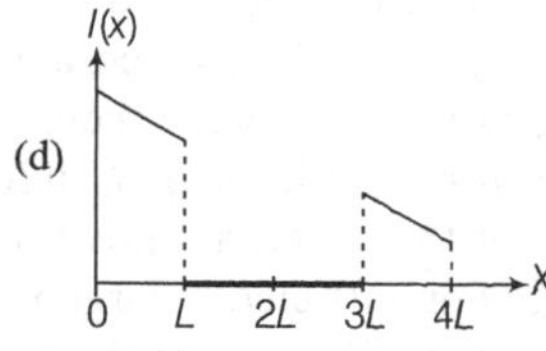


**10.** In an experiment to determine the acceleration due to gravity g, the formula used for the time period of a periodic motion is $T = 2\pi\sqrt{\frac{7(R-r)}{5g}}$. The values of R and r are measured to be $(60 \pm 1)$ mm and $(10 \pm 1)$ mm respectively. In five successive measurements, the time period is found to be 0.52 s, 0.56 , 0.57 s, 0.54 s and 0.59 s. The least count of the watch used for the measurement of time period is 0.01 s. Which of the following statement(s) is (are) true?

(a) The error in the measurement of $r$ is 10%

(b) The error in the measurement of $T$ is 3.57%

(c) The error in the measurement of $T$ is 2%

(d) The error in the determined of $g$ is 11%

**11.** A block with mass $M$ is connected by a massless spring with stiffness constant $k$ to a rigid wall and moves without friction on a horizontal surface. The block oscillates with small amplitude $A$ about an equilibrium position $x_0$. Consider two cases : (i) when the block is at $x_0$ and (ii) when the block is at $x = x_0 + A$. In both the cases, a particle with mass $m$ ($< M$) is softly placed on the block after which they stick to each other. Which of the following statement(s) is (are) true about the motion after the mass $m$ is placed on the mass $M$?

(a) The amplitude of oscillation in the first case changes by a factor of $\sqrt{\frac{M}{m+M}}$, whereas in the second case it remains unchanged

(b) The final time period of oscillation in both the cases is same

(c) The total energy decreases in both the cases

(d) The instantaneous speed at $x_0$ of the combined masses decreases in both the cases

**12.** While conducting the Young's double slit experiment, a student replaced the two slits with a large opaque plate in the $XY$-plane containing two small holes that act as two coherent point sources $(S_1, S_2)$ emitting light of wavelength 600 mm. The student mistakenly placed the screen parallel to the $XZ$-plane (for $z > 0$) at a distance $D = 3$ m from the mid-point of $S_1S_2$, as shown schematically in the figure. The distance between the source $d = 0.6003$ mm. The origin $O$ is at the intersection of the screen and the line joining $S_1S_2$.

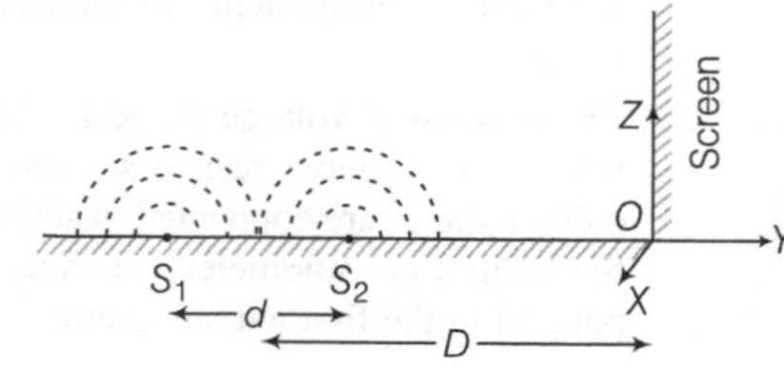

Which of the following is (are) true of the intensity pattern on the screen?

(a) Semi circular bright and dark bands centered at point $O$

(b) The region very close to the point $O$ will be dark

(c) Straight bright and dark bands parallel to the $X$-axis

(d) Hyperbolic bright and dark bands with foci symmetrically placed about $O$ in the $x$-direction

**13.** In the circuit shown below, the key is pressed at time $t = 0$. Which of the following statement(s) is (are) true?

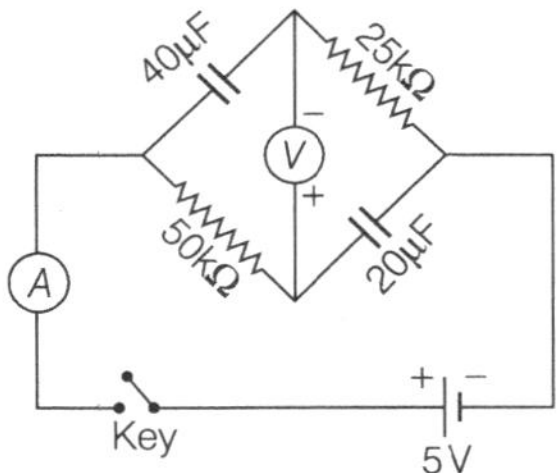


(a) The voltmeter display $-5$ V as soon as the key is pressed and displays $+5$ V after a long time

(b) The voltmeter will display 0 V at time $t = \ln 2$ seconds

(c) The current in the ammeter becomes $1/e$ of the initial value after 1 second

(d) The current in the ammeter becomes zero after a long time

**14.** Light of wavelength $\lambda_{ph}$ falls on a cathode plate inside a vacuum tube as shown in the figure. The work function of the cathode surface is $\phi$ and the anode is a wire mesh of conducting material kept at a distance $d$ from the cathode. A potential difference $V$ is maintained between the electrodes. If the minimum de-Broglie wavelength of the electrons passing through the anode is $\lambda_e$, which of the following statements(s) is (are) true?

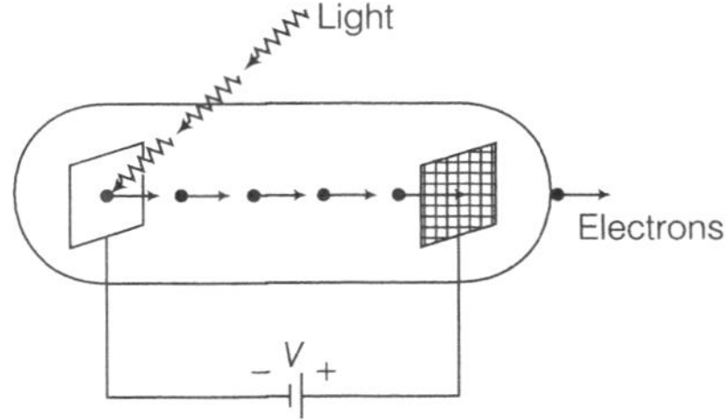


(a) $\lambda_e$ increases at the same rate as $\lambda_{ph}$ for $\lambda_{ph} < hc/\phi$

(b) $\lambda_e$ is approximately halved, if $d$ is doubled

(c) $\lambda_e$ decreases with increase in $\phi$ and 0 $\lambda_{ph}$

(d) For large potential difference $(V >> \phi/e)$, $\lambda_e$ is approximately halved if $V$ is made four times

## Section 3 (Maximum Marks : 16)

- This section contains **TWO** paragraphs.
- Based on each paragraph, there are **TWO** questions.
- Each question has **FOUR** options (a), (b) (c) and (d). **ONE ONLY** of these four options is correct.
- For each question, darken the bubble(s) corresponding to all the correct option(s) in the ORS.
- For each questions, marks will be awarded in one of the following categories.

| | | |
|---|---|---|
| Full Marks | : + 3 | If only the bubble corresponding to all the correct answer is darkened. |
| Zero Marks | : 0 | In all other cases. |

## Paragraph 1

A frame of the reference that is accelerated with respect to an inertial frame of reference is called a non-inertial frame of reference. A coordinate system fixed on a circular disc rotating about a fixed axis with a constant angular velocity $\omega$ is an example of a non-inertial frame of reference. The relationship between the force $\vec{\mathbf{F}}_{\text{rot}}$ experienced by a particle of mass $m$ moving on the rotating disc and the force $\vec{\mathbf{F}}_{\text{in}}$ experienced by the particle in an inertial frame of reference is,

$$\vec{\mathbf{F}}_{\text{rot}} = \vec{\mathbf{F}}_{\text{in}} + 2m(\vec{\mathbf{v}}_{\text{rot}} \times \vec{\omega}) + m(\vec{\omega} \times \vec{\mathbf{r}}) \times \vec{\omega},$$

where, $\mathbf{v}_{\text{rot}}$ is the velocity of the particle in the rotating frame of reference and $\mathbf{r}$ is the position vector of the particle with respect to the centre of the disc.

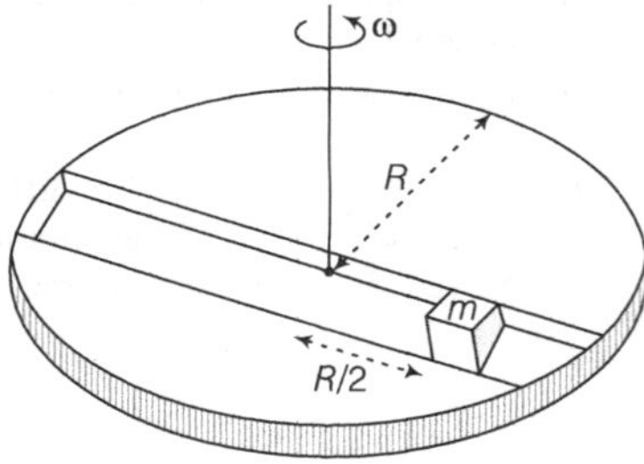


Now, consider a smooth slot along a diameter of a disc of radius $R$ rotating counter-clockwise with a constant angular speed $\omega$ about its vertical axis through its centre. We assign a coordinate system with the origin at the centre of the disc, the $X$-axis along the slot, the $Y$-axis perpendicular to the slot and the $Z$-axis along the rotation axis $(\omega = \omega\hat{\mathbf{k}})$. A small block of mass $m$ is gently placed in the slot at $\mathbf{r} = (R/2)\hat{\mathbf{i}}$ at $t = 0$ and is constrained to move only along the slot.

**15.** The distance $r$ of the block at time $t$ is

(a) $\frac{R}{2}\cos 2\omega t$ (b) $\frac{R}{2}\cos \omega t$

(c) $\frac{R}{4}(e^{\omega t} + e^{-\omega t})$ (d) $\frac{R}{4}(e^{2\omega t} + e^{-2\omega t})$

**16.** The net reaction of the disc on the block is

(a) $m\omega^2 R \sin \omega t\hat{\mathbf{j}} - mg\hat{\mathbf{k}}$

(b) $\frac{1}{2}m\omega^2 R\,(e^{\omega t} - e^{-\omega t})\hat{\mathbf{j}} + mg\hat{\mathbf{k}}$

(c) $\frac{1}{2}m\omega^2 R\,(e^{2\omega t} - e^{-2\omega t})\hat{\mathbf{j}} + mg\hat{\mathbf{k}}$

(d) $-m\omega^2 R \cos\omega t\hat{\mathbf{j}} - mg\hat{\mathbf{k}}$

## Paragraph 2

Consider an evacuated cylindrical chamber of height $h$ having rigid conducting plates at the ends and an insulating curved surface as shown in the figure. A number of spherical balls made of a light weight and soft material and coated with a conducting material are placed on the bottom plate. The balls have a radius $r << h$. Now, a high voltage source (HV) connected across the conducting plates such that the bottom plate is at $+V_0$ and the top plate at $-V_0$. Due to their conducting surface, the balls will get charge, will become equipotential with the plate and are repelled by it. The balls will eventually collide with the top plate, where the coefficient of restitution can be taken to be zero due to te soft nature of the material of the balls. The electric field in the chamber can be considered to be that of a parallel plate capacitor. Assume that there are no collisions between the balls and the interaction between them is negligible. (Ignore gravity)

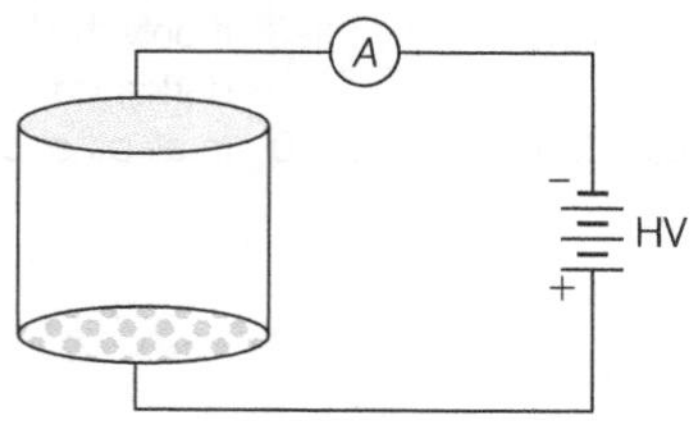

**17.** Which one of the following statement is correct?

(a) The balls will execute simple harmonic motion between the two plates

(b) The balls will bounce back to the bottom plate carrying the same charge they went up with

(c) The balls will stick to the top plate and remain there

(d) The balls will bounce back to the bottom plate carrying the opposite charge they went up with

**18.** The average current in the steady state registered by the ammeter in the circuit will be

(a) proportional to $V_0^2$
(b) proportional to the potential $V_0$
(c) zero
(d) proportions to $V_0^{1/2}$

# Chemistry

## Section 1 (Maximum Marks : 32)

- This section contains **SIX** questions.
- Each question has **FOUR** options (a), (b), (c) and (d). **ONLY ONE** of these four options is correct.
- Four each question, darken the bubble corresponding to the correct option in the ORS.
- For each question, marks will be awarded in one of the following categories.

| | | |
|---|---|---|
| Full Marks | : +3 | If only the bubble corresponding to the correct option is darkened. |
| Zero Marks | : 0 | If none of the bubbles is darkened. |
| Negative Marks | : −1 | In all other cases. |

**19.** The correct order of acidity for the following compounds is

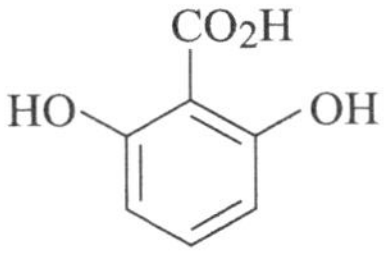

I: $CO_2H$, HO, OH (benzene ring)
II: $CO_2H$, OH (benzene ring)
III: $CO_2H$, OH (benzene ring)
IV: $CO_2H$, OH (benzene ring)

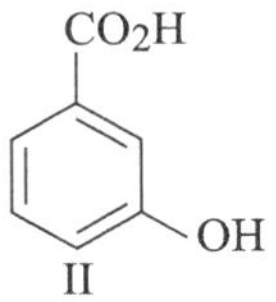

(a) I > II > III > IV
(b) III > I > II > IV
(c) III > IV > II > I
(d) I > III > IV > II

**20.** The major product of the following reaction sequence is

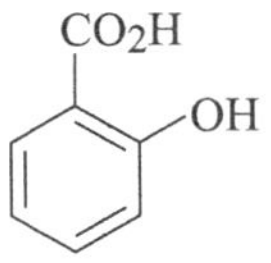

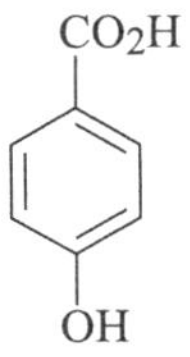

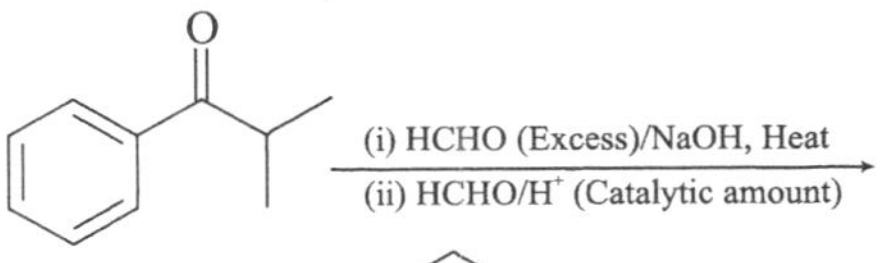

O (phenyl ketone) $\xrightarrow[\text{(ii) HCHO/H}^+\text{ (Catalytic amount)}]{\text{(i) HCHO (Excess)/NaOH, Heat}}$

(a) O, O (structure)

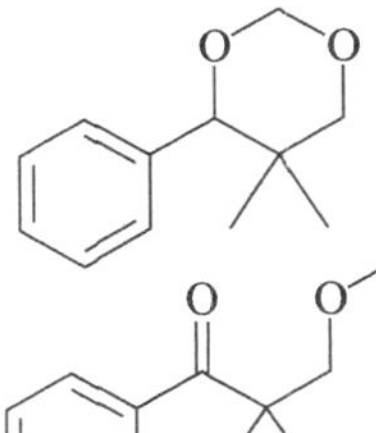

(b) O, O, OH (structure)

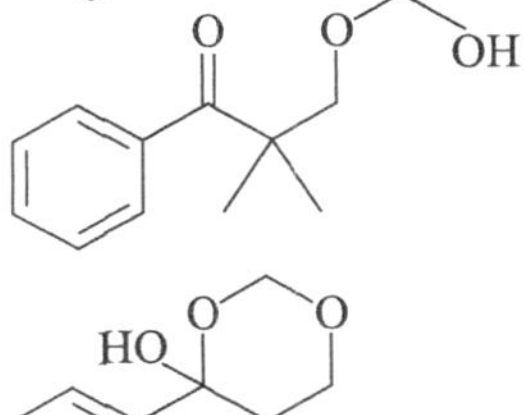

(c) HO, O, O (structure)

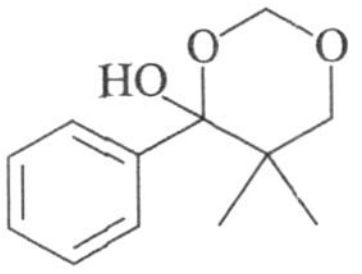

(d)

**21.** In the following reaction sequence in aqueous solution, the species *X*, *Y* and *Z*, respectively, are

$$S_2O_3^{2-} \xrightarrow{Ag^+} \underset{\text{(Clear solution)}}{X} \xrightarrow{Ag^+} \underset{\text{(White ppt.)}}{Y} \xrightarrow{\text{With time}} \underset{\text{(Black ppt.)}}{Z}$$

(a) $[Ag(S_2O_3)_2]^{3-}$, $Ag_2S_2O_3$, $Ag_2S$

(b) $[Ag(S_2O_3)_3]^{5-}$, $Ag_2SO_3$, $Ag_2S$

(c) $[Ag(SO_3)_2]^{3-}$, $Ag_2S_2O_3$, Ag

(d) $[Ag(SO_3)_3]^{3-}$, $Ag_2SO_4$, Ag

**22.** The qualitative sketches I, II and III given below show the variation of surface tension with molar concentration of three different aqueous solutions of KCl, $CH_3OH$ and $CH_3(CH_2)_{11}OSO_3^-Na^+$ at room temperature.

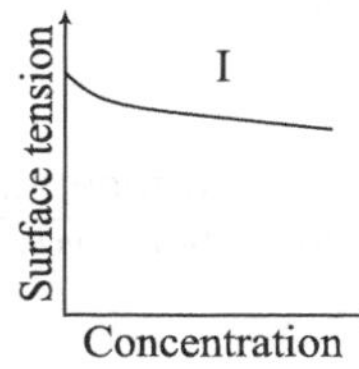


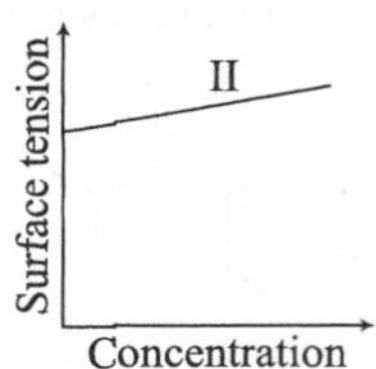


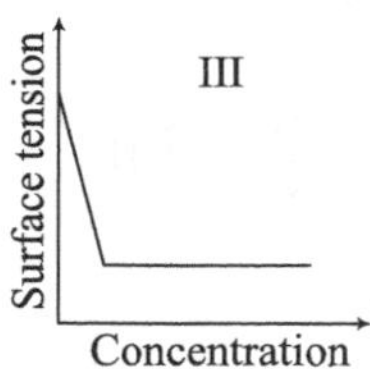


The correct assignment of the sketches is

(a) I KCl
II $CH_3OH$
III $CH_3(CH_2)_{11}OSO_3^-\ Na^+$

(b) I $CH_3(CH_2)_{11}OSO_3^-\ Na^+$
II $CH_3OH$
III KCl

(c) I KCl
II $CH_3(CH_2)_{11}OSO_3^-\ Na^+$
III $CH_3OH$

(d) I $CH_3OH$
II KCl
III $CH_3(CH_2)_{11}OSO_3^-\ Na^+$

**23.** The geometries of the ammonia complexes of $Ni^{2+}$, $Pt^{2+}$ and $Zn^{2+}$, respectively, are

(a) octahedral, square planar and tetrahedral

(b) square planar, octahedral and tetrahedral

(c) tetrahedral, square planar and octahedral

(d) octahedral, tetrahedral and square planar

**24.** For the following electrochemical cell at 298 K,

$$Pt(s)\,|\,H_2(g, 1\,\text{bar})\,|\,H^+(aq, 1\,M)\ ||\,M^{4+}(aq), M^{2+}(aq)\,|\,Pt(s)$$

$E_{cell} = 0.092$ V when $\dfrac{[M^{2+}(aq)]}{[M^{4+}(aq)]} = 10^x$

Given : $E^\circ_{M^{4+}/M^{2+}} = 0.151$ V; $2.303\dfrac{RT}{F} = 0.059$ V

The value of $x$ is

(a) – 2
(b) – 1
(c) 1
(d) 2

## Section 2 (Maximum Marks : 32)

- This section contains **EIGHT** questions.
- Each question has **FOUR** options (a), (b), (c) and (d). **ONE OR MORE THAN ONE** of these four options is (are) correct.
- For each question, darken the bubble(s) corresponding to all the correct option(s) in the ORS.
- For each question, marks will be awarded in **one of the following categories**:

Full Marks : + 4, If only the bubble corresponding to all the correct option(s) is (are) darkened.

Partial Marks : + 1, For darkening a bubble corresponding to **each correct option,** provided no incorrect option is darkened.

Zero Marks : 0, If none of the bubbles is darkened.

Negative Marks : −2, In all other cases.

For example, if (a), (c) and (d) are all the correct options for a question, darkening all these three will result in + 4 marks ; darkening only (a) and (d) will result in + 2 marks and darkening (a)and (b) will result in −2 marks, as a wrong option is also darkened.

**25.** According to molecular orbital theory, which of the following statements is(are) correct?

(a) $C_2^{2-}$ is expected to be diamagnetic

(b) $O_2^{2+}$ is expected to have a longer bond length than $O_2$

(c) $N_2^+$ and $N_2^-$ have the same bond order

(d) $He_2^+$ has the same energy as two isolated He atoms

**26.** The correct statement(s) for cubic close packed (ccp) three dimensional structure is (are)

(a) The number of the nearest neighbours of an atom present in the topmost layer is 12

(b) The packing efficiency of atom is 74%

(c) The number of octahedral and tetrahedral voids per atom are 1 and 2, respectively

(d) The unit cell edge length is $2\sqrt{2}$ times the radius of the atom

**27.** Reagent(s) which can be used to bring about the following transformation is (are)

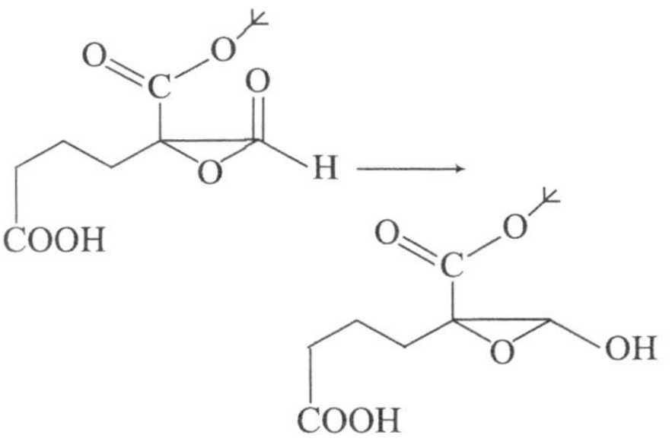

O C O O O H COOH → O C O O OH COOH

(a) $LiAlH_4$ in $(C_2H_5)_2O$

(b) $BH_3$ in THF

(c) $NaBH_4$ in $C_2H_5OH$

(d) Raney $Ni/H_2$ in THF

**28.** Extraction of copper from copper pyrite ($CuFeS_2$) involves

(a) crushing followed by concentration of the ore by froth-floatation

(b) removal of iron as slag

(c) self reduction step to produce 'blister copper' following evolution of $SO_2$

(d) refining of 'blister copper' by carbon reduction

**29.** The nitrogen containing compound produced in the reaction of $HNO_3$ with $P_4O_{10}$

(a) can also be prepared by reaction of $P_4$ and $HNO_3$
(b) is diamagnetic
(c) contains one N—N bond
(d) reacts with Na metal producing a brown gas

**30.** Mixture(s) showing positive deviation from Raoult's law at 35°C is (are)
(a) carbon tetrachloride + methanol
(b) carbon disulphide + acetone
(c) benzene + toluene
(d) phenol + aniline

**31.** For 'invert sugar', the correct statement(s) is (are)
(Given: specific rotations of (+) - sucrose, (+) - maltose, L-(−)-glucose and L-(+)-fructose in aqueous solution are +66°, +140°, − 52° and 92°, respectively)
(a) Invert sugar is prepared by acid catalysed hydrolysis of maltose
(b) Invert sugar is an equimolar mixture of D-(+) -glucose and D-(−)- fructose
(c) Specific rotation of invert sugar is − 20°
(d) On reaction with $Br_2$ water, invert sugar forms saccharic acid as one of the products

**32.** Among the following reactions(s), which gives(give) *tert*-butyl benzene as the major product?

(a) Benzene $\xrightarrow[\text{NaOC}_2\text{H}_5]{(CH_3)_3C\text{–Br}}$

(b) Benzene $\xrightarrow[\text{AlCl}_3]{(CH_3)_2CHCH_2\text{Cl}}$

(c) Benzene $\xrightarrow[\text{H}_2\text{SO}_4]{CH_2=C(CH_3)_2}$

(d) Benzene $\xrightarrow[\text{BF}_3\text{,OEt}_2]{(CH_3)_2CHCH_2\text{OH}}$

## Section 3 (Maximum Marks : 12)

- This section contains **FOUR** questions.
- Based on each paragraph, there are TWO questions.
- Each question has FOUR options (a), (b), (c) and (d). ONLY ONE of these four options is correct.
- Four each question, darken the bubble corresponding to the correct option in the ORS.
- For each question, marks will be awarded in **one of the following categories**:

Full Marks : +3, If only the bubble corresponding to the correct answer is darkened.

Zero Marks : 0, In all other cases.

### Paragraph for Questions 33-34

- Thermal decomposition of gaseous $X_2$ to gaseous $X$ at 298 K takes place according to the following equation:

$$X_2(g) \rightleftharpoons 2X(g)$$

- The standard reaction Gibbs energy, $\Delta_r G°$, of this reaction is positive. At the start of the reaction, there is one mole of $X_2$ and no $X$. As the reaction proceeds, the number of moles of $X$ formed is given by $\beta$. Thus, $\beta_{equilibrium}$ is the number of moles of $X$ formed at equilibrium. The reaction is carried out at a constant total pressure of 2 bar. Consider the gases to behave ideally. (Given, $R = 0.083$ L bar $K^{-1}$ $mol^{-1}$)

**33.** The equilibrium constant $K_p$ for this reaction at 298 K, in terms of $\beta_{\text{equilibrium}}$ is

(a) $\dfrac{8\beta^2_{\text{equilibrium}}}{2-\beta_{\text{equilibrium}}}$ (b) $\dfrac{8\beta^2_{\text{equilibrium}}}{4-\beta^2_{\text{equilibrium}}}$

(c) $\dfrac{4\beta^2_{\text{equilibrium}}}{2-\beta_{\text{equilibrium}}}$ (d) $\dfrac{4\beta^2_{\text{equilibrium}}}{4-\beta^2_{\text{equilibrium}}}$

**34.** The incorrect statement among the following for this reaction, is

(a) Decrease in the total pressure will result in the formation of more moles of gaseous $X$

(b) At the start of the reaction, dissociation of gaseous $X_2$ takes place spontaneously

(c) $\beta_{\text{equilibrium}} = 0.7$

(d) $K_C < 1$

## **Paragraph** for Questions 35-36

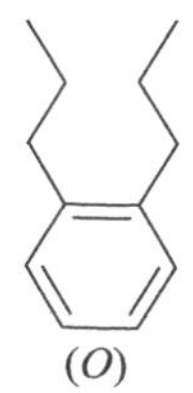

(O)

Treatment of compound $O$ with $KMnO_4/H^+$ gave $P$, which on heating with ammonia gave $Q$. The compound $Q$ on treatment with $Br_2/NaOH$ produced $R$. On strong heating, $Q$ gave $S$, which on further treatment with ethyl 2-bromopropanoate in the presence of KOH followed by acidification, gave a compound $T$.

**35.** The compound $R$ is

(a) $NH_2$ / $NH_2$

(b) O, Br / Br, O

(c) O, NHBr / NHBr, O

(d) O, NBr, O

**36.** The compound $T$ is

(a) glycine

(b) alanine

(c) valine

(d) serine

# Mathematics

## **Section 1** (Maximum Marks : 15)

- This section contains **SIX** questions.
- Each question has **FOUR** options (a), (b), (c) and (d). **ONLY ONE** of these four options is correct.
- For each question, darken the bubble corresponding to the correct option in the ORS.

For each question, marks will be awarded in one of the following categories :

Full Marks : +3 If only the bubble corresponding to the correct option is darkened.

Zero Marks : 0 If none of the bubbles is darkened.

Negative Marks : −1 In all other cases.

**37.** Let $P = \begin{bmatrix} 1 & 0 & 0 \\ 4 & 1 & 0 \\ 16 & 4 & 1 \end{bmatrix}$ and $I$ be the identity matrix of order 3. If $Q = [q_{ij}]$ is a matrix, such that $P^{50} - Q = I$, then $\dfrac{q_{31} + q_{32}}{q_{21}}$ equals

(a) 52 (b) 103 (c) 201 (d) 205

**38.** Area of the region $\{(x, y)\} \in R^2 : y \geq \sqrt{|x+3|}$, $5y \leq (x+9) \leq 15\}$ is equal to

(a) $\frac{1}{6}$ (b) $\frac{4}{3}$

(c) $\frac{3}{2}$ (d) $\frac{5}{3}$

**39.** The value of $\sum_{k=1}^{13} \frac{1}{\sin\left(\frac{\pi}{4} + \frac{(k-1)\pi}{6}\right)\sin\left(\frac{\pi}{4} + \frac{k\pi}{6}\right)}$ is equal to

(a) $3 - \sqrt{3}$

(b) $2(3 - \sqrt{3})$

(c) $2(\sqrt{3} - 1)$

(d) $2(2 + \sqrt{3})$

**40.** Let $b_i > 1$ for $i = 1, 2, \ldots, 101$. Suppose $\log_e b_1, \log_e b_2, \ldots, \log_e b_{101}$ are in AP with the common difference $\log_e 2$. Suppose $a_1, a_2, \ldots, a_{101}$ are in AP, such that $a_1 = b_1$ and $a_{51} = b_{51}$. If $t = b_1 + b_2 + \ldots + b_{51}$ and $s = a_1 + a_2 + \ldots + a_{51}$, then

(a) $s > t$ and $a_{101} > b_{101}$

(b) $s > t$ and $a_{101} < b_{101}$

(c) $s < t$ and $a_{101} > b_{101}$

(d) $s < t$ and $a_{101} < b_{101}$

**41.** The value of $\int_{-\pi/2}^{\pi/2} \frac{x^2 \cos x}{1 + e^x} dx$ is equal to

(a) $\frac{\pi^2}{4} - 2$ (b) $\frac{\pi^2}{4} + 2$

(c) $\pi^2 - e^{-\pi/2}$ (d) $\pi^2 + e^{\pi/2}$

**42.** Let $P$ be the image of the point (3, 1, 7) with respect to the plane $x - y + z = 3$. Then, the equation of the plane passing through $P$ and containing the straight line $\frac{x}{1} = \frac{y}{2} = \frac{z}{1}$ is

(a) $x + y - 3z = 0$

(b) $3x + z = 0$

(c) $x - 4y + 7z = 0$

(d) $2x - y = 0$

## Section 2 (Maximum Marks : 32)

- This section contains **EIGHT** questions.
- Each question has **FOUR** options (a), (b), (c) and (d). **ONE OR MORE THAN ONE** of these four option(s) is(are) correct.
- For each question, darken the bubble(s) corresponding to all the correct option(s) in the ORS.
- For each question, marks will be awarded in one of the following categories :

| | | |
|---|---|---|
| Full Marks | : +4 | If only the bubble(s) corresponding to all correct option(s) is (are) darkened. |
| Partial Marks | : +1 | For darkening a bubble corresponding **to each correct option**, provided no incorrect option is darkened. |
| Zero Marks | : 0 | If none of the dubbles is darkened. |
| Negative Marks | : –2 | In all other cases. |

**43.** Let $a, b \in R$ and $f : R \to R$ be defined by $f(x) = a\cos(|x^3 - x|) + b|x|\sin(|x^3 + x|)$. Then, $f$ is

(a) differentiable at $x = 0$, if $a = 0$ and $b = 1$

(b) differentiable at $x = 1$, if $a = 1$ and $b = 0$

(c) not differentiable at $x = 0$, if $a = 1$ and $b = 0$

(d) not differentiable at $x = 1$, if $a = 1$ and $b = 1$

**44.** Let $f(x)$

$$= \lim_{n \to \infty} \left[ \frac{n^n (x+n)\left(x+\frac{n}{2}\right)\cdots\left(x+\frac{n}{n}\right)}{n!(x^2+n^2)\left(x^2+\frac{n^2}{4}\right)\cdots\left(x^2+\frac{n^2}{n^2}\right)} \right]^{\frac{x}{n}},$$

for all $x = 0$. Then

(a) $f\left(\frac{1}{2}\right) \geq f(1)$ (b) $f\left(\frac{1}{3}\right) \leq f\left(\frac{2}{3}\right)$

(c) $f'(2) \leq 0$ (d) $\frac{f'(3)}{f(3)} \geq \frac{f'(2)}{f(2)}$

**45.** Let $f: R \to (0, \infty)$ and $g: R \to R$ be twice differentiable functions such that $f''$ and $g''$ are continuous functions on $R$.

Suppose $f'(2) = g(2) = 0$, $f''(2) \neq 0$ and $g'(2) \neq 0$. If $\lim_{x \to 2} \frac{f(x)g(x)}{f'(x)g'(x)} = 1$, then

(a) $f$ has a local minimum at $x = 2$
(b) $f$ has a local maximum at $x = 2$
(c) $f''(2) > f(2)$
(d) $f(x) - f''(x) = 0$, for atleast one $x \in R$

**46.** Let $\hat{u} = u_1\hat{i} + u_2\hat{j} + u_3\hat{k}$ be a unit vector in $R^3$ and $\hat{w} = \frac{1}{\sqrt{6}}(\hat{i} + \hat{j} + 2\hat{k})$. Given that there exists a vector $\vec{v}$ in $R^3$, such that $|\hat{u} + \vec{v}| = 1$ and $\hat{w} \cdot (\hat{u} + \vec{v}) = 1$.

Which of the following statement(s) is/are correct?

(a) There is exactly one choice for such $\vec{v}$
(b) There are infinitely many choices for such $\vec{v}$
(c) If $\hat{u}$ lies in the $XY$-plane, then $|u_1| = |u_2|$
(d) If $\hat{u}$ lies in the $XY$-plane, then $2|u_1| = |u_3|$

**47.** Let $P$ be the point on the parabola $y^2 = 4x$, which is at the shortest distance from the centre $S$ of the circle $x^2 + y^2 - 4x - 16y + 64 = 0$. Let $Q$ be the point on the circle dividing the line segment $SP$ internally. Then,

(a) $SP = 2\sqrt{5}$
(b) $SQ : QP = (\sqrt{5} + 1) : 2$
(c) the $x$-intercept of the normal to the parabola at $P$ is 6
(d) the slope of the tangent to the circle at $Q$ is $\frac{1}{2}$

**48.** Let $a, b \in R$ and $a^2 + b^2 \neq 0$. Suppose $S = \left\{ z \in C : z = \frac{1}{a + i\,bt}, t \in R, t \neq 0 \right\}$, where $i = \sqrt{-1}$. If $z = x + iy$ and $z \in S$, then $(x, y)$ lies on

(a) the circle with radius $\frac{1}{2a}$ and centre $\left(\frac{1}{2a}, 0\right)$ for $a > 0, b \neq 0$
(b) the circle with radius $-\frac{1}{2a}$ and centre $\left(-\frac{1}{2a}, 0\right)$ for $a < 0$, $b \neq 0$
(c) the $X$-axis for $a \neq 0, b = 0$
(d) the $Y$-axis for $a = 0, b \neq 0$

**49.** Let $a, \lambda, \mu \in R$. Consider the system of linear equations $ax + 2y = \lambda$ and $3x - 2y = \mu$. Which of the following statement(s) is/are correct?

(a) If $a = -3$, then the system has infinitely many solutions for all values of $\lambda$ and $\mu$
(b) If $a \neq -3$, then the system has a unique solution for all values of $\lambda$ and $\mu$
(c) If $\lambda + \mu = 0$, then the system has infinitely many solutions for $a = -3$
(d) If $\lambda + \mu \neq 0$, then the system has no solution for $a = -3$

**50.** Let $f:\left[-\frac{1}{2},2\right]\to R$ and $g:\left[-\frac{1}{2},2\right]\to R$ be functions defined by $f(x)=[x^2-3]$ and $g(x)=|x|\,f(x)+|4x-7|\,f(x)$, where $[y]$ denotes the greatest integer less than or equal to $y$ for $y\in R$. Then,

(a) $f$ is discontinuous exactly at three points in $\left[-\frac{1}{2},2\right]$

(b) $f$ is discontinuous exactly at four points in $\left[-\frac{1}{2},2\right]$

(c) $g$ is not differentiable exactly at four points in $\left(-\frac{1}{2},2\right)$

(d) $g$ is not differentiable exactly at five points in $\left(-\frac{1}{2},2\right)$

## Section 3 (Maximum Marks : 12)

- This section contains **TWO** paragraphs.
- Based on each paragraph, there are **TWO** questions.
- Each question has **FOUR** options (a), (b), (c) and (d). **ONLY ONE** of these four options is correct.
- For each question, darken the bubble corresponding to the correct option in the ORS.
- For each question, marks will be awarded in **one of the following categories**:

  Full Marks : + 3 If only the bubble corresponding to the correct option is darkened.
  
  Zero Marks : 0 In all other cases.

### Paragraph 1

Football teams $T_1$ and $T_2$ have to play two games against each other. It is assumed that the outcomes of the two games are independent. The probabilities of $T_1$ winning, drawing and losing a game against $T_2$ are $\frac{1}{2}$, $\frac{1}{6}$ and $\frac{1}{3}$, respectively. Each team gets 3 points for a win, 1 point for a draw and 0 point for a loss in a game. Let $X$ and $Y$ denote the total points scored by teams $T_1$ and $T_2$, respectively, after two games.

**51.** $P(X>Y)$ is

(a) $\frac{1}{4}$ (b) $\frac{5}{12}$ (c) $\frac{1}{2}$ (d) $\frac{7}{12}$

**52.** $P(X=Y)$ is

(a) $\frac{11}{36}$ (b) $\frac{1}{3}$ (c) $\frac{13}{36}$ (d) $\frac{1}{2}$

### Paragraph 2

Let $F_1(x_1, 0)$ and $F_2(x_2, 0)$, for $x_1<0$ and $x_2>0$, be the foci of the ellipse $\frac{x^2}{9}+\frac{y^2}{8}=1$. Suppose a parabola having vertex at the origin and focus at $F_2$ intersects the ellipse at point $M$ in the first quadrant and at point $N$ in the fourth quadrant.

**53.** The orthocentre of $\Delta F_1MN$ is

(a) $\left(-\frac{9}{10},0\right)$ (b) $\left(\frac{2}{3},0\right)$

(c) $\left(\frac{9}{10},0\right)$ (d) $\left(\frac{2}{3},\sqrt{6}\right)$

**54.** If the tangents to the ellipse at $M$ and $N$ meet at $R$ and the normal to the parabola at $M$ meets the $X$-axis at $Q$, then the ratio of area of $\Delta MQR$ to area of the quadrilateral $MF_1NF_2$ is

(a) 3 : 4 (b) 4 : 5 (c) 5 : 8 (d) 2 : 3

# Answers

## Paper 1

### Physics

1. (a) 2. (b) 3. (b) 4. (d) 5. (d) 6. (a, d) 7. (a,c)
8. (a,b,d) 9. (b,d) 10. (b,c,d) 11. (a,c,d) 12. (a,b,d) 13. (c, d) 14. (6)
15. (9) 16. (3) 17. (8) 18. (9)

### Chemistry

19. (c) 20. (c) 21. (b) 22. (b) 23. (a) 24. (b) 25. (b,c)
26. (a,c,d) 27. (a) 28. (b,d) 29. (a,b,c) 30. (b,c) 31. (b,c,d) 32. (5)
33. (9) 34. (5) 35. (6) 36. (4)

### Mathematics

37. (a) 38. (c) 39. (c) 40. (c) 41. (c) 42. (a, d) 43. (b,c,d)
44. (a,c,d) 45. (a,c) 46. (a,b,c) 47. (b, c) 48. (a) 49. (b, c) 50. (5)
51. (2) 52. (1) 53. (1)

## Paper 2

### Physics

1. (c) 2. (c) 3. (c) 4. (b) 5. (d) 6. (a) 7. (c,d)
8. (a,c) 9. (b,c) 10. (a,b,d) 11. (a,b,d) 12. (a,b) 13. (a,b,c,d) 14. (d)
15. (c) 16. (b) 17. (d) 18. (a)

### Chemistry

19. (a) 20. (a) 21. (a) 22. (d) 23. (a) 24. (d) 25. (a,c)
26. (b,c,d) 27. (c) 28. (a,b,c) 29. (b,d) 30. (a,b) 31. (b,c) 32. (b,c,d)
33. (b) 34. (c) 35. (a) 36. (b)

### Mathematics

37. (b) 38. (c) 39. (c) 40. (b) 41. (a) 42. (c) 43. (a,b)
44. (b,c) 45. (a,d) 46. (b,c) 47. (a,c,d) 48. (a,c,d) 49. (b,c,d) 50. (b,c)
51. (b) 52. (c) 53. (a) 54. (c)

# Answer *with* Explanations

## Paper 1

## Physics

**1.** *(a)*

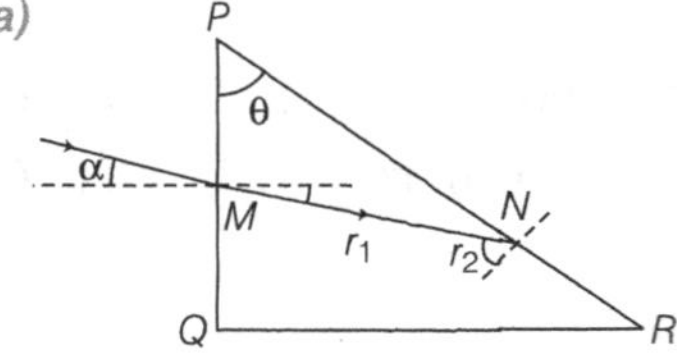


Applying Snell's law at $M$,

$$n = \frac{\sin\alpha}{\sin r_1} \Rightarrow \sqrt{2} = \frac{\sin 45^\circ}{\sin r_1}$$

$$\Rightarrow \quad \sin r_1 = \frac{\sin 45^\circ}{\sqrt{2}} = \frac{1/\sqrt{2}}{\sqrt{2}} = \frac{1}{2}$$

$$r_1 = 30^\circ$$

$$\sin\theta_c = \frac{1}{n} = \frac{1}{\sqrt{2}} \Rightarrow \theta_c = 45^\circ$$

Let us take $r_2 = \theta_c = 45^\circ$ for just satisfying the condition of TIR.

In $\Delta PNM$,

$$\theta + 90 + r_1 + 90 - r_2 = 180^\circ$$

or $\quad \theta = r_2 - r_1 = 45^\circ - 30^\circ = 15^\circ$

**Note** If $\alpha > 45^\circ$ (the given value). Then, $r_1 > 30^\circ$ (the obtained value)

$$r_2 > \theta_c \text{ (as } r_2 - r_1 = \theta \text{ or } r_2 = \theta + r_1)$$

or TIR will take place. So, for taking TIR under all conditions $\alpha$ should be greater than 45° or this is the minimum value of $\alpha$.

**2.** *(b)* $\frac{hc}{\lambda} - \phi = eV_0 \qquad (\phi = \text{work function})$

$$\frac{hc}{0.3\times10^{-6}} - \phi = 2e \qquad \text{...(i)}$$

$$\frac{hc}{0.4\times10^{-6}} - \phi = 1e \qquad \text{...(ii)}$$

Subtracting Eq. (ii) from Eq. (i)

$$hc\left(\frac{1}{0.3} - \frac{1}{0.4}\right)10^6 = e$$

$$hc\left(\frac{0.1}{0.12}\times10^6\right) = e$$

$$h = 0.64\times10^{-33}$$

$$= 6.4\times10^{-34}\,\text{J-s}$$

**3.** *(b)* Heat generated in device in 3 h

$$= \text{Time}\times\text{power}$$

$$= 3\times3600\times3\times10^3 = 324\times10^5\,\text{J}$$

Heat used to heat water $= ms\Delta\theta$

$$= 120\times1\times4.2\times10^3\times20\text{J}$$

Heat absorbed by coolant

$$Pt = 324\times10^5 - 120\times1\times4.2\times10^3\times20\text{J}$$

$$Pt = (325 - 100.8)\times10^5\,\text{J} = 223.2\times10^5\,\text{J}$$

$$P = \frac{223.2\times10^5}{3600} = 2067\text{ W}$$

**4.** *(d)*

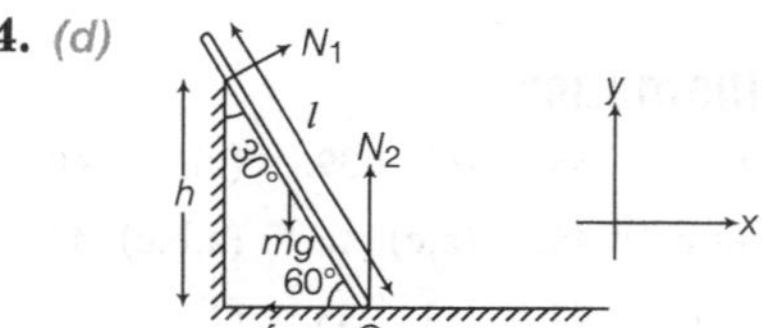


$$\Sigma F_x = 0, \quad N_1\cos 30^\circ - f = 0 \qquad \text{...(i)}$$

$$\Sigma F_y = 0, \quad N_1\sin 30^\circ + N_2 - mg = 0 \qquad \text{...(ii)}$$

$$\Sigma\tau_0 = 0$$

$$mg\frac{l}{2}\cos 60^\circ - N_1\frac{h}{\cos 30^\circ} = 0 \qquad \text{...(iii)}$$

Also, given $\quad N_1 = N_2 \qquad$ ...(iv)

Solving Eqs. (i), (ii), (iii) and (iv) we have

$$\frac{h}{l} = \frac{3\sqrt{3}}{16} \text{ and } f = \frac{16\sqrt{3}}{3}$$

**5.** *(d)* Suppose charger per unit length at any instant is $\lambda$. Initial value of $\lambda$ is suppose $\lambda_0$. Electric field at a distance $r$ at any instant is

$$E = \frac{\lambda}{2\pi\varepsilon r}$$

$$J = \sigma E = \sigma\frac{\lambda}{2\pi\varepsilon r}$$

$$i = \frac{dq}{dt} = J(A) = -J\sigma 2\pi rl$$

$$\frac{d\lambda l}{dt} = -\frac{\lambda}{2\pi\varepsilon r}\times\sigma 2\pi rl \qquad (q = \lambda l)$$

$$\int_{\lambda_o}^{\lambda}\frac{d\lambda}{\lambda} = -\frac{\sigma}{\varepsilon}\int_o^t dt \Rightarrow \lambda = \lambda_0 e^{-\frac{\sigma}{\varepsilon}t}$$

λ

$$J = \frac{\sigma}{2\pi\varepsilon r}\lambda = \frac{\sigma\lambda_0}{2\pi\varepsilon r}e^{-\frac{\sigma}{\varepsilon}t} = J_0 e^{-\frac{\sigma}{\varepsilon}t}$$

Here, $J_0 = \dfrac{\sigma\lambda_0}{2\pi\varepsilon r}$

$\therefore$ $J(t)$ decreases exponentially as shown in figure below.

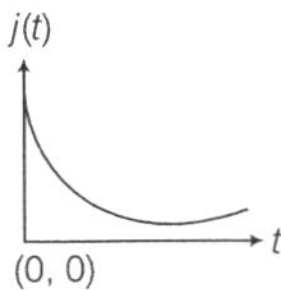


**6.** *(a,d)* **Case 1**

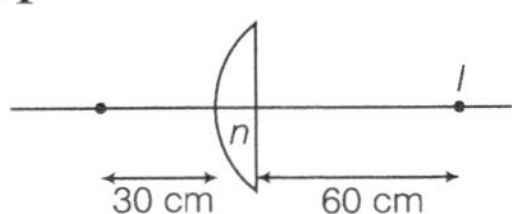


Using lens formula,

$$\frac{1}{60} + \frac{1}{30} = \frac{1}{f}$$

$$\Rightarrow \quad \frac{1}{f} = \frac{1}{60} + \frac{2}{60} \Rightarrow f = 20 \text{ cm}$$

Further, $\dfrac{1}{f_1} = (n-1)\left(\dfrac{1}{R} - \dfrac{1}{\infty}\right)$

$$\Rightarrow \quad f_1 = \frac{R}{n-1} = +20 \text{ cm}$$

**Case 2**

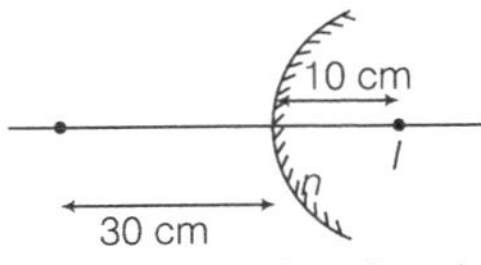


Using mirror formula, $\dfrac{1}{10} - \dfrac{1}{30} = \dfrac{1}{f_2}$

$$\frac{3}{30} - \frac{1}{30} = \frac{1}{f_2} = \frac{2}{30}$$

$$f_2 = 15 = \frac{R}{2} \Rightarrow R = 30$$

$$R = 30 \text{ cm}$$

$$\frac{R}{n-1} = +20 \text{ cm} = \frac{30}{n-1}$$

$$\Rightarrow \quad 2n - 2 = 3$$

$$\Rightarrow \quad f_1 = +20 \text{ cm}$$

Refractive index of lens is 2.5. Radius of curvature of convex surface is 30 cm. Faint image is erect and virtual. Focal length of lens is 20 cm.

**7.** *(a,c)* By reciprocity theorem of mutual induction, it can be assumed that current in infinite wire is varying at 10A/s and EMF is induced in triangular loop.

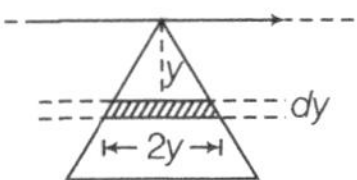


Flux of magnetic field through triangle loop, if current in infinite wire is $\phi$, can be calculated as follows:

$$d\phi = \frac{\mu_0 i}{2\pi y}\cdot 2y\,dy$$

$$d\phi = \frac{\mu_0 i}{\pi}dy \Rightarrow \phi = \frac{\mu_0 i}{\pi}\left(\frac{l}{\sqrt{2}}\right)$$

$$\Rightarrow \quad \text{EMF} = \left|\frac{d\phi}{dt}\right| = \frac{\mu_0}{\pi}\left(\frac{l}{\sqrt{2}}\right)\cdot\frac{di}{dt}$$

$$= \frac{\mu_0}{\pi}(10\,\text{cm})\left(10\frac{\text{A}}{\text{s}}\right) = \frac{\mu_0}{\pi}\text{ volt}$$

If we assume the current in the wire towards right then as the flux in the loop increases we know that the induced current in the wire is counter clockwise. Hence, the current in the wire is towards right. Field due to triangular loop at the location of infinite wire is into the paper. Hence, force on infinite wire is away from the loop. By cylindrical symmetry about infinite wire, rotation of triangular loop will not cause any additional EMF.

**8.** *(a,b,d)* $\mathbf{r} = \alpha t^3\hat{\mathbf{i}} + \beta t^2\hat{\mathbf{j}}$

$$\mathbf{v} = \frac{d\mathbf{r}}{dt} = 3\alpha t^2\hat{\mathbf{i}} + 2\beta t\hat{\mathbf{j}}$$

$$\mathbf{a} = \frac{d^2\mathbf{r}}{dt^2} = 6\alpha t\,\hat{\mathbf{i}} + 2\beta\hat{\mathbf{j}}$$

At $t = 1$ s,

(a) $\mathbf{v} = 3\times\dfrac{10}{3}\times 1\hat{\mathbf{i}} + 2\times 5\times 1\hat{\mathbf{j}}$

$$= (10\hat{\mathbf{i}} + 10\hat{\mathbf{j}})\text{ m/s}$$

(b) $\hat{\mathbf{L}} = \hat{\mathbf{r}}\times\hat{\mathbf{p}}$

$$= \left(\frac{10}{3}\times 1\hat{\mathbf{i}} + 5\times 1\hat{\mathbf{j}}\right)\times 0.1(10\hat{\mathbf{i}} + 10\hat{\mathbf{j}})$$

$$= \left(-\frac{5}{3}\hat{\mathbf{k}}\right)\text{N-ms}$$

(c) $\mathbf{F} = m\mathbf{a}$

$$= m\times\left(6\times\frac{10}{3}\times 1\hat{\mathbf{i}} + 2\times 5\hat{\mathbf{j}}\right) = (2\hat{\mathbf{i}} + \hat{\mathbf{j}})\text{N}$$

(d) $\tau = r \times \mathbf{F} = \left(\frac{10}{3}\hat{\mathbf{i}} + 5\hat{\mathbf{j}}\right) \times (2\hat{\mathbf{i}} + \hat{\mathbf{j}})$

$= +\frac{10}{3}\hat{\mathbf{k}} + 10(-\hat{\mathbf{k}}) = \left(-\frac{20}{3}\hat{\mathbf{k}}\right)$ N-m

**9.** *(b,d)* $[n] = [L^{-3}]$; $[q] = [AT]$

$[\varepsilon] = [M^{-1}L^{-3}A^2T^4]$

$[T] = [L]$

$[l] = [L]$

$[k_B] = [M^1L^2T^{-2}K^{-1}]$

(a) RHS

$= \sqrt{\frac{[L^{-3}A^2T^2]}{[M^{-1}L^{-3}T^4A^2][M^1L^2T^{-2}K^{-1}][K]}}$

$= \sqrt{\frac{[L^{-3}A^2T^2]}{[L^{-1}T^2A^2]}} = \sqrt{[L^{-2}]} = [L^{-1}]$ Wrong

(b) RHS

$= \sqrt{\frac{[M^{-1}L^{-3}T^4A^2][M^1L^2T^{-2}K^{-1}][K]}{[L^{-3}][A^2T^2]}}$

$= \sqrt{\frac{[L^{-1}T^2A^2]}{[L^{-3}T^2A^2]}} = [L]$ Correct

(c) RHS

$= \sqrt{\frac{[A^2T^2]}{[M^{-1}L^{-3}T^4A^2][L^{-2}][M^1L^2T^{-2}K^{-1}][K]}}$

$= \sqrt{[L^3]}$ Wrong

(d) RHS

$= \sqrt{\frac{[A^2T^2]}{[M^{-1}L^{-3}T^4A^2][L^{-1}][M^1L^2T^{-2}K^{-1}]}}$

$= \sqrt{\frac{[A^2T^2]}{[L^{-2}T^2A^2]}} = [L]$ Correct

**10.** *(b,c,d)*

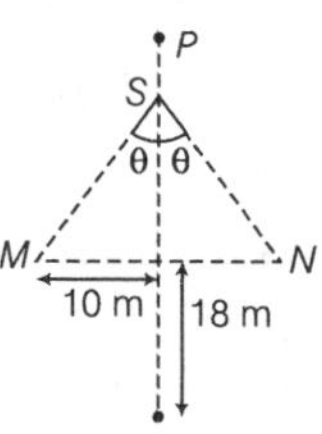


speed of car, $v = 60\,\text{km/h} = \frac{500}{3}$ m/s

At a point $S$, between $P$ and $Q$

$v'_M = v_M\left(\frac{C + v\cos\theta}{C}\right)$;

$v'_N = v_N\left(\frac{C + v\cos\theta}{C}\right)$

$\Rightarrow \Delta v = (v_N - v_M)\left(1 + \frac{v\cos\theta}{C}\right)$

Similarly, between $Q$ and $R$

$\Delta v = (v_N - v_M)\left(1 - \frac{v\cos\theta}{C}\right)$

$\frac{d(\Delta v)}{dt} = \pm(v_N - v_M)\frac{v}{C}\sin\theta\frac{d\theta}{dt}$

$\theta \approx 0°$ at $P$ and $R$ as they are large distance apart.

$\Rightarrow$ Slope of graph is zero.

at $Q$, $\theta = 90°$

$\sin\theta$ is maximum also value of $\frac{d\theta}{dt}$ is maximum as $\frac{d\theta}{dt} = \frac{v}{r}$, where $v$ is its velocity and $r$ is the length of the line joining $P$ and $S$. and $r$ is minimum at $Q$.

$\Rightarrow$ Slope is maximum at $Q$.

At $P$, $v_P = \Delta v = (v_N - v_M)\left(1 + \frac{v}{C}\right)$ $(\theta \approx 0°)$

At $R$, $v_R = \Delta v = (v_N - v_M)\left(1 - \frac{v}{C}\right)$ $(\theta \approx 0°)$

At $Q$, $v_Q = \Delta v = (v_N - v_M)$ $(\theta = 90°)$

From these equations, we can see that

$$v_P + v_R = 2v_Q$$

**11.** *(a,c,d)* From Snell's law,

$n\sin\theta = \text{constant}$

$\therefore\; n_1\sin\theta_i = n_2\sin\theta_f$

Further, $l$ will depend on $n_1$ and $n(z)$. But it will be independent of $n_2$.

**12.** *(a,b,d)* As radius $r \propto \frac{n}{z}$

$\Rightarrow \frac{\Delta r}{r} = \frac{\left(\frac{n+1}{z}\right)^2 - \left(\frac{n}{z}\right)^2}{\left(\frac{n}{z}\right)^2} = \frac{2n+1}{n^2} \approx \frac{2}{n} \propto \frac{1}{n}$

Energy $E \propto \frac{z^2}{n^2}$

$$\Rightarrow \quad \frac{\Delta E}{E} = \frac{\frac{z^2}{n^2} - \frac{z^2}{(n+1)^2}}{\frac{z^2}{(n+1)^2}}$$

$$= \frac{(n+1)^2 - n^2}{n^2 \cdot (n+1)^2} \cdot (n+1)^2$$

$$\Rightarrow \quad \frac{\Delta E}{E} = \frac{2n+1}{n^2} \simeq \frac{2n}{n^2} \propto \frac{1}{n}$$

Angular momentum, $L = \frac{nh}{2\pi}$

$$\Rightarrow \quad \frac{\Delta L}{L} = \frac{\frac{(n+1)h}{2\pi} - \frac{nh}{2\pi}}{\frac{nh}{2\pi}} = \frac{1}{n} \propto \frac{1}{n}$$

**13.** *(c,d)* Because of non-uniform evaporation at different section, area of cross-section would be different at different sections.

Region of highest evaporation rate would have rapidly reduced area and would become break up cross-section.

Resistance of the wire as whole increases with time.

Overall resistance increases hence power decreases.

$\left(P = \frac{V^2}{R} \text{ or } P \propto \frac{1}{R} \text{ as } V \text{ is constant}\right)$. At break up junction temperature would be highest, thus light of highest band frequency would be emitted at those cross-section.

**14.** **(6)** Energy of incident light (in eV)

$$E = \frac{12375}{970} = 12.7\,\text{eV}$$

After excitation, let the electron jumps to $n$th state, then $\frac{-13.6}{n^2} = -13.6 + 12.7$

Solving this equation, we get

$$n = 4$$

$\therefore$ Total number of lines in emission spectrum, $= \frac{n(n-1)}{2} = \frac{4(4-1)}{2} = 6$

**15.** **(9)** ${}^{12}_{5}B \longrightarrow {}^{12}_{6}C + {}^{0}_{-1}e + \bar{\nu}$

Mass of ${}^{12}_{6}\text{C} = 12.000$ u (by definition of 1 amu)

$Q$-value of reaction, $Q = (M_B - M_C) \times c^2$

$= (12.014 - 12.000) \times 931.5 = 13.041$ MeV

4.041MeV of energy is taken by ${}^{12}_{6}\text{C}$ *

$\Rightarrow$ Maximum KE of $\beta$-particle is $(13.041 - 4.041) = 9$ MeV

**16.** **(3)** Terminal velocity is given by

$$v_T = \frac{2}{9}\frac{r^2}{\eta}(d - \rho)g$$

$$\frac{v_P}{v_Q} = \frac{r_P^2}{r_Q^2} \times \frac{\eta_Q}{\eta_P} \times \frac{(d - \rho_P)}{(d - \rho_Q)}$$

$$= \left(\frac{1}{0.5}\right)^2 \times \left(\frac{2}{3}\right) \times \frac{(8 - 0.8)}{(8 - 1.6)} = 4 \times \frac{2}{3} \times \frac{7.2}{6.4} = 3$$

**17.** **(8)**

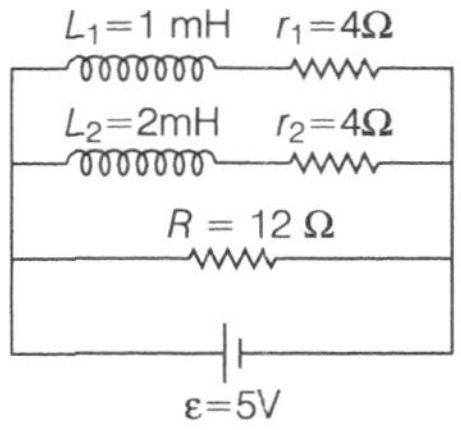

$$I_{\max} = \frac{\varepsilon}{R} = \frac{5}{12}\text{A} \qquad \text{(Initially at } t = 0\text{)}$$

$$I_{\min} = \frac{\varepsilon}{R_{eq}} = \varepsilon\left(\frac{1}{r_1} + \frac{1}{r_2} + \frac{1}{R}\right)$$

(finally in steady state)

$$= 5\left(\frac{1}{3} + \frac{1}{4} + \frac{1}{12}\right) = \frac{10}{3}\text{A}$$

$$\frac{I_{\max}}{I_{\min}} = 8$$

**18.** **(9)** $\log_2 \frac{p_1}{p_0} = 1$

Therefore, $\frac{p_1}{p_0} = 2$

According to Stefan's law, $p \propto T$

$$\Rightarrow \quad \frac{p_2}{p_1} = \left(\frac{T_2}{T_1}\right)^4 = \left(\frac{2767 + 273}{487 + 273}\right)^4 = 4^4$$

$$\frac{p_2}{p_1} = \frac{p_2}{2p_0} = 4^4 \quad \Rightarrow \quad \frac{p_2}{p_0} = 2 \times 4^4$$

$$\log_2 \frac{p_2}{p_0} = \log_2 [2 \times 4^4]$$

$$= \log_2 2 + \log_2 4^4$$

$$= 1 + \log_2 2^8 = 1 + 8 = 9$$

# Chemistry

**19.** *(c)* This graph shows the probability of finding the electron within shell at various distances from the nucleus (radial probability). The curve shows the maximum, which means that the radial probability is greatest for a given distance from the nucleus. This distance is equal to Bohr's radius = $a_0$

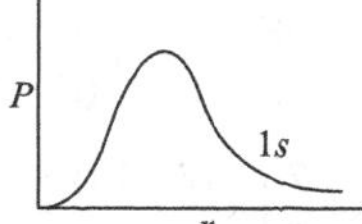


(a) It is for 2*s*-orbital.
(b) It is radial wave function for 1*s*.
(c) Correct
(d) Probability cannot be zero at a certain distance from nucleus.

**20.** *(c)* By first law, $\Delta E = Q + W$

For isothermal expansion, $\Delta E = 0$

$\therefore \quad Q = -W$

$$-Q_{\text{irrev}} = W_{\text{irrev}} = p\Delta V$$
$$= 3(2-1) = 3 \text{ L atm}$$

Also, $\Delta S_{\text{surr}} = \dfrac{Q_{\text{irrev}}}{T} = \dfrac{(-3 \times 101.3)\text{ J}}{300 \text{ K}}$

$$= -\frac{303.9}{300} = -1.013 \text{ JK}^{-1}$$

**21.** *(b)*

| Compounds | Hybridisation | Unpaired electron (s) | Magnetic character |
|---|---|---|---|
| $Ni(CO)_4$ | $sp^3$ | No | Diamagnetic |
| $[NiCl_4]^{2-}$ | $sp^3$ | two | Paramagnetic |
| $[Co(NH_3)_4Cl_2]Cl$ | $sp^3d^2$ | No | Diamagnetic |
| $Na_3[CoF_6]$ | $sp^3d^2$ | three | Paramagnetic |
| $Na_2O_2$ | — | No | Diamagnetic ($O_2^{2-}$) |
| $CsO_2$ | — | One | Paramagnetic |
| | | | $O_2^-$ (superoxide ion is paramagnetic) |

**22.** *(b)* Due to poor shielding of *d*-orbital in Ga, atomic radius of Ga is smaller than that of Al. Thus, Ga < Al < In < Tl.

**23.** *(a)* Natural rubber is formed by polymerisation of isoprene.

$$CH_2{=}\underset{\displaystyle CH_3}{\underset{|}{C}}{-}CH{=}CH_2 \longrightarrow \overset{\oplus}{C}H_2{-}\underset{\displaystyle CH_3}{\underset{|}{C}}{=}CH{-}\overset{\ominus}{C}H_2 \xrightarrow{\text{Polymerise}} \left[CH_2{-}\underset{\displaystyle CH_3}{\underset{|}{C}}{=}CH{-}CH_2\right]_n$$

Natural rubber

$$\left[CH_2{-}\underset{\displaystyle CH_3}{\underset{|}{C}}{=}CH{-}CH_2\right]_n \xrightarrow{H_2/Pt} \left[CH_2{-}\underset{\displaystyle CH_3}{\underset{|}{CH}}{-}CH_2{-}CH_2\right]_n$$

Propylene; Ethylene

Ethylene-propylene copolymer

This co-polymer is formed from propylene and ethylene.

$$nCH_2{=}\underset{\displaystyle CH_3}{\underset{|}{CH}} + nCH_2{=}CH_2 \longrightarrow \left[CH_2{-}\underset{\displaystyle CH_3}{\underset{|}{CH}}{-}CH_2{-}CH_2\right]_n$$

**24.** *(b)*

$$C_6H_5NH_2 \xrightarrow{Ac_2O/\text{Pyridine}} C_6H_5NHAc$$

$$KBrO_3 + HBr \longrightarrow Br_2$$

Ac is $CH_3CO$ (acetyl), it protects —$NH_2$ group from being oxidised.

$$C_6H_5NHAc \xrightarrow{Br_2 \text{ water}} p\text{-}Br{-}C_6H_4{-}NHAc \xrightarrow{H_3O^+} p\text{-}Br{-}C_6H_4{-}NH_2 + CH_3COOH$$

$NH_2$–C$_6$H$_4$–Br $\xrightarrow[\text{273-278 K (Diazotisation)}]{NaNO_2/HCl}$ $N_2Cl$–C$_6$H$_4$–Br $\xrightarrow[\text{(Sandmeyer reaction)}]{CuBr/HBr}$ Br–C$_6$H$_4$–Br

**25.** *(b, c)*

Cumene ($C_6H_5$–$CH(CH_3)_2$) $\xrightarrow{O_2}$ $C_6H_5$–O–O–CH($CH_3$)$_2$ $\xrightarrow{H_3O^+}$

Phenol (*P*) + $CH_3$–CO–$CH_3$ (Acetone) $\xrightarrow[\text{, Reimer-Tiemann reaction}]{CHCl_3/NaOH}$ salicylaldehyde (OH······O intramolecular H-bond) (*Q*) Major + *p*-hydroxybenzaldehyde (O—H, CHO) (*R*) Minor,

*o*-HO–C$_6$H$_4$–CHO (*Q*) Major $\xrightarrow{NaOH}$

*o*-(O–Na Br CH$_2$–Ph)–C$_6$H$_4$–CHO $\xrightarrow{\text{Williamson's synthesis}}$ *o*-(O–CH$_2$Ph)–C$_6$H$_4$–CHO (*S*)

(a) *R* is not steam volatile, but *Q* is steam volatile thus, incorrect.

(b) *Q* has enolic group thus, gives violet colour with 1% aqueous $FeCl_3$ solution thus, correct.

(c) *S* has carbonyl group hence, gives yellow precipitate with 2, 4-DNP thus, correct.

(d) *S* does not give colour with $FeCl_3$ thus, incorrect.

**26.** *(a, c, d)* $Na_2B_4O_7 \cdot 10H_2O$ (borax) is actually made of two tetrahedral and two triangular units, and is actually written as $Na_2[B_4O_5(OH)_4] \cdot 5H_2O$.

[Structure of $[B_4O_5(OH)_4]^{2-}$: HO—B, B—OH, B—OH (two tetrahedral B, two triangular B, bridged by O)]

(a) Thus, correct.

(b) Boron atoms are in different planes thus, incorrect.

(c) Two $sp^2$ and two $sp^3$-hybridised B atoms thus, correct.

(d) Each boron has one —OH group thus, correct.

**27.** *(a)* $S^{2-} + CuCl_2 \longrightarrow CuS\downarrow$ (black ppt.)

$SO_4^{2-} + CuCl_2 \longrightarrow$ Soluble, Thus

(a) $CuCl_2$ selectively precipitates $S^{2-}$.

(b) $S^{2-} + BaCl_2 \longrightarrow BaS\downarrow$ (soluble)

$SO_4^{2-} + BaCl_2 \longrightarrow BaSO_4\downarrow$ (white ppt.)

(b) precipitates $SO_4^{2-}$ and not $S^{2-}$.

(c) $S^{2-} + Pb^{2+} \longrightarrow PbS\downarrow$ (black ppt.)

$SO_4^{2-} + Pb^{2+} \longrightarrow PbSO_4\downarrow$ (white ppt.)

$S^{2-}$ and $SO_4^{2-}$, both are precipitated.

(d) $S^{2-} + Na_2[Fe(CN)]_5NO \longrightarrow$ (Sodium nitroprusside)

$Na_4[Fe(CN)]_5NOS]$ (Purple colour)

But no colour with $SO_4^{2-}$.

**28.** *(b, d)* For the elements with atomic number ($Z$) larger than 20,

Neutrons ($n$) > Protons ($p$); Thus, $n/p > 1$ Thus, there is upward deviation from linearity.

If $n < p$, Thus $n/p < 1$, then

(a) By $\beta^-$ - decay, ${}^{1}_{0}n \longrightarrow {}^{1}_{1}p + {}^{0}_{-1}e$ neutron changes to proton. Thus, $(n/p)$ ratio further decreases below 1. Thus, this decay is not allowed.

(b) By orbital or $K$- electron capture, ${}^{1}_{1}p + {}^{0}_{-1}e \longrightarrow {}^{1}_{0}n$ proton changes to neutron, hence, $(n/p)$ ratio increases. Thus stability increases. Thus correct.

(c) Neutron emission further decreases $n/p$ ratio.

(d) By $\beta^+$-emission, ${}^{1}_{1}p \longrightarrow {}^{1}_{0}n + {}^{0}_{+1}e$ proton changes to neutron. Hence, $n/p$ ratio increases. Thus correct.

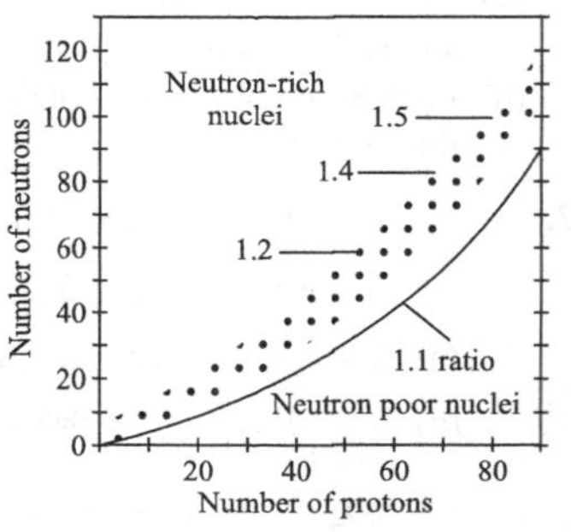


Plot of the number of neutrons against the number of protons in stable nuclei (shown by dots).

**29.** *(a, b, c)*

$$RCHO + \underset{\text{(Tollen's reagent)}}{Ag_2O} \xrightarrow{OH^-} RCOOH + \underset{\text{Silver mirror}}{2Ag}$$

Tollen's test is given by all aldehydes and all reducing sugars as glucose, fructose and α-hydroxy ketones

$$-\underset{\alpha}{\overset{\overset{OH}{|}}{CH}}-\overset{\overset{O}{||}}{C}-$$

(a) $H(Br)C{=}CH{-}CHO$ having an aldehyde group gives Tollen's test.

(b) $C_6H_5CHO$
Aldehyde

(c) $C_6H_5-\underset{\underset{OH}{|}}{CH}-\overset{\overset{O}{||}}{C}-C_6H_5$
α-hydroxy ketone

**30.** *(b,c)*

| Compounds | Hydridisation | Structures | Lone pair on central atom |
|---|---|---|---|
| $BrF_5$ | $sp^3d^2$ | | 1 |
| $ClF_3$ | $sp^3d$ | | 2 |
| $XeF_4$ | $sp^3d^2$ | | 2 |
| $SF_4$ | $sp^3d$ | | 1 |

**31.** *(b, c, d)*

Rate constant, $k = Ae^{-E_a/RT}$ where,

$E_a$ = activation energy and

$A$ = pre exponential factor

(a) If $E_a$ is high, it means lower value of $k$ hence, slow reaction. Thus, incorrect.

(b) On increasing temperature, molecules are raised to higher energy (greater than $E_a$), hence number of collisions increases. Thus, correct.

(c) $\log k = \log A - \dfrac{E_a}{RT}$

$$\Rightarrow \quad \frac{d(\log k)}{dT} = \frac{E_a}{RT^2}$$

Thus, when $E_a$ is high, stronger is the temperature dependence of the rate constant. Thus, correct.

(d) Pre-exponential factor ($A$) is a measure of rate at which collisions occur. Thus, correct.

**32. (5)** Given compound undergoes free-radical bromination under given conditions, replacing H by Br.

C* is chiral carbon.

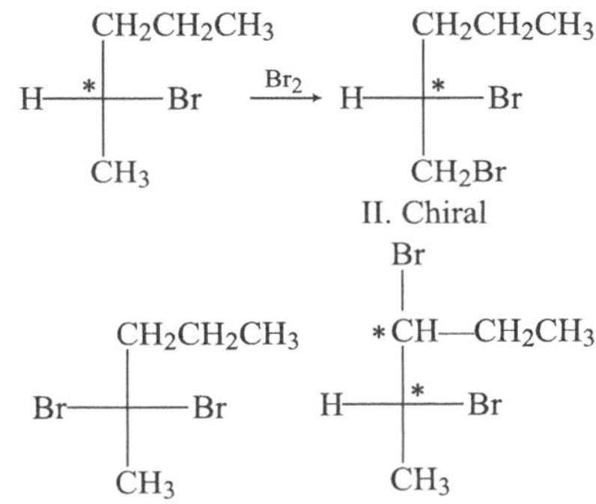

II. Chiral

II. Achiral

II. Chiral

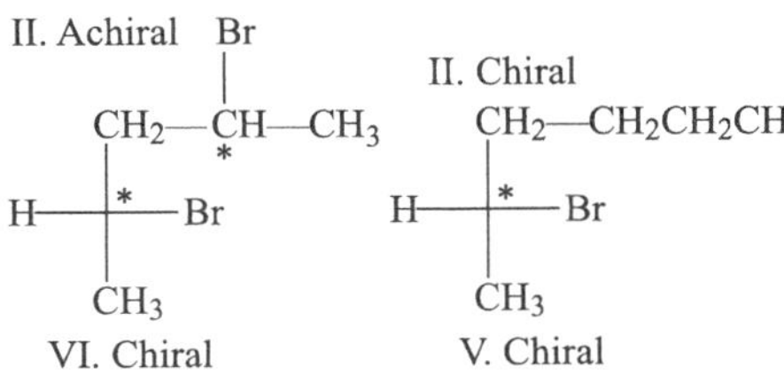

VI. Chiral

V. Chiral

(III) has two chiral centres and can have two structures.

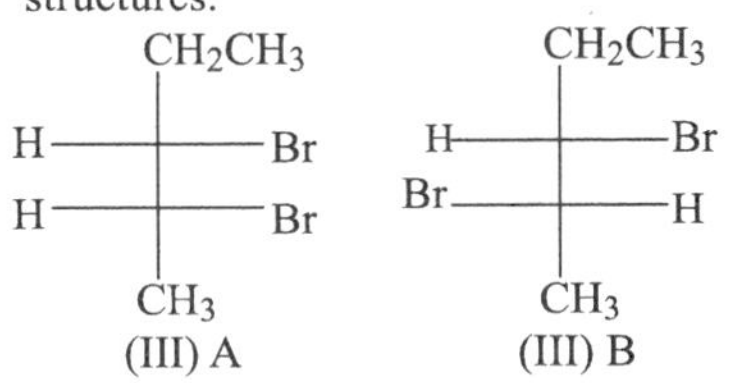

(III) A

(III) B

(IV) has also two chiral centres and can have two structures.

(IV)A

(IV)B

It has plane of symmetry thus, achiral.

Thus, chiral compounds are five. I, III A, III B, IV B and V.

**33. (9)** Moles of solute, $n_1 = \frac{w_1}{m_1}$; Moles of solvent, $n_2 = \frac{w_2}{m_2}$

$\chi_1$ (solute) = 0.1 and $\chi_2$ (solvent) = 0.9

$$\therefore \quad \frac{\chi_1}{\chi_2} = \frac{n_1}{n_2} = \frac{w_1}{m_1} \cdot \frac{m_2}{w_2} = \frac{1}{9}$$

$$\text{Molarity} = \frac{\text{Solute (moles)}}{\text{Volume (L)}} = \frac{w_1 \times 1000 \times 2}{m_1\,(w_1 + w_2)}$$

**Note** $\text{Volume} = \frac{\text{Total mass of solution}}{\text{Density}} = \left(\frac{w_1 + w_2}{2}\right) \text{mL}$

$$\text{Molality} = \frac{\text{Solute (moles)}}{\text{Solvent (kg)}} = \frac{w_1 \times 1000}{m_1 \times w_2}$$

Given, molarity = molality

hence, $\frac{2000 w_1}{m_1\,(w_1 + w_2)} = \frac{1000\, w_1}{m_1\, w_2}$

$\therefore \quad \frac{w_2}{w_1 + w_2} = \frac{1}{2}$

$\Rightarrow \quad w_1 = w_2 = 1$

$\therefore \quad \frac{w_1\, m_2}{m_1\, w_2} = \frac{1}{9}$

$\Rightarrow \quad \frac{m_1 \text{(solute)}}{m_2 \text{ (solvent)}} = 9$

**34. (5)** Ligand is $CH_2 - NH_2$ | $CH_2\bar{O}$

Geometrical isomers are

*Cis* *Trans* *Cis*

*Cis* *Trans*

**35.** **(6)** In neutral or faintly alkaline solution, $MnO_4^-$ is reduced to $MnO_2$ and $S_2O_3^{2-}$ is oxidised to $SO_4^{2-}$.

Change in ON = 4 units

$$\underset{+7}{Mn}O_4 + 1/2\ \underset{+2}{S_2}O_3^{2-} \longrightarrow \underset{+6}{S}O_4^{2-} + \underset{+4}{Mn}O_2$$

Change in ON = 3 units

Thus, $4MnO_4^- + \frac{3}{2}S_2O_3^{2-} \longrightarrow 3SO_4^{2-} + 4MnO_2$

or $8MnO_4^- + 3S_2O_3^{2-} \longrightarrow 6SO_4^{2-} + 8MnO_2$

Thus, moles of $SO_4^{2-}$ formed by 8 moles of $MnO_4^- = 6$

**36.** **(4)** (DC) Diffusion coefficient $\propto \lambda$ (mean free path) $\propto U_{mean}$, thus (DC) $\propto \lambda U_{mean}$

But, $\lambda = \frac{RT}{\sqrt{2}N_0\sigma p} \Rightarrow \lambda \propto \frac{T}{p}$

and $U_{mean} = \sqrt{\frac{8RT}{\pi M}} \Rightarrow U_{mean} \propto \sqrt{T}$

$\therefore$ $$\text{DC} \propto \frac{(T)^{3/2}}{p}$$

$$\frac{(DC)_2}{(DC)_1}(x) = \left(\frac{p_1}{p_2}\right)\left(\frac{T_2}{T_1}\right)^{3/2}$$

$$= \left(\frac{p_1}{2p_1}\right)\left(\frac{4T_1}{T_1}\right)^{3/2} = \left(\frac{1}{2}\right)(8) = 4$$

# Mathematics

**37.** *(a)* We have, 6 girls and 4 boys. To select 4 members (atmost one boy)
i.e. (1 boy and 3 girls) or (4 girls)

$$= {}^6C_3 \cdot {}^4C_1 + {}^6C_4 \quad \text{...(i)}$$

Now, selection of captain from 4 members = ${}^4C_1$ ...(ii)

$\therefore$ Number of ways to select 4 members (including the selection of a captain, from these 4 members)

$$= ({}^6C_3 \cdot {}^4C_1 + {}^6C_4)\,{}^4C_1$$
$$= (20 \times 4 + 15) \times 4$$
$$= 380$$

**38.** *(c)* Here, to find the least value of $\alpha \in R$, for which $4\alpha x^2 + \frac{1}{x} \geq 1$, for all $x > 0$.

i.e. to find the minimum value of $\alpha$ when $y = 4\alpha x^2 + \frac{1}{x}$; $x > 0$ attains minimum value of $\alpha$.

$\therefore$ $$\frac{dy}{dx} = 8\alpha x - \frac{1}{x^2} \quad \text{...(i)}$$

Now, $$\frac{d^2y}{dx^2} = 8\alpha + \frac{2}{x^3} \quad \text{...(ii)}$$

when $\frac{dy}{dx} = 0$,

then $8x^3\alpha = 1$

$\frac{d^2y}{dx^2} = 8\alpha + 16\alpha = 24\alpha$, Thus, $y$ attains minimum when

$$x = \left(\frac{1}{8\alpha}\right)^{1/3}; \alpha > 0.$$

$\therefore$ $y$ attains minimum when $x = \left(\frac{1}{8\alpha}\right)^{1/3}$.

i.e. $4\alpha\left(\frac{1}{8\alpha}\right)^{2/3} + (8\alpha)^{1/3} \geq 1$

$\Rightarrow \alpha^{1/3} + 2\alpha^{1/3} \geq 1 \Rightarrow 3\alpha^{1/3} \geq 1$

$\Rightarrow$ $$\alpha \geq \frac{1}{27}$$

Hence, the least value of $\alpha$ is $\frac{1}{27}$.

**39.** *(c)* Here, $x^2 - 2x\sec\theta + 1 = 0$ has roots $\alpha_1$ and $\beta_1$.

$$\therefore \alpha_1, \beta_1 = \frac{2\sec\theta \pm \sqrt{4\sec^2\theta - 4}}{2 \times 1}$$

$$= \frac{2\sec\theta \pm 2|\tan\theta|}{2}$$

Since, $\theta \in \left(-\frac{\pi}{6}, -\frac{\pi}{12}\right)$,

i.e. $\theta \in$ IV quadrant $= \frac{2\sec\theta \mp 2\tan\theta}{2}$

$\therefore \alpha_1 = \sec\theta - \tan\theta$ and $\beta_1 = \sec\theta + \tan\theta$ [as $\alpha_1 > \beta_1$]

and $x^2 + 2x \tan\theta - 1 = 0$ has roots $\alpha_2$ and $\beta_2$.

i.e. $\alpha_2, \beta_2 = \dfrac{-2\tan\theta \pm \sqrt{4\tan^2\theta + 4}}{2}$

$\therefore \quad \alpha_2 = -\tan\theta + \sec\theta$

and $\quad \beta_2 = -\tan\theta - \sec\theta \quad [\text{as } \alpha_2 > \beta_2]$

Thus, $\quad \alpha_1 + \beta_2 = -2\tan\theta$

**40.** *(c)* Given, $\sqrt{3}\sec x + \operatorname{cosec} x + 2(\tan x - \cot x) = 0,$

$(-\pi < x < \pi) - \{0, \pm \pi/2\}$

$\Rightarrow \quad \sqrt{3}\sin x + \cos x + 2(\sin^2 x - \cos^2 x) = 0$

$\Rightarrow \quad \sqrt{3}\sin x + \cos x - 2\cos 2x = 0$

Multiplying and dividing by $\sqrt{a^2 + b^2}$, i.e. $\sqrt{3+1} = 2$, we get

$$2\left(\frac{\sqrt{3}}{2}\sin x + \frac{1}{2}\cos x\right) - 2\cos 2x = 0$$

$$\Rightarrow \left(\cos x \cdot \cos\frac{\pi}{3} + \sin x \cdot \sin\frac{\pi}{3}\right) - \cos 2x = 0$$

$$\Rightarrow \quad \cos\left(x - \frac{\pi}{3}\right) = \cos 2x$$

$$\therefore \quad 2x = 2n\pi \pm \left(x - \frac{\pi}{3}\right)$$

[since, $\cos\theta = \cos\alpha \Rightarrow \theta = 2n\pi \pm \alpha$]

$$\Rightarrow \quad 2x = 2n\pi + x - \frac{\pi}{3}$$

or $\quad 2x = 2n\pi - x + \dfrac{\pi}{3}$

$$\Rightarrow \quad x = 2n\pi - \frac{\pi}{3} \text{ or } 3x = 2n\pi + \frac{\pi}{3}$$

$$\Rightarrow \quad x = 2n\pi - \frac{\pi}{3} \text{ or } x = \frac{2n\pi}{3} + \frac{\pi}{9}$$

$$\therefore \quad x = \frac{-\pi}{3} \text{ or } x = \frac{\pi}{9}, \frac{-5\pi}{9}, \frac{7\pi}{9}$$

Now, sum of all distinct solutions

$$= \frac{-\pi}{3} + \frac{\pi}{9} - \frac{5\pi}{9} + \frac{7\pi}{9} = 0$$

**41.** *(c)* Let $x = P$ (computer turns out to be defective, given that it is produced in plant $T_2$)

$$\Rightarrow \quad x = P\left(\frac{D}{T_2}\right) \quad \ldots\text{(i)}$$

where, $D =$ Defective computer

$\therefore$ $P$ (computer turns out to be defective given that is produced in plant $T_1$) $= 10x$

i.e. $\quad P\left(\dfrac{D}{T_1}\right) = 10x \quad \ldots\text{(ii)}$

Also, $P(T_1) = \dfrac{20}{100}$ and $P(T_2) = \dfrac{80}{100}$

Given, $P$ (defective computer) $= \dfrac{7}{100}$

i.e. $\quad P(D) = \dfrac{7}{100}$

Using law of total probability,

$$P(D) = 9(T_1) \cdot P\left(\frac{D}{T_1}\right) + P(T_2) \cdot P\left(\frac{D}{T_2}\right)$$

$$\therefore \quad \frac{7}{100} = \left(\frac{20}{100}\right) \cdot 10x + \left(\frac{80}{100}\right) \cdot x$$

$$\Rightarrow \quad 7 = (280)x$$

$$\Rightarrow \quad x = \frac{1}{40} \quad \ldots\text{(iii)}$$

$$\therefore \quad P\left(\frac{D}{T_2}\right) = \frac{1}{40} \text{ and } P\left(\frac{D}{T_1}\right) = \frac{10}{40}$$

$$\Rightarrow \quad P\left(\frac{\overline{D}}{T_2}\right) = 1 - \frac{1}{40} = \frac{39}{40}$$

and $\quad P\left(\dfrac{\overline{D}}{T_1}\right) = 1 - \dfrac{10}{40} = \dfrac{30}{40} \quad \ldots\text{(iv)}$

Using Baye's theorem,

$$P\left(\frac{T_2}{\overline{D}}\right) = \frac{P(T_2 \cap \overline{D})}{P(T_1 \cap \overline{D}) + P(T_2 \cap \overline{D})}$$

$$= \frac{P(T_2) \cdot P\left(\frac{\overline{D}}{T_2}\right)}{P(T_1) \cdot P\left(\frac{\overline{D}}{T_1}\right) + P(T_2) \cdot P\left(\frac{\overline{D}}{T_2}\right)}$$

$$= \frac{\frac{80}{100} \cdot \frac{39}{40}}{\frac{20}{100} \cdot \frac{30}{40} + \frac{80}{100} \cdot \frac{39}{40}} = \frac{78}{93}$$

**42.** *(a, d)* Given,

$$(x^2 + xy + 4x + 2y + 4)\frac{dy}{dx} - y^2 = 0$$

$$\Rightarrow \quad [(x^2 + 4x + 4) + y(x + 2)]\frac{dy}{dx} - y^2 = 0$$

$$\Rightarrow [(x + 2)^2 + y(x + 2)]\frac{dy}{dx} - y^2 = 0$$

Put $x + 2 = X$ and $y = Y$, then

$$(X^2 + XY)\frac{dY}{dX} - Y^2 = 0$$

$$\Rightarrow \quad X^2 dY + XYdY - Y^2 dX = 0$$

$\Rightarrow \quad X^2dY + Y(XdY - YdX) = 0$

$\Rightarrow \quad -\dfrac{dY}{Y} = \dfrac{XdY - YdX}{X^2}$

$\Rightarrow \quad -d(\log|Y|) = d\left(\dfrac{Y}{X}\right)$

On integrating both sides, we get

$$-\log|Y| = \frac{Y}{X} + C,$$

where $\quad x + 2 = X$ and $y = Y$

$\Rightarrow \quad -\log|y| = \dfrac{y}{x+2} + C \qquad \ldots(\text{i})$

Since, it passes through the point (1, 3).

$\therefore \quad -\log 3 = 1 + C$

$\Rightarrow \quad C = -1 - \log 3$

$\quad = -(\log e + \log 3)$

$\quad = -\log 3e$

$\therefore$ Eq. (i) becomes

$$\log|y| + \frac{y}{x+2} - \log(3e) = 0$$

$\Rightarrow \quad \log\left(\dfrac{|y|}{3e}\right) + \dfrac{y}{x+2} = 0 \qquad \ldots(\text{ii})$

Now, to check option (a), $y = x + 2$ intersects the curve.

$\Rightarrow \quad \log\left(\dfrac{|x+2|}{3e}\right) + \dfrac{x+2}{x+2} = 0$

$\Rightarrow \quad \log\left(\dfrac{|x+2|}{3e}\right) = -1$

$\Rightarrow \quad \dfrac{|x+2|}{3e} = e^{-1} = \dfrac{1}{e}$

$\Rightarrow \quad |x+2| = 3$ or $x + 2 = \pm 3$

$\therefore$ $x = 1, -5$ (rejected), as $x > 0$ [given]

$\therefore$ $x = 1$ only one solution.

Thus, (a) is the correct answer.

To check option (c), we have

$$y = (x+2)^2$$

and $\quad \log\left(\dfrac{|y|}{3e}\right) + \dfrac{y}{x+2} = 0$

$\Rightarrow \log\left[\dfrac{|x+2|^2}{3e}\right] + \dfrac{(x+2)^2}{x+2} = 0$

$\Rightarrow \quad \log\left[\dfrac{|x+2|^2}{3e}\right] = -(x+2)$

$\Rightarrow \quad \dfrac{(x+2)^2}{3e} = e^{-(x+2)}$

or $\quad (x+2)^2 \cdot e^{x+2} = 3e$

$\Rightarrow \quad e^{x+2} = \dfrac{3e}{(x+2)^2}$

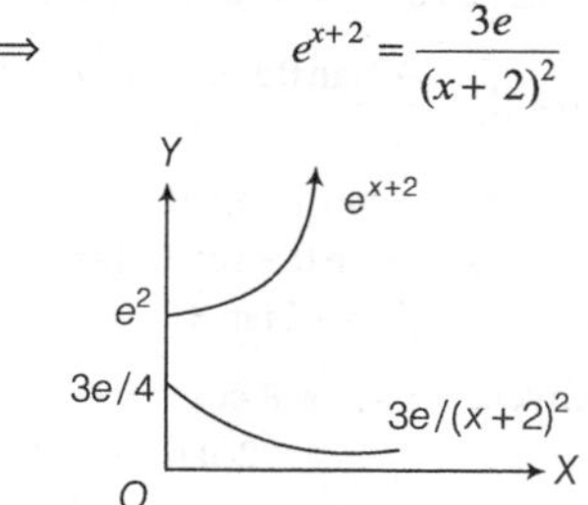


Clearly, they have no solution.

To check option (d),

$$y = (x+3)^2$$

i.e. $\log\left[\dfrac{|x+3|^2}{3e}\right] + \dfrac{(x+3)^2}{(x+2)} = 0$

To check the number of solutions.

Let $\quad g(x) = 2\log(x+3) + \dfrac{(x+3)^2}{(x+2)} - \log(3e)$

$\therefore \quad g'(x) = \dfrac{2}{x+3} + \left(\dfrac{(x+2)\cdot 2(x+3) - (x+3)^2 \cdot 1}{(x+2)^2}\right) - 0$

$= \dfrac{2}{x+3} + \dfrac{(x+3)(x+1)}{(x+2)^2}$

Clearly, when $x > 0$, then, $g'(x) > 0$

$\therefore$ $g(x)$ is increasing, when $x > 0$.

Thus, when $x > 0$, then $g(x) > g(0)$

$$g(x) > \log\left(\frac{3}{e}\right) + \frac{9}{4} > 0$$

Hence, there is no solution.

Thus, option (d) is true.

**43.** *(b, c, d)* Given, square base

$$OP = OR = 3$$

$\therefore \quad P(3,0,0), R = (0,3,0)$

Also, mid-point of $OQ$ is $T\left(\frac{3}{2}, \frac{3}{2}, 0\right)$.

Since, $S$ is directly above the mid-point $T$ of diagonal $OQ$ and $ST = 3$.

i.e. $S\left(\frac{3}{2}, \frac{3}{2}, 3\right)$

Here, DR's of $OQ$ (3, 3, 0) and DR's of $OS\left(\frac{3}{2}, \frac{3}{2}, 3\right)$.

$$\therefore \cos\theta = \frac{\frac{9}{2}+\frac{9}{2}}{\sqrt{9+9+0}\sqrt{\frac{9}{4}+\frac{9}{4}+9}}$$

$$= \frac{9}{\sqrt{18}\cdot\sqrt{\frac{27}{2}}} = \frac{1}{\sqrt{3}}$$

$\therefore$ Option (a) is incorrect.

Now, equation of the plane containing the $\Delta OQS$ is

$$\begin{vmatrix} x & y & z \\ 3 & 3 & 0 \\ 3/2 & 3/2 & 3 \end{vmatrix} = 0 \Rightarrow \begin{vmatrix} x & y & z \\ 1 & 1 & 0 \\ 1 & 1 & 2 \end{vmatrix} = 0$$

$\Rightarrow x(2-0) - y(2-0) + z(1-1) = 0$

$\Rightarrow \quad 2x - 2y = 0$ or $x - y = 0$

$\therefore$ Option (b) is correct.

Now, length of the perpendicular from $P(3, 0, 0)$ to the plane containing $\Delta OQS$ is

$$\frac{|3-0|}{\sqrt{1+1}} = \frac{3}{\sqrt{2}}$$

$\therefore$ Option (c) is correct.

Here, equation of $RS$ is

$$\frac{x-0}{3/2} = \frac{y-3}{-3/2} = \frac{z-0}{3} = \lambda$$

$$\Rightarrow \quad x = \frac{3}{2}\lambda,\ y = -\frac{3}{2}\lambda + 3,\ z = 3\lambda$$

To find the distance from $O(0, 0, 0)$ to $RS$.

Let $M$ be the foot of perpendicular.

O (0, 0, 0)
R (0,3,0)
M $\left(\frac{3\lambda}{2}, 3-\frac{3\lambda}{2}, 3\lambda\right)$
S $\left(\frac{3}{2}, \frac{3}{2}, 3\right)$

$\because \quad \overline{OM} \perp \overline{RS} \Rightarrow \overline{OM}\cdot\overline{RS} = 0$

$$\Rightarrow \frac{9\lambda}{4} - \frac{3}{2}\left(3 - \frac{3\lambda}{2}\right) + 3(3\lambda) = 0$$

$$\Rightarrow \quad \lambda = \frac{1}{3}$$

$$\therefore \quad M\left(\frac{1}{2}, \frac{5}{2}, 1\right)$$

$$\Rightarrow OM = \sqrt{\frac{1}{4} + \frac{25}{4} + 1} = \sqrt{\frac{30}{4}} = \sqrt{\frac{15}{2}}$$

$\therefore$ Option (d) is correct.

**44.** *(a, c, d)* Given a $\Delta XYZ$, where

$$2s = x + y + z$$

X
z
y
Y
Z
x

and $\quad \frac{s-x}{4} = \frac{s-y}{3} = \frac{s-z}{2}$

$$\therefore \quad \frac{s-x}{4} = \frac{s-y}{3} = \frac{s-z}{2}$$

$$= \frac{3s-(x+y+z)}{4+3+2} = \frac{s}{9}$$

or $\quad \frac{s-x}{4} = \frac{s-y}{3} = \frac{s-z}{2} = \frac{s}{9} = \lambda$ (let)

$\Rightarrow \quad s = 9\lambda,\ s = 4\lambda + x, s = 3\lambda + y$

and $\quad s = 2\lambda + z$

$\therefore \quad s = 9\lambda,\ x = 5\lambda,$

$\quad y = 6\lambda,\ z = 7\lambda$

Now, $\quad \Delta = \sqrt{s(s-x)(s-y)(s-z)}$ [heron's formula]

$$= \sqrt{9\lambda\cdot 4\lambda\cdot 3\lambda\cdot 2\lambda} = 6\sqrt{6}\lambda^2 \qquad \ldots\text{(i)}$$

Also, $\quad \pi r^2 = \frac{8\pi}{3}$

$$\Rightarrow \quad r^2 = \frac{8}{3} \qquad \ldots\text{(ii)}$$

and $\quad R = \frac{xyz}{4\Delta}$

$$= \frac{(5\lambda)(6\lambda)(7\lambda)}{4\cdot 6\sqrt{6}\lambda^2} = \frac{35\lambda}{4\sqrt{6}} \qquad \ldots\text{(iii)}$$

Now, $\quad r^2 = \frac{8}{3} = \frac{\Delta^2}{S^2} = \frac{216\lambda^4}{81\lambda^2}$

$$\Rightarrow \quad \frac{8}{3} = \frac{8}{3}\lambda^2 \qquad \text{[from Eq. (ii)]}$$

$$\Rightarrow \quad \lambda = 1$$

(a) $\Delta XYZ = 6\sqrt{6}\lambda^2 = 6\sqrt{6}$

$\therefore$ Option (a) is correct.

(b) Radius of circumcircle

$$(R) = \frac{35}{4\sqrt{6}}\lambda = \frac{35}{4\sqrt{6}}$$

$\therefore$ Option (b) is incorrect.

(c) Since,

$$r = 4R\sin\frac{X}{2}\cdot\sin\frac{Y}{2}\cdot\sin\frac{Z}{2}$$

$$\Rightarrow \frac{2\sqrt{2}}{\sqrt{3}} = 4\cdot\frac{35}{4\sqrt{6}}\sin\frac{X}{2}\cdot\sin\frac{Y}{2}\cdot\sin\frac{Z}{2}$$

$$\Rightarrow \frac{4}{35} = \sin\frac{X}{2}\cdot\sin\frac{Y}{2}\cdot\sin\frac{Z}{2}$$

$\therefore$ Option (c) is correct.

(d) $\sin^2\left(\frac{X+Y}{2}\right) = \cos^2\left(\frac{Z}{2}\right)$, as

$$\frac{X+Y}{2} = 90° - \frac{Z}{2} = \frac{s(s-z)}{xy} = \frac{9\times 2}{5\times 6} = \frac{3}{5}$$

$\therefore$ Option (d) is correct.

**45.** *(a, c)* Given, $RS$ is the diameter of $x^2 + y^2 = 1$.

Here, equation of the tangent at $P(\cos\theta, \sin\theta)$ is $x\cos\theta + y\sin\theta = 1$.

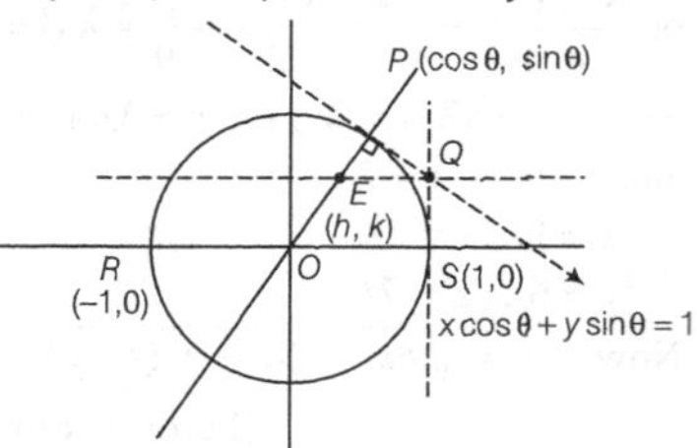


Intersecting with $x = 1$,

$$y = \frac{1-\cos\theta}{\sin\theta}$$

$$\therefore \quad Q\left(1, \frac{1-\cos\theta}{\sin\theta}\right)$$

$\therefore$ Equation of the line through $Q$ parallel to $RS$ is

$$y = \frac{1-\cos\theta}{\sin\theta} = \frac{2\sin^2\frac{\theta}{2}}{2\sin\frac{\theta}{2}\cos\frac{\theta}{2}} = \tan\frac{\theta}{2} \quad \dots\text{(i)}$$

Normal at $P$: $y = \frac{\sin\theta}{\cos\theta}\cdot x$

$$\Rightarrow \quad y = x\tan\theta \quad \dots\text{(ii)}$$

Let their point of intersection be $(h, k)$.

Then, $k = \tan\frac{\theta}{2}$

and $k = h\tan\theta$

$$\therefore \quad k = h\left(\frac{2\tan\frac{\theta}{2}}{1-\tan^2\frac{\theta}{2}}\right)$$

$$\Rightarrow \quad k = \frac{2h\cdot k}{1-k^2}$$

$$\Rightarrow k(1-k^2) = 2hk$$

$\therefore$ Locus for point $E: 2x = (1-y^2) \dots$(iii)

When $x = \frac{1}{3}$,

then $1 - y^2 = \frac{2}{3}$

$$\Rightarrow \quad y^2 = 1 - \frac{2}{3}$$

$$\Rightarrow \quad y = \pm\frac{1}{\sqrt{3}}$$

$\therefore \left(\frac{1}{3}, \pm\frac{1}{\sqrt{3}}\right)$ satisfy $2x = 1 - y^2$.

When $x = \frac{1}{4}$, then

$$1 - y^2 = \frac{2}{4}$$

$$\Rightarrow \quad y^2 = 1 - \frac{1}{2}$$

$$\Rightarrow \quad y = \pm\frac{1}{\sqrt{2}}$$

$\therefore \left(\frac{1}{4}, \pm\frac{1}{2}\right)$ does not satisfy $1 - y^2 = 2x$.

**46.** *(a, b, c)* Given,

$C_1 : x^2 + y^2 = 3$ intersects the parabola $x^2 = 2y$.

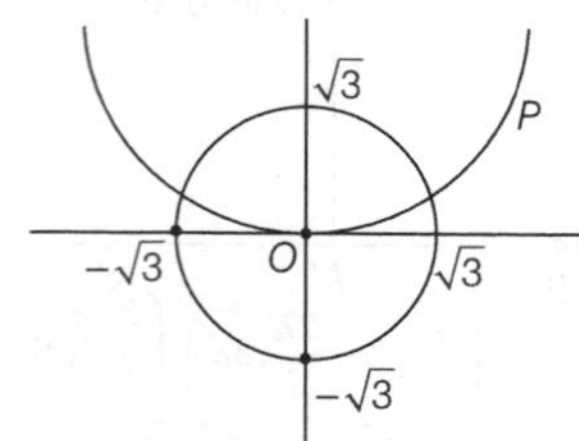

On solving $x^2 + y^2 = 3$ and $x^2 = 2y$, we get

$$y^2 + 2y = 3$$

$\Rightarrow$ $y^2 + 2y - 3 = 0$

$\Rightarrow$ $(y + 3)(y - 1) = 0$

$\therefore$ $y = 1, -3$

[neglecting $y = -3$, as $-\sqrt{3} \le y \le \sqrt{3}$]

$\therefore$ $y = 1$

$\Rightarrow$ $x = \pm\sqrt{2}$

$\Rightarrow$ $P(\sqrt{2}, 1) \in$ I quadrant

Equation of tangent at $P(\sqrt{2}, 1)$ to $C_1 : x^2 + y^2 = 3$ is

$$\sqrt{2}x + 1 \cdot y = 3 \quad \ldots\text{(i)}$$

Now, let the centres of $C_2$ and $C_3$ be $Q_2$ and $Q_3$, and tangent at $P$ touches $C_2$ and $C_3$ at $R_2$ and $R_3$ shown as below

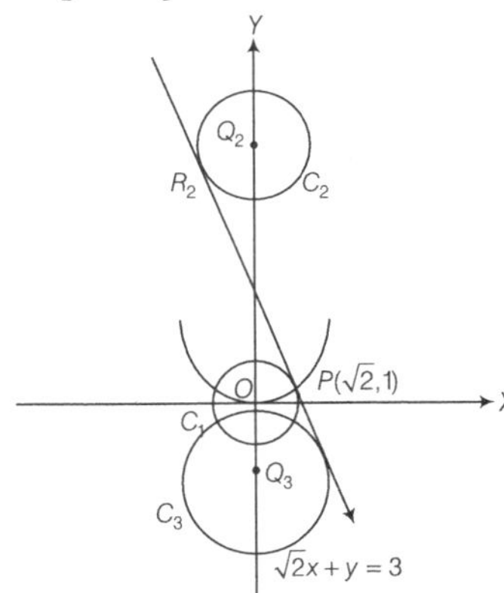


Let $Q_2$ be $(0, k)$ and radius is $2\sqrt{3}$.

$\therefore$ $\dfrac{|0 + k - 3|}{\sqrt{2 + 1}} = 2\sqrt{3}$

$\Rightarrow$ $|k - 3| = 6 \Rightarrow k = 9, -3$

$\therefore$ $Q_2(0, 9)$ and $Q_3(0, -3)$

Hence, $Q_2Q_3 = 12$

$\therefore$ Option (a) is correct.

Also, $R_2R_3$ is common internal tangent to $C_2$ and $C_3$,

and $r_2 = r_3 = 2\sqrt{3}$

$\therefore$ $R_2R_3 = \sqrt{d^2 - (r_1 + r_2)^2}$

$= \sqrt{12^2 - (4\sqrt{3})^2}$

$= \sqrt{144 - 48} = \sqrt{96} = 4\sqrt{6}$

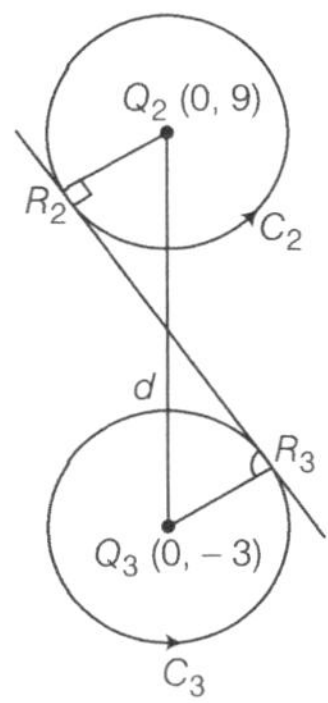


$\therefore$ Option (b) is correct.

$\because$ Length of perpendicular from $O(0, 0)$ to $R_2R_3$ is equal to radius of $C_1 = \sqrt{3}$.

$\therefore$ Area of $\Delta OR_2R_3 = \dfrac{1}{2} \times R_2R_3 \times \sqrt{3}$

$= \dfrac{1}{2} \times 4\sqrt{6} \times \sqrt{3}$

$= 6\sqrt{2}$

$\therefore$ Option (c) is correct.

Also, area of $\Delta PQ_2Q_3 = \dfrac{1}{2} Q_2Q_3 \times \sqrt{2}$

$= \dfrac{\sqrt{2}}{2} \times 12 = 6\sqrt{2}$

$\therefore$ Option (d) is incorrect.

**47.** *(b, c)* As, $g(f(x)) = x$

Thus, $g(x)$ is inverse of $f(x)$.

$\Rightarrow$ $g(f(x)) = x$

$\Rightarrow$ $g'(f(x)) \cdot f'(x) = 1$

$\therefore$ $g'(f(x)) = \dfrac{1}{f'(x)} \quad \ldots\text{(i)}$

[where, $f'(x) = 3x^2 + 3$]

When $f(x) = 2$, then

$x^3 + 3x + 2 = 2$

$\Rightarrow$ $x = 0$

i.e. when $x = 0$, then $f(x) = 2$

$\therefore$ $g'(f(x)) = \dfrac{1}{3x^2 + 3}$ at $(0, 2)$

$\Rightarrow$ $g'(2) = \dfrac{1}{3}$

$\therefore$ Option (a) is incorrect.

Now, $h(g(g(x))) = x$

$\Rightarrow$ $h(g(g(f(x))) = f(x)$

$\Rightarrow$ $h(g(x)) = f(x) \quad \ldots\text{(ii)}$

As $g(f(x)) = x$

$\therefore h(g(3)) = f(3) = 3^3 + 3(3) + 2 = 38$

$\therefore$ Option (d) is incorrect.

From Eq. (ii),

$$h(g(x)) = f(x)$$

$$\Rightarrow h(g(f(x))) = f(f(x))$$

$$\Rightarrow h(x) = f(f(x)) \quad \ldots\text{(iii)}$$

[using $g(f(x)) = x$]

$$\Rightarrow h'(x) = f'(f(x)) \cdot f'(x) \ldots\text{(iv)}$$

Putting $x = 1$, we get

$$h'(1) = f'(f(1)) \cdot f'(1)$$
$$= (3 \times 36 + 3) \times (6)$$
$$= 111 \times 6 = 666$$

$\therefore$ Option (b) is correct.

Putting $x = 0$ in Eq. (iii), we get

$$h(0) = f(f(0)) = f(2)$$
$$= 8 + 6 + 2 = 16$$

$\therefore$ Option (c) is correct.

**48.** *(a)* Here, $f'(x) = 2 - \frac{f(x)}{x}$

or $\frac{dy}{dx} + \frac{y}{x} = 2$

[i.e. linear differential equation in $y$]

Integrating Factor, IF

$$= e^{\int \frac{1}{x} dx} = e^{\log x} = x$$

$\therefore$ Required solution is

$$y \cdot (\text{IF}) = \int Q(\text{IF})dx + C$$

$$\Rightarrow y(x) = \int 2(x)\, dx + C$$

$$\Rightarrow yx = x^2 + C$$

$$\therefore y = x + \frac{C}{x} \quad [\because C \neq 0, \text{ as } f(1) \neq 1]$$

(a) $\lim_{x \to 0^+} f'\left(\frac{1}{x}\right) = \lim_{x \to 0^+} (1 - Cx^2) = 1$

$\therefore$ Option (a) is correct.

(b) $\lim_{x \to 0^+} x\, f\left(\frac{1}{x}\right) = \lim_{x \to 0^+} (1 + Cx^2) = 1$

$\therefore$ Option (b) is incorrect.

(c) $\lim_{x \to 0^+} x^2 f'(x)$

$$= \lim_{x \to 0^+} (x^2 - C) = -C \neq 0$$

$\therefore$ Option (c) is incorrect.

(d) $f(x) = x + \frac{C}{x}, C \neq 0$

For $C > 0$, $\lim_{x \to 0^+} f(x) = \infty$

$\therefore$ Function is not bounded in (0, 2).

$\therefore$ Option (d) is incorrect.

**49.** *(b, c)* Here, $P = \begin{bmatrix} 3 & -1 & -2 \\ 2 & 0 & \alpha \\ 3 & -5 & 0 \end{bmatrix}$

Now, $|P| = 3(5\alpha) + 1(-3\alpha) - 2(-10)$

$$= 12\alpha + 20 \quad \ldots\text{(i)}$$

$$\therefore \text{adj}(P) = \begin{bmatrix} 5\alpha & 2\alpha & -10 \\ -10 & 6 & 12 \\ -\alpha & -(3\alpha+4) & 2 \end{bmatrix}^T$$

$$= \begin{bmatrix} 5\alpha & -10 & -\alpha \\ 2\alpha & 6 & -3\alpha - 4 \\ -10 & 12 & 2 \end{bmatrix} \ldots\text{(ii)}$$

As, $PQ = kI$

$$\Rightarrow |P||Q| = |kI|$$

$$\Rightarrow |P||Q| = k^3$$

$$\Rightarrow |P|\left(\frac{k^2}{2}\right) = k^3 \quad \left[\text{given, } |Q| = \frac{k^2}{2}\right]$$

$$\Rightarrow |P| = 2k \quad \ldots\text{(iii)}$$

$$\because PQ = ki$$

$$\therefore Q = kp^{-1}I$$

$$= k \cdot \frac{\text{adj}P}{|P|} = \frac{k(\text{adj } P)}{2k} \quad \text{[from Eq. (iii)]}$$

$$= \frac{\text{adj } P}{2}$$

$$= \frac{1}{2}\begin{bmatrix} 5\alpha & -10 & -\alpha \\ 2\alpha & 6 & -3\alpha - 4 \\ -10 & 12 & 2 \end{bmatrix}$$

$$\therefore q_{23} = \frac{-3\alpha - 4}{2} \quad \left[\text{given, } q_{23} = -\frac{k}{8}\right]$$

$$\Rightarrow -\frac{(3\alpha + 4)}{2} = -\frac{k}{8}$$

$$\Rightarrow (3\alpha + 4) \times 4 = k$$

$$\Rightarrow 12\alpha + 16 = k \quad \ldots\text{(iv)}$$

From Eq. (iii), $|P| = 2k$

$$\Rightarrow 12\alpha + 20 = 2k$$

[from Eq. (i)] ...(v)

On solving Eqs. (iv) and (v), we get

$$\alpha = -1 \text{ and } k = 4 \quad \ldots\text{(vi)}$$

$\therefore 4\alpha - k + 8 = -4 - 4 + 8 = 0$

$\therefore$ Option (b) is correct.

Now, $|P \text{ adj } (Q)| = |P||\text{adj } Q|$

$$= 2k\left(\frac{k^2}{2}\right)^2 = \frac{k^5}{2} = \frac{2^{10}}{2} = 2^9$$

$\therefore$ Option (c) is correct.

**50.** (5) Coefficient of $x^2$ in the expansion of $\{(1+x)^2 + (1+x)^3 + \dots + (1+x)^{49} + (1+mx)^{50}\}$

$$\Rightarrow {}^2C_2 + {}^3C_2 + {}^4C_2 + \dots + {}^{49}C_2 + {}^{50}C_2 \cdot m^2 = (3n+1) \cdot {}^{51}C_3$$

$$\Rightarrow {}^{50}C_3 + {}^{50}C_2 m^2 = (3n+1) \cdot {}^{51}C_3$$

$$[\because {}^rC_r + {}^{r+1}C_r + \dots + {}^nC_r = {}^{n+1}C_{r+1}]$$

$$\Rightarrow \frac{50 \times 49 \times 48}{3 \times 2 \times 1} + \frac{50 \times 49}{2} \times m^2 = (3n+1)\frac{51 \times 50 \times 49}{3 \times 2 \times 1}$$

$$\Rightarrow m^2 = 51n + 1$$

$\therefore$ Minimum value of $m^2$ for which $(51n + 1)$ is integer (perfect square) for $n = 5$.

$\therefore \quad m^2 = 51 \times 5 + 1$

$\Rightarrow \quad m^2 = 256$

$\therefore \quad m = 16$

and $\quad n = 5$

Hence, the value of $n$ is 5.

**51.** (2) Given, $\begin{vmatrix} x & x^2 & 1+x^3 \\ 2x & 4x^2 & 1+8x^3 \\ 3x & 9x^2 & 1+27x^3 \end{vmatrix} = 10$

$$\Rightarrow x \cdot x^2 \begin{vmatrix} 1 & 1 & 1+x^3 \\ 2 & 4 & 1+8x^3 \\ 3 & 9 & 1+27x^3 \end{vmatrix} = 10$$

Apply $R_2 \to R_2 - 2R_1$ and $R_3 \to R_3 - 3R_1$, we get

$$x^3 \begin{vmatrix} 1 & 1 & 1+x^3 \\ 0 & 2 & -1+6x^3 \\ 0 & 6 & -2+24x^3 \end{vmatrix} = 10$$

$$\Rightarrow x^3 \cdot \begin{vmatrix} 2 & 6x^3 - 1 \\ 6 & 24x^3 - 2 \end{vmatrix} = 10$$

$$\Rightarrow x^3(48x^3 - 4 - 36x^3 + 6) = 10$$

$$\Rightarrow 12x^6 + 2x^3 = 10$$

$$\Rightarrow 6x^6 + x^3 - 5 = 0$$

$$\Rightarrow x^3 = \frac{5}{6}, -1$$

$$\therefore x = \left(\frac{5}{6}\right)^{1/3}, -1$$

Hence, the number of real solutions is 2.

**52.** (1) Here, $z = \dfrac{-1 + i\sqrt{3}}{2} = \omega$

$$\because P = \begin{bmatrix} (-\omega)^r & \omega^{2s} \\ \omega^{2s} & \omega^r \end{bmatrix}$$

$$P^2 = \begin{bmatrix} (-\omega)^r & \omega^{2s} \\ \omega^{2s} & \omega^r \end{bmatrix}\begin{bmatrix} (-\omega)^r & \omega^{2s} \\ \omega^{2s} & \omega^r \end{bmatrix}$$

$$= \begin{bmatrix} \omega^{2r} + \omega^{4s} & \omega^{r+2s}[(-1)^r + 1] \\ \omega^{r+2s}[(-1)^r + 1] & \omega^{4s} + \omega^{2r} \end{bmatrix}$$

Given, $\quad P^2 = -I$

$\therefore \quad \omega^{2r} + \omega^{4s} = -1$

and $\quad \omega^{r+2s}[(-1)^r + 1] = 0$

Since, $\quad r \in \{1, 2, 3\}$ and $(-1)^r + 1 = 0$

$\Rightarrow \quad r = \{1, 3\}$

Also, $\quad \omega^{2r} + \omega^{4s} = -1$

If $r = 1$, then $\omega^2 + \omega^{4s} = -1$

which is only possible, when $s = 1$.

As, $\quad \omega^2 + \omega^4 = -1$

$\therefore \quad r = 1, s = 1$

Again, if $r = 3$, then

$$\omega^6 + \omega^{4s} = -1$$

$\Rightarrow \quad \omega^{4s} = -2 \quad$ [never possible]

$\therefore \quad r \neq 3$

$\Rightarrow (r, s) = (1, 1)$ is the only solution.

Hence, the total number of ordered pairs is 1.

**53.** (1) Let $f(x) = \displaystyle\int_0^x \frac{t^2}{1+t^4}\, dt$

$$\Rightarrow f'(x) = \frac{x^2}{1+x^4} > 0, \text{ for all } x \in [0, 1]$$

$\therefore f(x)$ is increasing.

At $x = 0$, $f(0) = 0$ and at $x = 1$,

$$f(1) = \int_0^1 \frac{t^2}{1+t^4}\, dt$$

Because, $0 < \dfrac{t^2}{1+t^4} < \dfrac{1}{2}$

$$\Rightarrow \int_0^1 0 \cdot dt < \int_0^1 \frac{t^2}{1+t^4}\, dt < \int_0^1 \frac{1}{2} \cdot dt$$

$$\Rightarrow \quad 0 < f(1) < \frac{1}{2}$$

Thus, $f(x)$ can be plotted as

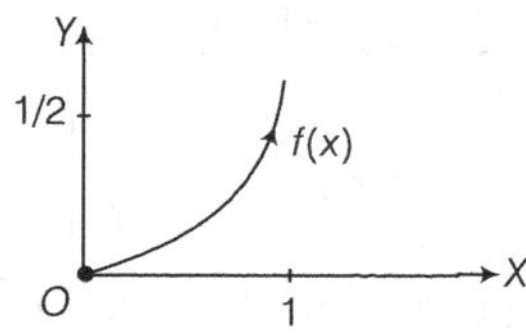


$\therefore y = f(x)$ and $y = 2x - 1$ can be shown as

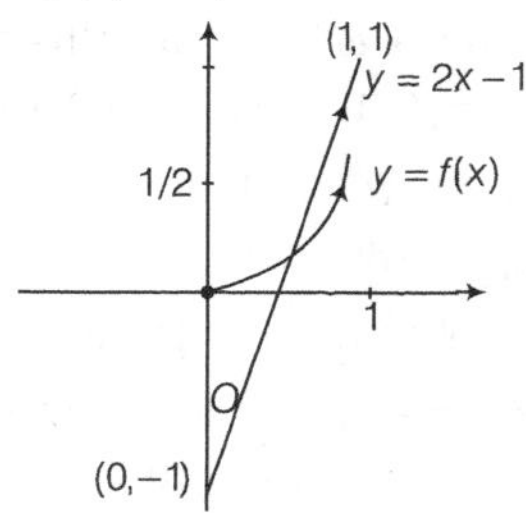


From the graph, the total number of distinct solutions for $x \in (0, 1] = 1$.

[as they intersect only at one point]

**54.** *(7)* Here, $\lim_{x \to 0} \dfrac{x^2 \sin(\beta x)}{\alpha x - \sin x} = 1$

$$\Rightarrow \lim_{x \to 0} \frac{x^2\left(\beta x - \frac{(\beta x)^3}{3!} + \frac{(\beta x)^5}{5!} - \dots\right)}{\alpha x - \left(x - \frac{x^3}{3!} + \frac{x^5}{5!} - \dots\right)} = 1$$

$$\Rightarrow \lim_{x \to 0} \frac{x^3\left(\beta - \frac{\beta^3 x^2}{3!} + \frac{\beta^5 x^4}{5!} - \dots\right)}{(\alpha - 1)x + \frac{x^3}{3!} + \frac{x^5}{5!} - \dots} = 1$$

Limit exists only,

when $\alpha - 1 = 0$

$$\Rightarrow \quad \alpha = 1 \quad \dots(\text{i})$$

$$\therefore \lim_{x \to 0} \frac{x^3\left(\beta - \frac{\beta^3 x^2}{3!} + \frac{\beta^5 x^4}{5!} - \dots\right)}{x^3\left(\frac{1}{3!} - \frac{x^2}{5!} - \dots\right)} = 1$$

$$\Rightarrow \quad 6\beta = 1 \quad \dots(\text{ii})$$

From Eqs. (i) and (ii),

we get

$$6(\alpha + \beta) = 6\alpha + 6\beta$$
$$= 6 + 1$$
$$= 7$$

# Paper 2

## Physics

**1.** *(c)* Using the relation

$$R = R_0\left(\frac{1}{2}\right)^n$$

Here, $R$ is activity of radioactive substance, $R_0$ initial activity and $n$ is number of half lives.

$$1 = 64\left(\frac{1}{2}\right)^n$$

Solving we get, $n = 6$

Now, $t = n(t_{1/2}) = 6(18 \text{ days}) = 108$ days

**2.** *(c)* Electrostatic energy

$$= \text{Binding energy of N} - \text{Binding energy of O}$$

$$= [(7M_H + 8M_n - M_N)] - [(8M_H + 7M_n - M_O)] \times c^2$$

$$= [-M_H + M_n + M_O - M_N]c^2$$

$$= [-1.007825 + 1.008665 + 15.003065 - 15.000109] \times 931.5$$

$$= +3.5359 \text{ MeV}$$

$$\Delta E = \frac{3}{5} \times \frac{1.44 \times 8 \times 7}{R} - \frac{3}{5} \times \frac{1.44 \times 7 \times 6}{R}$$

$$= 3.5359 \text{ MeV}$$

$$R = \frac{3 \times 1.44 \times 14}{5 \times 3.5359} = 3.42 \text{ fm}$$

**3.** *(c)* In the first process : $p_i V_i^\gamma = p_f V_f^\gamma$

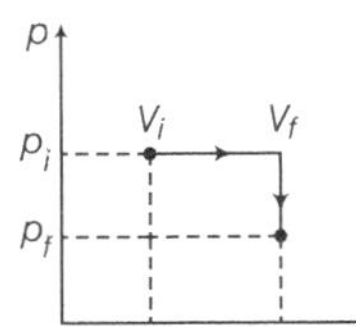


$$\Rightarrow \quad \frac{p_i}{p_f} = \left(\frac{V_f}{V_i}\right)^\gamma \Rightarrow 32 = 8^\gamma$$

$$\gamma = \frac{5}{3} \quad \ldots\text{(i)}$$

For the two step process,.

$$W = p_i(V_f - V_i) = 10^5(7 \times 10^{-3})$$

$$W = 7 \times 10^2 \text{J}$$

$$\Delta U = \frac{f}{2}(p_f V_f - p_i V_i)$$

$$= \frac{1}{\gamma - 1}\left(\frac{1}{4} \times 10^2 - 10^2\right)$$

$$\Delta U = -\frac{3}{2} \cdot \frac{3}{4} \times 10^2 = -\frac{9}{8} \times 10^2 \text{J}$$

$$Q - W = \Delta U$$

$$\Rightarrow \quad Q = 7 \times 10^2 - \frac{9}{8} \times 10^2$$

$$= \frac{47}{8} \times 10^2 \text{J} = 588 \text{ J}$$

**4.** *(b)* For vernier $C_1$

10 VSD = 9 MSD = 9 mm

1 VSD = 0.9 mm

$\Rightarrow$ LC = 1 MSD − 1 VSD

= 1 mm − 0.9 mm = 0.1 mm

Reading of $C_1$ = MSR + (VSR) (LC)

= 28 mm + (7)(0.1)

Reading of $C_1$ = 28.7 mm = 2.87 cm

For vernier $C_2$ : the vernier $C_2$ is abnormal, so we have to find the reading form basics.

The point where both of the marks are matching :

distance measured from main scale

= distance measured from vernier scale

28 mm + (1 mm) (8)

= (28 mm + $x$) + (1.1 mm) (7)

Solving we get, $x = 0.3$ mm

So, reading of $C_2$ = 28 mm + 0.3 mm

= 2.83 cm

**5.** *(d)*

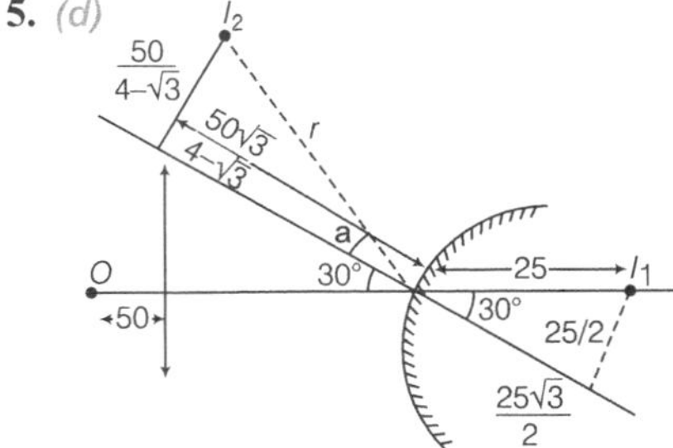


**For Lens**

$$\frac{1}{v} - \frac{1}{u} = \frac{1}{f} \Rightarrow v = \frac{uf}{u + f}$$

$$\Rightarrow \quad v = \frac{(-50)(30)}{-50 + 30} = 75 \text{ cm}$$

**For Mirror**

$$\frac{1}{v} + \frac{1}{u} = \frac{1}{f} \Rightarrow v = \frac{uf}{u - f}$$

$$\Rightarrow \quad v = \frac{\left(\frac{25\sqrt{3}}{2}\right)(50)}{\frac{25\sqrt{3}}{2} - 50} = \frac{-50\sqrt{3}}{4-\sqrt{3}} \text{ cm}$$

$$m = -\frac{v}{u} = \frac{h_2}{h_1}$$

$$\Rightarrow h_2 = -\left(\frac{\frac{-50\sqrt{3}}{4-\sqrt{3}}}{\frac{25\sqrt{3}}{2}}\right)\cdot\frac{25}{2} \Rightarrow h_2 = \frac{+50}{4-\sqrt{3}}$$

The $x$-coordinate of the images

$= 50 - v\cos 30 + h_2 \cos 60 \approx 25$

The $y$-coordinate of the images

$= v\sin 30 + h_2 \sin 60 \approx 25\sqrt{3}$

**6.** *(a)*

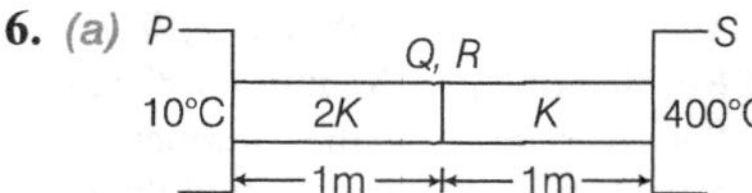


Rate of heat flow from $P$ to $Q$,

$$\frac{dQ}{dt} = \frac{2KA(T-10)}{1}$$

Rate of heat flow from $Q$ to $S$,

$$\frac{dQ}{dt} = \frac{KA(4000-T)}{1}$$

At steady state, state rate of heat flow is same

$$\therefore \quad \frac{2KA(T-10)}{1} = KA(400-T)$$

or $\quad 2T - 20 = 400 - T \quad$ or $\quad 3T = 420$

$\therefore \quad T = 140°$

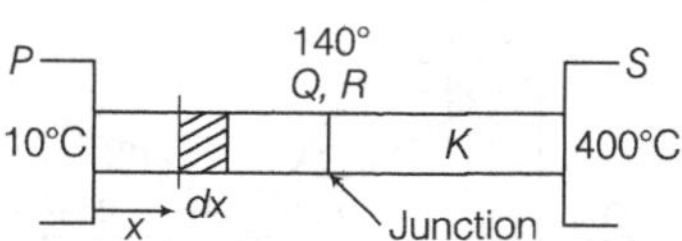


Temperature of junction is 140°C.

Temperature at a distance $x$ from end $P$ is $T_x = (130x + 10°)$

Change in length $dx$ is suppose $dy$.

Then, $dy = \alpha dx\,(T_x - 10)$

$$\int_0^{\Delta y} dy = \int_0^1 \alpha dx\,(130x + 10 - 10)$$

$$\Delta y = \left[\frac{\alpha x^2}{2} \times 130\right]_0^1$$

$$\Delta y = 1.2 \times 10^{-5} \times 65$$

$$\Delta y = 78.0 \times 10^{-5}\text{ m} = 0.78\text{ mm}$$

**7.** *(c,d)*

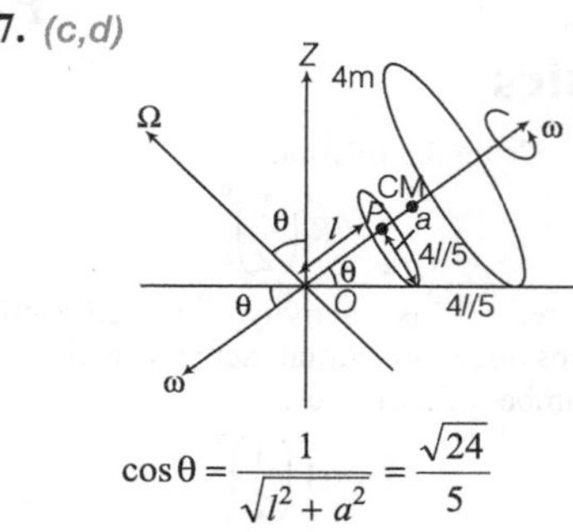


$$\cos\theta = \frac{1}{\sqrt{l^2 + a^2}} = \frac{\sqrt{24}}{5}$$

(c) Velocity of point $P$ : $a\omega = 1\,\Omega$ then

$\Omega = \frac{a\omega}{1}$ = Angular velocity of CM w.r.t. point $O$. Angular velocity fo CM w.r.t. $Z$-axis $= \Omega\cos\theta$

$$\omega_{CM-Z} = \frac{a\omega}{1}\,\frac{\sqrt{24}}{5} = \frac{a\omega}{\sqrt{24}\,a}\,\frac{\sqrt{24}}{5}$$

$$\omega_{CM-Z} = \frac{a\omega}{5}$$

(a) $L_z = L_{CM-O}\cos\theta - L_{D-CM}\sin\theta$

$$= \frac{81\sqrt{24}}{5}a^2 m\omega \times \frac{\sqrt{24}}{5} - \frac{17\,ma^2\omega}{2} \times \frac{1}{\sqrt{24}}$$

$$= \frac{81 \times 24 ma^2\omega}{25} - \frac{17ma^2\omega}{2\sqrt{24}}$$

(b) $L_{CM-O} = (5m)\left[\frac{9l}{5}\Omega\right]\frac{9l}{5} = \frac{81ml^2\omega}{5}$

$$= \frac{81ml^2}{5} \times \frac{a\omega}{l}$$

$$L_{CM-O} = \frac{81mla\omega}{5} = \frac{81\sqrt{24}\,a^2 m\omega}{5}$$

(d) $L_{D-CM} = \frac{ma^2}{2}\omega + \frac{4m(2a)^2}{2}\omega = \frac{17ma^2\omega}{2}$

**8.** *(a,c)* For maximum range of voltage resistance should be maximum. So, all four should be connected in series. For maximum range of current, net resistance should be least. Therefore, all four should be connected in parallel.

**9.** *(b,c)*

**When loop was entering ($x < L$)**

$$\phi = BLx \Rightarrow e = -\frac{d\phi}{dt} = -BL\frac{dx}{dt}$$

$$|e| = BLv$$

$$i = \frac{e}{R} = \frac{BLv}{R} \text{ (anti- clockwise)}$$

$$F = ilB \text{ (Left direction)}$$

$$= \frac{B^2L^2v}{R} \text{(in left direction)}$$

$$\Rightarrow \quad a = \frac{F}{m} = -\frac{B^2L^2v}{mR} \Rightarrow a = v\frac{dv}{dx}$$

$$v\frac{dv}{dx} = -\frac{B^2L^2v}{mR}$$

$$\Rightarrow \quad \int_{v_0}^{v} dv = -\frac{B^2L^2}{mR}\int_0^x dx$$

$$\Rightarrow \quad v = v_0 - \frac{B^2L^2v}{mR}x$$

(straight line of negative slope for $x < L$)

$$I = \frac{BL}{R}v$$

$\Rightarrow$ ($I$ *vs* $x$ will also be straight line of negative slope for $x < L$)

**$L \le x \le 3L$**

$$\frac{d\phi}{dt} = 0; \; e = 0, i = 0; F = 0$$

**$x > 4L$** $\quad e = Blv$

Force also will be in left direction.

$$i = \frac{BLv}{R} \qquad \text{(clockwise)}$$

$$a = -\frac{B^2L^2v}{mR} = v\frac{dv}{dx} \Rightarrow F = \frac{B^2L^2v}{R}$$

$$\int_L^x -\frac{B^2L^2}{mR}dx = \int_{v_i}^{v_f} dv$$

$$\Rightarrow \quad -\frac{B^2L^2}{mR}(x - L) = v_f - v_i$$

$$v_f = v_i - \frac{B^2L^2}{mR}(x - L)$$

(straight line of negative slope)

$$I = \frac{BLv}{R} \rightarrow \text{(clockwise)}$$

(straight line of negative slope)

**10.** (a,b,d) Mean time period

$$= \frac{0.52 + 0.56 + 0.57 + 0.54 + 0.59}{5}$$

$= 0.556 \cong 0.56$ sec as per significant figures

Error in reading $= |T_{mean} - T_1| = 0.04$

$|T_{mean} - T_2| = 0.00 \Rightarrow |T_{mean} - T_3| = 0.01$

$|T_{mean} - T_4| = 0.02 \Rightarrow |T_{mean} - T_5| = 0.03$

Mean error $= 0.1/5 = 0.02$

$$\% \text{ error in } T = \frac{\Delta T}{T} \times 100 = \frac{0.02}{0.56} \times 100 = 3.57\%$$

$$\% \text{ error in } r = \frac{0.001 \times 100}{0.010} = 10\%$$

$$\% \text{ error in } R = \frac{0.001 \times 100}{0.600} = 1.67\%$$

$$\% \text{ error in } \frac{\Delta g}{g} \times 100$$

$$= \frac{\Delta(R - r)}{R - r} \times 100 + 2 \times \frac{\Delta T}{T}$$

$$= \frac{0.002 \times 100}{0.05} + 2 \times 3.57$$

$$= 4\% + 7\% = 11\%$$

**11.** (a,b,d) **Case-I**

Just before $m$ is placed

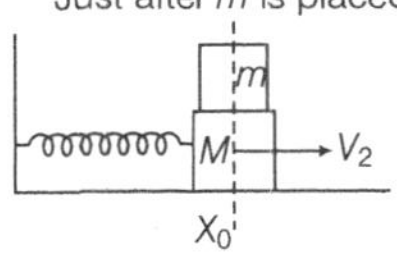


**Case-II**

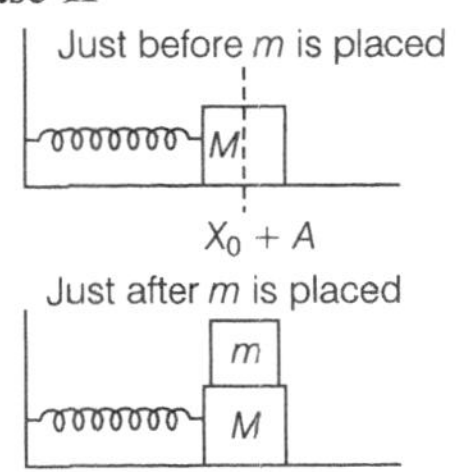

In case-1,

$$Mv_1 = (M + m)v_2 \Rightarrow v_2 = \left(\frac{M}{M+m}\right)v_1$$

$$\sqrt{\frac{k}{M+m}}\,A_2 = \left(\frac{M}{M+m}\right)\sqrt{\frac{k}{M}}\,A_1$$

$$A_2 = \sqrt{\frac{k}{M+m}}\,A_1$$

In case-2, $A_2 = A_1$

$$T = 2\pi\sqrt{\frac{M+m}{k}} \text{ in both cases.}$$

Total energy decreases in first case whereas remain same in 2nd case. Instantaneous speed at $x_0$ decreases in both cases.

**12.** (a,b)

Path difference at point $O = d = 0.6003$ mm $= 600300$ nm.

This path difference is equal to $\left(1000\lambda + \frac{\lambda}{2}\right)$.

$\Rightarrow$ Minima is formed at point $O$.

Line $S_1S_2$ and screen are perpendicular to each other, so fringe pattern is circular (semi-circular because only half of screen is available).

**13.** (a,b,c,d)

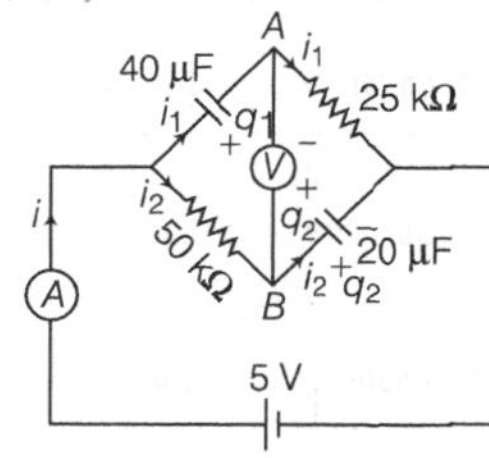


Just after pressing key,

$$5 - 25000i_1 = 0$$
$$5 - 50000i_2 = 0$$

(As charge in both capacitors = 0)

$\Rightarrow \quad i_1 = 0.2\text{mA} \Rightarrow i_2 = 0.1\text{ mA}$

And $V_B + 25000i_1 = V_A$

$\Rightarrow \quad V_B - V_A = -5\text{V}$

After a long time, $i_1$ and $i_2 = 0$ (steady state)

$$\Rightarrow \quad 5 - \frac{q_1}{40} = 0 \Rightarrow q_1 = 200\mu\text{C}$$

and $5 - \frac{q_2}{20} = 0 \Rightarrow q_2 = 100\mu\text{C}$

$$V_B - \frac{q_2}{20} = V_A \Rightarrow V_B - V_A = +5\text{V}$$

$\Rightarrow$ (a) is correct.

For capacitor 1,

$$q_1 = 200[1 - e^{-t/1}]\mu\text{C} \Rightarrow i_1 = \frac{1}{5}e^{-t/1}\text{mA}$$

For capacitor 2,

$$q_2 = 100[1 - e^{-t/1}]\mu\text{C} \Rightarrow i_2 = \frac{1}{10}e^{-t/1}\text{mA}$$

$$\Rightarrow \quad V_B - \frac{q_2}{20} + i_1 \times 25 = V_A$$

$$\Rightarrow V_B - V_A = 5[1 - e^{-t}] - 5e^{-t}$$
$$= -5[1 - 2e^{-t}]$$

At $t = \ln 2$,

$$V_B - V_A = 5[1 - 1] = 0$$

$\Rightarrow$ (b) is correct.

At $t = 1$,

$$i = i_1 + i_2 = \frac{1}{5}e^{-1} + \frac{1}{10}e^{-1} = \frac{3}{10}\cdot\frac{1}{e}$$

At $t = 0$,

$$i = i_1 + i_2 = \frac{1}{5} + \frac{1}{10} = \frac{3}{10}$$

$\Rightarrow$ (c) is correct.

After a long time, $i_1 = i_2 = 0$

$\Rightarrow$ (d) is correct.

**14.** (d) $K_{\max} = \dfrac{hc}{\lambda_{\text{ph}}} - \phi$

Kinetic energy of electron reaching the anode will be

$$K = \frac{hc}{\lambda_{\text{ph}}} - \phi + eV$$

Now, $\lambda_e = \dfrac{h}{\sqrt{2mK}}$

$$= \frac{h}{\sqrt{2m\left(\dfrac{hc}{\lambda_{\text{ph}}} - \phi + eV\right)}}$$

**If $eV >> \phi$,** $\lambda_e = \dfrac{h}{\sqrt{2m\left(\dfrac{hc}{\lambda_{\text{ph}}} + eV\right)}}$

**If $V_f = 4V_i$,** $(\lambda_e)_f \simeq \dfrac{(\lambda_e)_i}{}$

**15.** *(c)* Force on block along slot $= m\omega^2 r$

$$= ma = m\left(\frac{vdv}{dr}\right)$$

$$\int_0^v vdv = \int_{R/2}^r \omega^2 rdr$$

$$\Rightarrow \quad \frac{v^2}{2} = \frac{\omega^2}{2}\left(r^2 - \frac{R^2}{4}\right)$$

$$\Rightarrow \quad v = \omega\sqrt{r^2 - \frac{R^2}{4}} = \frac{dr}{dt}$$

$$\Rightarrow \quad \int_{R/4}^r \frac{dr}{\sqrt{r^2 - \frac{R^2}{4}}} = \int_0^t \omega dt$$

$$\ln\left(\frac{r + \sqrt{r^2 - \frac{R^2}{4}}}{\frac{R}{2}}\right) - \ln\left(\frac{\frac{R}{2} \text{ and } + \sqrt{\frac{R^2}{4} - \frac{R^2}{4}}}{\frac{R}{2}}\right) = \omega t$$

$$\Rightarrow \quad r + \sqrt{r^2 - \frac{R^2}{4}} = \frac{R}{2}e^{\omega t}$$

$$\Rightarrow \quad r^2 - \frac{R^2}{4} = \frac{R^2}{4}e^{2\omega t} + r^2 - 2r\frac{R}{2}e^{\omega t}$$

$$\Rightarrow \quad r = \frac{\frac{R^2}{4}e^{2\omega t} + \frac{R^2}{4}}{Re^{\omega t}} = \frac{R}{4}(e^{\omega t} + e^{-\omega t})$$

**16.** *(b)*

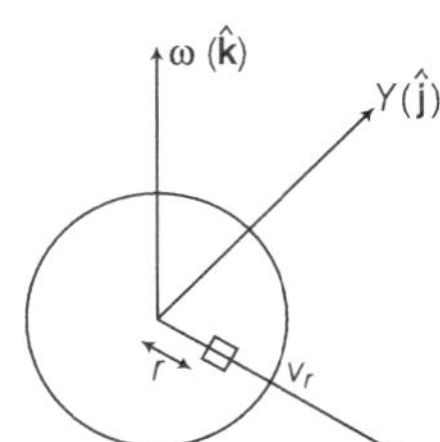


$$\mathbf{F}_{rot} = \mathbf{F}_{in} + 2m(v_{rot}\hat{\mathbf{i}}) \times \omega\hat{\mathbf{k}} + m(\omega\hat{\mathbf{k}} \times r\hat{\mathbf{i}}) \times \omega\hat{\mathbf{k}}$$

$$mr\omega^2\hat{\mathbf{i}} = \mathbf{F}_{in} + 2mv_{rot}\,\omega(-\hat{\mathbf{j}}) + m\omega^2 r\hat{\mathbf{j}}$$

$$\mathbf{F}_{in} = 2mv_r\omega\hat{\mathbf{j}},$$

$$r = \frac{R}{4}[e^{\omega t} + e^{-\omega t}]$$

$$\frac{dr}{dt} = v_r = \frac{R}{4}[\omega e^{\omega t} - \omega e^{-\omega t}]$$

$$\mathbf{F}_{in} = 2m\frac{R\omega}{4}[e^{\omega t} - e^{-\omega t}]\omega\hat{\mathbf{j}}$$

$$\mathbf{F}_{in} = \frac{mR\omega^2}{2}[e^{\omega t} - e^{-\omega t}]\hat{\mathbf{j}}$$

Also, reaction is due to disc surface, then

$$\mathbf{F}_{reaction} = \frac{mR\omega^2}{2}[e^{\omega t} - e^{-\omega t}]\hat{\mathbf{j}} + mg\hat{\mathbf{k}}$$

**17.** *(d)* Balls will gain positive charge and hence move towards negative plate.

On reaching negative plate, balls will attain negative charge and come back to positive plate.

And so on, balls will keep oscillating.

But oscillation is not SHM,

as force on balls is not $\propto x$.

$\Rightarrow$ (d) is correct.

**18.** *(a)* As the balls keep on carrying charge form one plate to another, current will keep on flowing even in steady state. When at bottom plate, if all balls attain charge $q$,

$$\frac{kq}{r} = V_0 \qquad \left(k = \frac{1}{4\pi\varepsilon_0}\right)$$

$$\Rightarrow \quad q = \frac{V_0 r}{k}$$

Inside cylinder, electric field,

$$E = [V_0 - (-V_0)]h = 2V_0 h.$$

$\Rightarrow$ Acceleration of each ball,

$$a = \frac{qE}{m} = \frac{2hr}{km}\cdot V_0^2$$

$\Rightarrow$ Time taken by balls to reach other plate,

$$t = \sqrt{\frac{2h}{a}} = \sqrt{\frac{2h.km}{2hrV_0^2}} = \frac{1}{V_0}\sqrt{\frac{km}{r}}$$

If there are $n$ balls,

then Average current,

$$i_{av} = \frac{nq}{t} = n \times \frac{V_0 r}{k} \times V_0\sqrt{\frac{r}{km}}$$

$$\Rightarrow \quad i_{av} \propto V_0^2$$

## Chemistry

**19.** *(a)* — OH group displays both kinds of effect;

an electron withdrawing acid-strengthening inductive effect from the *meta*-position and an electron-releasing acid weakening resonance effect from the *para*-position (at this position, resonance effect overweighs the inductive effect). Thus, III > IV.

*o*-hydroxybenzoic acid (II) is far stronger than the corresponding *meta* and *para* isomers as the carboxylate ion is stabilised by intramolecular H-bonding.

2,6-dihydroxybenzoic acid (I) forms carboxylate ion which is further stabilised by intramolecular H-bonding, Thus, correct order is

$$I > II > III > IV$$

II          I (Most stable)

**20.** (*a*)

H-atom at α carbon          $\xrightarrow{OH^-}$          Carbanion

Crossed aldol → $CH_2O^-$

$\xrightarrow{H_2O}$ $CH_2OH$ $\xrightarrow{HCHO,\ OH^-}$

α-carbon has no H atom hence, next reaction with HCHO is crossed Cannizzaro reaction

$CH_2OH + HCOO^-$

(By reduction)          (By oxidation)

$\xrightarrow[\text{Acetal formation}]{HCHO}$

**21.** (*a*) $2S_2O_3^{2-} + Ag^+ \longrightarrow [Ag(S_2O_3)_2]^{3-}$ (Clear solution)

$[Ag(S_2O_3)_2]^{3-} + 3Ag^+ \longrightarrow 2Ag_2S_2O_3$ (White ppt.)

$Ag_2S_2O_3 + H_2O \longrightarrow Ag_2S + H_2SO_4$ (Black ppt.)

**22.** (d) I ($CH_3OH$) : Surface tension decreases as concentration increases.

II (KCl) : Surface tension increases with concentration for ionic salt.

III [$CH_3(CH_2)_{11}OSO_3^- Na^+$] : It is an anionic detergent.

There is decrease in surface tension before micelle formation, and after CMC (Critical Micelle Concentration) is attained, no change in surface tension.

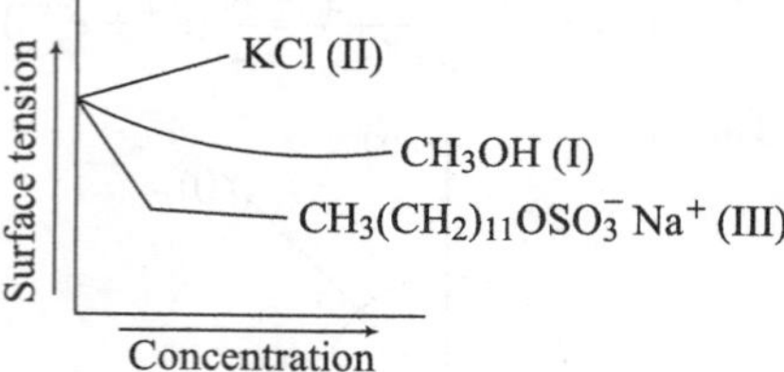


**23.** (a) $[Ni(NH_3)_6]^{2+}$ $sp^3d^2$ octahedral

$[Pt(NH_3)_4]^{2+}$ $dsp^2$ square planar

$[Zn(NH_3)_4]^{2+}$ $sp^3$ tetrahedral

**24.** (d) Oxidation at anode

$H_2(g) \longrightarrow 2H^+(aq) + 2e^-;\ E^\circ_{SHE} = 0.00\text{ V}$

Reduction at cathode

$M^{4+}(aq) + 2e^- \longrightarrow M^{2+}(aq);$

$E^\circ_{M^{4+}/M^{2+}} = 0.151\text{ V}$

Net: $M^{4+}(aq) + H_2(g)$

$\longrightarrow M^{2+}(aq) + 2H^+(aq);$

$$K = \frac{[M^{2+}][H^+]^2}{[M^{4+}]p_{H_2}} \quad (E^\circ_{cell} = 0.151 \text{ V})$$

$$= \frac{[M^{2+}]}{[M^{4+}]}$$

$$E_{cell} = E^\circ_{cell} - \frac{0.059}{2}\log K$$

$$0.092 = 0.151 - \frac{0.059}{2}\log\frac{[M^{2+}]}{[M^{4+}]}$$

$$0.059 = \frac{0.059}{2}\log 10^x$$

$\therefore \log 10^x = 2$

$\therefore \quad x = 2$

**25.** (a, c)

| Species | Electrons | MOEC | $N_B$ | $N_A$ | BO | Magnetic character |
|---|---|---|---|---|---|---|
| $C_2^{2-}$ | 14 | $\sigma 1s^2, \sigma^* 1s^2, \sigma 2s^2, \sigma^* 2s^2, \pi 2p_x^2 \approx -\pi 2p_y^2, \sigma 2p_z^2$ | 10 | 4 | 3 | Diamagnetic |
| $O_2^{2+}$ | 14 | As above according to number of electrons | 10 | 4 | 3 | Diamagnetic |
| $O_2$ | 16 | | 10 | 6 | 2 | Paramagnetic |
| $N_2^+$ | 13 | | 9 | 4 | 2.5 | Paramagnetic |
| $N_2^-$ | 15 | | 10 | 5 | 2.5 | Paramagnetic |
| $He_2^+$ | 3 | | 2 | 1 | 0.5 | Paramagnetic |

Thus, (a) is correct.

(b) Bond order $O_2^{2+} > O_2$ thus,
Bond length of $O_2^{2+} < O_2$ thus, incorrect.

(c) $N_2^+$ and $N_2^-$ have same bond order thus correct.

(d) $He_2^+$ with bond order $= 0.5$ is more stable thus, less energy than isolated He atoms. Thus, (d) is incorrect.

**26.** (b,c,d)

(a) Nearest neighbour in the topmost layer of ccp structure is 9 thus, incorrect.

(b) Packing efficiency is 74% thus, correct.

(c) Tetrahedral voids = 2
Octahedral voids = 1 per atom thus, correct.

(d) Edge length,

$$a = \frac{4}{\sqrt{2}}r = 2\sqrt{2}r$$

thus, correct explanation

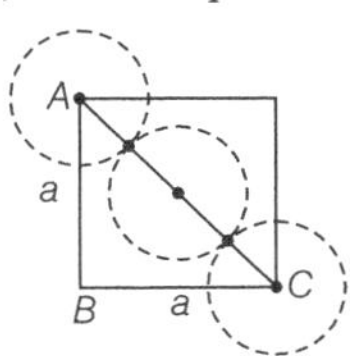


Edge length $= a$, Radius $= r$

$$AC^2 = AB^2 + BC^2$$

$$(4r)^2 = a^2 + a^2 = 2a^2$$

$$4r = \sqrt{2}a \Rightarrow r = \frac{\sqrt{2}}{4}a = \frac{a}{2\sqrt{2}}$$

$$\therefore \quad a = 2\sqrt{2}\,r$$

In ccp structure, the number of spheres is 4.

Hence, volume of 4 spheres $= 4\left(\frac{4}{3}\pi r^3\right)$

Total volume of unit cell $= a^3 = (2\sqrt{2}r)^3$

% of packing efficiency

$$= \frac{\text{Volume of 4 spheres}}{\text{Volume of unit cell}}$$

$$= \frac{4\left(\frac{4}{3}\pi r^3\right)}{[2(\sqrt{2}r)]^3} \times 100 = 74.05\% \simeq 74\%$$

**27.** (c) Only —CHO group is to be reduced to —$CH_2OH$.
It can be done using $NaBH_4$ in $C_2H_5OH$.

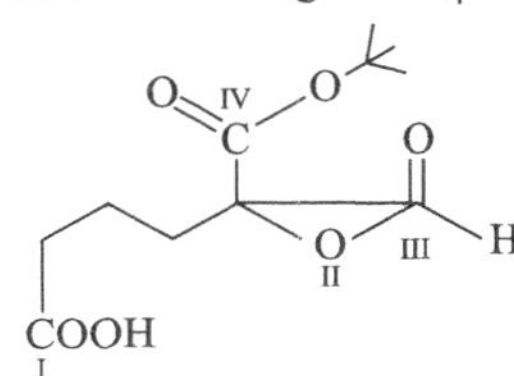

(a) $LiAlH_4 / (C_2H_5)_2O$ reduces I, II and III into —$CH_2OH$, and IV into diol.

(b) $BH_3$ / THF show same properties as (a).

(c) $NaBH_4 / C_2H_5OH$ reduces III into — $CH_2OH$.

(d) Raney nickel, same as (a) and (b), thus (c) is correct reagent.

**28.** (*a, b, c*) $CuFeS_2$ (copper pyrite) is converted into copper into following steps:

**Step I** Crushing (grinding ) followed by concentration by froth-floatation process.

**Step II** Roasting of ore in the presence of $SiO_2$ which removes iron as slag ($FeSiO_3$).

$$2CuFeS_2 + O_2 \longrightarrow Cu_2S + 2FeS + SO_2$$

$$2FeS + 3O_2 \longrightarrow 2SO_2 + 2FeO$$

$$FeO + SiO_2 \longrightarrow FeSiO_3(\text{Slag})$$

**Step III** Self-reduction in Bessemer converter

$$2Cu_2S + 3O_2 \longrightarrow 2Cu_2O + 2SO_2$$

$$2Cu_2O + Cu_2S \longrightarrow 6Cu + SO_2$$

Copper obtained is blister copper (98% pure).

**Step IV** Refining of blister copper is done by electrolysis

Impure copper—Anode

Pure copper— Cathode

At anode : $Cu \longrightarrow Cu^{2+} + 2e^-$

At cathode : $Cu^{2+} + 2e^- \longrightarrow Cu$

Carbon-reduction method is not used. Thus, (d) is incorrect.

**29.** (*b, d*) $P_4O_{10}$ is a dehydrating agent and converts $HNO_3$ into $N_2O_5$

$$2HNO_3 \longrightarrow N_2O_5 + H_2O$$

$$P_4O_{10} + 6H_2O \longrightarrow 4H_3PO_4$$

(a) $P_4 + 20HNO_3 \longrightarrow 4H_3PO_4 + 20NO_2 + 4H_2O$

Thus, (a) is incorrect.

(b) $N_2O_5$ has no unpaired electron and is thus, diamagnetic thus, (b) is correct.

(c) [Structure: O=N(→O)—O—N(=O)→O]

There is no N—N bond, thus, (c) is incorrect.

(d) $N_2O_5 + Na \longrightarrow NaNO_3 + NO_2$

$N_2O_5$ vapours are of brownish colour. Thus, (d) is correct.

**30.** (*a, b*) When intermolecular attraction between two components *A* and *B* in the mixture is same as between *A* and *A* or *B* and *B*, hence it is a case of ideal solution.

When intermolecular attraction between *A* and *B* in a mixture is smaller than that between A and A or B and B, then mixture is more vaporised, bp is lowered. It is a case of positive deviation from Raoult's law.

When intermolecular attraction between *A* and *B* is higher than that between *A* and *A* or *B* and *B*, then mixture is less vaporised, bp is increased. It is a case of negative deviation.

(a) Methanol molecules ($CH_3OH$) are hydrogen bonded. In a mixture of $CCl_4$ and $CH_3OH$, extent of H-bonding is decreased. Mixture is more vaporised thus, positive deviation from Raoult's law.

(b) Acetone molecules have higher intermolecular attraction due to dipole-dipole interaction. With $CS_2$, this interaction is decreased thus, positive deviation.

(c) Mixture of benzene and toluene forms ideal solution.

(d) Phenol and aniline have higher interaction due to intermolecular H-bonding. Hence, negative deviation.

**31.** (*b, c*) If there is inversion of specific rotation from (+) to (–), then invert sugar is formed.

(a) $\underset{\substack{(+)\text{Maltose} \\ 140^\circ}}{C_{12}H_{22}O_{11}} + H_2O \longrightarrow \underset{\substack{D(+) \\ 52^\circ}}{\text{Glucose}}$

(b) $\underset{\substack{(+)\text{Sucrose} \\ +66^\circ}}{C_{12}H_{22}O_{11}} + H_2O \longrightarrow \underset{\substack{D(+) \\ 52^\circ}}{\text{Glucose}} + \underset{\substack{L(-) \\ -92^\circ}}{\text{Fructose}}$

–40° for 2 moles mixture
–20° for 1 mole mixture

There is formation of invert sugar. Thus, correct.

(c) Specific rotation of invert sugar is –20° per mole. Thus, correct.

(d) $Br_2$ water is a weak oxidising agent. It oxidises —CHO to —COOH. —$CH_2OH$ group is not affected.

$$\underset{\text{Saccharic acid}}{COOH-(CHOH)_4-COOH} \xleftarrow{HNO_3} CHO-(CHOH)_4-CH_2OH$$

$$\xrightarrow{Br_2\text{ water}} \underset{\substack{\text{Gluconic acid}\\ \text{(one of the products)}}}{COOH-(CHOH)_4-CH_2OH}$$

$HNO_3$ (a strong oxidising agent) oxidises invert sugar to saccharic acid. Thus, incorrect.

**32.** (*b, c, d*)

(a) $(CH_3)_3C-Br \xrightarrow{NaOC_2H_5} (CH_3)_2C=CH_2$

$C_2H_5O^-$ (a strong nucleophile) causes E1 reaction to form isobutene as the major product.

(b)

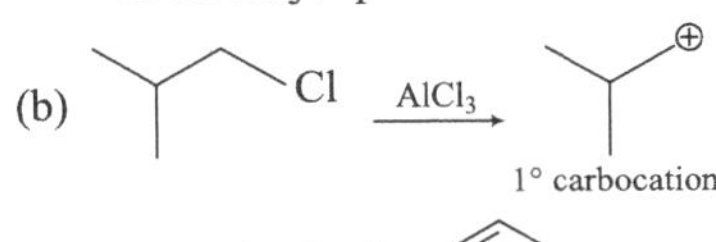


$\xrightarrow[\text{shift}]{1,\text{2-hydride}}$ $(CH_3)_3C^{\oplus}$ 3° carbocation (Major) $\xrightarrow[S_E]{C_6H_6}$ $C_6H_5C(CH_3)_3$ Tertiary butyl benzene($X$)

(c) $(CH_3)_2C=CH_2 \xrightarrow{H^+} (CH_3)_3C^{\oplus} \xrightarrow{C_6H_6} (X)$

(d) $(CH_3)_2CHCH_2OH + C_6H_6 \xrightarrow{BF_3,Et_2O} C_6H_5C(CH_3)_3$

**33.** (*b*) $X_2(g) \rightleftharpoons 2X(g)$

| | $X_2(g)$ | $2X(g)$ |
|---|---|---|
| At $t = 0$ | 1 | 0 |
| At equilibrium | $\left(1-\dfrac{x}{2}\right)$ | $x$ |

(where, $x = \beta_{eq}$)

Total moles $= \left(1+\dfrac{x}{2}\right)$ and

Mole fraction, $X_2(g) = \dfrac{\left(1-\dfrac{x}{2}\right)}{\left(1+\dfrac{x}{2}\right)}$

$X(g) = \left(\dfrac{x}{1+\dfrac{x}{2}}\right)$ and $p = 2$ bar

Partial pressure, $p_{X_2} = \left(\dfrac{1-\dfrac{x}{2}}{1+\dfrac{x}{2}}\right).p$

and $p_X = \dfrac{p\cdot x}{\left(1+\dfrac{x}{2}\right)}$

$$\therefore K_p = p_X^2 / p_{X_2} = \frac{\left[px/\left(1+\dfrac{x}{2}\right)\right]^2}{p\dfrac{(1-x/2)}{\left(1+\dfrac{x}{2}\right)}}$$

$$= \frac{4px^2}{(4-x^2)} = \frac{8\beta_{eq}^2}{(4-\beta_{eq}^2)}$$

**34.** (*c*)

(a) $K_p = \dfrac{4px^2}{(4-x^2)} = px^2 \quad (\because 4 >>> x)$

$\therefore \quad x \propto \sqrt{\dfrac{1}{p}}$

If $p$ decreases, $x$ increases. Equilibrium is shifted in the forward side. Thus, statement (a) is correct.

(b) At the start of the reaction, $Q = 0$ where, $Q$ is the reaction quotient

$$\Delta G = \Delta G^\circ + 2.303RT \log Q$$

Since, $\Delta G^\circ > 0$, thus $\Delta G$ is –ve.

Hence, dissociation takes place spontaneously.

Thus, (b) is correct.

(c) If we use $x = 0.7$ and $p = 2$ bar,

then $K_p = \dfrac{4\times 2(0.7)^2}{[4-(0.7)^2]} = 1.16 > 1$

Thus, (c) is incorrect.

(d) At equilibrium, $\Delta G = 0$

$\therefore \quad \Delta G^\circ = -2.303RT \log K_p$

Since, $\Delta G^\circ = +$ ve

Hence, $K_p < 1$, $K_C = \dfrac{K_p}{(RT)}$

Then $K_C < 1$. Thus, (d) is correct.

**35.** (*a*)

**36.** (*b*) Explanation

1,2-dipropylbenzene $\xrightarrow[\Delta]{MnO_4^-/H^+}$ phthalic acid (*P*) $\xrightarrow[\Delta]{NH_3}$ phthalamide (*Q*) $\xrightarrow[\Delta]{Br_2/NaOH}$ *o*-phenylenediamine (*R*)

*Q* $\xrightarrow{\Delta}$ phthalimide (*S*) $\xrightarrow{KOH}$

$S_1$ (potassium phthalimide, N—K) $\xrightarrow[-KBr]{Br-CH(CH_3)-COOC_2H_5}$ $S_2$ (N-substituted phthalimide with $CH(CH_3)COOC_2H_5$) $\xrightarrow{H_3O^+}$ phthalic acid (*P*) + $CH_3CH(NH_2)COOH$ (*T*)

## Mathematics

**37.** (***b***) Here, $P = \begin{bmatrix} 1 & 0 & 0 \\ 4 & 1 & 0 \\ 16 & 4 & 1 \end{bmatrix}$

$$\therefore \quad P^2 = \begin{bmatrix} 1 & 0 & 0 \\ 4 & 1 & 0 \\ 16 & 4 & 1 \end{bmatrix}\begin{bmatrix} 1 & 0 & 0 \\ 4 & 1 & 0 \\ 16 & 4 & 1 \end{bmatrix}$$

$$= \begin{bmatrix} 1 & 0 & 0 \\ 4+4 & 1 & 0 \\ 16+32 & 4+4 & 1 \end{bmatrix}$$

$$= \begin{bmatrix} 1 & 0 & 0 \\ 4 \times 2 & 1 & 0 \\ 16\,(1+2) & 4 \times 2 & 1 \end{bmatrix} \quad \ldots\text{(i)}$$

$$\text{and} \quad P^3 = \begin{bmatrix} 1 & 0 & 0 \\ 4 \times 2 & 1 & 0 \\ 16\,(1+2) & 4 \times 2 & 1 \end{bmatrix}\begin{bmatrix} 1 & 0 & 0 \\ 4 & 1 & 0 \\ 16 & 4 & 1 \end{bmatrix}$$

$$= \begin{bmatrix} 1 & 0 & 0 \\ 4 \times 3 & 1 & 0 \\ 16\,(1+2+3) & 4 \times 3 & 1 \end{bmatrix} \quad \ldots\text{(ii)}$$

From symmetry,

$$P^{50} = \begin{bmatrix} 1 & 0 & 0 \\ 4 \times 50 & 1 & 0 \\ 16\,(1+2+3+\ldots+50) & 4 \times 50 & 1 \end{bmatrix}$$

$$\because \quad P^{50} - Q = I \quad \text{[given]}$$

$$\therefore \quad \begin{bmatrix} 1-q_{11} & -q_{12} & -q_{13} \\ 200-q_{21} & 1-q_{22} & -q_{23} \\ 16 \times \frac{50}{2}(51) - q_{31} & 200-q_{32} & 1-q_{33} \end{bmatrix} = \begin{bmatrix} 1 & 0 & 0 \\ 0 & 1 & 0 \\ 0 & 0 & 1 \end{bmatrix}$$

$$\Rightarrow \quad 200 - q_{21} = 0,$$

$$\frac{16 \times 50 \times 51}{2} - q_{31} = 0,\ 200 - q_{32} = 0$$

$$\therefore \quad q_{21} = 200,\ q_{32} = 200,\ q_{31} = 20400$$

Thus, $\dfrac{q_{31}+q_{32}}{q_{21}} = \dfrac{20400+200}{200}$

$= \dfrac{20600}{200} = 103$

**38.** (*c*) Here, $\{(x, y) \in R^2 : y \geq \sqrt{|x+3|},\ 5y \leq (x+9) \leq 15\}$

$\therefore \quad y \geq \sqrt{x+3}$

$\Rightarrow \quad y \geq \begin{cases} \sqrt{x+3}, & \text{when } x \geq -3 \\ \sqrt{-x-3}, & \text{when } x \leq -3 \end{cases}$

or $\quad y^2 \geq \begin{cases} x+3, & \text{when } x \geq -3 \\ -3-x, & \text{when } x \leq -3 \end{cases}$

Shown as

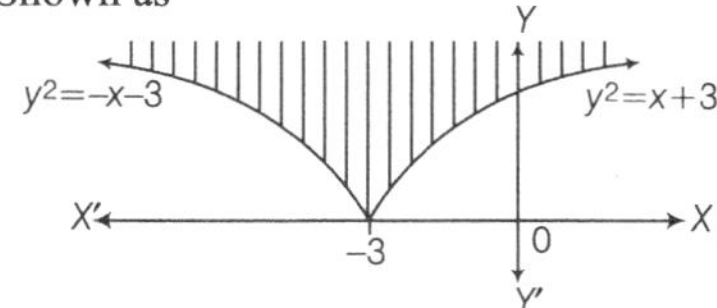


Also, $\quad 5y \leq (x+9) \leq 15$

$\Rightarrow \quad (x+9) \geq 5y$ and $x \leq 6$

Shown as

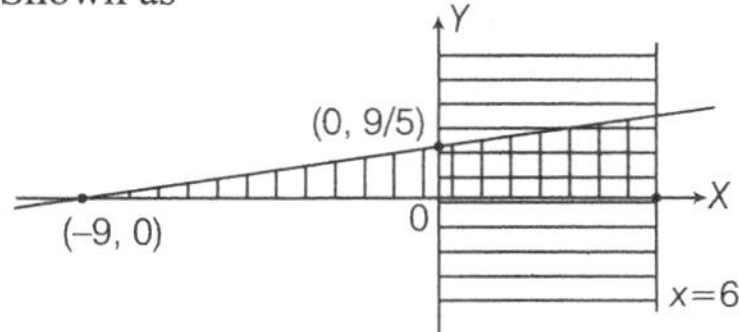


$\therefore \quad \{(x, y) \in R^2 : y \geq \sqrt{|x+3|},\ 5y \leq (x+9) \leq 15\}$

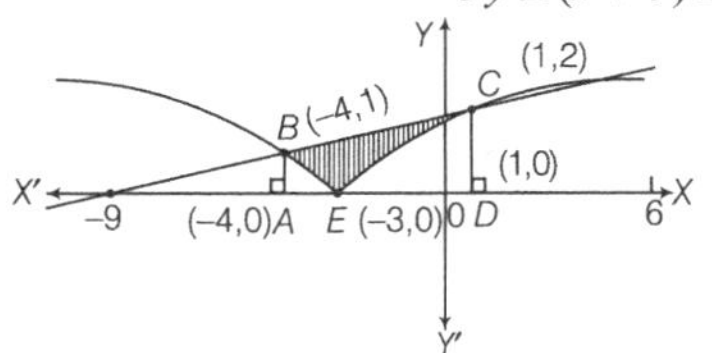


$\therefore$ Required area

= Area of trapezium $ABCD$

− Area of $ABE$ under parabola

− Area of $CDE$ under parabola

$$= \frac{1}{2}(1+2)(5) - \int_{-4}^{-3}\sqrt{-(x+3)}\,dx - \int_{-3}^{1}\sqrt{(x+3)}\,dx$$

$$= \frac{15}{2} - \left[\frac{(-3-x)^{3/2}}{-\frac{3}{2}}\right]_{-4}^{-3} - \left[\frac{(x+3)^{3/2}}{\frac{3}{2}}\right]_{-3}^{1}$$

$$= \frac{15}{2} + \frac{2}{3}[0-1] - \frac{2}{3}[8-0]$$

$$= \frac{15}{2} - \frac{2}{3} - \frac{16}{3} = \frac{15}{2} - \frac{18}{3} = \frac{3}{2}$$

**39.** (*c*) Here,

$$\sum_{k=1}^{13} \frac{1}{\sin\left\{\frac{\pi}{4} + \frac{(k-1)\pi}{6}\right\} \sin\left(\frac{\pi}{4} + \frac{k\pi}{6}\right)}$$

Converting into differences, by multiplying and dividing by

$$\sin\left[\left(\frac{\pi}{4} + \frac{k\pi}{6}\right) - \left\{\frac{\pi}{4} + \frac{(k-1)\pi}{6}\right\}\right],$$

i.e. $\sin\left(\frac{\pi}{6}\right)$.

$$\therefore \sum_{k=1}^{13} \frac{\sin\left[\left(\frac{\pi}{4} + k\frac{\pi}{6}\right) - \left\{\frac{\pi}{4} + (k-1)\frac{\pi}{6}\right\}\right]}{\sin\frac{\pi}{6}\left\{\sin\left\{\frac{\pi}{4} + (k-1)\frac{\pi}{6}\right\} \sin\left(\frac{\pi}{4} + k\frac{\pi}{6}\right)\right\}}$$

$$= 2\sum_{k=1}^{13} \frac{\left[\sin\left(\frac{\pi}{4} + \frac{k\pi}{6}\right)\cos\left\{\frac{\pi}{4} + (k-1)\frac{\pi}{6}\right\} - \sin\left\{\frac{\pi}{4} + (k-1)\frac{\pi}{6}\right\}\cos\left(\frac{\pi}{4} + \frac{k\pi}{6}\right)\right]}{\sin\left\{\frac{\pi}{4} + (k-1)\frac{\pi}{6}\right\}\sin\left(\frac{\pi}{4} + k\frac{\pi}{6}\right)}$$

$$= 2\sum_{k=1}^{13}\left[\cot\left\{\frac{\pi}{4} + (k-1)\frac{\pi}{6}\right\} - \cot\left(\frac{\pi}{4} + k\frac{\pi}{6}\right)\right]$$

$$= 2\left[\left\{\cot\left(\frac{\pi}{4}\right) - \cot\left(\frac{\pi}{4} + \frac{\pi}{6}\right)\right\} + \left\{\cot\left(\frac{\pi}{4} + \frac{\pi}{6}\right) - \cot\left(\frac{\pi}{4} + \frac{2\pi}{6}\right)\right\} + \ldots + \left\{\cot\left(\frac{\pi}{4} + 12\frac{\pi}{6}\right) - \cot\left(\frac{\pi}{4} + 13\frac{\pi}{6}\right)\right\}\right]$$

$$= 2\left\{\cot\frac{\pi}{4} - \cot\left(\frac{\pi}{4} + 13\frac{\pi}{6}\right)\right\}$$

$$= 2\left[1 - \cot\left(\frac{29\pi}{12}\right)\right] = 2\left[1 - \cot\left(2\pi + \frac{5\pi}{12}\right)\right]$$

$$= 2\left[1 - \cot\frac{5\pi}{12}\right] \quad \left[\because \cot\frac{5\pi}{12} = (2-\sqrt{3})\right]$$

$$= 2(1 - 2 + \sqrt{3}) = 2(\sqrt{3} - 1)$$

**40.** **(*b*)** If $\log b_1, \log b_2, \ldots, \log b_{101}$ are in AP, with common difference $\log_e 2$, then $b_1, b_2, \ldots, b_{101}$ are in GP, with common ratio 2.

$$\therefore \quad b_1 = 2^0 b_1$$
$$b_2 = 2^1 b_1$$
$$b_3 = 2^2 b_1$$
$$\vdots \quad \vdots \quad \vdots$$
$$b_{101} = 2^{100} b_1 \quad \ldots(\text{i})$$

Also, $a_1, a_2, \ldots, a_{101}$ are in AP.

Given, $a_1 = b_1$ and $a_{51} = b_{51}$

$$\Rightarrow \quad a_1 + 50D = 2^{50} b_1$$
$$\Rightarrow a_1 + 50\,D = 2^{50} a_1 \quad [\because a_1 = b_1] \ldots(\text{ii})$$

Now, $t = b_1 + b_2 + \ldots + b_{51}$

$$\Rightarrow \quad t = b_1 \frac{(2^{51} - 1)}{2 - 1} \quad \ldots(\text{iii})$$

and $s = a_1 + a_2 + \ldots + a_{51}$

$$= \frac{51}{2}(2a_1 + 50D) \ldots(\text{iv})$$

$$\therefore \quad t = a_1(2^{51} - 1) \ [\because a_1 = b_1]$$

or $\quad t = 2^{51} a_1 - a_1 < 2^{51} a_1 \quad \ldots(\text{v})$

and $s = \dfrac{51}{2}[a_1 + (a_1 + 50\,D)]$

[from Eq. (ii)]

$$= \frac{51}{2}[a_1 + 2^{50} a_1] = \frac{51}{2} a_1 + \frac{51}{2} 2^{50} a_1$$

$$\therefore \quad s > 2^{51} a_1 \quad \ldots(\text{vi})$$

From Eqs. (v) and (vi),

we get $\quad s > t$

Also, $\quad a_{101} = a_1 + 100\,D$

and $\quad b_{101} = 2^{100} b_1$

$$\therefore \quad a_{101} = a_1 + 100\left(\frac{2^{50} a_1 - a_1}{50}\right)$$

and $\quad b_{101} = 2^{100} a_1$

$$\Rightarrow a_{101} = a_1 + 2^{51} a_1 - 2a_1 = 2^{51} a_1 - a_1$$
$$\Rightarrow \quad a_{101} < 2^{51} a_1$$

and $b_{101} > 2^{51} a_1 \Rightarrow b_{101} > a_{101}$

**41.** **(*a*)** Let $I = \displaystyle\int_{-\pi/2}^{\pi/2} \frac{x^2 \cos x}{1 + e^x} dx \quad \ldots(\text{i})$

$$\left[\because \int_a^b f(x)dx = \int_a^b f(a + b - x)\,dx\right]$$

$$\Rightarrow \quad I = \int_{-\pi/2}^{\pi/2} \frac{x^2 \cos(-x)}{1 + e^{-x}} dx \quad \ldots(\text{ii})$$

On adding Eqs. (i) and (ii), we get

$$2I = \int_{-\pi/2}^{\pi/2} x^2 \cos x \left[\frac{1}{1 + e^x} + \frac{1}{1 + e^{-x}}\right] dx$$
$$= \int_{-\pi/2}^{\pi/2} x^2 \cos x \cdot (1)\,dx$$
$$\left[\begin{array}{l}\because \displaystyle\int_{-a}^{a} f(x)dx \\ = 2\displaystyle\int_0^a f(x)dx, \text{when } f(-x) = f(x)\end{array}\right]$$

$$\Rightarrow \ 2I = 2\int_0^{\pi/2} x^2 \cos x\,dx$$

Using integration by parts, we get

$$2I = 2\,[x^2(\sin x) - (2x)(-\cos x) + (2)(-\sin x)]_0^{\pi/2}$$

$$\Rightarrow 2I = 2\left[\frac{\pi^2}{4} - 2\right]$$

$$\therefore \quad I = \frac{\pi^2}{4} - 2$$

**42.** **(*c*)** Let image of $Q(3, 1, 7)$ w.r.t. $x - y + z = 3$ be $P(\alpha, \beta, \gamma)$.

$$\therefore \quad \frac{\alpha - 3}{1} = \frac{\beta - 1}{-1} = \frac{\gamma - 7}{1}$$
$$= \frac{-2(3 - 1 + 7 - 3)}{1^2 + (-1)^2 + (1)^2}$$
$$\Rightarrow \ \alpha - 3 = 1 - \beta = \gamma - 7 = -4$$
$$\therefore \quad \alpha = -1, \beta = 5, \gamma = 3$$

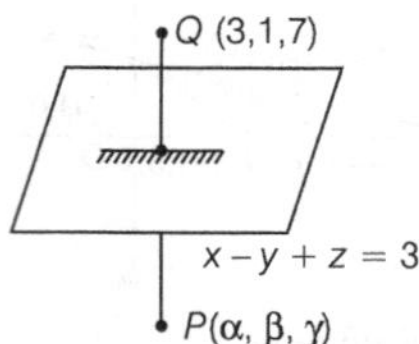


Hence, the image of $Q(3, 1, 7)$ is $P(-1, 5, 3)$.

To find equation of plane passing through $P(-1, 5, 3)$ and containing $\dfrac{x}{1} = \dfrac{y}{2} = \dfrac{z}{1}$

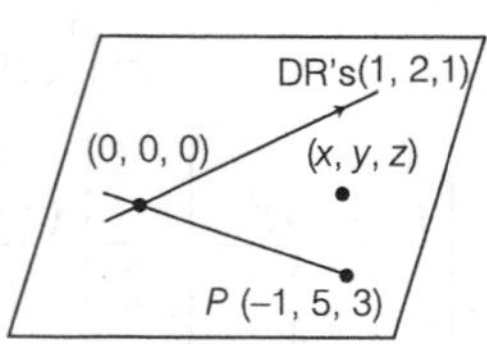

$$\Rightarrow \begin{vmatrix} x-0 & y-0 & z-0 \\ 1-0 & 2-0 & 1-0 \\ -1-0 & 5-0 & 3-0 \end{vmatrix} = 0$$

$\Rightarrow x(6-5) - y(3+1) + z(5+2) = 0$

$\therefore \quad x - 4y + 7z = 0$

**43.** **(*a*, *b*)** Here,

$f(x) = a\cos(|x^3 - x|) + b|x|\sin(|x^3 + x|)$

If $x^3 - x \geq 0$

$\Rightarrow \quad \cos|x^3 - x| = \cos(x^3 - x)$

$x^3 - x \leq 0$

$\Rightarrow \quad \cos|x^3 - x| = \cos(x^3 - x)$

$\therefore \quad \cos(|x^3 - x|) = \cos(x^3 - x),$

$\forall\, x \in R \quad \ldots\text{(i)}$

Again, if $x^3 + x \geq 0$

$\Rightarrow |x|\sin(|x^3 + x|) = x\sin(x^3 + x)$

$x^3 + x \leq 0$

$\Rightarrow |x|\sin(|x^3 + x|) = -x\sin\{-(x^3 + x)\}$

$\therefore \quad |x|\sin(|x^3 + x|) = x\sin(x^3 + x),$

$\forall\, x \in R \quad \ldots\text{(ii)}$

$\Rightarrow f(x) = a\cos(|x^3 - x|) + b|x|\sin(|x^3 + x|)$

$\therefore f(x) = a\cos(x^3 - x) + bx\sin(x^3 + x) \quad \ldots\text{(iii)}$

which is clearly sum and composition of differential functions.

Hence, $f(x)$ is always continuous and differentiable.

**44.** **(*b*, *c*)** Here, $f(x)$

$$= \lim_{n\to\infty}\left[\frac{n^n(x+n)\left(x+\frac{n}{2}\right)\ldots\left(x+\frac{n}{n}\right)}{n!\,(x^2+n^2)\left(x^2+\frac{n^2}{4}\right)\ldots\left(x^2+\frac{n^2}{n^2}\right)}\right]^{\frac{x}{n}}$$

$x > 0$

Taking log on both sides, we get $\log_e\{f(x)\}$

$$= \lim_{n\to\infty}\log\left[\frac{n^n(x+n)\left(x+\frac{n}{2}\right)\ldots\left(x+\frac{n}{n}\right)}{n!(x^2+n^2)\left(x^2+\frac{n^2}{4}\right)\ldots\left(x^2+\frac{n^2}{n^2}\right)}\right]^{\frac{x}{n}}$$

$$= \lim_{n\to\infty}\frac{x}{n}\cdot\log\left[\frac{\prod_{r=1}^{n}\left(x+\frac{1}{r/n}\right)}{\prod_{r=1}^{n}\left(x^2+\frac{1}{(r/n)^2}\right)\prod_{r=1}^{n}(r/n)}\right]$$

$$= x\lim_{n\to\infty}\frac{1}{n}\sum_{r=1}^{n}\log\left[\frac{x+\frac{n}{r}}{\left(x^2+\frac{n^2}{r^2}\right)\frac{r}{n}}\right]$$

$$= x\lim_{n\to\infty}\frac{1}{n}\sum_{r=1}^{n}\log\left(\frac{\frac{r}{n}\cdot x+1}{\frac{r^2}{n^2}\cdot x^2+1}\right)$$

Converting summation into definite integration, we get $\log_e\{f(x)\}$

$$= x\int_0^1\log\left(\frac{xt+1}{x^2t^2+1}\right)dt$$

Put $\quad tx = z \Rightarrow xdt = dz$

$$\therefore\ \log_e\{f(x)\} = x\int_0^x\log\left(\frac{1+z}{1+z^2}\right)\frac{dz}{x}$$

$$\Rightarrow \log_e\{f(x)\} = \int_0^x\log\left(\frac{1+z}{1+z^2}\right)dz$$

Using Newton-Leibnitz formula, we get

$$\frac{1}{f(x)}\cdot f'(x) = \log\left(\frac{1+x}{1+x^2}\right) \quad \ldots\text{(i)}$$

Here, at $x = 1$,

$$\frac{f'(1)}{f(1)} = \log(1) = 0$$

$\therefore \quad f'(1) = 0$

Now, sign scheme of $f'(x)$ is shown below

+ —————|————— −
x=1

$\therefore$ At $x = 1$, function attains maximum.

Since, $f(x)$ increases on $(0, 1)$.

$\therefore \quad f(1) > f(1/2)$

$\therefore$ Option (a) is incorrect.

$f(1/3) < f(2/3)$

$\therefore$ Option (b) is correct.

Also, $f'(x) < 0$, when $x > 1$

$\Rightarrow \quad f'(2) < 0$

$\therefore$ Option (c) is correct.

Also, $\dfrac{f'(x)}{f(x)} = \log\left(\dfrac{1+x}{1+x^2}\right)$

$$\therefore \frac{f'(3)}{f(3)} - \frac{f'(2)}{f(2)} = \log\left(\frac{4}{10}\right) - \log\left(\frac{3}{5}\right)$$
$$= \log(2/3) < 0$$
$$\Rightarrow \frac{f'(3)}{f(3)} < \frac{f'(2)}{f(2)}$$

$\therefore$ Option (d) is incorrect.

**45.** ***(a, d)*** Here, $\lim_{x \to 2} \frac{f(x) \cdot g(x)}{f'(x) \cdot g'(x)} = 1$

$$\Rightarrow \lim_{x \to 2} \frac{f(x)\, g'(x) + f'(x)\, g(x)}{f''(x)\, g'(x) + f'(x)\, g''(x)} = 1$$

[using L' Hospital's rule]

$$\Rightarrow \frac{f(2)\, g'(2) + f'(2)\, g(2)}{f''(2)\, g'(2) + f'(2)\, g''(2)} = 1$$

$$\Rightarrow \frac{f(2)\, g'(2)}{f''(2)\, g'(2)} = 1$$

$[\because f'(2) = g(2) = 0]$

$\Rightarrow \quad f(2) = f''(2) \quad \ldots \text{(i)}$

$\therefore \; f(x) - f''(x) = 0$, for atleast one $x \in R$.

$\Rightarrow$ Option (d) is correct.

Also, $f: R \to (0, \infty)$

$\Rightarrow \quad f(2) > 0$

$\therefore \quad f''(2) = f(2) > 0$ [from Eq. (i)]

Since, $\quad f'(2) = 0$ and $f''(2) > 0$

$\therefore \; f(x)$ attains local minimum at $x = 2$.

$\Rightarrow$ Option (a) is correct.

**46.** ***(b, c)*** Let $\theta$ be the angle between $\hat{u}$ and $\vec{v}$.

$\therefore \; |\hat{u} \times \vec{v}| = 1 \Rightarrow |\hat{u}||\vec{v}| \sin\theta = 1$

$\therefore \quad |\vec{v}| \sin\theta = 1 \qquad [\because |\hat{u}| = 1] \ldots \text{(i)}$

Clearly, there may be infinite vectors $\overrightarrow{OP} = \vec{v}$, such that $P$ is always 1 unit distance from $\hat{u}$.

$\therefore$ Option (b) is correct.

Again, let $\phi$ be the angle between $\hat{w}$ and $\hat{u} \times \vec{v}$.

$\therefore \; \hat{w} \cdot (\hat{u} \times \vec{v}) = 1 \Rightarrow |\hat{w}||\hat{u} \times \vec{v}| \cos\phi = 1$

$\Rightarrow \quad \cos\phi = 1 \Rightarrow \phi = 0$

Thus, $\quad \hat{w} = \hat{u} \times \vec{v}$

Now, if $\hat{u}$ lies in $XY$-plane, then

$$\hat{u} \times \vec{v} = \begin{vmatrix} \hat{i} & \hat{j} & \hat{k} \\ u_1 & u_2 & 0 \\ v_1 & v_2 & v_3 \end{vmatrix}$$

or $\quad \hat{u} = u\hat{i} + u_2\hat{j}$

$\therefore \; \hat{w} = (u_2v_3)\,\hat{i} - (u_1v_3)\hat{j} + (u_1v_2 - v_1u_2)\hat{k}$

$$= \frac{1}{\sqrt{6}}(\hat{i} + \hat{j} + 2\hat{k})$$

$$\therefore \quad u_2v_3 = \frac{1}{\sqrt{6}} \Rightarrow u_1v_3 = \frac{-1}{\sqrt{6}}$$

$$\Rightarrow \quad \frac{u_2v_3}{u_1v_3} = -1 \quad \text{or} \quad |u_1| = |u_2|$$

$\therefore$ Option (c) is correct.

Now, if $\hat{u}$ lies in $XZ$-plane, then $\hat{u} = u_1\hat{i} + u_3\hat{k}$

$$\therefore \quad \hat{u} \times \vec{v} = \begin{vmatrix} \hat{i} & \hat{j} & \hat{k} \\ u_1 & 0 & u_3 \\ v_1 & v_2 & v_3 \end{vmatrix}$$

$$\Rightarrow \hat{w} = (-v_2u_3)\,\hat{i} - (u_1v_3 - u_3v_1)\,\hat{j} + (u_1v_2)\,\hat{k}$$

$$\Rightarrow \quad \hat{w} = \frac{1}{\sqrt{6}}(\hat{i} + \hat{j} + 2\hat{k})$$

$$\Rightarrow \quad -v_2u_3 = \frac{1}{\sqrt{6}} \text{ and } u_1v_2 = \frac{2}{\sqrt{6}}$$

$\therefore \quad |u_2| = 2|u_3|$

$\therefore$ Option (d) is incorrect

**47.** ***(a, c, d)*** Tangent to $y^2 = 4x$ at $(t^2, 2t)$ is

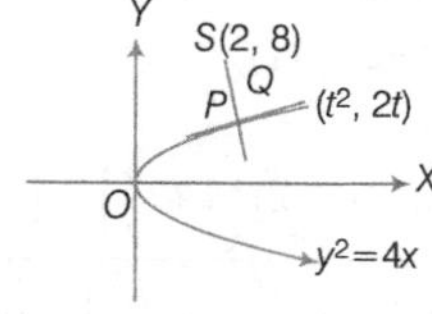


$$y(2t) = 2(x + t^2)$$

$\Rightarrow \quad yt = x + t^2 \quad \ldots \text{(i)}$

Equation of normal at $P(t^2, 2t)$ is

$$y + tx = 2t + t^3$$

Since, normal at $P$ passes through centre of circle $S(2, 8)$.

$\therefore \; 8 + 2t = 2t + t^3 \Rightarrow t = 2$, i.e. $P(4, 4)$

[since, shortest distance between two curves lie along their common normal and the common normal will pass through the centre of circle]

$\therefore \; SP = \sqrt{(4-2)^2 + (4-8)^2} = 2\sqrt{5}$

$\therefore$ Option (a) is correct.

Also, $SQ = 2$

$\therefore \quad PQ = SP - SQ = 2\sqrt{5} - 2$

Thus, $\dfrac{SQ}{QP} = \dfrac{1}{\sqrt{5}-1} = \dfrac{\sqrt{5}+1}{4}$

$\therefore$ Option (b) is incorrect.

Now, $x$-intercept of normal is

$$x = 2 + 2^2 = 6$$

$\therefore$ Option (c) is correct.

Slope of tangent $= \dfrac{1}{t} = \dfrac{1}{2}$

$\therefore$ Option (d) is incorrect

**48.** ***(a, c, d)*** Here, $x + iy = \dfrac{1}{a+ibt} \times \dfrac{a-ibt}{a-ibt}$

$$\therefore \quad x + iy = \frac{a - ibt}{a^2 + b^2t^2}$$

Let $a \neq 0, b \neq 0$

$\therefore \; x = \dfrac{a}{a^2+b^2t^2}$ and $y = \dfrac{-bt}{a^2+b^2t^2}$

$\Rightarrow \quad \dfrac{y}{x} = \dfrac{-bt}{a} \Rightarrow t = \dfrac{ay}{bx}$

On putting $x = \dfrac{a}{a^2+b^2t^2}$, we get

$$x\left(a^2 + b^2 \cdot \frac{a^2y^2}{b^2x^2}\right) = a$$

$\Rightarrow \quad a^2(x^2+y^2) = ax$

or $\quad x^2 + y^2 - \dfrac{x}{a} = 0 \quad \ldots$ (i)

or $\quad \left(x - \dfrac{1}{2a}\right)^2 + y^2 = \dfrac{1}{4a^2}$

$\therefore$ Option (a) is correct.

For $a \neq 0$ and $b = 0$,

$$x + iy = \frac{1}{a} \Rightarrow x = \frac{1}{a}, \; y = 0$$

$\Rightarrow z$ lies on $X$-axis.

$\therefore$ Option (c) is correct.

For $a = 0$ and $b \neq 0$,

$$x + iy = \frac{1}{ibt}$$

$\Rightarrow \quad x = 0, \; y = -\dfrac{1}{bt}$

$\Rightarrow z$ lies on $Y$-axis.

$\therefore$ Option (d) is correct.

**49.** ***(b, c, d)*** Here, $ax + 2y = \lambda$

and $\quad 3x - 2y = \mu$

For $a = -3$, above equations will be parallel or coincident, i.e. parallel for $\lambda + \mu \neq 0$ and coincident, if $\lambda + \mu = 0$ and if $a \neq -3$, equations are intersecting, i.e. unique solution.

**50.** ***(b, c)*** Here,

$$f(x) = [x^2 - 3] = [x^2] - 3$$

$$= \begin{cases} -3, & -1/2 \leq x < 1 \\ -2, & 1 \leq x < \sqrt{2} \\ -1, & \sqrt{2} \leq x < \sqrt{3} \\ 0, & \sqrt{3} \leq x < 2 \\ 1, & x = 2 \end{cases}$$

and $\; g(x) = |x|\,f(x) + |4x-7|\,f(x)$

$= (|x| + |4x-7|)\,f(x)$

$= (|x| + |4x-7|)[x^2-3]$

$$= \begin{cases} (-x-4x-7)(-3), & -1/2 \leq x < 0 \\ (x-4x+7)(-3), & 0 \leq x < 1 \\ (x-4x+7)(-2), & 1 \leq x < \sqrt{2} \\ (x-4x+7)(-1), & \sqrt{2} \leq x < \sqrt{3} \\ (x-4x+7)(0), & \sqrt{3} \leq x < 7/4 \\ (x+4x-7)(0), & 7/4 \leq x < 2 \\ (x+4x-7)(1), & x = 2 \end{cases}$$

$$\therefore \; g(x) = \begin{cases} 15x + 21, & -1/2 \leq x < 0 \\ 9x - 21, & 0 \leq x < 1 \\ 6x - 14, & 1 \leq x < \sqrt{2} \\ 3x - 7, & \sqrt{2} \leq x < \sqrt{3} \\ 0, & \sqrt{3} \leq x < 2 \\ 5x - 7, & x = 2 \end{cases}$$

Now, the graphs of $f(x)$ and $g(x)$ are shown below.

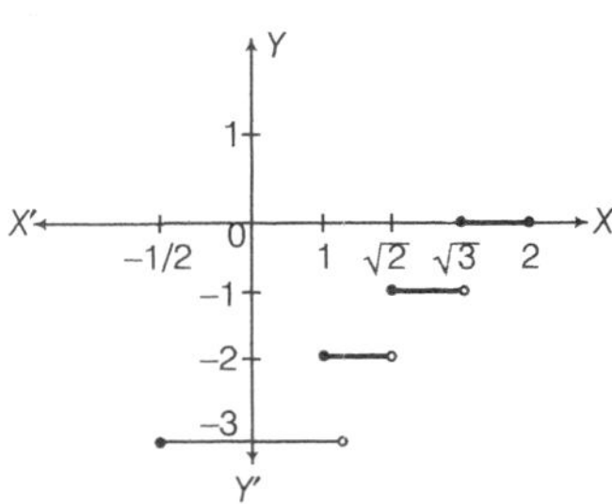


Graph for $f(x)$

Clearly, $f(x)$ is discontinuous at 4 points.

$\therefore$ Option (b) is correct.

Graph for $g(x)$

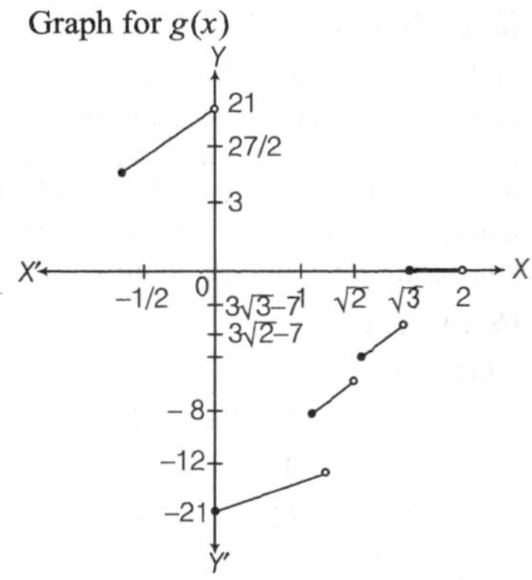


Clearly, $g(x)$ is not differentiable at 4 points, when

$$x \in (-1/2, 2).$$

$\therefore$ Option (c) is correct.

**51.** **(*b*)** Here, $P(X > Y) = P(T_1 \text{win})\, P(T_1 \text{ win})$

$$+ P(T_1 \text{ win})\, P(\text{draw}) + P(\text{draw})\, P(T_1 \text{ win})$$

$$= \left(\frac{1}{2}\times\frac{1}{2}\right) + \left(\frac{1}{2}\times\frac{1}{6}\right) + \left(\frac{1}{6}\times\frac{1}{2}\right) = \frac{5}{12}$$

**52.** **(*c*)** $P[X = Y] = P(\text{draw})\cdot P(\text{draw})$

$$+ P(T_1 \text{ win})\, P(T_2 \text{ win})$$

$$+ P(T_2 \text{ win})\cdot P(T_1 \text{ win})$$

$$= (1/6\times 1/6) + (1/2\times 1/3)$$

$$+ (1/3\times 1/2) = 13/36$$

**53.** **(*a*)** Here, $\quad \dfrac{x^2}{9} + \dfrac{y^2}{8} = 1 \quad$ ...(i)

has foci $(\pm\, ae, 0)$

where, $a^2e^2 = a^2 - b^2 \Rightarrow a^2e^2 = 9 - 8$

$\Rightarrow \quad ae = \pm\, 1$ i.e. $F_1, F_2 = (\pm\, 1, 0)$

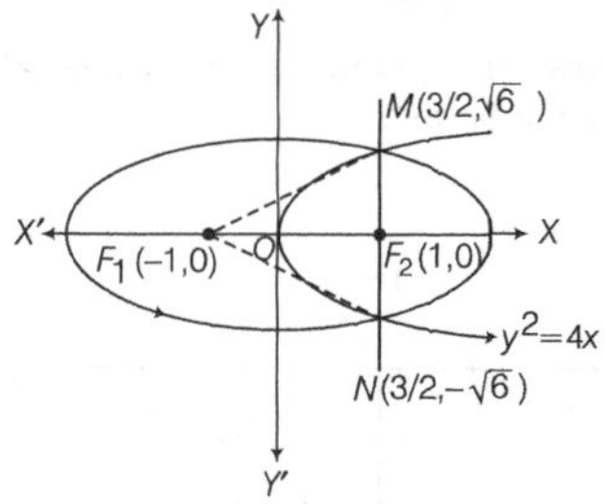


Equation of parabola having vertex $O(0,0)$ and $F_2(1, 0)$ (as, $x_2 > 0$)

$$y^2 = 4x \qquad \text{...(ii)}$$

On solving $\dfrac{x^2}{9} + \dfrac{y^2}{8} = 1$ and $y^2 = 4x$, we get

$$x = 3/2 \text{ and } y = \pm\,\sqrt{6}$$

Equation of altitude through $M$ on $NF_1$ is

$$\frac{y-\sqrt{6}}{x - 3/2} = \frac{5}{2\sqrt{6}}$$

$$\Rightarrow \quad (y-\sqrt{6}) = \frac{5}{2\sqrt{6}}(x - 3/2) \quad \text{...(iii)}$$

and equation of altitude through

$F_1$ is $y = 0 \qquad$ ...(iv)

On solving Eqs. (iii) and (iv), we get $\left(-\dfrac{9}{10}, 0\right)$ as orthocentre.

**54.** **(*c*)** Equation of tangent at $M(3/2, \sqrt{6})$ to $\dfrac{x^2}{9} + \dfrac{y^2}{8} = 1$ is

$$\frac{3}{2}\cdot\frac{x}{9} + \sqrt{6}\cdot\frac{y}{8} = 1 \qquad \text{...(i)}$$

which intersect $X$-axis at (6, 0).

Also, equation of tangent at $N(3/2, -\sqrt{6})$ is

$$\frac{3}{2}\cdot\frac{x}{9} - \sqrt{6}\,\frac{y}{8} = 1 \qquad \text{...(ii)}$$

Eqs. (i) and (ii) intersect on $X$-axis at $R(6, 0)$. ...(iii)

Also, normal at $M(3/2, \sqrt{6})$ is

$$y - \sqrt{6} = \frac{-\sqrt{6}}{2}\left(x - \frac{3}{2}\right)$$

On solving with $y = 0$,

we get $Q(7/2, 0)$ ...(iv)

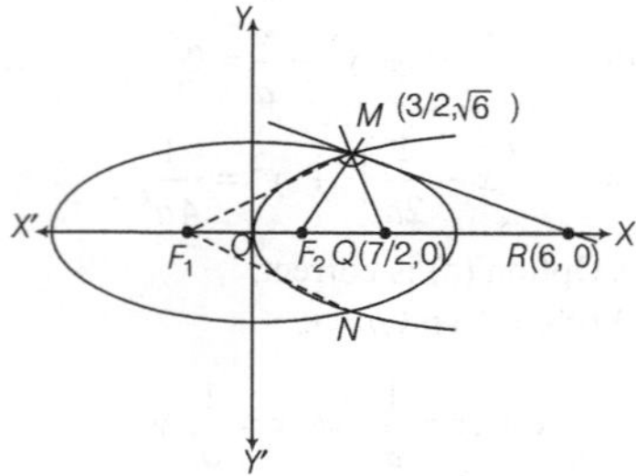


$\therefore$ Area of $\Delta MQR = \dfrac{1}{2}\left(6 - \dfrac{7}{2}\right)\sqrt{6} = \dfrac{5\sqrt{6}}{4}$ sq units

and area of quadrilateral $MF_1NF_2$

$$= 2\times\frac{1}{2}\,\{1-(-1)\}\,\sqrt{6} = 2\sqrt{6} \text{ sq units}$$

$$\therefore \quad \frac{\text{Area of } \Delta MQR}{\text{Area of quadrilateral } MF_1NF_2} = \frac{5}{8}$$

# Solved Paper 2015

# JEE Main

## Joint Entrance Examination

### Instructions

1. This test consists of 90 questions.
2. There are three parts in the question paper A, B, C consisting of Physics, Chemistry & Mathematics having 30 questions in each part of equal weightage. Each question is allotted 4 marks for correct response.
3. Candidates will be awarded marks as stated above in instruction no. 2 for correct response of each question. 1 marks will be deducted for indicating incorrect response of each question. No deduction from the total score will be made if no response is indicated for an item in the answer sheet.
4. There is only one correct response for each question. Filling up more than one response in any question will be treated as wrong response and marks for wrong response will be deducted according as per instructions.

## Physics

1. Two stones are thrown up simultaneously from the edge of a cliff 240 m high with initial speed of 10 m/s and 40 m/s respectively. Which of the following graph best represents the time variation of relative position of the second stone with respect to the first? Assume stones do not rebound after hitting the ground and neglect air resistance, take $g = 10\,\text{m/s}^2$ )

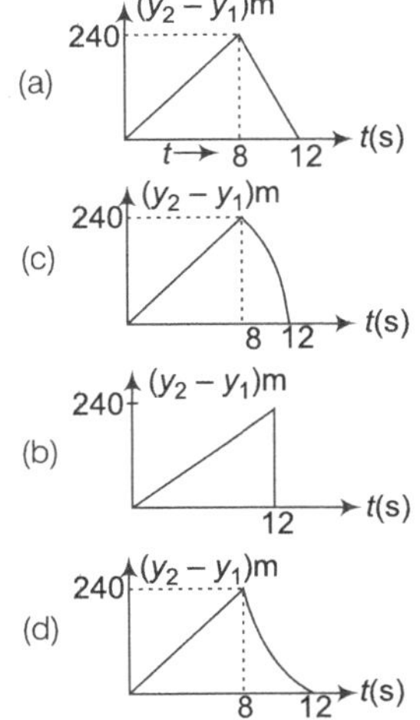


2. The period of oscillation of a simple pendulum is $T = 2\pi\sqrt{\frac{L}{g}}$. Measured value of $L$ is 20.0 cm known to 1 mm accuracy and time for 100 oscillations of the pendulum is found to be 90 s using a wrist watch of 1s resolution. The accuracy in the determination of $g$ is

(a) 2% (b) 3% (c) 1% (d) 5%

3. Given in the figure are two blocks $A$ and $B$ of weight 20 N and 100 N respectively. These are being pressed against a wall by a force $F$ as shown in figure. If the coefficient of friction between the blocks is 0.1 and between block $B$ and the wall is 0.15, the frictional force applied by the wall in block $B$ is

(a) 100 N
(b) 80 N
(c) 120 N
(d) 150 N

**4.** A particle of mass $m$ moving in the $x$-direction with speed $2v$ is hit by another particle of mass $2m$ moving in the $y$-direction with speed $v$. If the collision is perfectly inelastic, the percentage loss in the energy during the collision is close to

(a) 44% (b) 50% (c) 56% (d) 62%

**5.** Distance of the centre of mass of a solid uniform cone from its vertex is $z_0$. If the radius of its base is $R$ and its height is $h$, then $z_0$ is equal to

(a) $\frac{h^2}{4R}$ (b) $\frac{3h}{4}$

(c) $\frac{5h}{8}$ (d) $\frac{3h^2}{8R}$

**6.** From a solid sphere of mass $M$ and radius $R$, a cube of maximum possible volume is cut. Moment of inertia of cube about an axis passing through its centre and perpendicular to one of its faces is

(a) $\frac{MR^2}{32\sqrt{2}\pi}$ (b) $\frac{MR^2}{16\sqrt{2}\pi}$

(c) $\frac{4MR^2}{9\sqrt{3}\pi}$ (d) $\frac{4MR^2}{3\sqrt{3}\pi}$

**7.** From a solid sphere of mass $M$ and radius $R$, a spherical portion of radius $\left(\frac{R}{2}\right)$ is removed as shown in the figure. Taking gravitational potential $V = 0$ at $r = \infty$, the potential at the centre of the cavity thus formed is ($G$ = gravitational constant)

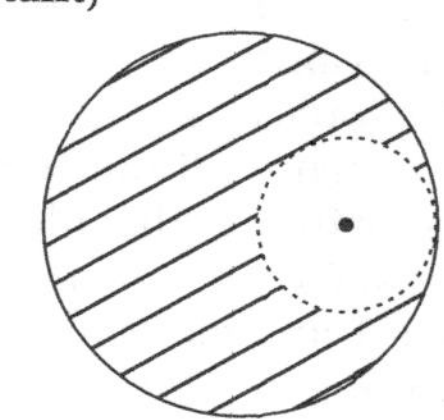

(a) $\frac{-GM}{2R}$ (b) $\frac{-GM}{R}$

(c) $\frac{-2GM}{3R}$ (d) $\frac{-2GM}{R}$

**8.** A pendulum made of a uniform wire of cross-sectional area $A$ has time period $T$. When an additional mass $M$ is added to its bob, the time period changes $T_M$. If the Young's modulus of the material of the wire is $Y$, then $\frac{1}{Y}$ is equal to ($g$ = gravitational acceleration)

(a) $\left[\left(\frac{T_M}{T}\right)^2 - 1\right]\frac{A}{Mg}$

(b) $\left[\left(\frac{T_M}{T}\right)^2 - 1\right]\frac{Mg}{A}$

(c) $\left[1-\left(\frac{T_M}{T}\right)^2\right]\frac{A}{Mg}$

(d) $\left[1-\left(\frac{T}{T_M}\right)^2\right]\frac{A}{Mg}$

**9.** Consider a spherical shell of radius $R$ at temperature $T$. The black body radiation inside it can be considered as an ideal gas of photons with internal energy per unit volume $u = \frac{U}{V} \propto T^4$ and pressure $p = \frac{1}{3}\left(\frac{U}{V}\right)$. If the shell now undergoes an adiabatic expansion, the relation between $T$ and $R$ is

(a) $T \propto e^{-R}$

(b) $T \propto e^{-3R}$

(c) $T \propto \frac{1}{R}$

(d) $T \propto \frac{1}{R^3}$

**10.** A solid body of constant heat capacity 1 J/°C is being heated by keeping it in contact with reservoirs in two ways

(i) Sequentially keeping in contact with 2 reservoirs such that each reservoir supplies same amount of heat.

(ii) Sequentially keeping in contact with 8 reservoirs such that each reservoir supplies same amount of heat.

In both the cases, body is brought from initial temperature 100°C to final temperature 200°C. Entropy change of the body in the two cases respectively, is

(a) ln2, 4ln2
(b) ln2, ln2
(c) ln2, 2ln2
(d) 2ln2, 8ln2

11. Consider an ideal gas confined in an isolated closed chamber. As the gas undergoes an adiabatic expansion, the average time of collision between molecules increases as $V^q$, where $V$ is the volume of the gas. The value of $q$ is $\left(\gamma = \frac{C_p}{C_V}\right)$

(a) $\frac{3\gamma + 5}{6}$
(b) $\frac{3\gamma - 5}{6}$
(c) $\frac{\gamma + 1}{2}$
(d) $\frac{\gamma - 1}{2}$

12. For a simple pendulum, a graph is plotted between its Kinetic Energy (KE) and Potential Energy (PE) against its displacement $d$. Which one of the following represents these correctly? (graphs are schematic and not drawn to scale)

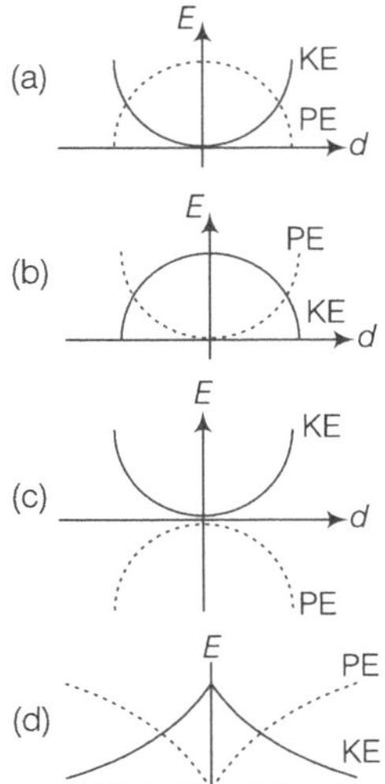


13. A train is moving on a straight track with speed 20 $ms^{-1}$. It is blowing its whistle at the frequency of 1000 Hz. The percentage change in the frequency heard by a person standing near the track as the train passes him is close to (speed of sound = 320 $ms^{-1}$)

(a) 6% (b) 12% (c) 18% (d) 24%

14. A long cylindrical shell carries positive surface charge $\sigma$ in the upper half and negative surface charge $-\sigma$ in the lower half. The electric field lines around the cylinder will look like figure given in (figures are schematic and not drawn to scale)

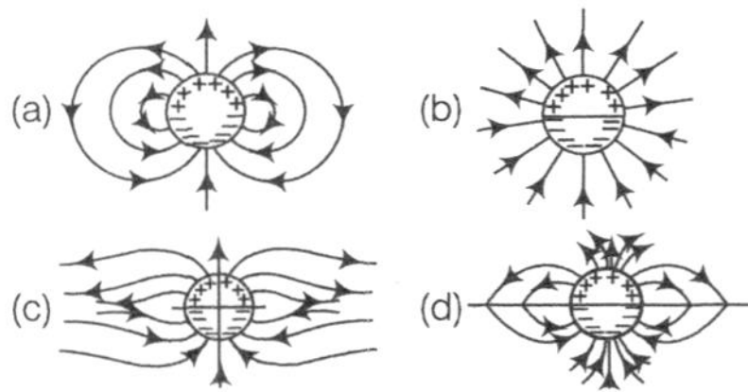


15. A uniformly charged solid sphere of radius $R$ has potential $V_0$ (measured with respect to $\infty$) on its surface. For this sphere, the equipotential surfaces with potentials $\frac{3V_0}{2}$, $\frac{5V_0}{4}$, $\frac{3V_0}{4}$ and $\frac{V_0}{4}$ have radius $R_1$, $R_2$, $R_3$, and $R_4$ respectively. Then,

(a) $R_1 = 0$ and $R_2 > (R_4 - R_3)$
(b) $R_1 \neq 0$ and $(R_2 - R_1) > (R_4 - R_3)$
(c) $R_1 = 0$ and $R_2 < (R_4 - R_3)$
(d) $2R < R_4$

16. In the given circuit, charge $Q_2$ on the 2 μF capacitor changes as $C$ is varied from 1 μF to 3μF. $Q_2$ as a function of $C$ is given properly by (figures are drawn schematically and are not to scale)

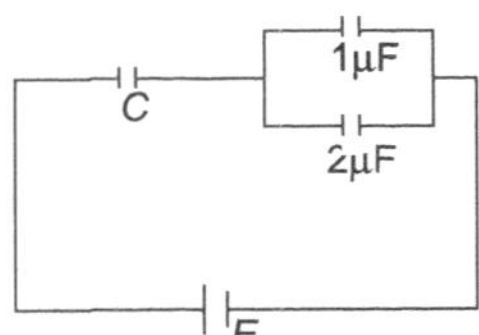

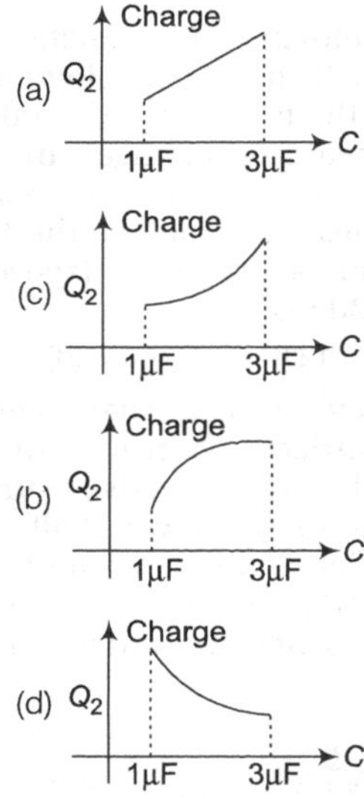


**17.** When 5V potential difference is applied across a wire of length 0.1m, the drift speed of electrons is $2.5 \times 10^{-4} \text{ms}^{-1}$. If the electron density in the wire is $8 \times 10^{28} \text{ m}^{-3}$ the resistivity of the material is close to

(a) $1.6 \times 10^{-8}\ \Omega\text{m}$ (b) $1.6 \times 10^{-7}\ \Omega\text{m}$
(c) $1.6 \times 10^{-6}\ \Omega\text{m}$ (d) $1.6 \times 10^{-5}\ \Omega\text{m}$

**18.** In the circuit shown below, the current in the $1\Omega$ resistor is

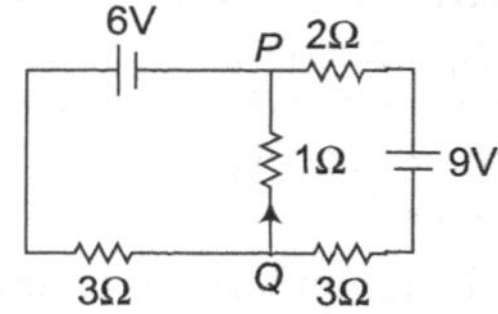


(a) 1.3 A, from $P$ to $Q$
(b) 0 A
(c) 0.13 A, from $Q$ to $P$
(d) 0.13 A, from $P$ to $Q$

**19.** Two coaxial solenoids of different radii carry current $I$ in the same direction. Let $\mathbf{F}_1$ be the magnetic force on the inner solenoid due to the outer one and $\mathbf{F}_2$ be the magnetic force on the outer solenoid due to the inner one. Then,

(a) $\mathbf{F}_1 = \mathbf{F}_2 = 0$
(b) $\mathbf{F}_1$ is radially inwards and $\mathbf{F}_2$ is radially outwards
(c) $\mathbf{F}_1$ is radially inwards and $\mathbf{F}_2 = 0$
(d) $\mathbf{F}_1$ is radially outwards and $\mathbf{F}_2 = 0$

**20.** Two long current carrying thin wires, both with current $I$, are held by insulating threads of length $L$ and are in equilibrium as shown in the figure, with threads making an angle $\theta$ with the vertical. If wires have mass $\lambda$ per unit length then, the value of $I$ is ($g$ = gravitational acceleration)

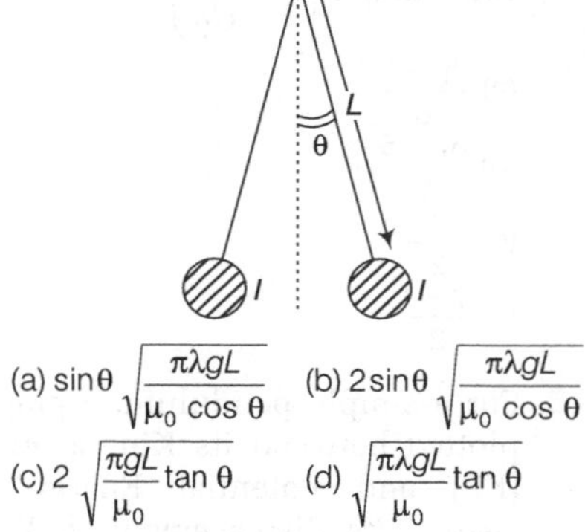


(a) $\sin\theta \sqrt{\dfrac{\pi\lambda gL}{\mu_0 \cos\theta}}$ (b) $2\sin\theta \sqrt{\dfrac{\pi\lambda gL}{\mu_0 \cos\theta}}$

(c) $2\sqrt{\dfrac{\pi gL}{\mu_0}\tan\theta}$ (d) $\sqrt{\dfrac{\pi\lambda gL}{\mu_0}\tan\theta}$

**21.** A rectangular loop of sides 10 cm and 5 cm carrying a current $I$ of 12 A is placed in different orientations as shown in the figures below.

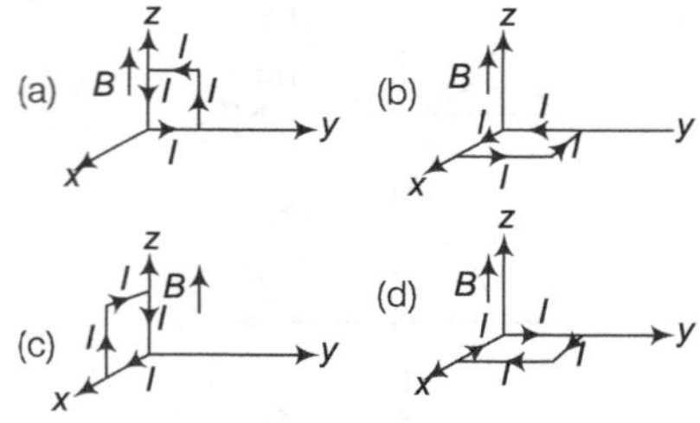


If there is a uniform magnetic field of 0.3 T in the positive $z$-direction in which orientations the loop would be in (i) stable equilibrium and (ii) unstable equilibrium?

(a) (a) and (b) respectively
(b) (a) and (c) respectively
(c) (b) and (d) respectively
(d) (b) and (c) respectively

22. An inductor ($L = 0.03$ H) and a resistor ($R = 0.15$ k$\Omega$) are connected in series to a battery of 15V EMF in a circuit shown below. The key $K_1$ has been kept closed for a long time. Then at $t = 0$, $K_1$ is opened and key $K_2$ is closed simultaneously. At $t = 1$ ms, the current in the circuit will be ($e^5 \cong 150$)

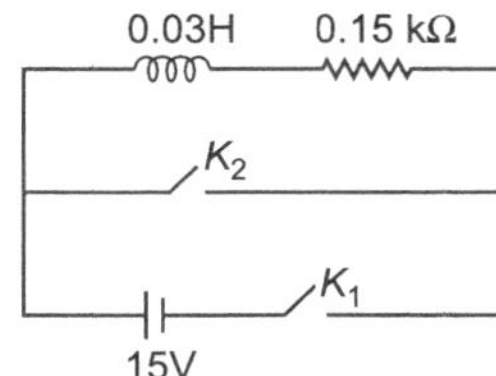


(a) 100 mA (b) 67 mA
(c) 6.7 mA (d) 0.67mA

23. A red LED emits light at 0.1 W uniformly around it. The amplitude of the electric field of the light at a distance of 1 m from the diode is

(a) 1.73 V/m
(b) 2.45 V/m
(c) 5.48 V/m
(d) 7.75 V/m

24. Monochromatic light is incident on a glass prism of angle $A$. If the refractive index of the material of the prism is $\mu$, a ray incident at an angle $\theta$, on the face $AB$ would get transmitted through the face $AC$ of the prism provided

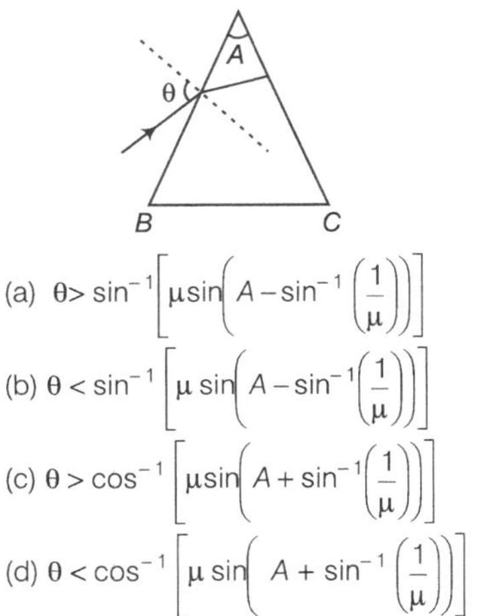


(a) $\theta > \sin^{-1}\left[\mu \sin\left(A - \sin^{-1}\left(\frac{1}{\mu}\right)\right)\right]$

(b) $\theta < \sin^{-1}\left[\mu \sin\left(A - \sin^{-1}\left(\frac{1}{\mu}\right)\right)\right]$

(c) $\theta > \cos^{-1}\left[\mu \sin\left(A + \sin^{-1}\left(\frac{1}{\mu}\right)\right)\right]$

(d) $\theta < \cos^{-1}\left[\mu \sin\left(A + \sin^{-1}\left(\frac{1}{\mu}\right)\right)\right]$

25. On a hot summer night, the refractive index of air is smallest near the ground and increases with height from the ground. When a light beam is directed horizontally, the Huygens principle leads us to conclude that as it travels, the light beam

(a) becomes narrower
(b) goes horizontally without any deflection
(c) bends downwards
(d) bends upwards

26. Assuming human pupil to have a radius of 0.25 cm and a comfortable viewing distance of 25 cm, the minimum separation between two objects that human eye can resolve at 500 nm wavelength is

(a) 1 $\mu$m (b) 30 $\mu$m (c) 100 $\mu$m (d) 300 $\mu$m

27. As an electron makes a transition from an excited state to the ground state of a hydrogen like atom/ion

(a) its kinetic energy increases but potential energy and total energy decrease
(b) kinetic energy, potential energy and total energy decrease
(c) kinetic energy decreases, potential energy increases but total energy remains same
(d) kinetic energy and total energy decrease but potential energy increases

28. Match List I (fundamental experiment) with List II (its conclusion) and select the correct option from the choices given below the list.

| List I | List II |
|---|---|
| A. Franck-Hertz experiment | 1. Particle nature of light |
| B. Photo-electric experiment | 2. Discrete energy levels of atom |
| C. Davisson-Germer experiment | 3. Wave nature of electron |
| | 4. Structure of atom |

| | A | B | C | | A | B | C |
|---|---|---|---|---|---|---|---|
| (a) | 1 | 4 | 3 | (b) | 2 | 4 | 3 |
| (c) | 2 | 1 | 3 | (d) | 4 | 3 | 2 |

**29.** A signal of 5 kHz frequency is amplitude modulated on a carrier wave of frequency 2MHz. The frequencies of the resultant signal is/are

(a) 2 MHz only
(b) 2005 kHz and 1995 kHz
(c) 2005 kHz 2000 kHz and 1995 kHz
(d) 2000 kHz and 1995 kHz

**30.** An *LCR* circuit is equivalent to a damped pendulum. In an *LCR* circuit, the capacitor is charged to $Q_0$ and then connected to the *L* and *R* as shown below.

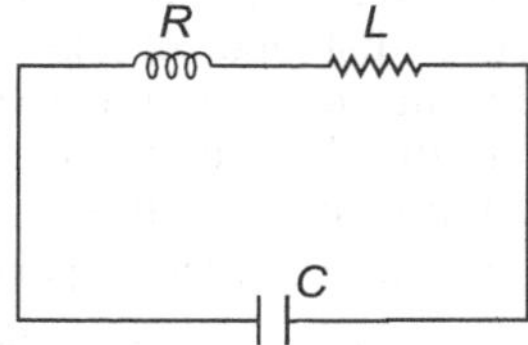


If a student plots graphs of the square of maximum charge ($Q^2_{Max}$) on the capacitor with time ($t$) for two different values $L_1$ and $L_2$ ($L_1 > L_2$) of $L$, then which of the following represents this graph correctly? (plots are schematic and not drawn to scale)

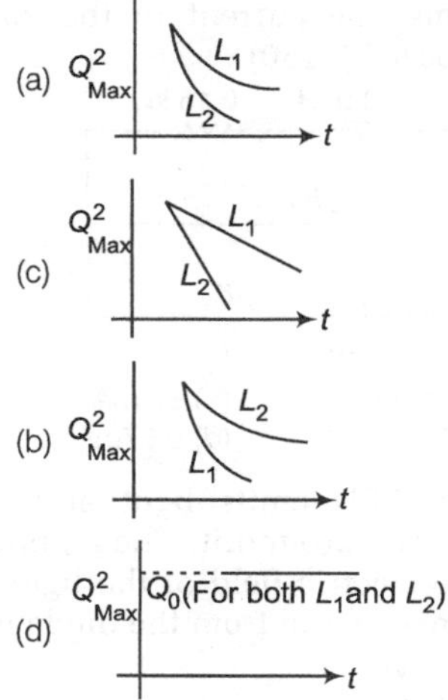


# Chemistry

**1.** The molecular formula of a commercial resin used for exchanging ions in water softening is $C_8H_7SO_3Na$ (mol. wt. = 206). What would be the maximum uptake of $Ca^{2+}$ ions by the resin when expressed in mole per gram resin?

(a) $\frac{1}{103}$
(b) $\frac{1}{206}$
(c) $\frac{2}{309}$
(d) $\frac{1}{412}$

**2.** Sodium metal crystallises in a body centred cubic lattice with a unit cell edge of 4.29 Å. The radius of sodium atom is approximately

(a) 1.86 Å (b) 3.22 Å (c) 5.72 Å (d) 0.93 Å

**3.** Which of the following is the energy of a possible excited state of hydrogen?

(a) +13.6 eV
(b) −6.8 eV
(c) −3.4 eV
(d) +6.8 eV

**4.** The intermolecular interaction that is dependent on the inverse cube of distance between the molecules is

(a) ion-ion interaction
(b) ion-dipole interaction
(c) London force
(d) hydrogen bond

**5.** The following reaction is performed at 298K

$$2NO(g) + O_2(g) \rightleftharpoons 2NO_2(g)$$

The standard free energy of formation of NO (*g*) is 86.6 kJ/mol at 298 K. What is the standard free energy of formation of $NO_2(g)$ at 298 K? ($K_p = 1.6 \times 10^{12}$)

(a) $R(298) \ln (1.6 \times 10^{12}) - 86600$
(b) $86600 + R(298) \ln (1.6 \times 10^{12})$
(c) $86600 - \frac{\ln (1.6 \times 10^{12})}{R(298)}$
(d) $0.5 [2 \times 86600 - R(298) \ln (1.6 \times 10^{12})]$

6. The vapour pressure of acetone at 20°C is 185 torr. When 1.2 g of a non-volatile substance was dissolved in 100 g of acetone at 20°C, its vapour pressure was 183 torr. The molar mass $(\text{g mol}^{-1})$ of the substance is

(a) 32 (b) 64 (c) 128 (d) 488

7. The standard Gibbs energy change at 300K for the reaction, $2A \rightleftharpoons B + C$ is 2494. 2 J. At a given time, the composition of the reaction mixture is $[A] = \frac{1}{2}$, $[B] = 2$ and $[C] = \frac{1}{2}$. The reaction proceeds in the $[R = 8.314 \text{ JK/mol}, e = 2.718]$

(a) forward direction because $Q > K_c$
(b) reverse direction because $Q > K_c$
(c) forward direction because $Q < K_c$
(d) reverse direction because $Q < K_c$

8. Two Faraday of electricity is passed through a solution of $CuSO_4$. The mass of copper deposited at the cathode is (at. mass of Cu = 63.5 u)

(a) 0g (b) 63.5g (c) 2g (d) 127g

9. Higher order (>3) reactions are rare due to

(a) low probability of simultaneous collision of all the reacting species
(b) increase in entropy and activation energy as more molecules are involved
(c) shifting of equilibrium towards reactants due to elastic collisions
(d) loss of active species on collision

10. 3 g of activated charcoal was added to 50 mL of acetic acid solution (0.06 N) in a flask. After an hour it was filtered and the strength of the filtrate was found to be 0.042 N. The amount of acetic acid adsorbed (per gram of charcoal) is

(a) 18 mg (b) 36 mg
(c) 42 mg (d) 54 mg

11. The ionic radii (in Å) of $N^{3-}$, $O^{2-}$ and $F^-$ respectively are

(a) 1.36, 1.40 and 1.71
(b) 1.36, 1.71 and 1.40
(c) 1.71, 1.40 and 1.36
(d) 1.71, 1.36 and 1.40

12. In the context of the Hall-Heroult process for the extraction of Al, which of the following statements is false?

(a) CO and $CO_2$ are produced in this process
(b) $Al_2O_3$ is mixed with $CaF_2$ which lowers the melting point of the mixture and brings conductivity
(c) $Al^{3+}$ is reduced at the cathode to form Al
(d) $Na_3AlF_6$ serves as the electrolyte

13. From the following statements regarding $H_2O_2$, choose the incorrect statement.

(a) It can act only as an oxidising agent
(b) It decomposed on exposure to light
(c) It has to be stored in plastic or wax lined glass bottles in dark
(d) It has to be kept away from dust

14. Which one of the following alkaline earth metal sulphates has its hydration enthalpy greater than its lattice enthalpy?

(a) $CaSO_4$ (b) $BeSO_4$ (c) $BaSO_4$ (d) $SrSO_4$

15. Which among the following is the most reactive?

(a) $Cl_2$ (b) $Br_2$ (c) $I_2$ (d) ICl

16. Match the catalysts to the correct processes.

| Catalyst | Process |
|---|---|
| (A) $TiCl_3$ | (i) Wacker process |
| (B) $PdCl_2$ | (ii) Ziegler- Natta polymerisation |
| (C) $CuCl_2$ | (iii) Contact process |
| (D) $V_2O_5$ | (iv) Deacon's process |

(a) (A)- (iii), (B) - (ii), (C) - (iv), (D) - (i)
(b) (A)- (ii), (B) - (i), (C) - (iv), (D) - (iii)
(c) (A)- (ii), (B) - (iii), (C) - (iv), (D) - (i)
(d) (A)- (iii), (B) - (i), (C) - (ii), (D) - (iv)

17. Which one has the highest boiling point?

(a) He (b) Ne (c) Kr (d) Xe

18. The number of geometric isomers that can exist for square planar $[Pt(Cl)(py)(NH_3)(NH_2OH)]^+$ is (py = pyridine)

(a) 2 (b) 3 (c) 4 (d) 6

**19.** The colour of $KMnO_4$ is due to

(a) $M \rightarrow L$ charge transfer transition
(b) $d-d$ transition
(c) $L \rightarrow M$ charge transfer transition
(d) $\sigma - \sigma^*$ transition

**20.** **Assertion (A)** Nitrogen and oxygen are the main components in the atmosphere but these do not react to form oxides of nitrogen.

**Reason (R)** The reaction between nitrogen and oxygen requires high temperature.

(a) Both Assertion and Reason are correct and the reason is the correct explanation for the Assertion.
(b) Both Assertion and Reason are correct but the reason is not the correct explanation for the Assertion.
(c) The Assertion is incorrect but the Reason is correct.
(d) Both the Assertion and Reason are incorrect.

**21.** In Carius method of estimation of halogens, 250 mg of an organic compound gave 141 mg of AgBr. The percentage of bromine in the compound is

(at. mass Ag = 108, Br = 80)

(a) 24 (b) 36 (c) 48 (d) 60

**22.** Which of the following compound will exhibit geometrical isomerism?

(a) 1-phenyl-2-butene
(b) 3-phenyl-1-butene
(c) 2-phenyl-1-butene
(d) 1, 1-diphenyl-1-propane

**23.** Which compound would give 5-keto-2-methyl hexanal upon ozonolysis?

(a) 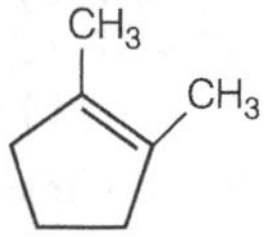 $CH_3$, $CH_3$ (b) 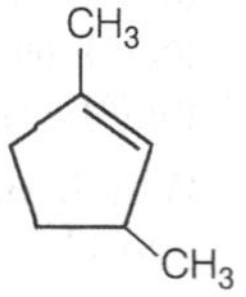 $CH_3$, $CH_3$

(c) 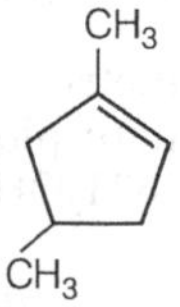 $CH_3$, $CH_3$ (d) 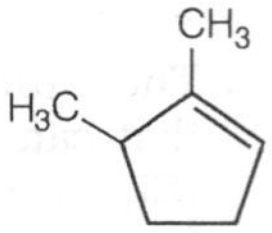 $CH_3$, $H_3C$

**24.** The synthesis of alkyl fluorides is best accomplished by

(a) free radical fluorination
(b) Sandmeyer's reaction
(c) Finkelstein reaction
(d) Swarts reaction

**25.** In the following sequence of reaction,

Toluene $\xrightarrow{KMnO_4} A \xrightarrow{SOCl_2} B \xrightarrow[BaSO_4]{H_2/Pd} C$

The product C is

(a) $C_6H_5COOH$
(b) $C_6H_5CH_3$
(c) $C_6H_5CH_2OH$
(d) $C_6H_5CHO$

**26.** In the reaction,

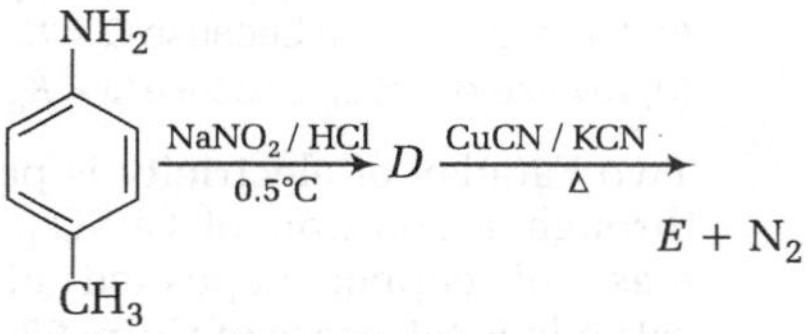

$NH_2$ / $CH_3$ (benzene ring) $\xrightarrow[0.5^\circ C]{NaNO_2/HCl} D \xrightarrow[\Delta]{CuCN/KCN} E + N_2$

The product $E$ is

(a) 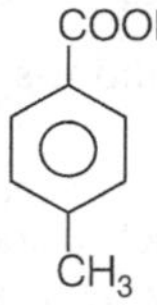 COOH, $CH_3$

(b) 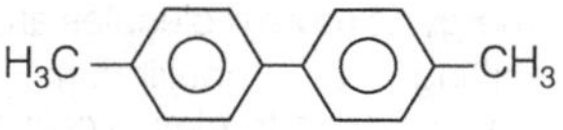 $H_3C$–(ring)–(ring)–$CH_3$

(c) 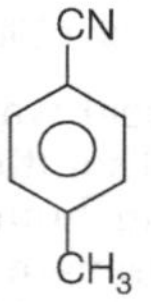 CN, $CH_3$

(d) 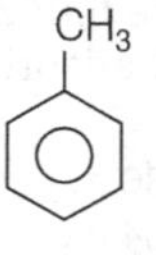 $CH_3$

**27.** Which polymer is used in the manufacture of paints and lacquers?

(a) Bakelite
(b) Glyptal
(c) Polypropene
(d) Polyvinyl chloride

28. Which of the vitamins given below is water soluble?
(a) Vitamin C (b) Vitamin D
(c) Vitamin E (d) Vitamin K

29. Which of the following compounds is not an antacid?
(a) Aluminium hydroxide
(b) Cimetidine
(c) Phenelzine
(d) Ranitidine

30. Which of the following compounds is not coloured yellow?
(a) $Zn_2[Fe(CN)_6]$
(b) $K_3[Co(NO_2)_6]$
(c) $(NH_4)_3[As(Mo_3O_{10})_4]$
(d) $BaCrO_4$

# Mathematics

1. Let $A$ and $B$ be two sets containing four and two elements respectively. Then, the number of subsets of the set $A \times B$, each having atleast three elements are
(a) 219
(b) 256
(c) 275
(d) 510

2. A complex number $z$ is said to be unimodular, if $|z| = 1$. Suppose $z_1$ and $z_2$ are complex numbers such that $\frac{z_1 - 2z_2}{2 - z_1\bar{z}_2}$ is unimodular and $z_2$ is not unimodular. Then, the point $z_1$ lies on a
(a) straight line parallel to $X$-axis
(b) straight line parallel to $Y$-axis
(c) circle of radius 2
(d) circle of radius $\sqrt{2}$

3. Let $\alpha$ and $\beta$ be the roots of equation $x^2 - 6x - 2 = 0$. If $a_n = \alpha^n - \beta^n$, for $n \geq 1$, then the value of $\frac{a_{10} - 2a_8}{2a_9}$ is equal to
(a) 6 (b) −6 (c) 3 (d) −3

4. If $A = \begin{bmatrix} 1 & 2 & 2 \\ 2 & 1 & -2 \\ a & 2 & b \end{bmatrix}$ is a matrix satisfying the equation $AA^T = 9I$, where $I$ is $3 \times 3$ identity matrix, then the ordered pair $(a, b)$ is equal to
(a) (2, −1) (b) (−2, 1)
(c) (2, 1) (d) (−2, −1)

5. The set of all values of $\lambda$ for which the system of linear equations $2x_1 - 2x_2 + x_3 = \lambda x_1$, $2x_1 - 3x_2 + 2x_3 = \lambda x_2$ and $-x_1 + 2x_2 = \lambda x_3$ has a non-trivial solution,
(a) is an empty set
(b) is a singleton set
(c) contains two elements
(d) contains more than two elements

6. The number of integers greater than 6000 that can be formed, using the digits 3, 5, 6, 7 and 8 without repetition, is
(a) 216 (b) 192 (c) 120 (d) 72

7. The sum of coefficients of integral powers of $x$ in the binomial expansion of $(1 - 2\sqrt{x})^{50}$ is
(a) $\frac{1}{2}(3^{50} + 1)$
(b) $\frac{1}{2}(3^{50})$
(c) $\frac{1}{2}(3^{50} - 1)$
(d) $\frac{1}{2}(2^{50} + 1)$

8. If $m$ is the AM of two distinct real numbers $l$ and $n(l, n > 1)$ and $G_1$, $G_2$ and $G_3$ are three geometric means between $l$ and $n$, then $G_1^4 + 2G_2^4 + G_3^4$ equals
(a) $4\,l^2mn$
(b) $4\,lm^2n$
(c) $4\,lmn^2$
(d) $4l^2m^2n^2$

**9.** The sum of first 9 terms of the series $\frac{1^3}{1}+\frac{1^3+2^3}{1+3}+\frac{1^3+2^3+3^3}{1+3+5}+\ldots$ is

(a) 71 (b) 96
(c) 142 (d) 192

**10.** $\lim_{x\to 0}\frac{(1-\cos 2x)(3+\cos x)}{x\tan 4x}$ is equal to

(a) 4 (b) 3
(c) 2 (d) $\frac{1}{2}$

**11.** If the function $g(x)=\begin{cases}k\sqrt{x+1}, & 0\le x\le 3\\ mx+2, & 3<x\le 5\end{cases}$ is differentiable, then the value of $k+m$ is

(a) 2 (b) $\frac{16}{5}$
(c) $\frac{10}{3}$ (d) 4

**12.** The normal to the curve $x^2+2xy-3y^2=0$ at $(1, 1)$

(a) does not meet the curve again
(b) meets the curve again in the second quadrant
(c) meets the curve again in the third quadrant
(d) meets the curve again in the fourth quadrant

**13.** Let $f(x)$ be a polynomial of degree four having extreme values at $x=1$ and $x=2$. If $\lim_{x\to 0}\left[1+\frac{f(x)}{x^2}\right]=3$, then $f(2)$ is equal to

(a) –8
(b) –4
(c) 0
(d) 4

**14.** The integral $\int\frac{dx}{x^2(x^4+1)^{\frac{3}{4}}}$ equals

(a) $\left(\frac{x^4+1}{x^4}\right)^{\frac{1}{4}}+c$ (b) $(x^4+1)^{\frac{1}{4}}+c$
(c) $-(x^4+1)^{\frac{1}{4}}+c$ (d) $-\left(\frac{x^4+1}{x^4}\right)^{\frac{1}{4}}+c$

**15.** The integral $\int_2^4\frac{\log x^2}{\log x^2+\log(36-12x+x^2)}dx$ is equal to

(a) 2 (b) 4
(c) 1 (d) 6

**16.** The area (in sq units) of the region described by $\{(x, y): y^2\le 2x$ and $y\ge 4x-1\}$ is

(a) $\frac{7}{32}$ (b) $\frac{5}{64}$ (c) $\frac{15}{64}$ (d) $\frac{9}{32}$

**17.** Let $y(x)$ be the solution of the differential equation $(x\log x)\frac{dy}{dx}+y=2x\log x,\quad (x\ge 1)$. Then, $y(e)$ is equal to

(a) $e$ (b) 0
(c) 2 (d) $2e$

**18.** The number of points having both coordinates as integers that lie in the interior of the triangle with vertices $(0, 0)$, $(0, 41)$ and $(41, 0)$ is

(a) 901 (b) 861
(c) 820 (d) 780

**19.** Locus of the image of the point $(2, 3)$ in the line $(2x-3y+4)+k(x-2y+3)=0$, $k\in R$, is a

(a) straight line parallel to $X$-axis
(b) straight line parallel to $Y$-axis
(c) circle of radius $\sqrt{2}$
(d) circle of radius $\sqrt{3}$

**20.** The number of common tangents to the circles $x^2+y^2-4x-6y-12=0$ and $x^2+y^2+6x+18y+26=0$ is

(a) 1 (b) 2 (c) 3 (d) 4

**21.** The area (in sq units) of the quadrilateral formed by the tangents at the end points of the latusrectum to the ellipse $\frac{x^2}{9}+\frac{y^2}{5}=1$ is

(a) $\frac{27}{4}$ (b) 18
(c) $\frac{27}{2}$ (d) 27

22. Let $O$ be the vertex and $Q$ be any point on the parabola $x^2 = 8y$. If the point $P$ divides the line segment $OQ$ internally in the ratio 1:3, then the locus of $P$ is

(a) $x^2 = y$
(b) $y^2 = x$
(c) $y^2 = 2x$
(d) $x^2 = 2y$

23. The distance of the point (1, 0, 2) from the point of intersection of the line $\frac{x-2}{3} = \frac{y+1}{4} = \frac{z-2}{12}$ and the plane $x - y + z = 16$ is

(a) $2\sqrt{14}$
(b) 8
(c) $3\sqrt{21}$
(d) 13

24. The equation of the plane containing the line $2x - 5y + z = 3$, $x + y + 4z = 5$ and parallel to the plane $x + 3y + 6z = 1$ is

(a) $2x + 6y + 12z = 13$
(b) $x + 3y + 6z = -7$
(c) $x + 3y + 6z = 7$
(d) $2x + 6y + 12z = -13$

25. Let **a**, **b** and **c** be three non-zero vectors such that no two of them are collinear and $(\mathbf{a} \times \mathbf{b}) \times \mathbf{c} = \frac{1}{3}|\mathbf{b}||\mathbf{c}|\mathbf{a}$. If $\theta$ is the angle between vectors **b** and **c**, then a value of $\sin\theta$ is

(a) $\frac{2\sqrt{2}}{3}$
(b) $\frac{-\sqrt{2}}{3}$
(c) $\frac{2}{3}$
(d) $\frac{-2\sqrt{3}}{3}$

26. If 12 identical balls are to be placed in 3 identical boxes, then the probability that one of the boxes contains exactly 3 balls, is

(a) $\frac{55}{3}\left(\frac{2}{3}\right)^{11}$ (b) $55\left(\frac{2}{3}\right)^{10}$
(c) $220\left(\frac{1}{3}\right)^{12}$ (d) $22\left(\frac{1}{3}\right)^{11}$

27. The mean of the data set comprising of 16 observations is 16. If one of the observation valued 16 is deleted and three new observations valued 3, 4 and 5 are added to the data, then the mean of the resultant data is

(a) 16.8 (b) 16.0 (c) 15.8 (d) 14.0

28. If the angles of elevation of the top of a tower from three collinear points $A$, $B$ and $C$ on a line leading to the foot of the tower are 30°, 45° and 60° respectively, then the ratio $AB : BC$ is

(a) $\sqrt{3} : 1$
(b) $\sqrt{3} : \sqrt{2}$
(c) $1 : \sqrt{3}$
(d) 2 : 3

29. Let

$$\tan^{-1} y = \tan^{-1} x + \tan^{-1}\left(\frac{2x}{1-x^2}\right),$$

where $|x| < \frac{1}{\sqrt{3}}$. Then, the value of $y$ is

(a) $\frac{3x - x^3}{1 - 3x^2}$ (b) $\frac{3x + x^3}{1 - 3x^2}$
(c) $\frac{3x - x^3}{1 + 3x^2}$ (d) $\frac{3x + x^3}{1 + 3x^2}$

30. The negation of $\sim s \vee (\sim r \wedge s)$ is equivalent to

(a) $s \wedge \sim r$ (b) $s \wedge (r \wedge \sim s)$
(c) $s \vee (r \vee \sim s)$ (d) $s \wedge r$

# Answer *with* Explanations

## Physics

1. *(c)* ***Central Idea*** *Concept of relative motion can be applied to predict the nature of motion of one particle with respect to the other.*

Consider the stones thrown up simultaneously as shown in the diagram below.

Considering motion of the second particle with respect to the first we have relative acceleration $|\mathbf{a}_{21}| = |\mathbf{a}_2 - \mathbf{a}_1| = g - g = 0$

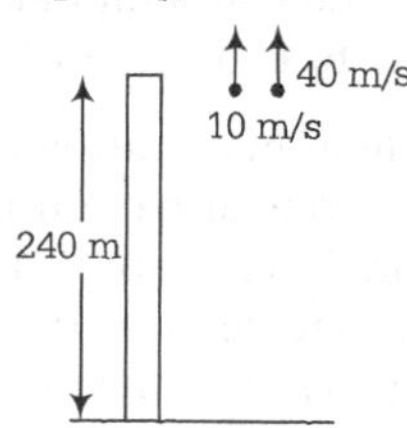


Thus, motion of first particle is straight line with respect to second particle till the first particle strikes ground at a time given by

$$-240 = 10t - \frac{1}{2} \times 10 \times t^2$$

or $t^2 - 2t - 48 = 0$

or $t^2 - 8t + 6t - 48 = 0$

or $t = 8, -6$ (not possible)

Thus, distance covered by second particle with respect to first particle in 8 s is

$$s_{12} = (v_{21})\, t = (40 - 10)\,(8\,\text{s})$$
$$= 30 \times 8 = 240 \text{ m}$$

Similarly, time taken by second particle to strike the ground is given by

$$-240 = 40t - \frac{1}{2} \times 10 \times t^2$$

or $-240 = 40t - 5t^2$

or $5t^2 - 40t - 240 = 0$

or $t^2 - 8t - 48 = 0$

$t^2 - 12t + 4t - 48 = 0$

or $t\,(t - 12) + 4\,(t - 12) = 0$

or $t = 12, -4$ (not possible)

Thus, after 8 s, magnitude of relative velocity will increase upto 12 s when second particle strikes the ground.

2. *(b)* ***Central Idea*** *Given time period* $T = 2\pi\sqrt{L/g}$

*Thus, changes can be expressed as*

$$\pm \frac{2\Delta T}{T} = \pm \frac{\Delta L}{L} \pm \frac{\Delta g}{g}$$

According to the question, we can write

$$\frac{\Delta L}{L} = \frac{0.1\text{cm}}{20.0\text{cm}} = \frac{1}{200}$$

Again time period

$$T = \frac{90}{100}\text{ s}$$

and $\Delta T = \frac{1}{100}\text{ s}$

$\Rightarrow \frac{\Delta T}{T} = \frac{1}{90}$

Now,

$\because T = 2\pi\sqrt{\frac{L}{g}}$

$\because g = 4\pi^2 \frac{L}{T^2}$

$\therefore \frac{\Delta g}{g} = \frac{\Delta L}{L} + \frac{2\Delta T}{T}$

or $\frac{\Delta g}{g} \times 100\% = \left(\frac{\Delta L}{L}\right) \times 100\% + \left(\frac{2\Delta T}{T}\right) \times 100\%$

$$= \left(\frac{1}{200} \times 100\right)\% + 2 \times \frac{1}{90} \times 100\%$$

$$\simeq 2.72\% \simeq 3\%$$

Thus, accuracy in the determination of $g$ is approx 3%.

3. *(c)* ***Central Idea*** *In vertical direction, weights are balanced by frictional forces.*

Consider FBD of block $A$ and $B$ as shown in diagram below.

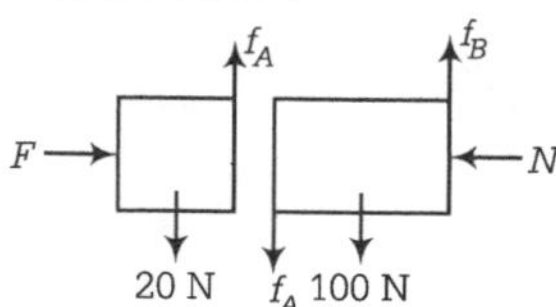


As the blocks are in equilibrium, balance forces are in horizontal and vertical direction.

For the system of blocks $(A + B)$.

$$F = N$$

For block $A$, $f_A = 20\text{N}$ and for block $B$,

$$f_B = f_A + 100$$
$$= 120\text{ N}$$

4. (c) ***Central Idea*** *Conservation of linear momentum can be applied but energy is not conserved.*

Consider the movement of two particles as shown below.

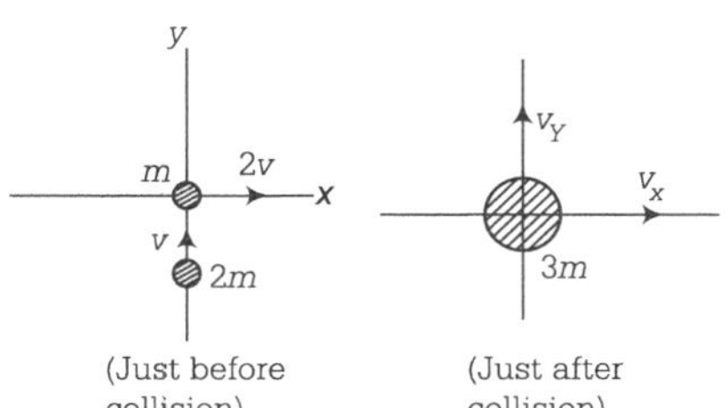


Conserving linear momentum in $x$-direction

$$(p_i)x = (p_f)x$$

or $$2mv = (2m + m)\, v_x$$

or $$v_x = \frac{2}{3}v$$

Conserving linear momentum in $y$-direction

$(p_i)y = (p_f)y$ or $2mv = (2m + m)\, v_y$

or $$v_y = \frac{2}{3}v$$

Initial kinetic energy of the two particles system is

$$E_i = \frac{1}{2}m\,(2v)^2 + \frac{1}{2}(2m)\,(v)^2$$
$$= \frac{1}{2} \times 4mv^2 + \frac{1}{2} \times 2mv^2$$
$$= 2mv^2 + mv^2 = 3mv^2$$

Final energy of the combined two particles system is

$$E_f = \frac{1}{2}(3m)\,(v_x^2 + v_y^2)$$
$$= \frac{1}{2}(3m)\left[\frac{4v^2}{9} + \frac{4v^2}{9}\right]$$
$$= \frac{3m}{2}\left[\frac{8\,v^2}{9}\right] = \frac{4mv^2}{3}$$

loss in the energy $\Delta E = E_i - E_f$

$$= mv^2\left[3 - \frac{4}{3}\right] = \frac{5}{3}mv^2$$

Percentage loss in the energy during the collision

$$\frac{\Delta E}{E_i} \times 100 = \frac{\frac{5}{3}mv^2}{3mv^2} \times 100 = \frac{5}{9} \times 100$$
$$\simeq 56\%$$

5. (b) We know that centre of mass of a uniform solid cone of height $h$ is at height $h/4$ from base, therefore

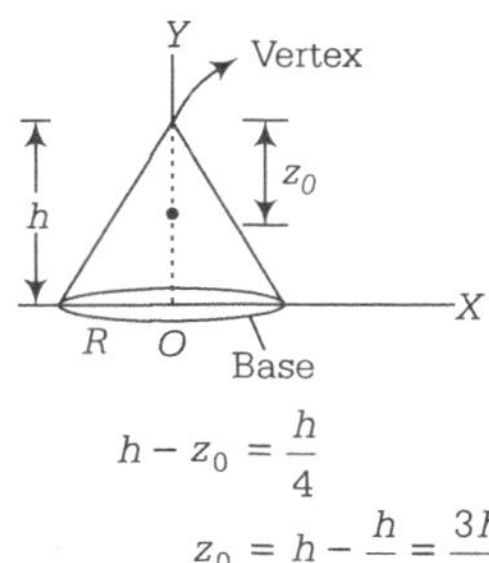


$$h - z_0 = \frac{h}{4}$$

or $$z_0 = h - \frac{h}{4} = \frac{3h}{4}$$

6. (c) ***Central Idea*** *Use geometry of the figure to calculate mass and side length of the cube interms of M and R respectively.*

Consider the cross-sectional view of a diametric plane as shown in the adjacent diagram.

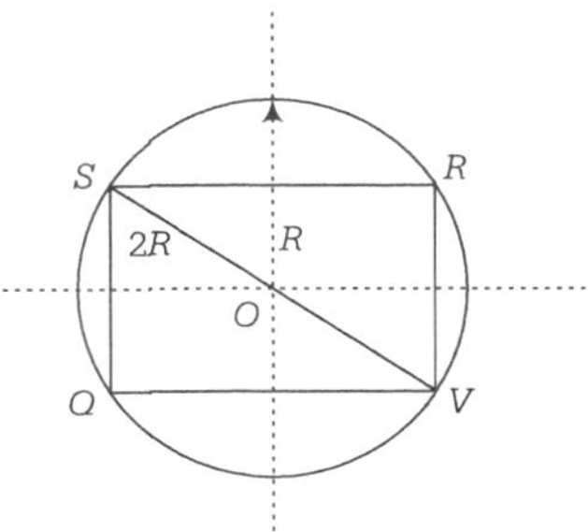


Cross-sectional view of the cube and sphere

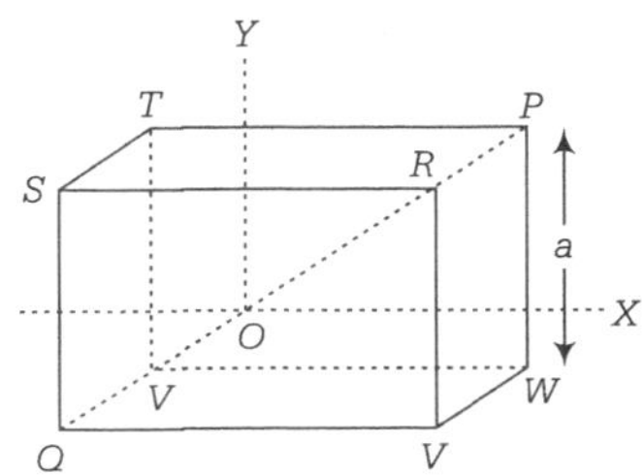


Using geometry of the cube

$$PQ = 2R = (\sqrt{3})\,a \quad \text{or} \quad a = \frac{2R}{\sqrt{3}}$$

Volume density of the solid sphere

$$\rho = \frac{M}{\frac{4}{3}\pi R^3} = \frac{3}{4\pi}\left(\frac{M}{R^3}\right)$$

Mass of cube $(m) = (\rho)(a)^3$

$$= \left(\frac{3}{4\pi} \times \frac{M}{R^3}\right)\left[\frac{2R}{\sqrt{3}}\right]^3$$

$$= \frac{3M}{4\pi R^3} \times \frac{8R^3}{3\sqrt{3}} = \frac{2M}{\sqrt{3}\pi}$$

Moment of inertia of the cube about given axis is

$$I_Y = \frac{ma^2}{12}(a^2 + a^2) = \frac{ma^2}{6}$$

$$\Rightarrow \quad I_Y = \frac{ma^2}{6} = \frac{2M}{\sqrt{3}\pi} \times \frac{1}{6} \times \frac{4R^2}{3}$$

$$= \frac{4MR^2}{9\sqrt{3}\pi}$$

7. *(b)* ***Central Idea*** *Consider cavity as negative mass and apply superposition of gravitational potential.*

Consider the cavity formed in a solid sphere as shown in figure.

$V(\infty) = 0$

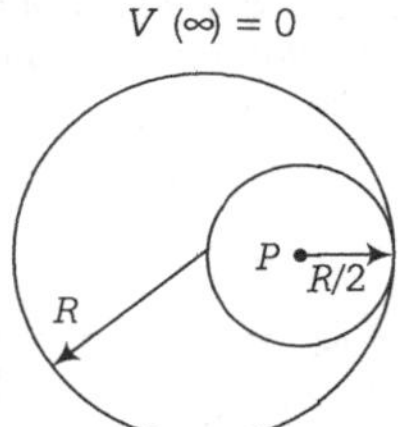


According to the question,we can write potential at an internal point $P$ due to complete solid sphere,

$$V_s = -\frac{GM}{2R^3}\left[3R^2 - \left(\frac{R}{2}\right)^2\right]$$

$$= \frac{-GM}{2R^3}\left[3R^2 - \frac{R^2}{4}\right]$$

$$= \frac{-GM}{2R^3}\left[\frac{11R^2}{4}\right] = \frac{-11GM}{8R}$$

Mass of removed part

$$= \frac{M}{\frac{4}{3} \times \pi R^3} \times \frac{4}{3}\pi\left(\frac{R}{2}\right)^3 = \frac{M}{8}$$

Potential at point $P$ due to removed part

$$V_c = \frac{-3}{2} \times \frac{GM/8}{\frac{R}{2}} = \frac{-3GM}{8R}$$

Thus, potential due to remaining part at point $P$,

$$V_P = V_s - V_c = \frac{-11GM}{8R} - \left(-\frac{3GM}{8R}\right)$$

$$= \frac{(-11+3)GM}{8R} = \frac{-GM}{R}$$

8. *(a)* We know that time period, $T = 2\pi\sqrt{\frac{L}{g}}$

When additional mass $M$ is added to its bob

$$T_M = 2\pi\sqrt{\frac{L + \Delta L}{g}}$$

where, $\Delta L$ is increase in length.

We know that

$$Y = \frac{Mg/A}{\Delta L/L} = \frac{MgL}{A\Delta L}$$

$$\Rightarrow \quad \Delta L = \frac{MgL}{AY}$$

$$\Rightarrow \therefore \quad T_M = 2\pi\sqrt{\frac{L + \frac{MgL}{AY}}{g}}$$

$$\Rightarrow \quad \left(\frac{T_M}{T}\right)^2 = 1 + \frac{Mg}{AY}$$

$$\text{or} \quad \frac{Mg}{AY} = \left(\frac{T_M}{T}\right)^2 - 1$$

$$\text{or} \quad \frac{1}{Y} = \frac{A}{Mg}\left[\left(\frac{T_M}{T}\right)^2 - 1\right]$$

9. *(c)* According to question,

$$p = \frac{1}{3}\left(\frac{U}{V}\right)$$

$$\Rightarrow \quad \frac{nRT}{V} = \frac{1}{3}\left(\frac{U}{V}\right) \quad [\because pV = nRT]$$

$$\text{or} \quad \frac{nRT}{V} \propto \frac{1}{3}T^4$$

$$\text{or} \quad VT^3 = \text{constant}$$

$$\text{or} \quad \frac{4}{3}\pi R^3 T^3 = \text{constant}$$

$$\text{or} \quad TR = \text{constant}$$

$$\Rightarrow \quad T \propto \frac{1}{R}$$

10. *(b)* Since, entropy is a state function, therefore change in entropy in both the processes must be same.Therefore, correct option should be (b).

11. *(c)* ***Central Idea*** *For an adiabatic process $TV^{\gamma - 1} = constant$.*

We know that average time of collision between molecules

$$\tau = \frac{1}{n\pi \sqrt{2}\, v_{\text{rms}}\, d^2}$$

where, $n$ = number of molecules per unit volume

$v_{\text{rms}}$ = rms velocity of molecules

As $\quad n \propto \frac{1}{V}$ and $v_{\text{rms}} \propto \sqrt{T}$

$$\tau \propto \frac{V}{\sqrt{T}}$$

Thus, we can write

$$n = K_1 V^{-1} \quad \text{and} \quad v_{\text{rms}} = K_2\, T^{1/2}$$

where, $K_1$ and $K_2$ are constants.

For adiabatic process, $TV^{\gamma - 1}$ = constant.

Thus, we can write

$$\tau \propto VT^{-1/2} \propto V\,(V^{1-\gamma})^{-1/2}$$

or $\quad \tau \propto V^{\frac{\gamma+1}{2}}$

**12.** *(b)* During oscillation, motion of a simple pendulum KE is maximum of mean position where PE is minimum. At extreme position, KE is minimum and PE is maximum. Thus, correct graph is depicted in option (b).

**13.** *(b)* Apparent frequency heard by the person before crossing the train.

$$f_1 = \left(\frac{c}{c - v_s}\right) f_0 = \left(\frac{320}{320 - 20}\right) 1000$$

Similarly, apparent frequency heard, after crossing the trains

$$f_2 = \left(\frac{c}{c + v_s}\right) f_0 = \left(\frac{320}{320 + 20}\right) 1000$$

[$c$ = speed of sound]

$$\Delta f = f_1 - f_2 = \left(\frac{2\,cv_s}{c^2 - v_s^2}\right) f_0$$

or $\quad \frac{\Delta f}{f_0} \times 100 = \left(\frac{2cv_s}{c^2 - v_s^2}\right) \times 100$

$$= \frac{2 \times 320 \times 20}{300 \times 340} \times 100$$

$$= \frac{2 \times 32 \times 20}{3 \times 34} = 12.54\% = 12\%$$

**14.** *(a)* Field lines should originate from positive charge and terminate to negative charge. Thus, (b) and (c) are not possible.

Electric field lines cannot form corners as shown in (d).

Thus, correct option is (a).

**15.** *(c, d)* Potential at the surface of the charged sphere

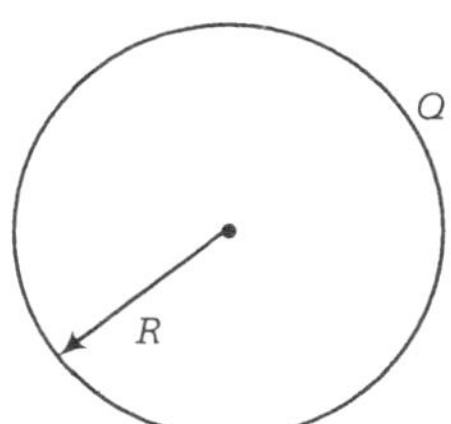


Charged sphere

$$V_0 = \frac{KQ}{R}$$

$$V = \frac{KQ}{r}, r \geq R$$

$$= \frac{KQ}{2R^3}(3R^2 - r^2);\ r \leq R$$

$$V_{\text{centre}} = V_c = \frac{KQ}{2R^3} \times 3R^2$$

$$= \frac{3KQ}{2R} = \frac{3V_0}{2}$$

$\Rightarrow \quad R_1 = 0$

As potential decreases for outside points.

Thus, according to the question, we can write

$$V_{R_2} = \frac{5V_0}{4} = \frac{KQ}{2R^3}(3R^2 - R_2^2)$$

$$\frac{5V_0}{4} = \frac{V_0}{2R^2}(3R^2 - R_2^2)$$

or $\quad \frac{5}{2} = 3 - \left(\frac{R_2}{R}\right)^2$

or $\quad \left(\frac{R_2}{R}\right)^2 = 3 - \frac{5}{2} = \frac{1}{2}$

or $\quad R_2 = \frac{R}{\sqrt{2}}$

Similarly,

$$V_{R_3} = \frac{3V_0}{4}$$

$\Rightarrow \quad \frac{KQ}{R_3} = \frac{3}{4} \times \frac{KQ}{R}$

or $\quad R_3 = \frac{4}{3}R$

$$V_{R4} = \frac{KQ}{R_4} = \frac{V_0}{4}$$

$\Rightarrow \quad \frac{KQ}{R_4} = \frac{1}{4} \times \frac{KQ}{R}$

or $\quad R_4 = 4R$

16. *(b)* ***Central Idea*** *Assume negative terminal of the battery as grounded* (0 V).

Suppose, potential of point $x$ is $V$.

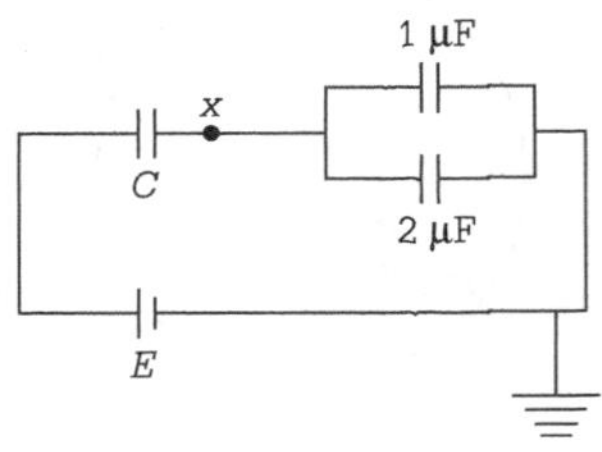


From the circuit diagram, we can write

$$Q_C = Q_1 + Q_2$$

or $C(E - V) = 1 \times V + 2 \times V$

or $V[C + 3] = CE$

or $V = \dfrac{CE}{3 + C}$

$\therefore$ $Q_2 = C_2 (V)$

$$= \frac{2\,CE}{3 + C}$$

$$= \frac{2E}{1 + 3/C}$$

As $C_1$ varied from 1 μF to 3 μF, charge increases with decreasing slope.

**Note** As $C \to \infty$, $Q_2 \to 2E =$ constant

17. *(d)* According to the question

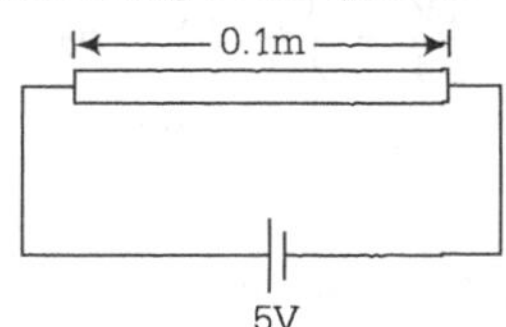


$$v_d = 2.5 \times 10^{-4}\,\text{m/s}$$

$\Rightarrow$ $n = 8 \times 10^{28}/\text{m}^3$

we know that

$$J = nev_d \text{ or } I = nev_d A$$

where, symbols have their usual meaning.

$\Rightarrow$ $\dfrac{V}{R} = nev_d A$

or $\dfrac{V}{\frac{\rho L}{A}} = nev_d A$

or $\dfrac{V}{\rho L} = nev_d$

or $\rho = \dfrac{V}{nev_d L}$

$$= \frac{5}{8 \times 10^{28} \times 1.6 \times 10^{-19} \times 2.5 \times 10^{-4} \times 0.1}$$

or $\rho = 1.6 \times 10^{-5}\ \Omega\text{m}$

18. *(c)* ***Central Idea*** *Connect point Q to ground and apply KCL.*

Consider the grounded circuit as shown below.

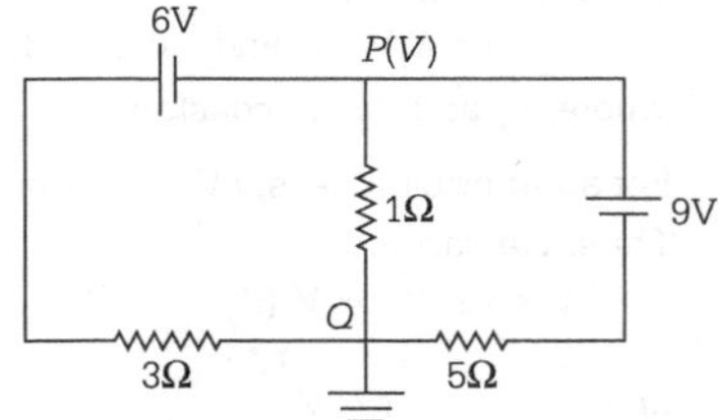


Applying KCL of point $Q$ we can write

Incoming current at $Q$ = outgoing current from $Q$

$\Rightarrow$ $\dfrac{V + 6}{3} + \dfrac{V}{1} = \dfrac{9 - V}{5}$

or $V\left[\dfrac{1}{3} + \dfrac{1}{5} + 1\right] = \dfrac{9}{5} - 2$

or $V\left[\dfrac{5 + 3 + 15}{15}\right] = \dfrac{9 - 10}{5}$

or $V\left[\dfrac{23}{15}\right] = \dfrac{-1}{5}$

or $V = \dfrac{-3}{23} = -0.13\,\text{V}$

Thus, current in the 1 Ω resistance is 0.13 A, from $Q$ to $P$.

19. *(a)* Consider the two coaxial solenoids. Due to one of the solenoids magnetic field at the centre of the other can be assumed to be constant.

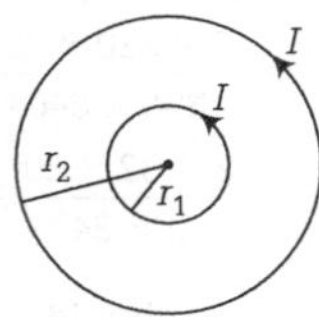


Due to symmetry, forces on upper and lower part of a solenoid will be equal and opposite and hence resultant is zero.

Therefore $F_1 = F_2 = 0$

20. *(b)* Consider free body diagram of the wire.

As the wires are in equilibrium, they must carry current in opposite direction.

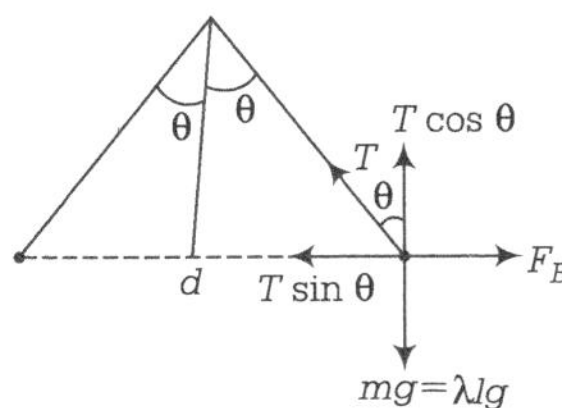


Here, $F_B = \frac{\mu_0 I^2 l}{2\pi d}$, where $l$ is length of each wire and $d$ is separation between wires.

From figure, $d = 2L \sin\theta$

$$T = \cos\theta = mg = \lambda lg$$

(in vertical direction)...(i)

$$T\sin\theta = F_B = \frac{\mu_0 I^2 l}{4\pi L \sin\theta}$$

(in horizontal direction)...(ii)

From Eqs. (i) and (ii),

$$\frac{T\sin\theta}{T\cos\theta} = \frac{\mu_0 I^2 l}{4\pi L \sin\theta \times \lambda lg}$$

$$\therefore \quad I = \sqrt{\frac{4\pi\lambda Lg \sin^2\theta}{\mu_0 \cos\theta}}$$

$$= 2\sin\theta\sqrt{\frac{\pi\lambda Lg}{\mu_0 \cos\theta}}$$

21. *(c) Since* **B** *is uniform only torque acts on a current carrying loop.*

As, $\tau = \mathbf{M} \times \mathbf{B}$

$\Rightarrow \quad |\tau| = |\mathbf{M}||\mathbf{B}| \sin\theta$

For orientation shown in (b) $\theta = 0^0$, $\tau = 0$ (stable equilibrium)

and for (d) $\theta = \pi$, $\tau = 0$ (unstable equilibrium)

22. *(d)* ***Central Idea*** *After long time inductor behave as short-circuit.*

At $t = 0$, the inductor behaves as short-circuited. The current

$$I_0 = \frac{E_0}{R} = \frac{15\text{V}}{0.15\text{ k}\Omega} = 100 \text{ mA}$$

As $K_2$ is closed, current through the inductor starts decay, which is given at any time $t$ as

$$I = I_0 e^{\frac{-tR}{L}} = (100\text{ mA}) e^{\frac{-t \times 15000}{3}}$$

At $\quad t = 1\text{ms}$

$$I = (100\text{mA}) e^{-\frac{1\times10^{-3}\times15\times10^3}{3}}$$

$$I = (100\text{ mA})\, e^{-5} = 0.6737 \text{ mA}$$

or $\quad I = 0.67$ mA

23. *(b)* Consider the LED as a point source of light.

Let power of the LED is $P$.

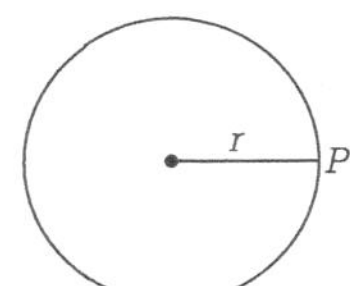


Intensity at $r$ from the source

$$I = \frac{P}{4\pi r^2} \quad \text{...(i)}$$

As we know that $I = \frac{1}{2}\varepsilon_0 E_0^2 c$ ...(ii)

From Eqs. (i) and (ii), we can write

$$\frac{P}{4\pi r^2} = \frac{1}{2}\varepsilon_0 E_0^2 c$$

$$\text{or} E_0^2 = \frac{2P}{4\pi\varepsilon_0 r^2 c} = \frac{2 \times 0.1 \times 9 \times 10^9}{1 \times 3 \times 10^8}$$

or $\quad E_0^2 = 6 \Rightarrow E_0 = \sqrt{6} = 2.45$ V/m

24. *(a)* ***Central Idea*** *The ray will get transmitted through face AC if* $i_{AC} < i_C$

Consider the ray diagram is shown below.

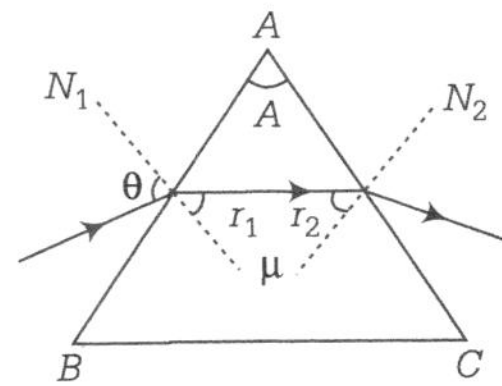


A ray of light incident on face $AB$ at an angle $\theta$.

$r_1$ = Angle of refraction on face $AB$

$r_2$ = Angle of incidence at face $AC$

For transmission of light through face $AC$

$$i_{AC} < i_C \quad \text{or} \quad A - r_1 < i_C$$

or $\quad \sin(A - r_1) < \sin i_C$

or $\quad \sin(A - r_1) < \frac{1}{\mu}$

$$A - r_1 < \sin^{-1}\left(\frac{1}{\mu}\right)$$

or $\quad \sin r_1 > \sin\left[A - \sin^{-1}\left(\frac{1}{\mu}\right)\right]$

Now, applying Snell's law at the face $AB$

$$1 \times \sin\theta = \mu \sin r_1 \quad \text{or} \quad \sin r_1 = \frac{\sin\theta}{\mu}$$

$$\Rightarrow \frac{\sin\theta}{\mu} > \sin\left[A - \sin^{-1}\left(\frac{1}{\mu}\right)\right]$$

$$\text{or} \quad \theta > \sin^{-1}\left[\mu \sin\left\{A - \sin^{-1}\left(\frac{1}{\mu}\right)\right\}\right]$$

**25.** *(d)* According to Snell's law,

$$\mu \sin\theta = \text{Constant}$$

$$\therefore \quad \sin\theta \propto \frac{1}{\mu}$$

As $\mu$ increases, $\theta$ decreases.

Hence, beam will bend upward.

**26.** *(b)* We can write resolving angle of necked eye as

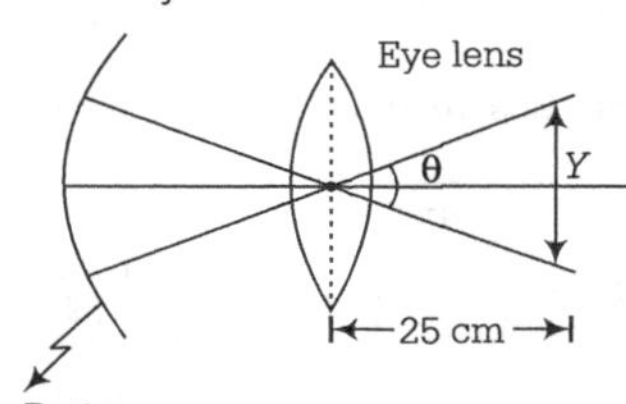


$$\theta = 1.22\frac{\lambda}{D}$$

Where, $D$ is the diameter of eye lens.

$$\text{or} \quad \frac{Y}{25 \times 10^{-2}} = \frac{1.22 \times 500 \times 10^{-9}}{0.25 \times 2 \times 10^{-2}}$$

$$Y = 30 \times 10^{-6} \text{ m} = 30 \mu\text{m}$$

**27.** *(a)* As we know that kinetic energy of an electron is

$$\text{KE} \propto \left(\frac{Z}{n}\right)^2$$

When the electron makes transition from an excited state to the ground state, then $n$ decreases and KE increases. We know that PE is lowest for ground state. As TE = – KE. TE also decreases.

**28.** *(c)* (A) Franck-Hertz experiments is associated with discrete energy levels of atom.

(B) Photo-electric experiment is associated with particle nature of light.

(C) Davisson-Germer experiment is associated with wave nature of electron.

**29.** *(c)* ***Central Idea*** *Frequency associated with AM are $f_c - f_m$, $f$, $f_c + f_m$*

According to the question

$$f_c = 2 \text{MHz} = 2000 \text{ kHz}$$

$$f_m = 5 \text{ kHz}$$

Thus, frequency of the resultant signal is/are carrier frequency $f_c = 2000$ kHz, LSB frequency $f_c - f_m = 2000 \text{ kHz} - 5 \text{ kHz} = 1995$ kHz and USB frequency $f_c + f_m = 2005$ kHz

**30.** *(a)* Considert the $LCR$ circuit at any time $t$

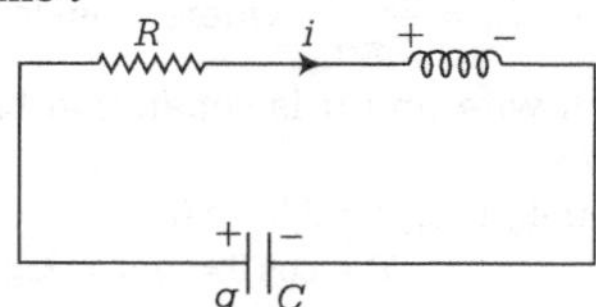


Now, applying KVL

We have $\quad \dfrac{q}{C} - iR - \dfrac{Ldi}{dt} = 0$

As current is decreasing with time we can write $i = -\dfrac{dq}{dt}$

$$\Rightarrow \quad \frac{q}{C} + \frac{dq}{dt}R + \frac{Ld^2q}{dt^2} = 0$$

$$\text{or} \quad \frac{d^2q}{dt^2} + \frac{R}{L}\frac{dq}{dt} + \frac{q}{LC} = 0$$

This equation is equivalent to that of a damped oscillator

Thus, we can write the solution as

$$Q_{\max}(t) = Q_0 \cdot e^{-Rt/2L}$$

$$\text{or} \quad Q_{\max}^2 = Q_0^2 \, e^{-\frac{Rt}{L}}$$

As $L_1 > L_2$ damping is faster for $L_2$

**Aliter**

Inductance is inertia of circuit. It means inductance opposes the flow of charge, more inductance means decay of charge is slow.

In option (a), in a given time to, $Q_1^2 > Q_2^2$.

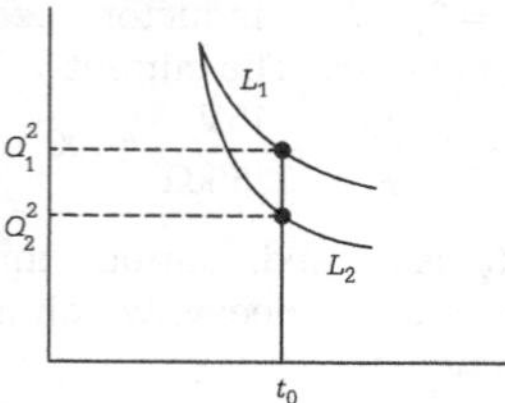


So, $L_1 > L_2$.

Hence option (a) is correct.

# Chemistry

1. *(d)* We know the molecular weight of $C_8H_7SO_3Na$

$= 12 \times 8 + 1 \times 7 + 32 + 16 \times 3 + 23 = 206$

We have to find, mole per gram of resin.

$\therefore$ 1 g of $C_8H_7SO_3Na$ has number of mole

$$= \frac{\text{Weight of given resin}}{\text{Molecular weight of resin}}$$

$$= \frac{1}{206} \text{ mol}$$

Now, reaction looks like

$$2C_8H_7SO_3Na + Ca^{2+} \longrightarrow (C_8H_7SO_3)_2Ca + 2Na^+$$

$\because$ 2 moles of $C_8H_7SO_3Na$ combines with 1 mol $Ca^{2+}$

$\therefore$ 1 mole of $C_8H_7SO_3Na$ will combine with $\frac{1}{2}$ mol $Ca^{2+}$

$\therefore \frac{1}{206}$ mole of $C_8H_7SO_3Na$ will combine with

$$\frac{1}{2} \times \frac{1}{206} \text{ mol } Ca^{2+} = \frac{1}{412} \text{ mol } Ca^{2+}$$

2. *(a)*

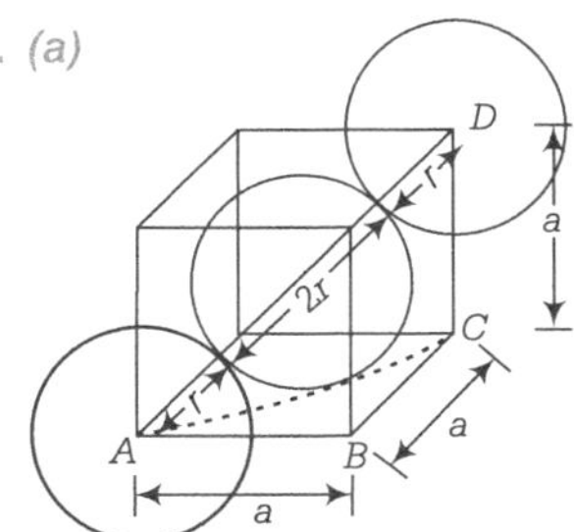


From this figure,

$$(AC)^2 = (AB)^2 + (BC)^2$$

$$(AC)^2 = a^2 + a^2 = 2a^2$$

Also, $(AD)^2 = (AC)^2 + (DC)^2$

$$(4r)^2 = 2a^2 + a^2$$

$$16r^2 = 3a^2$$

$$r = \frac{\sqrt{3}}{4} a$$

Now, when Na metal crystallises in bcc unit cell with unit cell edge, $a = 4.29$Å

We have the formula for radius,

i.e. $r = \frac{\sqrt{3}}{4} \times 4.29 \text{Å} = 1.86\text{Å}$

3. *(c)*

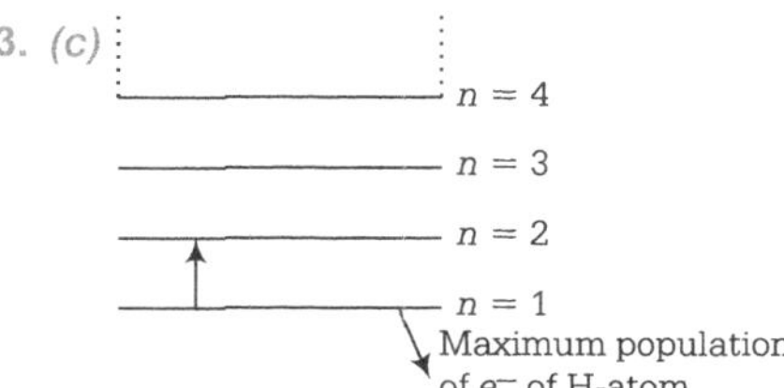


Since, at $n = 1$, the population of electrons is maximum *i.e.* at ground state. So, maximum excitation will take place from $n = 1$ to $n = 2$.

Hence, $n = 2$ is the possible excited state.

Now, we have the formula for energy of H-atom

$$(E_n)_H = -13.6 \frac{Z^2}{n^2} \text{ eV}$$

where, $Z$ = atomic number

$Z$ for H-atom = 1

$$\therefore \quad (E_n)_H = -13.6 \times \frac{1}{2^2} \text{ eV}$$

$$= -\frac{13.6}{4} \text{ eV}$$

$$= -3.4 \text{ eV}$$

4. *(b)* Ion-ion interaction is dependent on the square of distance,

i.e. ion-ion interaction $\propto \frac{1}{r^2}$

Similarly,

ion-dipole interaction $\propto \frac{1}{r^3}$

London forces $\propto \frac{1}{r^6}$

and dipole-dipole interactions $\propto \frac{1}{r^3}$

Superficially it seems as both ion dipole interaction and hydrogen bonding vary with the inverse cube of distance between the molecules but when we look at the exact expressions of field (force) created in two situations it comes as

$$|\mathbf{E}| \text{or } |\mathbf{F}| = \frac{2|\mathbf{P}|}{4\pi\varepsilon_0 r^3}$$

(In case of ion-dipole interaction)

and $$F = \frac{2q^2 r - 4q^2 a}{4\pi\varepsilon_0 r^3}$$

(In case of dipole-dipole interaction)

From the above it is very clear, the ion-dipole interaction is the better answer as compared to dipole-dipole interaction i.e. hydrogen bonding.

5. *(d)* For the given reaction,

$$2NO(g) + O_2(g) \rightleftharpoons 2NO_2(g)$$

Given, $\Delta G_f^\circ$ (NO) = 86.6 kJ/mol

$\Delta G_f^\circ$ $(NO_2)$ = ?

$K_p = 1.6 \times 10^{12}$

Now, we have,

$$\Delta G_f^\circ = 2\Delta G_{f(NO_2)}^\circ - [2\Delta G_{f(NO)}^\circ + \Delta G_{f(O_2)}^\circ]$$

$$= - RT \ln K_p = 2\Delta G_{f(NO_2)}^\circ - [2 \times 86{,}600 + 0]$$

$$\Delta G_{f(NO_2)}^\circ = \frac{1}{2} [2 \times 86{,}600 - R \times 298 \ln (1.6 \times 10^{12})]$$

$$\Delta G_{f(NO_2)}^\circ = 0.5 [2 \times 86{,}600 - R \times (298) \ln (1.6 \times 10^{12})]$$

6. *(b)* Given,

$p_o$ = 185 torr at 20°C

$p_s$ = 183 torr at 20°C

Mass of non-volatile substance,

$m$ = 1.2 g

Mass of acetone taken = 100 g

$M$ = ?

As , we have

$$\frac{p_o - p_s}{p_s} = \frac{n}{N}$$

Putting the values, we get,

$$\frac{185 - 183}{183} = \frac{\frac{1.2}{M}}{\frac{100}{58}} \Rightarrow \frac{2}{183} = \frac{1.2 \times 58}{100 \times M}$$

$$\therefore \quad M = \frac{183 \times 1.2 \times 58}{2 \times 100}$$

$M = 63.684 \approx 64$ g/mol

7. *(b)* Given, $\Delta G^\circ$ = 2494.2J

$$Q = \frac{[B][C]}{[A]^2} = \frac{2 \times \frac{1}{2}}{\left(\frac{1}{2}\right)^2} = 4$$

$\therefore$ We know,

$$\Delta G = \Delta G^\circ + RT \ln Q$$

$$= 2494.2 + 8.314 \times 300 \ln 4$$

$$= 28747.27\text{J}$$

= positive value

Also, we have

$$\Delta G = RT \ln \frac{Q}{K}$$

if $\Delta G$ is positive, $Q > K$.

Therefore, reaction shifts in reverse direction.

8. *(b)* Given, $Q = 2F$

Atomic mass of Cu = 63.5 u

Valency of the metal $Z$ = 2

We have, $CuSO_4 \longrightarrow Cu^{2+} + SO_4^{2-}$

$$\underset{1\text{mol}}{Cu^{2+}} + \underset{\substack{2\text{ mol}\\2F}}{2e^-} \longrightarrow \underset{1\text{mol} = 63.5\text{ g}}{Cu}$$

Alternatively, $W = ZQ$

$$= \frac{E}{F} \cdot 2F = 2E$$

$$= \frac{2 \times 63.5}{2} = 63.5$$

9. *(a)* The main conditions for the occurrence of a reaction is proper orientation and effective collision of the reactants.

Since the chances of simultaneous collision with proper orientation between more than 3 species is very rare, so reaction with order greater than 3 are rare.

10. *(a)* Given, initial strength of acetic acid

= 0.06 N

Final strength = 0.042 N

Volume given = 50 mL

$\therefore$ Initial $m$ moles of $CH_3COOH$

= 0.06 × 50 = 3

Final $m$ moles of $CH_3COOH$

= 0.042 × 50 = 2.1

$\therefore$ $m$ moles of $CH_3COOH$ adsorbed

= 3 − 2.1

= 0.9 m mol

Hence, mass of $CH_3COOH$ absorbed per gram of charcoal

$$= \frac{0.9 \times 60}{3}$$

($\because$ molar mass of $CH_3COOH = 60\text{ gmol}^{-1}$)

$$= \frac{54}{3} = 18\text{ mg.}$$

11. *(c)* Number of electrons in $N^{3-} = 7 + 3 = 10$

Number of electrons in $O^{2-} = 8 + 2 = 10$

Number of electrons in $F^{-} = 9 + 1 = 10$

Since, all the three species have each 10 electrons hence they are isoelectronic species.

It is considered that, in case of isoelectronic species as the negative charge increases, ionic radii increases and therefore the value of ionic radii are

$N^{3-} = 1.71$ (highest among the three)

$O^{2-} = 1.40$

$F^{-} = 1.36$ (lowest among the three)

***Time Saving Technique*** *There is no need to mug up the radius values for different ions. This particular question can be solved through following time saving.*

***Trick*** *The charges on the ions indicate the size as* $N^{3-} > O^{2-} > F^{-}$. *Thus, you have to look for the option in which the above trend is followed. Option (c) is the only one in which this trend is followed. Hence, it is the correct answer.*

12. *(d)* (a) In Hall-Heroult process for extraction of Al, carbon anode is oxidised to CO and $CO_2$.

(b) When $Al_2O_3$ is mixed with $CaF_2$, it lowers the melting point of the mixture and brings conductivity.

(c) $Al^{3+}$ is reduced at cathode to form Al.

(d) Here, $Al_2O_3$ is an electrolyte, undergoing the redox process. $Na_3AlF_6$ although is an electrolyte but serves as a solvent, not electrolyte.

13. *(a)* $H_2O_2$ acts as an oxidising as well as reducing agent, because oxidation number of oxygen in $H_2O_2$ is −1. So, it can be oxidised to oxidation state 0 or reduced to oxidation state −2.

$H_2O_2$ decomposes on exposure to light. So, it has to be stored in plastic or wax lined glass bottles in dark for the prevention of exposure. It also has to be kept away from dust.

14. *(b)* As we move down the group, size of metal increases. Be has lower size while $SO_4^{2-}$ has bigger size, that's why $BeSO_4$ breaks easily and lattice energy becomes smaller but due to lower size of Be, water molecules are gathered around and hence hydration energy increases.

On the other hand, rest metals i.e. Ca, Ba, Sr have bigger size and that's why lattice energy is greater than hydration energy.

***Time Saving Technique*** *In the question of finding hydration energy only check the size of atom. Smaller sized atom has more hydration energy.*

*Thus, in this question* Be *is placed upper most in the group has lesser size and not comparable with the size of sulphates. Hence,* $BeSO_4$ *is the right response.*

15. *(d)* $Cl_2$, $Br_2$ and $I_2$ are homonuclear diatomic molecule in which electronegativity of the combining atoms is same, so they are more stable and less reactive, whereas, I and Cl have different electronegativities and bond between them are polarised and hence, reactive. Therefore, interhalogen compounds are more reactive.

***Time Saving Technique*** *In this type of question of halogen, only go through the polarity of the molecule. As we know, diatomic molecule does not have polarity but molecules with dissimilar sizes have polarity resulting in more reactivity.*

16. *(b)* (a) $TiCl_3$ is used as Ziegler-Natta catalyst for the polymerisation of ethene.

(b) $PdCl_2$ is used in Wacker process, in which alkene changed into aldehyde *via* catalytic cyclic process initiated by $PdCl_2$.

(c) $CuCl_2$ is used in Deacon's process. (for $Cl_2$)

(d) $V_2O_5$ is used in contact process of manufacturing sulphuric acid.

***Time Saving Technique*** *This type of questions can also be solved through elimination technique. There is no need to know all the four matches to select the correct response. Even if you know (b) matches then also you can solve the problem. e.g. suppose you know the usage of* $V_2O_5$ *in contact process (i.e. D matches with (iii) and* $TiCl_3$ *is connected to Ziegler-Natta catalyst (i.e. A matches with ii). These two combinations are present only in option number (b). Likewise, for this question particularly if you know that* $V_2O_5$ *is used in* *contact process then this combination is present in option (b) only out of all the four option given. In this way you can eliminate wrong options to get the correct response.*

17. *(d)* As we move down the group of noble gases , molecular mass increases by which dipole produced for a moment and hence London forces increases from He to Xe. Therefore more amount of energy is required to break these forces, thus boiling point also increases from He to Xe.

18. *(b)* $[Pt(Cl)(py)(NH_3)(NH_2OH)]^+$ is square planar complex.

The structures are formed by fixing a group and then arranging all the groups.

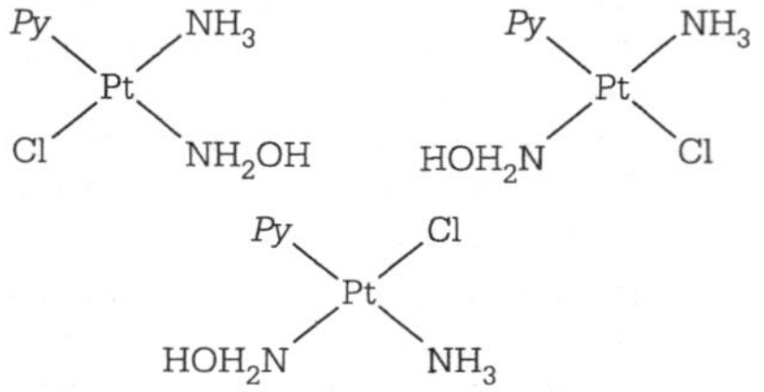

Py, $NH_3$, Pt, Cl, $NH_2OH$     Py, $NH_3$, Pt, $HOH_2N$, Cl     Py, Cl, Pt, $HOH_2N$, $NH_3$

Hence, this complex shows three geometrical isomers.

19. *(c)* $KMnO_4 \longrightarrow K^+ + MnO_4^-$

$\therefore$ In $MnO_4^-$, Mn has +7 oxidation state having no electron in *d*-orbitals.

It is considered that higher the oxidation state of metal, greater is the tendency to occur $L \rightarrow M$ charge transfer, because ligand is able to donate the electrons into the vacant *d*- orbital of metal.

Since, charge transfer is Laporte as well as spin allowed, therefore, it shows colour.

***Time Saving Technique*** *There is no need to check all the four options. Just find out the oxidation state of metal ion. If oxidation state is highest and ligand present there is of electron donating nature, gives LMCT, which shows more intense colour.*

20. *(a)* Nitrogen is an inert gas because of the presence of strong bond. That's why although there is 78% $N_2$ in the atmosphere but nitrogen oxide in not formed under ordinary conditions.

But when temperature is high enough i.e. $\approx$2000 K, it reacts with oxygen to form nitrogen oxide.

$$N_2 + O_2 \xrightarrow{\approx 2000\text{ K}} 2NO$$

Thus, Assertion and Reason are true and Reason is the correct explanation of the Assertion.

21. *(a)* Given,

Weight of organic compound = 250 mg

Weight of AgBr = 141 mg

$\therefore$ According to formula of % of bromine by Carius method

$$\% \text{ of Br} = \frac{\text{Atomic weight of Br}}{\text{Molecular weight of AgBr}} \times \frac{\text{Weight of AgBr}}{\text{Weight of organic bromide}} \times 100$$

$$\therefore \quad \% \text{ of Br} = \frac{80}{188} \times \frac{141}{250} \times 100$$

$$= \frac{1128000}{47000} = 24\ \%$$

22. *(a)* Alkene in which different groups are attached with the double bonded carbon atoms, exhibit geometrical isomerism.

1-phenyl-2-butene = (Ph, H, H) *trans*

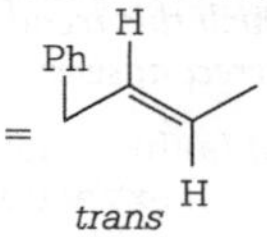

(Ph, H, H) *cis*

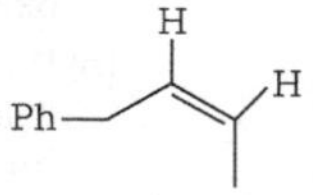

It will show geometrical isomerism.

3-phenyl-1-butene (H, H, H, Ph)

Same group, does not show geometrical isomerism.

2-phenyl-1-butene (Ph, H, H)

Same group, does not show geometrical isomerism.

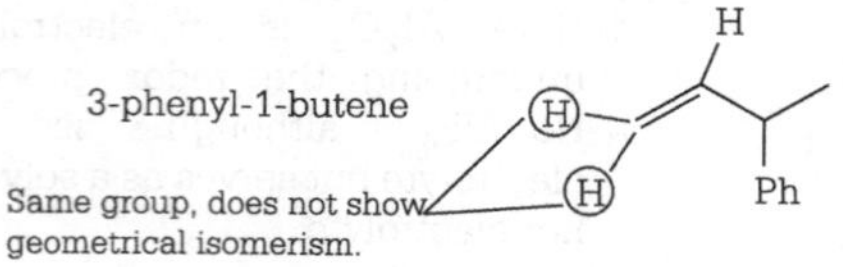

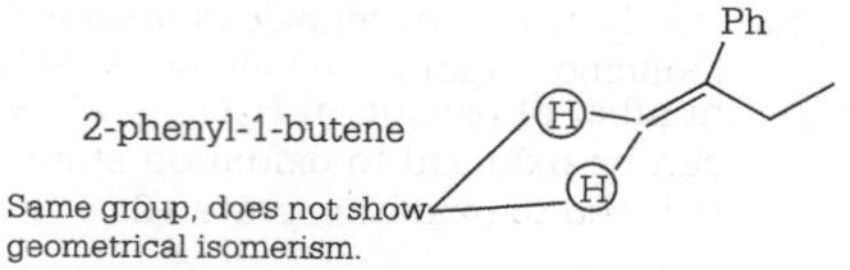

1,1-diphenyl-1-propane being an alkane (saturated compound) does not show geometrical isomerism.

***Time Saving Technique*** *We do not need the check all options, but it should remind that double bonded compounds show geometrical isomerism. Thus, (d) is eliminated, (i.e. propane). Now, eliminate the terminal alkene and get the correct response.*

**23.** *(b)* (a) 1,2-dimethylcyclopentene $\xrightarrow[\text{Zn, } H_2O_2]{O_3 \; \Delta}$

$$CH_3-\overset{O}{\overset{\|}{C}}-CH_2-CH_2-CH_2-\overset{O}{\overset{\|}{C}}-CH_3$$

Heptan-2,6-dione

(b) 1,5-dimethylcyclopentene $\xrightarrow[\text{Zn, } H_2O_2]{O_3 \; \Delta}$

$$\overset{6}{CH_3}-\overset{O}{\overset{\|}{C}}{}^5-\overset{4}{CH_2}-\overset{3}{CH_2}-\overset{2}{\underset{CH_3}{\underset{|}{CH}}}-\overset{1}{CHO}$$

5-keto-2-methyl hexanal

(c) 1,4-dimethylcyclopentene $\xrightarrow[\text{Zn, } H_2O_2]{O_3 \; \Delta}$

$$CH_3-\overset{O}{\overset{\|}{C}}-CH_2-\overset{CH_3}{\overset{|}{CH}}-CH_2-CHO$$

5-keto-3-methyl hexanal

(d) 1,5-dimethylcyclopentene ($H_3C$ on ring carbon adjacent to $C-CH_3$) $\xrightarrow[\text{Zn, } H_2O_2]{O_3 \; \Delta}$

$$CH_3-\overset{O}{\overset{\|}{C}}-\overset{CH_3}{\overset{|}{CH}}-CH_2-CH_2-CHO$$

5-keto-4-methyl hexanal

**24.** *(d)* Alkyl fluorides can be prepared by action of mercurous fluoride or antimony trifluorides (inorganic fluorides) on corresponding alkyl halide. This reaction is known as Swarts reaction.

$$CH_3Br + AgF \longrightarrow \underset{\text{Methyl fluoride}}{CH_3F} + AgBr$$

But, when action of NaI/acetone takes place on alkyl chloride or bromide, alkyl iodide forms. This reaction is called 'Finkelstein reaction'.

$$C_2H_5Cl \xrightarrow[\text{acetone}]{\text{NaI}} C_2H_5I + NaCl$$

Free radical fluorination is highly explosive reaction. so not preferred for the preparation of fluoride.

**25.** *(d)* Toluene undergoes oxidation with $KMnO_4$, forms benzoic acid. In this conversion, alkyl part of toluene converts into carboxylic group. Further, benzoic acid reacts with thionyl chloride ($SOCl_2$) to give benzoyl chloride which upon reduction with $H_2$/Pd or $BaSO_4$ forms benzaldehyde (Rosenmund Reduction).

The conversion look like,

$$\underset{\text{Toluene}}{C_6H_5CH_3} \xrightarrow{KMnO_4} \underset{\text{Benzoic acid}}{\underset{(A)}{C_6H_5COOH}} \xrightarrow{SOCl_2} \underset{\text{Benzoyl chloride}}{\underset{(B)}{C_6H_5COCl}} + SO_2 + HCl$$

$$\underset{(B)}{C_6H_5COCl} \xrightarrow[\text{(Rosenmund reaction)}]{H_2/Pd, \; BaSO_4} \underset{\text{Benzaldehyde}}{\underset{(C)}{C_6H_5CHO}} + HCl$$

**26.** *(c)*

$$\underset{\text{4- methyl aniline}}{CH_3C_6H_4NH_2} \xrightarrow[\text{0 – 5°C, diazotisation}]{NaNO_2/HCl} \underset{\text{4- methyl diazonium chloride}}{\underset{(D)}{CH_3C_6H_4\overset{+}{N_2}Cl^-}}$$

$$\xrightarrow[\Delta]{CuCN/KCN} \underset{\text{4- methyl benzocyanide}}{\underset{(E)}{CH_3C_6H_4CN}} + CuCl$$

27. *(b)* (a) Bakelite is used for making gears, protective coating and electrical fittings.

(b) Glyptal is used in the manufacture of paints and lacquers.

(c) PP is used in the manufacture of textile, packaging materials etc.

(d) Polyvinyl chloride (PVC) is used in the manufacture of rain coats, hand bags, leather clothes etc.

28. *(a)* Vitamin B and C are water soluble while vitamin A,D,E and K are fat soluble or water insoluble.

29. *(c)* Aluminium hydroxide $Al(OH)_3$, cimetidine and ranitidine are antacids while phenelzine is not.

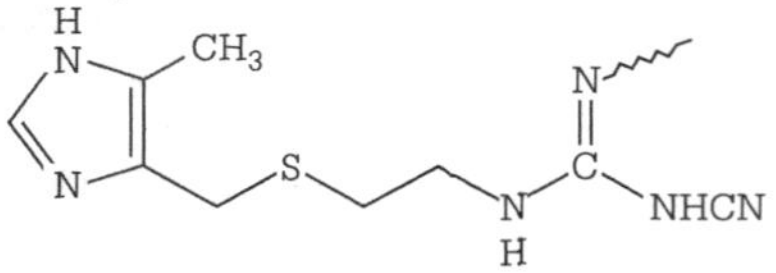

Cimetidine

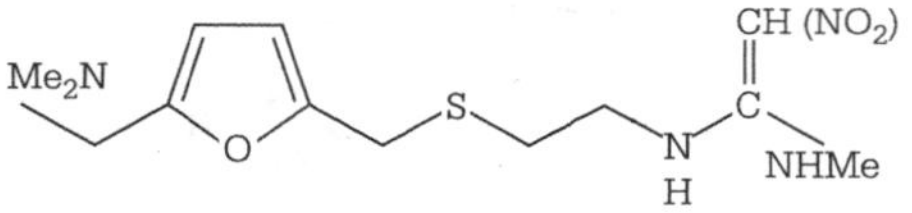

Ranitidine

Phenelzine is a tranquilizer, not an antacid.

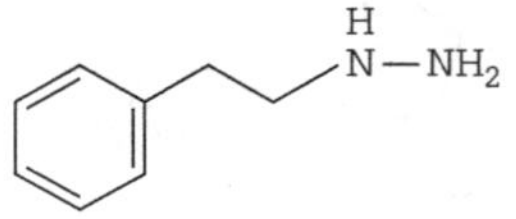

30. *(a)* $Zn_2[Fe(CN)_6]$, $K_3[Co(NO_2)_6]$ and $[(NH_4)_3As(Mo_3O_{10})_4]$ show colour due to *d-d* transition while $BaCrO_4$ is coloured due to charge transfer phenomenon.

Further, according to spectrochemical series the strong ligand possessing complex has higher energy and hence lower wavelength. Therefore, complexes containing $NO_2$, $NH_4^+$, $O^{2-}$ etc ligands show yellow colour while $CN^-$ forces the complex to impart white colour.

**Spectrochemical Series**

$$I^- < Br^- < S^{2-} < SCN^- < Cl^- < NO_3^- < N_3^- < F^- < OH^- < C_2O_4^{2-} \approx H_2O < NCS^- < CH_3CN < py < NH_3 < en < bipy < Phen < NO_2^- < PPh_3 < CN^- \approx CO$$

# Mathematics

1. *(a)* Given,

$$n(A) = 4, n(B) = 2$$

$$\Rightarrow \quad n(A \times B) = 8$$

Total number of subsets of set

$$(A \times B) = 2^8$$

Number of subsets of set $A \times B$ having no element (i.e. $\phi$) = 1

Number of subsets of set $A \times B$ having one element = ${}^8C_1$

Number of subsets of set $A \times B$ having two elements = ${}^8C_2$

$\therefore$ Number of subsets having atleast three elements

$$= 2^8 - (1 + {}^8C_1 + {}^8C_2)$$
$$= 2^8 - 1 - 8 - 28$$
$$= 2^8 - 37$$
$$= 256 - 37$$
$$= 219$$

2. *(c)* ***Central Idea*** *If z is unimodular, then $|z| = 1$. Also, use property of modulus i.e. $z\bar{z} = |z|^2$.*

Given, $z_2$ is not unimodular i.e. $|z_2| \neq 1$

and $\dfrac{z_1 - 2z_2}{2 - z_1\bar{z}_2}$ is unimodular

$$\Rightarrow \quad \left|\frac{z_1 - 2z_2}{2 - z_1\bar{z}_2}\right| = 1$$

$$\Rightarrow \quad |z_1 - 2z_2|^2 = |2 - z_1\bar{z}_2|^2$$

$$\Rightarrow (z_1 - 2z_2)(\bar{z}_1 - 2\bar{z}_2) = (2 - z_1\bar{z}_2)(2 - \bar{z}_1z_2) \quad (\because z\bar{z} = |z|^2)$$

$$\Rightarrow |z_1|^2 + 4|z_2|^2 - 2\bar{z}_1z_2 - 2z_1\bar{z}_2 = 4 + |z_1|^2|z_2|^2 - 2\bar{z}_1z_2 - 2z_1\bar{z}_2$$

$$\Rightarrow (|z_2|^2 - 1)(|z_1|^2 - 4) = 0$$

$$\because \quad |z_2| \neq 1$$

$$\therefore \quad |z_1| = 2$$

Let $z_1 = x + iy \Rightarrow x^2 + y^2 = (2)^2$

$\therefore$ Point $z_1$ lies on a circle of radius 2.

3. *(c)* Given, $\alpha$ and $\beta$ are the roots of the equation $x^2 - 6x - 2 = 0$.

$\because a_n = \alpha^n - \beta^n$ for $n \ge 1$

$$\therefore \quad a_{10} = \alpha^{10} - \beta^{10}$$
$$a_8 = \alpha^8 - \beta^8$$
$$a_9 = \alpha^9 - \beta^9$$

Now, consider

$$\frac{a_{10} - 2a_8}{2a_9} = \frac{\alpha^{10} - \beta^{10} - 2(\alpha^8 - \beta^8)}{2(\alpha^9 - \beta^9)}$$

$$= \frac{\alpha^8(\alpha^2 - 2) - \beta^8(\beta^2 - 2)}{2(\alpha^9 - \beta^9)}$$

$$= \frac{\alpha^8 \cdot 6\alpha - \beta^8 \cdot 6\beta}{2(\alpha^9 - \beta^9)}$$

$$= \frac{6\alpha^9 - 6\beta^9}{2(\alpha^9 - \beta^9)} = \frac{6}{2} = 3$$

[$\because \alpha$ and $\beta$ are the roots of $x^2 - 6x - 2 = 0$
or $x^2 = 6x + 2$
$\Rightarrow \alpha^2 = 6\alpha + 2$
$\Rightarrow \alpha^2 - 2 = 6\alpha$
and $\beta^2 = 6\beta + 2$
$\Rightarrow \beta^2 - 2 = 6\beta$]

**Aliter**

Since, $\alpha$ and $\beta$ are the roots of the equation

$$x^2 - 6x - 2 = 0$$

or $\quad x^2 = 6x + 2$

$\therefore \quad \alpha^2 = 6\alpha + 2$

$\Rightarrow \quad \alpha^{10} = 6\alpha^9 + 2\alpha^8 \quad \ldots(i)$

Similarly, $\beta^{10} = 6\beta^9 + 2\beta^8 \quad \ldots(ii)$

On subtracting Eq. (ii) from Eq. (i), we get

$$\alpha^{10} - \beta^{10} = 6(\alpha^9 - \beta^9) + 2(\alpha^8 - \beta^8)$$

$\Rightarrow \quad a_{10} = 6a_9 + 2a_8 \quad (\because a_n = \alpha^n - \beta^n)$

$\Rightarrow \quad a_{10} - 2a_8 = 6a_9 \Rightarrow \dfrac{a_{10} - 2a_8}{2a_9} = 3$

4. *(d)* Given, $A = \begin{bmatrix} 1 & 2 & 2 \\ 2 & 1 & -2 \\ a & 2 & b \end{bmatrix}$

$$A^T = \begin{bmatrix} 1 & 2 & a \\ 2 & 1 & 2 \\ 2 & -2 & b \end{bmatrix}$$

$$AA^T = \begin{bmatrix} 1 & 2 & 2 \\ 2 & 1 & -2 \\ a & 2 & b \end{bmatrix} \begin{bmatrix} 1 & 2 & a \\ 2 & 1 & 2 \\ 2 & -2 & b \end{bmatrix}$$

$$= \begin{bmatrix} 9 & 0 & a+4+2b \\ 0 & 9 & 2a+2-2b \\ a+4+2b & 2a+2-2b & a^2+4+b^2 \end{bmatrix}$$

It is given that $\quad AA^T = 9I$

$$\Rightarrow \begin{bmatrix} 9 & 0 & a+4+2b \\ 0 & 9 & 2a+2-2b \\ a+4+2b & 2a+2-2b & a^2+4+b^2 \end{bmatrix} = 9\begin{bmatrix} 1 & 0 & 0 \\ 0 & 1 & 0 \\ 0 & 0 & 1 \end{bmatrix}$$

$$\Rightarrow \begin{bmatrix} 9 & 0 & a+4+2b \\ 0 & 9 & 2a+2-2b \\ a+4+2b & 2a+2-2b & a^2+4+b^2 \end{bmatrix} = \begin{bmatrix} 9 & 0 & 0 \\ 0 & 9 & 0 \\ 0 & 0 & 9 \end{bmatrix}$$

On comparing, we get

$$a + 4 + 2b = 0$$

$\Rightarrow \quad a + 2b = -4 \quad \ldots(i)$

$$2a + 2 - 2b = 0$$

$\Rightarrow \quad a - b = -1 \quad \ldots(ii)$

and $\quad a^2 + 4 + b^2 = 9 \quad \ldots(iii)$

On solving Eqs. (i) and (ii), we get

$$a = -2, \; b = -1$$

This satisfies Eq. (iii).

Hence, $(a, b) \equiv (-2, -1)$

5. *(c)* Given system of linear equations

$$2x_1 - 2x_2 + x_3 = \lambda x_1$$

$\Rightarrow \quad (2 - \lambda)x_1 - 2x_2 + x_3 = 0 \quad \ldots(i)$

$$2x_1 - 3x_2 + 2x_3 = \lambda x_2$$

$\Rightarrow \quad 2x_1 - (3 + \lambda)x_2 + 2x_3 = 0 \quad \ldots(ii)$

$$-x_1 + 2x_2 = \lambda x_3$$

$\Rightarrow \quad -x_1 + 2x_2 - \lambda x_3 = 0 \quad \ldots(iii)$

Since, the system has non-trivial solution.

$$\therefore \quad \begin{vmatrix} 2-\lambda & -2 & 1 \\ 2 & -(3+\lambda) & 2 \\ -1 & 2 & -\lambda \end{vmatrix} = 0$$

$\Rightarrow (2 - \lambda)(3\lambda + \lambda^2 - 4) + 2(-2\lambda + 2) + 1(4 - 3 - \lambda) = 0$

$\Rightarrow (2 - \lambda)(\lambda^2 + 3\lambda - 4) + 4(1 - \lambda) + (1 - \lambda) = 0$

$\Rightarrow (2 - \lambda)(\lambda + 4)(\lambda - 1) + 5(1 - \lambda) = 0$

$\Rightarrow (\lambda - 1)[(2 - \lambda)(\lambda + 4) - 5] = 0$

$\Rightarrow (\lambda - 1)(\lambda^2 + 2\lambda - 3) = 0$

$\Rightarrow (\lambda - 1)[(\lambda - 1)(\lambda + 3)] = 0$

$\Rightarrow (\lambda - 1)^2(\lambda + 3) = 0$

$\Rightarrow \lambda = 1, 1, -3$

6. *(b)* The integer greater than 6000 may be of 4 digit or 5 digit. So, here two cases arise.

**Case I** When number is of 4 digit.

Four digit number can starts from 6, 7 or 8

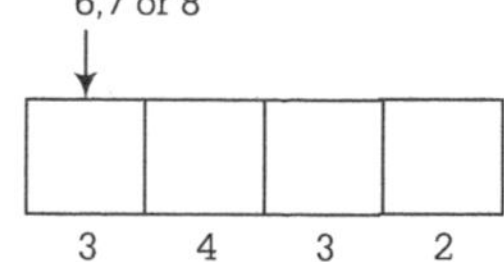


Thus, total number of 4 digit number, which are greater than 6000 $= 3 \times 4 \times 3 \times 2 = 72$

**Case II** When number is of 5 digit.

Total number of five digit number which are greater than 6000 = 5! = 120

$\therefore$ Total number of integers = 72 + 120 = 192

7. *(a)* Let $T_{r+1}$ be the general term in the expansion of $(1-2\sqrt{x})^{50}$

$$\therefore \quad T_{r+1} = {}^{50}C_r (1)^{50-r}(-2x^{1/2})^r = {}^{50}C_r 2^r x^{r/2}(-1)^r$$

For the integral power of $x$, $r$ should be even integer.

$$\therefore \text{Sum of coefficients} = \sum_{r=0}^{25} {}^{50}C_{2r}(2)^{2r}$$

$$= \frac{1}{2}[(1+2)^{50} + (1-2)^{50}]$$

$$= \frac{1}{2}[3^{50}+1]$$

**Aliter**

We have,

$$(1-2\sqrt{x})^{50} = C_0 - C_1 2\sqrt{x} + C_2(2\sqrt{x})^2 + \ldots + C_{50}(2\sqrt{x})^{50} \quad \ldots(i)$$

$$(1+2\sqrt{x})^{50} = C_0 + C_1 2\sqrt{x} + C_2(2\sqrt{x})^2 + \ldots + C_{50}(2\sqrt{x})^{50} \quad \ldots(ii)$$

On adding Eqs. (i) and (ii), we get

$$(1-2\sqrt{x})^{50} + (1+2\sqrt{x})^{50} = 2[C_0 + C_2(2\sqrt{x})^2 + \ldots + C_{50}(2\sqrt{x})^{50}]$$

$$\Rightarrow \frac{(1-2\sqrt{x})^{50} + (1+2\sqrt{x})^{50}}{2} = C_0 + C_2(2\sqrt{x})^2 + \ldots + C_{50}(2\sqrt{x})^{50}$$

On putting $x = 1$, we get

$$\frac{(1-2\sqrt{1})^{50} + (1+2\sqrt{1})^{50}}{2} = C_0 + C_2(2)^2 + \ldots + C_{50}(2)^{50}$$

$$\Rightarrow \frac{(-1)^{50} + (3)^{50}}{2} = C_0 + C_2(2)^2 + \ldots + C_{50}(2)^{50}$$

$$\Rightarrow \frac{1+3^{50}}{2} = C_0 + C_2(2)^2 + \ldots + C_{50}(2)^{50}$$

8. *(b)* Given, $m$ is the AM of $l$ and $n$

$$\therefore \quad l + n = 2m \quad \ldots(i)$$

and $G_1, G_2, G_3$ are geometric means between $l$ and $n$

$\therefore \quad l, G_1, G_2, G_3, n$ are in GP.

Let $r$ be the common ratio of this GP.

$$\therefore \quad G_1 = lr, \quad G_2 = lr^2, \quad G_3 = lr^3, \quad n = lr^4$$

$$\Rightarrow \quad r = \left(\frac{n}{l}\right)^{\frac{1}{4}}$$

Now, $G_1^4 + 2G_2^4 + G_3^4 = (lr)^4 + 2(lr^2)^4 + (lr^3)^4$

$$= l^4 \times r^4(1 + 2r^4 + r^8)$$

$$= l^4 \times r^4(r^4+1)^2$$

$$= l^4 \times \frac{n}{l}\left(\frac{n+l}{l}\right)^2$$

$$= ln \times 4m^2 = 4lm^2n$$

9. *(b)* **Central Idea** *Write the nth term of the given series and simplify it to get its lowest form. Then, apply,* $S_n = \Sigma T_n$.

Given series is

$$\frac{1^3}{1} + \frac{1^3+2^3}{1+3} + \frac{1^3+2^3+3^3}{1+3+5} + \ldots \infty$$

Let $T_n$ be the $n$th term of the given series.

$$\therefore \quad T_n = \frac{1^3+2^3+3^3+\ldots+n^3}{1+3+5+\ldots+ \text{to } n \text{ terms}}$$

$$= \frac{\left\{\frac{n(n+1)}{2}\right\}^2}{n^2}$$

$$= \frac{(n+1)^2}{4}$$

$$S_9 = \sum_{n=1}^{9} \frac{(n+1)^2}{4} = \frac{1}{4}[(2^2+3^2+\ldots+10^2) + 1^2 - 1^2]$$

$$= \frac{1}{4}\left[\frac{10(10+1)(20+1)}{6} - 1\right] = \frac{384}{4} = 96$$

**10.** *(c)* We have,

$$\lim_{x\to 0}\frac{(1-\cos 2x)(3+\cos x)}{x\tan 4x}$$

$$= \lim_{x\to 0}\frac{2\sin^2 x(3+\cos x)}{x\times\frac{\tan 4x}{4x}\times 4x}$$

$$= \lim_{x\to 0}\frac{2\sin^2 x}{x^2}\times\lim_{x\to 0}\frac{(3+\cos x)}{4}\times\frac{1}{\lim_{x\to 0}\frac{\tan 4x}{4x}}$$

$$= 2\times\frac{4}{4}\times 1 \quad \left(\because \lim_{\theta\to 0}\frac{\sin\theta}{\theta}=1 \text{ and } \lim_{\theta\to 0}\frac{\tan\theta}{\theta}=1\right)$$

$$= 2$$

**11.** *(a)* Since, $g(x)$ is differentiable $\Rightarrow g(x)$ must be continuous.

$$\therefore \quad g(x)=\begin{cases} k\sqrt{x+1}, & 0\le x\le 3\\ mx+2, & 3<x\le 5\end{cases}$$

At $x=3$, RHL $=3m+2$

and at $x=3$, LHL $=2k$

$$\therefore \quad 2k=3m+2 \quad \ldots\text{(i)}$$

$$\text{Also, } g'(x)=\begin{cases}\frac{k}{2\sqrt{x+1}}, & 0\le x<3\\ m, & 3<x\le 5\end{cases}$$

$\therefore L\{g'(3)\}=\frac{k}{4}$ and $R\{g'(3)\}=m$

$$\Rightarrow \quad \frac{k}{4}=m \text{ i.e. } k=4m \quad \ldots\text{(ii)}$$

On solving Eqs. (i) and (ii), we get

$$k=\frac{8}{5},\ m=\frac{2}{5}$$

$$\Rightarrow \quad k+m=2$$

**12.** *(d)* Given equation of curve is

$$x^2+2xy-3y^2=0 \quad \ldots\text{(i)}$$

On differentiating w.r.t. $x$, we get

$$2x+2xy'+2y-6yy'=0$$

$$\Rightarrow \quad y'=\frac{x+y}{3y-x}$$

At $x=1, y=1, y'=1$

i.e. $\left(\frac{dy}{dx}\right)_{(1,1)}=1$

Equation of normal at (1, 1) is

$$y-1=-\frac{1}{1}(x-1)$$

$$\Rightarrow \quad y-1=-(x-1)$$

$$\Rightarrow \quad x+y=2 \quad \ldots\text{(ii)}$$

On solving Eqs. (i) and (ii) simultaneously, we get

$$x^2+2x(2-x)-3(2-x)^2=0$$

$$\Rightarrow x^2+4x-2x^2-3(4+x^2-4x)=0$$

$$\Rightarrow -x^2+4x-12-3x^2+12x=0$$

$$\Rightarrow -4x^2+16x-12=0$$

$$\Rightarrow 4x^2-16x+12=0$$

$$\Rightarrow x^2-4x+3=0$$

$$\Rightarrow (x-1)(x-3)=0$$

$\Rightarrow x=1, 3$

Now, when $x=1$, then $y=1$

and when $x=3$, then $y=-1$

$\therefore$ $P=(1, 1)$ and $Q=(3,-1)$

Hence, normal meets the curve again at $(3,-1)$ in fourth quadrant.

**Aliter**

Given,

$$x^2+2xy-3y^2=0$$

$$\Rightarrow (x-y)(x+3y)=0$$

$$\Rightarrow x-y=0 \text{ or } x+3y=0$$

Equation of normal at (1, 1) is

$$y-1=-1(x-1)$$

$$\Rightarrow x+y-2=0$$

It intersects $x+3y=0$ at $(3,-1)$

and hence normal meet the curve in fourth quadrant.

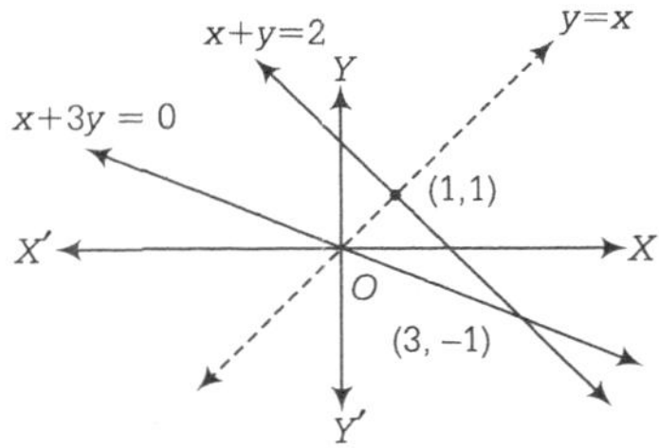


**13.** *(c)* ***Central Idea*** *Any function have extreme values (maximum or minimum) at its critical points, where $f'(x)=0$.*

Since, the function have extreme values at $x=1$ and $x=2$.

$$\therefore \quad f'(x)=0 \text{ at } x=1 \text{ and } x=2$$

$$\Rightarrow \quad f'(1)=0 \text{ and } f'(2)=0$$

Also it is given that

$$\lim_{x\to 0}\left[1+\frac{f(x)}{x^2}\right]=3$$

$$\Rightarrow \quad 1+\lim_{x\to 0}\frac{f(x)}{x^2}=3$$

$$\Rightarrow \quad \lim_{x\to 0}\frac{f(x)}{x^2}=2$$

$\Rightarrow$ $f(x)$ will be of the form

$$ax^4 + bx^3 + 2x^2$$

[$\because f(x)$ is of four degree polynomial]

Let $\quad f(x) = ax^4 + bx^3 + 2x^2$

$\Rightarrow \quad f'(x) = 4ax^3 + 3bx^2 + 4x$

$\Rightarrow \quad f'(1) = 4a + 3b + 4 = 0 \qquad \ldots(i)$

and $\quad f'(2) = 32a + 12b + 8 = 0$

$\Rightarrow \quad 8a + 3b + 2 = 0 \qquad \ldots(ii)$

On solving Eqs. (i) and (ii), we get

$$a = \frac{1}{2}, \; b = -2$$

$\therefore \quad f(x) = \frac{x^4}{2} - 2x^3 + 2x^2$

$\Rightarrow \quad f(2) = 8 - 16 + 8 = 0$

**14.** *(d)* $\displaystyle\int \frac{dx}{x^2(x^4+1)^{\frac{3}{4}}} = \int \frac{dx}{x^5\left(1 + \frac{1}{x^4}\right)^{\frac{3}{4}}}$

Put $\quad 1 + \frac{1}{x^4} = t^4$

$\Rightarrow \quad -\frac{4}{x^5}dx = 4t^3 dt$

$\Rightarrow \quad \frac{dx}{x^5} = -t^3 dt = \int \frac{-t^3 dt}{t^3}$

$$= -\int dt = -t + c = -\left(1 + \frac{1}{x^4}\right)^{1/4} + c$$

**15.** *(c)* ***Central Idea*** *Apply the property $\int_a^b f(x)dx = \int_a^b f(a+b-x)dx$ and then add.*

Let $\quad I = \displaystyle\int_2^4 \frac{\log x^2}{\log x^2 + \log(36 - 12x + x^2)} dx$

$$= \int_2^4 \frac{2\log x}{2\log x + \log(6-x)^2} dx$$

$$= \int_2^4 \frac{2\log x \, dx}{2[\log x + \log(6-x)]}$$

$\Rightarrow \quad I = \displaystyle\int_2^4 \frac{\log x \, dx}{[\log x + \log(6-x)]} \qquad \ldots(i)$

$\Rightarrow \quad I = \displaystyle\int_2^4 \frac{\log(6-x)}{\log(6-x) + \log x} dx \qquad \ldots(ii)$

$$\left[\because \int_a^b f(x)dx = \int_a^b f(a+b-x)dx\right]$$

On adding Eqs. (i) and (ii), we get

$$2I = \int_2^4 \frac{\log x + \log(6-x)}{\log x + \log(6-x)} dx$$

$\Rightarrow \quad 2I = \int_2^4 dx = [x]_2^4$

$\Rightarrow \quad 2I = 2 \Rightarrow I = 1$

**16.** *(d)* Given region is $\{(x, y) : y^2 \le 2x$ and $y \ge 4x - 1\}$

$y^2 \le 2x$ represents a region inside the parabola

$$y^2 = 2x \qquad \ldots(i)$$

and $y \ge 4x - 1$ represents a region to the left of the line

$$y = 4x - 1 \qquad \ldots(ii)$$

The point of intersection of the curve (i) and (ii) is

$$(4x-1)^2 = 2x$$

$\Rightarrow \quad 16x^2 + 1 - 8x = 2x$

$\Rightarrow \quad 16x^2 - 10x + 1 = 0$

$\Rightarrow \quad x = \frac{1}{2}, \frac{1}{8}$

$\therefore$ The points where these curves intersect, are $\left(\frac{1}{2}, 1\right)$ and $\left(\frac{1}{8}, -\frac{1}{2}\right)$.

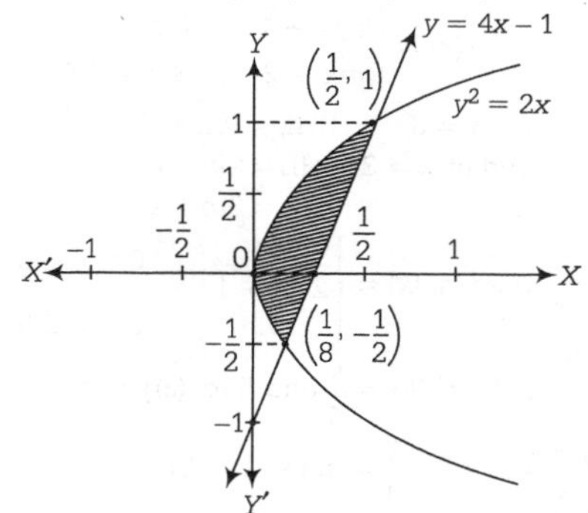


Hence, required area

$$= \int_{-1/2}^{1} \left(\frac{y+1}{4} - \frac{y^2}{2}\right) dy$$

$$= \frac{1}{4}\left(\frac{y^2}{2} + y\right)_{-1/2}^{1} - \frac{1}{6}(y^3)_{-1/2}^{1}$$

$$= \frac{1}{4}\left\{\left(\frac{1}{2} + 1\right) - \left(\frac{1}{8} - \frac{1}{2}\right)\right\} - \frac{1}{6}\left\{1 + \frac{1}{8}\right\}$$

$$= \frac{1}{4}\left\{\frac{3}{2} + \frac{3}{8}\right\} - \frac{1}{6}\left\{\frac{9}{8}\right\}$$

$$= \frac{1}{4} \times \frac{15}{8} - \frac{3}{16} = \frac{9}{32}$$

**17.** *(c)* Given differential equation is

$$(x\log x)\frac{dy}{dx} + y = 2x\log x, \qquad (x \ge 1)$$

$\Rightarrow \quad \frac{dy}{dx} + \frac{y}{x\log x} = 2$

This is a linear differential equation.

$$\therefore \text{IF} = e^{\int \frac{1}{x \log x} dx} = e^{\log(\log x)} = \log x$$

Now, the solution of given differential equation is given by

$$y \cdot \log x = \int \log x \cdot 2dx$$

$$\Rightarrow \quad y \cdot \log x = 2\int \log x \, dx$$

$$\Rightarrow \quad y \cdot \log x = 2[x \log x - x] + c$$

At $\quad x = 1, c = 2$

$$\Rightarrow \quad y \cdot \log x = 2[x \log x - x] + 2$$

At $\quad x = e,$

$$y = 2(e - e) + 2 \Rightarrow y = 2$$

18. (d) Required points $(x, y)$ are such that it satisfy $x + y < 41$

and $\quad x > 0, y > 0$

Number of positive integral solution of the equation $x + y + k = 41$ will be number of integral coordinates in the bounded region.

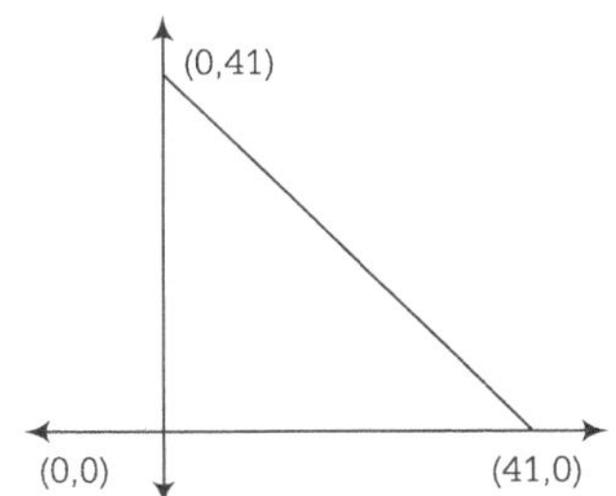


$\therefore$ Total number of integral coordinates

$$= {}^{41-1}C_{3-1} = {}^{40}C_2 = \frac{40!}{2!\,38!} = 780$$

**Aliter**

Consider the following figure :

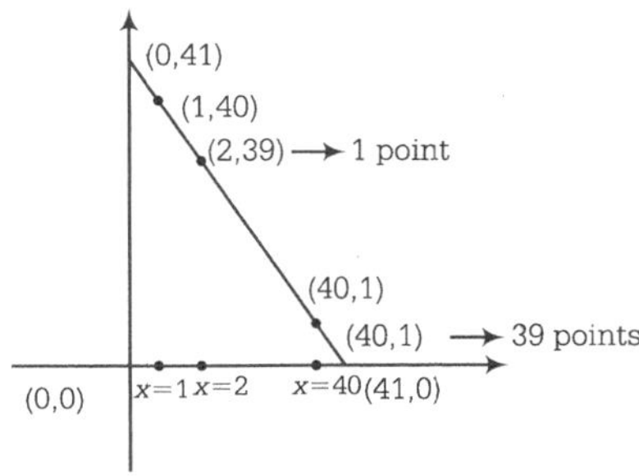


Clearly, the number of required points

$$= 1 + 2 + 3 + \ldots + 39$$

$$= \frac{39}{2}(39 + 1) = 780$$

19. (c) ***Central Idea*** *First of all find the point of intersection of the lines* $2x - 3y + 4 = 0$ *and* $x - 2y + 3 = 0$ *(say A). Now, the line* $(2x - 3y + 4) + k(x - 2y + 3) = 0$ *is the perpendicular bisector of the line joining points* $P(2, 3)$ *and image* $P'(h, k)$. *Now,* $AP = AP'$ *and simplify.*

Given line is

$$(2x - 3y + 4) + k(x - 2y + 3) = 0, k \in R \quad \ldots\text{(i)}$$

This line will pass through the point of intersection of the lines

$$2x - 3y + 4 = 0 \quad \ldots\text{(ii)}$$

and $$x - 2y + 3 = 0 \quad \ldots\text{(iii)}$$

On solving Eqs. (ii) and (iii), we get

$$x = 1, y = 2$$

$\therefore$ Point of intersection of lines (ii) and (iii) is (1, 2).

Let $M$ be the mid-point of $PP'$, then $AM$ is perpendicular bisector of $PP'$ (where, $A$ is the point of intersection of given lines).

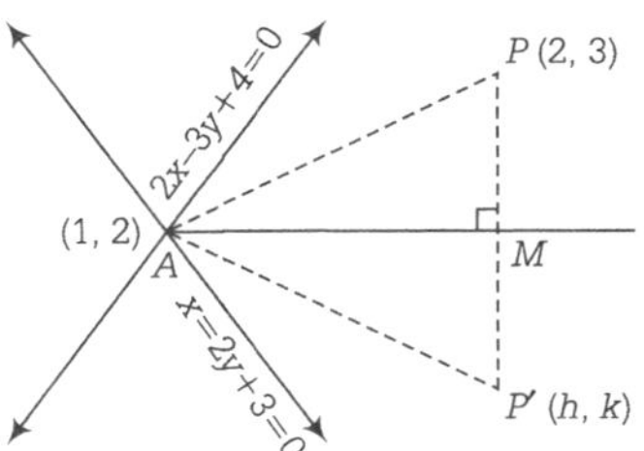


$$\therefore \quad AP = AP'$$

$$\Rightarrow$$

$$\sqrt{(2-1)^2 + (3-2)^2} = \sqrt{(h-1)^2 + (k-2)^2}$$

$$\Rightarrow \quad \sqrt{2} = \sqrt{h^2 + k^2 - 2h - 4k + 1 + 4}$$

$$\Rightarrow \quad \sqrt{2} = \sqrt{h^2 + k^2 - 2h - 4k + 5}$$

$$\Rightarrow \quad h^2 + k^2 - 2h - 4k + 5 = 2$$

$$\Rightarrow \quad h^2 + k^2 - 2h - 4k + 3 = 0$$

Thus, the required locus is

$$x^2 + y^2 - 2x - 4y + 3 = 0$$

which is a equation of circle with

$$\text{radius} = \sqrt{1 + 4 - 3} = \sqrt{2}$$

**Aliter**

$(2x - 3y + 4) + k(x - 2y + 3) = 0$ is family of lines passing through (1, 2). By congruency of triangles, we can prove that mirror image $(h, k)$ and the point (2, 3) will be equidistant from (1, 2).

$\therefore$ Locus of $(h, k)$ is $PR = PQ$

$\Rightarrow (h-1)^2 + (k-2)^2 = (2-1)^2 + (3-2)^2$

or $(x-1)^2 + (y-2)^2 = 2$

$\therefore$ Locus is a circle of radius $= \sqrt{2}$

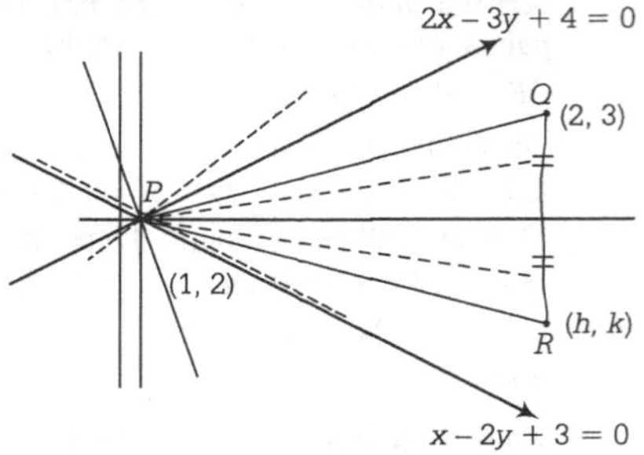


**20.** *(c)* ***Central Idea*** *Number of common tangents depend on the position of the circle with respect to each other.*

*(i) If circles touch externally $\Rightarrow$ $C_1C_2 = r_1 + r_2$, 3 common tangents*

*(ii) If circles touch internally $\Rightarrow$ $C_1C_2 = r_2 - r_1$, 1 common tangent*

*(iii) If circles do not touch each other, 4 common tangents*

Given equations of circles are

$$x^2 + y^2 - 4x - 6y - 12 = 0 \quad \dots(i)$$

$$x^2 + y^2 + 6x + 18y + 26 = 0 \quad \dots(ii)$$

Centre of circle (i) is $C_1$ (2, 3) and radius

$$= \sqrt{4 + 9 + 12} = 5(r_1) \quad \text{(say)}$$

Centre of circle (ii) is $C_2$ (–3, – 9) and radius

$$= \sqrt{9 + 81 - 26} = 8(r_2) \quad \text{(say)}$$

Now, $C_1C_2 = \sqrt{(2+3)^2 + (3+9)^2}$

$\Rightarrow \quad C_1C_2 = \sqrt{5^2 + 12^2}$

$\Rightarrow \quad C_1C_2 = \sqrt{25 + 144} = 13$

$\therefore \quad r_1 + r_2 = 5 + 8 = 13$

Also, $\quad C_1C_2 = r_1 + r_2$

Thus, both circles touch each other externally. Hence, there are three common tangents.

**21.** *(d)* Given equation of ellipse is

$$\frac{x^2}{9} + \frac{y^2}{5} = 1 \quad \dots(i)$$

$\therefore \quad a^2 = 9, b^2 = 5$

$\Rightarrow \quad a = 3, b = \sqrt{5}$

Now, $\quad e = \sqrt{1 - \frac{b^2}{a^2}} = \sqrt{1 - \frac{5}{9}} = \frac{2}{3}$

foci $= (\pm ae, 0) = (\pm\, 2, 0)$

and $\quad \frac{b^2}{a} = \frac{5}{3}$

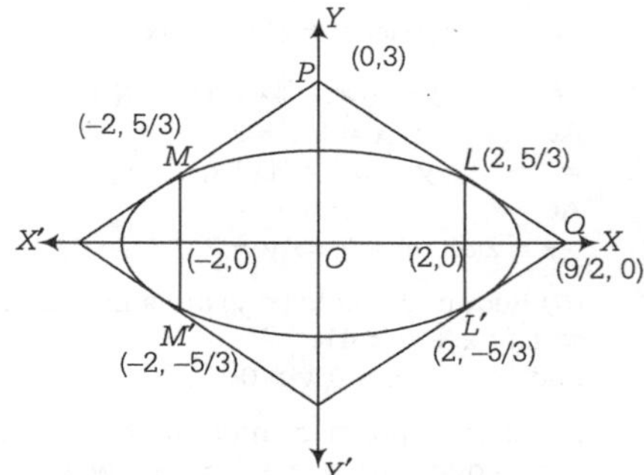


$\therefore$ Extremities of one of latusrectum are

$$\left(2, \frac{5}{3}\right)$$

and $\quad \left(2, \frac{-5}{3}\right).$

$\therefore$ Equation of tangent at $\left(2, \frac{5}{3}\right)$ is,

$$\frac{x(2)}{9} + \frac{y(5/3)}{5} = 1$$

or $\quad 2x + 3y = 9 \quad \dots(ii)$

Eq.(ii) intersects $X$ and $Y$-axes at $\left(\frac{9}{2}, 0\right)$ and $(0, 3)$, respectively.

$\therefore$ Area of quadrilateral

$= 4 \times$ Area of $\Delta POQ$

$= 4 \times \left(\frac{1}{2} \times \frac{9}{2} \times 3\right)$

$= 27$ sq units

**22.** *(d)* ***Central Idea*** *Any point on the parabola $x^2 = 8y$ is $(4t, 2t^2)$. Point P divides the line segment joining of $O(0, 0)$ and $Q(4t, 2t^2)$ in the ratio 1 : 3. Apply the section formula for internal division.*

Equation of parabola is

$$x^2 = 8y \quad \dots(i)$$

Let any point $Q$ on the parabola (i) is $(4t, 2t^2)$.

Let $P(h, k)$ be the point which divides the line segment joining (0,0) and $(4t, 2t^2)$ in the ratio 1 : 3.

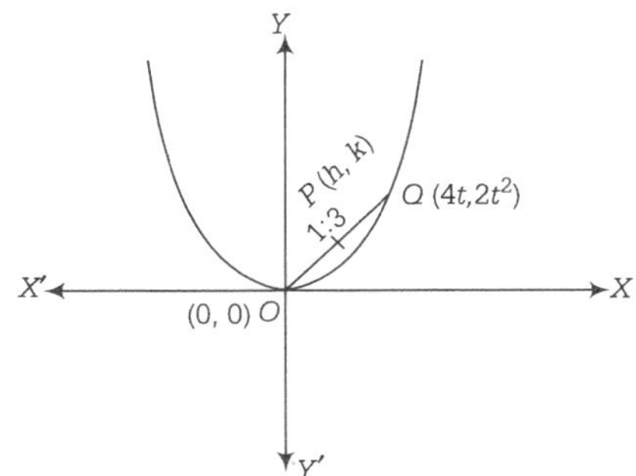


$\therefore \quad h = \dfrac{1 \times 4t + 3 \times 0}{4}$

$\Rightarrow \quad h = t$

and $\quad k = \dfrac{1 \times 2t^2 + 3 \times 0}{4}$

$\Rightarrow \quad k = \dfrac{t^2}{2}$

$\Rightarrow \quad k = \dfrac{1}{2} h^2 \qquad (\because t = h)$

$\Rightarrow 2k = h^2 \Rightarrow 2y = x^2$,

which is required locus.

**23.** *(d)* Given equation of line is

$$\frac{x-2}{3} = \frac{y+1}{4} = \frac{z-2}{12} = \lambda \quad \text{(say)} \ldots\text{(i)}$$

and equation of plane is

$$x - y + z = 16 \quad \ldots\text{(ii)}$$

Any point on the line (i) is

$$(3\lambda + 2, 4\lambda - 1, 12\lambda + 2)$$

Let this point be point of intersection of the line and plane.

$\therefore \ (3\lambda + 2) - (4\lambda - 1) + (12\lambda + 2) = 16$

$\Rightarrow \quad 11\lambda + 5 = 16$

$\Rightarrow \quad 11\lambda = 11$

$\Rightarrow \quad \lambda = 1$

$\therefore$ Point of intersection is (5, 3, 14).

Now, distance between the points (1, 0, 2) and (5, 3, 14)

$= \sqrt{(5-1)^2 + (3-0)^2 + (14-2)^2}$

$= \sqrt{16 + 9 + 144}$

$= \sqrt{169} = 13$

**24.** *(c)* Let equation of plane containing the lines $2x - 5y + z = 3$ and $x + y + 4z = 5$ be

$(2x - 5y + z - 3) + \lambda(x + y + 4z - 5) = 0$

$\Rightarrow (2 + \lambda)x + (\lambda - 5)y + (4\lambda + 1)z - 3 - 5\lambda = 0 \ldots\text{(i)}$

This plane is parallel to the plane $x + 3y + 6z = 1$.

$$\therefore \quad \frac{2+\lambda}{1} = \frac{\lambda - 5}{3} = \frac{4\lambda + 1}{6}$$

On taking first two equalities, we get

$6 + 3\lambda = \lambda - 5$

$\Rightarrow \quad 2\lambda = -11$

$\Rightarrow \quad \lambda = -\dfrac{11}{2}$

On taking last two equalities, we get

$6\lambda - 30 = 3 + 12\lambda$

$\Rightarrow \quad -6\lambda = 33$

$\Rightarrow \quad \lambda = -\dfrac{11}{2}$

So, the equation of required plane is

$$\left(2 - \frac{11}{2}\right)x + \left(\frac{-11}{2} - 5\right)y + \left(-\frac{44}{2} + 1\right)z - 3 + 5 \times \frac{11}{2} = 0$$

$\Rightarrow \quad -\dfrac{7}{2}x - \dfrac{21}{2}y - \dfrac{42}{2}z + \dfrac{49}{2} = 0$

$\Rightarrow \quad x + 3y + 6z - 7 = 0$

**25.** *(a)* Given,

$$(\mathbf{a} \times \mathbf{b}) \times \mathbf{c} = \frac{1}{3}|\mathbf{b}||\mathbf{c}|\mathbf{a}$$

$\Rightarrow \quad -\mathbf{c} \times (\mathbf{a} \times \mathbf{b}) = \dfrac{1}{3}|\mathbf{b}||\mathbf{c}|\mathbf{a}$

$\Rightarrow -(\mathbf{c} \cdot \mathbf{b}) \cdot \mathbf{a} + (\mathbf{c} \cdot \mathbf{a})\mathbf{b} = \dfrac{1}{3}|\mathbf{b}||\mathbf{c}|\mathbf{a}$

$$\left[\frac{1}{3}|\mathbf{b}||\mathbf{c}| + (\mathbf{c} \cdot \mathbf{b})\right]\mathbf{a} = (\mathbf{c} \cdot \mathbf{a})\mathbf{b}$$

Since, **a** and **b** are not collinear.

$$\mathbf{c} \cdot \mathbf{b} + \frac{1}{3}|\mathbf{b}||\mathbf{c}| = 0 \text{ and } \mathbf{c} \cdot \mathbf{a} = 0$$

$\Rightarrow |\mathbf{c}||\mathbf{b}|\cos\theta + \dfrac{1}{3}|\mathbf{b}||\mathbf{c}| = 0$

$\Rightarrow \quad |\mathbf{b}||\mathbf{c}|\left(\cos\theta + \dfrac{1}{3}\right) = 0$

$\Rightarrow \quad \cos\theta + \dfrac{1}{3} = 0 \quad (\because |\mathbf{b}| \neq 0, |\mathbf{c}| \neq 0)$

$\Rightarrow \quad \cos\theta = -\dfrac{1}{3}$

$\Rightarrow \quad \sin\theta = \dfrac{\sqrt{8}}{3} = \dfrac{2\sqrt{2}}{3}$

**26.** *(a)* There seems to be ambiguity in this question. It should be mentioned that boxes are different and one particular box has 3 balls.

Then, number of ways $= \dfrac{^{12}C_3 \times 2^9}{3^{12}}$

$= \dfrac{55}{3}\left(\dfrac{2}{3}\right)^{11}$

According to the question,

$$\dfrac{^3C_1 \times {^{12}C_3} 2^9 - {^3C_2} {^{12}C_3} {^9C_3} + \dfrac{12! \times 3!}{3!3!6!3!}}{3^{12}}$$

**27.** *(d)* Given, $\dfrac{x_1 + x_2 + x_3 + \ldots + x_{16}}{16} = 16$

$\Rightarrow \quad \sum_{i=1}^{16} x_i = 16 \times 16$

Sum of new observations

$= \sum_{i=1}^{18} y_i = (16 \times 16 - 16) + (3 + 4 + 5) = 252$

Number of observations = 18

$\therefore \quad$ New mean $= \dfrac{\sum_{i=1}^{18} y_i}{18} = \dfrac{252}{18} = 14$

**28.** *(a)* According to the given information, the figure should be as follows.
Let the height of tower = $h$

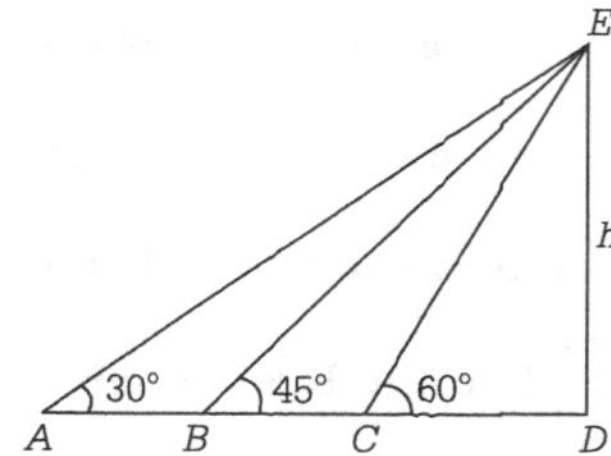


In $\Delta EDA$,

$$\tan 30° = \dfrac{ED}{AD}$$

$$\dfrac{1}{\sqrt{3}} = \dfrac{ED}{AD} = \dfrac{h}{AD}$$

$\Rightarrow \quad AD = h\sqrt{3}$

In $\Delta EDB$,

$\tan 45° = \dfrac{h}{BD} \quad \Rightarrow \quad BD = h$

In $\Delta EDC$,

$\tan 60° = \dfrac{h}{CD} \Rightarrow CD = \dfrac{h}{\sqrt{3}}$

Now, $\quad \dfrac{AB}{BC} = \dfrac{AD - BD}{BD - CD}$

$\Rightarrow \quad \dfrac{AB}{BC} = \dfrac{h\sqrt{3} - h}{h - \dfrac{h}{\sqrt{3}}}$

$\Rightarrow \quad \dfrac{AB}{BC} = \dfrac{h(\sqrt{3} - 1)}{\dfrac{h(\sqrt{3} - 1)}{\sqrt{3}}}$

$\Rightarrow \quad \dfrac{AB}{BC} = \dfrac{\sqrt{3} - 1}{(\sqrt{3} - 1)} \times \sqrt{3}$

$\Rightarrow \quad \dfrac{AB}{BC} = \dfrac{\sqrt{3}}{1}$

$\therefore \quad AB : BC = \sqrt{3} : 1$

**29.** *(a)* Given,

$$\tan^{-1} y = \tan^{-1} x + \tan^{-1}\left(\dfrac{2x}{1 - x^2}\right),$$

where $|x| < \dfrac{1}{\sqrt{3}}$

$\Rightarrow \quad \tan^{-1} y = \tan^{-1}\left\{\dfrac{x + \dfrac{2x}{1 - x^2}}{1 - x\left(\dfrac{2x}{1 - x^2}\right)}\right\}$

$\left[\because \tan^{-1} x + \tan^{-1} y = \tan^{-1}\left(\dfrac{x + y}{1 - xy}\right),\right.$
$\left. x > 0, y > 0, xy < 1\right]$

$$= \tan^{-1}\left(\dfrac{x - x^3 + 2x}{1 - x^2 - 2x^2}\right)$$

$$\tan^{-1} y = \tan^{-1}\left(\dfrac{3x - x^3}{1 - 3x^2}\right)$$

$\Rightarrow \quad y = \dfrac{3x - x^3}{1 - 3x^2}$

**Aliter**

$$|x| < \dfrac{1}{\sqrt{3}} \Rightarrow -\dfrac{1}{\sqrt{3}} < x < \dfrac{1}{\sqrt{3}}$$

Let $\quad x = \tan\theta$

$\Rightarrow \quad -\dfrac{\pi}{6} < \theta < \dfrac{\pi}{6}$

$\therefore \quad \tan^{-1} y = \theta + \tan^{-1}(\tan 2\theta)$

$= \theta + 2\theta = 3\theta$

$\Rightarrow \quad y = \tan 3\theta$

$\Rightarrow \quad y = \dfrac{3\tan\theta - \tan^3\theta}{1 - 3\tan^2\theta}$

$\Rightarrow \quad y = \dfrac{3x - x^3}{1 - 3x^2}$

**30.** *(d)* $\sim(\sim s \vee (\sim r \wedge s))$

$\equiv s \wedge (\sim(\sim r \wedge s))$

$\equiv s \wedge (r \vee \sim s)$

$\equiv (s \wedge r) \vee (s \wedge \sim s)$

$\equiv (s \wedge r) \vee \text{F} \quad (\because s \wedge \sim s$ is false)

$\equiv s \wedge r$

# Solved Paper 2015
# JEE Advanced

## Paper 1

## Physics

### Section 1 (Maximum Marks : 32)

- This section contains EIGHT questions.
- The answer to each question is a SINGLE DIGIT INTEGER ranging from 0 to 9, both inclusive.
- For each question, darken the bubble corresponding to the correct integer in ORS.
- Marking scheme :

  + 4 If the bubble corresponding to the answer is darkened

  0 in all other cases

**1.** A Young's double slit interference arrangement with slits $S_1$ and $S_2$ is immersed in water (refractive index $= 4/3$) as shown in the figure. The positions of maxima on the surface of water are given by $x^2 = p^2m^2\lambda^2 - d^2$, where $\lambda$ is the wavelength of light in air (refractive index $= 1$), $2d$ is the separation between the slits and $m$ is an integer. The value of $p$ is

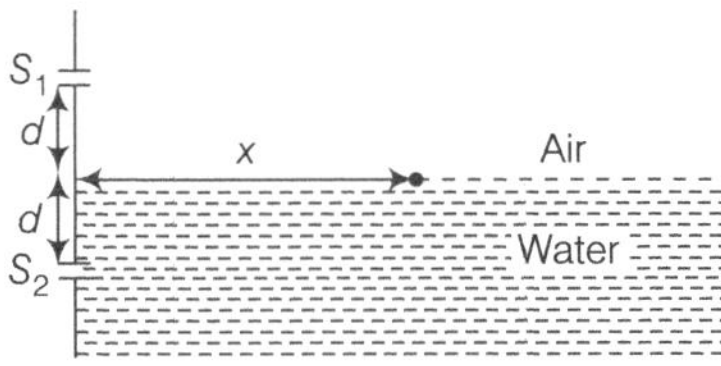


**2.** Consider a concave mirror and a convex lens (refractive index = 1.5) of focal length 10 cm each, separated by a distance of 50 cm in air (refractive index = 1) as shown in the figure. An object is placed at a distance of 15 cm from the mirror. Its erect image formed by this combination has magnification $M_1$. When the set-up is kept in a medium of refractive index $\frac{7}{6}$, the magnification becomes $M_2$. The magnitude $\left|\frac{M_2}{M_1}\right|$ is

**3.** An infinitely long uniform line charge distribution of charge per unit length $\lambda$ lies parallel to the $y$-axis in the $y$-$z$ plane at $z = \frac{\sqrt{3}}{2}a$ (see figure). If the magnitude of the flux of the electric field through the rectangular surface $ABCD$ lying in the $x$-$y$ plane with its centre at the origin is $\frac{\lambda L}{n\varepsilon_0}$ ($\varepsilon_0$ = permittivity of free space), then the value of $n$ is

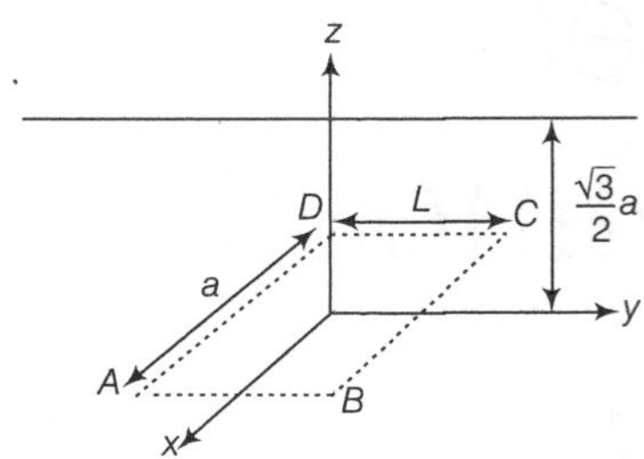


**4.** Consider a hydrogen atom with its electro in the $n^{\text{th}}$ orbital. An electromagnetic radiation of wavelength 90 nm is used to ionize the atom. If the kinetic energy of the ejected electron is 10.4 eV, then the value of $n$ is ($hc = 1242\,\text{eV nm}$)

**5.** A bullet is fired vertically upwards with velocity $v$ from the surface of a spherical planet. When it reaches its maximum height, its acceleration due to the planet's gravity is $\frac{1}{4}$th of its value at the surface of the planet. If the escape velocity from the planet is $v_{\text{sec}} = v\sqrt{N}$, then the value of $N$ is (ignore energy loss due to atmosphere)

**6.** Two identical uniform discs roll without slipping on two different surfaces $AB$ and $CD$ (see figure) starting at $A$ and $C$ with linear speeds $v_1$ and $v_2$, respectively, and always remain in contact with the surfaces. If they reach $B$ and $D$ with the same linear speed and $v_1 = 3$ m/s, then $v_2$ in m/s is($g = 10\,\text{m/s}^2$)

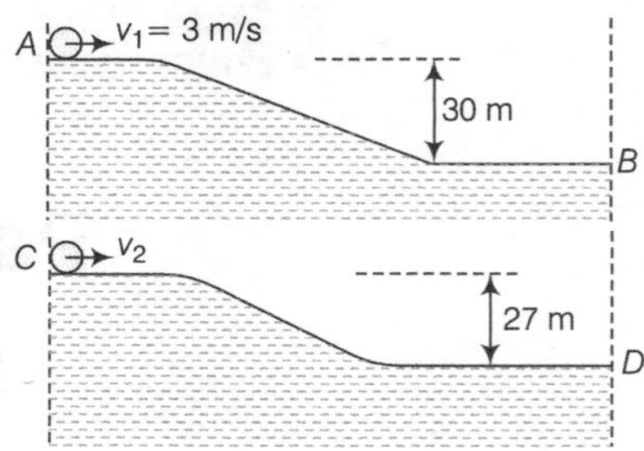


**7.** Two spherical stars $A$ and $B$ emit blackbody radiation. The radius of $A$ is 400 times that of $B$ and $A$ emits $10^4$ times the power emitted from $B$. The ratio $\left(\frac{\lambda_A}{\lambda_B}\right)$ of their wavelengths $\lambda_A$ and $\lambda_B$ at which the peaks occur in their respective radiation curves is

**8.** A nuclear power plant supplying electrical power to a village uses a radioactive material of half life $T$ years as the fuel.

The amount of fuel at the beginning is such that the total power requirement of the village is 12.5% of the electrical power available from the plant at that time. If the plant is able to meet the total power needs of the village for a maximum period of $nT$ years, then the value of $n$ is

## Section 2 (Maximum Marks : 40)

- This section contains TEN questions.
- Each question has FOUR options (a), (b), (c) and (d), ONE OR MORE THAN ONE of these four option(s) is(are) correct.
- For each question, darken the bubble(s) corresponding to all the correct option(s) in the ORS.
- Marking scheme :
  + 4 If only the bubble(s) corresponding to all the correct option(s) is(are) darkened.
  0 If none of the bubbles is darkened.
  −2 In all other cases.

**9.** A ring of mass $M$ and radius $R$ is rotating with angular speed $\omega$ about a fixed vertical axis passing through its centre $O$ with two point masses each of mass $\frac{M}{8}$ at rest at $O$. These masses can move radially outwards along two massless rods fixed on the ring as shown in the figure. At some instant, the angular speed of the system is $\frac{8}{9}\omega$ and one of the masses is at a distance of $\frac{3}{5}R$ from $O$. At this instant, the distance of the other mass from $O$ is

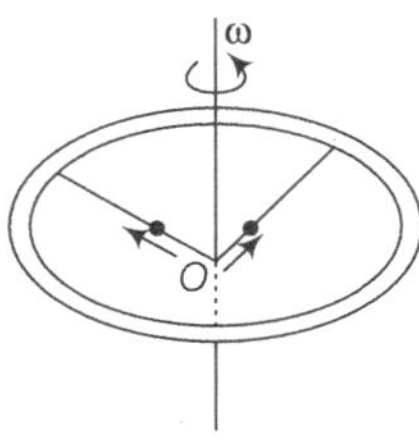


(a) $\frac{2}{3}R$ (b) $\frac{1}{3}R$
(c) $\frac{3}{5}R$ (d) $\frac{4}{5}R$

**10.** The figures below depict two situations in which two infinitely long static line charges of constant positive line charge density $\lambda$ are kept parallel to each other.

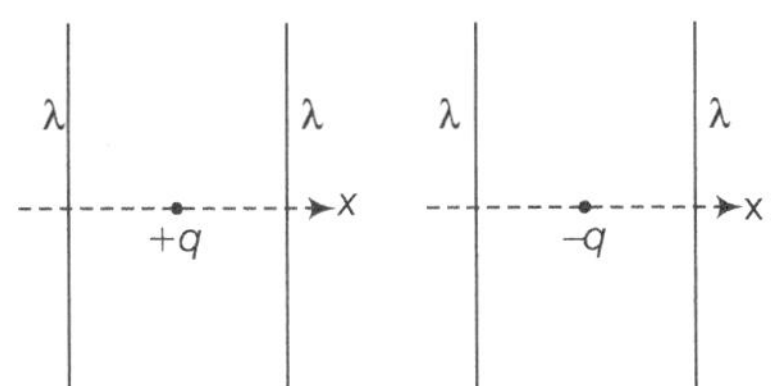


In their resulting electric field, point charges $q$ and $-q$ are kept in equilibrium between them. The point charges are confined to move in the $x$ direction only. If they are given a small displacement about their equilibrium positions, then the correct statements is/are

(a) both charges execute simple harmonic motion
(b) both charges will continue moving in the direction of their displacement
(c) charge $+q$ executes simple harmonic motion while charge $-q$ continues moving in the direction of its displacement
(d) charge $-q$ executes simple harmonic motion while charge $+q$ continues moving in the direction of its displacement

**11.** Two identical glass rods $S_1$ and $S_2$ (refractive index $=1.5$) have one convex end of radius of curvature 10 cm. They are placed with the curved surfaces at a distance $d$ as shown in the figure, with their axes (shown by the dashed line) aligned. When a point source of light $P$ is placed inside rod $S_1$ on its axis at a distance of 50 cm from the curved face, the light rays emanating from it are found to be parallel to the axis inside $S_2$. The distance $d$ is

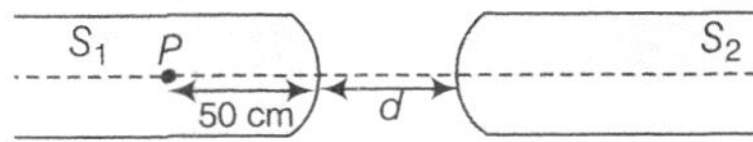


(a) 60 cm (b) 70 cm (c) 80 cm (d) 90 cm

**12.** A conductor (shown in the figure) carrying constant current $I$ is kept in the $x$-$y$ plane in a uniform magnetic field **B**. If $F$ is the magnitude of the total magnetic force acting on the conductor, then the correct statements is/are

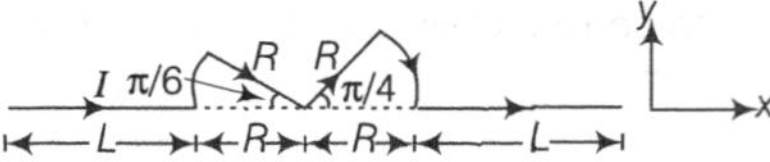


(a) if **B** is along $\hat{\mathbf{z}}$, $F \propto (L + R)$
(b) if **B** is along $\hat{\mathbf{x}}$, $F = 0$
(c) if **B** is along $\hat{\mathbf{y}}$, $F \propto (L + R)$
(d) if **B** is along $\hat{\mathbf{z}}$, $F = 0$

**13.** A container of fixed volume has a mixture of one mole of hydrogen and one mole of helium in equilibrium at temperature $T$. Assuming the gases are ideal, the correct statements is/are

(a) The average energy per mole of the gas mixture is $2RT$
(b) The ratio of speed of sound in the gas mixture to that in helium gas is $\sqrt{\frac{6}{5}}$
(c) The ratio of the rms speed of helium atoms to that of hydrogen molecules is $\frac{1}{2}$
(d) The ratio of the rms speed of helium atoms to that of hydrogen molecules is $\frac{1}{\sqrt{2}}$

**14.** In an aluminium (Al) bar of square cross section, a square hole is drilled and is filled with iron (Fe) as shown in the figure. The electrical resistivities of Al and Fe are $2.7 \times 10^{-8}\ \Omega$ m and $1.0 \times 10^{-7}\ \Omega$ m, respectively. The electrical resistance between the two faces $P$ and $Q$ of the composite bar is

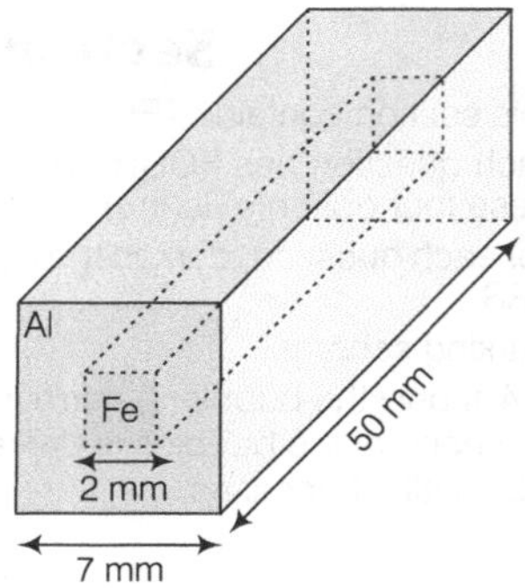


(a) $\frac{2475}{64}\ \mu\Omega$
(b) $\frac{1875}{64}\ \mu\Omega$
(c) $\frac{1875}{49}\ \mu\Omega$
(d) $\frac{2475}{132}\ \mu\Omega$

**15.** For photo-electric effect with incident photon wavelength $\lambda$, the stopping potential is $V_0$. Identify the correct variation(s) of $V_0$ with $\lambda$ and $\frac{1}{\lambda}$.

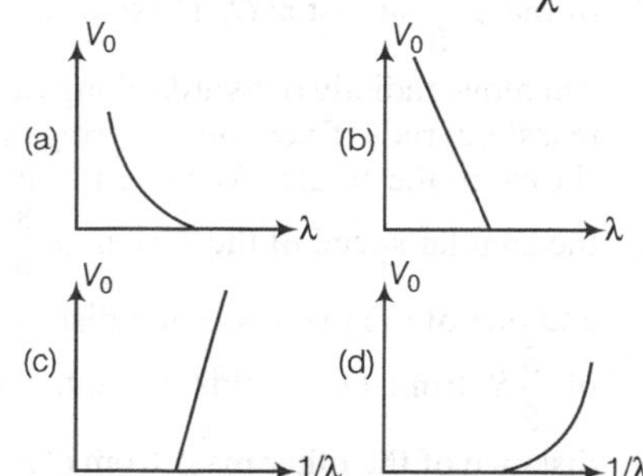


**16.** Consider a Vernier callipers in which each 1 cm on the main scale is divided into 8 equal divisions and a screw gauge with 100 divisions on its circular scale. In the Vernier callipers, 5 divisions of the Vernier scale coincide with 4 divisions on the main scale and in the screw gauge, one complete rotation of the circular scale moves it by two divisions on the linear scale. Then

(a) If the pitch of the screw gauge is twice the least count of the Vernier callipers, the least count of the screw gauge is 0.01 mm

(b) If the pitch of the screw gauge is twice the least count of the Vernier callipers, the least count of the screw gauge is 0.05 mm

(c) If the least count of the linear scale of the screw gauge is twice the least count of the Vernier callipers, the least count of the screw gauge is 0.01 mm

(d) If the least count of the linear scale of the screw gauge is twice the least count of the Vernier callipers, the least count of the screw gauge is 0.005 mm

**17.** Planck's constant $h$, speed of light $c$ and gravitational constant $G$ are used to form a unit of length $L$ and a unit of mass $M$. Then, the correct options is/are

(a) $M \propto \sqrt{c}$ (b) $M \propto \sqrt{G}$

(c) $L \propto \sqrt{h}$ (d) $L \propto \sqrt{G}$

**18.** Two independent harmonic oscillators of equal masses are oscillating about the origin with angular frequencies $\omega_1$ and $\omega_2$ and have total energies $E_1$ and $E_2$, respectively. The variations of their momenta $p$ with positions $x$ are shown in the figures. If $\frac{a}{b} = n^2$ and $\frac{a}{R} = n$, then the correct equations is/are

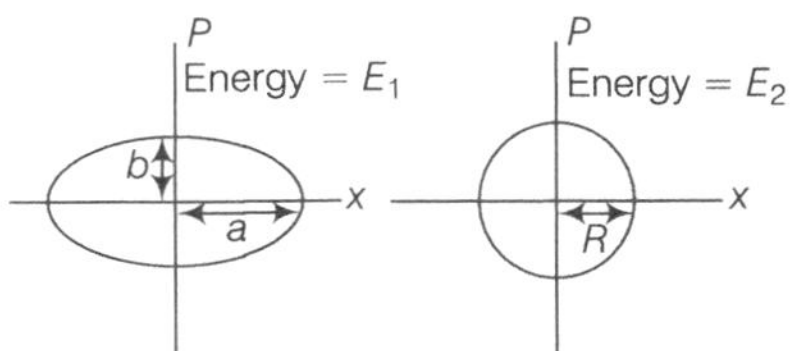


(a) $E_1\omega_1 = E_2\omega_2$ (b) $\frac{\omega_2}{\omega_1} = n^2$

(c) $\omega_1\omega_2 = n^2$ (d) $\frac{E_1}{\omega_1} = \frac{E_2}{\omega_2}$

## Section 3 (Maximum Marks : 16)

- This section contains TWO questions.
- Each question contains two columns, Column I and Column II.
- Column I has four entries (A), (B), (C) and (D).
- Column II has five entries (P), (Q), (R), (S) and (T).
- Match the entries in Column I with the entries in Column II.
- One or more entries in Column I may match with one or more entries in Column II.
- The ORS contains a 4 × 5 matrix whose layout will be similar to the one shown below :

| | | | | | |
|---|---|---|---|---|---|
| (A) | (P) | (Q) | (R) | (S) | (T) |
| (B) | (P) | (Q) | (R) | (S) | (T) |
| (C) | (P) | (Q) | (R) | (S) | (T) |
| (D) | (P) | (Q) | (R) | (S) | (T) |

- For each entry in Column I, darken the bubbles of all the matching entries. For example, if entry (A) in Column I matches with entries (Q), (R) and (T), then darken these three bubbles in the ORS. Similarly, for entries (B), (C) and (D).
- Marking schemes:

**For each entry in Column I**

+ 2 If only the bubble(s) corresponding to all the correct match(es) is (are) darkened.

0 If none of the bubbles is darkened.

– 1 In all other cases.

**19.** Match the nuclear processes given in Column I with the appropriate option(s) in Column II.

| | Column I | | Column II |
|---|---|---|---|
| A. | Nuclear fusion | P. | absorption of thermal neutrons by ${}^{235}_{92}U$ |
| B. | Fission in a nuclear reactor | Q. | ${}^{60}_{27}Co$ nucleus |
| C. | β-decay | R. | Energy production in stars via hydrogen conversion to helium |
| D. | γ-ray emission | S. | Heavy water |
| | | T. | Neutrino emission |

**20.** A particle of unit mass is moving along the $x$-axis under the influence of a force and its total energy is conserved. Four possible forms of the potential energy of the particle are given in Column I ($a$ and $U_0$ are constants). Match the potential energies in Column I to the corresponding statements in Column II.

| | Column I | | Column II |
|---|---|---|---|
| A. | $U_1(x) = \frac{U_0}{2}\left[1 - \left(\frac{x}{a}\right)^2\right]^2$ | P. | The force acting on the particle is zero at $x = a$ |
| B. | $U_2(x) = \frac{U_0}{2}\left(\frac{x}{a}\right)^2$ | Q. | The force acting on the particle is zero at $x = 0$ |
| C. | $U_3(x) = \frac{U_0}{2}\left(\frac{x}{a}\right)^2 \exp\left[-\left(\frac{x}{a}\right)^2\right]$ | R. | The force acting on the particle is zero at $x = -a$ |
| D. | $U_4(x) = \frac{U_0}{2}\left[\frac{x}{a} - \frac{1}{3}\left(\frac{x}{a}\right)^3\right]$ | S. | The particle experiences an attractive force towards $x = 0$ in the region $\lvert x\rvert < a$ |
| | | T. | The particle with total energy $\frac{U_0}{4}$ can oscillate about the point $x = -a$ |

# Chemistry

## Section 1 (Maximum Marks : 32)

- This section contains EIGHT questions.
- The answer to each question is a SINGLE DIGIT INTEGER ranging from 0 to 9, both inclusive.
- For each question, darken the bubble corresponding to the correct integer in ORS.
- Marking scheme :
  + 4 If the bubble corresponding to the answer is darkened.
  0 in all other cases.

**1.** The total number of stereoisomers that can exist for $M$ is

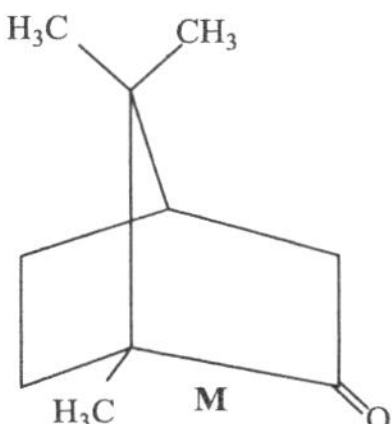


**2.** The number of resonance structures for $N$ is

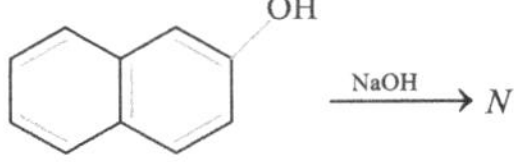


**3.** The total number of lone pair of electrons in $N_2O_3$ is

**4.** For the octahedral complexes of $Fe^{3+}$ in $SCN^-$ (thiocyanato-S) and in $CN^-$ ligand environments, the difference between the spin only magnetic moments in Bohr magnetons (when approximated to the nearest integer) is [atomic number of Fe = 26]

**5.** Among the triatomic molecules/ions $BeCl_2$, $N_3^-$, $N_2O$, $NO_2^+$, $O_3$, $SCl_2$, $ICl_2^-$, $I_3^-$ and $XeF_2$, the total number of linear molecules(s)/ion(s) where the hybridisation of the central atom does not have contribution from the $d$-orbital(s) is [atomic number of S = 16, Cl = 17, I = 53 and Xe = 54]

**6.** Not considering the electronic spin, the degeneracy of the second excited state ($n = 3$) of H-atom is 9, while the degeneracy of the second excited state of $H^-$ is

**7.** All the energy released from the reaction $X \rightarrow Y, \Delta_r G^\circ = -193\,\text{kJ mol}^{-1}$ is used for oxidising $M^+$ as $M^+ \rightarrow M^{3+} + 2e^-, E^\circ = -0.25\,\text{V}$.

Under standard conditions, the number of moles of $M^+$ oxidised when one mole of $X$ is converted to $Y$ is [$F = 96500\,\text{C mol}^-$]

**8.** If the freezing point of a 0.01 molal aqueous solution of a cobalt (III) chloride-ammonia complex (which behaves as a strong electrolyte) is $-0.0558°C$, the number of chloride(s) in the coordination sphere of the complex is

[$K_f$ of water $= 1.86\,\text{K kg mol}^{-1}$]

## Section 2 (Maximum Marks : 40)

- This section contains TEN questions.
- Each question has FOUR options (a), (b), (c) and (d), ONE OR MORE THAN ONE of these four option(s) is/are correct.
- For each question, darken the bubble(s) corresponding to the correct option(s) in the ORS.
- Marking scheme :
  + 4 If the bubble(s) corresponding to all the correct option(s) is/are darkened.
  0 If none of the bubble is darkened.
  −2 In all other cases.

**9.** Compound(s) that on hydrogenation produce(s) optically inactive compound (s) is/are

(a) $H_3C$, H, Br, $CH_3$

(b) $H_2C$, H, Br, $CH_3$

(c) $H_2C$, H, Br, $CH_3$, $CH_3$

(d) $H_2C$, Br, H, $CH_3$

**10.** The major product of the following reaction is

(a) $CH_3$, O

(b) $CH_3$, O

(c) O, $CH_3$

(d) O, $CH_3$

O, O, $CH_3$ $\xrightarrow[\text{(ii) } H^+, \text{ Heat}]{\text{(i) KOH, } H_2O}$

**11.** In the following reaction, the major product is

$CH_3$, $H_2C$, $CH_2$ $\xrightarrow{\text{1 equivalent HBr}}$

(a) $H_2C$, $CH_3$, $CH_3$, Br

(b) $H_3C$, $CH_3$, Br

(c) $H_2C$, $CH_3$, Br

(d) $H_3C$, $CH_3$, Br

**12.** The structure of D-(+)-glucose is

CHO
H—OH
HO—H
H—OH
H—OH
$CH_2OH$

The structure of L-(−)-glucose is

(a)
CHO
HO—H
H—OH
HO—H
HO—H
$CH_2OH$

(b)
CHO
H—OH
HO—H
H—OH
HO—H
$CH_2OH$

(c)

CHO
HO — H
HO — H
H — OH
HO — H
$CH_2OH$

(d)

CHO
HO — H
HO — H
HO — H
H — OH
$CH_2OH$

**13.** The major product of the reaction is

$H_3C$–CH($CH_3$)–$CH_2$–CH($NH_2$)–$CO_2H$ $\xrightarrow[0°C]{NaNO_2,\ aq.\ HCl}$

(a) $H_3C$–CH($CH_3$)–$CH_2$–CH(OH)–$NH_2$ (OH hashed)

(b) $H_3C$–CH($CH_3$)–$CH_2$–CH(OH)–$CO_2H$ (OH hashed)

(c) $H_3C$–CH($CH_3$)–$CH_2$–CH(OH)–$CO_2H$ (OH wedged)

(d) $H_3C$–CH($CH_3$)–$CH_2$–CH(OH)–$NH_2$ (OH wedged)

**14.** The correct statement(s) about $Cr^{2+}$ and $Mn^{3+}$ is/are [atomic number of Cr = 24 and Mn = 25]

(a) $Cr^{2+}$ is a reducing agent
(b) $Mn^{3+}$ is an oxidising agent
(c) both $Cr^{2+}$ and $Mn^{3+}$ exhibit $d^4$ electronic configuration
(d) when $Cr^{2+}$ is used as a reducing agent, the chromium ion attains $d^5$ electronic configuration

**15.** Copper is purified by electrolytic refining of blister copper. The correct statement(s) about this process is/are

(a) impure Cu strip is used as cathode
(b) acidified aqueous $CuSO_4$ is used as electrolyte
(c) pure Cu deposits at cathode
(d) impurities settle as anode-mud

**16.** $Fe^{3+}$ is reduced to $Fe^{2+}$ by using

(a) $H_2O_2$ in presence of NaOH
(b) $Na_2O_2$ in water
(c) $H_2O_2$ in presence of $H_2SO_4$
(d) $Na_2O_2$ in presence of $H_2SO_4$

**17.** The % yield of ammonia as a function of time in the reaction,

$N_2(g) + 3H_2(g) \rightleftharpoons 2NH_3(g)$;

$\Delta H < 0$

at $(p, T_1)$ is given below.

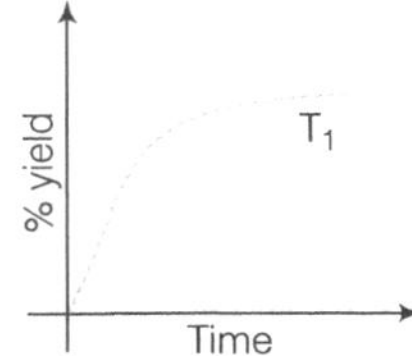


If this reaction is conducted at $(p, T_1)$, with $T_2 > T_1$ the % yield by of ammonia as a function of time is represented by

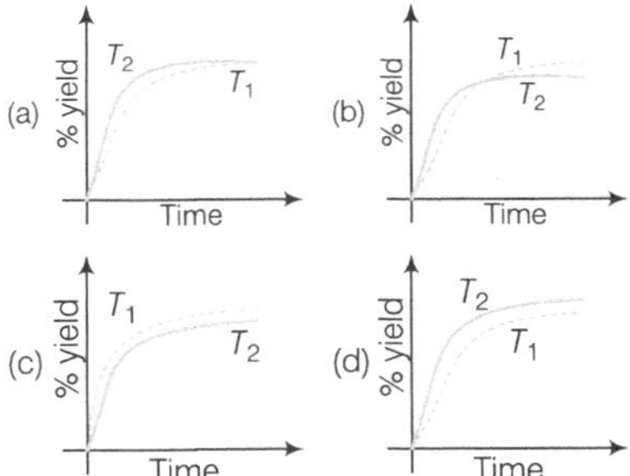


**18.** If the unit cell of a mineral has cubic close packed (ccp) array of oxygen atoms with $m$ fraction of octahedral holes occupied by aluminium ions and $n$ fraction of tetrahedral holes occupied by magnesium ions, $m$ and $n$ respectively, are

(a) $\frac{1}{2}, \frac{1}{8}$
(b) $1, \frac{1}{4}$
(c) $\frac{1}{2}, \frac{1}{2}$
(d) $\frac{1}{4}, \frac{1}{8}$

## Section 3 (Maximum Marks : 16)

- This section contains TWO questions.
- Each question contains two columns, Column I and Column II.
- Column I has four entries (A), (B), (C) and (D).
- Column II has five entries (P), (Q), (R), (S) and (T).
- Match the entries in Column I with the entries in Column II.
- One or more entries in Column I may match with one or more entries in Column II.
- The ORS contains a $4 \times 5$ matrix whose layout will be similar to the one shown below :
-

| | | | | | |
|---|---|---|---|---|---|
| (A) | (P) | (Q) | (R) | (S) | (T) |
| (B) | (P) | (Q) | (R) | (S) | (T) |
| (C) | (P) | (Q) | (R) | (S) | (T) |
| (D) | (P) | (Q) | (R) | (S) | (T) |

- For each entry in Column I, darken the bubbles of all the matching entries. For example, if entry (A) in Column I matches with entries (Q), (R) and (T), then darken these three bubbles in the ORS. Similarly, for entries (B), (C) and (D).
- Marking schemes:

**For each entry in Column I**

+ 2 If only the bubble(s) corresponding to all the correct match(es) is (are) darkened.

0 If none of the bubbles is darkened.

– 1 In all other cases.

**19.** Match the anionic species given in Column I that are present in the ore (s) given in Column II.

| Column I | Column II |
|---|---|
| A. Carbonate | P. Siderite |
| B. Sulphide | Q. Malachite |
| C. Hydroxide | R. Bauxite |
| D. Oxide | S. Calamine |
| | T. Argentite |

**20.** Match the thermodynamic processes given under Column I with the expressions given under Column II.

| Column I | Column II |
|---|---|
| A. Freezing of water at 273 K and 1 atm | P. $q = 0$ |
| B. Expansion of 1 mole of an ideal gas into a vacuum under isolated conditions | Q. $W = 0$ |
| C. Mixing of equal volumes of two ideal gases at constant temperature and pressure in an isolated container | R. $\Delta S_{sys} < 0$ |
| D. Reversible heating of $H_2$ (*g*) at 1 atm from 300 K to 600 K, followed by reversible cooling to 300 K at 1 atm | S. $\Delta U = 0$ |
| | T. $\Delta G = 0$ |

# Mathematics

## Section 1 (Maximum Marks : 32)

- This section contains EIGHT questions.
- The answer to each question is a SINGLE DIGIT INTEGER ranging from 0 to 9 both inclusive.
- For each question, darken the bubble corresponding to the correct integer in the ORS.
- Marking scheme :
  + 4 If the bubble corresponding to the answer is darkened.
  0 In all other cases.

**1.** Let the curve $C$ be the mirror image of the parabola $y^2 = 4x$ with respect to the line $x + y + 4 = 0$. If $A$ and $B$ are the points of intersection of $C$ with the line $y = -5$, then the distance between $A$ and $B$ is

**2.** The minimum number of times a fair coin needs to be tossed, so that the probability of getting atleast two heads is atleast 0.96, is

**3.** Let $n$ be the number of ways in which 5 boys and 5 girls can stand in a queue in such a way that all the girls stand consecutively in the queue. Let $m$ be the number of ways in which 5 boys and 5 girls can stand in a queue in such a way that exactly four girls stand consecutively in the queue. Then, the value of $\frac{m}{n}$ is

**4.** If the normals of the parabola $y^2 = 4x$ drawn at the end points of its latusrectum are tangents to the circle $(x-3)^2 + (y+2)^2 = r^2$, then the value of $r^2$ is

**5.** Let $f : R \to R$ be a function defined by $f(x) = \begin{cases} [x], & x \leq 2 \\ 0, & x > 2 \end{cases}$, where $[x]$ denotes the greatest integer less than or equal to $x$. If $I = \int_{-1}^{2} \frac{xf(x^2)}{2 + f(x+1)}\,dx$, then the value of $(4I - 1)$ is

**6.** A cylindrical container is to be made from certain solid material with the following constraints : It has a fixed inner volume of $V$ mm$^3$, has a 2 mm thick solid wall and is open at the top. The bottom of the container is a solid circular disc of thickness 2 mm and is of radius equal to the outer radius of the container.

If the volume of the material used to make the container is minimum, when the inner radius of the container is 10 mm, then the value of $\frac{V}{250\pi}$ is

**7.** Let $F(x) = \int_{x}^{x^2 + \frac{\pi}{6}} 2\cos^2 t\, dt$ for all $x \in R$ and $f : \left[0, \frac{1}{2}\right] \to [0, \infty)$ be a continuous function. For $a \in \left[0, \frac{1}{2}\right]$, if $F'(a) + 2$ is the area of the region bounded by $x = 0$, $y = 0$, $y = f(x)$ and $x = a$, then $f(0)$ is

**8.** The number of distinct solutions of the equation

$$\frac{5}{4}\cos^2 2x + \cos^4 x + \sin^4 x + \cos^6 x + \sin^6 x = 2$$

in the interval $[0, 2\pi]$ is

## Section 2 (Maximum Marks : 40)

- This section contains TEN questions.
- Each question has FOUR options (a), (b), (c) and (d). ONE OR MORE THAN ONE of these four options are correct.
- For each question, darken the bubble(s) corresponding to all the correct options in the ORS.
- Marking scheme :
  +4 If only the bubble(s) corresponding to all the correct options are darkened.
  0 If none of the bubbles is darkened.
  –2 In all the other cases.

**9.** Let $y(x)$ be a solution of the differential equation
$(1+e^x)y'+ye^x=1.$
If $y(0)=2$, then which of the following statement(s) is/are true?
(a) $y(-4)=0$
(b) $y(-2)=0$
(c) $y(x)$ has a critical point in the interval $(-1, 0)$
(d) $y(x)$ has no critical point in the interval $(-1, 0)$

**10.** Consider the family of all circles whose centres lie on the straight line $y=x$. If this family of circles is represented by the differential equation $Py''+Qy'+1=0$, where $P, Q$ are the functions of $x$, $y$ and $y'$ (here, $y'=\frac{dy}{dx}, y''=\frac{d^2y}{dx^2}$), then which of the following statement(s) is/are true?
(a) $P=y+x$
(b) $P=y-x$
(c) $P+Q=1-x+y+y'+(y')^2$
(d) $P-Q=x+y-y'-(y')^2$

**11.** Let $g: R \to R$ be a differentiable function with $g(0)=0, g'(0)=0$ and $g'(1)\neq 0$.

Let $f(x)=\begin{cases}\frac{x}{|x|}g(x), & x\neq 0\\ 0, & x=0\end{cases}$

and $h(x)=e^{|x|}$ for all $x\in R$. Let $(foh)(x)$ denotes $f\{h(x)\}$ and $(hof)(x)$ denotes $h\{f(x)\}$. Then, which of the following is/are true?
(a) $f$ is differentiable at $x=0$
(b) $h$ is differentiable at $x=0$
(c) $foh$ is differentiable at $x=0$
(d) $hof$ is differentiable at $x=0$

**12.** Let $f(x)=\sin\left[\frac{\pi}{6}\sin\left(\frac{\pi}{2}\sin x\right)\right]$ for all $x\in R$ and $g(x)=(\pi/2)\sin x$ for all $x\in R$. Let $(fog)(x)$ denotes $f\{g(x)\}$ and $(gof)(x)$ denotes $g\{f(x)\}$. Then, which of the following is/are true?
(a) Range of $f$ is $\left[-\frac{1}{2},\frac{1}{2}\right]$
(b) Range of $fog$ is $\left[-\frac{1}{2},\frac{1}{2}\right]$
(c) $\lim_{x\to 0}\frac{f(x)}{g(x)}=\frac{\pi}{6}$
(d) There is an $x\in R$ such that $(gof)(x)=1$

**13.** Let $\Delta PQR$ be a triangle. Let $\mathbf{a}=\mathbf{QR}$, $\mathbf{b}=\mathbf{RP}$ and $\mathbf{c}=\mathbf{PQ}$. If $|\mathbf{a}|=12$, $|\mathbf{b}|=4\sqrt{3}$ and $\mathbf{b}\cdot\mathbf{c}=24$, then which of the following is/are true?
(a) $\frac{|\mathbf{c}|^2}{2}-|\mathbf{a}|=12$
(b) $\frac{|\mathbf{c}|^2}{2}+|\mathbf{a}|=30$
(c) $|\mathbf{a}\times\mathbf{b}+\mathbf{c}\times\mathbf{a}|=48\sqrt{3}$
(d) $\mathbf{a}\cdot\mathbf{b}=-72$

**14.** Let $X$ and $Y$ be two arbitrary, $3\times 3$, non-zero, skew-symmetric matrices and $Z$ be an arbitrary, $3\times 3$, non-zero, symmetric matrix. Then, which of the following matrices is/are skew-symmetric?
(a) $Y^3Z^4-Z^4Y^3$ (b) $X^{44}+Y^{44}$
(c) $X^4Z^3-Z^3X^4$ (d) $X^{23}+Y^{23}$

**15.** Which of the following values of $\alpha$ satisfy the equation

$$\begin{vmatrix} (1+\alpha)^2 & (1+2\alpha)^2 & (1+3\alpha)^2 \\ (2+\alpha)^2 & (2+2\alpha)^2 & (2+3\alpha)^2 \\ (3+\alpha)^2 & (3+2\alpha)^2 & (3+3\alpha)^2 \end{vmatrix} = -648\alpha ?$$

(a) $-4$ (b) 9 (c) $-9$ (d) 4

**16.** In $R^3$, consider the planes $P_1 : y = 0$ and $P_2 : x + z = 1$. Let $P_3$ be a plane, different from $P_1$ and $P_2$, which passes through the intersection of $P_1$ and $P_2$. If the distance of the point (0, 1, 0) from $P_3$ is 1 and the distance of a point $(\alpha, \beta, \gamma)$ from $P_3$ is 2, then which of the following relation(s) is/are true?

(a) $2\alpha + \beta + 2\gamma + 2 = 0$
(b) $2\alpha - \beta + 2\gamma + 4 = 0$
(c) $2\alpha + \beta - 2\gamma - 10 = 0$
(d) $2\alpha - \beta + 2\gamma - 8 = 0$

**17.** In $R^3$, let $L$ be a straight line passing through the origin. Suppose that all the points on $L$ are at a constant distance from the two planes $P_1 : x + 2y - z + 1 = 0$ and $P_2 : 2x - y + z - 1 = 0$. Let $M$ be the locus of the foot of the perpendiculars drawn from the points on $L$ to the plane $P_1$. Which of the following point(s) lie(s) on $M$?

(a) $\left(0, -\frac{5}{6}, -\frac{2}{3}\right)$ (b) $\left(-\frac{1}{6}, -\frac{1}{3}, \frac{1}{6}\right)$
(c) $\left(-\frac{5}{6}, 0, \frac{1}{6}\right)$ (d) $\left(-\frac{1}{3}, 0, \frac{2}{3}\right)$

**18.** Let $P$ and $Q$ be distinct points on the parabola $y^2 = 2x$ such that a circle with $PQ$ as diameter passes through the vertex $O$ of the parabola. If $P$ lies in the first quadrant and the area of $\Delta OPQ$ is $3\sqrt{2}$, then which of the following is/are the coordinates of $P$?

(a) $(4, 2\sqrt{2})$
(b) $(9, 3\sqrt{2})$
(c) $\left(\frac{1}{4}, \frac{1}{\sqrt{2}}\right)$
(d) $(1, \sqrt{2})$

## Section 3 (Maximum Marks : 16)

- This section contains TWO questions.
- Each question contains two columns, Column I and Column II.
- Column I has four entries (A), (B), (C) and (D).
- Column II has five entries (P), (Q), (R), (S) and (T).
- Match the entries in Column I with the entries in Column II.
- One or more entries in Column I may match with one or more entries in Column II.
- The ORS contains a 4 × 5 matrix whose layout will be similar to the one shown below:

| (A) | (P) | (Q) | (R) | (S) | (T) |
|---|---|---|---|---|---|
| (B) | (P) | (Q) | (R) | (S) | (T) |
| (C) | (P) | (Q) | (R) | (S) | (T) |
| (D) | (P) | (Q) | (R) | (S) | (T) |

- For each entry in Column I, darken the bubbles of all the matching entries. For example, if entry (A) in Column I matches with entries (Q), (R) and (T), then darken these three bubbles in the ORS. Similarly, for entries (B), (C) and (D).
- Marking scheme :

**For each entry in Column I**

+2 If only the bubble(s) corresponding to all the correct matches are darkened.
0 If none of the bubbles is darkened.
−1 In all other cases.

**19.**

| | Column I | | Column II |
|---|---|---|---|
| A. | In $R^2$, if the magnitude of the projection vector of the vector $\alpha\hat{i} + \beta\hat{j}$ on $\sqrt{3}\,\hat{i} + \hat{j}$ is $\sqrt{3}$ and if $\alpha = 2 + \sqrt{3}\,\beta$, then possible value(s) of $\lvert\alpha\rvert$ is/are | P. | 1 |
| B. | Let $a$ and $b$ be real numbers such that the function $f(x) = \begin{cases} -3ax^2 - 2, & x < 1 \\ bx + a^2, & x \ge 1 \end{cases}$ is differentiable for all $x \in R$. Then, possible value(s) of $a$ is/are | Q | 2 |
| C. | Let $\omega\ (\neq 1)$ be a complex cube root of unity. If $(3 - 3\omega + 2\omega^2)^{4n+3} + (2 + 3\omega - 3\omega^2)^{4n+3} + (-3 + 2\omega + 3\omega^2)^{4n+3} = 0$, then the possible value(s) of $n$ is/are | R. | 3 |
| D. | Let the harmonic mean of two positive real numbers $a$ and $b$ be 4. If $q$ is a positive real number such that $a, 5, q, b$ is in arithmetic progression, then the value(s) of $\lvert q - 2a\rvert$ is/are | S. | 4 |
| | | T. | 5 |

**20.**

| | Column I | | Column II |
|---|---|---|---|
| A. | In $\Delta XYZ$, let $a$, $b$ and $c$ be the lengths of the sides opposite to the angles $X$, $Y$ and $Z$, respectively. If $2(a^2 - b^2) = c^2$ and $\lambda = \dfrac{\sin(X - Y)}{\sin Z}$, then possible value(s) of $n$ for which $\cos(n\pi\lambda) = 0$, is/are | P. | 1 |
| B. | In $\Delta XYZ$, let $a$, $b$ and $c$ be the lengths of the sides opposite to the angles $X$, $Y$ and $Z$, respectively. If $1 + \cos 2X - 2\cos 2Y = 2\sin X \sin Y$, then possible value(s) of $\dfrac{a}{b}$ is/are | Q. | 2 |
| C. | In $R^2$, let $\sqrt{3}\hat{i} + \hat{j}$, $\hat{i} + \sqrt{3}\hat{j}$ and $\beta\hat{i} + (1 - \beta)\hat{j}$ be the position vectors of $X$, $Y$ and $Z$ with respect to the origin $O$, respectively. If the distance of $Z$ from the bisector of the acute angle of **OX** with **OY** is $\dfrac{3}{\sqrt{2}}$, then possible value(s) of $\lvert\beta\rvert$ is/are | R. | 3 |
| D. | Suppose that $F(\alpha)$ denotes the area of the region bounded by $x = 0$, $x = 2$, $y^2 = 4x$ and $y = \lvert\alpha x - 1\rvert + \lvert\alpha x - 2\rvert + \alpha x$, where $\alpha \in \{0, 1\}$. Then, the value(s) of $F(\alpha) + \dfrac{8}{3}\sqrt{2}$, when $\alpha = 0$ and $\alpha = 1$, is/are | S. | 5 |
| | | T. | 6 |

# Paper 2

## Physics

### Section 1 (Maximum Marks : 32)

- This section contains EIGHT questions.
- The answer to each question is a SINGLE DIGIT INTEGER ranging from 0 to 9 both inclusive.
- For each question, darken the bubble corresponding to the correct integer in ORS.
- Marking scheme :
  +4 If the bubble corresponding to the answer is darkened.
  0 in all other cases.

**1.** Four harmonic waves of equal frequencies and equal intensities $I_0$ have phase angles 0, $\frac{\pi}{3}$, $\frac{2\pi}{3}$ and $\pi$. When they are superposed, the intensity of the resulting wave is $nI_0$. The value of $n$ is

**2.** For a radioactive material, its activity $A$ and rate of change of its activity $R$ are defined as $A = -\frac{dN}{dt}$ and $R = -\frac{dA}{dt}$, where $N(t)$ is the number of nuclei at time $t$. Two radioactive source $P$(mean life $\tau$) and $Q$ (mean life $2\tau$) have the same activity at $t = 0$. Their rate of change of activities at $t = 2\tau$ are $R_P$ and $R_Q$, respectively. If $\frac{R_P}{R_Q} = \frac{n}{e}$, then the value of $n$ is

**3.** A monochromatic beam of light is incident at 60° on one face of an equilateral prism of refractive index $n$ and emerges from the opposite face making an angle $\theta(n)$ with the normal (see figure). For $n = \sqrt{3}$ the value of $\theta$ is 60° and $\frac{d\theta}{dn} = m$. The value of $m$ is

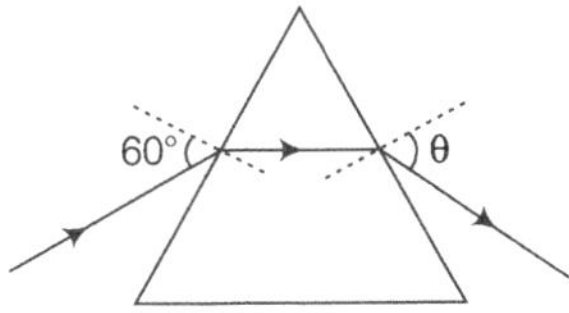


**4.** In the following circuit, the current through the resistor $R(= 2\Omega)$ is $I$ amperes. The value of $I$ is

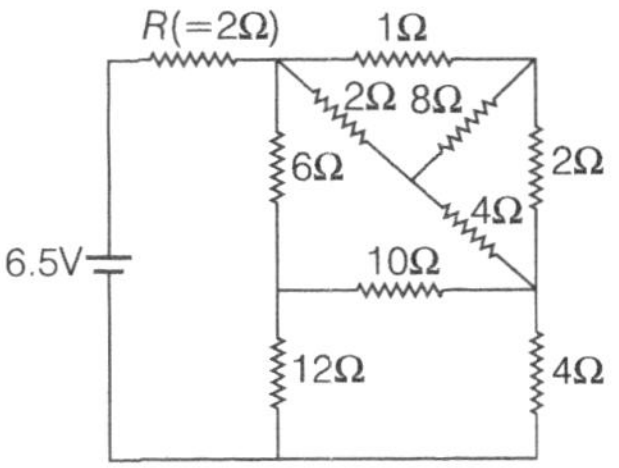


**5.** An electron in an excited state of $Li^{2+}$ ion has angular momentum $\frac{3h}{2\pi}$. The de Broglie wavelength of the electron in this state is $p\pi a_0$ (where $a_0$ is the Bohr radius). The value of $p$ is

**6.** A large spherical mass $M$ is fixed at one position and two identical masses $m$ are kept on a line passing through the centre of $M$ (see figure). The point masses are connected by a rigid massless rod of length $l$ and this assembly is free to move along the line connecting them.

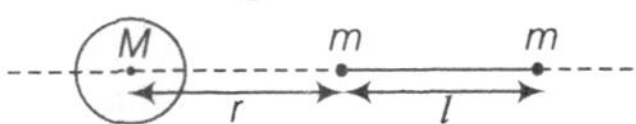


All three masses interact only through their mutual gravitational interaction. When the point mass nearer to $M$ is at a distance $r = 3l$ from $M$ the tension in the rod is zero for $m = k\left(\frac{M}{288}\right)$. The value of $k$ is

**7.** The energy of a system as a function of time $t$ is given as $E(t) = A^2 \exp(-\alpha t)$, where $\alpha = 0.2\,s^{-1}$. The measurement of $A$ has an error of 1.25%. If the error in the measurement of time is 1.50%, the percentage error in the value of $E(t)$ at $t = 5\,s$ is

**8.** The densities of two solids spheres $A$ and $B$ of the same radii $R$ vary with radial distance $r$ as $\rho_A(r) = k\left(\frac{r}{R}\right)$ and $\rho_B(r) = k\left(\frac{r}{R}\right)^5$, respectively, where $k$ is a constant. The moments of inertia of the individual spheres about axes passing through their centres are $I_A$ and $I_B$, respectively.

If $\frac{I_B}{I_A} = \frac{n}{10}$, the value of $n$ is

## Section 2 (Maximum Marks : 32)

- This section contains EIGHT questions.
- Each question has FOUR options (a), (b), (c) and (d), ONE OR MORE THAN ONE of these four option(s) is(are) correct.
- For each question, darken the bubble(s) corresponding to all the correct option(s) in the ORS.
- Marking scheme :

  \+ 4 If only the bubble(s) corresponding to all the correct optin(s) is(are) darkened.

  0 If none of the bubbles is darkened.

  −2 In all other cases.

**9.** Consider a uniform spherical charge distribution of radius $R_1$ centred at the origin $O$. In this distribution, a spherical cavity of radius $R_2$, centred at $P$ with distance $OP = a = R_1 - R_2$ (see figure) is made. If the electric field inside the cavity at position **r** is $\mathbf{E}(\mathbf{r})$, then the correct statements is/are

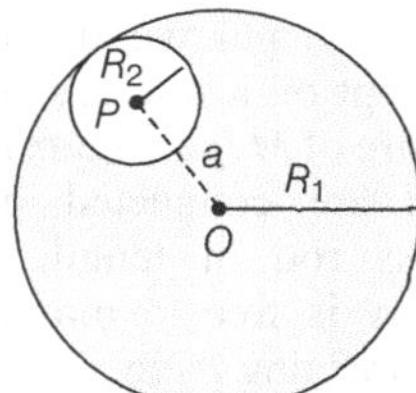


(a) **E** is uniform, its magnitude is independent of $R_2$ but its direction depends on **r**

(b) **E** is uniform, its magnitude depends on $R_2$ and its direction depends on **r**

(c) **E** is uniform, its magnitude is independent of '$a$' but its direction depends on **a**

(d) **E** is uniform and both its magnitude and direction depend on **a**

**10.** In plotting stress *versus* strain curves for two materials $P$ and $Q$, a student by mistake puts strain on the $y$-axis and stress on the $x$-axis as shown in the figure. Then, the correct statements is/are

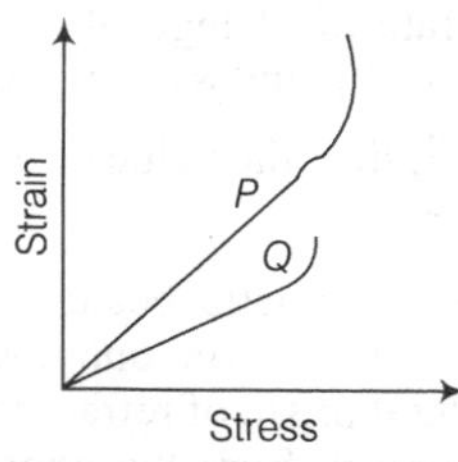


(a) $P$ has more tensile strength than $Q$

(b) $P$ is more ductile than $Q$

(c) $P$ is more brittle than $Q$

(d) The Young's modulus of $P$ is more than that of $Q$

**11.** A spherical body of radius $R$ consists of a fluid of constant density and is in equilibrium under its own gravity. If $P(r)$ is the pressure at $r (r < R)$, then the correct options is/are

(a) $P(r = 0) = 0$

(b) $\dfrac{P\left(r = \frac{3R}{4}\right)}{P\left(r = \frac{2R}{3}\right)} = \dfrac{63}{80}$

(c) $\dfrac{P\left(r = \frac{3R}{5}\right)}{P\left(r = \frac{2R}{5}\right)} = \dfrac{16}{21}$

(d) $\dfrac{P\left(r = \frac{R}{2}\right)}{P\left(r = \frac{R}{3}\right)} = \dfrac{20}{27}$

**12.** A parallel plate capacitor having plates of area $S$ and plate separation $d$, has capacitance $C_1$ in air. When two dielectrics of different relative permittivities ($\varepsilon_1 = 2$ and $\varepsilon_2 = 4$) are introduced between the two plates as shown in the figure, the capacitance becomes $C_2$. The ratio $\dfrac{C_2}{C_1}$ is

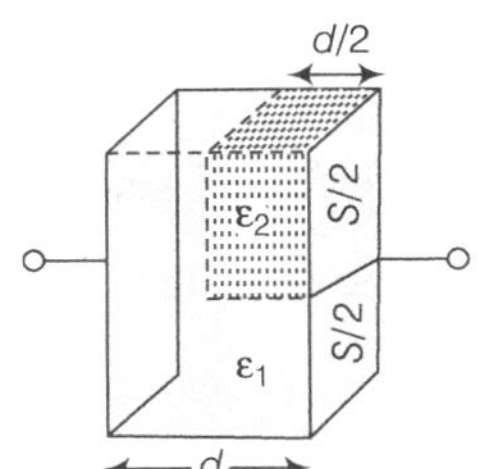


(a) $\frac{6}{5}$ (b) $\frac{5}{3}$

(c) $\frac{7}{5}$ (d) $\frac{7}{3}$

**13.** An ideal monoatomic gas is confined in a horizontal cylinder by a spring loaded piston (as shown in the figure). Initially the gas is at temperature $T_1$, pressure $P_1$ and volume $V_1$ and the spring is in its relaxed state. The gas is then heated very slowly to temperature $T_2$, pressure $P_2$ and volume $V_2$. During this process the piston moves out by a distance $x$.

Ignoring the friction between the piston and the cylinder, the correct statements is/are

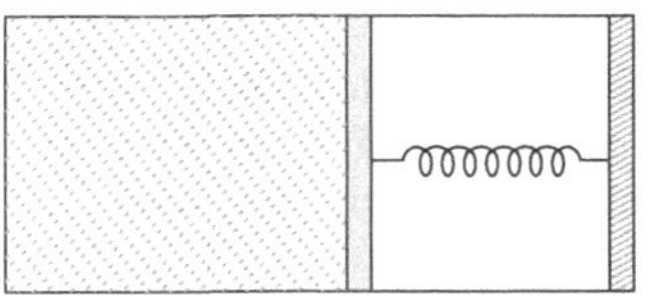

(a) If $V_2 = 2V_1$ and $T_2 = 3T_1$, then the energy stored in the spring is $\frac{1}{4}P_1V_1$

(b) If $V_2 = 2V_1$ and $T_2 = 3T_1$, then the change in internal energy is $3P_1V_1$

(c) If $V_2 = 3V_1$ and $T_2 = 4T_1$, then the work done by the gas is $\frac{7}{3}P_1V_1$

(d) If $V_2 = 3V_1$ and $T_2 = 4T_1$, then the heat supplied to the gas is $\frac{17}{6}P_1V_1$

**14.** A fission reaction is given by ${}^{236}_{92}\text{U} \rightarrow {}^{140}_{54}\text{Xe} + {}^{94}_{38}\text{Sr} + x + y$, where $x$ and $y$ are two particles. Considering ${}^{236}_{92}\text{U}$ to be at rest, the kinetic energies of the products are denoted by $K_{\text{Xe}}$, $K_{\text{Sr}}$, $K_x$(2 MeV) and $K_y$(2 MeV), respectively. Let the binding energies per nucleon of ${}^{236}_{92}\text{U}$, ${}^{140}_{54}\text{Xe}$ and ${}^{94}_{38}\text{Sr}$ be 7.5 MeV, 8.5 MeV and 8.5 MeV, respectively. Considering different conservation laws, the correct options is/are

(a) $x = n$, $y = n$, $K_{\text{Sr}} = 129$ MeV, $K_{\text{Xe}} = 86$ MeV

(b) $x = p$, $y = e^-$, $K_{\text{Sr}} = 129$ MeV, $K_{\text{Xe}} = 86$ MeV

(c) $x = p$, $y = n$, $K_{\text{Sr}} = 129$ MeV, $K_{\text{Xe}} = 86$ MeV

(d) $x = n$, $y = n$, $K_{\text{Sr}} = 86$ MeV, $K_{\text{Xe}} = 129$ MeV

**15.** Two spheres $P$ and $Q$ for equal radii have densities $\rho_1$ and $\rho_2$, respectively. The spheres are connected by a massless string and placed in liquids $L_1$ and $L_2$ of densities $\sigma_1$ and $\sigma_2$ and viscosities $\eta_1$ and $\eta_2$, respectively. They float in equilibrium with the sphere $P$ in $L_1$ and sphere $Q$ in $L_2$ and the string being taut (see figure).

If sphere $P$ alone in $L_2$ has terminal velocity $\mathbf{v}_P$ and $Q$ alone in $L_1$ has terminal velocity $\mathbf{v}_Q$, then

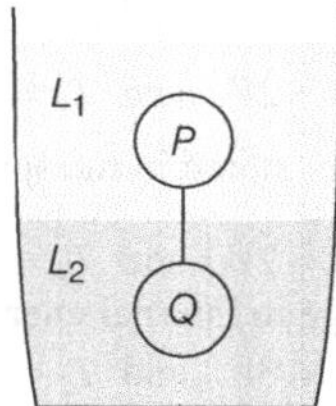


(a) $\frac{|\mathbf{v}_P|}{|\mathbf{v}_Q|} = \frac{\eta_1}{\eta_2}$ (b) $\frac{|\mathbf{v}_P|}{|\mathbf{v}_Q|} = \frac{\eta_2}{\eta_1}$

(c) $\mathbf{v}_P \cdot \mathbf{v}_Q > 0$ (d) $\mathbf{v}_P \cdot \mathbf{v}_Q < 0$

**16.** In terms of potential difference $V$, electric current $I$, permittivity $\varepsilon_0$, permeability $\mu_0$ and speed of light $c$, the dimensionally correct equations is/are

(a) $\mu_0 I^2 = \varepsilon_0 V^2$ (b) $\varepsilon_0 I = \mu_0 V$

(c) $I = \varepsilon_0 c V$ (d) $\mu_0 c I = \varepsilon_0 V$

## Section 3 (Maximum Marks : 16)

- This section contains TWO paragraphs.
- Based on each paragraph, there will be TWO questions.
- Each question has FOUR options (a), (b), (c) and (d). ONE OR MORE THAN ONE of these four option(s) is(are) correct.
- For each question, darken the bubble(s) corresponding to all the correct option(s) in the ORS.
- Marking scheme:
  + 4 If only the bubble(s) corresponding to all the corect option(s) is(are) darkened.
  0 If none of the bubbles is darkened.
  −2 In all other cases.

### Paragraph I

Light guidance in an optical fibre can be understood by considering a structure comprising of thin solid glass cylinder of refractive index $n_1$ surrounded by a medium of lower refractive index $n_2$. The light guidance in the structure takes place due to successive total internal reflections at the interface of the media $n_1$ and $n_2$ as shown in the figure. All rays with the angle of incidence $i$ less than a particular value $i_m$ are confined in the medium of refractive index $n_1$. The numerical aperture (NA) of the structure is defined as $\sin i_m$.

**17.** For two structures namely $S_1$ with $n_1 = \frac{\sqrt{45}}{4}$ and $n_2 = \frac{3}{2}$, and $S_2$ with $n_1 = \frac{8}{5}$ and $n_2 = \frac{7}{5}$ and taking the refractive index of water to be $\frac{4}{3}$ and that to air to be 1, the correct options is/are

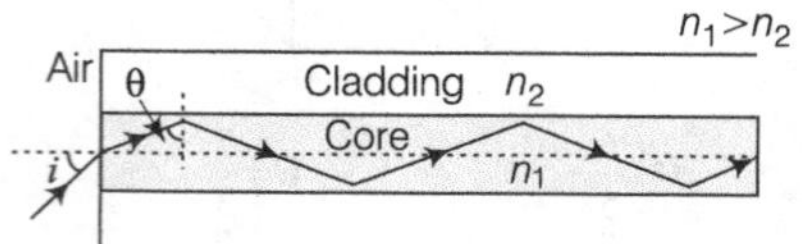


(a) $NA$ of $S_1$ immersed in water is the same as that of $S_2$ immersed in a liquid of refractive index $\frac{16}{3\sqrt{15}}$

(b) $NA$ of $S_1$ immersed in liquid of refractive index $\frac{6}{\sqrt{15}}$ is the same as that of $S_2$ immersed in water

(c) $NA$ of $S_1$ placed in air is the same as that $S_2$ immersed in liquid of refractive index $\frac{4}{\sqrt{15}}$

(d) $NA$ of $S_1$ placed in air is the same as that of $S_2$ placed in water

**18.** If two structures of same cross-sectional area, but different numerical apertures $NA_1$ and $NA_2 (NA_2 < NA_1)$ are joined longitudinally, the numerical aperture of the combined structure is

(a) $\dfrac{NA_1 NA_2}{NA_1 + NA_2}$ (b) $NA_1 + NA_2$

(c) $NA_1$ (d) $NA_2$

## Paragraph II

In a thin rectangular metallic strip a constant current $I$ flows along the positive $x$-direction, as shown in the figure. The length, width and thickness of the strip are $l$, $w$ and $d$, respectively. A uniform magnetic field **B** is applied on the strip along the positive $y$-direction. Due to this, the charge carriers experience a net deflection along the $z$-direction. This results in accumulation of charge carriers on the surface $PQRS$ and appearance of equal and opposite charges on the face opposite to $PQRS$. A potential difference along the $z$-direction is thus developed. Charge accumulation continues until the magnetic force is balanced by the electric force. The current is assumed to be uniformly distributed on the cross section of the strip and carried by electrons.

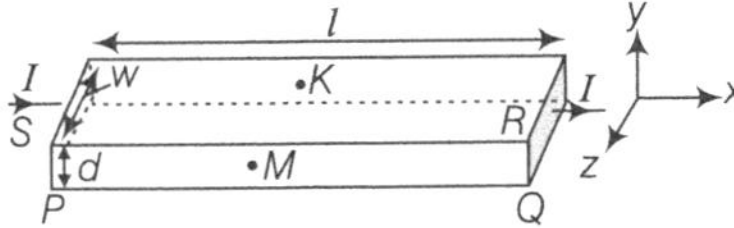


**19.** Consider two different metallic strips (1 and 2) of the same material. Their lengths are the same, widths are $w_1$ and $w_2$ and thicknesses are $d_1$ and $d_2$, respectively. Two points $K$ and $M$ are symmetrically located on the opposite faces parallel to the $x$-$y$ plane (see figure). $V_1$ and $V_2$ are the potential differences between $K$ and $M$ in strips 1 and 2, respectively. Then, for a given current $I$ flowing through them in a given magnetic field strength $B$, the correct statements is/are

(a) If $w_1 = w_2$ and $d_1 = 2d_2$, then $V_2 = 2V_1$

(b) If $w_1 = w_2$ and $d_1 = 2d_2$, then $V_2 = V_1$

(c) If $w_1 = 2w_2$ and $d_1 = d_2$, then $V_2 = 2V_1$

(d) If $w_1 = 2w_2$ and $d_1 = d_2$, then $V_2 = V_1$

**20.** Consider two different metallic strips (1 and 2) of same dimensions (length $l$, width $w$ and thickness $d$) with carrier densities $n_1$ and $n_2$, respectively. Strip 1 is placed in magnetic field $B_1$ and strip 2 is placed in magnetic field $B_2$, both along positive $y$-directions. Then $V_1$ and $V_2$ are the potential differences developed between $K$ and $M$ in strips 1 and 2, respectively. Assuming that the current $I$ is the same for both the strips, the correct options is/are

(a) If $B_1 = B_2$ and $n_1 = 2n_2$, then $V_2 = 2V_1$

(b) If $B_1 = B_2$ and $n_1 = 2n_2$, then $V_2 = V_1$

(c) If $B_1 = 2B_2$ and $n_1 = n_2$, then $V_2 = 0.5V_1$

(d) If $B_1 = 2B_2$ and $n_1 = n_2$, then $V_2 = V_1$

# Chemistry

## Section 1 (Maximum Marks : 32)

- This section contains EIGHT questions.
- The answer to each question is a SINGLE DIGIT INTEGER ranging from 0 to 9 both inclusive.
- For each question, darken the bubble corresponding to the correct integer in ORS.
- Marking scheme :
  +4 If the bubble corresponding to the answer is darkened.
  0 in all other cases.

**1.** In the complex acetylbromidodicarbonylbis (triethylphosphine) iron (II), the number of Fe—C bond (s) is

**2.** Among the complex ions, $[Co(NH_2—CH_2—CH_2—NH_2)_2Cl_2]^+$, $[CrCl_2(C_2O_4)_2]^{3-}$, $[Fe(H_2O)_4(OH)_2]^+$, $[Fe(NH_3)_2(CN)_4]^-$, $[Co(NH_2—CH_2—CH_2—NH_2)_2(NH_3)Cl]^{2+}$ and $[Co(NH_3)_4(H_2O)Cl]^{2+}$ the number of complex ion(s) that show(s) *cis-trans* isomerism is

**3.** Three moles of $B_2H_6$ are completely reacted with methanol. The number of moles of boron containing product formed is

**4.** The molar conductivity of a solution of a weak acid $HX$ (0.01 M) is 10 times smaller than the molar conductivity of a solution of a weak acid $HY$ (0.10 M). If $\lambda^0_{X^-} \approx \lambda^0_{Y^-}$, the difference in their $pK_a$ values, $pK_a(HX) - pK_a(HY)$, is (consider degree of ionisation of both acids to be <<1).

**5.** A closed vessel with rigid walls contains 1 mole of $^{238}_{92}U$ and 1 mole of air at 298 K. Considering complete decay of $^{238}_{92}U$ to $^{206}_{82}Pb$, the ratio of the final pressure to the initial pressure of the system at 298 K is

**6.** In dilute aqueous $H_2SO_4$ the complex diaquadioxalatoferrate (II) is oxidised by $MnO_4^-$. For this reaction, the ratio of the rate of change of $[H^+]$ to the rate of change of $[MnO_4^-]$ is

**7.** The number of hydroxyl group(s) in $Q$ is

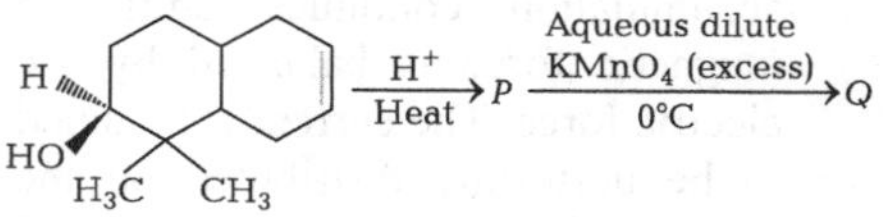

(Structure: bicyclic compound with H, HO, $H_3C$, $CH_3$) $\xrightarrow[\text{Heat}]{H^+} P \xrightarrow[0°C]{\text{Aqueous dilute } KMnO_4 \text{ (excess)}} Q$

**8.** Among the following the number of reaction(s) that produce(s) benzaldehyde is

I. 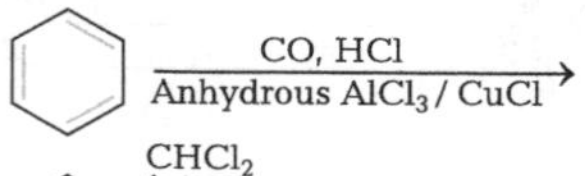 Benzene $\xrightarrow[\text{Anhydrous } AlCl_3/CuCl]{CO,\ HCl}$

II. 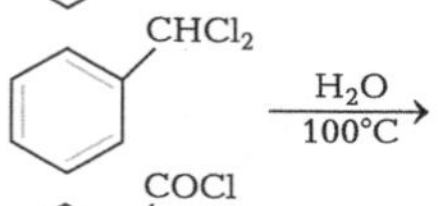 $C_6H_5CHCl_2$ $\xrightarrow[100°C]{H_2O}$

III. 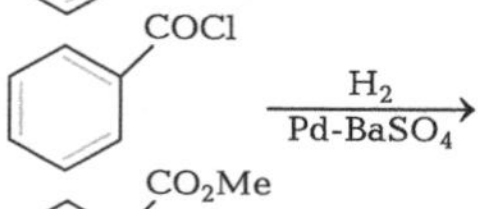 $C_6H_5COCl$ $\xrightarrow[Pd\text{-}BaSO_4]{H_2}$

IV. 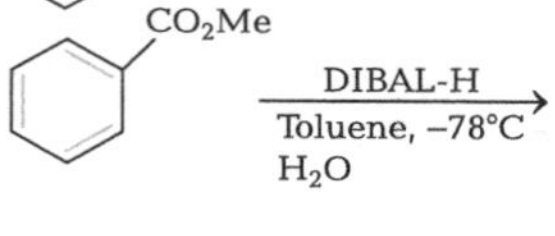 $C_6H_5CO_2Me$ $\xrightarrow[\text{Toluene, } -78°C,\ H_2O]{\text{DIBAL-H}}$

## Section 2 (Maximum Marks : 32)

- This section contains EIGHT questions.
- Each question has FOUR options (a), (b), (c) and (d), ONE OR MORE THAN ONE of these four option(s) is(are) correct.
- For each question, darken the bubble(s) corresponding to all the correct option(s) in the ORS.
- Marking scheme :
  + 4 If only the bubble(s) corresponding to all the correct option(s) is(are) darkened.
  0 If none of the bubbles is darkened.
  – 2 In all other cases.

**9.** Under hydrolysis conditions, the compounds used for preparation of linear polymer and for chain termination, respectively are

(a) $CH_3SiCl_3$ and $Si(CH_3)_4$

(b) $(CH_3)_2SiCl_2$ and $(CH_3)_3SiCl$

(c) $(CH_3)SiCl_2$ and $CH_3SiCl_3$

(d) $SiCl_4$ and $(CH_3)_3SiCl$

**10.** When $O_2$ is adsorbed on a metallic surface, electron transfer occurs from the metal to $O_2$. The true statement(s) regarding this adsorption is (are)

(a) $O_2$ is physisorbed

(b) heat is released

(c) occupancy of $^*\pi_{2p}$ of $O_2$ is increased

(d) bond length of $O_2$ is increased

**11.** One mole of a monoatomic real gas satisfies the equation $p(V - b) = RT$ where, $b$ is a constant. The relationship of interatomic potential $V(r)$ and interatomic distance $r$ for gas is given by

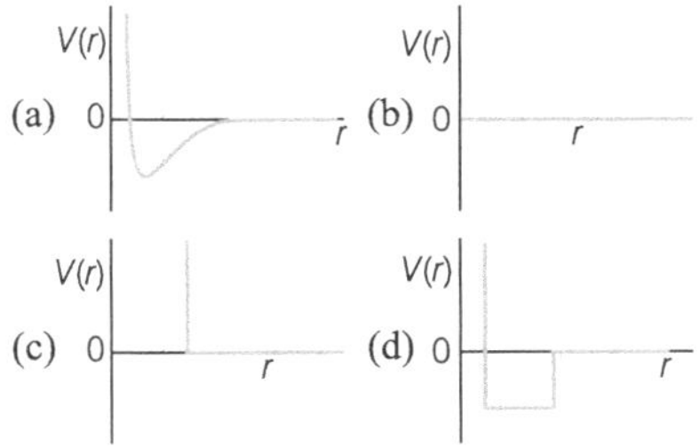


**12.** In the following reactions, the product $S$ is

$H_3C$ – (bicyclic compound) $\xrightarrow[\text{(ii) Zn, } H_2O]{\text{(i) } O_3} R \xrightarrow{\text{(i) } NH_3} S$

(a) $H_3C$, N (b) $H_3C$, N

(c) N, $H_3C$ (d) N, $H_3C$

**13.** The major product $U$ in the following reaction is

$\xrightarrow[\text{High pressure, Heat}]{CH_2{=}CH{-}CH_3, H^+} T \xrightarrow{\text{Radical initiator, } O_2} U$

(a) H, O, O, $CH_3$

(b) $H_3C$, $CH_3$, O, O, H

(c) H, O, O, $CH_2$

(d) $CH_2$, O, O, H

**14.** In the following reactions, the major product $W$ is

Aniline $\xrightarrow[0°C]{NaNO_2, HCl} V \xrightarrow{\text{2-naphthol}, NaOH} W$

(a) 1-(phenylazo)-2-naphthol (N=N at C-1, OH at C-2)

(b) N=N, OH

(c) N=N, OH

(d) HO, N=N

**15.** The correct statement(s) regarding, (i) HClO, (ii) $HClO_2$, (iii) $HClO_3$ and (iv) $HClO_4$ is (are)

(a) the number of Cl = O bonds in (ii) and (iii) together is two

(b) the number of lone pair of electrons on Cl in (ii) and (iii) together is three

(c) the hybridisation of Cl in (iv) is $sp^3$

(d) amongst (i) to (iv), the strongest acid is (i)

**16.** The pair(s) of ions where both the ions are precipitated upon passing $H_2S$ gas in presence of dilute HCl, is (are)

(a) $Ba^{2+}$, $Zn^{2+}$ (b) $Bi^{3+}$, $Fe^{3+}$

(c) $Cu^{2+}$, $Pb^{2+}$ (d) $Hg^{2+}$, $Bi^{3+}$

## Section 3 (Maximum Marks : 16)

- This section contains TWO paragraphs.
- Based on each paragraph, there will be TWO questions.
- Each equation has FOUR options (a), (b), (c) and (d). ONE OR MORE THAN ONE of these four option(s) is(are) correct.
- Four each question, darken the bubble(s) corresponding to all the correct option(s) in the ORS.
- Marking scheme:
  + 4 If only the bubble(s) corresponding to all the correct option(s) is(are) darkened.
  0 If none of the bubbles is darkened.
  −2 In all other cases.

### Paragraph 1

$C_8H_6 \xrightarrow[H_2]{Pd\text{-}BaSO_4} C_8H_8 \xrightarrow[(ii)\ H_2O_2, NaOH, H_2O]{(i)\ B_2H_6} X$

$C_8H_6 \xrightarrow{H_2O,\ HgSO_4, H_2SO_4} C_8H_8O \xrightarrow[(ii)\ H^+,\ Heat]{(i)\ EtMgBr,\ H_2O} Y$

**17.** Compound $X$ is

(a) $C_6H_5COCH_3$ (b) $C_6H_5CH(OH)CH_3$

(c) $C_6H_5CH_2CH_2OH$ (d) $C_6H_5CH_2CHO$

**18.** The major compound $Y$ is

(a) $C_6H_5CH_2CH=CHCH_3$

(b) $C_6H_5CH=CHCH_2CH_3$

(c) $C_6H_5C(=CH_2)CH_2CH_3$

(d) $C_6H_5C(CH_3)=CHCH_3$

**Paragraph 2**

When 100 mL of 1.0 M HCl was mixed with 100 mL of 1.0 M NaOH in an insulated beaker at constant pressure, a temperature increase of 5.7°C was measured for the beaker and its contents (Expt. 1). Because the enthalpy of neutralisation of a strong acid with a strong base is a constant $(-57.0 \text{ kJ mol}^{-1})$, this experiment could be used to measure the calorimeter constant. In a second experiment (Expt. 2), 100 mL of 2.0 M acetic acid $(K_a = 2.0 \times 10^{-5})$ was mixed with 100 mL of 1.0 M NaOH (under identical conditions to Expt. 1) where a temperature rise of 5.6°C was measured.

**19.** Enthalpy of dissociation (in $\text{kJ mol}^{-1}$) of acetic acid obtained from the Expt. 2 is

(a) 1.0
(b) 10.0
(c) 24.5
(d) 51.4

**20.** The pH of the solution after Expt. 2 is

(a) 2.8
(b) 4.7
(c) 5.0
(d) 7.0

# Mathematics

## Section 1 (Maximum Marks : 32)

- This section contains EIGHT questions.
- The answer to each question is a SINGLE DIGIT INTEGER ranging from 0 to 9 both inclusive.
- For each question, darken the bubble corresponding to the correct integer in the ORS.
- Marking scheme

  +4 If the bubble corresponding to the answer is darkened.

  0 In all other cases.

**1.** Suppose that all the terms of an arithmetic progression are natural numbers. If the ratio of the sum of the first seven terms to the sum of the first eleven terms is 6 : 11 and the seventh term lies in between 130 and 140, then the common difference of this AP is

**2.** The coefficient of $x^9$ in the expansion of $(1+x)(1+x^2)(1+x^3) \ldots (1+x^{100})$ is

**3.** Suppose that the foci of the ellipse $\frac{x^2}{9}+\frac{y^2}{5}=1$ are $(f_1, 0)$ and $(f_2, 0)$, where $f_1 > 0$ and $f_2 < 0$. Let $P_1$ and $P_2$ be two parabolas with a common vertex at (0, 0) with foci at $(f_1, 0)$ and $(2f_2, 0)$, respectively. Let $T_1$ be a tangent to $P_1$ which passes through $(2f_2, 0)$ and $T_2$ be a tangent to $P_2$ which passes through $(f_1, 0)$. If $m_1$ is the slope of $T_1$ and $m_2$ is the slope of $T_2$, then the value of $\left(\frac{1}{m_1^2}+m_2^2\right)$ is

**4.** Let $m$ and $n$ be two positive integers greater than 1. If $\lim_{\alpha \to 0}\left(\frac{e^{\cos(\alpha^n)}-e}{\alpha^m}\right) = -\left(\frac{e}{2}\right)$, then the value of $\frac{m}{n}$ is

**5.** If $\alpha = \int_0^1 (e^{9x+3\tan^{-1}x})\left(\frac{12+9x^2}{1+x^2}\right)dx$, where $\tan^{-1}x$ takes only principal values, then the value of $\left(\log_e|1+\alpha| - \frac{3\pi}{4}\right)$ is

**6.** Let $f: R \to R$ be a continuous odd function, which vanishes exactly at one point and $f(1)=\frac{1}{2}$. Suppose that $F(x)=\int_{-1}^{x} f(t)dt$ for all $x \in [-1, 2]$ and $G(x)=\int_{-1}^{x} t|f\{f(t)\}|\,dt$ for all $x \in [-1, 2]$. If $\lim_{x \to 1} \frac{F(x)}{G(x)} = \frac{1}{14}$, then the value of $f\left(\frac{1}{2}\right)$ is

**7.** Suppose that $\mathbf{p}, \mathbf{q}$ and $\mathbf{r}$ are three non-coplanar vectors in $R^3$. Let the components of a vector **s** along $\mathbf{p}, \mathbf{q}$ and **r** be 4, 3 and 5, respectively.

If the components of this vector **s** along $(-\mathbf{p}+\mathbf{q}+\mathbf{r})$, $(\mathbf{p}-\mathbf{q}+\mathbf{r})$ and $(-\mathbf{p}-\mathbf{q}+\mathbf{r})$ are $x, y$ and $z$ respectively, then the value of $2x+y+z$ is

**8.** For any integer $k$, let $\alpha_k = \cos\left(\frac{k\pi}{7}\right) + i\sin\left(\frac{k\pi}{7}\right)$, where $i=\sqrt{-1}$. The value of the expression $\frac{\sum_{k=1}^{12}|\alpha_{k+1}-\alpha_k|}{\sum_{k=1}^{3}|\alpha_{4k-1}-\alpha_{4k-2}|}$ is

## Section 2 (Maximum Marks : 32)

- This section contains EIGHT questions.
- Each question has four options (a), (b), (c) and (d). ONE OR MORE THAN ONE of these four options are correct.
- For each question, darken the bubble(s) corresponding to all the correct options in the ORS.
- Marking scheme :
  + 4 If only the bubble(s) corresponding to all the correct options are darkened.
  0 If none of the bubbles is darkened.
  – 2 In all other cases.

**9.** Let $f, g: [-1, 2] \to R$ be continuous functions which are twice differentiable on the interval $(-1, 2)$. Let the values of $f$ and $g$ at the points $-1$, 0 and 2 be as given in the following table:

| | $x=-1$ | $x=0$ | $x=2$ |
|---|---|---|---|
| $f(x)$ | 3 | 6 | 0 |
| $g(x)$ | 0 | 1 | –1 |

In each of the intervals $(-1, 0)$ and $(0, 2)$, the function $(f-3g)''$ never vanishes. Then, the correct statement(s) is/are

(a) $f'(x)-3g'(x)=0$ has exactly three solutions in $(-1, 0) \cup (0, 2)$

(b) $f'(x)-3g'(x)=0$ has exactly one solution in $(-1, 0)$

(c) $f'(x)-3g'(x)=0$ has exactly one solution in $(0, 2)$

(d) $f'(x)-3g'(x)=0$ has exactly two solutions in $(-1, 0)$ and exactly two solutions in $(0, 2)$

**10.** Let $f(x)=7\tan^8 x+7\tan^6 x-3\tan^4 x$ $x-3\tan^2 x$ for all $x \in \left(-\frac{\pi}{2}, \frac{\pi}{2}\right)$.

Then, the correct expression(s) is/are

(a) $\int_0^{\pi/4} x f(x)\,dx = \frac{1}{12}$

(b) $\int_0^{\pi/4} f(x)\,dx = 0$

(c) $\int_0^{\pi/4} x f(x)\,dx = \frac{1}{6}$

(d) $\int_0^{\pi/4} f(x)\,dx = 1$

**11.** Let $f'(x) = \dfrac{192x^3}{2 + \sin^4 \pi x}$ for all $x \in R$ with $f\left(\dfrac{1}{2}\right) = 0$. If $m \le \int_{1/2}^{1} f(x)\,dx \le M$, then the possible values of $m$ and $M$ are

(a) $m = 13, M = 24$
(b) $m = \dfrac{1}{4}, M = \dfrac{1}{2}$
(c) $m = -11, M = 0$
(d) $m = 1, M = 12$

**12.** Let $S$ be the set of all non-zero real numbers $\alpha$ such that the quadratic equation $\alpha x^2 - x + \alpha = 0$ has two distinct real roots $x_1$ and $x_2$ satisfying the inequality $|x_1 - x_2| < 1$. Which of the following interval(s) is/are a subset of $S$?

(a) $\left(-\dfrac{1}{2}, -\dfrac{1}{\sqrt{5}}\right)$ (b) $\left(-\dfrac{1}{\sqrt{5}}, 0\right)$
(c) $\left(0, \dfrac{1}{\sqrt{5}}\right)$ (d) $\left(\dfrac{1}{\sqrt{5}}, \dfrac{1}{2}\right)$

**13.** If $\alpha = 3 \sin^{-1}\left(\dfrac{6}{11}\right)$ and $\beta = 3\cos^{-1}\left(\dfrac{4}{9}\right)$, where the inverse trigonometric functions take only the principal values, then the correct option(s) is/are

(a) $\cos \beta > 0$
(b) $\sin \beta < 0$
(c) $\cos(\alpha + \beta) > 0$
(d) $\cos \alpha < 0$

**14.** Let $E_1$ and $E_2$ be two ellipses whose centres are at the origin. The major axes of $E_1$ and $E_2$ lie along the $X$-axis and $Y$-axis, respectively. Let $S$ be the circle $x^2 + (y-1)^2 = 2$. The straight line $x + y = 3$ touches the curves $S$, $E_1$ and $E_2$ at $P$, $Q$ and $R$, respectively.

Suppose that $PQ = PR = \dfrac{2\sqrt{2}}{3}$. If $e_1$ and $e_2$ are the eccentricities of $E_1$ and $E_2$ respectively, then the correct expression(s) is/are

(a) $e_1^2 + e_2^2 = \dfrac{43}{40}$
(b) $e_1 e_2 = \dfrac{\sqrt{7}}{2\sqrt{10}}$
(c) $|e_1^2 - e_2^2| = \dfrac{5}{8}$
(d) $e_1 e_2 = \dfrac{\sqrt{3}}{4}$

**15.** Consider the hyperbola $H$: $x^2 - y^2 = 1$ and a circle $S$ with centre $N(x_2, 0)$. Suppose that $H$ and $S$ touch each other at a point $P(x_1, y_1)$ with $x_1 > 1$ and $y_1 > 0$. The common tangent to $H$ and $S$ at $P$ intersects the $X$-axis at point $M$. If $(l, m)$ is the centroid of $\Delta PMN$, then the correct expression(s) is/are

(a) $\dfrac{dl}{dx_1} = 1 - \dfrac{1}{3x_1^2}$ for $x_1 > 1$

(b) $\dfrac{dm}{dx_1} = \dfrac{x_1}{3(\sqrt{x_1^2 - 1})}$ for $x_1 > 1$

(c) $\dfrac{dl}{dx_1} = 1 + \dfrac{1}{3x_1^2}$ for $x_1 > 1$

(d) $\dfrac{dm}{dy_1} = \dfrac{1}{3}$ for $y_1 > 0$

**16.** The option(s) with the values of $a$ and $L$ that satisfy the equation

$$\frac{\int_0^{4\pi} e^t (\sin^6 at + \cos^4 at)dt}{\int_0^{\pi} e^t (\sin^6 at + \cos^4 at)dt} = L,$$

is/are

(a) $a = 2, L = \dfrac{e^{4\pi} - 1}{e^{\pi} - 1}$

(b) $a = 2, L = \dfrac{e^{4\pi} + 1}{e^{\pi} + 1}$

(c) $a = 4, L = \dfrac{e^{4\pi} - 1}{e^{\pi} - 1}$

(d) $a = 4, L = \dfrac{e^{4\pi} + 1}{e^{\pi} + 1}$

## Section 3 (Maximum Marks : 16)

- This section contains TWO paragraphs.
- Based on each paragraph, there will be TWO questions.
- Each question has FOUR options (a), (b), (c) and (d). ONE OR MORE THAN ONE of these four options are correct.
- For each question, darken the bubbles(s) corresponding to all the correct options in the ORS.

Marking scheme :
+4 If only the bubble(s) corresponding to all the correct options are darkened.
0 If none of the bubbles is darkened.
–2 In all other cases.

### Paragraph 1

Let $F: R \to R$ be a thrice differentiable function. Suppose that $F(1) = 0, F(3) = -4$ and $F'(x) < 0$ for all $x \in (1, 3)$. Let $f(x) = xF(x)$ for all $x \in R$.

**17.** The correct statement(s) is/are
(a) $f'(1) < 0$
(b) $f(2) < 0$
(c) $f'(x) \neq 0$ for any $x \in (1, 3)$
(d) $f'(x) = 0$ for some $x \in (1, 3)$

**18.** If $\int_1^3 x^2 F'(x)\, dx = -12$ and $\int_1^3 x^3 F''(x)\, dx = 40$, then the correct expression(s) is/are
(a) $9f'(3) + f'(1) - 32 = 0$
(b) $\int_1^3 f(x)\, dx = 12$
(c) $9f'(3) - f'(1) + 32 = 0$
(d) $\int_1^3 f(x)\, dx = -12$

### Paragraph 2

Let $n_1$ and $n_2$ be the number of red and black balls, respectively in box I. Let $n_3$ and $n_4$ be the number of red and black balls, respectively in box II.

**19.** One of the two boxes, box I and box II was selected at random and a ball was drawn randomly out of this box. The ball was found to be red. If the probability that this red ball was drawn from box II, is $\frac{1}{3}$, then the correct option(s) with the possible values of $n_1, n_2, n_3$ and $n_4$ is/are
(a) $n_1 = 3, n_2 = 3, n_3 = 5, n_4 = 15$
(b) $n_1 = 3, n_2 = 6, n_3 = 10, n_4 = 50$
(c) $n_1 = 8, n_2 = 6, n_3 = 5, n_4 = 20$
(d) $n_1 = 6, n_2 = 12, n_3 = 5, n_4 = 20$

**20.** A ball is drawn at random from box I and transferred to box II. If the probability of drawing a red ball from box I, after this transfer, is $\frac{1}{3}$, then the correct option(s) with the possible values of $n_1$ and $n_2$ is/are
(a) $n_1 = 4$ and $n_2 = 6$
(b) $n_1 = 2$ and $n_2 = 3$
(c) $n_1 = 10$ and $n_2 = 20$
(d) $n_1 = 3$ and $n_2 = 6$

# Answers

## Paper 1

### Physics

**1.** (3) **2.** (7) **3.** (6) **4.** (2) **5.** (2) **6.** (7) **7.** (2)
**8.** (3) **9.** (d) **10.** (c) **11.** (b) **12.** (a,b,c) **13.** (a,b,d) **14.** (b)
**15.** (a,c) **16.** (b,c) **17.** (a,c,d) **18.** (b,d)

**19.** A → R or RT; B → P, S; C → Q, T; D → R

**20.** A → P,Q,R,T; B → Q, S; C → P,Q,R,S; D → P,R,T

### Chemistry

**1.** (2) **2.** (9) **3.** (8) **4.** (4) **5.** (4) **6.** (3) **7.** (4)
**8.** (1) **9.** (b,d) **10.** (a) **11.** (d) **12.** (a) **13.** (c) **14.** (a,b,c)
**15.** (b,c,d) **16.** (a,b) **17.** (b) **18.** (a)

**19.** A → P,Q,S; B → T; C → Q; D → R

**20.** A → R,T; B → P,Q,S; C → P,Q,S; D → P,Q,S,T

### Mathematics

**1.** (4) **2.** (8) **3.** (5) **4.** (2) **5.** (0) **6.** (4) **7.** (3)
**8.** (8) **9.** (a,c) **10.** (b,c) **11.** (a,d) **12.** (a,b,c) **13.** (a,c,d) **14.** (c,d)
**15.** (b,c) **16.** (b,d) **17.** (a,b) **18.** (a,d)

**19.** A → P,Q; B → P,Q; C → P,Q,S,T; D → Q,T

**20.** A → P,R,S; B → P,C; C → P,Q; D → S,T

## Paper 2

### Physics

**1.** (3) **2.** (2) **3.** (2) **4.** (1) **5.** (2) **6.** (7) **7.** (4)
**8.** (6) **9.** (d) **10.** (a,b) **11.** (b,c) **12.** (d) **13.** (a,b,c) **14.** (a)
**15.** (a,d) **16.** (a,c) **17.** (a) **18.** (d) **19.** (a,d) **20.** (a,c)

### Chemistry

**1.** (3) **2.** (6) **3.** (6) **4.** (3) **5.** (9) **6.** (8) **7.** (4)
**8.** (4) **9.** (b) **10.** (b,c,d) **11.** (c) **12.** (a) **13.** (b) **14.** (a)
**15.** (b,c) **16.** (c,d) **17.** (c) **18.** (d) **19.** (a) **20.** (b)

### Mathematics

**1.** (9) **2.** (8) **3.** (4) **4.** (2) **5.** (9) **6.** (7) **7.** (9)
**8.** (4) **9.** (b,c) **10.** (a,b) **11.** (b) **12.** (a,d) **13.** (b,c,d) **14.** (a,b)
**15.** (a,b,d) **16.** (a,c) **17.** (a,b,c) **18.** (a,b,c) **19.** (a,b) **20.** (c,d)

# Explanations

## Paper 1

### Physics

1. 

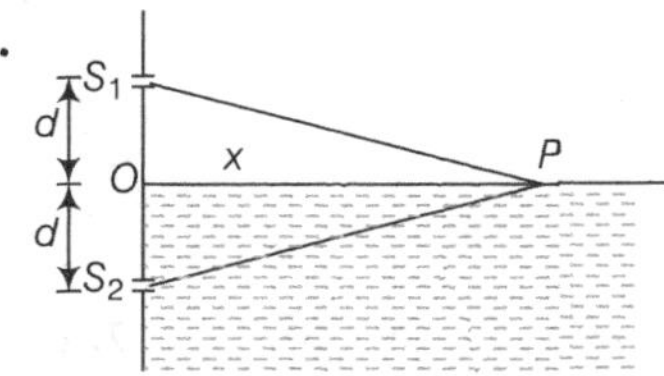


$$\mu(S_2P) - S_1P = m\lambda$$

$$\Rightarrow \quad \mu\sqrt{d^2 + x^2} - \sqrt{d^2 + x^2} = m\lambda$$

$$\Rightarrow \quad (\mu - 1)\sqrt{d^2 + x^2} = m\lambda$$

$$\Rightarrow \quad \left(\frac{4}{3} - 1\right)\sqrt{d^2 + x^2} = m\lambda$$

or
$$\sqrt{d^2 + x^2} = 3m\lambda$$

Squaring this equation we get,

$$x^2 = 9m^2\lambda^2 - d^2$$

$$\Rightarrow \quad p^2 = 9 \quad \text{or} \quad p = 3$$

2. ***Case* I**

Reflection from mirror

$$\frac{1}{f} = \frac{1}{v} + \frac{1}{u} \quad \Rightarrow \quad \frac{1}{-10} = \frac{1}{v} + \frac{1}{-15}$$

$$\Rightarrow \quad v = -30$$

For lens
$$\frac{1}{f} = \frac{1}{v} - \frac{1}{u}$$

$$\frac{1}{10} = \frac{1}{v} - \frac{1}{-20}$$

$$v = 20$$

$$|M_1| = \left|\frac{v_1}{u_1}\right|\left|\frac{v_2}{u_2}\right|$$

$$= \left(\frac{30}{15}\right)\left(\frac{20}{20}\right)$$

$$= 2 \times 1 = 2 \quad \text{(in air)}$$

***Case* II** For mirror, there is no change.

$$v = -30$$

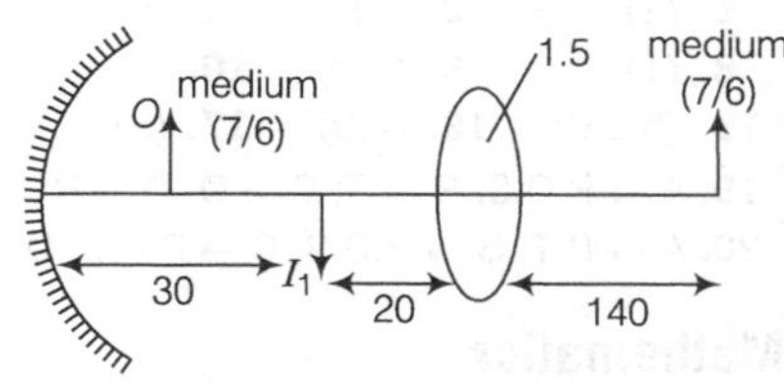


For lens,

$$\frac{1}{f_{\text{air}}} = \left(\frac{3/2}{1} - 1\right)\left(\frac{1}{R_1} - \frac{1}{R_2}\right)$$

$$\frac{1}{f_{\text{medium}}} = \left(\frac{3/2}{7/6} - 1\right)\left(\frac{1}{R_1} - \frac{1}{R_2}\right)$$

with
$$f_{\text{air}} = 10 \text{ cm}$$

We get
$$\frac{1}{f_{\text{medium}}} = \frac{4}{70} \text{ cm}^{-1}$$

$$\frac{1}{v} - \frac{1}{-20} = \frac{4}{70}$$

$$\frac{1}{v} + \frac{1}{20} = \left(\frac{2}{7}\right)\left(\frac{2}{10}\right) = \frac{4}{70}$$

$$\frac{1}{v} = \frac{4}{70} - \frac{1}{20}$$

$$v = 140,$$

$$|M_2| = \left|\frac{v_1}{u_1}\right|\left|\frac{v_2}{u_2}\right|$$

$$= \left(\frac{30}{15}\right)\left(\frac{140}{20}\right),$$

$$= (2)\left(\frac{140}{20}\right) = 14$$

$$\Rightarrow \quad \left|\frac{M_2}{M_1}\right| = \frac{14}{2} = 7$$

**3.** $ANBP$ is cross-section of a cylinder of length $L$. The line charge passes through the centre $O$ and perpendicular to paper.

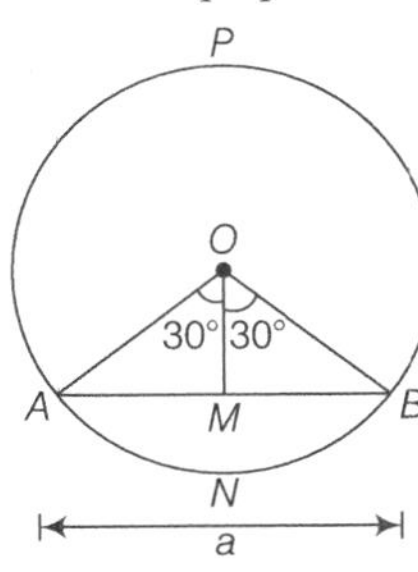


$$AM = \frac{a}{2}, \; MO = \frac{\sqrt{3}a}{2}$$

$$\therefore \quad \angle AOM = \tan^{-1}\left(\frac{AM}{OM}\right)$$

$$= \tan^{-1}\left(\frac{1}{\sqrt{3}}\right) = 30^\circ$$

Electric flux passing from the whole cylinder

$$\phi_1 = \frac{q_{\text{in}}}{\varepsilon_0} = \frac{\lambda L}{\varepsilon_0}$$

$\therefore$ Electric flux passing through $ABCD$ plane surface (shown only $AB$) = Electric flux passing through cylindrical surface $ANB$

$$= \left(\frac{60^\circ}{360^\circ}\right)(\phi_1) = \frac{\lambda L}{6\varepsilon_0}$$

$$\therefore \quad n = 6$$

**4.** Kinetic energy of ejected electron

= Energy of incident photon − energy required to ionize the electron from $n$th orbit (all in eV)

$$\therefore \quad 10.4 = \frac{1242}{90} - |E_n| = \frac{1242}{90} - \frac{13.6}{n^2}$$

$$\left(\text{as } E_n \propto \frac{1}{n^2} \text{ and } E_1 = -13.6 \text{ eV}\right)$$

Solving this equation, we get

$$n = 2$$

**5.** At height $h$

$$g' = \frac{g}{\left(1 + \frac{h}{R}\right)^2} \qquad \dots\text{(i)}$$

Given, $g' = \dfrac{g}{4}$

Substituting in Eq. (i) we get,

$$h = R$$

Now, from $A$ to $B$,

decrease in kinetic energy = increase in potential energy

$$\Rightarrow \quad \frac{1}{2}mv^2 = \frac{mgh}{1 + \frac{h}{R}}$$

$$\Rightarrow \quad \frac{v^2}{2} = \frac{gh}{1 + \frac{h}{R}} = \frac{1}{2}gR \qquad (h = R)$$

$$\Rightarrow \quad v^2 = gR \quad \text{or} \quad v = \sqrt{gR}$$

$$\text{Now,} \quad v_{\text{esc}} = \sqrt{2gR} = v\sqrt{2}$$

$$\Rightarrow \quad N = 2$$

**6.** In case of pure rolling, mechanical energy remains constant (as work-done by friction is zero). Further in case of a disc,

$$\frac{\text{translational kinetic energy}}{\text{rotational kinetic energy}} = \frac{K_T}{K_R}$$

$$= \frac{\frac{1}{2}mv^2}{\frac{1}{2}I\omega^2}$$

$$= \frac{mv^2}{\left(\frac{1}{2}mR^2\right)\left(\frac{v}{R}\right)^2} = \frac{2}{1}$$

$$\text{or,} \quad K_T = \frac{2}{3} \qquad \text{(Total kinetic energy)}$$

or, Total kinetic energy

$$K = \frac{3}{2}K_T = \frac{3}{2}\left(\frac{1}{2}mv^2\right) = \frac{3}{4}mv^2$$

Decrease in potential energy = increase in kinetic energy

$$\text{or} \quad mgh = \frac{3}{4}m\,(v_f^2 - v_i^2)$$

$$\text{or} \quad v_f = \sqrt{\frac{4}{3}gh + v_i^2}$$

As final velocity in both cases is same.

So, value of $\sqrt{\frac{4}{3}gh + v_i^2}$ should be same in both cases.

$$\therefore \quad \sqrt{\frac{4}{3} \times 10 \times 30 + (3)^2} = \sqrt{\frac{4}{3} \times 10 \times 27 + (v_2)^2}$$

Solving this equation we get,

$$v_2 = 7\,\text{m/s}$$

**7.** Power, $P = (\sigma T^4 A) = \sigma T^4 (4\pi R^2)$

or, $\quad P \propto T^4 R^2 \qquad \ldots(\text{i})$

According to Wien's law,

$$\lambda \propto \frac{1}{T}$$

($\lambda$ is the wavelength at which peak occurs)

$\therefore$ Eq. (i) will become,

$$P \propto \frac{R^2}{\lambda^4}$$

or $\quad \lambda \propto \left[\frac{R^2}{P}\right]^{1/4}$

$$\Rightarrow \quad \frac{\lambda_A}{\lambda_B} = \left[\frac{R_A}{R_B}\right]^{1/2}\left[\frac{P_B}{P_A}\right]^{1/4}$$

$$= [400]^{1/2}\left[\frac{1}{10^4}\right]^{1/4} = 2$$

**8.** Let initial power available from the plant is $P_0$. After time $t = nT$ or $n$ half lives, this will become $\left(\frac{1}{2}\right)^n P_0$. Now, it is given that,

$$\left(\frac{1}{2}\right)^n P_0 = 12.5\% \text{ of } P_0 = (0.125)\, P_0$$

Solving this equation we get,

$$n = 3$$

**9.** Let the other mass at this instant is at a distance of $x$ from the centre $O$.

Applying law of conservation of angular momentum, we have

$$I_1\omega_1 = I_2\omega_2$$

$$\therefore \quad (MR^2)(\omega) = \left[MR^2 + \frac{M}{8}\left(\frac{3}{5}R\right)^2 + \frac{M}{8}x^2\right]\left(\frac{8}{9}\omega\right)$$

Solving this equation, we get $x = \frac{4}{5}R.$

**Note** *If we take identical situations with both point masses, then answer will be (c). But in that case, angular momentum is not conserved.*

**10.** At the shown position, net force on both charges is zero. Hence they are in equilibrium. But equilibrium of $+q$ is stable equilibrium. So, it will start oscillations when displaced from this position. These small oscillations are simple harmonic in nature. While equilibrium of $-q$ is unstable. So, it continues to move in the direction of its displacement.

**11.** $R = 10\,\text{cm}$

Applying $\frac{\mu_2}{v} - \frac{\mu_1}{u} = \frac{\mu_2 - \mu_1}{R}$ two times

$$\frac{1}{v} - \frac{1.5}{-50} = \frac{1 - 1.5}{-10}$$

$$\frac{1}{v} + \frac{1.5}{50} = \frac{0.5}{10}$$

$$\frac{1}{v} = \frac{0.5}{10} - \frac{1.5}{50}$$

$$= \frac{2.5 - 1.5}{50}$$

$$\Rightarrow \quad v = 50$$

$MN = d$, $MI_1 = 50$ cm,

$NI_1 = (d - 50)$ cm



$$\text{Again, } \frac{1.5}{\infty} - \frac{1}{-(d-50)} = \frac{1.5 - 1}{10}$$

$$\frac{1}{d - 50} = \frac{1}{20}$$

$$d = 70$$

**12.**



Force on the complete wire = force on straight wire $PQ$ carrying a current $I$.

$$\mathbf{F} = I(\mathbf{PQ} \times \mathbf{B})$$

$$= I[\{2(L + R)\hat{\mathbf{i}}\} \times \mathbf{B}]$$

This force is zero if B is along $\hat{\mathbf{i}}$ direction or $x$-direction. If magnetic field is along $\hat{\mathbf{j}}$ direction or $\hat{\mathbf{k}}$ direction,

$$|\mathbf{F}| = F = (I)(2)(L + R)B \sin 90^\circ$$

or $\quad F = 2I(L + R)B$

or $\quad F \propto (L + R)$

∴ Options (a), (b) and (c) are correct.

**13.** (a) Total internal energy

$$U = \frac{f_1}{2}nRT + \frac{f_2}{2}nRT$$

$$(U_{\text{ave}})_{\text{per mole}} = \frac{U}{2n}$$

$$= \frac{1}{4}[5RT + 3RT] = 2RT$$

(b) $$\gamma_{\text{mix}} = \frac{n_1C_{p_1} + n_2C_{p_2}}{n_2C_{v_1} + n_2C_{v_2}}$$

$$= \frac{(1)\frac{7R}{2} + (1)\frac{5R}{2}}{(1)\frac{5R}{2} + (1)\frac{3R}{2}} = \frac{3}{2}$$

$$M_{\text{mix}} = \frac{n_1M_1 + n_2M_2}{n_1 + n_2}$$

$$= \frac{M_1 + M_2}{2} = \frac{2 + 4}{2} = 3$$

Speed of sound $V = \sqrt{\frac{\gamma RT}{M}}$

$$\Rightarrow \quad V \propto \sqrt{\frac{\gamma}{M}}$$

$$\frac{V_{\text{mix}}}{V_{\text{He}}} = \sqrt{\frac{\gamma_{\text{mix}}}{\gamma_{\text{He}}} \times \frac{M_{\text{He}}}{M_{\text{mix}}}}$$

$$= \sqrt{\frac{3/2}{5/3} \times \frac{4}{3}} = \sqrt{\frac{6}{5}}$$

(d) $$V_{\text{rms}} = \sqrt{\frac{3RT}{M}} \Rightarrow V_{\text{rms}} \propto \frac{1}{\sqrt{M}},$$

$$\frac{V_{\text{He}}}{V_{\text{H}}} = \sqrt{\frac{M_{\text{H}}}{M_{\text{He}}}} = \sqrt{\frac{2}{4}} = \frac{1}{\sqrt{2}}$$

**14.** *(b)* $$\frac{1}{R} = \frac{1}{R_{\text{Al}}} + \frac{1}{R_{\text{Fe}}} = \left(\frac{A_{\text{Al}}}{\rho_{\text{Al}}} + \frac{A_{\text{Fe}}}{\rho_{\text{Fe}}}\right)\frac{1}{\ell}$$

$$= \left[\frac{(7^2 - 2^2)}{2.7} + \frac{2^2}{10}\right]\frac{10^{-6}}{10^{-8}} \times \frac{1}{50 \times 10^{-3}}$$

Solving we get,

$$R = \frac{1875}{64} \times 10^{-6}\ \Omega$$

$$= \frac{1875}{64}\ \mu\Omega$$

**15.** $$eV_0 = \frac{hc}{\lambda} - W$$

$$V_0 = \left(\frac{hc}{e}\right)\left(\frac{1}{\lambda}\right) - \frac{W}{e}$$

$V_0$ versus $\frac{1}{\lambda}$ graph is in the form

$$y = mx - c$$

Therefore option (c) is correct.

Clearly, $V_0$ versus $\lambda$ graph is not a straight line but $V_0$ decreases with increase in $\lambda$ and $V_0$ becomes zero when $\frac{hc}{\lambda} = W$.

i.e. $\lambda = \lambda_0$ (Threshold wavelength)

∴ Option (a) is also correct.

**16. For Vernier callipers**

$$1 \text{ MSD} = \frac{1}{8} \text{ cm}$$

$$5 \text{ VSD} = 4 \text{ MSD}$$

$$\therefore \quad 1 \text{ VSD} = \frac{4}{5} \text{ MSD}$$

$$= \frac{4}{5} \times \frac{1}{8} = \frac{1}{10} \text{ cm}$$

Least count of Vernier callipers

$$= 1 \text{ MSD} - 1 \text{ VSD}$$

$$= \frac{1}{8} \text{ cm} - \frac{1}{10} \text{ cm} = 0.025 \text{ cm}$$

**(a) and (b)**

Pitch of screw gauge

$$= 2 \times 0.025 = 0.05 \text{ cm}$$

Least count of screw gauge

$$= \frac{0.05}{100} \text{ cm} = 0.005 \text{ mm}$$

**(c) and (d)**

Least count of linear scale of screw gauge = 0.05

Pitch = 0.05 × 2 = 0.1 cm

Least count of screw gauge

$$= \frac{0.1}{100} \text{ cm} = 0.01 \text{ mm}$$

**17.** $M \propto h^a c^b G^c$

$$M^1 \propto (ML^2T^{-1})^a (LT^{-1})^b (M^{-1}L^3T^{-2})^c$$

$$\propto M^{a-c} L^{2a+b+3c} T^{-a-b-2c}$$

$$a - c = 1 \quad \ldots(i)$$

$$2a + b + 3c = 0 \quad \ldots(ii)$$

$$a + b + 2c = 0 \quad \ldots(iii)$$

On solving (i), (ii), (iii), $a = \frac{1}{2}$, $b = +\frac{1}{2}$, $c = -\frac{1}{2}$

$\therefore M \propto \sqrt{c}$ only $\rightarrow$ (a) is correct.

In the same way we can find that,

$$L \propto h^{1/2} c^{-3/2} G^{1/2}$$

$L \propto \sqrt{h}$, $L \propto \sqrt{G} \rightarrow$ (c), (d) are also correct.

**18. Ist Particle**

$P = 0$ at $x = a \Rightarrow$ '$a$' is the amplitude of oscillation '$A_1$'.

At $x = 0$, $P = b$ (at mean position)

$$\Rightarrow mv_{max} = b \Rightarrow v_{max} = \frac{b}{m}$$

$$E_1 = \frac{1}{2}mv_{max}^2 = \frac{m}{2}\left[\frac{b}{m}\right]^2 = \frac{b^2}{2m}$$

$$A_1\omega_1 = v_{max} = \frac{b}{m}$$

$$\Rightarrow \omega_1 = \frac{b}{ma} = \frac{1}{mn^2} \quad \left(A_1 = a, \frac{a}{b} = n^2\right)$$

**IInd Particle**

$$P = 0 \text{ at } x = R \Rightarrow A_2 = R$$

At $x = 0, P = R \Rightarrow v_{max} = \frac{R}{m}$

$$E_2 = \frac{1}{2}mv_{max}^2 = \frac{m}{2}\left[\frac{R}{m}\right]^2 = \frac{R^2}{2m}$$

$$A_2\omega_2 = \frac{R}{m} \Rightarrow \omega_2 = \frac{R}{mR} = \frac{1}{m}$$

(b) $\frac{\omega_2}{\omega_1} = \frac{1/m}{1/mn^2} = n^2$

(c) $\omega_1\omega_2 = \frac{1}{mn^2} \times \frac{1}{m} = \frac{1}{m^2n^2}$

(d) $\frac{E_1}{\omega_1} = \frac{b^2/2m}{1/mn^2} = \frac{b^2n^2}{2} = \frac{a^2}{2n^2} = \frac{R^2}{2}$

$$\frac{E_2}{\omega_2} = \frac{R^2/2m}{1/m} = \frac{R^2}{2}$$

$$\Rightarrow \frac{E_1}{\omega_1} = \frac{E_2}{\omega_2}$$

**Note** *It is not given that the second figure is a circle. But from the figure and as per the requirement of question, we consider it is a circle.*

**19.** No solution is required.

**20.** (A) $F_x = \frac{-dU}{dx} = -\frac{2U_0}{a^3}[x-a][x][x+a]$

$F = 0$ at $x = 0$, $x = a, x = -a$ and $U = 0$ at $x = -a$ and $x = a$

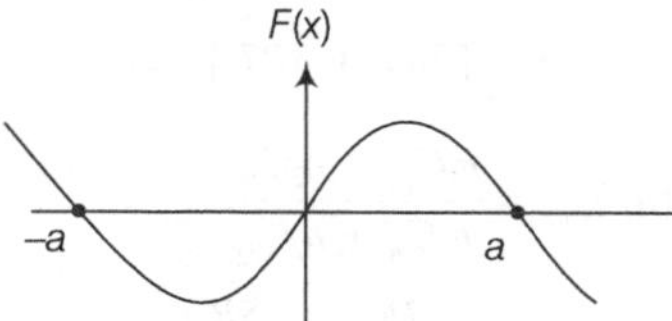


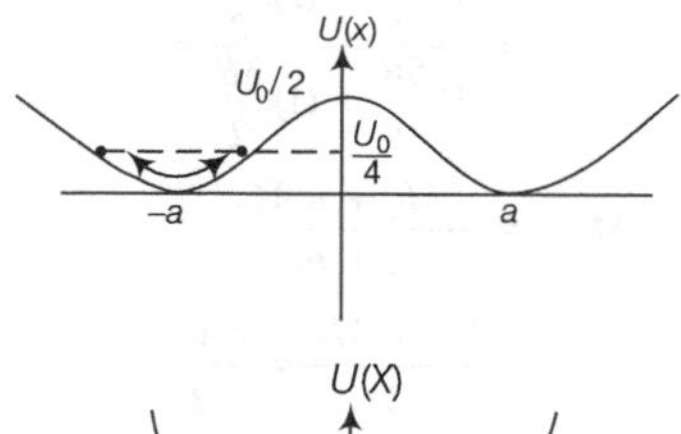


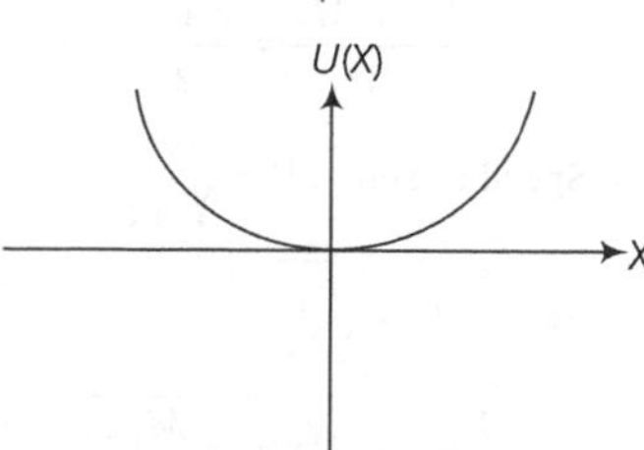


(B) $F_x = -\frac{dU}{dx} - U_0\left(\frac{x}{a}\right)$

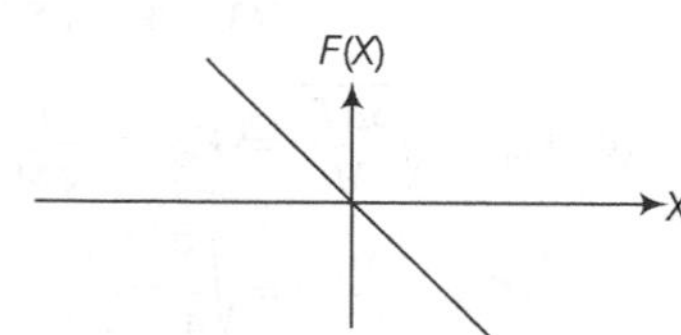


(C) $F_x = -\frac{dU}{dx}$

$$= U_0 \frac{e^{-x^2/x^2}}{a^3}[x][x-a][x+a]$$

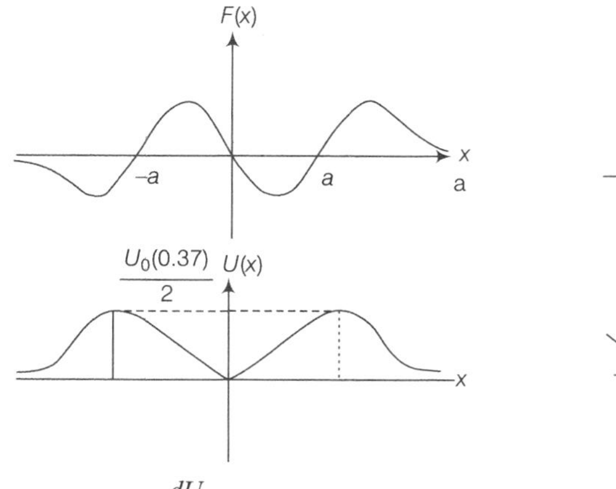


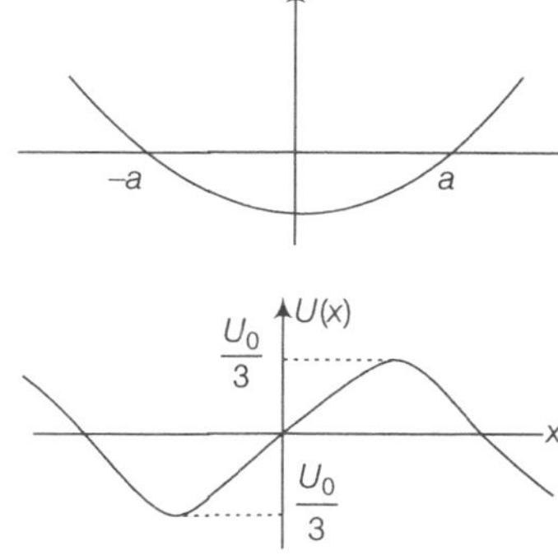


(D) $F_x = -\dfrac{dU}{dx}$

$$= -\frac{U_0}{2a^3}[(x-a)(x+a)]$$

# Chemistry

**1.** Although the compound has two chiral carbons (indicated by stars), it does not has four optically active isomers as expected. It is due to its existence in *cis*-form only.

$H_3C$, $CH_3$, $H_3C$, O — *cis*-form

$CH_3$, $CH_3$, $CH_3$, O — *trans*-form (only hypothetical)

The above shown transformation does not exist due to restricted rotation about the bridge head carbons, hence only *cis*-form and its mirror image exist.

**2.** OH $\xrightarrow{NaOH}$ N (I) ⟷ (II) ⟷ (III) ⟷ (IV) ⟷ (V) ⟷ (VI) ⟷ VII; (III) ⟷ (VIII) ⟷ IX

All the above shown nine resonance structures are different.

3. $N_2O_3$ has two proposed structures.

$O=N-O-N=O$

and

$O=N-N(\to O)=O$

In both cases, number of lone pair of electrons are eight.

4. When S is donor atom of $SCN^-$, it produces weak ligand field and forms high spin complex as

$$[Fe(SCN)_6]^{3-} : Fe^{3+} (3d^5) =$$

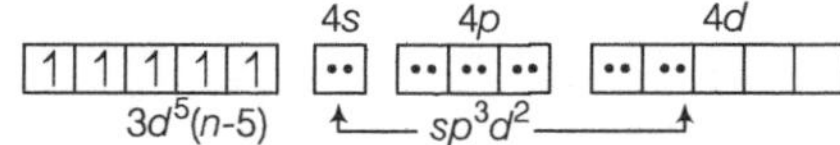


Spin only magnetic moment $(\mu_s) = \sqrt{5(5+2)}$ BM $= \sqrt{35}$ BM

In case of $CN^-$ ligand, carbon is the donor atom , it produces strong ligand field and forms low spin complex as

$$[Fe(CN)_6]^{3-} : Fe^{3+} (3d^5)$$

3d 4s 4p

$d^2sp^3$

Spin only magnetic moment $(\mu_s) = \sqrt{1(1+2)}$ BM

$= \sqrt{3}$ BM

Hence, difference in spin only magnetic moment

$$= \sqrt{35} - \sqrt{3} \approx 4 \text{ BM}$$

5. $Cl-\underset{\substack{\uparrow \\ sp}}{Be}-Cl$ $\quad N\equiv\underset{\substack{\uparrow \\ sp}}{N}\longrightarrow N^-$

$N\equiv\underset{\substack{\uparrow \\ sp}}{N}\to O$ $\quad O=\underset{\substack{\uparrow \\ sp}}{\overset{+}{N}}=O$

All the above mentioned molecules/ions have *sp*-hybridised central atom and no one pair at central atom, hence linear also.

Others are :

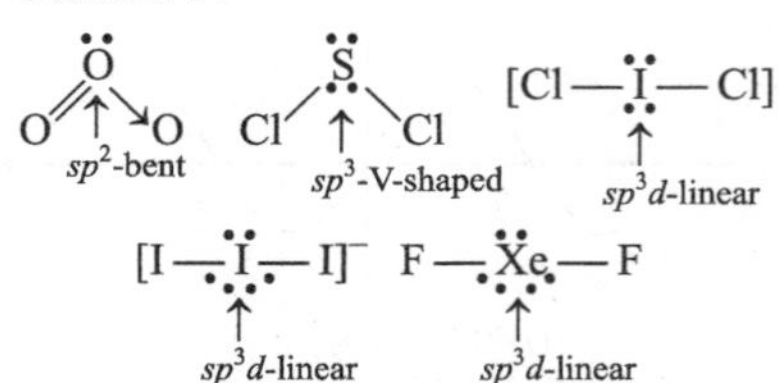


[Although $ICl_2^-$, $I_3^-$ and $XeF_2$ all also are linear but in them *d*-orbital contribute in hybridisation.]

6. In an one electron (hydrogenic) system, all orbitals of a shell remains degenerate, hence in second excited state, the degeneracy of H-atom is nine

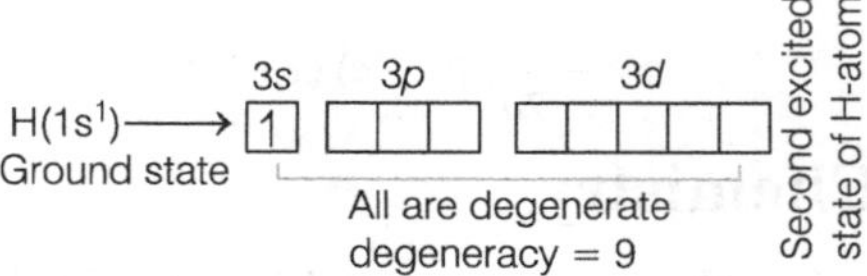


In case of many electrons system, different orbitals of a shell are non-degenerate. Hence,

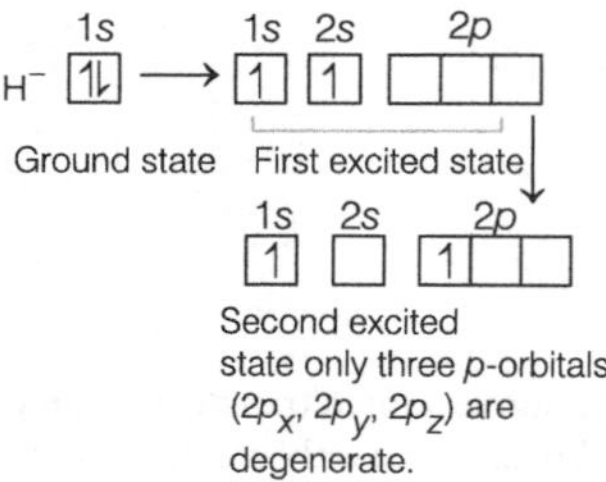


7. Energy obtained as one mole $X$ is converted into $Y$ is 193 KJ.

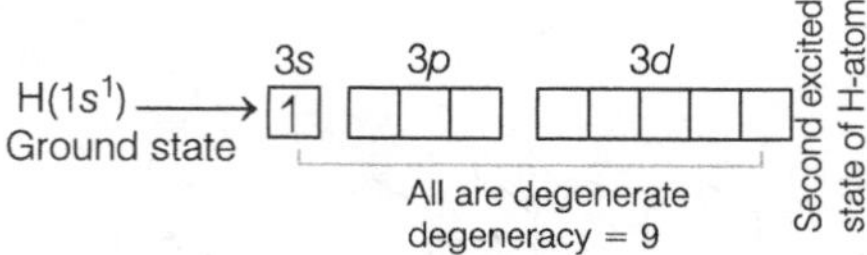


Energy consumed in converting one mole of $M^+$ to $M^{3+}$

$$= -nE^\circ F = 2 \times 96500 \times 0.25 \text{ J} = \frac{96500}{2} \text{ J}$$

$$\Rightarrow 193 \times 10^3 = n\left(\frac{96500}{2}\right)$$

$$\Rightarrow \qquad n = 4 \text{ mol}$$

**8.** + $\Delta T_f = iK_f m$

$$\Delta T_f = 0 - (-0.0558^\circ\text{ C}) = 0.0558^\circ\text{ C}$$

$$\Rightarrow i\text{ (vant Hoff's factor)} = \frac{0.0558}{1.86 \times 0.01} = 3$$

This indicates that complex upon ionisation produces three ions as:

$$[Co(NH_3)_5Cl]Cl_2 \rightarrow [Co(NH_3)_5Cl]^{2+}(aq) + 2Cl^-(aq)$$

Thus, only one Cl is inside the coordination sphere.

**9.** In both cases, hydrogenation of olefinic bond will render compound achiral as two identical ethyl group will come at the α-carbon which was earlier chiral carbon. However, in (*a*) and (*c*), chirality will be retained even after hydrogenation.

(a) 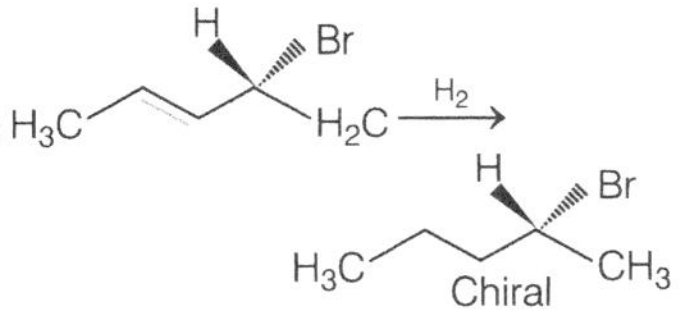 $H_3C$–CH=CH–C(H)(Br)–$H_2C$ $\xrightarrow{H_2}$ $H_3C$–$CH_2$–$CH_2$–C(H)(Br)–$CH_3$ Chiral

(b) 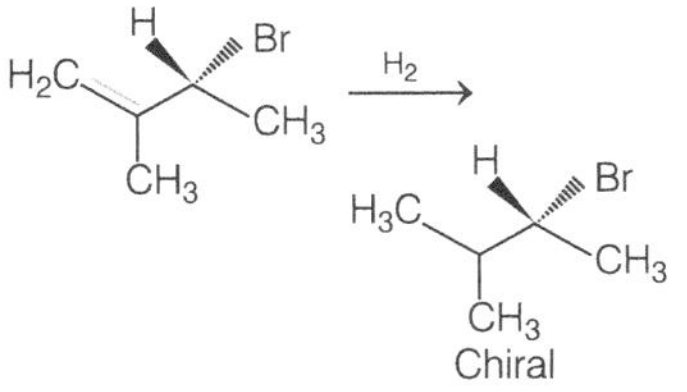 $H_2C$=C($CH_3$)–C(H)(Br)–$CH_3$ $\xrightarrow{H_2}$ $H_3C$–CH($CH_3$)–C(H)(Br)–$CH_3$ Chiral

**10.** 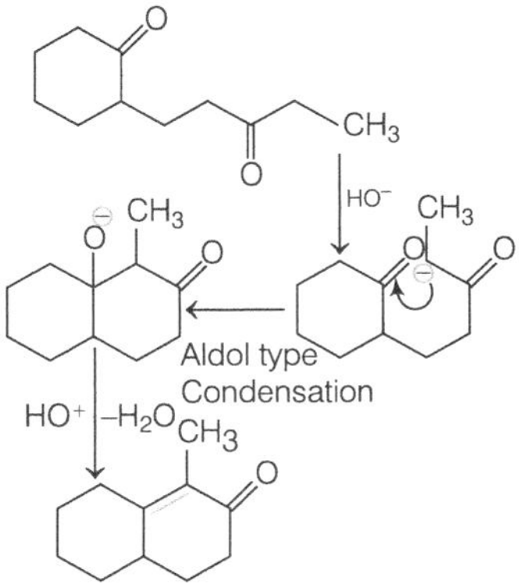 2-(3-oxopentyl)cyclohexanone $\xrightarrow{HO^-}$ enolate ($CH_3$) → Aldol type Condensation → alkoxide ($O^-$, $CH_3$) $\xrightarrow{HO^+,\ -H_2O}$ bicyclic enone ($CH_3$, O)

α, β-unsaturated ketone

**11.** Since, there is no mention of temperature, room temperature will be considered and thermodynamically controlled product would be the major product as:

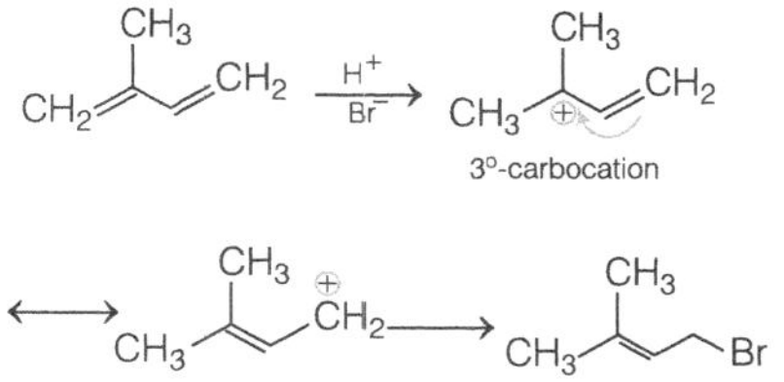

$CH_2$=C($CH_3$)–CH=$CH_2$ $\xrightarrow[Br^-]{H^+}$ $CH_3$–$\overset{\oplus}{C}$($CH_3$)–CH=$CH_2$ (3°-carbocation) ⟷ $CH_3$–C($CH_3$)=CH–$\overset{\oplus}{C}H_2$ ⟶ $CH_3$–C($CH_3$)=CH–$CH_2$–Br

**12.** L-(−)-glucose is enantiomer of D-(+)-glucose with hydroxy group on left of vertical at C-5 (L-configuration).

**13.** Reaction proceeds *via* diazonium salt with neighbouring group participation.

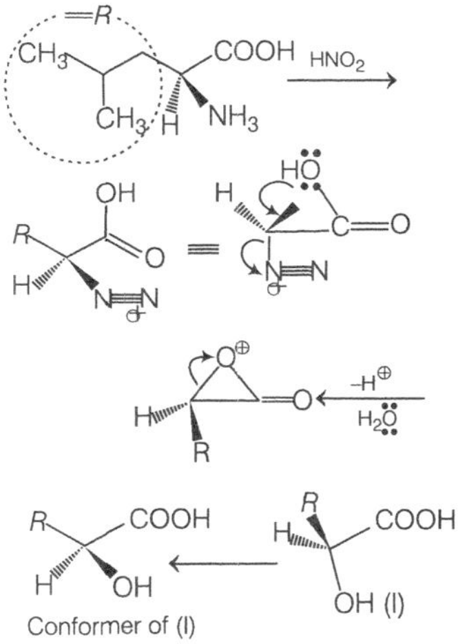

$CH_3$–CH($CH_3$)–$CH_2$–C(H)($NH_3$)–COOH (=R) $\xrightarrow{HNO_2}$ R–C(H)($N_2^+$)–C(OH)=O ≡ neighbouring participation ⟶ α-lactone (H, R) $\xleftarrow[H_2\ddot{O}]{-H^\oplus}$ ⟶ H–C(R)(OH)–COOH (I) ⟶ R–C(H)(OH)–COOH

Conformer of (I)

**14.** (*a*) In aqueous solution $Cr^{2+}(3d^4)$ acts as a reducing agent, oxidising itself to $Cr^{3+}(3d^3)$ that gives a completely half-field $t_{2g}$ level in octahedral ligand field of $H_2O$.

(*b*) $Mn^{3+}(3d^4)$ is an oxidising agent as it is reduced to $Mn^{2+}(3d^5)$, a completely half-filled stable configuration.

(*c*) Both $Cr^{2+}$ and $Mn^{3+}$ have $d^4$ configuration.

(*d*) $3d^4$ $Cr^{2+}(aq) \xrightarrow{R.A} Cr^{3+}(aq) + e^-$

Hence (*d*) is wrong statement.

**15.** (*a*) is wrong statement. Impure copper is set as anode where copper is oxidised to $Cu^{2+}$ and goes into electrolytic solutions.

(*b*) $CuSO_4$ is used as an electrolyte in purification process.

(*c*) Pure copper is deposited at cathode as:

$Cu^{2+} + 2e^- \longrightarrow Cu$ : (At cathode)

(*d*) Less active metals like Ag, Au etc settle down as anode mud.

**16.** $H_2O_2$ is alkaline medium acts as reducing agent, reduces $Fe^{3+}$ to $Fe^{2+}$. In acidic medium the same $H_2O_2$ oxidises $Fe^{2+}$ to $Fe^{3+}$.

**17.** Since, the reaction is exothermic, these will be less ammonia at equilibrium at higher temperature. However, rate of reaction increases with rise in temperature, $NH_3$ will be formed at faster rate in the initial stage when the temperature is high.

**18.** Oxide ions are at ccp positions, hence $4O^{2-}$ ions. Also, there are four octahedral voids and eight tetrahedral voids. Since '*m*' fraction of octahedral voids contain $Al^{3+}$ and '*n*' fraction of tetrahedral voids contain $Mg^{2+}$ ions, to maintain etectroneutrality $2(2Al^{3+} = +6\text{charge})$ and $(Mg^{2+} = +2\text{charge})$, will make unit cell neutral

Hence: $m = \frac{2}{4} = \frac{1}{2}$, $n = \frac{1}{8}$

**19.** Siderite $= FeCO_3$, Malachite $= CuCO_3 \cdot Cu(OH)_2$

Bauxite $= Al_2O_3 \cdot 2H_2O_2$ consisting some $Al(OH)_3$

Calamine $= ZnCO_3$, Argentite $= Ag_2S$

**20.** (A) $H_2O(l) \underset{1\,atm}{\overset{0°C}{\rightleftharpoons}} H_2O(s)$

$q<0$, $W<0$ (expansion)

$\Delta S_{sys}<0$ (solid state is more ordered than liquid state)

$\Delta U<0$; $\Delta G=0$ (At equilibrium)

(B) $q = 0$ (isolated), $W = 0$ ($p_{ext} = 0$)

$\Delta S_{sys} > 0 \because V_2 > V_1$

$\Delta U = 0 \because q = W = 0$

$\Delta G < 0 \because p_2 < p_1$

(C) $q = 0$ (isothermal mixing of ideal gases at constant $p$)

$W = 0 \because \Delta U = 0$; $q = 0$, $\Delta S_{sys} > 0$

$\because V_2 > V_1, \Delta U = 0$

$\because \Delta T = 0$

$\Delta G < 0 \because$ mixing is spontaneous.

(D) $q = 0$ (returning to same state and by same path)

$W = 0$

$\Delta S_{sys} = 0$ (same initial and final states)

$\Delta U = 0$

$\because T_i = T_f, \Delta G = 0$

## Mathematics

**1.** Let $P(t^2, 2t)$ be a point on the curve $y^2 = 4x$, whose image is $Q(x, y)$ on $x + y + 4 = 0$, then

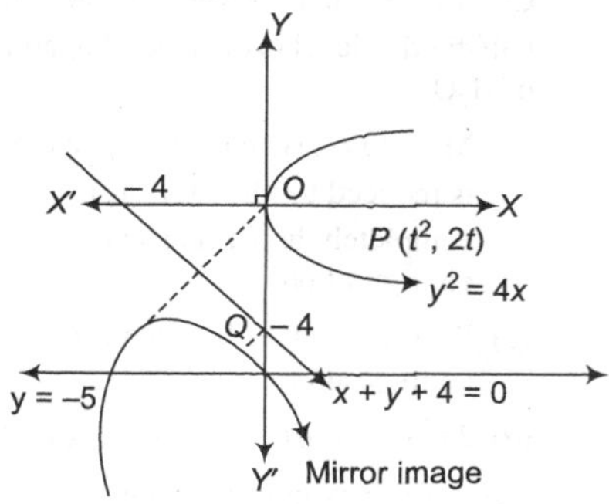


$$\frac{x - t^2}{1} = \frac{y - 2t}{1} = \frac{-2(t^2 + 2t + 4)}{1^2 + 1^2}$$

$\Rightarrow$ $x = -2t - 4$

and $y = -t^2 - 4$

Now, the straight line $y = -5$ meets the mirror image.

$\therefore$ $-t^2 - 4 = -5$

$\Rightarrow$ $t^2 = 1$

$\Rightarrow$ $t = \pm 1$

Thus, points of intersection of $A$ and $B$ are $(-6, -5)$ and $(-2, -5)$.

$\therefore$ Distance, $AB = \sqrt{(-2+6)^2 + (-5+5)^2}$

$= 4$

**2.** Using Binomial distribution,

$$P(X \geq 2) = 1 - P(X = 0) - P(X = 1)$$

$$= 1 - \left(\frac{1}{2}\right)^n - \left[{}^nC_1 \cdot \left(\frac{1}{2}\right) \cdot \left(\frac{1}{2}\right)^{n-1}\right]$$

$$= 1 - \frac{1}{2^n} - {}^nC_1 \cdot \frac{1}{2^n} = 1 - \left(\frac{1+n}{2^n}\right)$$

Given, $P(X \geq 2) \geq 0.96$

$$\therefore \quad 1 - \frac{(n+1)}{2^n} \geq \frac{24}{25} \Rightarrow \frac{n+1}{2^n} \leq \frac{1}{25}$$

$$\therefore \quad n = 8$$

**3.** Here, __ $B_1$ __ $B_2$ __ $B_3$ __ $B_4$ __ $B_5$ __

Out of 5 girls, 4 girls are together and 1 girl is separate. Now, to select 2 positions out of 6 positions between boys $= {}^6C_2$ ...(i)

4 girls are to be selected out of $5 = {}^5C_4$ ...(ii)

Now, 2 groups of girls can be arranged in 2!ways. ... (iii)

Also, the group of 4 girls and 5 boys is arranged in $4! \times 5!$ ways. ...(iv)

Now, total number of ways

$$= {}^6C_2 \times {}^5C_4 \times 2! \times 4! \times 5!$$

[from Eqs.(i), (ii), (iii) and (iv)]

$$\therefore \quad m = {}^6C_2 \times {}^5C_4 \times 2! \times 4! \times 5!$$

and $\quad n = 5! \times 6!$

$$\Rightarrow \quad \frac{m}{n} = \frac{{}^6C_2 \times {}^5C_4 \times 2! \times 4! \times 5!}{6! \times 5!}$$

$$= \frac{15 \times 5 \times 2 \times 4!}{6 \times 5 \times 4!} = 5$$

**4.** End points of latusrectum are $(a, \pm 2a)$ i.e. $(1, \pm 2)$.

Equation of normal at $(x_1, y_1)$ is

$$\frac{y - y_1}{x - x_1} = -\frac{y_1}{2a}$$

i.e. $$\frac{y-2}{x-1} = -\frac{2}{2}$$

and $$\frac{y+2}{x-1} = \frac{2}{2}$$

$\Rightarrow \quad x + y = 3$

and $\quad x - y = 3$

which is tangent to

$$(x-3)^2 + (y+2)^2 = r^2.$$

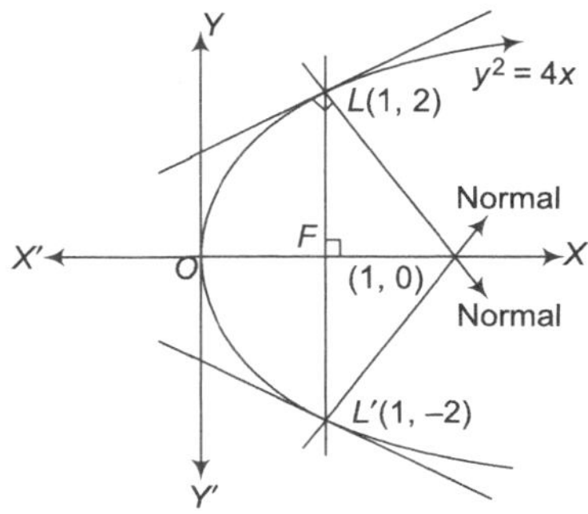


$\therefore$ Length of perpendicular from centre

= Radius

$$\Rightarrow \quad \frac{|3 - 2 - 3|}{\sqrt{1^2 + 1^2}} = r$$

$$\therefore \quad r^2 = 2$$

**5.** Here, $f(x) = \begin{cases} [x], & x \leq 2 \\ 0, & x > 2 \end{cases}$

$$\therefore \quad I = \int_{-1}^{2} \frac{x\,f(x^2)}{2 + f(x+1)}\,dx$$

$$= \int_{-1}^{0} \frac{x\,f(x^2)}{2 + f(x+1)}\,dx + \int_{0}^{1} \frac{x\,f(x^2)}{2 + f(x+1)}\,dx + \int_{1}^{\sqrt{2}} \frac{x\,f(x^2)}{2 + f(x+1)}\,dx + \int_{\sqrt{2}}^{\sqrt{3}} \frac{x\,f(x^2)}{2 + f(x+1)}\,dx + \int_{\sqrt{3}}^{2} \frac{x\,f(x^2)}{2 + f(x+1)}\,dx$$

$$= \int_{-1}^{0} 0\,dx + \int_{0}^{1} 0\,dx + \int_{1}^{\sqrt{2}} \frac{x \cdot 1}{2 + 0}\,dx + \int_{\sqrt{2}}^{\sqrt{3}} 0\,dx + \int_{\sqrt{3}}^{2} 0\,dx$$

$\because\ -1 < x < 0 \Rightarrow 0 < x^2 < 1 \Rightarrow [x^2] = 0,$

$0 < x < 1 \Rightarrow 0 < x^2 < 1 \Rightarrow [x^2] = 0,$

$1 < x < \sqrt{2} \Rightarrow$

$\begin{cases} 1 < x^2 < 2 & \Rightarrow [x^2] = 1 \\ 2 < x + 1 < 1 + \sqrt{2} & \Rightarrow f(x+1) = 0, \end{cases}$

$\sqrt{2} < x < \sqrt{3} \Rightarrow 2 < x^2 < 3$

$\Rightarrow \quad f(x^2) = 0,$

and $\sqrt{3} < x < 2 \Rightarrow 3 < x^2 < 4$

$\Rightarrow \quad f(x^2) = 0$

$$\Rightarrow \quad I = \int_1^{\sqrt{2}} \frac{x}{2} dx = \left[\frac{x^2}{4}\right]_1^{\sqrt{2}}$$

$$= \frac{1}{4}(2-1) = \frac{1}{4}$$

$\therefore \quad 4I = 1 \Rightarrow 4I - 1 = 0$

**6.** Here, volume of cylindrical container,

$$V = \pi r^2 h \quad \ldots\text{(i)}$$

and let volume of the material used be $T$.

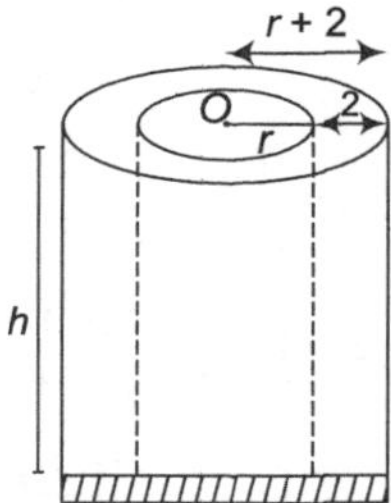


$$\therefore T = \pi[(r+2)^2 - r^2]h + \pi(r+2)^2 \times 2$$

$$T = \pi[(r+2)^2 - r^2]\cdot\frac{V}{\pi r^2} + 2\pi(r+2)^2$$

$$\left[\because V = \pi r^2 h \Rightarrow h = \frac{V}{\pi r^2}\right]$$

$$\Rightarrow \quad T = V\left(\frac{r+2}{r}\right)^2 + 2\pi(r+2)^2 - V$$

On differentiating w.r.t. $r$, we get

$$\frac{dT}{dr} = 2V\cdot\left(\frac{r+2}{r}\right)\cdot\left(\frac{-2}{r^2}\right) + 4\pi(r+2)$$

At $\quad r = 10, \dfrac{dT}{dr} = 0$

Now, $\quad 0 = (r+2)\cdot 4\left(\pi - \dfrac{V}{r^3}\right)$

$$\Rightarrow \quad \frac{V}{r^3} = \pi, \text{ where } r = 10$$

$$\Rightarrow \quad \frac{V}{1000} = \pi \text{ or } \frac{V}{250\pi} = 4$$

**7.** Since, $F'(a) + 2$ is the area bounded by $x = 0$, $y = 0$, $y = f(x)$ and $x = a$.

$$\therefore \quad \int_0^a f(x)dx = F'(a) + 2$$

Using Newton-Leibnitz formula,

$$f(a) = F''(a)$$

and $\quad f(0) = F''(0) \quad \ldots\text{(i)}$

Given, $\quad F(x) = \displaystyle\int_x^{x^2+\pi/6} 2\cos^2 t\, dt$

On differentiating,

$$F'(x) = 2\cos^2\left(x^2 + \frac{\pi}{6}\right)\cdot 2x - 2\cos^2 x \cdot 1$$

Again differentiating,

$$F''(x) = 4\left\{\cos^2\left(x^2 + \frac{\pi}{6}\right) - 2x\cos\left(x^2 + \frac{\pi}{6}\right)\sin\left(x^2 + \frac{\pi}{6}\right)2x\right\} + \{4\cos x\cdot\sin x\}$$

$$= 4\left\{\cos^2\left(x^2 + \frac{\pi}{6}\right) - 4x^2\cos\left(x^2 + \frac{\pi}{6}\right)\sin\left(x^2 + \frac{\pi}{6}\right)\right\} + 2\sin 2x$$

$$\therefore \quad F''(0) = 4\left\{\cos^2\left(\frac{\pi}{6}\right)\right\} = 3$$

$$\therefore \quad f(0) = 3$$

**8.** Here, $\dfrac{5}{4}\cos^2 2x + (\cos^4 x + \sin^4 x) + (\cos^6 x + \sin^6 x) = 2$

$$\Rightarrow \quad \frac{5}{4}\cot 2x + [(\cos^2 x + \sin^2 x)^2 - 2\sin^2 x\cos^2 x] + (\cos^2 x + \sin^2 x)[(\cos^2 x + \sin^2 x)^2 - 3\sin^2 x\cos^2 x] = 2$$

$$\Rightarrow \quad \frac{5}{4}\cos^2 2x + (1 - 2\sin^2 x\cos^2 x) + (1 - 3\cos^2 x\sin^2 x) = 2$$

$$\Rightarrow \quad \frac{5}{4}\cos^2 2x - 5\sin^2 x\cos^2 x = 0$$

$$\Rightarrow \quad \frac{5}{4}\cos^2 2x - \frac{5}{4}\sin^2 2x = 0$$

$$\Rightarrow \quad \frac{5}{4}\cos^2 2x - \frac{5}{4} + \frac{5}{4}\cos^2 2x = 0$$

$$\Rightarrow \quad \frac{5}{2}\cos^2 2x = \frac{5}{4}$$

$$\Rightarrow \quad \cos^2 2x = \frac{1}{2} \Rightarrow 2\cos^2 2x = 1$$

$$\Rightarrow \quad 1 + \cos 4x = 1$$

$$\Rightarrow \quad \cos 4x = 0, \text{ as } 0 \le x \le 2\pi$$

$\therefore$

$$4x = \left\{\frac{\pi}{2}, \frac{3\pi}{2}, \frac{5\pi}{2}, \frac{7\pi}{2}, \frac{9\pi}{2}, \frac{11\pi}{2}, \frac{13\pi}{2}, \frac{15\pi}{2}\right\}$$

as $\quad 0 \le 4x \le 8\pi$

$$\Rightarrow x = \left\{\frac{\pi}{8}, \frac{3\pi}{8}, \frac{5\pi}{8}, \frac{7\pi}{8}, \frac{9\pi}{8}, \frac{11\pi}{8}, \frac{13\pi}{8}, \frac{15\pi}{8}\right\}$$

Hence, the total number of solutions is 8.

**9.** Here, $(1 + e^x)y' + ye^x = 1$

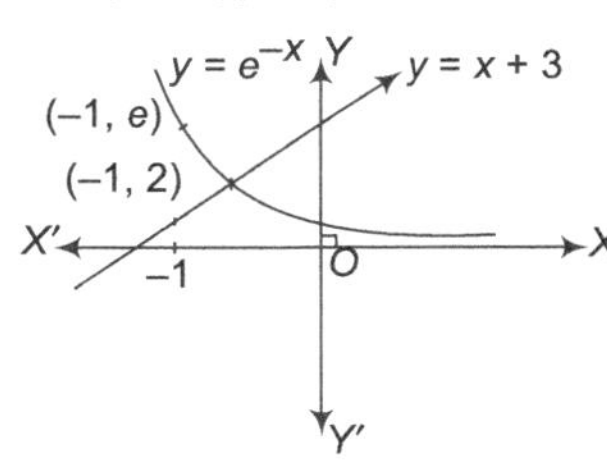


$$\Rightarrow \quad \frac{dy}{dx} + e^x \cdot \frac{dy}{dx} + ye^x = 1$$

$$\Rightarrow \quad dy + e^x dy + ye^x\,dx = dx$$

$$\Rightarrow \quad dy + d(e^x y) = dx$$

On integrating both sides, we get

$$y + e^x y = x + C$$

Given, $\quad y(0) = 2$

$$\Rightarrow \quad 2 + e^0 \cdot 2 = 0 + C$$

$$\Rightarrow \quad C = 4$$

$$\therefore \quad y(1 + e^x) = x + 4$$

$$\Rightarrow \quad y = \frac{x + 4}{1 + e^x}$$

Now at $x = -4$, $y = \dfrac{-4 + 4}{1 + e^{-4}} = 0$

$$\therefore \quad y(-4) = 0 \qquad \ldots\text{(i)}$$

For critical points, $\dfrac{dy}{dx} = 0$

i.e. $\quad \dfrac{dy}{dx} = \dfrac{(1 + e^x)\cdot 1 - (x + 4)e^x}{(1 + e^x)^2} = 0$

$$\Rightarrow \quad e^x(x + 3) - 1 = 0$$

or $\quad e^{-x} = (x + 3)$

Clearly, the intersection point lies between $(-1, 0)$.

$\therefore$ $y(x)$ has a critical point in the interval $(-1, 0)$.

**10.** Since, centre lies on $y = x$.

$\therefore$ Equation of circle is

$$x^2 + y^2 - 2ax - 2ay + c = 0$$

On differentiating, we get

$$2x + 2yy' - 2a - 2ay' = 0$$

$$\Rightarrow \quad x + yy' - a - ay' = 0$$

$$\Rightarrow \quad a = \frac{x + yy'}{1 + y'}$$

Again differentiating, we get

$$0 = \frac{(1 + y')[1 + yy'' + (y')^2] - (x + yy')\cdot(y'')}{(1 + y')^2}$$

$$\Rightarrow (1 + y')[1 + (y')^2 + yy''] - (x + yy')(y'') = 0$$

$$\Rightarrow 1 + y'[(y')^2 + y' + 1] + y''(y - x) = 0$$

On comparing with $Py'' + Qy' + 1 = 0$, we get

$$P = y - x \quad \text{and} \quad Q = (y')^2 + y' + 1$$

**11.** (a) Here, $f(x) = \begin{cases} g(x), & x > 0 \\ 0, & x = 0 \\ -g(x), & x < 0 \end{cases}$

$$f'(x) = \begin{cases} g'(x), & x \ge 0 \\ -g'(x), & x < 0 \end{cases}$$

$\therefore$ Option (a) is correct.

(b) $h(x) = e^{|x|} = \begin{cases} e^x, & x \ge 0 \\ e^{-x}, & x < 0 \end{cases}$

$$\Rightarrow \quad h'(x) = \begin{cases} e^x, & x \ge 0 \\ -e^{-x}, & x < 0 \end{cases}$$

$$\Rightarrow \quad h'(0^+) = 1$$

and $\quad h'(0^-) = -1$

So, $h(x)$ is not differentiable at $x = 0$.

$\therefore$ Option (b) is not correct.

(c) $(foh)(x) = f\{h(x)\}$, as $h(x) > 0$

$$= \begin{cases} g(e^x), & x \geq 0 \\ g(e^{-x}), & x < 0 \end{cases}$$

$$\Rightarrow \quad (foh)'(x) = \begin{cases} e^x g'(e^x), & x \geq 0 \\ -e^{-x} g'(e^{-x}), & x < 0 \end{cases}$$

$$\Rightarrow \quad (foh)'(0^+) = g'(1), (foh)'(0^-) = -g'(1)$$

So, $(foh)(x)$ is not differentiable at $x = 0$.

$\therefore$ Option (c) is not correct.

(d) $(hof)(x) = e^{|f(x)|} = \begin{cases} e^{|g(x)|}, & x \neq 0 \\ e^0 = 1, & x = 0 \end{cases}$

Now, $(hof)'(0) = \lim_{h \to 0} \dfrac{e^{|g(x)|} - 1}{x}$

$$= \lim_{h \to 0} \frac{e^{|g(x)|} - 1}{|g(x)|} \cdot \frac{|g(x)|}{x}$$

$$= \lim_{h \to 0} \frac{e^{|g(x)|} - 1}{|g(x)|} \cdot \lim_{h \to 0} \frac{|g(x) - 0|}{|x|} \cdot \lim_{h \to 0} \frac{|x|}{x}$$

$$= 1 \cdot g'(0) \cdot \lim_{h \to 0} \frac{|x|}{x}$$

$= 0$, as $g'(0) = 0$

$\therefore$ Option (d) is correct.

**12.** (a) $f(x) = \sin\left[\dfrac{\pi}{6} \sin\left(\dfrac{\pi}{2} \sin x\right)\right], x \in R$

$$= \sin\left(\frac{\pi}{6} \sin \theta\right), \theta \in \left[-\frac{\pi}{2}, \frac{\pi}{2}\right],$$

where $\theta = \dfrac{\pi}{2} \sin x$

$$= \sin \alpha, \alpha \in \left[-\frac{\pi}{6}, \frac{\pi}{6}\right],$$

where $\alpha = \dfrac{\pi}{6} \sin \theta$

$$\therefore \quad f(x) \in \left[-\frac{1}{2}, \frac{1}{2}\right]$$

Hence, range of $f(x) \in \left[-\dfrac{1}{2}, \dfrac{1}{2}\right]$

So, option (a) is correct.

(b) $f\{g(x)\} = f(t), t \in \left[-\dfrac{\pi}{2}, \dfrac{\pi}{2}\right]$

$$\Rightarrow \quad f(t) \in \left[-\frac{1}{2}, \frac{1}{2}\right]$$

$\therefore$ Option (b) is correct.

(c) $\lim_{x \to 0} \dfrac{f(x)}{g(x)}$

$$= \lim_{x \to 0} \frac{\sin\left[\frac{\pi}{6} \sin\left(\frac{\pi}{2} \sin x\right)\right]}{\frac{\pi}{2} (\sin x)}$$

$$= \lim_{x \to 0} \frac{\sin\left[\frac{\pi}{6} \sin\left(\frac{\pi}{2} \sin x\right)\right]}{\frac{\pi}{6} \sin\left(\frac{\pi}{2} \sin x\right)} \cdot \frac{\frac{\pi}{6} \sin\left(\frac{\pi}{2} \sin x\right)}{\left(\frac{\pi}{2} \sin x\right)}$$

$$= 1 \times \frac{\pi}{6} \times 1 = \frac{\pi}{6}$$

$\therefore$ Option (c) is correct.

(d) $g\{f(x)\} = 1$

$$\Rightarrow \quad \frac{\pi}{2} \sin\{f(x)\} = 1$$

$$\Rightarrow \quad \sin\{f(x)\} = \frac{2}{\pi} \quad \text{...(i)}$$

But $f(x) \in \left[-\dfrac{1}{2}, \dfrac{1}{2}\right] \subset \left[-\dfrac{\pi}{6}, \dfrac{\pi}{6}\right]$

$$\therefore \quad \sin\{f(x)\} \in \left[-\frac{1}{2}, \frac{1}{2}\right] \quad \text{...(ii)}$$

$$\Rightarrow \quad \sin\{f(x)\} \neq \frac{2}{\pi},$$

[from Eqs. (i) and (ii)]

i.e. No solution.

$\therefore$ Option (d) is not correct.

**13.** Given, $|\mathbf{a}| = 12$, $|\mathbf{b}| = 4\sqrt{3}$

$$\mathbf{a} + \mathbf{b} + \mathbf{c} = \mathbf{0}$$

$$\Rightarrow \quad \mathbf{a} = -(\mathbf{b} + \mathbf{c})$$

$\Rightarrow \quad |\mathbf{a}|^2 = |\mathbf{b}+\mathbf{c}|^2$

$\Rightarrow \quad |\mathbf{a}|^2 = |\mathbf{b}|^2 + |\mathbf{c}|^2 + 2\mathbf{b}\cdot\mathbf{c}$

$\Rightarrow \quad 144 = 48 + |\mathbf{c}|^2 + 48$

$\Rightarrow \quad |\mathbf{c}|^2 = 48 \Rightarrow |\mathbf{c}| = 4\sqrt{3}$

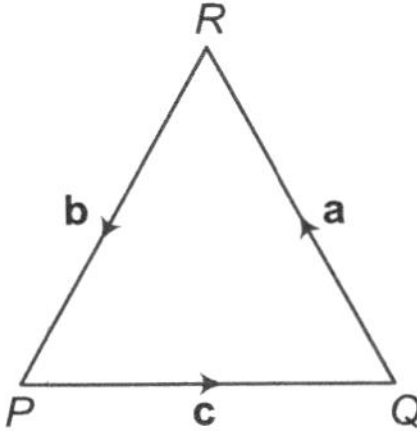


Also, $\quad |\mathbf{c}|^2 = |\mathbf{a}|^2 + |\mathbf{b}|^2 + 2\,\mathbf{a}\cdot\mathbf{b}$

$\Rightarrow \quad 48 = 144 + 48 + 2\,\mathbf{a}\cdot\mathbf{b}$

$\Rightarrow \quad \mathbf{a}\cdot\mathbf{b} = -72$

$\therefore$ Option (d) is correct.

Also, $\quad \mathbf{a}\times\mathbf{b} = \mathbf{c}\times\mathbf{a}$

$\Rightarrow \mathbf{a}\times\mathbf{b} + \mathbf{c}\times\mathbf{a} = 2\mathbf{a}\times\mathbf{b}$

$\Rightarrow |\mathbf{a}\times\mathbf{b} + \mathbf{c}\times\mathbf{a}| = 2\,|\mathbf{a}\times\mathbf{b}|$

$$= 2\sqrt{|\mathbf{a}|^2\,|\mathbf{b}|^2 - (\mathbf{a}\cdot\mathbf{b})^2}$$

$$= 2\sqrt{(144)\,(48) - (-72)^2}$$

$$= 2(12)\sqrt{48-36}$$

$$= 48\sqrt{3}$$

$\therefore$ Option (c) is correct.

Also, $\quad \dfrac{|\mathbf{c}|^2}{2} - |\mathbf{a}| = 24 - 12 = 12$

$\therefore$ Option (a) is correct.

and $\quad \dfrac{|\mathbf{c}|^2}{2} + |\mathbf{a}| = 24 + 12 = 36$

$\therefore$ Option (b) is not correct.

**14.** Given,

$$X^T = -X,\ Y^T = -Y,\ Z^T = Z$$

(a) Let $\quad P = Y^3Z^4 - Z^4Y^3$

Then, $\quad P^T = (Y^3\,Z^4)^T - (Z^4\,Y^3)^T$

$$= (Z^T)^4\,(Y^T)^3 - (Y^T)^3\,(Z^T)^4$$

$$= -Z^4Y^3 + Y^3Z^4 = P$$

$\therefore \quad P$ is symmetric matrix.

(b) Let $\quad P = X^{44} + Y^{44}$

Then, $\quad P^T = (X^T)^{44} + (Y^T)^{44}$

$$= X^{44} + Y^{44} = P$$

$\therefore \quad P$ is symmetric matrix.

(c) Let $\quad P = X^4Z^3 - Z^3X^4$

Then, $P^T = (X^4Z^3)^T - (Z^3X^4)^T$

$$= (Z^T)^3\,(X^T)^4 - (X^T)^4\,(Z^T)^3$$

$$= Z^3\,X^4 - X^4Z^3$$

$$= -P$$

$\therefore$ $P$ is skew-symmetric matrix.

(d) Let $\quad P = X^{23} + Y^{23}$

Then, $\quad P^T = (X^T)^{23} + (Y^T)^{23}$

$$= -X^{23} - Y^{23}$$

$$= -P$$

$\therefore \quad P$ is skew-symmetric matrix.

**15.** Given determinant could be expressed as product of two determinants.

i.e. $\begin{vmatrix} (1+\alpha)^2 & (1+2\alpha)^2 & (1+3\alpha)^2 \\ (2+\alpha)^2 & (2+2\alpha)^2 & (2+3\alpha)^2 \\ (3+\alpha)^2 & (3+2\alpha)^2 & (3+3\alpha)^2 \end{vmatrix}$

$$= -648\,\alpha$$

$$\Rightarrow \begin{vmatrix} 1+2\alpha+\alpha^2 & 1+4\alpha+4\alpha^2 & 1+6\alpha+9\alpha^2 \\ 4+4\alpha+\alpha^2 & 4+8\alpha+4\alpha^2 & 4+12\alpha+9\alpha^2 \\ 9+6\alpha+\alpha^2 & 9+12\alpha+4\alpha^2 & 9+18\alpha+9\alpha^2 \end{vmatrix}$$

$$= -648\,\alpha$$

$$\Rightarrow \begin{vmatrix} 1 & \alpha & \alpha^2 \\ 4 & 2\alpha & \alpha^2 \\ 9 & 3\alpha & \alpha^2 \end{vmatrix} \cdot \begin{vmatrix} 1 & 1 & 1 \\ 2 & 4 & 6 \\ 1 & 4 & 9 \end{vmatrix}$$

$$= -648\,\alpha$$

$$\Rightarrow \alpha^3 \begin{vmatrix} 1 & 1 & 1 \\ 4 & 2 & 1 \\ 9 & 3 & 1 \end{vmatrix} \cdot \begin{vmatrix} 1 & 1 & 1 \\ 2 & 4 & 6 \\ 1 & 4 & 9 \end{vmatrix}$$

$$= -648\,\alpha$$

$\Rightarrow \quad -8\alpha^3 = -648\alpha$

$\Rightarrow \quad \alpha^3 - 81\alpha = 0$

$\Rightarrow \quad \alpha\,(\alpha^2 - 81) = 0$

$\therefore \quad \alpha = 0, \pm 9$

**16.** Here, $P_3 : (x + z - 1) + \lambda y = 0$

i.e. $P_3 : x + \lambda y + z - 1 = 0$ ...(i)

whose distance from (0, 1, 0) is 1.

$$\therefore \quad \frac{|0 + \lambda + 0 - 1|}{\sqrt{1 + \lambda^2 + 1}} = 1$$

$$\Rightarrow \quad |\lambda - 1| = \sqrt{\lambda^2 + 2}$$

$$\Rightarrow \quad \lambda^2 - 2\lambda + 1 = \lambda^2 + 2 \Rightarrow \lambda = -\frac{1}{2}$$

$\therefore$ Equation of $P_3$ is $2x - y + 2z - 2 = 0$.

$\because$ Distance from $(\alpha, \beta, \gamma)$ is 2.

$$\therefore \quad \frac{|2\alpha - \beta + 2\gamma - 2|}{\sqrt{4 + 1 + 4}} = 2$$

$$\Rightarrow \quad 2\alpha - \beta + 2\gamma - 2 = \pm 6$$

$$\Rightarrow \quad 2\alpha - \beta + 2\gamma = 8$$

and $\quad 2\alpha - \beta + 2\gamma = -4$

**17.** Since, $L$ is at constant distance from two planes $P_1$ and $P_2$.

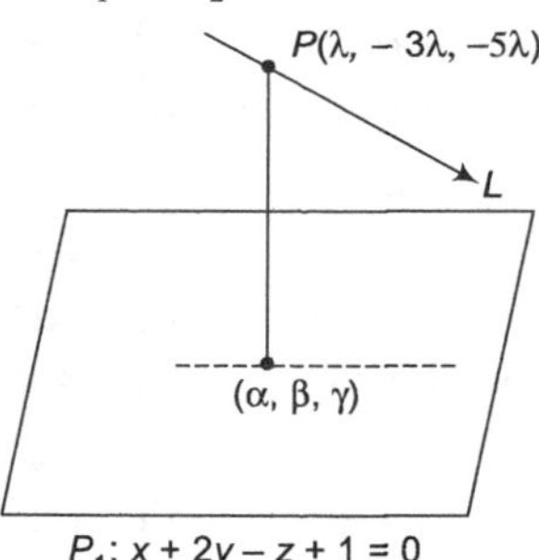


Therefore, $L$ is parallel to the line through intersection of $P_1$ and $P_2$.

$$\therefore \text{ DR's of } L = \begin{vmatrix} \hat{\mathbf{i}} & \hat{\mathbf{j}} & \hat{\mathbf{k}} \\ 1 & 2 & -1 \\ 2 & -1 & 1 \end{vmatrix}$$

$$= \hat{\mathbf{i}}(2 - 1) - \hat{\mathbf{j}}(1 + 2) + \hat{\mathbf{k}}(-1 - 4)$$

$$= \hat{\mathbf{i}} - 3\hat{\mathbf{j}} - 5\hat{\mathbf{k}}$$

$\therefore$ DR's of $L$ are (1, –3, –5) passing through (0, 0, 0).

Now, equation of $L$ is

$$\frac{x - 0}{1} = \frac{y - 0}{-3} = \frac{z - 0}{-5}$$

For any point on $L$,

$$\frac{x}{1} = \frac{y}{-3} = \frac{z}{-5} = \lambda \quad \text{[say]}$$

i.e. $P(\lambda, -3\lambda, -5\lambda)$

If $(\alpha, \beta, \gamma)$ is foot of perpendicular from $P$ on $P_1$, then

$$\frac{\alpha - \lambda}{1} = \frac{\beta + 3\lambda}{2} = \frac{\gamma + 5\lambda}{-1} = k \quad \text{[say]}$$

$$\Rightarrow \alpha = \lambda + k, \beta = 2k - 3\lambda, \gamma = -k - 5\lambda$$

which satisfy $P_1 : x + 2y - z + 1 = 0$

$$\Rightarrow (\lambda + k) + 2(2k - 3\lambda) - (-k - 5\lambda) + 1 = 0$$

$$\Rightarrow \quad k = -\frac{1}{6}$$

$$\therefore x = -\frac{1}{6} + \lambda, \; y = -\frac{1}{3} - 3\lambda, \; z = \frac{1}{6} - 5\lambda$$

which satisfy options (a) and (b).

**18.** Since, $\angle POQ = 90°$

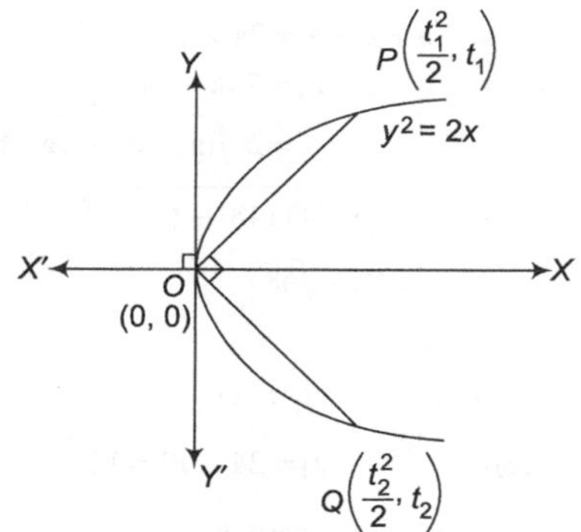


$$\Rightarrow \quad \frac{t_1 - 0}{\frac{t_1^2}{2} - 0} \cdot \frac{t_2 - 0}{\frac{t_2^2}{2} - 0} = -1$$

$$\Rightarrow \quad t_1 t_2 = -4 \quad \text{...(i)}$$

$$\because \quad \text{ar}(\Delta OPQ) = 3\sqrt{2}$$

$$\therefore \quad \frac{1}{2}\begin{vmatrix} 0 & 0 & 1 \\ t_1^2/2 & t_1 & 1 \\ t_2^2/2 & t_2 & 1 \end{vmatrix} = \pm 3\sqrt{2}$$

$$\Rightarrow \quad \frac{1}{2}\left(\frac{t_1^2 t_2}{2} - \frac{t_1 t_2^2}{2}\right) = \pm 3\sqrt{2}$$

$$\Rightarrow \quad \frac{1}{4}(-4t_1 + 4t_2) = \pm 3\sqrt{2}$$

$\Rightarrow t_1 + \frac{4}{t_1} = 3\sqrt{2}$ $\quad [\because t_1 > 0 \text{ for } P]$

$\Rightarrow \quad t_1^2 - 3\sqrt{2}t_1 + 4 = 0$

$\Rightarrow \quad (t_1 - 2\sqrt{2})(t_1 - \sqrt{2}) = 0$

$\Rightarrow \quad t_1 = \sqrt{2}$ or $2\sqrt{2}$

$\therefore \quad P(1, \sqrt{2})$ or $P(4, 2\sqrt{2})$

**19.** A. Projection of $(\alpha\hat{\mathbf{i}} + \beta\hat{\mathbf{j}})$ on $(\sqrt{3}\,\hat{\mathbf{i}} + \hat{\mathbf{j}}) = \sqrt{3}$

$\Rightarrow \quad \left|\frac{\sqrt{3}\alpha + \beta}{2}\right| = \sqrt{3}$

$\Rightarrow \quad \sqrt{3}\alpha + \beta = \pm 2\sqrt{3}$ ...(i)

and $\quad \alpha - \sqrt{3}\beta = 2$ ...(ii)

On solving Eqs. (i) and (ii), we get

$\alpha = 2, -1$

$\therefore \quad |\alpha| = 2, 1$

B. Since, if any function is differentiable, then it is continuous for all $x$.

By continuity at $x = 1$,

$-3a - 2 = b + a^2$ ...(i)

By differentiability,

$$f'(x) = \begin{cases} -6ax, & x < 1 \\ b, & x > 1 \end{cases}$$

$\therefore \quad -6a = b$ ... (ii)

From Eqs. (i) and (ii), we get

$-3a - 2 = -6a + a^2$

$\Rightarrow \quad a^2 - 3a + 2 = 0$

$\Rightarrow \quad a = 1, 2$

C. $(3 - 3\omega + 2\omega^2)^{4n+3}$

$+ (2 + 3\omega - 3\omega^2)^{4n+3}$

$+ (-3 + 2\omega + 3\omega^2)^{4n+3} = 0$

$\Rightarrow \{(-3 + 2\omega + 3\omega^2)\omega\}^{4n+3}$

$+ \{-3 + 2\omega + 3\omega^2)\omega^2\}^{4n+3}$

$+ \{(-3 + 2\omega + 3\omega^2)\}^{4n+3} = 0$

$\Rightarrow \quad (-3 + 2\omega + 3\omega^2)^{4n+3}\{\omega^{4n+3}$

$+ (\omega^2)^{4n+3} + 1\} = 0$

$\Rightarrow \quad \omega^n + \omega^{2n} + 1 = 0$

which is true only when $n$ is not a multiple of 3.

$\therefore \quad n = 1, 2, 4, 5$

D. Here, $\frac{2ab}{a+b} = 4$

and $\quad 2(5 - a) = b - 5$

$\Rightarrow \quad b = 15 - 2a$ ...(i)

Now, $\quad 2ab = 4(a + b)$

$\Rightarrow \quad 2a(15 - 2a) = 4(a + 15 - 2a)$

$\Rightarrow \quad 15a - 2a^2 = 30 - 2a$

$\Rightarrow 2a^2 - 17a + 30 = 0$

$\Rightarrow \quad a = 5/2, 6$

Hence, $|q - 2a| = |10 - 2a| = 5$ or $2$

**20.** A. Since, $2(a^2 - b^2) = c^2$

$\therefore \quad 2(\sin^2 x - \sin^2 y) = \sin^2 z$

[using sine law]

$\Rightarrow \quad 2\sin(x - y)\cdot\sin(x + y) = \sin^2 z$

$[\because x + y + z = \pi]$

$\Rightarrow \quad 2\sin(x - y)\cdot\sin z = \sin^2 z$

$\Rightarrow \quad \frac{\sin(x - y)}{\sin z} = \frac{1}{2} = \lambda$

Also, $\quad \cos(n\pi\lambda) = 0$

$\Rightarrow \quad \cos\left(\frac{n\pi}{2}\right) = 0$

$\therefore \quad n = 1, 3, 5$

B. $1 + \cos 2X - 2\cos 2Y = 2\sin X \cdot \sin Y$

$\Rightarrow \quad 1 + 1 - 2\sin^2 X - 2(1 - 2\sin^2 Y)$

$= 2\sin X \cdot \sin Y$

$\Rightarrow \quad -2a^2 + 4b^2 = 2ab$

$\Rightarrow \quad a^2 + ab - 2b^2 = 0$

$\Rightarrow \quad \left(\frac{a}{b}\right)^2 + \left(\frac{a}{b}\right) - 2 = 0$

$\Rightarrow \quad \left(\frac{a}{b} + 2\right)\left(\frac{a}{b} - 1\right) = 0$

$\Rightarrow \quad \frac{a}{b} = 1, -2$

C. $\mathbf{OX} = \sqrt{3}\,\hat{\mathbf{i}} + \hat{\mathbf{j}}, \mathbf{OY} = \hat{\mathbf{i}} + \sqrt{3}\,\hat{\mathbf{j}}$,

$\mathbf{OZ} = \beta\hat{\mathbf{i}} + (1 - \beta)\hat{\mathbf{j}}$

Angle bisector of **OX** and **OY** is along the line $y = x$ and its distance from $(\beta, 1 - \beta)$ is

$$\left|\frac{\beta-(1-\beta)}{\sqrt{2}}\right| = \frac{3}{\sqrt{2}}$$

$\Rightarrow$ $2\beta - 1 = \pm 3$

$\therefore$ $\beta = 2, -1$ or $|\beta| = 1, 2$

D. Area bounded by $x = 0, x = 2, y^2 = 4x$

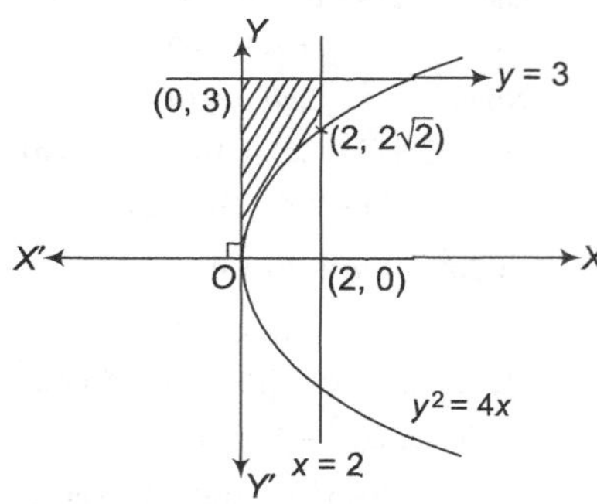


When $\alpha = 0$, then

Area bounded, $F(\alpha) = 2 \times 3 - \int_0^2 \sqrt{4}\ x\ dx$

$$= 6 - 2\left[\frac{x^{3/2}}{3/2}\right]_0^2$$

$$= 6 - \frac{4}{3} \cdot (2^{3/2} - 0)$$

$$= 6 - \frac{8}{3}\sqrt{2}$$

$\therefore$ $F(\alpha) + \frac{8\sqrt{2}}{3} = 6$ ...(i)

When $\alpha = 1$, then

$$y = |x-1| + |x-2| + x$$

$$= \begin{cases} 3x - 3, & x \geq 2 \\ x + 1, & 1 < x < 2 \\ 3 - x, & x \leq 1 \end{cases}$$

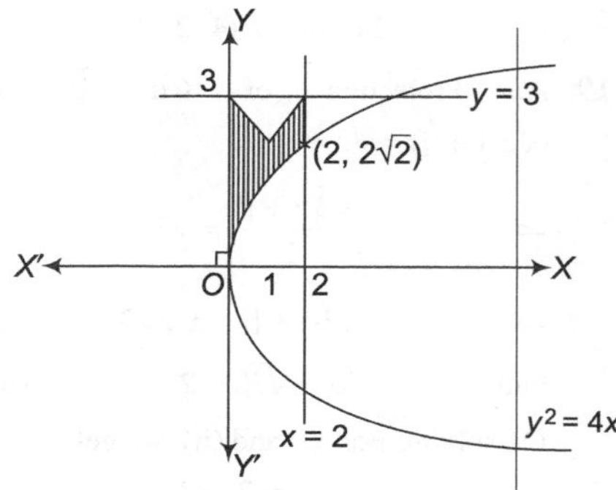


$\therefore$ Area bounded,

$$F(\alpha) = (6 - 1) - \int_0^2 2\sqrt{x}\ dx$$

$\Rightarrow$ $F(\alpha) = 5 - \frac{8}{3}\sqrt{2}$

$\therefore$ $F(\alpha) + \frac{8}{3}\sqrt{2} = 5$ ...(ii)

Hence, values of $F(\alpha) + \frac{8}{3}\sqrt{2}$ are 5 or 6.

[from Eqs. (i) and (ii)]

# Paper 2

## Physics

**1.** Let individual amplitudes are $A_0$ each. Amplitudes can be added by vector method.

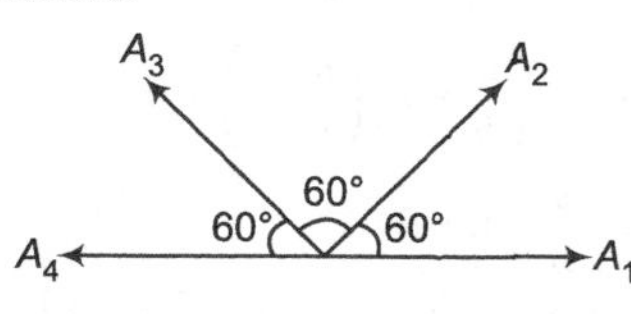


$$A_1 = A_2 = A_3 = A_4 = A_0$$

Resultant of $A_1$ and $A_4$ is zero. Resultant of $A_2$ and $A_3$ is

$$A = \sqrt{A_0^2 + A_0^2 + 2A_0A_0\cos 60^\circ} = \sqrt{3}A_0$$

This is also the net resultant.

Now, $I \propto A^2$

$\therefore$ Net intensity will become $3I_0$.

$\therefore$ Answer is 3.

**2.** Let initial numbers are $N_1$ and $N_2$.

$$\frac{\lambda_1}{\lambda_2} = \frac{\tau_2}{\tau_1} = \frac{2\tau}{\tau}$$

$$= 2 = \frac{T_2}{T_1} \quad (T = \text{Half life})$$

$$A = \frac{-dN}{dt} = \lambda N$$

Initial activity is same

$\therefore$ $\lambda_1 N_1 = \lambda_2 N_2$ ...(i)

Activity at time $t$,

$$A = \lambda N = \lambda N_0 e^{-\lambda t}$$

$$A_1 = \lambda_1 N_1 e^{-\lambda_1 t}$$

$$\Rightarrow \quad R_1 = -\frac{dA_1}{dt} = \lambda_1^2 N_1 e^{-\lambda_1 t}$$

Similarly, $R_2 = \lambda_2^2 N_2 e^{-\lambda_2 t}$

After $t = 2\tau$

$$\lambda_1 t = \frac{1}{\tau_1}(t) = \frac{1}{\tau}(2\tau) = 2$$

$$\lambda_2 t = \frac{1}{\tau_2}(t) = 1$$

$$= \frac{1}{2\tau}(2\tau) = 1$$

$$\frac{R_P}{R_Q} = \frac{\lambda_1^2 N_1 e^{-\lambda_1 t}}{\lambda_2^2 N_2 e^{-\lambda_2 t}}$$

$$\frac{R_P}{R_Q} = \frac{\lambda_1}{\lambda_1}\left(\frac{e^{-2}}{e^{-1}}\right) = \frac{2}{e}$$

**3.** Applying Snell's law at $M$ and $N$,

$$\sin 60° = n \sin r \quad \text{...(i)}$$

$$\sin \theta = n \sin (60 - r) \quad \text{...(ii)}$$

Differentiating we get

$$\cos \theta \frac{d\theta}{dn} = -n \cos (60 - r) \frac{dr}{dn} + \sin (60 - r)$$

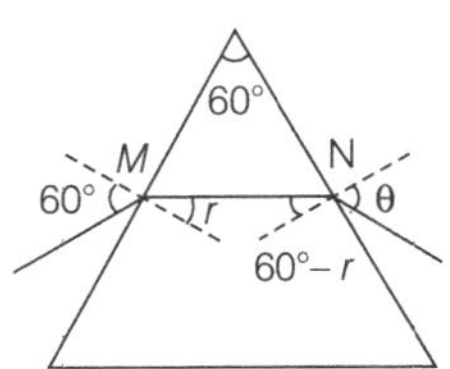


Differentiating Eq. (i),

$$n \cos r \frac{dr}{dn} + \sin r = 0$$

or $$\frac{dr}{dn} = -\frac{\sin r}{n \cos r} = \frac{-\tan r}{n}$$

$$\Rightarrow \quad \cos \theta \frac{d\theta}{dn} = -n \cos (60°-r)\left(\frac{-\tan r}{n}\right) + \sin (60°-r)$$

$$\frac{d\theta}{dn} = \frac{1}{\cos \theta}[\cos(60°-r)\tan r + \sin (60°-r)]$$

Form Eq. (i), $r = 30°$ for $n = \sqrt{3}$

$$\frac{d\theta}{dn} = \frac{1}{\cos 60}(\cos 30 \times \tan 30 + \sin 30)$$

$$= 2\left(\frac{1}{2} + \frac{1}{2}\right)$$

$$= 2$$

**4.**

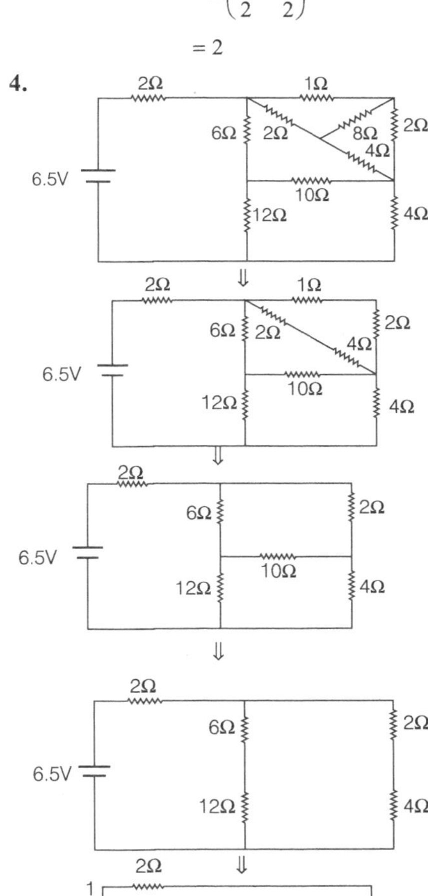


**5.** Angular momentum

$$= n\left(\frac{h}{2\pi}\right) = 3\left(\frac{h}{2\pi}\right)$$

$\therefore \quad n = 3$

Now, $\quad r_n \propto \frac{n^2}{z}$

$$\therefore \quad r_3 = \frac{(3)^2}{3}(a_0) = 3a_0$$

Now, $mv_3r_3 = 3\left(\frac{h}{2\pi}\right)$

$\therefore \quad mv_3(3a_0) = 3\left(\frac{h}{2\pi}\right)$

or $\quad \frac{h}{mv_3} = 2\pi a_0$

or $\quad \frac{h}{P_3} = 2\pi a_0 \qquad (P = mv)$

or $\quad \lambda_3 = 2\pi a_0 \qquad \left(\lambda = \frac{h}{P}\right)$

$\therefore$ Answer is 2.

**6.** For point mass at distance $r = 3l$

$$\frac{GMm}{(3l)^2} - \frac{Gm^2}{l^2} = ma \qquad \text{...(i)}$$

For point mass at distance $r = 4l$

$$\frac{GMm}{(4l)^2} + \frac{Gm^2}{l^2} = ma \qquad \text{...(ii)}$$

Equating the two equations we have,

$$\frac{GMm}{9l^2} - \frac{Gm^2}{l^2} = \frac{GMm}{16l^2} + \frac{Gm^2}{l^2}$$

$$\frac{7GMm}{144} = \frac{2Gm^2}{l^2}$$

$$m = \frac{7M}{288}$$

**7.** $E(t) = A^2e^{-\alpha t}$ ...(i)

$$\alpha = 0.2 \text{ s}^{-1}$$

$$\left(\frac{dA}{A}\right) \times 100 = 1.25\%$$

$$\left(\frac{dt}{t}\right) \times 100 = 1.50$$

$\Rightarrow \quad (dt \times 100) = 1.5t = 1.5 \times 5 = 7.5$

$$\therefore \quad \left(\frac{dE}{E}\right) \times 100 = \pm 2\left(\frac{dA}{A}\right) \times 100 \pm \alpha\,(dt \times 100)$$

Taking log on both sides of Eq. (i), we get

$$\log E = 2\log A - \alpha t$$

$$\frac{dE}{E} = \pm 2\frac{dA}{A} \pm \alpha dt$$

$$\therefore \quad \left(\frac{dE}{E}\right) \times 100 = \pm 2\left(\frac{dA}{A}\right) \times 100 \pm \alpha\,(dt \times 100)$$

$$= \pm 2\,(1.25) \pm 0.2\,(7.5)$$

$$= \pm 2.5 \pm 1.5$$

$$= \pm 4\%$$

**8.** Consider a shell of radius $r$ and thickness $dr$

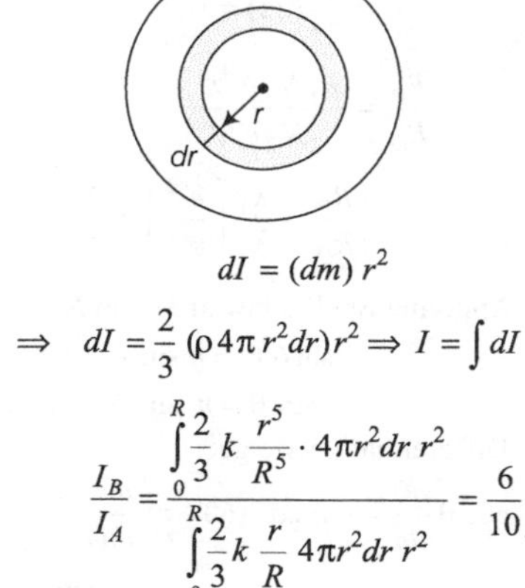


$$dI = (dm)\,r^2$$

$$\Rightarrow \quad dI = \frac{2}{3}(\rho 4\pi r^2 dr)r^2 \Rightarrow I = \int dI$$

$$\frac{I_B}{I_A} = \frac{\int_0^R \frac{2}{3} k \frac{r^5}{R^5} \cdot 4\pi r^2 dr\, r^2}{\int_0^R \frac{2}{3} k \frac{r}{R} 4\pi r^2 dr\, r^2} = \frac{6}{10}$$

**9.** The sphere with cavity can be assumed as a complete sphere with positive charge of radius $R_1$ + another complete sphere with negative charge and radius $R_2$.

$E_+ \rightarrow$ **E** due to total positive charge

$E_- \rightarrow$ **E** due to total negative charge.

$$\mathbf{E} = \mathbf{E}_+ + \mathbf{E}_-$$

If we calculate it at $P$, then $\mathbf{E}_-$ comes out to be zero.

$$\therefore \quad \mathbf{E} = \mathbf{E}_+$$

and $\mathbf{E}_+ = \frac{1}{4\pi\varepsilon_0}\frac{q}{R_1^3}(OP)$, in the direction of $OP$.

Here, $q$ is total positive charge on whole sphere.

It is in the direction of $OP$ or **a**.

Now, inside the cavity electric field comes out to be uniform at any point. This is a standard result.

**10.** $Y = \dfrac{\text{stress}}{\text{strain}}$ or $Y \propto \dfrac{1}{\text{strain}}$

(for same stress say $\sigma$)

$(\text{strain})_Q < (\text{strain})_P$

$\Rightarrow Y_Q > Y_P$

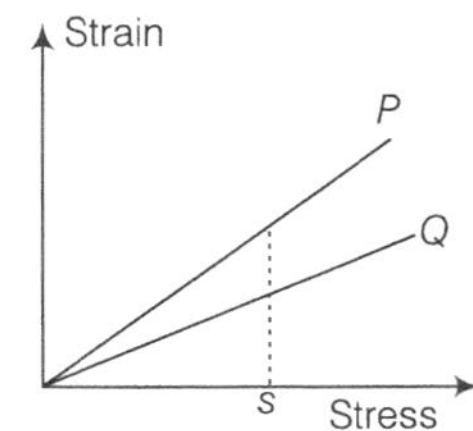


So, $P$ is more ductile than $Q$. Further, from the given figure we can also see that breaking stress of $P$ is more than $Q$. So, it has more tensile strength.

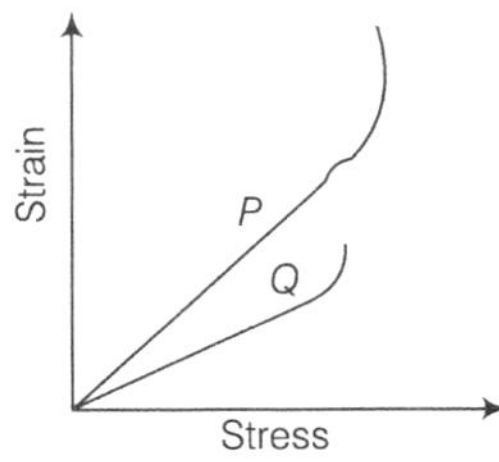


**11.** Gravitational field at a distance $r$ due to mass '$m$'

$$E = \frac{G\rho \frac{4}{3}\pi r^3}{r^2} = \frac{4G\rho\pi r}{3}$$

Consider a small element of width $dr$ and area $\Delta A$ at a distance $r$.

Pressure force on this element outwards = gravitational force on '$dm$' from '$m$' inwards

$$\Rightarrow \quad (dp)\Delta A = E(dm)$$

$$\Rightarrow \quad -dp \cdot \Delta A = \left(\frac{4}{3} G\pi\rho r\right)(\Delta A\, dr \cdot \rho)$$

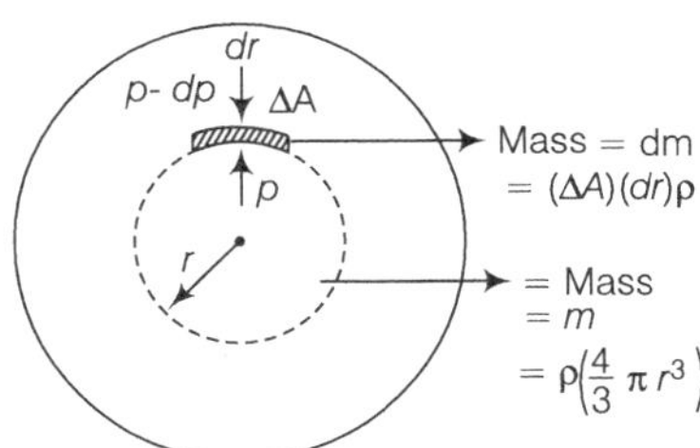


$$-\int_O^P dp = \int_R^r \left(\frac{4G\rho^2\pi}{3}\right) r\,dr$$

$$-p = \frac{4G\rho^2\pi}{3\times 2}[r^2 - R^2]$$

$$\Rightarrow \quad p = c(R^2 - r^2)$$

$$r = \frac{3R}{4},\ p_1 = c\left(R^2 - \frac{9R^2}{16}\right) = c\left(\frac{7R^2}{16}\right)$$

$$r = \frac{2R}{3},\ p_2 = c\left(R^2 - \frac{4R^2}{9}\right) = c\left(\frac{5R^2}{9}\right)$$

$$\frac{p_1}{p_2} = \frac{63}{80}$$

$$r = \frac{3R}{5},\ p_3 = c\left(R^2 - \frac{9}{25}R^2\right) = c\left(\frac{16R^2}{25}\right)$$

$$r = \frac{2R}{5},\ p_4 = c\left(R^2 - \frac{4R^2}{25}\right)$$

$$= c\left(\frac{21R^2}{25}\right) \quad \Rightarrow \frac{p_3}{p_4} = \frac{16}{21}$$

**12.**

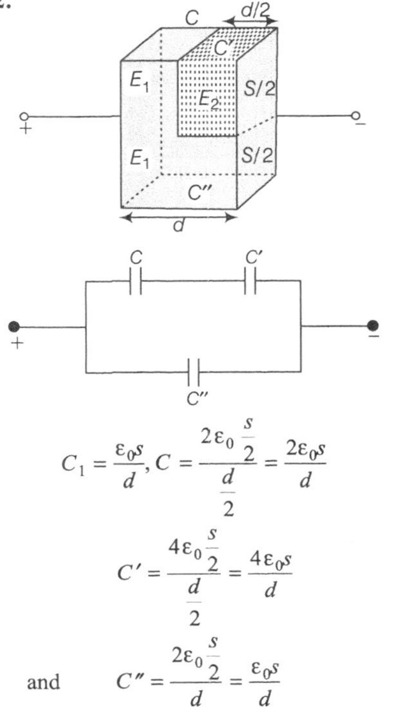


$$C_1 = \frac{\varepsilon_0 s}{d},\ C = \frac{2\varepsilon_0 \frac{s}{2}}{\frac{d}{2}} = \frac{2\varepsilon_0 s}{d}$$

$$C' = \frac{4\varepsilon_0 \frac{s}{2}}{\frac{d}{2}} = \frac{4\varepsilon_0 s}{d}$$

and $$C'' = \frac{2\varepsilon_0 \frac{s}{2}}{d} = \frac{\varepsilon_0 s}{d}$$

$$C_2 = \frac{CC'}{C + C'} + C'' = \frac{4}{3}\frac{\varepsilon_0 s}{d} + \frac{\varepsilon_0 s}{d}$$

$$= \frac{7}{3}\frac{\varepsilon_0 s}{d}\ \frac{C_2}{C_1} = \frac{7}{3}$$

**13.** **Note** *This question can be solved if right hand side chamber is assumed open, so that its pressure remains constant even if the piston shifts towards right.*

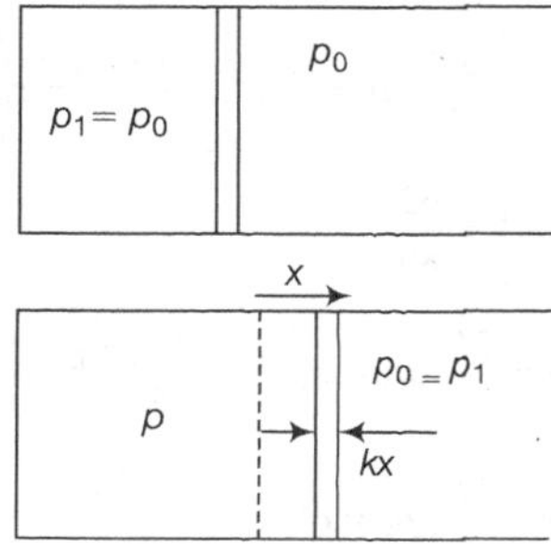


(a) $pV = nRT \Rightarrow p \propto \frac{T}{V}$

Temperature is made three times and volume is doubled

$$\Rightarrow \quad p_2 = \frac{3}{2} p_1$$

Further $x = \frac{\Delta V}{A} = \frac{V_2 - V_1}{A}$

$$= \frac{2V_1 - V_1}{A} = \frac{V_1}{A}$$

$$p_2 = \frac{3p_1}{2} = p_1 + \frac{kx}{A}$$

$$\Rightarrow \quad kx = \frac{p_1 A}{2}$$

Energy of spring

$$\frac{1}{2}kx^2 = \frac{p_1 A}{4}x = \frac{p_1 V_1}{4}$$

(b) $\Delta U = nc_V \Delta T = n\left(\frac{3}{2}R\right)\Delta T$

$$= \frac{3}{2}(p_2 V_2 - p_1 V_1)$$

$$= \frac{3}{2}\left[\left(\frac{3}{2}p_1\right)(2V_1) - p_1 V_1\right] = 3\,p_1 V_1$$

(c) $p_2 = \frac{4p_1}{3} \Rightarrow p_2 = \frac{4}{3}p_1 = p_1 + \frac{kx}{A}$

$$\Rightarrow kx = \frac{p_1 A}{3} \Rightarrow x = \frac{\Delta V}{A} = \frac{2V_1}{A}$$

$$W_{gas} = (p_0\,\Delta V + W_{spring})$$

$$= (p_1 Ax + \frac{1}{2}kx \cdot x)$$

$$= +\left(p_1 A \cdot \frac{2V_1}{A} + \frac{1}{2}\cdot\frac{p_1 A}{3}\cdot\frac{2V_1}{A}\right)$$

$$= 2p_1 V_1 + \frac{p_1 V_1}{3} = \frac{7p_1 V_1}{3}$$

(d) $\Delta Q = W + \Delta U$

$$= \frac{7p_1 V_1}{3} + \frac{3}{2}(p_2 V_2 - p_1 V_1)$$

$$= \frac{7p_1 V_1}{3} + \frac{3}{2}\left(\frac{4}{3}p_1 \cdot 3V_1 - p_1 V_1\right)$$

$$= \frac{7p_1 V_1}{3} + \frac{9}{2}p_1 V_1 = \frac{41 p_1 V_1}{6}$$

**Note** $\Delta U = \frac{3}{2}(p_2 V_2 - p_1 V_1)$ *has been obtained in part (b).*

**14.** From conservation laws of mass number and atomic number, we can say that $x = n$, $y = n$

$$(x = {}^1_0 n,\ y = {}^1_0 n)$$

$\therefore$ Only (a) and (d) options may be correct.

From conservation of momentum,

$$|P_{xe}| = |P_{sr}|$$

From $K = \frac{P^2}{2m} \Rightarrow K \propto \frac{1}{m}$

$$\frac{K_{sr}}{K_{xe}} = \frac{m_{xe}}{m_{sr}}$$

$\therefore$ $K_{sr} = 129\ \text{MeV}$,

$$K_{xe} = 86\ \text{MeV}$$

**Note** *There is no need of finding total energy released in the process.*

**15.** For floating, net weight of system = net upthrust

$\Rightarrow (\rho_1 + \rho_2)Vg = (\sigma_1 + \sigma_2)Vg$

Since string is taut, $\rho_1 < \sigma_1$ and $\rho_2 > \sigma_2$

$$v_P = \frac{2r^2 g}{2\eta_2}(\sigma_2 - \rho_1)$$

(upward terminal velocity)

$$v_Q = \frac{2r^2 g}{9\eta_1}(\rho_2 - \sigma_1)$$

(downward terminal velocity)

$$\left|\frac{v_P}{v_Q}\right| = \frac{\eta_1}{\eta_2}$$

Further, $\bar{v}_P \cdot \bar{v}_Q$ will be negative as they are opposite to each other.

**16.** (A) Energy of inductor

$$\frac{1}{2}LI^2 = \frac{1}{2}\frac{\mu_0 N^2 A}{l}I^2 e$$

Energy of capacitor

$$= \frac{1}{2}CV^2 = \frac{1}{2}\varepsilon_0\frac{A}{d}V^2 \mu_0\frac{A}{l}I^2 \text{ and}$$

$\varepsilon_0 \dfrac{A}{d}V^2$ have same dimensions.

So, $\mu_0 I^2$ and $\varepsilon_0 V^2$ have same dimensions.

(C) $Q = CV \Rightarrow \dfrac{Q}{t} = \dfrac{CV}{t}$

$$I = \varepsilon_0 \frac{A}{l}\frac{V}{t}$$

$\dfrac{A}{lt}$ have unit of speed.

So, $I = \varepsilon_0 cV$

**17.** $\dfrac{4}{3}\sin i = \dfrac{\sqrt{45}}{4}\sin(90 - \theta_c)$

$$= \frac{\sqrt{45}}{4}\cos\theta_c$$

$$\sin\theta_c = \frac{n_2}{n_1}$$

$$\therefore \quad \cos\theta_c = \sqrt{1 - \left(\frac{n_2}{n_1}\right)^2}$$

$$\Rightarrow \quad \frac{4}{3}\sin i = \frac{\sqrt{45}}{4}\frac{3}{\sqrt{45}}$$

$$\sin i = \frac{9}{16}$$

In second case,

$$\sin\theta_c = \frac{n_2}{n_1} = \frac{7}{8}$$

$$\Rightarrow \quad \cos\theta_c = \frac{\sqrt{15}}{8}$$

$$\frac{16}{3\sqrt{15}}\sin i = \frac{8}{5}\sin(90 - \theta_c)$$

Simplifying we get,

$$\sin i = \frac{9}{16}$$

(a) is correct.

Same approach can be adopted for other options. Correct answers are (a) and (c).

**18.** $\sin i_m = n_1 \sin(90 - \theta_c)$

$$\Rightarrow \quad \sin i_m = n_1 \cos\theta_c$$

$$\Rightarrow \quad NA = n_1\sqrt{1 - \sin^2\theta_c}$$

$$= n_1\sqrt{1 - \frac{n_2^2}{n_1^2}} = \sqrt{n_1^2 - n_2^2}$$

Substituting the values we get,

$$NA_1 = \frac{3}{4} \text{ and } NA_2 = \frac{\sqrt{15}}{5} = \sqrt{\frac{3}{4}}$$

$$NA_2 < NA_1$$

Therefore, the numerical aperture of combined structure is equal to the lesser of the two numerical aperture, which is $NA_2$.

**19.** $F_B = Bev = Be\dfrac{I}{nAe} = \dfrac{BI}{nA}$

$$F_e = eE$$

$$\Rightarrow \quad F_e = F_B$$

$$eE = \frac{BI}{nA}$$

$$\Rightarrow \quad E = \frac{B}{nAe}$$

$$V = Ed = \frac{BI}{nAe}\cdot w = \frac{BIw}{n(wd)e} = \frac{BI}{ned}$$

$$\frac{V_1}{V_2} = \frac{d_2}{d_1}$$

$\Rightarrow$ if $w_1 = 2w_2$ and $d_1 = d_2$

$$V_1 = V_2$$

$\therefore$ Correct answers are (a) and (d).

**20.** $V = \dfrac{BI}{ned}$

$$\Rightarrow \quad \frac{V_1}{V_2} = \frac{B_1}{B_2} \times \frac{n_2}{n_1}$$

If $B_1 = B_2$ and $n_1 = 2n_2$, then $V_2 = 2V_1$

If $B_1 = 2B_2$ and $n_1 = n_2$, then $V_2 = 0.5V_1$

$\therefore$ Correct answers are (a) and (c).

# Chemistry

**1.** acetylbromidodicarbonylbis (triethylphosphine)iron (II)

**2.** All six complex will show *cis-trans* isomerism

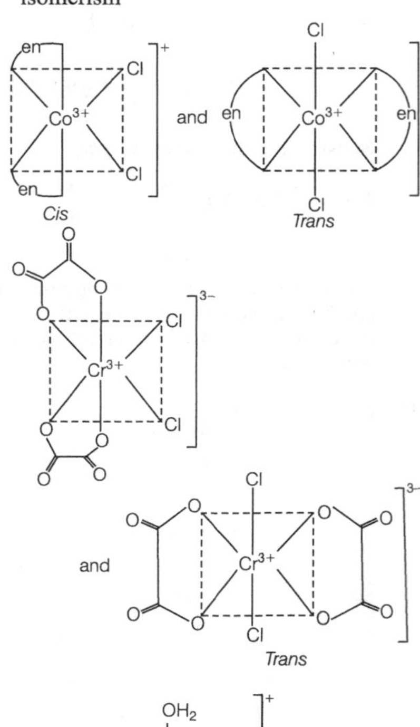

*Cis* and *Trans*

*Cis* and *Trans*

*Cis* and

*Cis* and

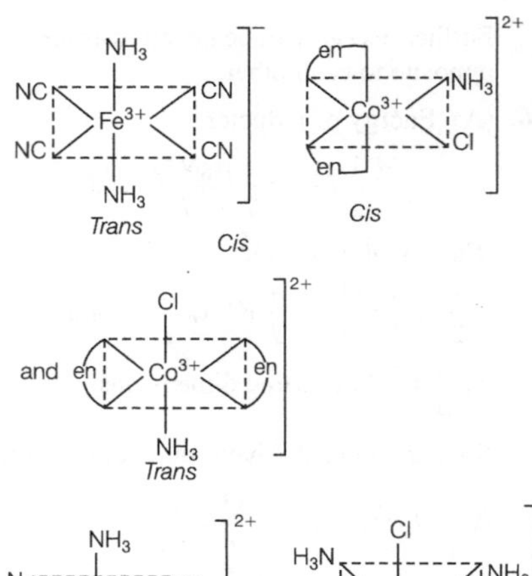

*Cis* and *Trans*

*Cis* and *Trans*

**3.** $B_2H_6 + 6\ CH_3OH \longrightarrow 2[B(OCH_3)_3] + 6H_2$

Therefore, from 3 moles of $B_2H_6$, 6 moles of $B(OCH_3)_3$ will be

**4.** Degree of ionisation $(\alpha) = \dfrac{\wedge m}{\wedge^{\infty}}$

Let $\wedge m\ (HY) = x$

$$\Rightarrow \quad \wedge m\ (HX) = \frac{X}{10}$$

$$\Rightarrow \quad \frac{\wedge m\ (HX)}{\wedge m\ (XY)} = \frac{1}{10} = \frac{\alpha\ (HX)}{\alpha\ (HY)}$$

$$[\because \wedge^{\infty}\ (HX) = \wedge^{\infty}\ (HY)]$$

Also : $K_a(HX) = (0.01)\,[\alpha(HX)]^2$ ...(i)

$$K_a(HY) = (0.10)\,[\alpha(HY)]^2$$

$$= 0.10\,[10\ \alpha\ (HX)]^2$$

$$= 10\,[\alpha\ (HX)]^2 \quad \text{... (ii)}$$

$$\Rightarrow \quad \frac{K_a(HX)}{K_a(HY)} = \frac{0.01}{10}$$

$$= \frac{1}{1000}$$

$$\Rightarrow \quad \log K_a(HX) - \log K_a(HY) = -3$$

$$\Rightarrow -\log K_a(HX) - [-\log K_a(HY)] = 3$$

$$\Rightarrow \quad pK_a(HX) - pK_a(HY) = 3$$

**5.** ${}_{92}U^{238} \longrightarrow {}_{82}Pb^{206} + 8\,{}_2He^4(g) + 6\,{}_{-1}\beta^0$

$n$(gas)[Initial] = 1 (air)

$n$(gas)[Final] = 8(He) + 1(air) = 9

⇒ At constant temperature and volume;

$p \propto n.$

So, $$\frac{p_f}{p_i} = \frac{n_f}{n_i} = \frac{9}{1} = 9$$

**6.** The balanced redox reaction is

$$MnO_4^- + [Fe(H_2O)_2(C_2O_4)_2]^{2-} + 8H^+ \longrightarrow Mn^{2+} + Fe^{3+} + 4CO_2 + 6H_2O$$

$$\Rightarrow \quad \frac{r[H^+]}{r[MnO_4^-]} = \frac{8}{1} = 81111$$

**7.**

H, HO, $H_3C$, $CH_3$ $\xrightarrow[\text{Heat}]{H^+}$

$H_3C$, $CH_3$ $\xrightarrow{\text{Methyl shift}}$ $H_3C$, ⊕, $CH_3$

$\xrightarrow[\text{Heat}]{-H^{\oplus}}$ $CH_3$, $CH_3$ P $\quad P \xrightarrow[0°C]{KMnO_4(\text{dil})}$

HO, OH, OH, $CH_3$, OH, $CH_3$

**8.** I. Gattermann-Koch reaction.

II. $C_6H_5CHCl_2 \xrightarrow[100°C\ (S_N1)]{H_2O} C_6H_5CH(OH)_2$ (*gem*-diol) $\xrightarrow{-H_2O} C_6H_5CHO$

III. Rosenmund's reduction.

IV. Acid chloride, anhydride and ester undergo controlled reduction with di-iso-butylaluminium hydride (DIBAL-H) at – 78°C to give aldehydes.

**9.**

$(CH_3)_2SiCl_2 + H_2O \longrightarrow HO{-}Si(CH_3)_2{-}OH$

$\xrightarrow{\text{Polymerisations}} HO{-}Si(CH_3)_2{-}O{-}[Si(CH_3)_2{-}O]_n{-}Si(CH_3)_2{-}OH \xrightarrow{(CH_3)_3XSiCl}$

Linear chain polymer

$CH_3{-}Si(CH_3)_2{-}O{-}Si(CH_3)_2{-}O{-}[Si(CH_3)_2{-}O]_n{-}Si(CH_3)_2{-}O{-}Si(CH_3)_2{-}CH_3$

Chain terminated linear chain silicone

**10.** Since, adsorption involves electron transfer from metal to $O_2$, it is chemical adsorption not physical adsorption, hence (*a*) is incorrect. Adsorption is spontaneous which involves some bonding between adsorbent and adsorbate, hence exothermic. The last occupied molecular orbital in $O_2$ is $\pi^* 2p$. Hence, electron transfer from metal to oxygen will increase occupancy of $\pi^* 2p$ molecular orbitals. Also increase in occupancy of $\pi^* 2p$ orbitals will decrease bond order and hence increase bond length of $O_2$.

**11.** Equation of state $p(V - b) = RT$ indicates absence of intermolecular attraction or repulsion, hence interatomic potential remains constant on increasing '$\pi$' in the beginning. As the molecules come very close, their electronic and nuclear repulsion increases abruptly.

**12.** $H_3C$– $\xrightarrow[\text{(ii) Zn–}H_2O]{\text{(i) }O_3}$ $H_3C$, $-\overset{O}{\overset{\|}{C}}-H$, CHO (R)

$\xrightarrow{NH_3}$ $H_3C$, CHO, CH=NH (X)

In the above reaction, $NH_3$ prefer to attack at aliphatic aldehyde group than an less reactive aromatic aldehyde group.

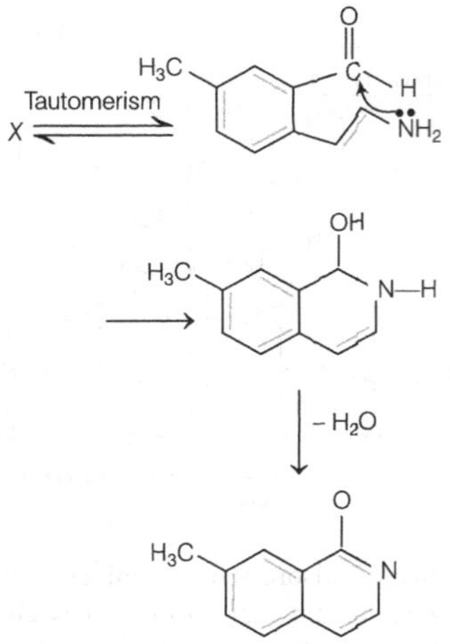

X ⇌ (Tautomerism) $H_3C$–C₆H₃(CHO)(CH=CH–$\ddot{N}H_2$) → $H_3C$–(cyclic, OH, N–H) $\xrightarrow{-H_2O}$ $H_3C$–(isoquinolinone) A

**13.** C₆H₆ + $CH_2{=}CH{-}CH_3$ $\xrightarrow{H^+}$ C₆H₅–CH($CH_3$)$CH_3$ (Cumene) $\xrightarrow{O_2}$ C₆H₅–C($CH_3$)₂–O–OH (Cumene hydroperoxide)

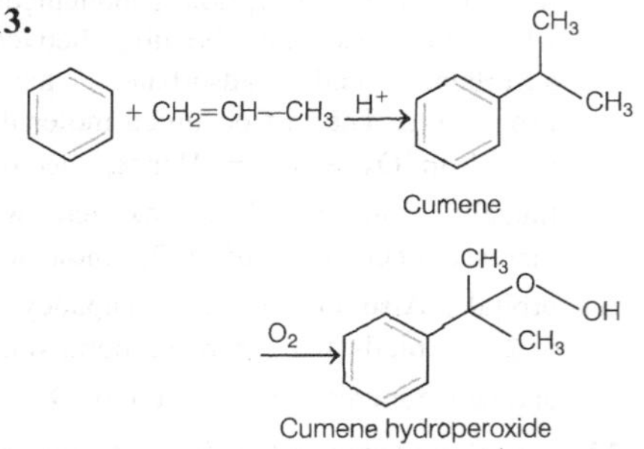

Cumene hydroperoxide formed above is an intermediate in the synthesis of phenol.

**14.** C₆H₅–$NH_2$ $\xrightarrow[0°C]{NaNO_2-HCl}$ C₆H₅–$\overset{+}{N_2}Cl^-$ (V) $\xrightarrow[NaOH]{\beta\text{-naphthol}}$ (N=N–Ph, OH naphthalene) W

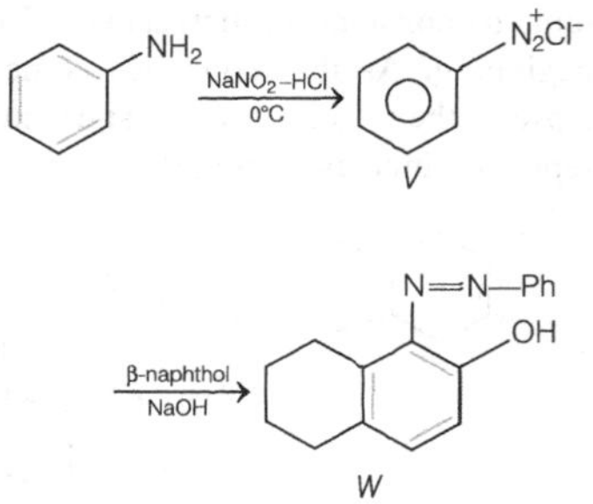

**15.** $H{-}\ddot{O}{-}\ddot{Cl}{:}$ (i) $\quad H{-}\ddot{O}{-}\ddot{Cl}{=}\ddot{O}$ (ii)

$H{-}\ddot{O}{-}Cl(=O){=}\ddot{O}$ (iii) $\quad H{-}\ddot{O}{-}Cl(=O)_2{=}\ddot{O}$ (iv)

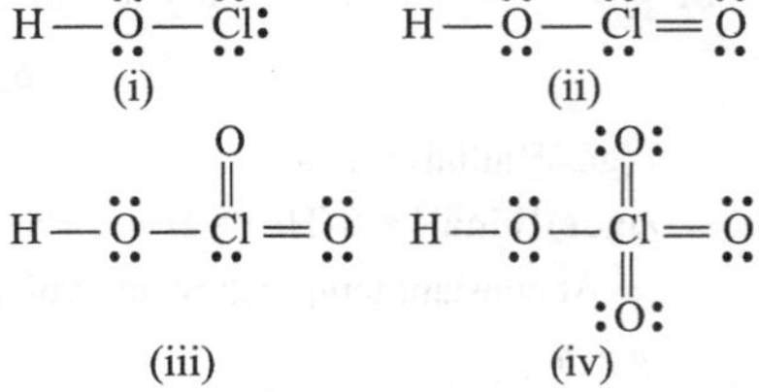

(*a*) Number of $Cl^- = O$ bonds in (ii) and (iii) together is three. Hence, wrong.

(*b*) Number of Lone Pair on Cl in (ii) and (iii) together is three. Hence, correct.

(*c*) In (iv), Cl is $sp^3$-hybridised. Hence, correct.

(*d*) Amongst (i) to (iv), the strongest acid is (iv). Hence, wrong.

**16.** Only radicals of I and II group of qualitative analysis get precipitated with $H_2S$ in the presence of dilute HCl.

(*c*) $Cu^{2+} + H_2S \xrightarrow{H^+} \underset{\text{Black}}{CuS\downarrow}$

$Pb^{2+} + H_2S \xrightarrow{H^+} \underset{\text{Black}}{PbS\downarrow}$

(*d*) $Hg^{2+} + H_2S \xrightarrow{H^+} \underset{\text{Black}}{HgS\downarrow}$

$Bi^{3+} + H_2S \xrightarrow{H^+} \underset{\text{Brown ppt}}{Bi_2S_3\downarrow}$

$Ba^{2+}$, $Zn^{2+}$ and $Fe^{3+}$ are not precipitated as sulphide.

## Paragraph 1 (17-18)

The reaction condition indicates that starting compound is phenyl acetylene.

**17.** C₆H₅–C≡C–H $\xrightarrow[H_2]{Pd/BaSO_4}$ C₆H₅–CH=$CH_2$ $\xrightarrow[\text{(ii) } H_2O_2,\ NaOH,\ H_2O]{\text{(i) } B_2H_6}$ C₆H₅–CH₂CH₂–OH (2-phenyl ethanol) (X)

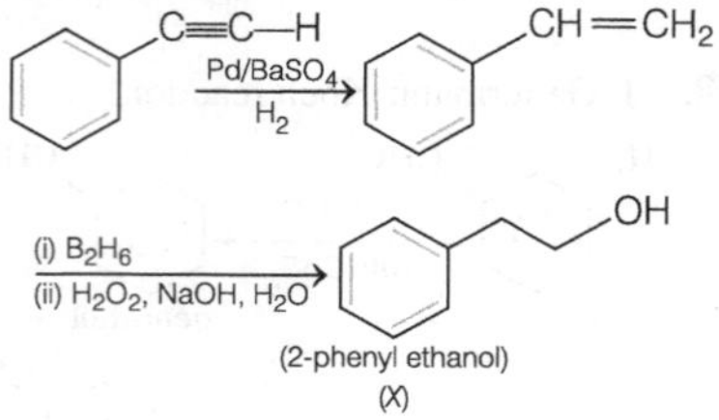

Hydroboration oxidation brings about anti-Markonikoff's hydration of alkene.

**18.** $\xrightarrow[H_2SO_4]{H_2O,\ HgSO_4}$ $\xrightarrow{C_2H_5MgBr,\ H_2O}$ $\xrightarrow{H^+/\Delta}$

**Paragraph 2** (19-20)

**19.** Let $C$ $JK^{-1}$ be the heat capacity of calorimeter.

Mass of solution

$= 200\text{ mL} \times 1\text{ g mL}^{-1} = 200\text{ g}$

Heat evolved in Expt.1

$= 57 \times 1000 \times 0.1(\text{mol}) = 5700\text{ J}$

$\Rightarrow 5700\text{ J} = (200 \times 4.2 + C) \times 5.7$

$\Rightarrow \quad 1000 = 200 \times 4.2 + C \qquad \ldots(\text{i})$

Let $x$ kJ/mol is heat evolved in neutralisation of acetic acid.

$\Rightarrow x \times 1000 \times 0.10$

$= (200 \times 4.2 + C) \times 5.6$

$\Rightarrow \dfrac{x \times 100}{5.6} = 200 \times 4.2 + C \qquad \ldots(\text{ii})$

From (i) and (ii) : $x = 56$ kJ/mol

$\Rightarrow$ Enthalpy of ionisation of acetic acid $= -56 - (-57) = 1$ kJ/mol

**20.**

$$CH_3COOH + NaOH \longrightarrow CH_3COONa + H_2O$$

| | $CH_3COOH$ | $NaOH$ | $CH_3COONa$ | $H_2O$ |
|---|---|---|---|---|
| | 200 mmol | 100 mol | 0 | 0 |
| | 100 mmol | 0 | 100 mmol | |

A buffer is now formed.

$$K_a = \frac{[H^+][CH_3COO^-]}{[CH_3COOH]} = [H^+]$$

$[\because [CH_3COOH] = [CH_3COO^-]]$

$\Rightarrow pH = pK_a = -\log(2 \times 10^{-5})$

$= 5 - \log 2 = 4.7$

# Mathematics

**1.** Given, $\dfrac{S_7}{S_{11}} = \dfrac{6}{11}$ and $130 < t_7 < 140$

$$\Rightarrow \frac{\frac{7}{2}[2a + 6d]}{\frac{11}{2}[2a + 10d]} = \frac{6}{11}$$

$$\Rightarrow \frac{7(2a + 6d)}{(2a + 10d)} = 6$$

$\Rightarrow \quad a = 9d \qquad \ldots(\text{i})$

Also, $\quad 130 < t_7 < 140$

$\Rightarrow \quad 130 < a + 6d < 140$

$\Rightarrow \quad 130 < 9d + 6d < 140$ [from Eq. (i)]

$\Rightarrow \quad 130 < 15d < 140$

$\Rightarrow \quad \dfrac{26}{3} < d < \dfrac{28}{3}$

[since, $d$ is a natural number]

$\therefore \quad d = 9$

**2.** Coefficient of $x^9$ in the expansion of $(1 + x)(1 + x^2)(1 + x^3) \ldots (1 + x^{100})$

= Terms having $x^9$

$= [1^{99} \cdot x^9,\ 1^{98} \cdot x \cdot x^8,\ 1^{98} \cdot x^2 \cdot x^7,\ 1^{98} \cdot x^3 \cdot x^6,$
$1^{98} \cdot x^4 \cdot x^5,\ 1^{97} \cdot x \cdot x^2 \cdot x^6,\ 1^{97} \cdot x \cdot x^3 \cdot x^5,$
$1^{97} \cdot x^2 \cdot x^3 \cdot x^4]$

$\therefore$ Coefficient of $x^9 = 8$

**3.**

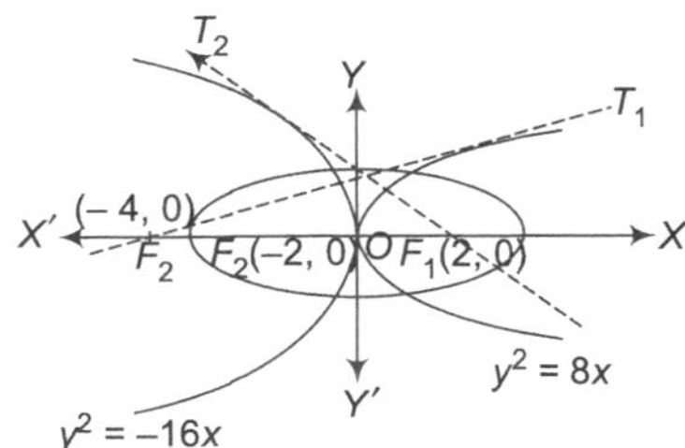


Tangent to $P_1$ passes through $(2f_2, 0)$ i.e. $(-4, 0)$.

$\therefore\ T_1 : y = m_1 x + \dfrac{2}{m_1}$

$\Rightarrow \quad 0 = -4m_1 + \dfrac{2}{m_1}$

$\Rightarrow \quad m_1^2 = 1/2 \qquad \ldots(\text{i})$

Also, tangent to $P_2$ passes through $(f_1, 0)$ i.e. (2, 0).

$$\Rightarrow T_2 : y = m_2 x + \frac{(-4)}{m_2}$$

$$\Rightarrow \quad 0 = 2m_2 - \frac{4}{m_2}$$

$$\Rightarrow \quad m_2^2 = 2 \quad \ldots(\text{ii})$$

$$\therefore \quad \frac{1}{m_1^2} + m_2^2 = 2 + 2 = 4$$

**4.** Given, $\lim_{\alpha \to 0}\left[\frac{e^{\cos(\alpha^n)} - e}{\alpha^m}\right] = -\frac{e}{2}$

$$\Rightarrow \lim_{\alpha \to 0} \frac{e\{e^{\cos(\alpha^n)-1} - 1\}}{\cos(\alpha^n) - 1} \cdot \frac{\cos(\alpha^n) - 1}{\alpha^m} = \frac{-e}{2}$$

$$\Rightarrow \lim_{\alpha \to 0} e\left\{\frac{e^{\cos(\alpha^n)-1} - 1}{\cos(\alpha^n) - 1}\right\} \cdot \lim_{\alpha \to 0} \frac{-2\sin^2 \frac{\alpha^n}{2}}{\alpha^m}$$

$$= -e/2$$

$$\Rightarrow e \times 1 \times (-2) \lim_{\alpha \to 0} \frac{\sin^2\left(\frac{\alpha^n}{2}\right)}{\frac{\alpha^{2n}}{4}} \cdot \frac{\alpha^{2n}}{4\alpha^m} = \frac{-e}{2}$$

$$\Rightarrow \quad e \times 1 \times -2 \times 1 \times \lim_{\alpha \to 0} \frac{\alpha^{2n-m}}{4} = \frac{-e}{2}$$

For this to be exists,

$$2n - m = 0 \Rightarrow \frac{m}{n} = 2$$

**5.** Here, $\alpha = \int_0^1 e^{(9x + 3\tan^{-1} x)}\left(\frac{12 + 9x^2}{1 + x^2}\right) dx$

Put $\quad 9x + 3\tan^{-1} x = t$

$$\Rightarrow \quad \left(9 + \frac{3}{1 + x^2}\right) dx = dt$$

$$\therefore \quad \alpha = \int_0^{9 + 3\pi/4} e^t \, dt = [e^t]_0^{9 + 3\pi/4}$$

$$= e^{9 + 3\pi/4} - 1$$

$$\Rightarrow \quad \log_e |1 + \alpha| = 9 + \frac{3\pi}{4}$$

$$\Rightarrow \quad \log_e |\alpha + 1| - \frac{3\pi}{4} = 9$$

**6.** Here, $\lim_{x \to 1} \frac{F(x)}{G(x)} = \frac{1}{14}$

$$\Rightarrow \quad \lim_{x \to 1} \frac{F'(x)}{G'(x)} = \frac{1}{14}$$

[using L' Hospital's rule] ...(i)

As $\quad F(x) = \int_{-1}^x f(t)\, dt$

$$\Rightarrow \quad F'(x) = f(x) \quad \ldots(\text{ii})$$

and $\quad G(x) = \int_{-1}^x t|f\{f(t)\}|\, dt$

$$\Rightarrow \quad G'(x) = x|f\{f(x)\}| \quad \ldots(\text{iii})$$

$$\therefore \lim_{x \to 1} \frac{F(x)}{G(x)} = \lim_{x \to 1} \frac{F'(x)}{G'(x)} = \lim_{x \to 1} \frac{f(x)}{x|f\{f(x)\}|}$$

$$= \frac{f(1)}{1|f\{f(1)\}|} = \frac{1/2}{|f(1/2)|} \ldots(\text{iv})$$

Given, $\quad \lim_{x \to 1} \frac{F(x)}{G(x)} = \frac{1}{14}$

$$\therefore \quad \frac{\frac{1}{2}}{\left|f\left(\frac{1}{2}\right)\right|} = \frac{1}{14} \Rightarrow \left|f\left(\frac{1}{2}\right)\right| = 7$$

**7.** Here, $\mathbf{s} = 4\mathbf{p} + 3\mathbf{q} + 5\mathbf{r} \quad \ldots(\text{i})$

and $\mathbf{s} = (-\mathbf{p} + \mathbf{q} + \mathbf{r})x + (\mathbf{p} - \mathbf{q} + \mathbf{r})y + (-\mathbf{p} - \mathbf{q} + \mathbf{r})z \quad \ldots(\text{ii})$

$$\therefore \quad 4\mathbf{p} + 3\mathbf{q} + 5\mathbf{r} = \mathbf{p}(-x + y - z) + \mathbf{q}(x - y - z) + \mathbf{r}(x + y + z)$$

On comparing both sides, we get

$$-x + y - z = 4, \; x - y - z = 3$$

and $\quad x + y + z = 5$

On solving above equations, we get

$$x = 4, \; y = \frac{9}{2}, \; z = \frac{-7}{2}$$

$$\therefore \; 2x + y + z = 8 + \frac{9}{2} - \frac{7}{2} = 9$$

**8.** Given, $\alpha_k = \cos\left(\frac{k\pi}{7}\right) + i\sin\left(\frac{k\pi}{7}\right)$

$$= \cos\left(\frac{2k\pi}{14}\right) + i\sin\left(\frac{2k\pi}{14}\right)$$

$\therefore$ $\alpha_k$ are vertices of regular polygon having 14 sides.

Let the side length of regular polygon be $a$.

$\therefore \quad |\alpha_{k+1} - \alpha_k|$ = length of a side of the regular polygon = $a$ ...(i) and $|\alpha_{4k-1} - \alpha_{4k-2}|$ = length of a side of the regular polygon $= a$ ...(ii)

$$\therefore \quad \frac{\sum_{k=1}^{12}|\alpha_{k+1}-\alpha_k|}{\sum_{k=1}^{3}|\alpha_{4k-1}-\alpha_{4k-2}|}=\frac{12(a)}{3(a)}=4$$

**9.** Let $F(x) = f(x) - 3g(x)$

$\therefore F(-1) = 3, F(0) = 3$ and $F(2) = 3$

So, $F'(x)$ will vanish atleast twice in $(-1, 0) \cup (0, 2)$.

$\because F''(x) > 0$ or $< 0, \forall x \in (-1, 0) \cup (0, 2)$

Hence, $f'(x) - 3g'(x) = 0$ has exactly one solution in $(-1, 0)$ and one solution in $(0,2)$.

**10.** Here, $f(x) = 7\tan^8 x + 7\tan^6 x - 3\tan^4 x - 3\tan^2 x$

for all $x \in \left(\frac{-\pi}{2}, \frac{\pi}{2}\right)$

$$\therefore \quad f(x) = 7\tan^6 x \sec^2 x - 3\tan^2 x \sec^2 x$$
$$= (7\tan^6 x - 3\tan^2 x)\sec^2 x$$

Now,

$$\int_0^{\pi/4} x\, f(x)dx = \int_0^{\pi/4} \underset{\text{I}}{x}\, \underset{\text{II}}{(7\tan^6 x - 3\tan^2 x)\sec^2 x}\, dx$$
$$= [x(\tan^7 x - \tan^3 x)]_0^{\pi/4} - \int_0^{\pi/4} 1\,(\tan^7 x - \tan^3 x)dx$$
$$= 0 - \int_0^{\pi/4} \tan^3 x(\tan^4 x - 1)dx$$
$$= -\int_0^{\pi/4} \tan^3 x\,(\tan^2 x - 1)\sec^2 x\, dx$$

Put $\tan x = t \Rightarrow \sec^2 x\, dx = dt$

$$\therefore \quad \int_0^{\pi/4} x\, f(x)dx = -\int_0^1 t^3(t^2-1)dt$$
$$= \int_0^1 (t^3 - t^5)dt = \left[\frac{t^4}{4} - \frac{t^5}{5}\right]_0^1 = \frac{1}{4} - \frac{1}{6} = \frac{1}{12}$$

Also,

$$\int_0^{\pi/4} f(x)dx = \int_0^{\pi/4} (7\tan^6 x - 3\tan^2 x)\sec^2 x\, dx$$
$$= \int_0^1 (7t^6 - 3t^2)dt = [t^7 - t^3]_0^1 = 0$$

**11.** (b) Here, $\quad f'(x) = \dfrac{192x^3}{2 + \sin^4 \pi x}$

$$\therefore \quad \frac{192x^3}{3} \le f'(x) \le \frac{192x^3}{2}$$

On integrating between the limits $\frac{1}{2}$ to $x$, we get

$$\int_{1/2}^{x} \frac{192x^3}{3}dx \le \int_{1/2}^{x} f'(x)dx \le \int_{1/2}^{x} \frac{192x^3}{2}dx$$
$$\Rightarrow \quad \frac{192}{12}\left(x^4 - \frac{1}{16}\right) \le f(x) - f(0) \le 24x^4 - \frac{3}{2}$$
$$\Rightarrow \quad 16x^4 - 1 \le f(x) \le 24x^4 - \frac{3}{2}$$

Again integrating between the limits $\frac{1}{2}$ to 1, we get

$$\int_{1/2}^{1} (16x^4 - 1)dx \le \int_{1/2}^{1} f(x)dx \le \int_{1/2}^{1}\left(24x^4 - \frac{3}{2}\right)dx$$
$$\Rightarrow \quad \left[\frac{16x^5}{5} - x\right]_{1/2}^{1} \le \int_{1/2}^{1} f(x)dx \le \left[\frac{24x^5}{5} - \frac{3}{2}x\right]_{1/2}^{1}$$
$$\Rightarrow \quad \left(\frac{11}{5} + \frac{2}{5}\right) \le \int_{1/2}^{1} f(x)dx \le \left(\frac{33}{10} + \frac{6}{10}\right)$$
$$\Rightarrow \quad 2.6 \le \int_{1/2}^{1} f(x)dx \le 3.9$$

**12.** Given, $x_1$ and $x_2$ are roots of $\alpha x^2 - x + \alpha = 0$.

$$\therefore \quad x_1 + x_2 = \frac{1}{\alpha} \text{ and } x_1 x_2 = 1$$

Also, $\quad |x_1 - x_2| < 1$

$\Rightarrow \quad |x_1 - x_2|^2 < 1 \Rightarrow (x_1 - x_2)^2 < 1$

or $\quad (x_1 + x_2)^2 - 4x_1x_2 < 1$

$$\Rightarrow \quad \frac{1}{\alpha^2} - 4 < 1 \text{ or } \frac{1}{\alpha^2} < 5$$
$$\Rightarrow \quad 5\alpha^2 - 1 > 0$$

or $\quad (\sqrt{5}\alpha - 1)(\sqrt{5}\alpha + 1) > 0$

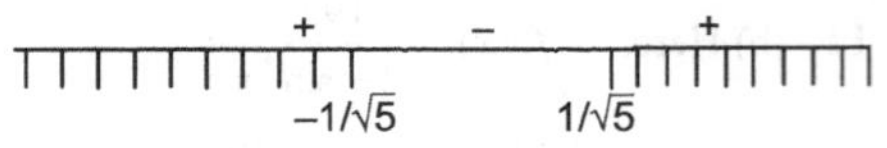


$\therefore \quad \alpha \in \left(-\infty, -\frac{1}{\sqrt{5}}\right) \cup \left(\frac{1}{\sqrt{5}}, \infty\right) \quad \ldots\text{(i)}$

Also, $\quad D > 0$

$\Rightarrow \quad 1 - 4\alpha^2 > 0 \quad \text{or} \quad \alpha \in \left(-\frac{1}{2}, \frac{1}{2}\right) \ldots\text{(ii)}$

From Eqs. (i) and (ii), we get

$$\alpha \in \left(-\frac{1}{2}, \frac{-1}{\sqrt{5}}\right) \cup \left(\frac{1}{\sqrt{5}}, \frac{1}{2}\right)$$

**13.** Here, $\alpha = 3\sin^{-1}\left(\frac{6}{11}\right)$

and $\beta = 3\cos^{-1}\left(\frac{4}{9}\right)$ as $\frac{6}{11} > \frac{1}{2}$

$\Rightarrow \quad \sin^{-1}\left(\frac{6}{11}\right) > \sin^{-1}\left(\frac{1}{2}\right) = \frac{\pi}{6}$

$\therefore \quad \alpha = 3\sin^{-1}\left(\frac{6}{11}\right) > \frac{\pi}{2} \quad \Rightarrow \quad \cos\alpha < 0$

Now, $\quad \beta = 3\cos^{-1}\left(\frac{4}{9}\right)$

As $\frac{4}{9} < \frac{1}{2} \Rightarrow \cos^{-1}\left(\frac{4}{9}\right) > \cos^{-1}\left(\frac{1}{2}\right) = \frac{\pi}{3}$

$\therefore \quad \beta = 3\cos^{-1}\left(\frac{4}{9}\right) > \pi$

$\therefore \quad \cos\beta < 0 \text{ and } \sin\beta < 0$

Now, $\alpha + \beta$ is slightly greater than $\frac{3\pi}{2}$.

$\therefore \quad \cos(\alpha + \beta) > 0$

**14.** Here, $E_1: \frac{x^2}{a^2} + \frac{y^2}{b^2} = 1, \ (a > b)$

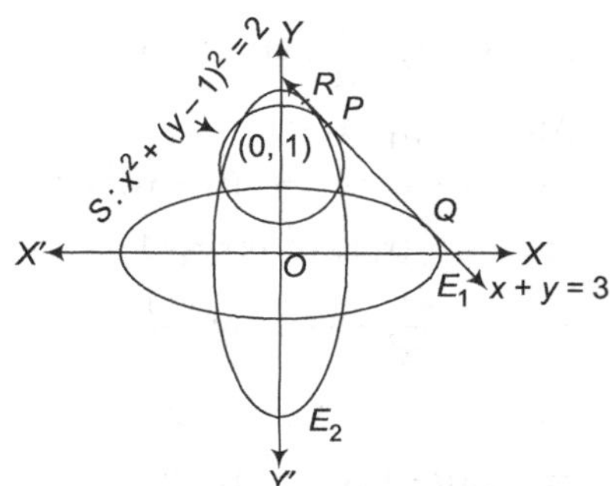


$E_2: \frac{x^2}{c^2} + \frac{y^2}{d^2} = 1, \ (c < d)$

and $\quad S: x^2 + (y-1)^2 = 2$

as tangent to $E_1$, $E_2$ and $S$ is $x + y = 3$.

Let the point of contact of tangent be $(x_1, y_1)$ to $S$.

$\therefore \quad x \cdot x_1 + y \cdot y_1 - (y + y_1) + 1 = 2$

or $xx_1 + yy_1 - y = (1 + y_1)$, same as $x + y = 3$.

$\Rightarrow \quad \frac{x_1}{1} = \frac{y_1 - 1}{1} = \frac{1 + y_1}{3}$

i.e. $\quad x_1 = 1 \text{ and } y_1 = 2$

$\therefore \quad P = (1, 2)$

Since, $PR = PQ = \frac{2\sqrt{2}}{3}$. Thus, by parametric form,

$$\frac{x-1}{-1/\sqrt{2}} = \frac{y-2}{1/\sqrt{2}} = \pm\frac{2\sqrt{2}}{3}$$

$\Rightarrow \quad \left(x = \frac{5}{3}, y = \frac{4}{3}\right)$

and $\quad \left(x = \frac{1}{3}, y = \frac{8}{3}\right)$

$\therefore \quad Q = \left(\frac{5}{3}, \frac{4}{3}\right) \quad \text{and} \quad R = \left(\frac{1}{3}, \frac{8}{3}\right)$

Now, equation of tangent at $Q$ on ellipse $E_1$ is

$$\frac{x \cdot 5}{a^2 \cdot 3} + \frac{y \cdot 4}{b^2 \cdot 3} = 1$$

On comparing with $x + y = 3$, we get

$$a^2 = 5 \text{ and } b^2 = 4$$

$\therefore \quad e_1^2 = 1 - \frac{b^2}{a^2} = 1 - \frac{4}{5} = \frac{1}{5} \quad \ldots\text{(i)}$

Also, equation of tangent at $R$ on ellipse $E_2$ is

$$\frac{x \cdot 1}{a^2 \cdot 3} + \frac{y \cdot 8}{b^2 \cdot 3} = 1$$

On comparing with $x + y = 3$, we get

$$a^2 = 1, b^2 = 8$$

$\therefore \quad e_2^2 = 1 - \frac{a^2}{b^2} = 1 - \frac{1}{8} = \frac{7}{8} \quad \ldots\text{(ii)}$

Now, $e_1^2 \cdot e_2^2 = \frac{7}{40} \Rightarrow e_1 e_2 = \frac{\sqrt{7}}{2\sqrt{10}}$

and $\quad e_1^2 + e_2^2 = \frac{1}{5} + \frac{7}{8} = \frac{43}{40}$

Also, $\quad |e_1^2 - e_2^2| = \left|\frac{1}{5} - \frac{7}{8}\right| = \frac{27}{40}$

**15.**

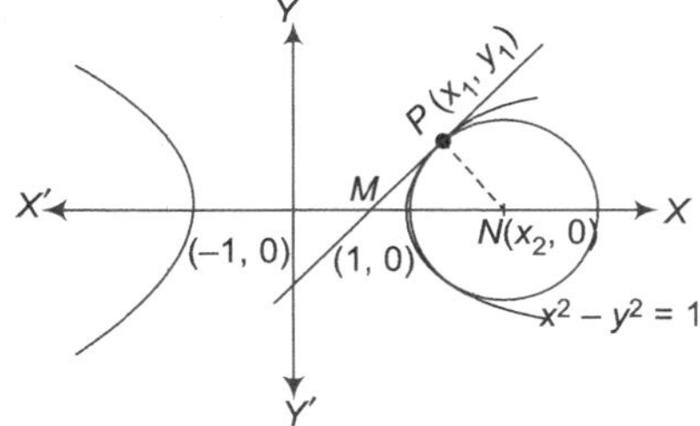


Equation of family of circles touching hyperbola at $(x_1, y_1)$ is

$(x - x_1)^2 + (y - y_1)^2 + \lambda(xx_1 - yy_1 - 1) = 0$

Now, its centre is $(x_2, 0)$.

$$\therefore \left[\frac{-(\lambda x_1 - 2x_1)}{2}, \frac{-(-2y_1 - \lambda y_1)}{2}\right] = (x_2, 0)$$

$\Rightarrow \quad 2y_1 + \lambda y_1 = 0 \Rightarrow \lambda = -2$

and $\quad 2x_1 - \lambda x_1 = 2x_2 \Rightarrow x_2 = 2x_1$

$\therefore \quad P(x_1, \sqrt{x_1^2 - 1})$

and $\quad N(x_2, 0) = (2x_1, 0)$

As tangent intersect $X$-axis at $M\left(\frac{1}{x}, 0\right)$.

Centroid of $\Delta PMN = (l, m)$

$$\Rightarrow \left(\frac{3x_1 + \frac{1}{x_1}}{3}, \frac{y_1 + 0 + 0}{3}\right) = (l, m)$$

$$\Rightarrow \quad l = \frac{3x_1 + \frac{1}{x_1}}{3}$$

On differentiating w.r.t. $x_1$, we get

$$\frac{dl}{dx_1} = \frac{3 - \frac{1}{x_1^2}}{3}$$

$$\Rightarrow \quad \frac{dl}{dx_1} = 1 - \frac{1}{3x_1^2}, \text{ for } x_1 > 1$$

and $\quad m = \dfrac{\sqrt{x_1^2 - 1}}{3}$

On differentiating w.r.t. $x_1$, we get

$$\frac{dm}{dx_1} = \frac{2x_1}{2 \times 3\sqrt{x_1^2 - 1}} = \frac{x_1}{3\sqrt{x_1^2 - 1}}, \text{ for } x_1 > 1$$

Also, $\quad m = \dfrac{y_1}{3}$

On differentiating w.r.t. $y_1$, we get

$$\frac{dm}{dy_1} = \frac{1}{3}, \text{ for } y_1 > 0$$

**16.** Let $I_1 = \int_0^{4\pi} e^t (\sin^6 at + \cos^6 at) dt$

$= \int_0^{\pi} e^t (\sin^6 at + \cos^6 at) dt$

$\quad + \int_{\pi}^{2\pi} e^t (\sin^6 at + \cos^6 at) dt$

$\quad + \int_{2\pi}^{3\pi} e^t (\sin^6 at + \cos^6 at) dt$

$\quad + \int_{3\pi}^{4\pi} e^t (\sin^6 at + \cos^6 at) dt$

$\therefore \quad I_1 = I_2 + I_3 + I_4 + I_5 \quad \ldots(\text{i})$

Now, $\quad I_3 = \int_{\pi}^{2\pi} e^t (\sin^6 at + \cos^6 at) dt$

Put $\quad t = \pi + t \Rightarrow dt = dt$

$\therefore \quad I_3 = \int_0^{\pi} e^{\pi + t} \cdot (\sin^6 at + \cos^6 at) dt$

$\quad = e^t \cdot I_2 \quad \ldots(\text{ii})$

Now, $\quad I_4 = \int_{2\pi}^{3\pi} e^t (\sin^6 at + \cos^6 at) dt$

Put $\quad t = 2\pi + t \Rightarrow dt = dt$

$\therefore \quad I_4 = \int_0^{\pi} e^{t + 2\pi} (\sin^6 at + \cos^6 at) dt$

$\quad = e^{2\pi} \cdot I_2 \quad \ldots(\text{iii})$

and $\quad I_5 = \int_{3\pi}^{4\pi} e^t (\sin^6 at + \cos^6 at) dt$

Put $\quad t = 3\pi + t$

$\therefore \quad I_5 = \int_0^{\pi} e^{3\pi + t} (\sin^6 at + \cos^6 at) dt$

$\quad = e^{3\pi} \cdot I_2 \quad \ldots(\text{iv})$

From Eqs. (i), (ii), (iii) and (iv), we get

$I_1 = I_2 + e^{\pi} \cdot I_2 + e^{2\pi} \cdot I_2 + e^{3\pi} \cdot I_2$

$\quad = (1 + e^{\pi} + e^{2\pi} + e^{3\pi}) I_2$

$$\therefore \quad L = \frac{\int_0^{4\pi} e^t (\sin^6 at + \cos^6 at) dt}{\int_0^{\pi} e^t (\sin^6 at + \cos^6 at) dt}$$

$\quad = (1 + e^{\pi} + e^{2\pi} + e^{3\pi})$

$\quad = \dfrac{1 \cdot (e^{4\pi} - 1)}{e^{\pi} - 1}$ for $a \in R$

**17.** According to the given data,

$F(x) < 0, \forall x \in (1, 3)$

We have, $f(x) = xF(x)$

$\Rightarrow \quad f'(x) = F(x) + xF'(x) \quad \ldots(\text{i})$

$\Rightarrow \quad f'(1) = F(1) + F'(1) < 0$ [given $F(1) = 0$ and $F'(x) < 0$]

Also, $f(2) = 2F(2) < 0$ [using $F(x) < 0, \forall x \in (1, 3)$]

Now, $f'(x) = F(x) + xF'(x) < 0$ [using $F(x) < 0, \forall x \in (1, 3)$]

$\Rightarrow \quad f'(x) < 0$

**18.** Given, $\int_1^3 x^2 F'(x)dx = -12$

$$\Rightarrow \quad [x^2F(x)]_1^3 - \int_1^3 2x \cdot F(x)dx = -12$$

$$\Rightarrow \quad 9F(3) - F(1) - 2\int_1^3 f(x)dx = -12$$

[$\because xF(x) = f(x)$, given]

$$\Rightarrow \quad -36 - 0 - 2\int_1^3 f(x)dx = -12$$

$$\therefore \quad \int_1^3 f(x)dx = -12$$

and $$\int_1^3 x^3 F''(x)dx = 40$$

$$\Rightarrow \quad [x^3F'(x)]_1^3 - \int_1^3 3x^2F'(x)dx = 40$$

$$\Rightarrow \quad [x^2(xF'(x)]_1^3 - 3 \times (-12) = 40$$

$$\Rightarrow \quad \{x^2 \cdot [f'(x) - F(x)]\}_1^3 = 4$$

$$\Rightarrow \quad 9[f'(3) - F(3)] - [f'(1) - F(1)] = 4$$

$$\Rightarrow \quad 9[f'(3) + 4] - [f'(1) - 0] = 4$$

$$\Rightarrow \quad 9f'(3) - f'(1) = -32$$

**19.**

| Box I | | Box II | |
|---|---|---|---|
| $n_1$ | Red | $n_3$ | Red |
| $n_2$ | Black | $n_4$ | Black |

Let $A$ = Drawing red ball

$$\therefore P(A) = P(B_1) \cdot P(A/B_1) + P(B_2) \cdot P(A/B_2)$$

$$= \frac{1}{2}\left(\frac{n_1}{n_1 + n_2}\right) + \frac{1}{2}\left(\frac{n_3}{n_3 + n_4}\right)$$

Given, $$P(B_2/A) = \frac{1}{3}$$

$$\Rightarrow \quad \frac{P(B_2) \cdot P(B_2 \cap A)}{P(A)} = \frac{1}{3}$$

$$\Rightarrow \quad \frac{\frac{1}{2}\left(\frac{n_3}{n_3 + n_4}\right)}{\frac{1}{2}\left(\frac{n_1}{n_1 + n_2}\right) + \frac{1}{2}\left(\frac{n_3}{n_3 + n_4}\right)} = \frac{1}{3}$$

$$\Rightarrow \quad \frac{n_3(n_1 + n_2)}{n_1(n_3 + n_4) + n_3(n_1 + n_2)} = \frac{1}{3}$$

Now, check options, then clearly options (a) and (b) satisfy.

**20.**

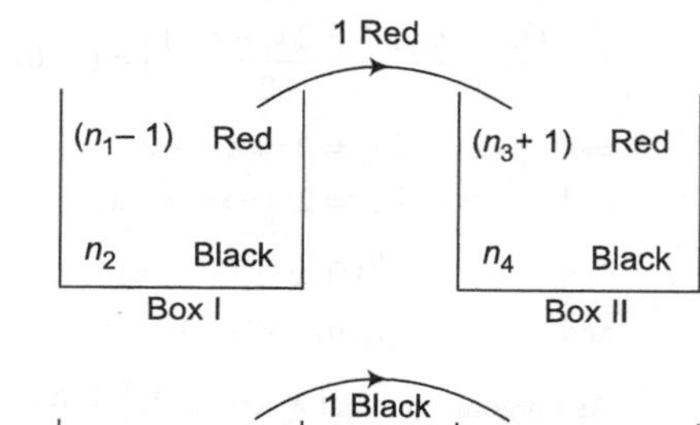


or

1 Black

| Box I | | Box II | |
|---|---|---|---|
| $n_1$ | Red | $n_3$ | Red |
| $(n_2 - 1)$ | Black | $(n_4 + 1)$ | Black |

$\therefore$ P (drawing red ball from $B_1$) $= \frac{1}{3}$

$$\Rightarrow \quad \left(\frac{n_1 - 1}{n_1 + n_2 - 1}\right)\left(\frac{n_1}{n_1 + n_2}\right) + \left(\frac{n_2}{n_1 + n_2}\right)\left(\frac{n_1}{n_1 + n_2 - 1}\right) = \frac{1}{3}$$

$$\Rightarrow \quad \frac{n_1^2 + n_1n_2 - n_1}{(n_1 + n_2)(n_1 + n_2 - 1)} = \frac{1}{3}$$

Clearly, options (c) and (d) satisfy.

# JEE Main
## Solved Paper 2014

M. Marks : 360 Time : 3 Hrs

### Instructions

- This test consists of 90 questions.
- There are three parts in the question paper A, B, C consisting of Physics, Chemistry and Mathematics having 30 questions in each part of equal weightage. Each question is allotetted 4 marks for correct response.
- Candidates will be awarded marks as stated above for correct response of each question. 1/4 marks will be deducted for indicating incorrect response of each question. No deduction from the total score will be made if no response is indicated for an item in the answer sheet.
- There is only one correct response for each question. Filling up more than one response in any question will be treated as wrong response and marks for wrong response will be deducted according as per instructions.

## Physics

**1.** The current voltage relation of diode is given by $I = (e^{1000V/T} - 1)$ mA, where the applied voltage $V$ is in volt and the temperature $T$ is in kelvin. If a student makes an error measuring $\pm 0.01$ V while measuring the current of 5 mA at 300K, what will be the error in the value of current in mA?

(a) 0.2 mA (b) 0.02 mA
(c) 0.5 mA (d) 0.05 mA

**2.** From a tower of height $H$, a particle is thrown vertically upwards with a speed $u$. The time taken by the particle to hit the ground, is $n$ times that taken by it to reach the highest point of its path. The relation between $H$, $u$ and $n$ is

(a) $2gH = n^2u^2$
(b) $gH = (n-2)^2u^2$
(c) $2gH = nu^2(n-2)$
(d) $gH = (n-2)^2u^2$

**3.** A mass $m$ supported by a massless string wound around a uniform hollow cylinder of mass $m$ and radius $R$. If the string does not slip on the cylinder, with what acceleration will the mass fall on release?

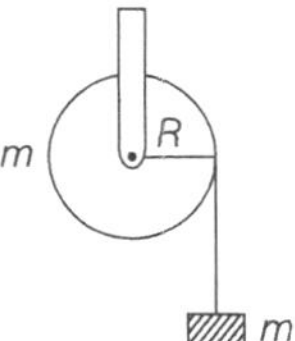


(a) $2g/3$ (b) $g/2$ (c) $5g/6$ (d) $g$

**4.** A block of mass $m$ is placed on a surface with a vertical cross-section given by $y = x^3/6$. If the coefficient of friction is 0.5, the maximum height above the ground at which the block can be placed without slipping is

(a) $\frac{1}{6}$ m (b) $\frac{2}{3}$ m (c) $\frac{1}{3}$ m (d) $\frac{1}{2}$ m

**5.** When a rubber band is stretched by a distance $x$, it exerts a restoring force of magnitude $F = a\,x + bx^2$, where $a$ and $b$ are constants. The work done in stretching the unstretched rubber-band by $L$ is

(a) $aL^2 + bL^3$

(b) $\frac{1}{2}(aL^2 + bL^3)$

(c) $\frac{aL^2}{2} + \frac{bL^3}{3}$

(d) $\frac{1}{2}\left(\frac{aL^2}{2} + \frac{bL^3}{3}\right)$

**6.** A bob of mass $m$ attached to an inextensible string of length $l$ is suspended from a vertical support. The bob rotates in a horizontal circle with an angular speed $\omega$ rad/s about the vertical support. About the point of suspension

(a) angular momentum is conserved

(b) angular momentum changes in magnitude but not in direction

(c) angular momentum changes in direction but not in magnitude

(d) angular momentum changes both in direction and magnitude

**7.** Four particles, each of mass $M$ and equidistant from each other, move along a circle of radius $R$ under the action of their mutual gravitational attraction, the speed of each particle is

(a) $\sqrt{\frac{GM}{R}}$ (b) $\sqrt{2\sqrt{2}\frac{GM}{R}}$

(c) $\sqrt{\frac{GM}{R}(1+2\sqrt{2})}$ (d) $\frac{1}{2}\sqrt{\frac{GM}{R}(1+2\sqrt{2})}$

**8.** The pressure that has to be applied to the ends of a steel wire of length 10 cm to keep its length constant when its temperature is raised by 100°C is (For steel, Young's modulus is $2 \times 10^{11}\text{Nm}^{-2}$ and coefficient of thermal expansion is $1.1 \times 10^{-5}\ \text{K}^{-1}$)

(a) $2.2 \times 10^8$ Pa (b) $2.2 \times 10^9$ Pa

(c) $2.2 \times 10^7$ Pa (d) $2.2 \times 10^6$ Pa

**9.** There is a circular tube in a vertical plane. Two liquids which do not mix and of densities $d_1$ and $d_2$ are filled in the tube. Each liquid subtends 90° angle at centre. Radius joining their interface makes an angle $\alpha$ with vertical. Ratio $d_1/d_2$ is

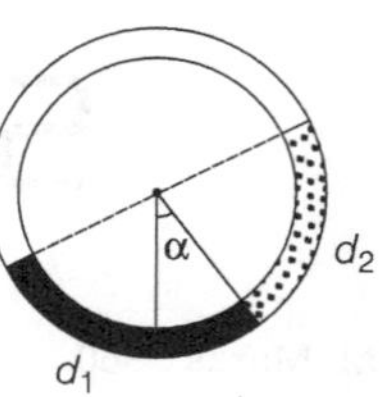


(a) $\frac{1+\sin\alpha}{1-\sin\alpha}$ (b) $\frac{1+\cos\alpha}{1-\cos\alpha}$

(c) $\frac{1+\tan\alpha}{1-\tan\alpha}$ (d) $\frac{1+\sin\alpha}{1-\cos\alpha}$

**10.** On heating water, bubbles beings formed at the bottom of the vessel detach and rise. Take the bubbles to be spheres of radius $R$ and making a circular contact of radius $r$ with the bottom of the vessel. If $r \ll R$ and the surface tension of water is $T$, value of $r$ just before bubbles detach is (density of water is $\rho$)

(a) $R^2\sqrt{\frac{\rho_w g}{3T}}$ (b) $R^2\sqrt{\frac{\rho_w g}{6T}}$

(c) $R^2\sqrt{\frac{\rho_w g}{T}}$ (d) $R^2\sqrt{\frac{3\rho_w g}{T}}$

**11.** Three rods of copper, brass and steel are welded together to form a $Y$-shaped structure. Area of cross-section of each rod is 4 cm$^2$. End of copper rod is maintained at 100°C whereas ends of brass and steel are kept at 0°C. Lengths of the copper, brass and steel rods are 46, 13 and 12 cm respectively.

The rods are thermally insulated from surroundings except at ends. Thermal conductivities of copper, brass and steel are 0.92, 0.26 and 0.12 in CGS units, respectively. Rate of heat flow through copper rod is

(a) 1.2 cal/s
(b) 2.4 cal/s
(c) 4.8 cal/s
(d) 6.0 cal/s

**12.** One mole of diatomic ideal gas undergoes a cyclic process $ABC$ as shown in figure. The process $BC$ is adiabatic. The temperatures at $A$, $B$ and $C$ are 400 K, 800 K and 600 K, respectively. Choose the correct statement.

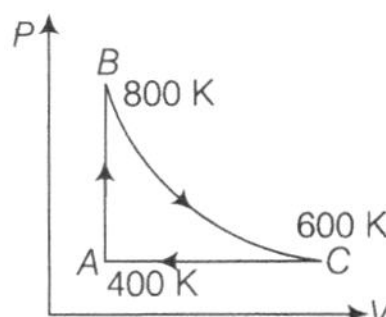


(a) The change in internal energy in whole cyclic process is 250 $R$
(b) The change in internal energy in the process $CA$ is 700 $R$
(c) The change in internal energy in the process $AB$ is $-350\,R$
(d) The change in internal energy in the process $BC$ is $-500\,R$

**13.** An open glass tube is immersed in mercury in such a way that a length of 8 cm extends above the mercury level. The open end of the tube is then closed and sealed and the tube is raised vertically up by additional 46 cm. What will be length of the air column above mercury in the tube now? (Atmospheric pressure = 76 cm of Hg)

(a) 16 cm (b) 22 cm
(c) 38 cm (d) 6 cm

**14.** A particle moves with simple harmonic motion in a straight line. In first $\tau$ sec, after starting from rest it travels a distance $a$ and in next $\tau$ sec, it travels $2a$, in same direction, then

(a) amplitude of motion is $3a$
(b) time period of oscillations is $8\pi$
(c) amplitude of motion is $4a$
(d) time period of oscillations is $6\pi$

**15.** A pipe of length 85 cm is closed from one end. Find the number of possible natural oscillations of air column in the pipe whose frequencies lie below 1250 Hz. The velocity of sound in air is 340 m/s.

(a) 12 (b) 8 (c) 6 (d) 4

**16.** Assume that an electric field $\mathbf{E} = 30x^2\,\hat{\mathbf{i}}$ exists in space. Then, the potential difference $V_A - V_O$, where $V_O$ is the potential at the origin and $V_A$ the potential at $x = 2$ m is

(a) 120 J (b) −120 J
(c) −80 J (d) 80 J

**17.** A parallel plate capacitor is made of two circular plates separated by a distance of 5 mm and with a dielectric of dielectric constant 2.2 between them. When the electric field in the dielectric is $3 \times 10^4$ V/m, the charge density of the positive plate will be close to

(a) $6 \times 10^{-7}$ C/m$^2$
(b) $3 \times 10^{-7}$ C/m$^2$
(c) $3 \times 10^{4}$ C/m$^2$
(d) $6 \times 10^{4}$ C/m$^2$

**18.** In a large building, there are 15 bulbs of 40 W, 5 bulbs of 100 W, 5 fans of 80 W and 1 heater of 1 kW. The voltage of the electric mains is 220 V. The minimum capacity of the main fuse of the building will be

(a) 8 A (b) 10 A (c) 12 A (d) 14 A

**19.** A conductor lies along the z-axis at $-1.5 \le z < 1.5$ m and carries a fixed current of 10.0 A in $-a_z$ direction (see figure). For a field $\mathbf{B} = 3.0 \times 10^{-4} e^{-0.2x} a_y$ T, find the power required to move the conductor at constant speed to $x = 2.0$ m, $y = 0$ in $5 \times 10^{-3}$ s. Assume parallel motion along the $x$-axis.

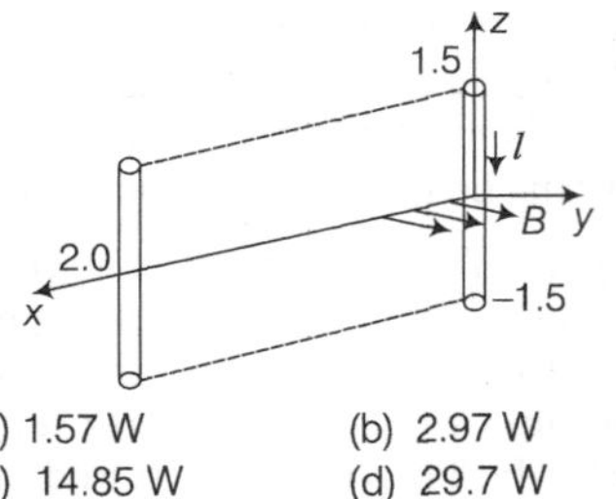


(a) 1.57 W (b) 2.97 W
(c) 14.85 W (d) 29.7 W

**20.** The coercivity of a small magnet where the ferromagnet gets demagnetised is $3 \times 10^3$ Am$^{-1}$. The current required to be passed in a solenoid of length 10 cm and number of turns 100, so that the magnet gets demagnetised when inside the solenoid is

(a) 30 mA (b) 60 mA
(c) 3 A (d) 6 A

**21.** In the circuit shown here, the point $C$ is kept connected to point $A$ till the current flowing through the circuit becomes constant. Afterward, suddenly point $C$ is disconnected from point $A$ and connected to point $B$ at time $t = 0$. Ratio of the voltage across resistance and the inductor at $t = L/R$ will be equal to

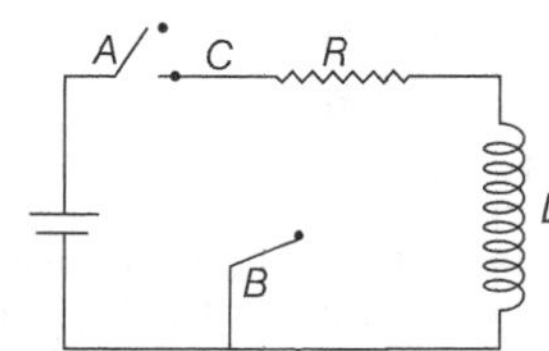


(a) $\frac{e}{1-e}$ (b) 1

(c) –1 (d) $\frac{1-e}{e}$

**22.** During the propagation of electromagnetic waves in a medium,

(a) electric energy density is double of the magnetic energy density
(b) electric energy density is half of the magnetic energy density
(c) electric energy density is equal to the magnetic energy density
(d) Both electric and magnetic energy densities are zero

**23.** A thin convex lens made from crown glass ($\mu = 3/2$) has focal length $f$. When it is measured in two different liquids having refractive indices 4/3 and 5/3. It has the focal lengths $f_1$ and $f_2$, respectively. The correct relation between the focal length is

(a) $f_1 = f_2 < f$
(b) $f_1 > f$ and $f_2$ becomes negative
(c) $f_2 > f$ and $f_1$ becomes negative
(d) $f_1$ and $f_2$ both become negative

**24.** A green light is incident from the water to the air-water interface at the critical angle ($\theta$). Select the correct statement.

(a) The entire spectrum of visible light will come out of the water at an angle of 90° to the normal
(b) The spectrum of visible light whose frequency is less than that of green light will come out of the air medium.
(c) The spectrum of visible light whose frequency is more than that of green light will come out to the air medium.
(d) The entire spectrum of visible light will come out of the water at various angles to the normal.

**25.** Two beams, $A$ and $B$, of plane polarised light with mutually perpendicular planes of polarisation are seen through a polaroid. From the position when the beam $A$ has maximum intensity (and beam $B$ has zero intensity), a rotation of polaroid through 30° makes the two beams appear equally bright. If the initial intensities of the two beams are $I_A$ and $I_B$ respectively, then $I_A / I_B$ equals

(a) 3 (b) 3/2
(c) 1 (d) 1/3

**26.** The radiation corresponding to $3 \rightarrow 2$ transition of hydrogen atom falls on a metal surface to produce photoelectrons. These electrons are made to enter a magnetic field of $3 \times 10^{-4}$ T. If the radius of the largest circular path followed by these electrons is 10.0 mm, the work function of the metal is close to

(a) 1.8 eV (b) 1.1 eV
(c) 0.8 eV (d) 1.6 eV

**27.** Hydrogen $({}_1H^1)$, deuterium $({}_1H^2)$, singly ionised helium $({}_2He^4)^+$ and doubly ionised lithium $({}_3Li^8)^{++}$ all have one electron around the nucleus. Consider an electron transition from $n = 2$ to $n = 1$. If the wavelengths of emitted radiation are $\lambda_1, \lambda_2, \lambda_3$ and $\lambda_4$, respectively for four elements, then approximately which one of the following is correct?

(a) $4\lambda_1 = 2\lambda_2 = 2\lambda_3 = \lambda_4$
(b) $\lambda_1 = 2\lambda_2 = 2\lambda_3 = \lambda_4$
(c) $\lambda_1 = \lambda_2 = 4\lambda_3 = 9\lambda_4$
(d) $\lambda_1 = 2\lambda_2 = 3\lambda_3 = 4\lambda_4$

**28.** The forward biased diode connection is

(a)  +2V —▷|—WWWW— −2V
(b)  −3V —▷|—WWWW— −3V
(c)  2V —▷|—WWWW— 4V
(d)  −2V —▷|—WWWW— +2V

**29.** A student measured the length of a rod and wrote it as 3.50 cm. Which instrument did he use to measure it?

(a) A meter scale
(b) A vernier calliper where the 10 divisions in vernier scale matches with 9 divisions in main scale and main scale has 10 divisions in 1 cm
(c) A screw gauge having 100 divisions in the circular scale and pitch as 1 mm
(d) A screw gauge having 50 divisions in the circular scale and pitch as 1 mm

**30.** Match List I (Electromagnetic wave type) with List II (Its association/application) and select the correct option from the choices given below the lists.

| List I | List II |
|---|---|
| A. Infrared waves | 1. To treat muscular strain |
| B. Radio waves | 2. For broadcasting |
| C. X-rays | 3. To detect fracture of bones |
| D. Ultraviolet | 4. Absorbed by the ozone layer of the atmosphere |

**Codes**

| | A | B | C | D |
|---|---|---|---|---|
| (a) | 4 | 3 | 2 | 1 |
| (b) | 1 | 2 | 4 | 3 |
| (c) | 3 | 2 | 1 | 4 |
| (d) | 1 | 2 | 3 | 4 |

# Chemistry

**1.** The correct set of four quantum numbers for the valence electrons of rubidium atom ($Z = 37$) is

(a) $5, 0, 0, +\frac{1}{2}$ (b) $5, 1, 0, +\frac{1}{2}$
(c) $5, 1, ,1, +\frac{1}{2}$ (d) $5, 0, 1, +\frac{1}{2}$

**2.** If $Z$ is a compressibility factor, van der Waals' equation at low pressure can be written as

(a) $Z = 1 + \frac{RT}{pb}$ (b) $Z = 1 - \frac{a}{VRT}$
(c) $Z = 1 - \frac{pb}{RT}$ (d) $Z = 1 + \frac{pb}{RT}$

**3.** CsCl crystallises in body centred cubic lattice. If $a$ its edge length, then which of the following expressions is correct?

(a) $r_{Cs^+} + r_{Cl^-} = 3a$
(b) $r_{Cs^+} + r_{Cl^-} = \frac{3a}{2}$
(c) $r_{Cs^+} + r_{Cl^-} = \frac{\sqrt{3}}{2}a$
(d) $r_{Cs^+} + r_{Cl^-} = \sqrt{3}a$

**4.** For the estimation of nitrogen, 1.4 g of an organic compound was digested by Kjeldahl's method and the evolved ammonia was absorbed in 60 mL of $\frac{M}{10}$ sulphuric acid. The unreacted acid required 20 mL of $\frac{M}{10}$ sodium hydroxide for complete neutralisation. The percentage of nitrogen in the compound is

(a) 6% (b) 10% (c) 3% (d) 5%

**5.** Resistance of 0.2 M solution of an electrolyte is 50 Ω. The specific conductance of the solution of 0.5 M solution of same electrolyte is 1.4 S $m^{-1}$ and resistance of same solution of the same electrolyte is 280 Ω. The molar conductivity of 0.5 M solution of the electrolyte in $S\ m^2\ mol^{-1}$ is

(a) $5 \times 10^{-4}$ (b) $5 \times 10^{-3}$
(c) $5 \times 10^{3}$ (d) $5 \times 10^{2}$

**6.** For the complete combustion of ethanol, $C_2H_5OH(l) + 3O_2(g) \longrightarrow 2CO_2(g) + 3H_2O(l)$, the amount of heat produced as measured in bomb calorimeter, is 1364.47 kJ $mol^{-1}$ at 25°C. Assuming ideality the enthalpy of combustion, $\Delta_C H$, for the reaction will be ($R$ = 8.314 J $K^{-1}mol^{-1}$)

(a) $-1366.95$ kJ $mol^{-1}$
(b) $-1361.95$ kJ $mol^{-1}$
(c) $-1460.50$ kJ $mol^{-1}$
(d) $-1350.50$ kJ $mol^{-1}$

**7.** The equivalent conductance of NaCl at concentration $C$ and at infinite dilution are $\lambda_C$ and $\lambda_\infty$, respectively. The correct relationship between $\lambda_C$ and $\lambda_\infty$ is given as (where, the constant $B$ is positive)

(a) $\lambda_C = \lambda_\infty + (B)C$
(b) $\lambda_C = \lambda_\infty - (B)C$
(c) $\lambda_C = \lambda_\infty - (B)\sqrt{C}$
(d) $\lambda_C = \lambda_\infty + (B)\sqrt{C}$

**8.** Consider separate solution of 0.500 M $C_2H_5OH\ (aq)$, 0.100 M $Mg_3(PO_4)_2\ (aq)$, 0.250 M KBr$(aq)$ and 0.125 M $Na_3PO_4(aq)$ at 25°C. Which statement is true about these solution, assuming all salts to be strong electrolytes?

(a) They all have the same osmotic pressure
(b) 0.100 M $Mg_3(PO_4)_2(aq)$ has the highest osmotic pressure
(c) 0.125 M $Na_3PO_4(aq)$ has the highest osmotic pressure
(d) 0.500 M $C_2H_5OH\ (aq)$ has the highest osmotic pressure

**9.** For the reaction,

$$SO_2(g) + \frac{1}{2}O_2(g) \rightleftharpoons SO_3(g)$$

if $K_p = K_C(RT)^x$

where, the symbols have usual meaning, then the value of $x$ is (assuming ideality)

(a) $-1$ (b) $-\frac{1}{2}$ (c) $\frac{1}{2}$ (d) 1

**10.** For the non-stoichiometric reaction $2A + B \rightarrow C + D$, the following kinetic data were obtained in three separate experiments, all at 298 K.

| | Initial concentration (*A*) | Initial concentration (*B*) | Initial rate of formation of *C* ($mol\ L^{-1}S^{-1}$) |
|---|---|---|---|
| (i) | 0.1 M | 0.1 M | $1.2 \times 10^{-3}$ |
| (ii) | 0.1 M | 0.2 M | $1.2 \times 10^{-3}$ |
| (iii) | 0.2 M | 0.1 M | $2.4 \times 10^{-3}$ |

The rate law for the formation of $C$ is

(a) $\frac{dC}{dt} = k[A][B]$ (b) $\frac{dC}{dt} = k[A]^2[B]$
(c) $\frac{dC}{dt} = k[A][B]^2$ (d) $\frac{dC}{dt} = k[A]$

**11.** Among the following oxoacids, the correct decreasing order of acid strength is

(a) $HOCl > HClO_2 > HClO_3 > HClO_4$
(b) $HClO_4 > HOCl > HClO_2 > HClO_3$
(c) $HClO_4 > HClO_3 > HClO_2 > HOCl$
(d) $HClO_2 > HClO_4 > HClO_3 > HOCl$

**12.** The metal that cannot be obtained by electrolysis of an aqueous solution of its salts is

(a) Ag (b) Ca (c) Cu (d) Cr

**13.** The octahedral complex of a metal ion $M^{3+}$ with four monodentate ligands $L_1$, $L_2$, $L_3$ and $L_4$ absorb wavelengths in the region of red, green, yellow and blue, respectively. The increasing order of ligand strength of the four ligands is

(a) $L_4 < L_3 , L_2 < L_1$
(b) $L_1 < L_3 < L_2 < L_4$
(c) $L_3 < L_2 < L_4 < L_1$
(d) $L_1 < L_2 < L_4 < L_3$

**14.** Which of the following properties is not shown by NO?

(a) It is diamagnetic in gaseous state
(b) It is a neutral oxide
(c) It combines with oxygen to form nitrogen dioxide
(d) Its bond order is 2.5

**15.** In which of the following reactions $H_2O_2$ acts as a reducing agent?

I. $H_2O_2 + 2H^+ + 2e^- \longrightarrow 2H_2O$
II. $H_2O_2 - 2e^- \longrightarrow O_2 + 2H^+$
III. $H_2O_2 + 2e^- \longrightarrow 2OH^-$
IV. $H_2O_2 + 2OH^- - 2e^- \rightarrow O_2 + 2H_2O$

(a) I and II (b) III and IV
(c) I and III (d) II and IV

**16.** The correct statement for the molecule, $CsI_3$ is

(a) it is a covalent molecule
(b) it contains $Cs^+$ and $I_3^-$
(c) it contains $Cs^{3+}$ and $I^-$ ions
(d) it contains $Cs^+$, $I^-$ and lattice $I_2$ molecule

**17.** The ratio of masses of oxygen and nitrogen of a particular gaseous mixture is 1 : 4. The ratio of number of their molecule is

(a) 1 : 4
(b) 7 : 32
(c) 1 : 8
(d) 3 : 16

**18.** Given below are the half-cell reactions

$Mn^{2+} + 2e^- \longrightarrow Mn \,; E° = -1.18 \text{ eV}$

$2(Mn^{3+} + e^- \longrightarrow Mn^{2+});$

$E° = +1.51 \text{ eV}$

The $E°$ for $3Mn^{2+} \rightarrow Mn + 2Mn^{3+}$ will be

(a) – 2.69 V; the reaction will not occur
(b) – 2.69 V; the reaction will occur
(c) – 0.33 V; the reaction will not occur
(d) – 0.33 V; the reaction will occur

**19.** Which series of reactions correctly represents chemical relations related to iron and its compound?

(a) $Fe \xrightarrow{\text{Dil. } H_2SO_4} FeSO_4 \xrightarrow{H_2SO_4, O_2} Fe_2(SO_4)_3 \xrightarrow{\text{Heat}} Fe$

(b) $Fe \xrightarrow{O_2, \text{ Heat}} FeO \xrightarrow{\text{Dil. } H_2SO_4} FeSO_4 \xrightarrow{\text{Heat}} Fe$

(c) $Fe \xrightarrow{Cl_2, \text{ Heat}} FeCl_3 \xrightarrow{\text{Heat, air}} FeCl_2 \xrightarrow{Zn} Fe$

(d) $Fe \xrightarrow{O_2, \text{ Heat}} Fe_3O_4 \xrightarrow{CO, 600°C} FeO \xrightarrow{CO, 700°C} Fe$

**20.** The equation which is balanced and represents the correct product(s) is

(a) $Li_2O + 2KCl \longrightarrow 2LiCl + K_2O$
(b) $[CoCl(NH_3)_5]^+ + 5H^+ \longrightarrow Co^{2+} + 5NH_4^+ + Cl^-$
(c) $[Mg(H_2O)_6]^{2+} + (EDTA)^{4-} \xrightarrow[\text{NaOH}]{\text{Excess}} [Mg(EDTA)]^{2+} + 6H_2O$
(d) $CuSO_4 + 4KCN \longrightarrow K_2[Cu(CN)_4] + K_2SO_4$

**21.** In $S_N2$ reactions, the correct order of reactivity for the following compounds $CH_3Cl, CH_3CH_2Cl$, $(CH_3)_2CHCl$ and $(CH_3)_3CCl$ is

(a) $CH_3Cl > (CH_3)_2CHCl > CH_3CH_2Cl > (CH_3)_3CCl$

(b) $CH_3Cl > CH_3CH_2Cl > (CH_3)_2CHCl > (CH_3)_3CCl$

(c) $CH_3CH_2Cl > CH_3Cl > (CH_3)_2CHCl > (CH_3)_3CCl$

(d) $(CH_3)_2CHCl > CH_3CH_2Cl > CH_3Cl > (CH_3)_3CCl$

**22.** On heating an aliphatic primary amine with chloroform and ethanolic potassium hydroxide, the organic compound formed is

(a) an alkanol

(b) an alkanediol

(c) an alkyl cyanide

(d) an alkyl isocyanide

**23.** The most suitable reagent for the conversion of $R—CH_2—OH \longrightarrow R—CHO$ is

(a) $KMnO_4$

(b) $K_2Cr_2O_7$

(c) $CrO_3$

(d) PCC (pyridinium chlorochromate)

**24.** The major organic compound formed by the reaction of 1,1,1-trichloroethane with silver powder is

(a) acetylene

(b) ethene

(c) 2-butyne

(d) 2-butene

**25.** Sodium phenoxide when heated with $CO_2$ under pressure at 125°C yields a product which on acetylation produces $C$.

ONa (on benzene ring) $+ CO_2 \xrightarrow[5\ atm]{125°} B \xrightarrow[Ac_2O]{H^+} C$

The major product $C$ would be

(a) OCOCH$_3$, COOH (benzene ring, ortho)

(b) OH, COCH$_3$, COCH$_3$ (benzene ring)

(c) OH, COOCH$_3$ (benzene ring, ortho)

(d) OCOCH$_3$, COOH (benzene ring, meta)

**26.** Considering the basic strength of amines in aqueous solution, which one has the smallest $pK_b$ value?

(a) $(CH_3)_2NH$ (b) $CH_3NH_2$

(c) $(CH_3)_3N$ (d) $C_6H_5NH_2$

**27.** For which of the following molecule significant $\mu \neq 0$?

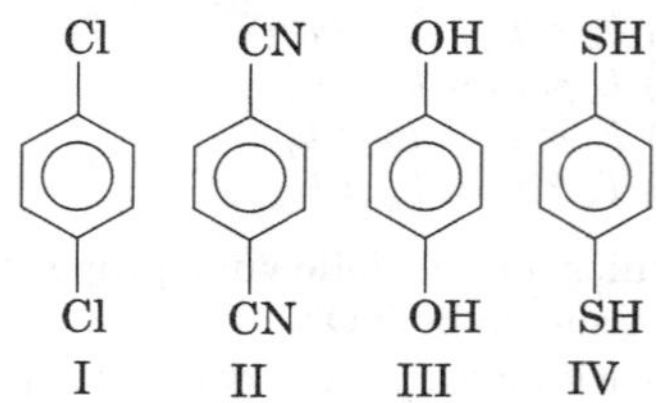


(a) Only I (b) I and II

(c) Only III (d) III and IV

**28.** Which one is classified as a condensation polymer?

(a) Dacron (b) Neoprene

(c) Teflon (d) Acrylonitrile

**29.** Which one of the following bases is not present in DNA?

(a) Quinoline (b) Adenine

(c) Cytosine (d) Thymine

**30.** In the reaction,

$CH_3COOH \xrightarrow{LiAlH_4} A \xrightarrow{PCl_5} B \xrightarrow{Alc.\ KOH} C$

The product $C$ is

(a) acetaldehyde

(b) acetylene

(c) ethylene

(d) acetyl chloride

# Mathematics

**1.** If $X = \{4^n - 3n - 1 : n \in N\}$ and $Y = \{9(n-1) : n \in N\}$; where $N$ is the set of natural numbers, then $X \cup Y$ is equal to

(a) $N$ (b) $Y - X$
(c) $X$ (d) $Y$

**2.** If $z$ is a complex number such that $|z| \geq 2$, then the minimum value of $\left| z + \frac{1}{2} \right|$

(a) is equal to 5/2
(b) lies in the interval (1, 2)
(c) is strictly greater than 5/2
(d) is strictly greater than 3/2 but less than 5/2

**3.** If $a \in R$ and the equation $-3(x - [x])^2 + 2(x - [x]) + a^2 = 0$ (where, $[x]$ denotes the greatest integer $\leq x$) has no integral solution, then all possible values of $a$ lie in the interval

(a) $(-1, 0) \cup (0, 1)$ (b) $(1, 2)$
(c) $(-2, -1)$ (d) $(-\infty, -2) \cup (2, \infty)$

**4.** Let $\alpha$ and $\beta$ be the roots of equation $px^2 + qx + r = 0$, $p \neq 0$. If $p, q$ and $r$ are in AP and $\frac{1}{\alpha} + \frac{1}{\beta} = 4$, then the value of $|\alpha - \beta|$ is

(a) $\frac{\sqrt{61}}{9}$ (b) $\frac{2\sqrt{17}}{9}$
(c) $\frac{\sqrt{34}}{9}$ (d) $\frac{2\sqrt{13}}{9}$

**5.** If $\alpha, \beta \neq 0$ and $f(n) = \alpha^n + \beta^n$ and

$$\begin{vmatrix} 3 & 1 + f(1) & 1 + f(2) \\ 1 + f(1) & 1 + f(2) & 1 + f(3) \\ 1 + f(2) & 1 + f(3) & 1 + f(4) \end{vmatrix}$$

$= K(1 - \alpha)^2 (1 - \beta)^2 (\alpha - \beta)^2$, then $K$ is equal to

(a) $\alpha\beta$ (b) $\frac{1}{\alpha\beta}$
(c) 1 (d) $-1$

**6.** If $A$ is a $3 \times 3$ non-singular matrix such that $AA^T = A^T A$ and $B = A^{-1} A^T$, then $BB^T$ is equal to

(a) $I + B$ (b) $I$ (c) $B^{-1}$ (d) $(B^{-1})^T$

**7.** If the coefficients of $x^3$ and $x^4$ in the expansion of $(1 + ax + bx^2)(1 - 2x)^{18}$ in powers of $x$ are both zero, then $(a, b)$ is equal to

(a) $\left(16, \frac{251}{3}\right)$ (b) $\left(14, \frac{251}{3}\right)$
(c) $\left(14, \frac{272}{3}\right)$ (d) $\left(16, \frac{272}{3}\right)$

**8.** The angle between the lines whose direction cosines satisfy the equations $l + m + n = 0$ and $l^2 = m^2 + n^2$ is

(a) $\frac{\pi}{3}$ (b) $\frac{\pi}{4}$
(c) $\frac{\pi}{6}$ (d) $\frac{\pi}{2}$

**9.** If $(10)^9 + 2(11)^1 (10)^8 + 3(11)^2 (10)^7 + \ldots + 10(11)^9 = k(10)^9$, then $k$ is equal to

(a) $\frac{121}{10}$ (b) $\frac{441}{100}$ (c) 100 (d) 110

**10.** Three positive numbers form an increasing GP. If the middle term in this GP is doubled, then new numbers are in AP. Then, the common ratio of the GP is

(a) $\sqrt{2} + \sqrt{3}$ (b) $3 + \sqrt{2}$
(c) $2 - \sqrt{3}$ (d) $2 + \sqrt{3}$

**11.** $\lim_{x \to 0} \frac{\sin(\pi \cos^2 x)}{x^2}$ is equal to

(a) $\frac{\pi}{2}$ (b) 1
(c) $-\pi$ (d) $\pi$

**12.** If $g$ is the inverse of a function $f$ and $f'(x) = \frac{1}{1 + x^5}$, then $g'(x)$ is equal to

(a) $1 + x^5$ (b) $5x^4$
(c) $\frac{1}{1 + \{g(x)\}^5}$ (d) $1 + \{g(x)\}^5$

**13.** If $f$ and $g$ are differentiable functions in $(0, 1)$ satisfying $f(0) = 2 = g(1)$, $g(0) = 0$ and $f(1) = 6$, then for some $c \in ]0, 1[$

(a) $2f'(c) = g'(c)$ (b) $2f'(c) = 3g'(c)$

(c) $f'(c) = g'(c)$ (d) $f'(c) = 2g'(c)$

**14.** If $x = -1$ and $x = 2$ are extreme points of $f(x) = \alpha \log|x| + \beta x^2 + x$, then

(a) $\alpha = -6, \beta = \frac{1}{2}$ (b) $\alpha = -6, \beta = -\frac{1}{2}$

(c) $\alpha = 2, \beta = -\frac{1}{2}$ (d) $\alpha = 2, \beta = \frac{1}{2}$

**15.** The integral $\int \left(1 + x - \frac{1}{x}\right) e^{x + \frac{1}{x}} dx$ is equal to

(a) $(x - 1) e^{x + \frac{1}{x}} + C$

(b) $x e^{x + \frac{1}{x}} + C$

(c) $(x + 1) e^{x + \frac{1}{x}} + C$

(d) $-x e^{x + \frac{1}{x}} + C$

**16.** The integral $\int_0^{\pi} \sqrt{1 + 4\sin^2 \frac{x}{2} - 4\sin \frac{x}{2}}\, dx$ is equal to

(a) $\pi - 4$ (b) $\frac{2\pi}{3} - 4 - 4\sqrt{3}$

(c) $4\sqrt{3} - 4$ (d) $4\sqrt{3} - 4 - \frac{\pi}{3}$

**17.** The area of the region described by $A = \{(x, y) : x^2 + y^2 \le 1 \text{ and } y^2 \le 1 - x\}$ is

(a) $\frac{\pi}{2} + \frac{4}{3}$ (b) $\frac{\pi}{2} - \frac{4}{3}$

(c) $\frac{\pi}{2} - \frac{2}{3}$ (d) $\frac{\pi}{2} + \frac{2}{3}$

**18.** Let the population of rabbits surviving at a time $t$ be governed by the differential equation $\frac{dp(t)}{dt} = \frac{1}{2} p(t) - 200$. If $p(0) = 100$, then $p(t)$ is equal to

(a) $400 - 300e^{\frac{t}{2}}$ (b) $300 - 200e^{-\frac{t}{2}}$

(c) $600 - 500e^{\frac{t}{2}}$ (d) $400 - 300e^{-\frac{t}{2}}$

**19.** If $PS$ is the median of the triangle with vertices $P(2, 2)$, $Q(6, -1)$ and $R(7, 3)$, then equation of the line passing through $(1, -1)$ and parallel to $PS$ is

(a) $4x - 7y - 11 = 0$

(b) $2x + 9y + 7 = 0$

(c) $4x + 7y + 3 = 0$

(d) $2x - 9y - 11 = 0$

**20.** Let $a$, $b$, $c$ and $d$ be non-zero numbers. If the point of intersection of the lines $4ax + 2ay + c = 0$ and $5bx + 2by + d = 0$ lies in the fourth quadrant and is equidistant from the two axes, then

(a) $2bc - 3ad = 0$ (b) $2bc + 3ad = 0$

(c) $2ad - 3bc = 0$ (d) $3bc + 2ad = 0$

**21.** The locus of the foot of perpendicular drawn from the centre of the ellipse $x^2 + 3y^2 = 6$ on any tangent to it is

(a) $(x^2 - y^2)^2 = 6x^2 + 2y^2$

(b) $(x^2 - y^2)^2 = 6x^2 - 2y^2$

(c) $(x^2 + y^2)^2 = 6x^2 + 2y^2$

(d) $(x^2 + y^2)^2 = 6x^2 - 2y^2$

**22.** Let $C$ be the circle with centre at $(1, 1)$ and radius 1. If $T$ is the circle centred at $(0, y)$ passing through origin and touching the circle $C$ externally, then the radius of $T$ is equal to

(a) $\frac{\sqrt{3}}{\sqrt{2}}$ (b) $\frac{\sqrt{3}}{2}$

(c) $\frac{1}{2}$ (d) $\frac{1}{4}$

**23.** The slope of the line touching both the parabolas $y^2 = 4x$ and $x^2 = -32y$ is

(a) $\frac{1}{2}$ (b) $\frac{3}{2}$ (c) $\frac{1}{8}$ (d) $\frac{2}{3}$

**24.** The image of the line $\frac{x-1}{3} = \frac{y-3}{1} = \frac{z-4}{-5}$ in the plane $2x - y + z + 3 = 0$ is the line

(a) $\frac{x+3}{3} = \frac{y-5}{1} = \frac{z-2}{-5}$

(b) $\frac{x+3}{-3} = \frac{y-5}{-1} = \frac{z+2}{5}$

(c) $\frac{x-3}{3} = \frac{y+5}{1} = \frac{z-2}{-5}$

(d) $\frac{x-3}{-3} = \frac{y+5}{-1} = \frac{z-2}{5}$

**25.** If $[\mathbf{a}\times\mathbf{b}\ \mathbf{b}\times\mathbf{c}\ \mathbf{c}\times\mathbf{a}] = \lambda\ [\mathbf{a}\ \mathbf{b}\ \mathbf{c}]^2$, then $\lambda$ is equal to

(a) 0 (b) 1
(c) 2 (d) 3

**26.** Let $A$ and $B$ be two events such that $P(\overline{A \cup B}) = \frac{1}{6}$, $P(A \cap B) = \frac{1}{4}$ and $P(\overline{A}) = \frac{1}{4}$, where $\overline{A}$ stands for the complement of the event $A$. Then , the events $A$ and $B$ are

(a) independent but not equally likely
(b) independent and equally likely
(c) mutually exclusive and independent
(d) equally likely but not independent

**27.** The variance of first 50 even natural numbers is

(a) $\frac{833}{4}$ (b) 833 (c) 437 (d) $\frac{437}{4}$

**28.** If $f_k(x) = 1/k\ (\sin^k x + \cos^k x)$, where $x \in R$ and $k \geq 1$, then $f_4(x) - f_6(x)$ is equal to

(a) 1/6 (b) 1/3 (c) 1/4 (d) 1/12

**29.** A bird is sitting on the top of a vertical pole 20 m high and its elevation from a point $O$ on the ground is 45°. It flies off horizontally straight away from the point $O$. After 1s, the elevation of the bird from $O$ is reduced to 30°. Then, the speed (in m/s) of the bird is

(a) $40(\sqrt{2} - 1)$ (b) $40(\sqrt{3} - \sqrt{2})$
(c) $20\sqrt{2}$ (d) $20(\sqrt{3} - 1)$

**30.** The statement $\sim (p \leftrightarrow \sim q)$ is

(a) equivalent to $p \leftrightarrow q$
(b) equivalent to $\sim p \leftrightarrow q$
(c) a tautology
(d) a fallacy

# Answers

**Physics**

| | | | | | | | | | |
|---|---|---|---|---|---|---|---|---|---|
| **1.** (a) | **2.** (c) | **3.** (b) | **4.** (a) | **5.** (c) | **6.** (c) | **7.** (d) | **8.** (a) | **9.** (c) | **10.** (*) |
| **11.** (c) | **12.** (d) | **13.** (a) | **14.** (d) | **15.** (c) | **16.** (c) | **17.** (a) | **18.** (c) | **19.** (b) | **20.** (c) |
| **21.** (c) | **22.** (c) | **23.** (b) | **24.** (d) | **25.** (d) | **26.** (b) | **27.** (c) | **28.** (a) | **29.** (b) | **30.** (d) |

**Note** '*' Means, none options are not correct.

**Chemistry**

| | | | | | | | | | |
|---|---|---|---|---|---|---|---|---|---|
| **1.** (a) | **2.** (b) | **3.** (c) | **4.** (b) | **5.** (a) | **6.** (a) | **7.** (c) | **8.** (a) | **9.** (b) | **10.** (d) |
| **11.** (c) | **12.** (b) | **13.** (b) | **14.** (a) | **15.** (d) | **16.** (b) | **17.** (b) | **18.** (a) | **19.** (d) | **20.** (b) |
| **21.** (b) | **22.** (d) | **23.** (d) | **24.** (c) | **25.** (a) | **26.** (a) | **27.** (d) | **28.** (a) | **29.** (a) | **30.** (c) |

**Mathematics**

| | | | | | | | | | |
|---|---|---|---|---|---|---|---|---|---|
| **1.** (d) | **2.** (b) | **3.** (a) | **4.** (d) | **5.** (c) | **6.** (b) | **7.** (d) | **8.** (a) | **9.** (c) | **10.** (d) |
| **11.** (d) | **12.** (d) | **13.** (d) | **14.** (c) | **15.** (b) | **16.** (d) | **17.** (a) | **18.** (a) | **19.** (b) | **20.** (c) |
| **21.** (c) | **22.** (d) | **23.** (a) | **24.** (a) | **25.** (b) | **26.** (a) | **27.** (b) | **28.** (d) | **29.** (d) | **30.** (a) |

# Hints & Solutions

## Physics

**1.** Given, $I = (e^{1000V/T} - 1)$ mA,

$$dV = \pm 0.01 \text{ V}$$
$$T = 300 \text{ K} \Rightarrow \text{mA}$$

So, $I = e^{1000V/T} - 1$

$$I + 1 = e^{1000V/T}$$

Taking log on both sides, we get

$$\log(I + 1) = \frac{1000\ V}{T}$$

On differentiating, $\frac{dI}{I+1} = \frac{1000}{T} dV$

$$dI = \frac{1000}{T} \times (I + 1)\, dV$$

$$\Rightarrow \quad dI = \frac{1000}{300} \times (5 + 1) \times 0.01$$
$$= 0.2 \text{ mA}$$

So, error in the value of current is 0.2 mA.

**2.** Time taken to reach the maximum height

$$t_1 = u/g$$

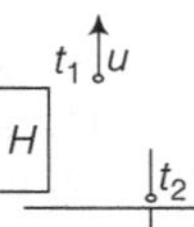


If $t_2$ is the time taken to hit the ground,

*i.e.*, $-H = ut_2 - \frac{1}{2}gt_2^2$

But $t_2 = nt_1$ [Given]

So, $-H = u\frac{nu}{g} - \frac{1}{2}g\frac{n^2u^2}{g^2}$

$$-H = \frac{nu^2}{g} - \frac{1}{2}\frac{n^2u^2}{g}$$

$$\Rightarrow \quad 2gH = nu^2(n-2)$$

**3.** For the mass $m$, $mg - T = ma$

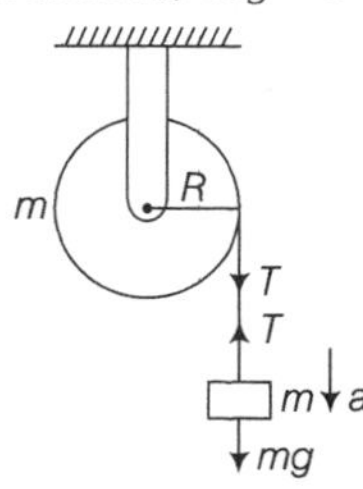


As we know, $a = R\alpha$

So, $mg - T = mR\alpha$ ...(i)

Torque about centre of pully

$$T \times R = mR^2\alpha \quad \text{...(ii)}$$

From Eqs. (i) and (ii), we get

$$a = \frac{g}{2}$$

Hence, the acceleration with the mass of a body fall is $\frac{g}{2}$.

**4.** A block of mass $m$ is placed on a surface with a vertical cross-section, then

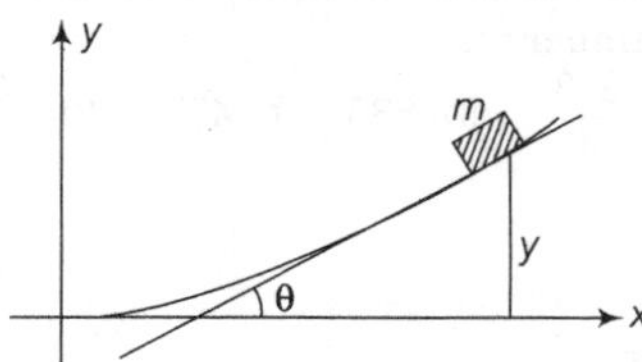


$$\tan\theta = \frac{dy}{dx} \frac{d\left(\frac{x^3}{6}\right)}{dx} = \frac{x^2}{2}$$

At limiting equilibrium, we get

$$\mu = \tan\theta$$
$$0.5 = \frac{x^2}{2}$$
$$\Rightarrow \quad x^2 = 1 \Rightarrow x = \pm 1$$

Now, putting the value of $x$ in $y = \frac{x^3}{6}$, we get

| When $x = 1$ | When $x = -1$ |
|---|---|
| $y = \frac{(1)^3}{6} = \frac{1}{6}$ | $y = \frac{(-1)^3}{6} = \frac{-1}{6}$ |

So, the maximum height above the ground at which the block can be placed without slipping is 1/6 m.

**5.** We know that change in potential energy of a system corresponding to a conservative internal force as

$$U_f - U_i = -W = -\int_i^f \mathbf{F} \cdot d\mathbf{r}$$

Given, $F = ax + bx^2$

We know that work done in stretching the rubber band by $L$ is

$$|dW| = |Fdx|$$

$$|W| = \int_O^L (ax + bx^2)dx$$

$$= \left[\frac{ax^2}{2}\right]_O^L + \left[\frac{bx^3}{3}\right]_O^L$$

$$= \left[\frac{aL^2}{2} - \frac{a \times (0)^2}{2}\right] + \left[\frac{b \times L^3}{3} - \frac{b \times (0)^3}{3}\right]$$

$$|W| = \frac{aL^2}{2} + \frac{bL^3}{3}$$

**6.** Angular momentum of the pendulum about the suspension point $O$ is

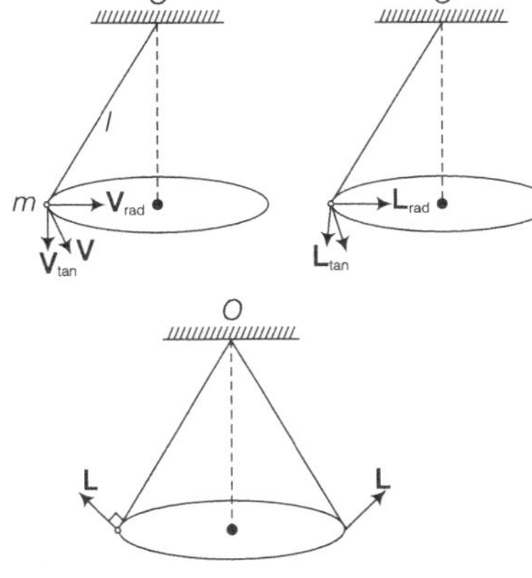


$$\mathbf{L} = m(\mathbf{r} \times \mathbf{v})$$

Then, $\mathbf{v}$ can be resolved into two components, radial component $r_{\text{rad}}$ and tangential component $r_{\tan}$. Due to $v_{\text{rad}}$, $L$ will be tangential and due to $v_{\tan}$, $L$ will be radially outwards as shown.

So, net angular momentum will be as shown in figure whose magnitude will be constant ($|L| = mvl$). But its direction will change as shown in the figure.

$$\mathbf{L} = m(\mathbf{r} \times \mathbf{v})$$

where, $r$ = radius of circle.

**7.** Net force acting on any one particle $M$,

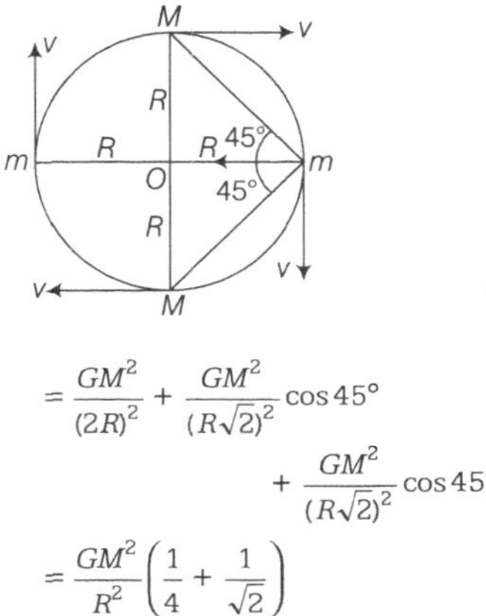


$$= \frac{GM^2}{(2R)^2} + \frac{GM^2}{(R\sqrt{2})^2}\cos 45^\circ + \frac{GM^2}{(R\sqrt{2})^2}\cos 45^\circ$$

$$= \frac{GM^2}{R^2}\left(\frac{1}{4} + \frac{1}{\sqrt{2}}\right)$$

This force will equal to centripetal force.

So, $$\frac{Mv^2}{R} = \frac{GM^2}{R^2}\left(\frac{1}{4} + \frac{1}{\sqrt{2}}\right)$$

$$v = \sqrt{\frac{GM}{4R}(1 + 2\sqrt{2})}$$

$$= \frac{1}{2}\sqrt{\frac{GM}{R}(2\sqrt{2} + 1)}$$

Hence, speed of each particle in a circular motion is

$$\frac{1}{2}\sqrt{\frac{GM}{R}(2\sqrt{2} + 1)}.$$

**8.** If the deformation is small, then the stress in a body is directly proportional to the corresponding strain.

According to Hooke's law *i.e.*,

$$\text{Young's modulus } (Y) = \frac{\text{Tensile stress}}{\text{Tensile strain}}$$

So, $$Y = \frac{F/A}{\Delta L/L} = \frac{FL}{A\Delta L}$$

If the rod is compressed, then compressive stress and strain appear. Their ratio $Y$ is same as that for tensile case.

Given, length of a steel wire ($L$) = 10 cm

Temperature ($\theta$) = 100°C

As length is constant.

$\therefore \quad \text{Strain} = \dfrac{\Delta L}{L} = \alpha \Delta\theta$

Now, pressure = stress

$= Y \times \text{strain}$

[Given, $Y = 2 \times 10^{11}\ \text{N/m}^2$ and

$\alpha = 1.1 \times 10^{-5}\ \text{K}^{-1}$]

$= 2 \times 10^{11} \times 1.1 \times 10^{-5} \times 100$

$= 2.2 \times 10^{8}\ \text{Pa}$

**9.** Equating pressure at $A$, we get

$R\sin\alpha\, d_2 + R\cos\alpha\, d_2 + R(1 - \cos\alpha)\, d_1 = R(1 - \sin\alpha)\, d_1$

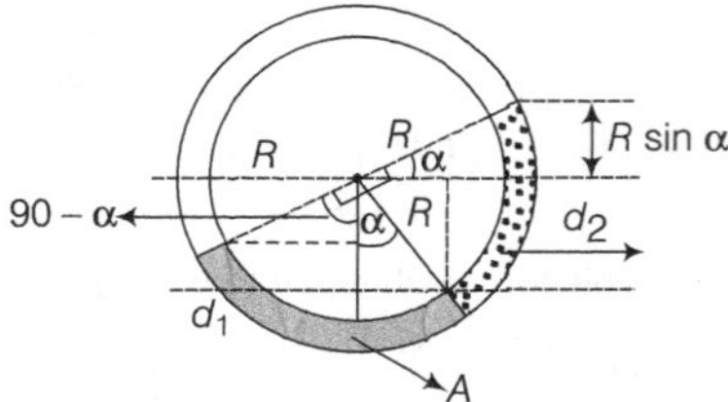


$(\sin\alpha + \cos\alpha)\, d_2 = d_1(\cos\alpha - \sin\alpha)$

$\Rightarrow \quad \dfrac{d_1}{d_2} = \dfrac{1 + \tan\alpha}{1 - \tan\alpha}$

**10.** The bubble will detach if,

Buoyant force ≥ Surface tension force

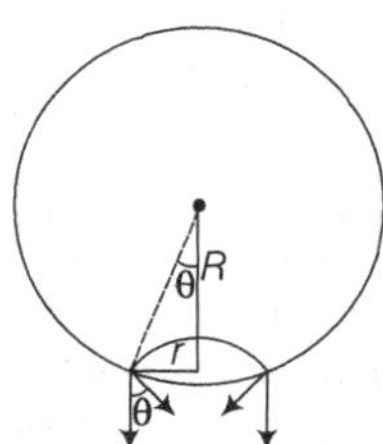


$\int \sin\theta\, T \times dl = T(2\pi r)\sin\theta$

$\dfrac{4}{3}\pi R^3 \rho_w g \geq \int T \times dl \sin\theta$

$(\rho_w)\left(\dfrac{4}{3}\pi R^3\right) g \geq (T)(2\pi r)\sin\theta$

$\Rightarrow \quad \sin\theta = \dfrac{r}{R}$

Solving, $r = \sqrt{\dfrac{2\rho_w R^4 g}{3T}} = R^2\sqrt{\dfrac{2\rho_w g}{3T}}$

No option matches with the correct answer.

**11.** In thermal conduction, it is found that in steady state the heat current is directly proportional to the area of cross-section $A$ which is proportional to the change in temperature $(T_1 - T_2)$.

Then, $\dfrac{\Delta Q}{\Delta t} = \dfrac{KA(T_1 - T_2)}{x}$

According to thermal conductivity, we get

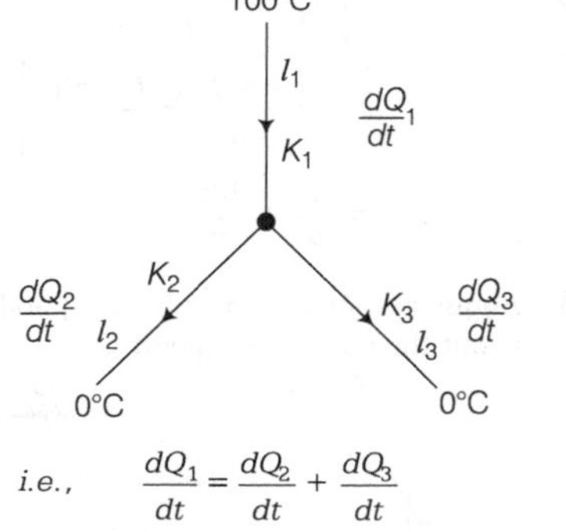


i.e., $\dfrac{dQ_1}{dt} = \dfrac{dQ_2}{dt} + \dfrac{dQ_3}{dt}$

$\dfrac{0.92(100 - T)}{46} = \dfrac{0.26(T - 0)}{13} + \dfrac{0.12(T - 0)}{12}$

$\Rightarrow \quad T = 40°\ \text{C}$

$\therefore \quad \dfrac{dQ_1}{dt} = \dfrac{0.92 \times 4(100 - 40)}{40}$

$= 4.8\ \text{cal/s}$

**12.** According to first law of thermodynamics, we get

(i) Change in internal energy from $A$ to $B$ i.e., $\Delta U_{AB}$

$\Delta U_{AB} = nC_V(T_B - T_A)$

$= 1 \times \dfrac{5R}{2}(800 - 400)$

$= 1000R$

(ii) Change in internal energy from $B$ to $C$

$\Delta U_{BC} = nC_V(T_C - T_B)$

$= 1 \times \dfrac{5R}{2}(600 - 800)$

$= -500\, R$

(iii) $\Delta U_{\text{isothermal}} = 0$

(iv) Change in internal energy from $C$ to $A$ *i.e.*, $\Delta U_{CA}$

$$\Delta U_{CA} = nC_V(T_A - T_C)$$
$$= 1 \times \frac{5R}{2}(400 - 600)$$
$$= -500\ R$$

**13.** In this question, the system is accelerating horizontally *i.e.*, no component of acceleration in vertical direction. Hence, the pressure in the vertical direction will remain unaffected.

*i.e.*, $p_1 = p_0 + \rho gh$

Again, we have to use the concept that the pressure in the same level will be same.

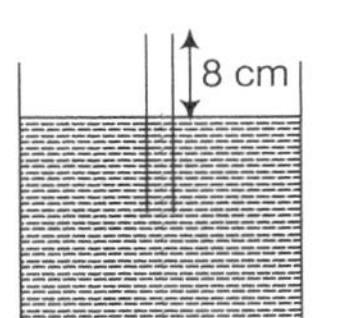


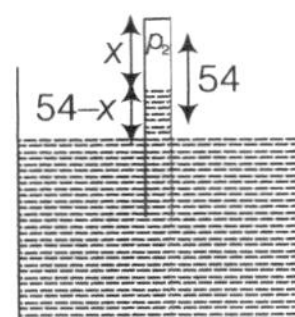


For air trapped in tube, $p_1V_1 = p_2V_2$

$$p_1 = p_{atm} = \rho g 76$$
$$V_1 = A \cdot 8$$
[$A$ = area of cross-section]
$$p_2 = p_{atm} - \rho g(54 - x)$$
$$= \rho g(22 + x)$$
$$V_2 = A \cdot x$$
$$\rho g 76 \times 8A = \rho g(22 + x)Ax$$
$$x^2 + 22x - 78 \times 8 = 0$$
$$\Rightarrow \quad x = 16 \text{ cm}$$

**14.** In SHM, a particle starts from rest, we have

*i.e.*, $x = A\cos \omega t$, at $t = 0$, $x = A$

When $t = \tau$, then $x = A - a$ ...(i)

When $t = 2\tau$, then $x = A - 3a$ ...(ii)

On comparing Eqs. (i) and (ii), we get

$$A - a = A\cos \omega\tau$$
$$A - 3a = A\cos 2\omega\tau$$

As $\cos 2\omega\tau = 2\cos^2 \omega\tau - 1$

$$\Rightarrow \quad \frac{A - 3a}{A} = 2\left(\frac{A - a}{A}\right)^2 - 1$$
$$\Rightarrow \quad \frac{A - 3a}{A} = \frac{2A^2 + 2a^2 - 4Aa - A^2}{A^2}$$
$$A^2 - 3aA = A^2 + 2a^2 - 4Aa$$
$$a^2 = 2aA$$
$$A = 2a$$

Now, $A - a = A\cos \omega\tau$

$$\Rightarrow \quad \cos \omega\tau = 1/2$$
$$\frac{2\pi}{T}\tau = \frac{\pi}{3}$$
$$\Rightarrow \quad T = 6\pi$$

**15.** For closed organ pipe

$$= \frac{(2n+1)v}{4l} \qquad [n = 0, 1, 2 \ldots]$$

85 cm

$$\frac{(2n+1)v}{4l} < 1250$$
$$(2n+1) < 1250 \times \frac{4 \times 0.85}{340}$$
$$(2n+1) < 1.25 \Rightarrow 2n < 11.50$$
$$n < 5.25$$

So, $n = 0, 1, 2, 3, \ldots 5$

So, we have 6 possibilities.

**Alternate method**

In closed organ pipe, fundamental node

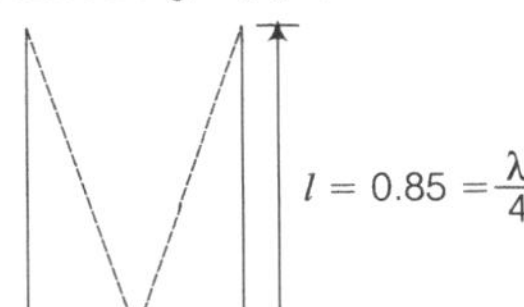


*i.e.*, $\frac{\lambda}{4} = 0.85 \Rightarrow \lambda = 4 \times 0.85$

As we know, $\nu = \frac{c}{\lambda}$

$$\Rightarrow \quad \frac{340}{4 \times 0.85} = 100 \text{ Hz}$$

$\therefore$ Possible frequencies = 100 Hz, 300 Hz, 500 Hz, 700 Hz, 900 Hz, 1100 Hz below 1250 Hz.

**16.** As we know, potential difference $V_A - V_O$ is

$$dV = -E dx$$
$$\int_{V_o}^{V_A} dV = -\int_0^2 30x^2 dx$$

$$V_A - V_O = -30 \times \left[\frac{x^3}{3}\right]_0^2$$
$$= -10 \times [2^3 - (0)^3]$$
$$= -10 \times 8 = -80 \text{ J}$$

**17.** When free space between parallel plates of capacitor,

$$E = \frac{\sigma}{\varepsilon_0}$$

When dielectric is introduced between parallel plates of capacitor, $E' = \frac{\sigma}{K\varepsilon_0}$

Electric field inside dielectric

$$\frac{\sigma}{K\varepsilon_0} = 3 \times 10^4$$

where, $K$ = dielectric constant of medium = 2.2

$\varepsilon_0$ = permitivity of free space $= 8.85 \times 10^{-12}$

$$\Rightarrow \quad \sigma = 2.2 \times 8.85 \times 10^{-12} \times 3 \times 10^4$$
$$= 6.6 \times 8.85 \times 10^{-8}$$
$$= 5.841 \times 10^{-7}$$
$$= 6 \times 10^{-7} \text{ C/m}^2$$

**18.** Total power ($P$) consumed

$$= (15 \times 40) + (5 \times 100) + (5 \times 80) + (1 \times 1000)$$
$$= 2500 \text{ W}$$

As we know,

Power *i.e.*, $P = VI$

$$\Rightarrow \quad I = \frac{2500}{220} \text{ A}$$
$$= \frac{125}{11} = 11.3 \text{ A}$$

Minimum capacity should be 12 A.

**19.** When force exerted on a current carrying conductor

$$F_{ext} = BIL$$

$$\text{Average power} = \frac{\text{Work done}}{\text{Time taken}}$$

$$P = \frac{1}{t}\int_0^2 F_{ext.} \, . \, dx$$
$$= \frac{1}{t}\int_0^2 B(x) IL \, dx$$
$$= \frac{1}{5 \times 10^{-3}}\int_0^2 3 \times 10^{-4} e^{-0.2x} \times 10 \times 3 \, dx$$
$$= 9\,[1 - e^{-0.4}]$$
$$= 9\left[1 - \frac{1}{e^{0.4}}\right] = 2.967 \approx 2.97 \text{ W}$$

**20.** For solenoid, the magnetic field needed to be magnetised the magnet.

$$B = \mu_0 n I$$

[where, $n = 100, l = 10 \text{ cm} = \frac{10}{100} \text{m} = 0.1 \text{ m}$]

$$\Rightarrow \quad 3 \times 10^3 = \frac{100}{0.1} \times I$$
$$I = 3\text{A}$$

**21.** After connecting $C$ to $B$ hanging the switch, the circuit will act like an $L$-$R$ discharging circuit.

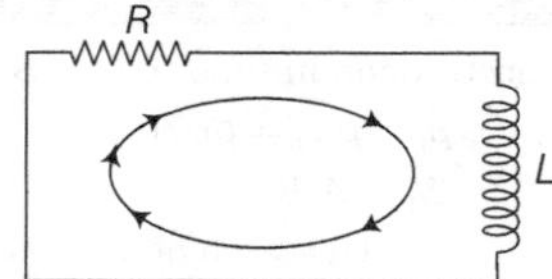


Applying Kirchhoff's loop equation,

$$V_R + V_L = 0 \quad \Rightarrow V_R = -V_L$$
$$\frac{V_R}{V_L} = -1$$

**22.** Both the energy densities are equal *i.e.*, energy is equally divided between electric and magnetic field.

**23.** It is based on lens maker's formula and its magnification.

*i.e.*, $\frac{1}{f} = (\mu - 1)\left(\frac{1}{R_1} - \frac{1}{R_2}\right)$

According to lens maker's formula, when the lens in the air.

$$\frac{1}{f} = \left(\frac{3}{2} - 1\right)\left(\frac{1}{R_1} - \frac{1}{R_2}\right)$$
$$\frac{1}{f} = \frac{1}{2x}$$
$$\Rightarrow \quad f = 2x$$

Here, $\left(\frac{1}{x} = \frac{1}{R_1} - \frac{1}{R_2}\right)$

In case of liquid, where refractive index is $\frac{4}{3}$ and $\frac{5}{3}$, we get

Focal length in first liquid

$$\frac{1}{f_1} = \left(\frac{\mu_s}{\mu_{l_1}} - 1\right)\left(\frac{1}{R_1} - \frac{1}{R_2}\right)$$

$$\Rightarrow \quad \frac{1}{f_1} = \left(\frac{\frac{3}{2}}{\frac{4}{3}} - 1\right)\frac{1}{x}$$

$\Rightarrow f_1$ is positive.

$$\frac{1}{f_1} = \frac{1}{8x} = \frac{1}{4(2x)} = \frac{1}{4f}$$

$$\Rightarrow \quad f_1 = 4f$$

Focal length in second liquid

$$\frac{1}{f_2} = \left(\frac{\mu_s}{\mu_{l_2}} - 1\right)\left(\frac{1}{R_1} - \frac{1}{R_2}\right)$$

$$\Rightarrow \quad \frac{1}{f_2} = \left(\frac{\frac{3}{2}}{\frac{5}{3}} - 1\right)\left(\frac{1}{x}\right)$$

$\Rightarrow f_2$ is negative.

**24.** For total internal reflection of light take place, following conditions must be obeyed.

(*i*) The ray must travel from denser to rarer medium.

(*ii*) Angle of incidence ($\theta$) must be greater than or equal to critical angle ($C$) *i.e.*,

$$C = \sin^{-1}\left[\frac{\mu_{\text{rarer}}}{\mu_{\text{denser}}}\right]$$

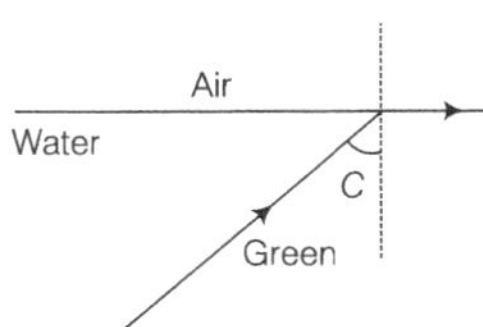


Here, $\quad \sin C = \frac{1}{n_{\text{water}}}$

and $\quad n_{\text{water}} = a + \frac{b}{\lambda^2}$

If frequency is less $\Rightarrow \lambda$ is greater and hence, RI $n_{(\text{water})}$ is less and therefore, critical angle increases. So, they do not suffer reflection and come out at angle less than 90°.

**25.** By law of Malus *i.e.*, $I = I_0 \cos^2\theta$

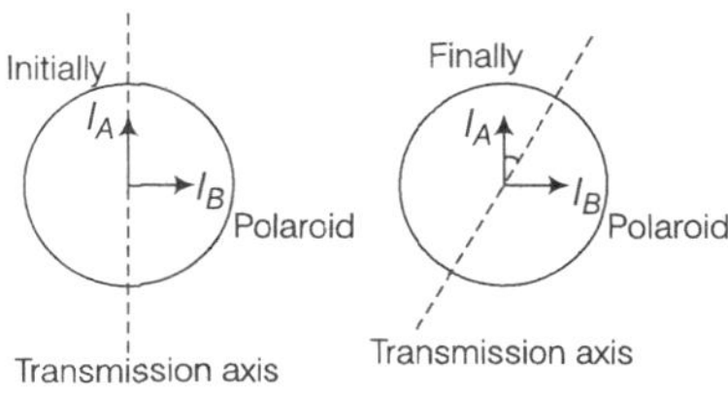


Now, $\quad I_{A'} = I_A \cos^2 30°$

$\quad I_{B'} = I_B \cos^2 60°$

As, $\quad I'_A = I'_B$

$$I_A \cos^2 30° = I_B \cos^2 60°$$

$$\Rightarrow \quad I_A \frac{3}{4} = I_B \frac{1}{4}$$

$$\Rightarrow \quad \frac{I_A}{I_B} = \frac{1}{3}$$

**26.** The problem is based on frequency dependence of photoelectric emission. When incident light with certain frequency (greater than on the threshold frequency is focus on a metal surface) then some electrons are emitted from the metal with substantial initial speed.

When an electron moves in a circular path, then

$$r = \frac{mv}{eB}$$

$$\Rightarrow \quad \frac{r^2e^2B^2}{2} = \frac{m^2v^2}{2}$$

$$\text{KE}_{\text{max}} = \frac{(mv)^2}{2m}$$

$$\Rightarrow \quad \frac{r^2e^2B^2}{2m} = (\text{KE})_{\text{max}}$$

Work function of the metal ($W$),

*i.e.*, $\quad W = h\nu - \text{KE}_{\text{max}}$

$$1.89 - \phi = \frac{r^2e^2B^2}{2m}\frac{1}{2}\text{eV}$$

$$= \frac{r^2eB^2}{2m}\text{eV}$$

[$h\nu \to 1.89$ eV, for the transition on from third to second orbit of H-atom]

$$= \frac{100 \times 10^{-6} \times 1.6 \times 10^{-19} \times 9 \times 10^{-8}}{2 \times 9.1 \times 10^{-31}}$$

$$\phi = 1.89 - \frac{1.6 \times 9}{2 \times 9.1}$$

$$= 1.89 - 0.79 = 1.1\,\text{eV}$$

**27.** For hydrogen atom, we get

$$\frac{1}{\lambda} = Rz^2\left(\frac{1}{1^2} - \frac{1}{2^2}\right) \Rightarrow \frac{1}{\lambda_1} = R(1)^2\left(\frac{3}{4}\right)$$

$$\frac{1}{\lambda_2} = R(1)^2\left(\frac{3}{4}\right)$$

$$\Rightarrow \quad \frac{1}{\lambda_3} = R(2)^2\left(\frac{3}{4}\right)$$

$$\frac{1}{\lambda_4} = R(3)^2\left(\frac{3}{4}\right)$$

$$\Rightarrow \quad \frac{1}{\lambda_1} = \frac{1}{4\lambda_3} = \frac{1}{9\lambda_4} = \frac{1}{\lambda_2}$$

**28.** For forward bias, *p*-side must be a higher potential than *n*-side.

+2 V —▷|— /\/\/\/\/\ — −2 V

So, is forward biased.

**29.** If student measures 3.50 cm, it means that there in an uncertainly of order 0.01 cm.

For vernier scale with 1 MSD = 1 mm

and 9 MSD = 10 VSD

LC of Vernier Calliper = 1MSD − 1VSD

$$= \frac{1}{10}\left(1 - \frac{9}{10}\right)$$

$$= \frac{1}{100} \text{ cm}$$

**30.** (a) Infrared rays are used to treat muscular strain.

(b) Radiowaves are used for broadcasting purposes.

(c) X-rays are used to detect fracture of bones.

(d) Ultraviolet rays are absorbed by ozone.

# Chemistry

**1.** Given, Atomic number of Rb, $Z = 37$

Thus, its electronic configuration is $[\text{Kr}]5s^1$. Since the last electron or valence electron enter in 5*s* subshell.

So, the quantum numbers are $n = 5$, $l = 0$, (for *s* orbital) $m = 0$ ($\because m = +l$ to $-l$), $s = +\frac{1}{2}$ or $-\frac{1}{2}$

**2.** To solve this problem, the stepwise approach required *i.e.*,

(*i*) Write the van der Waals' equation, then apply the condition that at low pressure, volume become high,

*i.e.*, $V - b \simeq V$

(*ii*) Now calculate the value of compressibility factor ($Z$). [$Z = pV / RT$]

According to van der Waals' equation,

$$\left(p + \frac{a}{V^2}\right)(V - b) = RT$$

At low pressure,

$$\left(p + \frac{a}{V^2}\right)V = RT$$

$$pV + \frac{a}{V} = RT$$

$$pV = RT - \frac{a}{V}$$

Divide both side by $RT$,

$$\frac{pV}{RT} = 1 - \frac{a}{RTV}$$

**3.** In CsCl, $Cl^-$ lie at corners of simple cube and $Cs^+$ at the body centre. Hence, along the body diagonal, $Cs^+$ and $Cl^-$ touch each other so

$$r_{Cs^+} + r_{Cl^-} = 2r$$

**Calculation of *r***

In $\Delta EDF$,

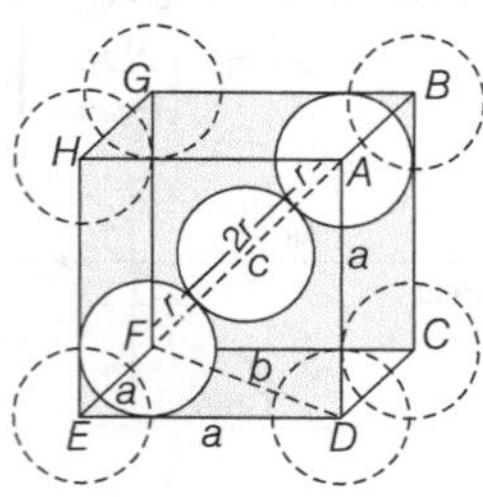


Body centred cubic unit cell

$$FD = b = \sqrt{a^2 + a^2} = \sqrt{2}a$$

In $\Delta AFD$,

$$c^2 = a^2 + b^2$$

$$= a^2 + (\sqrt{2}a)^2$$

$$= a^2 + 2a^2$$

$$c^2 = 3a^2$$
$$c = \sqrt{3}a$$

As $\Delta$ $AFD$ is an equilateral triangle.

$\therefore \quad \sqrt{3}a = 4r$

$\Rightarrow \quad r = \frac{\sqrt{3}a}{4}$

Hence, $r_{Cs^+} + r_{Cl} = 2r$

$$= 2 \times \frac{\sqrt{3}}{4}a = \frac{\sqrt{3}}{2}a$$

**4.** This problem is based on the estimation of percentage of N in organic compound using Kjeldahl's method. Use the concept of stoichiometry and follow the steps given below to solve the problem.

(*i*) Write the balanced chemical reaction for the conversion of N present in organic compound to ammonia, ammonia to ammonium sulphate and ammonium sulphate to sodium sulphate.

(*ii*) Calculate millimoles (m moles) of N present in organic compound followed by mass of N present in organic compound using the concept of stoichiometry.

(*iii*) At last, calculate % of N present in organic compound using formula

$$\% \text{ of N} = \frac{\textit{Mass of N} \times 100}{\textit{Mass of organic compound}}$$

Mass of organic compound = 1.4 g

Let it contain $x$ mmole of N atom.

$$\text{Organic compound} \longrightarrow \underset{x \text{ mmole}}{NH_3}$$

$$2NH_3 + \underset{\substack{6 \text{ m mole} \\ \text{initially taken}}}{H_2SO_4} \longrightarrow (NH_4)_2SO_4 \quad \ldots\text{(i)}$$

$$H_2SO_4 + 2\,NaOH \longrightarrow Na_2SO_4 + 2\,H_2O \quad \ldots\text{(ii)}$$

2 mmole NaOH reacted.

Hence, mmoles of $H_2SO_4$ reacted in Eq. (ii) = 1

$\Rightarrow$ mmoles of $H_2SO_4$ reacted from Eq. (i)

$$= 6 - 1 = 5 \text{ mmoles}$$

$\Rightarrow$ mmoles of $NH_3$ in Eq. (i)

$$= 2 \times 5 = 10 \text{ mmoles}$$

$\Rightarrow$ mmoles of N atom in the organic compound = 10 mmoles

$\Rightarrow$ Mass of N = $10 \times 10^{-3} \times 14$

$$= 0.14 \text{ g}$$

% of N

$$= \frac{\text{Mass of N present in organic compound}}{\text{Mass of organic compound}} \times 100$$

$$\Rightarrow \quad \% \text{ of N} = \frac{0.14}{1.4} \times 100 = 10\%$$

**5.** In order to solve the problem, calculate the value of cell constant of the first solution and then use this value of cell constant to calculate the value of *k* of second solution. Afterwards, finally calculate molar conductivity using value of *k* and *m*.

For first solution,

$$k = 1.4 \text{ Sm}^{-1}, \quad R = 50\ \Omega, \quad M = 0.2$$

$$\text{Specific conductance } (\kappa) = \frac{1}{R} \times \frac{l}{A}$$

$$1.4 \text{ Sm}^{-1} = \frac{1}{50} \times \frac{l}{A}$$

$$\frac{l}{A} = 50 \times 1.4 \text{ m}^{-1}$$

For second solution,

$$R = 280, \ \frac{l}{A} = 50 \times 1.4 \text{ m}^{-1}$$

$$\kappa = \frac{1}{280} \times 1.4 \times 50 = \frac{1}{4}$$

Now, molar conductivity

$$\lambda_m = \frac{\kappa}{1000 \times m} = \frac{1/4}{1000 \times 0.5} = \frac{1}{2000}$$

$$= 5 \times 10^{-4} \text{ S m}^2 \text{ mol}^{-1}$$

**6.** $C_2H_5OH(l) + 3O_2(g) \longrightarrow 2CO_2(g) + 3H_2O(l)$

$$\Delta U = -1364.47 \text{ kJ/mol}$$
$$\Delta H = \Delta U + \Delta n_g RT$$
$$\Delta n_g = -1$$
$$\Delta H = -1364.47 + \frac{-1 \times 8.314 \times 298}{1000}$$

[Here, value of $R$ in unit of J must be converted into kJ]

$$= -1364.47 - 2.4776$$
$$= -1366.94 \text{ kJ/mol}$$

During solving such problem, students are advised to keep much importance in unit conversion. As here, value of $R$ ($8.314 \text{ J K}^{-1} \text{ mol}^{-1}$) in $\text{JK}^{-1} \text{ mol}^{-1}$ must be converted into kJ by dividing the unit by 1000.

**7.** According to Debye Huckel Onsager equation,

$$\lambda_C = \lambda_\infty - B\sqrt{C}$$

where, $\lambda_C$ = limiting equivalent conductivity at concentration $C$

$\lambda_\infty$ = limiting equivalent conductivity at infinite dilution

$C$ = concentration

**8.** This problem includes concept of colligative properties (osmotic pressure here) and van't Hoff factor. Calculate the effective molarity of each solution.

*i.e.*, effective molarity = van't Hoff factor × molarity

0.5 M $C_2H_5OH\ (aq)$ $i = 1$

Effective molarity = 0.5

0.25 M KBr $(aq)$ $i = 2$

Effective molarity = 0.5 M

0.1 M $Mg_3(PO_4)_2(aq)$ $i = 5$

Effective molarity = 0.5 M

0.125 M $Na_3PO_4(aq)$ $i = 4$

Effective molarity = 0.5 M

Hence, all colligative properties are same.

**Note** *This equation is solved by assuming that the examiner has taken* $Mg_3(PO_4)_2$ *to be completely soluble. However, the fact is that it is insoluble (sparingly soluble).*

**9.** For the given reaction,

$$\Delta n_g = n_p - n_R$$

where, $n_p$ = number of moles of products

$n_R$ = number of moles of reactants

$$K_p = K_c(RT)^{\Delta n_g}$$

$$\Delta n_g = -(1/2)$$

**10.** This problem can be solved by determining the order of reaction w.r.t. each reactant and then writing rate law equation of the given equation accordingly as

$$r = \frac{dC}{dt} = k[A]^x[B]^y$$

where, $x$ = order of reaction w.r.t. $A$

$y$ = order of reaction w.r.t. $B$

$$1.2 \times 10^{-3} = k(0.1)^x(0.1)^y$$

$$1.2 \times 10^{-3} = k(0.1)^x(0.2)^y$$

$$2.4 \times 10^{-3} = k(0.2)^x(0.1)^y$$

$$R = k[A]^1[B]^0$$

As shown above, rate of reaction remains constant as the concentration of reactant ($B$) changes from 0.1 M to 0.2 M and becomes double when concentration of $A$ change from 0.1 to 0.2 (*i.e.*, doubled).

**11.** **Decreasing order of strength of oxoacids**

$$HClO_4 > HClO_3 > HClO_2 > HOCl$$

**Reason** Consider the structures of conjugate bases of each oxyacids of chlorine.

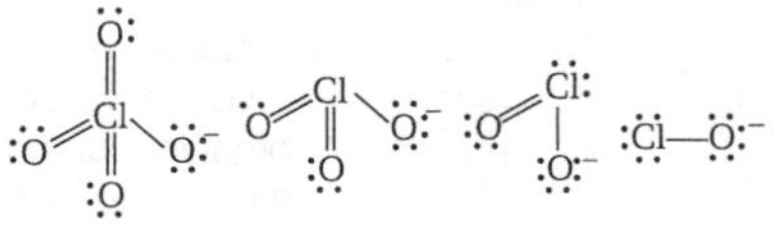

Negative charge is more delocalised on $ClO_4^-$ due to resonance, hence, $ClO_4^-$ is more stable (and less basic).

Hence, we can say as the number of oxygen atom(s) around Cl-atom increases as oxidation number of Cl-atom increases and thus, the ability of loose the $H^+$ increases.

**12.** Higher the position of element in the electro-chemical series more difficult is the reduction of its cations.

If $Ca^{2+}(aq)$ is electrolysed, water is reduced in preference to it. Hence, it cannot be reduced electrolytically from their aqueous solution.

$$Ca^{2+}(aq) + H_2O \longrightarrow Ca^{2+} + OH^- + H_2 \uparrow$$

**13.** Arrange the complex formed by different ligands $L_1, L_2, L_3$ and $L_4$, according to wavelength of their absorbed light, then use the following relation to answer the question.

Ligand field strength

$$\propto \text{Energy of light absorbed}$$

$$\propto \frac{1}{\text{Wavelength of light absorbed}}$$

| $\lambda$ Absorbed light | $L_1$ Red | $L_2$ Green | $L_3$ Yellow | $L_4$ Blue |
|---|---|---|---|---|

Wavelength of absorbed light decreases.

∴ Increasing order of energy of wavelengths absorbed reflect greater extent of crystal-field splitting, hence higher field strength of the ligand.

**Energy** Blue $(L_4)$ > green $(L_2)$
> yellow $(L_3)$
> red $(L_1)$

∴ $L_4 > L_2 > L_3 > L_1$ in field strength of ligands.

**14.** NO is paramagnetic in gaseous state because in gaseous state, it has one unpaired electron.

Total number of electron present
$$= 7 + 8$$
$$= 15\,e^-$$

Hence, there must be the presence of unpaired electron in gaseous state while in liquid state, it dimerises due to unpaired electron.

**15.** Release of electron is known as reduction. So, $H_2O_2$ acts as reducing agent when it releases electrons.

Here, in reaction (II) and (IV), $H_2O_2$ releases two electron hence, reaction (II) and (IV) is known as reduction.

In reaction (I) and (III), two electrons are being added, so (I) and (III) represents oxidation.

**16.** $I_3^-$ is an ion made up of $I_2$ and $I^-$ which has linear shape.

$CH_3Cl$ (Less crowded) > $CH_3CH_2Cl$ (1°) > $CH_3-CH(CH_3)-Cl$ (2°) > $CH_3-C(CH_3)_2-Cl$ (More crowded)

while $Cs^+$ is an alkali metal cation.

**17.** $$\frac{n_{O_2}}{n_{N_2}} = \frac{\dfrac{(m_{O_2})}{(M_{O_2})}}{\dfrac{(m_{N_2})}{(M_{N_2})}}$$

where, $m_{O_2}$ = given mass of $O_2$
$m_{N_2}$ = given mass of $N_2$
$M_{O_2}$ = molecular mass of $O_2$
$M_{N_2}$ = molecular mass of $N_2$
$n_{O_2}$ = number of moles of $O_2$
$n_{N_2}$ = number of moles of $N_2$

$$= \left[\frac{m_{O_2}}{m_{N_2}}\right]\frac{28}{32} = \frac{1}{4} \times \frac{28}{32}$$
$$= 7/32$$

**18.** Standard electrode potential of reaction $[E°]$ can be calculated as
$$E^\circ_{cell} = E_R - E_P$$
where, $E_R$ = SRP of reactant
$E_P$ = SRP of product

If $E^\circ_{cell}$ = +ve, then reaction is spontaneous otherwise non-spontaneous.

$$Mn^{3+} \xrightarrow{E_1^\circ = 1.51\,V} Mn^{2+}$$
$$Mn^{2+} \xrightarrow{E_2^\circ = -1.18\,V} Mn$$

∴ For $Mn^{2+}$ disproportionation,
$$E° = -1.51\,V - 1.18\,V$$
$$= -2.69\,V$$
$$< 0$$

**19.** Analyse each reaction given in the question and choose the correct answer on the basis of oxidation state and stability of iron compounds. Use the concept of Ellingham diagram to solve this problem.

*The correct reactions are as follows*

(a) $Fe + \text{dil } H_2SO_4 \longrightarrow FeSO_4 + H_2$

$$H_2SO_4 + 2FeSO_4 + \frac{1}{2}O_2 \longrightarrow Fe_2(SO_4)_3 + H_2O$$

$$Fe_2(SO_4)_3 \xrightarrow{\Delta} Fe_2O_3\,(s) + 3SO_3\uparrow$$

The given reaction is incorrect in question

(b) $Fe \xrightarrow[\Delta]{O_2} FeO$ [It could also be $Fe_2O_3$ or $Fe_3O_4$]

$$FeO + H_2SO_4 \longrightarrow FeSO_4 + H_2O$$
$$2FeSO_4 \xrightarrow{\Delta} Fe_2O_3 + SO_2 + SO_3$$

(c) $Fe \xrightarrow[Cl_2]{\Delta} FeCl_3 \xrightarrow[Air]{\Delta}$ No reaction

[It cannot give $FeCl_2$]

(d) $Fe \xrightarrow[\Delta]{O_2} Fe_3O_4 \xrightarrow[600°C]{CO} FeO \xrightarrow[700°C]{CO} Fe$

This is correct reaction.

**20.** This problem is based on conceptual mixing of properties of lithium oxide and preparation, properties of coordination compounds. To answer this question, keep in mind that on adding acid ammine complexes get destroyed.

(a) $Li_2O + KCl \longrightarrow 2LiCl + K_2O$

This is wrong equation, since a stronger base $K_2O$ cannot be generated by a weaker base $Li_2O$.

(b) $[CoCl(NH_3)_5]^+ + 5H^+ \longrightarrow Co^{2+}(aq) + 5\,NH_4^+ + Cl^-$

This is correct. All ammine complexes can be destroyed by adding $H^\oplus$. Hence, on adding acid to $[CoCl(NH_3)_5]$, It gets converted to $Co^{2+}(aq)^+$ $NH_4^+$ and $Cl^-$.

(c) $[Mg(H_2O)_6]^{2+} + EDTA^{4-} \xrightarrow[\text{excess}]{OH^-} [Mg(EDTA)]^{2+} + 6\,H_2O$

This is wrong, since the formula of complex must be $[Mg(EDTA)]^{2+}$ as EDT.

(d) The 4th reaction is incorrect. It can be correctly represented as

$$2CuSO_4 + 10KCN \longrightarrow 2K_3[Cu(CN)_4] + 2K_2SO_4 + (CN)_2 \uparrow$$

**21.** Steric hindrance (crowding) is the basis of $S_N2$ reaction, by using which we can arrange the reactant in correct order of their reactivity towards $S_N2$ reaction.

$$\text{Rate of } S_N2 \propto \frac{1}{\text{Steric crowding of 'C'}}$$

$$\underset{\text{(Less crowded)}}{CH_3Cl} > \underset{1°}{CH_3CH_2Cl} > \underset{2°}{CH_3-\underset{CH_3}{\underset{|}{CH}}-Cl} > \underset{\text{(More crowded)}}{CH_3-\overset{CH_3}{\overset{|}{\underset{CH_3}{\underset{|}{C}}}}H-Cl}$$

As steric hinderance (crowding) increases, rate of $S_N2$ reaction decreases.

**Note** *The order of reactivity towards $S_N2$ reaction for alkyl halides is*

$$\underset{(1°)}{\text{Primary halides}} > \underset{(2°)}{\text{Secondary halides}} > \underset{(3°)}{\text{Tertiary halides}}$$

**22.** This reaction is an example of carbylamine reaction which includes conversion of amine to isocyanide.

$$\underset{1°\text{ amine}}{R-\ddot{N}H_2} + CHCl_3 \xrightarrow[KOH]{C_2H_5OH} \underset{\text{Alkyl isocyanide}}{R-\overset{+}{N}\equiv\overset{-}{C}:}$$

**Note** *This reaction include rearrangement reaction of nitrene in which migration of alkyl group from carbon to nitrogen takes place.*

**23.** $R-CH_2OH \xrightarrow{PCC} R-CH=O$

Pyridinium chlorochromate is the mild oxidising agent which causes conversion of alcohol to aldehyde stage. While others causes conversion of alcohol to acid.

**24.** The reaction is $2CH_3-CCl_3 \xrightarrow[\Delta]{6Ag} \underset{\text{But-2-yne}}{CH_3-C\equiv C-CH_3} + 6\,AgCl$

**25.** It is a Kolbe Schmidt reaction.

$$C_6H_5ONa + CO_2 \xrightarrow[5\text{ atm}]{125°C} o\text{-}HOC_6H_4COONa \xrightarrow{H^+/Ac_2O} \underset{\text{(pain killer)}}{\underset{\text{Aspirin}}{o\text{-}CH_3COOC_6H_4COOH}}$$

The second step of the reaction is an example of acetylation reaction.

**26.** This problem can be solved by using the concept of effect of steric hindrance, hydration and H-bonding in basic strength of amines.

Order of basic strength of aliphatic amine in aqueous solution is as follows (order of $K_b$)

$$(CH_3)_2\ddot{N}H > CH_3\ddot{N}H_2 > (CH_3)_3\ddot{N} > C_6H_5\ddot{N}H_2$$

As we know, $pK_b = -\log K_b$. So, $(CH_3)_2\ddot{N}H$ will have smallest $pK_b$ value.

In case of phenyl amine, N is attached to $sp^2$ hybridised carbon, hence it has highest $pK_b$ and least basic strength.

**27.** Draw the structure of organic compounds indicating net dipole moment which includes lone pair and bond angle also.

In the quinol (p-$HO-C_6H_4-OH$) and thioquinol (p-$HS-C_6H_4-SH$)

—OH groups and —SH groups do not cancel their dipole moment as they exist in different confirmation.

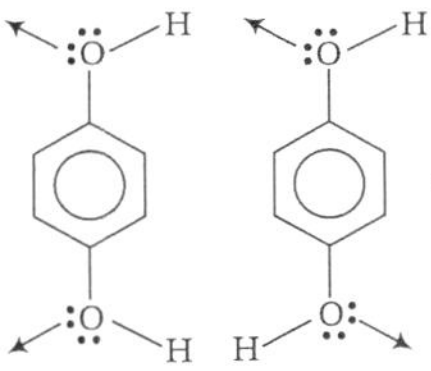

(Structures of quinol showing two conformations of —OH groups with lone pairs and dipole arrows) same as in thioquinol.

**28.** Dacron is a condensation polymer of ethylene glycol and methyl terepthalate. Formation of dacron can be shown as

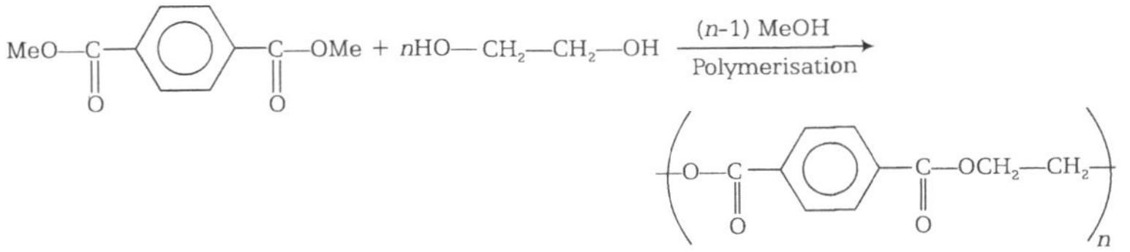

$$MeO-\underset{\underset{O}{\|}}{C}-C_6H_4-\underset{\underset{O}{\|}}{C}-OMe + nHO-CH_2-CH_2-OH \xrightarrow[\text{Polymerisation}]{(n-1)\ MeOH} \left(-O-\underset{\underset{O}{\|}}{C}-C_6H_4-\underset{\underset{O}{\|}}{C}-OCH_2-CH_2-\right)_n$$

Dacron

Here, elimination of MeOH occurs as a by product. So, this reaction is known as condensation polymerisation.

**29.** Quinoline is an alkaloid, it is not present in DNA. DNA has four nitrogen bases in adenine, guanine, cytosine and thymine.

**30.** This problem is based on successive reduction, chlorination and elimination reaction. To solve such problem, use the function of the given reagents.

(i) $LiAlH_4$ causes reduction

(ii) $PCl_5$ causes chlorination

(iii) Alc. KOH causes elimination reaction

$$CH_3COOH \xrightarrow{LiAlH_4} \underset{(A)}{CH_3CH_2OH} \xrightarrow{PCl_5} \underset{(B)}{CH_3CH_2Cl} \xrightarrow[-HCl]{Alc.KOH} \underset{\text{Ethylene}}{\underset{(C)}{CH_2=CH_2}}$$

# Mathematics

**1.** We have,

$$X = \{4^n - 3n - 1 : n \in N\}$$

$$X = \{0, 9, 54, 243, ...\}$$

[Put $n = 1, 2, 3, ...$]

$$Y = \{9(n-1) : n \in N\}$$

$$Y = \{0, 9, 18, 27, ...\}$$

[Put $n = 1, 2, 3, ...$]

It is clear that $X \subset Y$.

$\therefore \quad X \cup Y = Y$

**2.** $|z| \geq 2$ is the region on or outside circle whose centre is (0, 0) and radius is 2.

Minimum $\left|z + \frac{1}{2}\right|$ is distance of $z$, which lie on circle $|z| = 2$ from $\left(-\frac{1}{2}, 0\right)$.

$\therefore$ Minimum $\left|z + \frac{1}{2}\right|$ = Distance of $\left(-\frac{1}{2}, 0\right)$ from $(-2, 0)$

$$= \sqrt{\left(-2 + \frac{1}{2}\right)^2 + 0} = \frac{3}{2}$$

$$= \sqrt{\left(\frac{-1}{2} + 2\right)^2 + 0} = \frac{3}{2}$$

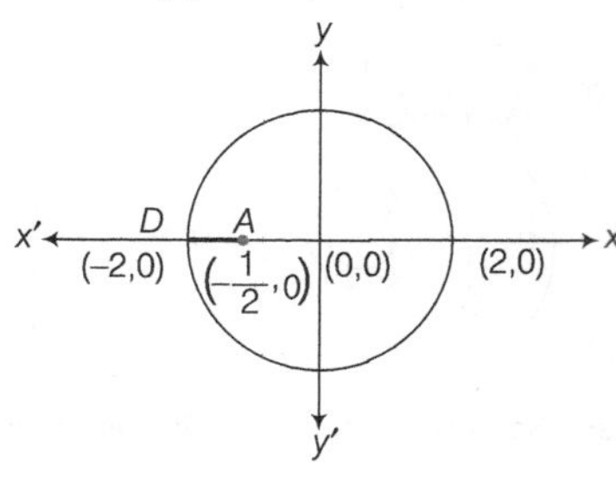


**Geometrically** Min $\left|z + \frac{1}{2}\right| = AD$

Hence, minimum value of $\left(z + \frac{1}{2}\right)$ lies in the interval (1, 2).

**3.** Put $t = x - [x] = \{X\}$, which is a fractional part function and lie between $0 \leq \{X\} < 1$ and then solve it.

Given, $a \in R$ and equation is

$$-3\{x - [x]\}^2 + 2\{x - [x]\} + a^2 = 0$$

Let $t = x - [x]$, then equation is

$$-3t^2 + 2t + a^2 = 0$$

$$\Rightarrow \quad t = \frac{1 \pm \sqrt{1 + 3a^2}}{3}$$

$$\because \quad t = x - [x] = \{X\}$$

[Fractional part]

$$\therefore \quad 0 \leq t \leq 1$$

$$0 \leq \frac{1 \pm \sqrt{1 + 3a^2}}{3} \leq 1$$

Taking positive sign, we get

$$0 \leq \frac{1 + \sqrt{1 + 3a^2}}{3} < 1 \qquad [\because \{x\} > 0]$$

$$\Rightarrow \quad \sqrt{1 + 3a^2} < 2 \Rightarrow 1 + 3a^2 < 4$$

$$\Rightarrow \quad a^2 - 1 < 0$$

$$\Rightarrow \quad (a + 1)(a - 1) < 0$$

+    +    (number line: –1, 1)

$$a \in (-1, 1)$$

For no integral solution of $a$, we consider the interval $(-1, 0) \cup (0, 1)$.

**Note** *Here, we figure out the integral solution, we get $a = 0$. This implies any interval excluding zero should be correct answer as it gives eitherwww no solution or no integral solution.*

**4.** If $ax^2 + bx + c = 0$ has roots $\alpha$ and $\beta$, then $\alpha + \beta = \frac{-b}{a}$ and $\alpha\beta = \frac{c}{a}$. Find the values of $\alpha + \beta$ and $\alpha\beta$ and then put in $(\alpha - \beta)^2 = (\alpha + \beta)^2 - 4\alpha\beta$ to get required value.

Given, $\alpha$ and $\beta$ are roots of

$$px^2 + qx + r = 0, p \neq 0$$

$$\therefore \quad \alpha + \beta = \frac{-q}{p}, \ \alpha\beta = \frac{r}{p} \qquad \text{...(i)}$$

Since, $p, q$ and $r$ are in AP.

$$\therefore \quad 2q = p + r \qquad \text{...(ii)}$$

Also, $\quad \frac{1}{\alpha} + \frac{1}{\beta} = 4$

$$\Rightarrow \quad \frac{\alpha + \beta}{\alpha\beta} = 4$$

$$\Rightarrow \quad \alpha + \beta = 4\alpha\beta$$

$$\Rightarrow \quad \frac{-q}{p} = \frac{4r}{p} \qquad \text{[From Eq. (i)]}$$

$$\Rightarrow \quad q = -4r$$

On putting the value of $q$ in Eq. (ii), we get $\Rightarrow 2(-4r) = p + r \Rightarrow p = -9r$

Now, $\alpha + \beta = \frac{-q}{p} = \frac{4r}{p} = \frac{4r}{-9r} = -\frac{4}{9}$ and $\alpha\beta = \frac{r}{p} = \frac{r}{-9r} = \frac{1}{-9}$

$\therefore$ $(\alpha - \beta)^2 = (\alpha + \beta)^2 - 4\alpha\beta = \frac{16}{81} + \frac{4}{9} = \frac{16 + 36}{81}$

$\Rightarrow$ $(\alpha - \beta)^2 = \frac{52}{81} \Rightarrow |\alpha - \beta| = \frac{2}{9}\sqrt{13}$

**5.** Use the property that, two determinants can be multiplied row-to-row or row-to-column, to write the given determinant as the product of two determinants and then expand.

Given, $f(n) = \alpha^n + \beta^n$, $f(1) = \alpha + \beta$, $f(2) = \alpha^2 + \beta^2$, $f(3) = \alpha^3 + \beta^3$, $f(4) = \alpha^4 + \beta^4$

Let $\Delta = \begin{vmatrix} 3 & 1 + f(1) & 1 + f(2) \\ 1 + f(1) & 1 + f(2) & 1 + f(3) \\ 1 + f(2) & 1 + f(3) & 1 + f(4) \end{vmatrix}$

$$\Rightarrow \Delta = \begin{vmatrix} 3 & 1 + \alpha + \beta & 1 + \alpha^2 + \beta^2 \\ 1 + \alpha + \beta & 1 + \alpha^2 + \beta^2 & 1 + \alpha^3 + \beta^3 \\ 1 + \alpha^2 + \beta^2 & 1 + \alpha^3 + \beta^3 & 1 + \alpha^4 + \beta^4 \end{vmatrix}$$

$$= \begin{vmatrix} 1\cdot 1 + 1\cdot 1 + 1\cdot 1 & 1\cdot 1 + 1\cdot\alpha + 1\cdot\beta & 1\cdot 1 + 1\cdot\alpha^2 + 1\cdot\beta^2 \\ 1\cdot 1 + 1\cdot\alpha + 1\cdot\beta & 1\cdot 1 + \alpha\cdot\alpha + \beta\cdot\beta & 1\cdot 1 + \alpha\cdot\alpha^2 + \beta\cdot\beta^2 \\ 1\cdot 1 + 1\cdot\alpha^2 + 1\cdot\beta^2 & 1\cdot 1 + \alpha^2\cdot\alpha + \beta^2\cdot\beta & 1\cdot 1 + \alpha^2\cdot\alpha^2 + \beta^2\cdot\beta^2 \end{vmatrix}$$

$$= \begin{vmatrix} 1 & 1 & 1 \\ 1 & \alpha & \beta \\ 1 & \alpha^2 & \beta^2 \end{vmatrix} \begin{vmatrix} 1 & 1 & 1 \\ 1 & \alpha & \beta \\ 1 & \alpha^2 & \beta^2 \end{vmatrix} = \begin{vmatrix} 1 & 1 & 1 \\ 1 & \alpha & \beta \\ 1 & \alpha^2 & \beta^2 \end{vmatrix}^2$$

On expanding, we get

$$\Delta = (1 - \alpha)^2 (1 - \beta)^2 (\alpha - \beta)^2$$

But given, $\Delta = K(1 - \alpha)^2 (1 - \beta)^2 (\alpha - \beta)^2$

Hence, $K(1 - \alpha)^2 (1 - \beta)^2 (\alpha - \beta)^2 = (1 - \alpha)^2 (1 - \beta)^2 (\alpha - \beta)^2$

$\therefore$ $K = 1$

**6.** Use the following properties of transpose $(AB)^T = B^T A^T$, $(A^T)^T = A$ and $A^{-1}A = I$ and simplify.

If $A$ is non-singular matrix, then $|A| \neq 0$

$AA^T = A^T A$ and $B = A^{-1}A^T$

$BB^T = (A^{-1}A^T)(A^{-1}A^T)^T = A^{-1}A^T A(A^{-1})^T$ $[\because (AB)^T = B^T A^T]$

$= A^{-1}AA^T (A^{-1})^T$ $[\because AA^T = A^T A]$

$= IA^T (A^{-1})^T$ $[\because A^{-1}A = I]$

$= A^T (A^{-1})^T = (A^{-1}A)^T$ $[\because (AB)^T = B^T A^T]$

$= I^T = I$

**7.** To find the coefficient of $x^3$ and $x^4$, use the formula of coefficient of $x^r$ in $(1 - x)^n$ is $(-1)^r\, {}^nC_r$ and then simplify.

In expansion of $(1 + ax + bx^2)(1 - 2x)^{18}$.

Coefficient of $x^3$ = Coefficient of $x^3$ in $(1 - 2x)^{18}$ + Coefficient of $x^2$ in $a(1 - 2x)^{18}$ + Coefficient of $x$ in $b(1 - 2x)^{18}$

$$= -{}^{18}C_3 \cdot 2^3 + a\,{}^{18}C_2 \cdot 2^2 - b\,{}^{18}C_1 \cdot 2$$

Given, coefficient of $x^3 = 0$

$$= {}^{18}C_3 \cdot 2^3 + a\,{}^{18}C_2 \cdot 2^2 - b\,{}^{18}C_1 \cdot 2 = 0$$

$$\Rightarrow -\frac{18\times17\times16}{3\times2}\cdot8+a\cdot\frac{18\times17}{2}\cdot2^2$$
$$-b\cdot18\cdot2=0$$
$$\Rightarrow \quad 17a-b=\frac{34\times16}{3} \quad ..(i)$$

Similarly, coefficient of $x^4$

$$^{18}C_4\cdot2^4-a\cdot{}^{18}C_3 2^3+b\cdot{}^{18}C_2\cdot2^2=0$$
$$\therefore \quad 32a-3b=240 \quad ...(ii)$$

On solving Eqs. (i) and (ii), we get

$$a=16,\ b=\frac{272}{3}$$

**8.** We know that angle between two lines is

$$\cos\theta=\frac{a_1a_2+b_1b_2+c_1c_2}{\sqrt{a_1^2+b_1^2+c_1^2}\sqrt{a_2^2+b_2^2+c_2^2}}$$
$$l+m+n=0$$
$$\Rightarrow \quad l=-(m+n)$$
$$\Rightarrow \quad (m+n)^2=l^2$$
$$\Rightarrow m^2+n^2+2mn=m^2+n^2$$
$$[\because l^2=m^2+n^2, \text{ given}]$$
$$\Rightarrow \quad 2mn=0$$

When $m=0 \Rightarrow l=-n$

Hence, $(l, m, n)$ is $(1, 0, -1)$.

When $n=0$, then $l=-m$

Hence, $(l, m, n)$ is $(1, 0, -1)$.

$$\therefore \quad \cos\theta=\frac{1+0+0}{\sqrt2\times\sqrt2}$$
$$=\frac12$$
$$\Rightarrow \quad \theta=\frac{\pi}{3}$$

**9.** Given, $k\cdot10^9=10^9+2(11)^1(10)^8$

$$+3(11)^2(10)^7+\ldots+10(11)^9$$
$$k=1+2\left(\frac{11}{10}\right)+3\left(\frac{11}{10}\right)^2$$
$$+\ldots+10\left(\frac{11}{10}\right)^9 \quad ...(i)$$
$$\left(\frac{11}{10}\right)k=1\left(\frac{11}{10}\right)+2\left(\frac{11}{10}\right)^2$$
$$+\ldots+9\left(\frac{11}{10}\right)^9+10\left(\frac{11}{10}\right)^{10} \quad ...(ii)$$

On subtracting Eq. (ii) from Eq. (i), we get

$$k\left(1-\frac{11}{10}\right)=1+\frac{11}{10}+\left(\frac{11}{10}\right)^2$$
$$+\ldots+\left(\frac{11}{10}\right)^9-10\left(\frac{11}{10}\right)^{10}$$
$$\Rightarrow k\left(\frac{10-11}{10}\right)=\frac{1\left[\left(\frac{11}{10}\right)^{10}-1\right]}{\left(\frac{11}{10}-1\right)}$$
$$-10\left(\frac{11}{10}\right)^{10}$$
$$\left[\because \text{In GP,sum of } n \text{ terms}=\frac{a(r^n-1)}{r-1}, \text{ when } r>1\right]$$
$$\Rightarrow -k$$
$$=10\left[10\left(\frac{11}{10}\right)^{10}-10-10\left(\frac{11}{10}\right)^{10}\right]$$
$$\therefore \quad k=100$$

**10.** Let $a, ar, ar^2$ are in GP $(r>1)$.

On multiplying middle term by 2, $a, 2ar, ar^2$ are in AP.

$$\Rightarrow \quad 4ar=a+ar^2$$
$$\Rightarrow \quad r^2-4r+1=0$$
$$\Rightarrow \quad r=\frac{4\pm\sqrt{16-4}}{2}$$
$$=2\pm\sqrt3$$
$$\Rightarrow \quad r=2+\sqrt3 \quad [\because \text{AP is increasing}]$$

**11.** $\lim\limits_{x\to0}\frac{\sin(\pi\cos^2x)}{x^2}$

$$=\lim_{x\to0}\frac{\sin\pi(1-\sin^2x)}{x^2}$$
$$=\lim_{x\to0}\frac{\sin(\pi-\pi\sin^2x)}{x^2}$$
$$=\lim_{x\to0}\frac{\sin(\pi\sin^2x)}{x^2}$$
$$[\because \sin(\pi-\theta)=\sin\theta]$$
$$=\lim_{x\to0}\frac{\sin\pi\sin^2x}{\pi\sin^2x}\times(\pi)\left(\frac{\sin^2x}{x^2}\right)$$
$$=\pi \quad \left[\because \lim_{\theta\to0}\frac{\sin\theta}{\theta}=1\right]$$

**12.** Here, '$g$' is the inverse of $f(x)$.

$\Rightarrow \quad fog(x) = x$

On differentiating w.r.t. $x$, we get

$$f'\{g(x)\} \times g'(x) = 1$$

$$g'(x) = \frac{1}{f'(g(x))} = \frac{1}{\frac{1}{1+\{g(x)\}^5}}$$

$$\left[\because f'(x) = \frac{1}{1+x^5}\right]$$

$$g'(x) = 1 + \{g(x)\}^5$$

**13.** Given, $f(0) = 2 = g(1), g(0) = 0$

and $\quad f(1) = 6$

$f$ and $g$ are differentiable in (0, 1).

Let $\quad h(x) = f(x) - 2g(x) \quad \ldots$(i)

$$h(0) = f(0) - 2g(0)$$
$$h(0) = 2 - 0$$
$$h(0) = 2$$

and $\quad h(1) = f(1) - 2g(1) = 6 - 2(2)$

$$h(1) = 2,\ h(0) = h(1) = 2$$

Hence, using Rolle's theorem,

$$h'(c) = 0, \text{ such that } c \in (0, 1)$$

Differentiating Eq. (i) at $c$, we get

$\Rightarrow \quad f'(c) - 2g'(c) = 0$

$\Rightarrow \quad f'(c) = 2g'(c)$

**14.** Here, $x = -1$ and $x = 2$ are extreme points of $f(x) = \alpha \log|x| + \beta x^2 + x$, then

$$f'(x) = \frac{\alpha}{x} + 2\beta x + 1$$

$$f'(-1) = -\alpha - 2\beta + 1 = 0 \quad \ldots\text{(i)}$$

[At extreme point, $f'(x) = 0$]

$$f'(2) = \frac{\alpha}{2} + 4\beta + 1 = 0 \quad \ldots\text{(ii)}$$

On solving Eqs. (i) and (ii), we get

$$\alpha = 2,\ \beta = -1/2$$

**15.** $\int \left(1 + x - \frac{1}{x}\right) e^{x + \frac{1}{x}} dx$

$$= \int e^{x+\frac{1}{x}} dx + \int x\left(1 - \frac{1}{x^2}\right) e^{x+\frac{1}{x}} dx$$

$$= \int e^{x+1/x} dx + x e^{x+1/x} - \int \frac{d}{dx}(x) e^{x+1/x} dx$$

$$= \int e^{x+\frac{1}{x}} dx + x e^{x+\frac{1}{x}} - \int e^{x+\frac{1}{x}} dx$$

$$\left[\because \int \left(1 - \frac{1}{x^2}\right) e^{x+\frac{1}{x}} dx = e^{x+\frac{1}{x}}\right]$$

$$= \int e^{x+1/x} dx + x e^{x+1/x} - \int ex^{+1/x} dx$$

$$= x e^{x+1/x} + C$$

**16.** Use the formula,

$$|x - a| = \begin{cases} x - a, & x \geq a \\ -(x-a), & x < a \end{cases}$$

to break given integral in two parts and then integrate separately.

$$\int_0^{\pi} \sqrt{\left(1 - 2\sin\frac{x}{2}\right)^2} dx = \int_0^{\pi} \left|1 - 2\sin\frac{x}{2}\right| dx$$

$$= \int_0^{\frac{\pi}{3}} \left(1 - 2\sin\frac{x}{2}\right) dx - \int_{\frac{\pi}{3}}^{\pi} \left(1 - 2\sin\frac{x}{2}\right) dx$$

$$= \left(x + 4\cos\frac{x}{2}\right)_0^{\pi/3} - \left(x + 4\cos\frac{x}{2}\right)_{\pi/3}^{\pi}$$

$$= 4\sqrt{3} - 4 - \frac{\pi}{3}$$

**17.** Given, $A = \{(x, y) : x^2 + y^2 \leq 1$ and $y^2 \leq 1 - x\}$

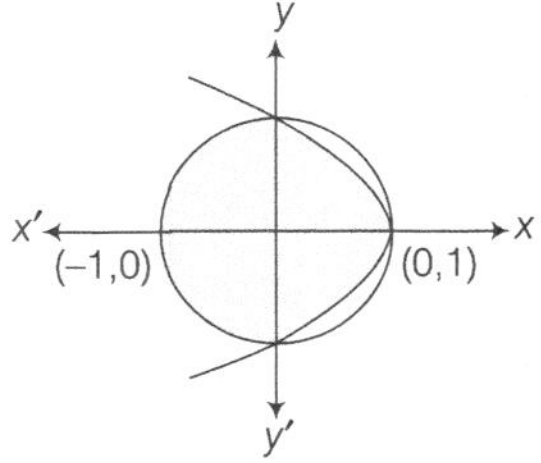


$$\text{Required area} = \frac{1}{2}\pi r^2 + 2\int_0^1 (1 - y^2) dy$$

$$= \frac{1}{2}\pi(1)^2 + 2\left(y - \frac{y^3}{3}\right)_0^1$$

$$= \frac{\pi}{2} + \frac{4}{3}$$

**18.** Given, differential equation $\frac{dp}{dt} - \frac{1}{2}p(t) = -200$ is a linear differential equation.

Here, $\quad p(t) = \frac{-1}{2}, Q(t) = -200$

$$\text{IF} = e^{\int -\left(\frac{1}{2}\right) dt} = e^{-t/2}$$

Hence, solution is

$$p(t)\cdot \text{IF} = \int Q(t)\,\text{IF}\,dt$$

$$p(t)\cdot e^{-\frac{t}{2}} = \int -200\cdot e^{-\frac{t}{2}}dt$$

$$p(t)\cdot e^{-\frac{t}{2}} = 400e^{-\frac{t}{2}} + K$$

$\Rightarrow \quad p(t) = 400 + ke^{-1/2}$

If $p(0) = 100$, then $k = -300$

$\Rightarrow \quad p(t) = 400 - 300e^{t/2}$

**19.** 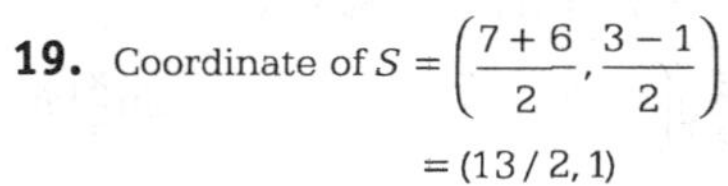

Coordinate of $S = \left(\frac{7+6}{2}, \frac{3-1}{2}\right)$

$= (13/2, 1)$

[$\because$ S is mid-point of line QR]

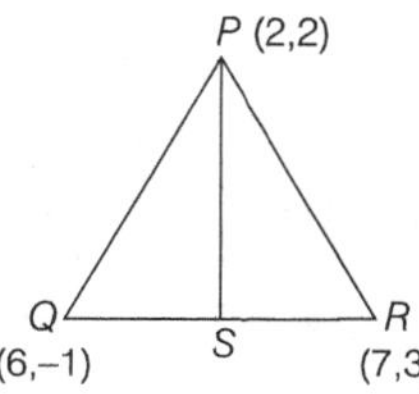


Slope of the line $PS$ is $\frac{-2}{9}$.

Required equation passes through $(1, -1)$ and parallel to $PS$ is

$$y + 1 = \frac{-2}{9}(x - 1)$$

$\Rightarrow \quad 2x + 9y + 7 = 0$

**20.** Let coordinate of the intersection point in fourth quadrant be $(\alpha, -\alpha)$.

Since, $(\alpha, -\alpha)$ lies on both lines

$$4ax + 2ay + c = 0$$

and $\quad 5bx + 2by + d = 0$

$\therefore \quad 4a\alpha - 2a\alpha + c = 0$

$\Rightarrow \quad \alpha = \frac{-c}{2a} \quad \ldots\text{(i)}$

and $\quad 5b\alpha - 2b\alpha + d = 0$

$\Rightarrow \quad \alpha = \frac{-d}{3b} \quad \ldots\text{(ii)}$

From Eqs. (i) and (ii), we get

$$\frac{-c}{2a} = \frac{-d}{3b}$$

$\Rightarrow \quad 3bc = 2ad$

$\Rightarrow \quad 2ad - 3bc = 0$

**21.** Equation of ellipse is $x^2 + 3y^2 = 6$ or $\frac{x^2}{6} + \frac{y^2}{2} = 1$.

Equation of the tangent is

$$\frac{x\cos\theta}{a} + \frac{y\sin\theta}{b} = 1$$

Let $(h, k)$ be any point on the locus.

$\therefore \quad \frac{h}{a}\cos\theta + \frac{k}{b}\sin\theta = 1 \quad \ldots\text{(i)}$

Slope of the tangent line is $\frac{-b}{a}\cot\theta$.

Slope of perpendicular drawn from centre (0,0) to $(h, k)$ is $k/h$.

Since, both the lines are perpendicular.

$\therefore \quad \left(\frac{k}{h}\right)\times\left(-\frac{b}{a}\cot\theta\right) = -1$

$\Rightarrow \quad \frac{\cos\theta}{ha} = \frac{\sin\theta}{kb} = \alpha \quad$ [Say]

$\Rightarrow \quad \cos\theta = \alpha ha,$

$\sin\theta = \alpha kb$

From Eq. (i), we get

$$\frac{h}{a}(\alpha ha) + \frac{k}{b}(\alpha kb) = 1$$

$\Rightarrow \quad h^2\alpha + k^2\alpha = 1$

$\Rightarrow \quad \alpha = \frac{1}{h^2 + k^2}$

Also, $\quad \sin^2\theta + \cos^2\theta = 1$

$\Rightarrow \quad (\alpha kb)^2 + (\alpha ha)^2 = 1$

$\Rightarrow \quad \alpha^2k^2b^2 + \alpha^2h^2a^2 = 1$

$\Rightarrow \quad \frac{k^2b^2}{(h^2+k^2)^2} + \frac{h^2a^2}{(h^2+k^2)^2} = 1$

$\Rightarrow \quad \frac{2k^2}{(h^2+k^2)^2} + \frac{6h^2}{(h^2+k^2)^2} = 1$

[$\because a^2 = 6, b^2 = 2$]

$\Rightarrow \quad 6x^2 + 2y^2 = (x^2 + y^2)^2$

[Replacing $k$ by $y$ and $h$ by $x$]

**22.** Use the property, when two circles touch each other externally, then distance between the centre is equal to sum of their radii, to get required radius.

Let the coordinate of the centre of $T$ be $(0, k)$.

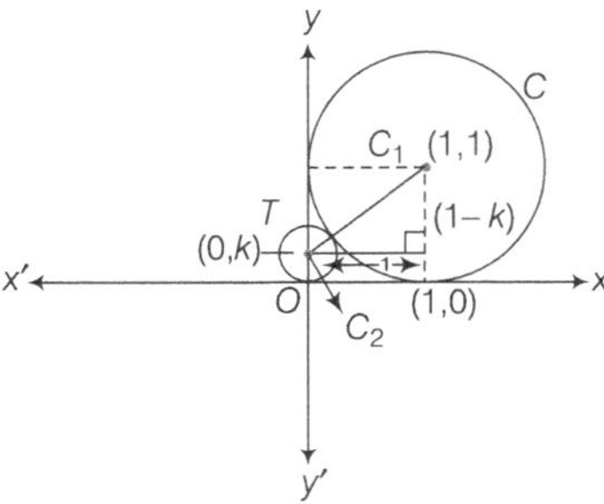


Distance between their centre

$k+1=\sqrt{1+(k-1)^2}$ $\quad [\because C_1C_2 = k+1]$

$\Rightarrow \quad k+1=\sqrt{1+k^2+1-2k}$

$\Rightarrow \quad k+1=\sqrt{k^2+2-2k}$

$\Rightarrow \quad k^2+1+2k = k^2+2-2k$

$\Rightarrow \quad k = 1/4$

So, the radius of circle $T$ is $k$ *i.e.*, $1/4$.

**23.** Let the tangent to parabola be $y = mx + a/m$, if it touches the other curve, then $D = 0$, to get the value of $m$.

For parabola, $y^2 = 4x$

Let $y = mx + 1/m$ be tangent line and it touches the parabola $x^2 = -32y$.

$\therefore \quad x^2 = -32\left(mx + \frac{1}{m}\right)$

$\Rightarrow \quad x^2 + 32mx + \frac{32}{m} = 0$

$\because \quad D = 0$

$\therefore \quad (32m)^2 - 4\cdot\left(\frac{32}{m}\right) = 0$

$\Rightarrow \quad m^3 = \frac{1}{8}$

$\therefore \quad m = \frac{1}{2}$

**24.** Here, plane, line and its image are parallel to each other. So, find any point on the normal to the plane from which the image line will be passed and then find equation of image line.

Here, plane and line are parallel to each other. Equation of normal to the plane through the point (1, 3, 4) is

$$\frac{x-1}{2} = \frac{y-3}{-1} = \frac{z-4}{1} = k \quad \text{[Say]}$$

Any point in this normal is

$$(2k+1, -k+3, 4+k).$$

Then,

$\left(\frac{2k+1+1}{2}, \frac{3-k+3}{2}, \frac{4+k+4}{2}\right)$ lies on plane. $\Rightarrow 2$

$(k+1) - \left(\frac{6-k}{2}\right) + \left(\frac{8+k}{2}\right) + 3 = 0$

$\Rightarrow \quad k = -2$

Hence, point through which this image pass is

$$(2k+1, 3-k, 4+k)$$

*i.e.*, $[2(-2)+1, 3+2, 4-2] = (-3, 5, 2)$

Hence, equation of image line is

$$\frac{x+3}{3} = \frac{y-5}{1} = \frac{z-2}{-5}$$

**25.** Use the following formula to simplify

$$\mathbf{a} \times (\mathbf{b} \times \mathbf{c}) = (\mathbf{a}\cdot\mathbf{c})\mathbf{b} - (\mathbf{a}\cdot\mathbf{b})\mathbf{c}$$

$$[\mathbf{a\ b\ c}] = [\mathbf{b\ c\ a}] = [\mathbf{c\ a\ b}]$$

and $[\mathbf{a\ a\ b}] = [\mathbf{a\ b\ b}] = [\mathbf{a\ c\ c}] = 0.$

Now, $[\mathbf{a\times b\ a\times c\ c\times a}]$

$= \mathbf{a}\times\mathbf{b}\cdot((\mathbf{b}\times\mathbf{c})\times(\mathbf{c}\times\mathbf{a}))$

$= \mathbf{a}\times\mathbf{b}\cdot(\mathbf{k}\times(\mathbf{c}\times\mathbf{a}))$

Let $\mathbf{k} = \mathbf{b}\times\mathbf{c}$

$= \mathbf{a}\times\mathbf{b}\cdot((\mathbf{k}\cdot\mathbf{a})\,\mathbf{c} - (\mathbf{k}\cdot\mathbf{c})\,\mathbf{a})$

$= (\mathbf{a}\times\mathbf{b})\cdot((\mathbf{b}\times\mathbf{c}\cdot\mathbf{a})\mathbf{c} - (\mathbf{b}\times\mathbf{c}\cdot\mathbf{c})\mathbf{a})$

$= (\mathbf{a}\times\mathbf{b})\cdot([\mathbf{b\ c\ a}]\,\mathbf{c}) - 0$

$\{\because [\mathbf{b}\times\mathbf{c}\cdot\mathbf{c}] = 0\}$

$= (\mathbf{a}\times\mathbf{b})\cdot\mathbf{c}\,[\mathbf{b\ c\ a}]$

$= [\mathbf{a\ bc}][\mathbf{bc\ a}] = [\mathbf{a\ bc}]^2$

$\{\because [\mathbf{a\ bc}] = [\mathbf{bc\ a}]\}$

But given,

$[\mathbf{a\times b\ b\times c\ c\times a}] = \lambda[\mathbf{a}\ \boldsymbol{b}\ \boldsymbol{c}]^2$

So, $\quad [\mathbf{abc}]^2 = \lambda[\mathbf{abc}]^2$

$\Rightarrow \quad \lambda = 1$

**26.** Given, $P(\overline{A\cup B}) = \frac{1}{6}$, $P(A\cap B) = \frac{1}{4}$,

$$P(\overline{A}) = \frac{1}{4}$$

$\therefore \quad P(A\cup B) = 1 - P(\overline{A\cup B})$

$= 1 - \frac{1}{6} = \frac{5}{6}$

and $\quad P(A) = 1 - P(\overline{A})$

$= 1 - \frac{1}{4} = \frac{3}{4}$

$$P(A\cup B) = P(A) + P(B) - P(A\cap B)$$

$\Rightarrow \quad \frac{5}{6} = \frac{3}{4} + P(B) - \frac{1}{4}$

$P(B) = \frac{1}{3} \Rightarrow A$ and $B$ are not equally likely.

$$P(A \cap B) = P(A) \cdot P(B) = \frac{1}{4}$$

So, events are independent.

**27.** Here, $\overline{X} = \frac{\sum X_i}{n}$

$$= \frac{2 + 4 + 6 + 8 + \dots + 100}{50}$$

$$= \frac{50 \times 51}{50}$$

$$= 51$$

$[\because \sum 2n = n(n+1)$, here $n = 50]$

Variance, $\sigma^2 = \frac{1}{n}\sum Xi^2 - (\overline{x})^2$

$$\sigma^2 = \frac{1}{50}(2^2 + 4^2 + \dots + 100^2) - (51)^2$$

$$= 833$$

**28.** Given, $f_k(x) = \frac{1}{k}(\sin^k x + \cos^k x)$,

where $x \in R$ and $k > 1$

$$f_4(x) - f_6(x) = \frac{1}{4}(\sin^4 x + \cos^4 x) - \frac{1}{6}(\sin^6 x + \cos^6 x)$$

$$= \frac{1}{4}(1 - 2\sin^2 x \cdot \cos^2 x) - \frac{1}{6}(1 - 3\sin^2 x \cdot \cos^2 x)$$

$$= \frac{1}{4} - \frac{1}{6} = \frac{1}{12}$$

**29.** In $\Delta OA_1B_1$,

$$\tan 45^\circ = \frac{A_1B_1}{OB_1} \Rightarrow \frac{20}{OB_1} = 1$$

$$\Rightarrow \quad OB_1 = 20$$

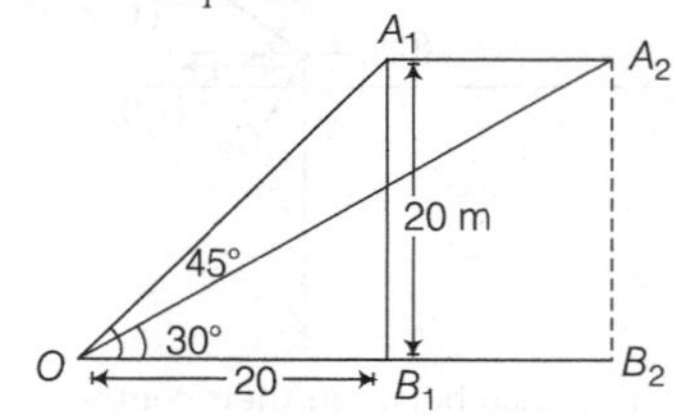


In $\Delta OA_2B_2$, $\tan 30^\circ = \frac{20}{OB_2}$

$$\Rightarrow \quad OB_2 = 20\sqrt{3}$$

$$\Rightarrow \quad B_1B_2 + OB_1 = 20\sqrt{3}$$

$$\Rightarrow \quad B_1B_2 = 20\sqrt{3} - 20$$

$$\Rightarrow \quad B_1B_2 = 20(\sqrt{3} - 1) \text{ m}$$

Now, speed $= \frac{\text{Distance}}{\text{Time}} = \frac{20(\sqrt{3} - 1)}{1}$

$$= 20(\sqrt{3} - 1) \text{ m/s}$$

**30.** $\sim(p \leftrightarrow \sim q)$ is equivalent to $(p \leftrightarrow q)$.

| $p$ | $q$ | $\sim p$ | $\sim q$ | $p \leftrightarrow q$ | $p \leftrightarrow \sim q$ | $\sim p \leftrightarrow q$ | $\sim(p \leftrightarrow \sim q)$ |
|---|---|---|---|---|---|---|---|
| T | F | F | T | F | T | T | F |
| F | T | T | F | F | T | T | F |
| T | T | F | F | T | F | F | T |
| F | F | T | T | T | F | F | T |

# JEE Advanced

## Solved Paper 2014

Time : 3 Hrs | M. Marks : 180

## Paper 1

### Instructions

- The question paper consists of two sections.
- Section 1 contains 10 multiple choice questions. Each question has four choices (a), (b), (c) and (d) out of which one or more than one are correct.
- Section 2 contains 10 questions. The answer to each of the question is a single-digit integer, ranging from 0 to 9 (both inclusive)
- For each question in Section 1, you will be awarded 3 marks if you darken all the bubble(s) corresponding to the correct answer(s) and zero mark if no bubbles are darkened. No negative marks will be awarded for incorrect answers in this section.
- For each question in Section 2, you will be awarded 3 marks if you darken only the bubble corresponding to the correct answer and zero mark if no bubble is darkened. No negative marks will be awarded for incorrect answer in this section.

## Physics

### Section 1

**Direction** (Q. No. 1-10) *This section contains 10 multiple choice questions. Each question has four choices (a), (b), (c) and (d) out of which one or more than one are correct.*

**1.** At time $t = 0$, terminal $A$ in the circuit shown in the figure is connected to $B$ by a key and an alternating current $I(t) = I_0 \cos(\omega t)$, with $I_0 = 1$ A and $\omega = 500$ rad s$^{-1}$ starts flowing in it with the initial direction shown in the figure. At $t = \frac{7\pi}{6\omega}$, the key is switched from $B$ to $D$. Now onwards only $A$ and $D$ are connected. A total charge $Q$ flows from the battery to charge the capacitor fully. If $C = 20\ \mu F$, $R = 10\ \Omega$ and the battery is ideal with emf of 50 V, identify the correct statement(s).

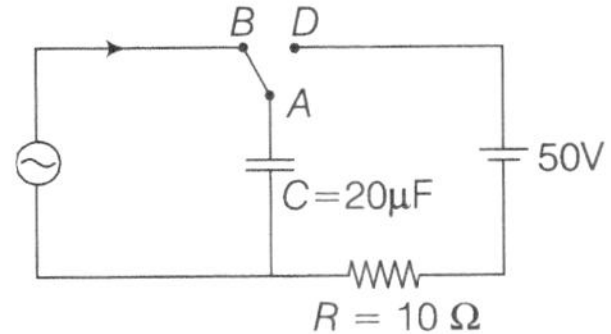

(a) Magnitude of the maximum charge on the capacitor before $t = \frac{7\pi}{6\omega}$ is $1 \times 10^{-3}$ C

(b) The current in the left part of the circuit just before $t = \frac{7\pi}{6\omega}$ is clockwise

(c) Immediately after $A$ is connected to $D$, the current in $R$ is 10 A

(d) $Q = 2 \times 10^{-3}$ C

**2.** A light source, which emits two wavelengths $\lambda_1 = 400$ nm and $\lambda_2 = 600$ nm, is used in a Young's double-slit experiment. If recorded fringe widths for $\lambda_1$ and $\lambda_2$ are $\beta_1$ and $\beta_2$ and the number of fringes for them within a distance $y$ on one side of the central maximum are $m_1$ and $m_2$, respectively, then

(a) $\beta_2 > \beta_1$

(b) $m_1 > m_2$

(c) from the central maximum, 3rd maximum of $\lambda_2$ overlaps with 5th minimum of $\lambda_1$

(d) the angular separation of fringes of $\lambda_1$ is greater than $\lambda_2$

**3.** One end of a taut string of length 3 m along the $x$-axis is fixed at $x = 0$. The speed of the waves in the string is 100 ms$^{-1}$. The other end of the string is vibrating in the $y$-direction so that stationary waves are set up in the string. The possible waveform(s) of these stationary wave is (are)

(a) $y(t) = A \sin \frac{\pi x}{6} \cos \frac{50\pi t}{3}$

(b) $y(t) = A \sin \frac{\pi x}{3} \cos \frac{100\pi t}{3}$

(c) $y(t) = A \sin \frac{5\pi x}{6} \cos \frac{250\pi t}{3}$

(d) $y(t) = A \sin \frac{5\pi x}{2} \cos 250\pi t$

**4.** A parallel plate capacitor has a dielectric slab of dielectric constant $K$ between its plates that covers 1/3 of the area of its plates, as shown in the figure. The total capacitance of the capacitor is $C$ while that of the portion with dielectric in between is $C_1$. When the capacitor is charged, the plate area covered by the dielectric gets charge $Q_1$ and the rest of the area gets charge $Q_2$. The electric field in the dielectric is $E_1$ and that in the other portion is $E_2$. Choose the correct option/options, ignoring edge effects.

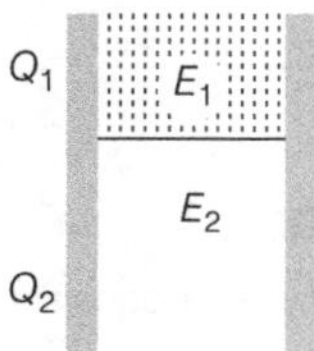


(a) $\frac{E_1}{E_2} = 1$

(b) $\frac{E_1}{E_2} = \frac{1}{K}$

(c) $\frac{Q_1}{Q_2} = \frac{3}{K}$

(d) $\frac{C}{C_1} = \frac{2 + K}{K}$

**5.** Let $E_1(r)$, $E_2(r)$ and $E_3(r)$ be the respective electric fields at a distance $r$ from a point charge $Q$, an infinitely long wire with constant linear charge density $\lambda$, and an infinite plane with uniform surface charge density $\sigma$.

If $E_1(r_0) = E_2(r_0) = E_3(r_0)$ at a given distance $r_0$, then

(a) $Q = 4\sigma\pi r_0^2$

(b) $r_0 = \frac{\lambda}{2\pi\sigma}$

(c) $E_1\left(\frac{r_0}{2}\right) = 2E_2\left(\frac{r_0}{2}\right)$

(d) $E_2\left(\frac{r_0}{2}\right) = 4E_3\left(\frac{r_0}{2}\right)$

**6.** A student is performing an experiment using a resonance column and a tuning fork of frequency 244 s$^{-1}$. He is told that the air in the tube has been replaced by another gas (assume that the column remains filled with the gas). If the minimum height at which resonance occurs is $(0.350 \pm 0.005)$ m, the gas in the tube is

**(Useful information**
$\sqrt{167RT} = 640\ J^{1/2}\ mole^{-1/2}$;
$\sqrt{140RT} = 590\ J^{1/2}\ mole^{-1/2}$. The molar masses $M$ in grams are given in the options. Take the value of $\sqrt{10/M}$ for each gas as given there.)

(a) Neon ($M = 20, \sqrt{10/20} = 7/10$)
(b) Nitrogen ($M = 28, \sqrt{10/28} = 3/5$)
(c) Oxygen ($M = 32, \sqrt{10/32} = 9/16$)
(d) Argon ($M = 36, \sqrt{10/36} = 17/32$)

**7.** Heater of an electric kettle is made of a wire of length $L$ and diameter $d$. It takes 4 minutes to raise the temperature of 0.5 kg water by 40 K. This heater is replaced by a new heater having two wires of the same material, each of length $L$ and diameter $2d$.

The way these wires are connected is given in the options. How much time in minutes will it take to raise the temperature of the same amount of water by 40 K?

(a) 4, if wires are in parallel
(b) 2, if wires are in series
(c) 1, if wires are in series
(d) 0.5, if wires are in parallel

**8.** In the figure, a ladder of mass $m$ is shown leaning against a wall. It is in static equilibrium making an angle $\theta$ with the horizontal floor. The coefficient of friction between the wall and the ladder is $\mu_1$ and that between the floor and the ladder is $\mu_2$. The normal reaction of the wall on the ladder is $N_1$ and that of the floor is $N_2$. If the ladder is about to slip, then

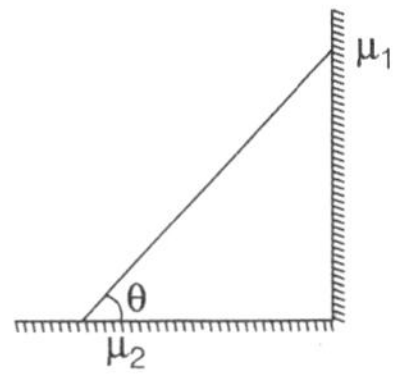


(a) $\mu_1 = 0, \mu_2 \neq 0$ and $N_2 \tan\theta = \frac{mg}{2}$
(b) $\mu_1 \neq 0, \mu_2 = 0$ and $N_1 \tan\theta = \frac{mg}{2}$
(c) $\mu_1 \neq 0, \mu_2 \neq 0$ and $N_2 = \frac{mg}{1 + \mu_1\mu_2}$
(d) $\mu_1 = 0, \mu_2 \neq 0$ and $N_1 \tan\theta = \frac{mg}{2}$

**9.** A transparent thin film of uniform thickness and refractive index $n_1 = 1.4$ is coated on the convex spherical surface of radius $R$ at one end of a long solid glass cylinder of refractive index $n_2 = 1.5$, as shown in the figure. Rays of light parallel to the axis of the cylinder traversing through the film from air to glass get focused at distance $f_1$ from the film, while rays of light traversing from glass to air get focused at distance $f_2$ from the film. Then

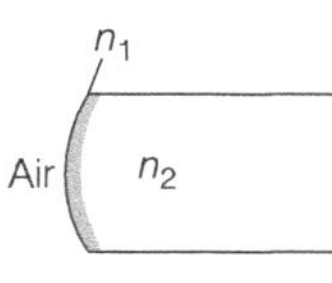


(a) $|f_1| = 3R$
(b) $|f_1| = 2.8R$
(c) $|f_2| = 2R$
(d) $|f_2| = 1.4R$

**10.** Two ideal batteries of emf $V_1$ and $V_2$ and three resistances $R_1$, $R_2$ and $R_3$ are connected as shown in the figure. The current in resistance $R_2$ would be zero if

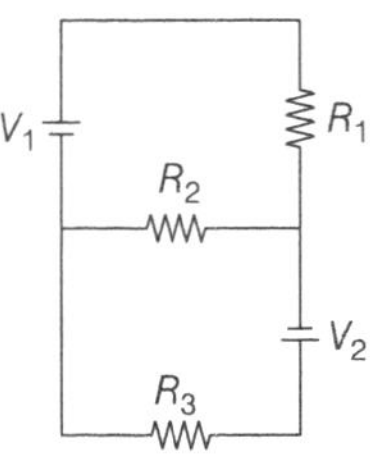


(a) $V_1 = V_2$ and $R_1 = R_2 = R_3$
(b) $V_1 = V_2$ and $R_1 = 2R_2 = R_3$
(c) $V_1 = 2V_2$ and $2R_1 = 2R_2 = R_3$
(d) $2V_1 = V_2$ and $2R_1 = R_2 = R_3$

## Section 2

**Direction** (Q. No. 11-20) *This section contains 10 questions. Each question, when worked out will result in one integer from 0 to 9 (both inclusive)*

**11.** Airplanes $A$ and $B$ are flying with constant velocity in the same vertical plane at angles 30° and 60° with respect to the horizontal respectively as shown in figure. The speed of $A$ is $100\sqrt{3}$ ms$^{-1}$. At time $t = 0$ s, an observer in $A$ finds $B$ at a distance of 500 m. This observer sees $B$ moving with a constant velocity perpendicular to the line of motion of $A$. If at $t = t_0$, $A$ just escapes being hit by $B$, $t_0$ in seconds is

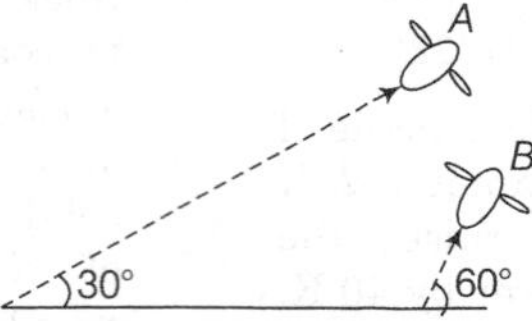


**12.** During Searle's experiment, zero of the vernier scale lies between $3.20 \times 10^{-2}$ m and $3.25 \times 10^{-2}$ m of the main scale. The 20th division of the vernier scale exactly coincides with one of the main scale divisions. When an additional load of 2 kg is applied to the wire, the zero of the vernier scale still lies between $3.20 \times 10^{-2}$ m and $3.25 \times 10^{-2}$ m of the main scale but now the 45th division of vernier scale coincides with one of the main scale divisions. The length of the thin metallic wire is 2 m and its cross-sectional area is $8 \times 10^{-7}$ m$^2$. The least count of the vernier scale is $1.0 \times 10^{-5}$ m. The maximum percentage error in the Young's modulus of the wire is

**13.** A uniform circular disc of mass 1.5 kg and radius 0.5 m is initially at rest on a horizontal frictionless surface. Three forces of equal magnitude $F = 0.5$ N are applied simultaneously along the three sides of an equilateral triangle $XYZ$ with its vertices on the perimeter of the disc (see figure). One second after applying the forces, the angular speed of the disc in rad s$^{-1}$ is

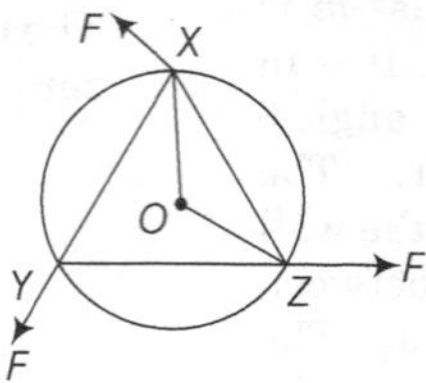


**14.** Two parallel wires in the plane of the paper are distance $X_0$ apart. A point charge is moving with speed $u$ between the wires in the same plane at a distance $X_1$ from one of the wires. When the wires carry current of magnitude $I$ in the same direction, the radius of curvature of the path of the point charge is $R_1$. In contrast, if the currents $I$ in the two wires have directions opposite to each other, the radius of curvature of the path is $R_2$. If $\frac{X_0}{X_1} = 3$, and value of $R_1/R_2$ is

**15.** To find the distance $d$ over which a signal can be seen clearly in foggy conditions, a railways engineer uses dimensional analysis and assumes that the distance depends on the mass density $\rho$ of the fog, intensity (power/area) $S$ of the light from the signal and its frequency $f$. The engineer finds that $d$ is proportional to $S^{1/n}$. The value of $n$ is

**16.** A galvanometer gives full scale deflection with 0.006 A current. By connecting it to a 4990 Ω resistance, it can be converted into a voltmeter of range 0-30 V. If connected to a $\frac{2n}{249}$ Ω resistance, it becomes an ammeter of range 0-1.5 A. The value of $n$ is

**17.** Consider an elliptically shaped rail $PQ$ in the vertical plane with $OP = 3$ m and $OQ = 4$ m. A block of mass 1 kg is pulled along the rail from $P$ to $Q$ with a force of 18 N, which is always parallel to line $PQ$ (see figure). Assuming no frictional losses, the kinetic energy of the block when it reaches $Q$ is $(n \times 10)$ J. The value of $n$ is (take acceleration due to gravity = 10 ms$^{-2}$)

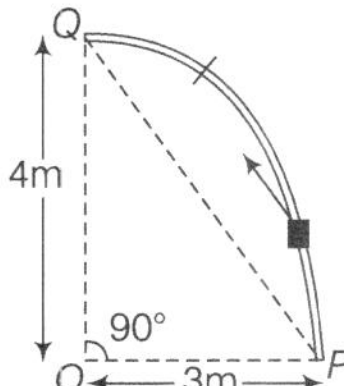


**18.** A rocket is moving in a gravity free space with a constant acceleration of 2 ms$^{-2}$ along $+x$ direction (see figure). The length of a chamber inside the rocket is 4 m. A ball is thrown from the left end of the chamber in $+x$ direction with a speed of 0.3 ms$^{-1}$ relative to the rocket. At the same time, another ball is thrown in $-x$ direction with a speed of 0.2 ms$^{-1}$ from its right end relative to the rocket. The time in seconds when the two balls hit each other is

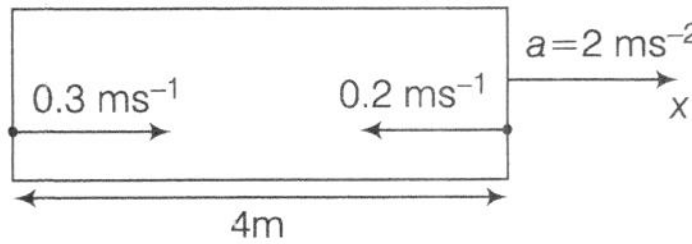


**19.** A horizontal circular platform of radius 0.5 m and mass 0.45 kg is free to rotate about its axis. Two massless spring toy-guns, each carrying a steel ball of mass 0.05 kg are attached to the platform at a distance 0.25 m from the centre on its either sides along its diameter (see figure). Each gun simultaneously fires the balls horizontally and perpendicular to the diameter in opposite directions. After leaving the platform, the balls have horizontal speed of 9 ms$^{-1}$ with respect to the ground. The rotational speed of the platform in rad s$^{-1}$ after the balls leave the platform is

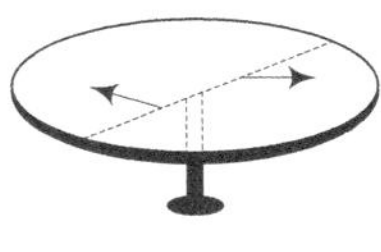

**20.** A thermodynamic system is taken from an initial state $i$ with internal energy $U_i = 100$ J to the final state $f$ along two different paths $iaf$ and $ibf$, as schematically shown in the figure. The work done by the system along the paths $af$, $ib$ and $bf$ are $W_{af} = 200$ J, $W_{ib} = 50$ J and $W_{bf} = 100$ J respectively. The heat supplied to the system along the path $iaf$, $ib$ and $bf$ are $Q_{iaf}$, $Q_{ib}$ and $Q_{bf}$ respectively. If the internal energy of the system in the state $b$ is $U_b = 200$ J and $Q_{iaf} = 500$ J, the ratio $Q_{bf} / Q_{ib}$ is

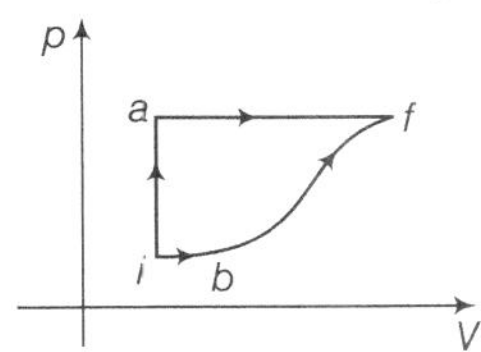

# Chemistry

## Section I

**Directions** (Q. Nos. 1-10) *This section contains 10 multiple choice questions. Each question has four choices (a), (b), (c) and (d) out of which one or more than one are correct.*

**1.** For the reaction,

$I^- + CiO_3^- + H_2SO_4 \rightarrow Cl^- + HSO_4^- + I_2$

the correct statement(s) in the balanced equation is/are

(a) stoichiometric coefficient of $HSO_4^-$ is 6
(b) iodide is oxidised
(c) sulphur is reduced
(d) $H_2O$ is one of the products

**2.** The pair(s) of reagents that yield paramagnetic species is/are

(a) Na and excess of $NH_3$
(b) K and excess of $O_2$
(c) Cu and dilute $HNO_3$
(d) $O_2$ and 2-ethylanthraquinol

**3.** In the reaction shown below, the major product(s) formed is/are

$NH_2$ ... $NH_2$ (O) $\xrightarrow[CH_2Cl_2]{\text{Acetic anhydride}}$ Product(s)

(a) N-H, $CH_3$, O ... $NH_2 + CH_3COOH$

(b) $NH_2$ ... H-N, $CH_3 + CH_3COOH$ (O, O)

(c) H-N, $CH_3$, O ... H-N, $CH_3 + H_2O$ (O, O)

(d) $\overset{\oplus}{N}H_3CH_2CO\overset{\ominus}{O}$ ... H-N, $CH_3$ (O, O)

**4.** In a galvanic cell, the salt bridge

(a) does not participate chemically in the cell reaction
(b) stops the diffusion of ions from one electrode to another
(c) is necessary for the occurrence of the cell reaction
(d) ensures mixing of the two electrolytic solutions

**5.** Upon heating with $Cu_2S$, the reagent(s) that give copper metal is/are

(a) $CuFeS_2$ (b) CuO (c) $Cu_2O$ (d) $CuSO_4$

**6.** Hydrogen bonding plays a central role in the following phenomena

(a) ice floats in water
(b) higher Lewis basicity of primary amines than tertiary amines in aqueous solutions
(c) formic acid is more acidic than acetic acid
(d) dimerisation of acetic acid in benzene

**7.** The reactivity of compound $Z$ with different halogens under appropriate conditions is given below

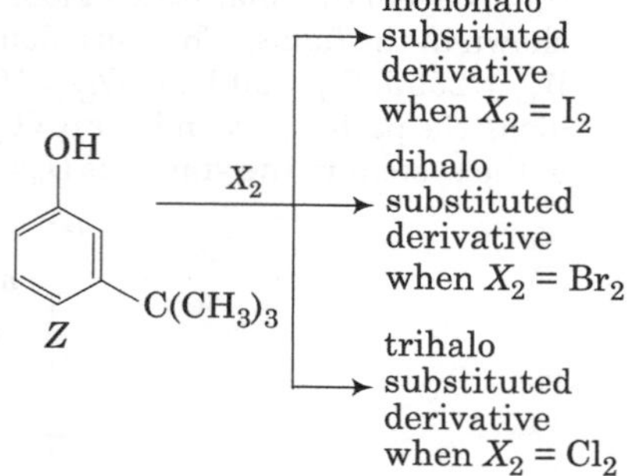

The observed pattern of electrophilic substitution can be explained by

(a) the steric effect on the halogen
(b) the steric effect of the *tert*-butyl group
(c) the electronic effect of the phenolic group
(d) the electronic effect of the *tert*-butyl group

**8.** The correct combination of names for isomeric alcohols with molecular formula $C_4H_{10}O$ is/are

(a) *tert*-butanol and 2-methylpropan-2-ol
(b) *tert*-butanol and 1,1-dimethylethan-1-ol
(c) *n*-butanol and butan-1-ol
(d) *iso*-butyl alcohol and 2-methylpropan-1-ol

**9.** An ideal gas in thermally insulated vessel at internal pressure $= p_1$, volume $= V_1$ and absolute temperature $= T_1$ expands irreversibly against zero external pressure, as shown in the diagram.

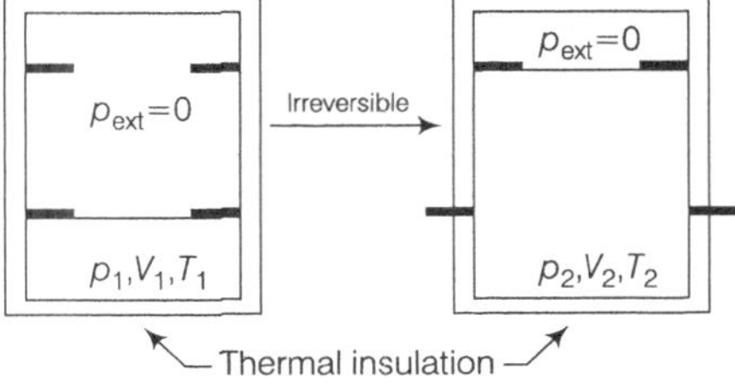


The final internal pressure, volume and absolute temperature of the gas are $p_2, V_2$ and $T_2$, respectively. For this expansion

(a) $q = 0$ (b) $T_2 = T_1$ (c) $p_2V_2 = p_1V_1$ (d) $p_2V_2^{\gamma} = p_1V_1^{\gamma}$

**10.** The correct statement(s) for orthoboric acid is/are

(a) it behaves as a weak acid in water due to self ionisation
(b) acidity of its aqueous solution increases upon addition of ethylene glycol
(c) it has a three dimensional structure due to hydrogen bonding
(d) it is a weak electrolyte in water

## Section II

**Directions** (Q. Nos. 11-20) *This section contains 10 question. Each question, when worked out will result in one integer from 0 to 9 (both inclusive)*

**11.** In an atom, the total number of electrons having quantum numbers $n = 4, |m_l| = 1$ and $m_s = -1/2$ is

**12.** The total number of distinct naturally occurring amino acids obtained by complete acidic hydrolysis of the peptide shown below is

**13.** If the value of Avogadro number is $6.023 \times 10^{23}$ mol$^{-1}$ and the value of Boltzmann constant is $1.380 \times 10^{-23}$ JK$^{-1}$, then the number of significant digits in the calculated value of the universal gas constant is

**14.** A compound $H_2X$ with molar weight of 80 g is dissolved in a solvent having density of 0.4 g mL$^{-1}$. Assuming no change in volume upon dissolution, the molality of a 3.2 molar solution is

**15.** $MX_2$ dissociates into $M^{2+}$ and $X^-$ ions in an aqueous solution, with a degree of dissociation ($\alpha$) of 0.5. The ratio of the observed depression of freezing point of the aqueous solution to the value of the depression of freezing point in the absence of ionic dissociation is

**16.** Consider the following list of reagents, acidified $K_2Cr_2O_7$, alkaline $KMnO_4$, $CuSO_4$, $H_2O_2$, $Cl_2$, $O_3$, $FeCl_3$, $HNO_3$ and $Na_2S_2O_3$. The total number of reagents that can oxidise aqueous iodide to iodine is

**17.** The total number(s) of stable conformers with non-zero dipole moment for the following compound is(are)

```
        Cl
  Br ───┼─── CH3
  Br ───┼─── Cl
        CH3
```

**18.** Among PbS, CuS, HgS, MnS, $Ag_2S$, NiS, CoS, $Bi_2S_3$ and $SnS_2$ the total number of black coloured sulphides is

**19.** Consider all possible isomeric ketones including stereoisomers of MW = 100. All these isomers are independently reacted with $NaBH_4$ (**Note** stereoisomers are also reacted separately). The total number of ketones that give a racemic product(s) is/are

**20.** A list of species having the formula $XZ_4$ is given below

$XeF_4$, $SF_4$, $SiF_4$, $BF_4^-$, $BrF_4^-$, $[Cu(NH_3)_4]^{2+}$, $[FeCl_4]^{2-}$, $[CoCl_4]^{2-}$ and $[PtCl_4]^{2-}$

Defining shape on the basis of the location of $X$ and $Z$ atoms, the total number of species having a square planar shape is

# Mathematics

## Section I

**Directions** (Q. Nos. 1-10) *This section contains 10 multiple choice questions. Each question has four choices (a), (b), (c) and (d) out of which one or more than one are correct.*

**1.** Let $f : [a, b] \to [1, \infty)$ be a continuous function and $g : R \to R$ be defined as

$$g(x) = \begin{cases} 0 & , \text{ if } \quad x < a \\ \int_a^x f(t)dt, & \text{if } \quad a \le x \le b. \\ \int_a^b f(t)dt, & \text{if } \quad x > b \end{cases}$$

Then,

(a) $g(x)$ is continuous but not differentiable at $a$

(b) $g(x)$ is differentiable on $R$

(c) $g(x)$ is continuous but not differentiable at $b$

(d) $g(x)$ is continuous and differentiable at either $a$ or $b$ but not both

**2.** For every pair of continuous function $f, g: [0,1] \to R$ such that max $\{f(x): x \in [0,1]\} = \max\{g(x): x \in [0,1]\}$.

The correct statement(s) is (are)

(a) $[f(c)]^2 + 3f(c) = [g(c)]^2 + 3g(c)$ for some $c \in [0,1]$

(b) $[f(c)]^2 + f(c) = [g(c)]^2 + 3g(c)$ for some $c \in [0,1]$

(c) $[f(c)]^2 + 3f(c) = [g(c)]^2 + g(c)$ for some $c \in [0,1]$

(d) $[f(c)]^2 = [g(c)]^2$ for some $c \in [0,1]$

**3.** Let $M$ be a $2 \times 2$ symmetric matrix with integer entries. Then, $M$ is invertible, if

(a) the first column of $M$ is the transpose of the second row of $M$

(b) the second row of $M$ is the transpose of the first column of $M$

(c) $M$ is a diagonal matrix with non-zero entries in the main diagonal

(d) the product of entries in the main diagonal of $M$ is not the square of an integer

**4.** Let $\mathbf{x}$, $\mathbf{y}$ and $\mathbf{z}$ be three vectors each of magnitude $\sqrt{2}$ and the angle between each pair of them is $\frac{\pi}{3}$. If $\mathbf{a}$ is a non-zero vector perpendicular to $\mathbf{x}$ and $\mathbf{y} \times \mathbf{z}$ and $\mathbf{b}$ is non-zero vector perpendicular to $\mathbf{y}$ and $\mathbf{z} \times \mathbf{x}$, then

(a) $\mathbf{b} = (\mathbf{b} \cdot \mathbf{z})(\mathbf{z} - \mathbf{x})$

(b) $\mathbf{a} = (\mathbf{a} \cdot \mathbf{y})(\mathbf{y} - \mathbf{z})$

(c) $\mathbf{a} \cdot \mathbf{b} = -(\mathbf{a} \cdot \mathbf{y})(\mathbf{b} \cdot \mathbf{z})$

(d) $\mathbf{a} = (\mathbf{a} \cdot \mathbf{y})(\mathbf{z} - \mathbf{y})$

**5.** From a point $P(\lambda, \lambda, \lambda)$, perpendiculars $PQ$ and $PR$ are drawn respectively on the lines $y = x, z = 1$ and $y = -x, z = -1$. If $P$ is such that $\angle QPR$ is a right angle, then the possible value(s) of $\lambda$ is (are)

(a) $\sqrt{2}$ (b) 1

(c) $-1$ (d) $-\sqrt{2}$

**6.** Let $M$ and $N$ be two $3 \times 3$ matrices such that $MN = NM$. Further, if $M \neq N^2$ and $M^2 = N^4$, then

(a) determinant of $(M^2 + MN^2)$ is 0

(b) there is a $3 \times 3$ non-zero matrix $U$ such that $(M^2 + MN^2)U$ is zero matrix

(c) determinant of $(M^2 + MN^2) \geq 1$

(d) for a $3 \times 3$ matrix $U$, if $(M^2 + MN^2)U$ equals the zero matrix, then $U$ is the zero matrix

**7.** Let $f: (0, \infty) \to R$ be given by

$$f(x) = \int_{1/x}^{x} e^{-\left(t + \frac{1}{t}\right)} \frac{dt}{t}.$$

Then,

(a) $f(x)$ is monotonically increasing on $[1, \infty)$

(b) $f(x)$ is monotonically decreasing on $[0, 1)$

(c) $f(x) + f\left(\frac{1}{x}\right) = 0$, for all $x \in (0, \infty)$

(d) $f(2^x)$ is an odd function of $x$ on $R$

**8.** Let $f: \left(-\frac{\pi}{2}, \frac{\pi}{2}\right) \to R$ be given by

$f(x) = [\log(\sec x + \tan x)]^3$.

Then,

(a) $f(x)$ is an odd function

(b) $f(x)$ is a one-one function

(c) $f(x)$ is an onto function

(d) $f(x)$ is an even function

**9.** A circle $S$ passes through the point $(0, 1)$ and is orthogonal to the circles $(x-1)^2 + y^2 = 16$ and $x^2 + y^2 = 1$. Then,

(a) radius of $S$ is 8

(b) radius of $S$ is 7

(c) centre of $S$ is $(-7, 1)$

(d) centre of $S$ is $(-8, 1)$

**10.** Let $a \in R$ and $f: R \to R$ be given by $f(x) = x^5 - 5x + a$.

Then,

(a) $f(x)$ has three real roots, if $a > 4$

(b) $f(x)$ has only one real root, if $a > 4$

(c) $f(x)$ has three real roots, if $a < -4$

(d) $f(x)$ has three real roots, if $-4 < a < 4$

## Section II

**Directions** (Q. Nos. 11-20) *This section contains 10 questions. Each question, when worked out will result in one integer from 0 to 9 (both inclusive).*

**11.** The slope of the tangent to the curve $(y-x^5)^2 = x(1+x^2)^2$ at the point (1, 3) is

**12.** Let $f: [0, 4\pi] \to [0, \pi]$ be defined by $f(x) = \cos^{-1}(\cos x)$. The number of points $x \in [0, 4\pi]$ satisfying the equation $f(x) = \dfrac{10-x}{10}$ is

**13.** The largest value of the non-negative integer a for which $\lim_{x\to 1}\left\{\dfrac{-ax+\sin(x-1)+a}{x+\sin(x-1)-1}\right\}^{\frac{1-x}{1-\sqrt{x}}} = \dfrac{1}{4}$ is

**14.** Let $f: R \to R$ and $g: R \to R$ be respectively given by $f(x) = |x| + 1$ and $g(x) = x^2 + 1$. Define $h: R \to R$ by

$$h(x) = \begin{cases} \max\{f(x), g(x)\}, & \text{if } x \le 0. \\ \min\{f(x), g(x)\}, & \text{if } x > 0. \end{cases}$$

The number of points at which $h(x)$ is not differentiable is

**15.** For a point $P$ in the plane, let $d_1(P)$ and $d_2(P)$ be the distances of the point $P$ from the lines $x-y=0$ and $x+y=0$, respectively. The area of the region $R$ consisting of all points $P$ lying in the first quadrant of the plane and satisfying $2 \le d_1(P) + d_2(P) \le 4$, is

**16.** Let $n_1 < n_2 < n_3 < n_4 < n_5$ be positive integers such that $n_1 + n_2 + n_3 + n_4 + n_5 = 20$. The number of such distinct arrangements $(n_1, n_2, n_3, n_4, n_5)$ is

**17.** The value of $\int_0^1 4x^3 \left\{\dfrac{d^2}{dx^2}(1-x^2)^5\right\} dx$ is

**18.** Let **a**, **b** and **c** be three non-coplanar unit vectors such that the angle between every pair of them is $\pi/3$. If $\mathbf{a} \times \mathbf{b} + \mathbf{b} \times \mathbf{c} = p\mathbf{a} + q\mathbf{b} + r\mathbf{c}$, where $p, q$ and $r$ are scalars, then the value of $\dfrac{p^2 + 2q^2 + r^2}{q^2}$ is

**19.** Let $a, b, c$ be positive integers such that $b/a$ is an integer. If $a, b, c$ are in geometric progression and the arithmetic mean of $a, b, c$ is $b + 2$, then the value of $\dfrac{a^2 + a - 14}{a+1}$ is

**20.** Let $n \ge 2$ be an integer. Take $n$ distinct points on a circle and join each pair of points by a line segment. Colour the line segment joining every pair of adjacent points by blue and the rest by red. If the number of red and blue line segments are equal, then the value of $n$ is

# Paper 2

## Instructions

- Section 1 contains 10 multiple choice questions. Each question has four choices (a), (b), (c) and (d) out of which one is correct.
- Section 2 contains 3 paragraphs each describing theory, experiment and data etc. Six questions relate to three paragraphs with two questions on each paragraph. Each question pertaining to a particular passage should have only one correct answer among the four given choices (a), (b), (c) and (d).
- Section 3 contains 4 multiple choice questions. Each question has two lists (**List I** P, Q, R and S; **List II** 1, 2, 3 and 4). The options for the correct match are provided as (a), (b), (c) and (d) out of which only one is correct.
- For each question in Section 1, 2 and 3 you will be awarded 3 marks if you darken the bubble corresponding to the correct answer and zero mark if no bubble is darkened. In all other cases, minus one (–1) mark will be awarded.

# Physics

## Section 1

**Direction** (Q. No. 1-10) *This section contains 10 multiple choice questions. Each question has four choices (a), (b), (c) and (d) out of which only one option is correct.*

**1.** A tennis ball is dropped on a horizontal smooth surface. It bounces back to its original position after hitting the surface. The force on the ball during the collision is proportional to the length of compression of the ball.

Which one of the following sketches describes the variation of its kinetic energy $K$ with time $t$ most appropriately? The figures are only illustrative and not to the scale.

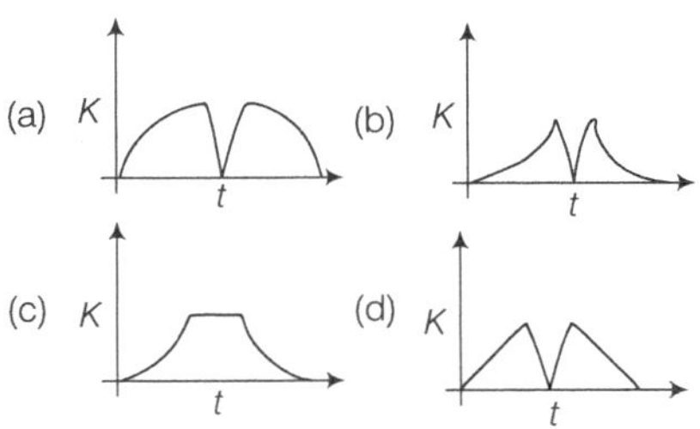


**2.** A wire, which passes through the hole in a small bead, is bent in the form of quarter of a circle. The wire is fixed vertically on ground as shown in the figure. The bead is released from near the top of the wire and it slides along the wire without friction. As the bead moves from $A$ to $B$, the force it applies on the wire is

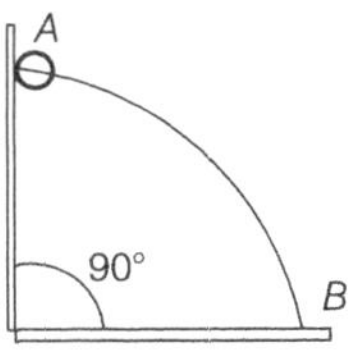


(a) always radially outwards
(b) always radially inwards
(c) radially outwards initially and radially inwards later
(d) radially inwards initially and radially outwards later

**3.** During an experiment with a metre bridge, the galvanometer shows a null point when the jockey is pressed at 40.0 cm using a standard resistance of 90 Ω, as shown in the figure. The least count of the scale used in the metre bridge is 1 mm. The unknown resistance is

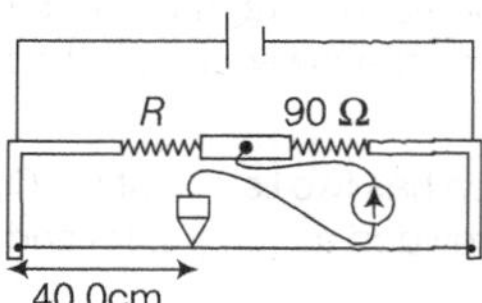


(a) 60 ± 0.15 Ω
(b) 135 ± 0.56 Ω
(c) 60 ± 0.25 Ω
(d) 135 ± 0.23 Ω

**4.** Charges $Q, 2Q$ and $4Q$ are uniformly distributed in three dielectric solid spheres 1, 2 and 3 of radii $R/2$, $R$ and $2R$ respectively, as shown in figure. If magnitudes of the electric fields at point $P$ at a distance $R$ from the centre of spheres 1, 2 and 3 are $E_1$, $E_2$ and $E_3$ respectively, then

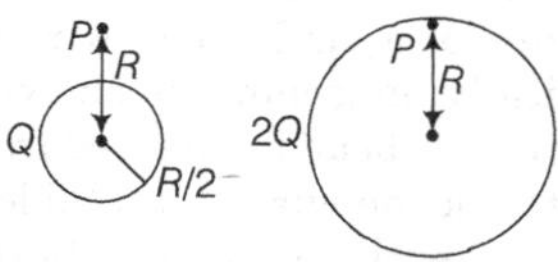


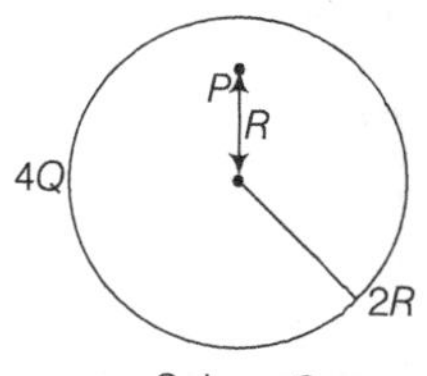


(a) $E_1 > E_2 > E_3$
(b) $E_3 > E_1 > E_2$
(c) $E_2 > E_1 > E_3$
(d) $E_3 > E_2 > E_1$

**5.** A point source $S$ is placed at the bottom of a transparent block of height 10 mm and refractive index 2.72. It is immersed in a lower refractive index liquid as shown in the figure. It is found that the light emerging from the block to the liquid forms a circular bright spot of diameter 11.54 mm on the top of the block. The refractive index of the liquid is

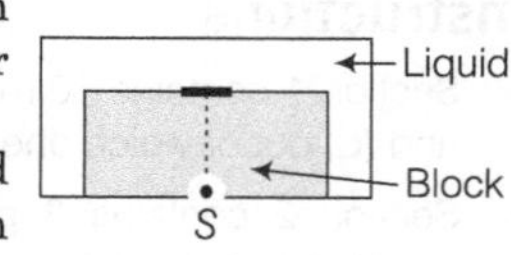


(a) 1.21 (b) 1.30 (c) 1.36 (d) 1.42

**6.** Parallel rays of light of intensity $I = 912 \text{ Wm}^{-2}$ are incident on a spherical black body kept in surroundings of temperature 300 K. Take Stefan constant $\sigma = 5.7 \times 10^{-8} \text{ Wm}^{-2} \text{ K}^{-4}$ and assume that the energy exchange with the surroundings is only through radiation. The final steady state temperature of the black body is close to

(a) 330 K (b) 660 K (c) 990 K (d) 1550 K

**7.** A metal surface is illuminated by light of two different wavelengths 248 nm and 310 nm. The maximum speeds of the photoelectrons corresponding to these wavelengths are $u_1$ and $u_2$, respectively. If the ratio $u_1 : u_2 = 2 : 1$ and $hc = 1240$ eV nm, the work function of the metal is nearly

(a) 3.7 eV (b) 3.2 eV
(c) 2.8 eV (d) 2.5 eV

**8.** If $\lambda_{Cu}$ is the wavelength of $K_\alpha$, X-ray line of copper (atomic number 29) and $\lambda_{MO}$ is the wavelength of the $K_\alpha$, X-ray line of molybdenum (atomic number 42), then the ratio $\lambda_{Cu}/\lambda_{Mo}$ is close to

(a) 1.99 (b) 2.14 (c) 0.50 (d) 0.48

**9.** A planet of radius $R = \frac{1}{10} \times$ (radius of earth) has the same mass density as earth. Scientists dig a well of depth $\frac{R}{5}$ on it and lower a wire of the same length and of linear mass density $10^{-3}$ $\text{kgm}^{-1}$ into it. If the wire is not touching anywhere, the force applied at the top of the wire by a person holding it in place is (take the radius of earth $= 6 \times 10^6$ m and the acceleration due to gravity of earth is $10\ \text{ms}^{-2}$)

(a) 96 N (b) 108 N
(c) 120 N (d) 150 N

**10.** A glass capillary tube is of the shape of truncated cone with an apex angle $\alpha$ so that its two ends have cross-sections of different radii. When dipped in water vertically, water rises in it to a height $h$, where the radius of its cross-section is $b$. If the surface tension of water is $S$, its density is $\rho$, and its contact angle with glass is $\theta$, the value of $h$ will be ($g$ is the acceleration due to gravity)

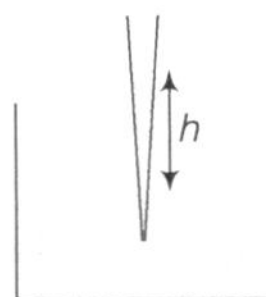


(a) $\frac{2S}{b\rho g} \cos(\theta - \alpha)$

(b) $\frac{2S}{b\rho g} \cos(\theta + \alpha)$

(c) $\frac{2S}{b\rho g} \cos(\theta - \alpha/2)$

(d) $\frac{2S}{b\rho g} \cos(\theta + \alpha/2)$

## Section 2

**Direction** (Q. No. 11-16) *This section contains 6 multiple choice questions relating to three paragraphs with two questions on each paragraph. Each question has four choices (a), (b), (c) and (d) out of which only one is correct.*

### Paragraph 1

In the figure a container is shown to have a movable (without friction) piston on top. The container and the piston are all made of perfectly insulating material allowing no heat transfer between outside and inside the container. The container is divided into two compartments by a rigid partition made of a thermally conducting material that allows slow transfer of heat. The lower compartment of the container is filled with 2 moles of an ideal monoatomic gas at 700 K and the upper compartment is filled with 2 moles of an ideal diatomic gas at 400 K. The heat capacities per mole of an ideal monoatomic gas are $C_V = \frac{3}{2}R$, $C_p = \frac{5}{2}R$, and those for an ideal diatomic gas are $C_V = \frac{5}{2}R, C_p = \frac{7}{2}R$.

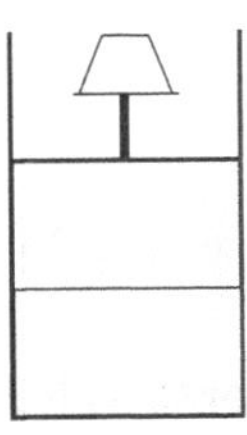

**11.** Consider the partition to be rigidly fixed so that it does not move. When equilibrium is achieved, the final temperature of the gases will be

(a) 550 K (b) 525 K (c) 513 K (d) 490 K

**12.** Now consider the partition to be free to move without friction so that the pressure of gases in both compartments is the same. Then total work done by the gases till the time they achieve equilibrium will be

(a) $250R$ (b) $200R$ (c) $100R$ (d) $-100R$

## Paragraph 2

A spray gun is shown in the figure where a piston pushes air out of nozzle. A thin tube of uniform cross-section is connected to the nozzle. The other end of the tube is in a small liquid container. As the piston pushes air through the nozzle, the liquid from the container rises into the nozzle and is sprayed out.

For the spray gun shown, the radii of the piston and the nozzle are 20 mm and 1 mm respectively. The upper end of the container is open to the atmosphere.

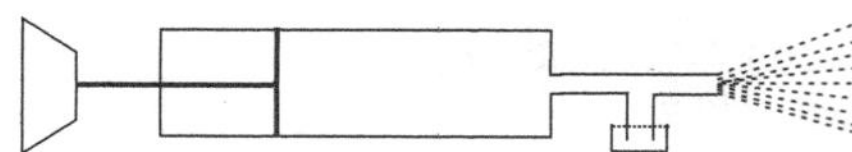

**13.** If the piston is pushed at a speed of 5 $mms^{-1}$, the air comes out of the nozzle with a speed of

(a) 0.1 $ms^{-1}$ (b) 1 $ms^{-1}$
(c) 2 $ms^{-1}$ (d) 8 $ms^{-1}$

**14.** If the density of air is $\rho_a$ and that of the liquid $\rho_l$, then for a given piston speed the rate (volume per unit time) at which the liquid is sprayed will be proportional to

(a) $\sqrt{\frac{\rho_a}{\rho_l}}$ (b) $\sqrt{\rho_a \rho_l}$
(c) $\sqrt{\frac{\rho_l}{\rho_a}}$ (d) $\rho_l$

## Paragraph 3

The figure shows a circular loop of radius $a$ with two long parallel wires (numbered 1 and 2) all in the plane of the paper. The distance of each wire from the centre of the loop is $d$. The loop and the wires are carrying the same current $I$. The current in the loop is in the counter-clockwise direction if seen from above.

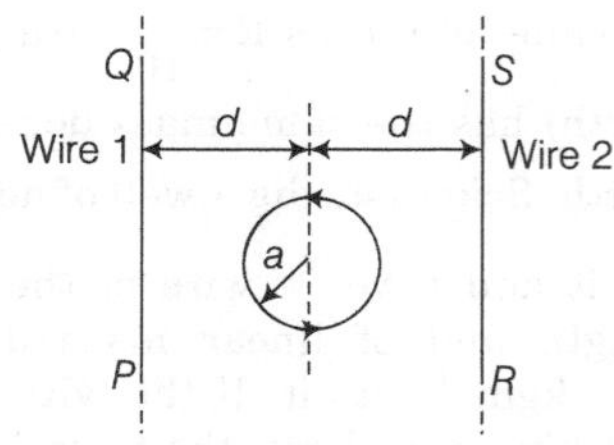


**15.** When $d \approx a$ but wires are not touching the loop, it is found that the net magnetic field on the axis of the loop is zero at a height $h$ above the loop. In that case

(a) current in wire 1 and wire 2 is the direction $PQ$ and $RS$, respectively and $h \approx a$
(b) current in wire 1 and wire 2 is the direction $PQ$ and $SR$, respectively and $h \approx a$
(c) current in wire 1 and wire 2 is the direction $PQ$ and $SR$, respectively and $h \approx 1.2a$
(d) current in wire 1 and wire 2 is the direction $PQ$ and $RS$, respectively and $h \approx 1.2a$

**16.** Consider $d >> a$, and the loop is rotated about its diameter parallel to the wires by 30° from the position shown in the figure. If the currents in the wires are in the opposite directions, the torque on the loop at its new position will be (assume that the net field due to the wires is constant over the loop)

(a) $\frac{\mu_0 I^2 a^2}{d}$
(b) $\frac{\mu_0 I^2 a^2}{2d}$
(c) $\frac{\sqrt{3} \mu_0 I^2 a^2}{d}$
(d) $\frac{\sqrt{3} \mu_0 I^2 a^2}{2d}$

## Section 3

**Direction** (Q. No. 17-20) *This section contains 4 multiple choice questions. Each question has matching lists. The codes for the lists have choices (a), (b), (c) and (d) out of which only one is correct.*

**17.** Four charges $Q_1, Q_2, Q_3$ and $Q_4$ of same magnitude are fixed along the $x$-axis at $x = -2a, -a, +a$ and $+2a$ respectively. A positive charge $q$ is placed on the positive $y$-axis at a distance $b > 0$. Four options of the signs of these charges are given in List I. The direction of the forces on the charge $q$ is given in List II. Match List I with List II and select the correct answer using the code given below the lists.

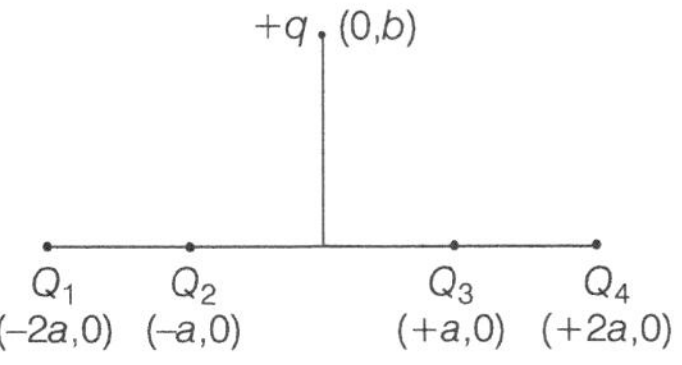


| | List I | | List II |
|---|---|---|---|
| P. | $Q_1, Q_2, Q_3, Q_4$ all positive | 1. | $+x$ |
| Q. | $Q_1, Q_2$ positive; $Q_3, Q_4$ negative | 2. | $-x$ |
| R. | $Q_1, Q_4$ positive; $Q_2, Q_3$ negative | 3. | $+y$ |
| S. | $Q_1, Q_3$ positive; $Q_2, Q_4$ negative | 4. | $-y$ |

**Codes**

| | P | Q | R | S |
|---|---|---|---|---|
| (a) | 3 | 1 | 4 | 2 |
| (b) | 4 | 2 | 3 | 1 |
| (c) | 3 | 1 | 2 | 4 |
| (d) | 4 | 2 | 1 | 3 |

**18.** Four combinations of two thin lenses are given in List I.The radius of curvature of all curved surfaces is $r$ and the refractive index of all the lenses is 1.5. Match lens combinations in List I with their focal length in List II and select the correct answer using the code given below the lists.

| | List I | | List II |
|---|---|---|---|
| P. | | 1. | $2r$ |
| Q. | | 2. | $r/2$ |
| R. | | 3. | $-r$ |
| S. | | 4. | $r$ |

**Codes**

| | P | Q | R | S |
|---|---|---|---|---|
| (a) | 1 | 2 | 3 | 4 |
| (b) | 2 | 4 | 3 | 1 |
| (c) | 4 | 1 | 2 | 3 |
| (d) | 2 | 1 | 3 | 4 |

**19.** A block of mass $m_1 = 1$ kg another mass $m_2 = 2$ kg are placed together (see figure) on an inclined plane with angle of inclination $\theta$. Various values of $\theta$ are given in List I.

The coefficient of friction between the block $m_1$ and the plane is always zero. The coefficient of static and dynamic friction between the block $m_2$ and the plane are equal to $\mu = 0.3$.

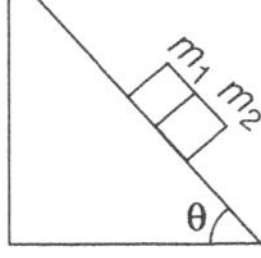


In List II expressions for the friction on the block $m_2$ are given. Match the correct expression of the friction in List II with the angles given in List I, and choose the correct option.

The acceleration due to gravity is denoted by $g$. **[Useful information** $\tan(5.5°) \approx 0.1$; **$\tan(11.5°) \approx 0.2$; $\tan(16.5°) \approx 0.3$]**

| | List I | | List II |
|---|---|---|---|
| P. | $\theta = 5°$ | 1. | $m_2 g \sin\theta$ |
| Q. | $\theta = 10°$ | 2. | $(m_1 + m_2) g \sin\theta$ |
| R. | $\theta = 15°$ | 3. | $\mu m_2 g \cos\theta$ |
| S. | $\theta = 20°$ | 4. | $\mu (m_1 + m_2) g \cos\theta$ |

**Codes**

| | P | Q | R | S |
|---|---|---|---|---|
| (a) | 1 | 1 | 1 | 3 |
| (b) | 2 | 2 | 2 | 3 |
| (c) | 2 | 2 | 2 | 4 |
| (d) | 2 | 2 | 3 | 3 |

**20.** A person in a lift is holding a water jar, which has a small hole at the lower end of its side. When the lift is at rest, the water jet coming out of the hole hits the floor of the lift at a distance $d$ of 1.2 m from the person.

In the following, state of the lift's motion is given in List I and the distance where the water jet hits the floor of the lift is given in List II. Match the statements from List I with those in List II and select the correct answer using the code given below the lists.

| | List I | | List II |
|---|---|---|---|
| P. | Lift is accelerating vertically up. | 1. | $d = 1.2$ m |
| Q. | Lift is accelerating vertically down with an acceleration less than the gravitational acceleration. | 2. | $d > 1.2$ m |
| R. | Lift is moving vertically up with constant speed. | 3. | $d < 1.2$ m |
| S. | Lift is falling freely. | 4. | No water leaks out of the jar |

**Codes**

| | P | Q | R | S |
|---|---|---|---|---|
| (a) | 2 | 3 | 2 | 4 |
| (b) | 2 | 3 | 1 | 4 |
| (c) | 1 | 1 | 1 | 4 |
| (d) | 2 | 3 | 1 | 1 |

# Chemistry

## Section I

**Directions** (Q. Nos. 1-10) *This section contains 10 multiple choice questions. Each question has four choices (a), (b), (c) and (d) out of which only one is correct.*

**1.** Assuming $2s$-$2p$ mixing is not operative, the paramagnetic species among the following is

(a) $Be_2$ (b) $B_2$
(c) $C_2$ (d) $N_2$

**2.** For the process,

$$H_2O(l) \longrightarrow H_2O(g)$$

at $T = 100°C$ and 1 atmosphere pressure, the correct choice is

(a) $\Delta S_{system} > 0$ and $\Delta S_{surrounding} > 0$
(b) $\Delta S_{system} > 0$ and $\Delta S_{surrounding} < 0$
(c) $\Delta S_{system} < 0$ and $\Delta S_{surrounding} > 0$
(d) $\Delta S_{system} < 0$ and $\Delta S_{surrounding} < 0$

**3.** For the elementary reaction $M \rightarrow N$, the rate of disappearance of $M$ increases by a factor of 8 upon doubling the concentration of $M$. The order of the reaction with respect to $M$ is

(a) 4 (b) 3 (c) 2 (d) 1

**4.** For the identification of β-naphthol using dye test, it is necessary to use

(a) dichloromethane solution of β-naphthol
(b) acidic solution of β-naphthol
(c) neutral solution of β-naphthol
(d) alkaline solution of β-naphthol

**5.** Isomers of hexane, based on their branching, can be divided into three distinct classes as shown in the figure.

I. 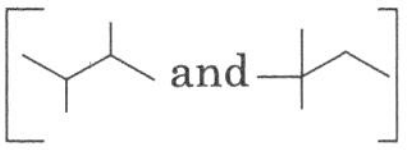 [ (2,3-dimethylbutane) and (2,2-dimethylbutane) ]

II. 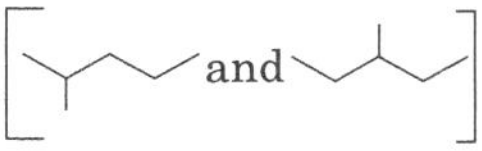 [ (2-methylpentane) and (3-methylpentane) ]

III. 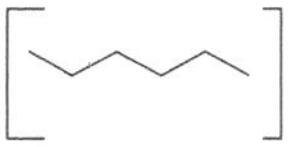 [ (n-hexane) ]

The correct order of their boiling point is

(a) I > II > III
(b) III > II > I
(c) II > III > I
(d) III > I > II

**6.** The major product in the following reaction is

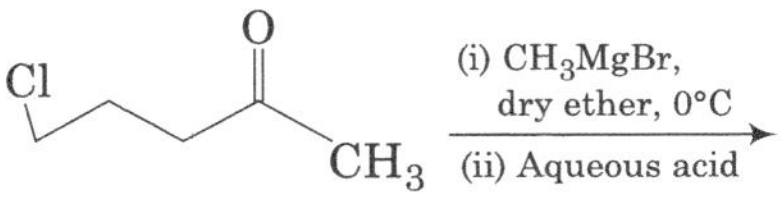

$Cl-CH_2CH_2CH_2-CO-CH_3 \xrightarrow[\text{(ii) Aqueous acid}]{\text{(i) } CH_3MgBr, \text{ dry ether, } 0°C}$

(a) 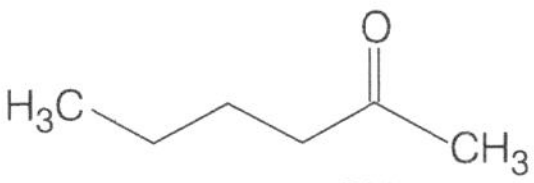 $H_3C-CH_2CH_2CH_2-CO-CH_3$

(b) 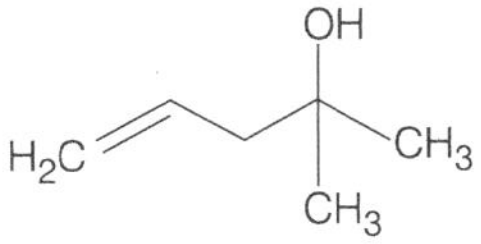 $H_2C=CH-CH_2-C(OH)(CH_3)-CH_3$

(c) (2-methylenetetrahydrofuran, $=CH_2$ on ring O)

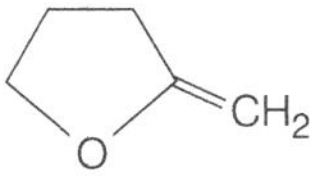

(d) 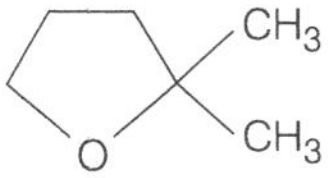 (2,2-dimethyltetrahydrofuran, $CH_3$, $CH_3$)

**7.** Under ambient conditions, the total number of gases released as products in the final step of the reaction scheme shown below is

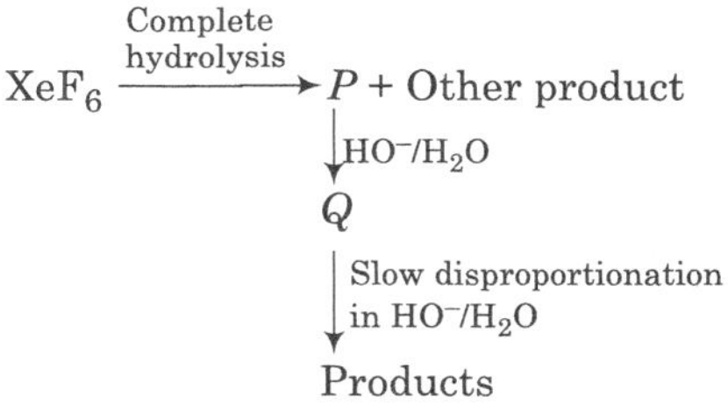

$$XeF_6 \xrightarrow[\text{hydrolysis}]{\text{Complete}} P + \text{Other product}$$

$\downarrow HO^-/H_2O$

$Q$

$\downarrow$ Slow disproportionation in $HO^-/H_2O$

Products

(a) 0 (b) 1 (c) 2 (d) 3

**8.** The product formed in the reaction of $SOCl_2$ with white phosphorus is

(a) $PCl_3$ (b) $SO_2Cl_2$ (c) $SCl_2$ (d) $POCl_3$

**9.** Hydrogen peroxide in its reaction with $KIO_4$ and $NH_2OH$ respectively, is acting as a

(a) reducing agent, oxidising agent
(b) reducing agent, reducing agent
(c) oxidising agent, oxidising agent
(d) oxidising agent, reducing agent

**10.** The acidic hydrolysis of ether ($X$) shown below is fastest when

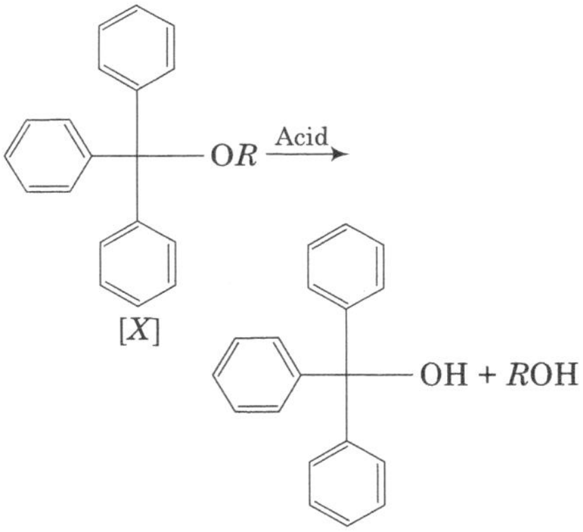

$(C_6H_5)_3C-OR \xrightarrow{\text{Acid}} (C_6H_5)_3C-OH + ROH$

[$X$]

(a) one phenyl group is replaced by a methyl group
(b) one phenyl group is replaced by a *para*-methoxyphenyl group
(c) two phenyl groups are replaced by two *para*-methoxyphenyl groups
(d) no structural change is made to $X$

## Section II

**Directions** (Q. Nos. 11-16) *This section contains 3 paragraphs, each describing theory, experiments, data etc. Six questions relate to the three paragraphs with two questions on each paragraph. Each question has only one correct answer among the four given options (a), (b), (c) and (d).*

### Paragraph 1

*X* and *Y* are two volatile liquids with molar weights of 10 g $mol^{-1}$ and 40 g $mol^{-1}$ respectively. Two cotton plugs, one soaked in *X* and the other soaked in *Y*, are simultaneously placed at the ends of a tube of length $L = 24$ cm, as shown in the figure. The tube is filled with an inert gas at 1 atm pressure and a temperature of 300 K. Vapours of *X* and *Y* react to form a product which is first observed at a distance *d* cm from the plug soaked in *X*. Take *X* and *Y* to have equal molecular diameters and assume ideal behaviour for the inert gas and the two vapours.

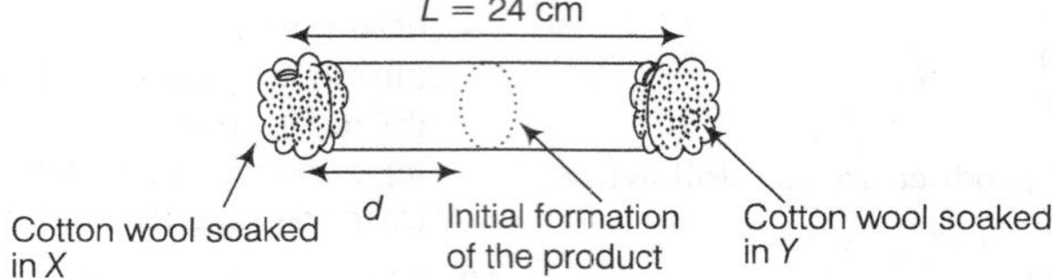


**11.** The value of *d* in cm (shown in the figure), as estimated from Graham's law, is
(a) 8 (b) 12 (c) 16 (d) 20

**12.** The experimental value of *d* is found to be smaller than the estimate obtained using Graham's law. This due to
(a) larger mean free path for *X* as a compared of that of *Y*
(b) larger mean free path for *Y* as compared to that of *X*
(c) increased collision frequency of *Y* with the inert gas as compared to that of *X* with the inert gas
(d) increased collision frequency of *X* with the inert gas as compared to that of *Y* with the inert gas

### Paragraph 2

Schemes 1 and 2 describe sequential transformation of alkynes *M* and *N*. Consider only the major products formed in each step for both schemes.

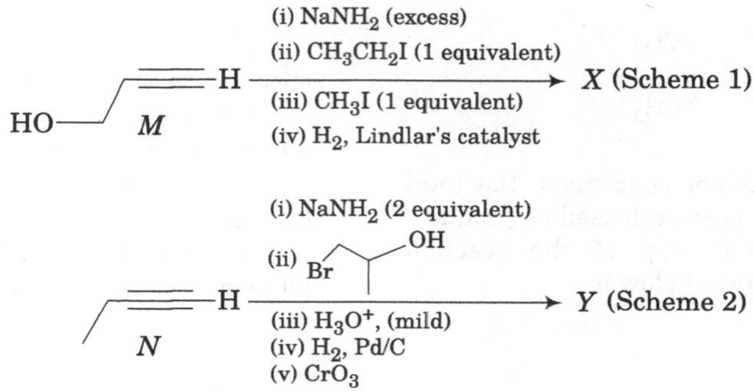

**13.** The product $X$ is

(a) $H_3CO$–$CH_2CH_2$–CH=CH–$CH_2CH_3$ (both H on the same side of the double bond)

(b) $H_3CO$–$CH_2CH_2$–CH=CH–$CH_2CH_3$ (H atoms on opposite sides of the double bond)

(c) $CH_3CH_2O$–$CH_2CH_2$–CH=CH–$CH_2CH_3$ (both H on the same side of the double bond)

(d) $CH_3CH_2O$–$CH_2CH_2$–CH=CH–$CH_2CH_3$ (H atoms on opposite sides of the double bond)

**14.** The correct statement with respect to product $Y$ is

(a) it gives a positive Tollen's test and is a functional isomer of $X$
(b) it gives a positive Tollen's test and is a geometrical isomer of $X$
(c) it gives a positive iodoform test and is a functional isomer of $X$
(d) it gives a positive iodoform test and is a geometrical isomer of $X$

## Paragraph 3

An aqueous solution of metal ion $M_1$ reacts separately with reagents $Q$ and $R$ in excess to give tetrahedral and square planar complexes, respectively. An aqueous solution of another metal ion $M_2$ always forms tetrahedral complexes with these reagents. Aqueous solution of $M_2$ on reaction with reagent S gives white precipitate which dissolves in excess of S. The reactions are summarized in the scheme given below

$$\text{Tetrahedral} \xleftarrow[\text{excess}]{Q} M_1 \xrightarrow[\text{excess}]{R} \text{Square planar}$$

$$\text{Tetrahedral} \xleftarrow[\text{excess}]{Q} M_2 \xrightarrow[\text{excess}]{R} \text{Tetrahedral}$$

$$M_2 \xrightarrow{S,\ \text{stoichiometric amount}} \text{White precipitate} \xrightarrow[\text{excess}]{S} \text{Precipitate dissolves}$$

**15.** $M_1$, $Q$ and $R$, respectively are

(a) $Zn^{2+}$, KCN and HCl
(b) $Ni^{2+}$, HCl and KCN
(c) $Cd^{2+}$, KCN and HCl
(d) $Co^{2+}$, HCl and KCN

**16.** Reagent S is

(a) $K_4[Fe(CN)_6]$
(b) $Na_2HPO_4$
(c) $K_2CrO_4$
(d) KOH

## Section III

**Directions** (Q. Nos. 17-20) *This section contains 4 multiple choice questions. Each question has four choices (a), (b), (c) and (d) out of which only one is correct.*

**17.** Match each coordination compound in List I with an appropriate pair of characteristics from List II and select the correct answer using the code given below the lists (en = $H_2NCH_2CH_2NH_2$; atomic numbers : Ti = 22; Cr = 24; Co = 27; Pt = 78)

| | List I | | List II |
|---|---|---|---|
| P. | $[Cr(NH_3)_4Cl_2]Cl$ | 1. | Paramagnetic and exhibits ionisation isomerism |
| Q. | $[Ti(H_2O)_5Cl](NO_3)_2$ | 2. | Diamagnetic and exhibits *cis-trans* isomerism |
| R. | $[Pt(en)(NH_3)Cl]NO_3$ | 3. | Paramagnetic and exhibits *cis-trans* isomerism |
| S. | $[Co(NH_3)_4(NO_3)_2]NO_3$ | 4. | Diamagnetic and exhibits ionisation isomerism |

**Codes**

| | P | Q | R | S | | P | Q | R | S |
|---|---|---|---|---|---|---|---|---|---|
| (a) | 4 | 2 | 3 | 1 | (b) | 3 | 1 | 4 | 2 |
| (c) | 2 | 1 | 3 | 4 | (d) | 1 | 3 | 4 | 2 |

**18.** Match the orbital overlap figures shown in List I with the description given in List II and select the correct answer using the code given below the lists.

| | List I | | List II |
|---|---|---|---|
| P. | | 1. | *p-d* π antibonding |
| Q. | | 2. | *d-d* σ bonding |
| R. | | 3. | *p-d* π bonding |
| S. | | 4. | *d-d* σ antibonding |

**Codes**

| | P | Q | R | S | | P | Q | R | S |
|---|---|---|---|---|---|---|---|---|---|
| (a) | 2 | 1 | 3 | 4 | (b) | 4 | 3 | 1 | 2 |
| (c) | 2 | 3 | 1 | 4 | (d) | 4 | 1 | 3 | 2 |

**19.** Different possible thermal decomposition pathways for peroxyesters are shown below. Match each pathway from List I with an appropriate structure from List II and select the correct answer using the code given below the lists.

(Peroxyester)

$P$: $\xrightarrow[-CO_2\uparrow]{}$ $R^{\bullet} + R'O^{\bullet}$

$Q$: $\xrightarrow[-CO_2\uparrow]{}$ $R^{\bullet} + R'O^{\bullet} \longrightarrow R^{\bullet} + X^{\bullet}$ + Carbonyl compound $\uparrow$

$R$: $\longrightarrow$ $RCO_2^{\bullet} + R'O^{\bullet} \xrightarrow[-CO_2]{} R^{\bullet} + X'^{\bullet}$ + Carbonyl compound $\uparrow$

$S$: $\longrightarrow$ $RCO_2^{\bullet} + R'O^{\bullet} \xrightarrow[-CO_2\uparrow]{} R^{\bullet} + R'O^{\bullet}$

| | List I | | List II |
|---|---|---|---|
| P. | Pathway *P* | 1. | $C_6H_5CH_2C(=O)-O-O-CH_3$ |
| Q. | Pathway *Q* | 2. | $C_6H_5C(=O)-O-O-CH_3$ |
| R. | Pathway *R* | 3. | $C_6H_5CH_2C(=O)-O-O-C(CH_3)_2CH_2C_6H_5$ |
| S. | Pathway *S* | 4. | $C_6H_5C(=O)-O-O-C(CH_3)_2C_6H_6$ |

**Codes**

| | P | Q | R | S | | P | Q | R | S |
|---|---|---|---|---|---|---|---|---|---|
| (a) | 1 | 3 | 4 | 2 | (b) | 2 | 4 | 3 | 1 |
| (c) | 4 | 1 | 2 | 3 | (d) | 3 | 2 | 1 | 4 |

**20.** Match the four starting materials (P, Q, R, S) given in List I with the corresponding reaction schemes (I, II, III, IV) provided in List II and select the correct answer using the code given below the lists.

| | List I | | List II |
|---|---|---|---|
| P. | $H-C\equiv C-H$ | 1. | **Scheme I**<br>$? \xrightarrow[\text{(iii) } SOCl_2 \text{ (iv) } NH_3]{\text{(i) } KMnO_4, HO^-, \text{ heat (ii) } H^+, H_2O} C_7H_6N_2O_3$ |
| Q. | OH, OH (benzene-1,3-diol) | 2. | Scheme II<br>$? \xrightarrow[\text{(iv) } HNO_3 \text{ (v) dil. } H_2SO_4, \text{ heat (vi) } HO^-]{\text{(i) Sn/HCl (ii) } CH_3COCl \text{ (iii) conc. } H_2SO_4} C_6H_6N_2O_2$ |
| R. | $NO_2$ (nitrobenzene) | 3. | **Scheme III**<br>$? \xrightarrow[\text{(iii) } H_2S.NH_3 \text{ (iv) } NaNO_2, H_2SO_4 \text{ (v) hydrolysis}]{\text{(i) red hot iron, 873 K (ii) fuming } HNO_3, H_2SO_4, \text{ heat}} C_6H_5NO_3$ |
| S. | $NO_2$, $CH_3$ (4-nitrotoluene) | 4. | **Scheme IV**<br>$? \xrightarrow[\text{(ii) conc. } HNO_3, \text{ conc. } H_2SO_4 \text{ (iii) dil.} H_2SO_4, \text{ heat}]{\text{(i) conc. } H_2SO_4, 60°C} C_6H_5NO_4$ |

**Codes**

| | P | Q | R | S | | P | Q | R | S |
|---|---|---|---|---|---|---|---|---|---|
| (a) | 1 | 4 | 2 | 3 | (b) | 3 | 1 | 4 | 2 |
| (c) | 3 | 4 | 2 | 1 | (d) | 4 | 1 | 3 | 2 |

# Mathematics

## Section I

**Directions** (Q. Nos. 1-10) *This section contains 10 multiple choice questions. Each question has four choices (a), (b), (c) and (d) out of which one is correct.*

**1.** Three boys and two girls stand in a queue. The probability that the number of boys ahead of every girl is atleast one more than the number of girls ahead of her, is

(a) 1/2 (b) 1/3 (c) 2/3 (d) 3/4

**2.** In a triangle, the sum of two sides is $x$ and the product of the same two sides is $y$. If $x^2 - c^2 = y$, where $c$ is the third side of the triangle, then the ratio of the inradius to the circumradius of the triangle is

(a) $\frac{3y}{2x(x+c)}$ (b) $\frac{3y}{2c(x+c)}$

(c) $\frac{3y}{4x(x+c)}$ (d) $\frac{3y}{4c(x+c)}$

**3.** Six cards and six envelopes are numbered 1, 2, 3, 4, 5, 6 and cards are to be placed in envelopes so that each envelope contains exactly one card and no card is placed in the envelope bearing the same number and moreover the card numbered 1 is always placed in envelope numbered 2. Then, the number of ways it can be done is

(a) 264 (b) 265 (c) 53 (d) 67

**4.** The common tangents to the circle $x^2 + y^2 = 2$ and the parabola $y^2 = 8x$ touch the circle at the points $P, Q$ and the parabola at the points $R, S$. Then, the area of the quadrilateral $PQRS$ is

(a) 3 (b) 6 (c) 9 (d) 15

**5.** The quadratic equation $p(x) = 0$ with real coefficients has purely imaginary roots. Then, the equation $p[p(x)] = 0$ has

(a) only purely imaginary roots
(b) all real roots
(c) two real and two purely imaginary roots
(d) neither real nor purely imaginary roots

**6.** The following integral $\int_{\pi/4}^{\pi/2} (2 \operatorname{cosec} x)^{17} dx$ is equal to

(a) $\int_0^{\log(1+\sqrt{2})} 2(e^u + e^{-u})^{16} du$

(b) $\int_0^{\log(1+\sqrt{2})} (e^u + e^{-u})^{17} du$

(c) $\int_0^{\log(1+\sqrt{2})} (e^u - e^{-u})^{17} du$

(d) $\int_0^{\log(1+\sqrt{2})} 2(e^u - e^{-u})^{16} du$

**7.** The function $y = f(x)$ is the solution of the differential equation $\frac{dy}{dx} + \frac{xy}{x^2-1} = \frac{x^4+2x}{\sqrt{1-x^2}}$ in $(-1, 1)$ satisfying $f(0) = 0$. Then, $\int_{-\frac{\sqrt{3}}{2}}^{\sqrt{3}/2} f(x)\, dx$ is

(a) $\frac{\pi}{3} - \frac{\sqrt{3}}{2}$ (b) $\frac{\pi}{3} - \frac{\sqrt{3}}{4}$

(c) $\frac{\pi}{6} - \frac{\sqrt{3}}{4}$ (d) $\frac{\pi}{6} - \frac{\sqrt{3}}{2}$

**8.** Let $f : [0, 2] \to R$ be a function which is continuous on $[0, 2]$ and is differentiable on $(0, 2)$ with $f(0) = 1$.

Let $F(x) = \int_0^{x^2} f(\sqrt{t})\, dt$, for $x \in [0, 2]$. If $F'(x) = f'(x)$, for all $x \in (0, 2)$, then $F(2)$ equals to

(a) $e^2 - 1$ (b) $e^4 - 1$
(c) $e - 1$ (d) $e^4$

**9.** Coefficient of $x^{11}$ in the expansion of $(1 + x^2)^4 (1 + x^3)^7 (1 + x^4)^{12}$ is

(a) 1051 (b) 1106
(c) 1113 (d) 1120

**10.** For $x \in (0, \pi)$, the equation $\sin x + 2 \sin 2x - \sin 3x = 3$ has

(a) infinitely many solutions
(b) three solutions
(c) one solution
(d) no solution

## Section II

**Directions** (Q. Nos. 11-16) *This section contains 3 paragraph, each describing theory, experiments, data etc. Six questions relate to the three paragraphs with two questions on each paragraph. Each question has only one correct answer among the four given options (a), (b), (c) and (d).*

### Paragraph 1

Box I contains three cards bearing numbers 1, 2, 3 ; box II contains five cards bearing numbers 1, 2, 3, 4, 5; and box III contains seven cards bearing numbers 1, 2, 3, 4, 5, 6, 7. A card is drawn from each of the boxes. Let $x_i$ be the number on the card drawn from the $i$th box $i = 1,2,3$.

**11.** The probability that $x_1 + x_2 + x_3$ is odd, is

(a) $\frac{29}{105}$ (b) $\frac{53}{105}$

(c) $\frac{57}{105}$ (d) $\frac{1}{2}$

**12.** The probability that $x_1, x_2, x_3$ are in an arithmetic progression, is

(a) $\frac{9}{105}$ (b) $\frac{10}{105}$

(c) $\frac{11}{105}$ (d) $\frac{7}{105}$

### Paragraph 2

Let $a, r, s, t$ be non-zero real numbers. Let $P(at^2, 2at), Q, R(ar^2, 2ar)$ and $S(as^2, 2as)$ be distinct point on the parabola $y^2 = 4ax$. Suppose that $PQ$ is the focal chord and lines $QR$ and $PK$ are parallel, where $K$ is the point $(2a, 0)$.

**13.** The value of $r$ is

(a) $-\frac{1}{t}$ (b) $\frac{t^2+1}{t}$

(c) $\frac{1}{t}$ (d) $\frac{t^2-1}{t}$

**14.** If $st = 1$, then the tangent at $P$ and the normal at $S$ to the parabola meet at a point whose ordinate is

(a) $\frac{(t^2+1)^2}{2t^3}$ (b) $\frac{a(t^2+1)^2}{2t^3}$

(c) $\frac{a(t^2+1)^2}{t^3}$ (d) $\frac{a(t^2+2)^2}{t^3}$

### Paragraph 3

Given that for each $a \in (0,1)$, $\lim_{h \to 0^+} \int_h^{1-h} t^{-a}(1-t)^{a-1} dt$ exists. Let this limit be $g(a)$. In addition, it is given that the function $g(a)$ is differentiable on $(0,1)$.

**15.** The value of $g\left(\frac{1}{2}\right)$ is

(a) $\pi$ (b) $2\pi$

(c) $\pi/2$ (d) $\pi/4$

**16.** The value of $g'(1/2)$ is

(a) $\frac{\pi}{2}$ (b) $\pi$

(c) $-\frac{\pi}{2}$ (d) 0

## Section III

**Directions** (Q. Nos. 17-20) *This section contains 4 multiple choice questions. Each question has four choices (a), (b), (c) and (d) out of which only one is correct.*

**17.** Match the following

| | List I | | List II |
|---|---|---|---|
| P. | The number of polynomials $f(x)$ with non-negative integer coefficients of degree $\le 2$, satisfying $f(0) = 0$ and $\int_0^1 f(x)dx = 1$, is | (i) | 8 |
| Q. | The number of points in the interval $[-\sqrt{13}, \sqrt{13}]$ at which $f(x) = \sin(x^2) + \cos(x^2)$ attains its maximum value, is | (ii) | 2 |
| R. | $\int_{-2}^{2} \frac{3x^2}{(1+e^x)} dx$ equals | (iii) | 4 |
| S. | $\frac{\left(\int_{-1/2}^{1/2} \cos 2x \log\left(\frac{1+x}{1-x}\right)dx\right)}{\left(\int_0^{1/2} \cos 2x \log\left(\frac{1+x}{1-x}\right)dx\right)}$ equals | (iv) | 0 |

**Codes**

| | P | Q | R | S | | P | Q | R | S |
|---|---|---|---|---|---|---|---|---|---|
| (a) | (iii) | (ii) | (iv) | (i) | (b) | (ii) | (iii) | (iv) | (i) |
| (c) | (iii) | (ii) | (i) | (iv) | (d) | (ii) | (iii) | (i) | (iv) |

**18.** Match List I with List II and select the correct answer using the codes given below the lists.

| | List I | | List II |
|---|---|---|---|
| P. | Let $y(x) = \cos(3\cos^{-1} x), x \in [-1,1], x \ne \pm\frac{\sqrt{3}}{2}$. Then, $\frac{1}{y(x)}\left\{(x^2-1)\frac{d^2y(x)}{dx^2} + \frac{xdy(x)}{dx}\right\}$ equals | (i) | 1 |
| Q. | Let $A_1, A_2, \ldots A_n, (n > 2)$ be the vertices of a regular polygon of $n$ sides with its centre at the origin. Let $a_k$ be the position vector of the point $A_k, k = 1,2,\ldots,n$. If $\left\lvert\sum_{k=1}^{n-1} (\mathbf{a}_k \cdot \mathbf{a}_{k+1})\right\rvert = \left\lvert\sum_{k=1}^{n-1} (\mathbf{a}_k \cdot \mathbf{a}_{k+1})\right\rvert$, then the minimum value of $n$ is | (ii) | 2 |
| R. | If the normal from the point $P(h,1)$ on the ellipse $\frac{x^2}{6} + \frac{y^2}{3} = 1$ is perpendicular to the line $x + y = 8$, then the value of $h$ is | (iii) | 8 |
| S. | Number of positive solutions satisfying the equation $\tan^{-1}\left(\frac{1}{2x+1}\right) + \tan^{-1}\left(\frac{1}{4x+1}\right) = \tan^{-1}(2/x^2)$ is | (iv) | 9 |

**Codes**

| | P | Q | R | S | | P | Q | R | S |
|---|---|---|---|---|---|---|---|---|---|
| (a) | (iv) | (iii) | (ii) | (i) | (b) | (ii) | (iv) | (iii) | (i) |
| (c) | (iv) | (iii) | (i) | (ii) | (d) | (ii) | (iv) | (i) | (iii) |

**19.** Let $f_1 : R \to R, f_2 : [0, \infty] \to R, f_3 : R \to R$ and $f_4 : R \to [0, \infty)$ be defined by

$$f_1(x) = \begin{cases} |x|, & \text{if } x < 0 \\ e^x, & \text{if } x \ge 0 \end{cases}; \; f_2(x) = x^2;$$

$$f_3(x) = \begin{cases} \sin x, & \text{if } x < 0 \\ x, & \text{if } x \ge 0 \end{cases}$$

and $f_4(x) = \begin{cases} f_2[f_1(x)], & \text{if } x < 0 \\ f_2[f_1(x)] - 1, & \text{if } x \ge 0 \end{cases}$

| | List I | | List II |
|---|---|---|---|
| P. | $f_4$ is | (i) | onto but not one-one |
| Q. | $f_3$ is | (ii) | neither continuous nor one-one |
| R. | $f_2 \, o \, f_1$ is | (iii) | differentiable but not one-one |
| S. | $f_2$ is | (iv) | continuous and one-one |

| | P | Q | R | S |
|---|---|---|---|---|
| (a) | (iii) | (i) | (iv) | (ii) |
| (b) | (i) | (iii) | (iv) | (ii) |
| (c) | (iii) | (i) | (ii) | (iv) |
| (d) | (i) | (iii) | (ii) | (iv) |

**20.** Let $z_k = \cos\left(\frac{2k\pi}{10}\right) + i \sin\left(\frac{2k\pi}{10}\right)$; $k = 1, 2, \ldots, 9$.

| | List I | | List II |
|---|---|---|---|
| P. | For each $z_k$, there exists a $z_j$ such that $z_k \cdot z_j = 1$ | (i) | True |
| Q. | There exists a $k \in \{1, 2, \ldots, 9\}$ such that $z_1 \cdot z = z_k$ has no solution $z$ in the set of complex numbers | (ii) | False |
| R. | $\frac{\lvert 1 - z_1 \rvert \lvert 1 - z_2 \rvert \ldots \lvert 1 - z_9 \rvert}{10}$ equals | (iii) | 1 |
| S. | $1 - \sum_{k=1}^{9} \cos\left(\frac{2k\pi}{10}\right)$ equals | (iv) | 2 |

**Codes**

| | P | Q | R | S |
|---|---|---|---|---|
| (a) | (i) | (ii) | (iv) | (iii) |
| (b) | (ii) | (i) | (iii) | (iv) |
| (c) | (i) | (ii) | (iv) | (iii) |
| (d) | (ii) | (i) | (iv) | (iii) |

# Answers

## Paper 1

**Physics**

**1.** (c,d) **2.** (a,b,c) **3.** (a,c,d) **4.** (a,d) **5.** (c) **6.** (d) **7.** (b,d) **8.** (c,d) **9.** (a,c) **10.** (a,b,d)
**11.** (5) **12.** (4) **13.** (2) **14.** (3) **15.** (3) **16.** (5) **17.** (5) **18.** (2 or 8) **19.** (4) **20.** (2)

**Chemistry**

**1.** (a,b,d) **2.** (a,b,c) **3.** (a) **4.** (a) **5.** (b,c,d) **6.** (a,b,d) **7.** (a,b,c) **8.** (a,c,d) **9.** (a,b,c) **10.** (b,d)
**11.** (6) **12.** (1) **13.** (4) **14.** (8) **15.** (2) **16.** (7) **17.** (3) **18.** (6/7) **19.** (5) **20.** (4)

**Mathematics**

**1.** (a,c) **2.** (a,d) **3.** (c,d) **4.** (a,b,c) **5.** (c) **6.** (a,b) **7.** (a,c,d) **8.** (a,b,c) **9.** (b,c) **10.** (b,d)
**11.** (8) **12.** (3) **13.** (0) **14.** (3) **15.** (6) **16.** (7) **17.** (2) **18.** (4) **19.** (4) **20.** (5)

## Paper 2

**Physics**

**1.** (b) **2.** (d) **3.** (c) **4.** (c) **5.** (c) **6.** (a) **7.** (a) **8.** (b) **9.** (b) **10.** (d)
**11.** (d) **12.** (d) **13.** (c) **14.** (a) **15.** (c) **16.** (b) **17.** (a) **18.** (b) **19.** (d) **20.** (c)

**Chemistry**

**1.** (c) **2.** (b) **3.** (b) **4.** (d) **5.** (b) **6.** (d) **7.** (c) **8.** (a) **9.** (a) **10.** (c)
**11.** (c) **12.** (d) **13.** (a) **14.** (c) **15.** (b) **16.** (d) **17.** (b) **18.** (c) **19.** (a) **20.** (c)

**Mathematics**

**1.** (a) **2.** (b) **3.** (c) **4.** (d) **5.** (d) **6.** (a) **7.** (b) **8.** (b) **9.** (c) **10.** (d)
**11.** (b) **12.** (c) **13.** (d) **14.** (b) **15.** (a) **16.** (d) **17.** (d) **18.** (a) **19.** (d) **20.** (c)

# Hints & Solutions

## Paper 1

### Physics

**1.** $\frac{dQ}{dt} = I \Rightarrow Q = \int I\, dt = \int (I_0 \cos \omega t)\, dt$

$\therefore \quad Q_{max} = \frac{I_0}{\omega} = \frac{1}{500}$

$= 2 \times 10^{-3}$ C

Just after switching

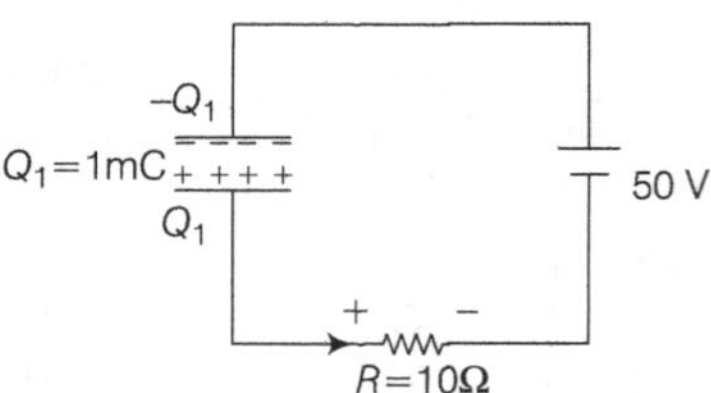


In steady state

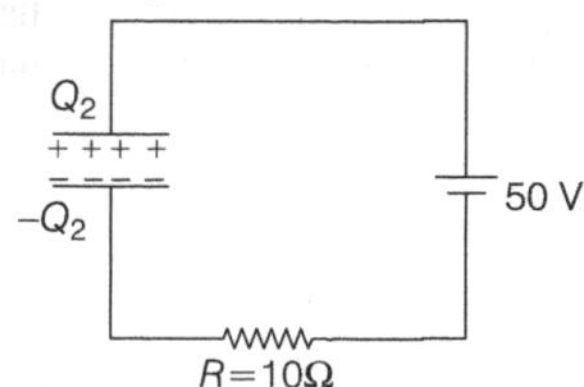


At $\quad t = \frac{7\pi}{6\omega}$ or $\omega t = \frac{7\pi}{6}$

Current comes out to be negative from the given expression. So, current is anti-clockwise.

Charge supplied by source from $t = 0$ to $t = \frac{7\pi}{6\omega}$

$$Q = \int_0^{\frac{7\pi}{6\omega}} \cos (500t)\, dt$$

$$= \left[\frac{\sin 500t}{500}\right]_0^{\frac{7\pi}{6\omega}} = \frac{\sin \frac{7\pi}{6}}{500}$$

$= -1$ mC

Apply Kirchhoff's loop law just after changing the switch to position $D$

$$50 + \frac{Q_1}{C} - IR = 0$$

Substituting the values of $Q_1$, $C$ and $R$ we get

$$I = 10 \text{ A}$$

In steady state $Q_2 = CV = 1$ mC

$\therefore$ Net charge flown from battery

$= 2$ mC

**2.** Fringe width $\beta = \frac{\lambda D}{d}$ or $\beta \propto \lambda$

$\therefore \quad \lambda_2 > \lambda_1$

So $\quad \beta_2 > \beta_1$

Number of fringes in a given width

$$m = \frac{y}{\beta} \quad \text{or} \quad m \propto \frac{1}{\beta}$$

$\Rightarrow \quad m_2 < m_1$ as $\beta_2 > \beta_1$

Distance of 3rd maximum of $\lambda_2$ from central maximum

$$= \frac{3\lambda_2 D}{d} = \frac{1800D}{d}$$

Distance of 5th minimum of $\lambda_1$ from central maximum

$$= \frac{9\lambda_1 D}{2d} = \frac{1800D}{d}$$

So, 3rd maximum of $\lambda_2$ will overlap with 5th minimum of $\lambda_1$

Angular separation (or angular fringe width)

$$= \frac{\lambda}{d} \propto \lambda$$

$\Rightarrow$ Angular separation for $\lambda_1$ will be lesser.

**3.** There should be a node at $x = 0$ and antinode at $x = 3$ m.

Also, $\quad v = \frac{\omega}{k} = 100$ m/s.

$\therefore \quad y = 0$ at $x = 0$

and $\quad y = \pm A$ at $x = 3$ m.

Only (a), (c) and (d) satisfy the condition.

**4.** $C = C_1 + C_2$

$$C_1 = \frac{K\varepsilon_0 A/3}{d},\ C_2 = \frac{\varepsilon_0 2A/3}{d}$$

$$\Rightarrow \quad C = \frac{(K+2)\varepsilon_0 A}{3d}$$

$$\Rightarrow \quad \frac{C}{C_1} = \frac{K+2}{K}$$

Also, $E_1 = E_2 = V/d$, where $V$ is potential difference between the plates.

**5.** $$\frac{Q}{4\pi\varepsilon_0 r_0^2} = \frac{\lambda}{2\pi\varepsilon_0 r_0} = \frac{\sigma}{2\varepsilon_0}$$

$$Q = 2\pi\sigma r_0^2$$

(a) is incorrect, $r_0 = \dfrac{\lambda}{\pi\sigma}$

(b) is incorrect, $E_1\left(\dfrac{r_0}{2}\right) = 4E_1(r_0)$

As $\quad E_1 \propto \dfrac{1}{r^2}$

$$E_2\left(\frac{r_0}{2}\right) = 2E_2(r_0) \text{ as } E_2 \propto \frac{1}{r}$$

$\Rightarrow$ (c) is correct

$$E_3\left(\frac{r_0}{2}\right) = E_3(r_0) = E_2(r_0)$$

as $\quad E_3 \propto r^0$

$\Rightarrow$ (d) is incorrect

**6.** Minimum length $= \dfrac{\lambda}{4} \Rightarrow \lambda = 4l$

Now, $\quad v = f\lambda = (244) \times 4 \times l$

as $\quad l = 0.350 \pm 0.005$

$\Rightarrow$ $v$ lies between 336.7 m/s to 346.5 m/s

Now, $v = \sqrt{\dfrac{\gamma RT}{M \times 10^{-3}}}$, here $M$ is molecular mass in gram

$$= \sqrt{100\,\gamma RT} \times \sqrt{\frac{10}{M}}.$$

For monoatomic gas,

$\gamma = 1.67$

$$\Rightarrow \quad v = 640 \times \sqrt{\frac{10}{M}}$$

For diatomic gas,

$$\gamma = 1.4 \Rightarrow v = 590 \times \sqrt{\frac{10}{M}}$$

$$\therefore \quad v_{Ne} = 640 \times \frac{7}{10} = 448 \text{ m/s}$$

$$v_{Ar} = 640 \times \frac{17}{32} = 340 \text{ m/s}$$

$$v_{O_2} = 590 \times \frac{9}{16} = 331.8 \text{ m/s}$$

$$v_{N_2} = 590 \times \frac{3}{5} = 354 \text{ m/s}$$

$\therefore$ Only possible answer is Argon.

**7.** Resistance of initially given kettle

$$R = \rho\frac{l}{A}$$

$$= \rho\frac{L}{\pi(d/2)^2} = \frac{4\rho L}{\pi d^2}$$

Power, $P = \dfrac{V^2}{R}$

or $\quad P \propto \dfrac{1}{R}$

Resistance of two replaced kettles

$$R_1 = \frac{\rho L}{\pi d^2}$$

and $\quad R_2 = \dfrac{\rho L}{\pi d^2}$

So $\quad R_1 = R_2 = \dfrac{R}{4}$

If wires are in parallel then equivalent resistance

$$R_P = \frac{R_1 R_2}{R_1 + R_2} = \frac{R}{8}$$

Then power $P_P = 8P \quad \left(\text{as } P \propto \dfrac{1}{R}\right)$

So, it will take 0.5 minute

(as $H = Pt$ or $t = \dfrac{H}{P}$ or $t \propto \dfrac{1}{P}$)

If wires are in series then equivalent resistance

$$R_S = R_1 + R_2 = \frac{R}{2}$$

Then power $P_S = 2P$

So, it will take 2 minutes.

**8.** $\mu_2$ can never be zero for maximum equilibrium.

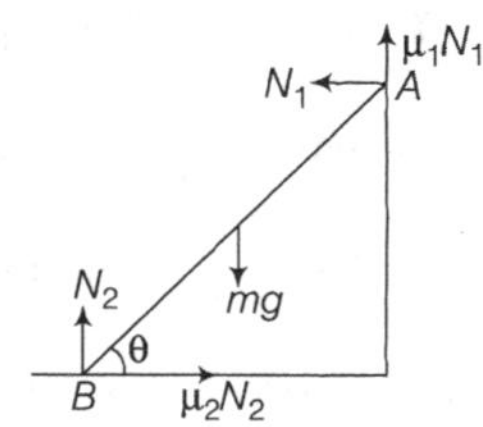


When $\mu_1 = 0$ we have

$$N_1 = \mu_2 N_2 \quad \ldots(i)$$

$$N_2 = m_2 g \quad \ldots(ii)$$

$$\tau_B = 0 \Rightarrow mg\frac{L}{2}\cos\theta = N_1 L \sin\theta$$

$$\Rightarrow \quad N_1 = \frac{mg\cot\theta}{2}$$

$$\Rightarrow \quad N_1 \tan\theta = \frac{mg}{2}$$

When $\mu_1 \neq 0$ we have

$$\mu_1 N_1 + N_2 = mg \quad \ldots(i)$$

$$\mu_2 N_2 = N_1 \quad \ldots(ii)$$

$$\Rightarrow \quad N_2 = \frac{mg}{1+\mu_1\mu_2}$$

**9.** $\dfrac{1}{f_{\text{film}}} = (n_1 - 1)\left(\dfrac{1}{R} - \dfrac{1}{R}\right)$

$$\Rightarrow \quad f_{\text{film}} = \infty \quad \text{(infinite)}$$

$\therefore$ There is no effect of presence of film.

**From Air to Glass**

Using the equation $\dfrac{n_2}{v} - \dfrac{1}{u} = \dfrac{n_2 - 1}{R}$

$$\frac{1.5}{v} - \frac{1}{\infty} = \frac{1.5-1}{R}$$

$$\Rightarrow \quad v = 3R$$

$$\therefore \quad f_1 = 3R$$

**From Glass to Air** Again using the same equation

$$\frac{1}{v} - \frac{n_2}{u} = \frac{1-n_2}{-R}$$

$$\Rightarrow \quad \frac{1}{v} - \frac{1.5}{\infty} = \frac{1-1.5}{-R}$$

$$\Rightarrow \quad v = 2R$$

$$\therefore \quad f_2 = 2R$$

**10.** Let us take $V_P = 0$. Then potentials across $R_1$, $R_2$ and $R_3$ are as shown in figure (ii)

In the same figure

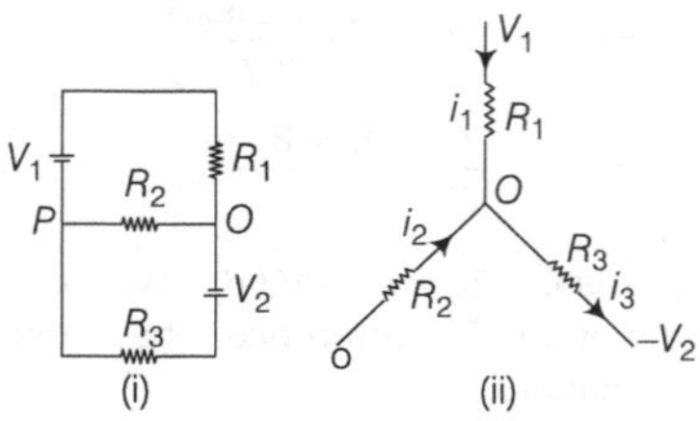


$$i_1 + i_2 = i_3$$

$$\therefore \quad \frac{V_1 - V_0}{R_1} + \frac{0 - V_0}{R_2} = \frac{V_0 - (-V_2)}{R_3}$$

Solving this equation we get

$$V_0 = \frac{\dfrac{V_1}{R_1} + 0 - \dfrac{V_2}{R_3}}{\dfrac{1}{R_1} + \dfrac{1}{R_2} + \dfrac{1}{R_3}}$$

Current through $R_2$ will be zero if

$$V_0 = 0 \Rightarrow \frac{V_1}{V_2} = \frac{R_1}{R_3}$$

In options (a), (b) and (d) this relation is satisfied.

**11.** Relative velocity of $B$ with respect to $A$ is perpendicular to line $PA$. Therefore, along line $PA$, velocity components of $A$ and $B$ should be same.

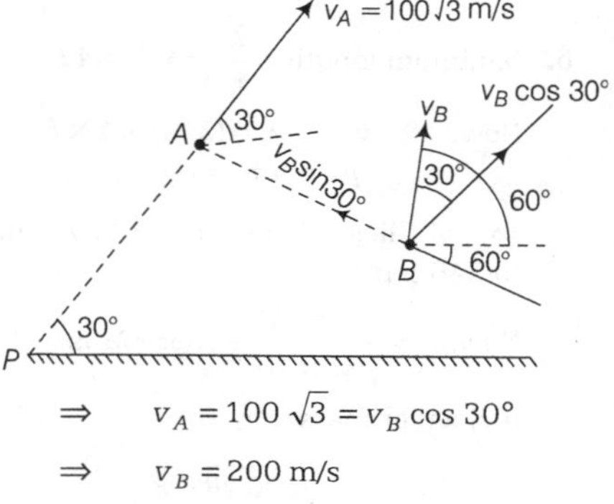


$$\Rightarrow \quad v_A = 100\sqrt{3} = v_B \cos 30°$$

$$\Rightarrow \quad v_B = 200 \text{ m/s}$$

$$t_0 = \frac{500}{v_B \sin 30°}$$

$$= \frac{500}{200 \times \dfrac{1}{2}} = 5 \text{ s}$$

**12.** $Y = \dfrac{F/A}{\dfrac{\Delta l}{l}}$, $\Delta l = 25 \times 10^{-50}$ m

$$Y = \frac{F}{A} \cdot \frac{l}{\Delta l}$$

$$\frac{\Delta Y}{Y} \times 100 = \frac{10^{-5}}{25 \times 10^{-5}} \times 100 = 4\%$$

**13.** Angular impulse = change in angular momentum

$$\therefore \quad \int \tau \, dt = I\omega$$

$$\Rightarrow \quad \omega = \frac{\int \tau \, dt}{I}$$

$$= \frac{\int_0^t 3F \sin 30^\circ \, R \, dt}{I}$$

Substituting the values, we have

$$\omega = \frac{3\,(0.5)\,(0.5)\,(0.5)\,(1)}{\dfrac{1.5\,(0.5)^2}{2}} = 2 \text{ rad/s}$$

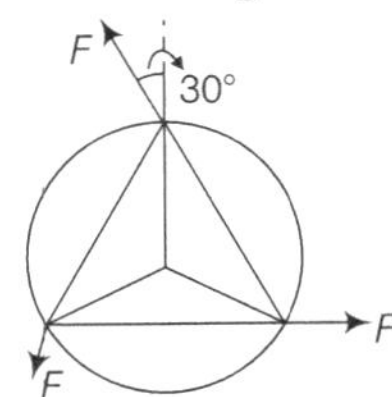


**14.** $B_2 = \dfrac{\mu_0 I}{2\pi x_1} + \dfrac{\mu_0 I}{2\pi\,(x_0 - x_1)}$

(when currents are in opposite directions)

$$B_1 = \frac{\mu_0 I}{2\pi x_1} - \frac{\mu_0 I}{2\pi\,(x_0 - x_1)}$$

I I

$\leftarrow x_1 \rightarrow \leftarrow x_0 - x_1 \rightarrow$

(when currents are in same direction)

Substituting $x_1 = \dfrac{x_0}{3}$ $\quad \left(\text{as } \dfrac{x_0}{x_1} = 3\right)$

$$B_1 = \frac{3\mu_0 I}{2\pi x_0} - \frac{3\mu_0 I}{4\pi x_0} = \frac{3\mu_0 I}{4\pi x_0}$$

$$R_1 = \frac{mv}{qB_1} \text{ and } B_2 = \frac{9\mu_0 I}{4\pi x_0}$$

$$R_2 = \frac{mv}{qB_2}$$

$$\frac{R_1}{R_2} = \frac{B_2}{B_1} = \frac{9}{3} = 3$$

**15.** Let $d = k(\rho)^a (s)^b (f)^c$

where, $k$ is a dimensionless. Then

$$[\text{L}] = \left[\frac{\text{M}}{\text{L}^3}\right]^a \left[\frac{\text{ML}^2\text{T}^{-2}}{\text{L}^2\text{T}}\right]^b \left[\frac{1}{\text{T}}\right]^c$$

Equating the powers of $M$ and $L$, we have

$$0 = a + b \quad \ldots\text{(i)}$$

$$1 = -3a \quad \ldots\text{(ii)}$$

Solving these two equations, we get

$$b = \frac{1}{3} \quad \therefore \quad n = 3$$

**16.** $i_g$ 4990 Ω (G) ← V →

$$i_g(G + 4990) = V$$

$$\Rightarrow \quad \frac{6}{1000}(G + 4990) = 30$$

$$\Rightarrow \quad G + 4990 = \frac{30{,}000}{6} = 5000$$

$$\Rightarrow \quad G = 10\ \Omega$$

$$V_{ab} = V_{cd}$$

$$\Rightarrow \quad i_g G = (1.5 - i_g)\,S$$

S, c, d, $(1.5 - i_g)$, 1.5 A, a, $i_g$, G, b

$$\Rightarrow \quad \frac{6}{1000} \times 10 = \left(1.5 - \frac{6}{1000}\right) S$$

$$\Rightarrow \quad S = \frac{60}{1494} = \frac{2n}{249}$$

$$\Rightarrow \quad n = \frac{249 \times 30}{1494} = \frac{2490}{498} = 5$$

**17.** From work-energy theorem.

Work done by all forces = change in kinetic energy

or $\quad W_F + W_{mg} = K_f - K_i$

$$18 \times 5 + (1 \times 10)(-4) = K_f$$

$$90 - 40 = K_f$$

$$K_f = 50 \text{ J} = 5 \times 10 \text{ J}$$

**18.**

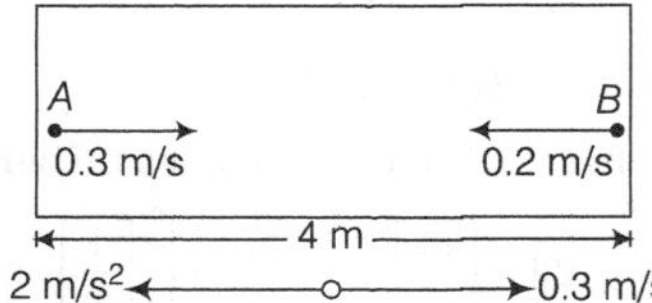


Motion of ball $A$ relative to rocket.

Consider motion of two balls with respect to rocket

Maximum distance of ball $A$ from left wall

$$= \frac{u^2}{2a} = \frac{0.3 \times 0.3}{2 \times 2}$$

$$= \frac{0.09}{4} \approx 0.02 \text{ m} \quad (\text{as } 0 = u^2 - 2aS)$$

So, collision of two balls will take place very near to left wall

← 2 m/s²

← O 0.2 m/s $\quad$ − ve ←→ + ve

Motion of ball $B$ relative to rocket

For $B$ $\quad S = ut + \frac{1}{2}at^2$

$$-4 = -0.2t - \left(\frac{1}{2}\right)2t^2$$

Solving this equation, we get,

$$t = 1.9 \text{ s}$$

$\therefore$ Nearest integer = 2 s

**Alternate**

$$S_1 = 0.2t + \frac{1}{2} \times 2 \times t^2$$

$$S_2 = 0.3t - \frac{1}{2} \times 2 \times t^2$$

$$S_1 + S_2 = 4$$

$$0.5t = 4$$

$\therefore \quad t = 8 \text{ s}$

**19.** Applying conservation of angular momentum

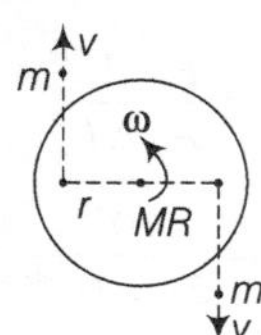


$$2mvr - \frac{MR^2}{2}\omega = 0$$

$$\omega = \frac{4mvr}{MR^2}$$

Substituting the values, we get

$$\omega = \frac{(4)(5 \times 10^{-2})(9)\left(\frac{1}{4}\right)}{45 \times 10^{-2} \times \frac{1}{4}}$$

$$\omega = 4 \text{ rad/s}$$

**20.**

$$W_{ibf} = W_{ib} + W_{bf}$$
$$= 50 \text{ J} + 100 \text{ J} = 150 \text{ J}$$
$$W_{iaf} = W_{ia} + W_{af}$$
$$= 0 + 200 \text{ J} = 200 \text{ J}$$
$$Q_{iaf} = 500 \text{ J}$$

So $\quad \Delta U_{iaf} = Q_{iaf} - W_{iaf}$
$$= 500 \text{ J} - 200 \text{ J} = 300 \text{ J}$$
$$= U_f - U_i$$

So, $\quad U_f = U_{iaf} + U_i$
$$= 300 \text{ J} + 100 \text{ J}$$
$$= 400 \text{ J}$$
$$\Delta U_{ib} = U_b - U_i$$
$$= 200 \text{ J} - 100 \text{ J}$$
$$= 100 \text{ J}$$
$$Q_{ib} = \Delta U_{ib} + W_{ib}$$
$$= 100 \text{ J} + 50 \text{ J} = 150 \text{ J}$$
$$Q_{ibf} = \Delta U_{ibf} + W_{ibf}$$
$$= \Delta U_{iaf} + W_{ibf}$$
$$= 300 \text{ J} + 150 \text{ J}$$
$$= 450 \text{ J}$$

So, the required ratio

$$\frac{Q_{bf}}{Q_{ib}} = \frac{Q_{ibf} - Q_{ib}}{Q_{ib}}$$

$$= \frac{450 - 150}{150} = 2$$

# Chemistry

**1. Plan** *This problem includes concept of redox reaction. A redox reaction consists of oxidation half-cell reaction and reduction half-cell reaction.*

*Write both half-cell reactions, i.e., oxidation half-cell reaction and reduction half-cell reaction.*

*Then balance both the equations.*

*Now determine the correct value of stoichiometry of* $H_2SO_4$.

➥ Oxidation half-reaction,

$$2I^- \longrightarrow I_2 + 2e^- \quad \ldots(i)$$

Here, $I^-$ is converted into $I_2$. Oxidation number of I is increasing from −1 to 0 hence, this is a type of oxidation reaction.

Reduction half-reaction

$$6H^+ + ClO_3^- + 6e^- \longrightarrow Cl^- + 3H_2O \quad \ldots(ii)$$

Here, $H_2O$ releases as a product. Hence, option (d) is correct.

Multiplying equation (i) by 3 and adding in equation (ii)

$$6I^- + ClO_3^- + 6H^+ \longrightarrow Cl^- + 3I_2 + 3H_2O$$

$$6I^- + ClO_3^- + 6H_2SO_4 \longrightarrow Cl^- + 3I_2 + 3H_2O + 6HSO_4^-$$

Stoichiometric coefficient of $HSO_4^-$ is 6.

Hence, option (a), (b) and (d) are correct.

**2. Plan** *Paramagnetic character of species can be easily explained on the basis of presence of unpaired electrons. The compounds containing unpaired electron(s) is/are paramagnetic.*

➥ Reaction of alkali metals with ammonia depends upon the physical state of ammonia whether it is in gaseous state or liquid state. If ammonia is considered as a gas then reaction will be

(a) $Na + \underset{\text{(Excess)}}{NH_3} \longrightarrow NaNH_2 + \frac{1}{2}H_2$

($NaNH_2$ + $1/2\,H_2$ are diamagnetic)

If ammonia is considered as a liquid then reaction will be

$$M + (x+y)NH_3 \longrightarrow [M(NH_3)_x]^+ + [e(NH_3)_y]^-$$

- ammoniated electron
- blue colour
- paramagnetic
- very strong reducing agent

(b) $K + \underset{\text{(Excess)}}{O_2} \longrightarrow KO_2(K^+, O_2^-)$
Potassium superoxide
Paramagnetic

(c) $3Cu + 8HNO_3 \longrightarrow 3Cu(NO_3)_2$ (Paramagnetic) $+ 2NO$ (Paramagnetic) $+ 4H_2O$

(d) 2-ethyl anthraquinol $\xrightarrow{O_2}$ 2-ethyl anthraquinone $+ H_2O_2$



2-ethyl anthraquinol

2-ethyl anthraquinone

Hence, option (a), (b) and (c) are correct choices.

**3. Plan** *This problem includes concept of acetylation reaction and regioselectivity of chemical reaction.*

➥ Regioselectivity means which group will react selectivily in the presence of two or more than two functional groups. Here, among two functional group —$NH_2$ and —$CONH_2$, $NH_2$ is more nucleophilic hence, $NH_2$ group will undergo reaction faster than $CONH_2$.

$$\text{(3-(aminomethyl)benzamide: } C_6H_4(CH_2-NH_2)(C(=O)-NH_2)\text{)} + CH_3-\overset{O}{\overset{\|}{C}}-O-\overset{O}{\overset{\|}{C}}-CH_3 \xrightarrow{CH_2Cl_2} C_6H_4(CH_2-NH-\overset{O}{\overset{\|}{C}}-CH_3)(C(=O)-NH_2)$$

$-CH_2-NH_2$ is more nucleophilic than $-\underset{O}{\underset{\|}{C}}-NH_2$.

Hence, correct choice is (a).

**4. Plan** *This problem is based on characteristics of salt bridge.*

➥ Functions of salt bridge

(i) It connects the two half-cells and completes the cell circuit.

(ii) It keeps the solutions of two half-cells and complete the cell circuit but does not participate chemically in the cell reaction.

(iii) It maintains the diffusion of ions from one electrode to another electrode.

(iv) A cell reaction may also occur in absence of salt bridge. Sometimes, both the electrodes dip in the same electrolyte solution and in such cases we do not require a salt bridge." So, option (c) is incorrect.

(v) This prevent mixing of two electrolytic solutions hence, option (d) is incorrect choice.

Hence, correct choice is (a)

**5.** (b) $4CuO \xrightarrow{1100°C} 2Cu_2O + O_2$

(c) $Cu_2S + 2Cu_2O \xrightarrow{\Delta} 6Cu + SO_2$

$$2Cu_2O + Cu_2S \xrightarrow{\Delta} 6Cu + SO_2$$

(d) $CuSO_4 \xrightarrow{720°C} CuO + SO_2 + \frac{1}{2}O_2$

$$4CuO \xrightarrow{1100°C} 2Cu_2O + O_2$$

$$2Cu_2O + Cu_2S \xrightarrow{\Delta} 6Cu + SO_2$$

Reaction is believed to proceed as

$$Cu_2S \rightleftharpoons 2Cu^+ + S^{2-}$$

$$2Cu_2O \rightleftharpoons 4Cu^+ + 2O^{2-}$$

$$S^{2-} + 2O^{2-} \longrightarrow SO_2 + 6e^-$$

$$6Cu^+ + 6e^- \longrightarrow 6Cu\,;\; E^\circ_{cell} = 0.52$$

Here, copper sulphide is reduced to copper metal. Solidified copper has blistered appearance due to evolution of $SO_2$ and thus obtained copper is known as blister copper.

Other compounds which give Cu are

(i) CuO as

$$4CuO \xrightarrow{1100°C} 2Cu_2O + O_2$$

$$2Cu_2O + Cu_2S \xrightarrow{\Delta} 6Cu + SO_2$$

(ii) $CuSO_4$ as

$$CuSO_4 \xrightarrow{720°C} CuO + SO_2 + \frac{1}{2}O_2$$

$$4CuO \xrightarrow{\Delta} 2Cu_2O + O_2$$

$$2Cu_2O + Cu_2S \xrightarrow{\Delta} 6Cu + SO_2$$

While $CuFeS_2$ will not give Cu on heating. The heating in the presence of $O_2$ gives $Cu_2S$ and FeS with the evolution of $SO_2$.

**6. Plan** *This problem can be solved by using concept of H-bonding and applications of H-bonding.*

(a) Ice floats in water due to the low density of ice as compare to water which is due to open cage like structure (formed by intermolecular H-bonding).

(b) Basic strength of $RNH_2 > R_3N$. It is also explained by hydrogen bonding.

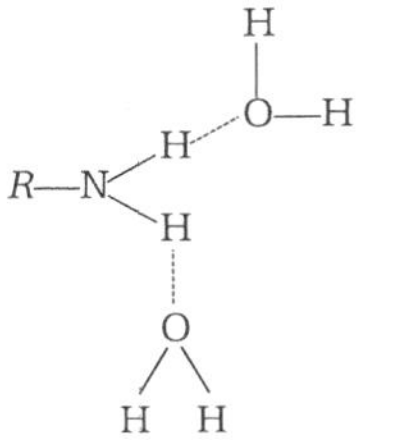

Two H-bonds are possible with water present in aqueous solution. (stabilize by solvation)

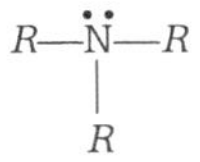

No H-bonding is possible with water present in aqueous solution. (stabilization by solvation is very less)

(c) HCOOH

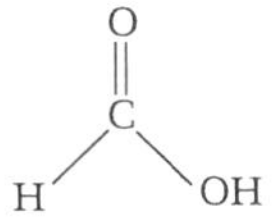

More acidic due to the presence of H. (Due to the absence of electron donating group)

$H_3C$—COOH

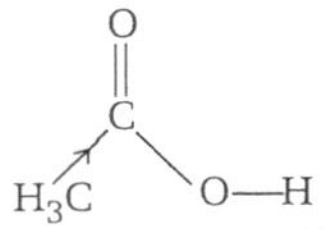

Less acidic than HCOOH due to presence of $CH_3$. (Electron donating group)

(d) Dimerisation of acetic acid in benzene is due to intermolecular hydrogen bonding.

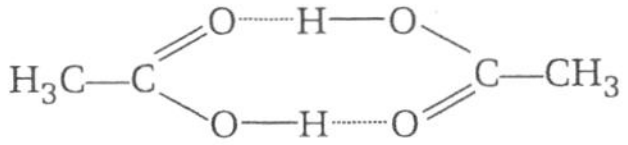

**7.** **Plan** *This problem includes concept of effect of steric and electronic effect on reactivity of organic compounds.*

*Steric effect of halogens are as follows $Cl_2 < Br_2 < I_2$*

*Electronic effect of phenolic group directs the approaching electrophile towards ortho and para positions. Tertiary butyl group has large size so it causes steric effect around aromatic nucleus.*

On the basis of above factors the products of the given reactions are as follows

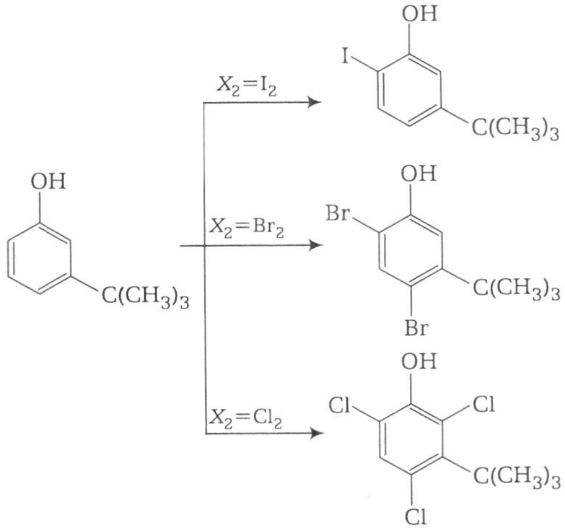

$X_2 = I_2$; $X_2 = Br_2$; $X_2 = Cl_2$

Hence, orientation in electrophilic substitution reaction is decided by

(a) The steric effect of the halogen

(b) The steric effect of the tert-butyl group

(c) The electronic effect of the phenolic group

So, (a), (b) and (c) are correct choices.

**8. Plan** *This problem is based on structure and nomenclature of organic compound.*

*Draw structure of each compound and write IUPAC name of the given compound.*

*Match the molecular formula of given compound with molecular formula of compound given in choices.*

➥ The combination of names for possible isomeric alcohols with molecular formula $C_4H_{10}O$ is/are

| Formula | Names |
|---|---|
| $CH_3CH_2CH_2CH_2OH$ | *n*-butyl alcohol / *n*-butanol / butan-1-ol |
| $CH_3—CH(CH_3)—CH_2—OH$ | Isobutyl alcohol / 2-methyl propan-1-ol |
| $CH_3—CH_2—CH(CH_3)—OH$ | Secondary butyl alcohol / butan-2-ol |
| $(CH_3)_3C—OH$ | Tertiary butyl alcohol / *tert* butanol / 2-methyl propan-2-ol |

Hence, all choices (a), (b) and (c) are correct.

**9. Plan** *This problem includes concept of Isothermal adiabatic irreversible expansion.*

➥ Process is adiabatic because of the use of thermal insolution therefore, $q = 0$

$$\because \quad p_{ext} = 0$$

$$w = p_{ext} \cdot \Delta V$$

$$= 0 \times \Delta V = 0$$

Internal energy can be written as

$$\Delta U = q + W = 0$$

The change in internal energy of an ideal gas depends only on temperature and change in internal energy $(\Delta U) = 0$ therefore, $\Delta T = 0$ hence, process is isothermal and

$$T_2 = T_1$$

and $$p_2V_2 = p_1V_1$$

(d) $p_2V_2^{\gamma} = p_1V_1^{\gamma}$ is incorrect, it is valid for adiabatic reversible process.

Hence, only (a), (b) and (c) are correct choices.

**10.** (a) It does not undergo self ionisation in water but accepts an electron pair from water, so it behaves as weak monobasic acid

$$H_3BO_3 + H_2O \rightleftharpoons B(OH)_4^- + H^+$$

Hence, (a) is incorrect.

(b) When treated with 1, 2-dihydroxy or polyhydroxy compounds, they form chelate (ring complex) which effectively remove $[B(OH)_4]^-$ species from solution and thereby produce maximum number of $H_3O^+$ or $H^+$ ions, *i.e.*, results in increased acidity.

(c) Baric acid crystallises in a layer structure in which planar triangular $BO_3^{3-}$ ions are bonded together through hydrogen bonds.

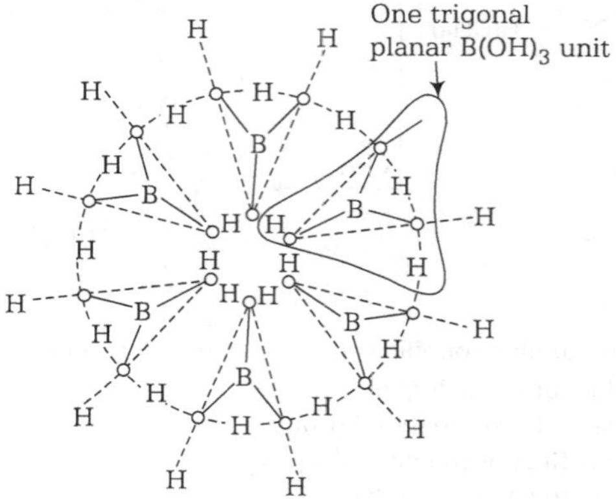

(d) In water the $pK_a$ value of $H_3BO_3$ is 9.25.

$$H_3BO_3 + H_2O \rightleftharpoons B(OH)_4^- + H^+; \; pK_a = 9.25$$

So, it is a weak electrolyte in water.

**11. Plan** *This problem is based on concept of quantum number. Follow the following steps to solve this problem.*

*Write all possible orbitals having combination of same principle, azimuthal, magnetic and spin quantum number.*

*Then count the all possible electrons having given set of quantum numbers.*

➥ For $n = 4$, the total number of possible orbitals are

| 4s | 4p | | | 4d | | | | | 4f | | | | | | |
|---|---|---|---|---|---|---|---|---|---|---|---|---|---|---|---|
| 0 | −1 | 0 | +1 | −2 | −1 | 0 | +1 | +2 | −3 | −2 | −1 | 0 | +1 | +2 | +3 |

According to question $|m_l| = 1$, *i.e.*, there are two possible values of $m_l$, *i.e.*, +1 and −1 and one orbital can contain maximum two electrons one having $s = +\frac{1}{2}$ and other having $s = -1/2$.

So, total number of orbitals having $\{|m_l| = 1\} = 6$

Total number of electrons having $\{|m_l| = 1 \text{ and } m_s = -\frac{1}{2}\} = 6$

**12. Plan** *This problem can be solved by performing hydrolysis of peptide and deciding the nature of product.*

➥ Chemical reaction and product formed after hydrolysis of given peptide can be represented as

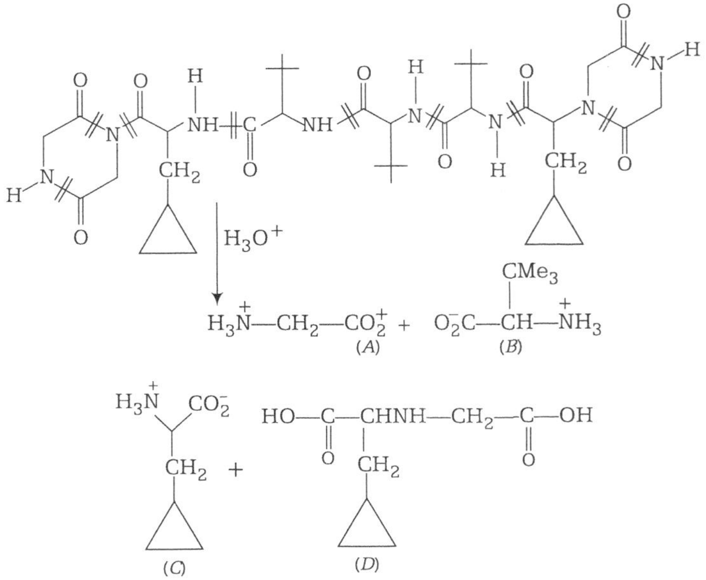

$H_3O^+$

$\overset{+}{H_3N}—CH_2—CO_2^+$ (A) + $O_2^-C—CH(CMe_3)—\overset{+}{N}H_3$ (B)

$\overset{+}{H_3N}$—CH($CO_2^-$)—$CH_2$—cyclopropyl (C) + HO—C(=O)—CH($CH_2$—cyclopropyl)NH—$CH_2$—C(=O)—OH (D)

(*A*) is glycine which is only naturally occurring amino acid.

While (*B*) , (*C*) and (*D*) are not the naturally occurring amino acids.

Hence, correct integer is (1).

**13.** **Plan** *This problem can be solved by using the concept involved in calculation of significant figure.*

➥ Universal gas constant $R = kN_A$

where, $k$ = Boltzman constant and $N_A$ = Avogadro number

$\therefore \quad R = 1.380 \times 10^{-23}$

$\times 6.023 \times 10^{23}$ J/K-mol

$= 8.31174 \cong 8.312$

Since, $k$ and $N_A$ both have four significant figures, so the value of $R$ is also rounded off upto 4 significant figures.

[When number is rounded off, the number of significant figure is reduced, the last digit is increased by 1 if following digits $\geq 5$ and is left as such if following digits is $\leq 4$.]

Hence, correct integer is (4).

**14.** **Plan** *This problem can be solved by using concept of conversion of molarity into molality.*

➥ Molarity = 3.2 M

Let, volume of solution

= 1000 mL = volume of solvent

Mass of solvent = $1000 \times 0.4 = 400$ g

Since, molarity of solution is 3.2 molar

$\therefore \quad n_{\text{solute}} = 3.2$ mol

$$\text{Molality } (m) = \frac{3.2}{400/1000} = 8$$

Hence, correct integer is (8).

**15.** $MX_2 \longrightarrow M^{2+} + 2X^-$

van't Hoff factor for any salt can be calculated using equation

$$i = 1 + \alpha\,(n - 1)$$

where, $n$ = number of constituent ions

$$\therefore \quad i(MX_2) = 1 + \alpha\,(3 - 1) = 1 + 2\alpha$$

$$\frac{(\Delta T_f)_{\text{observed}}}{(\Delta T_f)_{\text{theoretical}}} = i = 1 + 2\alpha$$

$$\therefore \quad i = 1 + 2 \times 0.5$$

$$i = 2$$

**16.** Acidified $K_2Cr_2O_7$, $CuSO_4$, $H_2O_2$, $Cl_2$, $O_3$, $FeCl_3$ and $HNO_3$ oxidise *aq.* iodide to iodine. Alkaline $KMnO_4$ oxidise *aq.* iodide to $IO_3^-$.

$Na_2S_2O_3$ is a strong reducing agent which on reaction with $I_2$ produces $I^-$.

$$Na_2S_2O_3 + I_2 \longrightarrow 2I^- + Na_2S_4O_6$$

Therefore, no reaction takes place between $Na_2S_2O_3$ and iodide ion.

Hence, correct integer is (7).

**17.** **Plan** *This problem can be solved by using concept of conformational analysis of given organic compound. To solve the question draw the stable conformational structures of organic compound and determine the net resultant dipole moment.*

➥ The conformations of the given compound are as follows

Cl
Br —+— $CH_3$
Br —+— Cl
$CH_3$

=

$CH_3$
Cl —+— Br
Br —+— Cl
$CH_3$

=

Cl, Br, Cl, Br, $H_3C$ $CH_3$ (Newman projection)

Stable conformer (with $\mu \neq 0$)

Cl, Br, Cl, $CH_3$, Br, $CH_3$

(Me-Me)gauche

$CH_3$, Br, Cl, Br, Cl, $CH_3$

(Br-Me)gauche

Br, Br, Cl, Cl, $CH_3$, $CH_3$

(Cl-Me)gauche

These three have non-zero dipole moment due to non-cancellation of all dipole moment created by C—Cl and C—Br bond.

**18.** From qualitative analysis of the different metal ions it is found that

PbS, CuS, HgS, $Ag_2S$, NiS, CoS are black coloured

MnS—dirty pink/buff coloured, $SnS_2$—yellow coloured

$Bi_2S_3$—brown/black (brownish black) coloured

Hence, correct integer is (7).

**19.** Molecular weight of the ketone is 100. So, molecular formula = $C_6H_{12}O$

Degree of unsaturation = $(6 + 1) - 12/2 = 1$

According to question, compound contains ketone group. Since, the compound which contain chiral centre lead to the formation of diastereomer while other produces enantiomers. Various isomers and their possible reduced product are as shown below.

(1) *n*-butyl–C(=O)—$CH_3$
(2) *Iso*-butyl–C(=O)—$CH_3$
(3) 3° butyl–C(=O)—$CH_3$

$\xrightarrow{NaBH_4}$

*n*-butyl–CH(OH)—$CH_3$
*Iso*-butyl–CH(OH)—$CH_3$
3° butyl–CH(OH)—$CH_3$

All are ± racemic mixture

(4) $CH_3—CH_2—\overset{*}{C}H(CH_3)—C(=O)—CH_3$
R(2°butyl)

(5) $CH_3—CH_2—\overset{*}{C}H(CH_3)—C(=O)—CH_3$
S(2°butyl)

$\xrightarrow{NaBH_4}$

$CH_3—CH_2—\overset{*}{C}H(CH_3)—\overset{*}{C}H(OH)—CH_3$
2-alcohols (*R,R*) and (*R,S*) diastereomeric pair

$CH_3—CH_2—\overset{*}{C}H(CH_3)—\overset{*}{C}H(OH)—CH_3$
2-alcohols (*S,S*) and (*S,R*) diastereomeric pair

(Here* represents chiral centre)

(6) *n*-propyl–C(=O)—Et
(7) *Iso*-propyl–C(=O)—Et

$\xrightarrow{NaBH_4}$

*n*-butyl–CH(OH)—Et (±)racemic mixture
*Iso*-propyl–CH(OH)—Et (±)racemic mixture

While in case of (4) and (5) they do not produce enantiomer due to presence of stereogenic centre on ketone.

**20. Plan** *This problem includes concept of hybridisation using VBT, VSEPR theory etc.*

$XeF_4$, $BrF_4^-$, $[Cu(NH_3)_4]^{2+}$, $[PtCl_4]^{2-}$ are square planar as shown below.

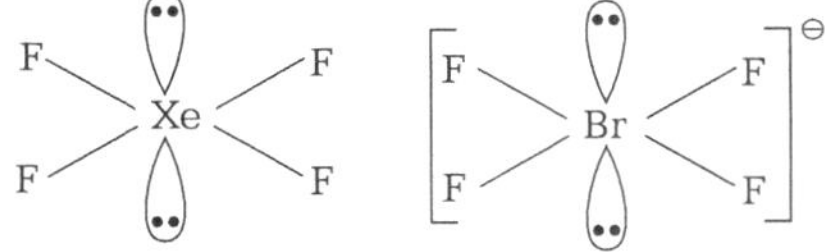

## Mathematics

**1. Plan** *A function f(x) is continuous at x = a, if* $\lim_{x\to a^-} f(x) = \lim_{x\to a^+} f(x) = f(a)$.

*Also, a function f(x) is differentiable at x = a, if* $\lim_{x\to a^-} \frac{f(x)-f(a)}{x-a} = \lim_{x\to a^+} \frac{f(x)-f(a)}{x-a}$

*i.e.,* $f'(a^-) = f'(a^+)$

Given that $f:[a, b] \to [1, \infty)$

and $g(x) = \begin{cases} 0 , & x < a \\ \int_a^x f(t)dt, & a \le x \le b \\ \int_a^b f(t)dt, & x > b \end{cases}$

Now, $g(a^-) = 0 = g(a^+) = g(a)$

[as $g(a^+) = \lim_{x\to a^+} \int_a^x f(t)dt = 0$

and $g(a) = \int_a^a f(t)dt = 0$]

$$g(b^-) = g(b^+) = g(b) = \int_a^b f(t)dt$$

$\Rightarrow g$ is continuous, $\forall x \in R$.

Now, $g'(x) = \begin{cases} 0 , & x < a \\ f(x) , & a < x < b \\ 0 , & x > b \end{cases}$

$g'(a^-) = 0$ but $g'(a^+) = f(a) \ge 1$

[$\because$ Range of $f(x)$ is $[1, \infty)$, $\forall x \in [a, b]$]

$\Rightarrow g$ is non-differentiable at $x = a$

and $g'(b^+) = 0$

but $g'(b^-) = f(b) \ge 1$

$\Rightarrow g$ is not differentiable at $x = b$.

**2. Plan** *If a continuous function has values of opposite sign inside an interval, then it has a root in that interval.*

$$f, g: [0,1] \to R$$

We take two cases.

Let $f$ and $g$ attain their common maximum value at $p$.

$\Rightarrow f(p) = g(p)$,

where $p \in [0, 1]$

Let $f$ and $g$ attain their common maximum value at different points.

$\Rightarrow f(a) = M$ and $g(b) = M$

$\Rightarrow f(a) - g(a) > 0$

and $f(b) - g(b) < 0$

$\Rightarrow f(c) - g(c) = 0$ for some $c \in [0, 1]$ as $f$ and $g$ are continuous functions.

$\Rightarrow f(c) - g(c) = 0$ for some $c \in [0, 1]$ for all cases. ...(i)

Option

(a) $\Rightarrow f^2(c) - g^2(c) + 3[f(c) - g(c)] = 0$ which is true from Eq. (i).

Option (d) $\Rightarrow f^2(c) - g^2(c) = 0$ which is true from Eq. (i)

Now, if we take

$$f(x) = 1 \text{ and } g(x) = 1, \ \forall x \in [0, 1]$$

Option (b) and (c) does not hold.

Hence, option (a) and (d) are correct.

**3. Plan** *A square matrix M is invertible iff* det$(M)$ *or* $|M| \ne 0$.

Let $M = \begin{bmatrix} a & b \\ b & c \end{bmatrix}$

(a) Given that $\begin{bmatrix} a \\ b \end{bmatrix} = \begin{bmatrix} b \\ c \end{bmatrix}$

$\Rightarrow a = b = c = \alpha$ (let)

$\Rightarrow M = \begin{bmatrix} \alpha & \alpha \\ \alpha & \alpha \end{bmatrix}$

$\Rightarrow |M| = 0$

$\Rightarrow M$ is non-invertible.

(b) Given that $[b \ c] = [a \ b]$

$\Rightarrow a = b = c = \alpha$ (let)

Again $|M| = 0$

$\Rightarrow M$ is non-invertible.

(c) As given $M = \begin{bmatrix} a & 0 \\ 0 & c \end{bmatrix}$

$\Rightarrow |M| = ac \ne 0$

($\because$ $a$ and $c$ are non-zero)

$\Rightarrow M$ is invertible.

(d) $M = \begin{bmatrix} a & b \\ b & c \end{bmatrix} \Rightarrow |M| = ac - b^2 \ne 0$

$\because$ $ac$ is not equal to square of an integer.

$\therefore$ $M$ is invertible.

**4. Plan**

(i) ***Dot product*** *let* **a** *and* **b** *be two non-zero vectors inclined at an angle* $\theta$. *Then,* $\mathbf{a} \cdot \mathbf{b} = |\mathbf{a}||\mathbf{b}|\cos\theta$

*(ii) The direction of* $\mathbf{a} \times \mathbf{b}$ *is perpendicular to the plane of* **a** *and* **b**.

*(iii)* ***Vector triple product***

$$\mathbf{a} \times (\mathbf{b} \times \mathbf{c}) = (\mathbf{a} \cdot \mathbf{c})\mathbf{b} - (\mathbf{a} \cdot \mathbf{b})\mathbf{c}$$

➥ Given that

$$|\mathbf{x}| = |\mathbf{y}| = |\mathbf{z}| = \sqrt{2}$$

and angle between each pair is $\frac{\pi}{3}$

$$\therefore \quad \mathbf{x} \cdot \mathbf{y} = |\mathbf{x}||\mathbf{y}| \cos \frac{\pi}{3}$$

$$= (\sqrt{2})\sqrt{2}\left(\frac{1}{2}\right) = 1$$

Similarly, $\mathbf{y} \cdot \mathbf{z} = \mathbf{z} \cdot \mathbf{x} = 1$

Also, $\mathbf{x} \cdot \mathbf{x} = |\mathbf{x}|^2 = 2$

Similarly, $\mathbf{y} \cdot \mathbf{y} = \mathbf{z} \cdot \mathbf{z} = 2$

Now, $\mathbf{a} \perp \mathbf{x}$ and $\mathbf{y} \times \mathbf{z}$

Let $\mathbf{a} = \lambda[\mathbf{x} \times (\mathbf{y} \times \mathbf{z})]$

$$= \lambda[(\mathbf{x} \cdot \mathbf{z})\mathbf{y} - (\mathbf{x} \cdot \mathbf{y})\mathbf{z}]$$

$$\mathbf{a} = \lambda(\mathbf{y} - \mathbf{z})$$

(as $\mathbf{x} \cdot \mathbf{z} = \mathbf{x} \cdot \mathbf{y} = 1$)

$$\mathbf{a} \cdot \mathbf{y} = \lambda(\mathbf{y} - \mathbf{z}) \cdot \mathbf{y}$$

$$\mathbf{a} \cdot \mathbf{y} = \lambda(\mathbf{y} \cdot \mathbf{y} - \mathbf{y} \cdot \mathbf{z}) = \lambda(2 - 1)$$

(as $\mathbf{y} \cdot \mathbf{y} = 2 = \mathbf{y} \cdot \mathbf{z} = 1$)

$$\mathbf{a} \cdot \mathbf{y} = \lambda$$

$$\Rightarrow \quad \mathbf{a} = (\mathbf{a} \cdot \mathbf{y})(\mathbf{y} - \mathbf{z}) \quad \text{...(i)}$$

Also, $\mathbf{b} \perp \mathbf{y}$ and $\mathbf{z} \times \mathbf{x}$

$\therefore$ Let $\mathbf{b} = \mu[\mathbf{y} \times (\mathbf{z} \times \mathbf{x})]$

$$= \mu[(\mathbf{y} \cdot \mathbf{x})\mathbf{z} - (\mathbf{y} \cdot \mathbf{z})\mathbf{x}]$$

$$= \mu(\mathbf{z} - \mathbf{x})$$

(as $\mathbf{y} \cdot \mathbf{x} = \mathbf{y} \cdot \mathbf{z} = 1$)

$$\mathbf{b} \cdot \mathbf{z} = \mu(\mathbf{z} - \mathbf{x}) \cdot \mathbf{z}$$

$$= \mu(\mathbf{z} \cdot \mathbf{z} - \mathbf{x} \cdot \mathbf{z})$$

$$= \mu(2 - 1) = \mu$$

$$\Rightarrow \quad \mathbf{b} = (\mathbf{b} \cdot \mathbf{z})(\mathbf{z} - \mathbf{x}) \quad \text{...(ii)}$$

Now, using Eqs. (i) and (ii), we get

$$\mathbf{a} \cdot \mathbf{b} = (\mathbf{a} \cdot \mathbf{y})(\mathbf{y} - \mathbf{z}) \cdot (\mathbf{b} \cdot \mathbf{z})(\mathbf{z} - \mathbf{x})$$

$$= (\mathbf{a} \cdot \mathbf{y})(\mathbf{b} \cdot \mathbf{z})(\mathbf{y} \cdot \mathbf{z} - \mathbf{y} \cdot \mathbf{x} - \mathbf{z} \cdot \mathbf{z} + \mathbf{z} \cdot \mathbf{x})$$

$$= (\mathbf{a} \cdot \mathbf{y})(\mathbf{b} \cdot \mathbf{z})(1 - 1 - 2 + 1)$$

$$\mathbf{a} \cdot \mathbf{b} = -(\mathbf{a} \cdot \mathbf{y})(\mathbf{b} \cdot \mathbf{z})$$

**5.** **Plan** *(i) Direction ratios of a line joining two points* $(x_1, y_1, z_1)$ *and* $(x_2, y_2, z_2)$ *are* $x_2 - x_1, y_2 - y_1, z_2 - z_1$.

*(ii) If the two lines with direction ratios* $a_1, b_1, c_1; a_2, b_2, c_2$ *are perpendicular, then* $a_1 a_2 + b_1 b_2 + c_1 c_2 = 0$.

➥ Line $L_1$ given by $y = x; z = 1$ can be expressed

$$L_1: \frac{x}{1} = \frac{y}{1} = \frac{z-1}{0}$$

$$\frac{x}{1} = \frac{y}{1} = \frac{z-1}{0} = \alpha \quad \text{(say)}$$

$$\Rightarrow \quad x = \alpha, y = \alpha, z = 1$$

Let the coordinates of $Q$ on $L_1$ be $(\alpha, \alpha, 1)$

Line $L_2$ given by $y = -x, z = -1$ can be expressed as

$$L_2: \frac{x}{1} = \frac{y}{-1} = \frac{z+1}{0}$$

$$\frac{x}{1} = \frac{y}{-1} = \frac{z+1}{0} = \beta \quad \text{(say)}$$

$$\Rightarrow \quad x = \beta, y = -\beta, z = -1$$

Let the coordinates of $R$ on $L_2$ be $(\beta, -\beta, -1)$.

Direction ratios of $PQ$ are $\lambda - \alpha, \lambda - \alpha, \lambda - 1$.

Now, $PQ \perp L_1$

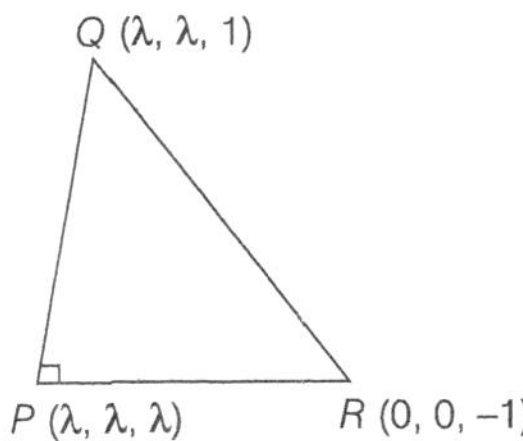


$$\therefore \quad 1(\lambda - \alpha) + 1 \cdot (\lambda - \alpha) + 0 \cdot (\lambda - 1) = 0$$

$$\Rightarrow \quad \lambda = \alpha$$

$$\therefore \quad Q(\lambda, \lambda, 1)$$

Direction ratio of $PR$ are

$$\lambda - \beta, \lambda + \beta, \lambda + 1.$$

Now, $PR \perp L_2$

$$\therefore \quad 1(\lambda - \beta) + (-1)(\lambda + \beta) + 0(\lambda + 1) = 0$$

$$\lambda - \beta - \lambda - \beta = 0$$

$$\Rightarrow \quad \beta = 0$$

$$\therefore \quad R(0, 0, -1)$$

Now, as $\angle QPR = 90^\circ$

(as $a_1a_2 + b_1b_2 + c_1c_2 = 0$,
if two lines with DR's
$a_1, b_1, c_1; a_2, b_2, c_2$ are perpendicular)

$\therefore (\lambda - \lambda)(\lambda - 0) + (\lambda - \lambda)(\lambda - 0) + (\lambda - 1)(\lambda + 1) = 0$

$\Rightarrow (\lambda - 1)(\lambda + 1) = 0$

$\Rightarrow \lambda = 1$ or $\lambda = -1$

$\lambda = 1$, rejected as $P$ and $Q$ are different points.

$\Rightarrow \lambda = -1$

**6. Plan**

*(i) If A and B are two non-zero matrices and $AB = BA$, then $(A-B)(A+B) = A^2 - B^2$.*

*(ii) The determinant of the product of the matrices is equal to product of their individual determinants i.e., $|AB| = |A||B|$.*

➥ Given, $M^2 = N^4$

$\Rightarrow M^2 - N^4 = 0$

$\Rightarrow (M - N^2)(M + N^2) = 0$ (as $MN = NM$)

Also, $M \neq N^2$

$\Rightarrow M + N^2 = 0$

$\Rightarrow \text{Det}(M + N^2) = 0$

Also, $\text{Det}(M^2 + MN^2)$

$= (\text{Det } M)(\text{Det}(M + N^2))$

$= (\text{Det } M)(0) = 0$

As, $\text{Det}(M^2 + MN^2) = 0$

Thus, there exists non-zero matrix $U$ such that $(M^2 + MN^2)U = 0$

**7. Plan** *(i) If $I(x) = \int_{\phi(x)}^{\psi(x)} f(t)\,dt$, then*

$$\frac{d}{dx}[I(x)] = f\{\psi(x)\}\left\{\frac{d}{dx}\psi(x)\right\} - f\{\phi(x)\}\left\{\frac{d}{dx}\phi(x)\right\}$$

*(ii) If $f'(x) > 0$, $\forall x \in [a, b]$, then $f(x)$ is monotonically increasing on $[a, b]$.*

*(iiii) $\int_a^b f(x)\,dx = -\int_b^a f(x)\,dx$*

*(iv) If $f(-x) = -f(x)$, $\forall x \in R$, then $f(x)$ is an odd function of x on R.*

➥ Given, $f(x) = \int_{\frac{1}{x}}^{x} \frac{e^{-\left(t+\frac{1}{t}\right)}}{t}\,dt$

$$f'(x) = 1 \cdot \frac{e^{-\left(x+\frac{1}{x}\right)}}{x} - \left(\frac{-1}{x^2}\right)\frac{e^{-\left(\frac{1}{x}+x\right)}}{1/x}$$

$$= \frac{e^{-\left(x+\frac{1}{x}\right)}}{x} + \frac{e^{-\left(x+\frac{1}{x}\right)}}{x}$$

$$= \frac{2e^{-\left(x+\frac{1}{x}\right)}}{x}$$

As, $f'(x) > 0$, $\forall x \in (0, \infty)$

$\therefore f(x)$ is monotonically increasing on $(0, \infty)$.

$\Rightarrow$ options (a) is correct and (b) is wrong.

$$\text{Now, } f(x) + f\left(\frac{1}{x}\right) = \int_{1/x}^{x} \frac{e^{-\left(t+\frac{1}{t}\right)}}{t}\,dt + \int_{x}^{1/x} \frac{e^{-\left(t+\frac{1}{t}\right)}}{t}\,dt$$

$= 0$, $\forall x \in (0, \infty)$

Now, let

$$g(x) = f(2^x) = \int_{2^{-x}}^{2^x} \frac{e^{-\left(t+\frac{1}{t}\right)}}{t}\,dt$$

$$g(-x) = f(2^{-x}) = \int_{2^x}^{2^{-x}} \frac{e^{-\left(t+\frac{1}{t}\right)}}{t}\,dt = -g(x)$$

$\therefore f(2^x)$ is an odd function.

**8. Plan**

*(i) If $f'(x) > 0$, $\forall x \in (a, b)$, then $f(x)$ is an increasing function in $(a, b)$ and thus $f(x)$ is one-one function in $(a, b)$.*

*(ii) If range of $f(x)$ = codomain of $f(x)$, then $f(x)$ is an onto function.*

*(iii) A function $f(x)$ is said to be odd function, if $f(-x) = -f(x)$, $\forall x \in R$ i.e., $f(-x) + f(x) = 0$, $\forall x \in R$*

➥ $f(x) = [\ln(\sec x + \tan x)]^3$

$$f'(x) = \frac{3\,[\ln(\sec x + \tan x)]^2 \dfrac{(\sec x \tan x + \sec^2 x)}{}}{(\sec x + \tan x)}$$

$$f'(x) = 3\sec x[\ln(\sec x + \tan x)]^2 > 0,$$

$$\forall x \in \left(\frac{-\pi}{2}, \frac{\pi}{2}\right)$$

$f(x)$ is an increasing function.

$\therefore$ $f(x)$ is an one-one function.

$$(\sec x + \tan x) = \tan\left(\frac{\pi}{4} + \frac{x}{2}\right),$$

as $x \in \left(-\frac{\pi}{2}, \frac{\pi}{2}\right)$, then

$$0 < \tan\left(\frac{\pi}{4} + \frac{x}{2}\right) < \infty$$

$$0 < \sec x + \tan x < \infty$$

$$\Rightarrow \quad -\infty < \ln(\sec x + \tan x) < \infty$$

$$-\infty < [\ln(\sec x + \tan x)]^3 < \infty$$

$$\Rightarrow \quad -\infty < f(x) < \infty$$

Range of $f(x)$ is $R$ and thus $f(x)$ is an onto function.

$$f(-x) = [\ln(\sec x - \tan x)]^3$$

$$= \left[\ln\left(\frac{1}{\sec x + \tan x}\right)\right]^3$$

$$f(-x) = -[\ln(\sec x + \tan x)]^3$$

$$f(x) + f(-x) = 0$$

$\Rightarrow f(x)$ is an odd function.

**9. Plan**

*(i) The general equation of a circle is $x^2 + y^2 + 2gx + 2fy + c = 0$ where, centre and radius are given by $(-g, -f)$ and $\sqrt{g^2 + f^2 - c}$, respectively.*

*(ii) If the two circles $x^2 + y^2 + 2g_1x + 2f_1y + c_1 = 0$ and $x^2 + y^2 + 2g_2x + 2f_2y + c_2 = 0$ are orthogonal, then $2g_1g_2 + 2f_1f_2 = c_1 + c_2$.*

➥ Let circle be

$$x^2 + y^2 + 2gx + 2fy + c = 0$$

It passes through (0, 1).

$$\therefore \quad 1 + 2f + c = 0 \qquad \ldots\text{(i)}$$

Orthogonal with

$$x^2 + y^2 - 2x - 15 = 0$$

$$2g(-1) = c - 15$$

$$\Rightarrow \quad c = 15 - 2g \qquad \ldots\text{(ii)}$$

Orthogonal with

$$x^2 + y^2 - 1 = 0$$

$$c = 1 \qquad \ldots\text{(iii)}$$

$$\Rightarrow \quad g = 7 \text{ and } f = -1$$

Centre is $(-g, -f) \equiv (-7, 1)$

$$\text{Radius} = \sqrt{g^2 + f^2 - c}$$

$$= \sqrt{49 + 1 - 1} = 7$$

**10. Plan**

*(i) Concepts of curve tracing are used in this question.*

*(ii) Number of roots are taken out from the curve traced.*

➥ Let $y = x^5 - 5x$

(i) As $x \to \infty, y \to \infty$ and as $x \to -\infty$, $y \to -\infty$

(ii) Also, at $x = 0, y = 0$, thus the curve passes through the origin.

(iii) $\dfrac{dy}{dx} = 5x^4 - 5 = 5(x^4 - 1)$

$$= 5(x^2 - 1)(x^2 + 1)$$

$$= 5(x - 1)(x + 1)(x^2 + 1)$$

+    −1    −    1    +

Now, $\dfrac{dy}{dx} > 0$ in $(-\infty, -1) \cup (1, \infty)$, thus $f(x)$ is increasing in these interval.

Also, $\dfrac{dy}{dx} < 0$ in $(-1, 1)$, thus decreasing in $(-1, 1)$.

(iv) Also, at $x = -1$, $\dfrac{dy}{dx}$ changes its sign from + ve to −ve.

$\therefore$ $x = -1$ is point of local maxima.

Similarly, $x = 1$ is point of local minima.

Local maximum value,

$$y = (-1)^5 - 5(-1) = 4$$

Local minimum value,

$$y = (1)^5 - 5(1) = -4$$

Now, let $y = -a$

As, evident from the graph, if $-a \in (-4, 4)$ *i.e.*, $a \in (-4, +4)$

Then, $f(x)$ has three real roots and if $-a > 4$ or $-a < -4$, then $f(x)$ has one real root.

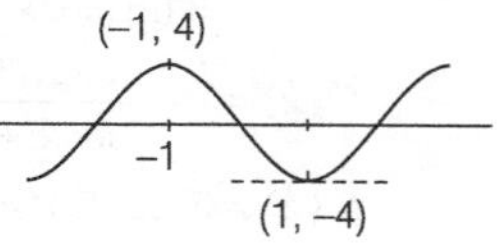


*i.e.*, for $a < -4$ or $a > 4$, $f(x)$ has one real root.

**11. Plan** *Slope of tangent at the point* $(x_1, y_1)$ *is* $\left(\frac{dy}{dx}\right)_{(x_1, y_1)}$.

➥ Given curve $\quad (y - x^5)^2 = x(1 + x^2)^2$

$$\Rightarrow \quad 2(y - x^5)\left(\frac{dy}{dx} - 5x^4\right) = (1 + x^2)^2 + 2x(1 + x^2)\cdot 2x$$

Put $x = 1, y = 3$

$$\therefore \quad \frac{dy}{dx} = 8$$

**12. Plan**

*(i) Using definition of* $f(x) = \cos^{-1}(x)$, *we trace the curve* $f(x) = \cos^{-1}(\cos x)$.

*(ii) The number of solutions of equations involving trigonometric functions and algebraic function, algebaric and algebraic functions are found using graphs of the curves.*

➥ We know, $\cos^{-1}(\cos x) = \begin{cases} x, \text{ if } x \in [0, \pi] \\ 2\pi - x, \text{ if } x \in [\pi, 2\pi] \\ -2\pi + x, \text{ if } x \in [2\pi, 3\pi] \\ 4\pi - x, \text{ if } x \in [3\pi, 4\pi] \end{cases}$

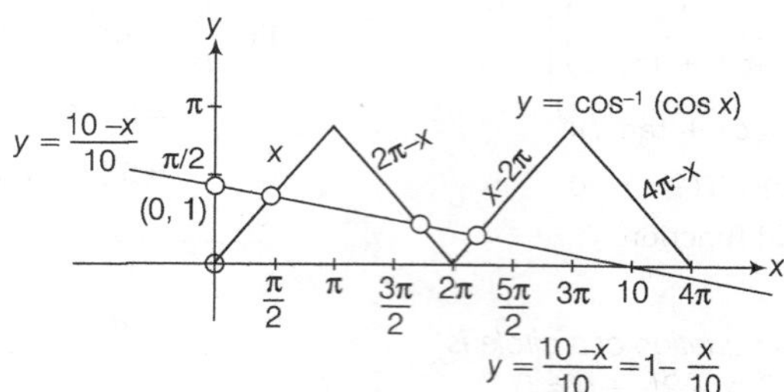


From above figure, it is clear that $y = \frac{10 - x}{10}$ and $y = \cos^{-1}(\cos x)$ intersect at three distinct points, so number of solutions is 3.

**13. Plan** $\lim_{x \to 0} \frac{\sin x}{x} = 1$

➥ Given, $\lim_{x \to 1} \left\{\frac{\sin(x - 1) + a(1 - x)}{(x - 1) + \sin(x - 1)}\right\}^{\frac{(1 + \sqrt{x})(1 - \sqrt{x})}{1 - \sqrt{x}}} = \frac{1}{4}$

$$\lim_{x \to 1} \left\{\frac{\frac{\sin(x - 1)}{(x - 1)} - a}{1 + \frac{\sin(x - 1)}{(x - 1)}}\right\}^{1 + \sqrt{x}} = \frac{1}{4}$$

$$\Rightarrow \quad \left(\frac{1-a}{2}\right)^2 = \frac{1}{4}$$

$$\Rightarrow \quad (a-1)^2 = 1$$

$$\Rightarrow \quad a = 2 \text{ or } 0$$

But for $a = 2$, base of above limit approaches $-1/2$ and exponent approaches to 2 and since base cannot be negative, hence limit does not exist.

**14. Plan**

*(i) In these type of questions, we draw the graph of the function.*

*(ii) The points at which the curve taken a sharp turn, are the points of non-differentiability.*

Curve of $f(x)$ and $g(x)$ are

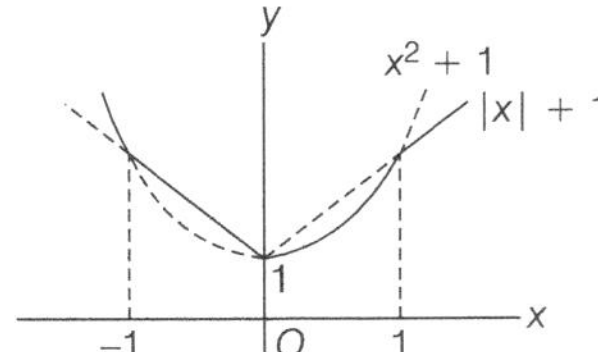


$h(x)$ is not differentiable at $x = \pm 1$ and 0.

As $h(x)$ take sharp turns at $x = \pm 1$ and 0.

Hence, number of points of non-differentiability of $h(x)$ is 3.

**15. Plan** *Distance of a point $(x_1, y_1)$ from $ax + by + c = 0$ is given by $\left|\frac{ax_1 + by_1 + c}{\sqrt{a^2 + b^2}}\right|$.*

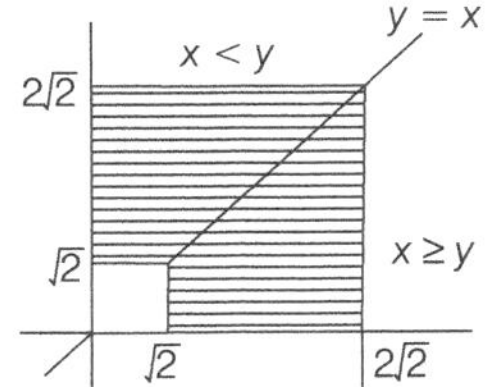


➥ Let $P(x, y)$ is the point in first quadrant.

Now, $2 \le \left|\frac{x-y}{\sqrt{2}}\right| + \left|\frac{x+y}{\sqrt{2}}\right| \le 4$

$$2\sqrt{2} \le |x-y| + |x+y| \le 4\sqrt{2}$$

*Case* I $x \ge y$

$$2\sqrt{2} \le (x-y) + (x+y) \le 4\sqrt{2}$$

$$\Rightarrow \quad x \in [\sqrt{2}, 2\sqrt{2}]$$

*Case* II $x < y$

$$2\sqrt{2} \le y - x + (x+y) \le 4\sqrt{2}$$

$$y \in [\sqrt{2}, 2\sqrt{2}]$$

$$\Rightarrow \quad A = (2\sqrt{2})^2 - (\sqrt{2})^2 = 6$$

**16. Plan** *Reducing the equation to a newer equation, where sum of variables is less. Thus, finding the number of arrangements becomes easier.*

➥ As, $n_1 \ge 1, n_2 \ge 2, n_3 \ge 3, n_4 \ge 4, n_5 \ge 5$

Let $n_1 - 1 = x_1 \ge 0, n_2 - 2 = x_2 \ge 0, \ldots, n_5 - 5 = x_5 \ge 0$

$\Rightarrow$ New equation will be

$$x_1 + 1 + x_2 + 2 + \ldots + x_5 + 5 = 20$$

$$\Rightarrow x_1 + x_2 + x_3 + x_4 + x_5 = 20 - 15 = 5$$

Now, $x_1 \le x_2 \le x_3 \le x_4 \le x_5$

| $x_1$ | $x_2$ | $x_3$ | $x_4$ | $x_5$ |
|---|---|---|---|---|
| 0 | 0 | 0 | 0 | 5 |
| 0 | 0 | 0 | 1 | 4 |
| 0 | 0 | 0 | 2 | 3 |
| 0 | 0 | 1 | 1 | 3 |
| 0 | 0 | 1 | 2 | 2 |
| 0 | 1 | 1 | 1 | 2 |
| 1 | 1 | 1 | 1 | 1 |

So, 7 possible cases will be there.

**17. Plan** *Integration by parts*

$$\int f(x)\, g(x)\, dx = f(x) \int g(x)\, dx - \int \left(\frac{d}{dx}[f(x)] \int g(x)\, dx\right) dx$$

➥ Given $I = \int_0^1 \underset{\text{I}}{4x^3} \frac{d^2}{dx^2} \underset{\text{II}}{(1-x^2)^5}\, dx$

$$= \left[4x^3 \frac{d}{dx}(1-x^2)^5\right]_0^1 - \int_0^1 12\, x^2 \frac{d}{dx}(1-x^2)^5\, dx$$

$$= \left[4x^3 \times 5(1-x^2)^4\ (-2x)\right]_0^1$$

$$- 12\left[[x^2\,(1-x^2)^5]_0^1 - \int_0^1 2x\,(1-x^2)^5 dx\right]$$

$$= 0 - 0 - 12(0 - 0) + 12\int_0^1 2x(1 - x^2)^5\, dx$$

$$= 12 \times \left[-\frac{(1 - x^2)^6}{6}\right]_0^1$$

$$= 12\left[0 + \frac{1}{6}\right] = 2$$

**18. Plan**

(i) $\mathbf{a} \cdot \mathbf{b} = |\mathbf{a}||\mathbf{b}|\cos\theta$

(ii) $[\mathbf{a}\ \mathbf{b}\ \mathbf{c}]^2 = \begin{vmatrix} \mathbf{a}\cdot\mathbf{a} & \mathbf{a}\cdot\mathbf{b} & \mathbf{a}\cdot\mathbf{c} \\ \mathbf{b}\cdot\mathbf{a} & \mathbf{b}\cdot\mathbf{b} & \mathbf{b}\cdot\mathbf{c} \\ \mathbf{c}\cdot\mathbf{a} & \mathbf{c}\cdot\mathbf{b} & \mathbf{c}\cdot\mathbf{c} \end{vmatrix}$

➥ $\mathbf{a}\cdot\mathbf{a} = |\mathbf{a}|^2 = 1$, similarly $\mathbf{b}\cdot\mathbf{b} = \mathbf{c}\cdot\mathbf{c} = 1$

$\mathbf{a}\cdot\mathbf{b} = |\mathbf{a}||\mathbf{b}|\cos\left(\frac{\pi}{3}\right) = \frac{1}{2}$, similarly

$\mathbf{b}\cdot\mathbf{c} = \mathbf{c}\cdot\mathbf{a} = \frac{1}{2}$

We know $[\mathbf{a}\ \ \mathbf{b}\ \ \mathbf{c}]^2 = \begin{vmatrix} \mathbf{a}\cdot\mathbf{a} & \mathbf{a}\cdot\mathbf{b} & \mathbf{a}\cdot\mathbf{c} \\ \mathbf{b}\cdot\mathbf{a} & \mathbf{b}\cdot\mathbf{b} & \mathbf{b}\cdot\mathbf{c} \\ \mathbf{c}\cdot\mathbf{a} & \mathbf{c}\cdot\mathbf{b} & \mathbf{c}\cdot\mathbf{c} \end{vmatrix}$

$$= \begin{vmatrix} 1 & \frac{1}{2} & \frac{1}{2} \\ \frac{1}{2} & 1 & \frac{1}{2} \\ \frac{1}{2} & \frac{1}{2} & 1 \end{vmatrix} = \frac{3}{4} - \frac{1}{4} = \frac{1}{2}$$

$\therefore \quad [\mathbf{a}\ \ \mathbf{b}\ \ \mathbf{c}] = \frac{1}{\sqrt{2}}$ ...(i)

As, given $\mathbf{a} \times \mathbf{b} + \mathbf{b} \times \mathbf{c} = p\mathbf{a} + q\mathbf{b} + r\mathbf{c}$

Take dot product with $\mathbf{a}$,

$$\mathbf{a}\cdot(\mathbf{a}\times\mathbf{b}) + \mathbf{a}\cdot(\mathbf{b}\times\mathbf{c}) = p\mathbf{a}^2 + q\mathbf{b}\cdot\mathbf{a} + r\mathbf{c}\cdot\mathbf{a}$$

$\Rightarrow \quad 0 + \frac{1}{\sqrt{2}} = p + \frac{q}{2} + \frac{r}{2}$ ...(ii)

$\because \quad [\mathbf{a}\ \mathbf{a}\ \mathbf{b}] = 0$

Now, take dot product with $\mathbf{b}$ and $\mathbf{c}$.

$0 = \frac{p}{2} + q + \frac{r}{2}$ ...(iii)

and $\frac{1}{\sqrt{2}} = \frac{p}{2} + \frac{q}{2} + r$ ...(iv)

On subtracting Eq.(ii) from Eq.(iv), we get

$$\frac{p}{2} - \frac{r}{2} = 0 \Rightarrow p = r$$

$\Rightarrow \quad p + q = 0$ [by Eq. (iii)]

$$\therefore \quad \frac{p^2 + 2q^2 + r^2}{q^2} = \frac{p^2 + 2p^2 + p^2}{p^2} = 4$$

**19. Plan**

(i) *If $a, b, c$ are in GP, then they can be taken as $a, ar, ar^2$ where $r$, $(r \neq 0)$ is the common ratio.*

(ii) *Arithmetic mean of $x_1, x_2, \ldots, x_n = \frac{x_1 + x_2 + \ldots + x_n}{n}$*

➥ Let $a, b, c$ are $a, ar, ar^2$, where $r \in N$

Also, $\frac{a + b + c}{3} = b + 2$

$\Rightarrow \quad a + ar + ar^2 = 3(ar) + 6$

$\Rightarrow \quad ar^2 - 2ar + a = 6$

$\Rightarrow \quad (r - 1)^2 = \frac{6}{a}$

$\because 6/a$ must be perfect square and $a \in N$.

$\therefore$ $a$ can be 6 only.

$\Rightarrow r - 1 = \pm 1 \Rightarrow r = 2$

and $\frac{a^2 + a - 14}{a + 1} = \frac{36 + 6 - 14}{7} = 4$

**20. Plan** *Number of line segment joining pair of adjacent point = $n$*

*Number of line segment obtained joining $n$ points on a circle = ${}^nC_2$*

➥ Number of red line segments $= {}^nC_2 - n$

Number of blue line segments $= n$

$\therefore \quad {}^nC_2 - n = n$

$\Rightarrow \quad \frac{n(n-1)}{2} = 2n$

$\Rightarrow \quad n = 5$

## Paper 2

### Physics

**1.**

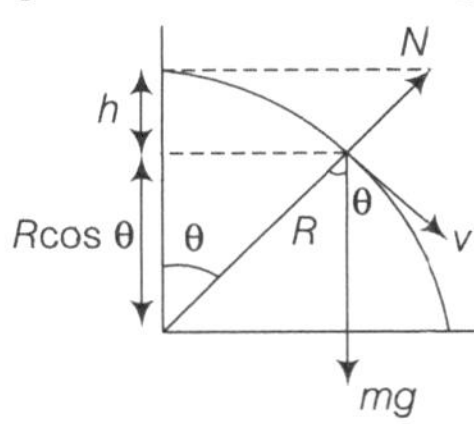


$$K = \frac{1}{2} mg^2t^2$$

$K \propto t^2$ Therefore, $K$-$t$ graph is parabola.

During collision retarding force is just like the spring force ($F \propto x$), therefore kinetic energy first decreases to elastic potential energy and then increases.

Most appropriate graph is therefore (b).

**2.** $h = R - R\cos\theta$

Using conservation of energy

$$mgR(1 - \cos\theta) = \frac{1}{2} mv^2$$

Radial force equation is

$$mg\cos\theta - N = \frac{mv^2}{R}$$

Here, $N$ = normal force on bead by wire

$$N = mg\cos\theta - \frac{mv^2}{R}$$

$$= mg(3\cos\theta - 2)$$

$N = 0$ at $\cos\theta = \frac{2}{3}$

$\Rightarrow$ Normal force act radially outward on bead if $\cos\theta > 2/3$ and normal force act radially inward on bead if $\cos\theta < 2/3$

$\therefore$ Force on ring is opposite to normal force on bead.

**3.** For balanced meter bridge

$$\frac{X}{R} = \frac{l}{(100 - l)} \quad \text{(where, } R = 90\,\Omega)$$

$$\therefore \quad \frac{X}{90} = \frac{40}{100 - 40}$$

$$\therefore \quad X = 60\,\Omega$$

$$X = R\frac{l}{(100 - l)}$$

$$\frac{\Delta X}{X} = \frac{\Delta l}{l} + \frac{\Delta l}{100 - l}$$

$$= \frac{0.1}{40} + \frac{0.1}{60}$$

$$\Delta X = 0.25$$

So, $X = (60 \pm 0.25)\,\Omega$

**4.** $E_1 = \frac{kQ}{R^2}$, where $k = \frac{1}{4\pi\varepsilon_0}$

$$E_2 = \frac{k(2Q)}{R^2} \Rightarrow E_2 = \frac{2kQ}{R^2}$$

$$E_3 = \frac{k(4Q)\,R}{(2R)^3} \Rightarrow E_3 = \frac{kQ}{2R^2}$$

$$E_3 < E_1 < E_2$$

**5.** At point $Q$ angle of incidence is critical angle $\theta_C$, where

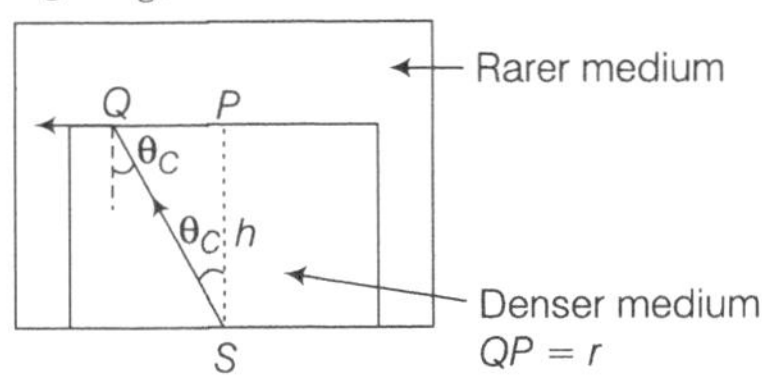


$$\sin\theta_C = \frac{\mu_l}{\mu_{\text{block}}}$$

In $\Delta PQS$, $\sin\theta_C = \frac{r}{\sqrt{r^2 + h^2}}$

$$\therefore \quad \frac{\mu_l}{\mu_{\text{block}}} = \frac{r}{\sqrt{r^2 + h^2}}$$

$$\Rightarrow \quad \mu_l = \frac{r}{\sqrt{r^2 + h^2}} \times 2.72$$

$$= \frac{5.77}{11.54} \times 2.72 = 1.36$$

**6.** In steady state

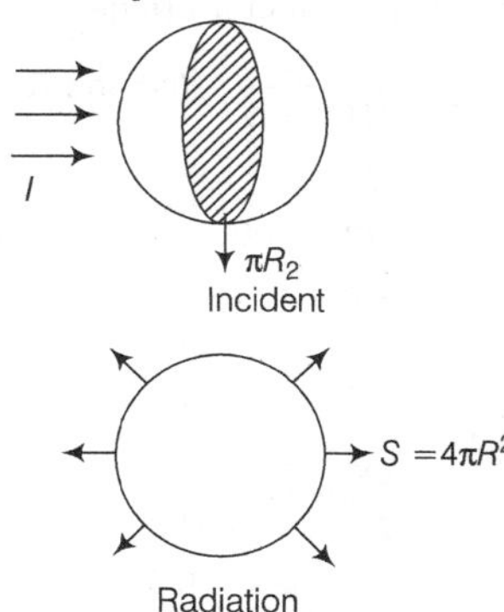


Energy incident per second = Energy radiated per second

$$\therefore \quad I\pi R^2 = \sigma (T^4 - T_0^4)\, 4\pi R^2$$

$$\Rightarrow \quad I = \sigma (T^4 - T_0^4)\, 4$$

$$\Rightarrow \quad T^4 - T_0^4 = 40 \times 10^8$$

$$\Rightarrow \quad T^4 - 81 \times 10^8 = 40 \times 10^8$$

$$\Rightarrow \quad T^4 = 121 \times 10^8$$

$$\Rightarrow \quad T \approx 330 \text{ K}$$

**7.** Energy corresponding to 248 nm wavelength

$$= \frac{1240}{248} \text{ eV} = 5 \text{ eV}$$

Energy corresponding to 310 nm wavelength

$$= \frac{1240}{310} \text{ eV} = 4 \text{ eV}$$

$$\frac{\text{KE}_1}{\text{KE}_2} = \frac{u_1^2}{u_2^2} = \frac{4}{1}$$

$$= \frac{5 \text{ eV} - W}{4 \text{ eV} - W}$$

$$\Rightarrow \quad 16 - 4W = 5 - W$$

$$\Rightarrow \quad 11 = 3W$$

$$\Rightarrow \quad W = \frac{11}{3} = 3.67 \text{ eV} \cong 3.7 \text{ eV}$$

**8.** $K_\alpha$ transition takes place from $n_1 = 2$ to $n_2 = 1$

$$\therefore \quad \frac{1}{\lambda} = R(Z - b)^2 \left[\frac{1}{(1)^2} - \frac{1}{(2)^2}\right]$$

For $K$-series, $b = 1$

$$\therefore \quad \frac{1}{\lambda} \propto (Z - 1)^2$$

$$\Rightarrow \quad \frac{\lambda_{\text{Cu}}}{\lambda_{\text{Mo}}} = \frac{(z_{\text{Mo}} - 1)^2}{(z_{\text{Cu}} - 1)^2}$$

$$= \frac{(42 - 1)^2}{(29 - 1)^2} = \frac{41 \times 41}{28 \times 28}$$

$$= \frac{1681}{784} = 2.144$$

**9.** Given, $R_{\text{planet}} = \dfrac{R_{\text{earth}}}{10}$

and density, $\rho = \dfrac{M_{\text{earth}}}{\frac{4}{3}\pi R_{\text{earth}}^3}$

$$= \frac{M_{\text{planet}}}{\frac{4}{3}\pi R_{\text{planet}}^3}$$

$$\Rightarrow \quad M_{\text{planet}} = \frac{M_{\text{earth}}}{10^3}$$

$$g_{\text{surface of planet}} = \frac{GM_{\text{planet}}}{R_{\text{planet}}^2}$$

$$= \frac{GM_e \cdot 10^2}{10^3 \cdot R_e^2}$$

$$= \frac{GM_e}{10 R_e^2}$$

$$= \frac{g_{\text{surface of earth}}}{10}$$

$$g_{\text{depth of planet}} = g_{\text{surface of planet}} \left(\frac{x}{R}\right)$$

where, $x$ = distance from centre of planet.

$\therefore$ Total force on wire

$$F = \int_{4R/5}^{R} \lambda \, dx \, g\left(\frac{x}{R}\right)$$

$$= \frac{\lambda g}{R}\left[\frac{x^2}{2}\right]_{4R/5}^{R}$$

Here, $g = g_{\text{surface of planet}}$

$R = R_{planet}$

Substituting the given values, we get

$F = 108$ N

**10.** Using geometry

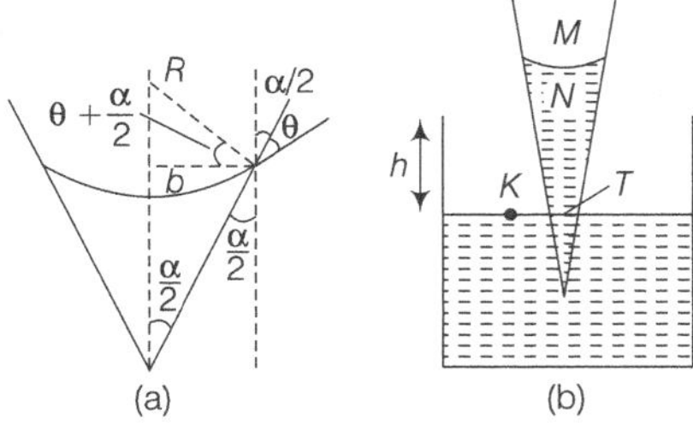


$$\frac{b}{R} = \cos\left(\theta + \frac{\alpha}{2}\right)$$

$$\Rightarrow \quad R = \frac{b}{\cos\left(\theta + \frac{\alpha}{2}\right)}$$

Using pressure equation along the path *MNTK*

$$p_0 - \frac{2S}{R} + h\rho g = p_0$$

Substituting the value of *R*, we get

$$h = \frac{2S}{R\rho g}$$

$$= \frac{2S}{b\rho g}\cos\left(\theta + \frac{\alpha}{2}\right)$$

**11.** Let final equilibrium temperature of gases is $T$

Heat rejected by gas by lower compartment

$$= nC_V\Delta T$$

$$= 2 \times \frac{3}{2} R(700 - T)$$

Heat received by the gas in above compartment

$$= nC_p\Delta T$$

$$= 2 \times \frac{7}{2} R(T - 400)$$

Equating the two, we get

$$2100 - 3T = 7T - 2800$$

$$\Rightarrow \quad T = 490 \text{ K}$$

**12.** $\Delta W_1 + \Delta U_1 = \Delta Q_1$ ...(i)

$\Delta W_2 + \Delta U_2 = \Delta Q_2$ ...(ii)

$\Delta Q_1 + \Delta Q_2 = 0$

$\therefore \quad (nC_p\Delta T)_1 + (nC_p\Delta T)_2 = 0$

But $n_1 = n_2 = 2$

$$\therefore \quad \frac{5}{2} R(T - 700) + \frac{7}{2} R(T - 400) = 0$$

Solving, we get $T = 525$ K

Now, from equations (i) and (ii), we get

$$\Delta W_1 + \Delta W_2 = -\Delta U_1 - \Delta U_2$$

as $\Delta Q_1 + \Delta Q_2 = 0$

$$\therefore \quad \Delta W_1 + \Delta W_2 = -[(nC_V\Delta T)_1 + (nC_V\Delta T)_2]$$

$$= -\left[2 \times \frac{3}{2} R \times (525 - 700) + 2 \times \frac{5}{2} R \times (525 - 400)\right]$$

$$= -100R$$

**13.** From continuity equation,

$$A_1 v_1 = A_2 v_2$$

Here, $A_1 = 400A_2$ because

$r_1 = 20r_2$ and $A = \pi r^2$

$$\therefore \quad v_2 = \frac{A_1}{A_2}(v_1) = 400\, v_1$$

$$= 400(5) \text{ mm/s}$$

$$= 2000 \text{ mm/s} = 2 \text{ m/s}$$

**14.** $p_1 - p_2 = \frac{1}{2}\rho_a v_a^2$

$$p_3 - p_2 = \frac{1}{2}\rho_l v_l^2$$

$$p_3 = p_1$$

$$\therefore \quad \frac{1}{2}\rho_l v_l^2 = \frac{1}{2}\rho_a v_a^2$$

$$\Rightarrow \quad v_l = \sqrt{\frac{\rho_a}{\rho_l}}\, v_a$$

$\therefore$ Volume flow rate $\propto \sqrt{\frac{\rho_a}{\rho_l}}$

**15.** $B_R = B$ due to ring

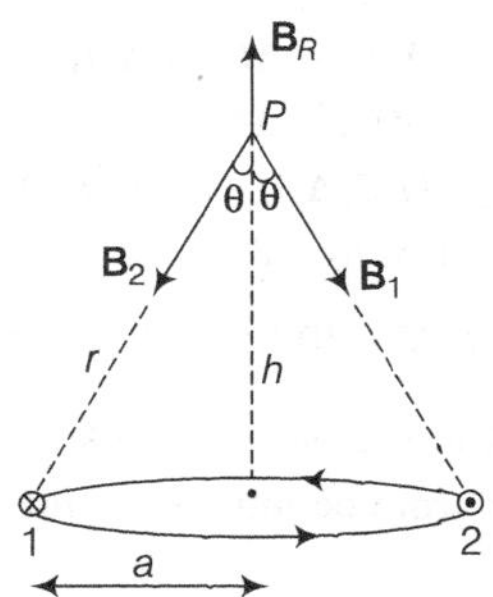


$B_1 = B$ due to wire-1

$B_2 = B$ due to wire-2

In magnitudes

$$B_1 = B_2 = \frac{\mu_0 I}{2\pi r}$$

Resultant of $B_1$ and $B_2$

$$= 2B_1 \cos\theta = 2\left(\frac{\mu_0 I}{2\pi r}\right)\left(\frac{h}{r}\right)$$

$$= \frac{\mu_0 I h}{\pi r^2}$$

$$B_R = \frac{\mu_0 I R^2}{2(R^2 + x^2)^{3/2}}$$

$$= \frac{2\mu_0 I \pi a^2}{4\pi r^3}$$

As, $R = a$, $x = h$ and $a^2 + h^2 = r^2$

For zero magnetic field at $P$,

$$\frac{\mu_0 I h}{\pi r^2} = \frac{2\mu_0 I \pi a^2}{4\pi r^3}$$

$$\Rightarrow \quad \pi a^2 = 2rh \Rightarrow h \approx 1.2a$$

**16.** Magnetic field at mid-point of two wires

= 2 (magnetic field due to one wire)

$$= 2\left[\frac{\mu_0}{2\pi}\frac{I}{d}\right] = \frac{\mu_0 I}{\pi d} \otimes$$

Magnetic moment of loop

$$M = IA = I\pi a^2$$

Torque on loop $= MB \sin 150°$

$$= \frac{\mu_0 I^2 a^2}{2d}$$

**17.** (P) Component of forces along $x$-axis will vanish. Net force along positive $y$-axis

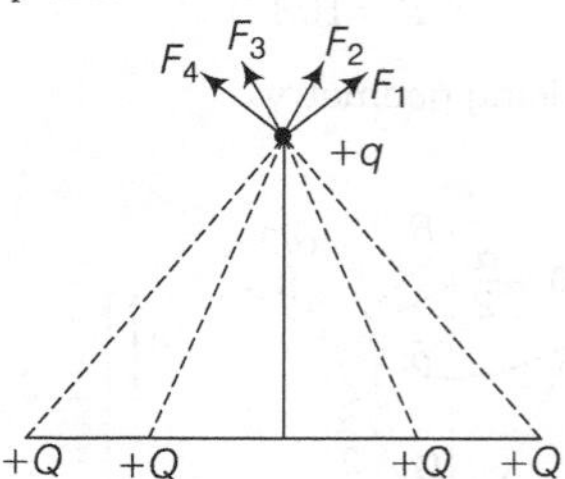


(Q) Component of forces along $y$-axis will vanish. Net force along positive $x$-axis

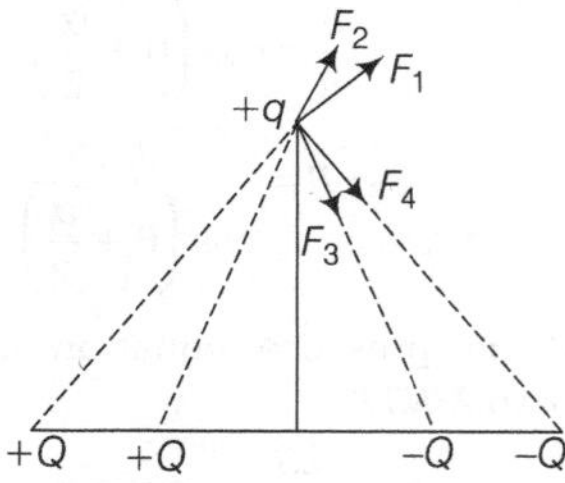


(R) Component of forces along $x$-axis will vanish. Net force along negative $y$-axis

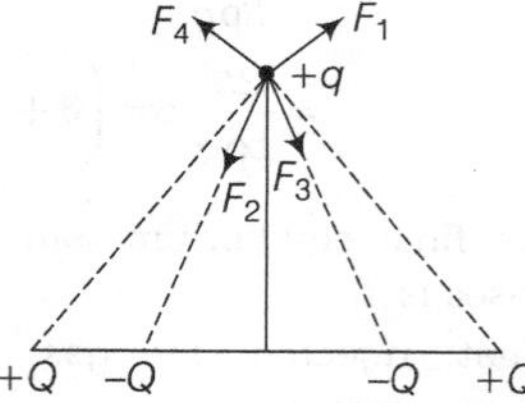


(S) Component of forces along $y$-axis will vanish. Net force along negative $x$-axis

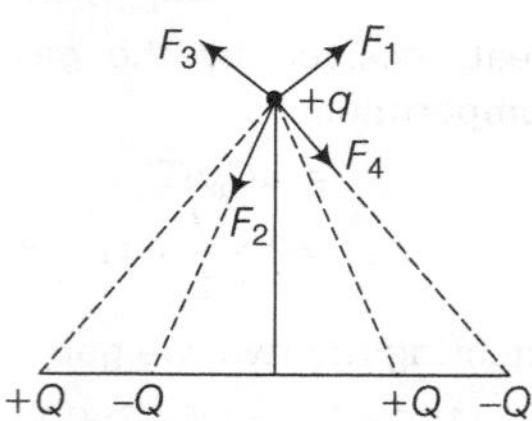

**18.** (P) $\frac{1}{f} = \left(\frac{3}{2} - 1\right)\left(\frac{1}{r} + \frac{1}{r}\right) = \frac{1}{r}$

$\Rightarrow \quad f = r$

$\Rightarrow \quad \frac{1}{f_{eq}} = \frac{1}{f} + \frac{1}{f} = \frac{2}{r}$

$\Rightarrow \quad f_{eq} = \frac{r}{2}$

(Q) $\frac{1}{f} = \left(\frac{3}{2} - 1\right)\left(\frac{1}{r}\right) \quad \Rightarrow f = 2r$

$\Rightarrow \quad \frac{1}{f} + \frac{1}{f} = \frac{2}{f} = \frac{1}{r}$

$\Rightarrow \quad f_{eq} = r$

(R) $\frac{1}{f} = \left(\frac{3}{2} - 1\right)\left(-\frac{1}{r}\right) = -\frac{1}{2r}$

$\Rightarrow \quad f = -2r$

$\Rightarrow \quad \frac{1}{f_{eq}} = \frac{1}{f} + \frac{1}{f} = -\frac{2}{2r}$

$\Rightarrow \quad f_{eq} = -r$

(S) $\frac{1}{f_{eq}} = \frac{1}{r} + \frac{1}{-2r} = \frac{1}{2r}$

$\Rightarrow \quad f_{eq} = 2r$

**19.** Block will not slip if

$(m_1 + m_2)\, g \sin\theta \le \mu m_2 g \cos\theta$

$\Rightarrow \quad 3 \sin\theta \le \left(\frac{3}{10}\right)(2) \cos\theta$

$\tan\theta \le \frac{1}{5} \quad \Rightarrow \quad \theta \le 11.5°$

(P) $\theta = 5°$ friction is static

$f = (m_1 + m_2)\, g \sin\theta$

(Q) $\theta = 10°$ friction is static

$f = (m_1 + m_2)\, g \sin\theta$

(R) $\theta = 15°$ friction is kinetic

$f = \mu m_2 g \cos\theta$

(S) $\theta = 20°$ friction is kinetic

$\Rightarrow f = \mu m_2 g \cos\theta$

**20.** $d = 2\sqrt{h_1 h_2} = \sqrt{4h_1 h_2}$

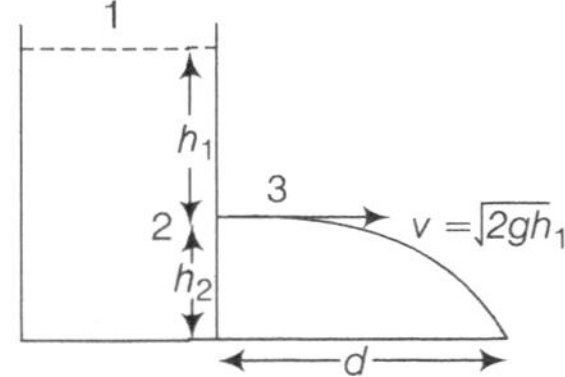


This is independent of the value of $g$.

(P) $g_{eff} > g \Rightarrow d = \sqrt{4h_1 h_2} = 1.2$ m

(Q) $g_{eff} < g \Rightarrow d = \sqrt{4h_1 h_2} = 1.2$ m

(R) $g_{eff} = g \Rightarrow d = \sqrt{4h_1 h_2} = 1.2$ m

(S) $g_{eff} = 0$

No water leaks out of jar. As there will be no pressure difference between top of the container and any other point. $p_1 = p_2 = p_3 = p_0$

## Chemistry

**1.** **Plan** *This problem can be solved by using the concept involved in molecular orbital theory. Write the molecular orbital electronic configuration keeping in mind that there is no 2s-2p mixing, then if highest occupied molecular orbital contain unpaired electron then molecule is paramagnetic otherwise diamagnetic.*

➥ Assuming that no $2s$-$2p$ mixing takes place the molecular orbital electronic configuration can be written in the following sequence of energy levels of molecular orbitals

$\sigma 1s, \sigma^* 1s, \sigma 2s, \sigma^* 2s, \sigma 2p_z, \pi 2p_x \equiv \pi 2p_y, \pi^* 2p_x \equiv \pi^* 2p_y, \sigma^* 2p_z$

(a) $Be_2 \rightarrow \sigma 1s^2, \sigma^* 1s^2, \sigma 2s^2, \sigma^* 2s^2$ (diamagnetic)

(b) $B_2 \rightarrow \sigma 1s^2, \sigma^* 1s^2, \sigma 2s^2, \sigma^* 2s^2, \sigma 2p_z^2, \begin{matrix} \pi 2p_x^0 \\ \pi 2p_y^0 \end{matrix}$

(diamagnetic)

(c) $C_2 \rightarrow \sigma 1s^2, \sigma^* 1s^2, \sigma 2s^2, \sigma^* 2s^2, \sigma 2p_z^2, \begin{matrix}\pi 2p_x^1\\ \pi 2p_y^1\end{matrix}, \begin{matrix}\pi^* 2p_x^0\\ \pi^* 2p_y^0\end{matrix}, \sigma^* 2p_z^0$

(paramagnetic)

(d) $N_2 \rightarrow \sigma 1s^2, \sigma^* 1s^2, \sigma 2s^2, \sigma^* 2s^2, \sigma 2p_z^2, \begin{matrix}\pi 2p_x^2\\ \pi 2p_y^2\end{matrix}, \begin{matrix}\pi^* 2p_x^0\\ \pi^* 2p_y^0\end{matrix}, \sigma^* 2p_z^0$

(diamagnetic)

Hence, (c) is the correct choice.

**2. Plan** *This problem is based on assumption that total entropy change of universe is zero.*

➡ At 100°C and 1 atmosphere pressure, $H_2O(l) \rightleftharpoons H_2O(g)$ is at equilibrium.

For equilibrium, $\Delta S_{total} = 0$ and $\Delta S_{system} + \Delta S_{surrounding} = 0$

As we know during conversion of liquid to gas entropy of system increases, in a similar manner entropy of surrounding decreases.

$\therefore \quad \Delta S_{system} > 0$

and $\Delta S_{surrounding} < 0$

Hence, (b) is the correct choice.

**3.** For the elementary reaction, $M \rightarrow N$

Rate law can be written as

$$\text{Rate} \propto [M]^n$$

$$\text{Rate} = k[M]^n \quad \ldots(i)$$

When we double the concentration of $[M]$, rate becomes 8 times, hence new rate law can be written as

$$8 \times \text{Rate} = k[2M]^n \quad \ldots(ii)$$

$$\frac{\text{Rate}}{8 \times \text{Rate}} = \frac{k[M]^n}{k[2M]^n}$$

$$\Rightarrow \quad \frac{1}{8} = \frac{1}{[2]^n} \Rightarrow [2]^n = 8 = [2]^3$$

$$n = 3$$

$\therefore$ Hence, (b) is the correct choice.

**4. Plan** *This problem can be solved by using the concept of synthesis of dye using electrophilic aromatic substitution reaction.*

➡ In basic (alkaline) solution naphthol exists as naphthoxide ion which is a strong *o, p*-directing group.

OH (2-naphthol) —KOH→ $O^{\ominus}$ (naphthoxide) ≡ resonance structure with -δ, -δ, -δ, -δ

Thus, formation of dye can be shown as

OH (2-naphthol) —$[Ph—\overset{\oplus}{N}\equiv N]Cl^-$ / alkaline solution→ 1-(N=N—Ph)-2-naphthol (OH)

Thus, (d) is the correct choice.

**5. Plan** *This problem is based on boiling point of isomeric alkanes.*

➡ As we know more the branching in an alkane, lesser will be its surface area and lesser will be the boiling point

**On moving left to right** (III to I)

- branching increases
- surface area decreases
- boiling point decreases

Hence, correct choice is (b).

**6. Plan** *This problem includes concept of nucleophilic addition reaction to carbonyl compound (ketone here) and intramolecular nucleophilic substitution reaction.*

➡ Complete reaction sequence is as shown below

$$Cl-CH_2CH_2CH_2-CO-CH_3 \xrightarrow[\text{Nucleophilic addition reaction}]{CH_3MgBr,\ \text{dry ether, } 0^\circ C} Cl-CH_2CH_2CH_2-C(OMgX)(CH_3)(CH_3)$$

$$\xrightarrow[\text{intramolecular nucleophilic substitution reaction}]{aq.\ \text{acid}} \text{2,2-dimethyltetrahydrofuran} + MgxCl$$

Hence, correct choice is (d).

**7. Plan** *This problem can be solved by using concept involved in chemical properties of xenon oxide and xenon fluoride.*

➥ $XeF_6$ on complete hydrolysis produces $XeO_3$.

$XeO_3$ on reaction with $OH^-$ produces $HXeO_4^-$ which on further treatment with $OH^-$. undergo slow disproportionation reaction and produces $XeO_6^{4-}$ along with $Xe(g)$, $H_2O(l)$ and $O_2(g)$ as a by-product.

Oxidation half-cell in basic aqueous solution

$$HXeO_4^- + 5OH^- \longrightarrow XeO_6^{4-} + 3H_2O + 2e$$

Reduction half-cell in basic aqueous solution

$$HXeO_4^- + 3H_2O + 6e^- \longrightarrow Xe + 7OH^-$$

Balanced overall disproportionation reaction is

$$4HXeO_4^- + 8OH^- \longrightarrow \underbrace{3XeO_6^{4-} + Xe}_{\text{2 products}} + 6H_2O$$

Complete sequence of reaction can be shown as

$$XeF_6 + 3H_2O \longrightarrow XeO_3 + 3H_2F_2 \xrightarrow{OH^-} HXeO_4^-$$

$$\downarrow OH^-/H_2O \text{ (disproportionation)}$$

$$XeO_6^{4-}(s) + Xe(g) + H_2O(l) + O_2(g)$$

**8. Plan** *This problem is based on chemical properties of phosphorus.*

➥ White phosphorus on reaction with thionyl chloride ($SOCl_2$) produces phosphorus trichloride

$$P_4(s) + 8SOCl_2(l) \longrightarrow 4PCl_3(l) + 4SO_2(g) + 2S_2Cl_2(g)$$

But if amount of thionyl chloride ($SOCl_2$) is in excess then it produces phosphorus pentachloride.

$$P_4 + 10SOCl_2(l) \longrightarrow 4PCl_5 + 10SO_2$$

Hence, (a) is the correct choice.

**9. Plan** *This problem can be solved by using concept of oxidant and reductant.*

***Oxidant*** *Oxidant increases the oxidation number of the species with which it is reacted.*

***Reductant*** *Reductant decreases the oxidation number of the species with which it is reacted.*

➥ $H_2O_2$ reacts with $KIO_4$ in the following manner

$$\overset{+7}{KIO_4} + H_2O_2 \longrightarrow \overset{+5}{KIO_3} + H_2O + O_2$$

On reaction of $KIO_4$ with $H_2O_2$, oxidation state of I varies from +7 to +5 *i.e.*, decreases, Thus, $KIO_4$ Thus, gets reduced hence, $H_2O_2$ is a reducing agent here.

With $NH_2OH$, it given following reaction

$$\overset{-1}{NH_2}OH + H_2O_2 \longrightarrow \overset{+3}{N_2}O_3 + H_2O$$

In the above reaction, oxidation state of N varies from −1 to +3. Here, oxidation number increases hence, $H_2O_2$ is acting as an oxidising agent here.
Hence, (a) is the correct choice.

**10.** **Plan** *This problem can be solved by using the concept of stability of carbocation and $S_N1$ reaction.*
When two phenyl groups are replaced by two *para* methoxy group, carbocation formed will be more stable. As the stability of carbocation formed increases, rate of acidic hydrolysis increases.

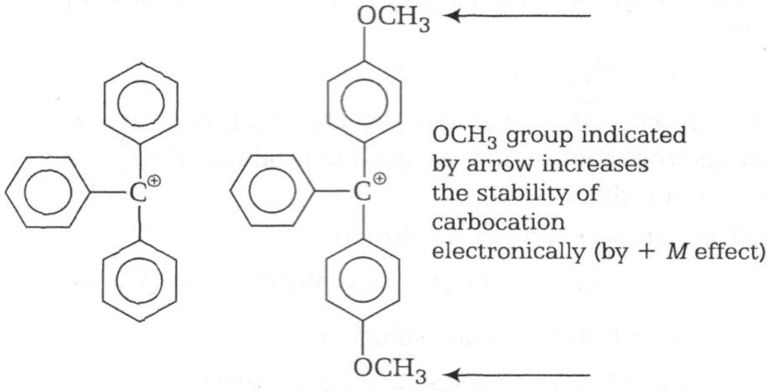

$OCH_3$

$C^{\oplus}$ $C^{\oplus}$

$OCH_3$ group indicated by arrow increases the stability of carbocation electronically (by + $M$ effect)

$OCH_3$

More stable carbocation

Hence, (c) is the correct choice.

**11.** **Plan** *This problem can be solved by using the concept of Graham's law of diffusion according to which rate of diffusion of non-reactive gases under similar conditions of temperature and pressure are inversely proportional to square root of their density.*

$$\text{Rate of diffusion} \propto \frac{1}{\sqrt{\text{molar weight of gas}}}$$

➥ Let distance covered by $X$ is $d$, then distance covered by $Y$ is 24-d.
If $r_X$ and $r_Y$ are the rate of diffusion of gases $X$ and $Y$,

$$\frac{r_X}{r_Y} = \frac{d}{24-d} = \sqrt{\frac{40}{10}} = 2 \qquad [\because \text{Rate of diffusion } \alpha \text{ distance travelled}]$$

$$d = 48 - 2d$$

$$\Rightarrow \qquad 3d = 48 \Rightarrow d = 16 \text{ cm}$$

Hences,(c) is the correct choice.

**12.** Collision frequency is the measurement of number of molecular collision taking place per second per unit volume of a gas.
As the collision frequency increases, molecular size increases and hence, covered distance decreases than the expected value.
Since, speed is higher for $x$ so it produces more collisions per second, *i.e.*, more collision frequency and hence.
covers lesser distance than that predicted by Graham's law.
Hence, (d) is the correct choice.

**13.** **Plan** *This problem can be solved by using the concept of nucleophilic substitution reaction, oxidation reaction and reduction reaction including strength of nucleophile and regioselectivity.*

Reaction of Scheme 1 can be completed as

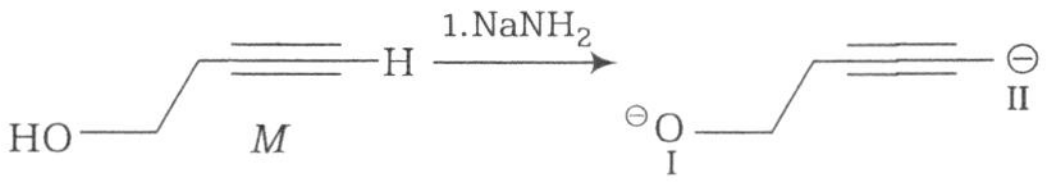

1.NaNH$_2$

HO M H I II

Among two nacked nucleophilic group I and II, II is more nucleophilic and then will react selectively as follows

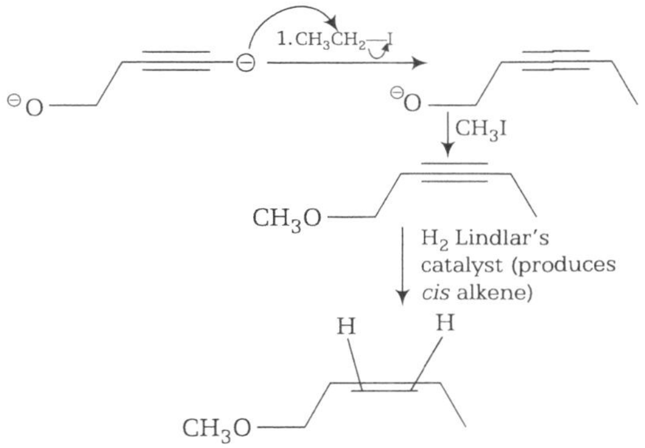

1.CH$_3$CH$_2$—I

CH$_3$I

CH$_3$O

H$_2$ Lindlar's catalyst (produces *cis* alkene)

H H

CH$_3$O

Hence, using the concept of regioselectivity we come on the conclusion that final product is correctly represented by structure (a).

**14. Plan** *This problem can be solved by using the concept of iodoform test and functional isomerism.*

**Iodoform test** The compound containing—COCH$_3$ or —CH(OH) group will undergo iodoform test.

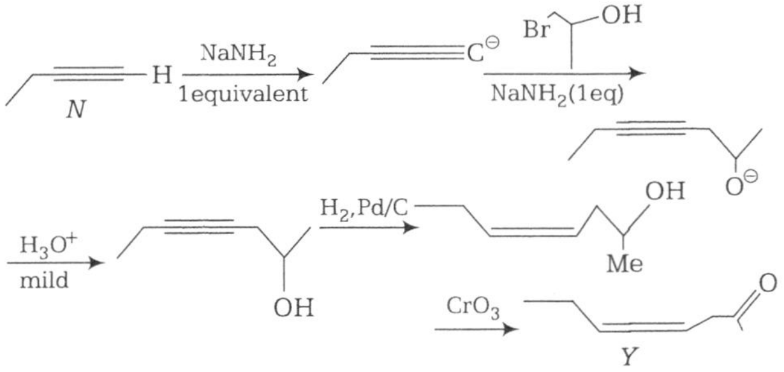

N H NaNH$_2$ 1equivalent C$^{\ominus}$ Br OH NaNH$_2$(1eq)

O$^{\ominus}$

H$_3$O$^+$ mild OH H$_2$,Pd/C OH Me

CrO$_3$ O Y

Thus, *X* and *Y* are functional isomers of each other and *Y* gives iodoform test due to the presence of CH$_3$CO group as Indicated.

Hence, correct choice is (c).

**15. Plan** *This problem can be solved by using concept of chemical reactions of transition metal ions (,) colour and structure of transition metal compounds.*

➡ Here, among given four option Ni$^{2+}$ and Zn$^{2+}$ has ability to form tetrahedral as well as square planar complex depending upon types of reagent used.

Ni$^{2+}$ on reaction with KCN forms square planar complex $[Ni(CN)_4]^{2-}$ due to strong field strength of CN.

$$Ni^{2+} + KCN \longrightarrow \underset{\text{Square planar}}{[Ni(CN)_4]^{2-}}$$

While on reaction with HCl, $Ni^{2+}$ forms stable tetrahedral complex $[Ni(Cl)_4]^{2-}$.

$Zn^{2+}$, on the other hand, on reaction with KCN as well as HCl produces tetrahedral complex because of its $d^{10}$ electronic configuration.

$$Zn^{2+} \xrightarrow{HCl} [ZnCl_4]^{2-}$$

$$Zn^{2+} \xrightarrow{KCN} [Zn(CN)_4]^{2-}$$

$$Zn^{2+} \xrightarrow[\text{excess}]{KOH} \underset{\text{White ppt}}{[Zn(OH)_4]^{2-}}$$

Complete reaction sequence can be shown as

$$\underset{\text{Tetrahedral}}{[NiCl_4]^{2-}} \xleftarrow[\text{excess}]{HCl(Q)} \underset{(M_1)}{Ni^{2+}} \xrightarrow[\text{excess}]{KCN(R)} \underset{\text{Square planar}}{[Ni(CN)_4]^{2-}}$$

$$\underset{\text{Tetrahedral}}{[ZnCl_4]^{2-}} \xleftarrow[\text{excess}]{HCl(Q)} \underset{(M_2)}{Zn^{2+}} \xrightarrow[\text{excess}]{KCN(R)} \underset{\text{Tetrahedral}}{[Zn(CN)_4]^{2-}}$$

$$Zn^{2+} \xrightarrow{KOH(s)} \underset{\text{White ppt.}}{Zn(OH)_2} \xrightarrow[\text{excess}]{KOH} \underset{\text{Soluble}}{[Zn(OH)_4]^{2-}}$$

**16.** $Zn^{2+}$ on treatment with excess of KOH produces $[Zn(OH)_4]^{2-}$ (Also refer to Q. 15).

**17. Plan** *This problem is based on concept of VBT and magnetic properties of coordination compound.*

*Draw VBT for each coordination compound.*

*If unpaired electron is present then coordination compound will be paramagnetic otherwise diamagnetic.*

*Coordination compounds of $[MA_4B_2]$ type show geometrical isomerism.*

➥ Molecular orbital electronic configuration (MOEC) for various coordination compound can be drawn using VBT as

MOEC for $[Cr(NH_3)_4Cl_2]Cl$ is

| 3d | | | | | 4s | 4p | | |
|---|---|---|---|---|---|---|---|---|
| 1 | 1 | 1 | ×× | ×× | ×× | ×× | ×× | ×× |
| | | | $NH_3$ | $NH_3$ | $NH_3$ | $NH_3$ | $NH_3$ | $NH_3$ |

Number of unpaired electrons $(n) = 3$

Magnetic properties = paramagnetic

Geometrical isomers of $[Cr(NH_3)_4Cl_2]^+$ are

cis: $[Cr(NH_3)_4Cl_2]^+$ — Cr with Cl (top), $H_3N$, $H_3N$ (left), Cl, $NH_3$ (right), $NH_3$ (bottom)

and

trans: $[Cr(NH_3)_4Cl_2]^+$ — Cr with Cl (top), $H_3N$, $H_3N$ (left), $NH_3$, $NH_3$ (right), Cl (bottom)

*cis* *trans*

MOEC of $[Ti(H_2O)_5Cl](NO_3)_2$ is

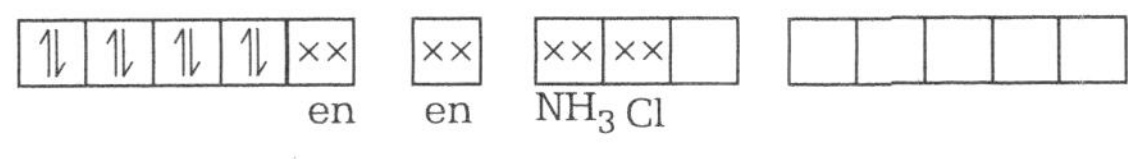


$$n = 1$$

Magnetic properties = paramagnetic

Ionisation isomers of $[Ti(H_2O)_5Cl](NO_3)_2$ are
$[Ti(H_2O)_5Cl](NO_3)_2$ and $[Ti(H_2O)_5(NO_3)]Cl(NO_3)$

MOEC of $[Pt(en)(NH_3)Cl]NO_3$ is

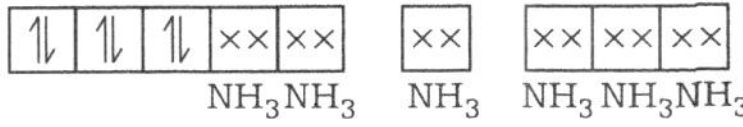


$$n = 0$$

Magnetic property = diamagnetic

Ionisation isomers are $[Pt(en)(NH_3)Cl]NO_3$ and $[Pt(en)NH_3(NO_3)]Cl$

MOEC of $[Co(NH_3)_4(NO_3)_2]NO_3$ is

NH3 NH3 NH3 NH3 NH3 NH3

$$n = 0$$

Magnetic property = Diamagnetic

Geometrical isomers are

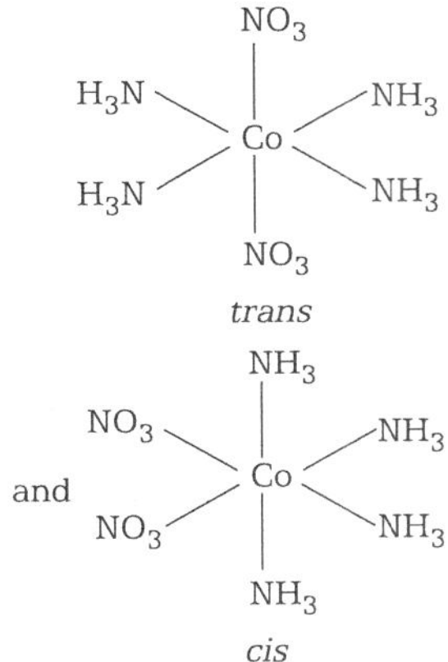


*trans*

and

NH3
NO3
NH3
Co
NO3
NH3
NH3

*cis*

Thus, magnetic property and isomerism in given coordination compound can be summarised as

(P) $[Cr(NH_3)_4Cl_2]Cl \rightarrow$ Paramagnetic and exhibits *cis-trans* isomerism (3)

(Q) $[Ti(H_2O)_5Cl](NO_3)_2 \rightarrow$ Paramagnetic and exhibits ionisation isomerism (1)

(R) $[Pt(en)(NH_3)Cl]NO_3 \rightarrow$ Diamagnetic and exhibits ionisation isomerism (4)

(S) $[Co(NH_3)_4(NO_3)_2]NO_3 \rightarrow$ Diamagnetic and exhibits *cis-trans* isomerism (2)

$\therefore$ P $\rightarrow$ 3, Q $\rightarrow$ 1, R $\rightarrow$ 4, S $\rightarrow$ 2

Hence, (b) is the correct choice.

**18. Plan** *This problem includes basic concept of bonding. It can be solved by using the concept of molecular orbital theory.*

Any orbital has two phase +ve and –ve. In the following diagram +ve phase is shown by darkening the lobes and –ve by without darkening the lobes.

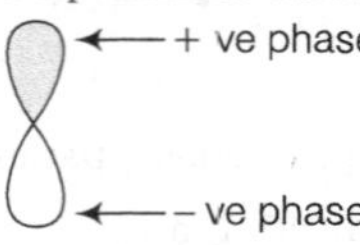


When two same phase overlap with each other it forms bonding molecular orbital otherwise antibonding.

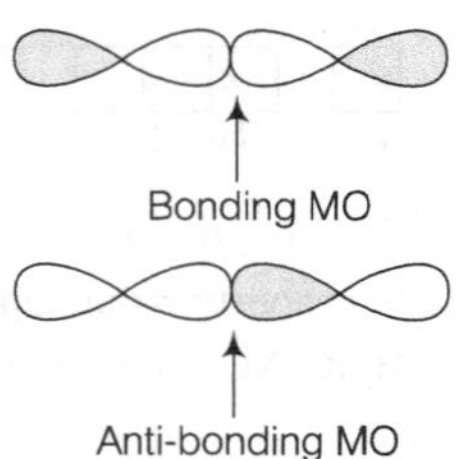


Axial overlapping leads to the formation of $\sigma$-bond and sideways overlapping leads to the formation of $\pi$-bond.

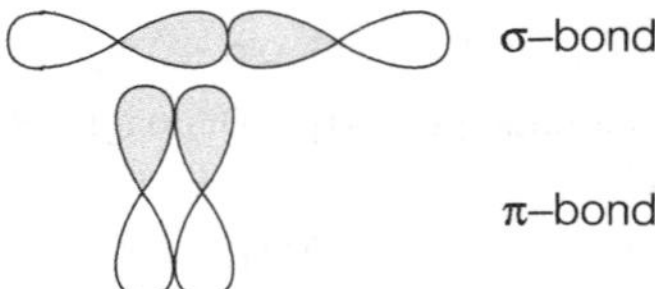


On the basis of above two concepts correct matching can be done as shown below

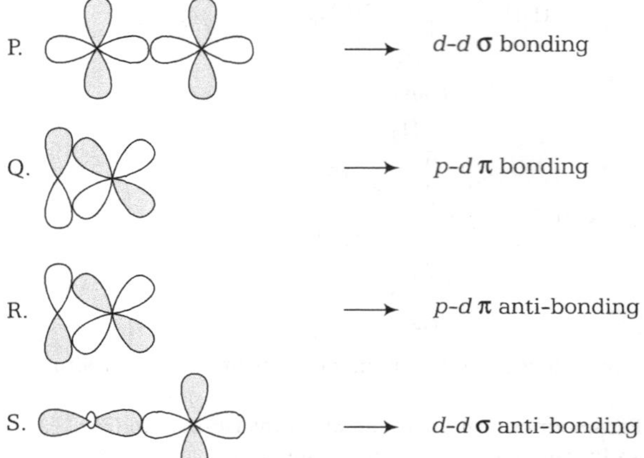


$\therefore$ P $\rightarrow$ 2, Q $\rightarrow$ 3, R $\rightarrow$ 1, S $\rightarrow$ 4

Hence, (c) is the correct option.

**19. Plan** *This problem can be solved by using the stability of radical obtained after fragmentation of peroxyester.*

*Allylic radical are more stable than alkyl radical, so when there is a possibility of formation of allyl radical, it will undergo fragmentation through formation of allyl radical. i.e., fragmentation produces stable radical.*

On the basis of stability of radical, fragmentation can be done as

| Column I | Column II | Explanation |
|---|---|---|
| P. | 1. $C_6H_5H_2C-C(=O)-O-O-CH_3$ | $C_6H_5-\dot{C}H_2+CO_2+CH_3\dot{O}$ |
| Q. | 3. $C_6H_5H_2C-C(=O)-O-O-C(CH_3)_2(CH_2C_6H_5)$ | $C_6H_5-\dot{C}H_2+CO_2+Ph-CH_2-C(O^{\bullet})(CH_3)-CH_3$ → $Ph-\dot{C}H_2+CH_3-CO-CH_3$ |
| R. | 4. $C_6H_5-C(=O)-O-O-C(CH_3)_2(C_6H_5)$ | $C_6H_5-\dot{C}O_2+CH_3-C(O^{\bullet})(C_6H_5)-CH_3$ $\xrightarrow{-CO_2}$ $\dot{P}h+CH_3-CO-Ph+\dot{C}H_3$ |
| S. | 2. $C_6H_5-C(=O)-O-O-CH_3$ | $C_6H_5-\dot{C}O_2+\dot{C}H_3O$; $C_6H_5\dot{C}O_2$ → $C_6\dot{H}_5+CO_2$ |

$P \to 1, Q \to 3, R \to 4, S \to 2$

Thus, (a) is the correct choice.

**20. Plan** *This problem can be solved by using the various concepts synthesis of benzene, electrophilic substitution reaction and directive influence of various substituents, including oxidation and reduction.*

$\Rightarrow$ —OH *and* —$NH_2$ *are o/p-directing groups.*

$\Rightarrow$ N-*acetylation is more favourable than* C-*acylation.*

$\Rightarrow$ N-*sulphonation is more favourable than* C-*sulphonation.*

$\Rightarrow$ $NO_2$ *is a meta-directing group.*

$\Rightarrow$ $H_2S \cdot NH_3$ *reduces only one* $NO_2$ *group selectively in the presence of two* $NO_2$ *groups.*

Using above concepts the correct sequence of reaction can be written as

P. $3CH \equiv CH \xrightarrow[\text{iron 873K}]{\text{Red hot}}$ benzene $\xrightarrow[\Delta]{\text{Fuming } HNO_3/H_2SO_4}$ 1,3-dinitrobenzene $\xrightarrow[\text{(selective reduction)}]{H_2S \cdot NH_3}$ 3-nitroaniline ($NH_2$, $NO_2$) $\xrightarrow[H_2SO_4]{NaNO_2}$ ($\overset{+}{N_2}HSO_4^-$, $NO_2$) $\xrightarrow{\text{Hydrolysis}}$ 3-nitrophenol ($OH$, $NO_2$)

Here N-sulphonation is more favourable in comparisons to C-sulphonation.

Q. Resorcinol ($OH$, $OH$) $\xrightarrow[60°C]{\text{Conc.}H_2SO_4}$ ($OH$, $OH$, $SO_3H$) $\xrightarrow{\overset{+}{N}O_2}$ ($O_2N$, $OH$, $OH$, $SO_3H$) $\xrightarrow[\Delta]{\text{Dil.}H_2SO_4}$ ($O_2N$, $OH$, $OH$)

Here nitration takes place at *o*-position of less hindered side.

R. Nitrobenzene ($NO_2$) $\xrightarrow{Sn/HCl}$ ($NH_2$) $\xrightarrow{CH_3COCl}$ ($NHCOCH_3$) $\xrightarrow{\text{Conc.}H_2SO_4}$ ($NHCOCH_3$, $SO_3H$) $\xrightarrow{HNO_3}$ ($NHCOCH_3$, $NO_2$, $SO_3H$) $\xrightarrow[\Delta]{\text{Dil.}H_2SO_4}$ ($NH_2$, $NO_2$) $\xleftarrow{OH^-}$ ($NHCOCH_3$, $NO_2$)

Here N-acetylation is more favourable than C-acetylation

S. ($NO_2$, $CH_3$) $\xrightarrow[-OH/\Delta]{KMnO_4}$ ($NO_2$, $COO^-$) $\xrightarrow{H_3O^+}$ ($NO_2$, $COOH$) $\xrightarrow{SOCl_2}$ ($NO_2$, $COCl$) $\xrightarrow{NH_3}$ ($NO_2$, $CONH_2$)

## Mathematics

**1. Plan** *(i) The number of arrangement of n distinct objects is given by n!.*

*(ii) Probability*

$$= \frac{\text{Number of favourable outcomes}}{\text{Number of total outcomes}}$$

➥ Total number of ways to arrange 3 boys and 2 girls are 5!.

According to given condition, following cases may arise.

B G G B B
G G B B B
G B G B B
G B B G B
B G B G B

So, number of favourable ways

$$= 5 \times 3! \times 2!$$
$$= 60$$

∴ Required probability

$$= \frac{60}{120} = \frac{1}{2}$$

**2. Plan**

*(i)* $\cos C = \dfrac{a^2 + b^2 - c^2}{2ab}$

*(ii)* $R = \dfrac{abc}{4\Delta}, r = \dfrac{\Delta}{s}$

*where, $R, r, \Delta$ denote the circumradius, inradius and area of triangle, respectively.*

➥ Let the sides of triangle be $a$, $b$ and $c$.

Given, $x = a + b$

$y = ab$

$x^2 - c^2 = y$

$\Rightarrow (a+b)^2 - c^2 = y$

$\Rightarrow a^2 + b^2 + 2ab - c^2 = ab$

$\Rightarrow a^2 + b^2 - c^2 = -ab$

$\Rightarrow \dfrac{a^2 + b^2 - c^2}{2ab} = -\dfrac{1}{2}$

$= \cos 120°$

$\Rightarrow \angle C = \dfrac{2\pi}{3}$

$\because R = \dfrac{abc}{4\Delta}, r = \dfrac{\Delta}{s}$

$\Rightarrow \dfrac{r}{R} = \dfrac{4\Delta^2}{s(abc)}$

$= \dfrac{4\left[\dfrac{1}{2} ab \sin\left(\dfrac{2\pi}{3}\right)\right]^2}{\dfrac{x+c}{2} \cdot y \cdot c}$

$\therefore \dfrac{r}{R} = \dfrac{3y}{2c(x+c)}$

**3. Plan** *If there are n different objects and n corresponding places, then the number of ways of putting these objects so that no object occupies its corresponding place* $= n!\left[\dfrac{1}{2!} - \dfrac{1}{3!} + \dots (-1)^n \dfrac{1}{n!}\right]$

➥ We have six cards $C_1, C_2, C_3, C_4, C_5, C_6$ and envelopes $E_1, E_2, E_3, E_4, E_5, E_6$.

Let the number of dearrangement when $C_1$ is put into $E_2 = X$

Similarly, the number of dearrangement when $C_1$ is put into $E_3 = X$

Similarly, the number of dearrangement when $C_1$ is put into $E_4 = X$

Similarly, the number of dearrangement when $C_1$ is put into $E_5 = X$

Similarly, the number of dearrangement when $C_1$ is put into $E_6 = X$

Thus, total number of dearrangements $= 5X = D_6$

$$D_6 = 6!\left(\frac{1}{2!} - \frac{1}{3!} + \frac{1}{4!} - \frac{1}{5!} + \frac{1}{6!}\right)$$

$$= 265$$

$\therefore X = 265/5 = 53$

**4. Plan**

*(i)* $y = mx + \dfrac{a}{m}$ *is equation of tangent to the parabola* $y^2 = 4ax$.

*(ii) A line is a tangent to circle, if distance of line from centre is equal to the radius of circle*

(iii) *Equation of chord drawn from exterior point $(x_1, y_1)$ to a circle/parabola is given by $T = 0$*

(iv) *Area of Trapezium $= \frac{1}{2}$ (sum of parallel sides) × (distance between them)*

Let equation of tangent to parabola be

$$y = mx + \frac{2}{m}$$

It also touches the circle $x^2 + y^2 = 2$.

$$\therefore \quad \left| \frac{2}{m\sqrt{1+m^2}} \right| = \sqrt{2}$$

$$\Rightarrow \quad m^4 + m^2 = 2$$

$$\Rightarrow \quad m^4 + m^2 - 2 = 0$$

$$\Rightarrow \quad (m^2 - 1)(m^2 + 2) = 0$$

$$\Rightarrow \quad m = \pm 1, m^2 = -2$$

$$(m^2 = -2 \text{ rejected})$$

So, tangents are $y = x + 2, y = -x - 2$

They, intersect at $(-2, 0)$.

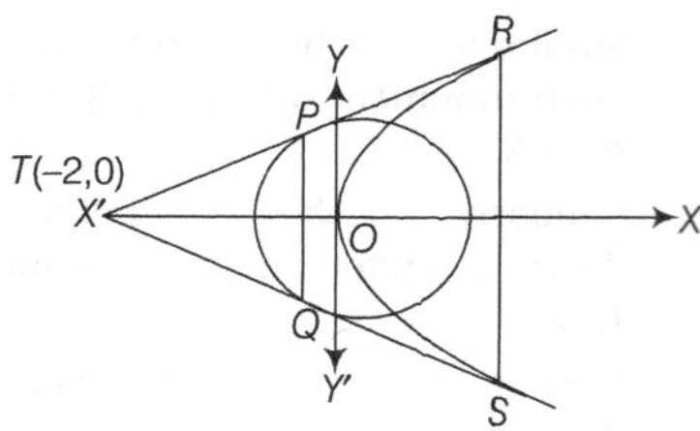


Equation of chord $PQ$ is, $-2x = 2$

$$\Rightarrow \quad x = -1$$

Equation of chord $RS$ is $O = 4(x - 2)$

$$\Rightarrow \quad x = 2$$

∴ Coordinates of $P, Q, R, S$ are

$P(-1, 1), Q(-1, -1), R(2, 4), S(2, -4)$

Hence, area of $PQRS = \frac{(2+8) \times 3}{2}$

$= 15$ sq units

**5. Plan** *If quadratic equation has purely imaginary roots, then coefficient of x must be equal to zero.*

➥ Let $p(x) = ax^2 + b$ with $a, b$ of same sign and $a, b \in R$.

Then, $p[p(x)] = a(ax^2 + b)^2 + b$

$p(x)$ has imaginary roots say $ix$.

Then, also $ax^2 + b \in R$

and $\quad (ax^2 + b)^2 > 0$

$\therefore \quad a(ax^2 + b)^2 + b \neq 0, \forall x$

Thus, $p[p(x)] \neq 0, \forall x$

**6. Plan** *This type of question can be done using appropriate substitution.*

➥ Given, $I = \int_{\pi/4}^{\pi/2} (2 \operatorname{cosec} x)^{17} dx$

$$= \int_{\pi/4}^{\pi/2} \frac{2^{17} (\operatorname{cosec} x)^{16} \operatorname{cosec} x \, (\operatorname{cosec} x + \cot x)}{(\operatorname{cosec} x + \cot x)} dx$$

Let $\operatorname{cosec} x + \cot x = t$

$\Rightarrow (-\operatorname{cosec} x \cdot \cot x - \operatorname{cosec}^2 x)\, dx = dt$

and $\operatorname{cosec} x - \cot x = 1/t$

$$\Rightarrow \quad 2 \operatorname{cosec} x = t + \frac{1}{t}$$

$$\therefore \quad I = -\int_{\sqrt{2}+1}^{1} 2^{17} \left( \frac{t + 1/t}{2} \right)^{16} \frac{dt}{t}$$

Let $t = e^u \Rightarrow dt = e^u du$. When $t = 1$, $e^u = 1 \Rightarrow u = 0$

and when $t = \sqrt{2} + 1, e^u = \sqrt{2} + 1$

$$\Rightarrow \quad u = \ln(\sqrt{2} + 1)$$

$$\Rightarrow \quad I = -\int_{\ln(\sqrt{2}+1)}^{0} 2(e^u + e^{-u})^{16} \frac{e^u du}{e^u}$$

$$= 2 \int_0^{\ln(\sqrt{2}+1)} (e^u + e^{-u})^{16} du$$

**7. Plan**

(i) *Solution of the differential equation $\frac{dy}{dx} + Py = Q$ is*

$$y \cdot (\text{IF}) = \int Q \cdot (\text{IF})\, dx + c$$

*where,* $\text{IF} = e^{\int P dx}$

(ii) $\int_{-a}^{a} f(x)\, dx = 2 \int_0^a f(x)\, dx$, *if* $f(-x) = f(x)$

➥ Given differential equation

$$\frac{dy}{dx} + \frac{x}{x^2 - 1} y = \frac{x^4 + 2x}{\sqrt{1 - x^2}}$$

This is a linear differential equation

$$\text{IF} = e^{\int \frac{x}{x^2-1}dx} = e^{\frac{1}{2}\ln|x^2-1|}$$

$$= \sqrt{1-x^2}$$

$\Rightarrow$ Solution is

$$y\sqrt{1-x^2} = \int \frac{x(x^3+2)}{\sqrt{1-x^2}} \cdot \sqrt{1-x^2}\, dx$$

or $y\sqrt{1-x^2} = \int (x^4+2x)\, dx$

$$= \frac{x^5}{5} + x^2 + c$$

$$f(0) = 0 \Rightarrow c = 0$$

$$\Rightarrow \quad f(x)\sqrt{1-x^2} = \frac{x^5}{5} + x^2$$

Now, $\int_{-\sqrt{3}/2}^{\sqrt{3}/2} f(x)\, dx = \int_{-\sqrt{3}/2}^{\sqrt{3}/2} \frac{x^2}{\sqrt{1-x^2}} dx$

[using property]

$$= 2\int_0^{\sqrt{3}/2} \frac{x^2}{\sqrt{1-x^2}} dx$$

$$= 2\int_0^{\pi/3} \frac{\sin^2\theta}{\cos\theta} \cos\theta\, d\theta$$

[taking $x = \sin\theta$]

$$= 2\int_0^{\pi/3} \sin^2\theta\, d\theta$$

$$= \int_0^{\pi/3} (1-\cos 2\theta)\, d\theta$$

$$= \left(\theta - \frac{\sin 2\theta}{2}\right)_0^{\pi/3}$$

$$= \frac{\pi}{3} - \frac{\sin 2\pi/3}{2}$$

$$= \frac{\pi}{3} - \frac{\sqrt{3}}{4}$$

**8.** **Plan** *Newton-Leibnitz's formula*

$$\frac{d}{dx}\left[\int_{\phi(x)}^{\psi(x)} f(t)\, dt\right] = f\{\psi(x)\}\left\{\frac{d}{dx}\psi(x)\right\} - f\{\phi(x)\}\left\{\frac{d}{dx}\phi(x)\right\}$$

➡ Given, $F(x) = \int_0^{x^2} f(\sqrt{t})\, dt$

$\therefore \quad F'(x) = 2x\, f(x)$

Also, $\quad F'(x) = f'(x)$

$$\Rightarrow \quad 2x\, f(x) = f'(x) \Rightarrow \frac{f'(x)}{f(x)} = 2x$$

$$\Rightarrow \quad \int \frac{f'(x)}{f(x)} dx = \int 2x\, dx$$

$$\Rightarrow \quad \ln f(x) = x^2 + c$$

$$\Rightarrow \quad f(x) = e^{x^2+c}$$

$$\Rightarrow \quad f(x) = K\, e^{x^2} \qquad [K = e^c]$$

Now, $\quad f(0) = 1$

$\therefore \quad 1 = K$

Hence, $\quad f(x) = e^{x^2}$

$$F(2) = \int_0^4 e^t dt$$

$$= [e^t]_0^4 = e^4 - 1$$

**9.** **Plan** *Coefficient of $x^r$ in $(1+x)^n$ is ${}^nC_r$.*

*In this type of questions, we find different composition of terms where product will give us $x^{11}$.*

➡ Coefficient of $x^{11}$ in $(1+x^2)^4(1+x^3)^7(1+x^4)^{12}$

Now, consider the following cases for $x^{11}$ in

$(1+x^2)^4(1+x^3)^7(1+x^4)^{12}$

Coefficient of $x^0\, x^3\, x^8$; Coefficient of $x^2\, x^9\, x^0$

Coefficient of $x^4\, x^3\, x^4$; Coefficient of $x^8\, x^3\, x^0$

$$= {}^4C_0 \times {}^7C_1 \times {}^{12}C_2 + {}^4C_1 \times {}^7C_3 \times {}^{12}C_0 + {}^4C_2 \times {}^7C_1 \times {}^{12}C_1 + {}^4C_4 \times {}^7C_1 \times {}^{12}C_0$$

$$= 462 + 140 + 504 + 7 = 1113$$

**10.** **Plan** *For solving this type of questions, obtain the LHS and RHS in equation and examine, the two are equal or not for a given interval.*

➡ Given, Trigonometrical Equation

$$(\sin x - \sin 3x) + 2\sin 2x = 3$$

$$\Rightarrow -2\cos 2x \sin x + 4\sin x \cos x = 3$$

[$\because$

$\sin C - \sin D = 2\cos\left(\frac{C+D}{2}\right)\sin\left(\frac{C-D}{2}\right)$

and $\sin 2\theta = 2\sin\theta\cos\theta$]

$$\Rightarrow \quad 2\sin x\,(2\cos x - \cos 2x) = 3$$

$$\Rightarrow 2\sin x\,(2\cos x - 2\cos^2 x + 1) = 3$$

$$\Rightarrow \quad 2\sin x\left(\frac{3}{2} - 2\left(\cos x - \frac{1}{2}\right)^2\right) = 3$$

$$\Rightarrow 3\sin x - 3 = 4\left(\cos x - \frac{1}{2}\right)^2 \sin x$$

As $x \in (0, \pi)$

LHS $\le 0$ and RHS $\ge 0$

For solution to exist, LHS = RHS = 0

Now, LHS = 0 $\Rightarrow$ $3\sin x - 3 = 0$

$$\Rightarrow \quad \sin x = 1$$

$$\Rightarrow \quad x = \frac{\pi}{2}$$

For $x = \frac{\pi}{2}$,

$$\text{RHS} = 4\left(\cos\frac{\pi}{2} - \frac{1}{2}\right)^2 \sin\frac{\pi}{2}$$

$$= 4\left(\frac{1}{4}\right)(1) = 1 \ne 0$$

$\therefore$ No solution of the equation exists.

**11. Plan**

*Probability*

$$= \frac{\textit{Number of favourable outcomes}}{\textit{Number of total outcomes}}$$

➥ As, $x_1 + x_2 + x_3$ is odd

So, all may be odd or one of them is odd and other two are even.

$\therefore$ Required probability

$$= \frac{{}^2C_1 \times {}^3C_1 \times {}^4C_1 + {}^1C_1 \times {}^2C_1 \times {}^4C_1 + {}^2C_1 \times {}^2C_1 \times {}^3C_1 + {}^1C_1 \times {}^3C_1 \times {}^3C_1}{{}^3C_1 \times {}^5C_1 \times {}^7C_1}$$

$$= \frac{24 + 8 + 12 + 9}{105} = \frac{53}{105}$$

**12. Plan** *If $x_1, x_2, x_3$ are in AP.*

*Then,* $x_2 - x_1 = x_3 - x_2$

$$\Rightarrow \quad 2x_2 = x_1 + x_3$$

$\because$ $x_1, x_2, x_3$ are in AP.

$$x_1 + x_3 = 2x_2$$

So, $x_1 + x_3$ should be even number

Either both $x_1$ and $x_3$ are odd or both are even.

$\therefore$ Required probability

$$= \frac{{}^2C_1 \times {}^4C_1 + {}^1C_1 \times {}^3C_1}{{}^3C_1 \times {}^5C_1 \times {}^7C_1}$$

$$= \frac{11}{105}$$

**13. Plan**

*(i) If $P(at^2, 2at)$ is one end point of focal chord of parabola $y^2 = 4ax$, then other end point is $\left(\frac{a}{t^2}, -\frac{2a}{t}\right)$.*

*(ii) Slope of line joining two points $(x_1, y_1)$ and $(x_2, y_2)$ is given by $\frac{y_2 - y_1}{x_2 - x_1}$.*

➥ If $PQ$ is focal chord, then coordinates of $Q$ will be $\left(\frac{a}{t^2}, -\frac{2a}{t}\right)$.

Now, slope of $QR$ = slope of $PK$

$$\frac{2ar + 2a/t}{ar^2 - a/t^2} = \frac{2at}{at^2 - 2a}$$

$$\Rightarrow \quad \frac{r + 1/t}{r^2 - 1/t^2} = \frac{t}{t^2 - 2}$$

$$\Rightarrow \quad \frac{1}{r - 1/t} = \frac{t}{t^2 - 2}$$

$$\Rightarrow \quad r - \frac{1}{t} = \frac{t^2 - 2}{t} = t - \frac{2}{t}$$

$$\Rightarrow \quad r = t - \frac{1}{t}$$

$$= \frac{t^2 - 1}{t}$$

**14. Plan** *Equation of tangent and normal at $(at^2, 2at)$ are given by $ty = x + at^2$ and $y + tx = 2at + at^3$, respectively.*

➥ Tangent at $P$ : $ty = x + at^2$

or $$y = \frac{x}{t} + at$$

Normal at $S$ :

$$y + \frac{x}{t} = \frac{2a}{t} + \frac{a}{t^3}$$

Solving, $$2y = at + \frac{2a}{t} + \frac{a}{t^3}$$

$$\Rightarrow \quad y = \frac{a(t^2 + 1)^2}{2t^3}$$

**15.** **Plan** $\int \frac{dx}{\sqrt{a^2 - x^2}} = \sin^{-1}\left(\frac{x}{a}\right) + c$

*As g(a) is defined in the question, first use the numerical value of 'a' given in question and then proved.*

➥ Given,

$$g(a) = \lim_{h \to 0^+} \int_h^{1-h} t^{-a}(1-t)^{a-1}\,dt$$

$\therefore g(1/2)$

$$= \lim_{h \to 0^+} \int_h^{1-h} t^{-1/2}(1-t)^{-1/2}\,dt$$

$$= \int_0^1 \frac{dt}{\sqrt{t-t^2}} = \int_0^1 \frac{dt}{\sqrt{\frac{1}{4} - \left(t - \frac{1}{2}\right)^2}}$$

$$= \sin^{-1}\left[\left(\frac{t-1/2}{1/2}\right)\right]_0^1$$

$$= \sin^{-1} 1 - \sin^{-1}(-1) = \pi$$

**16.** **Plan** $\int_0^a f(x)\,dx = \int_0^a f(a-x)\,dx$

*As g(a) is given, use the numerical value of 'a' given in the question and then proved.*

➥ Given that,

$$g(a) = \int_0^1 \frac{dt}{t^a(1-t)^{1-a}}$$

Clearly, $g(a) = g(1-a)$

[using $\int_0^a f(x)\,dx = \int_0^a f(a-x)\,dx$]

Now, differentiate w.r.t. '$a$', we get

$$g'(a) = g'(1-a)(-1)$$

Now, for $a = \frac{-1}{2}$, we have

$$-g'\left(\frac{1}{2}\right) = g'\left(\frac{1}{2}\right)$$

So, $g'\left(\frac{1}{2}\right) = 0$

**17.** (P) **Plan**

*(i) A polynomial satisfying the given conditions is taken.*

*(ii) The other conditions are also applied and the number of polynomial is taken out.*

➥ Let $f(x) = ax^2 + bx + c$

$$f(0) = 0 \Rightarrow c = 0$$

Now, $\int_0^1 f(x)\,dx = 1$

$$\Rightarrow \left(\frac{ax^3}{3} + \frac{bx^2}{2}\right)_0^1 = 1$$

$$\Rightarrow \frac{a}{3} + \frac{b}{2} = 1$$

$$\Rightarrow 2a + 3b = 6$$

As $a$, $b$ are non-negative integers.

So, $a = 0$, $b = 2$ or $a = 3$, $b = 0$

So, $f(x) = 2x$ or $f(x) = 3x^2$

(Q) **Plan** *Such type of questions are converted into only sine or cosine expression and then the number of points of maxima in given interval are obtained.*

➥ $f(x) = \sin(x^2) + \cos(x^2)$

$$= \sqrt{2}\left[\frac{1}{\sqrt{2}}\cos(x^2) + \frac{1}{\sqrt{2}}\sin(x^2)\right]$$

$$= \sqrt{2}\left[\cos x^2 \cos\frac{\pi}{4} + \sin\frac{\pi}{4}\sin(x^2)\right]$$

$$= \sqrt{2}\cos\left(x^2 - \frac{\pi}{4}\right)$$

For maximum value,

$$x^2 - \frac{\pi}{4} = 2n\pi$$

$$\Rightarrow x^2 = 2n\pi + \frac{\pi}{4}$$

$$\Rightarrow x = \pm\sqrt{\frac{\pi}{4}}, \text{for } n = 0$$

$$x = \pm\sqrt{\frac{9\pi}{4}}, \text{ for } n = 1$$

So, $f(x)$ attains maximum at 4 points in $[-\sqrt{13}, \sqrt{13}]$.

(R) **Plan**

*(i)* $\int_{-a}^a f(x)dx = \int_{-a}^a f(-x)dx$

*(ii)* $\int_{-a}^a f(x)dx = 2\int_0^a f(x)dx$, *if* $f(-x) = f(x)$, *i.e., f is an even function.*

➥ $$I = \int_{-2}^2 \frac{3x^2}{1+e^x}\,dx$$

$$I = \int_{-2}^2 \frac{3x^2}{1+e^{-x}}\,dx$$

$$2I = \int_{-2}^{2} \left( \frac{3x^2}{1+e^x} + \frac{3x^2(e^x)}{e^x+1} \right) dx$$

$$2I = \int_{-2}^{2} 3x^2 \, dx$$

$$\Rightarrow \quad 2I = 2\int_{0}^{2} 3x^2 \, dx$$

$$I = [x^3]_0^2 = 8$$

(S) **Plan** $\int_{-a}^{a} f(x)\,dx = 0$

*If* $f(-x) = -f(x)$

*i.e.,* $f(x)$ *is an odd function.*

➥ Let $f(x) = \cos 2x \log \left(\frac{1+x}{1-x}\right)$

$$f(-x) = \cos 2x \log \left(\frac{1-x}{1+x}\right)$$

$$= -f(x)$$

Hence, $f(x)$ is an odd function.

So, $\int_{-1/2}^{1/2} f(x)\,dx = 0$

(P) → (ii); (Q) → (iii); (R) → (i); (S) → (iv)

**18.** (P) $y = \cos(3\cos^{-1} x)$

$$y' = \frac{3\sin(3\cos^{-1}x)}{\sqrt{1-x^2}}$$

$$\sqrt{1-x^2}\, y' = 3\sin(3\cos^{-1}x)$$

$$\Rightarrow \frac{-x}{\sqrt{1-x^2}} y' + \sqrt{1-x^2}\, y'' = 3\cos(3\cos^{-1}x)\cdot \frac{-3}{\sqrt{1-x^2}}$$

$$\Rightarrow \quad -xy' + (1-x^2)\,y'' = -9y$$

$$\Rightarrow \quad \frac{1}{y}[(x^2-1)\,y'' + xy'] = 9$$

(Q) **Plan**

*(i) Angle subtended by a side of n sided regular polygon at the centre* $= \frac{2\pi}{n}$.

*(ii)* $|\mathbf{a} \times \mathbf{b}| = |\mathbf{a}||\mathbf{b}|\sin\theta$

*(iii)* $|\mathbf{a} \cdot \mathbf{b}| = |\mathbf{a}||\mathbf{b}|\cos\theta$

*(iv)* $\tan\theta = \tan\alpha \Rightarrow \theta = n\pi + \alpha, n \in Z$

Consider a polygon of (S) $n$ sides with centre at origin

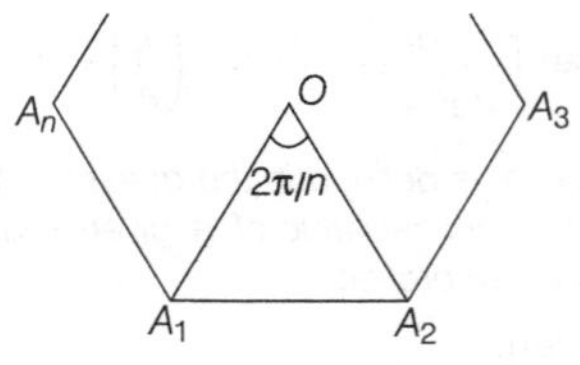


Let $|\mathbf{OA}_1| = |\mathbf{OA}_2| = \dots |\mathbf{OA}_n|$
$= r$ (say)

$$|\mathbf{a}_k \times \mathbf{a}_{k+1}| = r^2 \sin\frac{2\pi}{n}$$

$$|\mathbf{a}_k \cdot \mathbf{a}_{k+1}| = r^2 \cos\frac{2\pi}{n}$$

$$\Rightarrow \left|\sum_{k=1}^{n-1} \mathbf{a}_k \times \mathbf{a}_{k+1}\right| = \left|\sum_{k=1}^{n-1} \mathbf{a}_k \cdot \mathbf{a}_{k+1}\right|$$

$$\Rightarrow r^2(n-1)\sin\frac{2\pi}{n} = r^2(n-1)\cos\frac{2\pi}{n}$$

$$\tan\frac{2\pi}{n} = 1$$

$$\Rightarrow \quad \tan\frac{2\pi}{n} = \tan\frac{\pi}{4}$$

$$\Rightarrow \quad \frac{2\pi}{n} = t\pi + \frac{\pi}{4}, t \in Z$$

$$\Rightarrow \quad \frac{2}{n} = \frac{4t+1}{4}$$

$$\Rightarrow \quad n = \frac{8}{4t+1}, t \in Z$$

⇒ The minimum value of $n = 8$

(R) **Plan** *Equation of normal at the point* $(a\cos\theta, b\sin\theta)$ *of ellipse* $\frac{x^2}{a^2} + \frac{y^2}{b^2} = 1$ *is given by*

$$ax\sec\theta - by\,\text{cosec}\,\theta = a^2 - b^2$$

➥ Equation of normal is

$$\sqrt{6}\,x\sec\theta - \sqrt{3}\,y\,\text{cosec}\,\theta = 3$$

Its slope is $\frac{\sqrt{6}\sec\theta}{\sqrt{3}\,\text{cosec}\,\theta} = 1$

($\because$ slope of normal = slope of line perpendicular to

$$x + y = 8$$

$$\tan\theta = \frac{1}{\sqrt{2}}$$

So, normal is

$$\sqrt{6}\, x \frac{\sqrt{3}}{\sqrt{2}} - \sqrt{3} \times \sqrt{3}\, y = 3$$

$$3x - 3y = 3$$

$$\Rightarrow \quad x - y = 1$$

As it passes through $(h, 1)$.

So, $h - 1 = 1$

$h = 2$

(S) **Plan**

$$\tan^{-1} x + \tan^{-1} y = \tan^{-1}\left(\frac{x+y}{1-xy}\right)$$

Given equation

$$\tan^{-1}\left(\frac{1}{2x+1}\right) + \tan^{-1}\left(\frac{1}{4x+1}\right) = \tan^{-1}\left(\frac{2}{x^2}\right)$$

$$\Rightarrow \tan^{-1}\left(\frac{\frac{1}{2x+1} + \frac{1}{4x+1}}{1 - \frac{1}{(2x+1)(4x+1)}}\right) = \tan^{-1}\left(\frac{2}{x^2}\right)$$

$$\Rightarrow \tan^{-1}\left(\frac{6x+2}{(2x+1)(4x+1)-1}\right) = \tan^{-1}\left(\frac{2}{x^2}\right)$$

$$\Rightarrow \frac{3x+1}{4x^2+3x} = \frac{2}{x^2}$$

$$\Rightarrow 3x^3 + x^2 = 8x^2 + 6x$$

$$\Rightarrow x(3x^2 - 7x - 6) = 0$$

$$\Rightarrow x(x-3)(3x+2) = 0$$

$$\Rightarrow x = 0, \frac{-2}{3}, 3$$

So, only positive solution is $x = 3$.

(P) $\rightarrow$ (iv); (Q) $\rightarrow$ (iii); (R) $\rightarrow$ (ii); (S) $\rightarrow$ (i)

**19.** (P) **Plan**

*(i) For such questions, we need to properly define the functions and then we draw their graphs.*

*(ii) From the graphs, we can examine the function for continuity, differentiability, one-one and onto.*

$$f_1(x) = \begin{cases} -x, & x < 0 \\ e^x, & x \geq 0 \end{cases}$$

$$f_2(x) = x^2, \; x \geq 0$$

$$f_3(x) = \begin{cases} \sin x, & x < 0 \\ x, & x \geq 0 \end{cases}$$

$$f_4(x) = \begin{cases} f_2(f_1(x)), & x < 0 \\ f_2(f_1(x)) - 1, & x \geq 0 \end{cases}$$

Now, $f_2(f_1(x)) = \begin{cases} x^2, & x < 0 \\ e^{2x}, & x \geq 0 \end{cases}$

$$f_4 = \begin{cases} x^2, & x < 0 \\ e^{2x} - 1, & x \geq 0 \end{cases}$$

As $f_4(x)$ is continuous

$$f'_4(x) = \begin{cases} 2x, & x < 0 \\ 2e^{2x}, & x > 0 \end{cases}$$

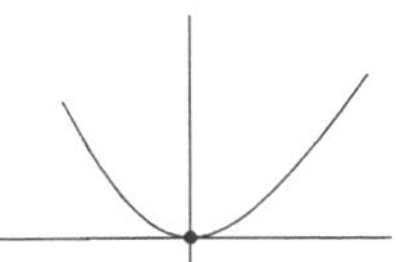

$f_4'(0)$ is not defined.

Its range is $[0, \infty)$.

Thus, range = codomain = $[0, \infty)$ thus, $f_4$ is onto.

Also, horizontal line (drawn parallel to $x$-axis) meets the curve more than once thus function is not one-one.

(Q) **Plan** $f_3(x)$

*differentiable at $x = 0$ and not one-one*

As evident, from the graph it is continuous and no sharp turn at $X = 0$ thus, $f(x)$ is differentiable at $X = 0$

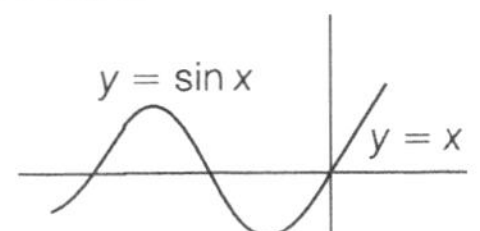


Also, a horizontal line intersects the graph more than once.

$\therefore$ It is not one-one.

(R) **Plan** $f_2(f_1(x))$

*It is neither continuous nor one-one.*

From the graph, it can be observed that the function is not continuous at $x = 0$.

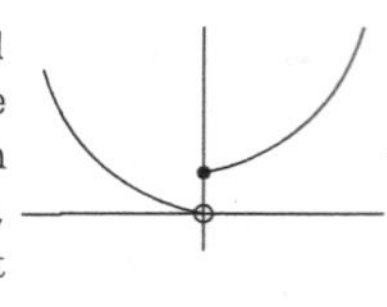

Also, the horizontal line intersects the curve at more than one point. So, $f_2[f_1(x)]$ is not one-one.

(S) **Plan** $f_2(x)$

*It is continuous and one-one.*

➥ As evident from graphs, the function is continuous.

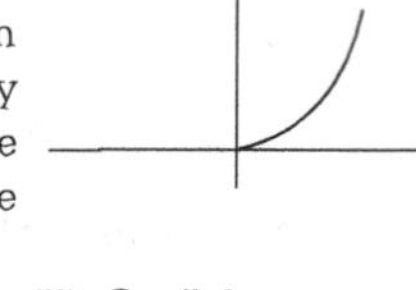

Also, the function is one-one, as any horizontal line will meet the graph only once.

P→(i), Q→(iii), R→(ii), S→(iv)

**20.** (P) **Plan** $e^{i\theta}\cdot e^{i\alpha}=e^{i(\theta+\alpha)}$

➥ Given $Z_k=e^{i\frac{2k\pi}{10}}$

$Z_k\cdot Z_j=e^{i\left(\frac{2\pi}{10}\right)(k+j)}$ $Z_k$ is 10th root of unity.

$\Rightarrow \bar{Z}_k$ will also be 10th root of unity.

Taking, $Z_j$ as $\bar{Z}_k$,

we have $Z_k\cdot Z_2=1$ (True)

(Q) **Plan** $\dfrac{e^{i\theta}}{e^{i\alpha}}=e^{i(\theta-\alpha)}$

➥ $z=\dfrac{z_k}{z_1}=e^{i\left(\frac{2k\pi}{10}-\frac{2\pi}{10}\right)}=e^{i\frac{\pi}{5}(k-1)}$

For $k=2$; $z=e^{i\frac{\pi}{5}}$ which is in the given set (False)

(R) **Plan**

*(i)* $1-\cos 2\theta=2\sin^2\theta$

*(ii)* $\sin 2\theta=2\sin\theta\cos\theta$ *and*

*(i)* $\cos 36°=\dfrac{\sqrt{5}-1}{4}$

*(ii)* $\cos 108°=\dfrac{\sqrt{5}+1}{4}$

$$\frac{|1-z_1||1-z_2|\ldots|1-z_9|}{10}$$

Note, $|1-z_k|$

$$=\left|1-\cos\frac{2\pi k}{10}-i\sin\frac{2\pi k}{10}\right|$$

$$=\left|2\sin\frac{\pi k}{10}\right|\left|\sin\frac{\pi k}{10}-i\cos\frac{\pi k}{10}\right|$$

$$=2\left|\sin\frac{\pi k}{10}\right|$$

Now, required product is

$$\frac{2^9\sin\frac{\pi}{10}\cdot\sin\frac{2\pi}{10}\cdot\sin\frac{3\pi}{10}\ldots\sin\frac{8\pi}{10}\cdot\sin\frac{9\pi}{10}}{10}$$

$$=\frac{2^9\left(\sin\frac{\pi}{10}\sin\frac{2\pi}{10}\sin\frac{3\pi}{10}\sin\frac{4\pi}{10}\right)^2\sin\frac{5\pi}{10}}{10}$$

$$=\frac{2^9\left(\sin\frac{\pi}{10}\cos\frac{\pi}{10}\cdot\sin\frac{2\pi}{10}\cos\frac{2\pi}{10}\right)^2\cdot 1}{10}$$

$$=\frac{2^9\left(\frac{1}{2}\sin\frac{\pi}{5}\cdot\frac{1}{2}\sin\frac{2\pi}{5}\right)^2}{10}$$

$$=\frac{2^5(\sin 36°\cdot\sin 72°)^2}{10}$$

$$=\frac{2^5}{2^2\times 10}(2\sin 36°\sin 72°)^2$$

$$=\frac{2^2}{5}(\cos 36°-\cos 108°)^2$$

$$=\frac{2^2}{5}\left[\left(\frac{\sqrt{5}-1}{4}\right)+\left(\frac{\sqrt{5}+1}{4}\right)\right]^2$$

$$=\frac{2^2}{5}\cdot\frac{5}{4}=1$$

(S) Sum of $n$th roots of unity $=0$

$$1+\alpha+\alpha^2+\alpha^3+\ldots+\alpha^9=0$$

$$1+\sum_{k=1}^{9}\alpha^k=0$$

$$1+\sum_{k=1}^{9}\left(\cos\frac{2k\pi}{10}+i\sin\frac{2k\pi}{10}\right)=0$$

$$1+\sum_{k=1}^{9}\cos\frac{2k\pi}{10}=0$$

So, $$1-\sum_{k=1}^{9}\cos\frac{2k\pi}{10}=2$$

(P) → (i), (Q) → (ii), (R) → (iii), (S) → (iv)

# JEE Main

# Solved Paper 2013

**M.Marks 360** **Time 3 Hrs**

## Instructions

- This test consists of 90 questions.
- There are three parts in the question paper A, B, C consisting of Physics, Chemistry and Mathematics having 30 questions in each part of equal weightage. Each question is allotted 4 marks for correct response.
- Candidates will be awarded marks as stated above for correct response of each question 1/4 marks will be deducted for indicating incorrect response of each question. No deduction from the total score will be made if no response is indicated for an item in the answer sheet.
- There is only one correct response for each question. Filling up more than one response in any question will be treated as wrong response and marks for wrong response will be deducted according as per instructions.

# PHYSICS

**1.** A uniform cylinder of length $L$ and mass $M$ having cross-sectional area $A$ is suspended, with its length vertical from a fixed point by a massless spring such that it is half submerged in a liquid of density $\sigma$ at equilibrium position. The extension $x_0$ of the spring when it is in equilibrium is

(a) $\frac{Mg}{k}$ (b) $\frac{Mg}{k}\left(1-\frac{LA\sigma}{M}\right)$

(c) $\frac{Mg}{k}\left(1-\frac{LA\sigma}{2M}\right)$ (d) $\frac{Mg}{k}\left(1+\frac{LA\sigma}{M}\right)$

**2.** A metallic rod of length $l$ is tied to a string of length $2l$ and made to rotate with angular speed $\omega$ on a horizontal table with one end of the string fixed. If there is a vertical magnetic field $B$ in the region, the emf induced across the ends of the rod is

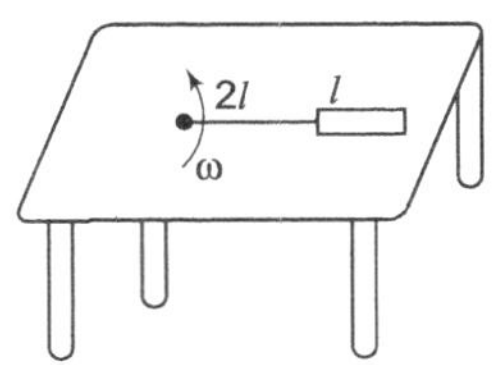


(a) $\frac{2B\,\omega\,l^3}{2}$ (b) $\frac{3B\,\omega\,l^3}{2}$

(c) $\frac{4B\omega\,l^2}{2}$ (d) $\frac{5B\omega\,l^2}{2}$

**3.** This question has statement I and statement II. Of the four choices given after the statements, choose the one that best describes the two statements.

**Statement I** A point particle of mass $m$ moving with speed $v$ collides with stationary point particle of mass $M$. If the maximum energy loss possible is given as $f\left(\frac{1}{2}mv^2\right)$, then $f = \left(\frac{m}{M+m}\right)$.

**Statement II** Maximum energy loss occurs when the particles get stuck together as a result of the collision.

(a) Statement I is true, Statement II is true, and Statement II is the correct explanation of Statement I
(b) Statement I is true, Statement II is true, but Statement II is not the correct explanation of Statement I
(c) Statement I is true, Statement II is false
(d) Statement I is false, Statement II is true

**4.** Let $[\varepsilon_0]$ denote the dimensional formula of the permittivity of vacuum. If $M$ = mass, $L$ = length, $T$ = Time and $A$ = electric current, then

(a) $[\varepsilon_0] = [M^{-1} L^{-3} T^2 A]$
(b) $[\varepsilon_0] = [M^{-1} L^{-3} T^4 A^2]$
(c) $[\varepsilon_0] = [M^{-2} L^2 T^{-1} A^{-2}]$
(d) $[\varepsilon_0] = [M^{-1} L^2 T^{-1} A^2]$

**5.** A projectile is given an initial velocity of $(\hat{\mathbf{i}} + 2\hat{\mathbf{j}})$ m/s, where $\hat{\mathbf{i}}$ is along the ground and $\hat{\mathbf{j}}$ is along the vertical. If $g = 10\,\text{m/s}^2$, the equation of its trajectory is

(a) $y = x - 5x^2$ (b) $y = 2x - 5x^2$
(c) $4y = 2x - 5x^2$ (d) $4y = 2x - 25x^2$

**6.** The amplitude of a damped oscillator decreases to 0.9 times its original magnitude is 5 s. In another 10 s it will decrease to $\alpha$ times its original magnitude, where $\alpha$ equals

(a) 0.7 (b) 0.81
(c) 0.729 (d) 0.6

**7.** Two capacitors $C_1$ and $C_2$ are charged to 120 V and 200 V respectively. It is found that by connecting them together the potential on each one can be made zero. Then

(a) $5C_1 = 3C_2$ (b) $3C_1 = 5C_2$
(c) $3C_1 + 5C_2 = 0$ (d) $9C_1 = 4C_2$

**8.** A sonometer wire of length 1.5 m is made of steel. The tension in it produces an elastic strain of 1%. What is the fundamental frequency of steel if density and elasticity of steel are $7.7 \times 10^3$ kg/m$^3$ and $2.2 \times 10^{11}$ N/m$^2$ respectively?

(a) 188.5 Hz (b) 178.2 Hz
(c) 200.5 Hz (d) 770 Hz

**9.** A circular loop of radius 0.3 cm lies parallel to a much bigger circular loop of radius 20 cm. The centre of the smaller loop is on the axis of the bigger loop. The distance between their centres is 15 cm. If a current of 2.0 A flows through the bigger loop, then the flux linked with smaller loop is

(a) $9.1 \times 10^{-11}$ Wb (b) $6 \times 10^{-11}$ Wb
(c) $3.3 \times 10^{-11}$ Wb (d) $6.6 \times 10^{-9}$ Wb

**10.** Diameter of a plano-convex lens is 6 cm and thickness at the centre is 3 mm. If speed of light in material of lens is $2 \times 10^8$ m/s, the focal length of the lens is

(a) 15 cm (b) 20 cm
(c) 30 cm (d) 10 cm

**11.** What is the minimum energy required to launch a satellite of mass $m$ from the surface of a planet of mass $M$ and radius $R$ in a circular orbit at an altitude of $2R$?

(a) $\frac{5GmM}{6R}$ (b) $\frac{2GmM}{3R}$
(c) $\frac{GmM}{2R}$ (d) $\frac{GmM}{3R}$

**12.** A diode detector is used to detect an amplitude modulated wave of 60% modulation by using a condenser of capacity 250 pico farad in parallel with a load resistance 100 kΩ. Find the maximum modulated frequency which could be detected by it.

(a) 10.62 MHz (b) 10.62 kHz
(c) 5.31 MHz (d) 5.31 kHz

**13.** A beam of unpolarized light of intensity $I_0$ is passed through a polaroid $A$ and then through another polaroid $B$ which is oriented so that its principal plane makes an angle of 45° relative to that of $A$. The intensity of the emergent light is

(a) $I_0$ (b) $I_0/2$
(c) $I_0/4$ (d) $I_0/8$

**14.** The supply voltage in a room is 120 V. The resistance of the lead wires is 6 Ω. A 60 W bulb is already switched on. What is the decrease of voltage across the bulb, when a 240 W heater is switched on in parallel to the bulb?

(a) zero (b) 2.9 V
(c) 13.3 V (d) 10.04 V

**15.** The shown $p$-$V$ diagram represents the thermodynamic cycle of an engine, operating with an ideal monoatomic gas. The amount of heat, extracted from the source in a single cycle is

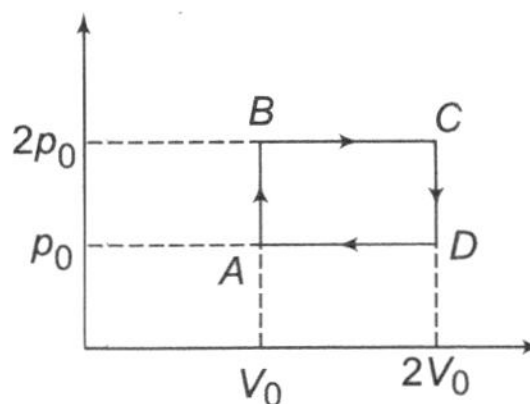


(a) $p_0V_0$ (b) $\left(\frac{13}{2}\right) p_0V_0$
(c) $\left(\frac{11}{2}\right) p_0V_0$ (d) $4p_0V_0$

**16.** A hoop of radius $r$ and mass $m$ rotating with an angular velocity $\omega_0$ is placed on a rough horizontal surface. The initial velocity of the centre of the hoop is zero. What will be the velocity of the centre of the hoop when it ceases to slip?

(a) $\frac{r\omega_0}{4}$ (b) $\frac{r\omega_0}{3}$
(c) $\frac{r\omega_0}{2}$ (d) $r\omega_0$

**17.** An ideal gas enclosed in a vertical cylindrical container supports a freely moving piston of mass $M$. The piston and the cylinder have equal cross-sectional area $A$. When the piston is in equilibrium, the volume of the gas is $V_0$ and its pressure is $p_0$. The piston is slightly displaced from the equilibrium position and released. Assuming that the system is completely isolated from its surrounding, the piston executes a simple harmonic motion with frequency

(a) $\frac{1}{2\pi}\frac{A_\gamma p_0}{V_0 M}$ (b) $\frac{1}{2\pi}\frac{V_0 M p_0}{A^2\gamma}$
(c) $\frac{1}{2\pi}\sqrt{\frac{A^2\gamma p_0}{MV_0}}$ (d) $\frac{1}{2\pi}\sqrt{\frac{MV_0}{A_\gamma p_0}}$

**18.** If a piece of metal is heated to temperature $\theta$ and then allowed to cool in a room which is at temperature $\theta_0$. The graph between the temperature $T$ of the metal and time $t$ will be closed to

(a) 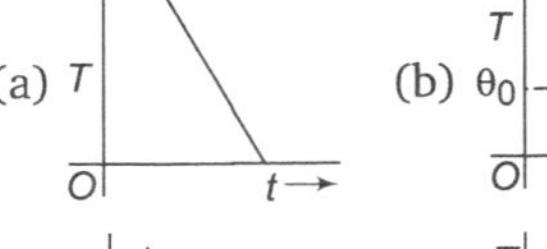


(b) 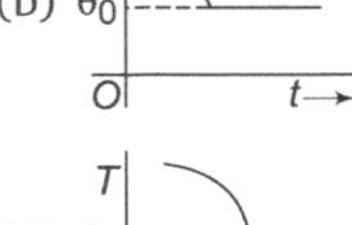


(c) 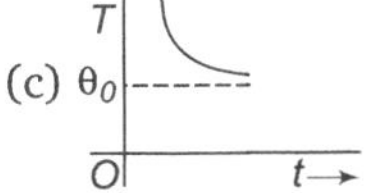


(d) T θ0 O t→

**19.** This question has Statement I and Statement II. Of the four choices given after the statements, choose the one that best describes the two statements.

**Statement I** Higher the range, greater is the resistance of ammeter.

**Statement II** To increase the range of ammeter, additional shunt needs to be used across it.

(a) Statement I is true, Statement II is true and Statement II is the correct explanation of Statement I
(b) Statement I is true, Statement II is true, but Statement II is not the correct explanation of Statement I
(c) Statement I is true, Statement II is false
(d) Statement I is false, Statement II is true

**20.** In a *L-C-R* circuit as shown below both switches are open initially. Now switch $S_1$ and $S_2$, are closed. ($q$ is charge on the capacitor and $\tau = RC$ is capacitance time constant). Which of the following statement is correct?

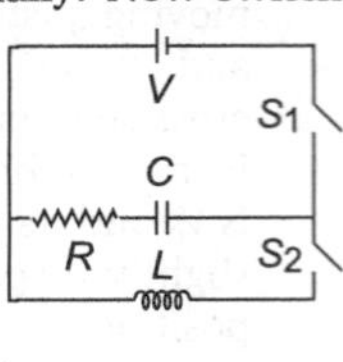


(a) Work done by the battery is half of the energy dissipated in the resistor
(b) At $t = \tau$, $q = CV/2$
(c) At $t = 2\tau$, $q = CV(1 - e^{-2})$
(d) At $t = \frac{\tau}{2}$, $q = CV(1 - e^{-1})$

**21.** Two coherent point sources $S_1$ and $S_2$ are separated by a small distance $d$ as shown. The fringes obtained on the screen will be

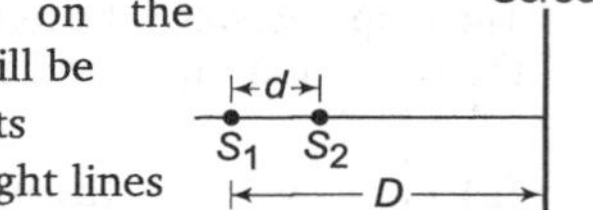


(a) points
(b) straight lines
(c) semi-circle
(d) concentric circles

**22.** The magnetic field in a travelling electromagnetic wave has a peak value of 20 nT. The peak value of electric field strength is

(a) 3 V/m (b) 6 V/m
(c) 9 V/m (d) 12 V/m

**23.** The anode voltage of a photocell is kept fixed. The wavelength $\lambda$ of the light falling on the cathode is gradually changed. The plate current $I$ of photocell varies as follows

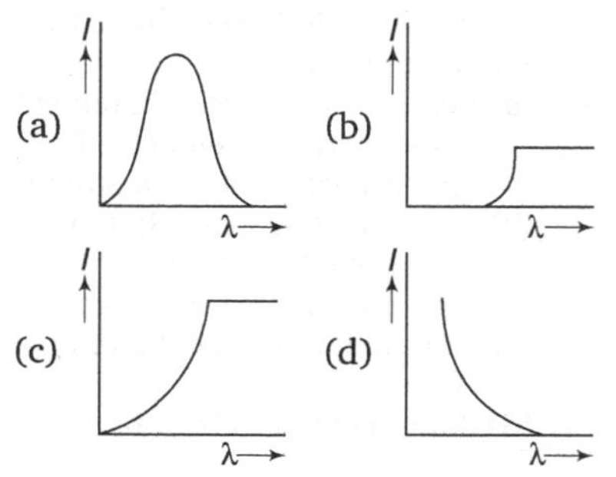


**24.** The *I-V* characteristic of an LED is

(a)
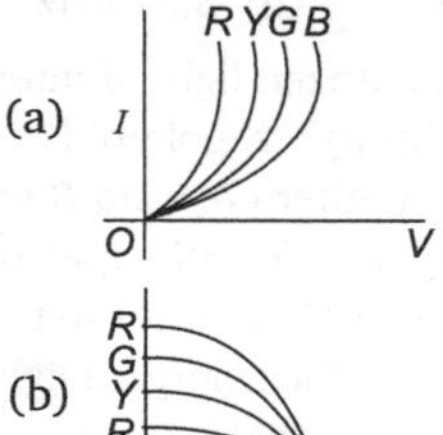


(b)
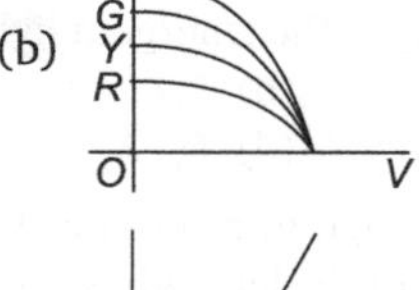


(c)
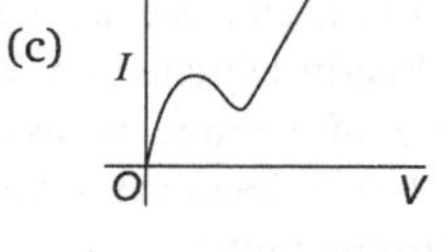


(d)
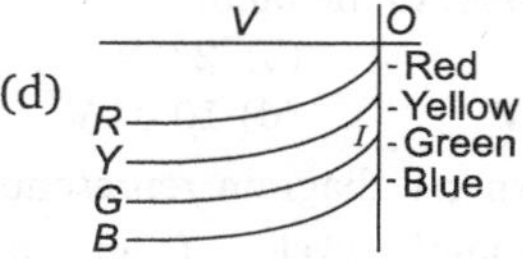


**25.** Assume that a drop of liquid evaporates by decrease in its surface energy, so that its temperature remains unchanged. What should be the minimum radius of the drop for this to be possible? The surface tension is $T$, denstiy of liquid is $\rho$ and $L$ is its latent heat of vaporization

(a) $\rho L/T$ (b) $\sqrt{T/\rho L}$
(c) $T/\rho L$ (d) $2T/\rho L$

**26.** In a hydrogen like atom electron makes transition from an energy level with quantum number $n$ to another with quantum number $(n - 1)$. If $n >> 1$, the frequency of radiation emitted is proportional to

(a) $\frac{1}{n}$ (b) $\frac{1}{n^2}$ (c) $\frac{1}{n^4}$ (d) $\frac{1}{n^3}$

**27.** The graph between angle of deviation ($\delta$) and angle of incidence ($i$) for a triangular prism is represented by

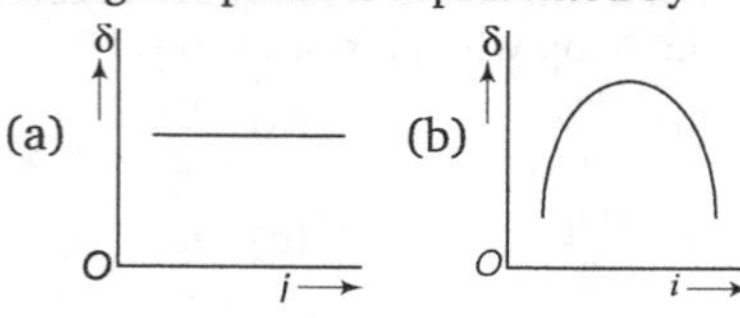

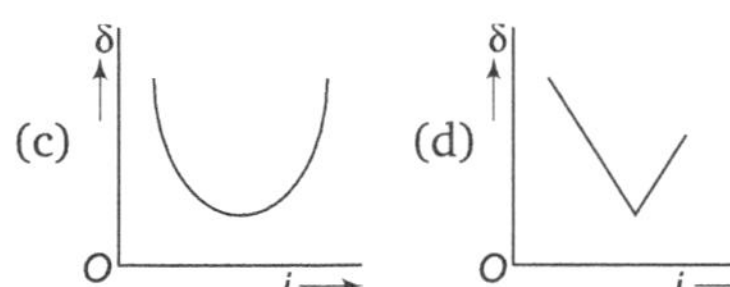


**28.** Two charges, each equal to $q$, are kept at $x = -a$ and $x = a$ on the $x$-axis. A particle of mass $m$ and charge $q_0 = \frac{q}{2}$ is placed at the origin. If charge $q_0$ is given a small displacement $y\ (y << a)$ along the $y$-axis, the net force acting on the particle is proportional to

(a) $y$ (b) $-y$ (c) $\frac{1}{y}$ (d) $-\frac{1}{y}$

**29.** Two short bar magnets of length 1 cm each have magnetic moments 1.20 $Am^2$ and 1.00 $Am^2$ respectively. They are placed on a hoirzontal table parallel to each other with their $N$ poles pointing towards the south. They have a common magnetic equator and are separated by a distance of 20.0 cm. The value of the resultant horizontal magnetic induction at the mid-point $O$ of the line joining their centres is close to (Horizontal component of the earth's magnetic induction is $3.6 \times 10^{-5}\ Wb/m^2$)

(a) $3.6 \times 10^{-5}\ Wb/m^2$
(b) $2.56 \times 10^{-4}\ Wb/m^2$
(c) $3.50 \times 10^{-4}\ Wb/m^2$
(d) $5.80 \times 10^{-4}\ Wb/m^2$

**30.** A charge $Q$ is uniformly distributed over a long rod $AB$ of length $L$ as shown in the figure. The electric potential at the point $O$ lying at distance $L$ from the end $A$ is

O ⟵ L ⟶ A [rod] B ⟵ L ⟶

(a) $\frac{Q}{8\pi\varepsilon_0 L}$ (b) $\frac{3Q}{4\pi\varepsilon_0 L}$
(c) $\frac{Q}{4\pi\varepsilon_0 L \ln 2}$ (d) $\frac{Q \ln 2}{4\pi\varepsilon_0 L}$

# CHEMISTRY

**1.** Which of the following complex species is not expected to exhibit optical isomerism?

(a) $[Co(en)_3]^{3+}$
(b) $[Co(en)_2Cl_2]^+$
(c) $[Co(NH_3)_3Cl_3]$
(d) $[Co(en)(NH_3)Cl_2]^+$

**2.** Which one of the following molecules is expected to exhibit diamagnetic behaviour?

(a) $C_2$ (b) $N_2$
(c) $O_2$ (d) $S_2$

**3.** A solution of $(-l)$ 1-chloro-1-phenylethane in toluene racemises slowly in the presence of a small amount of $SbCl_5$, due to the formation of

(a) carbanion (b) carbene
(c) carbocation (d) free radical

**4.** Given, $E^\circ_{Cr^{3+}/Cr} = -0.74$ V;

$E^\circ_{MnO_4^-/Mn^{2+}} = 1.51$ V

$E^\circ_{Cr_2O_7^{2-}/Cr^{3+}} = 1.33$ V;

$E^\circ_{Cl/Cl^-} = 1.36$ V

Based on the data given above strongest oxidising agent will be

(a) Cl (b) $Cr^{3+}$
(c) $Mn^{2+}$ (d) $MnO_4^-$

**5.** A piston filled with 0.04 mole of an ideal gas expands reversibly from 50.0 mL to 375 mL at a constant temperature of 37.0°C. As it does so, it absorbs 208 J of heat. The values of $q$ and $W$ for the process will be ($R = 8.314$ J/mol K, ln 7.5 = 2.01)

(a) $q = +208$ J, $W = -208$ J
(b) $q = -208$ J, $W = -208$ J
(c) $q = -208$ J, $W = +208$ J
(d) $q = +208$ J, $W = +208$ J

**6.** The molarity of a solution obtained by mixing 750 mL of 0.5 (M) HCl with 250 mL of 2(M) HCl will be

(a) 0.875 M (b) 1.00 M
(c) 1.75 M (d) 0.0975 M

**7.** Arrange the following compounds in the order of decreasing acidity

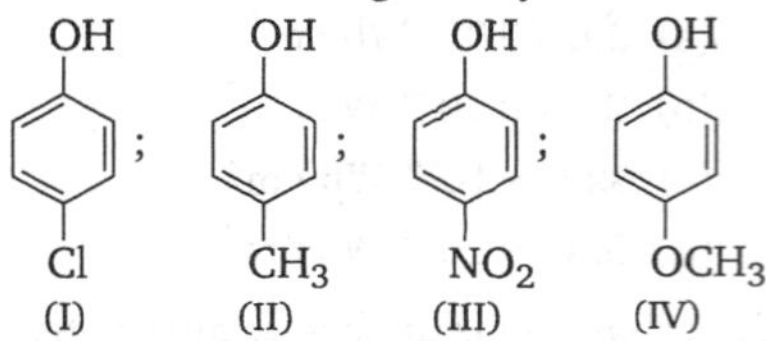


(a) II > IV > I > III (b) I > II > III > IV
(c) III > I > II > IV (d) IV > III > I > II

**8.** For gaseous state, if most probable speed is denoted by $C^*$, average speed by $\bar{C}$ and root square speed by $C$, then for a large number of molecules, the ratios of these speeds are

(a) $C^* : \bar{C} : C = 1.225 : 1.128 : 1$
(b) $C^* : \bar{C} : C = 1.128 : 1.225 : 1$
(c) $C^* : \bar{C} : C = 1 : 1.128 : 1.225$
(d) $C^* : \bar{C} : C = 1 : 1.225 : 1.128$

**9.** The rate of a reaction doubles when its temperature changes from 300 K to 310 K. Activation energy of such a reaction will be
($R = 8.314\ JK^{-1}\ mol^{-1}$ and
$\log 2 = 0.301$)

(a) 53.6 kJ $mol^{-1}$ (b) 48.6 kJ $mol^{-1}$
(c) 58.5 kJ $mol^{-1}$ (d) 60.5 kJ $mol^{-1}$

**10.** A compound with molecular mass 180 is acylated with $CH_3COCl$ to get a compound with molecular mass 390. The number of amino groups present per molecule of the former compound is

(a) 2 (b) 5
(c) 4 (d) 6

**11.** Which of the following arrangements does not represent the correct order of the property stated against it?

(a) $V^{2+} < Cr^{2+} < Mn^{2+} < Fe^{2+}$: paramagnetic behaviour
(b) $Ni^{2+} < Co^{2+} < Fe^{2+} < Mn^{2+}$ : ionic size
(c) $Co^{3+} < Fe^{3+} < Cr^{3+} < Sc^{3+}$ : stability in aqueous solution
(d) Sc < Ti < Cr < Mn : number of oxidation states

**12.** The order of stability of the following carbocations

$CH_2{=}CH{-}\overset{\oplus}{C}H_2$ ; $CH_3{-}CH_2{-}\overset{\oplus}{C}H_2$ ;
(I) (II)

$C_6H_5\overset{\oplus}{C}H_2$ is
(III)

(a) III > II > I (b) II > III > I
(c) I > II > III (d) III > I > II

**13.** Consider the following reaction,

$$xMnO_4^- + yC_2O_4^{2-} + zH^+ \longrightarrow xMn^{2+} + 2yCO_2 + \frac{z}{2}H_2O$$

The values of $x$, $y$ and $z$ in the reaction are, respectively

(a) 5, 2 and 16 (b) 2, 5 and 8
(c) 2, 5 and 16 (d) 5, 2 and 8

**14.** Which of the following is the wrong statement?

(a) ONCl and $ONO^-$ are not isoelectronic
(b) $O_3$ molecule is bent
(c) Ozone is violet-black in solid state
(d) Ozone is diamagnetic gas

**15.** A gaseous hydrocarbon gives upon combustion 0.72 g of water and 3.08 g of $CO_2$. The empirical formula of the hydrocarbon is

(a) $C_2H_4$ (b) $C_3H_4$ (c) $C_6H_5$ (d) $C_7H_8$

**16.** In which of the following pairs of molecules/ions both the species are not likely to exist?

(a) $H_2^+$, $He_2^{2-}$ (b) $H_2^-$, $He_2^{2-}$
(c) $H_2^{2+}$, $He_2$ (d) $H_2^-$, $He_2^{2+}$

**17.** Which of the following exists as covalent crystals in the solid state?

(a) Iodine (b) Silicon
(c) Sulphur (d) Phosphorus

**18.** Synthesis of each molecule of glucose in photosynthesis involves
(a) 18 molecules of ATP
(b) 10 molecules of ATP
(c) 8 molecules of ATP
(d) 6 molecules of ATP

**19.** The coagulating power of electrolytes having ions $Na^+$, $Al^{3+}$ and $Ba^{2+}$ for arsenic sulphide sol increases in the order
(a) $Al^{3+} < Ba^{2+} < Na^+$
(b) $Na^+ < Ba^{2+} < Al^{3+}$
(c) $Ba^{2+} < Na^{2+} < Al^{3+}$
(d) $Al^{3+} < Na^+ < Ba^{2+}$

**20.** Which of the following represents the correct order of increasing first ionization enthalpy for Ca, Ba, S, Se and Ar?
(a) Ca < S < Ba < Se < Ar
(b) S < Se < Ca < Ba < Ar
(c) Ba < Ca < Se < S < Ar
(d) Ca < Ba < S < Se < Ar

**21.** Energy of an electron is given by
$$E = -2.178 \times 10^{-18}\ \text{J}\left(\frac{Z^2}{n^2}\right)$$
Wavelength of light required to excite an electron in an hydrogen atom from level $n = 1$ to $n = 2$ will be
($h = 6.62 \times 10^{-34}$ Js and
$c = 3.0 \times 10^8\ \text{ms}^{-1}$)
(a) $1.214 \times 10^{-7}$ m
(b) $2.816 \times 10^{-7}$ m
(c) $6.500 \times 10^{-7}$ m
(d) $8.500 \times 10^{-7}$ m

**22.** Compound (*A*), $C_8H_9Br$ gives a white precipitate when warmed with alcoholic $AgNO_3$. Oxidation of (*A*) gives an acid (*B*), $C_8H_6O_4$. (*B*) easily forms anyhydride on heating. Identify the compound (*A*).

(a) $CH_2Br$, $CH_3$
(b) $C_2H_5$, Br
(c) $CH_2Br$, $CH_3$
(d) $CH_2Br$, $CH_3$

**23.** Four successive members of the first row transition elements listed below with atomic numbers. Which one of them is expected to have the highest $E^{\circ}_{M^{3+}/M^{2+}}$ value?
(a) Cr (Z = 24) (b) Mn (Z = 25)
(c) Fe (Z = 26) (d) Co (Z = 27)

**24.** How many litres of water must be added to 1 L of an aqueous solution of HCl with a pH of 1 to create an aqueous solution with pH of 2 ?
(a) 0.1 L (b) 0.9 L (c) 2.0 L (d) 9.0 L

**25.** The first ionisation potential of Na is 5.1 eV. The value of electron gain enthalpy of $Na^+$ will be
(a) – 2.55 eV (b) – 5.1 eV
(c) – 10.2 eV (d) + 2.55 eV

**26.** An organic compound *A* upon reacting with $NH_3$ gives *B*. On heating, *B* gives *C*. *C* in the presence of KOH reacts with $Br_2$ to give $CH_3CH_2NH_2$. *A* is
(a) $CH_3COOH$
(b) $CH_3CH_2CH_2COOH$
(c) $CH_3—CH(CH_3)—COOH$
(d) $CH_3CH_2COOH$

**27.** Stability of the species $Li_2$, $Li_2^-$ and $Li_2^+$ increases in the order of
(a) $Li_2 < Li_2^+ < Li_2^-$ (b) $Li_2^- < Li_2^+ < Li_2$
(c) $Li_2 < Li_2^- < Li_2^+$ (d) $Li_2^- < Li_2 < Li_2^+$

**28.** An unknown alcohol is treated with the "Lucas reagent" to determine whether the alcohol is primary, secondary or tertiary. Which alcohol reacts fastest and by what mechanism?
(a) Secondary alcohol by $S_N1$
(b) Tertiary alcohol by $S_N1$
(c) Secondary alcohol by $S_N2$
(d) Tertiary alcohol by $S_N2$

**29.** The gas leaked from a storage tank of the Union Carbide plant in Bhopal gas tragedy was
(a) Methyl isocyanate
(b) Methylamine
(c) Ammonia
(d) Phosgene

**30.** Experimentally it was found that a metal oxide has formula $M_{0.98}O$. Metal $M$, present as $M^{2+}$ and $M^{3+}$ in its oxide. Fraction of the metal which exists as $M^{3+}$ would be
(a) 7.01% (b) 4.08%
(c) 6.05% (d) 5.08%

# MATHEMATICS

**1.** Distance between two parallel planes $2x + y + 2z = 8$ and $4x + 2y + 4z + 5 = 0$ is
(a) $\frac{3}{2}$ (b) $\frac{5}{2}$ (c) $\frac{7}{2}$ (d) $\frac{9}{2}$

**2.** At present, a firm is manufacturing 2000 items. It is estimated that the rate of change of production $P$ with respect to additional number of workers $x$ is given by
$\frac{dP}{dx} = 100 - 12\sqrt{x}$. If the firm employees 25 more workers, then the new level of production of items is
(a) 2500 (b) 3000 (c) 3500 (d) 4500

**3.** Let $A$ and $B$ two sets containing 2 elements and 4 elements respectively. The number of subsets of $A \times B$ having 3 or more elements is
(a) 256 (b) 220 (c) 219 (d) 211

**4.** If the lines
$$\frac{x-2}{1} = \frac{y-3}{1} = \frac{z-4}{-k} \text{ and}$$
$$\frac{x-1}{k} = \frac{y-4}{2} = \frac{z-5}{1}$$
are coplanar, then $k$ can have
(a) any value
(b) exactly one value
(c) exactly two values
(d) exactly three values

**5.** If the vectors $\mathbf{AB} = 3\hat{\mathbf{i}} + 4\hat{\mathbf{k}}$ and $\mathbf{AC} = 5\hat{\mathbf{i}} - 2\hat{\mathbf{j}} + 4\hat{\mathbf{k}}$ are the sides of a $\Delta ABC$, then the length of the median through $A$ is
(a) $\sqrt{18}$ (b) $\sqrt{72}$
(c) $\sqrt{33}$ (d) $\sqrt{45}$

**6.** The real number $k$ for which the equation, $2x^3 + 3x + k = 0$ has two distinct real roots in $[0, 1]$
(a) lies between 1 and 2
(b) lies between 2 and 3
(c) lies between $-1$ and 0
(d) does not exist

**7.** The sum of first 20 terms of the sequence 0.7, 0.77, 0.777,..., is
(a) $\frac{7}{81}(179 - 10^{-20})$
(b) $\frac{7}{9}(99 - 10^{-20})$
(c) $\frac{7}{81}(179 + 10^{-20})$
(d) $\frac{7}{9}(99 + 10^{-20})$

**8.** A ray of light along $x + \sqrt{3}y = \sqrt{3}$ gets reflected upon reaching $x$-axis, the equation of the reflected ray is
(a) $y = x + \sqrt{3}$
(b) $\sqrt{3}y = x - \sqrt{3}$
(c) $y = \sqrt{3}x - \sqrt{3}$
(d) $\sqrt{3}y = x - 1$

**9.** The number of values of $k$, for which the system of equations
$$(k+1)x + 8y = 4k$$
$$kx + (k+3)y = 3k - 1$$
has no solution, is
(a) infinite (b) 1
(c) 2 (d) 3

**10.** If the equations $x^2 + 2x + 3 = 0$ and $ax^2 + bx + c = 0$, $a, b, c \in R$, have a common root, then $a : b : c$ is

(a) 1 : 2 : 3 (b) 3 : 2 : 1

(c) 1 : 3 : 2 (d) 3 : 1 : 2

**11.** The circle passing through (1, – 2) and touching the axis of $x$ at (3, 0) also passes through the point

(a) (– 5, 2) (b) (2, – 5)

(c) (5, – 2) (d) (– 2, 5)

**12.** If $x$, $y$ and $z$ are in AP and $\tan^{-1} x$, $\tan^{-1} y$ and $\tan^{-1} z$ are also in AP, then

(a) $x = y = z$ (b) $2x = 3y = 6z$

(c) $6x = 3y = 2z$ (d) $6x = 4y = 3z$

**13.** Consider

**Statement I**

$(p \wedge \sim q) \wedge (\sim p \wedge q)$ is a fallacy.

**Statement II**

$(p \rightarrow q) \leftrightarrow (\sim q \rightarrow \sim p)$ is a tautology.

(a) Statement I is true; Statement II is true; Statement II is a correct explanation for Statement I.

(b) Statement I is true; Statement II is true; Statement II is not a correct explanation for Statement I.

(c) Statement I is true; Statement II is false.

(d) Statement I is false; Statement II is true.

**14.** If $\int f(x)\, dx = \psi(x)$, then $\int x^5 f(x^3) dx$ is equal to

(a) $\frac{1}{3}[x^3\psi(x^3) - \int x^2\psi(x^3)dx] + C$

(b) $\frac{1}{3} x^3\psi(x^3) - 3\int x^3\psi(x^3)\, dx + C$

(c) $\frac{1}{3} x^3\psi(x^3) - \int x^2\psi(x^3)\, dx + C$

(d) $\frac{1}{3}[x^3\psi(x^3) - \int x^3\psi(x^3)\, dx] + C$

**15.** $\lim_{x \to 0} \frac{(1 - \cos 2x)(3 + \cos x)}{x \tan 4x}$ is equal to

(a) $-\frac{1}{4}$ (b) $\frac{1}{2}$ (c) 1 (d) 2

**16. Statement I** The value of the integral

$\int_{\pi/6}^{\pi/3} \frac{dx}{1 + \sqrt{\tan x}}$ is equal to $\pi/6$.

**Statement II**

$\int_a^b f(x)\, dx = \int_a^b f(a + b - x)\, dx$

(a) Statement I is true; Statement II is true; Statement II is a correct explanation for Statement I.

(b) Statement I is true; Statement II is true; Statement II is not a correct explanation for Statement I.

(c) Statement I is true; Statement II is false.

(d) Statement I is false; Statement II is true.

**17.** The equation of the circle passing through the foci of the ellipse $\frac{x^2}{16} + \frac{y^2}{9} = 1$ and having centre at (0, 3) is

(a) $x^2 + y^2 - 6y - 7 = 0$

(b) $x^2 + y^2 - 6y + 7 = 0$

(c) $x^2 + y^2 - 6y - 5 = 0$

(d) $x^2 + y^2 - 6y + 5 = 0$

**18.** A multiple choice examination has 5 questions. Each question has three alternative answers of which exactly one is correct. The probability that a student will get 4 or more correct answers just by guessing is

(a) $\frac{17}{3^5}$ (b) $\frac{13}{3^5}$

(c) $\frac{11}{3^5}$ (d) $\frac{10}{3^5}$

**19.** The $x$-coordinate of the incentre of the triangle that has the coordinates of mid-points of its sides as (0, 1), (1, 1) and (1, 0) is

(a) $2 + \sqrt{2}$

(b) $2 - \sqrt{2}$

(c) $1 + \sqrt{2}$

(d) $1 - \sqrt{2}$

**20.** The term independent of $x$ in expansion of $\left(\frac{x+1}{x^{2/3}-x^{1/3}+1}-\frac{x-1}{x-x^{1/2}}\right)^{10}$ is

(a) 4 (b) 120

(c) 210 (d) 310

**21.** The area (in square units) bounded by the curves
$y=\sqrt{x}$, $2y-x+3=0$, $x$-axis and lying in the first quadrant is

(a) 9 (b) 36

(c) 18 (d) $\frac{27}{4}$

**22.** Let $T_n$ be the number of all possible triangles formed by joining vertices of an $n$-sided regular polygon. If $T_{n+1}-T_n=10$, then the value of $n$ is

(a) 7 (b) 5

(c) 10 (d) 8

**23.** If $z$ is a complex number of unit modulus and argument $\theta$, then $\arg\left(\frac{1+z}{1+\bar{z}}\right)$ is equal to

(a) $-\theta$ (b) $\frac{\pi}{2}-\theta$

(c) $\theta$ (d) $\pi-\theta$

**24.** $ABCD$ is a trapezium such that $AB$ and $CD$ are parallel and $BC\perp CD$. If $\angle ADB=\theta$, $BC=p$ and $CD=q$, then $AB$ is equal to

(a) $\frac{(p^2+q^2)\sin\theta}{p\cos\theta+q\sin\theta}$

(b) $\frac{p^2+q^2\cos\theta}{p\cos\theta+q\sin\theta}$

(c) $\frac{p^2+q^2}{p^2\cos\theta+q^2\sin\theta}$

(d) $\frac{(p^2+q^2)\sin\theta}{(p\cos\theta+q\sin\theta)^2}$

**25.** If $P=\begin{bmatrix}1 & \alpha & 3\\ 1 & 3 & 3\\ 2 & 4 & 4\end{bmatrix}$ is the adjoint of a $3\times 3$ matrix $A$ and $|A|=4$, then $\alpha$ is equal to

(a) 4 (b) 11

(c) 5 (d) 0

**26.** The intercepts on $x$-axis made by tangents to the curve, $y=\int_0^x |t|\,dt$, $x\in R$, which are parallel to the line $y=2x$, are equal to

(a) $\pm 1$ (b) $\pm 2$ (c) $\pm 3$ (d) $\pm 4$

**27. Given** A circle, $2x^2+2y^2=5$ and a parabola, $y^2=4\sqrt{5}x$.

**Statement I** An equation of a common tangent to these curves is $y=x+\sqrt{5}$.

**Statement II** If the line, $y=mx+\frac{\sqrt{5}}{m}$ $(m\neq 0)$ is the common tangent, then $m$ satisfies $m^4-3m^2+2=0$.

(a) Statement I is true; Statement II is true; Statement II is a correct explanation for Statement I.

(b) Statement I is true; Statement II is true; Statement II is not a correct explanation for Statement I.

(c) Statement I is true; Statement II is false.

(d) Statement I is false; Statement II is true.

**28.** If $y=\sec(\tan^{-1}x)$, then $\frac{dy}{dx}$ at $x=1$ is equal to

(a) $\frac{1}{\sqrt{2}}$ (b) $\frac{1}{2}$ (c) 1 (d) $\sqrt{2}$

**29.** The expression $\frac{\tan A}{1-\cot A}+\frac{\cot A}{1-\tan A}$ can be written as

(a) $\sin A\cos A+1$

(b) $\sec A\operatorname{cosec} A+1$

(c) $\tan A+\cot A$

(d) $\sec A+\operatorname{cosec} A$

**30.** All the students of a class performed poorly in Mathematics. The teacher decided to give grace marks of 10 to each of the students. Which of the following statistical measures will not change even after the grace marks were given?

(a) Mean (b) Median

(c) Mode (d) Variance

# Answers

## Physics

| | | | | | | | | | |
|---|---|---|---|---|---|---|---|---|---|
| **1.** (c) | **2.** (d) | **3.** (d) | **4.** (b) | **5.** (b) | **6.** (c) | **7.** (b,c) | **8.** (b) | **9.** (a) | **10.** (c) |
| **11.** (a) | **12.** (b) | **13.** (c) | **14.** (d) | **15.** (b) | **16.** (c) | **17.** (c) | **18.** (c) | **19.** (d) | **20.** (c) |
| **21.** (d) | **22.** (b) | **23.** (d) | **24.** (a) | **25.** (d) | **26.** (d) | **27.** (c) | **28.** (a) | **29.** (b) | **30.** (d) |

## Chemistry

| | | | | | | | | | |
|---|---|---|---|---|---|---|---|---|---|
| **1.** (c) | **2.** (a.b) | **3.** (c) | **4.** (d) | **5.** (a) | **6.** (a) | **7.** (c) | **8.** (c) | **9.** (a) | **10.** (b) |
| **11.** (a) | **12.** (d) | **13.** (c) | **14.** (*) | **15.** (d) | **16.** (c) | **17.** (b) | **18.** (a) | **19.** (b) | **20.** (c) |
| **21.** (a) | **22.** (d) | **23.** (d) | **24.** (d) | **25.** (b) | **26.** (d) | **27.** (b) | **28.** (b) | **29.** (a) | **30.** (b) |

## Mathematics

| | | | | | | | | | |
|---|---|---|---|---|---|---|---|---|---|
| **1.** (c) | **2.** (c) | **3.** (c) | **4.** (c) | **5.** (c) | **6.** (d) | **7.** (c) | **8.** (b) | **9.** (b) | **10.** (a) |
| **11.** (c) | **12.** (a) | **13.** (b) | **14.** (c) | **15.** (d) | **16.** (d) | **17.** (a) | **18.** (c) | **19.** (b) | **20.** (c) |
| **21.** (a) | **22.** (b) | **23.** (c) | **24.** (a) | **25.** (b) | **26.** (a) | **27.** (b) | **28.** (a) | **29.** (b) | **30.** (d) |

# Hints & Solutions

## Physics

**1.** In equilibrium,

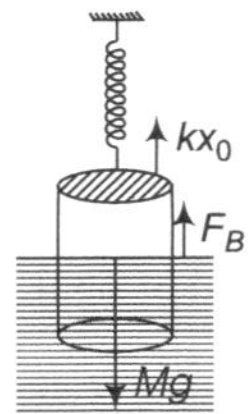


Upward force = Downward force

$$kx_0 + F_B = mg$$

Here, $kx_0$ is restoring force of spring and $F_B$ is buoyancy force.

$$kx_0 + \sigma \frac{L}{2} Ag = Mg$$

$$x_0 = \frac{Mg - \frac{\sigma LAg}{2}}{k}$$

$$= \frac{Mg}{k}\left(1 - \frac{\sigma LA}{2M}\right)$$

**2.** $e = \int_{2l}^{3l} (\omega x)\, B dx = B\omega \dfrac{[(3l)^2 - (2l)^2]}{2}$

$$= \frac{5Bl^2\omega}{2}$$

**3.** Maximum energy loss

$$= \frac{p^2}{2m} - \frac{p^2}{2(m+M)} \qquad \left(\because \text{KE} = \frac{p^2}{2m}\right)$$

Before collision the mass $m$ and after collision the mass is $m + M$

$$= \frac{p^2}{2m}\left[\frac{M}{(m+M)}\right]$$

$$= \frac{1}{2} mv^2 \left\{\frac{M}{m+M}\right\} \qquad \left(f = \frac{M}{m+M}\right)$$

**4.** From Coulomb's law $F = \frac{1}{4\pi\varepsilon_0} \frac{q_1 q_2}{R^2}$

$$\varepsilon_0 = \frac{q_1 q_2}{4\pi F R^2}$$

Substituting the units we have,

$$\varepsilon_0 = \frac{C^2}{N\text{-}m^2} = \frac{[AT]^2}{[MLT^{-2}]\,[L^2]}$$

$$= [M^{-1}L^{-3}T^4A^2]$$

**5.** Initial velocity $= (\mathbf{i} + 2\mathbf{j})$ m/s

Magnitude of initial velocity $u = \sqrt{(1)^2 + (2^2)} = \sqrt{5}$ m/s

Equation of trajectory of projectile is

$$y = x\tan\theta - \frac{gx^2}{2u^2}(1 + \tan^2\theta)$$

$$\left[\tan\theta = \frac{y}{x} = \frac{2}{1} = 2\right]$$

$$\therefore \quad y = x \times 2 - \frac{10\,(x)^2}{2(\sqrt{5})^2}[1 + (2)^2]$$

$$= 2x - \frac{10\,(x^2)}{2 \times 5}(1 + 4)$$

$$= 2x - 5x^2$$

**6.** Amplitude decreases exponentially. In 5 s it remains 0.9 times. Therefore in total 15 s it will remains (0.9) (0.9) (0.9) = 0.729 times its original value.

**7.** Polarity should be mentioned in the question. Potential on each of them can be zero if, $q_{net} = 0$

or $\quad q_1 \pm q_2 = 0$

or $\quad 120C_1 \pm 200C_2 = 0$

or $\quad 3C_1 \pm 5C_2 = 0$

**8.** Fundamental frequency of sonometer wire

$$f = \frac{v}{2l} = \frac{1}{2l}\sqrt{\frac{T}{\mu}}$$

$$= \frac{1}{2l}\sqrt{\frac{T}{Ad}}$$

Here, $\mu$ = mass per unit length of wire.

Also, Young's modulus of elasticity

$$Y = \frac{Tl}{A\Delta l}$$

$$\Rightarrow \quad \frac{T}{A} = \frac{Y\Delta l}{l} \Rightarrow f = \frac{1}{2l}\sqrt{\frac{Y\Delta l}{ld}}$$

$$l = 1.5 \text{ m}, \; d = 7.7 \times 10^3 \text{ kg/m}^3$$

$$Y = 2.2 \times 10^{11} \text{ N/m}^2$$

After substituting the values we get,

$$f \approx 178.2 \text{ Hz}$$

**9.** Magnetic field at the centre of smaller loop

$$B = \frac{\mu_0 i R_2^2}{2(R_2^2 + x^2)^{3/2}}$$

Area of smaller loop $S = \pi R_1^2$

$\therefore$ Flux through smaller loop $\phi = BS$

Substituting the values we get,

$$\phi \approx 9.1 \times 10^{-11} \text{ Wb}$$

**10.** By Pythagoras theorem

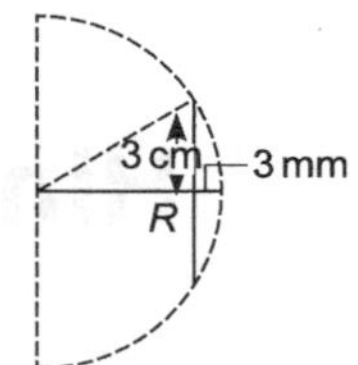


$$R^2 = (3)^2 + (R - 0.3)^2$$

$$\Rightarrow \quad R \approx 15 \text{ cm}$$

Refractive index of material of lens

$$\mu = \frac{c}{v}$$

Here $c$ = speed of light in vacuum $= 3 \times 10^8$ m/s

$v$ = speed of light in material of lens $= 2 \times 10^8$ m/s

$$= \frac{3 \times 10^8}{2 \times 10^8} = \frac{3}{2}$$

From lens maker's formula

$$\frac{1}{f} = (\mu - 1)\left(\frac{1}{R_1} - \frac{1}{R_2}\right)$$

Here, $R_1 = R$ and $R_2 = \infty$ (For plane surface)

$$\frac{1}{f} = \left(\frac{3}{2} - 1\right)\left(\frac{1}{15}\right)$$

$$\Rightarrow \quad f = 30 \text{ cm}$$

**11.** $E$ = Energy of satellite – energy of mass on the surface of planet

$$= -\frac{GMm}{2r} - \left(-\frac{GMm}{R}\right)$$

Here, $r = R + 2R = 3R$

Substituting in about equation we get,

$$E = \frac{5GMm}{6R}$$

**12.**

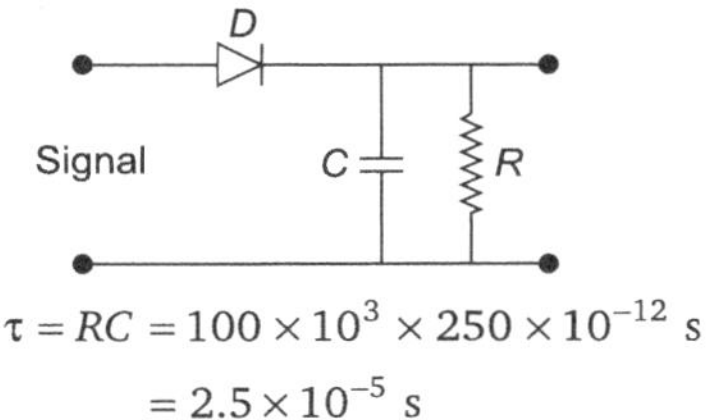


$$\tau = RC = 100 \times 10^3 \times 250 \times 10^{-12} \text{ s}$$
$$= 2.5 \times 10^{-5} \text{ s}$$

The higher frequency which can be detected with tolerable distortion is

$$f = \frac{1}{2\pi m_a RC}$$
$$= \frac{1}{2\pi \times 0.6 \times 2.5 \times 10^{-5}} \text{ Hz}$$
$$= \frac{100 \times 10^4}{25 \times 1.2\pi} \text{ Hz}$$
$$= \frac{4}{1.2\pi} \times 10^4 \text{ Hz}$$
$$= 10.61 \text{ kHz}$$

**13.** Relation between intensities is

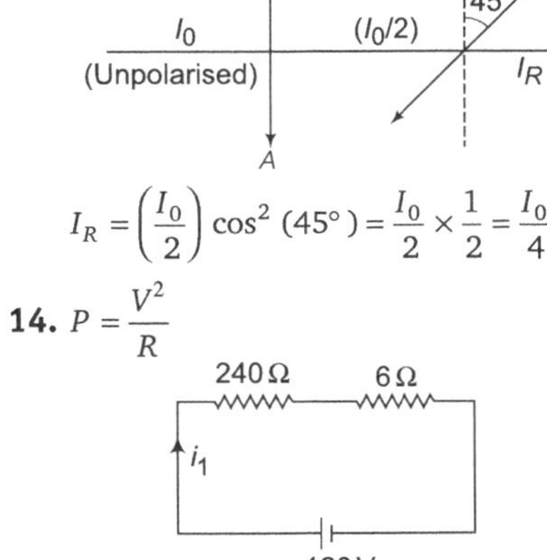


$$I_R = \left(\frac{I_0}{2}\right) \cos^2(45°) = \frac{I_0}{2} \times \frac{1}{2} = \frac{I_0}{4}$$

**14.** $P = \dfrac{V^2}{R}$

240 Ω 6 Ω
$i_1$
120 V

$$R = \frac{120 \times 120}{60} = 240\ \Omega$$
$$R_{eq} = 240 + 6 = 246\ \Omega$$
$$\Rightarrow \quad i_1 = \frac{V}{R_{eq}} = \frac{120}{246}$$

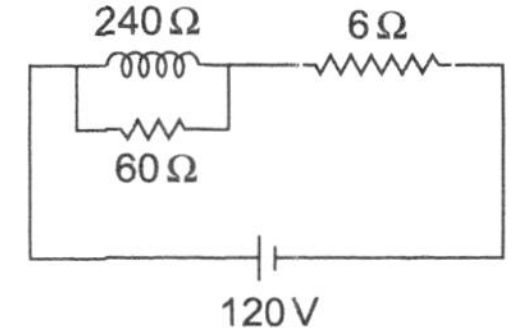


$$V_1 = \frac{240}{246} \times 120 = 117.073 \text{ V}$$
$$\Rightarrow \quad i_2 = \frac{120}{48 + 6}$$

48 Ω 6 Ω
$i_2$
120 V

$$V_2 = \frac{48}{54} \times 120$$
$$= 106.66 \text{ V}$$
$$V_1 - V_2 = 10.04 \text{ V}$$

**15.** Heat is extracted from the source means heat is given to the system (or gas). or $Q$ is positive. This is positive only along the path $ABC$.

Heat supplied

$$\therefore \quad Q_{ABC} = \Delta U_{ABC} + W_{ABC}$$
$$= nC_V (T_f - T_i) + \text{Area under } p\text{-}V \text{ graph}$$
$$= n\left(\frac{3}{2}R\right)(T_C - T_A) + 2p_0V_0$$
$$= \frac{3}{2}(nRT_C - nRT_A) + 2p_0V_0$$
$$= \frac{3}{2}(p_CV_C - p_AV_A) + 2p_0V_0$$
$$= \frac{3}{2}(4p_0V_0 - p_0V_0) + 2p_0V_0$$
$$= \frac{13}{2} p_0V_0$$

**16.**

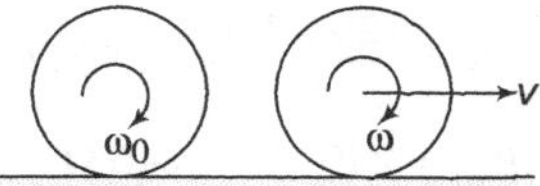


$$\omega = v/r$$

From conservation of angular momentum about bottommost point

$$mr^2\omega_0 = mvr + mr^2 \times \frac{v}{r}$$

$$\Rightarrow \quad v = \frac{\omega_0 r}{2}$$

**17.** In equilibrium,

$$p_0 A = Mg \qquad \ldots(\text{i})$$

when slightly displaced downwards,

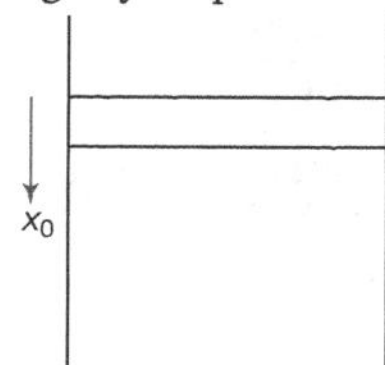


$$dp = -\gamma\left(\frac{p_0}{V_0}\right)dV$$

$\left(\text{As in adiabatic process, } \frac{dp}{dV} = -\gamma\,\frac{p}{V}\right)$

$\therefore$ Restoring force,

$$F = (dp)A = -\left(\frac{\gamma p_0}{V_0}\right)(A)(Ax)$$

$$F \propto -x$$

Therefore, motion is simple harmonic comparing with $F = -Kx$ we have :

$$K = \frac{\gamma p_0 A^2}{V_0}$$

$$\therefore \quad f = \frac{1}{2\pi}\sqrt{\frac{K}{m}} = \frac{1}{2\pi}\sqrt{\frac{\gamma p_0 A^2}{MV_0}}$$

**18.** According to Newton's cooling law, option (c) is correct answer.

**19.** Statement I is false and Statement II is true.

**20.** For charging of capacitor

$$q = CV\,(1 - e^{t/\tau})$$

at $t = 2\tau \Rightarrow q = CV\,(1 - e^{-2})$

**21.** It will be concentric circles.

**22.** Peak value of electric field

$$E_0 = B_0 c = 20 \times 10^{-9} \times 3 \times 10^8$$
$$= 6 \text{ V/m}$$

**23.** As $\lambda$ is increased, there will be a value of $\lambda$ above which photoelectron will cease to come out. So, photocurrent will be zero.

**24.** For same value of current higher value of voltage is required for higher frequency.

**25.** Decrease in surface energy = heat required in vaporization.

$$\therefore \quad T\,(dS) = L(dm)$$

$$\therefore T(2)(4\pi r)\,dr = L(4\pi r^2 dr)\rho$$

$$\therefore \quad r = \frac{2T}{\rho L}$$

**26.** $\Delta E = h\nu$

$$\nu = \frac{\Delta E}{h} = k\left[\frac{1}{(n-1)^2} - \frac{1}{n^2}\right]$$

$$= \frac{k2n}{n^2(n-1)^2}$$

$$\approx \frac{2k}{n^3} \propto \frac{1}{n^3}$$

**27.** We know that the angle of deviation depends upon the angle of incidence. If we determine experimentally, the angles of deviation corresponding to different angles of incidence and then plot $i$ (on-$x$-axis) and $\delta$ (on-$y$-axis), we get a curve as shown in figure.

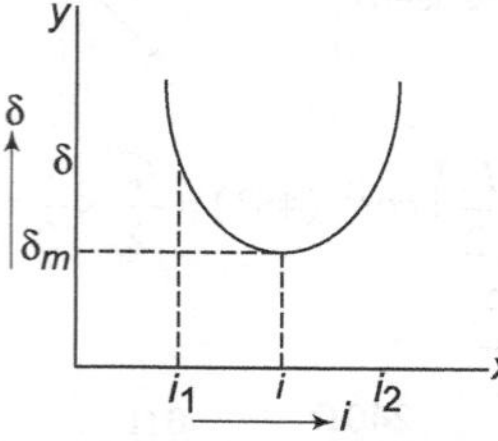


Clearly if angle of incidence is gradually increased, from a small value, the angle of deviation first decreases, becomes minimum for a particular angle of incidence and then begins to increase.

**28.**

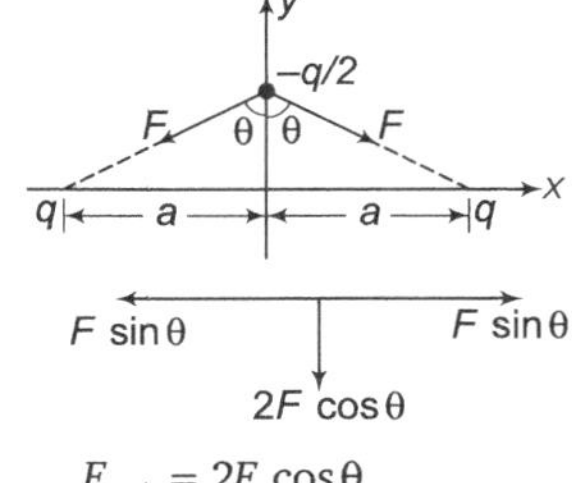


$$F_{net} = 2F\cos\theta$$

$$F_{net} = \frac{2kq\left(\frac{q}{2}\right)}{(\sqrt{y^2+a^2})^2}\cdot\frac{y}{\sqrt{y^2+a^2}}$$

$$F_{net} = \frac{2kq\left(\frac{q}{2}\right)y}{(y^2+a^2)^{3/2}} \Rightarrow \frac{kq^2y}{a^3} \propto y$$

**29.** $B_{net} = B_1 + B_2 + B_H$

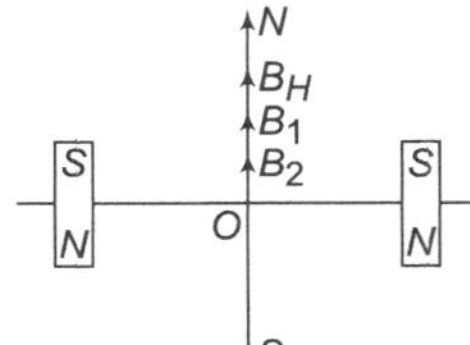


$$B_{net} = \frac{\mu_0}{4\pi}\frac{(M_1+M_2)}{r^3} + B_H$$

$$= \frac{10^{-7}(1.2+1)}{(0.1)^3} + 3.6\times10^{-5}$$

$$= 2.56\times10^{-4}\ \text{Wb/m}^2$$

**30.** O, A, B; $L$, $L$; $x$, $dx$

$$V = \int_L^{2L}\frac{k\,dQ}{x}$$

$$= \int_L^{2L}\frac{k\left(\frac{Q}{L}\right)dx}{x}$$

$$= \frac{Q}{4\pi\varepsilon_0 L}\int_L^{2L}\left(\frac{1}{x}\right)dx$$

$$= \frac{Q}{4\pi\varepsilon_0 L}[\log_e x]_L^{2L}$$

$$= \frac{Q}{4\pi\varepsilon_0 L}[\log_e 2L - \log_e L]$$

$$= \frac{Q}{4\pi\varepsilon_0 L}\ln(2)$$

# Chemistry

**1.** Optical isomerism is exhibited by only those complexes which lacks elements of symmetry.

$[Co(NH_3)_3Cl_3]$ shows facial as well as meridional isomerism. But both the forms contain plane of symmetry. Thus, this complex does not exhibit optical isomerism.

**2.** $C_2\,(6+6=12) = \sigma 1s^2, \overset{*}{\sigma}\,1s^2,$
$\sigma 2s^2, \overset{*}{\sigma}\,2s^2,\ \pi 2p_x^2 \approx \pi 2p_y^2$

Since, all the electrons are paired, it is a diamagnetic species.

$N_2\,(7+7=14) = \sigma 1s^2, \overset{*}{\sigma}\,1s^2, \sigma 2s^2,$
$\overset{*}{\sigma}\,2s^2,\ \pi 2p_x^2 \approx \pi 2p_y^2, \sigma 2p_z^2$

It is also a diamagnetic species because of the absence of unpaired electrons.

$O_2\,(8+8=16)$

or $S_2 = \sigma 1s^2, \overset{*}{\sigma}\,1s^2, \sigma 2s^2, \overset{*}{\sigma}\,2s^2,$
$\sigma 2p_z^2, \pi 2p_x^2 \approx \pi 2p_y^2$
$\overset{*}{\pi}\,2p_x^1 \approx \overset{*}{\pi}\,2p_y^1$

Due to the presence of two unpaired electrons $O_2$ and $S_2$ both are paramagnetic molecules.

**3.** $Cl—\underset{\underset{(-)}{Ph}}{C}H—CH_3 \xrightarrow[\text{Toluene}]{SbCl_5}$

$Ph—\overset{\oplus}{C}H—CH_3 + SbCl_6^-$
(Carbocation)
(planar)

$\longrightarrow Ph—\underset{Cl}{C}H—CH_3$
($d$ and $l$) mixture
$+ SbCl_5$

**4.** Higher the SRP, better is oxidising agent. Among the given $E^{\circ}_{MnO_4^-/Mn^{2+}}$ is highest, hence $MnO_4^-$ is the strongest oxidising agent.

**5.** The process is isothermal expansion, hence

$$Q = -W$$
$$\Delta E = 0$$
$$W = 2.303\, nRT \log \frac{V_2}{V_1}$$
$$= -2.303 \times 0.04 \times 8.314 \times 310 \times \log \frac{335}{50}$$
$$= -208 \text{ J}$$
$$Q = +208 \text{ J}$$
$$W = -208 \text{ J (expansion work)}$$

**6.** $$M_f = \frac{M_1V_1 + M_2V_2}{V_1 + V_2} = \frac{750 \times 0.5 + 250 \times 2}{750 + 250} = \frac{875}{1000} = 0.875 \text{ M}$$

OH-C6H4-$NO_2$ (−M, −I) > OH-C6H4-Cl (−I) > OH-C6H4-$CH_3$ (+I, hyper conjugation) > OH-C6H4-$OCH_3$ (+M)

Electron releasing group decreases while electron withdrawing group increases acidic strength by destabilising and stabilising the phenoxide ion formed respectively.

**8.** $C^* =$ most probable speed $= \sqrt{\frac{2RT}{M}}$

speed $= \sqrt{\frac{8RT}{\pi M}}$

$C =$ Mean square speed corrected as

rms $= \sqrt{\frac{3RT}{M}}$,

$$\overset{*}{C} < \bar{C} < C$$
$$\overset{*}{C} : \bar{C} : C = 1 : \sqrt{\frac{4}{\pi}} : \sqrt{\frac{3}{2}}$$
$$= 1 : 1.128 : 1.225$$

As no option correspond to mean square speed, it is understood as misprint. It should be root mean square speed.

**9.** From Arrhenius equation,

$$\log \frac{k_2}{k_1} = \frac{-E_a}{2.303\,R}\left(\frac{1}{T_2} - \frac{1}{T_1}\right)$$

Given, $\frac{k_2}{k_1} = 2; T_2 = 310$ K

$T_1 = 300$ K

On putting values,

$$\Rightarrow \log 2 = \frac{-E_a}{2.303 \times 8.314}\left(\frac{1}{310} - \frac{1}{300}\right)$$
$$\Rightarrow E_a = 53598.6 \text{ J/mol}$$
$$= 53.6 \text{ kJ/mol}$$

**10.** $R\text{—NH}_2 + CH_3\text{—}\overset{O}{\overset{\|}{C}}\text{—Cl} \xrightarrow[(-HCl)]{} R\text{—NH—}\overset{O}{\overset{\|}{C}}\text{—CH}_3$

Since each $\text{—COCH}_3$ group displace one H atom in the reaction of one mole of $CH_3\text{—}\overset{O}{\overset{\|}{C}}\text{—Cl}$ with one $\text{—NH}_2$ group, the molecular mass increases with 42 unit. Since the mass increases by $(390 - 180) = 210$ hence the number of $\text{—NH}_2$ group is $\frac{210}{42} = 5$.

**11.** (a) $V^{2+} = 3$ unpaired electrons

$Cr^{2+} = 4$ unpaired electrons

$Mn^{2+}$ unpaired electrons

$Fe^{2+} = 4$ unpaired electrons

Hence, the order of paramagnetic behaviour should be

$$V^{2+} < Cr^{2+} = Fe^{2+} < Mn^{2+}$$

(b) Ionic size decreases from left to right in the same period.

(c) (As per data from NCERT)

$Co^{3+}/Co^{2+} = 1.97;$

$Fe^{3+}/Fe^{2+} = 0.77;$

$Cr^{3+}/Cr^{2+} = -0.41$

$Sc^{3+}$ is highly stable (It does not show + 2).

(d) The oxidation states increases as we go from group 3 to group 7 in the same period.

**12.** The order of stability of carbocation will be

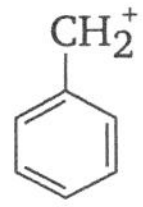

$C_6H_5\overset{+}{C}H_2 > CH_2{=}CH{-}\overset{+}{C}h_2$

Benzyl (more resonance stabilised)  Allyl (resonance stabilised)

$> CH_2{-}CH_2{-}\overset{+}{C}H_2$

Propyl (stabilised by $+I$ effect)

**13.** The half equations of the reaction are

$MnO_4^- \longrightarrow Mn^{2+}$

$C_2O_4^{2-} \longrightarrow CO_2$

The balanced half equations are

$MnO_4^- + 8H^+ + 5e^- \longrightarrow Mn^{2+} + 4H_2O$

$C_2O_4^{2-} \longrightarrow 2CO_2 + 2e^-$

On equating number of electrons, we get

$2MnO_4^- + 16H^+ + 10e^- \longrightarrow 2Mn^{2+} + 8H_2O$

$5C_2O_4^{2-} \longrightarrow 10CO_2 + 10e^-$

On adding both the equations, we get

$2MnO_4^- + 5C_2O_4^- + 16H^+ \longrightarrow 2Mn^{2+} + 2\times 5CO_2 + \frac{16}{2} H_2O$

Thus $x, y$ and $z$ are 2, 5 and 16 respectively.

**14.** (a) $ONCl = 8 + 7 + 17 = 32e^-$

$ONO^- = 8 + 7 + 8 + 1 = 24e^-$

(correct)

(b) 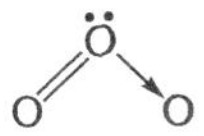 $\ddot{O}$ with $O{=}O$ double bond and coordinate bond $O \rightarrow O$ (bent)

Central O atom is $sp^2$ hybridised with 1 lone pair, so bent shape (correct).

(c) In solid state, ozone is violet-black ozone does not exist in solid state.

(d) $O_3$ has no unpaired electrons, so diamagnetic (correct).

*No option is correct.

**15.** 18 g $H_2O$ contains 2gH

$\therefore$ 0.72 g $H_2O$ contains 0.08 gH

44 g $CO_2$ contains 12 gC

$\therefore$ 3.08 g $CO_2$ contains 0.84 gC

$\therefore \quad C : H = \frac{0.84}{12} : \frac{0.08}{1}$

$= 0.07 : 0.08$

$= 7 : 8$

$\therefore$ Empirical formula = $C_7H_8$

**16.** Species having zero or negative bond order do not exist.

$H_2^{2+} (1 + 1 - 2 = 0) = \sigma 1s^0$

Bond order = 0

$He_2 \ (2 + 2 = 4) = \sigma 1s^2, \overset{*}{\sigma} 1s^2$

Bond order $= \frac{N_b - N_a}{2}$

$= \frac{2-2}{2} = 0$

So, both $H_2^{2+}$ and $He_2$ do not exist.

**17.** Silicon exists as covalent crystal in solid state. (Network like structure, like diamond).

**18.** $6CO_2 + 12NADPH + 18\ ATP \longrightarrow C_6H_{12}O_6 + 12NADP + 18\ ADP$

**19.** According to Hardy Schulze rule, greater the charge on oppositely charged ion, greater is its coagulating power. Since arsenic sulphide is a negatively charged sol, thus, the order of coagulating power is $Na^+ < Ba^{2+} < Al^{3+}$.

**20.** Ionisation energy increases along a period from left to right and decreases down a group. The position of given elements in the periodic table is as

| 2 | 16 | 18 |
|---|---|---|
| Ca | S | Ar |
| Ba | Se | |

Thus, the order of increasing $\Delta H_{IE_1}$ is Ba < Ca < Se < S < Ar

**21.** $\Delta E = 2.178 \times 10^{-18} \left(\frac{1}{1^2} - \frac{1}{2^2}\right) = \frac{hc}{\lambda}$

$$2.178 \times 10^{-18} \left(\frac{1}{1^2} - \frac{1}{2^2}\right) = \frac{6.62 \times 10^{-34} \times 3.0 \times 10^8}{\lambda}$$

$\therefore \lambda \approx 1.21 \times 10^{-7}$ m

**22.** Compound $A$ gives a precipitate with alcoholic $AgNO_3$, so it must contains Br in side chain. On oxidation, it gives $C_8H_6O_4$, which shows the presence of two alkyl chains attached directly with the benzene nucleus.

Since compound $B$ gives anhydride on heating, the two alkyl substituent must occupy adjacent (1, 2) position. Thus, $A$ must be

(benzene ring with $CH_2Br$ and $CH_3$ at adjacent positions)

and the reactions are as follows

(benzene with $C_2H_5$ and $CH_3$) $\xrightarrow[AgNO_3]{\text{Alcoholic}}$ (benzene with $CH_2$—$OR$ and $CH_3$) + AgBr↓

↓ Oxidation

(benzene with COOH and COOH) $\xrightarrow{\Delta}$ Phthalic anhydride

**23.** SRP value normally increases from left to right in the period of $d$-block elements. Some SRP value are exceptionally higher due to stability of product ion. *e.g.*,

$$E^{\circ}_{Mn^{3+}/Mn^{2+}} = +1.57 \text{ V};$$

$$E^{\circ}_{Co^{3+}/Co^{2+}} = +1.97 \text{ V}$$

Thus, $E^{\circ}_{M^{3+}/M^{2+}}$ is highest for Co.

**24.** $pH = 1 \therefore [H^+] = 10^{-1} = 0.1$ M

$pH = 2 \therefore [H^+] = 10^{-2} = 0.01$ M

For dilution of HCl $M_1V_1 = M_2V_2$

$0.1 \times 1 = 0.01 \times V_2$

$V_2 = 10$ L

Volume of water to be added $= 10 - 1 = 9$ L

**25.** $Na \longrightarrow Na^+ + e^-$ First IE

$Na^+ + e^- \longrightarrow Na$

Electron gain enthalpy of $Na^+$ is reverse of (IE)

Because reaction is reverse so

$\Delta H(eq) = -5.1$ eV

**26.** $\underset{(A)}{CH_3CH_2—\overset{O}{\overset{\|}{C}}—OH} \xrightarrow{NH_3} \underset{(B)}{CH_3CH_2COONH_4} \xrightarrow{\Delta}$

$\underset{(C)}{CH_3—CH_2—\overset{O}{\overset{\|}{C}}—NH_2} \xrightarrow[\text{Hofmann bromamide reaction}]{Br_2, KOH} CH_3—CH_2—NH_2$

**27.** $Li_2$ $(3 + 3 = 6) = \sigma 1s^2, \overset{*}{\sigma} 1s^2, \sigma 2s^2$

Bond order $= \frac{N_b - N_a}{2} = \frac{4-2}{2} = 1$

$Li_2^+$ $(3 + 3 - 1 = 5) = \sigma 1s^2, \overset{*}{\sigma} 1s^2, \sigma 2s^1$ B ond order $= \frac{3-2}{2} = \frac{1}{2} = 0.5$

$Li_2^-$ $(3 + 3 + 1 = 7) = \sigma 1s^2, \overset{*}{\sigma} 1s^2, \sigma 2s^2 \overset{*}{\sigma} 2s^1$

Bond order $= \frac{4-3}{2} = \frac{1}{2} = 0.5$

Stability order is $Li_2 > Li_2^+ > Li_2^-$

(because $Li_2^-$ has more number of electrons in antibonding orbitals which destabilises the species)

**28.** The reaction of alcohol with Lucas reagent is mostly an $S_N1$ reaction and the rate of reaction is directly proportional to the stability of carbocation formed in the reaction. Since 3° $R$—OH forms 3° carbocation (most stable) hence it will react fastest.

**29.** Methyl isocyanate $CH_3—N=C=O$ (MIC gas) gas leaked from the storage tank of the union carbide plant in Bhopal gas tragedy.

**30.** From the valency of $M^{2+}$ and $M^{3+}$, it is clear that three $M^{2+}$ ions will be replaced by $M^{3+}$ causing a loss of one $M^{3+}$ ion. Total loss of than from one molecule of Mo = 1 – 0.98 = 0.02

Total $M^{3+}$ present in one molecule of

$$Mo = 2 \times 0.02 = 0.04$$

That $M^{2+}$ and $M^{3+} = 0.98$

Thus $\quad \%$ of $M^{3+} = \dfrac{0.04 \times 100}{0.98}$

$$= 4.08\%$$

# Mathematics

**1.** Given planes are

$$2x + y + 2z - 8 = 0$$

and $2x + y + 2z + \dfrac{5}{2} = 0$

Distance between two planes

$$= \frac{|c_1 - c_2|}{\sqrt{a^2 + b^2 + c^2}} = \left| \frac{-8 - \frac{5}{2}}{\sqrt{2^2 + 1^2 + 2^2}} \right|$$

$$= \frac{\frac{21}{2}}{3} = \frac{7}{2}$$

**2.** Given, $\dfrac{dP}{dx} = (100 - 12\sqrt{x})$

$\Rightarrow \quad dP = (100 - 12\sqrt{x})\, dx$

On integrating both sides, we get

$$\int dP = \int (100 - 12\sqrt{x})\, dx$$

$$P = 100x - 8x^{3/2} + C$$

When $\quad x = 0$, then $P = 2000$

$\Rightarrow \quad C = 2000$

Now, when $x = 25$, then is

$$P = 100 \times 25 - 8 \times (25)^{3/2} + 2000$$

$$= 2500 - 8 \times 125 + 2000$$

$$= 4500 - 1000 = 3500$$

**3.** Given, $n(A) = 2$, $n(B) = 4$

$\therefore \quad n(A \times B) = 8$

The number of subsets of $A \times B$ having 3 or more elements

$$= {}^8C_3 + {}^8C_4 + \ldots + {}^8C_8$$

$$= ({}^8C_0 + {}^8C_1 + {}^8C_2 + {}^8C_3 + \ldots + {}^8C_8) - ({}^8C_0 + {}^8C_1 + {}^8C_2)$$

$\therefore \quad 2^n = {}^nC_0 + {}^nC_1 + \ldots + {}^nC_n$

$$= 2^8 - {}^8C_0 - {}^8C_1 - {}^8C_2$$

$$= 256 - 1 - 8 - 28 = 219$$

**4.** Condition for two lines are coplanar.

$$\begin{vmatrix} x_1 - x_2 & y_1 - y_2 & z_1 - z_2 \\ l_1 & m_1 & n_1 \\ l_2 & m_2 & n_2 \end{vmatrix} = 0$$

where, $(x_1, y_1, z_1)$ and $(x_2, y_2, z_2)$ are the points lie on a line (i) and (ii) respectively and $< l_1, m_1, n_1 >$ and $< l_2, m_2, n_2 >$ are the direction cosines of the line (i) and line (ii) respectively.

$$\therefore \quad \begin{vmatrix} 2-1 & 3-4 & 4-5 \\ 1 & 1 & -k \\ k & 2 & 1 \end{vmatrix} = 0$$

$$\Rightarrow \quad \begin{vmatrix} 1 & -1 & -1 \\ 1 & 1 & -k \\ k & 2 & 1 \end{vmatrix} = 0$$

$\Rightarrow \quad 1(1 + 2k) + (1 + k^2) - (2 - k) = 0$

$\Rightarrow \quad k^2 + 2k + k = 0$

$\Rightarrow \quad k^2 + 3k = 0 \Rightarrow k = 0, -3$

If 0 appears in the denominator, then the correct way of representing the equation of straight line is

$$\frac{x-2}{1} = \frac{y-3}{1};\ z = 4$$

**5.** We know that, the sum of three vectors of a triangle is zero.

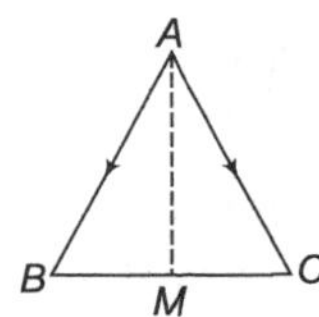


$\therefore\ \mathbf{AB} + \mathbf{BC} + \mathbf{CA} = 0$

$\Rightarrow \qquad \mathbf{BC} = \mathbf{AC} - \mathbf{AB}$

$\Rightarrow \qquad \mathbf{BM} = \dfrac{\mathbf{AC} - \mathbf{AB}}{2}$

($\because$ $M$ is a mid-point of $\mathbf{BC}$)

Also, $\quad \mathbf{AB} + \mathbf{BM} + \mathbf{MA} = 0$

(by properties of a triangle)

$\Rightarrow\ \mathbf{AB} + \dfrac{\mathbf{AC} - \mathbf{AB}}{2} = \mathbf{AM}$

$\Rightarrow\ \mathbf{AM} = \dfrac{\mathbf{AB} + \mathbf{AC}}{2}$

$= \dfrac{3\hat{\mathbf{i}} + 4\hat{\mathbf{k}} + 5\hat{\mathbf{i}} - 2\hat{\mathbf{k}} + 4\hat{\mathbf{k}}}{2}$

$= 4\hat{\mathbf{i}} - \hat{\mathbf{j}} + 4\hat{\mathbf{k}}$

$\Rightarrow |\mathbf{AM}| = \sqrt{4^2 + 1^2 + 4^2} = \sqrt{33}$

**6.** Let $f(x) = 2x^3 + 3x + k$

On differentiating w.r.t $x$, we get

$f'(x) = 6x^2 + 3 > 0,\ \forall\ x \in R$

$\Rightarrow f(x)$ is strictly increasing function.

$\Rightarrow f(x) = 0$ has only one real root, so two roots are not possible.

**7.** Let $S = 0.7 + 0.77 + 0.777 + \ldots$

$$= \frac{7}{10} + \frac{77}{10^2} + \frac{777}{10^3} + \ldots + \text{upto 20 terms}$$

$$= 7\left[\frac{1}{10} + \frac{11}{10^2} + \frac{111}{10^3} + \ldots + \text{upto 20 terms}\right]$$

$$= \frac{7}{9}\left[\frac{9}{10} + \frac{99}{100} + \frac{999}{1000} + \ldots + \text{upto 20 terms}\right]$$

$$= \frac{7}{9}\left[\left(1 - \frac{1}{10}\right) + \left(1 - \frac{1}{10^2}\right) + \left(1 - \frac{1}{10^3}\right) + \ldots + \text{upto 20 terms}\right]$$

$$= \frac{7}{9}\left[(1 + 1 + \ldots + \text{ upto 20 terms}) - \left(\frac{1}{10} + \frac{1}{10^2} + \frac{1}{10^3} + \ldots + \text{upto 20 terms}\right)\right]$$

$$= \frac{7}{9}\left[20 - \frac{\frac{1}{10}\left\{1 - \left(\frac{1}{10}\right)^{20}\right\}}{1 - \frac{1}{10}}\right]$$

$$\left[\because \sum_{i=1}^{20} = 20 \text{ and sum of } n \text{ terms of GP } S_n = \frac{a(1 - r^n)}{1 - r} \text{ when } (r < 1)\right]$$

$$= \frac{7}{9}\left[20 - \frac{1}{9}\left\{1 - \left(\frac{1}{10}\right)^{20}\right\}\right]$$

$$= \frac{7}{9}\left[\frac{179}{9} + \frac{1}{9}\left(\frac{1}{10}\right)^{20}\right]$$

$$= \frac{7}{81}[179 + (10)^{-20}]$$

**8.** Take any point $B\ (0, 1)$ on given line.

Equation of $AB'$

$$y - 0 = \frac{-1 - 0}{0 - \sqrt{3}}(x - \sqrt{3})$$

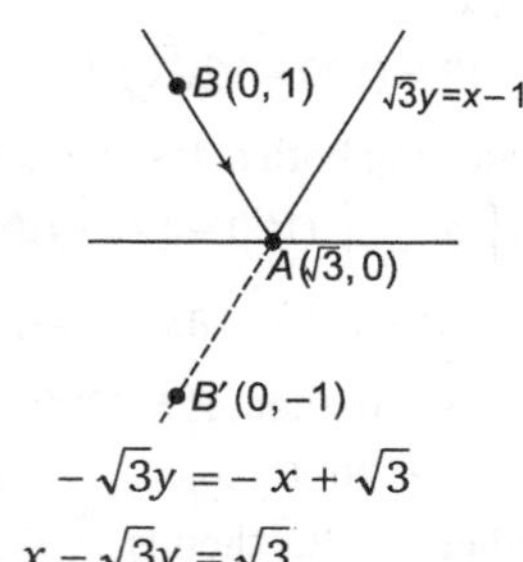


$\Rightarrow \quad -\sqrt{3}y = -x + \sqrt{3}$

$\Rightarrow x - \sqrt{3}y = \sqrt{3}$

$\Rightarrow \quad \sqrt{3}y = x - \sqrt{3}$

**9.** Given equations can be written in matrix form

$AX = B$

where, $A = \begin{bmatrix} k + 1 & 8 \\ k & k + 3 \end{bmatrix}, X = \begin{bmatrix} x \\ y \end{bmatrix}$

and $\quad B = \begin{bmatrix} 4k \\ 3k - 1 \end{bmatrix}$

For no solution, $|A| = 0$ and $(\text{adj } A) B \neq 0$

Now, $|A| = \begin{vmatrix} k+1 & 8 \\ k & k+3 \end{vmatrix} = 0$

$\Rightarrow \quad (k+1)(k+3) - 8k = 0$

$\Rightarrow \quad k^2 + 4k + 3 - 8k = 0$

$\Rightarrow \quad k^2 - 4k + 3 = 0$

$\Rightarrow \quad (k-1)(k-3) = 0$

$\Rightarrow \quad k = 1, k = 3$

Now, $\quad \text{adj } A = \begin{bmatrix} k+3 & -8 \\ -k & k+1 \end{bmatrix}$

Now, $(\text{adj } A) B = \begin{bmatrix} k+3 & -8 \\ -k & k+1 \end{bmatrix} \begin{bmatrix} 4 & k \\ 3 & k-1 \end{bmatrix}$

$= \begin{bmatrix} (k+3)(4k) - 8(3k-1) \\ -4k^2 + (k+1)(3k-1) \end{bmatrix}$

$= \begin{bmatrix} 4k^2 - 12k + 8 \\ -k^2 + 2k - 1 \end{bmatrix}$

Put $k = 1$

$(\text{adj } A) B = \begin{bmatrix} 4 - 12 + 8 \\ -1 + 2 - 1 \end{bmatrix}$

$= \begin{bmatrix} 0 \\ 0 \end{bmatrix}$ not true

Put $k = 3$

$(\text{adj } A) B = \begin{bmatrix} 36 - 36 + 8 \\ -9 + 6 - 1 \end{bmatrix}$

$= \begin{bmatrix} 8 \\ -4 \end{bmatrix} \neq 0$ true.

Hence, required value of $k$ is 3.

**Alternate Solution**

Condition for the system of equations has no solution,

$$\frac{a_1}{a_2} = \frac{b_1}{b_2} \neq \frac{c_1}{c_2}$$

$\therefore \quad \frac{k+1}{k} = \frac{8}{k+3} \neq \frac{4k}{3k-1}$

Take $\quad \frac{k+1}{k} = \frac{8}{k+3}$

$\Rightarrow \quad k^2 + 4k + 3 = 8k$

$\Rightarrow \quad k^2 - 4k + 3 = 0$

$\Rightarrow \quad (k-1)(k-3) = 0$

$k = 1, 3$

If $k = 1$, then $\frac{8}{1+3} \neq \frac{4 \cdot 1}{2}$, false

And if $k = 3$, then $\frac{8}{6} \neq \frac{4 \cdot 3}{9-1}$, true

Therefore $k = 3$

Hence, only one value of $k$ exist.

**10.** Given equations are

$$x^2 + 2x + 3 = 0 \quad \ldots\text{(i)}$$

and $\quad ax^2 + bx + c = 0 \quad \ldots\text{(ii)}$

Since, Eq. (i) has imaginary roots so, Eq. (ii) will also have both roots same as Eq. (i).

Thus, $\quad \frac{a}{1} = \frac{b}{2} = \frac{c}{3}$

Hence, $a : b : c$ is $1 : 2 : 3$

**11.** Let the equation of circle be

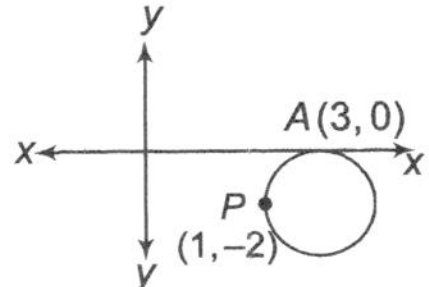


$$(x-3)^2 + (y-0)^2 + \lambda y = 0$$

As it passes through $(1, -2)$

$\therefore \quad (1-3)^2 + (-2)^2 + \lambda(-2) = 0$

$\Rightarrow \quad 4 + 4 - 2\lambda = 0 \Rightarrow \lambda = 4$

$\therefore$ Equation of circle is

$$(x-3)^2 + y^2 + 4y = 0$$

By hit and trial method, we see that point $(5, -2)$ satisfies equation of circle.

**12.** Since, $x, y$ and $z$ are in AP.

$\therefore \quad 2y = x + z$

Also, $\tan^{-1} x$, $\tan^{-1} y$ and $\tan^{-1} z$ are in AP.

$\therefore \quad 2\tan^{-1} y = \tan^{-1} x + \tan^{-1}(z)$

$\Rightarrow \quad \tan^{-1}\left(\frac{2y}{1-y^2}\right) = \tan^{-1}\left(\frac{x+z}{1-xz}\right)$

$\Rightarrow \quad \frac{x+z}{1-y^2} = \frac{x+z}{1-xz} \Rightarrow y^2 = xz$

Since $x$, $y$ and z are in AP as well as in GP

$\therefore \quad x = y = z$

**13.** **Statement II**

$(p \to q) \leftrightarrow (\sim q \to \sim p)$

$\equiv (p \to q) \leftrightarrow (p \to q)$

which is always true, so statement II is true.

**Statement I** $(p \wedge \sim q) \wedge (\sim p \wedge q)$

$\equiv p \wedge \sim q \wedge \sim p \wedge q$

$\equiv p \wedge \sim p \wedge \sim q \wedge q$

$\equiv f \wedge f \equiv f$

Hence, it is a fallacy statement.

So, statement I is true.

**Alternate Solution**

**Statement II**

$(p \to q) \leftrightarrow (\sim q \to \sim p)$

$\sim q \to \sim p$ is contrapositive of $p \to q$ hence $(p \to q) \leftrightarrow (p \to q)$ will be a tautology.

**Statement I** $(p \wedge \sim q) \wedge (\sim p \wedge q)$

| $p$ | $q$ | $\sim p$ | $\sim q$ | $p \wedge \sim q$ | $\sim p \wedge q$ | $(p \wedge \sim q) \wedge (\sim p \wedge q)$ |
|---|---|---|---|---|---|---|
| T | T | F | F | F | F | F |
| T | F | F | T | T | F | F |
| F | T | T | F | F | T | F |
| F | F | T | T | F | F | F |

Hence, it is a fallacy.

**14.** Given, $\int f(x)\,dx = \psi(x)$

Let $I = \int x^5 f(x^3)\,dx$

Put $x^3 = t$

$\Rightarrow \quad x^2 dx = \dfrac{dt}{3}$ ...(i)

$\therefore \quad I = \dfrac{1}{3}\int t\, f(t)\,dt$

$= \dfrac{1}{3}\left[t \cdot \int f(t)dt - \int \left\{\dfrac{d}{dt}(t) \int f(t)dt\right\}dt\right]$ (by parts)

$= \dfrac{1}{3}[t\,\psi(t) - \int \psi(t)\,dt]$

$= \dfrac{1}{3}[x^3\psi(x^3) - 3\int x^2\psi(x^3)\,dx] + C$ [$\because$ from Eq. (i)]

$= \dfrac{1}{3}x^3\psi(x^3) - \int x^2\psi(x^3)\,dx + C$

**15.** Let $I = \lim_{x \to 0} \dfrac{(1 - \cos 2x)}{x^2} \dfrac{(3 + \cos x)}{1} \cdot \dfrac{x}{\tan 4x}$

$= \lim_{x \to 0} \dfrac{2\sin^2 x}{x^2} \cdot \dfrac{3 + \cos x}{1} \cdot \dfrac{x}{\tan 4x}$

$= 2 \lim_{x \to 0} \left(\dfrac{\sin x}{x}\right)^2 \cdot \lim_{x \to 0} (3 + \cos x) \cdot \lim_{x \to 0} \dfrac{4x}{4\tan 4x}$

$= 2 \cdot (1)^2 \cdot (3 + \cos 0°) \cdot \dfrac{1}{4}(1)$

$= 2 \cdot 1 \cdot (3 + 1) \cdot \dfrac{1}{4} = 2 \cdot 4 \cdot \dfrac{1}{4} = 2$

**16.** Let $I = \int_{\pi/6}^{\pi/3} \dfrac{dx}{1 + \sqrt{\tan x}}$ ...(i)

$\therefore \quad I = \int_{\pi/6}^{\pi/3} \dfrac{dx}{1 + \sqrt{\tan\left(\dfrac{\pi}{2} - x\right)}}$

$= \int_{\pi/6}^{\pi/3} \dfrac{dx}{1 + \sqrt{\cot x}}$

$\Rightarrow \quad I = \int_{\pi/6}^{\pi/3} \dfrac{\sqrt{\tan x}\,dx}{1 + \sqrt{\tan x}}$ ...(ii)

On adding Eqs. (i) and (ii), we get

$2I = \int_{\pi/6}^{\pi/3} dx \Rightarrow 2I = [x]_{\pi/6}^{\pi/3}$

$\Rightarrow \quad I = \dfrac{1}{2}\left[\dfrac{\pi}{3} - \dfrac{\pi}{6}\right] = \dfrac{\pi}{12}$

Statement I is false.

But $\int_a^b f(x)\,dx = \int_a^b f(a + b - x)\,dx$ is a true statement by property of definite integrals.

**17.** Given equation of ellipse is $\dfrac{x^2}{16} + \dfrac{y^2}{9} = 1$

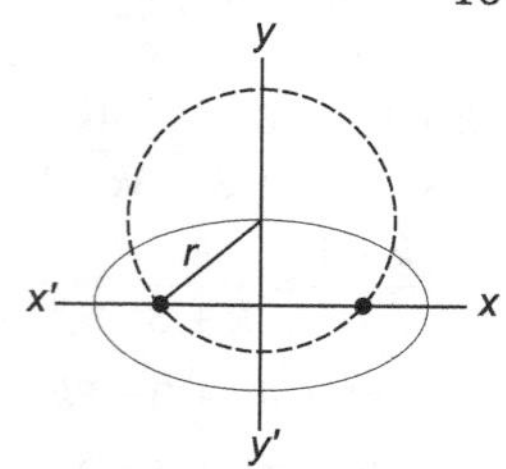

Here, $a = 4$, $b = 3$, $e = \sqrt{1 - \frac{9}{16}} \Rightarrow \frac{\sqrt{7}}{4}$

$\therefore$ Foci is $(\pm \text{ ae}, 0) = \left(\pm 4 \times \frac{\sqrt{7}}{4}, 0\right)$

$= (\pm \sqrt{7}, 0)$

$\therefore$ Radius of the circle,

$r = \sqrt{(ae)^2 + b^2} = \sqrt{7+9} = \sqrt{16} = 4$

Now, equation of circle is

$$(x-0)^2 + (y-3)^2 = 16$$

$$\therefore \quad x^2 + y^2 - 6y - 7 = 0$$

**18.** Probability of guessing a correct answer, $p = \frac{1}{3}$ and probability of guessing a wrong answer, $q = \frac{2}{3}$

$\therefore$ The probability of guessing a 4 or more correct answer

$$= {}^5C_4\left(\frac{1}{3}\right)^4 \cdot \frac{2}{3} + {}^5C_5\left(\frac{1}{3}\right)^5$$

$$= 5 \cdot \frac{2}{3^5} + \frac{1}{3^5} = \frac{11}{3^5}$$

**19.** Given mid-points of a triangle are (0, 1), (1, 1) and (1, 0). Plotting these points on a graph paper and make a triangle. So, the sides of a triangle will be 2, 2 and $\sqrt{2^2 + 2^2}$ *i.e.*, $2\sqrt{2}$

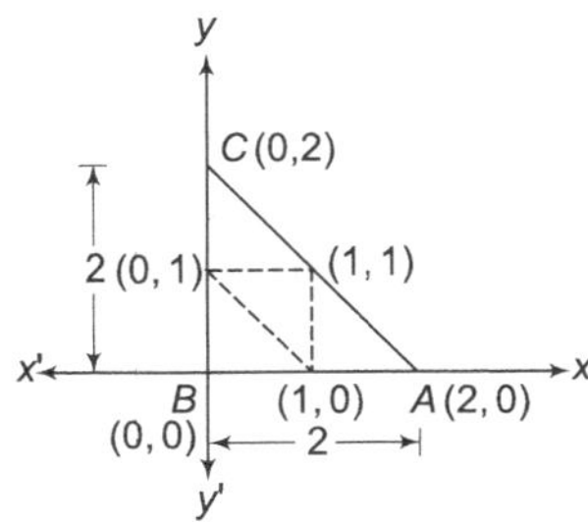


$x$-coordinate of incentre

$$= \frac{2 \times 0 + 2\sqrt{2} \cdot 0 + 2 \cdot 2}{2 + 2 + 2\sqrt{2}}$$

$$= \frac{2}{2 + \sqrt{2}} \times \frac{2 - \sqrt{2}}{2 - \sqrt{2}} = 2 - \sqrt{2}$$

**20.** $\therefore \left[\frac{x+1}{x^{2/3} - x^{1/3} + 1} - \frac{(x-1)}{x - x^{1/2}}\right]^{10}$

$$= \left[\frac{(x^{1/3})^3 + 1^3}{x^{2/3} - x^{1/3} + 1} - \frac{\{(\sqrt{x})^2 - 1\}}{\sqrt{x}\,(\sqrt{x} - 1)}\right]^{10}$$

$$= \left[\frac{(x^{1/3} + 1)(x^{2/3} + 1 - x^{1/3})}{x^{2/3} - x^{1/3} + 1} - \frac{\{(\sqrt{x})^2 - 1\}}{\sqrt{x}\,(\sqrt{x} - 1)}\right]^{10}$$

$$= \left[(x^{1/3} + 1) - \frac{(\sqrt{x} + 1)}{\sqrt{x}}\right]^{10}$$

$$= (x^{1/3} - x^{-1/2})^{10}$$

$\therefore$ The general term is

$$T_{r+1} = {}^{10}C_r (x^{1/3})^{10-r} (-x^{-1/2})^r$$

$$= {}^{10}C_r (-1)^r x^{\frac{10-r}{3} - \frac{r}{2}}$$

For independent of $x$, put

$$\frac{10-r}{3} - \frac{r}{2} = 0 \Rightarrow 20 - 2r - 3r = 0$$

$$\Rightarrow \quad 20 = 5r \Rightarrow r = 4$$

$$\therefore T_5 = {}^{10}C_4 = \frac{10 \times 9 \times 8 \times 7}{4 \times 3 \times 2 \times 1} = 210$$

**21.** Given curves are $y = \sqrt{x}$ ...(i)

and $2y - x + 3 = 0$ ...(ii)

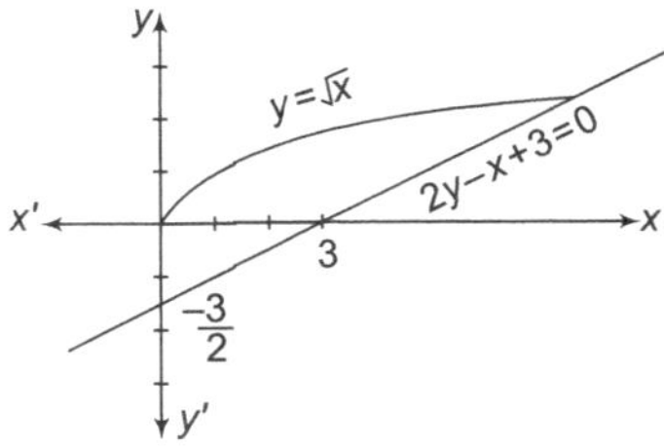


On solving Eqs. (i) and (ii), we get

$$2\sqrt{x} - (\sqrt{x})^2 + 3 = 0$$

$$\Rightarrow (\sqrt{x})^2 - 2\sqrt{x} - 3 = 0$$

$$\Rightarrow (\sqrt{x} - 3)(\sqrt{x} + 1) = 0$$

$$\Rightarrow \sqrt{x} = 3,$$

($\because \sqrt{x} = -1$ is not possible)

$$\therefore \quad y = 3$$

$\therefore$ Required area $= \int_0^3 (x_2 - x_1)\, dy$

$$= \int_0^3 \{(2y+3) - y^2\}\, dy$$

$$= \left[ y^2 + 3y - \frac{y^3}{3} \right]_0^3$$

$$= 9 + 9 - 9 = 9$$

**22.** Given, $T_n = {}^nC_3$

$$T_{n+1} = {}^{n+1}C_3$$

$\therefore\ T_{n+1} - T_n = {}^{n+1}C_3 - {}^nC_3 = 10$ (given)

$\Rightarrow\ {}^nC_2 + {}^nC_3 - {}^nC_3 = 10$

$(\because {}^nC_r + {}^nC_{r+1} = {}^{n+1}C_{r+1})$

$\Rightarrow\ {}^nC_2 = 10 \ \Rightarrow\ n = 5$

**23.** Given, $|z| = 1$, $\arg z = \theta \therefore z = e^{i\theta}$

But $\bar{z} = \dfrac{1}{z}$

$\therefore\ \arg\left(\dfrac{1+z}{1+\dfrac{1}{z}}\right) = \arg(z) = \theta$

**24.** Let $AB = x$

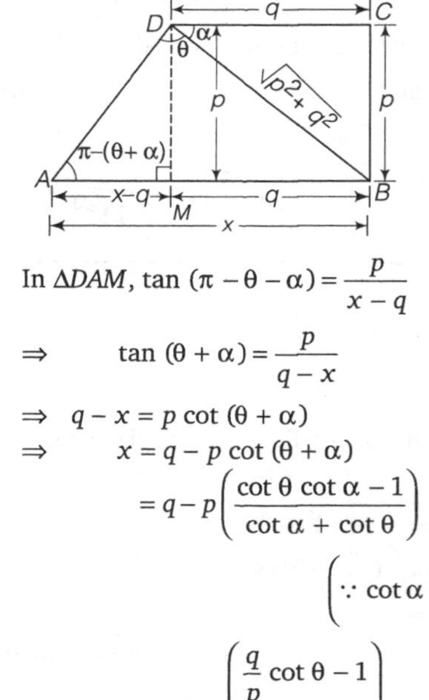


In $\Delta DAM$, $\tan(\pi - \theta - \alpha) = \dfrac{p}{x-q}$

$\Rightarrow\ \tan(\theta + \alpha) = \dfrac{p}{q-x}$

$\Rightarrow\ q - x = p\cot(\theta + \alpha)$

$\Rightarrow\ x = q - p\cot(\theta + \alpha)$

$$= q - p\left(\frac{\cot\theta\cot\alpha - 1}{\cot\alpha + \cot\theta}\right)$$

$\left(\because \cot\alpha = \dfrac{q}{p}\right)$

$$= q - p\left(\frac{\frac{q}{p}\cot\theta - 1}{\frac{q}{p} + \cot\theta}\right)$$

$$= q - p\left(\frac{q\cot\theta - p}{q + p\cot\theta}\right)$$

$$= q - p\left(\frac{q\cos\theta - p\sin\theta}{q\sin\theta + p\cos\theta}\right)$$

$$\Rightarrow\ x = \frac{q^2\sin\theta + pq\cos\theta - pq\cos\theta + p^2\sin\theta}{p\cos\theta + q\sin\theta}$$

$$\Rightarrow AB = \frac{(p^2+q^2)\sin\theta}{p\cos\theta + q\sin\theta}$$

**Alternate Solution**

Applying Sine rule in $\Delta ABD$,

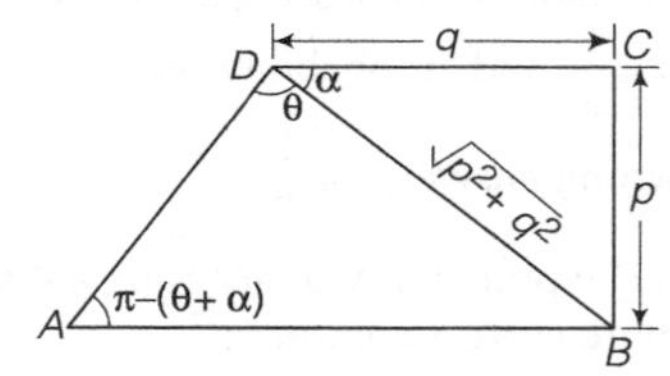


$$\frac{AB}{\sin\theta} = \frac{\sqrt{p^2+q^2}}{\sin\{\pi - (\theta+\alpha)\}}$$

$$\Rightarrow\ \frac{AB}{\sin\theta} = \frac{\sqrt{p^2+q^2}}{\sin(\theta+\alpha)}$$

$$\Rightarrow\ AB = \frac{\sqrt{p^2+q^2}\sin\theta}{\sin\theta\cos\alpha + \cos\theta\sin\alpha}$$

$$= \frac{(p^2+q^2)\sin\theta}{q\sin\theta + p\cos\theta}$$

$\left(\because \cos\alpha = \dfrac{q}{\sqrt{p^2+q^2}}\right.$

$$= \frac{(p^2+q^2)\sin\theta}{p\cos\theta + q\sin\theta}$$

and $\sin\alpha = \dfrac{p}{\sqrt{p^2+q^2}}\Bigg)$

**25.** Given, $P = \begin{bmatrix} 1 & \alpha & 3 \\ 1 & 3 & 3 \\ 2 & 4 & 4 \end{bmatrix}$

$\therefore\ |P| = 1(12-12) - \alpha(4-6) + 3(4-6)$

$$= 2\alpha - 6$$

$\because\ P = \text{adj}(A)$ (given)

$\therefore \quad |P| = |\text{adj } A| = |A|^2 = 16$

$(\because |\text{adj } A| = |A|^{n-1})$

$\therefore \quad 2\alpha - 6 = 16$

$\Rightarrow \quad 2\alpha = 22 \Rightarrow \alpha = 11$

**26.** Given, $y = \int_0^x |t|\, dt$

$\therefore \quad \frac{dy}{dx} = |x| \cdot 1 - 0 = |x|$

(by Leibnitz rule)

$\because$ Tangent to the curve

$y = \int_0^x (t)\, dt,\ x \in R$ are parallel to the line $y = 2x$

$\therefore$ Slope of both are equal

$\Rightarrow \quad x = \pm 2$

$\therefore$ Points, $y = \int_0^{\pm 2} |t|\, dt = \pm 2$

$\therefore$ Equation of tangent is

$$y - 2 = 2(x - 2)$$

and $\quad y + 2 = 2(x + 2)$

For $x$-intercept put $y = 0$, we get

$$0 - 2 = 2(x - 2)$$

and $\quad 0 + 2 = 2(x + 2) \Rightarrow x = \pm 1$

**27.** Equation of circle can be rewritten as

$$x^2 + y^2 = \frac{5}{2},$$

centre $\rightarrow$ (0, ) and radius $\rightarrow \sqrt{\frac{5}{2}}$

Let common tangent be

$$y = mx + \frac{\sqrt{5}}{m}$$

$\Rightarrow \quad m^2 x - my + \sqrt{5} = 0$

The perpendicular from centre to the tangent is equal to radius of the circle.

$$\therefore \quad \frac{\frac{\sqrt{5}}{m}}{\sqrt{1 + m^2}} = \sqrt{\frac{5}{2}}$$

$\Rightarrow \quad m\sqrt{1 + m^2} = \sqrt{2}$

$\Rightarrow \quad m^2(1 + m^2) = 2$

$\Rightarrow \quad m^4 + m^2 - 2 = 0$

$\Rightarrow \quad (m^2 + 2)(m^2 - 1) = 0$

$\Rightarrow \quad m = \pm 1$

$(\because m^2 + 2 \neq 0, \text{ as } m \in R)$

$\therefore y = \pm (x + \sqrt{5})$, both statements are correct as $m = \pm 1$ satisfies the given equation of statement II.

**28.** Given, $y = \sec(\tan^{-1} x)$

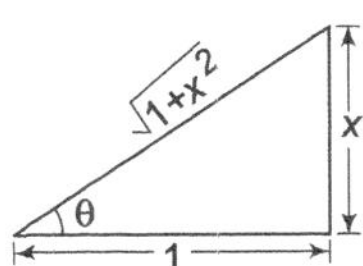


Let $\quad \tan^{-1} x = \theta \Rightarrow x = \tan\theta$

$\therefore \quad y = \sec\theta = \sqrt{1 + x^2}$

On differentiating w.r.t. $x$, we get

$$\frac{dy}{dx} = \frac{1}{2\sqrt{1 + x^2}} \cdot 2x$$

At $x = 1$, $\quad \frac{dy}{dx} = \frac{1}{\sqrt{2}}$

**29.** Given expression is

$$\frac{\tan A}{1 - \cot A} + \frac{\cot A}{1 - \tan A}$$

$$= \frac{\sin A}{\cos A} \times \frac{\sin A}{\sin A - \cos A} + \frac{\cos A}{\sin A} \times \frac{\cos A}{\cos A - \sin A}$$

$$= \frac{1}{\sin A - \cos A} \left\{ \frac{\sin^3 A - \cos^3 A}{\cos A \sin A} \right\}$$

$$= \frac{\sin^2 A + \sin A \cos A + \cos^2 A}{\sin A \cos A}$$

$$= \frac{1 + \sin A \cos A}{\sin A \cos A}$$

$$= 1 + \sec A \operatorname{cosec} A$$

**30.** If initially all marks were $x_i$, then

$$\sigma_1^2 = \frac{\sum (x_i - \bar{x})^2}{N}$$

Now, each is increased by 10

$$\therefore \sigma_2^2 = \frac{\sum (x_i + 10) - (\bar{x} + 10)]^2}{N} = \sigma_1^2$$

So, variance will not change whereas mean, median and mode will increase by 10.

# JEE Advanced

# Solved Paper 2013

## PAPER 1

- M. Marks 180 Time 3 Hrs

### Instructions

- The question paper consists of three sections.
- Section 1 contains 10 multiple choice questions each of 2 marks in this section there is no negative marking. Each question has four choices (a), (b), (c) and (d) out of which only one is correct.
- Section 2 contains 5 multiple choice questions each of 4 marks with negative marking of 1 mark. Each question has four choices (a), (b), (c) and (d) out of which one or more are correct.
- Section 3 contains 5 questions each of 4 marks with negative marking of 1 mark. The answer to each question is a single-digit integer, ranging from 0 to 9 (both inclusive)

## PHYSICS

### Section I

**Direction** (Q. No. 1-10) *This section contains* 10 *multiple choice questions. Each question has four choices (a), (b), (c) and (d) out of which only one is correct.*

**1.** A particle of mass $m$ is projected from the ground with an initial speed $u_0$ at an angle $\alpha$ with the horizontal. At the highest point of its trajectory, it makes a completely inelastic collision with another identical particle, which was thrown vertically upward from the ground with the same initial speed $u_0$. The angle that the composite system makes with the horizontal immediately after the collision is

(a) $\frac{\pi}{4}$ (b) $\frac{\pi}{4}+\alpha$

(c) $\frac{\pi}{4}-\alpha$ (d) $\frac{\pi}{2}$

**2.** The image of an object, formed by a plano-convex lens at a distance of 8 m behind the lens, is real and is one-third the size of the object. The wavelength of light inside the lens is $\frac{2}{3}$ times the wavelength in free space. The radius of the curved surface of the lens is

(a) 1 m (b) 2 m (c) 3 m (d) 6 m

**3.** The diameter of a cylinder is measured using a vernier callipers with no zero error. It is found that the zero of the vernier scale lies between 5.10 cm and 5.15 cm of the main scale. The vernier scale has 50 division equivalent to 2.45 cm. The 24th division of the vernier scale exactly coincides with one of the main scale divisions. The diameter of the cylinder is

(a) 5.112 cm (b) 5.124 cm
(c) 5.136 cm (d) 5.148 cm

**4.** The work done on a particle of mass $m$ by a force,

$$K\left[\frac{x}{(x^2+y^2)^{3/2}}\hat{\mathbf{i}}+\frac{y}{(x^2+y^2)^{3/2}}\hat{\mathbf{j}}\right]$$

($K$ being a constant of appropriate dimensions), when the particle is taken from the point $(a, 0)$ to the point $(0, a)$ along a circular path of radius $a$ about the origin in the $x$-$y$ plane is

(a) $\frac{2K\pi}{a}$ (b) $\frac{K\pi}{a}$ (c) $\frac{K\pi}{2a}$ (d) 0

**5.** One end of a horizontal thick copper wire of length $2L$ and radius $2R$ is welded to an end of another horizontal thin copper wire of length $L$ and radius $R$. When the arrangement is stretched by applying forces at two ends, the ratio of the elongation in the thin wire to that in the thick wire is

(a) 0.25 (b) 0.50
(c) 2.00 (d) 4.00

**6.** A ray of light travelling in the direction $\frac{1}{2}(\hat{\mathbf{i}}+\sqrt{3}\,\hat{\mathbf{j}})$ is incident on a plane mirror. After reflection, it travels along the direction $\frac{1}{2}(\hat{\mathbf{i}}-\sqrt{3}\,\hat{\mathbf{j}})$. The angle of incidence is

(a) 30° (b) 45°
(c) 60° (d) 75°

**7.** Two rectangular blocks, having indentical dimensions, can be arranged either in configuration I or in configuration II as shown in the figure. One of the blocks has thermal conductivity $K$ and the other $2K$. The temperature difference between the ends along the $x$-axis is the same in both the configurations. It takes 9 s to transport a certain amount of heat from the hot end to the cold end in the configuration I. The time to transport the same amount of heat in the configuration II is

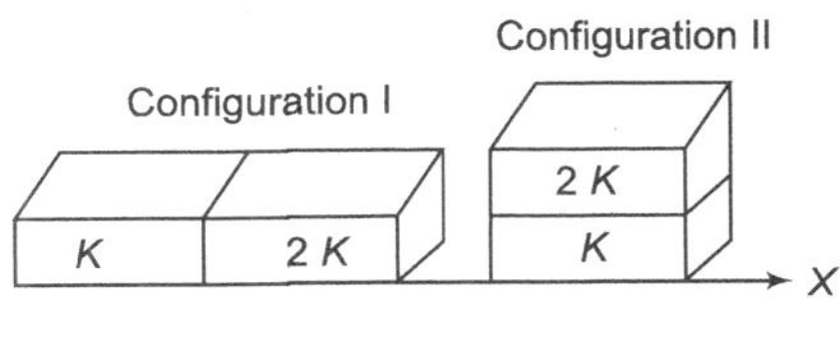


(a) 2.0 s (b) 3.0 s
(c) 4.5 s (d) 6.0 s

**8.** A pulse of light of duration 100 ns is absorbed completely by a small object initially at rest. Power of the pulse is 30 mV and the speed of light is $3\times10^8$ ms$^{-1}$. The final momentum of the object is

(a) $0.3\times10^{-17}$ kg-ms$^{-1}$
(b) $1.0\times10^{-17}$ kg-ms$^{-1}$
(c) $3.0\times10^{-17}$ kg-ms$^{-1}$
(d) $9.0\times10^{-17}$ kg-ms$^{-1}$

**9.** In the Young's double slit experiment using a monochromatic light of wavelength $\lambda$ the path difference (in terms of an integer $n$) corresponding to any point having half the peak intensity is

(a) $(2n+1)\frac{\lambda}{2}$ (b) $(2n+1)\frac{\lambda}{4}$
(c) $(2n+1)\frac{\lambda}{8}$ (d) $(2n+1)\frac{\lambda}{16}$

**10.** Two non-reactive monoatomic ideal gases have their atomic masses in the ratio 2 : 3. The ratio of their partial pressures, when enclosed in a vessel kept at a constant temperature, is 4 : 3. The ratio of their densities is

(a) 1 : 4 (b) 1 : 2
(c) 6 : 9 (d) 8 : 9

## Section II

**Direction** (Q. No. 11-15) *This section contains 5 multiple choice questions. Each question has four choices (a), (b), (c) and (d) out of which one or more are correct.*

**11.** Two non-conducting solid spheres of radii $R$ and $2R$, having uniform volume charge densities $\rho_1$ and $\rho_2$ respectively, touch each other. The net electric field at a distance $2R$ from the centre of the smaller sphere, along the line joining the centre of the spheres, is zero. The ratio $\frac{\rho_1}{\rho_2}$ can be

(a) $-4$ (b) $-\frac{32}{25}$ (c) $\frac{32}{25}$ (d) 4

**12.** A horizontal stretched string, fixed at two ends, is vibrating in its fifth harmonic according to the equation, $y(x, t) = (0.01\text{ m})[\sin(62.8\text{ m}^{-1})x]$ $\cos[(628\text{ s}^{-1})t]$.

Assuming $\pi = 3.14$, the correct statement(s) is (are)

(a) the number of nodes is 5
(b) the length of the string is 0.25 m
(c) the maximum displacement of the mid-point of the string from its equilibrium position is 0.01 m
(d) the fundamental frequency is 100 Hz

**13.** In the circuit shown in the figure, there are two parallel plate capacitors each of capacitance $C$. The switch $S_1$ is pressed first to fully charge the capacitor $C_1$ and then released. The switch $S_2$ is then pressed to charge the capacitor $C_2$. After some time, $S_2$ is released and then $S_3$ is pressed. After some time

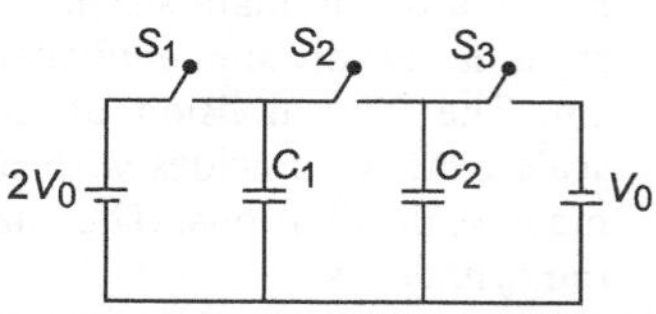


(a) the charge on the upper plate of $C_1$ is $2CV_0$
(b) the charge on the upper plate of $C_1$ is $CV_0$
(c) the charge on the upper plate of $C_2$ is 0
(d) the charge on the upper plate of $C_2$ is $-CV_0$

**14.** A particle of mass $M$ and positive charge $Q$, moving with a constant velocity $\mathbf{u}_1 = 4\hat{\mathbf{i}}\text{ ms}^{-1}$, enters a region of uniform static magnetic field normal to the $x$-$y$ plane. The region of the magnetic field extends from $x = 0$ to $x = L$ for all values of $y$. After passing through this region, the particle emerges on the other side after 10 milliseconds with a velocity $\mathbf{u}_2 = 2(\sqrt{3}\,\hat{\mathbf{i}} + \hat{\mathbf{j}})\text{ ms}^{-1}$. The correct statement(s) is (are)

(a) the direction of the magnetic field is $-z$ direction
(b) the direction of the magnetic field is $+z$ direction
(c) the magnitude of the magnetic field is $\frac{50\pi M}{3Q}$ units
(d) the magnitude of the magnetic field is $\frac{100\pi M}{3Q}$ units

**15.** A solid sphere of radius $R$ and density $\rho$ is attached to one end of a massless spring of force constant $k$. The other end of the spring is connected to another solid sphere of radius $R$ and density $3\rho$. The complete arrangement is placed in a liquid of density $2\rho$ and is allowed to reach equilibrium. The correct statement(s) is (are)

(a) the net elongation of the spring is $\frac{4\pi R^3 \rho g}{3k}$

(b) the net elongation of the spring is $\frac{8\pi R^3 \rho g}{3k}$

(c) the light sphere is partially submerged

(d) the light sphere is completely submerged

## Section III

**Direction** (Q. No. 16-20) *This section contains 5 questions. The answer to each question is a single digit integer, ranging from 0 to 9 (both inclusive)*

**16.** A uniform circular disc of mass 50 kg and radius 0.4 m is rotating with an angular velocity of 10 rad/s about its own axis, which is vertical. Two uniform circular rings, each of mass 6.25 kg and radius 0.2 m, are gently placed symmetrically on the disc in such a manner that they are touching each other along the axis of the disc and are horizontal. Assume that the friction is large enough such that the rings are at rest relative to the disc and the system rotates about the original axis. The new angular velocity (in rad $s^{-1}$) of the system is

**17.** The work functions of silver and sodium are 4.6 and 2.3 eV, respectively. The ratio of the slope of the stopping potential *versus* frequency plot for silver to that of sodium is

**18.** A bob of mass $m$, suspended by a string of length $l_1$, is given a minimum velocity required to complete a full circle in the vertical plane, At the highest point, it collides elastically with another bob of mass $m$ suspended by a string of length $l_2$, which is initially at rest. Both the strings are massless and inextensible. If the second bob, after collision acquires the minimum speed required to complete a full circle in the vertical plane, the ratio $\frac{l_1}{l_2}$ is

**19.** A particle of mass 0.2 kg is moving in one dimension under a force that delivers a constant power 0.5 W to the particle. If the initial speed (in $ms^{-1}$) of the particle is zero, the speed (in $ms^{-1}$) after 5s is

**20.** A freshly prepared sample of a radioisotope of half-life 1386 s has activity $10^3$ disintegrations per second. Given that $\ln 2 = 0.693$, the fraction of the initial number of nuclei (expressed in nearest integer percentage) that will decay in the first 80 s after preparation of the sample is

# CHEMISTRY

## Section I

**Direction** (Q. No. 1-10) *This section contains* 10 *multiple choice questions. Each question has four choices (a), (b), (c) and (d) out of which only one is correct.*

**1.** The standard enthalpies of formation of $CO_2(g)$, $H_2O(l)$ and glucose(s) at 25°C are – 400 kJ/mol, – 300 kJ/mol and – 1300 kJ/mol, respectively. The standard enthalpy of combustion per gram of glucose at 25°C is

(a) + 2900 kJ (b) – 2900 kJ
(c) – 16.11 kJ (d) + 16.11 kJ

**2.** KI in acetone, undergoes $S_N2$ reaction with each *P*, *Q*, *R* and *S*. The rates of the reaction vary as

$H_3C$—Cl (*P*)  (CH₃)₂CH—Cl (*Q*)  CH₂=CH—CH₂—Cl (*R*)

$C_6H_5$—CO—CH₂—Cl (*S*)

(a) $P > Q > R > S$
(b) $S > P > R > Q$
(c) $P > R > Q > S$
(d) $R > P > S > Q$

**3.** The compound that does not liberate $CO_2$, on treatment with aqueous sodium bicarbonate solution, is

(a) benzoic acid
(b) benzenesulphonic acid
(c) salicylic acid
(d) carbolic acid (Phenol)

**4.** Consider the following complex ions, *P*, *Q* and *R*.

$P = [FeF_6]^{3-}$, $Q = [V(H_2O)_6]^{2+}$ and $R = [Fe(H_2O)_6]^{2+}$

The correct order of the complex ions, according to their spin-only magnetic moment values (in BM) is

(a) $R < Q < P$ (b) $Q < R < P$
(c) $R < P < Q$ (d) $Q < P < R$

**5.** In the reaction,

$$P + Q \longrightarrow R + S$$

the time taken for 75% reaction of *P* is twice the time taken for 50% reaction of *P*. The concentration of *Q* varies with reaction time as shown in the figure. The overall order of the reaction is

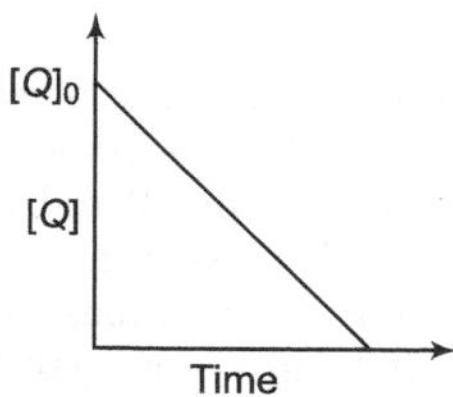


(a) 2 (b) 3 (c) 0 (d) 1

**6.** Concentrated nitric acid, upon long standing, turns yellow-brown due to the formation of

(a) NO (b) $NO_2$
(c) $N_2O$ (d) $N_2O_4$

**7.** The arrangement of $X^-$ ions around $A^+$ ion in solid *AX* is given in the figure (not drawn to scale). If the radius of $X^-$ is 250 pm, the radius of $A^+$ is

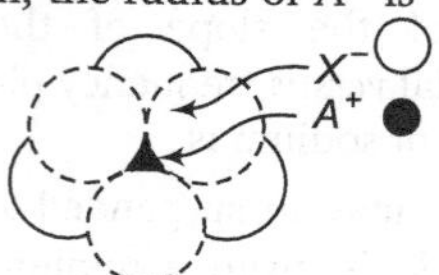


(a) 104 pm (b) 125 pm
(c) 183 pm (d) 57 pm

**8.** Upon treatment with ammoniacal $H_2S$, the metal ion that precipitates as a sulphide is

(a) Fe (III) (b) Al (III)
(c) Mg (II) (d) Zn (II)

**9.** Methylene blue, from its aqueous solution, is adsorbed on activated charcoal at 25°C. For this process, the correct statement is

(a) the adsorption requires activation at 25°C

(b) the adsorption is accompanied by a decreases in enthalpy

(c) the adsorption increases with increase of temperature

(d) the adsorption is irreversible

**10.** Sulphide ores are common for the metals

(a) Ag, Cu and Pb (b) Ag, Cu and Sn

(c) Ag, Mg and Pb (d) Al, Cu and Pb

## Section II

**Direction** (Q. No. 11-15) *This section contains 5 multiple choice questions. Each question has four choices (a), (b), (c) and (d) out of which one or more are correct.*

**11.** Benzene and naphthalene form an ideal solution at room temperature. For this process, the true statement(s) is (are)

(a) $\Delta G$ is positive

(b) $\Delta S_{\text{system}}$ is positive

(c) $\Delta S_{\text{surroundings}} = 0$

(d) $\Delta H = 0$

**12.** The pair(s) of coordination complexes/ions exhibiting the same kind of isomerism is(are)

(a) $[Cr(NH_3)_5Cl]Cl_2$ and $[Cr(NH_3)_4Cl_2]Cl$

(b) $[Co(NH_3)_4Cl_2]^+$ and $[Pt(NH_3)_2(H_2O)Cl]^+$

(c) $[CoBr_2Cl_2]^{2-}$ and $[PtBr_2Cl_2]^{2-}$

(d) $[Pt(NH_3)_3(NO_3)]Cl$ and $[Pt(NH_3)_3Cl]Br$

**13.** The initial rate of hydrolysis of methyl acetate (1 M) by a weak acid (HA, 1M) is 1/100th of that of a strong acid (HX, 1M), at 25°C. The $K_a$ (HA) is

(a) $1 \times 10^{-4}$ (b) $1 \times 10^{-5}$

(c) $1 \times 10^{-6}$ (d) $1 \times 10^{-3}$

**14.** The hyperconjugative stabilities of *tert*-butyl cation and 2-butene, respectively, are due to

(a) $\sigma \rightarrow p$ (empty) and $\sigma \rightarrow \pi^*$ electron delocalisations

(b) $\sigma \rightarrow \sigma^*$ and $\sigma \rightarrow \pi$ electron delocalisations

(c) $\sigma \rightarrow p$ (filled) and $\sigma \rightarrow \pi$ electron delocalisations

(d) $p$ (filled) $\rightarrow \sigma^*$ and $\sigma \rightarrow \pi^*$ electrons delocalisations

**15.** Among $P$, $Q$, $R$ and $S$, the aromatic compounds(s) is/are

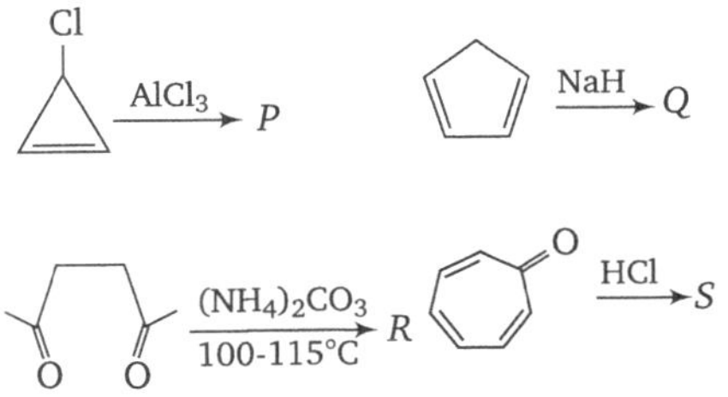

Cl
$\xrightarrow{AlCl_3} P$
$\xrightarrow{NaH} Q$
$\xrightarrow[100\text{-}115°C]{(NH_4)_2CO_3} R$
O O
O
$\xrightarrow{HCl} S$

(a) $P$ (b) $Q$

(c) $R$ (d) $S$

## Section III

**Direction** (Q. No. 16-20) *This section contains 5 questions. The answer to each question is a single digit integer, ranging from 0 to 9 (both inclusive).*

**16.** The total number of lone-pairs of electrons in melamine is

**17.** The total number of carboxylic acid groups in the product $P$ is

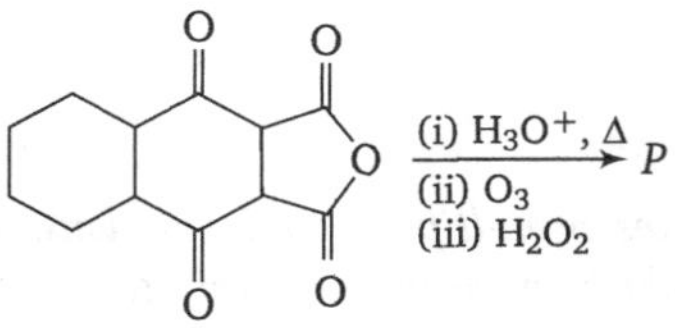

$\xrightarrow[\text{(ii) } O_3 \text{ (iii) } H_2O_2]{\text{(i) } H_3O^+, \Delta} P$

**18.** The atomic masses of He and Ne are 4 and 20 amu, respectively. The value of the de-Broglie wavelength of He gas at −73°C is "$M$" times that of the de-Broglie wavelength of Ne at 727°C. $M$ is

**19.** $EDTA^{4-}$ is ethylenediaminetetraacetate ion. The total number of N—Co—O bond angles in $[Co(EDTA)]^-$ complex ion is

**20.** A tetrapeptide has —COOH group on alanine. This produces glycine (Gly), valine (Val), phenyl alanine (Phe) and alanine (Ala), on complete hydrolysis. For this tetrapeptide, the number of possible sequences (primary structures) with $—NH_2$ group attached to a chiral center is

# MATHEMATICS

## Section I

**Directions** (Q. Nos. 1-10) *This section contains 10 multiple choice questions. Each question has four choices (a), (b), (c) and (d) out of which only one is correct.*

**1.** The value of

$$\cot\left\{\sum_{n=1}^{23} \cot^{-1}\left(1 + \sum_{k=1}^{n} 2k\right)\right\} \text{ is}$$

(a) $\frac{23}{25}$ (b) $\frac{25}{23}$

(c) $\frac{23}{24}$ (d) $\frac{24}{23}$

**2.** Let $\mathbf{PR} = 3\hat{\mathbf{i}} + \hat{\mathbf{j}} - 2\hat{\mathbf{k}}$ and $\mathbf{SQ} = \hat{\mathbf{i}} - 3\hat{\mathbf{j}} - 4\hat{\mathbf{k}}$ determine diagonals of a parallelogram $PQRS$ and $\mathbf{PT} = \hat{\mathbf{i}} + 2\hat{\mathbf{j}} + 3\hat{\mathbf{k}}$ be another vector. Then, the volume of the parallelopiped determined by the vectors **PT**, **PQ** and **PS** is

(a) 5 (b) 20

(c) 10 (d) 30

**3.** Let complex numbers $\alpha$ and $\frac{1}{\bar{\alpha}}$ lies on circles $(x - x_0)^2 + (y - y_0)^2 = r^2$ and $(x - x_0)^2 + (y - y_0)^2 = 4r^2$, respectively.

If $z_0 = x_0 + iy_0$ satisfies the equation $2|z_0|^2 = r^2 + 2$, then $|\alpha|$ is equal to

(a) $\frac{1}{\sqrt{2}}$ (b) $\frac{1}{2}$

(c) $\frac{1}{\sqrt{7}}$ (d) $\frac{1}{3}$

**4.** For $a > b > c > 0$, the distance between (1, 1) and the point of intersection of the lines $ax + by + c = 0$ and $bx + ay + c = 0$ is less than $2\sqrt{2}$. Then,

(a) $a + b - c > 0$ (b) $a - b + c < 0$

(c) $a - b + c > 0$ (d) $a + b - c < 0$

5. Perpendicular are drawn from points on the line $\frac{x+2}{2}=\frac{y+1}{-1}=\frac{z}{3}$ to the plane $x+y+z=3$. The feet of perpendiculars lie on the line

(a) $\frac{x}{5}=\frac{y-1}{8}=\frac{z-2}{-13}$

(b) $\frac{x}{2}=\frac{y-1}{3}=\frac{z-2}{-5}$

(c) $\frac{x}{4}=\frac{y-1}{3}=\frac{z-2}{-7}$

(d) $\frac{x}{2}=\frac{y-1}{-7}=\frac{z-2}{5}$

6. Four persons independently solve a certain problem correctly with probabilities $\frac{1}{2}, \frac{3}{4}, \frac{1}{4}, \frac{1}{8}$. Then, the probability that the problem is solved correctly by atleast one of them is

(a) $\frac{235}{256}$ (b) $\frac{21}{256}$ (c) $\frac{3}{256}$ (d) $\frac{253}{256}$

7. The area enclosed by the curves $y=\sin x+\cos x$ and $y=|\cos x-\sin x|$ over the interval $\left[0, \frac{\pi}{2}\right]$ is

(a) $4(\sqrt{2}-1)$ (b) $2\sqrt{2}(\sqrt{2}-1)$

(c) $2(\sqrt{2}+1)$ (d) $2\sqrt{2}(\sqrt{2}+1)$

8. A curve passes through the point $\left(1, \frac{\pi}{6}\right)$. Let the slope of the curve at each point $(x, y)$ be $\frac{y}{x}+\sec\left(\frac{y}{x}\right), x>0$.

Then, the equation of the curve is

(a) $\sin\left(\frac{y}{x}\right)=\log x+\frac{1}{2}$

(b) $\operatorname{cosec}\left(\frac{y}{x}\right)=\log x+2$

(c) $\sec\left(\frac{2y}{x}\right)=\log x+2$

(d) $\cos\left(\frac{2y}{x}\right)=\log x+\frac{1}{2}$

9. Let $f:\left[\frac{1}{2}, 1\right]\to R$ (the set of all real numbers) be a positive, non-constant and differentiable function such that $f'(x)<2f(x)$ and $f\left(\frac{1}{2}\right)=1$. Then, the value of $\int_{1/2}^{1} f(x)\,dx$ lies in the interval

(a) $(2e-1, 2e)$

(b) $(e-1, 2e-1)$

(c) $\left(\frac{e-1}{2}, e-1\right)$

(d) $\left(0, \frac{e-1}{2}\right)$

10. The number of points in $(-\infty, \infty)$, for which $x^2-x\sin x-\cos x=0$, is

(a) 6 (b) 4 (c) 2 (d) 0

## Section II

**Directions** (Q. Nos. 11-15) *This section contains 5 multiple choice questions. Each question has four choices (a), (b), (c) and (d) out of which one or more are correct.*

11. A rectangular sheet of fixed perimeter with sides having their lengths in the ratio 8 : 15 is converted into an open rectangular box by folding after removing squares of equal area from all four corners. If the total area of removed squares is 100, the resulting box has maximum volume. The lengths of the sides of the rectangular sheet are

(a) 24 (b) 32

(c) 45 (d) 60

12. Let $S_n=\sum_{2}^{4n}(-1)^{\frac{k(k+1)}{2}}k^2$. Then, $S_n$ can take value(s)

(a) 1056 (b) 1088

(c) 1120 (d) 1332

**13.** A line $l$ passing through the origin is perpendicular to the lines

$$l_1 : (3+t)\hat{\mathbf{i}} + (-1+2t)\hat{\mathbf{j}} + (4+2t)\hat{\mathbf{k}}, -\infty < t < \infty$$

$$l_2 : (3+2s)\hat{\mathbf{i}} + (3+2s)\hat{\mathbf{j}} + (2+s)\hat{\mathbf{k}}, -\infty < s < \infty$$

Then, the coordinate(s) of the point(s) on $l_2$ at a distance of $\sqrt{17}$ from the point of intersection of $l$ and $l_1$ is (are)

(a) $\left(\frac{7}{3}, \frac{7}{3}, \frac{5}{3}\right)$ (b) $(-1, -1, 0)$

(c) $(1, 1, 1)$ (d) $\left(\frac{7}{9}, \frac{7}{9}, \frac{8}{9}\right)$

**14.** Let $f(x) = x \sin \pi x$, $x > 0$. Then, for all natural numbers $n$, $f'(x)$ vanishes at

(a) a unique point in the interval $\left(n, n+\frac{1}{2}\right)$

(b) a unique point in the interval $\left(n+\frac{1}{2}, n+1\right)$

(c) a unique point in the interval $(n, n+1)$

(d) two points in the interval $(n, n+1)$

**15.** For $3 \times 3$ matrices $M$ and $N$, which of the following statement(s) is (are) not correct?

(a) $N^T M N$ is symmetric or skew-symmetric, according as $M$ is symmetric or skew-symmetric

(b) $MN - NM$ is symmetric for all symmetric matrices $M$ and $N$

(c) $M N$ is symmetric for all symmetric matrices $M$ and $N$

(d) $(\text{adj } M)\ (\text{adj } N) = \text{adj}(MN)$ for all invertible matrices $M$ and $N$

## Section III

**Directions** (Q. Nos. 16-20) *This section contains 5 questions. The answer to each question is a single digit integer, ranging from* 0 *to* 9 (*both inclusive.*)

**16.** A vertical line passing through the point $(h, 0)$ intersects the ellipse $\frac{x^2}{4} + \frac{y^2}{3} = 1$ at the points $P$ and $Q$. Let the tangents to the ellipse at $P$ and $Q$ meet at the point $R$. If $\Delta(h) =$ area of the $\Delta PQR$, $\Delta_1 = \max\limits_{1/2 \le h \le 1} \Delta(h)$ and $\Delta_2 = \min\limits_{1/2 \le h \le 1} \Delta(h)$, then $\frac{8}{\sqrt{5}}\Delta_1 - 8\Delta_2$ is equal to

**17.** The coefficients of three consecutive terms of $(1+x)^{n+5}$ are in the ratio $5 : 10 : 14$. Then, $n$ is equal to

**18.** Consider the set of eight vectors $V = [a\hat{\mathbf{i}} + b\hat{\mathbf{j}} + c\hat{\mathbf{k}} : a, b, c \in \{-1, 1\}]$. Three non-coplanar vectors can be chosen from $V$ in $2^p$ ways. Then, $p$ is

**19.** Of the three independent events $E_1$, $E_2$ and $E_3$, the probability that only $E_1$ occurs is $\alpha$, only $E_2$ occurs is $\beta$ and only $E_3$ occurs is $\gamma$. Let the probability $p$ that none of events $E_1$, $E_2$ or $E_3$ occurs satisfy the equations $(\alpha - 2\beta)$, $p = \alpha\beta$ and $(\beta - 3\gamma)$ $p = 2\beta\gamma$. All the given probabilities are assumed to lie in the interval $(0, 1)$.

Then, $\frac{\text{probability of occurrence of } E_1}{\text{probability of occurrence of } E_3}$ is equal to

**20.** A pack contains $n$ card numbered from 1 to $n$. Two consecutive numbered card are removed from the pack and the sum of the numbers on the remaining cards is 1224. If the smaller of the numbers on the removed cards is $k$, then $k - 20$ is equal to

# PAPER 2

## Instructions

- The question paper consists of three sections.
- Section 1 contains 8 multiple choice questions each of 3 marks with negative marking of 1 mark. Each question has four choices (a), (b), (c) and (d) out of which one or more are correct.
- Section 2 contains 4 paragraphs each describing theory, experiment, data etc. 8 question sections to four paragraphs with two questions on each paragraph, each of 3 marks with negative marking of 1 mark. Each question of paragraph has only one correct answer among the four choice (a), (b), (c) and (d).
- Section 3 contains 4 multiple choice questions each of 3 marks with negative marking of 1 mark. Each question has matching lists. The codes for the lists have choices (a), (b), (c) and (d) out of which only one is correct.

# PHYSICS

## Section I

**Direction** (Q. No. 1-8) *This section contains* 8 *multiple choice questions. Each question has four choices (a), (b), (c) and (d) out of which one or more than one are correct.*

**1.** Two bodies, each of mass $M$, are kept fixed with a separation $2L$. A particle of mass $m$ is projected from the mid-point of the line joining their centres, perpendicular to the line. The gravitational constant is $G$. The correct statement(s) is (are)

(a) The minimum initial velocity of the mass $m$ to escape the gravitational field of the two bodies is $4\sqrt{\frac{GM}{L}}$

(b) The minimum initial velocity of the mass $m$ to escape the gravitational field of the two bodies is $2\sqrt{\frac{GM}{L}}$

(c) The minimum initial velocity of the mass $m$ to escape the gravitational field of the two bodies is $\sqrt{\frac{2GM}{L}}$

(d) The energy of the mass $m$ remains constant

**2.** A particle of mass $m$ is attached to one end of a mass-less spring of force constant $k$, lying on a frictionless horizontal plane. The other end of the spring is fixed. The particle starts moving horizontally from its equilibrium position at time $t = 0$ with an initial velocity $u_0$. When the speed of the particle is $0.5u_0$, it collides elastically with a rigid wall. After this collision

(a) the speed of the particle when it returns to its equilibrium position is $u_0$

(b) the time at which the particle passes through the equilibrium position for the first time is $t = \pi\sqrt{\frac{m}{k}}$

(c) the time at which the maximum compression of the spring occurs is $t = \frac{4\pi}{3}\sqrt{\frac{m}{k}}$

(d) the time at which the particle passes through the equilibrium position for the second time is $t = \frac{5\pi}{3}\sqrt{\frac{m}{k}}$

**3.** A steady current $I$ flows along an infinitely long hollow cylindrical conductor of radius $R$. This cylinder is placed coaxially inside an infinite solenoid of radius $2R$. The solenoid has $n$ turns per unit length and carries a steady current $I$. Consider a point $P$ at a distance $r$ from the common axis. The correct statement(s) is (are)

(a) In the region $0 < r < R$, the magnetic field is non-zero

(b) In the region $R < r < 2R$, the magnetic field is along the common axis

(c) In the region $R < r < 2R$, the magnetic field is tangential to the circle of radius $r$, centered on the axis

(d) In the region $r > 2R$, the magnetic field is non-zero

**4.** Two vehicles, each moving with speed $u$ on the same horizontal straight road, are approaching each other. Wind blows along the road with velocity $w$. One of these vehicles blows a whistle of frequency $f_1$. An observer in the other vehicle hears the frequency of the whistle to be $f_2$. The speed of sound in still air is $v$. The correct statement(s) is (are)

(a) If the wind blows from the observer to the source, $f_2 > f_1$.

(b) If the wind blows from the source to the observer, $f_2 > f_1$

(c) If the wind blows from the observer to the source, $f_2 < f_1$

(d) If the wind blows from the source to the observer, $f_2 < f_1$

**5.** Using the expression $2d \sin\theta = \lambda$, one calculates the values of $d$ by measuring the corresponding angles $\theta$ in the range 0 to 90°. The wavelength $\lambda$ is exactly known and the error in $\theta$ is constant for all values of $\theta$. As $\theta$ increases from 0°

(a) the absolute error in $d$ remains constant

(b) the absolute error in $d$ increases

(c) the fractional error in $d$ remains constant

(d) the fractional error in $d$ decreases

**6.** Two non-conducting spheres of radii $R_1$ and $R_2$ and carrying uniform volume charge densities $+\rho$ and $-\rho$, respectively, are placed such that they partially overlap, as shown in the figure. At all points in the overlapping region

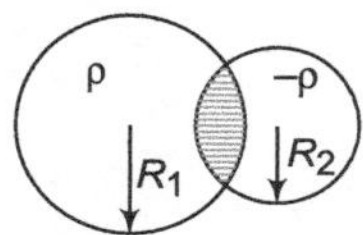


(a) the electrostatic field is zero

(b) the electrostatic potential is constant

(c) the electrostatic field is constant in magnitude

(d) the electrostatic field has same direction

**7.** The figure below shows the variation of specific heat capacity ($C$) of a solid as a function of temperature ($T$). The temperature is increased continuously from 0 to 500 K at a constant rate. Ignoring any volume change, the following statement(s) is (are) correct to reasonable approximation.

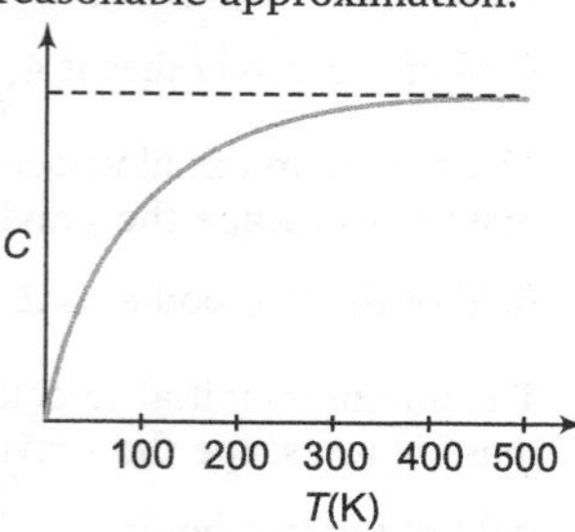


(a) the rate at which heat is absorbed in the range 0-100 K varies linearly with temperature $T$

(b) heat absorbed in increasing the temperature from 0-100 K is less than the heat required for increasing the temperature from 400-500 K

(c) there is no change in the rate of heat absorbtion in the range 400-500 K

(d) the rate of heat absorption increases in the range 200-300 K

**8.** The radius of the orbit of an electron in a Hydrogen-like atom is $4.5\, a_0$ where $a_0$ is the Bohr radius. Its orbital angular momentum is $\frac{3h}{2\pi}$. It is given that $h$ is Planck constant and $R$ is Rydberg constant. The possible wavelength(s), when the atom de-excites, is (are)

(a) $\frac{9}{32R}$ (b) $\frac{9}{16R}$ (c) $\frac{9}{5R}$ (d) $\frac{4}{3R}$

## Section II

**Direction** (Q. No. 9-16) *This section contains* 8 *multiple choice questions relating to four paragraphs with two questions on each paragraph. Each question has four choices (a), (b), (c) and (d) out of which only one is correct.*

### Paragraph I

A small block of mass 1 kg is released from rest at the top of a rough track. The track is a circular arc of radius 40 m. The block slides along the track without toppling and a frictional force acts on it in the direction opposite to the instantaneous velocity. The work done in overcoming the friction up to the point $Q$, as shown in the figure, is 150 J. (Take the acceleration due to gravity, $g = 10 \text{ ms}^{-2}$)

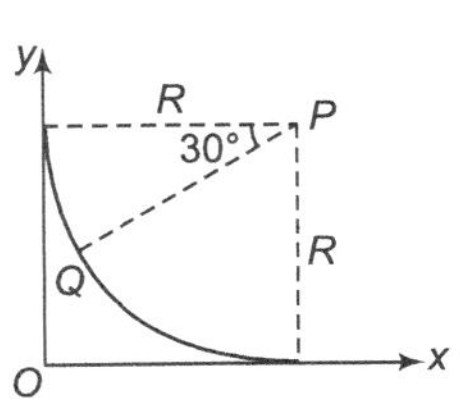


**9.** The speed of the block when it reaches the point $Q$ is

(a) $5 \text{ ms}^{-1}$ (b) $10 \text{ ms}^{-1}$

(c) $10\sqrt{3} \text{ ms}^{-1}$ (d) $20 \text{ ms}^{-1}$

**10.** The magnitude of the normal reaction that acts on the block at the point $Q$ is

(a) 7.5 N

(b) 8.6 N

(c) 11.5 N

(d) 22.5 N

### Paragraph II

A thermal power plant produces electric power of 600 kW at 4000 V, which is to be transported to a place 20 km away from the power plant for consumers' usage. It can be transported either directly with a cable of large current carrying capacity or by using a combination of step-up and step-down transformers at the two ends. The drawback of the direct transmission is the large energy dissipation. In the method using transformers, the dissipation is much smaller. In this method, a step-up transformer is used at the plant side so that the current is reduced to a smaller value. At the consumers' end a step-down transformer is used to supply power to the consumers at the specified lower voltage. It is reasonable to assume that the power cable is purely resistive and the transformers are ideal with a power factor unity. All the current and voltages mentioned are rms values.

**11.** If the direct transmission method with a cable of resistance $0.4\ \Omega\ \text{km}^{-1}$ is used, the power dissipation (in %) during transmission is

(a) 20 (b) 30

(c) 40 (d) 50

**12.** In the method using the transformers, assume that the ratio of the number of turns in the primary to that in the secondary in the step-up transformer is 1 : 10. If the power to the consumers has to be supplied at 200V, the ratio of the number of turns in the primary to that in the secondary in the step-down transformer is

(a) 200 : 1 (b) 150 : 1
(c) 100 : 1 (d) 50 : 1

## Paragraph III

A point charge $Q$ is moving in a circular orbit of radius $R$ in the $x$-$y$ plane with an angular velocity $\omega$. This can be considered as equivalent to a loop carrying a steady current $\frac{Q\omega}{2\pi}$. A uniform magnetic field along the positive $z$-axis is now switched on, which increases at a constant rate from 0 to B in one second. Assume that the radius of the orbit remains constant. The applications of the magnetic field induces an emf in the orbit. The induced emf is defined as the work done by an induced electric field in moving a unit positive charge around a closed loop. It is known that, for an orbiting charge, the magnetic dipole moment is proportional to the angular momentum with a proportionality constant $\gamma$.

**13.** The magnitude of the induced electric field in the orbit at any instant of time during the time interval of the magnetic field change is

(a) $\frac{BR}{4}$ (b) $\frac{-BR}{2}$ (c) $BR$ (d) $2BR$

**14.** The change in the magnetic dipole moment associated with the orbit, at the end of the time interval of the magnetic field change, is

(a) $\gamma BQR^2$ (b) $-\gamma\frac{BQR^2}{2}$
(c) $\gamma\frac{BQR^2}{2}$ (d) $\gamma BQR^2$

## Paragraph IV

The mass of a nucleus $^A_ZX$ is less that the sum of the masses of $(A-Z)$ number of neutrons and $Z$ number of protons in the nucleus. The energy equivalent to the corresponding mass difference is known as the binding energy of the nucleus. A heavy nucleus of mass $M$ can break into two light nuclei of masses $m_1$ and $m_2$ only if $(m_1 + m_2) < M$. Also two light nuclei of masses $m_3$ and $m_4$ can undergo complete fusion and form a heavy nucleus of mass $M'$ only if $(m_3 + m_4) > M'$. The masses of some neutral atoms are given in the table below :

| | | | | | |
|---|---|---|---|---|---|
| $^1_1H$ | 1.007825u | $^2_1H$ | 2.014102u | $^3_1H$ | 3.016050u |
| $^6_3Li$ | 6.01513u | $^7_3Li$ | 7.016004u | $^4_2He$ | 4.002603u |
| $^{152}_{64}Gd$ | 151.919803u | $^{206}_{82}Pb$ | 205.974455u | $^{70}_{30}Zn$ | 69.925325u |
| $^{82}_{34}Se$ | 81.916709u | $^{209}_{83}Bi$ | 208.980388u | $^{210}_{84}Po$ | 209.982876u |

**15.** The correct statement is

(a) The nucleus $^6_3Li$ can emit an alpha particle
(b) The nucleus $^{210}_{84}Po$ can emit a proton
(c) Deuteron and alpha particle can undergo complete fusion
(d) The nuclei $^{70}_{30}Zn$ and $^{82}_{34}Se$ can undergo complete fusion

**16.** The Kinetic energy (in keV) of the alpha particle, when the nucleus $^{210}_{84}Po$ at rest undergoes alpha decay, is

(a) 5319 (b) 5422
(c) 5707 (d) 5818

## Section III

**Direction** (Q. No. 17-20) *This section contains 4 multiple choice questions. Each question has matching lists. The codes for the lists have choices (a), (b), (c) and (d) out of which only one is correct.*

17. A right angled prism of refractive index $\mu_1$ is placed in a rectangular block of refractive index $\mu_2$, which is surrounded by a medium of refractive index $\mu_3$, as shown in the figure, A ray of light '*e*' enters the rectangular block at normal incidence. Depending upon the relationships between $\mu_1, \mu_2$ and $\mu_3$, it takes one of the four possible paths '*ef*', '*eg*', '*eh*' or '*ei*'.

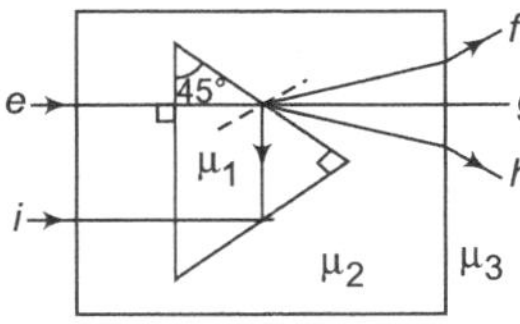

Match the paths in List I with conditions of refractive indices in List II and select the correct answer using the codes given below the lists.

| | List I | | List II |
|---|---|---|---|
| P. | $e \to f$ | 1. | $\mu_1 > \sqrt{2}\mu_2$ |
| Q. | $e \to g$ | 2. | $\mu_1 > \mu_1$ and $\mu_2 > \mu_3$ |
| R. | $e \to h$ | 3. | $\mu_1 = \mu_2$ |
| S. | $e \to i$ | 4. | $\mu_2 < \mu_1 < \sqrt{2}\mu_2$ and $\mu_2 > \mu_3$ |

**Codes**

| | P | Q | R | S | | P | Q | R | S |
|---|---|---|---|---|---|---|---|---|---|
| (a) | 2 | 3 | 1 | 4 | (b) | 1 | 2 | 4 | 3 |
| (c) | 4 | 1 | 2 | 3 | (d) | 2 | 3 | 4 | 1 |

18. Match List I with List II and select the correct answer using the codes given below the lists.

| | List I | | List II |
|---|---|---|---|
| P. | Boltzmann constant | 1. | $[ML^2T^{-1}]$ |
| Q. | Coefficient of viscosity | 2. | $[ML^{-1}T^{-1}]$ |
| R. | Planck constant | 3. | $[MLT^{-3}K^{-1}]$ |
| S. | Thermal conductivity | 4. | $[ML^2T^{-2}K^{-1}]$ |

**Codes**

| | P | Q | R | S | | P | Q | R | S |
|---|---|---|---|---|---|---|---|---|---|
| (a) | 3 | 1 | 2 | 4 | (b) | 3 | 2 | 1 | 4 |
| (c) | 4 | 2 | 1 | 3 | (d) | 4 | 1 | 2 | 3 |

19. One mole of a monatomic ideal gas is taken along two cyclic processes $E \to F \to G \to E$ and $E \to F \to H \to E$ as shown in the $p$-$V$ diagram.

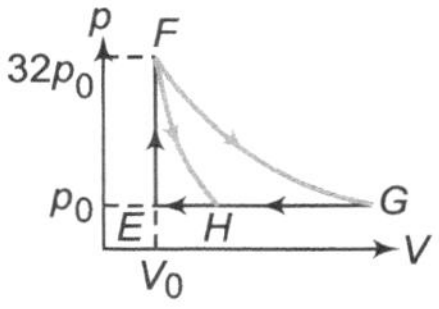

The processes involved are purely isochoric, isobaric, isothermal or adiabatic.

Match the paths in List I with the magnitudes of the work done in List II and select the correct answer using the codes given below the lists.

| | List I | | List II |
|---|---|---|---|
| P. | $G \to E$ | 1. | $160\ p_0V_0 \ln 2$ |
| Q. | $G \to H$ | 2. | $36\ p_0V_0$ |
| R. | $F \to H$ | 3. | $24\ p_0V_0$ |
| S. | $F \to G$ | 4. | $31\ p_0V_0$ |

**Codes**

| | P | Q | R | S | | P | Q | R | S |
|---|---|---|---|---|---|---|---|---|---|
| (a) | 4 | 3 | 2 | 1 | (b) | 4 | 3 | 1 | 2 |
| (c) | 3 | 1 | 2 | 4 | (d) | 1 | 3 | 2 | 4 |

20. Match List I of the nuclear process with List II containing parent nucleus and one of the end products of each process and then select the correct answer using the codes given below the lists.

| | List I | | List II |
|---|---|---|---|
| P. | Alpha decay | 1. | ${}^{15}_{8}O \to {}^{17}_{7}N + \ldots$ |
| Q. | $\beta^+$ decay | 2. | ${}^{238}_{92}U \to {}^{234}_{90}Th + \ldots$ |
| R. | Fission | 3. | ${}^{185}_{83}Bi \to {}^{184}_{82}Pb + \ldots$ |
| S. | Proton emission | 4. | ${}^{239}_{94}Pu \to {}^{140}_{57}La + \ldots$ |

**Codes**

| | P | Q | R | S | | P | Q | R | S |
|---|---|---|---|---|---|---|---|---|---|
| (a) | 4 | 2 | 1 | 3 | (b) | 1 | 3 | 2 | 4 |
| (c) | 2 | 1 | 4 | 3 | (d) | 4 | 3 | 2 | 1 |

# CHEMISTRY

## Section I

**Direction** (Q. No. 1-8) *This section contains* 8 *multiple choice questions. Each question has four choices (a), (b), (c) and (d) out of which one or more than one is correct.*

**1.** The $K_{sp}$ of $Ag_2CrO_4$ is $1.1 \times 10^{-12}$ at 298 K. The solubility (in mol/L) of $Ag_2CrO_4$ in a 0.1 M $AgNO_3$ solution is

(a) $1.1 \times 10^{-11}$ (b) $1.1 \times 10^{-10}$

(c) $1.1 \times 10^{-12}$ (d) $1.1 \times 10^{-9}$

**2.** In the following reaction, the product(s) formed is (are)

OH
CH₃
$\xrightarrow[OH^-]{CHCl_3}$ ?

OH, OHC, CHO, $CH_3$ (*P*)

O, $H_3C$, $CHCl_2$ (*Q*)

OH, $H_3C$, $CHCl_2$ (*R*)

OH, CHO, $CH_3$ (*S*)

(a) *P* (major)

(b) *Q* (minor)

(c) *R* (minor)

(d) *S* (major)

**3.** The major product(s) of the following reaction is (are)

OH, $SO_3H$ — Aqueous $Br_2$ (3.0 equivalents) →

OH, Br, Br, Br, $SO_3H$ (*P*)

OH, Br, Br, Br (*Q*)

OH, Br, Br, Br (*R*)

OH, Br, Br, Br, $SO_3H$ (*S*)

(a) *P* (b) *Q*

(c) *R* (d) *S*

**4.** After completion of the reactions (I and II), the organic compound(s) in the reaction mixtures is (are)

**Reaction I** $H_3C-CO-CH_3$ (1.0 mol) $\xrightarrow[\text{Aqueous/NaOH}]{Br_2 \text{ (1.0 mol)}}$

**Reaction II** $H_3C-CO-CH_3$ (1.0 mol) $\xrightarrow[CH_3COOH]{Br_2 \text{ (1.0 mol)}}$

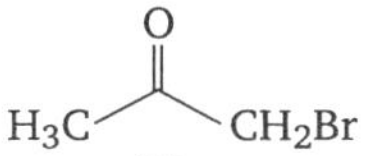

$H_3C-CO-CH_2Br$ (P) $H_3C-CO-CBr_3$ (Q)

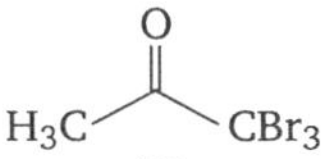

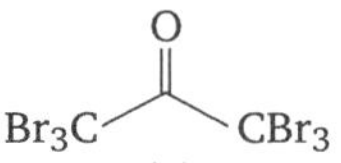

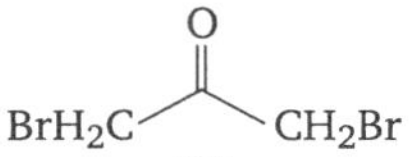

$Br_3C-CO-CBr_3$ (R) $BrH_2C-CO-CH_2Br$ (S)

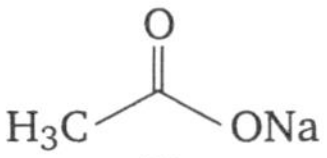

$H_3C-CO-ONa$ (T) $CHBr_3$ (U)

(a) Reaction I : *P* and Reaction II : *P*
(b) Reaction I : *U*, acetone and Reaction II : *Q*, acetone
(c) Reaction I : *T*, *U*, acetone and Reaction II : *P*
(d) Reaction I : *R*, acetone and Reaction II : *S*, acetone

**5.** The correct statement(s) about $O_3$ is (are)
(a) O—O bond lengths are equal
(b) thermal decomposition of $O_3$ is endothermic
(c) $O_3$ is diamagnetic in nature
(d) $O_3$ has a bent structure

**6.** In the nuclear transmutation
$^{9}_{4}Be + X \longrightarrow {}^{8}_{4}Be + Y$
$(X, Y)$ is (are)
(a) $(\gamma, n)$ (b) $(p, D)$
(c) $(n, D)$ (d) $(\gamma, p)$

**7.** The carbon-based reduction method is not used for the extraction of
(a) tin from $SnO_2$
(b) iron from $Fe_2O_3$
(c) aluminium from $Al_2O_3$
(d) magnesium from $MgCO_3$, $CaCO_3$

**8.** The thermal dissociation of equilibrium of $CaCO_3(s)$ is studied under different conditions.

$$CaCO_3(s) \rightleftharpoons CaO(s) + CO_2(g)$$

For this equilibrium, the correct statement(s) is (are)
(a) $\Delta H$ is dependent on $T$
(b) $K$ is independent of the initial amount of $CaCO_3$
(c) $K$ is dependent on the pressure of $CO_2$ at a given $T$
(d) $\Delta H$ is independent of the catalyst, if any

## Section II

**Direction** (Q. No. 9-16) *This section contains 8 multiple choice questions relating to four paragraphs with two questions on each paragraph. Each question has four choices (a), (b), (c) and (d) out of which only one is correct.*

### Paragraph I

An aqueous solution of a mixture of two inorganic salts,when treated with dilute HCl, gave a precipitate (P) and filtrate (Q). The precipitate P was found to dissolve in hot water. The filtrate (Q) remained unchanged, when treated with $H_2S$ in a dilute mineral acid medium. However, it gave a precipitate (R) with $H_2S$ in an ammoniacal medium. The precipitate R gave a coloured solution (S), when treated with $H_2O_2$ in an aqueous NaOH medium.

**9.** The precipitate *P* contains
(a) $Pb^{2+}$ (b) $Hg_2^{2+}$
(c) $Ag^{+}$ (d) $Hg^{2+}$

**10.** The coloured solution *S* contains
(a) $Fe_2(SO_4)_3$
(b) $CuSO_4$
(c) $ZnSO_4$
(d) $Na_2CrO_4$

## Paragraph II

$P$ and $Q$ are isomers of dicarboxylic acid $C_4H_4O_4$. Both decolourize $Br_2/H_2O$. On heating, $P$ forms the cyclic anhydride.

Upon treatment with dilute alkaline $KMnO_4$. $P$ as well as $Q$ could produce one or more than one form $S$, $T$ and $U$.

| | S | | | T | |
|---|---|---|---|---|---|
| | COOH | | | COOH | |
| H | | OH | H | | OH |
| H | | OH | HO | | H |
| | COOH | | | COOH | |

| | U | |
|---|---|---|
| | COOH | |
| HO | | H |
| H | | OH |
| | COOH | |

**11.** Compounds formed from $P$ and $Q$ are, respectively

(a) Optically active $S$ and optically active pair ($T$, $U$)

(b) Optically inactive $S$ and optically inactive pair ($T$, $U$)

(c) Optically active pair ($T$, $U$) and optically active $S$

(d) Optically inactive pair ($T$, $U$) and optically inactive $S$

**12.** In the following reaction sequences $V$ and $W$ are respectively

$$Q \xrightarrow[\Delta]{H_2/Ni} V$$

$$\text{Benzene} + V \xrightarrow{AlCl_3 \text{ (anhydrous)}} \xrightarrow[2.\ H_3PO_4]{1.\ Zn\text{—}Hg/HCl} W$$

(a) O, O, O and O (V) (W)

(b) $CH_2OH$, $CH_2OH$ and (V) (W)

(c) O, O, O and (V) (W)

(d) $HOH_2C$, $CH_2OH$ (V)

and $CH_2OH$ (W)

## Paragraph III

A fixed mass $m$ of a gas is subjected to transformation of states from $K$ to $L$ to $M$ to $N$ and back to $K$ as shown in the figure.

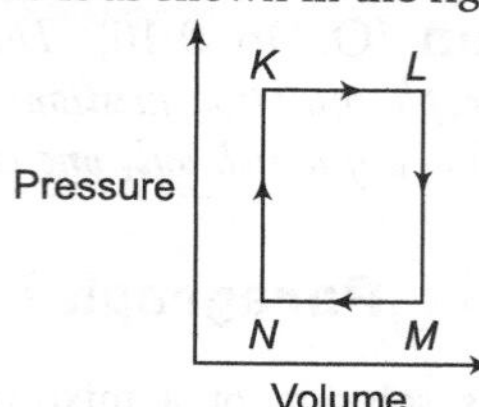


**13.** The succeeding operations that enable this transformation of states are

(a) heating, cooling, heating, cooling

(b) cooling, heating, cooling, heating

(c) heating, cooling, cooling, heating

(d) cooling, heating, heating, cooling

**14.** The pair of isochoric processes among the transormation of states is

(a) *K* to *L* and *L* to *M*
(b) *L* to *M* and *N* to *K*
(c) *L* to *M* and *M* to *N*
(d) *M* to *N* and *N* to *K*

## Paragraph IV

The reactions of $Cl_2$ gas with cold-dilute and hot-concentrated NaOH in water give sodium salts of two (different) oxoacids of chlorine, *P* and *Q*, respectively. The $Cl_2$ gas reacts with $SO_2$ gas, in the presence of charcoal, to give a product *R*. *R* reacts with white phosphorus to give a compound *S*. On hydrolysis, *S* gives an oxoacid of phosphorous *T*.

**15.** *P* and *Q* respectively, are the sodium salts of

(a) hypochlorous and chloric acids
(b) hypochlorous and chlorous acids
(c) chloric and perchloric acids
(d) chloric and hypochlorous acids

**16.** *R*, *S* and *T*, respectively, are

(a) $SO_2Cl_2$, $PCl_5$ and $H_3PO_4$
(b) $SO_2Cl_2$, $PCl_3$ and $H_3PO_3$
(c) $SOCl_2$, $PCl_3$ and $H_3PO_2$
(d) $SOCl_2$, $PCl_5$ and $H_3PO_4$

## Section III

**Direction** (Q. No. 17-20) *This section contains 4 multiple choice questions. Each question has four choices (a), (b), (c) and (d) out of which only one is correct.*

**17.** The unbalanced chemical reactions given in List I show missing reagent or condition (?) which are provided in List II. Match List I with List II and select the correct answer using the code given below the lists.

| | List I | | List II |
|---|---|---|---|
| P. | $PbO_2 + H_2SO_4 \xrightarrow{?} PbSO_4 + O_2 +$ other product | 1. | NO |
| Q. | $Na_2S_2O_3 + H_2O \xrightarrow{?} NaHSO_4 +$ other product | 2. | $I_2$ |
| R. | $N_2H_4 \xrightarrow{?} N_2 +$ other product | 3. | Warm |
| S. | $XeF_2 \xrightarrow{?} Xe +$ other product | 4. | $Cl_2$ |

**Codes**

| | P | Q | R | S |
|---|---|---|---|---|
| (a) | 4 | 2 | 3 | 1 |
| (b) | 3 | 2 | 1 | 4 |
| (c) | 1 | 4 | 2 | 3 |
| (d) | 3 | 4 | 2 | 1 |

**18.** Match the chemical conversion in List I with the appropriate reagents in List II and select the correct answer using the code given below the lists.

| | List I | | List II |
|---|---|---|---|
| P. | 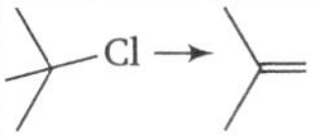 —Cl → (alkene) | 1. | (i) $Hg(OAc)_2$; (ii) $NaBH_4$ |
| Q. | 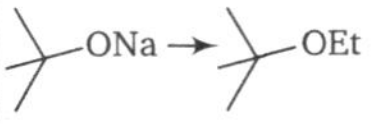 —ONa → —OEt | 2. | NaOEt |
| R. | 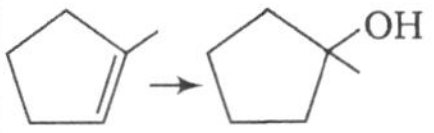 (methylcyclopentene) → (1-methylcyclopentanol, OH) | 3. | Et-Br |
| S. | 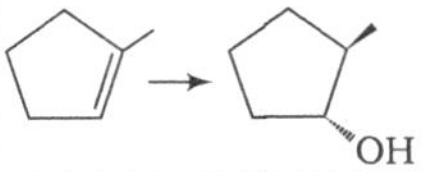 (methylcyclopentene) → (2-methylcyclopentanol, OH) | 4. | (i) $BH_3$; (ii) $H_2O_2$/NaOH |

**Codes**

| | P | Q | R | S |
|---|---|---|---|---|
| (a) | 2 | 3 | 1 | 4 |
| (b) | 3 | 2 | 1 | 4 |
| (c) | 2 | 3 | 4 | 1 |
| (d) | 3 | 2 | 4 | 1 |

**19.** An aqueous solution of $X$ is added slowly to an aqueous solution of $Y$ as shown in List I. The variation in conductivity of these reactions is given in List II. Match List I with List II and select the correct answer using the code given below the lists.

| | List I | | List II |
|---|---|---|---|
| P. | $(C_2H_5)_3N$ ($X$) + $CH_3COOH$ ($Y$) | 1. | Conductivity decreases and then increases |
| Q. | KI(0.1 M) ($X$) + $AgNO_3$(0.01 M) ($Y$) | 2. | Conductivity decreases and then does not change much |
| R. | $CH_3COOH$ ($X$) + KOH ($Y$) | 3. | Conductivity increases and then does not change much |
| S. | NaOH ($X$) + HI ($Y$) | 4. | Conductivity does not change much and then increases |

**Codes**

| | P | Q | R | S | | P | Q | R | S |
|---|---|---|---|---|---|---|---|---|---|
| (a) | 3 | 4 | 2 | 1 | (b) | 4 | 3 | 2 | 1 |
| (c) | 2 | 3 | 4 | 1 | (d) | 1 | 4 | 3 | 2 |

**20.** The standard reduction potential data at 25°C is given below.

$E°(Fe^{3+}/Fe^{2+}) = +0.77$ V;

$E°(Fe^{2}/Fe) = -0.44$ V;

$E°(Cu^{2+}/Cu) = +0.34$ V;

$E°(Cu^{+}/Cu) = +0.52$ V;

$E°(O_2(g) + 4H^+ + 4e^- \longrightarrow 2H_2O) = +1.23$ V;

$E°(O_2(g) + 2H_2O + 4e^- \longrightarrow 4OH) = +0.40$ V

$E°(Cr^{3+}/Cr) = -0.74$ V;

$E°(Cr^{2+}/Cr) = +0.91$ V

Match $E°$ of the rebox pair in List I with the values given in List II and select the correct answer using the code given below the lists.

| | List I | | List II |
|---|---|---|---|
| P. | $E°(Fe^{3+}/Fe)$ | 1. | − 0.18 V |
| Q. | $E°(4H_2O \rightleftharpoons 4H^+ + 4OH^-)$ | 2. | − 0.4 V |
| R. | $E°(Cu^{2+} + Cu \longrightarrow 2Cu^+)$ | 3. | − 0.04 V |
| S. | $E°(Cr^{3+}, Cr^{2+})$ | 4. | − 0.83 V |

**Codes**

| | P | Q | R | S | | P | Q | R | S |
|---|---|---|---|---|---|---|---|---|---|
| (a) | 4 | 1 | 2 | 3 | (b) | 2 | 3 | 4 | 1 |
| (c) | 1 | 2 | 3 | 4 | (d) | 3 | 4 | 1 | 2 |

# MATHEMATICS

## Section I

**Directions** (Q. Nos. 1-8) *This section contains 8 multiple choice questions. Each question has four choices (a), (b), (c) and (d) out of which one or more are correct.*

**1.** For $a \in R$ (the set of all real numbers), $a \neq -1$,

$$\lim_{n\to\infty} \frac{(1^a + 2^a + \ldots + n^a)}{(n+1)^{a-1}[(na+1)+(na+2)+\ldots+(na+n)]} = \frac{1}{60}.$$

Then, $a$ is equal to

(a) 5
(b) 7
(c) $\frac{-15}{2}$
(d) $\frac{-17}{2}$

**2.** Circle(s) touching $x$-axis at a distance 3 from the origin and having an intercept of length $2\sqrt{7}$ on $y$-axis is (are)

(a) $x^2 + y^2 - 6x + 8y + 9 = 0$

(b) $x^2 + y^2 - 6x + 7y + 9 = 0$

(c) $x^2 + y^2 - 6x - 8y + 9 = 0$

(d) $x^2 + y^2 - 6x - 7y + 9 = 0$

**3.** Two lines $L_1: x = 5, \frac{y}{3-\alpha} = \frac{z}{-2}$ and $L_2 : x = \alpha, \frac{y}{-1} = \frac{z}{2-\alpha}$ are coplanar. Then, $\alpha$ can take value(s)

(a) 1 (b) 2 (c) 3 (d) 4

**4.** In a $\Delta PQR$, $P$ is the largest angle and $\cos P = \frac{1}{3}$. Further in circle of the triangle touches the sides $PQ$, $QR$ and $RP$ at $N$, $L$ and $M$ respectively, such that the lengths of $PN$, $QL$ and $RM$ are consecutive even integers. Then, possible length(s) of the side(s) of the triangle is (are)

(a) 16 (b) 18

(c) 24 (d) 22

**5.** Let $w = \frac{\sqrt{3} + i}{2}$ and $P = \{w^n : n = 1, 2, 3, \dots\}$. Further $H_1 = \left\{z \in C : \text{Re } z > \frac{1}{2}\right\}$ and $H_2 = \left\{z \in C : \text{Re } z < -\frac{1}{2}\right\}$, where $C$ is the set of all complex numbers, if $z_1 \in P \cap H_1$, $z_2 \in P \cap H_2$ and $O$ represents the origin, then $\angle z_1 O z_2$ is equal to

(a) $\frac{\pi}{2}$ (b) $\frac{\pi}{6}$ (c) $\frac{2\pi}{3}$ (d) $\frac{5\pi}{6}$

**6.** If $3^x = 4^{x-1}$, then $x$ is equal to

(a) $\frac{2\log_3 2}{2\log_3 2 - 1}$ (b) $\frac{2}{2 - \log_2 3}$

(c) $\frac{1}{1 - \log_4 3}$ (d) $\frac{2\log_2 3}{2\log_2 3 - 1}$

**7.** Let $\omega$ be a complex cube root of unity with $\omega \neq 1$ and $P = [p_{ij}]$ be a $n \times n$ matrix with $p_{ij} = \omega^{i+j}$. Then, $P^2 \neq 0$, when $n$ is equal to

(a) 57 (b) 55 (c) 58 (d) 56

**8.** The function $f(x) = 2|x| + |x + 2| - ||x + 2| - 2|x||$ has a local minimum or a local maximum at $x$ is equal to

(a) $-2$ (b) $\frac{-2}{3}$ (c) 2 (d) $\frac{2}{3}$

## Section II

**Directions** (Q. Nos. 9-16) *This section contains 4 paragraphs each describing theory, experiment, data etc., eight questions relate to four paragraphs with two questions on each paragraph. Each question of a paragraph has only one correct answer among the four choices (a), (b), (c) and (d).*

### Paragraph I

Let $f : [0, 1] \to R$ (the set of all real numbers) be a function. Suppose the function $f$ is twice differentiable, $f(0) = f(1) = 0$ and satisfies $f''(x) - 2f'(x) + f(x) \geq e^x, x \in [0, 1]$.

**9.** Which of the following is true for $0 < x < 1$?

(a) $0 < f(x) < \infty$ (b) $-\frac{1}{2} < f(x) < \frac{1}{2}$

(c) $-\frac{1}{4} < f(x) < 1$ (d) $-\infty < f(x) < 0$

**10.** If the function $e^{-x}f(x)$ assumes its minimum in the interval [0, 1] a $x = \frac{1}{4}$, which of the following is true?

(a) $f'(x) < f(x), \frac{1}{4} < x < \frac{3}{4}$

(b) $f'(x) > f(x), 0 < x < \frac{1}{4}$

(c) $f'(x) < f(x), 0 < x < \frac{1}{4}$

(d) $f'(x) < f(x), \frac{3}{4} < x < 1$

## Paragraph II

Let $PQ$ be a focal chord of the parabola $y^2 = 4ax$. The tangents to the parabola at $P$ and $Q$ meet at a point lying on the line $y = 2x + a, a > 0$.

**11.** Length of chord $PQ$ is

(a) $7a$ (b) $5a$

(c) $2a$ (d) $3a$

**12.** If chord $PQ$ subtends an angle $\theta$ at the vertex of $y^2 = 4ax$, then $\tan\theta$ is equal to

(a) $\frac{2}{3}\sqrt{7}$ (b) $\frac{-2}{3}\sqrt{7}$

(c) $\frac{2}{3}\sqrt{5}$ (d) $\frac{-2}{3}\sqrt{5}$

## Paragraph III

Let $S = S_1 \cap S_2 \cap S_3$, where

$S_1 = \{z \in C : |z| < 4\}$, $S_2$

$$= \left\{z \in C : \text{Im}\left[\frac{z - 1 + \sqrt{3}i}{1 - \sqrt{3}i}\right] > 0\right\}$$

and $S_3 : \{z \in C : \text{Re}\, z > 0\}$

**13.** Area of $S$ is equal to

(a) $\frac{10\pi}{3}$ (b) $\frac{20\pi}{3}$

(c) $\frac{16\pi}{3}$ (d) $\frac{32\pi}{3}$

**14.** $\min\limits_{z \in S} |1 - 3i - z|$ is equal to

(a) $\frac{2 - \sqrt{3}}{2}$

(b) $\frac{2 + \sqrt{3}}{2}$

(c) $\frac{3 - \sqrt{3}}{2}$

(d) $\frac{3 + \sqrt{3}}{2}$

## Paragraph IV

A box $B_1$ contains 1 white ball, 3 red balls and 2 black balls. Another box $B_2$ contains 2 white balls, 3 red balls and 4 black balls. A third box $B_3$ contains 3 white balls, 4 red balls and 5 black balls.

**15.** If 1 ball is drawn from each of the boxes $B_1$, $B_2$ and $B_3$, the probability that all 3 drawn balls are of the same colour is

(a) $\frac{82}{648}$ (b) $\frac{90}{648}$

(c) $\frac{558}{648}$ (d) $\frac{566}{648}$

**16.** If 2 balls are drawn (without replacement) from a randomly selected box and one of the balls is white and the other ball is red, the probability that these 2 balls are drawn from box $B_2$ is

(a) $\frac{116}{181}$ (b) $\frac{126}{181}$

(c) $\frac{65}{181}$ (d) $\frac{55}{181}$

## Section III

**Directions** (Q. Nos. 17-20) *This section contains 4 multiple choice questions. Each question has four choices (a), (b), (c) and (d) out of which only one is correct.*

**17.** Match List I with List II and select the correct answer using the code given below the lists.

| | List I | | List II |
|---|---|---|---|
| P. | $\left[\frac{1}{y^2}\left\{\frac{\cos(\tan^{-1}y)+y\sin(\tan^{-1}y)}{\cot(\sin^{-1}y)+\tan(\sin^{-1}y)}\right\}^2+y^4\right]^{1/2}$ takes value | 1. | $\frac{1}{2}\sqrt{\frac{5}{3}}$ |
| Q. | If $\cos x+\cos y+\cos z=0=\sin x+\sin y+\sin z$, then possible value of $\cos\frac{x-y}{2}$ is | 2. | $\sqrt{2}$ |
| R. | If $\cos\left(\frac{\pi}{4}-x\right)\cos 2x+\sin x\sin 2x\sec x=\cos x\sin 2x\sec x+\cos\left(\frac{\pi}{4}+x\right)\cos 2x$, then possible value of $\sec x$ is | 3. | $\frac{1}{2}$ |
| S. | If $\cot\left(\sin^{-1}\sqrt{1-x^2}\right)=\sin[\tan^{-1}(x\sqrt{6})]$, $x\neq 0$. Then, possible value of $x$ is | 4. | 1 |

**Codes**

| | P | Q | R | S | | P | Q | R | S |
|---|---|---|---|---|---|---|---|---|---|
| (a) | 4 | 3 | 2 | 1 | (b) | 4 | 3 | 2 | 1 |
| (c) | 3 | 4 | 2 | 1 | (d) | 3 | 4 | 1 | 2 |

**18.** A line $L: y = mx + 3$ meets $y$-axis at $E(0, 3)$ and the arc of the parabola $y^2 = 16x, 0 \le y \le 6$ at the point $F(x_0, y_0)$. The tangent to the parabola at $F(x_0, y_0)$ intersects the $y$-axis at $G(0, y_1)$. The slope $m$ of the line $L$ is chosen such that the area of the $\Delta EFG$ has a local maximum

Match List I with List II and select the correct answer using the code given below the lists

| | List I | | List II |
|---|---|---|---|
| P. | $m =$ | 1. | 1/2 |
| Q. | Maximum area of $\Delta EFG$ is | 2. | 4 |
| R. | $y_0 =$ | 3. | 2 |
| S. | $y_1 =$ | 4. | 1 |

**Codes**

| | P | Q | R | S | | P | Q | R | S |
|---|---|---|---|---|---|---|---|---|---|
| (a) | 4 | 1 | 2 | 3 | (b) | 3 | 4 | 1 | 2 |
| (c) | 1 | 3 | 2 | 4 | (d) | 1 | 3 | 4 | 2 |

**19.** Match List I with List II and select the correct answer using the code given below the lists

| | List I | | List II |
|---|---|---|---|
| P. | Volume of parallelopiped determined by vectors **a**, **b** and **c** is 2. Then, the volume of the parallelopiped determined by vectors 2(**a** × **b**), 3(**b** × **c**) and (**c** × **a**) is | 1. | 100 |
| Q. | Volume of parallelopiped determined by vectors **a**, **b** and **c** is 5. Then, the volume of the parallelopiped determined by vectors 3(**a** + **b**), (**b** + **c**) and 2(**c** + **a**) is | 2. | 30 |
| R. | Area of a triangle with adjacent sides determined by vector **a** and **b** is 20. Then, the area of the triangle with adjacent sides determined by vectors (2**a** + 3**b**) and (**a** − **b**) is | 3. | 24 |
| S. | Area of a parallelogram with adjacent sides determined by vectors **a** and **b** is 30. Then, the area of the parallelogram with adjacent sides determined by vectors (**a** + **b**) and **a** is | 4. | 60 |

**Codes**

| | P | Q | R | S | | P | Q | R | S |
|---|---|---|---|---|---|---|---|---|---|
| (a) | 4 | 2 | 3 | 1 | (b) | 2 | 3 | 1 | 4 |
| (c) | 3 | 4 | 1 | 2 | (d) | 1 | 4 | 3 | 2 |

**20.** Consider the lines

$$L_1 : \frac{x-1}{2} = \frac{y}{-1} = \frac{z+3}{1},$$

$$L_2 : \frac{x-4}{1} = \frac{y+3}{1} = \frac{z+3}{2}$$

and the planes $P_1 : 7x + y + 2z = 3$, $P_2 : 3x + 5y - 6z = 4$. Let $ax + by + cz = d$ the equation of the plane passing through the point of intersection of lines $L_1$ and $L_2$ and perpendicular to planes $P_1$ and $P_2$.

Match List I with List II and select the correct answer using the code given below the lists.

| List I | | List II | |
|---|---|---|---|
| P. | $a =$ | 1. | 13 |
| Q. | $b =$ | 2. | – 13 |
| R. | $c =$ | 3. | 1 |
| S. | $d =$ | 4. | – 2 |

**Codes**

| | P | Q | R | S | | P | Q | R | S |
|---|---|---|---|---|---|---|---|---|---|
| (a) | 3 | 2 | 4 | 1 | (b) | 1 | 3 | 4 | 2 |
| (c) | 3 | 2 | 1 | 4 | (d) | 2 | 4 | 1 | 3 |

# Answers

## Paper 1

### Physics

**1.** (a) **2.** (c) **3.** (b) **4.** (d) **5.** (c) **6.** (a) **7.** (a) **8.** (b) **9.** (b) **10.** (d)
**11.** (b,d) **12.** (b,c) **13.** (b,d) **14.** (a,c) **15.** (a,d) **16.** (8) **17.** (1) **18.** (5) **19.** (8) **20.** (4)

### Chemistry

**1.** (c) **2.** (b) **3.** (d) **4.** (b) **5.** (d) **6.** (b) **7.** (a) **8.** (d) **9.** (b) **10.** (a)
**11.** (b,c,d) **12.** (b,d) **13.** (a) **14.** (a) **15.** (a,b,c,d) **16.** (6) **17.** (2) **18.** (5) **19.** (8) **20.** (4)

### Mathematics

**1.** (b) **2.** (c) **3.** (c) **4.** (a) **5.** (d) **6.** (a) **7.** (b) **8.** (a) **9.** (d) **10.** (c)
**11.** (a,c) **12.** (a,d) **13.** (b,d) **14.** (b,c) **15.** (c,d) **16.** (9) **17.** (6) **18.** (5) **19.** (6) **20.** (5)

## Paper 2

### Physics

**1.** (b,d) **2.** (a,d) **3.** (a,d) **4.** (a,b) **5.** (d) **6.** (c,d) **7.** (b,c,d) **8.** (a,c) **9.** (b) **10.** (a)
**11.** (b) **12.** (a) **13.** (b) **14.** (b) **15.** (c) **16.** (a) **17.** (d) **18.** (c) **19.** (a) **20.** (c)

### Chemistry

**1.** (b) **2.** (b,d) **3.** (b) **4.** (c) **5.** (a,c,d) **6.** (a,b) **7.** (c,d) **8.** (a,b,c,d) **9.** (a) **10.** (d)
**11.** (b) **12.** (a) **13.** (c) **14.** (b) **15.** (a) **16.** (a) **17.** (d) **18.** (a) **19.** (a) **20.** (d)

### Mathematics

**1.** (b,d) **2.** (a,c) **3.** (a,d) **4.** (b,d) **5.** (c,d) **6.** (a,b,c) **7.** (b,c,d) **8.** (a,b) **9.** (b) **10.** (c)
**11.** (b) **12.** (d) **13.** (b) **14.** (c) **15.** (a) **16.** (d) **17.** (b) **18.** (a) **19.** (c) **20.** (a)

# Hints & Solutions

## Paper 1

### Physics

**1.** From momentum conservation equation, we have,

$$\mathbf{p}_i = \mathbf{p}_f$$

$$\therefore m(u_0 \cos\alpha)\,\hat{\mathbf{i}} + m(\sqrt{u_0^2 - 2gH})\,\hat{\mathbf{j}} = (2m)\mathbf{v} \quad \ldots\text{(i)}$$

$$H = \frac{u_0^2 \sin^2\alpha}{2g} \quad \ldots\text{(ii)}$$

From Eqs. (i) and (ii)

$$\mathbf{v} = \frac{u_0 \cos\alpha}{2}\,\hat{\mathbf{i}} + \frac{u_0 \cos\alpha}{2}\,\hat{\mathbf{j}}$$

Since both components of **v** are equal. Therefore it is making 45° with horizontal.

**2.** $\mu = \dfrac{\lambda_{\text{air}}}{\lambda_{\text{medium}}} = \dfrac{1}{(2/3)} = \dfrac{3}{2}$

Further, $|m| = \dfrac{1}{3} = \left|\dfrac{v}{u}\right|$

$\therefore \quad |v| = \dfrac{|u|}{3}$

$u = -24$ m (Real object)

$\therefore \quad v = +8$ m (Real image)

Now, $\dfrac{1}{v} - \dfrac{1}{u} = \dfrac{1}{f} = (\mu - 1)\left(\dfrac{1}{+R} - \dfrac{1}{\infty}\right)$

$\therefore \quad \dfrac{1}{8} + \dfrac{1}{24} = \left(\dfrac{3}{2} - 1\right)\left(\dfrac{1}{R}\right)$

$\therefore \quad R = 3\text{ m}$

**3.** 1 MSD = 5.15 cm – 5.10 cm = 0.05 cm

$$1\text{ VSD} = \frac{2.45\text{ cm}}{50} = 0.049\text{ cm}$$

$\therefore$ LC = 1 MSD – 1 VSD = 0.01 cm

Hence, Diameter of cyclinder = (Main scale reading) + (Vernier scale reading) (LC)

= 5.10 + (24) (0.001) = 5.124 cm

**4.**

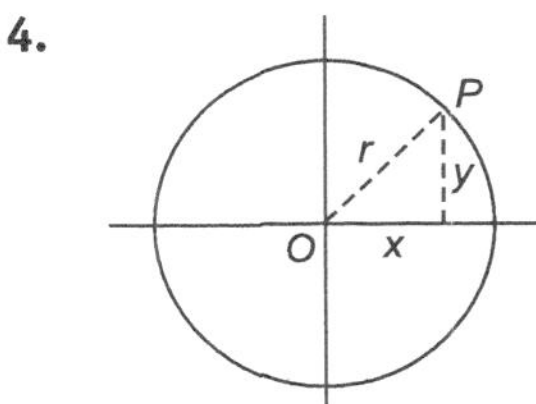


$$\mathbf{r} = \mathbf{OP} = x\,\hat{\mathbf{i}} + y\,\hat{\mathbf{j}}$$

$$\mathbf{F} = \frac{k}{(x^2 + y^2)^{3/2}}(x\,\hat{\mathbf{i}} + y\,\hat{\mathbf{j}}) = \frac{k}{r^3}(\mathbf{r})$$

Since **F** is along **r** or in radial direction. Therefore, work done is zero.

**5.** $\Delta l = \dfrac{FL}{AY} = \dfrac{FL}{(\pi r^2)Y}$

$\therefore \quad \Delta l \propto \dfrac{L}{r^2}$

$\therefore \quad \dfrac{\Delta l_1}{\Delta l_2} = \dfrac{L/R^2}{2L/(2R)^2} = 2$

**6.** Component along the plane $= \dfrac{1}{2}$

and component perpendicular to the plane $= \dfrac{\sqrt{3}}{2}$

$\therefore \quad \tan i = \dfrac{\left(\dfrac{1}{2}\right)}{\left(\dfrac{\sqrt{3}}{2}\right)} = \dfrac{1}{\sqrt{3}}$

$\therefore \quad i = 30° =$ angle of incidence

**7.** $R_I = R_1 + R_2 = \left(\frac{l}{KA}\right) + \left(\frac{l}{2KA}\right) = \frac{3}{2}\left(\frac{l}{KA}\right)$

$$\frac{1}{R_{II}} = \frac{1}{R_1} + \frac{1}{R_2} = \frac{KA}{l} + \frac{2KA}{l}$$

or $$R_{II} = \frac{l}{3KA} = \frac{R_I}{4.5}$$

Since thermal resistance $R_{II}$ is 4.5 times less than thermal resistance $R_I$.

$$\therefore \quad t_{II} = \frac{t_I}{4.5} = \frac{9}{4.5}\text{ s} = 2\text{ s}$$

**8.** Final momentum of object

$$= \frac{\text{Power} \times \text{time}}{\text{Speed of light}}$$

$$= \frac{30 \times 10^{-3} \times 100 \times 10^{-9}}{3 \times 10^8}$$

$$= 1.0 \times 10^{-17} \text{ kg-m/s}$$

**9.** $I = I_{max} \cos^2 \frac{\phi}{2}$ ...(i)

Given, $I = \frac{I_{max}}{2}$ ...(ii)

$\therefore$ From Eqs. (i) and (ii), we have

$$\phi = \frac{\pi}{2}, \frac{3\pi}{2}, \frac{5\pi}{2}$$

Or path difference, $\Delta x = \left(\frac{\lambda}{2\pi}\right) \cdot \phi$

$$\therefore \quad \Delta x = \frac{\lambda}{4}, \frac{3\lambda}{4}, \frac{5\lambda}{4} \dots \left(\frac{2n+1}{4}\right)\lambda$$

**10.** $\rho = \frac{pM}{RT}$

$\therefore \quad \rho \propto pM$

or $$\frac{\rho_1}{\rho_2} = \left(\frac{p_1}{p_2}\right)\left(\frac{M_1}{M_2}\right)$$

$$= \left(\frac{2}{3}\right)\left(\frac{4}{3}\right) = \frac{8}{9}$$

**11. At point *P***

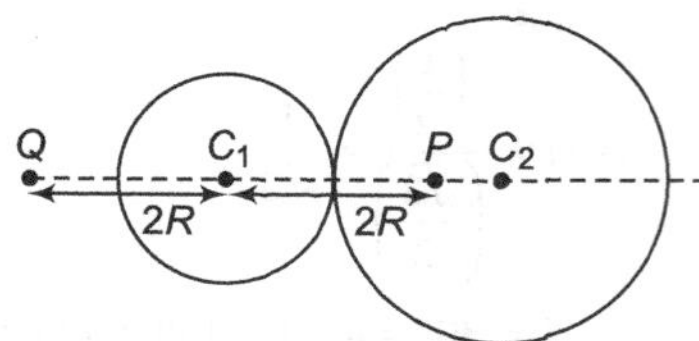


If resultant electric field is zero then

$$\frac{KQ_1}{4R^2} = \frac{KQ_2}{8R^3} R$$

$$\frac{\rho_1}{\rho_2} = 4$$

**At point *Q***

If resultant electric field is zero then

$$\frac{KQ_1}{4R^2} + \frac{KQ_2}{25R^2} = 0$$

$$\frac{\rho_1}{\rho_2} = -\frac{32}{25} \quad (\rho_1 \text{ must be negative})$$

**12.** [figure: string of length $l$ fixed at both ends, five loops, midpoint $P$]

Number of nodes = 6

From the given equation, we can see that

$$k = \frac{2\pi}{\lambda} = 62.8 \text{ m}^{-1}$$

$$\therefore \quad \lambda = \frac{2\pi}{62.8} \text{ m} = 0.1 \text{ m}$$

$$l = \frac{5\lambda}{2} = 0.25 \text{ m}$$

The mid-point of the string is $P$, an antinode

$\therefore$ maximum displacement = 0.01 m

$$\omega = 2\pi f = 628 \text{ s}^{-1}$$

$$\therefore \quad f = \frac{628}{2\pi} = 100 \text{ Hz}$$

But this is fifth harmonic frequency.

$\therefore$ Fundamental frequency $f_0$

$$= \frac{f}{5} = 20 \text{ Hz},$$

**13.** After pressing $S_1$ charge on upper plate of $C_1$ is $+ 2CV_0$.

After pressing $S_2$ this charge equally distributes in two capacitors. Therefore charge an upper plates of both capacitors will be $+ CV_0$.

When $S_2$ is released and $S_3$ is pressed, charge on upper plate of $C_1$ remains unchanged $(= + CV_0)$ but charge on upper plate of $C_2$ is according to new battery $(= -CV_0)$.

**14.** $\mathbf{u} = 4\hat{\mathbf{i}}$; $\mathbf{v} = 2(\sqrt{3}\,\hat{\mathbf{i}} + \hat{\mathbf{j}})$

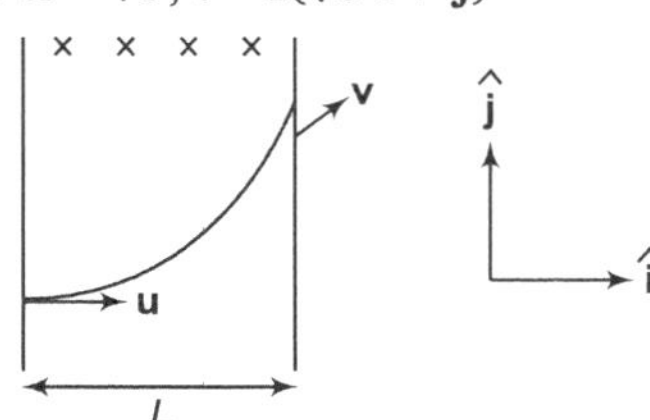


According to the figure, magnetic field should be in ⊗ direction, or along $-z$ direction.

Further, $\tan\theta = \dfrac{v_y}{v_x} = \dfrac{2}{2\sqrt{3}} = \dfrac{1}{\sqrt{3}}$

$\therefore \qquad \theta = 30°$ or $\dfrac{\pi}{6}$

= angle of **v** with $x$-axis

= angle rotated by the particle

$= Wt = \left(\dfrac{BQ}{M}\right)t$

$\therefore\ B = \dfrac{\pi M}{6Qt}$

$= \dfrac{50\pi M}{3Q}$ units (as $t = 10^{-3}$ second)

**15.** 

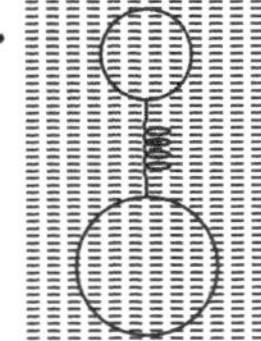

On small sphere

$$\frac{4}{3}\pi R^3(\rho)g + kx = \frac{4}{3}\pi R^3(2\rho)g \quad \dots\text{(i)}$$

On second sphere (large)

$$\frac{4}{3}\pi R^3(3\rho)g = \frac{4}{3}\pi R^3(2\rho)g + kx \quad \dots\text{(ii)}$$

By Eqs. (i) and (ii), we get

$$x = \frac{4\pi R^3 \rho g}{3k}$$

**16.** $I_1\omega_1 = I_2\omega_2$

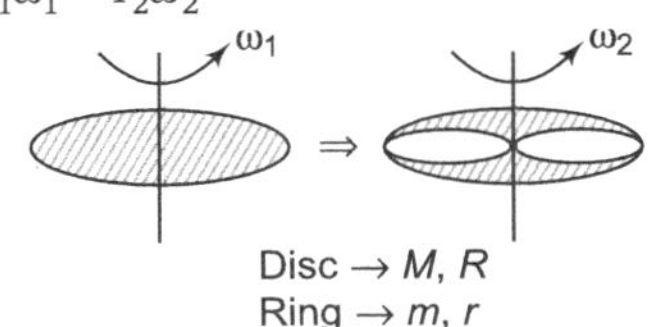


$\therefore \quad \omega_2 = \left(\dfrac{I_1}{I_2}\right)\omega_1$

$$= \left[\frac{\frac{1}{2}MR^2}{\frac{1}{2}MR^2 + 2(mr^2)}\right]\omega_1$$

$$= \left[\frac{50(0.4)^2}{50(0.4)^2 + 8 \times (6.25) \times (0.2)^2}\right](10)$$

= 8 rad/s

**17.** $eV_0 = hf - W$

$\therefore \qquad V_0 = \left(\dfrac{h}{e}\right)f - \dfrac{W}{e}$

$V_0$ *versus* $f$ graph is a straight line with slope $= \dfrac{h}{e}$ = a universal constant. Therefore the ratio of two slopes should be 1.

**18.** Velocity of first bob at highest point.

$v_1 = \sqrt{gR} = \sqrt{gl_1}$

(to just complete the vertical circle)

= velocity of second bob just after elastic collision.

= velocity of second bob at the bottommost point

$= \sqrt{5gl_2} \quad \therefore \quad \dfrac{l_1}{l_2} = 5$

**19.** $W = \dfrac{1}{2}mv^2$

$\therefore \quad Pt = \dfrac{1}{2}mv^2 \quad \therefore \quad v = \sqrt{\dfrac{2Pt}{m}}$

$= \sqrt{\dfrac{2 \times 0.5 \times 5}{0.2}} = 5$ m/s

**20.** Number of nuclei decayed in time $t$,

$N_d = N_0(1 - e^{-\lambda t})$

$\therefore$ % decayed $= \left(\dfrac{N_d}{N_0}\right) \times 100$

$= (1 - e^{-\lambda t}d) \times 100 \quad \dots\text{(i)}$

Here, $\lambda = \dfrac{0.693}{1386} = 5 \times 10^{-4}\ \text{s}^{-1}$

$\therefore \quad$ % decayed $\approx (\lambda t) \times 100$

$= (5 \times 10^{-4})(80)(100) = 4$

# Chemistry

**1. Plan** $\Delta_c H^\circ$ (Standard heat of combustion) is the standard enthalpy change when one mole of the substance is completely oxidised.

Also standard heat of formation ($\Delta_f H^\circ$) can be taken as the standard of that substance.

$$H^\circ_{H_2O} = \Delta_f H^\circ (H_2O) = -300 \text{ kJ mol}^{-1}$$

$$H^\circ_{\text{glucose}} = \Delta_f H^\circ \text{ (glucose)}$$
$$= -1300 \text{ kJ mol}^{-1}$$

$$H^\circ_{O_2} = \Delta_f H^\circ (O_2) = 0.00$$

$$C_6H_{12}O_6(s) + 6O_2(g) \longrightarrow 6CO_2(g) + 6H_2O(l)$$

$\Delta_c H^\circ$ (glucose)

$$= 6[\Delta_f H^\circ (CO_2) + \Delta_f H^0 (H_2O)] - [\Delta_f H^\circ (C_6H_{12}O_6) + 6\Delta_f H^\circ (O_2)]$$
$$= 6[-400 - 300] - [-1300 + 6 \times 0]$$
$$= -2900 \text{ kJ mol}^{-1}$$

Molar mass of $C_6H_{12}O_6 = 180 \text{ g mol}^{-1}$

Thus, standard heat of combustion of glucose per gram

$$= \frac{-2900}{180}$$
$$= -16.11 \text{ kJ g}^{-1}$$

**2. Plan** Acetone is an aprotic solvent and can dissolve both the nucleophile and the substrate and thus $S_N2$ reaction is favoured. Also

$S_N2 \xleftarrow{\quad 1^\circ \qquad 2^\circ \qquad 3^\circ \quad}$ Alkyl halides $\xrightarrow{\qquad\qquad\qquad} S_N1$

| | | |
|---|---|---|
| *S.* | (O, Cl) $C_6H_5COCH_2Cl$ | $1^\circ$ alkyl halide but (C—Cl). B.E. is decreased by electron withdrawing [$C_6H_5CO$] group, (a case of *I*-effect). Thus, maximum rate in $S_N2$ reaction |
| *Q.* | (Cl) | $2^\circ$ alkyl halide, rate is minimum |
| *P.* | $CH_3$—Cl | $1^\circ$ alkyl halide |
| R. | (Cl) | $1^\circ$ allylic halide but allylic $1^\circ$ carbocation is resonance stabilised in $S_N1$ reaction |

Thus, reactivity order is

$$S > P > R > Q$$

**3. Plan** $NaHCO_3 \rightleftharpoons Na^+ + HCO_3^-$

$HCO_3^-$ is decomposed by acid releasing $CO_2$

$$HCO_3^- + H^+ \longrightarrow H_2O + CO_2$$

If acid is stronger than $HCO_3^-$ then $CO_2$ is released.

Phenol is less acidic and thus does not liberate $CO_2$ with $NaHCO_3$.

**4. Plan** Spin only magnetic moment $= \sqrt{N(N+2)}$ BM. when *N* is the number of unpaired electrons. In presence of weak ligand (as $H_2O$, $Cl^-$, $F^-$) there is no pairing of electrons, and electrons donated by ligands are filled in outer vacant orbitals.

In presence of strong ligand (as $CN^-$, CO, $NH_3$, en) electrons are paired and electrons from ligands are filled in available inner orbitals

| Complex | Atomic no. of | O.N. | E.C. | Unpaired electrons | Magnetic moment |
|---|---|---|---|---|---|
| $P: [FeF_6]^{3-}$ weak ligand | 26 | +3 | [Ar] $3d^5$ | 5 | $\sqrt{35}$ BM |
| $Q: [V(H_2O)_6]^{2+}$ weak ligand | 23 | +2 | [Ar] | 3 | $\sqrt{15}$ BM |
| $R: [Fe(H_2O)_6]^{2+}$ | 26 | +2 | [Ar] $3d^6$ | 4 | $\sqrt{24}$ BM |

Thus, order of spin-only magnetic moment $= Q < R < P$

**5. Plan** Time of 75% reaction is twice the time taken for 50% reaction if it is first order reaction *w.r.t* $P$. From graph, $[Q]$ decreases linearly with time, thus it is zeroth order reaction *w.r.t.* $Q$

$$\frac{dx}{dt} = k[P]^a[Q]^b$$

Order *w.r.t.* $P = a = 1$

Order *w.r.t.* $Q = b = 0$

Thus, overall order of the reaction

$$= 1 + 0 = 1$$

**6. Plan** $NO_2$ is a brown coloured gas and imparts this colour to concentrated $HNO_3$ due to long standing.

$$4HNO_3 \longrightarrow 2H_2O + 2NO_2 + 3O_2$$

**7. Plan** Given arrangement represents octahedral void and for this

$$\frac{r_+ \text{ (cation)}}{r_- \text{ (anion)}} = 0.414$$

$$\frac{r(A^+)}{r(X^-)} = 0.414$$

$$r(A^+) = 0.414 \times r(X^-)$$

$$= 0.414 \times 250 \text{ pm} = 103.5 \text{ pm}$$

**8. Plan** $K_{sp}(ZnS)$ is very high and $Zn^{2+}$ is precipitated as ZnS by high concentration of $S^{2-}$ formed when $H_2S$ is passed in ammoniacal solution

$$H_2S \rightleftharpoons Zn^+ + S^{2-} \text{ (I)}$$

$$H^+ + OH^- \rightleftharpoons H_2O \text{ (II)}$$

Reaction (I) is favoured in forward side if $H^+$ is removed immediately by $OH^-$ $(NH_4OH)$

$$Zn^{2+} + S^{2-} \longrightarrow \underset{\text{White ppt}}{ZnS \downarrow}$$

$Fe^{3+}$ and $Al^{3+}$ are precipitated as hydroxide.

**9. Plan** Physical adsorption takes place with decrease in enthalpy thus exothermic change.

It is physical adsorption and does not require activation. Thus, (*a*) is incorrect.

Being physical adsorption $\Delta H < 0$ thus, (*b*) is correct. Exothermic reaction is favoured at low temperature thus (c) is incorrect.

Physical adsorption is always reversible, thus (d) is incorrect.

**10.**

| Element | Ores | Name |
|---|---|---|
| Ag | $Ag_2S$ | Argentite |
| Cu | $CuFeS_2$ | Copper pyrites |
| Pb | $PbS$ | Galena |
| Sn | $SnO_2$ | Cassiterite |
| Mg | $MgCO_3 \cdot CaCO_3$ | Dolomite |
| Al | $Al_2O_3 \cdot xH_2O$ | Bauxite |

**11. Plan** When an ideal solution is formed process is spontaneous thus

$$\Delta G < 0$$

$$\Delta S > 0$$

$$\Delta H = 0$$

$$\Delta S_{\text{surrounding}} = 0$$

**12. Plan** Depending on the structure of the complex,different types of isomerism are shown.

| | Complex | Isomerism |
|---|---|---|
| A. | $[Cr(NH_3)_5Cl]Cl_2$<br>$(C_2(NH_3)_4Cl_2)$ Cl | Neither of structural nor stereo isomerism |
| B. | $[Co(NH_3)_4Cl_2]^+$ | Cl, $H_3N$, $NH_3$, Co, $H_3N$, $NH_3$, Cl<br>*Trans w.r.t.* Cl |
| | $[Pt(H_2O)\cdot(NH_3)_2Cl]^+$ | $H_2O$, $NH_3$, Pt, Cl, $NH_3$<br>*Cis*<br>$H_2O$, $NH_3$, Pt, $H_3N$, Cl<br>*Trans* |
| C. | $[CoBr_2Cl_2]^{2-}$<br>$[PtBr_2Cl_2]^{2-}$ | $sp^3$tetrahedral<br>$dsp^2$-square planar |

| | Complex | Isomerism |
|---|---|---|
| D. | $[Pt(NH_3)_3(NO_3)]Cl$ | $[Pt(NH_3)_3(NO_3)]Cl$ <br> $[Pt(NH_3)_3Cl]NO_3$ <br> ionisation |
| | $[Pt(NH_3)_3Cl]Br$ | $[Pt(NH_3)_3Cl]Br$ <br> $[Pt(NH_3)_3Br]Cl$ <br> ionisation |

**13. Plan** $RCOOR' + H_2O \xrightarrow{H^+} RCOOH + R'OH$

Acid hydrolysis of ester is follows first order kinetics.

For same concentration of ester in each case rate is dependent on $[H^+]$ from acid.

$$\text{Rate} = k[RCOOR']$$

Also for weak acid $HA \rightleftharpoons H^+ + A^-$

$$K_a = \frac{[H^+][A^-]}{[HA]}$$

$$(\text{Rate})_{HA} = k[H^+]_{HA}$$

$$(\text{Rate})_{HX} = k[H^+]_{HX}$$

$$(\text{Rate})_{HX} = 100(\text{Rate})_{HA}$$

$\therefore$ Also in strong acid $[H^+] = [HX] = 1\text{ M}$

$$\frac{(\text{Rate})_{HX}}{(\text{Rate})_{HA}} = 100 = \frac{[H^+]_{HX}}{[H^+]_{HA}}$$

$$= \frac{1}{[H^+]_{HA}}$$

$$\therefore \quad [H^+]_{HA} = \frac{1}{100}$$

$$HA \rightleftharpoons H^+ + A^-$$

| | | |
|---|---|---|
| 1 | 0 | 0 |
| $(1-x)$ | $x$ | $x$ |

$x = 0\,0.01$

$$\therefore \quad K_a = \frac{[H^+][A^-]}{[HA]}$$

$$= \frac{0.01 \times 0.01}{0.99}$$

$$= 1.01 \times 10^{-4}$$

**14. Plan** Spreading out charge by the overlap of an empty *p*-orbital with an adjacent $\sigma$ bond is called hyperconjugation. This overlap (the hyperconjugation) delocalises the positive charge on the carbocation, spreading it over a larger volume, and this stabilises the carbocation.

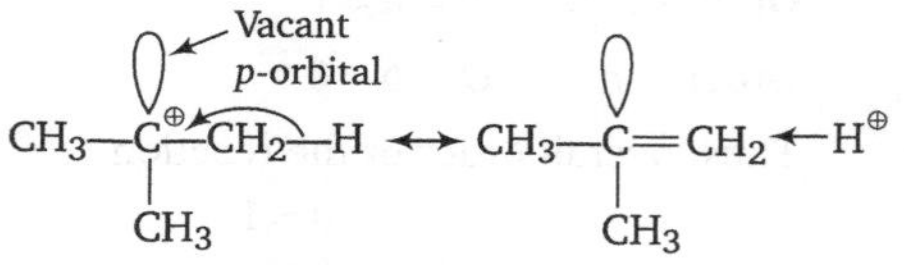

Vacant *p*-orbital

$$CH_3-\overset{\oplus}{C}(CH_3)-CH_2-H \longleftrightarrow CH_3-C(CH_3)=CH_2 \quad H^{\oplus}$$

tertiary butyl carbocation has one vacant *p*-orbital hence, it is stabilised by $\sigma$-*p* (empty) hyperconjugation.

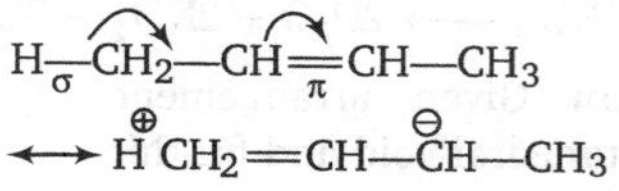

$$\underset{\sigma}{H-}CH_2-CH\underset{\pi}{=}CH-CH_3$$

$$\longleftrightarrow \overset{\oplus}{H}\ CH_2=CH-\overset{\ominus}{C}H-CH_3$$

In 2-butene, stabilisation is due to hyperconjugation between $\sigma - \pi^*$ electron delocalisation.

**15. Plan** A species is said to have aromatic character if

(a) ring is planar

(b) their is complete delocalisation of $\pi$ electrons

(c) Huckel rule *i.e.*, $(4n + 2)$ rule is followed.

where, *n* is the number of rings

$(4n + 2) = \pi$ electron delocalised.

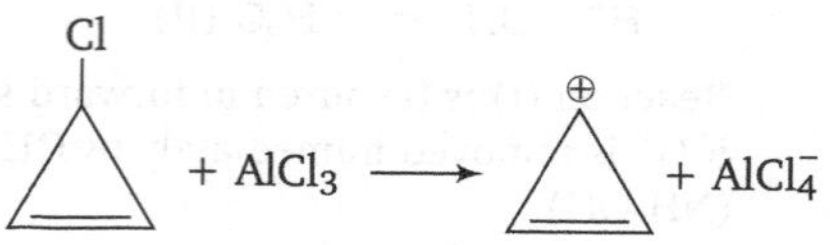

3-Chlorocyclopropene $+ AlCl_3 \longrightarrow$ cyclopropenyl cation $+ AlCl_4^-$

Aromatic (*P*)

Cyclopentadiene $+ NaH \longrightarrow$ cyclopentadienyl anion $+ H_2$

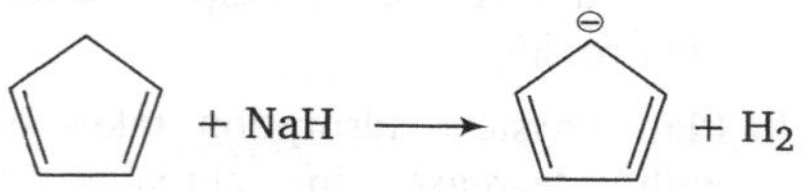

Aromatic (*Q*)

(Reaction scheme: 2,5-hexanedione $\xrightarrow[\Delta,\ 100–115°C]{(NH_4)_2CO_3}$ 2,5-dimethylpyrrole, Aromatic (R))

(Reaction scheme: tropone $\xrightarrow{HCl}$ hydroxytropylium ion, Aromatic (S))

| | $n$ | $(4n + 2)$ | $\pi$ electrons |
|---|---|---|---|
| P – | 0 | 2 | 2 |
| Q – | 1 | 6 | 6 (including lone pair) |
| R – | 1 | 6 | 6 (including lone pair on N) |
| S – | 1 | 6 | 6 |

In all cases there is complete delocalisation of $\pi$-electrons.

**16. Plan** Melamine is a heterocyclic compound.

(Structure of melamine with lone pairs on all six N atoms)

Each nitrogen atom has one pair of lone pair

Thus, in all **six** lone pairs.

**17. Plan** Reactant is cyclic anhydride and changes to dicarboxylic acid on hydrolysis.

Also there is decarboxylation on heating if there is keto group *w.r.t* —COOH group. Ozonolysis cleaves (C=C) bond and $H_2O_2$ oxidises —CHO to—COOH group.

(Reaction scheme: Anhydride $\xrightarrow{(i)\ H_3O^+}$ dicarboxylic acid (—OH, —OH) $\xrightarrow{\Delta\ \text{decarboxylation}}$ diketone $\xrightarrow{O_3}$ OHC, OHC dialdehyde $\xrightarrow{H_2O_2}$ HOOC, HOOC diacid (P))

Thus, number of —COOH groups in $P = 2$

**18. Plan** $KE = \frac{1}{2}mv^2 = \frac{3}{2}RT$

$\therefore \quad m^2v^2 = 2m\text{KE}$

$\therefore \quad mv = \sqrt{2m\text{KE}}$

$\lambda$ (wavelength)

$$= \frac{h}{mv} = \frac{h}{\sqrt{2m\text{KE}}} \propto \frac{h}{\sqrt{2m(T)}}$$

$$\lambda(\text{He at } -73°\text{C} = 200\text{ K}) = \frac{h}{\sqrt{2 \times 4 \times 200}}$$

$\lambda$ (Ne at 727°C = 1000 K)

$$= \frac{h}{\sqrt{2 \times 20 \times 1000}}$$

$$\therefore \quad \frac{\lambda(\text{He})}{\lambda(\text{Ne})} = M = \sqrt{\frac{2 \times 20 \times 1000}{2 \times 4 \times 200}} = 5$$

Thus, $\quad M = 5$

**19. Plan** EDTA is a multidentate ligand as it can donate six pairs of electrons – two pair from the two nitrogen atoms and four pair from the four terminal oxygens of the —$COO^-$ groups.

$$\begin{matrix} {}^-OOCCH_2 & & & & CH_2COO^- \\ & \ddot{N}{-}CH_2CH_2{-}\ddot{N} & & & \\ {}^-OOCCH_2 & & & & CH_2COO^- \end{matrix}$$

The structure of a chelate of a divalent $Co^{2+}$ with EDTA is shown as

CO, O, CH$_2$, CO—CH$_2$, O, N, CH$_2$, 2, 1, 3, Co, CH$_2$, O, N, 4, OC, CH$_2$, O, CH$_2$, CO

Each N has four N—Co—O bonds thus total eight N—Co—O bonds.

**20. Plan** A peptide linkage is hydrolysed to two free amino acids.

$$H_2N{-}\underset{\displaystyle R}{\underset{|}{CH}}{-}\overset{\displaystyle O}{\overset{\|}{C}}{-}NH{-}\underset{\displaystyle R'}{\underset{|}{CH}}{-}COOH$$

↑ peptide

$$H_2N{-}\underset{\displaystyle R}{\underset{|}{\overset{*}{C}H}}{-}\overset{\displaystyle O}{\overset{\|}{C}}{-}OH$$

$$+\ H_2N{-}\underset{\displaystyle R'}{\underset{|}{\overset{*}{C}H}}{-}COOH$$

$C^*$ is chiral carbon tetrapeptide has four amino acids joined by three peptide linkage.

—COOH group is on alanine part, thus it is at fixed C-terminal position in each combination.

Glycine is optically inactive thus it can't be on the N— terminal side.

Thus, possible combinations are

Phe-Gly-Val-Ala

Phe-Val-Gly-Ala

Val-Gly-Phe-Ala

Val-Phe-Gly-Ala

Thus, in all **four possible combinations.**

# Mathematics

**1. Concept Involved** Difference series, when ever, we have summation involving more than 3 terms we should always convert into differences.

*e.g.,* $$\sum_{r=1}^{5} \frac{1}{r(r+1)} = \frac{1}{1\cdot(2)} + \frac{1}{2\cdot(3)} + \frac{1}{3\cdot 4} + \frac{1}{4\cdot 5} + \frac{1}{5\cdot 6}$$

$$= \frac{2-1}{1\cdot(2)} + \frac{3-2}{2\cdot(3)} + \frac{4-3}{3\cdot(4)} + \frac{5-4}{4\cdot(5)} + \frac{6-5}{5\cdot(6)}$$

$$= \left(1-\frac{1}{2}\right) + \left(\frac{1}{2}-\frac{1}{3}\right) + \left(\frac{1}{3}-\frac{1}{4}\right) + \left(\frac{1}{4}-\frac{1}{5}\right) + \left(\frac{1}{5}-\frac{1}{6}\right)$$

$$= 1 - \frac{1}{6} = \frac{5}{6}$$

**Situation analysis**

convert into,

$$\tan^{-1} x - \tan^{-1} y = \tan^{-1}\left(\frac{x-y}{1+xy}\right)$$

$$\cot\left(\sum_{n=1}^{23} \cot^{-1}\left(1 + \sum_{k=1}^{n} 2k\right)\right)$$

$$\Rightarrow \cot\left(\sum_{n=1}^{23} \cot^{-1}(1 + 2 + 4 + 6 + 8 + \ldots + 2n)\right)$$

$$\Rightarrow \cot\left(\sum_{n=1}^{23} \cot^{-1}(1 + n(n+1))\right)$$

$$\Rightarrow \cot\left(\sum_{n=1}^{23} \tan^{-1} \frac{1}{1+n(n+1)}\right)$$

$$\Rightarrow \cot\left(\sum_{n=1}^{23} \tan^{-1} \left(\frac{(n+1)-n}{1+n(n+1)}\right)\right)$$

$$\Rightarrow \cot\left\{\sum_{n=1}^{23} (\tan^{-1}(n+1) - \tan^{-1} \ln n)\right\}$$

$$\Rightarrow \cot\{\tan^{-1} 2 - \tan^{-1} 1) + (\tan^{-1} 3 - \tan^{-1} 2) + (\tan^{-1} 4 - \tan^{-1} 3) + \ldots + (\tan^{-1} 24 - \tan^{-1} 23)\}$$

$$\Rightarrow \cot\{\tan^{-1} 24 - \tan^{-1} 1\}$$

$$\Rightarrow \cot\left(\tan^{-1} \frac{24-1}{1+24\cdot(1)}\right)$$

$$= \cot\left(\tan^{-1} \frac{23}{25}\right)$$

$$= \cot\left(\cot^{-1} \frac{25}{23}\right)$$

$$= \left(\frac{25}{23}\right)$$

**2. Concept Involved** It involves law of parallelogram and volume of parallelopiped. *i. e.*,

$\mathbf{a} + \mathbf{b} = \mathbf{p}$ and $\mathbf{b} - \mathbf{a} = \mathbf{q}$

$\therefore \quad \mathbf{a} = \dfrac{\mathbf{p} - \mathbf{q}}{2}$

and $\quad \mathbf{b} = \dfrac{\mathbf{p} + \mathbf{q}}{2}$

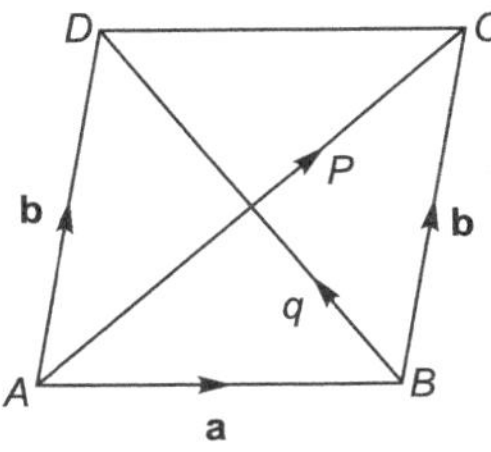


*i. e.*, If **p** and **q** are diagonals of parallelograms, then its sides are $\dfrac{\mathbf{p} - \mathbf{q}}{2}$ and $\dfrac{\mathbf{p} + \mathbf{q}}{2}$.

**Situation analysis**

After finding the sides of parallelogram we should find volume of parallelopiped *i. e.*, [**a b**]

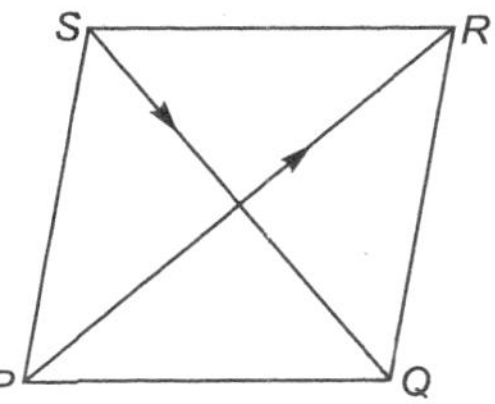


Here, sides of parallelogram are **PQ** and **PS**.

where, $\mathbf{PQ} = \dfrac{\mathbf{PR} + \mathbf{SQ}}{2}$

$$= \frac{(3\hat{\mathbf{i}} + \hat{\mathbf{j}} - 2\hat{\mathbf{k}}) + (\hat{\mathbf{i}} - 3\hat{\mathbf{j}} - 4\hat{\mathbf{k}})}{2}$$

$$\mathbf{PQ} = 2\hat{\mathbf{i}} - \hat{\mathbf{j}} - 3\hat{\mathbf{k}}$$

and $\mathbf{PS} = \dfrac{\mathbf{PR} - \mathbf{SQ}}{2}$

$$= \frac{(3\hat{\mathbf{i}} + \hat{\mathbf{j}} - 2\hat{\mathbf{k}}) - (\hat{\mathbf{i}} - 3\hat{\mathbf{j}} - 4\hat{\mathbf{k}})}{2}$$

$$\mathbf{PS} = \hat{\mathbf{i}} + 2\hat{\mathbf{j}} + \hat{\mathbf{k}}$$

$\therefore$ Volume of parallelopiped

$$= [\mathbf{PT}\ \mathbf{PQ}\ \mathbf{PS}]$$

$$= \begin{vmatrix} 1 & 2 & 3 \\ 2 & -1 & -3 \\ 1 & 2 & 1 \end{vmatrix}$$

$$= 1(-1+6) - 2(2+3) + 3(4+1)$$

$$= 5 - 10 + 15 = 10$$

**3. Concept Involved** Intersection of circles, the basic concept is to equations simultaneously and using properties of modules of complex numbers.

**Formula used** $|z|^2 = z \cdot \bar{z}$

and $|z_1 - z_2|^2 = (z_1 - z_2)(\bar{z}_1 - \bar{z}_2)$

$= |z_1|^2 - z_1\bar{z}_2 - z_2\bar{z}_1 + |z_2|^2$

Here, $(x - x_0)^2 + (y - y_0)^2 = r^2$

and $(x - x_0)^2 + (y - y_0)^2 = 4r^2$ could be written as,

$|z - z_0|^2 = r^2$ and $|z - z_0|^2 = 4r^2$

Since, $\alpha$ and $\frac{1}{\bar{\alpha}}$ lies on first and second respectively.

$\therefore \quad |\alpha - z_0|^2 = r^2$ and $\left|\frac{1}{\bar{\alpha}} - z_0\right|^2 = 4r^2$

$\Rightarrow \quad (\alpha - z_0)(\bar{\alpha} - \bar{z}_0) = r^2$

$\Rightarrow \quad |\alpha|^2 - z_0\bar{\alpha} - \bar{z}_0\alpha + |z_0|^2 = r^2 \quad \ldots(\text{i})$

and $\quad \left|\frac{1}{\bar{\alpha}} - z_0\right|^2 = 4r^2$

$\Rightarrow \quad \left(\frac{1}{\bar{\alpha}} - z_0\right)\left(\frac{1}{\alpha} - \bar{z}_0\right) = 4r^2$

$\Rightarrow \quad \frac{1}{|\alpha|^2} - \frac{z_0}{\alpha} - \frac{\bar{z}_0}{\bar{\alpha}} + |z_0|^2 = 4r^2,$

Since $\quad |\alpha|^2 = \alpha \cdot \alpha$

$\Rightarrow \quad \frac{1}{|\alpha|^2} - \frac{z_0 \cdot \bar{\alpha}}{|\alpha|^2} - \frac{\bar{z}_0}{|\alpha|^2} \cdot \alpha + |z_0|^2 = 4r^2$

$\Rightarrow \quad 1 - z_0\bar{\alpha} - \bar{z}_0\alpha + |\alpha|^2 |z_0|^2$

$= 4r^2|\alpha|^2 \quad \ldots(\text{ii})$

On subtracting Eqs. (i) and (ii), we get

$(|\alpha|^2 - 1) + |z_0|^2 (1 - |\alpha|^2)$

$= r^2 (1 - 4|\alpha|^2)$

$\Rightarrow \quad (|\alpha|^2 - 1)(1 - |z_0|^2) = r^2(1 - 4|\alpha|^2)$

$\Rightarrow \quad (|\alpha|^2 - 1)\left(1 - \frac{r^2 + 2}{2}\right)$

$= r^2(1 - 4|\alpha|^2)$

Given, $|z_0|^2 = \frac{r^2 + 2}{2}$

$\Rightarrow \quad (|\alpha|^2 - 1)\cdot\left(\frac{-r^2}{2}\right) = r^2(1 - 4|\alpha|^2)$

$\Rightarrow \quad |\alpha|^2 - 1 = -2 + 8|\alpha|^2$

$\Rightarrow \quad 7|\alpha|^2 = 1$

$\therefore \quad |\alpha| = \frac{1}{\sqrt{7}}$

**4. Concept involved** Application of inequality sum and differences, along with lengths of perpendicular. For this type of questions involving inequality we should always check all options.

**Situation analysis** Check all the inequalities according to options and use length of perpendicular from the point $(x_1, y_1)$ to $ax + by + c = 0$

*i. e.*, $\quad \frac{|ax_1 + by_1 + c|}{\sqrt{a^2 + b^2}}$

As, $\quad a > b > c > 0$

$a - c > 0$ and $b > 0$

$\Rightarrow \quad a + b - c > 0 \quad \ldots(\text{i})$

$a - b > 0$ and $c > 0$

$a + c - b > 0 \quad \ldots(\text{ii})$

$\therefore$ (a) and (c) is correct.

Also, the point of intersection for $ax + by + c = 0$ and $bx + ay + c = 0$

*i. e.*, $\quad \left(\frac{-c}{a + b}, \frac{-c}{a + b}\right)$

The distance between (1, 1) and $\left(\frac{-c}{a + b}, \frac{-c}{a + b}\right)$

*i. e.*, less than $2\sqrt{2}$

$\Rightarrow \quad \sqrt{\left(1 + \frac{c}{a + b}\right)^2 + \left(1 + \frac{c}{a + b}\right)^2} < 2\sqrt{2}$

$\Rightarrow \quad \left(\frac{a + b + c}{a + b}\right)\sqrt{2} < 2\sqrt{2}$

$\Rightarrow \quad a + b + c < 2a + 2b$

or $\quad a + b - c > 0 \quad \ldots(\text{ii})$

From Eqs. (i) and (ii), option (c) is correct.

**5. Concept Involved**

To find the foot of perpendiculars and find its locus.

**Formula used**

Foot of perpendicular from $(x_1, y_1, z_1)$ to $ax + by + cz + d = 0$ be $(x_2, y_2, z_2)$, then $\frac{x_2 - x_1}{a} = \frac{y_2 - y_1}{b} = \frac{z_2 - z_1}{c}$

$= \frac{-(ax_1 + by_1 + cz_1 + d)}{a^2 + b^2 + c^2}$

Any point on, $\quad \frac{x + 2}{2} = \frac{y + 1}{-1} = \frac{z}{3} = \lambda$

$\Rightarrow \quad x = 2\lambda - 2, y = -\lambda - 1, z = 3\lambda$

Let foot of perpendicular from $(2\lambda - 2, -\lambda - 1, 3\lambda)$ to $x + y + z = 3$ be $(x_2, y_2, z_2)$

$$\therefore \quad \frac{x_2 - (2\lambda - 2)}{1} = \frac{y_2 - (-\lambda - 1)}{1} = \frac{z_2 - (3\lambda)}{1}$$

$$= -\frac{(2\lambda - 2 - \lambda - 1 + 3\lambda - 3)}{1 + 1 + 1}$$

$$\Rightarrow \quad x_2 - 2\lambda + 2 = y_2 + \lambda + 1 = z_2 - 3\lambda = 2 - \frac{4\lambda}{3}$$

$$\therefore \quad x_2 = \frac{2\lambda}{3},\ y_2 = 1 - \frac{7\lambda}{3},\ z_2 = 2 + \frac{5\lambda}{3}$$

$$\Rightarrow \quad \lambda = \frac{x_2 - 0}{2/3} = \frac{y_2 - 1}{-7/3} = \frac{z_2 - 2}{5/3}$$

$\therefore$ Foot of perpendicular lie on.

$$\frac{x}{2/3} = \frac{y - 1}{-7/3} = \frac{z - 2}{5/3}$$

$$\Rightarrow \quad \frac{x}{2} = \frac{y - 1}{-7} = \frac{z - 2}{5}$$

**6. Concept Involved**

It is simple application of independent event, us to solve a certain problem or any type of compitition each event in independent of other.

**Formula used**

$P(A \cap B) = P(A) \cdot P(B)$, when $A$ and $B$ are independent events.

Probability that the problem is solved correctly by atleast one of them = 1 – (Problem is not solved by all)

$\therefore$ $P$ (Problem is solved) = 1 – $P$ (Problem is solved)

$$= 1 - P(\bar{A}) \cdot P(\bar{B}) \cdot P(\bar{C}) \cdot P(\bar{D})$$

$$= 1 - \left(\frac{1}{2} \cdot \frac{1}{4} \cdot \frac{3}{4} \cdot \frac{7}{8}\right)$$

$$= 1 - \frac{21}{256}$$

$$= \frac{235}{256}$$

**7. Concept Involved**

To find the bounded area between $y = f(x)$ and $y = g(x)$ between $x = a$ to $x = b$

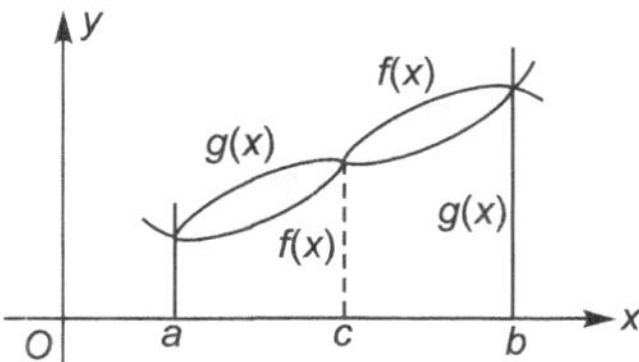


$\therefore$ Area bounded

$$= \int_a^c [g(x) - f(x)]\,dx + \int_c^b [f(x) - g(x)]\,dx$$

$$\Rightarrow \quad \int_a^b |f(x) - g(x)|\,dx$$

Here, $f(x) = y = \sin x + \cos x$, when

$$0 \le x \le \frac{\pi}{2}$$

and $g(x) = y = |\cos x - \sin x|$

$$= \begin{cases} \cos x - \sin x, & 0 \le x \le \frac{\pi}{4} \\ \sin x - \cos x, & \frac{\pi}{4} \le x \le \frac{\pi}{2} \end{cases}$$

Could be shown as

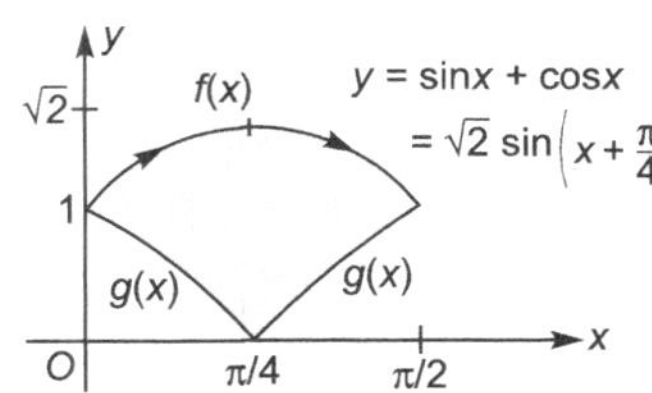


$\therefore$ Area bounded

$$= \int_0^{\pi/4} \{(\sin x + \cos x) - (\cos x - \sin x)\}\,dx$$

$$+ \int_{\pi/4}^{\pi/2} \{(\sin x + \cos x) - (\sin x - \cos x)\}\,dx$$

$$= \int_0^{\pi/4} 2\sin x\,dx + \int_{\pi/4}^{\pi/2} 2\cos x\,dx$$

$$= -2(\cos x)_0^{\pi/4} + 2(\sin x \cdot n)_{\pi/4}^{\pi/2}$$

$$= 4 - 2\sqrt{2} = 2\sqrt{2}(\sqrt{2} - 1)$$

**8. Concept Involved**

Solving of homogeneous differential equation *i. e.*, substitute $\frac{y}{x} = v$

$$\therefore \quad y = vx$$

$$\frac{dy}{dx} = v + x\frac{dv}{dx}$$

Here, slope of the curve at $(x, y)$ is

$$\frac{dy}{dx} = \frac{y}{x} + \sec\left(\frac{y}{x}\right)$$

Put $\quad \frac{y}{x} = v$

$\therefore \quad v + x\frac{dv}{dx} = v + \sec(v)$

$\Rightarrow \quad x\frac{dv}{dx} = \sec(v)$

$\Rightarrow \quad \int \frac{dv}{\sec v} = \int \frac{dx}{x}$

$\Rightarrow \quad \int \cos v\, dv = \int \frac{dx}{x}$

$\Rightarrow \quad \sin v = \log x + \log c$

$\Rightarrow \quad \sin\left(\frac{y}{x}\right) = \log(cx)$

As it passes through $\left(1, \frac{\pi}{6}\right)$

$\Rightarrow \quad \sin\left(\frac{\pi}{6}\right) = \log c$

$\Rightarrow \quad \log c = \frac{1}{2}$

$\therefore \quad \sin\left(\frac{y}{x}\right) = \log x + \frac{1}{2}$

**9. Concept Involved**

When ever we have linear differential equation containing inequality we should always check for increasing or decreasing

*i.e.*, for $\quad \frac{dy}{dx} + Py < 0$

$\Rightarrow \quad \frac{dy}{dx} + Py > 0$

Multiply by integrating factor *i.e.*, $e^{\int P dx}$ and convert into total differential equation.

Here, $f'(x) < 2f(x)$, multiplying by $e^{-\int 2dx}$

$$f'(x)\cdot e^{-2x} - 2e^{-2x}f(x) < 0$$

$\Rightarrow \quad \frac{d}{dx}(f(x)\cdot e^{-2x}) < 0$

$\therefore \phi(x) = f(x)e^{-2x}$ is decreasing for

$$x \in \left[\frac{1}{2}, 1\right]$$

Thus, when $x > \frac{1}{2}$

$$\phi(x) < \phi\left(\frac{1}{2}\right)$$

$\Rightarrow \quad e^{-2x}f(x) < e^{-1}\cdot f\left(\frac{1}{2}\right)$

$\Rightarrow \quad f(x) < e^{2x-1}\cdot 1$, given $f\left(\frac{1}{2}\right) = 1$

$\Rightarrow \quad 0 < \int_{1/2}^{1} f(x)\,dx < \int_{1/2}^{1} e^{2x-1}\,dx$

$\Rightarrow \quad 0 < \int_{1/2}^{1} f(x)\,dx < \left(\frac{e^{2x-1}}{2}\right)_{1/2}^{1}$

$\Rightarrow \quad 0 < \int_{1/2}^{1} f(x)\,dx < \frac{e-1}{2}$

**10. Concept Involved**

The given equation contains algebraic and trigonometric function called transcendal equation. To solve transcendal equations we should always plot the graph for LHS and RHS

Here, $\quad x^2 = x\sin x + \cos x$

Let $\quad f(x) = x^2$ and

$g(x) = x\sin x + \cos x$

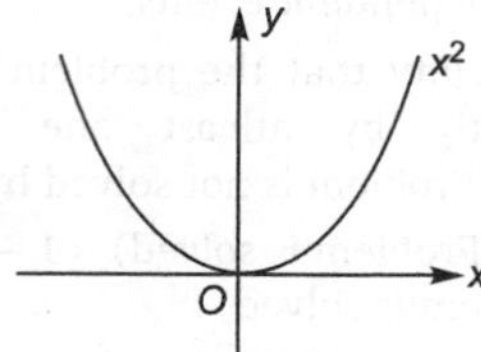


we know the graph for $f(x) = x^2$

To plot, $\quad g(x) = x\sin x + \cos x$

$g'(x) = x\cos x + \sin x - \sin x$

$$g'(x) = x\cos x \quad \ldots(i)$$

$$g''(x) = -x\sin x + \cos x \quad \ldots(ii)$$

Put $\quad g'(x) = 0$

$\Rightarrow \quad x\cos x = 0$

$\therefore \quad x = 0, \frac{\pi}{2}, \frac{3\pi}{2}, \frac{5\pi}{2}, \frac{7\pi}{2}, \ldots$

at $x = 0, \frac{3\pi}{2}, \frac{7\pi}{2}, \dots f''(x) > 0$

$\therefore$ minimum

at $x = \frac{\pi}{2}, \frac{5\pi}{2}, \frac{9\pi}{2}, \dots f''(x) < 0$

$\therefore$ maximum

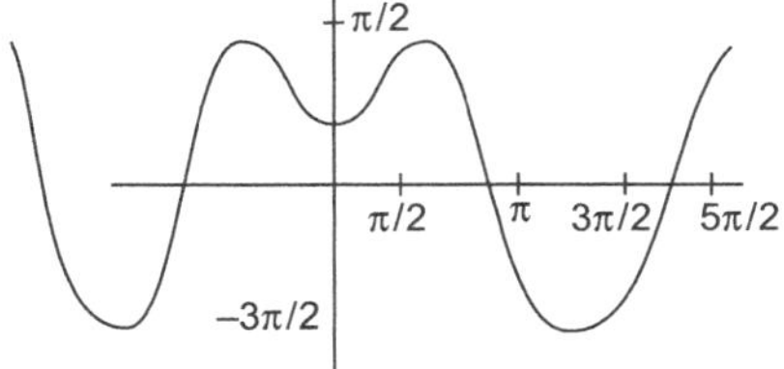


$\therefore f(x)$ and $g(x)$ are, shown as

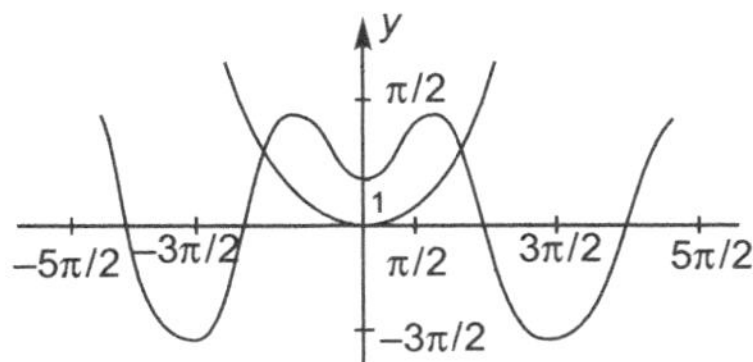


$\therefore$ Number of solutions are 2.

**11. Concept Involved**

The problem is based on the concept to maximise volume of cuboid.

*i.e.*, to form a function of volume, say $f(x)$ find $f'(x)$ and $f''(x)$.

Put $f'(x) = 0$ and check $f''(x)$ to be + ve or − ve for minimum and maximum, respectively.

Here, $l = 15x - 2a, b = 8x - 2a$
$h = a$

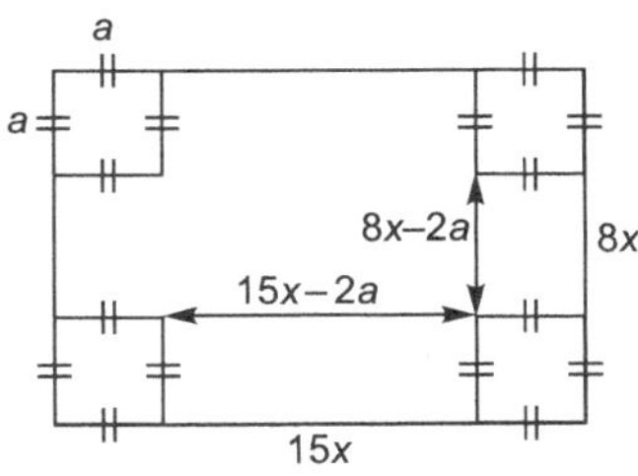


$\therefore$ Volume $= (8x - 2a)(15x - 2a)a$

$$V = 2a \cdot (4x - a)(15x - 2a) \quad \dots(i)$$

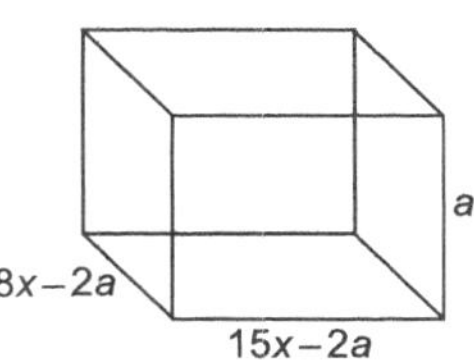


$$\frac{dv}{da} = 6a^2 - 46ax + 60x^2$$

$$\frac{d^2v}{da^2} = 12a - 46x$$

Here, $\left(\frac{dv}{da}\right) = 0$

$\Rightarrow \quad 6x^2 - 23x + 15 = 0$

at $\quad a = 5 \Rightarrow x = 3, \frac{5}{6}$

$$\left(\frac{d^2v}{da^2}\right) = 2(30 - 23x)$$

at $x = 3, \left(\frac{d^2v}{da^2}\right) = 2(30 - 69) < 0,$

$\therefore$ Maximum when $x = 3,$

also at $\quad x = \frac{5}{6} \Rightarrow \left(\frac{d^2v}{da^2}\right) > 0$

$\therefore$ at $x = \frac{5}{6}$, volume is minimum.

Thus, sides are $8x = 24$ and $15x = 45$

**12. Concept Involved**

It is to convert into differences and using sum of $n$ terms of AP.

*i.e.*, $\quad S_n = \frac{n}{2}[2a + (n-1)d]$

$$S_n = \sum_{k=1}^{4n} (-1)^{\frac{k(k+1)}{2}} \cdot k^2$$

$$= -(1)^2 - 2^2 + 3^2 + 4^2 - 5^2 - 6^2 + 7^2 + 8^2 + \dots$$

$$= (3^2 - 1^2) + (4^2 - 2^2) + (7^2 - 5^2) + (8^2 - 6^2) + \dots$$

$$= 2\{\underbrace{(4 + 6 + 12 + \dots)}_{n \text{ terms}} + \underbrace{(6 + 14 + 22 + \dots)}_{n \text{ terms}}\}$$

$$= 2\left[\frac{n}{2}\{2 \times 4 + (n-1)8\} + \frac{n}{2}\{2 \times 6 + (n-1)8\}\right]$$

$$= 2[n(4 + 4n - 4) + n(6 + 4n - 4)]$$

$$= 2[4n^2 + 4n^2 + 2n] = 4n(n+1)$$

Here, $1056 = 32 \times 33$, $1088 = 32 \times 34$,
$1120 = 32 \times 35$, $1332 = 36 \times 37$

1056 and 1332 are possible answers.

**13. Concept Involved**

Equation of straight line, is

$$l : \frac{x - x_1}{a} = \frac{y - y_1}{b} = \frac{z - z_1}{c}$$

Since, $l$ is perpendicular to $l_1$ and $l_2$.

$\therefore$ Its DR's are cross product of $l_1$ and $l_2$.

Now, to find a point on $l_2$ whose distance is given, assume a point and find its distance to obtain point.

Let $\quad l : \dfrac{x-0}{a} = \dfrac{y-0}{b} = \dfrac{z-0}{c}$

which is perpendicular to

$$l_1 : (3\hat{\mathbf{i}} - \hat{\mathbf{j}} + 4\hat{\mathbf{k}}) + t(\hat{\mathbf{i}} + 2\hat{\mathbf{j}} + 2\hat{\mathbf{k}})$$

$$l_2 : (3\hat{\mathbf{i}} + 3\hat{\mathbf{j}} + 2\hat{\mathbf{k}}) + s(2\hat{\mathbf{i}} + 2\hat{\mathbf{j}} + \hat{\mathbf{k}})$$

$\therefore$ DR's of $l$, is $\begin{vmatrix} \hat{\mathbf{i}} & \hat{\mathbf{j}} & \hat{\mathbf{k}} \\ 1 & 2 & 2 \\ 2 & 2 & 1 \end{vmatrix}$

$$= -2\hat{\mathbf{i}} + 3\hat{\mathbf{j}} - 2\hat{\mathbf{k}}$$

$\therefore \quad l : \dfrac{x}{-2} = \dfrac{y}{3} = \dfrac{z}{-2} = k_1, k_2$

Now, $\quad A(-2k_1, 3k_1, -2k_1)$

and $\quad B(-2k_2, 3k_2, -2k_2)$.

Since, $A$ lies on $l_1$

$$\therefore (-2k_1)\hat{\mathbf{i}} + (3k_1)\hat{\mathbf{j}} - (2k_1)\hat{\mathbf{k}} = (3+t)\hat{\mathbf{i}} + (-1+2t)\hat{\mathbf{j}} + (4+2t)\hat{\mathbf{k}}$$

$\Rightarrow \quad 3 + t = -2k_1, -1 + 2t = 3k_1,$
$4 + 2t = -2k_1$

$\therefore \quad k_1 = -1$

$\Rightarrow \quad A(2, -3, 2)$

Let any point on $l_2(3 + 2s, 3 + 2s, 2 + 5)$

$$\Rightarrow \sqrt{(2-3-2s)^2 + (-3-3-2s)^2 + (2-2-s)^2} = \sqrt{17}$$

$$\Rightarrow \quad 9s^2 + 28s + 37 = 17$$

$$\Rightarrow \quad 9s^2 + 28s + 20 = 0$$

$$\Rightarrow \quad 9s^2 + 18s + 10s + 20 = 0$$

$$\therefore \quad (9s + 10)(s + 2) = 0$$

$$\Rightarrow \quad s = -2, \frac{-10}{9}.$$

Hence, $(-1, -1, 0)$ and $\left(\frac{7}{9}, \frac{7}{9}, \frac{8}{9}\right)$ are required points.

**14. Concept Involved**

This question is based on the concept of transcendal equation and should be solved with the help of graph.

$$f(x) = x \sin \pi x, \; x > 0$$

$$f'(x) = \sin \pi x + \pi x \cos \pi x = 0$$

$$\Rightarrow \quad \tan \pi x = -\pi x$$

let $f(x) = \tan \pi x$ and $g(x) = -\pi x$

which could be plotted as,

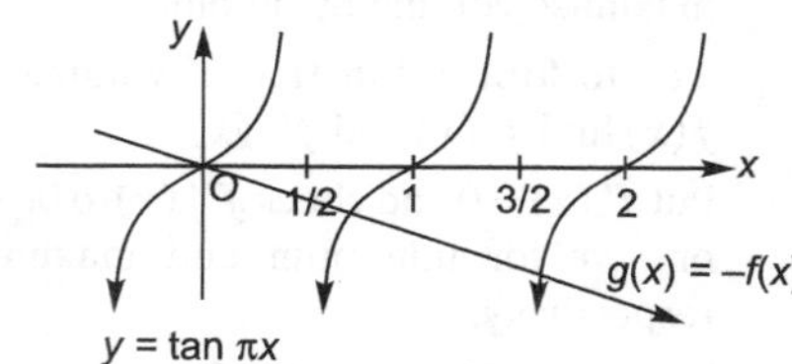


Clearly, $f(x)$ and $g(x)$ intersect when

$$\frac{1}{2} < x < 1.$$

or $\quad \dfrac{3}{2} < x < 2$

or $\quad \dfrac{5}{2} < x < 3$

$\therefore$ a unique point in $\left(n + \dfrac{1}{2}, n + 1\right)$

or $\quad (n, n+1)$

**15.** (a) $(N^T MN)^T = N^T M^T (N^T)^T$

$= N^T M^T N$, is symmetric is $M$ is symmetric and skew-symmetric is $M$ is skew-symmetric

(b) $(MN - NM)^T = (MN)^T - (NM)^T$

$= NM - MN = -(MN - NM)$

$\therefore$ Skew-symmetric, when $M$ and $N$ are symmetric.

(c) $(MN)^T = N^T M^T = NM \neq MN$

$\therefore$ not correct

(d) $(\text{adj}\, MN) = (\text{adj}\, N) \cdot (\text{adj}\, M)$

$\therefore$ not correct.

**16. Concept Involved**

As to maximise or minimise area of triangle we should find area is terms of parametric coordinate and we second derivative test.

Here, tangent at $P(2\cos\theta, \sqrt{3}\sin\theta)$ is

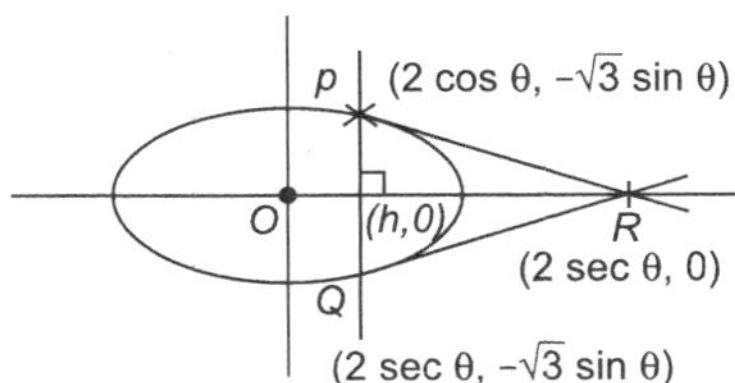


$$\frac{x}{2}\cos\theta + \frac{y}{\sqrt{3}}\sin\theta = 1$$

$\therefore \quad R(2\sec\theta, 0)$

$\Rightarrow \quad \Delta$ = area of $\Delta PQR$

$$= \frac{1}{2}(2\sqrt{3}\sin\theta)(2\sec\theta - 2\cos\theta)$$

$$= 2\sqrt{3} \cdot \sin^3\theta / \cos\theta \qquad \ldots(i)$$

Since, $\quad \frac{1}{2} \leq h \leq 1,$

$\therefore \quad \frac{1}{2} \leq 2\cos\theta \leq 1$

$\Rightarrow \quad \frac{1}{4} \leq \cos\theta \leq \frac{1}{2} \qquad \ldots(ii)$

$$\therefore \quad \frac{d\Delta}{d\theta} = \frac{2\sqrt{3}\{\cos\theta \cdot 3\sin^2\theta\cos\theta - \sin^3\theta(-\sin\theta)\}}{\cos^2\theta}$$

$$= \frac{2\sqrt{3}\cdot\sin^2\theta}{\cos^2\theta}[3\cos^2\theta + \sin^2\theta]$$

$$= \frac{2\sqrt{3}\sin^2\theta}{\cos^2\theta}\cdot[2\cos^2\theta + 1]$$

$$= 2\sqrt{3}\tan^2\theta\,(2\cos^2\theta + 1) > 0$$

when, $\frac{1}{4} \leq \cos\theta \leq \frac{1}{2}$

$\therefore \quad \Delta_1 = \Delta_{max}$ occurs at $\cos\theta = \frac{1}{4}$

$$= \left(\frac{2\sqrt{3}\cdot\sin^3\theta}{\cos\theta}\right)$$

When $\cos\theta = \frac{1}{4} = \frac{45\sqrt{5}}{8}$

$\Delta_2 = \Delta_{min}$ occurs at $\cos\theta = \frac{1}{2}$

$$= \left(\frac{2\sqrt{3}\sin^3\theta}{\cos\theta}\right),$$

when $\cos\theta = \frac{1}{2} = \frac{9}{2}$

$$\therefore \quad \frac{8}{\sqrt{5}}\Delta_1 - 8\Delta_2 = 45 - 36 = 9$$

**17.** Let the three consecutive terms in $(1+x)^{n+5}$ be $t_r, t_{r+1}, t_{r+2}$. Having coefficients ${}^{n+5}C_{r-1}, {}^{n+5}C_r, {}^{n+5}C_{r+1}$

given, ${}^{n+5}C_{r-1} : {}^{n+5}C_r : {}^{n+5}C_{r+1}$

$$= 5 : 10 : 14$$

$$\therefore \quad \frac{{}^{n+5}C_r}{{}^{n+5}C_{r-1}} = \frac{10}{5}$$

$$\text{and} \quad \frac{{}^{n+5}C_{r+1}}{{}^{n+5}C_r} = \frac{14}{10}$$

$$\Rightarrow \quad \frac{n+5-(r-1)}{r} = 2$$

$$\text{and} \quad \frac{n-r+5}{r+1} = \frac{7}{5}$$

$\Rightarrow \quad n - r + 6 = 2r$

and $\quad 5n - 5r + 25 = 7r + 7$

$\Rightarrow \quad n + 6 = 3r$

and $\quad 5n + 18 = 12r$

$$\therefore \quad \frac{n+6}{3} = \frac{5n+18}{12}$$

$\Rightarrow \quad 4n + 24 = 5n + 18$

$\Rightarrow \quad n = 6$

**18. Concept Involved**

The three vectors are coplanar, if volume is zero. Now, those vectors which are along diagonals of a cube, are non-coplanar.

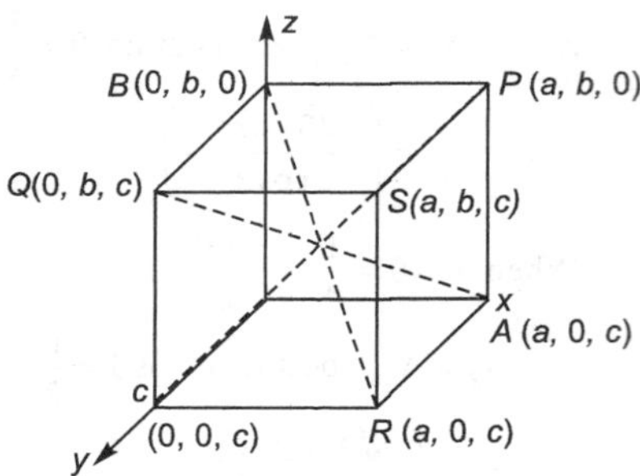


Here, the four diagonals are **OS**, **AQ**, **BR** and **CP**.

Among set of eight vectors four vectors form body diagonals of a cube, emagining four will be parallel (unlike) vectors.

$\therefore$ Number of ways of selecting three vector will be

$$^4C_3 \times 8 = 2^5 = 2^P$$

$$\Rightarrow \quad p = 5.$$

**19. Concept Involved**

For the events to be independent.

$$P(E_1 \cap E_2 \cap E_3) = P(E_1)\cdot P(E_2)\cdot P(E_3)$$

$$P(E_1 \cap \bar{E}_2 \cap \bar{E}_3) = P$$

(only $E_1$ occurs)

$$= P(E_1)\cdot(1 - P(E_2))(1 - P(E_3)).$$

let $x, y, z$ are probability of $E_1$, $E_2$ and $E_3$ respectively

$$\therefore \quad \alpha = x(1-y)(1-z) \quad \ldots\text{(i)}$$

$$\beta = (1-x)\cdot y(1-z) \quad \ldots\text{(ii)}$$

$$\gamma = (1-x)(1-y)z \quad \ldots\text{(iii)}$$

$$\Rightarrow \quad p = (1-x)(1-y)(1-z) \quad \ldots\text{(iv)}$$

Given, $(\alpha - 2\beta)p = \alpha\beta$

and $\quad (\beta - 3\gamma)p = 2\beta\gamma \quad \ldots$(v)

From above equations.

$$x = 2y \quad \text{and} \quad y = 3z$$

$$\therefore \quad x = 6z \quad \Rightarrow \quad \frac{x}{z} = 6$$

**20.** Let number of removed cards are $k$ and $(k+1)$

$$\therefore \quad \frac{n(n+1)}{2} - k - (k+1) = 1224$$

$$\Rightarrow \quad n^2 + n - 4k = 2450$$

$$\Rightarrow \quad n^2 + n - 2450 = 4k$$

$$\Rightarrow \quad (n+50)(n-49) = 4k$$

$$\therefore \quad n > 49 \quad \text{let } n = 50$$

$$\therefore \quad 100 = 4k, \; k = 25$$

$$\Rightarrow \quad k - 20 = 5$$

# Paper 2

## Physics

**1.**

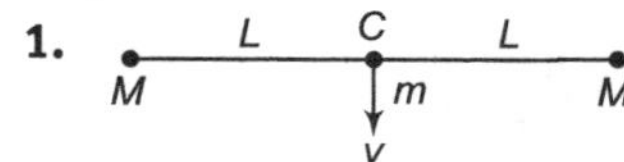


Let $v$ is the minimum velocity. From energy conservation,

$$U_c + K_c = U_\infty + K_\infty$$

$$\therefore \quad mV_c + \frac{1}{2}mv^2 = 0 + 0 \;\therefore\; v = \sqrt{-2V_c}$$

$$= \sqrt{(-2)\left(\frac{-2GM}{L}\right)} = 2\sqrt{\frac{GM}{L}}$$

**2.** (a) At equilibrium ($t = 0$) particle has maximum velocity $u_0$. Therefore velocity at time $t$ can be written as

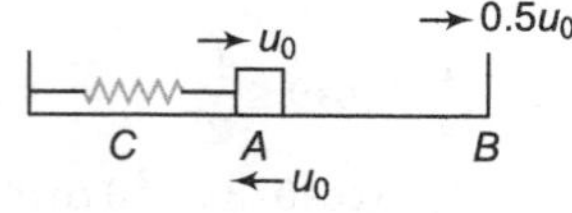


$$u = u_{max} \cos \omega t + u_0 \cos \omega t$$

Writing, $\quad u = 0.5u_0 = u_0 \cos \omega t$

$\therefore \quad \omega t = \dfrac{\pi}{3}$

$\therefore \quad \dfrac{2\pi}{T} t = \dfrac{\pi}{3} \quad \therefore\ t = \dfrac{T}{6}$

(b) $t = t_{AB} + t_{BA} = \dfrac{T}{6} + \dfrac{T}{6} = \dfrac{T}{3} = \dfrac{2\pi}{3}\sqrt{\dfrac{m}{k}}$

(c) $t = t_{AB} + t_{BA} + t_{AC}$

$$= \frac{T}{6} + \frac{T}{6} + \frac{T}{4} = \frac{7}{12}T = \frac{7\pi}{6}\sqrt{\frac{m}{k}}$$

(d) $t = t_{AB} + t_{BA} + t_{AC} + t_{CA}$

$$= \frac{T}{6} + \frac{T}{6} + \frac{T}{4} + \frac{T}{4} = \frac{5}{6}T = \frac{5\pi}{3}\sqrt{\frac{m}{k}}$$

**3.**

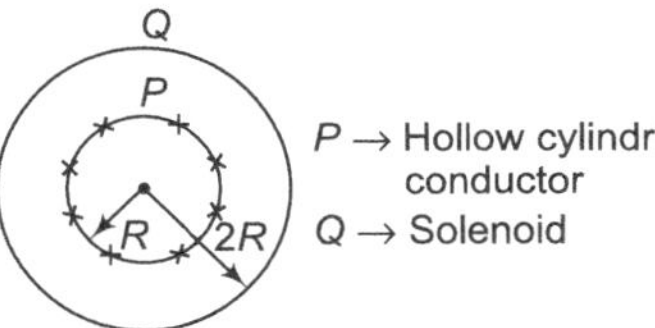


In the region, $0 < r < R$

$$B_P = 0,$$

$B_Q \neq 0$, along the axis

$\therefore \quad B_{net} \neq 0$

In the region, $\quad R < r < 2R$

$B_P \neq 0$, tangential to the circle of radius $r$, centred on the axis.

$B_Q \neq 0$, along the axis.

$\therefore B_{net} \neq 0$ neither in the directions mentioned in options (b) or (c).

In region, $r > 2R$

$$B_P \neq 0$$
$$B_Q \neq 0$$

$\therefore \quad B_{net} \neq 0$

**4.** When wind blows from $S$ to 0

$\rightarrow u$ [S] $\quad \leftarrow u$ [O]
$f_1 \quad f_2$

$$f_2 = f_1\left(\frac{v + w + u}{v + w - u}\right)$$

or $\quad f_2 > f_1$

When wind blows from 0 to $S$

$$f_2 = f_1\left(\frac{v - w + u}{v - w - u}\right)$$

$\therefore \quad f_2 > f_1$

**5.** $2d \sin\theta = \lambda$

$$d = \frac{\lambda}{2\sin\theta}$$

Differentiate $\partial(d) = \dfrac{\lambda}{2}\,\partial\,(\text{cosec}\theta)$

$$\partial(d) = \frac{\lambda}{2}(-\text{cosec}\,\theta\,\cot\theta)\,\partial\theta$$

$$\partial(d) = \frac{-\lambda\cos\theta}{2\sin^2\theta}$$

as $\theta$ = increases, $\dfrac{\lambda\cos\theta}{2\sin^2\theta}$ decreases

**Alternate Solution**

$$d = \frac{\lambda}{2\sin\theta}$$

$\ln d = \ln\lambda - \ln 2 - \ln\sin\theta$

$$\frac{\Delta(d)}{d} = 0 - 0 - \frac{1}{\sin\theta} \times \cos\theta\,(\Delta\theta)$$

Fractional error $|+(d)| = \cot\theta\,\Delta\theta\,|$

Absolute error $\Delta d = (d\cot\theta)\,\Delta\theta$

$$\frac{d}{2\sin\theta} \times \frac{\cos\theta}{\sin\theta}$$

$$\Delta d = \frac{\cos\theta}{\sin^2\theta}$$

**6.**

| $B_1$ | $B_2$ | $B_3$ |
|---|---|---|
| 1 whilte | 2 whilte | 3 whilte |
| 2 Red | 3 Red | 4 Red |
| 2 Black | 4 Black | 5 Black |

For electrostatic field,

$$\mathbf{E}_P = \mathbf{E}_1 + \mathbf{E}_2$$

$$= \frac{\rho}{3\varepsilon_0}\mathbf{C}_1\mathbf{P} + \frac{(-\rho)}{3\varepsilon_0}\mathbf{C}_1\mathbf{P}$$

$$= \frac{\rho}{3\varepsilon_0}(\mathbf{C}_1\mathbf{P} + \mathbf{PC}_2)$$

$$\mathbf{E}_P = \frac{\rho}{3\varepsilon_0}\mathbf{C}_1\mathbf{C}_2$$

For electrostatic potential. Since electric field is non zero so it is not equipotential.

**7.** $Q = mCT$

$$\frac{dQ}{dt} = mc\,\frac{dT}{dt}$$

$R$ = rate of absortion of heat $= \dfrac{dQ}{dt} \propto C$

(i) in 0 – 100 K
$C$ increase, so $R$ increases but not linearly

(ii) $\Delta Q = mC\Delta T$ as $C$ is more in (400 K – 500 K) then
(0 – 100 K) so heat is increasing

(iii) $C$ remains constant so there no change in $R$ from (400 K – 500 K)

(iv) $C$ is increases so $R$ increases in range(200 K – 300 K)

**8.** $L = 3\left(\dfrac{h}{2\pi}\right)$

$\therefore \quad n = 3$, as $L = n\left(\dfrac{h}{2\pi}\right)$

$$r_n \propto \frac{n^2}{z}$$

$$r_3 = 4.5a_0 \;\therefore\; z = 2$$

$$\frac{1}{\lambda_1} = Rz^2\left(\frac{1}{2^2} - \frac{1}{3^2}\right) = 4R\left(\frac{1}{4} - \frac{1}{9}\right)$$

$$\therefore \quad \lambda_1 = \frac{9}{5R}$$

$$\frac{1}{\lambda_2} = Rz^2\left(\frac{1}{1^2} - \frac{1}{3^2}\right) = 4R\left(1 - \frac{1}{9}\right)$$

$$\Rightarrow \quad \lambda_2 = \frac{9}{32R}$$

$$\frac{1}{\lambda_3} = Rz^2\left(\frac{1}{1^2} - \frac{1}{2^2}\right) = 4R\left(1 - \frac{1}{4}\right)$$

$$\Rightarrow \quad \lambda_3 = \frac{1}{3R}$$

**9.** Height fallen upto $Q$ is $R \sin 30° = \dfrac{R}{2} = 20$ m

$E_i - E_f$ = work done against friction.

$$\therefore \quad 0 - \left[-mgh + \frac{1}{2}mv^2\right] = 150$$

$$\therefore \quad (1)\,(10)\,(20) - \frac{1}{2} \times 1 \times v^2 = 150$$

or $\quad v = 10$ m/s

**10.** $N = mg \cos 60° = \dfrac{mv^2}{R}$

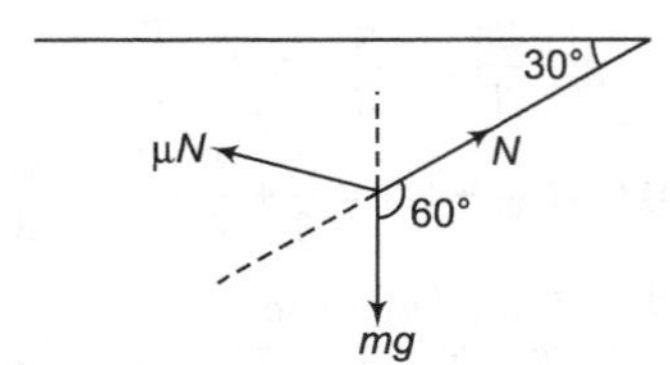


$$\therefore \quad N = \frac{mv^2}{R} + \frac{mg}{2}$$

$$= \frac{(1)\,(10)^2}{40} + \frac{(1)\,(10)}{2}$$

$$= 7.5 \text{ N}$$

**11.** $P = Vi$

$$\therefore \quad i = \frac{P}{V} = \frac{600 \times 10^3}{4000} = 150 \text{ A}$$

Total resistance of cables,

$$R = 0.4 \times 20 = 8\Omega$$

$\therefore$ Power loss in cables $= i^2R$

$$= (150)^2\,(8)$$

$$= 180000 \text{ W} = 180 \text{ kW}$$

This loss is 30% of 600 kW.

**12.** During step-up,

$$\frac{N_p}{N_s} = \frac{V_p}{V_s}$$

or $\quad \dfrac{1}{10} = \dfrac{4000}{V_s}$

or $\quad V_s = 40{,}000$ V

In step, down transformer,

$$\frac{N_p}{N_s} = \frac{V_p}{V_s} = \frac{40000}{200}$$

$$= \frac{200}{1}$$

**13.** The induced electric field is given by,

$$\oint \mathbf{E} \cdot \mathbf{dl} = -\frac{d\phi}{dt}$$

or $\quad El = -s\left(\dfrac{dB}{dt}\right)$

$\therefore \quad E(2\pi R) = -(\pi R^2)(B)$

or $\quad E = -\dfrac{BR}{2}$

**14.** $\frac{M}{L} = \frac{Q}{2m}$

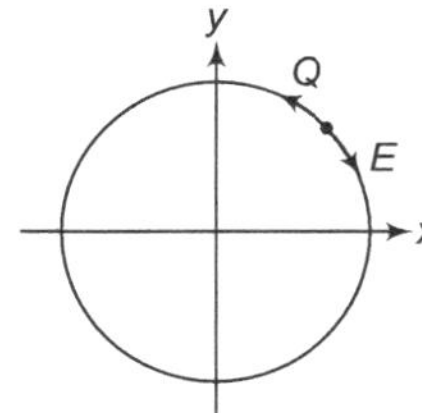


$\therefore \quad M = \left(\frac{Q}{2m}\right) L$

$\Rightarrow \quad M \propto L$, where $\gamma = \frac{Q}{2m}$

$= \left(\frac{Q}{2m}\right)(I\omega)$

$= \left(\frac{Q}{2m}\right)(mR^2\omega) = \frac{Q\omega R^2}{2}$

Induced electric field is opposite. Therefore,

$$\omega' = \omega - \alpha t$$

$$\alpha = \frac{\tau}{I} = \frac{(QE)R}{mR^2} = \frac{(Q)\left(\frac{BR}{2}\right)R}{mR^2} = \frac{QB}{2m}$$

$$\therefore \quad \omega' = \omega - \frac{QB}{2m}\cdot 1 = \omega - \frac{QB}{2m}$$

$$M_f = \frac{Q\omega' R^2}{2} = Q\left(\omega - \frac{QB}{2m}\right)\frac{R^2}{2}$$

$$\therefore \ \Delta M = M_f - M_i = -\frac{Q^2BR^2}{4m}$$

$$M = -\gamma \frac{QBR^2}{2} \qquad \left(\text{as } \gamma = \frac{Q}{2m}\right)$$

**15.** (a) ${}_3\text{Li}^7 \rightarrow {}_2\text{He}^4 + {}_1\text{H}^3$

$\Delta m = [\text{M}_{\text{Li}} - \text{M}_{\text{He}} - \text{M}_{\text{H}^3}]$
$= [6.01513 - 4.002603 - 3.016050]$
$= -1.003523\,\text{u}$

$\Delta m$ is negative so reaction is not possible.

(b) ${}_{84}\text{Po}^{210} \rightarrow {}_{83}\text{Bi}^{209} + {}_1\text{P}^1$

$\Delta m$ is negative so reaction is not possible.

(c) ${}_1\text{H}^2 \rightarrow {}_2\text{He}^4 + {}_3\text{Li}^6$

$\Delta m$ is positive so reaction is possible.

(d) ${}_{30}\text{Zn}^{70} + {}_{34}\text{Se}^{82} \rightarrow {}_{64}\text{Gd}^{152}$

$\Delta m$ is positive so reaction is not possible.

**16.** ${}_{84}\text{Po}^{210} \longrightarrow {}_2\text{He}^4 + {}_{82}\text{Pb}^{206}$

Mass defect $\Delta m = (m_{\text{Po}} - M_{\text{He}} - m_{\text{Pb}})$
$= 0.005818\,\text{u}$

$\therefore \quad Q = (\Delta m)(931.48)\,\text{MeV}$
$= 5.4193\,\text{MeV} = 5419\,\text{keV}$

$\alpha \longleftarrow \bullet \qquad \bullet \longrightarrow \text{Pb}$

From conservation of linear momentum,

$$p_{\text{Pb}} = p_\alpha$$

$$\therefore \quad \sqrt{2m_{\text{Pb}}\,k_{\text{Pb}}} = \sqrt{2m_\alpha\,k_\alpha}$$

$$\text{or} \quad \frac{k_\alpha}{k_{\text{Pb}}} = \frac{m_{\text{Pb}}}{m_\alpha} = \frac{206}{4}$$

$$\therefore \quad k_\alpha = \left(\frac{206}{206+4}\right)(k_{\text{total}})$$

$$= \left(\frac{206}{210}\right)(5419) = 5316\,\text{keV}$$

**17.** For $e \rightarrow i$

$$45° > \theta_c$$
$$\sin 45° > \sin\theta_c$$
$$\frac{1}{\sqrt{2}} > \frac{\mu_2}{\mu_1}$$
$$\mu_1 > \sqrt{2}\mu_2$$

For $e \rightarrow f$

angle of refraction is lesser than angle of incidence, so $\mu_2 > \mu_1$ and then $\mu_2 > \mu_3$

For $e \rightarrow g$, $\mu_1 = \mu_2$

for $e \rightarrow h$, $\mu_2 < \mu_1 < \sqrt{2}\mu_2$ and $\mu_2 > \mu_3$

**18.** (p) $U = \frac{1}{2}kT$

$\Rightarrow \quad [\text{ML}^2\text{T}^{-2}] = [k]\,K$

$\Rightarrow \quad [\text{K}] = [\text{ML}^2\text{T}^{-2}\text{K}^{-1}]$

(q) $F = \eta A \frac{dv}{dx}$

$\Rightarrow [\eta] = \frac{[\text{MLT}^{-2}]}{[\text{L}^2\text{LT}^{-1}\text{L}^{-1}]} = [\text{ML}^{-1}\,\text{T}^{-1}]$

(r) $E = h\nu$

$\Rightarrow \quad [\text{ML}^2\text{T}^2] = [h]\,[\text{T}^{-1}]$

$\Rightarrow \quad [h] = [\text{ML}^2\text{T}^{-1}]$

(s) $\frac{dQ}{dt} = \frac{kA\Delta\theta}{l}$

$\Rightarrow [k] = \frac{[ML^2T^{-3}L]}{[L^2K]} = [MLT^{-3}K^{-1}]$

**19.**

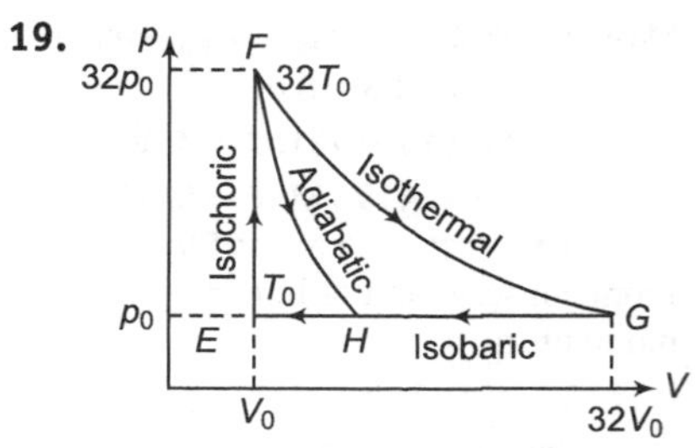


In $F \to G$ work done in isothermal process is

$nRT \ln\left(\frac{V_f}{V_i}\right) = 32\, p_0V_0 \ln\left(\frac{32V_0}{V_0}\right)$

$= 32\, p_0V_0 \ln 2^5 = 160p_0\, V_0 \ln 2$

In $G \to E$, $\Delta W = p_0\Delta V = p_0(31V_0)$

$= 31\, p_0V_0$

In $G \to H$ work done is less than $31\, p_0V_0$ *i.e.*, $24\, p_0V_0$

In $F \to H$ work done is $36\, p_0V_0$

**20.** (p) In α-decay mass number decreases by 4 and atomic number decreases by 2.

(q) In $\beta^+$-decay mass number remains unchanged while atomic number decreases by 1.

(r) In fission, parent nucleus breaks into all most two equal fragments.

(s) In proton emission both mass number and atomic number decreases by 1.

# Chemistry

**1. Plan** In presence of common ion (in this case $Ag^+$ ion) solubility of sparingly soluble salt is decreased.

Let solubility of $Ag_2CrO_4$ in presence of 0.1 M

$AgNO_3 = x$

$Ag_2CrO_4 \rightleftharpoons \underset{2x}{2Ag^+} + \underset{x}{CrO_4^{2-}}$

$AgNO_3 \rightleftharpoons \underset{0.1}{Ag^+} + \underset{0.1}{NO_3^-}$

Total $[Ag^+] = (2x + 0.1)\,M$

$\approx 0.1$ M as $x <<< 0.1$ M

$[CrO_4^{2-}] = x$ M

Thus, $[Ag^+]^2[CrO_4^{2-}] = K_{sp}$

$(0.1)^2(x) = 1.1 \times 10^{-12}$

$\because \quad x = 1.1 \times 10^{-10}$ M

**2. Plan** Phenolic compounds in alkaline solution react with chloroform ($CHCl_3$) at a temperature lower than that of $CHCl_3$ to form *ortho*-isomer as the major product (due to greater stability resulting from intramolecular hydrogen bonding)

$HO^- + H—CCl_3 \rightleftharpoons H_2O + \overset{\ominus}{C}Cl_3 \xrightarrow{-Cl^-} :CCl_2$ (dichloro carbene)

$O^\ominus$ (with $CH_3$ para) $+ :CCl_2 \longrightarrow$ (O, H, $\overset{\ominus}{C}Cl_2$, $CH_3$) $\xrightarrow[-H^+]{+H^+}$

OH, $CHCl_2$, $CH_3$ $\xrightarrow{OH^-}$ OH, CHO, $CH_3$ (*S*)

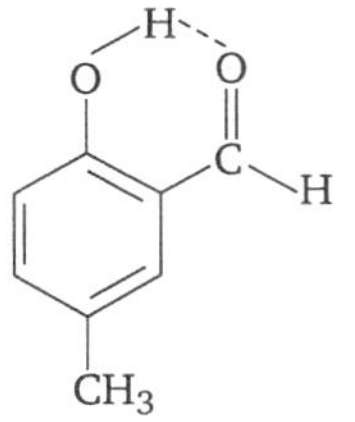

Major as stable due to intramolecular H-bonding.

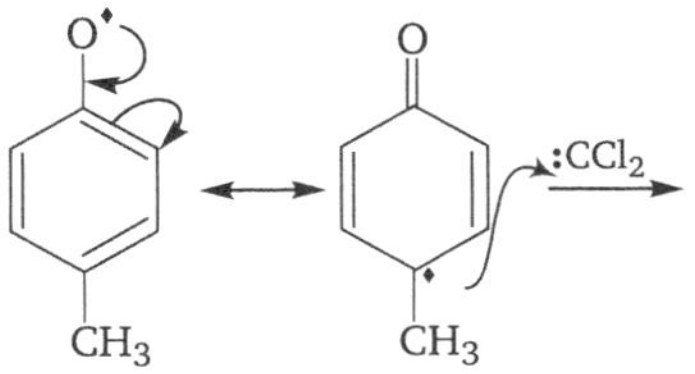

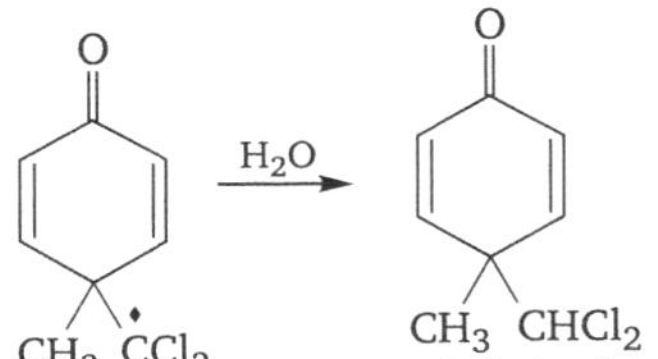

$\xrightarrow{:CCl_2}$

$\xrightarrow{H_2O}$ $CH_3$ $CHCl_2$

Q (minor)

**3. Plan** —OH group is activating group and is *o*- and *p*-directing.

Also $—SO_3H$ is a better leaving group and is knocked out by $Br^-$.

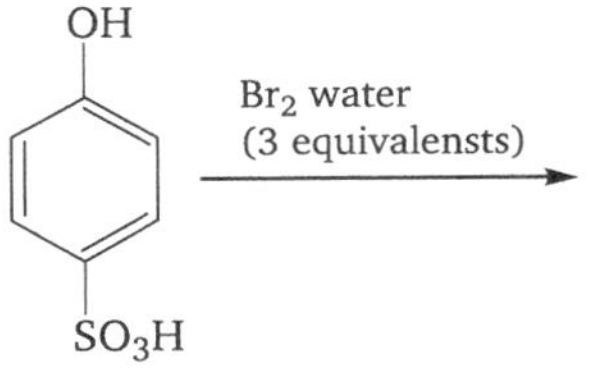

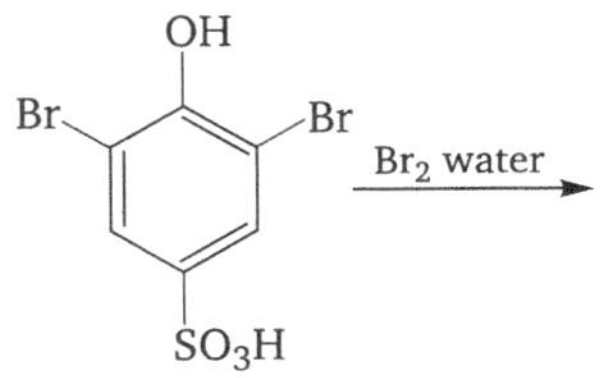

$\xrightarrow[\text{(3 equivalensts)}]{Br_2 \text{ water}}$

$\xrightarrow{Br_2 \text{ water}}$

*ortho*-attack

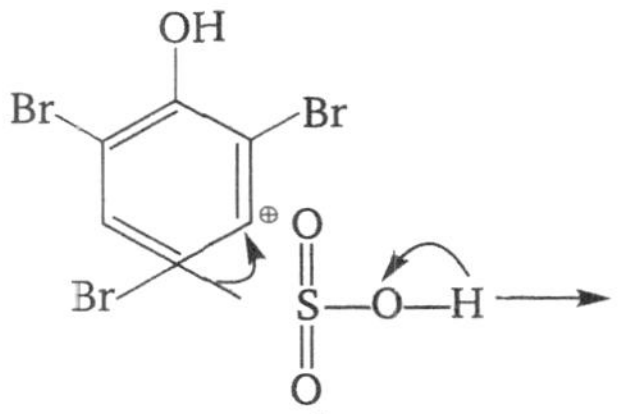

*para*-attack

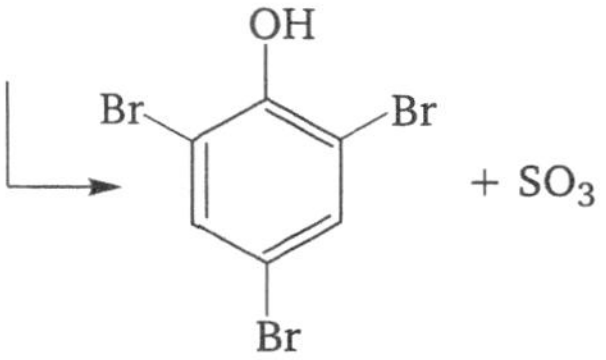

$+ SO_3$

(major)

**4. Plan** When acetone reacts with $Br_2$ in basic medium, bromoform is formed.

**Reaction I**

$$\underset{\substack{1 \text{ mol} \\ \frac{1}{3}\text{mol}}}{CH_3COCH_3} + \underset{\substack{3 \text{ mol} \\ 1 \text{ mol}}}{3\,Br_2} + 4\,NaOH$$

$$\rightarrow CH_3COONa + \underset{(U)}{CHBr_3} + 3NaBr + \underset{(T)}{3H_2O}$$

When $CH_3COCH_3$ and $Br_2$ are in equimolar quantity, all the $Br_2$ (limiting reactant) is converted into desired products and 2/3 mole of $CH_3COCH_3$ remains unreacted, being in excess.

When acetone reacts with $Br_2$ in acidic medium, there is monobromination of acetone.

**Reactions II**

$$\underset{1 \text{ mol}}{CH_3COCH_3} + \underset{1 \text{ mol}}{Br_2} \xrightarrow{CH_3COOH} \underset{(P)}{CH_3COCH_2Br} + HBr$$

$CH_3COCH_3$ and $Br_2$ react in 1 : 1 mole ratio and (*P*) is formed.

In reaction I, (*U*) and (*T*) are formed and acetone (reactant) remains unreacted.

In reaction II, (*P*) is formed.

**5. Plan** Due to resonance, bond-lengths between two atoms are equal. Species is said to be diamagnetic if all electrons are paired.

Process is endothermic if it takes place with absorption of heat.

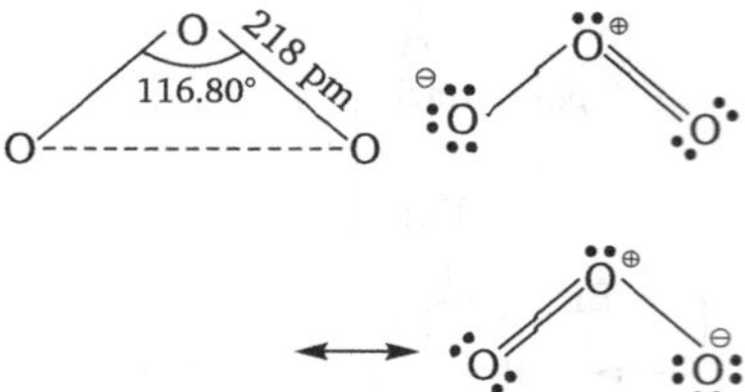


bent molecule

all electrons paired thus, diamagnetic

$$2O_3 \longrightarrow 3O_2 \quad \Delta H^\circ \approx -142 \text{ kJ mol}^{-1}$$

Exothermic

Thus, (b) is incorrect.

(a, c, d) are correct.

**6. Plan** ${}^{9}_{4}Be + {}^{b}_{a}X \longrightarrow {}^{8}_{4}Be + {}^{d}_{c}Y$

Atomic number same

$4 + a = 4 + c$

$9 + b = 8 + d$

If $X = {}^{0}_{0}\gamma$ — $a = 0$, $b = 0$

$Y = {}^{1}_{0}n$ — $c = 0$, $d = 1$

If $X = {}^{1}_{1}p$ — $a = 1$, $b = 1$

$Y = {}^{2}_{1}D$ — $c = 1$, $d = 2$

$${}^{9}_{4}Be + {}^{0}_{0}\gamma \longrightarrow {}^{8}_{4}Br + {}^{1}_{0}n$$

$${}^{9}_{4}Be + {}^{1}_{1}p \longrightarrow {}^{8}_{4}Be + {}^{2}_{1}D$$

**7.** Al has greater affinity for oxygen hence oxide is not reduced by carbon. MgO and CaO (formed in the calcination from carbonates) are stable species and not reduced by carbon.

**During Smelting**

$$SnO_2 + C \xrightarrow{1300°C} Sn + CO$$

$$2Fe_2O_3 + 3C \xrightarrow{\Delta} 4Fe + 3CO_2$$

**8. Plan** Heat of reaction is dependent on temperature (Kirchhoff's equation) in heterogeneous system, equilibrium constant is independent on the molar concentration of solid species.

Heat of reaction is not affected by catalyst. It lowers activation energy.

$$CaCO_3(s) \rightleftharpoons CaO(s) + CO_2(g)$$

**By Kirchhoff's equation,**

$$\Delta H_2^0 \text{ (at } T_2) = \Delta H_1^0 (\text{at } T_1) + \Delta C_p(T_2 - T_1)$$

$\Delta H^\circ$ varies with temperature.

Thus, (a) ... correct.

$$k = p_{CO_2}$$

$k$ is dependent on pressure of $CO_2$ but independent of molar concentration of $CaCO_3$.

Thus, (b) and (c) are correct.

At a given temperature, addition of catalysis lowers activation energy, $\Delta H$ remaining constant.

Thus, (d) also correct.

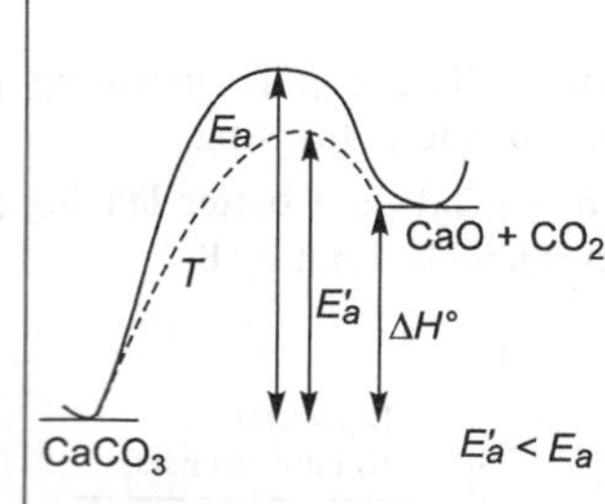


$E_a$ = Activation energy in absence of catalyst

$E'_a$ = Activation energy in presence of catalyst

**9, 10. Plan** $PbCl_2$ is soluble in hot water.

In ammoniacal medium, cations of group III and IV may be precipitated as hydroxide or sulphide and dissolved in $H_2O_2$ due to oxidation.

Mixture of two inorganic salts $\xrightarrow[\text{(ii) Filter}]{\text{(i) HCl}}$ Filtrate + ppt.

$\underset{\text{Yellow (S)}}{CrO_4^{2-}} \xleftarrow[\text{NaOH}]{H_2O_2} \underset{\text{ppt}}{Cr(OH)_3} \xleftarrow{NH_4OH/H_2S}$ Filtrate     ppt.

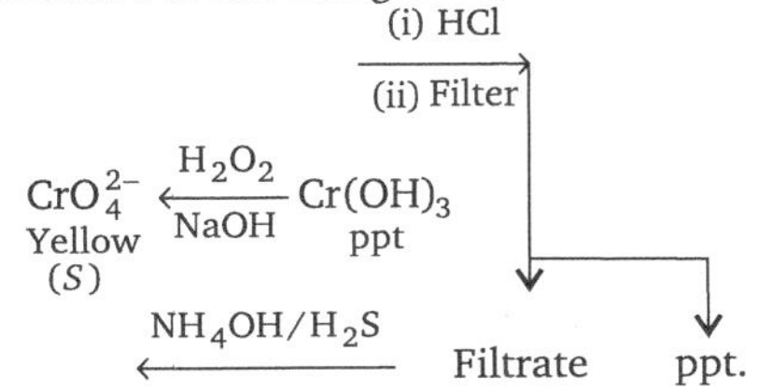

(as CH (NaCl) soluble in hot water)

Thus, **Q. 9.** *P* is $Pb^{2+}$

**Q. 10.** *S* is $Na_2CrO_4$

**11. Plan** alkenes decolourise $Br_2$ water

*cis*-isomer $\xrightarrow{\text{dil·KMnO}_4}$ *Meso* isomers by *syn* addition

*trans*-isomer $\xrightarrow{\text{dil·KMnO}_4}$ $d(+)$ and $l(-)$ isomers by *syn* addition

thus, reacemic mixture

Formation of anhydride from dicarboxylic acid indicates *cis*-isomer.

*P* and *Q* are isomers of dicarboxylic acids.

$$P, Q \xrightarrow{Br_2 \text{ water}} \text{decolourised}$$

*P* and *Q* have (C=C) bond

$$P \xrightarrow{\Delta} \text{anhydride}$$

Thus, *P* is *cis*-isomer.

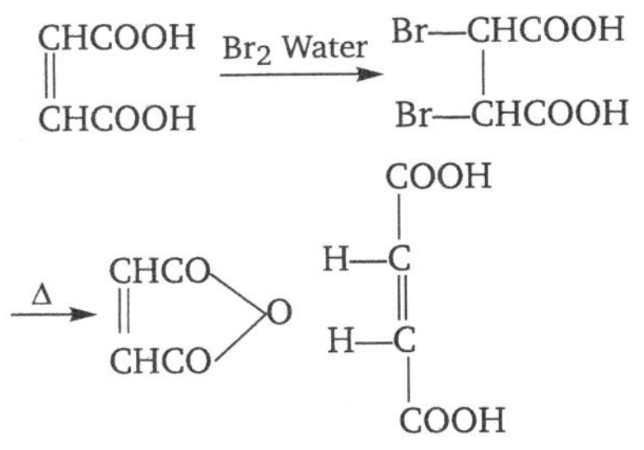

CHCOOH=CHCOOH $\xrightarrow{Br_2 \text{ Water}}$ Br—CHCOOH—Br—CHCOOH

$\xrightarrow{\Delta}$ CHCO=CHCO (cyclic anhydride, —O—)

COOH—C(H)=C(H)—COOH

(*P*) (maleic acid)

$\xrightarrow[\text{dilute alkaline KMnO}_4]{H_2O + O}$ COOH—C(H)(OH)—C(H)(OH)—COOH

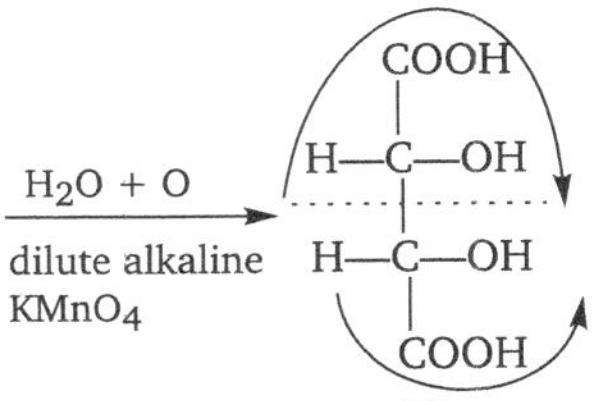

(*S*)

optically inactive due to internal cancellation of rotation (*meso*-isomer)

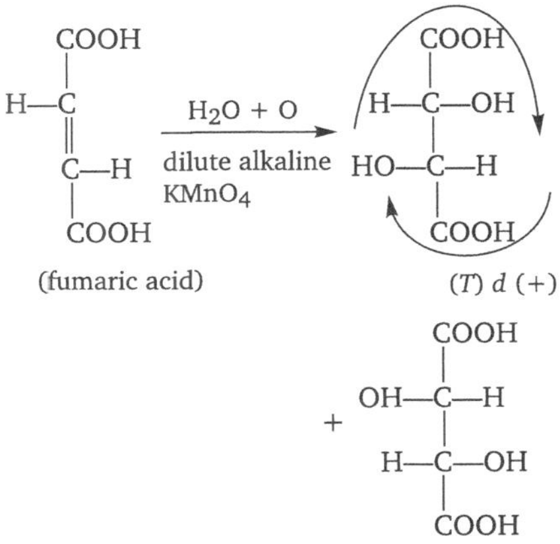

COOH—C(H)=C(H)—COOH (fumaric acid) $\xrightarrow[\text{dilute alkaline KMnO}_4]{H_2O + O}$ COOH—C(H)(OH)—C(HO)(H)—COOH

(*T*) *d* (+)

+ COOH—C(OH)(H)—C(H)(OH)—COOH

(*U*) *l* (–)

*T* and *U* (in 1 : 1 molar ratio) form optically inactive (racemic mixture) due to external cancellation.

**12. Plan** Ni/$H_2$ reduces (C=C) bond.

Benzene undergoes Friedel-Crafts reaction Zn-Hg/HCl reduces carbonyl group (Clemmensen reduction)

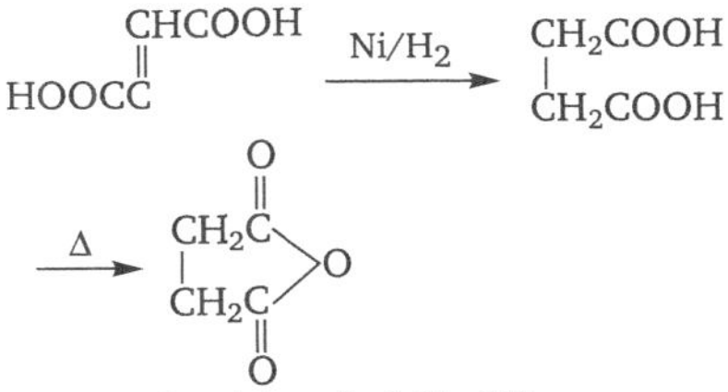

CHCOOH=HOOCC $\xrightarrow{Ni/H_2}$ $CH_2COOH$—$CH_2COOH$

$\xrightarrow{\Delta}$ $CH_2C(=O)$—$CH_2C(=O)$ (cyclic, —O—)

Succinic anhydride (*V*)

Benzene + succinic anhydride $\xrightarrow{\text{anhydrous } AlCl_3}$

$C_6H_5$—C(=O)—$CH_2$—$CH_2COOH$ $\xrightarrow[\text{Clemmensen}]{\text{Zn–Hg/HCl}}$

$C_6H_5$—$CH_2CH_2CH_2$—C(=O)OH $\xrightarrow{H_3PO_4}$ (tetralone, C=O)

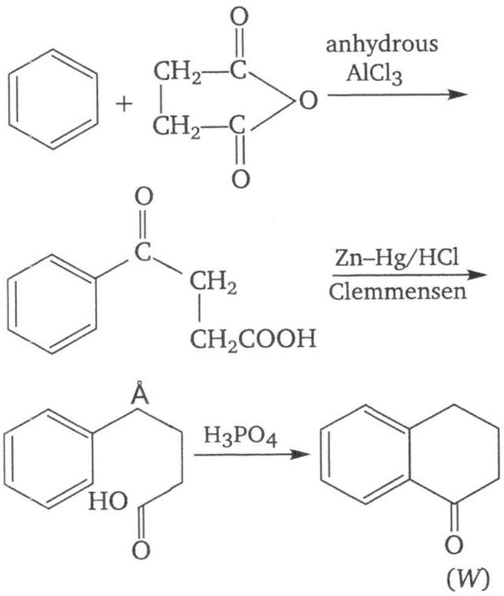

(*W*)

**13. Plan** By Boyle's law at constant temperature $p \propto \frac{1}{V}$

By Charles' law at constant pressure

$V \propto T$

Process taking place at

Constant temperature — isothermal

Constant pressure — isobaric

Constant volume — isochoric

Constant heat — adiabatic

| | | |
|---|---|---|
| $K \longrightarrow L$ | At constant $P$, volume increases | thus, heating |
| $L \longrightarrow M$ | At constant $V$, pressure decreases | thus, cooling |
| $M \longrightarrow N$ | At constant $P$, volume decreases | thus, cooling |
| $N \longrightarrow K$ | At constant $V$, pressure increases | thus, heating |

**14.** $L \longrightarrow M$ At constant $V$— isochoric

$N \longrightarrow K$

**15,16. Plan** $2NaOH + Cl_2 \xrightarrow{\text{cold}} NaCl + \underset{P}{NaOCl} + H_2O$

$$6NaOH + 3Cl_2 \xrightarrow{\text{hot}} 5NaCl + \underset{Q}{NaClO_3} + 3H_2O$$

$$\underset{\text{hypochlorous acid}}{HOCl} \xrightarrow{NaOH} \underset{P}{NaOCl}$$

$$\underset{\text{chloric acid}}{HClO_3} \xrightarrow{NaOH} \underset{Q}{NaClO_3}$$

$$Cl_2 + SO_2 \longrightarrow \underset{R}{SO_2Cl_2}$$

$$10SO_2Cl_2 + P_4 \longrightarrow \underset{S}{4PCl_5} + 10SO_2$$

$$PCl_5 + 4H_2O \longrightarrow \underset{T}{H_3PO_4} + 5HCl$$

**17.** (P) $2PbO_2 + 2H_2SO_4 \xrightarrow{\text{Warm (3)}} 2PbSO_4 + O_2 + 2H_2O$

(Q) $Na_2S_2O_3 + H_2O \xrightarrow{Cl_2 (4)} NaHSO_4 + HCl$

(R) $N_2H_4 \xrightarrow{I_2\ (2)} N_2 + HI$

(S) $XeF_2 \xrightarrow{NO(1)} Xe + NOF$

Thus,P —(3), Q—(4), R—(2), S — (1)

**18.**

| | List I | List II | Explanation |
|---|---|---|---|
| (P) | 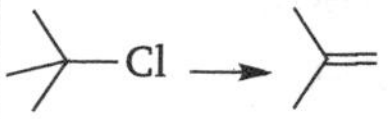 —Cl → | NaOEt(2) | $\bar{O}$Et (strong nucleophile) causes dehydrohaloge-nation of 3° alkyl halide |
| (Q) | 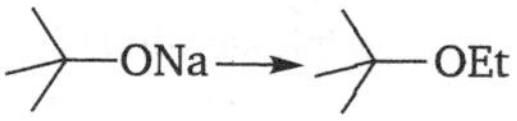 —ONa → —OEt | EtBr (3) | 3° butoxide undergoes $S_N$ reaction with 1° alkyl halide |
| (R) | 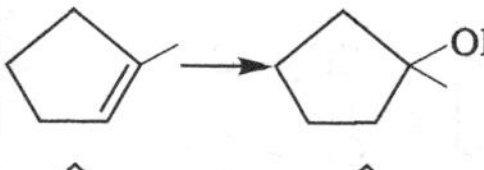 → OH | (i) $Hg(OAC)_2$ (ii) $NaBH_4$ (1) | Mercuration- demercuration adds $H_2O$ by Markownikoff's rule without rearrangement |
| (S) | 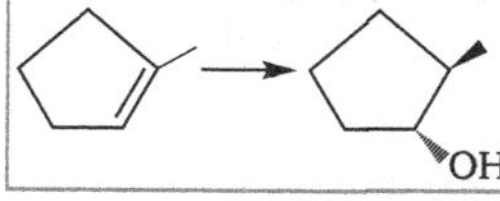 → OH | (i) $BH_3$ (ii) $H_2O_2$ / $OH^-$ (4) | Hydroboro-oxidation adds $H_2O$ by anti-Markowni-koff's rule |

Thus, P — (2), Q — (3), R — (1), S — (4)

**19.**

| | In burette Acid | In flask Base | Curve |
|---|---|---|---|
| I. | Strong | Strong | Conductance first decreases due to formation of $H_2O$ and then increases due to addition of strong electrolyte 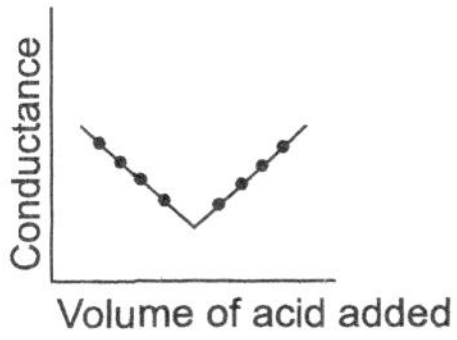 <br> Conductance / Volume of acid added |
| II. | Strong | Weak | Conductance increases slightly as $NH_4^+$ (salt) is hydrolysed forming HCl. After neutral point, it acid increases rapidly due to addition of strong 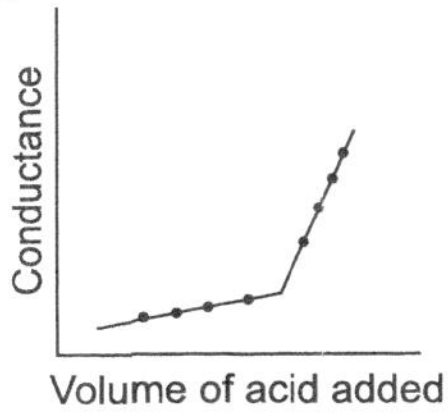 <br> Conductance / Volume of acid added |
| III. | Weak | Strong | Conductivity decreases due to neutralisation of conducting strong base and then remains constant due to addition of weak acid. 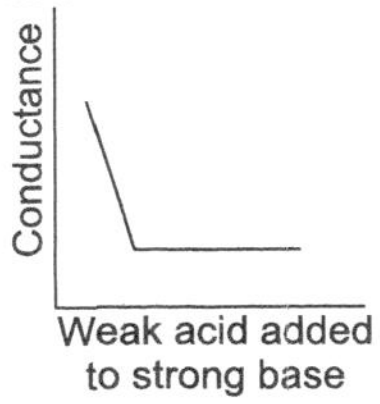 <br> Conductance / Weak acid added to strong base |
| IV. | Weak | Weak | Conductivity increases due to formation of ions and then remains constant due to addition of weak base. |
| V. | *KX* | $AgNO_3$ | Insoluble salt Ag*X* is formed, hence conductance remains constant. It increases due to addition of *KX*. 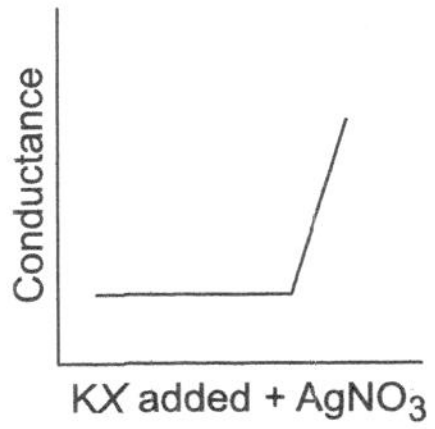 <br> Conductance / *KX* added + $AgNO_3$ |

P — (3), Q — (4), R — (2), S — (1)

**20. Plan** When different number of electrons are involved in a redox reaction

$$\Delta G^\circ_{net} = \Delta G^\circ_1 + \Delta G^\circ_2$$

$$-n_3FE^\circ_3 = -n_1FE^\circ_1 - n_2FE^\circ_2$$

$$\therefore \quad E^\circ_3 = \frac{n_1E^\circ_1 + n_2E^\circ_2}{n_3}$$

(P) $E^\circ_{3+}$ $Fe^{3+}/Fe$

**Net reaction** $Fe^{3+} \longrightarrow Fe$

| is obtained from | $n$ | $E°$ |
|---|---|---|
| $Fe^{3+} + e^- \rightarrow Fe^{2+}$ | $n_1 = 1$ | $E_1^{\circ d} = 0.77$ V |
| $Fe^{2+} + 2e^- \rightarrow Fe$ | $n_2 = 2$ | $E_2^\circ = -0.44$ V |
| $\because Fe^{3+} + 3e^- \rightarrow Fe$ | $n_3 = 3$ | $E_3^\circ = ?$ |

$$E°_3 = \frac{n_1E°_1 + n_2E°_2}{n_3}$$

$$= \frac{0.77 + 2(-0.44)}{3} = \frac{-0.11}{3} = -0.04 \text{ V}$$

Thus, P — (3)

**Net reaction**

$4H_2O \rightleftharpoons 4H^+ + 4OH^-$

| is obtained from | $n$ | $E°$ |
|---|---|---|
| $2H_2O \rightarrow O_2 + 4H^+ + 4e^-$ | $n_1 = 4$ | $-1.23$ V |
| $2H_2O + O_2 + 4e^- \rightarrow 4OH^-$ | $n_2 = 4$ | $+0.40$ V |
| $4H_2O \rightarrow 4H^+ + 4e^-$ | $n_3 = 4$ | ? |

$$E°_3 = \frac{n_1E°_1 + n_2E°_2}{n_3} = E°_1 + E°_2$$

$$= -1.23 + 0.40 = -0.83 \text{ V}$$

Thus, Q — (4)

(R) $Cu^{2+} + Cu \longrightarrow 2Cu^+$ (Oxidation; Reduction)

For thus $E°$ of $Cu^{2+} \longrightarrow Cu^+$

| is also required. | $n$ | $E°$ |
|---|---|---|
| $Cu^{2+} + 2e^- \longrightarrow Cu$ | 2 | 0.34 V |
| $Cu \longrightarrow Cu^+ + e^-$ | 1 | $-0.52$ V |
| $Cu^{2+} + e^- \longrightarrow Cu^+$ | $E°_3$ | ? |

$$E_3° = \frac{n_1E_1° + n_2E_2°}{n_3}$$

$$= \frac{2 \times 0.34 + 1 \times (-0.52)}{1} = 0.16 \text{ V}$$

| Thus, | $n$ | $E°$ |
|---|---|---|
| $Cu \rightarrow Cu^+ + e^-$ | $n_1 = 1$, | $-0.52$ V |
| $Cu^{2+} + e^- \rightarrow Cu^+$ | $n_2 = 1$ | 0.10 V |

$Cu^{2+} + Cu \rightarrow 2Cu^+$

$$E° = -0.52 + 0.16 = -0.36 \text{ V}$$

Thus, (R) — (1)

(S) $Cr^{3+} \longrightarrow Cr^{2+}$

| is obtained from | $n$ | $E°$ |
|---|---|---|
| $Cr^{3+} + 3e^- \rightarrow Cr$ | 3 | $-0.74$ V |
| $Cr \rightarrow Cr^{2+} + 2e^-$ | 2 | $+0.91$ V |
| $Cr^{3+} + e^- \rightarrow Cr^{2+}$ | 1 | ? |

$$E_3° = \frac{-0.74 \times 3 + 2 \times 0.91}{1} = -0.4 \text{ V}$$

Thus, $S = (2)$

P — (3), Q — (4), R — (1), S — (2)

# Mathematics

1. **Concept Involved**

Converting Infinite series into definite Integral

*i.e.*, $\lim_{n\to\infty} \frac{h(n)}{n}$

$$\lim_{n\to\infty} \frac{1}{n} \sum_{r=g(n)}^{h(n)} f\left(\frac{r}{n}\right) = \int f(x)\,dx$$

$$\lim_{n\to\infty} \frac{g(n)}{n}$$

where, $\frac{r}{n}$ is replaced with $x$.

$\Sigma$ is Replaced with Integral.

$$\text{Here, } \lim_{n\to\infty} \frac{1^a + 2^a + \ldots + n^a}{(n+1)^{a-1}\{(na+1) + (na+2) + \ldots + (na+n)\}} = \frac{1}{60}$$

$$\Rightarrow \lim_{n\to\infty} \frac{\sum_{r=1}^{n} r^a}{(n+1)^{a-1} \cdot \left[n^2a + \frac{n(n+1)}{2}\right]} = \frac{1}{60}$$

$$\Rightarrow \lim_{n\to\infty} \frac{2\sum_{r=1}^{n} \left(\frac{r}{n}\right)^a}{\left(1 + \frac{1}{n}\right)^{a-1} \cdot (2na + n + 1)}$$

$$\Rightarrow \lim_{n\to\infty} \frac{1}{n}\left(2\sum_{r=1}^{n}\left(\frac{r}{n}\right)^{a}\right)$$

$$\times \lim_{n\to\infty} \frac{1}{\left(1+\frac{1}{n}\right)^{a-1}\cdot\left(2a+1+\frac{1}{n}\right)}$$

$$\Rightarrow \quad 2\int_0^1 (x^a)\,dx\cdot\frac{1}{1\cdot(2a+1)}$$

$$\Rightarrow \quad \frac{2\cdot(x^{a+1})_0^1}{(2a+1)\cdot(a+1)} = \frac{2}{(2a+1)\cdot(a+1)}$$

$$\therefore \quad \frac{2}{(2a+1)(a+1)} = \frac{1}{60}$$

$$\Rightarrow \quad (2a+1)(a+1) = 120$$

$$\Rightarrow \quad 2a^2 + 3a + 1 - 120 = 0$$

$$\Rightarrow \quad 2a^2 + 3a - 119 = 0$$

$$\Rightarrow \quad (2a+17)(a-7) = 0$$

$$\Rightarrow \quad a = 7, \frac{-17}{2}$$

**2. Concept Involved**

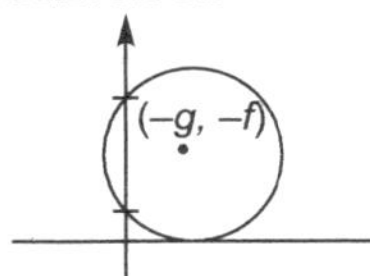


Here, the length of intercept on $y$-axis.

$$\Rightarrow \quad 2\sqrt{f^2 - c}$$

and if circle touches $x$-axis

$$\Rightarrow \quad g^2 = c$$

for $x^2 + y^2 + 2gx + 2fy + c = 0$

Here, $x^2 + y^2 + 2gx + 2fy + c = 0$

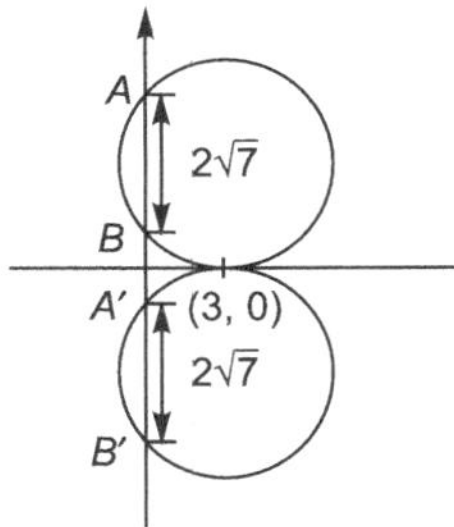


Passes through (3, 0)

$$\Rightarrow \quad 9 + 6g + c = 0 \quad \ldots(\text{i})$$

$$g^2 = c \quad \ldots(\text{ii})$$

and $\quad 2\sqrt{f^2 - c} = 2\sqrt{7}$

$$f^2 - c = 7 \quad \ldots(\text{iii})$$

From Eqs. (i) and (ii), we get

$$g^2 + 6g + 9 = 0$$

$$(g+3)^2 = 0$$

$$g = -3$$

and $\quad c = 9 \quad \therefore \quad f^2 = 16$

$$f = \pm 4$$

$$\therefore \quad x^2 + y^2 - 6x \pm 8y + 9 = 0$$

**3. Concept Involved**

If two straight lines are coplanar

*i.e.*, $\dfrac{x - x_1}{a_1} = \dfrac{y - y_1}{b_1} = \dfrac{z - z_1}{c_1}$ and $\dfrac{x - x_2}{a_2} = \dfrac{y - y_2}{b_2} = \dfrac{z - z_2}{c_2}$ are coplanar

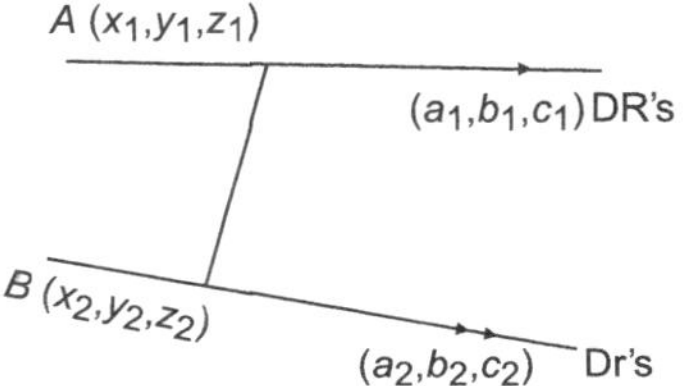


Then, $(x_2 - x_1, y_2 - y_1, z_2 - z_1)$, $(a_1, b_1, c_1)$ and $(a_2, b_2, c_2)$ are coplanar

*i.e.*, $\begin{vmatrix} x_2 - x_1 & y_2 - y_1 & z_2 - z_1 \\ a_1 & b_1 & c_1 \\ a_2 & b_2 & c_2 \end{vmatrix} = 0$

Here, $\quad x = 5, \dfrac{y}{3-\alpha} = \dfrac{z}{-2}$

$$\Rightarrow \quad \frac{x-5}{0} = \frac{y-0}{-(\alpha-3)} = \frac{z-0}{-2} \quad \ldots(\text{i})$$

and $\quad x = \alpha, \dfrac{y}{-1} = \dfrac{z}{2-\alpha}$

$$\Rightarrow \quad \frac{x-\alpha}{0} = \frac{y-0}{-1} = \frac{z-0}{2-\alpha} \quad \ldots(\text{ii})$$

$$\Rightarrow \quad \begin{vmatrix} 5-\alpha & 0 & 0 \\ 0 & 3-\alpha & -2 \\ 0 & -1 & 2-\alpha \end{vmatrix} = 0$$

$$\Rightarrow \quad (5-\alpha)[(3-\alpha)(2-\alpha) - 2] = 0$$

$$\Rightarrow \quad (5-\alpha)[\alpha^2 - 5\alpha + 4] = 0$$

$$\Rightarrow \quad (5-\alpha)(\alpha-1)(\alpha-4) = 0$$

$$\alpha = 1, 4, 5$$

**4. Concept Involved**

When ever cosine of angle and sides are given or to find out, we should always use cosine law.

*i.e.*, $\cos A = \dfrac{b^2 + c^2 - a^2}{2bc}$,

$\cos B = \dfrac{a^2 + c^2 - b^2}{2ac}$

and $\cos C = \dfrac{a^2 + b^2 - c^2}{2ab}$

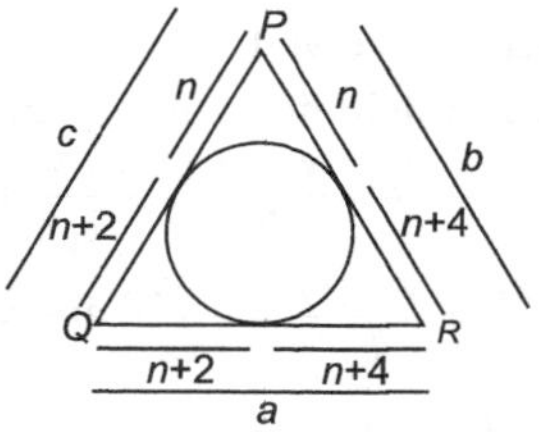

$\therefore \quad \cos P = \dfrac{b^2 + c^2 - a^2}{2bc}$

$\Rightarrow \dfrac{1}{3} = \dfrac{(2n+4)^2 + (2n+2)^2 - (2n+6)^2}{2(2n+4)(2n+2)}$

$\left(\because \cos p = \dfrac{1}{3}, \text{ given}\right)$

$\Rightarrow \dfrac{4n^2 - 16}{8(n+1)(n+2)} = \dfrac{1}{3}$

$\Rightarrow \dfrac{n^2 - 4}{2(n+1)(n+2)} = \dfrac{1}{3}$

$\Rightarrow \dfrac{(n-2)}{2(n+1)} = \dfrac{1}{3}$

$\Rightarrow 3n - 6 = 2n + 2 \Rightarrow n = 8$

$\therefore$ Sides are $2n + 2$, $2n + 4$, $2n + 6$

$\Rightarrow$ 18, 20, 22

**5. Concept Involved**

It is the simple representation of points on argand plane and to find the angle between the points.

Here, $P = W^n = \left(\cos\dfrac{\pi}{6} + i \sin\dfrac{\pi}{6}\right)^n$

$= \cos\dfrac{n\pi}{6} + i \sin\dfrac{n\pi}{6}$

$H_1 = \left\{Z \in C : \text{Re}(z) > \dfrac{1}{2}\right\}$

$\therefore P \cap H_1$ represents those points for which $\cos\dfrac{n\pi}{6}$ is + ve.

$\therefore$ It belongs to I or IV quadrant

$\Rightarrow \quad z_1 = P \cap H_1 = \cos\dfrac{\pi}{6} + i \sin\dfrac{\pi}{6}$

or $\quad \cos\dfrac{11\pi}{6} + i \sin\dfrac{11\pi}{6}$

$\therefore \quad z_1 = \dfrac{\sqrt{3}}{2} + \dfrac{i}{2} \text{ or } \dfrac{\sqrt{3}}{2} - \dfrac{i}{2} \quad \ldots\text{(i)}$

Similarly,

$z_2 = P \cap H_2$ *i.e.*, those points for which $\cos\dfrac{n\pi}{6} < 0$

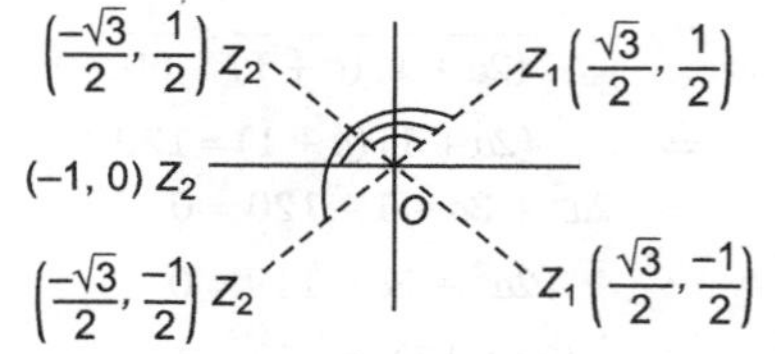

$\therefore z_2 = \cos\pi + i \sin\pi, \cos\dfrac{5\pi}{6} + i \sin\dfrac{5\pi}{6}, \cos\dfrac{7\pi}{6} + i \sin\dfrac{7\pi}{6}$

$\Rightarrow \quad z_2 = -1, \dfrac{-\sqrt{3}}{2} + \dfrac{i}{2}, \dfrac{-\sqrt{3}}{2} - \dfrac{i}{2}$

Thus, $\angle z_1 O z_2 = \dfrac{2\pi}{3}, \dfrac{5\pi}{6}, \pi$

**6.** $3^x = 4^{x-1}$, taking $\log_3$ on both sides

$\Rightarrow \quad x \log_3^3 = (x-1)\log_3^4$

$\Rightarrow \quad x = 2\log_3^2 \cdot x - \log_3^4$

$\Rightarrow \quad x(1 - 2\log_3^2) = -2\log_3^2$

$\Rightarrow \quad x = \dfrac{2\log_3^2}{2\log_3^2 - 1}$

$\Rightarrow x = \dfrac{1}{1 - \dfrac{1}{2\log_3^2}} = \dfrac{1}{1 - \dfrac{1}{\log_3^4}} = \dfrac{1}{1 - \log_4^3}$

$\Rightarrow x = \dfrac{2}{2 - \log_2^3}$

**7.** Here, $P = [p_{ij}]_{n \times n}$ with $p_{ij} = w^{i+j}$

$\therefore$ when $n = 1$

$P = [p_{ij}]_{1\times 1} = [w^2] \Rightarrow P^2 = [w^4] \neq 0$

$\therefore$ when $\quad n = 2$

$P = [p_{ij}]_{2\times 2} = \begin{bmatrix} p_{11} & p_{12} \\ p_{21} & p_{22} \end{bmatrix} = \begin{bmatrix} w^2 & w^3 \\ w^3 & w^4 \end{bmatrix}$

$= \begin{bmatrix} w^2 & 1 \\ 1 & w \end{bmatrix}$

$$P^2 = \begin{bmatrix} w^2 & 1 \\ 1 & w \end{bmatrix}\begin{bmatrix} w^2 & 1 \\ 1 & w \end{bmatrix}$$

$$P^2 = \begin{bmatrix} w^4+1 & w^2+w \\ w^2+w & 1+w^2 \end{bmatrix} \neq 0$$

when $n = 3$

$$P = [p_{ij}]_{3\times 3} = \begin{bmatrix} w^2 & w^3 & w^4 \\ w^3 & w^4 & w^5 \\ w^4 & w^5 & w^6 \end{bmatrix}$$

$$= \begin{bmatrix} w^2 & 1 & w \\ 1 & w & w^2 \\ w & w^2 & 1 \end{bmatrix}$$

$$P^2 = \begin{bmatrix} w^2 & 1 & w \\ 1 & w & w^2 \\ w & w^2 & 1 \end{bmatrix}\begin{bmatrix} w^2 & 1 & w \\ 1 & w & w^2 \\ w & w^2 & 1 \end{bmatrix}$$

$$= \begin{bmatrix} 0 & 0 & 0 \\ 0 & 0 & 0 \\ 0 & 0 & 0 \end{bmatrix} = 0$$

$\therefore$ $P^2 = 0$, when $n$ is multiple of 3.

$P^2 \neq 0$, when $n$ is not a multiple of 3.

$\Rightarrow$ $n = 57$ is not possible

$\therefore$ $n = 55, 58, 56$ is possible

**8. Concept Involved**

We know, $|x| = \begin{cases} x, & x \geq 0 \\ -x, & x < 0 \end{cases}$

$\Rightarrow$ $|x-a| = \begin{cases} x-a, & x \geq a \\ -(x-a), & x < a \end{cases}$

and for non-differentiable continuous function the maximum or minimum. Can be checked with graph as.

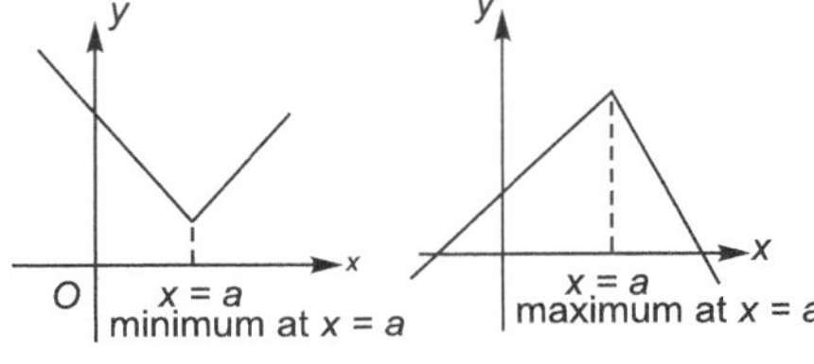


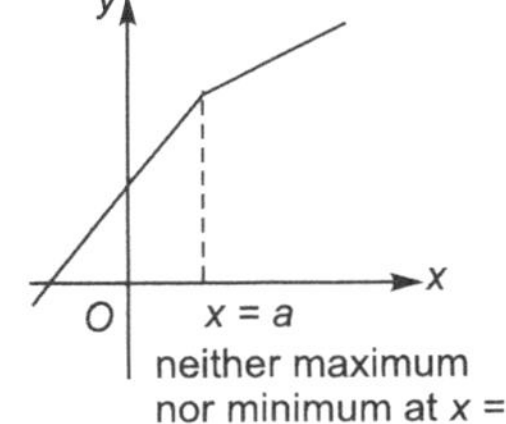


Here, $f(x) = 2|x| + |x+2| - ||x+2| - 2|x||$

$$= \begin{cases} -2x-(x+2)+(x-2), & \text{when } x \leq -2 \\ -2x+x+2+3x+2, & \text{when } -2 < x \leq -\frac{2}{3} \\ -4x, & \text{when } -\frac{2}{3} < x \leq 0 \\ 4x, & \text{when } 0 < x \leq 2 \\ 2x+4, & \text{when } x > 2 \end{cases}$$

$$= \begin{cases} -2x-4, & x \leq -2 \\ 2x+4, & -2 < x \leq -2/3 \\ -4x, & -\frac{2}{3} < x \leq 0 \\ 4x, & 0 < x \leq 2 \\ 2x+4, & x > 2 \end{cases}$$

Graph for $y = f(x)$ is shown as

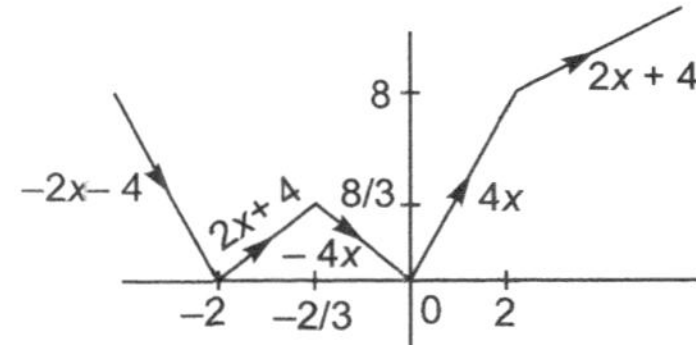


**Solutions** (Q. Nos. 9-10)

**Concept Involved** It is based on the concept of converting into total differential equation (*i.e.*, completing the equation into differential). So, as to check the function to be increasing or decreasing.

**9.** Here, $f''(x) - 2f'(x) + f(x) \geq e^x$

$\Rightarrow$ $f''(x)e^{-x} - f'(x)e^{-x} - f'(x)e^{-x} + f(x)e^{-x} \geq 0$

$\Rightarrow$ $\frac{d}{dx}(f'(x)e^{-x}) - \frac{d}{dx}(f(x)e^{-x}) \geq 1$

$\Rightarrow$ $\frac{d}{dx}(f'(x)e^{-x} - f(x)e^{-x}) \geq 1$

$\Rightarrow$ $\frac{d^2}{dx^2}(e^{-x}f(x)) \geq 1$ for all $x \in [0, 1]$

$\therefore$ $\phi(x) = e^{-x}f(x)$ is concave $n\phi$.

$f(0) = f(1) = 0$

$\Rightarrow$ $\phi(0) = 0 = \phi(1)$

$\Rightarrow$ $\phi(x) < 0$

$\Rightarrow$ $e^{-x}f(x) < 0$

$\therefore$ $f(x) < 0$

**10.** Here, $\phi'(x) < 0,\ x \in \left(0, \frac{1}{4}\right)$

and $\phi'(x) > 0,\ x \in \left(\frac{1}{4}, 1\right)$

$\Rightarrow e^{-x} f'(x) - e^{-x} f(x) < 0, x \in \left(0, \frac{1}{4}\right)$

$\Rightarrow \quad f'(x) < f(x),\ 0 < x < \frac{1}{4}\cdot$

**Solutions** (Q. Nos. 11-12)

**Concept Involved**

Intersection point of tangents at $(at_1^2, 2at_1)$ and $(at_2^2, 2at_2)$ is $(at_1t_2, a/t_1 + t_2))$, also tangents drawn at end point of focal chord are perpendicular and intersect on directrix.

**11.** Since, $R\left(-a, a\left(t - \frac{1}{t}\right)\right)$ lies on $y = 2x + a$

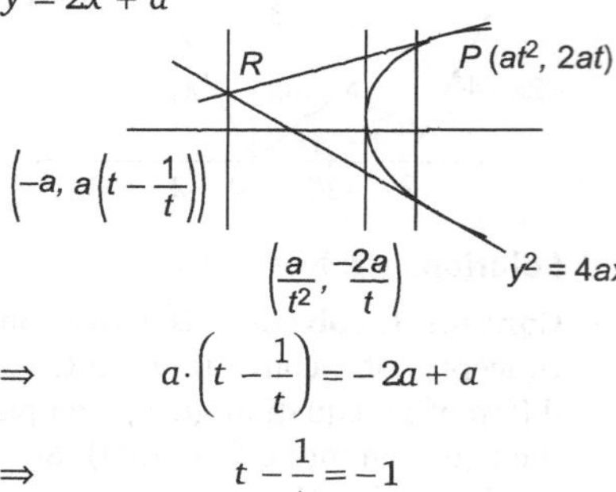


$\Rightarrow \quad a\cdot\left(t - \frac{1}{t}\right) = -2a + a$

$\Rightarrow \quad t - \frac{1}{t} = -1$

Thus, length of focal chord

$$= a\left(t + \frac{1}{t}\right)^2 = a\left\{\left(t - \frac{1}{t}\right)^2 + 4\right\} = 5a$$

**12.** $m_{OP} = \frac{2at - 0}{at^2 - 0} = \frac{2}{t}$

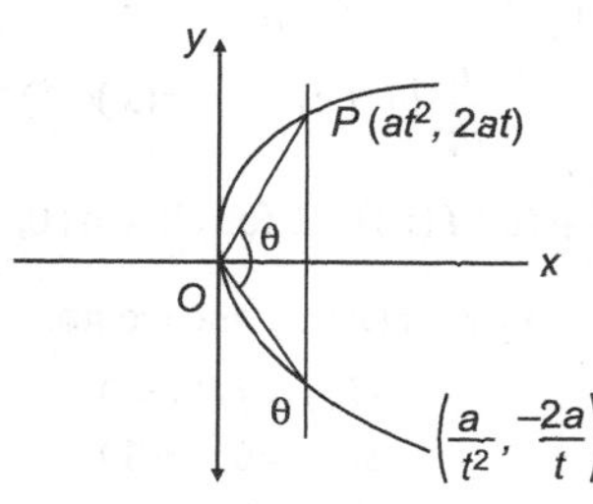


$$m_{OQ} = \frac{-2a/t}{a/t^2} = -2t$$

$\therefore \quad \tan\theta = \frac{\frac{2}{t} + 2t}{1 - \frac{2}{t}\cdot 2t} = \frac{2\left(t + \frac{1}{t}\right)}{1 - 4},$

where $\quad t + \frac{1}{t} = \sqrt{5} = \frac{2\sqrt{5}}{-3}$

**Solutions** (Q. Nos. 13-14)

Here, $\quad S = S_1 \cap S_2 \cap S_3$

$S_1 = \{Z \in C : |z| < 4\}$

$\therefore \quad S_1 : x^2 + y^2 < 4 \quad \ldots$(i)

and $S_2 = \left\{Z \in C : \text{In}\left(\frac{z - 1 + \sqrt{3}i}{1 - \sqrt{3}i}\right) > 0\right\}$

where,

$$\frac{z - 1 + i\sqrt{3}}{1 - i\sqrt{3}} = \frac{(x-1) + i(y + \sqrt{3})}{1 - i\sqrt{3}}$$

$$= \frac{(x-1)(1 + i\sqrt{3}) + i(y + \sqrt{3})(1 + i\sqrt{3})}{1 + 3}$$

$$= \frac{(x-1) + i\sqrt{3}(x-1) + i(y + \sqrt{3}) - \sqrt{3}(y + \sqrt{3})}{4}$$

$$= \frac{(x - 1 - \sqrt{3}y - 3)}{4} + \frac{i(\sqrt{3}x + y)}{4}$$

$\therefore \quad S_2 : \sqrt{3}x + y > 0 \quad \ldots$(ii)

$S_3 : \{Z \in C : \text{Re } Z > 0\}$

$\therefore \quad S_3 : x > 0 \quad \ldots$(iii)

**13.** Since, $S = S_1 \cap S_2 \cap S_3$

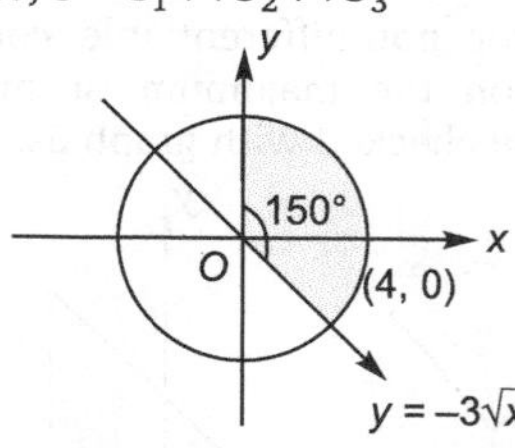


Clearly, the shaded region represents the area of sector

$$\therefore \quad S = \frac{1}{2} r^2\theta = \frac{1}{2} \times 4^2 \times \frac{5\pi}{6} = \frac{20\pi}{3}$$

**14.** $\min_{Z \in S} |1 - 3i - z|$

= perpendicular distance of point (1, − 3) From the line $\sqrt{3}x + y = 0$

$$\Rightarrow \quad \frac{|\sqrt{3} - 3|}{\sqrt{3 + 1}} = \frac{3 - \sqrt{3}}{2}$$

**15.**

| $B_1$ | $B_2$ | $B_3$ |
|---|---|---|
| 1 whilte | 2 whilte | 3 whilte |
| 2 Red | 3 Red | 4 Red |
| 2 Black | 4 Black | 5 Black |

$P$ (All 3 drawn balls are of the same colour)

$$= P(www) + P(RRR) + P(BBB)$$

$$= \left(\frac{1}{6} \times \frac{2}{9} \times \frac{3}{12}\right) + \left(\frac{3}{6} \times \frac{3}{9} \times \frac{4}{12}\right) + \left(\frac{2}{6} \times \frac{4}{9} \times \frac{5}{12}\right) = \frac{82}{648}$$

**16.** $P$(Ball drawn from box 2/one is white and one is red)

$$= \frac{P(A \cap B)}{P(B)}$$

$$= \frac{\frac{1}{3} \times \frac{2 \times 3}{{}^9C_2}}{\frac{1}{3}\left[\frac{1 \times 3}{{}^6C_2} + \frac{2 \times 3}{{}^9C_2} + \frac{3 \times 4}{{}^{12}C_2}\right]}$$

$$= \frac{\frac{1}{6}}{\frac{1}{5} + \frac{1}{6} + \frac{2}{11}} = \frac{\frac{1}{6}}{\frac{181}{55 \times 60}} = \frac{55}{181}$$

**17.** (P) Here, innermost function is inverse

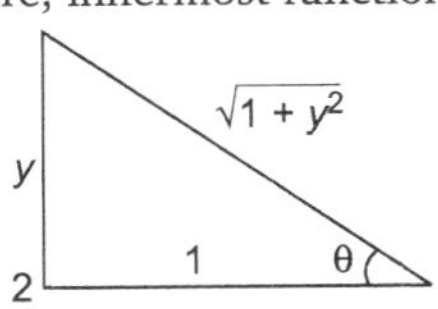


$\therefore$ put, $\tan^{-1} y = \theta \Rightarrow \tan\theta = y$

$$\Rightarrow \left[\frac{1}{y^2} \cdot \left(\frac{\cos(\tan^{-1} y) + y \sin(\tan^{-1} y)}{\cot(\tan^{-1} y) + \tan(\sin^{-1} y)}\right)^2 + y^4\right]^{1/2}$$

$$\Rightarrow \left[\frac{1}{y^2}\left\{\frac{\frac{1}{\sqrt{1+y^2}} + \frac{y^2}{\sqrt{1+y^2}}}{\frac{\sqrt{1-y^2}}{y} + \frac{y}{\sqrt{1-y^2}}}\right\}^2 + y^4\right]^{1/2}$$

$$\Rightarrow \left[\frac{1}{y^2} \cdot y^2(1 - y^4) + y^4\right]^{1/2} = 1$$

$$\cos x + \cos y = -\cos z$$

$$\sin x + \sin y = -\sin z$$

On squaring and adding, we get

(Q) $\cos^2 x + \sin^2 x + \cos^2 y + \sin^2 y + 2\cos x \cos y + 2\sin x \sin y = 1$

$$\Rightarrow 2 + 2(\cos(x - y)) = 1$$

$$\Rightarrow \cos(x - y) = -\frac{1}{2}$$

$$\Rightarrow 2\cos^2\left(\frac{x-y}{2}\right) - 1 = -\frac{1}{2}$$

$$\Rightarrow 2\cos^2\left(\frac{x-y}{2}\right) = \frac{1}{2}$$

$$\Rightarrow \cos\left(\frac{x-y}{2}\right) = \frac{1}{2}$$

(R) $\cos 2x \cdot \left(\cos\left(\frac{\pi}{4} - x\right) - \cos\left(\frac{\pi}{4} + x\right)\right) + 2\sin^2 x$

$$= 2\sin x \cdot \cos x$$

$$\Rightarrow \cos 2x \cdot (\sqrt{2} \sin x) + 2\sin^2 x = 2\sin x \cdot \cos x$$

$$\Rightarrow \sqrt{2}\sin x[\cos 2x + \sqrt{2}\sin x - \sqrt{2}\cos x] = 0$$

$$\Rightarrow \sin x = 0, (\cos x - \sin x)(\cos x + \sin x - \sqrt{2}) = 0$$

$$\Rightarrow \sec x = 1 \text{ or } \tan x = 1$$

$$\Rightarrow \sec x = 1 \text{ or } \frac{1}{\sqrt{2}}$$

(S) $\cot(\sin^{-1}\sqrt{1 - x^2}) = \sin(\tan^{-1}(x\sqrt{6}))$

$$\Rightarrow \frac{x}{\sqrt{1 - x^2}} = \frac{x\sqrt{6}}{\sqrt{1 + 6x^2}}$$

$$\Rightarrow 1 + 6x^2 = 6 - 6x^2$$

$$\Rightarrow 12x^2 = 5$$

$$x = \sqrt{\frac{5}{12}} = \frac{\sqrt{5}}{2\sqrt{3}}$$

(P)$\rightarrow$ 4, (Q) $\rightarrow$ 3, (R) $\rightarrow$ 2 or 4, (S) $\rightarrow$ 1

**18.** Here, $y^2 = 16x, 0 \le y \le 6$

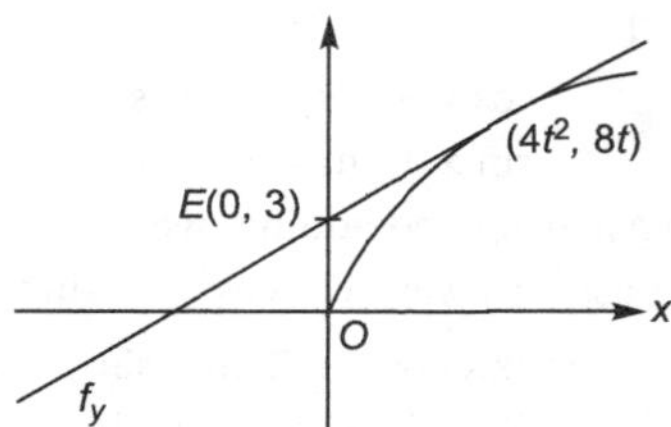


Tangent at $F$,

$$yt = x + at^2$$

at $\quad x = 0, y = at = 4t$

Also, $(4t^2, 8t)$ satisfy

$$y = mx + c$$

$$\Rightarrow \quad 8t = 4mt^2 + 3$$

$$\Rightarrow \quad 4mt^2 - 8t + 3 = 0$$

$$\therefore \quad \text{Area of } \Delta = \frac{1}{2}\begin{vmatrix} 0 & 3 & 1 \\ 0 & 4t & 1 \\ 4t^2 & 8t & 1 \end{vmatrix}$$

$$= \frac{1}{2} \cdot 4t^2(3 - 4t)$$

$$A = 2[3t^2 - 4t^3]$$

$$\therefore \quad \frac{dA}{dt} = 2[6t - 12t^2]$$

$$= -12t(2t - 1)$$

$-$ (left of 0), $+$ (between 0 and $\frac{1}{2}$), $-$ (right of $\frac{1}{2}$)

$\therefore$ Maximum at $t = \frac{1}{2}$

and $4mt^2 - 8t + 3 = 0$

$$\Rightarrow \quad m - 4 + 3 = 0$$

$$\Rightarrow \quad m = 1$$

$$G(0, 4t) \Rightarrow G(0, 2)$$

$$y_1 = 2$$

$$(x_0, y_0) = (4t^2, 8t) = (1, 4)$$

$$y_0 = 4$$

$$\text{Area } = 2\left(\frac{3}{4} - \frac{1}{2}\right) = \frac{1}{2}$$

**19.** (P) Given, $[\mathbf{a}\ \mathbf{b}\ \mathbf{c}] = 2$

$\therefore$ Volume of

$$2(\mathbf{a} \times \mathbf{b}), 3(\mathbf{b} \times \mathbf{c}), (\mathbf{c} \times \mathbf{a})$$

$$\Rightarrow \quad 6[\mathbf{a} \times \mathbf{b}\ \ \mathbf{b} \times \mathbf{c}\ \ \mathbf{c} \times \mathbf{a}]$$

$$\Rightarrow \quad 6[\mathbf{a}\ \mathbf{b}\ \mathbf{c}]^2$$

$$\Rightarrow \quad 6 \times 4 = 24$$

(Q) $[\mathbf{a}\ \mathbf{b}\ \mathbf{c}] = 5$, given

$\therefore$ Volume of

$$3(\mathbf{a} + \mathbf{b}), (\mathbf{b} + \mathbf{c}), 2(\mathbf{c} + \mathbf{a})$$

$$\Rightarrow \quad 6[\mathbf{a} + \mathbf{b}\ \ \mathbf{b} + \mathbf{c}\ \ \mathbf{c} + \mathbf{a}]$$

$$\Rightarrow \quad 6 \times 2[\mathbf{a}\ \mathbf{b}\ \mathbf{c}]$$

$$\Rightarrow \quad 12 \times 5 = 60$$

(R) $\frac{1}{2}|\mathbf{a} \times \mathbf{b}| = 20$, given

$$\therefore \quad \Delta_1 = \frac{1}{2}|(2\mathbf{a} + 3\mathbf{b}) \times (\mathbf{a} - \mathbf{b})|$$

$$= \frac{1}{2}|2\mathbf{a} \times \mathbf{a} - 2\mathbf{a} \times \mathbf{b} + 3\mathbf{b} \times \mathbf{a} - 3\mathbf{b} \times \mathbf{b}|$$

$$= \frac{1}{2}|2\mathbf{b} \times \mathbf{a} + 3\mathbf{b} \times \mathbf{a}| = \frac{5}{2}|\mathbf{a} \times \mathbf{b}|$$

$$= 5 \times 20 = 100$$

(S) Given, $|\mathbf{a} \times \mathbf{b}| = 30$

$$\therefore |(\mathbf{a} + \mathbf{b}) \times \mathbf{a}| = |\mathbf{a} \times \mathbf{b} + \mathbf{b} \times \mathbf{a}|$$

$$= |\mathbf{b} \times \mathbf{a}|$$

$$= |\mathbf{a} \times \mathbf{b}| = 30$$

**20.** $L_1 : \frac{x - 1}{2} = \frac{y - 0}{-1} = \frac{z - (-3)}{1}$

$$\text{Normal of plane } P : \mathbf{n} = \begin{vmatrix} \hat{\mathbf{i}} & \hat{\mathbf{j}} & \hat{\mathbf{k}} \\ 7 & 1 & 2 \\ 3 & 5 & -6 \end{vmatrix}$$

$$= \hat{\mathbf{i}}(-16) - \hat{\mathbf{j}}(-42 - 6) + \hat{\mathbf{k}}(32)$$

$$= -16\hat{\mathbf{i}} + 48\hat{\mathbf{j}} + 32\hat{\mathbf{k}}$$

DR's of normal $\mathbf{n} = \hat{\mathbf{i}} - 3\hat{\mathbf{j}} - 2\hat{\mathbf{k}}$

Point of intersection of $L_1$ and $L_2$.

$\Rightarrow 2K_1 + 1 = K_2 + 4$ and $-k_1 = k_2 - 3$

$\therefore \quad k_1 = 2$ and $k_2 = 1$

$\therefore$ Point of intersection $(5, -2, -1)$

$\therefore$ Equation of plane,

$$1 \cdot (x - 5) - 3(y + 2) - 2(z + 1) = 0$$

$$\Rightarrow \quad x - 3y - 2z - 13 = 0$$

$$\Rightarrow \quad x - 3y - 2z = 13$$

$$\therefore \quad a = 1, b = -3, c = -2, d = 13$$

# IIT JEE
## SOLVED PAPER 2012

### Physics

# Paper 1

## Section I

**Directions** (Q. Nos. 1-10) *This section contains* 10 *multiple choice questions. Each question has four choices (a), (b), (c) and (d) out of which only one is correct.*

1. Consider a thin spherical shell of radius $R$ with its centre at the origin, carrying uniform positive surface charge density. The variation of the magnitude of the electric field $|\mathbf{E}(r)|$ and the electric potential $V(r)$ with the distance $r$ from the centre, is best represented by which graph?

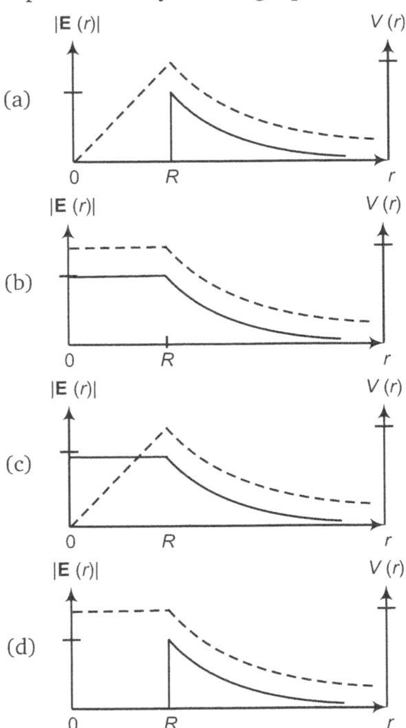


2. Young's double slit experiment is carried out by using green, red and blue light, one colour at a time. The fringe widths recorded are $\beta_G, \beta_R$ and $\beta_B$ respectively. Then,
(a) $\beta_G > \beta_B > \beta_R$
(b) $\beta_B > \beta_G > \beta_R$
(c) $\beta_R > \beta_B > \beta_G$
(d) $\beta_R > \beta_G > \beta_B$

3. Two large vertical and parallel metal plates having a separation of 1 cm are connected to a DC voltage source of potential difference $X$. A proton is released at rest midway between the two plates. It is found to move at 45° to the vertical just after release. Then $X$ is nearly
(a) $1 \times 10^{-5}$ V
(b) $1 \times 10^{-7}$ V
(c) $1 \times 10^{-9}$ V
(d) $1 \times 10^{-10}$ V

4. A bi-convex lens is formed with two thin plano-convex lenses as shown in the figure. Refractive index $n$ of the first lens is 1.5 and that of the second lens is 1.2. Both the curved surfaces are of the same radius of curvature $R = 14$ cm. For this bi-convex lens, for an object distance of 40 cm, the image distance will be

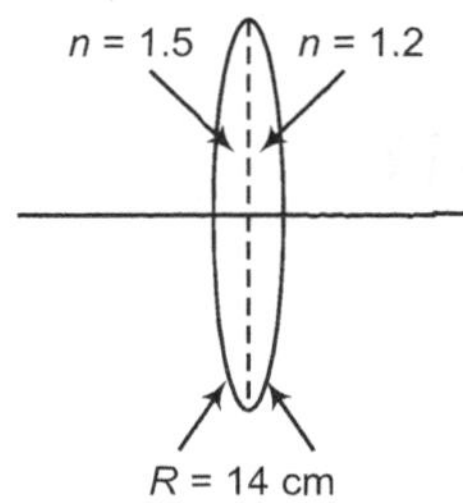


(a) – 280.0 cm (b) 40.0 cm
(c) 21.5 cm (d) 13.3 cm

**5.** A small mass $m$ is attached to a massless string whose other end is fixed at $P$ as shown in the figure. The mass is undergoing circular motion in the $x$-$y$ plane with centre at $O$ and constant angular speed $\omega$. If the angular momentum of the system, calculated about $O$ and $P$ are denoted by $\mathbf{L}_O$ and $\mathbf{L}_P$ respectively, then

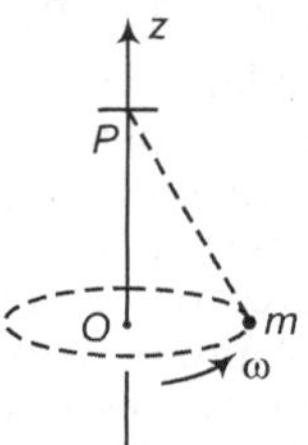


(a) $\mathbf{L}_O$ and $\mathbf{L}_P$ do not vary with time
(b) $\mathbf{L}_O$ varies with time while $\mathbf{L}_P$ remains constant
(c) $\mathbf{L}_O$ remains constant while $\mathbf{L}_P$ varies with time
(d) $\mathbf{L}_O$ and $\mathbf{L}_P$ both vary with time

**6.** A mixture of 2 moles of helium gas (atomic mass = 4 amu) and 1 mole of argon gas (atomic mass = 40 amu) is kept at 300 K in a container. The ratio of the rms speeds $\left(\frac{v_{rms}\text{ (helium)}}{v_{rms}\text{ (argon)}}\right)$ is

(a) 0.32
(b) 0.45
(c) 2.24
(d) 3.16

**7.** A thin uniform rod, pivoted at $O$, is rotating in the horizontal plane with constant angular speed $\omega$, as shown in the figure.

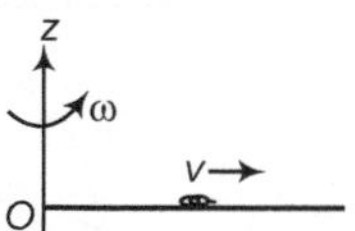


At time $t = 0$, a small insect starts from $O$ and moves with constant speed $v$ with respect to the rod towards the other end. It reaches the end of the rod at $t = T$ and stops. The angular speed of the system remains $\omega$ throughout. The magnitude of the torque $|\vec{\tau}|$ on the system about $O$, as a function of time is best represented by which plot?

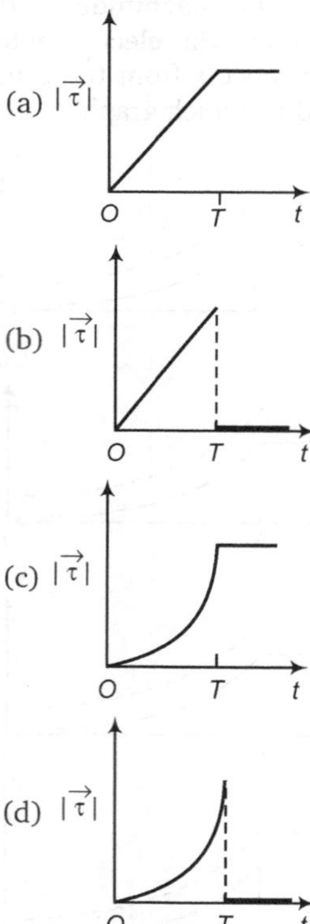

**8.** In the determination of Young's modulus $\left(Y = \frac{4MLg}{\pi l d^2}\right)$ by using Searle's method, a wire of length $L = 2$ m and diameter $d = 0.5$ mm is used. For a load $M = 2.5$ kg, an extension $l = 0.25$ mm in the length of the wire is observed. Quantities $d$ and $l$ are measured using a screw gauge and a micrometer, respectively. They have the same pitch of 0.5 mm. The number of divisions on their circular scale is 100. The contributions to the maximum probable error of the $Y$ measurement is

(a) due to the errors in the measurements of $d$ and $l$ are the same

(b) due to the error in the measurement of $d$ is twice that due to the error in the measurement of $l$.

(c) due to the error in the measurement of $l$ is twice that due to the error in the measurement of $d$.

(d) due to the error in the measurement of $d$ is four times that due to the error in the measurement of $l$.

**9.** A small block is connected to one end of a massless spring of unstretched length 4.9 m. The other end of the spring (see the figure) is fixed. The system lies on a horizontal frictionless surface. The block is stretched by 0.2 m and released from rest at $t = 0$. It then executes simple harmonic motion with angular frequency $\omega = \frac{\pi}{3}$ rad/s. Simultaneously at $t = 0$, a small pebble is projected with speed $v$ from point $P$ at an angle of 45° as shown in the figure. Point $P$ is at a horizontal distance of 10 m from $O$. If the pebble hits the block at $t = 1$ s, the value of $v$ is (take $g = 10$ m/s$^2$)

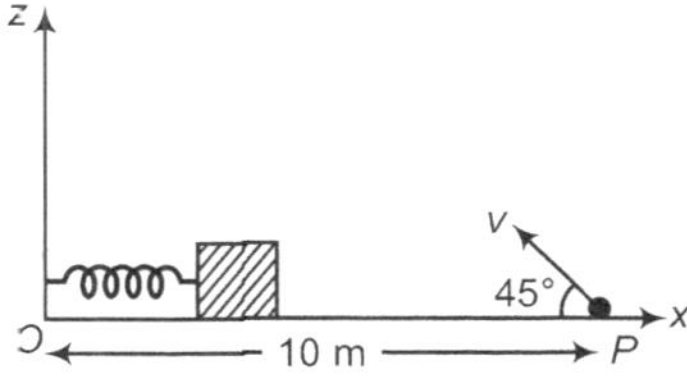


(a) $\sqrt{50}$ m/s

(b) $\sqrt{51}$ m/s

(c) $\sqrt{52}$ m/s

(d) $\sqrt{53}$ m/s

**10.** Three very large plates of same area are kept parallel and close to each other. They are considered as ideal black surfaces and have very high thermal conductivity. The first and third plates are maintained at temperatures $2T$ and $3T$ respectively. The temperature of the middle (*i.e.*, second) plate under steady state condition is

(a) $\left(\frac{65}{2}\right)^{\frac{1}{4}} T$

(b) $\left(\frac{97}{4}\right)^{\frac{1}{4}} T$

(c) $\left(\frac{97}{2}\right)^{\frac{1}{4}} T$

(d) $(97)^{\frac{1}{4}} T$

## Section II

**Directions** (Q. Nos. 11-15) *This section contains 5 multiple choice questions. Each question has four choices (a), (b), (c) and (d) out of which one or more are correct.*

**11.** For the resistance network shown in the figure, choose the correct option(s).

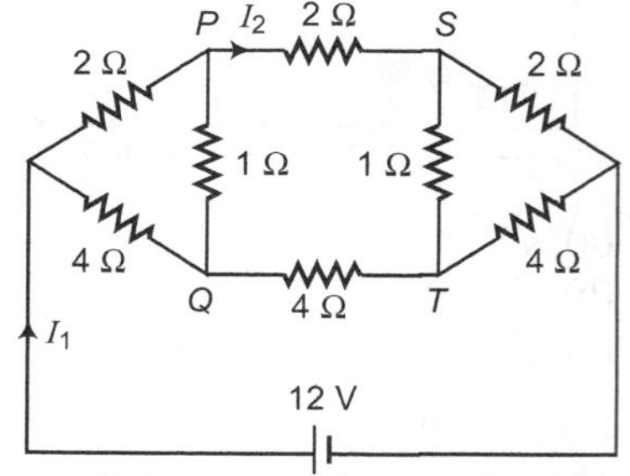


(a) The current through $PQ$ is zero
(b) $I_1 = 3\text{ A}$
(c) The potential at $S$ is less than that at $Q$
(d) $I_2 = 2\text{ A}$

**12.** A person blows into open-end of a long pipe. As a result, a high-pressure pulse of air travels down the pipe. When this pulse reaches the other end of the pipe,

(a) a high-pressure pulse starts travelling up the pipe, if the other end of the pipe is open
(b) a low-pressure pulse starts travelling up the pipe, if the other end of the pipe is open
(c) a low-pressure pulse starts travelling up the pipe, if the other end of the pipe is closed
(d) a high-pressure pulse starts travelling up the pipe, if the other end of the pipe is closed

**13.** Consider the motion of a positive point charge in a region where there are simultaneous uniform electric and magnetic fields $\mathbf{E} = E_0\,\hat{\mathbf{j}}$ and $\mathbf{B} = B_0\,\hat{\mathbf{j}}$. At time $t = 0$, this charge has velocity $\mathbf{v}$ in the $x$-$y$ plane, making an angle $\theta$ with the $x$-axis. Which of the following option(s) is(are) correct for time $t > 0$?

(a) If $\theta = 0°$, the charge moves in a circular path in the $x$-$z$ plane.
(b) If $\theta = 0°$, the charge undergoes helical motion with constant pitch along the $y$-axis
(c) If $\theta = 10°$, the charge undergoes helical motion with its pitch increasing with time, along the $y$-axis.
(d) If $\theta = 90°$, the charge undergoes linear but accelerated motion along the $y$-axis.

**14.** A cubical region of side $a$ has its centre at the origin. It encloses three fixed point charges, $-q$ at $(0, -a/4, 0)$, $+3q$ at $(0, 0, 0)$ and $-q$ at $(0, +a/4, 0)$. Choose the correct option(s).

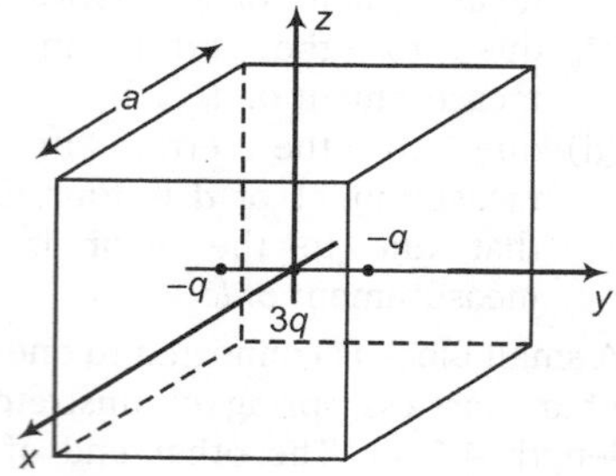


(a) The net electric flux crossing the plane $x = +a/2$ is equal to the net electric flux crossing the plane $x = -a/2$
(b) The net electric flux crossing the plane $y = +a/2$ is more than the net electric flux crossing the plane $y = -a/2$
(c) The net electric flux crossing the entire region is $q/\varepsilon_0$
(d) The net electric flux crossing the plane $z = +a/2$ is equal to the net electric flux crossing the plane $x = +a/2$

15. A small block of mass of 0.1 kg lies on a fixed inclined plane $PQ$ which makes an angle $\theta$ with the horizontal. A horizontal force of 1 N acts on the block through its centre of mass as shown in the figure. The block remains stationary if (take $g = 10\ m/s^2$)

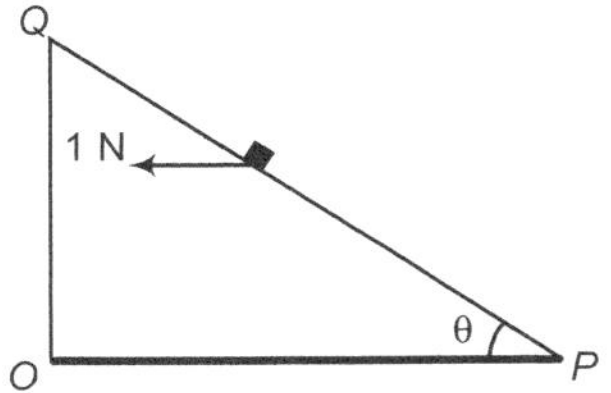


(a) $\theta = 45°$.
(b) $\theta > 45°$ and a frictional force acts on the block towards $P$
(c) $\theta > 45°$ and a frictional force acts on the block towards $Q$
(d) $\theta < 45°$ and a frictional force acts on the block towards $Q$

## Section III

**Directions** (Q. Nos. 16-20) *This section contains 5 questions. The answer to each question is a single digit integer, ranging from 0 to 9 (both inclusive).*

16. A cylindrical cavity of diameter $a$ exists inside a cylinder of diameter $2a$ as shown in the figure. Both the cylinder and the cavity are infinitely long. A uniform current density $J$ flows along the length. If the magnitude of the magnetic field at the point $P$ is given by $\frac{N}{12}\mu_0 aJ$, then the value of $N$ is

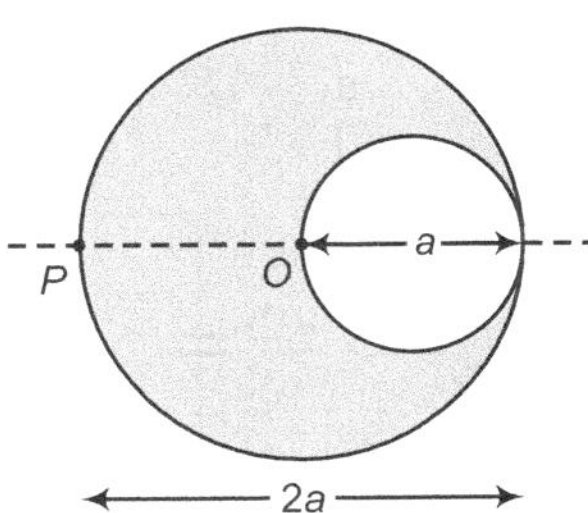


17. A proton is fired from very far away towards a nucleus with charge $Q = 120\ e$, where $e$ is the electronic charge. It makes a closest approach of 10 fm to the nucleus. The de-Broglie wavelength (in units of fm) of the proton at its start is

[Take the proton mass, $m_p = (5/3) \times 10^{-27}$ kg;
$h/e = 4.2 \times 10^{-15}$ J-s/C;
$\frac{1}{4\pi\varepsilon_0} = 9 \times 10^9$ m/F; 1 fm $= 10^{-15}$ m]

18. A circular wire loop of radius $R$ is placed in the $x$-$y$ plane centred at the origin $O$. A square loop of side $a$ ($a << R$) having two turns is placed with its centre at $z = \sqrt{3}\,R$ along the axis of the circular wire loop, as shown in figure. The plane of the square loop makes an angle of 45° with respect to the $z$-axis. If the mutual inductance between the loops is given by $\frac{\mu_0 a^2}{2^{p/2} R}$, then the value of $p$ is

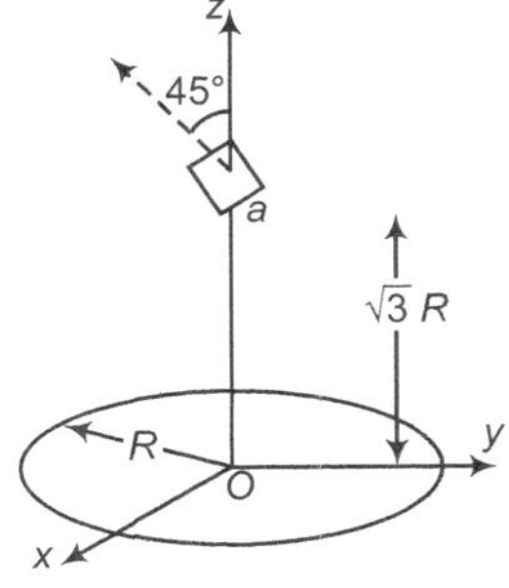

**19.** An infinitely long solid cylinder of radius $R$ has a uniform volume charge density $\rho$. It has a spherical cavity of radius $R/2$ with its centre on the axis of the cylinder, as shown in the figure. The magnitude of the electric field at the point $P$, which is at a distance $2R$ from the axis of the cylinder, is given by the expression $\frac{23\rho R}{16k\varepsilon_0}$. The value of $k$ is

**20.** A lamina is made by removing a small disc of diameter $2R$ from a bigger disc of uniform mass density and radius $2R$, as shown in the figure. The moment of inertia of this lamina about axes passing through $O$ and $P$ is $I_O$ and $I_P$, respectively. Both these axes are perpendicular to the plane of the lamina. The ratio $\frac{I_P}{I_O}$ to the nearest integer is

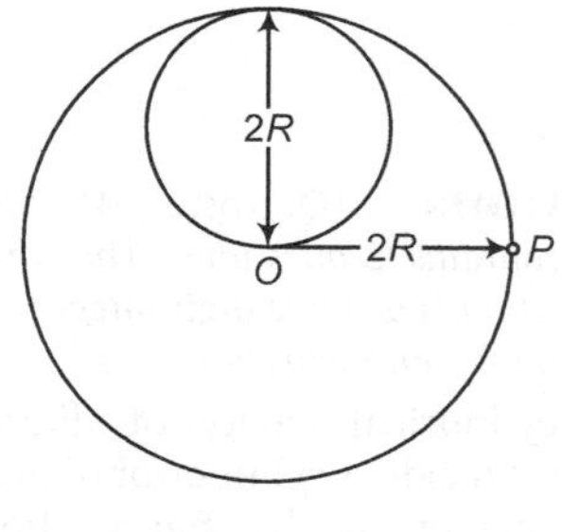


# Paper 2

## Section I

**Directions** (Q. Nos. 21-28) *This section contains 8 multiple choice questions. Each question has four choices (a), (b), (c) and (d) out of which only one is correct.*

**21.** A thin uniform cylindrical shell, closed at both ends, is partially filled with water. It is floating vertically in water in half-submerged state. If $\rho_c$ is the relative density of the material of the shell with respect to water, then the correct statement is that the shell is

(a) more than half-filled if $\rho_c$ is less than 0.5

(b) more than half-filled if $\rho_c$ is more than 1.0

(c) half-filled if $\rho_c$ is more than 0.5

(d) less than half-filled if $\rho_c$ is less than 0.5

**22.** In the given circuit, a charge of +80 μC is given to the upper plate of the 4 μF capacitor. Then in the steady state, the charge on the upper plate of the 3 μF capacitor is

(a) + 32 μC

(b) + 40 μC

(c) + 48 μC

(d) + 80 μC

**23.** Two moles of ideal helium gas are in a rubber balloon at 30°C. The balloon is fully expandable and can be assumed to require no energy in its expansion. The temperature of the gas in the balloon is slowly changed to 35°C. The amount of heat required in raising the temperature is nearly (take $R = 8.31$ J/mol-K)

(a) 62 J (b) 104 J
(c) 124 J (d) 208 J

**24.** Consider a disc rotating in the horizontal plane with a constant angular speed $\omega$ about its centre $O$. The disc has a shaded region on one side of the diameter and an unshaded region on the other side as shown in the figure. When the disc is in the orientation as shown, two pebbles $P$ and $Q$ are simultaneously projected at an angle towards $R$. The velocity of projection is in the $y$-$z$ plane and is same for both pebbles with respect to the disc. Assume that (i) they land back on the disc before the disc has completed $\frac{1}{8}$ rotation, (ii) their range is less than half the disc radius, and (iii) $\omega$ remains constant throughout. Then

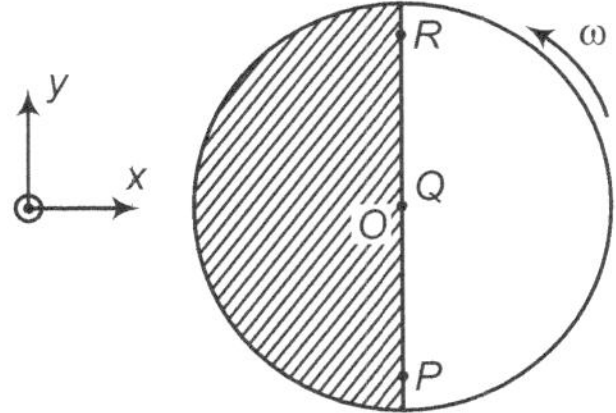


(a) $P$ lands in the shaded region and $Q$ in the unshaded region
(b) $P$ lands in the unshaded region and $Q$ in the shaded region
(c) both $P$ and $Q$ land in the unshaded region
(d) both $P$ and $Q$ land in the shaded region

**25.** A student is performing the experiment of resonance column. The diameter of the column tube is 4 cm. The frequency of the tuning fork is 512 Hz. The air temperature is 38° C in which the speed of sound is 336 m/s. The zero of the meter scale coincides with the top end of the resonance column tube. When the first resonance occurs, the reading of the water level in the column is

(a) 14.0 cm (b) 15.2 cm
(c) 16.4 cm (d) 17.6 cm

**26.** Two identical discs of same radius $R$ are rotating about their axes in opposite directions with the same constant angular speed $\omega$. The discs are in the same horizontal plane. At time $t = 0$, the points $P$ and $Q$ are facing each other as shown in the figure.

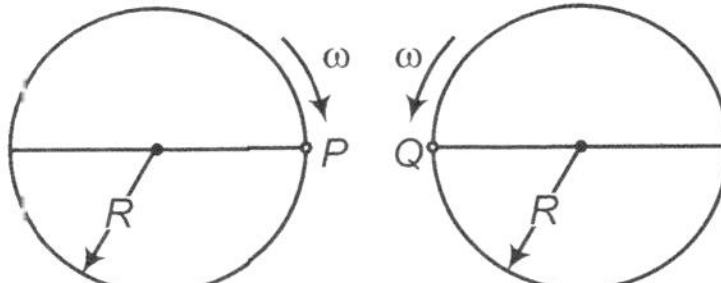


The relative speed between the two points $P$ and $Q$ is $v_r$. In one time period ($T$) of rotation of the discs, $v_r$ as a function of time is best represented by

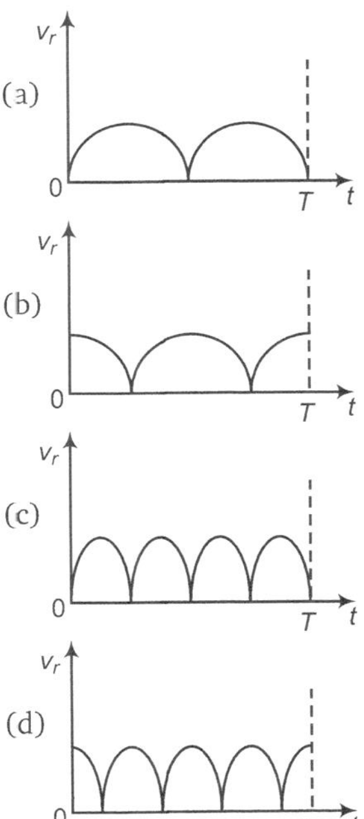

**27.** A loop carrying current $I$ lies in the $x$-$y$ plane as shown in the figure. The unit vector $\mathbf{k}$ is coming out of the plane of the paper. The magnetic moment of the current loop is

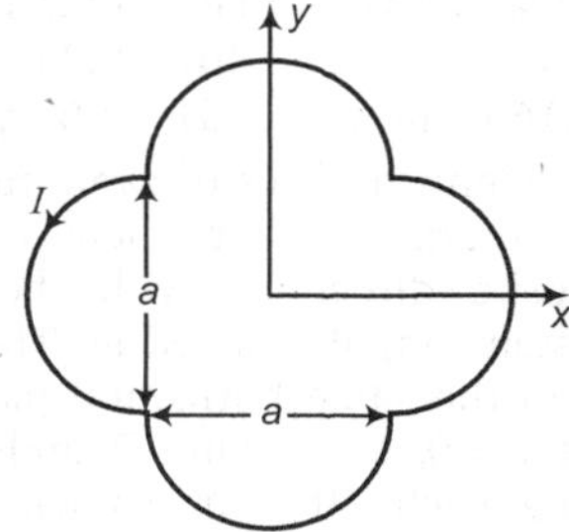


(a) $a^2 I\ \mathbf{k}$

(b) $\left(\frac{\pi}{2}+1\right) a^2\ I\ \mathbf{k}$

(c) $-\left(\frac{\pi}{2}+1\right) a^2\ I\ \mathbf{k}$

(d) $(2\pi + 1)\ a^2\ I\ \mathbf{k}$

**28.** An infinitely long hollow conducting cylinder with inner radius $R/2$ and outer radius $R$ carries a uniform current density along its length. The magnitude of the magnetic field, $|\mathbf{B}|$ as a function of the radial distance $r$ from the axis is best represented by

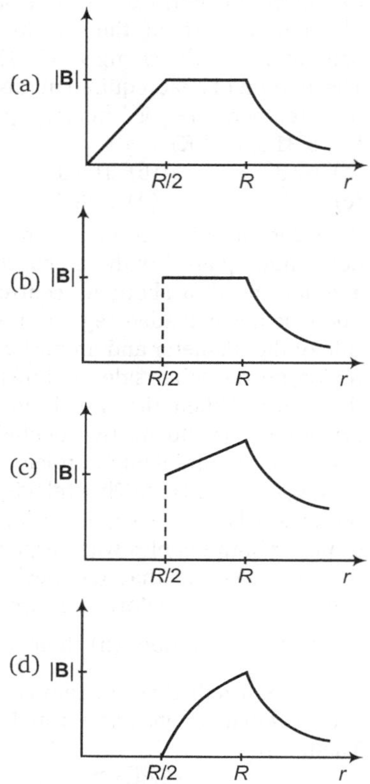


## Section II

**Directions** (Q. Nos. 29-34) *This section contains 6 multiple choice questions relating to three paragraphs with two questions on each paragraph. Each question has four choices (a), (b), (c) and (d) out of which only one is correct.*

### Paragraph

Most materials have the refractive index, $n > 1$. So, when a light ray from air enters a naturally occurring material, then by Snell's law, $\frac{\sin \theta_1}{\sin \theta_2} = \frac{n_2}{n_1}$, it is understood that the refracted ray bends towards the normal. But it never emerges on the same side of the normal as the incident ray. According to electromagnetism, the refractive index of the medium is given by the relation, $n = \left(\frac{c}{v}\right) = \pm \sqrt{\varepsilon_r \mu_r}$, where $c$ is the speed of electromagnetic waves in vacuum, $v$ its speed in the medium, $\varepsilon_r$ and $\mu_r$ are the relative permittivity and permeability of the medium respectively.

In normal materials, both $\varepsilon_r$ and $\mu_r$ are positive, implying positive $n$ for the medium. When both $\varepsilon_r$ and $\mu_r$ are negative, one must choose the negative root of $n$. Such negative refractive index materials can now be artificially prepared and are called meta-materials. They exhibit significantly different optical behaviour, without violating any physical laws. Since $n$ is negative, it results in a change in the direction of propagation of the refracted light. However, similar to normal materials, the frequency of light remains unchanged upon refraction even in meta-materials.

**29.** Choose the correct statement.

(a) The speed of light in the meta-material is $v = c|n|$

(b) The speed of light in the meta-material is $v = \frac{c}{|n|}$

(c) The speed of light in the meta-material is $v = c$

(d) The wavelength of the light in the meta-material $(\lambda_m)$ is given by $\lambda_m = \lambda_{air}|n|$, where $\lambda_{air}$ is the wavelength of the light in air.

**30.** For light incident from air on a meta-material, the appropriate ray diagram is

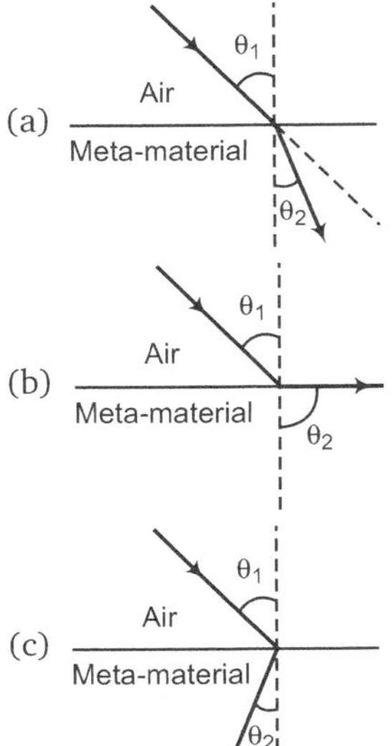


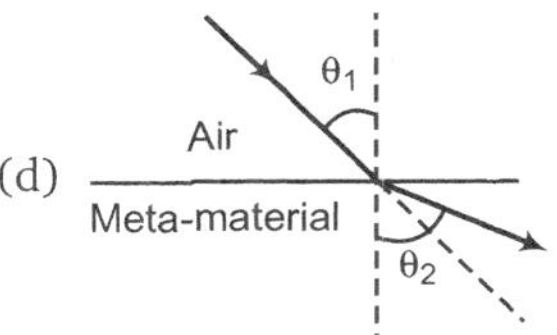


## Paragraph

The general motion of a rigid body can be considered to be a combination of (i) a motion of its centre of mass about an axis, and (ii) its motion about an instantaneous axis passing through the centre of mass. These axes need not be stationary. Consider, for example, a thin uniform disc welded (rigidly fixed) horizontally at its rim to a massless stick, as shown in the figure. When the disc-stick system is rotated about the origin on a horizontal frictionless plane with angular speed $\omega$, the motion at any instant can be taken as a combination of (i) a rotation of the centre of mass of the disc about the $z$-axis, and (ii) a rotation of the disc through an instantaneous vertical axis passing through its centre of mass (as is seen from the changed orientation of points $P$ and $Q$). Both these motions have the same angular speed $\omega$ in this case.

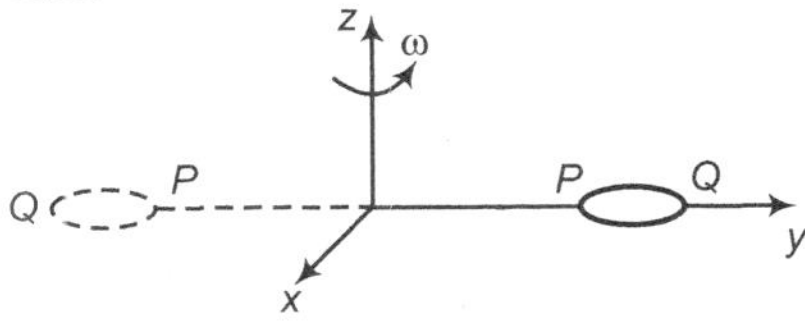


Now consider two similar systems as shown in the figure : Case (a) the disc with its face vertical and parallel to $x$-$z$ plane; Case (b) the disc with its face making an angle of 45° with $x$-$y$ plane and its horizontal diameter parallel to $x$-axis. In both the cases, the disc is welded at point $P$, and the systems are rotated with constant angular speed $\omega$ about the $z$-axis.

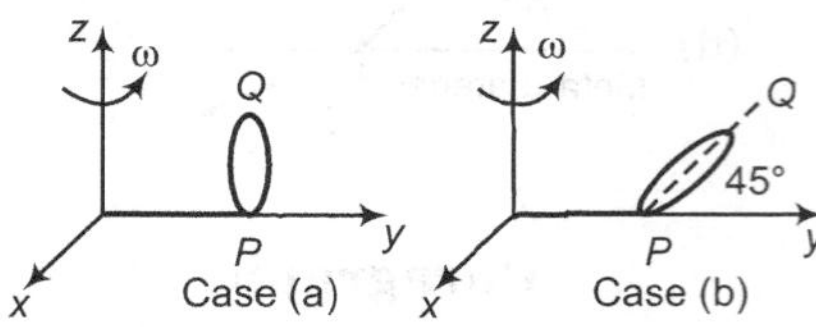


**31.** Which of the following statements regarding the angular speed about the instantaneous axis (passing through the centre of mass) is correct?

(a) It is $\sqrt{2}\omega$ for both the cases

(b) It is $\omega$ for case (a); and $\frac{\omega}{\sqrt{2}}$ for case (b)

(c) It is $\omega$ for case (a); and $\sqrt{2}\omega$ for case (b)

(d) It is $\omega$ for both the cases

**32.** Which of the following statements about the instantaneous axis (passing through the centre of mass) is correct?

(a) It is vertical for both the cases (a) and (b)

(b) It is vertical for case (a); and is at 45° to the $x$-$z$ plane and lies in the plane of the disc for case (b)

(c) It is horizontal for case (a); and is at 45° to the $x$-$z$ plane and is normal to the plane of the disc for case (b)

(d) It is vertical for case (a); and is at 45° to the $x$-$z$ plane and is normal to the plane of the disc for case (b)

## Paragraph

The β-decay process, discovered around 1900, is basically the decay of a neutron ($n$). In the laboratory, a proton ($p$) and an electron ($e^-$) are observed as the decay products of the neutron. Therefore, considering the decay of a neutron as a two-body decay process, it was predicted theoretically that the kinetic energy of the electron should be a constant. But experimentally, it was observed that the electron kinetic energy has a continuous spectrum. Considering a three-body decay process, *i.e.*, $n \rightarrow p + e^- + \bar{\nu}_e$, around 1930, Pauli explained the observed electron energy spectrum. Assuming the anti-neutrino ($\bar{\nu}_e$) to be massless and possessing negligible energy, and the neutron to be at rest, momentum and energy conservation principles are applied. From this calculation, the maximum kinetic energy of the electron is $0.8 \times 10^6$ eV. The kinetic energy carried by the proton is only the recoil energy.

**33.** If the anti-neutrino had a mass of 3 eV/$c^2$ (where $c$ is the speed of light) instead of zero mass, what should be the range of the kinetic energy $K$, of the electron?

(a) $0 \leq K \leq 0.8 \times 10^6$ eV

(b) $3.0 \text{ eV} \leq K \leq 0.8 \times 10^6$ eV

(c) $3.0 \text{ eV} \leq K < 0.8 \times 10^6$ eV

(d) $0 \leq K < 0.8 \times 10^6$ eV

**34.** What is the maximum energy of the anti-neutrino?

(a) Zero

(b) Much less than $0.8 \times 10^6$ eV

(c) Nearly $0.8 \times 10^6$ eV

(d) Much larger than $0.8 \times 10^6$ eV

## Section III

**Directions** (Q. Nos. 35-40) *This section contains 6 multiple choice questions. Each question has four choices (a), (b), (c) and (d) out of which one or more are correct.*

**35.** Six point charges are kept at the vertices of a regular hexagon of side $L$ and centre $O$ as shown in the figure. Given that $K = \frac{1}{4\pi\varepsilon_0}\frac{q}{L^2}$, which of the following statements(s) is(are) correct.

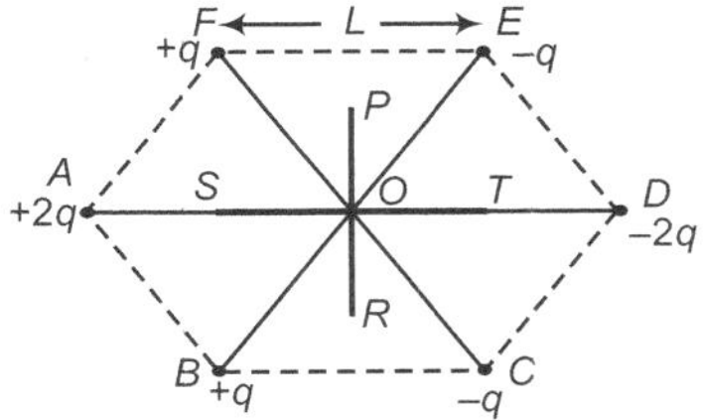


(a) The electric field at $O$ is $6K$ along $OD$
(b) The potential at $O$ is zero
(c) The potential at all points on the line $PR$ is same
(d) The potential at all points on the line $ST$ is same

**36.** Two solid cylinders $P$ and $Q$ of same mass and same radius start rolling down a fixed inclined plane from the same height at the same time. Cylinder $P$ has most of its mass concentrated near its surface, while $Q$ has most of its mass concentrated near the axis. Which statement(s) is(are) correct?

(a) Both cylinders $P$ and $Q$ reach the ground at the same time
(b) Cylinder $P$ has larger linear acceleration than cylinder $Q$
(c) Both cylinders reach the ground with same translational kinetic energy
(d) Cylinder $Q$ reaches the ground with larger angular speed

**37.** Two spherical planets $P$ and $Q$ have the same uniform density $\rho$, masses $M_P$ and $M_Q$, and surface areas $A$ and $4A$, respectively. A spherical planet $R$ also has uniform density $\rho$ and its mass is $(M_P + M_Q)$. The escape velocities from the planets $P$, $Q$ and $R$, are $V_P$, $V_Q$ and $V_R$, respectively. Then

(a) $V_Q > V_R > V_P$ (b) $V_R > V_Q > V_P$
(c) $V_R/V_P = 3$ (d) $V_P/V_Q = \frac{1}{2}$

**38.** The figure shows a system consisting of (i) a ring of outer radius $3R$ rolling clockwise without slipping on a horizontal surface with angular speed $\omega$ and (ii) an inner disc of radius $2R$ rotating anti-clockwise with angular speed $\omega/2$. The ring and disc are separated by frictionless ball bearings. The system is in the $x$-$z$ plane. The point $P$ on the inner disc is at a distance $R$ from the origin, where $OP$ makes an angle of 30° with the horizontal. Then with respect to the horizontal surface,

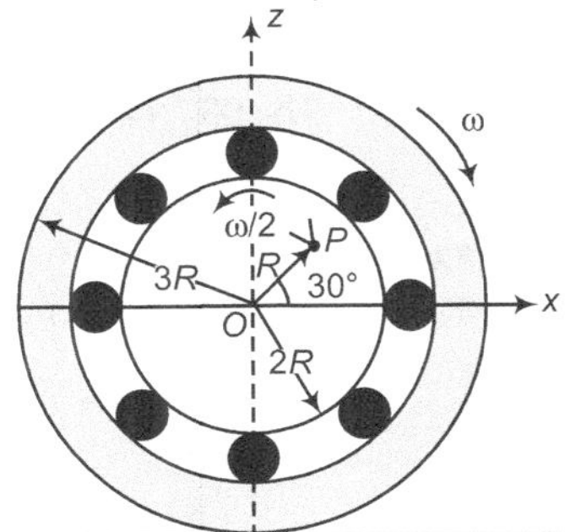


(a) the point $O$ has a linear velocity $3R\omega\,\mathbf{i}$
(b) the point $P$ has a linear velocity $\frac{11}{4}R\omega\,\mathbf{i} - \frac{\sqrt{3}}{4}R\omega\,\mathbf{k}$
(c) the point $P$ has a linear velocity $\frac{13}{4}R\omega\,\mathbf{i} - \frac{\sqrt{3}}{4}R\omega\,\mathbf{k}$

(d) the point $P$ has a linear velocity $\left(3 - \frac{\sqrt{3}}{4}\right) R\omega\, \mathbf{i} + \frac{1}{4} R\omega\, \mathbf{k}$

**39.** In the given circuit, the AC source has $\omega = 100$ rad/s. Considering the inductor and capacitor to be ideal, the correct choice(s) is(are)

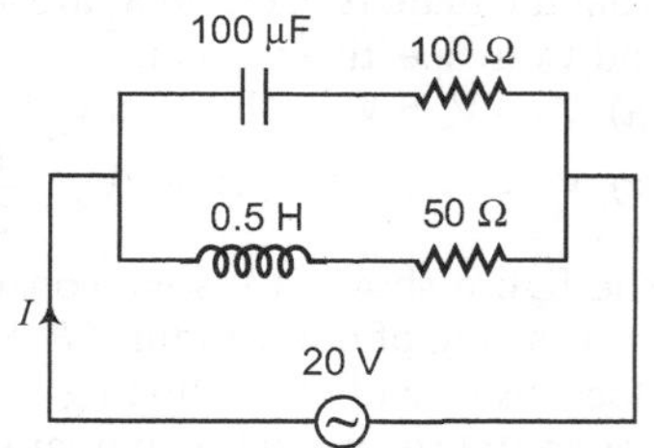


(a) The current through the circuit, $I$ is 0.3 A

(b) The current through the circuit, $I$ is $0.3\sqrt{2}$ A

(c) The voltage across 100 Ω resistor $= 10\sqrt{2}$ V

(d) The voltage across 50 Ω resistor = 10 V

**40.** A current carrying infinitely long wire is kept along the diameter of a circular wire loop, without touching it. The correct statement(s) is(are)

(a) The emf induced in the loop is zero if the current is constant

(b) The emf induced in the loop is finite if the current is constant

(c) The emf induced in the loop is zero if the current decreases at a steady rate

(d) The emf induced in the loop is finite if the current decreases at a steady rate

# Answers

## Paper 1

| | | | | | | | | | |
|---|---|---|---|---|---|---|---|---|---|
| **1.** (d) | **2.** (d) | **3.** (c) | **4.** (b) | **5.** (c) | **6.** (d) | **7.** (b) | **8.** (a) | **9.** (a) | **10.** (c) |
| **11.** (a,b,c,d) | | **12.** (b,d) | | **13.** (c,d) | | **14.** (a,c) | | **15.** (a,c) | |
| **16.** (5) | **17.** (7) | **18.** (7) | **19.** (6) | **20.** (3) | | | | | |

## Paper 2

| | | | | | | | | | |
|---|---|---|---|---|---|---|---|---|---|
| **21.** (a) | **22.** (c) | **23.** (d) | **24.** (*) | **25.** (b) | **26.** (a) | **27.** (b) | **28.** (d) | **29.** (b) | **30.** (c) |
| **31.** (d) | **32.** (a) | **33.** (d) | **34.** (c) | **35.** (a,b,c) | | **36.** (d) | **37.** (b,d) | | |
| **38.** (a,b) | | **39.** (a,c) | | **40.** (a,c) | | | | | |

**Note** : (*) No option is correct.

# Hints & Solutions

## Paper 1

1. **For inside points** $(r \le R)$

$E = 0$

$V = \text{constant} = \frac{1}{4\pi\varepsilon_0}\frac{q}{R}$

**For outside points** $(r \ge R)$

$$E = \frac{1}{4\pi\varepsilon_0} \cdot \frac{q}{r^2} \quad \text{or} \quad E \propto \frac{1}{r^2}$$

and $V = \frac{1}{4\pi\varepsilon_0}\frac{q}{r}$ or $V \propto \frac{1}{r}$

**On the surface** $(r = R)$

$$V = \frac{1}{4\pi\varepsilon_0}\frac{q}{R}$$

$$E = \frac{1}{4\pi\varepsilon_0} \cdot \frac{q}{R^2} = \frac{\sigma}{\varepsilon_0}$$

where, $\sigma = \frac{q}{4\pi R^2}$ = surface charge density corresponding to above equations the correct graphs are shown in option (d).

2. Fringe width $\beta = \frac{\lambda D}{d}$

or $\beta \propto \lambda$

Now, $\lambda_R > \lambda_G > \lambda_B$

$\therefore$ $\beta_R > \beta_G > \beta_B$

3. 
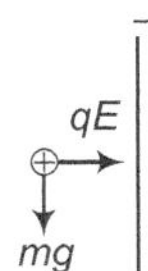


Net force is at 45° from vertical.

$\therefore$ $qE = mg$

or $\frac{qX}{d} = mg$ $\left(\because E = \frac{X}{d}\right)$

or $X = \frac{mgd}{q}$

$$= \frac{(1.67 \times 10^{-27})(9.8)(10^{-2})}{(1.6 \times 10^{-19})}$$

$$\approx 1 \times 10^{-9}\ \text{V}$$

4. Using the lens formula,

$$\frac{1}{v} - \frac{1}{u} = \frac{1}{F} = \frac{1}{f_1} + \frac{1}{f_2}$$

or $$\frac{1}{v} = \frac{1}{u} + \frac{1}{f_1} + \frac{1}{f_2}$$

$$= \frac{1}{u} + (n_1 - 1)\left(\frac{1}{R_1} - \frac{1}{R_2}\right) + (n_2 - 1)\left(\frac{1}{R_1'} - \frac{1}{R_2'}\right)$$

Substituting the values, we get

$$\frac{1}{v} = \frac{1}{-40} + (1.5 - 1)\left(\frac{1}{14} - \frac{1}{\infty}\right) + (1.2 - 1)\left(\frac{1}{\infty} - \frac{1}{-14}\right)$$

Solving this equations, we get

$$v = +40\ \text{cm}$$

5. Angular momentum of a particle about a point is given by :

$\mathbf{L} = \mathbf{r} \times \mathbf{p} = m\,(\mathbf{r} \times \mathbf{v})$

**For $\mathbf{L}_O$**

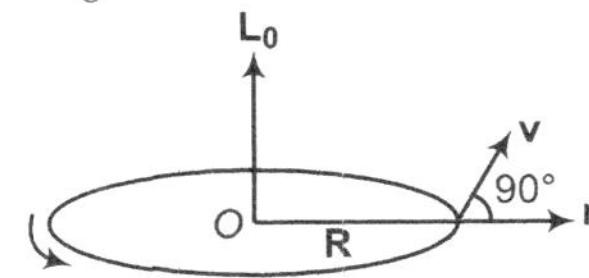


$$|\mathbf{L}| = (mvr \sin\theta) = m\,(R\omega)(R) \sin 90°$$
$$= mR^2\omega = \text{constant}$$

Direction of $\mathbf{L}_O$ is always upwards. Therefore, complete $\mathbf{L}_O$ is constant, both in magnitude as well as direction.

**For $\mathbf{L}_P$**

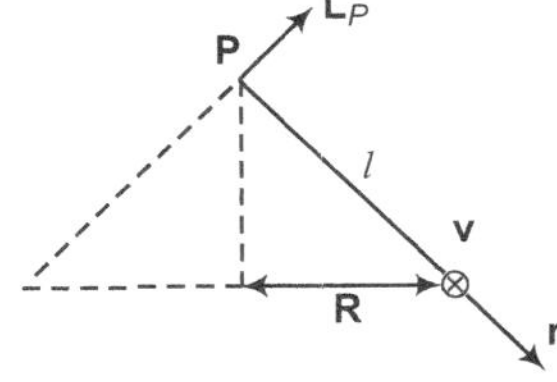

$$|\mathbf{L}_P| = (mvr \sin \theta)$$
$$= (m)(R\omega)(l) \sin 90^\circ$$
$$= (mRl\omega)$$

Magnitude of $\mathbf{L}_P$ will remain constant but direction of $\mathbf{L}_P$ keeps on changing.

**6.** $v_{\text{rms}} = \sqrt{\dfrac{3RT}{M}}$ or $v_{\text{rms}} \propto \dfrac{1}{\sqrt{M}}$

$$\therefore \quad \frac{(v_{\text{rms}})_{\text{He}}}{(v_{\text{rms}})_{\text{Ar}}} = \sqrt{\frac{M_{\text{Ar}}}{M_{\text{He}}}} = \sqrt{\frac{40}{4}}$$
$$= \sqrt{10} \approx 3.16$$

**7.** $|\mathbf{L}|$ or $L = I\omega$ (about axis of rod)

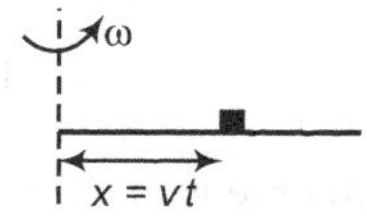


$$I = I_{\text{rod}} + mx^2 = I_{\text{rod}} + mv^2t^2$$

Here $m$ = mass of insect

$$\therefore \quad L = (I_{\text{rod}} + mv^2t^2)\,\omega$$

Now $|\vec{\tau}| = \dfrac{dL}{dt} = (2mv^2t\omega)$ or $|\vec{\tau}| \propto t$

*i.e.*, the graph is straight line passing through origin.

After time $T$,

$$L = \text{constant}$$
$$\therefore \quad |\vec{\tau}| \text{ or } \frac{dL}{dt} = 0$$

**8.** $\Delta d = \Delta l = \dfrac{0.5}{100}$ mm $= 0.005$ mm

$$Y = \frac{4MLg}{\pi l d^2}$$
$$\therefore \quad \left(\frac{\Delta Y}{Y}\right)_{\max} = \left(\frac{\Delta l}{l}\right) + 2\left(\frac{\Delta d}{d}\right)$$
$$\left(\frac{\Delta l}{l}\right) = \frac{0.5/100}{0.25} = 0.02$$

and $$\frac{2\Delta d}{d} = \frac{(2)(0.5/100)}{0.5} = 0.02$$

or $$\frac{\Delta l}{l} = 2 \cdot \frac{\Delta d}{d}$$

**9.** $t$ = time of flight of projectile

$$= \frac{2v \sin \theta}{g} \qquad (\theta = 45^\circ)$$
$$\therefore \quad v = \frac{gt}{2 \sin \theta} = \frac{10 \times 1}{2 \times 1/\sqrt{2}} = \sqrt{50} \text{ m/s}$$

**10.** Let temperature of middle plate in steady state is $T_0$

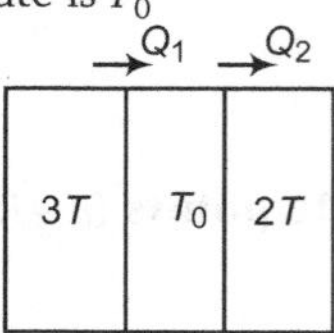


$$Q_1 = Q_2$$
$$Q = \text{net rate of heat flow}$$
$$\therefore \quad \sigma A (3T)^4 - \sigma A T_0^4 = \sigma A T_0^4 - \sigma A (2T)^4$$

Solving this equation, we get

$$T_0 = \left(\frac{97}{2}\right)^{1/4} T$$

**11.** Due to symmetry on upper side and lower side, points $P$ and $Q$ are at same potentials. Similarly, points $S$ and $T$ are at same potentials. Therefore, the simple circuit can be drawn as shown below :

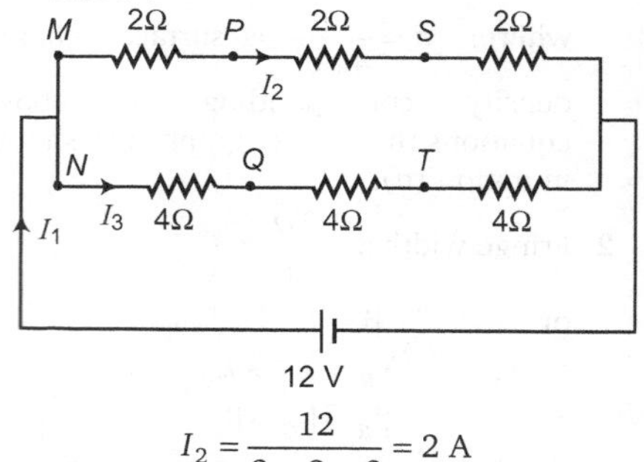


$$I_2 = \frac{12}{2+2+2} = 2 \text{ A}$$
$$I_3 = \frac{12}{4+4+4} = 1 \text{ A}$$
$$\therefore \quad I_1 = I_2 + I_3 = 3 \text{ A}$$

$I_{PQ} = 0$ because $V_P = V_Q$

Potential drop (from left to right) across each resistance is

$$\frac{12}{3} = 4 \text{V}$$
$$\therefore \quad V_{MS} = 2 \times 4 = 8 \text{ V}$$
$$V_{NQ} = 1 \times 4 = 4 \text{ V}$$

or $$V_S < V_Q$$

**12.** At open end phase of pressure wave changes by 180°. So, compression returns as rarefaction. At closed end there is no phase change. So, compression returns as compression and rarefaction as rarefaction.

13. (a) and (b)

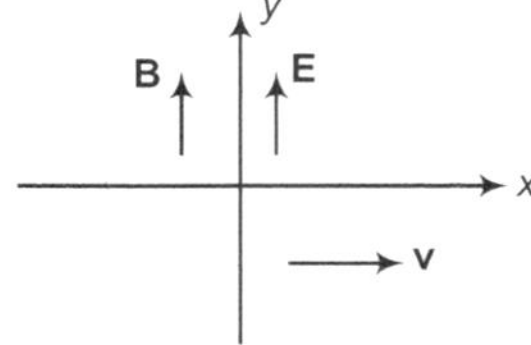


Magnetic field will rotate the particle in a circular path (in $x$-$z$ plane or perpendicular to **B**). Electric field will exert a constant force on the particle in positive $y$-direction. Therefore, resultant path is neither purely circular nor helical or the options (a) and (b) both are wrong.

(c)

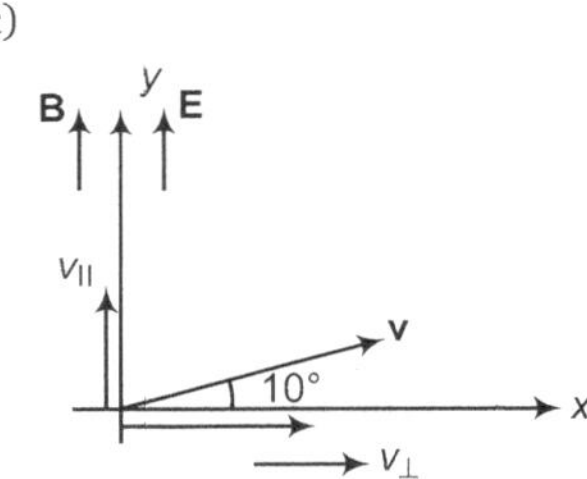


$v_\perp$ and **B** will rotate the particle in a circular path in $x$ - $z$ plane (or perpendicular to **B**). Further $v_{||}$ and **E** will move the particle (with increasing speed) along positive $y$-axis (or along the axis of above circular path). Therefore, the resultant path is helical with increasing pitch, along the $y$-axis (or along **B** and **E**). Therefore option (c) is correct.

(d)

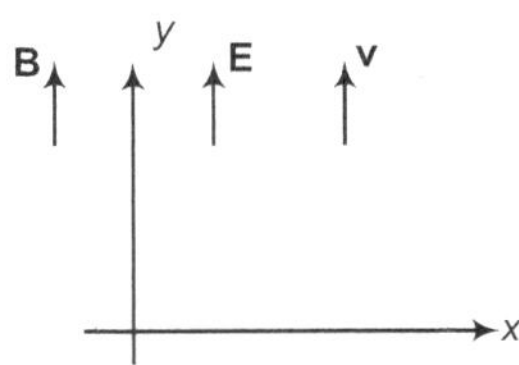


Magnetic force is zero, as θ between **B** and **v** is zero. But electric force will act in $y$-direction. Therefore, motion is 1-D and uniformly accelerated (towards positive $y$-direction).

14. Option (a) is correct due to symmetry.

Option (b) is wrong again due to symmetry.

Option (c) is correct because as per Gauss's theorem, net electric flux passing through any closed surface $= \dfrac{q_{in}}{\varepsilon_0}$

Here, $\quad q_{in} = 3q - q - q = q$

$\therefore$ Net electric flux $= \dfrac{q}{\varepsilon_0}$

Option (d) is wrong because there is no symmetry in two given planes.

15. $W = mg = 0.1 \times 10 = 1$ N

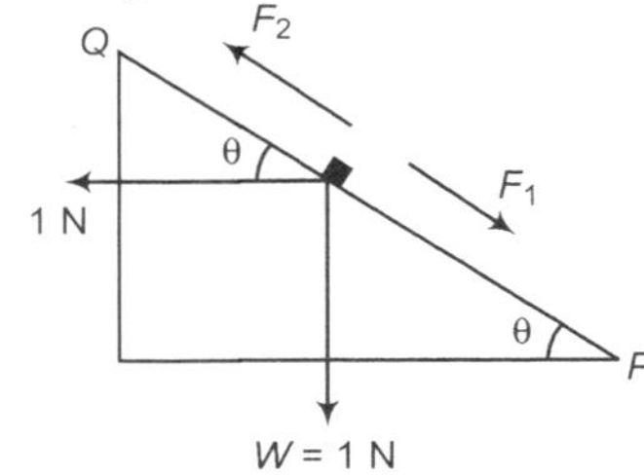


$F_1$ = component of weight
$= 1 \cdot \sin\theta = \sin\theta$

$F_2$ = component of applied force
$= 1 \cdot \cos\theta = \cos\theta$

Now, at $\theta = 45°$ : $F_1 = F_2$ and block remains stationary without the help of friction.

For $\theta > 45°$, $F_1 > F_2$, so friction will act towards $Q$.

For $\theta < 45°$, $F_2 > F_1$ and friction will act towards $P$.

16. $B_R = B_T - B_C$

$R$ = Remaining portion

$T$ = Total portion and

$C$ = cavity

$$B_R = \frac{\mu_0 I_T}{2a\pi} - \frac{\mu_0 I_C}{2\,(3a/2)\pi} \quad \ldots\text{(i)}$$

$$I_T = J\,(\pi a^2)$$

$$I_C = J\left(\frac{\pi a^2}{4}\right)$$

Substituting the values in Eq. (i), we have

$$B_R = \frac{\mu_o}{a\pi}\left[\frac{I_T}{2} - \frac{I_C}{3}\right]$$

$$= \frac{\mu_o}{a\pi}\left[\frac{\pi a^2 J}{2} - \frac{\pi a^2 J}{12}\right]$$

$$= \frac{5\mu_o aJ}{12}$$

$\therefore \quad N = 5$

**17.**

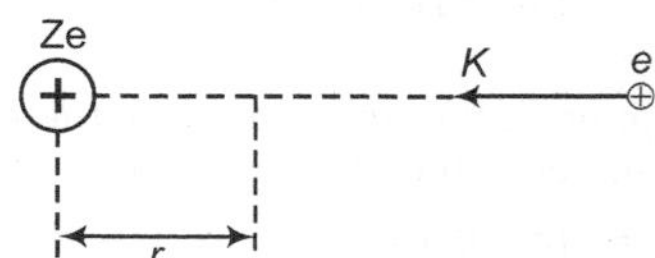


$r$ = closest distance = 10 fm

From energy conservation, we have

$$K_i + U_i = K_f + U_f$$

or $\quad K + 0 = 0 + \frac{1}{4\pi\varepsilon_0} \cdot \frac{q_1 q_2}{r}$

or $\quad K = \frac{1}{4\pi\varepsilon_0} \cdot \frac{(120\, e)(e)}{r} \quad \ldots\text{(i)}$

de-Broglie wavelength

$$\lambda = \frac{h}{\sqrt{2Km}} \quad \ldots\text{(ii)}$$

Substituting the given values in above two equations, we get

$$\lambda = 7 \times 10^{-15}\text{m} = 7 \text{ fm}$$

**18.** If $I$ current flows through the circular loop, then magnetic flux at the location of square loop is

$$B = \frac{\mu_0 I R^2}{2(R^2 + Z^2)^{3/2}}$$

Substituting the value of $Z$ $(= \sqrt{3}R)$, we have

$$B = \frac{\mu_0 I}{16R}$$

Now, total flux through the square loop is

$$\phi_T = NBS \cos\theta$$

$$= (2)\left(\frac{\mu_0 T}{16R}\right) a^2 \cos 45^\circ$$

Mutual inductance

$$M = \frac{\phi_T}{I} = \frac{\mu_0 a^2}{2^{7/2} R}$$

$\therefore \quad p = 7$

**19.** Volume of cylinder per unit length $(l = 1)$ is

$$V = \pi R^2 l = (\pi R^2)$$

$\therefore$ Charge per unit length,

$\lambda$ = (Volume per unit length) $\times$ (Volume charge density)

$= (\pi R^2 \rho)$

**Now at $P$**

$$E_R = E_T - E_C$$

$R$ = Remaining portion

$T$ = Total portion and

$C$ = cavity

$$\therefore \quad E_R = \frac{\lambda}{2\pi\varepsilon_0(2R)} - \frac{1}{4\pi\varepsilon_0}\frac{Q}{(2R)^2}$$

$Q$ = charge on sphere

$$= \frac{4}{3}\pi\left(\frac{R}{2}\right)^3 \rho = \frac{\pi R^3 \rho}{6}$$

Substituting the values, we have

$$E_R = \frac{(\pi R^2 \rho)}{4\pi\varepsilon_0 R} - \frac{1}{4\pi\varepsilon_0} \cdot \frac{(\pi R^3 \rho/6)}{4R^2}$$

$$= \frac{23\rho R}{96\varepsilon_0} = \frac{23\rho R}{(16)(6)\varepsilon_0}$$

$\therefore \quad k = 6$

**20.** $T$ = Total portion

$R$ = Remaining portion and

$C$ = Cavity and

let $\sigma$ = mass per unit area.

Then, $\quad m_T = \pi(2R)^2\sigma = 4\pi R^2\sigma$

$$m_C = \pi(R)^2\sigma = \pi R^2\sigma$$

**For $I_P$**

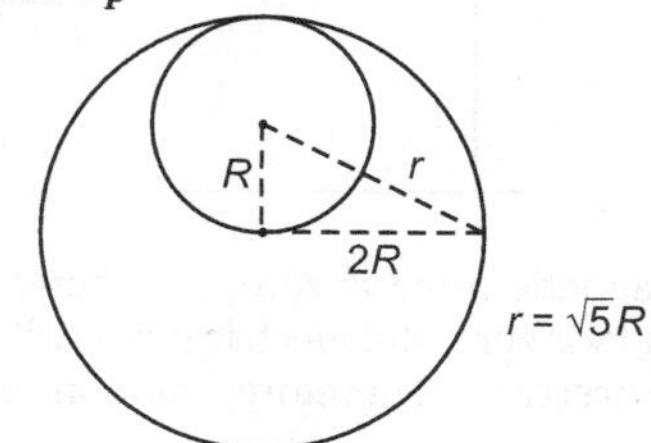

$$I_R = I_T - I_C$$
$$= \frac{3}{2} m_T (2R)^2 - \left[\frac{1}{2} m_C R^2 + m_C r^2\right]$$
$$= \frac{3}{2}(4\pi R^2 \sigma)(4R^2) - \left[\frac{1}{2}(\pi R^2 \sigma) + (\pi R^2 \sigma)(5R^2)\right]$$
$$= (18.5\,\pi R^4 \sigma)$$

**For $I_O$** $I_R = I_T - I_C$
$$= \frac{1}{2} m_T (2R)^2 - \frac{3}{2} m_C R^2$$
$$= \frac{1}{2}(4\pi R^2 \sigma)(4R^2) - \frac{3}{2}(\pi R^2 \sigma)(R^2)$$
$$= 6.5\,\pi R^4 \sigma$$
$$\therefore \quad \frac{I_P}{I_O} = \frac{18.5\pi R^4 \sigma}{6.5\pi R^4 \sigma} = 2.846$$

Therefore, the nearest integer is 3.

# Paper 2

**21.** Let $V_1$ = total material volume of shell
$V_2$ = total inside volume of shell and
$x$ = fraction of $V_2$ volume filled with water.

In floating condition,

Total weight = Upthrust

$$\therefore \quad V_1 \rho_c g + (xV_2)(1) g = \left(\frac{V_1 + V_2}{2}\right)(1) g$$

or $$x = 0.5 + (0.5 - \rho_c)\frac{V_1}{V_2}$$

From here we can see that,
$$x > 0.5 \text{ if } \rho_c < 0.5$$

**22.** Between $3\,\mu F$ and $2\,\mu F$ (in parallel) total charge of $80\,\mu C$ will distribute in direct ratio of capacity.

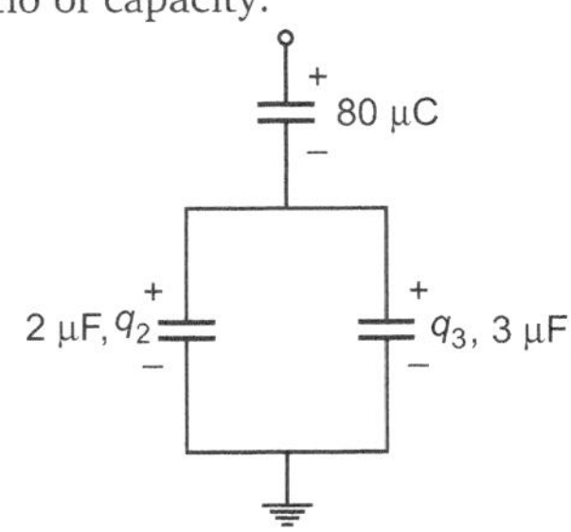


$$\therefore \quad \frac{q_3}{q_2} = \frac{3}{2}$$

or $$q_3 = \left(\frac{3}{3+2}\right)(80) = 48\,\mu C$$

**23.** The process may be assumed to be isobaric.

$$\therefore \quad Q = n C_p \Delta T = (2)\left(\frac{5}{2}R\right)(5)$$
$$= 5 \times 8.31 \times 5$$
$$= 207.75 \text{ J} \approx 208 \text{ J}$$

**24.** Language of question is not very clear. For example, disc is rotating. Its different points have different velocities. Relative velocity of pebble with respect to which point, it is not clear. Further, actual initial positions of $P$ and $Q$ are also not given.

**25.** With end correction,

$$f = n\left[\frac{v}{4(l+e)}\right], \quad \text{where } n = 1, 3, \ldots$$
$$= n\left[\frac{v}{4(l + 0.6\,r)}\right]$$

because, $e = 0.6\,r$, where $r$ is radius of pipe.

For first resonance, $n = 1$

$$\therefore \quad f = \frac{v}{4(l + 0.6\,r)}$$

or $$l = \frac{v}{4f} - 0.6r$$
$$= \left[\left(\frac{336 \times 100}{4 \times 512}\right) - 0.6 \times 2\right] \text{cm}$$
$$= 15.2 \text{ cm}$$

**26.** Language of question is wrong because relative speed is not the correct word. Relative speed between two is always zero. The correct word is magnitude of relative velocity.

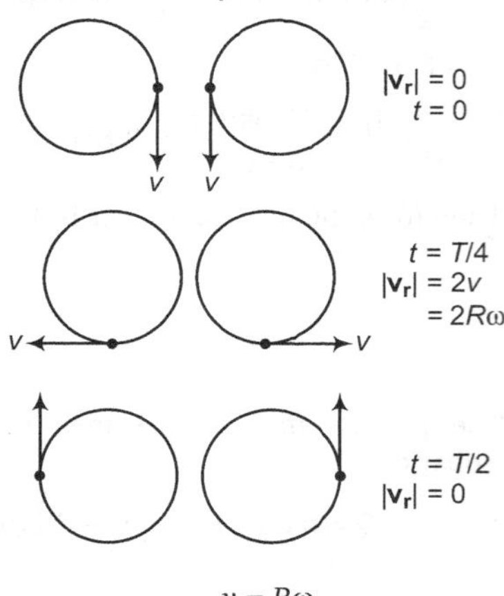


$$v = R\omega$$

Corresponding to above values, the correct graph is (a).

**27.** Area of the given loop is

$A$ = (area of two circles of radius $\frac{a}{2}$ and area of a square of side $a$)

$$= 2\pi\left(\frac{a}{2}\right)^2 + a^2$$

$$= \left(\frac{\pi}{2} + 1\right) a^2$$

$$|\mathbf{M}| = IA = \left(\frac{\pi}{2} + 1\right) a^2 I$$

From screw law direction of **M** is outwards or in positive $z$-direction.

$$\therefore \quad \mathbf{M} = \left(\frac{\pi}{2} + 1\right) a^2 I\, \hat{\mathbf{k}}$$

**28.** 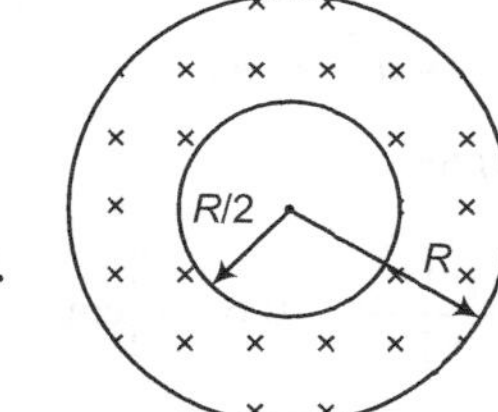


$r$ = distance of a point from centre.

**For $r \le R/2$** Using Ampere's circuital law,

$$\oint \mathbf{B} \cdot d\mathbf{l} \quad \text{or} \quad Bl = \mu_0 (I_{in})$$

$$\text{or} \quad B(2\pi r) = \mu_0 (I_{in})$$

$$\text{or} \quad B = \frac{\mu_0}{2\pi}\frac{I_{in}}{r} \quad \ldots\text{(i)}$$

Since $I_{in} = 0 \quad \therefore \quad B = 0$

**For $\frac{R}{2} \le r \le R$**

$$I_{in} = \left[\pi r^2 - \pi\left(\frac{R}{2}\right)^2\right]\sigma$$

Here $\sigma$ = current per unit area.

Substituting in Eq. (i), we have

$$B = \frac{\mu_0}{2\pi}\frac{\left[\pi r^2 - \pi\frac{R^2}{4}\right]\sigma}{r}$$

$$= \frac{\mu_0\sigma}{2r}\left(r^2 - \frac{R^2}{4}\right)$$

At $r = \frac{R}{2}$, $B = 0$

At $r = R$, $B = \frac{3\mu_0\sigma R}{8}$

**For $r \ge R$**

$$I_{in} = I_{Total} = I \text{ (say)}$$

Therefore, substituting in Eq. (i), we have

$$B = \frac{\mu_0}{2\pi}\cdot\frac{I}{r} \quad \text{or} \quad B \propto \frac{1}{r}$$

**29.** Since value of $n$ in meta-material is negative.

$$\therefore \quad v = \frac{c}{|n|}$$

**30.** According to the paragraph, refracted ray in meta-material should be on same side of normal.

**31-32.** (i) Every particle of the disc is rotating in a horizontal circle.

(ii) Actual velocity of any particle is horizontal.

(iii) Magnitude of velocity of any particle is

$$v = r\omega$$

where, $r$ is the perpendicular distance of that particle from actual axis of rotation ($z$-axis).

(iv) When it is broken into two parts then actual velocity of any particle is resultant of two velocities

$$v_1 = r_1\omega_1 \text{ and } v_2 = r_2\omega_2$$

Here,

$r_1$ = perpendicular distance of centre of mass from $z$-axis.

$\omega_1$ = angular speed of rotation of centre of mass from $z$-axis.

$r_2$ = distance of particle from centre of mass and

$\omega_2$ = angular speed of rotation of the disc about the axis passing through centre of mass.

(v) Net $v$ will be horizontal, if $v_1$ and $v_2$ both are horizontal. Further, $v_1$ is already horizontal, because centre of mass is rotating about a vertical $z$-axis. To make $v_2$ also horizontal, second axis should also be vertical.

**33-34.** Maximum kinetic energy of anti-neutrino is nearly $(0.8 \times 10^6)$ eV.

**35.** (a)

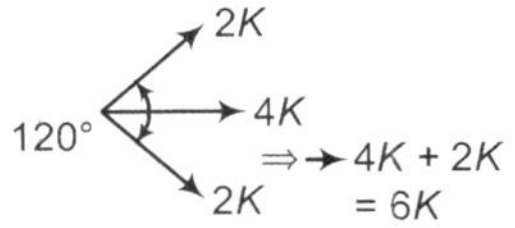

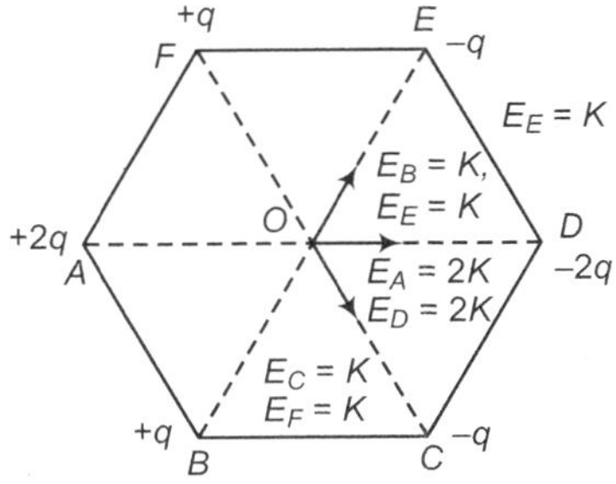

2K
120°
4K
⇒ 4K + 2K = 6K
2K

+q F, E −q, $E_E = K$, $E_B = K$, $E_E = K$, O, +2q A, D −2q, $E_A = 2K$, $E_D = 2K$, $E_C = K$, $E_F = K$, +q B, C −q

Resultant of 2K and 2K (at 120°) is also 2K towards 4K. Therefore, net electric field is 6K.

(b) $V_0 = \dfrac{1}{4\pi\varepsilon_0}\left[\dfrac{q_A}{L} + \dfrac{q_B}{L} + \dfrac{q_C}{L} + \dfrac{q_D}{L} + \dfrac{q_E}{L} + \dfrac{q_F}{L}\right]$

$$= \frac{1}{4\pi\varepsilon_0 L}(q_A + \ldots + q_F)$$

$$= 0$$

Because $q_A + q_B + q_C + q_D + q_E + q_F = 0$

(c) Only line PR, potential is same (= 0).

**36.** $I_P > I_Q$

In case of pure rolling,

$$a = \frac{g \sin\theta}{1 + I/mR^2}$$

$a_Q > a_P$ as its moment of inertia is less. Therefore, $Q$ reaches first with more linear speed and more translational kinetic energy.

Further, $\omega = \dfrac{v}{R}$

or $\omega \propto v$

$\therefore$ $\omega_Q > \omega_P$ as $v_P > v_Q$

**37.** Surface area of $Q$ is four times. Therefore, radius of $Q$ is two times. Volume is eight times. Therefore, mass of $Q$ is also eight times.

So, let $M_P = M$ and $R_P = r$

Then, $M_Q = 8$ M and $R_Q = 2r$

Now, mass of $R$ is $(M_P + M_Q)$ or $9\,M$. Therefore, radius of $R$ is $(9)^{1/3} r$. Now, escape velocity from the surface of a planet is given by

$$v = \sqrt{\frac{2GM}{r}}$$

($r$ = radius of that planet)

$$\therefore \quad v_P = \sqrt{\frac{2GM}{r}}$$

$$v_Q = \sqrt{\frac{2G(8M)}{(2r)}}$$

$$v_R = \sqrt{\frac{2G(9M)}{(9)^{1/3} r}}$$

From here we can see that,

$$\frac{v_P}{v_Q} = \frac{1}{2}$$

and $v_R > v_Q > v_P$

**38.** Velocity of point $O$ is

$$v_0 = (3R\omega)\,\hat{\mathbf{i}}$$

$\mathbf{v}_{PO}$ is $\frac{R \cdot \omega}{2}$ in the direction shown in figure. In vector form

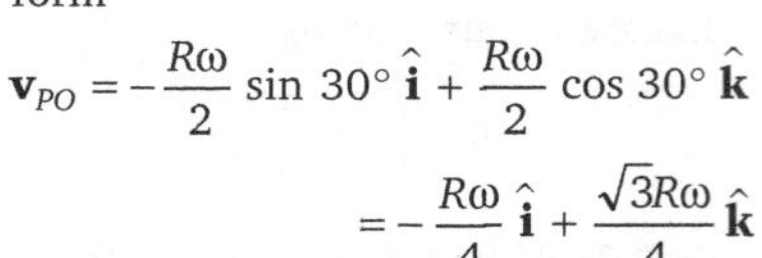


$$\mathbf{v}_{PO} = -\frac{R\omega}{2}\sin 30^\circ\,\hat{\mathbf{i}} + \frac{R\omega}{2}\cos 30^\circ\,\hat{\mathbf{k}}$$

$$= -\frac{R\omega}{4}\,\hat{\mathbf{i}} + \frac{\sqrt{3}R\omega}{4}\,\hat{\mathbf{k}}$$

But $\mathbf{v}_{PO} = \mathbf{v}_P - \mathbf{v}_O$

$\therefore \quad \mathbf{v}_P = \mathbf{v}_{PO} + \mathbf{v}_O$

$$= \left(-\frac{R\omega}{4}\,\hat{\mathbf{i}} + \frac{\sqrt{3}R\omega}{4}\,\hat{\mathbf{k}}\right) + 3R\omega\,\hat{\mathbf{i}}$$

$$= \frac{11}{4}R\omega\,\hat{\mathbf{i}} + \frac{\sqrt{3}}{4}R\omega\,\hat{\mathbf{k}}$$

**39.**

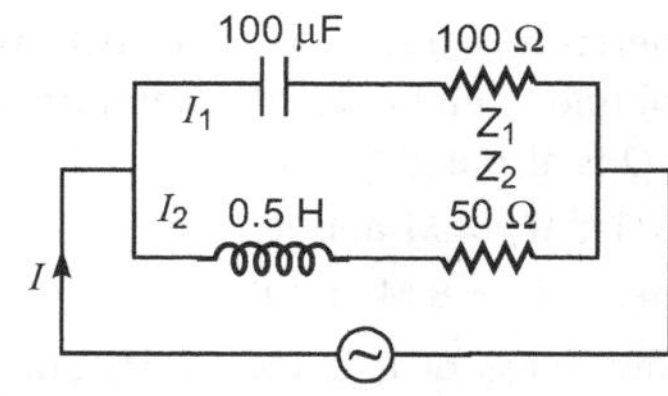


**Circuit 1**

$$X_C = \frac{1}{\omega C} = 100\ \Omega$$

$\therefore$

$$Z_1 = \sqrt{(100)^2 + (100)^2} = 100\sqrt{2}\ \Omega$$

$$\phi_1 = \cos^{-1}\left(\frac{R_1}{Z_1}\right) = 45^\circ$$

In this circuit current leads the voltage.

$$I_1 = \frac{V}{Z_1} = \frac{20}{100\sqrt{2}} = \frac{1}{5\sqrt{2}}\ \text{A}$$

$$V_{100\,\Omega} = (100)\,I_1 = (100)\frac{1}{5\sqrt{2}}\ \text{V}$$

$$= 10\sqrt{2}\ \text{V}$$

**Circuit 2**

$$X_L = \omega L = (100)\,(0.5) = 50\ \Omega$$

$$Z_2 = \sqrt{(50)^2 + (50)^2} = 50\sqrt{2}\ \Omega$$

$$\phi_2 = \cos^{-1}\left(\frac{R_2}{Z_2}\right) = 45^\circ$$

In this circuit voltage leads the current.

$$I_2 = \frac{V}{Z_2} = \frac{20}{50\sqrt{2}} = \frac{\sqrt{2}}{5}\ \text{A}$$

$$V_{50\,\Omega} = (50)\,I_2 = 50\left(\frac{\sqrt{2}}{5}\right) = 10\sqrt{2}\ \text{V}$$

Further, $I_1$ and $I_2$ have a mutual phase difference of 90°.

$$\therefore I = \sqrt{I_1^2 + I_2^2} = \sqrt{\frac{1}{50} + \frac{4}{50}} = \frac{1}{\sqrt{10}}\ \text{A}$$

$$\approx 0.3\ \text{A}$$

**40.**

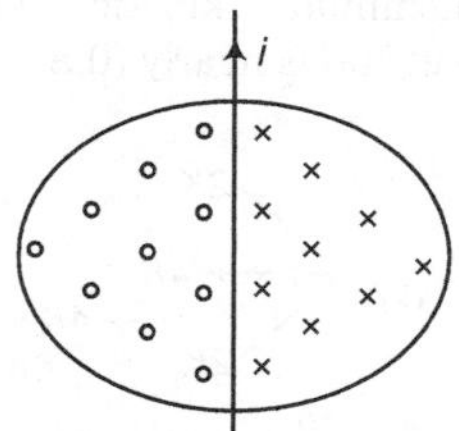


Due to the current in the straight wire, net magnetic flux from the circular loop is zero. Because in half of the circle magnetic field is inwards and in other half, magnetic field is outwards. Therefore, change in current will not cause any change in magnetic flux from the loop. Therefore, induced emf under all conditions through the circular loop is zero.

# IIT JEE
## SOLVED PAPER 2012

## Chemistry

## Paper 1

### Section I

**Direction** (Q. No. 1-10) *This section contains **10 multiple choice questions**, Each question has four choices, (a), (b), (c) and (d) out of which **ONLY ONE** is correct.*

1. Which ordering of compounds is according to the decreasing order of the oxidation state of nitrogen?
   (a) $HNO_3$, NO, $NH_4Cl$, $N_2$
   (b) $HNO_3$, NO, $N_2$, $NH_4Cl$
   (c) $HNO_3$, $NH_4Cl$, NO, $N_2$
   (d) NO, $HNO_3$, $NH_4Cl$, $N_2$

2. The kinetic energy of an electron in the second Bohr orbit of a hydrogen atom is [$a_0$ is Bohr radius]
   (a) $\dfrac{h^2}{4\pi^2 m a_0^2}$
   (b) $\dfrac{h^2}{16\pi^2 m a_0^2}$
   (c) $\dfrac{h^2}{32\pi^2 m a_0^2}$
   (d) $\dfrac{h^2}{64\pi^2 m a_0^2}$

3. The number of aldol reaction(s) that occurs in the given transformation is

$$CH_2CHO + 4HCHO \xrightarrow{\text{Conc. aq. NaOH}}$$

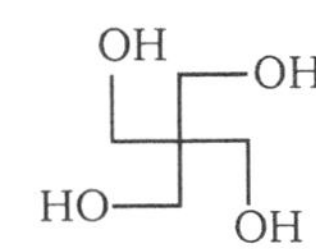


   (a) 1 (b) 2
   (c) 3 (d) 4

4. For one mole of a van der Waals' gas when $b = 0$ and $T = 300$ K, the $pV$ *vs.* $1/V$ plot is shown below. The value of the van der Waals' constant a (atm L $mol^{-2}$)

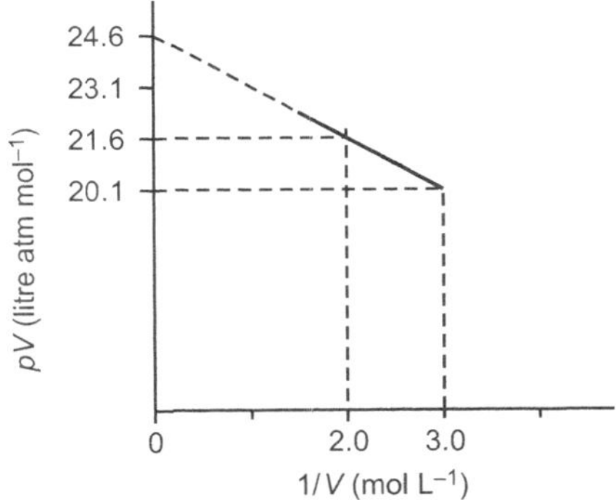


   (a) 1.0
   (b) 4.5
   (c) 1.5
   (d) 3.0

5. In allene ($C_3H_4$), the type(s) of hybridization of the carbon atoms, is (are)
   (a) $sp$ and $sp^3$
   (b) $sp$ and $sp^2$
   (c) only $sp^3$
   (d) $sp^2$ and $sp^3$

6. A compound $M_pX_q$ has cubic close packing (ccp) arrangement of $X$. Its unit cell structure is shown below. The empirical formula of the compound, is

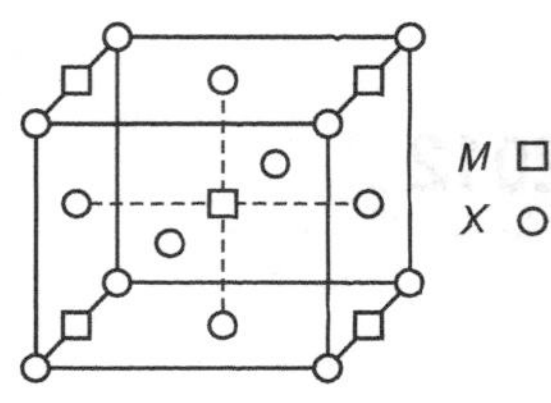


(a) $MX$ (b) $MX_2$

(c) $M_2X$ (d) $M_5X_{14}$

**7.** The number of optically active products obtained from the complete ozonolysis of the given compound, is

$CH_3$      H

$CH_3—CH= CH—C—CH=CH—C—CH=CH—CH_2$

C      $CH_3$

(a) 0 (b) 1

(c) 2 (d) 4

**8.** As per IUPAC nomenclature, the name of the complex $[Co(H_2O)_4(NH_3)_2]Cl_3$ is

(a) tetraaquadiaminecobalt (III) chloride

(b) tetraaquadiamminecobalt (III) chloride

(c) diaminetetraaquacobalt (III) chloride

(d) diamminetetraaquacobalt (III) chloride

**9.** The carboxyl functional group (—COOH) is present in

(a) picric acid (b) barbituric acid

(c) ascorbic acid (d) aspirin

**10.** The colour of light abosrbed by an aqueous solution of $CuSO_4$ is

(a) orange-red (b) blue-green

(c) yellow (d) violet

## Section II

**Direction** (Q. No. 11-15) *This section contains* **5 *multiple choice questions.*** *Each question has four choices (a), (b), (c) and (d) out of which* ***ONE or MORE*** *are correct.*

**11.** For an ideal gas, consider only *p-V* work in going from an initial state $X$ to the final state $Z$. The final state $Z$ can be reached by either of the two paths shown in the figure. Which of the following choice(s) is (are) correct?

[Take $\Delta S$ as change in entropy and $W$ as work done].

(a) $\Delta S_{X \to Z} = \Delta S_{X \to Y} + \Delta S_{Y \to Z}$

(b) $W_{X \to Z} = W_{X \to Y} + W_{Y \to Z}$

(c) $W_{X \to Z \to Z} = W_{X \to Y}$

(d) $\Delta S_{X \to Y \to Z} = \Delta S_{X \to Y}$

**12.** Which of the following molecules, in pure form, is (are) unstable at room temperature?

(a) 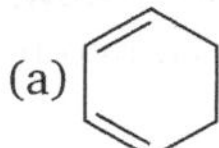 (b) 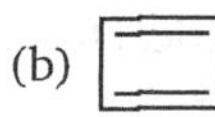

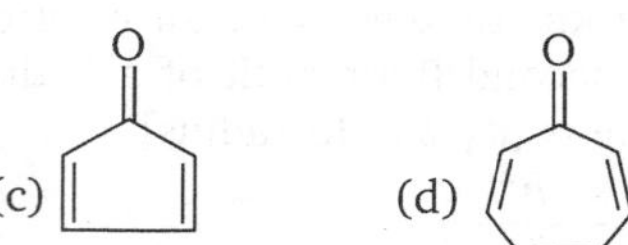


**13.** Identify the binary mixture(s) that can be separated into individual compounds, by differential extraction, as shown in the given scheme.

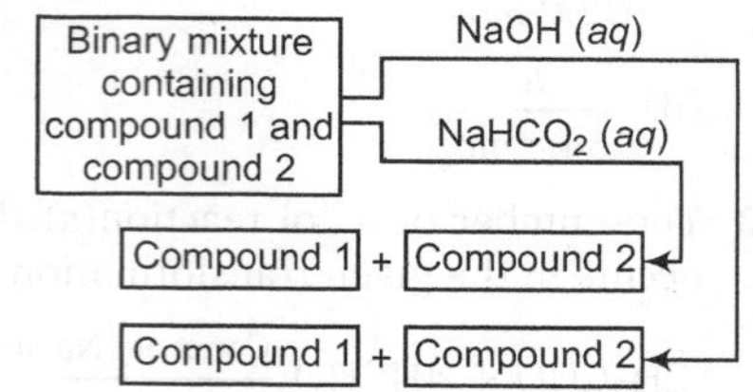


(a) $C_6H_5OH$ and $C_6H_5COOH$

(b) $C_6H_5COOH$ and $C_6H_5CH_2OH$

(c) $C_6H_5CH_2OH$ and $C_6H_5OH$

(d) $C_6H_5CH_2OH$ and $C_6H_5CH_2COOH$

**14.** Choose the correct reason(s) for the stability of the lyophobic colloidal particles.

(a) Preferential adsorption of ions on their surface from the solution.

(b) Preferential adsorption of solvent on their surface from the solution.

(c) Attraction between different particles having opposite charges on their surface.

(d) Potential difference between the fixed layer and the diffused layer of opposite charges around the colloidal particles.

**15.** Which of the following halides react(s) with $AgNO_3(aq)$ to give a precipitate that dissolves in $Na_2S_2O_3(aq)$?

(a) HCl (b) HF

(c) HBr (d) HI

## Section III

**Direction** (Q. No. 16-20) *This section contains* **5 *questions.*** *The answer to each question is a single-digit integer, ranging from 0 to 9 (both inclusive).*

**16.** An organic compound undergoes first order decomposition. The time taken for its decomposition to 1/8 and 1/10 of its initial concentration are $t_{1/8}$ and $t_{1/10}$ respectively. What is the value of $\frac{[t_{1/8}]}{[t_{1/10}]} \times 10$? ($\log_{10} 2 = 0.3$)

**17.** When the following aldohexose exists in its D-configuration, the total number of stereoisomers in its pyranose form, is

$$\begin{array}{c} CHO \\ | \\ CH_2 \\ | \\ CHOH \\ | \\ CHOH \\ | \\ CH_2OH \end{array}$$

**18.** The substituents $R_1$ and $R_2$ for nine peptides are listed in the table given below. How many of these peptides are positively charged at pH = 7.0 ?

$$\overset{\oplus}{H_3N}\text{—}\underset{\underset{H}{|}}{CH}CO\text{—}NH\text{—}\underset{\underset{R_1}{|}}{CH}\text{—}CONH\text{— ...}$$

$$\text{... —}\underset{\underset{R_2}{|}}{CH}\text{—}CONH\text{—}\underset{\underset{H}{|}}{CH}\text{—}CO\overset{\ominus}{O}$$

| Peptide | $R_1$ | $R_2$ |
|---|---|---|
| I | H | H |
| II | H | $CH_3$ |
| III | $CH_2COOH$ | H |
| IV | $CH_2CONH_2$ | $(CH_2)_4NH_2$ |
| V | $CH_2CONH_2$ | $CH_2CONH_2$ |
| VI | $(CH_2)_4NH_2$ | $(CH_2)_4NH_2$ |
| VII | $CH_2COOH$ | $CH_2CONH_2$ |
| VIII | $CH_2OH$ | $(CH_2)_4NH_2$ |
| IX | $(CH_2)_4NH_2$ | $CH_3$ |

**19.** The Periodic Table consists of 18 groups. An isotope of copper, on bombardment with protons, undergoes a nuclear reaction yielding element $X$ as shown below. To which group, element $X$ belongs in the Periodic Table?

$$^{63}_{29}Cu + ^{1}_{1}H — 6^{1}_{0}n + ^{4}_{2}\alpha + 2^{1}_{1}H + X$$

**20.** 29.2% ($w/W$) HCl stock solution has density of 1.25 g $mL^{-1}$. The molecular weight of HCl is 36.5 g $mol^{-1}$. The volume (mL) of stock solution required to prepare a 200 mL solution 0.4 M HCl is

# Paper 2

## Section I

**Direction** (Q. No. 21-28) *This section contains* **8** ***multiple choice questions.*** *Each question has four choices (a), (b), (c) and (d) out of which* ***ONLY ONE*** *is correct.*

**21.** $NiCl_2\{P(C_2H_5)_2(C_6H_5)\}_2$ exhibits temperature dependent magnetic behaviour (paramagnetic/diamagnetic) the coordination geometries of $Ni^{2+}$ in the paramagnetic and diamagnetic states are respectively

(a) tetrahedral and tetrahedral
(b) square planar and square planar
(c) tetrahedral and square planar
(d) square planar and tetrahedral

**22.** The reaction of white phosphorus with aqueous NaOH gives phosphine along with another phosphorus containing compound. The reaction type; the oxidation states of phosphorus in phosphine and the other product are respectively

(a) redox reaction; –3 and –5
(b) redox reaction; 3 and +5
(c) disproportionation reaction; –3 and +5
(d) disproportionation reaction; –3 and +3

**23.** In the cyanide extraction process of silver from argentite ore, the oxidizing and reducing agents used are

(a) $O_2$ and CO respectively
(b) $O_2$ and Zn dust respectively
(c) $HNO_3$ and Zn dust respectively
(d) $HNO_3$ and CO respectively

**24.** The compound that undergoes decarboxylation most readily under mild condition is

(a) 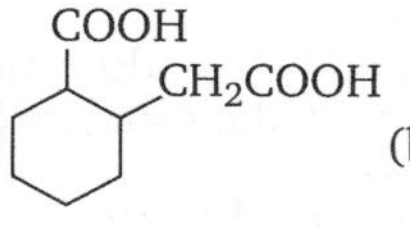 cyclohexane ring bearing COOH and, on the adjacent carbon, $CH_2COOH$
(b) 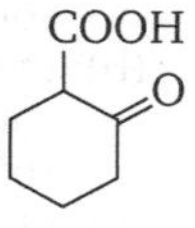 cyclohexane ring bearing COOH with a C=O on the adjacent ring carbon
(c) cyclohexane ring bearing COOH and, on the adjacent carbon, COOH
(d) 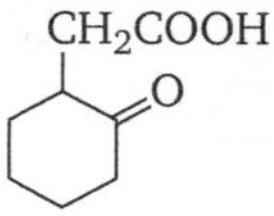 cyclohexane ring bearing $CH_2COOH$ with a C=O on the adjacent ring carbon

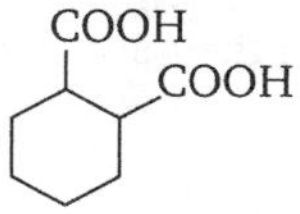

**25.** Using the data provided, calculate the multiple bond energy (kJ mol$^{-1}$) of a C≡C bond $C_2H_2$. That energy is (take the bond energy of a C—H bond as 350 kJ mol$^{-1}$)

$$2C(s) + H_2(g) \longrightarrow C_2H_2(g); \quad \Delta H = 225 \text{ kJ mol}^{-1}$$

$$2C(s) \longrightarrow 2C(g); \quad \Delta H = 1410 \text{ kJ mol}^{-1}$$

$$H_2(g) \longrightarrow 2H(g); \quad \Delta H = 330 \text{ kJ mol}^{-1}$$

**26.** The shape of $XeO_2F_2$ molecule is

(a) trigonal bipyramidal
(b) square planar
(c) tetrahedral
(d) see-saw

**27.** The major product H in the given reaction sequence is

$$CH_3—CH_2—CO—CH_3 \xrightarrow{^{\ominus}CN} G \xrightarrow{95\%\ H_2SO_4} H$$

(a) $CH_3—CH{=}C(CH_3)—COOH$

(b) $CH_3—CH{=}C(CH_3)—CN$

(c) $CH_3—CH_2—C(OH)(CH_3)—COOH$

(d) $CH_3—CH=C(CH_3)—CO—NH_2$

**28.** For a dilute solution containing 2.5 g of a non-volatile non-electrolyte solute in 100 g of water, the elevation in boiling point at 1 atm pressure is 2°C. Assuming concentration of solute is much lower than the concentration of solvent, the vapour pressure (mm of Hg) of the solution is (take $K_b = 0.76$ K kg mol$^{-1}$).

(a) 724
(b) 740
(c) 736
(d) 718

## Section II

**Direction** (Q. No. 29-34) *This section contains **6 multiple choice questions** relating to three paragraphs with two questions on each paragraph. Each question has four choices (a), (b), (c) and (d) out of which **ONLY ONE** is correct.*

### Paragraph

The electrochemical cell shown below is a concentration cell.

$M|M^{2+}$ (saturated solution of a sparingly soluble salt, $MX_2$)$||M^{2+}$ (0.001 mol dm$^{-3}$)$|M$.

The emf of the cell depends on the difference in concetration of $M^{2+}$ ions at the two electrodes. The emf of the cell at 298 is 0.059 V.

**29.** The solubility product ($K_{sp}$ ; mol$^3$dm$^{-9}$) of $MX_2$ at 298 based on the information available the given concentration cell is (take $2.303 \times R \times 298/F = 0.059$ V)

(a) $1 \times 10^{-15}$
(b) $4 \times 10^{-15}$
(c) $1 \times 10^{-12}$
(d) $4 \times 10^{-12}$

**30.** The value of $\Delta G$ (kJ mol$^{-1}$) for the given cell is (take 1 F = 96500 C mol$^{-1}$)

(a) – 5.7 (b) 5.7
(c) 11.4 (d) – 11.4

### Paragraph

Bleaching powder and bleach solution are produced on a large scale and used in several house hold products. The effectiveness of bleach solution is often measured by iodometry.

**31.** 25 mL of household bleach solution was mixed with 30 mL of 0.50 M KI and 10 mL of 4 N acetic acid. In the titration of the liberated iodine, 48 mL of 0.25 N $Na_2S_2O_3$ was used to reach the end point. The molarity of the household bleach solution, is

(a) 0.48 M (b) 0.96 M
(c) 0.24 M (d) 0.024 M

**32.** Bleaching powder contains a salt of an oxoacid as one of its components. The anhydride of that oxoacid, is

(a) $Cl_2O$ (b) $Cl_2O_7$
(c) $ClO_2$ (d) $Cl_2O_6$

### Paragraph

In the following reactions sequence, the compound $J$ is an intermediate.

$$I \xrightarrow[CH_3COONa]{(CH_3CO)_2O} J \xrightarrow[\text{(ii) } SOCl_2 \text{ (iii) anhyd. } AlCl_3]{\text{(i) } H_2, Pd/C} K$$

$J$ ($C_9H_8O_2$) gives effervescence on treatment with $NaHCO_3$ and positive Baeyer's test.

**33.** The compound $K$, is

(a) (b)

(c) (d)

**34.** The compound $I$, is

(a) (b)

(c) (d)

## Section III

**Direction** (Q. No. 35-40) *This section contains* **6** ***multiple choice questions.*** *Each question has four choices (a), (b), (c) and (d) out of which* ***ONE*** *or* ***MORE*** *are correct.*

**35.** With respect to graphite and diamond, which of the statement(s) given below is (are) correct?

(a) Graphite is harder than diamond.

(b) Graphite has higher electrical conductivity than diamond.

(c) Graphite has higher thermal conductivity than diamond.

(d) Graphite has higher C—C bond order than diamond.

**36.** The given graph/data I, II, III and IV represent general trends observed for different physisorption and chemisorption processes under mild conditions of temperature and pressure. Which of the following choice(s) about I, II, III and IV is (are) correct?

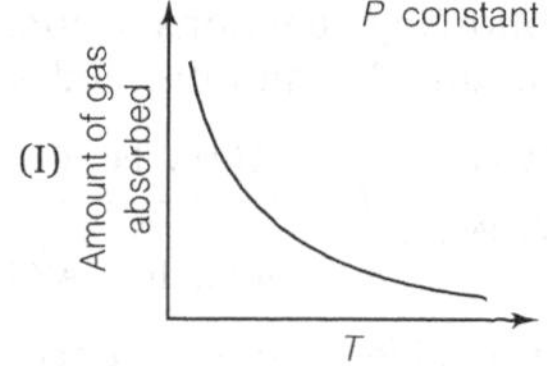


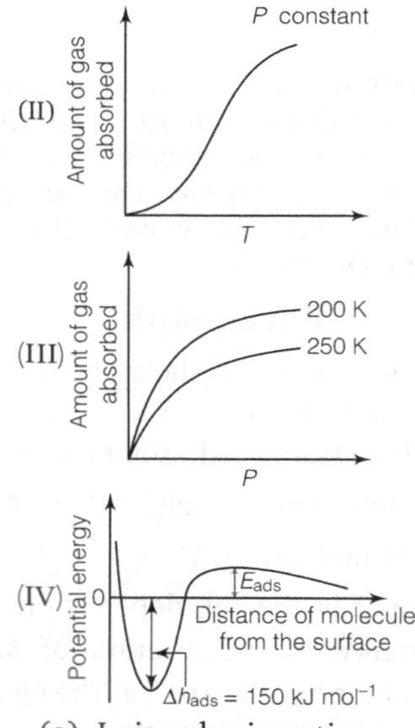


(a) I is physisorption and II is chemisorption

(b) I is physisorption and III is chemisorption

(c) IV is chemisorption and II is chemisorption

(d) IV is chemisorption and III is chemisorption

**37.** The reversible expansion on an ideal gas under adiabatic and isothermal conditions is shown in the figure. Which of the following statement(s) is (are) correct?

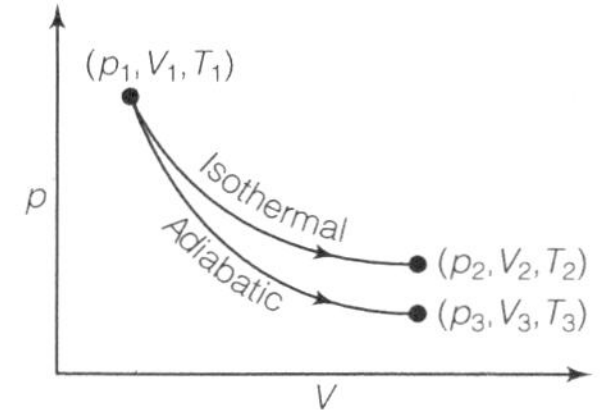


(a) $T_1 = T_2$
(b) $T_3 > T_1$
(c) $W_{\text{isothermal}} > W_{\text{adiabatic}}$
(d) $\Delta U_{\text{isothermal}} > \Delta U_{\text{adiabatic}}$

**38.** For the given aqueous reaction which of the statement(s) is (are) true?

Excess KI

$$+ K_3[Fe(CN)_6] \xrightarrow{\text{Dilute } H_2SO_4}$$

Brownish-yelllow solution

$\downarrow ZnSO_4$

(White precipitate + Brownish-yellow filtrate)

$\downarrow Na_2S_2O_3$

Colourless solution

(a) The first reaction is a redox reaction
(b) White precipitate is $Zn_3[Fe(CN)_6]_2$
(c) Addition of filtrate to starch solution gives blue colour
(d) White precipitate is soluble in NaOH solution

**39.** With reference to the scheme given, which of the given statement (s) about *T, U, V* and *W* is (are) correct?

$H_3C$ (lactone)

$\downarrow$ $LiAlH_4$ (excess)

$V \xleftarrow{CrO_3/H^+} U \xrightarrow{(CH_3CO)_2O} W$

(a) *T* is soluble in hot aqueous NaOH.
(b) *U* is optically active.
(c) Molecular formula of *W* is $C_{10}H_{18}O_4$.
(d) *V* gives effervescence on treatment with aqueous $NaHCO_3$.

**40.** Which of the given statement(s) about *N, O, P* and *Q* with respect to *M* is (are) correct?

M: HO, H, Cl, HO, H, $CH_3$
N: HO, H, $CH_3$, H, OH, $CH_3$
O: Cl, H, OH, HO, H, $CH_3$
P: $CH_3$, H, OH, HO, H, Cl
Q: $CH_3$, HO, H, HO, H, Cl

(a) *M* and *N* are non-mirror image stereoisomers
(b) *M* and *O* are identical
(c) *M* and *P* are enantiomers
(d) *M* and *Q* are identical

# Answers

## Paper 1

**1.** (b) **2.** (c) **3.** (c) **4.** (c) **5.** (b) **6.** (b) **7.** (a) **8.** (d) **9.** (d) **10.** (a)
**11.** (a,c) **12.** (b,c) **13.** (b,d) **14.** (a,d) **15.** (a,c,d)
**16.** (9) **17.** (8) **18.** (4) **19.** (8) **20.** (8)

## Paper 2

**21.** (c) **22.** (c) **23.** (b) **24.** (b) **25.** (d) **26.** (d) **27.** (a) **28.** (a) **29.** (b) **30.** (d)
**31.** (c) **32.** (a) **33.** (c) **34.** (a) **35.** (b,d) **36.** (a,c) **37.** (a,c,d)
**38.** (a,c,d) **39.** (a,c,d) **40.** (a,b,c)

# Hints & Solutions

## Paper 1

**1.** Let oxidation number of "N" be $x$.

In $HNO_3$ $+1 + x + 3(-2) = 0$

$\Rightarrow \quad x = +5$

In NO $\quad x - 2 = 0 \Rightarrow x = +2$

In $N_2$ $\quad x = 0$

In $NH_4Cl$ $\quad x + 4 - 1 = 0 \Rightarrow x = -3$

**2.** According to Bohr's model

$$mvr = \frac{nh}{2\pi}$$

$$\Rightarrow \quad (mv)^2 = \frac{n^2h^2}{4\pi^2r^2}$$

$$\Rightarrow \quad KE = \frac{1}{2}mv^2 = \frac{n^2h^2}{8\pi^2r^2m} \quad \ldots(i)$$

Also Bohr's radius for H-atom is $r = n^2a_0$

Substituting '$r$' in equation (i) gives

$$KE = \frac{h^2}{8\pi^2n^2a_0^2m}$$

when $n = 2$, $KE = \dfrac{h^2}{32\pi^2 a_0^2 m}$

**3.** The given reaction is an example of repeated aldol condensation followed by Cannizzaro reaction.

*Step I*

$CH_3CHO + OH^- \longrightarrow \bar{C}H_2—CHO + H_2O$

$H—C(=O)—H + \bar{C}H_2—CHO \rightleftharpoons H—C(O^-)(H)—CH_2—CHO \xrightarrow{H_2O} CH_2(OH)—CH_2—CHO$

*Step II*

$HOCH_2—CH_2—CHO + HO^- \rightleftharpoons HO—CH_2—\bar{C}H—CHO + H_2O$

$H—C(=O)—H + \bar{C}H(CH_2OH)—CHO \rightleftharpoons H—C(O^-)(H)—CH(CH_2OH)—CHO \xrightarrow{H_2O} CH_2(OH)—CH(CH_2OH)—CHO$

*Step III*

$HOCH_2—CH(CH_2OH)—CHO + HO^- \rightleftharpoons HOCH_2—\bar{C}(CH_2OH)—CHO + H_2O$

$H—C(=O)—H + {}^-CH(CH_2OH)(CH_2OH)—CHO \rightleftharpoons H—C(O^-)(H)—C(CH_2OH)(CH_2OH)—CHO \xrightarrow{H_2O} HOCH_2—C(CH_2OH)(CH_2OH)—CHO$

*Step IV* $HOCH_2—\underset{CH_2OH}{\overset{CH_2OH}{C}}—\overset{O}{\overset{\|}{C}}—H + H—\underset{H}{\overset{O^-}{C}}—OH$

$\xrightarrow{\text{Cannizzaro}}$ $C(CH_2OH)_4$ + $HCOO^-$

In the last step, formaldehyde is oxidized and the other aldehyde is reduced giving the desired products.

4. The van der Waals' equation of state is

$$\left(p + \frac{n^2a}{V^2}\right)(V - nb) = nRT$$

For one mole and when $b = 0$, the above equation condenses to

$$\left(p + \frac{a}{V^2}\right)V = RT$$

$$\Rightarrow \quad pV = RT - \frac{a}{V} \quad \ldots\text{(i)}$$

Equation (i) is a straight equation between $pV$ and $\frac{1}{V}$ whose slope is "$-a$".

Equating with slope of the straight line given in the graph.

$$-a = \frac{20.1 - 21.6}{3 - 2} = -1.5$$

$$\Rightarrow \quad a = 1.5$$

5. Allene is the name given to propdiene, $\underset{\text{Allene}}{H_2C{=}C{=}CH_2}$

Hybridization of an atom is determined by determining the number of hybrid orbitals at that atom which is equal to the number of sigma ($\sigma$) bonds plus number of lone pairs at the concerned atom.

**Note** Pi($\pi$) bonds are not formed by hybrid orbitals, therefore, not counted for hybridization.

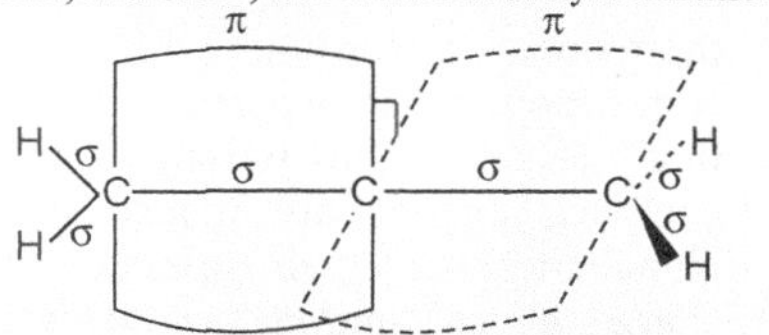


Three dimensional allene, the two consecutive pi-bonds are perpendicular to each other.

Here, the terminal carbons have only three sigma bonds associated with them, therefore, hybridization of terminal carbons is $sp^3$. The central carbon has only two sigma bonds associated, hence hybridization at central carbon is ***sp***.

6. Contribution of atom from the edge centre is 1/4. Therefore, number of

$M = \frac{1}{4} \times 4$ (from edge centre) + 1 (from body centre) = 2

Number of $X = \frac{1}{8} \times 8$ (from corners) $+ \frac{1}{2} \times 6$ (from face centre) = 4

$\Rightarrow$ Empirical formula $= M_2X_4 = MX_2$

7. Ozonolysis of the given triene occur as follows :

$$H_3C—CH{=}CH—\underset{H}{\overset{CH_3}{C}}—CH{=}CH—\underset{H}{\overset{CH_3}{C}}—CH{=}CH—CH_3$$

$$\xrightarrow{O_3}\ \xrightarrow[H_2O]{Zn} 2CH_3—CHO + \underset{CHO}{\overset{CHO}{|}} + 2\,HOC—\underset{H}{\overset{CH_3}{C}}—CHO$$

Since, none of the above dial is chiral, no optically active product is obtained.

**8.** First of all, the compound has complex positive part "$[Co(H_2O)_4(NH_3)_2]^{3+}$" therefore, according to IUPAC conventions, positive part will be named first. Secondly, in writing name of complex, ligands are named first in alphabetical order, irrespective of its charge, hence "ammine" will be written prior to "aqua".

**Note** In alphabetical order, original name of ligands are considered not the initials of prefixes. Also, special precaution should be taken in spelling the name of $NH_3$ ligand as it is ammine. Therefore, name of the complex is $[Co(H_2O)_4(NH_3)_2]Cl_3$
Diammine tetraaqua cobalt (III) chloride.

**9.** Structures of the various compounds are

Picric acid (A)

Barbituric acid (B)

Ascorbic acid (C)

Aspirin (D)

**10.** The aqueous solution of $CuSO_4$ consist of the complex $[Cu(H_2O)_4]^{2+}$ ion which adsorb in orange-red region and impart deep blue colouration to solution.

**11.** (a) Entropy is a state function, change in entropy in a cyclic process is zero.

Therefore,

$$\Delta S_{X \to Y} + \Delta S_{Y \to Z} + \Delta S_{Z \to X} = 0$$

$$\Rightarrow \quad -\Delta S_{Z \to X} = \Delta S_{X \to Y} + \Delta S_{Y \to Z}$$
$$= \Delta S_{X \to Z}$$

**Analysis of options (b) and (c).**

Work is a non-stable function, it does depends on the path followed. $W_{Y \to Z} = 0$ as $\Delta V = 0$. Therefore, $W_{X \to Y \to Z} = W_{X \to Y}$. Also, work is the area under the curve on *p-V* diagram.

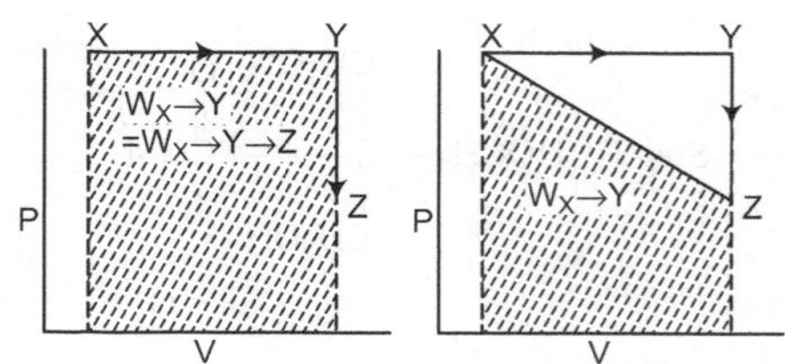


As shown above $W_{X \to Y} + W_{Y \to Z} = W_{X \to Y} = W_{X \to Y \to Z}$ but not equal to $W_{X \to Z}$.

**12.** According to Huckel rule, the compounds which have $4n$ ($n = 0, 1, 2, 3..$) delocalized $\pi$-electrons in a close-loop are anti-aromatic and characteristically unstable. Compound '*B*' satisfy the criteria of anti-aromaticity as :

Compound '*C*' is anti-aromatic in its resonance form :

anti-aromatic

Compound '*A*' has $4\pi$-electrons which are also delocalized but do not constitute close loop, hence, non-aromatic.

Compound *D* is atomatic, characteristically stable.

Tropyllium ion, aromatic

**13.** For separation be differential extraction one of the component must form salt with the given base so that the salt will be extracted in aqueous layer leaving other component in organic layer.

(a) Both phenol and benzoic acid forms salt with NaOH, hence this mixture can't be separated.

(b) Benzoic acid forms salt with NaOH while benzyl alcohol does not, hence the mixture can be separated using NaOH. Also benzoic acid forms salt with $NaHCO_3$ but benzyl alcohol does not, hence $NaHCO_3$ can be used for separation.

(c) Neither benzyl alcohol nor phenol forms salt with $NaHCO_3$, mixture cannot be separated using $NaHCO_3$.

(d) $C_6H_5CH_2COOH$ forms salt with NaOH, $C_6H_5CH_2OH$ does not, hence mixture can be separated using NaOH. $C_6H_5CH_2COOH$ forms salt with $NaHCO_3$. $C_6H_5CH_2OH$ does not, hence mixture can be separated using $NaHCO_3$.

**14.** Lyophobic sol, which is otherwise unstable, gets stabilized by preferential adsorption of ions on their surface developing a potential difference between the fixed layer and the diffused layer.

**15.** Solubilities of silver halides in water decreases from fluoride (AgF) to iodide (AgI). Silver fluoride is readily soluble in water hence when $AgNO_3$ solution is added to HF solution (HF being weak acid, its solution maintain very low concentration of $F^-$) no precipitate of AgF is formed.

HCl, HBr and HI being all strong acid, forms precipitates of AgCl, AgBr and AgI when $AgNO_3$ solution is added to their aqueous solution.

$$HCl(aq) + AgNO_3(aq) \longrightarrow \underset{\text{Curdy white}}{AgCl\,(s)} + HNO_3(aq)$$

$$HBr(aq) + AgNO_3(aq) \longrightarrow \underset{\text{Pale yellow}}{AgBr\,(s)} + HNO_3(aq)$$

$$HI\,(aq) + AgNO_3(aq) \longrightarrow \underset{\text{Yellow}}{AgI(s)} + HNO_3(aq)$$

The solubilities decreases from AgCl to AgI, AgCl dissolves in aqueous ammonia, AgBr dissolves only slightly in concentrated ammonia while AgI does not dissolve in ammonia solution.

$Na_2S_2O_3$ solution dissolve all three, AgCl, AgBr, AgI by forming complex $[Ag(S_2O_3)_2]^{3-}$ as $S_2O_3^{2-}$ is a stronger complexing agent than ammonia.

**16.** For a first order process $kt = \ln \frac{[A]_0}{[A]}$

where, $[A]_0$ = initial concentration.

$[A]$ = concentration of reactant remaining at time "$t$".

$$\Rightarrow \quad kt_{1/8} = \ln \frac{[A]_0}{[A]_0/8} = \ln 8 \quad \ldots(i)$$

$$\text{and } kt_{1/10} = \ln \frac{[A]_0}{[A]_0/10} = \ln 10 \quad \ldots(ii)$$

Therefore,

$$\frac{t_{1/8}}{t_{1/10}} = \frac{\ln 8}{\ln 10} = \log 8 = 3 \log 2$$

$$= 3 \times 0.3 = 0.9$$

$$\Rightarrow \quad \frac{t_{1/8}}{t_{1/10}} \times 10 = 0.9 \times 10 = 9$$

**17.** The D-from of given sugar is

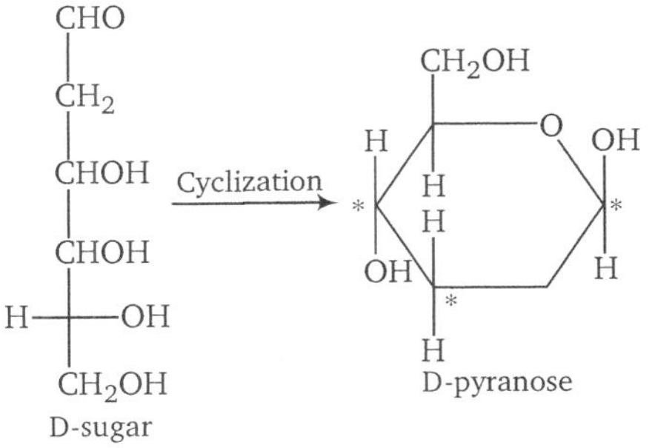

CHO
|
$CH_2$
|
CHOH
|
CHOH
|
H—C—OH
|
$CH_2OH$
D-sugar

—Cyclization→

D-pyranose (ring with $CH_2OH$, O, OH, H, H, H, OH, H substituents; chiral carbons starred *)

Configurations at the three chiral cabons (starred) can be changed maintaining D-configuration. Hence, the total number of steroisomers of D-pyranose = $2^3$ = 8.

**18.** The amino acid remain completely in Zwitter ionic form at its isoelectric point. Amino acids with additional acidic group have their isoelectric pH less than 7.0 and increasing pH above isoelectric point makes them anionic. On the other hand, amino acids with additional basic group have their isoelectric pH greater than 7.0 and decreasing pH below isoelectric point (by adding acid solution) makes them cationic. The given peptide with followings $R_1$ and $R_2$ are basic, will remain protonated (cationic) at pH = 7.0.

| Peptide | $R_1$ | $R_2$ |
|---|---|---|
| IV | $CH_2CONH_2$ | $(CH_2)_4NH_4$ |
| VI | $(CH_2)_4NH_2$ | $(CH_2)_4NH_4$ |
| VIII | $CH_2OH$ | $(CH_2)_4NH_4$ |
| IX | $(CH_2)_4NH_2$ | $CH_3$ |

**19.** Balancing the given nuclear reaction in terms of atomic number (charge) and mass number gives :

$$_{29}Cu^{63} + {}_1H^1 \longrightarrow 6\,{}_0n^1 + {}_2He^4(a) + 2\,{}_1H^1 + {}_2He^4 + {}_{26}X^{52}$$

The atomic number 26 correspond to transition metal "Fe" which belongs to $8^{th}$ group of modern Periodic Table.

**20.** Mass of HCl in 1.0 mL stock solution

$$= 1.25 \times \frac{29.2}{100}$$

$$= 0.365 \text{ g}$$

Mass of HCl required for 200 mL 0.4 M HCl

$$= \frac{200}{1000} \times 0.4 \times 36.5$$

$$= 0.08 \times 36.5 \text{ g}$$

∴ 0.365 g of HCl is present in 1.0 mL stock solution.

∴ $0.08 \times 36.5$ g HCl will be present in $\frac{0.08 \times 36.5}{0.365} = 8.0$ mL

## Paper 2

**21.** In the given comples, $NiCl_2\{P(C_2H_5)_2(C_6H_5)\}_2$ nickel is in + 2 oxidation state and the ground state electronic configuration of $Ni^{2+}$ ion in free gaseous state is

$Ni^{2+}$ [⇅|⇅|⇅|↑|↑] $3d^8$ [ ] $4s^0$ [ | | ] $4p^0$

For the given four-coordinated complex to be paramagnetic, it must possess unpaired electrons in the valence shell. To satisfy this condition, four lone pairs from the four ligands occupies the four $sp^3$ hybrid orbitals as :

$Ni^{2+}$ [⇅|⇅|⇅|↑|↑] [••] [••|••|••] ($sp^3$)

Therefore, geometry of paramagnetic complex must be tetrahedral. On the other hand, for complex to be diamagnetic, there should not be any unpaired electrons in the valence shell. This condition can be fullfilled by pairing electrons of 3*d* orbitals against Hund's rule as

$Ni^{2+}$ [⇅|⇅|⇅|⇅|••] $3d^8$ [••] [••|••| ] ($dsp^2$)

The above electronic arrangement gives $dsp^2$ hybridization and therefore, square planar geometry to the complex.

**22.** The reaction of white phosphorus with aqueous alkali is

$$P_4 + 3\,NaOH + 3H_2O \longrightarrow PH_3 + NaH_2PO_2$$

In the above reaction phosphorus is simultaneously oxidized $[P_4(0) \longrightarrow NaH_2\overset{+1}{P}O_2]$ as well as reduced $[P_4(0) \longrightarrow \overset{-3}{P}H_3]$. Therefore, this is an example of disproportionation reaction. Oxidation number of phosphorus in $PH_3$ is −3 and in $NaH_2PO_2$ is +1. However, + 1 oxidation no. is not given in any option, one might think that $NaH_2PO_2$ has gone to further decomposition on heating.

$$2NaH_2PO_2 \xrightarrow{\Delta} Na_2H\overset{+5}{P}O_4 + PH_3$$

**23.** The reactions involved in extraction of silver by cyanide process are

$$Ag_2S + CN^- + O_2 \longrightarrow [Ag(CN)_2]^- + SO_2 \quad \ldots(i)$$

$[Ag(CN)_2]^- + Zn \longrightarrow$
$[Zn(CN)_4]^{2-} + Ag$ ...(ii)

In reaction (i), sulphide is oxidized to $SO_2$ by oxygen. In the reaction (ii), silver ion ($Ag^+$) is reduced to Ag by Zn. Therefore, $O_2$ is oxidising agent and Zn is reducing agent.

**24.** It is a β-keto acid which undergo decarboxylation in very mild condtion *i.e.*, on simple heating. This occur through a six member cyclic transition state as

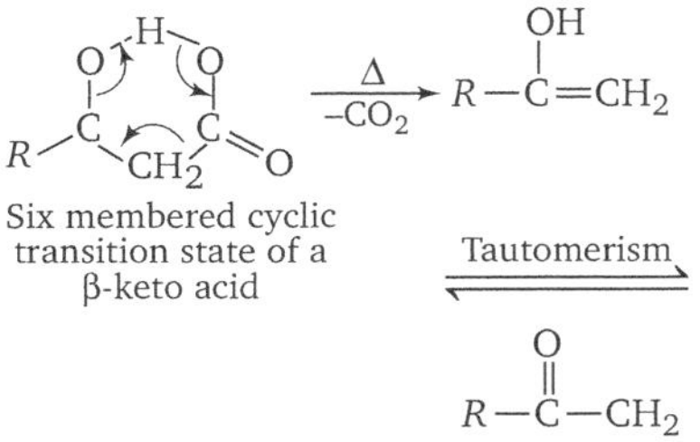

$$R-\overset{\overset{O\cdots H\cdots O}{|}}{C}(CH_2)C=O \xrightarrow[-CO_2]{\Delta} R-\overset{OH}{\overset{|}{C}}=CH_2$$

Six membered cyclic transition state of a β-keto acid

$$\overset{\text{Tautomerism}}{\rightleftharpoons} R-\overset{O}{\overset{\|}{C}}-CH_3$$

(i) Ordinary carboxylic acid require soda-lime catalyst for decarboxylation. (ii) Final step of decarboxylation in the above shown mechanism involve tautomerism, thererfore, for decarboxylation of β-ketro acid by above mechanism, the acid must contain an α-H].

**25.** For calculation of $C\equiv C$ bond energy, we must first calculate dissociation energy of $C_2H_2$ as

$C_2H_2(g) \longrightarrow 2C(g) + 2H(g)$ ...(i)

Using the given bond energies and enthalpies :

$C_2H_2(g) \longrightarrow 2C(g) + 2H(g)$; $\Delta H = -225$ kJ ...(ii)

$2C(s) \longrightarrow 2C(g)$; $\Delta H = 1410$ kJ ...(iii)

$H_2(g) \longrightarrow 2H(g)$; $\Delta H = 330$ kJ ...(iv)

Adding equation (ii), (iii) and (iv) gives equation (i).

$\Rightarrow$ $C_2H_2(g) \longrightarrow 2C(g) + 2H(g)$; $\Delta H = 1515$ kJ

$\Rightarrow$ 1515 kJ = 2 × (C—H) BE + ($C\equiv C$) BE
= 2 × 350 + ($C\equiv C$) BE

$\Rightarrow$ ($C\equiv C$) BE = 1515 − 700
= 815 kJ/mol

**26.** In $XeO_2F_2$, the bonding arrangement around the central atom Xe is

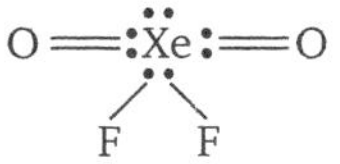

$$O=\ddot{X}e=O \text{ (with two F below)}$$

4σ-bonds + 1.0 lp = 5

$\Rightarrow$ Hybridization of Xe = $sp^3d$

$sp^3d$ hybridization corresponds to trigonal bipyramidal geometry. Also, in trigonal bipyramidal geometry, lone pairs remain present on equatorial positions in order to give less electronic repulsion.

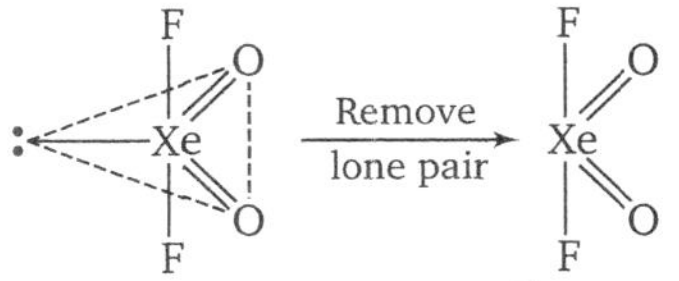

Remove lone pair →

See-saw shape

**Note** According to Bent's rule, the more electronegative atoms must be present on axial position. Hence, F are kept on axial positions.

**27.** The first step is cyanohydrin reaction

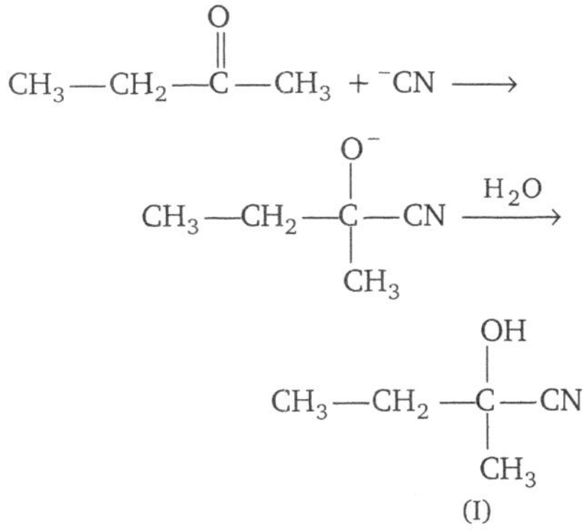

$$CH_3-CH_2-\overset{O}{\overset{\|}{C}}-CH_3 + {}^-CN \longrightarrow$$

$$CH_3-CH_2-\underset{CH_3}{\underset{|}{\overset{O^-}{\overset{|}{C}}}}-CN \xrightarrow{H_2O}$$

$$CH_3-CH_2-\underset{CH_3}{\underset{|}{\overset{OH}{\overset{|}{C}}}}-CN$$
(I)

In the second step the —CN of intermediate (I) is first hydrolysed and then dehydrated on heating in presence of conc. $H_2SO_4$.

$$CH_3—CH_2—\underset{CH_3}{\overset{OH}{\underset{|}{\overset{|}{C}}}}—CN \xrightarrow{H_2SO_4}$$
(I)

$$CH_3—CH_2—\underset{CH_3}{\overset{OH}{\underset{|}{\overset{|}{C}}}}—COOH$$

$$\xrightarrow{\Delta} CH_3—CH{=}\underset{CH_3}{\underset{|}{C}}—COOH$$

**28.** The elevation in boiling point is

$$\Delta T_b = K_b \cdot m : m = \text{molality} = \frac{n_2}{w_1} \times 1000$$

[$n_2$ = number of moles of solute, $w_1$ = weight of solvent in gram]

$$\Rightarrow \quad 2 = 0.76 \times \frac{n_2}{100} \times 1000$$

$$\Rightarrow \quad n_2 = \frac{5}{19}$$

Also, from Raoult's law of lowering of vapour pressure :

$$\frac{-\Delta p}{p_0} = x_2 = \frac{n_2}{n_1 + n_2} \approx \frac{n_2}{n_1} \; [\because n_1 >> n_2]$$

$$\Rightarrow \quad -\Delta p = 760 \times \frac{5}{19} \times \frac{18}{100}$$

$$= 36 \text{ mm of Hg}$$

$$\Rightarrow \quad p = 760 - 36 = 724 \text{ mm of Hg}$$

**Paragraph** (Q. No. 29 to 30) For the given concentration cell, the cell reactions are $M \longrightarrow M^{2+}$ at left hand electrode

$M^{2+} \longrightarrow M$ at right hand electrode

$\Rightarrow \quad M^{2+}$ (RHS electrode)

$\longrightarrow M^{2+}$ (LHS electrode)

$$E° = 0$$

Applying Nernst equation

$$E_{\text{cell}} = 0.059 = 0 - \frac{0.059}{2}$$

$$\log \frac{[M^{2+}] \text{ at LHS electrode}}{0.001}$$

$$\Rightarrow \quad \log \frac{[M^{2+}] \text{ at LHS electrode}}{0.001} = -2$$

$$\Rightarrow \quad [M^{2+}] \text{ at LHS electrode}$$

$$= 10^{-2} \times 0.001 = 10^{-5} \text{ M}$$

**29.** The solubility equilibria for $MX_2$ is

$$MX_2(s) \rightleftharpoons M^{2+}(aq) + 2X^-(aq)$$

Solubility product, $K_{sp} = [M^{2+}][X^-]^2$

$$= 10^{-5} \times (2 \times 10^{-5})^2 = 4 \times 10^{-15}$$

[$\because$ In saturated solution of $MX_2$, $[X^-] = 2[M^{2+}]$]

**30.** $\Delta G = -nEF = -\dfrac{2 \times 0.059 \times 96500}{1000}$ kJ

$= -11.4$ kJ

**Paragraph** (Q. No. 31 to 32)

**31.** The involved redox reactions are :

$$2H^+ + OCl^- + 2I^- \longrightarrow Cl^- + I_2 + H_2O \quad \ldots\text{(i)}$$

$$I_2 + 2S_2O_3^{2-} \longrightarrow 2I^- + S_4O_6^{2-} \quad \ldots\text{(ii)}$$

Also the $n$-factor of $S_2O_3^{2-}$ is one as

$$2S_2O_3^{2-} \longrightarrow S_4O_6^{2-} + 2e^-$$

[one '$e$' is produced per unit of $S_2O_3^{2-}$]

$\Rightarrow$ Molarity of $Na_2S_2O_3$ = 0.25 N × 1 = 0.25 M

$\Rightarrow m$ mol of $Na_2S_2O_3$ used up

$$= 0.25 \times 48 = 12$$

Now from stoichiometry of reaction (ii) 12 m mol of $S_2O_3^{2-}$ would have reduced 6 mmol of $I_2$.

From stoichiometry of reaction (i)

$m$ mol of $OCl^-$ reduced = $m$ mol in $I_2$ produced = 6

$\Rightarrow$ Molarity of household bleach solution

$$= \frac{6}{25} = \mathbf{0.24\ M}$$

**Shortcut Method**

Milliequivalent of $Na_2S_2O_3$

$= \text{milliequivalent of } OCl^-$
$= 0.25 \times 48 = 12$

Also $n$-factor of $OCl^- = 2[Cl^+ \longrightarrow Cl^-$, gain of $2e^-$

$\Rightarrow$ mmol of $OCl^- = \frac{12}{2} = 6$ mmol.

Remaining part is solved in the same manner.

**32.** Bleaching powder is $Ca(OCl)Cl$. Therefore, the oxoacid whose salt is present in bleaching powder is HOCl. Anhydride of HOCl is — $Cl_2O$ as

$$2\,HOCl \longrightarrow Cl_2O + H_2O$$

**Note** The oxidation number of element in anhydride and oxoacid remains the same].

**Paragraph** (Q. No. 33 to 34) The first step of reaction is Perkin's condensation.

$$\underset{I}{C_6H_5CHO} + (CH_3CO)_2O \xrightarrow{CH_3COONa} \underset{J}{C_6H_5—CH=CH—COOH}$$

$J$ being a carboxylic acid gives effervescence with $NaHCO_3$. Also, $J$ has olefinic bond, it wil decolourise Baeyer's reagent.

In the second step, $J$ on treatment with $H_2/Pd/C$ undergo hydrogenation at olefinic bond only as :

$$J + H_2 / Pd \longrightarrow C_6H_5—CH_2—CH_2—COOH$$

The hydrogenated acid, on treatment with $SOCl_2$ gives acid chloride.

$$C_6H_5—CH_2—CH_2—COOH + SOCl_2 \rightarrow C_6H_5—CH_2—CH_2—COCl + HCl + SO_2$$

In the final step, acid chloride formed above undergo intramolecular Friedel-Craft acylation as

$$C_6H_5CH_2CH_2COCl \xrightarrow{AlCl_3} \underset{\text{'K'}}{\text{1-indanone}}$$

**35.** Diamond has a three dimensional network structure, a hard substance where graphite is soft due to layered structure.

In graphite, only three valence electrons are involved in bonding and one electron remain free giving electrical conductivity. In diamond, all the four valence electrons are covalently bonded hence, insulator.

Diamond is better thermal conductor then graphite. Electrical conductivity is due to availability of free electrons, thermal conduction is due to transfer of thermal vibrational energy from one atom to another atom. A compact and precisely aligned crystals like diamond thus facilitate better movement of heat.

In graphite C—C bond acquire some double bond character, hence, higher bond order than in diamond.

**36.** Graph-I represents physiorption as in physisorption, absorbents are bonded to adsorbate through weak van der Waals' force.

Increasing temperature increases kinetic energy of adsorbed particles increasing the rate of desorption, hence amount of adsorption decreases.

Graph-III also represents physical adsorption as extent of adsorption increasing with pressure.

Graph-II represents chemisorption as it is simple activation energy diagram of a chemical reaction.

Graph-IV represents, chemisorption as it represents the potential energy diagram for the formation of a typical covalent bond.

**37.** (a) Since, change of state $(p_1, V_1, T_1)$ to $(p_2, V_2, T_2)$ is isothermal therefore, $T_1 = T_2$.

(b) Since, change of state $(p_1, V_1, T_1)$ to $(p_3, V_3, T_3)$ is an adiabatic expansion it brings about cooling of gas, therefore, $T_3 < T_1$.

(c) Work-done is the area under the curve of $p$-$V$ diagram. As obvious from the given diagram, magnitude of area under the isothermal curve is greater than the same under adiabatic curve, hence $W_{isothermal} > W_{adiabatic}$

**Note** Here only magnitudes of work is being considered otherwise both works have negative sign.

(d) $\Delta U = nC_v \Delta T$

In isothermal process, $\Delta U = 0$ as $\Delta T = 0$

In adiabatic process,

$$\Delta U = nC_v(T_3 - T_1) < 0 \text{ as } T_3 < T_1.$$

$$\Rightarrow \quad \Delta U_{isothermal} > \Delta U_{adiabatic}$$

**38.** $K_3[\overset{+3}{Fe}(CN)_6] + KI(\text{excess}) \longrightarrow K_4[\overset{+2}{Fe}(CN)_6] + KI_3$ (Redox)
Brownish yellow solution

$K_4[Fe(CN)_6] + ZnSO_4 \longrightarrow K_2Zn_3[Fe(CN)_6]_2$ or $K_2Zn[Fe(CN)_6]$
White ppt

$I_3^- + 2Na_2S_2O_3 \longrightarrow Na_2S_4O_6 + 2NaI + I_2$

$I_3^-$: Brownish yellow filtrate; $Na_2S_4O_6$: Clear solution; $I_2$: (Turns starch solution blue)

$K_2Zn[Fe(CN)_6]$ reacts with NaOH as

$K_2Zn[Fe(CN)_6] + NaOH \longrightarrow [Zn(OH)_4]^{2-} + [Fe(CN)_6]^{4-}$

**39.** (a) $T$ is an ester hydrolysis in hot aqueous alkali.

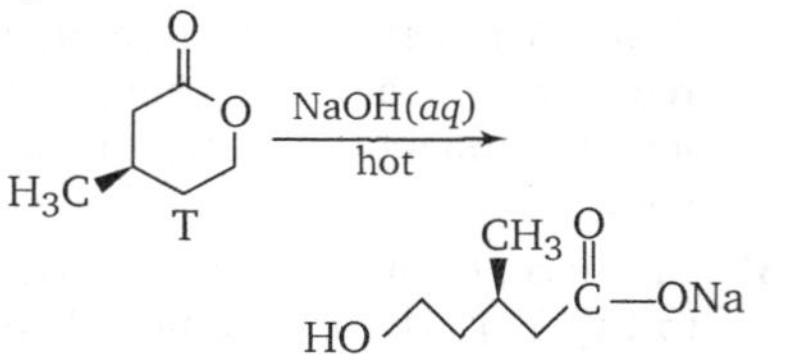

T (lactone with $H_3C$) $\xrightarrow[\text{hot}]{NaOH(aq)}$ HO–CH₂CH₂CH(CH₃)CH₂–C(=O)–ONa

Soluble in aqueous NaOH

(b) $LiAlH_4$ reduces ester to alcohol as

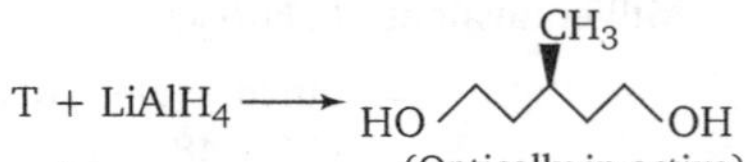

$T + LiAlH_4 \longrightarrow$ HO–CH₂CH₂CH(CH₃)CH₂CH₂–OH

(Optically in active)

"$U$" No chiral carbon optically inactive.

(c) $U$ on treatment with excess of acetic anhydride forms a diester as

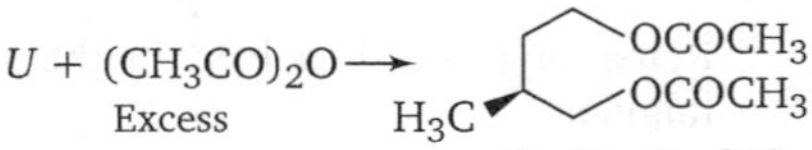

$U + (CH_3CO)_2O \longrightarrow$ (Excess) $H_3C$–CH(CH₂CH₂OCOCH₃)(CH₂OCOCH₃)

$C_{10}H_{18}O_4$ ($W$)

(d) $U$ on treatment with $CrO_3 | H^+$ undergo oxidation to diacid which gives effervescence with $NaHCO_3$.

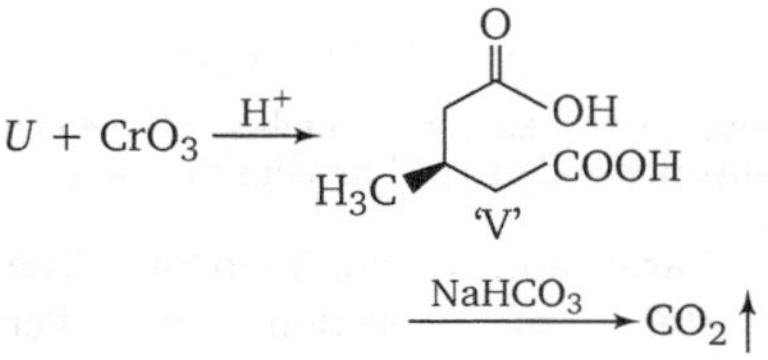

$U + CrO_3 \xrightarrow{H^+}$ $H_3C$–CH(CH₂COOH)(CH₂COOH) 'V' $\xrightarrow{NaHCO_3} CO_2\uparrow$

**40.** Converting all of them into Fischer projection.

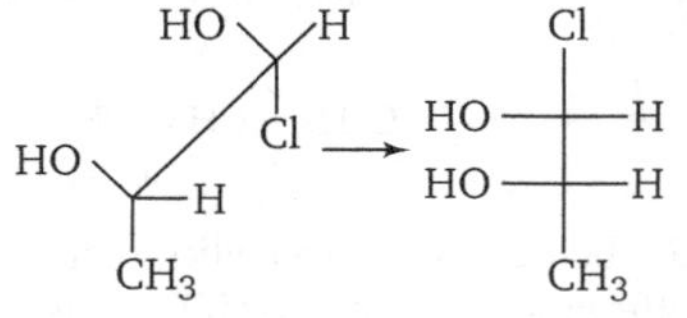

M: top Cl; HO—C—H; HO—C—H; bottom $CH_3$

N: top $CH_3$; HO—C—H; H—C—OH; bottom Cl

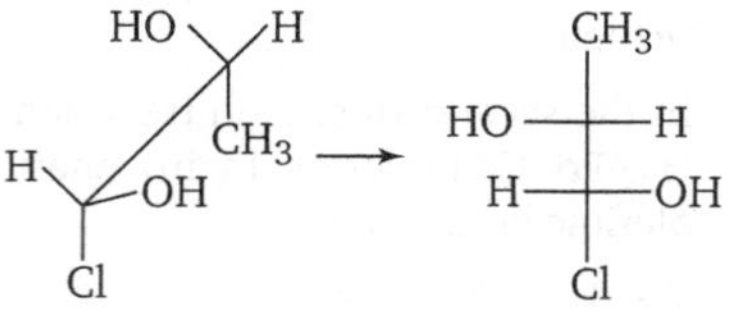

Since, '$M$' and '$N$' have —OH on same side and opposite side respectively, they cannot be miror image, they are diasteromers.

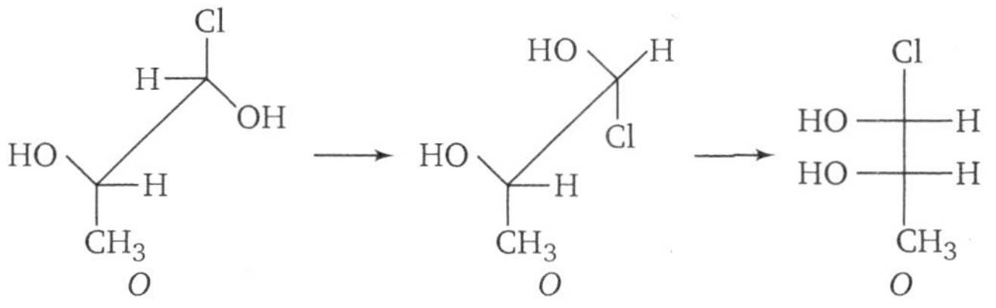

O O O

'M' and 'O' are identical.

**Note** Fischer projection represents eclipse form of sawhorse projection.
For comparision purpose, similar types of eclipse conformers must be drawn *i.e.*, both vertically up or both vertically down.

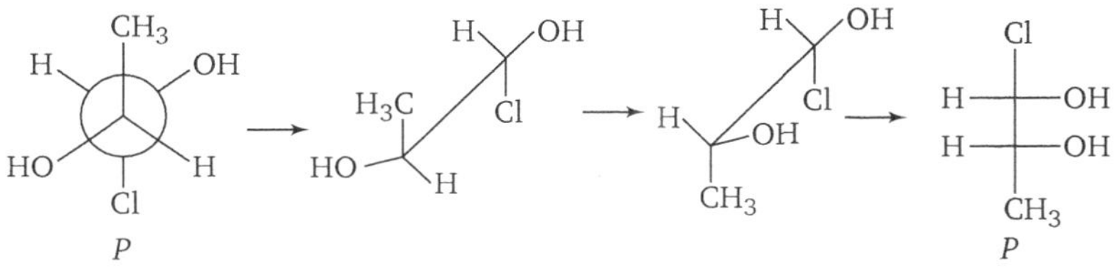

P P

'M' and 'P' are non-superimposable mirror images, hence, enantiomers.

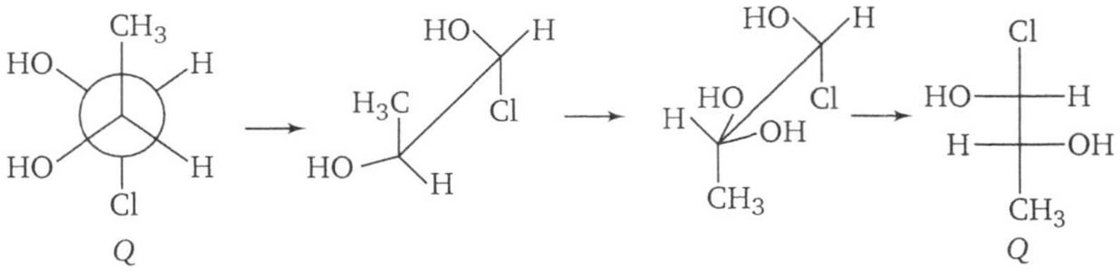

Q Q

'M' and 'Q' are non-identical they are distereomers.

# IIT JEE

## SOLVED PAPER 2012

## Mathematics

### Paper 1

#### Section I

**Direction** (Q. Nos. 1-10) *This section contains 10 multiple choice questions. Each question has four choices (a), (b), (c) and (d) out of which only one is correct.*

1. The point $P$ is the intersection of the straight line joining the points $Q(2, 3, 5)$ and $R(1, -1, 4)$ with the plane $5x - 4y - z = 1$. If $S$ is the foot of the perpendicular drawn from the point $T(2, 1, 4)$ to $QR$, then the length of the line segment $PS$ is

   (a) $\frac{1}{\sqrt{2}}$ (b) $\sqrt{2}$ (c) 2 (d) $2\sqrt{2}$

2. The integral $\int \frac{\sec^2 x}{(\sec x + \tan x)^{9/2}} dx$ equals to (for some arbitrary constant $K$)

   (a) $\frac{-1}{(\sec x + \tan x)^{11/2}} \left\{\frac{1}{11} - \frac{1}{7}(\sec x + \tan x)^2\right\} + K$

   (b) $\frac{1}{(\sec x + \tan x)^{11/2}} \left\{\frac{1}{11} - \frac{1}{7}(\sec x + \tan x)^2\right\} + K$

   (c) $\frac{-1}{(\sec x + \tan x)^{11/2}} \left\{\frac{1}{11} + \frac{1}{7}(\sec x + \tan x)^2\right\} + K$

   (d) $\frac{1}{(\sec x + \tan x)^{11/2}} \left\{\frac{1}{11} + \frac{1}{7}(\sec x + \tan x)^2\right\} + K$

3. Let $z$ be a complex number such that the imaginary part of $z$ is non-zero and $a = z^2 + z + 1$ is real. Then, $a$ cannot take the value

   (a) $-1$ (b) $\frac{1}{3}$ (c) $\frac{1}{2}$ (d) $\frac{3}{4}$

4. Let $f(x) = \begin{cases} x^2 \left|\cos \frac{\pi}{x}\right|, & x \neq 0, x \in R, \\ 0, & x = 0 \end{cases}$

   then $f$ is

   (a) differentiable both at $x = 0$ and at $x = 2$
   (b) differentiable at $x = 0$ but not differentiable at $x = 2$
   (c) not differentiable at $x = 0$ but differentiable at $x = 2$
   (d) differentiable neither at $x = 0$ nor at $x = 2$

5. The total number of ways in which 5 balls of different colours can be distributed among 3 persons so that each person gets atleast one ball is

   (a) 75 (b) 150 (c) 210 (d) 243

6. If $\lim_{x \to \infty} \left(\frac{x^2 + x + 1}{x + 1} - ax - b\right) = 4$, then

   (a) $a = 1, b = 4$ (b) $a = 1, b = -4$
   (c) $a = 2, b = -3$ (d) $a = 2, b = 3$

7. The function $f: [0, 3] \to [1, 29]$, defined by $f(x) = 2x^3 - 15x^2 + 36x + 1$, is

   (a) one-one and onto
   (b) onto but not one-one
   (c) one-one but not onto
   (d) neither one-one nor onto

**8.** The locus of the mid-point of the chord of contact of tangents drawn from points lying on the straight line $4x - 5y = 20$ to the circle $x^2 + y^2 = 9$ is

(a) $20(x^2 + y^2) - 36x + 45y = 0$

(b) $20(x^2 + y^2) + 36x - 45y = 0$

(c) $36(x^2 + y^2) - 20y + 45y = 0$

(d) $36(x^2 + y^2) + 20x - 45y = 0$

**9.** Let $P = [a_{ij}]$ be a $3 \times 3$ matrix and let $Q = [b_{ij}]$, where $b_{ij} = 2^{i+j} a_{ij}$ for $1 \le i, j \le 3$. If the determinant of $P$ is 2, then the determinant of the matrix $Q$ is

(a) $2^{10}$ (b) $2^{11}$ (c) $2^{12}$ (d) $2^{13}$

**10.** The ellipse $E_1 : \frac{x^2}{9} + \frac{y^2}{4} = 1$ is inscribed in a rectangle $R$ whose sides are parallel to the coordinate axes. Another ellipse $E_2$ passing through the point (0, 4) circumscribes the rectangle $R$. The eccentricity of the ellipse $E_2$ is

(a) $\frac{\sqrt{2}}{2}$ (b) $\frac{\sqrt{3}}{2}$

(c) $\frac{1}{2}$ (d) $\frac{3}{4}$

## Section II

**Direction** (Q. Nos. 11-15) *This section contains 5 multiple choice questions. Each question has four choices (a), (b), (c) and (d) out of which one or more are correct.*

**11.** If $y(x)$ satisfies the differential equation $y' - y \tan x = 2x \sec x$ and $y(0)$, then

(a) $y\left(\frac{\pi}{4}\right) = \frac{\pi^2}{8\sqrt{2}}$ (b) $y'\left(\frac{\pi}{4}\right) = \frac{\pi^2}{18}$

(c) $y\left(\frac{\pi}{3}\right) = \frac{\pi^2}{9}$

(d) $y'\left(\frac{\pi}{3}\right) = \frac{4\pi}{3} + \frac{2\pi^2}{3\sqrt{3}}$

**12.** A ship is fitted with three engines $E_1, E_2$ and $E_3$. The engines function independently of each other with respective probabilities 1/2, 1/4 and 1/4. For the ship to be operational atleast two of its engines must function. Let $X$ denote the event that the ship is operational and let $X_1, X_2$ and $X_3$ dentoes, respectively the events that the engines $E_1$, $E_2$ and $E_3$ are functioning. Which of the following is/are true?

(a) $P[X_1^c | X] = \frac{3}{16}$

(b) $P$ [Exactly two engines of the ship are functioning $x$] = 7/8

(c) $P[X | X_2] = \frac{5}{16}$

(d) $P[X | X_1] = \frac{7}{16}$

**13.** Let $\theta, \phi \in [0, 2\pi]$ be such that $2\cos\theta(1 - \sin\phi) = \sin^2\theta$

$\left(\tan\frac{\theta}{2} + \cot\frac{\theta}{2}\right)\cos\phi - 1,$

$\tan(2\pi - \theta) > 0$

and $-1 < \sin\theta < -\frac{\sqrt{3}}{2}$.

Then, $\phi$ cannot satisfy

(a) $0 < \phi < \frac{\pi}{2}$ (b) $\frac{\pi}{2} < \phi < \frac{4\pi}{3}$

(c) $\frac{4\pi}{3} < \phi < \frac{3\pi}{2}$ (d) $\frac{3\pi}{2} < \phi < 2\pi$

**14.** If $S$ be the area of the region enclosed by $y = e^{-x^2}$, $y = 0$, $x = 0$ and $x = 1$. Then,

(a) $S \ge \frac{1}{e}$

(b) $S \ge 1 - \frac{1}{e}$

(c) $S \le \frac{1}{4}\left(1 + \frac{1}{\sqrt{e}}\right)$

(d) $S \le \frac{1}{\sqrt{2}} + \frac{1}{\sqrt{e}}\left(1 - \frac{1}{\sqrt{2}}\right)$

**15.** Tangents are drawn to the hyperbola $\frac{x^2}{9} - \frac{y^2}{4} = 1$, parallel to the straight line $2x - y = 1$. The points of contacts of the tangents on the hyperbola are

(a) $\left(\frac{9}{2\sqrt{2}}, \frac{1}{\sqrt{2}}\right)$ (b) $\left(-\frac{9}{2\sqrt{2}}, -\frac{1}{\sqrt{2}}\right)$

(c) $(3\sqrt{3}, -2\sqrt{2})$

(d) $(-3\sqrt{3}, 2\sqrt{2})$

## Section III

**Direction** (Q. Nos. 16-20) *This section contains 5 multiple choice questions. Each question has four choices (a), (b), (c) and (d) out of which one or more are correct.*

**16.** Let $S$ be the focus of the parabola $y^2 = 8x$ and let $PQ$ be the common chord of the circle $x^2 + y^2 - 2x - 4y = 0$ and the given parabola. The area of the $\Delta OPS$ is

**17.** Let $p(x)$ be a real polynomial of least degree which has a local maximum at $x = 1$ and a local minimum at $x = 3$. If $p(1) = 6$ and $p(3) = 2$, then $p'(0)$ is

**18.** Let $f : R \to R$ be defined as $f(x) = |x| + |x^2 - 1|$. The total number of points at which $f$ attains either a local maximum or a local minimum is

**19.** The value of $6 + \log_{3/2}\left(\frac{1}{3\sqrt{2}}\sqrt{4 - \frac{1}{3\sqrt{2}}\sqrt{4 - \frac{1}{3\sqrt{2}}\sqrt{4 - \frac{1}{3\sqrt{2}}\cdots}}}\right)$ is

**20.** If $\mathbf{a}$, $\mathbf{b}$ and $\mathbf{c}$ are unit vectors satisfying $|\mathbf{a} - \mathbf{b}|^2 + |\mathbf{b} - \mathbf{c}|^2 + |\mathbf{c} - \mathbf{a}|^2 = 9$, then $|2\mathbf{a} + 5\mathbf{b} + 5\mathbf{c}|$, is

# Paper 2

## Section I

**Direction** (Q. Nos. 21-28) *This section contains 8 multiple choice questions. Each question has four choices (a), (b), (c) and (d) out of which only one is correct.*

**21.** The equation of a plane passing through the line of intersection of the planes $x + 2y + 3z = 2$ and $x - y + z = 3$ and at a distance $2/\sqrt{3}$ from the point $(3, 1, -1)$ is

(a) $5x - 11y + z = 17$

(b) $\sqrt{2}x + y = 3\sqrt{2} - 1$

(c) $x + y + z = \sqrt{3}$

(d) $x - \sqrt{2}y = 1 - \sqrt{2}$

**22.** If $\mathbf{a}$ and $\mathbf{b}$ are vectors such that $|\mathbf{a} + \mathbf{b}| = \sqrt{29}$ and $\mathbf{a} \times (2\mathbf{i} + 3\mathbf{j} + 4\mathbf{k}) = (2\mathbf{i} + 3\mathbf{j} + 4\mathbf{k}) \times \mathbf{b}$, then a possible value of $(\mathbf{a} + \mathbf{b}) \cdot (-7\mathbf{i} + 2\mathbf{j} + 3\mathbf{k})$ is

(a) 0 (b) 3 (c) 4 (d) 8

**23.** Let $PQR$ be a triangle of area $\Delta$ with $a = 2$, $b = \frac{7}{2}$ and $c = \frac{5}{2}$, where $a$, $b$ and $c$ are the lengths of the sides of the triangle opposite to the angles at $P$, $Q$ and $R$, respectively. Then, $\frac{2 \sin P - \sin 2P}{2 \sin P + \sin 2P}$ equals to

(a) $\frac{3}{4\Delta}$ (b) $\frac{45}{4\Delta}$

(c) $\left(\frac{3}{4\Delta}\right)^2$ (d) $\left(\frac{45}{4\Delta}\right)^5$

**24.** Four fair dice $D_1$, $D_2$, $D_3$ and $D_4$ each having six faces numbered 1, 2, 3, 4, 5 and 6 are rolled simultaneously. The probability that $D_4$ shows a number appearing on one of $D_1$, $D_2$ and $D_3$ is

(a) $\frac{91}{216}$ (b) $\frac{108}{216}$ (c) $\frac{125}{216}$ (d) $\frac{127}{216}$

**25.** The value of the integral $\int_{-\pi/2}^{\pi/2} \left( x^2 + \log \frac{\pi - x}{\pi + x} \right) \cos x \, dx$ is

(a) 0 (b) $\frac{\pi^2}{2} - 4$ (c) $\frac{\pi^2}{2} + 4$ (d) $\frac{\pi^2}{2}$

**26.** If $P$ is a $3 \times 3$ matrix such that $P^T = 2P + I$, where $P^T$ is the transpose of $P$ and $I$ is the $3 \times 3$ identity matrix, then there exists a column matrix, $X = \begin{bmatrix} x \\ y \\ z \end{bmatrix} \neq \begin{bmatrix} 0 \\ 0 \\ 0 \end{bmatrix}$ such that

(a) $PX = \begin{bmatrix} 0 \\ 0 \\ 0 \end{bmatrix}$ (b) $PX = X$

(c) $PX = 2X$ (d) $PX = -X$

**27.** Let $a_1, a_2, a_3, \ldots$ be in a harmonic progression with $a_1 = 5$ and $a_{20} = 25$. The least positive integer $n$ for which $a_n < 0$ is

(a) 22 (b) 23 (c) 24 (d) 25

**28.** Let $\alpha(a)$ and $\beta(a)$ be the roots of the equation $(\sqrt[3]{1+a} - 1)x^2 - (\sqrt{1+a} - 1)x + (\sqrt[6]{1+a} - 1) = 0$, where $a > -1$. Then, $\lim_{a \to 0^+} \alpha(a)$ and $\lim_{a \to 0^+} \beta(a)$ are

(a) $-\frac{5}{2}$ and 1 (b) $-\frac{1}{2}$ and $-1$

(c) $-\frac{7}{2}$ and 2 (d) $-\frac{9}{2}$ and 3

## Section II

**Direction** (Q. Nos. 29-34) *This section contains 6 multiple choice questions relating to three paragraphs with two questions on each paragraph. Each question has four choices (a), (b), (c) and (d) out of which only one is correct.*

### Paragraph

Let $f(x) = (1 - x)^2 \sin^2 x + x^2$ for all $x \in R$ and

let $g(x) = \int_1^x \left( \frac{2(t-1)}{t+1} - \ln t \right) f(t) \, dt$

for all $x \in (1, \infty)$.

**29.** Which of the following is true?

(a) $g$ is increasing on $(1, \infty)$

(b) $g$ is decreasing on $(1, \infty)$

(c) $g$ is increasing on $(1, 2)$ and decreasing on $(2, \infty)$

(d) $g$ is decreasing on $(1, 2)$ and increasing on $(2, \infty)$

**30.** Consider the statements

P : There exists some $x \in R$ such that $f(x) + 2x = 2(1 + x^2)$

Q : There exists some $x \in R$ such that $2f(x) + 1 = 2x(1 + x)$

Then,

(a) Both P and Q are true

(b) P is true and Q is false

(c) P is false and Q is true

(d) Both P and Q are false

### Paragraph

Let $a_n$ denote the number of all $n$-digit positive integers formed by the digits 0, 1 or both such that no consecutive digits in them are 0. Let $b_n$ = The number of such $n$-digit integers ending with digit 1 and $c_n$ = The number of such $n$-digit integers ending with digit 0.

**31.** Which of the following is correct?

(a) $a_{17} = a_{16} + a_{15}$ (b) $c_{17} \neq c_{16} + c_{15}$

(c) $b_{17} \neq b_{16} + c_{16}$ (d) $a_{17} = c_{17} + b_{16}$

**32.** The value of $b_6$ is

(a) 7 (b) 8 (c) 9 (d) 11

### Paragraph

A tangent $PT$ is drawn to the circle $x^2 + y^2 = 4$ at the point $P(\sqrt{3}, 1)$. A straight line $L$, perpendicular to $PT$ is a tangent to the circle $(x - 3)^2 + y^2 = 1$.

**33.** A common tangent of the two circles is

(a) $x = 4$ (b) $y = 2$

(c) $x+\sqrt{3}y=4$ (d) $x+2\sqrt{2}\,y=6$

**34.** A possible equation of $L$ is

(a) $x-\sqrt{3}y=1$ (b) $x+\sqrt{3}y=1$

(c) $x-\sqrt{3}y=-1$ (d) $x+\sqrt{3}y=5$

# Section III

**Direction** (Q. Nos. 35-40) *This section contains 6 multiple choice questions. Each question has four choices (a), (b), (c) and (d) out of which one or more are correct.*

**35.** Let $X$ and $Y$ be two events such that $P(X\mid Y)=\frac{1}{2}, P(Y\mid X)=\frac{1}{3}$ and $P(X\cap Y)\frac{1}{6}$. Which of the following is/are correct?

(a) $P(X\cup Y)=2/3$
(b) $X$ and $Y$ are independent
(c) $X$ and $Y$ are not independent
(d) $P(X^c\cap Y)=1/3$

**36.** If $f(x)=\int_0^x e^{t^2}(t-2)(t-3)\,dt,\ \forall$ $x\in(0,\infty)$, then

(a) $f$ has a local maximum at $x=2$
(b) $f$ is decreasing on (2, 3)
(c) there exists some $c\in(0,\infty)$ such that $f''(c)=0$
(d) $f$ has a local minimum at $x=3$

**37.** For every integer $n$, let $a_n$ and $b_n$ be real numbers. Let function $f:R\to R$ be given by

$$f(x)=\begin{cases}a_n+\sin\pi x, & \text{for } x\in[2n, 2n+1]\\ b_n+\cos\pi x, & \text{for } x\in(2n-1, 2n)\end{cases},$$

for all integers $n$.

If $f$ is continuous, then which of the following hold(s) for all $n$?

(a) $a_{n-1}-b_{n-1}=0$ (b) $a_n-b_n=1$
(c) $a_n-b_{n+1}=1$ (d) $a_{n-1}-b_n=-1$

**38.** If the straight lines $\frac{x-1}{2}=\frac{y+1}{K}=\frac{z}{2}$ and $\frac{x+1}{5}=\frac{y+1}{2}=\frac{z}{k}$ are coplanar, then the plane(s) containing these two lines is/are

(a) $y+2z=-1$ (b) $y+z=-1$
(c) $y-z=-1$ (d) $y-2z=-1$

**39.** If the adjoint of a $3\times 3$ matrix $P$ is $\begin{bmatrix}1&4&4\\2&1&7\\1&1&3\end{bmatrix}$, then the possible value(s) of the determinant of $P$ is/are

(a) $-2$ (b) $-1$ (c) 1 (d) 2

**40.** Let $f:(-1,1)\to R$ be such that $f(\cos 4\theta)=\frac{2}{2-\sec^2\theta}$ for $\theta\in\left(0,\frac{\pi}{4}\right)\cup\left(\frac{\pi}{4},\frac{\pi}{2}\right)$. Then, the value(s) of $f\left(\frac{1}{3}\right)$ is/are

(a) $1-\sqrt{\frac{3}{2}}$ (b) $1+\sqrt{\frac{3}{2}}$
(c) $1-\sqrt{\frac{2}{3}}$ (d) $1+\sqrt{\frac{2}{3}}$

# Answers

## Paper 1

**1.** (a) **2.** (c) **3.** (d) **4.** (b) **5.** (b) **6.** (b) **7.** (b) **8.** (a) **9.** (d) **10.** (c)
**11.** (a,d) **12.** (b,d) **13.** (a,c,d) **14.** (a,b,d) **15.** (a,b)
**16.** (4) **17.** (9) **18.** (5) **19.** (4) **20.** (3)

## Paper 2

**21.** (a) **22.** (c) **23.** (c) **24.** (a) **25.** (b) **26.** (d) **27.** (d) **28.** (b) **29.** (b) **30.** (c)
**31.** (a) **32.** (b) **33.** (d) **34.** (a) **35.** (a,b) **36.** (a,b,c,d) **37.** (b,d)
**38.** (b,c) **39.** (a,d) **40.** (a,b)

# Hints & Solutions

## Paper 1

1. ***Concept Involved*** It is based on two concept one is intersection of straight line and plane and other is the foot of perpendicular from a point to the straight line.

**Description of Situation**

(i) If the straight line

$$\frac{x-x_1}{a}=\frac{y-y_1}{b}=\frac{z-z_1}{c}=\lambda$$

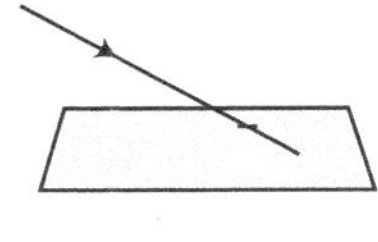

intersect the plane $Ax+By+Cz+d=0$.

Then, $(a\lambda+x_1, b\lambda+y_1+c\lambda+z_1)$ would satisfy

$Ax+By+Cz+d=0$

(ii) If $A$ is the foot of perpendicular from $P$ to $l$.

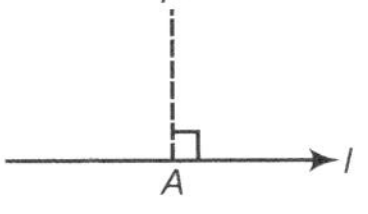


Then, (dr's of $PA$) is perpendicular to dr's of $l$.

$\Rightarrow \quad \overline{PA}\cdot\bar{l}=0$

**Sol.** Equation of straight line $QR$, is

$$\frac{x-2}{1-2}=\frac{y-3}{-1-3}=\frac{z-5}{4-5}$$

or $$\frac{x-2}{-1}=\frac{y-3}{-4}=\frac{z-5}{-1}$$

or $$\frac{x-2}{1}=\frac{y-3}{4}=\frac{z-5}{1}=\lambda$$

or $$\frac{x-2}{1}=\frac{y-3}{4}=\frac{z-5}{1}=\lambda \quad \ldots\text{(i)}$$

$\therefore$ $P(\lambda+2, 4\lambda+3, \lambda+5)$ must lie on $5x-4y-z=1$

$\Rightarrow 5(\lambda+2)-4(4\lambda+3)-(\lambda+5)=1$

$\Rightarrow \quad 5\lambda+10-16\lambda-12-\lambda-5=1$

$\Rightarrow \quad -7-12\lambda=1$

$\therefore \quad \lambda=\frac{-2}{3}$ or $P\left(\frac{4}{3},\frac{1}{3},\frac{13}{3}\right) \quad \ldots\text{(ii)}$

Again, we can assume $S$ from Eq. (i) as $S(\mu+2, 4\mu+3, \mu+5)$

$\therefore$ dr's of $TS=(\mu+2-2, 4\mu+3-1, \mu+5-4)$

$=(\mu, 4\mu+2, \mu+1)$

and dr's of $QR=(1, 4, 1)$

Since, perpendicular

$\therefore \; 1(\mu)+4(4\mu+2)+1(\mu+1)=0$

$\Rightarrow \quad \mu=-\frac{1}{2}$ and $S\left(\frac{3}{2},1,\frac{9}{2}\right) \quad \ldots\text{(iii)}$

$\therefore$ Length of

$$PS=\sqrt{\left(\frac{3}{2}-\frac{4}{3}\right)^2+\left(1-\frac{1}{3}\right)^2+\left(\frac{9}{2}-\frac{13}{3}\right)^2}$$

$$=\frac{1}{\sqrt{2}}$$

2. ***Concept Involved***

**Integration by Substitution**

*i.e.,* $\quad I=\int f(g(x))\cdot g'(x)\,dx$

Put, $g(x)=t \Rightarrow g'(x)\,dx=dt$

$\therefore \quad I=\int f(t)\,dt$

**Description of Situation** Generally, students gets confused after substitution *i.e.,* $\sec x+\tan x=t$.

Now, for $\sec x$ we should use

$$\sec^2 x-\tan^2 x=1$$

$\Rightarrow \quad (\sec x-\tan x)(\sec x+\tan x)=1$

$\Rightarrow \quad \sec x-\tan x=\frac{1}{t}$

**Sol.** Here, $I=\int\frac{\sec^2 dx}{(\sec x+\tan x)^{9/2}}$

Put, $\quad \sec x+\tan x=t$

$\Rightarrow \quad (\sec x\tan x+\sec^2 x)\,dx=dt$

$\Rightarrow \quad \sec x\cdot t\; dx=dt$

$\Rightarrow \quad \sec x\,dx=\frac{dt}{t}$

$\therefore \quad \sec x-\tan x=\frac{1}{t}$

$\Rightarrow \quad \sec x=\frac{1}{2}\left(t+\frac{1}{t}\right)$

$\therefore \quad I=\int\frac{\sec x\cdot\sec x\,dx}{(\sec x+\tan x)^{9/2}}$

$\Rightarrow \quad I=\int\frac{\frac{1}{2}\left(t+\frac{1}{t}\right)\cdot\frac{dt}{t}}{t^{9/2}}$

$$=\frac{1}{2}\int\left(\frac{1}{t^{9/2}}+\frac{1}{t^{13/2}}\right)dt$$

$$=-\frac{1}{2}\left\{\frac{2}{7t^{7/2}}+\frac{2}{11t^{11/2}}\right\}+K$$

$$=-\left[\frac{1}{7(\sec x+\tan x)^{7/2}}+\frac{1}{11(\sec x+\tan x)^{11/2}}\right]+K$$

$$=\frac{-1}{(\sec x+\tan x)^{11/2}}\left\{\frac{1}{11}+\frac{1}{7}(\sec x+\tan x)^2\right\}+K$$

**3.** ***Concept Involved*** If $ax^2+bx+c=0$ has roots $\alpha, \beta$ then,

$$\alpha, \beta=\frac{-b\pm\sqrt{b^2-4ac}}{2a}$$

For roots to be real $b^2-4ac\geq 0$.

**Description of Situation** As imaginary part of $z=x+iy$ is non-zero.

$\Rightarrow \quad y\neq 0$

**Sol.** **I Method** Let $z=x+iy$

$\therefore \quad a=(x+iy)^2+(x+iy)+1$

$\Rightarrow (x^2-y^2+x+1-a)+i(2xy+y)=0$

$\Rightarrow (x^2-y^2+x+1-a)+iy(2x+1)=0 \quad \ldots(i)$

It is purely real. If, $y(2x+1)=0$

but imaginary part of $z$, *i.e.*, $y$ is non-zero

$\Rightarrow \quad 2x+1=0$ or $x=-\frac{1}{2}$

$\therefore$ From Eq. (i) $\frac{1}{4}-y^2-\frac{1}{2}+1-a=0$

$\Rightarrow \quad a=-y^2+\frac{3}{4} \Rightarrow a<\frac{3}{4}$

**II Method** Here, $z^2+z+(1-a)=0$

$$\therefore \quad z=\frac{-1\pm\sqrt{1-4(1-a)}}{2\times 1}$$

$$\Rightarrow \quad z=\frac{-1\pm\sqrt{4a-3}}{2}$$

For $z$ do not have real roots,

$$4a-3<0 \Rightarrow a<\frac{3}{4}$$

**4.** ***Concept Involved*** To check differentiability at a point we use RHD and LHD at a point and if RHD = LHD, then $f(x)$ is differentiable at the point.

**Description of Situation**

As, $R\{f'(x)\}=\lim_{h\to 0}\frac{f(x+h)-f(x)}{h}$

and $L\{f'(x)\}=\lim_{h\to 0}\frac{f(x-h)-f(x)}{-h}$

Here, students generally gets confused in defining modulus.

**Sol. To check differentiable at $x=0$**

$$R\{f'(0)\}=\lim_{h\to 0}\frac{f(0+h)-f(0)}{h}$$

$$=\lim_{h\to 0}\frac{h^2\left|\cos\frac{\pi}{h}\right|-0}{h}$$

$$=\lim_{h\to 0}h\cdot\left|\cos\frac{\pi}{h}\right|=0$$

$$L\{f'(0)\}=\lim_{h\to 0}\frac{f(0-h)-f(0)}{-h}$$

$$=\lim_{h\to 0}\frac{h^2\left|\cos\left(-\frac{\pi}{h}\right)\right|-0}{-h}$$

$$=0$$

So, $f(x)$ is differentiable at $x=0$

**To check differentiability at $x=2$**

$$R\{f'(2)\}=\lim_{h\to 0}\frac{f(2+h)-f(2)}{h}$$

$$=\lim_{h\to 0}\frac{(2+h)^2\left|\cos\left(\frac{\pi}{2+h}\right)\right|-0}{h}$$

$$=\lim_{h\to 0}\frac{(2+h)^2\cdot\cos\left(\frac{\pi}{2+h}\right)}{h}$$

$$=\lim_{h\to 0}\frac{(2+h)^2\cdot\sin\left(\frac{\pi}{2}-\frac{\pi}{2+h}\right)}{h}$$

$$=\lim_{h\to 0}\frac{(2+h)^2\cdot\sin\left(\frac{\pi h}{2(2+h)}\right)}{h\cdot\frac{\pi}{2(2+h)}}\cdot\frac{\pi}{2(2+h)}$$

$\Rightarrow R(f'(2))=\pi$

$$L(f'(2)) = \lim_{h\to 0} \frac{f(2-h)-f(2)}{-h}$$

$$= \lim_{h\to 0} \frac{(2-h)^2 \cdot \left|\cos\frac{\pi}{2-h}\right| - 2^2 \cdot \left|\cos\frac{\pi}{2}\right|}{-h}$$

$$= \lim_{h\to 0} \frac{(2-h)^2 - \left(-\cos\frac{\pi}{2-h}\right) - 0}{-h}$$

$$= \lim_{h\to 0} \frac{-(2-h)^2 \cdot \sin\left(\frac{\pi}{2} - \frac{\pi}{2-h}\right)}{h}$$

$$= \lim_{h\to 0} \frac{(2-h)^2 \cdot \sin\left(-\frac{\pi h}{2(2-h)}\right)}{h \times \frac{-\pi}{2(2-h)}} \times \frac{-\pi}{2(2-h)}$$

$\Rightarrow \quad L(f'(2)) = -\pi$

Thus, $f(x)$ is differentiable at $x=0$ but not at $x=2$

5. ***Concept Involved***

Distribution of objects into groups

Here, the student should be particular about objects and groups. *i.e.,*

| **Objects** | **Groups** |
|---|---|
| Distinct | Distinct |
| Identical | Identical |
| Distinct | Identical |
| Identical | Distinct |

**Description of Situation** Here, 5 distinct balls are distributed amongst 3 persons so that each gets atleast one ball.

*i.e.,* Distinct $\rightarrow$ Distinct

So, we should make cases

***Case* I** $\left.\begin{matrix} A & B & C \\ 1 & 1 & 3 \end{matrix}\right\}$

***Case* II** $\left.\begin{matrix} A & B & C \\ 1 & 2 & 2 \end{matrix}\right\}$

**Sol.** Number of ways to distribute 5 balls is

$$\left({}^5C_1 \cdot {}^4C_1 \cdot {}^3C_3 \times \frac{3!}{2!}\right) + \left({}^5C_1 \cdot {}^4C_2 \cdot {}^2C_2 \times \frac{3!}{2!}\right)$$

$$= 60 + 90$$

$$= 150$$

6. ***Concept Involved*** $\left(\frac{\infty}{\infty}\right)$ form

$$\lim_{x\to\infty} \frac{a_0x^n + a_1x^{n-1} + \ldots + a_n}{b_0x^m + b_1x^{m-1} + \ldots + b_m} = \begin{cases} 0, & \text{If } n = m \\ \frac{a_0}{b_0}, & \text{If } n < m \\ +\infty, & \text{If } n > m \text{ and } a_0b_0 > 0 \\ -\infty, & \text{If } n > m \text{ and } a_0b_0 < 0 \end{cases}$$

**Description of Situation** As to make degree of numerator equal to degree of denominator.

**Sol.** $\lim_{x\to\infty}\left(\frac{x^2+x+1}{x+1} - ax - b\right) = 4$

$$\Rightarrow \lim_{x\to\infty} \frac{x^2 + x + 1 - ax^2 - ax - bx - b}{x+1} = 4$$

$$\Rightarrow \lim_{x\to\infty} \frac{x^2(1-a) + x(1-a-b) + (1-b)}{x+1} = 4$$

Here, we make degree of $N^r$ = degree of $D^r$

$\therefore \quad 1 - a = 0$

and $\lim_{x\to\infty} \frac{x(1-a-b)+(1-b)}{x+1} = 4$

$\Rightarrow \quad 1 - a - b = 4$

$\Rightarrow \quad b = -4 \quad [\because (-a) = 0]$

7. ***Concept Involved*** To check nature of function.

(i) **One-One** To check one-one, we must check whether

$f'(x) > 0$ or $f'(x) < 0$ in given domain.

(ii) **Onto** To check onto, we must check

Range = Co-domain

**Description of Situation** To find range in given domain $[a,b]$ put $f'(x) = 0$ and find $x = \alpha_1, \alpha_2, \ldots, \alpha_n \in [a,b]$

Now find,

$$\{f(a), f(\alpha_1), f(\alpha_2), \ldots, f(\alpha_n), f(b)\}$$

its the greatest and least gives you range.

**Sol.** $f : [0, 3] \rightarrow [1, 29]$

$$f(x) = 2x^3 - 15x^2 + 36x + 1$$

$$\therefore f'(x) = 6x^2 - 30x + 36$$

$$= 6(x^2 - 5x + 6) = 6(x-2)(x-3)$$

+   –   + (sign scheme with points 2, 3)

For given domain [0, 3], $f(x)$ is increasing as well as decreasing

$\Rightarrow$ **many-one**

Now, put $f'(x)=0 \Rightarrow x=2, 3$

Thus, for Range $f(0)=1$,
$f(2)=29$, $f(3)=28$

$\Rightarrow$ Range $\in [1, 29]$ $\therefore$ **onto.**

**8.** ***Concept Involved*** If

$$S: ax^2+2hxy+by^2+2gx+2fy+C$$

then equation of chord bisected at $P(x_1, y_1)$ is $T=S_1$

or $a\,xx_1+h(xy_1+yx_1)+b\,yy_1 + g(x+x_1)+f(y+y_1)+C = ax_1^2+2hx_1y_1+by_1^2+2gx_1+2fy_1+C$

**Description of Situation** As equation of chord of contact is $T=0$

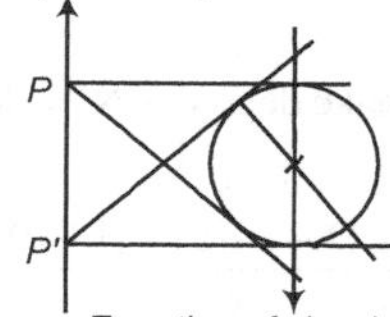


**Sol.** Here, equation of chord of contact *wrt P* is

$$x\lambda+y.\left(\frac{4\lambda-20}{5}\right)=9$$

$$5\lambda x+(4\lambda-20)y=45 \quad \ldots(\text{i})$$

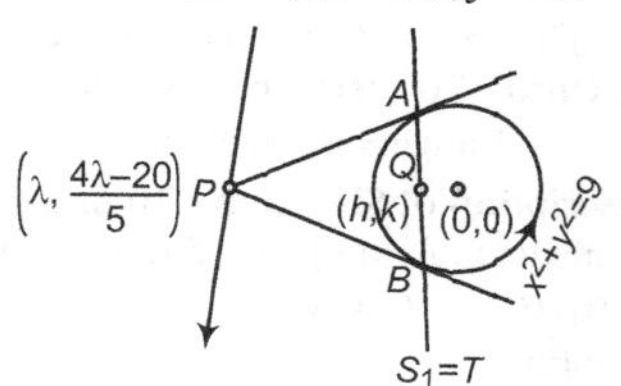


and equation of chord bisected at the point $Q(h, k)$ is

$$xh+yk-9=h^2+k^2-9$$

$$\Rightarrow \quad xh+ky=h^2+k^2 \quad \ldots(\text{ii})$$

From Eqs. (i) and (ii), we get

$$\frac{5\lambda}{h}=\frac{4\lambda-20}{k}=\frac{45}{h^2+k^2}$$

$$\therefore \quad \lambda=\frac{20h}{4h-5k} \text{ and } \lambda=\frac{9h}{h^2+k^2}$$

$$\Rightarrow \quad \frac{20h}{4h-5k}=\frac{9h}{h^2+k^2}$$

or $20(h^2+k^2)=9(4h-5k)$

or $20(x^2+y^2)=36x-45y$

**9.** ***Concept Involved*** It is a simple question on scalar multiplication *i.e.*,

$$\begin{vmatrix} ka_1 & ka_2 & ka_3 \\ b_1 & b_2 & b_3 \\ c_1 & c_2 & c_3 \end{vmatrix} = k\begin{vmatrix} a_1 & a_2 & a_3 \\ b_1 & b_2 & b_3 \\ c_1 & c_2 & c_3 \end{vmatrix}$$

**Description of Situation** Construction of Matrix

*i.e.*, If $A=[a_{ij}]_{3\times 3}=\begin{bmatrix} a_{11} & a_{12} & a_{13} \\ a_{21} & a_{22} & a_{23} \\ a_{31} & a_{32} & a_{33} \end{bmatrix}$

**Sol.** Here, $P=[a_{ij}]_{3\times 3}=\begin{bmatrix} a_{11} & a_{12} & a_{13} \\ a_{21} & a_{22} & a_{23} \\ a_{31} & a_{32} & a_{33} \end{bmatrix}$

$$Q=[b_{ij}]_{3\times 3}=\begin{bmatrix} b_{11} & b_{12} & b_{13} \\ b_{21} & b_{22} & b_{23} \\ b_{31} & b_{32} & b_{33} \end{bmatrix},$$

where, $b_{ij}=2^{i+j}a_{ij}$

$$\therefore \quad |Q|=\begin{vmatrix} 4a_{11} & 8a_{12} & 16a_{13} \\ 8a_{21} & 16a_{22} & 32a_{23} \\ 16a_{31} & 32a_{32} & 64a_{33} \end{vmatrix}$$

$$=4\times 8\times 16\begin{vmatrix} a_{11} & a_{12} & a_{13} \\ 2a_{21} & 2a_{22} & 2a_{23} \\ 4a_{31} & 4a_{32} & 4a_{33} \end{vmatrix}$$

$$=2^9\times 2\times 4\begin{vmatrix} a_{11} & a_{12} & a_{13} \\ a_{21} & a_{22} & a_{23} \\ a_{31} & a_{32} & a_{33} \end{vmatrix}$$

$$=2^{12}\cdot|P|=2^{12}\cdot 2=2^{13}$$

**10.** ***Concept Involved*** Equation of an ellipse is

$$\frac{x^2}{a^2}+\frac{y^2}{b^2}=1 \quad (a>b)$$

Eccentricity $e^2=1-\dfrac{b^2}{a^2}$ $(a>b)$

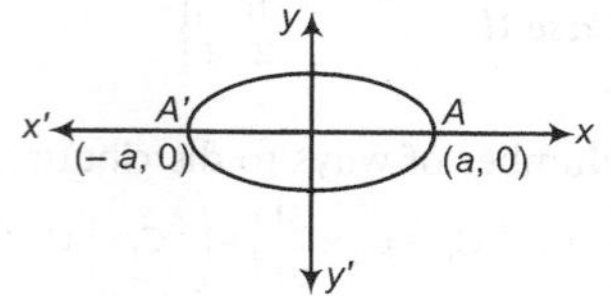


**Description of Situation** As ellipse circumscribes the rectangle, then it must pass through all four vertices.

**Sol.** Let the equation of Ellipse $E_2$ be

$$\frac{x^2}{a^2}+\frac{y^2}{b^2}=1, \text{ where } a<b \text{ and } b=4$$

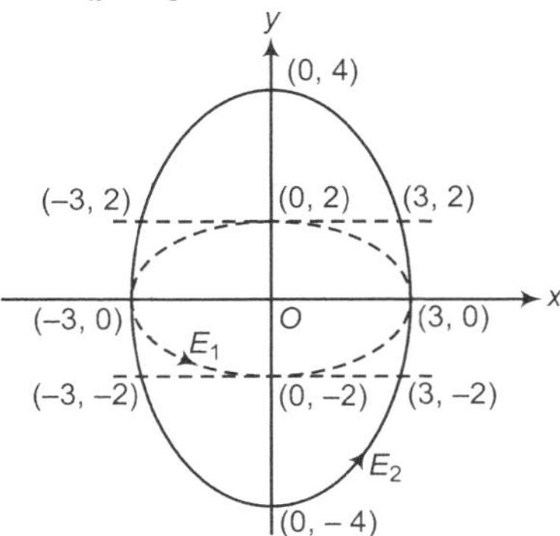


Also, it passes through (3, 2).

$$\Rightarrow \quad \frac{9}{a^2}+\frac{4}{b^2}=1 \qquad (\because b=4)$$

$$\Rightarrow \quad \frac{9}{a^2}+\frac{1}{4}=1 \quad \text{or} \quad a^2=12$$

Eccentricity of $E_2$

$$\Rightarrow \quad e^2=1-\frac{a^2}{b^2}=1-\frac{12}{16}=\frac{1}{4} \qquad (\because a<b)$$

$$\therefore \quad e=\frac{1}{2}$$

**11.** ***Concept Involved*** Linear differential equation under one variable.

$$\frac{dy}{dx}+Py=Q \quad ; \quad \text{IF}=e^{\int P\,dx}$$

$\therefore$ Solution is, $y\,(\text{IF})=\int Q\cdot(\text{IF})\,dx+C$

**Sol.** $y'-y\tan x=2x\sec x$ and $y(0)=0$

$$\Rightarrow \quad \frac{dy}{dx}-y\tan x=2x\sec x$$

$$\therefore \text{ IF}=\int e^{-\tan x}\,dx=e^{\log|\cos x|}=\cos x$$

Solution is

$$y\cdot\cos x=\int 2x\sec x\cdot\cos x\,dx+C$$

$$\Rightarrow \quad y\cdot\cos x=x^2+C,$$

As $\quad y(0)=0 \Rightarrow C=0$

$$\therefore \quad y=x^2\sec x$$

Now, $\quad y\left(\frac{\pi}{4}\right)=\frac{\pi^2}{8\sqrt{2}}$

$$\Rightarrow \quad y'\left(\frac{\pi}{4}\right)=\frac{\pi}{\sqrt{2}}+\frac{\pi^2}{8\sqrt{2}}$$

$$\Rightarrow \quad y\left(\frac{\pi}{3}\right)=\frac{2\pi^2}{9}$$

$$\Rightarrow \quad y'\left(\frac{\pi}{3}\right)=\frac{4\pi}{3}+\frac{2\pi^2}{3\sqrt{3}}$$

**12.** ***Concept Involved*** It is based on law of total probability and Bay's Law.

**Description of Situation** It is given that ship would work if atleast two of engines must work. If $X$ be event that the ship works. Then $X \Rightarrow$ either any two of $E_1, E_2, E_3$ works or all three engines $E_1, E_2, E_3$ works

**Sol.** Given $P(E_1)=\frac{1}{2}$, $P(E_2)=\frac{1}{4}$, $P(E_3)=\frac{1}{4}$

$$\therefore P(X)=\left\{\begin{array}{l} P(E_1\cap E_2\cap \bar{E}_3) \\ +P(E_1\cap \bar{E}_2\cap E_3) \\ +P(\bar{E}_1\cap E_2\cap E_3) \\ +P(E_1\cap E_2\cap E_3) \end{array}\right\}$$

$$=\left(\frac{1}{2}\cdot\frac{1}{4}\cdot\frac{3}{4}+\frac{1}{2}\cdot\frac{3}{4}\cdot\frac{1}{4}+\frac{1}{2}\cdot\frac{1}{4}\cdot\frac{1}{4}\right)+\left(\frac{1}{2}\cdot\frac{1}{4}\cdot\frac{1}{4}\right)$$

$$=\frac{1}{4}$$

Now, (a) $P(X_1^c / X)$

$$=P\left(\frac{X_1^c\cap X}{P(X)}\right)=\frac{P(\bar{E}_1\cap E_2\cap E_3)}{P(X)}$$

$$=\frac{\frac{1}{2}\cdot\frac{1}{4}\cdot\frac{1}{4}}{\frac{1}{4}}=\frac{1}{8}$$

(b) $P$ (Exactly two engines of the ship are functioning)

$$=\frac{P(E_1\cap E_2\cap \bar{E}_3)+P(E_1\cap \bar{E}_2\cap E_3)+P(\bar{E}_1\cap E_2\cap E_3)}{P(X)}$$

$$=\frac{\frac{1}{2}\cdot\frac{1}{4}\cdot\frac{3}{4}+\frac{1}{2}\cdot\frac{3}{4}\cdot\frac{1}{4}+\frac{1}{2}\cdot\frac{1}{4}\cdot\frac{1}{4}}{\frac{1}{4}}=\frac{7}{8}$$

(c) $P\left(\frac{X}{X_2}\right)=\frac{P(X\cap X_2)}{P(X_2)}$

$$= \frac{P(\text{Ship is operating with } E_2 \text{ function})}{P(X_2)}$$

$$= \frac{P(E_1 \cap E_2 \cap \bar{E}_3) + P(\bar{E}_1 \cap E_2 \cap E_3) + P(E_1 \cap E_2 \cap E_3)}{P(E_2)}$$

$$= \frac{\frac{1}{2}\cdot\frac{1}{4}\cdot\frac{3}{4} + \frac{1}{2}\cdot\frac{1}{4}\cdot\frac{1}{4} + \frac{1}{2}\cdot\frac{1}{4}\cdot\frac{1}{4}}{\frac{1}{4}} = \frac{5}{8}$$

(d) $P(X / X_1) = \dfrac{P(X \cap X_1)}{P(X_1)}$

$$= \frac{\frac{1}{2}\cdot\frac{1}{4}\cdot\frac{1}{4} + \frac{1}{2}\cdot\frac{3}{4}\cdot\frac{1}{4} + \frac{1}{2}\cdot\frac{1}{4}\cdot\frac{3}{4}}{\frac{1}{2}} = \frac{7}{16}$$

**13.** ***Concept Involved*** It is based on range of $\sin x$ *i.e.*, $[-1,1]$ and the internal for $a < x < b$.

**Description of Situation** As $\theta, \phi \in [0, 2\pi]$ and

$$\tan(2\pi - \theta) > 0, \ -1 < \sin\theta < -\frac{\sqrt{3}}{2}$$

$$\tan(2\pi - \theta) > 0 \Rightarrow -\tan\theta > 0$$

$\therefore$ $\theta \in$ II or IV quadrant.

Also, $\quad -1 < \sin\theta < -\dfrac{\sqrt{3}}{2}$

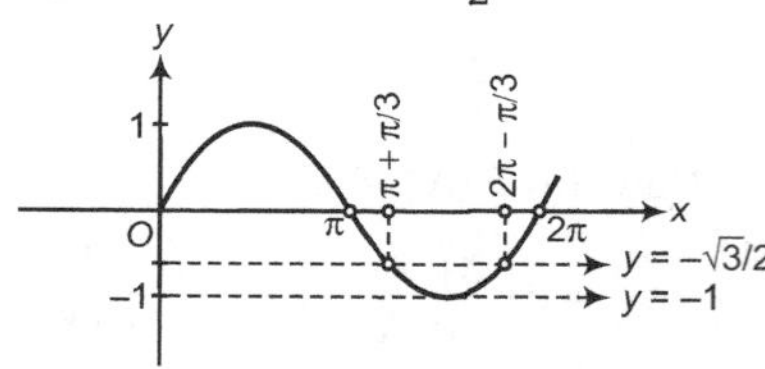


$\Rightarrow \dfrac{4\pi}{3} < \theta < \dfrac{5\pi}{3}$ but $\theta \in$ II or IV quadrant

$\Rightarrow \quad \dfrac{3\pi}{2} < \theta < \dfrac{5\pi}{3} \quad$ ...(i)

**Sol.** Here,

$$2\cos\theta(1 - \sin\phi) = \sin^2\theta\left(\tan\frac{\theta}{2} + \cot\frac{\theta}{2}\right)\cos\phi - 1$$

$$\Rightarrow 2\cos\theta - 2\cos\theta\sin\phi = \sin^2\theta\left(\frac{\sin^2\frac{\theta}{2} + \cos^2\frac{\theta}{2}}{\sin\frac{\theta}{2}\cos\frac{\theta}{2}}\right)\cos\phi - 1$$

$$\Rightarrow 2\cos\theta - 2\cos\theta\sin\phi = 2\sin^2\theta\left(\frac{1}{\sin\theta}\right)\cos\phi - 1$$

$$\Rightarrow 2\cos\theta + 1 = 2\sin\phi\cos\theta + 2\sin\theta\cos\phi$$

$$\Rightarrow \quad 2\cos\theta + 1 = 2\sin(\theta + \phi) \quad \text{...(ii)}$$

From Eq (i),

$$\frac{3\pi}{2} < \theta < \frac{5\pi}{3} \Rightarrow 2\cos\theta + 1 \in (1, 2)$$

$\therefore \quad 1 < 2\sin(\theta + \phi) < 2$

$\Rightarrow \quad \dfrac{1}{2} < \sin(\theta + \phi) < 1 \quad$ ...(iii)

$\Rightarrow \dfrac{\pi}{6} < \theta + \phi < \dfrac{5\pi}{6}$ or $\dfrac{13\pi}{6} < \theta + \phi < \dfrac{17\pi}{6}$

$\therefore \quad \dfrac{\pi}{6} - \theta < \phi < \dfrac{5\pi}{6} - \theta$

or $\quad \dfrac{13\pi}{6} - \theta < \phi < \left(\dfrac{17\pi}{6}\right) - \theta$

$\Rightarrow \phi \in \left(-\dfrac{3\pi}{2}, -\dfrac{2\pi}{3}\right)$ or $\left(\dfrac{2\pi}{3}, \dfrac{7\pi}{6}\right)$,

as $\theta \in \left(\dfrac{3\pi}{2}, \dfrac{5\pi}{3}\right)$

**14.** ***Concept Involved***

(i) Area of region $f(x)$ bounded between $x = a$ to $x = b$ is

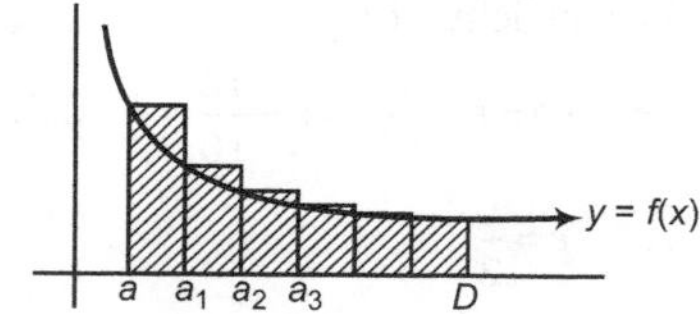


$\int_a^b f(x)\,dx =$ Sum of areas of rectangle shown in shaded part.

(ii) If $f(x) \geq g(x)$ when defined in $[a, b]$

$\Rightarrow \quad \int_a^b f(x)\,dx \geq \int_a^b g(x)\,dx$

**Description of Situation** As the given curve $y = e^{-x^2}$ cannot be integrated thus we have to bound this function by using above mentioned concept.

**Sol.** Graph for, $y = e^{-x^2}$

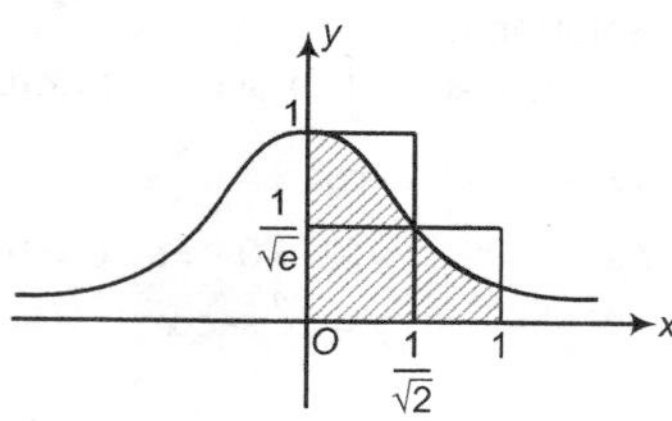


Since, $x^2 \leq x$ when $x \in [0, 1]$

$\Rightarrow \quad -x^2 \geq -x$ or $e^{-x^2} \geq e^{-x}$

$\therefore \quad \int_0^1 e^{-x^2} dx \ge \int_0^1 e^{-x} dx$

$\Rightarrow \quad S \ge -(e^{-x})_0^1 = 1 - \frac{1}{e} \quad \ldots(\text{i})$

Also, $\int_0^1 e^{-x^2} dx \le$ Area of two Rectangles.

$$\le \left(1 \times \frac{1}{\sqrt{2}}\right) + \left(1 - \frac{1}{\sqrt{2}}\right) \times \frac{1}{\sqrt{e}}$$

$$\le \frac{1}{\sqrt{2}} + \frac{1}{\sqrt{e}}\left(1 - \frac{1}{\sqrt{2}}\right) \quad \ldots(\text{ii})$$

$$\therefore \quad \frac{1}{\sqrt{2}} + \frac{1}{\sqrt{e}}\left(1 - \frac{1}{\sqrt{2}}\right) \ge S \ge 1 - \frac{1}{e}$$

[From Eqs. (i) and (ii)]

**15.** ***Concept Involved***

Equation of tangent to $\frac{x^2}{a^2} - \frac{y^2}{b^2} = 1$ is

$$y = mx \pm \sqrt{a^2m^2 - b^2}$$

**Description of Situation** If two straight lines

$$a_1x + b_1y + c_1 = 0$$

and $a_2x + b_2y + c_2 = 0$ are identical

$$\frac{a_1}{a_2} = \frac{b_1}{b_2} = \frac{c_1}{c_2}$$

**Sol.** Equation of tangent, parallel to

$$y = 2x - 1$$

$\Rightarrow \quad y = 2x \pm \sqrt{9(4) - 4}$

$\therefore \quad y = 2x \pm \sqrt{32} \quad \ldots(\text{i})$

The equation of tangent at $(x_1, y_1)$ is

$$\frac{xx_1}{9} - \frac{yy_1}{4} = 1 \quad \ldots(\text{ii})$$

From Eqs. (i) and (ii), we get

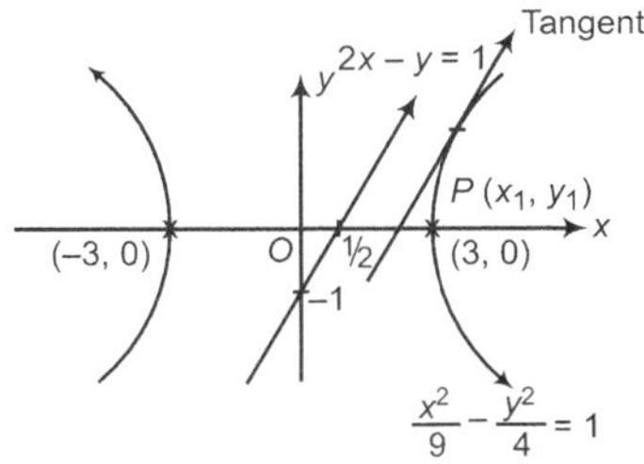


$$\frac{2}{\frac{x_1}{9}} = \frac{-1}{\frac{-y_1}{4}} = \frac{\pm\sqrt{32}}{1}$$

$\Rightarrow \quad x_1 = -\frac{9}{2\sqrt{2}}$ and $y_1 = -\frac{1}{\sqrt{2}}$

or $\quad x_1 = \frac{9}{2\sqrt{2}}, y_1 = \frac{1}{\sqrt{2}}$

**16. Concept Involved** Parametric coordinates for $y^2 = 4ax$ are $(at^2, 2at)$.

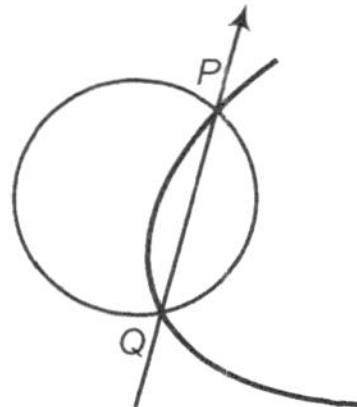


**Description of Situation** As the circle intersects the parabola at $P$ and $Q$. Thus, points $P$ and $Q$ should satisfy circle.

**Sol.** $P(2t^2, 4t)$ should lie on

$$x^2 + y^2 - 2x - 4y = 0$$

$\Rightarrow \quad 4t^4 + 16t^2 - 4t^2 - 16t = 0$

$\Rightarrow \quad 4t^4 + 12t^2 - 16t = 0$

$\Rightarrow \quad 4t(t^3 + 3t - 4) = 0$

$\Rightarrow \quad 4t(t-1)(t^2 + t + 4) = 0$

$\therefore \quad t = 0, 1 \quad \Rightarrow P(2, 4)$

Thus, area of $\Delta OPS = \frac{1}{2} \cdot OS \times PQ$

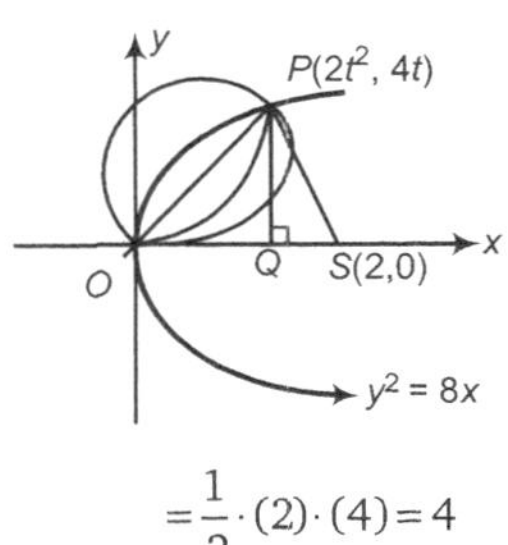


$$= \frac{1}{2} \cdot (2) \cdot (4) = 4$$

**17.** ***Concept Involved*** If $f(x)$ be least degree polynomial having local maximum and local minimum at $\alpha$ and $\beta$.

Then $\quad f'(x) = \lambda(x - \alpha)(x - \beta)$

**Sol.** Here, $p'(x) = \lambda(x-1)(x-3)$

$$= \lambda(x^2 - 4x + 3)$$

Integrating both sides between 1 to 3

$$\int_1^3 p'(x)\,dx = \int_1^3 \lambda(x^2-4x+3)\,dx$$

$$\Rightarrow \quad (p(x))_1^3 = \lambda\left(\frac{x^3}{3}-2x^2+3x\right)_1^3$$

$$\Rightarrow \quad p(3)-p(1)$$

$$= \lambda\left((9-18+9)-\left(\frac{1}{3}-2+3\right)\right)$$

$$\Rightarrow \quad 2-6 = \lambda\left\{\frac{-4}{3}\right\}$$

$$\Rightarrow \quad \lambda = 3$$

$$\therefore \quad p'(x) = 3(x-1)(x-3)$$

$$\therefore \quad p'(0) = 9$$

**18.** ***Concept Involved***

(i) Local maximum and local minimum are those point at which $f'(x)=0$ when, defined for all real numbers.

(ii) Local maximum and local minimum for piecewise function are also been checked at sharp edges.

**Description of Situation**

$$y = |x| = \begin{cases} x, & x \geq 0 \\ -x, & x < 0 \end{cases}$$

Also, $y = |x^2-1| = \begin{cases} (x^2-1), & x \leq -1 \text{ or } x \geq 1 \\ (1-x^2), & -1 \leq x \leq 1 \end{cases}$

**Sol.** $y = |x| + |x^2-1|$

$$= \begin{cases} -x+1-x^2, & x \leq -1 \\ -x+1-x^2, & -1 \leq x \leq 0 \\ x+1-x^2, & 0 \leq x \leq 1 \\ x+x^2-1, & x \geq 1 \end{cases}$$

$$= \begin{cases} -x^2-x+1, & x \leq -1 \\ -x^2-x+1, & -1 \leq x \leq 0 \\ -x^2+x+1, & 0 \leq x \leq 1 \\ x^2+x-1, & x \geq 1 \end{cases}$$

which could be graphically shown as

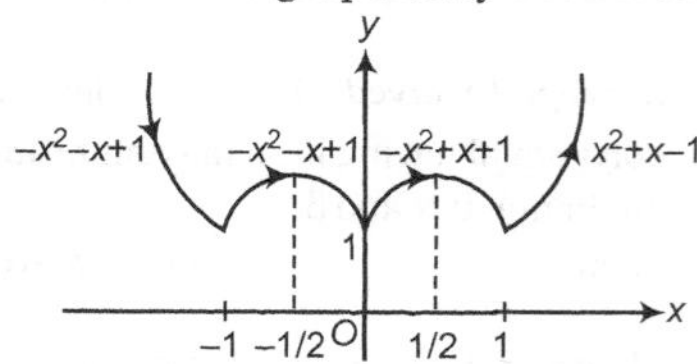


Thus, $f(x)$ attains maximum at $x = \frac{1}{2}, \frac{-1}{2}$ and $f(x)$ attains minimum at $x = -1, 0, 1$.

$\Rightarrow$ Total number of points = 5

**19.** ***Concept Involved***

Use of infinite series

*i.e.*, if $y = \sqrt{x\sqrt{x\sqrt{x\ldots\infty}}} \Rightarrow y = \sqrt{xy}$

**Sol.** $6+\log_{\frac{3}{2}}\left(\frac{1}{3\sqrt{2}}\sqrt{4-\frac{1}{3\sqrt{2}}\sqrt{4-\frac{1}{3\sqrt{2}}\sqrt{4-\frac{1}{3\sqrt{2}}\ldots}}}\right)$

Let $\sqrt{4-\frac{1}{3\sqrt{2}}\sqrt{4-\frac{1}{3\sqrt{2}}\sqrt{\ldots}}} = y$

$$\therefore \quad y = \sqrt{4-\frac{1}{3\sqrt{2}}y}$$

$$\Rightarrow \quad y^2 + \frac{1}{3\sqrt{2}}y - 4 = 0$$

$$\Rightarrow \quad 3\sqrt{2}y^2 + y - 12\sqrt{2} = 0$$

$$\therefore \quad y = \frac{-1 \pm 17}{6\sqrt{2}} \quad \text{or} \quad y = \frac{8}{3\sqrt{2}}$$

Now,

$$6+\log_{\frac{3}{2}}\left(\frac{1}{3\sqrt{2}}\cdot y\right) = 6+\log_{\frac{3}{2}}\left(\frac{1}{3\sqrt{2}}\cdot\frac{8}{3\sqrt{2}}\right)$$

$$= 6+\log_{\frac{3}{2}}\left(\frac{4}{9}\right) = 6+\log_{\frac{3}{2}}\left(\frac{3}{2}\right)^{-2}$$

$$= 6-2\cdot\log_{\frac{3}{2}}\left(\frac{3}{2}\right) = 4$$

**20.** ***Concept Involved*** If $\mathbf{a}, \mathbf{b}, \mathbf{c}$ are any three vectors

Then $|\mathbf{a}+\mathbf{b}+\mathbf{c}|^2 \geq 0$

$\Rightarrow |\mathbf{a}|^2+|\mathbf{b}|^2+|\mathbf{c}|^2+2(\mathbf{a}\cdot\mathbf{b}+\mathbf{b}\cdot\mathbf{c}+\mathbf{c}\cdot\mathbf{a}) \geq 0$

$\therefore \ \mathbf{a}\cdot\mathbf{b}+\mathbf{b}\cdot\mathbf{c}+\mathbf{c}\cdot\mathbf{a} \geq \frac{-1}{2}(|\mathbf{a}|^2+|\mathbf{b}|^2+|\mathbf{c}|^2)$

**Sol.** Given, $|\mathbf{a}-\mathbf{b}|^2+|\mathbf{b}-\mathbf{c}|^2+|\mathbf{c}-\mathbf{a}|^2 = 9$

$$\Rightarrow |\mathbf{a}|^2+|\mathbf{b}|^2-2\mathbf{a}\cdot\mathbf{b}+|\mathbf{b}|^2+|\mathbf{c}|^2 - 2\mathbf{b}\cdot\mathbf{c}+|\mathbf{c}|^2+|\mathbf{a}|^2-2\mathbf{c}\cdot\mathbf{a} = 9$$

$$\Rightarrow \quad 6-2(\mathbf{a}\cdot\mathbf{b}+\mathbf{b}\cdot\mathbf{c}+\mathbf{c}\cdot\mathbf{a}) = 9$$

$$(\because |\mathbf{a}|=|\mathbf{b}|=|\mathbf{c}|=1)$$

$$\Rightarrow \quad \mathbf{a}\cdot\mathbf{b}+\mathbf{b}\cdot\mathbf{c}+\mathbf{c}\cdot\mathbf{a}=-\frac{3}{2} \qquad \ldots\text{(i)}$$

Also, $\mathbf{a}\cdot\mathbf{b}+\mathbf{b}\cdot\mathbf{c}+\mathbf{c}\cdot\mathbf{a}$

$$\geq \frac{-1}{2}(|\mathbf{a}|^2+|\mathbf{b}|^2+|\mathbf{c}|^2)$$

$$\geq -\frac{3}{2} \qquad \ldots\text{(ii)}$$

From Eqs. (i) and (ii), we get

$$|\mathbf{a}+\mathbf{b}+\mathbf{c}|=0$$

as $\mathbf{a}\cdot\mathbf{b}+\mathbf{b}\cdot\mathbf{c}+\mathbf{c}\cdot\mathbf{a}$ is minimum when $|\mathbf{a}+\mathbf{b}+\mathbf{c}|=0$

$$\Rightarrow \quad \mathbf{a}+\mathbf{b}+\mathbf{c}=0$$

$$\therefore \quad |2\mathbf{a}+5\mathbf{b}+5\mathbf{c}|=|2\mathbf{a}+5(\mathbf{b}+\mathbf{c})|$$

$$=|2\mathbf{a}-5\mathbf{a}|=3$$

## Paper 2

**21.** ***Concept Invovled*** (i) Equation of plane through intersection of two planes.

*i.e.*, $(a_1x + b_1y + c_1z + d_1) + \lambda\,(a_2x + b_2y + c_2z + d_2) = 0$

(ii) Distance of a point $(x_1, y_1, z_1)$ from

$$ax + by + cz + d = 0$$

$$\Rightarrow \quad \frac{|ax_1 + by_1 + cz_1 + d|}{\sqrt{a^2 + b^2 + c^2}}$$

**Sol.** Equation of plane passing through intersection of two planes

$x + 2y + 3z = 2$ and $x - y + z = 3$ is

$$\Rightarrow \quad (x + 2y + 3z - 2) + \lambda\,(x - y + z - 3) = 0$$

$$\Rightarrow (1 + \lambda)\,x + (2 - \lambda)\,y + (3 + \lambda)\,z - (2 + 3\lambda) = 0$$

whose distance from $(3, 1, -1)$ is $\dfrac{2}{\sqrt{3}}$

$$\Rightarrow \frac{|3(1 + \lambda) + 1\cdot(2 - \lambda) - 1\,(3 + \lambda) - (2 + 3\lambda)|}{\sqrt{(1 + \lambda)^2 + (2 - \lambda)^2 + (3 + \lambda)^2}} = \frac{2}{\sqrt{3}}$$

$$\Rightarrow \quad \frac{|-2\lambda|}{\sqrt{3\lambda^2 + 4\lambda + 14}} = \frac{2}{\sqrt{3}}$$

$$\Rightarrow 3\lambda^2 = 3\lambda^2 + 4\lambda + 14 \Rightarrow \lambda = -\frac{7}{2}$$

$$\therefore \left(1 - \frac{7}{2}\right)x + \left(2 + \frac{7}{2}\right)y + \left(3 - \frac{7}{2}\right)z - \left(2 - \frac{21}{2}\right) = 0$$

$$\Rightarrow \quad -\frac{5x}{2} + \frac{11}{2}y - \frac{1}{2}z + \frac{17}{2} = 0$$

or $\quad 5x - 11y + z - 17 = 0$

**22.** **Concept Involved** If $\mathbf{a} \times \mathbf{b} = \mathbf{a} \times \mathbf{c}$

$$\Rightarrow \quad \mathbf{a} \times \mathbf{b} - \mathbf{a} \times \mathbf{c} = 0$$

$\Rightarrow \quad \mathbf{a} \times (\mathbf{b} - \mathbf{c}) = 0$ *i.e.*, $\mathbf{a} \,||\, (\mathbf{b} - \mathbf{c})$

or $\quad \mathbf{b} - \mathbf{c} = \lambda\mathbf{a}$

**Sol.** Here, $\mathbf{a} \times (2\mathbf{i} + 3\mathbf{j} + 4\mathbf{k}) = (2\mathbf{i} + 3\mathbf{j} + 4\mathbf{k}) \times \mathbf{b}$

$$\Rightarrow \mathbf{a} \times (2\mathbf{i} + 3\mathbf{j} + 4\mathbf{k}) - (2\mathbf{i} + 3\mathbf{j} + 4\mathbf{k}) \times \mathbf{b} = 0$$

$$\Rightarrow \quad (\mathbf{a} + \mathbf{b}) \times (2\mathbf{i} + 3\mathbf{j} + 4\mathbf{k}) = 0$$

$$\Rightarrow \quad \mathbf{a} + \mathbf{b} = \lambda\,(2\mathbf{i} + 3\mathbf{j} + 4\mathbf{k}) \qquad \ldots\text{(i)}$$

Since, $\quad |\mathbf{a} + \mathbf{b}| = \sqrt{29}$

$$\Rightarrow \quad \pm\lambda\sqrt{4 + 9 + 16} = \sqrt{29}$$

$$\Rightarrow \quad \lambda = \pm 1$$

$$\therefore \quad \mathbf{a} + \mathbf{b} = \pm\,(2\mathbf{i} + 3\mathbf{j} + 4\mathbf{k})$$

Now, $(\mathbf{a} + \mathbf{b})\cdot(-7\mathbf{i} + 2\mathbf{j} + 3\mathbf{k})$

$$= \pm\,(-14 + 6 + 12) = \pm 4$$

**23.** **Concept Involved** If $\Delta ABC$ has sides $a, b, c$. Then,

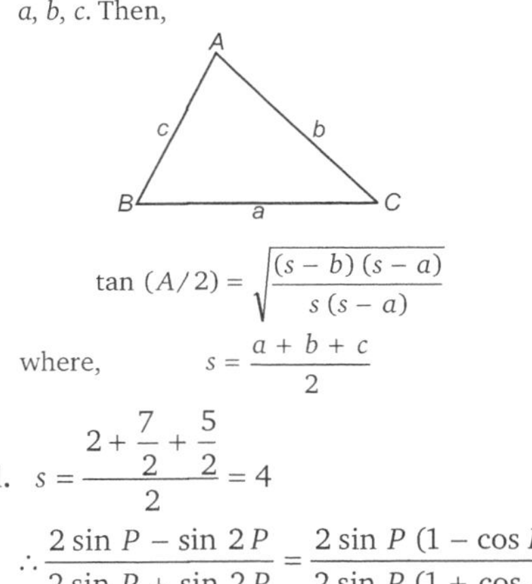


$$\tan(A/2) = \sqrt{\frac{(s - b)(s - a)}{s\,(s - a)}}$$

where, $\quad s = \dfrac{a + b + c}{2}$

**Sol.** $\quad s = \dfrac{2 + \dfrac{7}{2} + \dfrac{5}{2}}{2} = 4$

$$\therefore \frac{2\sin P - \sin 2P}{2\sin P + \sin 2P} = \frac{2\sin P\,(1 - \cos P)}{2\sin P\,(1 + \cos P)}$$

$$= \frac{2\sin^2(P/2)}{2\cos^2(P/2)} = \tan^2(P/2)$$



$$\Rightarrow \frac{(s-b)(s-c)}{s(s-a)} \times \frac{(s-b)(s-c)}{(s-b)(s-c)}$$

$$\Rightarrow \frac{((s-b)(s-c)^2)}{\Delta^2} = \frac{\left(4-\frac{7}{2}\right)^2\left(4-\frac{5}{2}\right)^2}{\Delta^2}$$

$$= \left(\frac{3}{4\Delta}\right)^2$$

**24.** ***Concept Involved*** As one of the dice shows a number appearing on one of $P_1$, $P_2$ and $P_3$. Thus, three cases arises.

(i) All shows same number.

(ii) Number appearing on $D_4$ appears on any one of $D_1, D_2, D_3$.

(iii) Number appearing on $D_4$ appears on any two of $D_1, D_2, D_3$.

**Sol.** Sample space $= 6 \times 6 \times 6 \times 6$

$= 6^4$ favourable events

= Case I or Case II or Case III

*Case* I First we should select one number for $D_4$ which appears on all *i.e.*, ${}^6C_1 \times 1$.

*Case* II For $D_4$ there are ${}^6C_1$ ways. Now, it appears on any one of $D_1, D_2, D_3$ *i.e.*, ${}^3C_1 \times 1$.

for other two there are $5 \times 5$ ways.

$$\Rightarrow \quad {}^6C_1 \times {}^3C_1 \times 1 \times 5 \times 5$$

*Case* **III** For $D_4$ there are ${}^6C_1$ ways now it appears on any two of $D_1, D_2, D_3 \Rightarrow {}^3C_2 \times 1^2$

for other one there are 5 ways.

$$\Rightarrow \quad {}^6C_1 \times {}^3C_2 \times 1^2 \times 5$$

Thus, probability

$$= \frac{{}^6C_1 + {}^6C_1 \times {}^3C_1 \times 5^2 + {}^6C_1 \times {}^3C_2 \times 5}{6^4}$$

$$= \frac{6(1+75+15)}{6^4} = \frac{91}{216}$$

**25.** $I = \int_{-\pi/2}^{\pi/2} \left(x^2 + \log\left(\frac{\pi - x}{\pi + x}\right)\right) \cos x \, dx$

As, $\int_{-a}^{a} f(x)\,dx = 0$,

when $f(-x) = -f(x)$

$$\therefore I = \int_{-\pi/2}^{\pi/2} x^2 \cos x \, dx + 0$$

$$= 2\int_0^{\pi/2} (x^2 \cos x)\,dx$$

$$= 2\{(x^2 \sin x)_0^{\pi/2} - \int_0^{\pi/2} 2x \cdot \sin x \, dx\}$$

$$= 2\left[\frac{\pi^2}{4} - 2\{(-x \cdot \cos x)_0^{\pi/2} - \int_0^{\pi/2} 1 \cdot (-\cos x)\,dx\}\right]$$

$$= 2\left[\frac{\pi^2}{4} - 2(\sin x)_0^{\pi/2}\right]$$

$$= 2\left[\frac{\pi^2}{4} - 2\right] = \left(\frac{\pi^2}{2} - 4\right)$$

**26.** Given, $\quad P^T = 2P + I \quad \ldots$(i)

$$\therefore \quad (P^T)^T = (2P+I)^T = 2P^T + I$$

$$\Rightarrow \quad P = 2P^T + I$$

$$\Rightarrow \quad P = 2(2P+I) + I$$

$$\Rightarrow \quad P = 4P + 3I \quad \text{or} \quad 3P = -3I$$

$$\Rightarrow \quad PX = -IX$$

**27.** **Concept Involved** $n^{\text{th}}$ term of HP $t_n = \dfrac{1}{a + (n-1)d}$

**Sol.** Here, $a_1 = 5$, $a_{20} = 25$ for HP

$$\therefore \quad \frac{1}{a} = 5 \text{ and } \frac{1}{a + 19d} = 25$$

$$\Rightarrow \quad \frac{1}{5} + 19d = \frac{1}{25}$$

$$\Rightarrow \quad 19d = \frac{1}{25} - \frac{1}{5} = -\frac{4}{25}$$

$$\therefore \quad d = \frac{-4}{19 \times 25}$$

Since, $\quad a_n < 0$

$$\Rightarrow \quad \frac{1}{5} + (n-1)d < 0$$

$$\Rightarrow \quad \frac{1}{5} - \frac{4}{19 \times 25}(n-1) < 0$$

$$\Rightarrow \quad (n-1) > \frac{95}{4}$$

$$\Rightarrow \quad n > 1 + \frac{95}{4} \text{ or } n > 24.75$$

least positive value of $n = 25$.

**28.** ***Concept Involved*** To make the quadratic into simple form we should eliminate radical sign.

**Description of Situation** As for given equation when $a \to 0$ the equation reduces to identity in $x$.

*i.e.,* $ax^2 + bx + c = 0$,

for all $x \in R$ or $a = b = c \to 0$

Thus, first we should make above equation independent from coefficients as 0.

**Sol.** Let $a + 1 = t^6$. Thus, when $a \to 0, t \to 1$.

$$\therefore \quad (t^2 - 1)x^2 + (t^3 - 1)x + (t - 1) = 0$$

$$\Rightarrow \quad (t-1)\{(t+1)x^2 + (t^2 + t + 1)x + 1\} = 0,$$

as $t \to 1$

$$2x^2 + 3x + 1 = 0$$

$$\Rightarrow \quad 2x^2 + 2x + x + 1 = 0$$

$$\Rightarrow \quad (2x+1)(x+1) = 0$$

Thus, $x = -1, -1/2$

or $\lim_{a \to 0^+} \alpha(a) = -\frac{1}{2}$

and $\lim_{a \to 0^+} \beta(a) = -1$

## Solutions (Q. Nos. 29 and 30)

**Concept Involved** Use of Newton leibnitz formula

*i.e.,* $\frac{d}{dx}\left\{\int_{g(x)}^{f(x)} \phi(t)\, dt\right\} = \phi(f(x)) \cdot f'(x) - \phi(g(x)) \cdot g'(x)$

**Sol.** Here, $f(x) = (1-x)^2 \sin^2 x + x^2$, for all $x$

$$g(x) = \int_1^x \left(\frac{2(t-1)}{t+1} - \log(t)\right) f(t)\, dt, \quad \forall \ x \in (1, \infty)$$

**29.** Here,

$$f(x) = (1-x)^2 \cdot \sin^2 x + x^2 \geq 0, \ \forall \ x$$

and $g(x) = \int_1^x \left(\frac{2(t-1)}{t+1} - \log t\right) f(t)\, dt$

$$\Rightarrow g'(x) = \left\{\frac{2(x-1)}{(x+1)} - \log x\right\} \cdot \underbrace{f(x)}_{+\text{ ve}} \quad \dots(\text{i})$$

For $g'(x)$ to be increasing or decreasing

Let $\phi(x) = \frac{2(x-1)}{(x+1)} - \log x$

$$\phi'(x) = \frac{4}{(x+1)^2} - \frac{1}{x} = \frac{-(x-1)^2}{x(x+1)^2}$$

$\phi'(x) < 0$, for $x > 1$

$$\Rightarrow \quad \phi(x) < \phi(1) \quad \Rightarrow \quad \phi(x) < 0 \quad \dots(\text{ii})$$

From Eqs. (i) and (ii), we get

$$g'(x) < 0 \text{ for } x \in (1, \infty)$$

$\therefore g(x)$ is decreasing for $x \in (1, \infty)$

**30.** Here, $f(x) + 2x = (1-x)^2 \cdot \sin^2 x + x^2 + 2x \quad \dots(\text{i})$

where, P : $f(x) + 2x = 2(1+x)^2 \quad \dots(\text{ii})$

$$\therefore 2(1 + x^2) = (1-x)^2 \sin^2 x + x^2 + 2x$$

$$\Rightarrow \quad (1-x)^2 \sin^2 x = x^2 - 2x + 2$$

$$\Rightarrow \quad (1-x)^2 \sin^2 x = (1-x)^2 + 1$$

$$\Rightarrow \quad (1-x)^2 \cos^2 x = -1$$

Which is never possible.

$\therefore$ P is false.

Again, let $h(x) = 2f(x) + 1 - 2x(1+x)$

where, $h(0) = 2f(0) + 1 - 0 = 1$

$h(1) = 2f(1) + 1 - 4 = -3$

as, $h(0)\,h(1) < 0$

$\Rightarrow h(x)$ must have a solution.

$\therefore$ $Q$ is true.

## Solution (Q. Nos. 31 and 32)

Since, $a_n$ be the $n$-digit positive integer formed by the digits 0, 1 or both such that no two consecutive digits are zero.

$b_n$ = numbers which are ending with 1

$c_n$ = numbers which are ending with 0

$$\therefore \quad a_n = b_n + c_n$$

*i.e.*, $a_n = 1\ (0 \text{ or } 1) \ldots\ldots (0 \text{ or } 1)$

$b_n = 1\ (0 \text{ or } 1) \ldots\ldots 1$

$n^{\text{th}}$place

$c_n = 1\ (0 \text{ or } 1) \ldots\ldots 0.$

**31.** As $a_n = b_n + c_n$

*i.e.*, $a_n = 1 \ldots\ldots\ldots (1 \text{ or } 0)$

$= \underbrace{1 \_\_\_\_\_\_\_\_\_}_{(n-1) \text{ places}}\boxed{1} = a_{n-1}$

or

$= \underbrace{1 \_\_\_\_\_\_}_{(n-2) \text{ places}}\boxed{1}\ \boxed{0} = a_{n-2}$

$\Rightarrow \quad a_n = a_{n-1} + a_{n-2}$

$\therefore \quad a_{17} = a_{16} + a_{15}$

**32.** $b_6$ = Six digit number ending with 1.

$\Rightarrow \boxed{1}\ \boxed{\ }\ \boxed{\ }\ \boxed{\ }\ \boxed{\ }\ \boxed{1}$

Now, the four places are to be filled.

*i.e.*, **Case I** $\_\ \_\ \_\ \boxed{1}$

**Case II** $\_\ \_\ \_\ \boxed{0}$

for **Case I** $\boxed{1}\ \_\ \_\ \_\ \boxed{1}\ \boxed{1}$

for 3 places, all 1's are used = 1 way

one zero is used $= {}^3C_1 = 3$

two zeros are used = 1 way

0 1 0 1.........

Total = 5 ways

**Case II** $\boxed{1}\ \_\ \_\ \_\ \boxed{0}\ \boxed{1}$

for 3 places,

all 1's are used = 1 way

one zero is used $= {}^2C_1 = 2$ ways

= 3 ways

Thus, $b_6 = 5 + 3 = 8$

**Solution (Q. Nos. 33 and 34)**

**Concept Involved** (i) Equation of tangent to at $(x_1, y_1)$ is

$x^2 + y^2 = r^2$

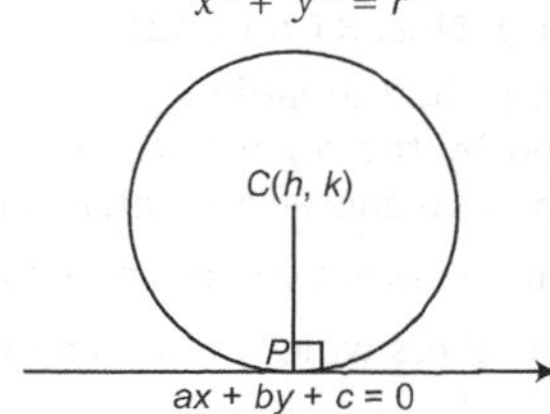


at $(x_1, y_1)$ is

$$xx_1 + yy_1 = r^2$$

(ii) If $ax + by + c = 0$ is tangent to

$$(x-h)^2 + (y-k)^2 = r^2$$

$\Rightarrow \quad |cp| = r$

**33.** Here, equation of common tangent be

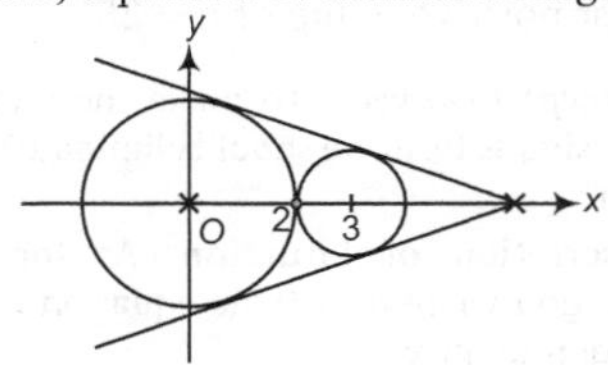


$$y = mx \pm 2\sqrt{1+m^2}$$

which is also the tangent to

$$(x-3)^2 + y^2 = 1$$

$\Rightarrow \quad \dfrac{|3m - 0 + 2\sqrt{1+m^2}|}{\sqrt{m^2+1}} = 1$

$\Rightarrow \quad 3m + 2\sqrt{1+m^2} = \pm\sqrt{1+m^2}$

$\Rightarrow \quad 3m = -3\sqrt{1+m^2}$

or $\quad 3m = -\sqrt{1+m^2}$

$\Rightarrow \quad m^2 = 1 + m^2$

or $\quad 9m^2 = 1 + m^2$

$\Rightarrow \quad m \in \phi \quad \text{or} \quad m = \pm\dfrac{1}{2\sqrt{2}}$

$\therefore \quad y = \pm\dfrac{1}{2\sqrt{2}}x \pm 2\sqrt{1+\dfrac{1}{8}}$

$\Rightarrow \quad y = \pm\dfrac{x}{2\sqrt{2}} \pm \dfrac{6}{2\sqrt{2}}$

$\Rightarrow \quad 2\sqrt{2}y = \pm(x+6)$

**34.** Here, tangent to $x^2 + y^2 = 4$ at $(\sqrt{3}, 1)$ is $\sqrt{3}x + y = 4$ ...(i)

As, $L$ is perpendicular to $\sqrt{3}x + y = 4$

$\Rightarrow \quad x - \sqrt{3}y = \lambda$

which is tangent to $(x-3)^2 + y^2 = 1$

$\Rightarrow \quad \dfrac{|3 - 0 - \lambda|}{\sqrt{1+3}} = 1$

$\Rightarrow \quad |3 - \lambda| = 2$

$\Rightarrow \quad 3 - \lambda = 2, -2$

$\therefore \quad \lambda = 1, 5$

$\Rightarrow \quad L : x - \sqrt{3}y = 1,\ x - \sqrt{3}y = 5$

**35. Concept Involved**

(i) Conditional probability *i.e.,*

$$P(A/B) = \frac{P(A \cap B)}{P(B)}$$

(ii) $P(A \cup B) = P(A) + P(B) - P(A \cap B)$

(iii) Independent event, then

$$P(A \cap B) = P(A) \cdot P(B)$$

**Sol.** Here, $P(X/Y) = \frac{1}{2}, P\left(\frac{Y}{X}\right) = \frac{1}{3}$

and $P(X \cap Y) = 6$

$(x_1, y_1, z_1)$ dr's$(a_1, b_1, c_1)$
dr's$(x_2 - x_1, y_2 - y_1, z_2 - z_1)$
dr's$(a_2, b_2, c_2)$
$(x_2, y_2, z_2)$

$$\therefore \quad P\left(\frac{X}{Y}\right) = \frac{P(X \cap Y)}{P(Y)}$$

$$\Rightarrow \quad \frac{1}{2} = \frac{1/6}{P(Y)}$$

$$\Rightarrow \quad P(Y) = \frac{1}{3} \qquad \ldots(\text{i})$$

$$P\left(\frac{Y}{X}\right) = \frac{1}{3}$$

$$\Rightarrow \quad \frac{P(X \cap Y)}{P(X)} = \frac{1}{3}$$

$$\Rightarrow \quad \frac{1}{6} = \frac{1}{3} P(X)$$

$$\therefore \quad P(X) = \frac{1}{2} \qquad \ldots(\text{ii})$$

$$P(X \cup Y) = P(X) + P(Y) - P(X \cap Y)$$

$$= \frac{1}{2} + \frac{1}{3} - \frac{1}{6} = \frac{2}{3} \qquad \ldots(\text{iii})$$

$$P(X \cap Y) = \frac{1}{6}$$

and $P(X) \cdot P(Y) = \frac{1}{2} \cdot \frac{1}{3} = \frac{1}{6}$

$\Rightarrow \quad P(X \cap Y) = P(X) \cdot P(Y)$

$\Rightarrow$ Independent events $\qquad \ldots(\text{iv})$

$$P(X^C \cap Y) = P(Y) - P(X \cap Y)$$

$$= \frac{1}{3} - \frac{1}{6} = \frac{1}{6} \qquad \ldots(\text{v})$$

**36. Concept Involved** Use of Newton Leibnitz formula

$$\frac{d}{dx}\left\{\int_{f(x)}^{g(x)} \phi(t)\,dt\right\} = \phi(g(x)) \cdot g'(x) - \phi(f(x)) \cdot f'(x)$$

**Sol.** Here, $f(x) = \int_0^x e^{t^2}(t-2)(t-3)\,dt$

$$\Rightarrow \quad f'(x) = e^{x^2}(x-2)(x-3)$$

+   2   −   3   +

$\therefore$ maximum at $x = 2$

minimum at $x = 3$

decreasing on $(2, 3)$

Also, $f'(x) = 0$ has two roots $x = 2$ and $x = 3$

*i.e.,* $\quad f'(2) = f'(3) = 0$

Thus, by Rolle's theorem

$f''(c) = 0$ must have atleast one root $\in (2, 3)$.

**37.** $f(2n) = a_n,\ f(2n^+) = a_n$

$$f(2n^-) = b_n + 1$$

$$\Rightarrow \quad a_n - b_n = 1$$

$$f(2n+1) = a_n$$

$$f((2n+1)^-) = a_n$$

$$f((2n+1)^+) = b_{n+1} - 1$$

$\Rightarrow \quad a_n = b_{n+1} - 1$ or $a_n - b_{n+1} = -1$

or $\quad a_{n-1} - b_n = -1$

**38.** ***Concept Involved*** If the straight lines are coplanar. They the should lie in same plane.

**Description of Situation** If straight lines are coplanar.

$$\Rightarrow \begin{vmatrix} x_2 - x_1 & y_2 - y_1 & z_2 - z_1 \\ a_1 & b_1 & c_1 \\ a_2 & b_2 & c_2 \end{vmatrix} = 0$$

**Sol.** Since, $\frac{x-1}{2} = \frac{y+1}{K} = \frac{z}{2}$

and $\frac{x+1}{5} = \frac{y+1}{2} = \frac{z}{k}$ are coplanar.

$$\Rightarrow \begin{vmatrix} 2 & 0 & 0 \\ 2 & K & 2 \\ 5 & 2 & K \end{vmatrix} = 0 \Rightarrow K^2 = 4$$

$$\Rightarrow \quad K = \pm 2$$

$\therefore \quad \mathbf{n}_1 = \mathbf{b}_1 \times \mathbf{d}_1 = 6\mathbf{j} - 6\mathbf{k}$, for $k = 2$

$\therefore\ \mathbf{n}_2 = \mathbf{b}_2 \times \mathbf{d}_2 = 14\mathbf{j} + 14\mathbf{k}$, for $k = -2$

So, equation of planes are $(\mathbf{r} - \mathbf{a}) \cdot \mathbf{n}_1 = 0$

$\Rightarrow\ y - z = -1$ and $(\mathbf{r} - \mathbf{a}) \cdot \mathbf{n}_2 = 0$

$\Rightarrow\ y + z = -1$

**39.** ***Concept Involved*** If $|A_{n \times n}| = \Delta$, then $|\text{adj}\, A| = \Delta^{n-1}$

**Sol.** Here, $\text{adj}\, P_{3 \times 3} = \begin{bmatrix} 1 & 4 & 4 \\ 2 & 1 & 7 \\ 1 & 1 & 3 \end{bmatrix}$

$\Rightarrow\ |\text{adj}\, P| = |P|^2$

$$\therefore\ |\text{adj}\, P| = \begin{vmatrix} 1 & 4 & 4 \\ 2 & 1 & 7 \\ 1 & 1 & 3 \end{vmatrix}$$

$$= 1(3-7) - 4(6-7) + 4(2-1)$$

$$= -4 + 4 + 4 = 4$$

$\Rightarrow\ |P| = \pm 2$

**40.** $f(\cos 4\theta) = \dfrac{2}{2 - \sec^2 \theta}$ ...(i)

At, $\cos 4\theta = \dfrac{1}{3}$

$\Rightarrow\ 2\cos^2 2\theta - 1 = \dfrac{1}{3}$

$\Rightarrow\ \cos^2 2\theta = \dfrac{2}{3}$

$\Rightarrow\ \cos 2\theta = \pm \sqrt{\dfrac{2}{3}}$ ...(ii)

$\therefore\ f(\cos 4\theta) = \dfrac{2 \cdot \cos^2 \theta}{2\cos^2 \theta - 1}$

$= \dfrac{1 + \cos 2\theta}{\cos 2\theta}$

$\Rightarrow\ f\left(\dfrac{1}{3}\right) = 1 \pm \sqrt{\dfrac{3}{2}}$ [From Eq. (ii)]

# IIT JEE

## SOLVED PAPER 2011

## Physics

## Paper I

### SECTION I

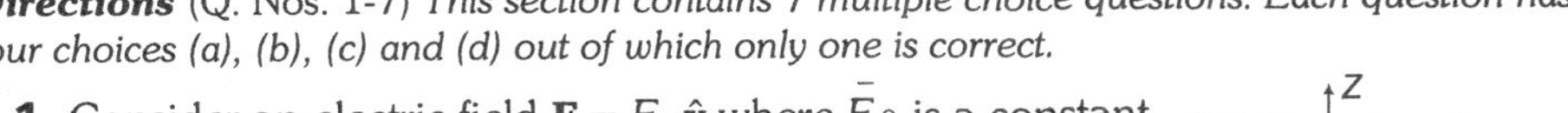

***Directions*** (Q. Nos. 1-7) *This section contains 7 multiple choice questions. Each question has four choices (a), (b), (c) and (d) out of which only one is correct.*

**1.** Consider an electric field $\mathbf{E} = E_0\hat{\mathbf{x}}$ where $\bar{E}_0$ is a constant. The flux through the shaded area ( as shown in the figure) due to this field is

(a) $2E_0 a^2$ (b) $\sqrt{2} E_0 a^2$

(c) $E_0 a^2$ (d) $\dfrac{E_0 a^2}{\sqrt{2}}$

**2.** A ball of mass $(m)$ 0.5 kg is attached to the end of a string having length $(L)$ 0.5 m. The ball is rotated on a horizontal circular path about vertical axis. The maximum tension that the string can bear is 324 N. The maximum possible value of angular velocity of ball (in rad/s) is

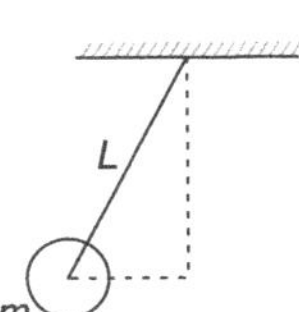

(a) 9 (b) 18 (c) 27 (d) 36

**3.** A 2μF capacitor is charged as shown in the figure. The percentage of its stored energy dissipated after the switch $S$ is turned to position 2, is

(a) 0% (b) 20%

(c) 75% (d) 80%

**4.** 5.6 L of helium gas at STP is adiabatically compressed to 0.7 L. Taking the initial temperature to be $T_1$, the work done in the process is

(a) $\dfrac{9}{8}RT_1$ (b) $\dfrac{3}{2}RT_1$ (c) $\dfrac{15}{8}RT_1$ (d) $\dfrac{9}{2}RT_1$

**5.** A police car with a siren of frequency 8 kHz is moving with uniform velocity 36 km/h towards a tall building which reflects the sound waves. The speed of sound in air is 320 m/s. The frequency of the siren heard by the car driver is

(a) 8.50 kHz (b) 8.25 kHz

(c) 7.75 kHz (d) 7.50 kHz

**6.** The wavelength of the first spectral line in the Balmer series of hydrogen atom is 6561 Å. The wavelength of the second spectral line in the Balmer series of singly ionized helium atom is

(a) 1215 Å (b) 1640 Å (c) 2430 Å (d) 4687 Å

**7.** A meter bridge is set up as shown in figure, to determine an unknown resistance $X$ using a standard 10 $\Omega$ resistor. The galvanometer shows null point when tapping key is at 52 cm mark. The end-corrections are 1 cm and 2 cm respectively for the ends $A$ and $B$. The determined value of $X$ is

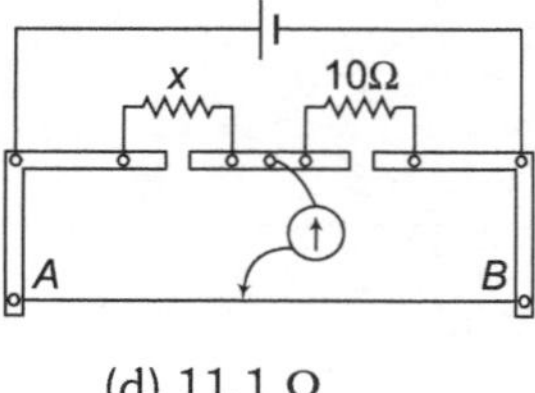


(a) 10.2 $\Omega$ (b) 10.6 $\Omega$ (c) 10.8 $\Omega$ (d) 11.1 $\Omega$

## SECTION II

**Directions** (Q. Nos. 8-11) *This section contains 4 multiple choice questions. Each question has four choices (a), (b), (c) and (d) out of which one or more may be correct.*

**8.** An electron and a proton are moving on straight parallel paths with same velocity. They enter a semi-infinite region of uniform magnetic field perpendicular to the velocity. Which of the following statement(s) is/are true?

(a) They will never come out of the magnetic field region
(b) They will come out travelling along parallel paths
(c) They will come out at the same time
(d) They will come out at different times

**9.** A composite block is made of slabs $A$, $B$, $C$, $D$ and $E$ of different thermal conductivities (given in terms of a constant, $K$) and sizes (given in terms of length, $L$) as shown in the figure. All slabs are of same width. Heat $Q$ flows only from left to right through the blocks. Then, in steady state

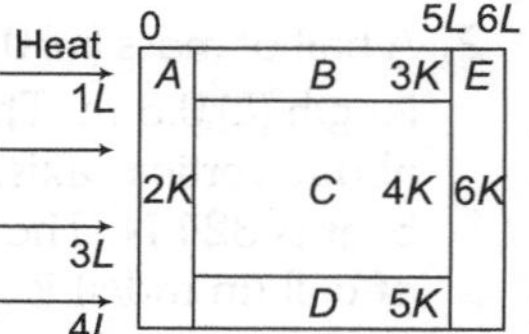


(a) heat flow through $A$ and $E$ slabs are same
(b) heat flow through slab $E$ is maximum
(c) temperature difference across slab $E$ is smallest
(d) heat flow through $C$ = heat flow through $B$ + heat flow through $D$

**10.** A spherical metal shell $A$ of radius $R_A$ and a solid metal sphere $B$ of radius $R_B < (R_A)$ are kept far apart and each is given charge $+Q$. Now, they are connected by a thin metal wire. Then

(a) $E_A^{\text{inside}} = 0$ (b) $Q_A > Q_B$
(c) $\dfrac{\sigma_A}{\sigma_B} = \dfrac{R_B}{R_A}$ (d) $E_A^{\text{on surface}} < E_B^{\text{on surface}}$

**11.** A metal rod of length $L$ and mass $m$ is pivoted at one end. A thin disc of mass $M$ and radius $R$ $(< L)$ is attached at its centre to the free end of the rod. Consider two ways the disc is attached. Case $A$—the disc is not free to rotate about its centre and case $B$—the disc is free to rotate about its centre. The rod-disc system performs SHM in vertical plane after being released from the same displaced position. Which of the following statement(s) is/are true?

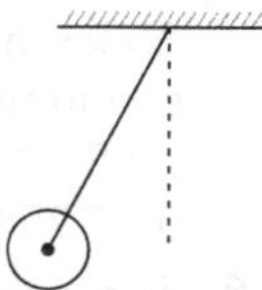

(a) Restoring torque in case $A =$ Restoring torque in case $B$
(b) Restoring torque in case $A <$ Restoring torque in case $B$
(c) Angular frequency for case $A >$ Angular frequency for case $B$
(d) Angular frequency for case $A <$ Angular frequency for case $B$

## SECTION III

***Directions*** (Q. Nos. 12-16) *This section contains 2 paragraphs. Based upon one of the paragraph 2 multiple choice questions and based on the other paragraph 3 multiple choice questions have to be answered. Each of these questions has four choices (a), (b), (c) and (d) out of which only one is correct.*

***Paragraph*** (Q. Nos. 12-13) *A dense collection of equal number of electrons and positive ions is called neutral plasma. Certain solids containing fixed positive ions surrounded by free electrons can be treated as neutral plasma. Let N be the number density of free electrons, each of mass m. When the electrons are subjected to an electric field, they are displaced relatively away from the heavy positive ions.If the electric field becomes zero, the electrons being to oscillate about the positive ions with a natural angular frequency $\omega_p$, which is called the plasma frequency. To sustain the oscillations, a time varying electric field needs to be applied that has an angular frequency $\omega$, where a part of the energy is absorbed and a part of it is reflected. As $\omega$ approaches $\omega_p$, all the free electrons are set to resonance together and all the energy is reflected. This is the explanation of high reflectivity of metals.*

**12.** Taking the electronic charge as $e$ and the permittivity as $\varepsilon_0$, use dimensional analysis to determine the correct expression for $\omega_p$.

(a) $\sqrt{\dfrac{Ne}{m\varepsilon_0}}$ (b) $\sqrt{\dfrac{m\varepsilon_0}{Ne}}$ (c) $\sqrt{\dfrac{Ne^2}{m\varepsilon_0}}$ (d) $\sqrt{\dfrac{m\varepsilon_0}{Ne^2}}$

**13.** Estimate the wavelength at which plasma reflection will occur for a metal having the density of electrons $N \approx 4 \times 10^{27}\ \text{m}^{-1}$. Take $\varepsilon_0 = 10^{-11}$ and $m = 10^{-30}$, where these quantities are in proper SI units.

(a) 800 nm (b) 600 nm (c) 300 nm (d) 200 nm

***Paragraph*** (Q. Nos. 14-16) *Phase space diagrams are useful tools in analyzing all kinds of dynamical problems. They are especially useful in studying the changes in motion as initial position and momentum are changed. Here we consider some simple dynamical systems in one-dimension. For such systems, phase space is a plane in which position is plotted along horizontal axis and momentum is plotted along vertical axis. The phase space diagram is x(t) vs p(t) curve in this plane. The arrow on the curve indicates the time flow. For example, the phase space diagram for a particle moving with constant velocity is a straight line as shown in the figure. We use the sign convention in which position or momentum upwards (or to right) is positive and downwards (or to left) is negative.*

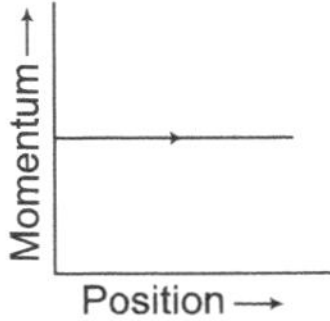


**14.** The phase space diagram for a ball thrown vertically up from ground is

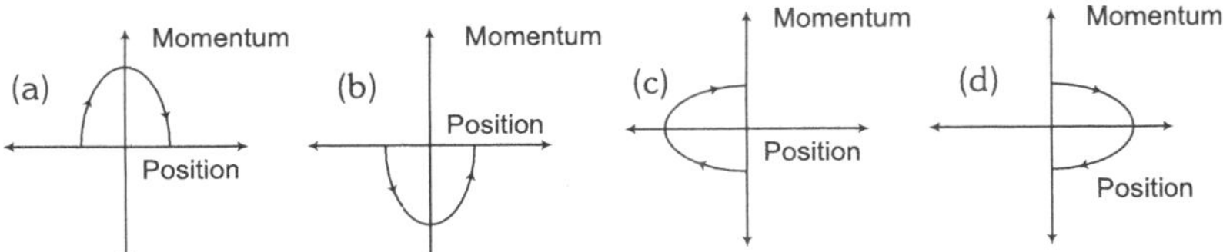

**15.** The phase space diagram for simple harmonic motion is a circle centered at the origin. In the figure, the two circles represent the same oscillator but for different initial conditions, and $E_1$ and $E_2$ are the total mechanical energies respectively. Then,

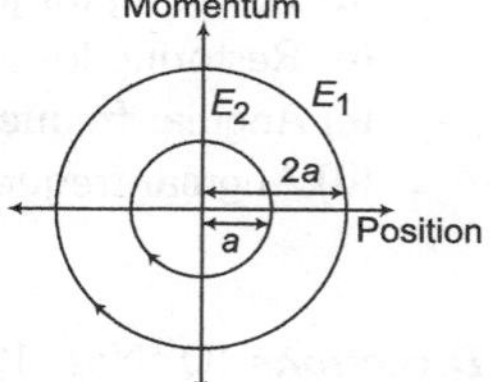


(a) $E_1 = \sqrt{2}E_2$ (b) $E_1 = 2\,E_2$

(c) $E_1 = 2E_2$ (d) $E_1 = 16\,E_2$

**16.** Consider the spring-mass system, with the mass submerged in water, as shown in the figure. The phase space diagram for one cycle of this system is

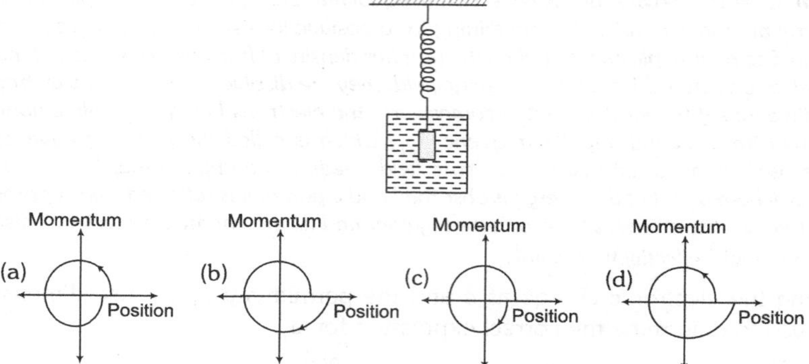


## SECTION IV

***Directions*** (Q. Nos. 17-23) *This section contains 7 questions. The answer to each question is a single digit integer, ranging from 0 to 9.*

**17.** Four point charges, each of $+\,q$, are rigidly fixed at the four corners of a square planar soap film of side $a$. The surface tension of the soap film is $\gamma$. The system of charges and planar film are in equilibrium, and $a = k\left[\dfrac{q^2}{\gamma}\right]^{1/N}$, where $k$ is a constant. Then, $N$ is

**18.** A block is moving on an inclined plane making an angle 45° with the horizontal and the coefficient of friction is $\mu$. The force required to just push it up the inclined plane is 3 times the force required to just prevent it from sliding down. If we define $N = 10\,\mu$, then $N$ is

**19.** A long circular tube of length 10 m and radius 0.3 m carries a current $I$ along its curved surface as shown. A wire loop of resistance 0.005 Ω and of radius 0.1 m is placed inside the tube with its axis coinciding with the axis of the tube. The current varies as $I = I_0 \cos 300\,t$, where $I_0$ is constant. If the magnetic moment of the loop is $N\mu_0 I_0 \sin 300\,t$, then $N$ is

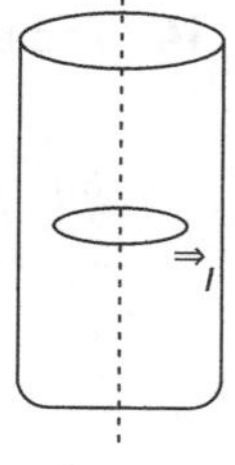

**20.** Steel wire of length $L$ at 40° C is suspended from the ceiling and then a mass $m$ is hung from its free end. The wire is cooled down from 40°C to 30°C to regain its original length $L$. The coefficient of linear thermal expansion of the steel is $10^{-5}/°C$, Young's modulus of steel is $10^{11}$ N/m$^2$ and radius of the wire is 1 mm. Assume that $L >>$ diameter of the wire. Then, the value of $m$ in kg is nearly

**21.** Four solid spheres each of diameter $\sqrt{5}$ cm and mass 0.5 kg ar placed with their centres at the corners of a square of side 4 cm. The moment of inertia of the system about the diagonal of the square is $N \times 10^{-4}$ kg-m$^2$, then $N$ is

**22.** A boy is pushing a ring of mass 2 kg and radius 0.5 m with a stick as shown in the figure. The stick applies a force of 2 N on the ring and rolls it without slipping with an acceleration of 0.3 m/s$^2$. The coefficient of friction between the ground and the ring is large enough that rolling always occurs and the coefficient of friction between the stick and the ring is $\frac{P}{10}$. The value of $P$ is

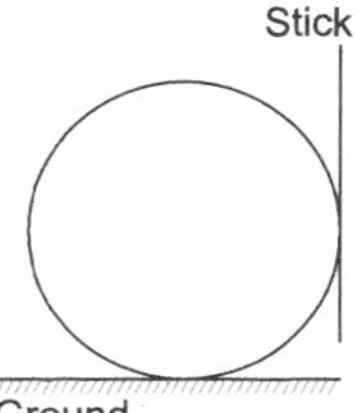


**23.** The activity of a freshly prepared radioactive sample is $10^{10}$ disintegrations per second, whose mean life is $10^{-9}$ s. The mass of an atom of this radioisotope is $10^{-25}$kg. The mass (in mg) of the radioactive sample is

# Paper II

## SECTION I

***Directions*** (Q. Nos. 1-8) *This section contains 8 multiple choice questions. Each question has four choices (a), (b), (c) and (d) out of which only one is correct.*

**1.** A light ray travelling in glass medium is incident on glass-air interface at an angle of incidence θ. The reflected ($R$) and transmitted ($T$) intensities, both as function of θ, are plotted. The correct sketch is

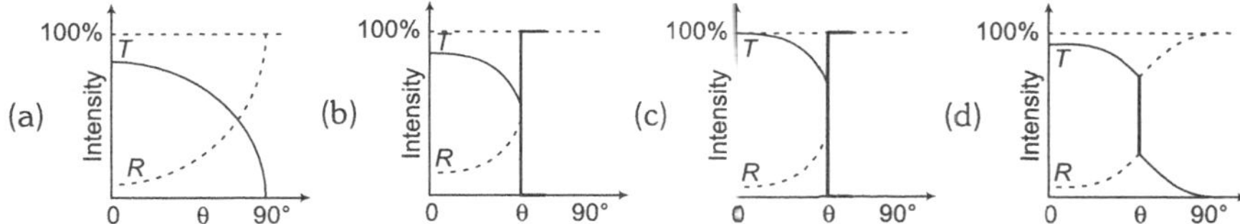


**2.** A wooden block performs SHM on a frictionless surface with frequency $\nu_0$. The block carries a charge $+Q$ cn its surface. If now a uniform electric field **E** is switched cn as shown, then the SHM of the block will be

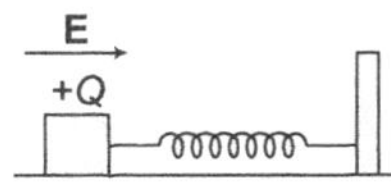


(a) of the dame frequency and with shifted mean position
(b) of the same frequency and with the same mean position
(c) of changed frequency and with shifted mean position
(d) of changed frequency and with the same mean position

**3.** The density of a solid ball is to be determined in an experiment. The diameter of the ball is measured with a screw gauge, whose pitch is 0.5 mm and there are 50 divisions on the circular scale. The reading on the main scale is 2.5 mm and that on the circular scale is 20 divisions. If the measured mass of the ball has a relative error of 2%, the relative percentage error in the density is

(a) 0.9% (b) 2.4% (c) 3.1% (d) 4.2%

**4.** A ball of mass 0.2 kg rests on a vertical post of height 5 m. A bullet of mass 0.01 kg, travelling with a velocity $v$ m/s in a horizontal direction, hits the centre of the ball. After the collision, the ball and bullet travel independently. The ball hits the ground at a distance of 20 m and the bullet at a distance of 100 m from the foot of the post. The initial velocity $v$ of the bullet is

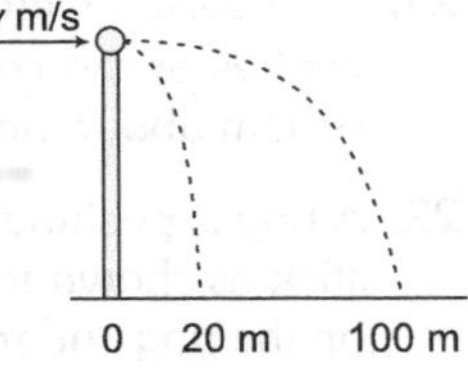


(a) 250 m/s (b) $250\sqrt{2}$ m/s (c) 400 m/s (d) 500 m/s

**5.** Which of the field patterns given in the figure is valid for electric field as well as for magnetic field?

(a) 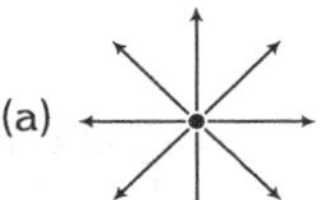 (b) 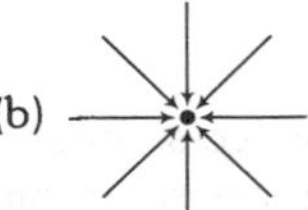 (c) 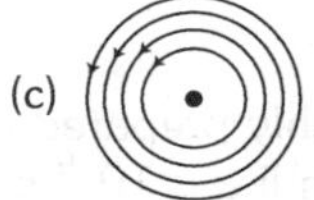 (d) 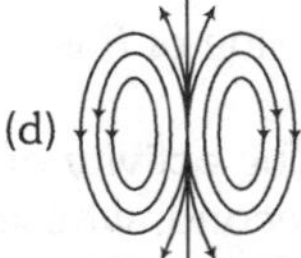

**6.** A point mass is subjected to two simultaneous sinusoidal displacements in $x$-direction, $x_1(t) = A \sin \omega t$ and $x_2(t) = A \sin\left(\omega t + \frac{2\pi}{3}\right)$. Adding a third sinusoidal displacement $x_3(t) = B \sin(\omega t + \phi)$ bring the mass to a complete rest. The values of $B$ and $\phi$ are

(a) $\sqrt{2}A, \frac{3\pi}{4}$ (b) $A, \frac{4\pi}{3}$ (c) $\sqrt{3}A, \frac{5\pi}{6}$ (d) $A, \frac{\pi}{3}$

**7.** A long insulated copper wire is closely wound as a spiral of $N$ turns. The spiral has inner radius $a$ and outer radius $b$. The spiral lies in the $XY$-plane and a steady current $I$ flows through the wire. The $Z$-component of the magnetic field at the centre of the spiral is

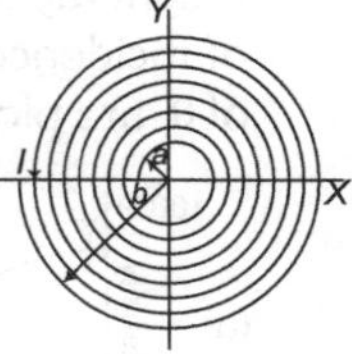


(a) $\frac{\mu_0 NI}{2(b-a)} \ln\left(\frac{b}{a}\right)$ (b) $\frac{\mu_0 NI}{2(b-a)} \ln\left(\frac{b+a}{b-a}\right)$

(c) $\frac{\mu_0 NI}{2b} \ln\left(\frac{b}{a}\right)$ (d) $\frac{\mu_0 NI}{2b} \ln\left(\frac{b+a}{b-a}\right)$

**8.** A satellite is moving with a constant speed $v$ in a circular orbit about the earth. An object of mass $m$ is ejected from the satellite such that it just escapes from the gravitational pull of the earth. At the time of its ejection, the kinetic energy of the object is

(a) $\frac{1}{2}mv^2$ (b) $mv^2$ (c) $\frac{3}{2}mv^2$ (d) $2mv^2$

## SECTION II

***Directions*** (Q. Nos. 9-12) *This section contains 4 multiple choice questions. Each question has four choices (a), (b), (c) and (d) out of which one or more may be correct.*

**9.** Two solid spheres $A$ and $B$ of equal volumes but of different densities $d_A$ and $d_B$ are connected by a string. They are fully immersed in a fluid of density $d_F$. They get arranged into an equilibrium state as shown in the figure with a tension in the string. The arrangement is possible only if

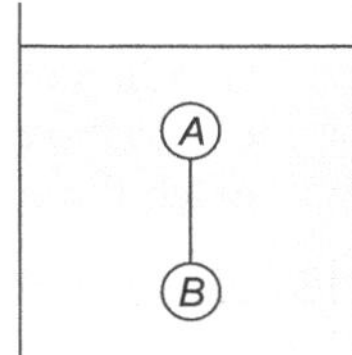


(a) $d_A < d_F$ (b) $d_B > d_F$

(c) $d_A > d_F$ (d) $d_A + d_B = 2d_F$

**10.** Which of the following statement(s) is/are correct?

(a) If the electric field due to a point charge varies as $r^{-2.5}$ instead of $r^{-2}$, then the Gauss's law will still be valid

(b) The Gauss's law can be used to calculate the field distribution around an electric dipole

(c) If the electric field between two point charges is zero somewhere, then the sign of the two charges is the same

(d) The work done by the external force in moving a unit positive charge from point $A$ at potential $V_A$ to point $B$ at potential $V_3$ is $(V_B - V_A)$

**11.** A series $R$-$C$ circuit is connected to AC voltage source. Consider two cases; ($A$) when $C$ is without a dielectric medium and ($B$) when $C$ is filled with dielectric of constant 4. The current $I_R$ through the resistor and voltage $V_C$ across the capacitor are compared in the two cases. Which of the following is/are true?

(a) $I_R^A > I_R^B$ (b) $I_R^A < I_R^B$ (c) $V_C^A > V_C^B$ (d) $V_C^A < V_C^B$

**12.** A thin ring of mass 2 kg and radius 0.5 m is rolling without slipping on a horizontal plane with velocity 1 m/s. A small ball of mass 0.1 kg, moving with velocity 20 m/s in the opposite directions, hits the ring at a height of 0.75 m and goes vertically up with velocity 10 m/s. Immediately after the collision

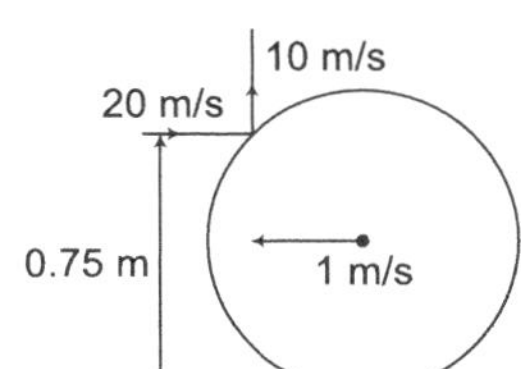


(a) the ring has pure rotation about its stationary CM

(b) the ring comes to a complete stop

(c) friction between the ring and the ground is to the left

(d) there is no friction between the ring and the ground

## SECTION III

***Directions*** (Q. Nos. 13-18) *This sections contain 5 questions. The answer to each of the question is a single-digit integer, ranging from 0 to 9.*

**13.** Two batteries of different emfs and different internal resistances are connected as shown. The voltage across $AB$ in volt is

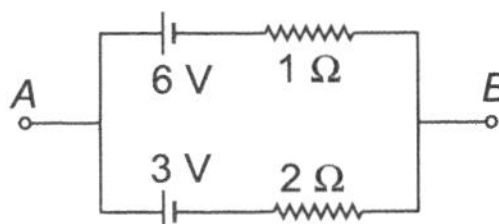

**14.** A series *R-C* combination is connected to an AC voltage of angular frequency $\omega = 500$ rad/s. If the impedance of the *R-C* circuit is $R\sqrt{1.25}$, the time constant (in millisecond) of the circuit is

**15.** A train is moving along a straight line with a constant acceleration *a*. A boy standing in the train throws a ball forward with a speed of 10 m/s, at an angle of 60° to the horizontal. The boy has to move forward by 1.15 m inside the train to catch the ball back at the initial height. The acceleration of the train in $m/s^2$, is

**16.** Water (with refractive index $= \frac{4}{3}$) in a tank is 18 cm deep. Oil of refractive index $\frac{7}{4}$ lies on water making a convex surface of radius of curvature $R = 6$ cm as shown. Consider oil to act as a thin lens. An object *S* is placed 24 cm above water surface. The location of its image is at *x* cm above the bottom of the tank. Then, *x* is

**17.** A block of mass 0.18 kg is attached to a spring of force constant 2 N/m. The coefficient of friction between the block and the floor is 0.1. Initially, the block is at rest and the spring is unstretched. An impulse is given to the block as shown in the figure. The block slides a distance of 0.06 m and comes to rest for the first time. The initial velocity of the block in m/s is $v = \frac{N}{10}$. Then, *N* is

**18.** A silver sphere of radius 1 cm and work function 4.7 eV is suspended from an insulating thread in free-space. It is under continuous illumination of 200 nm wavelength light. As photoelectrons are emitted, the sphere gets charged and acquires a potential. The maximum number of photoelectrons emitted from the sphere is $A \times 10^Z$ (where $1 < A < 10$). The value of *Z* is

## SECTION IV

***Directions*** (Q. Nos. 19-20) *This section contains 2 questions. Each question has four statements (A, B, C and D) given in Column I and five statements (p,q, r, s and t) in Column II. Any given statement in Column I can have correct matching with one or more statement(s) given in Column II.*

**19.** One mole of a monatomic ideal gas is taken through a cycle *ABCDA* as shown in the *p-V* diagram. Column II gives the characteristics involved in the cycle. Match them with each of the processes given in Column I.

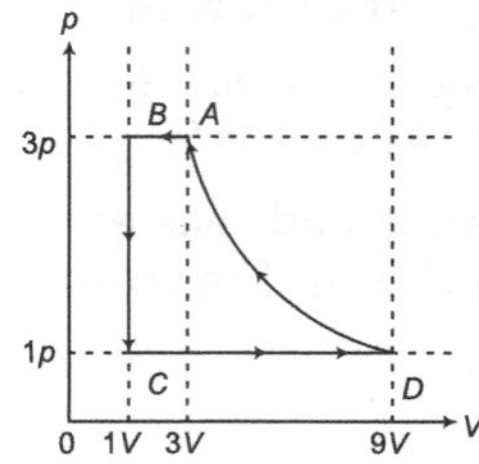

| Column I | Column II |
|---|---|
| (A) Process $A \to B$ | (p) Internal energy decreases |
| (B) Process $B \to C$ | (q) Internal energy increases |
| (C) Process $C \to D$ | (r) Heat is lost |
| (D) Process $D \to A$ | (s) Heat is gained |
| | (t) Work is done on the gas |

**20.** Column I showns four systems, each of the same length $L$, for producing standing waves. The lowest possible natural frequency of a system is called its fundamental frequency, whose wavelength is denoted as $\lambda_f$ . Match each system with statements given in Column II describing the nature and wave length of the standing waves.

| Column I | Column II |
|---|---|
| (A) Pipe closed at one end (O, L) | (p) Longitudinal waves |
| (B) Pipe open at both ends (O, L) | (q) Transverse waves |
| (C) Stretched wire clamped at both ends (O, L) | (r) $\lambda_f = L$ |
| (D) Stretched wire clamped at both ends and at mid-point (O, L/2, L) | (s) $\lambda_f = 2L$ |
| | (t) $\lambda_f = 4L$ |

## ANSWERS

### Paper I

| | | | | | |
|---|---|---|---|---|---|
| **1.** (c) | **2.** (d) | **3.** (d) | **4.** (a) | **5.** (a) | **6.** (a) |
| **7.** (b) | **8.** (b,d) | **9.** (a,c,d) | **10.** (a,b,c,d) | **11.** (a,d) | **12.** (c) |
| **13.** (b) | **14.** (d) | **15.** (c) | **16.** (b) | **17.** (3) | **18.** (5) |
| **19.** (6) | **20.** (3) | **21.** (9) | **22.** (4) | **23.** (1) | |

### Paper II

| | | | | | |
|---|---|---|---|---|---|
| **1.** (c) | **2.** (a) | **3.** (c) | **4.** (d) | **5.** (c) | **6.** (b) |
| **7.** (a) | **8.** (b) | **9.** (a,b,d) | **10.** (c,d) | **11.** (b,c) | **12.** (a,c) |
| **13.** (5) | **14.** (4) | **15.** (5) | **16.** (2) | **17.** (4) | **18.** (7) |

**19.** (A) → p,r,t; (B) → p,r; (C) → q, s; (D) → r **20.** (A) → p,t; (B) → p,s; (C) → q, s; (D) → q, r

# Hints & Solutions

## Paper I

**1.** Electric flux, $\mathbf{E} \cdot \mathbf{S}$, or $\phi = ES \cos \theta$

Here, $\theta$ is the angle between $\mathbf{E}$ and $\mathbf{S}$. In this question $\theta = 45°$, because $\mathbf{S}$ is perpendicular to the surface.

$$E = E_0$$

$$\Rightarrow \quad S = (\sqrt{2}a)\,(a) = \sqrt{2}a^2$$

$$\therefore \phi = (E_0)\,(\sqrt{2}a^2) \cos 45° = E_0\, a^2$$

$\therefore$ Correct option is (c).

**Analysis of Question**

(i) Question is moderately tough.

(ii) The given shaded area is a rectangle not a square. One side of this rectangle is $a$ and other side is $\sqrt{2}a$.

(iii) Electric field is uniform, whose magnitude is $E_0$ and direction is positive $x$. In uniform electric field we can use, $\phi = ES \cos \theta$

**2.**

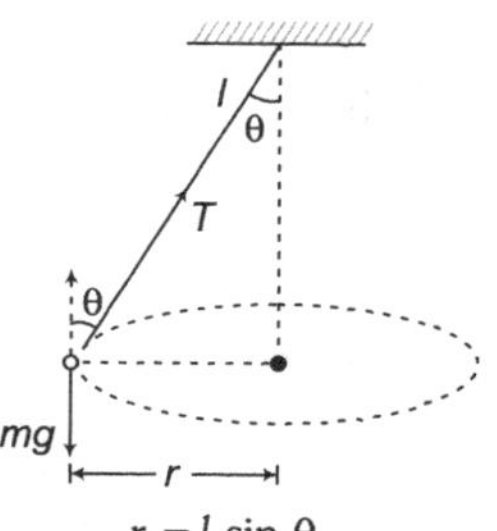


$$r = l \sin \theta$$

$T \cos \theta$ component will cancel $mg$.

$T \sin \theta$ component will provide necessary centripetal force to the ball towards centre $C$.

$$\therefore \quad T \sin \theta = mr\omega^2 = m(l \sin \theta)\,\omega^2$$

$$\text{or} \quad T = ml\omega^2$$

$$\therefore \quad \omega = \sqrt{\frac{T}{ml}}$$

$$\text{or} \quad \omega_{max} = \sqrt{\frac{T_{max}}{ml}} = \sqrt{\frac{324}{0.5 \times 0.5}}$$

$$= 36 \text{ rad/s}$$

$\therefore$ Correct option is (d).

**Analysis of Question**

(i) Question is simple.

(ii) This is called the conical pendulum.

(iii) The interesting fact in this problem is that $\omega$ or $T$ is independent of $\theta$.

$$\omega \propto \sqrt{T}$$

If $\omega$ is increased, $T$ will also increase.

**3.** $q_i = C_iV = 2V = q$ (say)

This charge will remain constant after switch is shifted from position 1 to position 2.

$$U_i = \frac{1}{2}\frac{q^2}{C_i} = \frac{q^2}{2 \times 2} = \frac{q^2}{4}$$

$$U_f = \frac{1}{2}\frac{q^2}{C_f} = \frac{q^2}{2 \times 10} = \frac{q^2}{20}$$

$\therefore$ Energy dissipated $= U_i - U_f = \dfrac{q^2}{5}$

This energy dissipated $\left(= \dfrac{q^2}{5}\right)$ is 80% of the initial stored energy $\left(= \dfrac{q^2}{4}\right)$.

**Analysis of Question**

(i) This question is moderately tough.

(ii) In a capacitor circuit, redistribution of charge takes place under following three conditions.

(a) A switch is closed.

(b) A closed switch is opened.

(c) A switch is shifted from one position to another position.

In the redistribution of charge, energy is dissipated.

**4.** At STP, 22.4 L of any gas is 1 mole.

$$\therefore \quad 5.6 \text{ L} = \frac{5.6}{22.4} = \frac{1}{4} \text{ moles} = n$$

In adiabatic process,

$$TV^{\gamma - 1} = \text{constant}$$

$$\therefore \quad T_2V_2^{\gamma - 1} = T_1V_1^{\gamma - 1}$$

$$\text{or} \quad T_2 = T_1 \left(\frac{V_1}{V_2}\right)^{\gamma - 1}$$

$$\gamma = \frac{C_p}{C_V} = \frac{5}{3} \text{ for monoatomic He gas.}$$

$$\therefore \quad T_2 = T_1 \left(\frac{5.6}{0.7}\right)^{\frac{5}{3}-1} = 4T_1$$

$\because$ Further in adiabatic process,

$$Q = 0$$

$$\therefore \quad W + \Delta U = 0$$

or

$$W = -\Delta U$$

$$= -nC_V \Delta T$$

$$= -n\left(\frac{R}{\gamma - 1}\right)(T_2 - T_1)$$

$$= -\frac{1}{4}\left(\frac{R}{\frac{5}{3} - 1}\right)(4T_1 - T_1)$$

$$= -\frac{9}{8} RT_1$$

$\therefore$ Correct option is (a).

**Analysis of Question**

(i) From calculation point of view question is moderately tough.

(ii) Volume of gas is decreasing. Therefore, work done by the gas should be negative.

**5.** $36 \text{ km/h}^{-1} = 36 \times \frac{5}{18} = 10 \text{ ms}^{-1}$

Observer

car —→ $10 \text{ ms}^{-1}$

$10 \text{ ms}^{-1}$

Building

←— car Source

For reflected sound

Apparent frequency of sound heard by car driver (observer) reflected from the building will be

$$f' = f\left(\frac{v + v_0}{v - v_s}\right)$$

$$= 8\left(\frac{320 + 10}{320 - 10}\right)$$

$$= 8.5 \text{ kHz}$$

$\therefore$ Correct option is (a).

**Analysis of Question**

(i) Question is simple.

(ii) Driver will listen two sounds, direct and reflected. Direct sound will be of 8 kHz as driver has no relative motion with the car. But reflected sound is of increased frequency because driver and image of car both are approaching towards each other.

**6.**

$n = 4$ ———— $n = 4$ ————

$n = 3$ ———— $n = 3$ ————

$n = 2$ ———— $n = 2$ ————

$n = 1$ ———— $n = 2$ ————

First line of Balmar series — Second line of Balmar series

For hydrogen or hydrogen type atoms

$$\frac{1}{\lambda} = RZ^2\left(\frac{1}{n_f^2} - \frac{1}{n_i^2}\right)$$

In the transition from $n_i \longrightarrow n_f$

$$\therefore \quad \lambda \propto \frac{1}{Z^2\left(\frac{1}{n_f^2} - \frac{1}{n_i^2}\right)}$$

$$\therefore \quad \frac{\lambda_2}{\lambda_1} = \frac{Z_1^2\left(\frac{1}{n_f^2} - \frac{1}{n_i^2}\right)_1}{Z_2^2\left(\frac{1}{n_f^2} - \frac{1}{n_i^2}\right)_2}$$

$$\lambda_2 = \frac{\lambda_1 Z_1^2\left(\frac{1}{n_f^2} - \frac{1}{n_i^2}\right)_1}{Z_2^2\left(\frac{1}{n_f^2} - \frac{1}{n_i^2}\right)_2}$$

Substituting the values, we have

$$= \frac{(6561 \text{ Å})(1)^2\left(\frac{1}{2^2} - \frac{1}{3^2}\right)}{(2)^2\left(\frac{1}{2^2} - \frac{1}{4^2}\right)} = 1215 \text{ Å}$$

$\therefore$ Correct option is (a).

**Analysis of Question**

(i) Question is simple.

(ii) In modern physics, mostly questions are asked on the emission of photon by the transition of electron from some higher energy state to some lower energy state.

(iii) For hydrogen and hydrogen like atoms which have only single electron, we can use the formula

$$\frac{1}{\lambda} = RZ^2\left(\frac{1}{n_f^2} - \frac{1}{n_i^2}\right)$$

(iv) Further, the student should also remember different series, as shown in figure below.

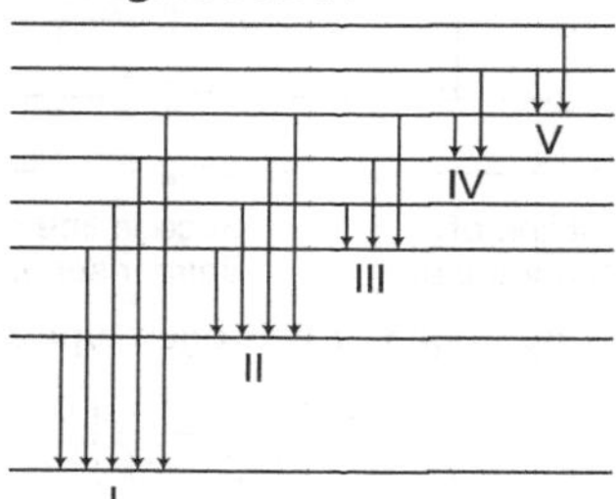


I → Lymen series
II → Balmer series
III → Paschen series
IV → Brackett series
V → Pfund series

**7.** Using the concept of balanced Wheatstone bridge, we have

$$\frac{P}{Q} = \frac{R}{S}$$

$$\therefore \quad \frac{X}{(52+1)} = \frac{10}{(48+2)}$$

$$\therefore \quad X = \frac{10 \times 53}{50} = 10.6\ \Omega$$

∴ Correct option is (b).

**Analysis of Question**

Question is moderately tough, because normally end corrections are not taught in the coaching/schools in this type of problem.

**8.** $$r = \frac{mv}{Bq}$$

or $r \propto m$

∴ $r_e < r_p$ as $m_e < m_p$

Further, $$T = \frac{2\pi m}{Bq}$$

or $T \propto m$

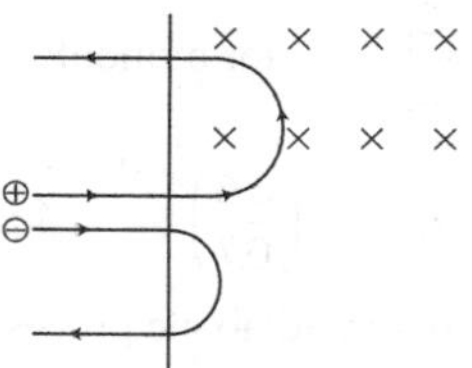

$$\therefore \quad T_e < T_p$$

$$t_e = \frac{T_e}{2}$$

and $$t_p = \frac{T_p}{2}$$

or $t_e < t_p$

∴ Correct options are (b) and (d).

**Note** In JEE 2011, official answer key, correct options were given bc, bd, bcd, here student should realise that both the electron and proton enter the magnetic field at the same time or at different times.

**Analysis of Question**

(i) Question is simple.

(ii) In the situation given above particle will never complete the circle unless magnetic field is all around.

**9.** Thermal resistance

$$R = \frac{l}{KA}$$

$$\therefore \quad R_A = \frac{L}{(2K)(4Lw)} = \frac{1}{8Kw}$$

(Here, $w$ = width)

$$R_B = \frac{4L}{3K(Lw)} = \frac{4}{3Kw}$$

$$R_C = \frac{4L}{(4K)(2Lw)} = \frac{1}{2Kw}$$

$$R_D = \frac{4L}{(5K)(Lw)} = \frac{4}{5Kw}$$

$$R_E = \frac{L}{(6K)(Lw)} = \frac{1}{6Kw}$$

$$R_A : R_B : R_C : R_D : R_E = 15 : 160 : 60 : 96 : 12$$

So, let us write, $R_A = 15\,R, R_B = 160\,R$ etc and draw a simple electrical circuit as shown in figure.

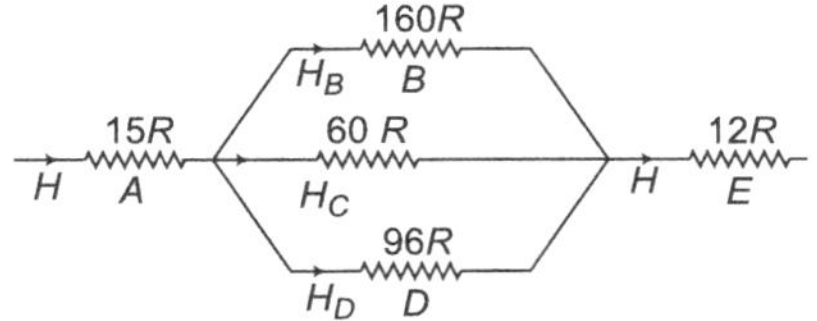


$H$ = Heat current = Rate of heat flow.

$$H_A = H_E = H \quad \text{[let]}$$

∴ Option (a) is correct.

In parallel, current distributes in inverse ratio of resistance.

$$\therefore \; H_B : H_C : H_D = \frac{1}{R_B} : \frac{1}{R_C} : \frac{1}{R_D}$$

$$= \frac{1}{160} : \frac{1}{60} : \frac{1}{96}$$

$$= 9 : 24 : 15$$

$$\therefore \quad H_B = \left(\frac{9}{9+24+15}\right) H = \frac{3}{16} H$$

$$H_C = \left(\frac{24}{9+24+15}\right) H = \frac{1}{2} H$$

and $$H_D = \left(\frac{15}{9+24+15}\right) H = \frac{5}{16} H$$

$$H_C = H_B + H_D$$

∴ Option (d) is correct.

Temperature difference (let us call it $T$)

= (Heat current) × (Thermal resistance)

$$T_A = H_A R_A = (H)(15R) = 15\,HR$$

$$T_B = H_B R_B = \left(\frac{3}{16} H\right)(160\,R) = 30\,HR$$

$$T_C = H_C R_C = \left(\frac{1}{2} H\right)(60\,R) = 30\,HR$$

$$T_D = H_D R_D = \left(\frac{5}{16} H\right)(96\,R) = 30\,HR$$

$$T_E = H_E R_E = (H)(12\,R) = 12\,HR$$

Here, $T_E$ is minimum. Therefore, option (c) is also correct.

∴ Correct options are (a), (c) and (d).

**Analysis of Question**

(i) From calculation point of view, question is difficult otherwise question is simple.

(ii) In heat transfer, questions are mainly asked from conduction and radiation topic.

**10.** Inside a conducting shell, electric field is always zero. Therefore, option (a) is correct. When the two are connected, their potentials become the same.

$$\therefore \quad V_A = V_B$$

or $$\frac{Q_A}{R_A} = \frac{Q_B}{R_B} \quad \left(\because V = \frac{1}{4\pi\varepsilon_0}\frac{Q}{R}\right)$$

Since, $R_A > R_B$

∴ $Q_A > Q_B$

∴ Option (b) is correct.

Potential is also equal to,

$$V = \frac{\sigma R}{\varepsilon_0}$$

$$V_A = V_B$$

$$\therefore \quad \sigma_A R_A = \sigma_B R_B$$

or $$\frac{\sigma_A}{\sigma_B} = \frac{R_B}{R_A}$$

∴ Option (c) is correct.

Electric field on surface,

$$E = \frac{\sigma}{E_0} \text{ or } E \propto \sigma$$

or $\sigma_A < \sigma_B$

Since, $\sigma_A < \sigma_B$

∴ $E_A < E_B$

∴ Option (d) is also correct.

∴ Correct options are (a), (b), (c) and (d).

**Analysis of Question**

(i) Question is simple.

(ii) $E = \dfrac{1}{4\pi\varepsilon_0}\dfrac{Q}{R^2} = \dfrac{\sigma}{\varepsilon_0}$ (On surface)

As, $\sigma$ = surface charge density

$$= \frac{Q}{4\pi R^2}$$

(iii) $V = \dfrac{1}{4\pi\varepsilon_0}\dfrac{Q}{R} = \dfrac{\sigma R}{\varepsilon_0}$

**11.** $\tau_A = \tau_B = mg\,\dfrac{L}{2}\sin\theta + MgL\sin\theta$

= Restoring torque about point $O$.

In case $A$, moment of inertia will be more. Hence, angular acceleration ($\alpha = \tau / I$) will be less. Therefore, angular frequency will be less.

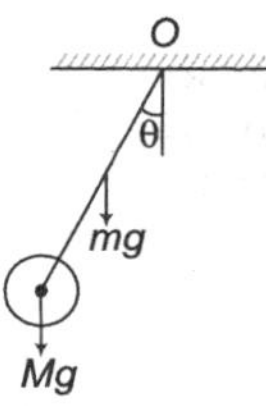


$\therefore$ Correct options are (a) and (d).

**Analysis of Question**

Question is difficult from my point of view. Because this type of SHM is rarely taught in the class and questions of this type are not given in standard books.

**12.** $N$ = Number of electrons per unit volume

$$\therefore \quad [N] = [L^{-3}], [e] = [q] = [It] = [AT]$$
$$[\varepsilon_0] = [M^{-1}L^{-3}T^4A^2]$$

Substituting the dimensions, we can see that,

$$\left[\sqrt{\frac{Ne^2}{m\varepsilon_0}}\right] = [T^{-1}]$$

Angular frequency has also the dimension $[T^{-1}]$.

$\therefore$ Correct option is (c).

**Analysis of Question**

(i) From calculation point of view, question is moderately difficult. Otherwise it is simple.

(ii) Students may commit a mistake in the dimensions of $N$. It is not dimensionless.

**13.** $$\omega = 2\pi f = \frac{2\pi c}{\lambda}$$

$$\therefore \quad \lambda = \frac{2\pi c}{\omega} = \frac{2\pi c}{\sqrt{Ne^2/m\varepsilon_0}}$$

Substituting the values, we get

$$\lambda = 600 \text{ nm}$$

$\therefore$ Correct option is (b).

**Analysis of Question**

(i) Paragraph does not make the solution very clear.

(ii) Solution is simple but based on error and trial method.

**14.** Momentum is first positive but decreasing. Displacement (or say position). is initially zero.

It will first increase. At highest point, momentum is zero and displacement is maximum. After that momentum is downwards (negative) and increasing but displacement is decreasing. Only (d) option satisfies these conditions.

**Analysis of Question**

(i) Question is simple.

(ii) Since all the graphs given in question are parabolic type, we need not to check the mathematical part of the question.

**15.** $$E = \frac{1}{2} m\omega^2 A^2$$

or $$E \propto A^2$$

$$\frac{E_2}{E_1} = \left(\frac{A_2}{A_1}\right)^2 = \left(\frac{a}{2a}\right)^2$$

or $$E_1 = 4E_2$$

$\therefore$ Correct option is (c).

**Analysis of Question**

(i) Question is simple.

(ii) Since, oscillator is same, $\omega_1 = \omega_2$

**16.** In all the given four figures, at mean position the position coordinate is zero. At the same time mass is starting from the extreme position in all fours cases. In figures (c) and (d), extreme position is more than the initial extreme position. But due to viscosity opposite should be the case. Hence, the answer should be either option (a) or option (b).

Correct answer is (b), because mass starts from positive extreme position (from uppermost position). Then, it will move downwards or momentum should be negative.

**Analysis of Question**

Question is simple but (*a*) and (*b*) options are confusing.

**17.**

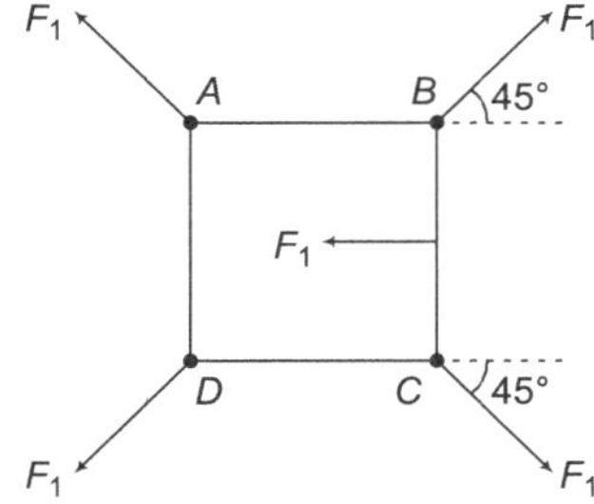


$F_1$ = Net electrostatic force on anyone charge due to rest of three charges

$$=\frac{1}{4\pi\varepsilon_0}\frac{q^2}{a^2}\left(\sqrt{2}+\frac{1}{2}\right)$$

$$F_2 = \text{Surface tension force} = \gamma\, a$$

If we see the equilibrium of line $BC$,

then, $\quad 2F_1 \cos 45° = F_2$

or $\quad \sqrt{2}\,F_1 = F_2$

or $\quad \frac{1}{4\pi\varepsilon_0}\frac{q^2}{a^2}\left(2+\frac{1}{\sqrt{2}}\right) = \gamma a$

$\therefore \quad a^3 = \frac{1}{4\pi\varepsilon_0}\left(2+\frac{1}{\sqrt{2}}\right)\frac{q^2}{\gamma}$

or $\quad a = \left\{\frac{1}{4\pi\varepsilon_0}\left(2+\frac{1}{\sqrt{2}}\right)\right\}^{1/3}\left[\frac{q^2}{\gamma}\right]^{1/3}$

$$= k\left[\frac{q^2}{\gamma}\right]^{1/3}$$

where, $\quad k = \left\{\frac{1}{4\pi\varepsilon_0}\left(2+\frac{1}{\sqrt{2}}\right)\right\}^{1/3}$

Therefore, $N = 3$

Answer is 3.

**Analysis of Question**

(i) Question is moderately difficult.

(ii) In my opinion problems of surface tension or viscosity are always not very simple questions.

**18.**

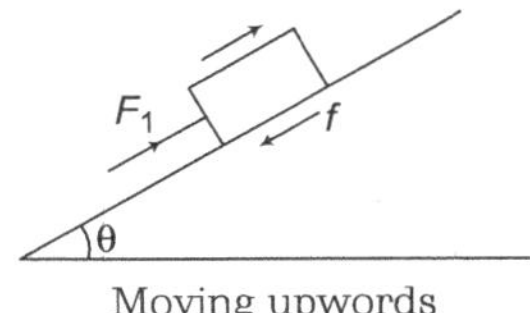


Moving upwords

F2 f θ = 45° θ

Just remains stationary

$$F_1 = mg \sin\theta + \mu mg \cos\theta$$
$$F_2 = mg \sin\theta - \mu mg \cos\theta$$

Given that, $F_1 = 3F_2$

or $\quad (\sin 45° + \mu \cos 45°)$
$$= 3(\sin 45° - \mu \cos 45°)$$

On solving, we get $\mu = 0.5$

$\therefore \quad N = 10\,\mu = 5$

$\therefore$ Answer is 5.

**Analysis of Question**

Question is simple. Only one has to take care of direction and magnitude of friction.

**19.** Take the circular tube as a long solenoid. The wires are closely wound. Magnetic field inside the solenoid is

$$B = \mu_0 ni$$

Here, $n$ = number of turns per unit length

$\therefore$ $ni$ = current per unit length

In the given problem,

$$ni = \frac{I}{L}$$

$\therefore \quad B = \frac{\mu_0 I}{L}$

Flux passing through the circular coil is

$$\phi = BS = \left(\frac{\mu_0 I}{L}\right)(\pi r^2)$$

Induced emf, $e = -\frac{d\phi}{dt} = -\left(\frac{\mu_0 \pi r^2}{LR}\right)\cdot\frac{dI}{dt}$

Induced current,

$$i = \frac{e}{R} = -\left(\frac{\mu_0 \pi r^2}{LR}\right)\frac{dI}{dt}$$

Magnetic moment $iA = i\pi r^2$

or $\quad M = -\left(\frac{\mu_0 \pi^2 r^4}{LR}\right)\cdot\frac{dI}{dt} \quad \ldots\text{(i)}$

Given, $\quad I = I_0 \cos 300\, t$

$\therefore \quad \frac{dI}{dt} = -\,300 I_0 \sin(300\, t)$

Substituting in Eq. (i), we get

$$M = \left(\frac{300\,\pi^2 r^4}{LR}\right)\mu_0 I_0 \sin 300\,t$$

$$\therefore \quad N = \frac{300\,\pi^2 r^4}{LR}$$

Substituting the values, we get

$$N = \frac{300\,(22/7)^2\,(0.1)^4}{(10)(0.005)}$$

$$= 5.926$$

or $N \simeq 6$

**Analysis of Question**

Question is difficult to understand inside the examination hall. But 5 to 10% question in IIT JEE are always difficult. Students should not panic. Because topper of IIT JEE scores approximately 80-90%.

**20.** $\Delta l_1 = \dfrac{FL}{AY} = \dfrac{mgL}{\pi r^2 Y}$ = Increase in length

$\Delta l_2 = L\,\alpha\,\Delta\theta$ = Decrease in length

To regain its original length,

$$\Delta l_1 = \Delta l_2$$

$$\therefore \quad \frac{mgL}{\pi r^2 Y} = L\,\alpha\,\Delta\theta$$

$$\therefore \quad m = \left(\frac{r^2 Y\,\alpha\,\Delta\theta}{g}\right)$$

Substituting the values, we get

$$m \simeq 3\text{ kg}$$

∴ Answer is 3.

**Analysis of Question**

Question is very simple. In my opinion any student who have gone through the syllabus once can solve this problem easily.

**21.**

$$r = \frac{d}{2} = \frac{\sqrt{5}}{2}\text{ cm}$$

$$= \frac{\sqrt{5}}{2} \times 10^{-2}\text{ m}$$

$$m = 0.5\text{ kg}$$

$$a = 4\text{ cm}$$

$$= 4 \times 10^{-2}\text{ m}$$

$$I_{XX} = I_1 + I_2 + I_3 + I_4$$

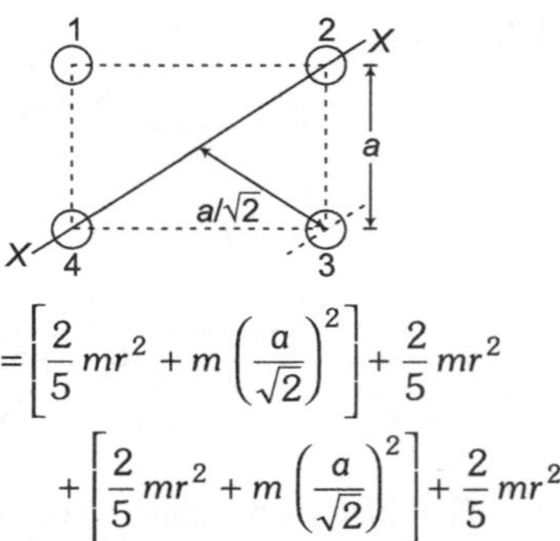


$$= \left[\frac{2}{5}mr^2 + m\left(\frac{a}{\sqrt{2}}\right)^2\right] + \frac{2}{5}mr^2 + \left[\frac{2}{5}mr^2 + m\left(\frac{a}{\sqrt{2}}\right)^2\right] + \frac{2}{5}mr^2$$

Substituting the values, we get

$$I_{XX} = 9 \times 10^{-4}\text{ Kgm}^{-2}$$

$$\therefore \quad N = 9$$

Answer is 9.

**Analysis of Question**

(i) Question is simple.

(ii) Only theorem of parallel axes is to be used properly.

**22.**

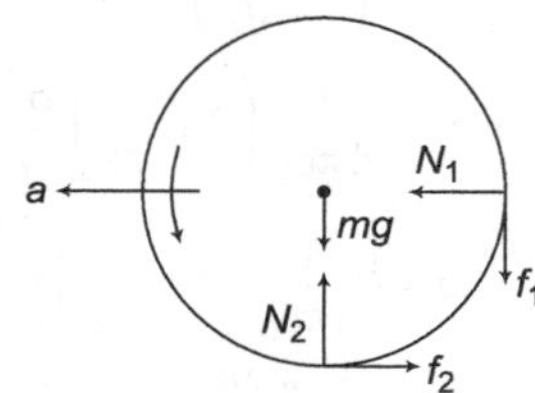


There is no slipping between ring and ground. Hence, $f_2$ is not maximum. But there is slipping between ring and stick. Therefore, $f_1$ is maximum. Now, let us write the equations.

$$I = mR^2 = (2)\,(0.5)^2$$

$$= \frac{1}{2}\text{ kgm}^{-2}$$

$$N_1 - F_2 = ma$$

or $N_1 - F_2 = (2)\,(0.3) = 0.6\text{ N}$ ...(i)

$$a = R\,\alpha = \frac{R\tau}{I}$$

$$= \frac{R(f_2 - f_1)R}{I} = \frac{R^2(f_2 - f_1)}{I}$$

$$\therefore \quad 0.3 = \frac{(0.5)^2\,(f_2 - f_1)}{(1/2)}$$

or $f_2 - f_1 = 0.6\text{ N}$ ...(ii)

$N_1^2 + f_1^2 = (2)^2 = 4$ ...(iii)

Further $F_1 = \mu N_1 = \left(\frac{P}{10}\right) N_1$ ...(iv)

Solving above four equation, we get

$P \simeq 3.6$

Therefore, the correct answer should be 4.

**Analysis of Question**

(i) Question is moderately tough from concept point of view. But calculations are lengthy.

(ii) One has to think about the two components of the force applied by the stick.

(iii) Answer comes an integer when you consider only $N_1$.

**23.** Activity $\left(-\frac{dN}{dt}\right) = \lambda N = \left(\frac{1}{t_{mean}}\right) \times N$

$\therefore \quad N = \left(-\frac{dN}{dt}\right) \times t_{mean}$

= Total number of atoms

Mass of one atom is $10^{-25}$ kg = $m$ (say)

$\therefore$ Total mass of radioactive substance

= (number of atoms) × (mass of one atom)

$= \left(-\frac{dN}{dt}\right)(t_{mean})(m)$

Substituting the values, we get

Total mass of radioactive substance

= 1 mg

$\therefore$ Answer is 1.

**Analysis of Question**

Question is very simple. Again in my opinion one can solve it.

# Paper II

**1.** After critical angle reflection will be 100% and transmission is 0%. Options (b) and (c) satisfy this condition. But option (c) is the correct option. Because in option (b) transmission is given 100% at $\theta = 0°$, which is not true.

**Analysis of Question**

From examination point of view question is moderately difficult. Because it takes little time to understand the given graphs clearly.

**2.** Frequency or time period of SHM depends on variable forces. It does not depend on constant external force. Constant external force can only change the mean position. For example, in the given question mean position is at natural length of spring in the absence of electric field. Whereas in the presence of electric field mean position will be obtained after a compression of $x_0$. Where $x_0$ is given by

$$kx_0 = QE$$

or $$x_0 = \frac{QE}{k}$$

$\therefore$ Correct answer is (a).

**Analysis of Question**

(i) Question is simple.

(ii) I think almost all the serious aspirants of JEE can solve this problem easily.

**3.** Least count of screw gauge $= \frac{0.5}{50}$

$= 0.01$ mm $= \Delta r$

Diameter, $r = 2.5 \text{ mm} + 20 \times \frac{0.5}{50}$

$= 2.70$ mm

$$\frac{\Delta r}{r} = \frac{0.01}{2.70}$$

or $$\frac{\Delta r}{r} \times 100 = \frac{1}{2.7}$$

Now, density $d = \frac{m}{V} = \frac{m}{\frac{4}{3}\pi\left(\frac{r}{2}\right)^3}$

Here, $r$ is the diameter.

$$\therefore \quad \frac{\Delta d}{d} \times 100 = \left\{\frac{\Delta m}{m} + 3\left(\frac{\Delta r}{r}\right)\right\} \times 100$$

$$= \frac{\Delta m}{m} \times 100 + 3 \times \left(\frac{\Delta r}{r}\right) \times 100$$

$= 2\% + 3 \times \frac{1}{27}$

$= 3.11\%$

**Analysis of Question**

(i) Question is moderately difficult.

(ii) In practical part, questions in JEE are asked from vernier callipers and screw gauge.

**4.** Time taken by the bullet and ball to strike the ground is

$$t = \sqrt{\frac{2h}{g}} = \sqrt{\frac{2 \times 5}{10}} = 1 \text{ s}$$

Let $v_1$ and $v_2$ are the velocities of ball and bullet after collision.

Then applying

$$x = vt$$

We have, $20 = v_1 \times 1$

or $v_1 = 20 \text{ ms}^{-1}$

$100 = v_2 \times 1$ or $v_2 = 100 \text{ m/s}^{-1}$

Now, from conservation of linear momentum before and after collision we have,

$$0.01\, v = (0.2 \times 20) + (0.01 \times 100)$$

On solving, we get

$$v = 500 \text{ ms}^{-1}$$

∴ Correct answer is (d).

**Analysis of Question**

Question is moderately lengthy from calculation point of view, otherwise it is simple.

**5.** Correct answer is (c), because induced electric field lines (produced by change in magnetic field) and magnetic field lines form closed loops.

**Analysis of Question**

Question is simple, provided you have an idea of induced electric field lines.

**6.**

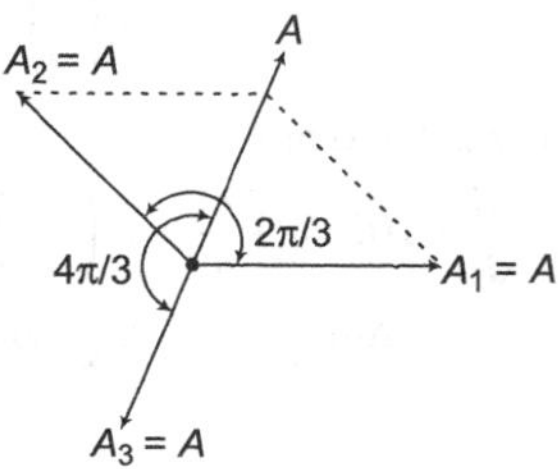


Resultant amplitude of $x_1$ and $x_2$ is $A$ at angle $\left(\frac{\pi}{3}\right)$ from $A_1$. To make resultant of $x_1, x_2$ and $x_3$ to be zero. $A_3$ should be equal to $A$ at angle $\phi = \frac{4\pi}{3}$ as shown in figure.

∴ Correct answer is (b).

**Alternate Solution**

If we substitute,

$$x_1 + x_2 + x_3 = 0$$

or $$A \sin \omega t + A \sin\left(\omega t + \frac{2\pi}{3}\right) + B \sin(\omega t + \phi) = 0$$

Then, by applying simple mathematics, we can prove that

$$B = A$$

and $$\phi = \frac{4\pi}{3}$$

**Analysis of Question**

(i) Question is simple.

(ii) Question can be solved by applying mathematics also.

(iii) Amplitudes of two or more sine or cosine functions of same frequency $\omega$ can be added by vector method.

**7.** If we take a small strip of $dr$ at distance $r$ from centre, then number of turns in this strip would be

$$dN = \frac{N}{b-a} dr$$

Magnetic field due to this element at the centre of the coil will be

$$dB = \frac{\mu_0 (dN) I}{2r} = \frac{\mu_0 NI}{2(b-a)} \frac{dr}{r}$$

$$\therefore \quad B = \int_{r=a}^{r=b} dB = \frac{\mu_0 NI}{2(b-a)} \ln \frac{b}{a}$$

∴ Correct answer is (a).

**Analysis of Question**

(i) If we see this problem independently, then I will rate this question moderately difficult. But the idea of this question is taken from question number 3.245 of IE Irodov.

(ii) Interestingly the same question was asked in IIT-JEE 2001 also.

**8.** In circular orbit of a satellite, potential energy

$$= -2 \times (\text{kinetic energy})$$
$$= -2 \times \frac{1}{2} mv^2 = -mv^2$$

Just to escape from the gravitational pull, its total mechanical energy should be zero. Therefore, its kinetic energy should be $+mv^2$.

$\therefore$ Correct answer is (b).

**Analysis of Question**

(i) Question is moderately difficult.

(ii) In circular orbit of satellite

$$U = -\frac{GMm}{r} \text{ and } K = \frac{GMm}{2r}$$

or $U = -2K$

Here, $U$ is potential energy and $K$ is kinetic energy.

**9.**

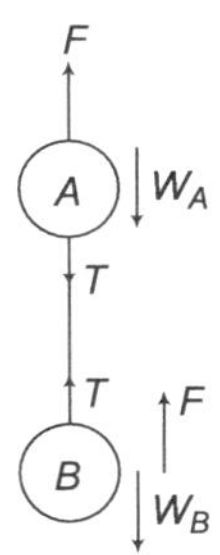


$$F = \text{Upthrust} = Vd_F g$$

Equilibrium of $A$

$$Vd_F g = T + W_A$$
$$= T + Vd_A g \quad \text{...(i)}$$

Equilibrium of $B$,

$$T + Vd_F g = Vd_B g \quad \text{...(ii)}$$

Adding Eqs. (i) and (ii), we get

$$2d_f = d_A + d_B$$

$\therefore$ Option (d) is correct.

From Eq. (i), we can see that

$$d_F > d_A \quad [\text{as } T > 0]$$

$\therefore$ Option (a) is correct.

From Eq. (ii) we can see that,

$$d_B > d_F$$

$\therefore$ Option (a) is correct.

$\therefore$ Correct options are (a), (b) and (d).

**Analysis of Question**

Question is moderately difficult but conceptwise it is good.

**10.** If charges are of opposite signs, then the two fields are along the same direction. So, they cannot be zero. Hence, the charges should be of same sign.

Therefore, option (c) is correct.

Further, work done by external force = change in potential energy

$$\therefore \quad W_{A \to B} = q\,(\Delta V)$$
$$= (+1)\,(V_B - V_A)$$

or $\quad W_{A \to B} = V_B - V_A$

Therefore, option (d) is also correct.

$\therefore$ Correct options are (c) and (d).

**Analysis of Question**

(i) Question is simple.

(ii) $W = \Delta U$ (by external force)

If charge is just moved, without changing the kinetic energy. If nothing is mentioned in the question regarding the kinetic energy then we take $W = \Delta U$.

**11.** $Z = \sqrt{R^2 + X_C^2} = \sqrt{R^2 + \left(\frac{1}{\omega C}\right)^2}$

In case (b) capacitance $C$ will be more.

Therefore, impedance $Z$ will be less.

Hence, current will be more.

$\therefore$ Option (b) is correct.

Further, $\quad V_C = \sqrt{V^2 - V_R^2}$

$$= \sqrt{V^2 - (IR)^2}$$

In case (b), since current $I$ is more.

Therefore, $V_C$ will be less.

$\therefore$ Option (c) is correct.

$\therefore$ Correct options are (b) and (c).

**Analysis of Question**

(i) Question is moderately difficult.

(ii) In my opinion problems of alternating currents are not very difficult.

(iii) Topic of $AC$ is small. One can feel comfortable in this topic by putting less efforts.

**12.** The data is incomplete. Let us assume that friction from ground on ring is not impulsive during impact.

From linear momentum conservation in horizontal direction, we have

$$(-2\times 1)+(0.1\times 20)$$
$$=(0.1\times 0)+(2\times v) \quad \overset{-\text{ve} \quad +\text{ve}}{\longleftarrow\!\!\longrightarrow}$$

Here, $v$ is the velocity of CM of ring after impact.

Solving the above equation, we have

$$v=0$$

Thus, CM becomes stationary.

$\therefore$ Correct answer is (a).

**Linear impulse during impact**

(i) In horizontal direction

$$J_1=\Delta P=0.1\times 20=2\,\text{Ns}$$

(ii) In vertical direction

$$J_2=\Delta P=0.1\times 10=1\text{ Ns}$$

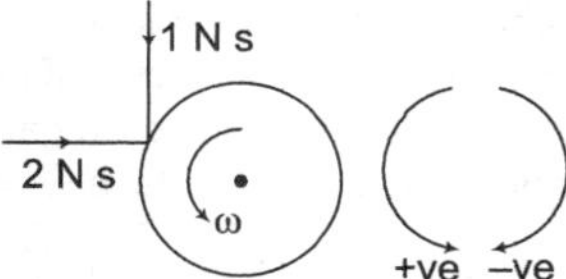


Writing the equation (about CM)

Angular impulse

= Change in angular momentum

$$1\times\left(\frac{\sqrt{3}}{2}\times\frac{1}{2}\right)-2\times 0.5\times\frac{1}{2}$$
$$=2\times(0.5)^2\left[\omega-\frac{1}{0.5}\right]$$

Solving this equation $\omega$ comes out to be positive or $\omega$ anti-clockwise. So just after collision rightwards slipping is taking place.

Hence, friction is leftwards.

Therefore, option (c) is also correct.

$\therefore$ Correct options are (a) and (c).

**Note** In JEE 2011 official answer key, correct option were given a, ac.

**Analysis of Question**

(i) Question is moderately difficult.

(ii) In such type of problems impulse due to friction during collision is ignored.

**13.** $V_{AB}$ = Equivalent emf of two batteries in parallel.

$$=\frac{E_1/r_1+E_1/r_2}{1/r_1+1/r_2}$$
$$=\frac{(6/1)+(3/2)}{(1/1)+(1/2)}=5\text{ V}$$

$\therefore$ Answer is 5.

**Analysis of Question**

(i) In my opinion this is the simplest question of JEE 2011.

(ii) When no current is drawn from this equivalent battery, then

$$V_{AB}=V=E$$

Otherwise, $V=E\pm ir$

(iii) Similar type of question was asked in JEE 1981 and cancelled paper of JEE 1997.

**14.** $Z=\sqrt{R^2+X_C^2}=R\sqrt{1.25}$

$\therefore \quad R^2+X_C^2=1.25\,R^2$

or $\quad X_C=\dfrac{R}{2}$

or $\quad \dfrac{1}{\omega C}=\dfrac{R}{2}$

$\therefore$ Time constant $=CR=\dfrac{2}{\omega}$

$$=\frac{2}{500}\text{ s}=4\text{ ms}$$

$\therefore$ Answer is 4.

**Analysis of Question**

(i) Question is very simple.

(ii) I think this is one of the simplest formula based question of this paper.

**15.** $t=T=\dfrac{2u\sin\theta}{g}$

$$=\frac{2\times 10\times\sin 60^\circ}{10}=\sqrt{3}\text{ s}$$

Displacement of train in time $t=\dfrac{1}{2}at^2$

Displacement of boy with respect to train

$$=1.15\text{ m}$$

$\therefore$ Displacement of boy with respect to ground $=\left(1.15+\dfrac{1}{2}at^2\right)$

Displacement of ball with respect to ground = $(u \cos 60°)t$

To catch the ball back at initial height,

$$1.15 + \frac{1}{2} at^2 = (u \cos 60°)\, t$$

$$\therefore\ 1.15 + \frac{1}{2} a (\sqrt{3})^2 = 10 \times \frac{1}{2} \times \sqrt{3}$$

Solving this equation, we get

$$a = 5 \text{ ms}^{-2}$$

$\therefore$ Answer is 5.

**Analysis of Question**

(i) Question is moderately tough.

(ii) Velocity of ball given in the question is with respect to ground.

**16.**

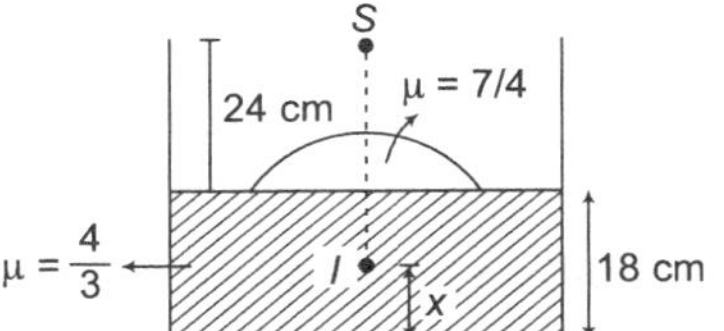


Two refractions will take place, first from spherical surface and the other from the plane surface.

So, applying

$$\frac{\mu_2}{v} - \frac{\mu_1}{u} = \frac{\mu_2 - \mu_1}{R}$$

two times with proper sign convention. Ray of light is travelling downwards. Therefore, downward direction is taken as positive direction.

$$\frac{7/4}{v} - \frac{1.0}{-24} = \frac{7/4 - 1.0}{+6} \quad \ldots(\text{i})$$

$$\frac{4/3}{(18 - x)} - \frac{7/4}{v} = \frac{4/3 - 7/4}{\infty} \quad \ldots(\text{ii})$$

Solving these equations, we get

$$x = 2 \text{ cm}$$

$\therefore$ Answer is 2.

**Analysis of Question**

(i) Question is moderately difficult from calculation point of view, otherwise it is simple.

(ii) $\frac{\mu_2}{v} - \frac{\mu_1}{u} = \frac{\mu_2 - \mu_1}{R}$ can be applied for plane surface also with $R = \infty$

**17.** Decrease in mechanical energy

= Work done against friction

$$\therefore\ \frac{1}{2} mv^2 - \frac{1}{2} kx^2 = \mu\, mgx$$

$$\text{or}\quad v = \sqrt{\frac{2\mu\, mgx + kx^2}{m}}$$

Substituting the values, we get

$$v = 0.4 \text{ ms}^{-1} = \left(\frac{4}{10}\right) \text{ms}^{-1}$$

$\therefore$ Answer is 4.

**Analysis of Question**

(i) Question is simple.

(ii) If $\mu_s$ and $\mu_k$ two values of coefficient of friction are given, then we will take $\mu_k$.

**18.** Photoemission will stop when potential on silver sphere becomes equal to the stopping potential.

$$\therefore\quad \frac{hc}{\lambda} - W = eV_0$$

$$\text{Here,}\quad V_0 = \frac{1}{4\pi\varepsilon_0} \frac{ne}{r}$$

$$\therefore\quad \left(\frac{1240}{1200} eV\right) - (4.7\ eV) = \frac{9 \times 10^9 \times n \times 1.6 \times 10^{-19}}{10^{-2}}$$

$$(6.2 - 4.7) = \frac{9 \times 10^9 \times n \times 1.6 \times 10^{-19}}{10^{-2}}$$

$$\text{or}\quad n = \frac{1.5 \times 10^{-2}}{9 \times 1.6 \times 10^{-10}} = 1.04 \times 10^7$$

$\therefore$ Answer is 7.

**Analysis of Question**

(i) Question is moderately difficult.

(ii) In $eV_0$ if we substitute the value of $e = 1.6 \times 10^{-19}$ C, then answer is in J. If we leave it then, answer is in eV.

**19.** Internal energy $\propto T \propto pV$

This is because

$$U = \frac{nf}{2} RT = \frac{f}{2} pV$$

Here, $n$ = number of moles

$f$ = degree of freedom

If the product $pV$ increases, then internal energy will increase and if product decreases, the internal energy will decrease.

Further, work is done on the gas, if volume of gas decreases. For heat exchange.

$$Q = W + \Delta U$$

Work done is area under $p$-$V$ graph. If volume increases work done by gas is positive and if volume decreases work done by gas is negative. Further $\Delta U$ is positive if product of $pV$ is increasing and $\Delta U$ is negative, if product of $pV$ is decreasing. If heat is taken by the gas $Q$ is positive and if heat is lost by the gas $Q$ is negative.

Keeping the above points in mind the answer to this question is as under.

(A) → (p, r, t)
(B) → (p, r)
(C) → (q, s)
(D) → (r, t)

**Analysis of Question**

(i) Calculation wise, question is slightly lengthy. Otherwise question is theory based and simple.

(ii) In process $DA$,

$$p_A V_A = p_D V_D$$

$\therefore \quad T_A = T_D$

or $\quad \Delta U = 0$

Further, volume of gas is decreasing. Therefore, work is done on the gas or work done by gas is negative. Therefore, $Q$ is negative or heat is lost.

(iii) This question covers almost all the concepts of first law of thermodynamics.

**20.** In organ pipes, longitudinal waves are formed. In string, transverse waves are formed. Open end of pipe is displacement antinode and closed end is displacement node. In case of string fixed end of a string is node.

Further, least distance between a node and an antinode is $\frac{\lambda}{4}$ and between two nodes is $\frac{\lambda}{2}$. Keeping these points in mind answer to this question is as under;

(A) → (p, t)
(B) → (p, s)
(C) → (q, s)
(D) → (q, r)

**Analysis of Question**

(i) Unlike the previous years's few match the column problems, this question seems to be more logical. If you know the subject, you can select correct matching without confusion.

(ii) In option (d) of column I, string is fixed from centre, therefore at lowest frequency two loops will be formed as shown below.

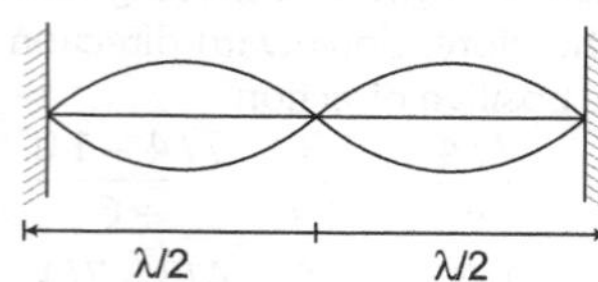

# IIT JEE

## UNSOLVED PAPER 2011

## Chemistry

### Paper I

#### SECTION I

***Directions*** (Q. Nos. 1 to 7) *This section contains 7 multiple choice questions. Each question has four choices (a), (b), (c), and (d), out of which only one is correct.*

**1.** Bombardment of aluminium by α-particle leads to its artificial disintegration in two ways, (i) and (ii) as shown. Products $X$, $Y$ and $Z$ respectively are

$$^{27}_{13}\text{Al} \xrightarrow{\text{(ii)}} {}^{30}_{15}\text{P} + Y$$

$$^{27}_{13}\text{Al} \xrightarrow{\text{(i)}} {}^{30}_{14}\text{Si} + X \qquad {}^{30}_{15}\text{P} \longrightarrow {}^{30}_{14}\text{Si} + Z$$

(a) proton, neutron, positron
(b) neutron, positron, proton
(c) proton, positron, neutron
(d) positron, proton, neutron

**2.** $AgNO_3(aq)$ was added to an aqueous KCl solution gradually and the conductivity of the solution was measured. The plot of conductance ($\Lambda$) *versus* the volume of $AgNO_3$ is

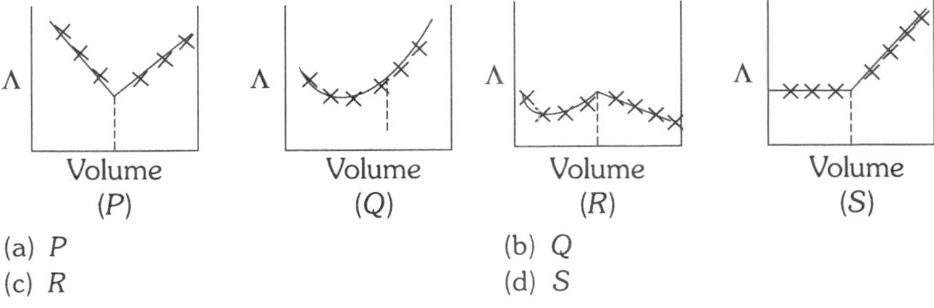


(a) $P$
(b) $Q$
(c) $R$
(d) $S$

**3.** The major product of the following reaction is

$$\text{Phthalimide (C}_6\text{H}_4\text{(CO)}_2\text{NH)} \xrightarrow[\text{(ii) Br}-\text{C}_6\text{H}_4-\text{CH}_2\text{Cl}]{\text{(i) KOH}}$$

(a) 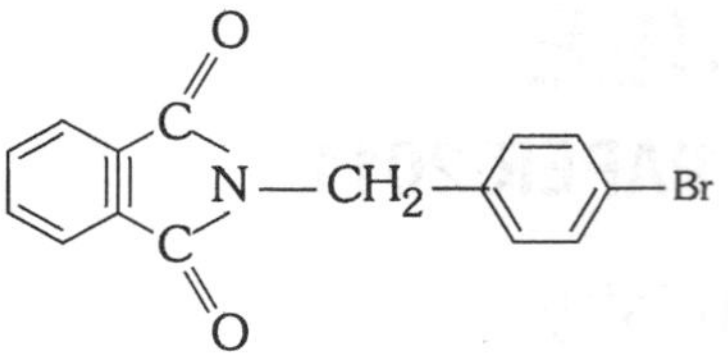 O, C, N — $CH_2$ — Br, C, O

(b) 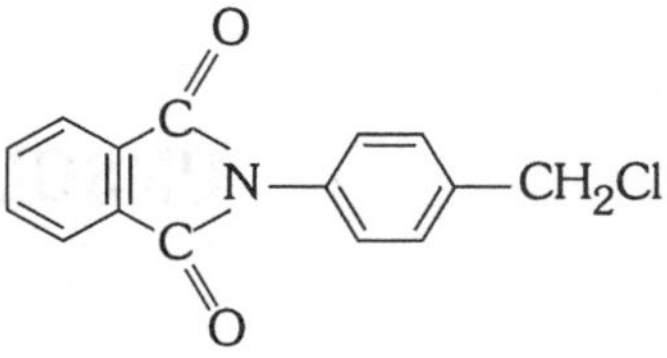 O, C, N — $CH_2Cl$, C, O

(c) O, C, N, C, O — $CH_2$ — Br

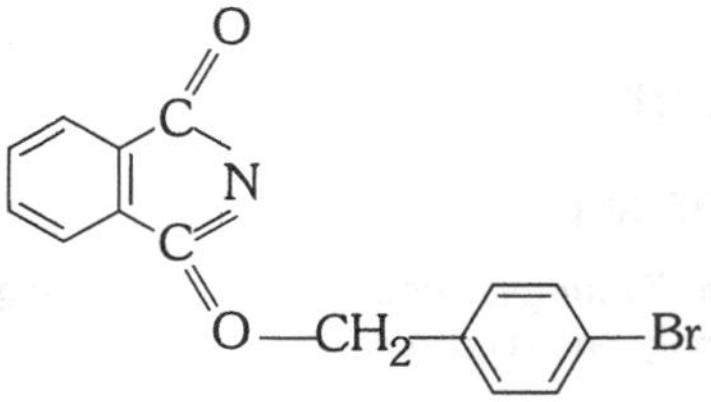

(d) 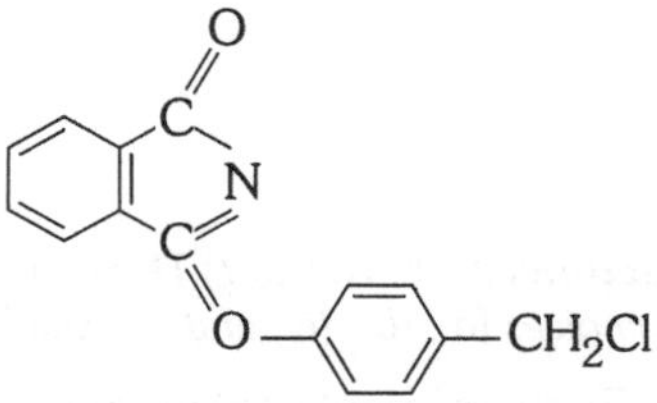 O, C, N, C, O — $CH_2Cl$

**4.** Geometrical shapes of the complexes formed by the reaction of $Ni^{2+}$ with $Cl^-$, $CN^-$ and $H_2O$, respectively, are

(a) octahedral, tetrahedral and square planar
(b) tetrahedral, square planar and octahedral
(c) square planar, tetrahedral and octahedral
(d) octahedral, square planar and octahedral

**5.** Extra pure $N_2$ can be obtained by heating

(a) $NH_3$ with CuO (b) $NH_4NO_4$ (c) $(NH_4)_2Cr_2O_7$ (d) $Ba(N_3)_2$

**6.** Dissolving 120 g of urea (molecular weight 60) in 1000 g of water gave a solution of density 1.15 g /mL. The molarity of the solution is

(a) 1.78 M (b) 2.00 M (c) 2.05 M (d) 2.22 M

**7.** Among the following compounds, the most acidic is

(a) *p*-nitrophenol (b) *p*-hydroxybenzoic acid
(c) *o*-hydroxybenzoic acid (d) *p*-toluic acid

## SECTION II

***Directions*** (Q. Nos. 8 to 11) *This section contains 4 multiple choice questions. Each question has four choices (a), (b), (c), and (d), out of which only one is correct.*

**8.** The correct statement(s) pertaining to the adsorption of a gas on a solid surface is(are)

(a) Adsorption is always exothermic
(b) Physisorption may transform into chemisorption at high temperature
(c) Physisorption increases with increasing temperature but chemisorption decreases with increasing temperature
(d) Chemisorption is more exothermic than physisorption, however it is very slow due to higher energy of activation

**9.** According to kinetic theory of gases

(a) collisions are always elastic
(b) heavier molecules transfer more momentum to the wall of the container

(c) only a small number of molecules have very high velocity
(d) between collisions, the molecules move in straight lines with constant velocities

**10.** Extraction of metal from the ore cassiterite involves
(a) carbon reduction of an oxide ore
(b) self-reduction of a sulphide ore
(c) removal of copper impurity
(d) removal of iron impurity

**11.** Amongst the given options, the compound (s) in which all the atoms are in one plane in all the possible conformations (if any), is (are)

(a) $H_2C{=}C(H){-}C(H){=}CH_2$
(b) $H{-}C{\equiv}C{-}C(H){=}CH_2$
(c) $H_2C{=}C{=}O$
(d) $H_2C{=}C{=}CH_2$

## SECTION III

***Directions*** (Q. Nos. 12 to 16) *This section contains 2 paragraphs. Based on the first paragraph 2 multiple choice questions and based on the paragraph 3 multiple choice questions have to be answered. Each of these questions has four choices (a), (b), (c) and (d) out of which only one is correct.*

*Paragraph 12-13 An acylic hydrocarbon P, having molecular formula $C_6H_{10}$, gave acetone as the only organic product through the following sequence of reactions, in which Q is an intermediate organic compounds.*

$$\underset{(C_6H_{10})}{P} \xrightarrow[\text{(iii) dil. acid}]{\text{(i) dil. } H_2SO_4/HgSO_4 \quad \text{(ii) } NaBH_4/\text{ethanol}} H_3C{-}\overset{O}{\overset{\|}{C}}{-}CH_3$$

**12.** The structure of compound $P$ is
(a) $CH_3CH_2CH_2CH_2{-}C{\equiv}C{-}H$
(b) $H_3CH_2C{-}C{\equiv}C{-}CH_2CH_3$
(c) $(H_3C)_2C(H){-}C{\equiv}C{-}CH_3$
(d) $(H_3C)_3C{-}C{\equiv}C{-}H$

**13.** The structure of the compound $Q$ is
(a) $(H_3C)_2C(H){-}C(OH)(H){-}CH_2CH_3$
(b) $(H_3C)_3C{-}C(OH)(H){-}CH_3$
(c) $(H_3C)_2C(H){-}C{-}CH_2CH(OH)CH_3$
(d) $CH_3CH_2CH_2CH(OH)CH_2CH_3$

***Paragraph* 14-16** *When a metal rod M is dipped into an aqueous colourless concentrated solution of compound N, the solution turns light blue. Addition of aqueous NaCl to the blue solution gives a white precipitate O. Addition of aqueous $NH_3$ dissolves O and gives an intense blue solution.*

**14.** The metal rod *M* is
(a) Fe (b) Cu (c) Ni (d) Co

**15.** The compound *N* is
(a) $AgNO_3$ (b) $Zn(NO_3)_2$ (c) $Al(NO_3)_3$ (d) $Pb(NO_3)_2$

**16.** The final solution contains
(a) $[Pb(NH_3)_4]^{2+}$ and $[CoCl_4]^{2-}$
(b) $[Al(NH_3)_4]^{3+}$ and $[Cu(NH_3)_4]^{2+}$
(c) $[Ag(NH_3)_2]^{+}$ and $[Co(NH_3)_4]^{2+}$
(d) $[Ag(NH_3)_2]^{+}$ and $[Ni(NH_3)_6]^{2+}$

## Section IV

***Directions*** (Q. Nos. 17 to 23) *This section contains 7 questions. The answer to each of the questions is a single-digit integer, ranging from 0 to 9.*

**17.** The maximum number of electrons that can have principal quantum number, $n = 3$ and spin quantum number, $m_s = -\frac{1}{2}$, is

**18.** Reaction of $Br_2$ with $Na_2CO_3$ in aqueous solution gives sodium bromide and sodium bromate with evolution of $CO_2$ gas. The number of sodium bromide molecules involved in the balanced chemical equation is

**19.** The total number of alkenes possible by dehydrobromination of 3-bromo-3-cyclopentylhexane using alcoholic KOH is

**20.** The work function ($\Phi$) of some metals is listed below. The number of metals which will show photoelectric effect when light of 300 nm wavelength falls on the metal is

| **Metal** | Li | Na | K | Mg | Cu | Ag | Fe | Pt | W |
|---|---|---|---|---|---|---|---|---|---|
| $\phi$ (eV) | 2.4 | 2.3 | 2.2 | 3.7 | 4.8 | 4.3 | 4.7 | 6.3 | 4.75 |

**21.** A decapeptide (molecular weight 796) on complete hydrolysis gives glycine (molecular weight 75), alanine and phenylalanine. Glycine contributes 47.0% to the total weight of the hydrolysed products. The number of glycine units present in the decapeptide is

**22.** The difference in the oxidation numbers of the two types of sulphur atoms in $Na_2S_4O_6$ is

**23.** To an evacuated vessel with movable piston under external pressure of 1 atm, 0.1 mole of He and 1.0 mole of an unknown compound (vapour pressure 0.68 atm at 0°C) are introduced. Considering the ideal gas behaviour, the total volume (in litre) of the gases at 0°C is close to

# Paper II

## SECTION I

***Directions*** (Q. Nos. 1 to 8) *This section contains 8 multiple choice questions. Each question has four choices (a), (b), (c) and (d) out of which only one is correct.*

**1.** Oxidation states of the metal in the minerals haematite and magnetite, respectively, are

(a) II, III in haematite and III in magnetite (b) II, III in haematite and II in magnetite
(c) II in haematite and II, III in magnetite (d) III in haematite and II, III in magnetite

**2.** The following carbohydrate is

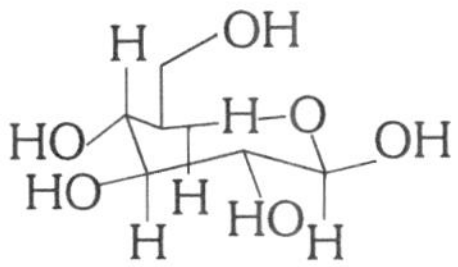

(a) a ketohexose (b) an aldohexose (c) an α-furanose (d) an α-pyranose

**3.** The major product of the following reaction is

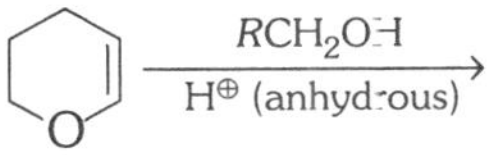

$$\xrightarrow[\text{H}^{\oplus}\text{ (anhydrous)}]{RCH_2OH}$$

(a) a hemiacetal (b) an acetal (c) an ether (d) an ester

**4.** Amongst the compounds given, the one that would form a brilliant coloured dye on treatment with $NaNO_2$ in dil. HCl followed by addition to an alkaline solution of β-naphthol is

(a) 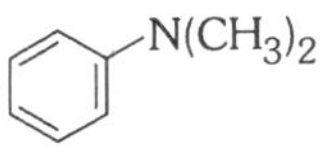 $C_6H_5N(CH_3)_2$ (b) $C_6H_5NHCH_3$ (c) $H_3C-C_6H_4-NH_2$ (d) 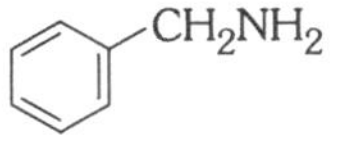 $C_6H_5CH_2NH_2$

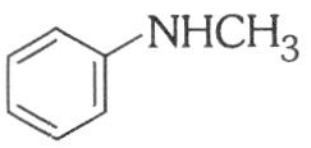

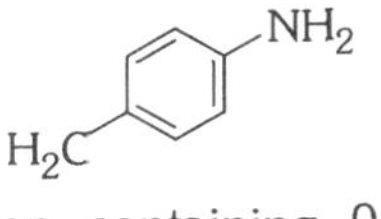

**5.** The freezing point (in °C) of solution containing 0.1 g of $K_3[Fe(CN)_6]$ (molecular weight 329) in 100 g of water ($K_f$ = 1.86 K kg mol$^{-1}$) is

(a) $-2.3\times10^{-2}$ (b) $-5.7\times10^{-2}$ (c) $-5.7\times10^{-3}$ (d) $-1.2\times10^{-2}$

**6.** Consider the following cell reaction,

$$2Fe\,(s)+O_2\,(g)+4H^+\,(aq)\longrightarrow 2Fe^{2+}\,(aq)+2H_2O\,(l),\ E^\circ=1.67\text{ V}$$

At $[Fe^{2+}]=10^{-3}$ M, P $(O_2)$ = 0.1 atm and pH = 3, the cell potential at 25°C is

(a) 1.47 V (b) 1.77 V (c) 1.87 V (d) 1.57 V

**7.** Passing $H_2S$ gas into a mixture of $Mn^{2+}$, $Ni^{2+}$, $Cu^{2+}$ and $Hg^{2+}$ ions in an acidified aqueous solution precipitates

(a) CuS and HgS (b) MnS and CuS (c) MnS and NiS (d) NiS and HgS

**8.** Among the following complexes $(K - P)$, $K_3[Fe(CN)_6]$ $(K)$,
$[Co(NH_3)_6]Cl_3$ $(L)$, $Na_3[Co(ox)_3]$ $(M)$,
$[Ni(H_2O)_6]Cl_2$ $(N)$; $K_2[Pt(CN)_4]$ $(O)$ and
$[Zn(H_2O)_6](NO_3)_2$ $(P)$
the diamagnetic complexes are
(a) $K, L, M, N$ (b) $K, M, O, P$ (c) $L, M, O, P$ (d) $L, M, N, O$

## SECTION II

**Directions** (Q. Nos. 9 to 12) *This section contains 4 multiple choice questions. Each question has four choices (a), (b), (c) and (d) out of which one or more may be correct.*

**9.** Reduction of the metal centre in aqueous permanganate ion involves
(a) three electrons in neutral medium (b) five electrons in neutral medium
(c) three electrons in alkaline medium (d) five electrons in acidic medium

**10.** For the first order reaction $2N_2O_5(g) \longrightarrow 4NO_2(g) + O_2(g)$
(a) the concentration of the reactant decreases exponentially with time.
(b) the half-life of the reaction decreases with increasing temperature
(c) the half-life of the reaction depends on the initial concentration of the reactant.
(d) the reaction proceeds of 99.6% completion in eight half-life duration.

**11.** The equilibrium $2Cu^{I} \rightleftharpoons Cu^{0} + Cu^{II}$ in aqueous medium at 25°C shifts towards the left in the presence of
(a) $NO_3^-$ (b) $Cl^-$ (c) $SCN^-$ (d) $CN^-$

**12.** The correct functional group $X$ and the reagent/reaction conditions $Y$ in the following schemes are $X-(CH_2)_4-X$

$$\xrightarrow{\text{(i) } Y} \text{Condensation polymer}$$

(i) $HO(O=)C-(CH_2)_4-C(=O)OH$, heat

(a) $X = COOCH_3$, $Y = H_2/Ni/heat$
(b) $X = CONH_2$, $Y = H_2/Ni/heat$
(c) $X = CONH_2$, $Y = Br_2/NaOH$
(d) $X = CN$, $Y = H_2/Ni/heat$

## SECTION III

**Directions** (Q.Nos. 13 to 18) *This section contains 6 questions. The answer to each of the question is a single digit integer, ranging from 0 to 9.*

**13.** In 1L saturated solution of AgCl [$K_{sp}$ AgCl $= 1.6 \times 10^{-10}$], 0.1 mole of CuCl [$K_{sp}$ (CuCl) $= 1.0 \times 10^{-6}$] is added. The resultant concentration of $Ag^+$ in the solution is $1.6 \times 10^{-x}$. The value of $x$ is

**14.** The maximum number of isomers (including stereoisomers) that are possible on monochlorination of the following compounds, is

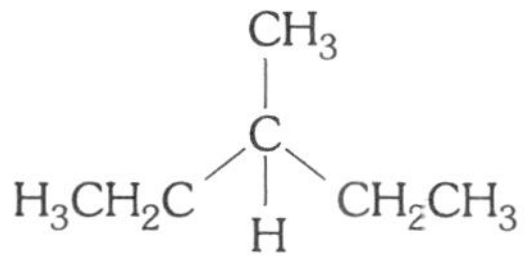

$CH_3$
C
$H_3CH_2C$ H $CH_2CH_3$

**15.** Among the following, the number of compounds than can react with $PCl_5$ to give $POCl_3$ is $O_2, CO_2, SO_2, H_2O, H_2SO_4, P_4O_{10}$.

**16.** The number of hexagonal faces that are present in a truncated octahedron is

**17.** The volume (in mL) of 0.1 M $AgNO_3$ required for complete precipitation of chloride ions present in 30 mL of 0.01 M solution of $[Cr(H_2O)_5Cl]Cl_2$, as silver chloride is close to

**18.** The total number of contributing structures showing hyperconjugation (involving C—H bonds) for the following carbocation is

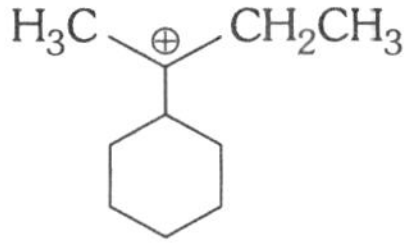

$H_3C$ $\oplus$ $CH_2CH_3$

## SECTION IV

***Directions*** (Q. Nos. 19 and 20) *This section contains 2 questions. Each question has four statements (A, B, C and D) given in Column I and five statements (p, q, r, s and t) in Column II. Any given statement in Column I can have correct matching with one or more statement (s) given in Column II.*

**19.** Match the transformation in Column I with appropriate options in Column II.

| | Column I | | Column II |
|---|---|---|---|
| A. | $CO_2(s) \rightarrow CO_2(g)$ | p. | Phase transition |
| B. | $CaCO_3(s) \rightarrow CaO(s) + CO_2(g)$ | q. | Allotropic change |
| C. | $2H \rightarrow H_2(g)$ | r. | $\Delta H$ is positive |
| D. | $P_{(white,\ solid)} \rightarrow P_{(red,\ solid)}$ | s. | $\Delta S$ is positive |
| | | t. | $\Delta S$ is negative |

**20.** Match the reactions in Column I with appropriate types of steps/reactive intermediate involved in these reactions as given in Column II.

| | Column I | Column II |
|---|---|---|
| A. | 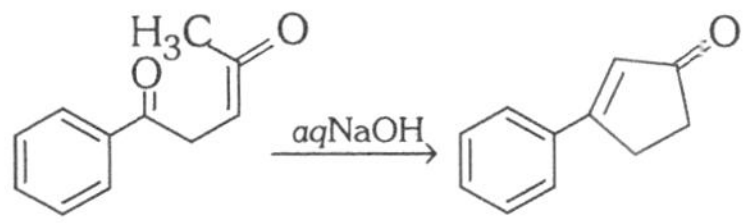 $H_3C$ O O $\xrightarrow{aq\,NaOH}$ O | p. Nucleophilic substitution |

| | Column I | Column II |
|---|---|---|
| B. | $C_6H_5COCH_2CH_2CH_2Cl \xrightarrow{CH_3MgI}$ 2-methyl-2-phenyltetrahydrofuran | q. Electrophilic substitution |
| C. | $C_6H_5COCH_2CH_2CH_2{}^{18}OH \xrightarrow{H_2SO_4}$ 2-phenylfuran ($^{18}O$) | r. Dehydration |
| D. | $C_6H_5CH_2CH_2CH_2C(CH_3)_2OH \xrightarrow{H_2SO_4}$ 1,1-dimethyltetralin ($H_3C$ $CH_3$) | s. Nucleophilic addition |
| | | t. Carbanion |

# ANSWERS

## Paper I

**1.** (a) **2.** (d) **3.** (a) **4.** (b) **5.** (d) **6.** (c)
**7.** (c) **8.** (a,b,d) **9.** (a,b,d) **10.** (a,d) **11.** (b,c) **12.** (d)
**13.** (b) **14.** (b) **15.** (a) **16.** (c) **17.** (9) **18.** (5)
**19.** (5) **20.** (4) **21.** (6) **22.** (5) **23.** (7)

## Paper II

**1.** (d) **2.** (b) **3.** (b) **4.** (c) **5.** (a) **6.** (d)
**7.** (a) **8.** (c) **9.** (a,c,d) **10.** (a,b,d) **11.** (b,c,d) **12.** (c,d)
**13.** (7) **14.** (8) **15.** (4) **16.** (8) **17.** (6) **18.** (6)
**19.** (A) → p, r, s; (B) → r, s, (C) → t; (D) → p, q, t
**20.** A → r, s, t; (B) → p, s; (C) → r, s; (D) → q, r

*According to IIT official answer key.

# Hints & Solutions

## Paper I

**1.** (i) $^{27}_{13}Al + ^{4}_{2}He\ (\alpha) \rightarrow ^{30}_{14}Si + ^{1}_{1}X \quad \Rightarrow X$ is proton $^{1}_{1}H$.

(ii) $^{27}_{13}Al + ^{4}_{2}He\ (\alpha) \rightarrow ^{30}_{15}P + ^{1}_{0}Y \quad \Rightarrow Y$ is neutron $^{1}_{0}n$.

$$^{30}_{15}P \rightarrow ^{30}_{14}Si + ^{0}_{1}Z$$

$\Rightarrow Z$ is positron $^{0}_{+1}e$.

**2.** As $AgNO_3$ is added to solution, KCl will be displaced accrding to following reaction

$$AgNO_3(aq) + KCl(aq) \longrightarrow AgCl(s) + KNO_3(aq)$$

For every mole of KCl displaced from solution, one mole of $KNO_3$ comes in solution resulting in almost constant conductivity. As the end point is reached, added $AgNO_3$ remain in solution increasing ionic concentration, hence conductivity increases.

**3.** It is the first step of Gabriel's phthalimide synthesis. The hydrogen bonded to nitrogen is sufficiently acidic due to two $\alpha$-carbonyls.

$$\text{Phthalimide N—H} + OH^- \longrightarrow H_2O + \text{phthalimide } N^- \longleftrightarrow N^- \longleftrightarrow O^-$$

The conjugate base formed above act as nucleophile in the subsequent step of reaction. As shown above, the nucleophile exist in three resonating form, one may think of oxygen being the donor atom in the nucleophilic attack. However, nitrogen act as donor as it is better donor than oxygen.

$$\text{Phthalimide } N^- + ClCH_2\text{—C}_6H_4\text{—Br} \xrightarrow{S_N2} \text{Phthalimide N—CH}_2\text{—C}_6H_4\text{—Br}$$

Bromine is not substituted in the above reaction as it is in resonance with benzene ring giving partial double bond character to C—Br bond, hence difficult to break.

$$ClCH_2\text{—C}_6H_4\text{—}\ddot{Br}: \longleftrightarrow ClCH_2\text{—C}_6H_4\text{=}\overset{+}{Br}$$

**4.** $Ni^{2+} + 4CN^- \longrightarrow [Ni(CN)_4]^{2-}$. Here $Ni^{2+}$ has $d^8$ configuration with $CN^-$ as strong ligand.

3d: 1l 1l 1l 1l •• | 4s: •• | 4p: •• •• _ — $dsp^2$

$d^8$-configuration in strong ligand field gives $dsp^2$-hybridisation, hence square planar geometry.

$$Ni^{2+} + 4Cl^- \longrightarrow [NiCl_4]^{2-}$$

Here $Ni^{2+}$ had $d^8$-configuration with $Cl^-$ as weak ligand.

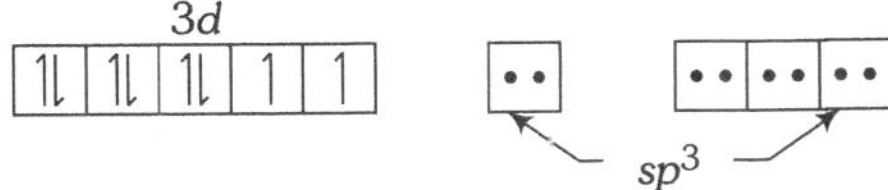

$d^8$-configuration in weak ligand field gives $sp^3$ hydridisation, hence tetrahedral geometry. $Ni^{2+}$ with $H_2O$ forms $[Ni(H_2O)_6]^{2+}$ complex and $H_2O$ is a weak ligand.

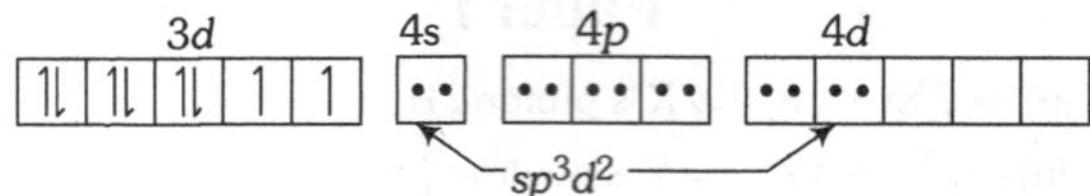


Therefore, $Ni(H_2O)_6^{2+}$ has octahedral geometry.

**5.** $$Ba(N_3)_2 \xrightarrow{\text{Heat}} Ba(s) + 3N_2(g)$$

Azide salt of barium can be obtained in purest form as well as the decomposition product contain solid Ba as byproduct along with gaseous nitrogen, hence no additional step of separation is required.

Other reactions are:

$$NH_4NO_3 \xrightarrow{\text{Heat}} N_2O + 2H_2O$$

$$2NH_3 + 3CuO \xrightarrow{\text{Heat}} 3Cu + 3H_2O + N_2$$

$$(NH_4)_2Cr_2O_7 \xrightarrow{\text{Heat}} Cr_2O_3 + 4H_2O + N_2$$

**6.** Molarity $= \dfrac{\text{Moles of solute}}{\text{Volume of solution (L)}}$. Moles of urea $= \dfrac{120}{60} = 2$

weight of solution = weight of solvent + weight of solute = 1000 + 120 = 1120 g

$\Rightarrow$ Volume $= \dfrac{1120\,g}{1.15\text{g}/\text{mL}} \times \dfrac{1}{1000\text{mL}/\text{L}} = 0.974\text{ L} \Rightarrow$ Molarity $= \dfrac{2.000}{0.974} = 2.05\text{ M}$

**7.** A monosubstituted benzoic acid is stronger than a monosubstituted phenol as former being a carboxylic acid. Among the given substituted benzoic acid, *ortho*-hydroxy acid is strongest acid although —OH cause electron donation by resonance effect which tends to decrease acid strength. It is due to a very high stabilization of conjugate base by intramolecular H-bond which outweight the electron donating resonance effect of —OH.

The overall order of acid-strength of given four acids is *ortho*-hydro *xy* benzoic acid ($pK_a = 2.98$) > Toluic acid $pK_a = 4.37$) > *p*-hydroxybenzoic acid ($pK_a = 4.58$) > *p*-nitrophenol ($pK_a = 7.15$).

**8.** (a) In the process of adsorption, a bond is formed between adsorbate and adsorbent, hence always exothermic.

(b) Physisorption require very low activation energy while chemisorption require high activation energy. Therefore a physisorption may transform into chemisorption but only at high temperature.

(c) It is wrong statement as at higher temperature, physically adsorbed substance starts desorbing.

(d) In physical adsorption, van der Waals' force hold the adsorbate and adsorbent together which is a weak electrostatic attraction. In chemisorption, strong chemical bond binds the adsorbate to the adsorbent. Therefore, chemisorption is more exothermic than physical adsorption.

**9.** (a) According to a postulate of kinetic theory of gases, collision between the molecules as well as with the wall of container is perfectly elastic in nature.

(b) If a gas molecule of man 'm' moving with speed 'u' collide to the wall of container, the change in momentum is $\Delta p = -2mu$. Therefore, heavier molecule will transfer more momentum to the wall as there will be greater change in momentum of the colliding gas malecule. However this was not postulated in kinetic theory of Gases.

(c) According to Maxwell-Boltzmann distribution of molecular speed, very few molecules have either very high or very low speeds. Most of the molecules moves in a specific, intermediate speed range.

(d) According to kinetic theory of gases, a gas molecule moves in straight line unless it collide with another molecular or to the wall of container and change in momentum is observed only after collision.

**10.** The important ore of tin is cassiterite ($SnO_2$). Tin is extracted from cassiterite ore by carbon reduction method in a blast furnace. $SnO_2 + 2C \rightarrow Sn + 2CO$

The product often contain traces of iron which is removed by blowing air through the melt to oxidise to FeO which then floats to the surface.

$$2Fe + O_2 \rightarrow 2FeO$$

**11.** In both (b) and (c), all the atoms are present in one single plane

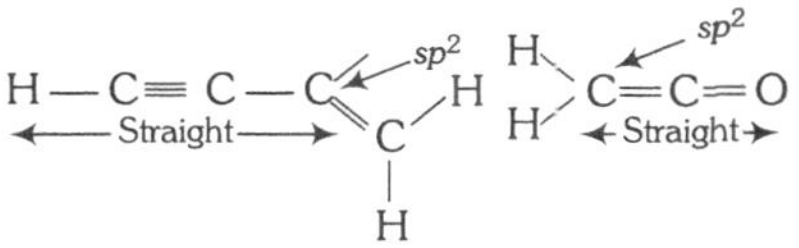

$$H—C\equiv C—CH{=}CH_2 \qquad H_2C{=}C{=}O$$

sp², sp²; ← Straight →; ← Straight →

In (a) 1, 3-butadiene, conformational change is possible between $C_2—C_3$ bond in which atoms will be present in more than one single plane.

In (d) allene, the terminals H—C—H planes are perpendicular to one another.

**Paragraph (12 and 13)** The final ozonolysis product indicates that the alkene before ozonolysis is

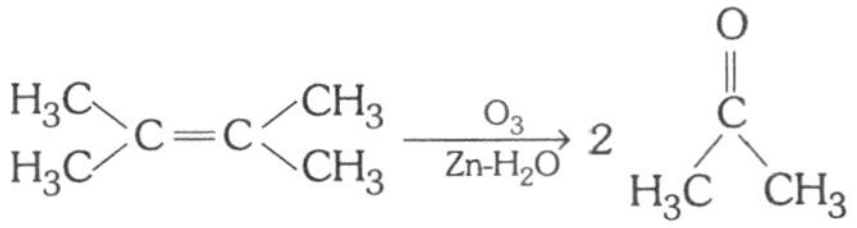

$$(H_3C)_2C{=}C(CH_3)_2 \xrightarrow[Zn\text{-}H_2O]{O_3} 2\ H_3C—\overset{\overset{\displaystyle O}{\|}}{C}—CH_3$$

Also $P(C_6H_{10})$ has two degree of unsaturation and oxymercuration demercuration hydration indicates that it is an alkyne. As alkyne, on hydration, gives a carbonyl compound which on reduction with $NaBH_4$ gives a 2° alcohol.

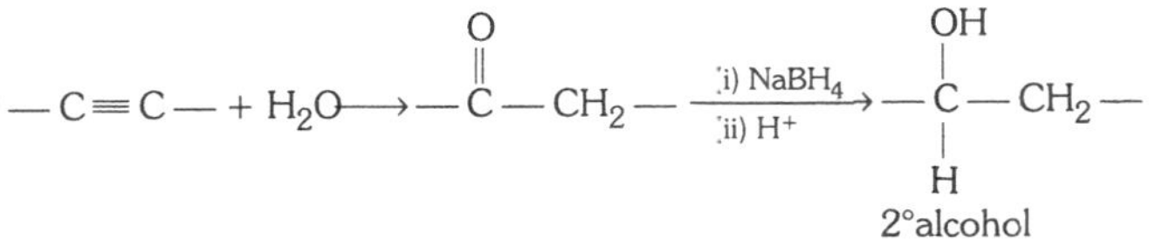

$$—C\equiv C— + H_2O \longrightarrow —\overset{\overset{\displaystyle O}{\|}}{C}—CH_2— \xrightarrow[(ii)\ H^+]{(i)\ NaBH_4} —\underset{\underset{\displaystyle H}{|}}{\overset{\overset{\displaystyle OH}{|}}{C}}—CH_2—$$

2°alcohol

The secondary alcohol that can give above shown alkene on acid catalysed dehydration is

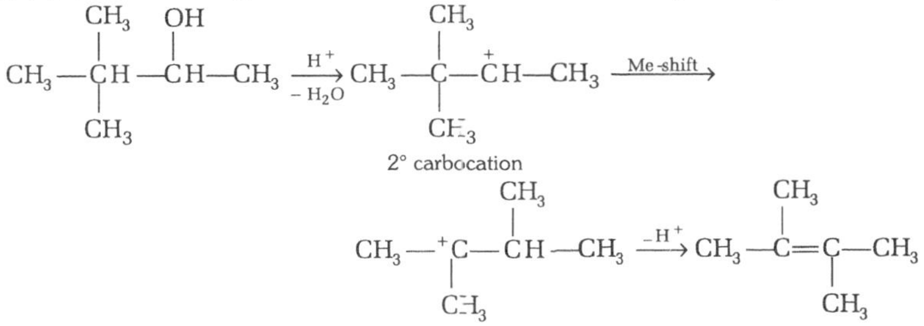

$$CH_3—\underset{\underset{\displaystyle CH_3}{|}}{\overset{\overset{\displaystyle CH_3}{|}}{C}H}—\overset{\overset{\displaystyle OH}{|}}{C}H—CH_3 \xrightarrow[-H_2O]{H^+} CH_3—\underset{\underset{\displaystyle CH_3}{|}}{\overset{\overset{\displaystyle CH_3}{|}}{C}}—\overset{+}{C}H—CH_3 \xrightarrow{Me\text{-}shift}$$

2° carbocation

$$CH_3—\overset{+}{\underset{\underset{\displaystyle CH_3}{|}}{C}}—\overset{\overset{\displaystyle CH_3}{|}}{C}H—CH_3 \xrightarrow{-H^+} CH_3—\overset{\overset{\displaystyle CH_3}{|}}{C}{=}\underset{\underset{\displaystyle CH_3}{|}}{C}—CH_3$$

**12.** $CH_3-\underset{CH_3}{\overset{CH_3}{\overset{|}{\underset{|}{C}}}}-C\equiv C-H \xrightarrow[H_2SO_4]{HgSO_4} CH_3-\underset{CH_3}{\overset{CH_3}{\overset{|}{\underset{|}{C}}}}-\overset{O}{\overset{\|}{C}}-CH_3 \xrightarrow[\text{(ii) } H^+]{\text{(i) } NaBH_4} Q$

**13.** Explained in the beginning.

**14.** Appearance of blue colour on addition of the metal rod $M$ to the solution of $N$ is an indication that metal may be Cu. Later formation of white precipitate on addition of NaCl and dissolution of this precipitate in aqueous ammonia confirm that while preipitate is of AgCl. This implies that the solution $N$ is of $AgNO_3$. This confirm that the metal $M$ is only Cu.

**15.** The compound $N$ is $AgNO_3$, explained above.

**16.** Since in the beginning, concentrated $AgNO_3$ solution was taken, some $AgNO_3$ remain in solution which gives white precipitate of AgCl on addition of NaCl. The precipitate finally dissolve in ammonia and $Cu^{2+}$ present in solution also forms complex with ammonia as

$$Cu(NO_3)_2 + 4NH_3(aq) \rightarrow \underset{\text{Blue}}{[Cu(NH_3)_4]^{2+}} + 2NO_3$$

$$AgCl(s) + 2NH_3(aq) \rightarrow [Ag(NH_3)_2]^+ + Cl^-$$

**17.** When $n = 3$, $l = 0, 1, 2$ i.e., there are $3s$, $3p$ and $3d$ orbitals. If all these orbitals are completely occupied as

| ⥮ |
|---|

| ⥮ | ⥮ | ⥮ |
|---|---|---|

| ⥮ | ⥮ | ⥮ | ⥮ | ⥮ |
|---|---|---|---|---|

Total 18 electrons, 9 electrons with $s = +\frac{1}{2}$ and 9 with $s = -\frac{1}{2}$.

***Alternatively*** *In any nth orbit, there can be a maximum of* $2n^2$ *electrons. Hence, when* $n = 3$, *number of maximum electrons* $= 18$. *Out of these 18 electrons, 9 can have spin* $-\frac{1}{2}$ *and remaining nine with spin* $= +\frac{1}{2}$.

**18.** $Br_2$ is disproportionated in basic medium as

$$3Br_2 + 3Na_2CO_3 \longrightarrow 5NaBr + NaBrO_3 + 3CO_2$$

**19.** The substrate has three different types of B—H, therefore, first, three structural isomers of alkenes are expected as

3 types of B-H $\xrightarrow{KOH}$ I + II + III

The last two alkenes II and III are also capable of showing geometrical isomerism hence two geometrical isomers for each of them will be counted giving a total of five isomers.

**20.** Energy of photon $= \frac{hc}{\lambda}\text{ J} = \frac{hc}{e\lambda}\text{ eV} = \frac{6.625 \times 10^{-34} \times 3 \times 10^8}{300 \times 10^{-9} \times 1.602 \times 10^{-19}} = 4.14\ eV$

For photoelectric effect to occur, energy of incident photons must be greater than work function of metal. Hence, only Li, Na, K and Mg have work functions less than 4.14 V.

**21.** A decapeptide has nine peptide (amide) linkage as

—CONH— bond

Therefore, no hydrolysis, it will absorb nine water molecules.

Hence, total mass of hydrolysis product = 796 + 18 × 9 = 958

$\Rightarrow$ mass of glycine in hydrolysis product = $\dfrac{958 \times 47}{100} = 450$

$\Rightarrow$ number of glycine molecule in one molecule of decapeptide = $\dfrac{450}{75} = 6$

**22.** $Na_2S_4O_6$ is a salt of $H_2S_4O_6$ which has the following structure

HO—S(=O)(=O)—S—S—S(=O)(=O)—OH (middle S: (0); terminal S: (v))

$\Rightarrow$ Difference in oxidation number of two types of sulphur = 5

**23.** Since, the external pressure is 1.0 atm, the gas pressure is also 1.0 atm as piston is movable. Out of this 1.0 atm partial pressure due to unknown compound is 0.68 atm.

Therefore, partial pressure of He = 1.00 − 0.68 = 0.32 atm.

$$\Rightarrow \quad \text{Volume} = \frac{n\,(\text{He})\,RT}{r\,(\text{He})} = \frac{0.1 \times 0.082 \times 273}{0.32} = 7\text{ L}$$

$\Rightarrow$ Volume of container = Volume of He.

# Paper II

**1.** Haematite is $Fe_2O_3$, in which oxidation number of iron is III. Magnetite is $Fe_3O_4$ which is infact a mixed oxide ($FeO \cdot Fe_2O_3$), hence, iron is present in both II and III oxidation state.

**2.** Here, the—OH of hemiacetal group is equatorial therefore, it is a β-pyranose of an aldohexose.

**3.** Dihydropyran $\xrightarrow{H^+}$ oxocarbenium ion $\xrightarrow[-H^+]{RCH_2OH}$ tetrahydropyranyl–$OCH_2R$ (Acetal)

Resonance stablisation

**4.** As we know, benzenediazonium salt forms brilliant coloured dye with β-naphthol, the compound under consideration must be *p*-toludine (c) as it is a primary aromatic amine. Primary aromatic amine, on treatment with $NaNO_2$ in dil. HCl forms the corresponding diazonium chloride salt.

$H_3C$–$C_6H_4$–$NH_2$ $\xrightarrow[HCl]{NaNO_2}$ $H_3C$–$C_6H_4$–$N_2^+Cl^-$ $\xrightarrow{\beta\text{-naphthol}}$ (OH)naphthyl–N=N–$C_6H_4$–$CH_3$

**5.** vant Hoff's factor ($i$) = 4 {$3K^+$ + $[Fe(CN)_6]^{3-}$}  Molality = $\dfrac{0.1}{329} \times \dfrac{1000}{100} = \dfrac{1}{329}$

$$\Rightarrow \quad -\Delta T_f = iK_f \cdot m = 4 \times 1.86 \times \frac{1}{329} = 2.3 \times 10^{-2} \Rightarrow T_f = -2.3 \times 10^{-2}\ ^\circ\text{C}$$

(As freezing point of water is 0°C)

**6.** The half reactions are

$$Fe(s) \longrightarrow Fe^{2+}(aq) + 2e^- \times 2$$
$$O_2(g) + 4H^+ + 4e^- \longrightarrow 2H_2O + 2Fe(s)$$
$$+ O_2(g) + 4H^+ \longrightarrow 2Fe^{2+}(aq) + 2H_2O(l)$$
$$E = E^\circ - \frac{0.059}{4} \log \frac{(10^{-3})^2}{(10^{-3})^4(0.1)} = 1.57 \text{ V}$$

**7.** In acidic medium, $H_2S$ is very feebly ionized giving very small concentration of sulphide ion for precipitation. Therefore, the most insoluble salts CuS and HgS are precipitated only.

**8.** For a diamagnetic complex, there should not be any unpaired electron in the valence shell of central metral. In $K_3[Fe(CN)_6]$, Fe (III) has $d^5$–configuration (odd electrons), hence it is paramagnetic. In $[Co(NH_3)_6]Cl_3$, Co (III) has $d^6$ configuration in a strong ligand field, hence all the electrons are paired and the complex is diamagnetic. In $Na_3[Co(ox)_3]$, Co (III) has $d^6$-configuration and oxalate being a chelating ligand, very strong ligand and all the six electrons remains paired in lower $t_{2g}$ level, diamagnetic. In $[Ni(H_2O)_6]Cl_2$, Ni (II) has $3d^8$-configuration and $H_2O$ is a weak ligand hence

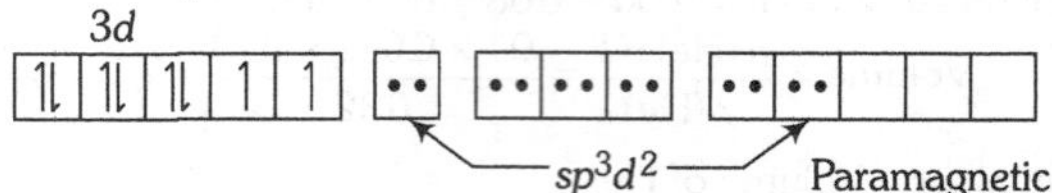


In $K_2[Pt(CN)_4]$, Pt (II) has $d^8$-configuration and $CN^-$ is a strong ligand, hence all the eight electrons are spin-paired. Therefore, complex is diamagnetic.
In $[Zn(H_2O)_6](NO_3)_2$, Zn (II) has $_3d^{10}$ configuration spin paired, hence diamagnetic.

**9.** In neutral medium : $MnO_4^- \rightarrow MnO_2$ $(Mn^{7+} + 3e^- \rightarrow Mn^{4+})$
In alkaline medium : $MnO_4^- \rightarrow MnO_2$ $(Mn^{7+} + 3\varepsilon^- \rightarrow Mn^{4+})$
In acidic medium : $Mn_4^- \rightarrow Mn^{2+}$ $(Mn^{7+} + 5e^- \rightarrow Mn^{2+})$

**10.** (a) For a first order reaction, the concentration of reactant remaining after time $t$ is given by $[A] = [A]_0 e^{-kt}$ Therefore, concentration of reactant decreases exponentially with time.

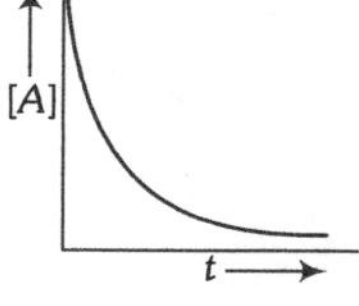


(b) Rise in temperature increases rate constant ($k$) and therefore decreases half-life ($t_{1/2}$) as

$$t_{1/2} = \frac{\text{In } 2}{k}$$

(d) For a first order reaction, if 100 moles of reactant is taken initially, after $n$ half-lives, reactant remaining is given by

$$\text{Percentage } A = 100\left(\frac{1}{2}\right)^n = 100\left(\frac{1}{2}\right)^8 = 0.3906$$

$$\Rightarrow A \text{reacted} = 100 - 0.3906 = 99.6\%$$

(c) Half-life of first order reaction is independent of initial concentration.

**11.** $Cl^-$, $CN^-$ and $SCN^-$ froms precipitate with Cu (I), remove Cu (I) ion from equilibrium and reaction shift in direction shift in backward direction according to Le-Chatelier's principle.

**12.** (a) When $X = COOCH_3$ $CH_3OOC—(CH_2)_4—COOCH_3$

$$\xrightarrow[\text{heat}]{H_2/Ni} HOCH_2—(CH_2)_4—CH_2OH + 2CH_3OH \xrightarrow[\text{heat}]{HOOC—(CH_2)_4—COOH}$$

$$\left[ O—(CH_2)_6—O—\overset{O}{\overset{||}{C}}—(CH_2)_4—\overset{O}{\overset{||}{C}} \right]_n$$

Ester, condensation polymer

(b) When $X = CONH_2$

$$H_2NOC—(CH_2)_4—CONH_2 \xrightarrow[\text{heat}]{H_2/Ni} H_2N—(CH_2)_6—NH_2$$

$$\xrightarrow{HOOC—(CH_2)_4—COOH} \left[ NH_2—(CH_2)_6—\overset{O}{\overset{||}{C}}—(CH_2)_4\overset{O}{\overset{||}{C}} \right]_n$$

Nylon, condensation polymer

(c) $$H_2NOC—(CH_2)_4—NONH_2 \xrightarrow[NaOH]{Br_2} H_2N—(CH_2)_4—NH_2$$

Hofmann's bromamide reaction

$$\xrightarrow[\text{heat}]{HOOC—(CH_2)_4—COOH} \left[ HN—(CH_2)_4—NH—\overset{O}{\overset{||}{C}}—(CH_2)_4\overset{O}{\overset{||}{C}} \right]_n$$

Nylon, condensation polymer

(d) When $X = CN$

$$NC—(CH_2)_4—CN \xrightarrow[\text{heat}]{H_2/Ni} H_2N—(CH_2)_6—NH_2 \xrightarrow[\text{heat}]{HOOC—(CH_2)_4—COOH}$$

$$\left[ HN—(CH_2)_6—NH—\overset{O}{\overset{||}{C}}—(CH_2)_4\overset{O}{\overset{||}{C}} \right]_n$$

Nylon, condensation polymer

In author's opinion (a) and (b) should also be the answer.

(a) $$MeO—\overset{O}{\overset{||}{C}}—(CH_2)_4—\overset{O}{\overset{||}{C}}—OMe \xrightarrow[\text{heat}]{H_2/Ni} HOCH_2—(CH_2)_4—CH_2OH$$

$$HOCH_2—(CH_2)_4—CH_2OH + HO—\overset{O}{\overset{||}{C}}—(CH_2)_4—\overset{O}{\overset{||}{C}}—OH \xrightarrow{\text{heat}}$$

$$\left[ O—(CH_2)_6—O—\overset{O}{\overset{||}{C}}—(CH_2)_4\overset{O}{\overset{||}{C}} \right]_n$$

Polyester, a condensation polymer

(b) $$H_2N—\overset{O}{\overset{||}{C}}—(CH_2)_4—\overset{O}{\overset{||}{C}}—NH_2 \xrightarrow[\text{heat}]{H_2/Ni} H_2NCH_2—(CH_2)_4—CH_2NH_2$$

$$H_2NCH_2—(CH_2)_4—CH_2NH_2 + HO—\overset{O}{\overset{||}{C}}—(CH_2)_4—OH \xrightarrow{\text{heat}}$$

$$\left[ NH—(CH_2)_6—NH—\overset{O}{\overset{||}{C}}—(CH_2)_4\overset{O}{\overset{||}{C}} \right]_n$$

Nylone, a condensation polymer

**13.** It is a case of simultaneous solubility of salts with a common ion. Here solubility product of CuCl is much greater than that of AgCl, it can be assumed that $Cl^-$ in solution comes mainly from CuCl.

$$\Rightarrow \qquad [Cl^-] = \sqrt{K_{sp}\ (CuCl)} = 10^{-3}\ M$$

Now for AgCl : $K_{sp} = 1.6 \times 10^{-10} = [Ag^+][Cl^-] = [Ag^+] \times 10^{-3} \Rightarrow [Ag^+] = 1.6 \times 10^{-7}$

**14.** I + II → III + IV

I has one chiral carbon = two isomers II has two chiral carbons and no symmetry = four isomers.

III and IV have no chiral carbon, no stereoisomers.

**15.** $PCl_5$ produces $POCl_3$ with the following reagents

$$PCl_5 + SO_2 \rightarrow POCl_3 + SOCl_2$$
$$PCl_5 + H_2O \rightarrow POCl_3 + 2HCl$$
$$PCl_5 + H_2SO_4 \longrightarrow SO_2Cl_2 + 2POCl_3 + 2HCl$$
$$6PCl_5 + P_4O_{10} \longrightarrow 10POCl_3$$

**16.** The truncated octahedron is the 14-faced Archimedean solid, with 14 total faces. 6 squares and 8 regular hexagons. The truncated octahedron is formed by removing the six right square pyramids one from each point of a regular octahedron as:

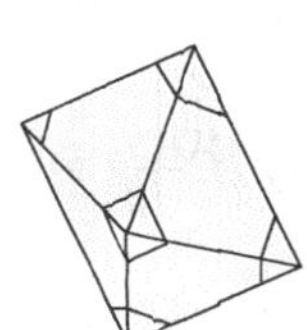

Truncated octahedron

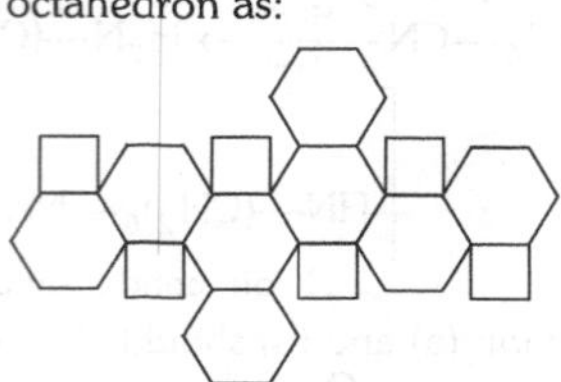

Truncated octahedron unfolded in two dimension

**17.** mmol of complex = $30 \times 0.01 = 0.3$ Also, 1 mole of complex $[Cr(H_2O)_5Cl]Cl_2$ gives only two moles of chloride ion when dissolved in solution

$$[Cr(H_2O)_5Cl]Cl_2 \longrightarrow [Cr(H_2O)_5Cl]^{2+} + 2Cl^-$$

$\Rightarrow$ mmol of $Cl^-$ ion produced from its 0.3 mmol = 0.6

Hence, 0.6 mmol of $Ag^+$ would be required for precipitation.

$\Rightarrow 0.60$ mmol of $Ag^+ = 0.1\text{ M} \times V$ (in mL)

$\Rightarrow \qquad V = 6$ mL.

**18.** These are total $6\alpha$—H to $sp^2$ carbon and they all can participate in hyperconjugation.

Three structures Two structures

**19.** (A) $CO_2(s) \longrightarrow CO_2(g)$

It is just a phase transition (sublimation) as no chemical change has occurred. Sublimation is always endothermic. Product is gas, more disordered, hence $\Delta s$ is positive.

(B) $CaCO_3(s) \longrightarrow CaO(s) + CO_2(g)$

It is a chemical decomposition, not a phase change. Thermal decomposition occur at the expense of energy, hence endothermic. Product contain a gaseous species, hence, $\Delta S > 0$.

(C) $2H \longrightarrow H_2\ (g)$

A new H—H covalent bond is being formed, hence, $\Delta H < 0$.

Also, product is less disordered than reactant, $\Delta S < 0$.

(D) Allotropes are considered as different phase, hence $P_{(white,\ solid)} \rightarrow P_{(red,\ solid)}$ is a phase transition as well as allotropic change.

Also, red phosphorus is more ordered than white phosphorus, $\Delta S < 0$.

**20.** (A) $Ph-\overset{O}{\overset{\|}{C}}-CH_2-CH_2-\overset{O}{\overset{\|}{C}}-CH_3 + OH^-$

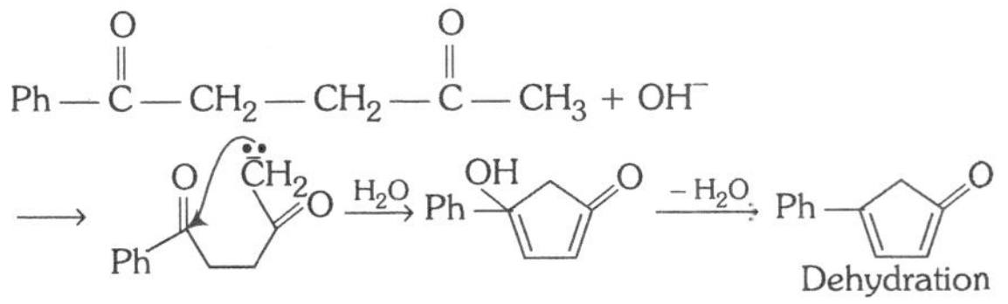

$\longrightarrow$ Carbonion $\xrightarrow{H_2O}$ $\xrightarrow{-H_2O}$ Dehydration

(B) $\xrightarrow{CH_3MgBr}$ $\xrightarrow{S_N2}$

(C) $\xrightarrow{H^+}$ Nucleophilic addition $\xrightarrow{-H^+}$ $\xrightarrow{-H_2O}$ dehydration

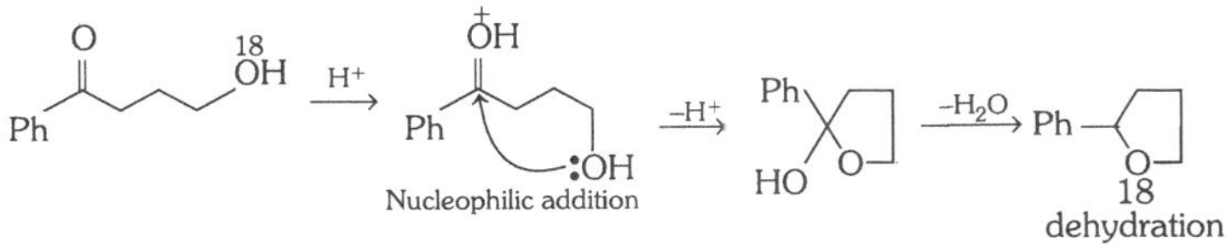

(D) $\xrightarrow[-H_2O]{H^+}$ $\xrightarrow{\text{Friedel-Craft alkylation}}$ Electrophilic subsitution

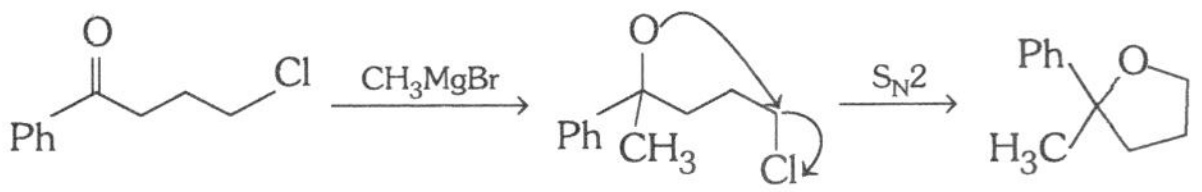

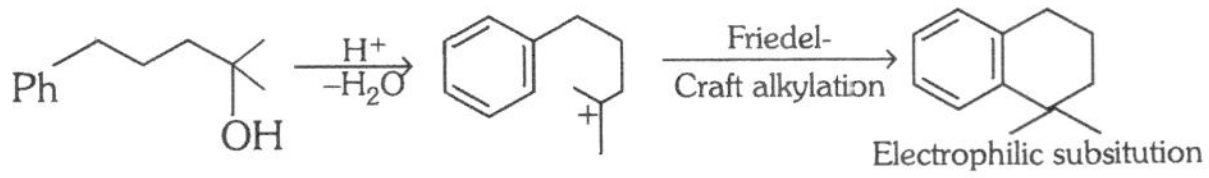

# IIT JEE
## SOLVED PAPER 2011

## Mathematics

## Paper I

### SECTION I

***Directions*** (Q. Nos. 1-7) *This section contains 7 multiple choice questions. Each question has four choices (a), (b), (c) and (d) out of which only one is correct.*

**1.** Let $\mathbf{a} = \hat{\mathbf{i}} + \hat{\mathbf{j}} + \hat{\mathbf{k}}$, $\mathbf{b} = \hat{\mathbf{i}} - \hat{\mathbf{j}} + \hat{\mathbf{k}}$ and $\mathbf{c} = \hat{\mathbf{i}} - \hat{\mathbf{j}} - \hat{\mathbf{k}}$ be three vectors. A vector $\mathbf{v}$ in the plane of $\mathbf{a}$ and $\mathbf{b}$ whose projection of $\mathbf{c}$ is $\frac{1}{\sqrt{3}}$, is given by

(a) $\hat{\mathbf{i}} - 3\hat{\mathbf{j}} + 3\hat{\mathbf{k}}$
(b) $-3\hat{\mathbf{i}} - 3\hat{\mathbf{j}} - \hat{\mathbf{k}}$
(c) $3\hat{\mathbf{i}} - \hat{\mathbf{j}} + 3\hat{\mathbf{k}}$
(d) $\hat{\mathbf{i}} + 3\hat{\mathbf{k}} - 3\hat{\mathbf{k}}$

**2.** Let the straight line $x = b$ divide the area enclosed by $y = (1-x)^2$, $y = 0$ and $x = 0$ into two parts $R_1(0 \le x \le b)$ and $R_2(b \le x \le 1)$ such that $R_1 - R_2 = \frac{1}{4}$. Then, $b$ equals to

(a) $\frac{3}{4}$ (b) $\frac{1}{2}$ (c) $\frac{1}{3}$ (d) $\frac{1}{4}$

**3.** A straight line $L$ through the point $(3, -2)$ is inclined at an angle $60°$ to the line $\sqrt{3}x + y = 1$. If $L$ also intersects the $X$-axis, then the equation of $L$ is

(a) $y + \sqrt{3}x + 2 - 3\sqrt{3} = 0$
(b) $y - \sqrt{3}x + 2 + 3\sqrt{3} = 0$
(c) $\sqrt{3}y - x + 3 + 2\sqrt{3} = 0$
(d) $\sqrt{3}y + x - 3 + 2\sqrt{3} = 0$

**4.** The value of $\int_{\sqrt{\log 2}}^{\sqrt{\log 3}} \frac{x \sin x^2}{\sin x^2 + \sin(\log 6 - x^2)} dx$ is

(a) $\frac{1}{4}\log\frac{3}{2}$ (b) $\frac{1}{2}\log\frac{3}{2}$ (c) $\log\frac{3}{2}$ (d) $\frac{1}{6}\log\frac{3}{2}$

**5.** Let $(x_0, y_0)$ be the solution of the following equations $(2x)^{\log 2} = (3y)^{\log 3}$, $3^{\log x} = 2^{\log y}$, then $x_0$ is equal to

(a) $\frac{1}{6}$ (b) $\frac{1}{2}$ (c) $\frac{1}{2}$ (d) 6

**6.** Let $P = \{\theta : \sin\theta - \cos\theta = \sqrt{2}\cos\theta\}$ and $Q = \{\theta : \sin\theta + \cos\theta = \sqrt{2}\sin\theta\}$ be two sets. Then,

(a) $P \subset Q$ and $Q - P \ne \Phi$
(b) $Q \not\subset P$
(c) $P \not\subset Q$
(d) $P = Q$

**7.** Let $\alpha$ and $\beta$ be the roots of $x^2 - 6x - 2 = 0$ with $\alpha > \beta$. If $a_n = \alpha^n - \beta^n$ for $n \geq 1$, then the value of $\frac{a_{10} - 2a_8}{2a_9}$ is

(a) 1 (b) 2 (c) 3 (d) 4

## SECTION II

***Directions*** (Q. Nos. 8-11) *This section contains 4 multiple choice questions. Each question has four choices (a), (b), (c) and (d) out of which one ore more may be correct.*

**8.** Let $M$ and $N$ be two $3 \times 3$ non-singular skew-symmetric matrices such that $MN = NM$. If $P^T$ denotes the transpose of $P$, then $M^2N^2(M^TN)^{-1}(MN^{-1})^T$ is equal to

(a) $M^2$ (b) $-N^2$ (c) $-M^2$ (d) $MN$

**9.** Let the eccentricity of the hyperbola $\frac{x^2}{a^2} - \frac{y^2}{b^2} = 1$ be reciprocal to that of the ellipse $x^2 + 4y^2 = 4$. If the hyperbola passes through a focus of the ellipse, then

(a) the equation of the hyperbola is $\frac{x^2}{3^2} - \frac{y^2}{2^2} = 1$

(b) a focus of the hyperbola is (2, 0)

(c) the eccentricity of the hyperbola is $\sqrt{\frac{5}{3}}$

(d) the equation of the hyperbola is $x^2 - 3y^2 = 3$

**10.** The vector(s) which is/are coplanar with vectors $\hat{\mathbf{i}} + \hat{\mathbf{j}} + 2\hat{\mathbf{k}}$ and $\hat{\mathbf{i}} + 2\hat{\mathbf{j}} + \hat{\mathbf{k}}$, are perpendicular to the vector $\hat{\mathbf{i}} + \hat{\mathbf{j}} + \hat{\mathbf{k}}$ is/are

(a) $\hat{\mathbf{j}} - \hat{\mathbf{k}}$ (b) $-\hat{\mathbf{i}} + \hat{\mathbf{j}}$ (c) $\hat{\mathbf{i}} - \hat{\mathbf{j}}$ (d) $-\hat{\mathbf{j}} + \hat{\mathbf{k}}$

**11.** Let $f : R \to R$ be a function such that $f(x + y) = f(x) + f(y), \forall x, y \in R$. If $f(x)$ is differentiable at $x = 0$, then

(a) $f(x)$ is differentiable only in a finite interval containing zero

(b) $f(x)$ is continuous, $\forall x \in R$

(c) $f'(x)$ is constant, $\forall x \in R$

(d) $f(x)$ is differentiable except at finitely many points

## SECTION III

***Directions*** (Q. Nos. 12-16) *This section contains 2 paragraphs. Based upon one of the paragraphs 2 multiple choice questions and based on the other paragraph 3 multiple choice questions have the be answered. Each of these questions has four choices (a), (b), (c) and (d) out of which only one is correct.*

***Paragraph*** (Q. Nos. 12-13)

*Let $U_1$ and $U_2$ be two urns such that $U_1$ contains 3 white and 2 red balls and $U_2$ contains only 1 white ball. A fair coin is tossed. If head appears then 1 ball is drawn at random from $U_1$ and put into $U_2$. However, if tail appears then 2 balls are drawn at random from $U_1$ and put into $U_2$. Now, 1 ball is drawn at random from $U_2$.*

**12.** The probability of the drawn ball from $U_2$ being white is

(a) $\frac{13}{30}$ (b) $\frac{23}{30}$ (c) $\frac{19}{30}$ (d) $\frac{11}{30}$

**13.** Given that the drawn ball from $U_2$ is white, the probability that head appeared on the coin is

(a) $\frac{17}{23}$ (b) $\frac{11}{23}$ (c) $\frac{15}{23}$ (d) $\frac{12}{23}$

***Paragraph*** (Q. Nos. 14-16)

*Let a, b and c be three real numbers satisfying* $[a \ b \ c]\begin{bmatrix} 1 & 9 & 7 \\ 8 & 2 & 7 \\ 7 & 3 & 7 \end{bmatrix} = [0 \ 0 \ 0]$ ...(E)

**14.** If the point $P(a, b, c)$, with reference to Eq. (E), lies on the plane $2x + y + z = 1$, then the value of $7a + b + c$ is

(a) 0 (b) 12 (c) 7 (d) 6

**15.** Let $\omega$ be a solution of $x^3 - 1 = 0$ with $\text{Im}(\omega) > 0$. If $a = 2$, with $b$ and $c$ satisfying Eq. (E), then the value of $\frac{3}{\omega^a} + \frac{1}{\omega^b} + \frac{3}{\omega^c}$

(a) $-2$ (b) 2 (c) 3 (d) $-3$

**16.** Let $b = 6$, with $a$ and $c$ satisfying Eq. (E). If $\alpha$ and $\beta$ are the roots of the quadratic equation $ax^2 + bx + c = 0$, then $\sum_{n=0}^{\infty}\left(\frac{1}{\alpha} + \frac{1}{\beta}\right)^n$ is

(a) 6 (b) 7 (c) $\frac{6}{7}$ (d) $\infty$

## SECTION IV

**Directions** (Q. Nos. 17-23) *This section contains 7 questions. The answer to each of the questions is a single digit integer, ranging from 0 to 9.*

**17.** The positive integer value of $n > 3$ satisfying the equation

$$\frac{1}{\sin\left(\frac{\pi}{n}\right)} = \frac{1}{\sin\left(\frac{2\pi}{n}\right)} + \frac{1}{\sin\left(\frac{3\pi}{n}\right)} \text{ is}$$

**18.** Let $a_1, a_2, a_3, \ldots, a_{100}$ be an arithmetic progression with $a_1 = 3$ and $S_p = \sum_{i=1}^{p} a_i$, $1 \le p \le 100$. For any integer $n$ with $1 \le n \le 20$, let $m = 5n$. If $\frac{S_m}{S_n}$ does not depend on $n$, then $a_2$ is

**19.** The minimum value of the sum of real numbers $a^{-5}$, $a^{-4}$, $3a^{-3}$, 1, $a^8$ and $a^{10}$ with $a > 0$ is

**20.** Let $f : [1, \infty) \to [2, \infty)$ be a differentiable function such that $f(1) = 2$. If $6\int_1^x f(t)dt = 3x\, f(x) - x^3$ for all $x \geq 1$, then the value of $f(2)$ is

**21.** Let $f(\theta) = \sin\left[\tan^{-1}\left(\frac{\sin\theta}{\sqrt{\cos 2\theta}}\right)\right]$, where $-\frac{\pi}{4} < \theta < \frac{\pi}{4}$. Then, the value of $\frac{d}{d(\tan\theta)}(f(\theta))$ is

**22.** Consider the parabola $y^2 = 8x$. Let $\Delta_1$ be the area of the triangle formed by the end points of its latusrectum and the point $P\left(\frac{1}{2}, 2\right)$ on the parabola and $\Delta_2$ be the area of the triangle formed by drawing tangents at $P$ and at the end points of the latusrectum. Then, $\frac{\Delta_1}{\Delta_2}$ is

**23.** If $z$ is any complex number satisfying $|z - 3 - 2i| \leq 2$, then the maximum value of $|2z - 6 + 5i|$ is

# Paper II

## SECTION I

**Directions** (Q. Nos. 1-8) *This section contains 8 multiple choice questions. Each question has four choices (a), (b), (c) and (d) out of which only one is correct.*

**1.** Let $P(6, 3)$ be a point on the hyperbola $\frac{x^2}{a^2} - \frac{y^2}{b^2} = 1$. If the normal at the point $P$ intersects the X-axis at (9, 0), then the eccentricity of the hyperbola is

(a) $\sqrt{\frac{5}{2}}$ (b) $\sqrt{\frac{3}{2}}$ (c) $\sqrt{2}$ (d) $\sqrt{3}$

**2.** Let $(x, y)$ be any point on the parabola $y^2 = 4x$. Let $P$ be the point that divides the line segment from (0, 0) to $(x, y)$ in the ratio 1 : 3. Then, the locus of $P$ is

(a) $x^2 = y$ (b) $y^2 = 2x$ (c) $y^2 = x$ (d) $x^2 = 2y$

**3.** Let $f(x) = x^2$ and $g(x) = \sin x$ for all $x \in R$. Then, the set of all $x$ satisfying $(fogogof)(x) = (gogof)(x)$, where $(fog)(x) = f(g(x))$ is

(a) $\pm\sqrt{n\pi}, n \in \{0, 1, 2, \ldots\}$ (b) $\pm\sqrt{n\pi}, n \in \{1, 2, \ldots.\}$

(c) $\frac{\pi}{2} + 2n\pi, n \in \{\ldots, -2, -1, 0, 1, 2, \ldots\}$ (d) $2n\pi, n \in \{\ldots, -2, -1, 0, 1, 2, \ldots\}$

**4.** Let $f : [-1, 2] \to [0, \infty)$ be a continuous function such that $f(x) = f(1 - x)$ for all $x \in [-1, 2]$. Let $R_1 = \int_{-1}^{2} x\, f(x)dx$ and $R_2$ be the area of the region bounded by $y = f(x)$, $x = -1$, $x = 2$ and the X-axis. Then,

(a) $R_1 = 2R_2$ (b) $R_1 = 3R_2$ (c) $2R_1 = R_2$ (d) $3R_1 = R_2$

**5.** If $\lim_{x \to 0} [1 + x \log(1 + b^2)]^{1/x} = 2b\sin^2\theta\, b > 0$ and $\theta \in (-\pi, \pi]$, then the value of $\theta$ is

(a) $\pm\frac{\pi}{4}$ (b) $\pm\frac{\pi}{3}$ (c) $\pm\frac{\pi}{6}$ (d) $\pm\frac{\pi}{2}$

**6.** The circle passing through the point $(-1, 0)$ and touching the $Y$-axis at $(0, 2)$, also passes through the point

(a) $\left(-\frac{3}{2}, 0\right)$ (b) $\left(-\frac{5}{2}, 2\right)$ (c) $\left(-\frac{3}{2}, \frac{5}{2}\right)$ (d) $(-1, -4)$

**7.** Let $\omega \neq 1$ be a cube root of unity and $S$ be the set of all non-singular matrices of the form $\begin{bmatrix} 1 & a & b \\ \omega & 1 & c \\ \omega^2 & \omega & 1 \end{bmatrix}$, where each of $a$, $b$ and $c$ is either $\omega$ or $\omega^2$. Then, the number of distinct matrices in the set $S$ is

(a) 2 (b) 6 (c) 4 (d) 8

**8.** The value of $b$ for which the equations $x^2 + bx - 1 = 0$, $x^2 + x + b = 0$ have one root in common is

(a) $-\sqrt{2}$ (b) $-i\sqrt{3}$ (c) $i\sqrt{5}$ (d) $\sqrt{2}$

## SECTION II

***Directions*** (Q. Nos. 9-12) *This section contains 4 multiple choice questions. Each question has four choices (a), (b), (c) and (d) out of which one or more may be correct.*

**9.** If $f(x) = \begin{cases} -x - \frac{\pi}{2}, & x \leq -\frac{\pi}{2} \\ -\cos x, & -\frac{\pi}{2} < x \leq 0 \\ x - 1, & 0 < x \leq 1 \\ \log x, & x > 1 \end{cases}$, then

(a) $f(x)$ is continuous at $x = -\frac{\pi}{2}$ (b) $f(x)$ is not differentiable at $x = 0$

(c) $f(x)$ is differentiable at $x = 1$ (d) $f(x)$ is differentiable at $x = -\frac{3}{2}$

**10.** Let $L$ be a normal to the parabola $y^2 = 4x$. If $L$ passes through the point $(9, 6)$, then $L$ is given by

(a) $y - x + 3 = 0$ (b) $y + 3x - 33 = 0$ (c) $y + x - 15 = 0$ (d) $y - 2x + 12 = 0$

**11.** Let $E$ and $F$ be two independent events. The probability that exactly one of them occurs is 11/25 and the probability of none of them occurring is 2/25. If $P(T)$ denotes the probability of occurrence of the event $T$, then

(a) $P(E) = \frac{4}{5}, P(F) = \frac{3}{5}$ (b) $P(E) = \frac{1}{5}, P(F) = \frac{2}{5}$

(c) $P(E) = \frac{2}{5}, P(F) = \frac{1}{5}$ (d) $P(E) = \frac{3}{5}, P(F) = \frac{4}{5}$

**12.** Let $f : (0, 1) \rightarrow R$ be defined by $f(x) = \dfrac{b - x}{1 - bx}$, where $b$ is a constant such that $0 < b < 1$. Then,

(a) $f$ if not invertible on (0, 1)

(b) $f \neq f^{-1}$ on (0, 1) and $f'(b) = \dfrac{1}{f'(0)}$

(c) $f = f^{-1}$ on (0, 1) and $f'(b) = \dfrac{1}{f'(0)}$

(d) $f^{-1}$ is differentiable on (0, 1)

## SECTION III

***Directions*** (Q. Nos. 13-18) *This section contains 6 questions. The answer to each of the questions is a single digit integer, ranging from 0 to 9.*

**13.** Let $y'(x) + y(x)g'(x) = g(x)g'(x)$ $y(0) = 0$, $x \in R$, where $f'(x)$ denotes $\dfrac{d\, f(x)}{dx}$ and $g(x)$ is a given non-constant differentiable function on $R$ with $g(0) = g(2) = 0$. Then, the value of $y(2)$ is...

**14.** Let $\mathbf{a} = -\hat{\mathbf{i}} - \hat{\mathbf{k}}$, $\mathbf{b} = -\hat{\mathbf{i}} + \hat{\mathbf{j}}$ and $\mathbf{c} = \hat{\mathbf{i}} + 2\hat{\mathbf{j}} + 3\hat{\mathbf{k}}$ be three given vectors. If $\mathbf{r}$ is a vector such that $\mathbf{r} \times \mathbf{b} = \mathbf{c} \times \mathbf{b}$ and $\mathbf{r} \cdot \mathbf{a} = 0$, then the value of $\mathbf{r} \cdot \mathbf{b}$ is

**15.** Let $\omega = e^{i\pi/3}$ and $a, b, c, x, y, z$ be non-zero complex numbers such that $a + b + c = x$, $a + b\omega + c\omega^2 = y$, $a + b\omega^2 + c\omega = z$.

Then, the value of $\dfrac{|x|^2 + |y|^2 + |z|^2}{|a|^2 + |b|^2 + |c|^2}$ is

**16.** Let $M$ be a $3 \times 3$ matrix satisfying

$$M\begin{bmatrix}0\\1\\0\end{bmatrix} = \begin{bmatrix}-1\\2\\3\end{bmatrix}, M\begin{bmatrix}1\\-1\\0\end{bmatrix} = \begin{bmatrix}1\\1\\-1\end{bmatrix} \text{ and } M\begin{bmatrix}1\\1\\1\end{bmatrix} = \begin{bmatrix}0\\0\\12\end{bmatrix}$$

Then, the sum of the diagonal entries of $M$ is

**17.** The number of distinct real roots of $x^4 - 4x^3 + 12x^2 + x - 1 = 0$ is

**18.** The straight line $2x - 3y = 1$ divides the circular region $x^2 + y^2 \leq 6$ into two parts. If $S = \left\{\left(2, \dfrac{3}{4}\right), \left(\dfrac{5}{2}, \dfrac{3}{4}\right), \left(\dfrac{1}{4}, -\dfrac{1}{4}\right), \left(\dfrac{1}{8}, \dfrac{1}{4}\right)\right\}$, then the number of point(s) in $S$ lying inside the smaller part is

## SECTION IV

***Directions*** (Q. Nos. 19-20) *This section contains 2 questions. Each question has four statements (A, B, C and D) given in Column I and five statements (p, q, r, s and t) in Column II. Any given statement in Column I can have correct matching with one or more statement(s) given in Column II.*

**19.** Match the statements given in Column I with the values given in Column II.

| Column I | Column II |
|---|---|
| (A) If $\mathbf{a} = \hat{\mathbf{j}} + \sqrt{3}\hat{\mathbf{k}}$, $\mathbf{b} = -\hat{\mathbf{j}} + \sqrt{3}\hat{\mathbf{k}}$ and $\mathbf{c} = 2\sqrt{3}\hat{\mathbf{k}}$ form a triangle, then the internal angle of the triangle between **a** and **b** is | (p) $\frac{\pi}{6}$ |
| (B) If $\int_a^b [f(x) - 3x]dx = a^2 - b^2$, then the value of $f\left(\frac{\pi}{6}\right)$ is | (q) $\frac{2\pi}{3}$ |
| (C) The value of $\frac{\pi^2}{\log 3} \int_{7/6}^{5/6} \sec(\pi x)dx$ is | (r) $\pi/3$ |
| (D) The maximum value of $\left|\arg\left(\frac{1}{1-z}\right)\right|$ for $\|z\| = 1, z \neq 1$ is given by | (s) $\pi$ |
| | (t) $\pi/2$ |

**20.** Match the statements given in Column I with the intervals/union of intervals given in Column II.

| Column I | Column II |
|---|---|
| (A) The set $\left\{\text{Re}\left(\frac{2iz}{1-z^2}\right) : z \text{ is a complex number}, \|z\| = 1, z \neq \pm 1\right\}$ | (p) $(-\infty, -1) \cup (1, \infty)$ |
| (B) The domain of the function $f(x) = \sin^{-1}\left[\frac{8(3)^{x-2}}{1-3^{2(x-1)}}\right]$ is | (q) $(-\infty, 0) \cup (0, \infty)$ |
| (C) If $f(\theta) = \begin{vmatrix} 1 & \tan\theta & 1 \\ -\tan\theta & 1 & \tan\theta \\ -1 & -\tan\theta & 1 \end{vmatrix}$, then the set $\{f(\theta) : 0 \leq \theta < \frac{\pi}{2}\}$ is | (r) $[2, \infty)$ |
| (D) If $f(x) = x^{3/2}(3x - 10)$, $x \geq 0$, then $f(x)$ is increasing in | (s) $(-\infty, -1] \cup [1, \infty)$ |
| | (t) $(-\infty, 0] \cup [2, \infty)$ |

## ANSWERS

### Paper I

| | | | | | |
|---|---|---|---|---|---|
| **1.** (c) | **2.** (b) | **3.** (b) | **4.** (a) | **5.** (c) | **6.** (d) |
| **7.** (c) | **8.** (c) | **9.** (b, d) | **10.** (a, d) | **11.** (b, c) | **12.** (b) |
| **13.** (d) | **14.** (d) | **15.** (a) | **16.** (b) | **17.** (7) | **18.** (3 or 9) |
| **19.** (8) | **20.** (8/3) | **21.** (1) | **22.** (2) | **23.** (5) | |

### Paper II

| | | | | | |
|---|---|---|---|---|---|
| **1.** (b) | **2.** (c) | **3.** (b) | **4.** (c) | **5.** (d) | **6.** (d) |
| **7.** (a) | **8.** (b) | **9.** (a, b, c, d) | **10.** (a, b, d) | **11.** (a, d) | **12.** (a) |
| **13.** (0) | **14.** (9) | **15.** (3) | **16.** (9) | **17.** (2) | **18.** (2) |

**19.** (A) → q; (B) → p; (C) → s; (D) → s

**20.** (A) → s; (B) → t; (C) → r; (D) → r

# Hints & Solutions

## Paper I

**1.** Let $\mathbf{v} = \mathbf{a} + \lambda\,\mathbf{b}$

$$\mathbf{v} = (1+\lambda)\,\hat{\mathbf{i}} + (1-\lambda)\,\hat{\mathbf{j}} + (1+\lambda)\hat{\mathbf{k}}$$

Projection of $\mathbf{v}$ on $\mathbf{c} = \dfrac{1}{\sqrt{3}}$

$$\Rightarrow \quad \frac{\mathbf{v}\cdot\mathbf{c}}{|\mathbf{c}|} = \frac{1}{\sqrt{3}}$$

$$\Rightarrow \quad \frac{(1+\lambda)-(1-\lambda)-(1+\lambda)}{\sqrt{3}} = \frac{1}{\sqrt{3}}$$

$$\Rightarrow \quad 1+\lambda-1+\lambda-1-\lambda = 1$$

$$\Rightarrow \quad \lambda - 1 = 1 \Rightarrow \lambda = 2$$

$$\therefore \quad \mathbf{v} = 3\hat{\mathbf{i}} - \hat{\mathbf{j}} + 3\hat{\mathbf{k}}$$

**2.** Here, area between 0 to $b$ is $R_1$ and $b$ to 1 is $R_2$.

$$\therefore \int_0^b (1-x)^2 dx - \int_b^1 (1-x)^2 dx = \frac{1}{4}$$

$$\Rightarrow \left(\frac{(1-x)^3}{-3}\right)_0^b - \left(\frac{(1-x)^3}{-3}\right)_b^1 = \frac{1}{4}$$

$$\Rightarrow -\frac{1}{3}\{(1-b)^3 - 1\} + \frac{1}{3}\{0-(1-b)^3\} = \frac{1}{4}$$

$$\Rightarrow \quad -\frac{2}{3}(1-b)^3 = -\frac{1}{3} + \frac{1}{4} = -\frac{1}{12}$$

$$\Rightarrow \quad (1-b)^3 = \frac{1}{8}$$

$$\Rightarrow \quad (1-b) = \frac{1}{2} \Rightarrow b = \frac{1}{2}$$

**3.** A straight line passing through $P$ and making an angle of $\alpha = 60°$, is given by

$$\frac{y-y_1}{x-x_1} = \tan(\theta \pm \alpha)$$

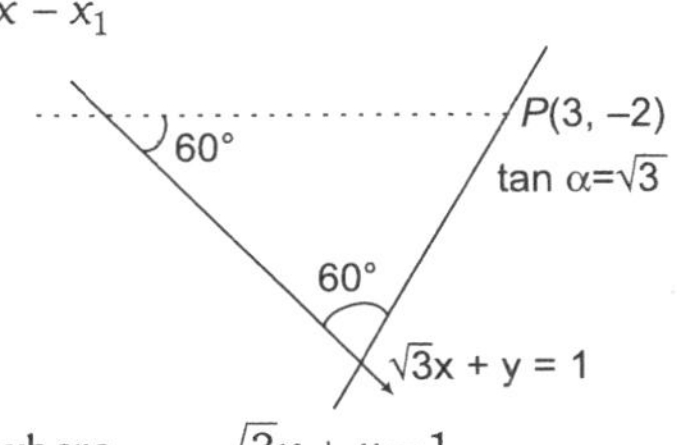


where, $\quad \sqrt{3}x + y = 1$

$$\Rightarrow \quad y = -\sqrt{3}x + 1$$

Then, $\quad \tan\theta = -\sqrt{3}$

$$\Rightarrow \quad \frac{y+2}{x-3} = \frac{\tan\theta \pm \tan\alpha}{1 \mp \tan\theta\tan\alpha}$$

$$\Rightarrow \quad \frac{y+2}{x-3} = \frac{-\sqrt{3}+\sqrt{3}}{1-(-\sqrt{3})(\sqrt{3})}$$

and $\quad \dfrac{y+2}{x-3} = \dfrac{-\sqrt{3}-\sqrt{3}}{1+(-\sqrt{3})(\sqrt{3})}$

$$\Rightarrow \quad y + 2 = 0$$

and $\quad \dfrac{y+2}{x-3} = \dfrac{-2\sqrt{3}}{1-3} = \sqrt{3}$

$$\Rightarrow \quad y + 2 = \sqrt{3}x - 3\sqrt{3}$$

Neglecting, $y + 2 = 0$ as it does not intersect $Y$-axis.

**4.** Put $x^2 = t \Rightarrow x\,dx = \dfrac{dt}{2}$

$$\therefore \quad I = \int_{\log 2}^{\log 3} \frac{\sin t \cdot \frac{dt}{2}}{\sin t + \sin(\log 6 - t)} \qquad \ldots(\text{i})$$

Using, $\quad \displaystyle\int_a^b f(x)dx = \int_a^b f(a+b-x)dx$

$$I = \frac{1}{2}\int_{\log 2}^{\log 3} \frac{\sin(\log 2 + \log 3 - t)}{\sin(\log 2 + \log 3 - t) + \sin(\log 6 - (\log 2 + \log 3 - t))} dt$$

$$I = \frac{1}{2}\int_{\log 2}^{\log 3} \frac{\sin(\log 6 - t)}{\sin(\log 6 - t) + \sin(t)} dt$$

$$\therefore I = \int_{\log 2}^{\log 3} \frac{\sin(\log 6 - t)}{\sin(\log 6 - t) + \sin t} dt \qquad \ldots(\text{ii})$$

On adding Eqs. (i) and (ii), we get

$$2I = \frac{1}{2}\int_{\log 2}^{\log 3} \frac{\sin t + \sin(\log 6 - t)}{\sin(\log 6 - t) + \sin t} dt$$

$$\therefore \quad 2I = \frac{1}{2}(t)_{\log 2}^{\log 3} = \frac{1}{2}(\log 3 - \log 2)$$

$$\Rightarrow I = \frac{1}{4}\log\left(\frac{3}{2}\right)$$

**5.** Taking log on both sides,

$$\log 2 \cdot \log(2x) = \log 3\,(\log 3y)$$

$\Rightarrow \quad \log 2\{\log 2+\log x\}$
$= \log 3\{\log 3+\log y\} \quad \ldots(i)$

and $\log x \cdot \log 3 = \log y \log 2$

$$\log y = \frac{\log x \cdot \log 3}{\log 2} \quad \ldots(ii)$$

From Eqs. (i) and (ii), we get

$$\log 2\{\log 2+\log x\} = \log 3 \cdot \left\{\log 3 + \frac{\log x \cdot \log 3}{\log 2}\right\}$$

$$\Rightarrow (\log 2)^2 + \log 2 \cdot \log x = (\log 3)^2 + \frac{(\log 3)^2}{(\log 2)} \cdot \log x$$

$$\Rightarrow \log x \left\{\frac{(\log 3)^2}{\log 2} - \log 2\right\} = (\log 2)^2 - (\log 3)^2$$

$$\Rightarrow \log x \left\{\frac{(\log 3)^2 - (\log 2)^2}{\log 2}\right\} = (\log 2)^2 - (\log 3)^2$$

$$\Rightarrow \log x = -\log 2 = \log 2^{-1}$$

$$\therefore \quad x = \frac{1}{2}$$

**6.** $P = \{\theta : \sin\theta - \cos\theta = \sqrt{2}\cos\theta\}$

$\Rightarrow \quad \cos\theta(\sqrt{2}+1) = \sin\theta$

$\Rightarrow \quad \tan\theta = \sqrt{2}+1 \quad \ldots(i)$

$Q = \{\theta : \sin\theta + \cos\theta = \sqrt{2}\sin\theta\}$

$\Rightarrow \quad \sin\theta(\sqrt{2}-1) = \cos\theta$

$$\Rightarrow \tan\theta = \frac{1}{\sqrt{2}-1} \times \frac{\sqrt{2}+1}{\sqrt{2}+1} = (\sqrt{2}+1) \quad \ldots(ii)$$

$\therefore \quad P = Q$

**7.** $$\frac{a_{10}-2a_8}{2a_9} = \frac{(\alpha^{10}-\beta^{10}) - 2(\alpha^8-\beta^8)}{2(\alpha^9-\beta^9)}$$

$$= \frac{\alpha^8(\alpha^2-2) - \beta^8(\beta^2-2)}{2(\alpha^9-\beta^9)}$$

[$\because \alpha$ is root of $x^2 - 6x - 2 = 0$
$\alpha^2 - 2 = 6\alpha$]

Also, $\beta$ is root of $x^2 - 6x - 2 = 0$

$\Rightarrow \quad \beta^2 - 2 = 6\beta$

$$= \frac{\alpha^8(6\alpha) - \beta^8(6\beta)}{2(\alpha^9-\beta^9)} = \frac{6(\alpha^9-\beta^9)}{2(\alpha^9-\beta^9)} = 3$$

**8.** Given, $M^T = -M, N^T = -N$

and $\quad MN = NM \quad \ldots(i)$

$\therefore \quad M^2N^2(M^TN)^{-1}(MN^{-1})^T$

$= M^2N^2N^{-1}(M^T)^{-1}(N^{-1})^T \cdot M^T$

$= M^2N(NN^{-1})(-M)^{-1}(N^T)^{-1}(-M)$

$= M^2NI(-M^{-1})(-N)^{-1}(-M)$

$= -M^2NM^{-1}N^{-1}M$

$= -M \cdot (MN)M^{-1}N^{-1}M$

$= -M(NM)M^{-1}N^{-1}M$

$= -MN(NM^{-1})N^{-1}M$

$= -M(NN^{-1})M = -M^2$

**Note** Here, non-singular word should not be used, since there is no non-singular $3 \times 3$ skew-symmetric matrix.

**9.** Here, equation of ellipse

$$\frac{x^2}{4} + \frac{y^2}{1} = 1$$

$$\Rightarrow e^2 = 1 - \frac{b^2}{a^2} = 1 - \frac{1}{4} = \frac{3}{4}$$

$$\therefore e = \frac{\sqrt{3}}{2} \text{ and focus } (\pm ae, 0)$$

$= (\pm\sqrt{3}, 0)$

For hyperbola $\dfrac{x^2}{a^2} - \dfrac{y^2}{b^2} = 1$,

$$e_1^2 = 1 + \frac{b^2}{a^2} \text{ where, } e_1^2 = \frac{1}{e^2} = \frac{4}{3}$$

$$\Rightarrow 1 + \frac{b^2}{a^2} = \frac{4}{3} \Rightarrow \frac{b^2}{a^2} = \frac{1}{3} \quad \ldots(i)$$

and hyperbola passes through $(\pm\sqrt{3}, 0)$.

Now, $\quad \dfrac{3}{a^2} = 1 \Rightarrow a^2 = 3 \quad \ldots(ii)$

From Eqs. (i) and (ii), we get

$b^2 = 1 \quad \ldots(iii)$

$\therefore$ Equation of hyperbola is

$$\frac{x^2}{3} - \frac{y^2}{1} = 1$$

Focus is $(\pm ae, 0)$.

Now, $\quad \left(\pm\sqrt{3} \cdot \dfrac{2}{\sqrt{3}}, 0\right) \Rightarrow (\pm 2, 0)$

Hence, both options (b) and (d) are correct.

**10.** Let $\mathbf{a} = \hat{\mathbf{i}} + \hat{\mathbf{j}} + 2\hat{\mathbf{k}}, \mathbf{b} = \hat{\mathbf{i}} + 2\hat{\mathbf{j}} + \hat{\mathbf{k}}$

and $\quad \mathbf{c} = \hat{\mathbf{i}} + \hat{\mathbf{j}} + \hat{\mathbf{k}}$

$\therefore$ A vector coplanar to **a** and **b**, and perpendicular to **c**.

Now, $\lambda\,(\mathbf{a}\times\mathbf{b})\times\mathbf{c}$

$\Rightarrow \lambda\,\{(\mathbf{a}\cdot\mathbf{c})\,\mathbf{b}-(\mathbf{b}\cdot\mathbf{c})\,\mathbf{a}\}$

$\Rightarrow \lambda\,\{(1+1+4)\,(\hat{\mathbf{i}}+2\hat{\mathbf{j}}+\hat{\mathbf{k}}) - (1+2+1)\,(\hat{\mathbf{i}}+\hat{\mathbf{j}}+2\hat{\mathbf{k}})\}$

$\Rightarrow \lambda\,\{6\hat{\mathbf{i}}+12\hat{\mathbf{j}}+6\hat{\mathbf{k}}-6\hat{\mathbf{i}}-6\hat{\mathbf{j}}-12\hat{\mathbf{k}}\}$

$\Rightarrow \lambda\,\{6\hat{\mathbf{j}}-6\hat{\mathbf{k}}\} \Rightarrow 6\lambda\,(\hat{\mathbf{j}}-\hat{\mathbf{k}})$

For $\lambda=\frac{1}{6} \Rightarrow$ Option (a) is correct.

For $\lambda=-\frac{1}{6} \Rightarrow$ Option (d) is correct.

**11.** $f(x+y)=f(x)+f(y)$, as $f(x)$ is differentiable at $x=0$.

$\Rightarrow \quad f'(0)=k \quad \ldots\text{(i)}$

Now, $f'(x)=\lim_{h\to 0}\frac{f(x+h)-f(x)}{h}$

$=\lim_{h\to 0}\frac{f(x)+f(h)-f(x)}{h}$

$=\lim_{h\to 0}\frac{f(h)}{h} \quad \left[\frac{0}{0}\text{ from}\right]$

Given, $f(x+y)=f(x)+f(y),\ \forall\, x, y$

$\therefore \quad f(0)=f(0)+f(0),$

when $\quad x=y=0 \Rightarrow f(0)=0$

Using L'Hospital's rule,

$\lim_{h\to 0}\frac{f'(h)}{1}=f'(0)=k \quad \ldots\text{(ii)}$

$\Rightarrow f'(x)=k$, on integrating both sides,

$f(x)=kx+C$, as $f(0)=0 \Rightarrow C=0$

So, $f(x)=kx$

$\therefore$ $f(x)$ is continuous for all $x\in R$ and $f'(x)=k$, *i.e.* constant for all $x\in R$.

Hence, both (b) and (c) are correct.

**Solutions** (Q. Nos. 12-13)

$u_1$ (3 W, 2 R), $u_2$ (1 W) — Initial

**Head appears.**

$u_1$ (2 W, 2 R) —1W→ $u_2$ (2 W)

or $u_1$ (3 W, 1 R) —1R→ $u_2$ (1 W, 1 R)

2 cases

**Tail appears.**

$u_1$ (1 W, 2 R) —2W→ $u_2$ (3 W)

$u_1$ (3 W, 0 R) → $u_2$ (1 W, 2 R)

$u_1$ (2 W, 1 R) → $u_2$ (2 W, 1 R)

3 cases

**12.** Now, probability of the drawn ball from $V_2$ being white is

$\Rightarrow P\,(\text{white}/V_2)$

$$=P(H)\cdot\left\{\frac{{}^3C_1}{{}^5C_1}\times\frac{{}^2C_1}{{}^2C_1}+\frac{{}^2C_1}{{}^5C_1}\times\frac{{}^1C_1}{{}^2C_1}\right\}+P(T)\left\{\frac{{}^3C_2}{{}^5C_2}\times\frac{{}^3C_2}{{}^3C_2}+\frac{{}^2C_2}{{}^5C_2}\times\frac{{}^1C_1}{{}^3C_2}+\frac{{}^3C_1\cdot{}^2C_1}{{}^5C_2}\times\frac{{}^2C_1}{{}^3C_2}\right\}$$

Now, $P\,(\text{white}/V_2)=\frac{1}{2}\left\{\frac{3}{5}\times 1+\frac{2}{5}\times\frac{1}{2}\right\}+\frac{1}{2}\left\{\frac{3}{10}\times 1+\frac{1}{10}\times\frac{1}{3}+\frac{6}{10}\times\frac{2}{3}\right\}=\frac{23}{30}$

**13.** $P$ (head appeared/white from $V_2$)

$$=P(H)\cdot\frac{\left\{\frac{{}^3C_1}{{}^5C_1}\times\frac{{}^2C_1}{{}^2C_1}+\frac{{}^2C_1}{{}^5C_1}\times\frac{{}^1C_1}{{}^2C_1}\right\}}{\frac{23}{30}}$$

$$=\frac{1}{2}\,\frac{\left\{\frac{3}{5}\times 1+\frac{2}{5}\times\frac{1}{2}\right\}}{\frac{23}{30}}=\frac{12}{23}$$

**Solutions** (Q. Nos. 14-16)

Given, $[a\ b\ c]_{1\times 3}\begin{bmatrix}1 & 9 & 7\\ 8 & 2 & 7\\ 7 & 3 & 7\end{bmatrix}_{3\times 3}=[0\,0\,0]$

$\Rightarrow \begin{bmatrix}a+8b+7c\\ 9a+2b+3c\\ 7a+7b+7c\end{bmatrix}=\begin{bmatrix}0\\0\\0\end{bmatrix}$

$\Rightarrow \quad a+8b+7c=0 \quad \ldots\text{(i)}$

$\Rightarrow \quad 9a+2b+3c=0 \quad \ldots\text{(ii)}$

$\Rightarrow \quad a+b+c=0 \quad \ldots\text{(iii)}$

On multiplying Eq. (iii) by 2, then subtract from Eq. (ii), we get

$$7a + c = 0 \quad \ldots(\text{iv})$$

Again multiplying Eq. (iii) by 3, then subtract from Eq. (ii), we get

$$6a - b = 0 \quad \ldots(\text{v})$$

$\therefore \quad b = 6a$ and $c = -7a$

**14.** As $(a, b, c)$ lies on $2x + y + z = 1$

$\Rightarrow \quad 2a + b + c = 1$

$\Rightarrow \quad 2a + 6a - 7a = 1$

$\Rightarrow \quad a = 1, b = 6$ and $c = -7$

$\therefore \quad 7a + b + c = 7 + 6 - 7 = 6$

**15.** If $a = 2$, $b = 12$ and $c = -14$

$\therefore \quad \dfrac{3}{\omega^a} + \dfrac{1}{\omega^b} + \dfrac{3}{\omega^c}$

$\Rightarrow \quad \dfrac{3}{\omega^2} + \dfrac{1}{\omega^{12}} + \dfrac{3}{\omega^{-14}} = \dfrac{3}{\omega^2} + 1 + 3\omega^2$

$= 3\omega + 1 + 3\omega^2$

$= 1 + 3(\omega + \omega^2)$

$= 1 - 3 = -2$

**16.** If $b = 6$, $a = 1$ and $c = -7$

$\therefore \quad ax^2 + bx + c = 0$

$\Rightarrow \quad x^2 + 6x - 7 = 0$

$\Rightarrow \quad (x + 7)(x - 1) = 0$

$\therefore \quad x = 1, -7$

$\Rightarrow \quad \sum_{n=0}^{\infty} \left(\dfrac{1}{1} - \dfrac{1}{7}\right)^n \Rightarrow \sum_{n=0}^{\infty} \left(\dfrac{6}{7}\right)^n$

$\Rightarrow \quad 1 + \dfrac{6}{7} + \left(\dfrac{6}{7}\right)^n + \ldots \infty$

$= \dfrac{1}{1 - \dfrac{6}{7}} = \dfrac{1}{1/7} = 7$

**17.** Given, $n > 3 \in$ Integer

and $\dfrac{1}{\sin\left(\dfrac{\pi}{n}\right)} = \dfrac{1}{\sin\left(\dfrac{2\pi}{n}\right)} + \dfrac{1}{\sin\left(\dfrac{3\pi}{n}\right)}$

$\Rightarrow \quad \dfrac{1}{\sin\dfrac{\pi}{n}} - \dfrac{1}{\sin\dfrac{3\pi}{n}} = \dfrac{1}{\sin\dfrac{2\pi}{n}}$

$\Rightarrow \quad \dfrac{\sin\dfrac{3\pi}{n} - \sin\dfrac{\pi}{n}}{\sin\dfrac{\pi}{n} \cdot \sin\dfrac{3\pi}{n}} = \dfrac{1}{\sin\dfrac{2\pi}{n}}$

$\Rightarrow \quad 2\cos\left(\dfrac{2\pi}{n}\right) \cdot \sin\dfrac{\pi}{n} = \dfrac{\sin\dfrac{\pi}{n} \cdot \sin\dfrac{3\pi}{n}}{\sin\dfrac{2\pi}{n}}$

$\Rightarrow \quad 2\sin\dfrac{2\pi}{n} \cdot \cos\dfrac{2\pi}{n} = \sin\dfrac{3\pi}{n}$

$\Rightarrow \quad \sin\dfrac{4\pi}{n} = \sin\dfrac{3\pi}{n}$

$\Rightarrow \quad \dfrac{4\pi}{n} = \pi - \dfrac{3\pi}{n} \Rightarrow \dfrac{7\pi}{n} = \pi \Rightarrow n = 7$

**18.** Given, $a_1 = 3, m = 5n$

and $\quad a_1, a_2, \ldots$ are in AP.

$\therefore \quad \dfrac{S_m}{S_n} = \dfrac{S_{5n}}{S_n}$ is independent of $n$.

Now, $\quad \dfrac{\dfrac{5n}{2}[2 \times 3 + (5n - 1)d]}{\dfrac{n}{2}[2 \times 3 + (n - 1)d]}$

$\Rightarrow \quad \dfrac{f\{(6 - d) + 5n\}}{(6 - d) + n}$

independent of $n$, if

$6 - d = 0 \Rightarrow d = 6$

$\therefore \quad a_2 = a_1 + d = 3 + 6 = 9$

*Or*

If $\quad d = 0$

$\dfrac{S_m}{S_n}$ is independent of $n$.

$\therefore \quad a_2 = 3$

**19.** Using AM $\geq$ GM,

$\dfrac{a^{-5} + a^{-4} + a^{-3} + a^{-3} + a^{-3} + 1 + a^8 + a^{10}}{8}$

$\geq (a^{-5} \cdot a^{-4} \cdot a^{-3} \cdot a^{-3} \cdot a^{-3} \cdot 1 \cdot a^8 \cdot a^{10})^{\frac{1}{8}}$

$\Rightarrow a^{-5} + a^{-4} + 3a^{-3} + 1 + a^8 + a^{10} \geq 8 \cdot 1$

Hence, minimum value is 8.

**20.** Given, $f(1) = \dfrac{1}{3}$ and $6\int_1^x f(t)\, dt$

$= 3x\, f(x) - x^3$, for all $x \geq 1$

Using (Newton-Leibnitz formula),

On Differentiating both sides,

$6f(x) \cdot 1 - 0 - 3f(x) + 3x\, f'(x) - 3x^2$

$\Rightarrow \quad 3x\, f'(x) - 3f(x) = 3x^2$

$\Rightarrow \quad f'(x) - \dfrac{1}{x} f(x) = x$

$$\Rightarrow \frac{x f'(x) - f(x)}{x^2} = 1 \Rightarrow \frac{d}{dx}\left\{\frac{f(x)}{x}\right\} = 1$$

On integrating both sides,

$$\frac{f(x)}{x} = x + C \qquad [\because f(1) = \frac{1}{3}]$$

$$\Rightarrow \frac{1}{3} = 1 + C \Rightarrow C = -\frac{2}{3}$$

Now, $f(x) = x^2 - \frac{2}{3}x$

$$\Rightarrow f(2) = 4 - \frac{4}{3} = \frac{8}{3}$$

**Note** Here, $f(1) = 2$ does not satisfy given function.

$$\therefore f(1) = \frac{1}{3}$$

For that, $f(x) = x^2 - \frac{2}{3}x$

and $f(2) = 4 - \frac{4}{3} = \frac{8}{3}$

**21.** $f(\theta) = \sin\left(\tan^{-1}\frac{\sin\theta}{\sqrt{\cos 2\theta}}\right), -\frac{\pi}{4} < \theta < \frac{\pi}{4}$

Let $\tan^{-1}\frac{\sin\theta}{\sqrt{\cos 2\theta}} = \phi$

$$\Rightarrow \tan\phi = \frac{\sin\theta}{\sqrt{\cos 2\theta}}$$

$\sqrt{\sin^2\theta + \cos 2\theta}$

$\sin\theta$

$\sqrt{\cos 2\theta}$

$$\therefore \sin\phi = \frac{\sin\theta}{\sqrt{\sin^2\theta + \cos 2\theta}}$$

$$= \frac{\sin\theta}{\sqrt{1 - \sin^2\theta}} = \frac{\sin\theta}{\cos\theta} = \tan\theta$$

$$\therefore f(\theta) = \sin\phi = \tan\theta$$

$$\Rightarrow \frac{d\, f(\theta)}{d(\tan\theta)} = 1$$

**22.** As, we know area of triangle formed by three points on parabola is twice the area of triangle formed by corresponding tangents, i.e. area of $\Delta PQR = 2$ area of $\Delta T_1T_2T_3$.

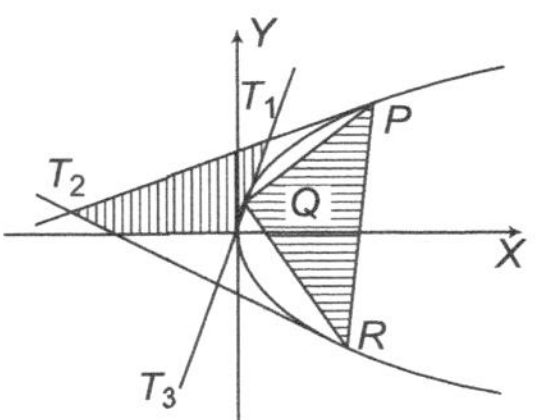


$$\therefore \Delta_1 = 2\Delta_2 \text{ or } \frac{\Delta_1}{\Delta_2} = 2$$

**23.** Given, $|z - 3 - 2i| \le 2$ ...(i)

To find minimum of $|2z - 6 + 5i|$

or $2\left|z - 3 + \frac{5}{2}i\right|$

Using triangle inequality,

i.e. $||z_1| - |z_2|| \le |z_1 + z_2|$

$$\therefore \left|z - 3 + \frac{5}{2}i\right|$$

$$= \left|z - 3 - 2i + 2i + \frac{5}{2}i\right|$$

$$= \left|(z - 3 - 2i) + \frac{9}{2}i\right| \ge \left||z - 3 - 2i| - \frac{9}{2}\right|$$

$$\ge \left|2 - \frac{9}{2}\right| \ge \frac{5}{2} \Rightarrow \left|z - 3 + \frac{5}{2}i\right| \ge \frac{5}{2}$$

or $|2z - 6 + 5i| \ge 5$

# Paper II

**1.** Equation of normal to hyperbola at $(x_1, y_1)$ is

$$\frac{a^2x}{x_1} + \frac{b^2y}{y_1} = a^2 + b^2$$

$$\therefore \text{At } (6, 3), \quad \frac{a^2x}{6} + \frac{b^2y}{3} = a^2 + b^2$$

It passes throught (9, 0).

Now, $\frac{a^2 \cdot 9}{6} = a^2 + b^2$

$$\Rightarrow \frac{3a^2}{2} - a^2 = b^2 \Rightarrow \frac{a^2}{b^2} = 2$$

$$\therefore e^2 = 1 + \frac{b^2}{a^2} = 1 + \frac{1}{2} \Rightarrow e = \sqrt{\frac{3}{2}}$$

**2.** By section formula,

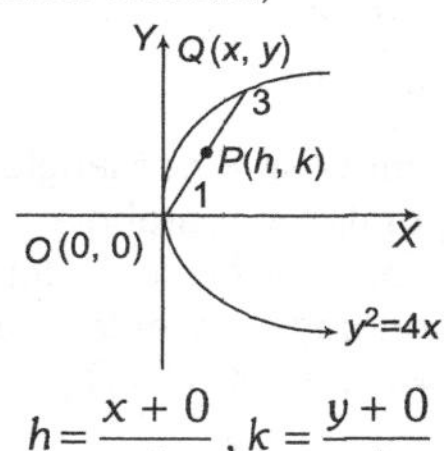


$$h=\frac{x+0}{4}, k=\frac{y+0}{4}$$

$\therefore \quad x = 4h$, and $y = 4k$

Substituting in $y^2 = 4x$,

$$(4k)^2 = 4(4h) \Rightarrow k^2 = h$$

or $y^2 = x$ is required locus.

**3.** $f(x) = x^2, g(x) = \sin x$

$$(gof)(x) = \sin x^2$$
$$go(gof)(x) = \sin(\sin x^2)$$
$$(fogogof)(x) = (\sin(\sin x^2))^2 \quad \ldots(i)$$

Again, $(gof)(x) = \sin x^2$

$$(gogof)(x) = \sin(\sin x^2) \quad \ldots(ii)$$

Given, $(fogogof)(x) = (gogof)(x)$

$\Rightarrow (\sin(\sin x^2))^2 = \sin(\sin x^2)$

$\Rightarrow \sin(\sin x^2)\{\sin(\sin x^2) - 1\} = 0$

$\Rightarrow \sin(\sin x^2) = 0$ or $\sin(\sin x^2) = 1$

$\Rightarrow \sin x^2 = 0$ or $\sin x^2 = \frac{\pi}{2}$

$\therefore \quad x^2 = n\pi$

(i.e. not possible as $-1 \le \sin\theta \le 1$)

$$x = \pm\sqrt{n\pi}$$

**4.** Here, $R_1 = \int_{-1}^{2} x\, f(x)\, dx \quad \ldots(i)$

Using, $\int_a^b f(x)\, dx = \int_a^b f(a+b-x)\, dx$

$$R_1 = \int_{-1}^{2}(1-x)\, f(1-x)dx,$$

$[\because f(x) = f(1-x)]$

$$\therefore \quad R_1 = \int_{-1}^{2}(1-x)\, f(x)\, dx \quad \ldots(ii)$$

Given, $R_2$ is area bounded by

$f(x), x-1$ and $x = 2$

$$\therefore \quad R_2 = \int_{-1}^{2} f(x)\, dx \quad \ldots(iii)$$

Adding Eqs. (i) and (ii), we get

$$2R_1 = \int_{-1}^{2} f(x)\, dx \quad \ldots(iv)$$

$\therefore$ From Eqs. (iii) and (iv), we get

$$2R_1 = R_2$$

**5.** Here, $\lim_{x\to 0}\{1 + x\log(1+b^2)\}^{1/x}$

[$1^\infty$ from]

$$\Rightarrow \quad e^{\lim_{x\to 0}\{x\log(1+b^2)\}\cdot\frac{1}{x}}$$

$$\Rightarrow \quad e^{\log(1+b^2)} = (1+b)^2 \quad \ldots(i)$$

Given,

$$\lim_{x\to 0}\{1 + x\log(1+b^2)\}^{1/x} = 2b\sin^2\theta$$

$$\Rightarrow \quad (1+b^2) = 2b\sin^2\theta$$

$$\therefore \quad \sin^2\theta = \frac{1+b^2}{2b} \quad \ldots(ii)$$

By AM $\ge$ GM,

$$\frac{b+\frac{1}{b}}{2} \ge \left(b\cdot\frac{1}{b}\right)^{1/2} \Rightarrow \frac{b^2+1}{2b} \ge 1 \quad \ldots(iii)$$

From Eqs. (ii) and (iii), we get $\sin^2\theta = 1$

$$\Rightarrow \quad \theta = \pm\frac{\pi}{2} \text{ as } \theta \in (-\pi, \pi]$$

**6.** Equation of circle passing through a point $(x_1, y_1)$ and touching the straight line $L$, is given by

$$(x-x_1)^2 + (y-y_1)^2 = \lambda L = 0$$

$\therefore$ Equation of circle passing through (0, 2) and touching $x = 0$.

Now, $(x-0)^2 + (y-2)^2 + \lambda x = 0 \quad \ldots(i)$

Also, it passes through $(-1, 0)$.

So, $1 + 4 - \lambda = 0 \Rightarrow \lambda = 5$

Eq. (i) becomes,

$$x^2 + y^2 - 4y + 4 + 5x = 0$$
$$\Rightarrow \quad x^2 + y^2 + 5x - 4y + 4 = 0$$

For $x$-intercept, put $y = 0$,

$$x^2 + 5x + 4 = 0$$
$$\Rightarrow \quad (x+1)(x+4) = 0$$
$$\therefore \quad x = -1, -4$$

**7.** $|A| \ne 0$, as non-singular.

$$\therefore \quad \begin{vmatrix} 1 & a & b \\ \omega & 1 & c \\ \omega^2 & \omega & 1 \end{vmatrix} \ne 0$$

$$\Rightarrow \quad 1(1-c\omega) - a(\omega - c\omega^2) + b(\omega^2 - \omega^2) \ne 0$$

$$\Rightarrow \quad 1 - c\omega - a\omega + ac\omega^2 \ne 0$$

$\Rightarrow \quad (1 - c\omega)(1 - a\omega) \neq 0$

$\Rightarrow \quad a \neq \frac{1}{\omega}, c \neq \frac{1}{\omega} \Rightarrow a = \omega, c = \omega$

and $\quad b \in \{\omega, \omega^2\} \Rightarrow 2$ solutions

**8.** If $\quad a_1x^2 + b_1x + c_1 = 0$

and $\quad a_2x^2 + b_2x + c_2 = 0$

have a common real root, then

$\Rightarrow \quad (a_1c_2 - a_2c_1)^2 = (b_1c_2 - b_2c_1)(a_1b_2 - a_2b_1)$

$\therefore \left.\begin{matrix} x^2 + bx - 1 = 0 \\ x^2 + x + b = 0 \end{matrix}\right\}$ have a common root.

$\Rightarrow \quad (1 + b)^2 = (b^2 + 1)(1 - b)$

$\Rightarrow \quad b^2 + 2b + 1 = b^2 - b^3 + 1 - b$

$\Rightarrow \quad b^3 + 3b = 0 \Rightarrow b(b^2 + 3) = 0$

$\Rightarrow \quad b = 0, \pm \sqrt{3}\, i$

**9.** $f(x) = \begin{cases} -x - \frac{\pi}{2}, & x \leq -\frac{\pi}{2} \\ -\cos x, & -\frac{\pi}{2} < x \leq 0 \\ x - 1, & 0 < x \leq 1 \\ \log x, & x > 1 \end{cases}$

Continuity at $x = -\frac{\pi}{2}$,

$f\left(-\frac{\pi}{2}\right) = -\left(-\frac{\pi}{2}\right) - \frac{\pi}{2} = 0$

RHL $\Rightarrow \lim_{h \to 0} -\cos\left(-\frac{\pi}{2} + h\right) = 0$

$\therefore$ Continuous at $x = 0$

Continuity at $x = 0 \Rightarrow f(0) = -1$

RHL $\Rightarrow \lim_{h \to 0}(0 + h) - 1 = -1$

$\therefore$ Continuous at $x = 0$

Continuity at $x = 1$; $f(1) = 0$

RHL $\Rightarrow \lim_{h \to 0} \log(1 + h) = 0$

$\therefore$ Continuous at $x = 1$

Here, $f'(x) = \begin{cases} -1, & x \leq -\frac{\pi}{2} \\ \sin x, & -\frac{\pi}{2} < x \leq 0 \\ 1, & 0 < x \leq 1 \\ \frac{1}{x}, & x > 1 \end{cases}$

Differentiable at $x = 0$,

LHD = 0, RHD = 1

$\therefore$ Not differentiable at $x = 0$

Differentiable at $x = 1$,

LHD = 1, RHD = 1

$\therefore$ Differentiable at $x = 1$

Also, for $x = -\frac{3}{2} \Rightarrow f(x) = -x - \frac{3}{2}$

$\therefore$ Differentiable at $x = -\frac{3}{2}$

**10.** Normal to $y^2 = 4x$, is

$y - mx - 2m - m^3$ which passes through (9, 6).

Now, $\quad 6 = 9m - 2m - m^3$

$\Rightarrow m^3 - 7m + 6 = 0 \Rightarrow m = 1, 2, -3$

$\therefore$ Equation of normals are

$y - x + 3 = 0,$

$y + 3x - 33 = 0$ and $y - 2x + 12 = 0$

**11.**

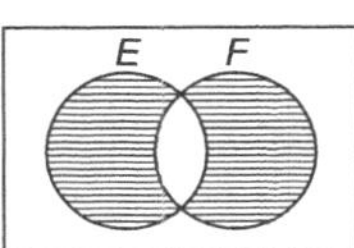


$P(E \cup F) - P(E \cap F) = \frac{11}{25}$ ...(i)

(i.e. only $E$ or only $F$)

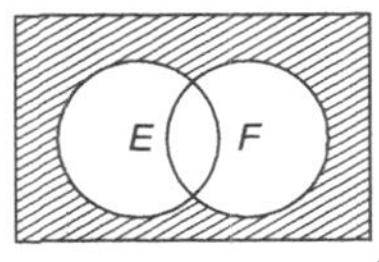


Neither of them occurs $= \frac{2}{25}$

$\Rightarrow \quad P(\bar{E} \cap \bar{F}) = \frac{2}{25}$ ...(ii)

From Eq. (i), we get

$P(E) + P(F) - 2P(E \cap F) = \frac{11}{25}$ ...(iii)

From Eq. (ii), we get

$(1 - P(E))(1 - P(F)) = \frac{2}{25}$

$\Rightarrow 1 - P(E) - P(F) + P(E) \cdot P(F) = \frac{2}{25}$ ...(iv)

From Eqs. (iii) and (iv), we get

$P(E) + P(F) = \frac{7}{5}$ and $P(E) \cdot P(F) = \frac{12}{25}$

$\therefore \quad P(E) \cdot \left\{\frac{7}{5} - P(E)\right\} = \frac{12}{25}$

$\Rightarrow \quad (P(E))^2 - \frac{7}{5}P(E) + \frac{12}{25} = 0$

$\Rightarrow \quad \left(P(E) - \frac{3}{5}\right)\left(P(E) - \frac{4}{5}\right) = 0$

$\therefore \quad P(E) = \frac{3}{4}$ or $\frac{4}{5} \Rightarrow P(F) = \frac{4}{5}$ or $\frac{3}{5}$

**12.** Here, $f(x) = \frac{b-x}{1-bx}$

where, $0 < b < 1, 0 < x < 1$

For function to be invertible it should be one-one onto.

$\therefore$ Check range :

Let $\quad f(x) = y \Rightarrow y = \frac{b-x}{1-bx}$

$\Rightarrow y - bxy = b - x \Rightarrow x(1 - by) = b - y$

$\Rightarrow \quad x = \frac{b-y}{1-by}$

where, $0 < x < 1$

$\therefore \quad 0 < \frac{b-y}{1-by} < 1$

$\frac{b-y}{1-by} > 0$ and $\frac{b-y}{1-by} < 1$

+ − + (sign scheme: $b$, $1/b$)

$\Rightarrow \quad y < b$ or $y > \frac{1}{b} \quad \ldots$(i)

$\frac{(b-1)(y+1)}{1-by} < -1 < y < \frac{1}{b} \quad \ldots$(ii)

From Eqs. (i) and (ii), we get

$y \in \left(-1, \frac{1}{b}\right) \subset$ Codomain

Thus, $f(x)$ is not invertible.

**13.** $\frac{dy}{dx} + y \cdot g'(x) = g(x)\, g'(x)$

$\text{IF} = e^{\int g'(x)\,dx} = e^{g(x)}$

$\therefore$ Solution is

$y(e^{g(x)}) = \int g(x) \cdot g'(x) \cdot e^{g(x)} dx + C$

Put $\quad g(x) = t, g'(x)dx = dt$

$y(e^{g(x)}) = \int t \cdot e^t\, dt + C$

$= t \cdot e^t - \int 1 \cdot e^t\, dt + C = t \cdot e^t - e^t + C$

$y\, e^{g(x)} = (g(x) - 1)\, e^{g(x)} + C \quad \ldots$(i)

Given, $y(0) = 0, g(0) = g(2) = 0$

$\therefore$ Eq. (i) becomes,

$y(0) \cdot e^{g(0)} = (g(0) - 1) \cdot e^{g(0)} + C$

$\Rightarrow \quad 0 = (-1) \cdot 1 + C \Rightarrow C = 1$

$\therefore \quad y(x) \cdot e^{g(x)} = (g(x) - 1)\, e^{g(x)} + 1$

$\Rightarrow y(2) \cdot e^{g(2)} = (g(2) - 1)\, e^{g(2)} + 1$

where, $\quad g(2) = 0$

$\Rightarrow \quad y(2) \cdot 1 = (-1) \cdot 1 + 1 \Rightarrow y(2) = 0$

**14.** $\mathbf{r} \times \mathbf{b} = \mathbf{c} \times \mathbf{b}$

$\Rightarrow \quad (\mathbf{r} - \mathbf{c}) \times \mathbf{b} = 0 \Rightarrow \mathbf{r} - \mathbf{c} + \lambda\, \mathbf{b}$

or $\quad \mathbf{r} = \mathbf{c} + \lambda\, \mathbf{b} \quad \ldots$(i)

Given, $\mathbf{r} \cdot \mathbf{a} = 0$, taking dot product with $\mathbf{a}$ for Eq. (i).

Now, $\quad \mathbf{r} \cdot \mathbf{a} = \mathbf{a} \cdot \mathbf{c} + \lambda\, \mathbf{a} \cdot \mathbf{b}$

$\therefore \quad \lambda = \frac{-\vec{a} \cdot \vec{c}}{\vec{a} \cdot \vec{b}} \quad [\because \vec{r} \cdot \vec{a} = 0] \ldots$(i)

From Eqs. (i) and (ii), we get

$\mathbf{r} = \mathbf{c} - \frac{\mathbf{a} \cdot \mathbf{c}}{\mathbf{a} \cdot \mathbf{b}}\mathbf{b}$

Taking dot with $\mathbf{b}$, we get

$\mathbf{r} \cdot \mathbf{b} = \mathbf{c} \cdot \mathbf{b} - \frac{\mathbf{a} \cdot \mathbf{c}}{\mathbf{a} \cdot \mathbf{b}}(\mathbf{b} \cdot \mathbf{b})$

where, $\begin{bmatrix} \mathbf{a} = -\hat{\mathbf{i}} - \hat{\mathbf{k}} \\ \mathbf{b} = -\hat{\mathbf{i}} + \hat{\mathbf{j}} \\ \mathbf{c} = \hat{\mathbf{i}} + 2\hat{\mathbf{j}} + 3\hat{\mathbf{k}} \end{bmatrix}$

$= (-1 + 2) - \frac{(-1-3)}{(1)}(1 + 1) = 1 + 8 = 9$

**15.** Here, $\omega = e^{i2\pi/3}$, then only integer solution exists.

Then, $\frac{|x^2| + |y^2| + |z^2|}{|a^2| + |b^2| + |c^2|} = 3$

**16.** Let $M = \begin{bmatrix} a_1 & a_2 & a_3 \\ b_1 & b_2 & b_3 \\ c_1 & c_2 & c_3 \end{bmatrix}$

$\therefore \quad M\begin{bmatrix} 0 \\ 1 \\ 0 \end{bmatrix} = \begin{bmatrix} -1 \\ 2 \\ 3 \end{bmatrix}, M\begin{bmatrix} 1 \\ -1 \\ 0 \end{bmatrix} = \begin{bmatrix} 1 \\ 1 \\ -1 \end{bmatrix},$

$M\begin{bmatrix} 1 \\ 1 \\ 1 \end{bmatrix} = \begin{bmatrix} 0 \\ 0 \\ 12 \end{bmatrix}$

$$\Rightarrow \begin{bmatrix} a_2 \\ b_2 \\ c_2 \end{bmatrix} = \begin{bmatrix} -1 \\ 2 \\ 3 \end{bmatrix}, \begin{bmatrix} a_1 - a_2 \\ b_1 - b_2 \\ c_1 - c_2 \end{bmatrix} = \begin{bmatrix} 1 \\ 1 \\ -1 \end{bmatrix},$$

$$\begin{bmatrix} a_1 + a_2 + a_3 \\ b_1 + b_2 + b_3 \\ c_1 + c_2 + c_3 \end{bmatrix} = \begin{bmatrix} 0 \\ 0 \\ 12 \end{bmatrix}$$

$\Rightarrow \quad a_2 = -1, b_2 = 2, c_2 = 3, a_1 - a_2 = 1,$
$b_1 - b_2 = 1, c_1 - c_2 = -1$

$\Rightarrow \quad a_1 + a_2 + a_3 = 0, b_1 + b_2 + b_3 = 0,$
$c_1 + c_2 + c_3 = 12$

$\therefore \quad a_1 = 0, b_2 = 2$ and $c_3 = 7$

Hence, sum of diagonal elements

$= 0 + 2 + 7 = 9$

**17.** $f(x) = x^4 - 4x^3 + 12x^2 + x - 1$

$f'(x) = 4x^3 - 12x^2 + 24x + 1$

$f''(x) = 12x^2 - 24x + 24$

$= 12(x^2 - 2x + 2)$

$= 12\{(x-1)^2 + 1\} > 0$, for all $x$

$\Rightarrow f'(x)$ is increasing.

Since, $f'(x)$ is cubic and increasing.

$\Rightarrow f'(x)$ has only one real root and two imaginary roots.

$\therefore \quad f(x)$ cannot have all distinct root.

$\Rightarrow$ Atmost 2 real roots.

Now, $\quad f(-1) = 15,$

$f(0) = -1$ and $f(1) = 9$

$\therefore \quad f(x)$ must have one root in $(-1, 0)$ and other in $(0, 1)$.

$\Rightarrow \quad$ 2 real roots.

**18.** $x^2 + y^2 \le 6$ and $2x - 3y = 1$ is shown as,

For the point to lie in the shaded part, origin and the point lie on opposite side of straight line $L$.

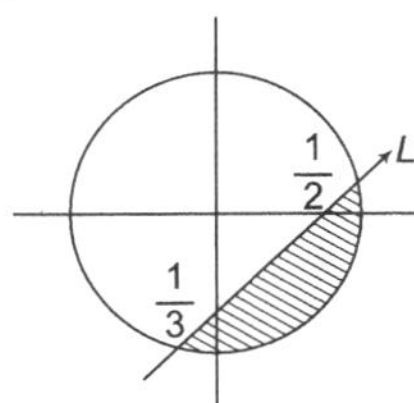


$\therefore$ For any point in shaded part $L > 0$ and for any point inside the circle $S < 0$.

Now, for $\left(2, \frac{3}{4}\right)$ $L : 2x - 3y - 1$

$L : 4 - \frac{9}{4} - 1 = \frac{3}{4} > 0$

and $\quad S : x^2 + y^2 - 6,$

$S : 4 + \frac{9}{16} - 6 < 0$

$\Rightarrow \quad \left(2, \frac{3}{4}\right)$ lies in shaded part.

For $\left(\frac{5}{2}, \frac{3}{4}\right) L : 5 - 9 - 1 < 0 \quad$ [neglect]

For $\left(\frac{1}{4}, -\frac{1}{4}\right) L : \frac{1}{2} + \frac{3}{4} - 1 > 0 \Big\}$

$\therefore \left(\frac{1}{4}, -\frac{1}{4}\right)$ lies in the shaded part.

For $\left(\frac{1}{8}, \frac{1}{4}\right) \; L : \frac{1}{4} - \frac{3}{4} - 1 < 0 \quad$ [neglect]

$\Rightarrow$ Only 2 points lie in the shaded part.

**19.** (A) $\therefore |\mathbf{a}| = \sqrt{1+3} = 2$

$|\mathbf{b}| = \sqrt{1+3} = 2$

$|\mathbf{c}| = \sqrt{12} = 2\sqrt{3}$

C
$\mathbf{b} = -\hat{\mathbf{j}} + \sqrt{3}\,\hat{\mathbf{k}}$ $\quad$ $\mathbf{a} = \hat{\mathbf{j}} + \sqrt{3}\,\hat{\mathbf{k}}$
A $\quad$ $\mathbf{c} = 2\sqrt{3}\,\hat{\mathbf{k}}$ $\quad$ B

Using cosine law,

$$\cos C = \frac{|\mathbf{a}|^2 + |\mathbf{b}|^2 - |\mathbf{c}|^2}{2|\mathbf{a}||\mathbf{b}|}$$

$$= \frac{4 + 4 - 12}{2 \times 2 \times 2} = \frac{-4}{8} = \frac{-1}{2}$$

$$\Rightarrow \quad \angle C = 120^\circ = \frac{2\pi}{3}$$

(B) $\int_a^b f(x)\,dx - 3\left(\frac{x^2}{2}\right)_a^b = (a^2 - b^2)$

$$\Rightarrow \quad \int_a^b f(x)\,dx - \frac{3}{2}(b^2 - a^2) = (a^2 - b^2)$$

$$\Rightarrow \quad \int_a^b f(x)dx = (a^2 - b^2) + \frac{3}{2}(b^2 - a^2)$$

$$= \frac{b^2 - a^2}{2}$$

$$\Rightarrow \quad \int_a^b f(x)\,dx = \frac{b^2 - a^2}{2}$$

$$f(x) = x \Rightarrow f\left(\frac{\pi}{6}\right) = \frac{\pi}{6}$$

(C) $\dfrac{\pi^2}{\log_e 3}\int_{7/6}^{5/6} \sec(\pi x)\,dx$

$$\Rightarrow \frac{\pi^2}{\log_e 3}\left\{\frac{\log|\sec \pi x + \tan \pi x|}{\pi}\right\}_{7/6}^{5/6}$$

$$\Rightarrow \frac{\pi}{\log 3}\left\{\log\left|\sec\frac{5\pi}{6} + \tan\frac{5\pi}{6}\right| - \log\left|\sec\frac{7\pi}{6} + \tan\frac{7\pi}{6}\right|\right\}$$

$$\Rightarrow \frac{\pi}{\log 3}\left\{\log|\sqrt{3}| - \log\left|\frac{1}{\sqrt{3}}\right|\right\}$$

$$\Rightarrow \frac{\pi}{\log 3}\{\log 3\} = \pi$$

(D) $\left|\arg\dfrac{1}{(1-z)}\right|$, for $|z| = 1$

$$\Rightarrow |\arg(1-z)^{-1}|$$

$$\Rightarrow |-\arg(1-z)| \Rightarrow |\arg(1-z)|$$



From figure, $\arg(z-1)$ is maximum $= \pi$.

**20.** (A) Given, $|z| = 1 \Rightarrow z \cdot \bar{z} = 1$

$$\therefore \quad \frac{2iz}{1-z^2} = \frac{2iz}{z\cdot\bar{z} - z^2} = \frac{2i}{\bar{z} - z},$$

Let $\quad z = x + iy$

$$\therefore z - \bar{z} = 2iy = \frac{2i}{-2iy} = -\frac{1}{y} \qquad \ldots\text{(i)}$$

where, $\quad y = \sqrt{1 - x^2}$

$$\therefore \quad -1 \le y \le 1 \Rightarrow -1 \le y$$

and $y \le 1 \Rightarrow -1 \ge \dfrac{1}{y}$ and $\dfrac{1}{y} \ge 1$

$$\Rightarrow \text{Re}\left(\frac{2iz}{1-z^2}\right) \in (-\infty, -1] \cup [1, \infty)$$

(B) $f(x) = \sin^{-1}\left(\dfrac{8(3^{x-2})}{1 - 3^{2(x-1)}}\right)$,

For domain, $-1 \le \dfrac{8(3^{x-2})}{1-3^{2(x-1)}} \le 1$

$$\Rightarrow -1 \le \frac{9\cdot(3^{x-2}) - (3^{x-2})}{1 - 3^{2(x-1)}} \le 1$$

$$\therefore \quad -1 \le \frac{3^x - 3^{x-2}}{1 - 3^x \cdot 3^{(x-2)}} \le 1$$

$$\frac{3^x - 3^{x-2}}{1 - 3^x \cdot 3^{x-2}} \ge -1$$

$$\Rightarrow \frac{(3^x - 1)(3^{x-2} - 1)}{(3^{x-1} + 1)(3^{x-1} - 1)} \ge 0$$



$$\Rightarrow x \in (-\infty, 0] \cup (1, \infty)$$

and $\dfrac{3^x - 3^{x-2}}{1 - 3^x \cdot 3^{x-2}} \le 1$

$$\Rightarrow \frac{(3^{x-2} - 1)(3^x + 1)}{(3^{x-1} + 1)(3^{x-1} - 1)} \ge 0$$



and $\quad x \in (-\infty, 1) \cup [2, \infty)$

$\therefore \quad x \in (-\infty, 0] \cup [2, \infty)$

(C) $f(\theta) = \begin{vmatrix} 1 & \tan\theta & 1 \\ -\tan\theta & 1 & \tan\theta \\ -1 & -\tan\theta & 1 \end{vmatrix}$

$R_1 \to R_1 + R_3$,

$$f(\theta) = \begin{vmatrix} 0 & 0 & 2 \\ -\tan\theta & 1 & \tan\theta \\ -1 & -\tan\theta & 1 \end{vmatrix}$$

$$= 2(\tan^2\theta + 1) = 2\sec^2\theta \ge 2$$

$f(\theta) \in [2, \infty)$

(D) $f(x) = x^{3/2}(3x - 10);\ x \ge 0$

$$f'(x) = x^{3/2}\cdot 3 + \frac{3}{2}\cdot x^{1/2}(3x - 10)$$

$$= 3x^{1/2}\left\{x + \frac{1}{2}(3x - 10)\right\}$$

$$= \frac{3}{2}x^{1/2}\{2x + 3x - 10\}$$

$$= \frac{15}{2}x^{1/2}(x - 2)$$



$\therefore \quad x \ge 2$

# IIT JEE

## SOLVED PAPER 2010

## Physics

## Paper I

### SECTION I — Single Correct Choice Type

*This section contains* **8 multiple choice questions**. *Each question has four choices (a), (b), (c) and (d) out of which* **ONLY ONE** *is correct.*

1. A thin uniform annular disc (*see figure*) of mass $M$ has outer radius $4R$ and inner radius $3R$. The work required to take a unit mass from point $P$ on its axis to infinity is

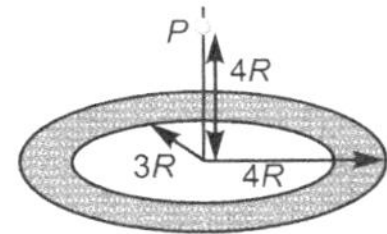


(a) $\dfrac{2GM}{7R}(4\sqrt{2}-5)$

(b) $-\dfrac{2GM}{7R}(4\sqrt{2}-5)$

(c) $\dfrac{GM}{4R}$

(d) $\dfrac{2GM}{5R}(\sqrt{2}-1)$

2. A block of mass $m$ is on an inclined plane of angle $\theta$. The coefficient of friction between the block and the plane is $\mu$ and $\tan\theta > \mu$. The block is held stationary by applying a force $P$ parallel to the plane. The direction of force pointing up the plane is taken to be positive. As $P$ is varied from $P_1 = mg(\sin\theta - \mu\cos\theta)$ to $P_2 = mg(\sin\theta + \mu\cos\theta)$, the frictional force $f$ *versus* $P$ graph will look like

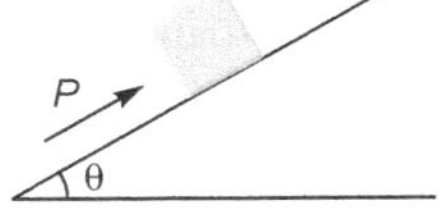


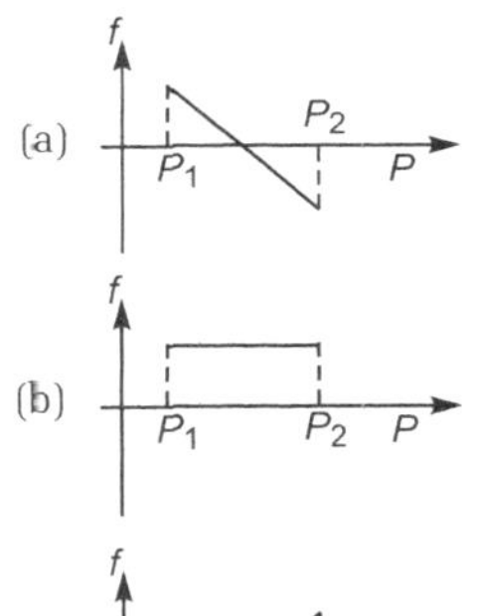


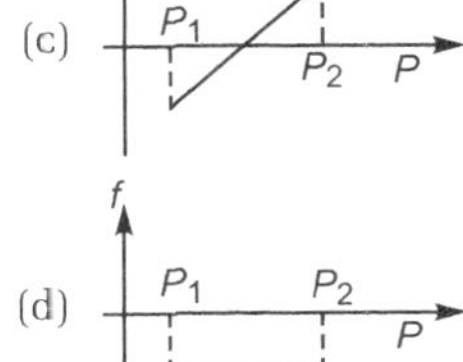


3. A real gas behaves like an ideal gas if its
   (a) pressure and temperature are both high
   (b) pressure and temperature are both low
   (c) pressure is high and temperature is low
   (d) pressure is low and temperature is high

4. Consider a thin square sheet of side $L$ and thickness $t$, made of a material of resistivity $\rho$. The resistance between

two opposite faces, shown by the shaded areas in the figure is

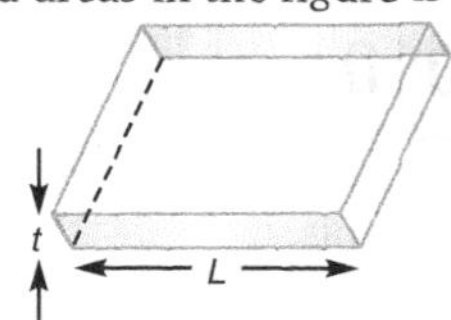


(a) directly proportional to $L$
(b) directly proportional to $t$
(c) independent of $L$
(d) independent of $t$

**5.** Incandescent bulbs are designed by keeping in mind that the resistance of their filament increases with the increase in temperature. If at room temperature, 100 W, 60 W and 40 W bulbs have filament resistances $R_{100}$, $R_{60}$ and $R_{40}$, respectively, the relation between these resistances is

(a) $\dfrac{1}{R_{100}} = \dfrac{1}{R_{40}} + \dfrac{1}{R_{60}}$
(b) $R_{100} = R_{40} + R_{60}$
(c) $R_{100} > R_{60} > R_{40}$
(d) $\dfrac{1}{R_{100}} > \dfrac{1}{R_{60}} > \dfrac{1}{R_{40}}$

**6.** To verify Ohm's law, a student is provided with a test resistor $R_T$, a high resistance $R_1$, a small resistance $R_2$, two identical galvanometers $G_1$ and $G_2$, and a variable voltage source $V$. The correct circuit to carry out the experiment is

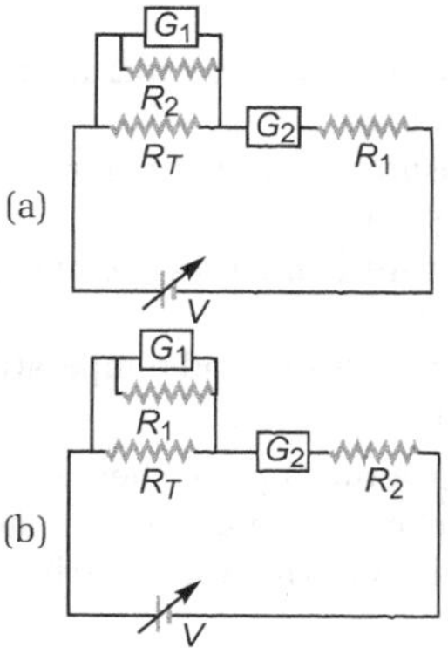


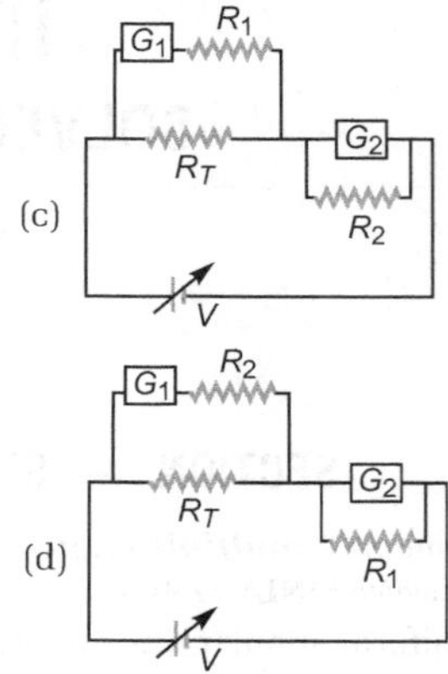


**7.** A thin flexible wire of length $L$ is connected to two adjacent fixed points and carries a current $I$ in the clockwise direction, as shown in the figure. When the system is put in a uniform magnetic field of strength $B$ going into the plane of the paper, the wire takes the shape of a circle. The tension in the wire is

(a) $IBL$
(b) $\dfrac{IBL}{\pi}$
(c) $\dfrac{IBL}{2\pi}$
(d) $\dfrac{IBL}{4\pi}$

**8.** An AC voltage source of variable angular frequency $\omega$ and fixed amplitude $V_0$ is connected in series with a capacitance $C$ and an electric bulb of resistance $R$ (*inductance zero*). When $\omega$ is increased

(a) the bulb glows dimmer
(b) the bulb glows brighter
(c) total impedance of the circuit is unchanged
(d) total impedance of the circuit increases

## SECTION II — Multiple Correct Choice Type

*This section contains* **5 multiple choice questions**. *Each question has four choices (a), (b), (c) and (d) out of which* **ONE OR MORE** *may be correct.*

**9.** A student uses a simple pendulum of exactly 1m length to determine $g$, the acceleration due to gravity. He uses a stop watch with the least count of 1 s for this and records 40 s for 20 oscillations. For this observation, which of the following statement(s) is/are true?

(a) Error $\Delta T$ in measuring $T$, the time period, is 0.05 s

(b) Error $\Delta T$ in measuring $T$, the time period, is 1 s

(c) Percentage error in the determination of $g$ is 5%

(d) Percentage error in the determination of $g$ is 2.5%

**10.** A few electric field lines for a system of two charges $Q_1$ and $Q_2$ fixed at two different points on the $x$-axis are shown in the figure. These lines suggest that

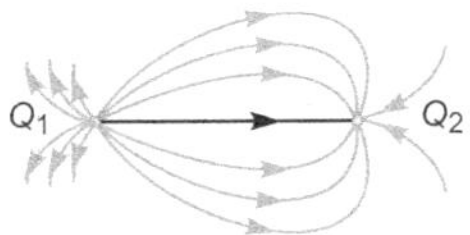


(a) $|Q_1| > |Q_2|$

(b) $|Q_1| < |Q_2|$

(c) at a finite distance to the left of $Q_1$ the electric field is zero

(d) at a finite distance to the right of $Q_2$ the electric field is zero

**11.** One mole of an ideal gas in initial state $A$ undergoes a cyclic process $ABCA$, as shown in the figure. Its pressure at $A$ is $p_0$. Choose the correct option($s$) from the following

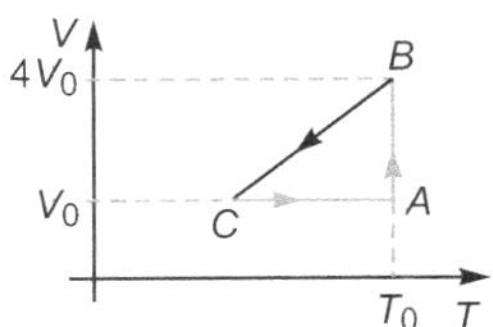


(a) Internal energies at $A$ and $B$ are the same

(b) Work done by the gas in process $AB$ is $p_0V_0 \ln 4$

(c) Pressure at $C$ is $\frac{p_0}{4}$

(d) Temperature at $C$ is $\frac{T_0}{4}$

**12.** A point mass of 1 kg collides elastically with a stationary point mass of 5 kg. After their collision, the 1 kg mass reverses its direction and moves with a speed of $2\ \text{ms}^{-1}$. Which of the following statement(s) is/are correct for the system of these two masses?

(a) Total momentum of the system is $3\ \text{kg-ms}^{-1}$

(b) Momentum of 5 kg mass after collision is $4\ \text{kg-ms}^{-1}$

(c) Kinetic energy of the centre of mass is 0.75 J

(d) Total kinetic energy of the system is 4 J

**13.** A ray $OP$ of monochromatic light is incident on the face $AB$ of prism $ABCD$ near vertex $B$ at an incident angle of 60° (*see figure*). If the refractive index of the material of the prism is $\sqrt{3}$, which of the following is (are) correct?

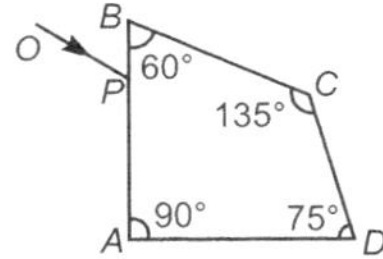


(a) The ray gets totally internally reflected at face $CD$

(b) The ray comes out through face $AD$

(c) The angle between the incident ray and the emergent ray is 90°

(d) The angle between the incident ray and the emergent ray is 120°

## SECTION III — Paragraph Type

*This section contains* **2 paragraphs.** *Based upon the first paragraph 2* **multiple choice questions** *and based upon the second paragraph 3* **multiple choice questions** *have to be answered. Each of these questions has four choices (a), (b), (c) and (d) out of* **WHICH ONLY ONE is CORRECT.**

$\boldsymbol{P_{14-15}}$ Electrical resistance of certain materials, known as superconductors, changes abruptly from a non-zero value to zero as their temperature is lowered below a critical temperature $T_C(0)$. An interesting property of superconductors is that their critical temperature becomes smaller than $T_C(0)$ if they are placed in a magnetic field *i.e.*, the critical temperature $T_C(B)$ is a function of the magnetic field strength $B$. The dependence of $T_C(B)$ on $B$ is shown in the figure.

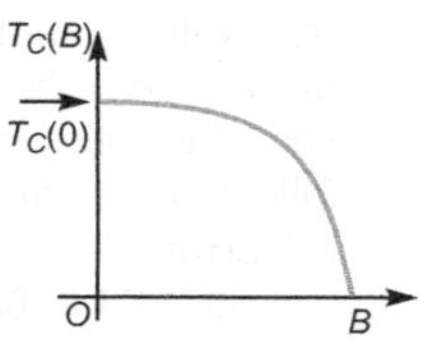


**14.** In the graphs below, the resistance $R$ of a superconductor is shown as a function of its temperature $T$ for two different magnetic fields $B_1$ (*solid line*) and $B_2$ (*dashed line*). If $B_2$ is larger than $B_1$, which of the following graphs shows the correct variation of $R$ with $T$ in these fields?

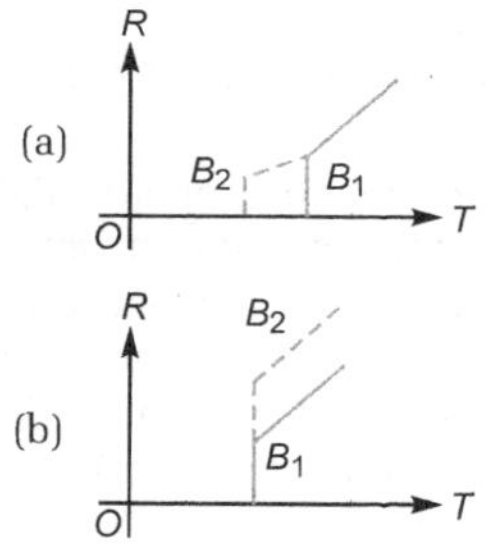


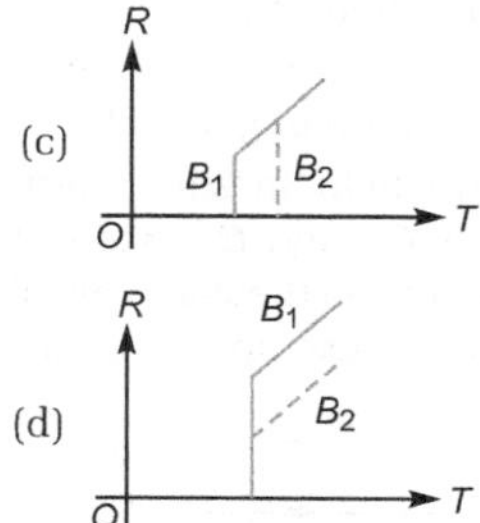


**15.** A superconductor has $T_C(0) = 100$ K. When a magnetic field of 7.5 Tesla is applied, its $T_C$ decreases to 75 K. For this material one can definitely say that when (Note : T = Tesla)

(a) $B = 5\,\text{T}, T_C(B) = 80$ K

(b) $B = 5\ \text{T}, 75\,\text{K} < T_C(B) < 100$ K

(c) $B = 10\,\text{T}, 75\,\text{K} < T_C\,(B) < 100$ K

(d) $B = 10\,\text{T}, T_C\,(B) = 70$ K

$\boldsymbol{P_{16-18}}$ When a particle of mass $m$ moves on the $x$-axis in a potential of the form $V(x) = kx^2$, it performs simple harmonic motion.

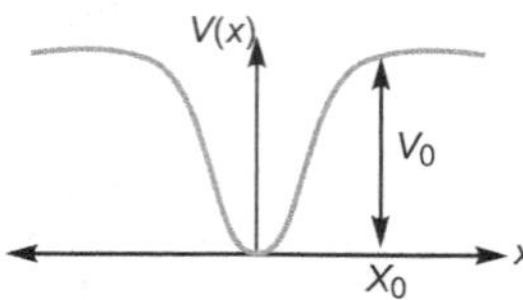


The corresponding time period is proportional to $\sqrt{\frac{m}{k}}$, as can be seen easily using dimensional analysis. However, the motion of a particle can be periodic even when its potential energy increases on both sides of $x = 0$ in a way different from $kx^2$ and its total energy is such that the particle does not escape to infinity. Consider a particle of mass $m$ moving on the $x$-axis. Its potential energy is $V(x) = \alpha x^4 (\alpha > 0)$ for $|x|$ near the origin and becomes a constant equal to $V_0$ for $|x| \ge X_0$ *(see figure)*.

**16.** If the total energy of the particle is $E$, it will perform periodic motion only if

(a) $E < 0$ (b) $E > 0$

(c) $V_0 > E > 0$ (d) $E > V_0$

**17.** For periodic motion of small amplitude $A$, the time period $T$ of this particle is proportional to

(a) $A\sqrt{\frac{m}{\alpha}}$ (b) $\frac{1}{A}\sqrt{\frac{m}{\alpha}}$

(c) $A\sqrt{\frac{\alpha}{m}}$ (d) $\frac{1}{A}\sqrt{\frac{\alpha}{m}}$

**18.** The acceleration of this particle for $|x| > X_0$ is

(a) proportional to $V_0$

(b) proportional to $\frac{V_0}{mX_0}$

(c) proportional to $\sqrt{\frac{V_0}{mX_0}}$

(d) Zero

## SECTION IV — Integer Type

*This section contains* ***TEN questions****. The answer to each question is a* ***Single-digit integer****, ranging from* 0 *to* 9. *The correct digit below the question number in the ORS is to be bubbled.*

**19.** A stationary source is emitting sound at a fixed frequency $f_0$, which is reflected by two cars approaching the source. The difference between the frequencies of sound reflected from the cars is 1.2% of $f_0$. What is the difference in the speeds of the cars *(in km per hour)* to the nearest integer ? The cars are moving at constant speeds much smaller than the speed of sound which is 330 $ms^{-1}$.

**20.** The focal length of a thin biconvex lens is 20 cm. When an object is moved from a distance of 25 cm in front of it to 50 cm, the magnification of its image changes from $m_{25}$ to $m_{50}$. The ratio $\frac{m_{25}}{m_{50}}$ is

**21.** An $\alpha$-particle and a proton are accelerated from rest by a potential difference of 100 V. After this, their de Broglie wavelengths are $\lambda_\alpha$ and $\lambda_p$ respectively. The ratio $\frac{\lambda_p}{\lambda_\alpha}$, to the nearest integer, is

**22.** When two identical batteries of internal resistance 1 Ω each are connected in series across a resistor $R$, the rate of heat produced in $R$ is $J_1$. When the same batteries are connected in parallel across $R$, the rate is $J_2$. If $J_1 = 2.25 J_2$ then the value of $R$ in Ω is

**23.** Two spherical bodies $A$ *(radius* 6 *cm* ) and $B$ *(radius* 18 *cm)* are at temperatures $T_1$ and $T_2$, respectively. The maximum intensity in the emission spectrum of $A$ is at 500 nm and in that of $B$ is at 1500 nm. Considering them to be black bodies, what will be the ratio of the rate of total energy radiated by $A$ to that of $B$?

**24.** When two progressive waves $y_1 = 4 \sin(2x - 6t)$ and $y_2 = 3 \sin\left(2x - 6t - \frac{\pi}{2}\right)$ are superimposed, the amplitude of the resultant wave is

**25.** A 0.1 kg mass is suspended from a wire of negligible mass. The length of the wire is 1 m and its cross-sectional area is $4.9 \times 10^{-7}$ m$^2$. If the mass is pulled a little in the vertically downward direction and released, it performs simple harmonic motion of angular frequency 140 rad s$^{-1}$. If the Young's modulus of the material of the wire is $n \times 10^9$ Nm$^{-2}$, the value of $n$ is

26. A binary star consists of two stars $A$ (*mass* $2.2M_S$) and $B$ (*mass* $11M_s$), where $M_S$ is the mass of the sun. They are separated by distance $d$ and are rotating about their centre of mass, which is stationary. The ratio of the total angular momentum of the binary star to the angular momentum of star $B$ about the centre of mass is

27. Gravitational acceleration on the surface of a planet is $\frac{\sqrt{6}}{11}g$, where $g$ is the gravitational acceleration on the surface of the earth. The average mass density of the planet is $\frac{2}{3}$ times that of the earth. If the escape speed on the surface of the earth is taken to be 11 kms$^{-1}$, the escape speed on the surface of the planet in kms$^{-1}$ will be

28. A piece of ice (*heat capacity* $= 2100\ J\ kg^{-1}\ °C^{-1}$ *and latent heat* $= 3.36 \times 10^5\ J\ kg^{-1}$) of mass $m$ gram is at –5°C at atmospheric pressure. It is given 420 J of heat so that the ice starts melting. Finally when the ice-water mixture is in equilibrium, it is found that 1 g of ice has melted. Assuming there is no other heat exchange in the process, the value of $m$ is

# Paper II

## SECTION I — Single Correct Choice Type

*This section contains* **6 multiple choice questions**. *Each question has four choices (a), (b), (c) and (d) out of which* ***ONLY ONE*** *is correct.*

1. A tiny spherical oil drop carrying a net charge $q$ is balanced in still air with a vertical uniform electric field of strength $\frac{81\pi}{7} \times 10^5$ Vm$^{-1}$. When the field is switched off, the drop is observed to fall with terminal velocity $2 \times 10^{-3}$ ms$^{-1}$. Given $g = 9.8$ ms$^{-2}$, viscosity of the air $= 1.8 \times 10^{-5}$ Ns m$^{-2}$ and the density of oil $= 900$ kg m$^{-3}$, the magnitude of $q$ is
   (a) $1.6 \times 10^{-19}$ C (b) $3.2 \times 10^{-19}$ C
   (c) $4.8 \times 10^{-19}$ C (d) $8.0 \times 10^{-19}$ C

2. A uniformly charged thin spherical shell of radius $R$ carries uniform surface charge density of $\sigma$ per unit area. It is made of two hemispherical shells, held together by pressing them with force $F$ (*see figure*). $F$ is proportional to

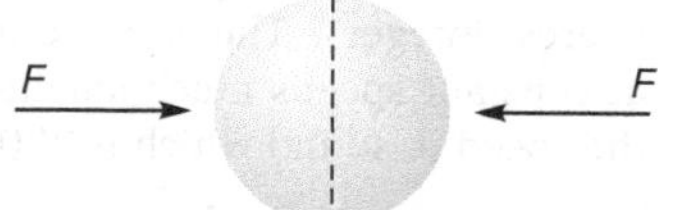


   (a) $\frac{1}{\varepsilon_0}\sigma^2 R^2$ (b) $\frac{1}{\varepsilon_0}\sigma^2 R$
   (c) $\frac{1}{\varepsilon_0}\frac{\sigma^2}{R}$ (d) $\frac{1}{\varepsilon_0}\frac{\sigma^2}{R^2}$

3. A block of mass 2 kg is free to move along the $x$-axis. It is at rest and from $t = 0$ onwards it is subjected to a time-dependent force $F(t)$ in the $x$ direction. The force $F(t)$ varies with $t$ as shown in the figure. The kinetic energy of the block after 4.5 s is

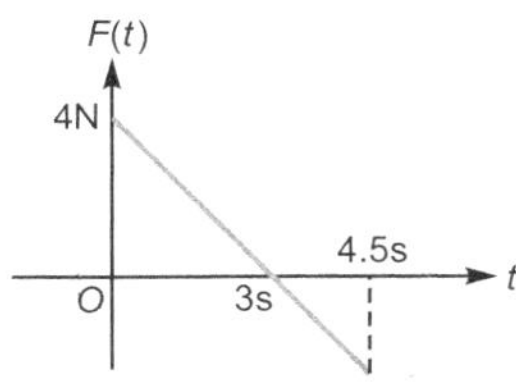


(a) 4.50 J (b) 7.50 J
(c) 5.06 J (d) 14.06 J

4. A biconvex lens of focal length 15 cm is in front of a plane mirror. The distance between the lens and the mirror is 10 cm. A small object is kept at a distance of 30 cm from the lens. The final image is
(a) virtual and at a distance of 16 cm from the mirror
(b) real and at a distance of 16 cm from the mirror
(c) virtual and at a distance of 20 cm from the mirror
(d) real and at a distance of 20 cm from the mirror

5. A vernier calipers has 1 mm marks on the main scale. It has 20 equal divisions on the vernier scale which match with 16 main scale divisions. For this vernier calipers, the least count is
(a) 0.02 mm
(b) 0.05 mm
(c) 0.1 mm
(d) 0.2 mm

6. A hollow pipe of length 0.8 m is closed at one end. At its open end a 0.5 m long uniform string is vibrating in its second harmonic and it resonates with the fundamental frequency of the pipe. If the tension in the wire is 50 N and the speed of sound is 320 $ms^{-1}$, the mass of the string is
(a) 5 g (b) 10 g
(c) 20 g (d) 40 g

## SECTION II — Integer Type

*This section contains* **5 questions**. *The answer to each of the questions is a* ***single-digit integer****, ranging from 0 to 9. The correct digit below the question number in the* **ORS** *is to be bubbled.*

7. To determine the half-life of a radioactive element, a student plots a graph of $\ln\left|\frac{dN(t)}{dt}\right|$ *versus* $t$. Here $\frac{dN(t)}{dt}$ is the rate of radioactive decay at time $t$. If the number of radioactive nuclei of this element decreases by a factor of $p$ after 4.16 yr, the value of $p$ is

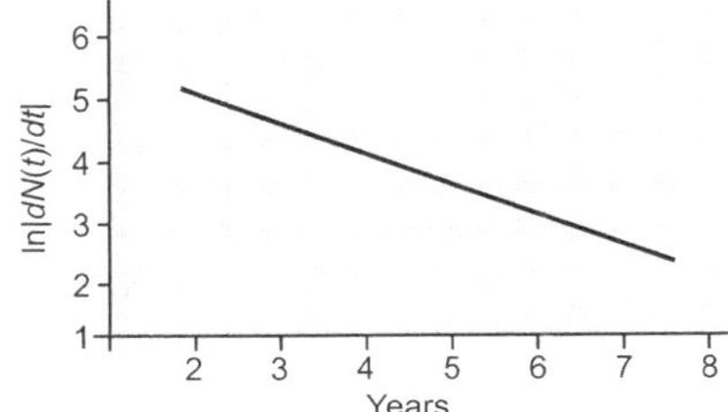


8. A diatomic ideal gas is compressed adiabatically to $\frac{1}{32}$ of its initial volume. If the initial temperature of the gas is $T_i$ (*in kelvin*) and the final temperature is $aT_i$, the value of $a$ is

9. At time $t = 0$, a battery of 10 V is connected across points $A$ and $B$ in the given circuit. If the capacitors have no charge initially, at what time (*in second*) does the voltage across them become 4 V?
[*Take* : $\ln 5 = 1.6$, $\ln 3 = 1.1$]

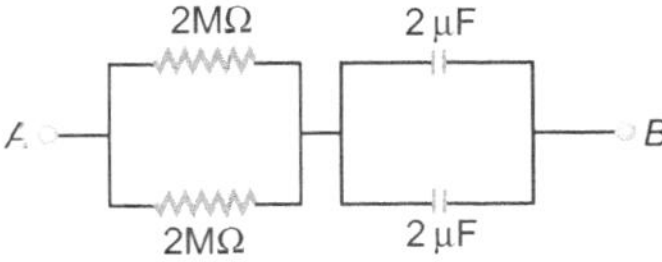


10. Image of an object approaching a convex mirror of radius of curvature 20 m along its optical axis is observed to move from $\frac{25}{3}$ m to $\frac{50}{7}$ m in 30 s. What is the speed of the object in km $h^{-1}$ ?

11. A large glass slab $\left(\mu = \frac{5}{3}\right)$ of thickness 8 cm is placed over a point source of light on a plane surface. It is seen that light emerges out of the top surface of the slab from a circular area of radius $R$ cm. What is the value of $R$?

## SECTION III — Paragraph Type

*This section contains* **2 paragraphs**. *Based upon each of the paragraphs* **3 multiple choice questions** *have to be answered. Each of these questions has four choices (a), (b), (c) and (d) out of which* **ONLY ONE** *is correct.*

$P_{12-14}$ The key feature of Bohr's theory of spectrum of hydrogen atom is the quantization of angular momentum when an electron is revolving around a proton. We will extend this to a general rotational motion to find quantized rotational energy of a diatomic molecule assuming it to be rigid. The rule to be applied is Bohr's quantization condition.

12. A diatomic molecule has moment of inertia $I$. By Bohr's quantization condition its rotational energy in the $n$th level ($n = 0$ *is not allowed*) is

(a) $\frac{1}{n^2}\left(\frac{h^2}{8\pi^2 I}\right)$ (b) $\frac{1}{n}\left(\frac{h^2}{8\pi^2 I}\right)$

(c) $n\left(\frac{h^2}{8\pi^2 I}\right)$ (d) $n^2\left(\frac{h^2}{8\pi^2 I}\right)$

13. It is found that the excitation frequency from ground to the first excited state of rotation for the CO molecule is close to $\frac{4}{\pi} \times 10^{11}$ Hz. Then the moment of inertia of CO molecule about its centre of mass is close to (*Take* $h = 2\pi \times 10^{-34}$ *J-s*)

(a) $2.76 \times 10^{-46}$ kg-m$^2$

(b) $1.87 \times 10^{-46}$ kg-m$^2$

(c) $4.67 \times 10^{-47}$ kg-m$^2$

(d) $1.17 \times 10^{-47}$ kg-m$^2$

14. In a CO molecule, the distance between C (*mass* = 12 *amu*) and O (*mass* = 16 *amu*), where 1 amu $= \frac{5}{3} \times 10^{-27}$ kg, is close to

(a) $2.4 \times 10^{-10}$ m (b) $1.9 \times 10^{-10}$ m

(c) $1.3 \times 10^{-10}$ m (d) $4.4 \times 10^{-11}$ m

$P_{15-17}$ When liquid medicine of density $\rho$ is to be put in the eye, it is done with the help of a dropper. As the bulb on the top of the dropper is pressed, a drop forms at the opening of the dropper. We wish to estimate the size of the drop.

We first assume that the drop formed at the opening is spherical because that requires a minimum increase in its surface energy. To determine the size, we calculate the net vertical force due to the surface tension $T$ when the radius of the drop is $R$. When this force becomes smaller than the weight of the drop, the drop gets detached from the dropper.

15. If the radius of the opening of the dropper is $r$, the vertical force due to the surface tension on the drop of radius $R$ (*assuming* $r << R$) is

(a) $2\pi rT$ (b) $2\pi RT$

(c) $\frac{2\pi r^2 T}{R}$ (d) $\frac{2\pi R^2 T}{r}$

16. If $r = 5 \times 10^{-4}$ m, $\rho = 10^3$ kg m$^{-3}$, $g = 10$ ms$^{-2}$, $T = 0.11$ Nm$^{-1}$, the radius of the drop when it detaches from the dropper is approximately

(a) $1.4 \times 10^{-3}$ m (b) $3.3 \times 10^{-3}$ m

(c) $2.0 \times 10^{-3}$ m (d) $4.1 \times 10^{-3}$ m

17. After the drop detaches, its surface energy is

(a) $1.4 \times 10^{-6}$ J (b) $2.7 \times 10^{-6}$ J

(c) $5.4 \times 10^{-6}$ J (d) $8.1 \times 10^{-9}$ J

## SECTION IV — Matrix Type

*This section contains* **2 questions**. *Each question has* ***four statements*** *(A, B, C and D) given in* ***Column I*** *and* ***five statements*** *(P, Q, R, S and T) in* ***Column II***. *Any given statement in* ***Column I*** *can have correct matching with one or more statement(s) given in* ***Column II***. *For example, if for a given question, statement B matches with the statements given in Q and R, then for that particular question, against statement B, darken the bubbles corresponding to Q and R in the* ***ORS***.

**18.** You are given many resistances, capacitors and inductors. These are connected to a variable DC voltage source (*the first two circuits*) or an AC voltage source of 50 Hz frequency (*the next three circuits*) in different ways as shown in **Column II**. When a current $I$ (*steady state for DC or rms for AC*) flows through the circuit, the corresponding voltage $V_1$ and $V_2$ (*indicated in circuits*) are related as shown in **Column I**. Match the two

| | Column I | | Column II |
|---|---|---|---|
| (A) | $I \neq 0$, $V_1$ is proportional to $I$ | (P) | $V_1$ $V_2$ 6 mH 3 μF $V$ |
| (B) | $I \neq 0$, $V_2 > V_1$ | (Q) | $V_1$ $V_2$ 6 mH 2 Ω $V$ |
| (C) | $V_1 = 0$, $V_2 = V$ | (R) | $V_1$ $V_2$ 6 mH 2 Ω $V$ |
| (D) | $I \neq 0$, $V_2$ is proportional to $I$ | (S) | $V_1$ $V_2$ 6 mH 3 μF $V$ |
| | | (T) | $V_1$ $V_2$ 1 kΩ 3 μF $V$ |

19. Two transparent media of refractive indices $\mu_1$ and $\mu_3$ have a solid lens shaped transparent material of refractive index $\mu_2$ between them as shown in figures in **Column II**. A ray traversing these media is also shown in the figures. In **Column I** different relationships between $\mu_1$, $\mu_2$ and $\mu_3$ are given. Match them to the ray diagram shown in **Column II**.

| | Column I | | Column II |
|---|---|---|---|
| (A) | $\mu_1 < \mu_2$ | (P) | 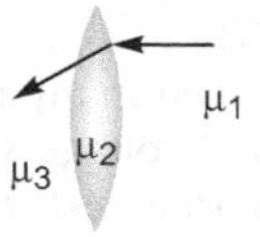 |
| (B) | $\mu_1 > \mu_2$ | (Q) | 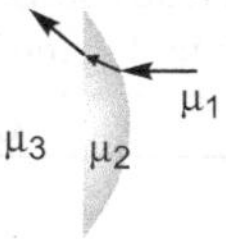 |
| (C) | $\mu_2 = \mu_3$ | (R) | 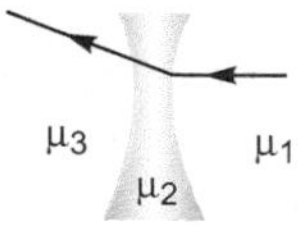 |
| (D) | $\mu_2 > \mu_3$ | (S) | 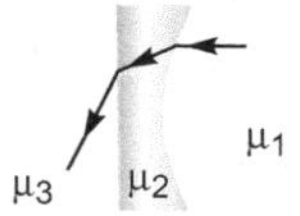 |
| | | (T) | 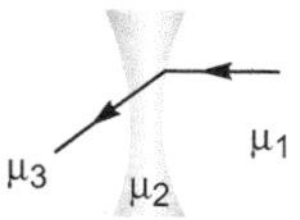 |

# Answers

## Paper I

| | | | | | | | |
|---|---|---|---|---|---|---|---|
| **1.** (a) | **2.** (a) | **3.** (d) | **4.** (c) | **5.** (d) | **6.** (c) | **7.** (c) | **8.** (b) |
| **9.** (a,c) | **10.** (a,d) | **11.** (a,b) | **12.** (a,c) | **13.** (a,b,c) | **14.** (a) | **15.** (b) | **16.** (c) |
| **17.** (b) | **18.** (d) | **19.** (7) | **20.** (6) | **21.** (3) | **22.** (4) | **23.** (9) | **24.** (5) |
| **25.** (4) | **26.** (6) | **27.** (3) | **28.** (8) | | | | |

## Paper II

| | | | | | | | |
|---|---|---|---|---|---|---|---|
| **1.** (d) | **2.** (a) | **3.** (c) | **4.** (b) | **5.** (d) | **6.** (b) | **7.** (8) | **8.** (4) |
| **9.** (2) | **10.** (3) | **11.** (6) | **12.** (d) | **13.** (b) | **14.** (c) | **15.** (c) | **16.** (a) |
| **17.** (b) | | | | | | | |

**18.** A– R, S, T; B–Q, R, S, T; C–Q, P, Q; D–Q, R, S, T

**19.** A– P, R; B–Q, S, T; C–P, R, T; D–Q, S

# Hints & Solutions

## Paper I

1. $W = \Delta U = U_f - U_i = U_\infty - U_p$

$$= -U_P$$
$$= -mV_P$$
$$= -V_P \qquad \text{(as } m = 1\text{)}$$

Potential at point $P$ will be obtained by integration as given below

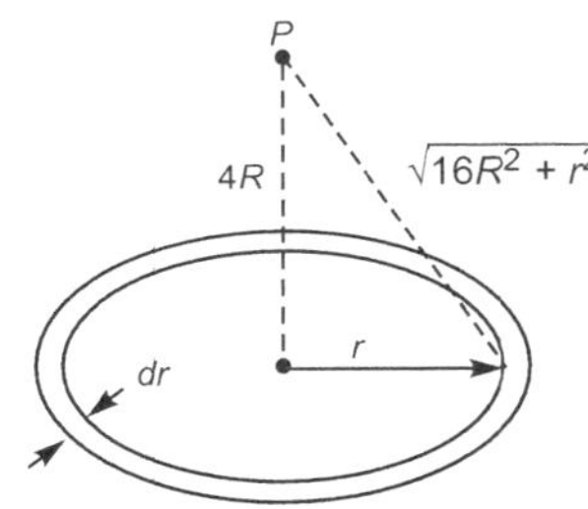


Let $dM$ be the mass of small ring as shown

$$dM = \frac{M}{\pi(4R)^2 - \pi(3R)^2}(2\pi r)\,dr$$
$$= \frac{2Mr\,dr}{7R^2}$$
$$dV_P = -\frac{G \cdot dM}{\sqrt{16R^2 + r^2}}$$
$$= -\frac{2GM}{7R^2}\int_{3R}^{4R}\frac{r}{\sqrt{16R^2 + r^2}} \cdot dr$$
$$= -\frac{2GM}{7R}(4\sqrt{2} - 5)$$
$$\therefore \quad W = +\frac{2GM}{7R}(4\sqrt{2} - 5)$$

$\therefore$ correct option is (a).

2. When

$P = mg(\sin\theta - \mu\cos\theta)$

$f = \mu\, mg\cos\theta$ (upwards)

when $P = mg\sin\theta$

$f = 0$

and when $P = mg(\sin\theta + \mu\cos\theta)$

$f = \mu mg\cos\theta$ (downwards)

Hence friction is first positive, then zero and then negative.

$\therefore$ correct option is (a).

3. A real gas behaves like an ideal gas at low pressure and high temperature.

$\therefore$ correct option is (d).

4. $R = \dfrac{\rho(L)}{A} = \dfrac{\rho L}{tL} = \dfrac{\rho}{t}$

*ie,* $R$ is independent of $L$.

Hence the correct option is (c).

5. $R = \dfrac{V^2}{P}$ or $R \propto \dfrac{1}{P}$

$$\therefore \quad \frac{1}{R_{100}} > \frac{1}{R_{60}} > \frac{1}{R_{40}}$$

Hence, the correct option is (d).

6. We will require a voltmeter, an ammeter, a test resistor and a variable battery to verify Ohm's law.

Voltmeter which is made by connecting a high resistance with a galvanometer is connected in parallel with the test resistor.

Further, an ammeter which is formed by connecting a low resistance in parallel with galvanometer is required to measure the current through test resistor.

The correct option is (c).

7.

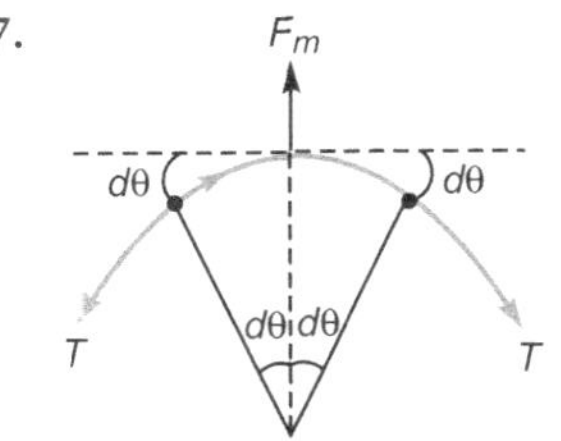


$$L = 2\pi R$$
$$\therefore \quad R = \frac{L}{2\pi}$$
$$2T\sin(d\theta) = F_m$$

for small angles, $\sin(d\theta) \approx d\theta$

$\therefore \; 2T\,(d\theta) = I\,(dL)\,B \sin 90^\circ$

$= I\,(2R \cdot d\theta) \cdot B$

$\therefore \quad T = IRB = \dfrac{ILB}{2\pi}$

$\therefore$ correct option is (c)

**8.** $Z = \sqrt{R^2 + X_C^2} : I_{\text{rms}} = \dfrac{V_{\text{rms}}}{Z} : P = I_{\text{rms}}^2 R$

where, $\quad X_C = \dfrac{1}{\omega C}$

As $\omega$ is increased, $X_C$ will decrease or $Z$ will decrease. Hence $I_{\text{rms}}$ or $P$ will increase. Therefore, bulb glows brighter.

Hence the correct option is (b).

**9.** $T = \dfrac{40\text{ s}}{20} = 2\text{ s}.$

Further, $t = nT = 20T$ or $\Delta t = 20\,\Delta T$

$\therefore \quad \dfrac{\Delta t}{t} = \dfrac{\Delta T}{T}$

or $\quad \Delta T = \dfrac{T}{t} \cdot \Delta t = \left(\dfrac{2}{40}\right)(1) = 0.05\text{ s}$

Further, $T = 2\pi\sqrt{\dfrac{l}{g}}$ or $T \propto g^{-1/2}$

$\therefore \dfrac{\Delta T}{T} \times 100 = -\dfrac{1}{2} \times \dfrac{\Delta g}{g} \times 100$

or % error in determination of $g$ is

$\dfrac{\Delta g}{g} \times 100 = -200 \times \dfrac{\Delta T}{T}$

$= -\dfrac{200 \times 0.05}{2} = -5\%$

$\therefore$ correct options are (a) and (c).

**10.** From the behaviour of electric lines, we can say that $Q_1$ is positive and $Q_2$ is negative. Further, $|Q_1| > |Q_2|$

At some finite distance to the right of $Q_2$, electric field will be zero. Because electric field due to $Q_1$ is towards right (*away from* $Q_1$) and due to $Q_2$ is towards left (*towards* $Q_2$). But since magnitude of $Q_1$ is more, the two fields may cancel each other because distance of that point from $Q_1$ will also be more.

$\therefore$ The correct options are (a) and (d).

**11.** $T_A = T_B \qquad \therefore U_A = U_B$

$W_{AB} = (1)(R)T_0 \ln\left(\dfrac{V_f}{V_i}\right)$

$= RT_0 \ln\left(\dfrac{4V_0}{V_0}\right)$

$= p_0 V_0 \ln(4)$

Information regarding $p$ and $T$ at $C$ can not be obtained from the given graph. Unless it is mentioned that line $BC$ passes through origin or not.

Hence, the correct options are (a) and (b).

**12.**

1 kg → 5 kg $\Rightarrow$ ← $2\text{ms}^{-1}$ 1 kg

$v_1' = \left(\dfrac{m_1 - m_5}{m_1 + m_5}\right) v_1 + \left(\dfrac{2m_5}{m_1 + m_5}\right) v_5$

$-2 = \left(\dfrac{1-5}{1+5}\right) v_1 + 0 \qquad (\text{as } v_5 = 0)$

$\therefore \quad v_1 = 3\text{ ms}^{-1}$

$v_5' = \left(\dfrac{m_5 - m_1}{m_1 + m_5}\right) v_5 + \left(\dfrac{2m_1}{m_1 + m_2}\right) v_1$

$= 0 + \left(\dfrac{2 \times 1}{6}\right)(3) = 1\text{ ms}^{-1}$

$P_{\text{CM}} = P_i = (1)(3)$

$= 3\,\dfrac{\text{kg-m}}{\text{s}}$

$P_5' = (5)(1) = 5\,\dfrac{\text{kg-m}}{5}$

$K_{\text{CM}} = \dfrac{P_{\text{CM}}^2}{2M_{\text{CM}}}$

$= \dfrac{9}{2 \times 6} = 0.75\text{ J}$

$K_{\text{total}} = \dfrac{1}{2} \times 1 \times (3)^2$

$= 4.5\text{ J}$

$\therefore$ correct options are (a) and (c).

**13.** $\sqrt{3} = \dfrac{\sin 60^\circ}{\sin r}$

$\therefore \; r = 30^\circ$

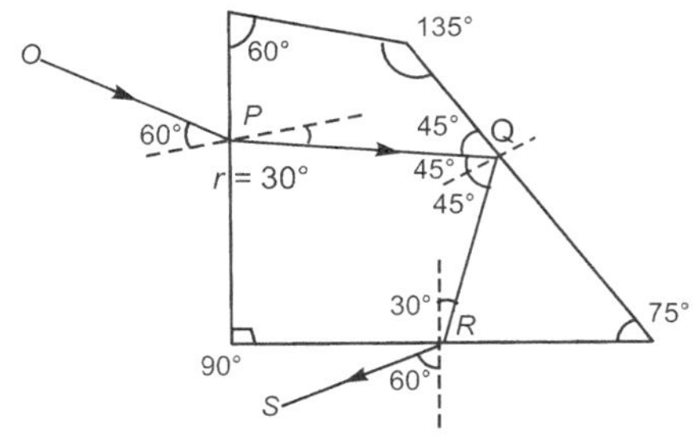


$$\theta_C = \sin^{-1}\left(\frac{1}{\sqrt{3}}\right)$$

or $\sin\theta_C = \frac{1}{\sqrt{3}} = 0.577$

At point $Q$, angle of incidence inside the prism is $i = 45°$.

Since $\sin i = \frac{1}{\sqrt{2}}$ is greater than $\sin\theta_C = \frac{1}{\sqrt{2}}$, ray gets totally internally reflected at face $CD$. Path of ray of light after point $Q$ is shown in figure.

From the figure, we can see that angle between incident ray $OP$ and emergent ray $RS$ is 90°.

Therefore, correct options are (a), (b) and (c).

**14.** If $B_2 > B_1$, critical temperature, (*at which resistance of semiconductors abruptly becomes zero*) in case–2 will be less than compared to case–1.

$\therefore$ The correct option is (a).

**15.** With increase in temperature, $T_C$ is decreasing.

$T_C(0) = 100$ K

$T_C = 75$ K at $B = 7.5$ T

Hence, at $B = 5$ T, $T_C$ should lie between 75K and 100K. Hence, the correct option should be (b).

**16.** If $E > V_0$, particle will escape. But simultaneously for oscillations, $E > 0$.

Hence, the correct answer is $V_0 > E > 0$ or the correct option is (c).

**17.** $[\alpha] = \left[\frac{\text{PE}}{x^4}\right] = \left[\frac{\text{ML}^2\text{T}^{-2}}{\text{L}^4}\right] = [\text{ML}^{-2}\text{T}^{-2}]$

$\therefore \quad \left[\frac{m}{\alpha}\right] = [\text{L}^2\text{T}^2]$

$\therefore \quad \left[\frac{1}{A}\sqrt{\frac{m}{\alpha}}\right] = [\text{T}]$

As dimensions of amplitude $A$ is [L].

Hence, the correct option is (b).

**18.** For $|x| > X_0$, potential energy is constant. Hence, kinetic energy, speed or velocity will also remain constant.

$\therefore$ Acceleration will be zero.

Hence, the correct option is (d).

**19.** Firstly, car will be treated as an observer which is approaching the source. Then, it will be treated as a source, which is moving in the direction of sound.

$C_1 \rightarrow v_1 \quad \bullet_S \quad v_2 \leftarrow C_2$

Hence, $f_1 = f_0\left(\frac{v + v_1}{v - v_1}\right)$

$f_2 = f_0\left(\frac{v + v_2}{v - v_2}\right)$

$\therefore \quad f_1 - f_2 = \left(\frac{1.2}{100}\right) f_0$

$= f_0\left[\frac{v + v_1}{v - v_1} - \frac{v + v_2}{v - v_2}\right]$

or $\left(\frac{1.2}{100}\right) f_0 = \frac{2v(v_1 - v_2)}{(v - v_1)(v - v_2)} f_0$

As $v_1$ and $v_2$ are very very less than $v$.

We can write, $(v - v_1)$ or $(v - v_2) \approx v$

$\therefore \quad \left(\frac{1.2}{100}\right) f_0 = \frac{2(v_1 - v_2)}{v} f_0$

or $(v_1 - v_2) = \frac{v \times 1.2}{200}$

$= \frac{330 \times 1.2}{200} = 1.98 \text{ ms}^{-1}$

$= 7.128 \text{ kmh}^{-1}$

$\therefore$ The nearest integer is 7.

**20.** $\frac{1}{v} - \frac{1}{u} = \frac{1}{f}$

or $\quad \frac{u}{v} - 1 = \frac{u}{f}$

or $\frac{u}{v} = \left(\frac{u+f}{f}\right)$

$\therefore$ $m = \frac{v}{u} = \left(\frac{f}{u+f}\right)$

$$\frac{m_{25}}{m_{50}} = \frac{\left(\frac{20}{-25+20}\right)}{\left(\frac{20}{-50+20}\right)} = 6$$

$\therefore$ answer is 6.

**21.** $\lambda = \frac{h}{P} = \frac{h}{\sqrt{2qqVm}}$ or $\lambda \propto \frac{1}{\sqrt{qm}}$

$$\frac{\lambda_p}{\lambda_\alpha} = \sqrt{\frac{q_\alpha}{q_p} \cdot \frac{m_\infty}{m_p}} = \sqrt{\frac{(2)(4)}{(1)(1)}} = 2.828$$

The nearest integer is 3.

$\therefore$ answer is 3.

**22.** In series, $i = \frac{2E}{2+R}$

$\therefore J_1 = i^2R = \left(\frac{2E}{2+R}\right)^2 \cdot R$

In parallel, $i = \frac{E}{0.5+R}$

$\therefore$ $J_2 = i^2R = \left(\frac{E}{0.5+R}\right)^2 \cdot R$

$$\frac{J_1}{J_2} = 2.25 = \frac{4(0.5+R)^2}{(2+R)^2}$$

or $1.5 = \frac{2(0.5+R)}{(2+R)}$

Solving we get, $R = 4\Omega$

$\therefore$ The answer is 4.

**23.** $\lambda_m \propto \frac{1}{T}$

$\therefore \frac{\lambda_A}{\lambda_B} = \frac{T_B}{T_A} = \frac{500}{1500} = \frac{1}{3}$

$E \propto T^4 A$ (where $A$ = surface area $= 4\pi R^2$)

$\therefore E \propto T^4 R^2$

$$\frac{E_A}{E_B} = \left(\frac{T_A}{T_B}\right)^4 \left(\frac{R_A}{R_B}\right)^2$$

$$= (3)^4 \left(\frac{6}{18}\right)^2 = 9$$

$\therefore$ Answer is 9.

**24.** Phase difference between the two waves is 90°. Amplitudes are added by vector method.

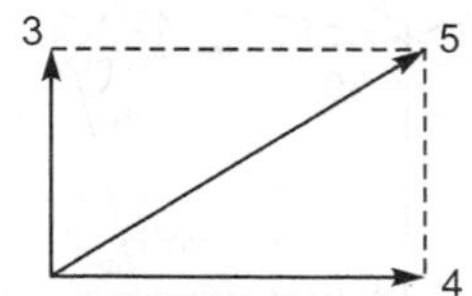


$\therefore$ Answer is 5.

**25.** $\omega = \sqrt{\frac{K}{m}} = \sqrt{\frac{YA}{lm}} = \sqrt{\frac{(n \times 10^9)(4.9 \times 10^{-7})}{1 \times 0.1}}$

Putting, $\omega = 140$ rad $s^{-1}$ in above equation we get,

$n = 4$

$\therefore$ Answer is 4.

**26.** $\frac{L_{Total}}{L_B} = \frac{(I_A + I_B)\omega}{I_B \cdot \omega}$ (as $\omega$ will be same in both cases)

$= \frac{I_A}{I_B} + 1 = \frac{m_A r_A^2}{m_B r_B^2} + 1$

$= \frac{r_A}{r_B} + 1$ (as $m_A r_A = m_B r_B$)

$= \frac{11}{2.2} + 1$ $\left(\text{as } r \propto \frac{1}{m}\right)$

$= 6$

$\therefore$ The correct answer is 6.

**27.** $g = \frac{GM}{R^2} = \frac{G\left(\frac{4}{3}\pi R^3\right)\rho}{R^2}$

or $g \propto \rho R$

or $R \propto \frac{g}{\rho}$

Now escape velocity, $v_e = \sqrt{2gR}$

or $v_e \propto \sqrt{gR}$

or $v_e \propto \sqrt{g \times \frac{g}{\rho}} \propto \sqrt{\frac{g^2}{\rho}}$

$\therefore (v_e)_{planet} = (11 \text{ km-s}^{-1})\sqrt{\frac{6}{121} \times \frac{3}{2}}$

$= 3 \text{ km-s}^{-1}$

$\therefore$ The correct answer is 3.

**28.** Language of question is slightly wrong. As heat capacity and specific heat are two different physical quantities. Unit of heat capacity is J - $kg^{-1}$ not J - $kg^{-1}$ - $°C^{-1}$. The heat capacity given in the question is really the specific heat. Now applying the heat exchange equation.

$$420 = (m \times 10^{-3})(2100)(5) + (1 \times 10^{-3})(3.36 \times 10^5)$$

Solving this equation we get,

$$m = 8\,g$$

∴ The correct answer is 8.

# Paper II

**1.** $qE = mg$ ...(i)

$$6\pi\eta rv = mg$$

$$\frac{4}{3}\pi r^3 \rho g = mg \quad \text{...(ii)}$$

$$\therefore \quad r = \left(\frac{3mg}{4\pi\rho g}\right)^{1/3} \quad \text{...(iii)}$$

Substituting the value of $r$ in Eq. (ii) we get,

$$6\pi\eta v\left(\frac{3mg}{4\pi\rho g}\right)^{1/3} = mg$$

or $$(6\pi\eta v)^3\left(\frac{3mg}{4\pi\rho g}\right) = (mg)^3$$

Again substituting $mg = qE$ we get,

$$(qE)^2 = \left(\frac{3}{4\pi\rho g}\right)(6\pi\eta v)^3$$

or $$qE = \left(\frac{3}{4\pi\rho g}\right)^{1/2}(6\pi\eta v)^{3/2}$$

$$\therefore \quad q = \frac{1}{E}\left(\frac{3}{4\pi\rho g}\right)^{1/2}(6\pi\eta v)^{3/2}$$

Substituting the values we get ,

$$q = \frac{7}{81\pi \times 10^5}\sqrt{\frac{3}{4\pi \times 900 \times 9.8} \times 216\pi^3} \times \sqrt{(1.8 \times 10^{-5} \times 2 \times 10^{-3})^3}$$

$$= 8.0 \times 10^{-19}\ C$$

∴ correct option is (d).

**2.** Electrical force per unit area $= \frac{1}{2}\varepsilon_0 E^2$

$$= \frac{1}{2}\varepsilon_0\left(\frac{\sigma}{\varepsilon_0}\right)^2 = \frac{\sigma^2}{2\varepsilon_0}$$

Projected area $= \pi R^2$

∴ net electrical force $= \left(\frac{\sigma^2}{2\varepsilon_0}\right)(\pi R^2)$

In equilibrium, this force should be equal to the applied force.

$$\therefore \quad F = \frac{\pi\sigma^2 R^2}{2\varepsilon_0}$$

or $$F \propto \frac{\sigma^2 R^2}{\varepsilon_0}$$

∴ The correct option is (a).

**3.** Area under $F$ - $t$ graph = momentum

$$= P = \sqrt{2km}$$

$$\therefore \quad k = \frac{A^2}{2m}$$

($A$ = net area of $F$ - $t$ graph)

$$= \frac{\left\{\left(\frac{4 \times 3}{2}\right) - \left(\frac{1.5 \times 2}{2}\right)\right\}^2}{2 \times 2} = 5.0625\ J$$

∴ The correct option is (c).

**4.** Object is placed at distance $2f$ from the lens. So first image $I_1$ will be formed at distance $2f$ on other side. This image $I_1$ will behave like a virtual object for mirror. The second image $I_2$ will be formed at distance 20 cm in front of the mirror, or at distance 10 cm to the left hand side of the lens.

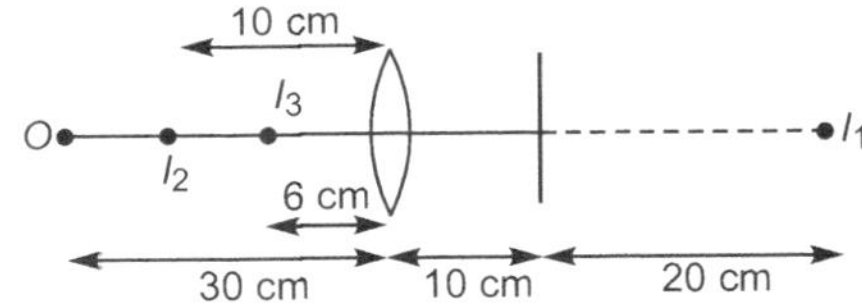

Now applying lens formula

$$\frac{1}{v}-\frac{1}{u}=\frac{1}{f}$$

$$\therefore \quad \frac{1}{v}-\frac{1}{+10}=\frac{1}{+15}$$

or $\quad v = 6$ cm

Therefore, the final image is at distance 16 cm from the mirror. But, this image will be real.

This is because ray of light is travelling from right to left.

$\therefore$The correct option is (b).

**5.** Least count of vernier calipers

LC = 1MSD – 1 VSD

$$=\frac{\text{Smallest division on main scale}}{\text{Number of divisions on vernier scale}}$$

20 divisions of vernier scale

= 16 divisions of main scale

$$\therefore \quad 1\text{ VSD} = \frac{16}{20}\text{ mm} = 0.8\text{ mm}$$

$$\therefore \quad \text{LC} = 1\text{MSD} - 1\text{VSD} = 1\text{ mm} - 0.8\text{mm}$$

$$\therefore \quad = 0.2\text{ mm}$$

$\therefore$The correct option is (d).

**6.** $2\left(\frac{v_1}{2l_1}\right)=\frac{v_2}{4l_2}$

$$\therefore \quad \frac{\sqrt{T/\mu}}{l_1}=\frac{320}{4l_2}$$

($\mu$ = mass per unit length of wire)

$$\text{or} \quad \frac{\sqrt{50/\mu}}{0.5}=\frac{320}{4\times 0.8}$$

Solving we get $\mu = 0.02\,\frac{\text{kg}}{\text{m}} = 20\,\frac{\text{g}}{\text{m}}$

$\therefore$Mass of string

$$=\left(20\,\frac{\text{g}}{\text{m}}\right)(0.5\,\text{m}) = 10\text{ g}$$

$\therefore$The correct option is (b).

**7.** $\left|\frac{dN}{dt}\right|$ = |Activity of radioactive substance|

$$= \lambda N = \lambda N_0\, e^{-\lambda t}$$

Taking log both sides

$$\ln\left|\frac{dN}{dt}\right| = \ln(\lambda N_0) - \lambda t$$

Hence, $\ln\left|\frac{dN}{dt}\right|$ *versus* $t$ graph is a straight line with slope $-\lambda$.

From the graph we can see that,

$$\lambda = \frac{1}{2} = 0.5\text{ yr}^{-1}$$

Now applying the equation,

$$N = N_0 e^{-\lambda t} = N_0\, e^{-0.5\times 4.16} = N_0\, e^{-2.08} = 0.125\, N_0 = \frac{N_0}{8}$$

*ie*, nuclei decreases by a factor of 8.

Hence the answer is 8.

**8.** In adiabatic process,

$$TV^{\gamma-1} = \text{constant}$$

$$\therefore \quad T_i\, V_i^{0.4} = T_f V_f^{0.4}$$

(as $\gamma = 1.4$ for diatomic gas)

$$\text{or} \quad T_i V_i^{0.4} = (\alpha\, T_i)\left(\frac{V_i}{32}\right)^{0.4}$$

$$\text{or} \quad \alpha\,(32)^{0.4} = 4$$

$\therefore$The correct answer is 4.

**9.** Voltage across the capacitors will increase from 0 to 10 V exponentially. The voltage at time $t$ will be given by

$V = 10\,(1 - e^{-t/\tau}\text{C})$

Here $\tau_c = C_{\text{net}}\, R_{\text{net}}$

$$= (1\times 10^6)(4\times 10^{-6}) = 4\text{ s}$$

$$\therefore \quad V = 10\,(1 - e^{-t/4})$$

Substituting $V = 4$ volt we have,

$$4 = 10\,(1 - e^{-t/4})$$

$$\text{or} \quad e^{-t/4} = 0.6 = \frac{3}{5}$$

Taking log both sides we have,

$$-\frac{t}{4} = \ln 3 - \ln 5$$

$$\text{or} \quad t = 4\,(\ln 5 - \ln 3) = 2\text{ s}.$$

Hence, the answer is 2 .

**10.** Using mirror formula twice,

$$\frac{1}{+25/3}+\frac{1}{-u_1}=\frac{1}{+10}$$

or $$\frac{1}{u_1}=\frac{3}{25}-\frac{1}{10} \text{ or } u_1 = 50\text{ m}$$

and, $$\frac{1}{(+50/7)}+\frac{1}{-u_2}=\frac{1}{+10}$$

$\therefore$ $$\frac{1}{u_2}=\frac{7}{50}-\frac{1}{10} \text{ or } u_2 = 25\text{ m}$$

$$\text{Speed of object} = \frac{u_1-u_2}{\text{time}} = \frac{25}{30}\text{ ms}^{-1} = 3\text{ kmh}^{-1}$$

$\therefore$ The answer is 3.

**11.**

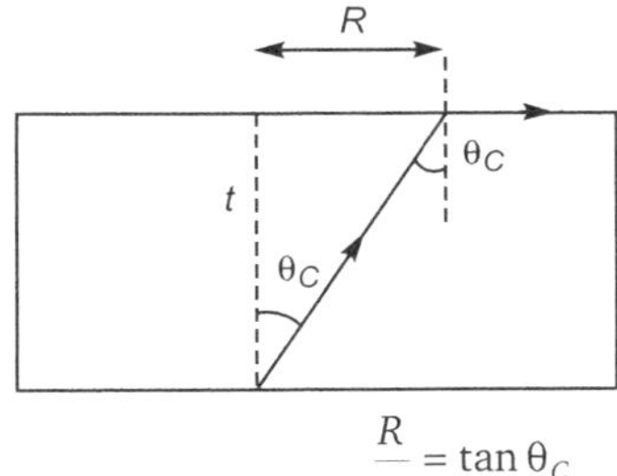


$$\frac{R}{t}=\tan\theta_C$$

or $R = t\,(\tan\theta_C)$

But, $\sin\theta_C = \frac{1}{\mu}=\frac{3}{5}$

$\therefore$ $\tan\theta_C = \frac{3}{4}$

$\therefore$ $R = \frac{3}{4}t = \frac{3}{4}(8\text{ cm}) = 6\text{ cm}.$

Hence the answer is 6.

**12.** $L = I\omega = \frac{nh}{2\pi} \therefore \omega = \frac{nh}{2\pi I}$

$$K = \frac{1}{2}I\omega^2 = \frac{1}{2}I\left(\frac{nh}{2\pi I}\right)^2 = \frac{n^2h^2}{8\pi^2 I}$$

$\therefore$ The correct answer is (d).

**13.** $$h\nu = K_2 - K_1 = \frac{3h^2}{8\pi^2 I}$$

$$\therefore I = \frac{3h}{8\pi^2 f} = \frac{3\times 2\pi\times 10^{-34}\times\pi}{8\times\pi^2\times 4\times 10^{11}}$$

$$= 1.87\times 10^{-46}\text{ kg - m}^2$$

$\therefore$ The correct answer is (b).

**14.** $I = \mu r^2$ ( where $\mu$ = reduced mass)

$$\mu = \frac{m_1 m_2}{m_1+m_2} = \frac{48}{7}\text{ amu}$$

$$= 11.43\times 10^{-27}\text{ kg}$$

Substituting in $I = \mu r^2$ we get,

$$r = \sqrt{\frac{I}{\mu}} = \sqrt{\frac{1.87\times 10^{-46}}{11.43\times 10^{-27}}}$$

$$= 1.28\times 10^{-10}\text{ m}$$

$\therefore$ The correct answer is (c).

**15.** Vertical force due to surface tension,

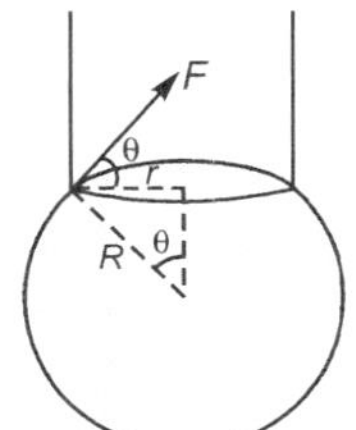


$$F_v = F\sin\theta = (T\cdot 2\pi r)(r/R) = \frac{2\pi r^2 T}{R}$$

$\therefore$ The correct option is (c).

**16.** $$\frac{2\pi r^2 T}{R} = mg = \frac{4}{3}\pi R^3\cdot\rho\cdot g$$

$$\therefore R^4 = \frac{3r^2T}{2\rho g} = \frac{3\times(5\times 10^{-4})^2\,(0.11)}{2\times 10^3\times 10}$$

$$= 4.125\times 10^{-12}\text{ m}^4$$

$\therefore$ $R = 1.425\times 10^{-3}\text{ m}\approx 1.4\times 10^{-3}\text{ m}.$

$\therefore$ The correct option is (a).

**17.** Surface energy,

$$E = (4\pi R^2)T = (4\pi)(1.4\times 10^{-3})^2\,(0.11) = 2.7\times 10^{-6}\text{ J}$$

$\therefore$ The correct option is (b).

**18.** **In circuit (P),** I can't be non zero in steady state.

**In circuit (Q) :**

$V_1 = 0$ and $V_2 = 2I = V$ (also)

**In circuit (R)** : $V_1 = X_L I = (2\pi f L) I$

$= (2\pi \times 50 \times 6 \times 10^{-3}) I$

$= 1.88\, I$

$V_2 = 2I$

**In circuit (S)**: $V_1 = X_L I = 1.88 I$

$$V_2 = X_C I = \left(\frac{1}{2\pi fC}\right) I$$

$$= \left(\frac{1}{2\pi \times 50 \times 3 \times 10^{-6}}\right) I = (1061) I$$

**In circuit (T)**:

$V_1 = IR = (1000) I$

$V_2 = X_C I = (1061) I$

Therefore the correct options are as under

(A) → R, S, T (B) → Q, R, S, T

(C) → Q or P, Q (D) → Q, R, S, T

**19.** (A)→ since $\mu_1 < \mu_2$, the ray of light will bend towards normal after first refraction.

(B) → $\mu_1 > \mu_2$, the ray of light will bend away from the normal after first refraction.

(C) → since $\mu_2 = \mu_3$ means in second refraction there will be no change in the path of ray of light.

(D) → Since $\mu_2 > \mu_3$, ray of light will bend away from the normal after second refraction.

Therefore the correct options are as under.

(A) → P, R

(B) → Q, S, T

(C) → P, R, T

(D) → Q, S

# IIT JEE

## SOLVED PAPER 2010

## Chemistry

## Paper I

### SECTION I—Single Correct Choice Type

*This section contains **8 multiple choice questions.** Each question has 4 choices (a), (b), (c), and (d), out of which **ONLY ONE** is correct.*

**1.** The correct statement about the following disaccharide is

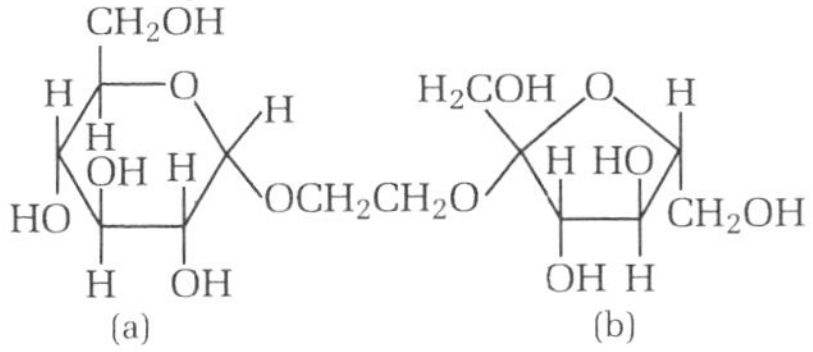

[Structure: two rings (a) and (b) joined by $OCH_2CH_2O$; ring (a) bears $CH_2OH$, H, O, H, OH, HO, H, OH; ring (b) bears $H_2COH$, O, H, H, HO, $CH_2OH$, OH, H]

(a) (b)

(a) Ring (a) is pyranose with α-glycosidic link

(b) Ring (a) is furanose with α-glycosidic link

(c) Ring (b) is furanose with α-glycosidic link

(d) Ring (b) is pyranose with β-glycosidic link

**2.** The synthesis of 3-octyne is achieved by adding a bromoalkane into a mixture of sodium amide and an alkyne. The bromoalkane and alkyne respectively are

(a) $BrCH_2CH_2CH_2CH_2CH_3$ and $CH_3CH_2C{\equiv}CH$

(b) $BrCH_2CH_2CH_3$ and $CH_3CH_2CH_2C{\equiv}CH$

(c) $BrCH_2CH_2CH_2CH_2CH_3$ and $CH_3C{\equiv}CH$

(d) $BrCH_2CH_2CH_2CH_3$ and $CH_3CH_2C{\equiv}CH$

**3.** The ionization isomer of $[Cr(H_2O)_4Cl(NO_2)]Cl$ is

(a) $[Cr(H_2O)_4(O_2N)]Cl_2$

(b) $[Cr(H_2O)_4Cl_2](NO_2)$

(c) $[Cr(H_2O)_4Cl(ONO)]Cl$

(d) $[Cr(H_2O)_4Cl_2(NO_2)]\cdot H_2O$

**4.** The correct structure of ethylenediaminetetraacetic acid (EDTA) is

(a) 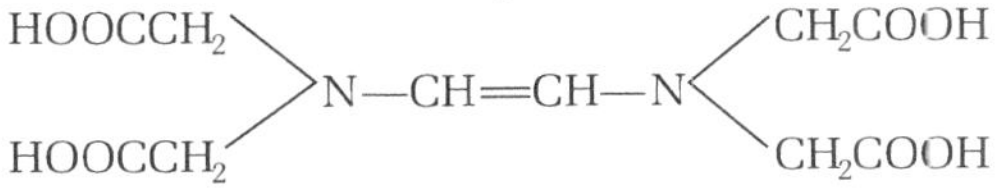 $(HOOCCH_2)_2N{-}CH{=}CH{-}N(CH_2COOH)_2$

(b) 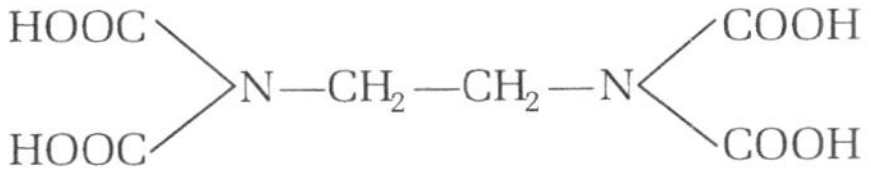 $(HOOC)_2N{-}CH_2{-}CH_2{-}N(COOH)_2$

(c) 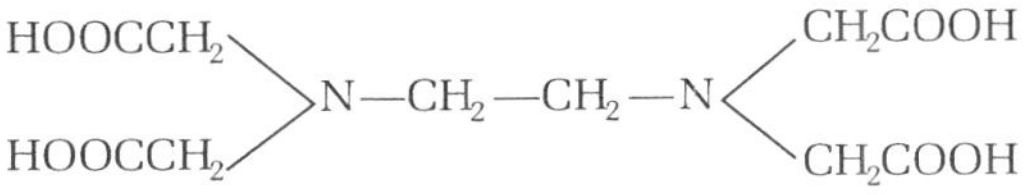 $(HOOCCH_2)_2N{-}CH_2{-}CH_2{-}N(CH_2COOH)_2$

(d) 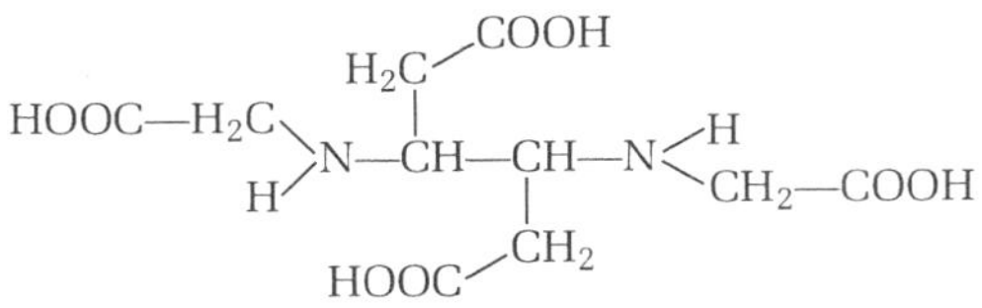 $HOOC{-}H_2C(H)N{-}CH(CH_2COOH){-}CH(CH_2COOH){-}NH(CH_2{-}COOH)$

**5.** The bond energy (in kcal $mol^{-1}$) of C—C single bond is approximately
(a) 1 (b) 10
(c) 100 (d) 1000

**6.** The species which by definition has zero standard molar enthalpy of formation at 298 K is
(a) $Br_2(g)$
(b) $Cl_2(g)$
(c) $H_2O(g)$
(d) $CH_4(g)$

**7.** Plots showing the variation of the rate constant ($k$) with temperature ($T$) are given below. The plot that follows Arrhenius equation is

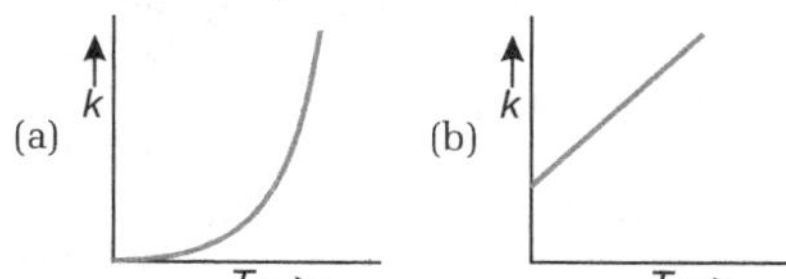


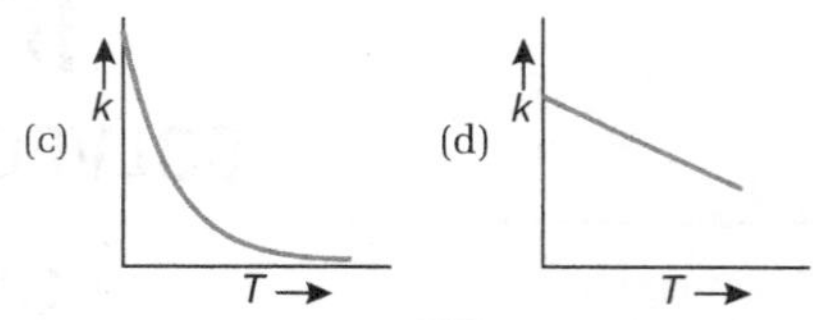


**8.** In the reaction, $C_6H_5$—$OCH_3$ $\xrightarrow{HBr}$ the products are,
(a) Br—$C_6H_4$—$OCH_3$ and $H_2$
(b) $C_6H_5$—Br and $CH_3Br$
(c) $C_6H_5$—Br and $CH_3OH$
(d) $C_6H_5$—OH and $CH_3Br$

## SECTION II–Multiple Correct Choice Type

*This section contains* **5 *multiple choice questions.*** *Each question has 4 choices (a), (b), (c), and (d), out of which* **one or more** *may be correct.*

**9.** Among the following, intensive property is (properties are)
(a) molar conductivity
(b) electromotive force
(c) resistance
(d) heat capacity

**10.** The reagent(s) used for softening the temporary hardness of water is(are)
(a) $Ca_3(PO_4)_2$ (b) $Ca(OH)_2$
(c) $Na_2CO_3$ (d) NaOCl

**11.** Aqueous solutions of $HNO_3$, KOH, $CH_3COOH$, and $CH_3COONa$ of identical concentrations are provided. The pair(s) of solutions which form a buffer upon mixing is(are)
(a) $HNO_3$ and $CH_3COOH$
(b) KOH and $CH_3COONa$
(c) $HNO_3$ and $CH_3COONa$
(d) $CH_3COOH$ and $CH_3COONa$

**12.** In the Newman projection for 2,2-dimethylbutane

(Newman projection: front groups X, $H_3C$, $CH_3$; back groups H, H, Y)

$X$ and $Y$ can respectively be
(a) H and H (b) H and $C_2H_5$
(c) $C_2H_5$ and H (d) $CH_3$ and $CH_3$

**13.** In the reaction, $C_6H_5OH$ $\xrightarrow{NaOH(aq)/Br_2}$ the intermediate(s) is (are)

(a)

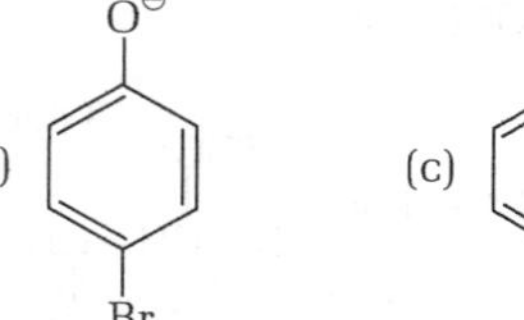


(b) 4,4-dibromocyclohexa-2,5-dienone (O, Br, Br)

(c) p-bromophenoxide ($O^{\ominus}$, Br)

(c) m-bromophenoxide ($O^{\ominus}$, Br)

## SECTION III—Paragraph Type

*This section contains **2 paragraphs**. Based upon the first paragraph 3 multiple choice questions and based upon the second paragraph **2 multiple choice questions** have to be answered. Each of these questions has 4 choices (a), (b), (c) and (d) out of **WHICH ONLY ONE CORRECT.***

***$C_{14-16}$*** Copper is the most noble of the first row transition metals and occurs in small deposits in several countries. Ores of copper include chalcanthite ($CuSO_4 \cdot 5H_2O$), atacamite ($Cu_2Cl(OH)_3$), cuprite ($Cu_2O$), copper glance ($Cu_2S$) and malachite ($Cu_2(OH)_2CO_3$). However, 80% of the world copper production comes from the ore chalcopyrite ($CuFeS_2$). The extraction of copper from chalcopyrite involves partial roasting, removal of iron and self-reduction.

**14.** Partial roasting of chalcopyrite produces
(a) $Cu_2S$ and FeO
(b) $Cu_2O$ and FeO
(c) CuS and $Fe_2O_3$
(d) $Cu_2O$ and $Fe_2O_3$

**15.** Iron is removed from chalcopyrite as
(a) FeO (b) FeS
(c) $Fe_2O_3$ (d) $FeSiO_3$

**16.** In self-reduction, the reducing species is
(a) S (b) $O^{2-}$ (c) $S^{2-}$ (d) $SO_2$

***$C_{17-18}$*** The concentration of potassium ions inside a biological cell is at least twenty times higher than the outside. The resulting potential difference across the cell is important in several processes such as transmission of nerve impulses and maintaining the ion balance. A simple model for such a concentration cell involving a metal M is :

$M(s) \mid M^+(aq; 0.05 \text{ molar}) \mid\mid M^+ (aq; 1 \text{ molar}) \mid M(s)$

For the above electrolytic cell the magnitude of the cell potential $|E_{cell}| = 70$ mV.

**17.** For the above cell
(a) $E_{cell} < 0; \Delta G > 0$
(b) $E_{cell} > 0; \Delta G < 0$
(c) $E_{cell} < 0; \Delta G° > 0$
(d) $E_{cell} > 0; \Delta G° < 0$

**18.** If the 0.05 molar solution of $M^+$ is replaced by a 0.0025 molar $M^+$ solution, then the magnitude of the cell potential would be
(a) 35 mV (b) 70 mV
(c) 140 mV (d) 700 mV

## SECTION IV—Integer Type

*This section contains **TEN questions**. The answer to each question is a **Single digit integer** ranging from 0 to 9. The correct digit below the question number in the ORS is to be bubbled.*

**19.** Amongst the following, the total number of compounds whose aqueous solution turns red litmus paper blue is

KCN $K_2SO_4$ $(NH_4)_2C_2O_4$ NaCl
$Zn(NO_3)_2$ $FeCl_3$ $K_2CO_3$ $NH_4NO_3$
LiCN

**20.** Based on VSEPR theory, the number of 90 degree F—Br—F angles in $BrF_5$ is

**21.** The value of $n$ in the molecular formula $Be_nAl_2Si_6O_{18}$ is

**22.** A student performs a titration with different burettes and finds titre values of 25.2 mL, 25.25 mL, and 25.0 mL. The number of significant figures in the average titre value is

**23.** The concentration of $R$ in the reaction $R \longrightarrow P$ was measured as a function of time and the following data is obtained :

| $[R]$ (molar) | 1.0 | 0.75 | 0.40 | 0.10 |
|---|---|---|---|---|
| $t$ (min) | 0.0 | 0.05 | 0.12 | 0.18 |

The order of the reaction is

**24.** The number of neutrons emitted when $^{235}_{92}U$ undergoes controlled nuclear fission to $^{142}_{54}Xe$ and $^{90}_{38}Sr$ is

**25.** The total number of basic groups in the following form of lysine is

$H_3\overset{\oplus}{N}-CH_2-CH_2-CH_2-H_2C-CH(NH_2)-COO^{\ominus}$

**26.** The total number of cyclic isomers possible for a hydrocarbon with the molecular formula $C_4H_6$ is

**27.** In the scheme given below, the total number of intramolecular aldol condensation products formed from '$Y$' is

Bicyclic alkene $\xrightarrow[2.\ Zn,\ H_2O]{1.\ O_3}$ $Y$ $\xrightarrow[2.\ heat]{1.\ NaOH(aq)}$

**28.** Amongst the following, the total number of compounds soluble in aqueous NaOH is

N,N-dimethylaniline ($H_3C-N-CH_3$ on benzene); cyclohexanecarboxylic acid (COOH); benzene ring with $OCH_2CH_3$ and $CH_2OH$; phenol (OH); nitrobenzene ($NO_2$); 4-(dimethylamino)phenol (OH, $H_3C-N-CH_3$); benzene ring with $CH_2CH_3$ and $CH_2CH_3$; naphthoic acid (COOH)

# Paper II

## SECTION I—Single Correct Choice Type

*This section contains* **6 *multiple choice questions***. *Each question has* 4 *choices* (*a*), (*b*), (*c*), and (*d*), *out of which* ***ONLY ONE*** *is correct.*

**1.** The compounds *P*, *Q* and *S*

*P*: $H_3C-C_6H_4-COOH$ *Q*: $H_3C-C_6H_4-OCH_3$ *S*: $H_3C-C_6H_4-C(=O)-O-C_6H_5$

where separately subjected to nitration using $HNO_3/H_2SO_4$ mixture. The major product formed in each case respectively, is

(a) HO, COOH, $NO_2$ ; $H_3C$, $OCH_3$, $NO_2$ ; $NO_2$, C(=O)–O–(benzene ring)

(b) HO, COOH, $NO_2$ ; $H_3C$, $OCH_3$, $NO_2$ ; C(=O)–O–(benzene ring), $NO_2$

(c) HO, COOH, $NO_2$ ; $H_3C$, $OCH_3$, $NO_2$ ; (benzene ring)–C(=O)–O–, $NO_2$

(d) HO, COOH, $NO_2$ ; $H_3C$, $OCH_3$, $NO_2$ ; (benzene ring)–C(=O)–O–, $NO_2$

**2.** Assuming that Hund's rule is violated, the bond order and magnetic nature of the diatomic molecule $B_2$ is

(a) 1 and diamagnetic
(b) 0 and diamagnetic
(c) 1 and paramagnetic
(d) 0 and paramagnetic

**3.** The packing efficiency of the two-dimensional square unit cell shown below is

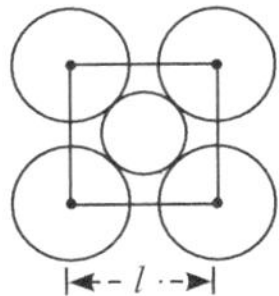


(a) 39.27% (b) 68.02%
(c) 74.05% (d) 78.54%

**4.** The complex showing a spin-only magnetic moment of 2.82 B.M. is

(a) $Ni(CO)_4$ (b) $[NiCl_4]^{2-}$
(c) $Ni(PPh_3)_4$ (d) $[Ni(CN)_4]^{2-}$

**5.** In the reaction $H_3C$–(benzene ring)–C(=O)$NH_2$ $\xrightarrow[\text{(2) } C_6H_5COCl]{\text{(1) NaOH/Br}_2}$ $T$

the structure of the product $T$ is

(a) $H_3C$–(benzene ring)–C(=O)–O–C(=O)–(benzene ring)

(b) $C_6H_5$–NH–C(=O)–$C_6H_4$–$CH_3$

(c) $H_3C$–$C_6H_4$–NH–C(=O)–$C_6H_5$

(d) $H_3C$–$C_6H_4$–C(=O)–NH–C(=O)–$C_6H_5$

6. The species having pyramidal shape is
(a) $SO_3$ (b) $BrF_3$
(c) $SiO_3^{2-}$ (d) $OSF_2$

## SECTION II—Integer Type

*This section contains* **5 questions**. *The answer to each question is a single-digit integer, ranging from* 0 *to* 9. *The correct digit below the question no. in the ORS is to be bubbled.*

7. Silver (atomic weight $= 108 \text{ g mol}^{-1}$) has a density of $10.5 \text{ g cm}^{-3}$. The number of silver atoms on a surface of area $10^{-12} \text{ m}^2$ can be expressed in scientific notation as $y \times 10^x$. The value of $x$ is

8. Among the following, the number of elements showing only one non-zero oxidation state is O, Cl, F, N, P, Sn, Tl, Na, Ti

9. One mole of an ideal gas is taken from $a$ to $b$ along two paths denoted by the solid and the dashed lines as shown in the graph below. If the work done along the solid line path is $W_s$ and that along the dotted line path is $W_d$, then the integer closest to the ratio $\frac{W_d}{W_s}$ is

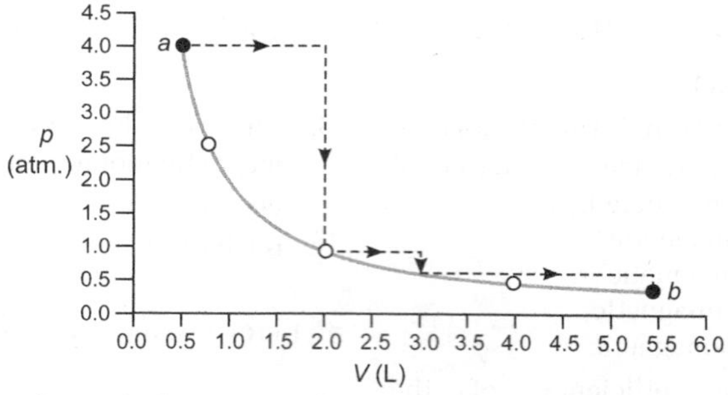


10. The total number of diprotic acids among the following is

| | | |
|---|---|---|
| $H_3PO_4$ | $H_2SO_4$ | $H_3PO_3$ |
| $H_2CO_3$ | $H_2S_2O_7$ | $H_3BO_3$ |
| $H_3PO_2$ | $H_2CrO_4$ | $H_2SO_3$ |

11. Total number of geometrical isomers for the complex $[RhCl(CO)(PPh_3)(NH_3)]$ is

## SECTION III— Paragraph Type

*This section contains* **2 paragraphs.** *Based upon each of the paragraphs* **3 multiple choice questions** *have to be answered. Each of these question has* 4 *choices (a), (b), (c) and (d) out of which* **ONLY ONE** *is correct.*

***C*$_{12-14}$** Two aliphatic aldehydes P and Q react in the presence of aqueous $K_2CO_3$ to give compound R, which upon treatment with HCN provides compound S. On acidification and heating, S gives the product shown below :

$H_3C$ $H_3C$ OH O O

**12.** The compounds *P* and *Q* respectively are

(a) $H_3C-CH(CH_3)-C(=O)-H$ and $H_3C-C(=O)-H$

(b) $H_3C-CH(CH_3)-C(=O)-H$ and $H-C(=O)-H$

(c) $H_3C-CH(CH_3)-CH_2-C(=O)-H$ and $H_3C-C(=O)-H$

(d) $H_3C-CH(CH_3)-CH_2-C(=O)-H$ and $H-C(=O)-H$

**13.** The compound *R* is

(a) $(H_3C)_2C(CH_2OH)-C(=O)-H$

(b) $(H_3C)_2C(CH(OH)CH_3)-C(=O)-H$

(c) $H_3C-CH(CH_3)-CH(CH_2OH)-C(=O)-H$

(d) $H_3C-CH(CH_3)-CH(CH(OH)CH_3)-C(=O)-H$

**14.** The compound *S* is

(a) $H_3C-CH(CH_3)-CH(CH_2CN)-C(=O)-H$

(b) $(H_3C)_2C(CH_2CN)-C(=O)-H$

(c) $H_3C-CH(CH_3)-CH(CH_2OH)-CH(CN)-OH$

(d) $(H_3C)_2C(CH_2OH)-CH(CN)-OH$

$C_{15-17}$ The hydrogen-like species $Li^{2+}$ is in a spherically symmetric state $S_1$ with one radial node. Upon absorbing light the ion undergoes transition to a state $S_2$. The state $S_2$ has one radial node and its energy is equal to the ground state energy of the hydrogen atom.

**15.** The state $S_1$ is
(a) 1 *s* (b) 2 *s*
(c) 2*p* (d) 3 *s*

**16.** Energy of the state $S_1$ in units of the hydrogen atom ground state energy is
(a) 0.75 (b) 1.50
(c) 2.25 (d) 4.50

**17.** The orbtial angular momentum quantum number of the state $S_2$ is
(a) 0
(b) 1
(c) 2
(d) 3

## SECTION IV—Matrix Type

*This section contains 2 questions. Each question has 4 statements (A), (B), (C) and (D) given in Column I and 5 statements (P), (Q), (R), (S) and (T) in Column II. Any given statement in Column I can have correct matching with one or more statement(s) given in Column II. For example, if for a given question, statement B matches with the statements given in (Q) and (R), then for that particular question, against statement B, darken the bubbles corresponding to Q and R in the* ***ORS****.*

**18.** Match the reaction in Column I with appropriate options in Column II.

| | Column I | | Column II |
|---|---|---|---|
| (A) | 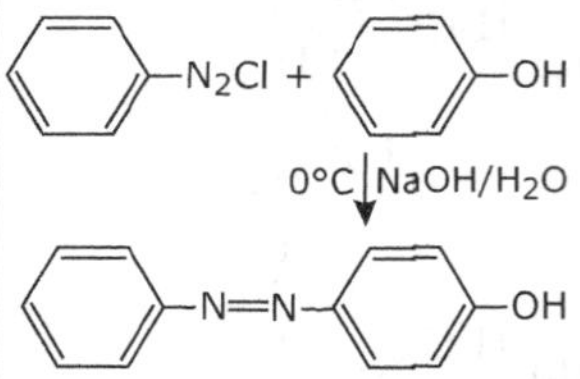 $C_6H_5-N_2Cl$ + $C_6H_5-OH$ $\xrightarrow{0°C,\ NaOH/H_2O}$ $C_6H_5-N{=}N-C_6H_4-OH$ | (P) | Racemic mixture |
| (B) | 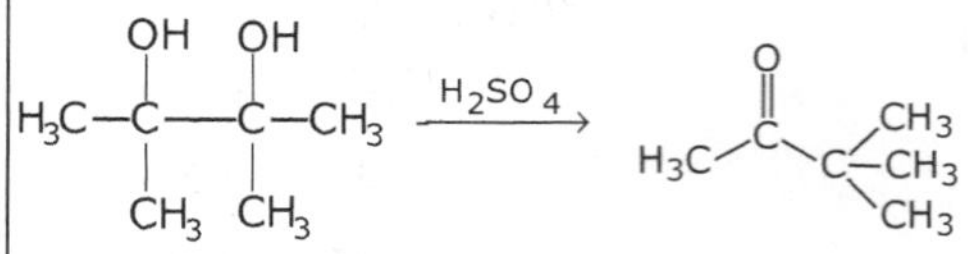 $H_3C-C(OH)(CH_3)-C(OH)(CH_3)-CH_3 \xrightarrow{H_2SO_4} H_3C-CO-C(CH_3)_3$ | (Q) | Addition reaction |
| (C) | 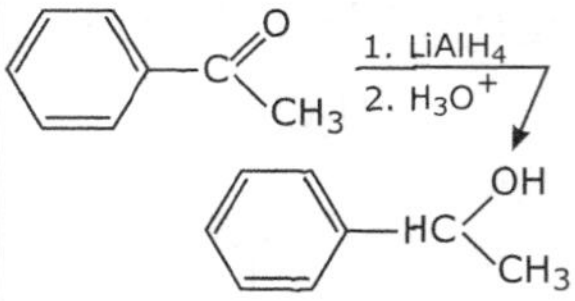 $C_6H_5-C(=O)-CH_3 \xrightarrow{1.\ LiAlH_4,\ 2.\ H_3O^+} C_6H_5-HC(OH)-CH_3$ | (R) | Substitution reaction |
| (D) | 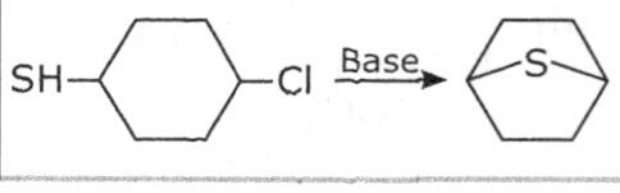 SH–(cyclohexane)–Cl $\xrightarrow{Base}$ (bicyclic sulfide, S-bridged) | (S) | Coupling reaction |
| | | (T) | Carbocation intermediate |

**19.** All the compounds listed in Column I react with water. Match the result of the respective reactions with the appropriate options listed in Column II.

| | Column I | | Column II |
|---|---|---|---|
| (A) | $(CH_3)_2SiCl_2$ | (P) | Hydrogen halide formation |
| (B) | $XeF_4$ | (Q) | Redox reaction |
| (C) | $Cl_2$ | (R) | Reacts with glass |
| (D) | $VCl_5$ | (S) | Polymerization |
| | | (T) | $O_2$ formation |

# Answers

## Paper I

| | | | | | | | | | |
|---|---|---|---|---|---|---|---|---|---|
| **1.** (a) | **2.** (d) | **3.** (b) | **4.** (c) | **5.** (c) | **6.** (b) | **7.** (a) | **8.** (d) | **9.** (a,b) | **10.** (b,c,d) |
| **11.** (c,d) | **12.** (b,d) | **13.** (a,c) | **14.** (a) | **15.** (d) | **16.** (c) | **17.** (b) | **18.** (c) | **19.** (3) | **20.** (0) |
| **21.** (3) | **22.** (3) | **23.** (0) | **24.** (3) | **25.** (2) | **26.** (5) | **27.** (1) | **28.** (4) | | |

## Paper II

| | | | | | | | | | |
|---|---|---|---|---|---|---|---|---|---|
| **1.** (c) | **2.** (a) | **3.** (d) | **4.** (b) | **5.** (b) | **6.** (d) | **7.** (7) | **8.** (2) | **9.** (2) | **10.** (6) |
| **11.** (3) | **12.** (b) | **13.** (a) | **14.** (d) | **15.** (b) | **16.** (c) | **17.** (b) | | | |

**18.** A – R, S, T; B – T; C – P, Q; D – R

**19.** A – P, S; B – P, Q, R, T; C – P, Q, T; D – P

# Hints & Solutions

## Paper I

1. Ring $A$ is pyranose (6 membered ring containing one O-atom) with $\alpha$-glycosidic linkage and ring $B$ is furanose with $\beta$-glycosidic linkage.

   Biomolecules

   Conceptual structural interpretation

   III

2. $CH_3—CH_2—C{\equiv}C—CH_2—CH_2—CH_2—CH_3$

   There are two (marked) possibility of halides

   Thus, in the given context

   $$CH_3CH_2C{\equiv}CH + NaNH_2 \longrightarrow CH_3CH_2—C{\equiv}\bar{C}Na^+$$

   $$CH_3CH_2C{\equiv}\bar{C}Na^+ + BrCH_2CH_2CH_2CH_3 \longrightarrow \underset{\text{3-octyne}}{CH_3CH_2C{\equiv}C—CH_2CH_2CH_2CH_3}$$

   Hydrocarbon

   Conceptual, reaction understanding

   II

3. The ionization isomer of $[Cr(H_2O)_4Cl(NO_2)]Cl$ is $[Cr(H_2O)_4Cl_2]NO_2$ because of exchanging of ligand and counter ions

   Coordination chemistry

   Straight conceptual

   II

**4.** EDTA structure is

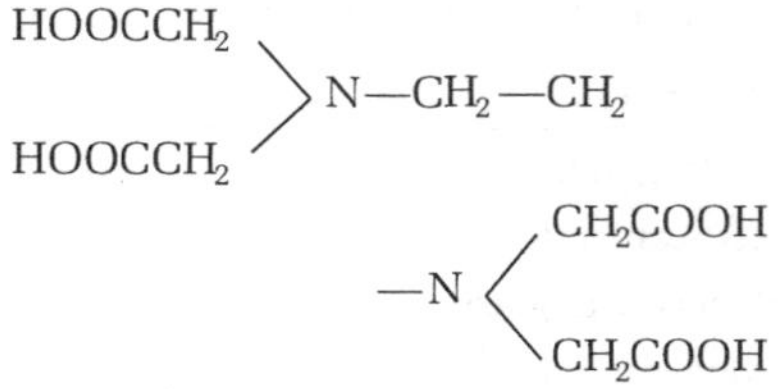

$$(HOOCCH_2)_2N-CH_2-CH_2-N(CH_2COOH)_2$$

Coordination chemistry
Factual
I

**5.** The bond energy of a C—C single bond is approximately 100 kcal $mol^{-1}$.
Energetics
Factual (usually it is not expected to be learnt)
III

**6.** $\Delta H_f^\circ(Cl_2, g) = 0$

$\Delta H_f^\circ$ of elements in their standard state is taken to be zero. $Cl_2$ is gas, $Br_2$ is liquid and $I_2$ is solid at room temperature.
Energetics
Conceptual, implication of rules/laws
III

**7.** $k = Ae^{-E_a/RT}$, As temperature increases $k$ increases exponentially.
Chemical kinetics
Conceptual mathematical analysis of data
I

**8.** $C_6H_5-OCH_3 \xrightarrow{HBr} C_6H_5-OH + CH_3Br$

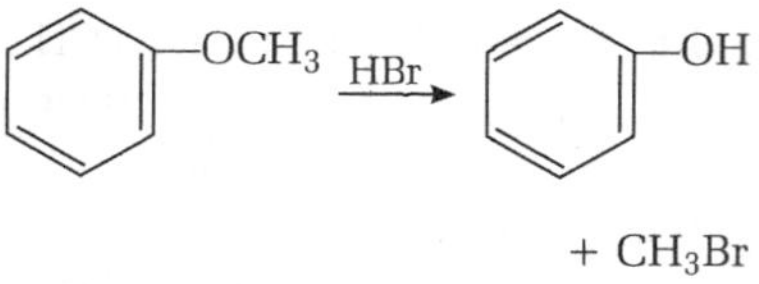

Alcohol-phenol-ether
Conceptual, understanding of mechanism
II (Given options raised the difficulty level)

**9.** Intensive property is one whose magnitude does not depend on the mass of the substance. Opposite to that is extensive property.
Energetics
Conceptual, concept analysis
III

**10.** Softening the temporary hardness of water means conversion of bicarbonate into calcium carbonate (precipitation)

$$Ca(HCO_3)_2 + Ca(OH)_2 \longrightarrow 2CaCO_3\downarrow + 2H_2O$$

$$Ca(HCO_3)_2 + Na_2CO_3 \longrightarrow CaCO_3\downarrow + 2NaHCO_3$$

Even NaOCl can be used, as in aqueous medium $\overset{-}{O}Cl$ gets hydrolysed and makes solution alkaline enough to cause precipitation of $CaCO_3$.

$$NaOCl + H_2O \rightleftharpoons HOCl + NaOH$$

$$OH^- + HCO_3^- \longrightarrow CO_3^{2-} + H_2O$$

Representative elements (Hydrogen and its compounds)
Conceptual
II

**11.** $HNO_3$ and $CH_3COONa$ can form buffer if the amount of $HNO_3$ is limited. However concentrations are same but the amount can differ with manipulation of volume.
Ionic equilibrium
Conceptual
II

**12.** $\underset{1}{CH_3}-\underset{2}{C}(CH_3)_2-\underset{3}{CH_2}-\underset{4}{CH_3}$

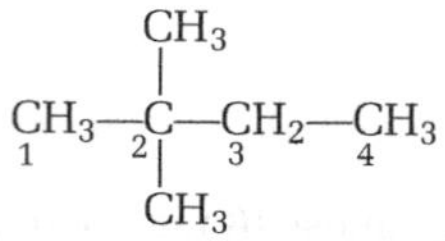

Conformation projection along $C_1-C_2$
$C_1$ contains all three Hs
So, $X = H$

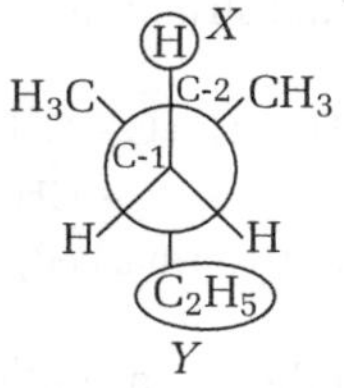

[Newman projection along C-1–C-2: front carbon C-1 bearing H (X), H, H; rear carbon C-2 bearing $H_3C$, $CH_3$, $C_2H_5$ (Y)]

$C_2$ contains two methyl and one ethyl group
So, $Y = C_2H_5$

Conformational projection along $C_2 - C_3$
$C_2$ contains three methyl groups
($C_2$ form back carbon in the given structure)
So, $Y = CH_3$

$C_3$ contains two Hs and one methyl group
($C_3$ form front carbon in the given structure)
So, $X = CH_3$
$C_3$ contains two Hs and one methyl group
($C_3$ form front carbon in the given structure)
So, $X = CH_3$
Isomerism (Stereochemistry)
Conceptual (Structural visualisation)
III

13. Phenol $\xrightarrow{NaOH(aq)/Br_2}$ 2,4,6-tribromophenol
(Final product)

—OH and in aqueous in alkaline medium $-\bar{O}$ are *ortho* and *para* directing. We expect the main intermediate products as mono *ortho*/*para* substituted product.
Remember! if we consider (b), which could be there in small amount then we need to consider even (d) because $-\bar{O}$ is strongest activating group that facilitates electrophilic substitution reaction overall on the ring. $-\bar{O}$ activates *ortho* and *para* position more but doesn't deactivate the *meta* position. We get a small amount of even the *meta* substituted product. But, as the question appears to be directly concerned answer should be only (a) and (c).
Reaction mechanism (Electrophilic substitution)
Conceptual
II

**C 14–16**

14. $2CuFeS_2 + 4O_2 \longrightarrow Cu_2S + 2FeO + 3SO_2$
15. Iron is removed in the form of slag of $FeSiO_3$
16. $S^{2-}$ acts as reducing species

$$2Cu_2O + Cu_2S^{2-} \longrightarrow 6Cu + \overset{+4}{S}O_2$$

Metallurgy
Conceptual
III II, II

**C 17–18**

17. The given cell is electrolytic concentration cell, So $E^\circ_{cell} = 0$

Anode :
$$M(s) \longrightarrow M^+(aq) + e^-$$
Cathode : $M^+(aq) + e^- \longrightarrow M(s)$

$$M^+(aq) \text{ cathode} \rightleftharpoons M^+(aq) \text{ anode}$$

$$E_{cell} = E^\circ_{cell} - \frac{0.0591}{1} \log \frac{[M^+]_{anodic}}{[M^+]_{cathodic}}$$

$$= 0 - \frac{0.0591}{1} \log \frac{0.05}{1}$$

$$E_{cell} = 70 \text{ mV } (+\text{ve})$$

Hence, $\Delta G = -nFE_{cell} = -\text{ve}$

18. $E_{cell} = E^\circ_{cell} - \frac{0.0591}{1} \log \frac{0.0025}{1}$

$E_{cell} = 140$ mV
Electrochemistry
Straight conceptual
III

19. The aqueous solution of KCN, $K_2CO_3$ and LiCN turn red litmus blue because of alkaline nature.

The anionic hydrolysis turns the solution alkaline.
Ionic equilibrium
Conceptual
III

**20.** In $BrF_5$ number of electrons = 6 (1 *lp* + 5 *bp*)
So, the structure is supposed to be square pyramidal but will be distorted because of additional *lp*-*bp* interaction.

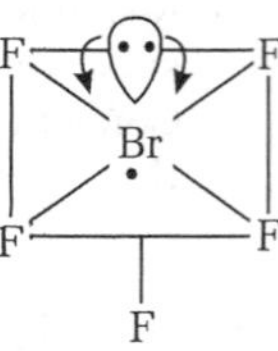


Additional *lp*-*bp* interaction reduces the all bond angle and do not let any angle to be 90°.
Chemical bonding
Conceptual
IV

**21.** For $Be_nAl_2Si_6O_{18}$ $n = 3$
$2n + 6 + 24 - 36 = 0 \quad n = 3$
Chemical bonding
Basic application
II

**22.** Average value $= \dfrac{25.2 + 25.25 + 25.0}{3} = \dfrac{75.45}{3}$
$= 25.15 = 25.2$ mL
Number of significant figure is 3.
Basics in chemistry
Rule application
III (Difficulty level is high because of shear ignorance of such topics)

**23.** $k_{\text{I}} = \dfrac{\Delta[R]}{\Delta t} = \dfrac{0.25}{0.05} = 5$
$k_{\text{II}} = \dfrac{\Delta[R]}{\Delta t} = \dfrac{0.60}{0.12} = 5$
So, reaction must be of zero order.
Chemical kinetics
Straight conceptual
II

**24.** ${}_{92}U^{235} \longrightarrow {}_{54}Xe^{142} + {}_{38}Sr^{90} + 3\,{}_{0}n^{1}$
Nuclear chemistry
Basic application
II

**25.** Lysine contains two basic groups, *e.g.*, $NH_2$
Biomolecules
Basic straight
I

**26.** $C_4H_6$ can have five cyclic isomers

Isomerism (organic)
Structural manipulation
III

**27.**

1. $O_3$
2. Zn, $H_2O$

5, 7, 4, 6, 8, 1, 3, 9, 2, 10

1. NaOH(*aq*)
2. Heat

For aldol condensation C-5 and C-7 can attack to C-1 similarly C-2 and C-10 can attack to C-6 but all give same product.
Carbonyl compounds
Conceptual
III

**28.** COOH, OH, OH, $H_3C$–N–$CH_3$ and COOH

are soluble in aqueous NaOH. Benzylic alcohol is less acidic than water so not soluble in aq. NaOH,
General organic chemistry
Conceptual (structural interpretation)
III

# Paper II

**1.** HO—C$_6$H$_4$—COOH ⟶ HO—C$_6$H$_3$(NO$_2$)—COOH

—OH controls the orientation, that is *ortho-para* directing

$H_3C$—C$_6$H$_4$—$OCH_3$ ⟶ $H_3C$—C$_6$H$_3$($NO_2$)—$OCH_3$

—$OCH_3$ controls the orientation (*o*/*p*-directing)

Ph(A)—C(=O)—O—Ph(B) ⟶ Ph—C(=O)—O—C$_6$H$_4$—$NO_2$

Ring B is more activated and —O—C(=O)—Ph is (*o*/*p*-directing)

General organic chemistry (GOC)
Conceptual
II

**2.** $B_2$ : Total electrons = 10

Configuration : $\sigma 1s^2\ \sigma^* 1s^2\ \sigma 2s^2 \sigma^* 2s^2\ \pi 2p_x^1 = \pi 2p_y^1$

If Hunds rule is violated, then

$$\sigma 1s^2\ \sigma^* 1s^2\ \sigma 2s^2\ \sigma^* 2s^2\ \ \pi 2p_x^2 = \pi 2p_y^0$$

So, diamagnetic

$$\text{Bond order} = \frac{6-4}{2} = 1$$

Chemical bonding
Conceptual
II

**3.** $a = (2\sqrt{2}r)$

$$\text{Packing fraction} = \frac{2 \times \pi r^2}{2(\sqrt{2}r)^2} = \frac{2\pi r^2}{8r^2} = \frac{\pi}{4}$$

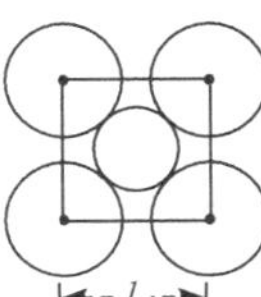


Solid state
Basic analysis (understanding)
II

4. Ni : $3d^8 4s^2$  $Ni^{2+}$ : $3d^8$

Since, Cl is a weak field ligand, it doesn't cause pairing of electron.

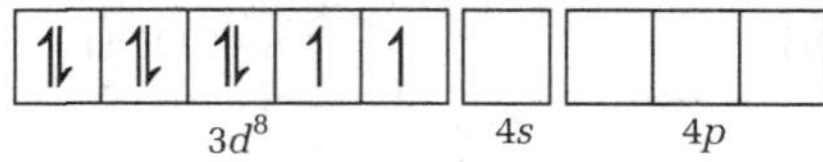


Number of unpaired electrons, $n = 2$

$$\mu = \sqrt{n(n+2)} = \sqrt{2(2+2)}\text{BM} = \sqrt{8}\text{ BM} = 2.82\text{ BM}$$

Coordination compound
Conceptual
III

5. $H_3C-C_6H_4-C(=O)NH_2 \xrightarrow{NaOH/Br_2} H_3C-C_6H_4-NH_2 \xrightarrow{C_6H_5COCl} H_3C-C_6H_4-NH-C(=O)-C_6H_5$

Acids and their derivatives
Conceptual (Reaction understanding)
II

6. $OSF_2$ has pyramidal shape
Chemical bonding
Basic understanding
II

F–S(=O)–F (with lone pair on S)

7. Volume of one mole of silver atoms $= \dfrac{108}{10.5}\text{ cm}^3/\text{mole}$

Volume of one silver atom $= \dfrac{108}{10.5} \times \dfrac{1}{6.022 \times 10^{23}}\text{ cm}^3$

So, $$\frac{4}{3}\pi R^3 = \frac{108}{10.5} \times \frac{1}{6.022 \times 10^{23}} = 1.708 \times 10^{-23}$$

$$R^3 = 0.407 \times 10^{-23}\text{ cm}^3$$
$$= 0.407 \times 10^{-29}\text{ m}^3$$

Area of each silver atom

$$\pi R^2 = \pi \times (0.407 \times 10^{-29}\text{ m}^3)^{2/3}$$

So number of silver atoms in given area

$$= \frac{10^{-12}}{(0.407 \times 10^{-29}\text{ m}^3)^{2/3}} = \frac{10^8}{(\pi \times 2)} = 1.6 \times 10^7 = y \times 10^x \quad \text{so,} \quad x = 7$$

Solid state
Mathematical understanding
III

**8.** F is the most electronegative element which cannot loose electron to other so only $-1$
Na is alkali metal which can loose only one electron so $+1$
Basics of general chemistry
Basic understanding
I

**9.** Solid line represents reversible isothermal process.

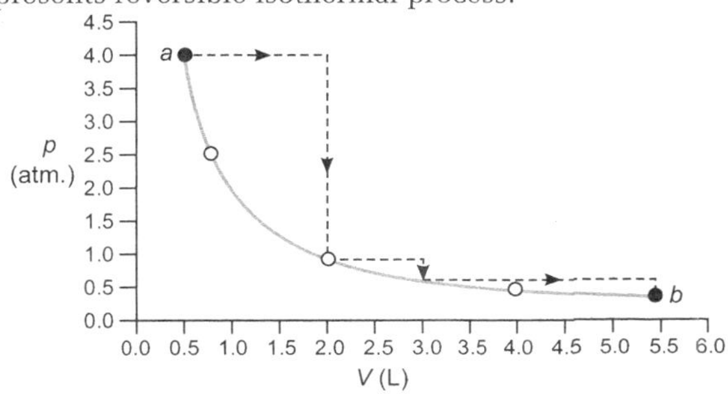


So, work $W_s = -4 \times 0.5 \ln\left(\frac{5.5}{0.5}\right) = -2 \ln 11$ L-atm

Dotted line represents irreversible process
So, work

$$W_d = -\left\{4 \times 1.5 + 1 \times 1 + \frac{2}{3} \times 2.5\right\} = -\left\{6 + 1 + \frac{5}{3}\right\} \text{ L-atm} = -\frac{26}{3} \text{ L-atm}$$

So $$\frac{W_d}{W_s} = \frac{26}{3 \times 2 \ln 11} \approx 2$$

Energetics
Conceptual
III

**10.** $H_2SO_4, H_2CO_3, H_2S_2O_7, H_2CrO_4, H_3PO_3, H_2SO_3$
Acid-base-salt
Basic structural understanding
II

**11.** $[M(abcd)]$ complex is square planar so will have 3 geometrical isomers.

$\left[\begin{matrix} a & & c \\ & M & \\ b & & d \end{matrix}\right]$, $\left[\begin{matrix} a & & b \\ & M & \\ d & & c \end{matrix}\right]$, $\left[\begin{matrix} a & & b \\ & M & \\ c & & d \end{matrix}\right]$

Coordination chemistry
Conceptual understanding of structure
II

**$C_{12-14}$**

$$\underset{\text{(Aliphatic aldehydes)}}{P + Q} \xrightarrow{K_2CO_3(aq)} R \xrightarrow{HCN} S$$

$\xrightarrow{H^+, \Delta}$ (lactone with OH, $H_3C$, $H_3C$, =O, O)

Some inferences from above data

- *aq* $K_2CO_3$ can facilitate aldol condensation to aliphatic aldehydes.
- HCN can lead to cyanohydrin formation
- $H^+$, $\Delta$ can lead to hydrolysis of cyanide to acid and following esterification.
  So *S* is cyanohydrin.

$CH_3$—$CH(CH_3)$—CH=O + HCH=O → $(H_3C)_2C(CH_2OH)$—CH=O → $(H_3C)_2C(CH_2OH)$—CH(OH)—C≡N ('S') → $(H_3C)_2C(CH_2OH)$—CH(OH)—COOH → lactone

Carbonyl compounds (Aldol condensation)
Conceptual reaction and mechanistic understanding
IV

**C** $_{15-17}$

Given : Hydrogen like species, $Li^{2+}$
Spherically symmetric state, $S_1 \Rightarrow$ *s*-orbital
$S_1$ has radial node $1 \Rightarrow 2s$ (No. of radial node $= 1 = n - l - 1$, so $n = 2$)
Species undergoes state $S_2$
$S_2$ has energy equal to the ground state energy of H-atom $= -13.6$ eV

So, $$-13.6 = -13.6\left(\frac{3^2}{n^2}\right)$$

$$n = 3$$

Again radial node $= 1 = n - l - 1$
or $1 = 3 - l - 1$ so $l = 1$
It suggests $S_2$ is $3p$

**16.** Energy of state, $S_1 = -13.6\left(\frac{3^2}{2^2}\right)$ eV/atom

$= \frac{9}{4}$ of energy of H-atom in ground state

$= 2.25$ of energy of H-atom in ground state

**17.** For *p*-orbital, orbital angular momentum quantum number = 1

Atomic structure
Conceptual data analysis
IV

**18.** (A) Phenol $\xrightarrow{NaOH}$ phenoxide ion ($C_6H_5O^-$)

$^-O-C_6H_4$ attacks $N\equiv\overset{+}{N}-C_6H_5$ $\longrightarrow$ $^-O-C_6H_4^{+}(H)-N=N-C_6H_5$ $\longleftrightarrow$ $O=C_6H_4(H)-N=N-C_6H_5$ $\longrightarrow$ $HO-C_6H_4-N=N-C_6H_5$

(B) $CH_3-C(OH)(CH_3)-C(OH)(CH_3)-CH_3 \xrightarrow[-H_2O]{H^+} CH_3-C(OH)(CH_3)-\overset{+}{C}(CH_3)-CH_3 \longrightarrow CH_3-CO-C(CH_3)_2-CH_3$

(C) $Ph-CO-CH_3 \xrightarrow{LiAlH_4}$ $Ph-C(O^-)(H)(CH_3)$ + $H-C(O^-)(Ph)(CH_3)$

(D) $HS-C_6H_{10}-Cl \xrightarrow[\text{attack}]{S_N2\text{-type}}$ bicyclic sulphide (S bridge)

Organic chemistry

Conceptual understanding of reaction mechanism

IV

**19.** $(CH_3)_2SiCl_2 + H_2O \longrightarrow (CH_3)_2Si(OH)_2 + HCl$ $-O-Si(CH_3)_2-O-Si(CH_3)_2-O-Si(CH_3)_2-O-$ (Polymer)

$$3XeF_4 + 6H_2O \longrightarrow XeO_3 + 2Xe + \frac{3}{2}O_2 + 12HF$$

$$Cl_2 + H_2O \longrightarrow 2HCl + \frac{1}{2}O_2$$

$$VCl_5 + H_2O \longrightarrow VOCl_3 + 2HCl$$

Representative elements

Factual

III

# IIT JEE

## SOLVED PAPER 2010

## Mathematics

## Paper I

### SECTION I

### (Straight objective Type) Only One Option Correct

*This section contains* **8 multiple choice questions**. *Each question has 4 choices (a), (b), (c) and (d) for its answer, out of which* **ONLY ONE** *is correct.*

1. The number of $3\times 3$ matrices $A$ whose entries are either 0 or 1 and for which the system $A\begin{bmatrix}x\\y\\z\end{bmatrix}=\begin{bmatrix}1\\0\\0\end{bmatrix}$ has exactly two distinct solutions, is

   (a) 0 (b) $2^9-1$

   (c) 168 (d) 2

2. The value of $\lim_{x\to 0}\frac{1}{x^3}\int_0^x \frac{t\log(1+t)}{t^4+4}\,dt$ is

   (a) 0 (b) $\frac{1}{12}$

   (c) $\frac{1}{24}$ (d) $\frac{1}{64}$

3. Let $p$ and $q$ be real numbers such that $p\neq 0$, $p^3\neq q$ and $p^3\neq -q$. If $\alpha$ and $\beta$ are non-zero complex numbers satisfying $\alpha+\beta=-p$ and $\alpha^3+\beta^3=q$, then a quadratic equation having $\frac{\alpha}{\beta}$ and $\frac{\beta}{\alpha}$ as its roots is

   (a) $(p^3+q)x^2-(p^3+2q)x+(p^3+q)=0$

   (b) $(p^3+q)x^2-(p^3-2q)x+(p^3+q)=0$

   (c) $(p^3-q)x^2-(5p^3-2q)x+(p^3-q)=0$

   (d) $(p^3-q)x^2-(5p^3+2q)x+(p^3-q)=0$

4. Let $f$, $g$ and $h$ be real-valued functions defined on the interval [0, 1] by $f(x)=e^{x^2}+e^{-x^2}$, $g(x)=xe^{x^2}+e^{-x^2}$ and $h(x)=x^2e^{x^2}+e^{-x^2}$. If $a$, $b$ and $c$ denote respectively, the absolute maximum of $f$, $g$ and $h$ on [0, 1], then

   (a) $a=b$ and $c\neq b$ (b) $a=c$ and $a\neq b$

   (c) $a\neq b$ and $c\neq b$ (d) $a=b=c$

5. If the angles $A$, $B$ and $C$ of a triangle are in an arithmetic progression and if $a$, $b$ and $c$ denote the lengths of the sides opposite to $A$, $B$ and $C$ respectively, then the value of the expression $\frac{a}{c}\sin 2C+\frac{c}{a}\sin 2A$ is

   (a) $\frac{1}{2}$ (b) $\frac{\sqrt{3}}{2}$

   (c) 1 (d) $\sqrt{3}$

6. Equation of the plane containing the straight line $\frac{x}{2}=\frac{y}{3}=\frac{z}{4}$ and perpendicular to the plane containing the straight lines $\frac{x}{3}=\frac{y}{4}=\frac{z}{2}$ and $\frac{x}{4}=\frac{y}{2}=\frac{z}{3}$ is

   (a) $x+2y-2z=0$ (b) $3x+2y-2z=0$

   (c) $x-2y+z=0$ (d) $5x+2y-4z=0$

7. Let $\omega$ be a complex cube root of unity with $\omega\neq 1$. A fair die is thrown three times. If $r_1$, $r_2$ and $r_3$ are the numbers obtained on the die, then the probability that $\omega^{r_1}+\omega^{r_2}+\omega^{r_3}=0$ is

(a) $\frac{1}{18}$ (b) $\frac{1}{9}$

(c) $\frac{2}{9}$ (d) $\frac{1}{36}$

**8.** Let $P$, $Q$, $R$ and $S$ be the points on the plane with position vectors $-2\hat{\mathbf{i}}-\hat{\mathbf{j}}$, $4\hat{\mathbf{i}}$, $3\hat{\mathbf{i}}+3\hat{\mathbf{j}}$ an d $-3\hat{\mathbf{i}}+2\hat{\mathbf{j}}$ respectively. The quadrilateral $PQRS$ must be a

(a) parallelogram, which is neither a rhombus nor a rectangle

(b) square

(c) rectangle, but not a square

(d) rhombus, but not a square

## SECTION II

### (Straight Objective Type) More than One Options are Correct

*This section contains* **5 multiple choice questions**. *Each question has 4 choices (a), (b), (c) and (d) for its answer, out of which* ***ONE or MORE*** *is/are correct.*

**9.** The value(s) of $\int_0^1 \frac{x^4(1-x)^4}{1+x^2}\,dx$ is (are)

(a) $\frac{22}{7}-\pi$ (b) $\frac{2}{105}$

(c) 0 (d) $\frac{71}{15}-\frac{3\pi}{2}$

**10.** Let $z_1$ and $z_2$ be two distinct complex numbers and let $z=(1-t)z_1+tz_2$ for some real number $t$ with $0<t<1$. If arg $(w)$ denotes the principal argument of a non-zero complex number $w$, then

(a) $|z-z_1|+|z-z_2|=|z_1-z_2|$

(b) $\arg(z-z_1)=\arg(z-z_2)$

(c) $\begin{vmatrix} z-z_1 & \bar{z}-\bar{z}_1 \\ z_2-z_1 & \bar{z}_2-\bar{z}_1 \end{vmatrix}=0$

(d) $\arg(z-z_1)=\arg(z_2-z_1)$

**11.** Let $f$ be a real-valued function defined on the interval $(0,\infty)$, by $f(x)=\ln x+\int_0^x \sqrt{1+\sin t}\,dt$. Then which of the following statement(s) is (are) true ?

(a) $f''(x)$ exists for all $x\in(0,\infty)$

(b) $f'(x)$ exists for all $x\in(0,\infty)$ and $f'$ is continuous on $(0,\infty)$, but not differentiable on $(0,\infty)$

(c) there exists $\alpha>1$ such that $|f'(x)|<|f(x)|$ for all $x\in(\alpha,\infty)$

(d) there exists $\beta>0$ such that $|f(x)|+|f'(x)|\le\beta$ from all $x\in(0,\infty)$

**12.** Let $ABC$ be a triangle such that $\angle ACB=\frac{\pi}{6}$ and let $a$, $b$ and $c$ denote the lengths of the sides opposite to $A$, $B$ and $C$ respectively. The value(s) of $x$ for which $a=x^2+x+1$, $b=x^2-1$ and $c=2x+1$ is (are)

(a) $-(2+\sqrt{3})$

(b) $1+\sqrt{3}$

(c) $2+\sqrt{3}$

(d) $4\sqrt{3}$

**13.** Let $A$ and $B$ be two distinct points on the parabola $y^2=4x$. If the axis of the parabola touches a circle of radius $r$ having $AB$ as its diameter, then the slope of the line joining $A$ and $B$ can be

(a) $-\frac{1}{r}$ (b) $\frac{1}{r}$

(c) $\frac{2}{r}$ (d) $-\frac{2}{r}$

## SECTION III

### (Linked Comprehension Type)

*This section contains* **2 paragraphs**. *Based upon the first paragraph* **3 multiple choice questions** *and based upon the second paragraph* **2 multiple choice questions** *have to answered. Each question has 4 choices (a), (b), (c) and (d) for its answer, out of* ***WHICH ONLY ONE.***

**Passage I for Q. Nos. 14 to 16 :** Let $p$ be an odd prime number and $T_p$ be the following set of $2\times 2$ matrices

$$T_p = \left\{ A = \begin{bmatrix} a & b \\ c & a \end{bmatrix}; a, b, c \in \{0, 1, 2, \dots, p-1\} \right\}$$

**14.** The number of $A$ in $T_p$ such that $A$ is either symmetric or skew-symmetric or both, and det $(A)$ is divisible by $p$ is

(a) $(p-1)^2$ (b) $2(p-1)$

(c) $(p-1)^2+1$ (d) $2p-1$

**15.** The number of $A$ in $T_p$ such that the trace of $A$ is not divisible by $p$ but det $(A)$ is divisible by $p$ is

[**Note :** The trace of a matrix is the sum of its diagonal entries.]

(a) $(p-1)(p^2-p+1)$

(b) $p^3-(p-1)^2$

(c) $(p-1)^2$

(d) $(p-1)(p^2-2)$

**16.** The number of $A$ in $T_p$ such that det $(A)$ is not divisible by $p$, is

(a) $2p^2$ (b) $p^3-5p$

(c) $p^3-3p$ (d) $p^3-p^2$

**Passage II for Q. Nos. 17 and 18 :** The circle $x^2+y^2-8x=0$ and hyperbola $\dfrac{x^2}{9}-\dfrac{y^2}{4}=1$ intersect at the points $A$ and $B$.

**17.** Equation of a common tangent with positive slope to the circle as well as to the hyperbola is

(a) $2x-\sqrt{5}y-20=0$

(b) $2x-\sqrt{5}y+4=0$

(c) $3x-4y+8=0$

(d) $4x-3y+4=0$

**18.** Equation of the circle with $AB$ as its diameter is

(a) $x^2+y^2-12x+24=0$

(b) $x^2+y^2+12x+24=0$

(c) $x^2+y^2+24x-12=0$

(d) $x^2+y^2-24x-12=0$

## SECTION IV

### (Numerical Value Based Questions)

*This section contains **TEN questions**. The answer to each question is a **Single-digit integer**, ranging from 0 to 9. The correct digit below the question number in the ORS is to be bubbled.*

**19.** Let $\omega$ be the complex number $\cos\dfrac{2\pi}{3}+i\sin\dfrac{2\pi}{3}$. Then the number of distinct complex number $z$ satisfying

$$\begin{vmatrix} z+1 & \omega & \omega^2 \\ \omega & z+\omega^2 & 1 \\ \omega^2 & 1 & z+\omega \end{vmatrix} = 0 \text{ is equal to}$$

**20.** Let $S_k$, $k=1, 2, \dots, 100$, denote the sum of the infinite geometric series whose first term is $\dfrac{k-1}{k!}$ and the common ratio is $\dfrac{1}{k}$. Then the value of

$$\frac{100^2}{100!}+\sum_{k=1}^{100}|(k^2-3k+1)S_k| \text{ is}$$

**21.** The number of all possible values of $\theta$, where $0<\theta<\pi$, for which the system of equations

$$(y+z)\cos 3\theta = (xyz)\sin 3\theta$$

$$x\sin 3\theta = \frac{2\cos 3\theta}{y}+\frac{2\sin 3\theta}{z}$$

and $(xyz)\sin 3\theta = (y+2z)\cos 3\theta + y\sin 3\theta$

have a solution $(x_0, y_0, z_0)$ with $y_0z_0 \neq 0$, is

**22.** Let $f$ be a real-valued differentiable function on $R$ (the set of all real numbers) such that $f(1)=1$. If the $y$-intercept of the tangent at any point

$P(x, y)$ on the curve $y = f(x)$ is equal to the cube of the abscissa of $P$, then the value of $f(-3)$ is equal to

**23.** The number of values of $\theta$ in the interval $\left(-\frac{\pi}{2}, \frac{\pi}{2}\right)$ such that $\theta \neq \frac{n\pi}{5}$ for $n = 0, \pm 1, \pm 2$ and $\tan\theta = \cot 5\theta$ as well as $\sin 2\theta = \cos 4\theta$ is

**24.** The maximum value of the expression $\frac{1}{\sin^2\theta + 3\sin\theta\cos\theta + 5\cos^2\theta}$ is

**25.** If $\vec{\mathbf{a}}$ and $\vec{\mathbf{b}}$ are vectors in space given by $\vec{\mathbf{a}} = \frac{\hat{\mathbf{i}} - 2\hat{\mathbf{j}}}{\sqrt{5}}$ and $\vec{\mathbf{b}} = \frac{2\hat{\mathbf{i}} + \hat{\mathbf{j}} + 3\hat{\mathbf{k}}}{\sqrt{14}}$, then the value of $(2\vec{\mathbf{a}} + \vec{\mathbf{b}}) \cdot [(\vec{\mathbf{a}} \times \vec{\mathbf{b}}) \times (\vec{\mathbf{a}} - 2\vec{\mathbf{b}})]$ is

**26.** The line $2x + y = 1$ is tangent to the hyperbola $\frac{x^2}{a^2} - \frac{y^2}{b^2} = 1$. If this line passes through the point of intersection of the nearest directrix and the $x$-axis, then the eccentricity of the hyperbola is

**27.** If the distance between the plane $Ax - 2y + z = d$ and the plane containing the lines $\frac{x-1}{2} = \frac{y-2}{3} = \frac{z-3}{4}$ and $\frac{x-2}{3} = \frac{y-3}{4} = \frac{z-4}{5}$ is $\sqrt{6}$, then $|d|$ is

**28.** For any real number $x$, let $[x]$ denotes the largest integer less than or equal to $x$. Let $f$ be a real valued function defined on the interval $[-10, 10]$ by

$$f(x) = \begin{cases} x - [x] & \text{if } [x] \text{ is odd} \\ 1 + [x] - x & \text{if } [x] \text{ is even} \end{cases}.$$

Then the value of $\frac{\pi^2}{10}\int_{-10}^{10} f(x)\cos \pi x \, dx$ is

# Paper II

## (Straight Objective Type) Only One Option Correct

*This section contains* 6 ***multiple choice questions****. Each question has* 4 *choices (a), (b), (c) and (d) for its answer, out of which* ***ONLY ONE*** *is correct.*

**1.** Let $S = \{1, 2, 3, 4\}$. The total number of unordered pairs of disjoint subsets of $S$ is equal to

(a) 25 (b) 34
(c) 42 (d) 41

**2.** Let $f$ be a real-valued function defined on the interval $(-1, 1)$ such that $e^{-x} f(x) = 2 + \int_0^x \sqrt{t^4 + 1}\, dt$, for all $x \in (-1, 1)$ and let $f^{-1}$ be the inverse function of $f$. Then $(f^{-1})'(2)$ is equal to

(a) 1 (b) $\frac{1}{3}$
(c) $\frac{1}{2}$ (d) $\frac{1}{e}$

**3.** For $r = 0, 1, \ldots, 10$, let $A_r$, $B_r$ and $C_r$ denote, respectively, the coefficient of $x^r$ in the expansions of $(1 + x)^{10}$, $(1 + x)^{20}$ and $(1 + x)^{30}$. Then $\sum_{r=1}^{10} A_r (B_{10}B_r - C_{10}A_r)$ is equal to

(a) $B_{10} - C_{10}$
(b) $A_{10}(B_{10}^2 - C_{10}A_{10})$
(c) 0
(d) $C_{10} - B_{10}$

**4.** Two adjacent sides of a parallelogram $ABCD$ are given by $\overrightarrow{\mathbf{AB}} = 2\hat{\mathbf{i}} + 10\hat{\mathbf{j}} + 11\hat{\mathbf{k}}$ and $\overrightarrow{\mathbf{AD}} = -\hat{\mathbf{i}} + 2\hat{\mathbf{j}} + 2\hat{\mathbf{k}}$. The side $AD$ is rotated by an acute angle $\alpha$ in the plane of the parallelogram so that $AD$ becomes $AD'$. If $AD'$ makes a right angle with the side $AB$, then the cosine of the angle $\alpha$ is given by

(a) $\frac{8}{9}$ (b) $\frac{\sqrt{17}}{9}$

(c) $\frac{1}{9}$ (d) $\frac{4\sqrt{5}}{9}$

**5.** A signal which can be green or red with probability $\frac{4}{5}$ and $\frac{1}{5}$ respectively, is received by station $A$ and then transmitted to station $B$. The probability of each station receiving the signal correctly is $\frac{3}{4}$. If the signal received at station $B$ is green, then the probability that the original signal green is

(a) $\frac{3}{5}$ (b) $\frac{6}{7}$

(c) $\frac{20}{23}$ (d) $\frac{9}{20}$

**6.** If the distance of the point $P(1, -2, 1)$ from the plane $x + 2y - 2z = \alpha$, where $\alpha > 0$, is 5, then the foot of the perpendicular from $P$ to the plane is

(a) $\left(\frac{8}{3}, \frac{4}{3}, -\frac{7}{3}\right)$ (b) $\left(\frac{4}{3}, -\frac{4}{3}, \frac{1}{3}\right)$

(c) $\left(\frac{1}{3}, \frac{2}{3}, \frac{10}{3}\right)$ (d) $\left(\frac{2}{3}, -\frac{1}{3}, \frac{5}{2}\right)$

## SECTION II — Integer Type

*This section contains* **5 questions**. *The answer to each question is a* ***single-digit integer,*** *ranging from* 0 *to* 9. The correct digit below the question number in the ORS is to be bubbled.

**7.** Let $a_1, a_2, a_3, \ldots, a_{11}$ be real numbers satisfying $a_1 = 15$, $27 - 2a_2 > 0$ and $a_k = 2a_{k-1} - a_{k-2}$ for $k = 3, 4, \ldots, 11$. If $\frac{a_1^2 + a_2^2 + \ldots + a_{11}^2}{11} = 90$, then the value of $\frac{a_1 + a_2 + \ldots + a_{11}}{11}$ is equal to

**8.** Let $k$ be a positive real number and let

$$A = \begin{bmatrix} 2k-1 & 2\sqrt{k} & 2\sqrt{k} \\ 2\sqrt{k} & 1 & -2k \\ -2\sqrt{k} & 2k & -1 \end{bmatrix} \text{ and}$$

$$B = \begin{bmatrix} 0 & 2k-1 & \sqrt{k} \\ 1-2k & 0 & 2\sqrt{k} \\ -\sqrt{k} & -2\sqrt{k} & 0 \end{bmatrix}$$

If $\det(\text{adj}\, A) + \det(\text{adj}\, B) = 10^6$, then $[k]$ is equal to

[**Note** : adj $M$ denotes the adjoint of a square matrix $M$ and $[k]$ denotes the largest integer less than or equal to $k$].

**9.** Consider a triangle $ABC$ and let $a$, $b$ and $c$ denote the lengths of the sides opposite to vertices $A$, $B$ and $C$ respectively. Suppose $a = 6$, $b = 10$ and the area of the triangle is $15\sqrt{3}$. If $\angle ACB$ is obtuse and if $r$ denotes the radius of the incircle of the triangle, then $r^2$ is equal to

**10.** Let $f$ be a function defined on $R$ (the set of all real numbers) such that $f'(x) = 2010(x - 2009)(x - 2010)^2(x - 2011)^3(x - 2012)^4$, for all $x \in R$. If $g$ is a function defined on $R$ with values in the interval $(0, \infty)$ such that $f(x) = \ln(g(x))$, for all $x \in R$, then the number of points in $R$ at which $g$ has a local maximum is

**11.** Two parallel chords of a circle of radius 2 are at a distance $\sqrt{3} + 1$ apart. If the chords subtend at the centre, angles of $\frac{\pi}{k}$ and $\frac{2\pi}{k}$, where $k > 0$, then the value of $[k]$ is

[**Note** : $[k]$ denotes the largest integer less than or equal to $k$]

## SECTION III

### (Linked Comprehension Type)

*This section contains 2 Passages. Each Passage has 3 multiple choice questions based on a paragraph. Each question has 4 choices (a), (b), (c) and (d) for its answer, out of which ONLY ONE is correct.*

**Passage I for Q. Nos. 12 to 14 :** Consider the polynomial $f(x) = 1 + 2x + 3x^2 + 4x^3$. Let $s$ be the sum of all distinct real roots of $f(x)$ and let $t = |s|$.

**12.** The real number $s$ lies in the interval

(a) $\left(-\frac{1}{4}, 0\right)$ (b) $\left(-11, -\frac{3}{4}\right)$

(c) $\left(-\frac{3}{4}, -\frac{1}{2}\right)$ (d) $\left(0, \frac{1}{4}\right)$

**13.** The area bounded by the curve $y = f(x)$ and the lines $x = 0$, $y = 0$ and $x = t$, lies in the interval

(a) $\left(\frac{3}{4}, 3\right)$ (b) $\left(\frac{21}{64}, \frac{11}{16}\right)$

(c) $(9, 10)$ (d) $\left(0, \frac{21}{64}\right)$

**14.** The function $f'(x)$ is

(a) increasing in $\left(-t, -\frac{1}{4}\right)$ and decreasing in $\left(-\frac{1}{4}, t\right)$

(b) decreasing in $\left(-t, -\frac{1}{4}\right)$ and increasing in $\left(-\frac{1}{4}, t\right)$

(c) increasing in $(-t, t)$

(d) decreasing in $(-t, t)$

**Passage II for Q. Nos. 15 to 17 :** Tangents are drawn from the point $P(3, 4)$ to the ellipse $\frac{x^2}{9} + \frac{y^2}{4} = 1$ touching the ellipse at points $A$ and $B$.

**15.** The coordinates of $A$ and $B$ are

(a) $(3, 0)$ and $(0, 2)$

(b) $\left(-\frac{8}{5}, \frac{2\sqrt{161}}{15}\right)$ and $\left(-\frac{9}{5}, \frac{8}{5}\right)$

(c) $\left(-\frac{8}{5}, \frac{2\sqrt{161}}{15}\right)$ and $(0, 2)$

(d) $(3, 0)$ and $\left(-\frac{9}{5}, \frac{8}{5}\right)$

**16.** The orthocentre of the triangle $PAB$ is

(a) $\left(5, \frac{8}{7}\right)$ (b) $\left(\frac{7}{5}, \frac{25}{8}\right)$

(c) $\left(\frac{11}{5}, \frac{8}{5}\right)$ (d) $\left(\frac{8}{25}, \frac{7}{5}\right)$

**17.** The equation of the locus of the point whose distance from the point $P$ and the line $AB$ are equal, is

(a) $9x^2 + y^2 - 6xy - 54x - 62y + 241 = 0$

(b) $x^2 + 9y^2 + 6xy - 54x + 62y - 241 = 0$

(c) $9x^2 + 9y^2 - 6xy - 54x - 62y - 241 = 0$

(d) $x^2 + y^2 - 2xy + 27x + 31y - 120 = 0$

## SECTION IV

### (Matrix Matches)

*This section contains 2 questions. Each question contains statements given in two columns, which have to be matched. The statements in Column I are labelled A, B, C and D, while the statements in Column II are labelled p, q, r, s and t. Any given statement in Column I can have correct matching with ONE OR MORE statement(s) in Column II.*

**18.** Match the statements of Column I with these in Column II.
[**Note** : Here $z$ takes values in the complex plane and $\operatorname{Im}(z)$ and $\operatorname{Re}(z)$ denote respectively, the imaginary part and the real part of $z$]

| | Column I | | Column II |
|---|---|---|---|
| (A) | The set of points $z$ satisfying $\|z - i\|z\|\| = \|z + i\|z\|\|$ is contained in or equal to | (p) | an ellipse with eccentricity 4/5 |
| (B) | The set of points $z$ satisfying $\|z+4\| + \|z-4\| = 0$ is contained in or equal to | (q) | the set of points $z$ satisfying $\operatorname{Im}(z) = 0$ |
| (C) | If $\|w\| = 2$, then the set of points $z = w - \frac{1}{w}$ is contained in or equal to | (r) | the set of points $z$ satisfying $\|\operatorname{Im} z\| \le 1$ |
| (D) | If $\|w\| = 1$, then the set of points $z = w + \frac{1}{w}$ is contained in or equal to | (s) | the set of points satisfying $\|\operatorname{Re} z\| \le 2$ |
| | | (t) | the set of points $z$ satisfying $\|z\| \le 3$ |

**19.** Match the statements of Column I with values of Column II.

| | Column I | | Column II |
|---|---|---|---|
| (A) | A line from the origin meets the lines $\frac{x-2}{1} = \frac{y-1}{2} = \frac{z+1}{1}$ and $\frac{x - \frac{8}{3}}{2} = \frac{y+3}{-1} = \frac{z-1}{1}$ at $P$ and $Q$ respectively. If length $PQ = d$, then $d^2$ is | (p) | −4 |
| (B) | The value of $x$ satisfying $\tan^{-1}(x+3) - \tan^{-1}(x-3) = \sin^{-1}\left(\frac{3}{5}\right)$ are | (q) | 0 |
| (C) | Non-zero vectors $\vec{\mathbf{a}}, \vec{\mathbf{b}}$ and $\vec{\mathbf{c}}$ satisfy $\vec{\mathbf{a}} \cdot \vec{\mathbf{b}} = 0$, $(\vec{\mathbf{b}} - \vec{\mathbf{a}}) \cdot (\vec{\mathbf{b}} + \vec{\mathbf{c}}) = 0$ and $2\|\vec{\mathbf{b}} + \vec{\mathbf{c}}\| = \|\vec{\mathbf{b}} - \vec{\mathbf{a}}\|$. If $\vec{\mathbf{a}} = \mu \vec{\mathbf{b}} + 4\vec{\mathbf{c}}$, then the possible value of $\mu$ are | (r) | 4 |
| (D) | Let $f$ be the function on $[-\pi, \pi]$ given by $f(0) = 9$ and $f(x) = \sin\left(\frac{9x}{2}\right) / \sin\left(\frac{x}{2}\right)$ for $x \ne 0$. The value of $\frac{2}{\pi}\int_{-\pi}^{\pi} f(x)\,dx$ is | (s) | 5 |
| | | (t) | 6 |

# Answers

## Paper I

1. (a) 2. (b) 3. (b) 4. (d) 5. (d) 6. (c) 7. (c) 8. (a) 9. (a) 10. (a,c,d)
11. (b,c) 12. (b) 13. (c,d) 14. (d) 15. (c) 16. (d) 17. (b) 18. (a) 19. 1 20. 4
21. 3 22. 9 23. 3 24. 2 25. 5 26. 2 27. 6 28. 4

## Paper II

1. (d) 2. (b) 3. (d) 4. (b) 5. (c) 6. (a) 7. 0 8. 4 9. 3 10. 1
11. 3 12. (c) 13. (a) 14. (b) 15. (d) 16. (c) 17. (a)
18. A → q,r; B → p; C → p,s,t; D → q,r,s,t 19. A → t; B → p,r; C → p; D → r

# Hints & Solutions

## Paper I

1. Since, $A\begin{bmatrix} x \\ y \\ z \end{bmatrix} = \begin{bmatrix} 1 \\ 0 \\ 0 \end{bmatrix}$ is linear equation in three variables and that could have only unique, no solution or infinitely many solution.

   $\therefore$ It is not possible to have two solutions.

   Hence, number of matrices $A$ is zero.

2. $\lim_{x \to 0} \frac{1}{x^3} \int_0^x \frac{t \log(1+t)}{4+t^4} dt$

   Using L' Hospital's rule,

   $$= \lim_{x \to 0} \frac{\frac{x \log(1+x)}{4+x^4}}{3x^2}$$

   $$= \lim_{x \to 0} \frac{\log(1+x)}{3x} \cdot \frac{1}{4+x^4}$$

   $$= \frac{1}{3} \cdot \frac{1}{4} = \frac{1}{12} \left[ \text{using, } \lim_{x \to 0} \frac{\log(1+x)}{x} = 1 \right]$$

3. Sum of roots $= \frac{\alpha^2 + \beta^2}{\alpha\beta}$ and product $= 1$

   Given, $\alpha + \beta = -p$ and $\alpha^3 + \beta^3 = q$

   $\Rightarrow (\alpha + \beta)(\alpha^2 - \alpha\beta + \beta^2) = q$

   $$\therefore \quad \alpha^2 + \beta^2 - \alpha\beta = \frac{-q}{p} \quad \text{...(i)}$$

   and $\quad (\alpha + \beta)^2 = p^2$

   $$\Rightarrow \quad \alpha^2 + \beta^2 + 2\alpha\beta = p^2 \quad \text{...(ii)}$$

   From Eqs. (i) and (ii), we get

   $$\alpha^2 + \beta^2 = \frac{p^3 - 2q}{3p}$$

   and $$\alpha\beta = \frac{p^3 + q}{3p}$$

   $\therefore$ Required equation is

   $$x^2 - \frac{(p^3 - 2q)x}{(p^3 + q)} + 1 = 0$$

   $$\Rightarrow (p^3 + q)x^2 - (p^3 - 2q)x + (p^3 + q) = 0$$

4. Given function,

   $f(x) = e^{x^2} + e^{-x^2}$,

   $g(x) = xe^{x^2} + e^{-x^2}$ and

   $h(x) = x^2 e^{x^2} + e^{-x^2}$ are strictly increasing on $[0, 1]$. Hence, at $x = 1$, the given function attains absolute maximum all equal to $e + \frac{1}{e}$.

   $\Rightarrow \quad a = b = c$

5. Since, $A, B, C$ are in AP

   $\Rightarrow \quad 2B = A + C$ ie, $\angle B = 60°$

   $$\therefore \frac{a}{c}(2\sin C \cos C) + \frac{c}{a}(2\sin A \cos A)$$

   $$= 2k(a\cos C + c\cos A)$$

   $$\left[\text{using, } \frac{a}{\sin A} = \frac{b}{\sin B} = \frac{c}{\sin C} = \frac{1}{k}\right]$$

   $= 2k(b)$

   $= 2\sin B$

   [using, $b = a\cos C + c\cos A$]

   $= \sqrt{3}$

6. The DR's of normal to the plane containing $\frac{x}{3} = \frac{y}{4} = \frac{z}{2}$ and $\frac{x}{4} = \frac{y}{2} = \frac{z}{3}$.

   $$\Rightarrow \quad \vec{\mathbf{n}}_1 = \begin{vmatrix} \hat{\mathbf{i}} & \hat{\mathbf{j}} & \hat{\mathbf{k}} \\ 3 & 4 & 2 \\ 4 & 2 & 3 \end{vmatrix} = (8\hat{\mathbf{i}} - \hat{\mathbf{j}} - 10\hat{\mathbf{k}})$$

   Also, equation of plane containing $\frac{x}{2} = \frac{y}{3} = \frac{z}{4}$ and DR's of normal to be $\vec{\mathbf{n}}_2 = a\hat{\mathbf{i}} + b\hat{\mathbf{j}} + c\hat{\mathbf{k}}$.

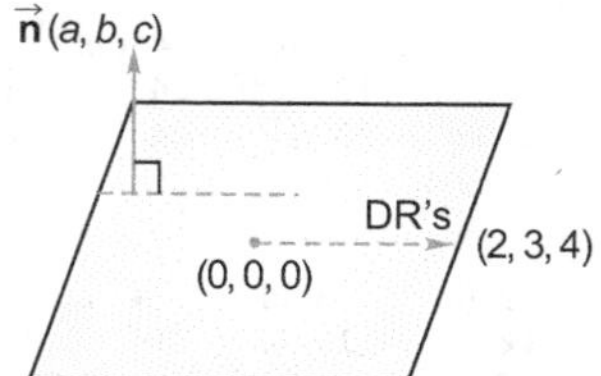

$\Rightarrow \quad ax + by + cz = 0 \quad \text{...(i)}$

where, $\vec{\mathbf{n}}_1 \cdot \vec{\mathbf{n}}_2 = 0$

$\Rightarrow \quad 8a - b - 10\,c = 0 \quad \text{...(ii)}$

and $\quad \vec{\mathbf{n}}_2 \perp (2\hat{\mathbf{i}} + 3\hat{\mathbf{j}} + 4\hat{\mathbf{k}})$

$\Rightarrow \quad 2a + 3\,b + 4c = 0 \quad \text{...(iii)}$

From Eqs. (ii) and (iii), we get

$$\frac{a}{\begin{vmatrix} -1 & -10 \\ 3 & 4 \end{vmatrix}} = \frac{b}{\begin{vmatrix} 8 \\ 2 \end{vmatrix}} = \frac{c}{\begin{vmatrix} -1 \\ 3 \end{vmatrix}}$$

$$\Rightarrow \quad \frac{a}{-4+30} = \frac{b}{-20-32} = \frac{c}{24+2}$$

$$\Rightarrow \quad \frac{a}{26} = \frac{b}{-52} = \frac{c}{26}$$

$$\Rightarrow \quad \frac{a}{1} = \frac{b}{-2} = \frac{c}{1} \quad \text{...(iv)}$$

From Eqs. (i) and (iv), required equation of plane, is $x - 2y + z = 0$.

7. **Sample space** A dice is thrown thrice, $n(s) = 6 \times 6 \times 6$.

**Favorable events** $\omega^{r_1} + \omega^{r_2} + \omega^{r_3} = 0$

*ie*, $(r_1, r_2, r_3)$ are ordered 3-triples which can take values,

$$\left.\begin{matrix} (1, 2, 3), & (1, 5, 3), & (4, 2, 3), & (4, 5, 3) \\ (1, 2, 6), & (1, 5, 6), & (4, 2, 6), & (4, 5, 6) \end{matrix}\right\}$$

*ie*, 8 ordered pairs and each can be arranged in 3! ways = 6

$\therefore \quad n(E) = 8 \times 6$

$$\Rightarrow \quad P(E) = \frac{8 \times 6}{6 \times 6 \times 6} = \frac{2}{9}$$

8. $m_{PQ} = \frac{1}{6}$, $m_{SR} = \frac{1}{6}$, $\quad m_{RQ} = -3$,

$m_{SP} = -3$

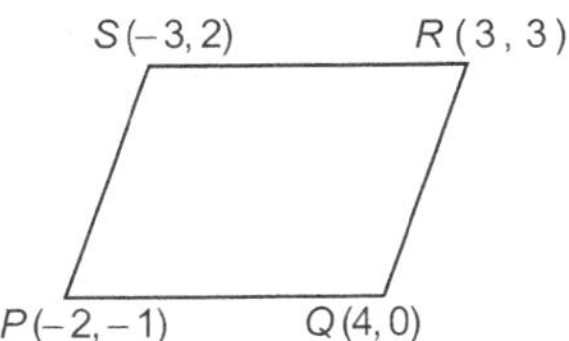


$\Rightarrow$ Parallelogram

But neither $PR = SQ$ nor $PR \perp SQ$.

$\therefore$ Parallelogram, which is neither a rhombus nor a rectangle.

9. Let $I = \int_0^1 \frac{x^4(1-x)^4}{1+x^2}\,dx$

$$= \int_0^1 \frac{(x^4-1)(1-x)^4 + (1-x)^4}{(1+x^2)}\,dx$$

$$= \int_0^1 (x^2-1)(1-x)^4\,dx + \int_0^1 \frac{(1+x^2-2x)^2}{(1+x^2)}\,dx$$

$$= \int_0^1 \left\{(x^2-1)(1-x)^4 + (1+x^2) - 4x + \frac{4x^2}{(1+x^2)}\right\} dx$$

$$= \int_0^1 \left((x^2-1)(1-x)^4 + (1+x^2) - 4x + 4 - \frac{4}{1-x^2}\right) dx$$

$$= \int_0^1 \left(x^6 - 4x^5 + 5x^4 - 4x^2 + 4 - \frac{4}{1+x^2}\right) dx$$

$$= \left[\frac{x^7}{7} - \frac{4x^6}{6} + \frac{5x^5}{5} - \frac{4x^3}{3} + 4x - 4\tan^{-1}x\right]_0^1$$

$$= \frac{1}{7} - \frac{4}{6} + \frac{5}{5} - \frac{4}{3} + 4 - 4\left(\frac{\pi}{4} - 0\right)$$

$$= \frac{22}{7} - \pi$$

10. Given, $z = \frac{(1-t)z_1 + t\,z_2}{(1-t)+t}$

$$\underset{z_1}{A} \text{ ———— } \underset{z}{P} \text{ ———— } \underset{z_2}{B}$$

$t : (1-t)$

Clearly, $z$ divides $z_1$ and $z_2$ in the ratio of $t : (1-t)$, $0 < t < 1$

$\Rightarrow \quad AP + BP = AB$

*ie*, $|z - z_1| + |z - z_2| = |z_1 - z_2|$

$\Rightarrow$ Option (a) is true.

and $\quad \arg(z - z_1) = \arg(z_2 - z)$

$\qquad = \arg(z_2 - z_1)$

$\Rightarrow$ (b) is false and (d) is true.

Also, $\arg(z - z_1) = \arg(z_2 - z_1)$

$\Rightarrow \quad \arg\left(\frac{z-z_1}{z_2-z_1}\right)=0$

$\therefore \frac{z-z_1}{z_2-z_1}$ is purely real.

$\Rightarrow \quad \frac{z-z_1}{z_2-z_1}=\frac{\bar{z}-\bar{z}_1}{\bar{z}_2-\bar{z}_1}$

or $\begin{vmatrix} z-z_1 & \bar{z}-\bar{z}_1 \\ z_2-z_1 & \bar{z}_2-\bar{z}_1 \end{vmatrix}=0$

$\therefore$ Option (c) is correct.

Hence, (a, c, d) is the correct option.

**11.** Here, $f'(x)=\frac{1}{x}+\sqrt{1+\sin x},\ x>0$ but $f(x)$ is not differentiable in $(0, \infty)$ as $\sin x$ may be $-1$ and then $f''(x)=-\frac{1}{x^2}+\frac{\cos x}{2\sqrt{1+\sin x}}$ will not exists.

$\Rightarrow f'(x)$ is continuous for all $x \in (0, \infty)$ but $f'(x)$ is not differentiable on $(0, \infty)$.

$\therefore$ Option (b) is true.

Also, $\quad f'(x)\le 3,\ \text{if } x>1$

and $\quad f(x)>3,\ \text{if } x>e^3$

$\therefore$ Let $\alpha=e^3$

$\Rightarrow$ Option (c) is true.

(d) is not possible as $f(x)\to\infty$ when $x\to\infty$.

Hence, (b, c) is the correct option.

**12.** Using, $\cos C=\frac{a^2+b^2-c^2}{2ab}$

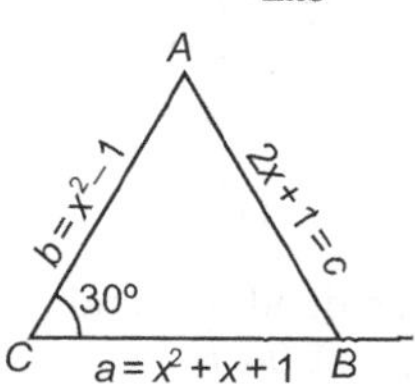


$$\Rightarrow \frac{\sqrt{3}}{2}=\frac{(x^2+x+1)^2+(x^2-1)^2-(2x+1)^2}{2(x^2+x+1)(x^2-1)}$$

$$\Rightarrow (x+2)(x+1)(x-1)x+(x^2-1)^2 = \sqrt{3}(x^2+x+1)(x^2-1)$$

$$\Rightarrow x^2+2x+(x^2-1)=\sqrt{3}(x^2+x+1)$$

$$\Rightarrow (2-\sqrt{3})x^2+(2-\sqrt{3})x-(\sqrt{3}+1)=0$$

$\Rightarrow x=-(2+\sqrt{3})$ and $x=1+\sqrt{3}$

But, $x=-(2+\sqrt{3}) \Rightarrow c$ is negative.

$\therefore \quad x=1+\sqrt{3}$ is the only solution.

Hence, (b) is the correct option.

**13.** Here, coordinate $M=\left(\frac{t_1^2+t_2^2}{2},\ t_1+t_2\right)$ *ie*, mid point of chord *AB*.

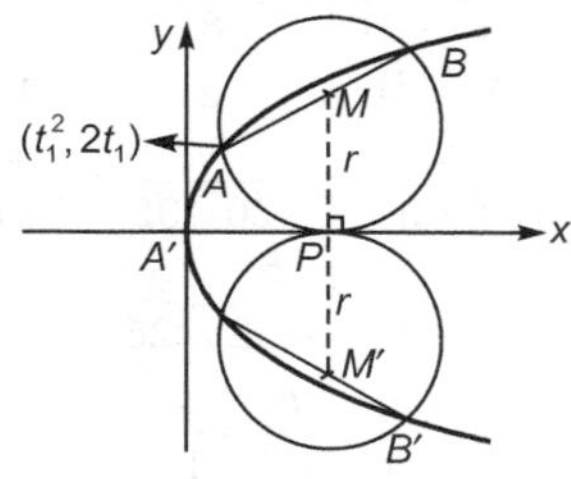


$$MP=t_1+t_2=r \quad \text{...(i)}$$

Also, $\quad m_{AB}=\frac{2t_2-2t_1}{t_2^2-t_1^2}=\frac{2}{t_2+t_1}$

(when *AB* is chord)

$\Rightarrow \quad m_{AB}=\frac{2}{r}$

[from Eq. (i)]

Also, $\quad m_{A'B'}=-\frac{2}{r}$

(when *A'B'* is chord)

Hence, (c, d) is the correct option.

**Solution for Q. Nos. 14 to 16**

**14.** Given, $A=\begin{bmatrix} a & b \\ c & a \end{bmatrix}$,

$a, b, c \in \{0, 1, 2, \ldots, p-1\}$

If $A$ is skew-symmetric matrix, then $a=0,\ b=-c$

$\therefore \quad |A|=-b^2.$

Thus, $P$ divides $|A|$ only when $b=0$ ...(i)

Again, if $A$ is symmetric matrix, then $b=c$ and $|A|=a^2-b^2$.

Thus, $p$ divides $|A|$ if either $p$ divides $(a-b)$ or $p$ divides $(a+b)$.

$p$ divides $(a-b)$, only when $a=b$

*ie*, $a=b\in\{0, 1, 2, \ldots, (p-1)\}$

*ie*, $p$ choices ...(ii)

$p$ divides $(a+b)$.

$\Rightarrow p$ choices, including $a=b=0$ included in (i)

$\therefore$ Total number of choices are $(p+p-1)=2p-1$.

Hence, (c) is the correct option.

**15.** Trace of $A=2a$, will be divisible by $p$ iff $a=0$.

$|A|=a^2-bc$, for $(a^2-bc)$ to be divisible by $p$. There are exactly $(p-1)$ ordered pairs $(b, c)$ for any value of $a$.

$\therefore$ Required number is $(p-1)^2$.

Hence, (c) is the correct option.

**16.** The number of matrices for which $p$ does not divide Tr $(A)=(p-1)p^2$ of these $(p-1)^2$ are such that $p$ divides $|A|$. The number of matrices for which $p$ divides Tr $(A)$ and $p$ does not divides $|A|$ are $(p-1)^2$.

$\therefore$Required number
$=(p-1)p^2-(p-1)^2+(p-1)^2$
$=p^3-p^2$.

**17.** Equation of tangent to hyperbola having slope $m$ is

$$y=mx+\sqrt{9m^2-4} \quad ...(i)$$

Equation of tangent to circle is

$$y=m(x-4)+\sqrt{16m^2+16} \quad ...(ii)$$

Eqs. (i) and (ii) will be identical for $m=\frac{2}{\sqrt{5}}$ satisfy.

$\therefore$ Equation of common tangent is $2x-\sqrt{5}\,y+4=0$.

**18.** The equation of the hyperbola is $\frac{x^2}{9}-\frac{y^2}{4}=1$ and that of circle is

$$x^2+y^2-8x=0$$

For their points of intersection

$$\frac{x^2}{9}+\frac{x^2-8x}{4}=1$$

$\Rightarrow \quad 4x^2+9x^2-72x=36$

$\Rightarrow \quad 13x^2-72x-36=0$

$\Rightarrow \quad 13x^2-78x+6x-36=0$

$\Rightarrow \quad 13x(x-6)+6(x-6)=0$

$\Rightarrow \quad x=6,\ x=-\frac{13}{6}$

$x=-\frac{13}{6}$ not acceptable

Now, for $x=6$, $y=\pm 2\sqrt{3}$

Required equation is
$(x-6)^2+(y+2\sqrt{3})(y-2\sqrt{3})=0$

$\Rightarrow \quad x^2-12x+y^2+24=0$

$\Rightarrow \quad x^2+y^2-12x+24=0$

**19.** Let $A=\begin{bmatrix}1 & \omega & \omega^2\\ \omega & \omega^2 & 1\\ \omega^2 & 1 & \omega\end{bmatrix}$

Now, $A^2=\begin{bmatrix}0&0&0\\0&0&0\\0&0&0\end{bmatrix}$

and Tr $(A)=0, |A|=0$

$\therefore \quad A^3=0$

$\Rightarrow \begin{vmatrix} z+1 & \omega & \omega^2\\ \omega & z+\omega^2 & 1\\ \omega^2 & 1 & z+\omega\end{vmatrix}=|A+zI|=0$

$\Rightarrow \quad z^3=0$

$\Rightarrow z=0$, the number of $z$ satisfying the given equation is 1.

**20.** We have, $S_k=\dfrac{\frac{k-1}{k!}}{1-\frac{1}{k}}=\dfrac{1}{(k-1)!}$

Now,
$(k^2-3k+1)S_k=\{(k-2)(k-1)-1\}\times\frac{1}{(k-1)!}$

$=\frac{1}{(k-3)!}-\frac{1}{(k-1)!}$

$\Rightarrow \sum_{k=1}^{100}|(k^2-3k+1)S_k|$

$=1+1+2-\left(\frac{1}{99!}+\frac{1}{98!}\right)=4-\frac{100^2}{100!}$

$\Rightarrow \quad \frac{100^2}{100!}+\sum_{k=1}^{100}|(k^2-3k+1)S_k|=4$

**21.** Given equations can be written as

$$x \sin 3\theta - \frac{\cos 3\theta}{y} - \frac{\cos 3\theta}{z} = 0 \quad \text{...(i)}$$

$$x \sin 3\theta - \frac{2\cos 3\theta}{y} - \frac{2\sin 3\theta}{z} = 0 \quad \text{...(ii)}$$

and $x \sin 3\theta - \frac{2}{y} \cos 3\theta$

$$- \frac{1}{z} (\cos 3\theta + \sin 3\theta) = 0 \quad \text{...(iii)}$$

Eqs. (ii) and (iii), implies

$$2 \sin 3\theta = \cos 3\theta + \sin 3\theta$$

$\Rightarrow \quad \sin 3\theta = \cos 3\theta$

$\therefore \quad \tan 3\theta = 1$

$\Rightarrow 3\theta = \frac{\pi}{4}, \frac{5\pi}{4}, \frac{9\pi}{4}$ or $\theta = \frac{\pi}{12}, \frac{5\pi}{12}, \frac{9\pi}{12}$.

**22.** The equation of the tangent at $(x, y)$ to the given curve $y = f(x)$ is

$$Y - y = \frac{dy}{dx}(X - x)$$

$$Y\text{-intercept} = y - x\frac{dy}{dx}$$

According to the question

$$x^3 = y - x\frac{dy}{dx}$$

$$\Rightarrow \quad \frac{dy}{dx} - \frac{y}{x} = -x^2$$

which is linear in $x$.

$$\text{IF} = e^{\int \frac{-1}{x} dx} = \frac{1}{x}$$

$\therefore$ Required solution is

$$y \cdot \frac{1}{x} = \int -x^2 \cdot \frac{1}{x} dx$$

$$\Rightarrow \quad \frac{y}{x} = \frac{-x^2}{2} + c$$

$$\Rightarrow \quad y = \frac{-x^3}{2} + cx$$

At $x = 1, y = 1$,

$$1 = \frac{-1}{2} + c$$

$$\Rightarrow \quad c = \frac{3}{2}$$

Now, $f(-3) = \frac{27}{2} + \frac{3}{2}(-3)$

$$= \frac{27 - 9}{2} = 9$$

**23.** Given, $\tan \theta = \cot 5\theta$

$$\Rightarrow \quad \tan \theta = \tan\left(\frac{\pi}{2} - 5\theta\right)$$

$$\Rightarrow \quad \frac{\pi}{2} - 5\theta = n\pi + \theta$$

$$\Rightarrow \quad 6\theta = \frac{\pi}{2} - n\pi$$

$$\Rightarrow \quad \theta = \frac{\pi}{12} - \frac{n\pi}{6}$$

Also, $\cos 4\theta = \sin 2\theta = \cos\left(\frac{\pi}{2} - 2\theta\right)$

$$\Rightarrow \quad 4\theta = 2n\pi \pm \left(\frac{\pi}{2} - 2\theta\right)$$

Taking positive

$$6\theta = 2n\pi + \frac{\pi}{2} \Rightarrow \theta = \frac{n\pi}{3} + \frac{\pi}{12}$$

Taking negative

$$2\theta = 2n\pi - \frac{\pi}{2} \Rightarrow \theta = n\pi - \frac{\pi}{4}$$

Above values of $\theta$ suggests that there are only 3 common solutions.

**24.** Let

$$f(\theta) = \frac{1}{\sin^2 \theta + 3 \sin \theta \cos \theta + 5 \cos^2 \theta}$$

Again let

$$g(\theta) = \sin^2 \theta + 3 \sin \theta \cos \theta + 5 \cos^2 \theta$$

$$= \frac{1 - \cos 2\theta}{2} + 5\left(\frac{1 + \cos 2\theta}{2}\right) + \frac{3}{2} \sin 2\theta$$

$$= 3 + 2 \cos 2\theta + \frac{3}{2} \sin 2\theta$$

$$\therefore \quad g(\theta)_{\min} = 3 - \sqrt{4 + \frac{9}{4}} = 3 - \frac{5}{2} = \frac{1}{2}$$

$$\therefore \quad f(\theta) = \frac{1}{g(\theta)_{\min}} = 2$$

**25.** From the given information, it is clear that

$$\vec{\mathbf{a}} = \frac{\hat{\mathbf{i}} - 2\hat{\mathbf{j}}}{\sqrt{5}}$$

$$\Rightarrow \quad |\vec{\mathbf{a}}| = 1, |\vec{\mathbf{b}}| = 1, \vec{\mathbf{a}} \cdot \vec{\mathbf{b}} = 0$$

Now, $(2\vec{\mathbf{a}} + \vec{\mathbf{b}}) \cdot [(\vec{\mathbf{a}} \times \vec{\mathbf{b}}) \times (\vec{\mathbf{a}} - 2\vec{\mathbf{b}})]$

$= (2\vec{\mathbf{a}} + \vec{\mathbf{b}})\cdot[a^2\,\vec{\mathbf{b}} - (\vec{\mathbf{a}}\cdot\vec{\mathbf{b}})\cdot\vec{\mathbf{a}} + 2\,b^2\cdot\vec{\mathbf{a}}$
$-2(\vec{\mathbf{b}}\cdot\vec{\mathbf{a}})\cdot a]$

$= [2\vec{\mathbf{a}} + \vec{\mathbf{b}}]\cdot[\vec{\mathbf{b}} + 2\vec{\mathbf{a}}] = 4\,\vec{\mathbf{a}}^{\,2} + \vec{\mathbf{b}}^{\,2}$

$= 4\cdot 1 + 1 = 5$ [as $a\cdot b = 0$]

**26.** On substituting $\left(\frac{a}{e}, 0\right)$ in $y = -2x + 1$,

we get

$$0 = -\frac{2a}{e} + 1$$

$$\Rightarrow \quad \frac{a}{e} = \frac{1}{2}$$

Also, $y = -2x + 1$ is tangent to hyperbola

$$\therefore \quad 1 = 4a^2 - b^2$$

$$\Rightarrow \quad \frac{1}{a^2} = 4 - (e^2 - 1)$$

$$\Rightarrow \quad \frac{4}{e^2} = 5 - e^2$$

$$\Rightarrow \quad e^4 - 5\,e^2 + 4 = 0$$

$$\Rightarrow \quad (e^2 - 4)(e^2 - 1) = 0$$

$$\Rightarrow \quad e = 2,\ e = 1$$

$e = 1$ gives the conic as parabola. But conic is given as hyperbola, hence $e = 2$

**27.** Equation of plane containing the given lines is $\begin{vmatrix} x-1 & y-2 & z-3 \\ 2 & 3 & 4 \\ 3 & 4 & 5 \end{vmatrix} = 0$

$\Rightarrow \quad (x-1)(-1) - (y-2)(-2) + (z-3)(-1) = 0$

$\Rightarrow \quad -x + 1 + 2y - 4 - z + 3 = 0$

$\Rightarrow \quad -x + 2y - z = 0$ ...(i)

Given plane is

$x - 2y + z = d$ ...(ii)

Eqs. (i) and (ii) are parallel.

Clearly, $A = 1$

Now, distance between plane

$$= \left|\frac{d}{\sqrt{1+4+1}}\right| = \sqrt{6}$$

$$\Rightarrow \quad |d| = 6$$

**28.** Given,

$$f(x) = \begin{cases} x - [x] & \text{if } [x] \text{ is odd} \\ 1 + [x] - x & \text{if } [x] \text{ is even} \end{cases}$$

$f(x)$ and $\cos \pi x$ both are periodic with period 2 and both are even.

$$\therefore \int_{-10}^{10} f(x)\cos \pi x\, dx$$

$$= 2\int_{0}^{10} f(x)\cos \pi x\, dx$$

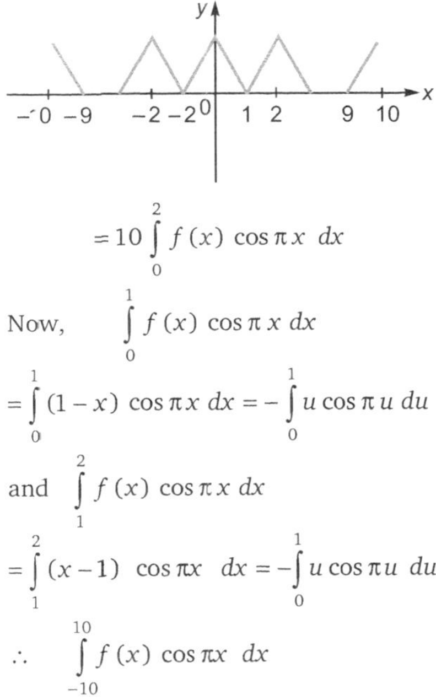


$$= 10\int_{0}^{2} f(x)\cos \pi x\, dx$$

Now, $\int_{0}^{1} f(x)\cos \pi x\, dx$

$$= \int_{0}^{1} (1-x)\cos \pi x\, dx = -\int_{0}^{1} u\cos \pi u\, du$$

and $\int_{1}^{2} f(x)\cos \pi x\, dx$

$$= \int_{1}^{2} (x-1)\cos \pi x\, dx = -\int_{0}^{1} u\cos \pi u\, du$$

$$\therefore \int_{-10}^{10} f(x)\cos \pi x\, dx$$

$$= -20\int_{0}^{1} u\cos \pi u\, du = \frac{40}{\pi^2}$$

$$\Rightarrow \quad \frac{\pi^2}{10}\int_{-10}^{10} f(x)\cos \pi x\, dx = 4$$

# Paper II

1. Let $A \cap B = \phi, A, B \subset S$

$$3^4 = \frac{81+1}{2} = 41$$

$$\Rightarrow \quad \frac{3^4+1}{2} = 41$$

2. We have,

$$e^{-x} f(x) = 2 + \int_0^x \sqrt{t^4+1}\, dt \quad x \in (-1, 1)$$

On differentiating w.r.t. $x$, we get

$$e^{-x}(f'(x) - f(x)) = \sqrt{x^4+1}$$

$$\Rightarrow \quad f'(x) = f(x) + \sqrt{x^4+1}\, e^x$$

$\because$ $f^{-1}$ is the inverse of $f$

$$\therefore \quad f^{-1}(f(x)) = x$$

$$\Rightarrow \quad f^{-1'}(f(x))\, f'(x) = 1$$

$$\Rightarrow \quad f^{-1'}(f(x)) = \frac{1}{f'(x)}$$

$$\Rightarrow \quad f^{-1'}(f(x)) = \frac{1}{f(x) + \sqrt{x^4+1}\, e^x}$$

At $x = 0, f(x) = 2$

$$f^{-1'}(2) = \frac{1}{2+1} = \frac{1}{3}$$

3. $A_r$ = Coefficient of $x^r$ in

$$(1+x)^{10} = {}^{10}C_r$$

$B_r$ = Coefficient of $x^r$ in

$$(1+x)^{20} = {}^{20}C_r$$

$C_r$ = Coefficient of $x^r$ in

$$(1+x)^{30} = {}^{30}C_r$$

$$\therefore \sum_{r=1}^{10} A_r (B_{10} B_r - C_{10} A_r)$$

$$= \sum_{r=1}^{10} A_r B_{10} B_r - \sum_{r=1}^{10} A_r C_{10} A_r$$

$$= \sum_{r=1}^{10} {}^{10}C_r\, {}^{20}C_{10}\, {}^{20}C_r - \sum_{r=1}^{10} {}^{10}C_r\, {}^{30}C_{10}\, {}^{10}C_r$$

$$= \sum_{r=1}^{10} {}^{10}C_{10-r}\, {}^{20}C_{10}\, {}^{20}C_r - \sum_{r=1}^{10} {}^{10}C_{10-r}\, {}^{30}C_{10}\, {}^{10}C_r$$

$$= {}^{20}C_{10} \sum_{r=1}^{10} {}^{10}C_{10-r} \cdot {}^{20}C_r - {}^{30}C_{10} \sum_{r=1}^{10} {}^{10}C_{10-r}\, {}^{10}C_r$$

$$= {}^{20}C_{10}({}^{30}C_{10} - 1) - {}^{30}C_{10}({}^{20}C_{10} - 1)$$

$$= {}^{30}C_{10} - {}^{20}C_{10} = C_{10} - B_{10}$$

4. $$\overrightarrow{\mathbf{AB}} = 2\hat{\mathbf{i}} + 10\hat{\mathbf{j}} + 11\hat{\mathbf{k}}$$

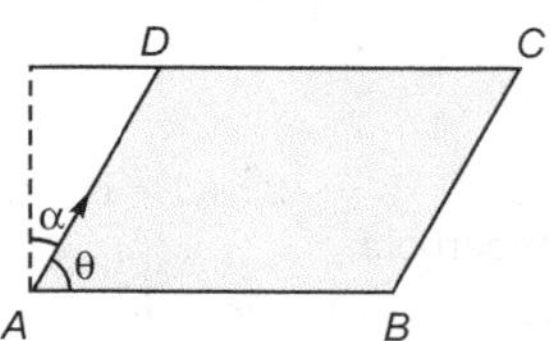


$$\overrightarrow{\mathbf{AD}} = -\hat{\mathbf{i}} + 2\hat{\mathbf{j}} + 2\hat{\mathbf{k}}$$

Angle 'θ' between $\overrightarrow{\mathbf{AB}}$ and $\overrightarrow{\mathbf{AD}}$ is

$$\cos(\theta) = \left|\frac{\overrightarrow{\mathbf{AB}} \cdot \overrightarrow{\mathbf{AD}}}{|\overrightarrow{\mathbf{AB}}||\overrightarrow{\mathbf{AD}}|}\right|$$

$$= \left|\frac{-2+20+22}{(15)(3)}\right| = \frac{8}{9}$$

$$\Rightarrow \quad \sin(\theta) = \frac{\sqrt{17}}{9}$$

Since, $\alpha + \theta = 90°$

$$\therefore \quad \cos(\alpha) = \cos(90° - \theta)$$

$$= \sin(\theta) = \frac{\sqrt{17}}{9}$$

5. From the tree-diagram it follows that

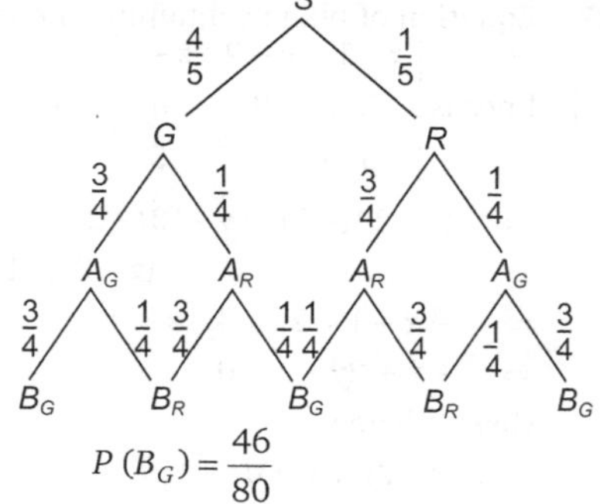


$$P(B_G) = \frac{46}{80}$$

$$P(B_G | G) = \frac{10}{16} = \frac{5}{8}$$

$$\therefore \quad P(B_G \cap G) = \frac{5}{8} \times \frac{4}{5} = \frac{1}{2}$$

$$P(G | B_G) = \frac{\frac{1}{2}}{P(B_G)} = \frac{1}{2} \times \frac{80}{46} = \frac{20}{23}$$

**6.** Distance of point $P$ from plane = 5

$\therefore$

$$5 = \left| \frac{1 - 4 - 2 - \alpha}{3} \right|$$

$$\alpha = 10$$

$P(1, -2, 1)$

$A(x, y, z)$

$x + 2y - 2z = \alpha$

Foot of perpendicular

$$\frac{x-1}{1} = \frac{y+2}{2} = \frac{z-1}{-2} = \frac{5}{3}$$

$$\Rightarrow \quad x = \frac{8}{3}, y = \frac{4}{3}, z = -\frac{7}{3}$$

Thus, the foot of the perpendicular is

$$A\left(\frac{8}{3}, \frac{4}{3}, -\frac{7}{3}\right)$$

**7.** $a_k = 2a_{k-1} - a_{k-2}$

$\Rightarrow$ $a_1, a_2, \ldots, a_{11}$ are in AP

$$\therefore \frac{a_1^2 + a_2^2 + \ldots + a_{11}^2}{11}$$

$$= \frac{11a^2 + 35 \times 11d^2 + 10ad}{11} = 90$$

$$\Rightarrow \quad 225 + 35d^2 + 150d = 90$$

$$35d^2 + 150d + 135 = 0$$

$$\Rightarrow \quad d = -3, -\frac{9}{7}$$

Given, $a_2 < \frac{27}{2} \therefore d = -3$ and $d \neq -\frac{9}{7}$

$$\Rightarrow \quad \frac{a_1 + a_2 + \ldots + a_{11}}{11}$$

$$= \frac{11}{2}[30 - 10 \times 3] = 0$$

**8.** $|A| = (2k+1)^3, |B| = 0$

But $\det(\text{adj} A) = \det(\text{adj} B) = 10^6$

$\Rightarrow (2k+1)^6 = 10^6$

$\Rightarrow k = \frac{9}{2} \Rightarrow [k] = 4$

**9.** $\sin C = \frac{\sqrt{3}}{2}$ and $C$ is given to be obtuse

$$\Rightarrow \quad C = \frac{2\pi}{3} = \sqrt{a^2 + b^2 - 2ab\cos C}$$

$$= \sqrt{6^2 + 10^2 - 2 \times 6 \times 10 \times \cos\frac{2\pi}{3}} = 14$$

$$\therefore \quad r = \frac{\Delta}{s} \Rightarrow r^2 = \frac{225 \times 3}{\left(\frac{6+10+14}{2}\right)^2} = 3$$

**10.** Let $g(x) = e^{f(x)}, \forall x \in R$

$\Rightarrow g'(x) = e^{f(x)} \cdot f'(x)$

$\Rightarrow f'(x)$ changes its sign from positive to negative in the neighbourhood of $x = 2009$

$\Rightarrow f(x)$ has local maxima at $x = 2009$

So, the number of local maximum is one.

**11.** Let $\theta = \frac{\pi}{2k}$

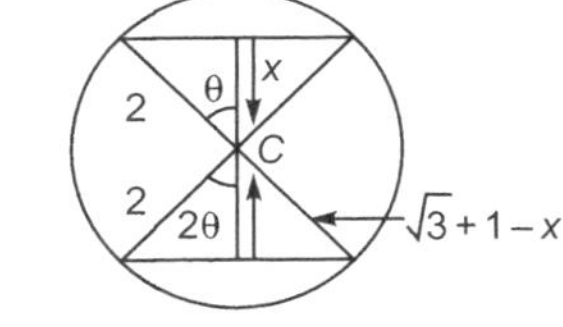


$$\cos\theta = \frac{x}{2}$$

$$\Rightarrow \quad \cos 2\theta = \frac{\sqrt{3} + 1 - x}{2}$$

$$\Rightarrow \quad 2\cos^2\theta - 1 = \frac{\sqrt{3} + 1 - x}{2}$$

$$\Rightarrow \quad 2\left(\frac{x^2}{4}\right) - 1 = \frac{\sqrt{3} + 1 - x}{2}$$

$$\Rightarrow \quad x^2 + x - 3 - \sqrt{3} = 0$$

$$\Rightarrow \quad x = \frac{-1 \pm \sqrt{1 + 12 + 4\sqrt{3}}}{2}$$

$$= \frac{-1 \pm \sqrt{13 + 4\sqrt{3}}}{2}$$

$$= \frac{-1 + 2\sqrt{3} + 1}{2} = \sqrt{3}$$

$$\therefore \quad \cos\theta = \frac{\sqrt{3}}{2} \Rightarrow \theta = \frac{\pi}{6}$$

$\therefore$ Required angle $= \frac{\pi}{k} = 2\theta = \frac{\pi}{3}$

$$\Rightarrow \quad k = 3$$

**Solution for Q. Nos. 12 to 14**

**12.** Given, $f(x) = 4x^3 + 3x^2 + 2x + 1$

$$f'(x) = 2(6x^2 + 3x + 1)$$

$$D = 9 - 24 < 0$$

Hence, $f(x) = 0$ has only one real root.

$$f\left(-\frac{1}{2}\right) = 1 - 1 + \frac{3}{4} - \frac{4}{8} > 0$$

$$f\left(-\frac{3}{4}\right) = 1 - \frac{6}{4} + \frac{27}{16} - \frac{108}{64}$$

$$= \frac{64 - 96 + 108 - 108}{64} < 0$$

$f(x)$ changes its sign in $\left(-\frac{3}{4}, \frac{-1}{2}\right)$, hence $f(x) = 0$ has a root in $\left(\frac{-3}{4}, \frac{-1}{2}\right)$.

**13.** $\int_0^{1/2} f(x)\,dx < \int_0^t f(x)\,dx < \int_0^{3/4} f(x)\,dx$

Now, $\int f(x)\,dx$

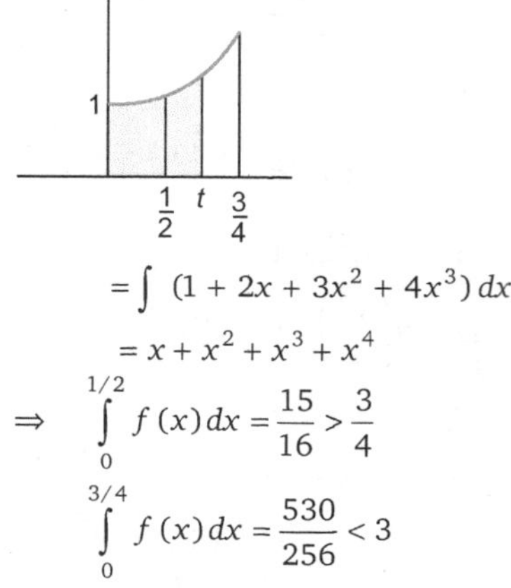


$$= \int (1 + 2x + 3x^2 + 4x^3)\,dx$$

$$= x + x^2 + x^3 + x^4$$

$$\Rightarrow \int_0^{1/2} f(x)\,dx = \frac{15}{16} > \frac{3}{4}$$

$$\int_0^{3/4} f(x)\,dx = \frac{530}{256} < 3$$

**14.** Figure is self explanatory

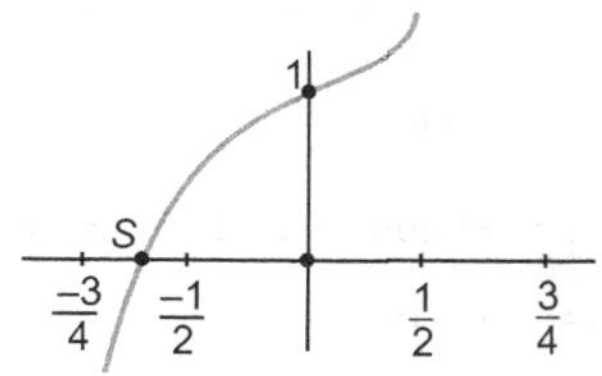


**Solution for Q. Nos. 15 to 17**

**15.** Figure is self explanatory

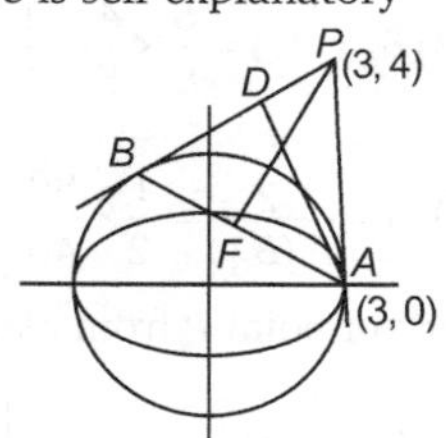


**16.** Equation of $AB$ is

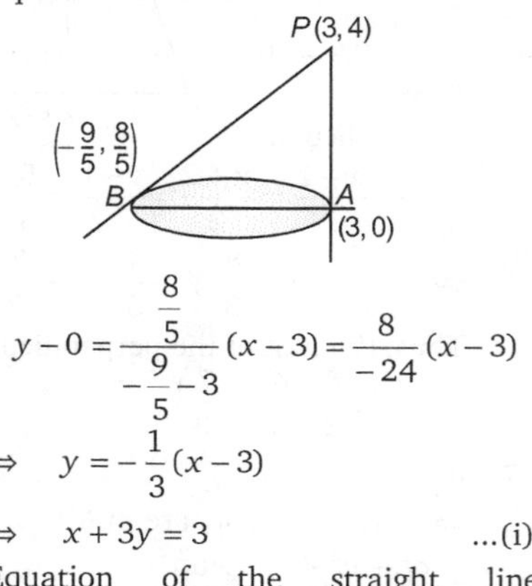


$$y - 0 = \frac{\frac{8}{5}}{-\frac{9}{5} - 3}(x - 3) = \frac{8}{-24}(x - 3)$$

$$\Rightarrow \quad y = -\frac{1}{3}(x - 3)$$

$$\Rightarrow \quad x + 3y = 3 \qquad \text{...(i)}$$

Equation of the straight line perpendicular to $AB$ through $P$ is $3x - y = 5$.

Equation of $PA$ is $x - 3 = 0$.

The equation of straight line perpendicular to $PA$ through $B\left(\frac{-9}{5}, \frac{8}{5}\right)$. is $y = \frac{8}{5}$.

Hence, the orthocentre is $\left(\frac{11}{5}, \frac{8}{5}\right)$.

**17.** Equation of $AB$ is $y - 0 = -\frac{1}{3}(x - 3)$

$$x + 3y - 3 = 0$$

$$|x + 3y - 3|^2 = 10[(x - 3)^2 + (y - 4)^2]$$

(Look at coefficient of $x^2$ and $y^2$ in the answers)

**18.** (A) $z$ is equidistant from the points $i|z|$ and $-i|z|$, whose perpendicular bisector is $\text{Im}(z) = 0$.

(B) Sum of distance of $z$ from (4, 0) and (– 4, 0) is a constant 10, hence locus of $z$ is ellipse with semi-major axis 5 and focus at $(\pm 4, 0)$, $ae = 4$.

$\therefore \quad e = \frac{4}{5}$

(C) $|z| \le |w| + \left|\frac{1}{w}\right| = \frac{5}{2} < 3$

(D) $|z| \le |w| + \left|\frac{1}{w}\right| = 2$

$\therefore \quad \text{Re}(z) \le |z| \le 2$

**19.** (A) Equation of the line passing through origin is

$$\frac{x}{a} = \frac{y}{b} = \frac{z}{c}$$

$$\therefore \quad \begin{vmatrix} 2 & 1 & -1 \\ 1 & -2 & 1 \\ a & b & c \end{vmatrix} = 0$$

$$\Rightarrow \quad a(-1) - b(3) + c(-5) = 0$$

$$\Rightarrow \quad -a - 3b - 5c = 0$$

$$\Rightarrow \quad a + 3b + 5c = 0 \qquad \text{...(i)}$$

Also, $\begin{vmatrix} \frac{8}{3} & -3 & 1 \\ 2 & -1 & 1 \\ a & b & c \end{vmatrix} = 0$

$$\therefore \quad a(-2) - b\left(\frac{2}{3}\right) + c\left(\frac{10}{3}\right) = 0$$

$$\Rightarrow \quad 2a + \frac{2b}{3} - \frac{10c}{3} = 0$$

$$3a + b - 5c = 0 \qquad \text{...(ii)}$$

From Eqs. (i) and (ii),

$$\frac{a}{-20} = \frac{b}{20} = \frac{c}{-8}$$

$$\frac{a}{5} = \frac{b}{-5} = \frac{c}{4}$$

Equation of line is

$$\frac{x}{5} = \frac{y}{-5} = \frac{z}{4} = \lambda \text{ (say)} \qquad \text{...(iii)}$$

Also, $\frac{x-2}{1} = \frac{y-1}{2} = \frac{z+1}{1} = k_1$ (say) ...(iv)

Now, $\frac{x - \frac{8}{3}}{2} = \frac{y+3}{-1} = \frac{z-1}{1} = k_2$ (say) ...(v)

Point on (iii) is $(5\lambda, -5\lambda, +4\lambda)$

Point on (iv) is

$(2 + k_1, 1 - 2k_1, -1 + k_1)$

Point on (v) is

$\left(\frac{8}{3} + 2k_2, -3 - k_2, 1 + k_2\right)$

On solving, $2 + k_1 + 1 - 2k_1 = 0$

$$-k_1 + 3 = 0$$

$k_1 = 3$

$P \equiv (5, -5, 2)$

Again, for $Q$

$$\frac{8}{3} + 2k_2 - 3 - k_2 = 0$$

$$k_2 - \frac{1}{3} = 0$$

$$k_2 = \frac{1}{3}$$

$$Q \equiv \left(\frac{10}{3}, \frac{-10}{3}, \frac{4}{3}\right)$$

Now, $PQ = \sqrt{\left(\frac{5}{3}\right)^2 + \left(\frac{5}{3}\right)^2 + \left(\frac{2}{3}\right)^2}$

$$= \frac{\sqrt{54}}{3}$$

$$PQ^2 = d^2 = \frac{54}{9} = 6$$

(B) $\tan^{-1}\left(\frac{x + 3 - x + 3}{1 + (x^2 - 9)}\right) = \tan^{-1}\left(\frac{3}{4}\right)$

$$\Rightarrow \quad \frac{6}{x^2 - 8} = \frac{3}{4}$$

$$\Rightarrow \quad 3x^2 = 48$$

$$\Rightarrow \quad x = \pm 4$$

(C) $(\vec{\mathbf{b}} - \vec{\mathbf{a}}) \cdot \left(\vec{\mathbf{b}} + \frac{\vec{\mathbf{a}} - \mu \vec{\mathbf{b}}}{4}\right) = 0$

$$\Rightarrow \quad (\vec{\mathbf{b}} - \vec{\mathbf{a}}) \cdot (4\vec{\mathbf{b}} + \vec{\mathbf{a}} - \mu \vec{\mathbf{b}}) = 0$$

$$(4 - \mu)\vec{\mathbf{b}}^2 - \vec{\mathbf{a}}^2 = 0 \qquad \text{...(i)}$$

Also, $2\left|\vec{\mathbf{b}} + \frac{\vec{\mathbf{a}} - \mu \vec{\mathbf{b}}}{4}\right| = |\vec{\mathbf{b}} - \vec{\mathbf{a}}|$

$$\Rightarrow \quad 2\left|\frac{(4 - \mu)\vec{\mathbf{b}} + \vec{\mathbf{a}}}{4}\right| = |\vec{\mathbf{b}} - \vec{\mathbf{a}}|$$

$$\Rightarrow \frac{(4-\mu)^2 \vec{\mathbf{b}}^2}{4} + \frac{\vec{\mathbf{a}}^2}{4} = \vec{\mathbf{b}}^2 + \vec{\mathbf{a}}^2$$

$$\Rightarrow \quad \frac{3\vec{\mathbf{a}}^2}{4} = \frac{(4-\mu)^2 - 4}{4} \cdot \vec{\mathbf{a}}^2$$

$$3\vec{\mathbf{a}}^2 = C\,(4-\mu)^2 - 4\vec{\mathbf{b}}^2 \qquad \text{...(ii)}$$

From Eqs. (i) and (ii)

$3(4-\mu) = (4-\mu)^2 - 4$

$(4-\mu)^2 - 3(4-\mu) - 4 = 0$

$\Rightarrow \qquad \mu = 0, 5$

$\mu = 5$ is not admissible.

(D) $f(0) = 9$,

$$f(x) = \frac{\sin\left(\frac{9x}{2}\right)}{\sin \frac{x}{2}}$$

$$= \left(3 - 4\sin^2 \frac{x}{2}\right)\left(3 - 4\sin^2 \frac{3x}{2}\right)$$

$$= 9 - 12\sin^2 \frac{x}{2} - 12\sin^2 \frac{3x}{2} + 16\sin^2 \frac{x}{2} \cdot \sin^2 \frac{3x}{2}$$

$$= 9 - 6(1-\cos x) - 6(1-\cos 3x) + 4(1-\cos x)(1-\cos 3x)$$

$$= 1 + 6\cos x + 6\cos 3x - 4\cos x - 4\cos 3x + 4\cos x \cos 3x$$

Let $\quad I = \dfrac{2}{\pi} \displaystyle\int_{-\pi}^{\pi} \frac{\sin \frac{9x}{2}}{\sin \frac{x}{2}}\, dx$

$$= \frac{4}{\pi} \int_0^{\pi} 1 + 2\cos x + 2\cos 3x + 2(\cos 4x + \cos x)dx$$

$$= \frac{4}{\pi} \times \pi$$

$$= 4$$